D0984089

The Fischer Indole Synthesis

Emil Fischer

The Fischer Indole Synthesis

Brian Robinson

Department of Pharmacy
University of Manchester

A Wiley Interscience Publication

JOHN WILEY & SONS

Chichester · New York · Brisbane · Toronto · Singapore

Library of Congress Cataloging in Publication Data:

Robinson, Brian.
The Fischer indole synthesis.
'A Wiley–Interscience publication.'
Includes index.
1. Indole. 2. Chemistry, Organic—Synthesis.
I. Title.
QD401.R598 547′.593 81-14749
ISBN 0 471 10009 9 AACR2

British Library Cataloguing in Publication Data:

Robinson, Brian
The Fischer indole synthesis.
1. Chemistry, Organic—Synthesis
2. Heterocyclic compounds
I. Title
547′.59 QD400.3

ISBN 0 471 10009 9

Filmset by Composition House Limited,
Salisbury, Wiltshire, England.

Printed by Pitman Press Limited, Bath, Avon

Contents

Preface

At approximately 10.30 a.m. on Friday, 30 September, 1960, during the course of my Ph.D. oral examination, I was asked to describe the (then accepted) mechanism of the Fischer indole synthesis. Thus were sown the seeds of this book.

The Fischer indole synthesis can be regarded as the elimination of the elements of ammonia from the arylhydrazone of an aldehyde or ketone, bearing at least one hydrogen atom on a C-2 atom, by treatment with either a Brønsted or Lewis acid, usually at an elevated temperature, or by refluxing in a high-boiling (b.p. *ca.* 200–250 °C) organic solvent. Since its discovery and recognition by Emil Fischer and two of his students in 1883 and 1884, it has become by far the most widely used and versatile of over thirty available syntheses of indoles. The reaction has been the subject of many reviews, each of which appears to have stimulated even further research effort. In view of the intense scientific interest in the synthesis, its wide ranging commercial application as evidenced by the large number of patents in which it is incorporated, the approach of the centenary of its discovery and the fact that the many previous reviews have been selective in their coverage, a fully comprehensive survey of the present state of our knowledge concerning the Fischer indole synthesis appears to be opportune.

This present study covers the investigations which have been reported upon the reaction since its discovery through to the 1981, volume 94 issue of *Chemical Abstracts.* Although an attempt has been made to refer to all known examples of the reaction which have been published, this ideal is impossible to accomplish since isolated examples of the reaction appear in literature which is not readily related to the synthesis and since the often extravagant claims of many patents are far from completely elucidated. However, it is hoped that all the salient aspects of the reaction have been included and that any omissions, for which one can only apologize, are of a minor nature. The work is intended to emphasize the three main features of the reaction, these being its development, the clarification of its mechanism, certain aspects of which are still far from rigorously established, and the utilizability of it and its extensions in the synthetic process. This policy has been achieved by surveying the reaction from seven different perspectives.

A conscious decision has been taken not to use the IUPAC system of nomenclature since this would have led to the introduction of unwieldy names for a large number of the compounds under investigation which are more easily identified using well established trivial root names and alternative methods of nomenclature. This approach is shared by a large number of contemporary authors of similar works.

Several institutions have provided me with invaluable help in putting at my disposal certain library facilities and literature which were not available to me within the University of Manchester. In this respect I express my gratitude to the University of Salford, especially Miss B. Marsden, A.L.A. and Mr G. N. Royds, B.A., A.L.A. of its library staff for their very willing help and co-operation, and to the Dyestuffs and Pharmaceuticals Divisions of Imperial Chemical Industries Ltd, especially Mr K. E. Howarth, B.Sc., M.I.Inf.Sc., of the latter Division for allowing me the use of modern information retrieval systems. My sincere thanks are also due to Mrs Shiela L. Trachsel for the monumental task of typing the manuscript, to Miss S. A. Meakin, B.Sc., M.P.S. for her assistance with the compilation of Table 23 and to two of my departmental colleagues, Dr P. A. Crooks and Dr R. D. Waigh for help with proof checking. While it is to be hoped that factual errors will be absent from this work, some typographical errors, for which I accept full responsibility, will, in spite of careful proof checking, doubtlessly still be present. However, the location of these errors should provide diversion for a reviewer.

February 1981 B. ROBINSON

The medal embossed on the cover was struck by the German Chemical Society in honour of Emil Fischer.

Hermann Emil Fischer (1852–1919)

On 9 October 1852 at Euskirchen, a small town in Rhenish Prussia some twenty-five miles from Cologne, a boy was born who in the scope and magnitude of his future scientific accomplishments was to become what many would regard as the greatest organic chemist in history. This boy was Hermann Emil Fischer who, preceded by five sisters, was the only surviving son of Julie Fischer (née Poensgen) and Laurenz Fischer. His mother came from a family which was well known in Rhenish industrial circles and his father, although with little formal education, was a successful business man, being a flourishing merchant involved in supplying the requirements of the peasant farmers inhabiting the Eifel and being associated with his brothers in other business ventures concerning a spinning mill, a brickworks and the foundation of a brewery at Dortmund. The family was Protestant and was well established in the Rhineland, having lived there since the end of the seventeenth century. The young Fischer had a very pleasant childhood and had numerous cousins who formed the nucleus of his playmates.

Because of their financially secure position, the Fischers were able to give Emil the very best of educational opportunities which were certainly not wasted on the boy. The first three years of his primary education were obtained by means of a private tutor, after which he spent four years at the local public school. Then followed two years at the Gymnasium in Wetzlar, succeeded by two more years at the Gymnasium in Bonn. During this time he proved to be an excellent pupil with a fine memory and the ability to quickly accept a wide range of subject matter. In the spring of 1869 he passed his final examinations in Bonn with great distinction and graduated as *primus omnium.*

Since at this time Fischer was his father's only son, it was reasonable for his father to wish him to train for and ultimately enter the family business enterprises in order that he might eventually become his successor. Accordingly, in 1869 at the age of seventeen years he was apprenticed to his brother-in-law, Ernst Friedrichs, a timber merchant. Emil Fischer apparently had little enthusiasm for, or ability in, the world of commerce and after two years in this work had ended in failure his father consented to his son's desire for a university education. Indeed, according to Fischer's posthumously published autobiography, his father said that Emil was 'too stupid to be a business man and had better be a student' and his uncle, a partner in the family lumber business, prophesied that 'the boy will never amount to anything'. The consent for a university education was however compromised by his father who saw to it that Emil's field of study was chemistry, since it was likely that at least this subject would provide him with a living (how times do change!). The commencement of his university education was unfortunately delayed when the eighteen-year-old youth contracted a persistent gastric catarrh but at last, in the spring of 1871, he enrolled as a student at the University of Bonn. It was here that Fischer came into contact with August von Kekulé and his active assistants Engelbach and Zincke. Although Kekulé delivered brilliantly inspiring lectures, he had practically no interest in laboratory instruction and this, together with the young Fischer's predilection for physics, nearly lost him to chemistry at this time. Indeed, had it not been for the persuasion of his cousin and fellow student, Otto Fischer (1852–1932) this might well have occurred. In line with the German student's tradition at this time in having a Wanderjahr, both young men transferred in the autumn of 1872 to the University of Strasbourg in Alsace. This had been newly established by the German authorities and provided with distinguished teachers and ample funding subsequent to the return of the area to Germany as a result of the Franco–German war. It was in Strasbourg that Fischer received both a theoretical and a practical up-to-date course of instruction in analytical chemistry from F. Rose and came under the magnetic teaching of Adolf von Baeyer (1835–1917). Under these influences, Fischer decided to devote the rest of his working life to organic chemistry. Indeed, he later declared that, next to his father, he owed to von Baeyer much of what he had accomplished in his life. According to contemporary custom, after Fischer had completed his training in analytical chemistry he commenced work upon his doctoral research. This was carried out under the supervision of von Baeyer and the degree of Doctor of Philosophy was duly conferred upon him in 1874 after the presentation of a thesis entitled 'Über Fluorescëin und Phtalëin-Orcin'. Some of the results embodied in this thesis are also published in the literature (Fischer, Emil, 1874). According to Fischer's own account, he did not fare too well in the accompanying oral examination but this was not too unexpected since all members of the Faculty at Strasbourg were allowed to ask questions in the examination and apparently a geologist present was not particularly impressed with the candidate's answers to his question on geology.

Fischer subsequently remained in Strasbourg as an independent research student and when, in the autumn of 1874, an assistantship in the organic chemistry section became vacant he accepted the position at the request of von Baeyer who obviously recognized his past student's chemical ability. In this capacity, Fischer was called upon to demonstrate to students in practical organic chemistry classes. Again according to his posthumously published autobiography, on one such class he was requested to assist a student who was attempting to synthesize 4,4′-biphenol via the diazotization of benzidine but was repeatedly obtaining only intractable 'muddy' products. After deciding that the formation of such products was the result of oxidation by the nitrous acid used in the diazotization, Fischer repeated this reaction for the student but added sodium sulphite to the reaction mixture to prevent this presumed oxidation. However, this addition resulted in the formation of a yellow precipitate, subsequent investigation of which showed it to be the salt of an arylhydrazine. This ultimately resulted in Fischer's discovery, by an analogous reduction of diazotized aniline, of phenylhydrazine (Fischer, Emil, 1875a, 1875b), a compound which was to prove to be the key to many of his future great discoveries. This latter reduction had, in fact, been investigated four years previously by Strecker and one of his students, Peter Römer, at the University of Würzburg when they had treated benzenediazonium nitrate with an excess of potassium hydrogen sulphite to afford a salt which they represented as $C_6H_5 \cdot NH(:NH) \cdot SO_3K$ (Strecker, 1871).

In 1875 von Baeyer accepted the invitation to succeed Justus von Liebig (1803–1873) at the University of Munich and Fischer accompanied him there in his capacity as an assistant in organic chemistry. At that time in Munich it was still customary to recognize the doctorate of another university only after another oral re-examination, the *nostrification.* This re-examination was made a pure formality for Fischer by von Baeyer who, not desiring to trouble his assistant too much, asked questions only pertaining to hydrazines which of course presented no difficulties to the candidate, the foremost expert in the subject. It was in Munich that Fischer began to investigate the chemistry of phenylhydrazine and other hydrazines and also worked in collaboration with his cousin, Otto Fischer, upon the rosaniline dyestuffs. These latter studies were an extension of his interest in the chemistry of dyestuffs emanating from his doctoral studies. In 1878 he qualified as a Privatdozent at Munich, the assigned theme of the qualifying lecture being 'The Present Problems of Chemistry', and at the beginning of 1879 he was appointed to an extraordinary professorship and made director of the analytical division in succession to Jacob Volhard (1834–1910) as a result of his earlier training in analytical chemistry under F. Rose at Strasbourg. In this same year, at the age of twenty-seven, he refused an offer of a full professorship in chemistry at the Technical University of Aix-la-Chapelle (Aachen) since he did not wish to leave the inspiring environment of the Munich department.

However, in 1882, at the age of thirty he once again succeeded Jacob Volhard, this time to the chair of chemistry at the University of Erlangen, where he was to stay for three years. It was here that Fischer began his studies in the purine and carbohydrate fields. It was also while he was at Erlangen that he discovered the synthesis of indoles which still carries his name. In 1883 the Badische Anilin-und Soda-Fabrik offered him the position as director of its scientific laboratory but since his father had now made him financially independent, it was possible to refuse this temptingly lucrative offer, preferring to remain in academic work. The final period of his stay in Erlangen was overshadowed by a return of the gastric catarrhal complaint from which he had suffered immediately prior to his undergraduate studies. This forced him to take a year's leave of absence and made him unable to accept the offer to succeed Viktor Meyer at the Federal Technical University of Zürich although soon after he had recovered from this illness and resumed his duties at Erlangen, he accepted, in 1885, in succession to Johannes Wislicenus (1835–1902), the chair of chemistry at the University of Würzburg where he was to remain for seven years. During this period he continued to develop his work upon purines and carbohydrates and also investigated the scope of his indole synthesis in collaboration with various research assistants. Fischer counted his seven years at Würzburg among the happiest of his life. The city and surrounding beautiful hilly countryside were much more attractive than Erlangen and the department contained a number of very able staff and offered a wide professional outlook. It was also during this period, in 1888, that he married Agnes Gerlach, the daughter of J. von Gerlach, Professor of Anatomy at Erlangen. He had initially accidentally met the Gerlachs several years earlier during a train journey between Munich and Erlangen. Although having himself once written that he had 'definitely decided to travel life's path alone', the friendship between the old professor's daughter and the young professor flourished and after several years of courtship they married. Unfortunately, after only seven years of happy marriage, Agnes Fischer died of a middle-ear inflammation. They had three sons but here again tragedy struck. One son was killed in the First World War and another committed suicide at the age of twenty-five as a result of compulsory military training. However, the third and eldest son, Hermann Otto Laurenz Fischer, who died in 1960, became Professor of Biochemistry in the University of California at Berkeley.

In 1892, August Wilhelm von Hofmann, professor and director of the Chemical Institute of the University of Berlin, died. The Faculty of the Institute suggested as possible successors, Baeyer, Fischer and Kekulé. Baeyer let it be known that he would not leave Munich which left the choice between Fischer and Kekulé, a decision ultimately being made in favour of the former. So, in 1892, at the age of forty, Emil Fischer was appointed to the most important chair of chemistry in Germany, a position that he held for the remaining twenty-seven years of his life. In Berlin the work upon purines was completed and the work upon carbohydrates was extended to include a study of glycosides, oligosaccharides and glycoside hydrolases. Also during this period these two fields were integrated in the synthesis of nucleosides. It was also in Berlin, in 1899, that Fischer made his initial studies upon proteins to which field, as with those of carbohydrates and purines, he would ultimately make such important and fundamental contributions. During this period were also effected significant studies upon enzymes and towards the end of his life he made significant contributions to the chemistry of the tannins.

Fischer's achievements were recognized by the award of many honours. He was made a Prussian Geheimrat (Excellenz), held honorary doctorates from the Universities of Brussels, Cambridge, Christiania and Manchester and was awarded the Prussian Order of Merit, the Maximilian Order for Arts and Sciences, the Elliott Crasson Gold Medal from the Franklin Institute of Philadelphia in 1913 and in 1890 the Davy Medal of the Royal Society of which he was elected a Foreign Member in 1899. He was elected a Member of the Berlin Academy of Sciences in 1893, an Honorary and Foreign Member of the Royal Institution of Great Britain in 1904 and earlier, in 1892, an Honorary and Foreign Member of the Chemical Society and delivered its Faraday Lecture in 1907 (Fischer, Emil, 1907). The title of this lecture was 'Synthetical Chemistry in its Relation to Biology' and it clearly illustrated Fischer's involvement in the chemistry of biological molecules and processes. In 1902 he was awarded the Nobel Prize for Chemistry for his studies upon purines and carbohydrates. This was the second award to a chemist but the first to an organic chemist, the initial award having been made in 1901 to Jacobus Henricus van't Hoff (1852–1911). It is interesting that Fischer, the ex-pupil, received this prize three years before von Baeyer, the ex-master, did so in 1905 for his research on organic dyestuffs and hydroaromatic compounds. The citation for Fischer's Nobel Prize reads 'in recognition of the extraordinary services he has rendered by his work on sugar and purine synthesis'. His Nobel Prize work on sugars was developed using, as one of the major tools, their reaction with phenylhydrazine to form osazones. He also applied the van't Hoff–Le Bel theory of the asymmetric carbon atom to their stereochemistry and succeeded in synthesizing a continuous sugar series containing from two to nine carbon atoms. Furthermore he recognized the nature of glycosides and by studying their enzymic hydrolysis he discovered the stereospecificity of enzyme reactivity. His simultaneous work in purines concerned the recognition that compounds such as xanthine, adenine, guanine, theobromine, theophylline, caffeine and uric acid were all closely related via the same parent ring structure, purine, work which involved their chemical interrelationships and total synthesis. Certainly, the combination of his work on sugars, purines, proteins and enzymes laid the chemical foundations for biochemistry.

Reference has already been made to the serious attacks of gastric catarrh from which Fischer suffered in the periods just prior to his entry as a student into the University of Bonn and towards the end of the period when he worked at the University of Erlangen. Furthermore, in the summer of 1881 he had suffered for three months from an attack of mercury poisoning as a result of inhalation of mercury diethyl produced during his study of the reaction between mercuric oxide and aliphatic hydrazines and then, ten years later, he fell victim to the toxic effects of phenylhydrazine, the vapours of which he had been inhaling over a long period of time. Just how much, if at all, these earlier illnesses contributed to his final illness we shall never know but on 15 July 1919, he died of cancer at his home in Wannsee, near Berlin, and it is in Wannsee that he is buried.

Biographical data relating to Fischer and his scientific achievements can be found in the following publications:

(i) *Aus Meinem Leben*, a posthumously-published autobiography (Fischer, Emil, 1922), which has been reviewed (Pope, 1922).

(ii) *Dem Andenken an Emil Fischer* consisting of five sections (Harries, Abderhalden, Weinberg, Trendelenburg and Lewin, 1919).

(iii) *Emil Fischer, sein Leben und sein Werk* (Hoesch, 1921).

(iv) *Dictionary of Scientific Biography* (Farber, 1972).

(v) B. Helferich (1961) in *Great Chemists.*

(vi) *German Nobel Prize Winners*, Heinz Moos Verlag, Munich, in collaboration with Inter Nationes, Bonn–Bad Godesberg, 1978, 2nd edition, p. 113.

(vii) *Nobel Prize Winners in Chemistry, 1901–1961* (Farber, 1963).

(viii) *Nobel Lectures – Chemistry – 1901–1921*, published by the Nobel Foundation and incorporating Fischer's presentation speech (Théel, 1966), his Nobel Lecture 'Syntheses in the Purine and Sugar Group' (Fischer, Emil, 1902) and a short biography on pages 36–39.

(ix) 'Address of Emil Fischer before the Berliner Akademie der Wissenschaften' (Fischer, Emil, 1894).

(x) *The Basic Work of Fischer and van't Hoff in Carbohydrate Chemistry* (Hudson, C. S., 1953) – includes personal biographical data.

(xi) Obituary notices (Duisberg, 1919; Forster, 1921–1922; Harrow, 1919; H.E.A., 1919; Hofmann, K. A., 1919; Jacobson, 1919; Knorr, 1919 and Wichelhaus, 1919).

(xii) 'Commemoration Speech on Emil Fischer' (Beckmann, 1920).

(xiii) 'Emil Fischer Memorial Lecture' (Forster, 1920, 1933).

(xiv) *Anniversaries in Chemistry and Physics* (Huntress, 1952).

(xv) *Emil Fischer zum 100. Geburtstag* (Helferich, 1953).

(xvi) *Five Greats born in 1852 (Becquerel, Fischer, Moissan, Ramsay, van't Hoff)* (Van de Velde, 1952).

(xvii) *Emil Fischer (1852–1952)* (Ulrich, 1952).

(xviii) *The Fischer Indole Synthesis* (Roussel, P. A., 1953) which incorporates a very brief review of the discovery of indole, the Fischer indole synthesis and a short biography.

(xix) *The Centennial of the Birth of Emil Fischer* (Okuno, 1952).

(xx) *J. van't Hoff and E. Fischer, Pioneers of Chemical Science* (Klare, 1980).

(xxi) Fischer's original scientific publications are collected in eight volumes, all published by Julius Springer, Berlin, in the years given in the respective brackets:

(a) *Untersuchungen über Aminosäuren, Polypeptide und Proteine, 1899–1906* (1906).

(b) *Untersuchungen in der Puringruppe, 1882–1906* (1907).
(c) *Untersuchungen über Kohlenhydrate und Fermente, 1884–1908* (1909).
(d) *Untersuchungen über Depside und Gerbstoffe, 1908–1919* (1920).
(e) *Untersuchungen über Kohlenhydrate und Fermente, II, 1908–1918* (1922).
(f) *Untersuchungen über Aminosäuren, Polypeptide und Proteine, II, 1907–1919* (1923) [see (Bergmann, M., 1924)].
(g) *Untersuchungen über Triphenylmethanfarbstoffe, Hydrazine und Indole* (*1924*) (pp. 1–28 of this volume relate to Fischer's Ph.D. thesis presented to the University of Strasbourg in 1874).
(h) *Untersuchungen aus verschiedenen Gebieten, Vorträge und Abhandlungen allgemeinen Inhalts* (*1924*) [see (Bergmann, M., 1924)].

CHAPTER I

Discovery of the Fischer Indole Synthesis, its Early Developments and Related Reactions

A. Early Developmental Investigations and Fischer's Studies on the Indole Synthesis

Some years before Fischer joined him as a research student at the University of Strasbourg, Baeyer was working at the Trade Academy in Berlin investigating the structure of indigo. It was here, in collaboration with Knop, that he found that reduction of isatin, an oxidation product of indigo (Erdmann, 1841; Laurent, 1841, 1842), with sodium amalgam afforded a compound named dioxindole ($C_8H_7NO_2$) and ultimately a compound named oxindole (C_8H_7NO). Although at the time the structures of these products were unknown, they were regarded as being oxygenated derivatives of a hypothetical parent compound, C_8H_7N, which was named indole (Baeyer and Knop, 1866). Subsequently, Baeyer himself obtained indole, referred to by him as 'die Muttersubstanz der Indigogruppe', by the reductive distillation of oxindole from its mixture with zinc dust (Baeyer, 1866) and later (Baeyer, 1868) by the reduction of indigo blue itself with tin and hydrochloric acid. It is interesting that this latter publication appears as the first paper in the inaugural volume of *Berichte der Deutschen Chemischen Gesellschaft*. The structure of indole was (Baeyer and Emmerling, 1869) confirmed ultimately by its synthesis involving fusion of 2-nitrocinnamic acid with a mixture of iron filings and potassium hydroxide. Subsequently, indole was again prepared from oxindole by heating it with a phosphorus pentachloride and phosphorus oxychloride mixture to afford a product whose molecular formula was established as $C_8H_5Cl_2N$. This was reductively dechlorinated by heating with zinc dust or a potassium hydroxide and iron filings mixture to give indole (Baeyer, 1879). Around this period other syntheses of indoles were also developed involving the cyclization of N-alkyl *ortho*-toluidines (Baeyer and Caro, H., 1877) or by reaction between bromomethyl phenyl ketone and aniline (Möhlau, 1881) but both these were of limited potential versatility and produced only low yields of product. It was Fischer who first developed a useful and versatile synthesis of indoles which furnished satisfactory yields of products and thereby opened up the field of indole chemistry.

In the decade following his discovery that diazotized aniline could be reduced with sodium sulphite to ultimately afford phenylhydrazine (Fischer, Emil, 1875a, 1875b), a method later modified to give much better results (Chattaway and Humphrey, 1927), Fischer and several of his research students, initially at the University of Strasbourg and subsequently at the Universities of Munich, Erlangen and Würzburg, respectively, carried out detailed studies of analogous syntheses from other aromatic primary amines [e.g. 1- and 2-hydrazinonaphthalene were prepared by reduction of diazotized 1- and 2-naphthylamine, respectively, using tin(II) chloride in concentrated hydrochloric acid (Fischer, Emil, 1886a), conditions which had been developed three years earlier at the Federal Technical University of Zürich for an alternative reduction of benzenediazonium chloride to phenylhydrazine (Meyer, V. and Lecco, 1883)] and developed a synthesis of N_α-alkyl- or N_α-arylphenylhydrazines by reducing the N-nitroso derivatives of the corresponding secondary amines [e.g. phenylmethylamine, phenylethylamine and diphenylamine were reacted with nitrous acid to produce the corresponding N-nitroso derivatives which were then reduced using zinc dust in acetic acid to yield N_α-methylphenylhydrazine (Fischer, Emil, 1886d), N_α-ethylphenylhydrazine (Fischer, Emil, 1875c) and N_α-phenylphenylhydrazine (Fischer, Emil, 1876b), respectively]. Zinc dust in acetic acid was also later (Altschul, 1892) introduced for accomplishing the reduction of aryldiazonium salts to the corresponding phenylhydrazines. A further synthesis of N_α-methylphenylhydrazine was published (Tafel, 1885) from the University of Erlangen in the year Fischer moved to Munich. This involved the N_α-methylation of N_β-benzoylphenylhydrazine, prepared previously (Fischer, Emil, 1878), by treatment with sodium methoxide and then methyl iodide, followed by hydrolysis of the benzoyl group. Fischer investigated in great detail not only the chemistry of arylhydrazines but also that of alkylhydrazines. A full account of his earlier studies in the area can be found in a paper 'Ueber die Hydrazinverbindungen' of one hundred and seventeen pages in length (Fischer, Emil, 1878) which encompasses a series of preliminary communications (Fischer, Emil, 1875a, 1875b, 1875c, 1876a, 1876b, 1877) upon the chemistry of aromatic hydrazines, the first two of these being published from the University of Strasbourg and the remainder from the University of Munich. For details of all his works in the area, Hoesch (1921) and the compilation of his published works by Julius Springer mentioned above can be consulted. It was soon found during these investigations that arylhydrazines react with aldehydes and ketones by eliminating the elements of water to form arylhydrazones and the reactivity of these compounds then came under scrutiny.

The first conversion of an arylhydrazone to an indole catalysed by acid was reported in 1883 from the University of Erlangen by Fischer and one of his

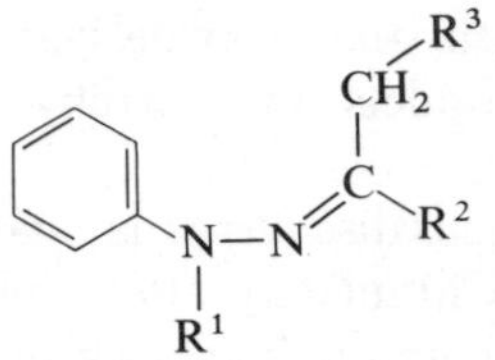

1

students, Friedrich Jourdan (Fischer, Emil and Jourdan, 1883). By reacting pyruvic acid with phenylhydrazine they prepared the corresponding hydrazone **1** ($R^1 = R^3 = H$, $R^2 = COOH$) which they found could be thermally decarboxylated to acetaldehyde phenylhydrazone, esterified to ethyl pyruvate phenylhydrazone and reduced with sodium amalgam to 2-(2-phenylhydrazino) propionic acid. They also found it to be soluble in acetic acid and dilute hydrochloric acid, but they did not heat these acidic solutions of the phenylhydrazone. The m.p. of **1** ($R^1 = R^3 = H$, $R^2 = COOH$) was originally quoted as 169 °C (Fischer, Emil and Jourdan, 1883). Later (Fischer, Emil, 1884) this was stated to be a misprint for 192 °C and subsequently (Japp and Klingemann, 1888a, 1888c) the m.p. was corrected to 185 °C. Similarly, pyruvic acid was reacted with N_α-methylphenylhydrazine to produce the hydrazone **1** ($R^1 = CH_3$, $R^2 = COOH$, $R^3 = H$) which was obtained as needles, m.p. = 78 °C (with softening at 70 °C) [This reaction was repeated seventy years later when a 71% yield of product was obtained (Shirley and Roussel, P. A., 1953)]. During further examination, the product was heated with 10% hydrochloric acid which converted it into a crystalline compound, m.p. 206 °C, of molecular formula $C_{10}H_9NO_2$. The loss of the elements of ammonia from the starting hydrazone in this reaction was clearly recognized and the equation (i) was written to account for it.

$$C_{10}H_{12}N_2O_2 = C_{10}H_9NO_2 + NH_3 \quad \text{(i)}$$

However, the reaction product could not be identified nor could its formation be accounted for although the fundamental importance of the transformation appears to have been realized by the two authors, who write as the last paragraph of their publication 'Dieser Vorgang ist so merkwürdig, dass wir einstweilen es nicht wagen, eine Erklärung desselben zu geben. Jedenfalls ist die neue Säure ein Repräsentant einer merkwürdigen Körperklasse, für welche Analogieen bis jetzt fehlen' (Fischer, Emil and Jourdan, 1883). Although reaction yields were not quoted by Fischer and Jourdan, it was reported seven decades later (reference 16 quoted in Roussel, P. A., 1953) that the reaction had been repeated several times with yields of between 71% and 78% and subsequently (Snyder and Cook, P. L., 1956) the reaction was again repeated to afford a 66% yield of product.

In the following year, Fischer and Otto Hess, another of his students at the University of Erlangen, identified (Fischer, Emil and Hess, 1884) the product from this above reaction as 1-methylindole-2-carboxylic acid (**2**; $R^1 = CH_3$, $R^2 = COOH$, $R^3 = H$) by effecting its thermal decarboxylation to produce 1-methylindole (**2**; $R^1 = CH_3$, $R^2 = R^3 = H$). This was recognized by being

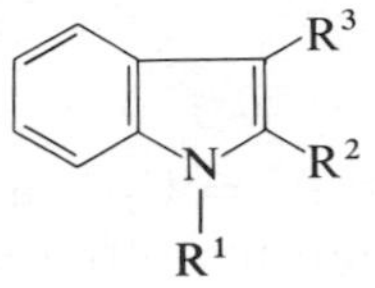

2

oxidized to a product formulated as methylpseudoisatin **3** ($R^1 = R^2 = H$, $R^3 = CH_3$) by analogy with the ethyl homologue **3** ($R^1 = R^2 = H$, $R^3 = C_2H_5$) which had been obtained the previous year by Adolf von Baeyer during his studies on indigo (Baeyer, 1883). Consequently, Fischer and Hess then prepared pyruvic acid N_α-ethylphenylhydrazone (**1**; $R^1 = C_2H_5$, $R^2 = COOH$, $R^3 = H$) which upon warming in 20% hydrochloric acid furnished 1-ethylindole-2-carboxylic acid (**2**, $R^1 = C_2H_5$, $R^2 = COOH$, $R^3 = H$), the identity of which they definitely established by its thermal decarboxylation to give **2** ($R^1 = C_2H_5$ $R^2 = R^3 = H$) which was oxidized to a compound identical with Baeyer's ethylpseudoisatin (**3**; $R^1 = R^2 = H$, $R^3 = C_2H_5$) (Baeyer, 1883).

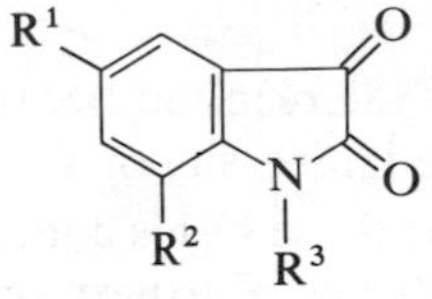

3

The synthetic sequence was also extended to the similar conversion of N_α-phenylphenylhydrazine and pyruvic acid into 1-phenylindole-2-carboxylic acid (**2**; $R^1 = C_6H_5$, $R^2 = COOH$, $R^3 = H$) but attempts to decarboxylate this product produced only an oil which was neither analysed nor characterized (Fischer, Emil and Hess, 1884). Three years later (Fischer, Emil, 1887) this decarboxylation was successfully accomplished and the product was oxidized to phenylpseudoisatin (**3**; $R^1 = R^2 = H$, $R^3 = C_6H_5$) and nearly seven decades later this indolization and decarboxylation were again effected in 20–25% overall yield (Shirley and Roussel, P. A., 1953). In 1885, Fischer's student, Otto Antrick, extended this synthetic sequence by heating pyruvic acid N_α-benzylphenylhydrazone with concentrated hydrochloric acid to afford 1-benzylindole-2-carboxylic acid which was subsequently thermally decarboxylated to give 1-benzylindole (Antrick, 1885). The results of yet further extensions by another of Fischer's students, S. Hegel, appeared in another publication from the University of Erlangen in 1886, the year after Fischer's departure to Würzburg. Pyruvic acid N_α,4-dimethylphenylhydrazone, N_α-ethyl-4-methylphenylhydrazone and N_α,2-dimethylphenylhydrazone (**4**; $R^1 = R^3 = CH_3$,

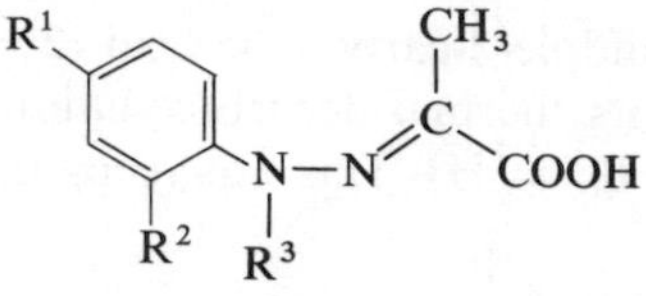

4 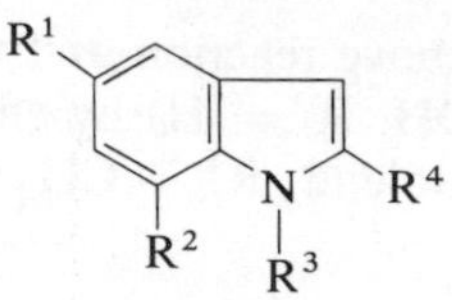 5

R^2 = H; R^1 = CH_3, R^2 = H, R^3 = C_2H_5 and R^1 = H, R^2 = R^3 = CH_3, respectively) were converted into the corresponding indoles **5** (R^1 = R^3 = CH_3, R^2 = H; R^1 = CH_3, R^2 = H, R^3 = C_2H_5 and R^1 = H, R^2 = R^3 = CH_3, respectively, R^4 = COOH) by heating with 10% hydrochloric acid. Thermal decarboxylation then afforded **5** (R^1 = R^3 = CH_3, R^2 = R^4 = H; R^1 = CH_3, R^2 = R^4 = H, R^3 = C_2H_5 and R^1 = R^4 = H, R^2 = R^3 = CH_3, respectively) which was oxidized into the corresponding pseudoisatins **3** (R^1 = R^3 = CH_3, R^2 = H; R^1 = CH_3, R^2 = H, R^3 = C_2H_5 and R^1 = H, R^2 = R^3 = CH_3, respectively) (Hegel, 1886). That Fischer and Hess recognized the potential of this synthetic route is apparent from the paragraph immediately preceding the experimental section of their paper (Fischer, Emil and Hess, 1884) which states 'Für die Synthese der Körper der Indigogruppe ist damit ein neues und, wie es scheint, recht fruchtbares Gebiet erschlossen'. However, little could they have known just how fertile the territory was that they had opened up.

The potential generality of the reaction became apparent in 1886 in a series of papers, published from the University of Würzburg, reporting the results of studies by Fischer himself and those of his colleagues and students Jos. Degen, Anton Roder and Adolf Schlieper. Furthermore, the use of sulphuric acid or phosphoric acid in place of hydrochloric acid to effect the indolization of pyruvic acid N_α-methylphenylhydrazone was investigated (Fischer, Emil, 1886b), as was the use of large quantities of zinc chloride which ultimately gave higher product yields (Fischer, Emil, 1886b). Thus, using this latter catalyst, Fischer himself converted acetone phenylhydrazone (**1**; R^1 = R^3 = H, R^2 = CH_3) into 2-methylindole (**2**; R^1 = R^3 = H, R^2 = CH_3) (59–62% yield), ethyl methyl ketone phenylhydrazone (**1**; R^1 = H, R^2 = R^3 = CH_3) into 2,3-dimethylindole (**2**, R^1 = H, R^2 = R^3 = CH_3) along with 2-ethylindole (**2**, R^1 = R^3 = H, R^2 = C_2H_5), these being separated by fractional crystallization of the total indolic reaction product (45% yield), methyl propyl ketone phenylhydrazone (**1**; R^1 = H, R^2 = CH_3, R^3 = C_2H_5) into 3-ethyl-2-methylindole (**2**; R^1 = H, R^2 = CH_3, R^3 = C_2H_5), acetophenone phenylhydrazone (**1**; R^1 = R^3 = H, R^2 = C_6H_5) into 2-phenylindole (**2**; R^1 = R^3 = H, R^2 = C_6H_5) (nearly quantitative yield), benzyl phenyl ketone phenylhydrazone (**1**; R^1 = H, R^2 = R^3 = C_6H_5) into 2,3-diphenylindole (**2**; R^1 = H, R^2 = R^3 = C_6H_5) (this reaction was performed by A. Schlieper), propionaldehyde phenylhydrazone (**1**; R^1 = R^2 = H, R^3 = H_3) into 3-methylindole (**2**; R^1 = R^2 = H, R^3 = CH_3) [34% yield, this being later

(Swaminathan and Ranganathan, 1957) increased to 64% by slight experimental modification], ethyl pyruvate phenylhydrazone (**1**; $R^1 = R^3 = H$, $R^2 = COOC_2H_5$) into indole-2-carboxylic acid (**2**; $R^1 = R^3 = H$, $R^2 = COOH$) (5–6% yield) along with the corresponding ethyl ester [when subsequently (Kermack, Perkin and Robinson, R., 1921) this route was again used to prepare indole-2-carboxylic acid from pyruvic acid and phenylhydrazone, the Reissert synthesis was preferred to synthesize this product which was required in large quantities] and laevulinic acid or ethyl laevulinate phenylhydrazone (**1**; $R^1 = H$, $R^2 = CH_3$, $R^3 = CH_2COOH$ or $CH_2COOC_2H_5$, respectively) into 2-methyl-3-indolylacetic acid or ethyl 2-methyl-3-indolylacetate (**2**; $R^1 = H$, $R^2 = CH_3$, $R^3 = CH_2COOH$ or $CH_2COOC_2H_5$, respectively), the latter product, without characterization, being hydrolysed to the corresponding acid. This acid was thermally decarboxylated to afford 2,3-dimethylindole, thereby affording an unambiguous synthesis of this latter compound (Fischer, Emil, 1886c, 1886e) [nearly fifty years later, other workers (Hoshino and Shimodaira, 1935) used this method to prepare 3-carboxymethyl-2-methylindole]. Likewise, acetone, acetophenone, ethylacetoacetate, ethyl laevulinate and propionaldehyde N_α-methylphenylhydrazone furnished 1,2-dimethylindole, 1-methyl-2-phenylindole (50% yield), ethyl 1,2-dimethylindole-3-carboxylate (75% yield) (characterized by hydrolysis to the corresponding acid), 3-carbethoxymethyl-1,2-dimethylindole (60% yield) (characterized by hydrolysis to the corresponding acid which was thermally decarboxylated to afford 1,2,3-trimethylindole) and 1,3-dimethylindole, respectively (Degen, 1886; see also Fischer, Emil, 1886e), ethyl pyruvate 3-carbethoxyphenylhydrazone (**6**) yielded, subsequent to hydrolysis of the indolization product, either **7** ($R^1 = H$, $R^2 = COOH$) or **7** ($R^1 = COOH$, $R^2 = H$) (Roder, 1886) and acetaldehyde, ethyl pyruvate and acetone 2-naphthylhydrazone (**8**; $R = H$, $COOC_2H_5$ and CH_3, respectively) afforded **9** ($R^1 = R^2 = H$)

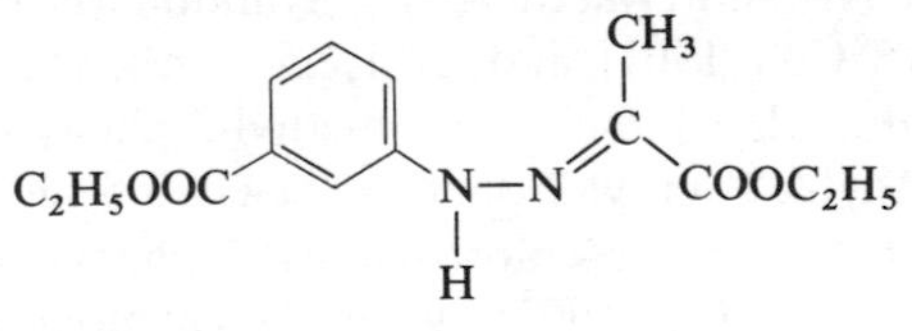

6 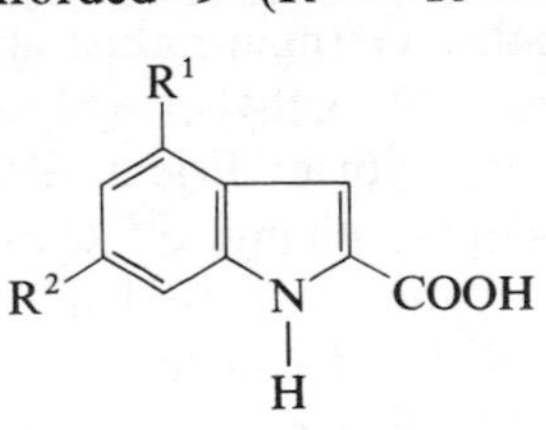 **7**

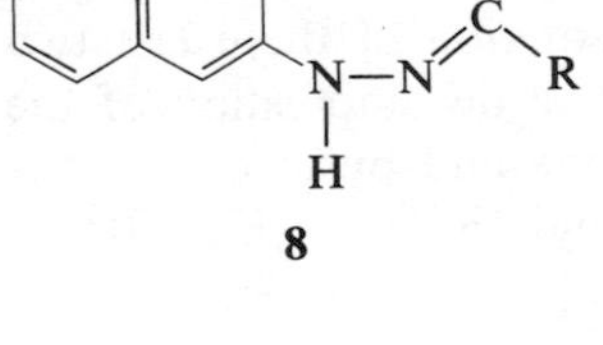

8 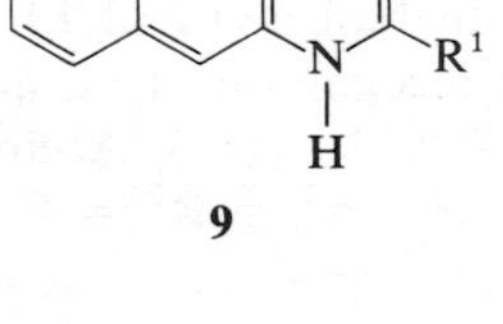 **9** 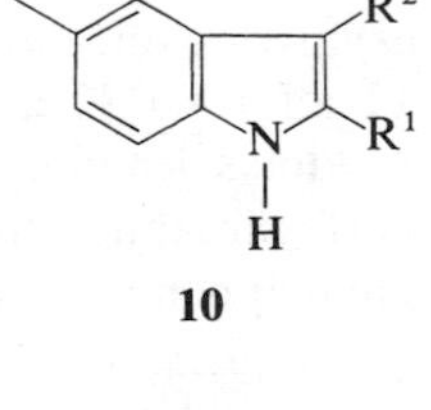 **10**

(a small amount), **9** (R^1 = COOH and H, R^2 = H) and **9** ($R^1 = CH_3$, R^2 = H), respectively, or **10** ($R^1 = R^2$ = H) (a small amount), **10** (R^1 = COOH and H, R^2 = H) and **10** ($R^1 = CH_3$, R^2 = H), respectively (Schlieper, 1886) [the product yield in the second of these indolizations was much later (Rydon and Siddappa, 1951) considerably improved although using similar catalytic conditions]. The results of these 1886 studies were summarized by Fischer (Fischer, Emil, 1886b, 1886e). A notable exception to the synthesis was observed (Fischer, Emil, 1886b, 1886e) when acetaldehyde phenylhydrazone failed to afford indole although this latter compound could be prepared by the thermal decarboxylation of indole-2-carboxylic acid which was prepared by indolization of pyruvic acid phenylhydrazone (Fischer, Emil, 1886e).

The indolizations reported prior to 1887 by Fischer and his co-workers were also covered by a patent, it being interesting that the patentee was an industrial concern (Farbwerke vorm. Meister, Lucius und Brüning, 1886a). This patent application included the use of hydrogen chloride, concentrated sulphuric acid, concentrated phosphoric acid and metal chlorides (e.g. tin(II) chloride and zinc chloride) as catalysts (fusing with zinc chloride was the preferred method in the specific examples cited) in the indolization of hydrazones derived from phenyl-, tolyl-, xylyl- and 1- and 2-naphthylhydrazine and (presumably N_α-) ethyl-, methyl-, and phenylphenylhydrazine and acetone, ethyl methyl ketone, diethyl ketone, methyl propyl ketone, acetophenone, benzyl methyl ketone, benzyl phenyl ketone, propionaldehyde, butyraldehyde, valeraldehyde, oenanthol (heptaldehyde), phenylacetaldehyde, pyruvic acid (and its esters), acetoacetic acid esters and laevulinic acid (and its esters). Among the specific examples cited, it is interesting that small quantities of **2** [$R^1 = R^3$ = H, $R^2 = (CH_2)_2COOH$] were isolated from the indolization of **1** (R^1 = H, $R^2 = CH_3$, $R^3 = CH_2COOH$), along with the major reaction product **2** (R^1 = H, $R^2 = CH_3$, $R^3 = CH_2COOH$) (see Chapter III, G). Another German industrial dyestuffs concern later patented the syntheses of 5-chloro-2-methyl-, 5-chloro-2-phenyl- and 5-methyl-2-phenylindole (Farbenfabriken vorm. Friedr. Bayer and Co., 1901) and of 1,2,5-trimethyl-, 1,5-dimethyl-2-phenyl-, 5-chloro-1,2-dimethyl-, 5-chloro-1-methyl-2-phenyl-, 1-ethyl-2-methyl-, 1-ethyl-2,5-dimethyl-, 1-ethyl-2-phenyl-, 1-ethyl-5-methyl-2-phenyl-, 5-chloro-1-ethyl-2-methyl- and 5-chloro-1-ethyl-2-phenylindole (Farbenfabriken vorm. Friedr. Bayer and Co., 1902) by methods analogous to those developed by Fischer and his colleagues involving heating mixtures of the appropriate arylhydrazone with an excess of zinc chloride.

In 1887, Fischer and his Würzburg team published further examples of arylhydrazone indolizations using a zinc chloride catalyst. Thus, the hydrazone **11** afforded **12** along with biphenyl and **13**, the isolation of these last two products leading to the recognition of an alternative decomposition of the arylhydrazone moiety (Arheidt, 1887). Acetophenone and pyruvic acid N_α-phenylphenylhydrazone (**1**; $R^1 = R^2 = C_6H_5$ and $R^1 = C_6H_5$, R^2 =

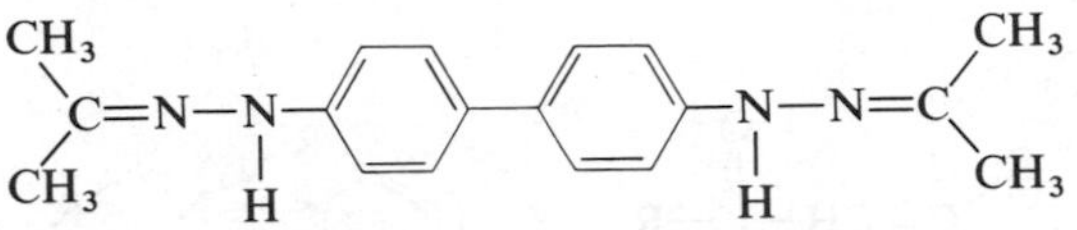

11

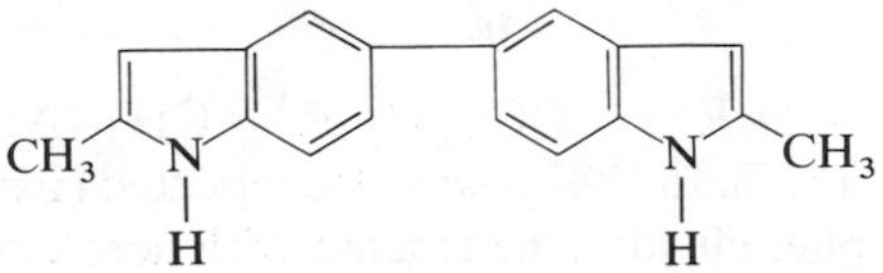

12

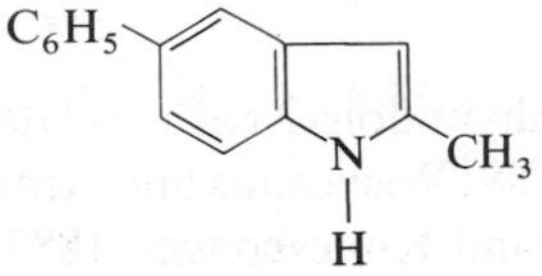

13

COOH, respectively, R^3 = H) furnished 1,2-diphenylindole (**2**; R^1 = R^2 = C_6H_5, R^3 = H) and 1-phenylindole-2-carboxylic acid (**2**; R^1 = C_6H_5, R^2 = COOH, R^3 = H), respectively, the latter being thermally decarboxylated into 1-phenylindole (Fischer, Emil, 1887). Ethyl pyruvate 4-methylphenylhydrazone afforded ethyl 5-methylindole-2-carboxylate (**5**; R^1 = CH_3, R^2 = R^3 = H, R^4 = $COOC_2H_5$) which after hydrolysis and decarboxylation gave 5-methylindole (**5**; R^1 = CH_3, R^2 – R^4 = H) (Raschen, 1887), acetone 4-methylphenylhydrazone gave 2,5-dimethylindole (**5**; R^1 = R^4 = CH_3, R^2 = R^3 = H) (Raschen, 1887), and ethyl pyruvate 2-methylphenylhydrazone gave ethyl 7-methylindole-2-carboxylate (**5**; R^1 = R^3 = H, R^2 = CH_3, R^4 = $COOC_2H_5$) which was hydrolysed to the corresponding acid but this was not very easily decarboxylated (Raschen, 1887). Ethyl pyruvate 1-naphthylhydrazone afforded **14** (R = $COOC_2H_5$) which was sequentially hydrolysed

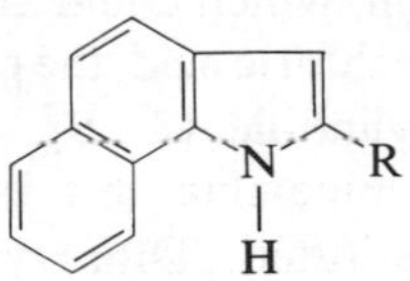

14

to yield **14** (R = COOH) and thermally decarboxylated to give **14** (R = H) (Schlieper, 1887) [this reaction sequence was much later (Rydon and Siddappa, 1951) repeated, using an ethanolic sulphuric acid indolization catalyst, in improved yield], acetone 1-naphthylhydrazone yielded **14** (R = CH_3) (Schlieper, 1887) and laevulinic acid 2-naphthylhydrazone [**8**; R = $(CH_2)_2COOH$] afforded one product, either **9** (R^1 = CH_3, R^2 = CH_2COOH) or **10** (R^1 = CH_3, R^2 = CH_2COOH) which was thermally decarboxylated to yield either **9** (R^1 = R^2 = CH_3) or **10** (R^1 = R^2 = CH_3) (Steche, 1887). At this period also, other colleagues of Fischer at Würzburg effected the indolization of ethyl 2-oxobutyrate phenylhydrazone (**1**; R^1 = H, R^2 = $COOC_2H_5$, R^3 = CH_3) and the corresponding acid **1** (R^1 = H, R^2 = COOH, R^3 = CH_3), by heating in ethanolic sulphuric acid, to produce ethyl

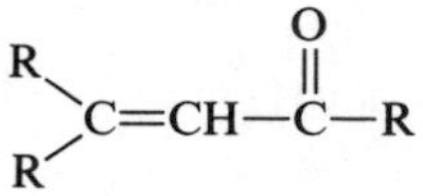

15 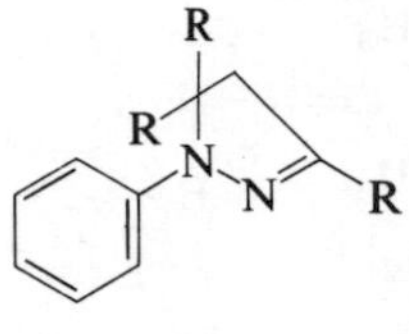 **16**

3-methylindole-2-carboxylate (**2**; $R^1 = H, R^2 = COOC_2H_5, R^3 = CH_3$) (Arnold, E., 1888; Wislicenus and Arnold, E. 1887). In 1887 it was also reported (Fischer, Emil and Knoevenagel, 1887) that phenylhydrazine reacted with acrolein (**15**; $R = H$) and mesityl oxide (**15**; $R = CH_3$) to afford the pyrazolines **16** ($R = H$ and CH_3, respectively) and not the corresponding phenylhydrazones. The results reported in some of these 1887 publications (Fischer, Emil and Knoevenagel, 1887; Raschen, 1887; Schlieper, 1887) are also embodied in two patents (Farbwerke vorm. Meister, Lucius und Brüning, 1886b, 1886c), the patentee being the same industrial concern similarly involved with the pre-1887 indolizations, as mentioned above.

Fischer and his students and colleagues continued to develop the indole synthesis at Würzburg and as a result published further papers in 1888. Thus, isovaleraldehyde, oenanthol (heptaldehyde) and benzyl methyl ketone phenylhydrazone [**1**; $R^1 = R^2 = H$, $R^3 = CH(CH_3)_2$; $R^1 = R^2 = H$, $R^3 = (CH_2)_4CH_3$ and $R^1 = H$, $R^2 = CH_3$, $R^3 = C_6H_5$, respectively] furnished 3-isopropylindole, 3-pentylindole and 2-methyl-3-phenylindole [**2**; $R^1 = R^2 = H$, $R^3 = CH(CH_3)_2$; $R^1 = R^2 = H$, $R^3 = (CH_2)_4CH_3$ and $R^1 = H$, $R^2 = CH_3$, $R^3 = C_6H_5$, respectively] when heated with zinc chloride. In the last indolization, which could also be effected by warming the hydrazone with alcoholic hydrochloric acid, the product was distinguished from the possible isomeric 2-benzylindole (**2**; $R^1 = R^3 = H$, $R^2 = CH_2C_6H_5$) since it did not give a positive 'pine-chip' test, indicating that both the indolic 2- and 3-positions were substituted. Dibenzylketone phenylhydrazone (**1**; $R^1 = H$, $R^2 = CH_2C_6H_5$, $R^3 = C_6H_5$) was also indolized using alcoholic hydrochloric acid to yield 2-benzyl-3-phenylindole (**2**; $R^1 = H$, $R^2 = CH_2C_6H_5$, $R^3 = C_6H_5$) (Trenkler, 1888).

Reaction of equimolar quantities of **17** ($R^1 = H$, $R^2 = OH$, $R^3 = C_6H_5$) and phenylhydrazine in the presence of sodium acetate afforded the hydrazone **18** which was indolized by heating with zinc chloride to produce a compound formulated as **19** (Laubmann, 1888b). This useful synthesis of the 3-hydroxy-

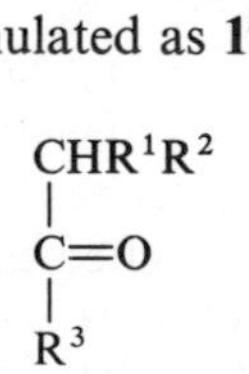

17 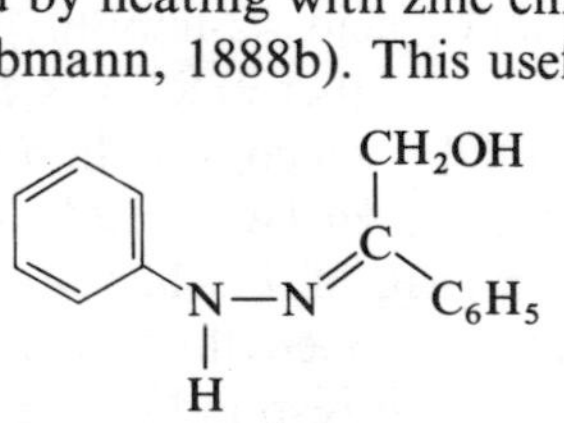 **18** 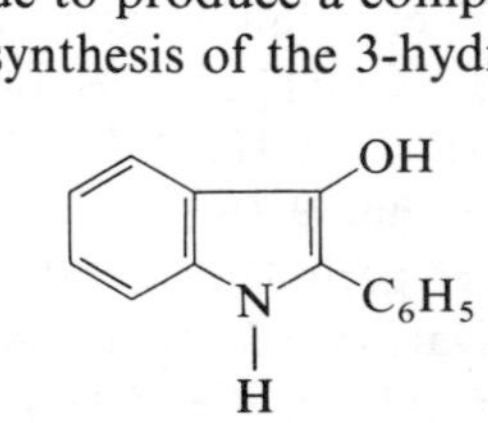 **19**

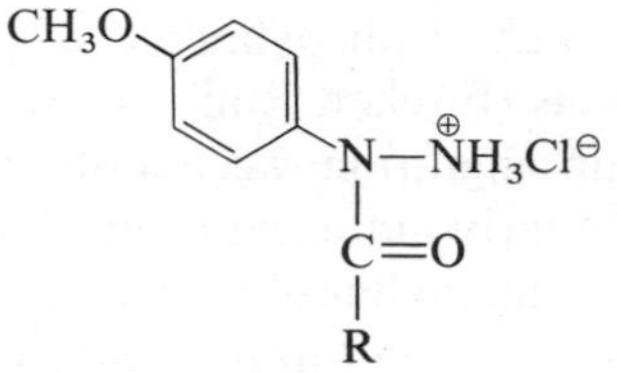

CH_3O N—$\overset{\oplus}{N}H_3Cl^{\ominus}$ C=O R

20

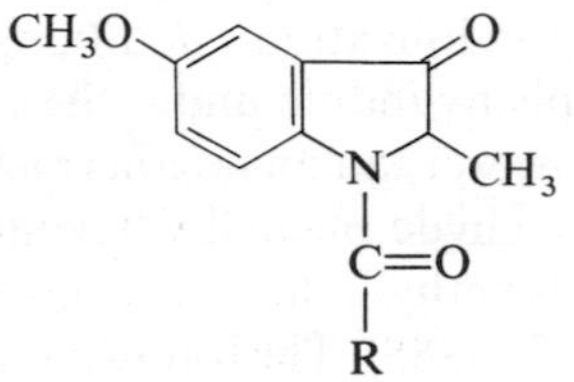

CH_3O O N CH_3 C=O R

21

CH=N—N(R^1)(C_6H_5)
C=N—N(R^1)(C_6H_5)
R^2

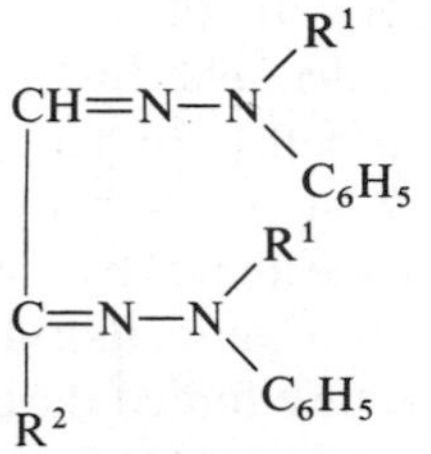

22

indole nucleus was used many decades later when **20** (R = $CH{=}CHC_6H_5$) was heated with acetol (**17**; R^1 = H, R^2 = OH, R^3 = CH_3) in dioxane to yield **21** (R = $CH{=}CHC_6H_5$) (Yamamoto, H., Nakamura, Y., Atami, Nakao and Kobayashi, T., 1970d) and when **20** (R = C_6H_5) and **17** (R^1 = H, R^2 = OH, R^3 = CH_3) under similar conditions afforded **21** (R = C_6H_5) (Yamamoto, H., Nakamura, Y., Atami, Nakao and Kobayashi, T., 1970c). (For other related uses of acetol in the Fischer indole synthesis, see also Yamamoto, H., Nakamura, Y. and Kobayashi, T., 1970.) It is interesting to note that reaction between **18** and phenylhydrazine furnished the osazone **22** (R^1 = H, R^2 = C_6H_5) (Laubmann, 1888b), two other of Fischer's co-workers also similarly finding that the osazones **22** (R^1 = CH_3, R^2 = C_6H_5) (Culmann, 1888) and **22** (R^1 = H, R^2 = COOH) (Nastvogel, 1888) were formed when bromoacetophenone (**17**; R^1 = H, R^2 = Br, R^3 = C_6H_5) reacted with a 3 × molar excess of N_α-methylphenylhydrazine and when dibromopyruvic acid (**17**; R^1 = R^2 = Br, R^3 = COOH) reacted with a 2 × molar excess of phenylhydrazine, respectively, and Fischer himself (Fischer, Emil, 1893) later finding that **17** (R^1 = NH_2, R^2 = R^3 = H) reacted with phenylhydrazone to form the osazone **22** (R^1 = R^2 = H), an observation extended by other of his colleagues in the University of Berlin (Gabriel and Pinkus, 1893) (see also Chapter VII, Q).

Fischer and another of his students, Theodor Schmitt (Schmidt?), found that whereas indolization of phenylacetaldehyde phenylhydrazone (**1**; R^1 = R^2 = H, R^3 = C_6H_5) with alcoholic hydrochloric acid afforded the expected 3-phenylindole (**2**; R^1 = R^2 = H, R^3 = C_6H_5) (Fischer, Emil and Schmidt, T., 1888), indolization of this hydrazone by heating with zinc chloride gave 2-phenylindole (**2**; R^1 = R^3 = H, R^2 = C_6H_5) (Fischer, Emil and Schmitt, T., 1888). This latter reaction involved the migration of the phenyl substituent under the indolization conditions and, in fact, 3-phenylindole was found to afford 2-phenylindole under these conditions (Fischer, Emil and Schmidt, T., 1888). However, an analogous methyl group migration was not observed when propionaldehyde phenylhydrazone was similarly indolized using zinc chloride to yield 3-methylindole with no trace of 2-methylindole (Fischer, Emil and Schmidt, T., 1888). The following year, this type of indolization with subsequent phenyl group migration was further investigated by another of Fischer's students in Würzburg (Ince, 1889). Phenylacetaldehyde N_α-methylphenylhydrazone was indolized with alcoholic hydrochloric acid, which does not effect the migration, to afford 1-methyl-3-phenylindole and this, upon heating with zinc chloride, produced 1-methyl-2-phenylindole which had previously been synthesized (Degen, 1886) (see above) by the indolization of acetophenone N_α-methylphenylhydrazone with a zinc chloride catalyst. Similarly, phenylacetaldehyde 2-naphthylhydrazone afforded the corresponding 3-phenylindole, upon treatment with alcoholic hydrochloric acid, which was isomerized, by heating with zinc chloride, into the 2-phenyl isomer which was also synthesized by the indolization of acetophenone 2-naphthylhydrazone (Ince, 1889) with a zinc chloride catalyst. As will be seen later, this type of migration and others subsequently found to be related to it, were to be the subjects of much further investigation in the decades to come (see Chapter V, A).

Also in 1889 another of Fischer's students published the results of indolizations and related studies using 1,3-diketones. Whereas benzoylacetone (**23**; R = C_6H_5, n = 1) reacted with phenylhydrazine to form a pyrazole derivative, the use of N_α-methylphenylhydrazine blocked pyrazole formation and a mono-N_α-methylphenylhydrazone was isolated. This was shown to have structure **24** (R = C_6H_5, X = O, n = 1) by indolization with heated zinc chloride to afford **2** ($R^1 = CH_3$, $R^2 = C_6H_5$, $R^3 = COCH_3$), the structure of which was established by hydrolytic deacetylation, using fuming hydrochloric acid, to the previously (Degen, 1886 – see above) synthesized 1-methyl-2-phenylindole (Kohlrausch, 1889). The alternative hydrazone formation would have ultimately resulted in hydrolytic debenzoylation yielding 1,2-dimethylindole in this reaction sequence. Hexan-2,5-dione (**23**; R = CH_3, n = 2) formed a bis-N_α-methylphenylhydrazone **24** (R = CH_3, X = $NN(CH_3)C_6H_5$, n = 2) which in acetic acid gave 2,5-dimethyl-1-(methylphenylamino)pyrrole (Kohlrausch, 1889).

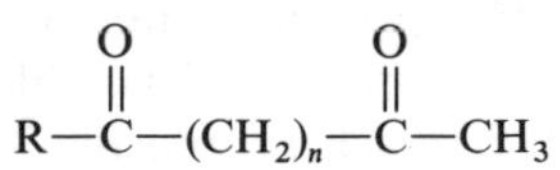

$$R{-}C(=O){-}(CH_2)_n{-}C(=O){-}CH_3$$

23

$$R{-}C(=N{-}N(CH_3){-}C_6H_5){-}(CH_2)_n{-}C(=X){-}CH_3$$

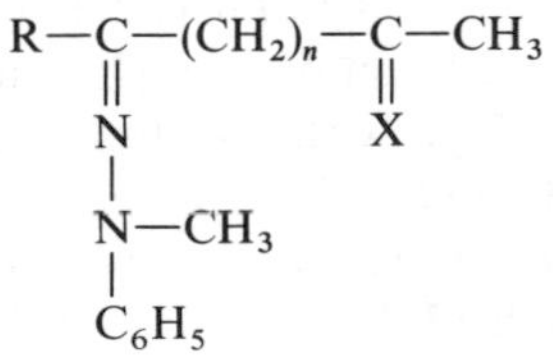

24

Many of the above studies by Fischer and his research students upon the development of his indole synthesis involved ambiguities with respect to the direction of indolization which were clearly recognized by these pioneers in the field. However, at the time, in all but one such reaction definite conclusions could not be reached concerning the preferred direction of indolization and thereby the structure of the products from such reactions. Indeed, as will be seen later, although some of these problems were solved after further experimentation over the following four or five decades, it is only comparatively recently that firm conclusions have been reached concerning others of them.

In 1887 Fischer published a book of fifty-eight preparative organic chemical exercises which were originally intended for the use of students at the University of Würzburg but soon acquired a much wider circulation. Two years later, an English translation of this work by Archibald Kling appeared. Included in it are the syntheses of phenylhydrazine by the reduction of benzenediazonium chloride with sodium sulphite (Fischer, Emil, 1889a) and of 2-methylindole by heating acetone phenylhydrazone with an excess of zinc chloride (Fischer, Emil, 1889b).

B. Other Early Applications of the Fischer Indole Synthesis; a Comparison with other Indole Syntheses; Reviews of the Reaction and Early Studies Related to its Catalysis

In 1889, the first two recognized syntheses of indoles from arylhydrazones in laboratories other than Fischer's were reported. One of these, published from the University of Berlin, involved the formation of **25** by heating 1-indanone phenylhydrazone in concentrated hydrochloric acid (Hausmann,

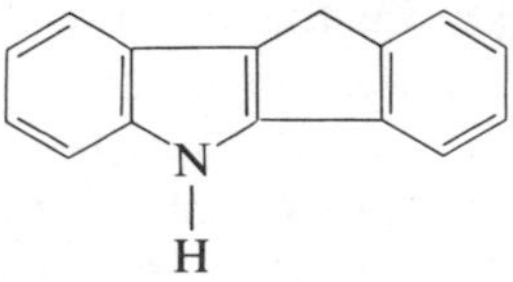

N
H

25

1889). The other was achieved by Baeyer and a co-worker at the University of Munich who reported the indolization of 3-carboxycyclohexanone phenylhydrazone (**26**; R = COOH) using warm mineral acid. The reaction yielded

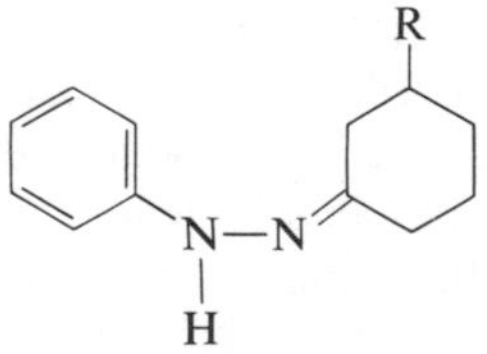

R
N—N
H

26

27

a product formulated as 4-carboxy-1,2,3,4-tetrahydrocarbazole (**27**; R = COOH) although the orientation of the substituent in the product was assigned without evidence to distinguish it from the possible correspondingly 2-substituted isomer (Baeyer and Tutein, 1889). Baeyer also later reported (Baeyer, 1894) the indolization of cyclohexanone phenylhydrazone (**26**; R = H) to afford 1,2,3,4-tetrahydrocarbazole (**27**; R = H) under similar catalytic conditions. The identification of this product, by means of a m.p. (120 °C) comparison, with that obtained (Graebe and Glaser, 1872) by reduction of carbazole with phosphorus and hydriodic acid to a basic hexahydro derivative, formation of the hydrochloride and subsequent pyrolytic dehydrogenation to a non-basic tetrahydrocarbazole, established the structure of the final product from this reaction sequence. Bayer definitely identified the structure of his above two reaction products (excepting the orientation of the COOH group in the former product) and recognized that he had effected Fischer indolizations. However, one year prior to the publication of the first of Baeyer's above studies, cyclohexanone (obtained by reduction of phenol) phenylhydrazone had been reacted by heating with dilute sulphuric acid (Drechsel, 1888). Although in this work it was recognized that the elements of ammonia had been eliminated from the starting hydrazone to form the reaction product, $C_{12}H_{13}N$, neither reference was made to nor analogy was drawn with Fischer's indolization studies which were published by that time. Furthermore, since the product obtained was reported to have a m.p. = 108 °C it was thought to be isomeric with the product, m.p. = 120 °C, obtained by Graebe and Glaser as already described, and the name 'hydrophenanilid' and structure **28** were proposed for it. It was

$N—C_6H_5$

28

only later (Baeyer and Tutein, 1889; Baeyer, 1894) that this product, in spite of its low m.p. of 108 °C, was recognized as 1,2,3,4-tetrahydrocarbazole and Drechsel's reaction was recognized as a Fischer indolization.

The first example of a Fischer indolization from a British laboratory was eported in 1891 (Dufton, 1891) when pyruvic acid 8-quinolylhydrazone was nverted into **29** (R^1 = COOH, R^2 = R^3 = H) by refluxing in concentrated rochloric acid solution. This work was effected in the Chemical Laboratories

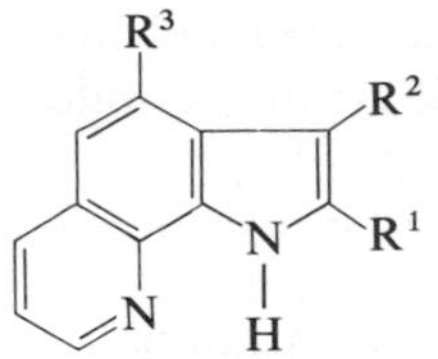

29

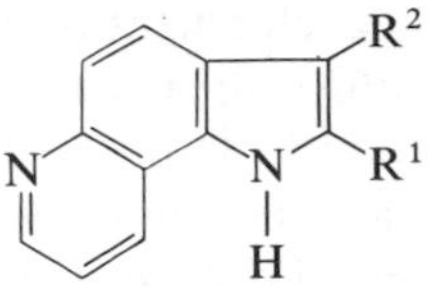

30

of the University of Cambridge and in the following year the results of a related indolization also carried out at Cambridge University, in which pyruvic acid 5-quinolylhydrazone was converted, by refluxing in concentrated hydrochloric acid, into **30** (R^1 = COOH, R^2 = H) were also reported by the same author (Dufton, 1892) after he had moved from the University of Cambridge to become the Head Science Master at Bradford Grammar School, England. However, in both of these papers no reference whatsoever was made to Fischer's previous extensive pioneering work in this area although it is difficult to believe that Dufton effected his reactions and so clearly formulated his products without a knowledge of Fischer's studies upon the indolization reaction.

By 1891 the Fischer indole synthesis had crossed the Atlantic Ocean. A paper published from Clark University, Worcester, Mass., USA referred to Fischer's earlier work and reported the indolization of ethyl acetoacetate phenylhydrazone by heating with concentrated sulphuric acid (Nef, 1891). Interestingly, this paper was published in the German literature (*Liebig's Annalen der Chemie*). Subsequently, in work also carried out in Worcester (Walker, C., 1892), the structure of the product obtained from this reaction was shown to be **2** (R^1 = H, R^2 = CH_3, R^3 = $COOC_2H_5$) rather than **2** (R^1 = H, R^2 = $CH_2COOC_2H_5$, R^3 = H) which would have resulted from the alternative direction of indolization.

A paper, the last associated with Fischer and his indole synthesis and somewhat isolated in time from his main studies in the area, appeared from the University of Würzburg as late as 1893. In this study, another of Fischer's pupils, Rudolf Brunck, reported further examples of the synthesis, involving the zinc chloride catalysed indolization of 2-acetylthiophen phenylhydrazone into **31** and what he assumed to be 1-acetylnaphthalene phenylhydrazone into **32** (Brunck, 1893). Later (Blades and Wilds, 1956) this latter product was reformulated correctly as **33**, Brunck's latter starting material having, in fact, been 2-acetylnaphthalene phenylhydrazone.

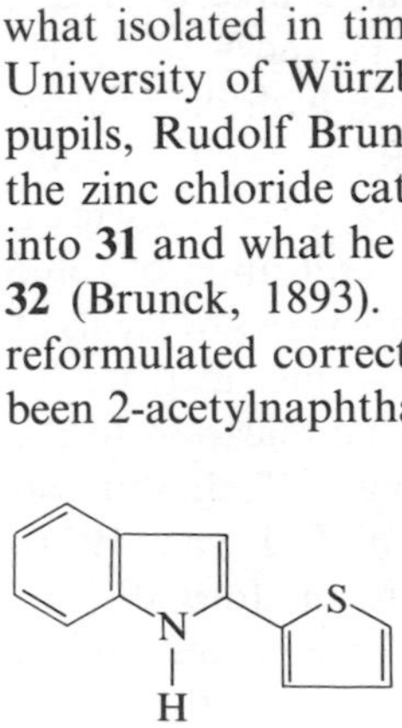

31

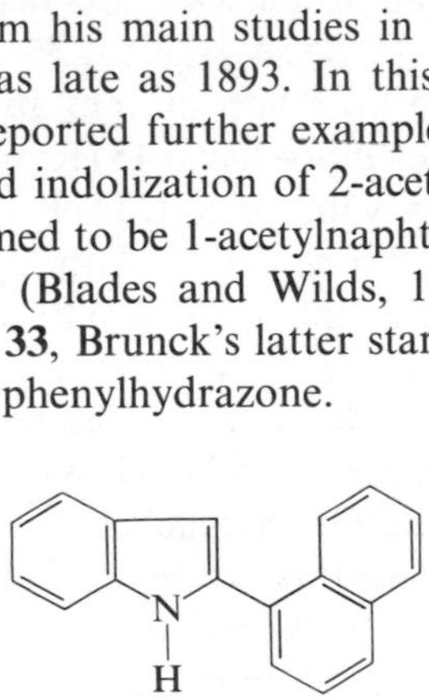

32

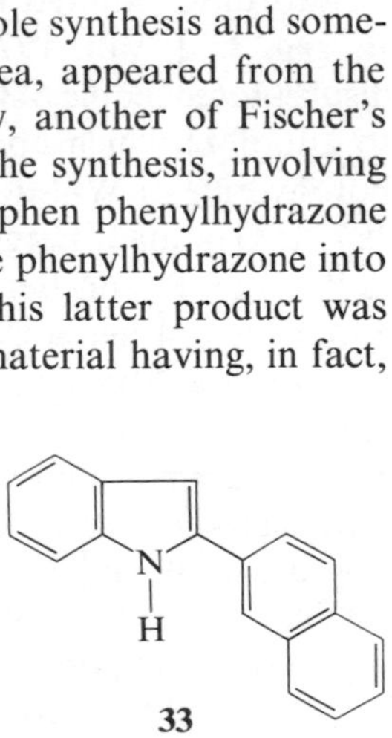

33

Fischer really spent only the years between 1883 and 1890 in the very active investigation and development of his indole synthesis. Indeed, in the accounts of his scientific studies already referred to, it is largely overshadowed, almost to the extent of obscurity, by his monumental investigations into the carbohydrates, purines and proteins. Nevertheless, the Fischer indole synthesis has since become by far, of over thirty available (Bard and Bunnett, 1980), the most frequently employed and versatile synthesis of indolic compounds although occasionally one of the other methods of synthesis has been preferred. Thus, for the preparation of 5-methoxy-6-methylindole-2-carboxylic acid, the Reissert synthesis was the method of choice (Allen, G. R., Poletto and Weiss, M. J., 1965) as it was also for the synthesis of 4-, 5-, 6- and 7-benzylmercaptoindoles since the preparation of mercaptophenylhydrazones was quite unsatisfactory (Piers, Haarsted, Cushley and Brown, R. K., 1962), the Madelung synthesis was preferred in another instance (Casnati, Langella, Piozzi, Ricca and Umani-Ronchi, 1964a, 1964b) and the oxidation of substituted 1,2,3,4-tetrahydroquinolin-3-ols with periodate (e.g. Merchant and Salgar, 1963; Pennington, Jellinek and Thurn, 1959; Pennington, Martin, L. J., Reid, R. E. and Lapp, 1959; Pennington, Tritle, Boyd, S. D., Bowersox and Aniline, 1965) may, in limited specific cases, be competitive in ease of preparation and yields of indole products. A recently introduced indole synthesis is outlined in Scheme 1 (Gassman and Bergen, T. J. van, 1973b; Gassman, Bergen, T. J. van, Gilbert, D. P. and Cue, 1974) and involves the sequential reaction of an aniline **34** (R^2 = H or alkyl) with *tert.*-butyl hypochlorite to afford **35**, with the 2-methylthio ketone **36** (R^3 = H) to form **37** (R^3 = H), with triethylamine to afford, via the ylide **38**, **39**, **40** and **41** and with Raney nickel to give **43** (R^3 = H) (Gassman, 1975a, 1976; Gassman and Bergen, T. J. van, 1973a; Gassman, Bergen, T. J. van, Gilbert, D. P. and Cue, 1974). Analogous use of **36** (R^4 = H) yielded unsubstituted indoles **43** (R^4 = H) (Gassman, 1975a; Gassman and Bergen, T. J. van, 1973b). This route was extended to the synthesis of 2,3-disubstituted indoles **43** (R^2 = H, $R^3 \neq$ H) using **35** (R^2 = H) and **36** ($R^3 \neq$ H) to afford **42** which upon reduction with lithium aluminium hydride, sodium borohydride or W-2 Raney nickel produced **43** (R^2 = H, $R^3 \neq$ H) (Gassman, 1975a; Gassman, Bergen, T. J. van, Gilbert, D. P. and Cue, 1974; Gassman, Gilbert, D. P. and Bergen, T. J. van, 1974). Although the utilization of some anilines (e.g. 4-methoxyaniline) in this synthesis afforded only very low yields of the corresponding indoles because of the instability of the intermediate N-chloroanilines **35**, this problem was later overcome by a modification to the route (Gassman and Gruetzmacher, 1976; Gassman, Bergen, T. J. van and Gruetzmacher, 1973; Gassman, Gruetzmacher and Bergen, T. J. van, 1974). The use of **36** (R^4 = OC_2H_5) in this synthetic route led to the formation of 3-methylthiooxindoles which with Raney nickel furnished the corresponding oxindoles and upon treatment with lithium aluminium hydride give the corresponding 3-substituted 2-unsubstituted indoles **43** (R^4 = H) (Gassman, 1975b;

34 **35**

$CH_3S{-}CH(R^3){-}C(=O){-}R^4$ **36**

38 **37**

39 **40**

(R^2 = H, R^3 ≠ H)

$-H_2O$ (R^3 = H)

42 **41**

43 (R^2 = H) **43** (R^3 = H)

Scheme 1

Gassman and Bergen, T. J. van, 1973c, 1974). The developers of this indole synthesis claimed (Gassman and Bergen, T. J. van, 1973a; Gassman, Bergen, T. J. van, Gilbert, D. P. and Cue, 1974; Gassman, Gilbert, D. P. and Bergen, T. J. van, 1974) that it possessed several advantages when compared with the Fischer indole synthesis. Two of these claims, involving superior or comparable product yields and the unambiguous synthesis of 2,3-disubstituted indoles from unsymmetrical ketones, appear to be invalid since many Fischer indolizations afford very high product yields and although Fischer indolizations using arylhydrazones of unsymmetrical ketones can often potentially, and indeed often do, form two isomeric indolic products (see Chapter III, G–L), this recent synthesis would also suffer from a similar analogous ambiguity in the 2-halogenation of such ketones, the product(s) from which are necessary for the synthesis of the required **36** by reaction with methyl mercaptide. The necessity for 2-halogeno ketones in the long established Bischler indole synthesis also limits the utility of this reaction. That this recent synthesis is applicable to the preparation of indole derivatives with functionalities sensitive to acidic, strongly alkaline or elevated thermal conditions, whereas such compounds are difficult to prepare by the Fischer indolization, would certainly be so, although it would be limited to the use of ketones whose functionalities were unaffected under conditions of 2-keto halogenation and which would not undergo concomitant halogenation at other sites. Furthermore, the preparation of indoles containing functional groups which would be reactive under the reductive conditions required to remove the methylthio groups would not be possible. Finally, although this synthesis utilizes anilines whereas the Fischer indolization requires that such anilines be converted to the corresponding arylhydrazones and thus one further reaction step, the Gassman synthesis requires that the ketone or aldehyde be converted into the corresponding **36** which thus involves two further reaction steps relative to the Fischer indolization.

The Fischer indolization has involved the use of an extremely wide range of aldehydes, ketones and arylhydrazones. However, the claim that a mixture of 2-carbethoxyacetanilide and phenylhydrazine, when refluxed in aqueous or alcoholic formic acid, undergoes a Fischer indolization to afford what is presumably 2-(2-carbethoxyphenylamino)indole (Saleha, Khan, N. H., Siddiqui and Kidwai, M. M., 1978) is difficult to understand although related syntheses of other 2-aminoindoles from amidrazones (amide 'phenylhydrazones') have also been claimed previously (Coxworth 1965). However, it is unfortunate, though perhaps significant, that experimental details relating to these earlier studies have not been published.

The synthesis has found considerable application in the preparation of pharmaceuticals and other compounds of industrial and commercial significance, indole alkaloids [e.g. it is used as the initial step in a synthesis of strychnine (Woodward, Cava, Ollis, Hunger, Daeniker and Schenker, 1954, 1963) and as the final step in synthesis of (±)-ibogamine (Ikezaki, Wakamatsu and

Ban, 1969; Sallay, 1967a)] and 5-hydroxytryptophan, 5-hydroxytryptamine and indole-3-acetic acid, together with many of their analogues and homologues. The elucidation of the mechanism of the reaction has involved many groups in an extremely interesting series of investigations extending over seven decades. The principles of the reaction have also been applied in the development of useful syntheses of 3*H*-indoles, pyrroles, oxindoles, benzofurans and 2-aminobenzothiazoles and the use of suitable dehydrogenations subsequent to the indolization of cyclic ketone arylhydrazones has furnished polycyclic aromatic compounds containing one or more indole nuclei. Investigations involving the direction of indolization where possible ambiguity exists (e.g. when starting with 3-substituted phenylhydrazines or appropriate unsymmetrical ketones), of rearrangements which occur concomitantly with indolization or involving indolizations of 2,6-disubstituted phenylhydrazones in which migration, elimination and/or substitution of these *ortho* substituents can occur have all provided fascinating scientific studies. The scope and usefulness of the reaction have been extended by the wide variety of Brønsted and Lewis acid catalysts that have been introduced for effecting it and during the last two decades the use of non-catalytic thermal indolization conditions has led to still further extensions. Recent studies (Ibrahim and Rippie, 1976) have investigated the effect of a lamellar lyotropic liquid crystal upon the indolizations of cyclohexanone and 2-methyl- and 2-propylcyclohexanone phenylhydrazone, reactions which were taken as models for biochemical transformations in proteinaceous liquid crystals. In view of the nature and properties of a liquid crystalline structure, it is not surprising that it was observed that the overall rate constants for the indolizations in a liquid crystal were reduced by comparison with the rate constants using an aqueous polyethylene glycol solution.

The Fischer indole synthesis and its extension to the synthesis of carbazoles developed by Borsche (see Chapter I,D) has received the attention of many previous reviewers. The first comprehensive reviews covered the literature up to the early 1920s (Hollins, 1924b, 1924c) and a few years earlier its specific application to the synthesis of 1,2,3,4-tetrahydrocarbazoles and carbazoles was thoroughly summarized (Cohn, 1919). In the 1940s three brief but useful reviews on the Fischer indole synthesis appeared (Meyer, H., 1940; Morton, 1946; Order and Lindwall, 1942) and the Borsche carbazole synthesis was reviewed (Campbell and Barclay, 1947). In the 1950s, two comprehensive treatises upon the chemistry of indoles and carbazoles included accounts of the Fischer (Julian, Meyer, E. W. and Printy, 1952; Sumpter and Miller, F. M., 1954) and Borsche (Freudenberg, 1952; Sumpter and Miller, F. M., 1954) syntheses. Later in the same decade, two very comprehensive reviews of the Fischer synthesis (Kitaev, 1959; Suvorov, Mamaev and Rodinov, 1959) and one concerning its mechanism as then understood (Arbuzov, A. E. and Kitaev, 1957d) appeared in the Russian literature and in 1963 the synthesis was subjected to similar comprehensive attention in the American literature (Robinson, B.,

1963a), this article being updated six years later (Robinson, B., 1969). The following year, significant coverage of the reaction was also included in a book published upon the chemistry of indoles (Sundberg, 1970). In a treatise which appeared in 1972 upon the synthesis of the indole nucleus (Brown, R. K., 1972a), the Fischer synthesis occupied a prominent position and some specific applications and extensions of the synthesis also formed part of a review of methods used for the cleavage of N—N bonds in organic compounds (Marshalkin and Yakhontov, 1973). Some other recent reviews (Grandberg, 1972; Grandberg and Sorokin, 1974; Kitaev and Troepol'skaya, 1978, 1979; Shine, 1967b) have tended to concentrate, although not exclusively so, upon the mechanistic aspects of the synthesis and another (Livingstone, 1973), though brief, included some interesting salient features.

Fischer obviously recognized the catalytic nature of the mineral acid or zinc chloride which he used to carry out his indole synthesis but he still employed these reagents in very large excesses. In a subsequent patent (Gesellschaft für Teerverwertung m.b.H., 1911) it was claimed that when the zinc chloride catalysed indolizations of the arylhydrazones of both aldehydes and ketones were effected in 'solvent naphtha' or 'neutral tar oils' with the reaction temperature kept as low as possible, the reactions proceeded more smoothly, yields were increased, resin formation was minimized and larger quantities of reactants could be used relative to the corresponding indolizations performed without a solvent. In 1910, A. E. Arbuzov, who was to become in later years one of the foremost figures in Russian organic chemistry (Pudovik, 1957; A. E. Arbuzov – birthday centenary, 1977), extended an earlier investigation reported by another investigator (Struthers, 1905) who had observed that thermal decomposition of phenylhydrazine in the presence of a mercury(II) cyanide or a copper(I) cyanide catalyst furnished aniline, ammonia, nitrogen and benzene. Changing to a copper(I) iodide catalyst in place of copper(I) cyanide in a similar reaction, it was found (Arbuzov, A. E. and Tikhvinskii, 1913a; Arbusow, A. E. and Tichwinsky, 1910a) that only the first three of these products were formed. The decomposition was also extended to the use of copper(I) bromide and copper(I) chloride catalysts (Arbuzov, A. E. and Tikhvinskii, 1913a; Arbusow, A. E. and Tichwinsky, 1910a), it being found (Arbuzov, A. E. and Tikhvinskii, 1913a) that copper(I) chloride was most active and copper(I) iodide was least active in catalysing the reaction. The investigation was also further extended (Arbusow, A. E., 1910; Arbuzov, A. E., 1913) to an investigation of the catalytic decomposition of several aldehyde phenylhydrazones under similar conditions when it was found that isovaleraldehyde, isobutyraldehyde and enanthole(?) phenylhydrazone, upon heating in the presence of a trace of copper(I) chloride, afforded isovaleronitrile, isobutyronitrile and enantho(?)nitrile, respectively, aniline also being identified as a product in the first transformation. This reaction was stated to be a valuable method of preparing aliphatic nitriles with at least four carbon atoms in their alkyl substituent (Arbuzov, A. E., 1913). Subsequently (Arbusow, A. E. and Tichwinsky, 1910b; Arbusov, A. E. and Tikhvinskii, 1913b) it was found that when ethyl methyl ketone phenylhydrazone (50 g.), propionaldehyde phenylhydrazone (50 g.) or propionaldehyde 4-methylphenylhydrazone (43 g.) were heated with copper(I) chloride (0.1 g.), evolution of ammonia occurred and 2,3-dimethylindole (*ca.* 60% yield), 3-methylindole (60% yield) and 3,5-dimethylindole, respectively, were isolated along with aniline and 4-methylaniline, respectively, from the second and third reactions. The corresponding alkyl nitriles were also formed in these reactions (Arbuzov, A. E., 1913). Similar small amounts of platinum(II) chloride or zinc chloride could also be substituted in place of copper(I) chloride in the second reaction. Contemporarily with this last report, Arbuzov and two more of his students reported that similar copper(I) chloride catalysed indolizations of methyl propyl ketone and dipropyl ketone phenylhydrazone produced, along with aniline in each case, 2-propylindole (Arbuzov, A. E. and Friauf, 1913) and 3-ethyl-2-propylindole (Arbuzov, A. E. and Wagner, 1913) respectively, although other workers (Braun and Bayer, 1925) preferred to use zinc chloride in alcohol rather than copper(I) chloride to catalyse the conversion of butyraldehyde phenylhydrazone to 3-ethylindole. Some twenty years later, along with other of his students, Arbuzov further used his technique to convert butyl methyl ketone and propiophenone phenylhydrazone into 2-methyl-3-propylindole and 3-methyl-2-phenylindole, respectively, and ethyl propylketone phenylhydrazone into 2,3-diethyl-and/or 3-methyl-2-propylindole (Arbusow, A. E., Saizew and Rasumow, 1935; Arbuzov, A. E. and Zaitzev, 1934) and, using similar trace quantities of zinc chloride in place of copper(I) chloride, converted diethyl ketone phenylhydrazone into 2-ethyl-3-methylindole (Arbuzov, A. E. and Rotermel, 1932). Because of these observations by Arbuzov and his co-workers that only trace amounts of the Lewis acids were required to achieve the indolization of arylhydrazones, thus experimentally recognizing the catalytic role of these acids in the reaction, the Fischer indole synthesis has occasionally (Janetzky and Verkade, 1945; Kitaev, 1971; Kitaev and Buzykin, 1972; Kitaev and Troepol'skaya, 1978; Verkade and Janetzky, 1943a, 1943b) been referred to as the Fischer–Arbuzov synthesis.

Another amalgamation of these two names was made since, eight decades after Fischer had prepared N_α-phenylphenylhydrazine by the reduction of N-nitrosodiphenylamine with zinc dust and glacial acetic acid (Fischer, Emil, 1876b), a route he had also used for the synthesis of other N_α-substituted phenylhydrazines, Arbuzov and one of his students found that higher yields from this particular reduction resulted if the zinc dust was added in portions during the course of the reaction rather than all being placed in the flask beforehand (Arbuzov, A. E. and Valitova, 1957). This slight modification led to this reaction being referred to in one instance (Bloss and Timberlake, 1963) as the Fischer–Arbuzov reduction. This reduction could also be accomplished

(Poierier and Benington, 1952) in 90% yield using lithium aluminium hydride (for other examples, see also Cross, A. D., King, F. E. and King, T. J., 1961; Heidt, Gombos and Tudos, 1966). However, whereas such a reduction using equimolar quantities of N-methyl-N-nitrosoaniline and lithium aluminium hydride afforded N_α-methylphenylhydrazine (77%) (Hanna and Schueler, 1952), the similar treatment of N-nitrosodiphenylamine using a 4 × molar excess of the reducing agent gave only diphenylamine (74%) (Schueler and Hanna, 1951).

More recently (Grandberg, Sibiryakova and Brovkin, 1969), the studies of a British and another Russian investigator in connection with the Fischer indole synthesis have led to it being referred to as the Robinson–Suvorov reaction.

A few years after Fischer's early developmental studies on the indole synthesis, its utility was extended by its application to the synthesis of 3*H*-indoles, 2-alkylidene indolines, indolin-2-ols and carbazoles and by the use of phenols in place of ketones.

C. Synthesis of 3*H*-Indoles, 2-Alkylidene Indolines and Indolin-2-ols

The treatment under indolization conditions of the arylhydrazones of ketones containing a tertiary C-2 atom, **44** (R^5 and $R^6 \neq$ H) affords the corresponding 3*H*-indoles, **45** (when R^3 = H in **44**). Furthermore, when starting with **44** (R^3, R^5 and $R^6 \neq$ H), either 2-alkylidene indolines or indolin-2-ols are formed, depending upon whether R^4 contains at least one hydrogen atom or no hydrogen atom (or when R^4 = H) on its carbon atom which is adjacent to the hydrazine group, respectively.

The first investigation of such an indolization involved the warming of **44** ($R^1 - R^3$ = H, R^4 = COOH, $R^5 = R^6 = CH_3$) with 15% alcoholic sulphuric acid. The product from this reaction was initially (Brunner, K., 1894) thought to be 3-methylindole but this was later (Brunner, K., 1895a) corrected to 2,3-dimethylindole. The formation of this product would involve decarboxylation followed by the now well established Plancher rearrangement of the immediate reaction product **45** (R^1–R^3 = H, $R^4 = R^5 = CH_3$) under the acidic indolization conditions (see Chapter V, A). Similarly, **44** (R^1–R^4 = H, $R^5 = R^6 = CH_3$) was indolized by heating with powdered zinc chloride to

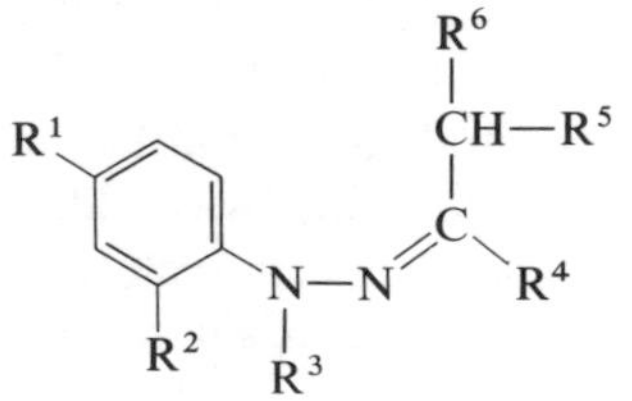

44

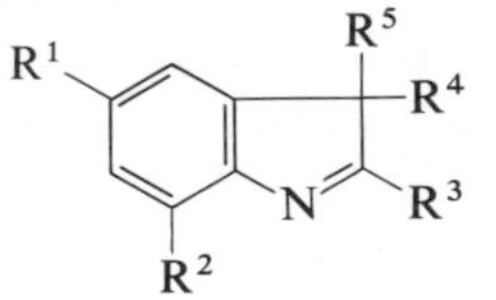

45

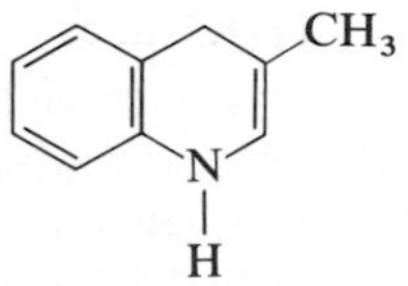

CH_3

N

H

46

again afford 2,3-dimethylindole. When this observation was reported (Brunner, K., 1895a), the reaction was suggested as proceeding via the 1,4-dihydro-3-methylquinoline (**46**), although a clear analogy was recognized between the reaction and the Fischer indole synthesis, with particular reference being made to the migration which occurred when phenylacetaldehyde phenylhydrazone was indolized by heating with zinc chloride to 2-phenylindole (Fischer, Emil and Schmidt, 1888; Fischer, Emil and Schmitt, 1888) (see earlier and Chapter V, A).

Subsequently (Robinson, R. and Suginome, 1932a), Brunner's initial reaction was repeated and once again 2,3-dimethylindole was isolated, along with ethyl 3,3-dimethyl-3*H*-indole-2-carboxylate (**45**; $R^1 = R^2 = H$, $R^3 = COOC_2H_5$, $R^4 = R^5 = CH_3$) and the corresponding phenylhydrazide **45** ($R^1 = R^2 = H$, $R^3 = CONHNHC_6H_5$, $R^4 = R^5 = CH_3$). When absolute ethanol saturated with dry hydrogen chloride was used as catalyst in place of the alcoholic sulphuric acid, the rearrangement of the 3*H*-indole into the indole was prevented and the products were 3,3-dimethyl-3*H*-indole-2-carboxylic acid (**45**; $R^1 = R^2 = H$, $R^3 = COOH$, $R^4 = R^5 = CH_3$), which upon pyrolysis at 135–140 °C gave 3,3-dimethyl-3*H*-indole (**45**; $R^1 = R^2 = R^3 = H$, $R^4 = R^5 = CH_3$), also isolated directly from the indolization, and ethyl 3,3-dimethyl-3*H*-indole-2-carboxylate. Extension of these observations led to the treatment of **44** ($R^1 = OC_2H_5$, $R^2 = R^3 = H$, $R^4 = COOH$, $R^5 = R^6 = CH_3$) with dry ethanolic hydrogen chloride which afforded 5-ethoxy-3,3-dimethyl-3*H*-indole-2-carboxylic acid (**45**; $R^1 = OC_2H_5$, $R^2 = H$, $R^3 = COOH$, $R^4 = R^5 = CH_3$) and 5-ethoxy-3,3-dimethyl-3*H*-indole (**45**, $R^1 = OC_2H_5$, $R^2 = R^3 = H$, $R^4 = R^5 = CH_3$), the latter product also being formed by heating the former product above its m.p. (Robinson, R. and Suginome, 1932a).

Returning to the earlier studies (Brunner, K., 1895b), the hydrazone **44** (R^1–$R^4 = H$, $R^5 = R^6 = CH_3$) was treated with zinc chloride in absolute ethanol (milder conditions than those without a solvent) and a product was isolated in the form of its dimer complex with zinc chloride, $(C_{10}H_{11}N)_2$ $ZnCl_2 + \frac{1}{2}C_2H_5OH$. Attempts to liberate the monomer, $C_{10}H_{11}N$, from this complex by basification led to the formation of a trimer, $(C_{10}H_{11}N)_3$ (see Chapter V, B), although a picrate salt of the monomeric product was prepared. However, no structure was proposed (Brunner, K., 1895b) for any of these compounds. Subsequently (Brunner, K., 1896a), the monomeric product was correctly formulated as **45** (R^1–$R^3 = H$, $R^4 = R^5 = CH_3$) and the synthesis was extended by the treatment of **44** ($R^1 = R^2 = R^4 = H, R^3 = R^5 = R^6 = CH_3$) with alcoholic zinc chloride which yielded a product, after reaction 'work-up' by basification, which was formulated as **47** but which was later corrected to **48** ($R^1 = CH_3, R_2 = H$) (Brunner, K., 1898b). At the time of its first isolation, this latter substance had been oxidized by using ammoniacal silver nitrate to afford a compound, the structure of which was correctly assigned as 1,3,3-trimethyloxindole (**49**; $R^1 = H, R^2–R^4 = CH_3$) (Brunner, K., 1896a). Later (Brunner, K., 1896b), this structural proposal was verified by synthesis by heating isobutyric acid N_α-methylphenylhydrazide (**50**; $R^1 = H, R^2–R^4 = CH_3$) with lime, a reaction which was to be the first example of a general synthesis of oxindoles which would carry Brunner's name (see Chapter VI, A). Starting with **44** ($R^1 = R^2 = R^4 = H, R^3 = C_6H_5, R^5 = R^6 = CH_3$), treatment with alcoholic zinc chloride or alcoholic hydriodic acid ultimately gave the indolin-2-ol **48** ($R^1 = C_6H_5, R^2 = H$) (Brunner, K., 1900).

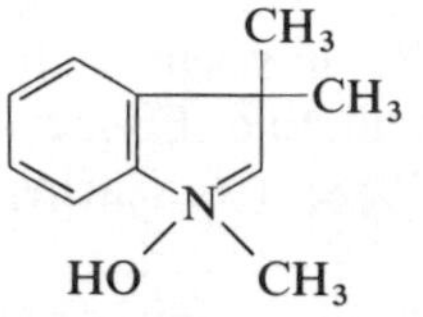

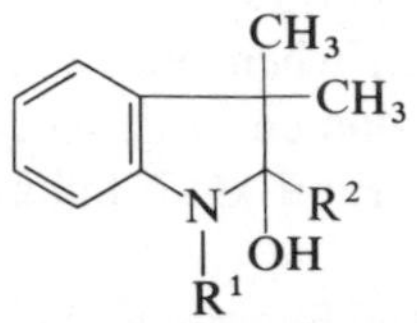

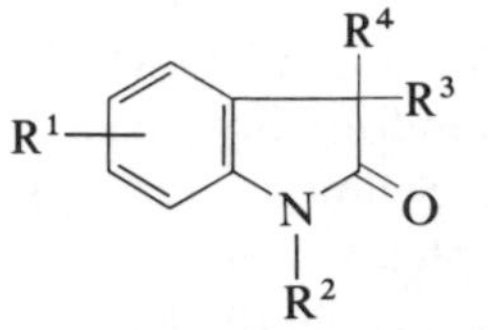

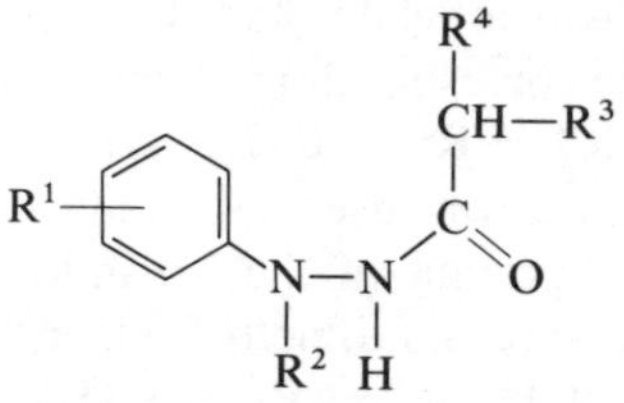

47 **48** **49**

50

Brunner continued to extend his investigations and after treatment of **44** ($R^1 = R^2 = H, R^3–R^6 = CH_3$) with zinc chloride in absolute alcohol, with subsequent basification of the reaction mixture, obtained a product which was formulated as **51** or **52** ($R^1 = R^3 = H, R^2 = R^4 = CH_3$) (Brunner, K., 1898a). Later (Brunner, K., 1898b, 1900), structure **52** ($R^1 = R^3 = H, R^2 = R^4 = CH_3$) was supported by the observation that **44** ($R^1 = R^2 = H$,

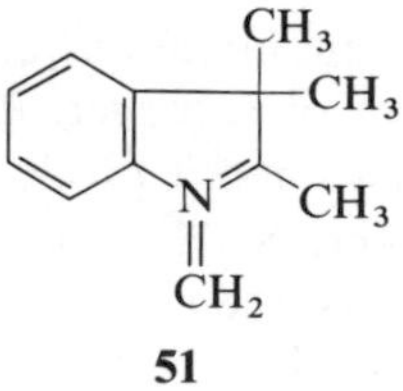

51 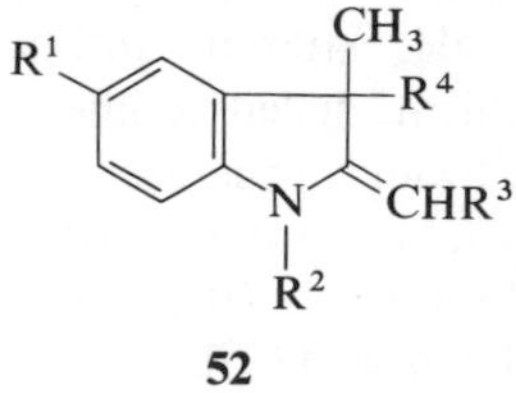 **52**

$R^3 = C_6H_5$, R^4–$R^6 = CH_3$) was cyclized by treatment with hydriodic acid in absolute ethanol to ultimately yield **52** ($R^1 = R^3 = H$, $R^2 = C_6H_5$, $R^4 = CH_3$). Over six decades later, a repeat of this indolization involving *in situ* hydrazone formation and using warm ethanolic dilute hydrochloric acid as the catalyst was the subject of a patent (National Cash Register Co., 1961), as were the related syntheses of **52** ($R^1 = R^3 = H$, $R^2 = CH_3$, C_2H_5 and C_6H_5, $R^4 = CH_3$) from the reaction of isopropyl methyl ketone with N_α-methyl-, N_α-ethyl- and N_α-phenylphenylhydrazine, respectively, in the presence of an acetic acid or 4-methylbenzene sulphonic acid catalyst in refluxing aqueous methanol or ethanol (Sumitomo, Hayakawa and Tsubojima, 1969) and the syntheses of **52** ($R^1 = H$, Cl and OCH_3, $R^2 = CH_3$, $R^3 = H$, $R^4 = CH_3$ and C_2H_5) from the corresponding hydrazones using a polyphosphate ester catalyst (Krutak, 1975).

Simultaneously with the publication of Brunner's above later papers, the results of a series of related studies by Plancher and his co-workers, involving similar extensions to Fischer's indole synthesis, began to appear. Using zinc chloride in absolute alcohol as catalyst, this group converted **44** (R^1–$R^3 = H$, R^4–$R^6 = CH_3$) into **45** ($R^1 = R^2 = H$, R^3–$R^5 = CH_3$) (Plancher, 1898a, 1898b, 1898c, 1898d, 1898g, 1901, 1905; Plancher and Bettinelli, 1899), via the isolation of an alkali labile complex $(C_{11}H_{13}N)_2ZnCl_2$ of the 3*H*-indole (Plancher, 1898b; Plancher and Bettinelli, 1899) [this reaction was later repeated under an atmosphere of nitrogen (Bajwa and Brown, R. K., 1968a) and also using refluxing glacial acetic acid as the catalyst (Hinman and Whipple, 1962)]. In one instance (Plancher and Bettinelli, 1899) a trace of 2,3-dimethylindole was also isolated from this reaction. Under similar conditions, **44** ($R^1 = R^2 = H$, R^3–$R^6 = CH_3$) afforded **52** ($R^1 = R^3 = H$, $R^2 = R^4 = CH_3$) (Plancher, 1898a, 1898b, 1898g), **44** (R^1–$R^3 = H$, $R^4 = R^5 = CH_3$, $R^6 = C_2H_5$) gave **45** ($R^1 = R^2 = H$, $R^3 = R^4 = CH_3$, $R^5 = C_2H_5$), via the alkali labile 3*H*-indole–zinc chloride complex $(C_{12}H_{15}N)_2\,ZnCl_2$ (Plancher, 1900a, 1902; Plancher and Bonavia, 1900), **44** ($R^1 = R^2 = H$, R^3–$R^5 = CH_3$, $R^6 = C_2H_5$) gave **52** ($R^1 = R^3 = H$, $R^2 = CH_3$, $R^4 = C_2H_5$) (Plancher, 1900a), **44** (R^1–$R^3 = H$, $R^4 = C_2H_5$, $R^5 = R^6 = CH_3$) yielded both **45** ($R^1 = R^2 = H$, $R^3 = C_2H_5$, $R^4 = R^5 = CH_3$) and 2-isopropyl-3-methylindole (Plancher, 1900a; Plancher and Bonavia, 1900, 1902), **44** [R^1–$R^3 = H$, $R^4 = CH(CH_3)_2$, $R^5 = R^6 = CH_3$] afforded **45** [$R^1 = R^2 = H$, $R^3 = CH(CH_3)_2$, $R^4 = R^5 = CH_3$] (Plancher, 1898a, 1898b, 1898g), **53** ($R^1 = R^3 = H$, $R^2 = CH_3$) led to both **54** ($R^1 = CH_3$, $R^2 = H$) and **55** ($R^1 = CH_3$,

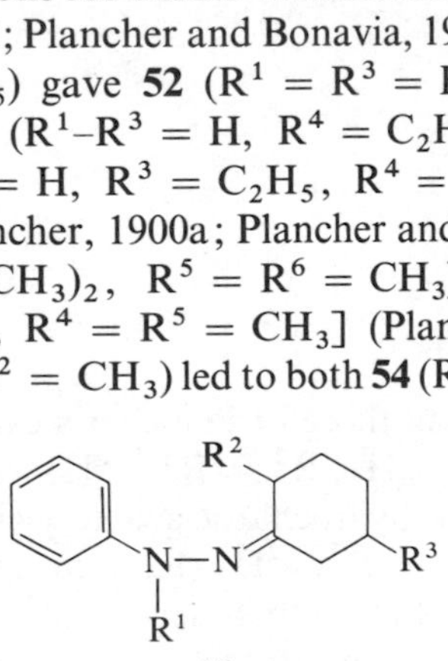

53

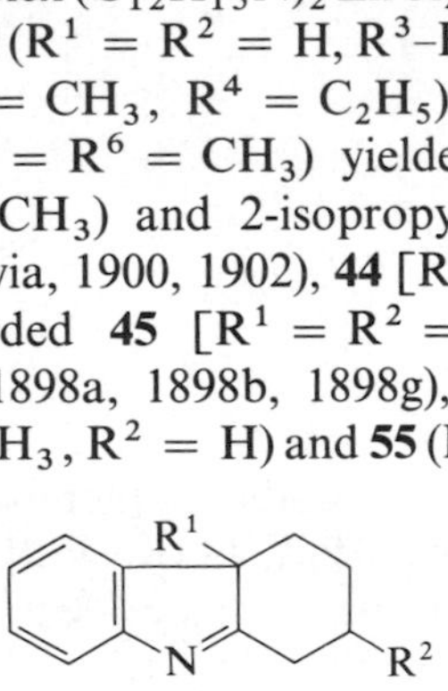

54

$R^2 = H$) (Plancher, 1900b; Plancher and Testoni, 1900) and **53** [$R^1 = H$, $R^2 = CH(CH_3)_2$, $R^3 = CH_3$] gave both **54** [$R^1 = CH(CH_3)_2$, $R^2 = CH_3$] and **55** [$R^1 = CH(CH_3)_2$, $R^2 = CH_3$] (Plancher and Carrasco, 1904). Indolization of **44** ($R^1 = R^2 = H$, $R^3 = R^5 = R^6 = CH_3$, $R^4 = C_2H_5$) and **44** ($R^1 = R^2 = H$, R^3–$R^5 = CH_3$, $R^6 = C_2H_5$) with an alcoholic hydriodic acid catalyst afforded **56** ($R^1 = C_2H_5$, $R^2 = CH_3$) (Plancher, 1900a; Plancher and Bonavia, 1902) and **56** ($R^1 = CH_3$, $R^2 = C_2H_5$) (Plancher, 1902), respectively.

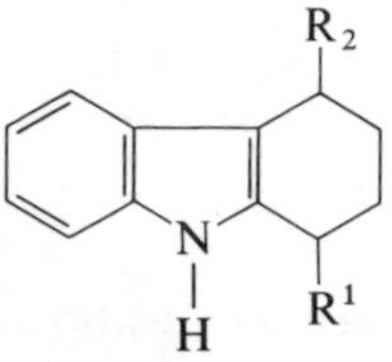

55 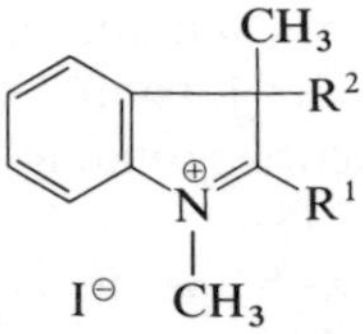 **56**

From the results of those early observations which were available, together with the observation that the zinc chloride catalysed indolization of methyl propyl ketone phenylhydrazone afforded 3-ethyl-2-methylindole (Plancher, 1898e), it was postulated that, upon indolization, arylhydrazones of ketones containing the ketonic moieties $CH_3{\cdot}CO{\cdot}CH\langle$, $-CH_2{\cdot}CO{\cdot}CH\langle$ and $CH_3{\cdot}CO{\cdot}CH_2-$ would produce, respectively, only the corresponding 3*H*-indole, a mixture of the corresponding indole and 3*H*-indole and mainly the corresponding 2-methylindole (Plancher and Bonavia, 1902).

Later, using silute sulphuric acid as the catalyst, **53** ($R^1 = R^3 = H$, $R^2 = C_2H_5$) formed both **54** ($R^1 = C_2H_5$, $R^2 = H$) and **55** ($R^1 = C_2H_5$, $R^2 = H$) and **53** ($R^1 = R^2 = C_2H_5$, $R^3 = H$) formed both 1,9-diethyl-1,2,3,4-tetrahydrocarbazole and **57** ($R = C_2H_5$) (Plancher and Ghigi, 1929) (see Chapter III, H).

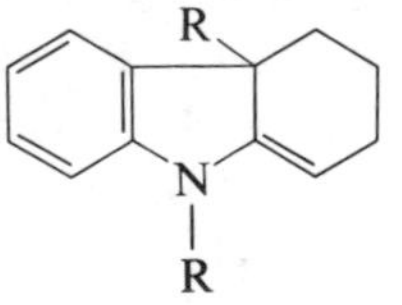

57

Other related studies contemporary with those of Plancher's earlier studies also appeared. Thus, **44** ($R^1 = R^4$–$R^6 = CH_3$, $R^2 = R^3 = H$) was indolized by treatment with zinc chloride in alcoholic hydrochloric acid to yield a product formulated as **52** ($R^1 = R^4 = CH_3$, $R^2 = R^3 = H$) (Konschegg, 1905), an erroneous assignment which also appeared in a subsequent publication by the same author (Konschegg, 1906) but which was later (Plancher and Carrasco,

1909) corrected to **45** (R^1 = R^3–R^5 = CH_3, R^2 = H). Compound **44** (R^1 = R^5 = R^6 = CH_3, R^2–R^4 = H) afforded **45** (R^1 = R^4 = R^5 = CH_3, R^2 = R^3 = H) [isolated by decomposition, by treatment with alkali, of the initially formed 3*H*-indole–zinc chloride complex, $(C_{11}H_{13}N)_2ZnCl_2$] upon treatment with zinc chloride in absolute alcohol (Grgin, 1906). Compound, **44** (R^1 = R^3 = H, R^2 = R^4–R^6 = CH_3) formed **45** (R^1 = H, R^2 = R^3–R^5 = CH_3) upon treatment with alcoholic hydriodic acid (Plangger, 1905) (the structure of the product was not given in this publication). Compound **44** (R^1 = R^2 = H, R^3 = R^5 = R^6 = CH_3, R^4 = C_6H_5) afforded **48** (R^1 = CH_3, R^2 = C_6H_5) after treatment with tin(II) chloride in alcoholic hydrochloric acid followed by basification (Jenisch, 1906), conditions which were also used in the much later (Carson, D. F. and Mann, 1965) indolization of **44** [R^1 = R^2 = H, R^3 = R^5 = R^6 = CH_3, R^4 = $C_6H_4N(CH_3)_2(4)$] into **48** [R^1 = CH_3, R^2 = $C_6H_4N(CH_3)_2(4)$]. Similarly, but using ethanolic hydrochloric acid as catalyst, **44** (R^1 = R^2 = H, R^3 = R^4 = C_6H_5, R^5 = R^6 = CH_3) was converted into **48** (R^1 = R^2 = C_6H_5) (Neber, Knöller, Herbst and Trissler, 1929).

Since these early developmental studies of Brunner, Plancher and the other contemporary workers, the syntheses of 3*H*-indoles, 2-alkylidene indolines and indolin-2-ols by these above routes have become well established and many examples appear in the appropriate sections of this present study.

D. The Borsche Synthesis of Carbazoles

In 1908 it was found that 1,2,3,4-tetrahydrocarbazoles, resulting from the indolization of various cyclohexanone arylhydrazones, could be dehydrogenated into the corresponding carbazoles by heating with lead(II) oxide (Borsche, Witte, A. and Bothe, 1908), a total reaction sequence which has often been referred to as the Borsche carbazole synthesis. Thus, cyclohexanone phenylhydrazone, 4-bromo-, 4-chloro-, 4-ethoxy-, 4-methoxy-, 4-methyl- and 4-nitrophenylhydrazone, 3-nitrophenylhydrazone, 2-chloro- and 2-nitrophenylhydrazone, 2,4-dimethylphenylhydrazone and 1- and 2-naphthylhydrazone, 3-methylcyclohexanone phenylhydrazone and 4-methyl-, 2,4,5-trimethyl- and 3-nitrophenylhydrazone and (−)-menthone[(−)-2-isopropyl-5-methylcyclohexanone] phenylhydrazone were converted, by warming their dilute sulphuric acid solutions, into the corresponding 1,2,3,4-tetrahydrocarbazoles **58** (R^1–R^7 = H; R^1 = R^3–R^7 = H; R^2 = Br, Cl, OC_2H_5, OCH_3, CH_3, NO_2; R^1 = R^2 = R^4–R^7 = H, R^3 = NO_2; R^1–R^3 = R^5–R^7 = H, R^4 = Cl, NO_2 and R^1 = R^3 = R^5–R^7 = H, R^2 = R^4 = CH_3), **59**, **60** and **58** [R^1–R^5 = R^7 = H, R^6 = CH_3; R^1 = R^2 = R^4 = R^6 = CH_3, R^3 = R^5 = R^7 = H; R^1 = R^2 = R^4 = R^5 = R^7 = H, R^3 = NO_2, R^6 = CH_3 and R^1–R^4 = R^6 = H, R^5 = $CH(CH_3)_2$, R^7 = CH_3], respectively. A selected number of these products were then pyrolysed with lead(II) oxide and pumice to convert them into the corresponding carbazoles **61** [R^1–R^5 = H;

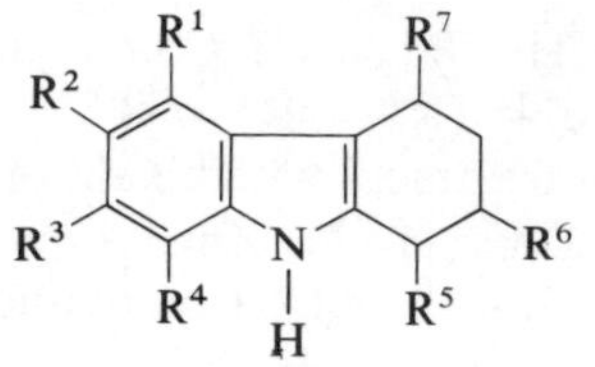

58 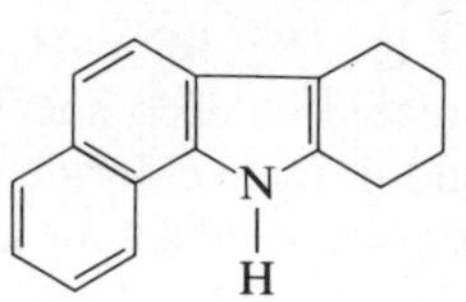 **59**

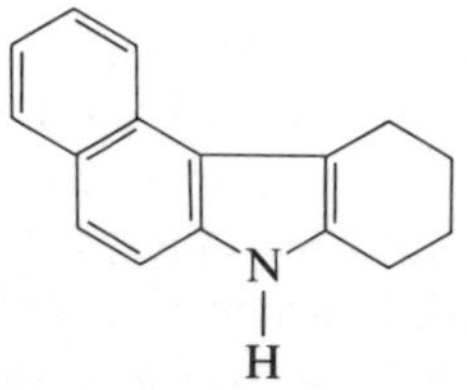

60 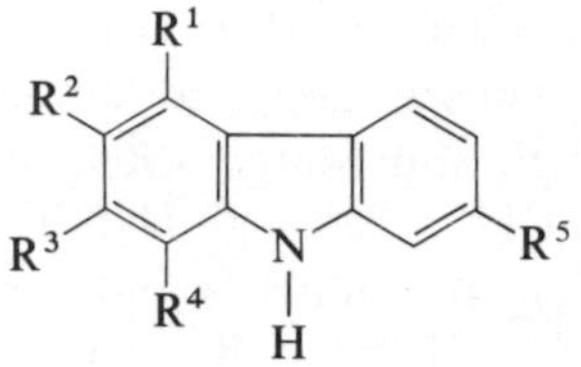 **61**

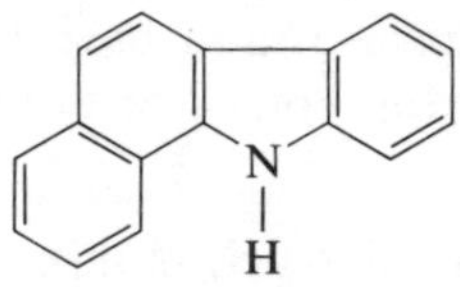

62

$R^1 = R^3$–$R^5 = H$, $R^2 = OCH_3$, CH_3; $R^1 = R^2 = R^4 = R^5 = H$, $R^3 = CH_3$; $R^1 = R^3 = R^5 = H$, $R^2 = R^4 = CH_3$; $R^1 = R^3 = R^4 = H$, $R^2 = R^5 = CH_3$; $R^1 = CH_3$, $R_2 = R_3 = R_5 = H$, $R^4 = CH(CH_3)_2$ and $R^1 = R^2 = R^4 = R^5 = CH_3$, $R^3 = H$] and **62**. Under these conditions, 6- and 8-chloro-1,2,3,4-tetrahydrocarbazole were reductively dechlorinated as well as dehydrogenated to afford carbazole in both cases. In this study several of the Fischer indolizations were ambiguous and in all such examples, except that using 3-methylcyclohexanone phenylhydrazone, product structures were assigned without verification. With the latter phenylhydrazone, the product was verified as being 1,2,3,4-tetrahydro-2-methylcarbazole by the above described dehydrogenation to 2-methylcarbazole, a compound already known from an unambiguous synthesis. For detailed discussions of the other types of ambiguous indolizations, including those referred to above, the appropriate sections of Chapter III in this book should be consulted. Other examples of the use of lead(II) oxide in Borsche carbazole syntheses were effected by heating the hydrocarbazole with the oxide (Ghigi, 1931), which at 400–600 °C also effected dealkylation of 2- and 3-methoxy-, 2- and 3-ethoxy-, 2- and 3-isopropoxy- and

2- and 3-isobutoxy-1,2,3,4-tetrahydrocarbazole to afford 2- and 3-hydroxycarbazole (Ballauf and Schmetzer, 1929; I. G. Farbenind. A.-G., 1930a) and with the oxide possibly associated with pumice (Ballauf and Schmetzer, 1929) and by distillation of a mixture of the hydrocarbazole and the oxide (Braun and Bayer, 1929), sometimes in a stream of carbon dioxide (Braun and Haensel, 1926).

Since it was apparent (Perkin and Plant, 1921) that the use of lead(II) oxide at high temperature would be inapplicable to the dehydrogenation of nitro- or halogeno derivatives of 1,2,3,4-tetrahydrocarbazole, other methods for achieving these dehydrogenations were sought. Following the oxidation of tetrahydroharmine into harmaline with potassium permanganate, attempted dehydrogenation of 1,2,3,4-tetrahydrocarbazole using this reagent was investigated. Although using potassium permanganate, in both dilute sulphuric acid and aqueous solutions, oxidation took place, as evidenced by the decoloration of the reaction mixture, no recognizable product could be isolated (Perkin and Plant, 1921). However, 1,2,3,4-tetrahydrocarbazole and its 9-methyl derivative were dehydrogenated using mercury(II) acetate [mercury(II) oxide in acetic acid] although the yields of carbazole and 9-methylcarbazole, respectively, were low. As evidenced by the precipitation of mercury(I) acetate during the reaction, the similar dehydrogenation of 6-nitro-1,2,3,4-tetrahydrocarbazole also occurred, although the reaction mixture was not investigated with a view to isolating organic products (Perkin and Plant, 1921). Later (Perkin and Plant, 1923a), sulphur in boiling quinoline was introduced as an alternative dehydrogenating agent and, using this, 1,2,3,4-tetrahydrocarbazole and its 9-methyl, 9-acetyl and 6-bromo derivatives were converted into carbazole (32%) and 9-methyl- (15%), 9-acetyl- (22%) and 3-bromo- (20%) carbazoles, respectively. It is important in these dehydrogenations that pure quinoline is used since, using crude coal tar quinoline, resinous products result from which only very small amounts of crystalline products can be isolated. Furthermore, using sulphur in refluxing quinoline, the attempted dehydrogenation of 6-nitro-1,2,3,4-tetrahydrocarbazole gave only a black resin and of ethyl 1,2,3,4-tetrahydrocarbazole-9-carboxylate gave carbazole (Perkin and Plant, 1923a). Other uses of sulphur in refluxing quinoline as a method of dehydrogenation in the Borsche carbazole synthesis can be found (Briscoe and Plant, 1928; Moggridge and Plant, 1937; Morgan and Walls, L. P., 1930; Oakeshott and Plant, 1926, 1928; Plant and Rogers, K. M., 1936; Plant, Rogers, K. M. and Williams, S. B. C., 1935; Plant and Tomlinson, M. L., 1932a; Plant and Williams, S. B. C., 1934) and direct pyrolysis in the presence of sulphur has also been similarly employed (Grandberg, Kost and Yaguzhinskii, 1960).

Pyrolysis with palladium-on-carbon in an atmosphere of hydrogen was introduced (Moggridge and Plant, 1937) as a method of effecting these dehydrogenations and, using this approach, methyl 1,2,3,4-tetrahydrocarbazole-6-, 7-, and 8-carboxylate were successfully dehydrogenated to ultimately furnish carbazole 3-, 2- and 1-carboxylic acid, respectively. However, attempts to similarly prepare carbazole-4-carboxylic acid from methyl 1,2,3,4-tetrahydrocarbazole-5-carboxylate failed and attempted dehydrogenation of 1,2,3,4-tetrahydrocarbazole-8-carboxylic acid under these conditions also effected decarboxylation to afford carbazole. Likewise, attempted dehydrogenation of 7-chloro-1,2,3,4-tetrahydrocarbazole by pyrolysis with palladium-on-carbon in an atmosphere of hydrogen also caused reductive dechlorination to yield carbazole, although 2-chlorocarbazole was produced when using sulphur in refluxing quinoline, conditions which also effected the conversion of 9-acetyl-5-chloro-1,2,3,4-tetrahydrocarbazole into 9-acetyl-4-chlorocarbazole (Moggridge and Plant, 1937). In these dehydrogenations, a mixture of the 1,2,3,4-tetrahydrocarbazole with 25% (w/w) of palladium-on-carbon was dry heated to *ca.* 300 °C. Subsequently (Horning, E. C., Horning, M. G. and Walker, G. N., 1948) it was found that these dehydrogenations could be achieved using a 5% palladium-on-carbon catalyst in refluxing trimethylbenzene (b.p. 168–172 °C) and by this method 1,2,3,4-tetrahydrocarbazole and its 6-, 7- and 9-methyl derivatives and its 2-isopropyl-4-methyl derivative were converted into carbazole (95%) and 3-methyl- (99%), 2-methyl- (90%) and 9-methyl- (86%) carbazoles and 2-isopropyl-4-methylcarbazole (90%), respectively, although the substituent orientation on this last compound was not established (see Chapter III, H). However, this method also caused simultaneous decarboxylation when 3-carboxy-1,2,3,4-tetrahydro-2,4-dimethylcarbazole afforded 2,4-dimethylcarbazole. Other uses of palladium-on-carbon in the Borsche carbazole synthesis have involved pyrolysis in a stream of hydrogen (Pausacker, 1950a; Pausacker and Schubert, 1949b), conditions which also caused concomitant reductive dechlorination when 8-chloro-1,2,3,4-tetrahydro-5-methylcarbazole yielded 4-methylcarbazole (Pausacker and Robinson, R., 1947), in a stream of nitrogen (Felton, 1952) or *in vacuo* (Clemo and Felton, 1951a). More specifically, pyrolysis after admixture with 5% palladium-on-carbon (Buu-Hoï, Périn and Jacquignon, 1960, 1966; Buu-Hoï and Saint-Ruf, 1965b) and 10% palladium-on-carbon (Govindachari, Rajappa and Sudarsanam, 1963; Maréchal, Christiaens, Renson and Jacquignon, 1978), which also caused defluorination when 1,2,3,4-tetrafluoro-5,6,7,8-tetrahydrocarbazole was converted into carbazole (Petrova, T. D., Mamaev and Yakobson, 1969), in an atmosphere of carbon dioxide (Carter, P. H., Plant and Tomlinson, M., 1957; Cummins, Kaye and Tomlinson, M. L., 1954; Cummins and Tomlinson, M. L., 1955; Milne and Tomlinson, M. L., 1952) [decarboxylation of carboxylic acid substituents can also occur if these conditions are applied too vigorously (Carter, P. H., Plant and Tomlinson, M., 1957)] or nitrogen (Cranwell and Saxton, 1962; Stillwell, 1964), which also effected N-demethylation when **63** was converted into **64** (Stillwell, 1964), have also been used. Sublimations *in vacuo* from palladium-on-carbon, which also effected N-debenzylation when **65** was converted into **66** (Roussel, O., Buu-Hoï and

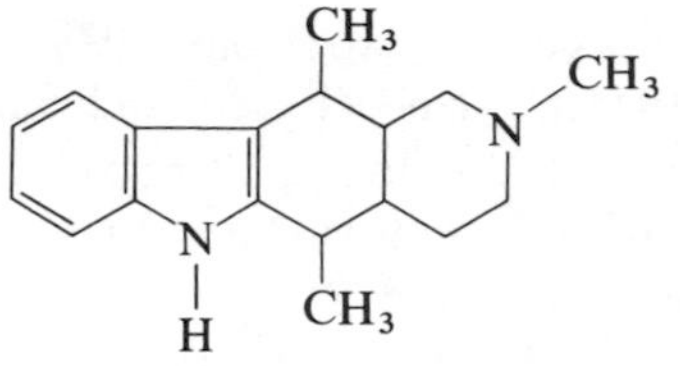

63

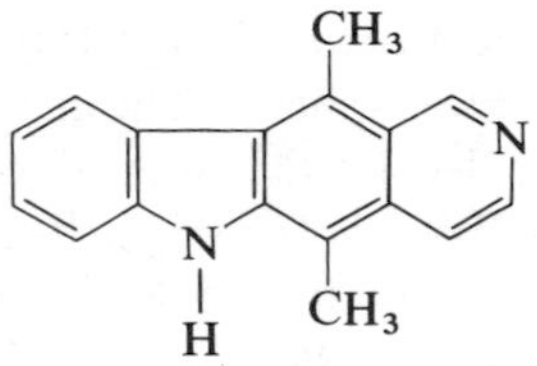

64

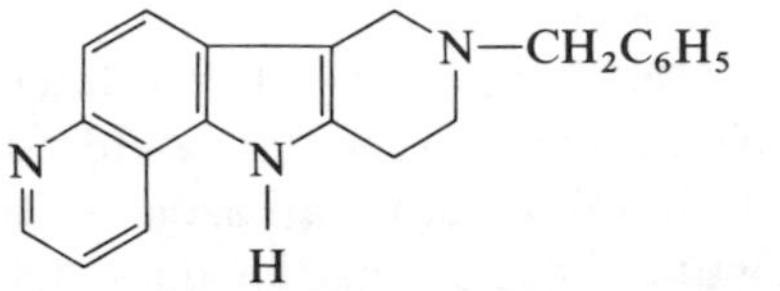

65

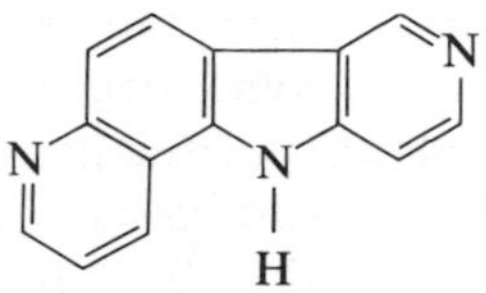

66

Jacquignon, 1965), and, more specifically, from 5% palladium-on-carbon (Buu-Hoï, Jacquignon and Hoeffinger, 1963; Buu-Hoï, Mangane and Jacquignon, 1967; Buu-Hoï, Martani, Ricci, Dufour, Jacquignon and Saint-Ruf, 1966; Buu-Hoï, Périn and Jacquignon, 1962, 1966; Buu-Hoï and Saint-Ruf, 1965a; Buu-Hoï, Saint-Ruf, Deschamps, Bigot and Hieu, 1971; Buu-Hoï Saint-Ruf and Dufour, 1964; Buu-Hoï, Saint-Ruf, Jacquignon and Marty, 1963), which also caused concomitant C-demethylations when **67**, **68** and **70** furnished **62**, **69** and **71** respectively (Buu-Hoï, Jacquignon and Ledésert, 1970), 10% palladium-on-carbon (Buu-Hoï, Croisy, Jacquignon and Martani, 1971) or 30% palladium-on-carbon (Govindachari, Rajappa and Sudarsanan, 1966) have also been used to effect such dehydrogenations. These have also been carried out by treatment with 5% palladium-on-carbon in refluxing mesitylene

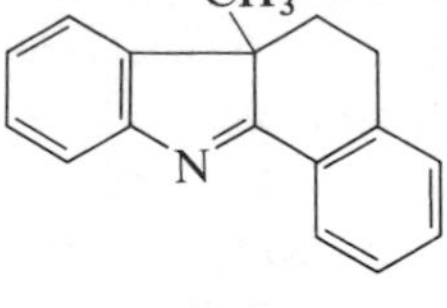

67

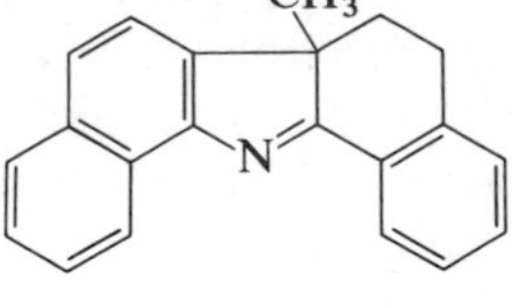

68

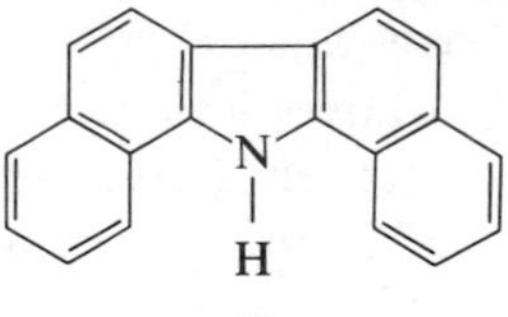

69

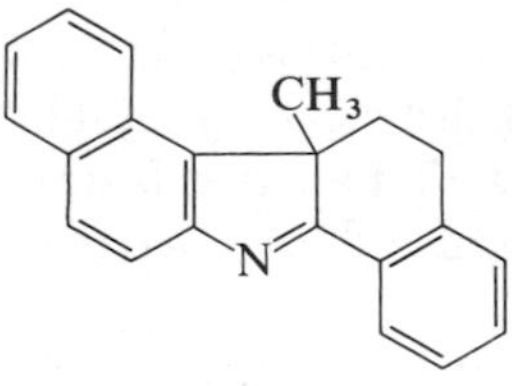

70

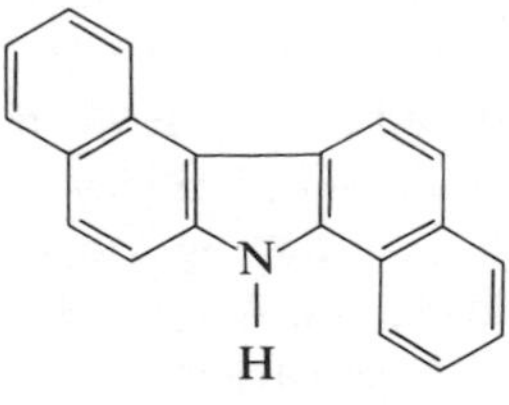

71

(Bajwa and Brown, R. K., 1968a, 1970; Fusco and Sannicolò, 1978c) under a nitrogen atmosphere (Thomas, T. J., 1975), xylene (Fusco and Sannicolò, 1978c) or decalin (Mosher, Crews, Acton and Goodman, 1966; Schmutz and Wittwer, 1960), with 10% palladium-on-carbon in refluxing decalin (Cohylakis, Hignett, Lichman and Joule, 1974; Winchester and Popp, 1975), quinoline (Winchester and Popp, 1975), xylene (Driver, Matthews and Sainsbury, 1979) and diphenylether, which does not remove nuclear fluorine substituents (Rastogi, Bindra, Rai and Anand, 1972) and with 30% palladium-on-carbon in refluxing decalin (Govindachari and Sudarsanam, 1967), xylene or *para*-cymene (Campaigne, Ergener, Hallum and Lake, R. D., 1959). Interestingly, this last catalyst in refluxing xylene effected the conversion of 8-chloro-1,2,3,4-tetrahydro-5-methylcarbazole into 1-chloro-4-methylcarbazole (Campaigne and Lake, R. D., 1959) (i.e. concomitant dechlorination does not occur). Distillation over palladium at 500–600 °C also caused, concomitantly with dehydrogenation, the dealkylation of 6-methoxy- and 7-ethoxy-substituents on the 1,2,3,4-tetrahydrocarbazole nucleus (I. G. Farbenind. A.-G., 1929). Pyrolysis with palladium-on-carbon has failed to effect dehydrogenation where treatment with sulphur in refluxing quinoline was ultimately successful (Forbes, E. J., Stacey, Tatlow and Wragg, R. T., 1960). Pyrolysis of 1-cyclohexyl-1,2,3,4-tetrahydrocarbazole with palladium-on-carbon afforded carbazole and another product which was not 1-phenylcarbazole (Berlin, K. D., Clark, P. E., Schroeder, J. T. and Hopper, 1968). The structure of this latter product remains to be established.

Heating 1,2,3,4-tetrahydrocarbazole with palladium black and an equimolar amount of safrole (acting as a hydrogen acceptor) in benzene solution gave carbazole. Maleic acid has also been similarly used as a hydrogen acceptor in related, although non-indolic, dehydrogenations (Akabori and Saito, K., 1930b). Likewise, pyrolysis with palladium black in molten cinnamic acid, acting both as a solvent and hydrogen acceptor, has also been used to effect the conversion of 1,2,3,4-tetrahydrocarbazole and its 6- and 7-(or 5?)-ethoxy derivatives into carbazole (89% yield) and 3- and 2-(or 4?)-ethoxycarbazole, respectively (Hoshino and Takiura, 1936). Palladium(II) chloride, in a refluxing aqueous medium containing just sufficient hydrochloric acid to retain the salt in solution, has also been used as a dehydrogenating agent although in connection with the present studies its use was limited to the conversion of 1,2,3,4-tetrahydrocarbazole into carbazole, although this was obtained in high (91%) yield (Cooke and Gulland, 1939). Likewise, 1,2,3,4-tetrahydrocarbazole produced carbazole when heated with a nickel catalyst in naphthalene (I. G. Farbenind. A.-G., 1935) or in phenol (I. G. Farbenindustrie A.-G., 1935) or with a catalyst containing copper, chromium and barium oxides and Florida earth (I. G. Farbenind. A.-G., 1933).

As a result of the earlier studies, it was apparent that a dehydrogenating agent of wide applicability which would afford high product yields without substituent elimination was required. These properties were found (Barclay and Campbell, 1945) in chloranil which in boiling xylene had previously (Arnold, R. T. and Collins, C. J., 1939; Arnold, R. T., Collins, C. and Zenk, 1940) been used to convert cyclohexyl nuclei into benzenoid nuclei. Under similar conditions, this reagent in sulphur free xylene was now used to convert 5-, 6-, 7- and 8-bromo-, 6- and 8-carbethoxy-, 6-carboxy-, 5-, 6-, 7- and 8-chloro-, 5,8-dichloro- and 6-ethoxy-1,2,3,4-tetrahydrocarbazole and 1,2,3,4-tetrahydro-2-, -6- and -8-methyl-, -5-, -6-, -7- and -8-nitro- and -9-phenylcarbazole into the corresponding carbazoles. Refluxing times were usually in the region of 18–24 h for complete dehydrogenation and yields of product varied between 50% and 95% depending upon the nature of the substituent(s). This reagent thus leaves intact carboxy, halogeno and nitro substituents although it is preferable to use the ethyl or methyl esters rather than the free carboxylic acids of 1,2,3,4-tetrahydrocarbazoles since the resulting carbazole carboxylic acid esters are more readily separated from the tetrachloroquinol, which is also formed during the reaction, than are the free acids. Since this introduction, chloranil has found very wide use in this synthetic approach to the carbazole nucleus. The reaction has been effected in refluxing xylene (Anderson, G., and Campbell, 1950; Buu-Hoï, Hoán and Khôi, 1950c; Buu-Hoï, Hoán, Khôi and Xuong, 1951; Buu-Hoï, Royer, Eckert and Jacquignon, 1952; Buu-Hoï and Saint-Ruf, 1965b; Carlin and Larson, 1957; Carlin and Moores, 1959, 1962; Clemo and Felton, 1952; Fusco and Sannicolò, 1978c; Nozoe, Sin, Yamane and Fujimori, K., 1975; Pausacker and Robinson, R., 1947; Wittig and Reichel, 1963), refluxing dry xylene (Barnes, Pausacker and Schubert, 1949; Buu-Hoï, Cagniant, Hoán and Khôi, 1950; Buu-Hoï and Hoán, 1951; Buu-Hoï, Hoán and Khôi, 1949, 1950b, 1950c; Buu-Hoï, Hoán, Khôi and Xuong, 1949, 1950; Buu-Hoï and Jacquignon, 1954, 1956; Buu-Hoï, Jacquignon and Lavit, 1956; Buu-Hoï, Khôi and Xuong, 1951), refluxing sulphur free xylene (Allen, F. L. and Suschitzky, 1953; Barclay, Campbell and Gow, 1946; Cummins and Tomlinson, M. L., 1955), refluxing pure, dry xylene (Green and Ritchie, 1949) and refluxing anhydrous tetrahydrofuran (Gilbert, J., Rousselle, Gansser and Viel, 1979) and with no solvent specified (Crum and Sprague, 1966; Miller, B. and Matjeka, 1977).

However, this use of chloranil has not been devoid of failures nor is it the method of choice in every case. Thus in some cases, pyrolysis with palladium-on-carbon (Buu-Hoï, Périn and Jacquignon, 1960), with 5% palladium-on-carbon under an atmosphere of carbon dioxide (Carter, P. H., Plant and Tomlinson, M., 1957), with selenium (the chloranil method failed) (Kulka and Manske, R. H. F., 1952), with sulphur in refluxing quinoline (the chloranil method failed) (Forbes, E. J., Stacey, Tatlow and Wragg, R. T., 1960), and with Raney nickel (Davidge, 1959) have been preferred, treatment with 30% palladium-on-carbon in refluxing xylene or *para*-cymene has been found to be more convenient (Campaigne, Ergener, Hallum and Lake, R. D., 1959) and repeated sublimation *in vacuo* over 5% palladium-on-carbon has been found to be superior with regard to both product yield and quality (Buu-Hoï, Jacquignon and Hoeffinger, 1963).

Extensions of this use of chloranil have employed bromanil in refluxing sulphur free xylene (Barnes, Pausacker and Schubert, 1949) and 2,3-dichloro-5,6-dicyano-1,4-benzoquinone in refluxing dry benzene (Nozoe, Sin, Yamane and Fujimori, K., 1975) and in stirred dioxane (Reed, G. W. B., 1975). However, the attempted dehydrogenations of **67** with either chloranil or 2,3-dichloro-5,6-dicyano-1,4-benzoquinone in xylene or of **68** with 2,3-dichloro-5,6-dicyano-1,4-benzoquinone in xylene were unsuccessful (Buu-Hoï, Jacquignon and Ledésert, 1970).

Other, less widely used methods of effecting the dehydrogenation step in the Borsche carbazole synthesis have used pyrolysis (Buu-Hoï, Hoán and Khôi, 1949; Buu-Hoï, Martani, Ricci, Dufour, Jacquignon and Saint-Ruf, 1966), refluxing in nitrobenzene (Braun and Bayer, 1929), distillation from a mixture with 2–3% copper powder (Braun and Bayer, 1929), pyrolysis with selenium (Kulka and Manske, R. H. F., 1952), pyrolysis with Raney nickel (Davidge, 1959; Miller, S. A., 1958), pyrolysis with Raney nickel in sulphur free xylene (Badcock and Pausacker, 1951), although this method failed where treatment with chloranil in refluxing xylene or with sulphur in refluxing quinoline was successful (Forbes, E. J., Stacey, Tatlow and Wragg, R. T., 1960) and was of limited applicability even when hydrogen acceptors were concomitantly employed (Badcock and Pausacker, 1951), heating with a nickel-on-nickel(II) chromite catalyst at 300 °C in benzene (Adkins and Lundsted, 1949) and treatment with a mixture of platinum in naphthalene in refluxing tetralin, which also effected reductive dechlorination when 5-acetyl-8-chloro-1,2,3,4-tetrahydrocarbazole was converted into 4-acetylcarbazole (Manske, R. H. F. and Kulka, 1950a), although pyrolysis with platinum has been unsuccessful (Kulka and Manske, R. H. F., 1952). The majority of these early developments in the Borsche synthesis of carbazoles have been briefly discussed in a review of the chemistry of carbazole (Campbell and Barclay, 1947).

Indolization of arylhydrazones of 1,2,3,4-tetrahydro-4-oxoquinolines or of ketones containing this ketonic moiety is often accompanied by spontaneous dehydrogenation, when using Brønsted acid catalysts, to yield fully aromatic polynuclear products. Analogous dehydrogenations do not occur using heterocyclic ketones having other hetero atoms in place of the nitrogen atom. The results of studies in this area are fully described in Chapter III, L. However, other dehydrogenations which occur concomitantly with indolization have also been observed, although often in these cases more drastic indolization conditions (i.e. zinc chloride at an elevated temperature) were employed. Thus, the 5-isoquinolyl-, 3-methyl-5-isoquinolyl- and 3-, 5-, 6- and 8- quinolylhydrazones of **72** were converted directly into **73** (R = H and CH_3) (Buu-Hoï, Jacquignon, Roussel, O., and Hoeffinger, 1964) and **74**, **75**, **76** and **77** (Buu-Hoï,

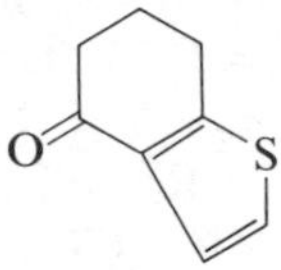

Jacquignon and Hoeffinger, 1963), respectively, and a series of 1- and 2-tetralone quinolylhydrazones were likewise directly converted into the corresponding series of benzopyridocarbazoles, whereas heating with a mixture of sulphuric and acetic acids afforded the dihydrobenzopyridocarbazoles (Buu-Hoï, Jacquignon and Hoeffinger, 1963).

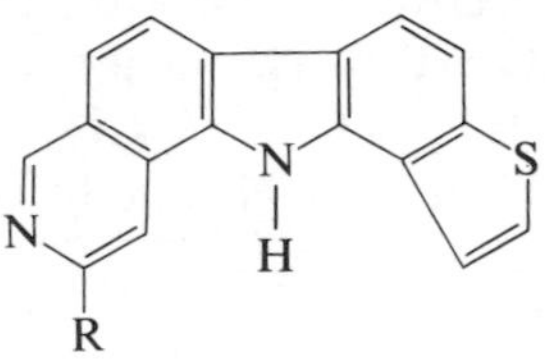

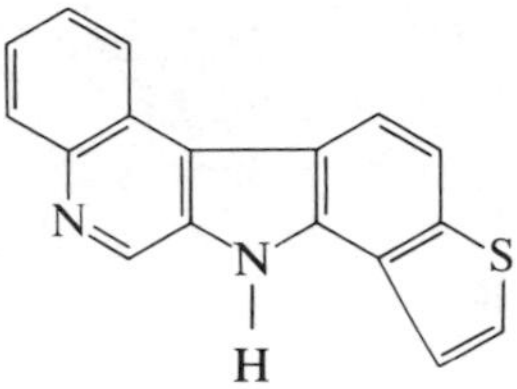

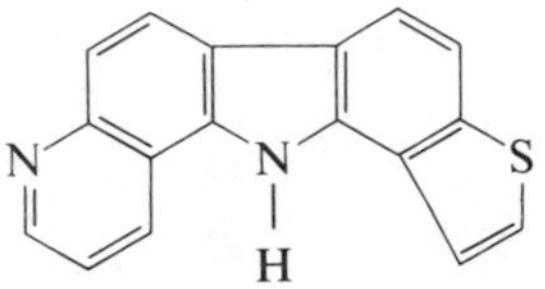

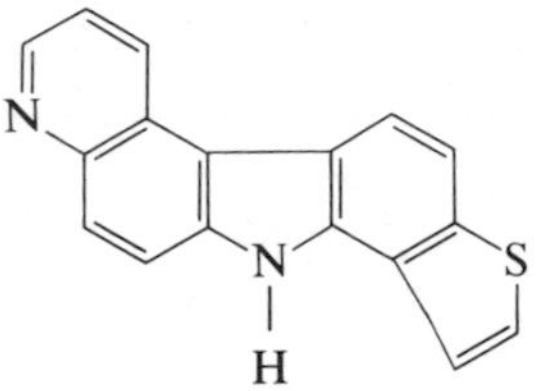

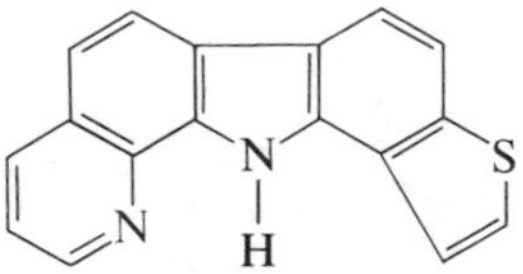

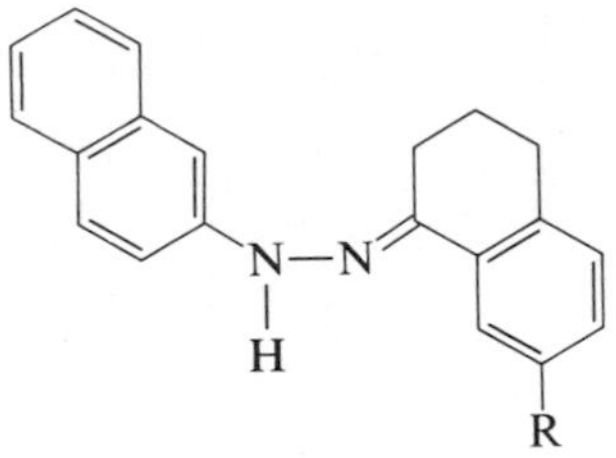

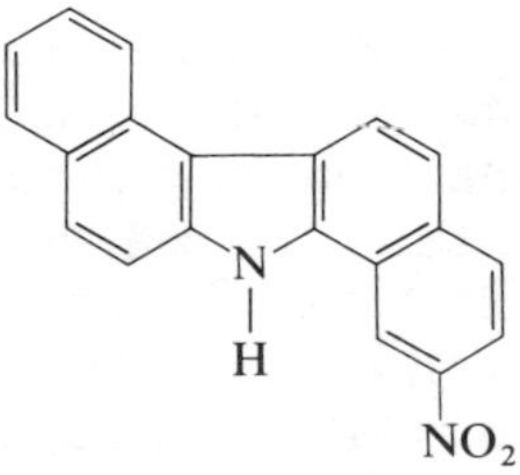

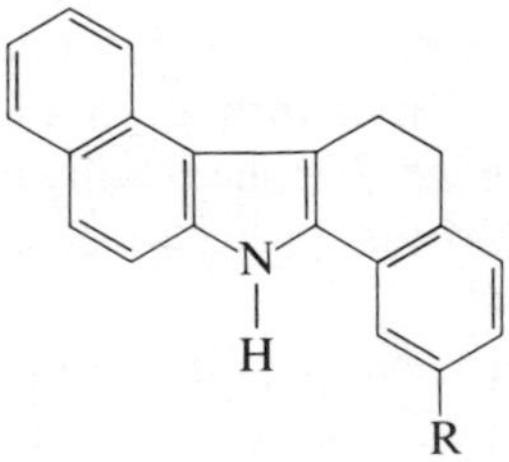

Related spontaneous dehydrogenations, which appear to be substituent specific, have been observed when using Brønsted acid indolization catalysts. Thus, whereas **78** (R = NO_2) was converted directly into **79** in the presence of dry hydrogen chloride in glacial acetic acid, the dihydro compounds **80** [R = H, CH_3, $C(CH_3)_3$ and OCH_3] were produced by similar treatment of **78** [R = H,

CH_3, $C(CH_3)_3$ and OCH_3, respectively] (Buu-Hoï, Hoán and Khôi, 1949). Similar catalytic conditions converted **81** ($R^1 = R^4 = CH_3$, $R^2 = R^3 = H$) into **82**, whereas from **81** (R^1–R^3 = H, $R^4 = CH_3$; $R^1 = R^2 = R^4 = H$, $R^3 = CH_3$; $R^1 = R^3 = R^4 = H$, $R^2 = CH_3$ and C_6H_5 and $R^1 = R^2 = H$, $R^3 = R^4 = CH_3$) the dihydro compounds **83** (R^1–R^3 = H, $R^4 = CH_3$;

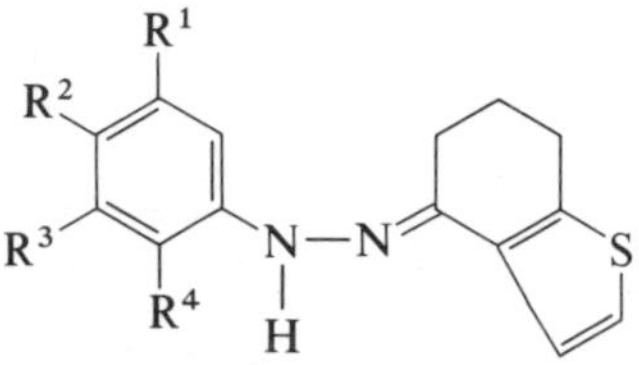

81 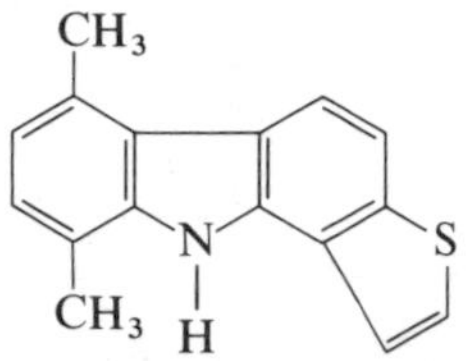 **82**

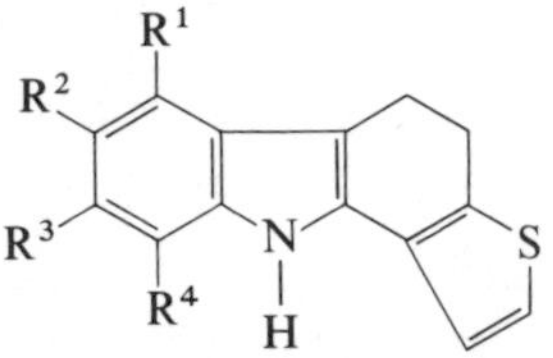

83

$R^1 = R^2 = R^4 = H$, $R^3 = CH_3$; $R^1 = R^3 = R^4 = H$, $R^2 = CH_3$ and C_6H_5 and $R^1 = R^2 = H$, $R^3 = R^4 = CH_3$, respectively) were produced Buu-Hoï, Hoán, Khôi and Xuong, 1949). A more recent related example involved the direct formation of 1,2-benzo-7-phenylcarbazole by treatment of a mixture of 3-biphenylylhydrazine with 1-tetralone with a solution of hydrogen chloride in acetic acid (Thang, D. C., Kossoff, Jacquignon and Dufour, 1976).

When cyclohexan-1,4-dione bisphenylhydrazone was indolized with either a sulphuric acid in acetic acid (Robinson, B., 1963d) or an aqueous sulphuric acid (Harley-Mason and Pavri, 1963; Teuber, Cornelius and Wolcke, 1966) catalyst, it afforded directly, along with other products (see Chapter III, K4), 5,11-dihydroindolo[3,2-*b*]carbazole which again resulted from dehydrogenation under indolization conditions. However, apart from the few examples above and those starting from arylhydrazones containing the 1,2,3,4-tetrahydro-4-oxoquinoline moiety, dehydrogenations did not usually accompany Fischer indolization when using a Brønsted acid catalyst in indolizations (Buu-Hoï, Jacquignon, Croisy and Ricci, 1968; Buu-Hoï, Mangane and Jacquignon, 1967; Buu-Hoï, Martani, Croisy, Jacquignon and Périn, 1966; Buu-Hoï, Martani, Ricci, Dufour, Jacquignon and Saint-Ruf, 1966; Buu-Hoï, Périn and Jacquignon, 1966; Buu-Hoï and Saint-Ruf, 1965a, 1965b) similar to those described above in which zinc chloride was used as the catalyst.

E. Synthesis of Benzocarbazoles by Reacting Aromatic Polycyclic Arylolic Compounds with Arylhydrazines

Fischer indole syntheses can be carried out by reacting aromatic polycyclic arylolic compounds, presumably acting as their keto tautomers, with arylhydrazines. The first such reaction was accomplished when 2-naphthol-3-carboxylic acid (**84**; $R^1 = R^2 = H$, $R^3 = COOH$) was refluxed with an equimolar amount of phenylhydrazine, ammonia being evolved and compound **85** ($R^1 = R^2 = R^4 = R^5 = H$, $R^3 = COOH$) being formed. This was subsequently decarboxylated to afford **85** (R^1–R^5 = H), m.p. = 120 °C (Schöpff, 1896). Upon later repetitions of this reaction sequence, the m.p. of the final product was given as 134.5 °C (Ullmann, 1898) and as 134–135 °C (Japp and Maitland, 1901) (see also Japp and Maitland, 1903a, 1903b).

The use of arylolic compounds as ketonic moieties in indolizations was extended by heating 9-hydroxyphenanthrene **86** with an excess of phenylhydrazine at 200 °C in an atmosphere of hydrogen to give **87** (R = H) (Japp and Findlay, A., 1897). Likewise, refluxing a mixture of **86** with 2-nitrophenylhydrazine in an acetic acid–concentrated hydrochloric acid mixture produced **87** ($R = NO_2$) (Kinsley and Plant, 1956). Upon attempting to apply the synthesis to the use of 2-naphthol as the phenolic moiety, no reaction occurred when this was heated with an excess of phenylhydrazine but upon addition of some dry phenylhydrazine hydrochloride, heating caused the evolution of ammonia and **85** (R^1–R^5 = H), identical with the product unambiguously synthesized earlier (Schöpff, 1896), was obtained (Japp and Maitland, 1901, 1903b). Likewise, 1-naphthol, when heated with a mixture of phenylhydrazine and its hydrochloride, afforded **62** but, relative to the analogous formation of

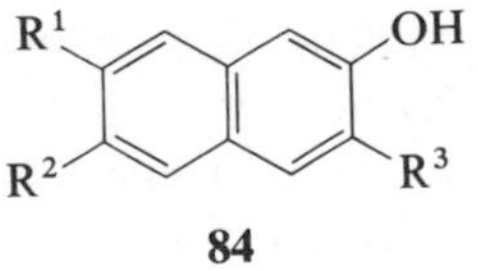

84

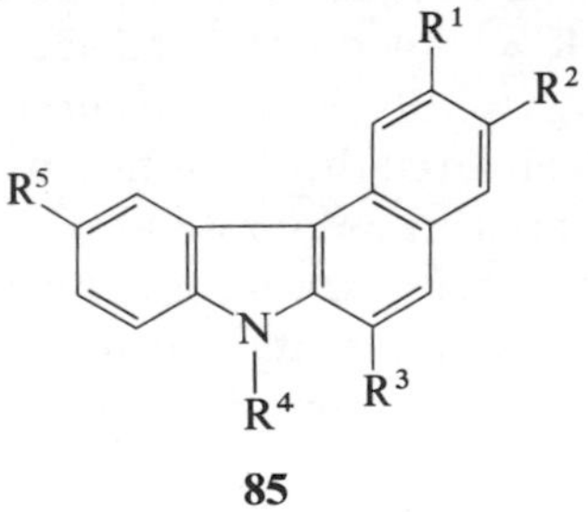

85

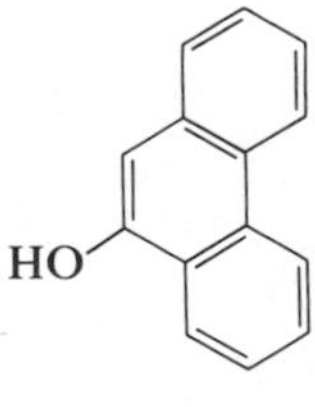

86

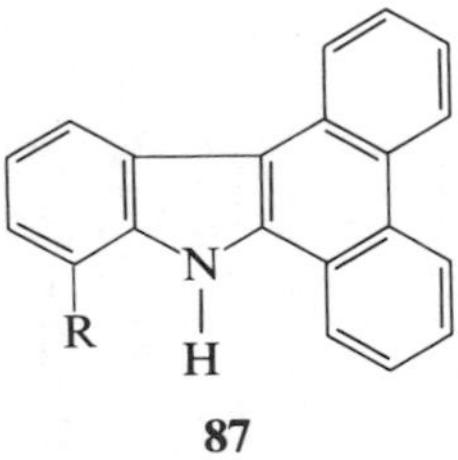

87

85 (R^1–R^5 = H), the reaction was more difficult and the yield was low (Japp and Maitland, 1901, 1903b). 2-Naphthol and 9-hydroxyphenanthrene (**86**) reacted similarly with 1- and 2-naphthylhydrazine in the presence of their corresponding hydrochlorides to yield **71**, **88**, **89** and **90**, respectively (Japp and Maitland, 1903a, 1903b).

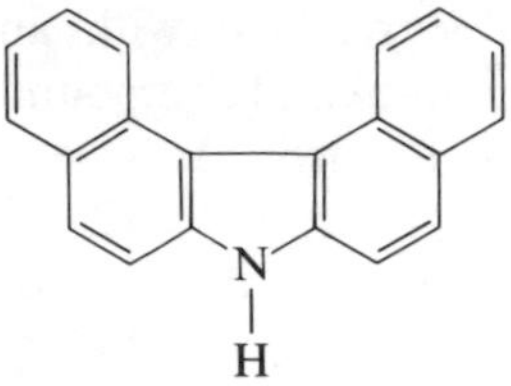

88 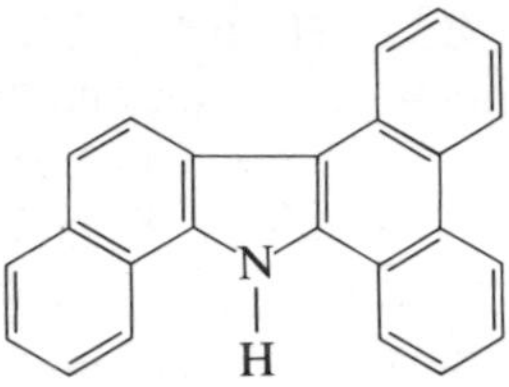 **89** 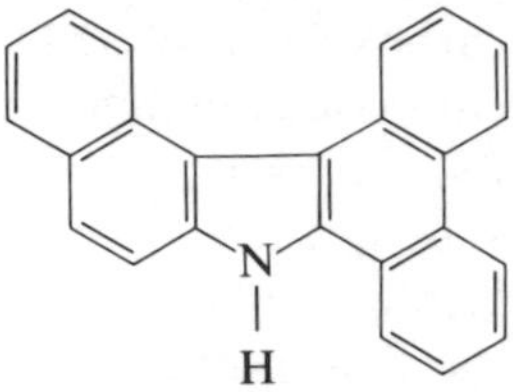 **90**

Several decades later, further uses of this indolization were reported when a mixture of 6-methyl-2-naphthol (**84**; R^1 = R^3 = H, R^2 = CH_3), phenylhydrazine and its hydrochloride were heated to form **85** (R^1 = R^3–R^5 = H, R^2 = CH_3) (Royer, 1946), when 6-ethyl-2-naphthol (**84**; R^1 = R^3 = H, R^2 = C_2H_5) was refluxed with a saturated solution of phenylhydrazine hydrogen sulphite to give **85** (R^1 = R^3–R^5 = H, R^2 = C_2H_5) (Buu-Hoï, Royer, Eckert and Jacquignon, 1952), when **84** (R^1 = H, R^2 = R^3 = CH_3) and **84** (R^1 = R^3 = CH_3, R^2 = H) were refluxed with a mixture of phenylhydrazine and its hydrochloride to form **85** (R^1 = R^4 = R^5 = H, R^2 = R^3 = CH_3 and R^1 = R^3 = CH_3, R^2 = R^4 = R^5 = H, respectively) (Buu-Hoï, Jacquignon and Long, C. T., 1957) and when 6-hydroxychrysene (**91**) was heated with mixtures of phenylhydrazine and its hydrochloride, 4-methylphenylhydrazine and its hydrochloride and 2-naphthylamine and its hydrochloride to afford **92** (R = H and CH_3) and **93**, respectively (Thang, D. C., Can, Buu-Hoï and Jacquignon, 1972). The direction of the last indolization was assumed to be that which involved the 1-position of the naphthalene nucleus, presumably by analogy with other indolizations of 2-naphthylhydrazones. Certainly only one product was isolated from the indolization although in this particular reaction the formation of the alternative isomeric product, formed by indolization involving the 3-position of the naphthalene nucleus, should not be excluded

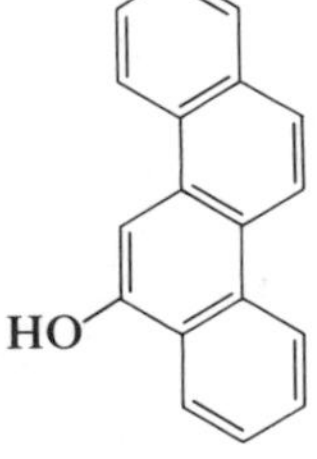

91 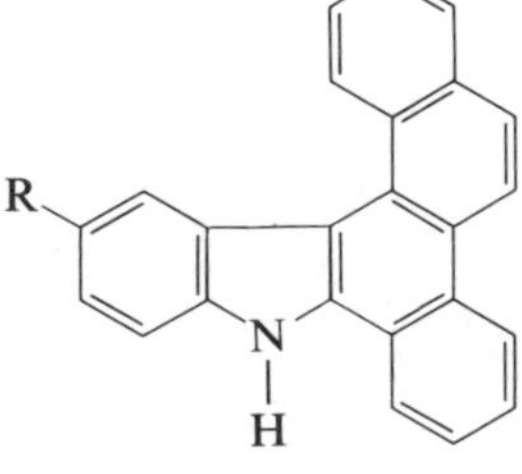 **92** 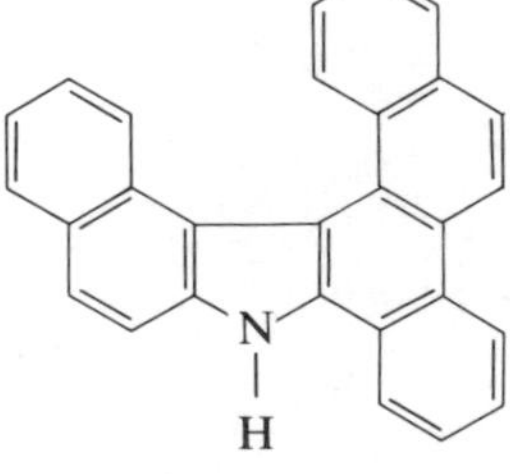 **93**

since it would be the sterically favoured product. Likewise, pyrolysis of a mixture of 3-biphenylylhydrazine and its hydrochloride with 2-naphthol afforded a product which was assumed to be 3,4-benzo-7-phenylcarbazole (Thang, D. C., Kossoff, Jacquignon and Dufour, 1976) although alternative possible isomeric formulations could not be excluded from the analytical data presented.

Neither phenol (Japp and Maitland, 1903a, 1903b) nor thymol (Japp and Maitland, 1903b) afforded the corresponding carbazoles when reacted with a mixture of phenylhydrazine and its hydrochloride, from which it was concluded (Japp and Maitland, 1903b) that monohydric monocyclic phenols in general do not take part in this reaction which is favoured by the degree of readiness with which the phenolic compound exhibits keto–enol tautomerism (Japp and Maitland, 1903b; Thang, D. C., Can, Buu-Hoï and Jacquignon, 1972). Furthermore, although resorcinol and phloroglucinol reacted violently when heated with a mixture of phenylhydrazine and phenylhydrazine hydrochloride, only intractable resinous products were formed (Japp and Maitland, 1903b – see also Baeyer and Kochendoerfer, 1899). Yields of indolic products were also extremely low when naphthols were heated with mixtures of naphthylhydrazines and their hydrochlorides since naphthylhydrazines and their salts are highly sensitive to heat (Buu-Hoï, Hoán and Khôi, 1949).

A modification of this reaction, the Bucherer carbazole synthesis [or more properly, as claimed by one worker (Friedländer, 1921), the Lepetit–Bucherer reaction in recognition of the related earlier studies by Lepetit], involves the treatment of an arylol, or the corresponding arylamine which under the reaction conditions is converted into the phenol by the Bucherer reaction, with an arylhydrazine in the presence of sodium bisulphite. The first of these reactions was achieved when a mixture of **84** ($R^1 = R^2 = H$, $R^3 = COOH$) with sodium hydroxide reacted with phenylhydrazine in the presence of sodium bisulphite to give **85** (R^1–$R^3 = R^5 = H$, $R^4 = SO_3^{\ominus}Na^{\oplus}$) which upon treatment with mineral acid furnished **85** (R^1–$R^5 = H$). Likewise, **84** ($R^1 = R^2 = H$, $R^3 = COOH$) reacted with 4-methylphenylhydrazine to afford **85** (R^1–$R^3 = H$, $R^4 = SO_3H$, $R^5 = CH_3$) which was converted, by treatment with mineral acid, into **85** (R^1–$R^4 = H$, $R^5 = CH_3$). Another product formed along with **85** (R^1–$R^3 = R^5 = H$, $R^4 = SO_3^{\ominus}Na^{\oplus}$) in the initial reaction was **94** and when this, or its hydrochloride, was heated, ammonia was evolved and **85** (R^1–$R^5 = H$) was formed (Bucherer and Seyde, 1908). Indeed, both prior and subsequent to this observation other examples of the formation of carbazoles by the acid

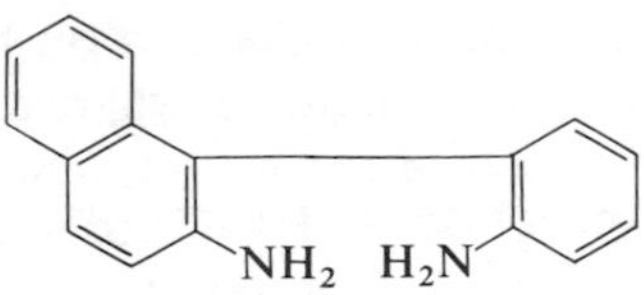

NH_2 H_2N

94

catalyzed cyclization of 2,2′-diaminobiaryls were reported (Carlin and Forshey, 1950; I. G. Farbenind. A.-G., 1930c; King, F. E. and King, T. J., 1945; Meisenheimer and Witte, K., 1903; Niementowskii, 1901; Nietzki and Goll, 1885; Poraĭ-Koshits and Salyamon, 1944; Roosmalen, 1934; Sako, 1934, 1936; Täuber, 1890, 1891, 1892; Täuber and Loewenherz, 1891; Warren, 1942). However, the utility of the reaction is limited by the inaccessibility of the necessary 2,2′-diaminobiaryls (Buu-Hoï, Hoán and Khôi, 1949). The mechanism of this deaminative cyclization is related to the latter stages of the mechanism of the Fischer indole synthesis (see Chapter II, A).

Other examples of the reaction between arylols or arylamines with arylhydrazines in the presence of sodium bisulphite have appeared (Bucherer and Schmidt, M., 1909; Bucherer and Seyde, 1909; Bucherer and Sonnenburg, 1910; Bucherer and Wahl, R., 1922; Bucherer and Zimmermann, 1922; Friedländer, 1916, 1921; Fuchs and Niszel, 1927; I. G. Farbenind. A.-G., 1929, 1930b, 1930e, 1931a, 1931b; König and Haller, 1920; Zander and Franke, W., 1963; Zander and Franke, W. H., 1969), the synthesis reported in the penultimate of these papers being claimed in part of a patent eleven years later (Murakami, Y. and Morimoto, 1974). However, its use is limited because of the inaccessibility of the necessary naphtholic starting materials (Buu-Hoï, Hoán and Khôi, 1949).

At its inception, the similarity between the conversion of a mixture of **84** (R^1–R^3 = H) with phenylhydrazine into **85** (R^1–R^5 = H) and the Fischer indole synthesis was recognized (Bucherer and Seyde, 1908) although the mechanism of the latter reaction was far from understood at this time and was only to become recognized many years later (see Chapter II). Thus, the mechanism initially proposed for the Bucherer carbazole synthesis, as exemplified by the reaction between 2-naphthol and phenylhydrazine in the presence of sodium bisulphite, was as shown in Scheme 2 in which the rearrangement of the intermediate **95** ultimately affords the carbazolic product (Bucherer and Seyde, 1908), a proposal which was later propagated (Bucherer and Sonnenburg, 1910). Subsequent criticism of this mechanistic scheme was made on the basis that it was without substantial experimental support and that it was improbable that an azo compound would form from a hydrazo compound in the presence of sodium bisulphite (Friedländer, 1921). Furthermore, compounds **85** (R^1–R^3 = R^5 = H, R^4 = CH_3 and C_2H_5) were formed when 2-naphthol was reacted with N_α-methylphenylhydrazine (Friedländer, 1921) and N_α-ethylphenylhydrazine (Friedländer, 1916), respectively, in the presence of bisulphite ion. These results clearly eliminated the proposal as outlined in Scheme 2 and led to the postulation of an alternative mechanism for the reaction in which the 2-naphthol, in the form of its keto tautomer **96**, reacts with the sodium bisulphite. The resulting adduct **97** then reacts with phenylhydrazine to afford **98** which ultimately yields **85** (R^1–R^5 = H) (Friedländer, 1921). A similar although slightly extended mechanistic proposal was reached independently by others with regard to a closely related reaction (König and Haller, 1920). The

$\mathbf{84}$ (R^1-R^3 = H) ⟶ [2-naphthyl $OSO_2^{\ominus}Na^{\oplus}$] ⟶ [2-naphthyl N(H)—N(H)—C_6H_5]

↓

95 [2-naphthyl N(H)—N($SO_3^{\ominus}Na^{\oplus}$)—$C_6H_5$] ⟵ [2-naphthyl N=N—$C_6H_5$]

↓

85 (R^1-R^3 = R^5 = H, R^4 = $SO_3^{\ominus}Na^{\oplus}$) ⟶ **85** ($R^1-R^5$ = H)

Scheme 2

proposal, as outlined, remained unchallenged for several decades (e.g. Sumpter and Miller, F. M., 1954) although the analogy of the reaction with the Fischer indole synthesis and of its mechanism with that proposed by G. M. and R. Robinson for the Fischer indolization (Robinson, G. M. and Robinson, R., 1918, 1924) (see Chapter II, A) was made during this period (Fuchs and Niszel, 1927).

As part of a programme of extensive studies, which have been reviewed by one of the investigators involved (Seeboth, 1967), upon the Bucherer reaction, it was established that sodium bisulphite reacted with 1- and 2-naphthol to form **99** (Rieche and Seeboth, 1958a, 1960a) and **100** (Rieche and Seeboth, 1960a, 1960b) respectively. These adducts reacted with phenylhydrazine under acidic conditions to ultimately afford 1,2-benzocarbazole and 3,4-benzocarbazole, respectively (Rieche and Seeboth, 1960c). The latter synthesis was

O

96

OH

$SO_3^{\ominus}Na^{\oplus}$

97

N—N(H)—C₆H₅ ring

98

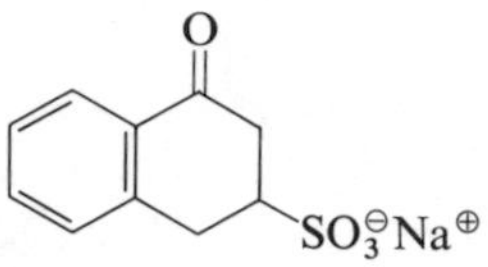

99

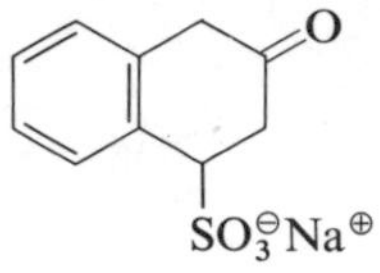

100

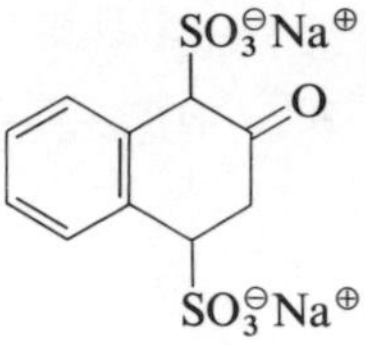

101

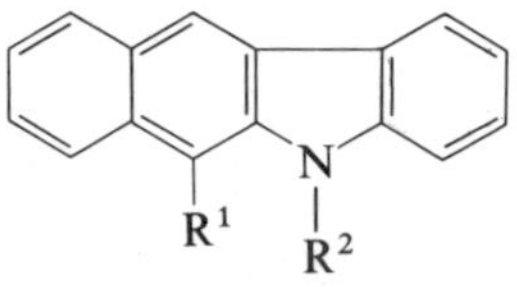

102

extended to the preparation of 3,4-benzo-9-butyl-, 3,4-benzo-9-methyl- and 3,4-benzo-6-bromocarbazole by reaction of **100** with N_α-butyl-, N_α-methyl- and 4-bromophenylhydrazine, respectively, under acidic conditions (Seeboth, Bärwolff and Becker, B., 1965). Another related reaction involved the conversion of **101** into **102** [$R^1 = H$, $R^2 = (CH_2)_3CH_3$] by reaction with N_α-butylphenylhydrazine in acetic acid and by a similar reaction with phenylhydrazine, with subsequent treatment of the product with hydrochloric acid, to afford **102** ($R^1 = SO_3H$, $R^2 = H$) (Seeboth, Neumann and Görsch, 1965). Starting with 1-naphthols and 1-naphthylamines, reaction with phenylhydrazine and sodium bisulphite yielded stable 1-tetralone phenylhydrazone-3-sulphonic acids which afforded 1,2-benzocarbazoles, together with other products, upon treatment with acids. In the corresponding 2-naphthalene series, the reaction did not usually stop at the hydrazones but afforded the 3,4-benzo-1,2-dihydrocarbazole-2-sulphonic acids which under the action of acids or bases furnished 3,4-benzocarbazoles (Seeboth, 1967). These studies led to the conclusion (Rieche and Seeboth, 1958b, 1960c; Seeboth, 1967) that the Bucherer carbazole synthesis, using 1- and 2-naphthol or 1- and 2-naphthylamine, proceeds via addition of the sodium bisulphite to afford **99** and **100**, respectively, which then form arylhydrazones. These then undergo Fischer indolization and finally loss of sodium bisulphite to form the corresponding benzocarbazoles. The mechanism proposed for these transformations was fully consistent with contemporary knowledge of the mechanism of Fischer indolization (see Chapter II) and with the isolation of intermediates.

F. Preparation of Arylhydrazines and Arylhydrazones

The scope of the Fischer indolization is further widened by the several versatile syntheses of arylhydrazines which have been developed. A brief but useful review of these has been given (Brown, R. K., 1972a). Arylhydrazines have been widely prepared by the reduction of the corresponding aryldiazonium salts. Several chemical reducing agents have been employed to carry out this change and a large number of examples can be found by reference to the appropriate sections of this present text. However, attempted reduction of 2-benzyloxyphenyldiazonium chloride with either tin(II) chloride in concentrated hydrochloric acid or with sodium sulphite failed, only 2-benzyloxyaniline hydrochloride being isolated from the reaction mixtures (Ek and Witkop, 1954). Electrochemical reductions of aryldiazonium salts to arylhydrazines have also been effected (Fioshin, Girina and Mamaev, 1956, and references therein quoted). Arylhydrazines which are unstable can be prepared and stored as their salts, usually their hydrochlorides. N_{α}-Alkylated or N_{α}-arylated phenylhydrazines are usually prepared by chemical reduction of the N-nitroso compounds of the corresponding secondary amines, recent examples using an aluminium–mercury amalgam in isopropanol as the reducing agent (Merck and Co., Inc., 1964a; Sarett and Shen, T.-Y., 1966a, 1966b). Lithium aluminium hydride, if used under carefully controlled conditions, is a very suitable reagent for these reductions of N-nitroso compounds (Poierier and Benington, 1952). Electrochemical reduction has also been used to achieve these transformations (Wells, J. E., Babcock and France, 1936). Again, unstable arylhydrazines thus produced can be isolated and stored as their salts or they can be prepared in the presence of a suitable aldehyde or ketone and converted, *in situ*, by the addition of an acid catalyst, into the corresponding indole (Clemo and Perkin, 1924a, 1924b; Clemo, Perkin and Robinson, R., 1924; Clifton and Plant, 1951; Linnell and Perkin, 1924; Plant and Rippon, 1928; Rogers, C. U. and Corson, 1947). Similarly, the solution of arylhydrazine salt resulting from reduction of an aryldiazonium salt or N-nitroso compound can be treated with an aldehyde or ketone and the resulting hydrazone, without isolation, converted directly into the indole (Ficken and Kendall, 1959, 1961; Khan, M. A. and Morley, 1978, 1979; Millson and Robinson, R., 1955), a procedure which is particularly useful when the arylhydrazine or arylhydrazone is unstable to heat or light. N_{α}-Alkylated or N_{α}-arylated arylhydrazines can be prepared by reacting the N_{α}-sodium salt of the arylhydrazine with alkyl halides or reactive aryl halides. The required sodium salts are prepared by reacting the arylhydrazine with sodium (Michaelis, 1886, 1889; Neuberg and Federer, 1905; Philips, 1889; Singh, B. K., 1913) [a method which ultimately also afforded traces of the N_{α},N_{β}-dialkylated arylhydrazine (Cornforth, J. W., Cornforth, R. H., Dalgliesh and Neuberger, 1951)], sodium ethoxide (Kratzl and Berger, K. P., 1958), sodamide (Audrieth, Weisiger and Carter, H. E., 1941; Crowther, Mann and Purdie, 1943; Grammaticakis, 1940a) or sodium hydride (Merck and Co. Inc., 1961; Palazzo and Baiocchi, 1965; Sterling Drug Inc., 1962).

Occasionally, the formation of hydrazines by reduction of the corresponding amine N-nitroso derivative either gave only very low product yield or failed altogether. Thus, chemical reduction of **103** (R = NO) caused cleavage of the ring system and catalytic hydrogenation caused denitrosation (Cattanach,

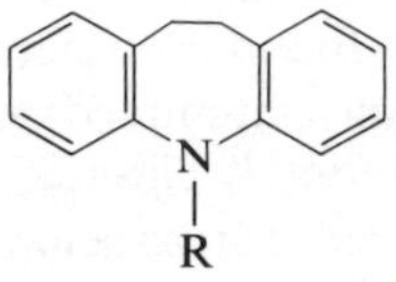

103

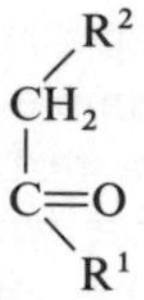

104

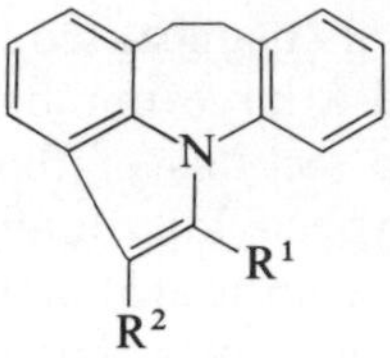

105

Cohen, A. and Heath-Brown, 1973). These failures were overcome by reaction of **103** (R = NO) with ethyl magnesium iodide which yielded **103** (R = N=$CHCH_3$), the acetaldehyde hydrazone of the required hydrazine. Since acetaldehyde arylhydrazones do not readily indolize (see Chapter IV, A), **103** (R = N=$CHCH_3$) could be reacted with **104** under indolization conditions, when transhydrazonation and subsequent indolization to give **105** occurred (Cattanach, Cohen, A. and Heath-Brown, 1973; Cohen, A., Heath-Brown and Cattanach, 1968; F. Hoffmann–La Roche and Co. A. G., 1965). This sequence has also been utilized by initially treating **103** (R = NO) with other alkyl magnesium iodides (Roche Products Ltd, 1965a, 1965c). Compound **103** (R = NO) has been satisfactorily reduced to **103** (R = NH_2) using an ethereal solution of chloramine (Cattanach, Cohen, A. and Heath-Brown, 1973; Roche Products Ltd, 1965b, 1968b).

Reduction of **106** (R = NO) into **106** (R = NH_2), with zinc dust in acetic acid, failed (Morton and Slaunwhite, 1949). However, it would appear that other methods of reduction were successful since **106** (R = NH_2) has been utilized in Fischer indolizations (Mooradian, 1975, 1976c). Similarly, whereas reduction of 2-cyanobenzenediazonium chloride with tin(II) chloride furnished only 3-aminoindazole (Aron and Elvidge, 1976), 2-cyanophenylhydrazones have been synthesized and indolized (Morooka, Tamoto and Matuura, 1976).

A method employed widely for the synthesis of arylhydrazones is the Japp–Klingemann reaction. When ethyl acetoacetate (**107**; $R^1 = R^2 = H, R^3 = C_2H_5$) was condensed initially with benzenediazonium nitrate in dilute potassium hydroxide solution, a product was isolated which was formulated as **108** (R = H)

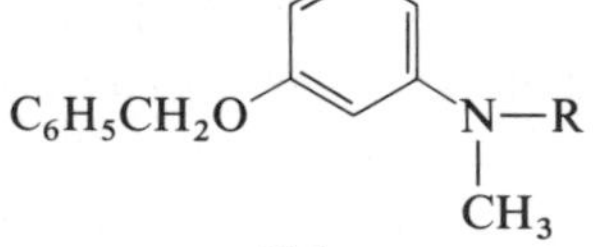

$$R^1—CH_2—C(=O)—CH(R^2)—COOR^3$$

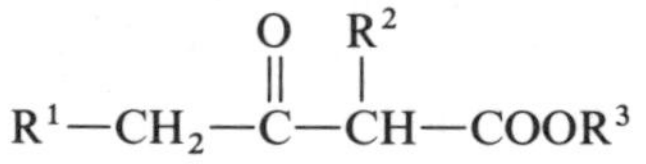

107

106

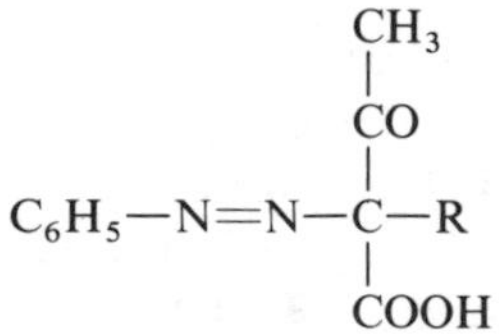

108

(Meyer, V., 1877). Ten years later, in an attempt to prepare **108** (R = CH_3) by treating **107** (R^1 = H, R^2 = CH_3, R^3 = C_2H_5) with benzenediazonium chloride, it was recognized that during the course of the reaction the acetyl group had been eliminated and a product was obtained which was formulated as **109** (R = C_2H_5) (Japp and Klingemann, 1887a) [many decades later (Heath-Brown and Philpott, P. G., 1965b) the intermediate azoester **110** in this

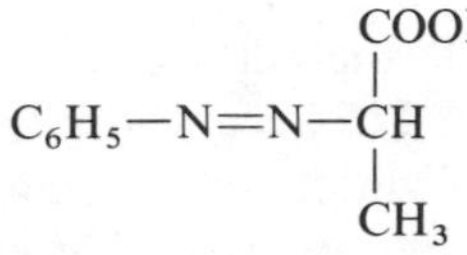

$$C_6H_5{-}N{=}N{-}CH(COOR)(CH_3)$$

109

$$C_6H_5{-}N{=}N{-}C(COOC_2H_5)(COCH_3){-}CH_3$$

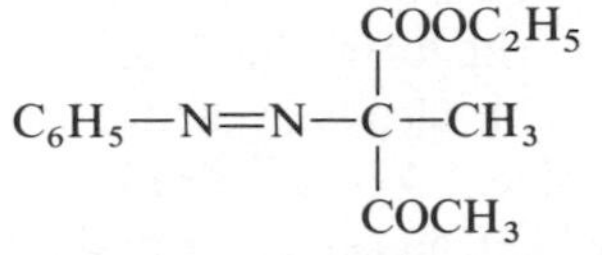

110

reaction was isolated as a distillable yellow oil]. Subsequently (Japp and Klingemann, 1887b, 1887d, 1888a, 1888c), the hydrolysis product of this substance was shown to be the same as the compound obtained earlier by condensing pyruvic acid with phenylhydrazine (Fischer, Emil and Jourdan, 1883) and in the earlier of these four publications the possibility of its formulation as the hydrazone (the term 'hydrazide' was used by Japp and Klingemann) **111** (R^1–R^3 = H) rather than **109** (R = H) was considered. In the later three publications (Japp and Klingemann, 1887d, 1888a, 1888c) this reformulation was verified and the structural postulation **109** (R = C_2H_5) was also corrected to **111** (R^1 = R^2 = H, R^3 = C_2H_5) and the reaction was extended to the preparation of **111** (R^1 = R^3 = H, R^2 = CH_3) [identical with the product obtained earlier by condensing 2-ketobutyric acid with phenylhydrazine (Wislicenus and Arnold, E., 1887)] and **111** (R^1 = 2-CH_3 and 4-CH_3, R^2 = R^3 = H) [identical with the products obtained earlier by condensing pyruvic acid with 2- and 4-methylphenylhydrazine hydrochloride, respectively (Raschen, 1887)]. Starting with the sodium salts of acetoacetic acid (**107**; R^1–R^3 = H) and its homologues **107** (R^1 = R^3 = H, R^2 = CH_3 and C_2H_5) rather than the corresponding esters, coupling with benzenediazonium chloride was found to involve concomitant decarboxylation, rather than deacetylation, to afford **112** (R = H) (Japp and Klingemann, 1888c – see also 1887c), **112**

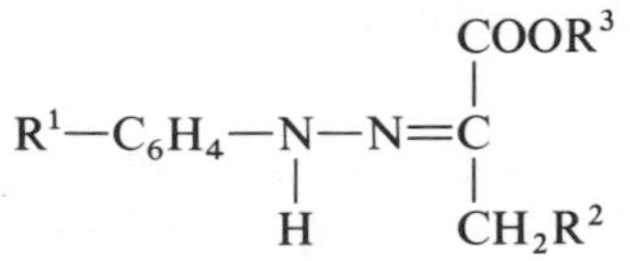

$$R^1{-}C_6H_4{-}N(H){-}N{=}C(COOR^3)(CH_2R^2)$$

111

$$C_6H_5{-}N(H){-}N{=}C(COCH_3)(R)$$

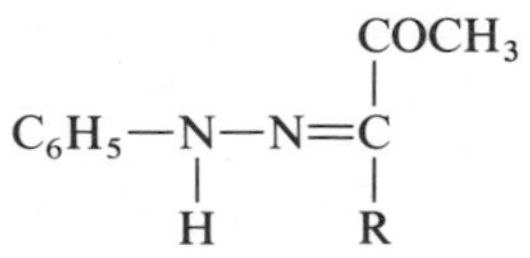

112

(R = CH_3) (Japp and Klingemann, 1881a, 1888b, 1888c, 1888d) and **112** (R = C_2H_5) (Japp and Klingemann, 1888a, 1888b, 1888c), respectively.

In connection with its use in the Fischer indole synthesis, the Japp–Klingemann reaction can be regarded as involving the electrophilic attack by an aryldiazonium salt **113** upon the anionic carbon atom of an active methylenyl or methinyl compound, **114**, under neutral, slightly acidic, but usually alkaline conditions. This forms the azo compound **115** which under thermal, alkaline, or, best of all, acidic conditions, is converted into **116** or **117**, depending upon the composition of the original active hydrogen compound **114** used. If **114** contains a free carboxylic acid group (R^2 = H), then synchronous decarboxylation occurs to afford **116**, the monoarylhydrazone of a 1,2-diketone. Upon indolization this will form the 2-acylindole **118**. If the carboxylic acid group is esterified in **114**, cleavage of the acyl group in **115** occurs to afford the arylhydrazone of a 2-ketoester, **117**. Upon indolization this forms **119** (R^5 = $COOR^2$) which upon hydrolysis followed by decarboxylation gives **119** (R^5 = H). A kinetic study of the reaction has been made (Genkina, Gordeev and Suvorov, 1976). Starting with diethyl 2-acetoglutarate (**114**; $R^2 = C_2H_5$, $R^3 = CH_2COOC_2H_5$, $R^4 = CH_3$) or ethyl 2-(2-cyanoethyl)acetoacetate (**114**; $R^2 = C_2H_5$, $R_3 = CH_2CN$, $R^4 = CH_3$), the reaction pathway has ultimately furnished 3-indolylacetic acids. Similarly, 3-indolyl-3-propionic acids and 3-indolyl-4-butyric acids are ultimately produced starting with 2-carbethoxycyclopentanone [**114**; $R^2 = C_2H_5$, $R^3 + R^4 = (CH_2)_2$], and 2-carbethoxycyclohexanone [**114**; $R^2 = C_2H_5$, $R^3 + R^4 = (CH_2)_3$], respectively. When the corresponding acids of these two cyclic 3-ketoacids are coupled with aryldiazonium salts, the corresponding monoarylhydrazones of cyclopentan-1,2-diones and cyclohexan-1,2-diones, respectively, are formed, as they are also from 2-formylcyclopentanone and 2-formylcyclohexanone, respectively. Piperidin-2-one-3-carboxylic acids likewise react with diazonium salts to yield piperidin-2, 3-dione 3-arylhydrazones which can be converted into tryptamines via the Fischer indolization. The uses of the Japp–Klingemann reaction in these above manners are to be found in the appropriate sections of this present study which are concerned with the indolizations of the arylhydrazones thus formed.

The Japp–Klingemann reaction was comprehensively reviewed up to 1959 with respect to its broader applications although specific reference was made to those arylhydrazones which were subsequently indolized (Phillips, 1959). A subsequent review of the reaction (Brown, R. K., 1972b) laid particular emphasis upon its use in the preparation of arylhydrazones which were ultimately subjected to Fischer indolization.

Under the conditions employed for most Japp–Klingemann reactions, the intermediate azo compounds **115** are unstable and the reactions give only the corresponding arylhydrazones **116** or **117**. However, several examples have now been reported in which the intermediate azo compounds have been isolated, these being referred to in the more recent review of the reaction (Brown, R. K., 1972b). Furthermore, such intermediate azo compounds have sometimes been directly subjected to indolization. Thus, treatment of **120** ($R^1 = OCH_3$ and $OCH_2C_6H_5$, R^2 = H) with ethanolic hydrogen chloride, of

R^1–C_6H_4–$N_2^{\oplus}X^{\ominus}$ + H–C(COR^4)(CH_2R^3)–$COOR^2$

113 **114**

↓

115

↓ ↓

116 **117**

↓↓↓ ↓↓↓

118 **119** (R^5 = $COOR^2$)

120 [R^1 = Cl, R^2 = C_6H_5 and $C_6H_4F(2)$] with ethanolic hydrogen chloride or isopropanolic sulphuric acid, of **120** [R^1 = Cl, R^2 = $C_6H_4Cl(2)$] with ethanolic hydrogen chloride, of **120** [R^1 = $C(CF_3)_2OH$, R^2 = C_6H_5] with concentrated hydrochloric acid in acetic acid, of **122** with dry hydrogen chloride in absolute ethanol, of **124** (R^1 = H and OCH_3, R^2 = H) with concentrated hydrochloric acid in acetic acid and of **126** with ethanolic hydrogen chloride afforded respectively **121** (R^1 = OCH_3 and $OCH_2C_6H_5$, R^2 = H) (Heath-Brown and Philpott, P. G., 1965b), **121** [R^1 = Cl, R^2 = C_6H_5 and $C_6H_4F(2)$] (Inaba, Ishizumi and Yamamoto, H., 1971; Yamamoto, H., Inaba, Hirohashi, T., Mori, Ishizumi and Maruyama, 1969; Yamamoto, H., Inaba, Okamoto, Hirohashi, T., Yamamoto, M., Mori, Ishiguro, Maruyama and Kobayashi, T., 1972c), **121** [R^1 = Cl, R^2 = $C_6H_4Cl(2)$] (Yamamoto, H., Inaba, Hirohashi,

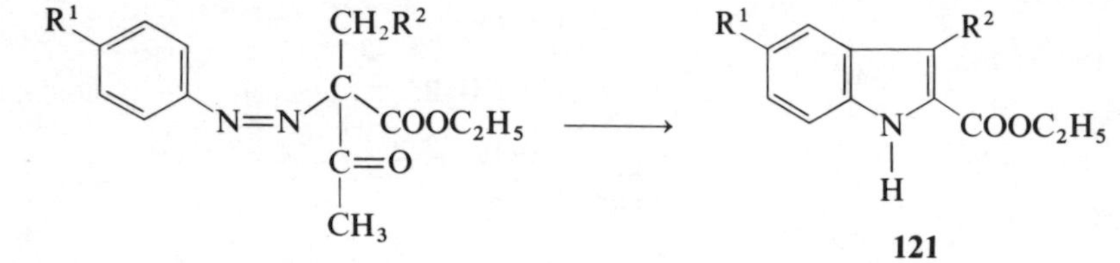

120 **121**

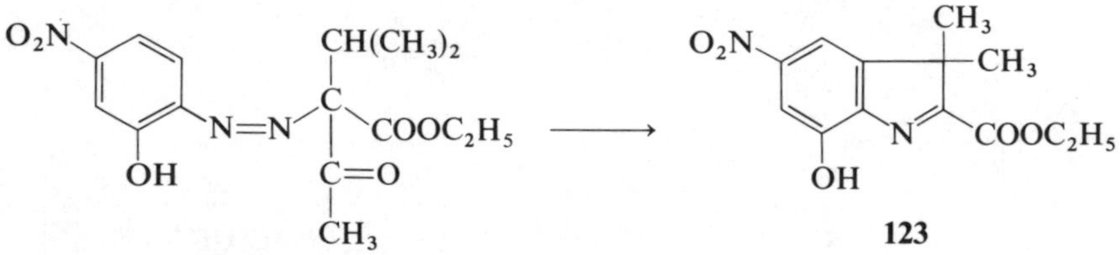

122 **123**

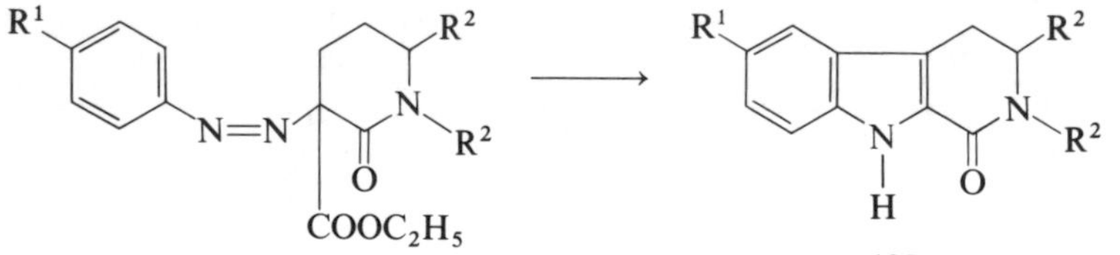

124 **125**

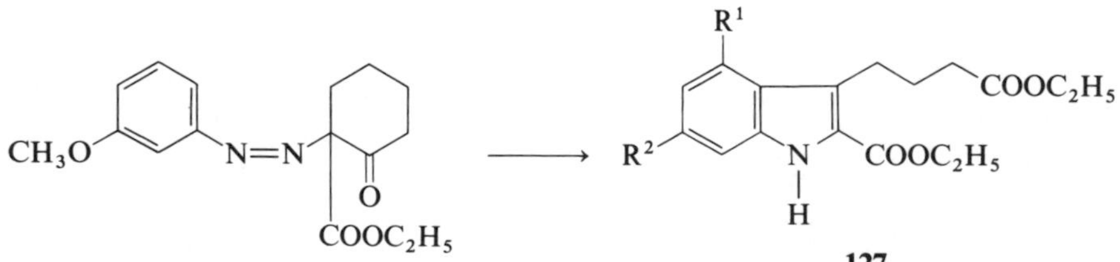

126 **127**

T., Mori, Ishizumi and Maruyama, 1969), **121** [R^1 = $C(CF_3)_2OH$, R^2 = C_6H_5] (Dalton, Fahrehnoltz and Silverzweig, 1979), **123** (Ried and Kleemann, 1968), **125** (R^1 = H and OCH_3, R^2 = H) (Henecka, Timmler, Lorenz and Geiger, 1957) and **127** (R^1 = OCH_3, R^2 = H and R^1 = H, R^2 = OCH_3) (Iyer, Jackson, A. H., Shannon and Naidoo, 1972, 1973). Other examples can be found in the patent literature (Henecka, Timmler and Lorenz, 1959; Reynolds, B. and Carson, J., 1969) and in the very recent literature (Alyab'eva, Khoshtariya, Vasil'ev, A. M., Tret'yakova, Efimova and Suvorov, 1979).

An interesting extension of the Japp–Klingemann reaction involves the coupling of aryldiazonium salts with enamines. The pioneering studies in this area were elaborated when **128** (R^1 = C_2H_5 and C_6H_5, R^2 = H) was treated with 4-chloro-, 4-methoxy- and 4-nitrobenzenediazonium chloride in cold dilute hydrochloric acid to give **129** (R^1 = C_2H_5 and C_6H_5, R^3 = 4-Cl, 4-OCH_3 and 4-NO_2, respectively) via a postulated mechanism shown in Scheme 3. Compound **129** (R^1 = C_6H_5, R^3 = 4-COOH) was also similarly synthesized. In support of this mechanism it was found that when **128** (R^1 = C_6H_5, R^2 = CH_3) and **128** (R^1 = R^2 = CH_3) were similarly reacted with

Scheme 3

4-carboxy-, 4-nitro- and 2,4-dinitrobenzenediazonium chloride and 4-chloro- and 4-nitrobenzenediazonium chloride, respectively, the products were acetophenone 4-carboxy-, 4-nitro- and 2,4-dinitrophenylhydrazone and acetone 4-chloro- and 4-nitrophenylhydrazone, respectively. In these reactions, the quaternary nature of the enamine C-3 atom blocks the mechanism shown in Scheme 3 by preventing the shift in the first intermediate which could thus undergo hydration to afford **130**, concerted cyclic rearrangement of which

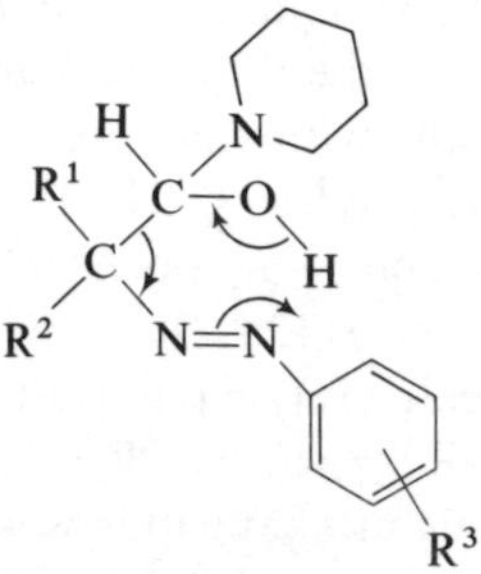

130

as shown would afford the arylhydrazones (Crary, Quayle and Lester, C. T., 1956). In a subsequent extension of this synthesis, the arylhydrazones **129** ($R^1 = CH_3$, $R^3 = H$, 2-NO_2, 3-Cl, 4-CH_3, 4-OCH_3 and 4-NO_2 and $R^1 = C_2H_5$, $R^3 = H$, 3-Cl, 4-CH_3 and 4-OCH_3) were prepared by reaction of the piperidine enamines of propionaldehyde and butyraldehyde, respectively, with the appropriate aryldiazonium salts (Shvedov, Altukhova and Grinev, A. N., 1966a). Further extensions involved the coupling of aryldiazonium salts with ketone enamines. Thus, cyclohexanone pyrrolidine enamine **131** ($R^3 = R^4 = H$, $m = n = 1$) reacted with benzenediazonium fluoroborate to produce, after hydrolysis of the intermediate iminium salt, cyclohexan-1,2-dione monophenylhydrazone (Kuehne, 1962), **132** ($R^1 = R^2 = H$; $R^1 = H$, $R^2 = CH_3$, OCH_3 and NO_2 and $R^1 = CH_3$, OCH_3 and NO_2, $R^2 = H$) reacted with **131** ($R^3 = R^4 = H$, $m = 2$, $n = 1$) to yield the corresponding **133** (the morpholine enamine did not react and the hexamethylene enamine furnished only low yields of arylhydrazones) (Shvedov, Altukhova and Grinev, A. N., 1964, 1965a), the syntheses of **133** [$R^1 = Cl$, COOH, $COOC_2H_5$, $CH_2COOC_2H_5$ and

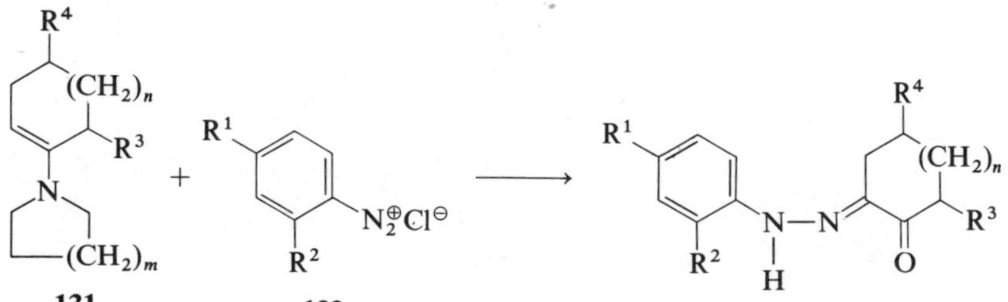

131 + **132** ⟶ **133**

$(CH_2)_2COOC_2H_5$, R^2–R^4 = H, $n = 1$; $R^1 = OCH_3$, $R^2 = NO_2$, $R^3 = R^4$ = H, $n = 1$; $R^1 = R^2 = R^4$ = H, $R^3 = CH_3$, $n = 1$; $R^1 = CH_3$, OCH_3 and Cl, $R^2 = R^4$ = H, $R^3 = CH_3$, $n = 1$; $R^1 = CH_3$, OCH_3 and Cl, $R^2 = R^4$ = H, $R^3 = CH_3$, $n = 1$; R^1–R^3 = H, $R^4 = OCH_3$, $n = 1$; $R^1 = CH_3$ and OCH_3, $R^2 = R^3$ = H, $R^4 = OCH_3$, $n = 1$; R^1–R^4 = H, $n = 2$ and $R^1 = CH_3$, R^2–R^4 = H, $n = 2$] were achieved by analogous reactions (Shvedov, Altukhova and Grinev, A. N., 1965b, 1966b) and **134** (R = H, X = CH_2, $m = 2$, $n = 1$ and 2 and R = H, X = O, $m = n = 2$) reacted with benzenediazonium chloride and 4-methoxybenzenediazonium chloride to afford

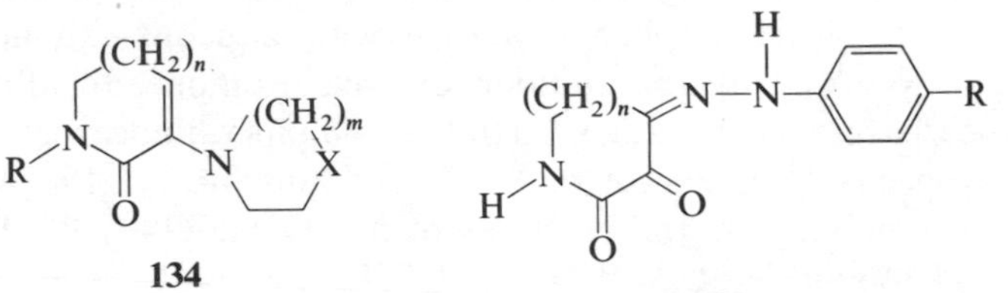

134

135

135 (R = H and OCH_3, $n = 1$ and 2) (Glushkov, Volskova, Smirnova and Magidson, 1969). Similarly, when **136** (R = H and C_2H_5) reacted with benzenediazonium chloride, **137** (R = H) (Jackson, A. and Joule, 1967; Jackson, A., Wilson, N. D. V., Gaskell and Joule, 1969) and **137** (R = C_2H_5)

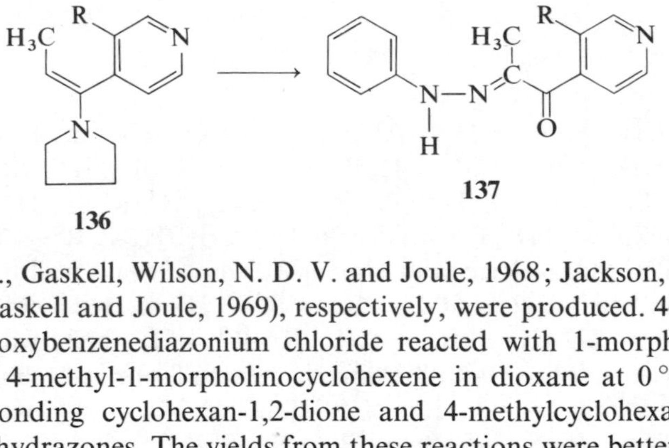

136

137

(Jackson, A., Gaskell, Wilson, N. D. V. and Joule, 1968; Jackson, A., Wilson, N. D. V., Gaskell and Joule, 1969), respectively, were produced. 4-Benzyloxy- and 4-methoxybenzenediazonium chloride reacted with 1-morpholinocyclohexene and 4-methyl-1-morpholinocyclohexene in dioxane at 0 °C to afford the corresponding cyclohexan-1,2-dione and 4-methylcyclohexan-1,2-dione 2-monoarylhydrazones. The yields from these reactions were better than those obtained by reacting the corresponding aryldiazonium chlorides with potassium 2-oxocyclohexane-1-carboxylate (Bisagni, Ducrocq, Lhoste, Rivalle and Civier, 1979). 1-Morpholinocyclohexene and 4-methyl-1-morpholinocyclohexene reacted analogously with other aryldiazonium salts (Bisagni, Ducrocq and Hung, 1980). A discussion of the earlier of these coupling reactions has appeared in a review of enamines in organic synthesis (Kuehne, 1969).

Other methods of considerable potential have been devised for the synthesis of arylhydrazines. Thus, 3-arylsydnones **138**, prepared from the

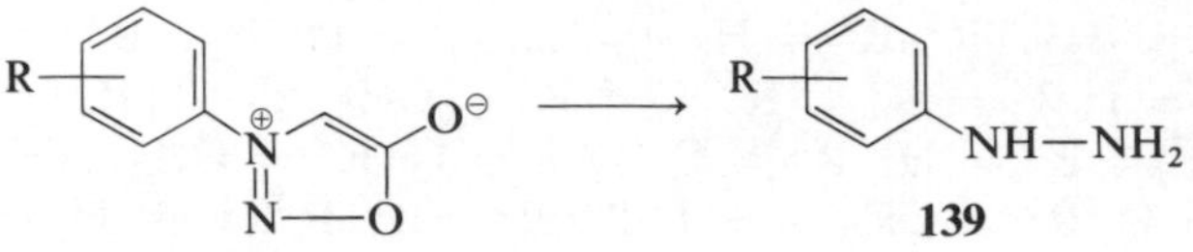

$$\text{138} \longrightarrow \text{139}$$

R—C$_6$H$_4$—$\overset{\oplus}{N}$(=N—O—C(O$^{\ominus}$)=CH) **138** ⟶ R—C$_6$H$_4$—NH—NH$_2$ **139**

corresponding arylamines, undergo acid catalysed conversion into the corresponding arylhydrazines **139** (Fugger, Tien and Hunsberger, 1955; Hadáček and Švehla, 1954; Tien and Hunsberger, 1955a, 1955b, 1955c). The *in situ* formation of arylhydrazines from such reactions has also been effected by treating 3-phenyl- or 3-(2-piperidinophenyl)sydnone with aqueous ethanolic hydrochloric acid, followed by the addition of cyclohexanone, to afford 1,2,3,4-tetrahydrocarbazole and 1,2,3,4-tetrahydro-8-piperidinocarbazole hydrochloride, respectively (Ainsworth, D. P. and Suschitzky, 1967a). Another method is exemplified by the reactions of mesitylene (**140**; R^1–R^3 = CH_3, R^4 = H) and **140** [R^1 = R^2 = $(CH_2)_2CH(CH_3)_2$, R^3 = R^4 = H] with diethyl azodicarboxylate (**141**) in the presence of boron trifluoride etherate to yield **142**

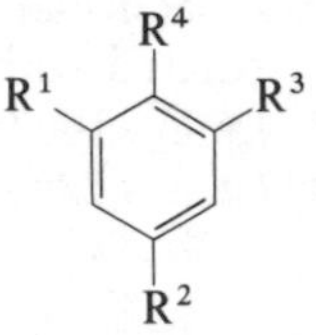

140 $H_5C_2OOC—N{=}N—COOC_2H_5$ **141**   **142**

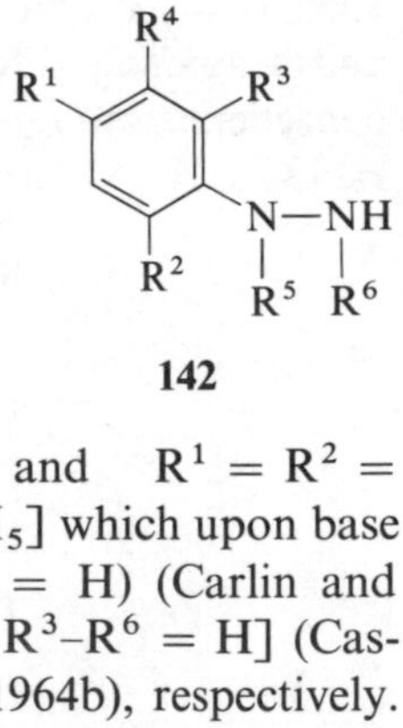

[R^1–R^3 = CH_3, R^4 = H, R^5 = R^6 = $COOC_2H_5$ and R^1 = R^2 = $(CH_2)_2CH(CH_3)_2$, R^3 = R^4 = H, R^5 = R^6 = $COOC_2H_5$] which upon base catalysed hydrolysis gave **142** (R^1–R^3 = CH_3, R^4–R^6 = H) (Carlin and Moores, 1962) and **142** [R^1 = R^2 = $(CH_2)_2CH(CH_3)_2$, R^3–R^6 = H] (Casnati, Langella, Piozzi, Ricci and Umani-Ronchi, 1964a, 1964b), respectively. Likewise, **140** (R^1 = R^2 = CH_3, R^3 = R^4 = H) reacted with **141** to ultimately produce **142** (R^1 = R^2 = CH_3, R^3–R^6 = H) which, without characterization, was indolized with 3,3-dimethylpentan-2-one in boiling monoethylene glycol to form a small yield of 5,7-dimethyl-2-*tert.*-butylindole (Casnati, Langella, Piozzi, Ricci and Umani-Ronchi, 1964b). Similarly, sulphuric acid can be used in place of boron trifluoride etherate as the catalyst but in such cases two molecules of **141** can also react with **140**. Thus **140** (R^1 = CH_3, R^2–R^4 = H) reacted with **141** to afford **142** (R^1 = CH_3, R^2–R^4 = H, R^5 = R^6 = $COOC_2H_5$) and **143** (R = H) and **140** (R^1 = R^3 = H, R^2 = R^4 = CH_3) similarly furnished **142** (R^1 = R^3 = H, R^2 = R^4 = CH_3, R^5 = R^6 = $COOC_2H_5$) and **143** (R = CH_3). Furthermore, selective reduction of **142** (R^1 = R^3 = H, R^2 = R^4 = CH_3, R^5 = R^6 = $COOC_2H_5$) with lithium aluminium hydride gave **142** (R^1 = R^3 = H, R^2 = R^4 = R^5 = CH_3, R^6 =

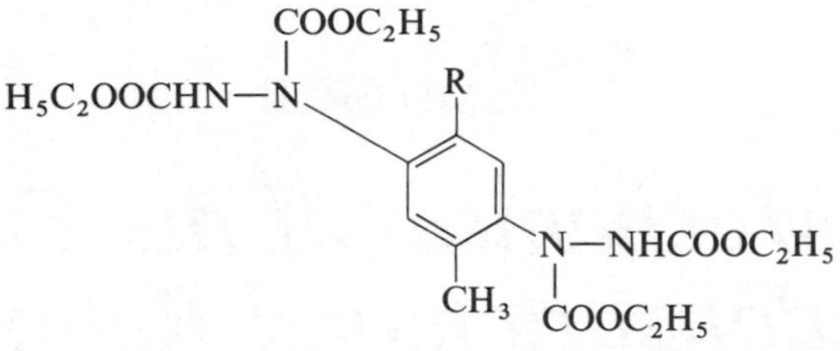
COOC2H5
H5C2OOCHN—N
R
N—NHCOOC2H5
CH3
COOC2H5

143

$COOC_2H_5$) which upon base catalysed hydrolysis gave **142** ($R^1 = R^3 = R^6 = H$, $R^2 = R^4 = R^5 = CH_3$) (Huisgen, Jakob, Siegel and Cadus, 1954). Contrary to these above acid catalysed reactions, pyrolysis of mixtures of **141** with alkylbenzenes effected addition of **141** to the C-2 atom of the alkyl substituent to yield the correspondingly substituted benzylhydrazines (Huisgen, Jakob, Siegel and Cadus, 1954).

CHAPTER II

The Mechanism of the Fischer Indolization and Related Studies

A. Mechanistic Postulations

Extensive investigations of this topic have been effected over the past several decades, during which time several mechanistic theories have been proposed.

Following the observation that the anil of acetophenone was oxidized, by phenylhydrazine or by phenylhydrazones, to 2-phenylindole, the following three-stage mechanism (Scheme 4) for the Fischer indolization was proposed (Reddelien, 1912). In this, the arylhydrazone **144** is reduced to an aniline and a ketoimine **145** (during the oxidation in the third stage), these two intermediates condense with the elimination of ammonia to give an anil **146** which is then subjected to oxidative ring closure to give the indole **147**.

Earlier, Brunner, K., (1898b) had made a comparison between the acid catalysed conversion of 2,2′-diaminobiphenyl (**148**) into carbazole and the acid catalysed indolization of methyl isopropyl ketone N_α-methylphenylhydrazone

144 $\xrightarrow{+2H}$ aniline + **145** $\xrightarrow{-NH_3}$ **146** $\xrightarrow{-2H}$ **147**

Scheme 4

(**149**) and had suggested that the latter reaction, leading to the product which he formulated as **151**, could proceed via the intermediate **150** which is analogous to the diamine **148**. However, this proposal remained unnoticed until 1957

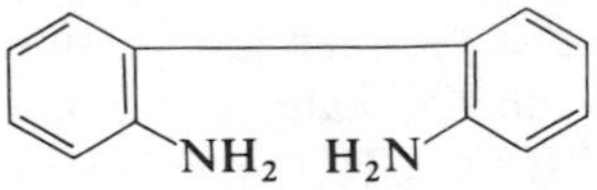

148

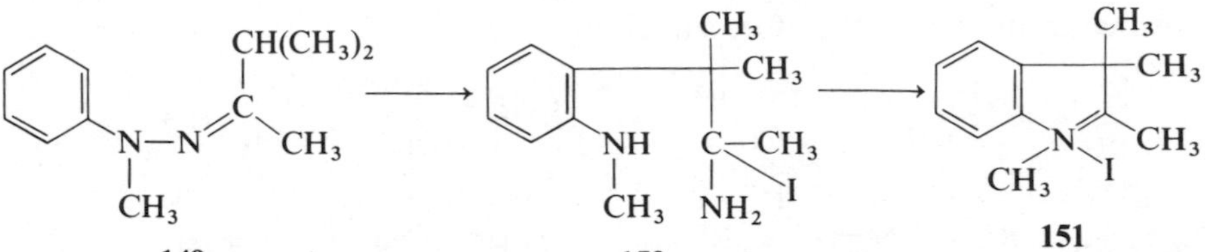

149 **150** **151**

(Arbuzov, A. E. and Kitaev, 1957d) and was not referred to by G. M. Robinson and R. Robinson when, in 1918 (Robinson, G. M. and Robinson, R., 1918), they proposed the following three-stage mechanistic scheme (Scheme 5) which essentially embodies Brunner's concept. The hydrazone **144** is transformed into the enehydrazine **152** which undergoes *ortho*-benzidine rearrangement (for reviews of the benzidine rearrangement, see Banthorpe, 1969; Ingold, 1969b; Shine, 1967a, 1969) into **153**, elimination of ammonia then affording **147**. The Robinsons gave chemical analogies for these three stages and, furthermore, extended their theory to the synthesis of 2,3,4,5-tetraphenylpyrrole

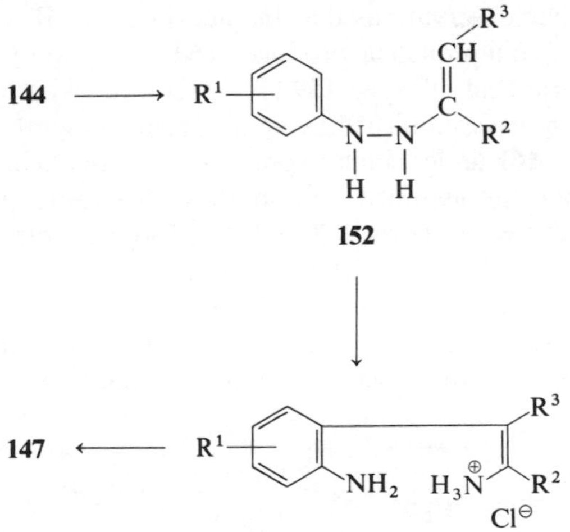

144 → **152**

147 ← **153**

Scheme 5

(**157**; $R^1 = R^2 = C_6H_5$, $R^3 = H$) from bisbenzyl phenyl ketone azine (**154**; $R^1 = R^2 = C_6H_5$), a reaction which they suggested would proceed via intermediates **155** and **156** (analogous to **152** and **153**, respectively). Unfortunately, as they appeared to be with Brunner's 1898 paper, the Robinsons were also apparently unaware of another German publication (Piloty, 1910) which eight years earlier had reported the acid catalysed conversion of bisdiethyl ketone azine (**154**; $R^1 = C_2H_5$, $R^2 = CH_3$) into **157** ($R^1 = C_2H_5$, $R^2 = CH_3$, $R^3 = H$). It is obvious from their publication that they had access to the German literature covering these two publications and their apparent unawareness of them is therefore surprising.

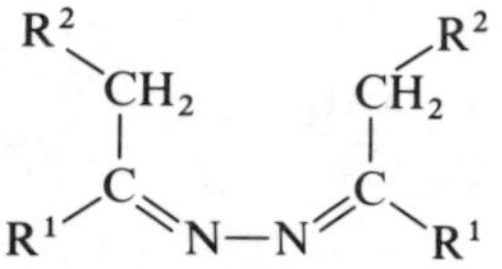

154

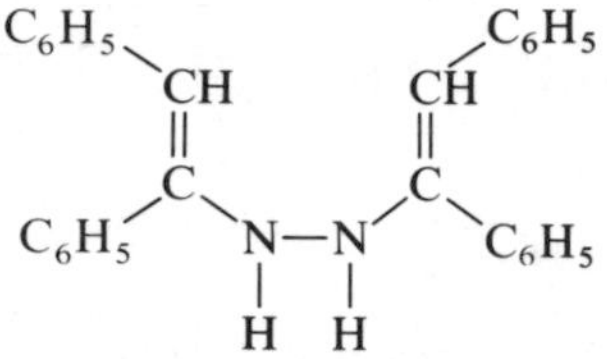

155

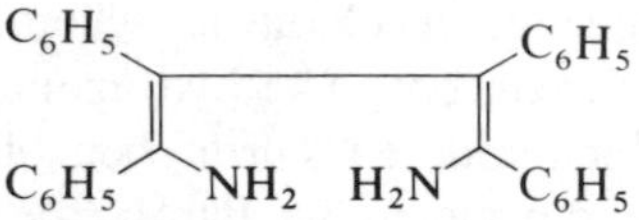

156

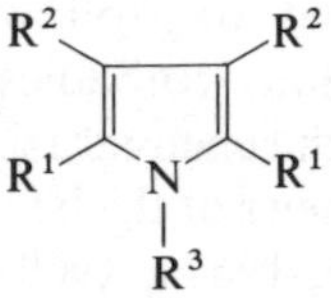

157

The Robinsons' analogy for the second and third stages of their mechanism involved the earlier observations that the action of tin(II) chloride in hydrochloric acid on 1-azonaphthalene produced **158** which upon boiling in concentrated hydrochloric acid afforded **159** (R = H) (Nietzki and Goll, 1885), the structure of which was established by an independent synthesis (Vesely, 1905) and that **160** gave **161** under similar conditions (Meisenheimer and Witte, K., 1903). The Robinsons assumed that in these reactions, **162** (R = H) was converted into **159** (R = H) via **158** and that **160** was converted into **161** via **163**. However, the following subsequent further studies of these and related reactions are at variance with this postulation.

Hydrazonaphthalenes can also rearrange to afford 2,2′-diaminobinaphthyls and dibenzocarbazoles, these being examples of reactions involving benzidine

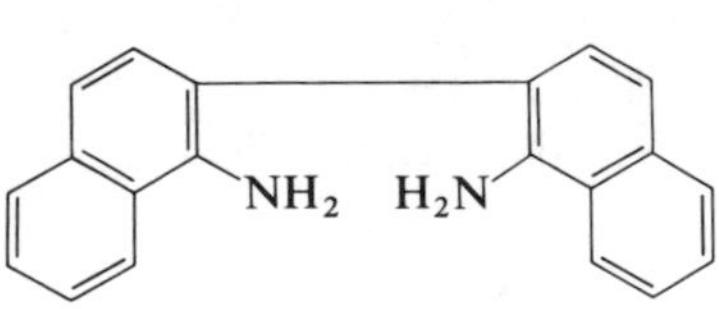

158

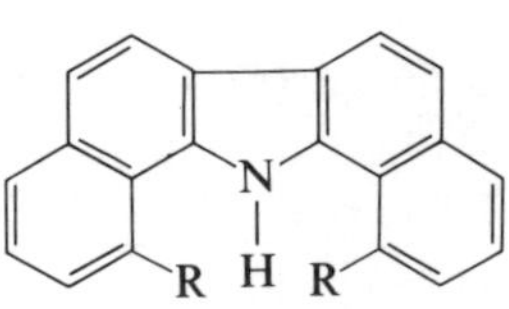

159

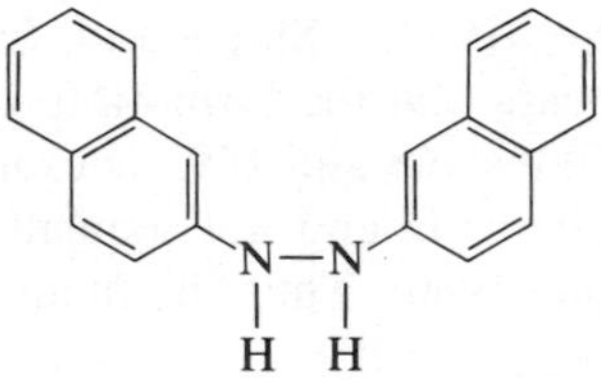

160

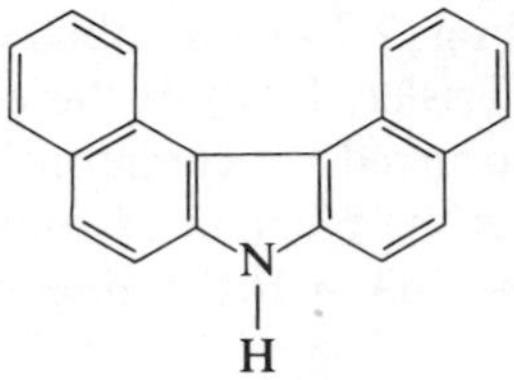

161

rearrangements (Banthorpe, 1972; Ingold, 1969b). Thus, pyrolysis of **162** (R = H) yielded **159** (R = H) together with traces of **158** and **164** (Shine and Snell, 1957), acid catalysed rearrangement of **165** yielded **166** and **71** (Shine, Huang and Snell, 1961), thermally induced rearrangement of **160** gave **161** and **163** (Shine and Trisler, 1960) and thermally induced rearrangement of **167** afforded **94** and only a very small amount of **85** (R^1–R^5 = H) (acid catalysed rearrangement of **167** afforded **94** as the only isolated product) (Shine, Huang and Snell, 1961). However, although the formation of these benzocarbazoles occurred along with the formation of the corresponding 2,2′-diaminobiaryls from the hydrazo reactant, these benzocarbazole formations occurred using conditions which were too mild to effect their formation by cyclization of the

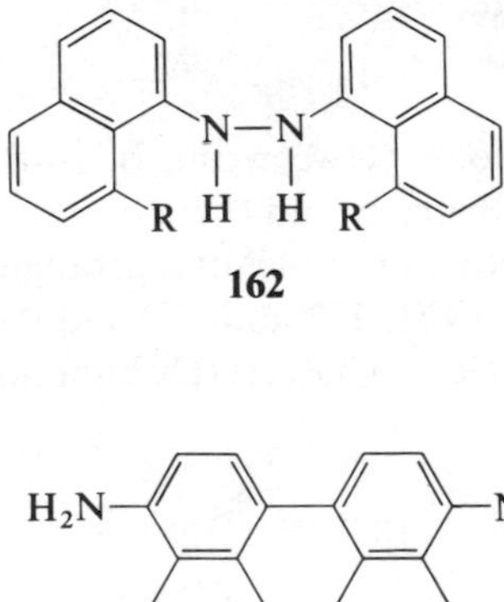

162

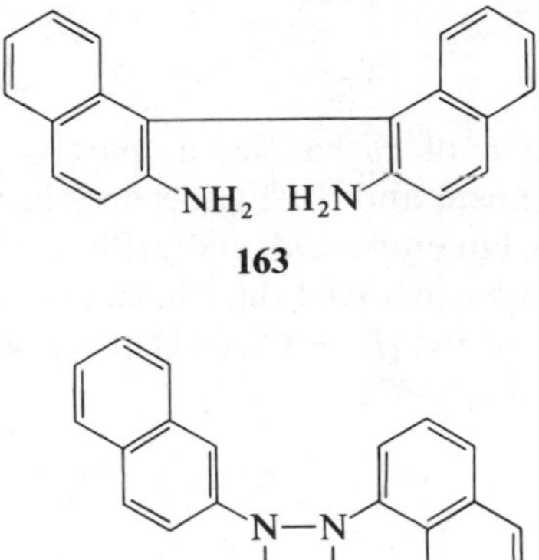

163

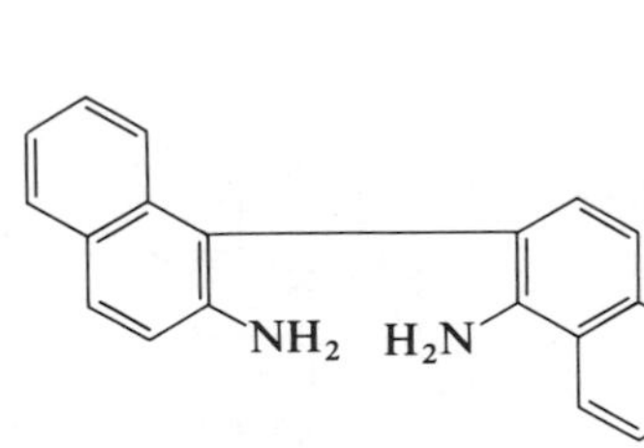

164

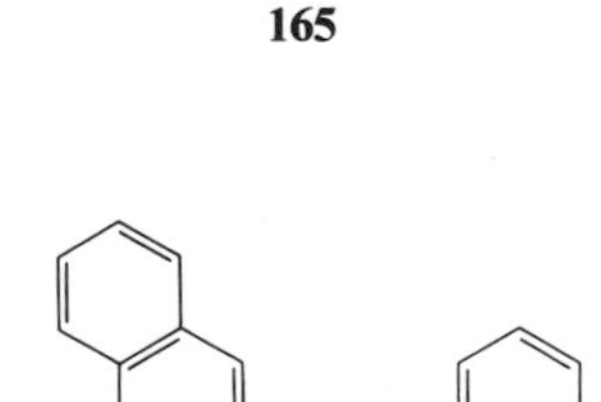

165

166

167

corresponding 2,2′-diaminobiaryls (Shine, 1967b; Shine and Snell, 1957; Shine and Trisler, 1960). It therefore appears that the formation of these two products occurred via common intermediates **168** and **169**, as exemplified by using **160** as the reactant as shown in Scheme 6, and not sequentially in the order **160** → **163** → **161** (Shine and Trisler, 1960). Thus the inclusion of this

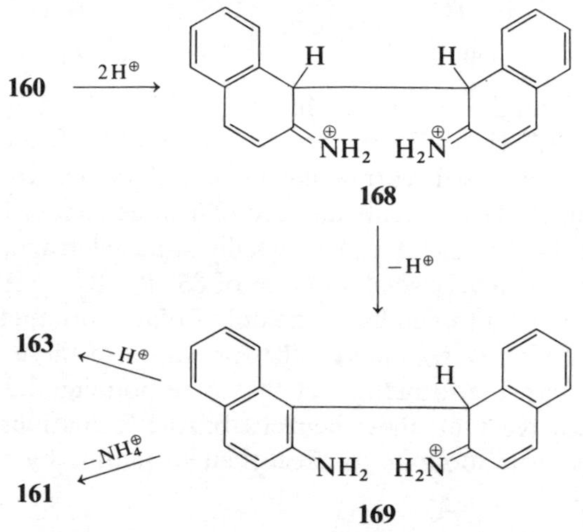

160 $\xrightarrow{2H^{\oplus}}$

H H

$\overset{\oplus}{N}H_2$ $H_2\overset{\oplus}{N}$

168

$-H^{\oplus}$

163 $\xleftarrow{-H^{\oplus}}$

H

NH_2 $H_2\overset{\oplus}{N}$

161 $\xleftarrow{-NH_4^{\oplus}}$

169

Scheme 6

formation of carbazoles as part of the analogy between the benzidine rearrangement and the Fischer indolization by Brunner and the Robinsons was correct, but only accidentally (Shine, 1967b). Other examples of rearrangements of this type involved the conversion of **170** into **171**, **172** and **173** and the conversion of **162** (R = COOH) into **174** and **159** (R = COOH) (Dokunikhin and Bystritskii, 1963).

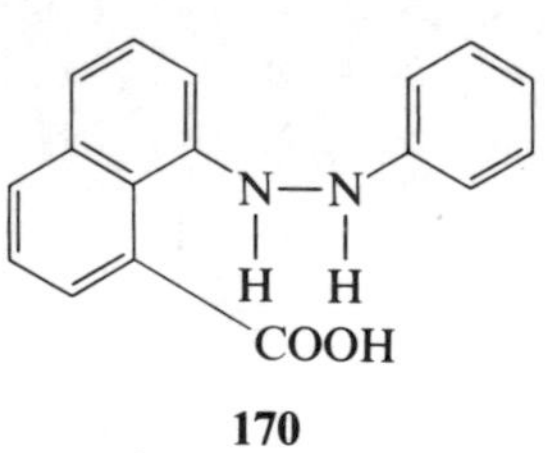

N—N
H H
COOH

170

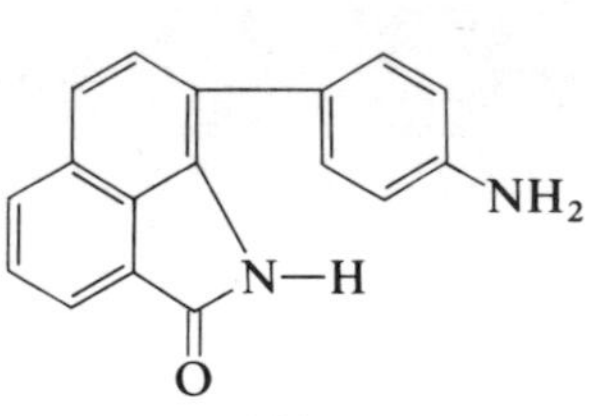

NH_2
N—H
O

171

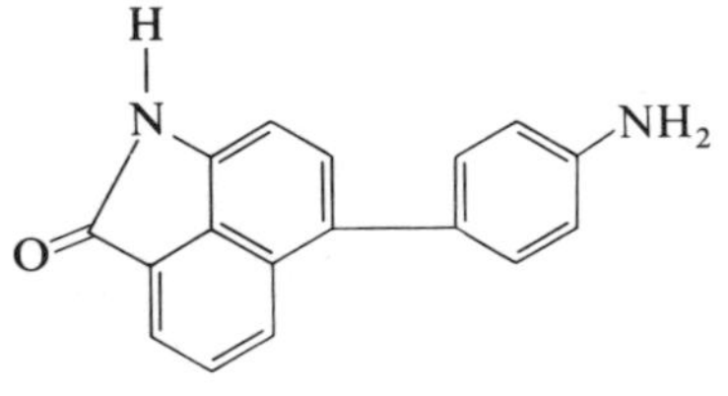

H
N
O
NH_2

172

N
H
COOH

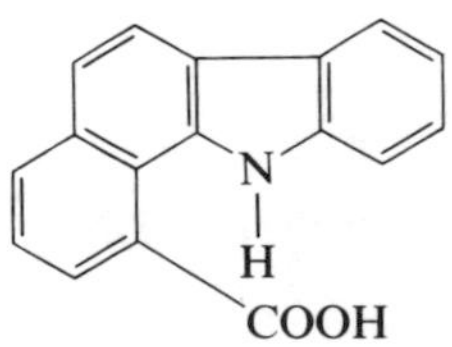

173

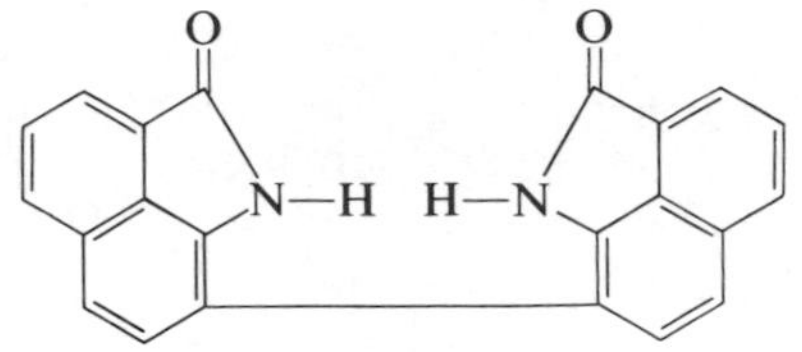

174

Another mechanistic proposal for the Fischer indolization (Cohn, 1919) involved the hydrazone **144** in an *ortho*-semidine rearrangement (Ingold, 1969c) to afford **175** which then cyclized with loss of ammonia to give the 3*H*-indole, **176**, and ultimately the indole, **147** (Scheme 7). Yet a further mechanistic scheme (Bamberger and Landau, 1919) was based upon the observation that phenylhydroxylamine (**177**) undergoes methylation to give N,N-dimethylaniline

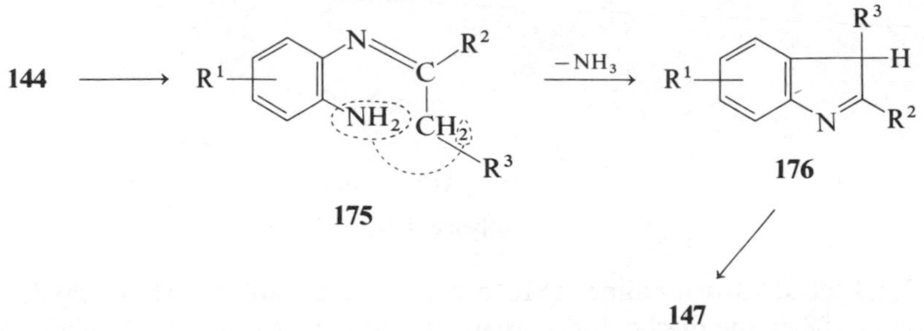

144 → 175 $\xrightarrow{-NH_3}$ 176 → 147

Scheme 7

oxide, a reaction which led to the suggestion that the tautomerism **177** ⇌ **178** (Scheme 8) for phenylhydroxylamine occurred, this being offered as one explanation for the formation of the oxidation product. By analogy with this, it was suggested that the Fischer indolization was effected as shown in Scheme 9. Hydrolysis of the hydrazone **144** occurred to produce its component hydrazino and ketonic moieties, the former of these then underwent tautomerism to give **179** which then condensed with the ketone to afford **180**. This was then deaminatively cyclized into **176** which then isomerized into **147**. The intermediacy

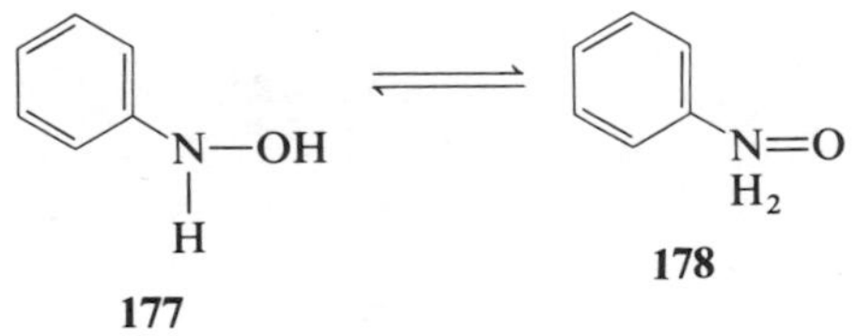

177 ⇌ 178

Scheme 8

144 ⟶ R^1-C$_6$H$_4$-NH—NH$_2$ + R^3CH$_2$C(=O)R^2

179

180

−NH_3

147 ⟵ **176**

Scheme 9

of 1,2,3,4-tetrahydrocinnolines **181**, formed by cyclization of the enehydrazine tautomer **152**, in the Fischer indolization has also been suggested (Neber, 1925; Burton and Duffield, 1949).

Of the above mechanistic proposals, that of Cohn was eliminated (Robinson, G. M. and Robinson, R., 1924) since it failed to account for the formation of 1-substituted indoles from N_α-alkylarylhydrazones and since it would also lead to an alteration in the substituent orientation on the benzenoid ring which it was established did not occur. Bamberger and Landau's theory was also eliminated since it also failed to account for the formation of 1-alkylindoles from N_α-alkylarylhydrazones and also because an explanation of the ammonia elimination was not clear (Hollins, 1922). Likewise, Neber's theory involving the intermediacy of 1,2,3,4-tetrahydrocinnolines subsequently failed to find

181

support (Neber, Knöller, Herbst and Trissler, 1929) and was also eliminated since such compounds are not transformed into indoles under the conditions used to effect the Fischer indolization (Allen, C. F. H. and Allen, J. A. van, 1951).

Although it was recognized that Reddelien's proposal (Scheme 4) would be inapplicable to the formation of 1-alkylindoles from N_α-alkylarylhydrazones, this difficulty was, at the same time as its recognition (Hollins, 1922, 1924b), overcome by suggesting that the ketoimine **145** might react as its tautomer **182** with the N-alkylaniline to give **183** which would ultimately afford the 1-alkylindole. Hollins was apparently unaware that this suggestion had been

145 ⇌ **182** [H_2N–C(R^2)=CH–R^3] + R^1–C$_6$H$_4$–NH–Alkyl ⟶ **183** [R^1–C$_6$H$_4$–N(Alkyl)–C(R^2)=CH–R^3]

made some ten years earlier (Wieland, H., 1913) with specific reference to the formation of 1,2-dimethylindole by Fischer indolization of acetone N_α-methylphenylhydrazone. Wieland commented at this time with regard to the mechanism of the indolization 'Es wäre wünschenswert, wenn die an sich sehr plausible Theorie über den Mechanismus der in ihren Wesen so lange rätselhaften Indolsynthese noch besser durch experimentelles Material gestützt würde. Dazu bietet sie die günstigsten Handhaben.' Little could he have realized what massive effort was to be concentrated upon this problem over the decades to come and that, even after a further seventy years, certain aspects of the reaction's mechanism would not be beyond debate. In this modified form, Hollins preferred the Reddelien theory rather than that of the Robinsons for both the Fischer indole synthesis (Hollins, 1922, 1924b) and the Piloty pyrrole synthesis (Hollins, 1924a). His main argument against the latter theory, as was that of other early critics (Neber, Knöller, Herbst and Trissler, 1929), was the absence of products resulting from *para* rearrangement in the second stage. Hollins's preference led to a subsequent somewhat personalized scientific attack upon him by the Robinsons (Robinson, G. M. and Robinson, R., 1924). After eliciting the intrinsic improbabilities of the Reddelien scheme, pointing out that it did not account for the necessary acid catalysis of the Fischer indolization and showing that added foreign aromatic amines did not ultimately produce 'mixed' indoles as would be expected, the Robinsons stated that 'It should be pointed out that Reddelien's paper was published in 1912 and that our criticism is mainly directed against the extension and advocacy of his suggestion by Hollins'. In a subsequent paper (Bodforss, 1925) the observation that added anilines did not produce 'mixed' indoles during Fischer indolization was again made and taken, along with other experimental data, as evidence of the invalidity of Reddelien's proposed mechanism. Later still (Campbell and Cooper, 1935), similar observations that fusion of both 'acetophenone-*p*- and *o*-toluidine' (anils) with 'phenylhydrazine zincichloride' afforded in each case only 2-phenylindole and not the 7- and 5-methyl homologues, respectively, further illustrated the invalidity of Reddelien's mechanistic proposal. In this latter work the identification of the reaction product was well established whereas in the similar earlier study (Bodforss, 1925) the identification was unsatisfactory (Campbell and Cooper, 1935). Surprisingly, Reddelien's proposal was resurrected some thirty years later (Kitaev, Troepol'skaya and Arbuzov, A. E., 1966a) to explain the formation of anils during Fischer indolization. However, such products could simply result from the condensation of the aniline and aldehyde or ketone formed by homolysis of the N—N bond and hydrolysis of the C=N bond, respectively, in the starting hydrazone.

Unlike all the other proposals, that of the Robinsons withstood the rigours of contemporary and later examination. Clear analogies were available for each stage, it offered an explanation for the necessity of an acid catalyst and its basis could be applied to the Brunner synthesis of oxindoles (see Chapter VI, A) (Robinson, G. M. and Robinson, R., 1924). Its one major criticism that *para* isomerization products (the *para*-benzidine counterparts) from stage 2 are absent in the reaction products was met by pointing out that such compounds, if formed, would be highly reactive 4-aminobenzyl alkyl ketones or 4-aminophenylacetaldehydes which would normally form tars under the vigorous indolization conditions (Robinson, G. M. and Robinson, R., 1924). In connection with this, the fact that the yields of the expected indoles from Fischer indolizations are often low and that many investigators refer to the isolation of these products from 'black, tarry reaction products' have also been referred to (Shine, 1967b).

However, products resulting from such *para* rearrangements have recently been isolated. Treatment of **184** with ethanolic hydrochloric acid or a hydrochloric acid–propionic acid mixture afforded **185** (R = H) (Boido and Boido Canu, 1973; Boido and Sparatore, 1968; Sparatore, Boido and Pirisino, 1974) and **185** (R = COC_2H_5) (Boido and Boido Canu, 1973), respectively, along with the expected indole **186**. Indolization of **187** produced the expected indole **188** (R = CH_3) along with another product formulated as **189**, although the

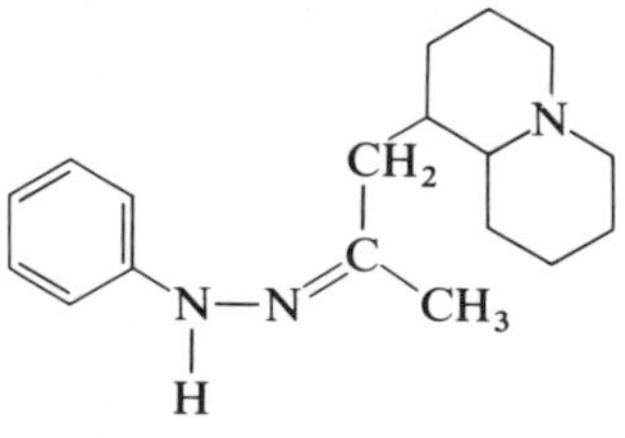

184

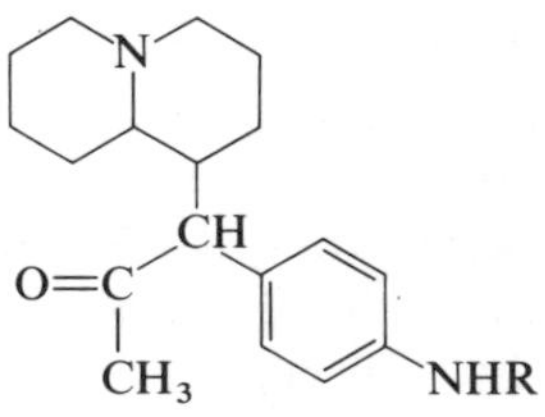

185

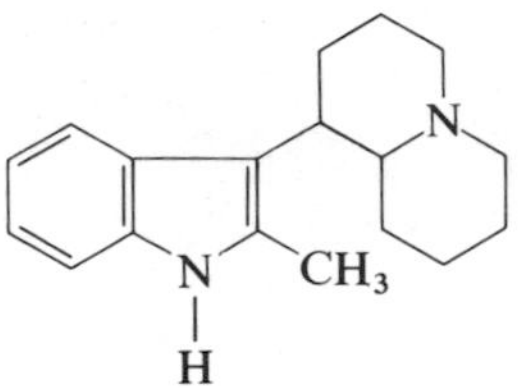

186 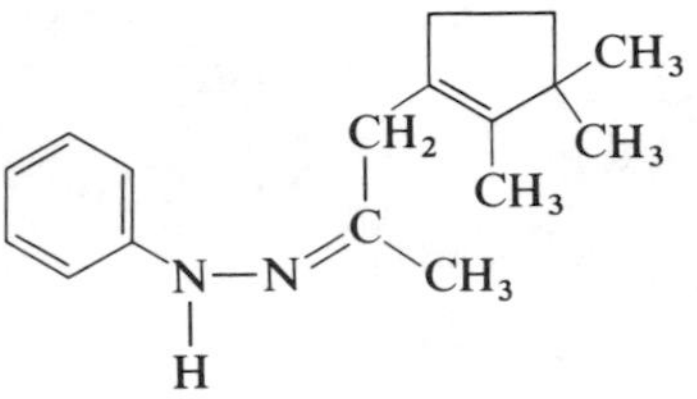 **187**

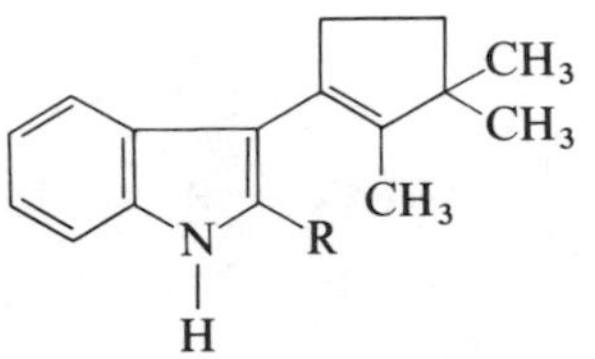

188 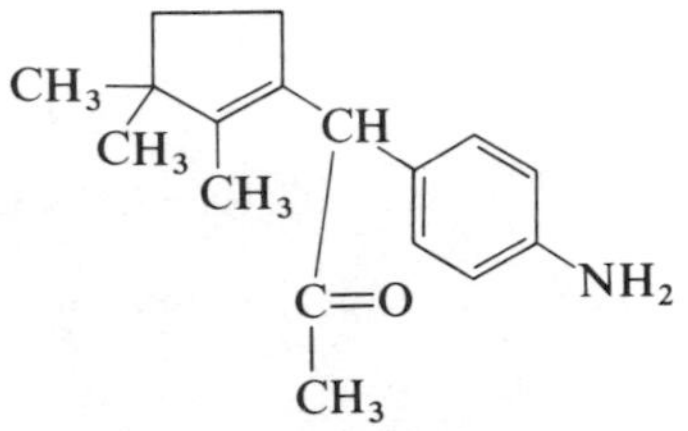 **189**

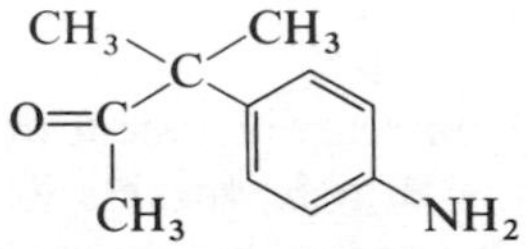

190

p.m.r. spectral data were not fully consistent with this structural proposal (Sparatore, Boido and Pirisino, 1974). The zinc chloride catalysed indolization of isopropyl methyl ketone phenylhydrazone yielded **190** along with the expected 2,3,3-trimethyl-3*H*-indole and other products (see Chapter V, A) (Boido and Boido Canu, 1977). When the aldehyde **192** was reacted with N_α,N_β-dimethylphenylhydrazine (**191**) at 20 °C the enehydrazine **193**, purified by chromatography on Alox at −10 °C, was formed. As will be seen later, such enehydrazines are far more easily converted into indoles than phenylhydrazones and in the present case, standing **193** at room temperature for 50 h produced a quantitative yield of a mixture of **194** (R = C_6H_5), resulting from mechanistic arrest after the new C—C bond closure in the indolization, and **195** resulting from the analogous *para* rearrangement. The structure of the latter product was based upon i.r. and p.m.r. spectroscopic analyses of the aldehyde obtained by its hydrolysis, this being also characterized as the 2,4-dinitrophenylhydrazone (Grieder, 1970; Grieder and Schiess, 1970). In particular, the p.m.r. spectrum indicate the presence of a 1,4-disubstituted benzene ring. During the course of the Bucherer carbazole synthesis (see Chapter I, E), *para* rearrangement products have also been isolated along with the expected benzocarbazoles (Bucherer and Sonnenburg, 1910, Bucherer and Zimmermann 1922; Rieche and Seeboth, 1958b, 1960b). The first and second of these reports would certainly have supported the Robinsons' mechanism against its criticism relating

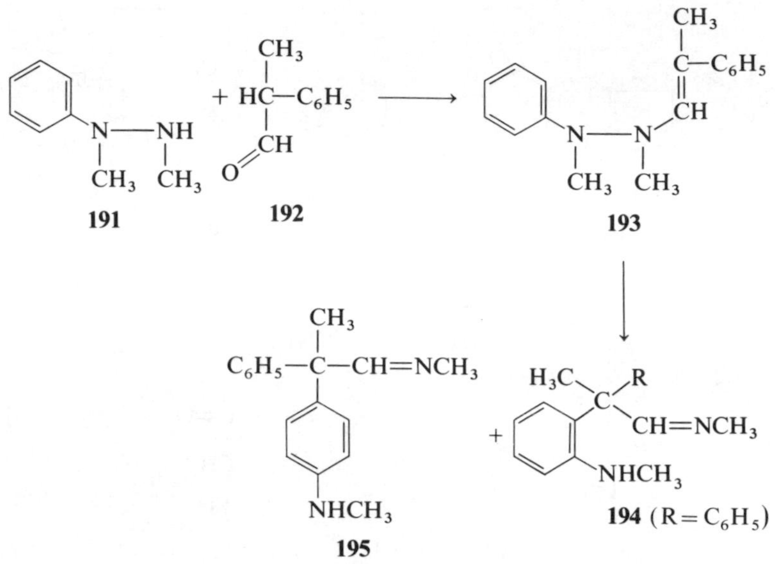

191 **192** **193**

195 **194** ($R = C_6H_5$)

to the lack of formation of *para* rearrangement products but it would appear that, as in other instances, the British workers were unaware of the German literature available to them. It appears highly likely that many other products of *para* rearrangements would be isolable upon careful analysis of the basic fractions of other Fischer indolization reaction mixtures. [5, 5]-Sigmatropic rearrangements of arylhydrazones under conditions of polyphosphoric acid catalysis, involving a *para* rearrangement within the hydrazino moiety, have also been observed (Fusco and Sannicolò, 1977, 1978a, 1980) (see Chapter VII, P).

The Robinsons' mechanism became generally accepted as representing that of the Fischer indolization. It was later subjected (Allen, C. F. H. and Wilson, C. V., 1943) to minor modification in that an imine structure **198** was preferred to **153**, the imine then cyclizing as shown (Scheme 10), either before (a) or after (b) hydrolysis, to give **147**. This possible ambiguity in the final stage of the Fischer indolization had been recognized by the Robinsons in 1924, as was a similar ambiguity later (Snyder, Merica, Force and White, E. G., 1958) during the reductive indolization of 2-nitrobenzyl cyanides. Although route (a) has been favoured (Allen, C. F. H. and Wilson, C. V., 1943), the reports that the indolizations of isopropyl phenyl ketone N_α-methylphenylhydrazone afforded **201** ($R = C_6H_5$) (Jenisch, 1906) and of isobutyraldehyde N_α-methylphenylhydrazone afforded **201** ($R = H$) (Brunner, K., 1896a, 1896b, 1900), appeared (Allen, C. F. H. and Wilson, C. V., 1943; Pausacker and Schubert, 1949b) to favour route (b). However, these compounds could have arisen via route (a) by attack of hydroxyl ion on the 3*H*-indolium cations **200** ($R = C_6H_5$ and H, respectively). Route (b) was favoured (Pausacker and Schubert, 1949b) by

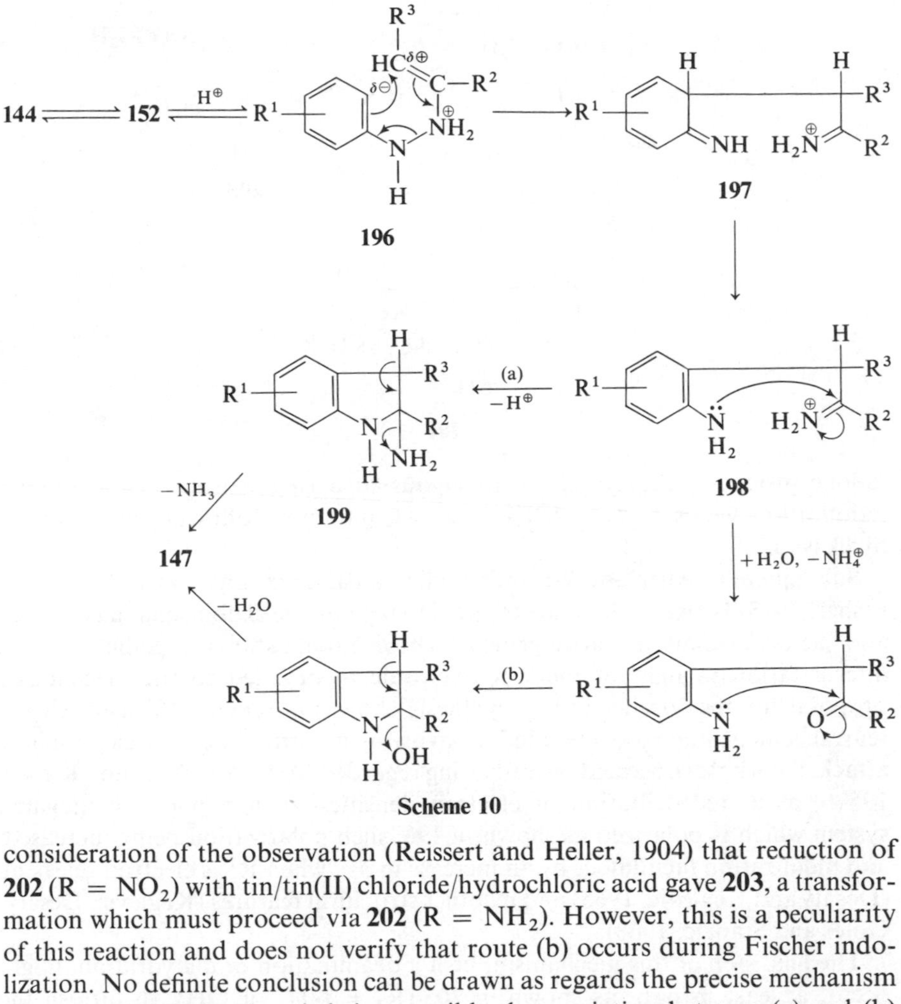

144 ⇌ 152 ⇌ ($H^{\oplus}$) 196 → 197 → 198 → (a) $-H^{\oplus}$ → 199 → ($-NH_3$) → 147

198 → ($+H_2O$, $-NH_4^{\oplus}$) → (b) → ($-H_2O$) → 147

Scheme 10

consideration of the observation (Reissert and Heller, 1904) that reduction of **202** (R = NO_2) with tin/tin(II) chloride/hydrochloric acid gave **203**, a transformation which must proceed via **202** (R = NH_2). However, this is a peculiarity of this reaction and does not verify that route (b) occurs during Fischer indolization. No definite conclusion can be drawn as regards the precise mechanism of ammonia elimination, it being possible that variation between (a) and (b) occurs depending upon the indolization conditions (Robinson, B., 1963a). It is interesting that in a recently published synthesis of indoles (Thyagarajan, Hillard, Reddy, K. V. and Majumdar, 1974), a mechanistic intermediate **204** was postulated which underwent ring closure as shown to form ultimately an

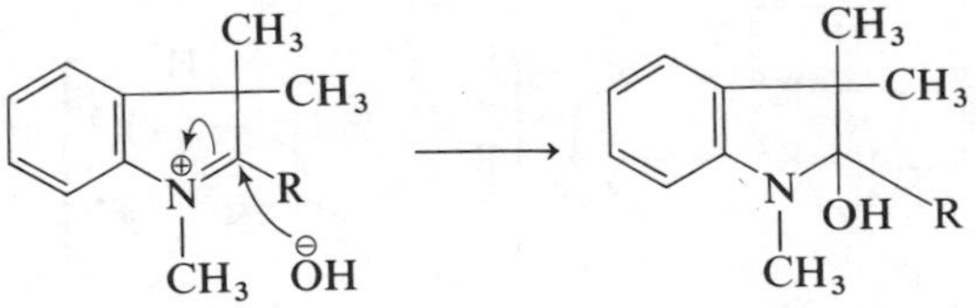

200 → **201**

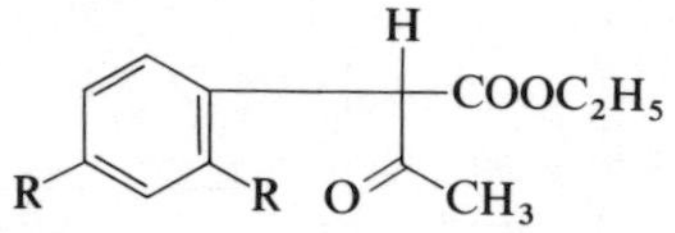

202

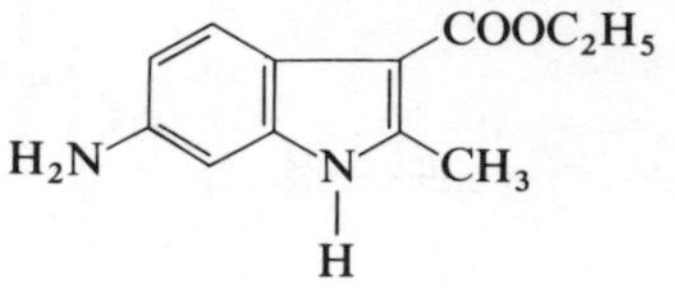

203

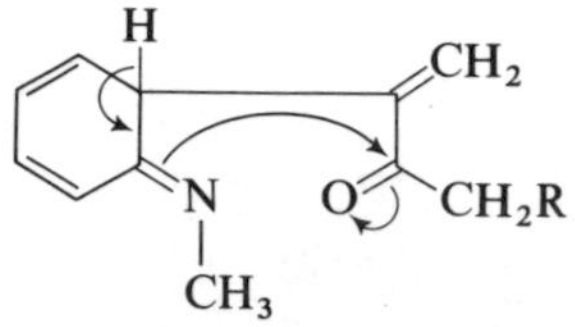

204

indolic product. Clearly this is analogous to a ring closure in the Fischer indolization which is preceded by loss of ammonia following imine group hydrolysis.

Subsequent to Allen and Wilson's studies, an analogy was drawn (Carlin and Fisher, 1948) between the rearrangement stage of the Robinsons' mechanism and the *ortho*-Claisen rearrangement. These modifications, together with an interpretation in light of modern electronic theory, led to the mechanistic proposal for the Fischer indole synthesis shown in Scheme 10 in which the rearrangement stage was regarded as involving an intramolecular electrophilic attack, the whole concerted process being regarded (Arbuzov, A. E. and Kitaev, 1957b) as a 'redistribution of electron densities' in a π–p–p–π conjugated system which is polarized as shown in **196**, such polarization being increased, and indolization facilitated, by an increase in $\delta^{\ominus}$, when R^1 is electron releasing (Desaty and Keglević, 1965) and by other structural features (Keglević, Desaty, Goleš and Stančić, 1968).

The final step of this mechanism, be it a deamination or dehydration, might occur, at least in part, as shown in **205** (R_4 = NH_2 or OH), to furnish the 3*H*-indole **206**, an eliminative route which occurs when 3,3-disubstituted 3*H*-indoles are synthesized by the Fischer method (see Chapter I, C). However, **206** contains a tertiary C-3 atom and would therefore spontaneously rearrange to afford **147**. Support for this proposal results from the observation that the

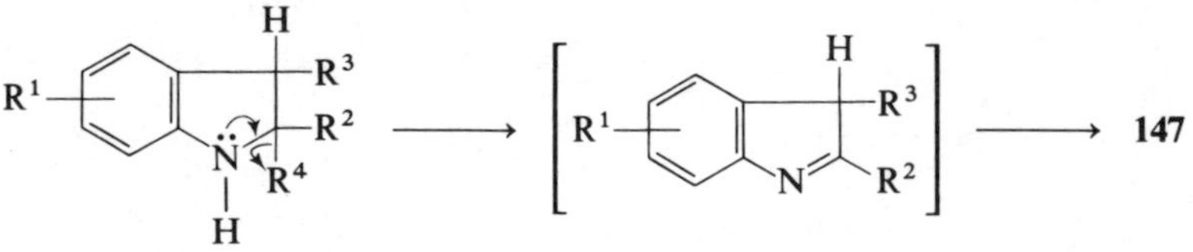

205 → 206 → 147

formic acid catalysed indolization of cyclopentanone phenylhydrazone produced **207** (R = H and CHO) (16% and 1% yields, respectively) and **208** (17% yields), it being suggested (Shimizu, J., Murakami, S., Oishi and Ban, 1971) that this latter product originated from the reductive formylation of **209** which arose from the intermediate **210** (R = NH_2 or OH) as shown. However, such a possible alternative route was limited by these workers to the indolizations of 2-unsubstituted cyclopentanone arylhydrazones, since in the subsequent intermediates containing two fused five membered rings, the formation of a double bond across the ring junction, would impose considerable steric strain relative to that in intermediates of type **209**.

A new mechanistic proposal for the rearrangement stage of the mechanism outlined in Scheme 10 has been made (Pausacker and Schubert, 1949a, 1949b). This resulted from the observation (Pausacker and Schubert, 1949a, 1949b) that indolization of a mixture of equal weights of cyclohexanone 2-methylphenylhydrazone and 2-methylcyclohexanone phenylhydrazone in glacial acetic acid yielded a basic product consisting of **211** (R = H) and **211** (R = CH_3) and a neutral product which after dehydrogenation gave the carbazoles **212** (R^1–R^3 = H) and **212** (R^1 = R^3 = H, R^2 = CH_3). In addition, an initial mixture of cyclohexanone 4-methylphenylhydrazone and 4-methylcyclohexanone phenylhydrazone gave, after dehydrogenation of the neutral reaction

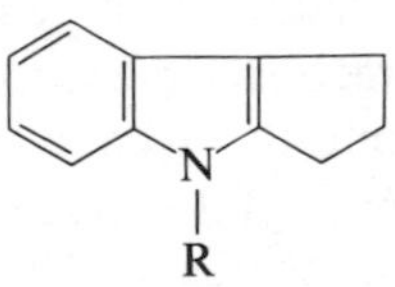

207 (structure: N–R)

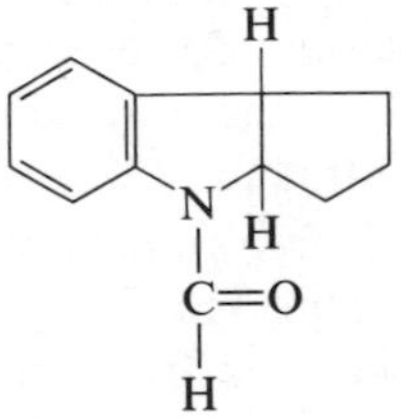

208 (structure: H, N, H, C=O, H)

209 (structure: H, N)

210 (structure: H, N̈, R, H)

211 (structure: CH_3, N, R)

212 (structure: R^1, R^3, N, H, R^2)

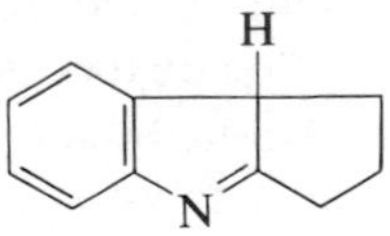

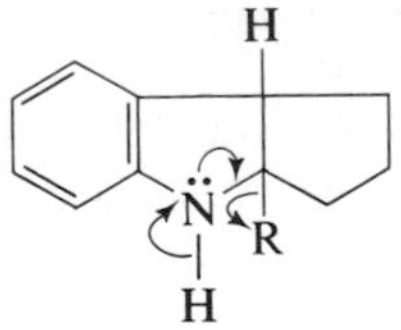

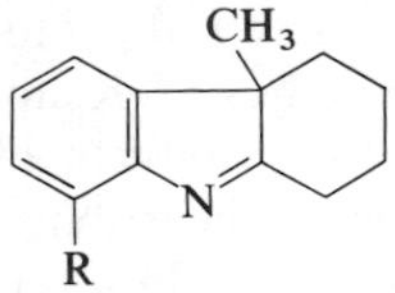

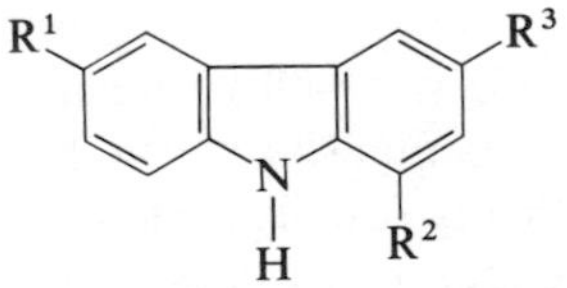

product, 3-methyl- and 3,6-dimethylcarbazole (**212**; $R^1 = CH_3$, $R^2 = R^3 = H$ and $R^1 = R^3 = CH_3$, $R^2 = H$, respectively), and carbazole itself was detected by i.r. spectral analysis (Pausacker and Schubert, 1949a). These results were interpreted (Pausacker and Schubert, 1949a, 1949b) as suggesting that the rearrangement stage of the mechanism was intermolecular in character and involved homolysis of the N—N bond, either before or after enehydrazine formation, to afford the radicals **213** and **214**. These, upon rearrangement, furnish **215** and **216**, respectively, which then combine to give **217** (**197** $- H^{\oplus}$) (Scheme 11), this then cyclizing and losing ammonia as previously described in Scheme 10. However, even at the time of the publication of these data and postulations (Robinson, R. in Pausacker and Schubert, 1949b), and again afterwards (Robinson, R. in Pausacker, 1949), it was suggested that these results could be simply caused by the hydrolysis of the initial hydrazones with subsequent condensation between the ketonic and hydrazino moieties followed by their indolization. This, in fact, was shown to be the case by three groups.

In the first report (Gore, Hughes, G. K. and Ritchie, 1949, 1950) it was shown that indolizations of mixtures of acetone phenylhydrazone and cyclohexanone, cyclohexanone 2,4-dinitrophenylhydrazone and phenylhydrazine, cyclohexanone 2,4-dinitrophenylhydrazone and acetone phenylhydrazone and benzaldehyde phenylhydrazone and cyclohexanone in boiling glacial acetic acid (under these conditions acetone phenylhydrazone and cyclohexanone 2,4-dinitrophenylhydrazone do not indolize) produced in each case 1,2,3,4-tetrahydrocarbazole as the only indolic product. Furthermore, the isolation of 2,4-dinitrophenylhydrazine (31% yield) from the second reaction mixture definitely excluded a free radical mechanism as did the failure to detect by-products typically expected from such a mechanism. Thus, although free radical mechanisms are far from 'clean', over 95% of 1,2,3,4-tetrahydrocarbazole could be obtained by the glacial acetic acid catalysed indolization of cyclohexanone phenylhydrazone, no hydrazobenzene or its acid catalysed rearrangement product, benzidine, were formed concomitantly with indolization (using tests sensitive to 3×10^{-5} g of benzidine), as would be expected by combination of phenylimino radicals, and large volumes of carbon dioxide were not evolved during indolizations involving acetic acid as the catalyst/solvent, as would be expected from its reaction with free radicals (Gore, Hughes, G. K. and Ritchie, 1950). Although attempts were made (Pausacker and Schubert, 1950) to rationalize the absence of these by-products in the event of the operation of a free radical mechanism, they are not very convincing. Other data which had also been used to support the operation of the free radical mechanism (Pausacker and Schubert, 1949b) were later (Gore, Hughes, G. K. and Ritchie, 1950) invalidated by being shown to be based upon products of unestablished structures, by being equally interpretable in terms of the Robinsons' mechanism or by being related to oxidation–reduction processes. In a related later study (Arbuzov, A. E. and Kitaev, 1957c) it was found that mixtures of acetone

144 ⇌ R^1–C_6H_4–$\dot{N}$H + R^3CH_2–C(R^2)=$\dot{N}$

213

152 ⇌ 213 + $R^3\dot{C}H$=C(R^2)–$\dot{N}$H

214

215 + 216

217 (197 − $H^{\oplus}$)

147

Scheme 11

phenylhydrazone with 2,4-dinitrophenylhydrazine and benzaldehyde, upon treatment with acetic acid, afforded acetone 2,4-dinitrophenylhydrazone and benzaldehyde phenylhydrazone, respectively, from which it was concluded that arylhydrazones are capable of transhydrazonation, a process which was probably responsible for Pausacker and Schubert's observations. However, a free radical mechanism involving initial homolysis of the hydrazone N—N bond was later (Sparatore, 1962) once again suggested to rationalize the formation of what were then thought to be the structures of the products resulting from the indolization of camphor phenylhydrazone. Others (Feofilaktov and Semenova, N. K., 1953b) also appeared to adopt, without good reason, the free radical mechanism in connection with their studies upon 2-ketoglutaric arylhydrazone indolizations. In a subsequent report (Geller and Skrunts, 1964), indolization of cyclohexanone phenylhydrazone in the presence of $^{15}N_{\beta}$-phenylhydrazine (1:1 molar ratio) and of an equilabelled mixture of cyclohexanone $^{15}N_{\alpha}$-phenylhydrazone and $^{15}N_{\beta}$-phenylhydrazine gave an approximate 2:3 distribution of the ^{15}N label between the ammonia produced and the phenylhydrazine remaining in the first reaction and an equidistribution of the ^{15}N label between the ammonia and tetrahydrocarbazole produced in the second reaction. Furthermore, indolization of cyclohexanone phenylhydrazone in the presence of ^{15}N-aniline gave 1,2,3,4-tetrahydrocarbazole with only the natural ^{15}N content and showed negligible ^{15}N loss from the aniline. If a process involving free radicals of type **213** had occurred, an equilibrium in this latter reaction would have become established, as shown in Scheme 12, which would have ultimately led to ^{15}N labelled 1,2,3,4-tetrahydrocarbazole and a reduction in the aniline ^{15}N label.

Scheme 12

Using a non-catalytic thermal indolization, which would be expected to favour any tendency to a free radical type mechanism, it was found that whereas cyclohexanone phenylhydrazone and 4-methylphenylhydrazone and cyclopentanone phenylhydrazone were all indolized normally by refluxing their diethylene glycol solutions, similar indolization of mixtures of cyclopentanone phenylhydrazone and cyclohexanone 4-methylphenylhydrazone and of cyclohexanone phenylhydrazone and ethyl methyl ketone afforded only the indolic products expected from the individual hydrazones, no 'crossed' indolization products being detected. Clearly the free radicals of the Pausacker proposal cannot be involved and, furthermore, under the non-catalytic thermal conditions it is apparent that transhydrazonation does not occur (Kelly, A. H., McLeod and Parrick, 1965).

However, in spite of these above data, it has been recently suggested (Kitaev and Troepol'skaya, 1978) that a free radical mechanism cannot be completely excluded from consideration since the radicals may exist briefly in a solvent cage. Furthermore, the mechanism of the reaction may be changed as the indolization conditions, which are very varied, are modified. Indeed, homolysis of arylhydrazones or their azo tautomers, with the ultimate involvement of free radical mechanisms, has been suggested to account for the formation of the corresponding anilines or substituted benzenes, respectively, concomitantly with indolization (see Chapter VII, J). Others (Grieder and Schiess, 1970; Heimgartner, Hansen and Schmid, H., 1979) have also suggested the occurrence of a radical process as opposed to a [3,3]-sigmatropic shift (see Chapter II, F) in certain cases.

The mechanism which is currently established for the Fischer indolization is basically that which was, as already mentioned, proposed by G. M. and R. Robinson and subsequently slightly enlarged by others to appear as shown in Scheme 10. This mechanism consists of four basic stages, these being (i) acid catalysed hydrazone–enehydrazine tautomerism, (ii) new C—C bond formation together with N—N bond cleavage, (iii) ring closure and (iv) deamination or dehydration. Apart from the contemporary and other evidence described above which has supported this mechanism, much further supportive data have been obtained which are now discussed.

B. Loss of the Nitrogen Atom

That it is the N_β-atom of the arylhydrazone which is eliminated during Fischer indolization was, in effect, established by Fischer himself as a result of his first indolization (Fischer, Emil and Jourdan, 1883) when the product resulting from the treatment of pyruvic acid N_α-methylphenylhydrazone with warm hydrochloric acid was recognized (Fischer, Emil and Hess, 1884) as being 1-methylindole-2-carboxylic acid. Since that time it has become well established that indolization of N_α-substituted arylhydrazones affords the corresponding 1-substituted indoles, a comprehensive list of all such examples published prior to 1924 having appeared (Hollins, 1924b) and a variety of later examples appearing throughout this present work and in Chapter IV (Table 31).

Further confirmation of the elimination of the N_β-atom was subsequently obtained using ^{15}N-tracers. An initial study (Allen, C. F. H. and Wilson, C. V., 1943) showed that indolization of acetophenone $^{15}N_\alpha$-phenylhydrazone furnished 2-phenyl-1-^{15}N-indole with complete retention of the ^{15}N label. The other nitrogen atom of a phenylhydrazone was labelled in later work (Clusius and Weisser, 1952) which showed that indolization of acetone $^{15}N_\beta$-phenylhydrazone gave rise to unlabelled 2-methylindole and ammonia which contained all the ^{15}N label of the original phenylhydrazone. Subsequently, similar results were obtained by indolization of cyclohexanone $^{15}N_\alpha$-phenylhydrazone

(Geller and Skrunts, 1964) and acetaldehyde $^{15}N_{\alpha}$-phenylhydrazone (Suvorov, Dmitrevskaya, Smushkevich and Pozdnyakov, 1972) to afford ^{15}N-1,2,3,4-tetrahydrocarbazole and ^{15}N-indole, respectively, with complete retention of the ^{15}N label in each case, and by indolization of acetaldehyde $^{15}N_{\beta}$-phenylhydrazone which yielded indole and ammonia (isolated as ammonium chloride), along with some aniline, with complete retention of the ^{15}N label in the ammonia (Suvorov, Dmitrevskaya, Smushkevich and Pozdnyakov, 1972).

Similar ^{15}N tracer studies have been effected upon the following two reactions, extensions of the Fischer indole synthesis which have already been discussed in Chapter I, E. Reaction of 2-naphthol-3-carboxylic acid (**84**; $R^1 = R^2 = H$, $R^3 = COOH$) with $^{15}N_{\beta}$-phenylhydrazine under thermal conditions afforded 3,4-benzocarbazole-1-carboxylic acid which contained 12% of the ^{15}N label as shown in **218** (R = COOH). These results were interpreted as involving an intermediate hydrazone **219** in an *ortho*-semidine rearrangement (Ingold, 1969c, Shine, 1969 and Chapter VII, M) to give **220** (R = COOH),

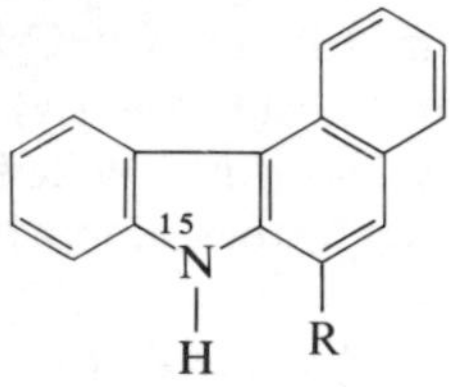

218

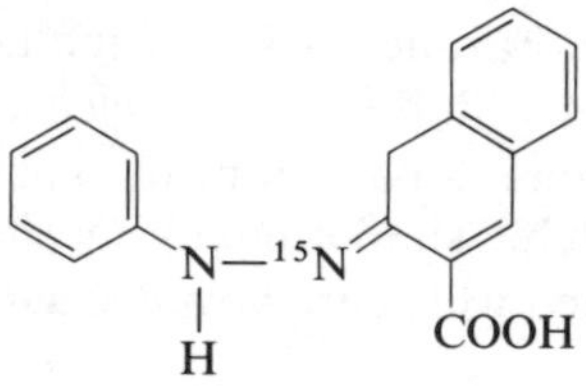

219

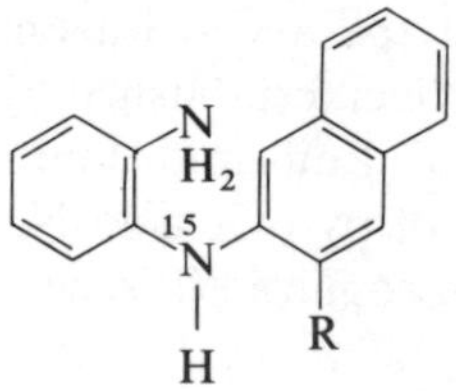

220

deaminative cyclization of which affords the above mentioned labelled product. Simultaneous reactions involving *para*-semidine and *ortho*- and *para*-benzidine rearrangements, from which products were not isolated, were postulated to account for the loss of the other 88% of the ^{15}N label (Clusius and Barsh, 1954). In another related study under the conditions of the Bucherer reaction (see Chapter I, E), 2-naphthol was reacted with $^{15}N_{\beta}$-phenylhydrazine in aqueous sulphur dioxide solution to yield 3,4-benzocarbazole which retained 6% of the ^{15}N label, the rest being located in the ammonia which was also formed. One explanation of these results again suggested the intermediacy of an *ortho*-semidine rearrangement to afford **220** (R = H). However, it was also recognized that these data were consistent with a mechanism shown in Scheme 13 which is analogous to the Fischer indolization. Support for this scheme was forthcoming from the observation that the diamine **223**, which was isolated from the reaction mixture, was converted under the reaction conditions into unlabelled **218** (R = H), in agreement with the known greater lability of a naphthyl amino group when compared with that of a phenyl amino group. The **218** (R = H) would then arise via intermediate **222** as shown. The formation of unlabelled **218** (R = H) could also occur by proton exchange in intermediate **221**, aromatization of the benzenoid moiety and ring closure as indicated in **224** (Holt,

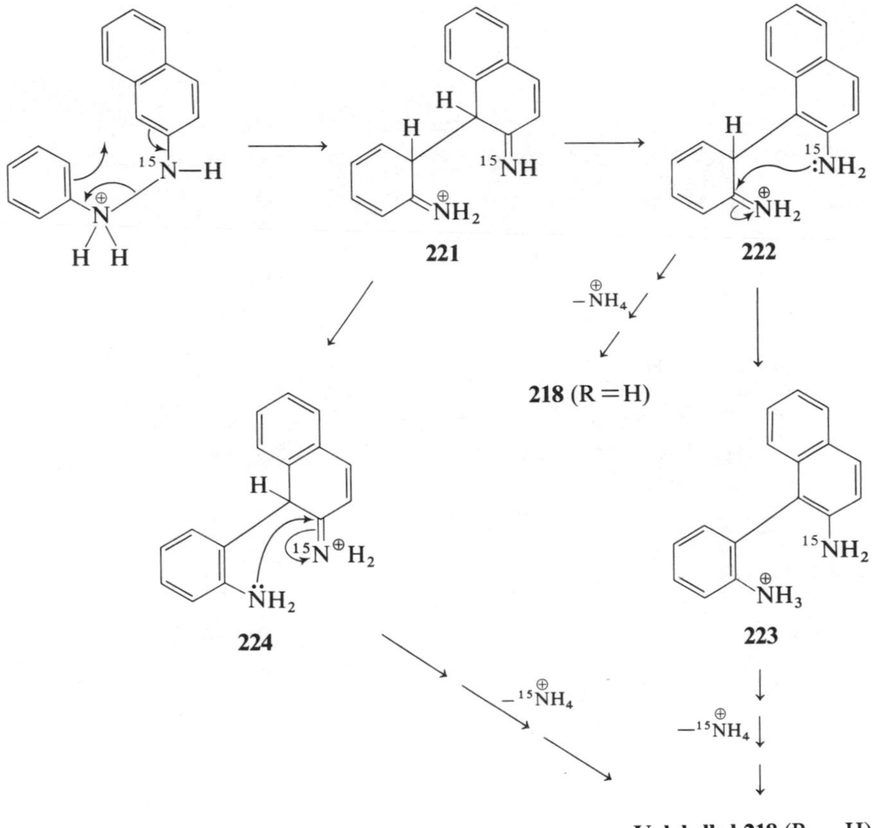

Scheme 13

P. F. and McNae, 1964). Application of Scheme 13 with or without concomitant operation of an *ortho*-semidine rearrangement would therefore also accommodate the above mentioned experimental data of Clusius and Barsh. Subsequent (Geller, 1978) criticisms of these mechanistic proposals are difficult to understand, especially since at the same time an essentially identical mechanistic explanation was postulated based upon the observation that when aniline and

$C_6H_5NH^{15}NH_2$

Na_2SO_3, SO_2, H_2O

223

$-^{15}NH_3$

$-HSO_3^{\ominus}$ Unlabelled **218** (R = H)

Scheme 14

1-naphthylamine reacted to afford 1-naphthylphenylamine, the ammonia which was also formed in the reaction originated from the 1-naphthylamine group. A reanalysis of these data in the latter publication in connection with the studies of Rieche and Seeboth upon the use of the Bucherer reaction in carbazole synthesis (see Chapter I, E) (Rieche and Seeboth, 1958a, 1958b, 1960a, 1960b, 1960c; Seeboth, 1967) led to the mechanistic proposal shown in Scheme 14 (Shine, 1967). Further support for the proposed involvement of **223** as shown in this scheme was derived from the observation that when unlabelled **223** was heated in 2M hydrochloric acid for 2.5 h, only a 15% yield of unlabelled **218** (R = H) was obtained whereas when unlabelled **223** was heated in sodium sulphite solution or in sulphurous acid, the yield of unlabelled **218** (R = H) obtained was 89% (Rieche and Seeboth, 1960c). Subsequently, the role of the sulphite ion in the Bucherer carbazole synthesis appeared to be completely ignored when it was suggested that the above example of this synthesis proceeded via the intermediate **224** as outlined in Scheme 13 (Geller, 1978).

C. Evidence for the Hydrazone–Enehydrazine Equilibrium

Investigations into the possible isomerizations of arylhydrazones among the hydrazone **225**, azo **226**, and enehydrazine **227** tautomers have occupied the attention of many groups and most of the studies effected in this area up to the end of the 1960s have been reviewed (Buckingham, 1969; Kitaev, Buzykin and Troepol'skaya, 1970).

R^1–C$_6$H$_4$–N=N–CH(R^2)–CH(R^3)–R^4

226

R^1–C$_6$H$_4$–NH–N=C(R^2)–CH(R^3)–R^4

225

R^1–C$_6$H$_4$–NH–NH–C(R^2)=C(R^3)–R^4

227

The first support for this initial stage of the indolization mechanism arose from the observation that the enolizability of aldehydes and ketones compared directly with the ease of indolization of their arylhydrazones (Robinson, G. M. and Robinson, R., 1918). In earlier studies (Freer, 1893, 1894) it had, in fact, been suggested that acetone and acetophenone phenylhydrazone might exist as their enehydrazine tautomers, suggestions which were later somewhat tenuously supported by oxidation and acetylation studies (Freer, 1899). Just prior to these studies, the reactions between ethyl acetoacetate and phenylhydrazine and between 2-carbethoxycyclopentanone and phenylhydrazine under mildly thermal conditions were reported to give the enehydrazine tautomers, **227** ($R^1 = R^3 = H$, $R^2 = CH_3$, $R^4 = COOC_2H_5$) (Nef, 1891) and **227** [$R^1 = H$, $R^2 = R^3 = (CH_2)_4$, $R^4 = COOC_2H_5$] (Dieckmann, 1901), respectively, of the corresponding hydrazones. However, although no evidence was presented to support these particular tautomeric formations, their extended conjugation would render them feasible. Indeed, recent studies (Ahlbrecht, 1971; Ahlbrecht and Henk, 1975, 1976) showed that hydrazones, including arylhydrazones, of 3-ketoesters and 3-ketonitriles tautomerize into their enehydrazine forms. Related to these latter observations are those in which the products formed when arylhydrazines reacted with diethyl acetylenedicarboxylate were shown by i.r. and p.m.r. spectroscopic analysis to be tautomeric mixtures of **225** ($R^2 = R^3 = COOCH_3$, $R^4 = H$) and **227** ($R^2 = R^3 = COOCH_3$, $R^4 = H$) (Heindel, Kennewell and Pfau, 1969, 1970; Sucrow and Slopianka, 1978). Crystallization of the reaction products from methanol yielded the enehydrazine tautomers whereas crystallization from benzene yielded the corresponding hydrazones, the more stable tautomers (Heindel, Kennewell and Pfau, 1969, 1970). However, contrary to these results it had been earlier observed (Acheson and Vernon, 1962) that the only product isolated by addition of N_α-methylphenylhydrazine to dimethyl acetylenedicarboxylate was the corresponding hydrazone, as evidenced by the absence of N—H absorption bands in its i.r. spectrum. Similarly, it appears that 2-(2-pyridyl)cyclohexanone phenylhydrazone and several of its analogues can also exist in the form of their enehydrazine tautomers, probably because of stabilization of these latter tautomeric forms by intramolecular hydrogen bonding as shown in **228** (Marchetti and Tosi, 1969a, 1969b).

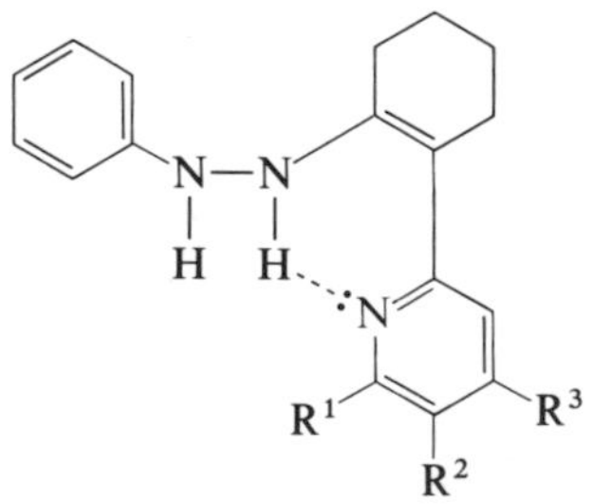

228

U.v. spectroscopic studies indicated that in neutral solution the 'normal' hydrazone structure was resistant to tautomeric change into the corresponding enehydrazine (or azo) structures (Auwers and Wunderling, 1931; Ramart-Lucas, Hoch and Martynoff, 1937). The possibility that arylhydrazones (**225**) might exist as their azo tautomers (**226**) had however been recognized soon after their first preparations (Fischer, Emil, 1896; Thiele and Heuser, 1896) and early u.v. spectroscopic studies led to the conclusion that acetaldehyde phenylhydrazone (Baly and Tuck, 1906; Chattaway, 1906), 4-bromophenylhydrazone and several other phenylhydrazones are converted into their azo tautomers upon exposure to sunlight (Baly and Tuck, 1906). Similar studies were later reported by others (Vémura and Inamura, 1935). Using 1-phenyl-1-phenylazocyclohexane as a reference compound, further u.v. and visible spectroscopic studies (Grammaticakis, 1947) led to the conclusion that, in solution, phenylhydrazones exist in equilibrium with small quantities of their corresponding azo tautomers. However, studies of spectra as a function of time were not reported and so the involvement of equilibria was not established. This work was later (Pausacker and Schubert, 1949b; Plieninger, 1950c) interpreted as being evidence in support of the existence of the enehydrazine tautomer under Fischer indolization conditions but the presence of no such tautomer had, in fact, been detected under the experimental conditions (in neutral organic solvents) used (Grammaticakis, 1947) although the probable existence of the enehydrazine tautomer had been postulated (Grammaticakis, 1946). Other spectroscopic studies using i.r., u.v. (O'Connor and Rosenbrook, 1961; O'Connor, 1961) and p.m.r. (O'Connor, 1961) also concluded that, in solution, arylhydrazones rapidly tautomerize to arylazoalkanes but these investigators were unable to detect the formation of the enehydrazine tautomer under the conditions employed (i.e. using neutral organic solvents). However, later u.v. (Bellamy and Guthrie, 1965a; Yao, 1964; Yao and Resnick, 1962 – see also Szmant and Planinsek, 1950), Raman (Arbuzov, A. E., Kitaev, Shagidullin and Petrova, L. E., 1967) and p.m.r. (Arbuzov, B. A., Samitov and Kitaev, 1966; Karabatsos, Graham and Vane, 1962; Karabatsos and Taller, 1963; Karabatsos, Vane, Taller and Hsi, 1964; Yao and Resnick, 1965) studies showed that, in solution in organic solvents or as pure liquids, arylhydrazones exist only in the hydrazone form (**225**) and that the tautomerism with the azo tautomer **226** claimed by the previous investigators as described above was caused by the oxidation of the hydrazones **225** to hydroperoxides (Bellamy and Guthrie, 1965a; Chernova, Shagidullin and Kitaev, 1967). I.r. spectral studies were initially thought (Shagidullin, Sattarova, Troepol'skaya and Kitaev, 1963a) to indicate a tautomerism between hydrazone (**225**) and azo (**226**) forms but a subsequent more careful analysis of the spectra (Shagidullin, Sattarova, Semenova, N. V., Troepol'skaya and Kitaev, 1963; Shagidullin, Satterova, Troepol'skaya and Kitaev, 1963b) indicated only the presence of the hydrazone **225** in solutions in organic solvents. These i.r. spectral data were also similarly reanalysed by a later group (Blair and Roberts, G. A. F., 1967).

The early u.v. spectral studies which claimed to have detected the conversion of arylhydrazones **225** into the corresponding azo structures **226** in the presence of sunlight (Baly and Tuck, 1906) were first shown to be erroneous nearly seventy years ago when it was found that the spectral changes were caused by oxidation of the arylhydrazones (Stobbe and Nowak, 1913), Fischer having first observed as far back as 1878 that benzaldehyde phenylhydrazone assumed a red coloration upon exposure to air (Fischer, Emil, 1878). Ultimately, the structures of the products resulting from such oxidations were postulated as **229** (Busch and Dietz, 1914). Later, these oxidation products were shown to be the hydroperoxides **230** (Bellamy and Guthrie, 1964, 1965a; Buckingham and Guthrie, 1967, 1968; Chaplin, Hey and Honeyman, 1959; Chernova, Shagidullin and Kitaev, 1964, 1967; Harvey, D. J., 1968; Karabatsos and Taller, 1963; O'Connor and Henderson, 1965; Schulz and Somogyi, 1967; Taylor, W. F., Weiss, H. A. and Wallace, T. J., 1968, 1969; Yao and Resnick, 1965) although earlier these hydroperoxides had been formulated as **230** ↔ **231** (Criegee and Lohaus, 1951), in which **231** comes close to **229** (Yao and Resnick, 1965), or as **230** or **232** in which **230** was favoured, although at the time it could not be unequivocally distinguished from **232** (Pausacker, 1950b). Clearly, the yellow coloured product, $C_{12}H_{17}N_3O_2$, which was obtained when 1-methylpiperidin-4-one phenylhydrazone was recrystallized from organic solvents under 'ordinary conditions' and, as stated at the time of its preparation, whose 'nature remains obscure' (Cook, A. H. and Reed, K. J., 1945), was the corresponding hydroperoxide.

The formation of **230** from **225** has been suggested (Bellamy and Guthrie, 1965a; Pausacker, 1950b) to occur via the route shown in Scheme 15. Consistent with this mechanism are the observations that N_α-methylphenylhydrazones did

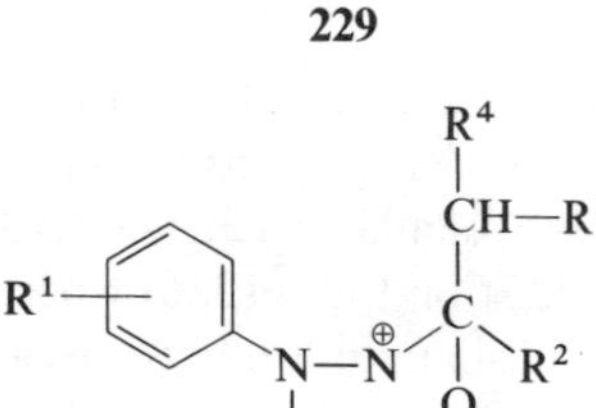

R¹ N—N C—R² CH—R³ R⁴ O O H

229

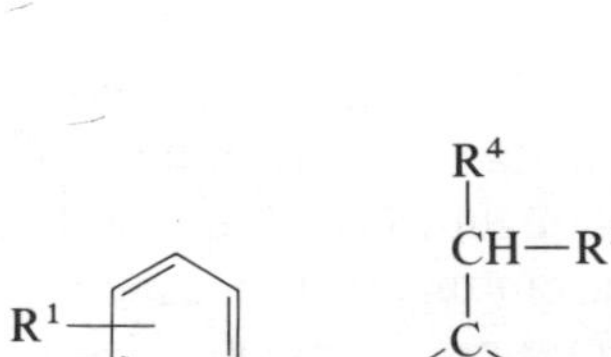

R¹ N=N C R² OOH CH—R³ R⁴

230

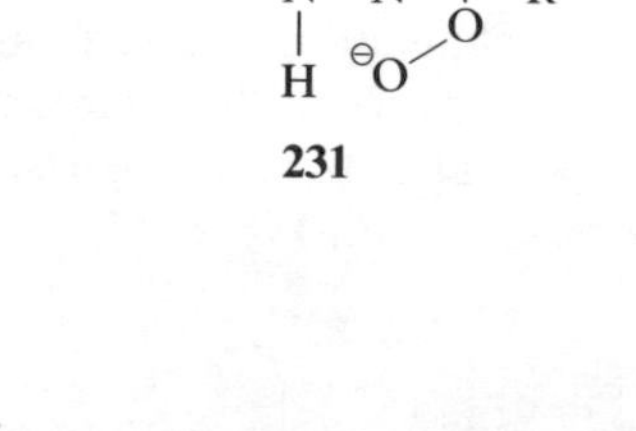

R¹ N—N⊕ C R² O ⊖O H CH—R³ R⁴

231

R¹ N—N=C R² OOH CH—R³ R⁴

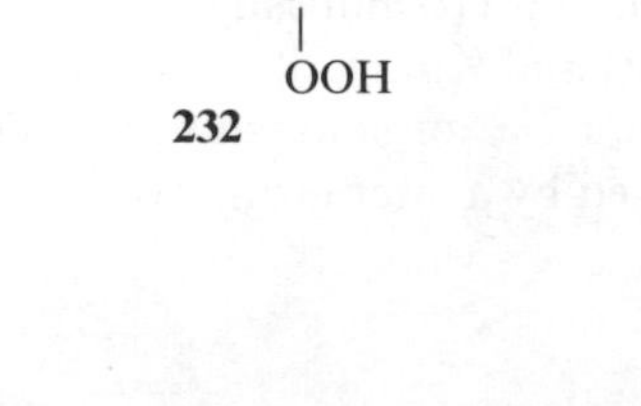

232

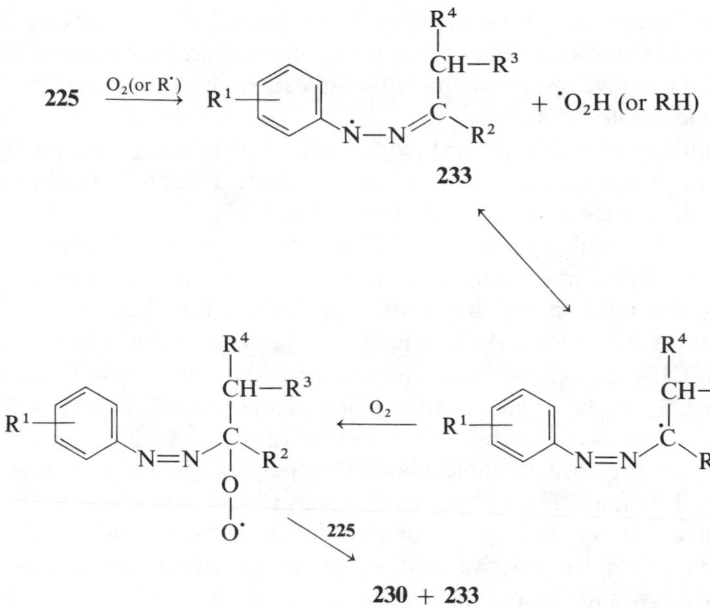

Scheme 15

not form peroxides (Bellamy and Guthrie, 1965a; Buckingham and Guthrie, 1968a; Pausacker, 1950b). The homolysis of the N_α—H bond in aryl (and alkyl) hydrazones has also been proposed as the initial stage of the mechanism of the reaction between these compounds and lead tetraacetate (Iffland, Salisbury and Schafer, 1961) although later evidence suggested that this reaction proceeds by an ionic mechanism (Harrison, M. J., Norman and Gladstone, 1967).

The suggestion that not only does oxygen react with arylhydrazones **225** to afford peroxides **230** but that it may also catalyse the formation of the azo tautomer **226** from **225** (O'Connor and Henderson, 1965) still requires verification. It was not supported by the observation that repeated attempts to use the method to convert cyclohexanone phenylhydrazone **225** [$R^1 = R^3 = H$, $R^2 + R^4 = (CH_2)_4$] into phenylazocyclohexane **226** [$R^1 = R^3 = H$, $R^2 + R^4 = (CH_2)_4$] were unsuccessful (Troepol'skaya, 1967).

From a study of the polarographic behaviour of a wide range of phenylhydrazones it has been concluded that, in their alcoholic solution, tautomerism between the hydrazone **225**, azo **226** and enehydrazine **227** forms occurred (Arbuzov, A. E. and Kitaev, 1957a; Kitaev and Arbuzov, A. E., 1957, 1960; Kitaev and Troepol'skaya, 1963a, 1963b). However, the results embodied in the earliest of these publications have been criticized and the conclusions drawn have been questioned in view of the reference compounds and experimental conditions used (O'Connor, 1961), comments which are indeed valid with regard to all five of these publications. Subsequently, two of their three authors

(Kitaev and Troepol'skaya, 1967) themselves retracted their earlier conclusions and indicated in particular that the apparent formation of the azo tautomers was, as in the above spectroscopic studies, caused by the formation of the hydroperoxides **230**.

Dipole moment studies of arylhydrazones also indicated the absence of tautomerism of hydrazones to either azo or enehydrazine forms in organic solvents (Kitaev, Flegontov and Troepol'skaya, 1966).

Adipaldehyde bisphenylhydrazone (**234**) was originally (Chittenden and Guthrie, 1964) thought to tautomerize into 1,6-bisphenylazohexane in chloroform–methanol solution but the product of this reaction was later (Bellamy, Guthrie and Chittenden, 1966) shown to be *trans*-1,2-bisphenylazocyclo hexane (**236**). It was suggested that this was formed by ring–chain tautomerism of **234** into **235**, this latter species then being oxidized to afford **236**. Although attempts to react two molecules of a monohydrazone together in a similar manner in the presence of molecular oxygen failed (Bellamy, Guthrie and Chittenden, 1966), aldehyde phenylhydrazones did afford oxidative dimerization products, along with other oxidation products depending upon the reaction conditions, when active manganese dioxide was used as the oxidizing agent (Bhatnagar and George, 1967).

$CH{=}NNHC_6H_5$ / $CH{=}NNHC_6H_5$ (**234**) $\rightleftharpoons$ $NHNHC_6H_5$ / $N{=}NC_6H_5$ (**235**) $\longrightarrow$ $N{=}NC_6H_5$ / $N{=}NC_6H_5$ (**236**)

It has been reported (Snyder and Smith, C. W., 1943) that an intermediate cherry red coloration was produced during the indolization of cyclohexanone phenylhydrazone catalysed by boron trifluoride etherate in glacial acetic acid, a coloration that was thought to be caused by the formation of a complex between the catalyst and the azo tautomer of the hydrazone since a similar reaction with cyclohexanone N_α-methylphenylhydrazone did not produce such a colour. However, this reason is unlikely since chromatographic attempts to isolate the azo tautomer were unsuccessful and when phenylazocyclohexane was treated with boron trifluoride etherate in glacial acetic acid, a cherry red colour only developed during ten minutes (Bellamy and Guthrie, 1965b). In fact, under basic, radical initiated or mildly acidic conditions, phenylazoalkanes were converted into the corresponding phenylhydrazones which are thus probably the more thermodynamically stable form. Under more vigorous acidic conditions, phenylazoalkanes were converted into indoles which were also obtained by similar treatment of the corresponding phenylhydrazones (Bellamy and Guthrie, 1965b). Other earlier investigators had also reported the base catalysed conversion of alkylazoalkanes into the corresponding alkylhydrazones (Fodor and Szarvas, 1943; Hutton and Steel, 1964) and of phenylazoalkanes into the corresponding phenylhydrazones under conditions of acid catalysis (Chaplin, Hey and Honeyman, 1959; Fischer, Emil, 1896).

Contrary to what might have been expected from the above studies, hydrazones have also been converted into the corresponding azo forms under conditions of base catalysis. Thus, studies of deuterium exchange of aldehyde phenylhydrazones in neutral methan[^{2}H]ol (Bellamy and Guthrie, 1968) and of aldehyde and ketone phenylhydrazones in neutral ethan[^{3}H]ol (Simon and Moldenhauer, 1967) furnished no evidence to suggest that under these conditions tautomeric equilibria exist between these hydrazones and the corresponding phenylazoalkanes and enehydrazines. However, similar studies (Simon and Moldenhauer, 1968) in alkaline media using ethan[^{3}H]ol showed that, under these conditions, hydroxyacetaldehyde phenylhydrazone did not exist in equilibrium with its enolhydrazine tautomer but did so with its azo tautomer and that 2-hydroxycyclohexanone phenylhydrazone was not in equilibrium with its enolhydrazine tautomer but was so with its enehydrazine tautomer. Furthermore, it has been observed that alkylhydrazones were converted into the corresponding azo compounds when heated in a solution of potassium hydroxide in diethylene glycol (Ioffe and Gershtein, 1969), distilled over granulated potassium hydroxide (Ioffe, Sergeeva, Z. I. and Stopskii, 1966) or heated in a dilute solution of potassium *tert.*-butoxide in *tert.*-butanol (Ioffe and Stopskii, 1968). Unfortunately, when arylhydrazones were heated with bases they underwent indolization, or N—N bond cleavage to afford nitriles and amines (Ioffe, Sergeeva, Z. I. and Stopskii, 1966). However, using formaldehyde phenylhydrazone at least removed the possibility of indolization and when this compound was heated in a 4% solution of potassium hydroxide in ethylene glycol, methylazobenzene was formed along with aniline and hydrogen cyanide, the products of N—N bond cleavage (Ioffe and Stopskii, 1967).

The above studies clearly indicate that, in neutral organic solvents or in the 'neat' state, arylhydrazones do not exist in equilibrium with measurable quantities of their possible azo or enehydrazine tautomers. Indeed, calculations using the LCAO method have shown that the enehydrazine form is 4 kcal/mol less favourable than the hydrazone from (Grandberg, Zuyanova, Przheval'skii and Minkin, 1970) and that the latter tautomer possesses π–p–π conjugation which should possibly stabilize it relative to the enehydrazine species (Arbuzov, A. E., Kitaev, Shagidullin and Petrova, L. E., 1967; Ioffe and Stopskii, 1968). However, it is believed (e.g. Arbuzov, A. E., Kitaev, Shagidullin and Petrova, L. E., 1967) that under Fischer indolization conditions, isomerization to the enehydrazine structure could occur under the influence of the acidic and/or thermal reaction conditions. Enehydrazine tautomers have also been postulated as transient intermediates in β-elimination reactions of arylhydrazones (Buckingham and Guthrie, 1966) and the formation of the enehydrazine tautomer of acetone phenylhydrazone has also been suggested (Le Fevre and Hamelin, 1979). Furthermore, although the existence of tautomerism between the hydrazone **225** and enehydrazine **227** forms has been doubted by one group (Kitaev and Troepol'skaya, 1978) it is possible that such an equilibration may still exist but may lie, as evidenced by the LCAO calculations, very much in favour of the hydrazone tautomer. Consequently, the techniques described in the above studies may have been too insensitive to detect the small quantities of the enehydrazine form present.

D. Kinetic Studies

By condensation of a number of aryl substituted phenylhydrazine hydrochlorides with a series of acetals in warm acetic acid (which generated the arylhydrazones *in situ*), it was found (Desaty and Keglević, 1964) that the subsequent indolization which occurred under these conditions took place readily (as measured by product yield) when the original arylhydrazine had an electron releasing substituent in the 4-position, happened less readily with such a substituent in the 2-position and failed with the isomeric 3-substituted analogues. Under similar conditions using phenylhydrazine hydrochloride and the corresponding acetals, only low yields of indoles resulted, whereas electron attracting substituents on the phenyl ring prevented indolization. These results were explained (Desaty and Keglević, 1964) by assuming that under the relatively mild indolization conditions used, the rate determining step of the indolization is enehydrazine formation. This is facilitated by electron release from electron releasing 4-(and less from 2-)substituents on the aromatic ring because of the increase in basicity of N_β, the protonation of which can be visualized as being the initial step in the conversion of the hydrazone to its enehydrazine tautomer. Such a mesomeric contribution from an electron releasing group substituted at the 3-position of the phenyl nucleus would not be possible. In support of this postulation are the earlier kinetic studies which showed that, under similar mild conditions, cyclohexanone 4-methoxyphenylhydrazone indolized much more rapidly than the corresponding 3-isomer (Pausacker and Schubert, 1950) and the observations that the basicities of the three isomeric methoxyphenylhydrazines are of the order $4 > 2 > 3$ (Stroh, H. H. and Westphal, 1963).

Kinetic studies upon the Fischer indole synthesis are far from conclusive. From the first investigation, which showed that the reaction rate is dependent upon both the concentration of the hydrazone and that of the acid catalyst, it was concluded that the rate determining step is the reaction between the hydrazone, or rearranged hydrazone, and a solvated proton (Pausacker and Schubert, 1950). However, these data from this early study were subsequently interpreted in terms of hydrazone protonation and enehydrazine formation being reversible and rapid in comparison with the step in which the formation of the new C—C bond occurs, which was assumed to be rate determining (Carlin, 1952). From further studies it was concluded that the rate determining step is the tautomerization between the protonated hydrazone and the protonated enehydrazine, although it was indicated that the slow stage of the mechanism may alter with a variation in the experimental conditions (Desaty and Keglević, 1964; McLean, J., McLean, S. and Reed, R. I., 1955). Fischer indolizations have been effected by reacting phenylhydrazones with methyl iodide, a reaction which can be effected at room temperature in a spectrophotometer sample tube. Since in these reactions intermediates could not be detected, it was concluded that the rate determining step of the indolization is that involving the reaction of the methyl iodide with the phenylhydrazone, equivalent to the protonation of the arylhydrazone in an acid catalysed indolization (Posvic, Dombro, Ito, H. and Telinski, 1974). Kinetic studies, in which an excess of cyclohexanone N_α-benzylphenylhydrazone was reacted with allyl bromide to afford 9-benzyl-1,2,3,4-tetrahydrocarbazole, showed the reaction to be first order (Grandberg, 1974; Grandberg, Przheval'skii, Ivanova, T. A., Zuyanova, Bobrova, Dashkevich, S. I., Nikitina, Shcherbina and Yaryshev, 1969). Two theses have also embodied the results of other studies upon the kinetics of the Fischer indolization (Elgersma, 1969; Scheltus, 1959). Unfortunately, these results have not been published in the international literature but the data from the latter thesis have been summarized in a review (Grandberg and Sorokin, 1974) as follows: '(i) On cyclisation of the phenylhydrazone of 2-deuterated cyclohexanone, a kinetic isotope effect was observed (K_D/K_H = 1.8); (ii) in the D_2O–CH_3OH–HCl solution, cyclohexanone phenylhydrazone reacts twice as fast as in the H_2O–CH_3OH–HCl system; (iii) the rate constants for the phenylhydrazones of cyclic ketones are satisfactorily correlated with the basicities of these compounds; (iv) an increase of the acidity of the medium leads to an increase of the rate of reaction.' These data and those contained in the other thesis (Elgersma, 1969), which again show that 2-deuteration in the ketonic moiety significantly retards indolization, do appear to support the monoprotonated hydrazone–enehydrazine tautomerization as being the rate determining step although in the later thesis the measured effects of benz substituents upon the rate of the reaction appear to be inconsistent with this postulation. It is also interesting that the reaction between arylhydrazines and 4-halogeno- or 4-tosylbutyraldehyde or propyl ketones was found to have an overall order of 1.28 (Grandberg and Przheval'skii, 1972) which could indicate the operation of more than one synchronous mechanism. In view of these above studies it would appear that the statement that 'it is well known that the slowest step of the Fischer reaction is tautomerisation of the hydrazone molecule' (Grandberg and Sorokin, 1973) may not be comprehensively correct. However, from the data and discussion presented in a recent review (Grandberg and Sorokin, 1974) it does appear that the tautomeric conversion of the hydrazone into the enehydrazine may be the rate determining step in the majority of indolizations.

Relevant to many of the above studies is the suggestion (Aksanova, Kucherova and Zagorevskii, 1964a) that in many cases, whereas electron accepting substituents may hinder indolization they may also minimize side reactions but electron donating substituents, although facilitating indolizations, may also promote side reactions and therefore lower the yields of indoles. The yields of indoles in such reactions, therefore, are not only related to the rates of the indolizations, irrespective of which stage in the mechanism is rate determining.

E. Isolation of Intermediates in the Fischer Indole Synthesis

1. Intermediates Corresponding to the Enehydrazine Tautomer **152** *in Scheme 10, and Related Studies*

The first claimed isolation of an N_{α}, N_{β}-diacetyl enehydrazine was reported when ethyl acetoacetate phenylhydrazone **237** ($R^1 = R^2 = H$, $R^3 = CH_3$, $R^4 = COOC_2H_5$) was heated with acetyl chloride in ethereal solution to produce an oily product formulated as **238** ($R^1 = R^3 = H$, $R^2 = CH_3$) (Nef, 1891) although the experimental analytical figures for the product were 2% high in carbon and, indeed, the product was subsequently (Walker, C., 1894) reformulated as **239**, resulting from the cyclodehydration of **237** ($R^1 = R^2 = H$, $R^3 = CH_3$, $R^4 = COOC_2H_5$). However, many decades later, treatment of the hydrazones **237** ($R^1 = R^2 = H$, $R^3 = R^4 = CH_3$) (Elgersma and Havinga, 1969; Palmer and McIntyre, 1969; Suvorov and Sorokina, N. P., 1961; Suvorov, Sorokina, N. P. and Sheinker, I. N., 1958; Suvorov, Sorokina, N. P. and Sheinker, Y. N., 1957; Grandberg and Przheval'skii, 1974), **237** [$R^1 = R^2 = H$, $R^3 + R^4 = (CH_2)_4$] (Elgersma, 1969; Elgersma and Havinga, 1969; Suvorov and Sorokina, N. P., 1961) and **237** [R^1–$R^3 = H$, $R^4 = CH_3$; $R^1 = R^2 = R^4 = H$, $R^3 = CH_3$; $R^1 = R^3 = R^4 = CH_3$, $R^2 = H$; $R^1 = OCH_3$, $R^2 = R^4 = H$, $R^3 = CH_3$; $R^1 = OCH_3$, $R^2 = H$, $R^3 = R^4 = CH_3$; $R^1 = NO_2$, $R^2 = H$, $R^3 = R^4 = CH_3$ and $R^1 = NO_2$, $R^2 = H$,

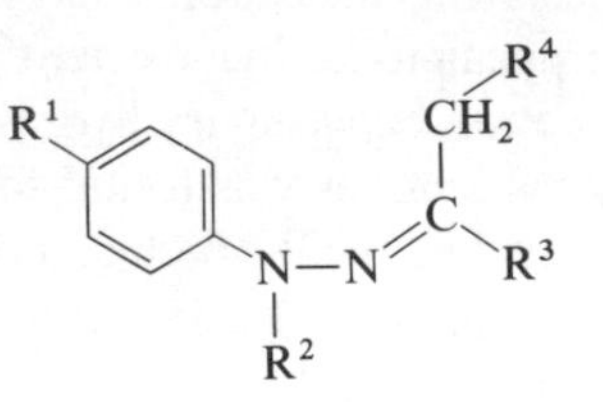

237

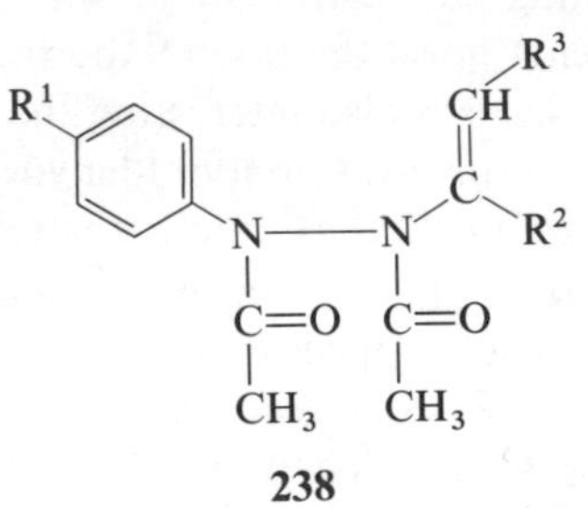

238

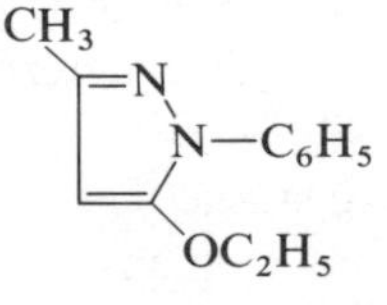

239

$R^3 + R^4 = (CH_2)_4$] (Suvorov and Sorokina, N. P., 1961) with acetic anhydride in the presence of 4-methylbenzene sulphonic acid yielded the corresponding diacetyl enehydrazines **238** [$R^1 = H, R^2 = R^3 = CH_3$; $R^1 = H$, $R^2 + R^3 = (CH_2)_4$; $R^1 = R^2 = H, R^3 = CH_3$; $R^1 = R^3 = H, R^2 = CH_3$; R^1–$R^3 = CH_3$; $R^1 = OCH_3$, $R^2 = H$, $R^3 = CH_3$; $R^1 = OCH_3$, $R^2 = R^3 = CH_3$; $R^1 = NO_2$, $R^2 = R^3 = CH_3$ and $R^1 = NO_2$, $R^2 + R^3 = (CH_2)_4$, respectively]. In an interesting related reaction it was found that whereas benzoylation of acetaldehyde 3- and 4-methoxyphenylhydrazone afforded the corresponding N_α-benzoyl derivatives, similar treatment of the corresponding 2-methoxyphenylhydrazone brought about C-benzoylation to give **240** (Ishii, H., Harada, K., Abe, K., Doki and Ikeda, 1971). The intermediate nature of these N-acetylated compounds was illustrated by the acid

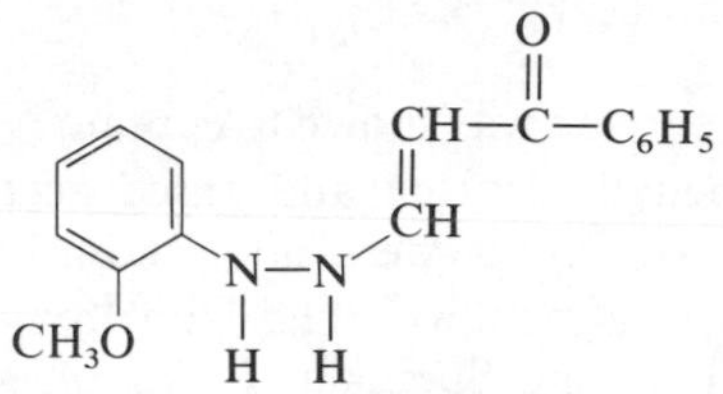

O
||
CH—C—C_6H_5
||
CH
N—N
CH_3O
H H

240

catalysed conversion of **238** ($R^1 = H$, $R^2 = R^3 = CH_3$) (Elgersma and Havinga, 1969; Suvorov and Sorokina, N. P., 1961; Suvorov, Sorokina, N. P. and Sheinker, Y. N., 1957; Suvorov, Sorokina, N. P. and Sheinker, Y. N., 1958) and **238** (R^1–$R^3 = CH_3$) (Suvorov and Sorokina, N. P., 1961) into 2,3-dimethyl- and 2,3,5-trimethylindole, respectively. The former transformation has also been effected by thermolysis (Grandberg and Przheval'skii, 1974). The above formation of **238** ($R^1 = H$, $R^2 = R^3 = CH_3$) was achieved in $>98\%$ yield and the formation of the possible alternative diacetyl derivative **238** ($R^1 = R^3 = H$, $R^2 = C_2H_5$) was undetectable. However, whereas treatment of the former diacetyl derivative with 40% (w/w) aqueous ethanolic sulphuric acid furnished only 2,3-dimethylindole, using 70% (w/w) acid the indolic product contained 25% 2-ethylindole, showing that conversion of the enehydrazine tautomer of **238** ($R^1 = H$, $R^2 = R^3 = CH_3$) into that of **238** ($R^1 = R^3 = H$, $R^2 = C_2H_5$) occurred under these latter conditions (Palmer and McIntyre, 1969).

Although earlier studies (Perkin and Plant, 1921) had shown that cyclohexanone N_α-acetylphenylhydrazone was indolized to 9-acetyl-1,2,3,4-tetrahydrocarbazole, repetition of this work afforded 1,2,3,4-tetrahydrocarbazole as the only product and it was suggested (Suvorov and Sorokina, N. P., 1961) that N_α-deacetylation in this cyclization and, likewise, N-deacetylation of the above diacetyl enehydrazines, were a necessary prelude to indolization. However, this conclusion was later (Elgersma, 1969) shown to be erroneous and it became well established that N_α-acylarylhydrazones form directly the corresponding 1-acylindoles upon indolization (see Chapter IV, C).

The corresponding monoacetyl enehydrazine intermediate **241** ($R^1 = R^2 = CH_3$) could not be isolated after reaction of ethyl methyl ketone N_α-methylphenylhydrazone with acetic anhydride and 4-methylbenzene sulphonic acid.

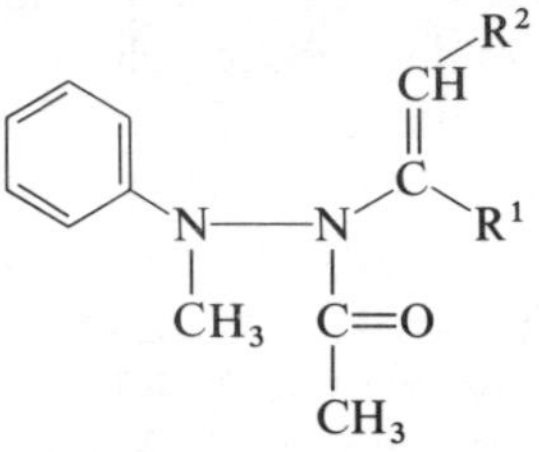

241

The product from this reaction was shown to contain N-acetyl-N-methylaniline, N_β-acetyl-N_α-methylphenylhydrazine and three acetyl derivatives of 1,2,3-trimethylindole whose structures were not established but which were also formed by analogous treatment of 1,2,3-trimethylindole. It was suggested (Suvorov, Sorokina, N. P. and Sheinker, I. N., 1959) that prevention of N_α-acetylation by the N_α-methyl group in this case allowed the N_α-p-electron pair to remain free and therefore spontaneous indolization of the enehydrazine tautomer occurred under the reaction conditions. Contrary to this failure to isolate an N_β-acetyl enehydrazine of type **241**, treatment of diethyl ketone, dibenzyl ketone, cyclopentanone, cyclohexanone, 2-tetralone, butyraldehyde and phenylacetaldehyde N_α-methylphenylhydrazone with acetyl chloride in pyridine at 0 °C afforded the corresponding **241** [$R^1 = C_2H_5$, $R^2 = CH_3$; $R^1 = CH_2C_6H_5$, $R^2 = C_6H_5$; $R^1 + R^2 = (CH_2)_{3 \text{ and } 4}$, $R^1 + R^2 = (CH_2)_2C_6H_4$ and $R^1 = H$, $R^2 = C_2H_5$ and C_6H_5, respectively]. All these products were converted, by heating at 170 °C, into the corresponding 1-methylindoles and acetamide, formed along with N-methylaniline, the isolation of which indicated that N—N bond homolysis is a major alternative reaction pathway. The N_β-acetyl enehydrazines also gave the corresponding indoles when treated with 0.5M dichloroacetic acid in anhydrous acetonitrile at room temperature although in this respect **241** ($R^1 = H$, $R^2 = C_2H_5$ and C_6H_5) reacted much slower than the analogous above compounds derived from ketones, a difference which was rationalized in terms of the sterically related increased basicity of this latter group of compounds relative to the former group (Schiess and Sendi, 1978).

Quaternization of cyclohexanone and 4-cyclohexylcyclohexanone N_α-methylphenylhydrazone (**242**; R = H and cyclohexyl, respectively) with methyl iodide furnished the corresponding indoles **246** (R = H and cyclohexyl) (Posvic, Dombro, Ito, H. and Telinski, 1974). No enehydrazine intermediate **244** nor any other intermediates could be isolated, even under these very mild

CH3 CH3 CH3
242 243
CH3 H CH3 CH3 CH3
245 244
CH3
246

conditions. However, under the appropriate conditions, **242** [R = $C(CH_3)_3$] reacted with methyl iodide to form **244** [R = $C(CH_3)_3$] and, indeed, it was then found that this was somewhat difficult to convert into **246** [R = $C(CH_3)_3$], a reaction eventually accomplished with boiling trifluoroacetic acid in dimethylformamide or acetic acid with ammonium chloride. I.r. spectroscopic monitoring of the second of these above three reactions led to the conclusion that in these reactions the formation of species **243** from **242** was the rate determining step and it was furthermore suggested (Posvic, Dombro, Ito, H. and Telinski, 1974) that whereas in moderately strong acids, N_β-protonation of **244** occurred to afford **245** which was then rapidly converted into **246**, in strong acids, C-protonation of the enamine system in **244** occurred to give **243** which cannot directly lead to indoles in the same manner. Although not referred to in the preceding studies (Posvic, Dombro, Ito, H. and Telinski, 1974), investigations reported five years earlier (Grandberg, Sibiryakova and Brovkin, 1969) had already shown that arylhydrazones reacted with alkylating agents to yield indoles. Thus, propionaldehyde, 4-oxopentan-1-ol and cyclohexanone N_α-methylphenylhydrazone reacted with equimolar quantities of benzyl chloride in refluxing ethanol to afford 1,3-dimethylindole, 1,2-dimethyltryptophol and 1,2,3,4-tetrahydro-9-methylcarbazole, respectively, and likewise cyclohexanone

N_{α}-methylphenylhydrazone reacted with dimethyl sulphate to afford 1,2,3,4-tetrahydro-9-methylcarbazole. Acetone and cyclohexanone N_{α}-benzylphenylhydrazone reacted with benzyl chloride to produce 1-benzyl-2-methylindole and 9-benzyl-1,2,3,4-tetrahydrocarbazole, respectively, and ethyl methyl ketone N_{α}-benzylphenylhydrazone reacted with allyl bromide to produce 1-benzyl-2,3-dimethylindole. It was proposed that N_{β}-alkylation formed a quaternary salt, the enehydrazine tautomeric form of which then underwent rearrangement with the ultimate formation of an indolic product. When analogous reactions were carried out with phenylhydrazones, both the 1-unsubstituted and 1-alkylated indoles were formed in each case. Thus, ethyl methyl ketone, dipropyl ketone and cyclohexanone phenylhydrazone reacted with benzyl chloride in refluxing ethanol to furnish 1-benzyl-2,3-dimethyl- and 2,3-dimethylindole, 9-benzyl-1,2,3,4-tetrahydro- and 1,2,3,4-tetrahydrocarbazole and 1-benzyl-3-ethyl-2-propyl- and 3-ethyl-2-propylindole, respectively. In each case the yield of the 1-benzylindolic product was greater than the yield of the corresponding 1-unsubstituted indolic product. It was suggested that the formation of the two products from such reactions was a result of initial differential alkylation of the phenylhydrazone at either N_{α} or N_{β} to give species **247** and **248**, respectively. These, after tautomerization to their enehydrazine forms, will ultimately afford the corresponding indoles (Grandberg, Sibiryakova and Brovkin, 1969).

R^2 CH_2 C R^1 N—$\overset{\oplus}{N}$ CH_2 H $Cl^{\ominus}$ C_6H_5

247

R^2 CH_2 C R^1 N—$\overset{\oplus}{N}$ H CH_2 $Cl^{\ominus}$ C_6H_5

248

Enehydrazines are also formed when N_{β}-alkyl or N_{α},N_{β}-dialkylarylhydrazines react with aldehydes and ketones. The first report of such a reaction involved that between 2,5,N_{α},N_{β}-tetramethylphenylhydrazine and dibenzylketone in acetic acid which afforded **249** ($R^1 = R^3$–$R^5 = CH_3$, $R^2 = R^7 = H$,

R^1 R^2 R^8 R^7 C C N—N R^6 R^3 R^4 R^5

249

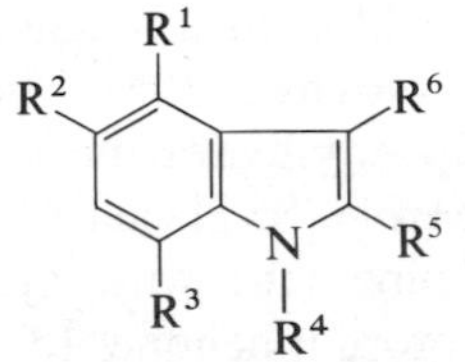

250

$R^6 = CH_2C_6H_5$, $R^8 = C_6H_5$). This yielded 2-benzyl-1,4,7-trimethyl-3-phenylindole (**250**; $R^1 = R^3 = R^4 = CH_3$, $R^2 = H$, $R^5 = CH_2C_6H_5$, $R^6 = C_6H_5$) upon treatment with concentrated hydrochloric acid in ethanol (Neber, Knöller, Herbst and Trissler, 1929). However, in a subsequent attempted repetition of the former reaction, the enehydrazine could not be isolated and the indole was the only product. Likewise, the reaction of N_α,N_β-dimethylphenylhydrazine with dibenzylketone, cyclopentanone or cyclohexanone under similar conditions produced only the indolic products **250** [R^1–R^3 = H, $R^4 = CH_3$ and $R^5 = CH_2C_6H_5$, $R^6 = C_6H_5$ and $R^5 + R^6 = (CH_2)_{3 \text{ and } 4}$, respectively] (Elgersma, 1969). The indolization of this type next reported involved the reaction between azobenzene and cyclohexanone in the presence of a boron trifluoride catalyst which produced 1,2,3,4-tetrahydrocarbazole together with aniline, as expected by cleavage of the N—N bond. The only product isolated when using concentrated sulphuric acid as the catalyst in this reaction was that resulting from a benzidine rearrangement of the hydrazobenzene formed from the azobenzene by reduction under the reaction conditions (Nesmeyanov and Golovnya, 1960). In a subsequent elaboration of this indolization, using a boron trifluoride–anisole complex as the catalyst, a mixture of azobenzene with cyclohexanone or 4-methylcyclohexanone afforded 1,2,3,4-tetrahydrocarbazole and 3-methyl-1,2,3,4-tetrahydrocarbazole, respectively, along with aniline in each case and benzidine (isolated as the sulphate) from the latter reaction. This last product, along with unchanged azobenzene, was the only compound isolated when using cyclopentanone, acetophenone, 1-tetralone or dipropyl ketone as the ketonic moieties in similar reactions with azobenzene. Despite the fact that the boron trifluoride–anisole complex readily catalysed the conversion of hydrazobenzene into benzidine, when a mixture of hydrazobenzene and cyclohexanone was treated with this catalyst, only 1,2,3,4-tetrahydrocarbazole and aniline were produced, indicating that the rate of indolization was apparently much greater than the rate of the benzidine rearrangement (Nesmeyanov and Golovnya, 1961). Since with strong acids in polar media the above reaction afforded only benzidine, the formation of which appeared to be dependent upon the initial diprotonation of the hydrazobenzene, a cation exchange resin in the acid form (Amberlite 1R-20) in refluxing toluene was used as the catalyst since this would not effect diprotonation of the hydrazobenzene. This, in fact, produced a 33% yield of 1,2,3,4-tetrahydrocarbazole (along with aniline as N-cyclohexylideneaniline) (Posvic, Dombro, Ito, H. and Telinski,

1974) as compared to a 30% yield using a boron trifluoride–anisole complex as catalyst (Nesmeyanov and Golovnya, 1961). In an extension of this use of Amberlite IR-20, N_β-methylphenylhydrazine reacted with cyclohexanone to afford 1,2,3,4-tetrahydrocarbazole (85% yield) (cf. a 20% yield using an acetic acid catalyst) and methylamine and with cyclopentanone, methyl pentyl ketone, heptaldehyde and propiophenone to give **250** [R^1–R^4 = H and $R^5 + R^6 = (CH_2)_3$, $R^5 = CH_3$, $R^6 = (CH_2)_3CH_3$, $R^5 = H$, $R^6 = (CH_2)_4CH_3$ and $R^5 = C_6H_5$, $R^6 = CH_3$, respectively]. Acetophenone did not react with N_β-methylphenylhydrazine under these conditions but yielded **250** (R^1–R^4 = R^6 = H, $R^5 = C_6H_5$) using 4-methylbenzene sulphonic acid as the catalyst. Surprisingly, since it is well established that N_α-methylation in phenylhydrazones facilitates their indolization, N_α,N_β-dimethylphenylhydrazine reacted only slowly (i.e. methylamine evolution was slow) with cyclohexanone, 4-cyclohexylcyclohexanone or methyl pentyl ketone in the presence of the Amberlite 1R-20 catalyst to produce only very low yields of **250** [R^1–R^3 = H, $R^4 = CH_3$ and $R^5 + R^6 = (CH_2)_4$, $R^5 + R^6 = (CH_2)_2CH(\text{cyclohexyl})CH_2$ and $R^5 = CH_3$, $R^6 = (CH_2)_2CH_3$, respectively] (Posvic, Dombro, Ito, H. and Telinski, 1974). When N_α,N_β-dimethylphenylhydrazine reacted with cyclohexanone, cyclopentanone, 2-tetralone, diethylketone, propionaldehyde and phenylacetaldehyde in the absence of an acid catalyst, the enehydrazines **249** [R^1–R^3 = R^7 = H, $R^4 = R^5 = CH_3$, $R^6 + R^8 = (CH_2)_4$] (Grandberg and Przheval'skii, 1969; Grieder and Schiess, 1970; Schiess and Grieder, 1969, 1974) and **249** [R^1–R^3 = R^7 = H, $R^4 = R^5 = CH_3$ and $R^6 + R^8 = (CH_2)_3$ and $(CH_2)_2C_6H_4$, $R^6 = C_2H_5$, $R^8 = CH_3$, $R^6 = H$, $R^8 = C_2H_5$ and $R^6 = H$, $R^8 = C_6H_5$] (Grieder and Schiess, 1970; Schiess and Grieder, 1969, 1974) were formed, respectively. These, upon treatment with acid or upon thermolysis, gave the corresponding indoles **250** [R^1–R^3 = H, $R^4 = CH_3$ and $R^5 + R^6 = (CH_2)_4$, $(CH_2)_3$ and $(CH_2)_2C_6H_4$, $R^5 = C_2H_5$, $R^6 = CH_3$, $R^5 = H$, $R^6 = C_2H_5$ and $R^5 = H$, $R^6 = C_6H_5$, respectively]. The relative rates of formation of these six products, 1.0, 0.05, 25, 0.5, 8.0 and 25, respectively {$T_{1/2}$ **250** [R^1–R^3 = H, $R^4 = CH_3$, $R^5 + R^6 = (CH_2)_4$] = 25 min} also supported the intermediacy of a monoprotonated enehydrazine tautomer in the mechanism of the Fischer indolization (Grieder and Schiess, 1970; Heimgartner, Hansen and Schmid, H., 1979). Similarly, when N_α,N_β-dimethylphenylhydrazine reacted with 2-phenylpropionaldehyde at 20 °C, **249** (R^1–R^3 = R^6 = H, $R^4 = R^5 = R^7 = CH_3$, $R^8 = C_6H_5$) was formed (Grieder and Schiess, 1970). In contrast to these results, it has been found that the corresponding enehydrazines were not isolable from the reactions between 2,6-dichloro-N_β-methylphenylhydrazine and cyclohexanone (Robinson, F. P. and Brown, R. K., 1964) and between 2,6-dimethylphenylhydrazine and cyclohexanone (Bajwa and Brown, R. K., 1968a), isobutyraldehyde (Bajwa and Brown, R. K., 1968b), propionaldehyde (Bajwa and Brown, R. K., 1969) and 2-methylcyclohexanone (Bajwa and Brown, R. K., 1970) in boiling benzene solution, these relatively mild conditions forcing the reaction surprisingly through the total indolization sequence. The significance of the very interesting indolic products obtained from these reactions is referred to in Chapter II, G2. In a similar manner, the reaction of hydrazobenzene with cyclohexanone, cyclopentanone and ethyl pyruvate in refluxing aromatic solvents (benzene, toluene or xylene) in the presence of a trace of 4-methylbenzene sulphonic acid gave the corresponding 1-substituted indolic products together with, in each case, aniline. The two cyclic ketones also similarly reacted with N_β-methylphenylhydrazine but a pyrazoline, formed by reaction between 2 molecules of cyclohexanone and 1 molecule of the hydrazine, and a linear condensation product, formed by reaction between 2 molecules of cyclopentanone and 2 molecules of the hydrazine were also formed, respectively, from these reactions. Only the corresponding pyrazoline and linear condensation product were isolated from the reaction between ethyl pyruvate and N_β-methylphenylhydrazine. Although the separation of water was observed when ethyl methyl ketone, benzyl methyl ketone and acetaldehyde reacted with N_β-methylphenylhydrazine in refluxing benzene, no other product could be isolated from these reactions (Chapelle, Elguero, Jacquier and Tarrago, 1970a). Clearly, however, the use of N_β-alkylated or N_β-arylated arylhydrazines in the synthesis of indoles may well succeed in some cases in which the corresponding arylhydrazones cannot be indolized or where the products are potentially unstable under acidic or strongly thermal conditions.

A series of studies, in principle analogous to those described above for the Fischer indolization, have been effected in connection with the Piloty synthesis of pyrroles. The results of these investigations are described in Chapter VI, D.

A synthesis of N_α-(4-chlorobenzyl)-4-methylphenylhydrazine has been achieved by reacting 4-methylphenylhydrazine with 4-chlorobenzyl chloride in the presence of triethylamine (Walton, Stammer, Nutt, Jenkins and Holly, 1965) and likewise a synthesis of N_α-(4-nitrobenzyl)-4-methoxyphenylhydrazine has been effected from 4-methoxyphenylhydrazine and 4-nitrobenzyl chloride (Sarett and Shen, T.-Y., 1966b). In the former study it was correctly recognized that N_β-alkylation would also occur during these reactions and, with regard to this, about a 20% yield of a by-product, presumably N_β-(4-chlorobenzyl)-4-methylphenylhydrazine, was produced. However, the statement that 'The presence of the by-product offers no problem as it is incapable of undergoing a Fischer-type ring closure . . .' (Walton, Stammer, Nutt, Jenkins and Holly, 1965) is obviously invalidated by the results of the above studies using N_β-alkylated and N_β-arylated arylhydrazines. Related to this it may be significant that only a 39.5% yield of methyl 1-(4-chlorobenzyl)-5-methylindole-3-acetate was obtained when the above mentioned hydrazine mixture (as the hydrochloride) was indolized with 3-carbomethoxypropionaldehyde in refluxing methanol (Walton, Stammer, Nutt, Jenkins and Holly, 1965). Presumably, the by-product could in this case also react to afford the intermediate **249**

[$R^1 = R^3 = R^4 = R^6 = R^7 = H$, $R^2 = CH_3$, $R^5 = CH_2C_6H_4Cl(4)$ and $R^8 = CH_2COOCH_3$] and ultimately methyl 5-methylindole-3-acetate (**250**; $R^1 = R^3$–$R^5 = H$, $R^2 = CH_3$ and $R^6 = CH_2COOCH_3$), together with 4-chlorobenzylamine.

Another enehydrazine isolation was effected (Eberle and Brzechffa, 1976) when 1-phenylpyrazolidin-5-one (**251**; R = H) reacted with 2,6-dichlorophenylacetaldehyde in warm toluene to afford **252**. This, upon recrystallization from hot aqueous ethanol lost water to produce **253** which was subsequently thermally indolized as described later in this chapter.

In all the above examples involving enehydrazine intermediates, the N_β-atoms were, of necessity, tertiary. What has been claimed (Wright and Gambino, 1979) to be the first isolation of enehydrazine intermediates, although this claim should more specifically also have referred to the secondary nature of the N_β-atoms, has been reported when 6-aminouracils or 6-amino-2-thiouracils reacted with arylhydrazines in acetic acid to form **254** (R^3 = H or CH_3, X = O and S, respectively) (Wright, 1976; Wright and Brown, N. C., 1974). The indolization of these compounds has been studied (Wright, 1976; Wright and Gambino, 1979) (see later in this chapter and Chapter IV, I). The isolation of **256** from the reaction between **255** and phenylhydrazine and their

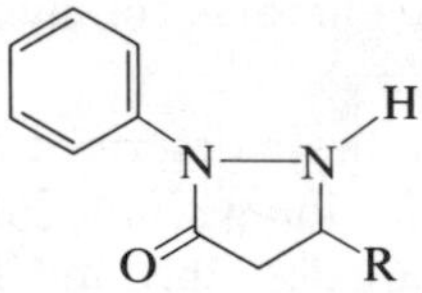

251

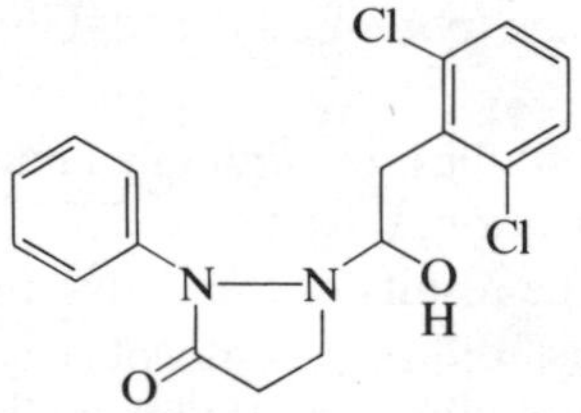

252

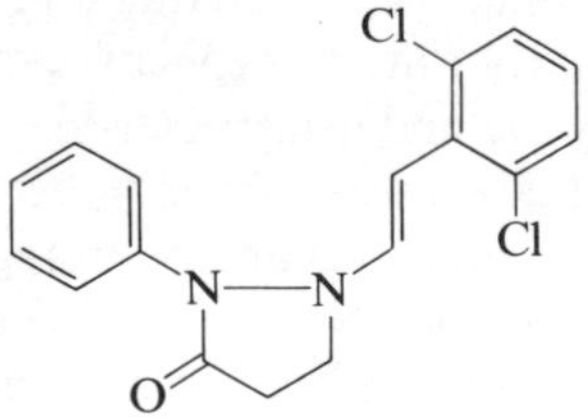

253

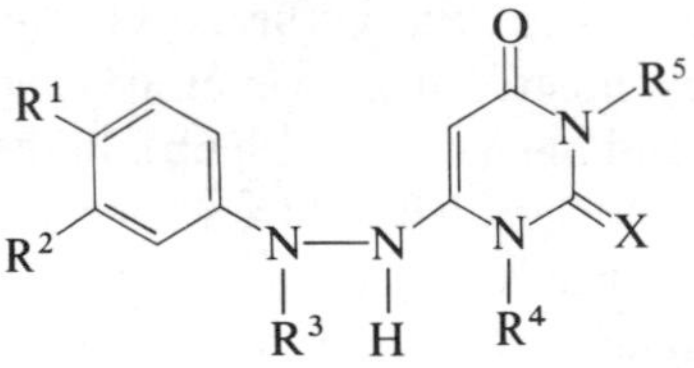

254

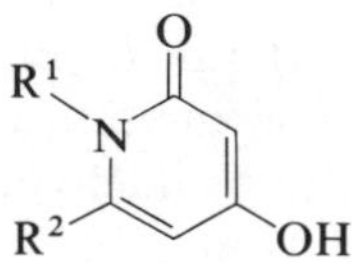

255

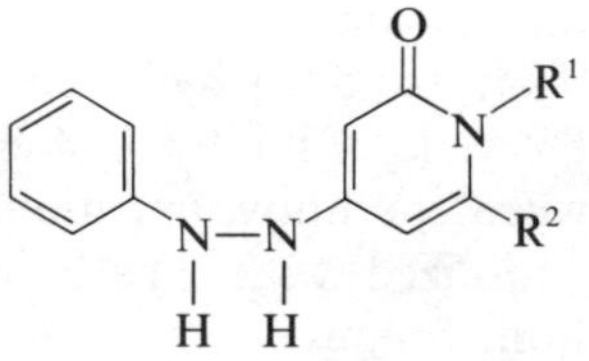

256

subsequent indolizations (see Chapter IV, I) (Ducrocq, Civier, André-Louisfert and Bisagni, 1975) represent isolations of enehydrazine intermediates analogous to those reported by Wright and his co-workers.

When **257** (R = H, CH_3 and C_2H_5) reacted with phenylhydrazine, when **257** (R = CH_3) reacted with 4-chlorophenylhydrazine or when **257** (R = $CH_2C_6H_5$) reacted with 4-chloro- or 4-methoxyphenylhydrazine, the corresponding arylhydrazones **258** were formed whereas when the ketones **257** (R = H, CH_3 and $CH_2C_6H_5$) reacted with N_α-benzyl- or N_α-methylphenylhydrazine or when **257** (R = CH_3) reacted with 4-chloro-N_α-methylphenylhydrazine, the condensation products existed as the enehydrazine tautomers, **259**, possibly because of their stabilization by intramolecular hydrogen bonding between the enehydrazine NH and the oxygen atom of either the 2-oxo or 4-carbamoyl groups (Yevich, Murphy, J. R., Dufresne and Southwick, 1978). These products were subjected to indolization as described later in this chapter.

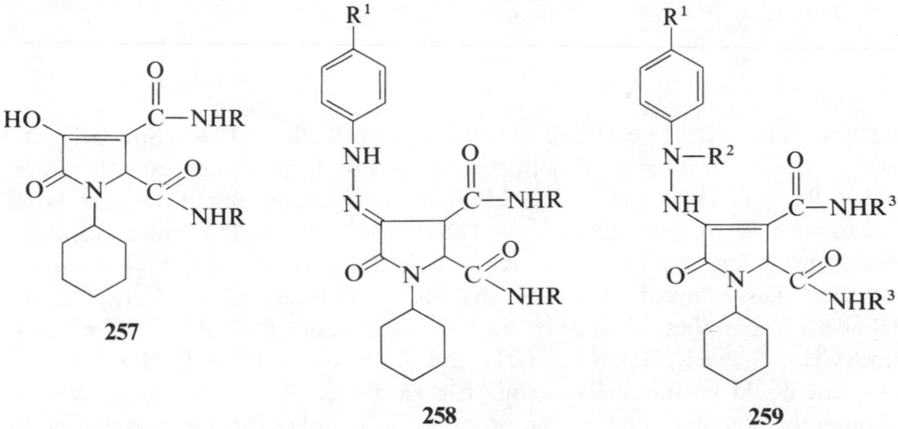

257 **258** **259**

An interesting series of thermal rearrangements has been effected upon the enehydrazines of general structure **260**. Thus, **260** [$R^1 = CH_3$, $R^2 = R^3 = C_6H_5$; $R^1 = C_2H_5$, $R^2 = R^3 = C_6H_5$; $R^1 = R^3 = CH_3$, $R^2 = C_6H_5$; $R^1 = CH_3$, $R^2 = C_6H_4CH_3(4)$, $R^3 = C_6H_5$; $R^1 = CH_2C_6H_5$, $R^2 = R^3 = C_6H_5$ and $R^1 = H$, $R^2 = R^3 = C_6H_5$] afforded **261** ($R^1 = CH_3$, $R^2 = R^3 = C_6H_5$; $R^1 = C_2H_5$, $R^2 = R^3 = C_6H_5$ and $R^1 = R^3 = CH_3$, $R^2 = C_6H_5$) (Kollenz, Ziegler, Eder and Prewedourakis, 1970), **261** [$R^1 = CH_3$, $R^2 = C_6H_4CH_3(4)$, $R^3 = C_6H_5$] (Kollenz and Labes, 1975); **261** ($R^1 = CH_2C_6H_5$, $R^2 = R^3 = C_6H_5$) (Kollenz, 1972b) and **262** (Kollenz, 1972b), respectively. Likewise, heating **260** ($R^1 = COC_6H_5$ and $COOC_2H_5$, $R^2 = R^3 = C_6H_5$) in decalin at 160 °C or in refluxing xylene, respectively, afforded **261** ($R^1 = COC_6H_5$ and $COOC_2H_5$, respectively, $R^2 = R^3 = C_6H_5$) [further heating of these products yielded **263** ($R^1 = CONHCOC_6H_5$ and OC_2H_5, respectively, $R^2 = C_6H_5$) (Kollenz, 1978)] and heating **264** (n = 1 and 2) in

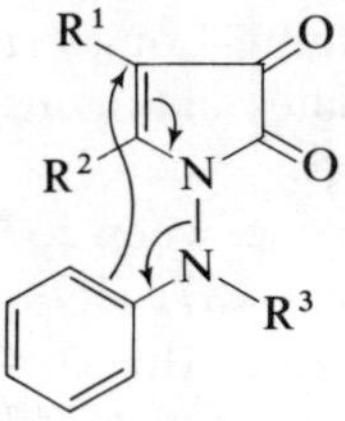

260

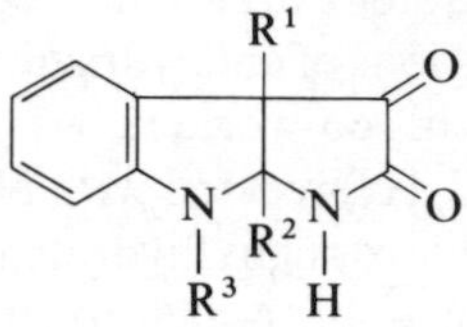

261

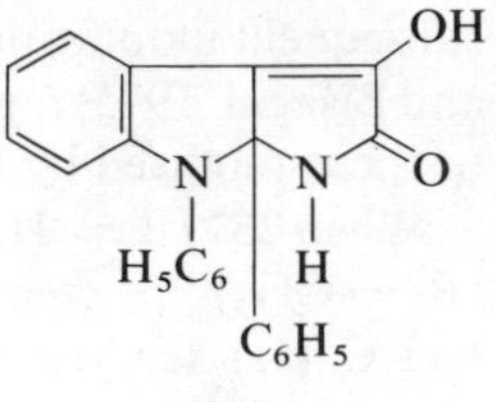

262

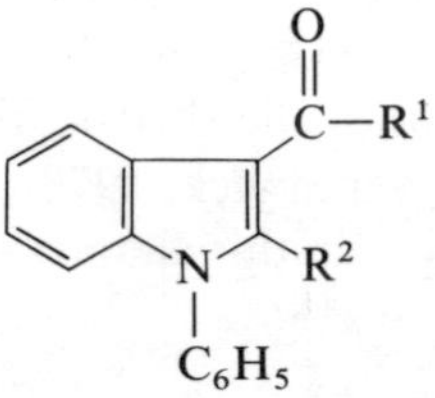

263

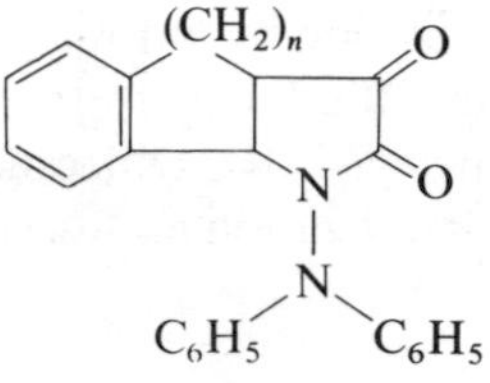

264

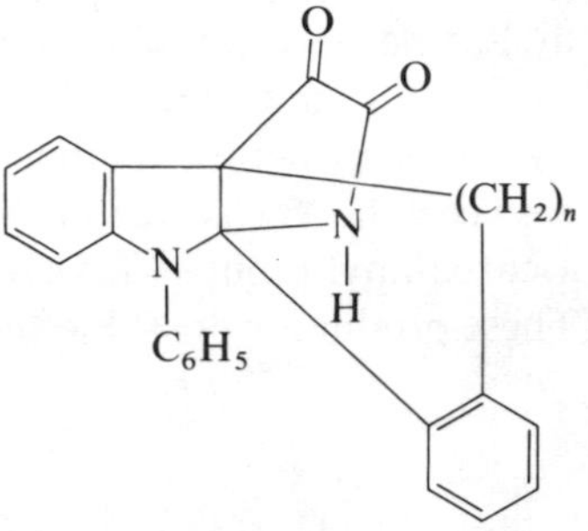

265

decalin yielded **265** (n = 1 and 2, respectively) (Kollenz, 1978). Similarly, **260** (R^1 = H, R^2 = R^3 = C_6H_5) in xylene at 140 °C produced **261** (R^1 = H, R^2 = R^3 = C_6H_5) which at 230–235 °C underwent opening of the pyrrolidin-2,3-dione ring to give **263** (R^1 = $CONH_2$, R^2 = C_6H_5), a similar cleavage occurring when **260** [R^1 = H, R^2 = $C_6H_4CH_3(4)$, R^3 = C_6H_5] in xylene solution was refluxed to afford **263** [R^1 = $CONH_2$, R^2 = $C_6H_4CH_3(4)$] (Kollenz and Labes, 1975). Of the four hydrazones **266** (R^1 = R^2 = C_6H_5 and CH_3; R^1 = C_6H_5, R^2 = CH_3 and R^1 = CH_3, R^2 = C_6H_5), only the last one could be thermally indolized (heating at 140 °C), affording **267**, although the potential indolization products of the other three were prepared by reaction of the appropriate **261** with the appropriate hydrazine (Kollenz, 1971). The mechanism of these indoline formations was recognized, when they were initially developed (Kollenz, Ziegler, Eder and Prwedourakis, 1970), in terms of that of the Fischer indolization current at the time and was postulated as involving new C—C bond formation through a 'concerted' electron movement

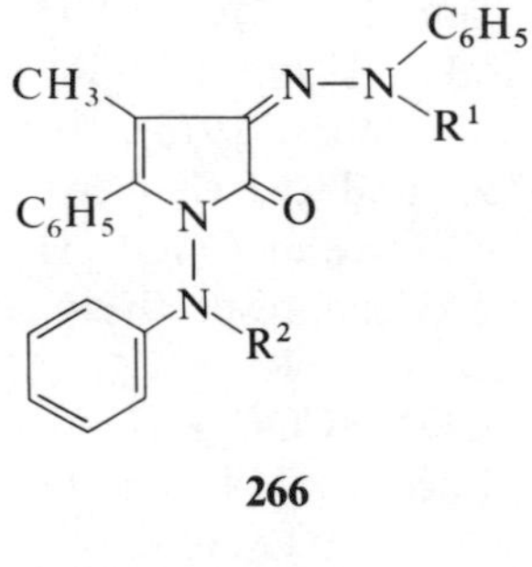

266

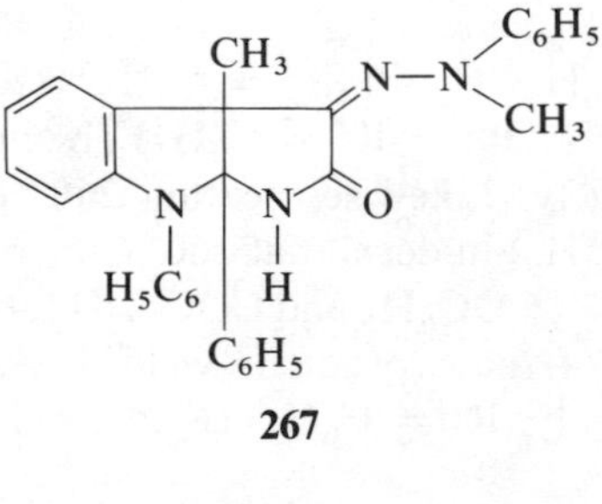

267

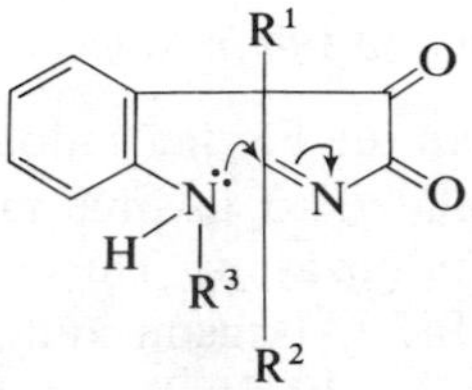

268

as shown in **260** to form **268** (after rearomatization) which underwent ring closure as shown to yield **261**. Subsequent studies (Kollenz and Labes, 1976) supported the intramolecular nature of the reaction mechanism when a mixture of **260** ($R^1 = CH_3$, $R^2 = R^3 = C_6H_5$) which was ^{14}C labelled in the benzenoid nuclei of the $(C_6H_5)_2N$-group and unlabelled **260** ($R^1 = H$, $R^2 = R^3 = C_6H_5$) was thermally indolized to furnish a mixture of products, heating of which at 220 °C afforded **263** ($R^1 = CONH_2$, $R^2 = C_6H_5$) which carried only 0.006% of the original activity present in the labelled **260** ($R^1 = CH_3$, $R^2 = R^3 = C_6H_5$). When $R^1 = CH_3$ in **261**, such a subsequent reaction could not occur – see above. Furthermore, kinetic studies which showed the reaction to be first order in reactant and investigation of the reaction activation parameters supported the involvement of a [3,3]-sigmatropic shift in the new C—C bond formation (Kollenz and Labes, 1976) (the involvement of a [3,3]-sigmatropic shift in the new C—C bond formation during Fischer indolization is discussed in detail later in this chapter). In an extension of this reaction, **269** was treated with polyphosphoric acid, when **271** was produced, presumably via **270** by the mechanism indicated in Scheme 16 (Kollenz, 1972a).

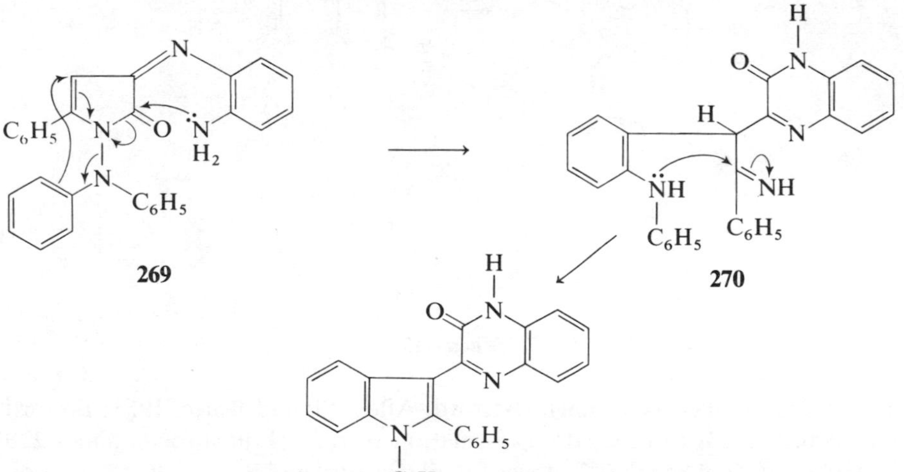

269

270

271

Scheme 16

2. Intermediates Corresponding to **198** *in Scheme 10*

Several reactions, other than the Fischer indolization, by which indoles can be formed, have been postulated to involve mechanisms analogous to the latter stages of the Fischer indolization. Thus, reductive ring contraction of cinnolines **272** into indoles **147** (Atkinson and Simpson, 1947; Baumgarten and Furnas, 1961; Brown, R. K., 1972f; Bruce, 1959; Neber, Knöller, Herbst and Trissler, 1929) has been postulated (Allen, C. F. H. and Allen, J. A. van, 1951; Ames, D. E., Novitt, Waite, D. and Lund, 1969; Besford and Bruce, 1964; Corbett and Holt, P. F., 1960; Jacobs, 1957b) to occur via intermediates **274** and **275** which subsequently form the indole **147** according to Scheme 17.

Scheme 17

This reduction has been shown (Besford, Allen, G. and Bruce, 1963; Besford and Bruce, 1964) to proceed via formation of the 1,4-dihydrocinnolines **273** and, in support of Scheme 17, it was found (Besford and Bruce, 1964) that upon reduction of 4-phenyl-2-^{15}N-cinnoline, the ^{15}N label was essentially eliminated in the ammonia produced. A similar intermediate, **277**, has also been postulated (Snyder, Merica, Force and White, E. G., 1958) in the reductive cyclization of

2-nitrobenzyl cyanides **276** into indoles **147** (R^2 = H) or 3*H*-indoles **278** (Brown, R. K., 1972d; Bourdais and Germain, 1970; Germain and Bourdais, 1976) and analogous intermediates have been postulated (Burton and Duffield, 1949) as being involved in the reductive cyclization of dinitrostyrenes **279** into indoles **147** (Brown, R. K., 1972e). 2-Nitrobenzyl ketones also undergo reductive cyclization to give indoles (Brown, R. K., 1972e), probably via the formation of analogous intermediates and the conversion of 2-(2-aminophenyl)ethanol into indole, by passage in the gaseous phase over a copper catalyst, also possibly occurred via a similar intermediate, 2-aminophenylacetaldehyde (Bakke, Heikman and Hellgren, 1974).

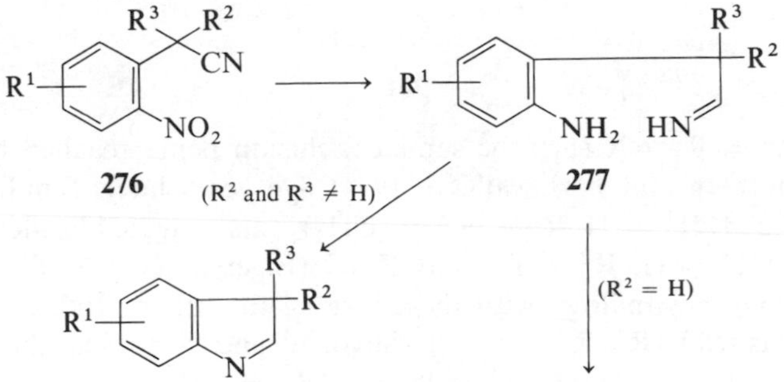

276 → **277**

(R^2 and R^3 ≠ H) → **278**

(R^2 = H) → **147** (R^2 = H)

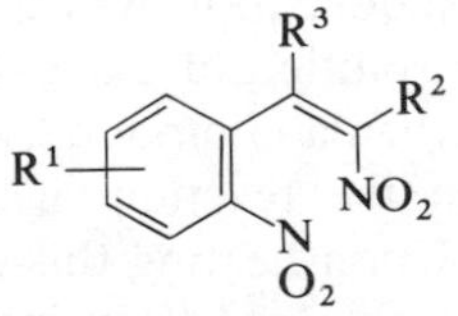

279

When the diacetyl enehydrazine intermediate **238** ($R^1 = OCH_3$, $R^2 = R^3 = CH_3$) was treated with alcoholic potassium hydroxide, 5-methoxy-2,3-dimethylindole was produced (Suvorov and Sorokina, N. P., 1961). However, similar treatment of **238** (R^1 = H and CH_3, $R^2 = R^3 = CH_3$) afforded compounds that were formulated initially as **280** (R^1 = H, $R^2 = R^3 = CH_3$) (Suvorov and Sorokina, N. P., 1961; Suvorov, Sorokina, N. P. and

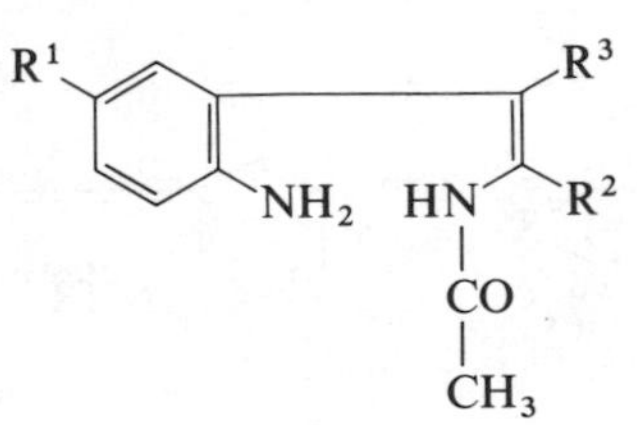

280

Sheinker, I. N., 1958) and **280** (R^1–R^3 = CH_3) (Suvorov and Sorokina, N. P., 1961), respectively. The structural assignment **280** (R^1 = H, R^2 = R^3 = CH_3) was later (Elgersma, 1969; Elgersma and Havinga, 1969) shown to be incorrect, the reaction simply involving hydrolysis of the N-acetyl group to yield **281**

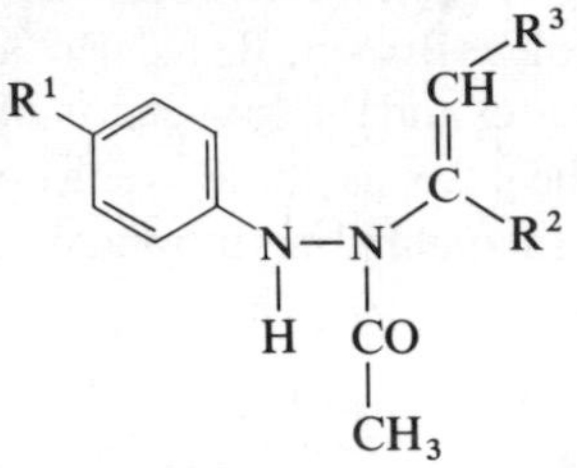

281

(R^1 = H, R^2 = R^3 = CH_3), the same conclusion being reached by another group (Grandberg and Przheval'skii, 1974) five years later. Similar alkaline treatment of **238** [R^1 = H, R^2 + R^3 = $(CH_2)_4$] also effected N_α-deacetylation to give **281** [R^1 = H, R^2 + R^3 = $(CH_2)_4$] (Elgersma, 1969; Elgersma and Havinga, 1969). By analogy with these results, the other product which was formulated as **280** (R^1–R^3 = CH_3) (Suvorov and Sorokina, N. P., 1961) should be structurally reassigned as **281** (R^1–R^3 = CH_3).

The first isolation of an identifiable intermediate corresponding to **198** in Scheme 10 for the mechanism of the Fischer indolization was achieved by passing dry hydrogen chloride into an ethanolic solution of 2-oxo-4-hydroxybutyrolactone phenylhydrazone (**282**; X = O) when a compound, isolated as its hydrochloride salt which precipitated from the reaction medium, was obtained and assigned structure **283** (R = H). Upon heating this compound with acid, the expected indole **284** (X = O) was obtained (Plieninger, 1950c; Plieninger and Nógradi, 1955b). Subsequently (Plieninger and Nógradi, 1955b),

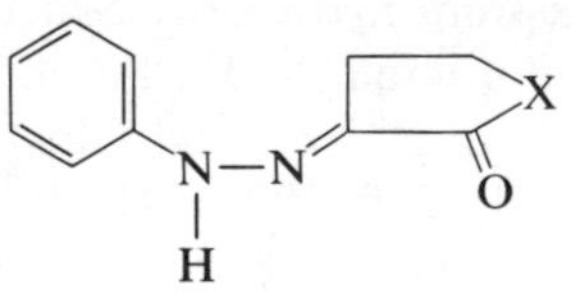

282

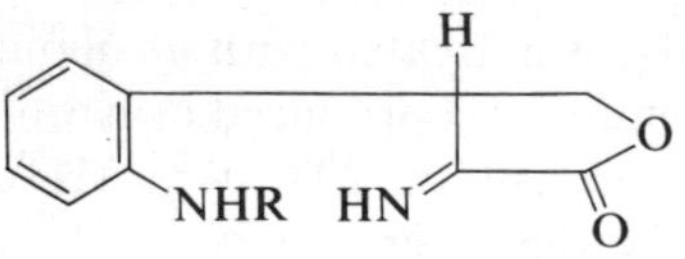

283

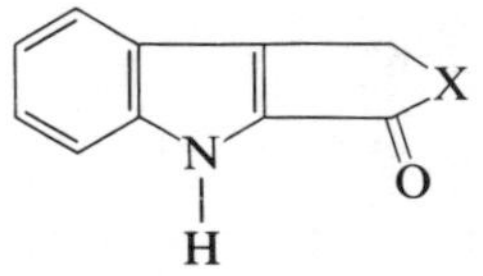

284

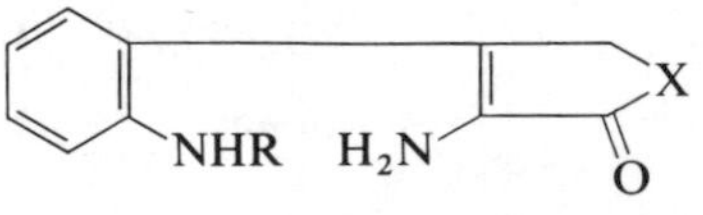

285

this product was acetylated to afford a derivative assigned the tautomeric structure **283** (R = $COCH_3$) $\rightleftharpoons$ **285** (R = $COCH_3$, X = O). Although these structural assignments were criticized (Suvorov, Sorokina, N. P. and Sheinker, I. N., 1958), later u.v. and p.m.r. spectroscopic investigations showed that essentially they were correct although the structure of the original amine was modified to **285** (R = H, X = O) (Owellen, Fitzgerald, J. A., Fitzgerald, B. M., Welsh, D. A., Walker, D. M. and Southwick, 1967; Elgersma, 1969), analogous to one of the tautomeric forms of the acetyl derivative. A similar compound, **287**, was also prepared from 2-oxo-4-hydroxybutyrolactone 2-naphthylhydrazone (**286**) (Elgersma, 1969). An analogous intermediate, **288** (R = H, X = S), was isolated when **254** (R^1 = CH_3, R^2–R^5 = H, X = S) was treated with refluxing 1M hydrochloric acid, this being converted into the corresponding indole **289** (R^1 = CH_3, R^2–R^5 = H, X = S) by refluxing in *N,N*-dimethylaniline (Wright and Gambino, 1979). Likewise, the major isolable product,

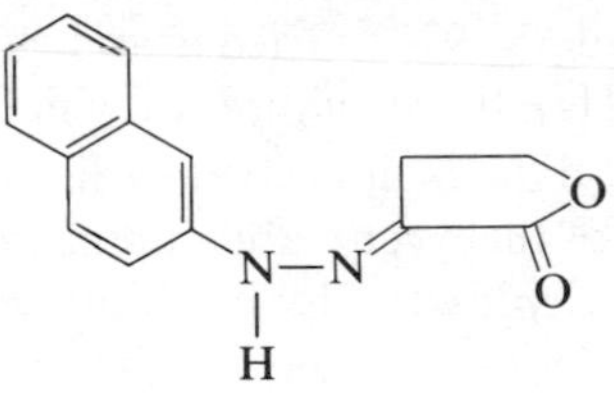

N—N
H
O
O
286

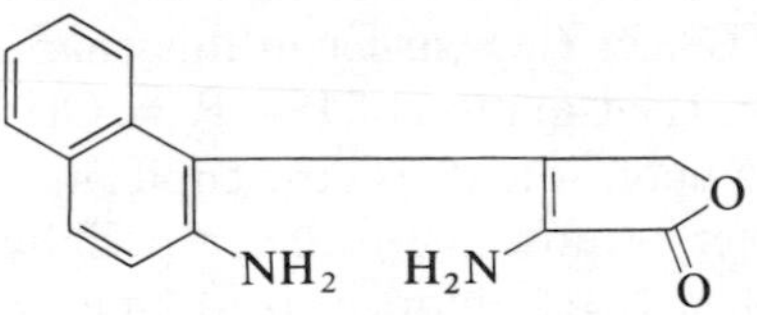

NH_2 H_2N
O
O
287

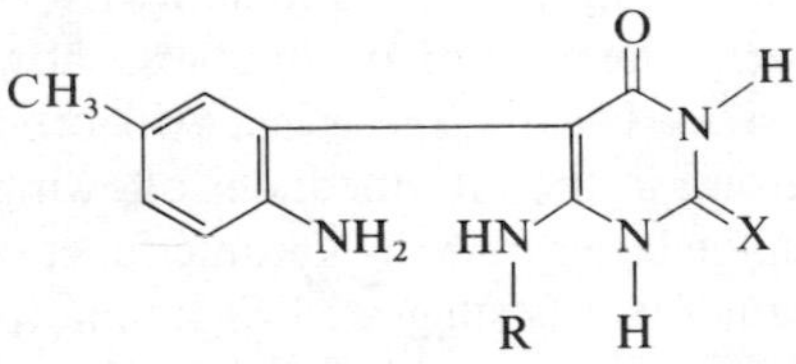

CH_3
O
H
N
NH_2 HN
N
X
R H
288

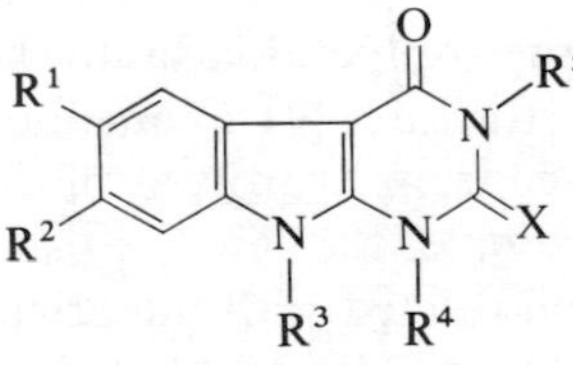

R^1
O
R^5
N
R^2
N
N
X
R^3 R^4
289

when using refluxing 98% formic acid in place of the 1M hydrochloric acid in this reaction, was **288** (R = CHO, X = S) which was hydrolysed by sodium hydroxide into **288** (R = H, X = S). Analogous intermediates were not isolated from similar treatments of **254** (X = O) which furnished the corresponding indoles **289** (X = O) (Wright, 1976) (see Chapter IV, I) and an attempt to prepare such an intermediate, **288** (R = H, X = O), by hydrolysis of **290**, formed by methylation of **288** (R = H, X = S) with methyl iodide in the

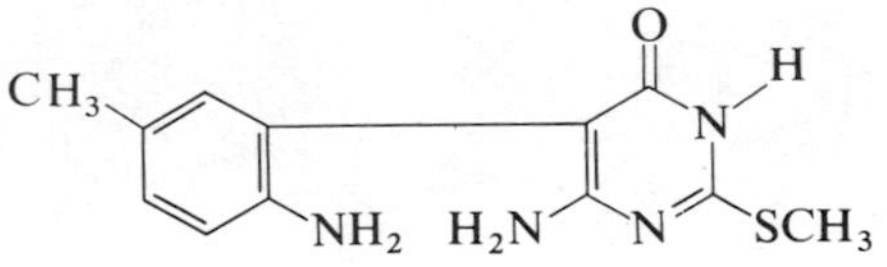

CH_3
O
H
N
NH_2 H_2N N SCH_3
290

presence of sodium hydroxide, afforded only **289** ($R^1 = CH_3$, R^2–R^5 = H, X = O), resulting from both hydrolysis and indolization (Wright and Gambino, 1979). Since the conversion of **254** ($R^1 = CH_3$, R^2–R^5 = H, X = S) into **288** (R = H, X = S) in effect isolated the [3,3]-sigmatropic rearrangement stage in the indolization mechanism by which the new C—C bond is formed (see later in this chapter), detailed kinetic studies to investigate the effect of hydrogen ion concentration upon this stage should now be possible (Wright and Gambino, 1979).

When a mixture of isobutyraldehyde with N_α,N_β-dimethylphenylhydrazine was heated with potassium hydroxide, a compound formulated as **194** (R = CH_3) was formed. Upon hydrolysis this yielded 1,3,3-trimethylindolinol. Similarly, when N_α,N_β-dimethylphenylhydrazine was reacted with 2-phenylpropionaldehyde, the enehydrazine **193** was formed. After purification by chromatography at −10 °C, this compound was stood at room temperature for 50 h, during which time it afforded **194** (R = C_6H_5) and the corresponding *para* isomer (see earlier in this chapter) (Grieder, 1970; Grieder and Schiess, 1970). The isolation of **194** (R = CH_3 and C_6H_5), the structures of which were based upon sound spectroscopic and chemical data as described earlier in this chapter, is most interesting, especially in view of their very labile imine groups.

When N_α-(4-chlorobenzyl)-4-methoxyphenylhydrazine hydrochloride (**291**) was heated with 2-methylcyclohexanone in acetic acid, one of the expected indolic products, **292**, was obtained along with a significant quantity of **293** (R = CH_3, X = O) (Yamamoto, H. and Atsumi, 1968a) and similarly, using 1-methyl-4-piperidone as the ketonic moiety, the reaction product was shown to have structure **294** (Yamamoto, H., 1967a). Clearly, these last two products resulted from hydrolysis of the imine group in the intermediate following the formation of the new C—C bond and in the latter case the ketonic function so formed reacted with the original hydrazine. It was suggested (Yamamoto, H. and Atsumi, 1968a) that these compounds were isolable because the amide nitrogen was not a strong enough nucleophile to react with the ketonic carbonyl function. Indeed, subsequent investigation of the indolization of the hydrazones formed by reacting **291** with cyclohexanone and laevulinic acid (**295**)

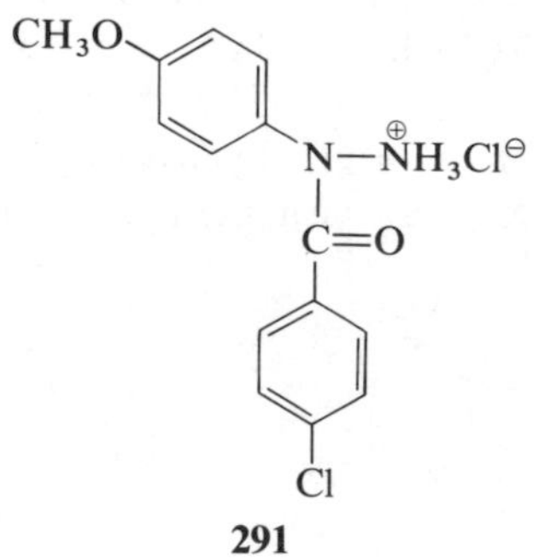

291

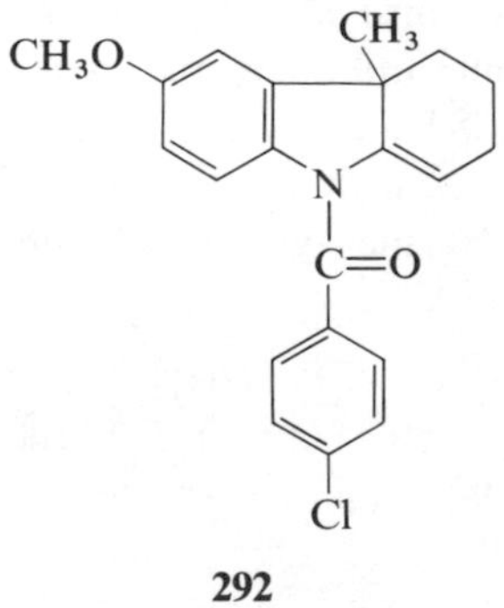

292

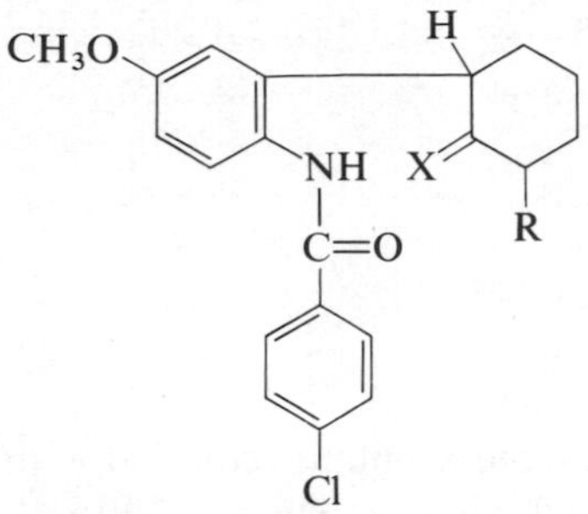

293

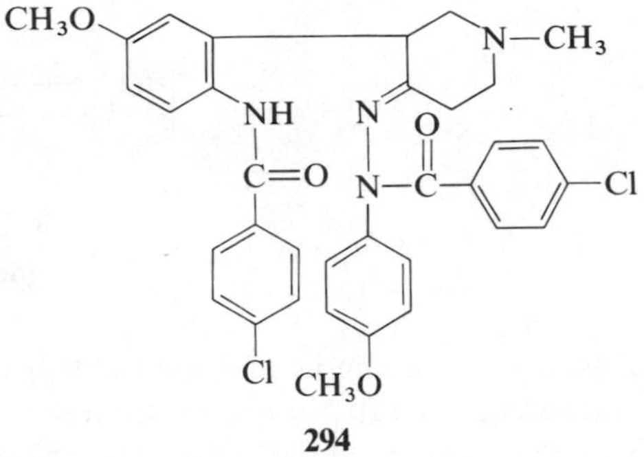

294

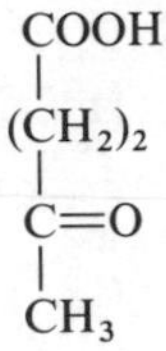

295

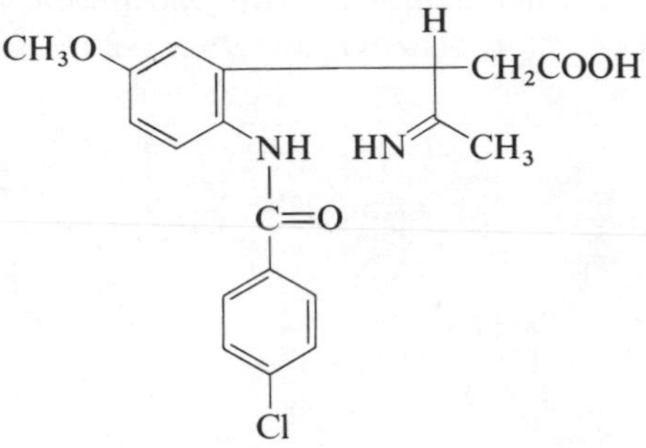

296

using ^{13}C-n.m.r. (Douglas, A. W., 1978) and ^{15}N-n.m.r. (Douglas, A. W., 1979) techniques clearly demonstrated the *in situ* intermediacy of the imines **293** (R = H, X = NH) and **296**, respectively.

Refluxing a solution of **297** (R = H and OCH_3) in tetralin afforded **298** (R = H and OCH_3, respectively), along with **299** (R = H), in each case and **299** [R = COC_6H_5 and $COC_6H_4OCH_3(4)$, respectively]. Upon treatment with mineral acids, **298** (R = H) gave the indolization product **299** (R = H) (Mills, Al Khawaja, Al-Saleh and Joule, 1981). The reaction sequence was presumed to be arrested because of the neutrality of both nitrogen atoms in

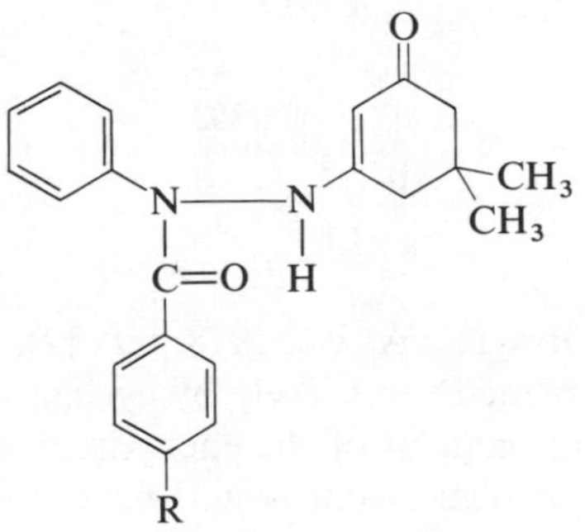

297

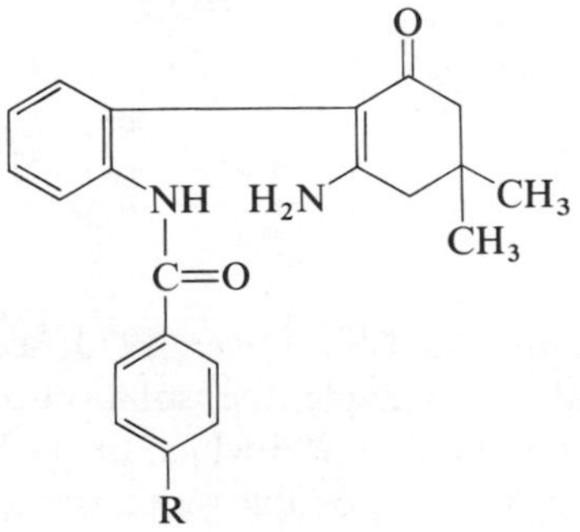

298

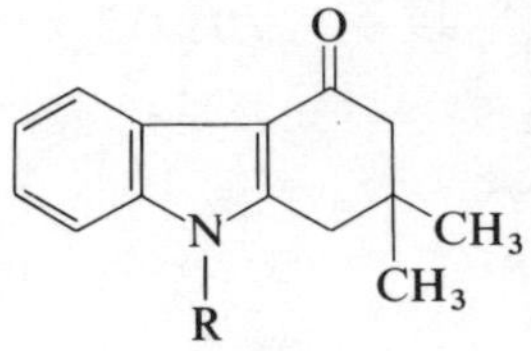

O
N
R
CH_3
CH_3

299

298, one being amidic and the other being vinylogously amidic, coupled with the absence of a strong proton source. Likewise, refluxing a mixture of **300** with laevulinic acid in toluene solution in the presence of phosphoric acid yielded **302** which upon further treatment with phosphoric acid in toluene furnished the expected 1-(4-chlorobenzoyl)-5-methoxy-2-methyl-3-indolylacetic acid (**303**)

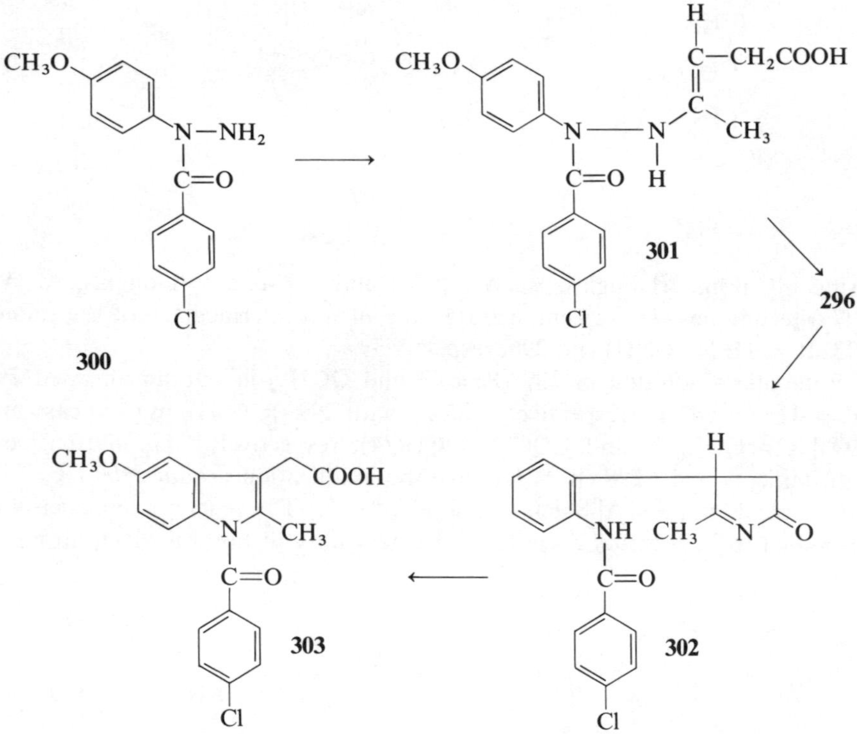

CH_3O
$N-NH_2$
$C=O$
Cl
300

H
$C-CH_2COOH$
C
CH_3
CH_3O
N — N
$C=O$ H
Cl
301

296

H
NH CH_3 N O
$C=O$
Cl
302

CH_3O
COOH
N CH_3
$C=O$
Cl
303

(Firestone and Sletzinger, 1970; see also Douglas, A. W., 1978, ref. 12). As in the above example, the isolation of **302**, formed most likely by cyclization of the intermediate **296** which arises by rearrangement of the enehydrazine **301**, was probably possible since the anilino nitrogen atom would have reduced basicity because of its aroylation.

When the hydrazone **304** was reacted with formic acid, three products were produced (Inoue and Ban, 1970). These were the expected indole **305**, the *N*-formyl indoline **306**, formed by formylation of the 3*H*-indolic alternative indolization product, and **307** which represents an intermediate in the pathway to the formation of the 3*H*-indolic product which has been trapped by formylation. In this case it was formylation which reduced the basicity of the nitrogen atoms and which was thereby a probable factor permitting the isolation of **307**.

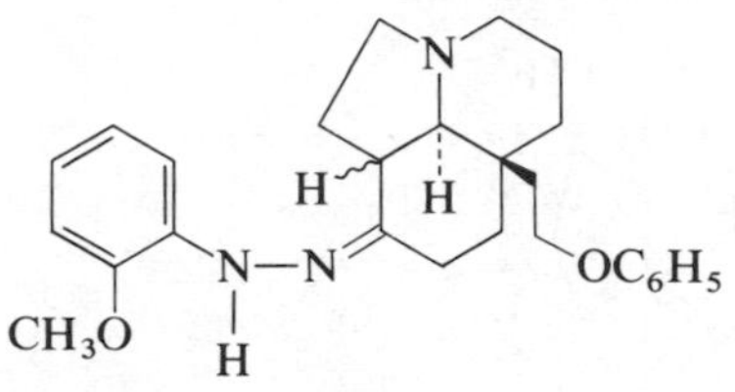

304 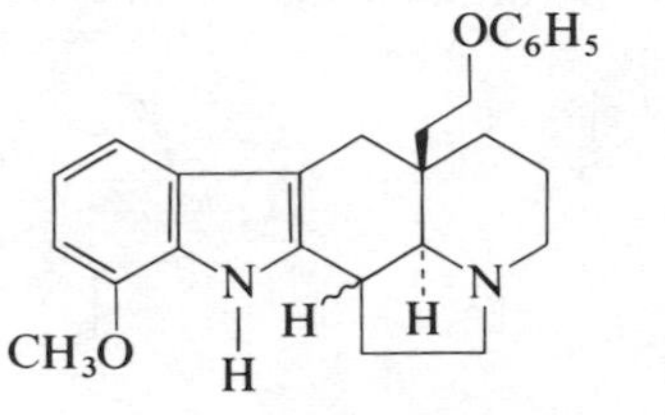 **305**

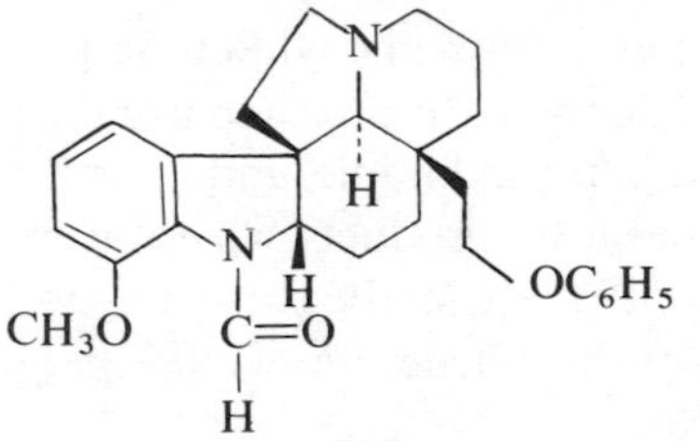

306 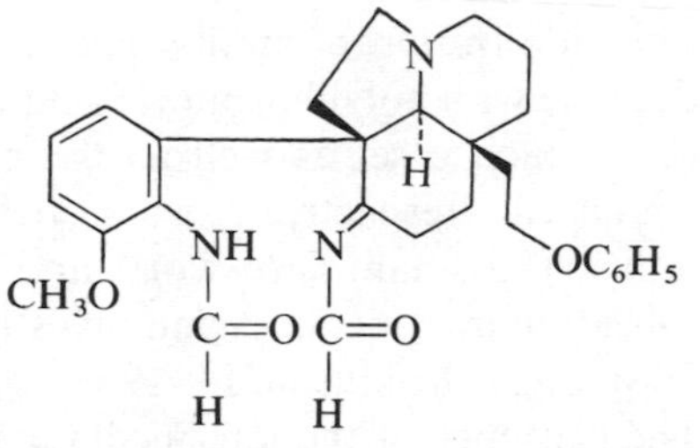 **307**

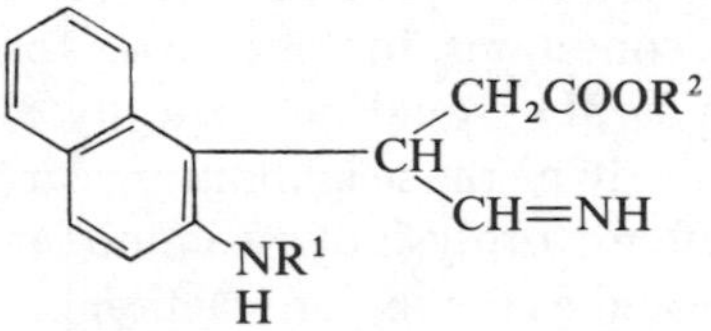

308

3-Carboxypropionaldehyde 2-naphthylhydrazone afforded two products upon treatment with ethanolic phosphoric acid. One of these upon hydrolysis gave 4,5(?) benzo-3-indolylacetic acid and the other similarly yielded a compound which, when diazotized and coupled with 2-naphthol, appeared to contain a primary aromatic amino function. Without further evidence, this product was formulated as **308** ($R^1 = H$, $R^2 = C_2H_5$). A mixture of 2-naphthylhydrazine hydrochloride with 3-carboxypropionaldehyde, upon successive treatment with sodium acetate and sulphuric acid, afforded an analogous product with the properties of an acetylated aromatic amine which was tentatively assigned structure **308** ($R^1 = COCH_3$, $R^2 = H$). Further investigation of this work (Borghero and Finsterle, 1955) and other work (Barnes, Pausacker and Badcock, 1951) in which camphenilone phenylhydrazone (**309**) furnished an isomeric basic product, $C_{15}H_{20}N_2$, characterized as a picrate, when treated with hydrogen chloride in dry ethanol, would be of interest (Owellen, Fitzgerald, J. A., Fitzgerald, B. M., Welsh, D. A., Walker, D. M. and Southwick, 1967; Robinson, B., 1963a). In connection with this, it should be noted that, in this latter work (Barnes, Pausacker and Badcock, 1951), enehydrazine tautomerization of **309** is impossible (Bredt's rule) unless methyl group migration or ring cleavage occurs (Robinson, B., 1963a).

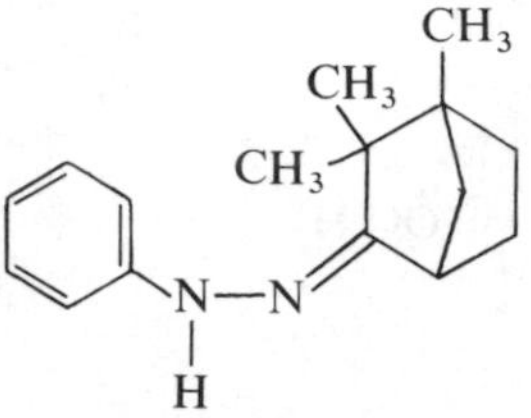

CH3 CH3 CH3 N—N H

309

The indolization of methyl phenyl ketone and ethyl phenyl ketone phenylhydrazone with polyphosphoric acid under milder conditions than usually used gave, in each case, as well as the expected 2-phenylindole and 3-methyl-2-phenylindole, respectively, small yields of basic by-products which were not identified (Kissman, Farnsworth and Witkop, 1952). Clearly, these may well be indolization intermediates and investigation of their structures would therefore be of interest (Robinson, B., 1963a).

The existence of the intermediate **198** was further demonstrated (Rapoport and Tretter, 1958) when the arylhydrazone **310** (R^1 = H, R^2 = C_2H_5) was subjected to indolization conditions. In this case the formation of an indole ring in an intermediate such as **311** would be sterically difficult and therefore an alternative cyclization, involving the neighbouring carbethoxy group, occurred to afford **312** (R = H), after hydrolysis of the imino function under the reaction conditions. This was isolated as the major reaction product along with a small

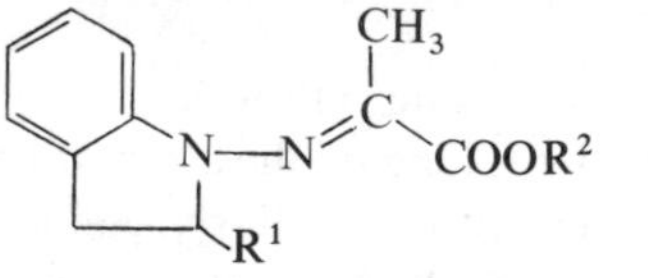

CH3 N—N=C COOR2 R1

310

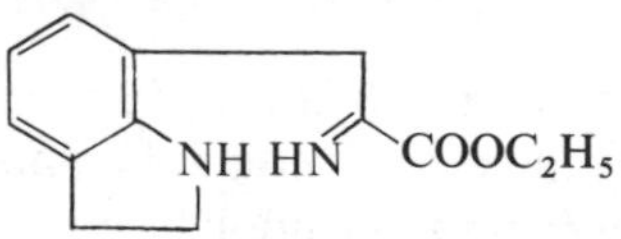

NH HN COOC2H5

311

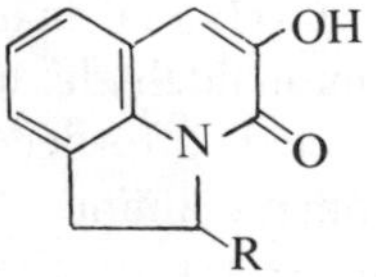

OH N O R

312

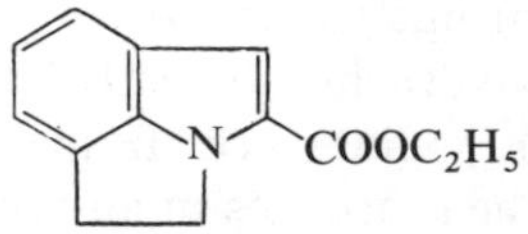

N COOC2H5

313

yield of the expected indole **313**. These observations were later (Yudin, Popravko and Kost, 1962) confirmed and extended when **310** (R^1 = H, CH_3 and C_6H_5, R^2 = H and C_2H_5) was converted into good yields of **312** (R = H, CH_3 and C_6H_5, respectively) using hydrochloric acid or sulphuric acid catalysts under varying conditions. Others (Sharkova, N. M., Kucherova and Zagorevskii, 1964) observed the formation of an analogous product from the ethanolic hydrogen chloride catalysed rearrangement of a related hydrazone of pyruvic acid. These formations of **312** and **313** are analogous to the earlier observations that **314** ($R^1 = R^2$ = H, X = NC_6H_5), upon heating in xylene solution, was converted into **315** ($R^1 = R^2$ = H) and aniline and upon pyrolysis or heating in quinoline or *N*,*N*-dimethylaniline solution afforded **316** ($R^1 = R^2$ = H,

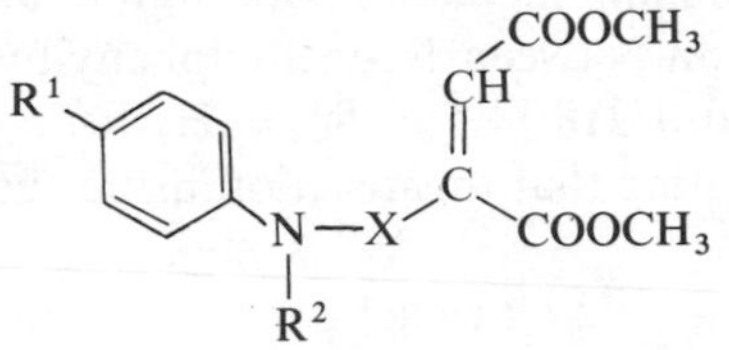

314

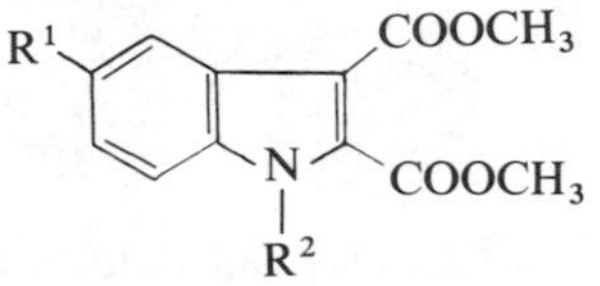

315

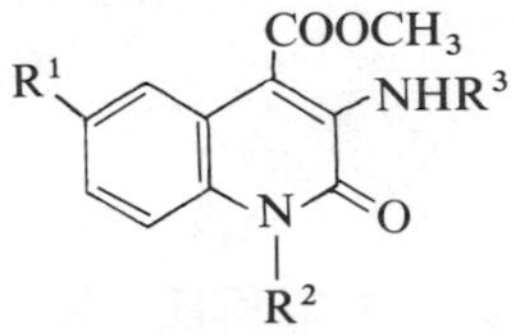

316

$R^3 = C_6H_5$) (Diels and Reese, 1934) and that **314** (R^1 = H, $R^2 = CH_2C_6H_5$, X = $NCH_2C_6H_5$) was converted into **315** (R^1 = H, $R^2 = CH_2C_6H_5$) and benzylamine and **316** (R^1 = H, $R^3 = R^3 = CH_2C_6H_5$) by refluxing in xylene and pyridine solutions, respectively (Diels and Reese, 1935). Likewise, **314** (R^1 = H, $R^2 = CH_2C_6H_5$, X = NH) was readily converted upon heating, under the conditions of its formation, into **315** (R^1 = H, $R^2 = CH_2C_6H_5$) (Diels and Reese, 1935). In the earlier paper (Diels and Reese, 1934) no serious attempts to mechanistically rationalize these transformations were made, suggestions that were presented involving a pentavalent carbon atom (misprint?) and a series of arbitrary hydrogen shifts and bond cleavages. In the later paper (Diels and Reese, 1935), the formations of **315** and **316** were clearly recognized as occurring by two possible different routes, also suggested later by others (Rapoport and Tretter, 1958), subsequent to an *ortho*-benzidine rearrangement of the enehydrazine. The above described transformations of **314** ($R^1 = R^2$ = H, X = NC_6H_5) were later confirmed (Huntress, Bornstein and Hearon, 1956; Huntress and Hearon, 1941) and attempts were made to extend the reactions. Thus, **314** [$R^1 = CH_3$, R^2 = H, X = $NC_6H_4CH_3(4)$] yielded only **316**

[R^1 = CH_3, R^2 = H, R^3 = $C_6H_4CH_3(4)$] when heated in either 2-methylpyridine or xylene solution, **314** [R^1 = Cl, R^2 = H, X = $NC_6H_4Cl(4)$] gave **315** (R^1 = Cl, R^2 = H) and 4-chloroaniline when refluxed in xylene solution **316** [R^1 = Cl, R^2 = H, R^3 = $C_6H_4Cl(4)$] when refluxed in 2-methylpyridine solution and **314** (R^1 = $OCOCH_3$, R^2 = H, X = NC_6H_5) furnished **316** (R^1 = $OCOCH_3$, R^2 = H, R^3 = C_6H_5) when refluxed in 2-methylpyridine solution but gave no recognizable products when refluxed in xylene, toluene, 1,3-diethylbenzene or tetralin solutions (Huntress, Bornstein and Hearon, 1956).

In the above studies, the starting products **314** were formed by reaction between the appropriate N_β-unsubstituted arylhydrazine and dimethyl acetylenedicarboxylate in refluxing methanol, although in another related indolization (see below), reaction between N_α-methylphenylhydrazine and **317** (R = C_2H_5) was used to afford **318** [R^1 = R^2 = H, R^3 = CH_3, R^4 = C_2H_5(?)] (Reif, 1909). It is interesting that this reaction using the hydrazine **319** formed

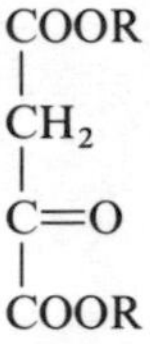

COOR–CH₂–C=O–COOR
317

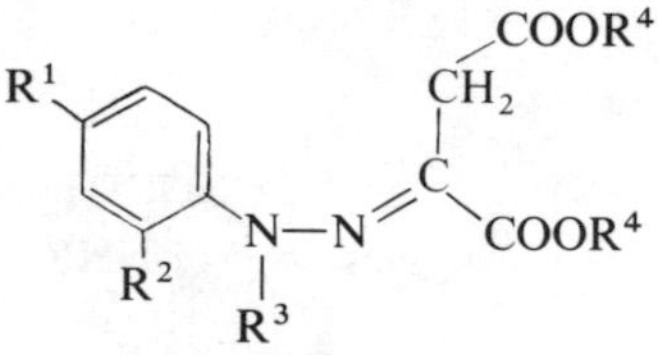

318

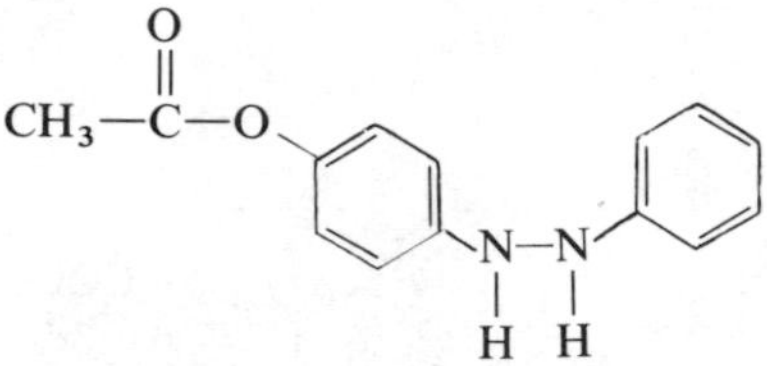

319

only **314** (R^1 = $OCOCH_3$, R^2 = H, X = NC_6H_5), as determined by the isolation of only **316** (R^1 = $OCOCH_3$, R^2 = H, R^3 = C_6H_5) from the subsequent reaction of the product in heated 2-methylpyridine (Huntress, Bornstein and Hearon, 1956). In an analogous reaction, phenylhydrazine reacted with **320** [R = C_2H_5 and $(CH_2)_3CH_3$] in refluxing toluene and by heating a mixture of the two reactants, respectively, by initial specifically orientated addition of the phenylhydrazine to the C≡C to afford intermediates formulated as **321** [R = C_2H_5 and $(CH_2)_3CH_3$, respectively]. These ultimately gave **322** [R = C_2H_5 and $(CH_2)_3CH_3$, respectively] after indolization. Alternatively, the intermediates **321** reacted further with phenylhydrazine to produce osazones (see Chapter VII, Q). The structure of **322** (R = C_2H_5) was verified by reductive desulphurization with Raney nickel in refluxing ethanol to afford ethyl indole-2-carboxylate (Bonnema and Arens, 1960a). Attempts to

320

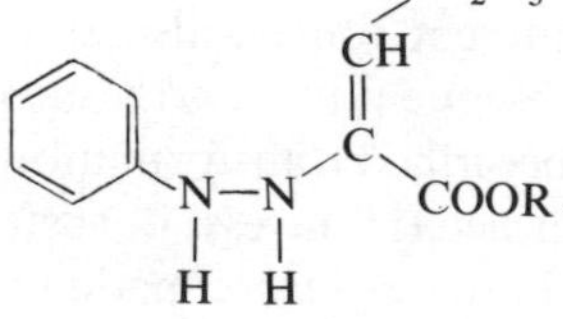

321

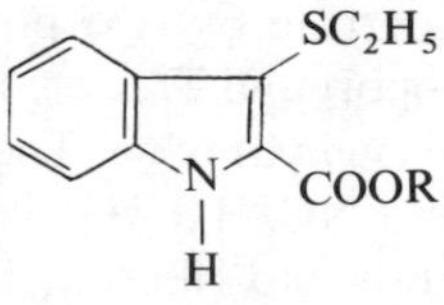

322

use other acetylenic compounds in the formation of analogous addition products were unsuccessful (Kost, Sviridova, Golubeva and Portnov, 1970a).

In these indolizations and comparable reactions, the use of acid catalysts is unnecessary since the starting materials, **314**, are already in the enehydrazine form (with $X \neq NH$). Subsequent acid catalysed indolizations converted the adducts **314** [$R^1 = H$, $R^2 + X = (CH_2)_3N$] and related substituted compounds into **315** [$R^1 = H$, $R^2 = (CH_2)_3NH_2$] and related substituted compounds (see also later in this chapter) using ethanolic hydrogen chloride (Kost, Sviridova, Golubeva and Portnov, 1970a), **318** ($R^1 = R^2 = H$, $R^3 = R^4 = CH_3$) into **315** ($R^1 = H$, $R^2 = CH_3$) using zinc chloride (Acheson and Vernon, 1962) (see also Reif, 1909; ester unspecified) and **318** ($R^1 = OCH_3$, $R^2 = H$, and $R^1 = H$, $R^2 = OCH_3$, $R^3 = CH_3$, $R^4 = C_2H_5$) into **323** ($R^1 = OCH_3$,

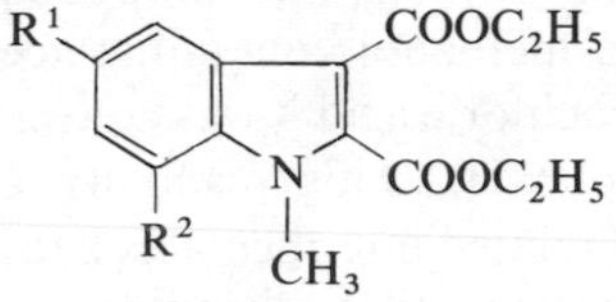

R^1 $COOC_2H_5$
N $COOC_2H_5$
R^2 CH_3
323

$R^2 = H$ and $R^1 = H$, $R^2 = OCH_3$, respectively) using sulphuric acid (Bell and Lindwall, 1948) although with this latter catalyst, **318** ($R^1 = OCH_3$, $R^2 = R^3 = H$, $R^4 = C_2H_5$) failed to indolize (Bell and Lindwall, 1948) but underwent an alternative cyclization, with concomitant ester hydrolysis, to afford 3-carboxy-1-(4-methoxyphenyl)-pyrazol-5-one. Similar alternative cyclizations occurred when the adducts **314** ($R^1 = H$ and CH_3, $R^2 = H$, $X = NH$) were treated with 4-methylbenzene sulphonic acid in methanol and when **314** ($R^1 = Cl$, $R^2 = H$, $X = NH$) was fused (Heindel, Kennewell and Pfau, 1970 – see also Heine, Hoye, T. R., Williard and Hoye, R. C., 1973; Sucrow and Slopianka, 1978), these being specific examples of a well established synthesis of the pyrazol-5-one system (Jacobs, 1957a; Johnson, A. W., 1950) (see Chapter VII, E).

Interesting observations akin to the above studies involved the reactions between N-phenyl- and N-(4-methylphenyl)hydroxylamine and dimethyl acetylene dicarboxylate which gave, along with other products, the monomethyl ester of indole-2,3-dicarboxylic acid and its 5-methyl homologue, respectively (Huntress, Lesslie and Hearon, 1956). Presumably the intermediate adducts **314** ($R^1 = H$ and CH_3, respectively, $R^2 = H$, $X = O$) could be involved in these reactions although it would appear that the initial addition occurred through the nitrogen atom (Agosta, 1961; Huntress, Lesslie and Hearon, 1956; Sheradsky, Nov, Segal and Frank, 1977; Winterfeldt, Krohn and Stracke, 1969). However, addition through the oxygen atom has been effected and treatment of the resulting products with basic catalysts has afforded the corresponding indoles as minor products (Sheradsky, Nov, Segal and Frank, 1977 (see Chapter VI, C).

3. Intermediates Corresponding to **199** *in Scheme 10*

In view of Plieninger's earlier mentioned studies concerning the indolization of **282** (X = O), a series of 1-substituted pyrrolidin-2,3-dionephenylhydrazones **282** [X = $N(CH_2)_2CH_3$, $N(CH_2)_3CH_3$, N-cyclohexyl, NC_6H_5, $N(CH_2)_2C_6H_5$, $NCH(CH_3)CH_2C_6H_5$ and $N(CH_2)_2C_6H_3(OCH_3)_2(3,4)$] was subjected to treatment with a hydrogen chloride–acetic acid catalyst. However, in all these cases no intermediates were isolated and the indolizations proceeded to completion to afford the corresponding indoles **284** [X = $N(CH_2)_2CH_3$, $N(CH_2)_3CH_3$, N-cyclohexyl, NC_6H_5, $N(CH_2)_2C_6H_5$, $NCH(CH_3)CH_2C_6H_5$ and $N(CH_2)_2C_6H_3(OCH_3)_2(3,4)$, respectively] (Southwick and Owellen, 1960). A subsequent synthesis, by an independent route, of **285** (R = H, X = N-cyclohexyl) was effected and although this compound was stable when heated in methanol, when heated in methanol containing acetic acid or when stood at room temperature in hydrochloric acid it was converted into the corresponding indole **284** (X = N-cyclohexyl). This reactivity accounted for the earlier observations that this and related intermediates had not been isolated from the indolizations. It was suggested that Plieninger's intermediate was isolable because its enamine nitrogen atom was less basic than that in the pyrrolidine analogues and that such protonation may be essential for the completion of the indolization (Owellen, Fitzgerald, J. A., Fitzgerald, B. M., Welsh, D. A., Walker, D. M. and Southwick, 1967). However, when the 1-substituted 4-benzylpyrrolidin-2,3-dione phenylhydrazones **324** [R = CH_3, $CH(CH_3)_2$, cyclohexyl and $CH_2C_6H_5$] were treated with methanolic hydrochloric acid, the indolization was arrested in every case, but at a later stage, corresponding to **199** in Scheme 10, when the 2-aminoindolines **325** [R = CH_3, $CH(CH_3)_2$, cyclohexyl and $CH_2C_6H_5$, respectively] were isolated. It was suggested (Southwick, McGrew, Engel, Millimen and Owellen, 1963) that the failures of these intermediates to undergo conversion into the corresponding 3*H*-indoles **326** might be caused by the ring strain that would be associated with these latter structures. This represented the first isolation of 2-aminoindolines from Fischer indolizations. It is interesting that it had been shown previously that the

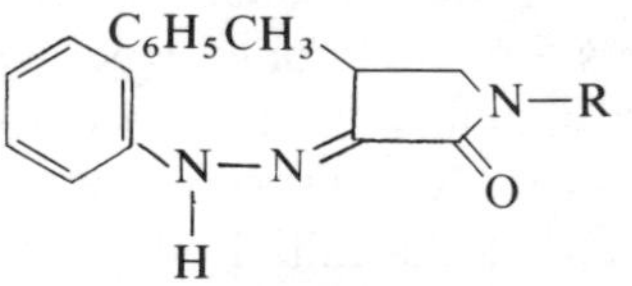

$C_6H_5CH_3$ N—R N—N O H

324

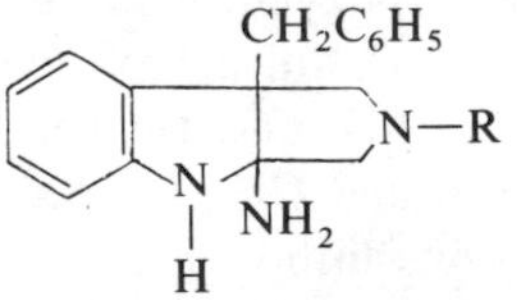

$CH_2C_6H_5$ N—R N NH_2 H

325

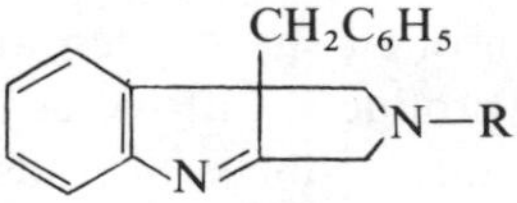

$CH_2C_6H_5$ N—R N

326

hydrazones **327** (R = C_2H_5 and C_6H_5, $n = 1$) and **327** [R = $CH_2C_6H_5$ and $(CH_2)_2C_6H_5$, $n = 2$], upon heating with zinc chloride in ethanol in the first case and by heating with concentrated hydrochloric acid in the other three cases, formed the 3*H*-indoles **328** (R = C_2H_5 and C_6H_5, $n = 1$) and **328** [R = $CH_2C_6H_5$ and $(CH_2)_2C_6H_5$, $n = 2$], respectively (Leuchs, Philpott, D., Sander, Heller, A. and Köhler, 1928). It may be argued that the steric strain in these 3*H*-indoles would not be as great as that in **326**. However, indolization of these hydrazones **327** under mild conditions may well lead to the isolation of reaction intermediates (cf. Brown, R. K., 1972a).

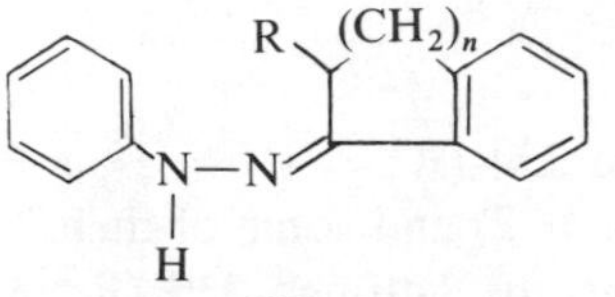

327

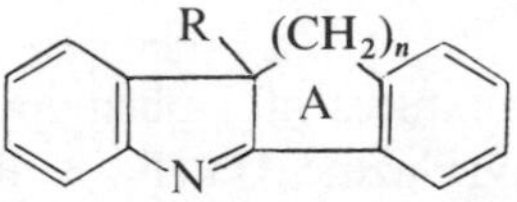

328

In a further development of these studies it was found that hydrochloric acid–acetic acid catalysed indolization of the 1-alkyl-4,5-di(N-substituted carbamoyl)-pyrrolidin-2,3-dionearylhydrazones **329** (R^1 = cyclohexyl, R^2 = H, CH_3, C_2H_5 and $CH_2C_6H_5$, R^3 = H, CH_3 and $CH_2C_6H_5$, R^4 = H, Cl and OCH_3) similarly afforded the corresponding 2-aminoindolines **330** (Southwick, Vida, Fitzgerald, B. M. and Lee, S. K., 1968; Yevich, Murphy, J. R., Dufresne and Southwick, 1978) and treatment of **331** in boiling dioxane gave **332** (Southwick, 1978). Once again, the elimination of the amino group was prevented, probably through the steric strain which would be present in the ultimate 3*H*-indoles or 3*H*-indolium cations if they were formed. However, pyrolysis of the hydrochlorides of **330** or of the free bases in ethylene glycol containing some of its sodium alcoholate caused elimination of the ring junction substituents to

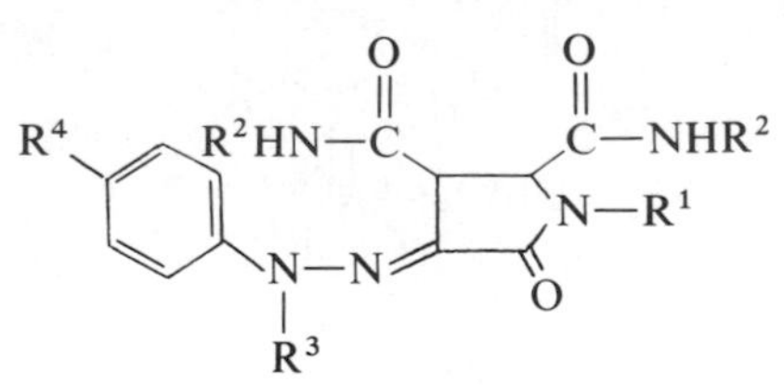

329

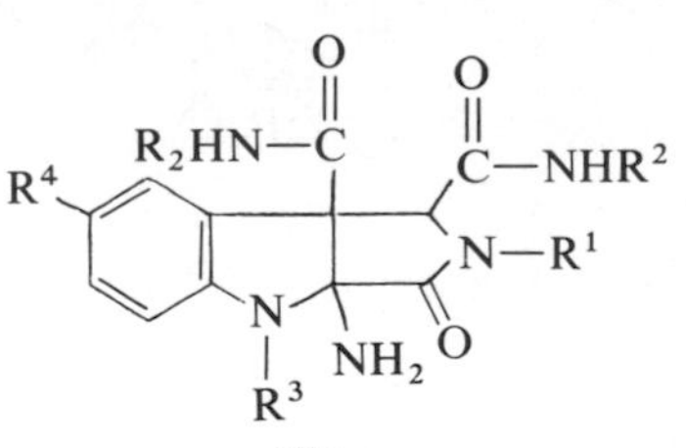

330

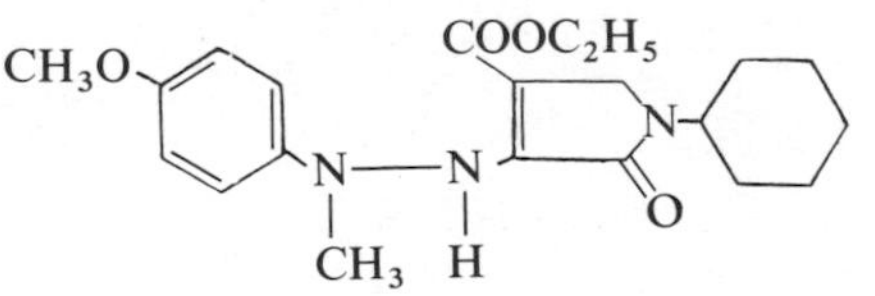

331

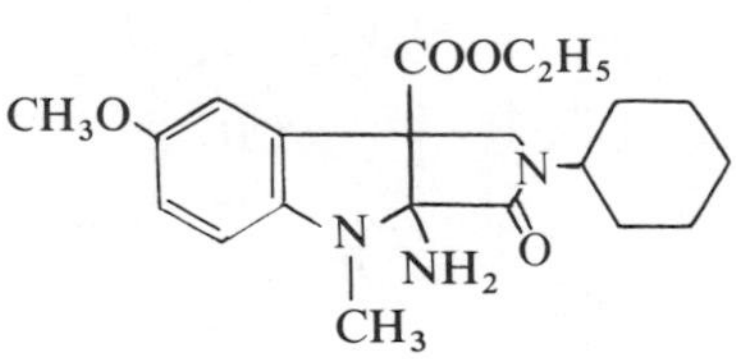

332

afford the indoles **333** (Southwick, Vida, Fitzgerald, B. M. and Lee, S. K., 1968; Yevich, Murphy, J. R., Dufresne and Southwick, 1978). Other similar studies relating to thermal cyclizations of 3-(2-arylhydrazino)-3-pyrroline derivatives have been effected (Dufresne, 1972).

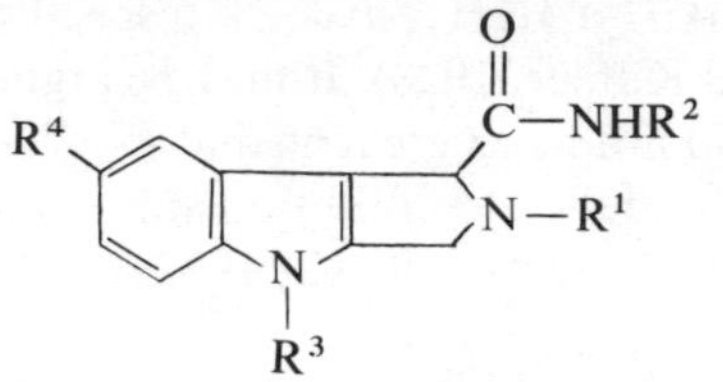

333

When mixtures of 1-phenylpyrazolidine **334** ($R^1 = R^2 = H$, $n = 1$) or of 1-phenylpyridazine **334** ($R^1 = R^2 = H$, $n = 2$) and some of their 3-alkylated and 3,4-dialkylated homologues with various ketones **335** ($R^5 = H$) were heated together, with or without acid catalysts, the hypothetical intermediate

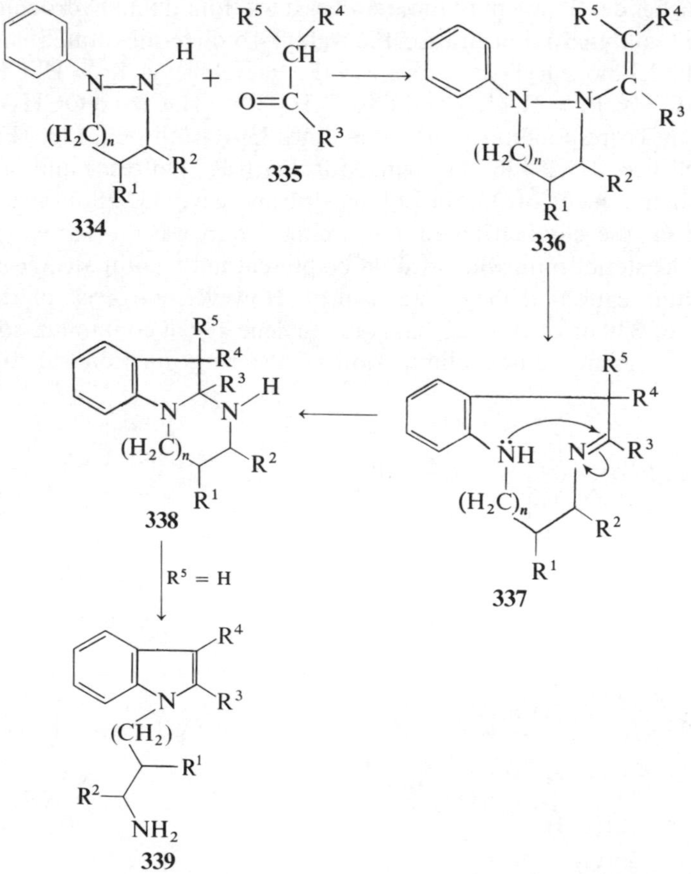

334 **335** **336** **337** **338** ($R^5 = H$) **339**

336 was presumably formed *in situ*. This underwent sequential C—C bond formation and N—N bond cleavage to give intermediate **337**, ring closure to form intermediate **338** and protonic loss of the benzylic hydrogen atom as shown to furnish the corresponding 1-aminoalkylindole **339** (Eberle, 1971; Eberle, Kahle and Talati, 1973; Kost, Sviridova, Golubeva and Portnov, 1970a, 1970b). Subsequent to these observations, attempts were made to product an intermediate analogous to **338** but in which the benzylic hydrogen atom was replaced by an alkyl group. Since this group, unlike the hydrogen atom, should not be readily lost, it was possible that such an intermediate, representing a blocked 2-aminoindoline intermediate in the Fischer indolization, would be isolable. To this effect, 2-methyl- and 2-carbethoxycyclohexanone were reacted with 1-phenylpyrazolidine (**334**; $R^1 = R^2 = H$, $n = 1$) but afforded only **339** [$R^1 = R^2 = H$, $R^3 + R^4 = CH(CH_3$ and $COOC_2H_5$, respectively) $(CH_2)_3$, $n = 1$] in high yields, no products resulting from indolization involving the tertiary carbon atom in the original ketones being isolated (Eberle, Kahle and Talati, 1973). However, in a similar reaction but starting with isobutyraldehyde (**335**; $R^3 = H$, $R^4 = R^5 = CH_3$) and cyclohexanecarboxyaldehyde [**335**; $R^3 = H$, $R^4 + R^5 = (CH_2)_5$] as the carbonyl moieties, the desired blocked intermediates **338** (R^1–$R^3 = H$, $R^4 = R^5 = CH_3$, $n = 1$) and **338** [R^1–$R^3 = H$, $R^4 + R^5 = (CH_2)_5$, $n = 1$], respectively, were obtained (Eberle, 1972a; Eberle and Kahle, 1973). Starting with methyl 4-formylcaproate [**335**; $R^3 = H$, $R^4 = C_2H_5$, $R^5 = (CH_2)_2COOCH_3$] and 5-norbornene-2-carboxaldehyde (**340**) as the carbonyl moieties, compounds **341** and **342**, respectively, were formed. Product **341** resulted from intramolecular lactamization of the intermediate **338** [R^1–$R^3 = H$, $R^4 = C_2H_5$, $R^5 = (CH_2)_2COOCH_3$, $n = 1$] (Eberle and Kahle, 1973). Treatment of **342** with maleic acid induced its rearrangement into **343** (Eberle and Kahle, 1973), a transformation which also occurred under the conditions of the thermal indolization (Eberle, 1972b).

OHC

340

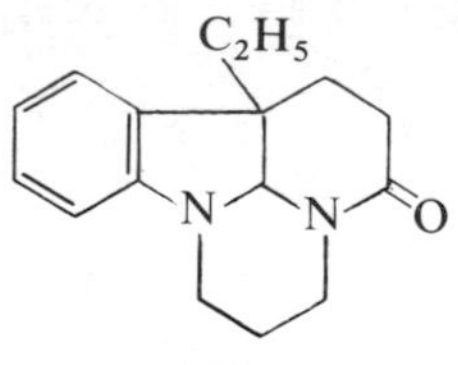

C_2H_5 N N O

341

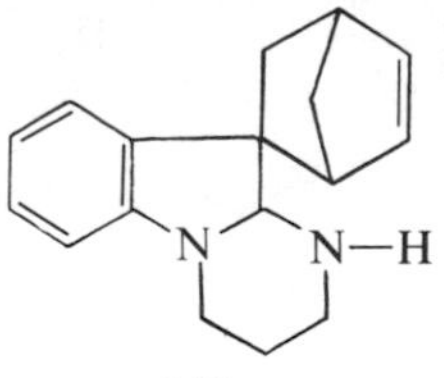

N N—H

342

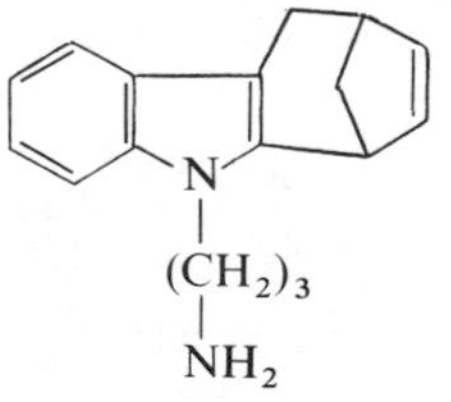

N $(CH_2)_3$ NH_2

343

In a parallel report from another laboratory (Croce, 1973), treatment of a mixture of **334** ($R^1 = R^2 = H, n = 1$) and **335** ($R^3 = CH_3, R^4 = COOCH_3, COC_6H_5$ and $COCH_3$, $R^5 = H$) with glacial acetic acid in benzene led to the isolation of the enehydrazine intermediates **336** ($R^1 = R^2 = R^5 = H$, $R^3 = CH_3$, $R^4 = COOCH_3$, COC_6H_5 and $COCH_3$, respectively, $n = 1$). These were hydrolysed back to the starting materials when treated with 5% hydrochloric acid but their treatment with dry hydrogen chloride in methanol furnished the corresponding **339** ($R^1 = R^2 = H$, $R^3 = CH_3$, $R^4 = COOCH_3$, COC_6H_5, $COCH_3$, respectively, $n = 1$). This same report included details of the formation of **339** [$R^1 = R^2 = H$, $R^3 + R^4 = (CH_2)_{3 \text{ and } 4}$, $n = 1$] from mixtures of **334** ($R^1 = R^2 = H$, $n = 1$) and **335** [$R^5 = H$, $R^3 + R^4 = (CH_2)_{3 \text{ and } 4}$] and, as described above by the other group, the formation of the blocked intermediate **338** (R^1–$R^3 = H$, $R^4 = R^5 = CH_3$, $n = 1$) from **334** ($R^1 = R^2 = H$, $n = 1$) and isobutyraldehyde (**335**: $R^3 = H$, $R^4 = R^5 = CH_3$).

This synthetic approach was later (Eberle and Brzechffa, 1976) extended by the use of the 1-phenylpyrazolidin-5-ones **344** (R = H and CH_3). Reaction with cyclohexanone, cycloheptanone, cyclooctanone and cyclododecanone afforded the final indolization products **345** (R = H and CH_3, $n = 4, 5, 6, 10$, respectively). However, with cyclopentanone the reaction was arrested at the 2-aminoindoline stage to yield **346**, the probable reason for this being associated with the steric strain in the potential indolic product. Likewise, heating the enehydrazine **253** in boiling 1,2-dichlorobenzene solution effected an arrested indolization and **347** was formed, probably isolable because, in the potential indole, this nucleus and the attached 2,6-dichlorophenyl nucleus were sterically unable to achieve coplanarity. An interesting analogy with these transformations is the base catalysed conversion of **348** into **349**, the *cis* stereochemistry of

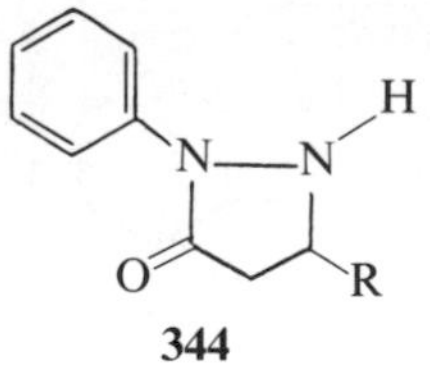

344 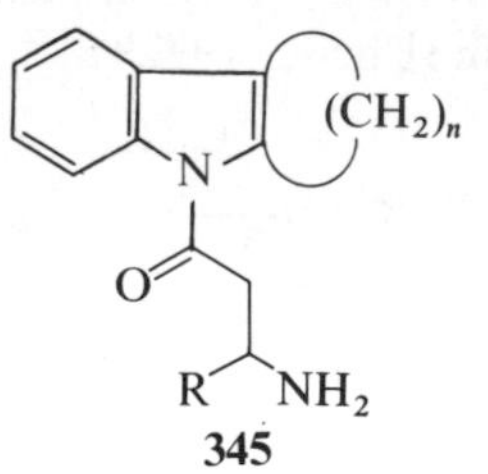 **345** 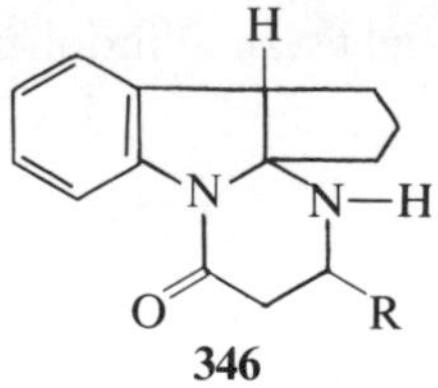 **346**

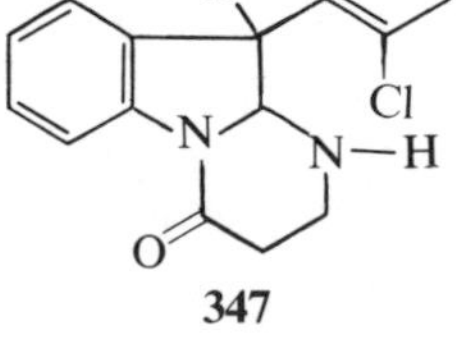

347 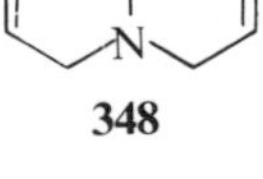 **348** 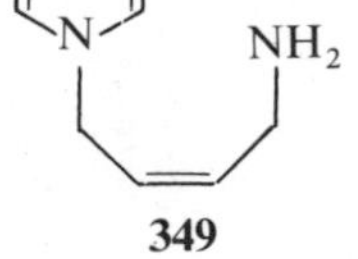 **349**

349 being established by the synthesis of the *trans* isomer **350** from succindialdehyde and *trans*-2-butene-1,4-diamine (Hinshaw, 1975).

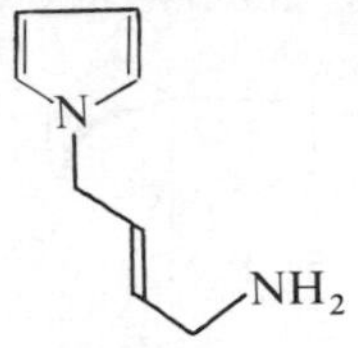

350

Indolization of **351** ($R^1 = R^3 = H$, $R^2 = CH_3$ and $CH_2C_6H_5$, $n = 2$) occurred in both possible directions and afforded the indoles **352** ($R^1 = H$, $R^2 = CH_3$ and $CH_2C_6H_5$, respectively, $n = 2$) together with the lactams **355** ($R^1 = H$, $R^2 = CH_3$ and $CH_2C_6H_5$, respectively, $X = NH$, $n = 2$). These latter products again resulted from an arrest of the final deamination stage of the indolization, in these cases because of intramolecular lactamization of the 2-aminoindole intermediates **354** ($R^1 = R^3 = H$, $R^2 = CH_3$ and $CH_2C_6H_5$, respectively, $n = 2$) which were formed via **353** as shown in Scheme 18 (Fritz and Losacker, 1967). Starting with **351** ($R^1 = R^3 = H$, $R^2 = CH_3$, $n = 1$), indolization again proceeded in both possible directions to form **352** ($R^1 = H$,

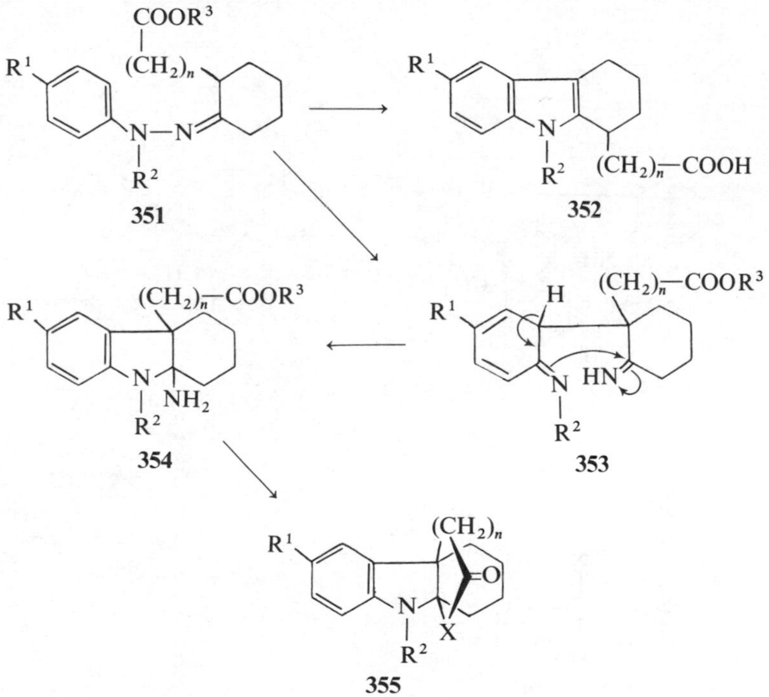

351 **352** **353** **354** **355**

Scheme 18

$R^2 = CH_3$, $n = 1$) together with the lactone **355** ($R^1 = H$, $R^2 = CH_3$, $X = O$, $n = 1$). In this case the intermediate **354** ($R^1 = R^3 = H$, $R^2 = CH_3$, $n = 1$), as its betaine **356**, underwent deamination as shown to give the

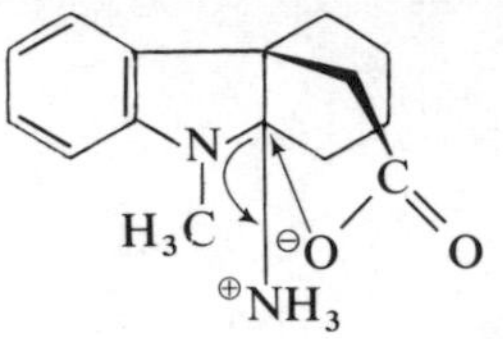

356

lactone (Fritz and Stock, 1970). However, indolization of the corresponding ethyl esters **351** ($R^1 = H$ and OCH_3, $R^2 = CH_3$, $R^3 = C_2H_5$, $n = 1$) again caused arrest of the indolization at the deamination stage when the new C—C bond formation involved the tertiary carbon atom of the ketonic moiety. Intermediate **354** ($R^1 = H$ and OCH_3, $R^2 = CH_3$, $R^3 = C_2H_5$, $n = 1$) underwent

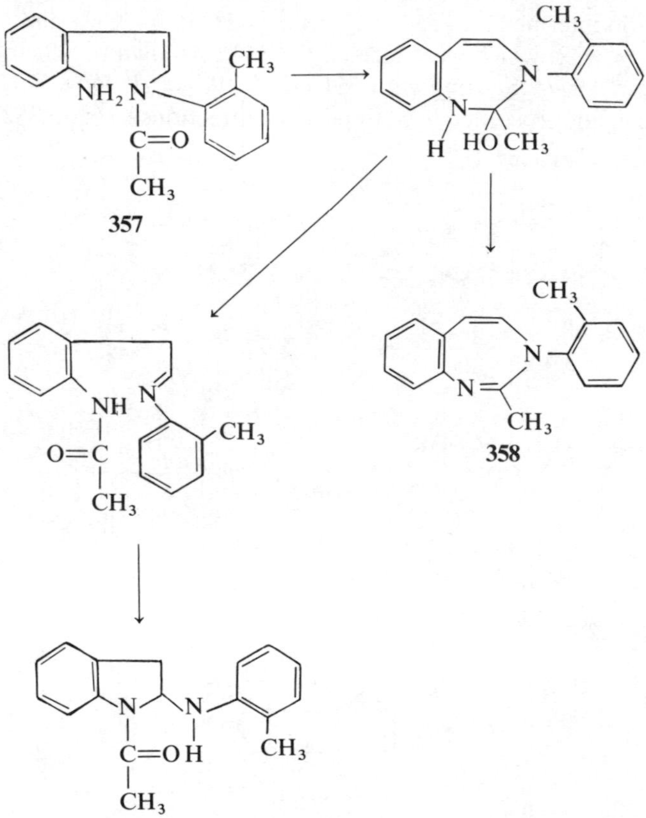

357

358

359

Scheme 19

intramolecular lactamization to afford **355** (R^1 = H and OCH_3, $R^2 = CH_3$, X = NH, n = 1) which was isolated along with the **352** (R^1 = H and OCH_3, $R^2 = CH_3$, n = 1) resulting from indolization in the alternative direction (Fritz and Stock, 1970).

The equivalent of the post C—C bond formation to 2-aminoindoline stages of the Fischer indolization has been effected in the transformation of **357** into **359**, catalyzed by molecular sieve in boiling benzene. This reaction also produced **358** and was suggested as proceeding as shown in Scheme 19. The elimination of 2-methylaniline from **359** to afford 1-acetylindole was subsequently carried out by its passage through a silica or alumina column (Chen, F. M. F. and Forrest, 1973; Forrest and Chen, F. M. F., 1972).

F. Mechanism of Formation of the New C—C Bond in the Fischer Indolization

Until the end of the 1960s, and sometimes even later, it was widely accepted (e.g. Brown, R. K., 1972a; Desaty and Keglević, 1965; Grandberg, Belyaeva and Dmitriev, L. B., 1971b, 1973; Grandberg, Zuyanova, Przheval'skii and Minkin, 1970; Ockenden and Schofield, 1957; Robinson, B., 1963a, Sundberg, 1970; Yakhontov and Marshalkin, 1972) that the formation of the new C—C bond in the Fischer indolization resulted from a concerted cyclic rearrangement involving an intramolecular electrophilic attack by the enehydrazine upon the aromatic nucleus of the enehydrazine tautomer of the arylhydrazone. This acceptance was based largely upon the observations that indolization of 3-substituted phenylhydrazones (see Chapter III, A) produced both the corresponding 4- and 6-substituted indoles in which the yield ratio of the 4-:6-isomer is generally <1 if the 3-substituent is *ortho–para* directly (electron donating) and is generally >1 if the 3-substituent is *ortho–meta* directing (electron attracting) and that electron donating substituents on the phenylhydrazone facilitate indolization and increase product yields (e.g. Bloink and Pausacker, 1950; Desaty and Keglević, 1965; Suvorov and Antonov, 1952; Suvorov, Antonov and Rokhlin, 1953), electron withdrawing groups having reverse effects (e.g. Da Settimo, Primofiore, Biagi and Santerini, 1976a; Feofilaktov and Semenova, N. K., 1953b). However, in these earlier studies the isomer separations and product isolations were often far from quantitative. Furthermore, as mentioned earlier in this chapter, no account was taken of possible rate modifications of side reactions with a change in substituent, the rate determining state of the indolization mechanism was not established and it is incorrect to assume that the yield of indolic product is directly proportional to the velocity of the rate determining step.

It was observed that indolization of a series of 2-, 3- and 4-nitrophenylhydrazones of several ketones gave, with the exception of ethyl methyl ketone, higher yields of indoles when using the 3-nitrophenylhydrazones than when using the corresponding 2- and 4-nitrophenylhydrazones. Since the nitro substituent deactivates the *ortho* and *para* positions towards electrophilic attack, these observations were interpreted in terms of the new C—C bond formation during indolization involving an intramolecular nucleophilic attack by the enehydrazine upon the aromatic nucleus (Frasca, 1962a). However, as in the preceding studies, these conclusions rely upon similar unsubstantiated assumptions concerning the rate determining step of the indolization and its relationship to product yields. Others have also claimed, without evidence, that the new C—C bond formation during Fischer indolization, using hydrazones derived from 6-hydrazino-1,3-dimethyluracil, occurs via an analogous intramolecular nucleophilic attack by the enehydrazine moiety (Senda and Hirota, 1972), a suggestion which was subsequently refuted (Duffy and Wibberley, 1974).

The similarity between the C—C bond formation stage in the Fischer indolization and the *ortho*-Claisen rearrangement has been recognized since 1948 (Carlin and Fisher, 1948). Further consideration of this analogy led to the suggestion (Hansen, Sutter and Schmid, H., 1968 – see also Schmid, M., Hansen and Schmid, H., 1971) that this stage of the indolization mechanism involves a [3,3]-sigmatropic shift. This postulation was also independently made the following year (Robinson, B., 1969), again with reference to the *ortho*-Claisen and other related rearrangements (Fukui and Fujimoto, H., 1966; Hoffmann, R. and Woodward, 1965a, 1965b; Jefferson and Scheinmann, 1968; Marvell, Stephenson, J. L. and Ong, J., 1965; Rhoads, 1963), but was cautiously limited to those indolizations which occur under thermal 'non-catalytic' conditions. This caution was later (Grandberg and Sorokin, 1974) criticized. Subsequent to these postulations, the role of a [3,3]-sigmatropic shift in the rearrangement stage in which the new C—C bond is formed during the Fischer indolization has become generally accepted (e.g. Duffy and Wibberley, 1974; GilChrist and Storr, 1972; Grandberg, 1972, 1974; Kollenz and Labes, 1976; Przheval'skii and Grandberg, 1974; Przheval'skii, Grandberg and Klyuev, N. A., 1976; Schiess and Grieder, 1974; Schiess and Sendi, 1978; Wright and Gambino, 1979).

A useful insight into the mechanism of this stage of the Fischer indolization has been gained by a study of the indolization of N_α, N_α-diarylhydrazones [the general unavailability of substituted N_α, N_α-diaryl hydrazines and problems associated with their stability, in particular those with electron with drawing substituents upon their aryl group(s), have been referred to (Harbert, Plattner, Welch, W. M., Weissman and Koe, 1980)] in which the two aryl groups are non-equivalent, although somewhat conflicting conclusions have been drawn from the results of these studies by some of the various groups who have effected them.

The first example of such a study appeared in a patent in which only the product **362** was reported from the indolization of a mixture of **360** and **361**

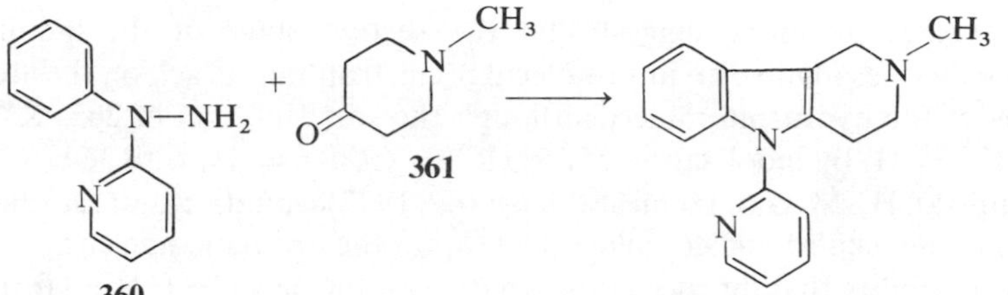

(Farbenfabriken Bayer A.-G., 1954). In a subsequent series of patents it was likewise reported that only those indoles resulting from indolization on to the most nucleophilic of the aryl nuclei were isolated. The results of these later studies, which involved heating mixtures of various N_α, N_α-diarylhydrazine hydrochlorides **363** with laevulinic acid or its ethyl ester **364** (R^3 = H or C_2H_5, respectively) to afford the corresponding **365** (R^3 = H or C_2H_5, respectively), are given in Table 1. That only the products indicated were reported as being isolated from these reactions does not eliminate the possibility that in some of these reactions the alternative isomeric products were also formed but not isolated. However, a quantitative yield of **365** ($R^1 = CH_3$, $R^2 = R^3$ = H) was claimed when **363** ($R^1 = CH_3$, R^2 = H) was heated with **364** (R^3 = H) (Yamamoto, H., Hirohashi, A., Izumi and Koshiba, 1973c).

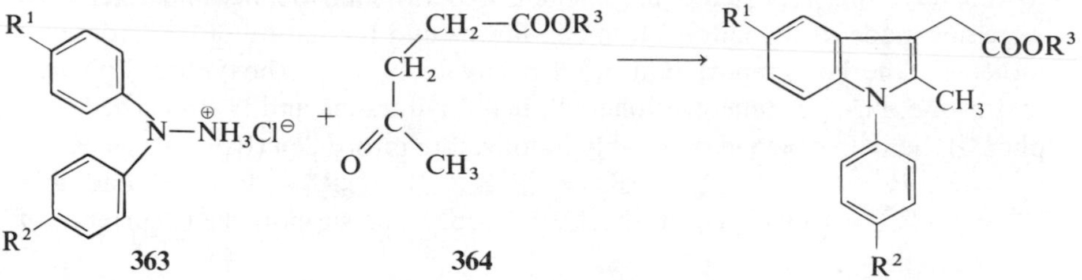

Table 1. Substituents Used in the Reaction between **363** and **364** to Afford **365**

R^1	R^2	R^3	Reference	R^1	R^2	R^3	Reference
CH_3	H	H	1–4	CH_3	H	C_2H_5	1
OCH_3	H	H	1–4	OCH_3	H	C_2H_5	1–3
CH_3	Cl	H	1, 3, 4	CH_3	Cl	C_2H_5	1
OCH_3	Cl	H	1, 2, 4	OCH_3	Cl	C_2H_5	1

References

1. Yamamoto, H., Hirohashi, A., Izumi and Koshiba, 1973a; 2. Yamamoto, H., Hirohashi, A., Izumi and Koshiba, 1973b; 3. Yamamoto, H., Hirohashi, A., Izumi and Koshiba, 1973c; 4. Yamamoto, H., Misaki, Izumi and Koshiba, 1972.

These studies therefore suggest that the second stage of the indolization mechanism does involve an intramolecular electrophilic attack on the aromatic nucleus of the hydrazino moiety although the sole isolation of **365** (R^1 = Cl, R^2 = R^3 = H) by indolization of **363** (R^1 = Cl, R^2 = H) with **364** (R^3 = H) (Yamamoto, H., Misaki, Izumi and Koshiba, 1972) would suggest the operation of an intramolecular nucleophilic attack upon the aromatic nucleus.

The possibility that the mechanistic nature of the new C—C bond formation during indolization may depend upon the indolization conditions is suggested by the observations that whereas indolization of ethyl pyruvate N_α-(substituted-phenyl)phenylhydrazones occurred mainly on to the electron rich aromatic nucleus, supporting the theory involving an intramolecular electrophilic attack, under sigmatropic conditions (presumably thermal in the absence of added acid) less substituent effect was observed during the indolization of the same hydrazones (Ishii, H., Murakami, Y., Takeda and Ikeda, 1974). Unfortunately, this report did not include experimental details and references were not made to the indolization conditions and the nature of the hydrazino moieties. The later report that ethyl pyruvate N_α-(4-methoxyphenyl)phenylhydrazone, N_α-(3,5-dimethoxyphenyl)phenylhydrazone and N_α-(4-carbethoxyphenyl)phenylhydrazone preferably indolized to afford **366** (R^1 = R^3 = R^4 = R^5 = H, R^2 = OCH_3; R^1 = R^3 = OCH_3, R^2 = R^4 = R^5 = H and R^1–R^4 = H, R^5 = $COOC_2H_5$, respectively) could also support the occurrence of

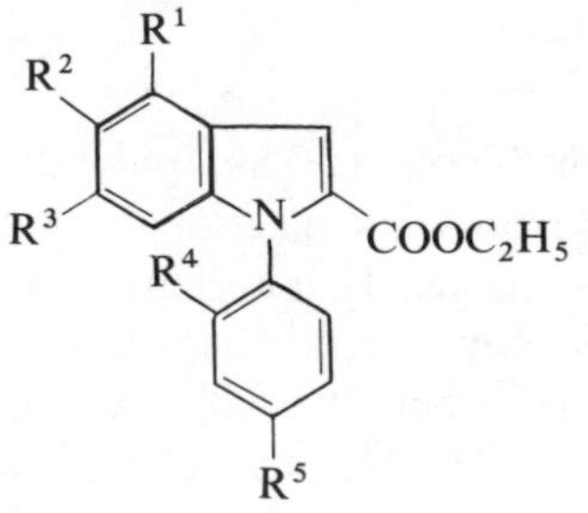

366

an intramolecular electrophilic attack. However, indolization of ethyl pyruvate N_α-(2-methoxyphenyl)phenylhydrazone using ethanolic hydrogen chloride afforded **366** (R^1–R^3 = R^5 = H, R^4 = OCH_3). Steric considerations of the two possible enehydrazine intermediates of this hydrazone led to the conclusion that only one, that ultimately furnishing **366** (R^1–R^3 = R^5 = H, R^4 = OCH_3), could undergo a [3,3]-sigmatropic shift. Likewise, similar steric and mechanistic considerations rationalized the observations that whereas ethyl pyruvate 2,6-dichlorophenylhydrazone, upon indolization with zinc chloride in acetic acid, gave ethyl 5,7-dichloroindole-2-carboxylate (the chlorine migration in such indolizations is discussed in detail later in this chapter), the N_α-methyl derivative of this hydrazone did not indolize under these conditions, although it is well established that N_α-methyl substituents normally facilitate arylhydrazone indolizations (Ishii, Hagiwara, Ishikawa, Ikeda and Murakami, Y., 1975).

The statement that 'it has been assumed that the cyclisation of N-phenyl-N-(3,5-dimethylphenyl)hydrazones takes place in the substituted ring' (Zagorevskii, Kucherova and Sharkova, N. M., 1977) is erroneous since the work referred to (Kricka and Vernon, 1974) only involved the indolization of N_α,2,5-trimethylphenylhydrazones and was therefore in this respect unambiguous.

The somewhat inconclusive work of the above mentioned two Japanese groups was followed by the far more convincing studies of Grandberg and his co-workers which clearly established that the new C—C bond was formed during indolization as a result of a [3,3]-sigmatropic shift. Indolization of **367** (R^1 = OCH_3, R^2 = R^3 = H, X = CH, Y = CH_2) with isopropanol saturated with hydrogen chloride, isopropanol containing sulphuric acid or in boiling diethylene glycol produced **368** (R^2 = OCH_3, R^1 = R^3–R^5 = H and R^1–R^3 = R^5 = H, R^4 = OCH_3, X = CH, Y = CH_2) in the percentage

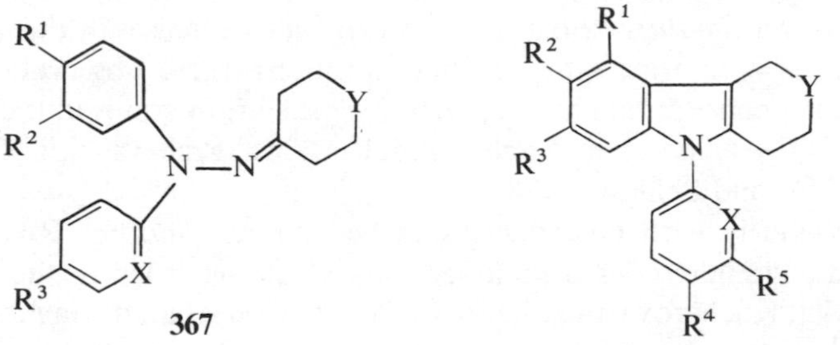

367

368

yield ratios of 80:20, 78:22 and 62:38, respectively. Hydrazone **367** (R^1 = Cl, R^2 = R^3 = H, X = CH, Y = CH_2) was similarly indolized to yield in each case **368** (R^1 = R^3–R^5 = H, R^2 = Cl and R^1–R^3 = R^5 = H, R^4 = Cl, X = CH, Y = CH_2) in the percentage yield ratios of 34:66, 32:68 and 32:68, respectively (Przheval'skii and Grandberg, 1974; Przheval'skii, Grandberg and Klyuev, N. A., 1976). Since it was established that the rates of electrophilic substitution in anisole are *ca.* $\times 10^{10}$ greater that the corresponding rates in chlorobenzene (Przheval'skii, Grandberg and Klyuev, N. A., 1976) and since in studies of the intramolecular electrophilic acylation of monosubstituted 3,3-diarylpropionic acids, only the product resulting from electrophilic attack at the most nucleophilic aryl nucleus was obtained in each case (Barltrop, Acheson, Philpot, MacPhee and Hunt, J. S., 1956; Braun, Manz and Reinsch, 1929; Johnston and Shotter, 1974), these present data were taken as confirmation that the second stage of the Fischer indolization involves a [3,3]-sigmatropic shift (Przheval'skii and Grandberg, 1974; Przheval'skii, Grandberg and Klyuev, N. A., 1976). Such a [3,3]-sigmatropic shift also operates in the *ortho*-Claisen rearrangement, a reaction which is likewise slightly but noticeably affected by the nature of aromatic substituents, electron releasing groups increasing the reaction rate and electron withdrawing groups decreasing it (Grandberg, 1972; GilChrist and Storr, 1972; Rhoads, 1963). In a subsequent extension of these studies, compounds **367** ($R^1 = R^3 = H$, $R^2 = OCH_3$ and Cl, X = CH, Y = CH_2) were indolized by heating with isopropanol saturated with dry hydrogen chloride, or with isopropanolic concentrated sulphuric acid or by refluxing in diethylene glycol. The products from these reactions were **368** ($R^1 = OCH_3$, R^2–R^5 = H, X = CH, Y = CH_2), **368** ($R^1 = R^2 = R^4 = R^5 = H$, $R^3 = OCH_3$, X = CH, Y = CH_2) and **368** (R^1–R^4 = H, $R^5 = OCH_3$, X = CH, Y = CH_2), the percentage yield ratio of the sum of the first two products to the third product being 43:57, 52:48 and 53:47, respectively, and **368** (R^1 = Cl, R^2–R^5 = H, X = CH, Y = CH_2), **368** ($R^1 = R^2 = R^4 = R^5 = H$, R^3 = Cl, X = CH, Y = CH_2) and **368** (R^1–R^4 = H, R^5 = Cl, X = CH, Y = CH_2), the corresponding percentage yield ratio being respectively 17:83, 27:73 and 35:65, respectively. Since the rate of electrophilic substitution in a benzene ring increases by three to four orders of magnitude when there is an *ortho* or *para* methoxy group present and decreases by the same degree when there is a chlorine atom in these positions, these observations further supported the concept that a [3,3]-sigmatropic shift operates in the formation of the new C—C bond in Fischer indolizations (Przheval'skii, Grandberg, Klyuev, N. A. and Belikov, 1978).

Simultaneously with Grandberg's earlier studies, another Russian group performed a similar series of indolizations which again led to the conclusion (Zagorevskii, Kucherova and Sharkova, N. M., 1977) that, during indolization, the new C—C bond was formed as a result of a [3,3]-sigmatropic shift. Thus, **367** ($R^1 = CH_3$, $R^2 = R^3 = H$; $R^1 = OCH_3$, $R^2 = R^3 = H$ and $R^1 = CH_3$, $R^2 = H$, $R^3 = OCH_3$, X = CH, Y = CH_2) was indolized by refluxing a mixture of cyclohexanone with the corresponding hydrazine hydrochloride in ethanol or in some cases ethanolic hydrogen chloride, to yield in each case mixtures of both possible indoles (in the percentage yield ratios indicated), **368** ($R^1 = R^3$–R^5 = H, $R^2 = CH_3$ and R^1–$R^3 = R^5$ = H, $R^4 = CH_3$; X = CH, Y = CH_2) (60:40), ($R^1 = R^3$–R^5 = H, $R^2 = OCH_3$, and R^1–$R^3 = R^5$ = H, $R^4 = OCH_3$; X = CH, Y = CH_2) (82:18) and ($R^1 = R^3 = R^5$ = H, $R^2 = OCH_3$, $R^4 = CH_3$ and $R^1 = R^3 = R^5$ = H, $R^2 = CH_3$, $R^4 = OCH_3$; X = CH, Y = CH_2) (76:24), respectively (Sharkova, N. M., Kucherova and Zagorevskii, 1974; Zagorevskii, Kucherova, Sharkova, N. M., Ivanova, T. I. and Klyuev, S. M., 1975). Compounds **367** ($R^1 = CH_3$, $R^2 = R^3 = H$, X = CH, Y = S and NCH_3) were similarly indolized to give a mixture of **368** ($R^1 = R^3$–R^5 = H, $R^2 = CH_3$ and $R^1 = R^3 = R^5$ = H, $R^4 = CH_3$, X = CH, Y = S) (75:25) and ($R^1 = R^3$–R^5 = H, $R^2 = CH_3$ and R^1–$R^3 = R^5$ = H, $R^4 = CH_3$, X = CH, Y = NCH_3) (67:33), respectively (Zagorevskii, Kucherova, Sharkova, N. M., Ivanova, T. I. and Klyuev, S. M., 1975) and

367 (R^1 = $COOC_2H_5$, R^2 = R^3 = H, X = CH, Y = CH_2) furnished both **368** (R^1 = R^3–R^5 = H, R^2 = $COOC_2H_5$ and R^1–R^3 = R^5 = H, R^4 = $COOC_2H_5$, X = CH, Y = CH_2) (15:85) (Zagorevskii, Kucherova and Sharkova, N. M., 1977). Here again in these studies, indolization slightly favoured the involvement of the aromatic ring bearing the most nucleophilic substituents. With **367** (R^1 = R^3 = H, R^2 = OCH_3, X = CH, Y = CH_2) indolization gave **368** (R^1–R^3 = R^4 = H, R^5 = OCH_3; R^1 = OCH_3, R^2–R^5 = H and R^1 = R^2 = R^4 = R^5 = H, R^3 = OCH_3; X = CH, Y = CH_2) in 54%, 20% and 26% relative percentage yields, results in accord with the earlier observations (Elgersma, 1969; Pausacker and Schubert, 1950) that cyclohexanone 3-methoxyphenylhydrazone is indolized more slowly than the corresponding phenylhydrazone. Similarly, **369** afforded both **370** and **371**

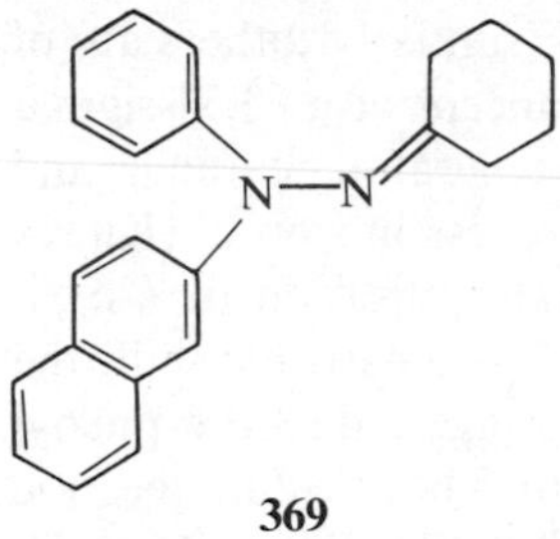

369 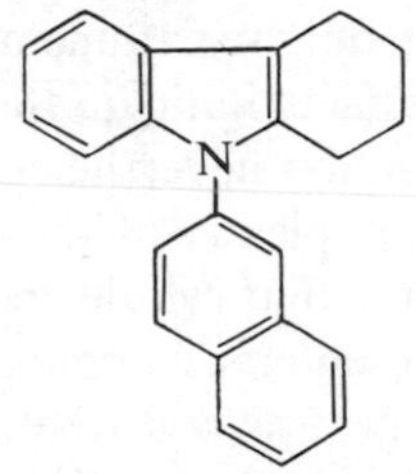 **370** 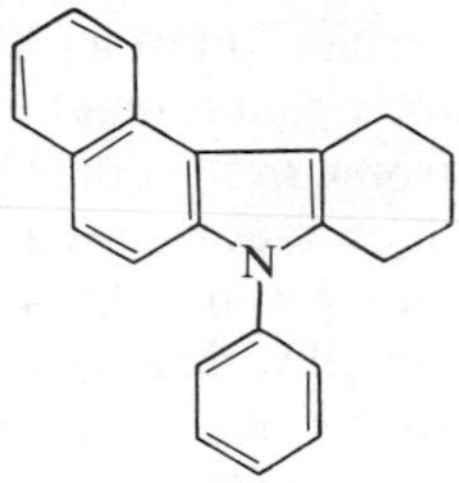 **371**

in the relative yields of 33% to 67%, respectively, no product resulting from indolization at the 3-position of the naphthalene nucleus (see Chapter III, D) being detected (Zagorevskii, Kucherova and Sharkova, N. M., 1977). When one of the aryl nuclei was much less nucleophilic than the other, indolization appeared to involve exclusively the latter. Thus, **367** (R^1–R^3 = H, X = N, Y = CH_2 and S) gave **368** (R^1–R^5 = H, X = N, Y = CH_2 and S), isolated in 74% and 61% yields, respectively, as the only products, **372** reacted with cyclohexanone to yield only **373** in 64% yield (Sharkova, N. M., Kucherova and Zagorevskii, 1974; Zagorevskii, Kucherova and Sharkova, N. M., 1977) and **374** (R = H and OCH_3) and **376** reacted with laevulinic acid to afford **375** (R = H and OCH_3) and **377**, respectively, as the only isolated products (Birchall, G. R., Hepworth and Smith, S. C., 1973). Unlike Grandberg and his

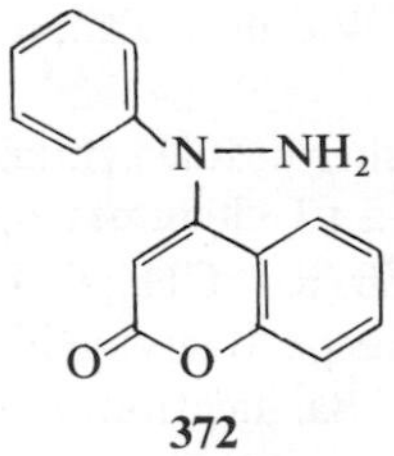

372 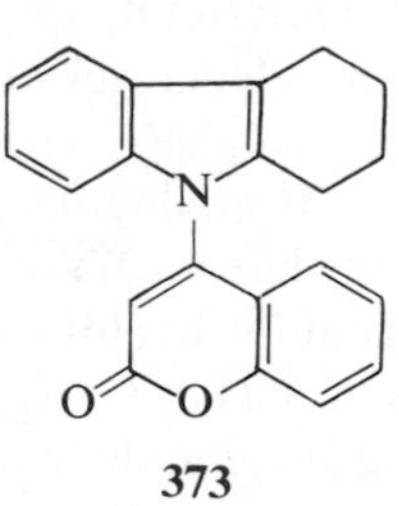 **373** 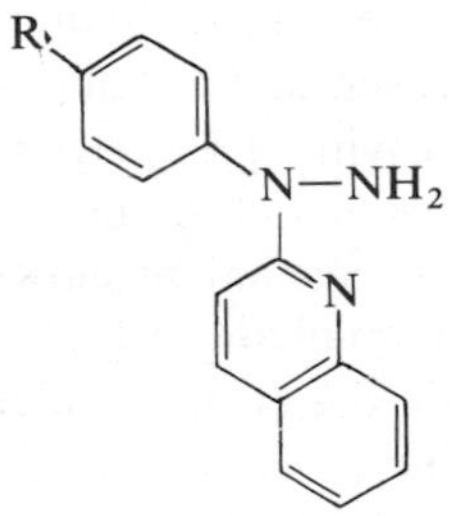 **374**

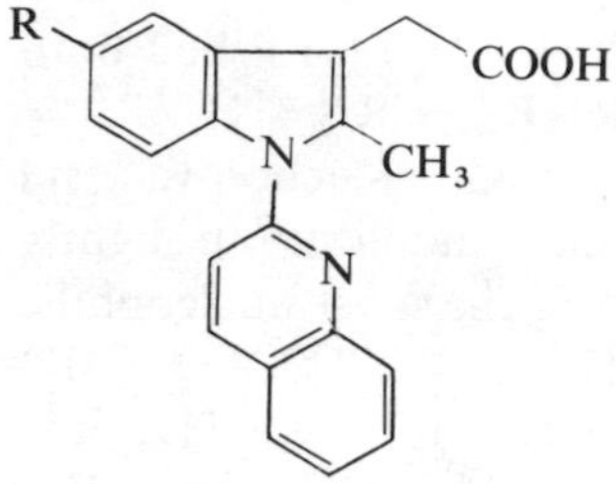

375 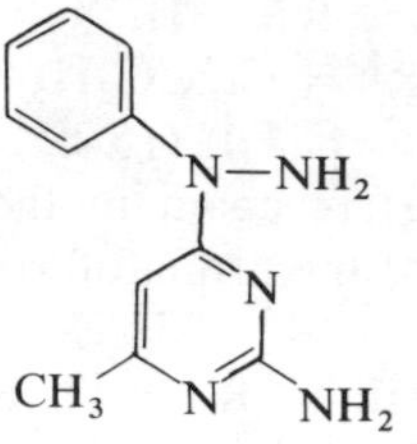 **376** 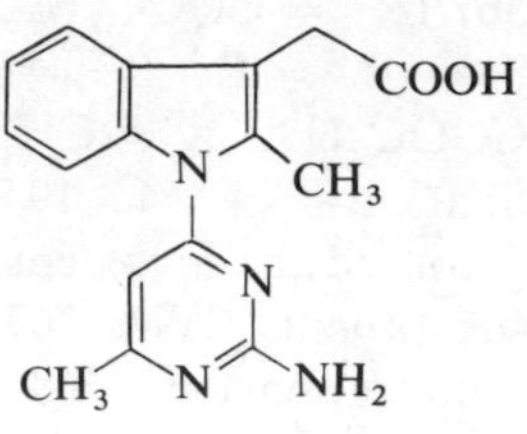 **377**

co-workers, this other Russian group (Sharkova, N. M., Kucherova and Zagorevskii, 1974; Zagorevskii, Kucherova and Sharkova, N. M., 1977) did not draw from their observations categorical conclusions regarding the direct verification of a [3,3]-sigmatropic rearrangement and, similarly, others are of the opinion that 'all of these facts indicate that the concept of a [3,3]-sigmatropic shift in the Fischer reaction nevertheless is too general in character and does not exhaust all of the complexity of the cyclic process in step b' (Kitaev and Troepol'skaya, 1978) and that 'while the thermal cyclisation of 4-pyrimidylhydrazones is best represented as a concerted [3,3]-sigmatropic shift, the effects of electron releasing groups described above suggest that the phenyl ring provides the attacking electron pair' (Wright, 1976). The effect in question involved the hydrochloric acid catalysed indolizations of **254** ($R^1 = R^2 = H$, $R^1 = CH_3$, $R^2 = H$ and $R^1 = R^2 = CH_3$, respectively, R^3–$R^5 = H$, $X = O$) into **289** ($R^1 = R^2 = H$, $R^1 = CH_3$, $R^2 = H$ and $R^1 = R^2 = CH_3$, respectively, R^3–$R^5 = H$, $X = O$) in 21%, 64% and 84% yields, respectively. However, as already pointed out, such an interpretation of product yields is not necessarily valid.

It would appear from these above data that the exclusive operation of a [3,3]-sigmatropic shift is unlikely in new C—C bond formation during Fischer indolization and that in some cases an intramolecular electrophilic attack upon the aryl nucleus of the arylenehydrazine may operate alternatively or concomitantly. Perhaps the early original caution when the possible operation of such a [3,3]-sigmatropic shift was suggested (Robinson, B., 1969) is justified.

Interesting analogies have been made between this stage of the mechanism of the Fischer indolization, a 3,4-diaza-Cope rearrangement (Heimgartner, Hansen and Schmid, H., 1979; Winterfeldt, 1970), and other rearrangements involving [3,3]-sigmatropic shifts (Hendrickson, 1974; Widmer, Zsindely, Hansen and Schmid, H., 1973).

A mechanism shown in Scheme 20, analogous to the Fischer indole synthesis but involving a [3,7]-sigmatropic shift in **379** in place of a [3,3]-sigmatropic shift, has been postulated to account for the formation of **380** ($R = CH_3$, C_2H_5, C_6H_5) by treatment of **379** ($R = CH_3$, C_2H_5, C_6H_5, respectively) with a refluxing solution of 40% hydrogen bromide in acetic acid. Basification of **380** afforded the corresponding **381** (Fritz and Schenk, 1972).

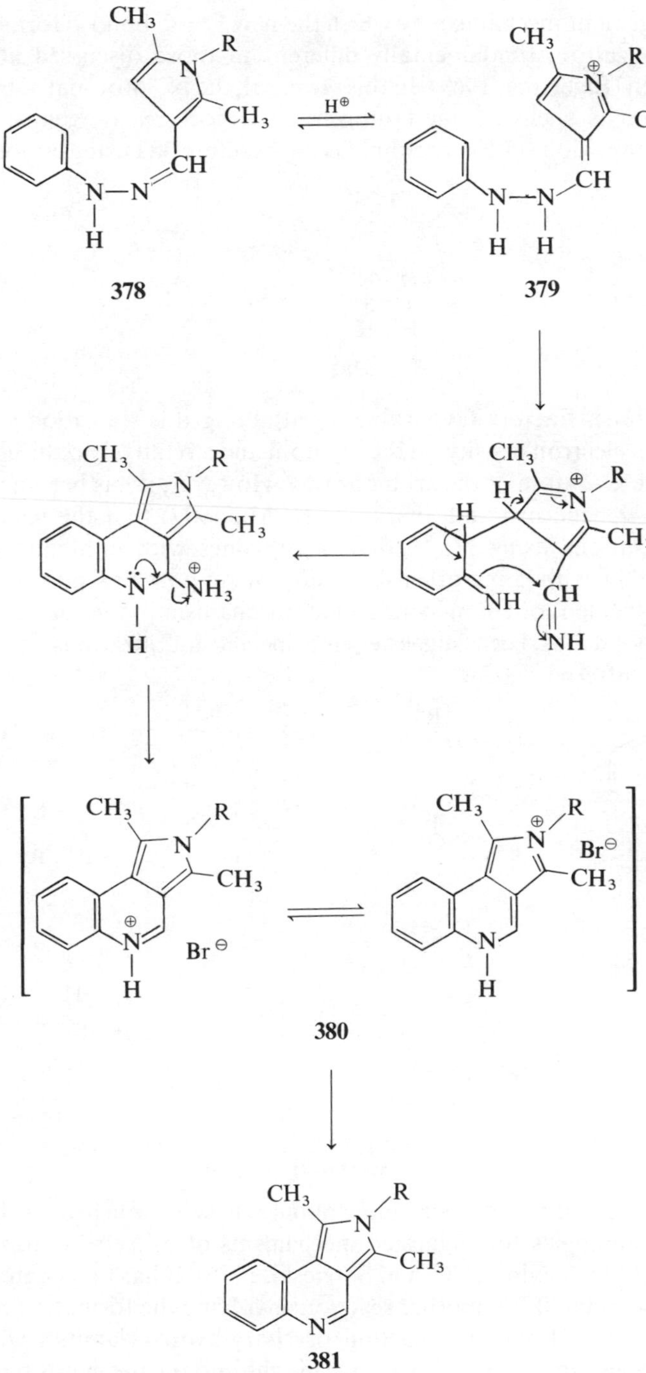

378

379

380

381

Scheme 20

A rearrangement mechanism by which the new C—C bond is formed during Fischer indolization, fundamentally different to those discussed above, has been proposed (Elgersma, 1969). In this proposal, the N_β-protonated hydrazone form, and not its enehydrazine tautomer, undergoes rearrangement through a non-cyclic transition state, as exemplified in structure **382** using cyclohexanone

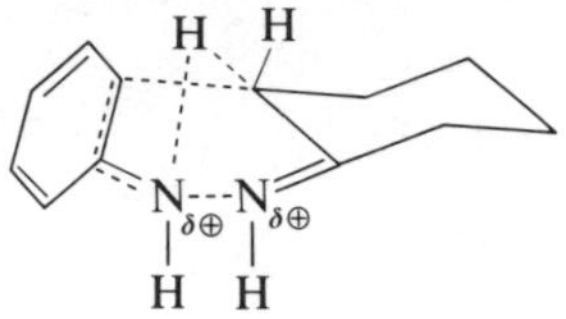

382

phenylhydrazone. Factors favourable for attaining this transition state are a relatively high electron density on the N_α-atom and a relatively reduced electron density on the C-2 atom of the aromatic ring. However, it has been pointed out (Zagorevskii, Kucherova and Sharkova, N. M., 1977) that the results of the above indolizations using N_α,N_α-diarylhydrazones with non-equivalent aryl groups, together with some other facts, do not support this postulation. The possible involvement of a non-synchronous mechanism in the formation of the new C—C bond has been suggested for specific indolizations as shown in Scheme 21 (Dufresne, 1972).

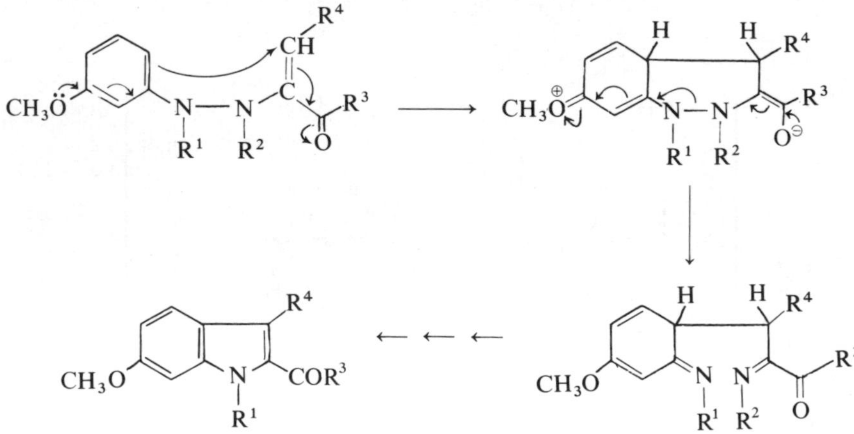

Scheme 21

The mechanism of the Fischer indolization as now established has been used in providing analogies for suggested mechanisms of other reactions (Haynes and Hewgill, 1972; Taylor, E. C. and Sowinski, 1975). It has been stated that the mechanism postulated for another reaction, involving the formation of 1,2,3,4-tetrahydrocarbazol-4-ones by reaction of N-arylhydroxylamines with cyclohexan-1,3-diones, 'may have a suggestive insight into the mechanism of Fischer indole synthesis' (Okamoto and Shudo, 1973), although it is far from clear what this insight could involve.

G. Group Migrations, Substitutions and Eliminations during the Indolization of 2-Monosubstituted and 2,6-Disubstituted Phenylhydrazones

The Robinsons' mechanism, with its modifications as shown in Scheme 10, has been further supported by the extensive indolization investigations which have been carried out using a variety of 2-monosubstituted and 2,6-disubstituted phenylhydrazones, the results of which can only be rationalized by considering the intermediacy of a dienone imine structure **197** or a corresponding cyclized structure.

1. 2-Halogeno- and 2,6-Dihalogenophenylhydrazones

Following the recognition of the similarity between the second stage of the Fischer indolization, during which the formation of the new C—C bond occurs, and the *ortho*-Claisen rearrangement and with particular reference to the previous studies which had been effected upon the rearrangements of 2,6-dibromo- and 2,6-dichlorophenyl allyl ether, several 2,6-dichlorophenylhydrazones, **383**, were subjected to indolization by heating with zinc chloride with or without added solvent (Carlin and Fisher, 1948). Isolated from the reaction were small amounts, up to a maximum of 33% yield, of the corresponding 5,7-dichloroindoles **384**. Although at the time (Carlin and Fisher, 1948) no analogy involving a 1,3-migration of a halogen atom in either a benzidine or a Claisen rearrangement was available, later studies (Piers and Brown, R. K., 1963) showed that such rearrangements do occur in zinc chloride catalysed Claisen rearrangements of allyl 2,6-dichlorophenyl ethers. The chlorine migrations during the indolizations were shown (Carlin and Fisher, 1948) to occur before formation of the indole ring and to be specific for dichlorophenylhydrazones with a 2,6-orientation. A similar migration was later observed during the sulphuric acid catalysed indolization of cyclohexanone 2,6-dichlorophenylhydrazone which gave 6,8-dichloro-1,2,3,4-tetrahydrocarbazole (Barnes, Pausacker and Schubert, 1949), with cyclohexanone 2-chloro-1-naphthylhydrazone which afforded 6-chloro-1,2,3,4-tetrahydro-7,8-benzocarbazole (Barnes, Pausacker and Badcock, 1951) and during the zinc chloride catalysed indolization of ethyl pyruvate 2,6-dichlorophenylhydrazone in acetic acid which yielded, in 82.3% yield, ethyl 5,7-dichloroindole-2-carboxylate (Ishii, Hagiwara, Ishikawa, Ikeda and Murakami, Y., 1975). Further studies (Carlin, Wallace, J. G. and Fisher, 1952 – see also Carlin and Amoros-Marin, 1959) furnished no evidence to suggest that the chlorine migration was intermolecular [indeed, in the absence of later studies, another (Dewar, 1963) also favoured an intramolecular mechanism], other than possibly between rearranging arylhydrazones, but strongly suggested that the chlorine migrated as a positive ion in which case it could form a π-electron complex with the aromatic system during its migration. On the basis of these above data, the following mechanism (Scheme 22) for the reaction was proposed (Carlin, Wallace, J. G. and Fisher, 1952). An alternative proposal that involved the

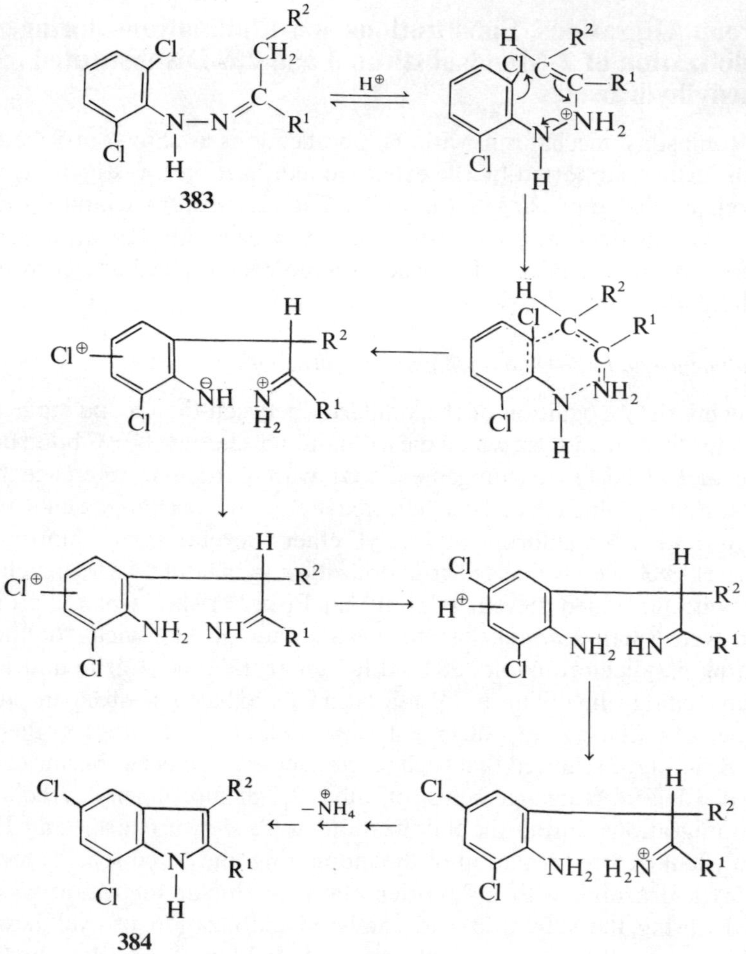

Scheme 22

intermediacy of the species **385**, in which chlorine migrates as shown to give **386** (Southwick, quoted in Carlin, Wallace, J. G. and Fisher, 1952), had the additional advantage that it would explain, by reductive removal of the allylic chlorine in either **385** or **386**, the formation of 7-chloroindoles in the above type of reaction using tin(II) chloride as catalyst (Carlin, Wallace, J. G. and Fisher, 1952). Another mechanistic proposal (Arbuzov, A. E. and Kitaev, 1957b,

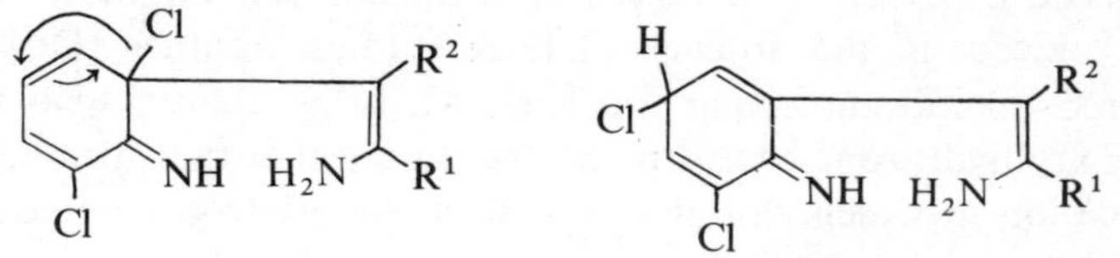

385 **386**

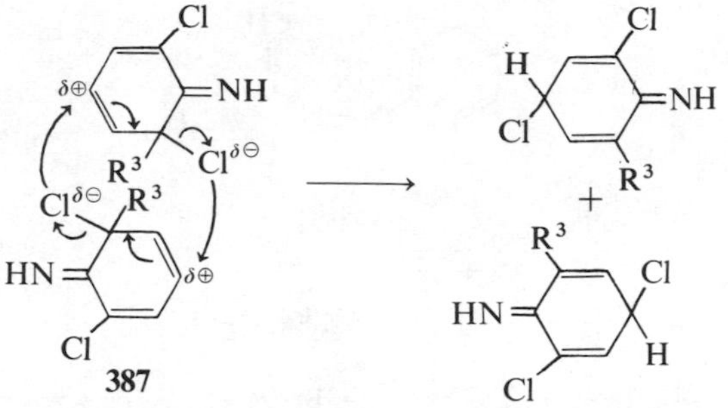

387

1957d) for these halogen migrations involved the polarization in the intermediate **385** which could lead to the chlorine 1,3-migration as shown in **387** ($R^3 = CR^2{=}CR^1NH_2$). However, such a mechanism would not explain the following observations involving halogen exchange. The zinc chloride catalysed indolization of acetophenone 2,6-dibromophenylhydrazone (**388**) afforded a mixture containing approximately equal amounts of 5,7-dibromo-2-phenylindole (**389**) and 7-bromo-5-chloro-2-phenylindole (**390**) and a trace of 7-bromo-2-phenylindole (**391**) and the zinc bromide catalysed indolization of acetophenone 2,6-dichlorophenylhydrazone (**392**) formed a mixture of 5,7-dichloro-2-phenylindole (**393**) and 5-bromo-7-chloro-2-phenylindole (**394**) (Carlin and Larson, 1957). From these results it was apparent that the previously proposed mechanism (Carlin, Wallace, J. G. and Fisher, 1952) for these halogen migrations, involving a 'positive halogen', was invalid since it was incompatible with the relative oxidation potentials of bromide and chloride ions (i.e. although exchange was feasible between $Cl^{\oplus}$ and $ZnBr_2$ which would ultimately give rise to **394**, it was unlikely between $Br^{\oplus}$ and $ZnCl_2$ and so **390** would not be an expected product in the above reaction). To account for these results, the

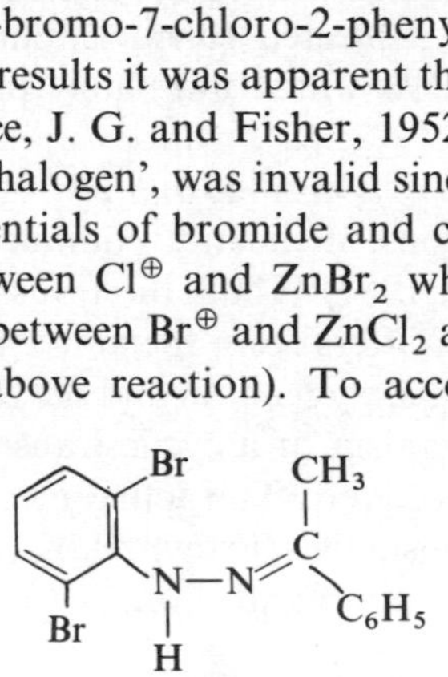

388

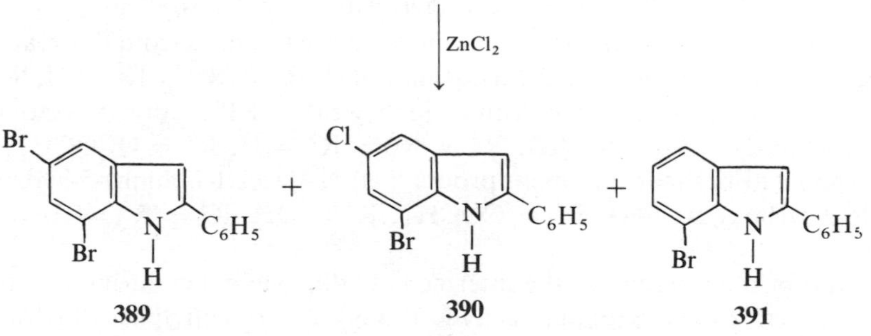

$ZnCl_2$

389 + 390 + 391

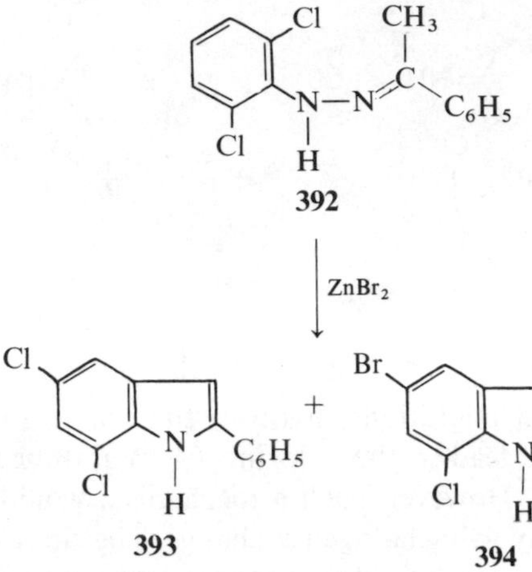

392

$\xrightarrow{ZnBr_2}$

393 + 394

following mechanistic proposal (Scheme 23) was advanced (Carlin and Larson, 1957) in which the halogen migrates as its more normal anionic ion. This is in full agreement with the Robinsons' mechanism, as modified in Scheme 10, involving the formation of a dienone–imine intermediate **395**. Removal of $X^{\ominus}$ affords **396** which is then subjected to nucleophilic attack by the complex $(ZnY_2X)^{\ominus}$ to give **397** or **398** which ultimately produces the indoles **399** or **400**, respectively.

Although this conversion of **395** → **396** → **397** or **398** was in agreement with the experimental observations, it did not eliminate other mechanistic possibilities (Carlin and Larson, 1957). Thus, the transformation could involve an S_N1 mechanism which only differs from the above mechanism in the degree of separation of the migrating halogen anion and the cation in the transition state. Alternatively, an S_N2 mechanism or a cyclic transition state might operate as shown in **401** and **402**, respectively. The actual mechanism operating may well depend upon reaction circumstances (Brown, R. K., 1972a). The S_N2 mechanism was later (Shine, 1967b) favoured and thought to be reminiscent of the Bamberger rearrangement.

Similar displacement of halogen was observed (Ishii, Murakami, Y., Furuse, Hosoya and Ikeda, 1973) when ethyl pyruvate 2,6-dichlorophenylhydrazone was indolized using zinc chloride in acetic acid, the products from this reaction being ethyl 5,7-dichloroindole-2-carboxylate (**403**; $R^1 = R^3 = R^5 = H, R^2 = R^4 = Cl$), isolated in the significantly high yield of 82%, ethyl 5-acetoxy-7-chloroindole-2-carboxylate (**403**; $R^1 = R^3 = R^5 = H$, $R^2 = OCOCH_3$, $R^4 = Cl$) and its Fries rearrangement product, ethyl 4-acetyl-7-chloro-5-hydroxyindole-2-carboxylate (**403**; $R^1 = COCH_3$, $R^2 = OH$, $R^3 = R^5 = H$, $R^4 = Cl$).

In view of the isolation of the intermediate dienone-imine during the indolization of a mixture of propionaldehyde and N_β,2,6-trimethylphenylhydrazine

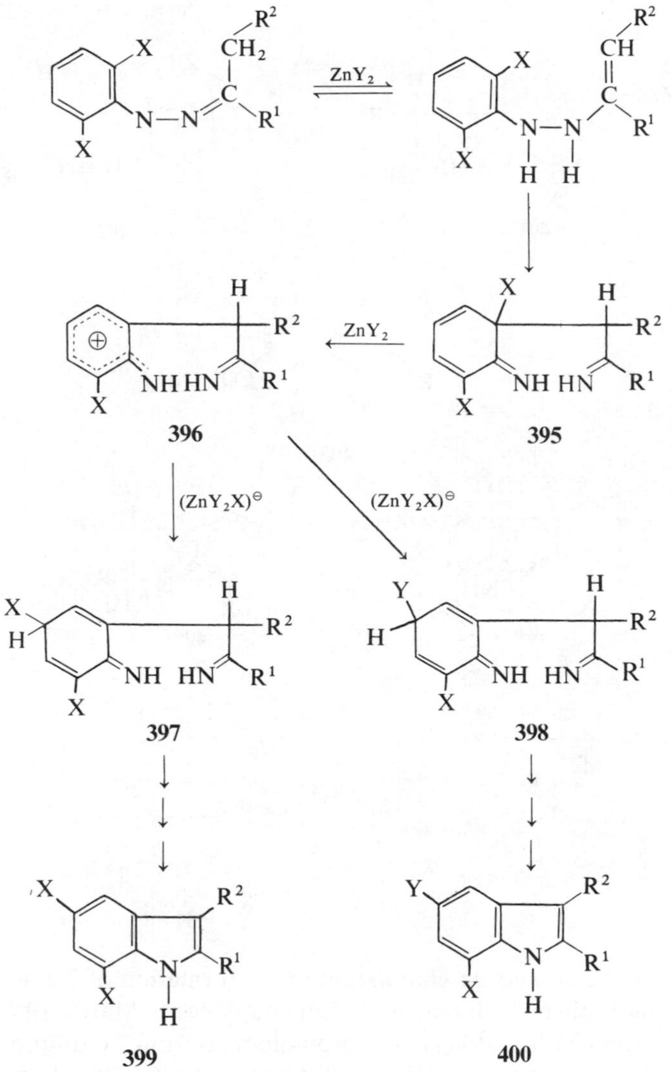

Scheme 23

hydrochloride as described later in this chapter (Bajwa and Brown, R. K., 1969), it was suggested (Brown, R. K., 1972a) that the above halogen exchanges could operate on an analogous intermediate **404** which would thus give **405** and ultimately the indole. Alternatively, by analogy with a suggestion concerning similar methyl group migrations, halogen migration might accompany the ring closure to **404** (Carlin and Moores, 1962) although evidence is available which suggests that such methyl group migrations precede the ring closure (Carlin and Harrison, J. W., 1965).

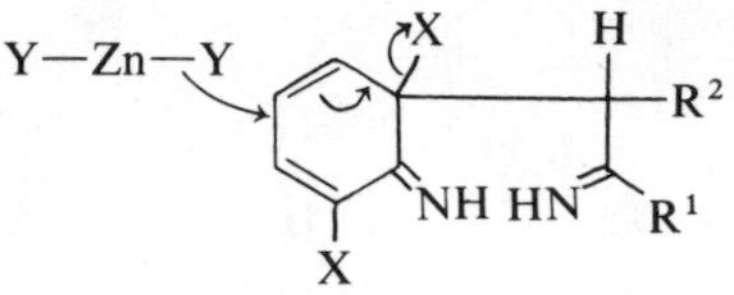

401

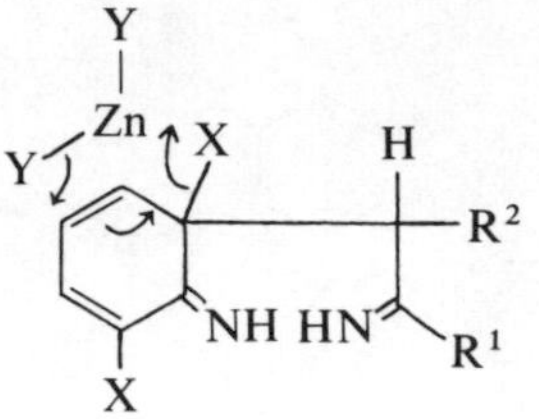

402

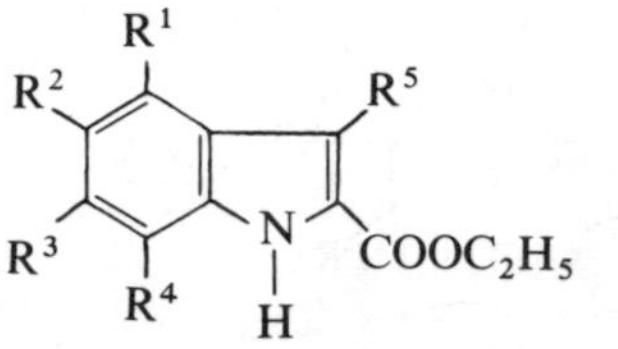

403

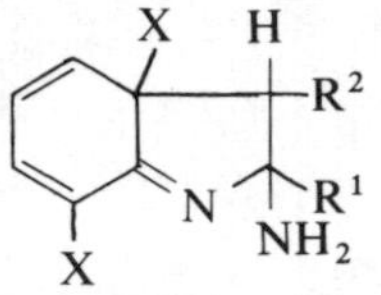

404

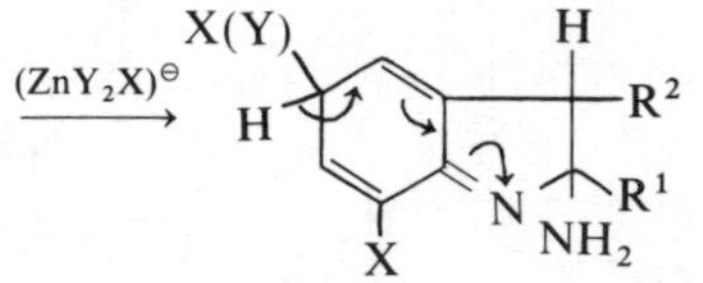

405

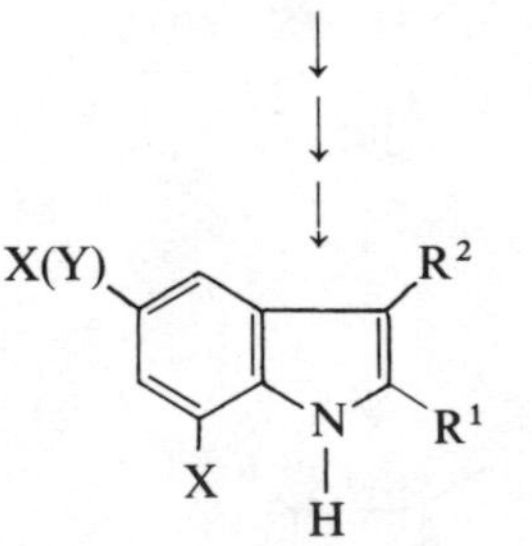

The mechanism of halogen elimination in the formation of 7-chloroindoles from 2,6-dichlorophenylhydrazones (Carlin and Amoros-Marin, 1959; Carlin, Wallace, J. G. and Fisher, 1952) and 7-bromoindoles from 2,6-dibromophenylhydrazones (Carlin and Larson, 1957) still requires clarification. Likewise does the halogen elimination from the reaction between 2,6-dichloro-N_β-methylphenylhydrazine (**406**) and cyclohexanone in benzene solution at room temperature which afforded 5-amino-6-chloro-1,2,3,4-tetrahydro-9-methylcarbazole (**409**), 8-chloro-1,2,3,4-tetrahydrocarbazole (**408**) and methylamine hydrochloride by a proposed mechanism involving the intermediate **407** as shown in Scheme 24 (Robinson, F. P. and Brown, R. K., 1964). It was suggested (Robinson, F. P. and Brown, R. K., 1964) that a readily oxidizable substance in the reaction mixture reductively removed the allylic halogen, it being possible (Carlin, 1952) that traces of water in the reaction mixture [via the formation

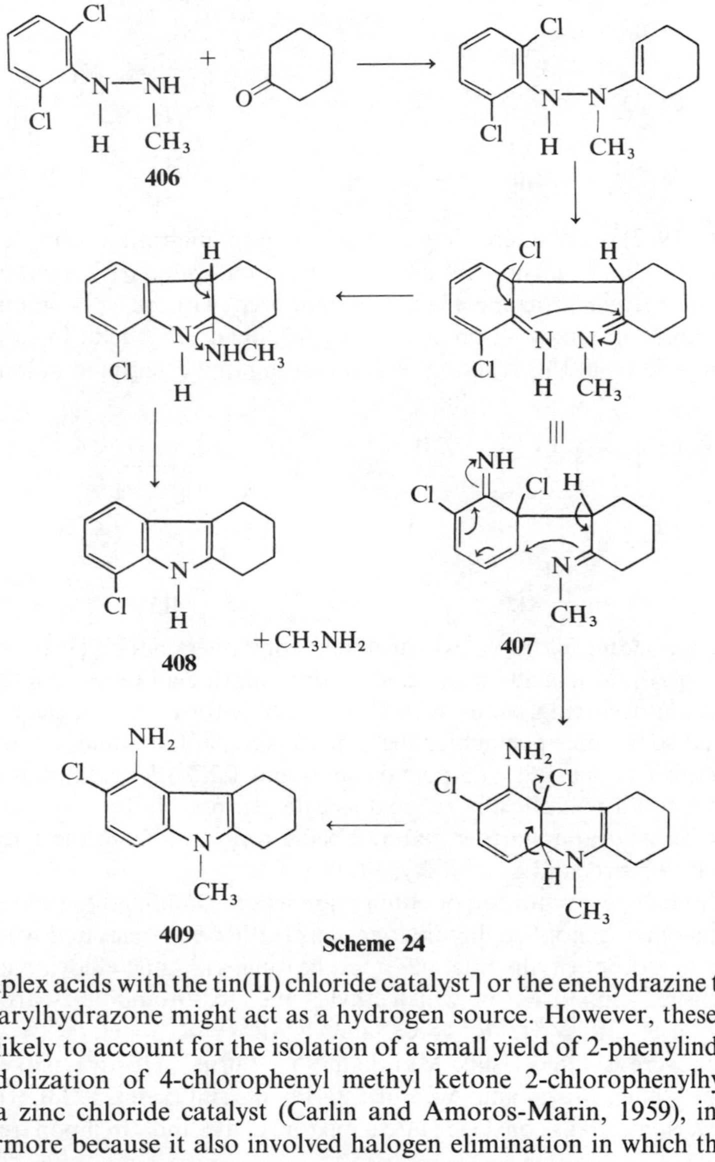

Scheme 24

of complex acids with the tin(II) chloride catalyst] or the enehydrazine tautomer of the arylhydrazone might act as a hydrogen source. However, these theories are unlikely to account for the isolation of a small yield of 2-phenylindole from the indolization of 4-chlorophenyl methyl ketone 2-chlorophenylhydrazone using a zinc chloride catalyst (Carlin and Amoros-Marin, 1959), interesting furthermore because it also involved halogen elimination in which there is an unsubstituted *ortho* position in the original arylhydrazone.

Mechanisms analogous to that proposed above explain more recent examples of halogen migration and elimination. Thus, the indolization of the 2-bromo-5-methoxyphenylhydrazone of piperidin-2,3-dione, using an acetic acid–hydrochloric acid mixture as catalyst, afforded a mixture of **410** and **411** (Morozovskaya, Ogareva and Suvorov, 1969) [in an earlier study of this reaction, only the latter product was isolated (Suvorov, Fedotova, M. V., Orlova, L. M. and

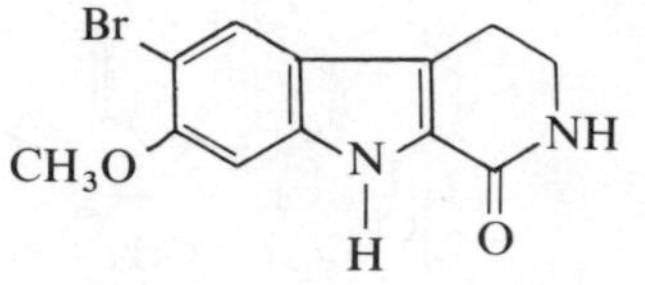

410

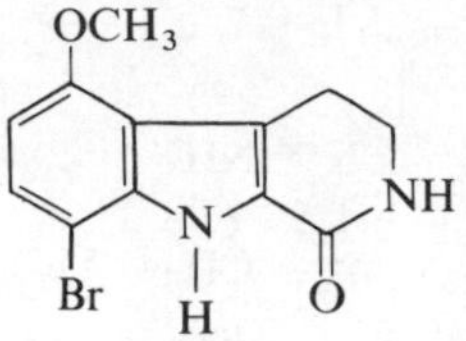

411

Orgareva, 1962)], an interesting case of halogen migration using an arylhydrazone with an unsubstituted *ortho* position. Similarly, acetophenone pentachlorophenylhydrazone (**412**) was indolized with polyphosphoric acid to give 4,5,6,7-tetrachloro-2-phenylindole (**413**) in an 81% yield. In view of the very high yield from this reaction, it is surprising that attempted indolizations

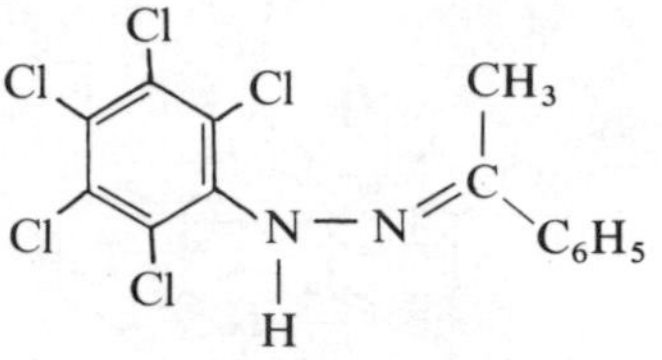

412

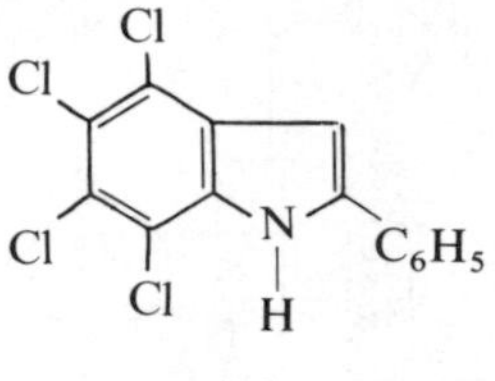

413

of the corresponding hydrazones of acetone, propiophenone and cyclohexanone using polyphosphoric acid, acetic acid or zinc chloride all failed, considerable tar formation occurring, along with the invariable formation of pentachloroaniline and sometimes pentachlorobenzene. Likewise, the attempted indolization of acetophenone 1,2,3,4,5-pentabromo- and 1,2,3,4,5-pentafluorophenylhydrazone, using polyphosphoric acid as catalyst, both failed to yield indolic products, only tar and starting material being recovered from these reactions (Collins, I., Roberts and Suschitzky, 1971).

That this halogen migration or elimination was not confined to indolizations of only 2,6-dihalogenophenylhydrazones was further demonstrated when ethyl pyruvate 2-chlorophenylhydrazone, upon heating with zinc chloride in acetic acid, furnished a mixture from which ethyl 5- and 7-chloroindole-2-carboxylate (**403**; $R^1 = R^3\text{–}R^5 = H$, $R^2 = Cl$ and $R^1\text{–}R^3 = R^5 = H$, $R^4 = Cl$, respectively) were isolated (Ishii, Murakami, Y., Furuse, Hosoya, Takeda and Ikeda, 1971; Ishii, Murakami, Y., Furuse, Hosoya and Ikeda, 1973), when **414** (*syn* or *anti* isomer) was converted into a mixture of **415** and **416**, upon treatment

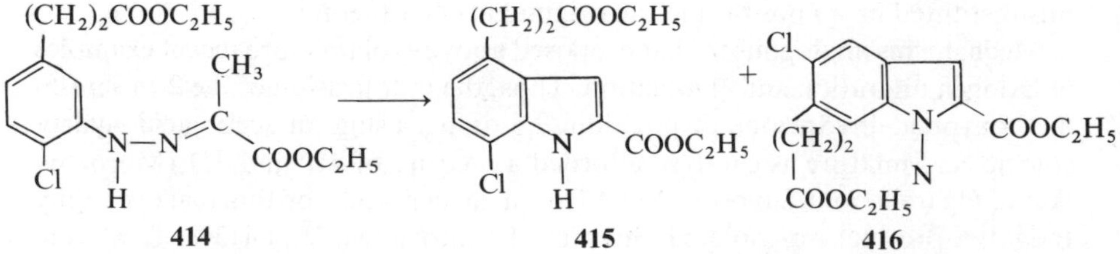

414

415

416

with zinc chloride in refluxing acetic acid (Nagasaka, Yuge and Ohki, 1977) and when **417** afforded both **418** and **419** upon treatment with a mixture of glacial acetic acid and concentrated hydrochloric acid (Morozovskaya, Ogareva and Suvorov, 1970). Similarly, halogen elimination occurred when indolization of cyclohexanone 5,8-dichloro-6-quinolylhydrazone took place at the 5-position of the quinoline nucleus (Kulka and Manske, R. H. F., 1952) (see Chapter III, E).

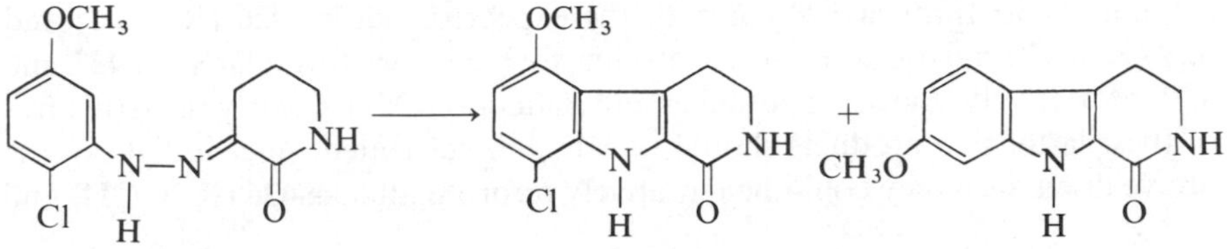

OCH_3 NH N—N H Cl O **417** → OCH_3 NH N H Cl O **418** + NH CH_3O N H O **419**

It is now possible to rationalize the formation of 1,2,3,4-tetrahydro-6-hydroxycarbazole from the dilute sulphuric acid catalysed indolization of cyclohexanone 2-chlorophenylhydrazone. The products from this reaction were found to be the expected 8-chloro-1,2,3,4-tetrahydrocarbazole and a second compound to which structure **420** was initially assigned (Barnes, Pausacker and Schubert,

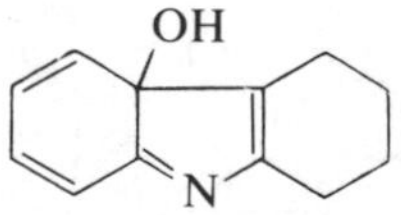

OH N **420**

1949) but this was later (Milne and Tomlinson, M. L., 1952) corrected to 1,2,3,4-tetrahydro-6-hydroxycarbazole. Clearly, the formation of this compound involved the nucleophilic displacement of the allylic chlorine atom in structures analogous to **395–397**, **401**, **404** or **405**, after the rearrangement of the enehydrazine tautomer of the starting hydrazone has occurred at the C-2 atom of the benzenoid ring. A similar rearrangement and halogen displacement seemed to occur when indolization of cyclohexanone 2-fluorophenylhydrazone afforded, along with the 'normal' indolization product, another product which appeared to be 1,2,3,4-tetrahydro-6-hydroxycarbazole (Allen, F. L. and Suschitzky, 1953). Likewise, the indolization of cyclohexanone 2-chloro-5-methylphenylhydrazone (**421**) yielded 8-chloro-1,2,3,4-tetrahydro-5-methylcarbazole (**422**) together with a product which was originally (Pausacker and Robinson, R., 1947) formulated as **423**. The structure of this compound was later (Cummins, Kaye and Tomlinson, M. L., 1954) reassigned as **424**, although this reassignment was based upon its synthesis via the indolization of cyclohexanone

4-methoxy-3-methylphenylhydrazone, the direction of which is ambiguous (see Chapter III, A). However, structure **424** is most likely to be correct and is certainly in agreement with a mechanism of formation based upon the above analogies although in the earlier paper (Pausacker and Robinson, R., 1947) the compound was stated to be devoid of phenolic character (cf. Milne and Tomlinson, M. L., 1952).

The indolization of cyclohexanone 2-chloro-1-naphthylhydrazone (**425**) with acetic acid and cyclohexanone 2,6-dichlorophenylhydrazone (**428**) with dilute sulphuric acid furnished along with the expected indoles **426** (R = Cl) and **429** (R = Cl), respectively, other products which were formulated as **427** and **430**, respectively (Barnes, Pausacker and Badcock, 1951). Clearly the structures of these latter two products should now be reinvestigated, since in light of the above discussion they could be tentatively reformulated as **426** (R = OH) and

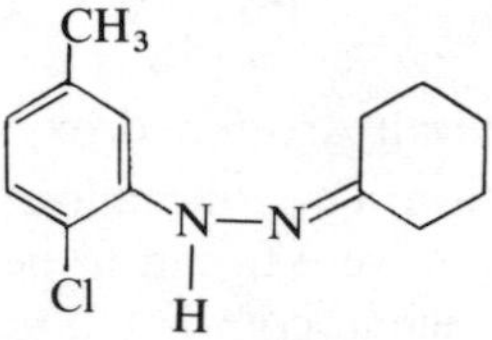

421 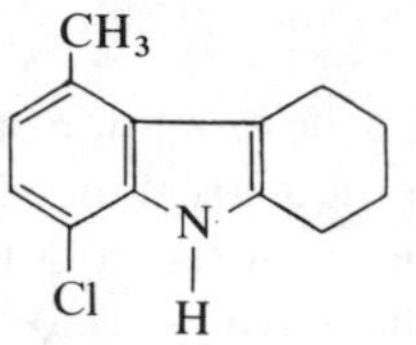 **422** 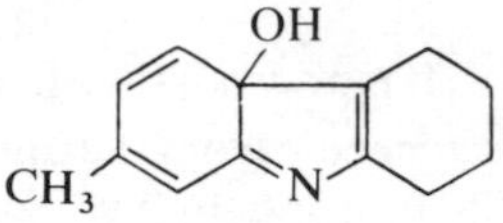 **423**

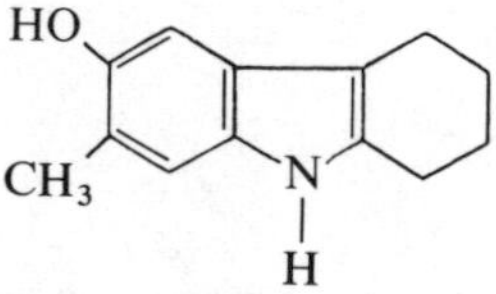

424 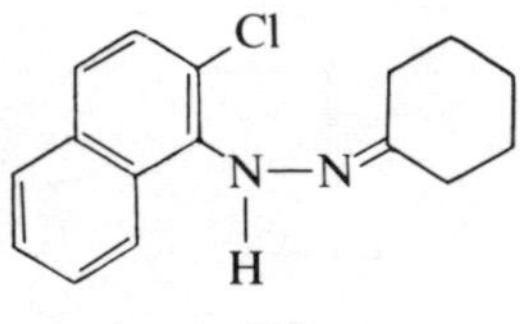 **425** 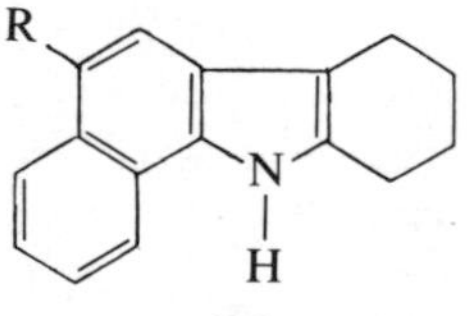 **426**

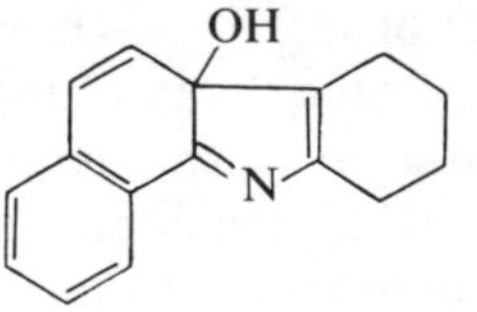

427 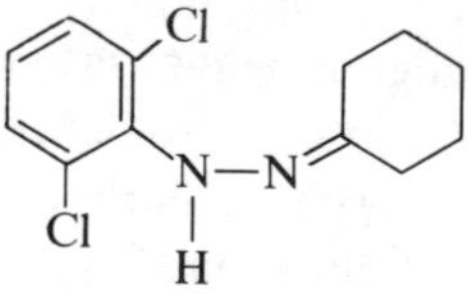 **428**

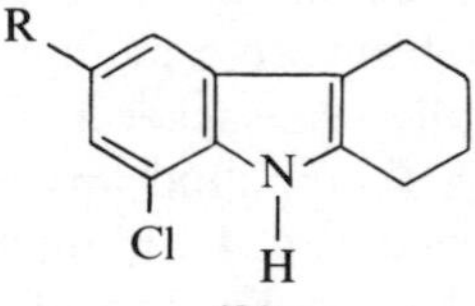

429 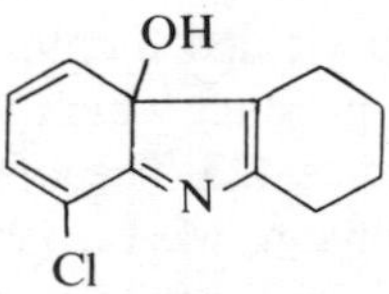 **430**

429 (R = OH), respectively. Likewise, the structures of the products resulting from the indolization of cyclohexanone 2,4,6-trichloro- and 2,4,6-tribromophenylhydrazone (**431**; X = Cl and Br, respectively) and which have been (Barnes, Pausacker and Schubert, 1949) formulated as **432** (X = Cl and Br, respectively) are worthy of reinvestigation.

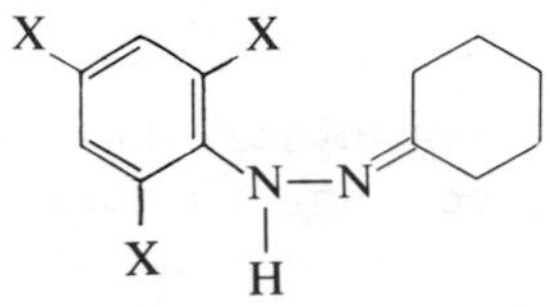

431 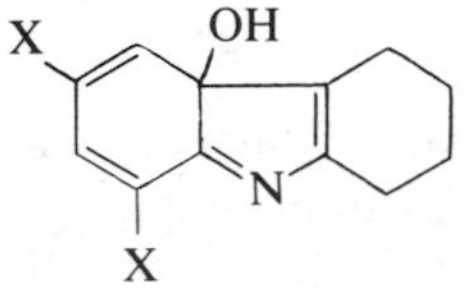 **432**

2. 2-Alkyl- and 2,6-Dialkylphenylhydrazones

The first losses of an alkyl group during Fischer indolization of 2-alkylated arylhydrazones were noted (Huisgen, 1948a) when cyclohexanone 1-methyl-2-naphthylhydrazone (**433**; R = H, X = CH), cyclohexanone 5,8-dimethyl-6-quinolylhydrazone (**433**; R = CH_3, X = N) and cyclohexanone 6-methyl-5-quinolylhydrazone (**435**) were found to give **434** (R = H, X = CH), **434** (R = CH_3, X = N) and **436**, respectively. It is interesting to note that, in the case of **433**, the indolizations were directed towards the methyl substituted *ortho* carbon atom in spite of there being an alternative unsubstituted *ortho* position. This is in accord with the direction of indolization of 2-naphthylhydrazones which occurs towards the 1-position of the naphthalene nucleus (see Chapter III, D). However, attempted indolization of **437** (R = CH_3) was unsuccessful although **437** (R = H) indolized in very high yield to afford **438** (Huisgen, 1948a).

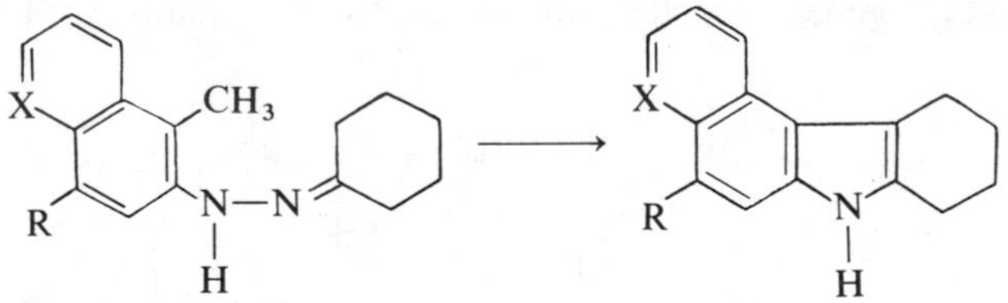

433 **434**

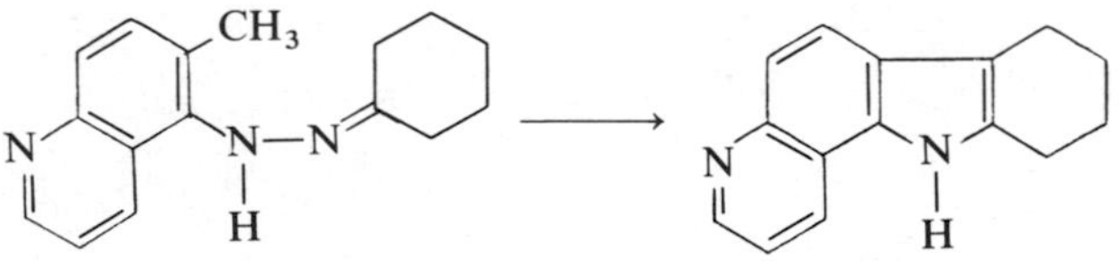

435 **436**

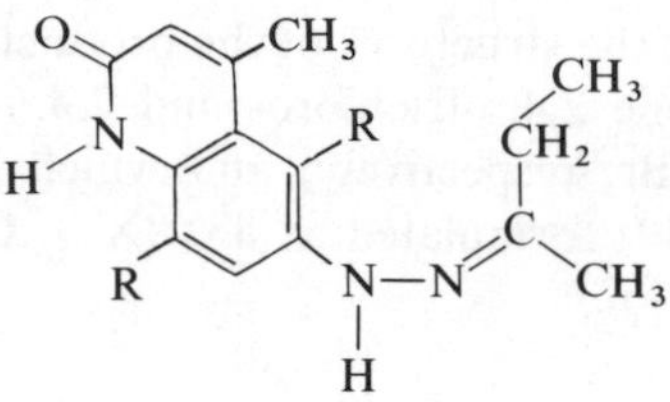

437

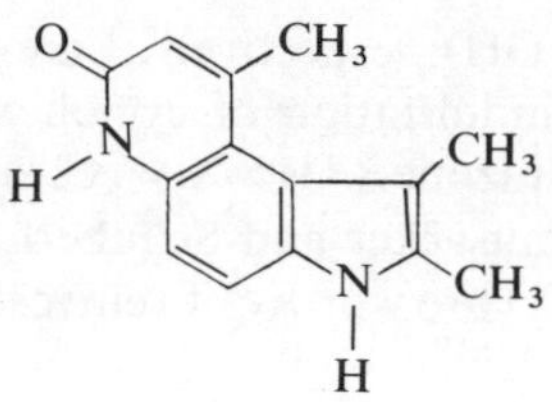

438

Investigation of the behaviour of 2,6-dimethylphenylhydrazones [for a synthesis of 2,6-dimethylphenylhydrazine, see Franzen, Onsager and Faerden (1918)] under Fischer indolization conditions was suggested when these above reactions were investigated (Huisgen, 1948a). However, the first systematic investigation of this subject was reported in a series of papers by Carlin and several of his co-workers. During the indolization of ethyl pyruvate 2,6-dimethylphenylhydrazone (**439**; $R^1 = R^3 = H$, $R^2 = COOC_2H_5$) by pyrolysis with a zinc chloride catalyst it was found that, similar to the 1,3-migration which occurs upon indolization of 2,6-dihalogenophenylhydrazones, a 1,2-migration of a methyl group occurred to produce ethyl 4,7-dimethylindole-2-carboxylate (**440**; $R^1 = R^3 = H$, $R^2 = COOC_2H_5$) and in one instance, which could not be repeated or explained, the methyl group migrated to the indolic 3-position to give ethyl 3,7-dimethylindole-2-carboxylate (Carlin, Henley and Carlson, 1957). These results were corroborated when acetophenone 2,6-dimethylphenylhydrazone (**439**; $R^1 = R^3 = H$, $R^2 = C_6H_5$) was heated with zinc chloride in nitrobenzene, a reaction which led to a reaction product from which could be isolated a trace of a solid $C_{24}H_{25}NO_2$ of unassigned structure, 2,6-dimethylaniline (**441**) (8%), acetophenone (trace), 4,7-dimethyl-2-phenylindole (**440**; $R^1 = R^3 = H$, $R^2 = C_6H_5$) (4%) and **442** (33%) (Carlin and Carlson, 1957, 1959).

To explain the formations of **440** ($R^1 = R^3 = H$, $R^2 = COOC_2H_5$ and C_6H_5) and **442**, a general mechanism was postulated, analogous to that earlier

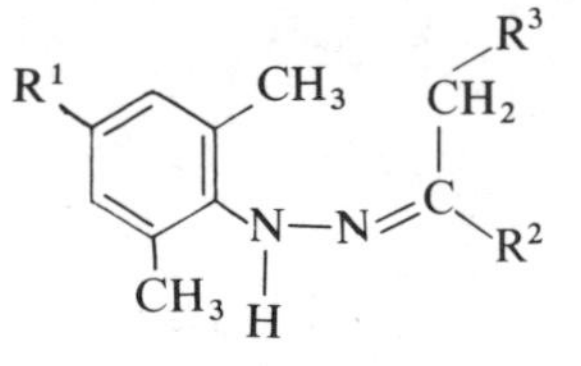

439

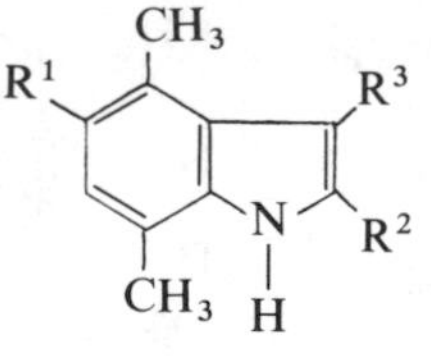

440

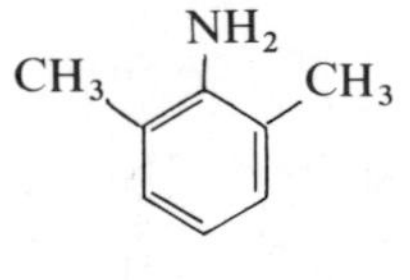

441

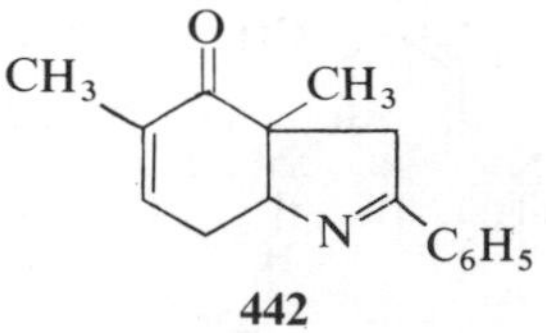

442

postulated and already described to explain halogen migrations. The hydrazone **439**, via its enehydrazine tautomer **443**, forms the key intermediate in the mechanism (Scheme 25), the dienone-imine **444**, which can react either by route (i), involving 1,2-migration of a methyl group, to afford the intermediate **446**, or by route (ii), respectively. This proposal was supported by the later observation (Carlin, Magistro and Mains, 1964) that, when using acetophenone $^{15}N_{\beta}$-2,6-dimethylphenylhydrazone, the product **447** ($R^1 = R^3 = H$, $R^2 = C_6H_5$) retained the full ^{15}N label, whereas only a trace of the label was found in **440** ($R^1 = R^3 = H$, $R^2 = C_6H_5$).

When the indolization of cyclohexanone 2,4,6-trimethylphenylhydrazone [**439**; $R^1 = CH_3$, $R^2 + R^3 = (CH_2)_4$] [for syntheses of 2,4,6-trimethylphenylhydrazine, see Franzen, Onsager and Faerden (1918) and Hunsberger, Shaw, E. R., Fugger, Ketcham and Lednicer (1956)] was first investigated using boiling acetic acid as catalyst, structure **448** was assigned to the product (Barnes, Paucacker and Badcock, 1951). However, it was later found that this product was 1,2,3,4-tetrahydro-6,7,8-trimethylcarbazole (**449**; $R^1 = H$, $R^2 = CH_3$) by dehydrogenation to afford 1,2,3-trimethylcarbazole which was unambiguously synthesized (Carlin and Moores, 1959, 1962; Miller, B. and Matjeka, 1980). Although the formation of **449** ($R^1 = H$, $R^2 = CH_3$) could result from three 1,2-shifts in the intermediate **445** [$R^1 = CH_3$, $R^2 + R^3 = (CH_2)_4$], the possibility of a direct 1,4-migration via a transition state **450** was favoured but nevertheless admitted to be 'unconventional' (Carlin and Moores, 1959, 1962). Such a direct 1,4-shift was, indeed, later (Dewar, 1963; Shine, 1967b) thought to be improbable, mainly because of the unlikelihood of the formation of the transition state. The main reason for favouring a direct 1,4-migration of the methyl group had been that no 1,2,3,4-tetrahydro-5,6,8-trimethylcarbazole (**449**; $R^1 = CH_3$, $R^2 = H$) could be isolated from the indolization. This was expected to be formed as a result of loss of proton from **446** [$R^1 = CH_3$, $R^2 + R^3 = (CH_2)_4$], the intermediate which would result from the first 1,2-methyl group shift in a mechanism involving three such consecutive shifts (Carlin and Moores, 1962). However, it was later (Dewar, 1963) suggested that the steric interaction between the cyclohexyl group and the migrating methyl group might well encourage further migration rather than aromatization. Such steric interactions would be absent in **446** ($R^1 = R^3 = H$, $R^2 = C_6H_5$) which could therefore lose a proton to ultimately give 4,7-dimethyl-2-phenylindole as already described. Such a suggestion would also invalidate the argument (Thomas, T. J., 1975) that the failure to detect any 1,2,3,4-tetrahydro-5,8-dimethylcarbazole along with the 11–14% of **451** and 80–85% of **452** from the indolization of cyclohexanone 2,6-dimethylphenylhydrazone (a reaction which is discussed below in detail) renders unlikely the possibility that the 1,4-migration of methyl groups occurs by a series of three 1,2-shifts. Furthermore, the extra methyl group in **446** [$R^1 = CH_3$, $R^2 + R^3 = (CH_2)_4$] as compared to **446** ($R^1 = R^3 = H$, $R^2 = C_6H_5$) will stabilize the

Scheme 25

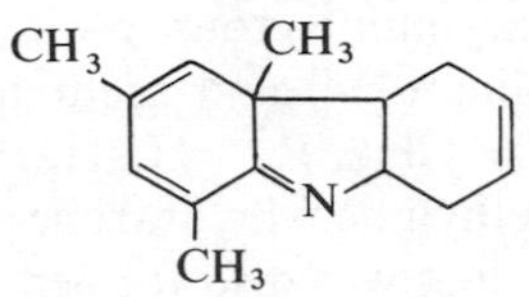

448

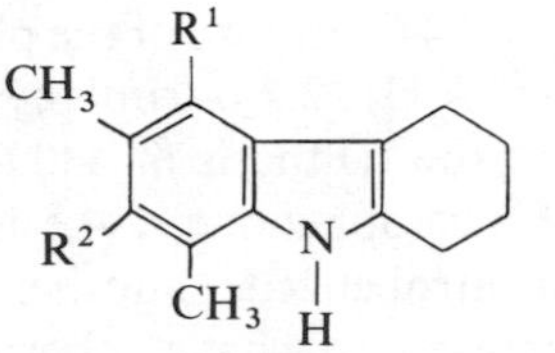

449

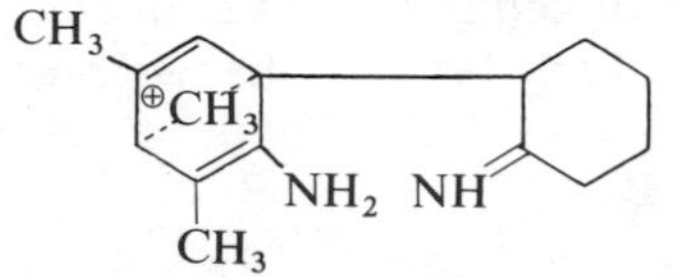

450

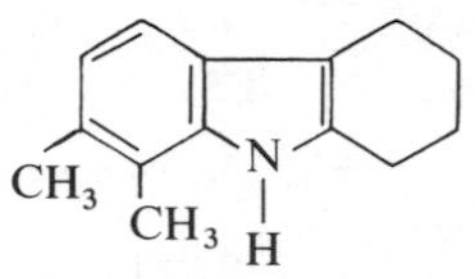

451

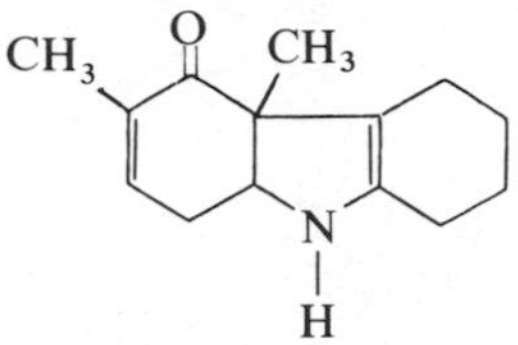

452

benzenonium ion and therefore discourage proton removal and the very different catalytic conditions under which the two indolizations involving these proposed intermediates were effected could well have differential influences upon the rates of rearrangement of and proton loss from these intermediates (Dewar, 1963). Subsequent work also supported the occurrence of three 1,2-shifts (Miller, B. and Matjeka, 1977) (see later in this chapter).

Contrary to the above observation, acetophenone 2,4,6-trimethylphenylhydrazone (**439**; $R^1 = CH_3$, $R^2 = C_6H_5$, $R^3 = H$) yielded no product resulting from a 1,4-migration of one of the *ortho* methyl groups when it was indolized using zinc chloride in nitrobenzene. The products isolated from this reaction (Carlin and Harrison, 1965) were 4,5,7-trimethyl-2-phenylindole (**440**; $R^1 = CH_3$, $R^2 = C_6H_5$, $R_3 = H$), resulting from a 1,2-migration of the methyl group in the intermediate dienone-imine **444** ($R^1 = CH_3$, $R^2 = C_6H_5$, $R^3 = H$), 2,4,6-trimethyl-3-phenacylaniline (**453**), resulting from a

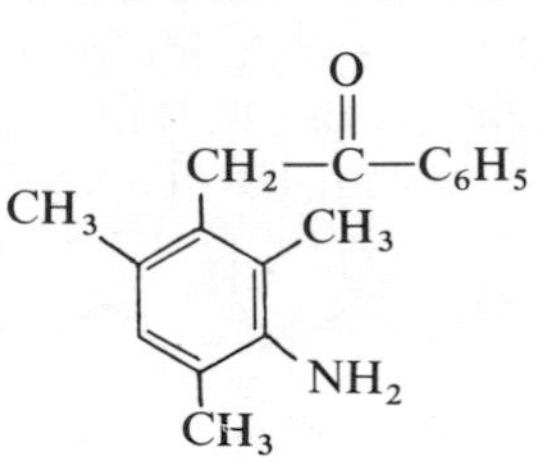

453

1,2- (or possibly 1,4-) migration of the phenacylimino group in **444** ($R^1 = CH_3$, $R^2 = C_6H_5$, $R^3 = H$), 2,4,6-trimethylaniline (**454**) and acetophenone. Furthermore, during the synthesis of **440** ($R^1 = CH_3$, $R^2 = C_6H_5$, $R^3 = H$) by indolization of acetophenone 2,4,5-trimethylphenylhydrazone (**455**) using zinc chloride in nitrobenzene, another product was also formed, the spectral data of which showed it to have either structure **456** or **457**, both of which are feasible with regard to Scheme 25 (Carlin and Harrison, J. W., 1965). Unfortunately, a p.m.r. spectral comparison was not made with **442**, since this should have differentiated clearly between the two possible structures.

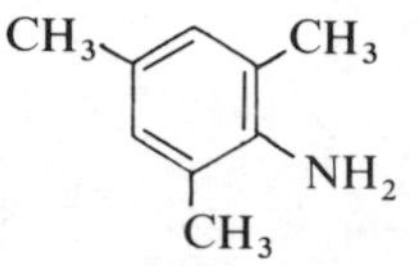

454 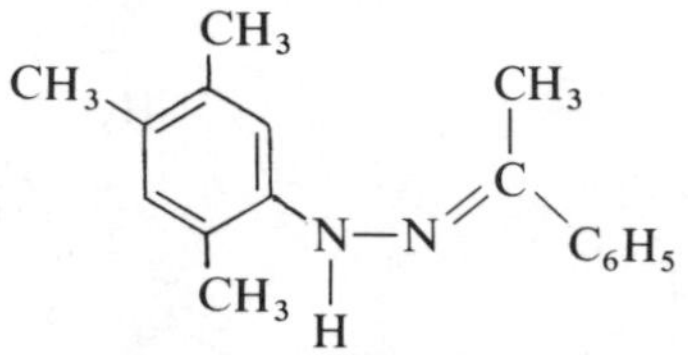 **455**

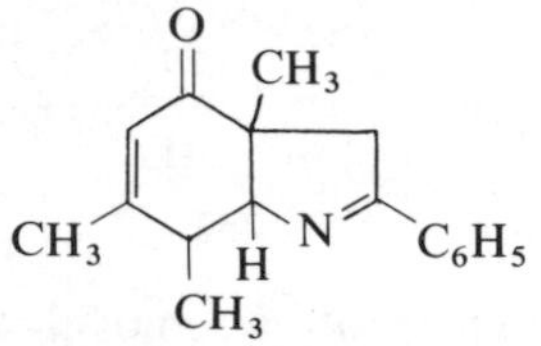

456 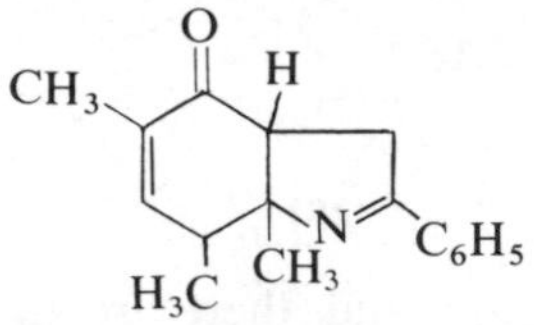 **457**

The indolization of cyclohexanone 2,6-dimethylphenylhydrazone [**439**; $R^1 = H$, $R^2 + R^3 = (CH_2)_4$] has been the subject of extensive investigation by Carlin and various of his students. Initially (Harrison, 1962) the only product isolated was 1,2,3,4-tetrahydro-7,8-dimethylcarbazole (**451**) (1.5% yield). Using a 25% polyphosphoric acid–75% glacial acetic acid catalyst, the products **451** (<1%), **452** (*ca.* 1%) and 2,6-dimethylaniline (**441**) (13.5%) (Goel, 1966) were isolated. Attempts to optimize the reaction conditions led to the isolation of a maximum yield of only 3.5% of **451** after warming the hydrazone in acetic acid at 87 °C (boiling trichloroethylene) for 1 h. Ultimately, following the possibility that aerial oxidation may be responsible for these low product yields, the indolization was carried out by heating at 87 °C in glacial acetic acid containing a suspension of sodium bisulphite as an oxygen and/or peroxide scavenger. This led to the isolation of **451** (11–14%) and **452** (80–85%), spectroscopic investigation of the latter product indicating that it existed in tautomeric equilibrium with **458** (Thomas, T. J., 1975). The observed

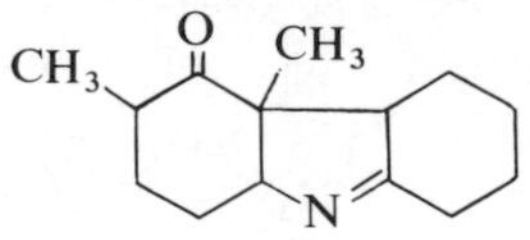

458

sensitivity of **451** and **452** to oxygen was probably responsible for the low product yields reported in many of the related above indolizations (Thomas, T. J., 1975). The effects of reaction medium and structure of the ketonic moiety on the course of the indolizations of 2,6-dimethylphenylhydrazones have been studied (Goel, 1966) and summarized (Thomas, T. J., 1975). A zinc chloride catalyst in nitrobenzene favours 1,2-methyl migration, glacial acetic acid favours 1,4-methyl migration, hydrochloric acid favours methyl group elimination and polyphosphoric acid appears to be largely non-specific in these respects. Using acetophenone and cyclohexanone, 1,2- and 1,4-methyl migration, respectively, are promoted. However, acetophenone and cyclohexanone hydrazones have frequently been indolized using zinc chloride in nitrobenzene and acetic acid, respectively, so either structural or solvent effects, or both, may be involved.

Carlin and his co-workers (Carlin, Henley and Carlson, 1957) attempted to trap, as the Diels–Alder adduct, the postulated dienone-imine intermediate during the zinc chloride catalysed indolization of ethyl pyruvate 2,6-dimethylphenylhydrazone by adding maleic anhydride to the reaction mixture. This technique had previously (Conroy and Firestone, 1953, 1956; Curtin and Crawford, 1957) been successful in isolating the corresponding intermediate in the Claisen rearrangement but no material indicative of such an adduct was isolated from the indolization under these conditions although the expected ethyl-4,7-dimethylindole-2-carboxylate was formed. It is possible that the intermediate is too reactive in the indolization pathway to form such an adduct, especially under the vigorous catalytic conditions used. In view of this failure, R. K. Brown and his co-workers performed the indolization of several enehydrazines which were formed by reacting 2,6-dichloro- or 2,6-dimethyl-N_β-methylphenylhydrazine with various ketones. It was hoped that by so locking the enehydrazine–hydrazone tautomerism in favour of the enehydrazine, the first intermediate in the mechanism, it might be possible to use very mild indolization conditions under which the 2,6-substituents may block the synthesis at the dienone-imine stage. The results of their work using chlorine atoms as the 2,6-substituents and effecting indolization in benzene at room temperature with no catalyst have been discussed earlier. It did not lead to the isolation of the dienone-imine intermediates but the structures of the products which were isolated could only be rationalized on the basis of the intervention of such an intermediate in their formation (Robinson, F. P. and Brown, R. K., 1964).

Since the halogen atom was so easily lost in the above investigation, the reactions between N_β,2,6-trimethylphenylhydrazine hydrochloride (the free base is somewhat unstable) and various ketones with subsequent reaction of the corresponding enehydrazines in boiling benzene were investigated. Using cyclohexanone as the ketonic moiety, the products from the reaction were identified as ammonium chloride and 1,2,3,4-tetrahydro-8,9-dimethylcarbazole

(**459**; $R^1 = R^2 = CH_3$). Similarly, when cyclohexanone and 2,6-diethyl-N_β-methylphenylhydrazine were reacted together the reaction product yielded ammonium chloride and 8-ethyl-1,2,3,4-tetrahydro-9-methylcarbazole (**459**; $R^1 = C_2H_5$, $R^2 = CH_3$), together with a trace of methylammonium chloride, and when cyclohexanone was reacted with N_β-ethyl-2,6-dimethylphenylhydrazine hydrochloride there were formed ammonium chloride, 9-ethyl-1,2,3,4-tetrahydro-8-methylcarbazole (**459**; $R^1 = CH_3$, $R^2 = C_2H_5$) and a trace of

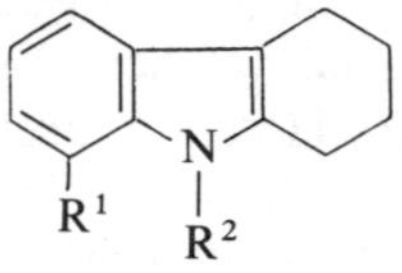

N R¹ R²

459

ethyl ammonium chloride (Bajwa and Brown, R. K., 1968a). Clearly, one of the original *ortho* alkyl groups was lost during these reactions, although its fate was unknown. Although the corresponding proposed dienone-imine intermediates **460** were not isolated, their formation was implied in the mechanism (Scheme 26) suggested in order to account for the reaction products (Bajwa and Brown, R. K., 1968a). Structure **461** could, after protonation, lose the angular methyl group as $R^{1\oplus}$ to give rise to the more stable 8-alkyl-1,2,3,4-tetrahydrocarbazole **462** (Brown, R. K., 1972a). The major reaction pathway is (i) which is in agreement both with the experimental results and with the facts that it involves the loss of the labile hydrogen atom in **460** in effecting ring closure and that the final loss of $R^{1\oplus}$ gives rise to an aromatic system (Bajwa and Brown, R. K., 1968a). Although in the published studies of Carlin and his co-workers, elimination of methyl groups was not observed, analogous methyl group eliminations had been observed (Huisgen, 1948a) (i.e. **433** → **434** and **435** → **436**). Further, one of Carlin's students reported that the indolization of **463** in ethanolic hydrochloric acid gave 1,2,3,4-tetrahydro-8,9-dimethylcarbazole (**459**; $R^1 = R^2 = CH_3$) in 64–83% yield (Laufer, 1958). When isobutyraldehyde N_β,2,6-trimethylphenylhydrazone hydrochloride (**464**; $R = CH_3$) was heated in benzene it afforded **466** ($R = CH_3$) which represents an arrest of the indolization pathway immediately succeeding the dienone-imine **465** ($R = CH_3$) formation (Bajwa and Brown, R. K., 1968b). Using the propionaldehyde analogue, a similar arrest occurred and the reaction furnished **466** ($R = H$) after 10–12 min reaction time. When this product was heated under reflux in benzene for 20h, a 3:1 mixture of ammonium chloride and methylamine hydrochloride, along with a 36% yield of 1,3,7-trimethylindole (**469**; $R = CH_3$), was obtained. Spectral investigation of the reaction residues also indicated the presence of 3,7-dimethylindole (**469**; $R = H$) but this could not be isolated (Bajwa and Brown, R. K., 1969). Scheme 27 was suggested (Bajwa and Brown, R. K., 1969) to account for these observations. Since there is considerable steric

·HCl

·HCl

460

(i)

(ii)

$(-NH_3 \rightarrow NH_4Cl)$

$(-NH_2R^2 \rightarrow R^2\overset{\oplus}{N}H_3Cl^{\ominus})$

461

459

262

Scheme 26

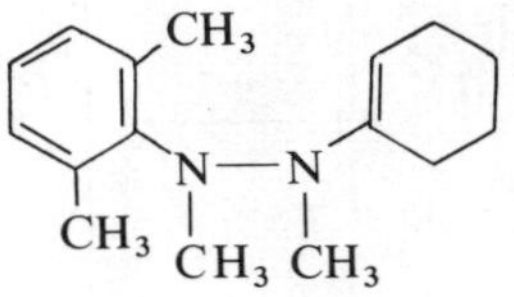

crowding around the ring junction in **468**, relative to **467**, it undergoes elimination more readily than species **467**, this accounting for the relative yields of products from the reaction (Bajwa and Brown, R. K., 1969). Reaction of 2-methylcyclohexanone with N_{β},2,6-trimethylphenylhydrazine hydrochloride in boiling benzene afforded approximately equal quantities of ammonium chloride and methylamine hydrochloride and 1,2,3,4-tetrahydro-1,8,9-trimethylcarbazole (**472**; R = CH_3). These results were mechanistically rationalized (Bajwa and Brown, R. K., 1970) as shown in Scheme 28. The isolation of methylamine hydrochloride suggests that the alternative direction of cyclization in **471** may have occurred, although no **472** (R = H) could be isolated. Furthermore, no products resulting from the alternative enehydrazine **470** could be isolated, possibly because the rearrangement of this enehydrazine to form the dienone-imine would be sterically inhibited. In support of this suggestion it was found (Bajwa and Brown, R. K., 1970) that a mixture of 2,6-dimethylcyclohexanone and N_{β},2,6-trimethylphenylhydrazine hydrochloride failed to react at all when boiled for 41 h in benzene.

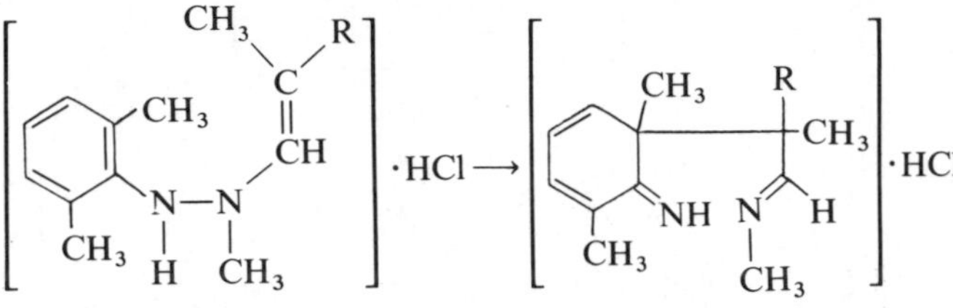

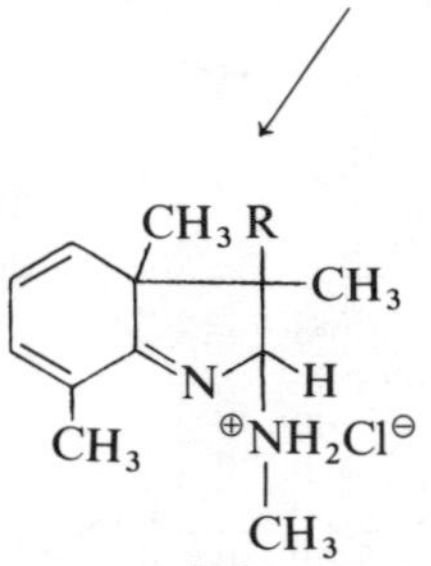

It has been suggested (Geller, 1978) that the 1-methylindoles produced in the above studies of R. K. Brown and Bajwa could arise, for example when starting with **464** (R = H), via a 1,3-migration of the angular methyl group in **466**

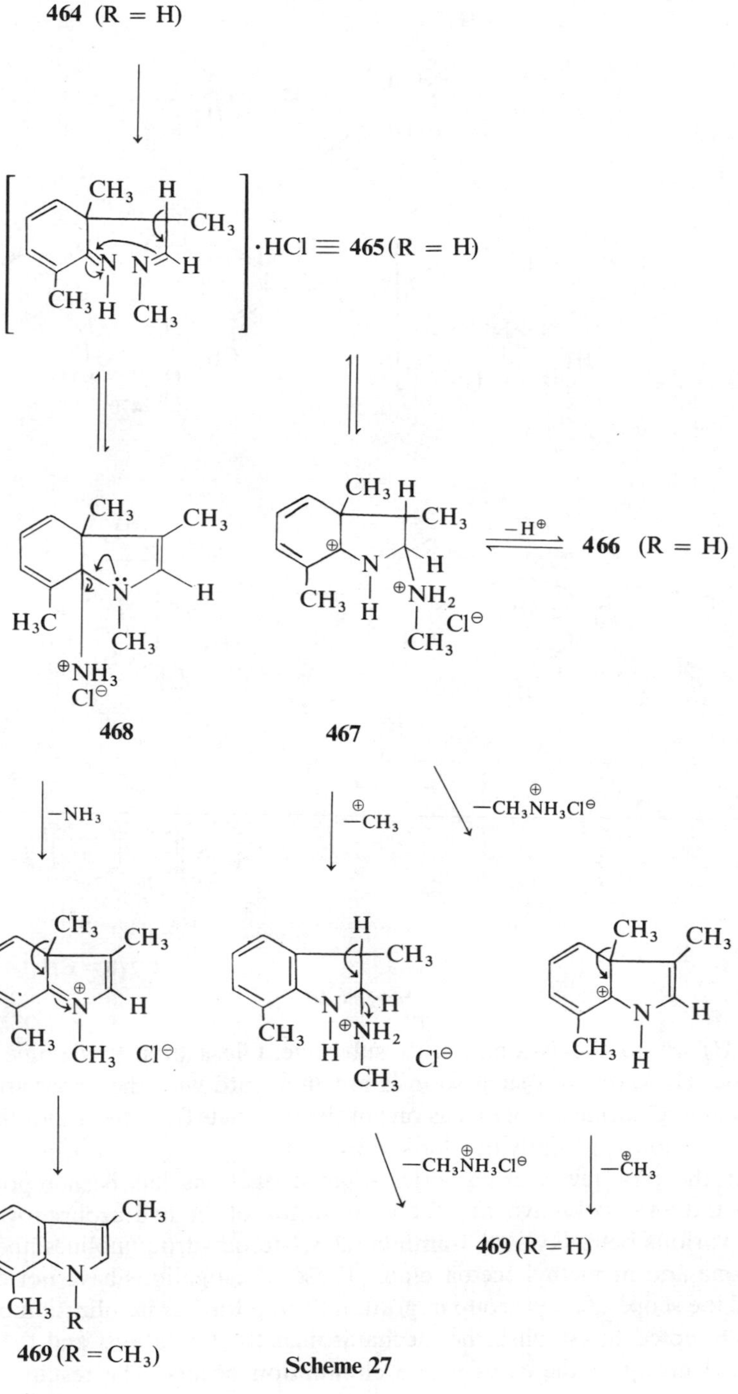

464 (R = H)

·HCl ≡ **465** (R = H)

$-H^{\oplus}$ **466** (R = H)

468

467

$-NH_3$

$-\overset{\oplus}{C}H_3$

$-CH_3\overset{\oplus}{N}H_3Cl^{\ominus}$

$-CH_3\overset{\oplus}{N}H_3Cl^{\ominus}$

$-\overset{\oplus}{C}H_3$

469 (R = H)

469 ($R = CH_3$)

Scheme 27

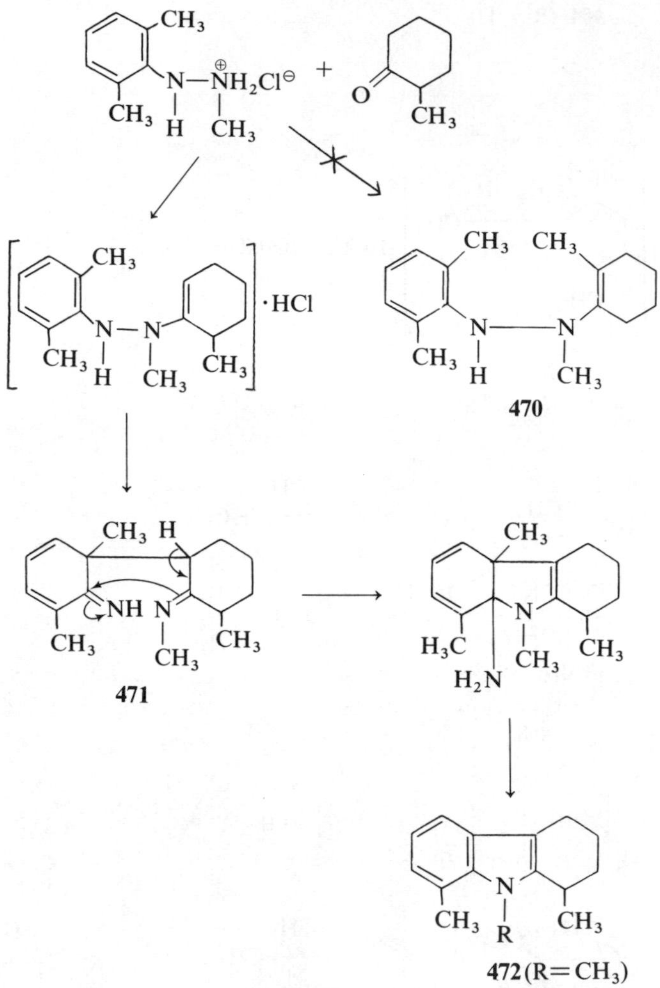

Scheme 28

(R = H) on to the N-atom, with subsequent loss of methylamine hydrochloride. However, this suggestion is not in accord with the observation that ammonium chloride is isolated as one of the products from these indolizations, an observation apparently overlooked by Geller.

Over the past few years a series of investigations has been reported by Fusco and Sannicolò involving the indolization of the hydrazones formed between various benzalkylated 1-amino-1,2,3,4-tetrahydroquinolines and cyclohexanone and/or methyl acetoacetate. These investigations have not only extended the scope of alkyl group migration during Fischer indolization but have also attempted to establish the mechanism of the migrations and the fate of the alkyl group in the cases where elimination occurs. The results of these indolizations are partially summarized in Table 2, brief refluxing with hydrochloric or sulphuric acid in methanol and brief heating in acetic acid being used to indolize the methyl acetoacetate and cyclohexanone hydrazones, respectively.

Table 2. The Indolic Products Resulting from the Acid Catalysed Transformations of the Hydrazones formed between Cyclohexanone and/or Methyl Acetoacetate and Various Benz-alkylated 1-Amino-1,2,3,4-tetrahydroquinolines

Starting Hydrazone	Product(s)	Route of Formation (see Scheme 29)	Reference
473 [$R^1 = CH_3$, $R^2 = R^3 = H$, $R^4 + R^5 = (CH_2)_4$]	**474** [$R^1 = CH_3$, $R^2 = R^3 = H$, $R^4 + R^5 = (CH_2)_4$]	(ii)	1
	475 [$R^1 = CH_3$, $R^2 = R^3 = H$, $R^4 + R^5 = (CH_2)_4$]	(i)[b]	1
473 ($R^1 = R^4 = CH_3$, $R^2 = R^3 = H$, $R^5 = COOCH_3$)	**474** ($R^1 = R^4 = CH_3$, $R^2 = R^3 = H$, $R^5 = COOCH_3$)	(ii)	1
	474 ($R^1 = R^4 = CH_3$, $R^2 = R^3 = R^5 = H$)	(ii)[c]	1
	474 ($R^1 - R^3 = H$, $R^4 = CH_3$, $R^5 = COOCH_3$)	(iii)[d]	1
	475 ($R^1 = R^4 = CH_3$, $R^2 = R^3 = H$, $R^5 = COOCH_3$)	(i)[b]	1
	475 [$R^1 = (CH_2)_3NH_2$, $R^2 = R^3 = H$, $R^4 = CH_3$, $R^5 = COOCH_3$]	(v)	1
473 [$R^1 = R^2 = CH_3$, $R^3 = H$, $R^4 = R^5 = (CH_2)_4$]	**474** [$R^1 = R^2 = CH_3$, $R^3 = H$, $R^4 + R^5 = (CH_2)_4$]	(iv)[e]	1
	474 [$R^1 = R^3 = CH_3$, $R^2 = H$, $R^4 + R^5 = (CH_2)_4$]	(vi)	1
	475 [$R^1 = R^2 = CH_3$, $R^3 = H$, $R^4 + R^5 = (CH_2)_4$]	(i)[b]	1
473 ($R^1 = R^2 = R^4 = CH_3$, $R^3 = H$, $R^5 = COOCH_3$	**474** ($R^1 = R^2 = R^4 = CH_3$, $R^3 = H$, $R^5 = COOCH_3$)	(iv)[e]	2
473 [$R^1 = R^3 = CH_3$, $R^2 = H$, $R^4 + R^5 = (CH_2)_4$]	**474** [$R^1 = R^2 = CH_3$, $R^3 = H$, $R^4 + R^5 = (CH_2)_4$]	(ii)	2

(continued)

Table 2 (*continued*)

Starting Hydrazone	Product(s)	Route of Formation (see Scheme 29)	Reference
473 [$R^1 = R^3 = CH_3$, $R^2 = H$, $R^4 + R^5 = (CH_2)_4$]	**474** [$R^1 = H$, $R^2 = R^3 = CH_3$, $R^4 + R^5 = (CH_2)_4$]	(vi)	1
	475 [$R^1 = R^3 = CH_3$, $R^2 = H$, $R^4 + R^5 = (CH_2)_4$]	(i)[b]	1
473 ($R^1 = R^3 = R^4 = CH_3$, $R^2 = H$, $R^5 = COOCH_3$)	**474** ($R^1 = R^2 = R^4 = CH_3$, $R^3 = H$, $R^5 = COOCH_3$)	(ii)	1
	474 $R^1 = R^3 = H$, $R^2 = R^4 = CH_3$, $R^5 = COOCH_3$)	(iii)	1
	475 ($R^1 = R^3 = R^4 = CH_3$, $R^2 = H$, $R^5 = COOCH_3$)	(i)[b]	1
473 [$R^1 + R^2 = (CH_2)_4$, $R^3 = H$, $R^4 = CH_3$, $R^5 = COOCH_3$]	**474** [$R^1 + R^2 = (CH_2)_4$, $R^3 = H$, $R^4 = CH_3$, $R^5 = COOCH_3$]	(iv)[e]	1
	474 [$R^1 + R^2 = (CH_2)_4$, $R^3 = R^5 = H$, $R^4 = CH_3$]	(iv)[e,f]	3
	475 [$R^1 + R^2 = (CH_2)_4$, $R^3 = H$, $R^4 = CH_3$, $R^5 = COOCH_3$][a]	(i)[b]	3
473 [$R^1 + R^2 = (CH_2)_3$, $R_3 = H$, $R^4 = CH_3$, $R^5 = COOCH_3$]	**474** [$R^1 + R^2 = (CH_2)_3$, $R^3 = H$, $R^4 = CH_3$, $R^5 = COOCH_3$]	(iv)[e]	3
	474 [$R^1 = (CH_2)_3OCH_3$, $R^2 = R^3 = H$, $R^4 = CH_3$, $R^5 = COOCH_3$]	[g]	4
473 ($R^1 = C_6H_5$, $R^2 = R^3 = H$, $R^4 = CH_3$, $R^5 = COOCH_3$)	**474** ($R^1 = C_6H_5$, $R^2 = R^3 = R^5 = H$, $R^4 = CH_3$)	(ii)[h,i]	5

Table 2 (*continued*)

473 [R^1 = $CH_2N(C_2H_5)_2$, $R^2 = R^3$ = H, $R^4 = CH_3$, $R^5 = COOCH_3$]	**474** ($R^1 - R^3$ = H, $R^4 = CH_3$, $R^5 = COOCH_3$	(iii)[j]	5
473 ($R^1 = CH_2C_6H_5$, $R^2 = R^3$ = H, $R^4 = CH_3$, $R^5 = COOCH_3$)	**474** ($R^1 - R^3$ = H, $R^4 = CH_3$, $R^5 = COOCH_3$)[l]	(iii)[k]	1

Notes

a. The corresponding aniline, **476** [$R^1 + R^2 = (CH_2)_4$], was also isolated; b. Loss of the C_3H_7N fragment could occur via formation, by ring closure of **481**, of the propellane **477** which could then undergo acid catalysed ring opening to form **478** from which loss of the 1-aminopropyl group could occur, possibly involving attack by $Cl^{\ominus}$ ion, when using hydrochloric acid as indolization catalyst, as shown in **478**. This postulation was supported when $^{15}N_{\beta}$–**473** [$R^1 + R^2 = (CH_2)_4$, R^3 = H, $R^4 = CH_3$, $R^5 = COOCH_3$] was found to yield **475** [$R^1 + R^2 = (CH_2)_4$, R^3 = H, $R^4 = CH_3$, $R^5 = COOCH_3$] containing the total ^{15}N label when treated with methanolic hydrochloric acid and also by the isolation of 1-amino-3-chloropropane (as its benzoyl derivative) from a similar reaction utilizing unlabelled hydrazone (Fusco and Sannicolò, 1974); c. Since **474** ($R^1 = R^4 = CH_3$, $R^2 = R^3$ = H, $R^5 = COOCH_3$) was extremely resistant to base or acid catalysed hydrolysis, decarboxymethylation probably preceded indolization in the formation of **474** ($R' = R^4 = CH_3$, $R^2 = R^3 = R^5$ = H) (Fusco and Sannicolò, 1978c); d. Repetition of this indolization using hydrogen chloride in benzene as the indolization catalyst furnished four of the five indolic products [**474** ($R^1 = R^4 = CH_3$, $R^2 = R^3 = R^5$ = H) was not isolated] isolated from the methanolic hydrochloric acid catalysed reaction, along with methyl chloride formed from the trapping of the eliminated methyl group. Similar results were obtained when the corresponding ethyl ester hydrazone was treated with hydrogen chloride in benzene (Fusco and Sannicolò, 1978c); e. Evidence for the occurrence of two 1,2-shifts rather than one 1,3-shift was obtained by indolization of **473** ($R^1 = R^4 = CH_3$, $R^2 = CD_3$, R^3 = H, $R^5 = COOCH_3$) into an approximately equimolar mixture of **474** ($R^1 = R^4 = CH_3$, $R^2 = CD_3$, R^3 = H, $R^5 = COOCH_3$) and **474** ($R^1 = CD_3$, $R^2 = R^4 = CH_3$, R^3 = H, $R^5 = COOCH_3$) (Fusco and Sannicolò, 1975a, 1975b) whereas a direct 1,3-shift would have afforded only the latter product; f. Since **474** [$R^1 + R^2 = (CH_2)_4$, R^3 = H, $R^4 = CH_3$, $R^5 = COOCH_3$] was resistant to base catalysed, and therefore presumably acid catalysed hydrolysis, decarboxymethylation probably preceded indolization in the formation of **474** [$R^1 + R^2 = (CH_2)_4$, $R^3 = R^5$ = H, $R^4 = CH_3$] (Fusco and Sannicolò, 1973); g. Results from solvolysis (indolization effected in methanolic sulphuric acid) of the strained intermediate **479** which arises from the first 1,2-shift (Fusco and Sannicolò, 1975c); h. Using 0.1M perchloric acid solution in anhydrous acetic acid as catalyst; i. Since **474** ($R^1 = C_6H_5$, $R^2 = R^3$ = H, $R^4 = CH_3$, $R^5 = COOCH_3$) was very resistant to hydrolysis, decarboxymethylation probably preceded indolization (Fusco and Sannicolò, 1976a); j. Anticipated since the eliminated cation, $^{\oplus}CH_2N(C_2H_5)_2$, would be resonance stabilized. It is also probable that this cation is a precursor of the pyridine **480** which was also isolated from this reaction (Fusco and Sannicolò, 1976a, 1978c); k. The eliminated benzyl cation would be resonance stabilized (Fusco and Sannicolò, 1978c); l. 8-Benzyl-1,2,3,4-tetrahydroquinoline (**476**; R^1 = H, $R^2 = CH_2C_6H_5$) was also isolated.

References

1. Fusco and Sannicolò, 1978c; 2. Fusco and Sannicolò, 1975a; 3. Fusco and Sannicolò, 1973; 4. Fusco and Sannicolò, 1975c; 5. Fusco and Sannicolò, 1976a).

The results in Table 2 and those discussed immediately below show that five different transformations of the cyclohexadienoneimine intermediate, **482**, which results from a [3,3]-sigmatropic shift involving the 8-position of the 1,2,3,4-tetrahydroquinoline nucleus in **473**, can occur (Scheme 29), whereas when the [3,3]-sigmatropic shift involves the other *ortho* carbon atom, the intermediate **481** can indolize by loss of the propylamine chain (Scheme 29) (Fusco and Sannicolò, 1978c). These studies have been reviewed (Fusco and Sannicolò, 1980).

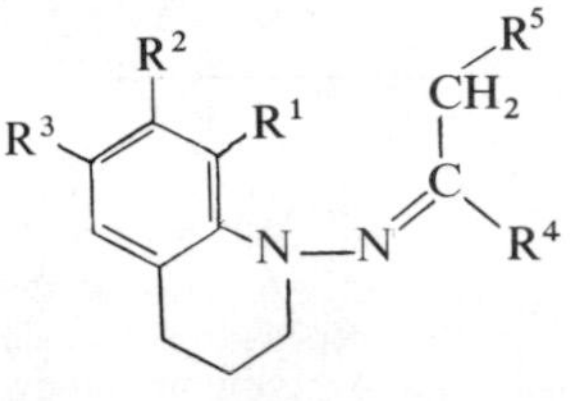

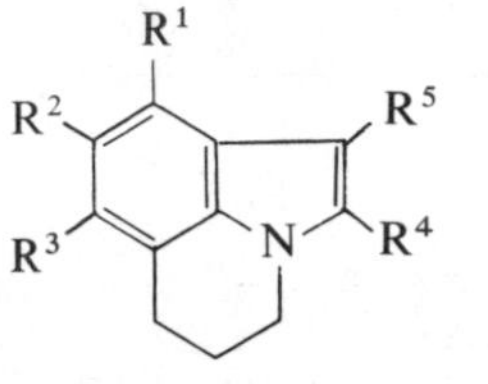

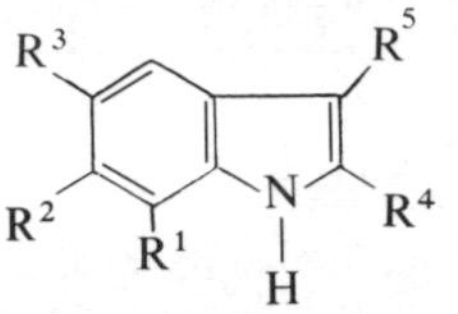

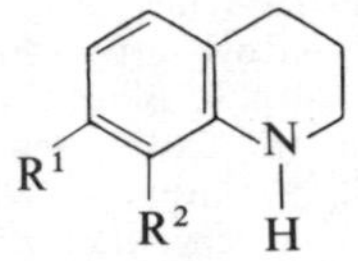

In order to investigate the nature of the 1,4-migration of the alkyl group during indolization, the hydrazone **473** [$R^1 = CH_3$, $R^2 = H$, $R^3 = CD_3$, $R^4 + R^5 = (CH_2)_4$] was subjected to indolization followed by chloranil dehydrogenation. This resulted in the formation of only **483** which led to the suggestion (Fusco and Sannicolò, 1975b) that this migration is not the result of three 1,2-shifts as earlier suggested (Dewar, 1963). However, the invalidity of this conclusion was later (Fusco and Sannicolò, 1978b) recognized when it was realized that the only conclusion that could be drawn from this observation was that the migration did not involve at any stage the attachment of the migrating CH^3 group to the carbon atom bearing the CD^3 group. Indeed, [1,4]-sigmatropic shifts of alkyl groups in carbonium ions have been indicated

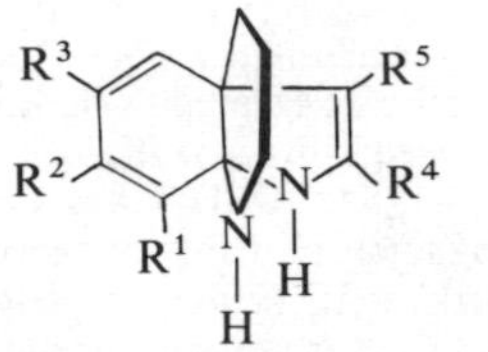

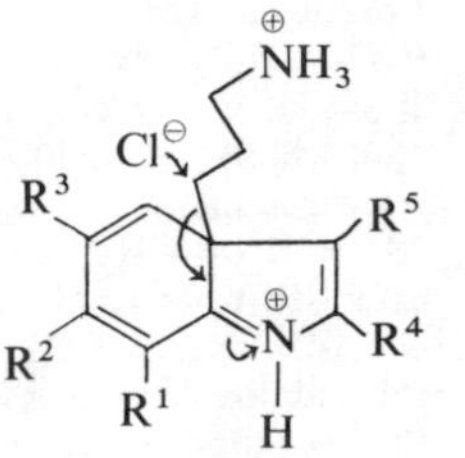

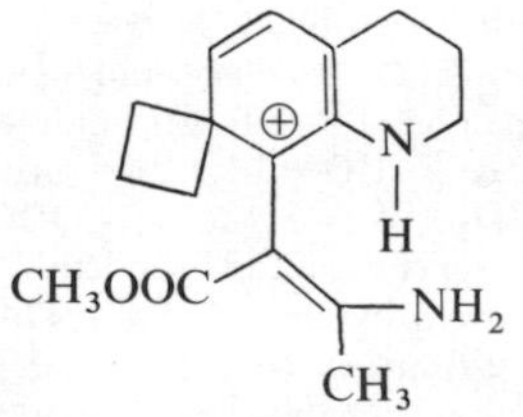

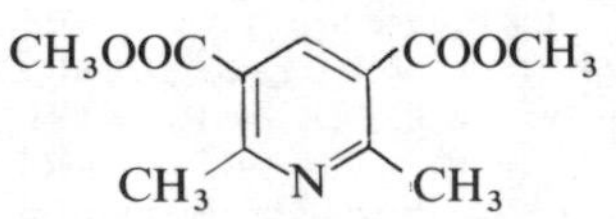

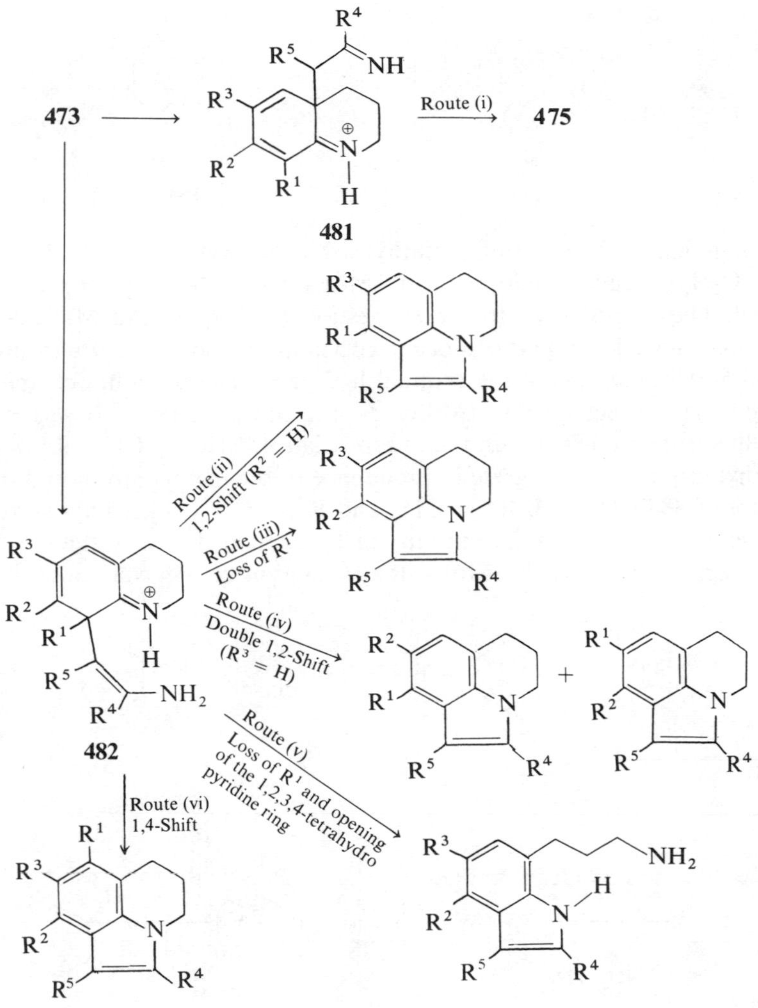

Scheme 29

to be improbable (Dewar, 1963; Shine, 1967) (see earlier in this chapter) or even theoretically forbidden (Miller, B. and Matjeka, 1977). Evidence that the alkyl group 1,4-migrations during Fischer indolization do occur via three 1,2-shifts was obtained by indolization of cyclohexanone 2-ethyl-6-methylphenylhydrazone followed by chloranil dehydrogenation, which afforded the carbazoles **484** ($R^1 = H$, $R^2 = C_2H_5$, $R^3 = CH_3$; $R^1 = CH_3$, $R^2 = H$, $R^3 = C_2H_5$ and $R^1 = C_2H_5$, $R^2 = H$, $R^3 = CH_3$), resulting from alkyl migration, and **484** ($R^1 = R^2 = H$, $R^3 = CH_3$ and C_2H_5), resulting from alkyl elimination.

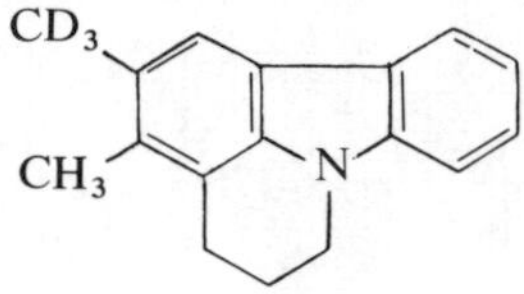

483

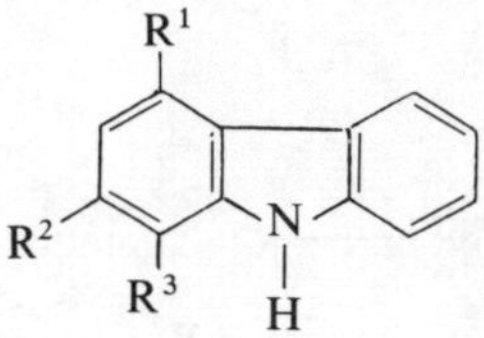

484

Very significantly, no 1-ethyl-2-methylcarbazole (**484**; $R^1 = H$, $R^2 = CH_3$, $R^3 = C_2H_5$), which would result from a direct 1,4-methyl migration, was formed. These results led to the suggestion (Miller, B. and Matjeka, 1977, 1980) that these 1,4-migrations occurred via a series of 1,2-shifts through the 1- and 6-carbon atoms of the original hydrazone as shown in Scheme 30, or possibly via a direct 1,5-shift (Miller, B. and Matjeka, 1980). It was expected from literature models that in the intermediate **486** ($R^1 = CH_3$, $R^2 = C_2H_5$) the ethyl group would migrate in preference to the methyl group and thus the absence of **484** ($R^1 = H$, $R^2 = CH_3$ and $R^3 = C_2H_5$) in the above reaction product is explicable. What remains to be explained is why these 1,2-shifts should apparently initially involve the formation of **485** when a shift in the

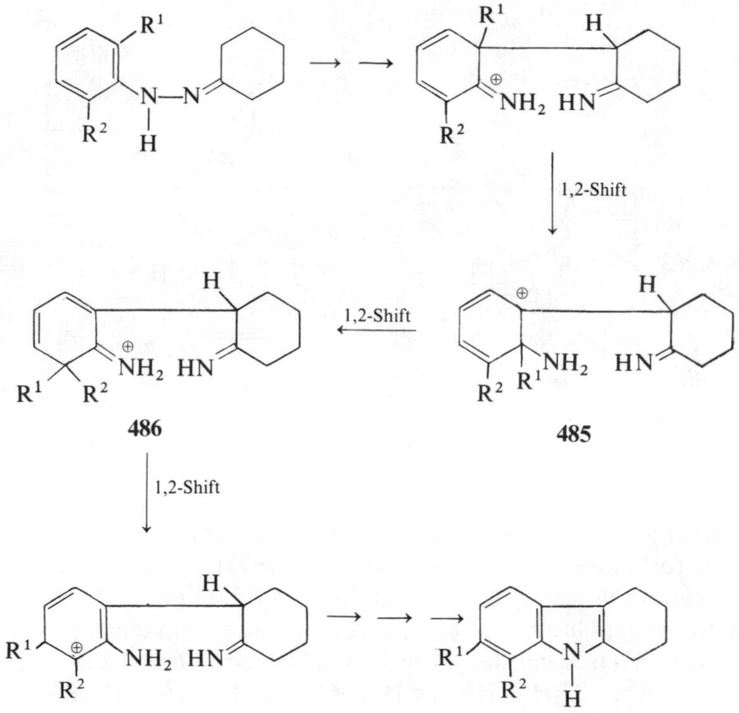

1,2-Shift

1,2-Shift

486

485

1,2-Shift

Scheme 30

alternative direction would give **487** which would be expected to be much more stable (Miller, B. and Matjeka, 1977, 1980). However, this theory does offer an explanation why overall 1,4-methyl migrations, which not only occur with cyclohexanone but also when 2,3-diethylketone 2,6-dimethylphenylhydrazone affords 2-ethyl-3,5,6,7-tetramethylindole, occur, whereas using the methyl ketonic (e.g. acetone, acetophenone and ethyl acetoacetate) and aldehydic

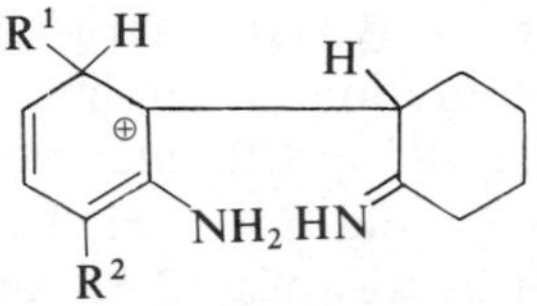

487

moieties referred to above, such migrations did not occur. What appears to be necessary for the occurrence of ultimate 1,4-migration is that the benzenoid ring should interact in the [3,3]-sigmatropic shift with an enehydrazine carbon atom which is not derived from a methyl group in the original hydrazone. The intermediate then resulting from a 1,2-shift onto the vinyl carbon atom leading to 1,2- and 1,3-migrations would then have structure **488**. In this there is much more steric crowding than in the intermediate **489** which results from a 1,2-shift on to the imino carbon atom and which leads to ultimate 1,4-migration (Miller, B. and Matjeka, 1977, 1980). However other factors are probably also involved and their elucidation awaits further study (Miller, B. and Matjeka, 1980).

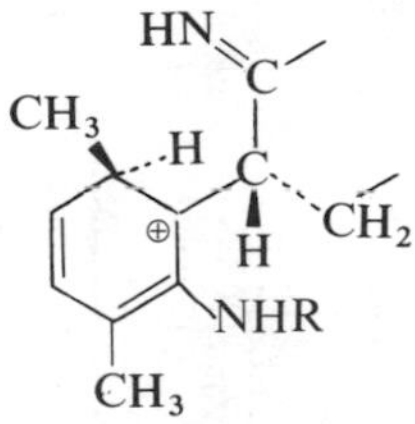

488 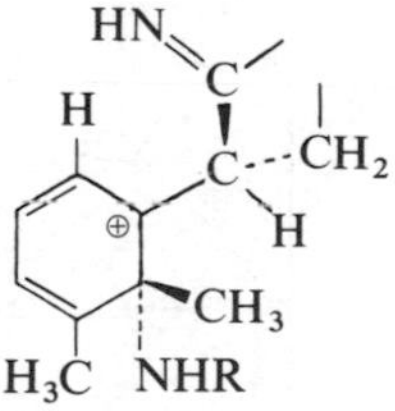 **489**

The mechanism in Scheme 30 is consistent with the recent observations (Fusco and Sannicolò, 1978b) that the indolization of cyclohexanone 2,3-dimethylphenylhydrazone and cyclohexanone 2,5-dimethylphenylhydrazone produced, after dehydrogenation over palladium, 1,2-, 3,4- and 1,4-dimethylcarbazole and 1,4-, 2,4- and 1,2-dimethylcarbazole, respectively. These three products from each reaction obviously result from 'normal' indolization and a 1,2- and 1,5-migration of methyl group, respectively, the latter resulting from either two 1,2-shifts according to Scheme 29 or a direct [1,5]-sigmatropic shift. Significantly, no 2,4-dimethylcarbazole was isolable from the dehydrogenation product of the former indolization product. Such a product would be derived from a direct 1,4-shift (or three 1,2-shifts) (Fusco and Sannicolò,

1978b). These studies of Fusco and Sannicolò have been reviewed (Fusco and Sannicolò, 1980).

In accordance with the above observations, indolization of **490** involved either of the benzenoid *ortho* carbon atoms in the [3,3]-sigmatropic shift and proceeded presumably via the intermediates **491** or **492**. In boiling aqueous sulphuric acid, **491** was either dealkenylated or underwent a 1,2-migration of this group, with subsequent hydration, to afford very low yields of **493** [$R^1 = R^2 = H$, $R^3 = Cl$ and $R^1 = (CH_2)_2C(OH)(CH_3)_2$, $R^2 = H$, $R^3 = Cl$, respectively]. Thermal indolization in boiling 1,2-dichlorobenzene again yielded **493** ($R^1 = R^2 = H$, $R^3 = Cl$), along with **493** [$R^1 = H$, $R^2 = C(CH_3)_2CH{=}CH_2$, $R^3 = Cl$ and $R^1 = H$, $R^2 = CH_2{-}CH{=}C(CH_3)_2$, $R^3 = Cl$], formed from **491** by way of a [3,3]- and a [1,3]-sigmatropic shift, respectively, and **493** [$R^1 = R^2 = H$, $R^3 = CH^2{-}CH{=}C(CH_3)_2$], formed by elimination of the chlorine atom from **492** (Baldwin and Tzodikov, 1977).

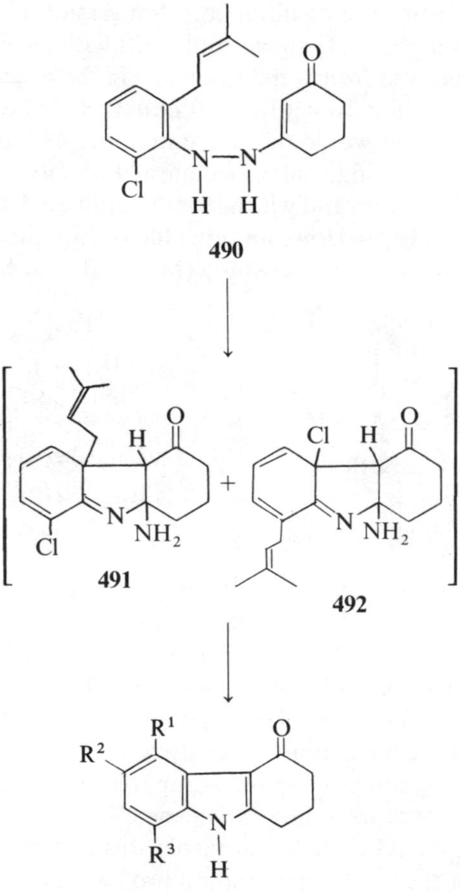

Cl N—N H H O

490

H O NH$_2$ N Cl + Cl H O N NH$_2$

491 **492**

R^1 O R^2 N R^3 H

493

Indolization of a mixture of N_α,2,5-trimethylphenylhydrazine and benzyl phenyl ketone with zinc chloride afforded a 2% yield of 1,4,7-trimethyl-2,3-diphenylindole as the only isolated product. The same hydrazine, when mixed with ethyl methyl ketone and then treated with polyphosphoric acid gave a 10% yield of 1,2,3,4,7-pentamethylindole as the only isolated product (Kricka and Vernon, 1974). It would be interesting to reinvestigate these reaction mixtures with a view to isolating possible products resulting from indolization involving the methyl substituted *ortho* carbon atom of the benzenoid nucleus, especially in view of the low yields of 'normal' indolization products obtained. Related to this is the observation that the polyphosphoric acid catalysed indolization of ethyl pyruvate 2-methylphenylhydrazone afforded ethyl 7-methylindole-2-carboxylate as the major product along with some ethyl 4-methylindole-2-carboxylate (Heath-Brown and Philpott, P. G., 1965b). In another recent example, 4-methoxyacetophenone 2,6-dimethylphenylhydrazone was indolized by treatment with polyphosphoric acid at 100–110 °C to produce a small yield of 2-(4-methoxyphenyl)-4,7-dimethylindole, along with the major product which was formed by an alternative reaction (Fusco and Sannicolò, 1978a) (see Chapter VII, P).

3. 2-Methoxy- and 2,6-Dimethoxyphenylhydrazones

The indolization of hydrazone **494** (R^1–R^3 = H) into ethyl 7-methoxyindole-2-carboxylate (**495**; R^1–R^3 = R^5 = H, R^4 = OCH_3) in low yield under unspecified conditions was reported in 1948 (Bell and Lindwall, 1948) when the corresponding N_α-methylphenylhydrazone of pyruvic acid was also indolized, again in very low yield, using an acetic acid–hydrochloric acid catalyst, into 7-methoxy-1-methylindole-2-carboxylic acid (**496**). The analogous preparation of **495** (R^1–R^3 = R^5 = H, R^4 = OC_2H_5) from ethyl pyruvate 2-ethoxyphenylhydrazone has also been described (Hughes, G. K. and Lions, 1937–1938a). Subsequently, however, indolization of **494** (R^1–R^3 = H) with ethanolic

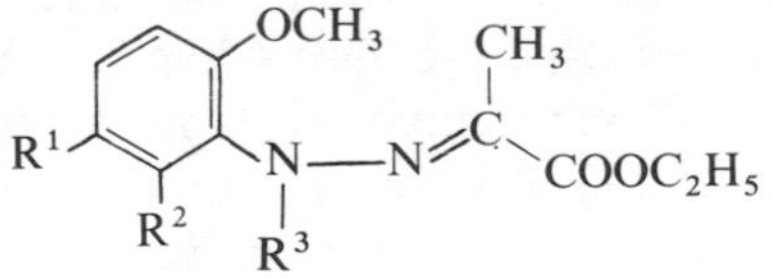

OCH_3 CH_3 R^1 N—N=C $COOC_2H_5$ R^2 R^3

494

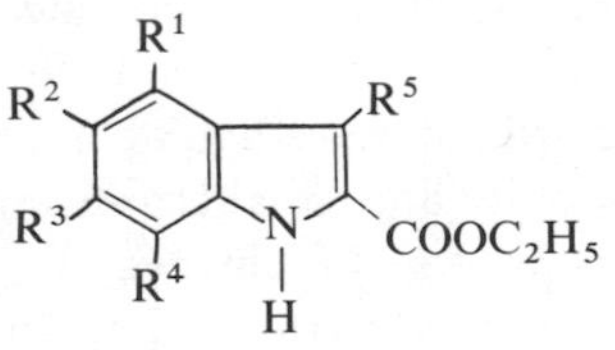

R^1 R^2 R^5 R^3 N $COOC_2H_5$ R^4 H

495

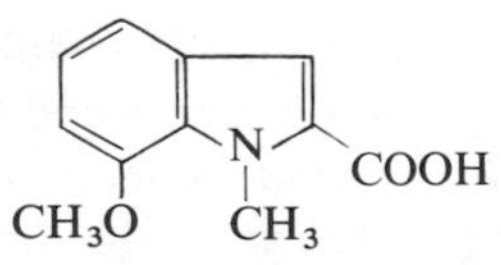

N COOH CH_3O CH_3

496

hydrogen chloride produced an unidentified indole which was not **495** (R^1–R^3 = R^5 = H, R^4 = OCH_3) (Pappalardo and Vitali, 1958b), later studies (Gannon, Benigni, Dickson and Minnis, 1969) showing that this product was most probably ethyl 6-chloroindole-2-carboxylate (**495**; R^1 = R^2 = R^4 = R^5 = H, R^3 = Cl).

A more detailed investigation (Gannon, Benigni, Dickson and Minnis, 1969) of the indolization of **494** (R^1–R^3 = H) showed that, with ethanolic hydrogen chloride as the catalyst, five products were produced. Three of these were shown to be ethyl 7-methoxyindole-2-carboxylate (**495**; R^1–R^3 = R^5 = H, R^4 = OCH_3), ethyl 3-chloroindole-2-carboxylate (**495**; R^1–R^4 = H, R^5 = Cl) and ethyl 6-chloroindole-2-carboxylate (**495**; R^1 = R^2 = R^4 = R^5 = H, R^3 = Cl), respectively, spectroscopic evidence suggested that the fourth compound was ethyl 6-ethoxyindole-2-carboxylate (**495**; R^1 = R^2 = R^4 = R^5 = H, R^3 = OC_2H_5) and from mass spectral data it was deduced that the fifth compound was an indole dimer but no structure was established. Using acetic acid–sulphuric acid as catalyst, the only recognizable product was **495** (R^1–R^3 = R^5 = H, R^4 = OCH_3) and using polyphosphoric acid no recognizable products could be isolated. Ethanolic hydrogen chloride catalysed indolization of ethyl pyruvate 2-benzyloxyphenylhydrazone produced a complex mixture of products in which **495** (R^1 = R^2 = R^4 = R^5 = H, R^3 = Cl) could be identified. Since **495** (R^1–R^3 = R^5 = H, R^4 = OCH_3) was stable in refluxing ethanolic hydrogen chloride, the indolization conditions, a mechanism of formation of **495** (R^1 = R^2 = R^4 = R^5 = H, R^3 = Cl) was suggested (Gannon, Benigni, Dickson and Minnis, 1969) involving the substitution of the methoxy group by a chlorine atom in the dienone-imine intermediate **497** and proceeding via **498**, **499** and **500** (Scheme 31). Presumably **495** (R^1 = R^2 = R^4 = R^5 = H, R^3 = OC_2H_5) could be formed in a similar manner with ethanol rather than chloride anion nucleophilically attacking **499**. Furthermore, the possibility of an additional shift involving the second double bond in **498** was recognized but attempts to detect ethyl 4-chloroindole-2-carboxylate (**495**; R^1 = Cl, R^2–R^5 = H) in the reaction mixture were unsuccessful. The mechanism of formation of ethyl 3-chloroindole-2-carboxylate was suggested to be as shown in Scheme 32 which involves, as have some other similar mechanisms already discussed in this chapter, the elimination of the N_α-atom of the original hydrazone during indolization. However, the later (Ishii, Murakami, Y., Furuse, Hosoya, Takeda and Ikeda, 1971; Ishii, Murakami, Y., Hosoya, Takeda, Suzuki, Y. and Ikeda, 1973) observation that ethyl pyruvate 2-methoxy-N_α-methylphenylhydrazone (**494**; R^1 = R^2 = H, R^3 = CH_3) afforded, upon indolization, a mixture of products, one of which was ethyl 3-chloro-1-methylindole-2-carboxylate eliminated this possibility and an alternative mechanism involving the 'normal' N_β-elimination was proposed (Scheme 33).

Over the past ten years the indolization of ethyl pyruvate 2-methoxyphenylhydrazone has been intensively studied by Ishii and his co-workers using a

494 ($R^1 - R^3 = H$)

OCH_3 NH HN $COOC_2H_5$

497

$H^{\oplus}$

OCH_3 $\overset{\oplus}{N}H_2$ HN $COOC_2H_5$

498

OCH_3 H $\oplus$ NH_2 HN $COOC_2H_5$

499

$Cl^{\ominus}$

OCH_3 H Cl NH_2 HN $COOC_2H_5$

500

Cl NH_2 HN $COOC_2H_5$ ⟶ ⟶ ⟶ **495** ($R^1 = R^2 = R^4 = R^5 = H, R^3 = Cl$)

Scheme 31

OCH_3 H H $\overset{\oplus}{N}H_2$ HN $COOC_2H_5$

498

OCH_3 H_2N N H $COOC_2H_5$

OCH_3 $Cl^{\ominus}$ N $COOC_2H_5$

H Cl N $COOC_2H_5$

495 ($R^1 - R^4 = H, R^5 = Cl$)

Scheme 32

494 ($R^1 - R^3 = H$) → **497**

OCH_3 H H NH HN $COOC_2H_5$ ≡ NH OCH_3 H H HN $COOC_2H_5$

495 ($R^1 - R^3 = R^5 = H$, $R^4 = OCH_3$)

1,2-Migration of OCH_3

495 ($R^1 = OCH_3 R^2 - R^5 = H$)

1,3-Migration of OCH_3

495 ($R^1 = R^3 - R^5 = H$, $R^2 = OCH_3$)

$+H^{\oplus}$ $-NH_3$

NH_2 OCH_3 H N H $COOC_2H_5$

495 ($R^1 = NHCOCH_2COOC_2H_5$, $R^2 - R^5 = H$)

OCH_3 $\overset{\oplus}{N}$ H $COOC_2H_5$

Nucleophile (Nu) at C_6

Nu H OCH_3 N H $COOC_2H_5$

$Cl^{\ominus}$ at C_4

Cl H OCH_3 N H $COOC_2H_5$

495 ($R^1 = R^2 = R^4 = R^5 = H$, $R^3 = Cl$)($Nu = Cl^{\ominus}$)
and **495** ($R^1 = R^2 = R^4 = R^5 = H$, $R^3 = OC_2H_5$)($Nu = C_2H_5OH$)
and **501** ($R^1 = R^2 = R^4 = H$, $R^3 = Cl$)[Nu = **495** ($R^1 = R^2 = R^4 = R^5 = H$, $R^3 = Cl$)]
and **501** (R^1–$R^3 = H$, $R^4 = OCH_3$)[Nu = **495** (R^1–$R^3 = R^5 = H$, $R^4 = OCH_3$)]
and **501** ($R^1 = NHCOCH_2COOC_2H_5$, R^2–$R^4 = H$)[Nu = **495** ($R^1 = NHCOCH_2COOC_2H_5$, R^2–$R^5 = H$)]
and **501** ($R^1 = R^3 = R^4 = H$, $R^2 = OH$)[Nu = **495** ($R^1 = R^3$–$R^5 = H$, $R^2 = OH$)]

495 ($R^1 = Cl, R^2–R^5 = H$)

$Cl^{\ominus}$ at C_3 ($-{}^{\ominus}OCH_3$) → ... → **495** ($R^1–R^4 = H, R^5 = Cl$)

$Cl^{\ominus}$ at C_5 ($-{}^{\ominus}OCH_3$) → ... → **495** ($R^1 = R^3 - R^5 = H, R^2 = Cl$)

C_2H_5OH at C_7 ($-{}^{\ominus}OCH_3$) → ... → **495** ($R^1 - R^3 = R^5 = H, R^4 = OC_2H_5$)

Tosyl-O at C_5 ($-{}^{\ominus}OCH_3$) → ... → **495** ($R^1 = R^3–R^5 = H, R^2 =$ O-Tosyl)

↓

495 ($R^1 = R^3 - R^5 = H, R^2 = OH$)

Scheme 33

Table 3. Products Resulting from the Indolization of Ethyl Pyruvate 2-Methoxyphenylhydrazone

Catalyst	Products (Unless otherwise stated,		
C_2H_5OH–3M HCl	**495** ($R^4 = OCH_3$)[a,b,c]	**495** ($R^3 = Cl$)[a,b,c]	**495** ($R^3 = OC_2H_5$)[a,b,c]
C_2H_5OH–HCl gas	**495** ($R^4 = OCH_3$)[b,c]	**495** ($R^4 = OC_2H_5$)[b,c]	**495** ($R^3 = Cl$)[b,c,d]
C_2H_5OH–H_2SO_4	**495** ($R^4 = OCH_3$)[a,b,c]	**495** ($R^3 = OC_2H_5$)[a,b,c]	**495**[a,b,c]
HCl–CH_3COOH	**495** ($R^4 = OCH_3$)[b,c]	**495** ($R^3 = Cl$)[b,c]	**495** ($R^5 = Cl$)[b,c]
$ZnCl_2$–CH_3COOH	**495** ($R^4 = OCH_3$)[c,d,e]	**495** ($R^2 = Cl$)[c,d,e]	**495** ($R^2 = OCH_3$)[c,d,e]
BF_3–CH_3COOH	**495** ($R^4 = OCH_3$)[a,c]	**495**[a,c,e]	**495** ($R^2 = OCH_3$)[a,c,e]
BF_3–$CH_3COOC_2H_5$	**495** ($R^4 = OCH_3$)[a,c e]	**495**[a,c,e]	**495** ($R^2 = OCH_3$)[a,c,e]
H_2SO_4–CH_3COOH	**495** ($R^4 = OCH_3$)[a,c,e]	**495**[a,c,e]	**495** ($R^2 = OCH_3$)[a,c,e]
$HO_3SC_6H_4CH_3(4)$ /$CH_2(COOC_2H_5)_2$	**495** ($R^4 = OCH_3$)[f]	**495** ($R^2 = OCH_3$)[f]	**495** ($R^2 = $ OTosyl)[f]

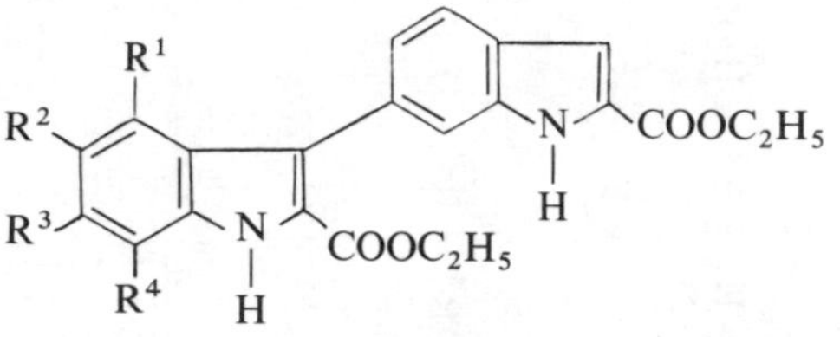

501

variety of protic and Lewis acids. The results of their investigations are given in Table 3. These were rationalized (Ishii, Murakami, Y., Furuse, Hosoya, Takeda and Ikeda, 1971; Ishii, Murakami, Y., Takeda and Furuse, 1974; Ishii, Murakami, Y., Hosoya, Takeda, Suzuki, Y. and Ikeda, 1973) in terms of the mechanisms shown in Scheme 33. These mechanisms are similar to those previously suggested for the halogen and methyl group migrations and eliminations as already discussed and, like these, Scheme 33 emphasizes the key position of the dienone-imine intermediate **497**, or its immediate cyclization product, and the potential lability of the 2-substituent upon indolization of 2-substituted phenylhydrazones. However, Scheme 33 also includes a mechanistic route whereby 4-, 5-, 6- and 7-substitution, with elimination of the methoxy group, can occur.

The formation of indole (3% yield) along with 7-methoxyindole during the indolization of acetaldehyde 2-methoxyphenylhydrazone using heated alumina as catalyst (Suvorov, Bykhovskii and Podkhalyuzina, 1977) can be accommodated by a mechanism analogous to that in Scheme 33 in which hydride ion is the attacking nucleophile involved in the elimination of the methoxy group. It is mechanistically significant that only the expected methoxyindoles are formed by corresponding indolizations of acetaldehyde 3- and 4-methoxyphenylhydrazone (Suvorov, Bykhovskii and Podkhalyuzina, 1977). As with the earlier elimination of halide (Carlin, 1962; Robinson, F. P. and Brown, R. K.,

using Various Catalysts (see Ishii, H., Murakami, Y. and Ishikawa, 1977a)

R^1–R^5 = H in **495** and R^1–R^4 = H in **501**)				
495 ($R^4 = OC_2H_5$)[b,c] **495** ($R^3 = OC_2H_5$)[b,c,d]	**495** ($R^1 = OCH_3$)[b,c] **495** (R^5 = Cl)[b,c]	**495** (R^5 = Cl)[b,c] **495** (R^1 = Cl)[b,c]	**495** (R^1 = Cl)[b,c] **495** (R^2 = Cl)[b,c]	**495** (R^2 = Cl)[b,c] **501** (R^3 = Cl)[b,c]
495 (R^1 = Cl)[b,c] **495** ($R^1 = OCH_3$)[c,d,e]	**495** (R^2 = Cl)[b,c] **495**[c,d,e]			
501 ($R^4 = OCH_3$)[f]	**495** (R^1 = NHCO– $-CH_2COOC_2H_5$)[f]	**501** (R^1 = NHCO– $-CH_2COOC_2H_5$)[f]	**501** (R^2 = OH)[f]	

Note References

a. Ishii, Murakami, Y., Suzuki, Y. and Ikeda, 1970; b. Ishii, Murakami, Y., Hosoya, Takeda, Suzuki, Y. and Ikeda, 1973; c. Ishii, Murakami, Y., and Ishikawa, 1977a; d. Ishii, Murakami, Y., Furuse, Hosoya, Takeda and Ikeda, 1971 (Chromatography on silicic acid of the reaction mixture resulting from the ethanolic hydrogen chloride catalysed reaction gave five indolic compounds and a mixture of three indoles, each bearing a halogen atom, which were detected by v.p.c.); e. Ishii, Murakami, Y., Furuse, Hosoya and Ikeda, 1973; f. Ishii, Murakami, Y., Takeda and Furuse, 1974.

1964), mechanistic clarification of the loss of the methoxy group [to afford **495** (R^1–R^5 = H)] is still required.

Zinc chloride in acetic acid catalysed the conversion of ethyl pyruvate 5-chloro-2-methoxyphenylhydrazone (**494**; R^1 = Cl, $R^2 = R^3$ = H) into a mixture of ethyl 5,6-dichloro-, 4-chloro-7-methoxy-, 6-chloro-5-methoxy and 4-amino-6-chloroindole-2-carboxylate (Ishii, Murakami, Y., Takeda and Furuse, 1974), ethanolic hydrogen chloride catalysed the indolization of ethyl pyruvate 2-methoxy-N_α-methylphenylhydrazone (**494**; $R^1 = R^2$ = H, R^3 = CH_3) into a mixture of ethyl 3-chloro-, 6-chloro-, 7-methoxy- and an ethoxy-1-methylindole-2-carboxylate, ethyl 1-methylindole-2-carboxylate, the dimers **502** (R^1 = H, $R^2 = OCH_3$, n = 1) and **502** (R^1 = Cl, R^2 = H, n = 1), the trimer **502** (R^1 = Cl, R^2 = H, n = 2) and ethyl 3-, 4-, 5- and 6-chloroindole-2-carboxylate (Ishii, Murakami, Y., Hosoya, Takeda, Suzuki, Y. and Ikeda,

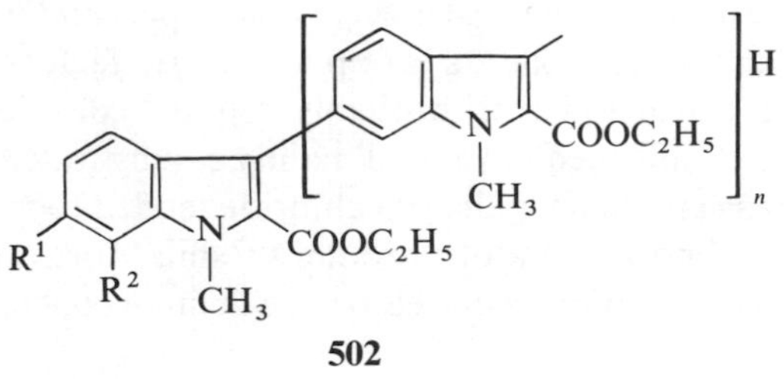

502

1973) and ethanolic hydrogen chloride or zinc chloride in acetic acid catalysed the indolization of ethyl pyruvate 2,6-dimethoxyphenylhydrazone (**494**; $R^1 = R^3 = H$, $R^2 = OCH_3$) into ethyl 5-chloro-7-methoxyindole-2-carboxylate or this indole along with ethyl 6-chloro-7-methoxyindole-2-carboxylate, respectively (Ishii, Murakami, Y., Furuse, Hosoya and Ikeda, 1973). All these reactions are mechanistically rationalizable with regards to Scheme 33.

Earlier (Barnes, Pausacker and Schubert, 1949) it was reported that cyclohexanone 2-methoxyphenylhydrazone, when treated with dilute sulphuric acid, afforded the 'normal' indolization product, 1,2,3,4-tetrahydro-8-methoxycarbazole and a second product which was assigned structure **503** [when using polyphosphate ester as indolization catalyst, the formation of only the former product could be detected (Kanaoka, Ban, Miyashita, Irie and Yonemitsu, 1966)]. Later, following further studies upon the indolization of ethyl pyruvate 2-methoxyphenylhydrazone as already described (Gannon, Benigni, Dickson and Minnis, 1969) it was suggested that the latter product, which was shown to be dimeric, consisted of two 1,2,3,4-tetrahydro-8-methoxycarbazole units linked with loss of a methoxyl group. This suggestion was subsequently (Ishii, Murakami, Y., Furuse, Hosoya, Takeda and Ikeda, 1973) further elaborated to the proposal of structure **504** for this product, based upon the above observations of similar indolizations together with mass and p.m.r. spectral data.

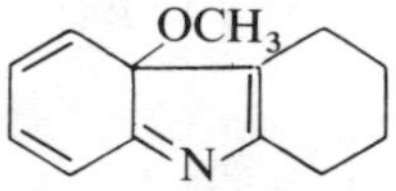

OCH_3 N

503

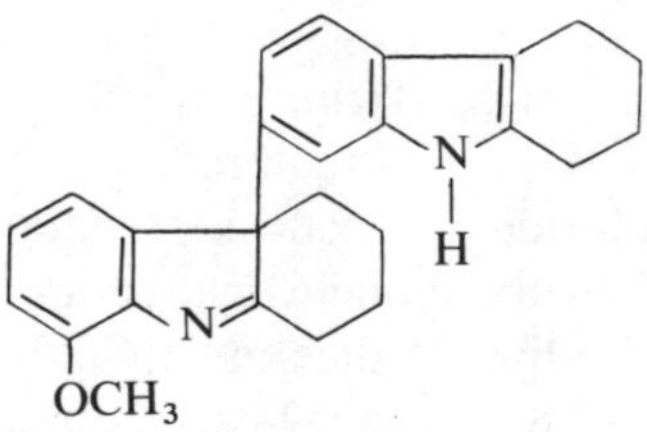

N H N OCH_3

504

Determinant factors which are involved in these reactions with regards to both the qualitative and quantitative aspects of product formation have been discussed (Ishii, Murakami, Y., Furuse, Hosoya and Ikeda, 1973). Acid strength and type of acid catalyst, electron density and distribution within the starting hydrazone, the nature of the substituents on the benzenoid ring of the phenylhydrazino moiety and the nucleophilicity of the nucleophiles present in the reaction mixture have all been discussed in this respect and other factors, such as the nucleophilicity of the leaving group and steric factors of additional substituents in the benzenoid ring probably require consideration. Using zinc chloride as catalyst afforded fewer and mainly 5-substituted 'abnormal' indolic products, whereas ethanolic hydrogen chloride tended to lead to the formation of 6-substituted 'abnormal' indolic products (Ishii, Murakami, Y., Takeda and Furuse, 1974). More work is required before definite conclusions can be drawn in this area.

The indolization of ethyl pyruvate 2-methoxyphenylhydrazone in the presence of nucleophilic reagents has been introduced as a route to the synthesis of the corresponding 6-substituted indoles. Thus, by indolization of ethyl pyruvate 2-methoxyphenylhydrazone, using 4-methylbenzene sulphonic acid in boiling benzene as catalyst, in the presence of ethyl acetoacetate (Ishii, Ikeda and Murakami, Y., 1972; Ishii and Murakami, Y., 1975; Ishii, Murakami, Y., Furuse, Hosoya, Takeda and Ikeda, 1971; Ishii, Murakami, Y., Furuse, Takeda and Ikeda, 1973; Ishii, Murakami, Y., Hosoya, Furuse, Takeda and Ikeda, 1972; Ishii, Murakami, Y. and Ishikawa, 1977a), acetylacetone (Ishii, Ikeda and Murakami, Y., 1972; Ishii, Murakami, Y., Furuse, Hosoya, Takeda and Ikeda, 1971; Ishii, Murakami, Y., Hosoya, Furuse, Takeda and Ikeda, 1972; Ishii, Murakami, Y., Furuse, Hosoya, Takeda and Ikeda, 1973; Ishii, Murakami, Y., and Ishikawa, 1977a), ethyl indole-2-carboxylate (Ishii, Ikeda and Murakami, Y., 1972; Ishii, Murakami, Y., Furuse and Hosoya, 1979), ethyl 6-chloroindole-2-carboxylate (Ishii, Ikeda and Murakami, Y., 1972; Ishii, Murakami, Y., Furuse, Hosoya, Takeda and Ikeda, 1971, 1973; Ishii, Murakami, Y., Hosoya, Furuse, Takeda and Ikeda, 1972) and ethyl 7-methoxyindole-2-carboxylate (Ishii, Murakami, Y., Furuse, Hosoya, Takeda and Ikeda, 1971, 1973; Ishii, Murakami, Y., Hosoya, Furuse, Takeda and Ikeda, 1972) there were obtained, along with other indolization products, compounds **495** [$R^1 = R^2 = R^4 = R^5 = H$, $R^3 = CH(COCH_3)COOC_2H_5$ and $CH(COCH_3)_2$], **501** (R^1–R^4 = H) [along with traces of **501** (R^1–R^3 = H, $R^4 = OCH_3$) and **505** (Ishii, Murakami, Y., Furuse and Hosoya, 1979)], **501** ($R^1 = R^2 = R^4 = H$, $R^3 = Cl$) and **501** (R^1–R^3 = H, $R^4 = OCH_3$) and **505**, respectively, the last two products being obtained

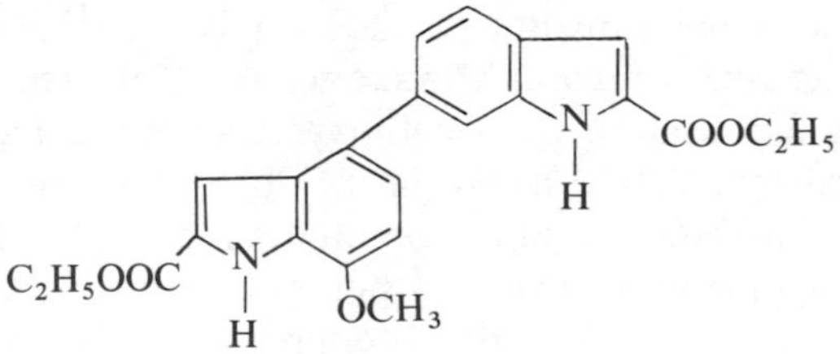

$COOC_2H_5$ H C_2H_5OOC N H OCH_3

505

from the last indolization. Compound **501** (R^1–R^4 = H) could be hydrolysed with ethanolic potassium hydroxide into the dicarboxylic acid which upon heating in quinoline with copper[copper(II)?] chromite afforded 3,6′-bisindolyl (Ishii, Murakami, Y., Furuse and Hosoya, 1979). However, apart from the product **495** [$R^1 = R^2 = R^4 = R^5 = H$, $R^3 = CH(COCH_3)COOC_2H_5$], which was isolated in 56% yield, the yields in the other indolizations were low and the desired 6-substituted indoles have to be separated from other reaction products formed according to Scheme 33. Despite these drawbacks, these reactions still provide a useful synthetic source of the appropriate bisindolyls which would be difficult to prepare by other known routes.

An attempt to use indole as the nucleophile in the above synthetic route did not produce bisindoles. Thus, treatment of *anti*-ethyl pyruvate 2-methoxyphenylhydrazone (see Chapter VII, I) with 4-methylbenzene sulphonic acid in boiling benzene in the presence of indole gave a mixture of the geometrical isomer of the starting hydrazone (see Chapter VII, I), the product **506** which was formed by condensation of indole with ethyl pyruvate and the indole trimer **507** (Ishii, Murakami, Y., Furuse and Hosoya, 1979). The structure of **507**, which was also formed when indole alone was reacted in refluxing benzene solution in the presence of 4-methylbenzene sulphonic acid (Ishii, Murakami, Y., Hosoya, Furuse, Takeda and Ikeda, 1972; Ishii, Murakami, K., Murakami, Y. and Hosoya, 1977), had earlier (Ishii, Murakami, K., Murakami, Y. and Hosoya, 1977) been established by synthesis.

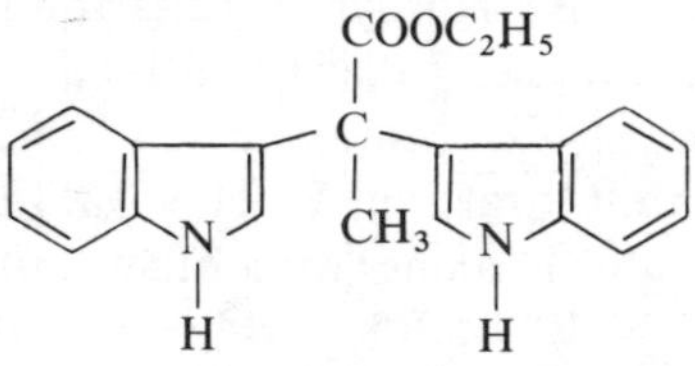

$COOC_2H_5$ C CH_3 N H N H

506

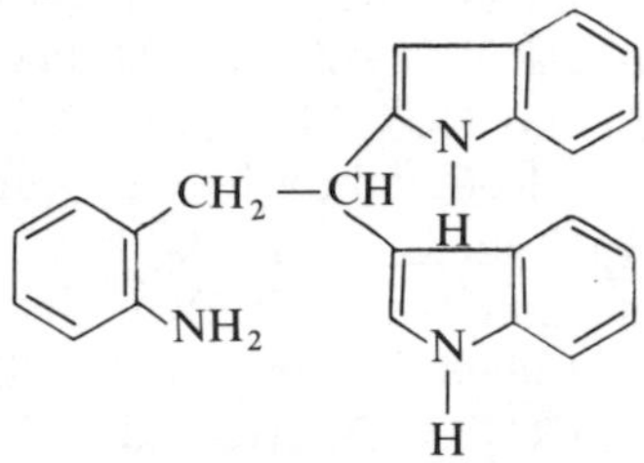

CH_2-CH N H NH_2 N H

507

The p.m.r. spectra of several of the above ethyl indole-2-carboxylates in the presence of the shift reagent [Eu(DPM)$_3$] have been measured and the relationship between the indolic substituents and the position of the coordinated Eu atom has been discussed (Ishii and Murakami, Y., 1979). The earlier of the above studies of the indolization of 2-methoxyphenylhydrazones have been reviewed (Ishii, Murakami, Y. and Ishikawa, 1977b).

Substituent group migration and/or elimination during indolization of 2-substituted phenylhydrazones obviously poses problems in the design of syntheses of 7-substituted indoles. For example, since such phenomena occurred during the synthesis of ethyl 7-benzyloxy-5-chloroindole-2-carboxylate (Regis Chemical Company, 1973), which was obtained only in very low yield using the appropriate Fischer synthesis, alternative approaches were utilized in the synthesis of 5,7- and 6,7-dibenzyloxyindole (Lee, F. G. H., Dickson, Suzuki, J., Zirnis and Manian, 1973).

Two reviews, which also include new experimental observations, upon group migration, elimination and substitution during the indolization of 2,6-disubstituted arylhydrazones, with particular emphasis upon their authors' contributions to this subject, have recently appeared (Fusco, 1978; Fusco and Sannicolò, 1978c).

It would certainly be of interest to extend these studies upon *ortho* group migration, elimination and/or substitution during indolization of *ortho* substituted phenylhydrazones to the use of 2-alkyl-6-halogeno-, 2-alkoxy-6-alkyl- and 2-alkoxy-6-halogenophenylhydrazones. Indeed, an example of such a study has already been discussed with respect to the indolization of **490** and a related investigation has recently been reported (Holla, S. and Ambekar, 1979b) in which ethyl pyruvate 2-chloro-6-methylphenylhydrazone was indolized by heating with polyphosphoric acid. Chromatography of the reaction product afforded ethyl 7-chloro-4-methylindole-2-carboxylate (10% yield) as the only isolated product.

CHAPTER III

Fischer Indolizations of Potential Directional Ambiguity and Closely Related Indolizations

A. Indolization of 3-Substituted Phenylhydrazones and Related Compounds and Methods Used to Establish the Orientation of the Benzenoid Substituent in the Resulting Indolic Product(s)

Indolization of 3-substituted phenylhydrazones, **508**, can possibly give rise to two isomeric products, a 4-substituted indole **509** and a 6-substituted indole **510**. Over fifty years ago, in relation to this problem, it was stated (Hollins, 1924b) that 'of the two *ortho* positions in *meta* substituted phenylhydrazones the reactive one is probably that which is *para* to the substituent group'.

508 **509** **510**

However, this early hypothesis was not supported by the results of the many reactions of this type which were subsequently effected and which are given in Table 4. From this table it can be seen that both possible isomeric indolic products were usually formed, although apart from some examples in which $R^1 = NO_2$, the yield of **510** usually exceeded that of **509**. However, in many of the examples quoted in Table 4, total yields of the indolic products were low and isomer separations were far from quantitative so that the **509** : **510** yield ratios given may not truly reflect the actual yield ratios. Furthermore, apart from the nature of R^1, it is possible that the **509** : **510** yield ratio from these indolizations may also depend upon the structure of the carbonyl moiety of the hydrazone and also upon the cyclization conditions (Grandberg, Belyaeva and Dmitriev, L. B., 1971a). Indeed, cases are known in which a change in indolization catalyst can reverse the **509** : **510** yield ratio (Ockenden and Schofield, 1957) (see Table 4).

Table 4. Products Resulting from the Potentially Ambiguous Indolizations of 3-Substituted Phenylhydrazones

Arylhydrazone **508**				Indolization Product Yield Ratio			
R^1	R^3 (unless otherwise stated, R^2 = H)	R^4	Catalyst	4-isomer	6-isomer	Separation Method	Reference(s)
CH_3	H	CH_3	$ZnCl_2$	a	a	b	1
CH_3	H	SC_6H_5	HCl/C_2H_5OH	0	c,d	b	2
CH_3	H	$CH_2C(COOC_2H_5)_2$ with $NHCOCH_3$ on the C	Dilute H_2SO_4	60–63[f]	55–60[f]	e	3
CH_3	CH_3	H	$ZnCl_2$	c,g		b	4
CH_3	CH_3	CH_3	$ZnCl_2$		c,g	b	1
CH_3	CH_3	CH_3	BF_3 etherate/CH_3COOH or CH_3COOH	0	18[f]	h	5
CH_3	CH_3	CH_2COOH [R^2 = $COC_6H_4Cl(4)$]	i	1	1	j	6
CH_3	CH_3	$(CH_2)_nCOOH$[k] (n = 1,2,3)	CH_3COOH		c,g		7
CH_3	CH_3	$SCH_2COOC_2H_5$	HCl/C_2H_5OH	0	c	b	2
CH_3	CH_3	SC_6H_5	HCl/C_2H_5OH	0	c	b	2
CH_3	COOH	CH_2COOH	H_2SO_4/C_2H_5OH	c,g		b	8
CH_3	$COOC_2H_5$	H	j	1	1	j	9
CH_3	C_6H_5	H	$ZnCl_2$	0	c	b	10
CH_3	C_6H_5	C_6H_5	$ZnCl_2$	2.3	9.9	m	11
CH_3	C_6H_5	C_6H_5	BF_3 etherate/CH_3COOH	n	n	h	5
CH_3	2-Pyridyl	C_6H_5	PPA	c,g	c,g	j	12
CH_3	4-Pyridyl	C_6H_5	PPA	c.g	c.g	j	12
CH_3	$(CH_2)_4$[o]		H_2SO_4	c(?),g	c(?),g	j	13
CH_3	$(CH_2)_4$ (R^2 = CH_3)		CH_3COOH	p	p	q	14
CH_3	$(CH_2)_2C(CH_3)_2CH_2$		H_2SO_4/C_2H_5OH	r	r	r	15
CH_3	$CO(CH_2)_3$		PPA		c	j	16
CH_3	$(CH_2)_2N(CH_3)CH_2$ [R^2 = $(CH_2)_2$–(6-methyl-3-pyridyl), i.e. $(CH_2)_2$–C_5H_3N–CH_3]		HCl/C_2H_5OH		c,g	j	17
CH_3	$(CH_2)_2N(C_2H_5)(CH_2)_2$[s]		j	1	1	j	18
CH_3	$C{=}C(CH_3){-}S$ with $COOC_2H_5$ on the C		CH_3COOH/H_2O		c,g	b	19

(continued)

Table 4 (*continued*)

Arylhydrazone **508**			Indolization Product Yield Ratio			
R^1	R^3 R^4 (unless otherwise stated, $R^2 = H$)	Catalyst	4-isomer	6-isomer	Separation Method	Reference(s)
CH_3	$CO(CH_2)_3$	HCl/CH_3COOH	l,g	l,g	m	20
CH_3	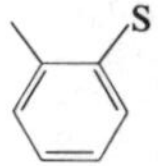 S	CH_3COOH	t	t	t	21
CH_3	$(CH_2)_2$ CH_3 CH_3	HCl/CH_3COOH		c,g	b	22
CH_3	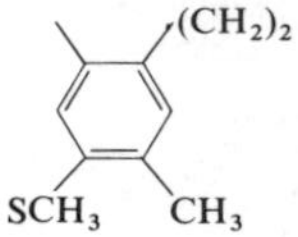 $(CH_2)_2$ SCH_3 CH_3	HCl/CH_3COOH		c,g	b	23
CH_3	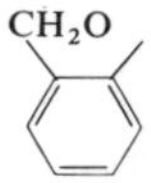 CH_2O	HCl/CH_3COOH	l	l	h	24
CH_3	$(CH_2)_2$ S	CH_3COOH/H_2O		c,g	b	19
CH_3	$(CH_2)_2$ S CH_3	HCl/CH_3COOH		c,g	b	25
CH_3	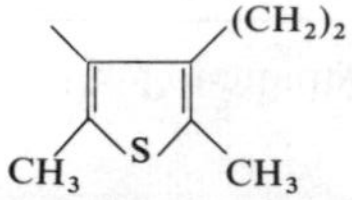 $(CH_2)_2$ CH_3 S CH_3	HCl/CH_3COOH		c,g	b	25

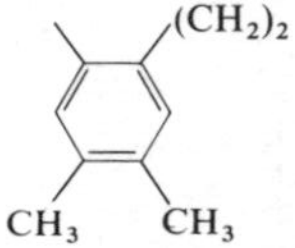

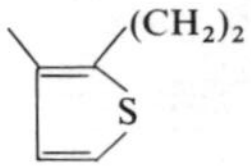

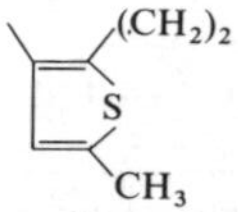

C_2H_5	H	$CH_2C(COOC_2H_5)_2$ (with $NHCOCH_3$)	Dilute H_2SO_4	u	u	u	26
$CH_2COOC_2H_5$	$COOC_2H_5$	$CH_2COOC_2H_5$	HCl/C_2H_5OH	l	2	e	27
C_6H_5		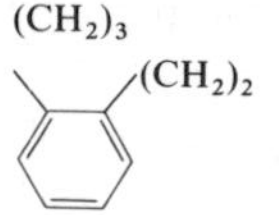 $(CH_2)_3$	$H_2SO_4(?)/CH_3OH$	v	v	b	28
C_6H_5		[structure] $(CH_2)_2$	HCl/CH_3COOH		c,g	b	29
C_6H_5		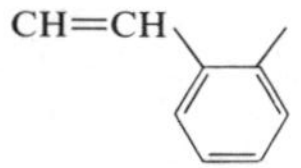 $CH{=}CH$ [structure]	HCl/CH_3COOH		c,g	b	29
w	H	$(CH_2)_2SCH_2C_6H_5$	$CH_3COOH/C_2H_5OH/H_2O$	2	3	h	30
w	CH_3	$(CH_2)_nCOOH$[k] (n = 1,2,3)	CH_3COOH		c,g	j	7
w		$(CH_2)_2N(CH_3)CH_2$ [$R^2 = (CH_2)_2$-3-(4-methylpyridyl)]	HCl/C_2H_5OH	l	l	j	17
w		$(CH_2)_2C(CH_2C_6H_5)CH_2$	H_2SO_4/CH_3COOH		c,g	j	31
w		$CO(CH_2)_3$	HCl/CH_3COOH	l	l	x	32
w		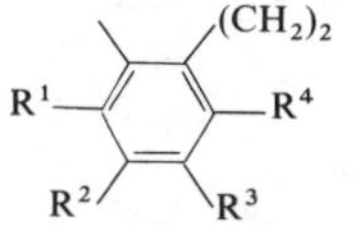 [structure] $(CH_2)_2$, R^1, R^2, R^3, R^4 ($R^1 = R^3 = R^4 = H$, $R^2 = CH_3$; $R^1 = R^4 = H$, $R^2 = R^3 = CH_3$ and $R^1 = R^4 = CH_3$, $R^2 = R^3 = H$)	HCl/CH_3COOH		c,g	b	22
w		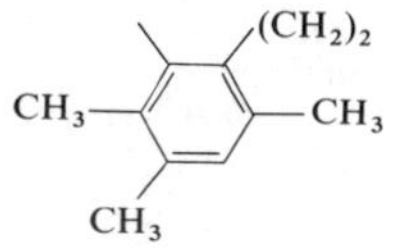 [structure] $(CH_2)_2$, CH_3, CH_3, CH_3	HCl/CH_3COOH		c,g	b	33
w		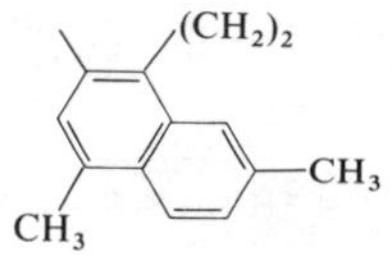 [structure] $(CH_2)_2$, CH_3, CH_3	HCl/CH_3COOH		c,g	b	34
w		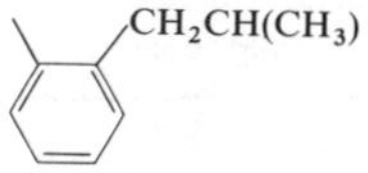 [structure] $CH_2CH(CH_3)$	HCl/CH_3COOH		c,g	b	35

(*continued*)

Table 4 (*continued*)

Arylhydrazone **508**				Indolization Product Yield Ratio			
R^1	R^3 (unless otherwise stated, R^2 = H)	R^4	Catalyst	4-isomer	6-isomer	Separation Method	Reference(s)
w	S; OCH_3		H_2SO_4/CH_3COOH		c,g	b	36
w	NHCONHCO		Dilute HCl	1	1	e	37
y	CH_3	$(CH_2)_nCOOH^k$ (n = 1,2,3)	CH_3COOH		c,g	j	7
y	$COCH_3$	H	HCl/CH_3OH		c	h	38
y	$(CH_2)_4$		Dilute H_2SO_4		c	b	39
z	$COOCH_3$	H	HCl/CH_3OH or H_2SO_4/CH_3OH	aa	aa	b	40
z	$(CH_2)_n$; R; R [R = H, n = 1 and 2; R + R = $(CH_2)_4$, n = 2]		HCl/CH_3COOH		c,g	b	41
bb	$(CH_2)_2N(CH_3)CH_2$ [R^2 = $(CH_2)_2$–(N; CH_3)]		HCl/C_2H_5OH		c,g	j	17
cc	CH_3	$(CH_2)_nCOOH^k$ (n = 1,2,3)	CH_3COOH		c,g	j	7
dd	CH_3	CH_2COOH	HCl	1	1	j	42
dd	CH_3	$CH(C_2H_5)COOH$	HCl	1	1	j	43
dd	CH_3	$CH[C(CH_3)_3]COOH$	HCl	1	1	j	42, 43
ee	CH_3	CH_2COOH	HCl	1	1	j	42, 43
ff	CH_3	$CH[C(CH_3)_3]COOH$	HCl	1	1	j	42, 43
gg	$NHCH_2$; Cl		HCl/C_2H_5OH	0	45[f]	b	44

hh		$(CH_2)_4$		HCl/CH_3COOH	5	3	e	45
OCH_3	H		H	Al_2O_3	1	1	h	46
OCH_3	H		CH_2COOH [$R^2 = COC_6H_4Cl(4)$]	HCl/C_2H_5OH	1	1	j	47
OCH_3	H		$(CH_2)_2NH_2$	$ZnCl_2$	ii	ii	ii	48
OCH_3	H		SC_6H_5	HCl/C_2H_5OH	0	c,jj	b	2
OCH_3	CH_3		H	$ZnCl_2$		c,g	b	49
OCH_3	CH_3		CH_3	CH_3COOH	21 (or 50)[kk]	50 (or 21)[kk]	e	5
OCH_3	CH_3		CH_3	CH_3COOH	kk	kk	e	50
OCH_3	CH_3		CH_3	HCl		c	b	51
OCH_3	CH_3		C_2H_5	CH_3COOH		kk	e	50
OCH_3	CH_3		$SCH_2COOC_2H_5$	HCl/C_2H_5OH		c	b	2
OCH_3	CH_3		SC_6H_5	HCl/C_2H_5OH		c	b	2
OCH_3	$COCH_3$		$(CH_2)_2$-phthalimido	$HCl/CH_3(CH_2)_3OH$		c	b	52
OCH_3	COOH		H[ll]	HCl/C_2H_5OH	a	a	b	53
OCH_3	COOH		CH_2COOH	HCl/C_2H_5OH		c	b	54
OCH_3	COOH		$(CH_2)_2COOH$	H_2SO_4/C_2H_5OH		c	b	55
OCH_3	$COOCH_3$		CH_3 H —C CH_2COOH_3	HCl/C_2H_5OH		c	j	56
OCH_3	$COOCH_3$		CH_3 H —C CH_2COOH_3	j		c	j	57
OCH_3	$COOC_2H_5$		H	j		c	j	9
OCH_3	$COOC_2H_5$		H	$ZnCl_2/CH_3COOH$	22	107	h	58
OCH_3	$COOC_2H_5$		CH_3	H_2SO_4/C_2H_5OH		c	b	59
OCH_3	$COOC_2H_5$		$CH_2COOC_2H_5$	HCl/C_2H_5OH		c,g	b	60
OCH_3	$COOC_2H_5$		$(CH_2)_3COOC_2H_5$[mm]	HCl/C_2H_5OH	nn	nn	h	61
OCH_3	$COOC_2H_5$		C_6H_5	HCl/C_2H_5OH		c,g	b	62
OCH_3	C_6H_5		C_6H_5	HCl/C_2H_5OH	oo	oo	h	5
OCH_3	$C_6H_4OCH_3(4)$		CH_3	$ZnCl_2$		c,g	b	63
OCH_3	$C_6H_4OCH_3(4)$		$C_6H_4OCH_3(4)$	HCl/C_2H_5OH	1.06	11.08	h	64
OCH_3	$C_6H_4OCH_3(4)$		$C_6H_4OCH_3(4)$	HCl	1	1	j	65
OCH_3		$(CH_2)_4$		CH_3COOH	1	6	m	66
OCH_3		$(CH_2)_4$		$HCl/(CH_3)_2CHOH$	1	1	h	67
OCH_3		pp ($R^2 = CH_3$)		H_2SO_4/CH_3COOH	qq	qq	h	68

(continued)

Table 4 (*continued*)

Arylhydrazone **508**				Indolization Product Yield Ratio			
R^1	R^3	R^4 (unless otherwise stated, $R^2 = H$)	Catalyst	4-isomer	6-isomer	Separation Method	Reference(s)
OCH_3	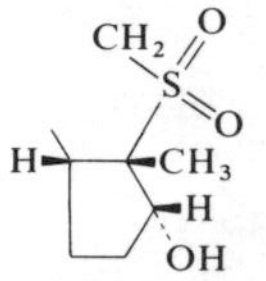		HCl/C_2H_5OH		c	h	69
OCH_3		$(CH_2)_2N(R)CH_2$ [$R = H, CH_3, (CH_2)_2CH_3, CH(CH_3)_2$ and $(CH_2)_3CH_3$]	HCl/C_2H_5OH		c	b	70
OCH_3		$(CH_2)_2N(CH_3)CH_2$	Dilute H_2SO_4		*ca.* 45,[f,g]	rr	71
OCH_3		$(CH_2)_2N(CH_2C_6H_5)CH_2$	HCOOH		c,g	j	72
OCH_3		$(CH_2)_2N(C_2H_5)(CH_2)_2$[s]	j		c,g	j	73
OCH_3		$(CH_2)_2N(COC_6H_5)(CH_2)_2$	HCOOH	6.82	23.6	h	74
OCH_3		$CO(CH_2)_3$	Dilute H_2SO_4		c	b	75
OCH_3		$CO(CH_2)_3$	HCl/CH_3COOH	0.49	1.2	h	76
OCH_3		$CONH(CH_2)_2$	HCOOH	3	5	e	76, 77
OCH_3		$CONH(CH_2)_2$	HCl/C_2H_5OH		c	e	76
OCH_3		$CONHCH_2CH(CH_3)$	j		c	j	57
OCH_3	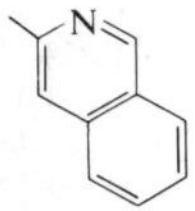	CH_3	HCl/C_2H_5OH		c,g	b	78
OCH_3		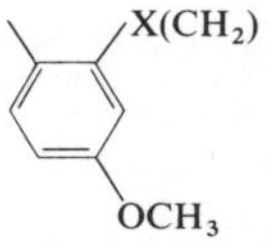 ($X = CH_2$ and NH)	HCl/C_2H_5OH		c,g	b	79
OCH_3	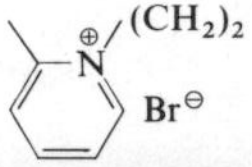		HBr/CH_3OH		c,g	b	80

OCH_3		$\overset{\oplus}{N}$—$(CH_2)_2$ $Br^{\ominus}$		HBr/CH_3OH		c,g	b	80
OC_2H_5		$(CH_2)_4$		Dilute H_2SO_4		c,g	b	81
OC_2H_5		$(CH_2)_2N(CH_3)CH_2$		HCl/C_2H_5OH		c	b	70
$O(CH_2)_2CH_3$		$(CH_2)_2N(CH_3)CH_2$		HCl/C_2H_5OH		c	b	70
$O(CH_2)_2CH_3$		$(CH_2)_2N(CH_3)CH_2$		HCl/C_2H_5OH		c	j	82
$O(CH_2)_3CH_3$		$(CH_2)_2N(CH_3)CH_2$		HCl/C_2H_5OH		c	b	70
$OCH_2C_6H_5$	H		CH_3	j		c,g	j	83
$OCH_2H_6H_5$	H		CH_2COOCH_3	HCl/C_2H_5OH	c,g		j	47
$OCH_2C_6H_5$	H		$(CH_2)_2SCH_2C_6H_5$	$CH_3COOH/C_2H_5OH/H_2O$	2	5	h	30
$OCH_2C_6H_5$	COOH		H[s]	HCl/C_2H_5OH		c	b	84
$OCH_2C_6H_5$	$COOC_2H_5$		$(CH_2)_2$-phthalimido	HCl/C_2H_5OH		c.g	j	85
$OCH_2C_6H_5$	$COOC_2H_5$		C_6H_5	HCl/C_2H_5OH		17[f],g	b	62
$OCH_2C_6H_5$		$(CH_2)_2CHCH_2$ with $N(CH_3)_2$[ll] on CH		j		c	j	86
$OCH_2C_6H_5$		$(CH_2)_2CHCH_2$ with piperidino on CH		j		c	j	86
$OC_6H_4CN(4)$	$C_6H_4CN(4)$		H	$ZnCl_2$	r	r	r	87
OH		$(CH_2)_4$		Naphthalene-1,5-disulphonic acid		c,g	b	88
OH		$(CH_2)_4$		HCl/CH_3COOH	ss	c	j	89
tt		$CO(CH_2)_3$		HCl/CH_3COOH		c	h	90
uu	CH_3		CH_3	HCl/C_2H_5OH		c,g	b	91
uu	CH_3		CH_2COOH	H_2SO_4/C_2H_5OH		c	h	92
uu	$COCH_3$		C_6H_5	HCl/C_2H_5OH		c,g	b	93
uu	COOH		$(CH_2)_2OH$	HCl/C_2H_5OH		c	b	94
uu	$COOC_2H_5$		H	HCl/C_2H_5OH		c,g	b	93, 95
uu	$COOC_2H_5$		CH_3	HCl/C_2H_5OH		c,g	b	93
uu	$COOC_2H_5$		$CH_2COOC_2H_5$	HCl/C_2H_5OH		c,g	b	60
uu	$COOC_2H_5$		$(CH_2)_3COOH$	HCl/C_2H_5OH		c,g	b	93
uu	$COOC_2H_5$		C_6H_5	HCl/C_2H_5OH		c,g	b	93
uu		$(CH_2)_4$		HCl/C_2H_5OH		c,g	b	95
uu		$CO(CH_2)_3$		Dilute H_2SO_4		c	b	75
uu		$COCH_2CH(CH_3)CH_2$		HCl/CH_3COOH		c,g	j	96
vv	CH_3		CH_3	HCl/CH_3COOH	oo	oo	b	97

(*continued*)

Table 4 (*continued*)

Arylhydrazone **508**					Indolization Product Yield Ratio			
R^1	R^3 (unless otherwise stated, R^2 = H)		R^4	Catalyst	4-isomer	6-isomer	Separation Method	Reference(s)
vv	CH_3		$(CH_2)_2N$–(piperidine-4-yl)–$(CH_2)_2$–OH	HCl/C_2H_5OH		c	b	98
vv		$(CH_2)_3$		H_2SO_4		c, ww	j	99
vv		$(CH_2)_4$		CH_3COOH		oo	b	97
xx	H		CH_2COOH [R^2 = $COC_6H_4Cl(4)$]	HCl/C_2H_5OH		c, g	j	83
xx		$CONH(CH_2)_2$		j		c, g	j	100
yy	CH_3		$CH(C_2H_5)COOH$	HCl	1	1	j	42, 43
zz	CH_3		$(CH_2)_5CH_3$–CHCOOH	HCl	1	1	j	42, 43
aaa	CH_3		$(CH_2)_3CH_3$–CHCOOH	HCl	1	1	j	42, 43
bbb		$CO(CH_2)_3$		HCl/CH_3COOH	1.4[f]	72[f]	e, h	101
ccc		$(CH_2)_4$		HCl/CH_3COOH		c	j	89
ddd		$(CH_2)_4$		HCl/CH_3COOH		c	j	89
eee		$(CH_2)_4$		HCl		70[f]	b	45
CF_3	H		CH_2COOH (R^2 = $CH_2C_6H_5$)	HCl/C_2H_5OH	c		j	83
CF_3	H		CH_2COOCH_3	$ZnCl_2/CH_3COOH$	1	1	j	102, 103
CF_3	H		$CH_2COOC_3H_7$ (R^2 = $CH_2C_6H_5$)	HCl/C_2H_5OH	c, g		j	47
CF_3	CH_3		CH_2COOH	H_2SO_4/C_2H_5OH	1	1	h	102, 103
CF_3	$COOC_2H_5$		H	$ZnCl_2$	9.7	5.8	e	104
CF_3	$C_6H_4OCH_3(4)$		$C_6H_4OCH_3(4)$	Heat in monoethylene glycol	4.4	1.7	j	105
CF_3	$(CH_2)_4$			H_2SO_4/CH_3COOH	0	70[f]	b	106, 107
CF_3	$(CH_2)_4$		(fff)	HCl/C_2H_5OH	oo	oo	b	108
CF_3		$(CH_2)_2C(CH_3)_2CH_2$		H_2SO_4/CH_3COOH	0.683	0.578	h	15
CF_3		$(CH_2)_2N(CH_3)CH_2$[fff]		H_2SO_4/CH_3COOH	oo	oo	b	108
CF_3		$(CH_2)_2NRCH_2$ (R = H and CH_3)		H_2SO_4/CH_3COOH	oo	oo	b	70
CF_3		$(CH_2)_2N(CH_3)CH_2$ [R^2 = $(CH_2)_2$-3-(4-methylpyridyl)]		HCl/C_2H_5OH		c, g	j	17
CF_3		$(CH_2)_2SCH_2$[fff]		HCl/C_2H_5OH	oo	oo	b	108

CF_3	[$R^2 = (CH_2)_2N(CH_3)_2$]		HCl/C_2H_5OH	oo	oo	b	108
CF_3	SCH_2[fff]		HCl/C_2H_5OH	oo	oo	b	108
ggg	H	CH_2COOH ($R^2 = CH_2C_6H_5$)	HCl/C_2H_5OH	c, g		j	47
F	H	CH_2COOH ($R^2 = CH_2C_6H_5$)	HCl/C_2H_5OH	c, g		j	47
F	H	CH_2COOCH_3	HCl/C_2H_5OH	l	l	j	102, 103
F	H	$(CH_2)_3OH$	i	l	3	j	109
F	$COOC_2H_5$	H	H_2SO_4/CH_3COOH	hhh	hhh	e, h	110
F	$CH_2C_6H_4COOC_2H_5(4)$	H	PPA or HCl/C_2H_5OH	l	l	j	111
F	$C_6H_4F(4)$	H	PPA		c, g	b	112
F	$(CH_2)_4$		CH_3COOH	6	5	h	113
iii	$(CH_2)_2CH(NHCH_3)CH_2$		j		c	j	114
iii	$(CH_2)_2CH\ N(CH_3)_2\ CH_2$		j		c	j	114
jjj	$CONH(CH_2)_2$		HCOOH		c	b	115
kkk	CH_3	CHRCOOH [$R = CH_3$ and $CH(CH_3)_2$]	HCl	l	l	j	42, 43
Cl	H	CH_2COOH	H_2SO_4/C_2H_5OH	lll	lll	e	116
Cl	H	$(CH_2)_3COOCH_3$	H_2SO_4/CH_3OH	ss	c	ss	117
Cl	H	$(CH_2)_3COOC_2H_5$	H_2SO_4/C_2H_5OH	mmm	mmm	mmm	117
Cl	H	$(CH_2)_3COOC_2H_5$	H_2SO_4/C_2H_5OH	l	l	h	118, 119, 120, 121
Cl	H	SC_6H_5	H_2SO_4/C_2H_5OH		c, nnn	b	2
Cl	CH_3	H	$ZnCl_2$	l	l	m	122
Cl	CH_3	H (ooo)	$ZnCl_2$	l, g	l, g	j	13
Cl	CH_3	CH_3	BF_3 etherate/CH_3COOH or $ZnCl_2$		c	e	5
Cl	CH_3	$CH_2COOC_2H_5$	$ZnCl_2$	ppp	ppp	b	123
Cl	CH_3	C_6H_5	$ZnCl_2$(?)		c, g	j	124
Cl	CH_3	SC_6H_5	HCl/C_2H_5OH		c	b	2
Cl	COOH	CH_2COOH	H_2SO_4/C_2H_5OH		c, g	b	8
Cl	$COOC_2H_5$	H	j		c	j	9
Cl	$COOC_2H_5$	$(CH_2)_2COOC_2H_5$	j	l	l	j	118, 119, 125
Cl	C_6H_5	C_6H_5	BF_3 etherate/CH_3COOH or $ZnCl_2$	c		h	5
Cl	$CH_2C_6H_4COOC_2H_5(4)$	H	PPA or HCl/C_2H_5OH	l	l	j	111
Cl	$(CH_2)_4$		Dilute H_2SO_4	l	l	e	126, 127
Cl	$(CH_2)_4$		$HCl/(CH_3)_2CHOH$	l	l	d	67
Cl	$(CH_2)_4$ (o, ooo)		$ZnCl_2$		c, g	j	13
Cl	$(CH_2)_2N(CH_3)CH_2$ ($R^2 = CH_2C_6H_5$)		HCl/C_2H_5OH	l	l	e	128

(*continued*)

Table 4 (*continued*)

Arylhydrazone **508**				Indolization Product Yield Ratio			
R^1	R^3 (unless otherwise stated, R^2 = H)	R^4	Catalyst	4-isomer	6-isomer	Separation Method	Reference(s)
Cl		$(CH_2)_2N(CH_3)CH_2$ ($R^2 = CH_2C_6H_5$)	Dilute H_2SO_4	2–4, g	9–11, g	qqq	71
Cl		$(CH_2)_2N(CH_3)CH_2$ [$R^2 = CH_2C_6H_4Cl(4)$]	Dilute H_2SO_4		*ca.* 55[f], g	rrr	71
Cl		$(CH_2)_2N(CH_3)CH_2$	HCl/C_2H_5OH		c, g	b	128
Cl		$(CH_2)_2N(CH_3)CH_2$ [$R^2 = (CH_2)_2$-4-pyridyl]	HCl/C_2H_5OH		c, g	j	129
Cl		$(CH_2)_2N(CH_3)CH_2$ [$R^2 = (CH_2)_2$-3-(4-methylpyridyl)]	HCl/C_2H_5OH		c, g	j	17
Cl		$(CH_2)_2NR(CH_2)_2$ R = CH_3($R^2 = CH_2C_6H_5$) R = C_2H_5(R^2 = H and $CH_2C_6H_5$)	j	1	1	j	18
Cl		S	CH_3COOH	5	6	e	130
Cl		S	CH_3COOH	1	1	e	131
Cl		S, Cl	CH_3COOH		c	j	131
sss	CH_3	CH_2COOH	H_2SO_4/C_2H_5OH	1	1	h	92
sss	$COOC_2H_5$	H	$ZnCl_2/CH_3COOH$	69	111	h	132
sss		$(CH_2)_2N(CH_3)CH_2$ [$R^2 = (CH_2)_2$-3-(4-methylpyridyl)]	HCl/C_2H_5OH		c, g	j	17
ttt		$(CH_2)_2N(CH_3)CH_2$	HCl/C_2H_5OH		c, g	j	17
uuu	CH_3	CHRCOOH (R = C_2H_5 and C_5H_{11})	HCl	1	1	j	42, 43
Br	$COOC_2H_5$	H	j	1	1	j	9
Br	C_6H_5	C_6H_5	HCl/C_2H_5OH	1, g	1, g	e	133

Br	C_6H_5	$C_6H_4Br(4)$	PPA	r	r	r	87
Br	$C_6H_4Br(3)$	H	PPA		c,g	b	134
Br	$C_6H_4Br(4)$	H	PPA		c,g	b	134
Br	$C_6H_4Br(4)$	CH_3	PPA		c,g	b	134
Br	$C_6H_4Br(4)$	C_6H_5	PPA	r	r	r	87
Br	$C_6H_4[O(CH_2)_2$-$OC_6H_4Br(4)](4)$	H	$ZnCl_2$	r	r	r	87
Br	2-(5-Bromobenzofuranyl)	H	PPA		c, g	b	135
Br	$(CH_2)_4$		$H_2SO_4/C_2H_5OH/H_2O$	4	1	h	127
Br	$(CH_2)_4$		Dilute H_2SO_4	1	1	vvv	136
Br	$(CH_2)_2N(CH_3)CH_2$ ($R^2 = CH_2C_6H_5$)		HCl/C_2H_5OH		c, g	b	128
Br	$(CH_2)_2N(CH_3)CH_2$		Dilute H_2SO_4	*ca.* 4	*ca.* 8–9	qqq	71
Br	$CO(CH_2)_3$		HCl/CH_3COOH	l, g	l, g	vvv	137
Br	CH_2		Dilute H_2SO_4	l, g	l, g	vvv	138
Br	S		CH_3COOH		c, g	b	135
www	$CO(CH_2)_3$		HCl/CH_3COOH	l, g	l, g	x	137
xxx	$COOC_2H_5$	2-Pyridyl	j		c	j	137a
NO_2	H	C_2H_5	HCl	3	1	h	139
NO_2	H	$(CH_2)_2Cl$	HCl	7	8	h	139
NO_2	CH_3	H	PPA	4	1	h	140
NO_2	CH_3	CH_3	HCl	oo	oo	b	141
NO_2	CH_3	CH_3	HCl/CH_3COOH	l, g	l, g	e	133, 142
NO_2	CH_3	CH_3	HCl	5	8	h	143
NO_2	CH_3	CH_3	HCl	8.5	23.7	h	144
NO_2	CH_3	CH_3	HCl	5	7	h	145
NO_2	CH_3	CH_3	HCl or HCl/CH_3COOH	2	3	h	5
NO_2	CH_3	CH_3	BF_3 etherate/CH_3COOH	1.88	1.47	h	5
NO_2	CH_3	C_2H_5	HCl	2.27	7.48	h	146
NO_2	CH_3	$CH(CH_3)_2$	HCl	25	32	h	147
NO_2	CH_3	$(CH_2)_3CH_3$	HCl/CH_3COOH	1	1	h	147
NO_2	CH_3	$(CH_2)_4CH_3$	HCl/CH_3COOH	31	35.5	h	147
NO_2	CH_3	C_6H_5	HCl/CH_3COOH	45	16.5	h	147
NO_2	CH_3	C_6H_5	HCl/CH_3COOH	c		h	146

(*continued*)

Table 4 (*continued*)

Arylhydrazone **508**				Indolization Product Yield Ratio			
R^1	R^3 (unless otherwise stated, R^2 = H)	R^4	Catalyst	4-isomer	6-isomer	Separation Method	Reference(s)
NO_2	COOH	C_6H_5	HCOOH	oo	oo	b	148
NO_2	$COOC_2H_5$	H	PPA	19	8.2	e	149
NO_2	$COOC_2H_5$	H	PPA	34	23	j	150
NO_2	$COOC_2H_5$	C_6H_5	H_2SO_4/C_2H_5OH	1.5	1.7	e	151
NO_2	C_6H_5	H	PPA	c		h	152
NO_2	C_6H_5	CH_3	HCl	1	1	h	147
NO_2	C_6H_5	C_6H_6	HCl/CH_3COOH	l, g	l, g	e	153
NO_2	C_6H_5	C_6H_6	HCl/CH_3COOH	2.93	1.38	e, h	146
NO_2	C_6H_5	C_6H_6	BF_3 etherate/CH_3COOH or $ZnCl_2$	1.87	1.12	h	5
NO_2	$C_6H_4CH_3(4)$	H	PPA	1, yyy	1, yyy	h	152
NO_2	$C_6H_4Cl(4)$	H	PPA	2, yyy	1, yyy	h	152
NO_2	$(CH_2)_3$		H_2SO_4	3	1	e	142, 154
NO_2	$(CH_2)_4$		H_2SO_4		c, g	b	155, 156
NO_2	$(CH_2)_4$		H_2SO_4	2	1	vvv	157
NO_2	$(CH_2)_4$		H_2SO_4	1	1	h	127
NO_2	$(CH_2)_4$		HCl	19	36	h	147
NO_2	$(CH_2)_4$		PPA	13	16	h	147
NO_2	$CH_2CH(CH_3)(CH_2)_2$		H_2SO_4		c, g	b	155
NO_2	$(CH_2)_2CH(CH_3)CH_2$		Dilute H_2SO_4	*ca.* 1	*ca.* 1	vvv	126, 158
NO_2	$(CH_2)_5$		HCl/CH_3COOH	27	34	h	147
NO_2	$(CH_2)_6$		HCl/CH_3COOH	19	36	h	159
NO_2	$(CH_2)_7$		HCl/CH_3COOH	27	34	h	159
NO_2	$(CH_2)_8$		HCl/CH_3COOH	15	22	h	159
NO_2	$(CH_2)_9$		HCl/CH_3COOH	15	18	h	159
NO_2	$(CH_2)_{10}$		HCl/CH_3COOH	11	21	h	159
NO_2	$(CH_2)_{11}$		HCl/CH_3COOH	15	16	h	159
NO_2	$(CH_2)_{12}$		HCl/CH_3COOH	20	21	h	159
NO_2	$(CH_2)_{13}$		$HCl(CH_3COOH$	17	19	h	159
NO_2	$(CH_2)_{14}$		HCl/CH_3COOH	7	8	h	159
NO_2	$(CH_2)_{15}$		HCl/CH_3COOH	3	4	h	159
NO_2			HCl	oo	oo	b	160

NO_2		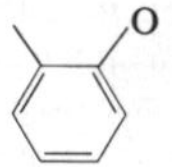 (ring structure) O		j		c, g	j	161
zzz	$COOC_2H_5$		H	PPA		45[f]	b	162
zzz		$(CH_2)_4$		H_2SO_4		c	b	126
zzz		$CO(CH_2)_3$		HCl/CH_3COOH	l, g	l, g	e	137
aaaa		$(CH_2)_3$		H_2SO_4	l	l	e	142, 163
aaaa		$(CH_2)_4$		H_2SO_4	1	1	e	126, 158
aaaa		$(CH_2)_4$		BF etherate/CH_3COOH	1	4	e	164
bbbb	$COOC_2H_5$		C_6H_5	H_2SO_4/C_2H_5OH	1	l	e	151
cccc	$COOC_2H_5$		C_6H_5	H_2SO_4/C_2H_5OH	1	l	e	151
COOH	CH_3		CH_3	HCl/CH_3COOH	5.5, g	16, g	e	165, 166
COOH	C_6H_5		C_6H_5	HCl/CH_3COOH	5.6	10.9	e	11
COOH		$(CH_2)_4$		H_2SO_4	3[dddd]	8[dddd]	e	126, 167, 168
COOH		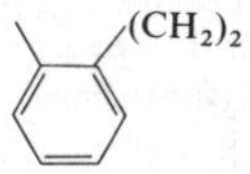 (ring structure) $(CH_2)_2$		j		c, g	j	169[eeee]
$COOC_2H_5$	$COOC_2H_5$		H	$ZnCl_2$	oo	oo	b	170
$COOC_2H_5$	$COOC_2H_5$		$(CH_2)_2COOH$	H_2SO_4/C_2H_5OH	29	20	e	171
$COOC_2H_5$		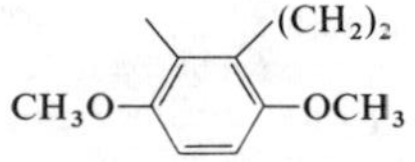 CH_3O– (ring structure) –OCH_3, $(CH_2)_2$		H_2SO_4/C_2H_5OH	1	1	j	172
$COCH_3$		$CONH(CH_2)_2$		HCOOH	1	3	e	173, 174
$COCH_3$		$CONHCH_2CH(CH_3)$		HCOOH	l	l	e	174
COC_6H_5		$CONHCH_2CH(CH_3)$		HCOOH	37	52	e	173
CN	COOH		H	CH_3COOH, HCl or $ZnCl_2$	ffff	ffff	ffff	175
CN	$C_6H_4[(OC_6H_4CN(4)](4)$		H	$ZnCl_2$	r	r	r	87
SO_3H[gggg]	CH_3		H	$ZnCl_2$		c, g	j	176
SO_3H[gggg]	C_6H_5		H	H_2SO_4		c, g	j	176
SO_3H[gggg]	$C_6H_4Cl(4)$		H	$ZnCl_2$		c, g	j	176
SO_3H[gggg]		$(CH_2)_4$		$ZnCl_2$		c, g	j	177

Notes

a. Only one isomer was isolated which was thought to be the 6-substituted indole by analogy with previous related indolizations (Kermack, Perkin and Robinson, R., 1921). However, no structural verification was presented; b. Fractional crystallization or crystallization; c. Only isomer isolated; d. Other 3-methylphenylhydrazones likewise afforded the corresponding 6-methylindoles as the only isolated products (Istituto Luso Farmaco d'Italia S.r.l., 1967; Laskowski, 1968); e. Fractional crystallization; f. % yield; g. Orientation not established from published data; h. Chromatography; i. Heat the arylhydrazine hydrochloride with the carbonyl compound; j. Not specified; k. R^2 = Benzoyl, 4-bromo-, 4-chloro-, 4-fluoro-, 4-methoxy-, 4-methyl- and 4-methylthiobenzoyl, 2-furanylcarbonyl, isonicotinoyl, 2-naphthoyl, nicotinoyl and 2-thienylcarbonyl; l. Both isomers were isolated but yields not specified; m. Fractional crystallization, in part of the picrates; n. The 6-isomer (isolated pure) was the major product and the 4-isomer (isolated crude) was the minor product; o. Also with $R^2 = (CH_2)_2N(CH_3)_2$; p. Both isomers were probably formed but they were only partially separated; q. High vacuum fractional sublimation; r. The formation of both isomers was detected by t.l.c. but a preparative separation was not effected; s. Also with $R^2 = CH_2C_6H_5$; t. Sharp melting product which could be a mixture of isomers; u. Both isomers were apparently formed but these could not be separated; v. Both isomers were formed but only the 4-isomer was isolated by crystallization. Subsequent chromatography on silica gel afforded a eutectic mixture of the two isomers; w. 3,4-Dimethylphenylhydrazone used; x. Manual separation followed by recrystallization; y. 4-Methoxy-3-methylphenylhydrazone used; z. 3,4-Tetramethylenephenylhydrazone used; aa. Both isomers were formed but only the 4-isomer was isolated; bb. 4-Chloro-3-methylphenylhydrazone used; cc. 4-Ethoxy- or 4-fluoro-3-methylphenylhydrazone used; dd. 3-Methyl-4-nitrophenylhydrazone used; ee. 3-Ethyl-4-nitrophenylhydrazone used; ff. 4-Nitro-3-propylphenylhydrazone used; gg. 3-Diethylaminomethyl-4-methoxyphenylhydrazone used; hh. 1-Benzoyl-1,2,3,4-tetrahydroquinolylhydrazone used; ii. Both isomeric products were produced but only the 6-isomer was isolated (as the aminoethyl N-acetyl derivative) by fractional crystallization of the reaction mixture subsequent to acetylation; jj. Other 3-methoxy- (Istituto Luso Farmaco d'Italia S.r.l., 1967; Laskowski, 1968) and 3-ethoxy- (Laskowski, 1968) phenylhydrazones likewise gave the corresponding 6-alkoxyindoles as the only reported products; kk. Both isomers were isolated but the orientations of the methoxy substituents were not established; ll. Also with $R^2 = CH_3$; mm. Starting with the azo compound **126**; nn. Only the 6-isomer was initially isolated but later, after further chemical transformations, traces of the 4-isomer were detected; oo. Only one product was isolated but the orientation of the benz substituent(s) was not established; pp. Using 2-methylcyclohexanone as the ketonic moiety; qq. From the neutral fraction of the reaction product, only one compound was isolated but the orientation of the methoxy group at the 5- or 7-position was not established. From the basic component of the reaction product, both isomers were isolated, after catalytic hydrogenation to the 1,2,3,4,4a,9a-hexahydrocarbazole. No product yields were quoted; rr. Crystallization of the naphthalene-1,5-disulphonate; ss. 4-Isomer remained in solution during 'work-up'; tt. 3-Hydroxy-4-methylphenylhydrazone and 3-hydroxy-4-hydroxymethylphenylhydrazone used, respectively; uu. 3,4-Dimethoxyphenylhydrazone used. Other 3,4-dimethoxyphenylhydrazones were also indolized to yield the corresponding 5,6-dimethoxyindoles as the only reported products (Archer, S. 1970; Istituto Luso Farmaco d'Italia S.r.l., 1967; Laskowski, 1968; Zenitz, 1966); vv. 3,4-Methylenedioxyphenylhydrazone used; ww. Other 3,4-methylenedioxyphenylhydrazones and 3,4-diethoxyphenylhydrazones (Istituto Luso Farmaco d'Italia S.r.l., 1967; Laskowski, 1968) and 3,4-ethylene-1,2-dioxyphenylhydrazones and 3-ethoxy-4-methoxyphenylhydrazones (Laskowski, 1968) similarly afforded the corresponding 5,6-disubstituted indoles as the only reported products; xx. 4-Chloro-3-methoxyphenylhydrazone used; yy. 3-Methoxy-4-nitrophenylhydrazone used; zz. 3-Ethoxy-4-nitrophenylhydrazone used; aaa. 4-Nitro-3-propoxyphenylhydrazone used; bbb. 3-Hydroxy-4-methylphenylhydrazone used; ccc. 4-Carboxy-3-hydroxyphenylhydrazone used; ddd. 4-CONH $[C_6H_4Cl(4)]$, 3-OH; eee. 1-Acetyl-1,2,3,4-tetrahydro-7-quinolylhydrazone used; fff. Also with $R^2 = N(CH_2)_2N(CH_3)_3$ and $N(CH_2)_3N\langle\rangle NCH_3$; ggg. Using 4-dimethylamino-3-trifluoromethylphenylhydrazone and 4-hydroxy-3-trifluoromethylphenylhydrazone; hhh. Both isomers were isolated but the yield of the 6-isomer $\gg$ the yield of the 4-isomer; iii. 3,4-Difluorophenylhydrazone used; jjj. 3-Fluoro-4-methoxyphenylhydrazone used; kkk. 3-Fluoro-4-nitrophenylhydrazone used; lll. The 6-isomer was isolated, along with a eutectic mixture of the 4- and 6-isomers; mmm. Both isomers were probably formed but they were only partially separated; nnn. Other 3-chlorophenylhydrazones afforded only the corresponding 6- (Istituto Luso Farmaco d'Italia S.r.l., 1967; Laskowski, 1968) or 4- (Laskowski,

1968) substituted indoles as the reported products;   ooo. Also with $R^2 = (CH_2)_2N(C_2H_5)_2$;   ppp. It was initially (Stevens, F. J. and Higginbotham, 1954) thought that only one isomer, separated by crystallization, was formed. However, (Piper and Stevens, F. J., 1962; Stevens, F. J., 1962) it was subsequently shown that the reaction product was a eutectic mixture of both possible isomeric indoles in which the 6-isomer was present in about 5–6% predominance;   qqq. Fractional crystallization furnished the 6-isomer, the 4-isomer being subsequently isolated as its naphthalene-1,5-disulphonate;   rrr. Crystallization of the hydrochloride;   sss. 3,4-Dichlorophenylhydrazone used;   ttt. 3-Chloro-4-methylphenylhydrazone used;   uuu. 3-Chloro-4-nitrophenylhydrazone used;   vvv. Fractional crystallization, totally or in part of the N-acetyl derivatives;   www. 3-Bromo-4-methylphenylhydrazone used;   xxx. 3-Bromo-4-chlorophenylhydrazone used;   yyy. Indazoles were also simultaneously produced (see Chapter VII, O);   zzz. 4-Methyl-3-nitrophenylhydrazone used;   aaaa. 4-Chloro-3-nitrophenylhydrazone used;   bbbb. 4-Carboxy-3-nitrophenylhydrazone used;   cccc. 4-Carbethoxy-3-nitrophenylhydrazone used;   dddd. When this reaction was initially performed using a mineral acid as catalyst, the orientation(s) of the substituent on the 4- or 6-position of the indole nucleus in the reaction product(s) was(were) not established (Baeyer and Tutein, 1889);   eeee. The indolizations of 1-tetralone and 6,7-dimethoxy-1-tetralone 3-carboxyphenylhydrazones also appear to have been effected but details in the published abstract of these studies are very brief (Avanesova, Astvatsatryan, Sarkisyan, Garibyan and Tatevosyan, 1975);   ffff. No recognizable product was formed from this reaction although the product gave a positive Ehrlich colour test;   gggg. For the synthesis of 3-hydrazinobenzene sulphonic acid, see Johnson, M. (1921) and Limpricht (1888). A useful report (Hunsberger, Shaw, E. R., Fugger, Ketcham and Lednicer, 1956) also described the synthesis of thirty-five different arylhydrazines by the reduction of the corresponding aryldiazonium salts and included the preparation of salts of **142** [R^1–$R^3 = R^5 = R^6 = H$, $R^4 = CH_3$, $C(CH_3)_3$, $CH{=}CHCOOH$, OCH_3, OC_2H_5, Br and I]. Syntheses of **142** (R^1–$R^3 = R^5 = R^6 = H$, $R^4 = OC_2H_5$) (Franzen and Schmidt, M., 1917), **142** (R^1–$R^3 = R^5 = R^6 = H$, $R^4 = F$) (Suschitzky, 1953) and **142** ($R^1 = R^4 = CH_3$, $R^2 = R^3 = R^5 = R^6 = H$) (Franzen, Onsager and Faerden, 1918) have also appeared and references to the syntheses of some of these and other 3-substituted phenylhydrazines can be located in the references presented in connection with Table 4.

References

1. Mendlik and Wibaut, 1931;   2. Wieland, T. and Rühl, 1963;   3. Snyder, Beilfuss and Williams, J. K., 1953;   4. Plancher and Ciusa, 1906;   5. Ockenden and Schofield, 1957;   6. Yamamoto, H. and Nakao, 1969c;   7. Yamamoto, H. and Nakao, 1971b;   8. Feofilaktov and Semenova, N. K., 1953c;   9. Kotov, Sagitullin and Gorbunov, 1970;   10. Campbell and Cooper, 1935;   11. Coldham, Lewis, J. W. and Plant, 1954;   12. Lakshmanan, 1960;   13. Farbenfabriken Bayer A.-G., 1955b;   14. Bloss and Timberlake, 1963;   15. Bergman and Erdtman, 1969;   16. Shah, G. D. and Patel, B. P. J., 1979;   17. Berger, L. and Corraz, 1968;   18. Geigy, A.-G., 1966, 1968;   19. Buu-Hoï, Hoán Khôi and Xuong, 1949;   20. Kent and McNeil, 1938;   21. Dalgleish and Mann, 1947;   22. Buu-Hoï, Saint-Ruf, Jacquignon and Marty, 1963;   23. Buu-Hoï and Hoán, 1951;   24. Bistochi, De Meo, Ricci, Croisy and Jacquignon, 1978;   25. Buu-Hoï, Hoán and Khôi, 1950b;   26. Lingens and Weiler, 1963;   27. Plieninger, 1954;   28. Jones, G. and Tringham, 1975;   29. Thang, D. C., Kosoff, Jacquignon and Dufour, 1976;   30. Keglević and Goleš, 1970;   31. Buu-Hoï, Roussel, O., and Jacquignon, 1964;   32. Anderson, G. and Campbell, 1950;   33. Buu-Hoï and Saint-Ruf, 1962;   34. Buu-Hoï and Saint-Ruf, 1965a;   35. Buu-Hoï, Mangane and Jacquignon, 1967;   36. Buu-Hoï and Saint-Ruf, 1963;   37. Wright, 1976;   38. Remers, Roth and Weiss, M. J., 1964;   39.Cummins, Kaye and Tomlinson, M. L., 1954;   40. Shagalov, Ostapchuk, Zlobina, Eraksina, Babushkina, T. A., Vasil'ev, A. M., Ogorodnikova and Suvorov, 1978 – see also Shagalov, Ostapchuk, Zlobina, Eraksina and Suvorov, 1977;   41. Buu-Hoï and Jacquignon, 1951;   42. Carey, Gal and Sletzinger, 1968;   43. Carey, Gal and Sletzinger, 1967;   44. Marquez, Cranston, Ruddon, Kier and Burckhalter, 1972;   45. Kulka and Manske, R. H. F., 1952;   46. Suvorov, Bykhovskii and Podkhalyuzina, 1977;   47. Shen, T.-Y., 1964;   48. Späth and Lederer, 1930a;   49. Späth and Brunner, O., 1925;   50. Vejdělek, 1957;   51. Neuss, Boaz and Forbes, J. W., 1954;   52. Manske, R. H. F., Perkin and

References (*continued*)

Robinson, R., 1927; 53. Kermack, Perkin and Robinson, R., 1922; 54. Kermack, Perkin and Robinson, R., 1921; 55. Barrett, H. S. B., Perkin and Robinson, R., 1929; 56. Sandoz Patents Ltd., 1965; 57. Frey, A. J., Ott, Bruderer and Stadler, 1960; 58. Ishii, Murakami, Y., Hosoya, Takeda, Suzuki, Y. and Ikeda, 1973; 59. Wieland, T. and Grimm, 1965; 60. Findlay, S. P. and Dougherty, 1948; 61. Iyer, Jackson, A. H., Shannon and Naidoo, 1972, 1973; Iyer, Jackson, A. H. and Shannon, 1973; 62. Morton and Slaunwhite, 1949; 63. Mentzer, 1946; 64. Szmuszkovicz, Glenn, Heinzelman, Hester and Youngdale, 1966; 65. Upjohn Co., 1966; 66. Chalmers, J. R., Openshaw and Smith, G. F., 1957; 67. Przheval'skii, Grandberg, Klyuev, N. A. and Belikov, 1978; 68. Millson and Robinson, R., 1955; 69. Jogdeo and Bhide, 1980; 70. Cattanach, Cohen, A. and Heath-Brown, 1968; 71. Hörlein, U., 1954; 72. Hester, 1968; 73. Geigy, A.-G., 1966, 1968; 74. Hester, Tang, Keasling and Veldkamp, 1968; 75. Douglas, B., Kirkpatrick, Moore, B. P. and Weisbach, 1964; 76. Bhide, Tikotkar and Tilak, 1960; 77. Abramovitch, 1956; Morozovskaya, Ogareva and Suvorov, 1969; 78. Elderfield and Wythe, 1954; 79. Cross, P. E. and Jones, E. R. H., 1964; 80. Elderfield, Lagowski, McCurdy and Wythe, 1958; 81. Hoshino and Takiura, 1936; 82. Cohen, A. and Cattanach, 1967; 83. Horning, E. C., Sweeley, Dalgliesh and Kelly, W., 1959; 84. Cortes and Walls, F., 1964; 85. Gaimster, 1960; 86. Mooradian, 1975, 1976c; 87. Dann, Fick, Pietzner, Walkenhorst, Fernbach and Zeh, 1975; 88. Herdieckerhoff and Tschunkur, 1933; 89. Long, R. S., 1956; 90. Chakraborty, Chatterji, Ganguly, 1969; Chakraborty, Islam and Bhattacharyya, P., 1973; 91. Beer, McGrath, Robertson and Woodier (with Holker), 1949; 92. Lanzilotti, Littell, Fanshawe, McKenzie and Lovell, 1979; 93. Lions and Spruson, 1932; 94. Sterling Drug Inc., 1970; 95. Perkin and Rubenstein, 1926; 96. Bhattacharyya, P., Basak, Islam and Chakraborty, 1975; 97. Clemo and Weiss, J., 1945; 98. Zenitz, 1965; 99. Jones, G., 1975; 100. Roussell UCLAF, 1963a; 101. Mester, Choudhury and Reisch, 1980; 102. Shen, T.-Y., 1965; 103. Shen, T.-Y., 1967b; 104. Bornstein, Leone, Sullivan and Bennett, O. F., 1957; 105. Szmuszkovicz, 1970; 106. Forbes, E. J., Stacey, Tatlow and Wragg, R. T., 1960; 107. Forbes, E. J., Tatlow and Wragg, R. T., 1960; 108. Aksanova, Sharkova, N. M., Baranova, Kucherova and Zagorevskii, 1966; 109. Katsube, Sasajima, Ono, Nakao, Muruyama, Takayama, Katayama, Tanaka, Y., Inaba and Yamamoto, H., 1974; 110. Allen, F. L., Brunton and Suschitzky, 1955; 111. Duncan and Boswell, 1973; 112. Joshi, K. C., Pathak and Chand, 1978b; 113. Allen, F. L. and Suschitzky, 1953; 114. Mooradian, 1973; 115. Kirk, 1976; 116. Fox and Bullock, 1951b – see also Marumo, Abe, H., Hattori and Munakata, 1968; 117. Bullock and Hand, J. J., 1956b; 118. Suvorov, Mamaev and Shagalov, 1953; 119. Shagalov, Sorokina, N. P. and Suvorov, 1964; 120. Suvorov, Fedatova, M. V., Orlova, L. M. and Ogareva, 1962; 121. Suvorov, 1957; 122. Piper and Stevens, F. J., 1966; 123. Stevens, F. J. and Higginbotham, 1954; 124. Aktieselskabet Dumex (Dumex Ltd), 1964; 125. Yoshitomi Pharmaceutical Industries Ltd, 1967; 126. Moggridge and Plant, 1937; 127. Barclay and Campbell, 1945; 128. Hoerlein, 1956; 129. Berger, L. and Corraz, 1970a, 1972a; 130. Werner, L. H., Schroeder, D. C. and Ricca, 1957; 131. Ciba Ltd, 1960; 132. Ishii, Murakami, Y., Takeda and Furuse, 1974; 133. Plant and Tomlinson, M. L., 1933; 134. Dann, Bergen, G., Demant and Volz, 1971; 135. Dann, Volz, Demant, Pfeifer, Bergen, G., Fick and Walkenhorst, 1973; 136. Plant and Wilson, A. E. J., 1939; 137. Mears, Oakeshott and Plant, 1934; 137a. Steinman and Tahbaz, 1980; 138. Plant and Tomlinson, M. L., 1931; 139. McKay, Parkhurst, Silverstein and Skinner, 1963; 140. Noland Smith, L. R. and Rush, 1965; 141. Bauer and Strauss, 1932; 142. Plant and Whitaker, 1940; 143. Schofield and Theobald, 1949; 144. Atkinson, Simpson and Taylor, A., 1954; 145. Pappalardo and Vitali, 1958a; 146. Schofield and Theobald, 1950 (see also Vejdělek, 1957); 147. Frasca, 1962a; 148. Kidwai, A. R. and Khan, N. H., 1963; 149. Parmerter, Cook, A. G. and Dixon, 1958; 150. Scriven, Suschitzky, Thomas, D. R. and Newton, 1979; 151. Sato, Y., 1963b; 152. Dennler and Frasca, 1966b; 153. Fennell and Plant, 1932; 154. Plant, 1929; 155. Borsche, Witte, A. and Bothe, 1908; 156. Perkin and Plant, 1921; 157. Plant, 1936; 158. Plant and Rosser, 1928; 159. Dennler and Frasca, 1966a; 160. Bannister and Plant, 1948; 161. Schroeder, D. C., Corcoran, Holden and Mulligan, 1962; 162. Gadaginamath, 1976; 163. Massey and Plant, 1931; 164. Robinson, F. P. and Brown, R. K., 1964; 165. Brown, U. M., Carter, P. H. and Tomlinson, M., 1958; 166. Brimblecombe, Downing and Hunt, R. R., 1966; 167. Collar and Plant, 1926; 168. Plant and Williams, S. B. C., 1934; 169. Avanesova, Musaelyan and Tatevosyan, 1972; 170. Roder, 1886; 171. Koelsch, 1943; 172. Avanesova and Tatevosyan, 1970; 173. Strandtmann, Cohen, M. P. and Shavel, 1963; 174. Shavel, Strandtmann and Cohen, M. P., 1965; 175. Acheson and Vernon, 1962; 176. I. G. Farbenindustrie Akt.-Ges., 1932; 177. General Aniline Works Inc., 1932.

These possible dependencies have been further and more thoroughly investigated, particular attention being paid in these studies to effect near quantitative isomer isolation/separation by the use of preparative g.l.c. (Grandberg, Belyaeva and Dmitriev, L. B., 1971a, 1971b, 1973) or isomer detection by quantitative p.m.r. analysis (Grandberg, Belyaeva and Dmitriev, L. B., 1971b, 1973). Indolization of cyclohexanone and diethyl ketone 3-methylphenylhydrazone [**508**; $R^1 = CH_3$, $R^2 = H$, $R^3 + R^4 = (CH_2)_4$ and $R^2 = C_2H_5$, $R^4 = CH_3$, respectively] using zinc chloride, copper(I) chloride, ethanolic sulphuric acid or hydrogen chloride, polyphosphoric acid, sulphosalicylic acid or boron trifluoride etherate in acetic acid as catalysts, or by boiling a mixture of the hydrazine hydrochloride and ketone in dimethylformamide, furnished in each case both possible isomeric indolic products, the **509** : **510** yield ratio being within the range of 1 : 1–5 : 9, depending upon the catalyst (Grandberg, Belyaeva and Dmitriev, L. B., 1971a, 1971c). A similar investigation using a similar catalytic spectrum in the indolizations of diethyl ketone 3-chloro-, 3-ethyl- and 3-methoxyphenylhydrazone again led to the formation of both possible isomeric products in all cases, the **509** : **510** yield ratio being within the range 1 : 0.8–1 : 1.8 except in one reaction, the indolization of diethyl ketone 3-methoxyphenylhydrazone using ethanolic hydrogen chloride, in which the yield ratio was **509** : **510** = 1 : 5.4 (Grandberg, Belyaeva and Dmitriev, L. B., 1971b). Thus, with only two exceptions, the indolization of diethylketone 3-chlorophenylhydrazone with a catalytic or a molar quantity of zinc chloride which gave **509** : **510** yield ratios = 1 : 0.9 and 1 : 0.8, respectively, the above indolizations usually afforded slightly greater yields of the corresponding 6- rather than 4-substituted indoles, although in several examples the two isomers were formed in equal amounts. Extension of these investigations to the indolization of diethyl ketone 3-nitrophenylhydrazone using zinc chloride, 30% sulphuric acid, concentrated hydrochloric acid, polyphosphoric acid or boron trifluoride etherate as catalyst reversed this situation and produced the corresponding 2-ethyl-3-methyl-4- and -6-nitroindole in a yield ratio ranging between 3 : 1 and 12 : 13, respectively, depending upon the catalyst used (Grandberg, Belyaeva and Dmitriev, L. B., 1973).

These data were, quite surprisingly in view of contemporary opinion, interpreted in terms of the formation of the C—C bond formation during indolization involving an intramolecular electrophilic attack upon the benzenoid nucleus, *ortho–para* directing 3-substituents (e.g. alkyl, alkoxy, halogeno) directing predominantly *para* and thus ultimately affording 6-substituted indoles as the major products and *ortho–meta* directing 3-substituents (e.g. nitro) yielding 4-substituted indoles predominantly (Grandberg, Belyaeva and Dmitriev, L. B., 1971b, 1973) although some of the data presented in Table 4 had been likewise but earlier interpreted (Ockenden and Schofield, 1957 – see also Szmuszkovicz, Glenn, Heinzelmann, Hester and Youngdale, 1966). However, it is now well established (see Chapter II, F) that the new C—C bond formation during indolization most likely occurs by means of a [3,3]-sigmatropic shift, a process largely independent of the electronic nature of substituents on the benzenoid ring. Indeed, the **509** : **510** yield ratio, when reliably quantitatively investigated as above, does not differ greatly from unity and is therefore in accord with this mechanistic interpretation, substituents which increase the electron density predominantly at the *para* position leading to a **509** : **510** yield ratio $\leqslant 1$ and substituents which decrease the electron density predominantly at the *para* position leading to a **509** : **510** yield ratio $\geqslant 1$. The apparent independence of the Fischer indolization of the electronic characteristics of substituents in the aryl nucleus of the arylhydrazone had, in fact, been noted as long ago as 1949 (Schofield and Theobald, 1949).

Regardless of the electronic nature of the benzenoid substituent, steric considerations will always favour the formation of the 6-substituted product, **510**, over the 4-substituted isomer, **509** (Grandberg, Belyaeva and Dmitriev, L. B., 1971a, 1971b, 1973; Ockenden and Schofield, 1957; Schofield and Theobald, 1949; Szmuszkovicz, Glenn, Heinzelman Hester and Youngdale, 1966). It was suggested that this steric effect was clearly illustrated by the observation (Grandberg, Belyaeva and Dmitriev, L. B., 1973) that the hydrochloric acid catalysed indolization of isopropyl methyl ketone 3-nitrophenylhydrazone **511** ($R = NO_2$) yielded **512** ($R^1 = H$, $R^2 = NO_2$) as the only product. However,

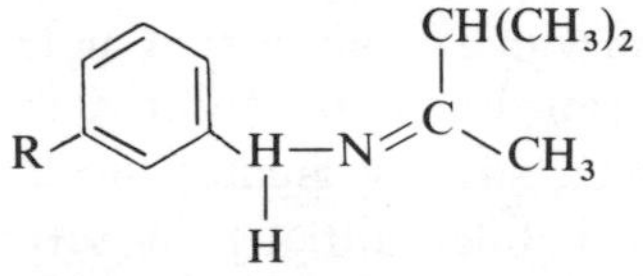

511 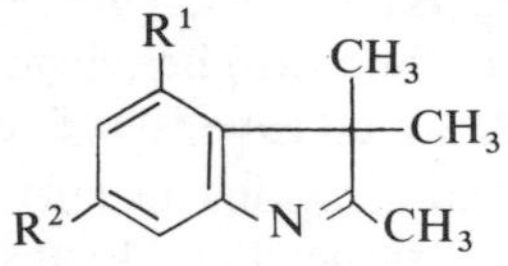 **512**

earlier (Sych, 1953) it had been reported that **511** ($R = NO_2$) was indolized, using glacial acetic acid as catalyst, to afford **512** ($R^1 = NO_2$, $R^2 = H$) with no reference being made to the formation of the 6-isomer. It is possible that these observations represent a complete change in the direction of indolization with a change in catalyst. However, since the former product had m.p. = 135 °C (Grandberg, Belyaeva and Dmitriev, L. B., 1973) and the latter product had m.p. = 136–137 °C (Sych, 1953), it is possible that they are identical and therefore one of the formulations is erroneous. The structure **512** ($R^1 = H$, $R^2 = NO_2$) appeared to be supported by spectroscopic data (Grandberg, Belyaeva and Dmitriev, L. B., 1973). The hydrazone **511** ($R = CH_3$) has also been indolized, using zinc chloride in ethanol as catalyst, to give a product which was formulated as **512** ($R^1 = H$, $R^2 = CH_3$) without experimental verification for the orientation of the methyl substituent (Ghigi, 1933b). Indolization of isobutyraldehyde 3-nitrophenylhydrazone using zinc chloride in alcohol as the catalyst appears to produce both possible isomeric 3*H*-indolic products but these were neither characterized nor differentiated (Brunner, K.,

Wiedner and Kling, 1931). Further investigation of all these reactions and the structures of their products would certainly be of interest, as would the indolization of other 3-substituted phenylhydrazones of isopropyl methyl ketone and other ketones or aldehydes which would afford 3*H*-indoles with potential steric interaction between the alkyl substituents at the 3-position and the possible 4-substituent.

It would also be of interest to examine the relative yields of both possible indolic products obtained upon indolization of phenylhydrazones substituted in both *meta* positions with dissimilar substituents. From the above discussion it might be expected that the relative electronic natures of the two *meta* substituents would have little effect upon which *ortho* carbon atom became involved in the new C—C bond formation but that this may be far more dependent upon the relative steric effects of the two substituents. Two published examples of such a reaction involved the indolizations of **513** (R = H and CH_3) which in each case furnished only one product which was formulated, without experimental verification, as **514** (R = H and CH_3, respectively) (Beer, Brown, J. P. and Robertson, 1951). U.v. spectral investigation should distinguish between these structures and their possible isomers.

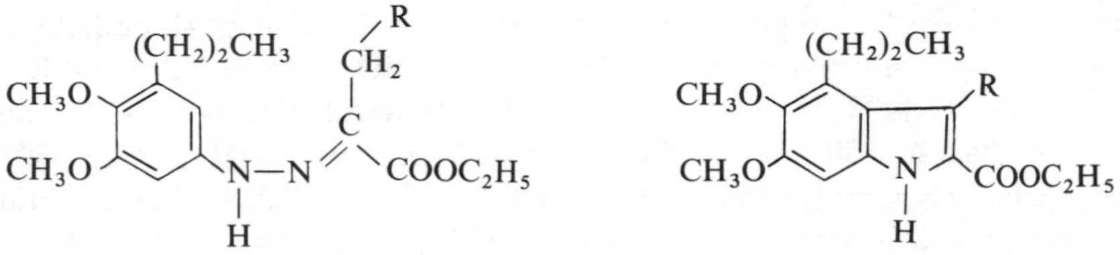

$(CH_2)_2CH_3$ R CH_2 CH_3O C CH_3O N—N $COOC_2H_5$ H

513

$(CH_2)_2CH_3$ CH_3O R CH_3O N $COOC_2H_5$ H

514

Several methods have been employed for distinguishing between the 4- and/or 6-substituted indoles formed upon indolization of 3-substituted phenylhydrazones.

1. *Directed Indolizations*

The use of a suitable 2-substituted phenylhydrazone **515** directs the indolization unambiguously to give **516**, subsequent removal of the original 2-substituent then affording the corresponding 4-substituted indole **516** (R^2 = H).

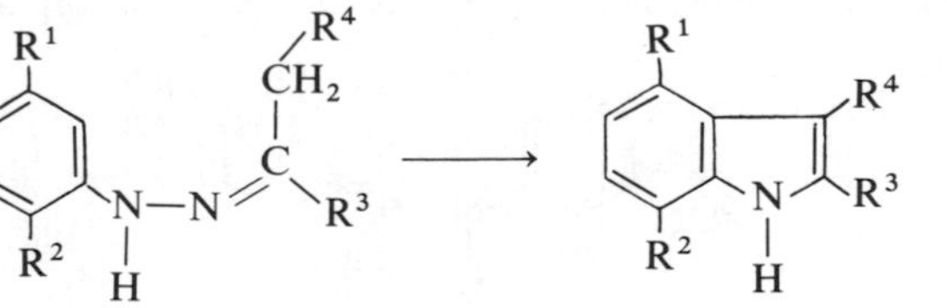

R^1 R^4 CH_2 C N—N R^3 R^2 H

515

R^1 R^4 N R^3 R^2 H

516

Although this approach has been so far used mainly for the unambiguous synthesis of 4-substituted indoles, an obvious modification in the substitution pattern of the original hydrazone would unambiguously furnish 6-substituted indoles.

Often R^2 is halogen in these directed syntheses. Thus, contrary to the earlier reports (Borsche, Witte, A. and Bothe, 1908; Perkin and Plant, 1921) that sulphuric acid catalysed indolization of cyclohexanone 3-nitrophenylhydrazone [**508**; $R^1 = NO_2$, $R^2 = H$, $R^3 + R^4 = (CH_2)_4$] produced only one product which was, without evidence, assigned structure **510** [$R^1 = NO_2$, $R^2 = H$, $R^3 + R^4 = (CH_2)_4$], it was later found that this reaction afforded a mixture of both possible indolic products, 1,2,3,4-tetrahydro-5- and -7-nitrocarbazole **509** and **510** [$R^1 = NO_2$, $R^2 = H$, $R^3 + R^4 = (CH_2)_4$], respectively (Plant, 1936). The structures of these products were assigned by the reduction of the former to **509** [$R^1 = NH_2$, $R^2 = H$, $R^3 + R^4 = (CH_2)_4$] which was unambiguously synthesized by the sulphuric acid catalysed indolization of **515** [$R^1 = NO_2$, $R^2 = Cl$, $R^3 + R^4 = (CH_2)_4$] into **516** [$R^1 = NO_2$, $R^2 = Cl$, $R^3 + R^4 = (CH_2)_4$] (Hall and Plant, 1953; Perkin and Plant, 1921) followed by reduction of the nitro group, with simultaneous reductive removal of the chloro group, with tin and hydrochloric acid (Hall and Plant, 1953; Plant, 1936). Likewise, the sulphuric acid catalysed indolization of 4-methylcyclohexanone 3-nitrophenylhydrazone [**508**; $R^1 = NO_2$, $R^2 = H$, $R^3 + R^4 = (CH_2)_2CH(CH_3)CH_2$] afforded both 1,2,3,4-tetrahydro-3-methyl-5- and -7-nitrocarbazole **509** and **510** [$R^1 = NO_2$, $R^2 = H$, $R^3 + R^4 = (CH_2)_2CH(CH_3)CH_2$], respectively (Plant and Rosser, 1928). The structures of these products were later established (Moggridge and Plant, 1937) by reduction of the former to afford **509** [$R^1 = NH_2$, $R^2 = H$, $R^3 + R^4 = (CH_2)_2CH(CH_3)CH_2$]. This product was then unambiguously synthesized by the indolization of **515** [$R^1 = NO_2$, $R^2 = Cl$, $R^3 + R^4 = (CH_2)_2CH(CH_3)CH_2$] into **516** [$R^1 = NO_2$, $R^2 = Cl$, $R^3 + R^4 = (CH_2)_2CH(CH_3)CH_2$] which, upon treatment with tin and hydrochloric acid, underwent reduction of the nitro group and reductive dechlorination to form the desired product. However, a similar attempt to establish the orientation of the nitro substituent in the products resulting from the indolization of ethyl methyl ketone 3-nitrophenylhydrazone was unsuccessful (Plant and Whitaker, 1940). Several piperidin-4-one 3-alkoxyphenylhydrazones **517** [$R^1 = CH_3$, $R^2 = R^3 = H$; $R^1 = (CH_2)_2CH_3$ and $(CH_2)_3CH_3$, $R^2 = H$, $R^3 = CH_3$ and $R^1 = CH_3$, $R^2 = H$, $R^3 = (CH_2)_2CH_3$, $CH(CH_3)_2$ and $(CH_2)_3CH_3$]

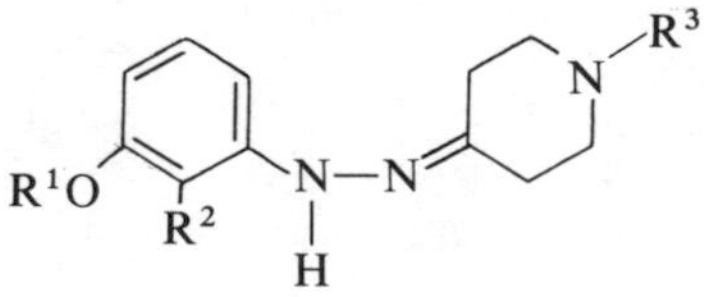

517

have been indolized to afford in each case only one isolated product, **518** [$R^1 = CH_3$, $R^2 = R^3 = H$; $R^1 = (CH_2)_2CH_3$ and $(CH_2)_3CH_3$, $R^2 = H$, $R^3 = CH_3$ and $R^1 = CH_3$, $R^2 = H$, $R^3 = (CH_2)_2CH_3$, $CH(CH_3)_2$ and $(CH_2)_3CH_3$, respectively]. The orientation of **518** ($R^1 = R^3 = CH_3, R^2 = H$) was established by its unambiguous synthesis via the indolization of **517** ($R^1 = R^3 = CH_3$, $R^2 = Cl$) to give **518** ($R^1 = R^3 = CH_3$, $R^2 = Cl$) which

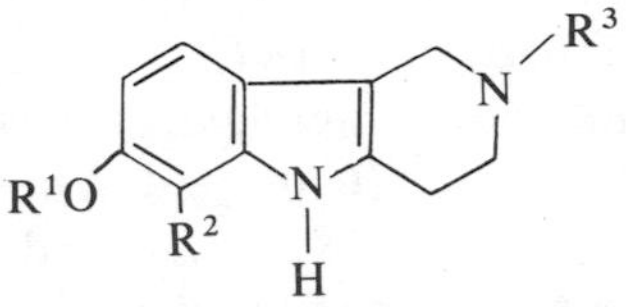

R³ N R¹O R² N H

518

was then dehalogenated by treatment with sodium in liquid ammonia. The orientations of the corresponding propoxy and butoxy homologues were then related to this unambiguously synthesized product by O-dealkylation and the structural orientations of other products were assumed by analogy (Cattanach, Cohen, A. and Heath-Brown, 1968). Indolization of **519** (R = H) produced a mixture of **520** ($R^1 = CH_3$, $R^2 = R^3 = H$) and **520** ($R^1 = R^3 = H$, $R^2 = CH_3$). These structures were differentiated by the reduction of the former isomer to the corresponding tetrahydrocarbazole **521** [$R^1 = R^2 = CH_3$,

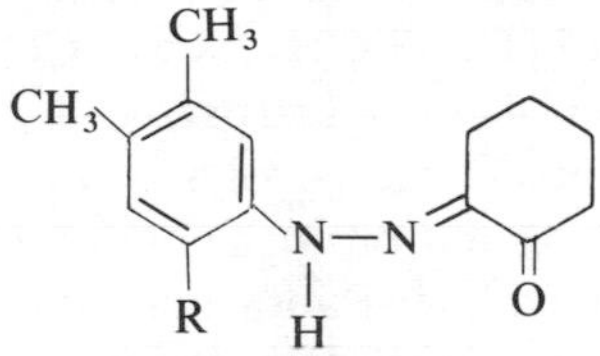

CH₃ CH₃ N—N R H O

519

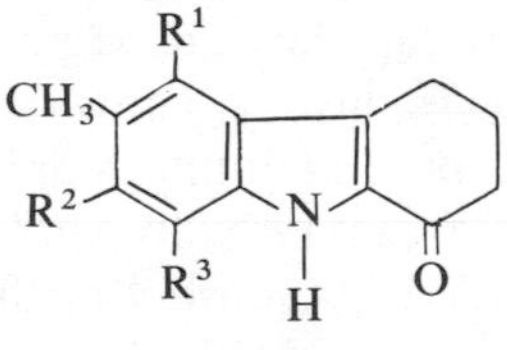

R¹ CH₃ R² N R³ H O

520

R¹ R² R⁵ R³ N R⁴ H

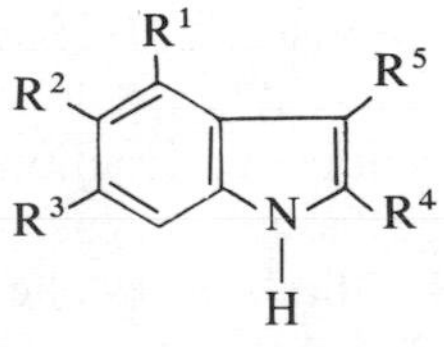

521

$R^3 = H$, $R^4 + R^5 = (CH_2)_4$] which was dehydrogenated to afford 3,4-dimethylcarbazole. This was unambiguously synthesized by indolization of **519** (R = Br) to yield **520** ($R^1 = CH_3$, $R^2 = H$, $R^3 = Br$) which upon successive Clemmensen reduction, chloranil dehydrogenation and reductive debromination with red phosphorus and hydrobromic acid gave 3,4-dimethylcarbazole (Anderson, G. and Campbell, 1950). Indolization of cyclohexanone 4-methyl-3-nitrophenylhydrazone afforded only one product which was shown to be 1,2,3,4-tetrahydro-6-methyl-7-nitrocarbazole. That the indolization product was not the corresponding 5-nitro isomer was established by the synthesis of this isomer by nitration of 9-acetyl-1,2,3,4-tetrahydro-6-methylcarbazole with subsequent hydrolysis of the acetyl group. The position of the nitro substituent in this nitration product was established by its reduction with tin and alcoholic hydrochloric acid followed by benzoylation to furnish 5-benzamido-9-benzoyl-1,2,3,4,4a,9a-hexahydro-6-methylcarbazole which was unambiguously synthesized by indolization of cyclohexanone 2-bromo-4-methyl-5-nitrophenylhydrazone to afford 1-bromo-5,6,7,8-tetrahydro-3-methyl-4-nitrocarbazole which was then sequentially reduced with tin and alcoholic hydrochloric acid and benzoylated (Moggridge and Plant, 1937).

Many other examples of this directed synthetic approach have appeared and, indeed, it has become a recognized route for the synthesis of 4-substituted indoles (Nagasaka, Yuge and Ohki, 1977). Thus, **515** ($R^1 = NO_2$, $R^2 = Cl$, $R^3 = R^4 = CH_3$) was indolized into **516** ($R^1 = NO_2$, $R^2 = Cl$, $R^3 = R^4 = CH_3$) which upon reduction with tin and hydrochloric acid yielded **516** ($R^1 = NH_2$, $R^2 = H$, $R^3 = R^4 = CH_3$) (Plant and Whitaker, 1940). Compounds **515** [$R^1 = COOH$, $R^2 = Cl$, $R^3 = COOC_2H_5$ and H, $R^4 = (CH_2)_2COOH$] were indolized using boron trifluoride-in-acetic acid and 2.25M sulphuric acid, respectively, into **516** [$R^1 = COOH$, $R^2 = Cl$, $R^3 = COOC_2H_5$ and H, respectively, $R^4 = (CH_2)_2COOH$] which afforded **516** [$R^1 = COOH$, $R^2 = H$, $R^3 = COOH$ and H, respectively, $R^4 = (CH_2)_2COOH$] upon hydrogenolysis in the presence of 10% palladium-on-carbon in alkaline solution (Bowman, Goodburn and Reynolds, A. A., 1972 – see also Campaigne and Lake, 1959; Cummins and Tomlinson, M. L., 1955). Compound **515** [$R^1 = COCH_3$, $R^2 = Cl$, $R^3 + R^4 = (CH_2)_4$] in 17% sulphuric acid gave **516** [$R^1 = COCH_3$, $R^2 = Cl$, $R^3 + R^4 = (CH_2)_4$] which upon heating with distilled naphthalene and Adam's platinum catalyst in tetralin underwent both reductive dechlorination and dehydrogenation to afford 4-acetylcarbazole (Manske, R. H. F. and Kulka, 1950a). Compound **515** [$R^1 = CH_3$, $R^2 = Cl$, $R^3 + R^4 = (CH_2)_4$] was similarly converted into 4-methylcarbazole by sequential reaction with boiling dilute sulphuric acid and pyrolysis with palladium-on-carbon in a stream of hydrogen (dehydrogenation with chloranil in boiling xylene afforded 1-chloro-4-methylcarbazole) (Pausacker and Robinson, R., 1947). Compound **515** [$R^1 = (CH_2)_2COOC_2H_5$, $R^2 = Cl$, $R^3 = COOC_2H_5$, $R^4 = H$] was converted into **516** [$R^1 = (CH_2)_2COOC_2H_5$, R^2–$R^4 = H$] via **516** [$R^1 = (CH_2)_2COOC_2H_5$, $R^2 = Cl$, $R^3 = COOH$, $R^4 = H$] and **516** [$R^1 = (CH_2)_2COOC_2H_5$, $R^2 = Cl$, $R^3 = R^4 = H$] (Nagasaka, Yuge and Ohki, 1977). In this last indolization, the [3,3]-sigmatropic shift also involved the benzenoid carbon atom carrying the chloro substituent, with subsequent 1,3-migration of the chlorine atom, to afford **416** (see Chapter II, G1).

Certainly, when using halogeno (and alkyl – see below) *ortho* substituents to unambiguously direct Fischer indolizations, the possibility of the [3,3]-sigmatropic shift involving the substituted *ortho* carbon atom with subsequent migration, elimination or substitution of the original *ortho* substituent (see Chapter II, G) must be considered. For these reasons, unambiguous syntheses alternative to those using Fischer indolizations directed by *ortho* methyl groups have been employed (Miller, B. and Matjeka, 1980). Such a substitution of an *ortho* chlorine atom has been made use of in directional studies upon ambiguous indolizations. Thus, the only product isolated from the indolization of **522** [$R^1 = OCH_3$, $R^2 = CH_3$, $R^3 + R^4 = (CH_2)_4$] was shown to have structure **521** [$R^1 = H$, $R^2 = OCH_3$, $R^3 = CH_3$, $R^4 + R^5 = (CH_2)_4$] by its dehydrogenation and O-demethylation into 3-hydroxy-2-methylcarbazole.

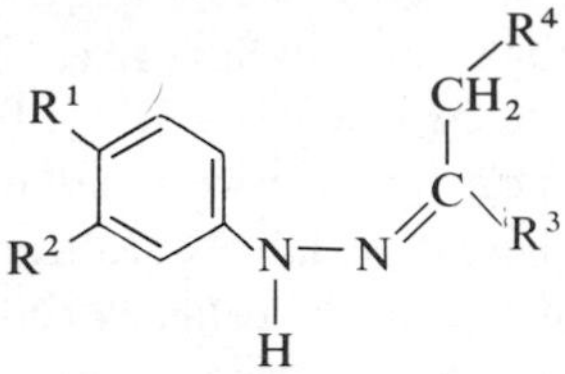

522

The structure of this product was established by indolization of cyclohexanone 2-chloro-5-methylphenylhydrazone with a dilute sulphuric acid catalyst. This afforded 8-chloro-1,2,3,4-tetrahydro-5-methylcarbazole as the expected major product, along with a minor product, **521** [$R^1 = H$, $R^2 = OH$, $R^3 = CH_3$, $R^4 + R^5 = (CH_2)_4$], in which the chloro substituent was replaced by a hydroxyl group after new C—C bond formation involving the chlorine-bearing *ortho* carbon atom (see Chapter II, G1). Dehydrogenation of this minor product then produced 3-hydroxy-2-methylcarbazole **523** (Cummins, Kaye and Tomlinson, M. L., 1954). When the indolization of cyclohexanone 2-chloro-5-methylphenylhydrazone was first performed (Pausacker and Robinson, R., 1947), the structure of the minor product was not established, it being tentatively formulated as **423**. Surprisingly, in view of the later (Cummins, Kaye and Tomlinson, M. L., 1954) establishment of its structure as 1,2,3,4-tetrahydro-

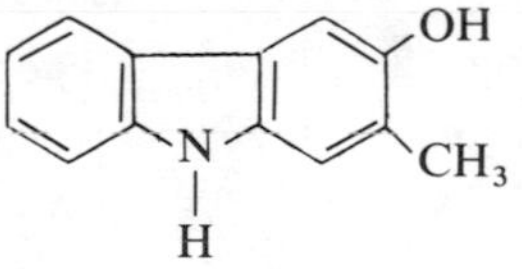

523

6-hydroxy-7-methylcarbazole, it was in the earlier paper (Pausacker and Robinson, R., 1947) reported to be devoid of phenolic character, although 1,2,3,4-tetrahydro-6-hydroxycarbazole was later (Milne and Tomlinson, M. L., 1952) found to possess normal phenolic properties.

Compound **524** ($R^1 = OCH_3$, $R^2 = Br$) was indolized into **525** ($R^1 = OCH_3$, $R^2 = Br$), using a hydrochloric acid–acetic acid mixture, which

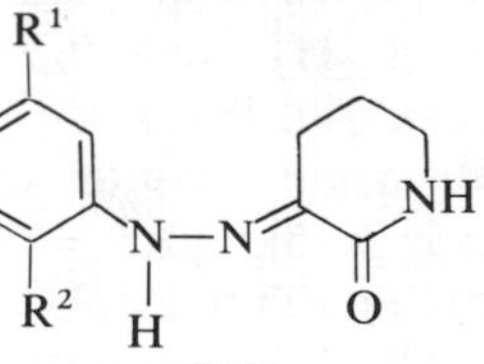

524 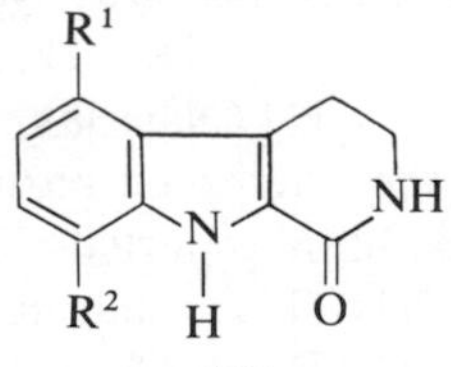 **525**

subsequently afforded **525** (R^1 = OCH_3, R^2 = H) [the m.p. reported for this product differs considerably from that previously reported (Abramovitch, 1956)] upon treatment with Raney nickel and hydrazine hydrate in refluxing ethanol (Suvorov, Fedotova, M. V., Orlova, L. M. and Ogareva, 1962). This indolization was later repeated (Morozovskaya, Ogareva and Suvorov, 1969) and then (Morozovskaya, Ogareva and Suvorov, 1970) extended to that of **524** (R^1 = OCH_3, R^2 = Cl) into **525** (R^1 = OCH_3, R^2 = Cl) which yielded **525** (R^1 = OCH_3, R^2 = H) upon heating with palladium-on-carbon in ethanol containing hydrazine. In these two later indolizations, the new C—C bond formation also occurred at the other *ortho* carbon atom, with the ultimate elimination of the bromine and chlorine atoms, respectively, to produce a 0.5% yield of the corresponding 6-methoxy indolic product (Morozovskaya, Ogareva and Suvorov, 1969, 1970) (see Chapter II, G1). Treatment of **524** (R^1 = $COCH_3$, R^2 = Cl) with refluxing formic acid afforded **525** (R^1 = $COCH_3$, R^2 = Cl). This product, upon alkaline hydrolysis and subsequent refluxing in a hydrochloric acid–acetic acid mixture, furnished **526** (R^1 = CH_3, R^2 = Cl, R^3 = H and COOH, R^4 = H). The former product, upon treatment with hydrogen in the presence of 10% palladium-on-carbon at room temperature and atmospheric pressure gave **526** (R^1 = CH_3, R^2–R^4 = H) (after the uptake of 1 mole of hydrogen) and ultimately **527** (R = H). Compound **526** (R^1 = CH_3, R^2 = Cl, R^3 = $COOC_2H_5$, R^4 = H) likewise afforded **527** (R = $COOC_2H_5$) (Strandtmann, Cohen, M. P. and Shavel, 1965).

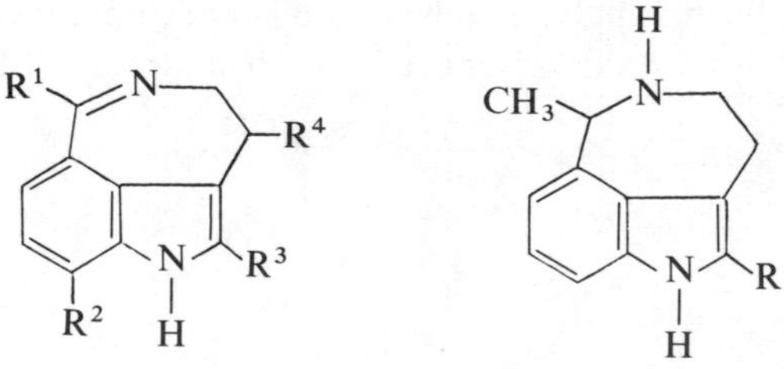

526 **527**

Syn- or *anti-* **524** [R^1 = $COOCH_3$, $CH_2COOC_2H_5$ and $(CH_2)_2COOC_2H_5$, R^2 = Cl], upon treatment with boron trifluoride etherate in acetic acid, yielded **525** [R^1 = $COOCH_3$, $CH_2COOC_2H_5$ and $(CH_2)_2COOC_2H_5$, respectively, R^2 = Cl] which upon catalytic hydrogenolysis in 60% methanol with a 5% palladium-on-carbon catalyst in the presence of ammonium acetate afforded **525** [R^1 = $COOCH_3$, $CH_2COOC_2H_5$ and $(CH_2)_2COOC_2H_5$, respectively, R^2 = H] (Nagasaka, Yuge and Ohki, 1977).

Failures at the attempted reductive hydrogenolysis stage in this synthetic approach have been reported. Thus, although **515** [R^1 = NO_2, R^2 = Cl, $R^3 + R^4$ = $(CH_2)_3$] was indolized in refluxing dilute sulphuric acid to yield **516** [R^1 = NO_2, R^2 = Cl, $R^3 + R^4$ = $(CH_2)_3$], attempts to reductively dechlorinate this product into **516** [$R^1 = NO_2, R^2 = H, R^3 + R^4 = (CH_2)_3$] failed (Plant and Whitaker, 1940). Likewise, **515** ($R^1 = CH_3$, $R^2 = Cl$, $R^3 = R^4 = C_6H_5$) was indolized with a hydrochloric acid–acetic acid mixture to produce **516** ($R^1 = CH_3$, $R^2 = Cl$, $R^3 = R^4 = C_6H_5$) but heating this with palladium-on-carbon in a hydrogen atmosphere failed to give any recognizable product, although hydrogen chloride was evolved (Coldham, Lewis, J. W. and Plant, 1954). The desired product from this last reaction, **516** ($R^1 = CH_3$, $R^2 = H$, $R^3 = R^4 = C_6H_5$), was, however, synthesized via the indolization of **515** ($R^1 = CH_3$, $R^2 = COOH$, $R^3 = R^4 = C_6H_5$), using a hydrochloric acid–acetic acid catalyst, to afford **516** ($R^1 = CH_3$, $R^2 = COOH$, $R^3 = R^4 = COOH$) which was subsequently decarboxylated by pyrolysis of its sodium salt with soda-lime (a similar reaction sequence using cyclohexanone in place of benzyl phenyl ketone afforded the hydroperoxide **528**). The product obtained from this reaction sequence, **516** ($R^1 = CH_3$, $R^2 = H$, $R^3 = R^4 = C_6H_5$), was identical with one of the two isomeric products obtained by indolization of benzyl phenyl ketone 3-methylphenylhydrazone using zinc chloride as the catalyst (Coldham, Lewis, J. W. and Plant, 1954).

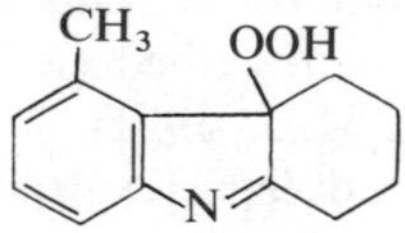

528

As well as undergoing hydrogenolysis, 7-chloro substituents in 4-nitroindoles can be nucleophilically displaced with secondary amines. When a 2-carboxylic acid or ester grouping is also present, this reaction occurs when the reactants are refluxed. Thus, for example, ethyl pyruvate 2-chloro-5-nitrophenylhydrazone (**515**; $R^1 = NO_2$, $R^2 = Cl$, $R^3 = COOC_2H_5$, $R^4 = H$) was indolized by treatment with polyphosphoric acid to furnish **516** ($R^1 = NO_2$, $R^2 = Cl$, $R^3 = COOC_2H_5$, $R^4 = H$) which upon reaction, under these conditions, with the appropriate secondary amine [$(CH_2)_{4-6}NH$, morpholine or 4-methylpiperazine] afforded the corresponding aminoindoles **516** [$R^1 = NO_2$, $R^2 = (CH_2)_{4-6}N$, morpholino and 4-methyl-1-piperazinyl, respectively, $R^3 = COOC_2H_5$, $R^4 = H$], along with the corresponding amides formed by reaction between the ester function and the secondary amines (Ainsworth, D. P. and Suschitzky, 1967b). When using indoles not carrying a 2-carboxylic acid or ester function, such nucleophilic displacements could only be effected when the reactants were heated together in a sealed tube at 140 °C (Ainsworth, D. P. and Suschitzky, 1967a, 1967b).

Use of a carboxylic acid group in directing indolization involved the conversion of **515** [$R^1 = Cl$, $R^2 = COOH$, $R^3 + R^4 = (CH_2)_4$] into **516** [$R^1 = Cl$, $R^2 = COOH$, $R^3 + R^4 = (CH_2)_4$]. Although this product could not be decarboxylated, its N-acetyl derivative was readily decarboxylated to produce 9-acetyl-5-chloro-1,2,3,4-tetrahydracarbazole, identical (Plant and Wilson, A. E. J., 1939) with the product obtained by acetylation of one of the two indolization products of cyclohexanone 3-chlorophenylhydrazone (Moggridge and Plant, 1937). In this study, the use of an *ortho* halogen as a blocking group was obviously not possible.

An *ortho* methyl group has also been used to direct Fischer indolization. Thus, indolization of cyclohexanone 3-fluorophenylhydrazone [**508**; $R^1 = F$, $R^2 = H$, $R^3 + R^4 = (CH_2)_4$] in refluxing glacial acetic acid afforded a mixture of **509** and **510** [$R^1 = F$, $R^2 = H$, $R^3 + R^4 = (CH_2)_4$]. These were dehydrogenated to the corresponding 4- and 2-fluorocarbazole, respectively, and the substituent orientation of the former isomer was established by its unambiguous synthesis by indolization of **515** [$R^1 = F$, $R^2 = CH_3$, $R^3 + R^4 = (CH_2)_4$] into **516** [$R^1 = F$, $R^2 = CH_3$, $R^3 + R^4 = (CH_2)_4$], followed by dehydrogenation to the corresponding carbazole. This was demethylated via selenium dioxide oxidation to the carboxylic acid with subsequent decarboxylation by treatment with copper[copper(II)?] chromite in boiling quinoline (Allen, F. L. and Suschitzky, 1953 – see also Suschitzky, 1953).

Another unambiguous synthetic route to 4-substituted indoles from 4,5,6,7-tetrahydroindol-4-ones has been described (Remers, Roth, Gibs and Weiss, M. J., 1971; Remers and Weiss, M. J., 1971). This may also be of use in establishing the structure(s) of the indole(s) resulting from the indolization of 3-substituted phenylhydrazones.

2. *Oxidation Degradations*

Benzyl phenyl ketone 3-carboxyphenylhydrazone (**508**; $R^1 = COOH$, $R^2 = H$, $R^3 = R^4 = C_6H_5$) gave both **509** and **510** ($R^1 = COOH$, $R^2 = H$, $R^3 = R^4 = C_6H_5$). These structures were differentiated by oxidation of the indolic 2,3-double bond in the latter product, with chromium(VI) oxide in acetic acid, to afford **529** ($R^1 = R^2 = H$, $R^3 = COOH$, $R^4 = COC_6H_5$, $R^5 = C_6H_6$) (Coldham, Lewis, J. W. and Plant, 1954). The hydrazone **508** [$R^1 = COOH$, $R^2 = H$, $R^3 = COOC_2H_5$, $R^4 = (CH_2)_2COOH$] was indolized using ethanolic sulphuric acid to yield **509** and **510** [$R^1 = R^3 = COOC_2H_5$, $R^2 = H$, $R^4 = (CH_2)_2COOC_2H_5$]. Structural assignment of these products resulted from their oxidation with chromium(VI) oxide in acetic acid to furnish **529** [$R^1 = COOC_2H_5$, $R^2 = R^3 = H$, $R^4 = COCOOC_2H_5$, $R^5 = (CH_2)_2COOC_2H_5$] and **529** [$R^1 = R^2 = H$, $R^3 = COOC_2H_5$, $R^4 = COCOOC_2H_5$, $R^5 = (CH_2)_2COOC_2H_5$], respectively.

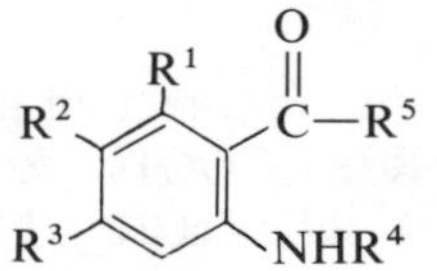

O
R¹ ‖
R² C—R⁵
R³ NHR⁴

529

These products, upon treatment with hot aqueous sodium hydroxide, afforded the intramolecular Claisen condensation product **530**, which had to arise from the original 4-carbethoxyindolic product, along with the simple hydrolysis product, and the hydrolysis product **529** [$R^1 = R^2 = R^4 = H, R^3 = COOH, R^5 = (CH_2)_2COOH$], respectively (Koelsch, 1943). Although it was initially

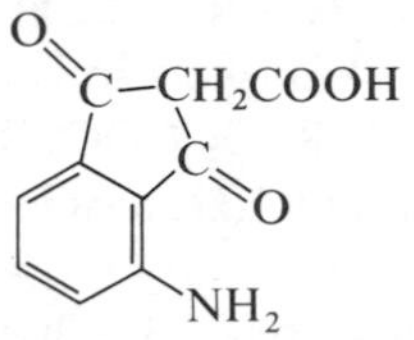

O
C—CH_2COOH
C
O
NH_2
350

claimed (Bauer and Strauss, 1932) that indolization of ethyl methyl ketone 3-nitrophenylhydrazone (**508**; $R^1 = NO_2$, $R^2 = H$, $R^3 = R^4 = CH_3$) produced only one indolic product, it was later found that both possible isomeric indolic products were produced. The structures of these products were distinguished by their chromium(VI) oxide in acetic acid oxidation which yielded **529** ($R^1 = NO_2$, $R^2 = R^3 = H$, $R^4 = COCH_3$, $R^5 = CH_3$) and **529** ($R^1 = R^2 = H$, $R^3 = NO_2$, $R^4 = COCH_3$, $R^5 = CH_3$), hydrolysis of which afforded the corresponding primary amines (Schofield and Theobald, 1949). [A similar degradative sequence was subsequently performed upon the analogous 1-acetylindoles (Atkinson, Simpson and Taylor, A., 1954).] The latter amine was diazotized and then reacted with hypophosphorus acid to afford 4-nitroacetophenone which upon oxidation with hypochlorite yielded 4-nitrobenzoic acid. The identification of this product established the orientation of the nitro group as being in the 6-position on the original indole nucleus (Schofield and Theobald, 1949). Likewise, methyl propyl ketone and benzyl methyl ketone 3-nitrophenylhydrazone were indolized to give **509** and **510** ($R^1 = NO_2$, $R^2 = H$; $R^3 = CH_3$, $R^4 = C_2H_5$ and $R^3 = R^4 = C_6H_5$, respectively). These were oxidized with chromium(VI) oxide in acetic acid, followed by hydrolysis, to give 2-amino-6- and -4-nitropropiophenone and 6- and 4-nitro-2-aminobenzophenone, respectively. The second of these four products was sequentially deaminated, reduced and acetylated to afford the known 4-acetamidopropiophenone and similarly the fourth product was converted into the known 4-aminobenzophenone, these correlations establishing the orientation of the nitro groups as being in the 6-positions of the original indoles (Schofield and Theobald, 1950). From the indolization of benzyl methyl ketone 3-nitrophenylhydrazone, only **509** ($R^1 = NO_2$, $R_2 = H$, $R^3 = CH_3$, $R^4 = C_6H_5$) was isolated, the structure of which was established by chromium(VI) oxide in acetic acid oxidation, followed by hydrolysis, to form the known 2-amino-6-nitrobenzophenone. Other unidentified products were also formed (Schofield and Theobald, 1950). The products obtained from the indolization of **508** ($R^1 = NO_2$, $R^2 = H$, $R^3 = COOC_2H_5$, $R^4 = C_6H_5$) were identified as

509 and **510** ($R^1 = NO_2$, $R^2 = H$, $R^3 = COOC_2H_5$, $R^4 = C_6H_5$) by sequential chromium(VI) oxide in acetic acid oxidation and hydrolysis into **529** ($R^1 = NO_2$, R^2–$R^4 = H$, $R^5 = C_6H_5$) and **529** ($R^1 = R^2 = R^4 = H$, $R^3 = NO_2$, $R^5 = C_6H_5$), respectively (Sato, Y., 1963b). Indolization of **508** ($R^1 = NO_2$, $R^2 = H$, $R^3 = C_6H_5$, $R^4 = CH_3$) afforded both **509** and **510** ($R^1 = NO_2$, $R^2 = H$, $R^3 = C_6H_5$, $R^4 = CH_3$), the structures of which were established by their chromium(VI) oxide in acetic acid oxidations to produce **529** ($R^1 = NO_2$, $R^2 = R^3 = H$, $R^4 = COC_6H_5$, $R^5 = CH_3$) and **529** ($R^1 = R^2 = H$, $R^3 = NO_2$, $R^4 = COC_6H_5$, $R^5 = CH_3$), respectively, known compounds (Frasca, 1962a).

Indolization of ethyl methyl ketone 3-methylphenylhydrazone (**508**; R^1–$R^3 = CH_3$) afforded only **510** ($R^1 = R^3 = R^4 = CH_3$, $R^2 = H$) (Ockenden and Schofield, 1957), the structure of which was again established by oxidative scission of the indolic 2,3-double bond but now using ozone, to yield 2-acetamido-4-methylacetophenone (Ockenden and Schofield, 1953a). Similar ozonolyses were subsequently (Ockenden and Schofield, 1957) used to establish substituent orientations in the products resulting from other indolizations with 3-substituted phenylhydrazones although the method failed when applied to the indolic products resulting from the indolization of benzyl phenyl ketone and ethyl methyl ketone 3-methoxyphenylhydrazone, their treatment with ozone resulting only in tar formation.

Indolization of ethyl pyruvate 3-nitrophenylhydrazone (**508**; $R^1 = NO_2$, $R^2 = R^4 = H$, $R^3 = COOC_2H_5$) afforded both **509** and **510** ($R^1 = NO_2$, $R^2 = R^4 = H$, $R^3 = COOC_2H_5$). The structure of the latter isomer was established by its hydrolysis and decarboxylation to 6-nitroindole (Parmerter, Cook, A. G. and Dixon, 1958), which had previously (Majima and Kotake, M., 1930) been prepared by nitration of indole-3-carboxylic acid with subsequent decarboxylation of the product. The orientation of the nitro group in the nitration product had been established by its oxidation, with potassium permanganate, into 2-amino-4-nitrobenzoic acid (Majima and Kotake, M., 1930). Indolization of **508** [$R^1 = Br$, $R^2 = H$, $R^3 + R^4 = (CH_2)_4$] with boiling sulphuric acid gave both **509** and **510** [$R^1 = Br$, $R^2 = H$, $R^3 + R^4 = (CH_2)_4$]. The structure of the latter isomer was established by its 9-acetylation and subsequent reaction with nitric acid–acetic acid to give **531** which furnished **532** ($R^1 = H$, $R^2 = Br$, $R^3 = COCH_3$) upon boiling in acetic anhydride. This

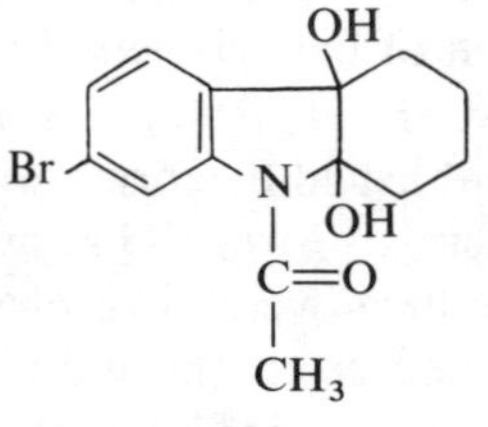

Br OH OH N C=O CH3

531

O R1 R2 R1 N R3

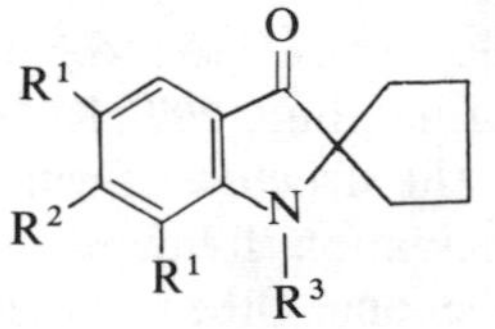

532

compound was then hydrolysed and nitrated to afford **532** ($R^1 = NO_2$, $R^2 = Br$, $R^3 = H$). This reacted with aniline to yield **532** ($R^1 = NO_2$, $R^2 = NHC_6H_5$, $R^3 = H$) (Plant and Wilson, A. E. J., 1939) which was similarly prepared from the known (Moggridge and Plant, 1937) 9-acetyl-7-chloro-1,2,3,4-tetrahydrocarbazole. Indolization of **508** [$R^1 = NO_2$, $R^2 = H$, $R^3 + R^4 = (CH_2)_3$] afforded both **509** and **510** [$R^1 = NO_2$, $R^2 = H$, $R^3 + R^4 = (CH_2)_3$] (Plant, 1929). The structure of the 1-acetyl derivative of the latter isomer (Perkin and Plant, 1923b) was established by treating it with nitric acid to give **533** (R = H) which was degraded with alkali to yield **529** [$R^1 = R^2 = H$, $R^3 = NO_2$, $R^4 = COCH_3$, $R^5 = (CH_2)_3COOH$], oxidation then affording

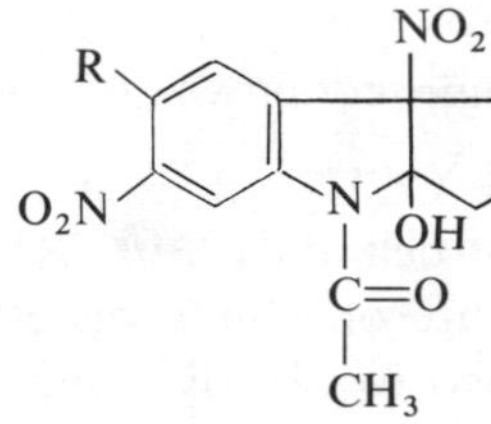

533

2-acetamido-4-nitrobenzoic acid (Plant and Whitaker, 1940). This product was also obtained by a similar degradation of one of the two indolic products resulting from the indolization of ethyl methyl ketone 3-nitrophenylhydrazone, thereby again establishing the orientation of the nitro substituent (Plant and Whitaker, 1940). Cyclopentanone 4-chloro-3-nitrophenylhydrazone [**522**; $R^1 = Cl$, $R^2 = NO_2$, $R^3 + R^4 = (CH_2)_3$] was indolized to both possible isomeric products, **521** [$R^1 = NO_2$, $R^2 = Cl$, $R^3 = H$, $R^4 + R^5 = (CH_2)_3$] and **521** [$R^1 = H$, $R^2 = Cl$, $R^3 = NO_2$, $R^4 + R^5 = (CH_2)_3$] (Massey and Plant, 1931). The structures of these isomeric products were differentiated by establishing that of the latter by N-acetylation followed by treatment with nitric acid in boiling acetic acid to afford **533** (R = Cl). This was treated with aqueous potassium hydroxide to afford **529** [$R^1 = H$, $R^2 = Cl$, $R^3 = NO_2$, $R^4 = COCH_3$, $R^5 = (CH_2)_3COOH$] which upon oxidation with potassium permanganate gave **529** ($R^1 = H$, $R^2 = Cl$, $R^3 = NO_2$, $R^4 = COCH_3$, $R^5 = OH$), characterized and identified as the methyl ester of its amide hydrolysis product, methyl 2-amino-5-chloro-4-nitrobenzoate (Plant and Whitaker, 1940). Indolization of cyclohexanone 3-methoxyphenylhydrazone [**508**; $R^1 = OCH_3$, $R^2 = H$, $R^3 + R^4 = (CH_2)_4$] afforded **509** and **510** [$R^1 = OCH_3$, $R^2 = H$, $R^3 + R^4 = (CH_2)_4$] (Chalmers, J. R., Openshaw and Smith, G. F., 1957) whose structures were assigned by melting point comparisons with previously synthesized authentic specimens (Cummins and Tomlinson, M. L., 1955). The structure of the latter product was further verified by its electrolytic reduction, followed by acetylation and nitration to give **534** which upon oxidation with potassium permanganate furnished a 2% yield of a

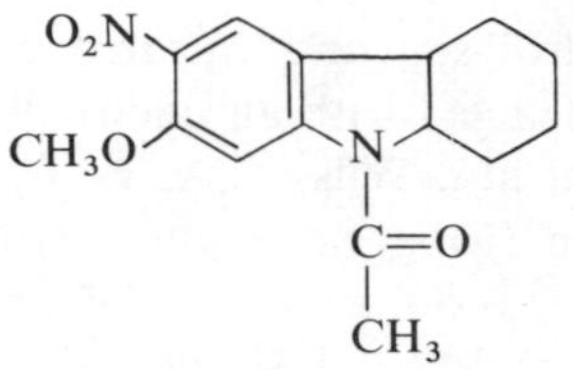

O_2N CH_3O N C=O CH_3

534

compound which very likely had structure **529** (R^1 = H, R^2 = NO_2, R^3 = OCH_3, R^4 = $COCH_3$, R^5 = OH), it being identified as the nitration product of **529** (R^1 = R^2 = H, R^3 = OCH_3, R^4 = $COCH_3$, R^5 = OH) (Chalmers, J. R., Openshaw and Smith, G. F., 1957).

3. Other Degradations to Compounds of Known Structures

Indolization of **508** (R^1 = OCH_3, R^2 = R^4 = H, R^3 = $COOC_2H_5$) in ethanolic sulphuric acid afforded only **510** (R^1 = OCH_3, R^2 = R^4 = H, R^3 = $COOC_2H_5$), the structure of which was established (Wieland, T. and Grimm, 1965) by its hydrolysis to **510** (R^1 = OCH_3, R^2 = R^4 = H, R^3 = COOH), followed by decarboxylation, to afford **510** (R^1 = OCH_3, R^2–R^4 = H). Both these last compounds were available (Kermack, Perkin and Robinson, R., 1921). The arylhydrazone **508** [R^1 = COOH, R^2 = H, R^3 + R^4 = $(CH_2)_4$], upon indolization in boiling sulphuric acid, gave both **509** and **510** [R^1 = COOH, R^2 = H, R^3 + R^4 = $(CH_2)_4$] (Collar and Plant, 1926). The structure of the latter isomer was established by dehydrogenation of its methyl ester, with subsequent ester hydrolysis, to yield carbazole-2-carboxylic acid (Moggridge and Plant, 1937), a known compound (Plant and Williams, S. B. C., 1934). The arylhydrazone **508** [R^1 = Cl, R^2 = H, R^3 + R^4 = $(CH_2)_4$] was similarly indolized into both **509** and **510** [R^1 = Cl, R^2 = H, R^3 + R^4 = $(CH_2)_4$], the structure of the latter isomer being determined (Moggridge and Plant, 1937) by its dehydrogenation to give 2-chlorocarbazole (Ullmann, 1904). Indolization of **508** (R^1 = OCH_3, R^2 = H, R^3 = $COOC_2H_5$, R^4 = $CH_2COOC_2H_5$), using ethanolic hydrogen chloride, afforded only **510** (R^1 = OCH_3, R^2 = H, R^3 = $COOC_2H_5$, R^4 = $CH_2COOC_2H_5$), the structure of which was verified by its hydrolysis into the corresponding dicarboxylic acid (Findlay, S. P. and Dougherty, 1948). Indolization of **508** [R^1 = OCH_3, R^2 = H, R^3 + R^4 = $CONH(CH_2)_2$] led to the isolation of both **509** and **510** [R^1 = OCH_3, R^2 = H, R^3 + R^4 = $CONH(CH_2)_2$] (Abramovitch, 1956). The structure of the latter product was based upon the similarity of its m.p. (199–200 °C) with that of a previously synthesized specimen (m.p. = 198 °C) (Barrett, H. S. B., Perkin and Robinson, R., 1929) and upon its hydrolysis and decarboxylation to 6-methoxytryptamine (Akabori and Saito, K., 1930b). Indolization of **508** (R^1 = R^4 = $CH_2COOC_2H_5$, R^2 = H, R^3 = $COOC_2H_5$) afforded both **509** and **510** (R^1 = R^4 = $CH_2COOC_2H_5$, R^2 = H, R^3 = $COOC_2H_5$). These were separated and differentiated by their hydrolyses to the corresponding tricarboxylic acids which were decarboxylated, by heating with resorcinol, to give 3,4- and 3,6-dimethylindole, respectively. The former product had physical constants consistent with those recorded for a previously unambiguously synthesized specimen (Plieninger, 1954). Indolization of **508** ($R^1 = OCH_3$, $R^2 = H$, $R^3 = H$ and CH_3, $R^4 = SC_6H_5$) and **508** ($R^1 = Cl$, $R^2 = R^3 = H$, $R^4 = SC_6H_5$) gave **510** ($R^1 = OCH_3$, $R^2 = H$, $R^3 = H$ and CH_3, $R^4 = SC_6H_5$) and **510** ($R^1 = Cl$, $R^2 = R^3 = H$, $R^4 = SC_6H_5$), respectively, as the only isolated products. The structures of these were verified (Wieland, T. and Rühl, 1963) by desulphurization with Raney nickel to afford the previously unambiguously synthesized 6-methoxyindole (**510**; $R^1 = OCH_3$, R^2–$R^4 = H$) (Harvey, D. G. and Robson, 1938), 6-methoxy-2-methylindole (**510**; $R^1 = OCH_3$, $R^2 = R^4 = H$, $R^3 = CH_3$) (Späth and Brunner, O., 1925) and 6-chloroindole (**510**; $R^1 = Cl$, R^2–$R^4 = H$) (Uhle, 1949), respectively. Both **509** and **510** [$R^1 = CH_3$, $R^2 = R^3 = H$, $R^4 = CH_2C(NCOCH_3)(COOC_2H_5)_2$] were formed when **508** [$R^1 = CH_3$, $R^2 = R^3 = H$, $R^4 = CH_2C(NCOCH_3)(COOC_2H_5)_2$] was indolized (Snyder, Beilfuss and Williams, J. K., 1953), the structures of the products being supported by a comparison of their m.p.s. with previously unambiguously synthesized specimens (Rydon, 1948). Indolization of compound **126** with ethanolic hydrogen chloride was initially thought to give only **510** [$R^1 = OCH_3$, $R^2 = H$, $R^3 = COOC_2H_5$, $R^4 = (CH_2)_3COOC_2H_5$] (Iyer, Jackson, A. H., Shannon and Naidoo, 1972, 1973). However, from the subsequent (Iyer, Jackson and Shannon, 1973) conversion of this product into 1,2,3,4-tetrahydro-7-methoxycarbazol-1-one there was also isolated a small (1%) yield of the corresponding 5-methoxy isomer, the structure of which was confirmed by its reduction with diborane into the known (Chalmers, J. R., Openshaw and Smith, G. F., 1957) 1,2,3,4-tetrahydro-5-methoxycarbazole. This showed that the original indolization product contained **509** [$R^1 = OCH_3$, $R^2 = H$, $R^3 = COOC_2H_5$, $R^4 = (CH_2)_3COOC_2H_5$] as a minor component. Treatment of a mixture of 3-methoxyphenylhydrazine and 4-aminobutyraldehyde diethylacetal with zinc chloride afforded a mixture of 4- and 6-methoxytryptamine [**509** and **510**, respectively, $R^1 = OCH_3$, $R^2 = R^3 = H$, $R^4 = (CH_2)_2NH_2$]. The 6-methoxy isomer was separated as its N-acetyl derivative **510** [$R^1 = OCH_3$, $R^2 = R^3 = H$, $R^4 = (CH_2)_2NHCOCH_3$], the orientation of the methoxyl group being established by conversion of this derivative into harmaline (**535**) by treatment with phosphorus pentoxide (Späth and Lederer, 1930a). The formation of **535** and ultimately its dehydrogenation product, harman, from the indolic product

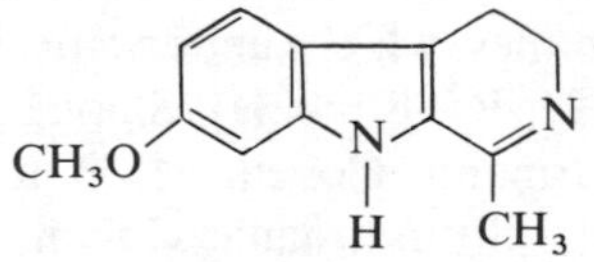

CH_3O N N H CH_3

535

resulting from the indolization of **508** [R^1 = OCH_3, R^2 = H, R^3 = $COCH_3$, R^4 = $(CH_2)_2$-phthalimido], established that this product was methoxy substituted at the indole 6-position (Manske, R. H. F., Perkin and Robinson, R., 1927). Indolization of **508** [R^1 = OCH_3, R^2 = H, R^3 = $COOC_2H_5$, R^4 = $(CH_2)_2COOH$], by treatment with ethanolic concentrated sulphuric acid, afforded only **510** [R^1 = OCH_3, R^2 = H, R^3 = $COOC_2H_5$, R^4 = $(CH_2)_2COOC_2H_5$]. The structure of this product was established by its sequential hydrolysis, pyrolytic 2-decarboxylation, conversion to the hydrazide, formation of the azide and acid treatment which yielded the known 1,2,3,4-tetrahydro-7-methoxy-β-carbolin-1-one (Barrett, H. S. B., Perkin and Robinson, R., 1929).

When cyclohexanone 3-trifluoromethylphenylhydrazone [**508**; R^1 = CF_3, R^2 = H, R^3 + R^4 = $(CH_2)_4$] was indolized, only one product was obtained (Forbes, E. J., Stacey, Tatlow and Wragg, R. T., 1960). This was dehydrogenated into 2-trifluoromethylcarbazole (**536**; R^1 = R^3 = H, R^2 = CF_3), the structure of which was established (Forbes, E. J., Tatlow and Wragg, R. T., 1960) by the following unambiguous synthesis. 2-Nitro-4-trifluoromethyldiphenyl (**537**;

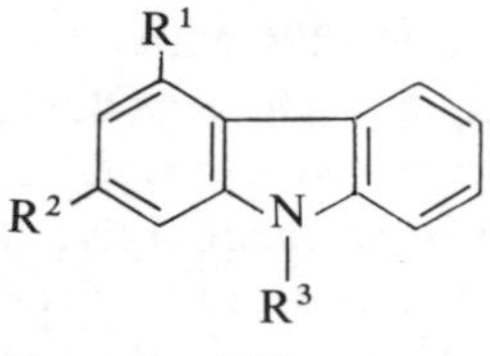

536 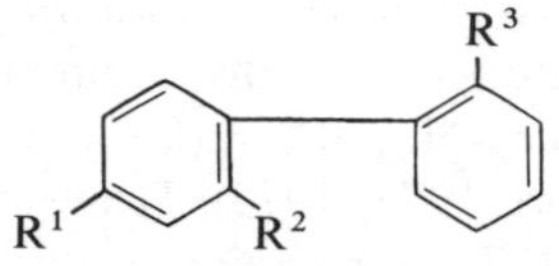 **537**

R^1 = CF_3, R^2 = NO_2, R^3 = H) was catalytically reduced to yield **537** (R^1 = CF_3, R^2 = NH_2, R^3 = H). This was diazotized, treated with sodium azide and the resulting 2-azido compound cyclized by irradiation to afford **536** (R^1 = R^3 = H, R^2 = CF_3). The isomeric carbazole, as its 9-acetyl derivative **536** (R^1 = CF_3, R^2 = H, R^3 = $COOCH_3$) was similarly synthesized starting from **537** (R^1 = H, R^2 = NO_2, R^3 = CF_3) (Forbes, E. J., Tatlow and Wragg, R. T., 1960).

The arylhydrazones **508** [R^1 = $COCH_3$ and COC_6H_5, R^2 = H, R^3 + R^4 = $CONH(CH_2)_2$] each formed both possible isomeric indoles upon indolization. The 4-substituted products, **509** [R^1 = $COCH_3$ and COC_6H_5, R^2 = H, R^3 + R^4 = $CONH(CH_2)_2$], were distinguished from the corresponding 6-substituted isomers by hydrolysis of the former's amide bonds, followed by decarboxylation of the indole-2-carboxylic acid moiety to afford intermediate 4-acyltryptamines which underwent intramolecular condensation to give **526** (R^1 = CH_3, R^2–R^4 = H) (Shavel, Strandtmann and Cohen, M. P., 1962, 1965; Strandtmann, Cohen, M. P. and Shavel, 1963) and **526** (R^1 = C_6H_5, R^2–R^4 = H) (Strandtmann, Cohen, M. P. and Shavel, 1963), respectively. A similar sequence of reactions starting from **508** [R^1 = $COCH_3$,

$R^2 = H$, $R^3 + R^4 = CONHCH_2CH(CH_3)$] afforded **509** [$R^1 = COCH_3$, $R^2 = H$, $R^3 + R^4 = CONHCH_2CH(CH_3)$] and **526** ($R^1 = R^4 = CH_3$, $R^2 = R^3 = H$) (Shavel, Strandtmann and Cohen, M. P., 1965).

Indolization of **508** ($R^1 = CH_3$, $R^2 = R^4 = H$, $R^3 = C_6H_5$) yielded **510** ($R^1 = CH_3$, $R^2 = R^4 = H$, $R^3 = C_6H_6$) as the only isolated product, this structure being confirmed by its unambiguous synthesis from **538** using the Madelung synthesis (Campbell and Cooper, 1935).

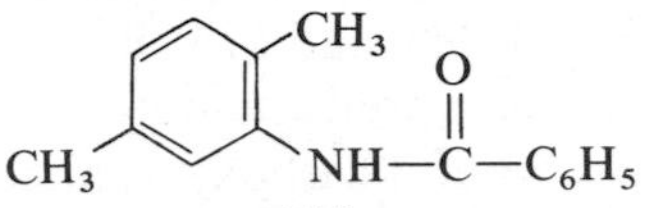

538

Hydrazone **508** ($R^1 = CF_3$, $R^2 = R^4 = H$, $R^3 = COOC_2H_5$) formed both **509** and **510** ($R^1 = CF_3$, $R^2 = R^4 = H$, $R^3 = COOC_2H_5$) upon indolization. The orientation of the CF_3 substituent in these products was established (Bornstein, Leone, Sullivan and Bennett, O. F., 1957) by their hydrolyses with 10M sodium hydroxide into **509** and **510** ($R^1 = R^3 = COOH$, $R^2 = R^4 = H$), the diethyl esters of which, **509** and **510** ($R^1 = R^3 = COOC_2H_5$, $R^2 = R^4 = H$), were identified by comparison with authentic specimens (Roder, 1886; Kermack, 1924, respectively). However, although the orientation of the carbethoxy group in the latter work (Kermack, 1924) was beyond doubt [since an unambiguous Reissert synthesis involving the reduction of **539** ($R^1 = R^3 = H$, $R^2 = CN$) affording the correspondingly 6-substituted intermediate **540** was

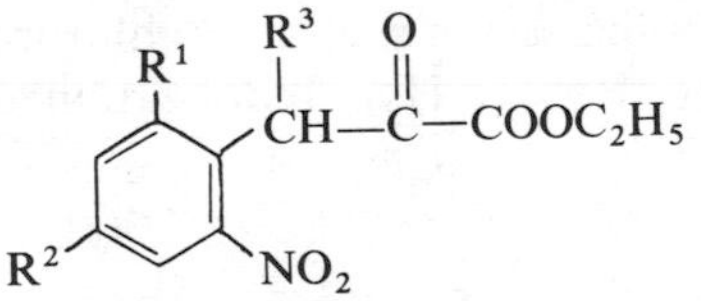

539 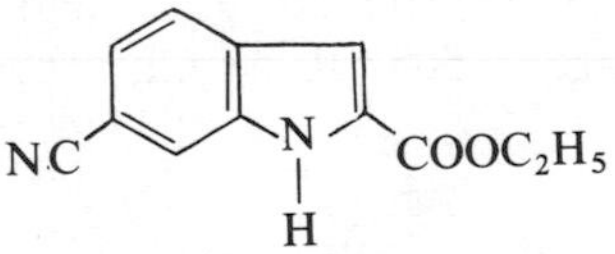 **540**

involved in its preparation] the former (Roder, 1886) synthesis involved the indolization of ethyl pyruvate 3-carbethoxyphenylhydrazone, the direction of which was not established. However, these latter studies (Bornstein, Leone, Sullivan and Bennett, O. F., 1957; Kermack, 1924) now established the orientation of the carbethoxy group as being at the 4-position of the indole nucleus in this early study (Roder, 1886). The hydrazone **508** ($R^1 = OCH_3$, $R^2 = H$, $R^3 = COOH$, $R^4 = CH_2COOH$) afforded only **510** ($R^1 = OCH_3$, $R^2 = H$, $R^3 = COOH$, $R^4 = CH_2COOH$) upon indolization with ethanolic hydrogen chloride followed by hydrolysis with methanolic potassium hydroxide. The structure of this product was established by its thermal decarboxylation into 6-methoxy-3-methylindole which was unambiguously synthesized by the Reissert method, starting with **539** ($R^1 = H$, $R^2 = OCH_3$, $R^3 = CH_3$) (Kermack, Perkin and Robinson, R., 1921). Reissert syntheses starting from **539**

(R^1 = F, R^2 = R^3 = H and R^1 = R^3 = H, R^2 = F) yielded ethyl 4- and 6-fluoroindole-2-carboxylate, respectively, which were used to identify the products resulting from the indolization of ethyl pyruvate 3-fluorophenylhydrazone (**508**; R^1 = F, R^2 = $COOC_2H_5$, R^3 = H) (Allen, F. L., Brunton and Suschitzky, 1955). Likewise, 4- and 6-chloroindole were unambiguously prepared (Fox and Bullock, 1951b; Uhle, 1949) and, by reaction of their magnesium iodide complexes with chloroacetonitrile, were converted into their 3-cyanomethyl derivatives, hydrolyses of which furnished the corresponding 4- and 6-chloroindole acetic acid (**509** and **510**; R^1 = Cl, R^2 = R^3 = H, R^4 = CH_2COOH), respectively. These products were used to identify the products resulting from the indolization of 3-carboxypropionaldehyde 3-chlorophenylhydrazone (**508**; R^1 = Cl, R^2 = R^3 = H, R^4 = CH_2COOH) (Fox and Bullock, 1951b). Indolization of acetaldehyde 3-methoxyphenylhydrazone (**508**; R^1 = OCH_3, R^2–R^4 = H) gave both 4- and 6-methoxyindole (Suvorov, Bykhovskii and Podkhalyuzina, 1977) which were identified by m.p. comparison with samples earlier unambiguously synthesized (Blaikie and Perkin, 1924; Kermack, Perkin and Robinson, R., 1921, 1922) by the Reissert method. A Reissert synthesis starting from 4-methyl-3-nitroanisole afforded ethyl 6-methoxyindole-2-carboxylate, one of the two isomeric indoles resulting from treatment of ethyl pyruvate 3-methoxyphenylhydrazone with zinc chloride in acetic acid (Ishii, Murakami, Y., Hosoya, Takeda, Suzuki, Y. and Ikeda, 1973). Related syntheses, involving the reduction of **541** (R^1 = Cl, R^2 = H and R^1 = H, R^2 = Cl) into **509** and **510** (R^1 = Cl, R^2 = R^4 = H, R^3 = CH_3), respectively, have also been used (Piper and Stevens, F. J., 1966) to establish the structures of the products resulting from the indolization of acetone 3-chlorophenylhydrazone **508** (R^1 = Cl, R^2 = R^4 = H, R^3 = CH_3). Indolization of **508**

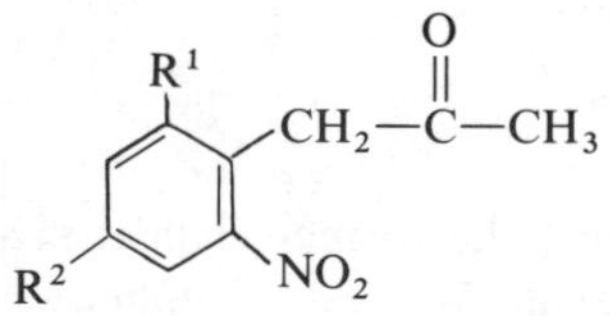

R^1, CH_2—C(=O)—CH_3, NO_2, R^2

541

[R^1 = Cl, R^2 = H, R^3 = $COOC_2H_5$, R^4 = $(CH_2)_3COOC_2H_5$] afforded both possible isomeric indolic products, **509** and **510** [R^1 = Cl, R^2 = H, R^3 = COOH, R^4 = $(CH_2)_3COOH$], which were 2-decarboxylated to the corresponding 3-indolylbutyric acids (Suvorov, Mamaev and Shagalov, 1953). The structures of these latter compounds were subsequently differentiated by their unambiguous synthesis by reaction of 4- and 6-chloroindole with methyl magnesium iodide followed by treatment with 4-butyrolactone (Shagalov, Sorokina and Suvorov, 1964). The 6-chloroindole had previously (Suvorov, Fedotova, M. V., Orlova, L. M. and Ogareva, 1962) been unambiguously synthesized using a Reissert synthesis whereas, for the synthesis of the 4-chloroindole, an alternative unambiguous synthesis involving the reductive deaminative cyclization of **542** was used (Shagalov, Sorokina and Suvorov, 1964) since in this case the Reissert synthetic approach was unsatisfactory.

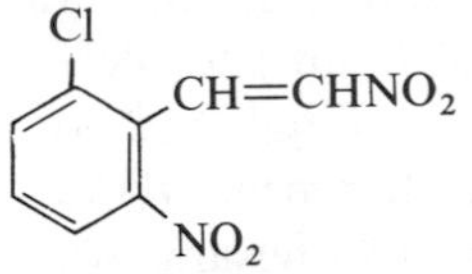

Cl

$CH{=}CHNO_2$

NO_2

542

Indolization of **508** [R^1 = $NHCOCH_3$, R^2 = H, R^3 + R^4 = $(CH_2)_4$] gave a poor yield of only one product which was formulated as **510** [R^1 = $NHCOCH_3$, R^2 = H, R^3 + R^4 = $(CH_2)_4$] since it was hydrolysed to an amino-1,2,3,4-tetrahydrocarbazole prepared by the reduction of what was thought to be 1,2,3,4-tetrahydro-7-nitrocarbazole (Perkin and Plant, 1921, 1923b). However, since it was later (Plant, 1936) shown that the compound thought to have been 1,2,3,4-tetrahydro-5-nitrocarbazole in these earlier two papers (Perkin and Plant, 1921, 1923b) was the 7-nitro isomer, it appears that what was thought to be the 7-nitro isomer might be the 5-nitro isomer. Thus, the acetylamino group in the above indolization product should apparently be substituted at the 5-position of the 1,2,3,4-tetrahydrocarbazole nucleus. Likewise, the product resulting from the indolization of **508** [R^1 = NHCOCOOH, R^2 = H, R^3 + R^4 = $(CH_2)_4$] with acetic acid was formulated as **510** [R^1 = NHCOCOOH, R^2 = H, R^3 + R^4 = $(CH_2)_4$] since it was hydrolysed, using concentrated hydrochloric acid, to an amino-1,2,3,4-tetrahydrocarbazole which was again also obtained by the reduction of the supposed 1,2,3,4-tetrahydro-7-nitrocarbazole (Edwards and Plant, 1923). Again it appears that this product should be reformulated as the 5-isomer. However, since indolization yields were poor, this does not exclude the formation of the 7-isomer. Reinvestigation of these indolizations using spectroscopic techniques (see next section) would certainly be of interest. Indolization of acetone 3-methylphenylhydrazone (**508**; R^1 = R^3 = CH_3, R^2 = R^4 = H) afforded only one product which was formulated, without verification, as 2,4-dimethylindole (**509**; R^1 = R^3 = CH_3, R^2 = R^4 = H) (Plancher and Ciusa, 1906). Subsequently, the alternative possible formulation of this product as 2,6-dimethylindole (**510**; R^1 = R^2 = CH_3, R^2 = R^4 = H) was excluded by a claimed unambiguous synthesis of this latter product (Allen, C. F. H., Young, D. M. and Gilbert, M. R., 1937). However, the experimental data given in this latter paper do not appear to substantiate this claim and further work is necessary to clarify this problem. Certainly the exclusive formation of **509** (R^1 = R^3 = CH_3, R^2 = R^4 = H) in this indolization is very unlikely and, indeed, its formation as only the minor possible isomeric indolic product might be expected.

Indolization of cyclohexanone 4-chloro-3-nitrophenylhydrazone afforded both possible isomeric products **521** [$R^1 = NO_2$, R^2 = Cl, R^3 = H, $R^4 + R^5 = (CH_2)_4$] and **521** [R^1 = H, R^2 = Cl, $R^3 = NO_2$, $R^4 + R^5 = (CH_2)_4$] (Plant and Rosser, 1928 – see also Robinson, F. P. and Brown, R. K., 1964). The structure of the former isomer product was established (Moggridge and Plant, 1937) by treatment with tin and alcoholic hydrochloric acid, which effected reduction of the nitro and indolic 2,3-double bond groups and reductive dechlorination, to give, after benzoylation, the known (Plant, 1936) 5-benzamido-9-benzoyl-1,2,3,4,4a,9a-hexahydrocarbazole.

Both possible isomeric products, **521** [$R^1 + R^2 = (CH_2)_4$, $R^3 = R^5$ = H, $R^4 = COOCH_3$ and $R^1 = R^5$ = H, $R^2 + R^3 = (CH_2)_4$, $R^4 = COOCH_3$] were formed by the indolization of **543** [$R^1 + R^2 = (CH_2)_4$, R^3 = COOH,

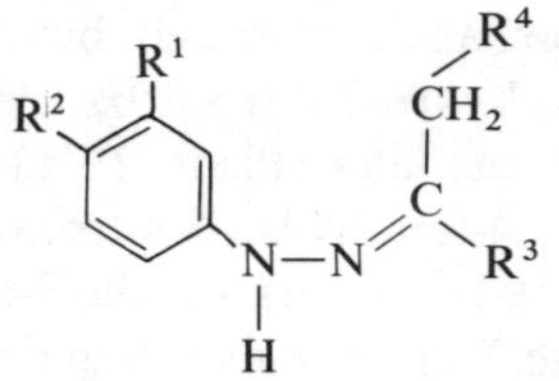

543

R^4 = H] by treatment with either hydrogen chloride in methanol or with refluxing methanolic sulphuric acid. However, from the reaction mixture only the former isomer could be isolated and this was then sequentially hydrolysed into **521** [$R^1 + R^2 = (CH_2)_4$, $R^3 = R^5$ = H, R^4 = COOH], thermally decarboxylated into **521** [$R^1 + R^2 = (CH_2)_4$, R^3–R^5 = H] and dehydrogenated, by treatment with 2,3-dichloro-5,6-dicyano-1,4-benzoquinone in benzene at room temperature, to afford the known **544** (R = H). The first two stages

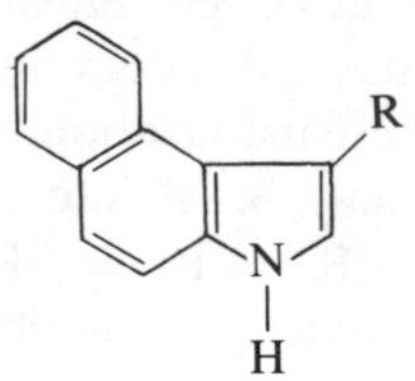

544

of this reaction sequence were repeated with the isomeric mixture of products resulting from the indolization but, as with the mixture of the starting methyl esters, neither the acid mixture nor the final resulting mixture of **521** [$R^1 + R^2 = (CH_2)_4$, R^3–R^5 = H] and **521** [$R^1 = R^4 = R^5$ = H, $R^2 + R^3 = (CH_2)_4$] could be resolved (Shagalov, Ostapchuk, Zlobina, Eraksina, Babushkina, T. A., Vasil'ev, A. M., Ogorodnikova and Suvorov, 1978).

Indolization of **545** afforded a mixture of **546** and **547**. After separation, these products were dehydrogenated to yield **548** and **549**, respectively. The structure

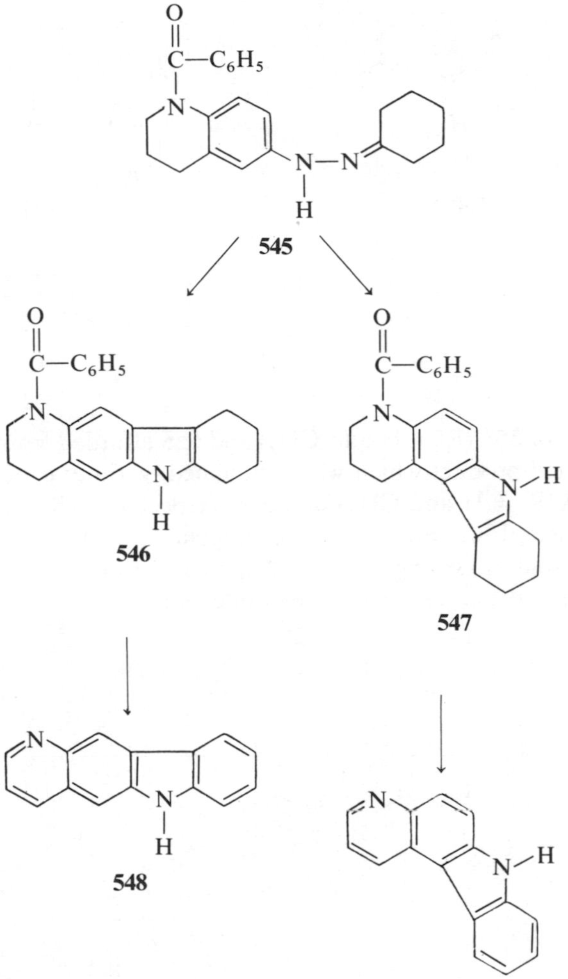

545

546

547

548

549

of the last compound was established by its synthesis from the indolization of cyclohexanone 6-quinolylhydrazone, which was known to occur towards the 5- and not the 7-position of the quinoline nucleus (see later in this chapter), followed by dehydrogenation (Kulka and Manske, R. H. F., 1952). Contrary to the above observations, indolization of **550** afforded only one product. This was shown to have structure **551** by its deacetylation and subsequent dehydrogenation to **552**, the structure of which was known since the product was not identical with the compound obtained by dehydrogenation of the product resulting from the indolization of cyclohexanone 7-quinolylhydrazone, a reaction which was known to occur towards the 8-position of the quinoline nucleus (see Chapter III, E) (Kulka and Manske, R. H. F., 1952).

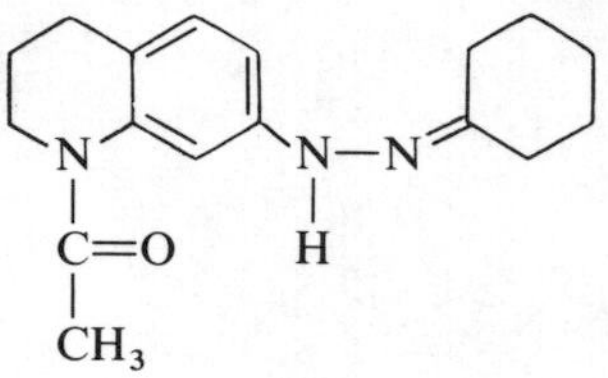

550

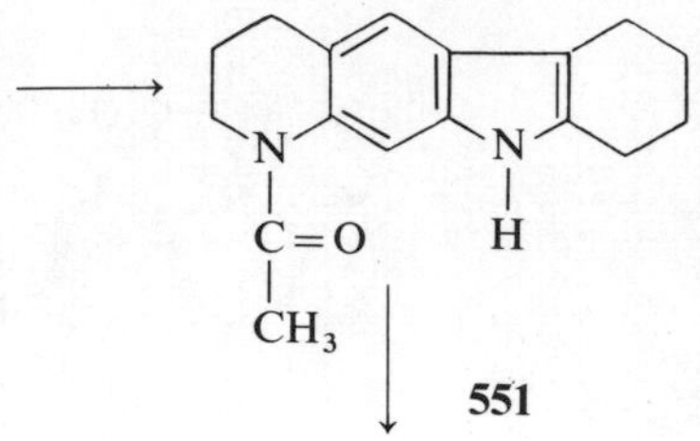

551

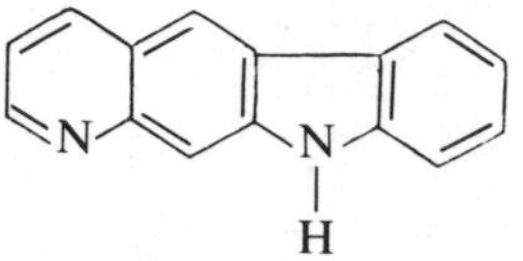

552

Indolization of **553** (R = H and CH_3) and **555** afforded from each reaction only one isolated product which was formulated, without experimental verification, as **554** (R = H and CH_3) and **556**, respectively (Kempter, Schwalba, Stoss and Walter, 1962) rather than the angular isomer. Degradative and/or spectroscopic studies (see next section) upon these products, with a view to establishing their structures, would be of interest.

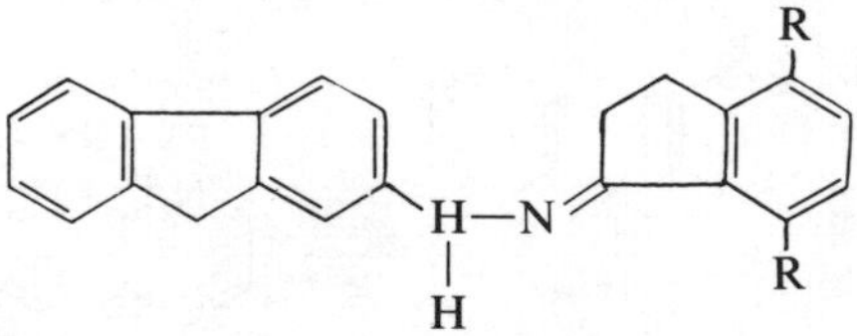

553

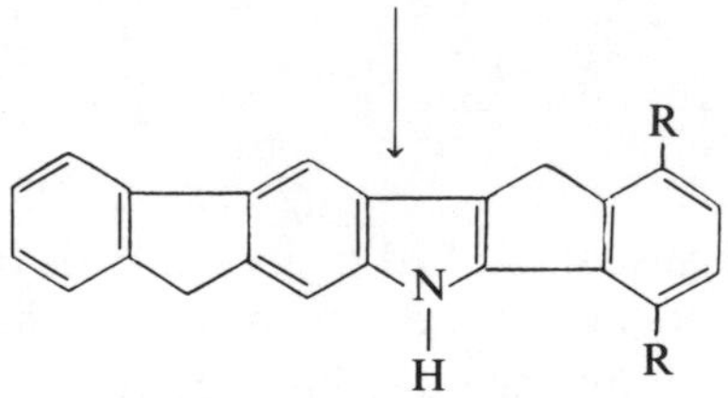

554

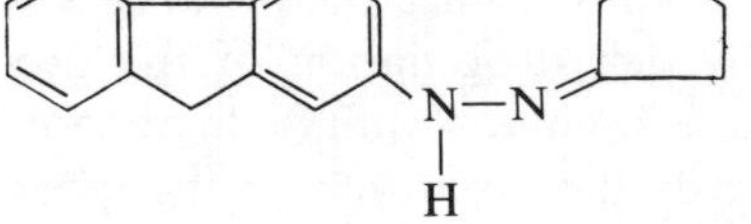

555

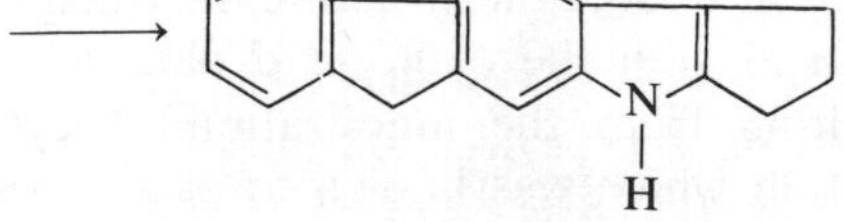

556

4. Spectroscopic Techniques

Indolization of phenylpyruvic acid 3-nitrophenylhydrazone (**508**; $R^1 = NO_2$, $R^2 = H$, $R^3 = COOH$, $R^4 = C_6H_5$) afforded both possible isomeric indolic products which were subsequently reduced to the corresponding amino acids **509** and **510** ($R^1 = NH_2$, $R^2 = H$, $R^3 = COOH$, $R^4 = C_6H_5$). These two structures were differentiated by i.r. examination of their ethyl esters which indicated in the latter isomer absorption bands attributed to the out-of-plane deformation vibrations of an isolated hydrogen atom and two adjacent hydrogen atoms on the benzenoid nucleus (Sato, Y., 1963b). In a similar manner, i.r. spectral examination differentiated between the two indolic products resulting from the indolization of **557** (Werner, L. H., Schroeder, D. C. and Ricca, 1957). The absence of absorption between 810 and 730 cm^{-1} in the i.r. spectrum of the product resulting from the indolization of **558** indicated the absence of 3 adjacent protons on a benzenoid ring and showed that it had structure **559** ($R^1 = H$, $R^2 = OCH_3$) and not **559** ($R^1 = OCH_3$, $R^2 = H$) (Cross, P. E. and Jones, E. R. H., 1964). The detection of bands in their i.r. spectra characteristic of

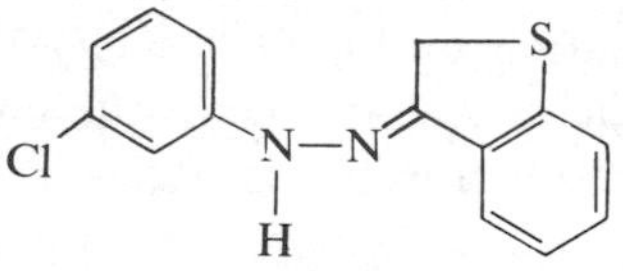

557

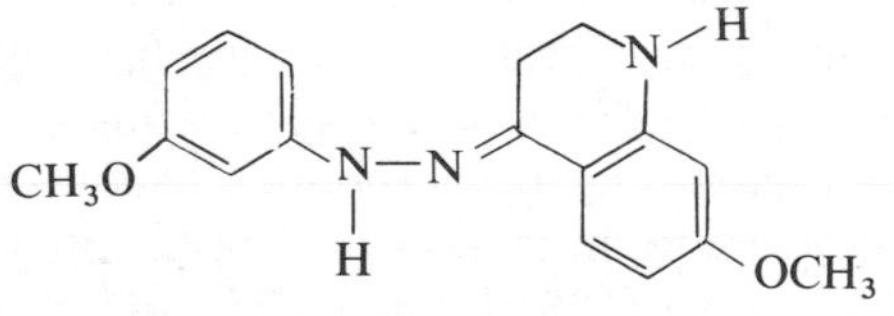

558

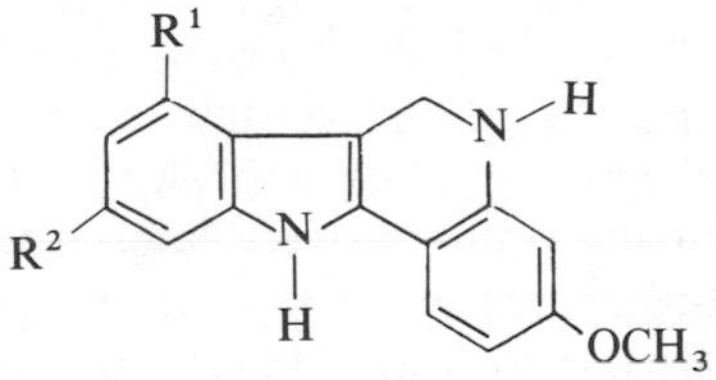

559

1,2,3- and 1,2,4-substituted benzenoid nuclei differentiated between structures **509** and **510** [$R^1 = CF_3$, $R^2 = H$, $R^3 + R^4 = (CH_2)_2C(CH_3)_2CH_2$], respectively, the products formed by indolization of 4,4-dimethylcyclohexanone 3-trifluoromethylphenylhydrazone [**508**; $R^1 = CF_3$, $R^2 = H$, $R^3 + R^4 = (CH_2)_2C(CH_3)_2CH_2$] (Bergman and Erdtman, 1969). The structures of the two isomeric indoles resulting in each case from the indolization of **508** [$R^1 = COCH_3$ and COC_6H_5, $R^2 = H$, $R^3 + R^4 = CONH(CH_2)_2$] were similarly distinguished (Strandtmann, Cohen, M. P. and Shavel, 1963). In connection with all these i.r. spectral studies, the spectra of unambiguously synthesized 4- and 6-chloro-2-methylindole-3-carboxylic acid showed strong absorption bands at 740 and 773 cm^{-1} and at 800 cm^{-1}, respectively (Piper and Stevens, F. J., 1962). Indolization of ethyl laevulinate 3-chlorophenylhydrazone with a zinc chloride catalyst yielded a eutectic mixture of 4- and 6-chloro-2-methylindole-3-acetic acid. Comparison of the i.r. spectrum of this mixture with those formed by mixing unambiguously synthesized specimens of the 4- and 6-chloro isomers in varying proportions indicated that in the eutectic mixture the 6-chloro isomer was present in about 5–6% predominance (Piper and Stevens, F. J., 1962). Indolization of **543** ($R^1 = NO_2$, $R^2 = R^3 = COOC_2H_5$, $R^4 = C_6H_5$) gave both **521** ($R^1 = NO_2$, $R^2 = R^4 = COOC_2H_5$, $R^3 = H$, $R^5 = C_6H_5$) and **521** ($R^1 = H$, $R^2 = R^4 = COOC_2H_5$, $R^3 = NO_2$, $R^5 = C_6H_5$). These structures were differentiated by their reduction to the corresponding amino compounds, the i.r. spectra of which showed absorption bands attributed to the out-of-plane deformation vibrations of two adjacent hydrogen atoms and isolated hydrogen atoms on the benzenoid nucleii, respectively. Confirmation of these assignments was obtained by the chromium(VI) oxide in acetic acid oxidation of **521** ($R^1 = NO_2$, $R^2 = R^4 = COOC_2H_5$, $R^3 = H$, $R^5 = C_6H_5$) into **529** ($R^1 = NO_2$, $R^2 = COOC_2H_5$, $R^3 = H$, $R^4 = COCOOC_2H_5$, $R^5 = C_6H_5$) which, after hydrolysis and decarboxylation, afforded **529** ($R^1 = NO_2$, $R^2 = COOH$, $R^3 = R^4 = H$, $R^5 = C_6H_5$) (Sato, Y., 1963b).

Both **509** and **510** [$R^1 = OCH_3$, $R^2 = H$, $R^3 = R^4 = C_6H_4OCH_3(4)$] were produced by indolization of **508** [$R^1 = OCH_3$, $R^2 = H$, $R^3 = R^4 = C_6H_4OCH_3(4)$], these structures again being differentiated by examination of these out-of-plane deformation absorption bands in their i.r. spectra but now coupled with the analysis of their p.m.r. spectra, that of the former isomer clearly exhibiting the pair of doublets (J = 2 and 7 Hz) caused by the H-5 proton of the indole nucleus (Szmuszkovicz, Glenn, Heinzelmanm Hester and Youngdale, 1966). Similar uses of i.r. and p.m.r. spectroscopic investigations, involving complete analyses of the signals in the p.m.r. spectra caused by aromatic protons, were later made in establishing the structures of the two indolic products formed in each case by the indolization of 1-benzoylhexahydroazepin-4-one 3-methoxyphenylhydrazone [**508**; $R^1 = OCH_3$, $R^2 = H$, $R^3 + R^4 = (CH_2)_2N(COC_6H_5)(CH_2)_2$] (Hester, Tang, Keasling and Veldkamp, 1968), of cyclohexanone 3-methylphenylhydrazone [**508**; $R^1 = CH_3$, $R^2 = H$, $R^3 + R^4 = (CH_2)_4$] (Grandberg, Belyaeva and Dmitriev, L. B., 1971a), of cyclohexanone 3-chloro- and 3-methoxyphenylhydrazone (Przhenval'skii, Grandberg, Klyuev, N. A. and Belikov, 1978), of diethyl ketone 3-chloro-, 3-ethyl- and 3-methoxyphenylhydrazone (**508**; $R^1 = Cl$, C_2H_5 and OCH_3, respectively, $R^2 = H$, $R^3 = C_2H_5$, $R^4 = CH_3$) (Grandberg, Belyaeva and Dmitriev, L. B., 1971b) and of diethylketone 3-nitrophenylhydrazone (**508**; $R^1 = NO_2$, $R^2 = H$, $R^3 = C_2H_5$, $R^4 = CH_3$) (Grandberg, Belyaeva and Dmitriev, L. B., 1973). In the p.m.r. spectra of the products from these reactions, *ortho* and *meta* coupling between the aromatic protons was clearly observed but *para* coupling was apparently too small to be observed under the investigational conditions. This was also the case in the p.m.r. spectra of the major product **510** [$R^1 = OCH_3$, $R^2 = H$, $R^3 = COOC_2H_5$, $R^4 = (CH_2)_3COOC_2H_5$], resulting from the indolization of **126** (Iyer, Jackson, A. H.,

Shannon and Naidoo, 1973) and in the two products resulting in each case from the indolization of butyraldehyde and 4-chlorobutyraldehyde 3-nitrophenylhydrazones, the structures of which were distinguished by p.m.r. spectral analysis (McKay, Parkhurst, Silverstein and Skinner, 1963). A recent example established the orientation of a 6-methoxyindole obtained by Fischer indolization, as distinct from the possible 4-isomer, by p.m.r. spectroscopic comparison with the aromatic proton signal patterns of authentic specimens of 1,2,3,4-tetrahydro-5- and -7-methoxycarbazole (Jogdeo and Bhide, 1980). The detection of the H-6 and H-7 indolic protons as doublets (J = 8 Hz) centred at $\tau = 2.63$ and $\tau = 2.09$, respectively, permitted the formulation of the only product isolated from the indolization of **543** ($R^1 = NO_2$, $R^2 = CH_3$, $R^3 = COOC_2H_5$, $R^4 = H$) as **521** ($R^1 = NO_2$, $R^2 = CH_3$, $R^3 = R^5 = H$, $R^4 = COOC_2H_5$) rather than the alternative possible isomer (Gadaginamath, 1976). Indolization of ethyl pyruvate 3,4-dichlorophenylhydrazone (**543**; $R^1 = R^2 = Cl$, $R^3 = COOC_2H_5$, $R^4 = H$) afforded both **521** ($R^1 = R^2 = Cl$, $R^3 = R^5 = H$, $R^4 = COOC_2H_5$) and **521** ($R^1 = R^5 = H$, $R^2 = R^3 = Cl$, $R^4 = COOC_2H_5$), the latter product exhibiting two doublets (J = 9.0 Hz) in its p.m.r. spectrum indicative of the *ortho* coupled aromatic protons whereas the aromatic proton signals in the former compound appeared as two singlets (Ishii, Murakami, Y., Takeda and Furuse, 1974), *para* coupling constants again presumably being too small for observation. Similarly, the p.m.r. spectrum of the only product obtained by indolization of **560** exhibited two 1-proton aromatic singlets, confirming structure **561** [$R^1 = H$, $R^2 = CH_2N(C_2H_5)_2$], rather than the alternative possible isomer **561** [$R^1 = CH_2N(C_2H_5)_2$, $R^2 = H$] (Marquez, Cranston, Ruddon, Kier and Burckhalter, 1972). The p.m.r. spectra

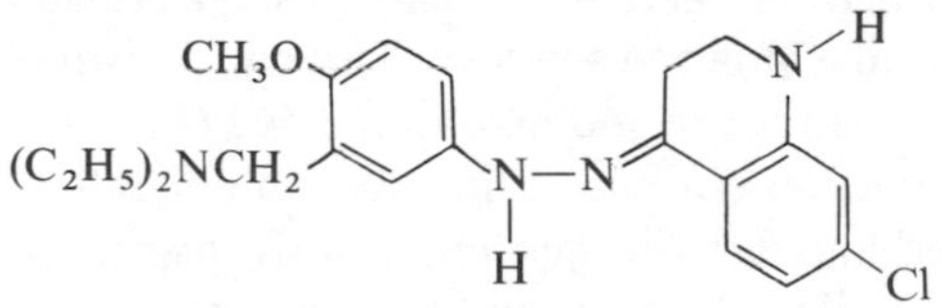

560 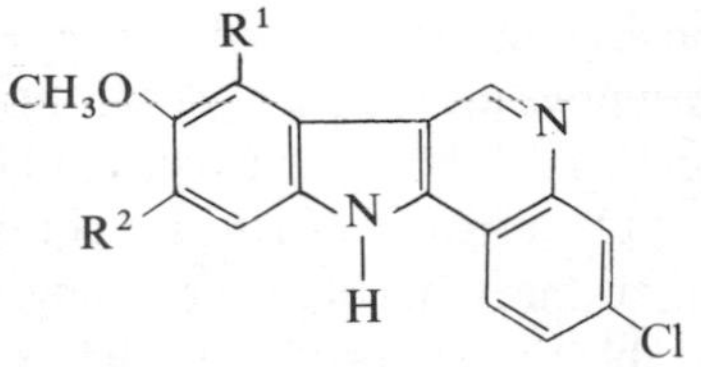 **561**

of the two indolic products resulting from the indolization of a mixture of 3,4-dimethylphenylhydrazine hydrochloride and 4-benzylthiobutyraldehyde diethyl acetal showed them to have structures **521** [$R^1 = R^2 = CH_3$, $R^3 = R^4 = H$, $R^5 = (CH_2)_2SCH_2C_6H_5$] and **521** [$R^1 = R^4 = H$, $R^2 = R^3 = CH_3$, $R^5 = (CH_2)_2SCH_2C_6H_5$]. The spectrum of the latter isomer exhibited two aromatic 1-proton singlets, indicative of the *para* protons (again no *para* coupling was observed) and a 6-proton singlet at $\tau = 7.67$ ($CDCl_3$) or $\tau = 7.74$ (CCl_4), indicating that the protons of the two CH_3 groups are fortuitously magnetically equivalent. The aromatic protons in the former isomer are also fortuitously magnetically equivalent and give rise to a 2-proton singlet whereas one of the methyl group signals is now moved downfield [$\tau = 7.50$ ($CDCl_3$) and $\tau = 7.62$ (CCl_4)], indicating the presence of a methyl group at the 4-position of the indole nucleus (Keglević and Goleš, 1970). Indolization of cyclohexanone 4-chloro-3-nitrophenylhydrazone [**543**; $R^1 = NO_2$, $R^2 = Cl$, $R^3 + R^4 = (CH_2)_4$] yielded both possible isomeric products, **521** [$R^1 = NO_2$, $R^2 = Cl$, $R^3 = H$, $R^4 + R^5 = (CH_2)_4$] and **521** [$R^1 = H$, $R^2 = Cl$, $R^3 = NO_2$, $R^4 + R^5 = (CH_2)_4$]. The former isomer was 9-methylated and reduced to afford 5-amino-6-chloro-1,2,3,4-tetrahydro-9-methylcarbazole, the p.m.r. spectrum of which exhibited a characteristic AB aromatic proton signal (Robinson, F. P. and Brown, R. K., 1964). This result was in agreement with the earlier (Moggridge and Plant, 1937; Plant and Rosser, 1928) structural assignment of the two isomeric indolization products based upon chemical evidence which has been discussed above. From the indolization of **562** ($R = H$, $X = H_2$), using saturated ethanolic hydrogen chloride, a 70% yield of **563** ($R = H$, $X = H_2$) was obtained, the structure of which was distinguished from its

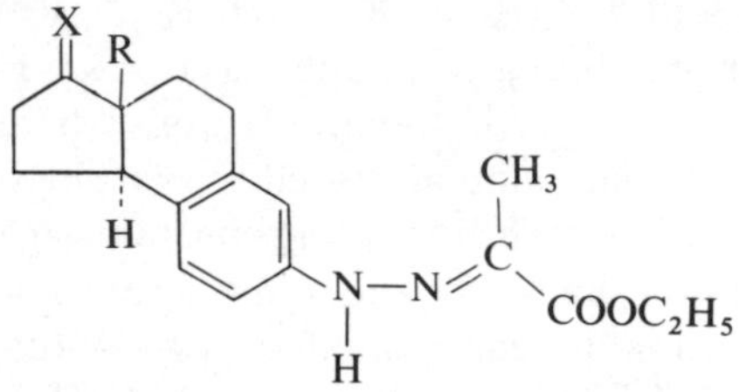

562

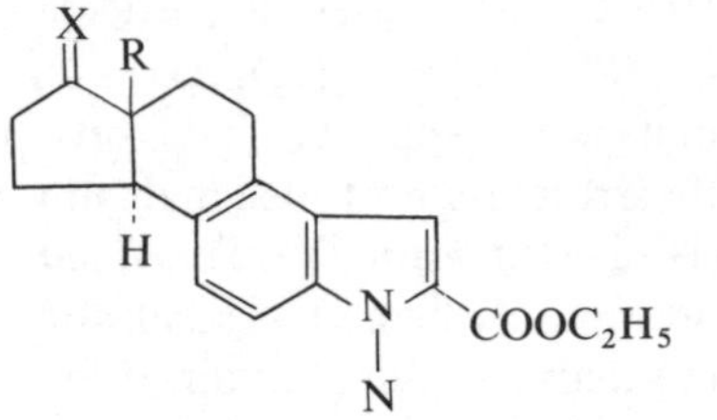

563

possible isomer using p.m.r. spectral techniques which clearly indicated the AB aromatic proton system. Small amounts of the corresponding alternative isomeric indole may have been formed but were not isolated (Sladkov, Shner, Anisimova and Suvorov, 1972; Sladkov, Shner, Alekseeva, Turchin, Anisimova, Sheinker, Y. N. and Suvorov, 1971). However, indolization of **562** ($R = CH_3$, $X = O$), formed *in situ* by coupling the corresponding diazonium salt with 2-methylacetoacetic ester, using ethanolic hydrogen chloride furnished both possible indolic products, **563** ($R = CH_3$, $X = O$) and **564**, in a yield ratio of 3 : 2, respectively, the structures being distinguished by p.m.r. spectral data which confirmed the 1,2- and 1,4-proton systems on the benzenoid nuclei,

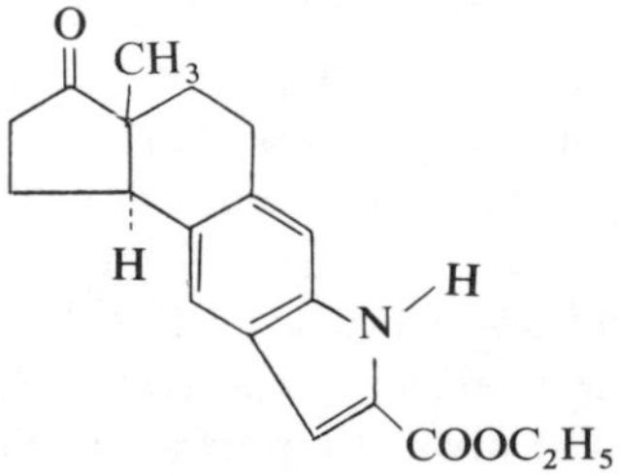

564

respectively (Sladkov, Shner, Anisimova, Alekseeva, Lisitsa, Terekhina and Suvorov, 1974). Similarly, alcoholic sulphuric acid catalysed indolization of **565** afforded both possible isomeric indolic products, **566** and **567**, in a yield ratio of 17 : 3, respectively. Again, the structures were distinguished by the detection of the 1,2- and 1,4-proton systems, respectively, on the benzenoid nuclei by p.m.r. spectral investigation (Sladkov, Anisimova and Suvorov, 1977), as were the structures of the two possible products resulting in each case from the indolizations of **522** [$R^1 + R^2$ = $NHCOCH_2S$ and SCH_2CONH, $R^3 + R^4$ = $CO(CH_2)_3$] (Brunelli, Fravolini, Grandolini and Strappaghetti, 1980). The p.m.r. spectra of indoles have been reviewed (Hiremath and Hosane, 1973) and in connection with these present studies it is relevant that the signal caused by

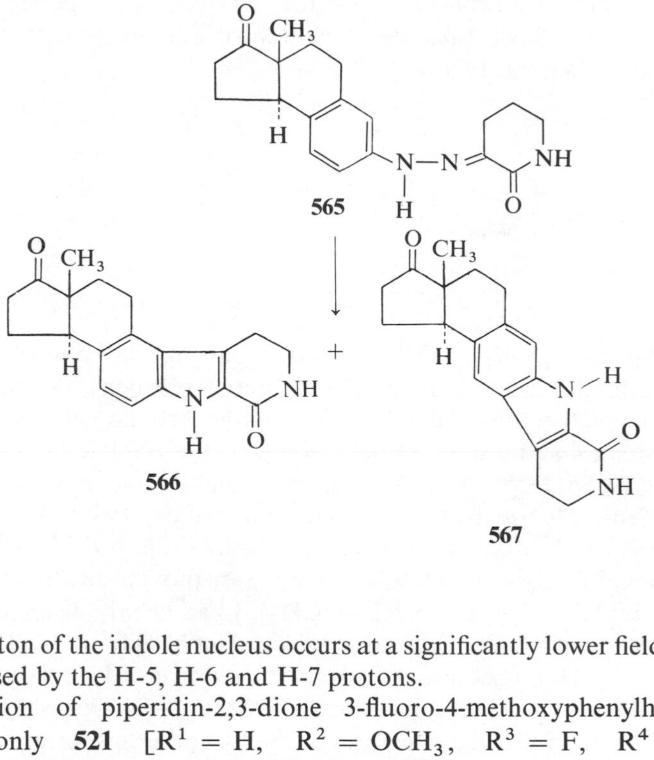

565

566

567

the H-4 proton of the indole nucleus occurs at a significantly lower field than the signals caused by the H-5, H-6 and H-7 protons.

Indolization of piperidin-2,3-dione 3-fluoro-4-methoxyphenylhydrazone afforded only **521** [R^1 = H, R^2 = OCH_3, R^3 = F, $R^4 + R^5$ = $CONH(CH_2)_2$], the structure of which was differentiated from the alternative possible isomeric product by p.m.r. analysis which showed *ortho* and *meta* proton–fluorine spin–spin interactions, again without discernible *para* proton–proton coupling (Kirk, 1976).

Indolization of **508** [R^1 = C_6H_5, R^2 = H, $R^3 + R^4$ = $(CH_2)_3$] produced a mixture of both possible isomeric products, **509** and **510** [R^1 = C_6H_5, R^2 = H, $R^3 + R^4$ = $(CH_2)_3$] from which only the former isomer was isolated in a pure form. The structures of the two products were differentiated by sodium periodate oxidation of the indolic 2,3-double bond in the mixture of the two indolic products which gave after fractional crystallization, **568** ($R^1 = C_6H_5$, $R^2 = H$) and **568** ($R^1 = H$, $R^2 = C_6H_5$), respectively. The latter isomer was clearly distinguished from the former by its p.m.r. spectrum which exhibited a low field aromatic doublet ($\tau = 1.85$, $J = 8$ Hz) characteristic of the H-7 proton which was deshielded by the adjacent carbonyl group (Jones, G. and Tringham, 1975). Similar deshielding enabled the structures of the two products resulting from the indolization of **254** ($R^1 = R^2 = CH_3$, R^3–R^5 = H, X = S) to be distinguished. Whereas in one of the products, **378** ($R^1 = R^2 = CH_3$, R^3–R^5 = H, X = S), the protons of both methyl groups appeared as a singlet at $\tau = 7.73$, in the other product the signal of one of the methyl groups was shifted downfield ($\tau = 7.16$), as would be expected of the C-5 methyl protons in **569** because of the close proximity of the carbonyl group at the 4-position (Wright, 1976).

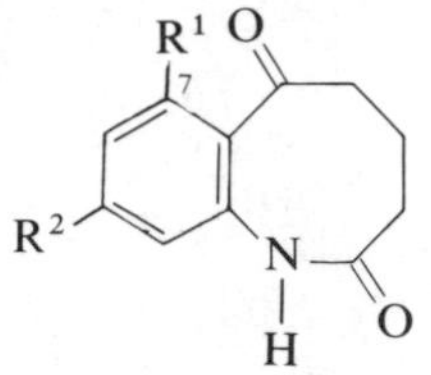

568 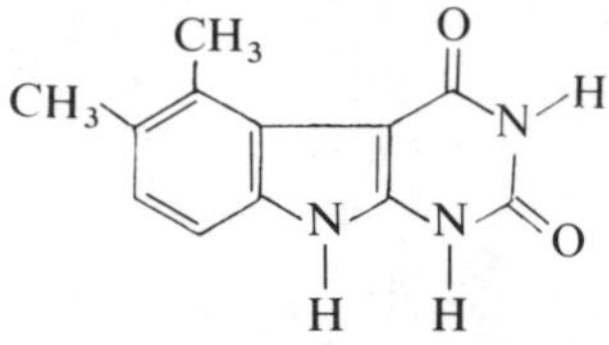 **569**

U.v. spectroscopy has also been used to distinguish between the isomers formed from the indolization of 3-substituted phenylhydrazones. The u.v. spectra of 6-substituted indoles show a bathochromic shift relative to the corresponding 4-substituted isomers (Keglevič and Goleš, 1970; Pappalardo and Vitali, 1958a, 1958b; Stoll, Troxler, Peyer and Hofmann, A., 1955; Szmuszkovicz, Glenn, Heinzelman, Hester and Youngdale, 1966; Wieland, T. and Grimm, 1965 and other references quoted below). Thus, the products formed from the indolization of cyclohexanone 3-methoxyphenylhydrazone [**508**; $R^1 = OCH_3$, $R^2 = H$, $R^3 + R^4 = (CH_2)_4$] had clearly distinguishable u.v. spectra, one of which was assigned to the product **510** [$R^1 = OCH_3$, $R^2 = H$, $R^3 + R^4 = (CH_2)_4$] because of its similarity to that of 1,2,3,4-tetrahydroharmine, a 7-methoxyindole chromophore (Chalmers, J. R., Openshaw and Smith, G. F., 1957). Likewise, the structures of the two products formed in each case from the indolization of various ketonic 3-nitrophenylhydrazones were assigned by comparison of their u.v. spectra with those of model 4- and 6-nitroindolic chromophores (Frasca, 1962a – see also Berti, Da Settimo and Segnini, 1960; Pappalardo and Vitali, 1958a), as were the two isomeric indolic products resulting in each case from the treatment with polyphosphoric acid of 4-chloro- and 4-methylacetophenone 3-nitrophenylhydrazone [**508**; $R^1 = NO_2$, $R^2 = R^4 = H$, $R^3 = C_6H_4Cl(4)$ and $C_6H_4CH_3(4)$, respectively] and the one indolic product, **509** ($R^1 = NO_2$, $R^2 = R^4 = H$, $R^3 = C_6H_5$), resulting from the similar treatment of **508** (R^1 = NO_2, R^2 = R^4 = H, R^3 = C_6H_5) (Dennler, and Frasca, 1966b – see also McKay, Parkhurst, Silverstein and Skinner, 1963). Indolization of a mixture of 3-methoxy-N_α-methylphenylhydrazine with 2-methylcyclohexanone afforded either 1,2,3,4-tetrahydro-5- or -7-methoxy-1,9-dimethylcarbazole as the neutral product and a mixture of **570** (R^1 = OCH_3, R^2 = H) and **570** (R^1 = H, R^2 = OCH_3) as the basic product. The structures of these two isomeric basic products were established by catalytic hydrogenation to, and subsequent u.v. spectral examination of, the corresponding 1,2,3,4,4a,9a-hexahydrocarbazoles **571** (R^1 = OCH_3, R^2 = H, R^3 = CH_3) and **571**

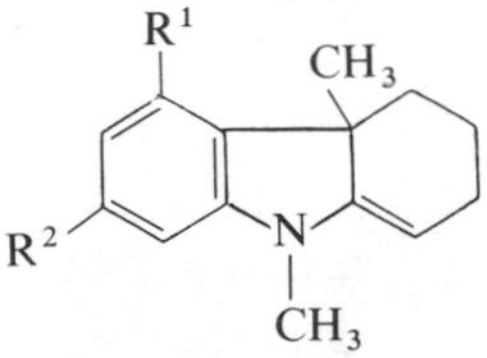

R1 CH3 R2 N CH3

570

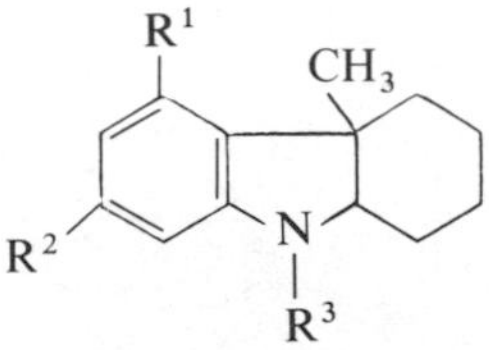

R1 CH3 R2 N R3

571

(R^1 = H, R^2 = OCH_3, R^3 = CH_3) (Millson and Robinson, R., 1955 – see also Chalmers, J. R., Openshaw and Smith, G. F., 1957). In a related study, a mixture of 3-methoxyphenylhydrazine and 2-methylcyclohexanone was refluxed in acetic acid, the basic reaction product was isolated and reduced with sodium borohydride to give **571** (R^1 = OCH_3, R^2 = R^3 = H and R^1 = R^3 = H, R^2 = OCH_3) in a yield ratio of *ca.* 1 : 3, respectively (Sakai, Wakabayashi and Nishina, 1969). Unfortunately, the non-basic product from this indolization was not examined.

The use of u.v., i.r. and p.m.r. spectroscopic techniques in combination allowed the assignment of structures to the two indolic products, **509** and **510** [R^1 = $OCH_2C_6H_5$, R^2 = R^3 = H, R^4 = $(CH_2)_2SCH_2C_6H_5$], obtained by treatment of a mixture of 3-benzyloxyphenylhydrazine hydrochloride with 4-benzylthiobutyraldehyde diethyl acetal in ethanolic aqueous acetic acid (Keglevič and Goleš, 1970).

B. Indolization of *meta*- and *para*-Phenylenedihydrazones

The indolization of biscyclohexanone *meta*-phenylenedihydrazone [**572**; R^1 + R^2 = $(CH_2)_4$] [for the synthesis of *meta*-phenylenedihydrazine dihydrochloride from *meta*-phenylenediamine, see Rull and Le Strat (1975) and Schoutissen (1935)] afforded only one of the two possible isomeric products, **573** [R^1 + R^2 = $(CH_2)_4$] and **574** [R^1 + R^2 = $(CH_2)_4$, R^3 = H]. The product was assigned the 'angular' structure **574** [R^1 + R^2 = $(CH_2)_4$, R^3 = H] (Tomlinson, M. L., 1951) from theoretical considerations and by analogy with reactions such as the Skraup reaction with *meta*-phenylenediamine.

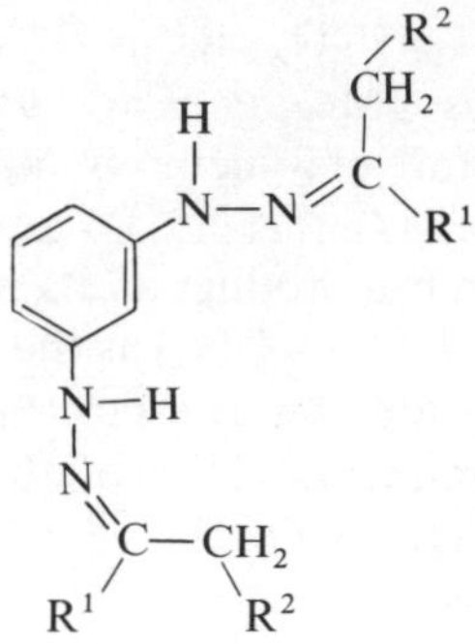

572

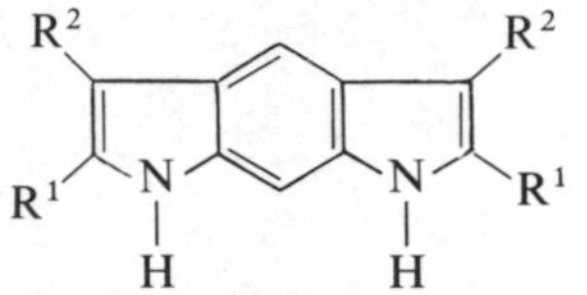

573

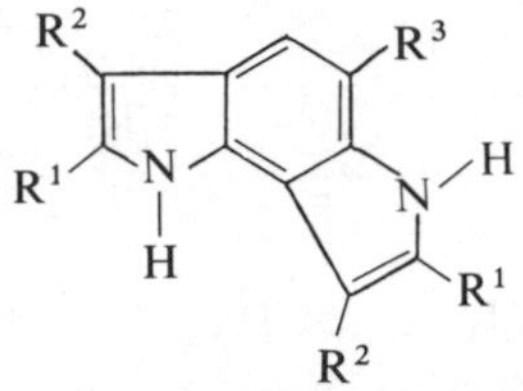

574

Support for this structural assignment was certainly evident from earlier (Smith, C. R., 1930) and later (Kulka and Manske, R. H. F., 1952; Jones, N. A. and Tomlinson, M. L., 1953) studies which showed that in related reactions, 'angular' products are formed in preference to 'linear' ones. Confirmation of structure **574** [$R^1 + R^2 = (CH_2)_4$, $R^3 = H$] for the indolization product was obtained by its dehydrogenation, by pyrolysis with palladium-on-carbon under nitrogen, to give **575** (Tomlinson, M. L., 1951) which was later unambiguously synthesized by dehydrogenation of **574** [$R^1 + R^2 = (CH_2)_4$,

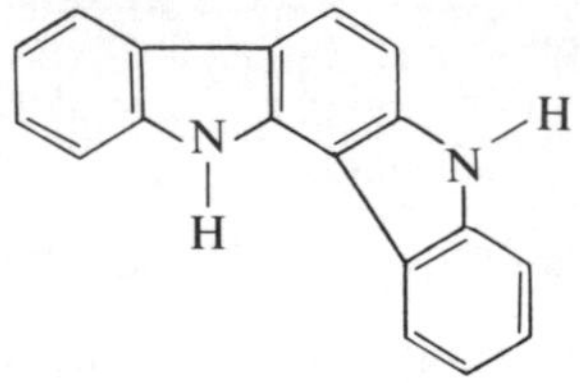

575

$R^3 = H$], prepared, but unfortunately not purified and characterized, by heating 5-amino-1,2,3,4-tetrahydrocarbazole with 2-chlorocyclohexanone (Hall and Plant, 1953). Alternatively, indolization of 1,2,3,4-tetrahydrocarbazol-4-one phenylhydrazone produced **575**, spontaneous dehydrogenation (see Chapter I, D) also occurring under the indolization conditions (Mann and

Willcox, 1958). It is unfortunate that, in this study, the origin of the 1,2,3,4-tetrahydrocarbazol-4-one was not quoted although presumably it was synthesized by indolization of cyclohexan-1,3-dione monophenylhydrazone as published earlier (Clemo and Felton, 1951b) (see Chapter III, K2). In accordance with these observations, the indolization of bisethyl pyruvate *meta*-phenylenedihydrazone (**572**; $R^1 = COOC_2H_5$, $R^2 = H$) gave only the 'angular' product **574** ($R^1 = COOC_2H_5$, $R^2 = R^3 = H$). The structure of this product was distinguished from that of the corresponding 'linear' isomer **573** ($R^1 = COOC_2H_5$, $R^2 = H$) by examination of the p.m.r. spectra of it and its hydrolysis product **574** ($R^1 = COOH$, $R^2 = R^3 = H$) and this latter's decarboxylation product **574** (R^1–$R^3 = H$). All these spectra indicated the presence of the *ortho* benzenoid protons ($J = 8.5$–8.8 Hz) and the two non-equivalent pyrrolic 1- and 3-protons and in the third compound the two non-equivalent pyrrolic 2-protons were evident (Samsoniya, Targamadze, Tret'yakova, Efimova, Turchin, Gverdtsiteli, I. M. and Suvorov, 1977). A more recent (Samsoniya, Targamadze and Suvorov, 1980) study has found that both **574** ($R^1 = COOC_2H_5$, $R^2 = R^3 = H$) (60% yield) and **573** ($R^1 = COOC_2H_5$, $R^2 = H$) (8% yield) are formed from the indolization of **572** ($R^1 = COOC_2H_5$, $R^2 = H$). Prior to these above studies, the 'linear' structure **573** ($R^1 = R^2 = C_6H_5$) had been postulated for the product resulting from the indolization of **572** ($R^1 = R^2 = C_6H_5$) (Ruggli and Petitjean, 1936) and likewise, but later than and contrary to the results of the above studies, the 'linear' structures **573** ($R^1 = C_6H_5$, $R^2 = H$; $R^1 = COOC_2H_5$, $R^2 = H$ and $R^1 = H$, $R^2 = CH_3$) were proposed for the indolization products of **572** ($R^1 = C_6H_5$, $R^2 = H$; $R^1 = COOC_2H_5$, $R^2 = H$ and $R^1 = H$, $R^2 = CH_3$, respectively) [**572** ($R^1 = CH_3$, $R^2 = CH_2COOC_2H_5$) could not be indolized] (Giuliano and Leonardi, 1957). In the former of these studies, the structure of the product was based upon its identification with the compound obtained earlier by condensation of *meta*-phenylenediamine with benzoin. However, in this earlier work (Japp and Meldrum, 1899) it was rightly stated that the structure of the product could be either the corresponding 'linear' or 'angular' isomer. Clearly, contemporary experimental verification for the former structural postulation was lacking (Tomlinson, M. L., 1951). Indeed, it was later shown (Jones, N. A. and Tomlinson, M. L., 1953) to be incorrect when structure **574** ($R^1 = R^2 = C_6H_5$, $R^3 = H$) was established for the compound obtained by Japp and Meldrum (1899), and hence for the indolization product of Ruggli and Petitjean (1936), by acetylation of **574** ($R^1 = R^2 = C_6H_5$, $R^3 = H$) to afford **574** ($R^1 = R^2 = C_6H_5$, $R^3 = COCH_3$) which was unambiguously synthesized by condensation of 2,4-diaminoacetophenone with benzoin. In the latter of the above studies (Giuliano and Leonardi, 1957), the 'linear' structures for all three products were based, by analogy, upon the claim that the indolization of **572** ($R^1 = C_6H_5$, $R^2 = H$) produced a compound with all the characteristics of that obtained (Ruggli and Straub, 1938) by catalytic hydrogenation, with subsequent cyclization, of **576**. However, a direct comparison of these products was not effected and, furthermore, no actual physical properties of the indolization product were quoted (Giuliano and Leonardi, 1957). These latter

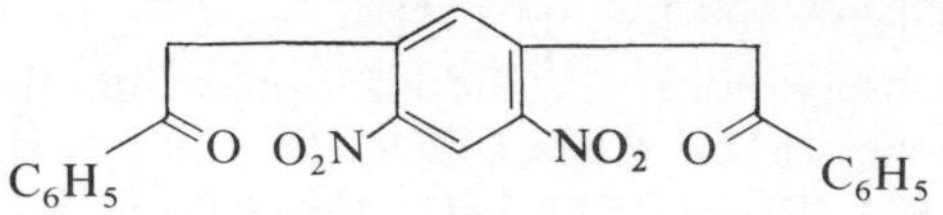

576

indolization products should now be subjected to a p.m.r. spectral examination which, especially in the last two products because of the absence of other aromatic protons, will readily differentiate between the 1,4- and 1,2-aromatic proton systems of **573** ($R^1 = COOC_2H_5$, $R^2 = H$ and $R^1 = H$, $R^2 = CH_3$) and **574** ($R^1 = COOC_2H_5$, $R^2 = R^3 = H$ and $R^1 = R^3 = H$, $R^2 = CH_3$), respectively.

Attempts to indolize biscyclohexanone *para*-phenylenedihydrazone [**577**; $R^1 = H$, $R^2 = N{=}C(CH_2)_5$, which could give rise to two possible products, **578** and/or **579** [$2R = (CH_2)_4$], have been unsuccessful. However, in one case

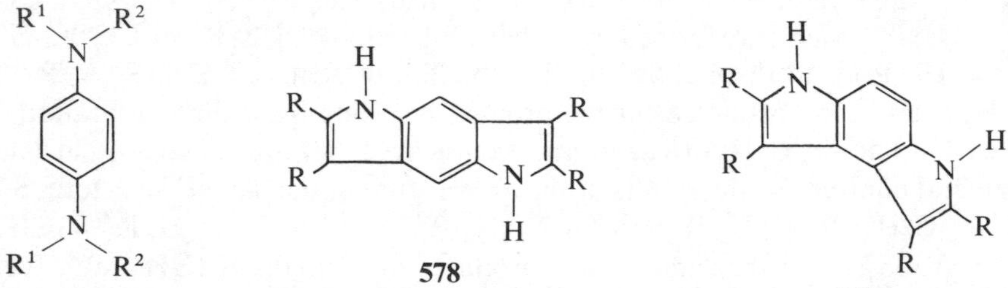

577 **578** **579**

(Clifton and Plant, 1951) no specific catalysts nor experimental conditions were given and in the other case (Tomlinson, M. L., 1951), whereas decomposition was noted in cold glacial acetic acid, using alcoholic sulphuric acid, ammonium sulphate was isolated which indicated the possible occurrence of indolization. In view of this, it would certainly be worth while to attempt further this indolization and that of other bisketone *para*-phenylenedihydrazones. In connection with this it should be noted that the two possible products, **578** and **579** ($R = C_6H_5$), which could result from the indolization of the *para*-phenylenedihydrazone [**577**; $R^1 = H$, $R^2 = N{=}C(CH_2C_6H_5)C_6H_5$] have already been synthesized by alternative routes and their structures have been established (Kinsley and Plant, 1958) and *para*-phenylenedihydrazine dihydrochloride has been readily synthesized from *para*-phenylenediamine (Lee, W. Y., 1974; Lee, W. Y. and Lee, Y. Y., 1969; Rull and Le Strat, 1975; Schoutissen, 1933, 1934). Attempts to produce the dihydrazone **577** [$R^1 = CH_3$, $R^2 = N{=}C(CH_2)_5$] *in situ* by the reduction of the dinitroso compound **577** ($R^1 = CH_3$, $R^2 = NO$) in the presence of cyclohexanone and acid have produced neither the dihydrazone nor indolic products (Clifton and Plant, 1951; Manjunath, 1927).

C. Indolization of 3-Pyridyl- and 3-Quinolylhydrazones and Related Compounds

Whereas the indolization of 2- and 4-pyridylhydrazones, in the latter case without additional substituents on the 2-position of the pyridine ring, are unambiguous with regard to the formation of the new C—C bond, this aspect of the orientation of indolization in 3-pyridylhydrazones is ambiguous. It has been assumed that indolization of acetone and propionaldehyde 2-chloro-5-pyridylhydrazone (**580**; R^1 = Cl, R^2 = R^3 = R^5 = R^6 = H, R^4 = CH_3 and R^1 = Cl, R^2–R^5 = H, R^6 = CH_3, respectively) by heating with a zinc

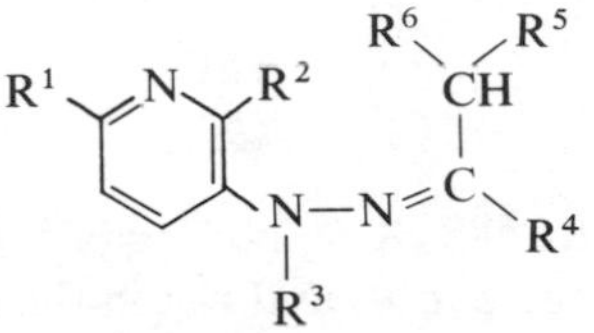

580

chloride catalyst (Deutsche Gold- und Silberschiedeanstalt, 1925) and that the reaction between acetone and 2-chloro-5-pyridylhydrazine (no catalyst specified) (Takahashi, T., Saikachi, Goto and Shimamura, 1944) effected the new C—C bond formation only at the 6-position, rather than the 4-position, of the pyridine nucleus, to give 5-chloro-2-methyl-4-azaindole (**581**; R^1 = Cl, R^2 = R^3 = R^5 = H, R^4 = CH_3, X = N, Y = CH) and 5-chloro-3-methyl-4-azaindole (**581**; R^1 = Cl, R^2 = R^4 = H, R^5 = CH_3, X = N, Y = CH), respectively. Similarly, the indolization of isopropyl methyl ketone 3-pyridylhydrazone by heating with zinc chloride yielded only one product, **582** (X = CH, Y = N) or **582** (X = N, Y = CH) (Ficken and Kendall, 1961). Somewhat

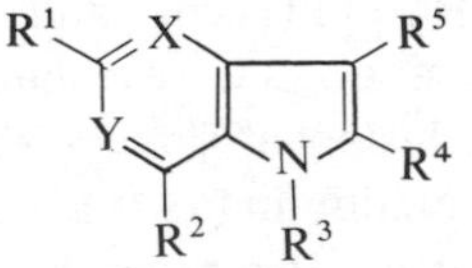

581 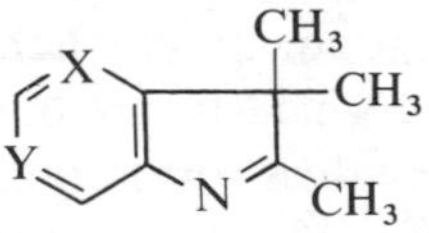 **582**

tentative support for the latter structure was obtained from i.r. spectroscopic investigation, using model compounds for comparative purposes, by consideration of the aromatic C—H out-of-plane deformation vibrations which were thought to give rise to strong absorption in the 900–700 cm^{-1} region. Since at the time the new C—C bond formation during indolization was thought to occur by an intramolecular electrophilic attack, the sole formation of **582** (X = N, Y = CH) was also stated (Ficken and Kendall, 1961) to be in agreement with the limited evidence then available that pyridines possessing an *ortho–para* directing 3-substituent underwent electrophilic attack at the 2- rather than the 4-position. The application of p.m.r. spectroscopic analysis, unfortunately not widely available in 1961, would have readily solved this structural problem. From the indolization of isopropyl methyl ketone N_α-methyl-3-pyridylhydrazone with zinc chloride, only one compound was again isolated which, since it was related to **582** (X = N, Y = CH) obtained in the previous indolization through their reaction with methyl iodide and hydrogen iodide, respectively, was formulated as **583** (Ficken and Kendall, 1961).

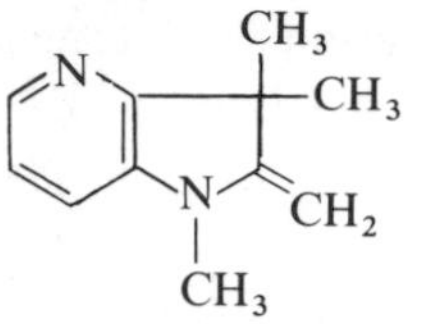

583

Furthermore, the structure **582** (X = N, Y = CH) was supported by this exclusive N_1-quaternization, since in it, and not in the corresponding 6-aza isomer, the quaternization of the pyridino N-atom might have well been expected to be difficult because of the steric hindrance by the neighbouring *gem* dimethyl group. In accordance with these above observations, the only products isolated from the zinc chloride catalysed indolizations of piperidin-2,3-dione 3-pyridylhydrazone [**580**; R^1–R^3 = R^5 = H, R^4 + R^6 = $CONH(CH_2)_2$] and its 1-oxide were shown to have structures **581** [R^1–R^3 = H, R^4 + R^5 = $CONH(CH_2)_2$, X = N and N→O, respectively, Y = CH] since the former product exhibited a characteristic ABX aromatic proton signal in its p.m.r. spectrum (see also Pietra and Tacconi, 1964) and also was not identical with the product obtained by the catalytic hydrogenolysis of the chlorine atom from **581** [R^1 = R^3 = H, R^2 = Cl, R^4 + R^5 = $CONH(CH_2)_2$, X = CH, Y = N] which was unambiguously synthesized by indolization of **580** [R^1 = R^3 = R^5 = H, R^2 = Cl, R^4 + R^6 = $CONH(CH_2)_2$] (Tacconi and Perotti, 1965).

In the above indolizations where only 4-azaindoles were isolated, it is unlikely that they were formed to the complete exclusion of the corresponding 6-aza isomers although their formation may well predominate. It is significant in this respect that the yields of the 4-aza isomers from these reactions were low and that indolization of **580** [R^1 = R^3 = R^5 = H, R^2 = CH_3, R^4 = R^6 = $(CH_2)_4$], in which the formation of a 4-azaindole is prevented (unless methyl group migration or elimination occurs – see Chapter II, G2) afforded **581** [R^1 = R^3 = H, R^2 = CH_3, R^4 = R^5 = $(CH_2)_4$, X = CH, Y = N]. The corresponding ethyl pyruvate hydrazone **580** (R^1 = R^3 = R^5 = R^6 = H, R^2 = CH_3, R^4 = $COOC_2H_5$) was also prepared but was not indolized (Clemo and Holt, R. J. W., 1953). Confirmation of the formation of both 4- and 6-azaindoles from indolization of 3-pyridylhydrazones was forthcoming in the observation (Abramovitch and Adams, K. A. H., 1962) that cyclohexanone 3-pyridylhydrazone [**580**; R^1–R^3 = R^5 = H, R^4 + R^6 = $(CH_2)_4$], upon heating with zinc chloride, produced in high yield an indolic product from which were separated, by column chromatography, both **581** [$R^1 = R^3 = H$, $R^4 + R^5 = (CH_2)_4$, X = N, Y = CH] and **581** [$R^1 = R^3 = H$, $R^4 + R^5 = (CH_2)_4$, X = CH, Y = N], the yield of the former compound greatly exceeding that of the latter compound. The structures of these two products were confirmed by their dehydrogenation, with 10% palladium-on-carbon in boiling 1,3,5-trimethylbenzene, to give the known δ- and β-carbolines, respectively (Abramovitch and Adams, 1962). In further related examples, propionaldehyde, ethyl methyl ketone, cyclopentanone, cyclohexanone, phenylacetaldehyde and benzyl phenyl ketone 3-pyridylhydrazones were all indolized, by refluxing in either mono- or diethylene glycol, to afford the corresponding 4- and 6-azaindoles in which the yield of the former isomer greatly predominated in all cases (see Table 29 in Chapter IV, I). Only the corresponding 4-azaindoles were isolated from the similar indolizations of butyraldehyde, acetophenone and 1-tetralone 3-pyridylhydrazones although traces of the corresponding 6-aza isomers may also have been formed (Kelly, A. H. and Parrick, 1970). Now that model 4- and 6-azaindoles are available, i.r. and u.v. spectroscopy can be reliably used to establish the orientation of the indolization in the product(s) obtained from 3-pyridylhydrazones (Kelly, A. H. and Parrick, 1970). Indolization of piperidin-2,3-dione 4-(2-methylpyridyl)hydrazone-1-oxide (**584**, $R^1 = CH_3$, $R^2 = R^3 = H$) furnished only one of the two possible isomeric indoles, **585** ($R^1 = R^3 = H$, $R^2 = CH_3$ or $R^1 = CH_3$, $R^2 = R^3 = H$) but the position of the CH_3 group was not established (Tacconi and Pietra, 1965). Clearly, a p.m.r. spectroscopic investigation, as used below, would have accomplished this.

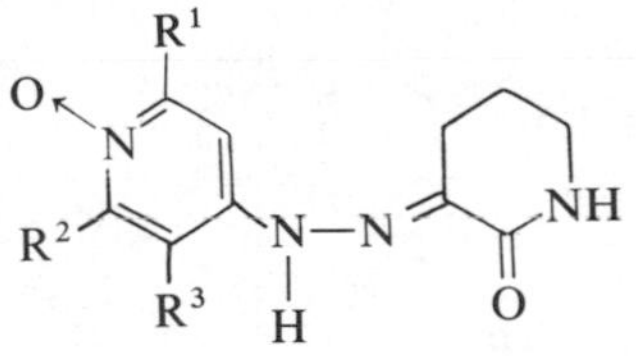

584

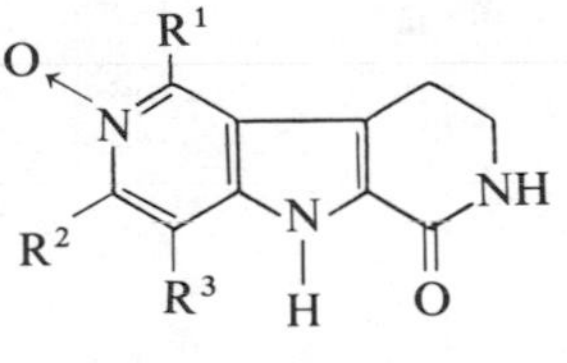

585

It has recently been reported (Bisagni, Ducrocq and Civier, 1976) that upon boiling under reflux in diphenylether solution, the hydrazones **586** [$R^1 = CH_3$, $R^2 = H$; $R^1 = R^2 = CH_3$; $R^1 = C_6H_5$, $R^2 = CH_3$; $R^1 = CH_3$, $R^2 = C_6H_5$; $R^1 = C_6H_5$, $R^2 = C_2H_5$ and $R^1 + R^2 = (CH_2)_{3-6}$] afforded the corresponding 4,5-dihydro-5-azaindol-4-ones **587** [$R^1 = CH_3$, $R^2 = H$;

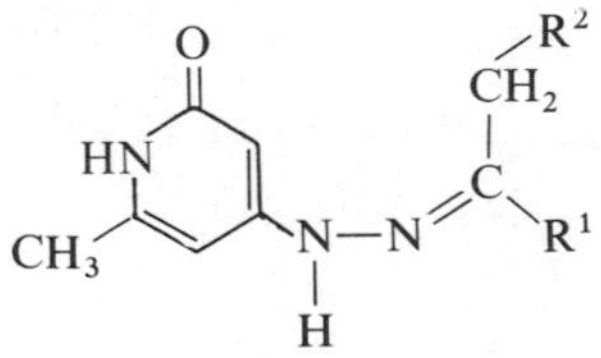

586

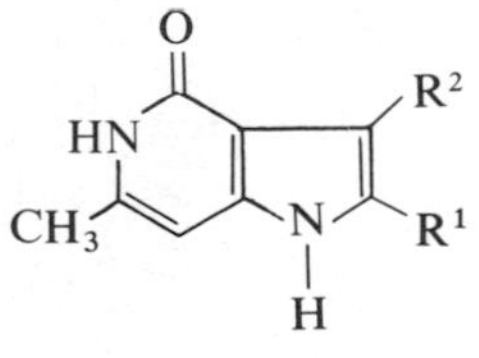

587

$R^1 = R^2 = CH_3$; $R^1 = C_6H_5$, $R^2 = CH_3$; $R^1 = CH_3$, $R^2 = C_6H_5$; $R^1 = C_6H_5$, $R^2 = C_2H_5$ and $R^1 + R^2 = (CH_2)_{3-6}$, respectively] and likewise the corresponding 4,5,6,7-tetrahydrobenzofuran-4-one and 1- and 2-tetralone hydrazones were converted into **588**, **589** and **590**, respectively. The reaction only gave a resin when similar indolization attempts were made using 1-benzylpiperidin-4-one as the ketonic moiety. That the orientation of indolization was as indicated by these structures was established in one case, and inferred by analogy in the others, by treatment of **587** [$R^1 + R^2 = (CH_2)_4$] with phosphorus oxychloride to afford **591** (R = Cl) which was hydrogenolysed to give **591** (R = H), the p.m.r. spectrum of which clearly indicated the 1,4-protons in the aromatic ring. If the indolization had occurred in the alternative direction, the corresponding product would have contained the two aromatic protons on adjacent carbon atoms. In a related indolization (Ducrocq, Civier, André-Louisfert amd Bisagni, 1975) in which the mechanistic role of the 2-pyridone moiety was reversed relative to that above, compounds **255** [$R^1 = H$, $R^2 = CH_3$; $R^1 = H$, $R^2 = (CH_2)_2CH_3$ and $R^1 = R^2 = CH_3$] were reacted with phenylhydrazine under a Dean and Stark trap and the corresponding intermediate enehydrazines **256** were then reacted in boiling diphenylether to afford **592** [$R^1 = H$, $R^2 = CH_3$; $R^1 = H$, $R^2 = (CH_2)_2CH_3$ and $R^1 = R^2 = CH_3$, respectively]. That indolization had occurred in the direction as shown was demonstrated by utilizing a method analogous to that used in the aforementioned study.

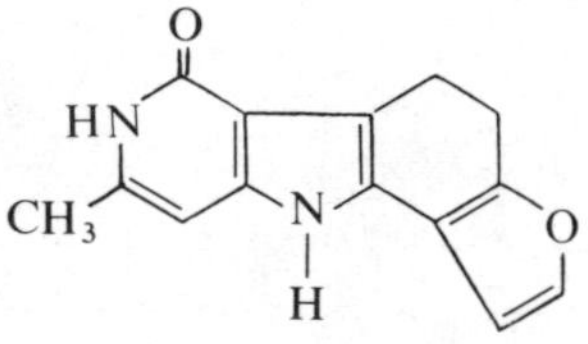

588

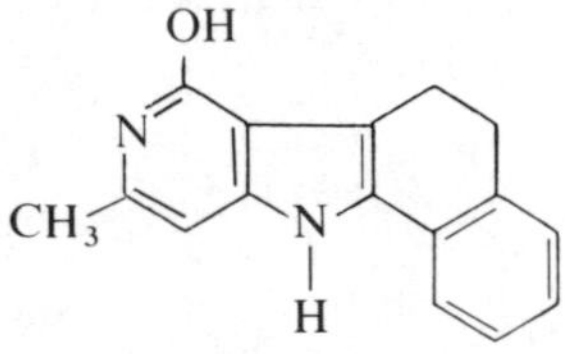

589

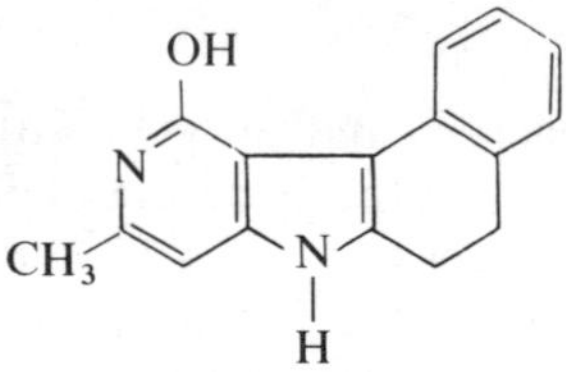

590

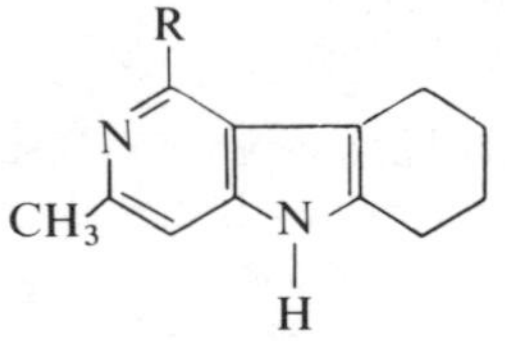

591

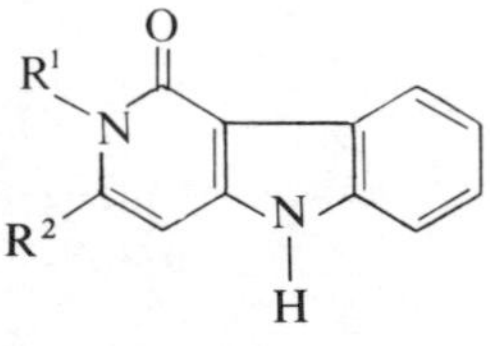

592

As with 3-pyridylhydrazones, the indolization of 3-quinolylhydrazones can also theoretically yield two products. However, in all the examples so far reported, only products resulting from indolization towards the C-4 atom of the quinoline ring have been isolated. These observations might be expected on the basis of data relating to the relative reactivities of the 2- and 4-positions of quinoline. Thus, indolization of cyclohexanone 3-quinolylhydrazone with an acetic acid–sulphuric acid catalyst afforded **593** [R^1 = H, $R^2 + R^3$ = $(CH_2)_4$]. The structure of this product was established by its dehydrogenation to the known compound **594** (R = H) (Clemo and Felton, 1951a) which was unambiguously synthesized by sequential indolization of **595** to **596**, reduction

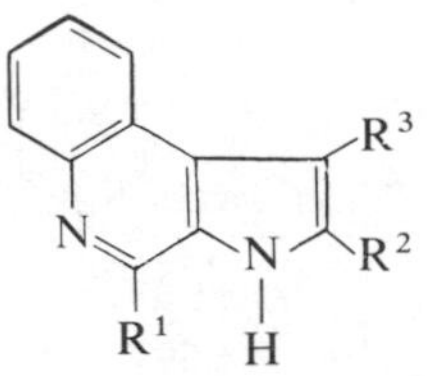

593

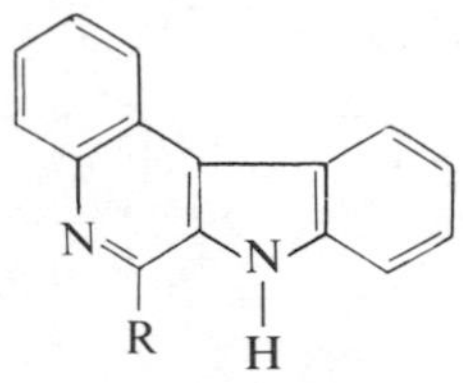

594

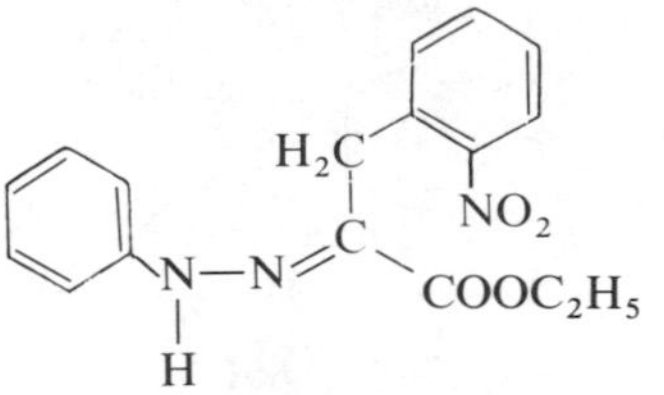

595

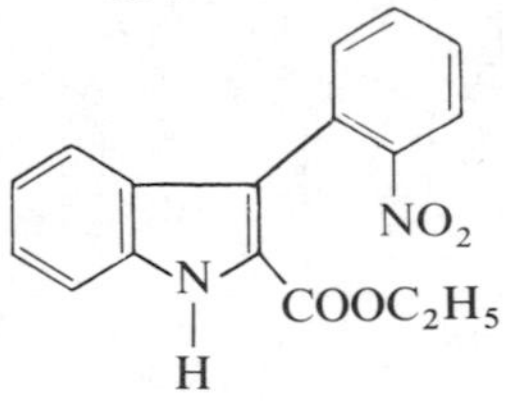

596

of the nitro group and spontaneous lactamization to **594** (R = OH) and conversion into **594** (R = H) via **594** (R = Cl) [this reaction sequence was first investigated by Kermack and Slater (1928) and again used later by Kermack and Tebrich (1940) – see also Freedman and Judd, 1972, Neber, Knöller, Herbst and Trissler, 1929]. Similarly, ethyl pyruvate 3-quinolylhydrazone furnished **593** (R^1–R^3 = H), ester hydrolysis and decarboxylation also occurring under the indolization conditions (fusion with zinc chloride) which was followed by sublimation by heating *in vacuo* (Govindachari, Rajappa and Sudarsanam, 1961). The identity of this product was based upon the close similarity of its m.p. to that reported previously for an authentic sample (Eiter and Nagy, 1949). Indolization of 2-tetralone 3-quinolylhydrazone (**597**;

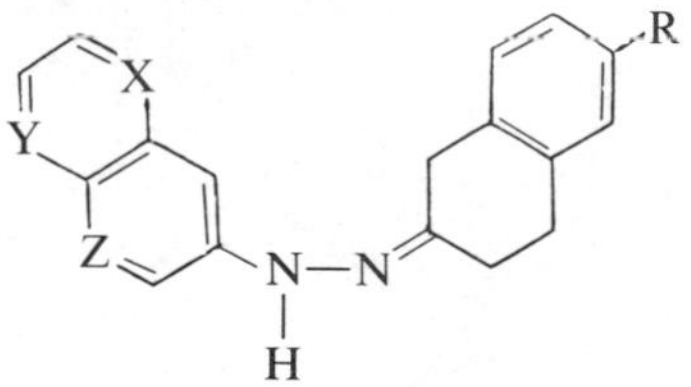

597

X = Y = CH, Z = N, R = H) in an acetic acid–sulphuric acid mixture gave a product which was assumed to have structure **598** since it was dehydrogenated to a compound which was formulated as **599** (X = Y = CH, Z = N, R = H) because its u.v. spectrum was closely similar to that of the benzenoid isostere **599** (X–Z = CH, R = H) (Buu-Hoï, Périn and Jacquignon, 1962). Likewise, the products resulting from the indolizations of **597** (R = CH_3, X = Y = CH, Z = N), **600** (R = CH_3, W = X = CH, Y = N, Z = CH=CH) and **600** (R = H, W = X = CH, Y = N, Z = S), with subsequent dehydrogenations, were formulated as **599** (R = CH_3, X = Y = CH, Z = N), **601** (R = CH_3,

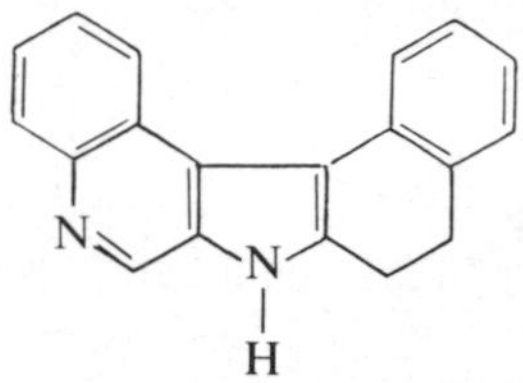

598 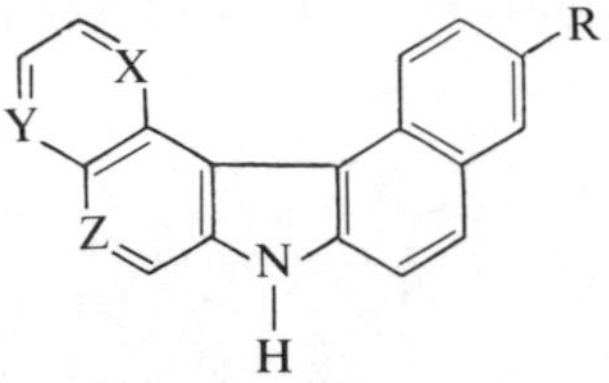 **599**

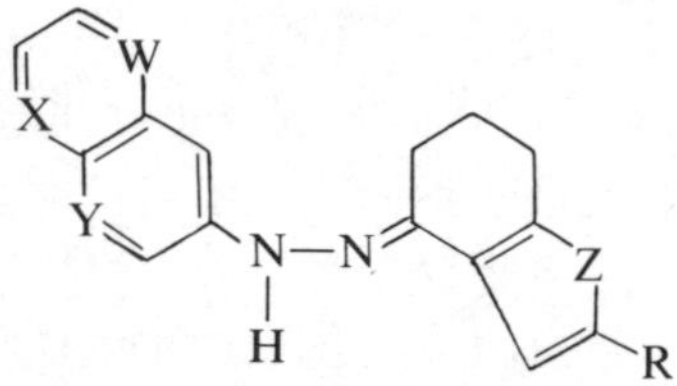

600 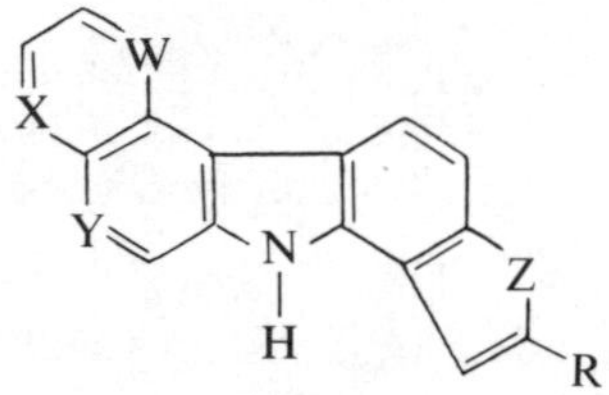 **601**

W = X = CH, Y = N, Z = CH=CH) and **601** (R = H, W = X = CH, Y = N, Z = S), respectively (Buu-Hoï, Jacquignon and Hoeffinger, 1963) since their u.v. spectra closely resembled those of their benzenoid isosteres. Without any experimental verification, the new C—C bond formation was assumed to involve the 4- rather than the 2-position of the quinoline nucleus when **600** (R = H, W = X = CH, Y = N, Z = CH=CH), **602** (X = CH, Y = N) and **604** (X = CH, Y = N), upon indolization with a zinc chloride catalyst (attempts to use an acetic acid–sulphuric acid catalyst were unsuccessful) and subsequent dehydrogenation, were reported to afford **601** (R = H, W = X = CH, Y = N, Z = CH=CH), **603** (X = CH, Y = N) and **605** (X = CH, Y = N), respectively (Buu-Hoï, Périn and Jacquignon, 1962). Acetone, ethyl

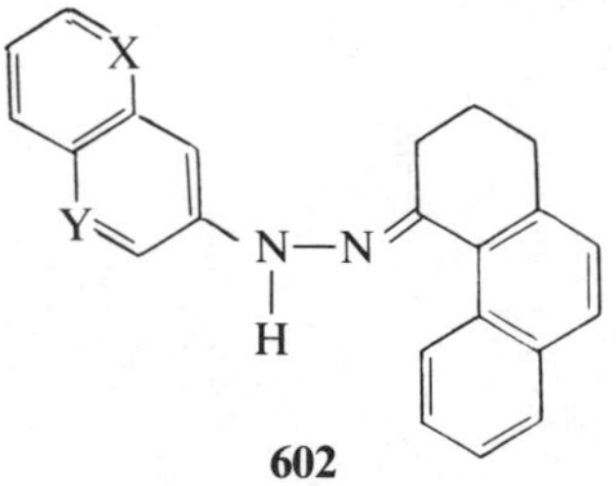

602 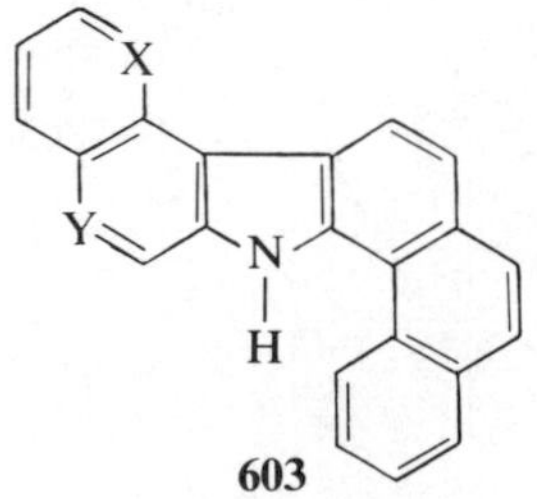 **603**

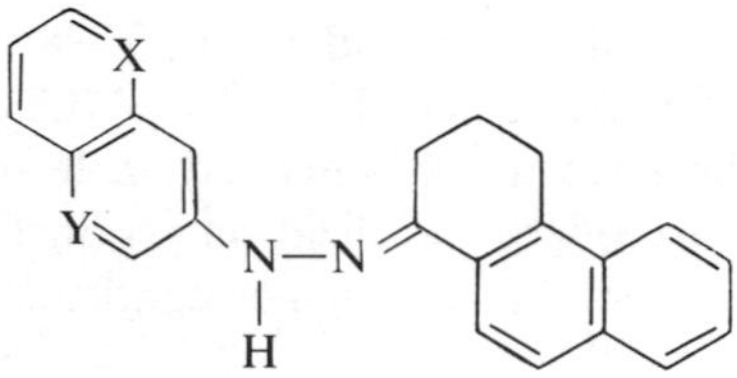

604 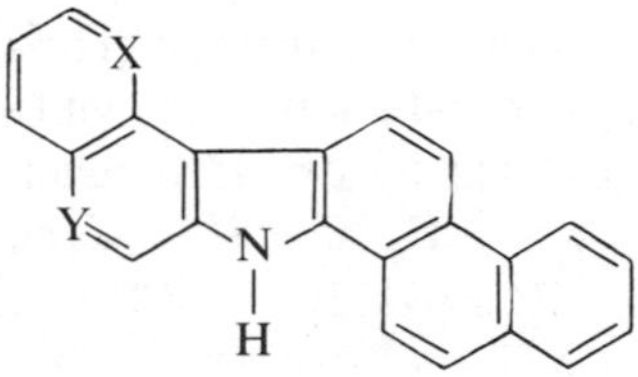 **605**

methyl ketone, diethyl ketone, acetophenone, propiophenone, benzyl phenyl ketone and 1-tetralone 3-quinolylhydrazone have all been indolized, using a zinc chloride catalyst, to compounds which were assigned structures **593** ($R^1 = R^3 = H$, $R^2 = CH_3$; $R^1 = H$, $R^2 = R^3 = CH_3$; $R^1 = H$, $R^2 = C_2H_5$, $R^3 = CH_3$; $R^1 = R^3 = H$, $R^2 = C_6H_5$; $R^1 = H$, $R^2 = C_6H_5$, $R^3 = CH_3$ and $R^1 = H$, $R^2 = R^3 = C_6H_5$) and **601** (R = H, W = X = CH, Y = N, Z = CH=CH), respectively (Govindachari, Rajappa and Sudarsanam, 1961). However, no evidence was offered to support these formulations in preference to the alternative isomeric structures for these products, although u.v. spectral data were quoted (Govindachari, Rajappa and Sudarsanam, 1961) from which it could be seen that the spectra of the first three products were similar to that of **593** (R^1–R^3 = H), the structure of which had been established as already described. Furthermore, analogy with the above described related studies would suggest that the structures given for these seven products are probably correct.

D. Indolization of 2-Naphthylhydrazones and Related Compounds

Indolization of cyclohexanone 2-naphthylhydrazone [**606**; R^1–R^3 = H, $R^4 + R^5 = (CH_2)_4$, X–Z = CH] could possibly afford two isomeric products, **607** [R^1 = H, $R^2 + R^3 = (CH_2)_4$, X = CH] and/or **608** [$R^1 = R^2$ = H, $R^3 + R^4 = (CH_2)_4$, X–Z = CH]. It has been found in practice that only

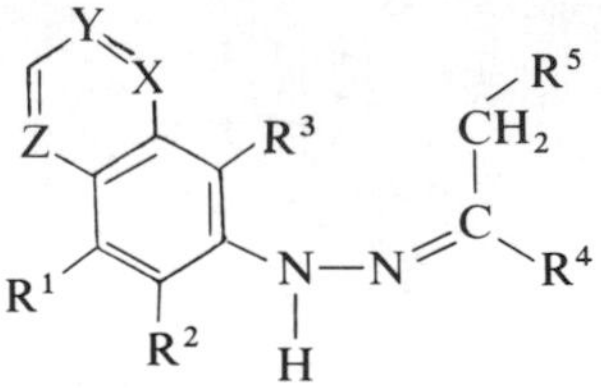

606 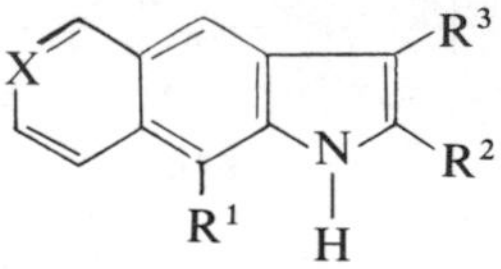 **607**

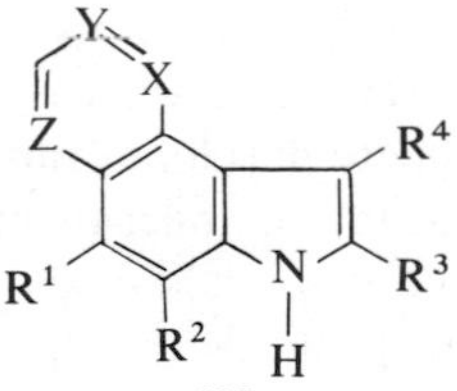

608

one product was formed (Huisgen, 1948a; Oakeshott and Plant, 1928), the structure of which was shown to be **608** [$R^1 = R^2 = H$, $R^3 + R^4 = (CH_2)_4$, X–Z = CH] by its dehydrogenation to the known 3,4-benzocarbazole (**609**; $R^1 = R^2 = H$, X = Y = CH). The structure of the indolization product **608** [$R^1 = R^2 = H$, $R^3 + R^4 = (CH_2)_4$, X–Z = CH] was further confirmed by

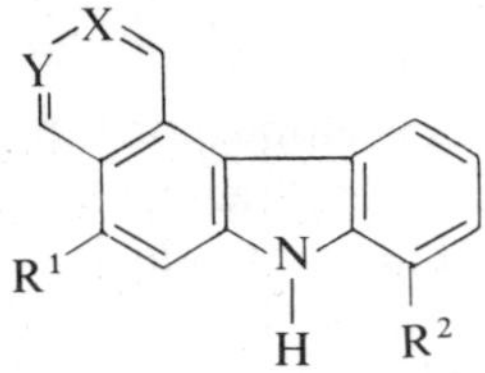

609

its formation from the indolization of **606** [$R^1 = R^2 = H$, $R^3 = CH_3$, $R^4 + R^5 = (CH_2)_4$, X–Z = CH] using nickel(II) chloride in acetic acid as catalyst. The involvement of the naphthyl 1-position in the indolization process is so strongly favoured that even when it is quaternary it can still be involved in the new C—C bond formation, in this instance the original methyl substituent ultimately being eliminated (Huisgen, 1948a) (see Chapter II, G2). Without verification, the two products obtained by the dilute sulphuric acid catalysed indolization of 2-methylcyclohexanone 2-naphthylhydrazone [**606**; R^1–R^3 = H, $R^4 + R^5 = CH(CH_3)(CH_2)_3$, X–Z = CH] were formulated as **607** [$R^1 = H$, $R^2 + R^3 = CH(CH_3)(CH_2)_3$, X = CH] and **610** (Cecchetti and Ghigi, 1930). However, the former product was later (Bryant and Plant, 1931) shown to have structure **608** [$R^1 = R^2 = H$, $R^3 + R^4 = CH(CH_3)(CH_2)_3$, X–Z = CH] by its dehydrogenation into **609** ($R^1 = H$, $R^2 = CH_3$, X = Y = CH) which was unambiguously synthesized and, by analogy, the latter indolization product was then formulated as **611**. It was later reported (Campaigne,

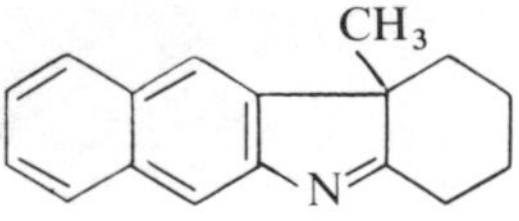

610

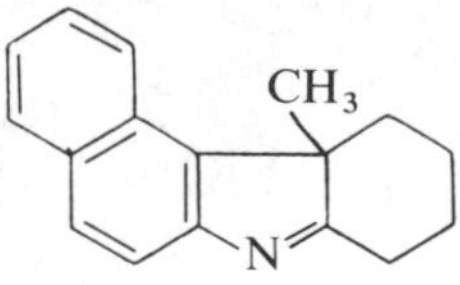

611

Ergener, Hallum and Lake, 1959) that a mixture of 2-methylcyclohexanone with 2-naphthylhydrazine hydrochloride in aqueous methanol at room temperature failed to indolize. This failure was attributed to steric hindrance since mixtures of cyclohexanone or 3- or 4-methylcyclohexanone with 2-naphthylhydrazine hydrochloride underwent indolization under these conditions. Certainly, under such mild conditions steric effects could become manifest. Cyclohexanone 1-bromo-3-naphthylhydrazone [**606**; $R^1 = Br$, $R^2 = R^3 = H$, $R^4 + R^5 = (CH_2)_4$, X–Z = CH], upon indolization in refluxing dilute sulphuric acid, gave only one product which was shown to have structure **609** (R^1 = Br, R^2 = H, X–Z = CH) by being structurally related to **609** (R^1 = R^2 = H, R^3 + R^4 = $(CH_2)_4$, X–Z = CH) by the latter's bromination (Plant and Tomlinson, M. L., 1932b). Indolization of ethyl pyruvate and acetone 2-naphthylhydrazone **606** (R^1–R^3 = R^5 = H, R^4 = COOH and CH_3, respectively, X–Z = CH) by fusion with zinc chloride afforded in each case products resulting from reaction in one direction only (Schlieper, 1886). The products obtained from the former reaction were found to be **608** (R^1 = R^2 = R^4 = H, R^3 = COOH, X–Z = CH) and its decarboxylation product **608** (R^1–R^4 = H, X–Z = CH) since the latter compound was found to be different from 5,6-benzindole, the product which would have ultimately resulted from indolization at the C-3 atom of the 2-naphthylhydrazone nucleus (Rydon and Siddappa, 1951). The product resulting from the latter indolization is, by analogy, most likely to have structure **608** (R^1 = R^2 = R^4 = H, R^3 = CH_3, X–Z = CH) rather than the alternative 'linear' isometric structure. The properties of **608** (R^1 = R^2 = R^4 = H, R^3 = COOH, X–Z = CH) reported by three different groups (Goldsmith and Lindwall, 1953; Rydon and Siddappa, 1951; Schlieper, 1886) were in close agreement whereas those reported by a fourth group (Hughes, G. K. and Lions, 1937–1938a) were widely divergent from these other three for apparently the same product formed by similar indolization. It has been suggested that this could have been caused by the existence of polymorphic forms of this product, although no evidence for this could be found (Goldsmith and Lindwall, 1953).

Synthetic entry into the linear series corresponding to **607** (X = CH) has been accomplished in several ways. Fischer indolization of ethyl pyruvate 1-methoxy-2-naphthylhydrazone **606** (R^1 = R^2 = R^5 = H, R^3 = OCH_3, R^4 = $COOC_2H_5$, X–Z = CH) with dry hydrogen chloride in ethanol afforded **607** (R^1 = OCH_3, R^2 = $COOC_2H_5$, R^3 = H, X = CH) (Goldsmith and Lindwall, 1953) and alternative syntheses have given 5,6-benzindole (Eraksina, Maslennikova, Shagalov and Suvorov, 1979; Tret'yakova, Suvorov, Efimova, Vasil'ev, A. M., Shagalov and Babushkina, T. A., 1978) and 2,3-benzocarbazole (Buu-Hoï, Hoán and Khôi, 1950a; Grotta, Riggle and Bearse, 1961). A further attempted synthetic entry into this series by a Fischer indolization was only partially successful when pyruvic acid 1,2,3,4-tetrahydro-6-naphthylhydrazone was indolized with either methanolic hydrogen chloride or methanolic sulphuric acid. Although both possible isomeric products (as their methyl esters) were formed, only that resulting from indolization involving the 5-position of the hydrazone nucleus could be isolated and no homogeneous compound could be separated from the mixture resulting after the original mixture of isomeric products was sequentially hydrolysed and decarboxylated (Shagalov, Ostapchuk, Zlobina, Eraksina, Babushkina, T. A., Vasil'ev, A. M., Ogorodnikova and Suvorov, 1978). However, in a related study, a separation of these products, **521** [R^1 + R^2 = $(CH_2)_4$, R^3–R^5 = H and R^1 = R^4 = R^5 = H, R^2 + R^3 = $(CH_2)_4$], appears to have been achieved (Shagalov,

Ostapchuk, Zlobina, Eraksina and Suvorov, 1977). Using the Bucherer carbazole synthesis, **101** reacted with N_α-butylphenylhydrazine in acetic acid to afford **102** [R^1 = H, R^2 = $(CH_2)_3CH_3$] and similarly reacted with phenylhydrazine, after treatment of the reaction product with hydrochloric acid, to yield **102** (R^1 = SO_3H, R^2 = H). In related syntheses, **612** (R = H) reacted

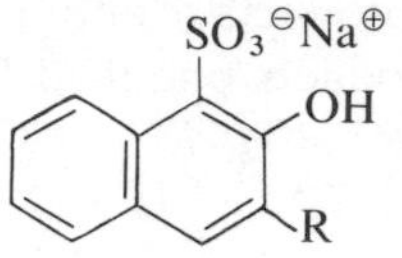

$SO_3^{\ominus}Na^{\oplus}$ OH R

612

with phenylhydrazine and N_α-butylphenylhydrazine in the presence of sodium bisulphite to yield **102** [R^1 = H, R^2 = H and $(CH_2)_3CH_3$, respectively] and **612** (R = COOH) reacted with N_α-butyl- and N_α-methylphenylhydrazine under similar conditions to give **102** [R^1 = H, R^2 = $(CH_2)_3CH_3$ and CH_3, respectively] (Seeboth, Neumann and Görsch, 1965). Indolization apparently occurred in both possible directions when **606** [R^1–R^3 = H, R^4 = $COOC_2H_5$, R^5 = C_6H_4Br (2), X–Z = CH] was treated with alcoholic hydrogen chloride. Fractional crystallization of the reaction product afforded two isomeric products, $C_{21}H_{16}O_2NBr$, for which structures **607** [R^1 = H, R^2 = $COOC_2H_5$, R^3 = $C_6H_4Br(2)$, X = CH] and **608** [R^1 = R^2 = H, R^3 = $COOC_2H_5$, R^4 = $C_6H_4Br(2)$, X–Z = CH] were postulated (Chalmers, A. J. and Lions, 1933). Obvious steric crowding in **608** [R^1 = R^2 = H, R^3 = $COOC_2H_5$, R^4 = $C_6H_4Br(2)$, X–Z = CH] could have been responsible for the alternative partial formation of its linear isomer. The much later apparent criticism (Shagalov, Eraksina, Turchin and Suvorov, 1970) of this suggested formation of both possible isomeric products is not well founded. Indeed, it relies upon the false assumption that Chalmers, A. J. and Lions (1933) implied that the formation of both possible isomeric products from 2-naphthylhydrazone indolizations is of a more general occurrence.

Acetophenone and phenylacetaldehyde 2-naphthylhydrazone have both been indolized (Ince, 1889). Although it might be expected, by analogy with the above results, that the product formed in each case would have the corresponding angular structure **608** (R^1 = R^2 = R^4 = H, R^3 = C_6H_5, X–Z = CH and R^1–R^3 = H, R^4 = C_6H_5, X–Z = CH), it is possible that the latter indolization product could have structure **607** (R^1 = R^2 = H, R^3 = C_6H_5, X = CH) if steric effects, as postulated in the previous paragraph, were to be manifest in the reaction.

From these above studies it can be concluded that, in the absence of occasional possible steric effects, the formation of the new C—C bond during the indolization of 2-naphthylhydrazones occurs at the 1-position rather than at the 3-position of the naphthalene nucleus. This is in accordance with the known chemistry of naphthalene which would direct the [3,3]-sigmatropic shift in this manner. This conclusion has become generally accepted in the many examples in which 2-naphthylhydrazones have been indolized (Aksanova, Kucherova and Zagorevskii, 1964a, 1964b; Aksanova, Sharkova, L. M. and Kucherova, 1970; Beyts and Plant, 1939; Bistochi, De Meo, Ricci, Croisy and Jacquignon, 1978; Borghero and Finsterle, 1955; Borsche, Witte, A. and Bothe, 1908; Bremner and Browne, 1975; Bryant and Plant, 1931; Buu-Hoï, Bellavita, Ricci, Hoeffinger and Balucani, 1966; Buu-Hoï, Cagniant, Hoán and Khôi, 1950; Buu-Hoï, Croisy, Jacquignon, Hien, Martani and Ricci, 1972; Buu-Hoï, Croisy, Jacquignon and Martani, 1971; Buu-Hoï, Croisy, Jacquignon, Renson and Ruwet, 1970; Buu-Hoï and Hoán, 1949, 1951, 1952; Buu-Hoï, Hoán and Khôi, 1949, 1950a, 1950b, 1950c; Buu-Hoï, Hoán, Khôi and Xuong, 1949, 1950, 1951; Buu-Hoï and Jacquignon, 1954; Buu-Hoï, Jacquignon, Croisy, Loiseau, Périn, Ricci and Martani, 1969; Buu-Hoï, Jacquignon, Croisy and Ricci, 1968; Buu-Hoï, Jacquignon and Ledésert, 1970; Buu-Hoï, Khôi and Xuong, 1950, 1951; Buu-Hoï, Mangane and Jacquignon, 1966, 1967; Buu-Hoï, Martani, Croisy, Jacquignon and Périn, 1966; Buu-Hoï, Roussel, O. and Jacquignon, 1964; Buu-Hoï, Royer, Eckert and Jacquignon, 1952; Buu-Hoï and Saint-Ruf, 1962, 1963, 1965a, 1965b; Buu-Hoï, Saint-Ruf and Dufour, 1964; Buu-Hoï, Saint-Ruf, Martani, Ricci and Balucani, 1968; Buu-Hoï and Xuong, 1952; Campaigne, Ergener, Hallum and Lake, 1959; Cawley and Plant, 1938; Clemo and Felton, 1952; Cornforth, J. W., Hughes, G. K., Lions and Harradence, 1937–1938; Croisy, Jacquignon and Fravolini, 1974; Dufour, Buu-Hoï, Jacquignon and Hien, 1972; Felton, 1952; Feofilaktov, 1947a; Feofilaktov and Semenova, N. K., 1953b; Findlay, S. P. and Dougherty, 1948; Goldsmith and Lindwall, 1953; Hughes, G. K. and Lions, 1937–1938a; Jacquignon, Croisy-Delcey and Croisy, 1972; Jacquignon, Fravolini, Feron and Croisy, 1974; Kakurina, Kucherova and Zagorevskii, 1965a, 1965b; Komzolova, Kucherova and Zagorevskii, 1964; Korczynski, Brydowna and Kierzek, 1926; Kucherova, Petruchenko and Zagorevskii, 1962; Lions, 1932; Maréchal, Christiaens, Rénson and Jacquignon, 1978; Nominé and Pénasse, 1959b; Oliver, G. L., 1968; Patel, H. P. and Tedder, 1963; Pellicciari, Natalini, Ricci, Alunni-Bistocchi and De Meo, 1978; Périn-Roussel, Buu-Hoï and Jacquignon, 1972; Plant and Thompson, 1950; Plant and Tomlinson, M. L., 1932b; Roussel, O. Buu-Hoï and Jacquignon, 1965; Rydon and Siddappa, 1951; Shagalov, Eraksina, Tkachenko and Suvorov, 1973; Shagalov, Eraksina, Turchin and Suvorov, 1970; Shagalov, Tkachenko, Eraksina and Suvorov, 1973; Sharkova, L. M., Aksanova and Kucherova, 1971, Sharkova, N. M., Kucherova, Aksanova and Zagorevskii, 1969; Sharkova, N. M., Kucherova, Portnova and Zagorevskii, 1968; Sharkova, N. M., Kucherova and Zagorevskii, 1962; Stevens, F. J., Ashby and Downey, 1957; Suvorov, Preobrazhenskaya and Uvarova, 1963; Suvorov, Preobrazhenskaya, Uvarova and Sheinker, 1963). By further analogy, the product obtained by indolization of laevulinic acid 2-naphthylhydrazone (Steche, 1887) should most likely be formulated as **608**

($R^1 = R^2 = H$, $R^3 = CH_3$, $R^4 = CH_2COOH$, X–Z = CH) rather than the corresponding linear isomer.

2-Anthrylhydrazine hydrochloride (**613**; R = $\overset{\oplus}{N}H_3Cl^{\ominus}$, X = CH), upon treatment with **614** in the presence of refluxing ethanolic phosphoric acid, furnished only one product which was assumed to have structure **615** (R = $CH_2COOC_2H_5$, X = CH) rather than the alternative linear structure (Finsterle, 1955). Indolization of cyclohexanone 2-phenanthrylhydrazone (**616**) with hydrogen chloride in acetic acid likewise afforded only one isolated product which, without experimental evidence, was assumed to have structure **617** rather than the alternative **618** (Buu-Hoï, Saint-Ruf, Deschamps, Bigot and Hieu, 1971).

Boiling alcoholic solutions of ethyl methyl ketone (Sharkova, N. M., Kucherova and Zagorevskii, 1972b, 1972c), methyl propyl ketone, cyclohexanone, tetrahydro-4*H*-thiopyran-4-one and 1-methylpiperidin-4-one (Sharkova, N. M. Kucherova and Zagorevskii, 1972c) with **619** ($R^1 = H$, $R^2 = CH_3$, X = O) yielded **620** [$R^1 = H$, $R^2 = CH_3$; $R^3 = R^4 = CH_3$, $R^3 = CH_3$, $R^4 = C_2H_5$, $R^3 + R^4 = (CH_2)_4$, $(CH_2)_2SCH_2$ and $(CH_2)_2N(CH_3)CH_2$, respectively] as the only isolated products. The structures of **620** ($R^1 = H$, $R^2 = CH_3$; $R^3 = R^4 = CH_3$ and $R^3 = CH_3$, $R^4 = C_2H_5$) were distinguished

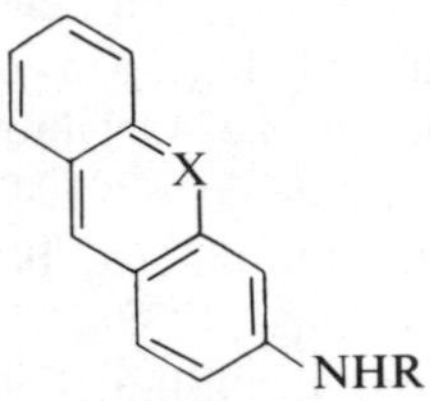

613 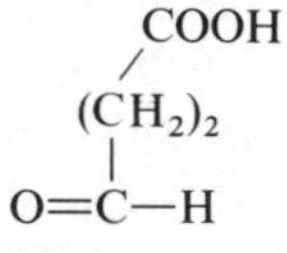 **614** 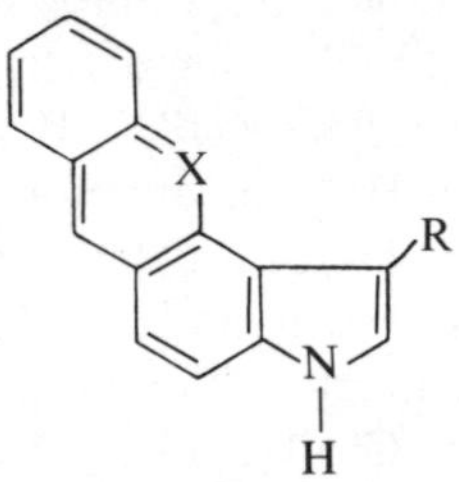 **615**

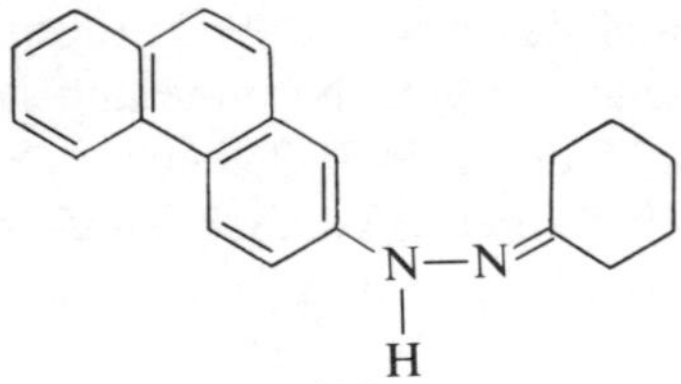

616 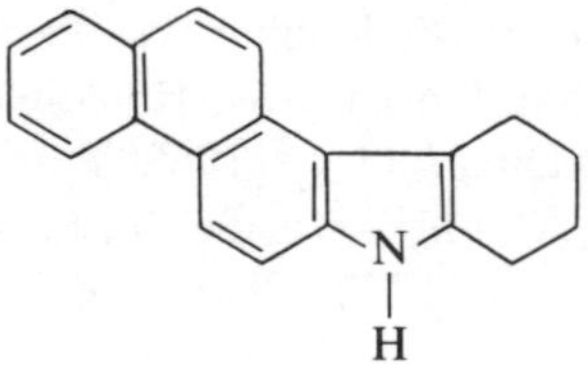 **617**

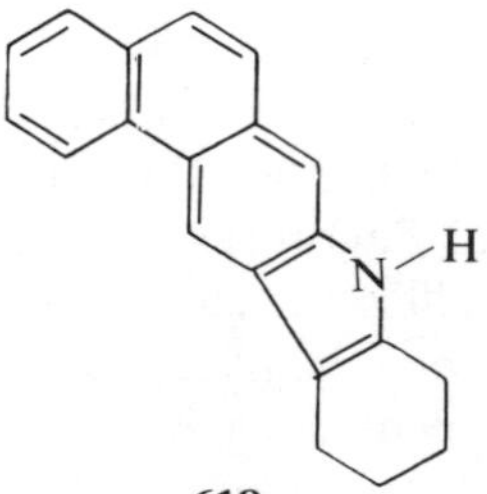

618 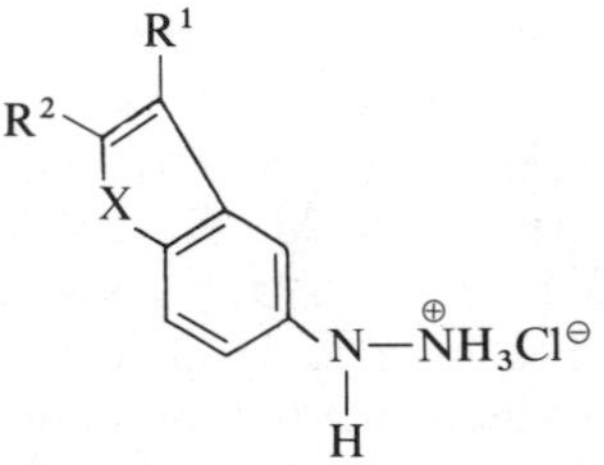 **619** 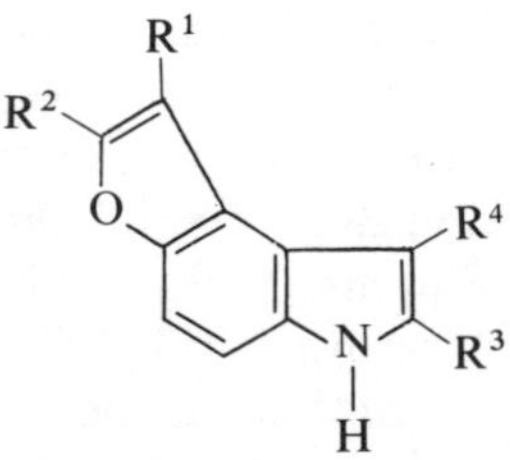 **620**

from the possible linear isomers by p.m.r. spectral investigations which confirmed the *ortho* protons on the benzenoid ring as two doublets (J = 8–9 Hz) (Sharkova, N. M., Kucherova and Zagorevskii, 1972b, 1972c). Similarly, **619** ($R^1 = R^2 = CH_3$, X = O) reacted with tetrahydro-4*H*-thiopyran-4-one to afford **620** [$R^1 = R^2 = CH_3$, $R^3 + R^4 = (CH_2)_2SCH_2$] (Sharkova, N. M., Kucherova and Zagorevskii, 1972b, 1972c) and **619** [$R^1 + R^2 = CH_2N(CH_3)(CH_2)_2$, X = O] reacted with ethyl methyl ketone, tetrahydro-4*H*-thiopyran-4-one, 1-methylpiperidin-4-one (Sharkova, N. M., Kucherova and Zagorevskii, 1972c) and cyclohexanone (Sharkova, N. M., Kucherova and Zagorevskii, 1972b, 1972c) to give **620** [$R^1 + R^2 = CH_2N(CH_3)(CH_2)_2$; $R^3 = R^4 = CH_3$, $R^3 + R^4 = (CH_2)_2SCH_2$, $(CH_2)_2N(CH_3)CH_2$ and $(CH_2)_4$, respectively]. Reaction between **619** ($R^1 = R^2$ = H, X = S) and isopropyl methyl ketone in the presence of 8% sulphuric acid furnished a product which was, without any verification, assumed to have structure **621** rather than the linear isomeric alternative (Abramenko and Zhiryakov, 1971). Likewise, the structures of the single products resulting from the indolization of the

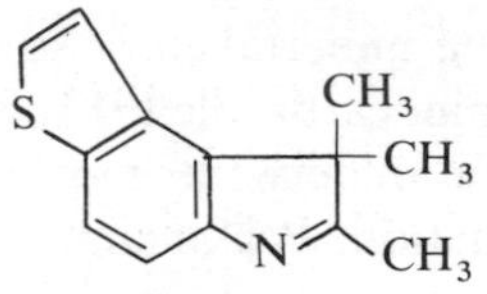

621

hydrazones formed by reacting **619** ($R^1 = R^2$ = H, X = Se) and the ketones **622** (X = CH=CH, S and Se) were formulated as **623** (X = CH=CH, S and Se, respectively) (Maréchal, Christiaens, Renson and Jacquignon, 1978) and the products resulting from the indolizations of the hydrazones formed between **624** (R = H and CH_3) and **625** and various ketones were assumed to be the angular rather than the corresponding linear isomers (Buu-Hoï, Martani, Ricci, Dufour, Jacquignon and Saint-Ruf, 1966; Buu-Hoï, Saint-Ruf, Martani,

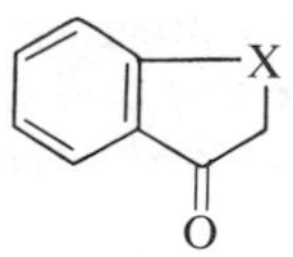

622 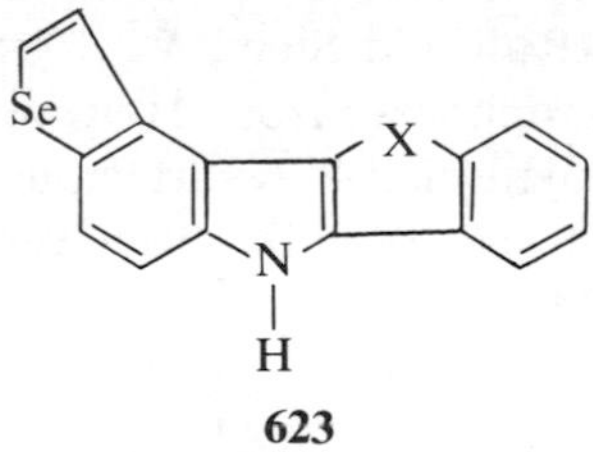 **623**

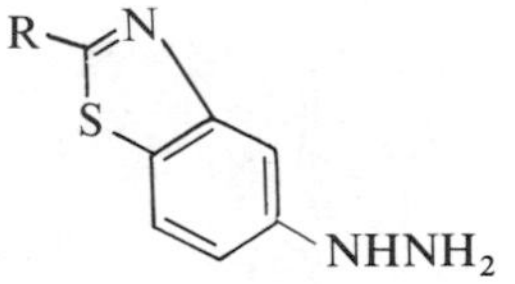

624 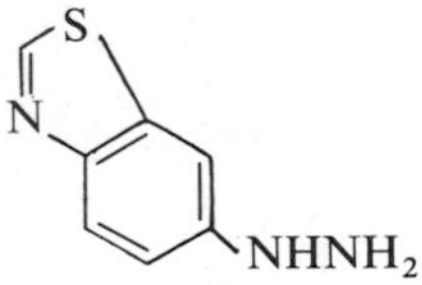 **625**

Ricci and Balucani, 1968), as were the products resulting from the indolizations of hydrazones derived from the hydrazines **626** and **627** (R = H, alkyl and aryl) (Maksimov, Chetverikov and Kost, 1979).

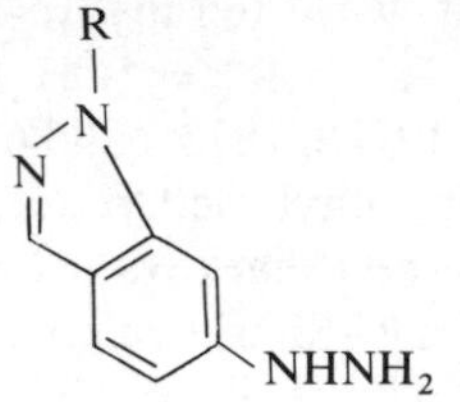

626 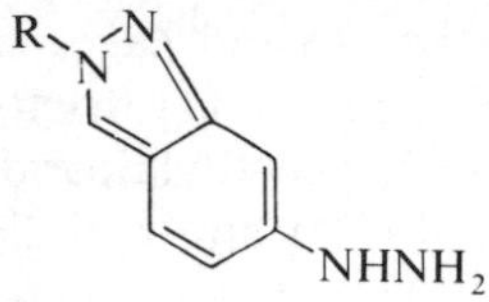 **627**

Apparently contrary to what would be expected from the above observations, the product obtained by heating a mixture of 2-hydrazinocarbazole (**628**; R^1 = H, R^2 = $NHNH_2$) with diethyl ketone in acetic acid was formulated as **629** rather than the angular isomer (I. G. Farbenindustrie Akt.-Ges., 1934). However, no structural verification was presented. Further investigation of the direction of indolization of hydrazones derived from **628** (R^1 = H, R^2 = $NHNH_2$) should therefore be undertaken, a study which may be facilitated by a recent synthesis of 2-aminocarbazole (**628**; R^1 = H, R^2 = NH_2) in high yield from carbazole which underwent sequential 9-acetylation, Friedel–Crafts 2-acetylation, oxamination, Beckmann rearrangement, and hydrolysis (Kyziol and Lyzniak, 1980).

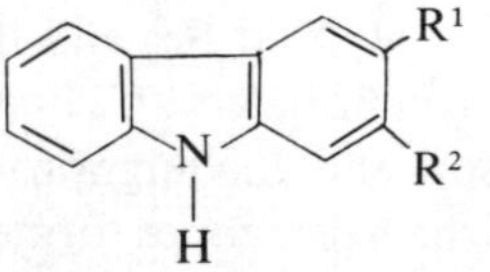

628 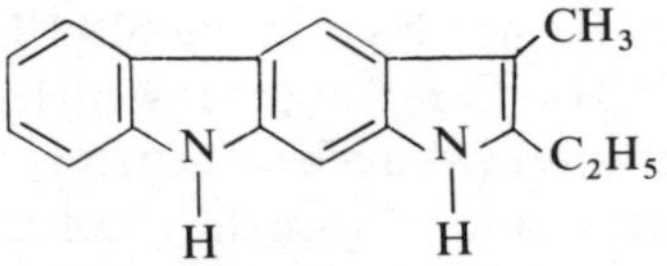 **629**

Routes to the indoloindole system(s) **630** and/or **631** via 3-hydrazinocarbazole (**628**; R^1 = $NHNH_2$, R^2 = H) initially appeared not to be possible since it was reported (Ruff and Stein, V., 1901) that this hydrazine could not be prepared from 3-aminocarbazole. However, other synthetic approaches to this required intermediate should be attempted before this possible route is eliminated. Indeed, it has recently been reported (Khoshtariya, Sikharulidze,

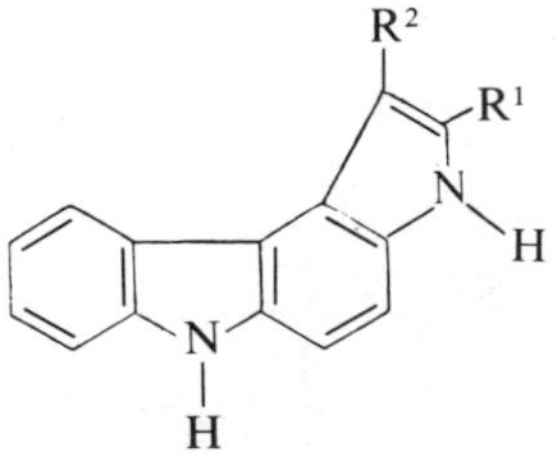

630 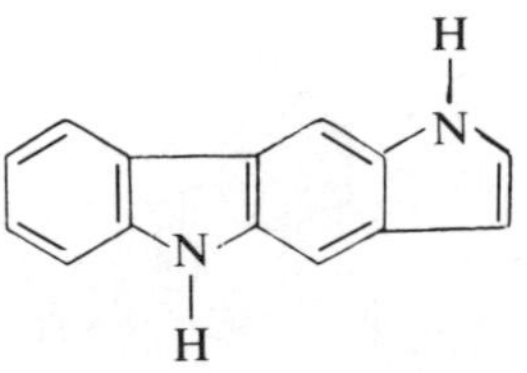 **631**

Tret'yakova, Efimova and Suvorov, 1979) that ethyl pyruvate 3-carbazolylhydrazone [**628**; $R^1 = NHN{=}C(CH_3)COOC_2H_5$, $R^2 = H$], prepared by a Japp–Klingemann reaction between diazotized 3-aminocarbazole and ethyl 2-methylacetoacetate, was indolized in refluxing ethanolic hydrogen chloride to afford **630** ($R^1 = COOC_2H_5$, $R^2 = H$).

Indolization of mixtures of **632** (formed *in situ*) with ethyl methyl ketone, benzyl methyl ketone, acetophenone, propiophenone, cyclohexanone and propionaldehyde produced **633** [$R^1 = R^2 = CH_3$, $R^1 = CH_3$, $R^2 = C_6H_5$, $R^1 = C_6H_5$, $R^2 = H$ and CH_3, $R^1 + R^2 = (CH_2)_4$ and $R^1 = H$, $R^2 = CH_3$, respectively] (a pure product could not be obtained using pyruvic acid as the ketonic moiety). Similarly, mixtures of **634** (formed *in situ*) with ethyl methyl ketone, benzyl methyl ketone and cyclohexanone afforded **635** [$R^1 = R^2 = CH_3$, $R^1 = CH_3$, $R^2 = C_6H_5$ and $R^1 + R^2 = (CH_2)_4$, respectively].

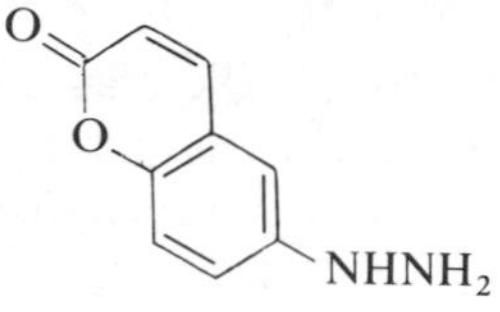

O O NHNH$_2$ **632**

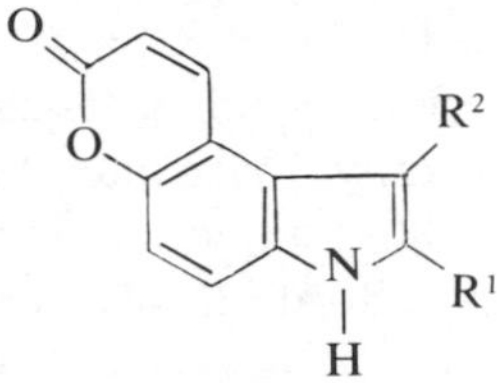

O O R^2 R^1 N H **633**

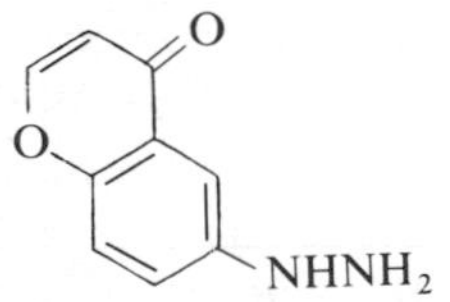

O O NHNH$_2$ **634**

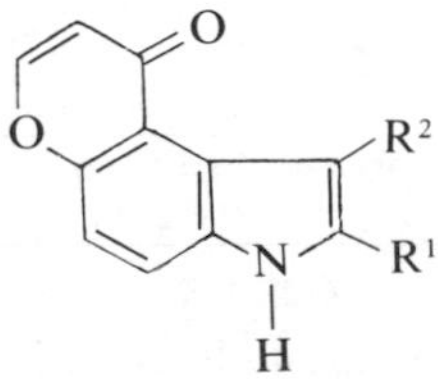

O O R^2 R^1 N H **635**

The structures of these products were distinguished from the possible linear isomers by the appearance of the 1,2-benzenoid proton systems as pairs of doublets (J = 9 Hz) in their p.m.r. spectra (Khan, M. A. and Morley, 1979).

E. Indolization of 6- and 7-Quinolylhydrazones

For syntheses of 6-hydrazinoquinoline, see Hunsberger, Shaw, E. R., Fugger, Ketcham and Lednicer (1956) and Knueppel (1900). Indolization of ethyl pyruvate 6-quinolylhydrazone (**606**; R^1–$R^3 = R^5 = H$, $R^4 = COOC_2H_5$, X = Y = CH, Z = N) with a zinc chloride catalyst led to the isolation of only one product. The structure of this was verified as **608** ($R^1 = R^2 = R^4 = H$, $R^3 = COOC_2H_5$, X = Y = CH, Z = N) by its sequential hydrolysis and decarboxylation to **608** (R^1–$R^4 = H$, X = Y = CH, Z = N) which had a u.v. spectrum similar to that of **608** ($R^1 = R^3 = H$, $R^2 = R^4 = C_2H_5$, X = Y = CH, Z = N). This was synthesized by indolization of **606** ($R^1 = R^3 = H$, $R^2 = R^5 = C_2H_5$, R^4 = COOH, X = Y = CH, Z = N), followed by decarboxylation (Wieland, H. and Horner, L., 1938). The former sequence of reactions, ultimately affording **608** (R^1–R^4 = H, X = Y = CH, Z = N), was repeated thirty-nine years later (Gryaznov, Akhvlediani, Volodina, Vasil'ev, A. M., Babushkina, T. A. and Suvorov, 1977) but using an acetic acid–sulphuric acid catalyst and also starting with pyruvic acid 6-quinolylhydrazone (Suvorov, Sergeeva, Gryaznev, Shabunova, Tret'yakova, Efimova, Volodina, Morozova, Akhvlediani *et al.*, 1977). By analogy with the above earlier results, indolization of **606** ($R^1 = R^5 = C_2H_5$, $R^2 = R^3 = H$, R^4 = COOH, X = Y = CH, Z = N) was again found to occur at the 5- rather than at the 7-position of the quinoline nucleus since the product obtained by decarboxylation of the indolization product had a u.v. spectrum similar to that of **608** ($R^1 = R^3 = H$, $R^2 = R^4 = C_2H_5$, X = Y = CH, Z = N) (Horner, L., 1939). In later related studies (Huisgen, 1948b), the zinc chloride catalysed indolizations of ethyl pyruvate 4-ethyl-7-methyl-6-quinolylhydrazone and ethyl pyruvate 7,8-dimethyl-6-quinolylhydrazone were effected, subsequent decarboxylations affording **608** ($R^1 = R^3 = R^4 = H$, $R^2 = CH_3$, X = CC_2H_5, Y = CH, Z = N) and **608** ($R^1 = R^2 = CH_3$, $R^3 = R^4 = H$, X = Y = CH, Z = N), respectively. Comparison of the u.v. spectroscopic data from these products with that referred to above further supported the above structural assignments and those of other potentially ambiguous indolizations of 6-quinolylhydrazones (Huisgen, 1948b). Without experimental verification, the product obtained by indolization of cyclohexanone 6-quinolylhydrazone [**606**; R^1–R^3 = H, $R^4 + R^5 = (CH_2)_4$, X = Y = CH, Z = N] was formulated as **608** [$R^1 = R^2 = H$, $R^3 + R^4 = (CH_2)_4$, X = Y = CH, Z = N] (Clemo and Felton, 1951a; Dewar, 1944). This postulation was subsequently confirmed (Kulka and Manske, R. H. F., 1952) by unambiguous synthesis via the indolization of cyclohexanone 5,8-dichloro-6-quinolylhydrazone [**606**; $R^1 = R^3$ = Cl, R^2 = H, $R^4 + R^5 = (CH_2)_4$, X = Y = CH, Z = N] using hydrogen chloride in refluxing butanol as the catalyst. The product from this reaction was **608** [R^1 = Cl, R^2 = H, $R^3 + R^4 = (CH_2)_4$, X = Y = CH, Z = N], the loss of the chlorine atom verifying the indolization at the 5-position of the quinoline nucleus. Reductive dechlorination then formed **608** [$R^1 = R^2 = H$, $R^3 + R^4 = (CH_2)_4$, X = Y = CH, Z = N]. In a similar manner, indolization of **606** [$R^1 = R^3 = CH_3$, R^2 = H, $R^4 + R^5 = (CH_2)_4$, X = Y = CH, Z = N] occurred at the C-5 atom of the quinoline nucleus and ultimately effected demethylation to afford **608** [$R^1 = CH_3$, R^2 = H, $R^3 + R^4 = (CH_2)_4$, X = Y = CH, Z = N] which was also formed by indolization of **606** [$R^1 = CH_3$, $R^2 = R^3 = H$, $R^4 + R^5 = (CH_2)_4$, X = Y = CH, Z = N]. The direction of this latter indolization was therefore established (Huisgen, 1948a). As already mentioned in this chapter, the angular mode of indolization is so preferred in these ambiguous indolizations that, where necessary, a quaternary aromatic carbon atom may be involved in the new C—C bond formation, with ultimate elimination of the original aromatic substituent, rather than formation of the linear indolic product (see also Chapter II, G). The indolizations of 1- and 2-tetralone 6-quinolylhydrazone (**600**; R = H, W = Y = CH, X = N, Z = CH=CH) and **597** (R = H, X = Z = CH, Y = N), respectively, have also been found to occur at the 5- rather than the 7-position of the quinoline nucleus (see Chapter III, I for a discussion of the direction of indolization of 2-tetralone arylhydrazones) since dehydrogenation of the original indolization products afforded compounds which, from the observations that their u.v. spectra resembled those of their benzenoid isosteres, were formulated as **636** (R = H, V = N, W–Z = CH) and **599** (R = H, X = Z = CH, Y = N), respectively (Buu-Hoï, Jacquignon and Périn, 1962; Buu-Hoï, Périn and Jacquignon, 1960 – see also Buu-Hoï, Saint-Ruf, Jacquignon and Barrett, G. C., 1958). Likewise, the u.v. spectrum of **636** (R = H, V = N, W–Z = CH) was very similar to those of **636** (R = H, V–X = Z = CH, Y = N) and **636** (R = H, V = W = Y = Z = CH, X = N) which were synthesized via the indolization of 2-tetralone

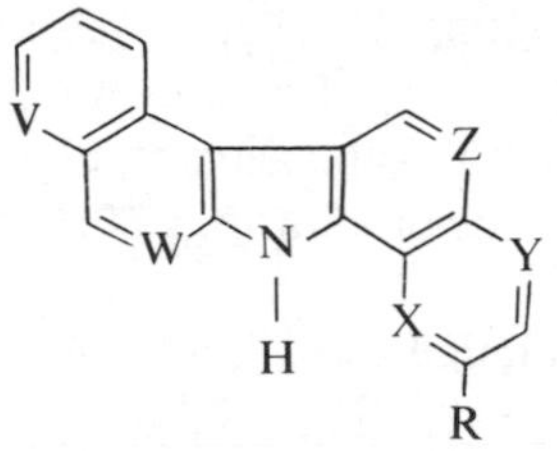

636

5- and 8-quinolylhydrazone, respectively (Buu-Hoï, Jacquignon and Périn, 1962), the products resulting from the indolization of **597** (R = CH_3, X = Z = CH, Y = N) and **600** (R = CH_3, W = Y = CH, X = N, Z = CH=CH), with subsequent dehydrogenation, were formulated as **599** (R = CH_3, X = Z = CH, Y = N) and **636** (R = CH_3, V = N, W–Z = CH), respectively, since their u.v. spectra were similar to those of their desmethylbenzenoid isosteres, and **600** (R = H, X = N, W = Y = CH, Z = S), upon sequential indolization and dehydrogenation, afforded a product which was formulated as **601** (R = H, X = N, W = Y = CH, Z = S) since its u.v. spectrum resembled that of the benzenoid isostere **636** (R = H, V–Z = CH) (Buu-Hoï, Jacquignon and Hoeffinger, 1963).

Without experimental verification it has been assumed that the products resulting from the indolizations of acetone, ethyl methyl ketone, 4-methylcyclohexanone, 1,2,3,4-tetrahydroquinolin-4-one, benzothiophen-3(2*H*)-one and 6-methoxybenzothiophen-3(2*H*)-one 6-quinolylhydrazone have structures **608** (R^1 = R^2 = R^4 = H, R^3 = CH_3, X = Y = CH, Z = N) (Dewar,

1944), **608** ($R^1 = R^2 = H$, $R^3 = R^4 = CH_3$, X = Y = CH, Z = N) (Verkade, Janetzky, Werner, E. C. G. and Lieste, 1943), **608** [$R^1 = R^2 = H$, $R^3 + R^4 = (CH_2)_2CH(CH_3)CH_2$, X = Y = CH, Z = N] (Clemo and Felton, 1951), **636** (R = H, V = Z = N, W–Y = CH) (Roussel, O., Buu-Hoï and Jacquignon, 1965), **637** (R = H) (Buu-Hoï, Bellavita, Ricci, Hoeffinger and Balucani, 1966) and **637** ($R = OCH_3$) (Buu-Hoï and Saint-Ruf, 1963), respectively. Similarly, indolization of **437** (R = H) afforded a product which has been assumed to be **438** rather than the corresponding linear isomer

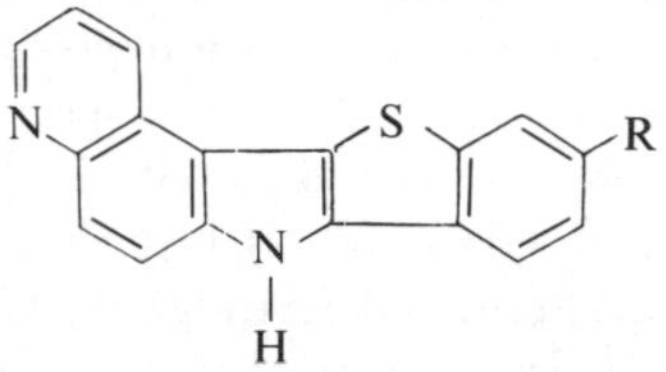

637

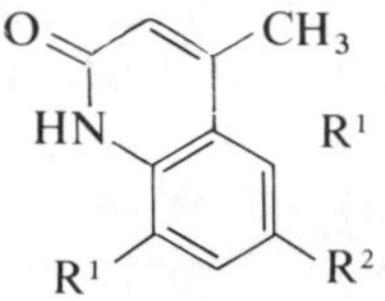

638

(Huisgen, 1948a) and the product resulting in each case from the indolization of 1,2,3,4-tetrahydrophenanthren-1-and -4-one 6-quinolylhydrazone have been similarly formulated (Buu-Hoï, Périn and Jacquignon, 1960), as has the product resulting from the indolization of cyclooctanone 6-quinolylhydrazone (Buu-Hoï, Saint-Ruf, Jacquignon and Barrett, G. C., 1958). By analogy with the results of the above studies, the orientations of these structural postulations are most probably correct.

Only one product, formulated as **639**, was isolated when a mixture of ethyl methyl ketone and **638** ($R^1 = H$, $R^2 = NHNH_2$) was indolized in acetic acid. Unfortunately, attempts to verify the structure of the product were unsuccessful when it was found that **638** [$R^1 = CH_3$, $R^2 = NHN{=}C(CH_3)C_2H_5$] could not be indolized (Huisgen, 1948a).

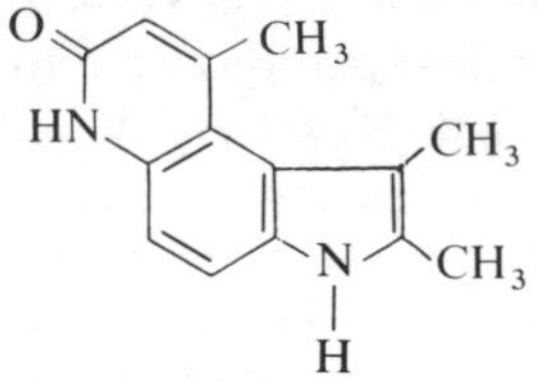

639

Cyclohexanone 7-quinolylhydrazone [**606**; R^1–R^3 = H, $R^4 + R^5 = (CH_2)_4$, X = N, Y = Z = CH] indolized towards the C-8 atom of the quinoline nucleus to yield **608** [$R^1 = R^2 = H$, $R^3 + R^4 = (CH_2)_4$, X = N, Y = Z = CH] as the only isolated product. The structure of this product was verified by its synthesis involving reductive dechlorination of **608** [$R^1 = H$, $R^2 = Cl$, $R^3 + R^4 = (CH_2)_4$, X = N, Y = Z = CH] which was prepared by indolization of **606** [$R^1 = R^3 = H$, $R^2 = Cl$, $R^4 + R^5 = (CH_2)_4$, X = N, Y = Z = CH] (Kulka and Manske, R. H. F., 1952). Indolization of 1- and 2-tetralone 7-quinolylhydrazone, **600** (R = H, W = N, X = Y = CH, Z = CH=CH) and **597** (R = H, X = N, Y = Z = CH), respectively, afforded in each case only one product. These were assigned structures resulting from new C—C bond formation at the C-8 atom of the quinoline nucleus since upon dehydrogenation they furnished the products **601** (R = H, W = N, X = Y = CH, Z = CH=CH) and **599** (R = H, X = N, Y = Z = CH), respectively, the structures of which were distinguished from their possible isomers by the similarity of the u.v. spectra of these products with the corresponding spectra of their benzenoid isosteres and, in the case of the former, also with the u.v. spectrum of **636** (R = H, V = W = Y = Z = CH, X = N) (Buu-Hoï, Jacquignon and Périn, 1962). Likewise, sequential indolization and dehydrogenation of **600** ($R = CH_3$, W = N, X = Y = CH, Z = CH=CH) and **597** ($R = CH_3$, X = N, Y = Z = CH) afforded in each case only one product, formulated as **601** ($R = CH_3$, W = N, X = Y = CH, Z = CH=CH) and **599** ($R = CH_3$, X = N, Y = Z = CH), respectively, because of the similarity of their u.v. spectra with those of the corresponding benzenoid isosteres (Buu-Hoï, Jacquignon and Hoeffinger, 1963). Indolization of **602** (X = N, Y = CH) and **604** (X = N, Y = CH), with subsequent dehydrogenation of the products, afforded in each case one compound, formulated as **603** (X = N, Y = CH) and **605** (X = N, Y = CH), respectively (Buu-Hoï, Périn, and Jacquignon, 1962). Although no evidence was presented to support these formulations in preference to the alternative isomeric structures, the suggested structures would be expected by analogy with the above observations.

The hydrazone **613** [$R = N{=}C(CH_3)COOC_2H_5$, X = N] was indolized by heating with zinc chloride followed by aqueous 'work up', which also caused hydrolysis and decarboxylation [the latter presumably occurred subsequent to indolization since the corresponding acetaldehyde hydrazone would not be expected to indolize (see Chapter IV, A)], to give **615** (R = H, X = N). This structure was distinguished from the linear isomer by p.m.r. spectral analysis (Suvorov, Alyab'eva and Khoshtariya, 1978; Alyabeva, Khoshtariya, Vasil'ev, A. M., Tret'vakova, Efimova and Suvorov, 1979).

F. Indolization of 6- and 7-Isoquinolylhydrazones

None of the desired product could be isolated from an attempt to convert 6-aminoisoquinoline into 6-hydrazinoisoquinoline, probably because of the instability of the hydrazine under the reaction conditions. Because of this failure, the indolization of 6-isoquinolylhydrazones was not investigated (Manske, R. H. F. and Kulka, 1950b), although attempts were not made to generate 6-isoquinolylhydrazine *in situ* in the presence of ketones or aldehydes, to afford the necessary hydrazones, or to generate the hydrazone *in situ* in the presence of indolization catalysts. In connection with these latter possibilities, both feasible isomeric products, **609** ($R^1 = R^2 = H, X = CH, Y = N$) and **640** ($R^1$–$R^4 = H, X = N, Y = CH$), which could result from such indolizations using cyclohexanone as the ketonic reactant, after dehydrogenative

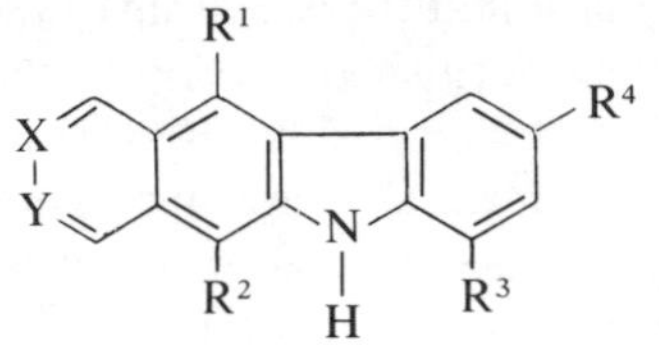

640

aromatization, have been unambiguously synthesized by alternative routes (Manske, R. H. F. and Kulka, 1950b). Comparison with these two products will allow ready identification of the direction of indolization in cyclohexanone 6-isoquinolylhydrazone if it can be effected. The reaction might be expected to involve the C-5 rather than the C-7 atom of the isoquinoline nucleus (Robinson, 1963a) to ultimately yield **609** ($R^1 = R^2 = H, X = CH, Y = N$).

Cyclohexanone 7-isoquinolylhydrazone [**606**; R^1–$R^3 = H$, $R^4 + R^5 = (CH_2)_4$, $X = Z = CH$, $Y = N$] has been indolized to produce only one isolated product which was subsequently dehydrogenated to give a product of structure **609** ($R^1 = R^2 = H, X = N, Y = CH$) or **640** ($R^1$–$R^4 = H, X = CH, Y = N$) (Manske, R. H. F. and Kulka, 1949). Although attempts (Manske, R. H. F. and Kulka, 1949, 1950b) to distinguish between these possible isomers by unambiguous synthesis failed, indolization to **608** [$R^1 = R^2 = H$, $R^3 + R^4 = (CH_2)_4$, $X = Z = CH$, $Y = N$] and ultimately to **609** ($R^1 = R^2 = H$, $X = N$, $Y = CH$) was favoured (Manske, R. H. F. and Kulka, 1949, 1950b). Of the tried unambiguous synthetic approaches, one involved an attempted indolization of cyclohexanone 8-chloro-7-isoquinolylhydrazone using either 17% sulphuric acid or dry hydrogen chloride in organic solvents as catalyst but these afforded only a small amount of an unidentified and chlorine free product when the reaction temperature was 100 °C or higher, or low-boiling products, only one of which was identified (as 8-chloroisoquinoline), when the reaction was attempted at room temperature over a long period (Manske, R. H. F. and Kulka, 1949). Another synthetic approach also involved a Fischer indolization when **641** ($R = CH_3$) reacted in warm 17% sulphuric acid to afford a neutral product **642** ($R = CH_3$) which was sequentially hydrolysed and

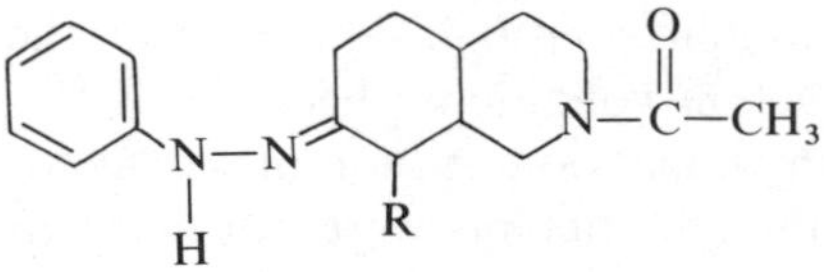

641

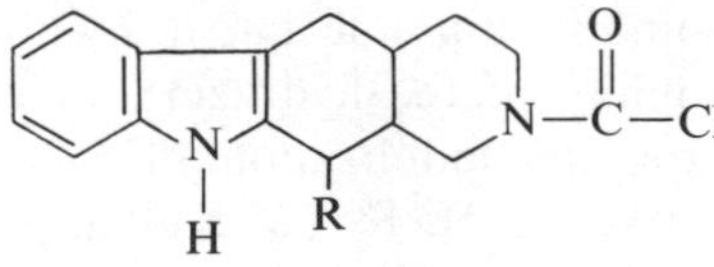

642

dehydrogenated to yield **640** ($R^1 = R^3 = R^4 = H$, $R^2 = CH_3$, X = CH, Y = N). However, attempts to demethylate this product via oxidation followed by decarboxylation failed at the initial stage (Manske, R. H. F. and Kulka, 1949).

G. Indolization of Arylhydrazones with Ambiguous Acyclic Ketonic Moieties

Three years after his discovery of the indole synthesis, Emil Fischer (1886b, 1886c, 1886e) clearly recognized that the formation of two indolic products, **644** and **645**, was possible from the indolization of arylhydrazones of methyl alkyl ketones of structure **643**. Some years later, consideration of this problem was extended and the following general rules governing the direction of indolization of asymmetrical dialkyl ketone arylhydrazones were published (Plancher and Bonavia, 1902).

1. Methyl alkyl ketone arylhydrazones of the type **643** afford the corresponding 2-methylindoles **644** as the major products, an observation later (Yamamoto, H., Nakao and Kobayashi, A., 1968) attributed, at least in part, to the relative hyperconjugative stabilities of the two possible enehydrazine tautomeric intermediates.

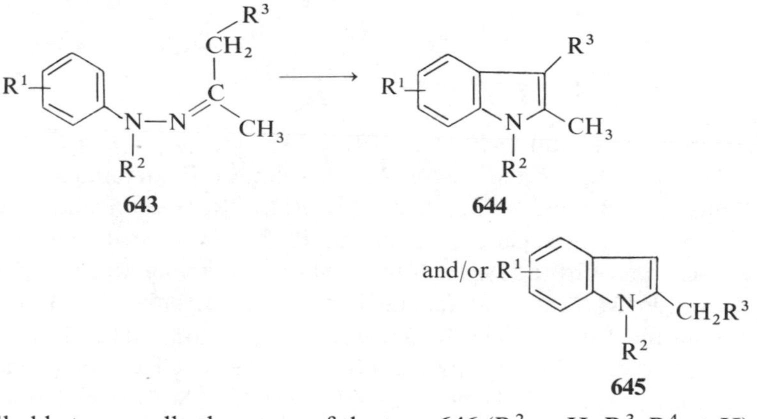

R^3, CH_2, R^1, C, N—N, CH_3, R^2

643

R^3, R^1, N, CH_3, R^2

644

and/or R^1, N, CH_2R^3, R^2

645

2. Methyl alkyl ketone arylhydrazones of the type **646** ($R^2 = H$, $R^3, R^4 \neq H$) afford only the corresponding 2-methyl-3*H*-indoles **647**.

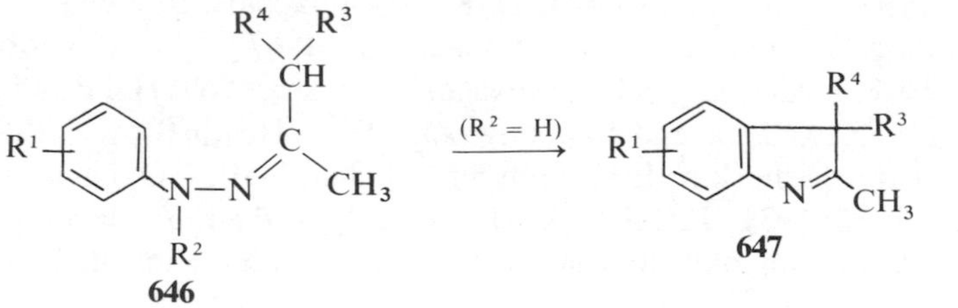

R^4, R^3, CH, R^1, C, N—N, CH_3, R^2

646

($R^2 = H$)

R^4, R^1, R^3, N, CH_3

647

3. Dialkyl ketone arylhydrazones of the type **648** give both indoles **649** and **650** (if R^4 = H in **648**) or **650** and **651** (if R^3 and $R^4 \neq$ H in **648**).

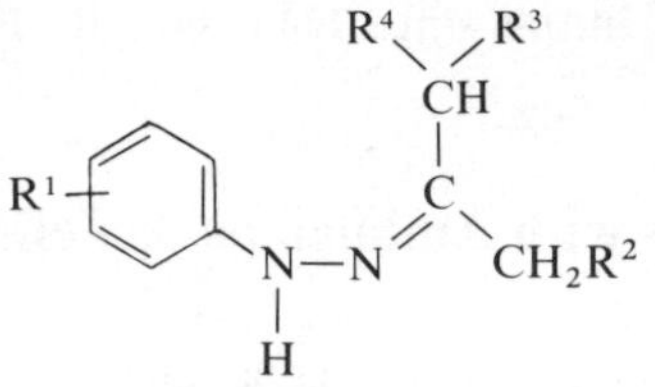

648 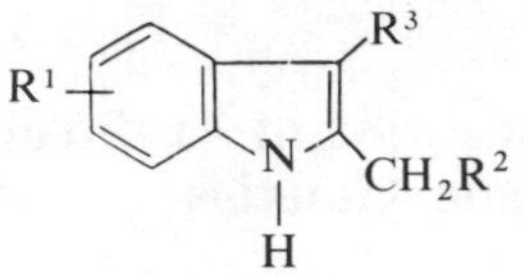 **649**

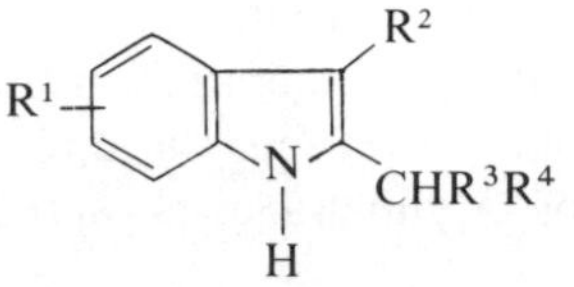

650 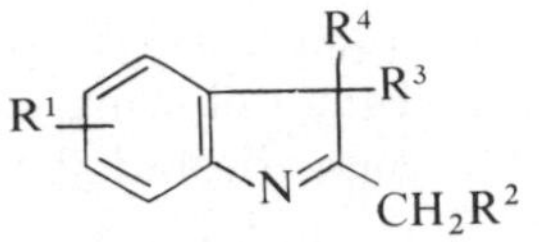 **651**

Many experimental data have been accumulated to support rule 1. Thus, the corresponding 3-substituted 2-methylindoles have been obtained starting from arylhydrazones of ketones **652** with R = CH_3 (Ainsworth, D. P. and Suschitzky, 1967a; Aksanova, Pidevich, Sharkova, L. M. and Kucherova, 1968;

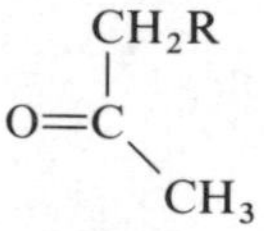

652

Arbusow, A. E. and Tichwinsky, 1910b; Arbuzov, A. E. and Tikhvinski, 1913b; Atkinson, Simpson and Taylor, A., 1954; Bailey, Baldry and Scott, P. W., 1979; Bauer and Strauss, 1932; Beer, McGrath, Robertson and Woodier, (with Holker), 1949; Bogat-skii, Ivanova, R. Y., Andronati and Zhilina, 1979; Borsche and Groth, 1941; Brimblecombe, Downing and Hunt, R. R., 1966; Brown, U. M., Carter, P. H. and Tomlinson, M. 1958; Brown, R. K., Nelson, Sandin and Tanner, 1952; Buu-Hoï and Jacquignon, 1960; Clemo and Weiss, J., 1945; Colwell, Horner, J. K. and Skinner, 1964; CRC Compagnia di Ricerca Chimica S.A., 1979; Dave, V., 1976; Degen, 1886; Deorha and Joshi, S. S., 1961; Duuren, 1961; Fischer, Emil, 1886c, 1886e; Fitzpatrick, J. T. and Hiser, 1957; Frasca, 1962a; Gardner, J. H. and Stevens, J. R., 1947; Gaudion, Hook and Plant, 1947; Glish and Cooks, 1978; Govindachari, Rajappa and Sudarsanam, 1966; Govindachari, and Sudarsanam, 1967, 1971; Grandberg and Bobrova, 1978; Grandberg, Sibiryakova and Brovkin, 1969; Hahn and Bartnik, 1972; Hahn, Zawadzka and Szwedowska, 1969; Heath-Brown, 1975; Herdieckerhoff and Tschurkur, 1933; Huisgen, 1948a, 1948b; I. G. Farbenindustrie Akt. Ges., 1932, 1934; Jackson, A. H. and Smith, P., 1968; Jacquignon and Buu-Hoï, 1961; Kanaoka, Ban, Miyashita, Irie and Yonemitsu, 1966; Kelly,

A. H., McLeod and Parrick, 1965; Kesswani, 1967, 1970; Khan, M. A. and Morley, 1978, 1979; Khan, M. A. and Rocha, 1978; Kimura, Inaba and Yamamoto, H., 1973a; Kissman, Farnsworth and Witkop, 1952; Kost, Sugorova and Yakubov, 1965; Kost, Yudin, Berlin and Terent'ev, A. P., 1959; Kost, Yudin, Budylin and Yaryshev, 1965; Kost, Yudin, Dmitriev, B. A. and Terent'ev, A. P., 1959; Kost, Yudin and Terent'ev, A. N., 1959; Marion and Oldfield, 1947; Nagaraja and Sunthankar, 1958; Namis, Cortes, Collera and Walls, F., 1966; Neuss, Boaz and Forbes, J. W., 1954; Ockenden and Schofield, 1953a, 1953b, 1957; Orlova, E. K., Sharkova, N. M., Meshcheryakova, Zagorevskii and Kucherova, 1975; Pappalardo and Vitali, 1958a; Parrick and Wilcox, 1976; Plant and Thompson, 1950; Plant and Tomlinson, M. L., 1933; Pretka and Lindwall, 1954; Quadbeck and Röhm, 1954; Roche Products Ltd, 1974; Rothstein and Feitelson, 1956; Saleha, Siddiqi and Khan, N. H., 1979; Sandoz Ltd, 1980; Sato, Y. and Sunagawa, 1967; Schofield and Theobald, 1949, 1950; Sellstedt and Wolf, M., 1974; Sergeeva, Akhvlediani, Shabunova, Kovolev, Vasil'ev, A. M. Babushkina, T. N. and Suvorov, 1975; Sharkova, N. M., Kucherova and Zagorevskii, 1964, 1972b, 1972c; Shvedov, Trofimkin, Vasil'eva, and Grinev, 1975; Sibiryakova, Brovkin, Belyakova and Grandberg, 1969; Snyder and Smith, C. W., 1943; Sturm, Tritschler and Zeidler, 1972; Sunagawa and Sato, Y., 1967; Sunagawa, Soma, Nakano, H. and Matsumoto, 1961; Suvorov, Avramenko, Shkil'kova and Zamyshlyaeva, 1974; Suvorov and Sorokina, N. P., 1960; Suvorov, Sorokina, N. P. and Sheinker, I. N., 1959, Suzuki, N., and Sato, Y., 1955; Terzyan and Tatevosyan, 1960; Vejdělek, 1957; Verkade, Janetzky, Werner, E. C. G. and Lieste, 1943; Verley and Beduwé, 1925; Vinogradova, Daut, Kost and Terent'ev, A. P., 1962; Watanabe, Y., Yamamoto, M., Shim, Miyanaga and Mitsudo, 1980; Witkop, 1944; Yakhontov, Suvorov, Pronina, E. V., Kanterov, Podkhalyuzina, Starostenko and Shkil'kova, 1972; Yamada, S., Chibata and Tsurui; 1953; Yamada, S. *et al.*, 1954; Yamamoto, H. and Nakao, 1968a; Yudin, Budylin and Kost, 1964; Yudin, Kost and Berlin, 1958), $R = C_2H_5$ (Aksanova, Pidevich, Sharkova, L. M. and Kucherova, 1968; Dave, V., 1976; Deorha and Joshi, S. S., 1961; Fischer, Emil, 1886b, 1886c, 1886e; Jackson, A. H. and Smith, P., 1968; Julian and Pikl, 1935; Kimura, Inaba and Yamamoto, H., 1973a; Namis, Cortes, Collera and Walls, F., 1966; Ockenden and Schofield, 1953b; Plancher, 1898e; Rothstein and Feitelson, 1956; Sandoz Ltd., 1980; Schlittler, Burckhardt and Gellert, 1953; Schofield and Theobald, 1950; Sellstedt and Wolf, M., 1974; Sharkova, N. M., Kucherova and Zagorevskii, 1964, 1972c; Shaw, E., 1954; Shvedov, Trofimkin, Vasil'eva and Grinev, 1975; Sugasawa and Nakamura, S., 1953; Vejdělek, 1957; Watanabe, Y., Yamamoto, M., Shim, Miyanaga and Mitsudo, 1980; Yamamoto, H. and Nakao, 1968a), $R = (CH_2)_2CH_3$ (Arbusow, A. E., Saizew and Rasumov, 1935; Dave, V., 1976; Kost, Yudin, Berlin and Terent'ev, A. P., 1959; Kost, Yudin, Dmitriev, B. A. and Terent'ev, A. P., 1959; Kost, Yudin and Terent'ev, A. N., 1959; Kuroda, 1923; Rothstein and Feitelson, 1956; Sellstedt and Wolf, M., 1974; Terzyan, Akopyan and Tatevosyan, 1961; Vinogradova, Daut, Kost and Terent'ev, A. P., 1962; Yamamoto, H. and Nakao, 1968a; Yamamoto, H., Nakao and Kobayashi, T., 1969), R = $CH(CH_3)_2$ (Akopyan and Tatevosyan, 1974; Colwell, Horner, J. K. and Skinner, 1964; Dave, V., 1976; Frasca, 1962a; Jackson, A. H. and Smith, P., 1968; Kuroda, 1923; Rothstein and Feitelson, 1956; Sandoz Ltd, 1980), R = $(CH_2)_3CH_3$ (Dave, V., 1976; Esayan, Terzyan, Astratyan, Dzhanpoladyan and Tatevosyan, 1968; Frasca, 1962a; Posvic, Dombro, Ito, H. and Telinski, 1974; Shaw, E. and Woolley, D. W., 1953), R = $CH_2CH(CH_3)_2$ (Akopyan and Tatevosyan, 1974), R = $CH(CH_3)C_2H_5$ (Kimura, Inaba and Yamamoto, H., 1973a), R = $C(CH_3)_3$ (David and Régent, 1964), R = $(CH_2)_4CH_3$ (Buu-Hoï and Royer, 1947; Dave, V., 1976; Esayan, Terzyan, Astratyan, Dzhanpoladyan and Tatevosyan, 1968; Frasca, 1962a), R = $(CH_2)_5CH_3$ (Buu-Hoï and Jacquignon, 1960; Jacquignon and Buu-Hoï, 1961), R = $(CH_2)_7CH_3$ (Buu-Hoï and Royer, 1947; Kuroda, 1923), R = $(CH_2)_{9,\,13\text{ and }15}CH_3$ (Buu-Hoï and Jacquignon, 1960; Jacquignon and Buu-Hoï, 1961), R = $CH_2CH{=}C(CH_3)_2$ (Buu-Hoï and Royer, 1947; Chemerda and Sletzinger, 1968e), R = $CH_2CH{=}CClCH_3$ (Boyakhchyan, Rashidyan and Tatevosyan, 1966; Chemerda and Sletzinger, 1968e), R = $CH_2CCl{=}CH_2$, $CH_2CH{=}CHCl$, $CH_2C(OH){=}CHCH_3$, $CH_2C{\equiv}CH$, $CH_2C{\equiv}CCH_3$ and $CH_2CH(OH)CH_3$ (Chemerda and Sletzinger, 1968e), R = $(CH_2)_2OH$ (Brenner, Clamkowski, Hinkley and Gal, 1969; Glamkowski, Gal and Sletzinger, 1973; Grandberg, Kost and Terentyev, 1957; Grandberg, Sibiryakova and Brovkin, 1969; Kost, Vinogradova, Daut and Terent'ev, A. P., 1962; Kost, Yudin, Dmitriev, B. A. and Terent'ev, A. P., 1959; Kost, Yudin and Popravko, 1962; Shvedov, Trofimkin, Vasil'eva and Grinev, A. N., 1975; Yamamoto, H., Nakamura, Y., Nakao and Kobayashi, T., 1970a; Yamamoto, H. and Nakao, 1968c), R = $(CH_2)_2OC_6H_5$ (Boyd-Barrett and Robinson, R., 1932; King, F. E. and Robinson, R., 1932, 1933), R = $CH_2N(C_2H_5)_2$ and $(CH_2)_2NHC_6H_5$ (Yamamoto, H., Nakao and Atsumi, 1968b), R = $(CH_2)_2N(C_2H_5)_2$ (I. G. Farbenind, A.-G., 1930d; Steck, Fletcher and Carabateas, 1974), R = $(CH_2)_2NO_2$ (Chemerda and Sletzinger, 1969a), R = $(CH_2)_2Cl$ (Gaines, Sletzinger and Ruyle, W., 1961; McCrea, 1966; Shaw, E. and Woolley, D. W., 1953; Shvedov, Trofimkin, Vasil'eva and Grinev, A. N., 1975; Sletzinger, Gaines and Ruyle, W. V., 1957; Sletzinger, Ruyle, W. V. and Gaines, 1961; Steck, Fletcher and Carabateas, 1974), R = $CH_2CH(CH_3)Cl$ (Gaines, Sletzinger and Ruyle, W., 1961; Sletzinger, Ruyle and Gaines, 1961), R = $(CH_2)_{2-4}Si(CH_3)_3$ and $(CH_2)_{2-4}Si(C_2H_5)_3$ (Shostakovskii, Komarov and Roman, 1968), R = $COOCH_3$ (Fusco and Sannicolò, 1973, 1975a, 1975b, 1976a; Thyagarajan, Hillard, Reddy, K. V. and Majumdar, 1974), R = $COOC_2H_5$ (Bauer and Strauss, 1932; Binder, Habison and Noe, 1977; Blake, Tretter and Rapoport, 1965; Degen, 1886; Fischer, Emil, 1886b, 1886e; King, F. E. and L'Ecuyer, 1934; Kornet, Thio and Tolbert, 1980; Kost, Vinogradova,

Daut and Terent'ev, A. P., 1962; Kost, Yudin, Dmitriev, B. A. and Terent'ev, A. P., 1959; Kost, Yudin and Popravko, 1962; Kost, Yudin and Zinchenko, 1973; Michaelis and Luxembourg, 1893; Nef, 1891; Steck, Fletcher and Carabateas, 1974, Sumitomo Chemical Co. Ltd., 1966; Walker, C., 1892, 1894), R = CH_2COOH (Amorosa, 1956; Bertazzoni, Bartoletti and Perlotto, 1970; Bikova, Vitev, Dyankova and Ilarionov, 1972; Bikova, Vitev and Ilarionov, 1974; Biniecki and Jakubowski, 1974; Birchall, G. R., Hepworth and Smith, S. C., 1973; Boltze, Opitz, Raddatz, Seidel, Jacobi, Dell and Schoellnhammer, 1979; Brenner, Clamkowski, Hinkley and Gal, 1969; Bullock and Fox, 1951; Carey, Gal and Sletzinger, 1967; Carey, Gal, Sletzinger and Reinhold, 1968; Carry, Gal and Sletzinger, 1968; Cassebaum, Dierbach and Hilger, 1970; CRC Compagnia di Ricerca Chimica S.A., 1979; De Bellis and Stein, M. L., 1961; Degen, 1886; Doyle and Smith, S. C., 1974; Ellsworth, Gatto, Meriwether and Mertel, 1978; Fabbrica Italiana Sintetici S.p.A., 1968; Firestone and Sletzinger, 1970; Fischer, Emil, 1886b, 1886c, 1886e; Fisnerova and Nemecek, 1976; Francia Barra and Carmelo Marin Moga, 1979; Hoshino and Shimodaira, 1935; I.C.I. Ltd, 1975; Kimura, Inaba and Yamamoto, H., 1972, 1973b, 1973c, 1973d; Kögl and Kostermans, 1935; Kosa and Kovacs, 1970, 1972, 1975; Lanzilotti, Littell, Fanshawe, McKenzie and Lovell, 1979; Merck and Co. Inc., 1962, 1964b; Nakao, Takahashi, K. and Yamamoto, H., 1971; Nakatsuka, Hazue, Makari, Kawahara, Endo and Yoshitake, 1976; Okamoto, Niizaki, S., Kobayashi, T., Izumi and Yamamoto, H., 1972; Okamoto, Niizaki, M., Kobayashi, T., Izumi, Yamamoto, H., Inaba, Nakamura, Y. and Nakao, 1971; Pakula, Wojciehowski, Poslinska, Pichnej, Ptaszynski, Przepalkowski and Logwinienko, 1969, 1971, 1975; Poslinska, Pakula, Wojciechowski and Pichnej, 1972; Saleha, Khan, N. H., Siddiqui and Kidwai, M. M., 1978; Saleha, Siddiqi and Khan, N. H., 1979; Sarett and Shen, T.-Y., 1966a; Scherrer, 1969; Shen, T.-Y., 1964, 1966, 1967a, 1967b, 1967c; Shen, T.-Y., Gal and Utne, 1970; Shen, T.-Y. and Sarett, 1966; Shvedov, Trofimkin, Vasil'eva and Grinev, A. N., 1975; Sletzinger and Gal, 1968b; Sletzinger, Gal and Chemerda, 1968; Snyder and Smith, C. W., 1943; Steche, 1887; Steck, Fletcher and Carabateas, 1974; Stevens, F. J., Ashby and Downey, 1957; Sumitomo Chemical Co. Ltd, 1966, 1967, 1968a, 1968c, 1969b; Tani, H., Otani, Mizutani and Mashimo, 1969; Toth, Szaba, Eibel and Somfai, 1972, 1973; Walton, Stammer, Nutt, Jenkins and Holly, 1965; Woolley, D. W. and Shaw, E., 1955; Yamamoto, H., 1967b, 1968; Yamamoto, H., and Atami, 1969; Yamamoto, H., Atsuko and Takuhiro, 1970; Yamamoto, H., Inaba, Nakamura, Y., Nakao, Niizaki, M., and Atami, 1970; Yamamoto, H., Inaba, Nakao and Niizaki, M., 1970; Yamamoto, H. and Kimura, 1974; Yamamoto, H., Misaki and Izumi, 1968, 1970; Yamamoto, H., Nakamura, Y., Atami, Kobayashi, T. and Nakao, 1971; Yamamoto, H., Nakamura, Atami, Nakao and Kobayashi, T., 1970a, 1970f; Yamamoto, H., Nakamura, Y., Atami, Nakao, Kobayashi, T., Saito, C. and Awata, 1970; Yamamoto, H., Nakamura,

Y., Atami, Nakao, Kobayashi, T., Saito, C. and Kurita, 1969; Yamamoto, H., Nakamura, Y. and Kobayashi, T., 1970; Yamamoto, H., Nakamura, Y., Nakao and Kobayashi, T., 1969a, 1969b, 1970b; Yamamoto, H. and Nakao, 1968b, 1968f, 1968g, 1969a, 1969c, 1969d, 1969e, 1970b, 1970c, 1970d, 1970g, 1970i, 1971a, 1971b, 1973, 1974a, 1974b; Yamamoto, H., Nakao and Kobayashi, A., 1968; Yamamoto, H., Nakao and Kurita, 1968a; Yamamoto, H., Nakao and Okamoto, 1968; Yamamoto, H., Saito, C., Okamoto, Awata, Inukai, Hirohashi, A. and Yukawa, 1969), R = $CH_2{}^{14}COOH$ (Nakatsuka, Hazue, Makari, Kawahara, Endo and Yoshitake, 1976), R = CH_2COOCH_3 (Kosa and Kovacs, 1975; Okamoto, Niizuki, M., Kobayashi, T., Izumi, Yamamoto, H., Inaba, Nakamura, Y. and Nakao, 1971; Sarett and Shen, T.-Y., 1966a; Shaw, E., 1955; Shen, T.-Y., 1964, 1965, 1966, 1967c; Sumitomo Chemical Co. Ltd, 1966, 1967; Woolley, D. and Shaw, E. N., 1959; Yamamoto, H., Inaba, Nakao and Niizuki, M., 1970; Yamamoto, H. and Nakao, 1969e, 1971a, 1974a; Yamamoto, H., Nakao and Okamoto, 1968), R = $CH_2COOC_2H_5$ (Beitz, Stroh, H.-H. and Fiebig, 1967; Bullock and Hand, J. J., 1956a; Carlin and Larson, 1957; De Bellis and Stein, M. L., 1961; Degen, 1886; Fischer, Emil, 1886b, 1886c; Fox and Bullock, 1955a; Hansch and Muir, 1950; Hoshino and Shimodaira, 1935; Kimura, Inaba and Yamamoto, H., 1973c, 1973d; Kosa and Kovacs, 1975; Kost, Vinogradova, Daut and Terent'ev, A. P., 1962; Merck and Co. Inc., 1964a; Newberry, 1974; Piper and Stevens, F. J., 1962; Saleha, Khan, N. H., Siddiqui and Kidwai, M. M., 1978; Sarett and Shen, T.-Y., 1966b; Shen, T.-Y., 1967a, 1967c; Shvedov, Trofimkin, Vasil'eva and Grinev, A. N., 1975; Snyder and Smith, C. W., 1943; Steche, 1887; Stevens, F. J. and Fox, 1948; Stevens, F. J. and Higginbotham, 1954; Sumitomo Chemical Co. Ltd, 1966; Walton, Stammer, Nutt, Jenkins and Holly, 1965; Yamamoto, H., Inaba, Nakamura, Y., Nakao, Niizaki, M., and Atami, 1970; Yamamoto, H., Inaba, Nakao and Koshiba, 1972; Yamamoto, H., Nakamura, Y., Atami, Nakao, Kobayashi, T., Saito, C. and Kurita, 1969; Yamamoto, H. and Nakao, 1969e, 1974a; Yamamoto, H., Nakao and Kobayashi, A., 1968), R = $CH_2COOCH(CH_3)_2$ and $CH_2COO(CH_2)_3CH_3$ (Kosa and Kovacs, 1970; Stevens, F. J., Ashby and Downey, 1957), R = $CH_2COOC(CH_3)_3$ (Kosa and Kovacs, 1972, 1975; Merck and Co. Inc., 1965, 1967b; Shen, T.-Y., 1966; Sletzinger and Gal, 1968a; Sumitomo Chemical Co. Ltd, 1966, 1968a, 1969a; Yamamoto, H. and Nakao, 1974a), R = $CH_2COOCH_2C_6H_5$ (Shen, T.-Y., 1966; Sumitomo Chemical Co. Ltd, 1966; Yamamoto, H. and Nakao, 1971a, 1974a), R = $CH_2COOC(C_6H_5)_3$ (Merck and Co. Inc., 1965), R = $CH_2COOCH_2COOCH_2C_6H_5$ (Boltze, Brendler, Dell and Jacobi, 1974), R = $CH_2COO(CH_2)_2N(CH_3)_2$, $CH_2COO(CH_2)_2N(C_2H_5)_2$ and $CH_2COOCH_2CH(CH_3)N(CH_3)_2$ (Yamamoto, H., Okamoto and Kobayashi, T., 1970) (for other esters, see Noda, Nakagawa, Miyata, Nakajima and Ide, 1979), R = CH_2CN (Shen, T.-Y., 1966), R = CH_2CONH_2 (Okamoto, Kobayashi, T. and Yamamoto, H., 1974; Shen, T.-Y., 1964, 1966; Sumitomo

Chemical Co. Ltd, 1966; Yamamoto, H. and Nakao, 1968e), R = $CH_2CON(CH_3)_2$, $CH_2CONH(CH_2)_2OH$ and CH_2CO-morpholino (Okamoto, Kobayashi, T. and Yamamoto, H., 1970, 1974), R = $CH_2CONHCH_3$, $CH_2CONH(CH_2)_3N(CH_3)_2$ and CH_2CONH-2-pyrimidyl (Okamoto, Kobayashi, T. and Yamamoto, H., 1970), R = $(CH_2)_2COOH$ (Akopyan and Tatevosyan, 1971; Kimura, Inaba and Yamamoto, H., 1973d; Mndzhoyan, Tatevosyan and Ekmekdzhayan, 1957; Ogandzhanyan, Avanesova and Tatevosyan, 1968; Ogandzhanyan and Tatevosyan, 1970; Sumitomo Chemical Co. Ltd, 1966; Teuber, Cornelius and Worbs, 1964; Yamamoto, H., Atsuko and Takuhiro, 1970; Yamamoto, H., Inaba, Nakamura, Y., Nakao, Niizaki, M. and Atami, 1970; Yamamoto, H., Inaba, Nakao and Niizaki, M., 1970; Yamamoto, H., Misaki and Izumi, 1968, 1970; Yamamoto, H. and Nakao, 1968b, 1968g, 1969e, 1970g, 1971a, 1971b, 1974a), R = $(CH_2)_2COOCH_3$ (Yamamoto, H. and Nakao, 1971a, 1974a), R = $(CH_2)_2COOC_2H_5$ (Sumitomo Chemical Co. Ltd, 1966; Yamamoto, H. and Nakao, 1974a; Yoshitomi Pharmaceutical Industries Ltd, 1967), R = $(CH_2)_2COOC(CH_3)_3$ (Yamamoto, H. and Nakao, 1974a), R = $(CH_2)_2COOCH_2C_6H_5$ (Yamamoto, H. and Nakao, 1971a, 1974a), R = $(CH_2)_2CN$ (Kost, Yudin and Yü-Chou, 1964; R = $(CH_2)_3COOH$ (Sumitomo Chemical Co. Ltd, 1966, 1968a; Yamamoto, H., Inaba, Nakao and Niizaki, M., 1970; Yamamoto, H., Nakamura, Y., Atami, Nakao and Kobayashi, T., 1970a; Yamamoto, H., Nakamura, Y., Atami, Nakao, Kobayashi, T., Saito, C. and Kurita, 1969; Yamamoto, H., Nakamura, Y., Nakao and Kobayashi, T., 1969a; Yamamoto, H. and Nakao, 1968b; 1968g, 1969e, 1970g, 1971b, 1973, 1974a; Yamamoto, H., Nakao and Kobayashi, A., 1968; Yamamoto, H., Nakao and Kurita, 1968b), R = $(CH_2)_3COOCH_3$ (Sumitomo Chemical Co. Ltd, 1966; Yamamoto, H. and Nakao, 1974a), R = $(CH_2)_3COOC_2H_5$ (Suvorov and Antonov, 1952 – see also Suvorov, 1957; Yamamoto, H. and Nakao, 1974a), R = $(CH_2)_3COOC(CH_3)_3$ and $(CH_2)_3COOCH_2C_6H_5$ (Yamamoto, H. and Nakao, 1974a), R = $CH_2CH{=}CHCOOH$ (Yamamoto, H., Nakamura, Y., Atami, Nakao and Kobayashi, 1970h), R = $CH(CH_3)COOH$ (Carey, Gal and Sletzinger, 1967; Carry, Gal and Sletzinger, 1968; Kimura, Inaba and Yamamoto, H., 1972, 1973b, 1973d; Sarett and Shen, T.-Y., 1966a, 1966b; Shen, T.-Y., 1964; Shen, T.-Y. and Sarett, 1966; Yamamoto, H. and Nakao, 1968g, 1974a), R = $CH(CH_3)COOCH_3$ (Shen, T.-Y., 1964, 1967c; Yamamoto, H. and Nakao, 1974a), R = $CH(CH_3)COOC_2H_5$ (Kimura, Inaba and Yamamoto, H., 1973b, 1973c, 1973d; Merck and Co. Inc., 1962, 1963, 1964a, 1964b; Sarett and Shen, T.-Y., 1965, 1966a, 1966b; Shen, T.-Y., 1964, 1965, 1966, 1967b; Sumitomo Chemical Co. Ltd, 1966; Yamamoto, H. and Nakao, 1974a), R = $CH(CH_3)COOC(CH_3)_3$ (Merck and Co. Inc., 1965, 1967b; Yamamoto and Nakao, 1974a), R = $CH(CH_3)COOCH_2C_6H_5$ (Yamamoto and Nakao, 1974a), R = $CH(C_2H_5)COOH$ (Carey, Gal and Sletzinger, 1967; Carry, Gal and Sletzinger, 1968; Shen, T.-Y., 1964), R = $CH(C_2H_5)COOC_2H_5$ (Merck and Co. Inc.,

1962; Sarett and Shen, T.-Y., 1966b), R = $CH(C_2H_5)COOC(CH_3)_3$ (Merck and Co. Inc., 1967b), R = $CH[(CH_2)_2CH_3]COOH$ (Carry, Gal and Sletzinger, 1968), R = $CH[(CH_2)_2CH_3]COOC(CH_3)_3$ (Merck and Co. Inc., 1967b), R = $CH[CH(CH_3)_2]COOH$ (Carey, Gal and Sletzinger, 1967; Carry, Gal, and Sletzinger, 1968), R = $CH[(CH_2)_3CH_3]COOH$ (Carey, Gal and Sletzinger, 1967; Carry, Gal and Sletzinger, 1968), R = $CH[(CH_2)_3CH_3]COOC(CH_3)_3$ (Merck and Co. Inc., 1967b), R = $CH[CH_2CH(CH_3)_2]COOH$ (Carey, Gal and Sletzinger, 1967; Carry, Gal and Sletzinger, 1968), R = $CH[C(CH_3)_3]COOH$ (Carey, Gal and Sletzinger, 1967; Carry, Gal and Sletzinger, 1968), R = $CH(C_5H_{11})COOH$ and $CH(C_6H_{13})COOH$ (Carey, Gal and Sletzinger, 1967; Carry, Gal and Sletzinger, 1968), R = CH(Alkyl)COOH (alkyl group unspecified in patent abstract) (Carey, Gal, Sletzinger and Reinhold, 1968), R = $CH(C_6H_5)COOH$ (Yamamoto, H., Nakamura, Y., Atami, Nakao and Kobayashi, T., 1970g), R = $CH(COOH)_2$ (Sumitomo Chemical Co. Ltd, 1966; Yamamoto, H. and Nakao, 1968g, 1969c), R = $CH[COOC(CH_3)_3]_2$ (Chemerda and Sletzinger, 1968d, 1968f, 1969c), R = CHBrCOOH (Carey, Gal and Sletzinger, 1967), R = $CH_2CHR^1COOH[R^1 = CH_3, C_2H_5, (CH_2)_2CH_3, (CH_2)_3CH_3, C_6H_5$ and $CH_2C_6H_5]$ (Mndzhoyan, Tatevosyan, Terzyan and Ekmekdzhyan, 1958; Ogandzhanyan, Avenesova and Tatevosyan, 1968; Ogandzhanyan and Tatevosyan, 1970), R = $CH_2CH(\overset{\oplus}{N}H_3)COO^{\ominus}$ (Brenner, Clamkowski, Hinkley and Gal, 1969), R = $CH_2CH(COOCH_3)_2$ and $CH_2CH(COOC_2H_5)_2$ (Sumitomo Chemical Co. Ltd, 1966), R = CH_2CHO, $CH_2CH(OCH_2C_6H_5)_2$

and CH_2CH (O, O ring fused to benzene) (Brenner, Glamkowski, Hinkley and Russ, 1969),

R = $CH_2CH(OC_2H_5)_2$ (Yamamoto, H. and Nakao, 1968g, 1970f), R = C_6H_5 (Aktieselskabet Dumex (Dumex Ltd), 1963, 1964; Atkinson, Simpson and Taylor, A., 1954; Beer, Donavanik and Robertson, 1954; Frasca, 1962a; Govindachari, Rajappa and Sudarsanam, 1966; Govindachari and Sudarsanam, 1967; 1971; Ishizumi, Mori, Inaba and Yamamoto, H., 1973; Khan, M. A. and Morley, 1978, 1979; Khan, M. A. and Rocha, 1978; Kost, Sugorova and Yakubov, 1965; Kost and Yudin, 1957; Kost, Yudin, Berlin and Terent'ev, A. P., 1959; Kost, Yudin, Dmitriev and Terent'ev, A. P., 1959; Kost, Yudin and Terent'ev, A. N., 1959; Nagazaki, 1960a; Ockenden and Schofield, 1953a; Schofield and Theobald, 1950; Sellstedt and Wolf, M., 1974; Shvedov, Trofimkin, Vasil'eva and Grinev, A. N., 1975; Teotino, 1959; Trenkler, 1888; Verkade and Janetsky, 1943b; Vinogradova, Daut, Kost and Terent'ev, A. P., 1962; Welstead and Chen, Y.-H., 1971, 1976; Yakhontov and Pronina, E. V., 1968; Yamamoto, H., Inaba, Hirohashi, T., Ishigura, Maruyama and Mori, 1971; Yamamoto, H., Inaba, Ishigura, Maruyama, Hirohashi, T. and Mori, 1971; Yamamoto, H. and Nakao, 1968a), R = $C_6H_4OCH_3(4)$ (Szmuszkovicz,

Glenn, Heinzelman, Hester and Youngdale, 1966), R = $C_6H_3(NO_2)_2$(2,4) (Joshi, S. S. and Gambhir, 1956), R = $CH_2C_6H_5$ (Aktieselskabet Dumex (Dumex Ltd), 1963; CRC Compagnia di Ricerca Chimica S.A., 1979; Dave, V., 1976; Hoshino, 1933; Janetsky and Verkade, 1945; Kuroda, 1923; Ockenden and Schofield, 1953a, 1953b; Yamamoto, H. and Nakao, 1968a), R = $CH_2C_6H_4OCH_3$(4) (Szmuszkovicz, Glenn, Heinzelman, Hester and Youngdale, 1966), R = $CH_2C_6H_3(OCH_3)_2$(3,4) (Julian, 1931), R = $(CH_2)_2C_6H_5$ (Esayan, Terzyan, Astratyan, Dzhanpoladyan and Tatevosyan, 1968), R = CH_2-1-naphthyl (Ockenden and Schofield, 1953b), R = 2-naphthyl (CRC Compagnia di Ricerca Chimica S.A., 1979); R = 3-pyridyl (Schut, Ward, Lorenzetti and Hong, 1970), R = OH (Yamamoto, H., Nakamura, Y., Atami, Nakao and Kobayashi, T., 1970c, 1970d), R = $N(CH_3)_2$, $N(C_2H_5)_2$, piperidino and morpholino (Yoneda, Miyamae and Nitta, 1967) and R = SCH_2COOCH_3, $SCH_2COOC_2H_5$ and SC_6H_5 (Wieland, T. and Rühl, 1963). Other analogous examples are known with **652** in which R = **653** (Ehrhart, 1953), **654** (R^1-R^3 = H, n = 2) (Ehrhart, 1953), **654** [R^1 = R^2 = H, R^3 = C_6H_5; R^1 = H, R^2 = OH, R^3 = C_6H_5 and C_6H_4Cl(4), n = 3] (Yamamoto,

—C(CH$_3$)CH$_2$—N(pyrrolidine)

653

—(CH$_2$)$_n$—N(piperidine, R^1, R^2, R^3)

654

H., Okamoto, Sasajima, Nakao, Maruyama and Katayama, 1971), **654** [R^1 = R^2 = H, R^3 = $(CH_2)_2OH$, n = 2] (Zenitz, 1965), **654** (R^1 = CH_3, R^2 = R^3 = H, n = 2) (Zenitz, 1977), **654** [R^1 = $(CH_2)_{0 \text{ and } 3}$-cyclohexyl, R^2 = R^3 = H, n = 2] (Zenitz, 1977), **654** (R^1 = CH_2-cyclohexyl, R^2 = R^3 = H, n = 2) (Zenitz, 1966, 1974, 1977), **654** [R^1 = R^2 = H, R^3 = $(CH_2)_{0 \text{ and } 2}$-cyclohexyl, n = 2] (Zenitz, 1977), **654** (R^1 = R^2 = H, R^3 = CH_2-cyclohexyl, n = 2) (Zenitz, 1966), **655** [R = H, alkyl or aryl (for details see Chapter IV, F), n = 1–4] (Archer, S., 1967, 1970; Istituto Luso Farmaco d'Italia S.r.l., 1967; Laskowski, 1968; Steck, Fletcher and Carabateas, 1974), **656** (Archer, S., 1963; McCrea, 1966; Merck and Co. Inc., 1961; Sankyo Co. Ltd, 1964; Sato, Y. and Sunagawa, 1967; Sletzinger, Gaines and Ruyle, W. V., 1957; Sletzinger, Ruyle, W. V. and Gaines, 1961; Steck, Fletcher and Carabateas, 1974; Sterling Drug Inc., 1962; Sunagawa and Sato, Y., 1967), **657** [R = H, CH_3, C_2H_5,

—(CH$_2$)$_n$—N(piperazine)N—R

655

—(CH$_2$)$_2$—N(phthalimide)

656

—CHR—(1H-benzimidazol-2-yl)

657

$(CH_2)_2CH_3$ and $(CH_2)_3CH_3$] (Esayan and Tatevosyan, 1972), **658** [R^1 = H and OH, R^2 = H, C_6H_5, $C_6H_4Cl(4)$ and $C_6H_4CF_3(4)$ (Yamamoto, H., Okamoto, Sasajima, Nakao, Maruyama and Katayama, 1975), **659** (Dickel and DeStevens, 1970, 1972), **660** (n = 0 and 1) (Boido and Boido Canu, 1973 – see also Sparatore, Boido and Pirisino, 1974) and **661** (CRC Compagnia di Ricerca Chimica S.A., 1979; Kajfez and Mihalic, 1978).

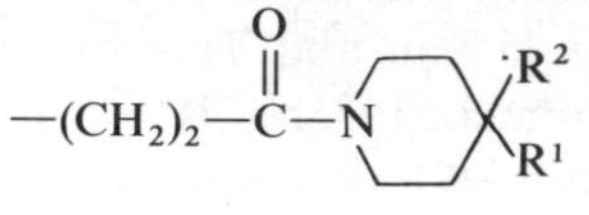

658

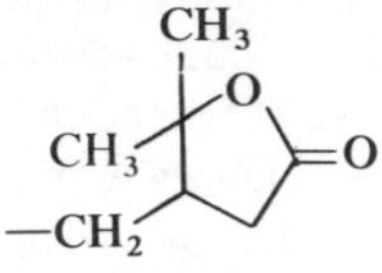

659

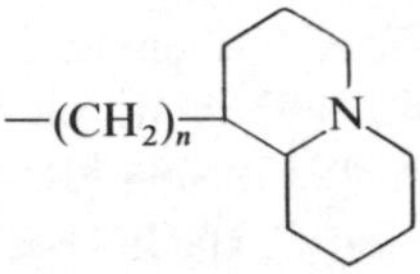

660

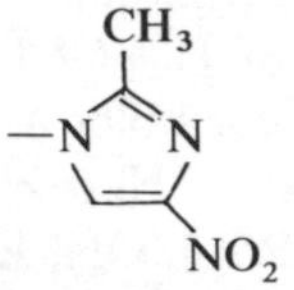

661

Exceptions to rule 1 (see page 241) have, however, been observed. Thus, an initial copper(I) chloride catalysed indolization of methyl propyl ketone phenylhydrazone was reported (Arbuzov, A. E. and Friauf, 1913) to afford 2-propylindole. However, structural verification of the product was lacking and, indeed, later (Julian and Pikl, 1935 – see also Janetsky and Verkade, 1946) the principal product from this reaction was shown to be 3-ethyl-2-methylindole. Traces of the corresponding 2-benzylindoles may have been formed, along with the 2-methyl-3-phenylindoles, upon indolization of benzyl methyl ketone arylhydrazones (Ockenden and Schofield, 1953a; Schofield and Theobald, 1950). Ethyl methyl ketone phenylhydrazone, upon treatment with zinc chloride (Fischer, Emil, 1886b) or nickel(II) chloride (Korczynski, Brydowna and Kierzek, 1926), furnished both 2,3-dimethylindole and 2-ethylindole [for an unambiguous synthesis of 2-ethylindole, see Leete (1961a) and Verley and Beduvé (1925)]. In the latter paper, the identity of the 2-ethylindole, obtained in 20% yield, was established by coupling it with 4-nitrobenzenediazonium salt to give **662**, although the structure of this product, but not its molecular formula, was given in error as **663** in the original paper (Korczynski,

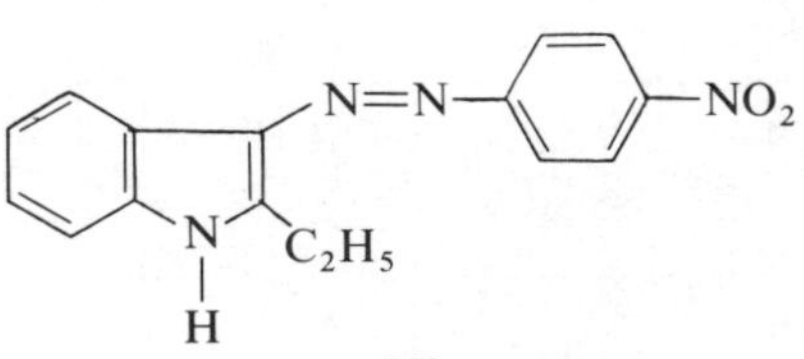

662

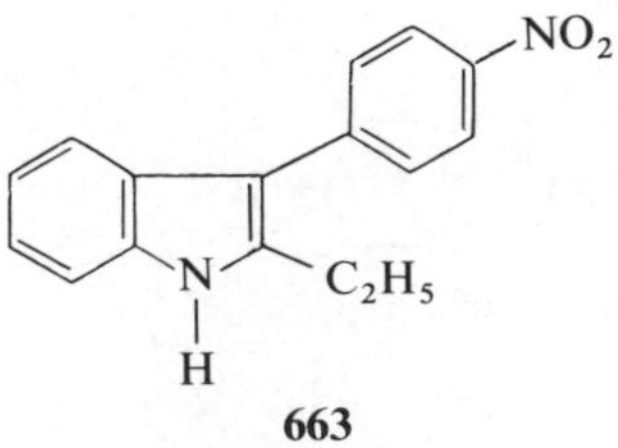

663

Brydowna and Kierzek, 1926): 2,3-dimethylindole will not react with aryldiazonium salts. Similarly, laevulinic acid phenylhydrazone (**643**; $R^1 = R^2 = H$, $R^3 = CH_2COOH$), upon indolization by fusion with zinc chloride, afforded **644** ($R^1 = R^2 = H$, $R^3 = CH_2COOH$) as the major reaction product, along with a small amount of **645** ($R^1 = R^2 = H$, $R^3 = CH_2COOH$) (Farbwerke vorm. Meister, Lucius and Brüning, 1886a). Although when **643** ($R^1 = 3\text{-}CH_3$, $R^2 = H$, $R^3 = SCH_2COOC_2H_5$) was indolized using alcoholic hydrochloric acid, only **644** ($R^1 = 6\text{-}CH_3$, $R^2 = H$, $R^3 = SCH_2COOC_2H_5$) was isolated from this reaction, the p.m.r. spectrum of the total indolic reaction mixture exhibited a signal at $\tau = 6.65$ attributed to the $—SCH_2—$ protons in **645** ($R^1 = 6\text{-}CH_3$, $R^2 = H$, $R^3 = SCH_2COOC_2H_5$). A signal at $\tau = 7.6$ was assigned to the 2-methyl group in the isolated product (Wieland, T. and Rühl, 1963) but unfortunately the relative integral intensities of these above two peaks, presumably singlets, in the total reaction product were not quoted so the relative yields of the two isomeric products cannot be estimated.

Several other methods have been used to distinguish between the isomeric structures **644** and **645** possible for the products from the above type of indolization. An approach related to that above, employing coupling with aryldiazonium salts, involved the reaction with Ehrlich's reagent. Thus, hydrazone **643** [$R^1 = 4\text{-}OCH_3$, $R^2 = H$, $R^3 = (CH_2)_2OC_6H_5$] with boiling alcoholic sulphuric acid afforded only one isolated product, either **644** [$R^1 = 5\text{-}OCH_3$, $R^2 = H$, $R^3 = (CH_2)_2OC_6H_5$] or **645** [$R^1 = 5\text{-}OCH_3$, $R^2 = H$, $R^3 = (CH_2)_2OC_6H_5$], the former structure being favoured since no color developed when the product was treated with Ehrlich's reagent. As a test standard, **664** was similarly indolized, unambiguously, to yield **665** ($R = COOC_2H_5$) which

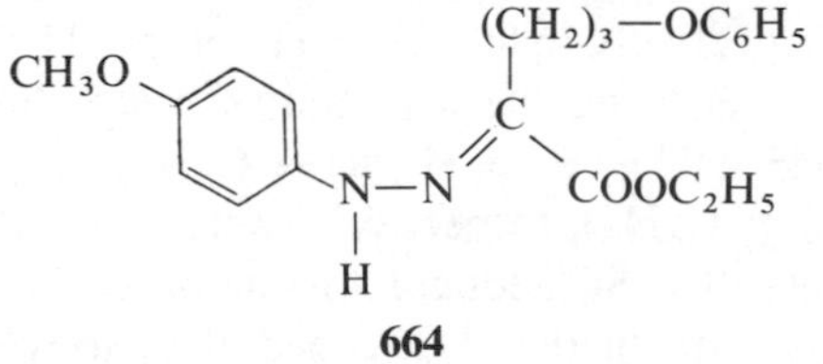

CH_3O $(CH_2)_3—OC_6H_5$ C $COOC_2H_5$ N—N H

664

CH_3O $(CH_2)_2OC_6H_5$ N R H

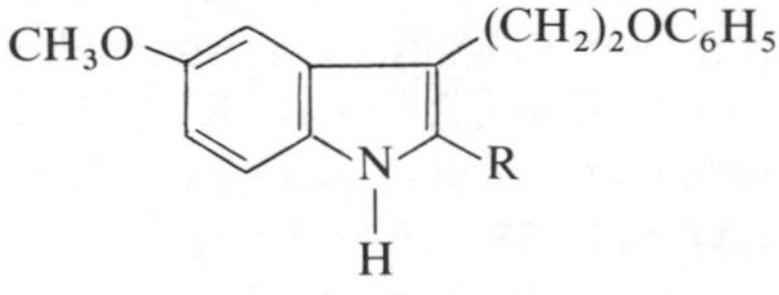

665

after hydrolysis and decarboxylation produced **665** ($R = H$). This product developed an intense red-purple color with Ehrlich's reagent (King, F. E. and Robinson, R., 1932). Other chemical methods for structural differentiation have involved acetylation with acetic anhydride which afforded a 1-acetyl or 1,3-diacetyl derivative, respectively (when $R^2 = H$ in **644** and **645**) (Buu-Hoï and Royer, 1947), C-3 methylation with boiling methanolic methyl iodide and sodium acetate (again when $R^2 = H$ in **644** and **645**) which afforded the basic 3*H*-indole **666** or the neutral 3-methylindole **667** derivatives, respectively (Deorha and Joshi, S. S., 1961), oxidative scission of the indolic 2,3-double bond with ozone (Ockenden and Schofield, 1953a, 1953b) or with chromium(VI) oxide in acetic acid (Schofield and Theobald, 1949, 1950) which led to known

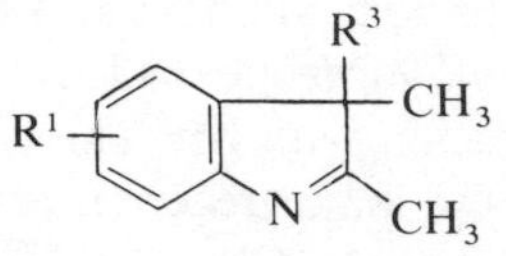

666 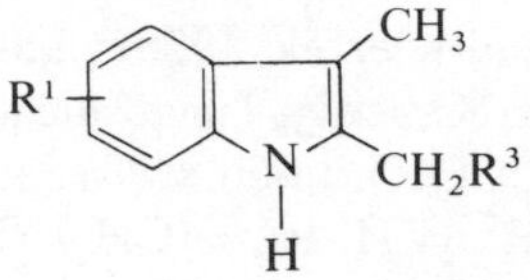 **667**

substituted benzene derivatives and alternative unambiguous synthesis of the product before (Janetsky and Verkade, 1945) [in this paper reference is incorrectly made to the benzylation of skatole magnesium iodide, when the reaction referred to involved the benzylation of 2-methylindole magnesium iodide (Hoshino, 1933)], or after (Deorha and Joshi, S. S., 1961) further chemical modification. Spectroscopic methods involved u.v. analysis when R^3 was functionalized, in which case **645** exhibited indolic absorption whereas **644** did not do so because of conjugation with the functionalized R^3 group (Yoneda, Miyamae and Nitta, 1967), and p.m.r. analysis which, in particular, identified the 3H singlet of the indolic 2-CH_3 group in **644** and the indolic 3-H signal in **645** which was well upfield from the other aromatic proton signals (Bailey and Seager, 1974; Palmer and McIntyre, 1969; Yamamoto, H., Misaki and Imanaka, 1968; Yoneda, Miyamae and Nitta, 1967).

The first attempts (Buu-Hoï, Jacquignon and Périn-Roussel, 1965) to investigate the effect of the nature of the indolization catalyst upon the direction of indolization in alkyl methyl ketone arylhydrazones involved the indolization of benzyl methyl ketone phenyl- and 4-methylphenylhydrazone **643** (R^1 = H and 4-CH_3, respectively, R^2 = H, R^3 = C_6H_5) using hydrogen chloride in acetic acid, zinc chloride and polyphosphoric acid as catalysts. Using the first two catalysts, the only product isolated in each case was 2-methyl-3-phenylindole and 2,5-dimethyl-3-phenylindole, respectively (**644**; R^1 = H and 5-CH_3, respectively, R^2 = H, R^3 = C_6H_5) whereas using the last catalyst both the possible indolic products, **644** and **645** ($R^1 = R^2$ = H, R^3 = C_6H_5) and **644** and **645** (R^1 = 5-CH_3, R^2 = H, R^3 = C_6H_5), respectively, were isolated via fractional crystallization of their picrates. Subsequent (Yamamoto, H., Misaki and Imanaka, 1968) investigations confirmed and extended these above conclusions. Using acetic acid solutions, hydrogen chloride, boron trifluoride etherate, zinc chloride and 85% *ortho*-phosphoric acid as catalysts it was found that indolization of ethyl methyl ketone and benzyl methyl ketone phenylhydrazone (**643**; $R^1 = R^2$ = H, R^3 = CH_3 and C_6H_5, respectively) furnished **644** ($R^1 = R^2$ = H, R^3 = CH_3 and C_6H_5, respectively) as the only isolated products. However, with polyphosphoric acid as catalyst, both these products and **645** ($R^1 = R^2$ = H, R^3 = CH_3 and C_6H_5, respectively) were obtained, the ratio of **644** : **645** in both cases increasing with increasing concentrations of phosphorus pentoxide to *ortho*-phosphoric acid in the polyphosphoric acid (Table 5) (Yamamoto, H., Misaki and Imanaka, 1968). Later (Yamamoto, H., Misaki, Izumi and Koshiba, 1971), 2-ethyl- and 2-benzylindole, respectively,

Table 5. Products Obtained by Indolization of Ethyl Methyl Ketone and Benzyl Methyl Ketone Phenylhydrazone using Polyphosphoric Acid of Varying Compositions as the Catalysts

	Yield (%) from **643** ($R^1 = R^2 = H, R^3 = CH_3$) of:		Yield (%) from **643** ($R^1 = R^2 = H, R^3 = C_6H_5$) of:	
Catalyst Composition ($P_2O_5 : H_3PO_4$)	**644** ($R^1 = R^2 = H$, $R^2 = CH_3$)	**645** ($R^1 = R^2 = H$, $R^3 = CH_3$)	**644** ($R^1 = R^2 = H$, $R^3 = C_6H_5$)	**645** ($R^1 = R^2 = H$, $R^3 = C_6H_5$)
0.5 : 1	60.0	19.8	62.3	9.4
1 : 1	36.8	30.4	28.9	36.3
1.5 : 1	33.4	32.6	18.3	40.7
2 : 1	25.7	29.6	14.7	44.8

were claimed as the only products from these two indolizations using polyphosphoric acid as the catalyst. Similar results were also obtained by indolization of ethyl methyl ketone and benzyl methyl ketone N_α-(4-chlorobenzoyl) phenylhydrazone with acetic acid solutions, hydrogen chloride, boron trifluoride etherate and zinc chloride as catalysts which afforded **644** [$R^1 = H$, $R^2 = COC_6H_4Cl(4)$, $R^3 = CH_3$ and C_6H_5, respectively] as the only products. However, starting with the latter hydrazone and using polyphosphoric acid ($P_2O_5 : H_3PO_4 = 1:1$) as catalyst, only a 0.4% yield of **645** ($R^1 = R^2 = H$, $R^3 = C_6H_5$) was obtained, together with **644** ($R^1 = R^2 = H$, $R^3 = C_6H_5$) (29.2% yield) and **644** [$R^1 = H$, $R^2 = COC_6H_4Cl(4)$, $R^3 = C_6H_5$] (26.4%). It was suggested that in this case, even though a polyphosphoric acid catalyst would be expected to effect considerable indolization with the new C—C bond formation involving the methyl group rather than the methylene group, the steric interaction between the benzyl and 4-chlorobenzoyl groups would lead the hydrazone to adopt the conformation **668** and thereby ultimately lead to the observed indolic products, de-4-chlorobenzoylation occurring subsequent to indolization (Yamamoto, H., Misaki and Imanaka, 1968).

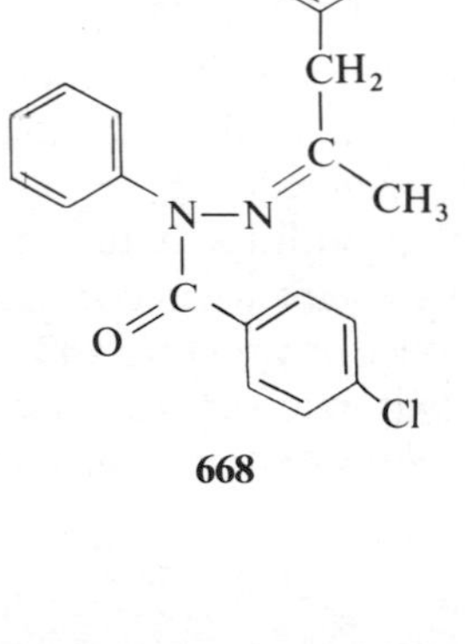

668

The product yields and/or ratios quoted in the above studies may be questionable because of the methods of analysis of the total reaction products which may afford less than 100% product recovery. This problem was overcome in a later study (Palmer and McIntyre, 1969) when analysis of the crude reaction mixtures by p.m.r. spectroscopy readily determined the ratio of **644** : **645** by comparison of the integrals of the indolic 3-H signal from **645**, which is well upfield from the other aromatic proton signals, and the indolic 2-CH_3 signal from **644**. The indolizations studied were those of ethyl methyl ketone, benzyl methyl ketone and methyl 4-nitrobenzyl ketone phenylhydrazone [**643**; $R^1 = R^2 = H$, $R^3 = CH_3$, C_6H_5 and $C_6H_4NO_2(4)$, respectively]. Using polyphosphoric acids of varying compositions with all three hydrazones, the yields of the respective **644** and **645** were determined (Table 6). It can be seen that, in accord with earlier studies, as the concentration of phosphorus pentoxide in the catalyst increased, so did the relative yield of **645** to **644** in the product. A similar result was observed when ethyl methyl ketone phenylhydrazone was treated with aqueous ethanolic (1 : 1 molar) sulphuric acid when, using 40% (w/w) sulphuric acid, only **644** ($R^1 = R^2 = H$, $R^3 = CH_3$) was formed but with 70% (w/w) sulphuric acid a mixture of **644** ($R^1 = R^2 = H$, $R^3 = CH_3$) (49%) and **645** ($R^1 = R^2 = H$, $R^3 = CH_3$) (51%) resulted. Similarly, indolization of benzyl methyl phenylhydrazone in 80% (w/w) aqueous sulphuric acid afforded both **644** ($R^1 = R^2 = H$, $R^3 = C_6H_5$) (79%) and **645** ($R^1 = R^2 = H$, $R^3 = C_6H_5$) (21%) although only the former product was formed when using boron trifluoride etherate in acetic acid or 32% (w/w) ethanolic sulphuric acid as catalysts. Since these indolic products were stable under indolization conditions, the involvement of intramolecular rearrangements subsequent to indolization in the formation of the isomeric indolic products was eliminated. These above observations were rationalized by assuming that, in weak acid media, the hydrazone **643** was monoprotonated to give **669** which could then lose a proton by an E_1 mechanism to afford the enehydrazine **671** (analogous to the Saytzeff olefin, i.e. the most alkylated olefin) which would ultimately lead to the formation of **644**. In strong acid media, the hydrazone **643** was diprotonated to give **670** which, because of the additional inductive electron withdrawal from the alkyl groups, would then lose a proton by an E_2 type mechanism, possibly involving a concerted attack by the solvent upon the C—H bond. This, by analogy with the Hofmann elimination which also occurs by an E_2 mechanism, would then form the enehydrazine **672** (i.e. equivalent to the least alkylated olefin formed during Hofmann elimination) and ultimately **645**.

It is therefore apparent that both the nature and concentration of the acid catalyst must be taken into account when considering the direction of indolization in ketone arylhydrazones of structure **643**. However, steric effects which may favour enehydrazine tautomerism, and hence indolization, in one particular direction, as already discussed in connection with structure **668** and as

Table 6. Product Ratios Resulting from the Indolization of Ethyl Methyl Ketone, Benzyl Methyl Ketone and Methyl 4-Nitrobenzyl Ketone Phenylhydrazone using Polyphosphoric Acid Catalyst of varying Concentrations

Concentration of Phosphorus Pentoxide[a]	Product Ratios ($R^1 = R^2 = H$ in all cases)					
	644 ($R^3 = CH_3$)	**645** ($R^3 = CH_3$)	**644** ($R^3 = C_6H_5$)	**645** ($R^3 = C_6H_5$)	**644** [$R^3 = C_6H_4NO_2(4)$]	**645** [$R^3 = C_6H_4NO_2(4)$]
66.4[b]	100	0	100	0	100	0
73.2	92	8	98	2	—	—
79.8	60	40	91	9	—	—
83.0	49	51	71	29	—	—
86.4	50	50	44	56	66	34

Notes

a. Percentage (w/w) P_2O_5 content in $P_2O_5 + H_2O$ system; b. 90% (w/w) *ortho*-phosphoric acid.

643 ⟶ 669 ⟶ 670

669 ↓ 671 ↓↓↓ 644

670 ↓ 672 ↓↓↓ 245

discussed later in connection with the nature of the acid catalyst (Lyle, R. E. and Skarlos, 1966), must be considered. Furthermore, enehydrazine formation in **643** with R^3 = amino, ester, phenyl, thio, etc. would be favoured in the direction of the methylene group since stabilizing extended conjugation involving the p- or π-electrons of the R^3 group would then be possible.

Consideration of the results of some of the above studies allowed the indolization of **643** [$R^1 = H$, $R^2 = CH_3$, $R^3 = (CH_2)_2CN$] to be carried out, using a dilute sulphuric acid catalyst, to give **644** [$R^1 = H$, $R^2 = CH_3$, $R^3 = (CH_2)_2CN$] as the only isolated product whereas, with a polyphosphoric acid catalyst, both **644** [$R^1 = H$, $R^2 = CH_3$, $R^3 = (CH_2)_2CN$] and **645** [$R^1 = H$, $R^2 = CH_3$, $R^3 = (CH_2)_2CN$] were formed along with the corresponding amides **644** [$R^1 = H$, $R^2 = CH_3$, $R^3 = (CH_2)_2CONH_2$] and **645** [$R^1 = H$, $R^2 = CH_3$, $R^3 = (CH_2)_2CONH_2$], respectively (Julia and Lenzi, 1971). Although the nitrile mixture was chromatographically separated from the amide mixture, the isomeric indolic products were not themselves separated, their formation being based upon p.m.r. studies of products resulting from further transformations. Similar considerations led to the indolization of **643** [$R^1 = R^2 = H$, $R^3 = (CH_2)_3COOH$], using concentrated sulphuric acid in aqueous ethanol, to afford **645** [$R^1 = R^2 = H$, $R^3 = (CH_2)_3COOC_2H_5$] as the major product, along with equal quantities of **645** [$R^1 = R^2 = H$, $R^3 = (CH_2)_3COOH$] and **644** [$R^1 = R^2 = H$, $R^3 = (CH_2)_3COOC_2H_5$] (Bailey and Seager, 1974).

Polyphosphoric acid is certainly the catalyst of choice when indolization of a methyl ketone involving the methyl group carbon atom in the new C—C bond formations is required, even when ambiguities in indolization direction do not exist. Thus, using this catalyst, **673** (R^1 = H, R^2 = CH_3) afforded **674** (R^1 = H, R^2 = CH_3) (Casnati, Langella, Piozzi, Ricca and Umani-Ronchi, 1964b; Colle, M.-A. and David, 1960), **673** (R^1 = R^2 = CH_3) yielded **674** (R^1 = R^2 = CH_3) (Casnati, Langella, Piozzi, Ricca and Umani-Ronchi, 1946b), **673** [R^1 = $(CH_2)_2CH(CH_3)_2$, R^2 = C_2H_5] gave **674** [R^1 = $(CH_2)_2CH(CH_3)_2$, R^2 = C_2H_5] (Casnati, Langella, Piozzi, Ricca and Umani-Ronchi, 1964a, 1964b), **675** (R^1–R^3 = H, R^4 = R^5 = OCH_3, X = CH) afforded **676** (R^1–R^3 = H, R^4 = R^5 = OCH_3, X = CH) (Woodward, Cava, Ollis, Hunger, Daeniker and Schenker, 1963), **675** [R^1–R^3 = R^5 = H, R^4 = I, CH_3, C_2H_5 and $(CH_2)_2CH_3$, X = CH; R^1 = R^3–R^5 = H, R^2 = OH, X = CH and R^1 = R^2 = R^4 = R^5 = H, R^3 = NO_2, X = CH] afforded **676** [R^1–R^3 = R^5 = H, R^4 = I, CH_3, C_2H_5 and $(CH_2)_2CH_3$, X = CH; R^1 = R^3–R^5 = H, R^2 = OH, X = CH and R^1 = R^2 = R^4 = R^5 = H, R^3 = NO_2, X = CH, respectively] (Calvaire and Pallaud, 1960), **675** (R^1 = R^2 = R^5 = H, R^3 = R^4 = F, X = CH; R^1 = R^4 = F, R^2 = R^3 = R^5 = H, X = CH; R^1 = R^4 = F, R^2 = R^5 = H, R^3 = CH_3, X = CH; R^1 = NO_2, R^2 = R^3 = R^5 = H, R^4 = F, X = CH; R^1 = NO_2, R^2 = R^5 = H, R^3 = CH_3, R^4 = F, X = CH; R^1 = NO_2, R^2 = CH_3, R^3 = R^5 = H, R^4 = F, X = CH; R^1 = NO_2, R^2 = F, R^3 = R^4 = H, R^5 = CH_3, X = CH and R^1 = NO_2, R^2 = R^5 = H, R^3 = Cl, R^4 = F, X = CH) gave **676** (R^1 = R^2 = R^5 = H, R^3 = R^4 = F, X = CH; R^1 = R^4 = F, R^2 = R^3 = R^5 = H, X = CH; R^1 = R^4 = F, R^2 = R^5 = H, R^3 = CH_3, X = CH; R^1 = NO_2, R^2 = R^3 = R^5 = H, R^4 = F, X = CH; R^1 = NO_2, R^2 = R^5 = H, R^3 = CH_3, R^4 = F, X = CH; R^1 = NO_2, R^2 = CH_3, R^3 = R^5 = H, R^4 = F, X = CH; R^1 = NO_2, R^2 = F, R^3 = R^4 = H, R^5 = CH_3, X = CH and R^1 = NO_2, R^2 = R^5 = H, R^3 = Cl, R^4 = F, X = CH, respectively) (Joshi, K. C.,

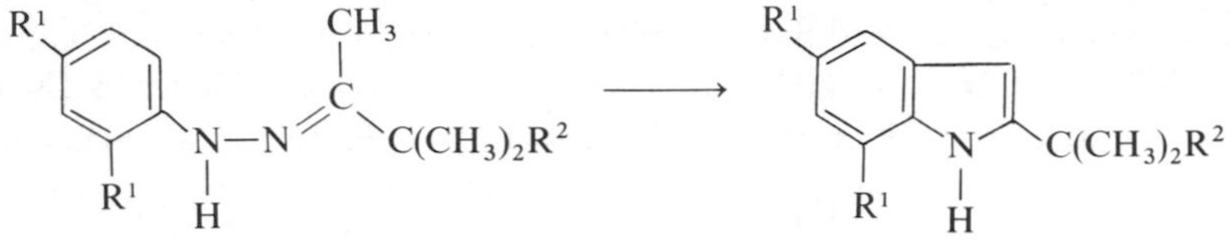

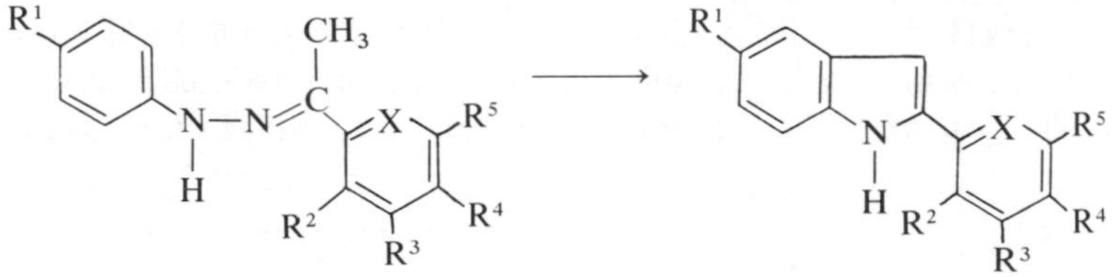

Pathak and Chand, 1978b), **675** [R^1–R^3 = H, R^4 + R^5 = $(CH_2)_4$, X = CH] yielded **676** [R^1–R^3 = H, R^4 = R^5 = $(CH_2)_4$, X = CH] (Schindler and Häfligers, 1957), **675** (R^1–R^4 = H, R^5 = OC_2H_5, X = N and R^1–R^3 = R^5 = H, R^4 = C_2H_5, X = N) afforded **676** (R^1–R^4 = H, R^5 = OC_2H_5, X = N and R^1–R^3 = R^5 = H, R^4 = C_2H_5, X = N, respectively) (Buchmann, Rehor and Wegwart, 1968), **675** (R^1–R^5 = H, X = N) afforded **676** (R^1–R^5 = H, X = N) (Al-Azawe and Sarkis, 1973; Sugasawa, Terashima and Kanaoka, 1956), **677** (X = N, Y = CH and X = CH, Y = N) afforded **678** (X = N, Y = CH and X = CH, Y = N, respectively) (Al-Azawe and Sarkis, 1973;

CH_3
N—N=C
H
X
Y
→
N
H
X
Y

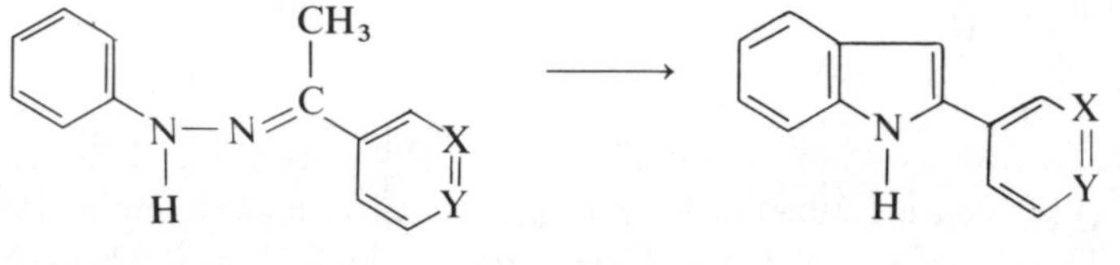

677 **678**

Gray and Archer, W. L., 1957; Huffman, 1962; Sugasawa, Terashima and Kanaoka, 1956), **679** (R = H, X = O) furnished **680** (X = O) (Calvaire and Pallaud, 1960; Holla, B. S. and Ambekar, 1976), **679** (R = H, X = S) afforded **680** (X = S) (Holla, S. and Ambekar, 1974), **681** afforded **682** (Al-Azawe and Sarkis, 1973; Schindler and Häfligers, 1957) and **683**, **685**, **687**, **689** and **691** formed **684**, **686**, **688**, **690** and **692**, respectively (Buu-Hoï, Delcey, Jacquignon

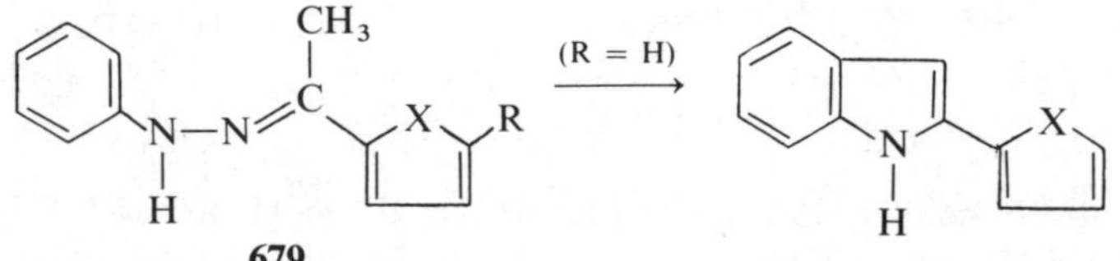

CH_3
N—N=C
H
X
R
(R = H)
→
N
H
X

679 **680**

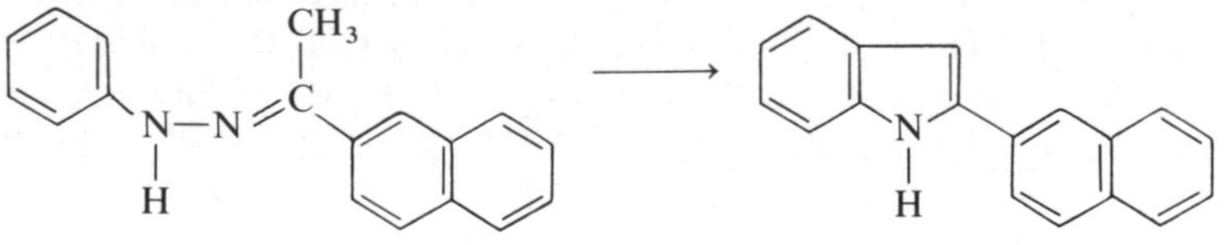

CH_3
N—N=C
H
→
N
H

681 **682**

CH_3
N—N=C
H
→
N
H

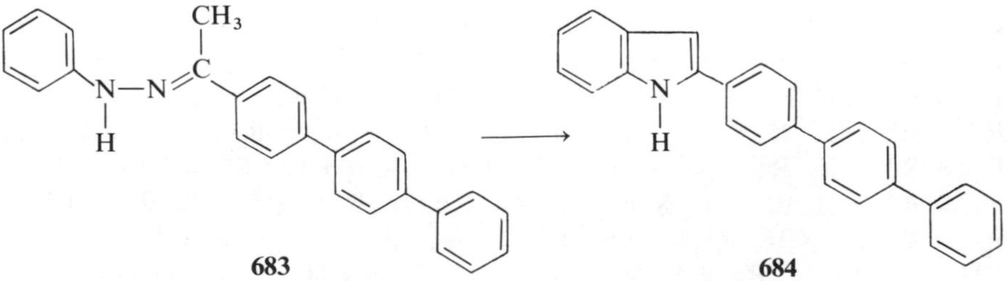

683 **684**

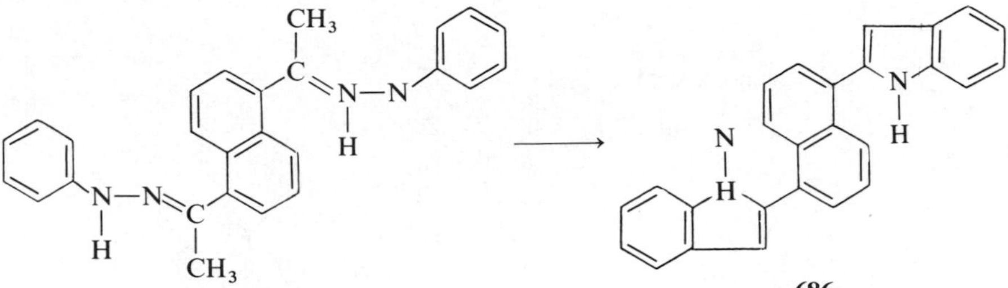

685 686

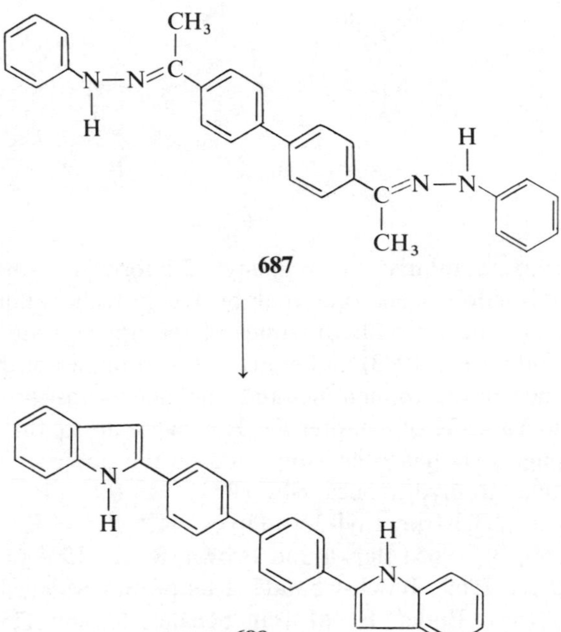

687

688

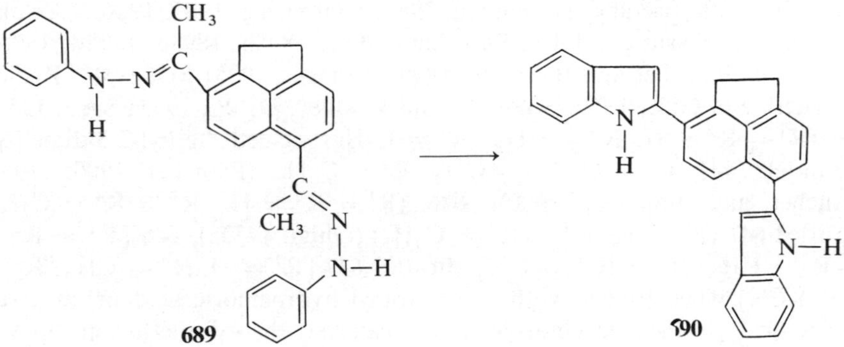

689 690

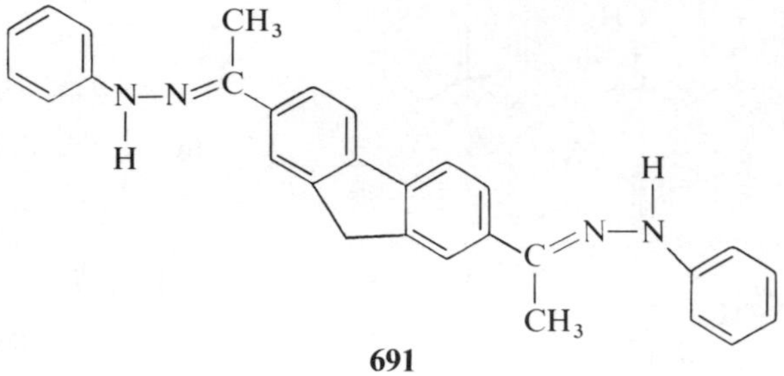

691

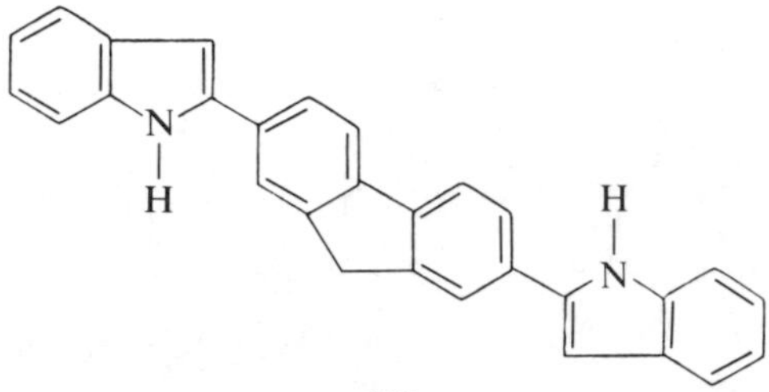

692

and Périn, 1968). Similarly, 2-(4-biphenyl-, 2-fluorenyl-, 5-indanyl-, 5-indolinyl and 3-indolyl-)indole were prepared by the polyphosphoric acid catalysed indolizations of the phenylhydrazones of the appropriate acetyl derivatives (Al-Azawe and Sarkis, 1973) and many of the examples of the indolizations of arylhydrazones of acetophenone and the benz substituted acetophenones referred to in Table 32 of Chapter IV, K likewise utilize this catalyst.

Rule 2 (page 241) has been supported by the observations that isopropyl methyl ketone arylhydrazones **646** ($R^2 = H$, $R^3 = R^4 = CH_3$) gave the corresponding 2,3,3-trimethyl-3*H*-indoles **647** ($R^2 = R^3 = CH_3$) (Ahmed and Robinson, B., 1965; Bajwa and Brown, R. K., 1968a; Berti, Da Settimo and Nannipieri, 1969; Bloch-Chaudé, Rumpf and Sadet, 1955; Deorha and Joshi, S. S., 1961; Ender, F., Moisar, Schäfer, K. and Teuber, 1959; Ghigi, 1933b; Ficken and Kendall, 1959, 1961; Grammaticakis, 1940b; Grandberg, Belyaeva and Dmitriev, L. B., 1973; Hinman and Whipple, 1962; Kanaoka, Ban, Miyashita, Irie and Yonemitsu, 1966; Konschegg, 1905, 1906, – see also Plancher and Carrasco, 1909; Plancher, 1898a, 1898b, 1898c, 1898d, 1898g, 1901, 1905; Plancher and Bettinelli, 1899; Plangger, 1905; Redies, F., Redies, B., Tuerk and Gille, 1969; Shaw, E. and Woolley, D. W., 1953; Sych, 1953), **646** ($R^1 = R^2 = H$, $R^3 = CH_3$, $R^4 = C_2H_5$) yielded 3-ethyl-2,3-dimethyl-3*H*-indole (**647**; $R^1 = H$, $R^3 = CH_3$, $R^4 = C_2H_5$) (Plancher, 1900a, 1902; Plancher and Bonavia, 1900), **646** ($R^1 = R^2 = H$, $R^3 = R^4 = C_2H_5$) afforded **647** ($R^1 = H$, $R^3 = R^4 = C_2H_5$) (Ghigi, 1933a), **646** [$R^1 = R^2 = H$, $R^3 = CH_3$, $R^4 = (CH_2)_2CN$] afforded **647** [$R^1 = H$, $R^3 = CH_3$, $R^4 = (CH_2)_2CN$] when treated with concentrated hydrochloric acid or 20% sulphuric acid {using zinc chloride as the catalyst, the cyanoethyl group was eliminated to afford 2,3-dimethylindole which was also formed by treatment of **647** [R^1 = H, R^3 = CH_3, R^4 = $(CH_2)_2CN$] with zinc chloride} (Kost, Yudin and Yü-Chou, 1964; Yudin, Kost and Chernyshova, 1970), **646** (R^1 = R^2 = H, R^3 = CH_3, R^4 = CH_2COOH) formed **647** (R^1 = H, R^3 = CH_3, R^4 = CH_2COOH) (methyl ester formation also occurred under the appropriate indolization conditions) (Rosenstock, 1966), **646** [R^1 = R^2 = H, R^3 = CH_3, R^4 = $CH_2CH(CH_3)COOH$] formed **647** [R^1 = H, R^3 = CH_3, R^4 = $CH_2CH(CH_3)COOH$] (Teuber and Cornelius, 1965), **646** [R^1 = R^2 = H, R^3 + R^4 = $(CH_2)_4$] afforded **647** [R^1 = H, R^3 + R^4 = $(CH_2)_4$] using an acetic acid catalyst (Witkop and Patrick, 1951a, 1953), **646** (R^1 = R^2 = H, R^3 = R^4 = $CH_2C_6H_5$) afforded **647** (R^1 = H, R^3 = R^4 = $CH_2C_6H_5$) using a zinc chloride catalyst (Leuchs, Heller, A. and Hoffmann, A., 1929) and **646** (R^1 = R^2 = H, R^3 = CH_3, R^4 = C_6H_5) furnished **647** (R^1 = H, R^3 = CH_3, R^4 = C_6H_5) (Evans, F. J. and Lyle, R. E., 1960; Evans, F. J., Lyle, G. G., Watkins and Lyle, R. E., 1962).

An interesting exception to this generalization seemed to occur when **646** (R_1 = R^2 = H, R^3 = CH_3, R^4 = $COOC_2H_5$) was treated with concentrated sulphuric acid when, along with the main reaction product which is a pyrazolonesulphonic acid (see Chapter VII, E), an indolic product was isolated which was formulated as **693** (Walker, C., 1894). Certainly, this study was

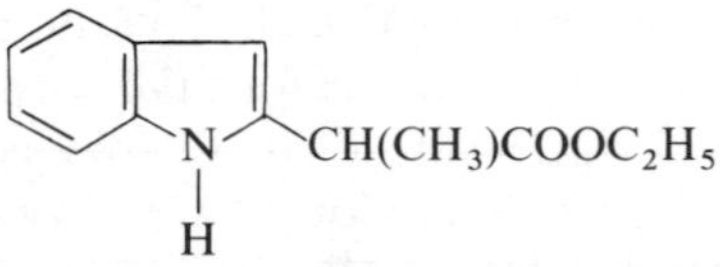

[Structure: indole (N–H) with 2-$CH(CH_3)COOC_2H_5$ substituent]

693

effected long before the mechanism of the indolization was understood and before the formation of 3*H*-indoles from such reactions had been considered. However, this product did appear to be indolic rather than 3*H*-indolic since it was apparently extracted into ether from an acidic reaction mixture and it gave a positive indole colour test ('splinter-reaction'). However, two attempted repetitions of this reaction led in only one case to the isolation of a 2% yield of a compound whose u.v., p.m.r. and mass spectral properties supported its formulation as **647** (R^1 = H, R^3 = CH_3, R^4 = $COOC_2H_5$) (Mills, 1969). Furthermore, when **646** (R^1 = R^2 = H, R^3 = C_2H_5, R^4 = $COOC_2H_5$) was treated with concentrated sulphuric acid, the ethyl group appeared to be eliminated concomitantly with indolization since the product apparently had structure **644** (R^1 = R^2 = H, R^3 = $COOC_2H_5$), an alternative synthesis of it being effected by treatment of **643** (R^1 = R^2 = H, R^3 = $COOC_2H_5$) with concentrated sulphuric acid (Walker, C., 1894). At the time, this loss of the elements of ethylamine led to the suggestion that the mechanism of this indolization is 'quite different' from that of those indolizations in which the elements of ammonia are lost (Walker, C., 1894). Compound **644** ($R^1 = R^2 = H, R^3 = COOC_2H_5$) could, however, be formed in this reaction by elimination of the ethyl group from the intermediate 3*H*-indole **647** ($R^1 = H$, $R^3 = C_2H_5$, $R^4 = COOC_2H_5$), analogous to the above mentioned elimination of the cyanoethyl group from **647** [$R^1 = H$, $R^3 = CH_3$, $R^4 = (CH_2)_2CN$] (see also Chapter V, A). A further study of these reactions and of their products should be undertaken.

As with the observations in connection with rule 1, the use of a polyphosphoric acid catalyst in indolizations of **646** (R^3, $R^4 \neq H$) again favoured the involvement of the carbon atom of the methyl group in new C—C bond formation. Thus, under these conditions. **694** furnished **695** (Langlois, Y., Langlois, N. and Potier, 1975).

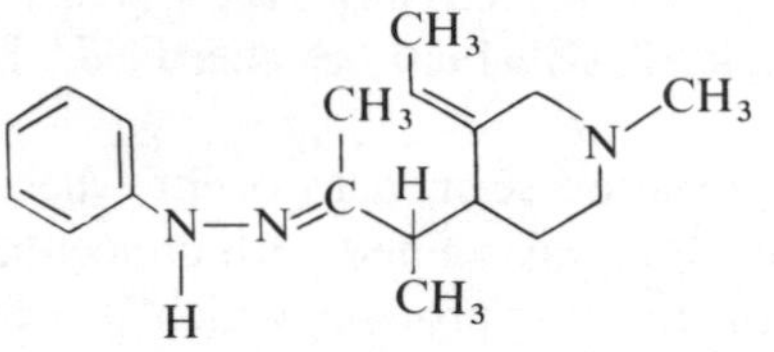

694

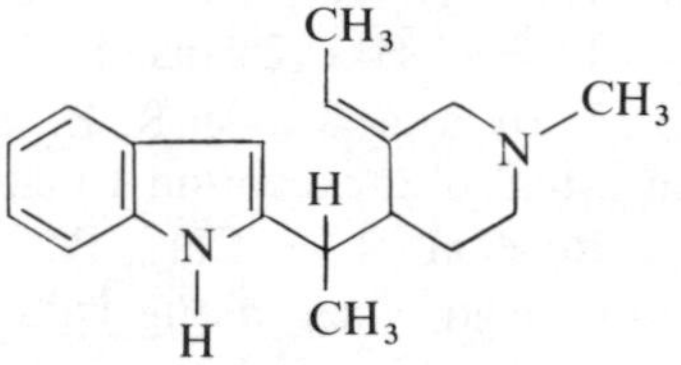

695

When indolizing N_α-substituted derivatives of **646** (i.e. $R^2 \neq H$), new C—C bond formation again appeared to involve the 2-carbon atom of the 2-methinyl rather than that of the methyl group to afford the corresponding 2-methyleneindolines **696** (Brunner, 1898a, 1898b, 1900; Drapkina, Grigor'eva and Doroshina, 1974; Gal'bertshtam and Samoilova, 1973; Gordian and Gal'bershtam, 1971; Krutak, 1975; National Cash Register Co., 1961; Plancher, 1898a, 1898b, 1898g, 1900a, 1902; Sumitomo, Hayakawa and Tsubojima, 1969).

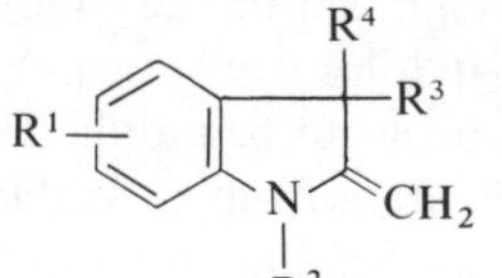

696

In many of the above examples in which only the basic 3*H*-indoles or 2-methyleneindoles were isolated, the reaction mixtures were not examined for the presence of non-basic components which would include indoles.

It was initially reported (Hughes, G. R. and Lions, 1937–1938b) that, using refluxing glacial acetic acid as the catalyst, indolization of **697** ($X = CH_2$) afforded **698** ($X = CH_2$) as the only isolated product (using boiling 10% sulphuric acid instead of glacial acetic acid only caused hydrolysis of the hydrazone). This formation of **698** ($X = CH_2$) is in agreement with the many observations and generalizations mentioned previously. However, it was later

697

698

reported (Adkins and Coonradt, 1941) that using presumably zinc chloride as catalyst, since the indolization method referred to was that of Emil Fischer (1886c), the reaction furnished **699** (X = CH_2) (18% yield) as the only isolated product. Subsequently, further investigations of this indolization and that of **697** (X = NCH_3) under a variety of catalytic conditions were carried out

699

(Lyle, R. E. and Skarlos, 1966), the results of which are given in Table 7. From this it can be seen that a polyphosphoric acid catalyst, in accordance with earlier observations in this chapter, favoured indolization involving the methyl carbon atom rather than the methinyl carbon atom of the arylhydrazone.

Table 7. Variation of Yields with Catalyst in the Fischer Indolization of **697** (X = CH_2 and NCH_3)

		Percentage Yields with Varying Catalysts		
Phenylhydrazone	Product	Polyphosphoric Acid	Acetic Acid	Zinc Chloride
697 (X = CH_2)	**698** (X = CH_2)	5.2	73	67
	699 (X = CH_2)	80	—	—
697 (X = NCH_3)	**698** (X = NCH_3)	2.3	4.9	91
	699 (X = NCH_3)	64	—	—

Although to some extent these data in Table 7 may be analysed in terms of the mono-/diprotonation theory which was subsequently proposed (Illy and Funderburk, 1968; Palmer and McIntyre, 1969) (see below), steric considerations of the two possible enehydrazine tautomeric intermediates in each of these indolizations were offered in explanation at the time these data were reported. Thus, using a 'small' acid (i.e. a proton from polyphosphoric acid) the preferred enehydrazine intermediate would be **700** in which there is no steric interaction between the benzenoid and the cyclohexanoid rings. This would ultimately produce **699** ($X = CH_2$). However, using a 'large' acid [i.e. zinc chloride or acetic acid (the whole molecule of acetic acid was assumed to be involved, probably as an acetylating agent)], the steric repulsion between A ($\equiv ZnCl_2$ or $OCOCH_3$) and the cyclohexanoid ring favours the formation of the enehydrazine **701**, which would ultimately lead to **698** ($X = CH_2$), even though in **701**

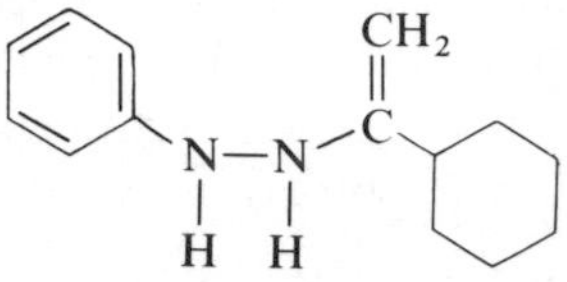

700 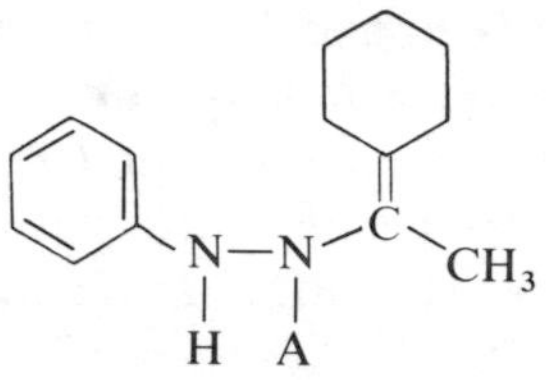 **701**

there would be steric hindrance between either the 2- and 6- or 3- and 5-axial hydrogen atoms on the cyclohexanoid ring and the benzenoid ring when the latter approaches the π-lobes of the double bond of the enehydrazine tautomer, necessary to effect the new C—C bond formation. Thus, attempted prediction of the direction of ambiguous indolizations of ketone arylhydrazones must include, when applicable, considerations of the relative steric strain in the two possible enehydrazines involved and the relative steric hindrance to their subsequent reaction (see Grandberg and Sorokin, 1974), as well as considerations of the stabilization effected by multiple alkyl substitution of the double bond of the enehydrazine tautomer (Lyle, R. E. and Skarlos, 1966). The relative stabilities of the transition states from the enehydrazine intermediates have been estimated (Lyle, R. E. and Skarlos, 1966) by considering the principles of $A^{1,2}$ strain as elaborated with enamines of cyclohexane derivatives (Johnson, F. and Malhotra, 1965; Malhotra and Johnson, F., 1965) and the relative stabilities of **700** and **701** have also (Grandberg and Sorokin, 1974) been related to those of enamines of 2-substituted cyclohexanones (Gurowitz and Joseph, 1965, 1967; Johnson, F. and Whitehead, 1964). However, studies upon the indolization of isopropyl methyl ketone phenylhydrazone (**646**; $R^1 = R^2 = H$, $R^3 = R^4 = CH_3$) showed that, even with 'small' acids as catalysts, 2,3,3-trimethyl-3*H*-indole (**647**; $R^1 = H$, $R^2 = R^3 = CH_3$) could be formed almost exclusively and only when using concentrated strong 'small' acids in considerable excess were appreciable quantities of 2-isopropylindole (**650**; $R^1 = R^2 = H$, $R^3 = R^4 = CH_3$) formed (Tables 8 and 9) (Illy and Funderburk, 1968). From these data it was suggested (Illy and Funderburk, 1968) that in weak acids, monoprotonation of the hydrazone occurred to give the enehydrazine **702** which then ultimately afforded **647** ($R^1 = H$, $R^2 = R^3 = CH_3$) whereas in strong acids [i.e. with Hammett acidity function ($-H_0$) of at least 3.38 when using sulphuric acid as catalyst] diprotonation could occur to form **703** which would ultimately give

Table 8. The Effect of Concentration and Amount of Sulphuric Acid upon the Product Yield from the Indolization of Isopropyl Methyl Ketone Phenylhydrazone (Reaction Temperature = 90 °C)

Percentage Sulphuric Acid	$-H_0$[a]	Mole ratio Sulphuric Acid : **646** ($R^1 = R^2 = H$, $R^3 = R^4 = CH_3$)	Percentage Product Yields **647** ($R^1 = H$, $R^2 = R^3 = CH_3$)	**650** ($R^1 = R^2 = H$, $R^3 = R^4 = CH_3$)
10	0.31	1 : 1	94.8	1.1
		5 : 1	88.5	2.1
20	1.01	1 : 1	93.9	1.4
		5 : 1	93.0	1.8
30	1.72	1 : 1	97.7	<1.0
		5 : 1	91.4	<1.0
40	2.41	1 : 1	96.4	1.0
		5 : 1	91.2	<1.0
50	3.38	1 : 1	97.8	<1.0
		5 : 1	85.3	4.4
60	4.46	1 : 1	98.1	<1.0
		5 : 1	56.1	30.3
70	5.65	1 : 1	96.5	<1.0
		5 : 1	27.1	67.1
78	6.71	1 : 1	91.5	2.1
		5 : 1	20.2	79.0
		6 : 1	12.9	84.0

Note

a. Acidity function.

rise to **650** ($R^1 = R^2 = H$, $R^3 = R^4 = CH_3$). Subsequently (Palmer and McIntyre, 1969), indolization of isopropyl methyl ketone phenylhydrazone was effected with polyphosphoric acid of varying concentrations as the catalyst (Table 10). These results were again interpreted in terms of a mono/diprotonation theory. The formation of the two possible enehydrazine intermediates (≡**702** and **703**) was likened to an E_1 (Saytzeff type → most alkylated olefin) elimination of proton from the monoprotonated hydrazone and an E_2 (Hofmann elimination type → least alkylated olefin) elimination of proton from the diprotonated hydrazone, respectively, as discussed earlier in this chapter.

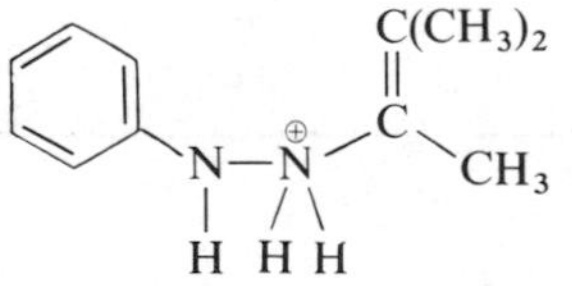

$C_6H_5-NH-\overset{\oplus}{N}H_2-C(=C(CH_3)_2)-CH_3$

702

$[C_6H_5-NH-NH_2-C(=CH_2)-CH(CH_3)_2]^{2\oplus}$ —H

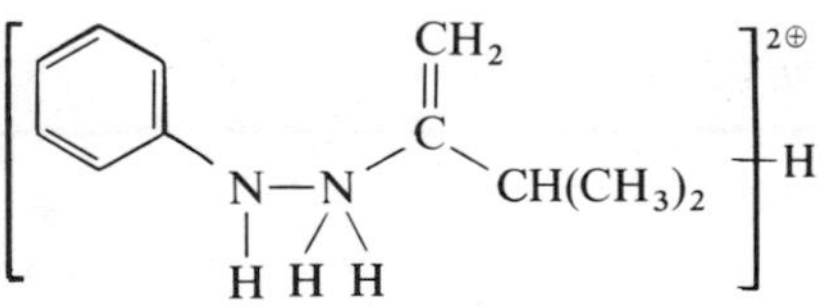

703

Table 9. Effect of Catalyst upon the Product Yield from the Indolization of Isopropyl Methyl Ketone Phenylhydrazone

Catalyst	$-H_0$[a]	Mole Ratio Catalyst : Hydrazone	Percentage Product Yields: **647** ($R^1 = H$, $R^2 = R^3 = CH_3$)	Percentage Product Yields: **650** ($R^1 = R^2 = H$, $R^3 = R^4 = CH_3$)
$ZnCl_2$		1 : 1	87.3	—
		5 : 1	90.4	—
100% CH_3COOH		6 : 1	90.3	—
50% $KHSO_4$	−0.4	5 : 1	98.0	—
75% $KHSO_4$	−0.4	5 : 1	95.2	—
85% H_3PO_4	3.7	5 : 1	97.2	—
PPA	4.80	1 : 1	73.5	14.3
		5 : 1	8.4	72.2
10% HCl	1.00	1 : 1	59.0	—
		5 : 1	91.5	2.1
37% HCl	4.41	1 : 1	81.2	—
		2 : 1	95.4	—
		5 : 1	81.7	10.1
		17 : 1	69.9	23.8

Note

a. Acidity function.

Table 10. Product Ratios from the Indolization of Isopropyl Methyl Ketone Phenylhydrazone with Polyphosphoric Acid of varying Concentrations

Concentration of Phosphorus Pentoxide	**647** ($R^1 = H, R^2 = R^3 = CH_3$)	**650** ($R^1 = R^2 = H, R^3 = R^4 = CH_3$)
66.4[a]	100	0
73.2	74	26
79.8	45	55
83.0	21	79
86.4	17	83

Note

a. 90% (w/w) *ortho*-phosphoric acid.

A limited number of studies have been effected with respect to rule 3 (page 242). Indolization of ethyl propyl ketone phenylhydrazone (**648**; $R^1 = R^3 =$ H, $R^2 = CH_3$, $R^4 = C_2H_5$) would be expected to give rise to both 2,3-diethylindole and 3-methyl-2-propylindole. However, using copper(I) chloride as the indolization catalyst, an oily indolic product was produced from which an apparently homogeneous picrate, m.p. = 144 °C, was obtained (Arbuzov, A. E. and Zaitzev, 1934; Arbusow, A. E., Saizew and Rasumow, 1935). The structure of this picrate was not established although it was later (Leete, 1961a) suggested that it was not identical with unambiguously synthesized 2,3-diethylindole picrate, m.p. = 135–136 °C (Leete, 1961a), an earlier claimed alternative synthesis having furnished a picrate, m.p. = 121–122 °C (Martynov and Martynova, 1954). Even if the picrate of the Fischer indolization product was homogeneous, the formation of both possible isomeric indolic products is by no means eliminated by the experimental data presented. Indeed, the zinc chloride catalysed indolization of ethyl hexyl ketone and hexyl propyl ketone phenylhydrazones [**648**: $R^1 = R^3 =$ H, $R^2 = (CH_2)_4CH_3$, $R^4 = CH_3$ and C_2H_5, respectively] appeared to produce both possible indolic products in each case, 2-hexyl-3-methyl- and 2-ethyl-3-pentylindole and 3-ethyl-2-hexyl- and 3-pentyl-2-propylindole, respectively. However, although one apparently homogeneous picrate was obtained from each of the two reaction products, no separation of isomers was effected in either case and no structural studies were performed upon the isolated picrates (Buu-Hoï and Royer, 1947). The zinc chloride catalysed indolizations of ethyl isobutyl ketone and isobutyl propyl ketone phenylhydrazones (**648**; $R^1 = R^3 =$ H, $R^2 = CH(CH_3)_2$, $R^4 = CH_3$ and C_2H_5, respectively) were claimed (Buu-Hoï and Royer, 1947) to form only one indolic product in each case since only one homogeneous picrate was formed from each reaction mixture. Since the original indolic products were oily liquids and not solids, the products from these reactions were very tenuously formulated as 2-isobutyl-3-methyl- and 3-ethyl-2-isobutylindole, respectively. The only products isolated from indolizations involving arylhydrazone **648** ($R^2 = CH_3$, $R^3 =$ H, $R^4 = CH_2COOH$) (Kögl and Kostermans, 1935; Sarett and Shen, T.-Y., 1966a) and **648** [$R^2 = CH_3$, C_2H_5 and $(CH_2)_2CH_3$, $R^3 =$ H, $R^4 = CH_2COOC(CH_3)_3$] (Merck and Co. Inc., 1965) were the corresponding 3-indolylacetic acids and the *tert*.-butyl esters, respectively **649** [$R^2 = CH_3$, $R^3 = CH_2COOH$; $R^2 = CH_3$, C_2H_5 and $(CH_2)_2CH_3$, $R_3 = CH_2COOC(CH_3)_3$, respectively]. Contrary to these results, reaction of a mixture of N_α-(4-chlorobenzyl)-4-methoxyphenylhydrazine hydrochloride with ethyl 3-propionylpropionate in refluxing ethanol yielded essentially only one product (t.l.c. analysis) which by p.m.r. spectral investigation was shown to be ethyl 3-{2-[1-(4-chlorobenzyl)-5-methoxy-3-methylindolyl]}propionate (Walton, Jenkins, Nutt and Holly, 1968) and the reaction of **648** [$R^1 = R^3 =$ H, $R^2 = CH_3$, $R^4 = (CH_2)_2COOH$] in warm acetic acid afforded products **649** [$R^1 =$ H, $R^2 = CH_3$, $R^3 = (CH_2)_2COOH$] and **704** ($R = CH_3$), these resulting from indolization

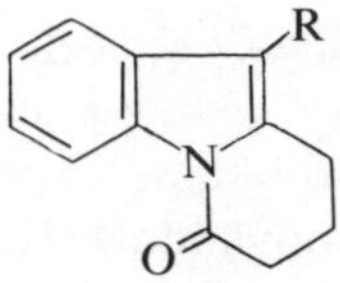

704

in both possible directions. The latter product was produced by cyclization, under the reaction conditions, of the initially formed **649** [R^1 = H, R^2 = $(CH_2)_2COOH$, R^3 = CH_3] (Teuber, Cornelius and Worbs, 1964). In relation to this last indolization, it is interesting that **648** [R^1–R^3 = H, R^4 = $(CH_2)_2COOH$], when similarly treated by warming in acetic acid, yielded only **649** [R^1 = R^2 = H, R^3 = $(CH_2)_2COOH$] (Teuber, Cornelius and Worbs, 1964). It would be interesting to attempt the indolization of this hydrazone using a polyphosphoric acid catalyst in which case indolization might be expected to occur, at least partially, in the alternative direction (i.e. involving new C—C bond formation at the methyl carbon atom). Such catalytic conditions might also be applied, with similar results, to other methyl alkyl ketone arylhydrazones which have as mentioned above, afforded the corresponding 3-substituted 2-methylindoles as the only isolated products subsequent to their indolization. Copper(I) chloride catalysed indolization of benzyl ethyl ketone phenylhydrazone, 4-methylphenylhydrazone and N_α-methylphenylhydrazone afforded 2-benzyl-3-methyl, 2-benzyl-3,5-dimethyl- and 2-benzyl-1,3-dimethylindole, respectively, as the only isolated product in each case (Janetzky and Verkade, 1945). Although the structures of these products were verified by alternative unambiguous syntheses, the yields from the indolizations were low so the concomitant formation of the alternative isomeric indole in each reaction is not precluded. The formation of such products might be expected to be preferred over those actually isolated since the intermediate enehydrazine in their formation would involve the benzenoid ring of the benzyl group in an extended conjugation. In accordance with this postulation, other examples are known in which benzyl ethyl ketone arylhydrazones gave corresponding 2-ethyl-3-phenylindoles as the only isolated products [Aktieselskabet Dumex (Dumex Ltd), 1963; Shvedov, Trofimkin, Vasil'eva and Grinev, A. N., 1975]. Similarly, heating a mixture of 4-methoxyphenylhydrazine hydrochloride and **705** (R^1 = $COOCH_3$, R^2 = Cl, R^3 = CH_3, $n = 2$) in dry methanol gave **706** (R^1 = $COOCH_3$, R^2 = Cl, R^3 = CH_3, R^4 = H, R^5 = OCH_3, $n = 2$) as the only

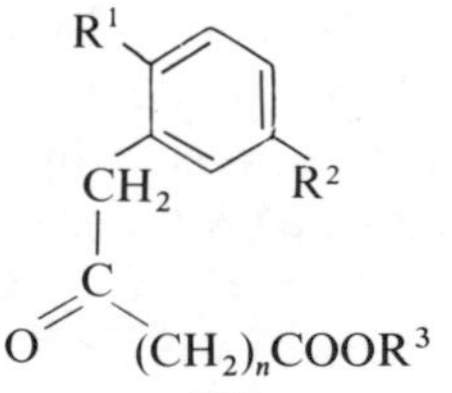

705 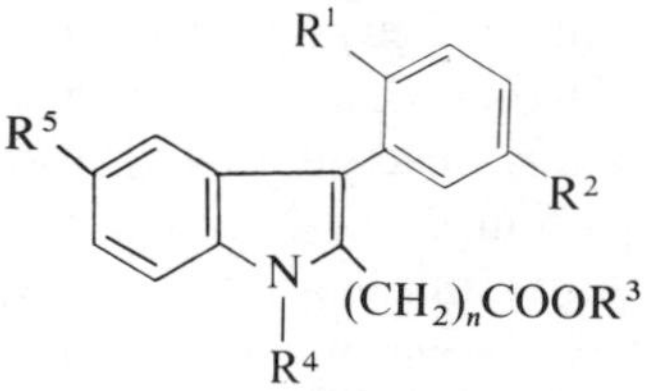 **706**

isolated product [when the indolization of this mixture was attempted using polyphosphoric acid as the catalyst, the only product isolated was **707**, resulting from the intramolecular cyclization of **705** (R^1 = $COOCH_3$, R^2 = Cl, R^3 = CH_3, n = 2)] (Wheeler, 1970) and when mixtures of **705** (R^1–R^3 = H, n = 2)

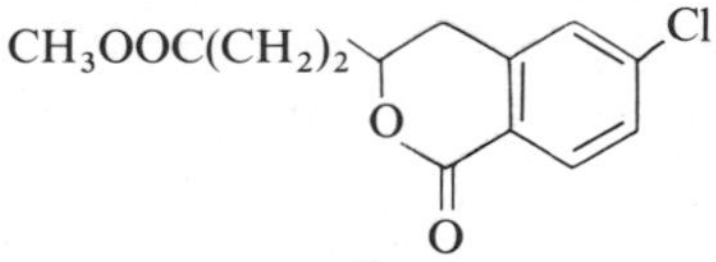

$CH_3OOC(CH_2)_2$ ... Cl

707

and N_α-(4-chlorobenzoyl)-4-methoxyphenylhydrazine hydrochloride or 4-methoxy-N_α-nicotinoylphenylhydrazine were heated in acetic acid, the only products isolated were **706** [R^1–R^3 = H, R^4 = $COC_6H_4Cl(4)$ and nicotinoyl, respectively, R^5 = OCH_3, n = 2] (Nakao, Takahashi, K. and Yamamoto, H., 1971). However, contrary to these observations, the corresponding 2-benzylindoles have been isolated from the indolization of arylhydrazones of the ketones **708** (R^1 = R^2 = H, n = 1) (Shen, T.-Y., 1967c), **708** (R^1 = CH_3, R^2 = H, n = 1) (Sarett and Shen, T.-Y., 1966a) and **708** (R^1 = CH_3, R^2 = C_2H_5, n = 1) (Sarett and Shen, T.-Y., 1966b). Indolization of a mixture of N_α-methylphenylhydrazine hemisulphate and **708** (R^1 = H, R^2 = C_2H_5, n = 1)

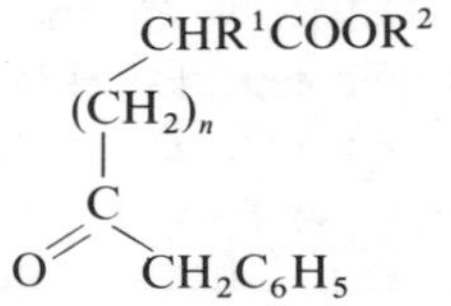

CHR^1COOR^2

$(CH_2)_n$

$O=C-CH_2C_6H_5$

708

afforded **706** (R^1 = R^2 = R^5 = H, R^3 = C_2H_5, R^4 = CH_3, n = 1) (Teotino, 1959; Teotino and Maffii, 1962). Indolizations occurred in similar directions when the N_α-methylphenylhydrazone, N_α-benzylphenylhydrazone and 4-methoxy-N_α-methylphenylhydrazone of **705** (R^1 = R^2 = H, R^3 = CH_3, n = 1) were treated with methanolic sulphuric acid, the major products (77–92% yields) being **706** (R^1 = R^2 = R^5 = H, R^3 = R^4 = CH_3, n = 1; R^1 = R^2 = R^5 = H, R^3 = CH_3, R^4 = $CH_2C_6H_5$, n = 1 and R^1 = R^2 = H, R^3 = R^4 = CH_3, R^5 = OCH_3, n = 1, respectively) and only from the last reaction was the isolation of the alternative isomeric product, **709** (R^1 = OCH_3, R^2 = CH_3), effected, and then in only a 2% yield. However, when using

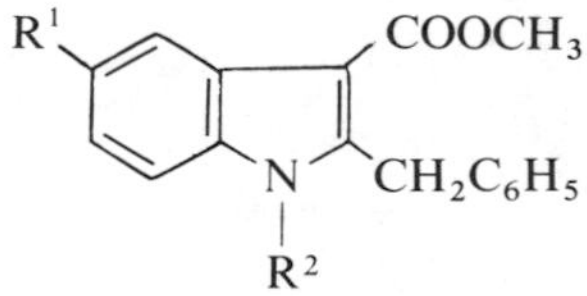

R^1 $COOCH_3$ N $CH_2C_6H_5$ R^2

709

polyphosphoric acid as the catalyst in the indolizations of these three hydrazones, the only isolated products were **710** (R^1 = H, R^2 = CH_3 and $CH_2C_6H_5$ and R^1 = OCH_3, R^2 = CH_3, respectively), formed by cyclization of the

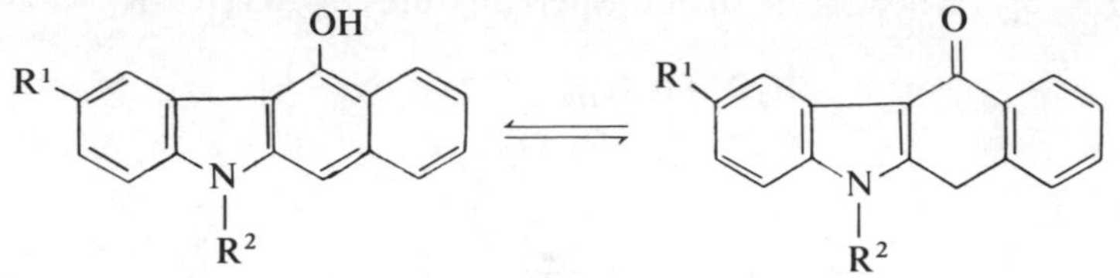

710

corresponding intermediate indolic products **709** (R^1 = H, R^2 = CH_3 and $CH_2C_6H_5$ and R^1 = OCH_3, R^2 = CH_3, respectively) under the reaction conditions. Non-catalytic indolization of the last hydrazone (experimental details were not given) formed predominantly **709** (R^1 = OCH_3, R^2 = CH_3), along with a 1.8% yield of **706** (R^1 = R^2 = H, R^3 = R^4 = CH_3, R^5 = OCH_3, n = 1) (Shvedov, Kurilo and Grinev, A. N., 1972). Other analogous indolizations of arylhydrazones with a similar ketonic moiety are known in which the carbon atom of the methylene group adjacent to the carbomethoxy group is involved in the formation of the new C—C bond (Shvedov, Trofimkin, Vasil'eva and Grinev, A. N., 1975). Mixtures of 4-chloro- and 4-nitrophenylhydrazine with **711** afforded **712** (R = Cl and NO_2, respectively) when refluxed in 18% ethanolic hydrogen chloride (Ishizumi and Mori, 1976). Indolization of mixtures of **713** (R^1 = alkyl, alkoxy, halogen, nitro and carbalkoxy, R^2 = alkyl, aryl, arylalkyl and acyl) with **714** (R^3 = unsubstituted or substituted alkyl, R^4 = alkyl or aryl) afforded the corresponding indoles **715**

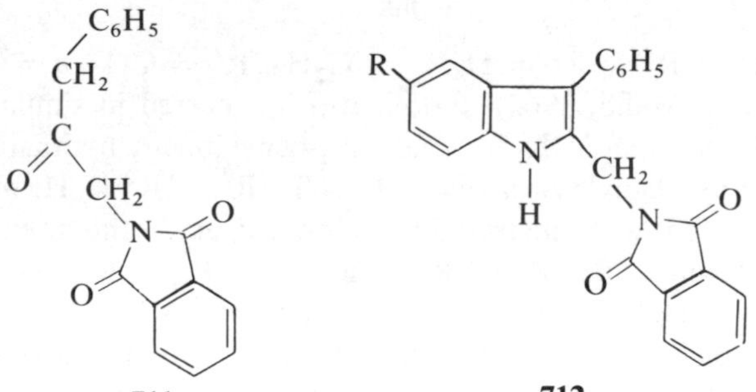

711 **712**

713 **714** **715**

which could be desulphurized using Raney nickel to give the corresponding 3-unsubstituted 2-indolylacetate esters (Trofimov, Tsyshkova and Grenev, 1975). These products would not usually be formed, or would be formed as only the very minor products by comparison with the formation of the other possible isomers, by direct indolization of the corresponding 3-oxobutyric ester arylhydrazones, although yields might be maximized in favour of the 2-indolylacetate ester by using polyphosphoric acid as the indolization catalyst. In more specific examples relating to the above observations, the indolizations of **716** (R^1 = H, R^2 = CH_3 and C_6H_5 and R^1 = OCH_3, R^2 = C_6H_5) in

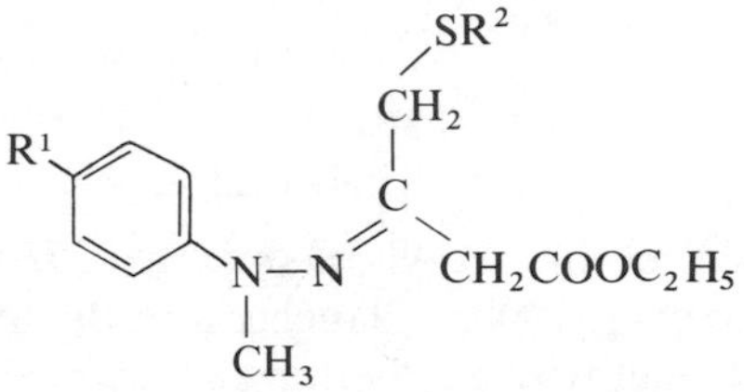

716

alcoholic solutions of formic or sulphuric acids gave **715** (R^1 = H, R^2 = CH_3, R^3 = C_2H_5, R^4 = CH_3 and C_6H_5 and R^1 = 5–OCH_3, R^2 = CH_3, R^3 = C_2H_5, R^4 = C_6H_5, respectively), as the only isolated products. As with the above generalized examples, these could be likewise desulphurized to yield the corresponding 3-unsubstituted indoles (Trofimov, Tsyshkova, Garnova and Grinev, A. N., 1975 – see also Trofimov, Garnova, Grinev, A. N. and Tsyshkova, 1979). However, when the last of these hydrazones was indolized with polyphosphoric acid at 180–200 °C, considerable resinification occurred but the indolic product **717**, resulting from the alternative direction of indolization, was chromatographically detectable in the reaction product (Trofimov, Tsyshkova, Garnova and Grinev, A. N., 1975). The only product reported after reacting **718** with 4-methoxyphenylhydrazine hydrochloride in refluxing dry benzene was **719** (Chemerda, and Sletzinger, 1968a, 1968b). Compounds **721** (R = H and CH_3) were reported as the only products from the indolization of **720** (R = H and CH_3, respectively) (Akopyan and Tatevosyan, 1976). Indolization of ethyl isopropyl ketone phenylhydrazone with alcoholic zinc

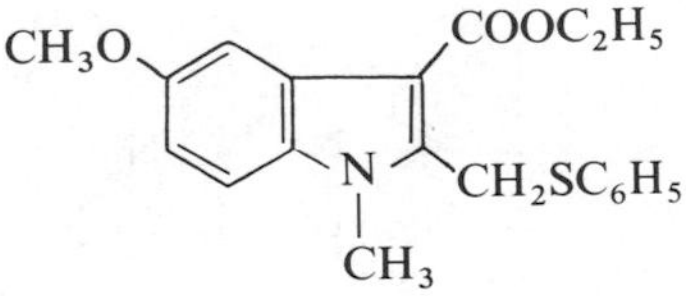

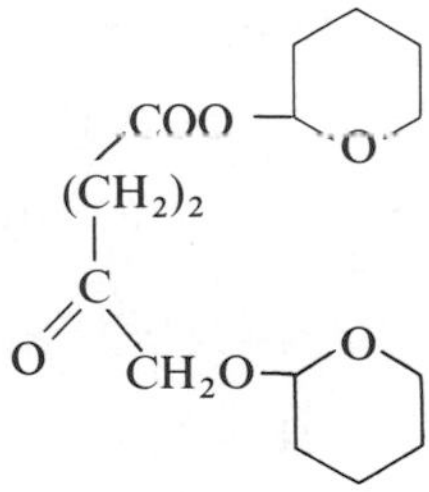

718

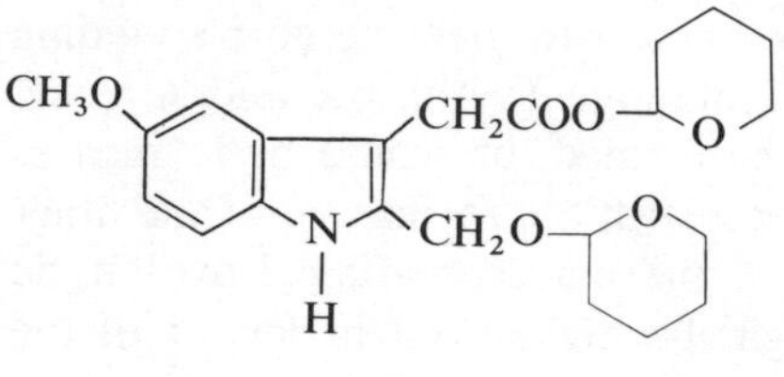

719

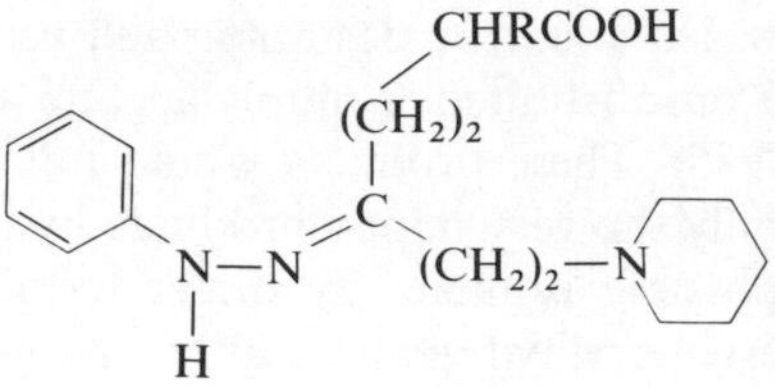

720

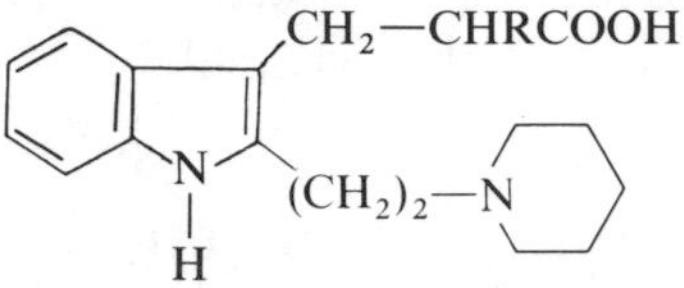

721

chloride produced both 2-ethyl-3,3-dimethyl-3*H*-indole and 2-isopropyl-3-methylindole (Plancher, 1900a; Plancher and Bonavia, 1900, 1902). Using the corresponding N_α-methylphenylhydrazone with alcoholic hydriodic acid as catalyst, the reaction yielded **722** (Plancher, 1900a; Plancher and Bonavia,

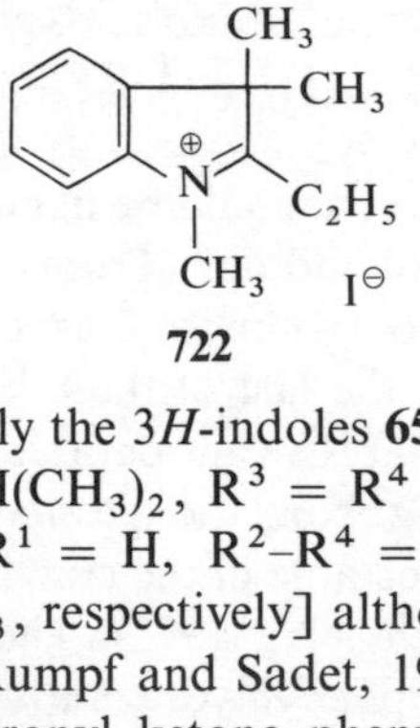

722

1902). It was inferred that only the 3*H*-indoles **651** (R^1 = H, R^2–R^4 = CH_3) and **651** [R^1 = H, R^2 = $CH(CH_3)_2$, R^3 = R^4 = CH_3] were isolated from the indolizations of **648** [R^1 = H, R^2–R^4 = CH_3 and R^1 = H, R^2 = $CH(CH_3)_2$, R^3 = R^4 = CH_3, respectively] although no experimental details were given (Bloch-Chaudé, Rumpf and Sadet, 1955). Zinc chloride catalysed indolization of isobutyl isopropyl ketone phenylhydrazone was claimed to form only 2,3-diisopropylindole but structural verification for the product was lacking (Buu-Hoï and Royer, 1947) even though in this case the product resulting from the alternative direction of indolization would have been the basic 3*H*-indole **651** [R^1 = H, R^2 = $CH(CH_3)_2$, R^3 = R^4 = CH_3]. Similarly, the only compound isolated from the indolization of **723** was **724**

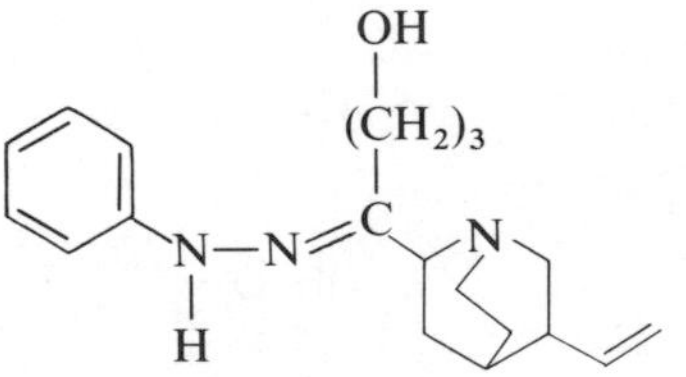

723

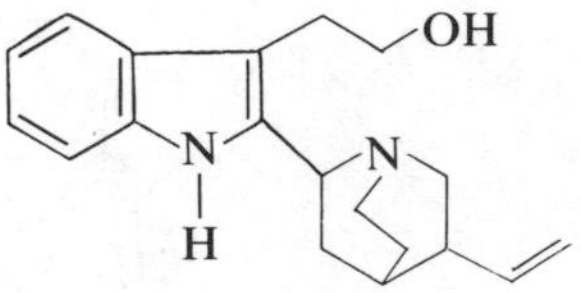

724

(Ch'ang-pai, Evstigneeva and Preobrazhenskii, N., 1958; Ch'ang-pai, Evstigneeva and Preobrazhenskii, N. A., 1960) although it might be expected that a careful investigation of the total reaction mixture by chromatography would also indicate the presence of the corresponding 3*H*-indolic product **725**. Likewise, indolization of **726** or of a mixture of its component ketonic and hydrazino moieties, afforded **727** as the only isolated product (Rodionov and Suvorov, 1950). The formation of the corresponding 3*H*-indolic product in this reaction may however be unlikely because of considerable steric crowding around the C-3 atom in this potential product. Only **651** [R^1 = H, R^2 = $(CH_2)_2OCOCH_3$, $R^3 = R^4 = CH_3$] and **728** were isolated from the indolization of mixtures of **729** with phenyl- and 2-naphthylhydrazine, respectively

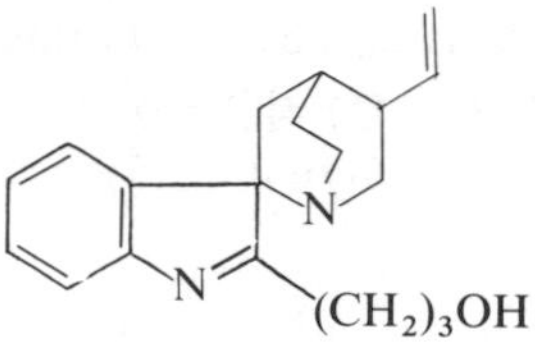

$(CH_2)_3OH$

725

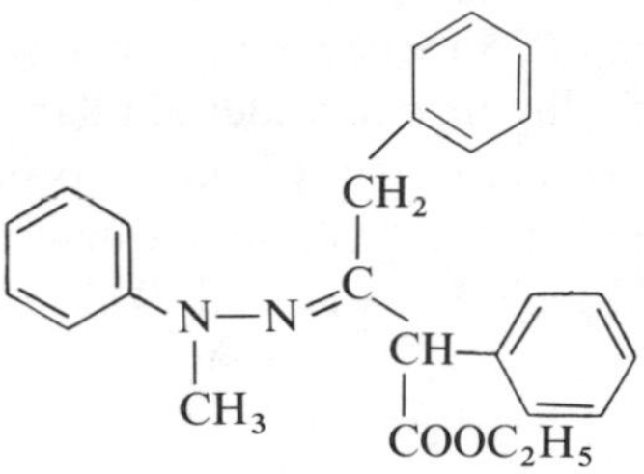

CH_2 C N—N CH CH_3 $COOC_2H_5$

726

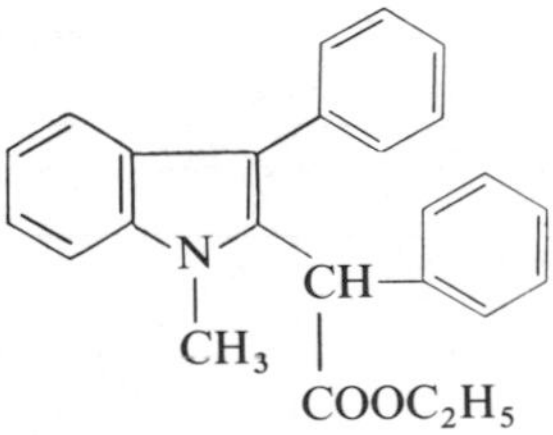

N CH_3 CH $COOC_2H_5$

727

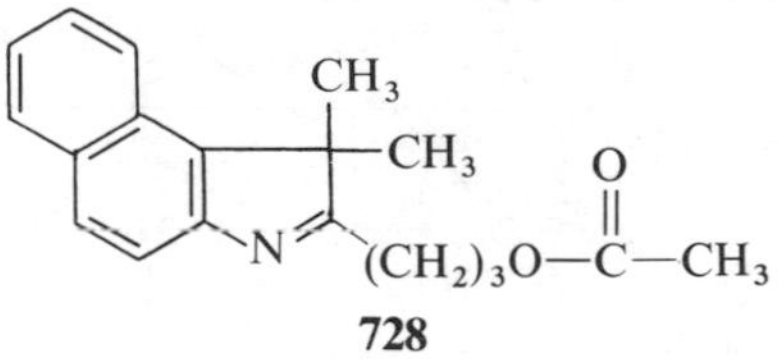

CH_3 CH_3 O N $(CH_2)_3O—C—CH_3$

728

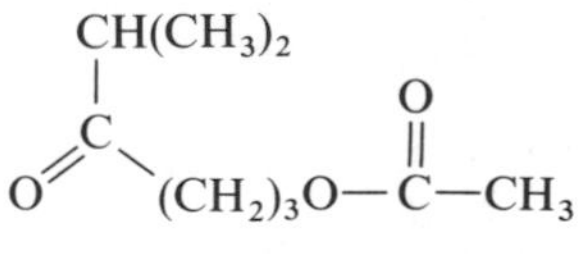

$CH(CH_3)_2$ C O O $(CH_2)_3O—C—CH_3$

729

(Oliver, G. L., 1968). Clearly, this aspect of the Fischer indolization requires further study. Total reaction products should be carefully analysed using chromatographic and/or spectroscopic techniques and product structures should be firmly established. Although in the majority of such reactions both possible products might be expected to be formed, the relationship between the nature of the indolization catalyst upon the yield ratio of the two products should be investigated.

Heating mixtures of phenylhydrazine and **730** [$R^1 = CH_3$, $R^2 = C_2H_5$, $(CH_2)_2CH_3$ and $(CH_2)_5CH_3$ and $R^1 = R^2 = C_2H_5$] initiated exothermic reactions and resulted in each case in the formation of two products, **734** and **735** [$R^1 = CH_3$, $R^2 = C_2H_5$, $(CH_2)_2CH_3$ and $(CH_2)_5CH_3$ and $R^1 = R^2 = C_2H_5$, respectively], both isolated in approximately equal yields. Presumably the initial adducts existed as the corresponding tautomeric structures **731** and **732** [$R^1 = CH_3$, $R^2 = C_2H_5$, $(CH_2)_2CH_3$ and $(CH_2)_5CH_3$ and $R^1 = R^2 = C_2H_5$, respectively]. The former underwent intramolecular nucleophilic attack as shown, to yield the corresponding **734**, rather than an apparently energetically considerably less favourable [3,3]-sigmatropic rearrangement which would have ultimately yielded the corresponding 3-cyanoindoles **733**. The latter tautomer, however, underwent normal Fischer indolization to afford the corresponding **735** (Scheme 34). Reaction of **730** [$R^1 = H$, $R^2 = (CH_2)_2CH_3$] with phenylhydrazine under similar conditions afforded **734** [$R^1 = H$, $R^2 = (CH_2)_2CH_3$] as the only isolated product, presumably because in this case the unconjugated intermediate adduct **732** rearranged very quickly to the conjugated adduct **731** (Landor, S. R., Landor, P. D., Fomum and Mpango, 1977).

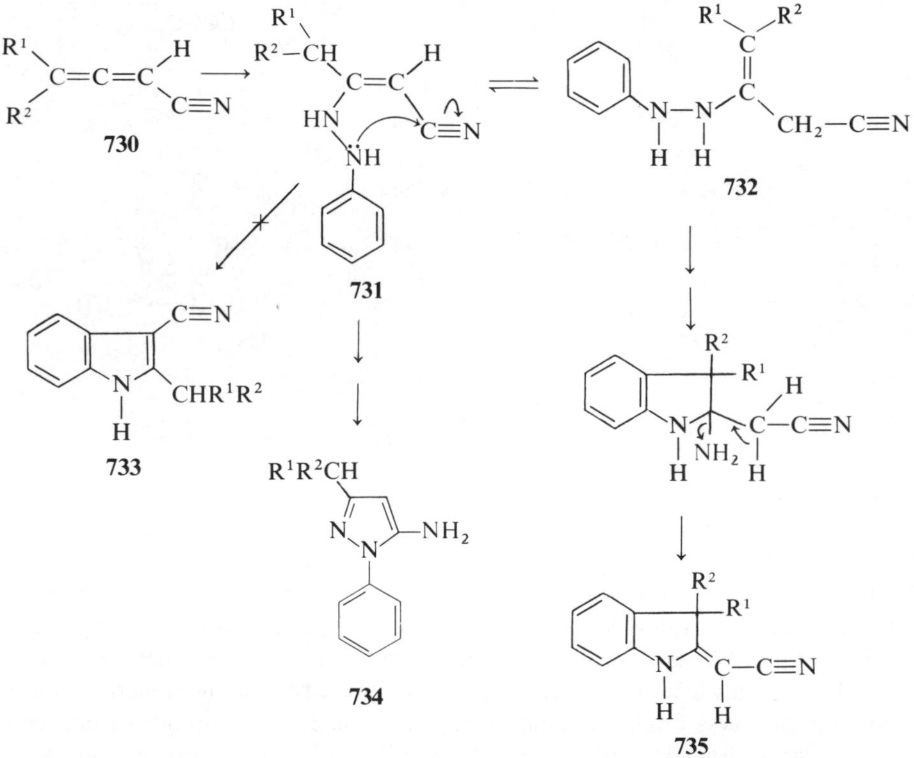

Scheme 34

H. Indolization of 2-Substituted Cycloalkanone Arylhydrazones

Indolizations of these compounds, **736** ($R^2 \neq H$) can give rise to neutral indoles **738** ($R^2 \neq H$), basic 3*H*-indoles **737** and their further transformation products, the composition of the reaction product depending upon the nature of the indolization catalyst, the reaction conditions and the nature of the cycloalkanoid 2-substituent, R^2. Furthermore, the appropriately functionalized R^2 group in **738** can further react with the indolic N atom under the indolization conditions (Scheme 35).

736

737

Elimination of R^2 and other further reactions

738

Interaction between N and R^2

Scheme 35

Early examples of indolizations in this area involved the following reactions. Use of alcoholic zinc chloride (Plancher, 1900b; Plancher and Testoni, 1900), either methyl magnesium iodide or phenyl magnesium bromide (Grammaticakis, 1939, 1940b) and sulphuric acid (Grammaticakis, 1940b; Grandberg, Kost and Yaguzhinskii, 1960; Pausacker and Schubert, 1949a; Plancher and Ghigi, 1929) as catalysts in the indolization of 2-methylcyclohexanone phenylhydrazone (**736**, $R^1 = R^3 = H$, $R^2 = CH_3$, $n = 2$) afforded in each case a mixture of **737** ($R^1 = R^3 = H$, $R^2 = CH_3$, $n = 2$) and **738** ($R^1 = R^3 = H$, $R^2 = CH_3$, $n = 2$), the former of these products being later (Fritz and Stock, 1969) resolved and the absolute configuration of each of the optical isomers being established. Use of glacial acetic acid (Lions, 1937–1938), polyphosphate ester (Kanaoka, Ban, Miyashita, Irie and Yonemitsu, 1966) and sulphuric acid (Plancher and Ghigi, 1929) as catalysts in the indolization of 2-ethylcyclohexanone phenylhydrazone (**736**, $R^1 = R^3 = H$, $R^2 = C_2H_5$, $n = 2$) yielded both **737** (R^1 = R^3 = H, R^2 = C_2H_5, n = 2) and **738** (R^1 = R^3 = H, R^2 = C_2H_5, n = 2) in each case, indolization of the corresponding N_α-ethylphenylhydrazone with a sulphuric acid catalyst giving a mixture of 1,9-diethyl-1,2,3,4-tetrahydrocarbazole and **57** (R = C_2H_5) (Plancher and Ghigi, 1929) [for a preparation of 2-ethylcyclohexanone, see Witkop and Patrick (1953)]. Use of sulphuric acid as catalyst converted 2-methylcyclohexanone 2-methylphenylhydrazone into a mixture of **737** (R^1 = 8-CH_3, R^2 = CH_3, R^3 = H, n = 2) and **738** (R^1 = 8-CH_3, R^2 = CH_3, R^3 = H, n = 2) (Bajwa and Brown, R. K., 1970; Pausacker and Schubert, 1949b). Use of ethanolic hydrogen chloride converted 2-methylcyclohexanone and 2,5-dimethylcyclohexanone 4-methoxyphenylhydrazone into mixtures of **737** (R^1 = 6-OCH_3, R^2 = CH_3, R^3 = H, n = 2) and **738** (R^1 = 6-OCH_3, R^2 = CH_3, R^3 = H, n = 2) and **737** (R^1 = 6-OCH_3, R^2 = R^3 = CH_3, n = 2) and **738** (R^1 = 6-OCH_3, R^2 = R^3 = CH_3, n = 2), respectively (Gilbert, J., Rousselle, Ganser and Viel, 1979; Rousselle, Gilbert, J. and Viel, 1977). Use of glacial acetic acid effected the conversion of 2-phenylcyclohexanone phenylhydrazone (**736**; R^1 = R^3 = H, R^2 = C_6H_5, n = 2) into a mixture of **737** (R^1 = R^3 = H, R^2 = C_6H_5, n = 2) and **738** (R^1 = R^3 = H, R^2 = C_6H_5, n = 2) (Green and Ritchie, 1949). Use of alcoholic zinc chloride converted menthone[(−)-2-isopropyl-5-methylcyclohexanone] phenylhydrazone [**736**, R^1 = H, R^2 = $CH(CH_3)_2$, R^3 = CH_3, n = 2] into a mixture of **737** (R^1 = H, R^2 = $CH(CH_3)_2$, R^3 = CH_3, n = 2], b.p. = 170–171 °C/14 mm (picrate m.p. = 166–167 °C and methiodide, m.p. = 209–21 °C) and **738** [R^1 = H, R^2 = $CH(CH_3)_2$, R^3 = CH_3, n = 2], b.p. = 202–204 °C/14 mm (picrate m.p. = 164–165 °C) (Plancher and Carrasco, 1904). Subsequent to this indolization of menthone phenylhydrazone, indolization using a dilute sulphuric acid catalyst furnished only **738** [R^1 = H, R^2 = $CH(CH_3)_2$, R^3 = CH_3, n = 2], isolated as a crystalline solid, m.p. = 114.5 °C (Borsche, Witte, A. and Bothe, 1908). An even later (Kuroda, 1923) repetition of this indolization, again using zinc chloride as the catalyst, afforded a non-basic oil, 'menthoindole' ($C_{16}H_{21}N$), b.p. = 213 °C/20 mm (presumably 1,2,3,4-tetrahydro-1-isopropyl-4-methylcarbazole) and a basic oil, $C_{16}H_{23}N$, b.p. = 181 °C/18 mm which is likely to be **737** [R^1 = H, R^2 = $CH(CH_3)_2$, R^3 = CH_3, n = 2], the elemental analysis probably being unsatisfactory. Further clarification of the indolization of menthone phenylhydrazone is certainly required. Likewise, treatment of 2-methylcyclohexanone 1-naphthylhydrazone (**739**) with dilute sulphuric acid afforded a mixture of **740** and **741** (Cecchetti and Ghigi, 1930) and the

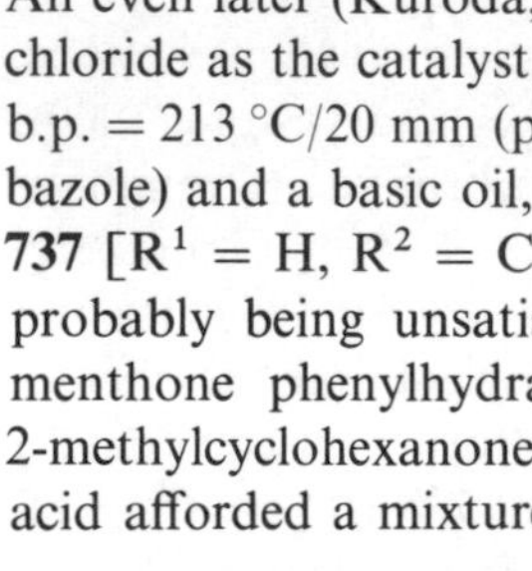

CH_3 N—N H

739

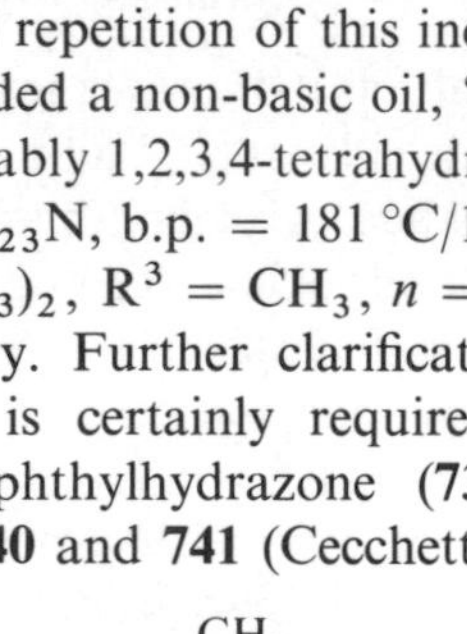

CH_3 N

740

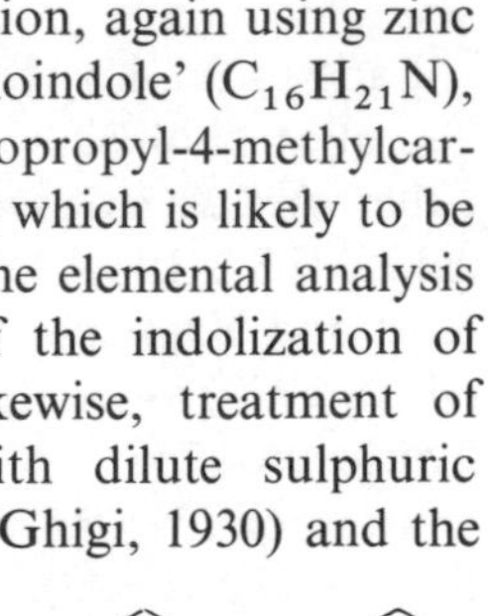

N H CH_3

741

corresponding 2-naphthylhydrazone [**606**; R^1–R^3 = H, $R^4 + R^5$ = $CH(CH_3)(CH_2)_3$, X–Z = CH] afforded a basic and a non-basic product which were initially formulated as **610** and **607** [R^1 = H, $R^2 + R^3$ = $CH(CH_3)(CH_2)_3$, X = CH], respectively (Cecchetti and Ghigi, 1930). Later (Bryant and Plant, 1931), the structure of the latter product from the second reaction was corrected to **608** [$R^1 = R^2$ = H, $R^3 + R^4$ = $CH(CH_3)(CH_2)_3$, X–Z = CH] and, by analogy, **610** was reformulated as **611** (see earlier in this chapter).

In other examples, only the corresponding **737** (n = 2) was isolated starting from arylhydrazones of 2-methylcyclohexanone (Buu-Hoï, Jacquignon and Loc, 1958; Deorha and Joshi, S. S., 1961; Ender, F., Moisar, Schäfer, K. and Teuber, 1959; Sakai, S., Wakabayashi and Nishina, 1969), 2-ethylcyclohexanone (Kanaoka, Ban, Yonemitsu, Irie and Miyashita, 1965), 2-propyl, 2-butyl-, 2-hexyl, 2-octyl- and 2-cyclohexylcyclohexanone (Buu-Hoï, Jacquignon and Loc, 1958), 2-alkylcyclohexanones (Plancher, Testoni and Olivari, 1920), 2-phenylcyclohexanone (Nakazaki, Yamamoto, K. and Yamagami, 1960) and 2-benzylcyclohexanone (Nakazaki and Isoe, 1955) or only the corresponding **738** (n = 2) were isolated starting from arylhydrazones of 2-methylcyclohexanone (Esayan, Terzyan, Astratyan, Dzhanpoladyan and Tatevosyan, 1968; Gupta, Jetley, Rani and Malik, 1975; Kost, Yudin and Terent'ev, A. N., 1959), 2-ethylcyclohexanone (Witkop and Patrick, 1951a), 2,4-dimethylcyclohexanone (Gupta, Jetley, Rani and Malik, 1975), 2-cyclohexylcyclohexanone (Berlin, Clark, P. E., Schroeder, J. T. and Hopper, 1968; Buu-Hoï, Binh, Loc, Xuong and Jacquignon, 1957) [later (Buu-Hoï, Jacquignon and Loc, 1958), **737** ($R^1 = R^3$ = H, R^2 = cyclohexyl, n = 2) was also claimed as a by-product from this reaction], 2-phenyl-4-*tert*.-butyl- and 4-methyl-2-(2-methylphenyl)-cyclohexanone (Bozzini, Gratton, Pellizer, Risaliti and Stener, 1979), 2-ethyl-3-methyl- and 3-ethyl-2-methylcyclohexanone (Miller, B. and Matjeka, 1977, 1980) and 3-carboxy-4-ethyl-2-methylcyclohexanone (Buchi and Warnhoff, 1959). Similarly, only **738** [$R^1 = R^3$ = H, R^2 = $(CH_2)_3OH$, n = 2] was isolated from the indolization of a mixture of **742** with phenylhydrazine using

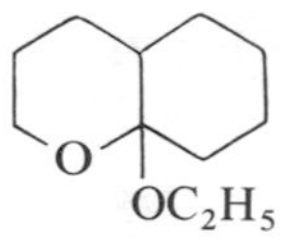

O OC_2H_5

742

an aqueous sulphuric acid catalyst (Dolby and Esfandiari, 1972), although closely-related reactions using refluxing acetic acid as the catalyst led to the initial formation of the corresponding 3*H*-indolium cations (Britten, A. Z., Bardsley and Hill, C. M., 1971) (see Chapter IV. G). and treatment of **641** (R = CH_3) with dilute sulphuric acid gave **642** (R = CH_3) (Manske, R. H. F. and Kulka, 1949). Although only one of the two possible isomeric products was isolated from these indolizations, it is reasonable to assume that both were formed but that 'work-up' conditions afforded only one of them. In fact, the basic and neutral properties of **737** and **738**, respectively, facilitate their ready separation. Thus, after the indolizations of **736** ($R^1 = R^3 = H$, $R^2 = C_6H_5$, $n = 2$) and **743** using ethanol saturated with hydrogen chloride as catalyst, the non-basic products, which would include the corresponding indoles, were removed by ether extraction of the reaction mixture and only the corresponding basic products **737** ($R^1 = R^3 = H$, $R^2 = C_6H_5$, $n = 2$) and **744**, respectively, were isolated (Nakazaki, Yamamoto, K. and Yamagami,

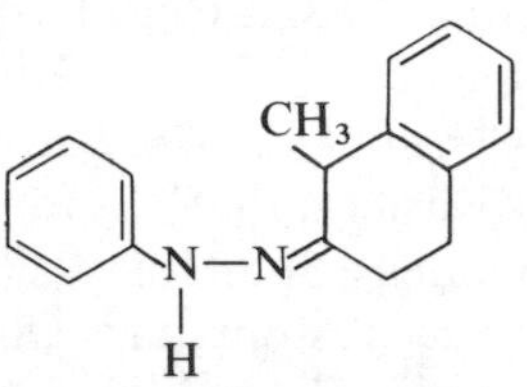

CH₃ N—N H **743**   CH₃ N **744**

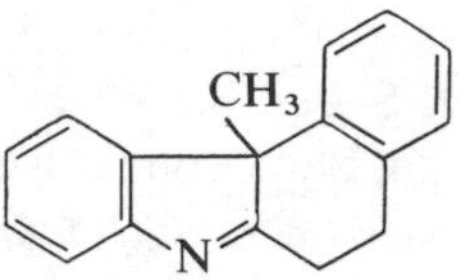

1960). A similar separation allowed the isolation of 1,2,3,4-tetrahydro-1,1-dimethylcarbazole as the neutral product resulting from the indolization of a mixture of 2,2- and 2,6-dimethylcyclohexanone phenylhydrazone (Robinson, B. and Smith, G. F., 1960). Likewise, only the corresponding furano- or pyrano-indoline ring systems, formed by intramolecular cyclisation of the corresponding 3*H*-indolium cations, were isolated after mixtures of 2-(2-hydroxyethyl)cyclohexanone diethyl ketal with phenylhydrazine, 2-(2-hydroxy-ethyl(cyclohexanone ethylene ketal with N_α-methylphenylhydrazine and 2-(2-hydroxypropyl)cyclohexanone diethyl ketal with phenylhydrazine or 2-(3-hydroxypropyl)cyclohexanone diethyl ketal with phenylhydrazine or N_α-methylphenylhydrazine, respectively, were refluxed in glacial acetic acid (Britten, A. Z., Bardsley and Hill, C. M., 1971) (see Chapter IV, G).

Surprisingly, attempted indolizations of mixtures of 2-methylcyclohexanone and menthone with **745** were unsuccessful, it being concluded that the presence

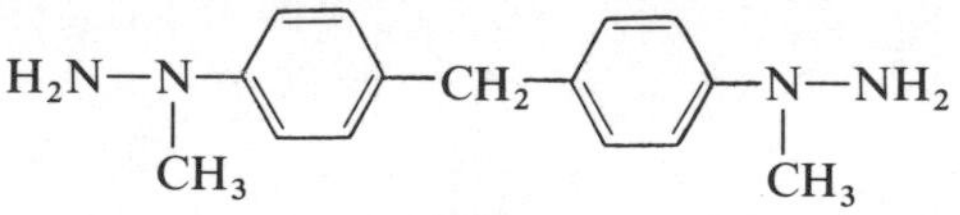

H_2N—N—⟨⟩—CH_2—⟨⟩—N—NH_2 with CH_3 on each N

745

of a CH_2COCH_2 group in the ketonic moiety was necessary for the occurrence of indolization using this particular arylhydrazine (Braun, 1908b). The reinvestigation of these indolizations should certainly be undertaken since, with the appropriate catalyst(s), they should be successful.

Several investigations have been focused upon the relationship between the nature of the indolization catalyst or the nature of the 2-substituent in the cyclohexanone moiety and the ratio of **737** ($n = 2$) : **738** ($n = 2$) formed as in

Scheme 35 or in closely related reactions. Thus, the formation of the corresponding indolic products was favoured when the indolizations of 2-methylcyclohexanone 2-, 3- and 4-methoxy-N_α-methylphenylhydrazone were effected using sulphuric acid as the catalyst whereas, using a glacial acetic acid catalyst, the corresponding compounds **746** were predominantly formed although

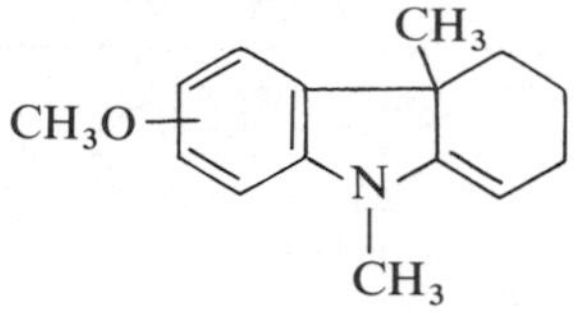

746

these were then reduced, without purification, to afford the corresponding 1,2,3,4,4a,9a-hexahydrocarbazoles (Millson and Robinson, R., 1955). Using concentrated hydrochloric acid in glacial acetic acid as catalyst, indolization of **736** [$R^1 = R^3 = H$, $R^2 = (CH_2)_{1-4,6,8}H$, $n = 2$] afforded as the only detectable products the corresponding 3*H*-indoles **737** [$R^1 = R^3 = H$, $R^2 = (CH_2)_{1-4,6,8}H$, $n = 2$, respectively] (Buu-Hoï, Jacquignon and Loc, 1958). Likewise, formic acid yielded predominantly the 3*H*-indole **737** ($R^1 = R^3 = H$, $R^2 = C_2H_5$, $n = 2$), although this was only isolated in low yield (3.3%) since it reacted with the formic acid under the indolization conditions to give 4a-ethyl-9-formyl-1,2,3,4,4a,9a-hexahydrocarbazole (**747**) (63.7%). The

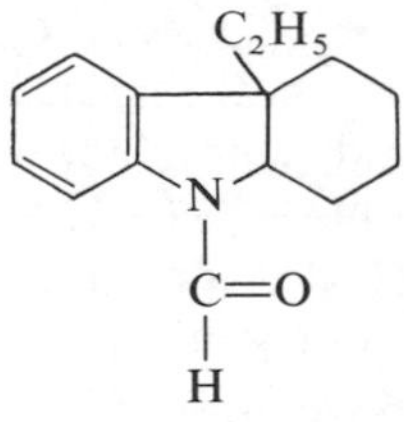

747

formation of a small amount of **738** ($R^1 = R^3 = H$, $R^2 = C_2H_5$, $n = 2$) from this indolization was detected by u.v. spectral analysis (Ban, Oishi, Kishio and Iijima, I., 1967) but this compound could not be isolated from the reaction mixture. Starting with **736** [$R^1 = R^3 = H$, $R^2 = CH_3$, C_2H_5, $CH(CH_3)_2$, cyclohexyl and C_6H_5] and using dilute sulphuric acid and glacial acetic acid as indolization catalysts, it was found that the corresponding **737** : **738** yield ratios were <1 and >1, respectively (Table 11) (Pausacker, 1950a). Also included in Table 11 are similar comparative results obtained starting from some of these hydrazones and using the non-catalytic thermal technique (i.e. refluxing the hydrazone in a solvent such as monoethylene glycol, diethylene glycol or tetralin) (Kelly, A. H., McLeod and Parrick, 1965).

The investigation of the **737** ($R^1 = R^3 = H$, $R^2 = CH_3$, $n = 2$) : **738** ($R^1 = R^3 = H$, $R^2 = CH_3$, $n = 2$) yield ratio in the product obtained by

Table 11. Directional Effects of the 2-Substituent upon the Indolization of 2-Substituted Cyclohexanone Phenylhydrazones under Three Different Cyclization Conditions

Starting Hydrazone **736** ($R^1 = R^3 = H, n = 2$) R^2	Corresponding Yield (%) or Isomer Ratios **737** : **738** ($R^1 = R^3 = H, n = 2$) from the Product		
	Catalyst Used		
	20% Sulphuric Acid[a]	Acetic Acid[a]	Thermal Cyclization[b]
CH_3	21 : 45 (28 : 50)[c]	61 : 6 (61 : 7)[c]	51 : 49
C_2H_5	28 : 44[d]	76 : 13 (15 : 2)[e]	—
$(CH_2)_2CH_3$	—	—	28 : 72
$CH(CH_3)_2$	18 : 44	70 : 16	—
$CH(CH_2)_5$	17 : 29	79 : 9	4 : 96
C_6H_5	19 : 54	69 : 27	13 : 87

Notes

a. Yield (%) ratios; b. Isomer ratios; c. Ratio of percentage yields obtained by McLean, J., McLean, S. and Reed, R. I. (1955); d. Using polyphosphate ester as indolization catalyst, the **737** : **738** yield ratio = 48 : 21, as percentage product yields (Kanaoka, Ban, Miyashita, Irie and Yonemitsu, 1966); e. Ratio of yields obtained by Lions, (1937–1938).

indolization of **736** ($R^1 = R^3 = H, R^2 = CH_3, n = 2$) under varying catalytic conditions has also been effected (Table 12) (Pausacker, 1950a). A similar comparative study using **736** [$R^1 = R^3 = H, R^2 = (CH_2)_3CH_3, n = 2$] and hydrogen chloride in acetic acid, zinc chloride and polyphosphoric acid as catalysts afforded in each case both **737** [$R^1 = R^3 = H, R^2 = (CH_2)_3CH_3, n = 2$] and **738** [$R^1 = R^3 = H, R^2 = (CH_2)_3CH_3, n = 2$] in the ratios 9 : 7, 1.3 : 2.8 and 5 : 6, respectively (Buu-Hoï, Jacquignon and Béranger, 1968).

From the results of these above studies it can be seen that several factors govern the direction of indolization of **736** into **737** and/or **738**. Thus, the use of 'large' acids, in particular acetic acid, mono-, di- and trichloroacetic acids and an acetic acid–hydrochloric acid mixture as catalysts favours the formation of **737**, often exclusively (Table 12), whereas using 'small' acids such as aqueous hydrochloric acid and sulphuric acid as catalysts affords both **737** and **738** with the latter product predominating. These results are as expected considering acid 'size' in relation to the stability of the two possible enamine intermediates as discussed earlier (Lyle, R. E. and Skarlos, 1966). The steric effect of the R^2 group in determining the direction of the indolization is also apparent from the data presented in Table 11, the neighbouring presence of a bulky R^2 group probably sterically retarding the new C—C bond formation in the indolization. The suggestion (Grandberg and Sorokin, 1974) that the data presented in Tables 11 and 12 are not in accord with these steric considerations does not appear to be valid. The strength of the acid catalyst is also likely to be involved

Table 12. Directional Effects of the Acid Catalyst upon the Indolization of 2-Methylcyclohexanone Phenylhydrazone

Catalyst	Solvent	% Yield of **737** ($R^1 = R^3 = H, R^2 = CH_3$, $n = 2$)	% Yield of **738** ($R^1 = R^3 = H, R^2 = CH_3$, $n = 2$)
CH_3COOH	CH_3COOH	61	6
CH_3COOH	H_2O	56–40[a]	0
$ClCH_2COOH$	H_2O	42	0
$Cl_2CHCOOH$	H_2O	51	0
Cl_3CCOOH	H_2O	53	0
H_2SO_4	H_2O[b]	21	45
H_2SO_4	C_2H_5OH	36	33
HCl(conc.)	H_2O	38	40
HCl (gas)	C_2H_5OH	44	19
HCl (gas)	C_6H_6[c]	65	4

Notes

a. The yields varied with the concentration of the aqueous acetic acid solution; b. Using the same catalyst at the same concentration (10%) under similar reflux conditions it was later reported by other workers (Grandberg, Kost and Yaguzhinskii, 1960) that this reaction yielded **737** ($R^1 = R^3 = H$, $R^2 = CH_3$, $n = 2$) (54%) and **738** ($R^1 = R^3 = H$, $R^2 = CH_3$, $n = 2$) (19%) (i.e. a marked reversal of the tabulated observation); c. Continuously saturated with dry hydrogen chloride.

in determining the relative yields of **737** and **738**, by the effect that it would have upon the formation of the two possible enehydrazine intermediates via mono- or diprotonation of the arylhydrazone as described earlier (Illy and Funderburk, 1968; Palmer and McIntyre, 1969).

From the above studies upon the indolization of both asymmetric cyclic and acyclic ketone arylhydrazones it can be concluded that, when using weaker acids or low acid concentrations, indolization towards the more branched C-2 atom of the ketonic moiety predominated whereas with strong acids or high acid concentrations, indolization predominantly involved the less branched C-2 atom. Further studies concerning 2-alkylcyclohexanone phenylhydrazone indolizations with subsequent g.c. analyses of the total reaction product (Miller, F. M. and Schinske, 1978) have corroborated this conclusion. Thus, the **737** ($R^1 = R^3 = H, R^2 = CH_3, n = 2$) : **738** ($R^1 = R^3 = H, R^2 = CH_3, n = 2$) yield ratios observed when 2-methylcyclohexanone phenylhydrazone was indolized with various acids at 80 °C have been determined and are presented in Table 13. Using various concentrations of sulphuric acid in ethanol as catalyst, these ratios were also determined and are given in Table 14, the reason why the indole formation was maximal when using *ca.* 40% ethanolic sulphuric acid remaining obscure. Using 10% sulphuric acid in various alcohols as catalyst, these product ratios varied as shown in Table 15. The increasing size of the

Table 13. The Effect of the Indolization Catalyst at 80 °C upon the Product Ratio from the Indolization of 2-Methylcyclohexanone Phenylhydrazone

Catalyst	$-H_0$[a]	Ratio[b]
CH_3COOH	*ca.* −2.5	40
10% (w/w) H_2SO_4 in C_2H_5OH	0.43	1.8
BF_3/C_2H_5OH	7–10	0.9
$ZnCl_2/C_2H_5OH$	—	0.3
PPA	*ca.* 7	0.2

Notes

a. Acidity function; b. **737** ($R^1 = R^3 = H$, $R^2 = CH_3$, $n = 2$) : **738** ($R^1 = R^3 = H$, $R^2 = CH_3$, $n = 2$) yield ratio.

Table 14. The Effect of Acid Concentration at 80 °C upon the Product Ratio from the Indolization of 2-Methylcyclohexanone Phenylhydrazone

% (w/w) H_2SO_4 in C_2H_5OH	$-H_0$[a]	Ratio[a]
10	0.43	1.8
20	1.10	0.8
40	2.54	0.4
60	4.51	0.5
80	7.52	0.7

Note

a. See footnotes in Table 13.

Table 15. The Effect of Alcoholic Solvent at 80 °C upon the Product Ratio from the 10% (w/w) Sulphuric Acid Catalysed Indolization of 2-Methylcyclohexanone Phenylhydrazone

Solvent Alcohol	Ratio[a]
CH_3OH	1.4
C_2H_5OH	1.8
$(CH_3)_2CHOH$	2.2
$(CH_3)_3COH$	3.1

Note

a. See note b in Table 13.

alkyl group in the protonated alcohol, $R—\overset{\oplus}{O}H_2$, reduced the effectiveness of the acid in protonating the intermediate in indole formation and therefore, as expected by analogy with the above observations, the ratio increased as the size of R increased. The effect of reaction temperature upon this ratio was investigated using a polyphosphoric acid catalyst and the results are given in Table 16. These results are in contrast to the variations of this ratio observed under thermal non-catalytic conditions (heating in diethylene glycol) (Table 17).

Table 16. Temperature Effects upon the Product Ratio obtained from the Polyphosphoric Acid Catalyzed Indolization of 2-Methylcyclohexanone Phenylhydrazone

Temperature (°C)	Ratio[a]
80	0.2
125	0.4
160	1.0

Note

a. See note b in Table 13.

Table 17. Temperature Effects upon the Product Ratio from the Thermal Indolization of 2-Methylcyclohexanone Phenylhydrazone

Temperature (°C)	Ratio[a]
155	2.0
200	0.5
245	0.2

Note

a. See note b in Table 13.

These opposite temperature effects are not inconsistent with the operation of two competing pathways, each being favoured by different factors. Finally, steric control of the reaction was investigated by determining the corresponding **737** : **738** product ratios from the indolizations of 2-ethyl-, 2-isopropyl- and 2-*tert.*-butylcyclohexanone phenylhydrazone with various acid catalysts (Table 18). It is apparent from this study that increasing the size of the 2-alkyl substituent in the 2-alkylcyclohexanone moiety tends to lower the ratio [i.e. favour the formation of the corresponding **738** ($R^1 = R^3 = H$, $n = 2$) which is formed exclusively using 2-*tert.*-butylcyclohexanone phenylhydrazone].

Table 18. Product Ratios from the Indolization of 2-Alkylcyclohexanone Phenylhydrazones with Various Catalysts at 80 °C

Catalyst	Ratios[a]			
	2-Methyl	2-Ethyl	2-Isopropyl	2-*tert*.-Butyl
CH_3COOH	40	12	6.5	0
10% (w/w) H_2SO_4 in C_2H_5OH	1.8	4.1	0.8	0
BF_3/C_2H_5OH	0.9	2.7	0.1	0
$ZnCl_2/C_2H_5OH$	0.3	3.0	0.5	0
PPA	0.2	4.4	0.6	0

Note

a. Ratios of the corresponding **737** ($R^1 = R^3 = H, n = 2$) : **738** ($R^1 = R^3 = H, n = 2$).

Following the establishment that, under the indolization conditions, the products **737** did not rearrange to the corresponding products **738**, the data presented in Tables 13–18 were interpreted in terms of the following mechanism (Scheme 36). The two possible enehydrazine tautomers **749** and **750** are produced from the hydrazone **748** by a rearrangement which, although not essentially acid catalysed, is facilitated by such conditions. The formation of the more highly substituted double bond isomer, **749**, should predominate over the formation of **750** and, indeed, at low reaction temperatures in the non-catalysed reaction (Table 17) this factor apparently prevails (kinetic control). Under these conditions at higher temperatures, more of **750** is produced (thermodynamic control) and this, for steric reasons, can undergo more ready [3,3]-sigmatropic rearrangement to **754** than can **749** to **751**. At low acidities, in the absence of severe steric effects, the large proportions of 3*H*-indoles produced suggest that the acid is serving only to catalyse enehydrazine formation. At higher acid concentrations, the formations of **752** and **753** are suggested which, in view of earlier studies (Schiess and Grieder, 1974), should rearrange into **756** and **757**, respectively, far more readily than the unprotonated species **749** and **750** rearrange into **751** and **754**, respectively. Therefore, at higher acid concentrations, routes B and C should be followed in preference to routes A and D and the fact that route C is preferred over B could reflect the greater stability of the indolic product **758** compared with that of the 3*H*-indolic product **755**. Alternatively, since it had been earlier (Hinman, 1968) shown that 3-substitution in enamines lowers their basicity, then **750** would be more readily protonated than **749**, the formation of **753** would therefore predominate over that of **752** and the formation of **758** would therefore be favoured (Miller, F. M. and Schinske, 1978).

Scheme 36

During the formation of the enehydrazine tautomer of a 2-alkylcyclohexanone arylhydrazone, the loss of the axial proton from the 2- or 6-positions should be favoured over the loss of the equatorial protons since the axial bond electrons are coplanar with the developing double bond (see **759** → **760**).

759 **760**

Therefore, if the 2-alkyl substituent was in the axial position, then, neglecting other factors, the enehydrazine formation would have to involve loss of a proton from the 6-position. However, because of the conformational flexibility of the cyclohexane ring, axial and equatorial positions can be interconverted (Lohr, 1977). Therefore, for such an effect to influence the product ratio of **737** ($n = 2$) : **738** ($n = 2$), cyclohexanones of rigid conformation were employed. Using various catalysts, 2-methylcyclohexanone, 2,4-dimethylcyclohexanone and 2-methyl-4-*tert.*-butylcyclohexanone phenylhydrazone were indolized, the results of these reactions being given in Table 19. From this it can be seen that whereas the 4-methyl substituent has little effect upon the

Table 19. Product Ratios of 737 ($n = 2$) : 738 ($n = 2$) resulting from the Indolizations of 2-Methyl-2, 4-Dimethyl- and 2-Methyl-4-*tert.*-butylcyclohexanone Phenylhydrazone with various Catalysts

		Ratios using the Phenylhydrazones of:		
Catalyst	$-H_0$[a]	2-Methylcyclo-hexanone	2,4-Dimethyl-cyclohexanone	2-Methyl-4-*tert.*-butylcyclo-hexanone
CH_3COOH	−2.5	40.0	27.4	1.83
10% (w/w) H_2SO_4/C_2H_5OH	0.31	1.78	—	1.96
BF_3/C_2H_5OH	—	0.86	0.97	1.75
$ZnCl_2$	—	0.28	0.30	2.50
PPA	4.8	0.19	0.16	2.16

Note

a. Acidity function.

indolization direction, the 4-*tert*.-butyl group, by conformationally locking the cyclohexane ring, renders the product ratio largely independent of catalyst. This ratio then depends largely upon the fact that the most substituted enehydrazine [which would ultimately afford the corresponding **737** ($n = 2$)] would be formed in every case, since the 2-methyl substituent would be predominantly equatorial. The results of thermal indolization of 2-methyl-4-*tert*.-butylcyclohexanone phenylhydrazone in diethylene glycol at varying temperatures also showed that the **737** ($n = 2$) : **738** ($n = 2$) yield ratio was almost constant (Table 20) although in all these reactions, which are thermodynamically

Table 20. Product Ratios of 737 ($n = 2$) : 738 ($n = 2$) resulting from the Thermal Indolization of 2-Methyl-4-*tert*.-butylcyclohexanone Phenylhydrazone in Diethylene Glycol at varying Temperatures

Temperature (°C)	Ratio
150	0.34
200	0.22
245	0.27

controlled, the formation of the more stable indolic product **738** ($n = 2$) predominated (Lohr, 1977). In a related study, isopinocamphone (**761**) phenylhydrazone was indolized using acetic acid, sulphuric acid, boron trifluoride, zinc chloride and polyphosphoric acid as the catalysts and in each case only the corresponding product **762** was formed, this being in accordance with the above theory (Lohr, 1977).

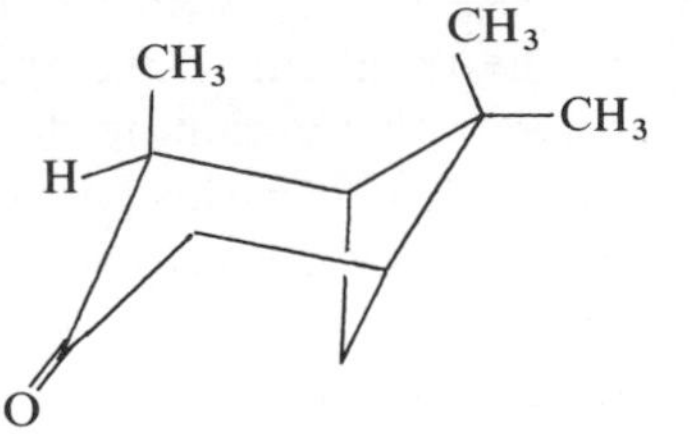

CH_3 CH_3 CH_3 H O

761

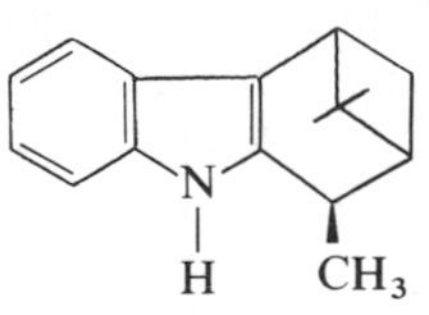

N H CH_3

762

The exclusive formation of 1-substituted 1,2,3,4-tetrahydrocarbazoles by Fischer indolization can be achieved starting with 2-substituted adipoin arylhydrazones, **763** (R^1 = OH), and using formic acid as the catalyst. Thus, compounds **763** (R^1 = OH, R^2 = CH_3, C_2H_5 and $CH_2C_6H_5$) were converted into **764** (R = CH_3, C_2H_5 and $CH_2C_6H_5$, respectively) and **765** (R = CH_3, C_2H_5 and $CH_2C_6H_5$, respectively), the formation of these two products in

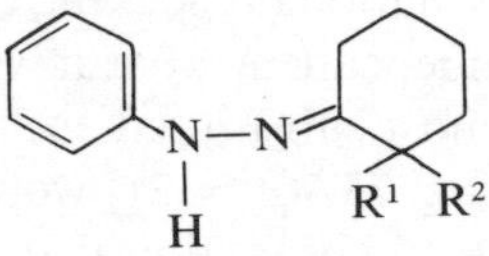

763

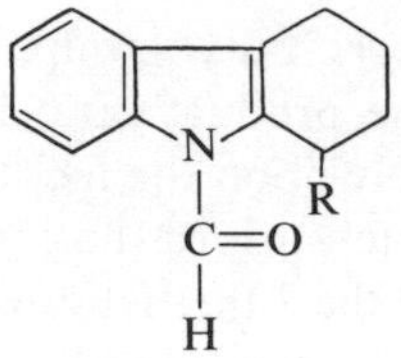

764

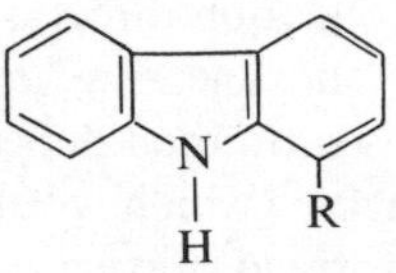

765

each case probably resulting from disproportionation of intermediate 3,4-dihydrocarbazoles (Wakamatsu, Hara and Ban, 1977). However, the yields from these reactions were low.

Other 2,2-disubstituted cyclohexanone phenylhydrazones **763** [$R^1 + R^2 = (CH_2)_5$] (Rice, Sheth and Wheeler, J. W., 9171), **763** ($R^1 = CH_3$, $R^2 = COOCH_3$) (Asselin, Humber and Dobson, 1977), **763** [$R^1 = CH_3$, $R^2 = (CH_2)_nCOOCH_3$] (Asselin, Humber and Dobson, 1976) and related cyclopentanone and cycloheptanone analogues (Asselin, Humber and Dobson, 1976) have also been indolized into the corresponding 1,1-disubstituted-1,2,3,4-tetrahydrocarbazoles and related systems (see also Robinson, B. and Smith, G. F., 1960).

When 2-methyl- and 2,5-dimethylcyclohexanone 4-methoxyphenylhydrazone were treated with hydrochloric acid, indolization occurred, as expected, in both possible directions to afford **737** ($R^1 = 6\text{-}OCH_3$, $R^2 = CH_3$, $R^3 = H$, $n = 2$) and **738** ($R^1 = 6\text{-}OCH_3$, $R^2 = CH_3$, $R^3 = H$, $n = 2$) and **737** ($R^1 = 6\text{-}OCH_3$, $R^2 = R^3 = CH_3$, $n = 2$) and **738** ($R^1 = OCH_3$, $R^2 = R^3 = CH_3$, $n = 2$), respectively. The 1,2,3,4-tetrahydrocarbazolic components of these products were subsequently dehydrogenated to furnish 6-methoxy-1-methylcarbazole and 6-methoxy-1,4-dimethylcarbazole, respectively. An alternative synthesis of these two products was also accomplished via unambiguous Fischer indolizations when 4-methoxycyclohexanone 2-methyl- and 2,5-dimethylphenylhydrazone, respectively, were treated with ethanolic hydrochloric acid and the corresponding 1,2,3,4-tetrahydro-3-methoxycarbazoles were, without purification, dehydrogenated with chloranil in dry tetrahydrofuran (Gilbert, J., Rousselle, Gansser and Viel, 1979). This type of alternative synthetic approach to the appropriate carbazoles is certainly worthy of further applications.

Reaction of **766** with methanolic hydrogen chloride at 0 °C afforded **767** as the only isolated product (Stillwell, 1964). It is interesting that dehydrogenation

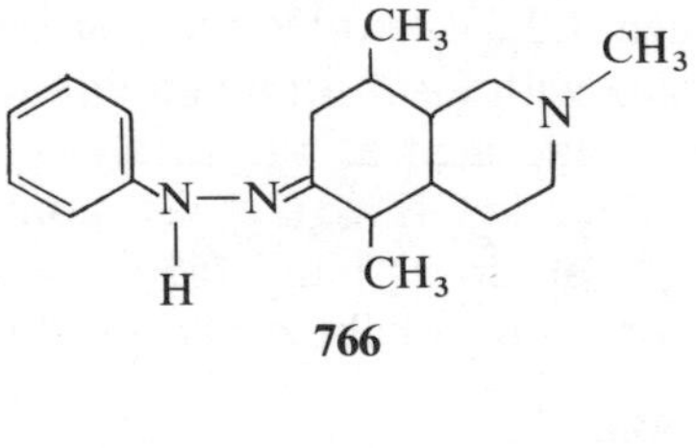

766

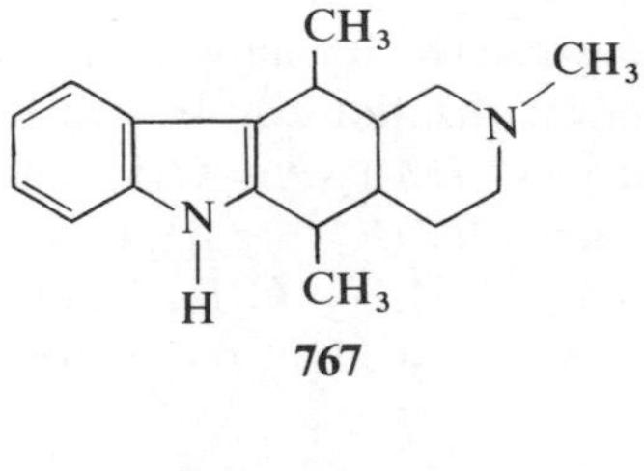

767

of this product, by pyrolysis with 10% palladium-on-carbon under nitrogen in a sealed tube, caused concomitant N-demethylation to afford ellipticine (**640**; $R^1 = R^2 = CH_3$, $R^3 = R^4 = H$, X = N, Y = CH) (Stillwell, 1964).

The only product isolated from the indolization of a mixture of phenylhydrazine and dihydrocodeinone (**768**) was **769** (Ekmekdzhyan and Tatevosyan, 1960).

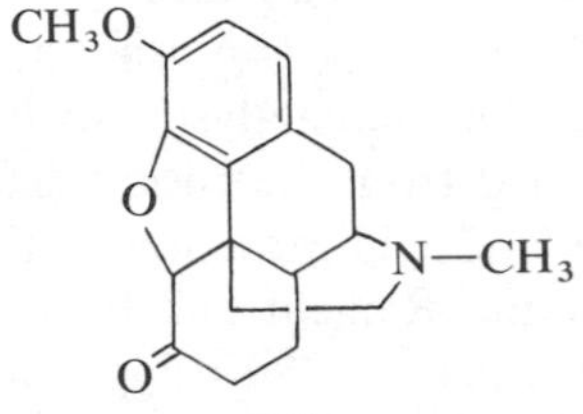

768

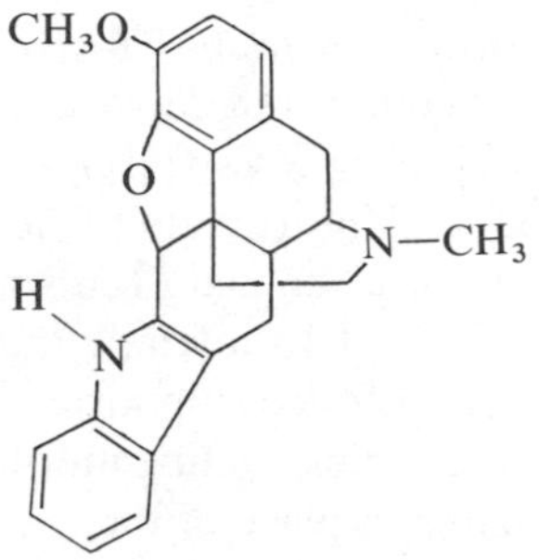

769

Indolization of *trans*-**770** (R = H and OCH_3, X = Y = CH_2), **770** (R = H, X = $NCOCH_3$, Y = CH_2) (probably the *trans* isomer) and **770** (R = H, X = CH_2, Y = $NCOCH_3$) with an acetic acid catalyst afforded **771** (R = H, and OCH_3, X = Y = CH_2) (Georgian, 1957, 1958, 1962), **771** (R = H, X = $NCOCH_3$, Y = CH_2) (Georgian, 1957, 1958, 1962) and **771** (R = H, X = CH_2, Y = $NCOCH_3$) (Georgian, 1958), respectively. In the earlier of

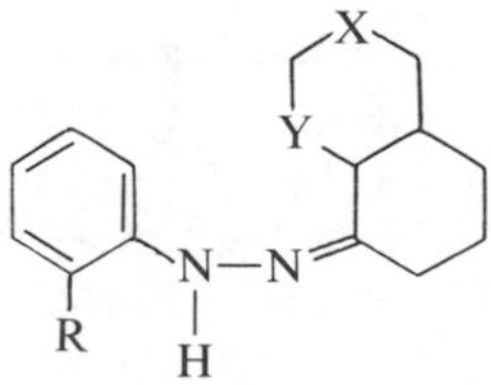

770

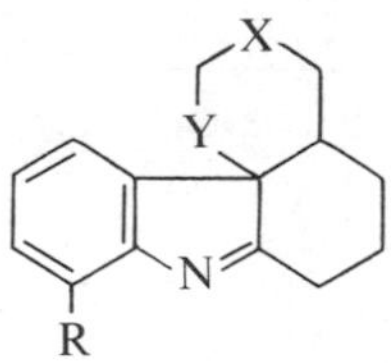

771

these three publications, only minimum amounts (not characterized) of the corresponding indoles were reported as being formed using the carbocyclic ketone (Georgian, 1957) and none was reported as being formed in any of the three publications when using the heterocyclic ketones. These results are not surprising since it is well established that acetic acid greatly favours, often exclusively, the indolization of 2-substituted cyclohexanones to 3*H*-indoles (see Tables 11 and 12). However, in the two later reports (Georgian, 1958, 1962), examination of the neutral fractions as well as the basic fractions of the reaction mixtures resulting from the acetic acid catalysed indolizations of *trans*-**770** (R = H and OCH_3, X = Y = CH_2) did give reasonable yields of the indoles **772** (R = H and OCH_3, respectively), although these were both less

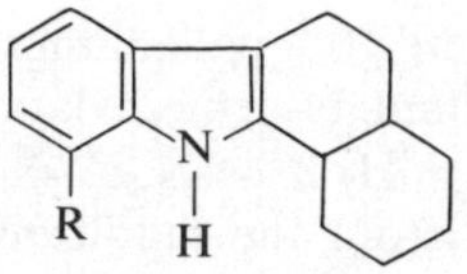

772

than the yields of the corresponding 3*H*-indoles simultaneously produced. In a recent indolization of *trans*-**770** (R = H, X = Y = CH_2) using a mixture of acetic acid and concentrated hydrochloric acid as the catalyst, only a 23% yield of **771** (R = H, X = Y = CH_2) was obtained from the basic fraction of the reaction product. Unfortunately, the neutral and acidic fractions of the reaction product were not examined (Teuber, Gholami, Reinehr and Bader, 1979). It would be of interest to attempt the indolization of **770** using, for example, moderately concentrated sulphuric acid as the catalyst in attempts to maximize the yields of the corresponding indolic products. Indolization of *cis*-1-decalone phenylhydrazone in place of the *trans* isomer also afforded the 3*H*-indole **771** (R = H, X = Y = CH_2) (Georgian, 1958). Contrary to these observations, however, acetic acid treatment of **773** produced only a 7–9% yield of **774**, the predominant product (about 90% yield) being **775** (Fritz and Rubach, 1968). Starting with **776**, indolization after admixture with phenylhydrazine or 2-methoxyphenylhydrazine is unambiguous, although the expected products **777** (R = H and OCH_3, respectively) were only isolated in low yields, considerable quantities of starting ketone being recovered (Georgian, 1958, 1962). Many other hydrocarbazoles were obtained in an analogous manner (Georgian,

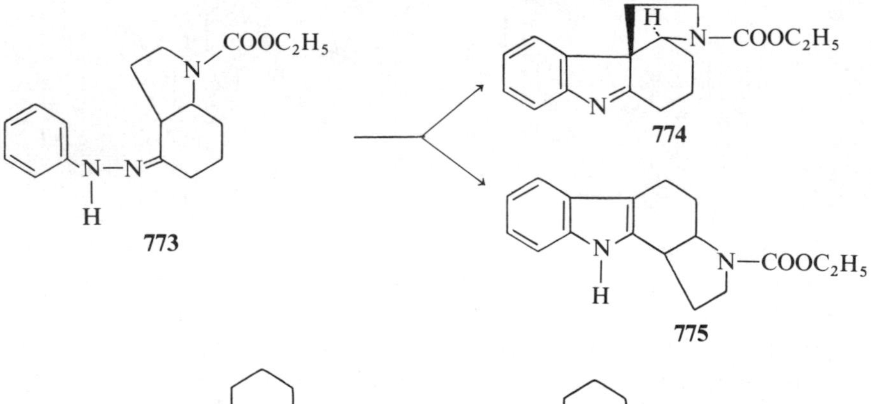

773 **774** **775**

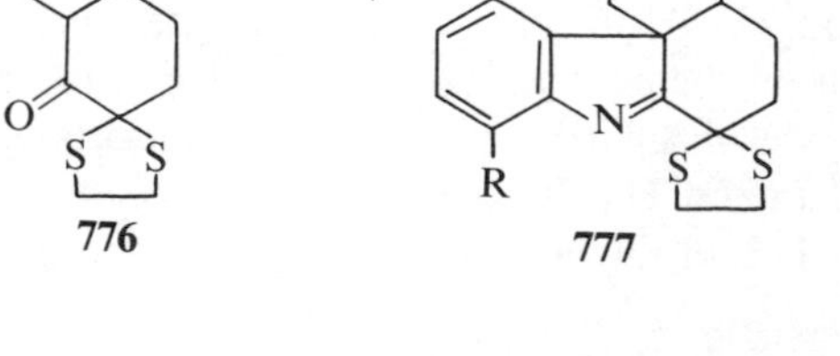

776 **777**

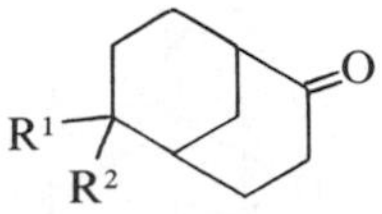

778

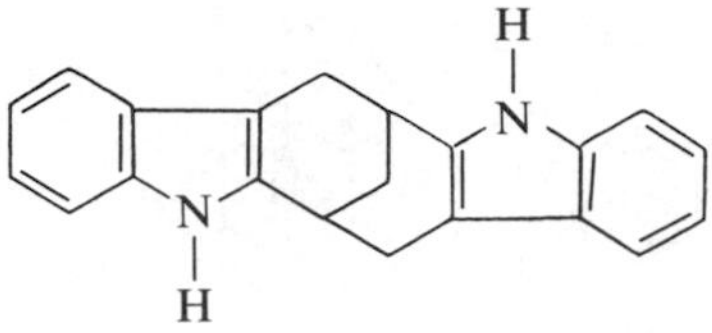

779

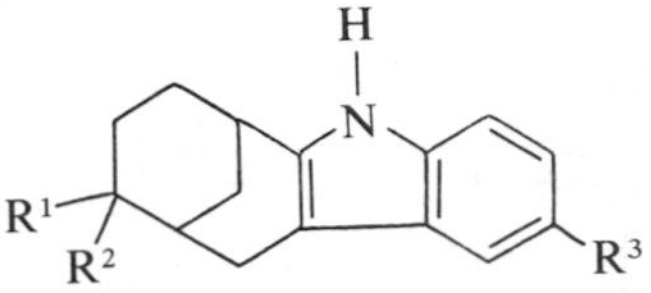

780

1958). Only the product **779** was isolated when **778** ($R^1 + R^2 = O$) was reacted with phenylhydrazine in refluxing ethanolic hydrochloric acid (65% yield) or when the bisphenylhydrazone of **778** ($R^1 + R^2 = O$) was added to polyphosphoric acid (96% yield). Only **780** [$R^1 + R^2 = O(CH_2)_2O$, $R^3 = H$] was isolated when a mixture of **778** [$R^1 + R^2 = O(CH_2)_2O$] with phenylhydrazine was heated in alcoholic acetic acid (55% yield) [this product could be hydrolysed in warm 1 M sulphuric acid to afford **780** ($R^1 + R^2 = O$, $R^3 = H$)]. Only **780** ($R^1 = H$, $R^2 = OH$, $R^3 = NO_2$) was isolated when **778** ($R^1 = H$, $R^2 = OH$) was reacted with 4-nitrophenylhydrazine in refluxing ethanolic phosphoric acid (53% yield), in methanolic sulphuric acid at room temperature (*ca.* 20% yield) [the corresponding 4-nitrophenylhydrazone (*ca.* 15% yield) was also isolated from this reaction] or when the 4-nitrophenylhydrazone of **778** ($R^1 = H$, $R^2 = OH$) was refluxed with polyphosphoric esters in chloroform (*ca.* 30% yield). When a mixture of **778** ($R^1 + R^2 = O$) with 4-nitrophenylhydrazine was refluxed in alcoholic phosphoric acid, the occurrence of indolization was not detected and only the corresponding bis-4-nitrophenylhydrazone was isolated and likewise only the 2-nitrophenylhydrazone was isolated when **778** ($R^1 = H$, $R^2 = OH$) was reacted with 2-nitrophenylhydrazine under similar conditions (Kadzyauskas, Butkus, Vasyulite, Averina and Zefirov, 1979). It is unfortunate that in these studies the reaction mixtures were not examined for basic products. In view of the low yields of indolic products usually obtained from these reactions it is very likely that the corresponding 3*H*-indolic products were also formed.

Indolizations of **781** ($R = C_2H_5$) (Ban, Sato, Y., Inoue, Nagai, M., Oishi, Terashima, Yonemitsu and Kanaoka, 1965 – see also Gilbert, B., 1968), **782** ($R^1 = OCH_3$, $R^2 = C_2H_5$) (Ban and Iijima, I., 1969 – see also Stevens, R. V., Fitzpatrick, J. M., Kaplan and Zimmerman, 1971), **783** ($R^1 = C_2H_5$, $R^2 = H$) [Kuehne and Bayha, 1966; Stork and Dolfini, 1963 – the stereochemistry of the ketonic moiety used in this latter study was established to be as shown in **783** by a later revision (Ban and Iijima, I., 1969; Ban, Iijima, I.,

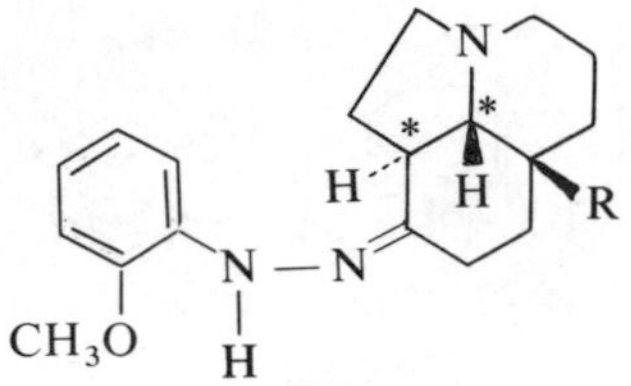

781

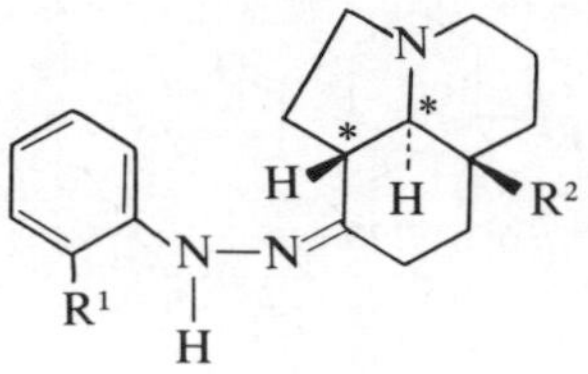

782

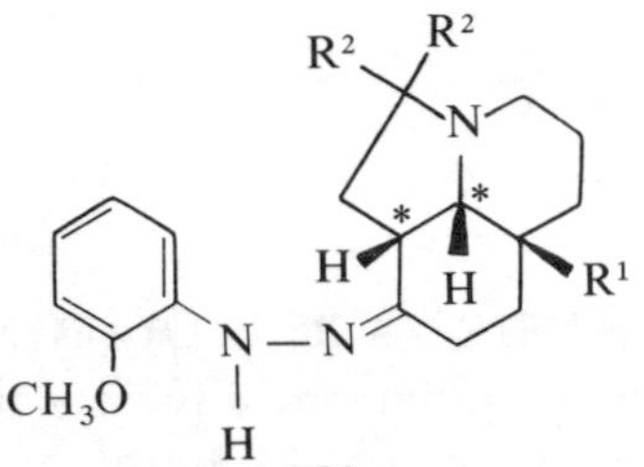

783

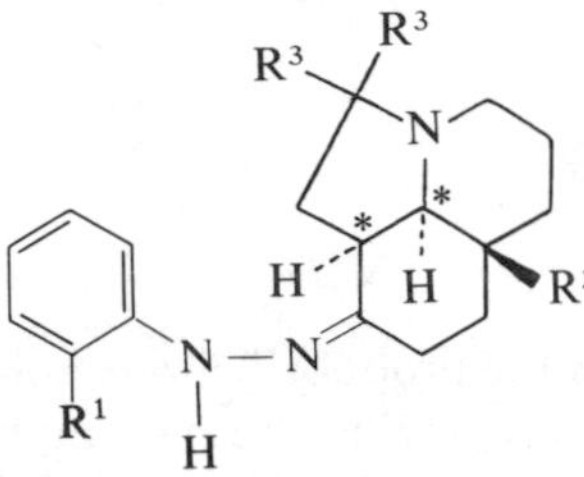

784

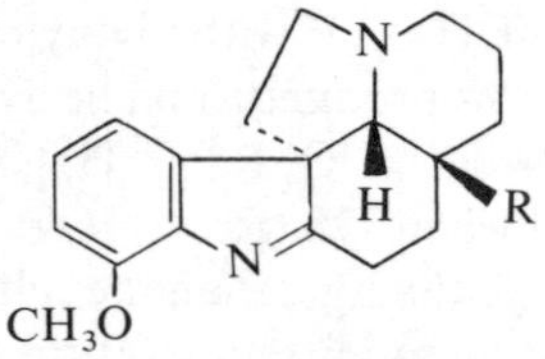

785

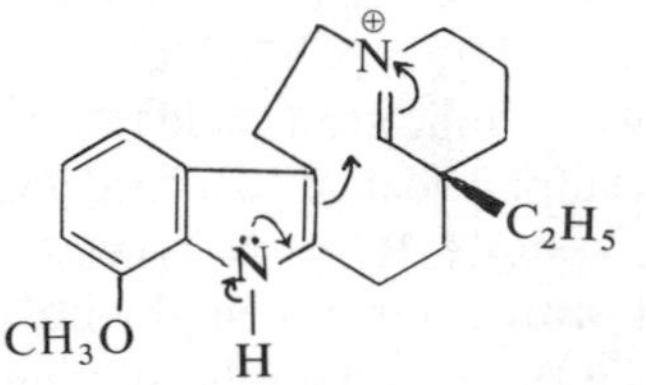

786

Inoue, Akagi and Oishi, 1969)] and **784** ($R^1 = OCH_3$, $R^2 = C_2H_5$, $R^3 = H$) (Kuehne and Bayha, 1966) have all been effected using acetic acid as catalyst to give in each case **785** ($R = C_2H_5$). The formation of this product involved in each case the formation of the new C—C bond at the ring junction carbon atom and also involved equilibration at the asymmetric *C. This equilibration arose through the intermediacy of the species **786**, formed by retro Mannich condensation of the 3*H*-indoles initially formed, which then cyclized as shown to the thermodynamically most stable isomer **785** ($R = C_2H_5$) (Gilbert, B., 1965, 1968; Inoue and Ban, 1970; Klioze and Darmory, 1975; Smith, G. F. and Wróbel, 1960; Stork, 1964; Stork and Dolfini, 1963). The earlier investigations of these indolizations have been reviewed (Gilbert, B., 1965, 1968) but the structural conformations used in this review require reconsideration in the light of later studies (Ban and Iijima, I., 1969; Ban, Iijima, I., Inoue, Akagi and Oishi, 1969). Under the above indolization conditions, the non-equilibrated product **787** ($R = C_2H_5$) was also obtained, along with **785** ($R = C_2H_5$), using **782** ($R^1 = OCH_3$, $R^2 = C_2H_5$) as the hydrazone reactant (Ban and Iijima, I., 1969). When attempting the indolization of this same hydrazone using formic acid as the catalyst, the equilibration was prevented altogether and **787** ($R = C_2H_5$), the thermodynamically less stable isomer, but not **785** ($R = C_2H_5$), was isolated,

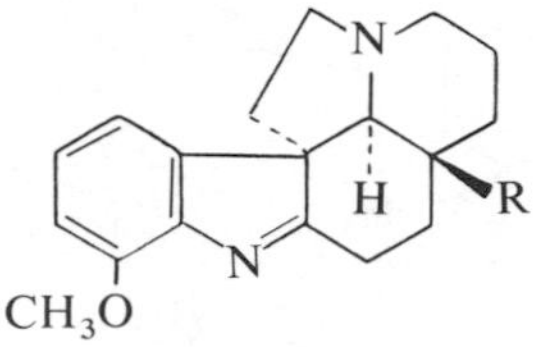

787

along with the indole resulting from indolization in the alternative possible direction (Ban and Iijima, I., 1969). An analogous alternative indolization occurred concomitantly with the above mentioned conversion of **781** (R = C_2H_5) into **785** (R = C_2H_5) (Ban, Sato, Y., Inoue, Nagai, M., Oishi, Terashima, Yonemitsu and Kanaoka, 1965). Similar observations have been made with related reactions. Thus, indolization of **304** with acetic acid or polyphosphate ester or with formic acid as catalyst afforded the thermodynamically less stable 3*H*-indole or **306**, respectively [the highest yield (16%) of **306** was obtained using formic acid dried over anhydrous copper(II) sulphate as catalyst] along with in each case the indole **305** resulting from indolization in the alternative possible direction, but no **785** [R = $(CH_2)_2OC_6H_5$] was detected in any of these reactions (Inoue and Ban, 1970). Treatment of **782** (R^1 = R^2 = H)/**784** (R^1–R^3 = H) with formic acid yielded only the diastereoisomers **788** (R^1 = R^2 = H) and **789** (Akagi, Oishi and Ban, 1969), similar treatment of **784** (R^1 = R^2 = H, R^3 + R^3 = O) gave **788** (R^1 = CHO, R^2 + R^2 = O)

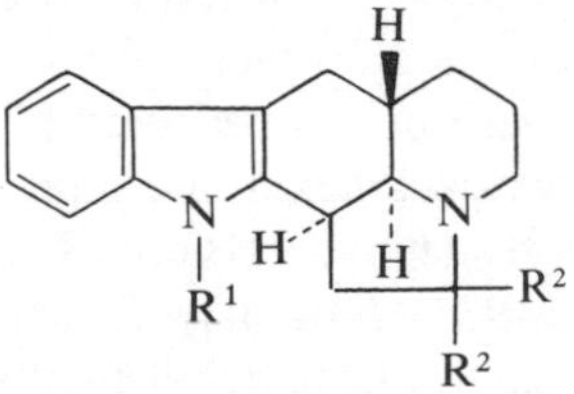

788

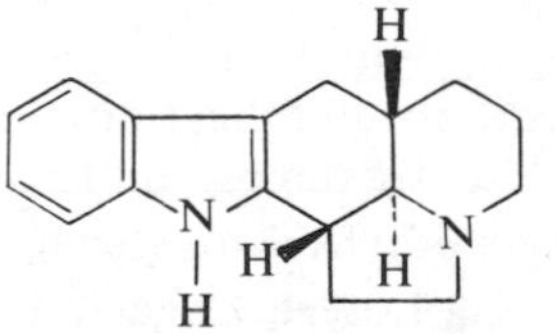

789

(Akagi, Oishi and Ban, 1969) and acetic acid catalysed indolization of **783** (R^1 = C_2H_5, R^2 + R^2 = O) yielded only neutral indolic products (Stork and Dolfini, 1963). With the latter two reactions, the presence of the amide carbonyl group led to a five membered ring in the transition state, which would ultimately have afforded 3*H*-indoles containing three trigonal atoms, an unstable situation which is rendered much more favourable when the number of such atoms is reduced to one by prior reduction of the amide carbonyl function (Stork and Dolfini, 1963). Indolization of **781** (R = $CH_2CH{=}CH_2$)/**783** (R^1 = $CH_2CH{=}CH_2$, R^2 = H) using an acetic acid catalyst afforded **785** (R = $CH_2CH{=}CH_2$) and under similar conditions **782** (R^1 = OCH_3, R^2 = $CH_2CH{=}CH_2$)/**784** (R^1 = OCH_3, R^2 = $CH_2CH{=}CH_2$, R^3 = H) afforded a mixture of **785** (R = $CH_2CH{=}CH_2$) and **787** (R = $CH_2CH{=}CH_2$). As might be expected by analogy with the above observations, the formation

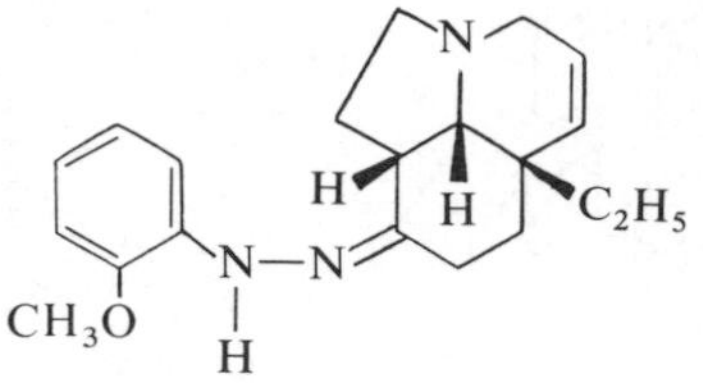

790

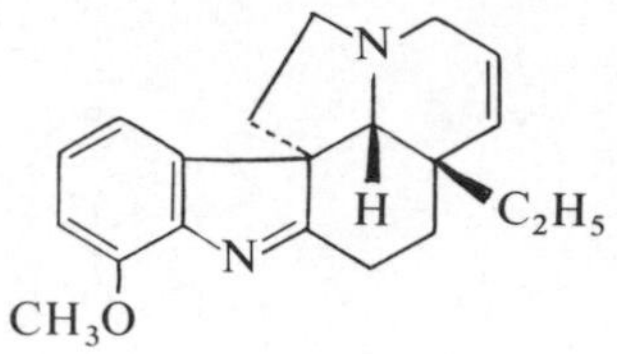

791

of only the latter product was detected when this last indolization was effected using formic acid as the catalyst (Saxton, Smith, A. J. and Lawton, 1975). Indolization of **790** using an acetic acid catalyst produced **791** (Klioze and Darmory, 1975). It should be noted that in the above reactions, the 3*H*-indoles are rarely characterized, being either immediately reduced with lithium aluminium hydride to the corresponding indolines (Ban, Sato, Y., Inoue, Nagai, M., Oishi, Terashima, Yonemitsu and Kanaoka, 1965; Inoue and Ban, 1970; Klioze and Darmory, 1975) or formylated under the indolization conditions, when using formic acid as the catalyst, to furnish the corresponding 1-formylindolines (Ban and Iijima, I., 1969; Inoue and Ban, 1970; Saxton, Smith, A. J. and Lawton, 1975).

Indolizations of **736** ($R^1 = R^3 = H$, $n = 2$) containing functionalized R^2 groups have also been investigated. Reactions in which R^2 = 2-hydroxyethyl or 3-hydroxypropyl are discussed later (see Chapter IV, G). Only the corresponding 1-carbethoxy- and 1-cyanomethyl-1,2,3,4-tetrahydrocarbazoles and 2-, 3- or 4-monomethyl substituted homologues were reported as being isolated from the indolization of a series of variously substituted phenylhydrazones of 2-carbethoxy- and 2-cyanomethylcyclohexanone and their corresponding methyl substituted homologues, respectively (Biere, Rufer, Ahrens, Schroeder, E., Losert, Loge and Schillinge, 1973). Similarly, 2-carbomethoxycyclohexanone afforded 1-carbomethoxy-1,2,3,4-tetrahydro-9-methylcarbazole (Bailey, Peach and Vandrevala, 1978), presumably by a Fischer indolization, and the only products reported from the indolization of various 2-carboxycyclohexanone arylhydrazones were the corresponding 1-carboxy-1,2,3,4-tetrahydrocarbazoles (Berger, L. and Corraz, 1973).

Indolization of **792** ($R^1 = R^3 = H$, $R^2 = CH_3$, $m = 1$, $n = 2$) produced a mixture of **793** ($R^1 = H$, $R^2 = CH_3$, $X = O$, $m = 1$, $n = 2$) and **794** ($R^1 = R^3 = H$, $R^2 = CH_3$, $m = 1$, $n = 2$), the former product probably arising via deaminative cyclization of the intermediate betaine as shown in **795** (Fritz,

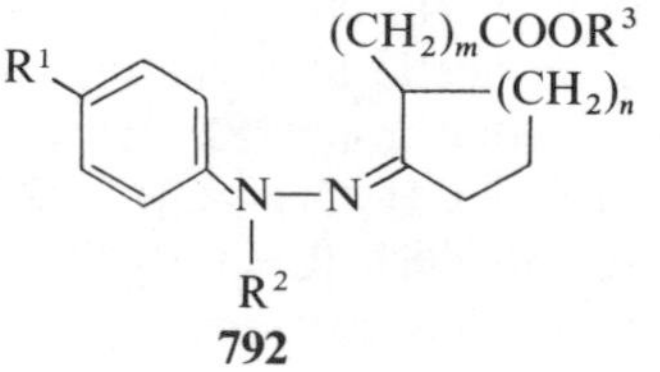

792

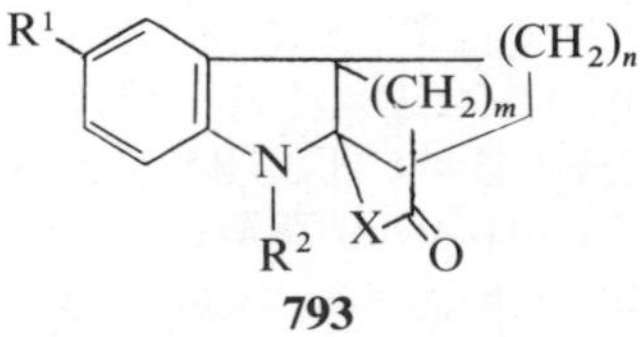

793

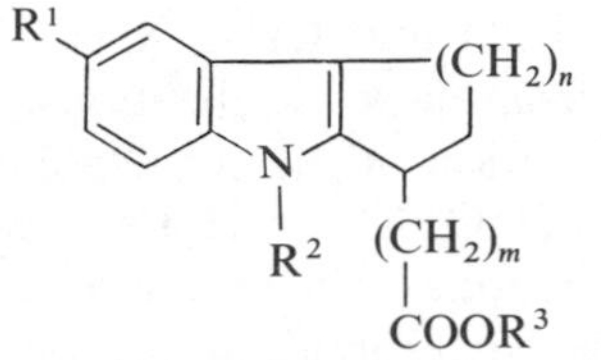

794

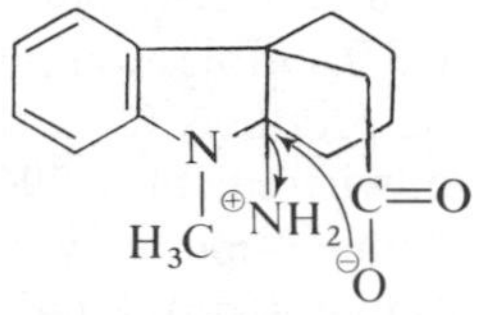

795

Losacker, Stock and Gerber, 1971; Losacker, 1966). Practically none of the lactam **793** (R^1 = H, R^2 = CH_3, X = NH, m = 1, n = 2) was formed from this indolization although indolization of the corresponding ethyl ester and its methoxylated analogue **792** (R^1 = H and OCH_3, respectively, R^2 = CH_3, R^3 = C_2H_5, m = 1, n = 2) gave the lactams **793** (R^1 = H and OCH_3, respectively, R^2 = CH_3, X = NH, m = 1, n = 2) (Losacker, 1966). Similarly, the acetic acid catalysed indolization of **792** (R^1 = OCH_3, R^2 = CH_3, R^3 = C_2H_5, $m = n = 1$) gave both **793** (R^1 = OCH_3, R^2 = CH_3, X = NH, $m = n = 1$) and **794** (R^1 = OCH_3, R^2 = CH_3, R^3 = H, $m = n = 1$) (Fritz and Stock, 1970). Only **794** (R^1 = R^2 = H, R^3 = C_2H_5, m = 1, n = 2) (75% yield) was isolated from the indolization of **792** (R^1 = R^2 = H, R^3 = C_2H_5, m = 1, n = 2) using 20% sulphuric acid as the catalyst (Bailey, Scott, P. W. and Vandrevala, 1980) although the basic product(s) of the reaction remained unexamined. The 20% aqueous sulphuric acid catalysed indolization of **792** (R^1–R^3 = H, $m = n = 2$) afforded **737** [R^1 = R^3 = H, R^2 = $(CH_2)_2COOH$, n = 2] and **796** (R = H, X = CH_2) in a yield ratio of 9 : 20, respectively, the latter product being formed by intramolecular lactamization of the intermediate **794** (R^1–R^3 = H, $m = n = 2$) (Openshaw and Robinson, R., 1937). Surprisingly, a later repetition of this reaction again produced **737** [R^1 = R^3 = H, R^2 = $(CH_2)_2COOH$, n = 2] and **796** (R = H, X = CH_2) but in a yield ratio of 42 : 17, respectively (Kost, Yudin and Yü-Chou, 1964).

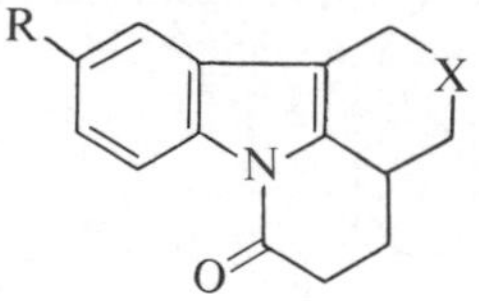

796

In yet a further repetition, both products were again isolated but the relative yields were not stated (Fritz and Fischer, O., 1964) and in a later related indolization, **792** (R^1 = OCH_3, R^2 = R^3 = H, $m = n = 2$) was converted into **737** [R^1 = 6-OCH_3, R^2 = $(CH_2)_2COOH$, R^3 = H, n = 2] and **796** (R = OCH_3, X = CH_2) (Fritz and Losacker, 1967). Starting with the appropriate arylhydrazones, compounds **796** (R = H, CH_3, NO_2 and $COOC_2H_5$, X = S) have likewise been prepared, in the last case along with the corresponding 3*H*-indole (Kakurina, Kucherova and Zagorevskii, 1964, 1965b) (see later in this chapter). In a later study, both **796** (R = H and CH_3, X = CH_2) and the corresponding 3*H*-indoles were isolated from these indolizations, as were both similar possible products from the indolization of the corresponding 2-naphthylhydrazone, whereas only **796** (R = CH_3O, X = CH_2) was isolated when the corresponding 4-methoxyphenylhydrazone was indolized and only the 3*H*-indolic product was isolated when the corresponding 4-carbethoxyphenylhydrazone was indolized (Borisova, Kucherova and Zagorevskii, 1970a). Similarly, the indolization of 5-oxoheptanoic acid afforded 2-(2-ethyl-3-indolyl)propionic acid (22% yield) and **704** (R = CH_3) (23% yield), resulting from initial indolization in both possible directions, although indolization of 5-oxocaproic acid phenylhydrazone yielded only 3-(2-methyl-3-indolyl)propionic acid (Teuber, Cornelius and Worbs, 1964) as discussed earlier. Using as the catalyst glacial acetic acid, a weak and 'large' acid, **792** ($R^1 = R^2$ = H, $R^3 = C_2H_5$, $m = n = 2$) afforded an almost quantitative yield of **737** [$R^1 = R^3$ = H, $R^2 = (CH_2)_2COOC_2H_5$, $n = 2$], no **794** ($R^1 = R^2$ = H, $R^3 = C_2H_5$, $m = n = 2$) nor its derived lactam **796** (R = H, X = CH_2) being obtained (Lions, 1937–1938). Significantly, using polyphosphoric acid, a strong and 'small' (Lyle, R. E. and Skarlos, 1966) acid, the corresponding N_α-methylphenylhydrazone **792** (R^1 = H, $R^2 = CH_3$, $R^3 = C_2H_5$, $m = n = 2$) afforded, as the only isolated product, a 48% yield of **794** (R^1 = H, $R^2 = CH_3$, $R^3 = C_2H_5$, $m = n = 2$) (Dolby and Esfandiari, 1972). Indolization of **792** ($R^1 = R^3$ = H, $R^2 = CH_3$ and $CH_2C_6H_5$, $m = n = 2$) furnished both **793** (R^1 = H, $R^2 = CH_3$ and $CH_2C_6H_5$, respectively, X = NH, $m = n = 2$) and **794** (R^1 = H, $R^2 = CH_3$ and $CH_2C_6H_5$, respectively, R^3 = H, $m = n = 2$) (Fritz and Losacker, 1967). Similarly, indolization of a mixture of **797** with N_α-methylphenylhydrazine gave **798** (Avanesova and Tatevosyan, 1979) although **799** (R = H and COOH) were produced when mixtures of **797** with phenylhydrazine and 4-carboxyphenylhydrazine, respectively, were indolized (Avanesova and Tatevosyan, 1974). Indolization of **792** (R^1–R^3 = H, m = 3

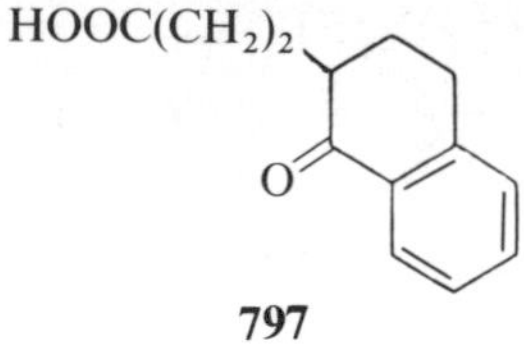

797

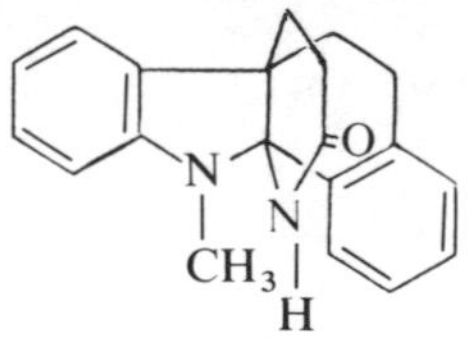

798

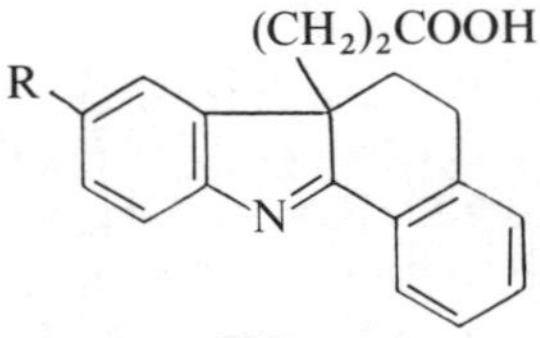

799

and 4, $n = 2$) produced **737** [$R^1 = R^3 = H$, $R^2 = (CH_2)_3COOH$ and $(CH_2)_4COOH$, respectively, $n = 2$] and **794** (R^1–$R^3 = H$, $m = 3$ and 4, respectively, $n = 2$) (Fritz and Losacker, 1967). The lactam carbonyl group of the above indolization products has been reduced to a methylene unit (Avanesova and Tatevosyan, 1979; Fritz and Losacker, 1967; Fritz and Stock, 1970) to yield the corresponding cyclic secondary amines. When starting with **793** ($R^1 = OCH_3$, $R^2 = CH_3$, X = NH, $m = n = 1$), further sequential transformations of this reduction product led to **800**, an interesting analogue of the anticholinesterase drug, physostigmine (Fritz and Stock, 1970). The product formed by indolization of **801** could not be purified but after its reduction the lactam **802** could be isolated (Openshaw and Robinson, R., 1937).

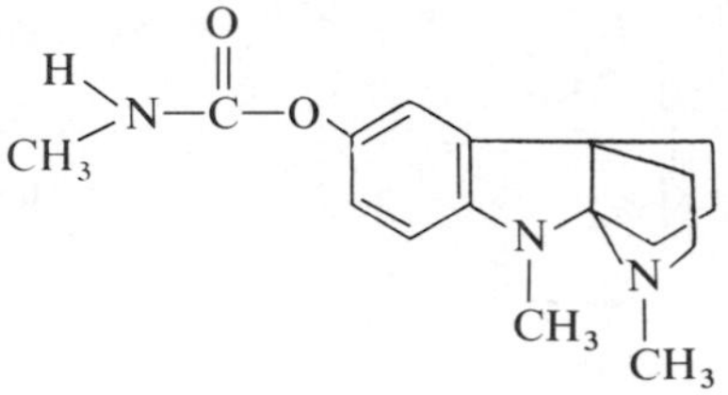

800

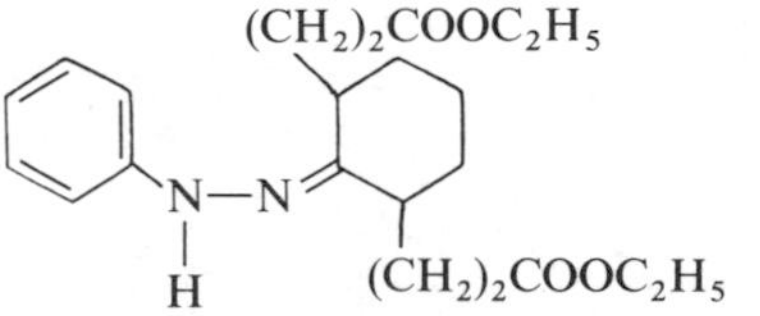

801

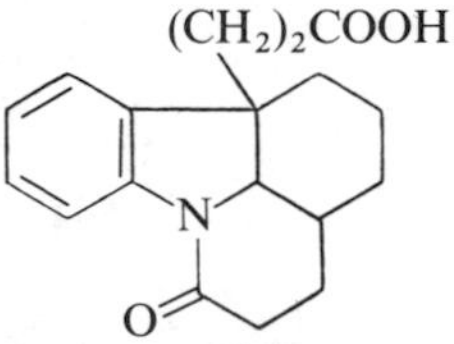

802

The zinc chloride catalysed indolization of **736** [$R^1 = R^3 = H$, $R^2 = (CH_2)_2CN$, $n = 2$] at 200–220 °C afforded only 1,2,3,4-tetrahydrocarbazole (44%) and at 170–190 °C afforded 1,2,3,4-tetrahydracarbazole (13.5%) and **738** [$R^1 = R^3 = H$, $R^2 = (CH_2)_2CN$, $n = 2$] (32%) and the sulphuric acid catalysed indolization of the same hydrazone gave **737** [$R^1 = R^3 = H$, $R^2 = (CH_2)_2CN$, $n = 2$] (18%) and **738** [$R^1 = R^3 = H$, $R^2 = (CH_2)_2CN$, $n = 2$] (50%). Since treatment of the latter compound with zinc chloride at 170–190 °C yielded only **796** (R = H, X = CH_2) whereas similar treatment of the former compound yielded 1,2,3,4-tetrahydrocarbazole, it was suggested that the 1,2,3,4-tetrahydrocarbazole formed during these indolizations arose in a similar manner (Kost, Yudin and Yü-Chou, 1964). It is interesting that related eliminations have been observed (Nakazaki, Isoe and Tanno, 1955) when **737** [$R^1 = R^3 = H$, $R^2 = C(CH_3)_3$, $n = 2$] picrate was boiled in ethanol and when **737** [$R^1 = R^3 = H$, $R^2 = CH(CH_3)_2$ and $CH_2C_6H_5$, $n = 2$] was heated in 4 M hydrochloric acid to form 1,2,3,4-tetrahydrocarbazole in each case (see Chapter V, A). Another product, reported later in 4% yield from the above sulphuric acid catalysed indolization and in 21.5% yield using a glacial acetic acid catalyst,

736 [$R^1 = R^3 = H, R^2 = (CH_2)_2CN, n = 2$]

803

804

Scheme 37

was shown to have structure **804** by an independent synthesis and was suggested as being formed as shown in Scheme 37 via intermediate **803** which underwent a Fischer type indolization followed by aromatization (Kost, Yudin and Chernyshova, 1968). Support for this postulated mechanism was later (Yudin, Kost and Chernyshova, 1970) derived from the observation that when the hydrazone **805** ($R^1 = R^2 = H$) was refluxed with glacial acetic acid, the α-carboline **809** ($R^1 = R^2 = H$) (8% yield) was formed, presumably via intermediates **806**–**808** ($R^1 = R^2 = H$) as shown in Scheme 38. Likewise, **805** ($R^1 = CH_3$, $R^2 = H$; $R^1 = H$, $R^2 = CH_3$ and $R^1 = R^2 = CH_3$) was converted into **809** ($R^1 = CH_3$, $R^2 = H$; $R^1 = H$, $R^2 = CH_3$ and $R^1 = R^2 = CH_3$, respectively) (7%, 32.5% and 22.5% yields, respectively). In the case of **805** ($R^1 = H, R^2 = CH_3$), the effect of variation of indolization catalyst upon the reaction products was investigated. Using 95% acetic acid, glacial acetic acid at room temperature for two months, polyphosphoric acid or concentrated sulphuric acid at room temperature, the yields of **809** ($R^1 = H, R^2 = CH_3$) obtained were 4–5%, 0%, 8% and 0%, respectively. No other products were reported as being isolated from these reactions. Heating with 20% sulphuric acid or concentrated hydrochloric acid formed respectively 4% and 3–4%

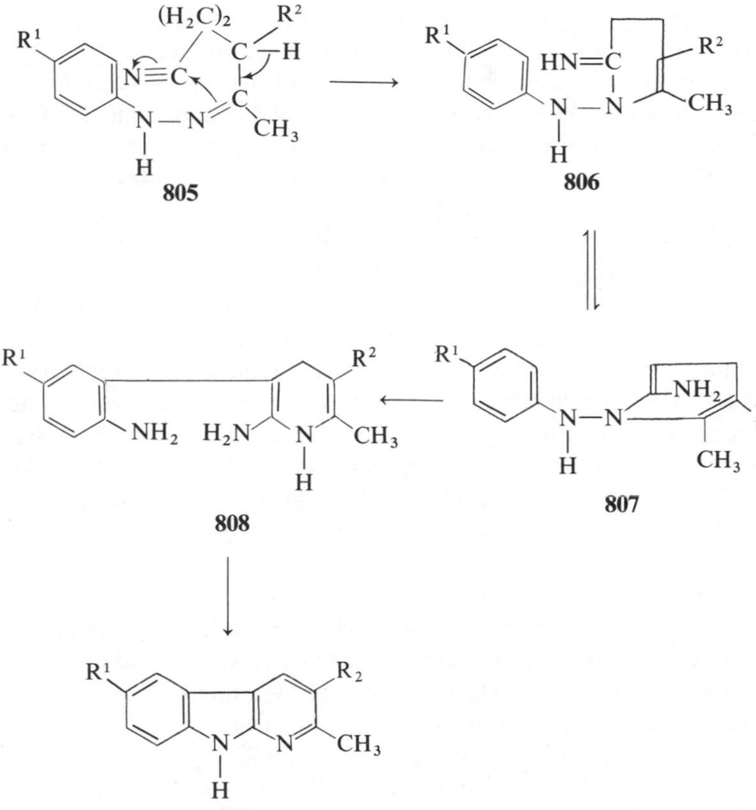

Scheme 38

yields of this product together with in each case *ca.* 50% yield of one of the possible 'normal' indolization products, 3-(2-cyanoethyl)-2,3-dimethyl-3*H*-indole (Yudin, Kost, and Chernyshova, 1970). This route to substituted α-carbolines, by the treatment of 5-oxocapronitrile arylhydrazones with boiling glacial acetic acid, has also been the subject of a patent (Yudin, Chernyshova and Kost, 1968).

Passage of dry hydrogen chloride into ethanolic solutions of **736** [R^1 = R^3 = H, R^2 = $CH_2N(CH_3)_2$, $CH_2N(C_2H_5)_2$, CH_2-piperidino and CH_2-morpholino, $n = 2$] yielded products which were originally formulated as **737** [R^1 = R^3 = H, R^2 = $CH_2N(CH_3)_2$, $CH_2N(C_2H_5)_2$, CH_2-piperidino and CH_2–morpholino, respectively, $n = 2$] (Harradence and Lions, 1939). Later, the first of these products was dehydrogenated, by heating with sulphur, to form 1-methylcarbazole and it was therefore reformulated as **738** [R^1 = R^3 = H, R^2 = $CH_2N(CH_3)_2$, $n = 2$] (Grandberg, Kost and Yaguzhinskii, 1960). This reformulation is also in accord with the acid 'size' hypothesis for the catalyst in relation to the direction of indolization as discussed earlier (Lyle, R. E. and Skarlos, 1966). By analogy, the other reaction products should now be reformulated as the corresponding indoles, rather than the 3*H*-indoles as originally proposed (Harradence and Lions, 1939) and, indeed, such a reformulation has been supported by a recent u.v. spectral examination of these products (Fritz, Losacker, Stock and Gerber, 1971). Likewise, only **738** ($R^1 = R^3 = H$, $R^2 = CH_2$-piperidino, $n = 2$) has been isolated after treatment of a mixture of the ketonic and hydrazino components of **736** ($R^1 = R^3 = H$, $R^2 = CH_2$-piperidino, $n = 2$) with ethanolic hydrochloric acid (Hahn, W. E. and Zawadzka, 1969). Further investigation of the indolization of **736** [$R^1 = R^3 = H$, $R^2 = CH_2N(CH_3)_2$, $n = 2$] by heating the hydrochloride of its ketonic moiety with phenylhydrazine or phenylhydrazine hydrochloride or by pyrolysis of the phenylhydrazone itself (non-catalytic indolization) afforded 1,2,3,4-tetrahydrocarbazole (46.5%), 1,2,3,4-tetrahydrocarbazole (6%) and **738** [$R^1 = R^3 = H$, $R^2 = CH_2N(CH_3)_2$, $n = 2$] (41%) and 1,2,3,4-tetrahydrocarbazole (12%) and **738** [$R^1 = R^3 = H$, $R^2 = CH_2N(CH_3)_2$, $n = 2$] (20%), respectively. These observations were in accord with the earlier report that when a mixture of phenylhydrazine with 2-piperidinomethylcyclohexanone was heated, one of the products isolated was 1,2,3,4-tetrahydrocarbazole (along with ammonia, aniline and piperidine) (Mannich and Hönig, 1927). It was suggested that the 1,2,3,4-tetrahydrocarbazole was formed by indolization of cyclohexanone phenylhydrazone produced by a retro Mannich reaction of the starting hydrazone (Grandberg, Kost and Yaguzhinskii, 1960 – see also Mannich and Hönig, 1927). However, it is equally likely that it resulted from a retro Mannich reaction involving the initially formed **737** [$R^1 = R^3 = H$, $R^2 = CH_2N(CH_3)_2$, $n = 2$] (i.e. post indolization) as shown, a reaction for which many analogies are known (e.g. Fritz, Losacker, Stock and Gerber, 1971). A similar reaction could also account for the origin of the morpholine hydrochloride, formed along with ammonium chloride during the treatment of 2-methyl-1-morpholinopentan-3-one phenylhydrazone with dry ethanolic hydrogen chloride, although no other recognizable products could be isolated from this reaction (Harradence and Lions, 1939). The claim (Grandberg, Kost and Yaguzhinskii, 1960) that when, after its formation by Fischer indolization, **738** [$R^1 = R^3 = H$, $R^2 = CH_2N(CH_3)_2$, $n = 2$] was boiled in water, dilute acid and 2 M alkali, it was completely converted back to 'the starting material' is difficult to understand.

When a mixture of **810** and phenylhydrazine was warmed in acetic acid, the only product isolated was 1,2,3,4-tetrahydrocarbazole, presumably formed via the sequential formation of **811** and **812**, the latter then undergoing elimination of the 4a-substitutent (Teuber, Worbs and Cornelius, 1968).

Several attempts to effect a Fischer indolization of **813** were unsuccessful, although no experimental details were recorded (Katritzky, Dennis, Sabongi and Turker, 1979).

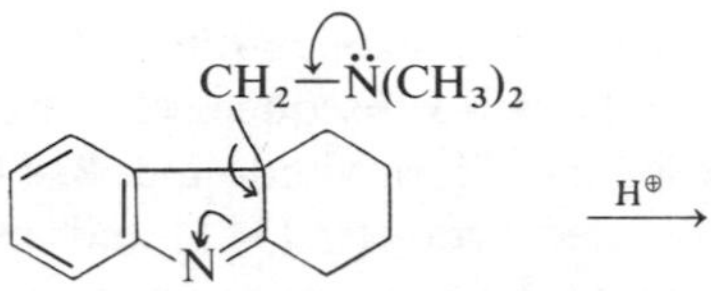

CH_2—$\ddot{N}(CH_3)_2$

$\xrightarrow{H^{\oplus}}$ **738** ($R^1 - R^3 = H$, $n = 2$)

737 [$R^1 = R^3 = H$, $R^2 = (CH_3)_2NCH_2$, $n = 2$]

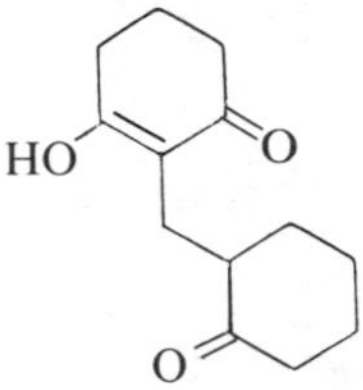

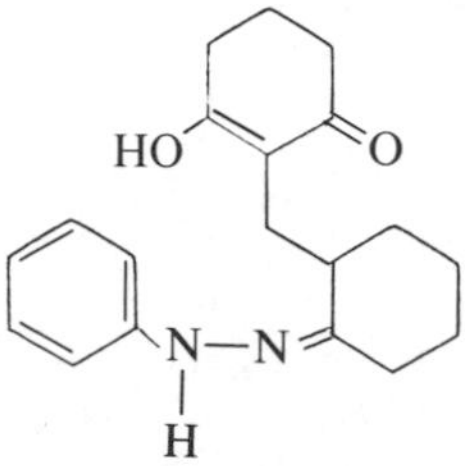

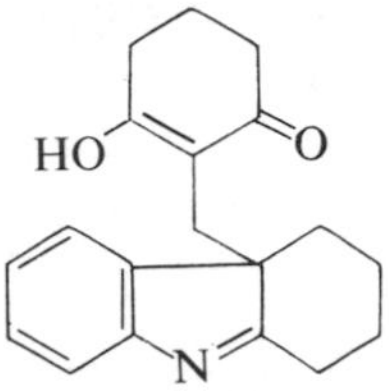

HO O O **810**

HO O N—N H **811**

HO O N **812**

CH_3 N N CH_3 N H N—N O **813**

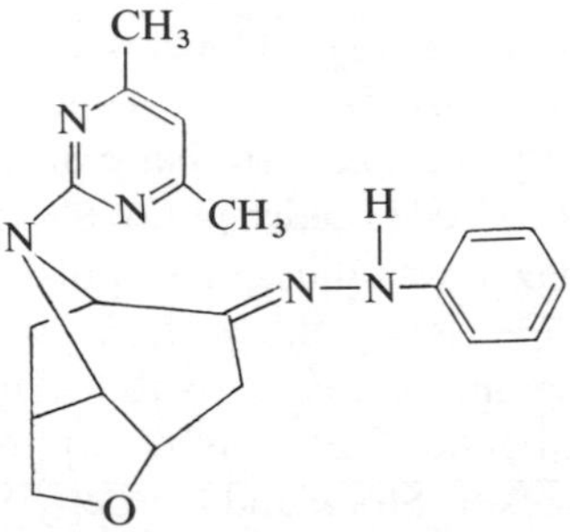

By comparison with the wide applicability of the indolizations of 2-substituted cyclohexanone arylhydrazones, indolization of 2-substituted cyclopentanone arylhydrazones has met with very limited success. Thus, attempted indolization of **736** ($R^1 = R^3 = H$, $R^2 = C_2H_5$, $n = 1$) using hydrogen chloride in acetic acid produced only resins (Buu-Hoï, Jacquignon and Loc, 1958). Similar intractable materials also resulted from attempts (Kelly, A. H., McLeod and Parrick, 1965) to indolize **736** ($R^1 = R^3 = H$, $R^2 = CH_3$, $n = 1$) using glacial acetic acid, dilute sulphuric acid or zinc chloride as catalysts. However, by refluxing the hydrazone in diethylene glycol, indolization was effected to furnish both **737** ($R^1 = R^3 = H$, $R^2 = CH_3$, $n = 1$) and **738** ($R^1 = R^3 = H$, $R^2 = CH_3$, $n = 1$) in an approximately 5 : 1 yield ratio, respectively, although in low percentage yields (Kelly, A. H., McLeod and Parrick, 1965). Likewise, 2-methylcyclopentanone 4-benzyloxyphenylhydrazone (**736**; $R^1 = 4\text{-}OCH_2C_6H_5$, $R^2 = CH_3$, $R^3 = H$, $n = 1$) was indolized in boiling monoethylene glycol but the only isolated product (52% yield) was the indole **738** ($R^1 = 5\text{-}OCH_2C_6H_5$, $R^2 = CH_3$, $R^3 = H$, $n = 1$) (Ahmed, 1966). Contrary to these observations, formic acid catalysed indolization of

2-methyl- and 2-ethylcyclopentanone phenylhydrazone (**736**; $R^1 = R^3 = H$, $R^2 = CH_3$ and C_2H_5, respectively, $n = 1$) with subsequent chromatographic 'work-up' of the reaction mixture afforded the 1-formylindolines **814** (R = CH_3 and C_2H_5, respectively) as the only isolated products (34% and 33% yields, respectively). These products were formed by reductive formylation of the initially formed indolization products **737** ($R^1 = R^3 = H$, $R^3 = CH_3$ and C_2H_5, respectively, $n = 1$) (Shimizu, J., Murakami, S., Oishi and Ban, 1971).

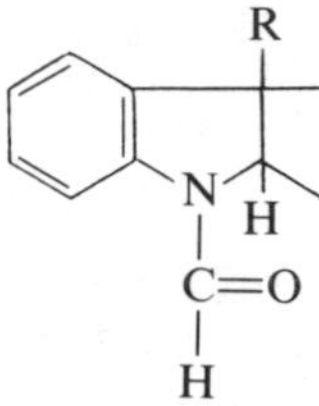

R

N H

C=O

H

814

Treatment of 2-morpholinomethylcyclopentanone phenylhydrazone (**736**; $R^1 = R^3 = H$, $R^2 = CH_2$-morpholino, $n = 1$) with dry ethanolic hydrogen chloride gave an apparently homogeneous basic product which was formulated as **737** ($R^1 = R^3 = H$, $R^2 = CH_2$-morphilino, $n = 1$) (Harradence and Lions, 1939). However, no evidence was presented to distinguish this structure from the possible alternative, **738** ($R^1 = R^3 = H$, $R^2 = CH_2$-morpholino, $n = 1$), the latter formulation being analogous to the structures established for the products obtained from the indolization of the cyclohexanone homologue and its analogues (Fritz, Losacker, Stock and Gerber, 1971; Grandberg, Kost and Yaguzhinskii, 1960) as described above. Indolization of 2-carbomethoxycyclopentanone phenylhydrazone afforded **738** ($R^1 = R^3 = H$, $R^2 = COOCH_3$, $n = 1$) (Lacoume, Milcent and Olivier, 1972). When a solution of **815** in

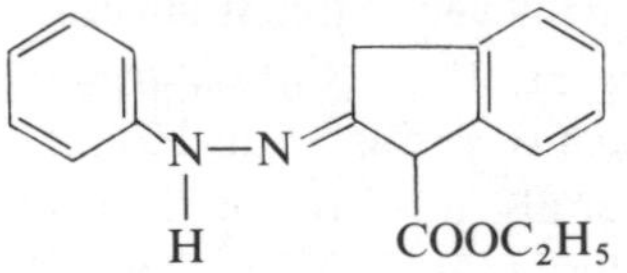

N—N

H $COOC_2H_5$

815

acetic acid was refluxed, then along with the main reaction product, an uncrystallizable resin, two other products were isolated. One of these showed a positive 'pine-chip' reaction, indicating the presence of an indole nucleus, and on the basis of this and an empirical formula determination it was formulated, although incorrectly numbered in the original text, as **816**. The other product, which gave a negative 'pine-chip' reaction, was isomeric with **816** and was thought to be polymeric (Perkin and Titley, 1922). This product is possibly a trimer of the 3*H*-indole **817** (see Chapter V, B) which would be formed from the alternative direction of indolization of **815**. Further study of this reaction and

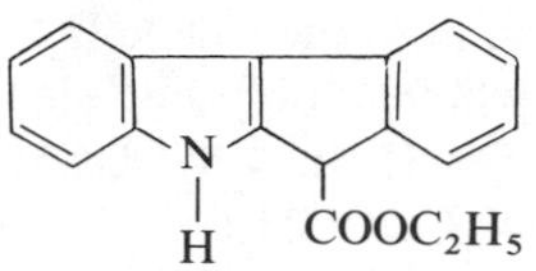

$COOC_2H_5$

N H $COOC_2H_5$

816

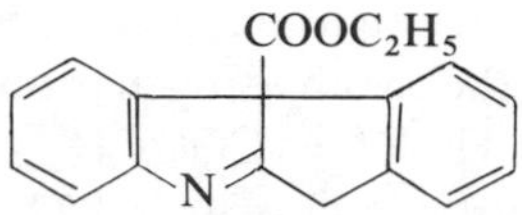

N

817

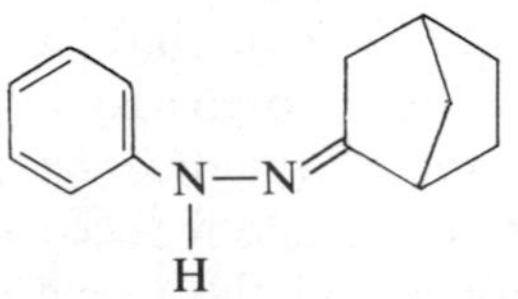

O

N—N H

818

H CH_3 CH_3

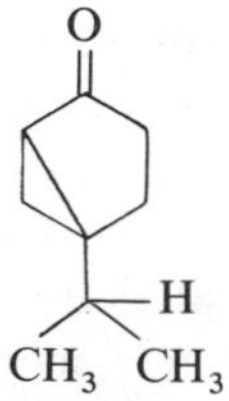

819

the structure of the products would be of interest. Attempted indolization of norcamphor phenylhydrazone (**818**) was unsuccessful, treatment with hydrogen chloride in acetic acid yielding only starting material, with polyphosphoric acid causing decomposition and with zinc chloride in acetic acid giving unidentified resins (Buu-Hoï, Jacquignon and Béranger, 1968). Likewise, the attempted use of **819** as the ketonic moiety in Fischer indolizations was unsuccessful (Braun, 1908b).

I. Indolization of Unsymmetrically Substituted (other than 2-Substituted) Cycloalkanone Arylhydrazones

Fischer indolization of 3-substituted cyclopentanone arylhydrazones **820** ($n = 1$) could give rise to **821** ($n = 1$) and/or **822** ($n = 1$). Starting with **820** ($R^1 = R^2 = H$, $R^3 = COOH$, $n = 1$), only one product was obtained which was assigned structure **821** ($R^1 = R^2 = H$, $R^3 = COOH$, $n = 1$) on the basis of a p.m.r. investigation of its methyl ester (Lacoume, Milcent and Olivier,

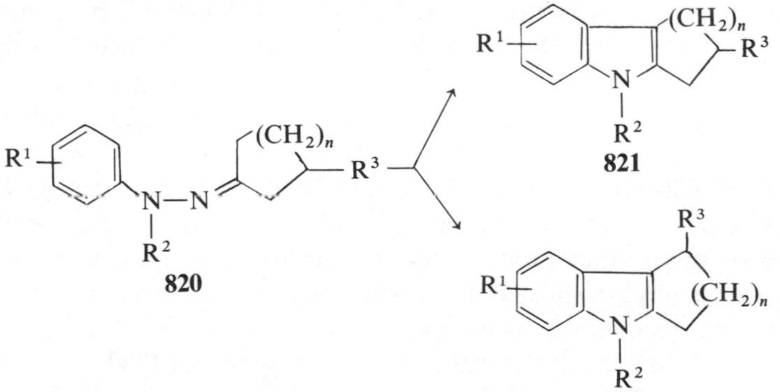

R^1 $(CH_2)_n$ R^3 N R^2

821

R^1 $(CH_2)_n$ R^3 N—N R^2

820

R^3 R^1 $(CH_2)_n$ N R^2

822

1972). This investigation used $Eu(DPM)_3$ which induced shifts said to be indicative of the two hydrogen atoms *cis* vicinal to the $COOCH_3$ group in **821** ($R^1 = R^2 = H$, $R^3 = COOCH_3$, $n = 1$) whereas only one such *cis* vicinal hydrogen atom would exist in **822** ($R^1 = R^2 = H$, $R^3 = COOCH_3$, $n = 1$). A model compound **738** ($R^1 = R^3 = H$, $R^2 = COOCH_3$, $n = 1$), having such a single *cis* vicinal hydrogen atom, was synthesized via indolization of 2-carbomethoxycyclopentanone phenylhydrazone and comparatively investigated. Further support for structure **821** ($R^1 = R^2 = H$, $R^3 = COOCH_3$, $n = 1$) was obtained when a negligible lanthanide induced shift was observed in the indolic 4-H atom of its 5-methoxy analogue, prepared by indolization of 3-carbomethoxycyclopentanone 4-methoxyphenylhydrazone. The use of model compounds indicated that a significant lanthanide induced shift would have been expected if this indolization product had the isomeric structure corresponding to **822** ($R^1 = OCH_3$, $R^2 = H$, $R^3 = COOCH_3$, $n = 1$). However, neglect of the angle term (Cockerill, Davies, G. L. O., Harden and Rackham, 1973) in this study is not justified and the conclusions may therefore by open to some doubt although, in accordance with the above conclusions, several arylhydrazones of 3-carboxycyclopentanone (Berger, L. and Corraz, 1977) have been found to indolize 'away' from the 3-substituted carbon atom of the cyclic ketonic moiety to afford the corresponding **821** ($n = 1$) as the only isolated products. On the contrary, indolization of **820** [$R^1 = 4\text{-}OCH_3$, $R^2 = COC_6H_4Cl(4)$, $R^3 = COOH$, $n = 1$] gave a mixture of **821** [$R^1 = 5\text{-}OCH_3$, $R^2 = COC_6H_4Cl(4)$, $R^3 = COOH$, $n = 1$] and **822** [$R^1 = 5\text{-}OCH_3$, $R^2 = COC_6H_4Cl(4)$, $R^3 = COOH$, $n = 1$] (Sumitomo Chemical Co. Ltd, 1968b).

The first 3-substituted cyclohexanone arylhydrazone to be subjected to indolization was **820** ($R^1 = R^2 = H$, $R^3 = COOH$, $n = 2$). When this reaction was initially performed (Baeyer and Tutein, 1889), only one of the two possible isomeric indolic products was obtained but the orientation of the carboxylic acid group was not established although the product was formulated as 4-carboxy-1,2,3,4-tetrahydrocarbazole. The formation of only one product from this reaction was later (Allen, G. R., 1970) confirmed and its structure was established as **821** ($R^1 = R^2 = H$, $R^3 = COOH$, $n = 2$) by dehydrogenation of its derived methyl ester to afford the known 2-carbomethoxycarbazole. The corresponding 4-methoxyphenylhydrazone and N_α-(4-chlorobenzoyl)-4-methoxyphenylhydrazone were also indolized to give **821** [$R^1 = OCH_3$, $R^2 = H$ and $COC_6H_4Cl(4)$, respectively, $R^3 = COOCH_3$, $n = 2$] as the only isolated products. Upon hydrolysis, the latter product gave the former product, the structure of this former product being established by its conversion into 6-methoxycarbazole-2-carboxaldehyde which was distinguished from the possible isomeric 4-carboxaldehyde by its u.v. spectrum (Allen, G. R., 1970). When the indolization of 3-methylcyclohexanone phenylhydrazone (**820**; $R^1 = R^2 = H$, $R^3 = CH_3$, $n = 2$) was initially investigated it was claimed (Plancher and

Carrasco, 1904) that only one of the two possible isomeric products was formed although the orientation of the methyl group was not established. In subsequent studies (Barclay and Campbell, 1945; Borsche, Witte, A. and Bothe, 1908; Grammaticakis, 1939; Rogers, C. U. and Corson, 1947) only one product could again be isolated from this reaction although the fact that this was isolated in low, unspecified, unspecified, and 65% yields, respectively, does not exclude the concomitant formation of the other possible isomer. The structure of the isolated product was established as **821** ($R^1 = R^2 = H$, $R^3 = CH_3$, $n = 2$) by its dehydrogenation to afford the known 2-methylcarbazole (Barclay and Campbell, 1945; Borsche, Witte, A. and Bothe, 1908). Related to these investigations, other studies (Morgan and Walls, L. P., 1930), of which Barclay and Campbell (1945) were apparently unaware, had shown that both (±)- and (+)-3-methylcyclohexanone 4-methylphenylhydrazone indolized upon warming in glacial acetic acid to afford in each case only one isolated product. These were both shown to be the corresponding 1,2,3,4-tetrahydro-2,6-dimethylcarbazoles by their dehydrogenation into 2,6-dimethylcarbazole which was unambiguously synthesized by two alternative routes. An analogous indolization orientation was subsequently shown to occur when 3-methylcyclohexanone 2-naphthylhydrazone was indolized. The only product isolated from this reaction was shown to be 5,6-benzo-1,2,3,4-tetrahydro-2-methylcarbazole (**823**) by its synthesis via an alternative unambiguous route. (Campaigne,

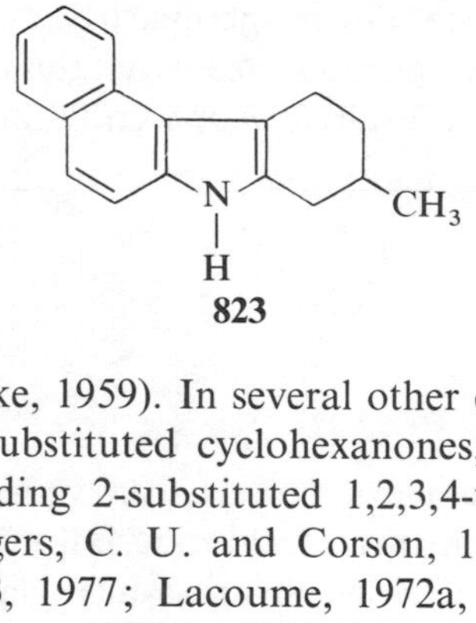

N
H
CH_3

823

Ergener, Hallum and Lake, 1959). In several other examples of indolizations of arylhydrazones of 3-substituted cyclohexanones, the only isolated products were the corresponding 2-substituted 1,2,3,4-tetrahydrocarbazoles **821** [$n = 2$, $R^3 = CH_3$ (Rogers, C. U. and Corson, 1947), COOH (Berger, L. and Corraz, 1972c, 1973, 1977; Lacoume, 1972a, 1972b; Organon, 1972), $(CH_2)_{1-3}COOH$ (Lacoume, 1972a), $COOC_2H_5$, CN and CH_2OH (Berger, L. and Corraz, 1977), $CH(CH_3)COOH$ (Berger, L. and Corraz, 1974), $CCH_3(COOC_2H_5)_2$ (Gurien and Teitel, 1979, 1980) and NH_2 (from 3-acetylaminocyclohexanone) (Fliedner, 1979)]. Indolization of **824** afforded only **825**, the structure of which was distinguished from the possible isomeric **826** by the appearance of a 4H-allylic signal in the p.m.r. spectrum of the product (Rice, Sheth and Wheeler, J. W., 1971).

Examples of indolizations of 3-substitued cyclohexanone arylhydrazones are known in which the orientation of the substituent in the product has not

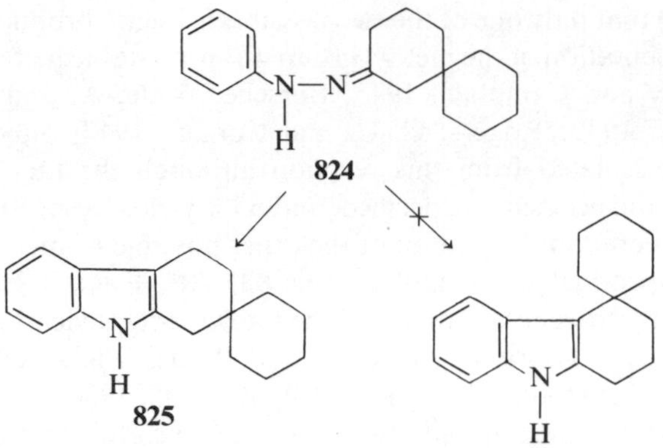

824

825

826

been established. Thus, the product resulting from each of the indolizations of 3-methylcyclohexanone 4-methyl-, 2,4-dimethyl-, 2,4,5-trimethyl- and 3-nitrophenylhydrazone (Borsche, Witte, A. and Bothe, 1908) was arbitrarily assigned the corresponding 1,2,3,4-tetrahydro-2-methylcarbazole structure. Indolization of 3-methylcyclohexanone 1-naphthylhydrazone (**827**) with a dilute sulphuric acid catalyst gave, in accordance with the results of the above studies, only one product (Cecchetti and Ghigi, 1930). Although no structure was proposed for this product, it was clearly recognized that one of the two possible 7,8-benzo-1,2,3,4-tetrahydromethylcarbazoles had been formed. By analogy with the above examples, it is likely that this product would have structure **828** but verification is awaited. Apparently at variance with the above

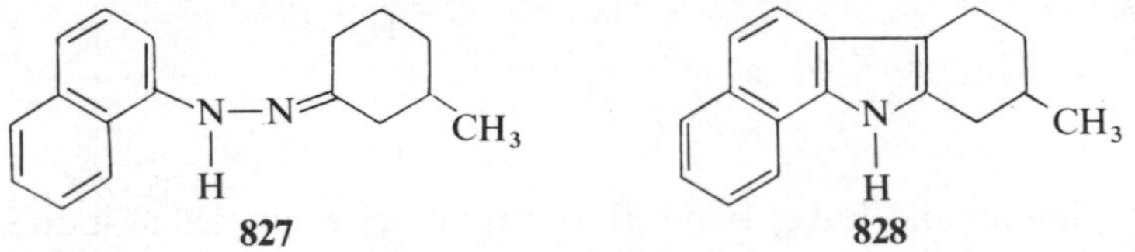

827

828

observations, 3-methylcyclohexanone N_α-benzylphenylhydrazone, upon indolization with alcoholic hydrogen chloride, afforded a product which was formulated as 9-benzyl-1,2,3,4-tetrahydro-4-methylcarbazole (Namis, Cortes, Collera and Walls, F., 1966). However, evidence for the orientation of the methyl substituent was lacking. Indolization of 3-isopropyl-5-methylcyclohexanone phenylhydrazone (**829**) with an acetic acid catalyst was reported (Horning, E. C., Horning, M. G. and Walker, G. N., 1948) to yield an apparently homogeneous product which for steric reasons was assigned structure **830**. However, no verification of this postulation was given and further investigation of the structure of this product would be of interest.

It might be expected that both possible indolic products should be formed from 3-substituted cyclohexanone arylhydrazones and no doubt in many of the

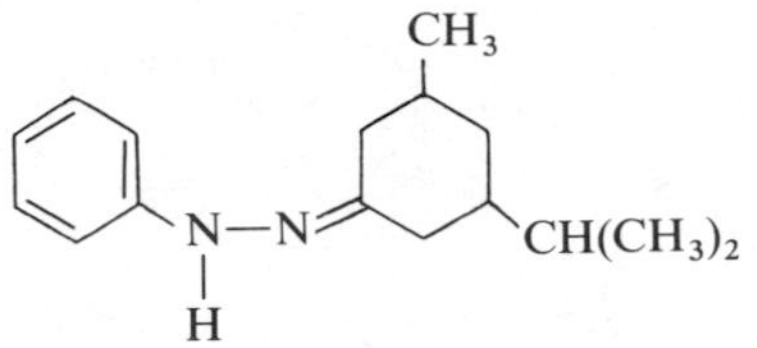

829

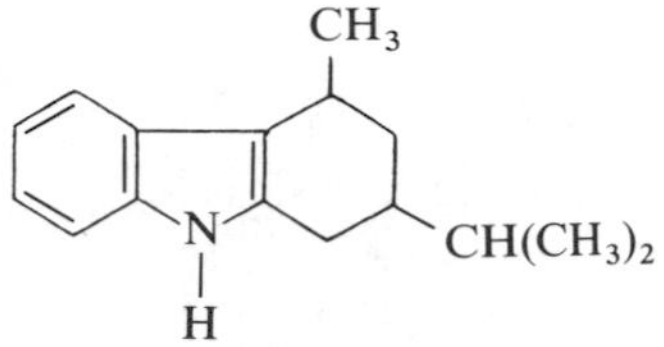

830

above mentioned indolizations from which only one product was isolated, the alternative isomeric product was simultaneously formed but was not isolated.

Indolization of a mixture of **831** [R = $CH_2CON(CH_3)_2$, $CH_2CON(C_2H_5)_2$, CH_2CO-1-pyrrolidinyl and $CH_2C_6H_5$] with phenylhydrazine hydrochloride using 7% ethanolic sulphuric acid as catalyst yielded only **832** [R^1 = H, R^2 = $CH_2CON(CH_3)_2$, $CH_2CON(C_2H_5)_2$, CH_2CO-1-pyrrolidinyl and $CH_2C_6H_5$,

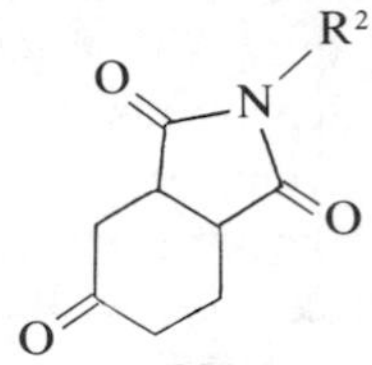

831

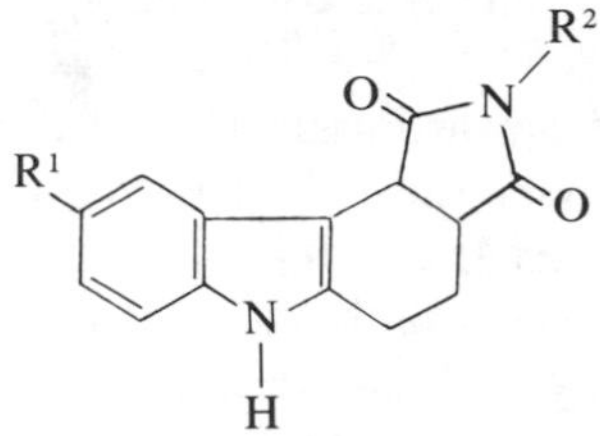

832

respectively] and likewise the indolization of a mixture of **831** (R = CH_2COOH) with 4-carboxyphenylhydrazine hydrochloride under similar conditions afforded only **832** (R^1 = $COOC_2H_5$, R^2 = $CH_2COOC_2H_5$) (Rashidyan, Asratyan, Karagezyan, Mkrtchyan, Sedrakyan and Tatevosyan, 1968). However, contrary to these observations in which only one of the two possible isomeric indolization products was isolated in each case, indolization of the phenylhydrazone of **831** (R = CH_3) using 7% ethanolic sulphuric acid, 15% methanolic sulphuric acid or a saturated solution of hydrogen chloride in acetic acid as catalyst afforded in each case a separable mixture of **832** (R^1 = H, R^2 = CH_3) and **833** in a yield ratio of approximately 3 : 1, respectively (Reed, G. W. B., 1975). The direction of indolization of the structurally related hydrazone **834** depended upon the indolization catalyst used. Thus, the use of a saturated solution of hydrogen chloride in acetic acid as catalyst led to the

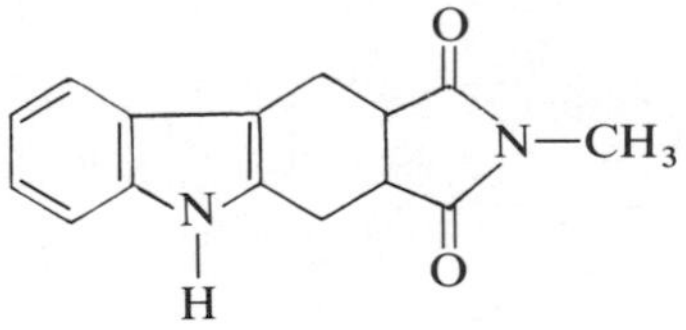

833

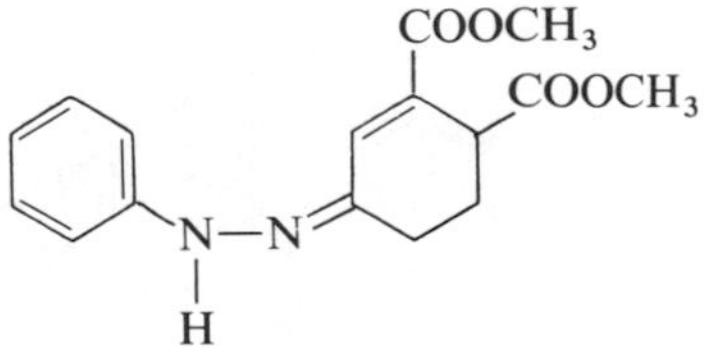

834

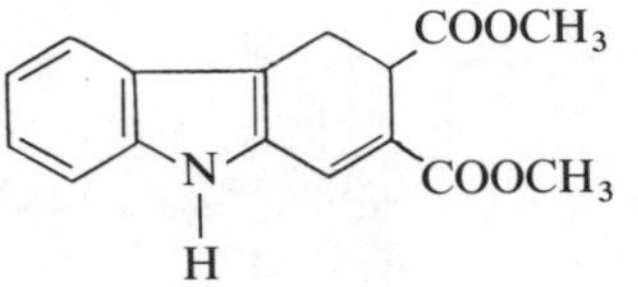

835

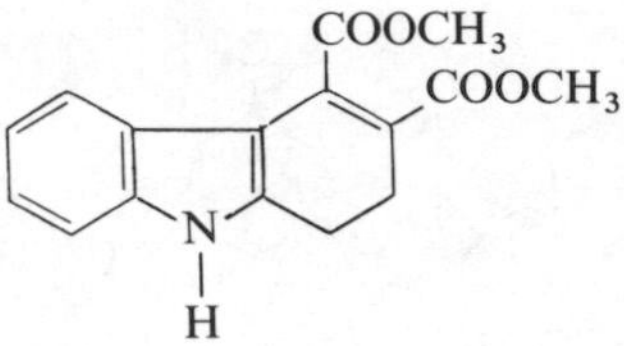

836

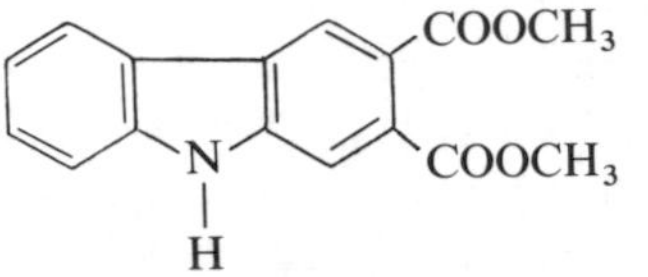

837

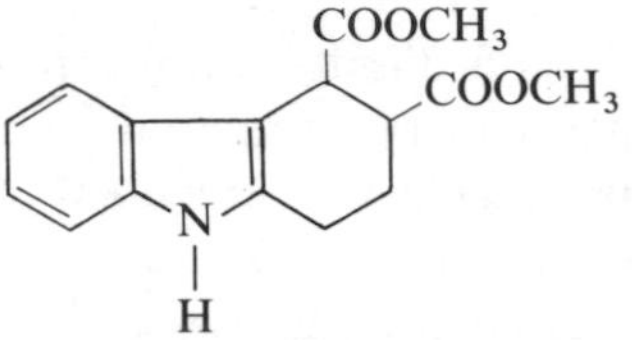

838

isolation of only **835** whereas using 7% or 15% methanolic sulphuric acid gave **835** and **836** in a 1 : 2 yield ratio, respectively, along with traces of **837** and **838** resulting from aromatization of **835** with synchronous reduction of **836**, respectively, under the indolization conditions. Upon using 60% methanolic sulphuric acid as the indolization catalyst, **837** became the major isolated product, being formed along with traces of **838** (Reed, G. W. B., 1975).

The indolization of *cis*-2-decalone phenylhydrazone (**839**; R = H), by boiling in acetic acid containing hydrogen chloride, produced a mixture, separable by fractional crystallization, of **840** (R = H) and **841**, in a yield ratio of approximately 2 : 1. The structures of these products were established by their dehydrogenation, using chloranil, into the known 2,3- and 3,4-benzocarbazole, respectively (Buu-Hoï, Jacquignon and Lavit, 1956). The authentic specimen of the latter dehydrogenation product referred to in this work had been synthesized (Japp and Maitland, 1901, 1903b) by indolization of a mixture of 2-naphthol with phenylhydrazine/phenylhydrazine hydrochloride (see Chapter

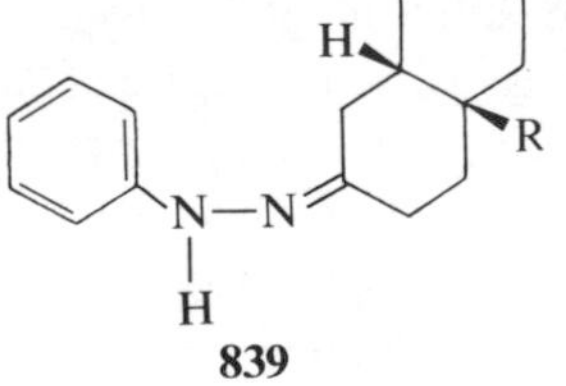

839

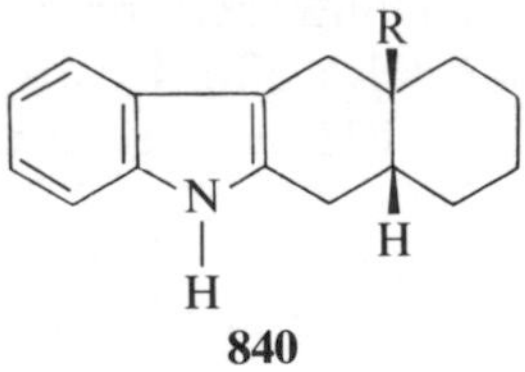

840

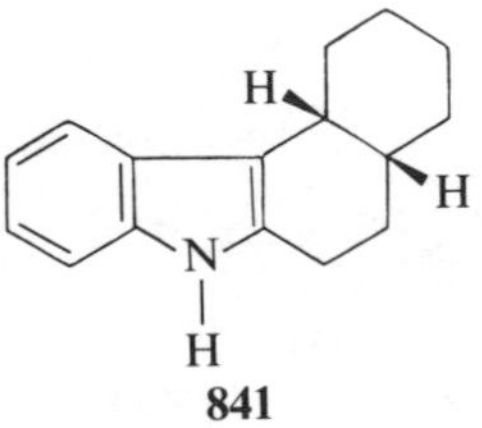

841

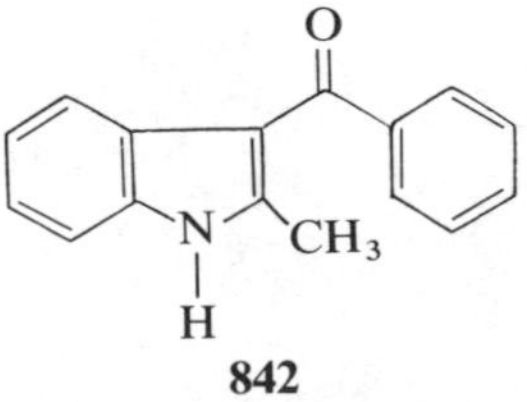

842

I, E), a synthesis which is itself potentially ambiguous and could afford **85** (R^1–R^5 = H) and/or **102** (R^1 = R^2 = H). However, the only product obtained in this latter mentioned work was shown to be 3,4-benzocarbazole by comparison with a specimen obtained earlier by an unambiguous synthesis involving indolization of a mixture of 2-hydroxynaphthalene-3-carboxylic acid with phenylhydrazine, followed by decarboxylation (Schöpff, 1896; Ullmann, 1898). The authentic sample of 2,3-benzocarbazole had been obtained (Buu-Hoï, Hoán and Khôi, 1950a) by an unambiguous Elbs reaction involving thermal cyclization of 3-benzoyl-2-methylindole (**842**). Indolization of both *cis*- and *trans*-9-methyl-3-decalone phenylhydrazone (**839**; R = CH_3) and (**843**), using zinc chloride in refluxing ethanol or acetic acid as catalyst, furnished in both cases only the linear isomeric products, **840** (R = CH_3) and **844**,

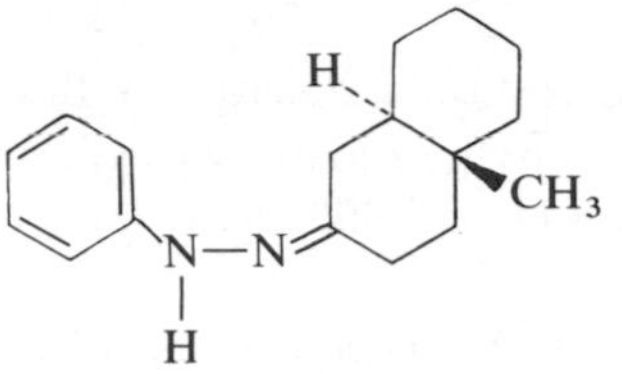

843 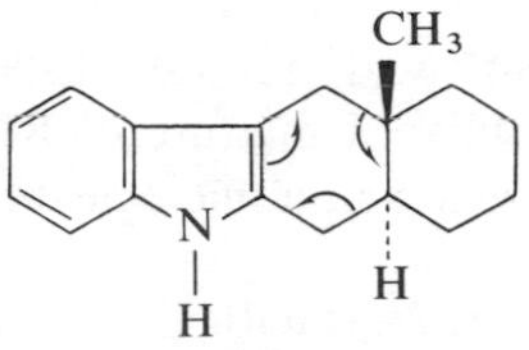 **844**

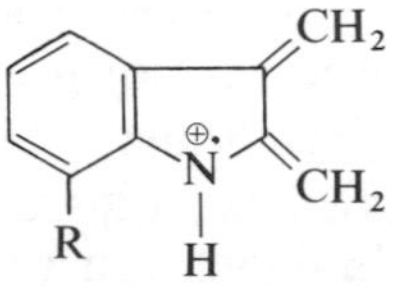

845

respectively. As in earlier related studies (Stork and Dolfini, 1963 – see below), the structures of these products were differentiated from their possible angular isomers by their mass spectral fragmentation pattern which indicated the occurrence of a retro Diels–Alder reaction, as illustrated in **844**, to afford a radical ion **845** (R = H) (m/e = 143) (Miller, F. M. and Lohr, 1978). The formation of only the linear isomeric product in each of these indolizations is analogous to the formation of the 2-bromo derivative from both *cis*- and *trans*-9-methyl-3-decalone (Yanagita and Yamakawa, 1956, 1957). Both these results and those of the indolizations were considered in terms of a conformational analysis of the 9-methyl-3-decalone system, with particular attention to the elimination of non-bonded hydrogen–hydrogen interactions when enolization occurs towards the C-2 atom in both the *cis* and *trans* isomers (Lohr, 1977; Miller, F. M. and Lohr, 1978). Indolization of **846** with a zinc chloride catalyst in benzene afforded **847**. The structure of this product was distinguished from the alternative angular isomer by its mass spectrum which again exhibited a major peak resulting from a retro Diels–Alder reaction as shown in **847** (Britten, A. and Lockwood, 1974). The stereochemistry of the ring junction in

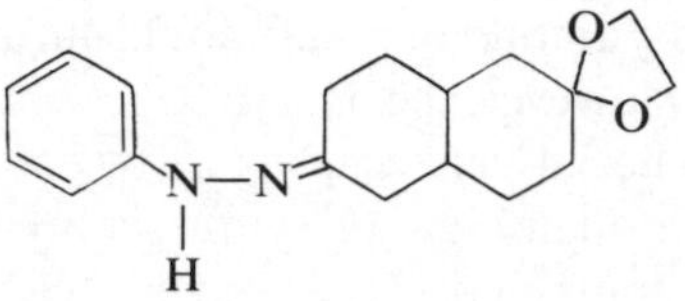

846 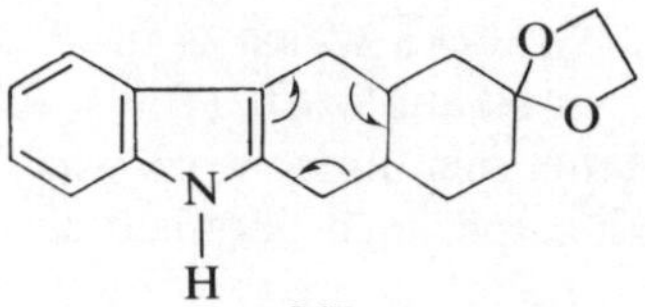 **847**

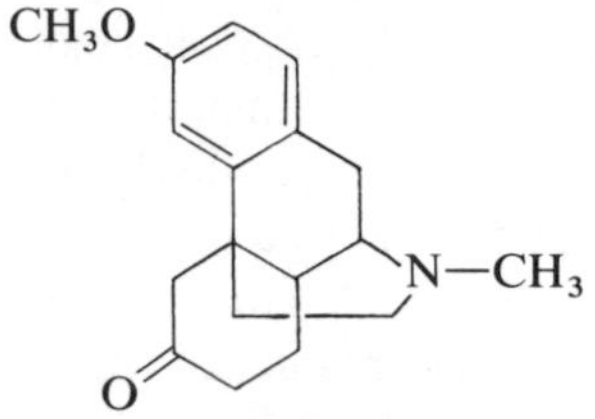

848

846 and **847** was not established. Indolization occurred in both possible directions when a mixture of **848** and phenylhydrazine was treated with 4-methylbenzene sulphonic acid to furnish **849** and **850** (Sawa and Miyamoto, 1970).

Treatment of **851** with acetic acid afforded 15% yields of each of the products shown to be the corresponding stereoisomers **852** and **853**. These structures were readily distinguished from the possible angular isomers by their mass spectra which in both cases showed a major peak at m/e = 173 corresponding to the radical ion **845** (R = OCH_3) formed by a retro Diels–Alder reaction as illustrated in **852** (Stork and Dolfini, 1963). Likewise, the indolization of *trans*-1-benzyl-decahydroquinolin-7-one phenylhydrazone occurred at the 6-position of the quinoline nucleus to give the linear isomeric indole as the only

849 **850**

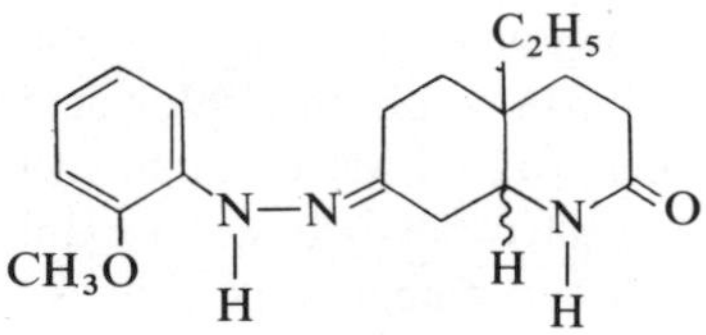

851 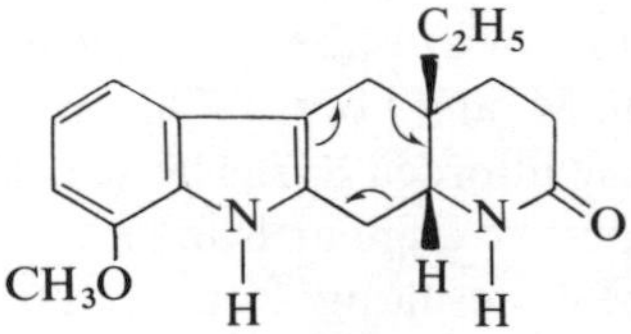 **852**

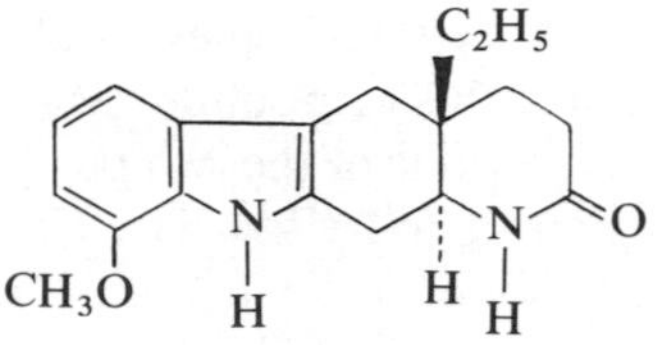

853

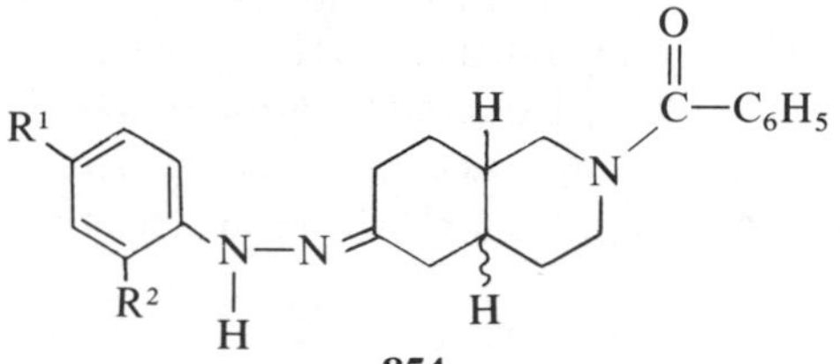

854

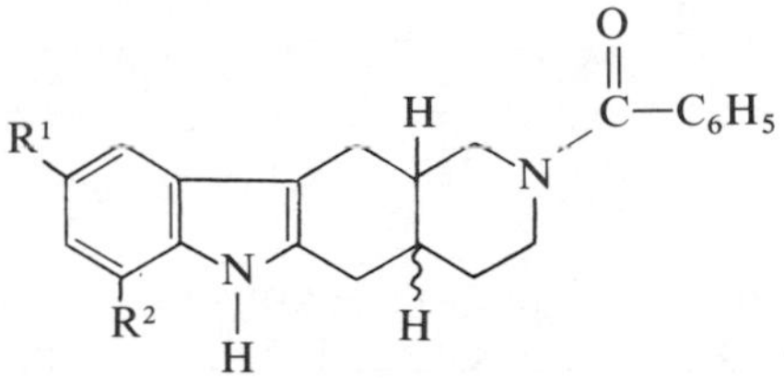

855

$(R^1{=}R^2{=}H)$

640 (R^1-R^4 = H, X = N, Y = CH)

isolated product (Kumar and Jain, 1979). Similar catalytic treatment of *cis*- and *trans*-**854** (R^1 = R^2 = H, R^1 = H, R^2 = F and CH_3 and R^1 = F and OCH_3, R^2 = H) (these indolizations could be more conveniently achieved by heating mixtures of the appropriate hydrazine and ketone in ethanolic hydrochloric acid) afforded, as the only isolated products, the corresponding *cis* and *trans* linear indoles **855** (R^1 = R^2 = H, R^1 = H, R^2 = F and CH_3 and R^1 = F and OCH_3, R^2 = H, respectively). The structures of the products were established (Rastogi, Bindra, Rai and Anand, 1972) by their i.r., u.v., p.m.r. and mass spectral properties and those of their corresponding dehydrogenation products **640** (R^1–R^4 = H, R^1 = R^2 = R^4 = H, R^3 = F and CH_3 and R^1–R^3 = H, R^4 = F and OCH_3, respectively, X = N, Y = CH), formed by heating with 10% palladium-on-carbon in diphenylether but, unfortunately, these spectral data were not presented. The product formulated as **640** (R^1–R^4 = H, X = N, Y = CH) in these studies also had a m.p. close to that quoted in the literature for this compound (Manske, R. H. F. and Kulka, 1949, 1950b) (see page 240). Only one of the two possible isomeric products was isolated from the indolization of **641** (R = H). Upon dehydrogenation, this afforded the same product as obtained by sequential indolization and dehydrogenation of cyclohexanone 7-isoquinolylhydrazone (Manske, R. H. F. and Kulka, 1949). Since this latter product most likely has structure **609** (R^1 = R^2 = H, X = N, Y = CH) (Manske, R. H. F. and Kulka, 1949, 1950b) it is likely that the product isolated from indolization of **641** (R = H) has structure **856**. Structural verification by mass spectral investigation should now be effected. However, since the pure dehydrogenation product of **856** was only

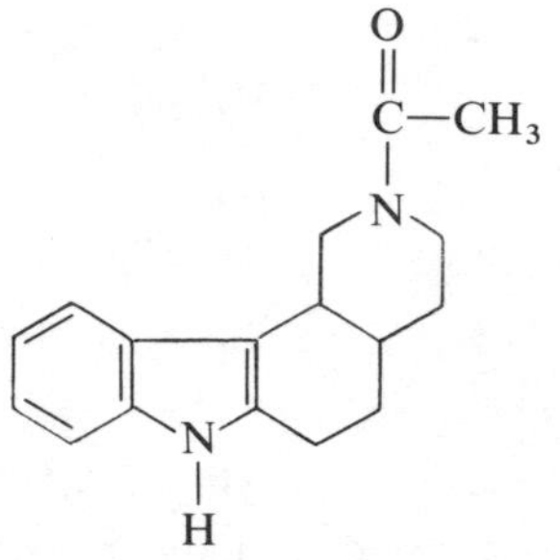

856

isolated in low yield from the dehydrogenation of the total indolization reaction mixture (Manske, R. H. F. and Kulka, 1949), the concomitant formation of **642** (R = H) is by no means eliminated. Indeed, sulphuric acid catalysed indolization of **641** (R = CH_3) gave a non-basic product (Manske, R. H. F. and Kulka, 1949) which must have structure **642** (R = CH_3) since indolization in the alternative possible direction would have given a basic 3*H*-indole, the simultaneous formation of which is not eliminated since **642** (R = CH_3) was only isolated in low yield and the basic reaction products were not investigated.

Indolization of 3-cyanoheptanone phenylhydrazone (**857**; R = H) and 4-methoxyphenylhydrazone (**857**; R = OCH_3) afforded in each case only one product, **858** (R^1 = H and OCH_3, respectively, R^2 = CN). The structures of these products were distinguished (Lacoume, Milcent and Olivier, 1972) from the possible isomers **859** (R = H and OCH_3, respectively) by conversion into the corresponding methyl esters **858** (R^1 = H and OCH_3, respectively, R^2 = $COOCH_3$). P.m.r. spectral examination of these esters showed that, in each

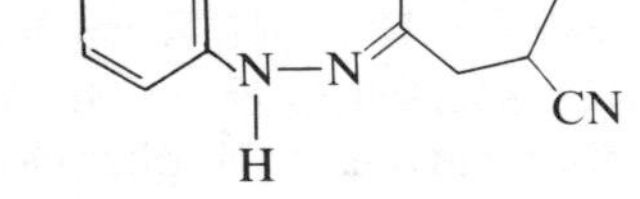

857

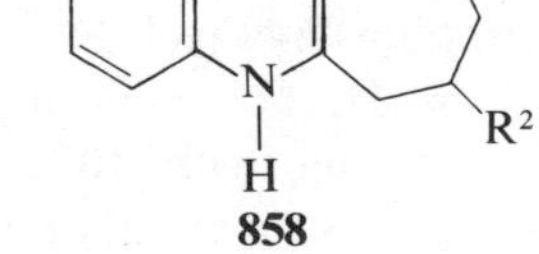

858

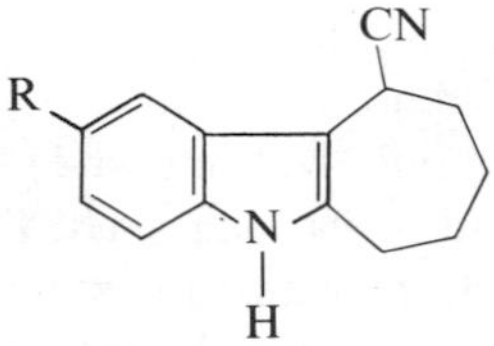

859

case, addition of $Eu(DPM)_3$ caused a downfield shift of the signals caused by two of the four allylic protons. The structure of the analogous esters from **859** (R = H and OCH_3) would not be compatible with these observations although, once again, neglect of the angle term, as already referred to earlier in this section, may cast some doubt upon this interpretation.

It can be concluded from these above results that indolization of 3-substituted cycloalkanones afforded only indoles resulting from reaction at the cycloalkanone carbon atom 'away' from the 3-substituent, resulting from the formation during indolization of the enehydrazine tautomer **860** rather than the formation of **861** from **820**. Supporting this hypothesis were the experimental observations that bromination of 3-carboxycyclohexanone primarily formed **862** and that the enol acetate obtained from 3-carbomethoxycyclohexanone was mainly **863** and the calculated conclusion, using the LCAO method, that equatorial 4-methylcyclohexene is 0.8 kcal/mol more stable than pseudo equatorial 3-methylcyclohexene (Lacoume, Milcent and Olivier, 1972). Other similar small energy differences had also earlier been calculated to exist between related isomers [Malhotra, Moackley and Johnson, F., 1967; Charles, Descotes, Martin, J. C. and Querou, 1968 (quoted by Lacoume, Milcent and Olivier,

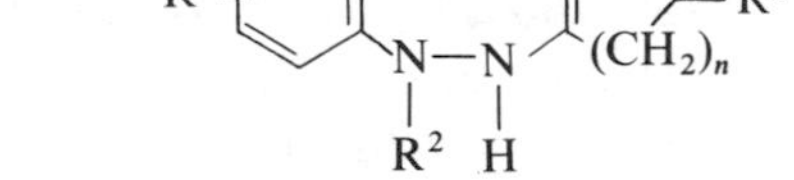

820

860

861

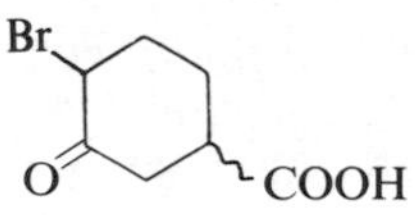

862

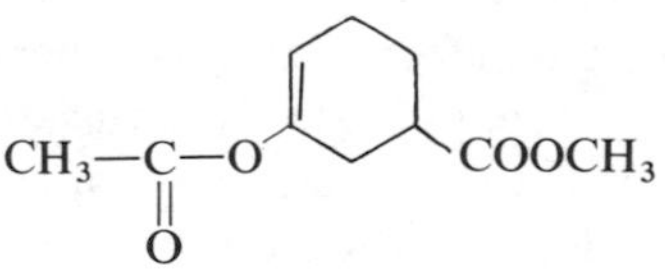

863

1972)]. However, it was later (Grandberg and Sorokin, 1974) suggested that these calculated energy differences, being only small, may not exert large directive influences upon such indolizations when they are effected under fairly severe reaction conditions and it was suggested that it was much more likely that steric factors operating in the transition state in the formation of the new C—C bond were the directive influence. A combination of the two factors may be most likely.

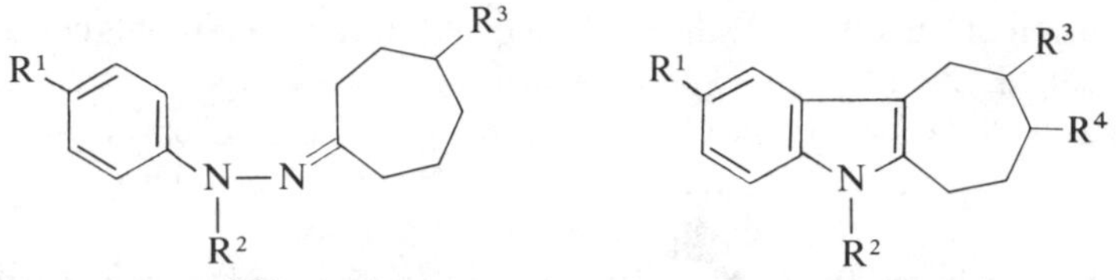

864 **865**

Indolization of 4-carboxycycloheptanone 4-methoxyphenylhydrazone (**864**; $R^1 = OCH_3$, $R^2 = H$, $R^3 = COOH$) yielded both possible isomeric indolic products, **865** ($R^1 = OCH_3$, $R^2 = R^3 = H$, $R^4 = COOH$ and $R^1 = OCH_3$, $R^2 = R^4 = H$, $R^3 = COOH$) which were separated by fractional crystallization. The structures of these two products were distinguished from a study of the effect of $Eu(DPM)_3$ upon the p.m.r. spectra of their methyl esters. The product having the p.m.r spectrum in which two allylic and two alicyclic protons were sensitive towards the addition of the shift reagent was assigned structure **865** ($R^1 = OCH_3$, $R^2 = R^4 = H$, $R^3 = COOCH_3$) and that having the p.m.r. spectrum in which analogous sensitivity was exhibited by four alicyclic, non-allylic protons was assigned structure **865** ($R^1 = OCH_3$, $R^2 = R^3 = H$, $R^4 = COOCH_3$) (Lacoume, Milcent and Olivier, 1972). Similarly, 4-benzoyloxycycloheptanone N_α-methylphenylhydrazone (**864**; $R^1 = H$, $R^2 = CH_3$, $R^3 = OCOC_6H_5$) afforded both possible isomeric indolic products, **865** ($R^1 = R^3 = H$, $R^2 = CH_3$, $R^4 = OCOC_6H_5$ and $R^1 = R^4 = H$, $R^2 = CH_3$, $R^3 = OCOC_6H_5$) which were separated by chromatography. The structural differentiation of these products was based upon p.m.r. spectral comparisons which indicated the higher degree of symmetry in the former isomer (Julia and Lenzi, 1971).

2-Tetralone phenylhydrazone (**866**; R^1–R^3 = H) was indolized, using dilute sulphuric acid as a catalyst [indolization using a zinc chloride catalyst in refluxing absolute ethanol with subsequent distillation of the product under reduced pressure also caused isomerization and dehydrogenation (see Chapter V, A) to yield 1,2-benzocarbazole (**62**) (Ghigi, 1931)], to give only one product which was formulated as either **867** or **868** (R^1–R^3 = H) (Ghigi, 1930). Subsequently (Ghigi, 1931), the latter structure was verified by dehydrogenation of the product, by distillation from lead(II) oxide-on-pumice in a current of carbon dioxide, to form the known 3,4-benzocarbazole (**609**; $R^1 = R^2 = H$, X = Y = CH). The preferential formation of **868** (R^1–R^3 = H) from the

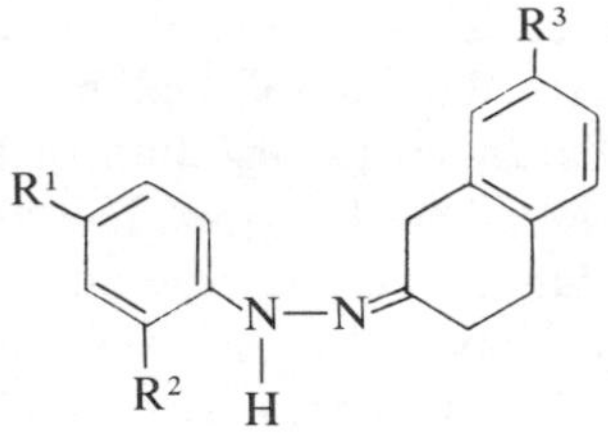

indolization was rationalized (Robinson, B., 1963a) by the suggestion that the enehydrazine tautomer of **866** (R^1–R^3 = H), **869** (or its protonated form) which would ultimately lead to **868** (R^1–R^3 = H), would be preferentially formed since it contains greater conjugative stabilization than the alternative enehydrazine tautomer **870** which would utlimately lead to **867**. However, this interpretation was questioned (Brown, R. K., 1972a) since it was not compatible with the earlier observation (Janetsky and Verkade, 1945) that the only product isolated from the copper(I) chloride catalysed indolization of benzyl ethyl ketone phenylhydrazone was 2-benzyl-3-methylindole. Other 2-tetralone arylhydrazones, **866** (R^1 = F, Cl, Br, I, CH_3 and OCH_3, R^2 = R^3 = H and R^1 = R^3 = H, R^2 = F, Cl, Br, I, CH_3 and OCH_3), have been indolized (Pecca and Albonico, 1971) using hydrogen chloride in acetic acid as the catalyst. The use of this catalyst, which also greatly improved the yield of **868** (R^1–R^3 = H) obtained from **866** (R^1–R^3 = H) by comparison with the sulphuric acid catalyst used earlier (Ghigi, 1930), furnished very good yields of **868** (R^1 = F, Cl, Br, I, CH_3 and OCH_3, respectively, R^2 = R^3 = H) (50–96%) but only moderate yields of **868** (R^1 = R^3 = H, R^2 = F, Cl, Br, I, CH_3 and OCH_3, respectively) (40–60%). In these examples, the angular rather than the corresponding linear structures for the products were confirmed by spectral

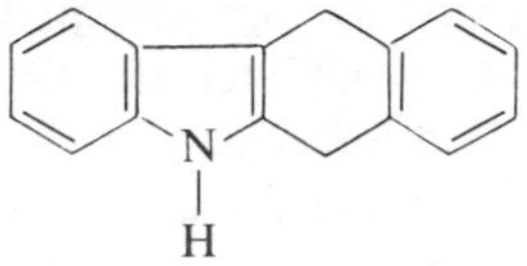

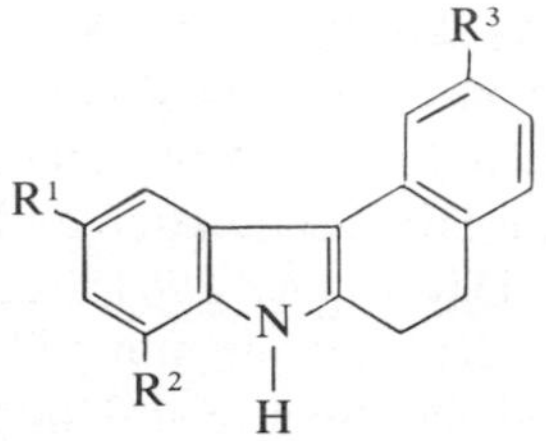

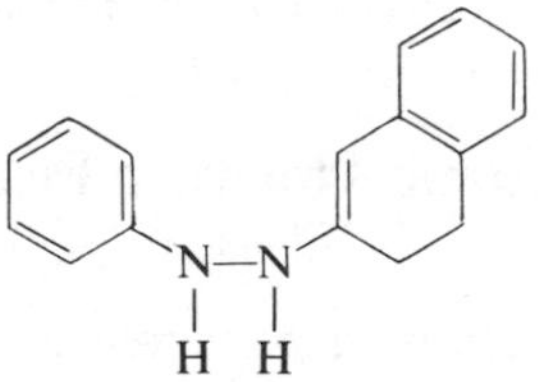

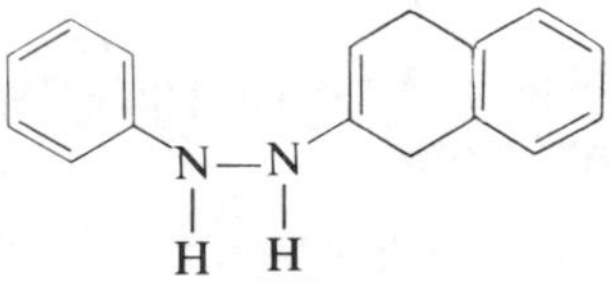

methods and although details were not presented, u.v. and p.m.r. data would be conclusive in this respect. Indeed, u.v. spectral data established that the product obtained from the polyphosphate ester catalysed indolization of **866** ($R^1 = R^2 = H$, $R^3 = OCH_3$) was **868** ($R^1 = R^2 = H$, $R^3 = OCH_3$) and not the corresponding linear isomer (Kanaoka, Ban, Miyashita, Irie and Yonemitsu, 1966). 2-Tetralone 2-, 3- and 4-carboxyphenylhydrazone have all been indolized to the corresponding 3,4-benzo-1,2-dihydrocarbazoles (Avanesova, Musaelyan and Tatevosyan, 1972), as have the 2-, 3- and 4-carbethoxyphenylhydrazone of 5,8-dimethoxy-2-tetralone (Avanesova and Tatevosyan, 1970). By analogy with these above results it has been assumed that the new C—C bond formation during the indolization of other 2-tetralone (Buu-Hoï, Hoán and Khôi, 1949, 1950b; Buu-Hoï, Jacquignon and Hoeffinger, 1963; Buu-Hoï, Jacquignon, Roussel, O. and Hoeffinger, 1964; Buu-Hoï, Périn and Jacquignon, 1960, 1962; Buu-Hoï, Saint-Ruf, Jacquignon and Barrett, G. C., 1958; Nakazaki, 1960a) and 6-methyl-2-tetralone (Buu-Hoï, Jacquignon and Hoeffinger, 1963; Buu-Hoï, Jacquignon, Roussel, O. and Hoeffinger, 1964) arylhydrazones occurred at the C-1 and not at the C-3 atom of the 2-tetralone moiety.

J. The Use of Ketosteroids as the Ketonic Moiety in Fischer Indolizations

The early studies which have led to the synthesis of indolosteroids by this route have been briefly referred to in reviews upon the chemistry of steroidal heterocycles (Zhungietu and Dorofeenko, 1967) and upon derivatives of steroids with condensed heterocycles (Akhrem and Titov, 1967).

1. 2-Ketosteroids

Work on this subject has been limited to the polyphosphoric acid catalysed indolization of a mixture of N_α-methylphenylhydrazine and 5α-cholestan-2-one (**871**). This reaction afforded a product which was formulated as **872** rather than the angular isomer (Warnhoff and NaNonggai, 1962) since **871** was known (Djerassi and Nakano, T., 1960) to give the Δ^2-enol and so, by analogy, enehydrazine formation, the initial stage of the indolization mechanism, would

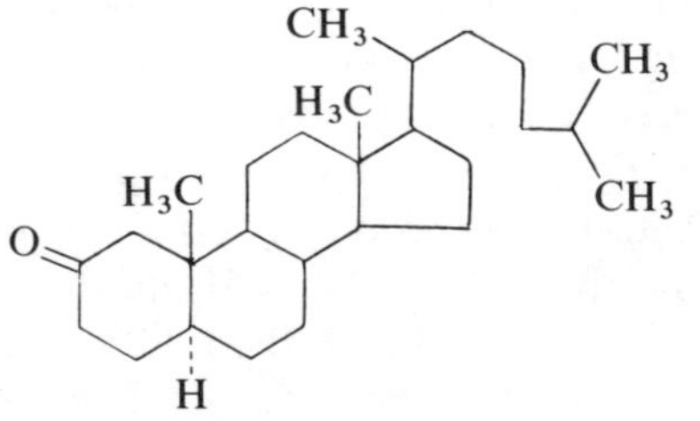

CH_3 CH_3 H_3C H_3C CH_3 O H

871

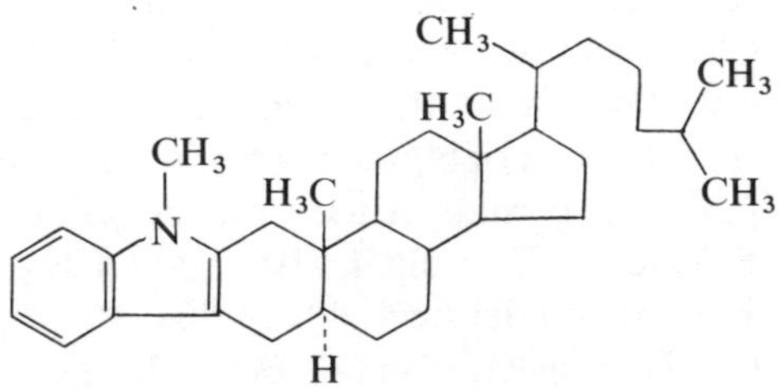

CH_3 CH_3 H_3C CH_3 N H_3C CH_3 H

872

be expected to occur in the same direction. Furthermore, formation of the angular isomer would involve the introduction of considerable steric crowding relative to the formation of **872** (Buckingham and Guthrie, 1968b).

2. 3-Ketosteroids

Many detailed studies have been carried out upon the indolization of 3-ketosteroid arylhydrazones. The first reaction of this type was reported in 1908 (Dorée and Gardner, J. A.) when 5β-cholestan-3-one [**873**; R^1–R^5 = H, R^6 = $CH(CH_3)(CH_2)_3CH(CH_3)_2$], at that time of unknown structure but recognized

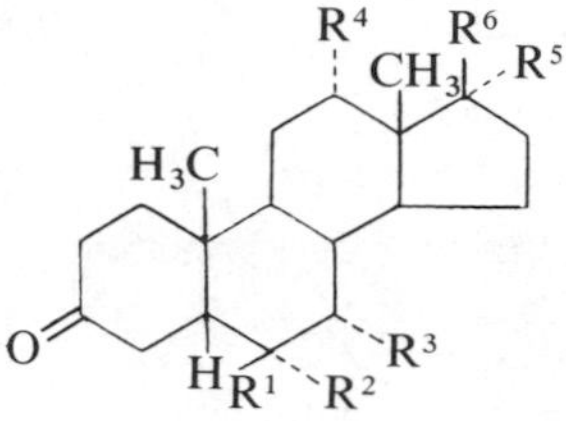

873

as a ketone, was boiled with phenylhydrazine or reacted with phenylhydrazine in warm acetic acid. Elemental analytical data clearly indicated that the product of this reaction was not the expected phenylhydrazone although the derived empirical formula of $C_{33}H_{51}N$ threw no light upon its chemical structure. In the following year (Dorée, 1909) these analytical data were reassessed, resulting in the empirical formula being modified to $C_{33}H_{49}N$ which, assuming it to represent the molecular formula, indicated a loss of the elements of water and ammonia from a combination of the reactants. This, together with an awareness of earlier studies (Dreschsel, 1888; Borsche, Witte, A. and Bothe, 1908) upon the application of the Fischer indolization to cyclohexanone arylhydrazones, led to the recognition (Dorée, 1909) of the indolic nature of the reaction product. It is interesting that this product was formed simply by boiling a mixture of the ketone with phenylhydrazine (Dorée and Gardner, J. A., 1908), this representing the earliest known example of a non-catalytic thermal indolization (see Chapter IV, L).

Twenty-six and twenty-eight years later (Dorée and Petrow, V. A., 1935; Schwenk and Whitman, 1937, respectively, a related synthesis, using refluxing glacial acetic acid as the catalyst and using **874** [R^1–R^7 = H, R^8 = $CH(CH_3)(CH_2)_3CH(CH_3)_2$] in place of the 5$\beta$-steroidal reactant, was effected. The product, which was also obtained by reacting 'monobromocholestanone' with phenylhydrazine (Schwenk and Whitman, 1937), was formulated as **875** [5α-H, R^1–R^7 = H, R^8 = $CH(CH_3)(CH_2)_3CH(CH_3)_2$, X = H_2] rather than the linear isomer **876** [5α-H, R^1–R^7 = H, R^8 = $CH(CH_3)(CH_2)_3CH(CH_3)_2$] on the basis of surface film measurements and chemical analogies available at

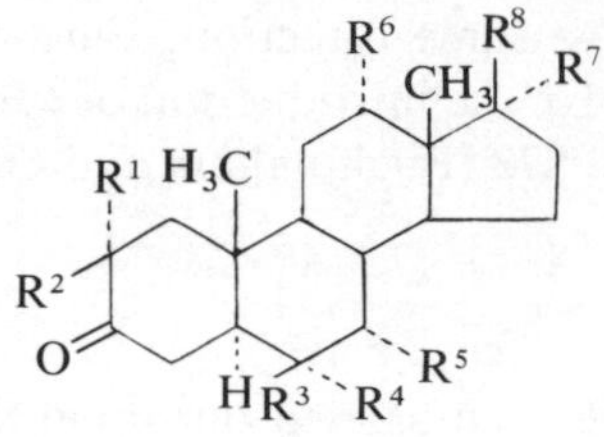

874

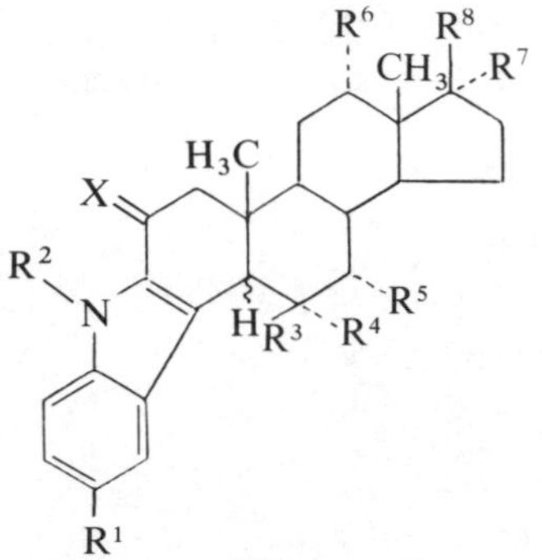

875

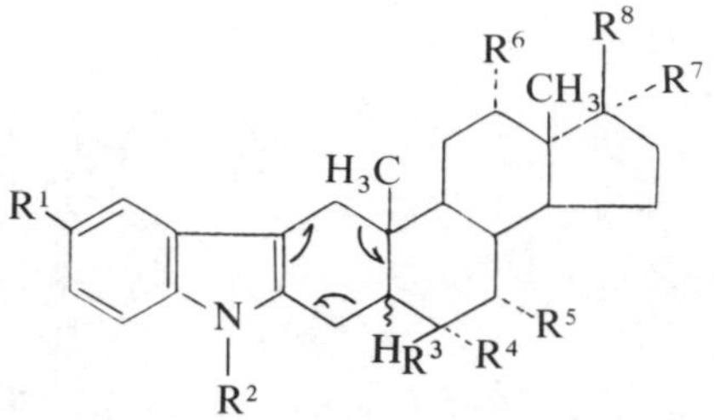

876

the time. A later group (Buckingham and Guthrie, 1968a), when repeating the indolization of this 5α-isomer in refluxing acetic acid, were apparently, from their literature citations, unaware of the earlier (Ban and Sato, Y., 1965 – see below) verification of the linear isomeric structure for the product, when they formulated it as such but only making reference to the studies (Dorée and Petrow, V. A., 1935) in which it had been formulated as the corresponding angular isomer as already mentioned. Another investigator (Rossner, 1937) did not distinguish between these two possible structures. This structural assignment was subsequently (Antaki and Petrow, 1951) reversed, although without further reason or evidence, the product from the reaction of the 5α-isomer **874** [R^1–R^7 = H, R^8 = $CH(CH_3)(CH_2)_3CH(CH_3)_2$] with phenylhydrazine in acetic acid now being assigned structure **876** [5α-H, R^1–R^7 = H, R^8 = $CH(CH_3)(CH_2)_3CH(CH_3)_2$], the angular structure **875** [5β-H, R^1–R^7 = H, R^8 = $CH(CH_3)(CH_2)_3CH(CH_3)_2$, X = H_2)] being assigned to the product obtained by reaction of 5β-cholestan-3-one **873** [R^1–R^5 = H, R^6 = $CH(CH_3)(CH_2)_3CH(CH_3)_2$] with phenylhydrazine in acetic acid (Dorée, 1909 and Dorée and Gardner, J. A., 1908).

Structure **875** [5β-H, R^1–R^7 = H, R^8 = $CH(CH_3)(CH_2)_3CH(CH_3)_2$] for the above mentioned product resulting from the indolization of 5β-cholestan-3-one phenylhydrazone has been verified (Ban and Sato, Y., 1965) by subjecting it to ozonolysis to yield **877**. This product was reduced with sodium borohydride to form the corresponding alcohol which underwent spontaneous dehydration to yield **878**, sequential ozonolysis and base catalysed hydrolysis of which

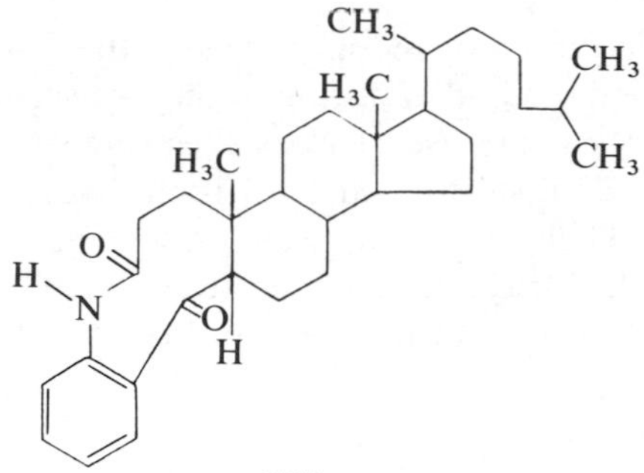

877 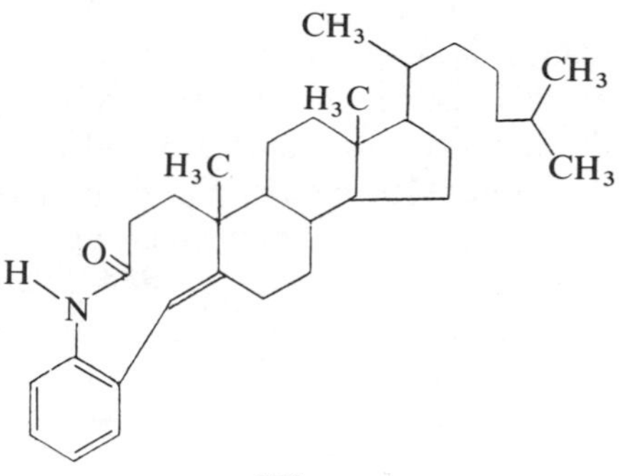 **878**

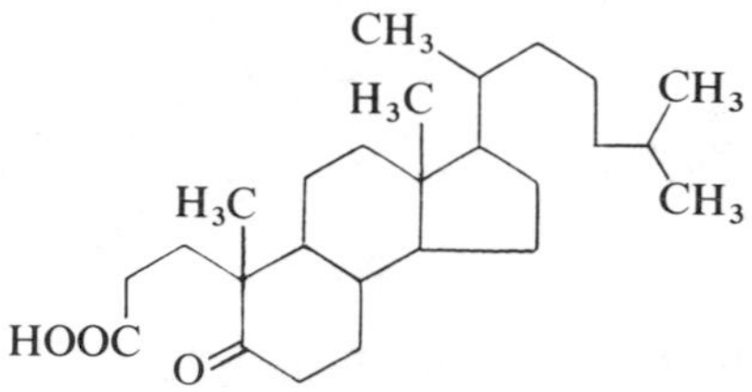

879 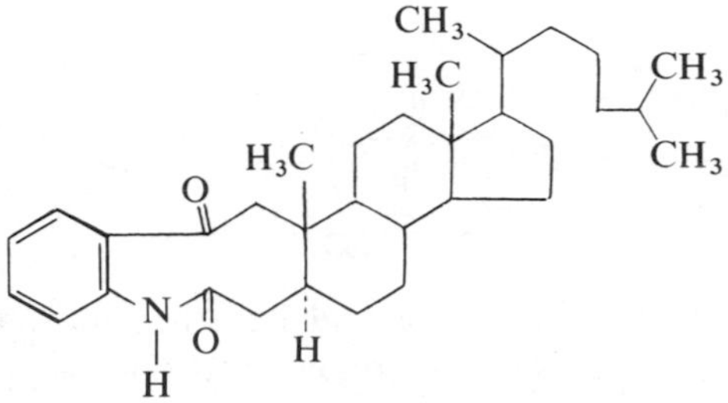 **880**

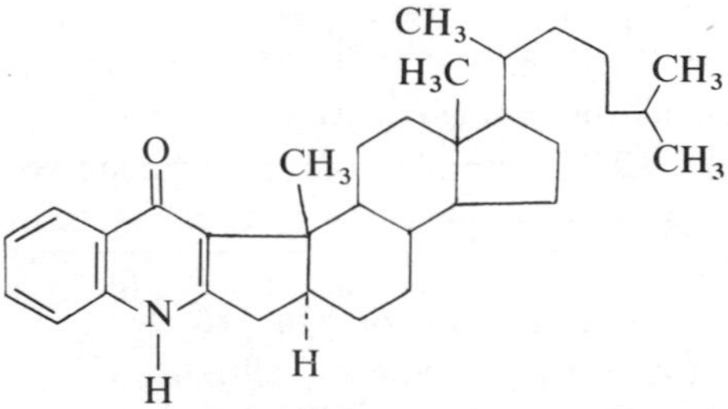

881 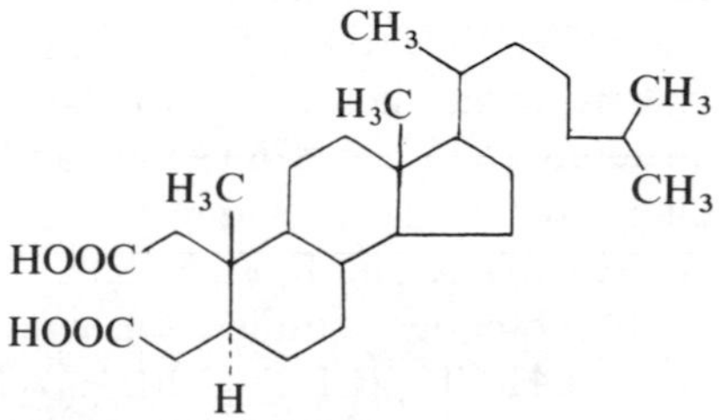 **882**

furnished the known compound **879**. The product resulting from the indolization of 5α-cholestan-3-one phenylhydrazone was also subjected to ozonolysis to afford, after chromatography of the crude product on alumina, compounds **880** and **881**. The latter product was formed by intramolecular condensation of the former product under the chromatographic alkaline conditions. The former product, upon sequential oxidation with hydrogen peroxide in acetic acid and hydrolysis produced the known compound **882**. These results confirmed the verification (Warnhoff and NaNonggai, 1962) that the product obtained by the polyphosphoric acid catalysed indolization of a mixture of 5β-cholestan-3-one and N_α-methylphenylhydrazine had structure **875** [5β-H, $R^1 = R^3$–R^7 = H, $R^2 = CH_3$, $R^8 = CH(CH_3)(CH_2)_3CH(CH_3)_2$, $X = H_2$] since it was shown to be the C-5 epimer of the compound obtained by catalytic hydrogenation of the product resulting from the indolization of a mixture of 5α-cholest-1-en-3-one

(**883**) and N_α-methylphenylhydrazine which can only indolize towards the C-4 atom of the steroidal nucleus. A more recent method of distinguishing between the isomeric structures **875** and **876** involved the use of mass spectrometry. Structure **876**, containing as it does a 2,3-disubstituted-1,2,3,4-tetrahydrocarbazole moiety, fragmented by a retro Diels–Alder pathway, as shown, to give rise to a radical ion **884** whereas the fragmentation of structure **875** (X = H_2) does not give rise to such a radical ion (Britten, A. and Lockwood,

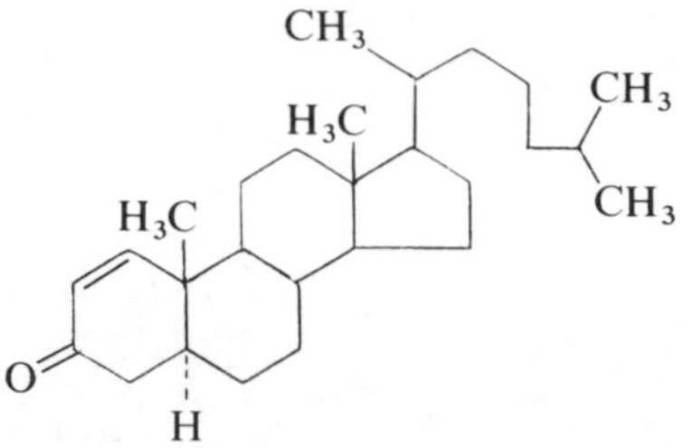

883

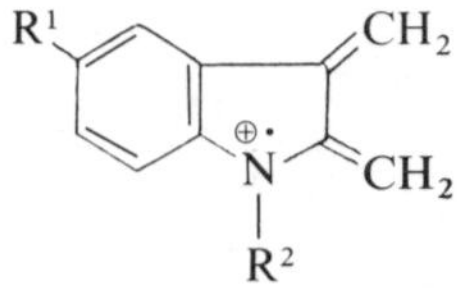

884

1974; Catsoulacos and Papadopoulos, 1976; Jacquignon, Croisy-Delçey and Croisy, 1972). In practice, the mass spectra of the linear isomers **876** had very intense ion peaks corresponding to **884** whereas such intense fragmentation ion peaks were absent in the mass spectra of the angular isomers **875** (X = CH_2) (Harvey, D. J., Laurie and Reed, R. I., 1971a, 1971b). P.m.r. spectroscopic data have also been used to distinguish between the isomeric structures **875** and **876** (Akiba and Ohki, 1970) and also to verify the stereochemistry of the A/B ring junction of the steroidal moiety in these systems (Harvey, D. J. and Reid, S. T., 1972).

In subsequent studies, the conversion of a mixture of **874** [R^1–R^7 = H, R^8 = $CH(CH_3)(CH_2)_3CH(CH_3)_2$] with phenylhydrazine in refluxing glacial acetic acid into **876** [5α-H, R^1–R^7 = H, R^8 = $CH(CH_3)(CH_2)_3CH(CH_3)_2$] was repeated (Harvey, D. J., 1968) and under similar conditions, mixtures of this ketone with N_α-methylphenylhydrazine, 4-methyl-, 4-methoxy-, 4-benzyloxy-, 4-benzoyloxy-, 4-chloro- and 4-bromophenylhydrazine, N_α-benzylphenylhydrazine and N_α-benzyl-5-benzyloxyphenylhydrazine afforded **876** [5α-H, R^1 = R^3–R^7 = H, R^2 = CH_3, R^8 = $CH(CH_3)(CH_2)_3CH(CH_3)_2$] (Doorenbos and Wu, 1968; Harvey, D. J., 1968), **876** [5α-H, R^1 = CH_3 and OCH_3, R^2–R^7 = H, R^8 = $CH(CH_3)(CH_2)_3CH(CH_3)_2$] (Harvey, D. J., 1968), **876** [5α-H, R^1 = $OCH_2C_6H_5$, R^2–R^7 = H, R^8 = $CH(CH_3)(CH_2)_3CH(CH_3)_2$] (Harvey, D. J., 1968 and Harvey, D. J.; Reid, S. T., 1972), **876** [5α-H, R^1 = $OCOC_6H_5$, Cl and Br, R^2–R^7 = H, R^8 = $CH(CH_3)(CH_2)_3CH(CH_3)_2$] and **876** [5α-H, R^1 = R^3–R^7 = H, R^2 = $CH_2C_6H_5$ and R^1 = $OCH_2C_6H_5$, R^2 = $CH_2C_6H_5$, R^3–R^7 = H, R^8 = $CH(CH_3)(CH_2)_3CH(CH_3)_2$] (Harvey, D. J., 1968), respectively. Likewise, the conversion of a mixture of **873** [R^1–R^5 = H, R^6 = $CH(CH_3)(CH_2)_3CH(CH_3)_2$] with phenylhydrazine in refluxing acetic acid into **875** [5β-H, R^1–R^7 = H, R^8 = $CH(CH_3)(CH_2)_3CH(CH_3)_2$,

$X = H_2$] was repeated (Harvey, D. J., 1968), a mixture of this ketone with N_α-methylphenylhydrazine was converted into **875** [5β-H, $R^1 = R^3$–R^7 = H, $R^2 = CH_3$, $R^8 = CH(CH_3((CH_2)_3CH(CH_3)_2$, $X = H_2$] by refluxing in acetic acid (Doorenbos and Wu, 1968; Harvey, D. J., 1968) [this reaction had previously (Warnhoff and NaNonggai, 1962) been effected using a polyphosphoric acid catalyst], a mixture of this ketone with 4-benzyloxyphenylhydrazine in refluxing acetic acid afforded **875** [5β-H, $R^1 = OCH_2C_6H_5$, R^2–R^7 = H, $R^8 = CH(CH_3)(CH_2)_3CH(CH_3)_2$, $X = H_2$] (Harvey, D. J., 1968; Harvey, D. J. and Reid, S. T., 1972) and mixtures of this ketone with 4-methylphenylhydrazine and N_α-benzylphenylhydrazine were likewise converted into **875** [5β-H, $R^1 = CH_3$, R^2–R^7 = H, $R^8 = CH(CH_3)(CH_2)_3CH(CH_3)_2$, $X = H_2$ and 5β-H, $R^1 = R^3$–R^7 = H, $R^2 = CH_2C_6H_5$, $R^8 = CH(CH_3)(CH_2)_3CH(CH_3)_2$, $X = H_2$, respectively] (Harvey, D. J., 1968). Products from these reactions were, in general, more difficult to crystallize than were the corresponding products obtained using **874** [R^1–R^7 = H, $R^8 = CH(CH_3)(CH_2)_3CH(CH_3)_2$] (Harvey, D. J., 1968).

By analogy with these above observations, and although the reaction between **873** [$R^1 = R^2 = R^5$ = H, $R^3 = R^4$ = OH, $R^6 = CH(CH_3)(CH_2)_2COOCH_3$] and phenylhydrazine in refluxing acetic acid gave only a non-crystalline indolic product (Chaplin, Hey and Honeyman, 1959; Harvey, D. J., 1968), use of N_α-methylphenylhydrazine in place of phenylhydrazine in this reaction furnished crystalline **875** [5β-H, $R^1 = R^3 = R^4 = R^7$ = H, $R^2 = CH_3$, $R^5 = R^6$ = OH, $R^8 = CH(CH_3)(CH_2)_2COOCH_3$, $X = H_2$] (Harvey, D. J., 1968). The products resulting from the acetic acid catalysed reaction between phenylhydrazine and the three 5β-3-ketosteroids **873** [R^1–R^5 = H, R^6 = $CH(CH_3)(CH_2)_2COOCH_3$ and $CH(CH_3)(CH_2)_2CONH_2$ and $R^1 = R^2 = R^5$ = H, $R^3 = R^4 = OCOCH_3$, $R^6 = CH(CH_3)(CH_2)_2COOCH_3$] were similarly formulated as **875** [5β-H, R^1–R^7 = H, $R^8 = CH(CH_3)(CH_2)_2COOCH_3$ and $CH(CH_3)(CH_2)_2CONH_2$, $X = H_2$ and R^1–$R^4 = R^7$ = H, $R^5 = R^6 = OCOCH_3$, $R^8 = CH(CH_3)(CH_2)_2COOCH_3$, $X = H_2$, respectively] (Chaplin, Hey and Honeyman, 1959). Likewise, mixtures of 5α- or 5β-androstan-3-one with various arylhydrazines have been indolized in hot or boiling acetic acid and the reaction products formulated accordingly (Table 21) and 5α-spirostan-3-one (**885**) underwent Fischer indole syntheses with phenylhydrazine and

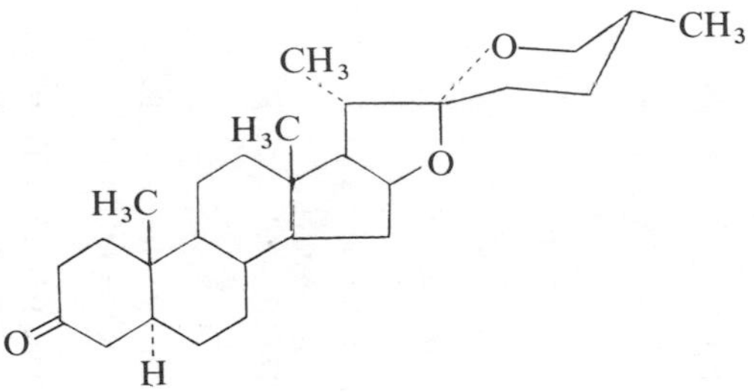

885

Table 21. Products Resulting from the Indolization of Various 5α- and 5β-Androstan-3-one Arylhydrazones

Androstane Derivative	Arylhydrazine	Product	Reference	Catalyst
873 (R^1–R^5 = H, R^6 = $OCOCH_3$)	Phenylhydrazine	**875** (5β-H, R^1–R^7 = H, R^8 = $OCOCH_3$, X = H_2)	1	CH_3COOH
873 (R^1–R^5 = H, R^6 = $OCOCH_3$)	4-Methoxyphenylhydrazine	**875** (5β-H, R^1 = OCH_3, R^2–R^7 = H, R^8 = $OCOCH_3$, X = H_2)[a]	2	CH_3COOH
873 (R^1–R^5 = H, R^6 = $OCOCH_3$)	4-Bromophenylhydrazine	**875** (5β-H, R^1 = Br, R^2–R^7 = H, R^8 = $OCOCH_3$, X = H_2)[a]	2	CH_3COOH
874 (R^1–R^6 = H, R^7 = CH_3, R^8 = OH)	Phenylhydrazine	**876** (5α-H, R^1–R^6 = H, R^7 = CH_3, R^8 = OH)	3[c]	CH_3COOH
874 (R^1–R^6 = H, R^7 = CH_3, R^8 = OH	N_α-Methylphenylhydrazine	**876** (5α-H, R^1 = R^3–R^6 = H, R^2 = R^7 = CH_3, R^8 = OH)	3	CH_3COOH
874 (R^1–R^6 = H, R^7 = CH_3, R^8 = OH	4-Benzyloxyphenylhydrazine	**876** (5α-H, R^1 = $OCH_2C_6H_5$, R^2–R^6 = H, R^7 = CH_3, R^8 = $OCOCH_3$)[b]	4	CH_3COOH
874 (R^1–R^6 = H, R^7 = CH_3, R^8 = OH	4-Benzoyloxyphenylhydrazine	**876** (5α-H, R^1 = $OCOC_6H_5$, R^2–R^6 = H, R^7 = CH_3, R^8 = OH)	4	CH_3COOH
874 (R^1–R^6 = H, R^7 = CH_3, R^8 = OH	4-Benzyloxy-N_α-(4-chlorobenzyl)phenylhydrazine	**876** (5α-H, R^1 = $OCH_2C_6H_5$, R^2 = $CH_2C_6H_4Cl(4)$, R^3–R^6 = H, R^7 = CH_3, R^8 = OH)	4	CH_3COOH
874 (R^1–R^6 = H, R^7 = CH_3, R^8 = OH	N_α-Benzoyl-4-hydroxyphenylhydrazine	**876** (5α-H, R^1 = R^8 = OH, R^2 = COC_6H_5, R^3–R^6 = H, R^7 = CH_3	4	CH_3COOH
874 (R^1–R^7 = H, R^8 = OH)	Phenylhydrazine	**876** (5α-H, R^1–R^7 = H, R^8 = OH)	5	CH_3COOH
874 (R^1–R^7 = H, R^8 = OH)	4-Methoxyphenylhydrazine	**876** (5α-H, R^1 = OCH_3, R^2–R^7 = H, R^8 = $OCOCH_3$)[b]	4	CH_3COOH
874 (R^1–R^7 = H, R^8 = OH)	4-Nitrophenylhydrazine	**876** (5α-H, R^1 = NO_2, R^2–R^7 = H, R^8 = $OCOCH_3$)[b]	4	CH_3COOH
874 (R^1–R^7 = H, R^8 = $OCOCH_3$)	Phenylhydrazine	**876** (5α-H, R^1–R^7 = H, R^8 = $OCOCH_3$) R^8 = $OCOCH_3$)	1	CH_3COOH
874 (R^1–R^7 = H, R^8 = $OCOCH_3$)	Phenylhydrazine	**876** (5α-H, R^1–R^7 = H, R^8 = $OCOCH_3$)[a]	6	CH_3COOH saturated with HCl

874 (R^1–R^7 = H, R^8 = $OCOCH_3$)	4-Methylphenylhydrazine	**876** (5α-H, R^1 = CH_3, R^2–R^7 = H, R^8 = $OCOCH_3$)[a]	6	CH_3COOH saturated with HCl
874 (R^1–R^7 = H, R^8 = $OCOCH_3$)	4-Methoxyphenylhydrazine	**876** (5α-H, R^1 = OCH_3, R^2–R^7 = H, R^8 = $OCOCH_3$)[a]	2	CH_3COOH
874 (R^1–R^7 = H, R^8 = $OCOCH_3$)	4-Bromophenylhydrazine	**876** (5α-H, R^1 = Br, R^2–R^7 = H, R^8 = $OCOCH_3$)[a]	2	CH_3COOH
874 (R^1–R^7 = H, R^8 = $OCOCH_3$)	1-Naphthylhydrazine	[structure: H_3C, H_3C, $OCOCH_3$[a], N–H, H]	6	CH_3COOH saturated with HCl
874 (R^1–R^7 = H, R^8 = $OCOCH_3$)	2-Naphthylhydrazine	[structure: H_3C, H_3C, $OCOCH_3$[a], N–H, H]	6	CH_3COOH saturated with HCl

Notes

a. Structures were confirmed by mass spectral investigation; b. Acetylation of the 17β-OH group occurred under the reaction conditions; c. This reaction has also been achieved under unspecified conditions (Carelli, Marchini, Cardellini, Micheletti Moracci and Liso, 1969).

References

1. Akiba and Ohki, 1970; 2. Catsoulacos and Papadopoulos, 1976; 3. Doorenbos and Wu, 1968; 4. Lester, M. G., Petrow, V. and Stephenson, O., 1965; 5. Orr and Bowers, 1962 (although it was not specified in this publication that the 5α- and not the 5β-3-ketosteroid was used, the use of the former stereoisomer is at present assumed because of the direction of indolization; 6. Jacquignon, Croisy-Delcey and Croisy, 1972.

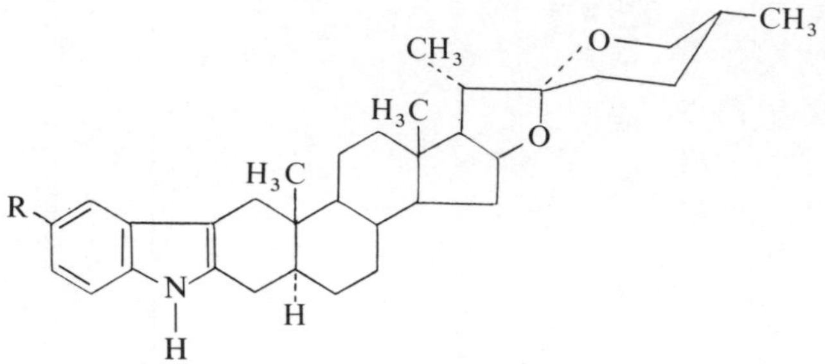

886

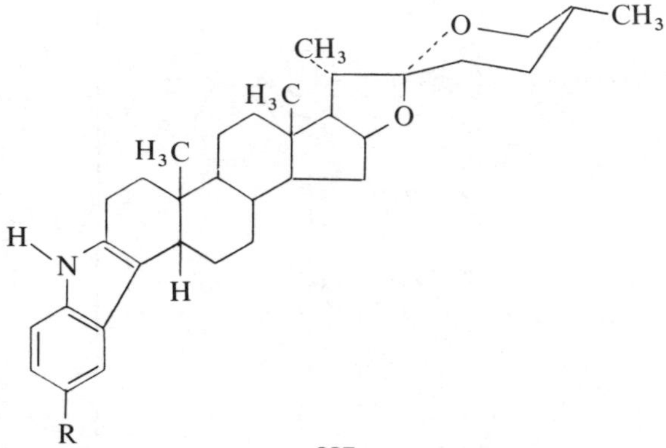

887

4-methyl- and 4-nitrophenylhydrazines to afford **886** (R = H, CH_3 and NO_2, respectively) whereas the 5β-isomer was indolized with similar arylhydrazines to produce **887** (R = H, CH_3 and NO_2, respectively) (Irismetov, Goryaev and Rivkina, 1979).

From these above studies it could be concluded that indolization of the arylhydrazones of 5α- and 5β-3-ketosteroids gives the isomeric linear and angular indolosteroids (**876** and **875**), respectively, and, indeed, this would be in accord with the first stage of the mechanism of the indolization and the known (Hart, 1963) direction of enolization of 5α- and 5β-cholestan-3-one. Furthermore, the direction of cyclization in the limited studies which have been effected upon the Piloty pyrrolization using 3-ketosteroids also complies with these observations (Sucrow and Chondromatidis, 1970) (see Chapter VI, D). However, it has been shown (Liston, 1966; Liston and Howarth, 1967) that 3-ketosteroids yield both 2- and 3-enols and therefore it is not surprising that a more thorough investigation of the reaction products resulting from the indolization of their arylhydrazones using g.l.c. techniques has led to the isolation of both linear and angular isomers. Thus, 5α- and 5β-cholestan-3-one were reacted with phenylhydrazine and N_α-methylphenylhydrazine in boiling acetic acid to afford products which upon g.l.c. analysis were shown to consist of compounds

876 [5α-H, R^1–R^7 = H, R^8 = $CH(CH_3)(CH_2)_3CH(CH_3)_2$] (88.5%) along with 11.5% of the corresponding angular isomer **875** [5α-H, R^1–R^7 = H, R^8 = $CH(CH_3)(CH_2)_3CH(CH_3)_2$, X = H_2], **876** [5α-H, R^1 = R^3–R^7 = H, R^2 = CH_3, R^8 = $CH(CH_3)(CH_2)_3CH(CH_3)_2$] (89.5%) along with 10.5% of the corresponding angular isomer **875** [5α-H, R^1 = R^3–R^7 = H, R^2 = CH_3, R^8 = $CH(CH_3)(CH_2)_3CH(CH_3)_2$, X = H_2], **875** [5β-H, R^1–R^7 = H, R^8 = $CH(CH_3)(CH_2)_3CH(CH_3)_2$, X = H_2] (92%) along with 8% of the corresponding linear isomer **876** [5β-H, R^1–R^7 = H, R^8 = $CH(CH_3)(CH_2)_3CH(CH_3)_2$] and **875** [5β-H, R^1 = R^3–R^7 = H, R^2 = CH_3, R^8 = $CH(CH_3)(CH_2)_3CH(CH_3)_2$, X = H_2] (81.2%) along with 10.8% of the corresponding linear isomer **876** [5β-H, R^1 = R^3–R^7 = H, R^2 = CH_3, R^8 = $CH(CH_3)(CH_2)_3CH(CH_3)_2$], respectively (Harvey, D. J. and Reid, 1972). It is also interesting that from this latter reaction a 4% yield of compound **888** (R^1 = CH_3, R^2 = H) was isolated, a small yield of this compound,

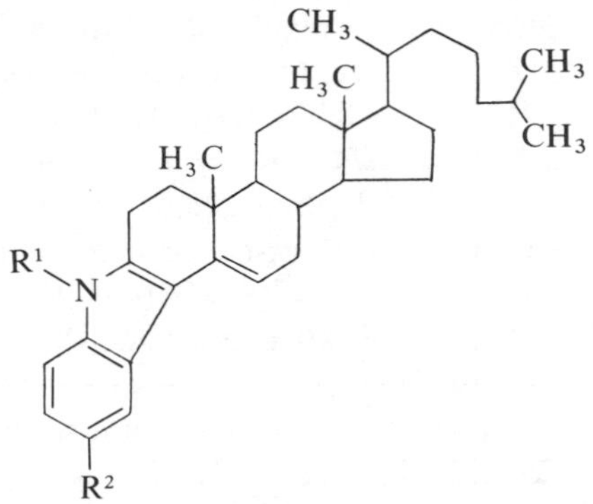

888

resulting from 5,6-dehydrogenation under indolization conditions, having previously (Warnhoff and NaNonggai, 1962) been isolated, along with the major reaction product **875** [5β-H, R^1 = R^3–R^7 = H, R^2 = CH_3, R^8 = $CH(CH_3)(CH_2)_3CH(CH_3)_2$, X = H_2], from the same reaction but using polyphosphoric acid instead of acetic acid as catalyst. From these data it would appear that the indolizations of 5α- and 5β-3-ketosteroids form predominantly the corresponding linear (**876**) and angular (**875**) indolosteroids, respectively, along with small amounts (*ca.* 10%) of the corresponding isomers **875** and **876**, respectively. However, exceptions to these findings are known for whereas 5α-androstan-3,17-dione (**874**; R^1–R^6 = H, R^7 + R^8 = O) reacted with phenylhydrazine hydrochloride in refluxing aqueous ethanol to furnish an almost quantitative yield of 5α-androst-2-ene[3,2-b]indol-17-one (**876**; 5α-H, R^1–R^6 = H, R^7 + R^8 = O) (Britten, A. and Lockwood, 1974) [this reaction was also earlier effected by reacting the ketone and phenylhydrazine in refluxing acetic acid (Harvey, D. J., 1968)], as expected by analogy with the above conclusions, the 5β-isomer similarly indolized to 5β-androst-2-ene-[3,2-b]indol-17-one (**876**; 5β-H, R^1–R^6 = H, R^7 + R^8 = O) (Britten, A. and Lockwood, 1974). The structures of these two products were established by their mass spectra but the formation in almost quantitative yield of the latter product is very surprising since such a formation of the linear indolosteroid from a 5β-3-ketosteroid arylhydrazone is contrary to all the above examples. In these several previous studies, the indolizations were performed by warming or boiling either the arylhydrazone or a mixture of the arylhydrazine with the 3-ketosteroid in acetic acid whereas in this later study (Britten, A. and Lockwood, 1974) the indolizations were achieved (selectively involving the 3-keto group – see page 335) by dropwise addition of an aqueous solution of phenylhydrazine hydrochloride into a refluxing ethanolic solution of the ketosteroid. These differences could possibly account for the differences in direction of indolization. The presence of the 17-keto group may also influence the direction of these indolizations.

Another synthetic entry into the angular structures **875** (5α-H) starting with a 5α-cholestan-3-one has been achieved by treating 2α-bromo-5α-cholestan-3-one [**874**; R^1 = Br, R^2–R^7 = H, R^8 = $CH(CH_3)(CH_2)_3CH(CH_3)_2$] or 2,2-dibromo-5$\alpha$-cholestan-3-one [**874**; R^1 = R^2 = Br, R^3–R^7 = H, R^8 = $CH(CH_3)(CH_2)_3CH(CH_3)_2$] in boiling ethanol with N_α-methylphenylhydrazine in acetic acid to afford in each case a product formulated as **875** [5α-H, R^1 = R^3–R^7 = H, R^2 = CH_3, R^8 = $CH(CH_3)(CH_2)_3CH(CH_3)_2$, X = $NN(CH_3)C_6H_5$]. Since 1,2,3,4-tetrahydrocarbazol-1-one could not be induced to indolize with phenylhydrazine or N_α-methylphenylhydrazine, it was suggested that these indolizations involving bromosteroids did not involve the intermediacy of **875** [5α-H, R^1 = R^3–R^7 = H, R^2 = CH_3, R^8 = $CH(CH_3)(CH_2)_3CH(CH_3)_2$, X = O] but rather involved the indolization of cholestan-2,3-dione bis-N_α-methylphenylhydrazone (Buckingham and Guthrie, 1968b), the formation of such an intermediate being an example of osazone formation from α-halogenoketones (see Chapter VII, Q). Structure **875** [5α-H, R^1 = R^3–R^7 = H, R^2 = CH_3, R^8 = $CH(CH_3)(CH_2)_3CH(CH_3)_2$, X = $NN(CH_3)C_6H_5$] was preferred to the possible alternative structure in which indolization had occurred on to the steroidal C-1 atom because of the steric hindrance at this position (Buckingham and Guthrie, 1968b).

The influence of a 6-substituent upon the direction of indolization of 5α- and 5β-3-ketosteroid arylhydrazones has been investigated (Harvey, D. J. and Reid, S. T., 1972). When 5α-cholestan-3,6-dione [**874**; R^1 = R^2 = R^5–R^7 = H, R^3 + R^4 = O, R^8 = $CH(CH_3)(CH_2)_3CH(CH_3)_2$] was reacted with various arylhydrazines in glacial acetic acid, hydrazone formation occurred preferentially at C-3 (the more reactive of the two carbonyl functions – see page 335) and the subsequent indolizations gave between 70% and 90% yields of the corresponding linear isomers **876** [5α-H, R^1, R^2 = H, CH_3; H, $CH_2C_6H_5$; $OCH_2C_6H_5$, H; CH_3, H; Cl, H; $OCOC_6H_5$, H; $OCH_2C_6H_5$,

$CH_2C_6H_5$ and OCH_3, $CH_2C_6H_5$, $R^3 + R^4 = O$, R^5–$R^7 = H$, $R^8 = CH(CH_3)(CH_2)_3CH(CH_3)_2$]. Similar reaction of the 5β-isomer afforded products resulting from indolization in both possible directions. Thus, whereas only cholest-4-ene-3,6-dione 3-monophenylhydrazone could be isolated from the attempted indolization of a mixture of 5β-cholestan-3,6-dione [**873**; $R^1 + R^2 = O$, R^3–$R^5 = H$, $R^6 = CH(CH_3)(CH_2)_3CH(CH_3)_2$] and phenylhydrazine in acetic acid, the use of N_α-methylphenylhydrazine in place of phenylhydrazine led to the formation of products resulting from indolization in both possible directions, with the linear products, resulting from indolization on to the steroidal C-2 atom, greatly predominating. Thus, after refluxing in acetic acid, there were isolated, by fractional crystallization or by column chromatography, **876** [5α-H, $R^1 = R^5$–$R^7 = H$, $R^2 = CH_3$, $R^3 + R^4 = O$, $R^8 = CH(CH_3)(CH_2)_3CH(CH_3)_2$] [epimerization most likely occurred either during or subsequent to the indolization since 5β-cholestan-3,6-dione was stable in acetic acid under the indolization conditions (Harvey, D. J. and Reid, S. T., 1972) and the 5β-isomeric product appeared to slowly epimerize in hot acetic acid (Harvey, D. J., 1968)] (16% yield), **876** [5β-H, $R^1 = R^5$–$R^7 =$ H, $R^2 = CH_3$, $R^3 + R^4 = O$, $R^8 = CH(CH_3)(CH_2)_3CH(CH_3)_2$] (33% yield) (epimerized to the 5α-isomer by treatment with hydrochloric acid in acetic acid) and **889**, but no products having the angular structure, resulting

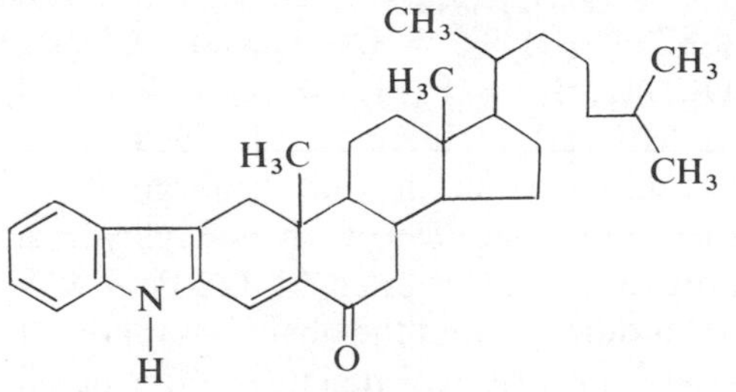

889

from new C—C bond formation during indolization at the steroidal C-4 atom, could be detected. However, repetition of the indolization, with subsequent epimerization of the total reaction product, in hydrochloric acid followed by column chromatography, led to the isolation of **876** [5α-H, $R^1 =$ R^5–$R^7 = H$, $R^2 = CH_3$, $R^3 + R^4 = O$, $R^8 = CH(CH_3)(CH_2)_3CH(CH_3)_2$] (65% yield), **889** (12%) and **875** (5β-H, $R^1 = R^5$–$R^7 = H$, $R^2 = CH_3$, $R^3 + R^4 = O$, $R^8 - CH(CH_3)(CH_2)_3)CH(CH_3)_2$, $X = H_2$] (10%) (Harvey, D. J. and Reid, S. T., 1972). Whereas reaction of 6β-hydroxy-5β-cholestan-3-one [**873**; $R^1 = OH$, R^2–$R^5 = H$, $R^6 = CH(CH_3)(CH_2)_3CH(CH_3)_2$] with phenylhydrazine in refluxing acetic acid yielded only a non-crystalline product, although its i.r. spectral examination suggested that indolization had occurred, reaction between 6β-hydroxy-5β-cholestan-3-one and N_α-methylphenylhydrazine in refluxing acetic acid furnished, as the major product, the angular indolosteroid **875** [5β-H, R^1 = R^4–R^7 = H, R^2 = CH_3, R^3 = OH, R^8 = $CH(CH_3)(CH_2)_3CH(CH_3)_2$, X = H_2] together with small amounts of the corresponding linear isomer **876** [5β-H, R^1 = R^4–R^7 = H, R^2 = CH_3, R^3 = OH, R^8 = $CH(CH_3)(CH_2)_3CH(CH_3)_2$] and, likewise, the 6$\beta$-acetoxy derivative afforded an 87% yield of **875** [5β-H, R^1 = R^4–R^7 = H, R^2 = CH_3, R^3 = $OCOCH_3$, R^8 = $CH(CH_3)(CH_2)_3CH(CH_3)_2$, X = H_2]. Unlike the corresponding 6β-hydroxy compound, 6β-acetoxy-5β-cholestan-3-one also underwent a Fischer indolization when reacted with phenylhydrazine in acetic acid to afford **875** [5β-H, R^1 = R^2 = R^4–R^7 = H, R^3 = $OCOCH_3$, R^8 = $CH(CH_3)(CH_2)_3CH(CH_3)_2$, X = H_2] (44% yield) as a crystalline product. When a mixture of 6β-hydroxy-5α-cholestan-3-one [**874**; R^1 = R^2 = R^4–R^7 = H, R^3 = OH, R^8 = $CH(CH_3)(CH_2)_3CH(CH_3)_2$] and N_α-methylphenylhydrazine reacted in refluxing acetic acid, the linear indolosteroid **876** [5α-H, R^1 = R^4–R^7 = H, R^2 = CH_3, R^3 = OH, R^8 = $CH(CH_3)(CH_2)_3CH(CH_3)_2$] was produced (Harvey, D. J. and Reid, S. T., 1972). Reaction of 6α-nitro-5α-cholestan-3-one [**874**; R^1–R^3 = R^5–R^7 = H, R^4 = NO_2, R^8 = $CH(CH_3)(CH_2)_3CH(CH_3)_2$] with phenylhydrazine and a series of substituted phenylhydrazines in refluxing acetic acid gave the corresponding indolosteroids **876** [5α-H, R^1 = H, CH_3, OCH_3, $OCH_2C_6H_5$, Cl, Br and I, R^2 = H and R^1 = H, R^2 = CH_3 and $CH_2C_6H_5$, R^3 = R^5–R^7 = H, R^4 = NO_2, R^8 = $CH(CH_3)(CH_2)_3CH(CH_3)_2$] (Harvey, D. J., 1968; Harvey, D. J. and Reid, S. T., 1972) and **876** [5α-H, R^1 = OCH_3 and $OCH_2C_6H_5$, R^2 = $CH_2C_6H_5$ and R^1 = $OCOC_6H_5$, R^2 = H, R^3 = R^5 = R^7 = H, R^4 = NO_2, R^8 = $CH(CH_3)(CH_2)_3CH(CH_3)_2$] (Harvey, D. J., 1968).

The product resulting from the indolization of a mixture of 2α,17α-dimethylandrostan-17β-ol-3-one (**890**) with phenylhydrazine in refluxing acetic acid was arbitrarily formulated (Lester, M. G., Petrow, V. and Stephenson, O., 1965) as **891**, a 3*H*-indole, without the obvious support of u.v. spectral evidence which would have readily distinguished it from the possible angular isomer, an indole, formed by indolization at the C-4 atom of the steroid nucleus. However, the above studies relating to the indolizations of 2-unsubstituted 3-ketosteroid arylhydrazones do, by analogy, support structure **891**, at least for the major product from this reaction. Likewise, the obvious u.v. spectral confirmation was not quoted when the products resulting from the indolizations of friedelin

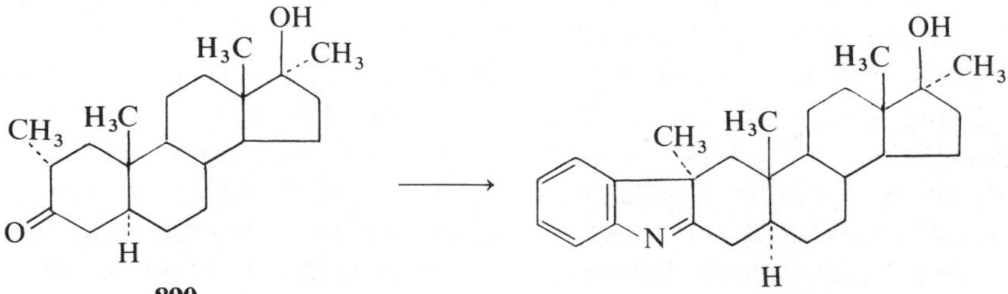

OH
H_3C CH_3
CH_3 H_3C
O
H
890

OH
H_3C CH_3
CH_3 H_3C
N
H
891

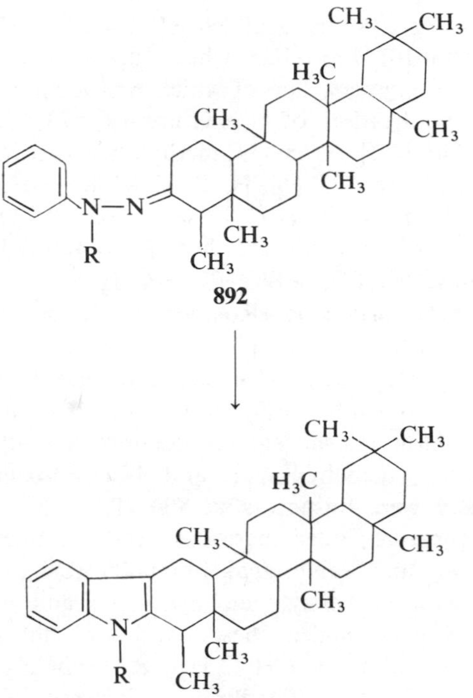

phenylhydrazone (**892**; R = H) and N_{α}-methylphenylhydrazone (**892**; R = CH_3) were formulated as the indoles **893** (R = H and CH_3, respectively) rather than the possible alternative 3*H*-indoles (Buu-Hoï and Jacquignon, 1956).

3. Δ^4-3-Ketosteroids

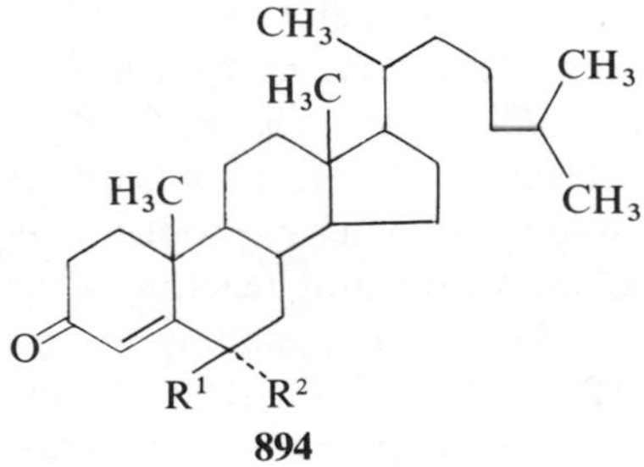

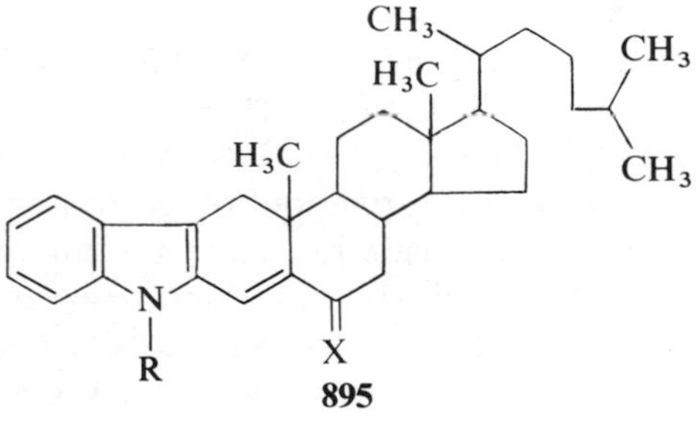

Unlike simpler α,β-unsaturated ketone arylhydrazones, attempted indolization of which failed (see Chapter VII, F), indolization of a mixture of **894** ($R^1 = R^2 = H$) with phenylhydrazine led to a product which was formulated, without evidence, as **895** (R = H, X = H_2) (Rossner, 1937). However, the related indolization of a mixture of **894** ($R^1 = R^2 = H$) with N_α-methylphenylhydrazine afforded a product which, upon catalytic hydrogenation, yielded a pair of C-5 epimers, one of which was identical with the product obtained by the indolization of a mixture of **873** [R^1–R^5 = H, R^6 = $CH(CH_3)(CH_2)_3CH(CH_3)$] with N_α-methylphenylhydrazine, which was known to have structure **875** [5β-H, $R^1 = R^3$–R^7 = H, $R^2 = CH_3$, R^8 = $CH(CH_3)(CH_2)_3CH(CH_3)_2$, X = H_2] (see above). The structure of the indolization product from a mixture of **894** ($R^1 = R^2 = H$) with N_α-methylphenylhydrazine must therefore be **888** ($R^1 = CH_3$, $R^2 = H$) and the structure **895** (R = H, X = H_2) originally (Rossner, 1937) assigned to the product obtained from the corresponding phenylhydrazone must, by analogy, be revised to **888** ($R^1 = R^2 = H$). These structures are as expected, since **894** ($R^1 = R^2 = H$) was known to enolize towards the C-6 atom (Warnhoff and NaNonggai, 1962). Although subsequent attempts to indolize mixtures of **894** ($R^1 = R^2 = H$) with phenylhydrazine and 4-benzyloxyphenylhydrazine in refluxing acetic acid were unsuccessful, **894** ($R^1 = R^2 = H$) reacted with N_α-methylphenylhydrazine under these conditions to furnish the previously synthesized (Warnhoff and NaNonggai, 1962) **888** ($R^1 = CH_3$, $R^2 = H$) and similarly reacted with N_α-benzylphenylhydrazine and 4-methoxy- and 4-benzyloxyphenylhydrazine under these conditions to afford **888** (R^1 = $CH_2C_6H_5$, R^2 = H and $R^1 = CH_2C_6H_5$, $R^2 = OCH_3$ and $OCH_2C_6H_5$, respectively) (Harvey, D. J., 1968). Products identical with these were also isolated when mixtures of N_α-substituted phenylhydrazines with **896** were indolized in refluxing acetic acid (Harvey, D. J., 1968). When a mixture of **894** (R^1 = OH, R^2 = H) with phenylhydrazine was reacted in refluxing acetic acid, only the phenylhydrazone of **897** was obtained, whereas when a mixture

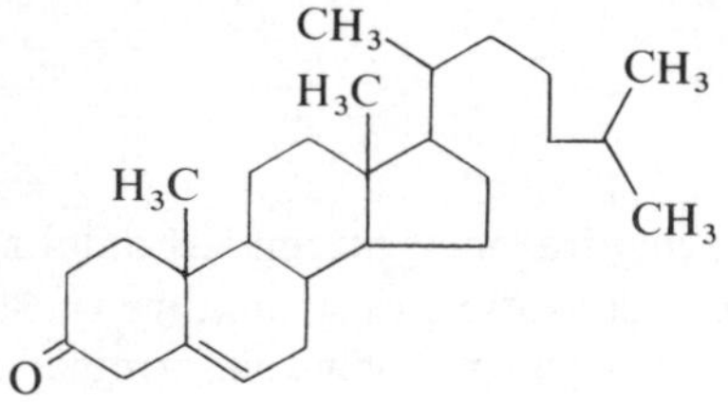

896

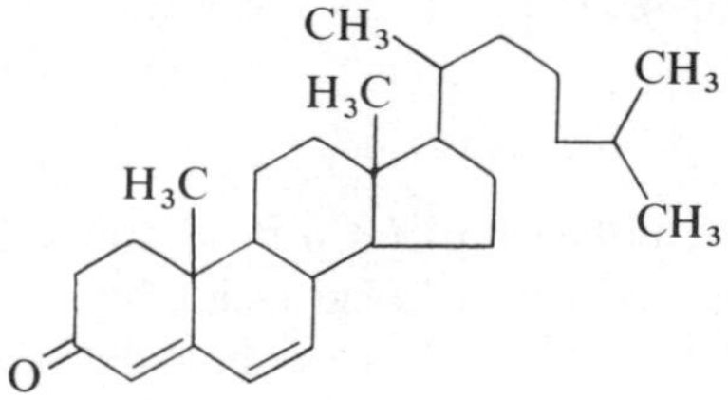

897

of **894** (R^1 = OH, R^2 = H) with N_α-methylphenylhydrazine was similarly treated, the products were **895** (R = CH_3, X = H_2) together with a small amount of **876** [5α-H, $R^1 = R^5$–R^7 = H, $R^2 = CH_3$, $R^3 + R^4$ = O, R^8 = $CH(CH_3)(CH_2)_3CH(CH_3)$]. Although no evidence was found to indicate the simultaneous formation of corresponding angular isomeric products, no attempt was made to isolate any minor product(s) from this reaction which therefore merits further study (Harvey, D. J., 1968). No crystalline product was isolated from the reaction between **894** ($R^1 = OCOCH_3$, R^2 = H) and N_α-methylphenylhydrazine in refluxing acetic acid (Harvey, D. J., 1968).

Indolization of the 3-monophenylhydrazone of **894** ($R^1 + R^2 = O$) by heating with polyphosphoric acid and of a mixture of **894** ($R^1 + R^2 = O$) with N_α-methylphenylhydrazine in refluxing acetic acid gave **895** (R = H and CH_3, respectively, X = O) (Harvey, D. J. and Reid, S. T., 1972).

Reaction of a mixture of **897** with N_α-methylphenylhydrazine gave **898**. The possibility that the new C—C bond formation had occurred at the steroidal C-2 atom, in accordance with previous evidence (Heilbron, Kennedy, Spring and Swain, 1938) regarding the enolization of **897**, was established by catalytic hydrogenation of **898** to afford the known **876** [5α-H, $R^1 = R^3\text{–}R^7 = H$, $R^2 = CH_3$, $R^8 = CH(CH_3)(CH_2)_3CH(CH_3)_2$] (see above) along with a smaller amount of the 5β-isomer (Harvey, D. J. and Reid, S. T., 1972).

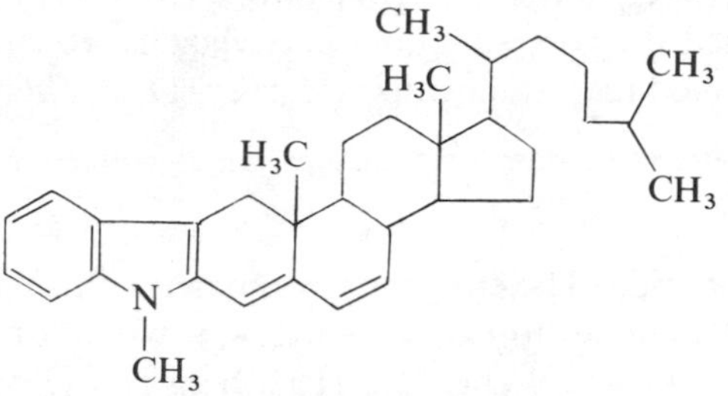

898

4. *7-Ketosteroids*

No indolic material could be detected in the products resulting from the reaction of **899** ($R = OCOCH_3$) with either phenylhydrazine or N_α-methylphenylhydrazine in acetic acid. However, the use of either hydrochloric or sulphuric acid–acetic acid as catalysts gave products, i.r. spectroscopic examination of which indicated the presence of **900** ($R^1 = OCOCH_3$ and OH, $R^2 = H$ and $R^1 = OCOCH_3$ and OH, $R^2 = CH_3$, respectively). Evidence was also obtained that the former indolization product also contained the 3*H*-indole **901** but none of this could be isolated. Similarly, the reaction between **899** (R = H) and N_α-methylphenylhydrazine furnished a product, the

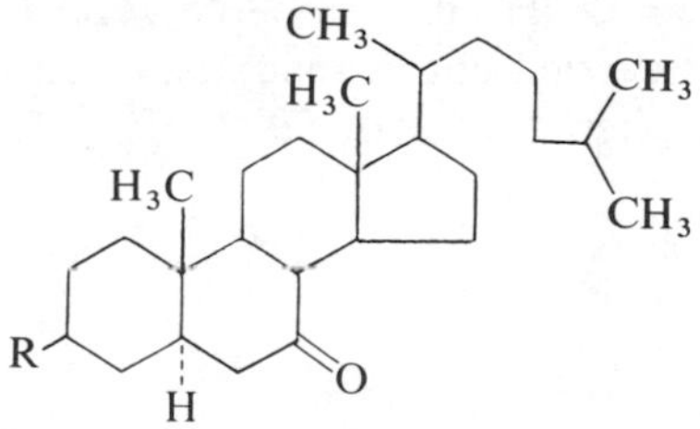

899

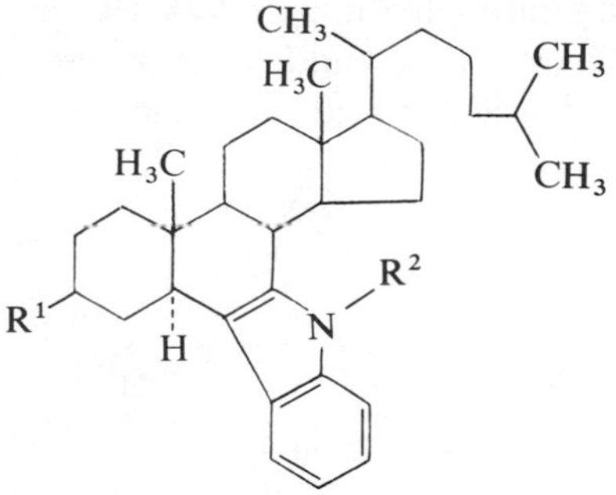

900

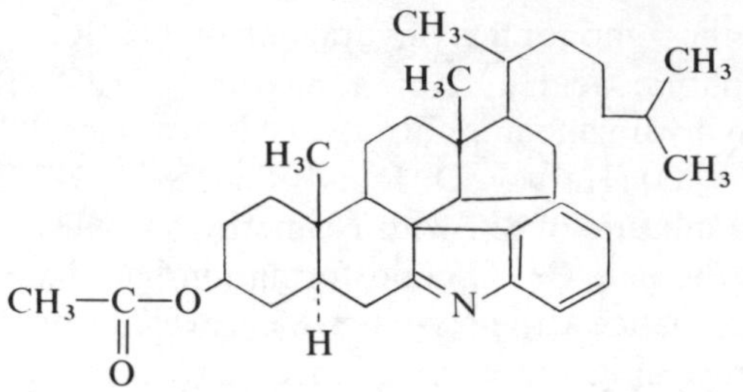

901

ir. spectrum of which indicated the presence of an indole ring system but recognizable products could not be isolated. Attempts to effect indolization via the isolation of the phenylhydrazone of **899** (R = H) were not made since this product formed the corresponding phenylazohydroperoxide (see Chapter II, C) upon attempted recrystallization (Harvey, D. J., 1968).

5. *11-Ketosteroids*

Although the steroidal 11-keto group will react with small ketonic reagents (e.g. hydroxylamine) under forcing conditions, it will not react with the larger ketonic reagents (Rausser, Weber, L., Hershberg and Oliveto, 1966 and refs. cited therein). Becuase of this latter observation, attempts were not made to utilize 11-ketosteroids as the ketonic moieties in Fischer indolizations (Harvey, D. J., 1968).

6. *17-Ketosteroids*

Several studies involving the use of 17-ketosteroids in Fischer indolizations have been reported although early attempts were unsuccessful. Under the conditions normally used for the synthesis of the corresponding indolosteroids from 3-ketosteroids, 17-ketosteroids did not react analogously. Thus, when **902** (R^1 = OH, $R^2 + R^3$ = O) was reacted with phenylhydrazine in warmed acetic acid, the only product isolated was the hydroperoxide of the corresponding phenylhydrazone **902** (R^1 = OH, R^2 = OOH, R^3 = $N{=}NC_6H_5$) (see Chapter II, C) and, when the reaction was carried out in refluxing ethanolic hydrochloric acid, only the corresponding phenylhydrazone **902** (R^1 = OH,

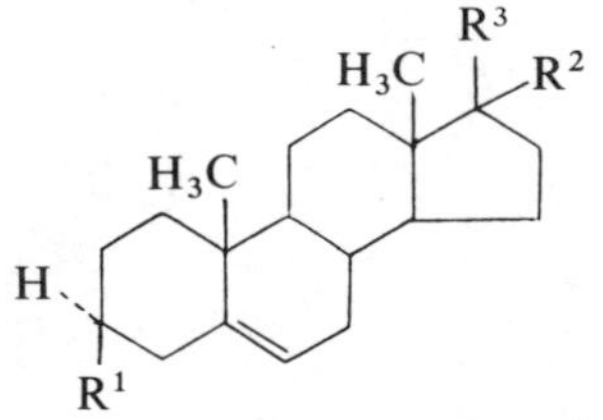

902

$R^2 + R^3 = NNHC_6H_5$) was formed which again readily formed the hydroperoxide **902** ($R^1 = OH$, $R^2 = OOH$, $R^3 = N{=}NC_6H_5$). Likewise, when **902** ($R^1 = OCOCH_3$, $R^2 + R^3 = O$) reacted with phenylhydrazine in ethanolic hydrochloric acid, the corresponding phenylhydrazone **902** ($R^1 = OCOCH_3$, $R^2 + R^3 = NNHC_6H_5$) was formed. This was recovered unchanged after boiling in acetic acid or after shaking at room temperature with hydrochloric acid but was hydrolysed to **902** ($R^1 = OH$, $R^2 + R^3 = O$) upon refluxing in hydrochloric acid (Chaplin, Hey and Honeyman, 1959). This unreactivity of the steroidal 17-keto group relative to the reactivity of the steroidal 3-keto group was also apparent from the previously mentioned observations that when **874** (R^1–R^6 = H, $R^7 + R^8 = O$) reacted with one equivalent of phenylhydrazine in refluxing acetic acid (Harvey, D. J., 1968) or when an aqueous solution of phenylhydrazine hydrochloride was added dropwise to a refluxing ethanolic solution of **874** (R^1–R^6 = H, $R^7 + R^8 = O$) (Britten, A. and Lockwood, 1974), only **876** (5α-H, R^1–R^6 = H, $R^7 + R^8 = O$) was isolated, in almost quantitative yield from the latter reaction. Likewise, when an aqueous solution of phenylhydrazine hydrochloride was added to a solution of **873** (R^1–R^4 = H, $R^5 + R^6 = O$), only **876** (5β-H, R^1–R^6 = H, $R^7 + R^8 = O$) was isolated in almost quantitative yield (Britten, A. and Lockwood, 1974). Similar reactions using 4-benzyloxy- and N_α-benzyl- and N_α-methylphenylhydrazine in place of phenylhydrazine with **874** (R^1–R^6 = H, $R^7 + R^8 = O$) in refluxing acetic acid afforded **876** (5α-H, $R^1 = OCH_2C_6H_5$, R^2–R^6 = H, $R^7 = R^8 = O$ and $R^1 = R^3$–R^6 = H, $R^2 = CH_2C_6H_5$ and CH_3, $R^7 = R^8 = O$, respectively) (Harvey, D. J., 1968). The last of these three reactions also produced a small yield of **903** ($R = CH_3$), a product which was isolated exclusively after **874** (R^1–R^6 = H, $R^7 + R^8 = O$) reacted with a 2 × molar equivalent of N_α-methylphenylhydrazine in refluxing acetic acid. N_α-Benzylphenylhydrazine similarly reacted with **874** (R^1–R^6 = H, $R^7 + R^8 = O$) to yield **903** ($R = CH_2C_6H_5$) (Harvey, D. J., 1968) but when **874** (R^1–R^6 = H, $R^7 + R^8 = O$) reacted with a 2 × molar equivalent of phenylhydrazine, the only product isolated was **876** (5α-H, R^1–R^6 = H, $R^7 + R^8 = NNHC_6H_5$) which upon further heating in acetic acid solution was hydrolysed

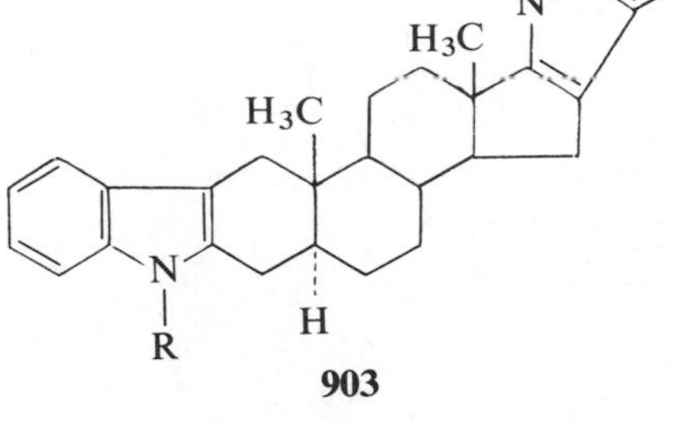

903

to afford **876** (5α-H, R^1–R^6 = H, $R^7 + R^8$ = O) (Harvey, D. J., 1968). Clearly, these results indicate facilitation of indolization by hydrazone N_α-alkylation, a well established phenomenon (Chalmers, A. J. and Lions, 1933; Chastrette, 1970; Diels and Köllisch, 1911; Ishii, Murakami, Y., Hosoya, Takeda, Suzuki, Y. and Ikeda, 1973; Fischer, Emil, 1886b; Kermack, Perkin and Robinson, R., 1921; Kermack and Slater, 1928; Mann and Wilkinson, 1957; Padfield and Tomlinson, M. L., 1950; Perkin and Plant, 1923b; Robinson, R. and Thornley, 1924; Woolley, D. W. and Shaw, E., 1955). In an extension of these studies, **904** (R = OH) was reacted with N_α-benzyl- and N_α-methylphenylhydrazine in refluxing acetic acid to afford **905** (R^1 = H, R^2 = $OCOCH_3$, R^3 = $CH_2C_6H_5$ and CH_3, respectively, R^4 = H) and likewise **904** (R = $OCOCH_3$) reacted with N_α-methylphenylhydrazine to form **905** (R^1 = H, R^2 = $OCOCH_3$, R^3 = CH_3, R^4 = H). Subsequent sequential hydrolysis of the acetyl group using dilute sodium hydroxide and Oppenauer oxidation of these products afforded **905** ($R^1 + R^2$ = O, R^3 = $CH_2C_6H_5$ and CH_3, respectively, R^4 = H) (Harvey, D. J., 1968). However, more recent studies have shown that it is not necessary to use N_α-alkylated arylhydrazones of 17-ketosteroids in order to achieve indolization. Thus, the 4-bromophenylhydrazone of **904** (R = $OCOCH_3$) and the hydrazones **906** (R = Br and OCH_3), upon refluxing in acetic acid, afforded **905** (R^1 = R^3 = H, R^2 = $OCOCH_3$, R^4 = Br) and **907** (R = Br and OCH_3, respectively), respectively

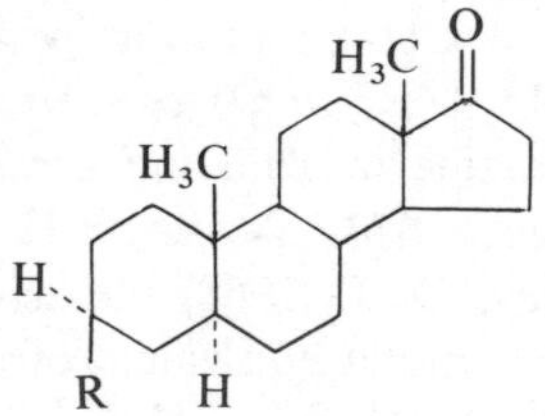

904

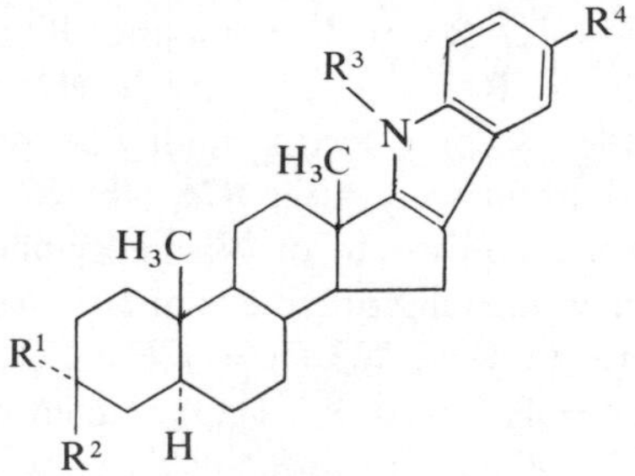

905

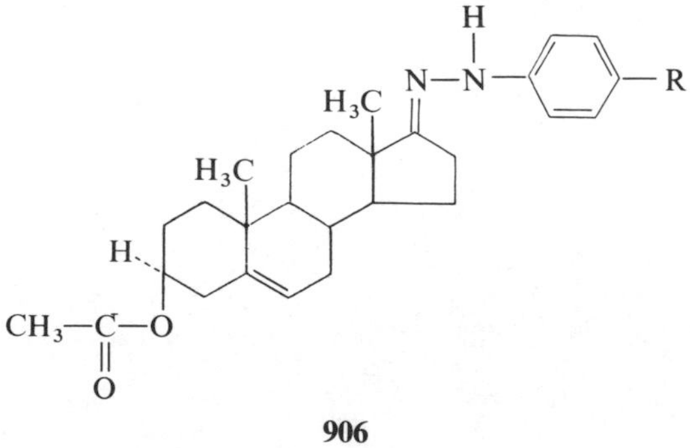

906

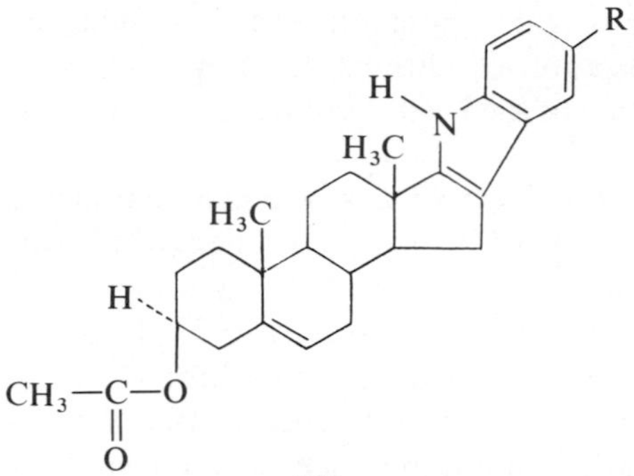

907

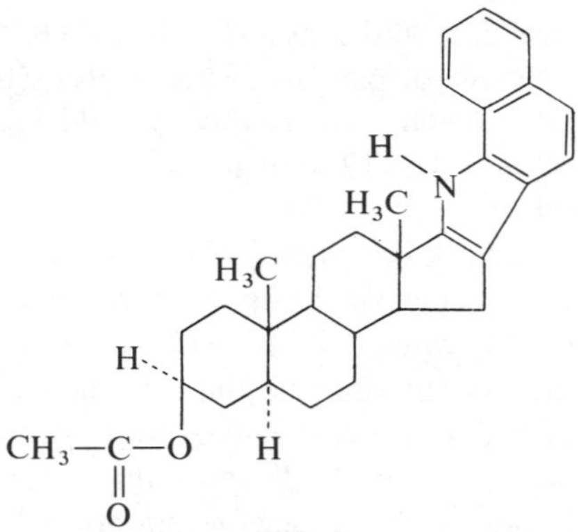

908

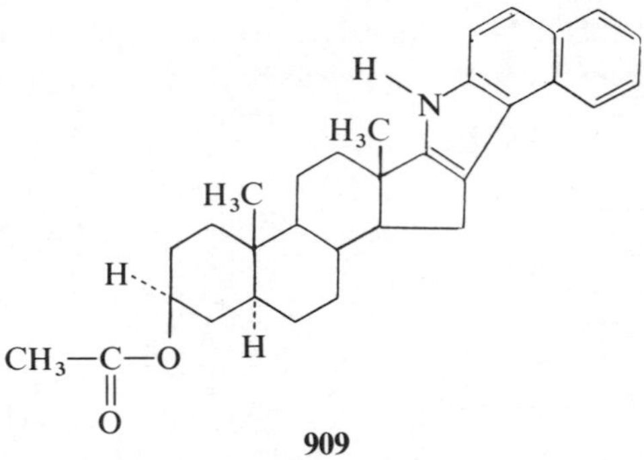

909

(Catsoulacos and Papadopoulos, 1976) and the phenylhydrazone, 4-methylphenylhydrazone and 1- and 2-naphthylhydrazone of **904** (R = $OCOCH_3$) afforded **905** ($R^1 = R^3 = H$, $R^2 = OCOCH_3$, $R^4 = H$ and CH_3, respectively), **908** and **909**, respectively, upon warming in a solution of acetic acid saturated with hydrogen chloride (Jacquignon, Croisy–Delcey and Croisy, 1972).

7. Selective Indolizations using Polyketosteroids

Selective indolizations using polyketosteroids have been observed. Such indolizations have already been referred to which involved only the 3-keto group in 5α- and 5β-cholestan-3,6-dione (Harvey, D. J. and Reid, S. T., 1972), in accordance with the observation that only the 3-keto group of 3,6-diketosteroids reacted with 2,4-dinitrophenylhydrazine to afford the 3-mono-2,4-dinitrophenylhydrazone (Fieser, L. F. and Fieser, M., 1959), and in 5α- and 5β-androstan-3,17-dione (Britten, A. and Lockwood, 1974; Harvey, D. J., 1968). The 3,7,12-triketosteroid **910** ($R^1 + R^1 = O$, $R^2 = H$) reacted with phenylhydrazine and several substituted phenylhydrazines in refluxing acetic acid to furnish **911** (R = H) and the corresponding substituted analogues and, under similar conditions, **910** (R^1 = OCH_3, R^2 = CH_3) reacted with N_α-methylphenylhydrazine to furnish **911** (R = CH_3). These observations indicated that, under such relatively mild indolization conditions, the low reactivities of the 7- and 12-keto groups relative to that of the 3-keto group become manifest (Harvey, D. J., 1968).

A route by which 3,17-diketosteroido[6,7-*b*]indoles can be prepared utilizing the Fischer synthesis has been developed. Androst-4-ene-3,17-dione (**912**) was readily converted into **913**, in nearly quantitative yield, by reacting with pyrrolidine in refluxing thiophen-free benzene solution (Heyl and Herr, 1953). Reaction of **913** with benzenediazonium fluoroborate and with 2- and 4-bromo-, 2-, 3- and 4-chloro, 2-, 3- and 4-fluoro- and 3- and 4-iodobenzenediazonium fluoroborate afforded the corresponding **914** which upon treatment with phosphorus

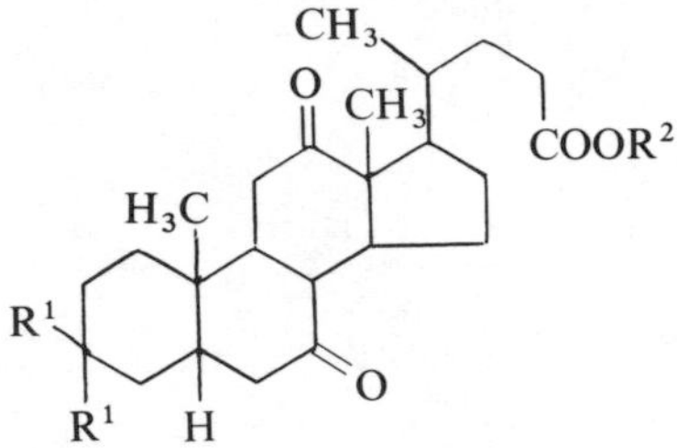

910

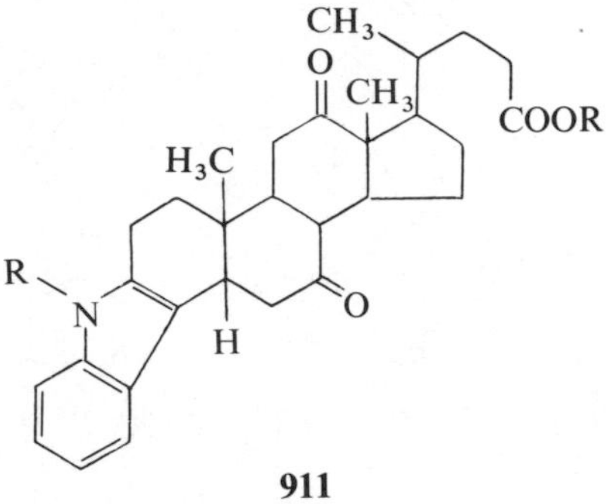

911

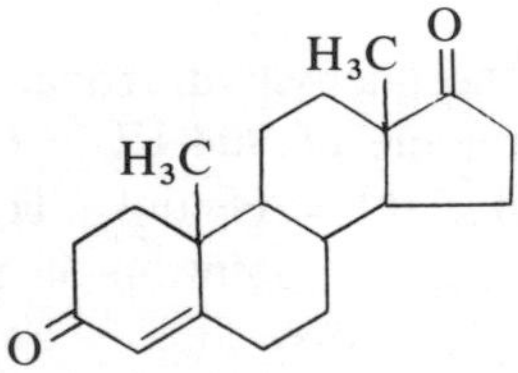

912

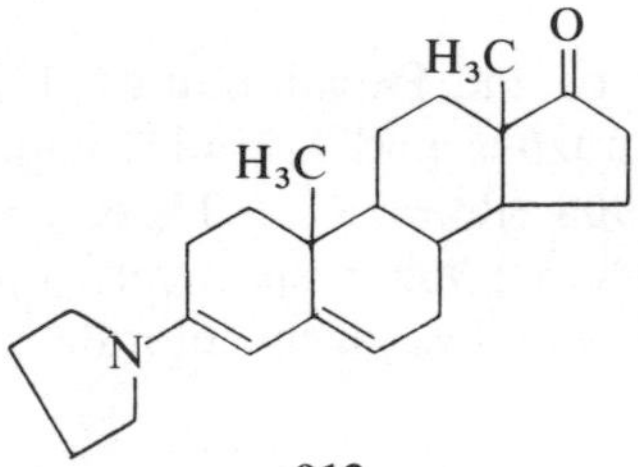

913

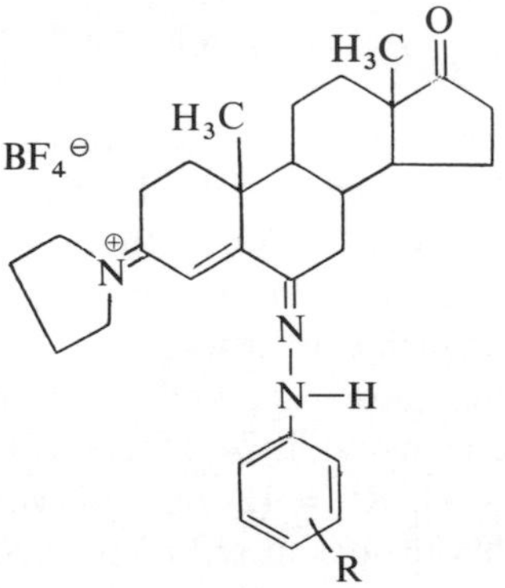

914

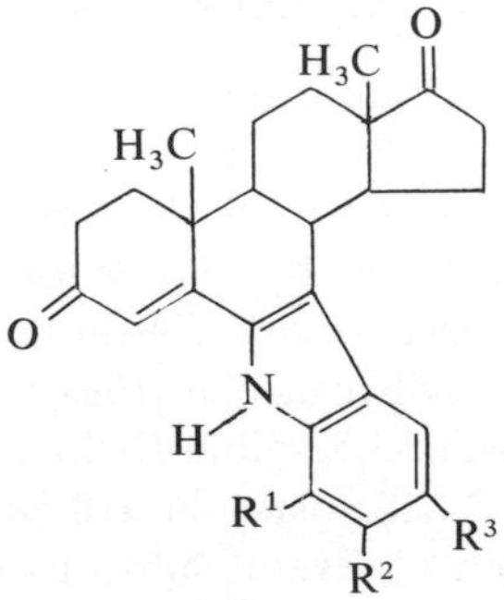

915

oxychloride at room temperature underwent indolization to give products which, without purification, were hydrolysed in 2% methanolic sodium hydroxide to yield **915** (R^1–R^3 = H; R^1 = Br, R^2 = R^3 = H; R^1 = R^2 = H, R^3 = Br; R^1 = Cl, R^2 = R^3 = H; R^1 = R^3 = H, R^2 = Cl; R^1 = R^2 = H, R^3 = Cl; R^1 = F, R^2 = R^3 = H, R^1 = R^3 = H, R^2 = F; R^1 = R^2 = H, R^3 = F; R^1 = R^3 = H, R^2 = I and R^2 = H, R^3 = I, respectively) (Manhas, Brown, J. W. and Pandit, 1975; Manhas, Brown, J. W., Pandit and Houdewind,

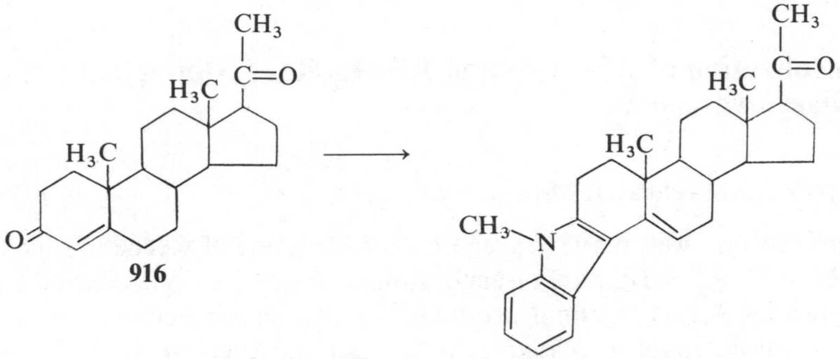

CH_3 $C{=}O$ H_3C H_3C O

916

CH_3 $C{=}O$ H_3C H_3C CH_3 N

917

CH_3 $C{=}O$ H_3C H_3C O H

918

CH_3 $C{=}O$ H_3C H_3C N CH_3 H

919

O $CH_2O{-}C{-}CH_3$ $C{=}O$ H_3C OH O H_3C O

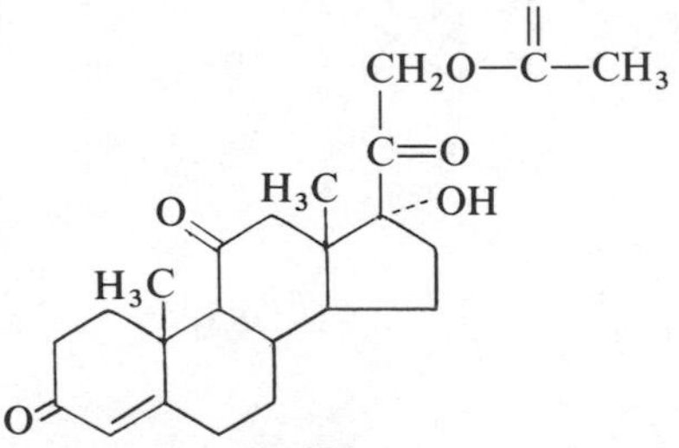

920

O $CH_2O{-}C{-}CH_3$ $C{=}O$ H_3C OH O H_3C CH_3 N

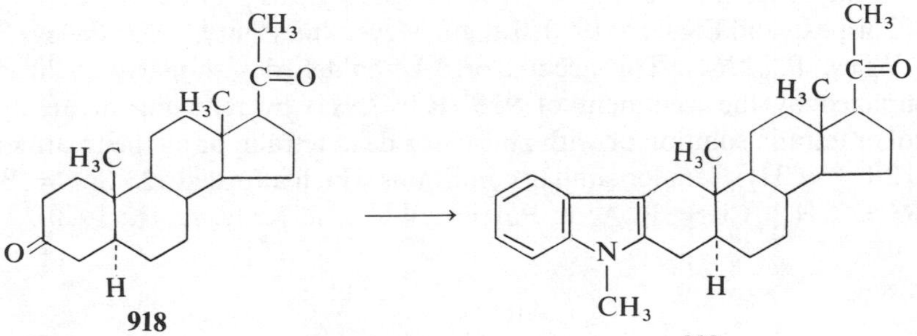

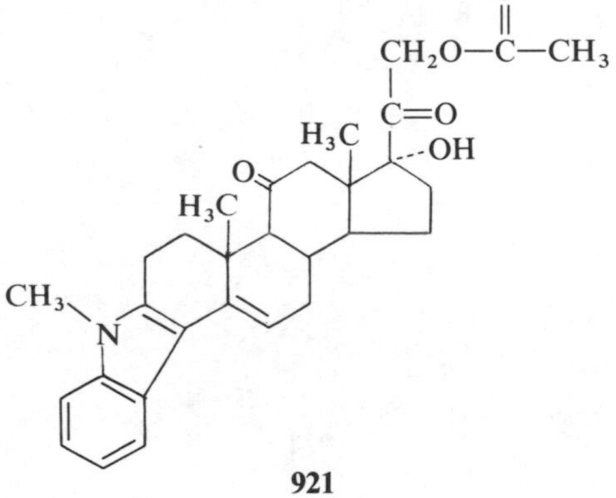

921

1975). Verification was not given for the assigned orientation of the halogen substituent in the **915** obtained from the 3-halogenophenylhydrazones (see Chapter III, A).

Other examples of selective carbonyl reactivities in ketosteroid arylhydrazone indolizations were apparent when mixtures of progesterone (**916**), 5α-pregn-3,20-dione (**918**) and cortisone acetate (**920**) with N_α-methylphenylhydrazine reacted in refluxing acetic acid to afford **917**, **919** and **921**, respectively (Harvey, D. J., 1968).

K. Indolization of 1,3-, 1,4- and 1,5-Diketone Mono- and Diarylhydrazones

1. Acyclic 1,3-Diketone Arylhydrazones

When attempts were made to prepare arylhydrazones of acyclic 1,3-diketones **922** (R^2 = H, R^4 = H, alkyl or aryl), subsequent cyclization occurred to give the pyrazoles **923** as the final products. Indeed, such reactions constitute a major synthetic route to pyrazoles (Coispeau and Elguero, 1970; Kost and Grandberg, 1966; Phillips, 1959). Similar reactions also occurred with 3-keto-ester arylhydrazones **922** (R^2 = H, R^4 = O-alkyl) to produce pyrazol-3-ones **924** (Coispeau and Elguero, 1970; Julian, Meyer and Printy, 1952; Wiley, R. H. and Wiley, P., 1964). The acetals of 3-ketoaldehydes similarly cyclized, as exemplified by the treatment of **925** (R = CH_3) by refluxing in diethylene glycol or tetralin solution or with zinc chloride in tetralin or by the treatment of **925** [2R = $(CH_2)_2$] under similar conditions which afforded **923** (R^1 = R^4 = H, R^3 = CH_3) (Clark, B. A. J., Parrick, West and Kelly, A. H., 1970). These

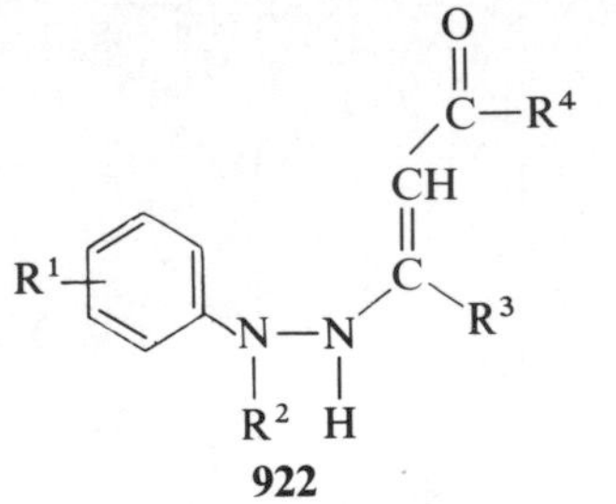

922

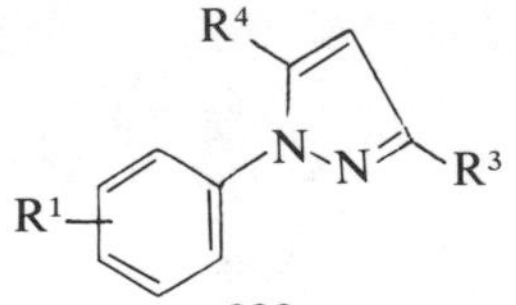

923

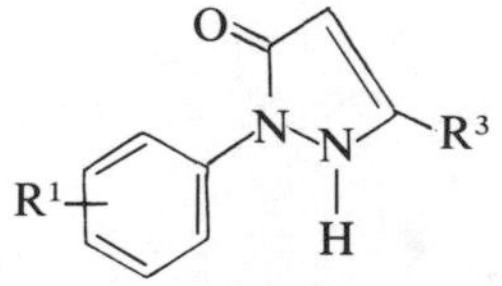

924

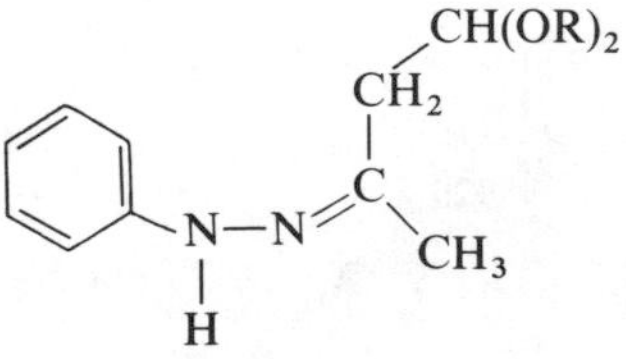

925

cyclizations, as alternatives to indolization, of arylhydrazones, along with related reactions are discussed in more detail in Chapter VII, D and E.

Using N_α-substituted arylhydrazones of 1,3-diketones, pyrazole formation is prevented and the anticipated indolizations can be achieved. Thus, a mixture of pentan-2,4-dione with N_α-methylphenylhydrazine was indolized to furnish 3-acetyl-1,2-dimethylindole (**2**; $R^1 = R^2 = CH_3$, $R^3 = COCH_3$), C—C bond formation occurring during indolization at the C-3 atom, as might be expected, and not the C-1 atom of the ketonic moiety (Robinson, B., 1969). Similarly, indolization of **926** (R = H and C_6H_5) with a 20% sulphuric acid catalyst gave **927** (R = H and C_6H_5, respectively) (Kost, Yudin and Popravko, 1962). An

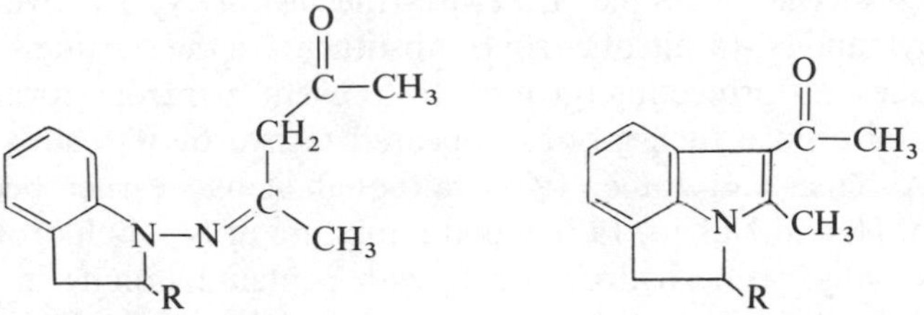

926

927

interesting related reaction involved that between 1-phenylbutan-1,3-dione and N_α-methylphenylhydrazone which apparently afforded only the mono-arylhydrazone **922** (R^1 = H, $R^2 = R^4 = CH_3$, $R^3 = C_6H_5$), by selective reaction at the 1-keto group. The structure of this product was established by its indolization to afford **2** ($R^1 = CH_3$, $R^2 = C_6H_5$, $R^3 = COCH_3$) which was subjected to acid catalysed hydrolysis to give the known 1-methyl-2-phenylindole (**2**; $R^1 = CH_3$, $R^2 = C_6H_5$, R^3 = H) (Kohlrausch, 1889). Other examples of similar selective hydrazone formations are known. Thus, when N_α-methylphenylhydrazine was heated with ethyl benzoylpyruvate it formed an uncharacterized hydrazone which was indolized, by heating with zinc chloride, to afford 3-benzoyl-1-methylindole-2-carboxylic acid (**2**; $R^1 = CH_3$, R^2 = COOH, $R^3 = COC_6H_5$). The structure of this product was confirmed by its alternative synthesis by reaction of 1-methylindole-2,3-dicarboxylic acid anhydride with diphenylcadmium (Staunton and Topham, 1953). A mixture of acetylacetaldehyde with N_α-benzyl-4-methoxyphenyl-hydrazine indolized in warm acetic acid to yield 3-acetyl-1-benzyl-5-methoxy-indole, isolated as its N_α-benzyl-4-methoxyphenylhydrazone (**928**) which was

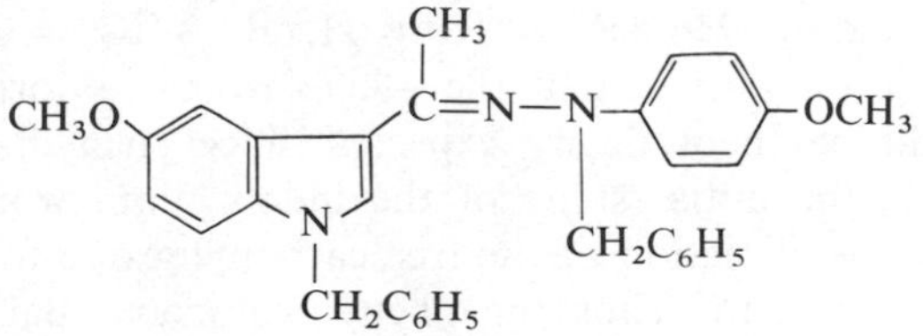

928

formed under the reaction conditions (Cross, A. D., King, F. E. and King, T. J., 1961). A further product of this reaction was benzaldehyde N_α-benzyl-4-methoxyphenylhydrazone, the benzaldehyde probably being derived from the N_α-benzyl group in the original hydrazine or hydrazone by a disproportionation reaction (Cross, A. D., King, F. E. and King, T. J., 1961). When the related indolization, from the reaction of a mixture of pentan-2,4-dione with N_α-benzylphenylhydrazine in refluxing ethylene glycol, was effected, 3-acetyl-1-benzyl-2-methylindole (**2**; $R^1 = CH_2C_6H_5$, $R^2 = CH_3$, $R^3 = COCH_3$) was obtained. However, attempts to debenzylate related 1-benzylindoles were not successful so the application of this route to the synthesis of 1-unsubstituted 3-acylindoles was not possible. Likewise, the use of N_α-benzoyl, N_α-(4-methoxybenzoyl) and N_α-(4-nitrobenzoyl) substituents in the starting 1,3-diketone arylhydrazones as protecting groups to prevent pyrazole formation was without promise since such groups appeared to hydrolyse readily (Mills, Al Khawaja, Al-Saleh and Joule, 1981), although it had earlier been claimed (Yamamoto, H. and Nakao, 1971c) that a mixture of N_α-(4-chlorobenzoyl)-4-methoxyphenylhydrazine hydrochloride with pentan-2,4-dione in acetic acid at 70 °C afforded 3-acetyl-1-(4-chlorobenzoyl)-5-methoxy-2-methylindole. However, doubt has been cast upon this claim following the observation that the reaction of N_α-(4-methoxybenzoyl)phenylhydrazine hydrochloride with pentan-2,4-dione under these conditions led only to the isolation of the carbonyl conjugated enehydrazine tautomer of the corresponding monoarylhydrazone (Mills, Al Khawaja, Al-Saleh and Joule, 1981).

2. Cyclic 1,3-Diketone Arylhydrazones

N_α-Unsubstituted arylhydrazones of these diketones can be prepared and indolized since, unlike their acyclic analogues, they cannot undergo pyrazole formation. Thus, 2,2-dimethylcyclopentan-1,3-dione monophenylhydrazone, upon refluxing in monoethylene glycol under an atmosphere of nitrogen, afforded **929** which was also prepared by reacting 2,2-dimethylcyclopentan-1,3-dione with phenylhydrazine under indolization conditions (Dashkevick, 1978).

In the above formation of **929**, the direction of indolization was unambiguous. However, indolization of cyclohexan-1,3-dione monoarylhydrazones could give rise to 1,2,3,4-tetrahydrocarbazol-2 and/or 4-ones (**930**; $R^3 + R^4 = O$, $R^5 = R^6 = H$) and/or **930** ($R^3 = R^4 = H$, $R^5 + R^6 = O$), respectively, although it is well established that the 4-keto isomer is formed exclusively. This is as would be theoretically expected since enehydrazine formation in the hydrazone, the initial stage of the indolization, would be expected to be favoured in the direction of the free carbonyl group to produce an extended conjugated system. Thus the phenylhydrazone (Ballantine, Barrett, C. B., Beer, Boggiano, Eardley, Jennings and Robertson, 1957; Clemo and

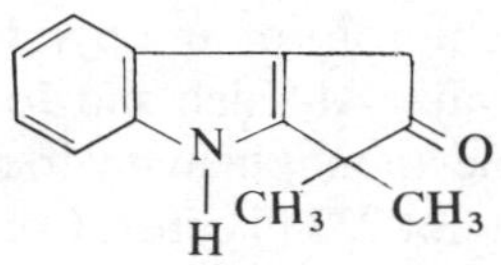

929 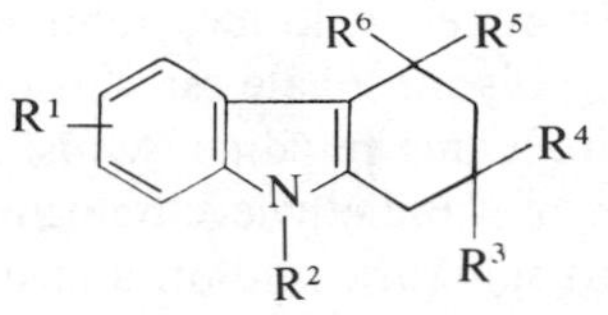 **930**

Felton, 1951b; Hester, 1967, 1969a; Newell, R., 1975; Zinnes, 1975), 4-ethoxyphenylhydrazone (Teuber, Cornelius and Wölcke, 1966) and N_α-methylphenylhydrazone (Ballantine, Barrett, C. B., Beer, Boggiano, Eardley, Jennings and Robertson, 1957) of cyclohexan-1,3-dione gave rise to compounds **930** ($R^1 = R^2 = H$, $R^1 = 6\text{-}OC_2H_5$, $R^2 = H$ and $R^1 = H$, $R^2 = CH_3$, respectively, $R^3 = R^4 = H$, $R^5 + R^6 = O$) [for another similar example, see (Baldwin and Tzodikov, 1977)]. The orientation of the carbonyl group in these products was based upon u.v. spectral data (Teuber, Cornelius and Wölcke, 1966) and upon the observation (Clemo and Felton, 1951b) that the analogous indolization of 5-methylcyclohexan-1,3-dione monophenylhydrazone afforded **930** (R^1–$R^3 = H$, $R^4 = CH_3$, $R^5 + R^6 = O$), as shown by reduction of its carbonyl group and indolic 2,3-double bond to furnish the corresponding hexahydromethylcarbazole which was identical to the product obtained by reduction of the indolic 2,3-double bond in 1,2,3,4-tetrahydro-2-methylcarbazole, the structure of which had been established earlier (Borsche, Witte, A. and Bothe, 1908). Furthermore, compound **930** (R^1–$R^4 = H$, $R^5 + R^6 = O$) has been recently (Oikawa and Yonemitsu, 1977) synthesized by oxidation of 1,2,3,4-tetrahydrocarbazole with 2,3-dichloro-5,6-dicyano-1,4-benzoquinone and a m.p. = 219–221 °C quoted for it, this comparing favourably with that of 223 °C quoted (Clemo and Felton, 1951b) for the product resulting from the above indolization.

Other similar indolizations have been effected. Thus, a mixture of 5,5-dimethylcyclohexan-1,3-dione with phenylhydrazine, using a 4-methylbenzene sulphonic acid catalyst in refluxing toluene under a Dean and Stark trap, produced 1,2,3,4-tetrahydro-2,2-dimethylcarbazol-4-one (**930**; $R^1 = R^2 = H$, $R^3 = R^4 = CH_3$, $R^5 + R^6 = O$) (Borch and Newell, R. G., 1973; Newell, R., 1975). Indolizations of 5,5-dimethylcyclohexan-1,3-dione mono-N_α-benzoyl- and N_α-(4-methoxybenzoyl)phenylhydrazone (**297**; R = H and OCH_3, respectively), but not the mono-N_α-(4-nitrobenzoyl)phenylhydrazone (**297**; $R = NO_2$), were effected in hot 1.75M sulphuric acid, with boron trifluoride in refluxing acetic acid or in refluxing monoethylene glycol to afford **930** ($R^1 = R^2 = H$, $R^3 = R^4 = CH_3$, $R^5 + R^6 = O$). The stage at which the N_α-acyl groups were hydrolysed remains unestablished but the last reaction represents the most efficient way of synthesizing **930** ($R^1 = R^2 = H$, $R^3 = R^4 = CH_3$, $R^5 + R^6 = O$). All three of these above hydrazones were unchanged upon treatment with zinc chloride in refluxing acetic acid, refluxing ethanolic hydrogen chloride, refluxing formic acid, dichloroacetic acid in acetonitrile, polyphosphate ester in refluxing chloroform and pyridine hydrochloride in refluxing pyridine (Mills, Al Khawaja, Al-Saleh and Joule, 1981). Indolization of 4,6-diethylcyclohexan-1,3-dione monophenylhydrazone (**931**) with aqueous sulphuric acid produced **932** and not **933** (Teuber, Cornelius and Wölcke, 1966).

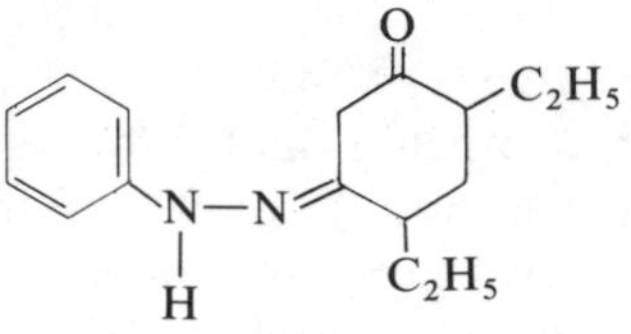

931

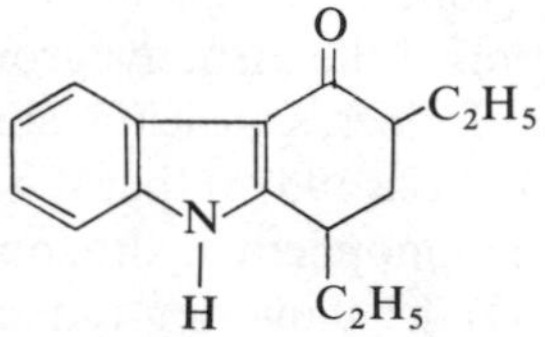

932

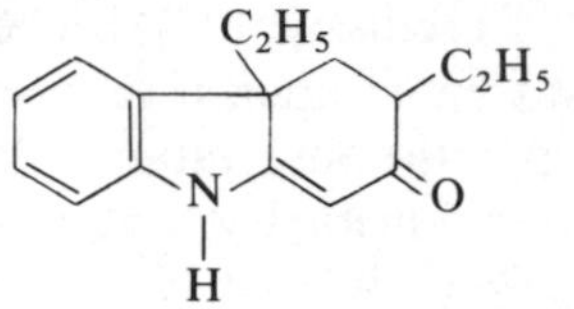

933

Although 1,2,3,4-tetrahydrocarbazol-2-ones **930** ($R^3 + R^4 = O$, $R^5 = R^6 = H$) can be made by other routes (Teuber and Cornelius, 1964), they can also be prepared from cyclohexan-1,3-dione monoarylhydrazones via the formation of the corresponding monoketal with ethylene glycol. Thus, cyclohexan-1,3-dione monophenylhydrazone was converted into the ketal **930** [$R^1 = R^2 = R^5 = R^6 = H$, $R^3 + R^4 = O(CH_2)_2O$] by reaction with ethylene glycol and concomitant indolization in the presence of a 4-methylbenzene sulphonic acid catalyst in refluxing toluene using a Dean and Stark trap. The ketal was then hydrolyzed with 10% sulphuric acid to afford 1,2,3,4-tetrahydrocarbazol-2-one (**930**; $R^1 = R^2 = R^5 = R^6 = H$, $R^3 = R^4 = O$) (Borch and Newell, R. G., 1973). Likewise, **934** ($R^1 = CH_3$, $R^2 = H$) was indolized into **930** ($R^1 = R^2 = R^5 = H$, $R^3 + R^4 = O$, $R^6 = CH_3$) whereas the 5,5-dimethyl analogue **934** ($R^1 = R^2 = CH_3$), when similarly indolized,

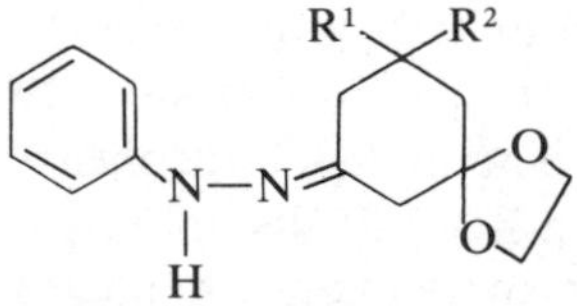

934

but using concentrated sulphuric acid in place of 4-methylbenzene sulphonic sulphonic acid, gave only small yields of both **930** ($R^1 = R^2 = H$, $R^5 = R^6 = CH_3$, $R^3 + R^4 = O$) (6%) and **930** ($R^1 = R^2 = H$, $R^3 = R^4 = CH_3$, $R^5 + R^6 = O$) (5%). It was suggested (Borch and Newell, R. G., 1973) that on steric grounds the enehydrazination of **934** ($R^1 = R^2 = H$ and $R^1 = H$, $R^2 = CH_3$) was favoured in the direction away from the ketal grouping, to give **935** ($R^1 = R^2 = H$ and $R^1 = H$, $R^2 = CH_3$, respectively), which ultimately gave rise to the 1,2,3,4-tetrahydrocarbazol-2-ones, whereas the formation of **935** ($R^1 = R^2 = CH_3$) was energetically unfavourable and thus both the 1,2,3,4-tetrahydrocarbazol-2- and 4-ones were formed but in low yield. In agreement with

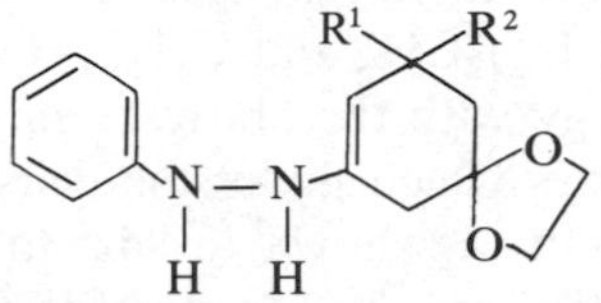

935

these above observations, a mixture of 4-chlorophenylhydrazine hydrochloride with cyclohexan-1,3-dione was converted into 6-chloro-1,2,3,4-tetrahydrocarbazol-2-one (**930**; R^1 = 6-Cl, $R^2 = R^5 = R^6 = H$, $R^3 + R^4 = O$) via the ketal **930** [R^1 = 6-Cl, $R^2 = R^5 = R^6 = H$, $R^3 + R^4 = O(CH_2)_2O$] (Berger, L. and Scott, J. W., 1979). Other ketalizing reagents, namely triethyl orthoformate and triethyl orthoacetate with 4-methylbenzene sulphonic acid were investigated for use in this synthetic approach but, starting with cyclohexan-1,3-dione monophenylhydrazone, these led to the formation of the tetrahydroindazoles **936** (R = H and CH_3, respectively) (Borsch and Newell, R. G., 1973 – see also Newell, R., 1975).

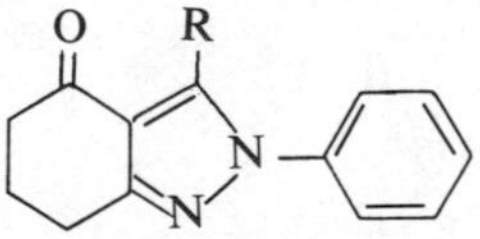

936

The indolization of 1,2,3,4-tetrahydro-9-methylcarbazol-4-one phenylhydrazone **930** ($R^1 = R^3 = R^4 = H$, $R^2 = CH_3$, $R^5 + R^6 = NNHC_6H_5$) failed using a variety of conditions (Bhide, Tikotkar and Tilak, 1957) but vacuum pyrolysis of its 9-demethyl hydrochloride **937** caused, consecutively, indolization and dehydrogenation to afford 5,12-dihydroindolo(3,2-*a*)carbazole (**938**)

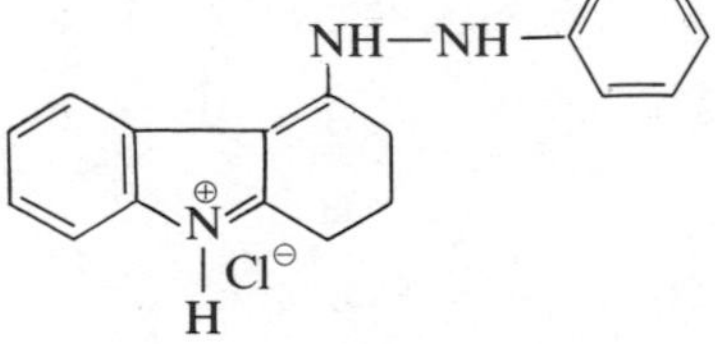

937

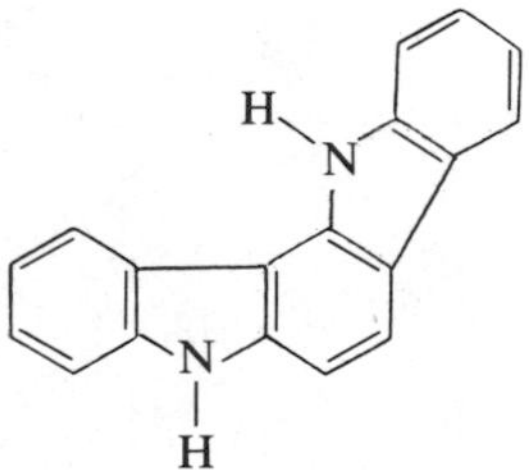

938

(Mann and Willcox, 1958). Although an earlier report (Bhide, Tikotkar and Tilak, 1957) had claimed that **930** (R^1–R^4 = H, $R^5 + R^6$ = O) was unaffected when boiled with phenylhydrazine in acetic acid, **937** was prepared by reacting the ketone with phenylhydrazine in refluxing ethanol containing a trace of acetic acid followed by treatment of the reaction mixture with hydrogen chloride (Mann and Willcox, 1958).

Indolization of several 2-alkylcyclohexan-1,3-dione monoarylhydrazones **939** [R = CH_3, C_2H_5, $(CH_2)_2CH_3$ and $CH_2C_6H_5$)] afforded, via the enehydrazine **940** and by analogy with the above findings, the corresponding 3*H*-indoles **941** as intermediates which under the reaction conditions underwent cleavage and rearrangement, as shown in **942**, to afford the corresponding compounds **943** and **704** (Scheme 39). None of the indoles **944**, which would result from indolization after initial enehydrazine formation away from the carbonyl group, were isolated from these reactions (Teuber, Cornelius and Worbs, 1964; Teuber, Worbs and Cornelius, 1968 – see also Schlittler and Weber, N., 1972).

Scheme 39

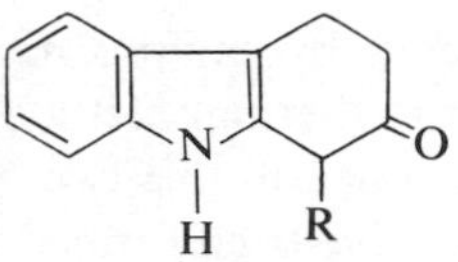

944

3. *Acyclic 1,4-Diketone Arylhydrazones*

Acyclic 1,4-diketones react with arylhydrazines in several ways. Thus, depending upon the conditions, hexan-2,5-dione reacted with N_α-methylphenylhydrazone to form either the bisarylhydrazone **945** (R = CH_3) or the pyrrole **946** ($R^1 = R^2 = R^4 = R^5 = H$, $R^3 = CH_3$) (Kohlrausch, 1889).

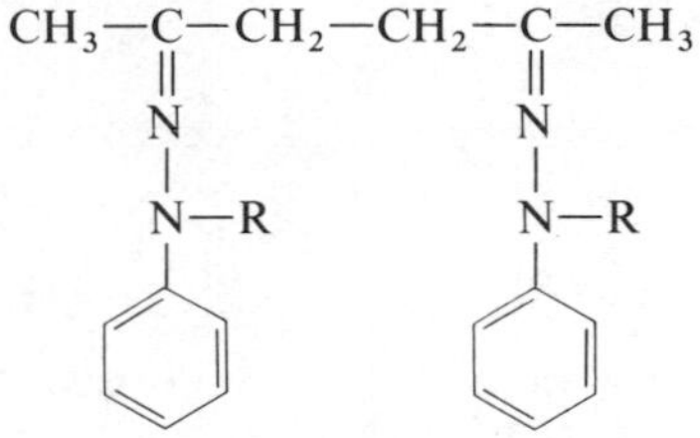

945

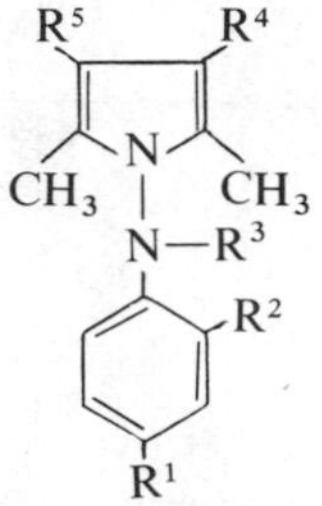

946

Related to this, the bisphenylhydrazone **945** (R = H) (Smith, A., 1896; Smith, A. and McCoy, 1902), or the corresponding hexan-2,5-dione monophenylhydrazone (Arbusow, A. and Chrutzki, 1913; Arbuzov, A. E. and Khrutzkii, 1913), upon treatment under mineral acidic conditions or with copper(I) chloride, respectively, afforded **946** (R^1–R^5 = H). Similarly, hexan-2,5-dione reacted with 2,4-dinitrophenylhydrazine in methanolic hydrochloric acid to give **946** ($R^1 = R^2 = NO_2$. R^3–R^5 = H) (reaction in pyridine solution gave the corresponding mono-2,4-dinitrophenylhydrazone) and ethyl 2-acetonylacetoacetate reacted with 2,4-dinitrophenylhydrazine in ethanolic sulphuric acid to afford the bis-2,4-dinitrophenylhydrazone and **946** ($R^1 = R^2 = NO_2$, $R^3 = R^4 = H$, $R^5 = COOC_2H_5$) (the former of these products yielded the latter product upon standing in ethanolic sulphuric acid at room temperature). Diethyl 2,3-diacetylsuccinate reacted with 4-nitrophenylhydrazine and 2,4-dinitrophenylhydrazine in aqueous acetic acid and ethanolic sulphuric acid, respectively, to yield **946** ($R^1 = NO_2$, $R^2 = R^3 = H$, $R^4 = R^5 = COOC_2H_5$ and $R^1 = R^2 = NO_2$, $R^3 = H$, $R^4 = R^5 = COOC_2H_5$, respectively) (Binns and Brettle, 1966) and 2,4-dichlorophenylhydrazine reacted with diethyl 2,3-diacetylsuccinate in aqueous acetic acid to afford **946** ($R^1 = R^2 = Cl$, $R^3 = H$, $R^4 = R^5 = COOC_2H_5$) (Bülow, 1918). Although the indolization of **945** (R = CH_3) was not attempted, indolization of the bisphenylhydrazone

945 (R = H) could be effected by boiling in solution with monoethylene glycol; i.e. via the non-catalytic thermal method, to afford the bisindolyl **947** ($R^1 = R^2 = H$) rather than one of the two possible alternative isomeric products, although these may have been formed in traces. This preferential or exclusive formation of compound **947** ($R^1 = R^2 = H$) was not surprising since in the bisenehydrazine **948** leading to its formation there is a fully conjugated system, whereas the analogous intermediates which would ultimately

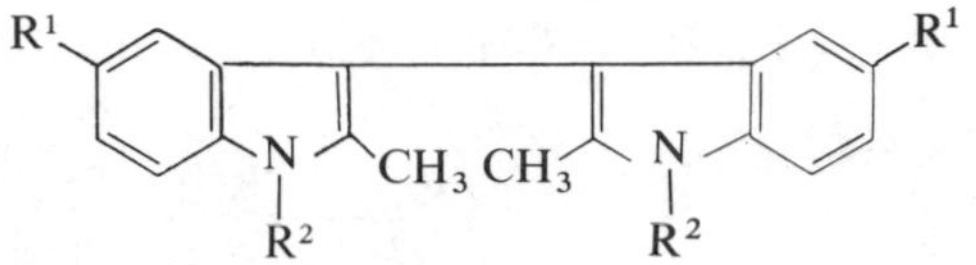

947

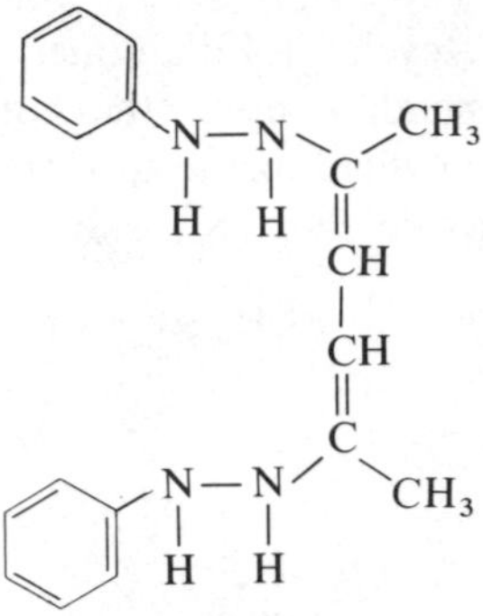

948

afford the other two possible products contained two unconjugated monoenehydrazine systems (Robinson, B., 1964b). When a mixture of N_α-(4-chlorobenzoyl-4-methoxyphenylhydrazine hydrochloride with hexan-2,5-dione was heated in acetic acid it was similarly indolized into **947** [$R^1 = OCH_3$, $R^2 = COC_6H_4Cl(4)$] (Yamamoto, H. and Atami, 1970). In another related study, mixtures of hexan-2,5-dione with 4-methoxy- or 4-nitrophenylhydrazine (Chemerda and Sletzinger, 1969b) or their hydrochlorides (Chermerda and Sletzinger, 1968c) were boiled under reflux in *tert.*-butanol to afford the indoles **949** (R = OCH_3 and NO_2, respectively) (Chemerda and Sletzinger, 1968c, 1969b). This indolization only utilized one of the carbonyl moieties and also appeared to employ a non-catalytic thermal technique in the former reactions.

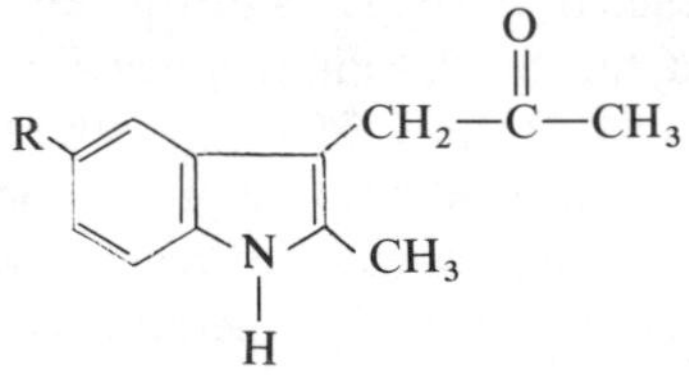

949

4. Cyclic 1,4-Diketone Arylhydrazones

Whereas all attempts to prepare cyclohexan-1,4-dione monophenylhydrazone (**950**; $R^1 = H$, $R^2 + R^3 = NNHC_6H_5$, X = O) have failed (Harley-Mason and Pavri, 1963), 3,5-dimethylcyclohexan-1,4-dione 1-phenylhydrazone

(**950**; $R^1 = CH_3$, $R^2 + R^3 = NNHC_6H_5$, $X = O$) has been prepared (Teuber, Cornelius and Wölcke, 1966), steric hindrance by the two adjacent methyl groups no doubt hindering phenylhydrazone formation at C-4. This product was indolized by warming in dilute sulphuric acid to yield **951** ($R^1 = H$, $R^2 = R^5 = CH_3$, $R^3 + R^4 = O$) (Teuber, Cornelius and Wölcke, 1966).

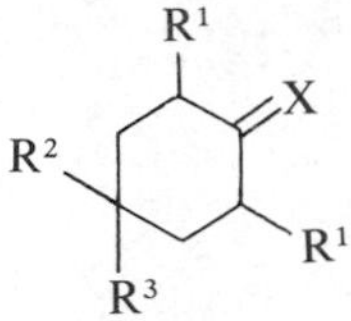

950

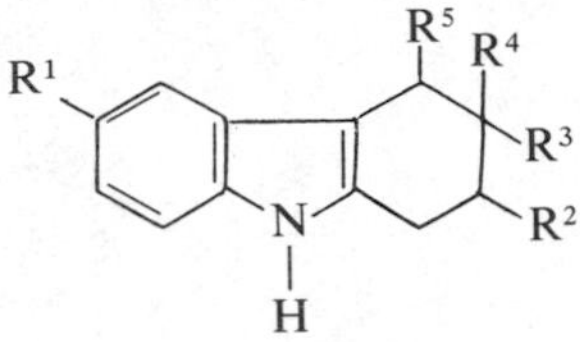

951

Several investigations have been effected upon the indolization of cyclohexan-1,4-dione bisphenylhydrazone (**950**; $R^1 = H$, $R^2 + R^3 = X = NNHC_6H_5$) and the nature of the products has been found to depend upon the catalyst used. Using 50% aqueous sulphuric acid at 100 °C followed by vacuum sublimation, 1,2,3,4-tetrahydrocarbazol-3-one (**951**; $R^1 = R^2 = R^5 = H$, $R^3 + R^4 = O$) and 5,11-dihydroindolo[3,2-*b*]carbazole (**952**) were obtained (Harley-Mason and

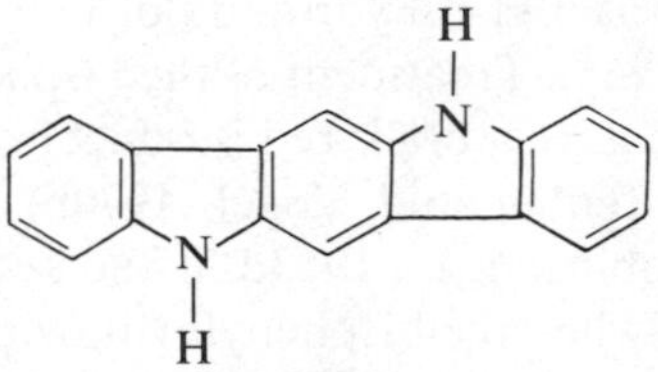

952

Pavri, 1963), compound **952** being formed by dehydrogenation of the dihydro precursor under the reaction conditions used. Heating the bisphenylhydrazone with a 1:5 (v/v) mixture of concentrated sulphuric acid with glacial acetic acid afforded compound **952**, a synthesis of this product which was later (Hünig and Steinmetzer, 1976) preferred over other published syntheses, and a compound thought to be 3-phenylhydrazinocarbazole (**953**; $R = NNHC_6H_5$) (Robinson,

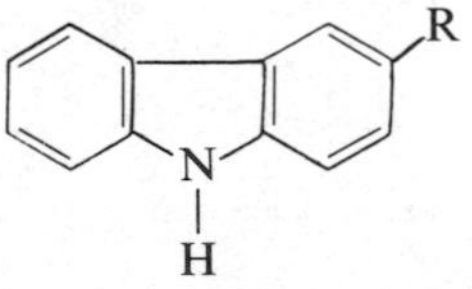

953

B., 1963d). However, the structure of this latter product was later (Teuber and Vogel, 1970a) corrected to 3-aminocarbazole (**953**; $R = NH_2$). Refluxing with ethanolic 1.5 M aqueous sulphuric acid afforded (Teuber, Cornelius and Wölcke, 1966) 3-aminocarbazole (**953**; $R = NH_2$), at the time thought to be 3-phenylhydrazinocarbazole (**953**; $R = NNHC_6H_5$) but later (Teuber and Vogel,

1970a) corrected to **953** (R = NH_2), 3-hydroxycarbazole (**953**; R = OH) and a compound $C_{20}H_{17}N_3$ which was subsequently (Teuber and Vogel, 1970a) shown to have structure **954** (R = CH_3) and to be formed by condensation of the hypothetical intermediate **951** ($R^1 = R^2 = R^5 = H$, $R^3 + R^4 = NNHC_6H_5$), or its conjugated enehydrazine tautomer, with acetaldehyde

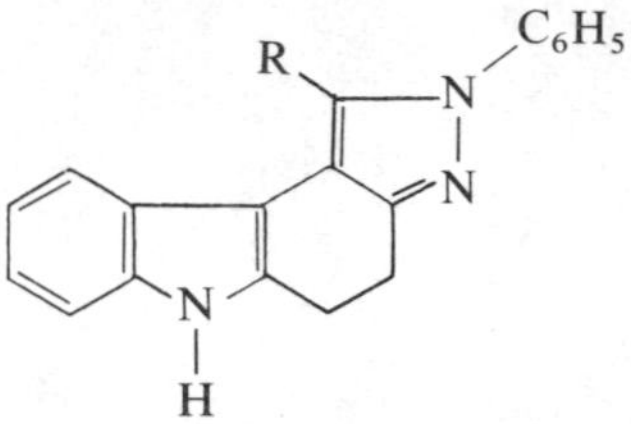

954

which was formed by concomitant oxidation of the reaction solvent, ethanol, by phenylhydrazinium ion. It is interesting in connection with this latter suggestion that when ethanol was replaced by methanol as the reaction solvent, product **954** (R = CH_3) was replaced by **954** (R = H) (Teuber and Vogel, 1970a). The only product isolated by heating the bisphenylhydrazone with 28% aqueous sulphuric acid was 5,11-dihydroindolo[3,2-*b*]carbazole (**952**) (Teuber, Cornelius and Wölcke, 1966). Treatment of the bisphenylhydrazone with ethanolic 0.5 M sulphuric acid gave a brick-red product from which compounds **955** and **956** were isolated (Teuber and Vogel, 1970b). It is interesting to note that in earlier studies (Robinson, B., 1963d) it had been noted that a red colour had immediately formed when the bisphenylhydrazone was added to a mixture of concentrated sulphuric acid and glacial acetic acid.

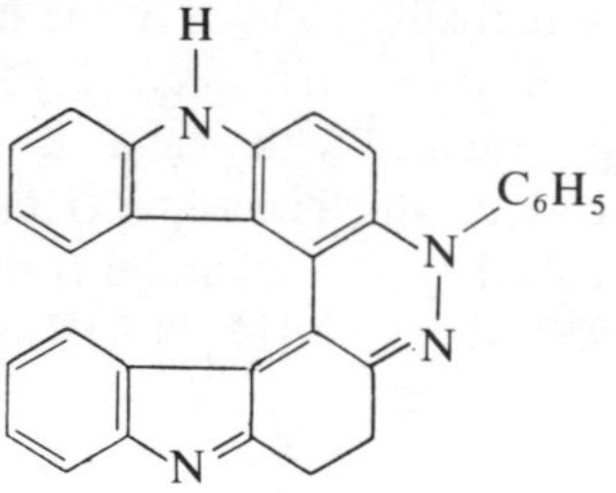

955

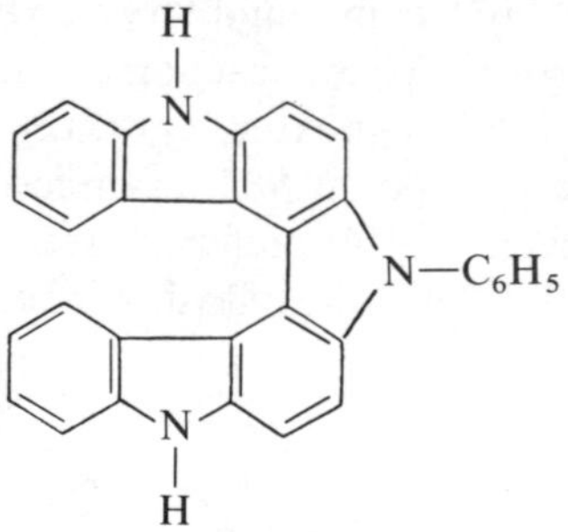

956

The formation of all the above reaction products has been rationalized (Teuber and Vogel, 1970b) as occurring via the common intermediate **951** ($R^1 = R^2 = R^5 = H$, $R^3 + R^4 = NNHC_6H_5$) which was formed by monoindolization of the bisphenylhydrazone. This could then undergo disproportionation, with or without hydrolysis, to form 3-amino- or 3-hydroxycarbazole

(**953**; R = NH_2 and OH, respectively), respectively, hydrolysis to afford 1,2,3,4-tetrahydrocarbazol-3-one (**951**; $R^1 = R^2 = R^5 = H$, $R^3 + R^4 = O$), indolization followed by dehydrogenation to afford 5,11-dihydroindolo[3,2-*b*]-carbazole (**952**), reaction with concomitantly formed aldehydes to give the pyrazolocarbazoles **954** or reaction with 1,2,3,4-tetrahydrocarbazol-3-one (**951**; $R^1 = R^2 = R^5 = H$, $R^3 + R^4 = O$) to give the helicene **955** which would act as a precursor of the helicene **956** under the reaction conditions. It is interesting that no 5,8-dihydroindolo[2,3-*c*]carbazole nor its tetrahydro intermediate was isolated from any of these reactions.

Probably the most synthetically useful product from the above reaction was 1,2,3,4-tetrahydrocarbazol-3-one (**951**; $R^1 = R^2 = R^5 = H$, $R^3 + R^4 = O$) but this was either not produced in very large yields (Harley-Mason and Pavri, 1963) or was not produced at all (Britten, A. and Lockwood, 1974). Although an alternative synthesis for its preparation is available (Teuber and Cornelius, 1964), it can also be synthesized in good yields by utilizing the appropriate Fischer synthesis. Thus, the cyclohexan-1,4-dione ketal phenylhydrazone **950** [$R^1 = H$, $R^2 + R^3 = O(CH_2)_2O$, $X = NNHC_6H_2$] was indolized in anhydrous benzene solution with freshly fused zinc chloride as catalyst to afford **951** [$R^1 = R^2 = R^5 = H$, $R^3 + R^4 = O(CH_2)_2O$] (74% yield) which was quantitatively hydrolysed to the required product (Britten, A. and Lockwood, 1974). Refluxing an acetic acid solution of 4-benzyloxyphenylhydrazine hydrochloride with 4-acetoxycyclohexanone (**950**; $R^1 = R^2 = H$, $R^3 = OCOCH_3$, $X = O$) in the presence of sodium acetate furnished 3-acetoxy-6-benzyloxy-1,2,3,4-tetrahydrocarbazole (**951**; $R^1 = OCH_2C_6H_5$, $R^2 = R^3 = R^5 = H$, $R^4 = OCOCH_3$) in 84% yield. This was hydrolytically deacetylated in 95% yield and the resulting secondary alcohol was subjected to Oppenauer oxidation to afford 6-benzyloxy-1,2,3,4-tetrahydrocarbazol-3-one (**951**; $R^1 = OCH_2C_6H_5$, $R^2 = R^5 = H$, $R^3 + R^4 = O$) in 63% yield. Also prepared by this route were **951** ($R^1 = R^2 = R^5 = H$, $R^3 + R^4 = O$) and its 9-methyl derivative (Coombes, Harvey, D. J. and Reid, S. T. 1970). Upon refluxing a solution of 4-benzoyloxy-2,6-dimethylcyclohexanone (**950**; $R^1 = CH_3$, $R^2 = H$, $R^3 = OCOC_6H_5$, $X = O$) with phenylhydrazine in glacial acetic acid, 3-benzoyloxy-

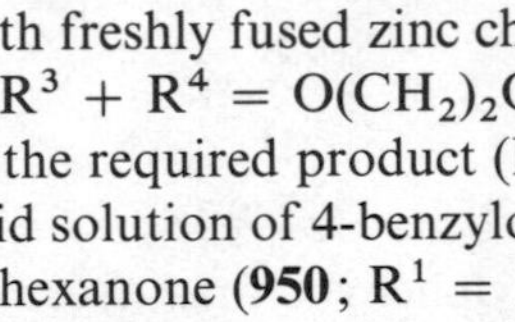

957

1,2,3,4-tetrahydro-1,4a-dimethyl-4a*H*-carbazole (**957**; $R^1 = H$, $R^2 = OCOC_6H_5$) was produced. This was hydrolysed with base to the secondary alcohol **957** ($R^1 = H$, $R^2 = OH$) which was dehydrogenated with Raney nickel and cyclohexanone to yield **957** ($R^1 + R^2 = O$) (Teuber and Cornelius, 1965).

In a related synthetic sequence (Harley-Mason and Pavri, 1963; Teuber, Cornelius and Wölcke, 1966), a mixture of 4-benzoyloxycyclohexanone (**950**; $R^1 = R^2 = H$, $R^3 = OCOC_6H_5$, $X = O$) with phenylhydrazine was indolized in glacial acetic acid and in ethanolic hydrochloric acid to yield 3-benzoyloxy-1,2,3,4-tetrahydrocarbazole (**951**; R^1–$R^3 = R^5 = H$, $R^4 = OCOC_6H_5$) which was subjected to base catalysed hydrolysis to afford the corresponding secondary alcohol. Unfortunately, and surprisingly in view of the above described analogous reactions, attempts to oxidize this alcohol to afford **951** ($R^1 = R^2 = R^5 = H$, $R^3 + R^4 = O$) were unsuccessful.

The monophenylhydrazone **958** ($X = NNHC_6H_5$) has been prepared from the *cis* 1,4-diketone **958** ($X = O$) but attempts to effect its indolization with acetic acid failed and with sulphuric acid only a small yield of the *trans* compound, **959**, was obtained, along with considerable quantities of the *trans* 1,4-diketone **960** (Harvey, D. J. and Reid, S. T., 1970). An attempt to react

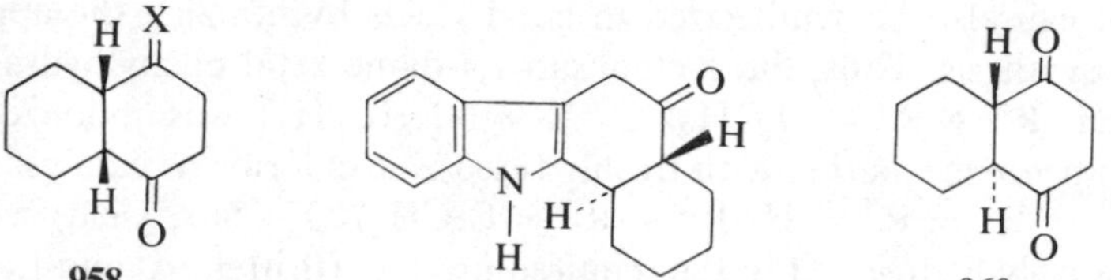

958 **959** **960**

958 ($X = O$) with N_α-methylphenylhydrazine led only to a rapid decomposition of the reaction mixture (Harvey, D. J., 1968). Phenylhydrazine reacted with **960** to produce the bisphenylhydrazone but all attempts to purify this or to prepare the corresponding monophenylhydrazone resulted in failure, so this potential approach to the synthesis of **959** was not pursued further (Harvey, D. J., 1968). Indolization similar to those described above has been reported to occur when **961** ($X = NNHC_6H_5$) was converted into **962** ($R = H$) [which was later (Harvey, D. J., 1968) shown by conformational analysis to be a racemic mixture of **962** ($R = H$) and **963** ($R = H$)] using an acetic acid–hydrochloric acid mixture (Sallay, 1964, 1967b) or 10% sulphuric acid (Harvey,

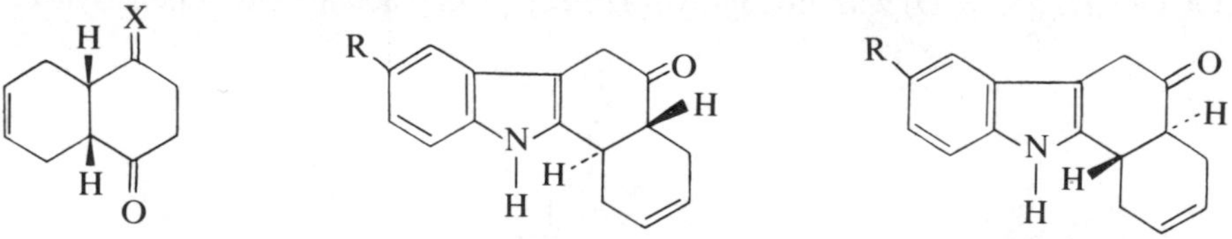

961 **962** **963**

D. J. and Reid, S. T., 1970) as the catalyst. In this reaction, epimerization at the ring junction again occurred to afford the racemic mixture of the *trans* isomer **962** ($R = H$)/**963** ($R = H$) (Harvey, D. J., 1968) which could also be obtained by similar indolization of the *trans* analogue of **961** ($X = NNHC_6H_5$) (Sallay, 1964). Similarly, **961** [$X = NNHC_6H_4OCH_2C_6H_5(4)$] afforded a racemic mixture (Harvey, D. J., 1968) of **962** (R = $OCH_2C_6H_5$) and **963** (R = $OCH_2C_6H_5$) (35% yield) by refluxing in 1 M sulphuric acid (Harvey, D. J. and Reid, S. T., 1970). However, an attempt to react **961** (X = O) with N_α-methylphenylhydrazine led only to a rapid decomposition of the reaction mixture (Harvey, D. J., 1968). In all these reactions the products isolated were non-basic and resulted from indolization occurring in the direction away from the ring junction. Since the yields were sometimes low, it would be interesting to reinvestigate these reactions with a view to isolating the possible isomeric basic 3*H*-indoles. In connection with this, indolization of **961** (X = $NNHC_6H_5$) has been attempted in refluxing acetic acid using various reaction times but only unidentified dark tarry oils were produced. Not only might such reaction conditions have favoured 3*H*-indole formation but the above observed acid catalysed epimerization at the ring junction during indolization may also have been prevented (Harvey, D. J., 1968; Harvey, D. J. and Reid, S. T., 1970).

Indolization of **964** with alcoholic hydrogen chloride led to the isolation of only one neutral product, **965** (ring junction stereochemistry unspecified), although the reaction mixture was not examined for basic products (Hudson,

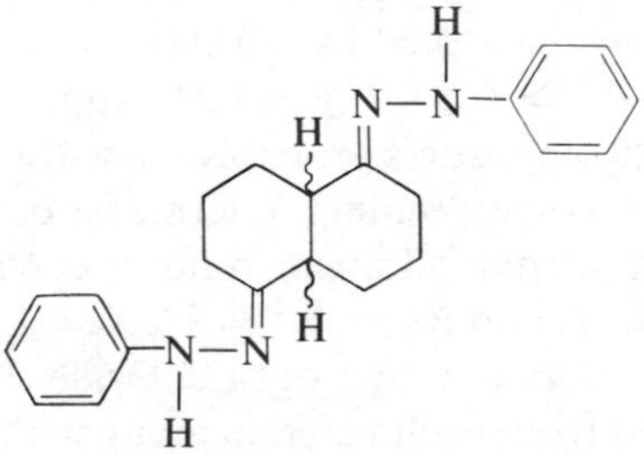

964

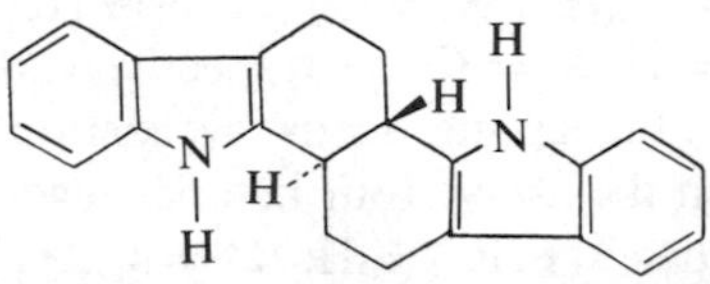

965

B. J. F. and Robinson, R., 1942). Concomitantly with this investigation, attempted indolization of the *trans* isomer of **964** again afforded only one neutral, crystalline product. However, the molecular formula of this product, m.p. 292–296 °C, appeared to contain oxygen and its structure remained unestablished (Hudson, B. J. F. and Robinson, R., 1942). Further investigation of this latter indolization, effected by boiling a mixture of phenylhydrazine with the *trans* diketone in acetic acid–hydrochloric acid resulted in the isolation of unreacted *trans* diketone, **965**, **966**, **967** and **968** (Bhide, Pai, Tikotkar and Tilak, 1958). The melting points quoted (Bhide, Pai, Tikotkar and Tilak, 1958), for the last three products were 195 °C, 232–235 °C (softens at 225 °C) and

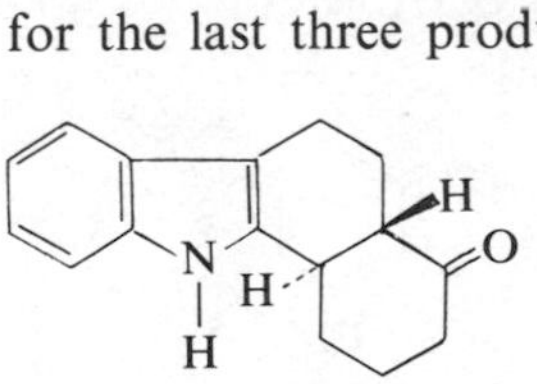

966

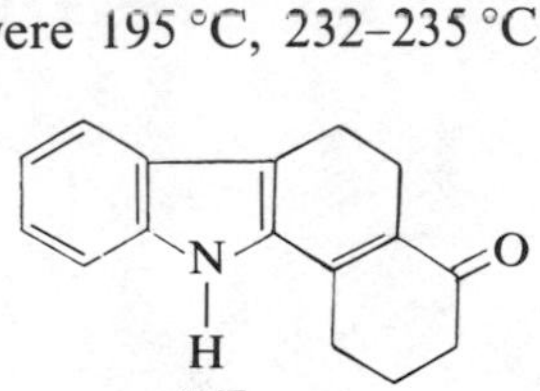

967

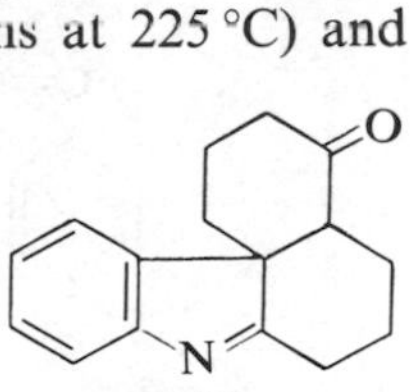

968

168 °C (shrinks at 158 °C), respectively, so it would appear that none of these products was identical with the product, m.p. = 292–296 °C, obtained by the earlier investigators (Hudson, B. J. F. and Robinson, R., 1942) as mentioned above. Further work to establish the structure of this product would be of interest.

5. Acyclic 1,5-Diketone Arylhydrazones

Indolization of a mixture of **969** with heptan-2,6-dione, by heating in acetic acid, afforded **970** (Yamamoto, H. and Atami, 1970). Another study (Moskovkina and Tilichenko, 1976) used **971** [$R^1 = H$, $R^2 = C_6H_5$, $X = CH_2$; $R^1 = H$, $R^2 = C_6H_4OCH_3(4)$, $X = CH_2$; $R^1 = R^2 = H$, $X = CH_2$ and $R^1 = CH_3$, $R^2 = C_6H_5$, $X = O$] as the ketonic moieties. These reacted with phenylhydrazine in the presence of acetic acid at room temperature to produce small amounts of the monophenylhydrazones **974** [$R^1 = H$, $R^2 = C_6H_5$, $X = CH_2$; $R^1 = H$, $R^2 = C_6H_4OCH_3(4)$, $X = CH_2$ and $R^1 = R^2 = H$, $X = CH_2$) and **972** ($R^1 = CH_3$, $R^2 = C_6H_5$, $X = O$), respectively. Heating **972** ($R^1 = CH_3$, $R^2 = C_6H_5$, $X = O$) and **974** ($R^1 = H$, $R^2 = C_6H_5$, $X = CH_2$ and $R^1 = R^2 = H$, $X = CH_2$) in acetic acid gave **973** ($R^1 = CH_3$, $R^2 = C_6H_5$, $X = O$) and **976** ($R^1 = H$, $R^2 = C_6H_5$, $X = CH_2$ and $R^1 = R^2 = H$, $X = CH_2$), respectively, the last two products probably being formed from the spontaneous cyclization of the corresponding intermediate **975**. When the above four ketones reacted with phenylhydrazine in hot acetic acid or when **971** ($R^1 = H$, $R^2 = C_6H_5$, $X = CH_2$ and $R^1 = R^2 = H$, $X = CH_2$) reacted with phenylhydrazine in refluxing alcoholic hydrogen chloride, both the corresponding **973** and **976** were formed from each reaction along with the corresponding **977**. It was suggested that the **977** arose via the corresponding **978**, these being formed when **971** reacted with phenylhydrazine, with subsequent elimination of aniline which was also formed in the reactions (Moskovkina and Tilichenko, 1976). However, the formation of the corresponding anilines as side products during the indolization of arylhydrazones is widely known and very general (see Chapter VII, J) and so, in these reactions the aniline need not result from the postulated intermediates **978**. Indeed, compounds **977**

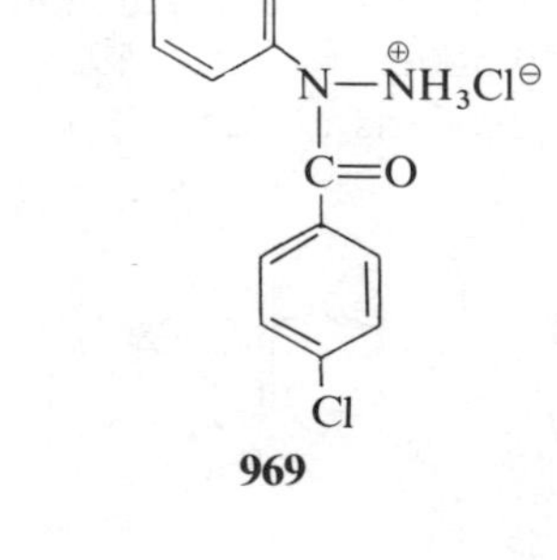

969

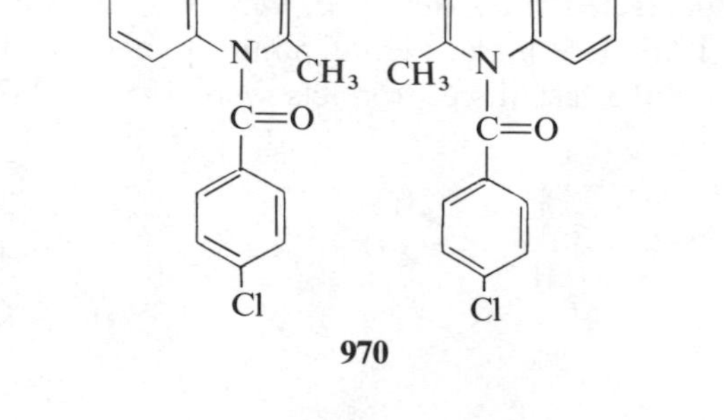

970

971 **972** **973**

974 **975**

976

977 **978**

could arise via the reaction of **971** with the ammonium ion formed from the indolization. Surprisingly, no corresponding products **979** or **980**, which could result from the indolization of **974** in the alternative direction, were isolated from these reactions.

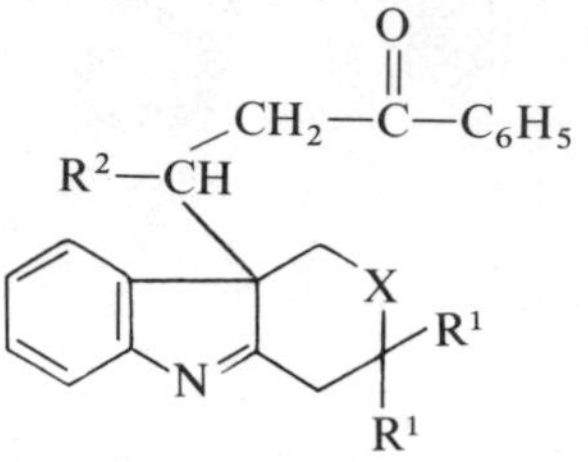

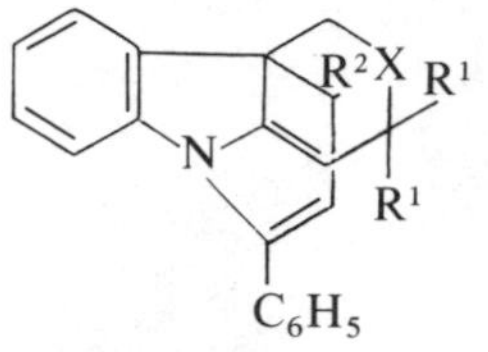

6. Cyclic 1,5-Diketone Arylhydrazones

The indolization of cyclooctan-1,5-dione bisphenylhydrazone (**981**) afforded **982** which could be converted into **983**, although in very low yield (Maiti, Thomson and Mahendran, 1978). It is unfortunate that reaction conditions and the evidence supporting the structural assignment **983**, as opposed to its possible isomer, was not given although this assignment must be based upon p.m.r. spectral data. Indolization of mixtures of the appropriate arylhydrazine hydrochlorides **984** with the appropriate diketones **985** in acetic acid has given products formulated as **986** [R^1–R^3 = H; R^1 = H, CH_3, OCH_3 and Cl, R^2 = $COC_6H_4Cl(4)$, R^3 = H; and R^1 = H, R^2 = $COC_6H_4Cl(4)$, R^3 = CH_3] rather than the alternative isomers (Morita, Noguchi, Kishimoto, Agata and Otsuka, 1972). Although structural verification appeared to be lacking, steric considerations would support these formulations as **986**.

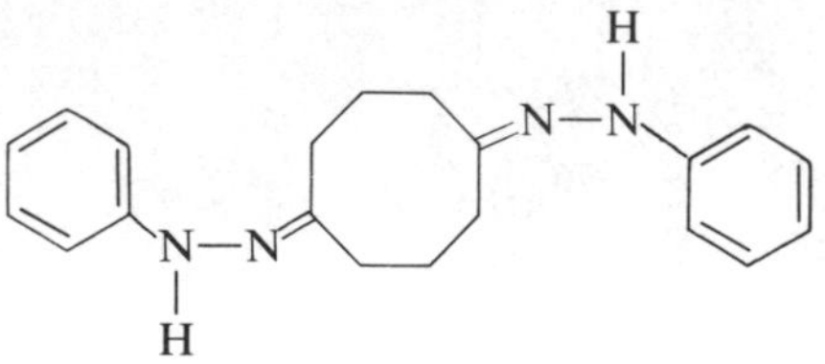

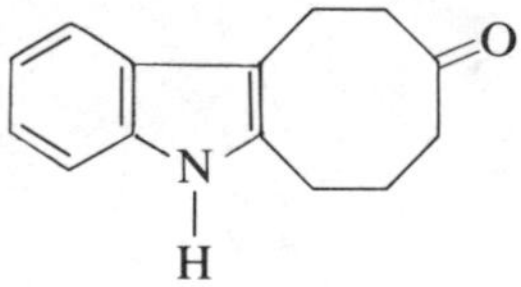

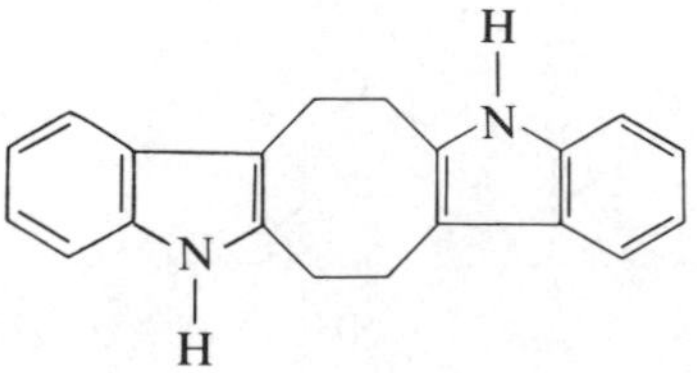

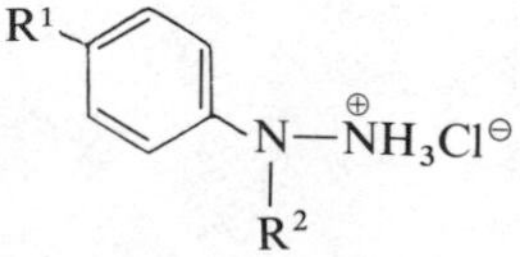

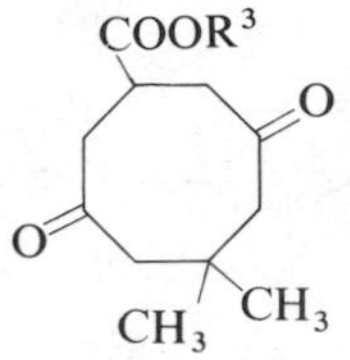

L. Indolization of Heterocyclic Ketone Arylhydrazones

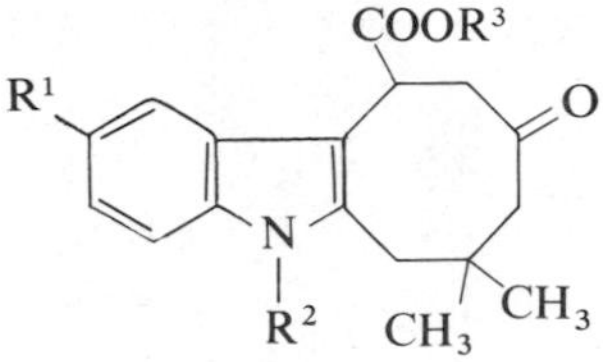

1. Pyrrolidin-3-one Arylhydrazones

The only products isolated from the indolization of these hydrazones result from new C—C bond formation at the C-4 atom of the pyrrolidine nucleus (i.e. indolization takes place in a direction 'away' from the hetero atom). Thus, **987** [$R^1 = Cl$, $R^2 = CH(CH_3)_2$, $R^3 = H$] (Welch, W. M. and Harbert, 1977), **987** ($R^1 = R^3 = H$, $R^2 = COOC_2H_5$) (Welch, W. M., 1977a, 1977b), **987** ($R^1 = F$, $R^2 = COOC_2H_5$, $R^3 = H$) (Plattner, Harbert, Tretter and Welch, W. M., 1975; Welch, W. H. and Harbert, 1976; Welch, W. M. and Harbert, 1977, 1980) **987** ($R^1 = Cl$, $R^2 = COOC_2H_5$, $R^3 = H$) (Plattner, Harbert, Tretter and Welch, W. M., 1975; Welch, W. M. and Harbert, 1977) and **987** ($R^1 = H$ and Br, $R^2 = COOC_2H_5$, $R^3 = H$) (Plattner, Harbert, Tretter and Welch, W. M., 1975) afforded the corresponding indoles **988** and not **989**. Likewise, only the indoles **988** [$R^1 = CH_3$, $R^2 = (CH_2)_2CH_3$; $R^1 = OCH_3$, $R^2 = (CH_2)_3CH_3$; $R^1 = H$, $R^2 = (CH_2)_2CH_3$; $R^1 = CH_3$, $R^2 =$ cyclohexyl; $R^1 = H$, $R^2 =$ cyclohexyl; $R^1 = OCH_3$, $R^2 =$ cyclohexyl; $R^1 = CH_3$, $R^2 = CH_2C_6H_5$ and $R^1 = OCH_3$, $R^2 = CH_2C_6H_5$; $R^3 = H$] and **991** [$R^1 =$ cyclohexyl, $CH_2C_6H_5$ and $(CH_2)_3CH_3$, $R^2 = H$] were produced by indolization of the corresponding hydrazones **987** and **990** (Sharkova, N. M., Kucherova, Aksanova and Zagorevskii, 1969). In the latter study, none of the possible isomeric indolic products could be detected in the reaction mixture and the structures **988** and **991** were confirmed by u.v. and p.m.r. spectral measurements. Further confirmation of structure **988** ($R^1 = R^3 = H$, $R^2 =$ cyclohexyl) was obtained by identifying it with the compound previously (Southwick and Owellen, 1960) obtained by lithium aluminium hydride reduction of the indolization product **284** (X = N-cyclohexyl) from 1-cyclohexylpyrrolidin-2,3-dione phenylhydrazone. Using **987** ($R^1 = R^3 = CH_3$, $R^2 = H$) and **990** ($R^1 = H$, $R^2 = CH_3$), indolizations were unambiguous and the products from such transformations were **988** ($R^1 = R^3 = CH_3$, $R^2 = H$) and **991** ($R^1 = H$, $R^2 = CH_3$), respectively (Aksanova, Kucherova and Sharkova, L. M., 1969).

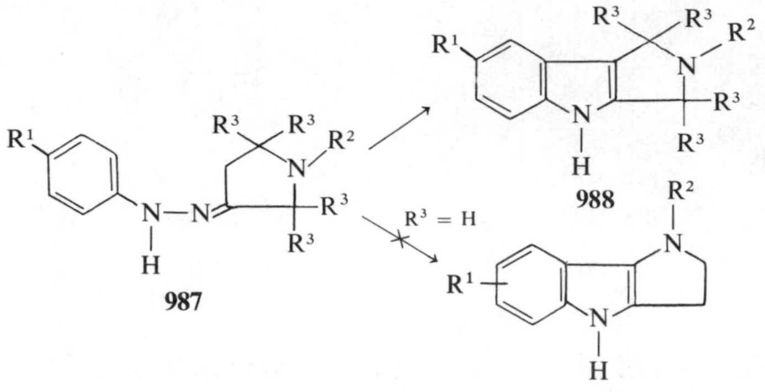

987 988 989

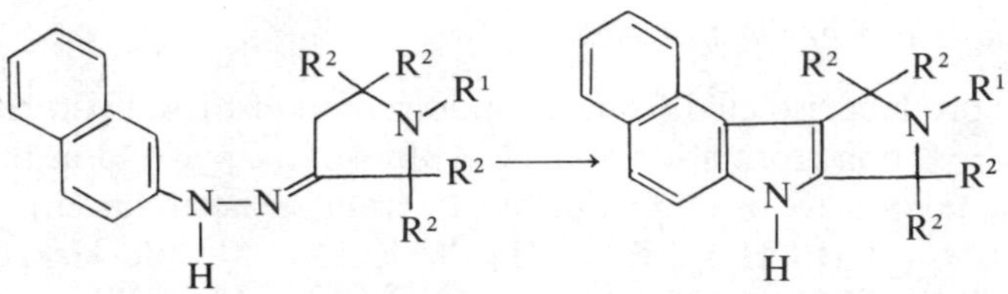

990 991

Unlike the benzothiophen-3(2*H*)-ones **992** (X = S) and benzofuran-3(2*H*)-ones **992** (X = O) (see later in this chapter), the possible indolization of mixtures of indoxyls **992** (X = NR^2) with arylhydrazines has not been investigated although the parent product from such a reaction, 5,10-dihydroindolo-[3,2-*b*]indole, has been prepared by other methods (Cadogan, Cameron-Wood, Mackie and Searle, 1965; Heller, G., 1917; Meyer, K., 1964; Ruggli, 1917). How-

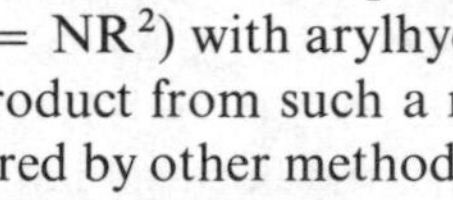

992

ever, closely related reactions have been effected. Thus, reduction of **993** (R^1–R^3 = H; $R^1 = R^3$ = H, $R^2 = OCH_3$ and $R^1 = R^2$ = H, $R^3 = CH_3$ and Cl) with zinc in acetic acid in the presence of acetic anhydride and sodium acetate afforded **994** ($R^2 = R^3$ = H, $R^4 = CH_3$; $R^2 = OCH_3$, R^3 = H, $R^4 = CH_3$ and R^2 = H, $R^3 = CH_3$ and Cl, $R^4 = CH_3$, respectively) apparently through the intermediates depicted in Scheme 40 (Shvedov, Kurilo, Cherkasova and Grinev, A. N., 1975) in which the first two stages of the Fischer indolization mechanism are clearly represented. It would be of interest to attempt these reductive rearrangements in the absence of acetylating agents. By similar routes and under similar conditions, **993** ($R^1 = COCH_3$, $R^2 = R^3$ = H; $R^1 = COCH_3$, $R^2 = CH_3$ and $COOC_2H_5$, R^3 = H; $R^1 = COCH_3$, R^2 = H, $R^3 = CH_3$; $R^1 = COC_6H_5$ and $COCH_2C_6H_5$, $R^2 = R^3$ = H) afforded

N=N, R[3], N, R[2], R[1]

993

H H, N—N, R[3], N, R[2], R[1]

H O, N—C—CH[3], N, R[3], R[1], HN, R[2], C=O, CH[3]

NH[2], N, R[3], R[1], H[2]N, R[2]

H O, N—C—CH[3], N, R[3], R[1], N, R[4]

994

Scheme 40

994 ($R^2 = R^3 = H$, $R^4 = CH_3$; $R^2 = CH_3$ and $COOC_2H_5$, $R^3 = H$, $R^4 = CH_3$; $R^2 = H$, $R^3 = R^4 = CH_3$ and $R^2 = R^3 = H$, $R^4 = C_6H_5$ and $CH_2C_6H_5$, respectively) (Shvedov, Kurilo, Cherkasova and Grinev, A. N., 1977) and **995** ($R^1 = R^2 = H$) reacted with 4-chloro- and 4-nitrophenylhydrazine, **995** ($R^1 = Br$, $R^2 = H$) reacted with phenyl- and 4-chlorophenylhydrazine and **995** ($R^1 = H$, $R^2 = NO_2$) reacted with phenyl-, 4-chlorophenyl- and 4-nitrophenylhydrazine in refluxing acetic acid to furnish **996** ($R^1 = R^2 = H$, $R^3 = Cl$ and NO_2, $R^4 = NH_2$; $R^1 = Br$, $R^2 = H$, $R^3 = H$ and Cl, $R^4 = NH_2$ and $R^1 = H$, $R^2 = NO_2$, $R^3 = H$, Cl and NO_2, $R^4 = NH_2$, respectively) (Kurilo, Ryabova and Grinev, A. N., 1979a, 1979b).

The synthesis of the as yet rarely synthesized 1,4-dihydropyrrolo[3,2-*b*]indole system has not been attempted by reacting 3-hydroxypyrroles with arylhydrazines.

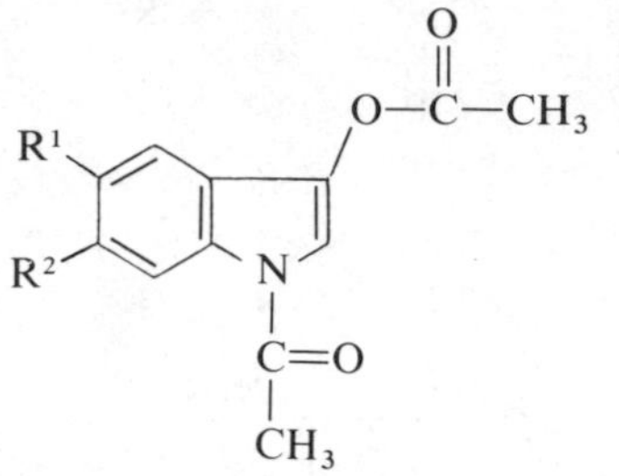

995

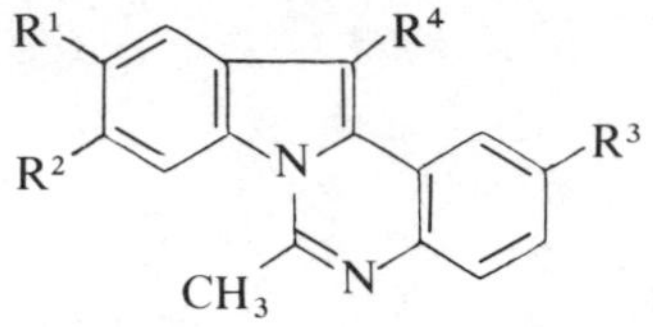

996

2. Piperidin-3-one Arylhydrazones

Although the possibility that both 1,2,3,4-tetrahydro-β- and δ-carbolines, **998** and **999**, respectively, might be produced by indolization of these hydrazones, **997** ($X = H_2$), in practice only the 1,2,3,4-tetrahydro-β-carbolines were isolated, often in good yields, from the reaction mixtures using **997** (R^1 = H,

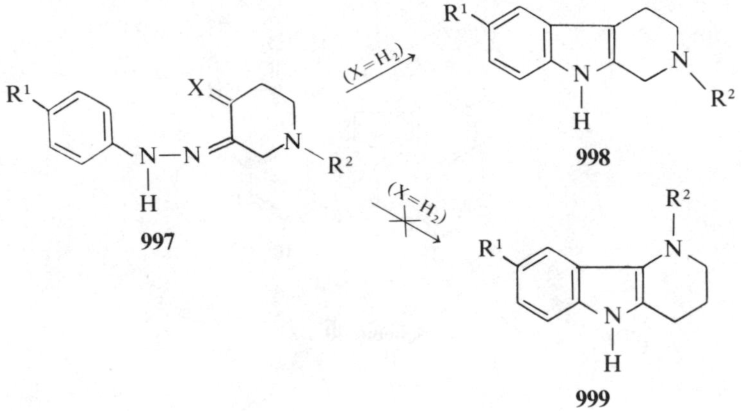

997

998

999

CH_3 and OCH_3, $R^2 = CH_2C_6H_5$, $X = H_2$) and the corresponding 1- and 2-naphthylhydrazone (Périn-Roussel, Buu-Hoï and Jacquignon, 1972) and **997** ($R^1 = CH_3$ and OCH_3, $R^2 = CH_3$, $X = H_2$) (Sharkova, N. M., Kucherova and Zagorevskii, 1972a) although attempts to indolize **997** (R^1 = H Br and $COOC_2H_5$, $R^2 = CH_3$, $X = H_2$) were unsuccessful (Sharkova, N. M., Kucherova and Zagorevskii, 1972a). The structures **998** were distinguished from the corresponding possible **999** by their u.v. (Sharkova, N. M., Kucherova and Zagorevskii, 1972a) and p.m.r. (Périn-Roussel, Buu-Hoï and Jacquignon, 1972; Sharkova, N. M., Kucherova and Zagorevskii, 1972a) spectral properties and by alternative unambiguous synthesis (Périn-Roussel, Buu-Hoï and Jacquignon,

1972). Thus, by analogy with pyrrolidin-3-one arylhydrazones, indolization took place 'away' from the hetero nitrogen atom in **997** ($X = H_2$). Pyrolysis of the above synthesized **998** ($R^2 = CH_2C_6H_5$) with 5% palladium-on-carbon, followed by sublimation, effected debenzylation and dehydrogenation to give the corresponding β-carbolines, the total sequence constituting a useful synthetic route to these compounds (Périn-Roussel, Buu-Hoï and Jacquignon, 1972) which were only obtained as minor products via the Fischer indolization of cyclohexanone 3-pyridylhydrazones (Abramovitch and Adams, K. A. H., 1962) (see Chapter III, C). Attempts to indolize **997** ($R^1 = H$, $R^2 = COOC_2H_5$, $X = O$) were unsuccessful (Morosawa, 1960).

Using aryl hydrazones of piperidin-3-ones which were fused to other rings, indolization again occurred away from the heterocyclic nitrogen atom. Thus, racemic **1000**, upon treatment with ethanolic hydrogen chloride, produced **1001** (Keufer, 1950 and Reckhow and Tarbell, 1952) and under similar conditions

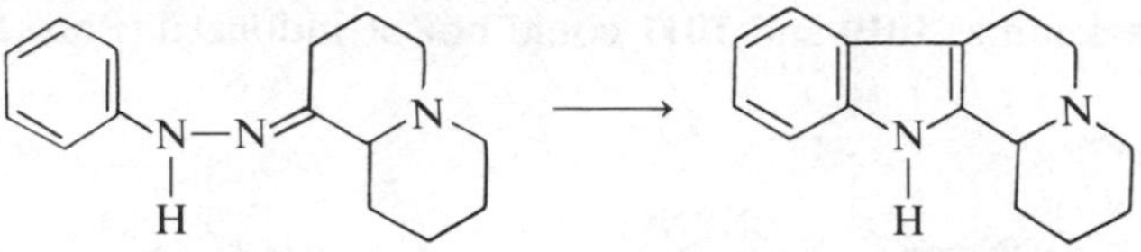

N—N N H N N H

1000 **1001**

R(−)-**1000** and S(+)-**1000** produced R(+)- and (±)-**1001** and S(−)- and (±)-**1001**, respectively (Yamada, S. and Kunieda, 1967). Likewise, the suggestion was made (Clemo and Swan, 1949) that indolization of **1002** ($R^1 = R^2 = H$, $R^3 + R^3 = O$) would afford **1003** ($R^1 = R^2 = H$, $R^3 + R^3 = O$)

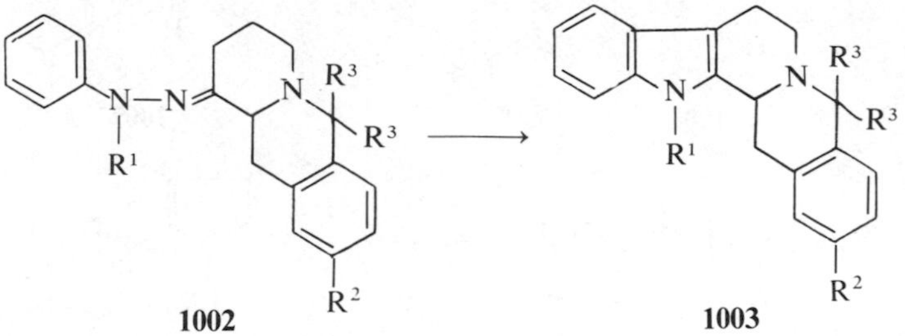

N—N N R^3 R^3 R^1 R^2 N N R^3 R^3 R^1 R^2

1002 **1003**

and experimentally it was shown that the hydrazones **1002** (R^1–R^3 = H; $R^1 = CH_3$, $R^2 = R^3 = H$ and $R^1 = R^3 = H$, $R^2 = OCH_3$) gave **1003** [R^1–R^3 = H (Clemo and Swan, 1946); $R^1 = CH_3$, $R_2 = R^3 = H$ (Julian and Magnani, 1949) and $R^1 = R^3 = H$, $R^2 = OCH_3$ (Swan, 1950), respectively] using ethanolic hydrogen chloride, ethanolic hydrogen chloride, and 5% sulphuric acid, respectively, as catalysts. Starting with R(+)- and S(−)-**1002** (R^1–R^3 = H), indolization using ethanolic hydrogen chloride as catalyst afforded in each case (±)-**1003** (R^1–R^3 = H) along with R(+)- and S(−)-**1003**

(R^1–R^3 = H), respectively (Yamada, S. and Kunieda, 1967). It would be of interest to reinvestigate these indolizations using a 'large' weak acid (e.g. acetic acid) catalyst with a view to effecting indolization involving the ring junction carbon atom, forming the corresponding 3*H*-indoles (see Chapter III, H). Furthermore, since the protonation of the heterocyclic nitrogen atom may be the decisive factor in determining the direction of indolization, which leads to **1001** and **1003**, indolization of these hydrazones using the non-catalytic thermal technique (see Chapter IV, L) should also be investigated.

The molecular skeletons of **1001** and **1003** have been unambiguously synthesised by ethanolic hydrogen chloride or hydrogen bromide catalysed indolizations of hydrazones of type **1004**, **1006** and **1008** which afforded **1005** (Elderfield, Lagowski, McCurdy and Wythe, 1958; Glover and Jones, G., 1958; Kaneko, 1960; Prasad and Swan, 1958; Swan and Thomas, P. R., 1963), **1007** (Elderfield, Lagowski, McCurdy and Wythe, 1958; Glover and Jones, G., 1958; Swan, 1958) and **1009** (Swan, 1958), respectively. However, the phenylhydrazones of the ketones **1010** and **1011** could not be indolized (Prasad and Swan, 1958).

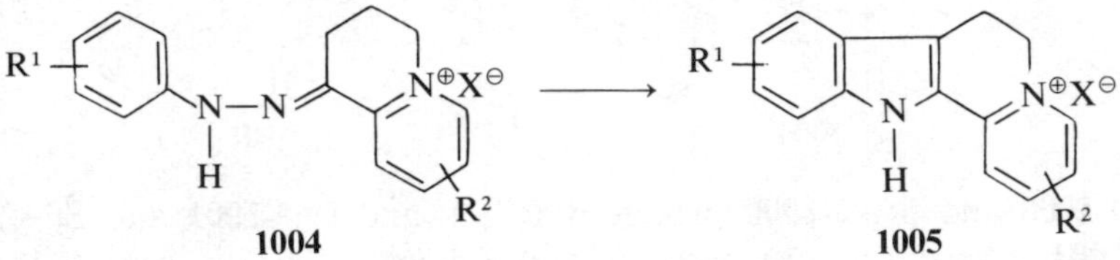

1004 → **1005**

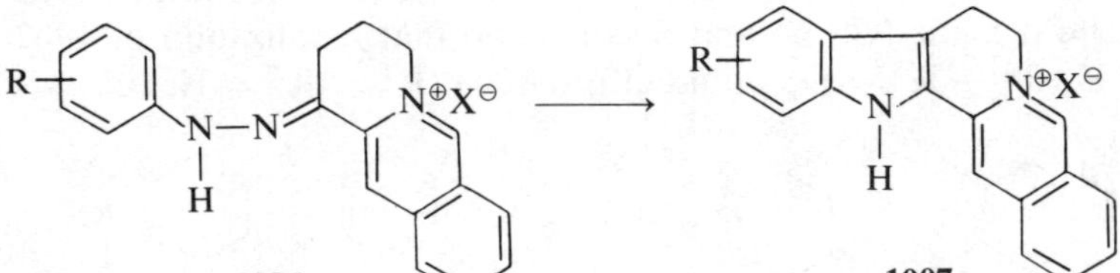

1006 → **1007**

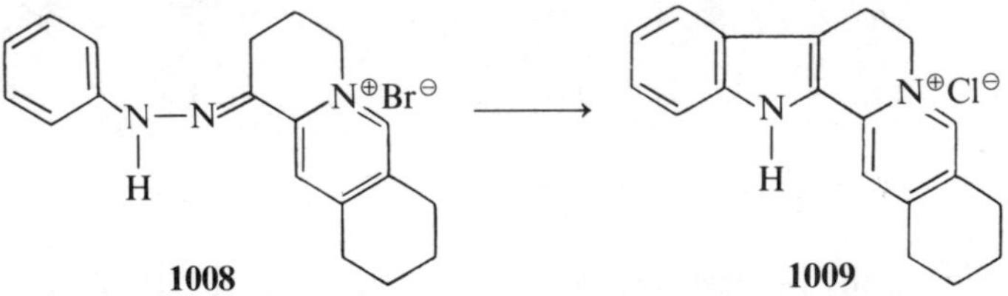

1008 → **1009**

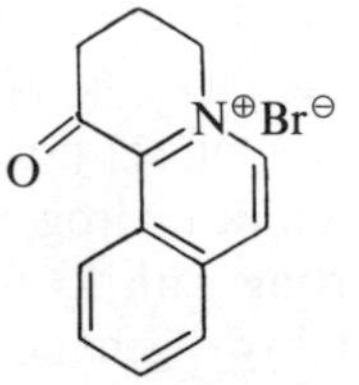

1010 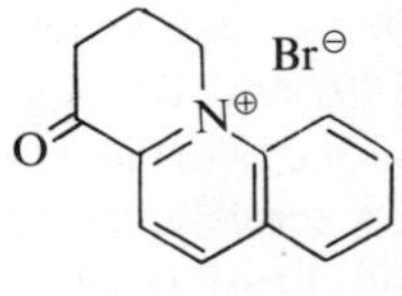 **1011**

Attempted indolization of quinuclidone phenylhydrazone (**1012**) failed. With hydrogen chloride in acetic acid starting material was recovered, with polyphosphoric acid considerable carbonization occurred and with zinc chloride in acetic acid unidentifiable resins were obtained (Buu-Hoï, Jacquignon and Béranger, 1968). In this case, indolization 'away' from the nitrogen atom would be inhibited by Bredt's rule which would prevent the formation of the necessary enehydrazine tautomer.

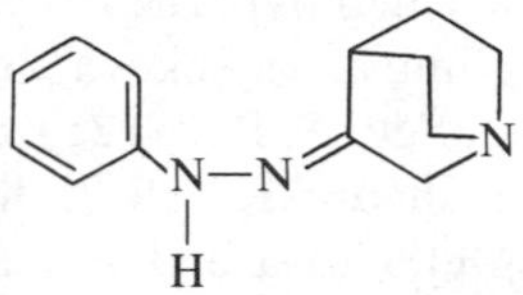

1012

3. Piperidin-4-one Arylhydrazones

The indolization of these arylhydrazones has been used extensively in the synthesis of 1,2,3,4-tetrahydro-γ-carbolines for which the reaction is the major synthetic source (Sharkova, N. M., Kucherova and Zagorevskii, 1972a). A discussion of this reaction has appeared in a review upon 1,2,3,4-tetrahydro-γ-carbolines (Kost, Yurovskaya and Trofimov, 1973).

The hydrazones were usually formed *in situ* when mixtures of their component hydrazine and carbonyl [or ketal (Abbott Laboratories, 1967; Johnson, R. P. and Oswald, 1968; Nagai, Y., Uno, Shimizu, M. and Karasawa, 1977)] compounds were subjected directly to indolization conditions, comparative studies of which have shown that optimal results were obtained using refluxing 7–10% ethanolic hydrogen chloride (Kucherova and Kochetkov, 1956). Thus, the piperidin-4-one arylhydrazones **1013** (R^3–R^6 = H) (Abbott Laboratories,

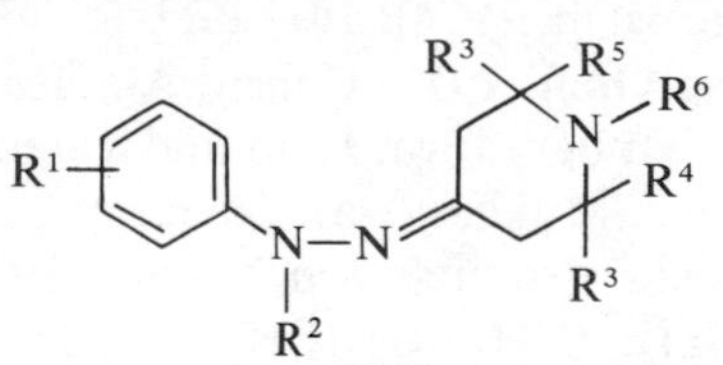

1013

1967; Berger, L. and Corraz, 1972a; Cattanach, Cohen, A. and Heath-Brown, 1968, 1973; F. Hoffman-La Roche, 1963; Hahn, W. E., Bartnik and Zawadzka, 1966; Johnson, R. P. and Oswald, 1968; Nagai, Y., Uno, Shimizu, M. and Karasawa, 1977; Orlova, E. K., Sharkova, N. M., Meshcheryakova, Zagorevskii and Kucherova, 1975; Pachter, 1967; Rajagopalan, 1974; Roche Products Ltd, 1968a), **1013** (R^3–R^5 = H, R^6 = CH_3) (Adams, C. DeW., 1976; Aksanova, Pidevich, Sharkova, L. M. and Kucherova, 1968; Aksanova, Sharkova, N. M., Baronova, Kucherova and Zagorevskii, 1966; Barbulescu,

Bornaz and Greff, 1971; Berger, L. and Corraz, 1968, 1970a, 1970b, 1972a, 1972b; Boekelheide and Ainsworth, C., 1950a; Cattanach, Cohen, A. and Heath-Brown, 1968, 1973; Cohen, A. and Cattanach, 1967; Cohen, A., Heath-Brown and Cattanach, 1968; Cohen, A., Heath-Brown, Smithen and Cattanach, 1968; Cook, A. H. and Reed, K. J., 1945; Farbenfabriken Bayer A.-G., 1954, 1955a; F. Hoffmann-LaRoche, 1963, 1965; Hahn, W. E., Bartnik and Zawadzka, 1966; Heath-Brown, 1975; Hoerlein, 1956; Hörlein, H. U., 1955, 1957; Hörlein, U., 1956; Kochetkov, Kucherova, Pronina, L. P. and Petruchenko, 1959; Kost, Trofimov, Tsyshkova and Shadurskii, 1969; Kost, Vinogradova, Daut and Terent'ev, A. P., 1962; Kost, Vinogradova, Trofimov, Mukhanova, Nozdrich and Shadurskii, 1967; Kost, Yudin and Popravko, 1962; Kost, Yurovskaya, Mel'nikova and Potanina, 1973; Kucherova and Kochetkov, 1956; Orlova, E. K., Sharkova, N. M., Meshcheryakova, Zagorevskii and Kucherova, 1975; Pachter, 1967; Roche Products Ltd, 1965a, 1965c, 1968a; Sellstedt and Wolf, M., 1974; Sharkova, N. M., Kucherova and Zagorevskii, 1964, 1972b, 1972c; Shvedov, Trofimkim, Vasil'eva and Grinev, A. N., 1975; Spickett, 1966; Sumitomo Chemical Co. Ltd., 1968b; Sunagawa and Sato, Y., 1962a; Yamamoto, H. and Atami, 1976; Yamamoto, H., Atsumi, Aono and Kuwazima, 1970; Yamamoto, H., Nakamura, Y., Atami, Nakao and Kobayashi, T., 1970b), **1013** (R^3–R^5 = H, R^6 = C_2H_5) (Adams, C. DeW., 1976; Barbulescu, Bornaz and Greff, 1971; Cohen, A., Heath-Brown, Smithen and Cattanach, 1968; F. Hoffmann-La Roche, 1963, 1965; Roche Products Ltd, 1965c, 1968a; Rosnati and Palazzo, 1954; Spickett, 1966), **1013** [R^3–R^5 = H, R^6 = $(CH_2)_2CH_3$] (Barbulescu, Bornaz and Greff, 1971; Cohen, A., Heath-Brown, Smithen and Cattanach, 1968; Kost, Vinogradova, Trofimov, Mukhanova, Nozdrich and Shadurskii, 1967; Roche Products Ltd, 1968a), **1013** [R^3–R^5 = H, R^6 = $CH(CH_3)_2$] (Adams, C. DeW., 1976; F. Hoffmann-La Roche, 1963), **1013** (R^3–R^5 = H, R^6 = $CH_2CH{=}CH_2$) (Barbulescu, Bornaz and Greff, 1971; Rajagopalan, 1974), **1013** [R^3–R^5 = H, R^6 = $(CH_2)_3CH_3$] (Barbulescu, Bornaz and Greff, 1971; Cohen, A., Heath-Brown, Smithen and Cattanach, 1968; Kost, Vinogradova, Daut and Terent'ev, A. P., 1962; Kost, Vinogradova, Trofimov, Mukhanova, Nozdrich and Shadurskii, 1967; Rajagopalan, 1974; Roche Products Ltd, 1968a), **1013** [R^3–R^5 = H, R^6 = $CH_2CH(CH_3)_2$, $C(CH_3)_3$, $(CH_2)_2CH(CH_3)_2$, $(CH_2)_6CH_3$ and cyclohexyl] (Kost, Vinogradova, Trofimov, Mukhanova, Nozdrich and Shadurskii, 1967), **1013** (R^3–R^5 = H, R^6 = alkyl) (Cattanach, Cohen, A. and Heath-Brown, 1973), **1013** (R^3–R^5 = H, R^6 = lower alkyl) (Pachter, 1967), **1013** (R^3–R^5 = H, R^6 = cycloalkylmethyl and exo-7-norcarylmethyl (Adams, C. DeW., 1976), **1013** (R^3–R^5 = H, R^6 = alkyl, aralkyl and $COCF_3$) (Rajagopalan, 1974), **1013** (R^3–R^5 = H, R^6 = $CH_2C_6H_5$) (Abbott Laboratories, 1967; Adams, C. DeW., 1976; Berger, L. and Corraz, 1968, 1972a; Buu-Hoï, Roussel, O. and Jacquignon, 1964; Johnson, R. P. and Oswald, 1968; Nozoe, Sin, Yamane and Fujimori, K., 1975; Pachter, 1967; Roussel, O., Buu-Hoï and Jacquignon,

1965; Spickett, 1966), **1013** [R^3–R^5 = H, R^6 = $(CH_2)_2C_6H_5$] (Berger, L. and Corraz, 1972a; Spickett, 1966) and **1013** (R^3–R^5 = H, R^6 = $COOC_2H_5$) (Harbert, Plattner, Welch, W. M., Weissmann and Koe, 1980; Plattner, Harbert, Tretter and Welch, W. M., 1975; Rajagopalan, 1974; Welch, W. M., 1977a, 1977b) were all indolized to afford the corresponding 1,2,3,4-tetrahydro-γ-carbolines **1014**. The indolization of these last mentioned arylhydrazones, in

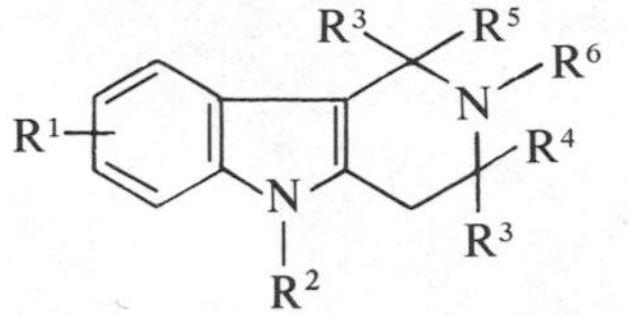

1014

which the piperidin-4-one nitrogen atom was non-basic, proceeded in higher yields than when using a piperidin-4-one moiety containing a basic nitrogen atom. Removal of this basic property probably suppressed the retro Michael reaction which would have led to polymerization in acidic media. Subsequent to indolization, the protecting carbethoxy group could readily be removed by hydrolysis with ethanolic potassium hydroxide (Harbert, Plattner, Welch, W. M., Weissmann and Koe, 1980). The attempted indolization of **1013** (R^1–R^5 = H, R^6 = C_6H_5) has been reported to fail (Gallagher and Mann, 1962). Arylhydrazones **1013** (R^3–R^5 = CH_3, R^6 = H) (Komzolova, Kucherova and Zagorevskii, 1964, 1967; Robinson, R. and Thornley, 1924; Rosnati and Palazzo, 1954) furnished the corresponding **1014** upon indolization but attempts to indolize **1013** (R^1–R^3 = H, R^4–R^6 = CH_3) and the structurally related tropinone phenylhydrazone [**1013**; R^1–R^3 = H, R^4 + R^5 = $(CH_2)_2$, R^6 = CH_3] were unsuccessful (Rosnati and Palazzo, 1954).

An arylhydrazone of an unsymmetrically substituted piperidin-4-one could give rise to two isomeric products upon indolization. Thus, when compounds **1013** (R^1 = H, 4-CH_3, 4-Cl and 4-Br, R^2 = R^5 = R^6 = H, R^3 = R^4 = CH_3) were indolized, both the 1,2,3,4-tetrahydro-2,2,4-trimethyl-γ-carbolines **1014** (R^1 = H, 6-CH_3, 6-Cl and 6-Br, R^2 = R^5 = R^6 = H, R^3 = R^4 = CH_3) and 1,2,3,4-tetrahydro-2,4,4-trimethyl-γ-carbolines **1014** (R^1 = H, 4-CH_3, 4-Cl and 4-Br, R^2 = R^4 = R^6 = H, R^3 = R^5 = CH_3) were formed, the yield of the former isomer being predominant in each case. This predominance became exclusive when starting with the corresponding 4-methoxyphenylhydrazone, only **1014** (R^1 = 6-OCH_3, R^2 = R^5 = R^6 = H, R^3 = R^4 = CH_3) being isolated from the reaction mixture (Novikova, Silenko, Kucherova, Rozenberg and Zagorevskii, 1971).

Several groups have shown that indolization of 2-alkylpiperidin-4-one arylhydrazones **1015** produced, in good yields, compounds of structure **1019** (Borisova, Kucherova, Kartashova and Zagorevskii, 1972; Borisova, Kucherova and Zagorevskii, 1970a; Cattanach, Cohen, A. and Heath-Brown, 1971;

1015 → 1016

1015 → 1017 → (H⊕) 1018 → 1019

1018 → 1020

Scheme 41

Ebnöther, Niklaus and Süess, 1968, 1969; Kucherova, Borisova, Sharkova, N. M. and Zagorevskii, 1970) along with, in some cases, low yields of the corresponding indoles **1016** (Cattanach, Cohen, A. and Heath-Brown, 1971; Ebnöther, Niklaus and Süess, 1969). It was suggested (Ebnöther, Niklaus and Süess, 1969) that the compounds **1019** were formed via the 3*H*-indoles **1017** which underwent a retro Mannich reaction to afford **1018** and then cyclized (Scheme 41). Other products from these reactions were the isotryptamines **1020** which, since compounds **1019** were stable under the indolization conditions used, were thought to also arise from the intermediate **1018** by the elimination of the methylene group by hydrolysis of the quaternary imino functional group (Cattanach, Cohen, A. and Heath-Brown, 1971). Prior to the recognition of this rearrangement, other similar indolizations had been effected. Thus, treatment of **1021** ($R^1 = R^2 = R^6 = H$, $R^3 = R^5 = C_2H_5$, $R^4 = CH_3$) with alcoholic hydrogen chloride yielded a product which was formulated as **1022** (Rosnati

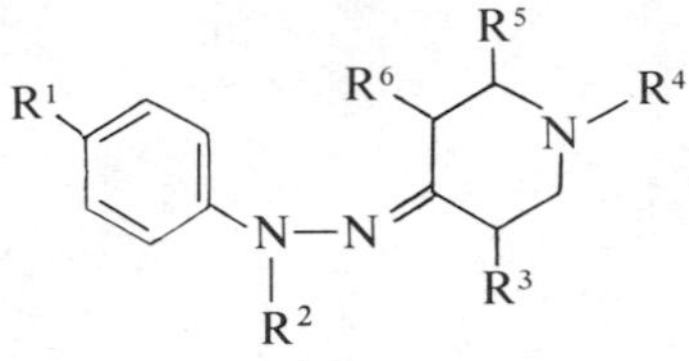

1021

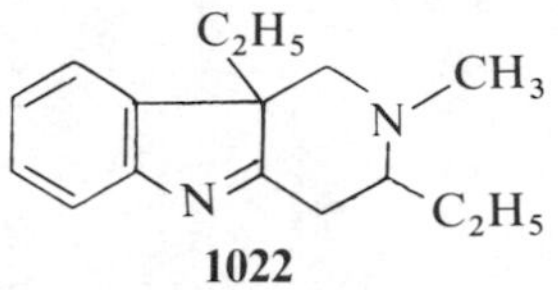

1022

and Palazzo, 1954) and similar treatment of **1021** ($R^1 = R^2 = H$, $R^3 = R^4 = CH_3$, $R^5 = H$ and CH_3, $R^6 = H$ and $R^1 = OC_2H_5$, $R^2 = H$ and $CH_2C_6H_5$, $R^3 = R^4 = CH_3$, $R^5 = R^6 = H$) gave products which were formulated as **1023** ($R^1 = R^2 = H$, $R^3 = CH_3$, $R^4 = H$ and CH_3 and $R^1 = OC_2H_5$, $R^2 = H$ and $CH_2C_6H_5$, $R^3 = R^4 = CH_3$, respectively) (Kucherova and Kochetkov, 1956). However, it has been independently recognized that the products

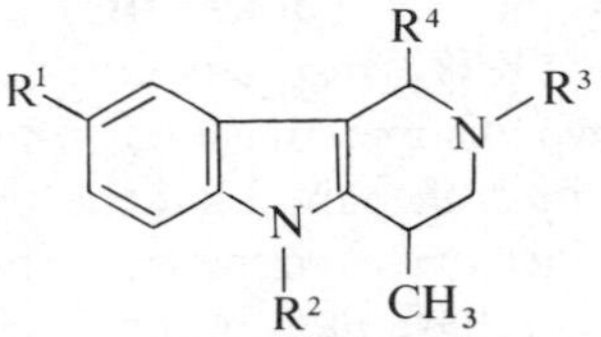

1023

from both these studies (Cattanach, Cohen, A. and Heath-Brown, 1971) and from the later study (Kucherova, Borisova, Sharkova, N. M. and Zagorevskii, 1970) should probably be reformulated with regard to the rearrangement depicted in Scheme 41. This would then require the corresponding reformulation of the further transformation products of some of the initial indolization products obtained in the latter investigation (Kochetkov, Kucherova and Zhukova, 1961; Kucherova, Zhukova, Kamzolova, Petruchenko, Sharkova, N. M. and Kochetkov, 1961). Likewise, the product resulting from the indolization of **1021** [$R^1 = R^5 = R^6 = H$, $R^2 = (CH_2)_2$-(4-methyl-3-pyridyl), $R^3 = CH_3$, $R^4 = (CH_2)_3CH_3$] and formulated as **1023** [$R^1 = R^4 = H$, $R^2 = (CH_2)_2$-(4-methyl-3-pyridyl), $R^3 = (CH_2)_3CH_3$] (Kost, Vinogradova, Trofimov, Mukhanova, Nozdrich and Shadurskii, 1967) should also probably be reformulated accordingly. The indolizations of the four phenylhydrazones **1024** ($R^1 = H$ and CH_3, $R^2 = OCH_3$ and $R^1 = C_2H_5$, $R^2 = H$ and OCH_3) have been investigated (Gerszberg, Cueva and Frasca, 1972). Using the initial compound, both possible indolic products, **1025** and **1026** ($R^1 = H$, $R^2 = OCH_3$), were isolated and the indolization of the other three hydrazones

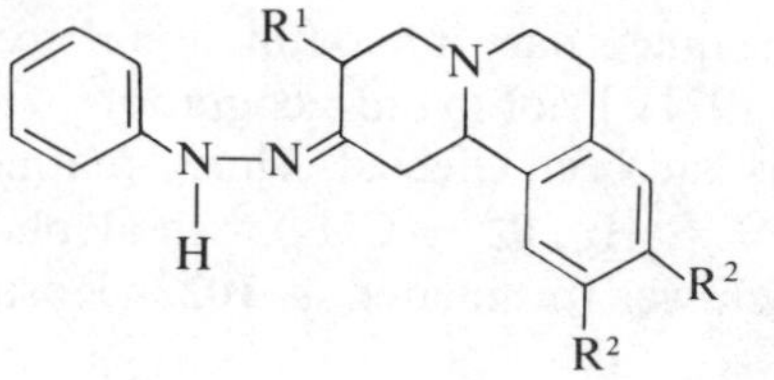

1024

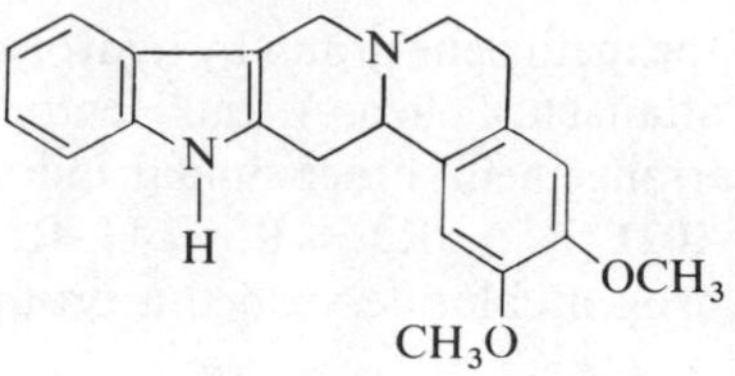

1025

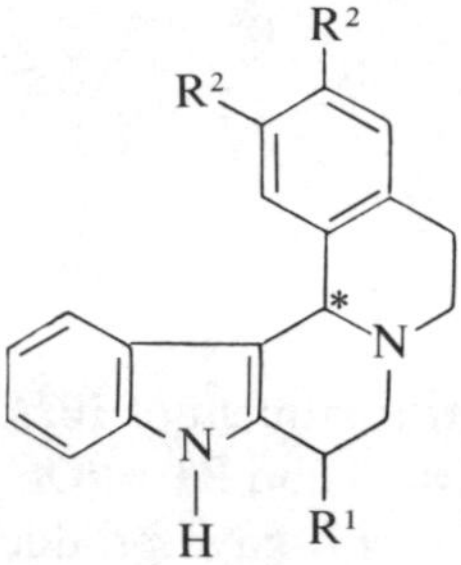

1026

similarly occurred in both possible directions to afford the indoles **1026** ($R^1 = CH_3$, $R^2 = OCH_3$; $R^1 = C_2H_5$, $R^2 = H$ and OCH_3, respectively) and products formulated as the 3*H*-indoles **1027** ($R^1 = CH_3$, $R^2 = OCH_3$; $R^1 = C_2H_5$, $R^2 = H$ and OCH_3, respectively). The last three indolic products were each isolated as two distinct diastereoisomers whose stereochemistry differed at the *C atom in each pair. Like the indoles **1026**, the 3*H*-indolic structures **1027** also contained two asymmetric carbon atoms but these products were not isolated as pairs of diastereoisomers. The possibility that these latter products had structures **1028** ($R^1 = CH_3$, $R^2 = OCH_3$; $R^1 = C_2H_5$, $R^2 = H$ and OCH_3, respectively), being formed from the corresponding **1027** by a

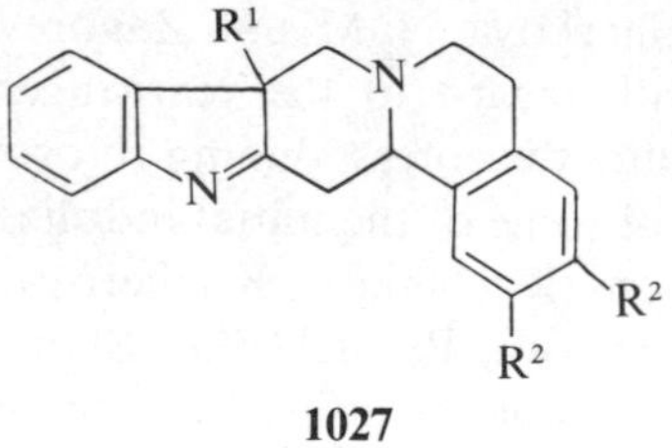

1027

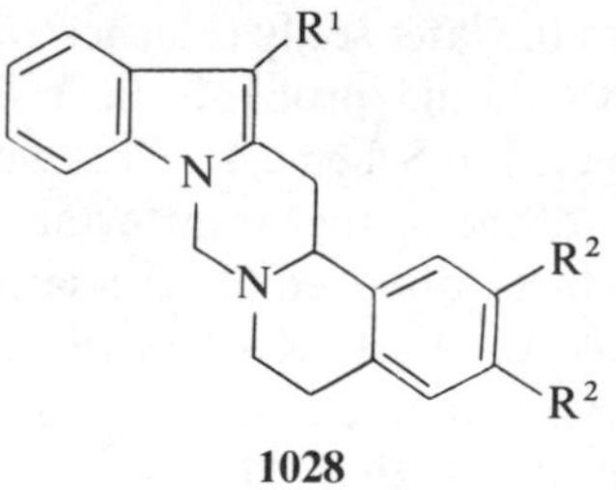

1028

rearrangement analogous to that shown in Scheme 41, was not considered (Gerszberg, Cueva and Frasca, 1972). Such reformulations, containing only one asymmetric carbon atom, would thus remove the possibility of diastereoisomer formation in these latter products and, furthermore, are not at variance with the spectroscopic data quoted (Gerszberg, Cueva and Frasca, 1972) for these products. Indeed, their u.v. spectra do appear to indicate the presence of indolic rather than 3*H*-indolic chromophores. Clearly, a more detailed spectroscopic examination of these products should be effected.

Although in 1966 (Hahn, W. E., Bartnik and Zawadzka, 1966) and 1968 (Hahn, W., Nowaczyk, M., Bartnik and Zawadska, 1968) it was reported that the indolization of **1029** [R^1–R^5 = H and $R^1 = R^2 = R^4$ = H, R^3 = $(CH_2)_2CN$, R^5 = $COCH_3$] in ethanol saturated with hydrogen chloride formed **1030** [$R^1 = R^2$ = H and R^1 = $(CH_2)_2CN$, R^2 = $COCH_3$, respectively], in 1924 (Clemo and Perkin, 1924a) it had been observed that the indolization of the former hydrazone using a dilute sulphuric acid catalyst gave **1031** (R^1–R^4 = H), formed by dehydrogenation of the expected indole **1030** ($R^1 = R^2$ = H) under the indolization conditions. This observation was not

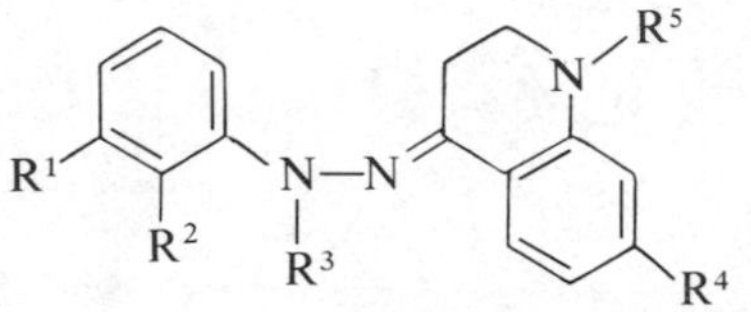

1029

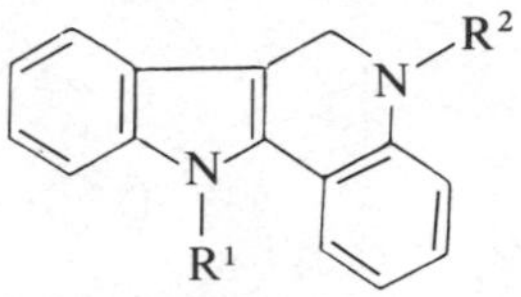

1030

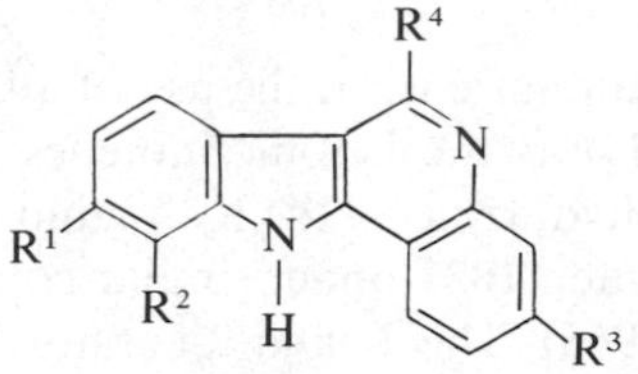

1031

pursued until 1949 (Cookson and Mann, 1949 and Mann, 1949) when it was found that the treatment of 1,2,3,4-tetrahydro-1-phenylquinolin-4-one phenylhydrazone (**1029**; R^1–R^4 = H, R^5 = C_6H_5) with refluxing ethanolic hydrogen chloride again caused sequential indolization and dehydrogenation to give **1032** (R = C_6H_5) and likewise **1029** (R^1–R^4 = H, R^5 = CH_3) was later (Braunholtz and Mann, 1955a) converted into **1032** (R = CH_3). Indolization of **1029** [$R^1 = R^2 = R^4$ = H; R^3 = CH_3, R^5 = C_6H_5; R^3 = C_6H_5, R^5 = CH_3

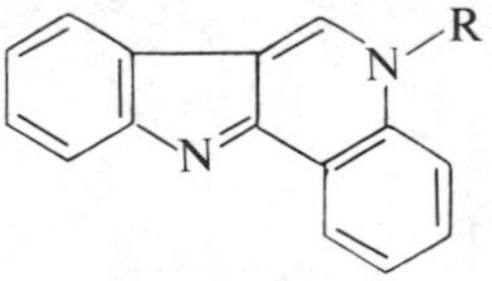

1032

and R^3 = $(CH_2)_2CN$, R^5 = $COCH_3$] under similar conditions afforded **1030** (R^1 = CH_3, R^2 = C_6H_5) (Mann, 1949), **1030** (R^1 = C_6H_5, R^2 = CH_3) (Braunholtz and Mann, 1955a) and **1030** [R^1 = $(CH_2)_2CN$, R^2 = $COCH_3$] (Hahn, W., Nowaczyk, M., Bartnik and Zawadzka, 1968), respectively. In these reactions, dehydrogenative aromatization was prevented by the presence of the two N-substituents. Further examples of similar indolizations in which subsequent dehydrogenation occurred have been observed. Compound **1029** ($R^1 = R^4 = OCH_3$, $R^2 = R^3 = R^5 = H$ and $R^1 = R^3 = R^5 = H$, $R^2 = R^4 = OCH_3$) yielded **1031** ($R^1 = R^3 = OCH_3$, $R^2 = R^4 = H$ and $R^1 = R^4 = H$, $R^2 = R^3 = OCH_3$, respectively) upon refluxing in ethanolic hydrochloric acid (Cross, P. E. and Jones, E. R. H., 1964). Compound **560** was converted into **561** [$R^1 = H$, $R^2 = CH_2N(C_2H_5)_2$] by refluxing in ethanolic hydrochloric acid (Marquez, Cranston, Ruddon, Kier and Burckhalter, 1972). The phenylhydrazone of **1033** ($R^1 = R^2 = H$) gave **1031** (R^1–$R^4 = H$) by

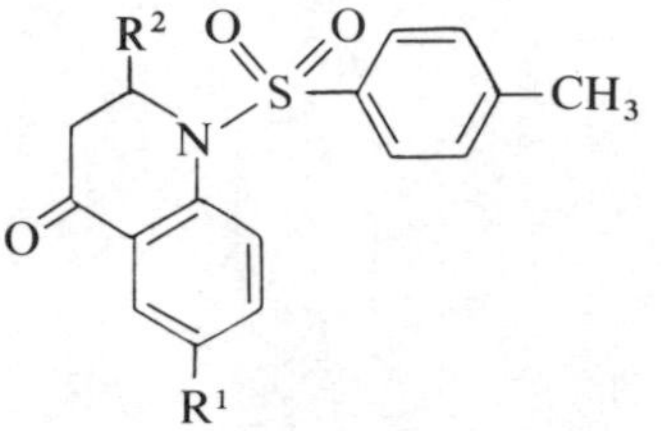

R2 O O S CH3 N O R1

1033

heating in acetic acid–sulphuric acid, the use of **1033** ($R^1 = OCH_3$, $R^2 = H$ and $R^1 = Cl$, $R^2 = CH_3$) as the ketonic moieties and of 4-methyl- and 2,3- and 3,4-dimethylphenylhydrazine as the hydrazino moieties also affording the correspondingly substituted **1031** under similar conditions (Roussel, O., Buu-Hoï and Jacquignon, 1965). The 1- and 2-naphthylhydrazone of **1033** ($R^1 = R^2 = H$; $R^1 = OCH_3$, $R^2 = H$ and $R^1 = Cl$, $R^2 = CH_3$) and the 6-quinolylhydrazone of **1033** ($R^1 = R^2 = H$) gave **1034** ($R^1 = R^2 = H$, $R^1 = OCH_3$, $R^2 = H$ and $R^1 = Cl$, $R^2 = CH_3$), **1035** ($R^1 = R^2 = H$, $R^1 = OCH_3$, $R^2 = H$ and $R^1 = Cl$, $R^2 = CH_3$, $X = CH$) and **1035** ($R^1 = R^2 = H$, $X = N$), respectively, by heating in acetic acid–sulphuric acid (Roussel, O., Buu-Hoï and Jacquignon, 1965). Compound **1036** gave **1037** by refluxing

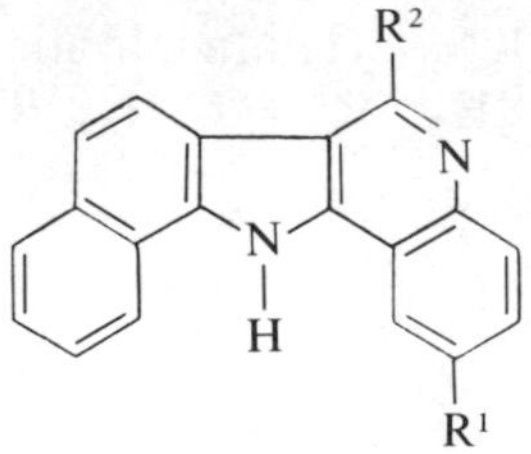

R2 N N H R1

1034

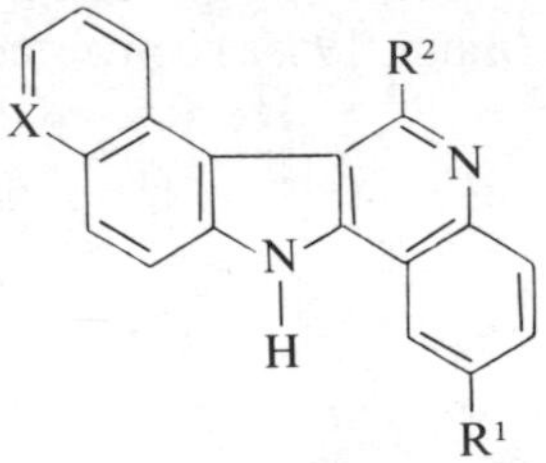

X R2 N N H R1

1035

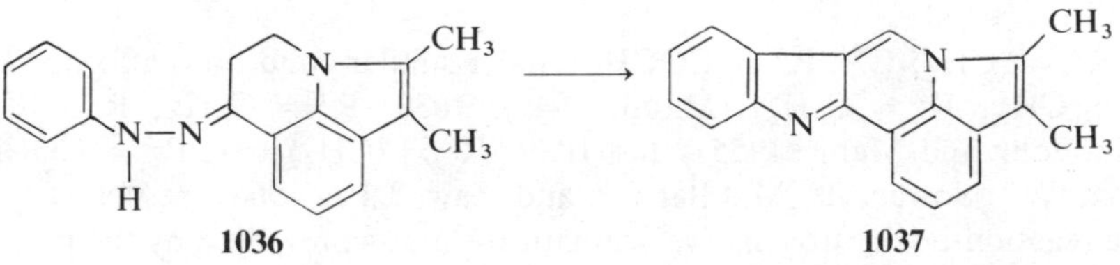

CH3 N N—N CH3 H

1036

→

CH3 N N CH3

1037

in ethanolic hydrogen chloride (Almond and Mann, 1952). Compound **1038** (R = H) gave **1039** by refluxing in dilute sulphuric acid or aqueous ethanolic hydrogen chloride, the latter reaction also affording a small amount of another product, the structure of which remained unestablished (Mann and Smith, B. B., 1951). Compounds **1040** ($R^2 = CH_3, NH_2, OC_2H_5$ and SH) were converted into **1041** ($R^2 = CH_3$) (Da Settimo, Primofiore, Biagi and Santerini, 1976a) using dilute sulphuric acid or polyphosphoric acid as the catalyst (when mixtures of the separate ketonic moiety with the appropriate arylhydrazine

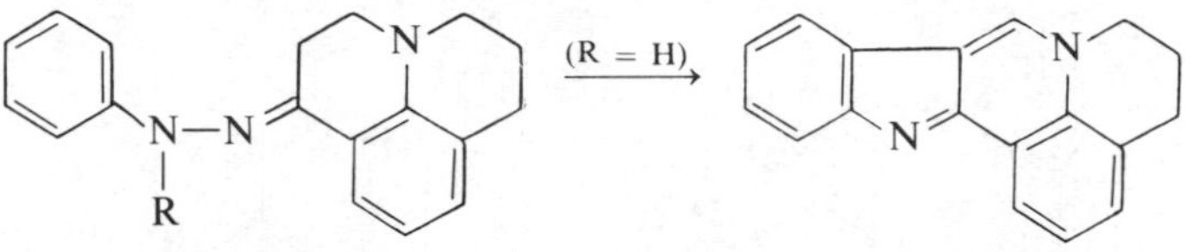

(R = H)

1038 **1039**

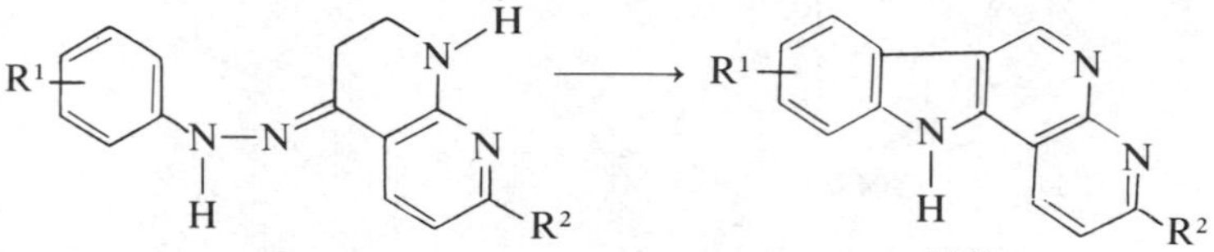

1040 **1041**

were heated in ethanolic concentrated hydrochloric acid, indolization proceeded without concomitant dehydrogenation, although upon attempted recrystallization of the products, aromatization did occur), **1041** ($R = NH_2$) (Da Settimo, Primofiore, Biagi and Santerini, 1976b) using ethanolic hydrochloric acid as catalyst (dilute sulphuric acid or polyphosphoric acid were unsatisfactory) and **1041** ($R^2 = OC_2H_5$ and SH) (Da Settimo, Biagi, Primofiore, Ferrarini and Livi, 1978) using ethanolic hydrochloric acid as catalyst (dilute sulphuric acid or polyphosphoric acid were again unsatisfactory in this respect), respectively. Related to these transformations is the observation that indolization of **1038** ($R = CH_3$) in refluxing dilute sulphuric acid afforded a mixture of products from which a crystalline substance was isolated for which structure **1042** (R + R = O) was suggested. This is probably formed by oxidation of the initially produced **1042** (R = H) (Mann and Smith, B. B., 1951). Analogous simultaneous dehydrogenations failed to accompany the

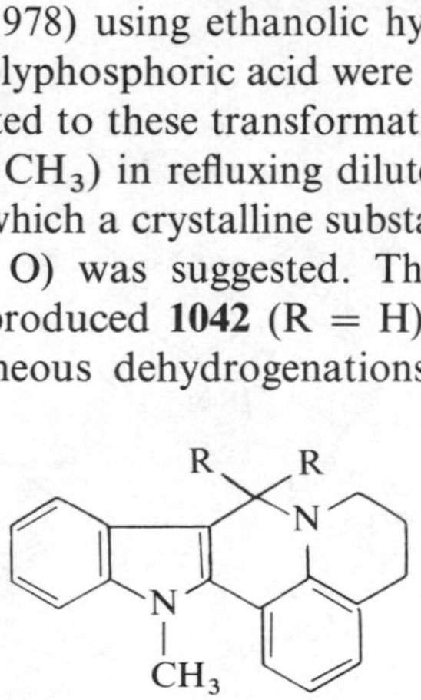

1042

indolizations of **1043** into **1044** (Almond and Mann, 1951), of **1045** (R = H and C_6H_5) into **1046** (R = H and C_6H_5, respectively) (Braunholtz and Mann, 1955b), of **1047** (R = H and CH_3) into **1048** (R = H and CH_3, respectively) (in these cases, evidence suggested that the indoles initially formed had isomerized to afford the products shown) (Mann and Tetlow, 1957) and of **1049** into **1050** (Braunholtz and Mann, 1958), all these reactions being effected in refluxing ethanolic hydrogen chloride. These simultaneous dehydrogenations

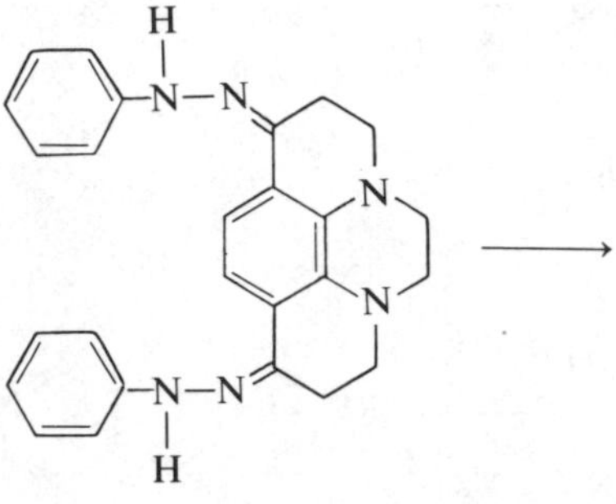

1043 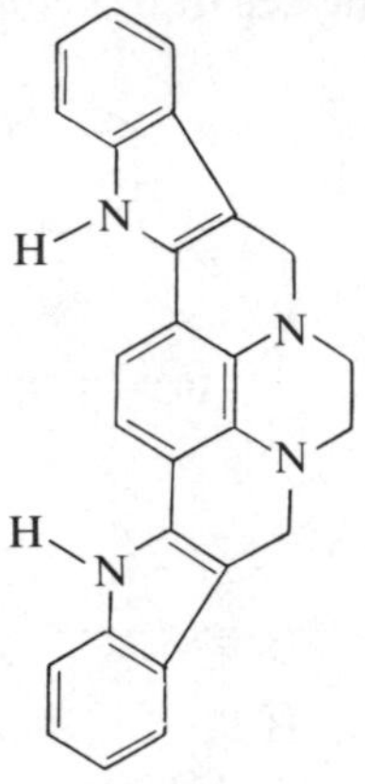 → **1044**

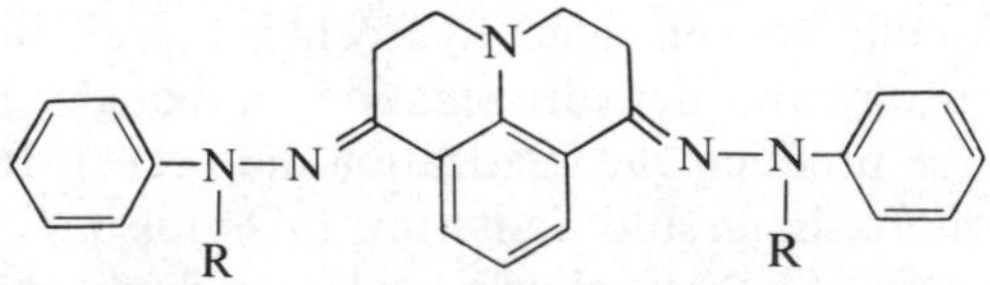

1045 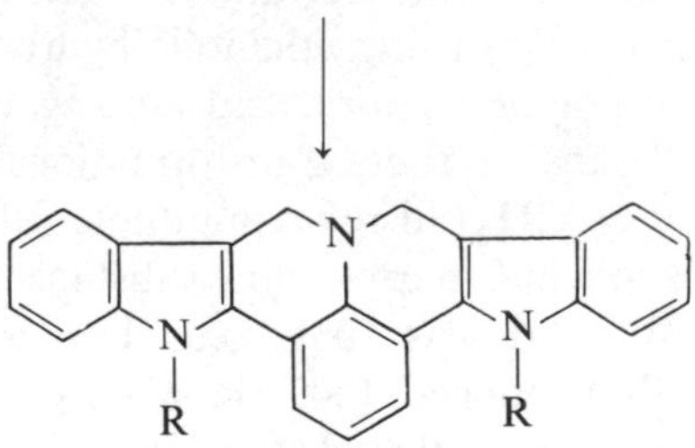 → **1046**

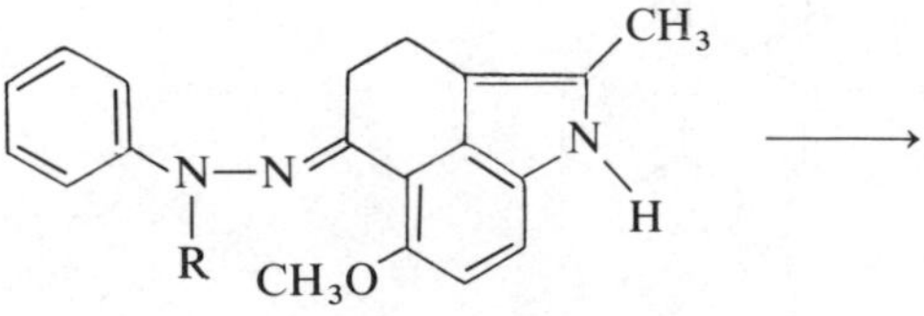

1047 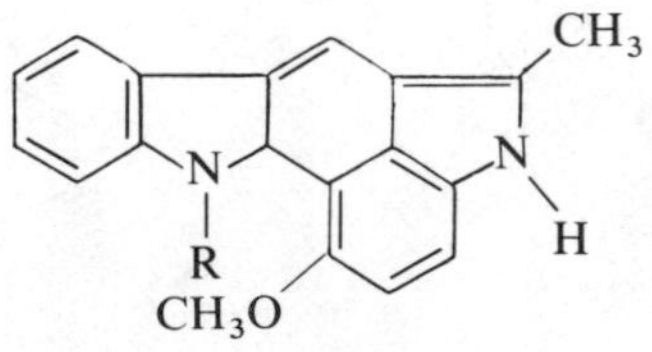 → **1048**

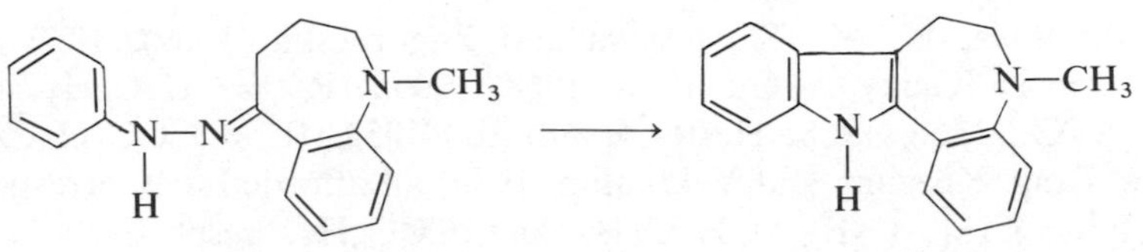

1049 → 1050

which accompanied indolizations promoted in refluxing mineral acids have been ascribed to atmospheric oxidation of the initial indolization products (Da Settimo, Biagi, Primofiore, Ferrarini and Livi, 1978; Da Settimo, Primofiore, Biagi and Santerini, 1976a, 1976b). They appear to be limited to reactions involving specific nitrogen heterocyclic ketone arylhydrazones since, of the examples so far studied (see below), the corresponding arsenic, oxygen, phosphorus, selenium and sulphur analogues did not ultimately undergo similar dehydrogenation subsequent to indolization. It has been concluded (Braunholtz and Mann, 1955a, 1955b, 1958; Mann and Wilkinson, 1957) that for this type of dehydrogenation to accompany indolization, the following conditions have to be satisfied: (i) conjugation must exist between the nitrogen atom of the aminoketone and the potential indole nitrogen atom (i.e. between the nitrogen atom and ketone group in the original ketone), (ii) the ring fusion formed through the indolization has to be such that dehydrogenation will further conjugate the two hetero nitrogen atoms and (iii) aromatization must produce fully aromatic cations such as **1051** (Cookson and Mann, 1949; Mann, 1949) which are the initial products resulting from indolization and dehydrogenation prior to basification of the reaction mixtures during 'work up'.

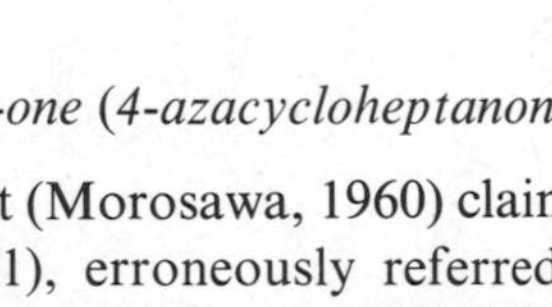

1051

4. *Hexahydro-4H-azepin-4-one* (*4-azacycloheptanone*) *Arylhydrazones*

Although an initial report (Morosawa, 1960) claimed that **1052** ($R^1 = R^2 =$ H, $R^3 = CH_2C_6H_5$, $n = 1$), erroneously referred to by others (Sharkova, N. M., Kucherova, Portnova and Zagorevskii, 1968) as the phenylhydrazone of 1-aza-1-benzylcyclopentan-4-one, and **1052** ($R^1 = R^2 =$ H, $R^3 =$ $COOC_2H_5$, $n = 1$) did not participate in the Fischer indolization, these failures appear to have been caused by an incorrect choice of catalyst since in many subsequent studies such indolizations have been effected. Thus, a large number of benz substituted phenylhydrazones **1052** ($R^3 = CH_3$, $n = 1$) (Geigy A.-G., 1966, 1968; Sharkova, N. M., Kucherova, Portnova and Zagorevskii,

1968; Sharkova, N. M., Kucherova and Zagorevskii, 1969), **1052** ($R^3 = C_2H_5$, $n = 1$) (Geigy A.-G., 1966, 1968), **1052** ($R^3 = CH_2C_6H_5$, $n = 1$) (Geigy A.-G., 1966, 1968; Hester, 1968) and **1052** ($R^3 = COC_6H_5$, $n = 1$) (Hester, Tang, Keasling and Veldkamp, 1968) all afforded the corresponding **1053** ($R^3 = CH_3$, C_2H_5, $CH_2C_6H_5$ and COC_6H_5, respectively, $n = 1$). No products **1054**, resulting from indolization in the alternative possible direction on the ketonic moiety, were detected. In some cases the indolic nature of the products was confirmed from u.v. and i.r. spectroscopic data and structures **1053** ($n = 1$) were distinguished from the possible **1054** using p.m.r. spectroscopic evidence. Thus, signals at $\tau = 6.16$ and $\tau = 7.03$ were attributed to the methylene protons adjacent to the nitrogen atom and the indole 2- and 3-position, respectively, in **1053** ($R^1 = R^2 = H$, $R^3 = COC_6H_5$, $n = 1$) (Hester, Tang, Keasling and Veldkamp, 1968) whereas in **1053** ($R^1 = R^2 = H$, $R^3 = CH_3$, $n = 1$) these eight protons were far more equivalent and gave rise to a single peak at $\tau = 7.2$ (Sharkova, N. M., Kucherova, Portnova and Zagorevskii, 1968). These p.m.r. spectral data eliminated the corresponding **1054** as the possible product structures, containing as they do a methylene group adjacent to two sp^3-hybridized carbon atoms and another methylene group situated between an aromatic ring and a nitrogen atom. Furthermore, structures **1053** ($R^1 = R^2 = H$, $R^3 = CH_3$, $n = 1$ and $R^1 = H$, $R^2 = R^3 = CH_3$, $n = 1$), obtained by indolizations of the respective phenyl- and N_α-methylphenylhydrazone, were confirmed by their Emde degradation, via their methiodides, to furnish **1055** (R = H and CH_3, respectively) (Sharkova, N. M., Kucherova, Portnova and Zagorevskii, 1968).

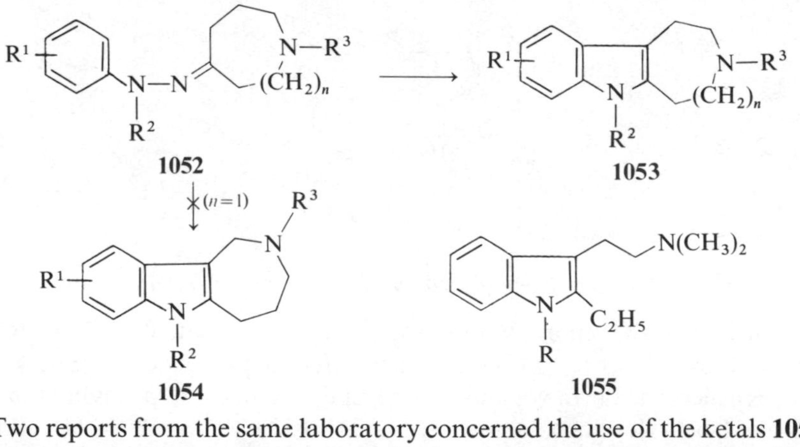

R^1 N—N N—R^3 $(CH_2)_n$ R^2 **1052** ⟶ R^1 N N—R^3 $(CH_2)_n$ R^2 **1053**

1052 ↛ ($n=1$) **1054**: R^1 N R^2 N R^3

1055: $N(CH_3)_2$ N C_2H_5 R

Two reports from the same laboratory concerned the use of the ketals **1056** as the ketonic moieties in Fischer indolizations. One of these, a patent (Sallay, 1966), claimed the indolization of a mixture of **1056** ($R^1 = R^2 = H$) (with the stereochemistry of the ring junction unspecified) with phenylhydrazine in refluxing dilute sulphuric acid to afford **1057** (R^1–$R^3 = H$) (with the stereochemistry of the ring junction unspecified) and extended the claim to include an

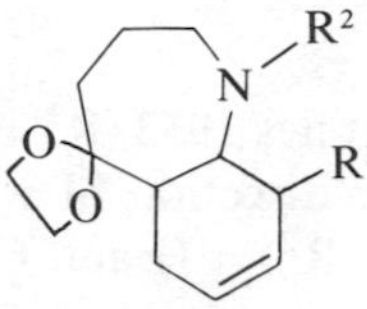

1056

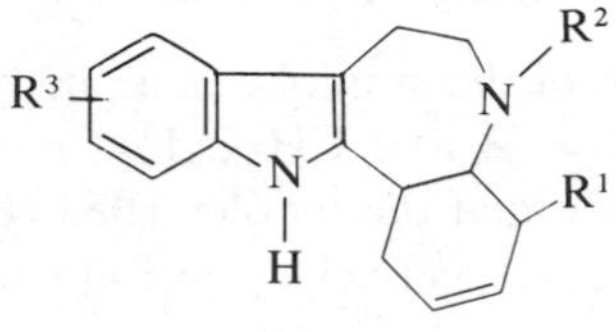

1057

unspecified range of R^1, R^2 and R^3 substituents. The other (Sallay, 1964), while limiting the nature of R^1–R^3, specified the stereochemistry of the ring junction in the starting ketal **1056** ($R^1 = R^2 = H$). Thus, using a sulphuric acid catalyst, a mixture of *cis*-**1056** ($R^1 = R^2 = H$) with phenylhydrazine afforded *cis*-**1057** (R^1–$R^3 = H$) (70–78% yield), together with traces of **1058**

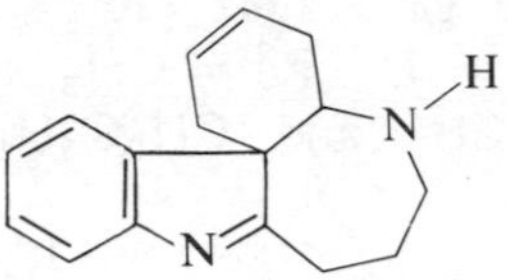

1058

and *trans*-**1057** (R^1–$R^3 = H$). This last product was also produced by indolization of a mixture of *trans*-**1056** ($R^1 = R^2 = H$) with phenylhydrazine. It would be interesting to investigate the possibility of increasing the relative yield of **1058** from these reactions by using a 'large' weak acid which should favour 3*H*-indole rather than indole formation.

Indolizations of **1059** ($R^1 + R^2 = O$, $R^3 = R^4 = H$) (Augustine and Pierson, 1969), **1059** ($R^1 = R^2 = R^4 = H$, $R^3 = C_2H_5$) (Sallay, 1967a and Ikezaki, Wakamatsu and Ban, 1969), **1059** (R^1–$R^3 = H$, $R^4 = C_2H_5$) (Ikezaki, Wakamatsu and Ban, 1969) and **1059** ($R^1 = R^2 = H$, R^3, $R^4 = H$, C_2H_5) (Ban, Wakamatsu, Fujimoto, Y. and Oishi, 1968) have all been effected. In all cases the initial enehydrazine formation involved the secondary and not the tertiary carbon atom (as might be expected since the generation of an sp^2-hybridized carbon at the tertiary centre would be difficult), ultimately giving **1060** ($R^1 + R^2 = O$, $R^3 = R^4 = H$; $R^1 = R^2 = R^4 = H$, $R^3 = C_2H_5$; R^1–$R^3 = H$, $R^4 = C_2H_5$ and $R^1 = R^2 = H$, R^3, $R^4 = H$, C_2H_5, respectively) as the sole products.

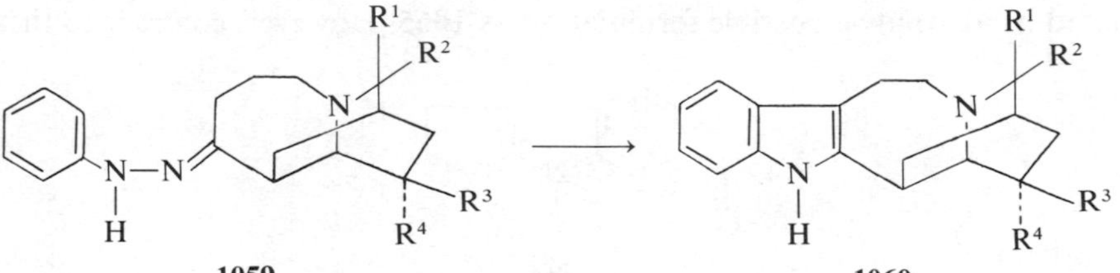

1059 → **1060**

5. 5-Azacyclooctanone Arylhydrazones

Indolization of the aminoketone phenylhydrazones **1052** (R^1 = H, R^2 = H and CH_3, R^3 = H and $CH_2C_6H_5$, $n = 2$), the direction of which was unambiguous, afforded the indoles **1053** (R^1 = H, R^2 = H and CH_3, R^3 = H and $CH_2C_6H_5$, respectively, $n = 2$) (Yardley and Smith, H., 1972).

6. Dihydrothiophen-3(2H)-one Arylhydrazones

Contrary to the direction of indolization of pyrrolidin-3-one arylhydrazones which has been already discussed, indolization of dihydrothiophen-3(2*H*)-one arylhydrazones **1061** [R^1 = H, CH_3, $OCH_2C_6H_5$ and $COOC_2H_5$, R^2 = H; R^1 = H, R^2 = CH_3 and $CH_2C_6H_5$ and R^1 = $COOC_2H_5$, R^2 = $(CH_2)_3CH_3$ and $CH_2C_6H_5$] and **1063** gave **1062** [R^1 = H, CH_3, $OCH_2C_6H_5$ and $COOC_2H_5$, R^2 = H; R^1 = H, R^2 = CH_3 and $CH_2C_6H_5$ and R^1 = $COOC_2H_5$, R^2 = $(CH_2)_3CH_3$ and $CH_2C_6H_5$, respectively] and **1064**,

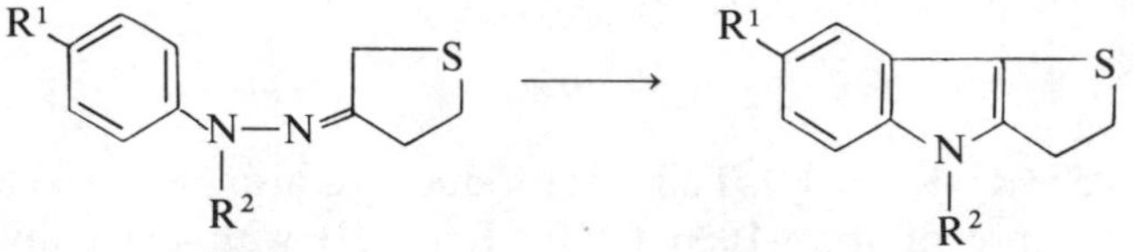

1061 **1062**

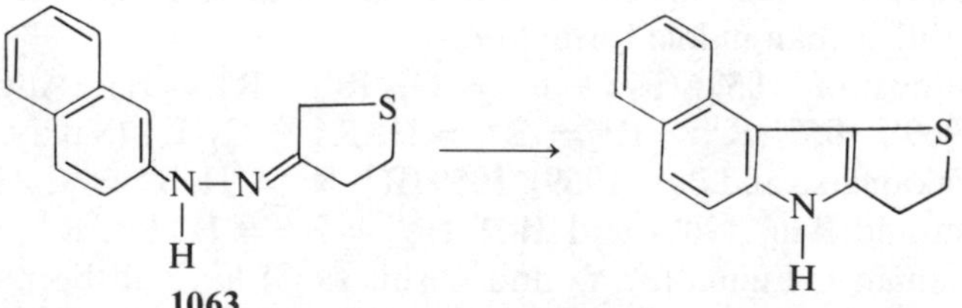

1063 **1064**

respectively (Aksanova, Kucherova and Zagorevskii, 1964a). The structure of **1062** ($R^1 = R^2$ = H), and, by analogy, the other indolization products, was established by its reductive desulphurization with nickel and hydrazine hydrate to yield 2-ethylindole. Furthermore, its u.v. spectrum was unlike those of cyclopent[2,3]indole and 1,2,3,4-tetrahydrocarbazole, which therefore eliminated its alternative possible formulation as **1065**. However, contrary to these

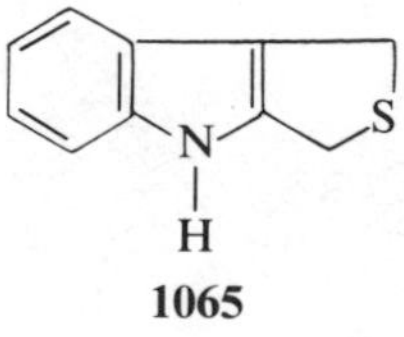

1065

observations, the acetic acid catalysed indolizations of mixtures of **1066** (R = H) with **1067** (R = H) and of **1066** (R = Cl) with **1067** (R = H and CH_3) furnished **1068** ($R^1 = R^2 = H$, $R^1 = Cl$, $R^2 = H$ and CH_3, respectively), the structures of the products in each case being established by reductive desulphurization with Raney nickel in ethanol to afford the corresponding 2-methylindole-3-acetic acid derivatives **1069** ($R^1 = R^2 = H$, $R^1 = Cl$, $R^2 =$ H and CH_3, respectively) (Sorrentino, 1968).

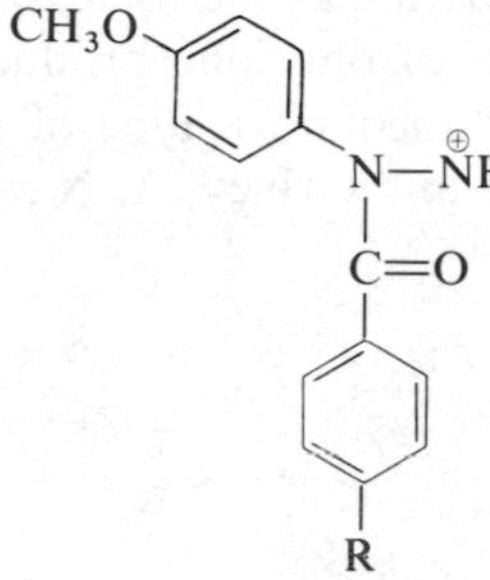

1066

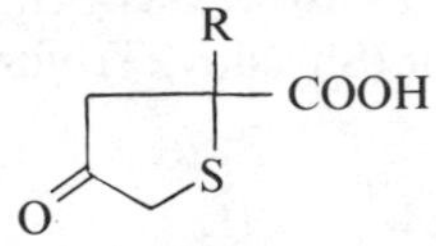

1067

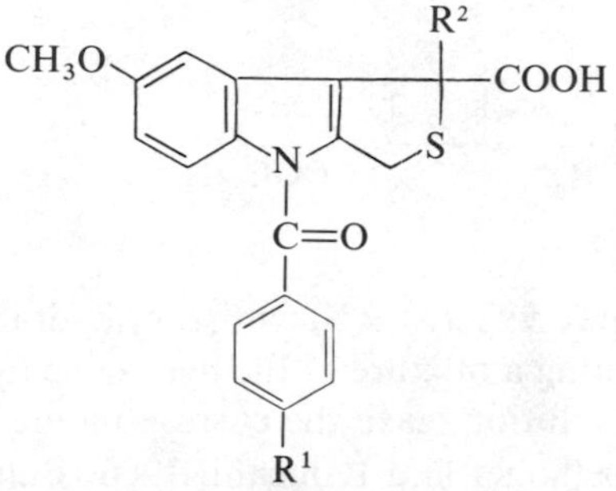

1068

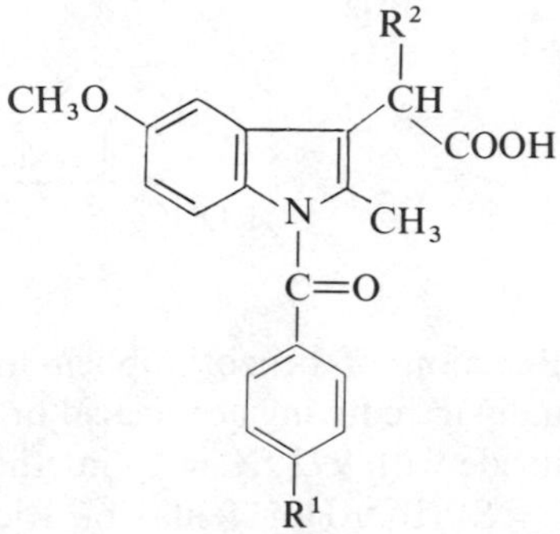

1069

When a mixture of **1070** (R = OH) with phenylhydrazine was refluxed in acetic acid it produced **1071** (R = H) (Benary and Baravian, 1915). This synthetic approach was subsequently (Buu-Hoï, Hoán, Khôi and Xuong, 1949) extended by similarly indolizing mixtures of **1070** (R = OH) with 2-, 3- and 4-methyl-, 2,3-dimethyl-, 4-bromo- and 4-chlorophenylhydrazine and 4-biphenylylhydrazine, and with 1- or 2-naphthylhydrazine, again using acetic acid as catalyst, to afford the corresponding substituted derivatives of **1071**. The

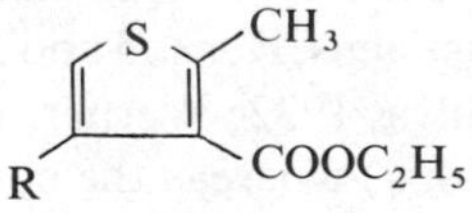

1070

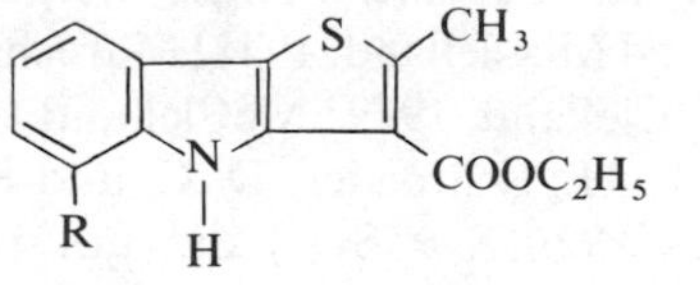

1071

product resulting from the analogous treatment of **1070** (R = OH) with 2-biphenylylhydrazine was formulated as either **1071** (R = C_6H_5) or **1072**, although the former structure was preferred (Buu-Hoï, Hoán, Khôi and Xuong, 1949). The products resulting from other indolizations of 2-biphenylylhydrazones have been similarly formulated (Buu-Hoï and Hoán, 1951 and Buu-Hoï, Hoán and Khôi, 1950b). These possible alternative formulations to indoles are discussed further in Chapter VII, P). Attempts to synthesize **1073** by reaction of **1070** (R = OH) with **1070** (R = NHNHCHO) (acting as the hydrazine precursor) under acidic conditions led to the isolation of only one product which was formulated as **1074**, this being probably formed via a type of benzidine rearrangement (Shvedov, Trofimkin, Vasil'eva and Grinev, A. N., 1975).

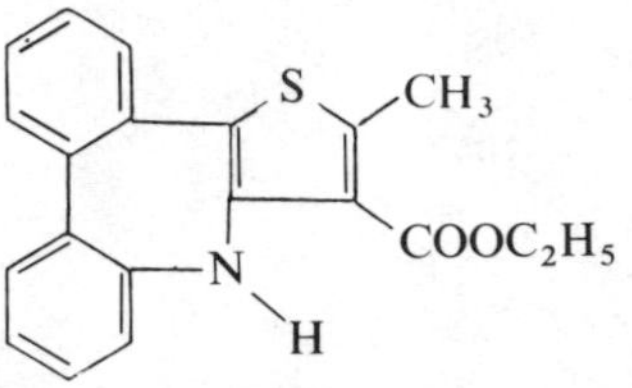

S CH_3 $COOC_2H_5$ N H

1072

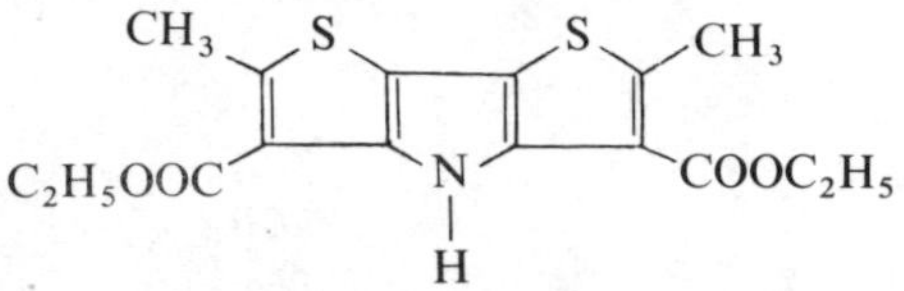

CH_3 S S CH_3 C_2H_5OOC N $COOC_2H_5$ H

1073

CH_3 S S CH_3 C_2H_5OOC NH_2 H_2N $COOC_2H_5$

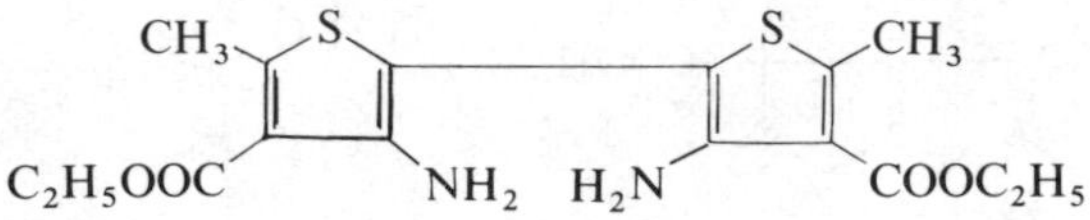

1074

Reaction of benzothiophen-3(2*H*)-ones **992** (X = S) with arylhydrazines, usually in refluxing acetic acid or by heating a mixture of the hydrazine hydrochloride with **992** (X = S) in ethanolic solution, gave the corresponding **1075** (X = S) (Buu-Hoï, Bellavita, Ricci, Hoeffinger and Balucani, 1966; Buu-Hoï and Saint-Ruf, 1963; Buu-Hoï, Saint-Ruf, Martani, Ricci and Balucani, 1968;

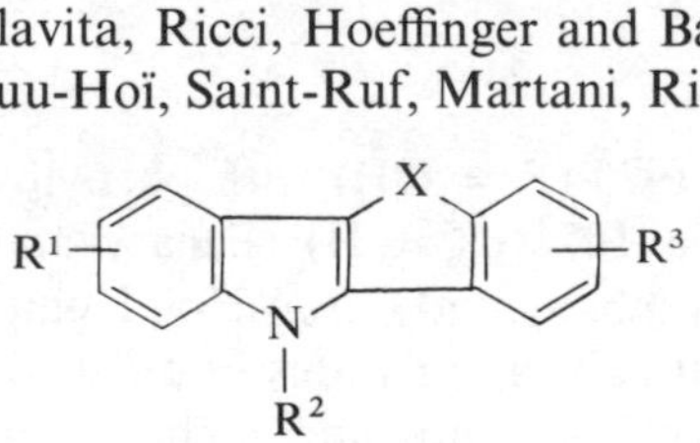

X R^1 R^3 N R^2

1075

Buu-Hoï and Xuong, 1952; Ciba Ltd., 1960; Dalgliesh and Mann, 1947; Dann, Volz, Demant, Pfeifer, Bergen, G., Fick and Walkenhorst, 1973; Fowkes and McClelland, 1941; Maréchal, Christiaens, Renson and Jacquignon, 1978; McClelland, 1929; McClelland and D'Silva, 1932; Werner, L. H., 1962; Werner, L. H., Schroeder, D. C. and Ricca, 1957) whereas the corresponding sulphones **992** (X = SO_2) did not participate in indolization (Aksanova, Sharkova, L. M., Kucherova and Zagorevskii, 1973; Fowkes and McClelland,

1941; McClelland and D'Silva, 1932 – see also Cornforth, J. W., Hughes, G. K., Lions and Harradence, 1937–1938). Substituents at the 4-position in **992** (X = S) and at the 2-position of the phenylhydrazine moiety also inhibited these indolizations (Dalgliesh and Mann, 1947). In an analogous manner, **1076** and **1078** reacted with phenylhydrazine to afford **1077** and **1079**, respectively, and **992** (R^1 = H, X = S), **1076** and **1078** reacted with 1-amino-1,2,3,4-tetrahydroquinoline to produce **1080**, **1081** and **1082**, respectively (Dalgliesh and Mann, 1947). Attempts to achieve the indolization of a mixture of **1083** with phenylhydrazine, by heating in acetic acid solution, failed (Dalgliesh and Mann, 1947).

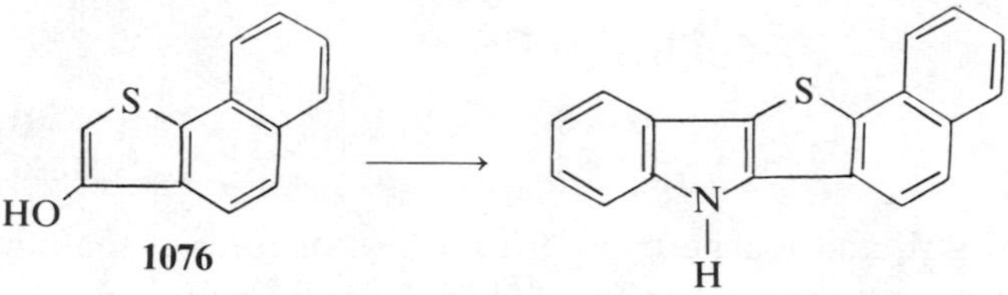

1076 → **1077**

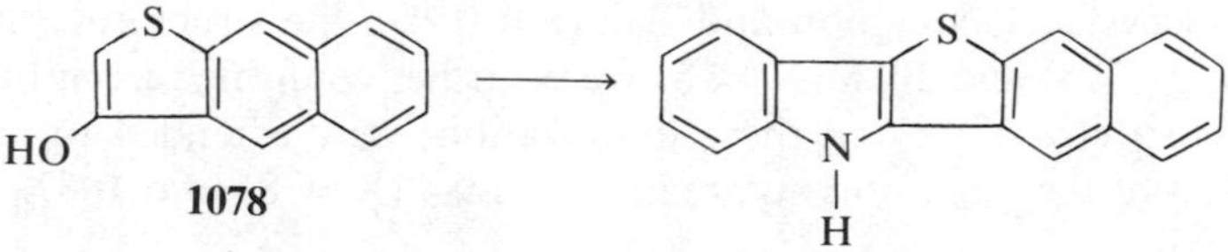

1078 → **1079**

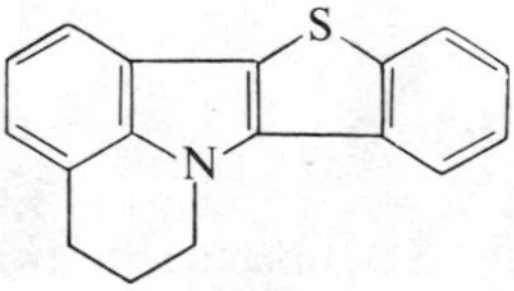

1080 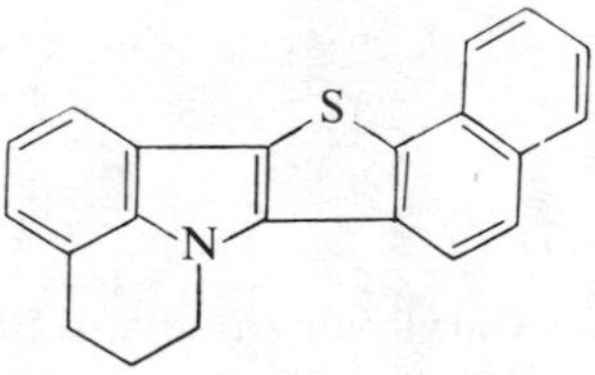 **1081**

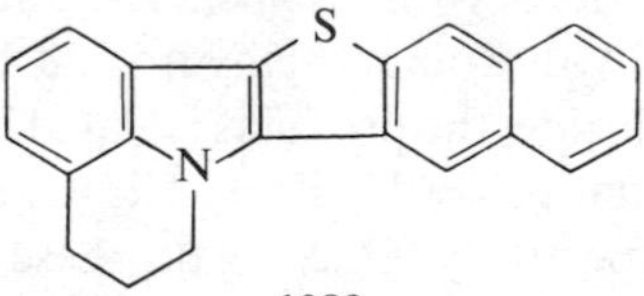

1082 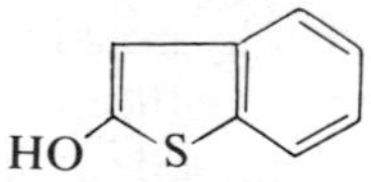 **1083**

7. *Dihydro-2H-thiopyran-3(4H)-one Arylhydrazones*

Indolization of these hydrazones occurred 'towards' the sulphur atom, as with the dihydrothiophen-3(2*H*)-one arylhydrazones but again contrary to direction of indolization in the analogous aminoketones already discussed.

Thus, indolization of mixtures of **1084** ($X = S$) with phenylhydrazine hydrochloride, 2-naphthylhydrazine hydrochloride, and 4-methylphenylhydrazine hydrochloride in refluxing ethanol, ethanol, and methanol, respectively, and with N_{α}-methylphenylhydrazine in refluxing ethanolic hydrogen chloride yielded **1085** ($R^1 = R^2 = H$, $X = S$), **1086** ($X = S$) and **1085** ($R^1 = CH_3$, $R^2 = H$ and $R^1 = H$, $R^2 = CH_3$, $X = S$), respectively, although the corresponding 4-carbethoxyphenylhydrazone could not be indolized (Aksanova,

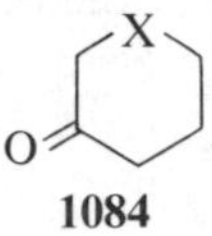

1084

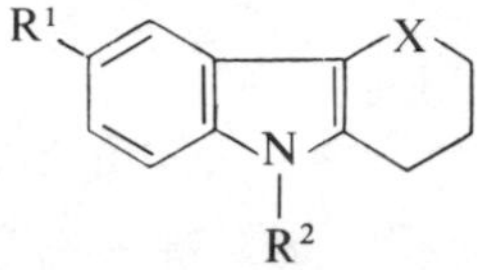

1085

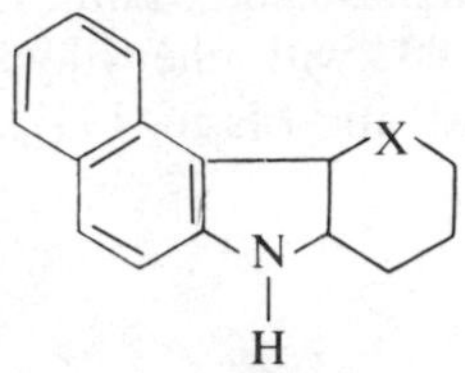

1086

Sharkova, L. M. and Kucherova, 1970). The structures of the other products were assigned by analogy with that of **1085** ($R^1 = R^2 = H$, $X = S$), the structure of which was confirmed by its p.m.r. spectral properties. Subsequently (Croisy, Ricci, Janćevska, Jacquignon and Balucani, 1976) the structures **1085** ($R^1 = R^2 = H$, $X = S$) and **1086** ($X = S$) were further confirmed from both p.m.r. and mass spectral data and these indolizations were extended to include the conversion of the 1-naphthylhydrazone of **1084** ($X = S$) into **1087**. However,

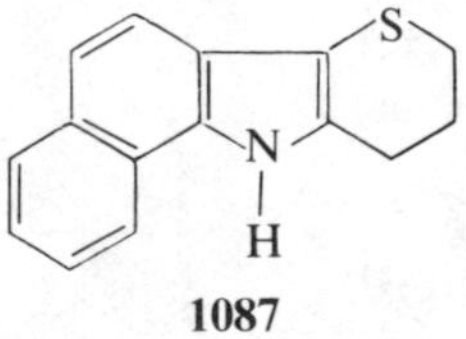

1087

indolization of arylhydrazones of **1084** ($X = SO_2$) occurred 'away' from the sulphur atom. Thus, when mixtures of this ketone with phenylhydrazine and 4-methylphenylhydrazine hydrochloride, with N_{α}-benzyl- and N_{α}-methylphenylhydrazine hydrochloride and with 2-naphthylhydrazine hydrochloride were heated in ethanolic hydrogen chloride, the products isolated had structures **1088** ($R^1 = H$ and CH_3, $R^2 = H$ and $R^1 = H$, $R^2 = CH_2C_6H_5$ and CH_3, $X = SO_2$) and **1089** ($X = SO_2$), respectively. Once again these formulations

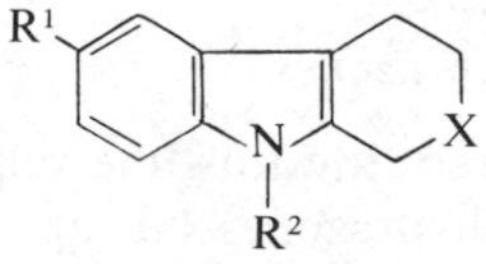

1088

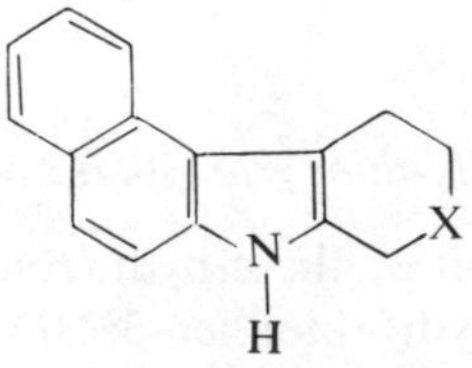

1089

were based, by analogy, upon that of the parent compound **1088** ($R^1 = R^2 = H$, $X = SO_2$), the structure of which was verified from p.m.r. spectral data (Aksanova, Sharkova, L. M. and Kucherova, 1970).

Indolization of **1090** with ethanolic hydrogen chloride furnished **1091** (Kiang and Mann, 1951). Mixtures of **1092** (X = S) with phenylhydrazine and 1- and 2-naphthylhydrazine in acetic acid saturated with hydrogen chloride afforded **1093** (R^1–R^4 = H, X = O), **1094** (X = S) and **1095** (X = S), respectively, the structures of which were distinguished from the possible isomeric formulations using p.m.r. and mass spectral data (Mispelter, Croisy, Jacquignon, Ricci, Rossi and Schiaffella, 1977). Refluxing solutions of the phenylhydrazones of **1096** (R = H and CH_3) in acetic acid saturated with hydrogen chloride and of the 8-quinolylhydrazone of **1096** (R = H) in acetic acid–sulphuric acid gave **1097** (R = H and CH_3) and **1098**, respectively (Buu-Hoï, Jacquignon, Ledésert, Ricci and Balucani, 1969).

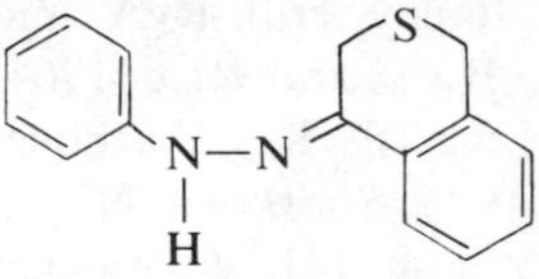

1090 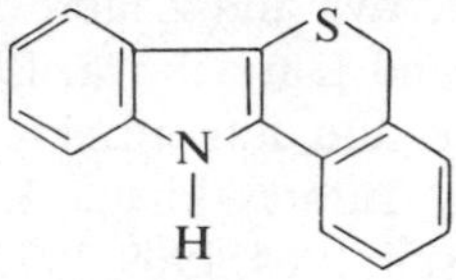 **1091** 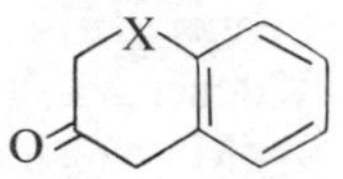 **1092**

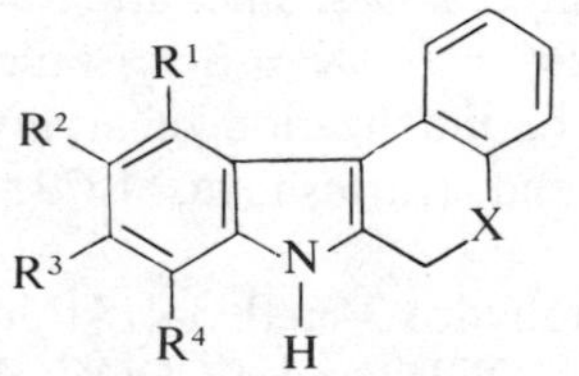

1093 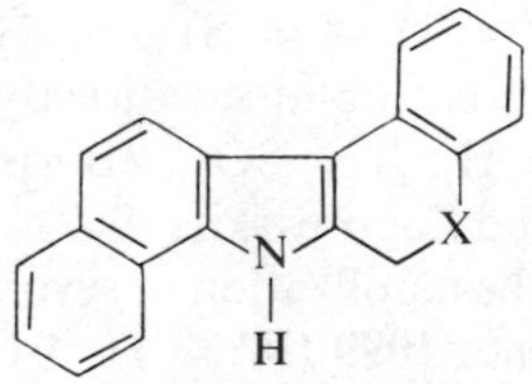 **1094**

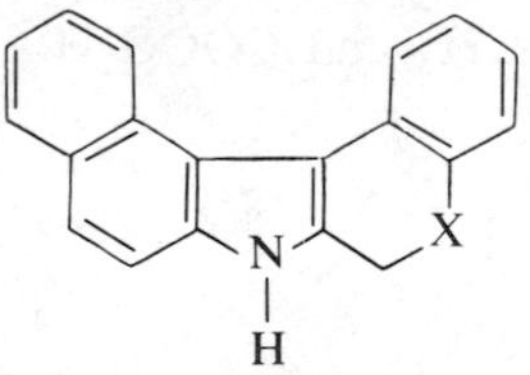

1095 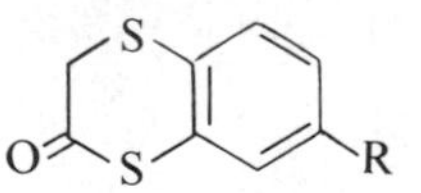 **1096**

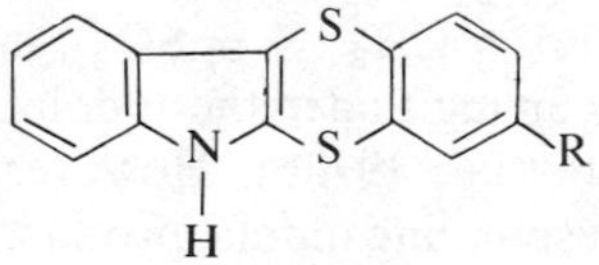

1097 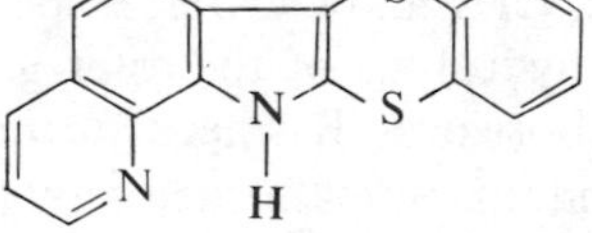 **1098**

8. Tetrahydro-4H-thiopyran-4-one Arylhydrazones

The first report (Bennett, G. and Waddington, 1929) of the indolization of hydrazones of this type involved the refluxing acetic acid catalysed conversion of the phenylhydrazone and 4-bromophenylhydrazone **1099** (R^1 = H and 4-Br, $R^2 = R^3$ = H, X = S) into **1100** (R^1 = H and 6-Br, respectively, $R^2 = R^3$ = H, X = S). Under similar conditions the corresponding 4-nitrophenylhydrazone failed to indolize. Subsequently, many other groups (Aksanova, Pidevich, Sharkova, L. M. and Kucherova, 1968; Aksanova, Sharkova,

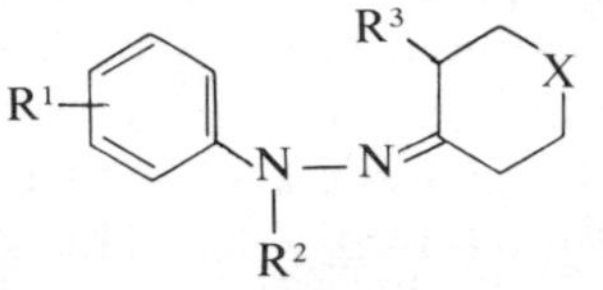

1099 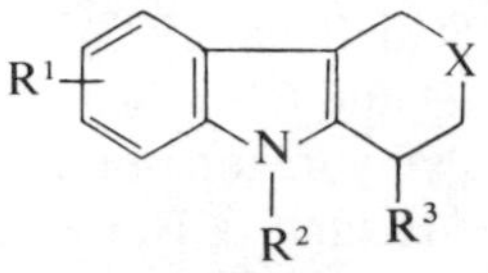 **1100**

N. M., Baranova, Kucherova and Zagorevskii, 1966; Bailey, Hill, P. A. and Seager, 1974; Borisova and Kartashova, 1979; Freed, Hertz and Rice, 1964; Kucherova, Petruchenko and Zagorevskii, 1962; Orlova, E. K., Sharkova, N. M., Meshcheryakova, Zagorevskii and Kucherova, 1975; Sharkova, N. M., Kucherova and Zagorevskii, 1964, 1969, 1972b, 1972c; Young, T. E., Ohnmacht and Hamel, 1967; Young, T. E., and Scott, P. H., 1968a) extended this synthesis to include the use of a large variety of arylhydrazines. Since attempted oxidation of **1100** (R^2 = H, X = S) into **1100** (R^2 = H, X = SO_2) failed, the latter compounds were obtained directly by the indolization of arylhydrazones of **1099** (R^3 = H, X = SO_2) (Borisova and Kartashova, 1979; Kucherova, Aksanova and Zagorevskii, 1963).

Whereas the indolization of several tetrahydro-3-methyl-4*H*-thiopyran-4-one arylhydrazones **1099** (R^1 = H, 4-CH_3, 4-OCH_3 and 4-$COOC_2H_5$, R^2 = H; R^3 = CH_3, X = S) afforded both the corresponding indoles **1100** (R^1 = H, 6-CH_3, 6-OCH_3 and 6-$COOC_2H_5$, respectively, R^2 = H, R^3 = CH_3, X = S) and 3*H*-indoles **1101** (R^1 = H, CH_3, OCH_3 and $COOC_2H_5$, respectively,

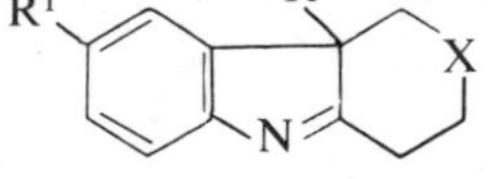

1101

R^2 = CH_3, X = S), indolization of the sulphones **1099** (R^1 = H, 4-CH_3 and 4-$COOC_2H_5$, R^2 = H, R^3 = CH_3, X = SO_2) gave only the 3*H*-indoles **1101** (R^1 = H, CH_3 and COOH, respectively, R^2 = CH_3, X = SO_2), in the latter reaction hydrolysis of the ester group occurring under the indolization conditions (Borisova, Kucherova and Zagorevskii, 1970b). However, a recent related indolization afforded only the corresponding indolic product, although the basic component(s) of the reaction were not investigated (Jogdeo and

Bhide, 1980). The isolation of the 3*H*-indoles **1101** (X = S) is in contrast to the corresponding aza analogues which, as described above, underwent further rearrangement under the indolization conditions to afford **1019** (Scheme 41). Products isolated after the replacement of the 3-methyl group by a 3-(2-carbethoxyethyl) group in **1099** (R^1 = H, 4-CH_3, 4-NO_2 and 4-$COOC_2H_5$, R^2 = H, R^3 = CH_3, X = S) and in the corresponding 2-naphthylhydrazone, followed by indolization, were the indoles **796** (R = H, CH_3, NO_2 and $COOC_2H_5$, respectively, X = S) and **1102**, respectively. It was suggested that these were formed via ring closure of the intermediate indoles **1100** [R^1 = H, CH_3, NO_2 and $COOC_2H_5$, respectively, R^2 = H, R^3 = $(CH_2)_2COOC_2H_5$, X = S] and **1103**, respectively, under the indolization conditions (Kakurina, Kucherova and Zagorevskii, 1964, 1965b), analogous to the observations with the carbocyclic analogues discussed earlier in this chapter. In only one of these present reactions was the corresponding 3*H*-indole {i.e. **1101** [R^1 = $COOC_2H_5$, R^2 = $(CH_2)_2COOC_2H_5$, X = S]} isolated.

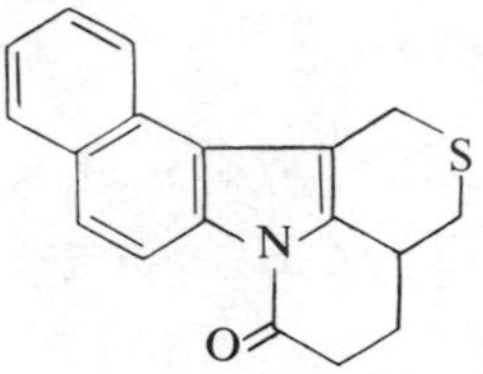

S N O

1102

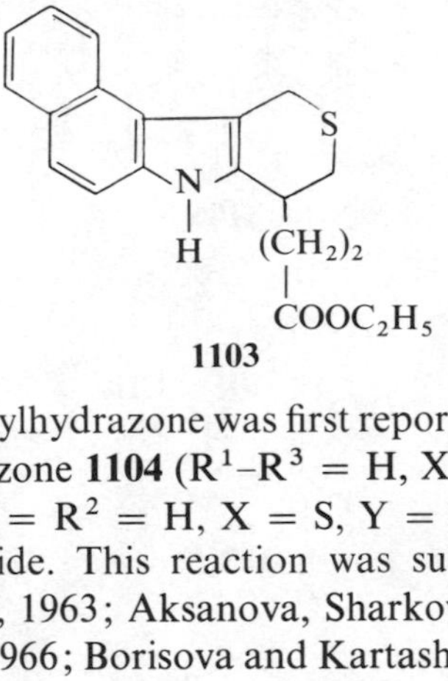

S N H $(CH_2)_2$ $COOC_2H_5$

1103

Indolization of a thiochroman-4-one arylhydrazone was first reported (Kiang and Mann, 1951) when the parent hydrazone **1104** (R^1–R^3 = H, X = S, Y = CH=CH) was converted into **1105** (R^1 = R^2 = H, X = S, Y = CH=CH) using boiling ethanolic hydrogen chloride. This reaction was subsequently (Aksanova, Kucherova and Zagorevskii, 1963; Aksanova, Sharkova, N. M., Baranova, Kucherova and Zagorevskii, 1966; Borisova and Kartashova, 1979;

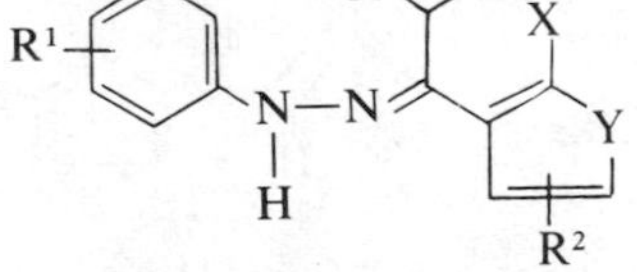

R^3 R^1 X N—N Y H R^2

1104

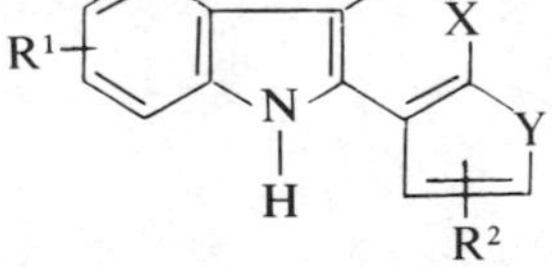

R^1 X N Y H R^2

1105

Buu-Hoï, Croisy, Jacquignon, Hien, Martani and Ricci, 1972; Buu-Hoï, Jacquignon, Croisy, Loiseau, Périn, Ricci and Martani, 1969; Buu-Hoï, Martani, Croisy, Jacquignon and Périn, 1966; Croisy, Jacquignon and Fravolini, 1974; Fravolini, Croisy and Jacquignon, 1976; Jacquignon, Fravolini, Feron and Croisy, 1974; Kucherova, Aksanova, Sharkova, N. M. and Zagorevskii, 1963; Sharkova, N. M., Kucherova and Zagorevskii, 1964; Young, T. E.

and Scott, P. H., 1965, 1966, 1968a, 1968b) extended to the synthesis of a variety of **1105** (X = S, Y = CH=CH) starting with variously benz substituted phenylhydrazones, 1- and 2-naphthylhydrazone and other hydrazones. Furthermore, the indolizations of **1104** (R^1–R^3 = H, X = Y = S) into **1105** (R^1 = R^2 = H, X = Y = S), of the corresponding 1- and 2-naphthylhydrazone into **1106** and **1107**, respectively, and of the phenyl- and 1- and 2-naphthylhydrazones of **1108** into **1109**, **1110** and **1111**, respectively, have been effected (Buu-Hoï, Jacquignon, Croisy and Ricci, 1968). Similarly, the phenyl- and 1- and 2-naphthylhydrazone of **1112** afforded **1113**, **1114** and **1115**, respectively, but the bisphenylhydrazone of **1116** could not be indolized (Buu-Hoï, Jacquignon, Croisy, Loiseau, Périn, Ricci and Martani, 1969). This synthetic approach has been extended to the synthesis of the sulphones **1105** (R^2 = H, X = SO_2, Y = CH=CH) starting from the corresponding hydrazones **1104** (R^2 = R^3 = H, X = SO_2, Y = CH=CH) (Aksanova, Sharkova,

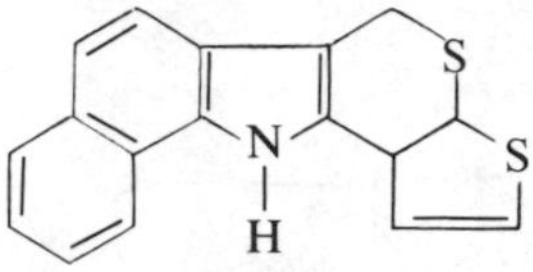

1106 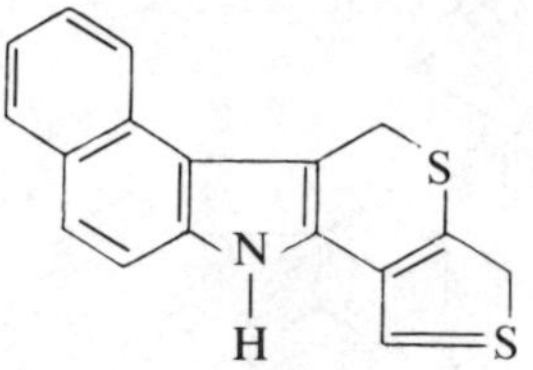 **1107**

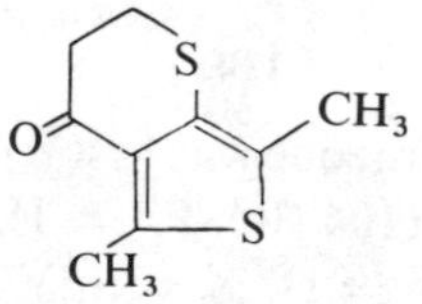

1108 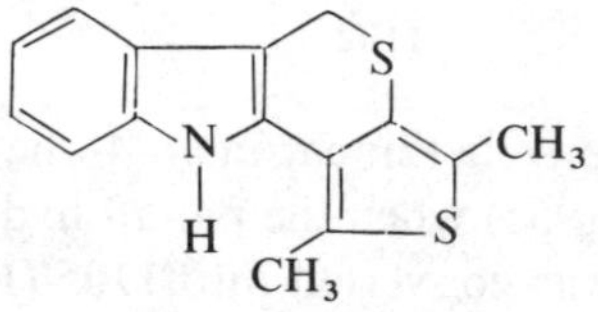 **1109**

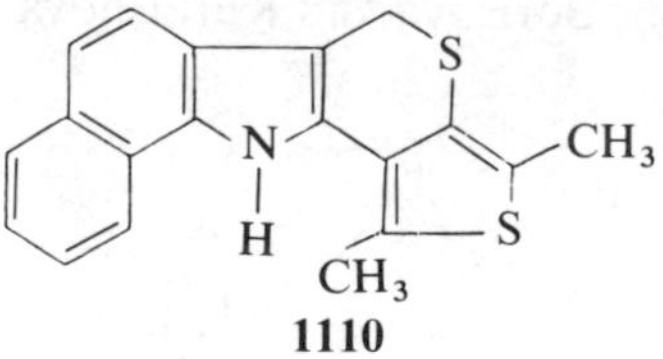

1110 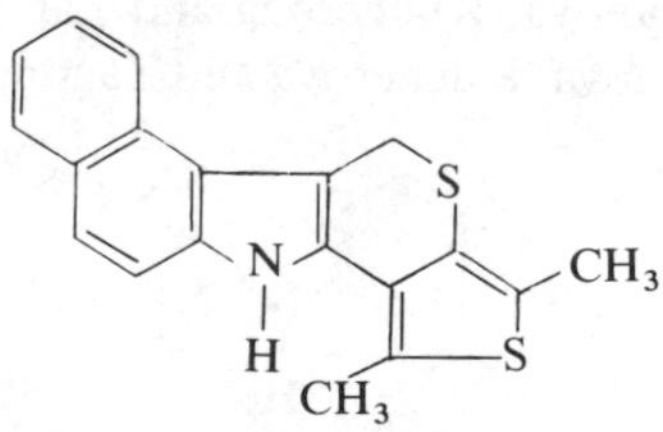 **1111**

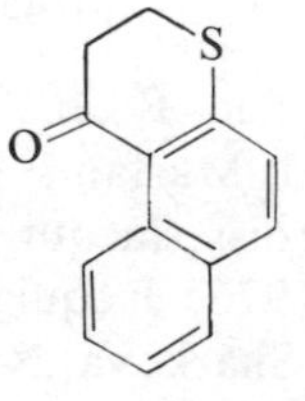

1112 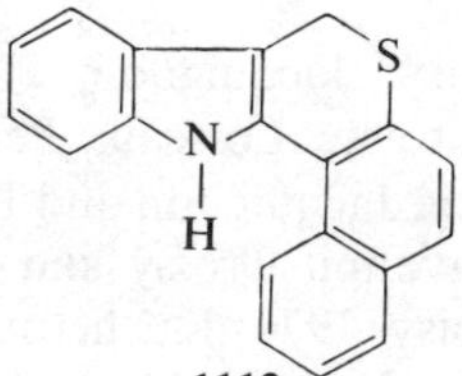 **1113**

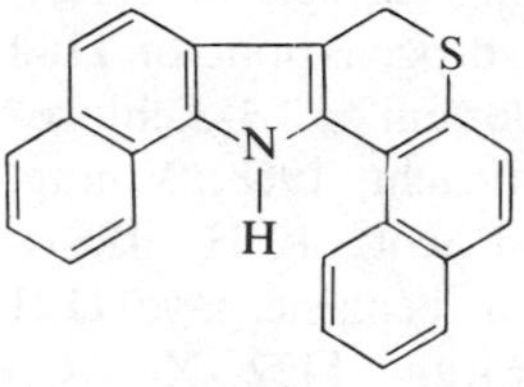

1114

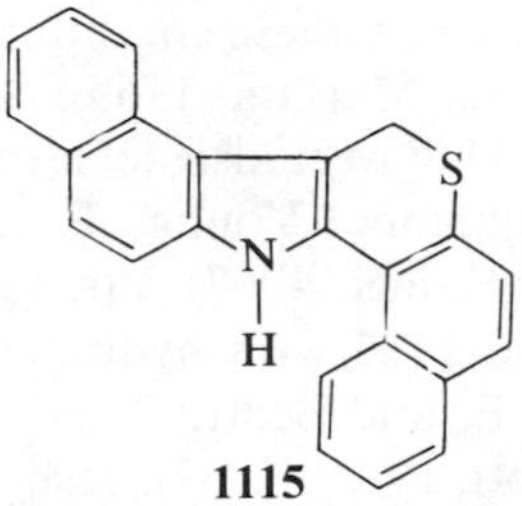

1115

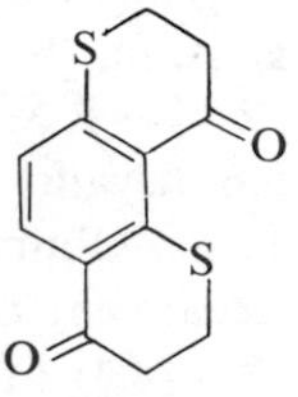

1116

N. M., Baranova, Kucherova and Zagorevskii, 1966; Borisova and Kartashova, 1979; Buu-Hoï, Jacquignon, Croisy, Loiseau, Périn, Ricci and Martani, 1969; Kucherova, Aksanova and Zagorevskii, 1963). Indolization of 3-methylthiochroman-4-one arylhydrazone (**1104**; $R^1 = 4\text{-}COOC_2H_5$, $R^2 = H$, $R^3 = CH_3$, X = S, Y = CH=CH) yielded the corresponding 3*H*-indole **1117** (R = $COOC_2H_5$, X = S), although the reaction failed with the corresponding 4-nitrophenyl- and phenylhydrazone. Likewise, the sulphones **1104** (R^1 = H, 4-CH_3 and 4-$COOC_2H_5$, R^2 = H, $R^3 = CH_3$, X = SO_2, Y = CH=CH) and the corresponding 2-naphthylhydrazone afforded **1117** (R = H, CH_3 and COOH, respectively, X = SO_2) and **1118**, respectively (Kakurina, Kucherova and Zagorevskii, 1965a).

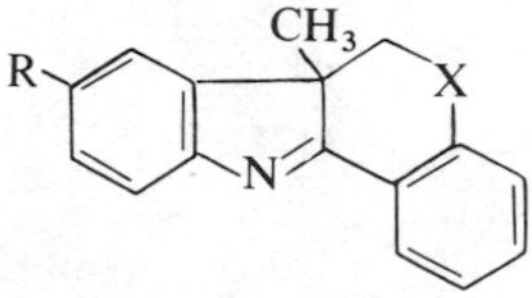

1117

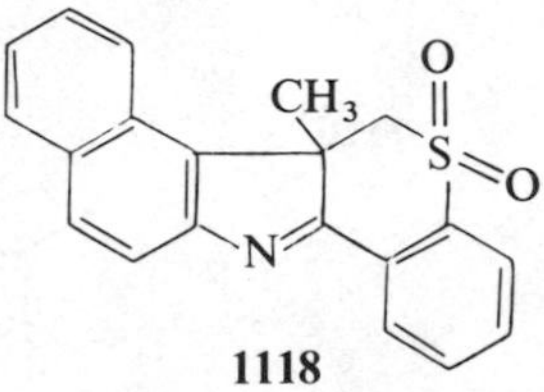

1118

Several methods have been developed for dehydrogenating the **1100** (R^2 = H, X = S) and **1105** (X = S, Y = CH=CH) systems produced in the above indolizations. Thus, treatment of **1105** (X = S, Y = CH=CH) with picric acid in a refluxing solvent such as ethanol or acetic acid afforded **1119** (X = S) (Buu-Hoï, Croisy, Jacquignon, Hien, Martani and Ricci, 1972; Buu-Hoï, Croisy, Ricci, Jacquignon and Périn, 1966; Buu-Hoï, Jacquignon, Croisy, Loiseau, Périn, Ricci and Martani, 1969; Buu-Hoï, Jacquignon, Croisy and Ricci, 1968; Buu-Hoï, Martani, Croisy, Jacquignon and Périn, 1966; Croisy, Jacquignon and Fravolini, 1974; Fravolini, Croisy and Jacquignon, 1976; Jacquignon, Fravolini, Feron and Croisy, 1974) although the corresponding

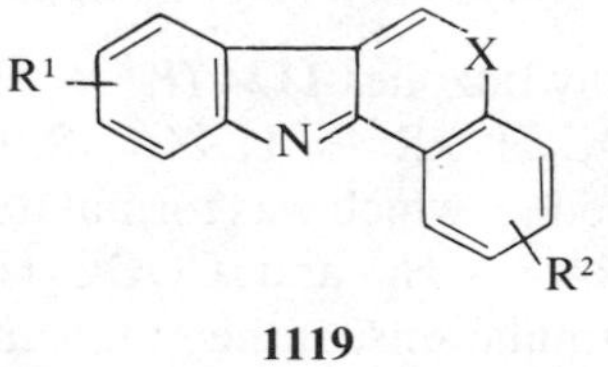

1119

sulphones were unaffected by these conditions (Buu-Hoï, Jacquignon, Croisy, Loiseau, Périn, Ricci and Martani, 1969). Similar dehydrogenations of **1100** (R^2 = H, X = S) into **1120** were also achieved using chloranil or 2,3-dichloro-5,6-dicyano-1,4-benzoquinone (Young, T. E. and Ohnmacht, 1968; Young, T. E., Ohnmacht and Hamel, 1967; Young, T. E. and Scott, P. H., 1967). Treatment of **1119** and **1120** with hydrogen chloride in benzene gave **1121** (X = Cl) (Young, T. E. and Scott, P. H., 1965, 1966) and **1122** (X = Cl) (Young, T. E. and Scott, P. H., 1967), respectively, reactions which could be reversed upon basification (Young, T. E. and Scott, P. H., 1966). This attempted salt formation was sometimes unsuccessful since extensive decomposition occurred under the acidic conditions employed (Young, T. E. and Ohnmacht, 1968). However, **1121** (X = ClO_4) could be prepared directly from **1105** (X = S, Y = CH=CH) by reaction with trityl perchlorate in acetic acid (Young, T. E. and Soctt, P. H., 1965, 1966, 1968a, 1968b), similar treatment of **1100** (R^1–R^3 = H, X = S) producing **1123** which upon reaction with 2,3-dichloro-5,6-dicyano-1,4-benzoquinone in perchloric acid gave **1122** (R^1 = R^3 = H, X = ClO_4) (Young, T. E. and Ohnmacht, 1968).

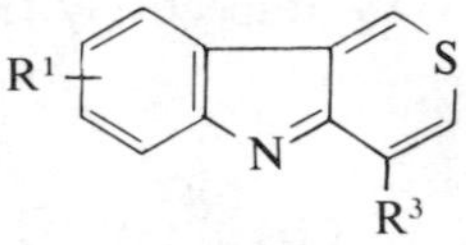

1120

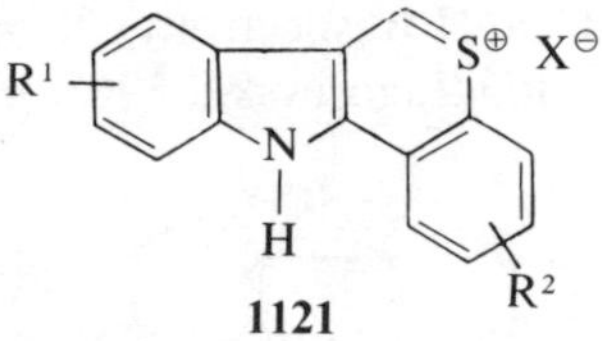

1121

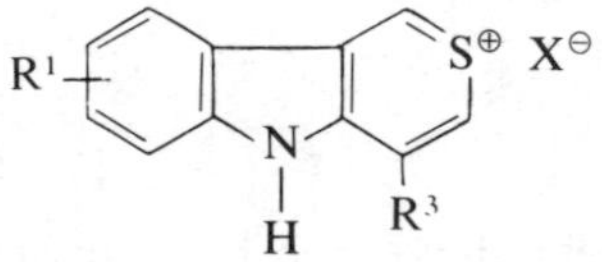

1122

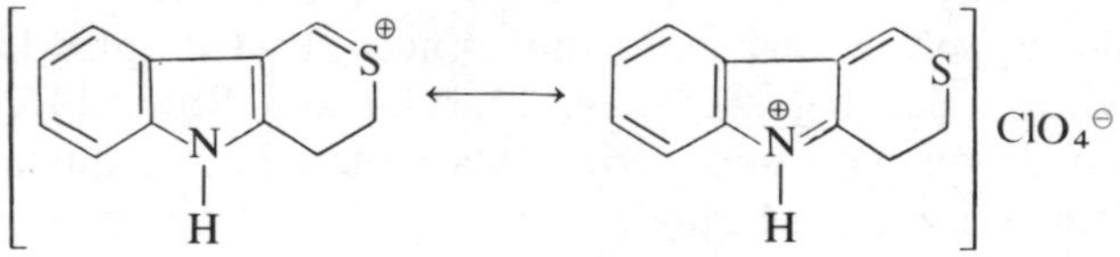

1123

9. 4-Thiepanone (4-Thiacycloheptanone) and Homologous Arylhydrazones

Indolization of the arylhydrazones **1124** (R^1 = H, R^2 = H and $CH_2C_6H_5$ and R^1 = CH_3 and $COOC_2H_5$, R^2 = H, X = S, n = 1) led to the isolation in each case of only one product which was formulated as **1125** (R^1 = H, R^2 = H and $CH_2C_6H_5$ and R^1 = CH_3 and $COOC_2H_5$, R^2 = H, respectively, X = S, n = 1). These formulations, rather than the corresponding possible

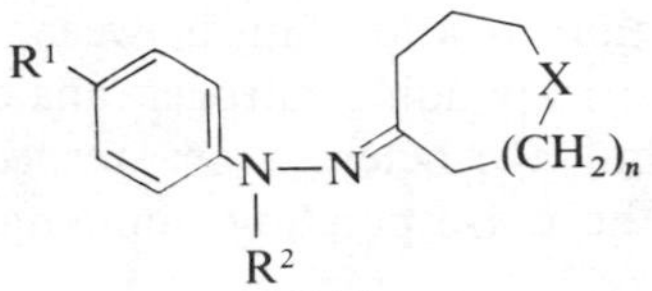

1124 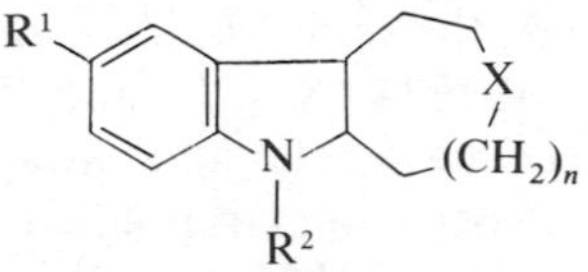 1125

isomers, were based upon the seemingly sound evidence that the product obtained from the N_α-benzylphenylhydrazone, when subjected to reductive desulphurization, furnished 1-benzyl-2,3-diethylindole whose identity was apparently established by comparison of its i.r. spectrum and the R_f (t.l.c.) value with an authentic sample which was synthesized by 1-benzylation of 2,3-diethylindole which was unambiguously synthesized by an alternative (Leete, 1961) route. The other indolization products were then formulated by analogy (Aksanova, Kucherova and Zagorevskii, 1965). However, later (Aksanova, Kucherova, Portnova and Zagorevskii, 1967) these observations and conclusions were refuted when, on the basis of a p.m.r. spectral analysis of the methylene groups, structure **1126** (R^1 = H, R^2 = $CH_2C_6H_5$, X = S) was

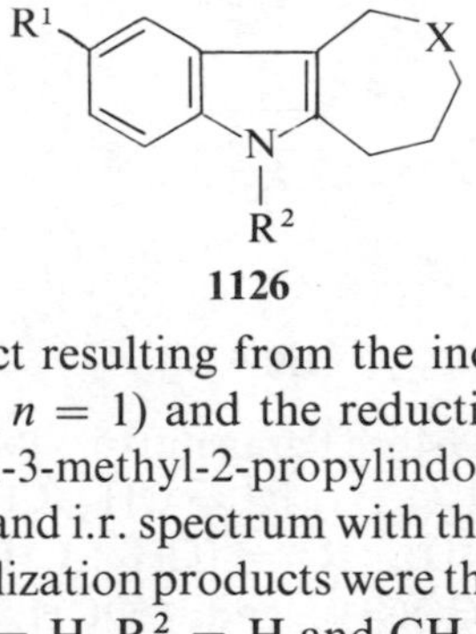

1126

established for the product resulting from the indolization of **1124** (R^1 = H, R^2 = $CH_2C_6H_5$, X = S, n = 1) and the reductive desulphurization product was shown to be 1-benzyl-3-methyl-2-propylindole, again based upon a comparison of R_f (t.l.c.) value and i.r. spectrum with those of an authentic specimen. By analogy, the other indolization products were thus reformulated accordingly. Indolization of **1124** [R^1 = H, R^2 = H and $CH_2C_6H_5$; R^1 = CH_3, R^2 = H and R^1 = $COOC_2H_5$, R^2 = H and $(CH_2)_3CH_3$, X = SO_2, n = 1] afforded in each case only one isolated product which were formulated, without any evidence, as **1125** [R^1 = H, R^2 = H and $CH_2C_6H_5$; R^1 = CH_3, R^2 = H and R^1 = $COOC_2H_5$, R^2 = H and $(CH_2)_3CH_3$, respectively, X = SO_2, n = 1] rather than the corresponding isomeric formulations **1126** [R^1 = H, R^2 = H and $CH_2C_6H_5$; R^1 = CH_3, R^2 = H and R^1 = $COOC_2H_5$, R^2 = H and $(CH_2)_3CH_3$, respectively, X = SO_2] (Aksanova, Kucherova and Zagorevskii, 1965). Later (Aksanova, Kucherova, Portnova and Zagorevskii, 1967), p.m.r. spectroscopic analysis of the methylene groups in the product resulting from the indolization of **1124** (R^1 = H, R^2 = $CH_2C_6H_5$, X = SO_2, n = 1) supported its formulation as **1125** (R^1 = H, R^2 = $CH_2C_6H_5$, X = SO_2). Thus, in these indolizations, the difference in indolization orientation between the sulphide and sulphone is analogous to that found in the indolizations of the arylhydrazones of **1084** (X = S) and **1084** (X = SO_2) already discussed.

Furthermore, the indolization–orientation relationship between the arylhydrazones of **1084** (X = S) and the corresponding nitrogen analogues **997** (X = H_2) (i.e. 'towards' and 'away from' the hetero atom, respectively, as already discussed) is extrapolated to the corresponding homologues, **1124** (X = S, *n* = 1) and **1052** (*n* = 1).

Indolization of **1127** [R^1 = H, R^2 = H, CH_3 and $CH_2C_6H_5$; R^1 = CH_3, $COOC_2H_5$ and $NHCOCH_3$, R^2 = H and R^1 = $COOC_2H_5$, R^2 = $(CH_2)_3CH_3$, X = S], **1127** [R^1 = R^2 = H; R^1 = CH_3, $NHCOCH_3$ and $COOC_2H_5$, R^2 = H and R^1 = $COOC_2H_5$, R^2 = $(CH_2)_3CH_3$ and $CH_2C_6H_5$, X = SO_2] and **1128** (X = S and SO_2) gave **1129** [R^1 = R^2 = H, X = S) (Aksanova,

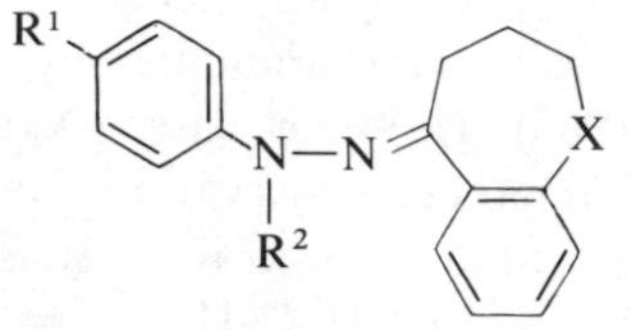

R^1 N—N X R^2 **1127**

N—N X H **1128**

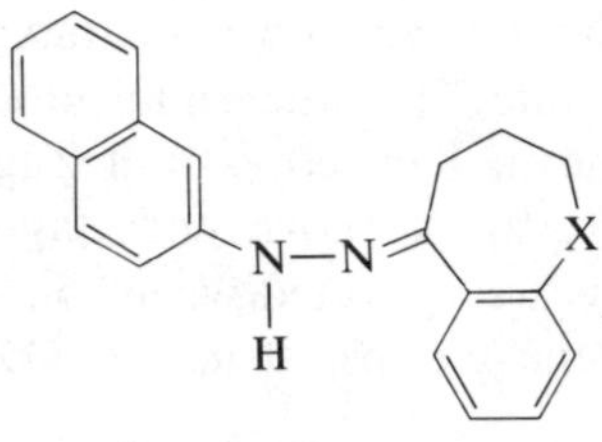

R^1 X N R^2 **1129**

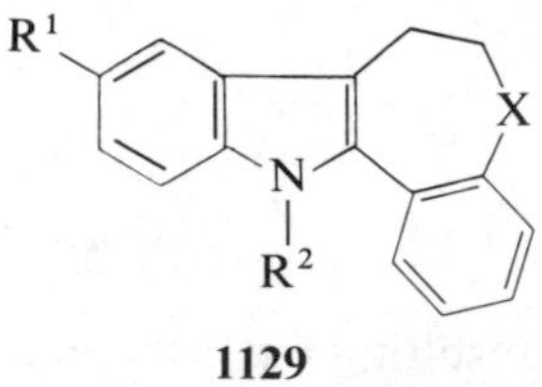

Kucherova, Zagorevskii, 1964b; Paragamian, 1973, 1974a, 1974b), **1129** [R^1 = H, R^2 = CH_3 and $CH_2C_6H_5$; R^1 = CH_3, $COOC_2H_5$ and $NHCOCH_3$, R^2 = H and R^1 = $COOC_2H_5$, R^2 = $(CH_2)_3CH_3$, X = S], **1129** [R^1 = R^2 = H; R^1 = CH_3, $NHCOCH_3$ and COOH, R^2 = H; R^1 = COOH, R^2 = $(CH_2)_3CH_3$ and R^1 = $COOC_2H_5$, R^2 = $CH_2C_6H_5$, X = SO_2] and **1130** (X = S and SO_2) (Aksanova, Kucherova and Zagorevskii, 1964b), respectively. The phenyl- and 1-naphthylhydrazone of **1131** produced exclusively **1132** and **1133**, respectively, resulting from the expected initial enehydrazine formation towards the benzenoid ring. However, with the 2-naphthylhydrazone, both possible isomeric indoles, **1134** and **1135**, were formed

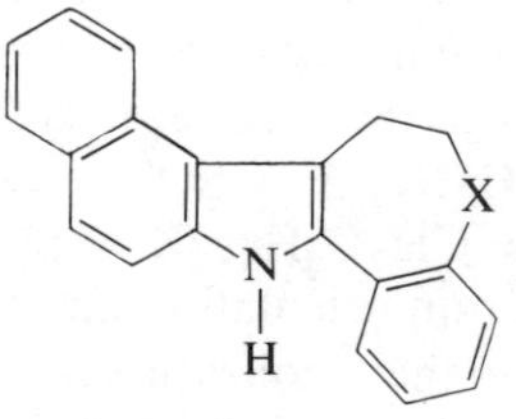

X N H **1130**

S O **1131**

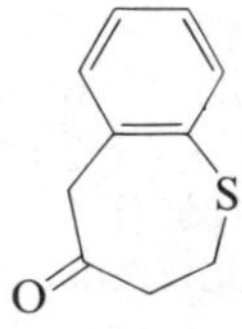

S N H **1132**

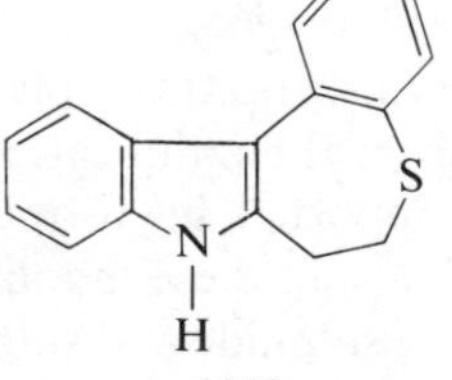

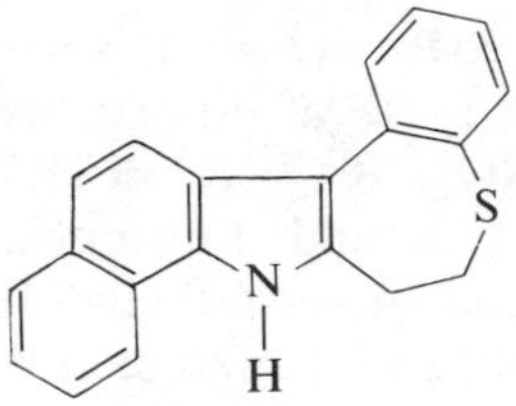

1133

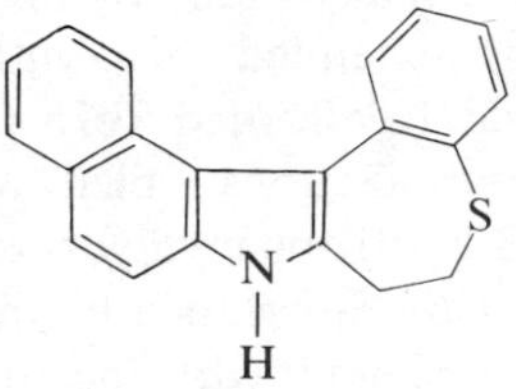

1134

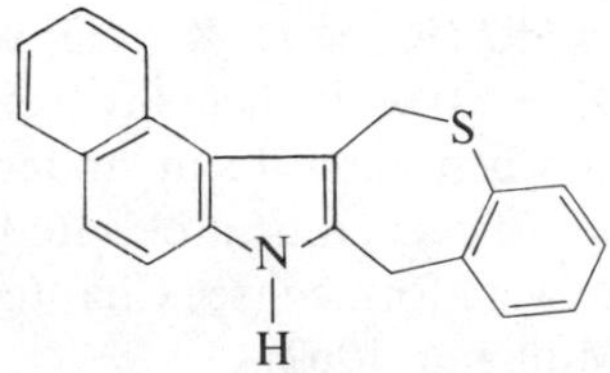

1135

(Pellicciari, Natalini, Ricci, Alunni-Bistocchi and De Meo, 1978), possibly because of the steric interaction between the terminal benzenoid nuclei in **1134** favouring the formation of some **1135**, even though the formation of the latter product would involve the initial enehydrazine formation in the less favoured direction.

Indolization of **1124** (R^1 = H, C_{1-6} alkyl, C_{1-6} alkoxy and halogen, R^2 = H, X = SO_2, $n = 2$), by refluxing mixtures of the ketone and the appropriate phenylhydrazine in acetic acid, afforded **1125** (R^1 = H, C_{1-6} alkyl, C_{1-6} alkoxy and halogen, respectively, R^2 = H, X = SO_2, $n = 2$) (Cetenko and Morrison, 1979).

10. Benzofuran-3(2H)-one (Coumaran-3-one) Arylhydrazones

Benzofuran-3(2*H*)one (**992**; R^1 = H, X = O) reacted with arylhydrazines (i.e. phenylhydrazine, N_α-methyl-, 4-methyl- and 4-bromophenylhydrazine and 2-naphthylhydrazine) to afford the corresponding hydrazones which upon treatment with acetic acid furnished the indolic products **1075** (R^1–R^3 = H; $R^1 = R^3$ = H, R^2 = CH_3 and R^1 = 5-CH_3 and 5-Br, respectively, R^2 = R^3 = H, X = O) and **1136** (X = O), respectively (Cornforth, J. W., Hughes, G. K., Lions and Harradence, 1937–1938). A contemporary report (Cawley

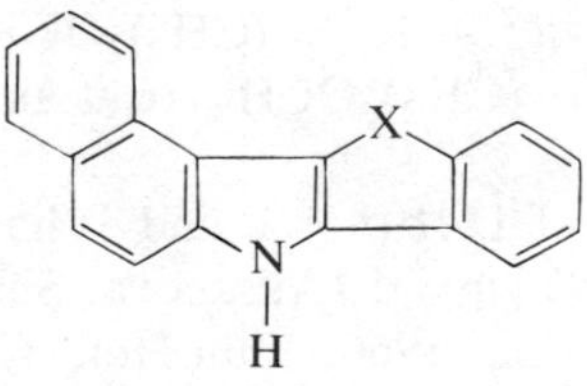

1136

and Plant, 1938) also similarly prepared **1075** (R^1–R^3 = H, X = O) and **1136** (X = O) and extended this synthesis to the use of 4-nitrophenylhydrazine which ultimately afforded **1075** (R^1 = 5-NO_2, R^2 = R^3 = H, X = O) but analogous indolizations failed when using 2- and 3-nitrophenylhydrazine although the 2-nitrophenylhydrazone was later successfully indolized (Kinsley and Plant, 1956). Subsequently, the indolization of a mixture of **992** (R^1 = H, X = O) with phenylhydrazine to form **1075** (R^1–R^3 = H, X = O) was yet again repeated and the synthesis was extended to include the analogous indolizations of mixtures of **992** (R^1 = H, X = O) with 3-nitrophenylhydrazine and of **992** (R^1 = 5-Cl, X = O) with phenylhydrazine. It was found that these reactions proceeded well when carried out under controlled conditions but that using large quantities of reactants at elevated temperatures, by-products of unestablished structure were formed (see Chapter VII, N) (Schroeder, D. C., Corcoran, Holden and Mulligan, 1962).

The possible indolization of mixtures of 3-hydroxyfurans with arylhydrazines does not appear to have been studied.

11. Tetrahydro-4H-pyran-4-one Arylhydrazones

All attempts which have been made to indolize the phenylhydrazone of tetrahydro-4*H*-pyran-4-one (**1137**; R = H) have failed (Cawley and Plant, 1938; Sharkova, N. M., Kucherova and Zagorevskii, 1962), as did attempts to indolize the 4-nitrophenylhydrazone (Cawley and Plant, 1938). However, the 2-naphthylhydrazone could be indolized to afford **1138** (R^1 = R^2 = H) and,

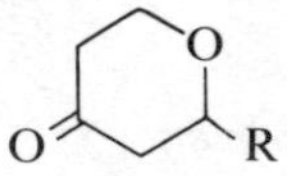

O R O
1137

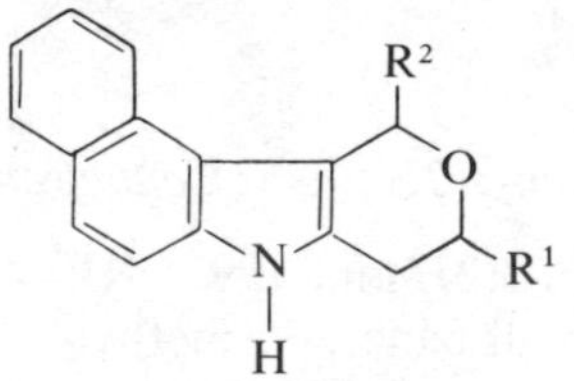

R^2 O N R^1 H
1138

likewise, the 2-naphthylhydrazone of **1137** (R = CH_3) gave apparently only one of the two possible isomeric products **1138** (R^1 = H, R^2 = CH_3 or R^1 = CH_3, R^2 = H) but the orientation of the methyl group was not established (Sharkova, N. M., Kucherova and Zagorevskii, 1962). Likewise, the hydrazone **1139** [R^1 = H, R^2 + R^3 = $(CH_2)_2OCH_2$] has been indolized into **1140** [R^1 = H, R^2 + R^3 = $(CH_2)_2OCH_2$] (Cattanach, Cohen, A. and Heath-Brown, 1973).

Several arylhydrazones of **1141** (n = 2 and 3) have been converted into the corresponding indoles **1142** [n = 1 (Aksanova, Sharkova, N. M., Baronova, Kucherova and Zagorevskii, 1966; Buu-Hoï, Croisy, Jacquignon, Hien, Martani and Ricci, 1972; Buu-Hoï, Martani, Croisy, Jacquignon and Périn,

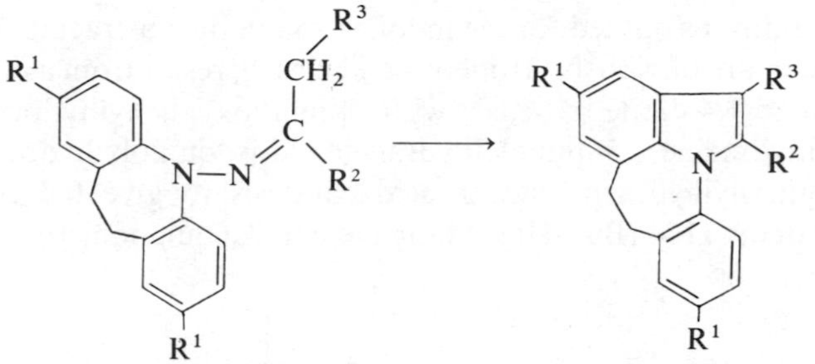

1139

1140

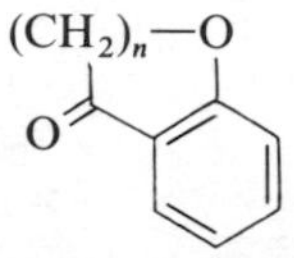

1141

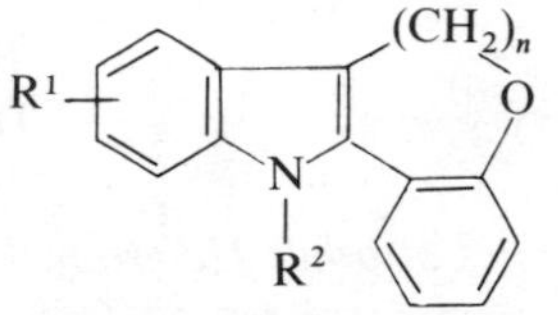

1142

1966; Sharkova, N. M., Kucherova and Zagorevskii, 1962, 1969) and $n = 2$ (Sharkova, L. M., Aksanova and Kucherova, 1971), respectively]. However, attempts to indolize the phenylhydrazone (Harradence, Hughes, G. K. and Lions, 1938) and 4-nitrophenylhydrazone (Padfield and Tomlinson, M. L., 1950) of **1141** ($n = 2$) were unsuccessful although the corresponding N_α-methylphenylhydrazone readily afforded **1142** ($R^1 = H$, $R^2 = CH_3$, $n = 1$) upon treatment with dilute sulphuric acid (Padfield and Tomlinson, M. L., 1950). Attempts to dehydrogenate **1142** ($n = 1$) with picric acid, successful with the sulphur analogues (see above), failed (Buu-Hoï, Croisy, Jacquignon, Hien, Martani and Ricci, 1972).

From the indolizations of the phenylhydrazone, 2-, 3- and 4-methylphenylhydrazone and 1- and 2-naphthylhydrazone of **1092** (X = O), only **1093** (R^1–R^4 = H; R^1–R^3 = H, $R^4 = CH_3$; $R^1 = CH_3$, R^2–R^4 = H; $R^1 = R^2 = R^4 = H$, $R^3 = CH_3$ and $R^1 = R^3 = R^4 = H$, $R^2 = CH_3$, X = O), **1094** (X = O) and **1095** (X = O), respectively, were isolated (Bistocchi, De Meo, Ricci, Croisy and Jacquignon, 1978).

The product **1143** was obtained by the indolization of the corresponding phenylhydrazone (Kametani, Fukumoto and Masuko, 1963). However, the

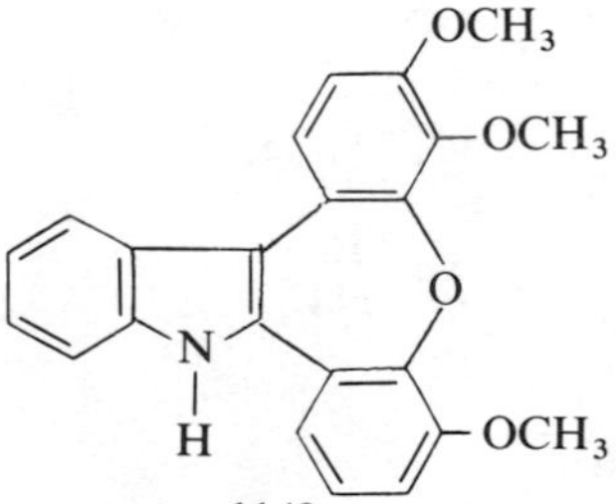

1143

catalytic conditions quoted for the indolization in the abstract in English of the original paper are obviously erroneous and clearly result from a mistranslation.

Isochroman-1,4-dione reacted with 4-methoxyphenylhydrazine, 1- and 2-naphthylhydrazine, 5-quinolylhydrazine, 5-isoquinolylhydrazine and 6-methyl-8-quinolylhydrazine under acid catalysis to give the corresponding indolic products **1144** (Buu-Hoï, Mangane and Jacquignon, 1966).

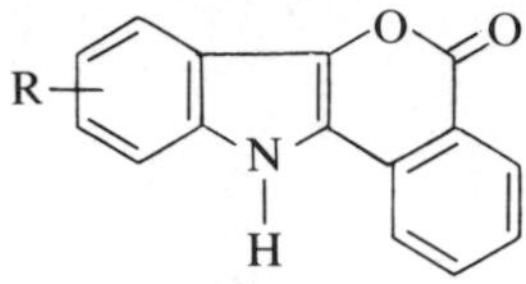

1144

12. Arylhydrazones of other Heterocyclic Ketones

When the phenylhydrazone **1145** (R = H) was treated with hydrogen chloride, hydrolysis of the hydrazone group was the only observed reaction. However, when the hydrazone was heated with a large excess of polyphosphoric acid, indolization occurred to afford **1146** (R = H) (Loev and Snader, 1967). The hydrazones **1145** (R = H, Cl and CF_3) were analogously converted into **1146** (R = H, Cl and CF_3, respectively) using acetic acid as the catalyst (Loev, 1969).

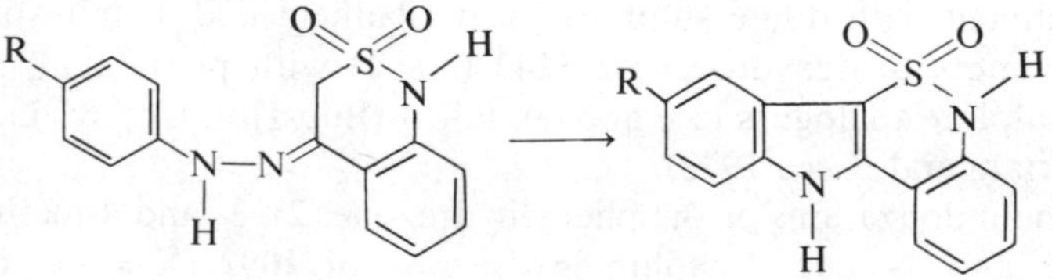

1145 → **1146**

Attempted indolization of **1147** (R = H, X = AsC_6H_5) was unsuccessful, yielding only intractable tars (Gallagher and Mann, 1962). Heating with zinc chloride was required to indolize **1148** (R^1 = H, R^2 = OCH_3, X = $AsCH_3$ into **1149** (R^1 = H, R^2 = OCH_3, X = $AsCH_3$) whereas compounds **1148**

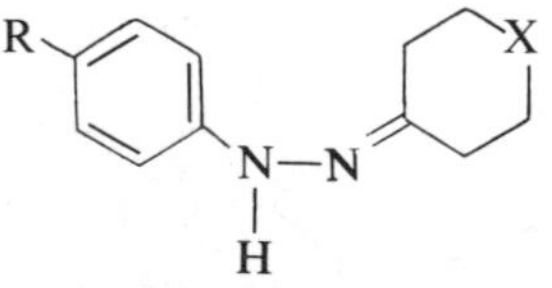

1147

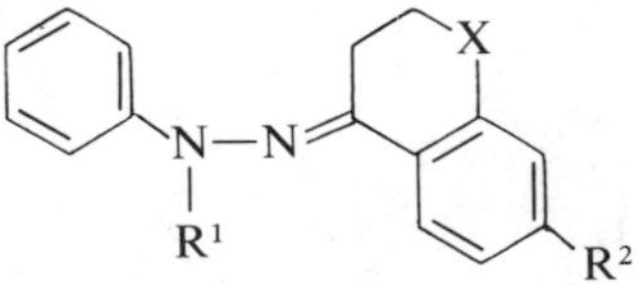

1148

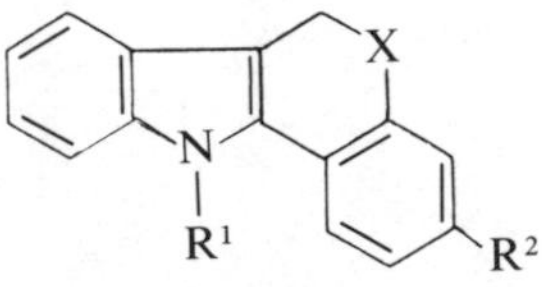

1149

(R^1 = CH_3 and C_6H_5, R^2 = OCH_3, X = $AsCH_3$) were readily converted into **1149** (R^1 = CH_3 and C_6H_5, respectively, R^2 = OCH_3, X = $AsCH_3$) by simply warming in ethanolic acetic acid (Mann and Wilkinson, 1957). The bisphenylhydrazone **1150**, when heated in acetic acid containing zinc chloride, gave a product formulated as **1151** (R^1 = H, R^2 = OCH_3), one hydrazone moiety having undergone indolization and the other hydrolysis (Mann and Wilkinson, 1957). However, no evidence was presented to eliminate the other possible structure, **1151** (R^1 = OCH_3, R^2 = H), for the product.

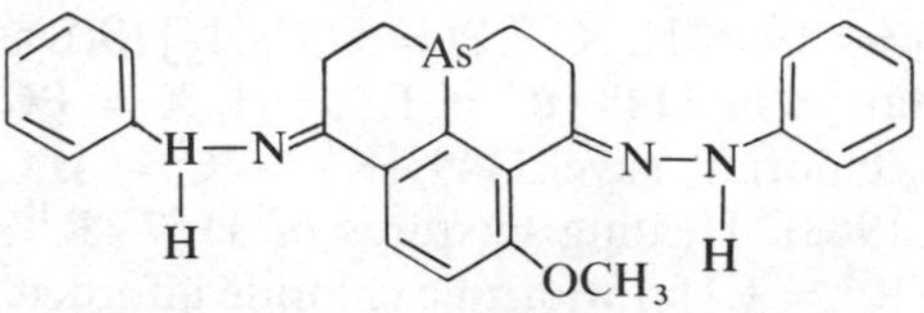

1150 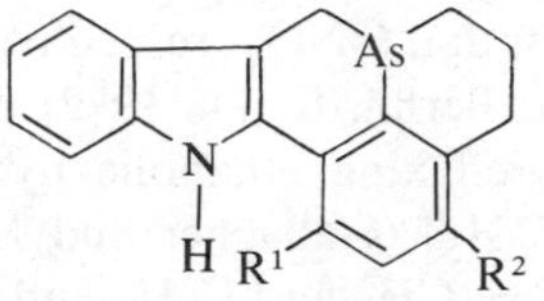 **1151**

Indolization of mixtures of **992** (R^1 = H, X = Se) with phenylhydrazine and 2-naphthylhydrazine, using hydrogen chloride in acetic acid as catalyst, afforded **1075** (R^1–R^3 = H, X = Se) and **1136** (X = Se), respectively, other arylhydrazines having also been used in analogous indolizations with the same ketone (Buu-Hoï, Saint-Ruf, Martani, Ricci and Balucani, 1968; Maréchal, Christiaens, Renson and Jacquignon, 1978). Indolization of the phenylhydrazone and 1- and 2-naphthylhydrazone of **1152** (X = Se) afforded **1105** (R^1 = R^2 = H, X = Se, Y = CH=CH), **1153** and **1154**, respectively. These products were, similarly to the sulphur analogues discussed earlier, dehydrogenated by boiling with picric acid in acetic acid solution to yield **1119** (R^1 = R^2 = H, X = Se), **1155** and **1156**, respectively (Buu-Hoï, Croisy, Jacquignon, Renson and Ruwet, 1970).

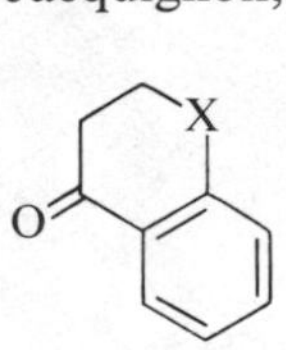

1152 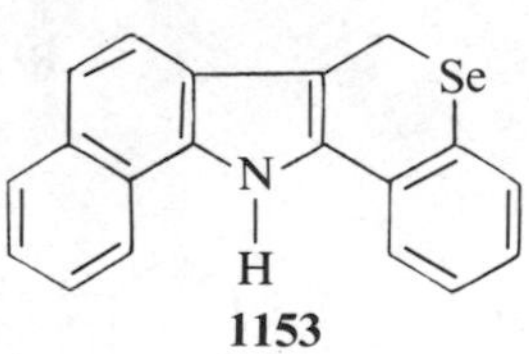 **1153** 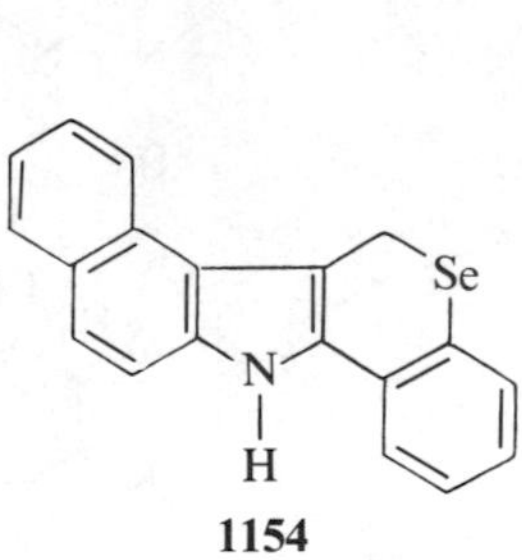 **1154**

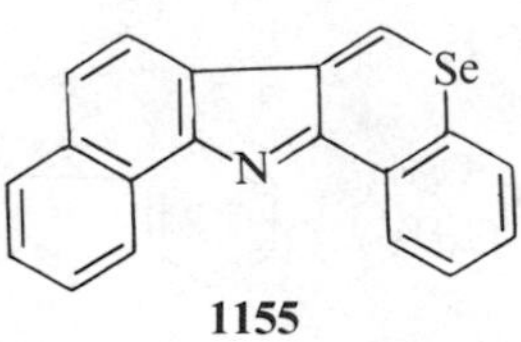

1155 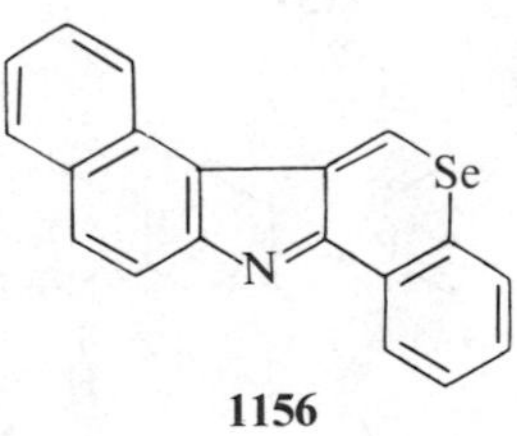 **1156**

Arylhydrazones **1147** (R = H, CH_3, OH and F, X = PC_6H_5) were indolized *in situ*, using acetic acid–concentrated hydrochloric acid, to give **1100** (R^1–R^3 = H, X = PC_6H_5) (Gallagher and Mann, 1962; Srivastava and Berlin, K. D., 1972) and **1100** (R^1 = 6-CH_3, 6-OH and 6-F, respectively, R^2 = R^3 = H, X = PC_6H_5) (Srivastava and Berlin, K. D., 1972). These products were readily oxidized into **1100** [R^1 = H, 6-CH_3, 6-OH and 6-F, R^2 = R^3 = H, X = P(= O)C_6H_5] and, indeed, from the indolizations of **1147** (R = Cl, Br and NO_2, X = PC_6H_5) using an acetic acid–concentrated hydrochloric acid catalyst, the only products which could be isolated were **1100** [R^1 = 6-Cl, 6-Br and 6-NO_2, respectively, R^2 = R^3 = H, X = P(= O)C_6H_5] (Srivastava and Berlin, K. D., 1972). Indolization of **1148** (R^1 = R^2 = H, X = PC_6H_5) in refluxing ethanolic hydrogen chloride gave **1149** (R^1 = R^2 = H, X = PC_6H_5) (Gallagher and Mann, 1963). Heating mixtures of **1157** (R^1 = H, R^2 = CH_3 and C_6H_5 and R^1 = R^2 = CH_3) with zinc chloride afforded **1158** (R^1 = H, R^2 = CH_3 and C_6H_5) and **1159**, respectively, as the only isolated products. Reaction of **1158** (R^1 = H, R^2 = C_6H_5) with triethyl- and trimethyloxonium fluoroborate yielded **1160** (R^1 = H, R^2 = C_2H_5 and CH_3, respectively). These two fluoroborates also effected the indolization of **1157** (R^1 = H, R^2 = C_6H_5) to afford **1160** (R^1 = H and CH_3, R^2 = CH_3) and **1158** (R^1 = CH_3, R^2 = C_6H_5) and **1160** (R^1 = R^2 = C_2H_5), respectively. Pyrolysis of **1160** (R^1 = R^2 = CH_3 and C_2H_5) at 300 °C furnished **1161** (R = CH_3 and C_2H_5, respectively). Only **1163** was isolated after heating a mixture of **1162** with zinc chloride (Märkl, Habel and Baier, 1979).

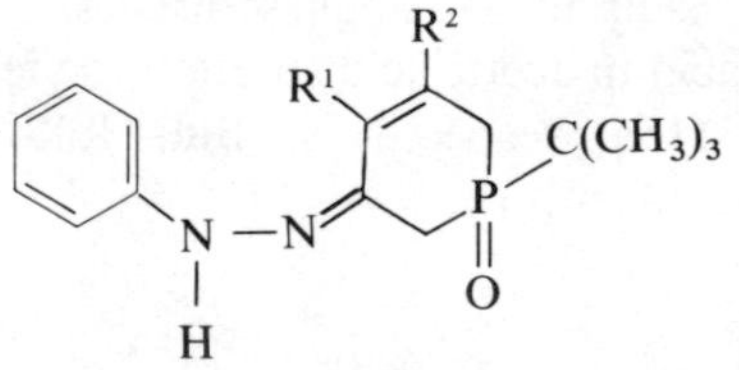

1157 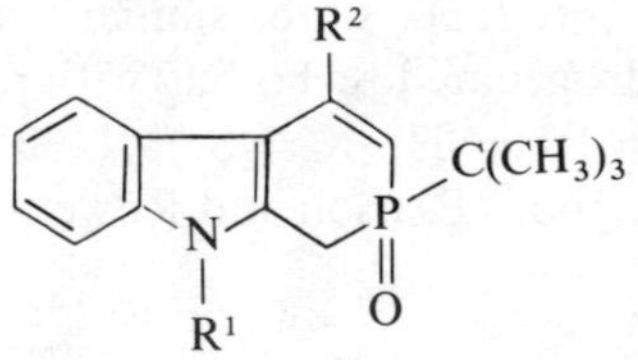 **1158**

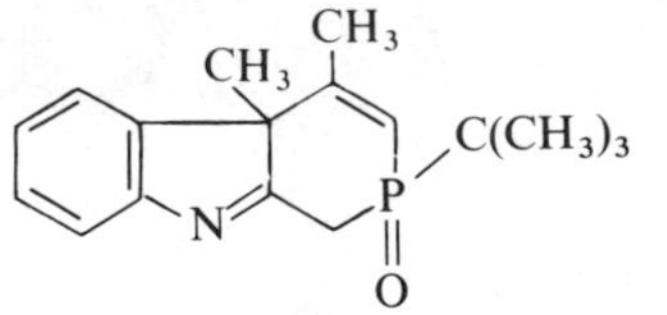

1159 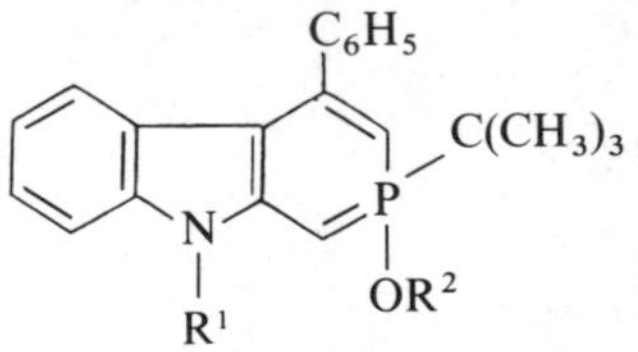 **1160**

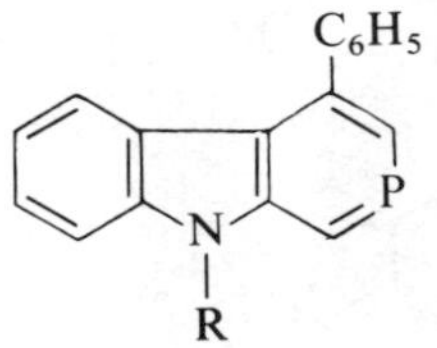

1161 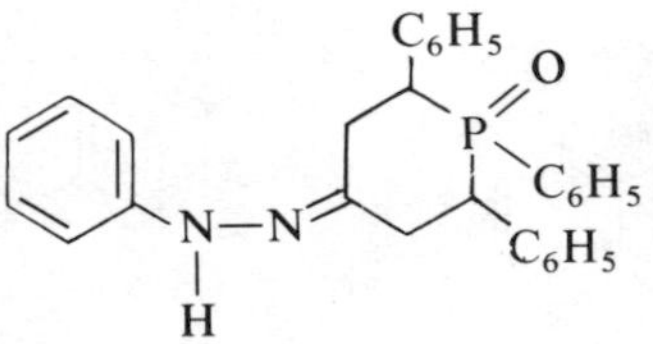 **1162** 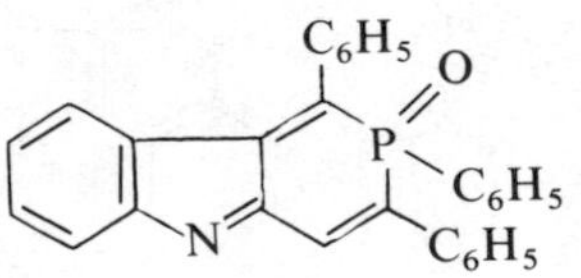 **1163**

CHAPTER IV

Selected Examples and Uses of the Fischer Indole Synthesis

A. Indolization of Acetaldehyde Phenylhydrazone

For many decades (see, for examples, Suvorov, Mamaev and Rodinov, 1959; Suvorov and Sorokina, N. P., 1961; Terent'ev, A. P. and Preobrazhenskaia, 1958) the failure to convert acetaldehyde phenylhydrazone into indole remained a notable exception to the Fischer indole synthesis. Thus, Fischer himself (Fischer, Emil, 1886b, 1886e) was unable to isolate indole after heating a mixture of acetaldehyde phenylhydrazone with zinc chloride and subsequent studies likewise failed to yield indole when the hydrazone was treated with tin(IV) chloride in hydrochloric acid (Brunner, K., 1900), boron trifluoride etherate in acetic acid and other unspecified acid catalysts (Snyder and Smith, C. W., 1943), formic acid (Kidwai, A. R. and Khan, N. H., 1963) or polyphosphate ester in boiling chloroform (Kanaoka, Ban, Miyashita, Irie and Yonemitsu, 1966; Kanaoka, Ban, Yonemitsu, Irie and Miyashita, 1965) or after refluxing in monoethylene glycol (Robinson, B., 1963a) or diethylene glycol (Kelly, A. H., McLeod and Parrick, 1965) solutions. Although indole could not be isolated from these non-catalytic thermal indolizations, the evolution of ammonia was detected in both cases and, in the latter case, aniline, N-ethylaniline and a non-basic compound of empirical formula C_6H_6N were also isolated from the reaction mixture. It has been suggested (Kitaev, and Troepol'skaya, 1978) that this last product might be 5-methyl-1-phenylpyrazoline (see Chapter VII, R) containing a small amount of indole. Certainly, this pyrazoline was obtained by heating acetaldehyde phenylhydrazone with a copper(I) chloride catalyst (Kitaev and Troepol'skaya, 1978) but the non-basic nature of the product C_6H_6N obtained from the attempted thermal indolization would not be in accord with a pyrazoline formulation.

Attempts to indolize several other acetaldehyde arylhydrazones have also been unsuccessful. Thus, acetaldehyde N_α-methylphenylhydrazone could not be indolized using a tin(IV) chloride in hydrochloric acid catalyst (Brunner, K., 1900), even though it is now well established (Chalmers, A. J. and Lions, 1933; Chastrette, 1970; Diels and Köllisch, 1911; Fischer, Emil, 1886b; Harvey, D. J. and Reid, S. T., 1972; Ishii, Murakami, Y., Hosoya, Takeda, Suzuki, Y. and Ikeda, 1973; Kermack, Perkin and Robinson, R., 1921; Kermack and

Slater, 1928; Mann and Wilkinson, 1957; Padfield and Tomlinson, M. L., 1950; Perkin and Plant, 1923b; Robinson, R. and Thornley, 1924 and Woolley, D. W. and Shaw, E., 1955) that N_α-methylation facilitates the indolization of arylhydrazones. Failure also attended attempts to indolize acetaldehyde 4-bromo- and 4-nitrophenylhydrazone (and acetophenone 4-bromophenylhydrazone) by heating with a nickel(II) chloride catalyst, although these observations were ascribed to the electron withdrawing influences of the bromo and nitro substituents, respectively (Korczynski, Brydowna and Kierzek, 1926). Likewise, acetaldehyde 2-pyridylhydrazone failed to afford 7-azaindole upon treatment with zinc chloride or hydrochloric acid (Fargher and Furness, 1915) or with polyphosphoric acid (Okuda and Robison, 1959) and acetaldehyde 2,5-dichloro-4-pyridylhydrazone did not indolize upon heating with zinc chloride (Yakhantov, Marshalkin, Anisimova and Kostyuchenko, 1973) although these failures could be caused, in part, by the presence of the pyridino N-atom (see Chapter IV,I).

Other related observations have also been made. Thus, acetaldehyde arylhydrazines have been reacted with ketones and other aldehydes under Fischer indolization conditions, when transhydrazonization occurred followed by indolization to give only the corresponding 2- and/or 3-substituted indole (Cattanach, Cohen, A. and Heath-Brown, 1973; F. Hoffmann-La Roche and Co. A. G., 1965; Cohen, A., Heath-Brown and Cattanach, 1968; Yamamoto, H. and Nakao, 1969a; Yamamoto, H., Nakao and Kobayashi, A., 1968; Yamamoto, H., Nakao and Okamoto, 1968). Attempted indolization of pyruvic acid 6-quinolylhydrazone (**606**; R^1–R^3 = R^5 = H, R^4 = COOH, X = Y = CH, Z = N), presumably by heating after admixture with zinc chloride, failed, since under the indolization conditions, decarboxylation of the hydrazone occurred to afford acetaldehyde 6-quinolylhydrazone (Wieland, H. and Horner, L., 1938). However, the use of ethyl pyruvate 6-quinolylhydrazone (**606**; R^1–R^3 = R^5 = H, R^4 = $COOC_2H_5$, X = Y = CH, Z = N) prevented this decarboxylation and indolization could thus be achieved to furnish **608** (R^1 = R^2 = R^4 = H, R^3 = $COOC_2H_5$, X = Y = CH, Z = N). This was then hydrolysed to the corresponding acid which was decarboxylated to produce **608** (R^1–R^4 = H, X = Y = CH, Z = N) (Wieland, H. and Horner, L., 1938). Indeed, this synthetic route has found a wide application in the synthesis of 2,3-diunsubstituted indoles from the appropriate ethyl pyruvate arylhydrazones (see Chapter IV, B). Advantage has been taken of the lack of reactivity towards indolization of acetaldehyde arylhydrazones in the development of an unambiguous synthesis of N_α-acylaryl- or N_α-aroylarylhydrazones. Thus, the arylhydrazine was converted into the acetaldehyde arylhydrazone which reacted with acyl or aroyl chlorides to afford the N_α-acyl or N_α-aroyl derivatives. These were then selectively hydrolysed with alcoholic hydrogen chloride at low temperature to afford the corresponding N_α-acylaryl- or N_α-aroylarylhydrazines (CRC Compagnia di Ricerca Chimica S. A., 1979; Yamamoto, H., 1967a; Yamamoto, H. and Nakao, 1969e,

1974a, 1974b; Yamamoto, H., Nakao and Kobayashi, A., 1968; Yamamoto, H., Nakao and Okamoto, 1968; Yamamoto, H., Saito, C., Okamoto, Awata, Inukai, Hirohashi, A. and Yukawa, 1969).

Contrary to these above apparent indolization failures of acetaldehyde arylhydrazones, it was noted during an early development of the indolization by one of Fischer's students (Schlieper, 1886) that a small yield of 4,5-benzindole (**10**; $R^1 = R^2 = H$) was formed when a mixture of acetaldehyde 2-naphthylhydrazone with an equal quantity of zinc chloride was fused, although at the time the structure of this product was not distinguished from the possible 5,6-benzindole (**9**; $R^1 = R^2 = H$). Other successful indolizations of acetaldehyde arylhydrazones were observed twenty years ago when compounds **1164** (R^1–$R^3 = R^5 = R^6 = H$, $R^4 = H$ and Br) were reacted with either a polyphosphoric or sulphuric acid catalyst to yield **1165** (R^1–$R^3 = R^5 = R^6 = H$, $R^4 = H$ and Br, respectively) (Sunagawa, Soma, Nakano, H. and Matsumoto, 1961), although acetaldehyde 2-methylaminotropone azine **1166** (R^1–$R^3 = H$) could not be cyclized (Sato, Y. and Sunagawa, 1967).

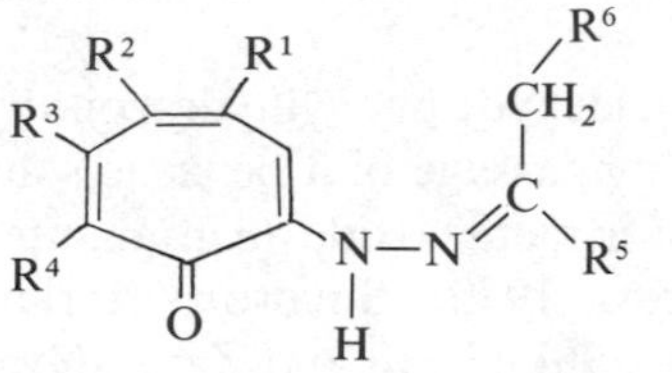

1164

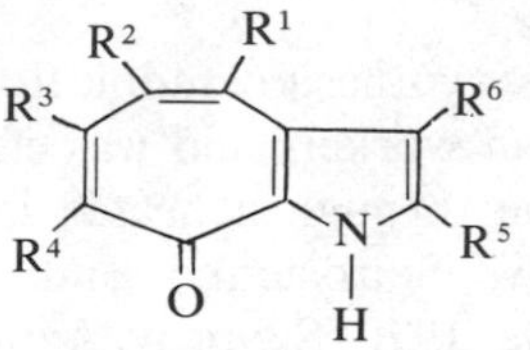

1165

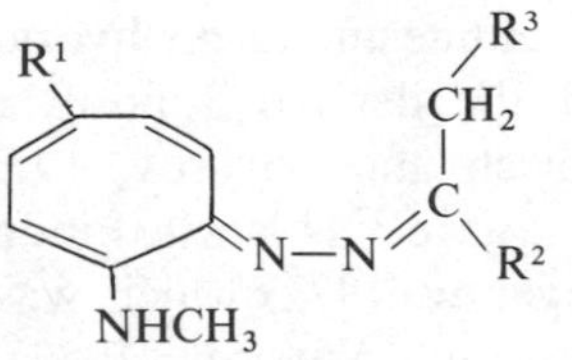

1166

Since the validity of polarographic studies (Arbuzov, A. E. and Kitaev, 1957a; Kitaev and Arbuzov, A. E., 1957, 1960; Kitaev and Troepol'skaya, 1963a, 1963b) which claimed to have detected the formation of the enehydrazine tautomer of acetaldehyde phenylhydrazone has been questioned (O'Connor, 1961) and since attempts (Suvorov and Sorokina, N. P., 1961) to trap this tautomer as its diacetyl derivative **238** (R^1–$R^3 = H$), successful with other arylhydrazones (Suvorov and Sorokina, N. P., 1961; Suvorov, Sorokina, N. P. and Sheinker, I. N., 1958; Suvorov, Sorokina, N. P. and Sheinker, Y. N., 1957 – see also Chapter II, E1), have failed, it was suggested (Robinson, B., 1963a) that the failure to indolize acetaldehyde phenylhydrazone might be caused by a failure to establish the hydrazone–enehydrazine tautomerism. However, that enehydrazine formation, however unfavoured, does occur with acetaldehyde phenylhydrazone was demonstrated by the following recent reports from three groups who achieved the indolization of this hydrazone using flow techniques which immediately removed the indole, once it was formed, from contact with the catalyst. It appears that indole may have been formed during some of the attempted indolizations mentioned earlier but that it immediately reacted with unreacted hydrazone under the drastic conditions of the indolizations and therefore its formation remained undetected (Nakazaki and Yamamoto, K., 1976). Alternatively, polymerization of indole under strongly acidic indolization conditions could well account for its absence in some of the products resulting from these earlier attempted indolizations. In connection with this, only traces of 3-ethylindole could be isolated after treatment of butyraldehyde phenylhydrazone with strong acids (e.g. ethanolic sulphuric acid or ethanolic hydrochloric acid), possibly because of the acid catalysed polymerization of this product, whereas when using boron trifluoride etherate in refluxing benzene as the catalyst, a 15% yield of 3-ethylindole was obtained (Bullock and Fox, 1951).

The first synthesis of indole from acetaldehyde phenylhydrazone was reported by Russian workers and was effected by passage of a benzene solution of the hydrazone through a heated layer of a metal oxide (e.g. aluminium oxide) (Kanterov, Starostenko and Suvorov, 1970; Suvorov, Avramenko and Shkil'kova, 1970; Suvorov, Avramenko, Shkil'kova and Zamyshlyaeva, 1969a, 1969b, 1974; Suvorov, Shkil'kova, Avramenko and Zamyshlyaeva, 1970). Starting with ^{14}C labelled nitrobenzene this reaction was, via the sequential formation of ^{14}C labelled aniline and phenylhydrazine, used to synthesize benz ^{14}C labelled indole and the derived 3-indolylacetic acid and tryptamine (Dmitrevskaya, Smushkevich and Suvorov, 1973) and starting from $^{15}N_{\alpha}$ labelled phenylhydrazine it gave ^{15}N labelled indole (Suvorov, Dmitrevskaya, Smushkevich and Pozdnyakov, 1972) which was ultimately (Dmitrevskaya, Smushkevich, Pozdnyakov and Suvorov, 1973) converted into the correspondingly labelled 3-indolylacetic acid, tryptamine and N,N-dimethyltryptamine. Indolizations under similar catalytic conditions have also been extended to the use of other hydrazones (Suvorov, Shkil'kova, Avramenko and Zamyshlyaeva, 1970). Thus, using the appropriate substituted phenylhydrazone of acetaldehyde, 1-methylindole (Suvorov, Avramenko, Shkil'kova and Zamyshlyaeva, 1969b, 1974), 5-bromo- and 5-chloroindole (Suvorov, Avramenko, Shkil'kova and Zamyshlyaeva, 1969a, 1969b, 1974), 5-methylindole (Suvorov, Avramenko, Shkil'kova and Zamyshlyaeva, 1969b, 1974), 5-methoxyindole (Suvorov, Avramenko, Shkil'kova and Zamyshlyaeva, 1974; Suvorov, Bykhovskii and Podkhalyuzina, 1977) and 4-, 6- and 7-methoxyindole (Suvorov, Bykhovskii and Podkhalyuzina, 1977) have been synthesized. Acetone and butyraldehyde phenylhydrazone (Suvorov, Avramenko, Shkil'kova and

Zamyshlyaeva, 1969a, 1969b, 1974), acetone 4-methoxy- and 4-methylphenylhydrazone (Suvorov, Avramenko, Shkil'kova and Zamyshlyaeva, 1969a) and methyl propyl ketone phenylhydrazone (Suvorov, Avramenko, Shkil'kova and Zamyshlyaeva, 1974) have also been similarly indolized into the expected indoles. Application of these indolization conditions to acetaldehyde 2-pyridylhydrazone afforded 7-azaindole, along with considerable quantities of 2-aminopyridine and **1167** (R = H) and under similar conditions, acetone and ethyl methyl ketone 2-pyridylhydrazone gave 2-methyl- and -2,3-dimethyl-7-azaindole, respectively, along with 2-aminopyridine and **1167** (R = H) in each case

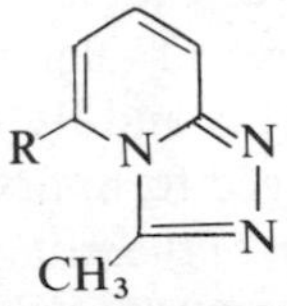

1167

(Yakhontov, Suvorov, Kanterov, Podkhalyuzina, Pronina, E. V., Starostenko and Shkil'kova, 1972; Yakhontov, Suvorov, Pronina, E. V., Kanterov, Podkhalyuzina, Starostenko and Shkil'kova, 1972). Similarly, as well as indole and ammonia, considerable quantities of aniline, together with traces (< 1 %) of nitrogen, methane and ethane were also formed during the above indolizations of acetaldehyde phenylhydrazone (Kanterov, Starostenko, Oleinikov and Suvorov, 1970). The spent catalyst from the above gas phase indolizations of acetaldehyde anylhydrazones has been regenerated by pyrolysis in a stream of gas containing oxygen which removes resinous coatings on the catalytic surfaces (Ushakov, Timofeev and Tyuryaev, 1979).

Starting with 1-tritioacetaldehyde phenylhydrazone (Suvorov, Dmitrevskaya, Smushkevich and Przhiyalgovskaya, 1976) and 2,2,2-trideuterio- or 2,2,2-tritritioacetaldehyde phenylhydrazone (Suvorov, Dmitrevskaya, Smushkevich and Petrovskaya, 1975), the isotopic label was found to be distributed throughout the three major reaction products (i.e. indole, ammonia and aniline) after indolization, these distributions being rationalized in terms of the various equilibria established during the reactions. Of several metallic oxides which have been employed in this transformation (Suvorov, Bykhovskii and Podkhalyuzina, 1977), aluminium oxide has been most frequently used. Using this catalyst it has been found that (*a*) a maximum yield of indole (65 %) was formed at an optimum reaction temperature of 310 °C, the yield of aniline, formed by an apparently purely thermal process, increasing very little over the temperature range investigated (i.e. 270–350 °C) (Kanterov, Suvorov, Starostenko and Oleinikov, 1970), (*b*) a maximum yield of indole (*ca.* 45 %) was reached at a constant reaction temperature of 279 °C when the benzene solution of acetaldehyde phenylhydrazone reached a concentration of 25 %, further concentration increases causing no further change in yields (Kanterov, Suvorov,

Starostenko and Oleinikov, 1970), (*c*) the acidity and catalytic activity of the aluminium oxide were linked, both reaching a maximum when heat treatment prior to reaction was effected at a temperature of 600 °C (Suvorov, Starostenko, Antipina, Kanterov, Podkhalyuzina and Bulgakov, 1972), (*d*) fluorination of the aluminium oxide maximized its activity at a 4.8 % fluorine content, this catalyst affording a maximum yield of indole (*ca.* 60 %) at a reaction temperature of 250 °C (i.e. 60 °C lower than with the unfluorinated aluminium oxide) (Suvorov, Starostenko, Antipina, Kanterov, Podkhalyuzina and Bulgakov, 1972), (*e*) the addition of alkali to the catalyst retarded the decomposition of the hydrazone, decreased the yield of indole but increased the yield of aniline (Kanterov, Suvorov, Starostenko and Oleinikov, 1970; Suvorov, Starostenko, Antipina, Kanterov, Podkhalyuzina and Bulgakov, 1972), and (*f*) the reaction kinetics of the indolization changed from first order to zero order, a transition related to the degree of saturation of the catalyst surface, and where the reaction was of first order kinetics its activation energy was 35.5 $\pm$ 1.5 kcal/mol (Suvorov, Bykhovskii, Pozdnyakov and Sadovnikov, 1975), the activation energy of the thermal decomposition of the hydrazone into aniline being 41 kcal/mol (Kanterov, Starostenko, Oleinikov and Suvorov, 1970).

A Japanese group next reported the synthesis of indole (in 68.3 % yield) by a technique closely related to that of the Russians except that, in place of acetaldehyde phenylhydrazone in benzene solution, a mixture of phenylhydrazine and paraldehyde was passed over an aluminium oxide catalyst at 290 °C using nitrogen as the carrier gas (Yamauchi, Yada and Kudo, 1974).

A third indolization of acetaldehyde phenylhydrazone was reported from both Russian and Japanese sources. This indolization was achieved by passage of the hydrazone over heated zinc chloride [a similar compound was one of Emil Fischer's original indolization catalysts (Fischer, Emil, 1886b)], indole being produced in 24 % (Suvorov, Smushkevich, Pozdnyakov and Shteinpress, 1974) and 36 % (Nakazaki and Yamamoto, K., 1976) yields, respectively. It is interesting that in this latter work, a 46 % yield of aniline was also produced along with a 10 % yield of acetonitrile, both of which are also produced by a purely thermal decomposition of the hydrazone. This latter reaction also produced traces of indole, as exemplified by the formation of indole, along with aniline and acetonitrile, during the v.p.c. of acetaldehyde phenylhydrazone at an injection chamber temperature of 250 °C (Nakazaki and Yamamoto, K., 1976). These reactions showed that acetaldehyde phenylhydrazone can be thermally indolized under non-catalytic conditions. Subsequently (Glish and Cooks, 1978), it was also shown to be indolized in the gas phase under mass spectral investigational conditions.

Studies which have been effected upon Fischer indolizations under conditions of heterogeneous catalysis have been reviewed (Suvorov, Bykhonskii, Dmitrevskaya, Starostenko, Smushkevich and Shkil'kova, 1977 – see also Suvorov, Starostenko, Kanterov and Podkhalyuzina, 1976).

Relevant to studies upon acetaldehyde arylhydrazone indolizations are the observations that when cooled acetaldehyde was added dropwise, with stirring, to solutions of phenylhydrazine and 4-bromo-, 4-chloro-, 4-iodo-, 4-methoxy- and 4-methylphenylhydrazine in anhydrous ether at 2–3 °C, the corresponding hydrazones were formed whereas when phenylhydrazine and 4-chloro-, 4-methoxy- and 4-methylphenylhydrazine were added to ethanolic solutions of acetaldehyde at 0 °C, the only products isolated were **1168** (R = H, Cl, OCH_3 and CH_3, respectively) (Bokii, Babushkina, T. A., Vasil'ev, A. M., Volodina, Kozik, Struchkov and Suvorov, 1975). Care is therefore required when undertaking the synthesis of acetaldehyde arylhydrazones.

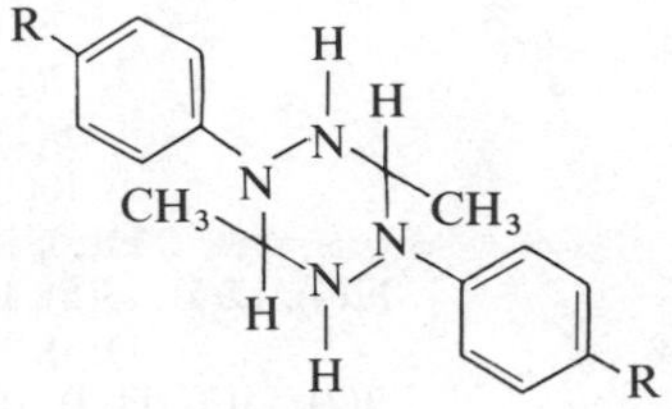

1168

It is unfortunate that attempts were not made to indolize acetaldehyde phenylhydrazone when flow-through indolization techniques were first investigated, using heated Zeolite catalysts to convert acetone and cyclohexanone phenylhydrazone into fairly good yields of 2-methylindole and 1,2,3,4-tetrahydrocarbazole, respectively (Venuto and Landis, 1968).

B. Synthesis of 2,3-Diunsubstituted Indoles via Indolization of Pyruvic Ester or Pyruvic Acid Arylhydrazones

In view of the failures already mentioned which for decades attended the attempted indolizations of acetaldehyde arylhydrazones, a widely applied route was developed for the synthesis of 2,3-diunsubstituted indoles from pyruvic ester (usually ethyl) or pyruvic acid arylhydrazones. Upon indolization these produced the corresponding indole-2-carboxylic esters **1169** (R^3 = COOalkyl) or indole-2-carboxylic acids **1169** (R^3 = COOH). Alkaline hydrolysis of the former, followed by acidification, afforded the latter which upon decarboxylation under a variety of conditions gave the desired **1169** (R^3 = H). Simple indoles which have been synthesized by this route are given in Table 22. Application of this sequence to the synthesis of indole itself (Table 22), starting from

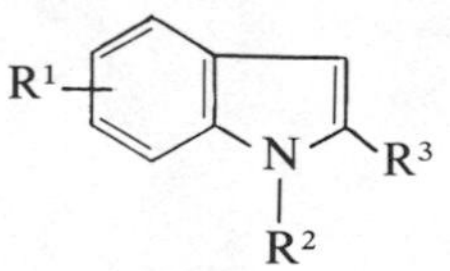

1169

Table 22. Simple 2,3-Diunsubstituted Indoles which have been prepared via the Fischer Indolization of Pyruvic Acid or Pyruvic Ester Phenylhydrazones or Substituted Phenylhydrazones

Substituent on the Indole Produced	Reference (Decarboxylation Catalyst)
—	1(1), 2(1)
1-CH_3	3(2), 4(1), 5(1), 6(1)
1-C_2H_5	4(1), 7(1)
1-$(CH_2)_2CH_3$	7(1)
1-$CH_2CH{=}CH_2$	8(1)
1-$CH(CH_3)_2$	7(1)
1-$CH_2CH(CH_3)_2$	7(1)
1-$(CH_2)_2CH(CH_3)_2$	7(1)
1-C_6H_5	9(1)
1-$CH_2C_6H_5$	10(1)
4-CH_3	11(3), 12(4)
5-CH_3	11(4), 12(5), 13(5), 14(5), 15(1), 16(1)
6-CH_3	11(3), 11(6)
7-CH_3	3(2), 11(3), 11(4), 16(1), 17(2), 18(5)
7-C_6H_5	19(7)
1,5-$(CH_3)_2$	20(1), 21(1)
1,7-$(CH_3)_2$	3(2), 20(1)
4,7-$(CH_3)_2$	3(2), 22(1), 23(6)
6,7-$(CH_3)_2$	3(2)
1-C_2H_5, 5-CH_3	20(1)
1-CH_3, 7-C_6H_5	24(1)
5-OCH_3	17(6), 25(1), 26(6), 27(8), 28(1), 29(6), 30(1)
6-OCH_3	31(5)
7-OCH_3	29(1), 30(6)
5-OC_2H_5	32(9), 33(1), 34(1)
5-$OCH_2C_6H_5$	35(1), 36(6)
5-OCH_3, 1-CH_3	37(1), 38(1)
6-OCH_3, 7-CH_3	39(10)
5-OC_2H_5, 1-CH_3	40(1)
5-$OCH_2C_6H_5$, 7-CH_3	17(6)
4,5,6-$(OCH_3)_3$	41(11)
4-F	42(1)
5-F	42(1), 43(5)
6-F	42(1)
7-F	42(1)
5-Cl	17(6), 44(6), 45(7)
7-Cl	44(6), 45(7)
5-Br	45(7), 46(6), 47(6)
7-Br	45(7), 48(6)
5-Cl, 7-CH_3	49(6)
7-Cl, 5-CH_3	50(6)
7-Cl, 5-$OCH_2C_6H_5$	51(6)
4,6-Cl_2	45(6)
4,7-Cl_2	45(6)
5,7-Cl_2	17(6), 45(6)

Table 22. (*continued*)

5-Br, 1-CH_3	52(12)
5-Br, 7-CH_3	49(6)
7-Br, 5-CH_3	49(6)
4-NO_2	53(13)
5-NO_2	53(13), 54(6)
6-NO_2	53(13)
7-NO_2	53(13), 55(14), 56(6), 57(15)
4-NO_2, 5-CH_3	58(13)
4-NO_2, 7-CH_3	59(13)
5-NO_2, 7-CH_3	59(13)
7-NO_2, 5-CH_3	59(13)
7-NO_2, 5-OCH_3	59(13), 60(6)
4-NO_2, 7-Cl	61(13)
5-$COCH_3$	62(4)
5-SO_2NH_2	63(16)

Decarboxylation Method

1. Pyrolysis [in the synthesis of indole itself by this method, the use of pure indole-2-carboxylic acid is particularly important since attempts to pyrolytically decarboxylate the impure acid provided only very low yields of indole (Elks, Elliott and Hems, 1944c)]; 2. Heat with copper-bronze powder in quinoline; 3. Heat with phosphorus pentoxide; 4. Heat with copper(I) chloride in quinoline; 5. Pyrolysis of the ammonium salt of the indole-2-carboxylic acid, a technique which appeared to be preferable to method 1. (Boyd, W. J. and Robson, 1935; Kermack and Slater, 1928 – see also Bergmann, E. D. and Pelchowicz, 1959; Harvey, D. G., 1955; Harvey, D. G. and Robson, 1938); 6. Heat with 'copper' chromite (Lazier and Arnold, H. R., 1943; Vogel, 1951) in quinoline [when using this method, the use of sulphate-free carboxylic acid reactants is advisable (Harvey, D. G., 1958, 1959; Rydon and Tweddle, 1955) and the use of synthetic quinoline has also been specified (Leggetter and Brown, R. K., 1960)]; 7. Heat *in vacuo*; 8. Heat with 'copper' acetate in quinoline; 9. Heat with bronze powder; 10. Heat with copper-bronze in 2-benzylpyridine; 11. No catalyst specified; 12. Heat with quinoline; 13. Heat with copper(II) oxide in quinoline [for other examples using this method, see Sladkov, Shner, Alekseeva, Turchin, Anisimova, Sheinker and Suvorov (1971); Sladkov, Shner, Anisimova, Alekseeva, Lisitsa, Terekhina and Suvorov (1974) and Sladkov, Shner, Anisimova and Suvorov (1972)]; 14. Heat in glycerol; 15. The carboxylic acid, completely free from inorganic salts, was heated with one-tenth of its weight of its 'copper' salt in refluxing dimethylacetamide under an atmosphere of nitrogen; 16. Reflux in a mixture of 'copper' chromite, quinoline and Dowtherm.

References

1. Fischer, Emil, 1886c, 1886e – see also Andrisano and Vitali, 1957; Ciamician and Zatti, 1889; Porcher and Hervieux, 1907; Shorygin and Polyakova, 1939, 1940; 2. Elks, Elliott and Hems, 1944c; 3. Marion and Oldfield, 1947; 4. Fischer, Emil and Hess, 1884; 5. Snyder and Eliel, 1948; 6. Kögl and Kostermans, 1935; 7. Michaelis, 1897; 8. Michaelis and Luxembourg, 1893; 9. Fischer, Emil, 1887; Fischer, Emil and Hess, 1884 – see also Shirley and Roussel, P. A., 1953; 10. Antrick, 1885; 11. Andrisano and Vitali, 1957;

(*continued*)

Table 22. *References continued*

12. Robson, 1924–1925 – see also Jackman and Archer, S., 1946; 13. Rydon, 1948; 14. Gordon and Jackson, R. W., 1935; 15. Raschen, 1887; 16. Kruber, 1926; 17. Heath-Brown and Philpott, P. G., 1965b; 18. Boyd, W. J. and Robson, 1935; 19. Kost, Sugorova and Yakubov, 1965; 20. Hegel, 1886; 21. Braun and Kruber, 1912; 22. Plancher and Caravaggi, 1905; 23. Carlin, Henley and Carlson, 1957; 24. Little, Taylor, W. I. and Thomas, B. R., 1954; 25. Amorosa, 1955; 26. Kralt, Asma, Haeck and Moed, 1961; 27. Julia and Manoury, 1965; 28. Blaikie and Perkin, 1924; 29. Pappalardo and Vitali, 1958b; 30. Sagitullin and Koronelli, 1964; 31. Kermack, Perkin and Robinson, R., 1922; 32. Murphy, H. W., 1964; 33. Rydon and Siddappa, 1951; 34. Deulofeu, 1938 (an analogous attempt to synthesize 5-methoxyindole was unsuccessful); 35. Boehme, 1953; 36. Ash and Wragg, W. R., 1958 – see also Stoll, Troxler, Peyer and Hofmann, A., 1955 [dimorphic forms of 5-benzyloxyindole are known (Burton and Leong, 1953)]; 37. Bell and Lindwall, 1948; 38. Cook, J. W., Loudon and McCloskey, 1951; 39. Troxler, Bormann and Seeman, 1968; 40. Kolosov and Preobrazhensky, 1953; 41. Zee and Ho, Y.-S. 1970; 42. Allen, F. L., Brunton and Suschitzky, 1955; 43. Quadbeck and Röhm, 1954; 44. Rydon and Tweddle, 1955 – see also Rydon and Long, C. A., 1949; 45. Pappalardo and Vitali, 1958c [the decarboxylations of 4- and 6-bromoindole-2-carboxylic acids into 4- and 6-bromoindoles, respectively, have been effected by heating the reactants in 2.5% acetic acid solution, with stirring, in an autoclave at 230 °C (Anderson, J. A. and Kohler, 1974)]; 46. Cavallini, Ravenna and Grasso, 1958; 47. Harvey, D. G., 1959; 48. Leggetter and Brown, R. K., 1960; 49. Ambekar and Siddappa, 1967; 50. Ambekar and Siddappa, 1964–1965; 51. Lee, F. G. H., Dickson, Suzuki, J., Zirnis and Manian, 1973; 52. Kunoni, 1962; 53. Parmerter, Cook, A. G. and Dixon, 1958; 54. Cavallini and Ravenna, 1958; 55. Hughes, G. K., Lions and Ritchie, 1939 – however, see Rydon and Siddappa, (1951); 56. Singer and Shive, 1957; 57. Casini and Goodman, 1964; 58. Gadaginamath, 1976; 59. Hiremath and Siddappa, 1964; 60. Julia and Nickel, 1966; 61. Ainsworth, D. P. and Suschitzky, 1967b; 62. Avramenko, Mosina and Suvorov, 1970; 63. De Bellis and Stein, M. L., 1961.

pyruvic acid phenylhydrazone, afforded good product yields when used on a small scale (*ca.* 10 g of hydrazone) but was difficult to perform when using large quantities of reactants (Elks, Elliott and Hems, 1944c). Indolization of purified ethyl pyruvate phenylhydrazone using a boron trifluoride etherate catalyst gave a 55% yield of ethyl indole-2-carboxylate whereas this product was obtained in only poor yield when using unpurified hydrazone or a formic acid catalyst (Bailey, Scott, P. W. and Vandrevala, 1980). Presumably, the 5-chloro-, 4,6-dichloro-, 5-ethoxy-, 5-fluoro-, 5-methoxy-, 5-methyl-, 4,7- and 5,7-dimethyl- and 5-methylthioindole used in a study of these compounds (Yoshitomi Pharmaceutical Industries Ltd, 1967) were synthesized by applications of this sequence, although preparative details were not given.

Other indoles, which have similarly been synthesized by sequential indolizations of the appropriate 2-ketoester or 2-ketoacid arylhydrazone, hydrolysis (where necessary) and decarboxylation, include 5-methoxy-3-methyl- (Blaikie and Perkin, 1924), 6-methoxy-3-methyl- (Wieland, T. and Grimm, 1965),

7-methoxy-3-methyl- (Blaikie and Perkin, 1924), 3-ethyl-5-methoxy- (Blaikie and Perkin, 1924; Goutarel, Janot, Le Hir, Corrodi and Prelog, 1954), 5-ethoxy-3-methyl- (Keimatsu and Sugasawa, 1928a; Kobayashi, T., 1939), 5-benzyloxy-3-ethyl- (Shaw, E., 1955), 5,6-dihydroxy-7-propyl- and 5,6-dihydroxy-3-methyl-7-propyl- (Beer, Brown, J. P. and Robertson, 1951) (in all these cases, decarboxylation was effected by pyrolysis), 3-(2-phenoxyethyl)- (decarboxylation was achieved by heating with finely divided copper) (Manske, R. H. F., 1931a), 5-methoxy-3-(2-phenoxyethyl)- (decarboxylation was effected by pyrolysis) (King, F. E. and Robinson, R., 1932), 3-phenyl- and 5-methyl-3-phenyl- (Borsche and Klein, 1941) [in both cases, decarboxylation occurred under the indolization conditions, (i.e. heating in ethanolic hydrogen chloride)], 1-methyl-3-phenyl- [decarboxylation was carried out by heating in concentrated sulphuric acid (cf. Chalmers, A. J. and Lions, 1933)] (Borsche and Klein, 1941), 3-(2-nitrophenyl)- (decarboxylation was best effected by pyrolysis of the ammonium salt of the acid) (Kermack and Slater, 1928), 1-methyl-3-(2-nitrophenyl)- (decarboxylation was effected by pyrolysis) (Kermack and Slater, 1928), 3-(4-nitrophenyl)- (decarboxylation was effected by pyrolysis in the presence of a trace of the 'copper' salt of the acid reactant) (Hino, Suzuki, T. and Nakagawa, 1973 – see also Wislicenus and Schultz, F., 1924), 3-(3,4-dimethoxyphenyl)-1-methyl- (decarboxylation was achieved by treatment with ethanolic hydrogen chloride) (Chalmers, A. J. and Lions, 1933), 3-(5-chloro-2-nitrophenyl)- (the decarboxylation method was not quoted in the patent abstract) (Freedman and Judd, 1972), 3-(2-pyridylmethyl)- (decarboxylation was performed by pyrolysis with copper powder in resorcinol) (Finkelstein and Lee, J., 1956) and 3-(2-naphthyl-)- (decarboxylation was achieved by pyrolysis) (Sempronj, 1938) indole. Several hydrazones **1170** ($R^1 = NO_2$, $R^2 = CH_3$ and OCH_3) have been indolized with a polyphosphoric acid catalyst to the corresponding **1171** ($R^3 = COOC_2H_5$) which upon hydrolysis followed by decarboxylation by

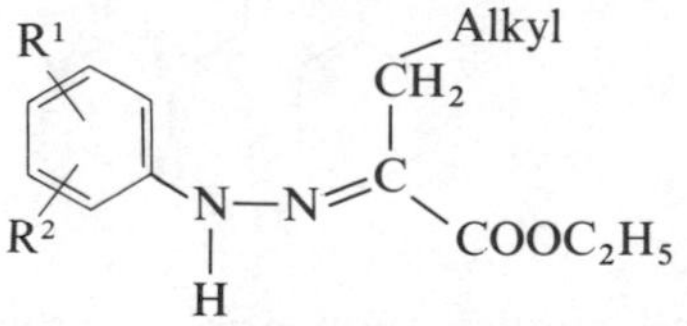

1170 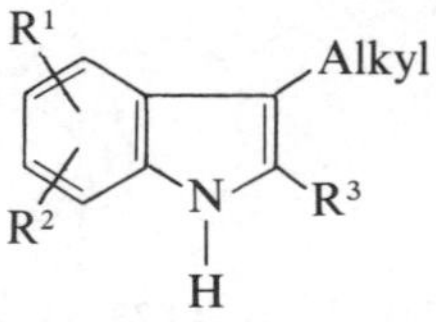 **1171**

heating in quinoline containing copper(II) oxide gave **1171** (R^3 = H). Unfortunately, neither the orientation of the benz substituents nor details of the alkyl group were given in the abstract (Weng, Wang, C.-T. and Wang, T., 1962). Further examples of this synthetic sequence leading to the formation of 2,3-diunsubstituted indoles include the conversion of pyruvic acid 1-naphthylhydrazone into 6,7-benzindole (**14**; R = H) (Shagalov, Eraksina, Tkachenko and Suvorov, 1973; Shagalov, Tkachenko, Eraksina and Suvorov, 1973), of ethyl pyruvate 1-naphthylhydrazone into 6,7-benzindole (**14**; R = H) (Rydon and Siddappa, 1951; Schlieper, 1887) [whereas the decarboxylations reported in these studies were effected by pyrolysis of the acid, it was subsequently (Hosmane, Hiremath and Schneller, 1973) achieved by heating a mixture of the acid with 'copper' chromite in quinoline], this sequence having earlier been established as far as the 6,7-benzindole-2-carboxylic acid by others (Hughes, G. K. and Lions, 1937–1938a). Likewise, pyruvic acid 2-naphthylhydrazone gave 4,5-benzindole (**544**; R = H) (Shagalov, Eraksina, Tkachenko and Suvorov, 1973; Shagalov, Tkachenko, Eraksina and Suvorov, 1973), as did ethyl pyruvate 2-naphthylhydrazone, decarboxylation being carried out by pyrolysis of the acid (Schlieper, 1886) or of the ammonium salt of the acid (Rydon and Siddappa, 1951), this sequence also having earlier been established as far as the 4,5-benzindole-2-carboxylic acid by others (Hughes, G. K. and Lions, 1937–1938a). It is interesting that when the indolization of ethyl pyruvate 2-naphthylhydrazone was effected by heating with zinc chloride, ester group hydrolysis with subsequent decarboxylation also occurred under the indolization conditions to yield 4,5-benzindole (**544**; R = H) (Schlieper, 1886). Since treatment of acetaldehyde 2-naphthylhydrazone under similar conditions afforded a small yield of 4,5-benzindole (**544**; R = H) (Schlieper, 1886), these reactions which are additional to indolization may, at least in part, have preceded the indolization stage. In a related sequence, **544** (R = CH_3) was prepared by thermal decarboxylation of the corresponding indole-2-carboxylic acid (Hosmane, Hiremath and Schneller, 1974) which had been prepared earlier (Hughes, G. K. and Lions, 1937–1938a) by the appropriate Fischer indolization followed by hydrolysis of the ethyl ester so formed. Indolization of **1172** in refluxing methanolic sulphuric acid afforded **1173** (R = $COOCH_3$) which was sequentially hydrolysed to **1173** (R = COOH) and thermally

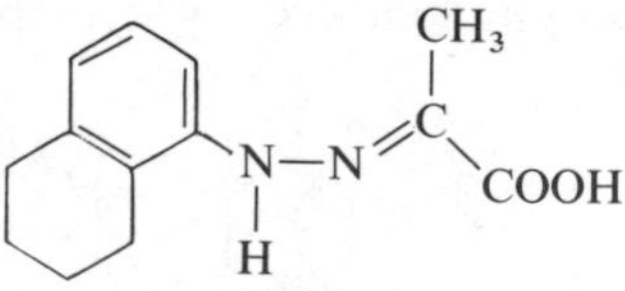

1172 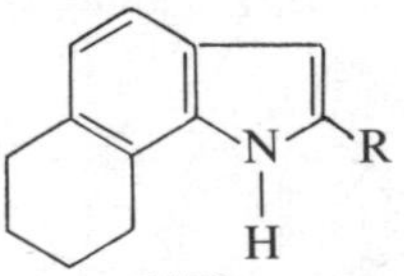 **1173**

decarboxylated to **1173** (R = H) and similar sequential treatment of the isomeric pyruvic acid 1,2,3,4-tetrahydro-6-naphthylhydrazone [**543**; $R^1 + R^2 = (CH_2)_4$, R^3 = COOH, R^4 = H] furnished both possible isomeric products **521** [$R^1 + R^2 = (CH_2)_4$, R^3–R^5 = H and $R^1 = R^4 = R^5$ = H, $R^2 + R^3 = (CH_2)_4$] (Shagalov, Ostapchuk, Zlobina, Eraksina, Babushkina, T. A., Vasil'ev, A. M., Ogoradnikova and Suvorov, 1978; Shagalov, Ostapchuk, Zlobina, Eraksina and Suvorov, 1977). The studies effected upon the indolization of pyruvic acid and ethyl pyruvate 6-quinolylhydrazone and related ethyl substituted hydrazones, with subsequent ester hydrolysis, where necessary, followed by decarboxylation, have already been discussed in Chapter III, E).

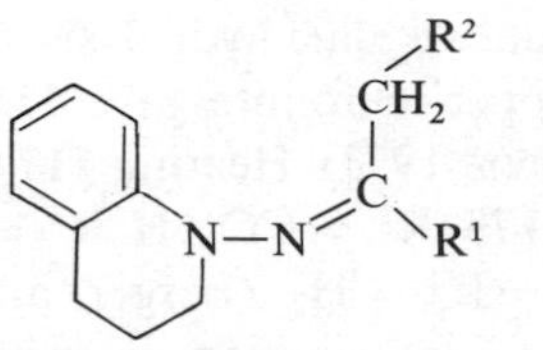

1174

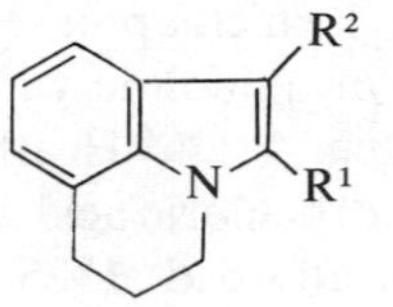

1175

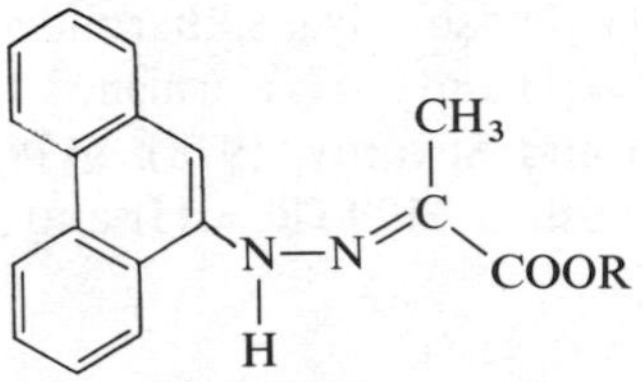

1176

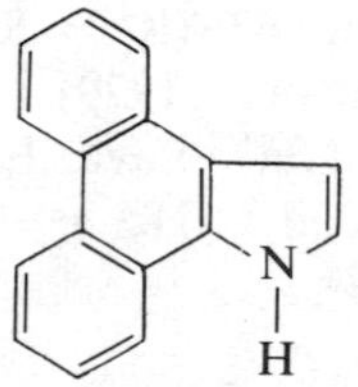

1177

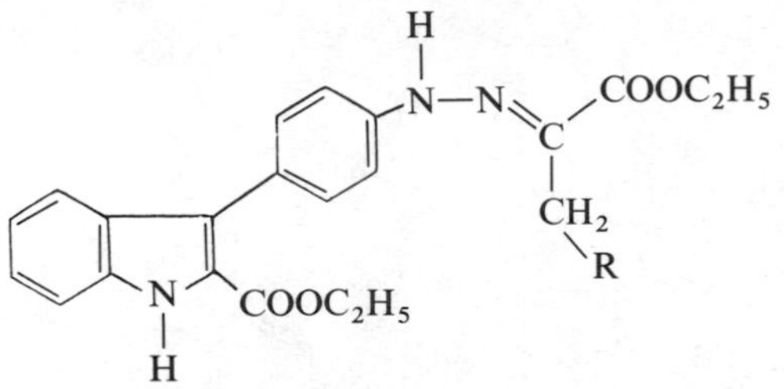

1178

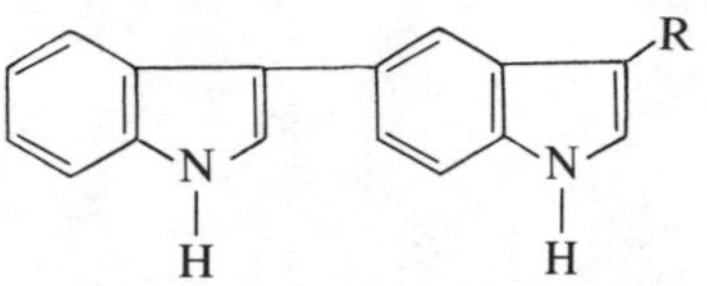

1179

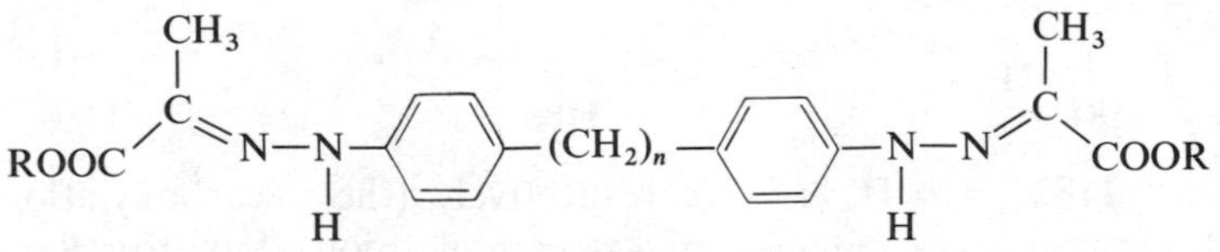

1180

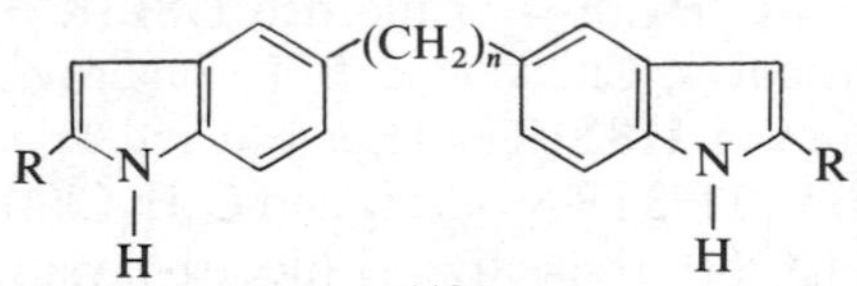

1181

The pyruvic acid or ethyl pyruvate 5- and 8-quinolylhydrazone have likewise been converted into **30** ($R^1 = R^2 =$ H) and **29** (R^1–R^3 = H) via the carboxylic acids **30** (R^1 = COOH, R^2 = H) and **29** (R^1 = COOH, $R^2 = R^3$ = H), respectively (Gryaznov, Akhvlediani, Volodina, Vasil'ev, A. M., Babushkina, T. A. and Suvorov, 1977; Suvorov, Sergeeva, Z. F. Gryaznov, Shabunova, Tret'yakova, Efimova, Volodina, Morozova, Akhvlediani, *et al.*, 1977). Indolization of **1139** ($R^1 = R^3$ = H, R^2 = COOH) yielded **1140** ($R^1 = R^3$ = H,

$R^2 = COOC_2H_5$) which upon sequential alkaline hydrolysis and decarboxylation by heating in quinoline with 'copper' chromite gave **1140** (R^1–R^3 = H) (Cattanach, Cohen, A. and Heath-Brown, 1973). Heating **1174** (R^1 = COOH, R^2 = H) with hydrochloric acid gave **1175** (R^1 = COOH, R^2 = H) which upon decarboxylation afforded **1175** ($R^1 = R^2$ = H) (Barger and Dyer, 1938). Following a similar reaction sequence of indolization, ester hydrolysis (where necessary) and decarboxylation, **628** [$R^1 = NHN{=}C(CH_3)COOC_2H_5$, R^2 = H] afforded **630** ($R^1 = R^2$ = H) (Khoshtariya, Sikharulidze, Tret'yakova, Efimova and Suvorov, 1979), **1176** (R = H and C_2H_5) furnished **1177** (Buyanov, Mirzametova, Tret'yakova, Efimova and Suvorov, 1978), **1178** (R = H) and **1180** (R = H and C_2H_5, $n = 0$) furnished **1179** (R = H) and **1181** (R = H,

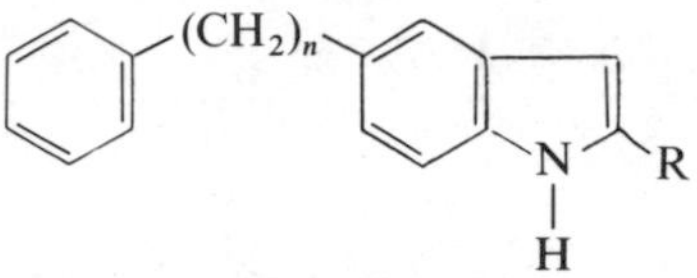

1182 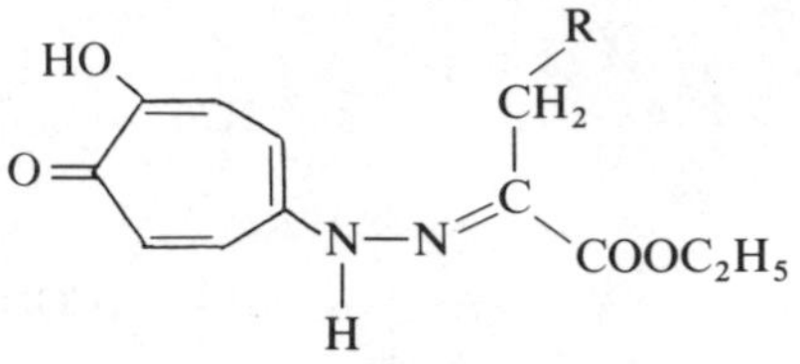 **1183**

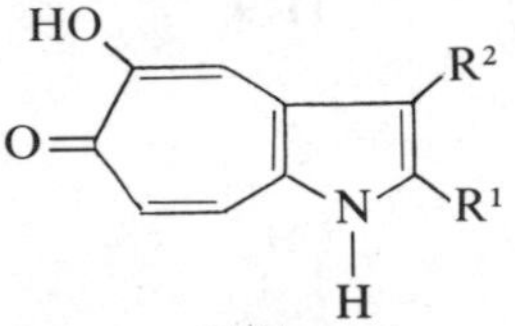

1184 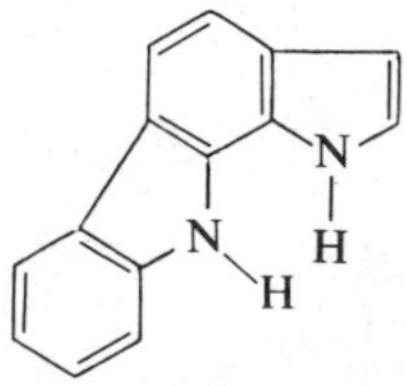 **1185** 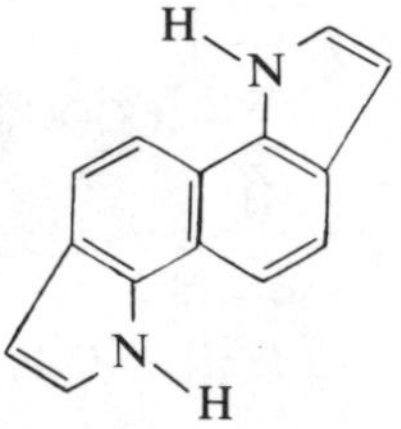 **1186**

$n = 0$) and **1182** (R = H, $n = 0$), respectively (the decarboxylations were achieved by pyrolysis of the acids in a stream of argon) (Suvorov, Samsoniya, Chilikin, Chikvaidze, I. S., Turchin, Efimova, Tret'yakova and Gverdtsiteli, I. M., 1978), **1180** (R = C_2H_5, $n = 1$) afforded **1181** (R = H, $n = 1$) and **1182** (R = H, $n = 1$) (Samsoniya, Chikvaidze, S. I., Suvorov and Gverdtsiteli, N., 1978) [these formations of **1182** (R = H, $n = 0$ and 1) are discussed in more detail in Chapter VII, J], **1183** [R = C_6H_5 and $C_6H_4Cl(4)$ gave **1184** [R^1 = H, $R^2 = C_6H_5$ and $C_6H_4Cl(4)$, respectively] (decarboxylations were effected by heating with copper powder in quinoline) (Sato, Y., 1963b), ethyl pyruvate 1-carbazolylhydrazone afforded **1185** (Sikharulidze, Khoshtariya, Kurkovskaya, Tret'yakova, Efimova and Suvorov, 1979), bispyruvic acid naphthyl-1,5-dihydrazone gave **1186** (Samsoniya, Trapaidze, Gverdtsiteli, I. M., and Suvorov, 1977; Samsoniya, Trapaidze, Suvorov and Gverdtsiteli, I. M., 1978) and bisethyl pyruvate *meta*-phenylenedihydrazone (**572**; $R^1 = COOC_2H_5$, R^2 = H) gave **574** (R^1–R^3 = H) (Samsoniya, Targamadze, Tret'yakova, Efimova, Turchin, Gverdtsiteli, I. M. and Suvorov, 1979). Analogously with the indolization of ethyl pyruvate 2-naphthylhydrazone (Schlieper, 1886) mentioned above, pyrolysis of mixtures of ethyl pyruvate 3-quinolylhydrazone and **613** [R = N=C(CH$_3$)COOC$_2$H$_5$, X = N] with zinc chloride gave **593** (R^1–R^3 = H) (Govindachari, Rajappa and Sudarsanam, 1961) and **615** (R = H, X = N) (Suvorov, Alyab'eva and Khoshtariya, 1978), respectively, ester hydrolysis with subsequent decarboxylation occurring under these indolization conditions.

The Reissert indole synthesis also yields indole-2-carboxylic acids, many of which have been analogously decarboxylated. Apart from many examples which use the decarboxylation conditions given in Table 22 in connection with the Fischer indole synthesis, in relation to the Reissert indole synthesis, 3-arylindole-2-carboxylic acids have been decarboxylated when treated with halogenic acid (Muramatsu, Ishizumi and Katsube, 1978), copper(I) bromide in boiling quinoline has been employed to decarboxylate 4- and 6-bromoindole-2-carboxylic acid (Plieninger, Suehiro, Suhr and Decker, 1955) as has a copper(I) iodide–copper powder mixture in boiling quinoline in the decarboxylation of 4-iodoindole-2-carboxylic acid (Hardegger and Corrodi, 1956). The use of copper(I) cyanide in boiling quinoline also caused this decarboxylation but, under these conditions, nucleophilic displacement of halide ion by cyanide ion also occurs. Thus, such treatment of 4- and 7-chloro- and 5-bromoindole-2-carboxylic acid afforded 4- (Uhle, 1949), 7- (Singer and Shive, 1955b) and 5- (Singer and Shive, 1955a) cyanoindole, respectively. Such a transformation was effected in two separate stages when 5-benzyloxy-4-chloroindole-2-carboxylic acid was heated with 'copper' chromite in quinoline to afford 5-benzyloxy-4-chloroindole which upon heating with copper(I) cyanide in quinoline gave 5-benzyloxy-4-cyanoindole (Robinson, P. and Slaytor, 1961). Indole-2-decarboxylations can also occur under indolization conditions, not only when using heated zinc chloride as the catalyst as mentioned above (Govindachari, Rajappa and Sudarsanam, 1961; Schlieper, 1886; Suvorov, Alyab'eva and Khoshtariya, 1978) but also when using refluxing ethanol saturated with dry hydrogen chloride to bring about indolization (Kermack and Slater, 1928).

Several failures in this approach to the synthesis of 2,3-diunsubstituted indoles are known. Thus, the indolization stage failed when using ethyl pyruvate 4-iodophenylhydrazone, although no experimental details were given (Harvey, D. G., 1958), pyruvic acid 4-methoxyphenylhydrazone could not be indolized (again no experimental conditions were given) (Deulofeu, 1938), although 5-methoxyindole has been prepared by many other investigators using this approach (see Table 22) as has 5-chloroindole (see Table 22) although attempts to indolize ethyl pyruvate 4-chlorophenylhydrazone using either acetic acid or ethanolic hydrogen chloride as catalyst were unsuccessful (Young, E. H. P., 1958). Treatment of **1187** with polyphosphoric acid afforded a paste which was insoluble in most organic solvents and from which **1188** could not be isolated (Khan, M. A. and Morley, 1978). The decarboxylation stage has also been found to fail with **1190** (R^1 = R^2 = H, R^3 = C$_2$H$_5$, n = 1) which was prepared by

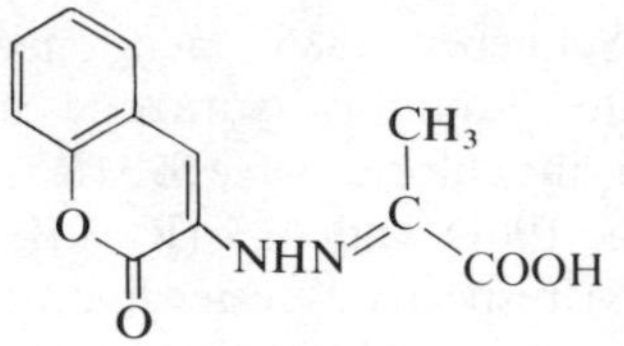

1187

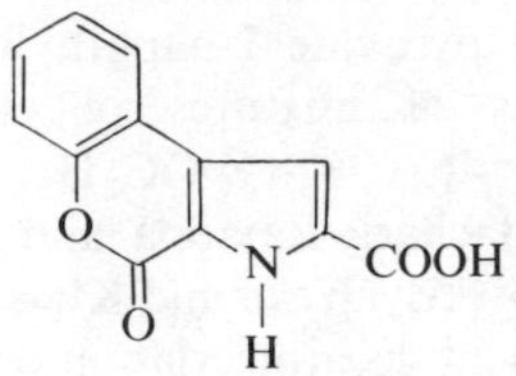

1188

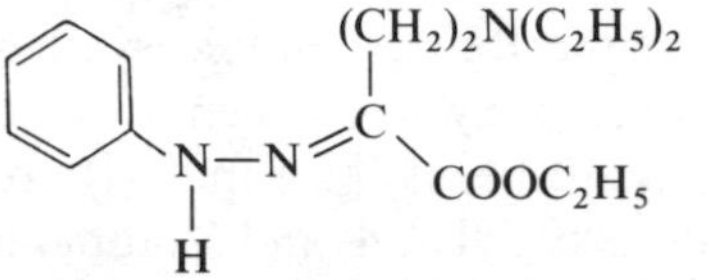

1189

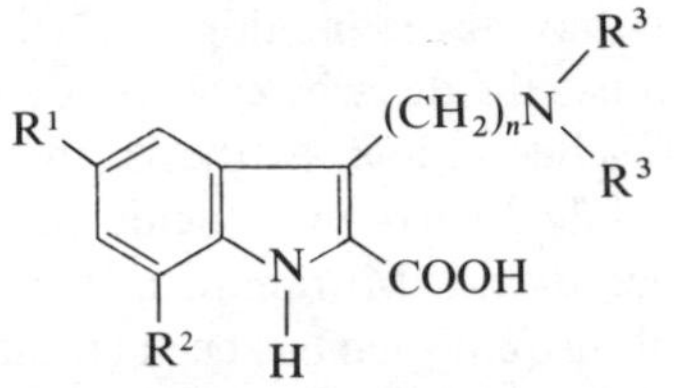

1190

sequential indolization and ester hydrolysis of **1189** (Hegedüs, 1946) (see also below). Furthermore, the decarboxylation stage is very dependent upon the reaction conditions used. Thus, pyrolysis, refluxing quinoline, alone or with copper(I) bromide or oxide or copper powder, or refluxing resorcinol all failed to decarboxylate 4- and 6-bromoindole-2-carboxylic acid (Barltrop and Taylor, D. A. H., 1954) whereas treatment in the molten state with copper(I) bromide followed by boiling quinoline furnished *ca.* 60% yields of the desired products (Plieninger, Suehiro, Suhr and Desker, 1955). Heating with glycerol was the method of choice in decarboxylations of 6-hydroxy-7-methoxyindole-2-carboxylic acid (Beer, Clarke, Davenport and Robertson, 1951) and 6-benzyloxyindole-2-carboxylic acid (Burton and Stoves, 1937), as it was in the decarboxylation of 5-, 6- and 7-hydroxyindole-2-carboxylic acid, when all other conditions and catalysts which were tried were unsuccessful. However, in the decarboxylations of these last three isomers, the yields varied considerably between the three isomers (Beer, Clarke, K., Khorana and Robertson, 1948). Heating with 'copper' chromite in refluxing quinoline, widely used decarboxylating conditions, has been found to have limitations and in such cases simple fusion of the acid in a stream of nitrogen was the preferred method of decarboxylation (Bretherick, Gaimster and Wragg, W. R., 1961). Whereas pyrolysis in glycerol or in quinoline solutions or of the sodium or potassium salts of some indole-2-carboxylic acids failed to effect their decarboxylation, such a reaction was achieved in high yield by pyrolysis of the ammonium salt (Kermack and Slater, 1928). Attempts to decarboxylate 4-trifluoromethylindole-2-carboxylic acid by pyrolysis of its ammonium or silver salts and by heating in quinoline, with or without copper powder, failed to afford any recognizable product. No attempt was made to decarboxylate the corresponding 6-isomer since it was only available in small quantities (Bornstein, Leone, Sullivan and Bennett, O. F., 1957). A useful comparative study has been made of the methods available for effecting the decarboxylation of both indole-2- and 3-carboxylic acids, the 'copper' salt of the reactant acid being introduced as a useful alternative catalyst in such reactions (Piers and Brown, R. K., 1962). Pyrolysis of the 'copper' salt of the acid in dimethylacetamide has also been used to achieve the decarboxylation of indole-2-carboxylic acids (Suvorov, Gordeev and Vasin, 1974).

A study (Abramovitch, 1956) using benz substituted tryptamine 2-carboxylic acids has concluded that the ease of decarboxylation decreased in the order $OCH_3 > CH_3 > H > Cl > NO_2$ of substituents on the benzenoid ring, the reverse of the order expected (Brown, B. R., 1951) for other decarboxylations, and that 1-methylation facilitated decarboxylation. As a result of these observations, it was suggested (Abramovitch, 1956) that indole-2-carboxylic acids **1169** (R^3 = COOH) may undergo decarboxylation via their 1-protonated form **1191** as shown. The formation of **1191** would be facilitated by 1-methylation and

1169 (R^3 = COOH) $\rightleftharpoons$ R^1 ... $N^{\oplus}$ ... R^2 ... H ... C=O ... O—H

1191

by electron releasing groups R^1 and the $N^{\oplus}$ would attract the electrons from the C—C bond as shown. This suggestion was later supported (Strandtmann, Puchalski and Shavel, 1964) by the observation that decarboxylation of 5-acetylgramine-2-carboxylic acid (**1190**; R^1 = $COCH_3$, R^2 = H, R^3 = CH_3, $n = 1$) under acidic conditions was very difficult, only proceeding in refluxing acetic acid–hydrochloric acid and then only in a 2.5% product yield. Likewise was the earlier observation that **1190** ($R^1 = R^2 = H$, $R^3 = C_2H_5$, $n = 1$) could not be decarboxylated (Hegedüs, 1946) (see above). It was suggested that the positive charge carried by the side chain nitrogen atom under the acidic reaction conditions retarded the necessary protonation of the N-1 atom which was supposed to be the rate determining step in the decarboxylation. Significantly, the decarboxylation of 5-acetyltryptamine-2-carboxylic acid (**1190**; R^1 = $COCH_3$, $R^2 = R^3$ = H, $n = 2$), in which the protonated side chain was now further removed from the N-1 atom, proceeded readily (Strandtmann, Puchalski and Shavel, 1964), although the introduction of a 7-chloro substituent, to give 5-acetyl-7-chlorotryptamine-2-carboxylic acid (**1190**; R^1 = $COCH_3$, R^2 = Cl, R^3 = H, $n = 2$), retarded the rate of decarboxylation (Strandtmann, Cohen, M. P. and Shavel, 1963, 1965). Significant, too, was the observation that attempts to decarboxylate 5-nitrotryptamine-2-carboxylic acid were unsuccessful whereas the 1-methyl homologue, under vigorous conditions, produced a small yield of 1-methyl-5-nitrotryptamine (Abramovitch, 1956).

Pyruvamide reacted with phenylhydrazine in dilute alcoholic solution to afford the phenylhydrazone **1192** (Abramovitch, 1956) of which both the *syn* and *anti* isomers (see Chapter VII, I) were later prepared by variation of the reaction conditions (Abramovitch and Spenser, 1957). However, no attempt was made in either case to effect the indolization of these hydrazones.

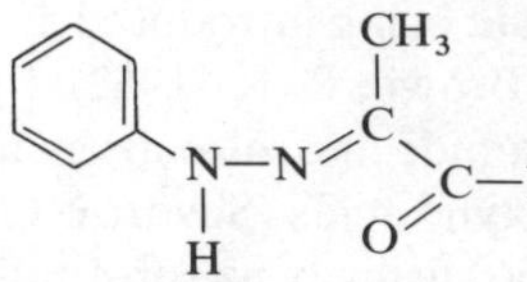

Indolization of the arylhydrazones of 2-oxoglutaric acid (or its mono- or diester) and its homologues (see Table 32 in this chapter) gives the corresponding 2-carboxy-3-indolylacetic acids and its homologues. These products could be bisdecarboxylated into the corresponding 3-alkylindoles. Thus, pyrolysis of **1193** ($R^1 = R^2 = R^4 = H$, $R^3 = OCH_3$, $n = 1$) at an unspecified temperature yielded **1194** ($R^1 = R^2 = R^4 = H$, $R^3 = OCH_3$, $n = 1$) (Kermack, Perkin and Robinson, R., 1921), of **1193** (R^1–$R^3 = H$, $R^4 = CH_3$, $n = 1$) at its m.p. *in vacuo* or in diphenylamine or quinoline solutions yielded **1194** (R^1–$R^3 = H$, $R^4 = CH_3$, $n = 1$) (King, F. E. and L'Ecuyer, 1934), of **1193** ($R^1 = R^3 = H$, $R^2 = OC_2H_5$, $R^4 = CH_3$, $n = 1$) at 250 °C gave **1194** ($R^1 = R^3 = H$, $R^2 = OC_2H_5$, $R^4 = CH_3$, $n = 1$) (Stedman, 1924) and of **1193** ($R^1 = CH_2COOH$, R^2–$R^4 = H$, $n = 1$ and $R^1 = R^2 = R^4 = H$, $R^3 = CH_2COOH$, $n = 1$) in resorcinol afforded **1194** ($R^1 = CH_3$, R^2–$R^4 = H$, $n = 1$ and $R^1 = R^2 = R^4 = H$, $R^3 = CH_3$, $n = 1$, respectively) (Plieninger, 1954). Similarly, pyrolytic decarboxylation of 1-methyl-3-indolylacetic acid, 2-methyl-3-indolylacetic acid, 1,2-dimethyl-3-indolylacetic acid and 4,5-benzo-3-indolylacetic acid afforded 1,3-dimethylindole (Snyder and Eliel, 1948), 2,3-dimethylindole (Fischer, Emil, 1886c, 1886e), 1,2,3-trimethylindole (Degen, 1886) and 4,5-benzo-2,3-dimethylindole (Steche, 1887), respectively, a route which was preferred (Fischer, Emil, 1886c) for the synthesis of 3-methylindoles via the Fischer indole synthesis, and 2-indolylacetic acids could be decarboxylated to yield the corresponding 2-methylindoles (Teotino, 1959). However, far more examples are known in which such dicarboxylic acids were specifically monodecarboxylated to lose the carboxylic group at the indolic 2-position. Thus, pyrolysis of **1195** (R^1–$R^5 = R^7 = R^8 = H$, $R^6 = COOH$, $n = 0$), at 225–230 °C (Feofilaktov, 1952; Kalb, Schweizer and Schimpf, 1926; Kalb, Schweizer, Zellner and Berthold, 1926; Unanyan and Tatevosyan, 1961) or with copper powder in quinoline (Feofilaktov and Semenova, N. K., 1952a; King, F. E. and L'Ecuyer, 1934), afforded **1195** (R^1–$R^8 = H$, $n = 0$) [in the earlier of these studies (King, F. E. and L'Ecuyer, 1934) the decarboxylation was arrested after the

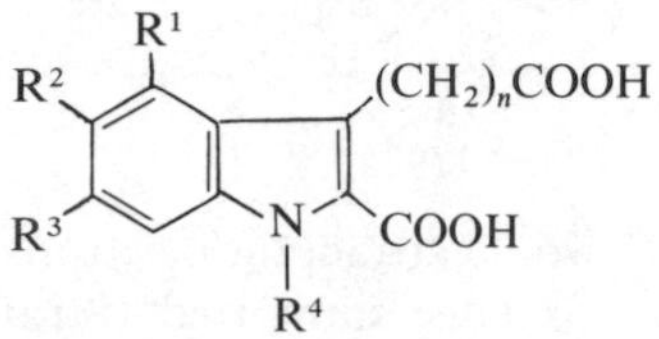

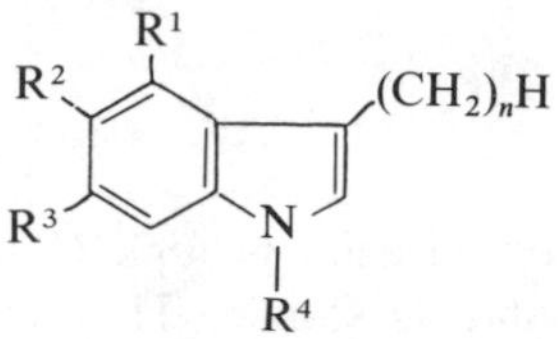

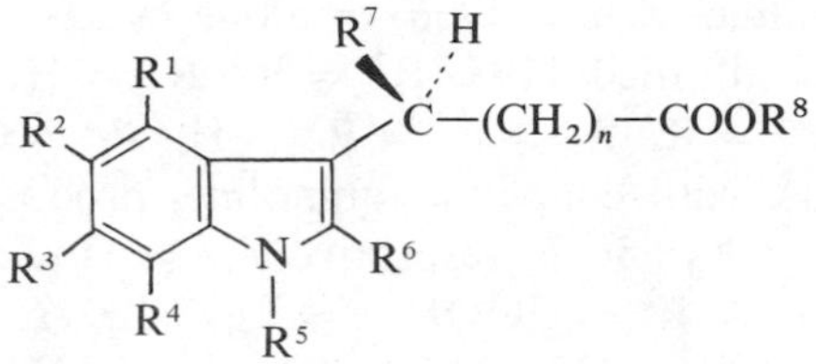

1195

evolution of one mole equivalent of carbon dioxide], of 7a-^{14}C-**1195** (R^1–R^5 = R^7 = R^8 = H, R^6 = COOH, $n = 0$), *in vacuo* followed by sublimation, afforded 7a-^{14}C-**1195** (R^1–R^8 = H, $n = 0$) (Robinson, J. R., 1957), of **1195** (R^1–R^4 = R^7 = R^8 = H, R^5 = CH_3, R^6 = COOH, $n = 0$) at its m.p. *in vacuo* or in quinoline with a trace of copper powder gave **1195** (R^1–R^4 = R^6 = R^8 = H, R^5 = CH_3, $n = 0$) (King, F. E. and L'Ecuyer, 1934) (again the decarboxylation was arrested after the evolution of one mole equivalent of carbon dioxide), of **1195** (R^1–R^4 = R^7 = R^8 = H, R^5 = CH_3, C_2H_5, $CH_2{=}CHCH_2$ and $CH_2C_6H_5$, R^6 = COOH, $n = 0$), in an acetic acid–water–concentrated hydrochloric acid mixture (13 : 5 : 5 v/v) at 112–115 °C, afforded high yields of **1195** (R^1–R^4 = R^6–R^8 = H, R^5 = CH_3, C_2H_5, $CH_2{=}CHCH_2$ and $CH_2C_6H_5$, respectively, $n = 0$) (Rosenmund, Meyer, G. and Hansel, 1975), of **1195** (R^1–R^5 = R^7 = R^8 = H, R^6 = COOH, $n = 1$) at an unspecified temperature (Manske, R. H. F. and Robinson, R., 1927; Yoshitomi Pharmaceutical Industries Ltd, 1967), at 225–230 °C (Feofilaktov, 1952) or at 220–230 °C (Keimatsu and Sugasawa, 1928b), yielded **1195** (R^1–R^8 = H, $n = 1$), of **1195** (R^1 = R^3–R^5 = R^7 = R^8 = H, R^2 = CH_3, R^6 = COOH, $n = 1$), with a small amount of copper powder at 235–240 °C, yielded **1195** (R^1 = R^3–R^8 = H, R^2 = CH_3, $n = 1$) (Manske, R. H. F. and Leitch, 1936), of **1195** (R^1 = R^3–R^5 = R^7 = R^8 = H, R^2 = OCH_3, R^6 = COOH, $n = 1$), at 230 °C (Rensen, 1959), at an unspecified temperature (Unanyan and Tatevosyan, 1961) or in diphenylamine solution (Barrett, H. S. B., Perkin and Robinson, R., 1929), gave **1195** (R^1 = R^3–R^8 = H, R^2 = OCH_3, $n = 1$), of **1195** (R^1 = R^2 = R^4 = R^5 = R^7 = R^8 = H, R^3 = OCH_3, R^6 = COOH, $n = 1$), at 220 °C, gave **1195** (R^1 = R^2 = R^4 = R^5–R^8 = H, R^3 = OCH_3, $n = 1$) (Barrett, H. S. B., Perkin and Robinson, R., 1929), of **1195** (R^1 = R^2 = R^4 = R^5 = R^8 = H, R^3 = OCH_3, R^6 = COOH, R^7 = CH_3, $n = 1$), with copper powder in quinoline solution, afforded **1195** (R^1 = R^2 = R^4–R^6 = R^8 = H, R^3 = OCH_3, R^7 = CH_3, $n = 1$) (Sandoz Patents Ltd, 1965 – see also Frey, A. J., Ott, Bruderer and Stadler, 1960), of **1195** (R^1–R^3 = R^5 = R^7 = R^8 = H, R^4 = OCH_3, R^6 = COOH, $n = 1$), with a small amount of copper powder at 235–240 °C, produced **1195** (R^1–R^3 = R^5–R^8 = H, R^4 = OCH_3, $n = 1$) (Manske, R. H. F., 1931a), of **1195** (R^1 = R^3–R^5 = R^7 = R^8 = H, R^2 = $OCH_2C_6H_5$, R^6 = COOH, $n = 1$), in refluxing tetralin (Justoni and Pessina, 1955), in paraffin oil or tetrahydro- or decahydronaphthalene [Drogas Vacunas y Sueros, S. A. (Drovyssa), 1957], with copper in quinoline (Zenitz, 1966) or alone in *ca.* 20 g portions (Justoni and Pessina, 1957b), afforded **1195** ($R^1 = R^3$–R^8 = H, $R^2 = OCH_2C_6H_5$, $n = 1$), of **1195** ($R^1 = R^3$–$R^5 = R^7 = R^8$ = H, $R^2 = CH_3$, CF_3 and F, R^6 = COOH, $n = 1$), with copper in quinoline, produced **1195** ($R^1 = R^3$–R^8 = H, $R^2 = CH_3$, CF_3 and F, respectively, $n = 1$) (Zenitz, 1966), of **1195** (R^1–$R^5 = R^7 = R^8$ = H, R^6 = COOH, $n = 2$), at 220 °C (Jackson, R. W. and Manske, R. H. F., 1930), at 220–240 °C (Pacheco, 1951), at an unspecified temperature (Feofilaktov, 1947b), *in vacuo* (Polaczkowa and Porowska, 1950), or with copper powder (Feofilaktov, and Semenova, N. K., 1952b), yielded **1195** (R^1–R^8 = H, $n = 2$) [during the purification of this product via the formation of its methyl ester, a trace of 1,2,3,4-tetrahydrocarbazol-1-one was also isolated (Jackson, R. W. and Manske, R. H., 1930)], of **1195** ($R^1 = R^2 = R^4 = R^5 = R^7 = R^8$ = H, $R^3 = OCH_3$, R^6 = COOH, $n = 2$), by application of gentle heating to portions of acid of *ca.* 1 g, afforded **1195** ($R^1 = R^2 = R^4$–R^8 = H, $R^3 = OCH_3$, $n = 2$) (Iyer, Jackson, A. H., Shannon and Naidoo, 1972, 1973), of **1195** ($R^1 = R^3$–$R^5 = R^7 = R^8$ = H, $R^2 = CF_3$, F and $OCH_2C_6H_5$, R^6 = COOH, $n = 2$), with copper in quinoline, yielded **1195** ($R^1 = R^3$–R^8 = H, $R^2 = CF_3$, F and $OCH_2C_6H_5$, respectively, $n = 2$) (Zenitz, 1966) and of **1195** (R^1–$R^5 = R^7 = R^8$ = H, R^6 = COOH, $n = 3$), at 210 °C, furnished **1195** (R^1–R^8 = H, $n = 3$), purified via its methyl ester (Manske, R. H. F. and Leitch, 1936). Similarly, **1196** (R = COOH, $n = 2$) afforded **1196** (R = H, $n = 2$) upon pyrolysis at 220–230 °C (Feofilaktov, 1947a) and **1197** (R = COOH) afforded

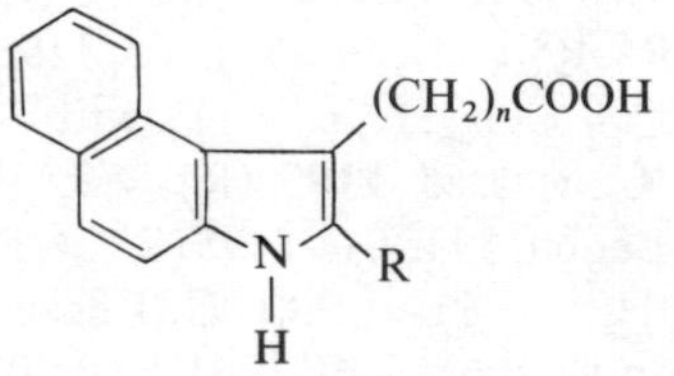

1196 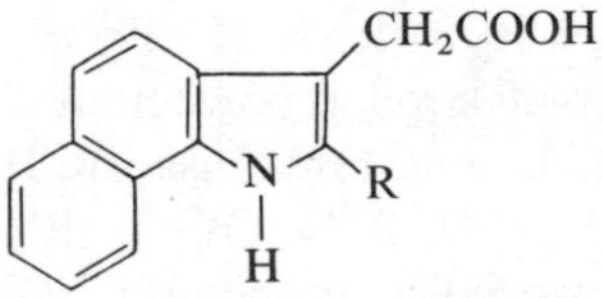 **1197**

1197 (R = H) upon heating with copper powder in quinoline (Feofilaktov and Semenova, N. K., 1953b) and selective decarboxylation also occurred when **516** [$R^1 = (CH_2)_2COOH$, R^2 = Cl, R^3 = COOH, R^4 = H] produced **516** [$R^1 = (CH_2)_2COOH$, R^2 = Cl, $R^3 = R^4$ = H] upon heating with copper powder in quinoline (Nagasaka, Yuge and Ohki, 1977). In these selective decarboxylations there was also a tendency for the other carboxyl group to be removed during the reaction, as described earlier. As a result, the yields of the required 3-indolylalkanoic acids were often low. However, selective decarboxylation of the dicarboxylic acids **1193** has been achieved by reacting them in the presence of metal ions in aqueous basic media at elevated temperatures and pressures. Thus, **1193** (R^1–R^4 = H, $n = 1$), when treated with sodium bicarbonate, triethylamine or sodium hydroxide afforded **1195** (R^1–R^8 = H, $n = 0$) in 86%, 76% and 93% yields, respectively, and **1193** (R^1–R^4 = H, $n = 2$

and 3) and indole-2-carboxylic acid, upon similar treatment with sodium hydroxide furnished **1195** (R^1–R^8 = H, n = 1 and 2) and indole, respectively, in 76% 86% and 70% yields, respectively (Bowman and Islip, 1971). An application of this method followed the observation that, although **1195** (R^1 = COOH, R^2–R^5 = R^7 = R^8 = H, R^6 = COOH, n = 1), obtained by sequential hydrolysis and reductive dechlorination of **1195** (R^1 = COOH, R^2 = R^3 = R^5 = R^7 = R^8 = H, R^4 = Cl, R^6 = $COOC_2H_5$, n = 1), could be converted into **1195** (R^1 = COOH, R^2–R^8 = H, n = 1) in boiling 4-methylquinoline, on a large scale this reaction was both unpleasant and capricious. However, **1195** (R^1 = COOH, R^2 = R^3 = R^5 = R^7 = R^8 = H, R^4 = Cl, R^6 = $COOC_2H_5$, n = 1) could be converted into **1195** (R^1 = COOH, R^2–R^8 = H, n = 1) in a 'one-pot' reaction involving sequential refluxing with aqueous sodium hydroxide, hydrogenolysis in the presence of 10% palladium-on-carbon under pressure and heating at 240 °C in an autoclave (Bowman, Goodburn and Reynolds, A. A., 1972).

Selective decarboxylation of the dicarboxylic acids **1193** has also been attained via monoesterification. Thus, **1195** (R^1–R^5 = R^8 = H, R^6 = COOH, n = 0) was converted into **1195** (R^1–R^5 = R^7 = H, R^6 = COOH, R^8 = C_2H_5, n = 0) which was decarboxylated upon heating with quinoline and copper(I) oxide in diethylene glycol, subsequent hydrolysis giving **1195** (R^1–R^8 = H, n = 0) (Robinson, J. R. and Good, 1957). This reaction sequence had earlier (Tanaka, Z., 1940a) been carried out via the corresponding monomethyl ester, formed by treatment of the dicarboxylic acid with methanolic hydrogen chloride, and decarboxylation involving pyrolysis with 'copper' chromite in quinoline. Similarly, treatment of **1195** (R^1–R^4 = R^7 = R^8 = H, R^5 = CH_3, R^6 = COOH, n = 0) with heated ethanolic hydrogen chloride produced **1195** (R^1–R^4 = R^7 = H, R^5 = CH_3, R^6 = COOH, R^8 = C_2H_5, n = 0) which upon sequential decarboxylative pyrolysis and hydrolysis furnished **1195** (R^1–R^4 = R^6–R^8 = H, R^5 = CH_3, n = 0) (King, F. E. and L'Ecuyer, 1934) and the 5-, 6- and 7-methoxy- and 4,5- and 6,7-benzo derivatives of **1195** (R^1–R^5 = R^7 = R^8 = H, R^6 = COOH, n = 0) were likewise converted into the corresponding substituted **1195** (R^1–R^8 = H, n = 0) via their methyl esters, although attempted application of this sequence to the 5,6-dimethoxy analogue failed at the decarboxylation stage (Findlay, S. P. and Dougherty, 1948). Monoesterification of **1198** (R^1 = H, R^2 = COOH, R^3 = H and CH_3) with ethanolic hydrogen chloride afforded **1198** (R^1 = C_2H_5, R^2 = COOH, R^3 = H and CH_3). Decarboxylative pyrolysis of these products with a 'copper–chromium oxide'

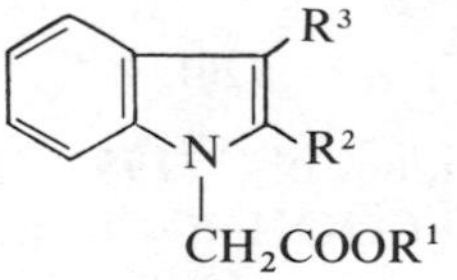

R^3 / N / R^2 / CH_2COOR^1

1198

catalyst (for the preparation of this catalyst, see Connor, Folkers and Adkins, 1932) was effected in a large volume of quinoline which minimized reactant decomposition. However, small amounts of aniline present in the quinoline resulted in simultaneous anilide formation to yield 1-indolylacetic acid anilide and 3-methyl-1-indolylacetic acid anilide, respectively, along with the corresponding ethyl esters. Hydrolysis of either esters or anilides afforded the corresponding 1-indolylacetic acids (Smith, W. S. and Moir, 1952).

Selective hydrolysis of **1199** ($R^1 = COOC_2H_5$, $R^2 = C_2H_5$) with ethanolic potassium hydroxide at room temperature yielded **1199** ($R^1 = COOH$, $R^2 = C_2H_5$) which upon pyrolysis at 180–190 °C, with subsequent hydrolysis, yielded **1199** ($R^1 = R^2 = H$) (King, F. E. and L'Ecuyer, 1934).

$COOR^2$ CH_2R^1 N CH_3

1199

Indolization of 2-oxoglutaric acid phenylhydrazone furnished a mixture of **1193** (R^1–R^4 = H, $n = 1$) and **1195** (R^1–R^8 = H, $n = 0$) which could be separated by fractional distillation (Fox and Bullock, 1951a) or by fractional crystallization of their di- (insoluble) and mono- (soluble) sodium salts, respectively (Fox and Bullock, 1951b). A simultaneous decarboxylation also occurred under the indolization conditions when **1200** was refluxed in ethanol saturated with hydrogen chloride to give both **1201** (R = COOH) and **1201** (R = H) (Millson and Robinson, R., 1955). However, under similar conditions, **122** and its derived 2-ketoester 2-hydroxy-4-nitrophenylhydrazone afforded **123** (Ried and Kleemann, 1968).

CH_3O $CH(CH_3)_2$ N—N=C COOH H

1200

CH_3O CH_3 CH_3 N R

1201

Attempted decarboxylation of **1195** ($R^1 = R^3$–$R^5 = R^7 = R^8 = H$, $R^2 = SO_2NH_2$, R^6 = H and COOH, $n = 0$) was unsuccessful (De Bellis and Stein, M. L., 1961).

C. Indolization of N_α-Acylarylhydrazones and N_α-Aroylarylhydrazones

The first of these indolizations was reported in 1921 when cyclohexanone N_α-acetylphenylhydrazone in boiling dilute sulphuric acid furnished 9-acetyl-1,2,3,4-tetrahydrocarbazole [**1202**; $R^1 = R^4 = H$, $R^5 = COCH_3$, $R^6 + R^7 = (CH_2)_4$], although, interestingly, the indolization only occurred with difficulty

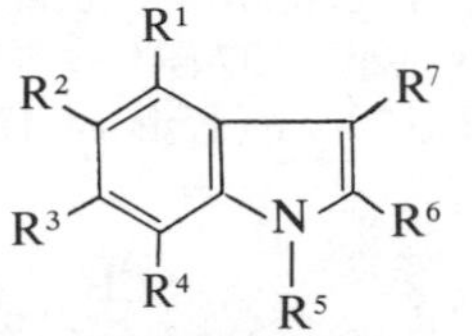

R1 R2 R7 R3 N R6 R4 R5

1202

in boiling acetic acid, conditions which readily converted cyclohexanone phenylhydrazone into 1,2,3,4-tetrahydrocarbazole (Perkin and Plant, 1921). However, forty years later, other investigators (Suvorov and Sorokina, N. P., 1961) failed to repeat the former indolization, obtaining only 1,2,3,4-tetrahydrocarbazole, a result which led to the suggestion that an acyl or aroyl substituent on the N_α-atom of the hydrazone inhibits the stage in the indolization mechanism involving the formation of the new C—C bond and therefore needs to be lost, thus freeing the N_α p-electron pair, prior to indolization. However, these latter experimental observations and related conclusions were invalidated (Yamamoto, H., 1967a) when cyclohexanone N_α-acetylphenylhydrazone was found to afford a mixture of 9-acetyl-1,2,3,4-tetrahydrocarbazole and 1,2,3,4-tetrahydrocarbazole after refluxing in 20% aqueous sulphuric acid. Furthermore, in refluxing 20% aqueous sulphuric acid, 9-acetyl-1,2,3,4-tetrahydrocarbazole was hardly deacetylated, no debenzoylation of 9-benzoyl-1,2,3,4-tetrahydrocarbazole could be detected, benzaldehyde N_α-acetylphenylhydrazone was deacetylated at a rate much slower than was the rate of formation of 1,2,3,4-tetrahydrocarbazole from cyclohexanone N_α-acetylphenylhydrazone and acetanilide was readily deacetylated to give aniline. These observations led to the conclusion that the major route by which deacetylation occurred during the above indolization involved the loss of the acetyl group after the formation of the new C—C bond but before the formation of the indole nucleus, probably by deacetylation of intermediate **1203** ($R = COCH_3$) into **1203** ($R = H$)

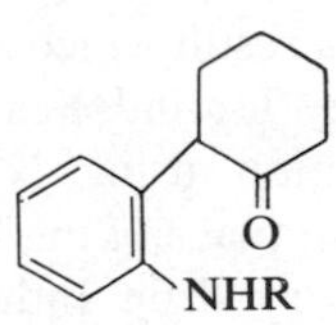

O NHR

1203

(Yamamoto, H. and Atsumi, 1968b). Contrary to the earlier observation (Perkin and Plant, 1921) referred to above, glacial acetic acid under reflux was also found to effect the indolization readily and produced only 1-acetyl-1,2,3,4-tetrahydrocarbazole from a mixture of N_α-acetylphenylhydrazine hydrochloride and cyclohexanone (Yamamoto, H., 1967a; Yamamoto, H. and Atsumi, 1968b).

When a mixture of N_α-acetylphenylhydrazine hydrochloride and 2-methylcyclohexanone was refluxed in glacial acetic acid, compound **1204** (R^1 = H, R^2 = CH_3) was formed along with **737** (R^1 = R^3 = H, R^2 = CH_3, n = 2), the latter product being formed by hydrolysis of **1204** (R^1 = H, R^2 = CH_3). No

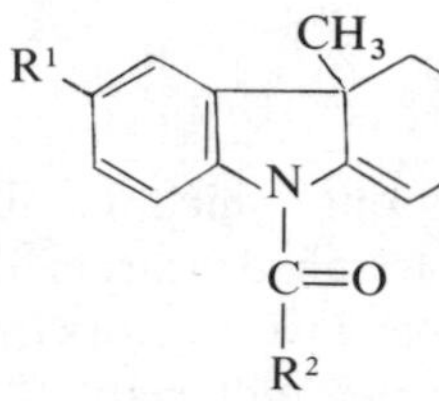

R^1 CH_3 N C=O R^2

1204

indolic product was formed resulting from indolization involving the C-6 atom of the 2-methylcyclohexanone moiety. Likewise, a mixture of N_α-benzoylphenylhydrazine hydrochloride and 2-methyl cyclohexanone afforded only **1204** (R^1 = H, R^2 = C_6H_5) which, unlike **1204** (R^1 = H, R^2 = CH_3), did not undergo hydrolysis under the indolization conditions (Yamamoto, H. and Atsumi, 1968a), and a mixture of N_α-(4-chlorobenzoyl)-4-methoxyphenylhydrazine hydrochloride and 2-methylcyclohexanone afforded **1204** [R^1 = OCH_3, R^2 = $C_6H_4Cl(4)$] (Atami, Izumi and Yamamoto, H., 1971; Yamamoto, H. and Atsumi, 1968a) and a product **293** (R = CH_3, X = O), suggested as being formed by arrest of the indolization mechanistic reaction sequence subsequent to new C—C bond formation involving the C-6 atom of the 2-methylcyclohexanone moiety (Yamamoto, H. and Atsumi, 1968a). Other similar intermediates have also been isolated (Yamamoto, H., 1967a) (see also Chapter II, E2).

Concomitantly with these above studies it was found that a mixture of laevulinic acid and N_α-acetyl-4-methoxyphenylhydrazine in acetic acid at 75–80 °C produced **1202** (R^1 = R^3 = R^4 = H, R^2 = OCH_3, R^5 = $COCH_3$, R^6 = CH_3, R^7 = CH_2COOH), a result which again showed that deacetylation of the N_α-atom was unnecessary for indolization to occur (Yamamoto, H., 1967b). Extension of this reaction using N_α-(4-chlorobenzoyl)-4-methoxyphenylhydrazine hydrochloride in place of the N_α-acetyl analogue afforded the well established anti-inflammatory drug Indomethacin [**1202**; R^1 = R^3 = R^4 = H, R^2 = OCH_3, R^5 = $COC_6H_4Cl(4)$, R^6 = CH_3, R^7 = CH_2COOH]

(Yamamoto, H., 1968). This drug had earlier been synthesized from 5-methoxy-2-methyl-3-indolylacetic acid [prepared by a Fischer indolization or later via 1-benzyl-5-methoxy-2-methyl-3-indolylacetonitrile which was readily available from a general 'one-pot' synthesis of indole-3-acetonitriles (Makisumi and Takada, 1976)]. This, via its anhydride, was converted into the *tert.*-butyl ester which was then sequentially 4-chlorobenzoylated at the nitrogen atom and pyrolytically de-*tert.*-butylated (Shen, T. Y., Windholz, Rosegay, Witzel, Wilson, A. N., Willett, Holtz, Ellis, Matzuk, Lucas, Stammer, Holly, Sarett, Risley, Nuss and Winter, C. A., 1963). However, the later of the above syntheses, subsequently repeated by many groups (see Table 23), involving a direct Fischer indolization was far more readily achieved than that in 1963 and gave almost quantitative product yields. This later synthesis has also been applied to the preparation of Indomethacin labelled with ^{14}C at the indolic C-2 position (Ellsworth, Gatto, Meriwether and Mertel, 1978).

Because of the therapeutic importance of Indomethacin and the readiness with which N_α-acylated or N_α-aroylated arylhydrazones could be indolized to the corresponding 1-acylated indoles, a large number of 1-acylated or 1-aroylated indoles have been synthesized by this route, usually using heated glacial acetic acid as the catalyst and mainly by the route's pioneer, H. Yamamoto, and his co-workers at the Sumitomo Chemical Co. in Japan. Often, however, a mixture of the N_α-acyl- or N_α-aroylarylhydrazine hydrochloride and ketonic moiety or the N_α-acyl- or N_α-aroylarylhydrazine and ketoacid moiety are merely heated in ethanolic solution to effect indolization and, in other cases, catalysts such as hydrogen chloride (Bikova, Vitev, Dyankova and Ilarionov, 1972; Sletzinger, Gal and Chemerda, 1968; Yamamoto, H., Nakamura, Y., Atami, Nakao, Kobayashi, T., Saito, C. and Awata, 1970; Yamamoto, H. and Nakao, 1969a, 1969d), phosphoric acid (Firestone and Sletzinger, 1970; Fisnerova and Nemecek, 1976; Sletzinger and Gal, 1968b), polyphosphoric acid (Ishizumi, Mori, Inaba and Yamamoto, H., 1973; Sletzinger and Gal, 1968a) or zinc chloride (Ishizumi, Mori, Inaba and Yamamoto, H., 1973; Yamamoto, H., Inaba, Izumi, Okamoto, Hirohashi, T., Ishizumi, Maruyama, Kobayashi, T., Yamamoto, M. and Mori, 1971) have been employed. The results of the Japanese studies and the related studies of others are given in Table 23. Apart from these examples, many others are mentioned in patents but the abstracts of these are not comprehensive (Alexander and Mooradian, 1972, 1975; Takayama, Nakao, Kimura, Inaba and Yamamoto, H., 1974; Yamamoto, H., Nakamura, Y. and Kobayashi, T., 1970) and in other patent abstracts not all examples are specified (Alexander and Mooradian, 1976; Boltze, Opitz, Raddatz, Seidel, Jacobi, Dell and Schoellnhammer, 1979; Kimura, Inaba and Yamamoto, H., 1973b; Noda, Nakagawa, Miyata, Nakajima and Ide, 1979; Pakula, Wojciechowski, Poslinska, Pichnej, Ptaszynski, Przepalkowski and Logwinienko, 1971; Poslinska, Pakula, Wojciechowski and Pichnej, 1972; Sumitomo Chemical Co. Ltd, 1969a; Takayama, Nakao, Kimura, Inaba and

Table 23. 1-Acylindoles which have been Synthesized Directly by Indolization of the Corresponding N_α-Acyl- and N_α-Aroylarylhydrazones **1202** (Unless otherwise stated, R^1, R^3 and R^4 = H)

R^5	R^6	R^7	R^2	References
Acetyl	CH^3	H	OCH_3	1
Acetyl	CH^3	CH_2COOH	OCH_3	2–5
Acetyl	CH^3	$(CH_2)_2COOH$	OCH_3	4
Acetyl	$C_6H_4OCH_3(4)$	CH_2COOH	OCH_3	6
Acetyl	$C_6H_4OCH_3(4)$	$C_6H_4OCH_3(4)$	H	7
Acetyl	$(CH_2)_4$		H	8–10
Chloroacetyl	CH_3	CH_2COOH	OCH_3	3, 4
Chloroacetyl	CH_3	$(CH_2)_2COOH$	OCH_3	4
N,N-Dimethylaminoacetyl	CH_3	C_2H_5	OCH_3	11
N,N-Dimethylaminoacetyl	C_6H_5	CH_3	OCH_3	11
3-Chloropropionyl	$C_6H_4OH(4)$	CH_3	Cl	11
4-Chlorobutyryl	CH_3	CH_2COOH	OCH_3	3, 4, 12–14
4-Chlorobutyryl	CH_3	$(CH_2)_2COOH$	OCH_3	4
But-2-enoyl	CH_3	CH_2COOH	OCH_3	3, 4
But-2-enoyl	CH_3	$(CH_2)_2COOH$	OCH_3	4
4-Methylpentanoyl	CH_3	CH_2COOH	OCH_3	3, 4
4-Methylpentanoyl	CH_3	$(CH_2)_2COOH$	OCH_3	4
Hexanoyl	CH_3	CH_2COOH	OCH_3	3, 4, 13, 14
Hexanoyl	CH_3	CH_2COOH	a	15
Hexanoyl	CH_3	$(CH_2)_2COOH$	OCH_3	4
2,4-Hexadienoyl	H	CH_3	OCH_3	16
2,4-Hexadienoyl	CH_3	CH_2COOH	OCH_3	3, 4, 12–14, 17
2,4-Hexadienoyl	CH_3	$(CH_2)_2COOH$	OCH_3	3, 4
2,4-Hexadienoyl	CH_3	$(CH_2)_3COOH$	OCH_3	3
2,4-Hexadienoyl	COOH	CH_2COOH	OCH_3	18
Octanoyl	CH_3	CH_2COOH	OCH_3	4
Octanoyl	CH_3	$(CH_2)_2COOH$	OCH_3	4
Decanoyl	CH_3	CH_2COOH	OCH_3	4, 14

Decanoyl	CH_3	$(CH_2)_2COOH$	OCH_3	4
Linoleoyl	CH_3	CH_2COOH	OCH_3	19
Cyclopropylcarbonyl	CH_3	CH_2COOH	OCH_3	20, 21
Cyclobutylcarbonyl	CH_3	CH_2COOH	OCH_3	20, 21
Cyclopentylcarbonyl	CH_3	CH_2COOH	OCH_3	20, 21
Cyclohexylcarbonyl	CH_3	CH_2COOH	OCH_3	20, 21
Cyclohexylcarbonyl	CH_3	$(CH_2)_3COOH$	OCH_3	20
1-Cyclohexenylcarbonyl	CH_3	CH_2COOH	OCH_3	20
Tetrahydropyranylcarbonyl	CH_3	CH_2COOH	OCH_3	21
Phenoxyacetyl	CH_3	CH_2COOH	OCH_3	22–25
Phenoxyacetyl	CH_3	$(CH_2)_2OH$	OCH_3	26
Phenoxyacetyl	CH_3	$(CH_2)_3COOH$	OCH_3	23, 24
2,3-Dimethylphenoxyacetyl	CH_3	CH_2COOH	OCH_3	27
2,4-Dimethylphenoxyacetyl	CH_3	CH_2COOH	OCH_3	27
2,6-Dimethylphenoxyacetyl	CH_3	CH_2COOH	OCH_3	27
4-*tert*.-Butylphenoxyacetyl	CH_3	CH_2COOH	OCH_3	23–25
4-*tert*.-Butylphenoxyacetyl	CH_3	$(CH_2)_2OH$	OCH_3	26
2-Methoxyphenoxyacetyl	CH_3	CH_2COOH	OCH_3	27
4-Methoxyphenoxyacetyl	CH_3	CH_2COOH	OCH_3	27
4-Chlorophenoxyacetyl	CH_3	CH_2COOH	OCH_3	22–25
4-Chloro-3-methylphenoxyacetyl	CH_3	CH_2COOH	OCH_3	27
2,6-Dichlorophenoxyacetyl	CH_3	CH_2COOH	OCH_3	27
2,4,6-Trichlorophenoxyacetyl	CH_3	CH_2COOH	OCH_3	27
4-Bromophenoxyacetyl	CH_3	CH_2COOH	OCH_3	22
2-Nitrophenoxyacetyl	CH_3	CH_2COOH	OCH_3	27
4-Nitrophenoxyacetyl	CH_3	CH_2COOH	OCH_3	27
1-Naphthyloxyacetyl	CH_3	CH_2COOH	OCH_3	27
2-Naphthyloxyacetyl	CH_3	CH_2COOH	OCH_3	23, 28
2-Naphthyloxyacetyl	CH_3	$(CH_2)_2OH$	OCH_3	26
2-(4-Chlorophenoxy)-2-methyl-propionyl	CH_3	CH_2COOH	OCH_3	23, 24

(*continued*)

Table 23 (*continued*)

R^5	R^6	R^7	R^2	References
3-Phenylthiopropionyl	CH_3	CH_2COOH	OCH_3	23–25
3-Phenylthiopropionyl	CH_3	$(CH_2)_2OH$	OCH_3	26
Benzoyl	H	CH_2COOH	H, CH_3, OCH_3, OC_2H_5, SCH_3, F, Cl, NO_2	29(b)
Benzoyl	H	$(CH_2)_2COOH$	H, CH_3, OCH_3, OC_2H_5, SCH_3, F, Cl, NO_2	29
Benzoyl	H	$CH(CH_3)COOH$	H, CH_3, OCH_3, OC_2H_5, SCH_3, F, Cl, NO_2	29
Benzoyl	H	$(CH_2)_3COOH$	H, CH_3, OCH_3, OC_2H_5, SCH_3, F, Cl, NO_2	29
Benzoyl	CH_3	H	OCH_3	1
Benzoyl	CH_3	CH_2COOH	H	29–34
Benzoyl	CH_3	CH_2COOH	CH_3	29, 31–33, 35
Benzoyl	CH_3	CH_2COOH	OCH_3	29, 32–38
Benzoyl	CH_3	CH_2COOH	OC_2H_5	29, 32, 33
Benzoyl	CH_3	CH_2COOH	$OCH_2C_6H_5$	37
Benzoyl	CH_3	CH_2COOH	SCH_3	29, 33
Benzoyl	CH_3	CH_2COOH	$N(CH_3)_2$	35
Benzoyl	CH_3	CH_2COOH	F	29, 32, 33
Benzoyl	CH_3	CH_2COOH	Cl	29, 31, 33, 38
Benzoyl	CH_3	CH_2COOH	NO_2	29, 33
Benzoyl	CH_3	CH_2COOH	H, CH_3, OCH_3, OC_2H_5, F(c)	29
Benzoyl	CH_3	CH_2COOCH_3	CH_3, OCH_3, $N(CH_3)_2$	35
Benzoyl	CH_3	$CH_2COOC_2H_5$	CH_3	35
Benzoyl	CH_3	$CH_2COOC_2H_5$	OCH_3	35, 38
Benzoyl	CH_3	$CH_2COOC_2H_5$	$N(CH_3)_2$	35
Benzoyl	CH_3	$CH_2COOC(CH_3)_3$	CH_3, OCH_3, $N(CH_3)_2$	35
Benzoyl	CH_3	$CH_2COOCH_2C_6H_5$	OCH_3	31
Benzoyl	CH_3	$(CH_2)_2COOH$	H, CH_3, OCH_3, OC_2H_5, F(b)	29, 32

Benzoyl	CH_3	$(CH_2)_2COOH$	SCH_3, Cl, NO_2	29
Benzoyl	CH_3	$CH(CH_3)COOH$	H, CH_3, OCH_3, OC_2H_5, SCH_3, F, Cl, NO_2	29
Benzoyl	CH_3	$(CH_2)_3COOH$	H	29, 32, 34
Benzoyl	CH_3	$(CH_2)_3COOH$	CH_3, OC_2H_5, F	29, 32
Benzoyl	CH_3	$(CH_2)_3COOH$	OCH_3	29, 32, 34
Benzoyl	CH_3	$(CH_2)_3COOH$	SCH_3, Cl, NO_2	29
Benzoyl	CH_3	$(CH_2)_3COOH$	H, CH_3, OCH_3, OC_2H_5, F(c)	32
Benzoyl	CH_3	$COOC_2H_5$	H	31
Benzoyl	$C_6H_4OCH_3(4)$	$COOC_2H_5$	OCH_3	6, 39(d)
Benzoyl	$C_6H_4Cl(4)$	$COOC_2H_5$	OCH_3	6, 39(d)
Benzoyl	$(CH_2)_4$		H	8–10, 40
Benzoyl	$(CH_2)_2CH(CH_2OH)CH_2$		H	41
Benzoyl	$(CH_2)_2C(COOH)_2CH_2$		H	42, 43
Benzoyl	$(CH_2)_2CH(COOH)CH_2$		H	43
Benzoyl	$CH_2SCH(COOH)$		OCH_3	44
4-Methylbenzoyl	H	$COOC_2H_5$	H, CH_3, OCH_3, OC_2H_5, SCH_3, F, Cl, NO_2	29
4-Methylbenzoyl	H	$(CH_2)_2COOH$	H, CH_3, OCH_3, OC_2H_5, SCH_3, F, Cl, NO_2	29
4-Methylbenzoyl	H	$CH(CH_3)COOH$	H, CH_3, OCH_3, OC_2H_5, SCH_3, F, Cl, NO_2	29
4-Methylbenzoyl	H	$(CH_2)_3COOH$	H, CH_3, OCH_3, OC_2H_5, SCH_3, F, Cl, NO_2	29
4-Methylbenzoyl	CH_3	CH_2COOH	H	29, 31–33, 45
4-Methylbenzoyl	CH_3	CH_2COOH	CH_3	29, 32, 33, 45
4-Methylbenzoyl	CH_3	CH_2COOH	OCH_3	29–33, 38, 46
4-Methylbenzoyl	CH_3	CH_2COOH	OC_2H_5, F	29, 32, 33
4-Methylbenzoyl	CH_3	CH_2COOH	SCH_3, Cl, NO_2	29, 33
4-Methylbenzoyl	CH_3	CH_2COOH	$N(CH_3)_2$	45
4-Methylbenzoyl	CH_3	CH_2COOH	H, CH_3, OCH_3, OC_2H_5, F(c)	32

(continued)

Table 23 (*continued*)

R^5	R^6	R^7	R^2	References
4-Methylbenzoyl	CH_3	CH_2COOH	a	47
4-Methylbenzoyl	CH_3	$CH_2COOC(CH_3)_3$	H, CH_3, $N(CH_3)_2$	45
4-Methylbenzoyl	CH_3	$(CH_2)_2COOH$	H, CH_3, OCH_3, OC_2H_5, F	29, 32
4-Methylbenzoyl	CH_3	$(CH_2)_2COOH$	SCH_3, Cl, NO_2	29
4-Methylbenzoyl	CH_3	$(CH_2)_2COOH$	H, CH_3, OCH_3, OC_2H_5, F(c)	32
4-Methylbenzoyl	CH_3	$CH(CH_3)COOH$	H, CH_3, OCH_3, OC_2H_5, SCH_3, F, Cl, NO_2	29
4-Methylbenzoyl	CH_3	$(CH_2)_3COOH$	H, CH_3, OCH_3, OC_2H_5, SCH_3, F, Cl, NO_2	29, 32
4-Methylbenzoyl	CH_3	$(CH_2)_3COOH$	H, CH_3, OCH_3, OC_2H_5, F(c)	32
4-Methylbenzoyl	CH_3	$(CH_2)_2NHC_6H_5$	OCH_3	48
4-Methylbenzoyl	C_6H_5	CH_2COOH	CH_3	6
4-Methylbenzoyl (d)	C_6H_5	CH_2COOH	OCH_3	39
4-Methoxybenzoyl	H	CH_2COOH	H, CH_3, OCH_3, OC_2H_5, SCH_3, F, Cl, NO_2	29
4-Methoxybenzoyl	H	$(CH_2)_2COOH$	H, CH_3, OCH_3, OC_2H_5, SCH_3, F, Cl, NO_2	29
4-Methoxybenzoyl	H	$CH(CH_3)COOH$	H, CH_3, OCH_3, OC_2H_5, SCH_3, F, Cl, NO_2	29, 32
4-Methoxybenzoyl	H	$(CH_2)_3COOH$	H, CH_3, OCH_3, OC_2H_5, SCH_3, F, Cl, NO_2	29
4-Methoxybenzoyl	CH_3	CH_2COOH	H	29, 32, 33, 45
4-Methoxybenzoyl	CH_3	CH_2COOH	CH_3	25, 29, 33, 35, 45
4-Methoxybenzoyl	CH_3	CH_2COOH	OCH_3	29–33, 35
4-Methoxybenzoyl	CH_3	CH_2COOH	OC_2H_5, F	29, 32, 33
4-Methoxybenzoyl	CH_3	CH_2COOH	SCH_3, Cl, NO_2	29, 33

4-Methoxybenzoyl	CH_3	CH_2COOH	$N(CH_3)_2$	35, 45
4-Methoxybenzoyl	CH_3	CH_2COOH	H, CH_3, OCH_3, OC_2H_5, F(c)	32
4-Methoxybenzoyl	CH_3	CH_2COOCH_3	CH_3, OCH_3, $N(CH_3)_2$	35
4-Methoxybenzoyl	CH_3	$CH_2COOC_2H_5$	CH_3, OCH_3, $N(CH_3)_2$	35
4-Methoxybenzoyl	CH_3	$CH_2COOC(CH_3)_3$	CH_3, OCH_3, $N(CH_3)_2$	35, 45
4-Methoxybenzoyl	CH_3	$(CH_2)_2COOH$	H, CH_3, OCH_3,OC_2H_5, F	29, 32
4-Methoxybenzoyl	CH_3	$(CH_2)_2COOH$	SCH_3, Cl, NO_2	29
4-Methoxybenzoyl	CH_3	$(CH_2)_2COOH$	H, CH_3, OCH_3, OC_2H_5, F(c)	32
4-Methoxybenzoyl	CH_3	$CH(CH_3)COOH$	H, CH_3, OCH_3, OC_2H_5, SCH_3, F, Cl, NO_2	29
4-Methoxybenzoyl	CH_3	$(CH_2)_3COOH$	H, CH_3, OCH_3, OC_2H_5, F	29, 32
4-Methoxybenzoyl	CH_3	$(CH_2)_3COOH$	SCH_3, Cl, NO_2	29
4-Methoxybenzoyl	CH_3	$(CH_2)_3COOH$	H, CH_3, OCH_3, OC_2H_5, F(c)	32
4-Methylthiobenzoyl	H	CH_2COOH	H, CH_3, OCH_3, OC_2H_5, SCH_3, F, Cl, NO_2	29
4-Methylthiobenzoyl	H	$(CH_2)_2COOH$	H, CH_3, OCH_3, OC_2H_5, SCH_3, F, Cl, NO_2	29
4-Methylthiobenzoyl	H	$CH(CH_3)COOH$	H, CH_3, OCH_3, OC_2H_5, SCH_3, F, Cl, NO_2	29
4-Methylthiobenzoyl	H	$(CH_2)_3COOH$	H, CH_3, OCH_3, OC_2H_5, SCH_3, F, Cl, NO_2	29
4-Methylthiobenzoyl	CH_3	CH_2COOH	H, CH_3, OCH_3, OC_2H_5, F	29, 32, 33
4-Methylthiobenzoyl	CH_3	CH_2COOH	SCH_3, Cl, NO_2	29, 33
4-Methylthiobenzoyl	CH_3	CH_2COOH	H, CH_3, OCH_3, OC_2H_5, F(c)	32
4-Methylthiobenzoyl	CH_3	$CH_2COOC(CH_3)_3$	OCH_3	31
4-Methylthiobenzoyl	CH_3	$(CH_2)_2COOH$	OCH_3	29, 31, 32
4-Methylthiobenzoyl	CH_3	$(CH_2)_2COOH$	H, CH_3, OC_2H_5, F	29, 32
4-Methylthiobenzoyl	CH_3	$(CH_2)_2COOH$	SCH_3, Cl, NO_2	29

(*continued*)

Table 23 (*continued*)

R^5	R^6	R^7	R^2	References
4-Methylthiobenzoyl	CH_3	$(CH_2)_2COOH$	H, CH_3, OCH_3, OC_2H_5, F(c)	32
4-Methylthiobenzoyl	CH_3	$CH(CH_3)COOH$	H, CH_3, OCH_3, OC_2H_5, SCH_3, F, Cl, NO_2	29
4-Methylthiobenzoyl	CH_3	$(CH_2)_3COOH$	H, CH_3, OCH_3, OC_2H_5, F	29, 32
4-Methylthiobenzoyl	CH_3	$(CH_2)_3COOH$	SCH_3, Cl, NO_2	29
4-Methylthiobenzoyl	CH_3	$(CH_2)_3COOH$	H, CH_3, OCH_3, OC_2H_5, F(c)	32
2-Hydroxybenzoyl	CH_3	CH_2COOH	OCH_3	49
3,4-Methylenedioxybenzoyl	CH_3	CH_2COOH	OCH_3	50(e), 51–53
3,4-Methylenedioxybenzoyl	CH_3	CH_2COOH	Cl	52, 53
3,4-Methylenedioxybenzoyl	CH_3	CH_2COOH	a	47
3,4-Methylenedioxybenzoyl	CH_3	CH_2CONH_2	OCH_3	54
3,4-Methylenedioxybenzoyl	CH_3	$CH_2CON(CH_3)_2$	OCH_3	54, 55
3,4-Methylenedioxybenzoyl	CH_3	$CH_2CONH(CH_2)_2OH$	OCH_3	54, 55
3,4-Methylenedioxybenzoyl	CH_3	$CH_2COO(CH_2)_2N(CH_3)_2$	OCH_3	56
3,4-Methylenedioxybenzoyl	CH_3	CH_2CO-morpholino	OCH_3	54, 55
3,4-Methylenedioxybenzoyl	CH_3	$(CH_2)_2COOH$	OCH_3	52, 53
4-Trifluoromethylbenzoyl	H	CH_2COOH	H, CH_3, OCH_3, OC_2H_5, SCH_3, F, Cl, NO_2	29
4-Trifluoromethylbenzoyl	H	$(CH_2)_2COOH$	H, CH_3, OCH_3, OC_2H_5, SCH_3, F, Cl, NO_2	29
4-Trifluoromethylbenzoyl	H	$CH(CH_3)COOH$	H, CH_3, OCH_3, OC_2H_5, SCH_3, F, Cl, NO_2	29
4-Trifluoromethylbenzoyl	CH_3	CH_2COOH	H	29, 31, 33, 34
4-Trifluoromethylbenzoyl	CH_3	CH_2COOH	CH_3, OC_2H_5, SCH_3, F, Cl, NO_2	29, 33
4-Trifluoromethylbenzoyl	CH_3	CH_2COOH	OCH_3	29, 33, 34
4-Trifluoromethylbenzoyl	CH_3	$(CH_2)_2COOH$	H, CH_3, OCH_3, OC_2H_5, SCH_3, F, Cl, NO_2	29

4-Trifluoromethylbenzoyl	CH_3	$CH(CH_3)COOH$	H, CH_3, OCH_3, OC_2H_5, SCH_3, F, Cl, NO_2	29
4-Trifluoromethylbenzoyl	CH_3	$(CH_2)_3COOH$	H, OCH_3	29, 34
4-Trifluoromethylbenzoyl	CH_3	$(CH_2)_3COOH$	CH_3, OC_2H_5, SCH_3, F, Cl, NO_2	29
4-Fluorobenzoyl	H	CH_2COOH	H, CH_3, OCH_3, OC_2H_5, SCH_3, F, Cl, NO_2	29
4-Fluorobenzoyl	H	$(CH_2)_2COOH$	H, CH_3, OCH_3, OC_2H_5, SCH_3, F, Cl, NO_2	29
4-Fluorobenzoyl	H	$CH(CH_3)COOH$	H, CH_3, OCH_3, OC_2H_5, SCH_3, F, Cl, NO_2	29
4-Fluorobenzoyl	H	$(CH_2)_3COOH$	H, CH_3, OCH_3, OC_2H_5, SCH_3, F, Cl, NO_2	29
4-Fluorobenzoyl	CH_3	CH_2COOH	H, CH_3, OC_2H_5, F	29, 32, 33
4-Fluorobenzoyl	CH_3	CH_2COOH	OCH_3	29, 31–33, 57
4-Fluorobenzoyl	CH_3	CH_2COOH	SCH_3, Cl, NO_2	29, 33
4-Fluorobenzoyl	CH_3	CH_2COOH	H, CH_3, OCH_3, OC_2H_5, F(c)	32
4-Fluorobenzoyl	CH_3	CH_2COOH	a	47
4-Fluorobenzoyl	CH_3	$(CH_2)_2COOH$	H, CH_3, OCH_3, OC_2H_5, F	29, 32
4-Fluorobenzoyl	CH_3	$(CH_2)_2COOH$	SCH_3, Cl, NO_2	29
4-Fluorobenzoyl	CH_3	$(CH_2)_2COOH$	H, CH_3, OCH_3, OC_2H_5, F(c)	32
4-Fluorobenzoyl	CH_3	$CH(CH_3)COOH$	H, CH_3, OCH_3, OC_2H_5, SCH_3, F, Cl, NO_2	29
4-Fluorobenzoyl	CH_3	$(CH_2)_3COOH$	H, CH_3, OCH_3, OC_2H_5, F	29, 32
4-Fluorobenzoyl	CH_3	$(CH_2)_3COOH$	SCH_3, Cl, NO_2	29
4-Fluorobenzoyl	CH_3	$(CH_2)_3COOH$	H, CH_3, OCH_3, OC_2H_5, F(c)	32
4-Chlorobenzoyl	H	CH_3	OCH_3	16
4-Chlorobenzoyl	H	C_2H_5	OCH_3	16
4-Chlorobenzoyl	H	C_6H_5	OCH_3	16
4-Chlorobenzoyl	H	CH_2COOH	H, CH_3, OCH_3, OC_2H_5, SCH_3, F, Cl, NO_2	29

(*continued*)

Table 23 (*continued*)

R^5	R^6	R^7	R^2	References
4-Chlorobenzoyl	H	$(CH_2)_2COOH$	H, CH_3, OCH_3, OC_2H_5, SCH_3, F, Cl, NO_2	29
4-Chlorobenzoyl	H	$CH(CH_3)COOH$	H, CH_3, OCH_3, OC_2H_5, SCH_3, F, Cl, NO_2	29
4-Chlorobenzoyl	H	$(CH_2)_3OH$	OCH_3	58
4-Chlorobenzoyl	H	$(CH_2)_3COOH$	H, CH_3, OCH_3, OC_2H_5, SCH_3, F, Cl, NO_2	29
4-Chlorobenzoyl	CH_3	CH_3	H	59
4-Chlorobenzoyl	CH_3	CH_3	OCH_3	16
4-Chlorobenzoyl	CH_2	CH_3	a	60
4-Chlorobenzoyl	CH_3	C_2H_5	H	16
4-Chlorobenzoyl	CH_3	C_2H_5	OCH_3	16
4-Chlorobenzoyl	CH_3	C_2H_5	a	60
4-Chlorobenzoyl	CH_3	C_6H_5	H	16, 59
4-Chlorobenzoyl	CH_3	$CH_2C_6H_5$	OCH_3	16
4-Chlorobenzoyl	CH_3	CH_2COOH	H	29, 31–34, 45
4-Chlorobenzoyl	CH_3	CH_2COOH	CH_3	29, 31–33, 35 45, 61
4-Chlorobenzoyl	CH_3	CH_2COOH	OCH_3(f)	29–37, 44, 57, 62–69, 70(e)
4-Chlorobenzoyl	CH_3	CH_2COOH	OC_2H_5	29, 31–33, 57
4-Chlorobenzoyl	CH_3	CH_2COOH	OC_6H_5	71
4-Chlorobenzoyl	CH_3	CH_2COOH	$OCH_2C_6H_5$	36, 37
4-Chlorobenzoyl	CH_3	CH_2COOH	SCH_3	29, 33
4-Chlorobenzoyl	CH_3	CH_2COOH	$N(CH_3)_2$	35, 45
4-Chlorobenzoyl	CH_3	CH_2COOH	F	29, 31–33
4-Chlorobenzoyl	CH_3	CH_2COOH	Cl	29, 31, 33, 72
4-Chlorobenzoyl	CH_3	CH_2COOH	NO_2	29, 33
4-Chlorobenzoyl	CH_3	CH_2COOH	H(c)	31, 32, 73

4-Chlorobenzoyl	CH_3	CH_2COOH	H(g)	73
4-Chlorobenzoyl	CH_3	CH_2COOH	CH_3, OCH_3, OC_2H_5, F(c)	32
4-Chlorobenzoyl	CH_3	CH_2COOH	a	74
4-Chlorobenzoyl	CH_3	CH_2COOH	h	47
4-Chlorobenzoyl	CH_3	CH_2COOH	i	75
4-Chlorobenzoyl	CH_3	CH_2COOCH_3	H	57
4-Chlorobenzoyl	CH_3	CH_2COOCH_3	CH_3	31, 35
4-Chlorobenzoyl	CH_3	CH_2COOCH_3	OCH_3, $N(CH_3)_2$	35
4-Chlorobenzoyl	CH_3	CH_2COOCH_3	NO_2	31
4-Chlorobenzoyl	CH_3	$CH_2COOC_2H_5$	CH_3	35
4-Chlorobenzoyl	CH_3	$CH_2COOC_2H_5$	$N(CH_3)_2$	35, 76
4-Chlorobenzoyl	CH_3	$CH_2COOC_2H_5$	OCH_3	31, 35
4-Chlorobenzoyl	CH_3	$CH_2COO(CH_2)_3CH_3$	CH_3	30
4-Chlorobenzoyl	CH_3	$CH_2COOC(CH_3)_3$	H	45
4-Chlorobenzoyl	CH_3	$CH_2COOC(CH_3)_3$	CH_3	35, 45
4-Chlorobenzoyl	CH_3	$CH_2COOC(CH_3)_3$	OCH_3	35
4-Chlorobenzoyl	CH_3	$CH_2COOC(CH_3)_3$	$N(CH_3)_2$	35, 45, 77
4-Chlorobenzoyl	CH_3	$CH_2COOC(CH_3)_3$	F	31
4-Chlorobenzoyl	CH_3	$CH_2COO(CH_2)_2OH$	OCH_3	78
4-Chlorobenzoyl	CH_3	$CH_2COO(CH_2)_2OAlkyl$	OCH_3	78
4-Chlorobenzoyl	CH_3	$CH_2COO(CH_2)_2OCOCH_3$	OCH_3	78
4-Chlorobenzoyl	CH_3	$CH_2COOCH_2CF_3$	OCH_3	78
4-Chlorobenzoyl	CH_3	$CH_2COO(CH_2)_2CF_3$	OCH_3	78
4-Chlorobenzoyl	CH_3	$CH_2COOCH_2C_6H_4CH_3(4)$	OCH_3	78
4-Chlorobenzoyl	CH_3	$CH_2COOCH_2COOCH_2C_6H_5$	OCH_3	79
4-Chlorobenzoyl	CH_3	$(CH_2)_2OH$	OCH_3	58, 63, 80
4-Chlorobenzoyl	CH_3	$(CH_2)_2OH$	Cl	58
4-Chlorobenzoyl	CH_3	$(CH_2)_2COOH$	H, CH_3, OC_2H_5, F	29, 32
4-Chlorobenzoyl	CH_3	$(CH_2)_2COOH$	OCH_3	29, 31, 32
4-Chlorobenzoyl	CH_3	$(CH_2)_2COOH$	SCH_3, Cl, NO_2	29
4-Chlorobenzoyl	CH_3	$(CH_2)_2COOH$	H, CH_3, OCH_3, OC_2H_5 F(c)	32

(*continued*)

Table 23 (*continued*)

R^5	R^6	R^7	R^2	References
4-Chlorobenzoyl	CH_3	$(CH_2)_2COOH$	a	47
4-Chlorobenzoyl	CH_3	$CH(CH_3)COOH$	H, CH_3, OC_2H_5, SCH_3, F, Cl, NO_2	29
4-Chlorobenzoyl	CH_3	$CH(CH_3)COOH$	OCH_3	29, 44
4-Chlorobenzoyl	CH_3	$CH(CH_3)COOH$	a	47
4-Chlorobenzoyl	CH_3	$CH_2CH(NH_2)COOH$	OCH_3	63
4-Chlorobenzoyl	CH_3	$(CH_2)_3COOH$	H	29, 32, 34, 38
4-Chlorobenzoyl	CH_3	$(CH_2)_3COOH$	CH_3	29, 32, 81
4-Chlorobenzoyl	CH_3	$(CH_2)_3COOH$	OCH_3	29, 31, 32, 34, 81
4-Chlorobenzoyl	CH_3	$(CH_2)_3COOH$	OC_2H_5, F	29, 32
4-Chlorobenzoyl	CH_3	$(CH_2)_3COOH$	SCH_3, Cl, NO_2	29
4-Chlorobenzoyl	CH_3	$(CH_2)_3COOH$	H, CH_3, OCH_3, OC_2H_5, F(c)	32
4-Chlorobenzoyl	CH_3	$(CH_2)_3COOCH_3$	H	31
4-Chlorobenzoyl	CH_3	$CH_2CH{=}CHCOOH$	OCH_3	82
4-Chlorobenzoyl	CH_3	$CH_2CH(COOC_2H_5)_2$	OCH_3	31
4-Chlorobenzoyl	CH_3	$CH(C_6H_5)COOH$	OCH_3	83
4-Chlorobenzoyl	CH_3	CH_2CHO	OCH_3	84
4-Chlorobenzoyl	CH_3	$CH_2CH(OCH_2C_6H_5)_2$	OCH_3	84
4-Chlorobenzoyl	CH_3	CH_2CH	OCH_3	84
4-Chlorobenzoyl	CH_3	$COCH_3$	OCH_3	85
4-Chlorobenzoyl	CH_3	$CH_2N(C_2H_5)_2$	OCH_3	48
4-Chlorobenzoyl	CH_3	CH_2CONH_2	OCH_3	31, 85
4-Chlorobenzoyl	CH_3	OH(keto tautomer)	OCH_3	87

4-Chlorobenzoyl	CH_3	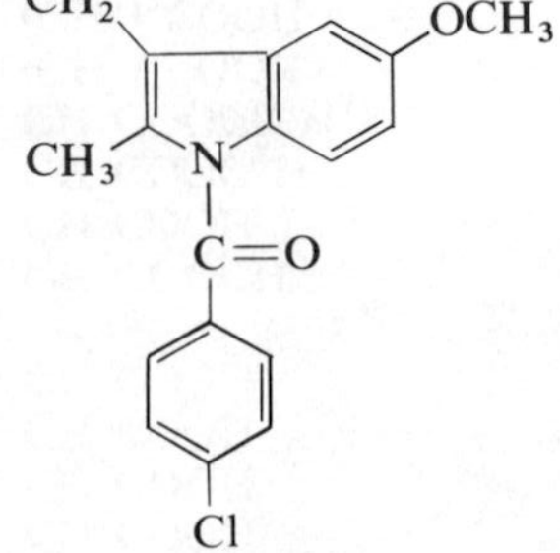 CH_2 OCH_3 CH_3 N C=O Cl	OCH_3	88
4-Chlorobenzoyl	CH_3	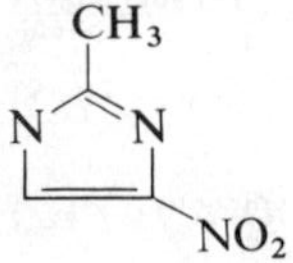 CH_3 N N NO_2	OCH_3	89
4-Chlorobenzoyl	C_2H_5	H	H	59
4-Chlorobenzoyl	$CH_2C_6H_5$	H	H	59
4-Chlorobenzoyl	$(CH_2)_4$		OCH_3	8, 9, 40, 90
4-Chlorobenzoyl	$(CH_2)_2CH(CH_3)CH_3$		OCH_3	8, 9, 40
4-Chlorobenzoyl	$CH_2CH(CH_3)(CH_2)_2$		OCH_3	8
4-Chlorobenzoyl	$(CH_2)_3CH(CH_3)$		OCH_3	8
4-Chlorobenzoyl	$(CH_2)_2CH(COOH)$		OCH_3	8
4-Chlorobenzoyl	$CH_2CH(COOH)CH_2$		OCH_3	8
4-Chlorobenzoyl	$(CH_2)_2NHCH_2$		H	40
4-Chlorobenzoyl	$(CH_2)_2{}^{\oplus}NH(CH_3)CH_2$ $Cl^{\ominus}$		OCH_3	9
4-Chlorobenzoyl	$(CH_2)_2N(CH_3)CH_2$		OCH_3	8
4-Chlorobenzoyl	$CH_2C(CH_3)_2CH_2COCH_2CH(COOH)$		H, CH_3, OCH_3, Cl	91

(*continued*)

Table 23 *(continued)*

R^5	R^6	R^7	R^2	References
4-Chlorobenzoyl	$CH_2C(CH_3)_2CH_2COCH_2CH(COOCH_3)$		H	91
4-Chlorobenzoyl	$CH_2SCH(COOH)$		OCH_3	44
4-Chlorobenzoyl	$CH_2SC(CH_3)(COOH)$		OCH_3	44
4-Chlorobenzoyl	COOH	CH_2COOH	OCH_3	92
4-Chlorobenzoyl	$(CH_2)_2COOH$	C_6H_5	OCH_3	93
4-Chlorobenzoyl	C_6H_5	CH_2COOH	CH_3, OCH_3	6, 39(d)
4-Chlorobenzoyl	$C_6H_4Cl(4)$	CH_2COOH	OCH_3	6, 39(d)
4-Chlorobenzoyl	$C_6H_4OH(4)$	CH_3	OCH_3	39(d)
4-Chlorobenzoyl	$C_6H_4OCH_3(4)$	$C_6H_4OCH_3(4)$	H, OCH_3	7
4-Bromobenzoyl	H	CH_2COOH	H, CH_3, OCH_3, OC_2H_5, SCH_3, F, Cl, NO_2	29
4-Bromobenzoyl	H	$(CH_2)_2COOH$	H, CH_3, OCH_3, OC_2H_5, SCH_3, F, Cl, NO_2	29
4-Bromobenzoyl	H	$CH(CH_3)COOH$	H, CH_3, OCH_3, OC_2H_5, SCH_3, F, Cl, NO_2	29
4-Bromobenzoyl	H	$(CH_2)_3COOH$	H, CH_3, OCH_3, OC_2H_5, SCH_3, F, Cl, NO_2	29
4-Bromobenzoyl	CH_3	CH_2COOH	H, OC_2H_5, F	29, 32, 33
4-Bromobenzoyl	CH_3	CH_2COOH	CH_3	29, 32, 33, 35
4-Bromobenzoyl	CH_3	CH_2COOH	SCH_3, Cl, NO_2	29, 33
4-Bromobenzoyl	CH_3	CH_2COOH	$N(CH_3)_2$	35
4-Bromobenzoyl	CH_3	CH_2COOH	H, CH_3, OCH_3, OC_2H_5, F(c)	32
4-Bromobenzoyl	CH_3	CH_2COOCH_3	OCH_3	31, 35
4-Bromobenzoyl	CH_3	CH_2COOH	CH_3, $N(CH_3)_2$	35
4-Bromobenzoyl	CH_3	$CH_2COOC_2H_5$	CH_3, OCH_3, $N(CH_3)_2$	35
4-Bromobenzoyl	CH_3	$CH_2COOC(CH_3)_3$	OC_2H_5, $N(CH_3)_2$	35
4-Bromobenzoyl	CH_3	$(CH_2)_2COOH$	H, CH_3, OCH_3, OC_2H_5, F	29, 32
4-Bromobenzoyl	CH_3	$(CH_2)_2COOH$	SCH_3, Cl, NO_2	29

4-Bromobenzoyl	CH_3	$(CH_2)_2COOH$	H, CH_3, OCH_3, OC_2H_5, F(c)	32
4-Bromobenzoyl	CH_3	$CH(CH_3)COOH$	H, CH_3, OCH_3, OC_2H_5, SCH_3, F, Cl, NO_2	29
4-Bromobenzoyl	CH_3	$(CH_2)_3COOH$	H, CH_3, OCH_3, OC_2H_5, F	29, 32
4-Bromobenzoyl	CH_3	$(CH_2)_3COOH$	SCH_3, Cl, NO_2	29
4-Bromobenzoyl	CH_3	$(CH_2)_3COOH$	H, CH_3, OCH_3, OC_2H_5, F(c)	32
3,5-Dinitrobenzoyl	CH_3	CH_2COOH	OCH_3	22
Phenylacetyl	CH_3	CH_2COOH	OCH_3	12, 14, 22, 94–96
Phenylacetyl	CH_3	$CH_2CH(OC_2H_5)_2$	OCH_3	97
4-Chlorophenylacetyl	CH_3	CH_2COOH	OCH_3	12, 94
4-Chlorophenylacetyl	CH_3	CH_2COOH	OC_2H_5	17
4-Chlorophenylacetyl	CH_3	CH_2COOH	j	47, 74
4-Nitrophenylacetyl	CH_3	CH_2COOH	OCH_3	94
3,4-Dimethoxyphenylacetyl	CH_3	CH_2COOH	OCH_3	14, 94
Chlorophenylacetyl	CH_3	CH_2COOH	OCH_3	14, 94, 96
Diphenylacetyl	CH_3	CH_2COOH	OCH_3	22, 94
3-Phenylpropionyl	CH_3	CH_2COOH	OCH_3	94
3,3-Diphenylpropionyl	CH_3	CH_2COOH	OCH_3	22
2-Phenylbutyryl	CH_3	CH_2COOH	OCH_3	14, 17, 94
4-Phenylbutyryl	CH_3	CH_2COOH	OCH_3	94
2-Phenylcyclopropylcarbonyl	CH_3	CH_2COOH	OCH_3	98
4-(4-Methoxyphenyl)butyryl	CH_3	CH_2COOH	OCH_3	94
1-Naphthylacetyl	CH_3	CH_2COOH	OCH_3	12, 13, 93
Cinnamoyl	CH_3	CH_3	h	60
Cinnamoyl	CH_3	CH_2COOH	CH_3	99
Cinnamoyl	CH_3	CH_2COOH	OCH_3	12–14, 17, 22, 94, 99, 100–103
Cinnamoyl	CH_3	CH_2COOH	$N(CH_3)_2$	104
Cinnamoyl	CH_3	CH_2COOH	a	47, 74, 105
Cinnamoyl	CH_3	CH_2COOH	j	47, 105
Cinnamoyl	CH_3	CH_2COOCH_3	CH_3	99

(continued)

Table 23 *(continued)*

R^5	R^6	R^7	R^2	References
Cinnamoyl	CH_3	CH_2COOCH_3	OCH_3	99, 102
Cinnamoyl	CH_3	CH_2COOCH_3	j	74
Cinnamoyl	CH_3	$CH_2COOC_2H_5$	CH_3	17, 94
Cinnamoyl	CH_3	$CH_2COOC_2H_5$	OCH_3	14, 94, 102
Cinnamoyl	CH_3	$CH_2COOC_2H_5$	$N(CH_3)_2$	76
Cinnamoyl	CH_3	$CH_2COOC(CH_3)_3$	OCH_3, OC_2H_5	106
Cinnamoyl	CH_3	$CH_2COOCH_2CH(CH_3)N(CH_3)_2$	OCH_3, OC_2H_5	56
Cinnamoyl	CH_3	$CH_2COO(CH_2)_2N(CH_3)_2$	OCH_3, OC_2H_5	56
Cinnamoyl	CH_3	$CH_2COO(CH_2)_2N(C_2H_5)_2$	OCH_3, OC_2H_5	56
Cinnamoyl	CH_3	$CH_2COOCH_2CH(CH_3)N(CH_3)_2$	OCH_3, OC_2H_5	56
Cinnamoyl	CH_3	$(CH_2)_2COOH$	OCH_3, OC_2H_5	14, 17, 94, 102
Cinnamoyl	CH_3	$CH(CH_3)COOH$	OCH_3, OC_2H_5	94
Cinnamoyl	CH_3	$CH(CH_3)COOH$	a, j	105
Cinnamoyl	CH_3	$(CH_2)_3COOH$	OCH_3	14, 94, 102
Cinnamoyl	CH_3	$CH_2CONHCH_3$	OCH_3	55
Cinnamoyl	CH_3	$CH_2CON(CH_3)_2$	OCH_3	55
Cinnamoyl	CH_3	$CH_2CONH(CH_2)_2OH$	OCH_3	55
Cinnamoyl	CH_3	$CH_2CONH(CH_2)_3N(CH_3)_2$	OCH_3	55
Cinnamoyl	CH_3	CH_2CO-morpholino	OCH_3	55
Cinnamoyl	CH_3	CH_2CONH-2-pyrimidyl	OCH_3	55
Cinnamoyl	CH_3	$CH_2CH(OC_2H_5)_2$	OCH_3	97
Cinnamoyl	CH_3	OH(keto tautomer)	OCH_3	107
Cinnamoyl	COOH	CH_3	OCH_3	108, 109
Cinnamoyl	COOH	CH_2COOH	OCH_3	18, 110
4-Methylcinnamoyl	CH_3	CH_2COOH	OCH_3	12, 14, 102
4-Methoxycinnamoyl	CH_3	CH_2COOH	OCH_3	13, 14, 17, 94, 102
4-Chlorocinnamoyl	CH_3	CH_2COOH	OCH_3	13, 94, 102
3-Nitrocinnamoyl	CH_3	CH_2COOH	OCH_3	94, 102
4-Nitrocinnamoyl	CH_3	CH_2COOH	OCH_3	102
3,4-Dimethylcinnamoyl	CH_3	CH_2COOH	OCH_3	111

3,4-Methylenedioxycinnamoyl	CH_3	CH_2COOH	OCH_3	52, 53
2-Methylcinnamoyl	CH_3	CH_2COOH	OCH_3	94, 102
2-Phenylcinnamoyl	CH_3	CH_2COOH	OCH_3	94, 102
2-Naphthoyl	H	CH_2COOH	H, CH_3, OCH_3, OC_2H_5, SCH_3, F, Cl, NO_2	29
2-Naphthoyl	H	$(CH_2)_2COOH$	H, CH_3, OCH_3, OC_2H_5, SCH_3, F, Cl, NO_2	29
2-Naphthoyl	H	$CH(CH_3)COOH$	H, CH_3, OCH_3, OC_2H_5, SCH_3, F, Cl, NO_2	29
2-Naphthoyl	H	$(CH_2)_3COOH$	H, CH_3, OCH_3, OC_2H_5, SCH_3, F, Cl, NO_2	29
2-Naphthoyl	CH_3	CH_2COOH	H, CH_3, OCH_3, OC_2H_5, F	29, 32, 33
2-Naphthoyl	CH_3	CH_2COOH	SCH_3, Cl, NO_2	29, 33
2-Naphthoyl	CH_3	CH_2COOH	H, CH_3, OCH_3, OC_2H_5, F(c)	32
2-Naphthoyl	CH_3	CH_2COOCH_3	OCH_3	31
2-Naphthoyl	CH_3	$(CH_2)_2COOH$	H, CH_3, OCH_3, OC_2H_5, F(c)	29, 32
2-Naphthoyl	CH_3	$(CH_2)_2COOH$	SCH_3, Cl, NO_2	29
2-Naphthoyl	CH_3	$CH(CH_3)COOH$	H, CH_3, OCH_3, OC_2H_5, SCH_3, F, Cl, NO_2	29
2-Naphthoyl	CH_3	$(CH_2)_3COOH$	H, CH_3, OCH_3, OC_2H_5, F	29, 32
2-Naphthoyl	CH_3	$(CH_2)_3COOH$	SCH_3, Cl, NO_2	29
2-Naphthoyl	CH_3	$(CH_2)_3COOH$	H, CH_3, OCH_3, OC_2H_5, F(c)	32
1-Indanylcarbonyl	CH_3	CH_3	k	60
1-Indanylcarbonyl	CH_3	CH_2COOH	a, j	105
1-Indanylcarbonyl	CH_3	$CH(CH_3)COOH$	a, j	105
2-Indanylcarbonyl	CH_3	CH_2COOH	OCH_3	20, 21
2-Indanylcarbonyl	CH_3	$CH_2COOC_2H_5$	OCH_3	20
2-Indanylcarbonyl	CH_3	$(CH_2)_3COOH$	OCH_3	20
5-Indanylcarbonyl	CH_3	CH_2COOH	a	15, 105
5-Indanylcarbonyl	CH_3	CH_2COOH	a, j	105

(*continued*)

Table 23 (*continued*)

R^5	R^6	R^7	R^2	References
5-Indanylcarbonyl	CH_3	$CH(CH_3)COOH$	a, j	105
2-Indenylcarbonyl	CH_3	CH_2COOH	OCH_3	20
Picolinoyl	CH_3	CH_2COOH	$OCH_2C_6H_5$	36
Picolinoyl	CH_3	CH_2COOH	a, j	105
Picolinoyl	CH_3	$CH(CH_3)COOH$	a, j	105
Nicotinoyl	H	CH_2COOH	H, CH_3, OCH_3, OC_2H_5, SCH_3, F, Cl, NO_2	29
Nicotinoyl	H	$(CH_2)_2COOH$	H, CH_3, OCH_3, OC_2H_5, SCH_3, F, Cl, NO_2	29
Nicotinoyl	H	$CH(CH_3)COOH$	H, CH_3, OCH_3, OC_2H_5, SCH_3, F, Cl, NO_2	29
Nicotinoyl	H	$(CH_2)_3COOH$	H, CH_3, OCH_3, OC_2H_5, SCH_3, F, Cl, NO_2	29
Nicotinoyl	CH_3	CH_3	a	60
Nicotinoyl	CH_3	CH_2COOH	H	29, 32–34
Nicotinoyl	CH_3	CH_2COOH	CH_3, OC_2H_5, F	29, 32, 33
Nicotinoyl	CH_3	CH_3COOH	OCH_3	29, 32–34, 36, 37, 112, 113
Nicotinoyl	CH_3	CH_3COOH	$OCH_2C_6H_5$	37
Nicotinoyl	CH_3	CH_3COOH	SCH_3, Cl, NO_2	29, 33
Nicotinoyl	CH_3	CH_3COOH	H, CH_3, OCH_3, OC_2H_5, F(c)	32
Nicotinoyl	CH_3	CH_3COOH	a	15
Nicotinoyl	CH_3	CH_2COOCH_3	OCH_3	31, 112, 113
Nicotinoyl	CH_3	$CH_2COOC(CH_3)_3$	OCH_3, OC_2H_5	106
Nicotinoyl	CH_3	$(CH_2)_2COOH$	H, CH_3, OCH_3, OC_2H_5, F	29, 32
Nicotinoyl	CH_3	$(CH_2)_2COOH$	SCH_3, Cl, NO_2	29
Nicotinoyl	CH_3	$(CH_2)_2COOH$	H, CH_3, OCH_3, OC_2H_5, F(c)	32

Nicotinoyl	CH_3	$CH(CH_3)COOH$	H, CH_3, OCH_3, OC_2H_5, SCH_3, F, Cl, NO_2	29
Nicotinoyl	CH_3	$(CH_2)_3COOH$	H, OCH_3	29, 32, 34
Nicotinoyl	CH_3	$(CH_2)_3COOH$	CH_3, OC_2H_5, F	29, 32
Nicotinoyl	CH_3	$(CH_2)_3COOH$	SCH_3, Cl, NO_2	29
Nicotinoyl	CH_3	$(CH_2)_3COOH$	H, CH_3, OCH_3, OC_2H_5, F(c)	32
Nicotinoyl	$(CH_2)_2CH(CH_3)CH_2$		OCH_3	8
Nicotinoyl	CH_3	CH_2CONH_2	OCH_3, OC_2H_5	86
Nicotinoyl	COOH	CH_2COOH	OCH_3, OC_2H_5	114
Nicotinoyl	$(CH_2)_2COOH$	C_6H_5	OCH_3, OC_2H_5	93
Nicotinoyl	C_6H_5	CH_2COOH	OCH_3, OC_2H_5	6
Isonicotinoyl	H	CH_2COOH	H, CH_3, OCH_3, OC_2H_5, SCH_3, F, Cl, NO_2	29
Isonicotinoyl	H	$(CH_2)_2COOH$	H, CH_3, OCH_3, OC_2H_5, SCH_3, F, Cl, NO_2	29
Isonicotinoyl	H	$CH(CH_3)COOH$	H, CH_3, OCH_3, OC_2H_5, SCH_3, F, Cl, NO_2	29
Isonicotinoyl	H	$(CH_2)_3COOH$	H, CH_3, OCH_3, OC_2H_5, SCH_3, F, Cl, NO_2	29
Isonicotinoyl	CH_3	CH_2COOH	H	29, 32–34
Isonicotinoyl	CH_3	CH_2COOH	CH_3, OC_2H_5, F	29, 32, 33
Isonicotinoyl	CH_3	CH_2COOH	OCH_3	29, 31–34, 36, 37, 113
Isonicotinoyl	CH_3	CH_2COOH	$OCH_2C_6H_5$	37
Isonicotinoyl	CH_3	CH_2COOH	SCH_3, Cl, NO_2	29, 33
Isonicotinoyl	CH_3	CH_2COOH	H, CH_3, OCH_3, OC_2H_5, F(c)	32
Isonicotinoyl	CH_3	$CH_2COOC(CH_3)_3$	OCH_3, OC_2H_5	106
Isonicotinoyl	CH_3	$(CH_2)_2COOH$	H, CH_3, OCH_3, OC_2H_5, F	29, 32
Isonicotinoyl	CH_3	$(CH_2)_2COOH$	SCH_3, Cl, NO_2	29
Isonicotinoyl	CH_3	$(CH_2)_2COOH$	H, CH_3, OCH_3, OC_2H_5, F(c)	32

(continued)

Table 23 (*continued*)

R^5	R^6	R^7	R^2	References
Isonicotinoyl	CH_3	$CH(CH_3)COOH$	H, CH_3, OCH_3, OC_2H_5, SCH_3, F, Cl, NO_2	29
Isonicotinoyl	CH_3	$(CH_2)_3COOH$	H, OCH_3	29, 32, 34
Isonicotinoyl	CH_3	$(CH_2)_3COOH$	CH_3, OC_2H_5, F	29, 32
Isonicotinoyl	CH_3	$(CH_2)_3COOH$	H, CH_3, OCH_3, OC_2H_5, F(c)	32
Isonicotinoyl	COOH	CH_2COOH	OCH_3	114
2-Furylcarbonyl	H	CH_2COOH	H, CH_3, OCH_3, OC_2H_5, SCH_3, F, Cl, NO_2	29
2-Furylcarbonyl	H	$(CH_2)_2COOH$	H, CH_3, OCH_3, OC_2H_5, SCH_3, F, Cl, NO_2	29
2-Furylcarbonyl	H	$CH(CH)_3COOH$	H, CH_3, OCH_3, OC_2H_5, SCH_3, F, Cl, NO_2	29
2-Furylcarbonyl	H	$(CH_2)_3COOH$	H, CH_3, OCH_3, OC_2H_5, SCH_3, F, Cl, NO_2	29
2-Furylcarbonyl	CH_3	CH_3	CH_3	16
2-Furylcarbonyl	CH_3	CH_2COOH	H, CH_3, OCH_3, F	29, 32, 33
2-Furylcarbonyl	CH_3	CH_2COOH	CH_3	29, 31–33, 113
2-Furylcarbonyl	CH_3	CH_2COOH	OCH_3	29, 32, 33, 36, 115, 116
2-Furylcarbonyl	CH_3	CH_2COOH	SCH_3, Cl, NO_2	29, 33
2-Furylcarbonyl	CH_3	CH_2COOH	H, CH_3, OCH_3, OC_2H_5, F(c)	32
2-Furylcarbonyl	CH_3	CH_2COOCH_3	OCH_3	31
2-Furylcarbonyl	CH_3	$(CH_2)_2COOH$	CH_3	29, 31, 32, 113
2-Furylcarbonyl	CH_3	$(CH_2)_2COOH$	H, OCH_3, OC_2H_5, F	29, 32
2-Furylcarbonyl	CH_3	$(CH_2)_2COOH$	SCH_3, Cl, NO_2	29
2-Furylcarbonyl	CH_3	$(CH_2)_2COOH$	H, CH_3, OCH_3, OC_2H_5, F(c)	32

2-Furylcarbonyl	CH_3	$CH(CH_3)COOH$	H, CH_3, OCH_3, OC_2H_5, SCH_3, F, Cl, NO_2	29
2-Furylcarbonyl	CH_3	$(CH_2)_3COOH$	H, CH_3, OCH_3, OC_2H_5, F	29, 32
2-Furylcarbonyl	CH_3	$(CH_2)_3COOH$	SCH_3, Cl, NO_2	29
2-Furylcarbonyl	CH_3	$(CH_2)_3COOH$	H, CH_3, OCH_3, OC_2H_5, F(c)	32
2-Furylcarbonyl	CH_3	$CH_2CH(COOCH_3)_2$	Cl	31
2-(5-Nitrofuryl)carbonyl	CH_3	CH_2COOH	OCH_3	115
3-(2-Furyl)acryloyl	CH_3	CH_2COOH	OCH_3	12, 14, 93, 101
3-(2-Furyl)acryloyl	CH_3	CH_2COOH	$N(CH_3)_2$	104
3-(2-Furyl)acryloyl	CH_3	CH_2COOH	a	15
3-(2-Furyl)acryloyl	CH_3	$CH_2COOC_2H_5$	OCH_3	17
3-(2-Furyl)acryloyl	CH_3	$CH_2COOC_2H_5$	j	15, 47
3-(2-Furyl)acryloyl	CH_3	$CH(CH_3)COOC_2H_5$	j	15, 74
3-(2-Furyl)acryloyl	CH_3	$CH(CH_3)C_2H_5$	k	60
2-Thienylcarbonyl	H	CH_2COOH	H, CH_3, OCH_3, OC_2H_5, SCH_3, F, Cl, NO_2	29
2-Thienylcarbonyl	H	$(CH_2)_2COOH$	H, CH_3, OCH_3, OC_2H_5, SCH_3, F, Cl, NO_2	29
2-Thienylcarbonyl	H	$CH(CH_3)COOH$	H, CH_3, OCH_3, OC_2H_5, SCH_3, F, Cl, NO_2	29
2-Thienylcarbonyl	H	$(CH_2)_3COOH$	H, CH_3, OCH_3, OC_2H_5, SCH_3, F, Cl, NO_2	29
2-Thienylcarbonyl	CH_3	CH_2COOH	H, CH_3, OC_2H_5, F	29, 32, 33
2-Thienylcarbonyl	CH_3	CH_2COOH	OCH_3	29, 31–33, 36, 113
2-Thienylcarbonyl	CH_3	CH_2COOH	SCH_3, Cl, NO_2	29, 33
2-Thienylcarbonyl	CH_3	CH_2COOH	H, CH_3, OCH_3, OC_2H_5, F(c)	32
2-Thienylcarbonyl	CH_3	$CH_2COOC_2H_5$	OCH_3	31
2-Thienylcarbonyl	CH_3	$(CH_2)_2COOH$	H, CH_3, OCH_3, OC_2H_5, F	29, 32
2-Thienylcarbonyl	CH_3	$(CH_2)_2COOH$	SCH_3, Cl, NO_2	29

(continued)

Table 23 *(continued)*

R^5	R^6	R^7	R^2	References
2-Thienylcarbonyl	CH_3	$(CH_2)_2COOH$	H, CH_3, OCH_3, OC_2H_5, F(c)	32
2-Thienylcarbonyl	CH_3	$CH(CH_3)COOH$	H, CH_3, OCH_3, OC_2H_5, SCH_3, F, Cl, NO_2	29
2-Thienylcarbonyl	CH_3	$(CH_2)_3COOH$	CH_3	29, 32, 113
2-Thienylcarbonyl	CH_3	$(CH_2)_3COOH$	OCH_3	29, 31, 32
2-Thienylcarbonyl	CH_3	$(CH_2)_3COOH$	H, OC_2H_5, F	29, 32
2-Thienylcarbonyl	CH_3	$(CH_2)_3COOH$	SCH_3, Cl, NO_2	29
2-Thienylcarbonyl	CH_3	$(CH_2)_3COOH$	H, CH_3, OCH_3, OC_2H_5, F(c)	32
2-(5-Chlorothienyl)carbonyl	H	CH_2COOH	H, CH_3, OCH_3, OC_2H_5, SCH_3, F, Cl, NO_2	29
2-(5-Chlorothienyl)carbonyl	H	$(CH_2)_2COOH$	H, CH_3, OCH_3, OC_2H_5, SCH_3, F, Cl, NO_2	29
2-(5-Chlorothienyl)carbonyl	H	$CH(CH_3)COOH$	H, CH_3, OCH_3, OC_2H_5, SCH_3, F, Cl, NO_2	29
2-(5-Chlorothienyl)carbonyl	H	$(CH_2)_3COOH$	H, CH_3, OCH_3, OC_2H_5, SCH_3, F, Cl, NO_2	29
2-(5-Chlorothienyl)carbonyl	CH_3	CH_3COOH	H, OCH_3, OC_2H_5, SCH_3, F, Cl, NO_2	29, 33
2-(5-Chlorothienyl)carbonyl	CH_3	CH_3COOH	CH_3	29, 33, 113
2-(5-Chlorothienyl)carbonyl	CH_3	CH_2COOCH_3	CH_3, OCH_3	31
2-(5-Chlorothienyl)carbonyl	CH_3	$(CH_2)_2COOH$	H, CH_3, OCH_3, OC_2H_5, SCH_3, F, Cl, NO_2	29
2-(5-Chlorothienyl)carbonyl	CH_3	$CH(CH_3)COOH$	H, CH_3, OCH_3, OC_2H_5, SCH_3, F, Cl, NO_2	29
2-(5-Chlorothienyl)carbonyl	CH_3	$(CH_2)_3COOH$	H, CH_3, OCH_3, OC_2H_5, SCH_3, F, Cl, NO_2	29
3-(2-Thienyl)acryloyl	CH_3	CH_2COOH	OCH_3	14, 101
3-(2-Thienyl)acryloyl	CH_3	CH_2COOH	$N(CH_3)_2$	104

3-(2-Thienyl)acryloyl	CH_3	$CH_2COOC_2H_5$	OCH_3	17
4-(2-Chlorothienyl)carbonyl	CH_3	CH_2COOH	OCH_3	115
4-(2-Bromothienyl)carbonyl	CH_3	CH_2COOH	OCH_3	115
2-Quinolylcarbonyl	CH_3	CH_2COOH	OCH_3	36
Phthalimidoacetyl	H	C_6H_5	Cl	117, 118
Phthalimidoacetyl	CH_3	CH_2COOH	OCH_3	119
3-Phthalimidopropionyl	CH_3	$(CH_2)_3COOH$	OCH_3	119
CH_3O, $CH_2C(=O)$, N, CH_3, R (R=H and CH_3)	CH_3	CH_2COOH	OCH_3	120
Phenylsulphonyl	CH_3	CH_2COOH	H, OCH_3	121
4-Chlorophenylsulphonyl	CH_3	CH_2COOH	CH_3, OCH_3	121
4-Methylphenylsulphonyl	CH_3	CH_2COOH	OCH_3	122

Notes

a. 5,6-Methylenedioxy; b. In reference 29, the methyl, ethyl, *tert.*-butyl and benzyl esters were also prepared in every case by use of the corresponding ketoester in the indolizations; c. With $R^3 = CH_3$; d. The location(s) of the aromatic substituent(s) at position 3 appear to be incorrectly quoted in the abstract, since, from the nature of the reactants, their actual location(s) would be at position 4; e. Product has ^{14}C at the indolic C-2 position; f. Indomethacin; g. With $R^4 = CH_3$; h. 5,6-$O(CH_2)_2O$; i. 4,5-CH=CH—C(OCH_3)=CH—; j. 5,6-Trimethylene; k. 5,6-Pentamethylene.
Direct syntheses of other 1-(4-halogeno-, 4-methoxy- and 3,5-dinitrobenzoyl)indoles have recently been effected by this route (CRC Compagnia di Ricerca Chimica S.A., 1979).

Table 23. (*continued*)

References

1. Yamamoto, H., Nakamura, Y., Atami, Nakao and Kobayashi, T., 1970e; 2. Yamamoto, H., 1967b; 3. Yamamoto, H. and Nakao, 1968b; 4. Yamamoto, H. and Nakao, 1971a; 5. Yamamoto, H. and Nakao, 1969a; 6. Sumitomo Chemical Co. Ltd, 1968c; 7. Yamamoto, H., Nakao and Atami, 1969; 8. Sumitomo Chemical Co. Ltd, 1968b; 9. Yamamoto, H., 1967a; 10. Yamamoto, H. and Atsumi, 1968b; 11. Yamamoto, H., Nakao and Kobayashi, T., 1969; 12. Yamamoto, H. and Nakao, 1970b; 13. Yamamoto, H. and Nakao, 1970c; 14. Yamamoto, H. and Nakao, 1970g; 15. Kimura, Inaba and Yamatoto, H., 1973c; 16. Yamamoto, H. and Nakao, 1968a; 17. Yamamoto, H., Inaba, Nakamura, Y., Nakao, Niizaki, M. and Atami, 1970; 18. Yamamoto, H. and Nakao, 1970h; 19. Yamamoto, H. and Atami, 1969; 20. Yamamoto, H., Nakamura, Y., Atami, Nakao, Kobayashi, T., Saito, C. and Kurita, 1969; 21. Yamamoto, H., Nakamura, Y., Atami, Nakao, Kobayashi, T., Saito, C. and Awata, 1970; 22. Bikova, Vitev, Dyankova and Ilarionov, 1972; 23. Sumitomo Chemical Co. Ltd, 1968a; 24. Yamamoto, H., Nakamura, Y., Nakao and Kobayashi, T., 1969a; 25. Yamamoto, H., Nakamura, Y., Nakao and Kobayashi, T., 1970b; 26. Yamamoto, H., Nakamura, Y., Nakao and Kobayashi, T., 1970a; 27. Bikova, Vitev and Ilarionov, 1974; 28. Sumitomo Chemical Co. Ltd, 1969b; 29. Yamamoto, H. and Nakao, 1974a; 30. Kosa and Kovacs, 1970; 31. Sumitomo Chemical Co. Ltd, 1966; 32. Yamamoto, H., and Nakao, 1971b; 33. Yamamoto, H. and Nakao, 1974b; 34. Yamamoto, H. and Nakao, 1973; 35. Kosa and Kovacs, 1975; 36. Pakula, Wojciehowski, Poslinska, Pichnej, Ptaszynski, Przepalkowski and Logwinienko, 1969; 37. Pakula, Wojciechowski, Poslinska, Pichnej, Ptaszynski, Przepalkowski and Logwinienko, 1975; 38. Yamamoto, H., Nakao and Kobayashi, T., 1968; 39. Yamamoto, H., Nakao and Atsumi, 1968a; 40. Yamamoto, H., Nakamura, Y., Atami, Nakao and Kobayashi, T., 1970b; 41. Alexander and Mooradian, 1977a; 42. Alexander and Mooradian, 1976; 43. Alexander and Mooradian, 1975; 44. Sorrentino, 1968; 45. Kosa and Kovacs, 1972; 46. Yamamoto, H., Nakao and Kurita, 1968a; 47. Kimura, Inaba and Yamamoto, H., 1973d; 48. Yamamoto, H., Nakao and Atsumi, 1968b; 49. Biniecki and Jakubowskii, 1974; 50. Nakatsuka, Hazue, Makari, Kawahara, Endo and Yoshitake, 1976; 51. Yamamoto, H., Saito, C., Okamoto, Awata, Inukai, Hirohashi, A. and Yukawa, 1969; 52. Yamamoto, H., Atsuko and Takuhiro, 1970; 53. Yamamoto, H., Misaki and Izumi, 1968, 1970; 54. Okamoto, Kobayashi, T. and Yamamoto, H., 1974; 55. Okamoto, Kobayashi, T. and Yamamoto, H., 1970; 56. Yamamoto, H., Okamoto and Kobayashi, T., 1970; 57. Sumitomo Chemical Co. Ltd, 1967; 58. Yamamoto, H. and Nakao, 1968c; 59. Yamamoto, H., Misaki and Imanaka, 1968; 60. Kimura, Inaba and Yamamoto, H., 1973a; 61. Sletzinger, Gal and Chemerda, 1968; 62. Bertazzoni, Bartoletti and Perlotto, 1970; 63. Brenner, Clamkowski, Hinkley and Gal, 1969; 64. Firestone and Sletzinger, 1970; 65. Fisnerova and Nemecek, 1976; 66. Sletzinger and Gal, 1968b; 67. Tani, H., Otani, Mizutani and Mashimo, 1969; 68. Yamamoto, H., 1968; 69. Yamamoto, H. and Nakao, 1970i; 70. Ellsworth, Gatto, Meriwether and Mertel, 1978; 71. Yamamoto, H., Nakamura, Y., Atami, Nakao and Kobayashi, T., 1970f; 72. Cassebaum, Dierbach and Hilger, 1970; 73. Yamamoto, H. and Nakao, 1969c; 74. Kimura, Inaba and Yamamoto, H., 1973b; 75. Boltze, Opitz, Raddatz, Seidel, Jacobi, Dell and Schoellnhammer, 1979; 76. Yamamoto, H., Inaba, Nakao and Koshiba, 1972; 77. Sletzinger and Gal, 1968a; 78. Noda, Nakagawa, Miyata, Nakajima and Ide, 1979; 79. Boltze, Brendler, Dell and Jacobi, 1974; 80. Glamkowski, Gal and Sletzinger, 1973; 81. Yamamoto, H., Nakao and Kurita, 1968b; 82. Yamamoto, H., Nakamura, Y., Atami, Nakao and Kobayashi, T., 1970h; 83. Yamamoto, H., Nakamura, Y., Atami, Nakao and Kobayashi, T., 1970g; 84. Brenner, Glamkowski, Hinkely and Russ, 1969; 85. Yamamoto, H. and Nakao, 1971c; 86. Yamamoto, H. and Nakao, 1968e; 87. Yamamoto, H., Nakamura, Y., Atami, Nakao and Kobayashi, T., 1970c; 88. Yamamoto, H. and Atami, 1970; 89. Kajfez and Mihalic, 1978; 90. Yamamoto, H., Nakamura, Y., Nakao, Atsumi and Kobayashi, T., 1970; 91. Morita, Naguchi, Kishimoto, Agata and Otsuka, 1972; 92. Yamamoto, H. and Nakao, 1968d; 93. Nakao, Takahashi, K. and Yamamoto, H., 1971; 94. Yamamoto, H. and Nakao, 1968g; 95. Yamamoto, H. and Nakao, 1969d; 96. Yamamoto, H. and Nakao, 1970d; 97. Yamamoto, H. and Nakao, 1970f; 98. Yamamoto, H. and Kimura, 1974; 99. Okamoto, Niizaki, M., Kobayashi, T., Izumi, Yamamoto, H., Inaba, Nakamura, Y. and Nakao, 1971; 100. Yamamoto, H., Nakamura, Y., Atami, Kobayashi, T. and Nakao, 1971; 101. Okamoto, Niizaki, S., Kobayashi, T., Izumi and Yamamoto, H., 1972; 102. Yamamoto, H. and Nakao, 1969e; 103. Yamamoto, H. and Nakao, 1969b; 104. Shen, T.-Y., Gal and Utne, 1970; 105. Kimura, Inaba and Yamamoto, H., 1972; 106. Sumimoto Chemical Co. Ltd, 1969a; 107. Yamamoto, H., Nakamura, Y., Atami, Nakao and Kobayashi, T., 1970d; 108. Nakao, Katayama and Yamamoto, H., 1972a; 109. Nakao, Katayama and Yamamoto, H., 1972b; 110. Yamamoto, H. and Nakao, 1970e; 111. Yamamoto, H., Misaki and Izumi, 1970; 112. Yamamoto, H., Nakao and Okamoto, 1968; 113. Yamamoto, H., Inaba, Nakao and Niizaki, M., 1970; 114. Yamamoto, H. and Nakao, 1970a; 115. Toth, Szabo, Eibel and Somfai, 1972; 116. Toth, Szabo, Eibel and Somfai, 1973; 117. Ishizumi, Mori, Inaba, and Yamamoto, H., 1973; 118. Yamamoto, H., Inaba, Izumi, Okamoto, Hirohashi, T., Ishizumi, Maruyama, Kobayashi, T., Yamamoto, M. and Mori, 1971; 119. Yamamoto, H., Nakamura, Y., Atami, [illegible] 121. Yamamoto, H. and Nakao

Yamamoto, H., 1974; Yamamoto, H., Inaba, Izumi, Okamoto, Hirohashi, T., Ishizumi, Maruyama, Kobayashi, T., Yamamoto, M. and Mori, 1971; Yamamoto, H., Nakamura, Y., Atami, Nakao and Kobayashi, T., 1970e; Yamamoto, H., Nakamura, Y. and Kobayashi, T., 1970; Yamamoto, H., Nakamura, Y., Nakao, Atsumi and Kobayashi, T., 1970; Yamamoto, H. and Nakao, 1969a, 1969b, 1970i).

An interesting synthesis of lower alkyl 1-acyl-2-methyl-3-indolylacetates has also been effected by reacting N_α-acylarylhydrazines with acetol (**17**; $R^1 = H$, $R^2 = OH$, $R^3 = CH_3$) and the appropriate esters of 2-haloacetic acid (Yamamoto, H., Nakamura, Y. and Kobayashi, T., 1970).

In view of these intensive studies upon the application of this synthetic route to 1-acyl- or 1-aroylindoles and especially because of its many earlier applications to the synthesis of **1202** (R^1–$R^4 = H$, $R^5 = COC_6H_5$, $R^6 = CH_3$, $R^7 = CH_2COOH$) (see Table 23), the recent patenting (Francia Barra and Carmelo Marin Moga, 1979) of yet another synthesis of this product by this route, involving heating a mixture of N_α-benzoylphenylhydrazine hydrochloride with laevulinic acid at 130–140 °C, appears to be superfluous.

The early studies of H. Yamamoto and his co-workers in this area included a study of the synthesis of N_α-acyl- and N_α-aroylarylhydrazines and the corresponding hydrazones (Yamamoto, H., Nakao and Kobayashi, A., 1968). 4-Methoxyphenylhydrazine hydrochloride was aroylated on both N_α and N_β when it was reacted with benzoyl chloride in the presence of a hydrogen chloride acceptor. However, prior formation of the acetaldehyde or benzaldehyde 4-methoxyphenylhydrazone directed the benzoylation specifically to N_α, the hydrazone C=N bond in both products subsequently being hydrolysed with alcoholic hydrogen chloride to afford N_α-benzoyl-4-methoxyphenylhydrazine hydrochloride. Alternatively, the acetaldehyde or benzaldehyde N_α-benzoyl-4-methoxyphenylhydrazone could be reacted with a potentially indolizable ketone (e.g. laevulinic acid) under acidic conditions which effected the establishment of ketone–hydrazone equilibria from which the laevulinic acid N_α-benzoyl-4-methoxyphenylhydrazone was irreversibly removed by indolization. The ethyl ester of this last compound could also be prepared by N_α-benzoylation of ethyl laevulinate 4-methoxyphenylhydrazone in the presence of a hydrogen chloride acceptor. Another synthesis of N_α-acyl- and N_α-aroylarylhydrazines was achieved by treating sodium arylhydrazine-β-sulphonates, **1205**, with the appropriate acid chloride in pyridine or aqueous *tert.*-butanol, followed by acid catalysed hydrolysis of the sulphonic acid group (Karady, Ly, Pines, Chemerda

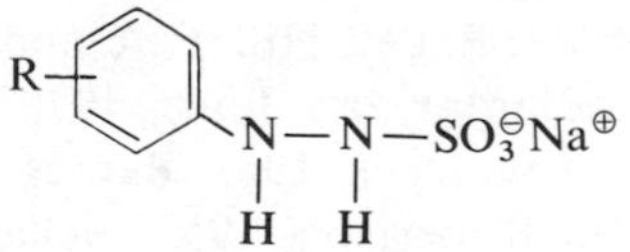

R–C_6H_4–N(H)–N(H)–$SO_3^{\ominus}Na^{\oplus}$

1205

and Sletzinger, 1973, Sletzinger and Gal, 1968b; Toth, Szabo, Eibel and Somfai, 1973).

The indolizations referred to in Table 23 utilized preformed N_α-acyl- or N_α-aroylarylhydrazones, or their component carbonyl and hydrazine moieties. However, when methyl laevulinate 4-methoxyphenylhydrazone was heated in acetic acid with acetyl chloride and 2,4-hexadienoyl chloride and when ethyl laevulinate 4-methoxyphenylhydrazone was similarly treated with 2,4-hexadienoyl chloride, the corresponding 1-acylated indoles **1202** [$R^1 = R^3 = R^4 = H$, $R^2 = OCH_3$, $R^5 = COCH_3$ and $CO(CH = CH)_2CH_3$, $R^6 = CH_3$, $R^7 = CH_2COOCH_3$ and $R^1 = R^3 = R^4 = H$, $R^2 = OCH_3$, $R^5 = CO(CH = CH)_2CH_3$, $R^6 = CH_3$, $R^7 = CH_2COOC_2H_5$, respectively] were produced (Yamamoto, H. and Nakao, 1970j). It remains to be established at which stage the N-acylations occurred in these reactions which certainly appear to be worthy of further application since they circumvent the isolation of the N_α-acyl- or N_α-aroylarylhydrazines or the corresponding hydrazones.

D. Synthesis of 2,3-Polymethylene Indoles and Related Compounds

Indolizations of cycloalkanone arylhydrazones **1206** afford **1207**, examples of such transformations being given in Table 24. As well as these examples, the hydrazones formed between cyclopentanone and N_α-[(2-methyl-5-pyridyl)-ethyl]phenylhydrazine (Vinogradova, Daut, Kost and Terent'ev, A. P., 1962),

R^1-C$_6$H$_4$-N(R^2)-N=C(CH$_2$)(CH$_2$)$_n$ ⟶ R^1-indole(N-R^2)-(CH$_2$)$_n$

1206 **1207**

6- and 8-chloro-4-quinolylhydrazine (Khan, M. A. and Rocha, 1978), 3-methyl-5-isoquinolylhydrazine (Buu-Hoï, Jacquignon, Roussel, O. and Hoeffinger, 1964), **624** ($R = CH_3$) (Buu-Hoï, Saint-Ruf, Martani, Ricci and Balucani, 1968) and 3-carbethoxy-2-methyl-4-thienylhydrazine (Shvedov, Trofimkin, Vasil'eva and Grinev, A. N., 1975), the hydrazones **1139** [$R^1 = H$, $R^2 + R^3 = (CH_2)_3$] (Cattanach, Cohen, A. and Heath-Brown) and **555** (Kempter, Schwalba, Stoss and Walter, 1962), the hydrazones formed between cyclohexanone and a wide range of arylhydrazines (Ainsworth, D. P. and Suschitzky, 1967a; Baeyer, 1894; Baeyer and Tutein, 1889; Bailey, Hill, P. A. and Seager, 1974; Bajwa and Brown, R. K., 1968a; Baradarani and Joule, 1978; Barclay and Campbell, 1945; Barnes, Pausacker and Badcock, 1951; Barnes, Pausacker and Schubert, 1949; Beer, Broadhurst and Robertson, 1952; Bellamy and Guthrie, 1965b; Binder, Habison and Noe, 1977; Bloss and Timberlake, 1963; Borsche and

Table 24. Products Resulting from the Indolization of Various Cycloalkanone (Excluding Cyclohexanone) Arylhydrazones

Product **1207** (Unless otherwise stated, R^1 and R^2 = H)

R^1	R^2	*n*	Reference(s)
		3	1–9
	CH_3	3	1, 10
	$(CH_2)_2CN$	3	11, 12
	$CH_2C_6H_5$	3	13
5-CH_3		3	4
5-C_2H_5		3	14
5-$(CH_2)_2CH_3$		3	14
5-$CH(CH_3)_2$		3	3, 14
4-C_6H_5		3	3
5-C_6H_5		3	3, 14
6-C_6H_5		3	3
5-Cl		3	15
5-Br		3	16
5-OCH_3		3	17
5-OC_2H_5		3	17
5-$OCH(CH_3)_2$		3	14
5-$OCH_2C_6H_5$		3	3
5,6-OCH_2O		3	14
4-NO_2		3	18, 19
5-NO_2		3	18
6-NO_2		3	18, 19
7-NO_2		3	18
7-$N(CH_3)_2$		3	9
7-Piperidino		3	9
7-Morpholino		3	9
7-[1-(4-Methylpiperazinyl)]		3	9
5-Cl, 4-NO_2		3	15
5-Cl, 6-NO_2		3	15
		5	2, 6, 9, 20–24
	CH_3	5	10, 20
	$(CH_2)_2CN$	5	12
5-CH_3		5	22
5-$COOC_2H_5$		5	25
5-F		5	22
5-Cl		5	22
5-Br		5	21
5-OCH_3		5	22
4-NO_2		5	26
5-NO_2		5	22, 26
6-NO_2		5	26
7-NO_2		5	27
7-$N(CH_3)_2$		5	9
7-Piperidino		5	9
7-Morpholino		5	9

(*continued*)

Table 24. (*continued*)

R^1	R^2	*n*	Reference(s)
7-[1-(4-Methylpiperazinyl)]		5	9
		6	22, 28, 29
	CH_3	6	2, 28
$4\text{-}NO_2$		6	30
$5\text{-}NO_2$		6	22, 30
$6\text{-}NO_2$		6	30
$7\text{-}NO_2$		6	27, 30
		7	22, 31
$5\text{-}CH_3$		7	31
$4\text{-}NO_2$		7	30
$5\text{-}NO_2$		7	30
$6\text{-}NO_2$		7	30
$7\text{-}NO_2$		7	30
		8	22, 31
$5\text{-}CH_3$		8	31
$4\text{-}NO_2$		8	30
$5\text{-}NO_2$		8	30
$6\text{-}NO_2$		8	30
$7\text{-}NO_2$		8	30
		9	22
$4\text{-}NO_2$		9	30
$5\text{-}NO_2$		9	30
$6\text{-}NO_2$		9	30
$7\text{-}NO_2$		9	30
		10	22, 32
$5\text{-}CH_3$		10	32
$4\text{-}NO_2$		10	30
$5\text{-}NO_2$		10	30
$6\text{-}NO_2$		10	30
$7\text{-}NO_2$		10	30
		11	22, 33
$4\text{-}NO_2$		11	30
$5\text{-}NO_2$		11	30
$6\text{-}NO_2$		11	30
$7\text{-}NO_2$		11	30
$4\text{-}NO_2$		12	30
$5\text{-}NO_2$		12	30
$6\text{-}NO_2$		12	30
$7\text{-}NO_2$		12	30
		13	22, 33
	CH_3	13	28
$4\text{-}NO_2$		13	30
$5\text{-}NO_2$		13	30
$6\text{-}NO_2$		13	30
$7\text{-}NO_2$		13	30
		14	33

Table 24. (*continued*)

References

1. Perkin and Plant, 1923b; Snyder and Smith, C. W., 1943; Watanabe, Y., Yamamoto, M., Shim, Miyanaga and Matsudo, 1980; 2. Gale and Wilshire, 1974; 3. Jones, G. and Tringham, 1975; 4. Kelly, A. H., McLeod and Parrick, 1965; 5. Grammaticakis, 1939 (see note); 6. Kanaoka, Ban, Miyashita, Irie and Yonemitsu, 1966; 7. Bailey, Scottergood and Warr, 1971; 8. Shimizu, J., Murakami, S., Oishi and Ban, 1971; 9. Ainsworth, D. P. and Suschitzky, 1967a; 10. Gale, Lin and Wilshire, 1976; 11. Hahn, W. E., Bartnik, Zawadzka and Nowaczyk, W., 1968; 12. Hahn, W. E., Nowaczyk, M. and Bartnik, 1968; 13. Namis, Cortes, Collera and Walls, F., 1966; 14. Jones, G., 1975; 15. Massey and Plant, 1931; 16. Plant and Tomlinson, M. L., 1931; 17. Kempter, Schwalba, Stoss and Walter, 1962; 18. Plant, 1929; 19. Plant and Whitaker, 1940; 20. Plancher, Cecchetti and Ghigi, 1929; Plancher and Ghigi, 1929; 21. Perkin and Plant, 1928; 22. Rice, Hertz and Freed, 1964; 23. Witkop, 1953a; 24. Anderson, A. G. and Tazuma, 1952; 25. Aksanova, Pidevich, Sharkova, L. M. and Kucherova, 1968; 26. Frasca, 1962a; 27. Sparatore and Cerri, 1968; 28. Jacquignon and Buu-Hoï, 1956; 29. Witkop, Patrick and Rosenblum, 1951; 30. Dennler and Frasca, 1966a; 31. Buu-Hoï, Jacquignon and Loc, 1958; 32. Jacquignon, Buu-Hoï and Dufour, 1966; 33. Buu-Hoï, 1949a.

Note

The reaction product was erronesously formulated by this author as **209** although it was correctly named (i.e. as 2,3-trimethyleneindole) in his text.

Kienitz, 1910; Borsche, Witte, A. and Bothe, 1908; Braun, 1908b; Braun and Schörnig, 1925; Bruck, 1970; Bülow, 1918; Buu-Hoï, Jacquignon, Roussel, O. and Hoeffinger, 1964; Buu-Hoï, Saint-Ruf, Deschamps, Bigot and Hieu, 1971; Campaigne, Ergener, Hallum and Lake, 1959; Carlin and Fisher, 1948; Carlin and Larson, 1957; Carlin and Moores, 1959, 1962; Carlin, Wallace, J. G. and Fisher, 1952; Cattanach, Cohen, A and Heath-Brown, 1973; Chalmers, J. R., Openshaw and Smith, G. F., 1957; Clemo and Felton, 1951a, 1952; Clemo and Perkin, 1924b; Clemo, Perkin and Robinson, R., 1924; Cohylakis, Hignett, Lichman and Joule, 1974; Coldham, Lewis, J. W. and Plant, 1954; Collar and Plant, 1926; Cranwell and Saxton, 1962, Crum and Sprague, 1966; Davidge, 1959; Dewar, 1944; Edwards and Plant, 1923; Elgersma, 1969 Esayan, Terzyan, Astratyan, Dzhanpoladyan and Tatevosyan, 1968; Farbenfabriken Bayer A.-G., 1955b; Gale, Lin and Wilshire, 1976; General Aniline Works Inc., 1932; Govindachari, Rajappa and Sundarsanam, 1963; Govindachari and Sudarsanam, 1967; Grammaticakis, 1939, 1960; Grandberg, Sibiryakova and Brovkin, 1969; Hahn, W. E., Bartnik, Zawadzka and Nowaczyk, W., 1968; Hahn, W. E., Kryczka and Bartnik, 1974; Hahn, W. E., Nowaczyk, M. and Bartnik, 1968; Hahn, W. E. and Zawadzka, 1967, 1969; Hahn, W., Kryczka and Bartnik, 1975; Hahn, W., Nowaczyk, M. Bartnik and Zawadzka, 1968; Herdieckerhoff and Tschunkur, 1933; Horning, E. C., Horning, M. G. and Walker, G. N., 1948; Hoshino and Takiura, 1936; Huisgen, 1948a; Kanaoka,

Ban, Miyashita, Irie and Yonemitsu, 1966; Khan, M. A. and Morley, 1978, 1979; Khan, M. A. and Rocha, 1978; Kochetkov, Kucherova and Evdakov, 1956, 1957; Kost, Sugorova and Yakubov, 1965; Kost, Terent'ev, A. P., Vinogradova, Terent'ev, P. B. and Ershov, 1960; Kost, Vinogradova, Trofimov, Mukhanova, Nozdrich and Shadurskii, 1967; Kost, Yudin, Berlin, Y. A. and Terent'ev, A. P., 1959; Kost, Yudin, Dmitriev, B. A. and Terent'ev, A. P., 1959; Kost, Yudin and Terent'ev, A. N., 1959; Linnell and Perkin, 1924; Lions and Ritchie, 1939a, 1939b; Long, R. S., 1956; Manske, R. H. F. and Kulka, 1947, 1949; Marshalkin and Yakhontov, 1972b; Milne and Tomlinson, M. L., 1952; Namis, Cortes, Collera and Walls, F., 1966; Nandi, 1940; Noland and Baude, 1966; Nozoe, Sin, Yamane and Fujimori, K., 1975; Oakeshott and Plant, 1928; Orlova, E. K., Sharkova, N. M., Meshcheryakova, Zagorevskii and Kucherova, 1975; Pausacker and Schubert, 1949b; Pecca and Albonico, 1970; Perkin and Plant, 1921; Petrova, T. D., Mamaev and Yakobson, 1967, 1969; Plant, 1936; Plant and Rosser, 1928; Plant and Tomlinson, M. L., 1932a, 1932b; Plant and Williams, S. B. C., 1934; Rogers, C. U. and Corson, 1947, 1950, 1963; Schmutz and Wittwer, 1960; Sharkova, N. M., Kucherova and Zagorevskii, 1964, 1972b, 1972c; Shaw, E. and Woolley, D. W., 1957; Shvedov, Trofimkin, Vasil'eva and Grinev, A. N., 1975; Sibiryakova, Brovkin, Belyakova and Grandberg, 1969; Smith, A. and Utley, 1970; Snyder and Smith, C. W., 1943; Sumitomo Chemical Co. Ltd, 1968b; Thomas, R. C., 1975; Venuto and Landis, 1968; Vinogradova, Daut, Kost and Terent'ev, A. P., 1962; Welch, W. M., 1977a, 1977b; Winchester and Popp, 1975; Yamada, S., Chibata and Tsurui, 1953; Yamada, S. *et al.*, 1955; Yamamoto, H., Nakamura, Y., Atami, Nakao and Kobayashi, T., 1970b; Yamamoto, H., Nakamura, Y., Nakao, Atsumi and Kobayashi, T., 1970), between cycloheptanone and 6-quinolylhydrazine (Buu-Hoï, Périn and Jacquignon, 1960), 3-methyl-5-isoquinolylhydrazine (Buu-Hoï, Jacquignon, Roussel, O. and Hoeffinger, 1964), 3-carbethoxy-2-methyl-4-thienylhydrazine (Shvedov, Trofimkin, Vasil'eva and Grinev, A. N., 1975), 1-aminoindoline (Bailey, Hill, P. A. and Seager, 1974) and 1-amino-1,2,3,4-tetrahydroquinoline (Bahadur, Bailey and Baldry, 1977), the hydrazone **1139** [$R^1 = H, R^2 + R^3 = (CH_2)_5$] (Cattanach, Cohen, A. and Heath-Brown, 1973) and cyclooctanone 6-quinolylhydrazone (**1208**) (indolization presumably occurring at the 5-position of the quinolyl nucleus – see Chapter III, E) (Buu-Hoï, Saint-Ruf, Jacquignon and Barrett, G. C., 1958) have also been successfully indolized. Similarly, compounds **1209** (n = 5, 7, 8 and 10) have been converted

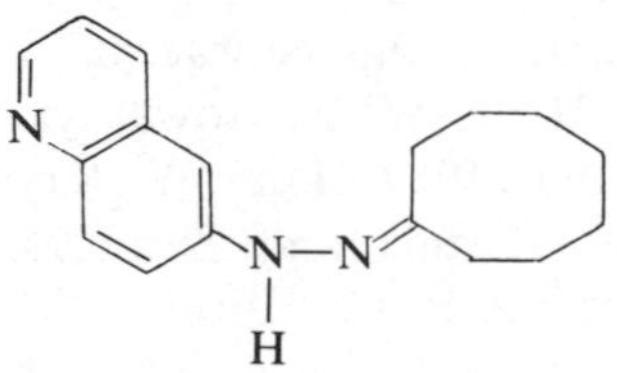

N
N—N
H

1208

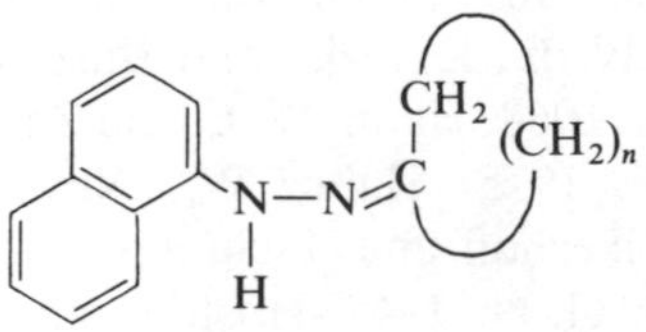

CH_2
$(CH_2)_n$
N—N=C
H

1209

into **1210** [$n = 5$ (Freed, Hertz and Rice, 1964), 7, 8 (Buu-Hoï, Jacquignon and Loc, 1958) and 10 (Jacquignon, Buu-Hoï and Dufour, 1966), respectively], **1211** furnished **1212** (Jacquignon, Buu-Hoï and Dufour, 1966) and **1213** furnished a product formulated as **1214** (Buu-Hoï, 1949a). Other arylhydrazones of cyclopentanone, cycloheptanone and cyclooctanone have also been indolized (Sellstedt and Wolf, M., 1974).

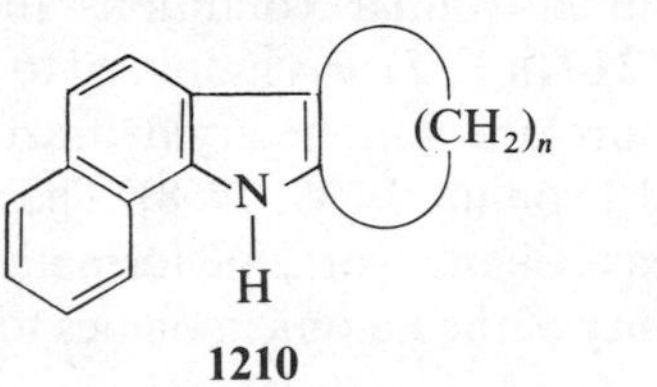

1210

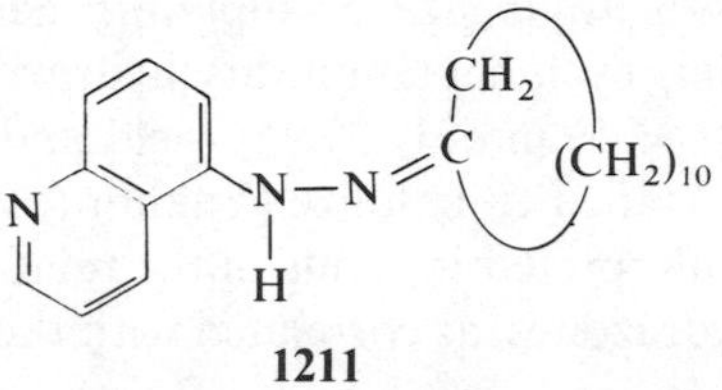

1211

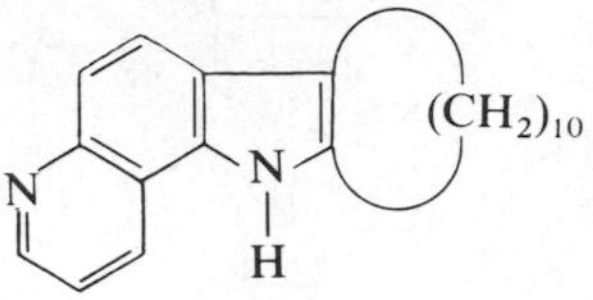

1212

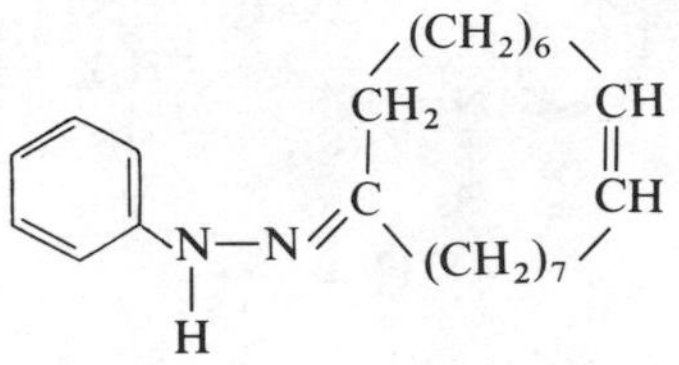

1213

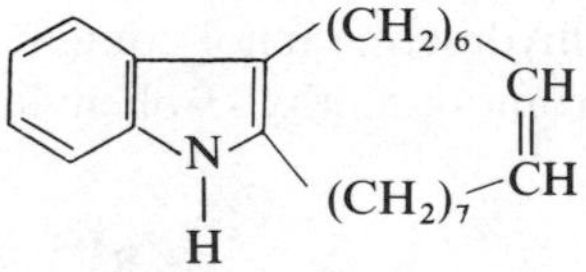

1214

Although **1215** ($R^1 = R^2 = H$, $n = 1$ and 3) (Borsche and Kienitz, 1910) and **1215** ($R^1 = H$ and CH_3, $R^2 = CH_3$, $n = 2$) (Braun, 1908b) were indolized to afford **1216** ($R^1 = R^2 = H$, $n = 1$ and 3) and **1216** ($R^1 = H$ and CH_3,

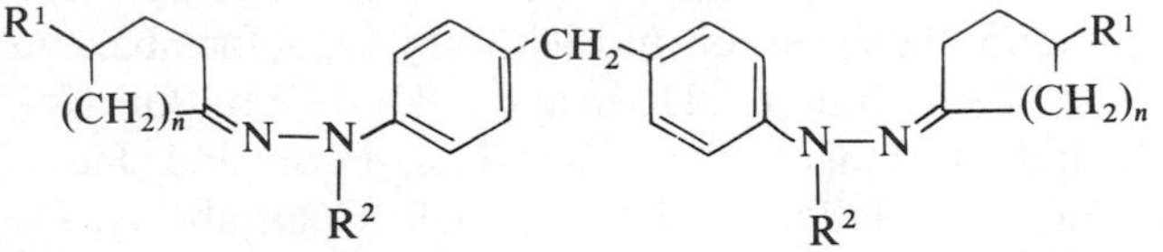

1215

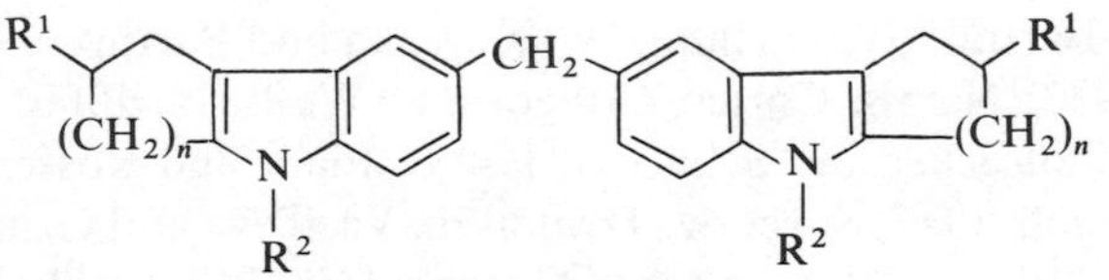

1216

$R^2 = CH_3$, $n = 2$), respectively, no definite product could be obtained from the cyclopentanone hydrazone **1215** ($R^1 = H$, $R^2 = CH_3$, $n = 1$) (Braun, 1908b). A similar failure attended another attempt to indolize a cyclopentanone arylhydrazone. Thus, whereas cyclohexanone and cyclooctanone 6-bromo-2-pyridylhydrazones (**1217**; $n = 4$ and 6, respectively) were indolized into **1218** ($n = 4$ and 6, respectively) in 93.6% and 100% (?) yields, respectively, by heating with 4-methylbenzene sulphonic acid, under similar conditions the corresponding cycloheptanone arylhydrazone **1217** ($n = 5$) was indolized to produce **1218** ($n = 5$) in only 29.9% yield and the cyclopentanone arylhydrazone **1217** ($n = 3$) failed to indolize (Yakhontov and Pronina, E. V., 1968). These results were interpreted in terms of the relative ease of enehydrazine formation in the arylhydrazones, as correlated with the ability of the ketonic moieties to enolize,

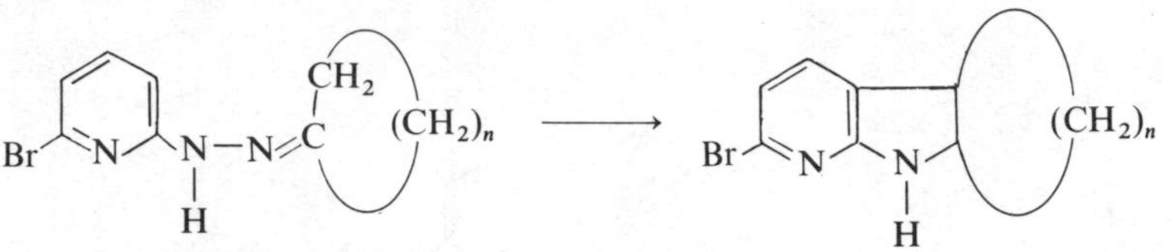

1217 → **1218**

this latter being of the order cyclooctanone > cyclohexanone > cycloheptanone ≫ cyclopentanone. Similarly, acetone 6-bromo-2-pyridylhydrazone (enol content in acetone = 1.5×10^{-4}%) did not indolize whereas benzyl methyl ketone 6-bromo-2-pyridylhydrazone (enol content in benzyl methyl ketone = 2.9%) afforded 7-aza-6-bromo-2-methyl-3-phenylindole (Yakhontov and Pronina, E. V., 1968).

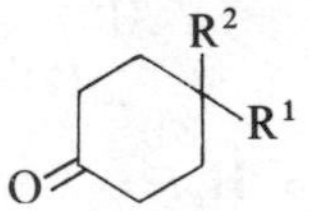

1219

Arylhydrazones of the 4-substituted cyclohexanones **1219** ($R^1 = H$, $R^2 = CH_3$) (Braun, 1908b; Braun and Schörnig, 1925; Brimblecombe, Downing and Hunt, R. R., 1966; Buu-Hoï, Hoán and Khôi, 1950b; Buu-Hoï, Jacquignon, Roussel, O. and Hoeffinger, 1964; Buu-Hoï, Périn and Jacquignon, 1960; Campaigne, Ergener, Hallum and Lake, 1959; Chakraborty, Das, K. C. and Chowdhury, 1966a, 1966b, 1969; Clemo and Felton, 1951a; Cranwell and Saxton, 1962; Deorha and Joshi, S. S. 1961; Grammaticakis, 1939; Hahn, W. E., Kryczka and Bartnik, 1974; Hahn, W., Kryczka and Bartnik, 1975; Moggridge and Plant, 1937; Namis, Cortes, Collera and Walls, F., 1966; Oakeshott and Plant, 1926; Pausacker and Schubert, 1949b; Plant and Rosser, 1928; Rogers, C. U. and Corson, 1947; Shvedov, Trofimkin, Vasil'eva and Grinev, A. N., 1975; Smith, A. and Utley, 1970; Sumitomo Chemical Co. Ltd, 1968b; Winchester and

Popp, 1975; Yamamoto, H., Nakamura, Y., Atami, Nakao and Kobayashi, T., 1970b), **1219** (R^1 = H, R^2 = C_2H_5) (Plant, Rogers, K. M. and Williams, S. B. C., 1935), **1219** [R^1 = H, R^2 = $C(CH_3)_3$] (Berlin, K. D., Clark, P. E., Schroeder, J. T. and Hopper, 1968; Smith, A. and Utley, 1970), **1219** (R^1 = H, R^2 = cyclohexyl) (Buu-Hoï, Binh, Loc, Xuong and Jacquignon, 1957), **1219** (R^1 = H, R^2 = CH_2OH) (ketal used) (Alexander and Mooradian, 1977a, 1977b), **1219** (R^1 = H, R^2 = CF_3) (Forbes, E. J., Stacey, Tatlow and Wragg, R. T., 1960), **1219** (R^1 = H, R^2 = COOH) (Alexander and Mooradian, 1975; Berger, L. and Corraz, 1973; Perkin, 1904a, 1904b), **1219** (R^1 = H, R^2 = $COOC_2H_5$) (Rice and Scott, K. R., 1970; Utley and Yeboah, 1978), **1219** (R^1 = H, R^2 = OCH_3) (Gilbert, J., Rousselle, Gansser and Viel, 1979; Smith, A. and Utley, 1970), **1219** (R^1 = H, R^2 = $OCOCH_3$) (Coombes, Harvey, D. J. and Reid, S. T., 1970; Harvey, D. J., 1968), **1219** (R^1 = H, R^2 = $OCOC_6H_5$) (Julia and Lenzi, 1971; Maki, Masugi, Hiramitsu and Ogiso, 1973; Teuber, Cornelius and Wölcke, 1966), **1219** (R^1 = H, R^2 = $NHCH_3$) (Mooradian, 1973, 1976b; Schut, 1968), **1219** [R^1 = H, R^2 = $N(CH_3)_2$] (Mooradian, 1973, 1975, 1976a, 1976b, 1976c; Sterling Drug Inc., 1975), **1219** [R^1 = H, R^2 = NHC_2H_5, $NH(CH_2)_3CH_3$, $NHCH_2C_6H_5$ and $NH(CH_2)_2N(C_2H_5)_2$] (Mooradian, 1976b), **1219** (R^1 = H, R^2 = 1-pyrrolidinyl (Mooradian, 1975, 1976b, 1976c), **1219** (R^1 = H, R^2 = piperidino, morpholino and 1-(4-methylpiperazinyl) (Mooradian, 1976b), **1219** (R^1 = R^2 = CH_3) (Bergman and Erdtman, 1969; Rice, Sheth and Wheeler, J. W. 1971; Smith and Utley, 1970), **1219** (R^1 = R^2 = C_2H_5) (Rice, Sheth and Wheeler, J. W., 1971), **1219** (R^1 = R^2 = COOH) (Alexander and Mooradian, 1972, 1975, 1976), **1219** (R^1 = $COO^{\ominus}$, R^2 = $\overset{\oplus}{N}H_3$) (Britten, A. and Lockwood, 1974), **1219** [$R^1 + R^2$ = $(CH_2)_{4\text{ and }5}$ and $(CH_2)_2CHCH_3(CH_2)_2$] (Rice, Sheth and Wheeler, J. W., 1971), **1219** [R^1 + R^2 = $O(CH_2)_2O$] (Britten, A. and Lockwood, 1974) and **1219** ($R^1 + R^2$ = CONHCONH) (Maki, Masugi, Hiramitsu and Ogiso, 1973) (in this case, hydrolysis of the indolization product gave the 2-aminoacid, 3-amino-1,2,3,4-tetrahydrocarbazole-3-carboxylic acid) and **1220** (R^1 = R^3 = H, R^2 = CH_3) (Braun and Haensel, 1926; Rogers, C. U. and Corson, 1947), **1220** (R^1 = H, R^2 = CH_3, R^3 = COOH) (Horning, E. C., Horning, M. G. and Walker, G. N., 1948) and **1220** [$R^1 + R^1$ = $(CH_2)_5$, R^2 = R_3 = H], a somewhat sterically hindered ketone (Rice, Sheth and Wheeler, J. W., 1971), have also been indolized. Treatment of **1221** with acetic acid afforded **1222** (R^1 = H, R^2 = $OCOC_6H_5$) which was then sequentially converted into **1222** (R^1 = H, R^2 = OH) and **1222** ($R^1 + R^2$ = O) (Teuber and Cornelius, 1965).

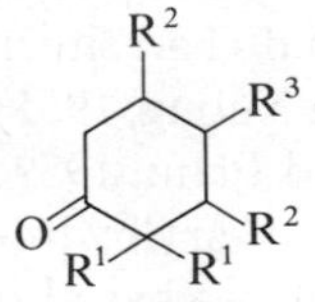

1220 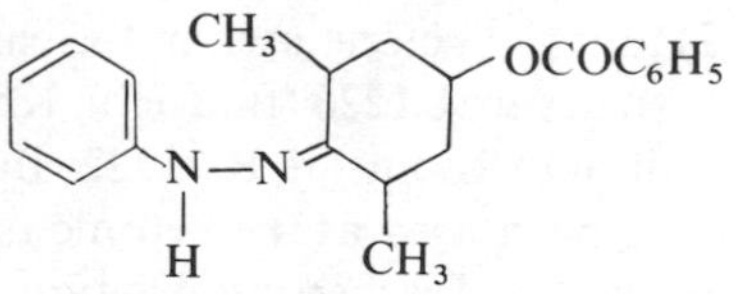 **1221** 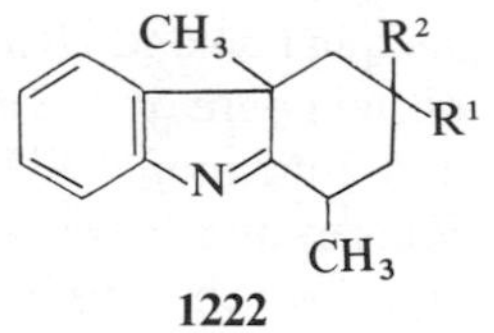 **1222**

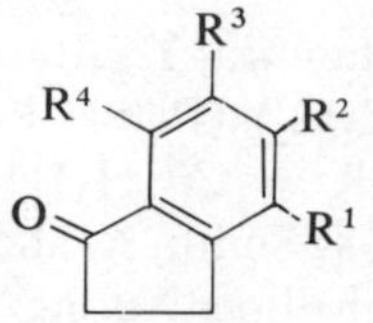

R3, R4, R2, O, R1

1223

Related to these above examples, arylhydrazones of 1-indanone (**1223**; R^1–R^4 = H) (Armit and Robinson, R., 1922; Bryant and Plant, 1931; Buu-Hoï, Hoán and Khôi, 1950a; Buu-Hoï and Jacquignon, 1951; Hausmann, 1889; Kanaoka, Ban, Miyashita, Irie and Yonemitsu, 1966; Kempter, Schwalba, Stoss and Walter, 1962; Kipping, 1894; Kollenz, 1978; Leuchs and Kowalski, 1925; Plant and Tomlinson, M. L., 1931; Seka and Kellermann, 1942; Titley, 1928; Treibs, 1959) and its substituted analogues **1223** ($R^1 = R^2 = R^4$ = H, R^3 = CH_3) (Buu-Hoï, Hoán and Khôi, 1950a; Buu-Hoï and Xuong, 1952; Seka and Kellermann, 1942), **1223** ($R^1 = R^2 = R^4$ = H, R^3 = C_2H_5) (Buu-Hoï and Xuong, 1952), **1223** [$R^1 = R^2 = R^4$ = H, R^3 = $CH(CH_3)_2$] (Buu-Hoï and Xuong, 1952), **1223** ($R^1 = R^3$ = CH_3, $R^2 = R^4$ = H) (Buu-Hoï and Xuong, 1952), **1223** ($R^1 = R^4$ = CH_3, $R^2 = R^3$ = H) (Kempter, Schwalba, Stoss and Walter, 1962), **1223** [$R^1 = R^4$ = H, $R^2 + R^3$ = $(CH_2)_{3\ \mathrm{and}\ 4}$] (Dufour, Buu-Hoï, Jacquignon and Hien, 1972), **1223** ($R^1 = R^2 = R^4$ = H, R^3 = Cl) (Buu-Hoï and Xuong, 1952; Seka and Kellermann, 1942) have been indolized. Likewise, **1224** (X = H_2, Y = O) (Dufour, Buu-Hoï, Jacquignon and Hien, 1972), **1224** (X = O, Y = H_2) (Buu-Hoï and Xuong, 1952; Dufour, Buu-Hoï, Jacquignon and Hien, 1972), **1225** (Bannister and Plant, 1948; Dufour, Buu-Hoï,

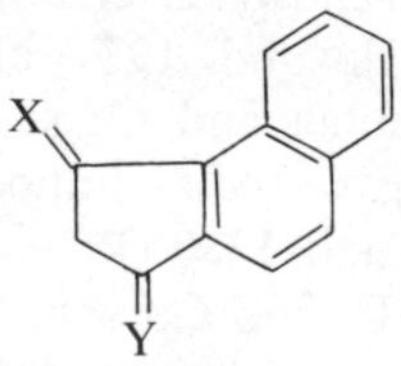

X, Y, O

1224 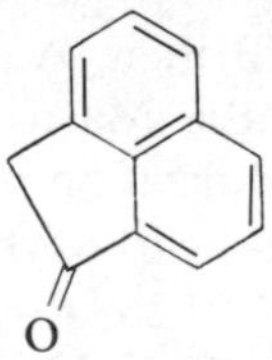 **1225**

Jacquignon and Hien, 1972; Sircar and Gopalan, 1932) [the phenylhydrazone and 1-naphthylhydrazone of this ketone were also indolized by others (Korczynski, Brydowna and Kierzek, 1926) but the products were different from specimens prepared later (Sircar and Gopalan, 1932; Dufour, Buu-Hoï, Jacquignon and Hien, 1972) which led to the suggestion (Dufour, Buu-Hoï, Jacquignon and Hien, 1972) that the ketone used in the earlier study had either been impure or had another structure], **1226** (Buu-Hoï, Khôi and Xuong, 1951) and 2-indanone (**1227**) (Armit and Robinson, R., 1922; Beyts and Plant, 1939; Kinsley and Plant, 1956) have been used as the ketonic moieties of arylhydrazones which have been subjected to Fischer indolization. Similar use has also

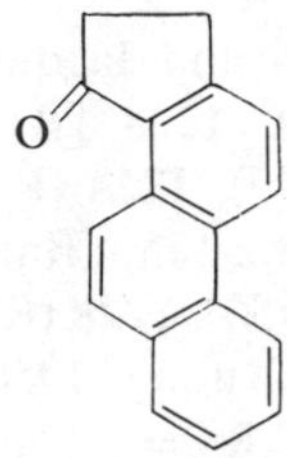

1226

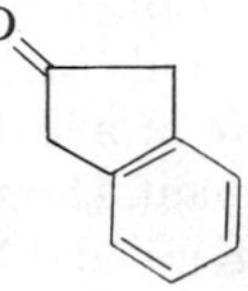

1227

been made of 1-tetralone (**1228**; R^1–R^6 = H) (Ainsworth, D. P. and Suschitzky, 1967a; Avanesova, Astvatsatryan, Sarkisyan, Garibyan and Tatevosyan, 1975; Bryant and Plant, 1931; Buu-Hoï, Hoán and Khôi, 1949, 1950b; Buu-Hoï and Jacquignon, 1951, 1956; Buu-Hoï, Jacquignon and Hoeffinger, 1963; Buu-Hoï, Jacquignon, Roussel, O. and Hoeffinger, 1964; Buu-Hoï, Mangane and Jacquignon, 1967; Buu-Hoï, Martani, Ricci, Dufour, Jacquignon and Saint-Ruf, 1966; Buu-Hoï, Périn and Jacquignon, 1960, 1962; Buu-Hoï, Saint-Ruf, Jacquignon and Marty, 1963; Buu-Hoï, Saint-Ruf, Martani, Ricci and Balucani, 1968; Driver, Matthews and Sainsbury, 1979; Felton, 1952; Freed, Hertz and Rice, 1964; Ghigi, 1930; Hahn, W. E., Kryczka and Bartnik, 1974; Hahn, W. E., Nowaczyk, M. and Bartnik, 1968; Hahn, W. E. and Zawadzka, 1967; Hahn, W., Kryczka and Bartnik, 1975; Hahn, W., Nowaczyk, M., Bartnik and Zawadzka, 1968; Kipping and Hill, A., 1899; Kollenz, 1978; Maréchal, Christiaens, Renson and Jacquignon, 1978; Mosher, Crews, Acton and Goodman, 1966; Nakazaki, 1960a; Ried and Kleemann, 1968; Rogers, C. U. and Corson, 1947, 1950; Shvedov, Trofimkin, Vasil'eva and Grinev, A. N., 1975; Thang, D. C., Kossoff, Jacquignon and Dufour, 1976; Winchester and Popp, 1975; Wittig and Reichel, 1963) and its substituted analogues **1228** (R^1–R^5 = H, R^6 = CH_3) (Buu-Hoï, Mangane and Jacquignon, 1967), **1228** (R^1–R^4 = R^6 = H, R^5 = CH_3) (Hahn, W., Kryczka and Bartnik, 1975), **1228**

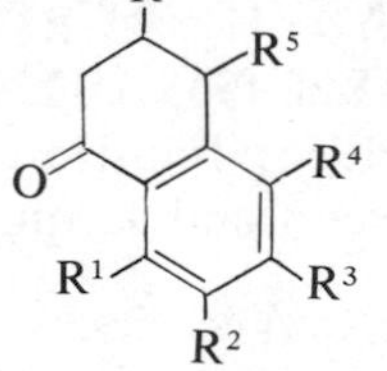

1228

(R^1 = R^3 − R^6 = H, R^2 = CH_3) (Buu-Hoï, Hoán and Khôi, 1949; Buu-Hoï, Jacquignon and Hoeffinger, 1963; Buu-Hoï, Jacquignon, Roussel, O. and Hoeffinger, 1964; Buu-Hoï, Saint-Ruf, Jacquignon and Marty, 1963; Buu-Hoï, Saint-Ruf, Martani, Ricci and Balucani, 1968), **1228** (R^1 = R^3–R^6 = H, R^2 = C_2H_5) (Buu-Hoï, Hoán and Khôi, 1950c), **1228** [R^1 = R^3–R^6 = H,

$R^2 = CH(CH_3)_2$] (Buu-Hoï, Royer, Eckert and Jacquignon, 1952; Buu-Hoï, Saint-Ruf, Jacquignon and Marty, 1963), **1228** [$R^1 = R^3$–$R^6 = H$, $R^2 = C(CH_3)_3$] (Buu-Hoï, Hoán and Khôi, 1949), **1228** ($R^1 = R^3$–$R^6 = H$, $R^2 = OCH_3$) (Buu-Hoï, Cagniant, Hoán and Khôi, 1950; Buu-Hoï, Hoán and Khôi, 1949; Buu-Hoï, Hoán, Khôi and Xuong, 1950), **1228** ($R^1 = R^2 = R^4$–$R^6 = H$, $R^3 = OCH_3$) (Buu-Hoï, Hoán, Khôi and Xuong, 1950; Buu-Hoï, Périn and Jacquignon, 1966), **1228** ($R^1 = R^3$–$R^6 = H$, $R^2 = Cl$ and Br) (Buu-Hoï, Hoán, Khôi and Xuong, 1951), **1228** ($R^1 = R^3$–$R^6 = H$, $R^2 = NO_2$) (Buu-Hoï, Hoán and Khôi, 1949, 1950b), **1228** ($R^1 = R^4$–$R^6 = H$, $R^2 = R^3 = CH_3$) (Buu-Hoï, Cagniant, Hoán and Khôi, 1950; Buu-Hoï, Martani, Ricci, Dufour, Jacquignon and Saint-Ruf, 1966; Buu-Hoï, Saint-Ruf, Jacquignon and Marty, 1963; Buu-Hoï, Saint-Ruf, Martani, Ricci and Balucani, 1968), **1228** ($R^1 = R^3 = R^5 = R^6 = H$, $R^2 = R^4 = CH_3$) (Buu-Hoï, Cagniant, Hoán and Khôi, 1950; Buu-Hoï, Saint-Ruf, Jacquignon and Marty, 1963), **1228** ($R^1 = R^4 = CH_3$, $R^2 = R^3 = R^5 = R^6 = H$) (Buu-Hoï, Hoán and Khôi, 1950b; Buu-Hoï, Saint-Ruf, Jacquignon and Marty, 1963; Buu-Hoï, Saint-Ruf, Martani, Ricci and Balucani, 1968), **1228** [$R^1 = R^4$–$R^6 = H$, $R^2 + R^3 = (CH_2)_4$] (Buu-Hoï and Jacquignon, 1951, 1954, 1956), **1228** (R^1–$R^3 = CH_3$, R^4–$R^6 = H$) (Buu-Hoï, Saint-Ruf and Dufour, 1964), **1228** ($R^1 = R^2 = R^4 = CH_3$, $R^3 = R^5 = R^6 = H$) (Buu-Hoï and Saint-Ruf, 1962), **1228** ($R^1 = R^5 = R^6 = H$, R^2–$R^4 = CH_3$) (Buu-Hoï, Saint-Ruf and Dufour, 1964), **1228** ($R^1 = R^3$–$R^6 = H$, $R^2 = SCH_3$) (Buu-Hoï, Khôi and Xuong, 1950) **1228** ($R^1 = R^4$–$R^6 = H$, $R^2 = SCH_3$, $R^3 = CH_3$) (Buu-Hoï and Hoán, 1951), **1228** ($R^1 = R^4$–$R^6 = H$, $R^2 = OCH_3$, $R^3 = CH_3$) (Buu-Hoï, Cagniant, Hoán and Khôi, 1950), **1228** ($R^1 = CH_3$, $R^2 = R^3 = R^5 = R^6 = H$, $R^4 = OCH_3$) (Buu-Hoï, Cagniant, Hoán and Khôi, 1950), **1228** ($R^1 = R^4$–$R^6 = H$, $R^2 = OCH_3$, $R^3 = Cl$) (Buu-Hoï, Hoán, Khôi and Xuong, 1951) and **1228** ($R^1 = R^4$–$R^6 = H$, $R^2 = R^3 = OCH_3$) (Avanesova, Astvatsatryan, Sarkisyan, Garibyan and Tatevosyan, 1975; Buu-Hoï, Cagniant, Hoán and Khôi, 1950). The ketones **1229** (R^1–$R^5 = H$, $X = O$, $Y = H_2$) (Buu-Hoï, Périn and Jacquignon, 1960, 1962), **1229** (R^1–$R^3 = R^5 = H$, $R^4 = CH_3$, $X = O$, $Y = H_2$; R^1–$R^3 = H$, $R^4 = R^5 = CH_3$, $X = O$, $Y = H_2$) (Buu-Hoï and Saint-Ruf, 1965a), **1229** (R^1–$R^5 = H$, $X = H_2$, $Y = O$) (Buu-Hoï and Jacquignon, 1956; Buu-Hoï, Jacquignon, Roussel, O. and Hoeffinger, 1964; Buu-Hoï, Périn and Jacquignon, 1960, 1962), **1229** ($R^1 =$

Y, X, R1, R2, R3, R4, R5

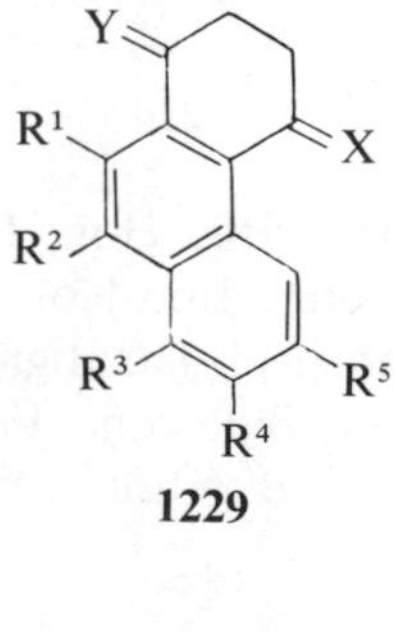

1229

$R^2 = CH_3$, R^3–R^5 = H, X = H_2, Y = O; $R^1 = R^3 = R^4$ = H, $R^2 = R^5 = CH_3$, X = H_2, Y = O and $R^1 = R^3 = R^5$ = H, $R^2 = R^4 = CH_3$, X = H_2, Y = O) (Buu-Hoï and Saint-Ruf, 1965a), **1229** [$R^1 = R^4 = R^5$ = H, $R^2 + R^3 = (CH_2)_2$, X = H_2, Y = O] (Buu-Hoï, Khôi and Xuong, 1951), **1230** (Braun and Bayer, 1929), **1231** (R^1–R^3 = H; $R^1 = R^2$ = H, $R^3 = CH_3$; $R^1 = R^3$ = H, $R^2 = CH_3$; $R^1 = CH_3$, $R^2 = R^3$ = H and R^1 = H, $R^2 = R^3 = CH_3$) (Buu-Hoï and Saint-Ruf, 1965a), **1232** (Buu-Hoï and Jacquignon, 1954), **1233** (Buu-Hoï and Saint-Ruf, 1965a), **1234** (X = NC_2H_5) (Buu-Hoï and Saint-Ruf, 1965b), **1234** (X = Se) (Buu-Hoï and Hoán, 1952), **1235** (R = H) (Buu-Hoï, Hoán, Khôi and Xuong, 1949; Buu-Hoï, Jacquignon and Hoeffinger, 1963; Buu-Hoï, Jacquignon, Roussel, O. and Hoeffinger, 1964; Buu-Hoï, Saint-Ruf, Martani, Ricci and Balucani, 1968), **1235** (R = CH_3) (Buu-Hoï,

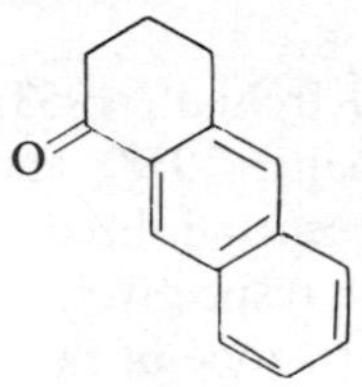

1230

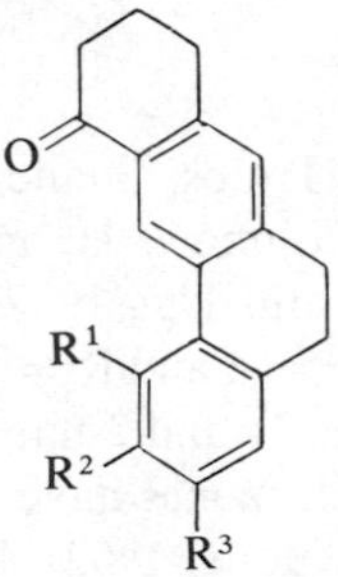

1231

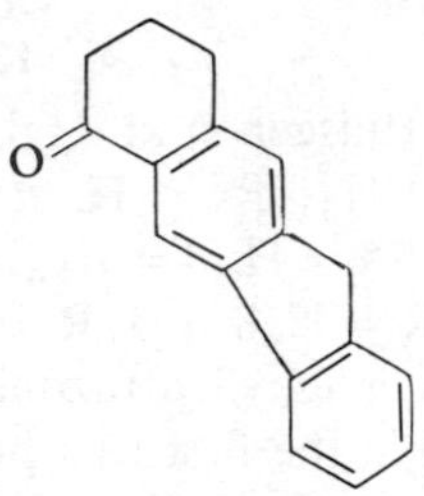

1232

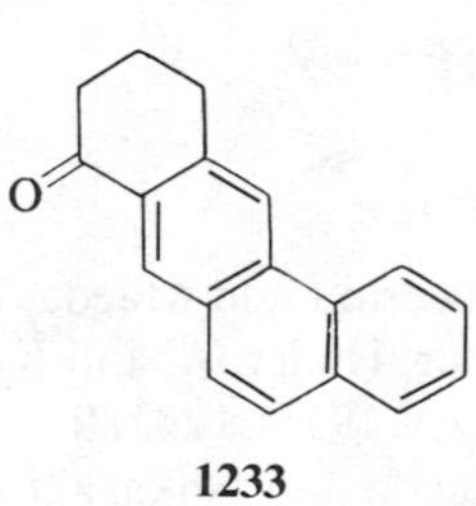

1233

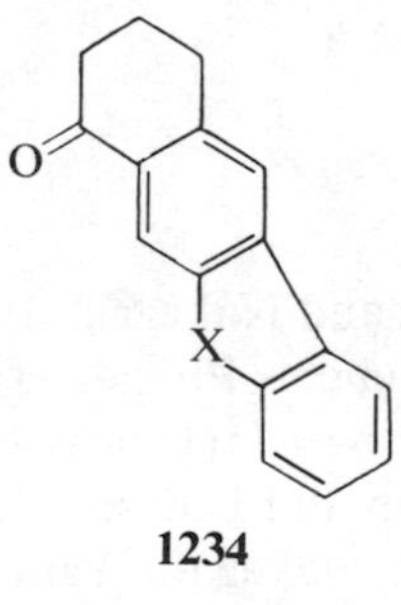

1234

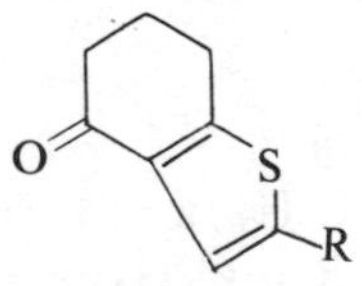

1235

Hoán and Khôi, 1950b), **1235** (R = C_2H_5) (Buu-Hoï, Hoán and Khôi, 1950c), **1235** (R = Cl) (Buu-Hoï, Hoán, Khôi and Xuong, 1951), **1236** (Buu-Hoï and Jacquignon, 1954, 1956), **1237** (Buu-Hoï, Hoán and Khôi, 1950b), **1238** (Buu-Hoï, Croisy, Jacquignon and Martani, 1971) and **1239** (Bremner and Browne, 1975) have also been used in indolizations. The phenylhydrazones **1240** ($R^1 = R^2$ = H, n = 3) (Buu-Hoï and Xuong, 1952; Fujimori, H. and Yamane,

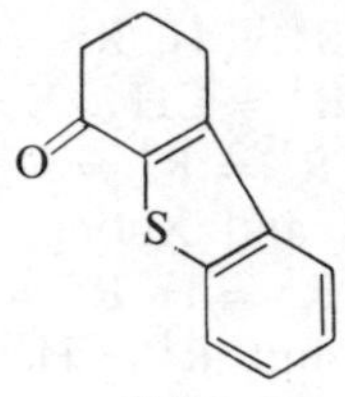

1236 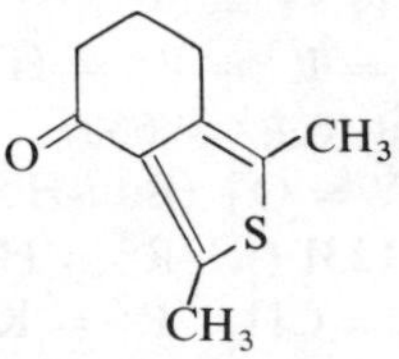 **1237** 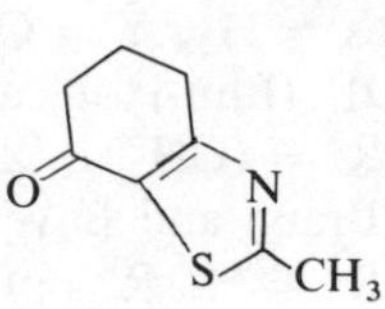 **1238**

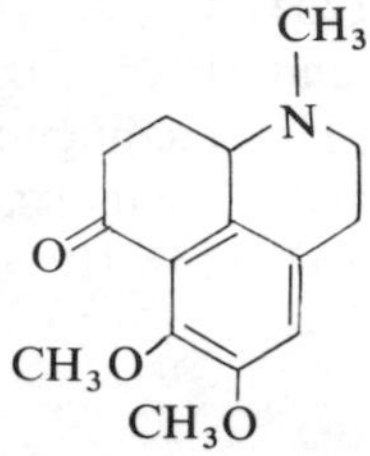

1239 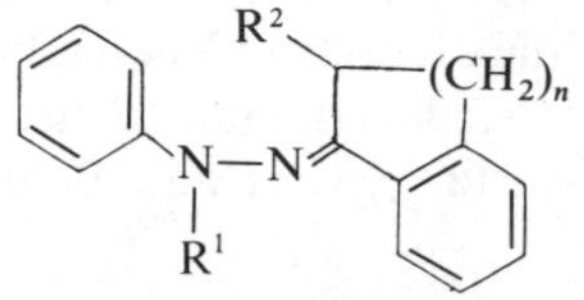 **1240**

1978; Huisgen and Ugi, 1957; Treibs, Steinert and Kirchhof, 1953), **1240** ($R^1 = CH_3$, $R^2 = H$, $n = 3$) (Fujimori, H. and Yamane, 1978) and **1240** ($R^1 = R^2 = H$, $n = 4$) (Huisgen and Ugi, 1957) have been indolized to yield **1241** ($R = H, n = 3$; $R = CH_3, n = 3$ and $R = H, n = 4$, respectively) and the 4-bromophenylhydrazone and the 1- and 2-naphthylhydrazone of the ketonic moiety in the first two phenylhydrazones have likewise afforded the expected indolic products (Buu-Hoï and Xuong, 1952). Indolization of **1240** [$R^1 = H$, $R^2 = C_2H_5, C_6H_6$ and $CH_2C_6H_5$, $n = 1$; $R^1 = H$, $R^2 = CH_3, CH_2C_6H_5$ and $(CH_2)_2C_6H_5$, $n = 2$ and $R^1 = H$, $R^2 = CH_3$, $n = 3$ and 4] gave **1242**

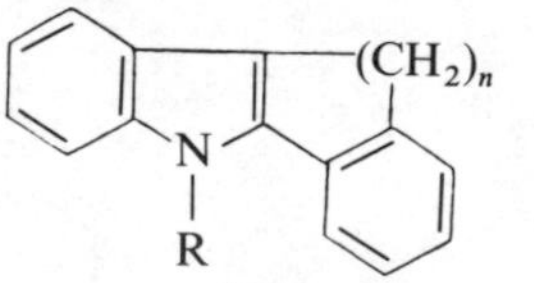

1241 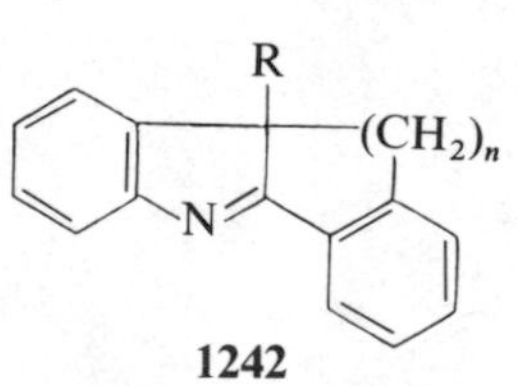 **1242**

($R = C_2H_5, n = 1$) (Maeda and Nakazaki, 1960; Nakazaki and Maeda, 1962), **1242** ($R = C_6H_5, n = 1$) (Leuchs, Philpott, D., Sander, Heller, A. and Köhler, 1928), **1242** ($R = CH_2C_6H_5, n = 1$) (Leuchs and Winzer, 1925), **1242** ($R = CH_3$, $n = 2$) (Buu-Hoï, Jacquignon and Ledésert, 1970; Maeda and Nakazaki, 1960; Nakazaki and Maeda, 1962; Nakazaki, Yamamoto, K. and Yamagami, 1960), **1242** [$R = CH_2C_6H_5$ and $(CH_2)_2C_6H_5$, $n = 2$] (Leuchs, Philpott, D., Sander, Heller, A. and Köhler, 1928) and **1242** ($R = CH_3$, $n = 3$ and 4) (Maeda and Nakazaki, 1960; Nakazaki and Maeda, 1962), respectively, although several attempts to prepare **1242** ($R = CH_3, n = 1$) by indolization of **1240** ($R^1 = H$, $R^2 = CH_3, n = 1$) were unsuccessful (Nakazaki and Maeda, 1962). The 1- and 2-naphthylhydrazone of the ketonic moiety of **1240** ($R^2 = CH_3$, $n = 2$) afforded **1243** and **1244**, respectively, upon indolization (Buu-Hoï, Jacquignon and Ledésert, 1970). Indolization of the phenylhydrazone of **1245** produced **1246** (Muth, Steiniger and Papanastassiou, 1955), of the phenylhydrazone of **1247** produced **1248** (Muth and Hoyle, 1964) and of the phenylhydrazone and 1-naphthylhydrazone of **1249** yielded **1250** and **1251**, respectively (Menichi, Bisagni and Royer, 1964).

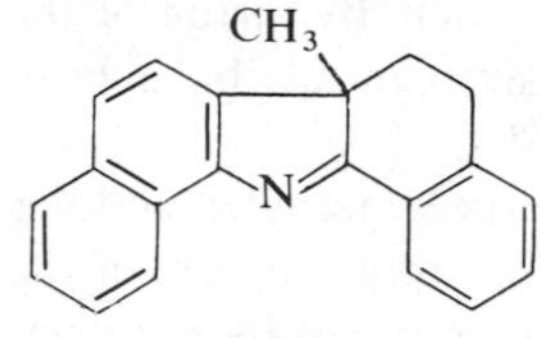

CH_3 N **1243**

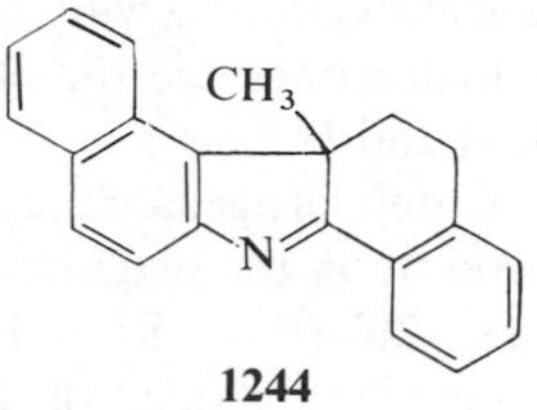

CH_3 N **1244**

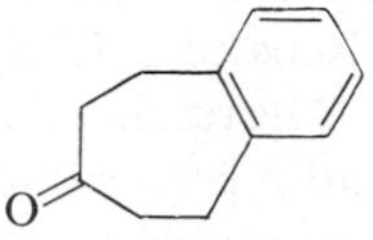

O **1245**

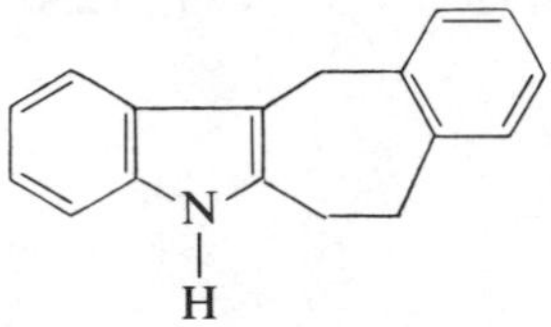

N H **1246**

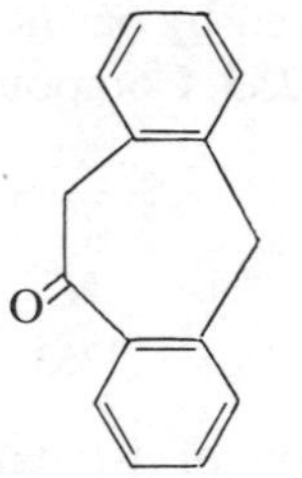

O **1247**

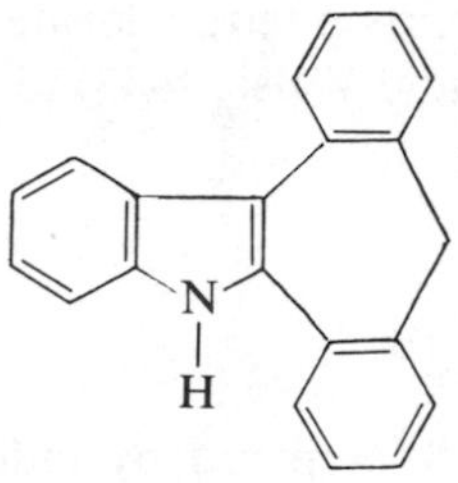

N H **1248**

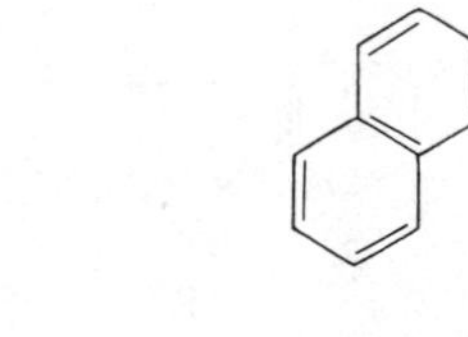

C_2H_5 O O **1249**

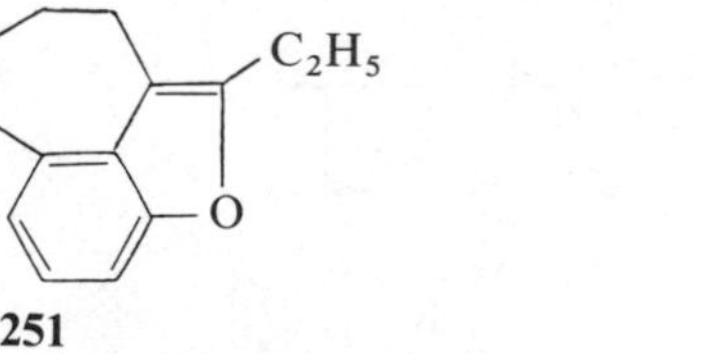

C_2H_5 N H O **1250**

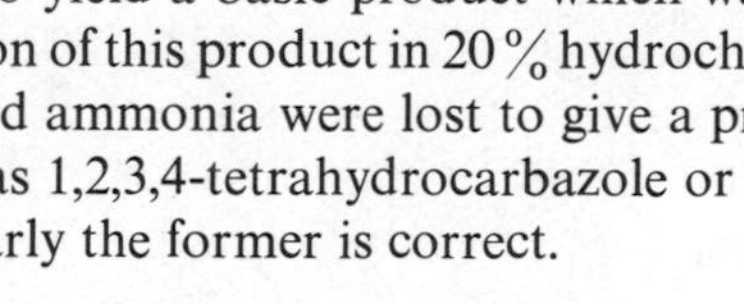

C_2H_5 N H O **1251**

An interesting reaction occurred when cyclohexanone cyanohydrin reacted with phenylhydrazine to yield a basic product which was formulated as **1252**. Upon refluxing a solution of this product in 20% hydrochloric acid, the elements of hydrogen cyanide and ammonia were lost to give a product, m.p. = 116 °C, which was formulated as 1,2,3,4-tetrahydrocarbazole or **57** (R = H) (Bucherer and Brandt, 1934). Clearly the former is correct.

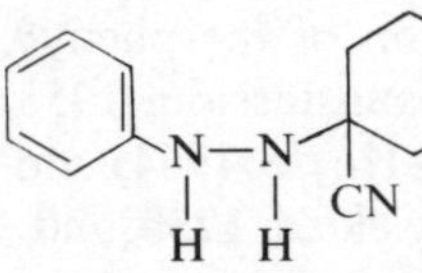

1252

The indolizations of cycloalkanone arylhydrazones which, by virtue of the ketonic moieties of the hydrazones are directionally ambiguous, have been discussed in Chapter III, H and I).

The results of attempts to dehydrogenate the above 2,3-polymethyleneindoles and related systems depend upon the magnitude of *n* (i.e. upon the size of the polymethylene ring). Thus, **1207** ($R^1 = R^2 = H$, $n = 3$), prepared by indolization of cyclopentanone phenylhydrazone, was dehydrogenated under unstated conditions (Treibs, 1959) or by treatment with N-bromosuccinimide or bromine in carbon tetrachloride followed by basification of the reaction product (Paul, H. and Weise, 1963) to give **1253**, Compounds **1254** ($R^1 = R^2 = H$), **1256** and

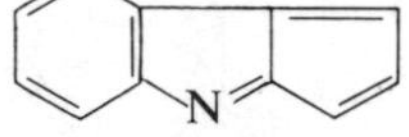

1253

1258, prepared by indolization of 1-indanone phenylhydrazone and 1- and 2-naphthylhydrazone, respectively, were dehydrogenated to afford **1255** ($R^1 = R^2 = H$), **1257** and **1259**, respectively (Treibs, 1959), and compounds **1254** ($R^1 = CH_3$, OCH_3 and C_6H_5, $R^2 = H$ and CH_3) and **554** (R = H and CH_3), prepared by indolization of the appropriate 1-indanone and 4,7-dimethyl-1-indanone arylhydrazones, yielded **1255** ($R^1 = CH_3$, OCH_3 and C_6H_5, $R^2 =$ H and CH_3), and **1260** (R = H and CH_3), respectively, when treated with flowers of sulphur in refluxing quinoline (Kempter, Schwalba, Stoss and Walter, 1962). However, **1261**, prepared by indolization of the hydrazone of cyclopentanone with **624** ($R = CH_3$), remained unchanged when heated with

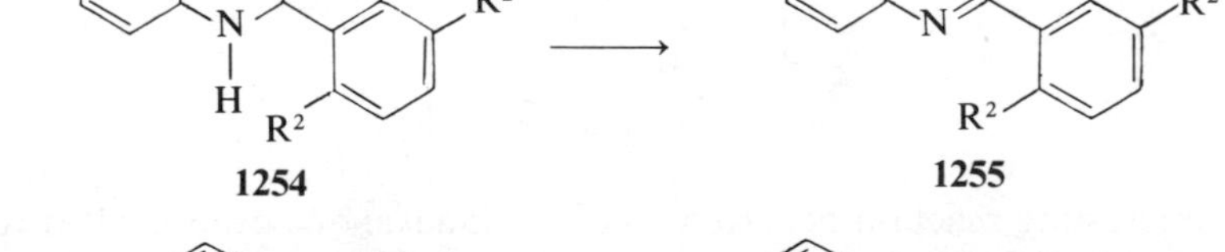

1254 1255

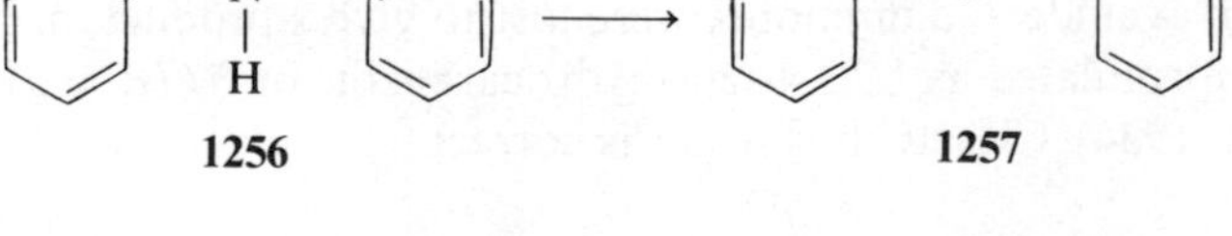

1256 1257

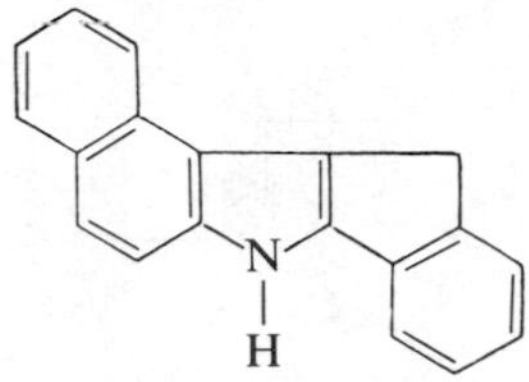

1258

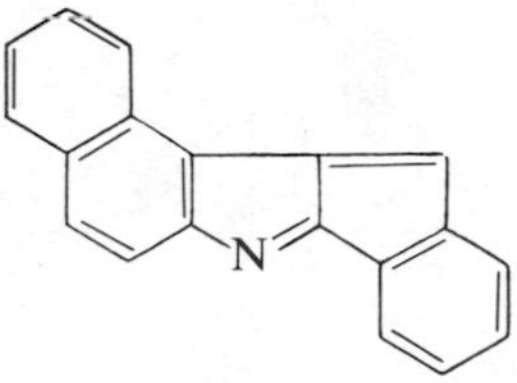

1259

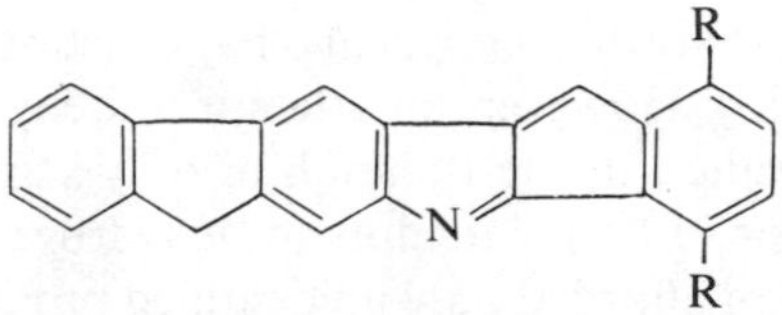

1260

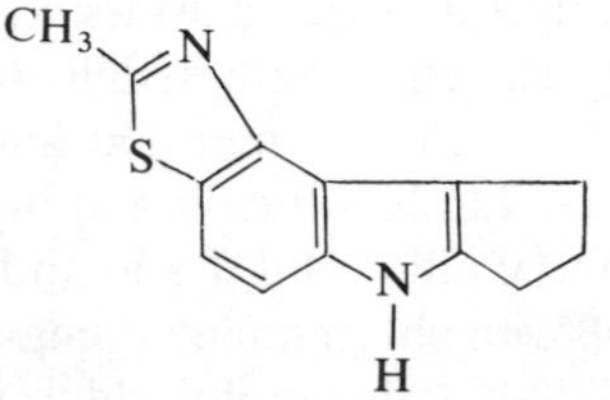

1261

palladium-on-carbon (Buu-Hoï, Saint-Ruf, Martani, Ricci and Balucani, 1968). The large number of studies which have been carried out upon the dehydrogenation of 1,2,3,4-tetrahydrocarbazoles into carbazoles have been discussed in Chapter I, D. The homologous **1207** ($R^1 = R^2 = H$, $n = 5$) was dehydrogenated by heating with iodine in nitrobenzene (Treibs, 1952), by treatment with chloranil in refluxing amyl alcohol (Treibs, Steinert and Kirchhof, 1953) or by passage in the vapour phase over heated 5% palladium-on-carbon on magnesium oxide (Anderson, A. G. and Tazuma, 1952; Muth, Steiniger and Papanastassiou, 1955) to afford **1262** (R = H). The spectroscopic properties of **1262** ($R = CH_3$) supported this structural assignment, rather than the alternative formulation **1263**, and hence, by analogy supported structure

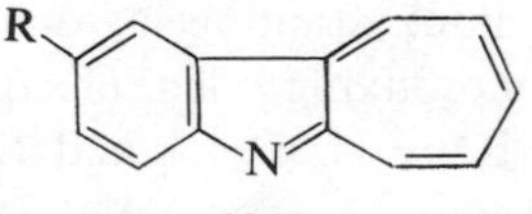

1262

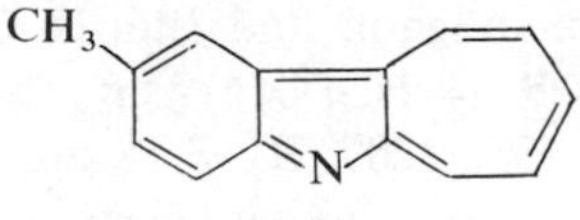

1263

1262 (R = H) (Takase, Asao and Hirata, 1968). Chloranil has also been used to convert **1241** (R = H, $n = 3$) into **1264** (De Jong and Boyer, 1972; Muth, Steiniger and Papanastassiou, 1955; Treibs, Steinert and Kirchhof, 1953), a dehydrogenation which has also been effected by heating with palladium-on-carbon (Muth, Steiniger and Papanastassiou, 1955). When compounds **1241** (R = H and CH_3, $n = 3$) were treated with 2,3-dichloro-5,6-dicyano-1,4-benzoquinone in wet dioxan at 5 °C, compounds **1265** (R = H and CH_3) were formed (Fujimori, H., and Yamane, 1978), similar oxidations also being effected upon 2,3-polymethylene indoles (Oikawa and Yonemitsu, 1977; Yamane and Fujimori, K., 1976). Dehydrogenation of **1246** using chloranil in

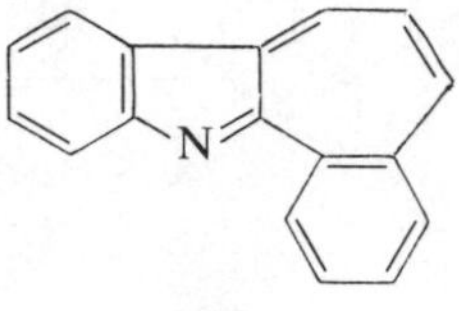

1264 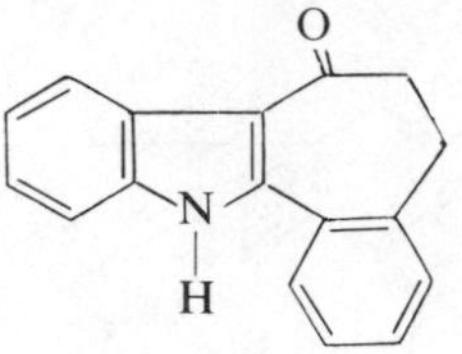 1265

refluxing benzene or by heating with palladium-on-carbon yielded **1266** (Muth, Steiniger and Papanastassiou, 1955). Chloranil appeared to be the reagent of choice for effecting these dehydrogenations (Muth, Steiniger and Papanastassiou, 1955; Treibs, Steinert and Kirchhof, 1953). However, an attempt to dehydrogenate **1267** with chloranil in xylene resulted in the isolation of only starting material (Buu-Hoï, Périn and Jacquignon, 1960). Attempts to dehydrogenate **1248** with chloranil in boiling amyl alcohol afforded only unidentified products and also failed using chloranil in boiling benzene and upon heating with palladium-on-carbon on magnesium oxide at 350–370 °C/25 mm. A study on molecular models suggested that the potential product, **1268**, from this dehydrogenation may be unstable and may be incapable of existence (Muth and

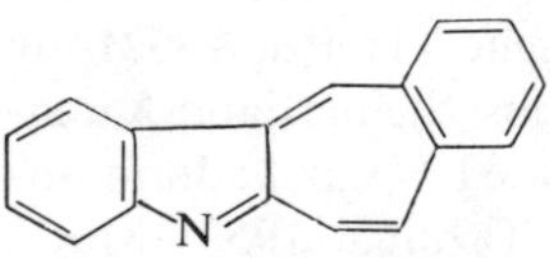

1266 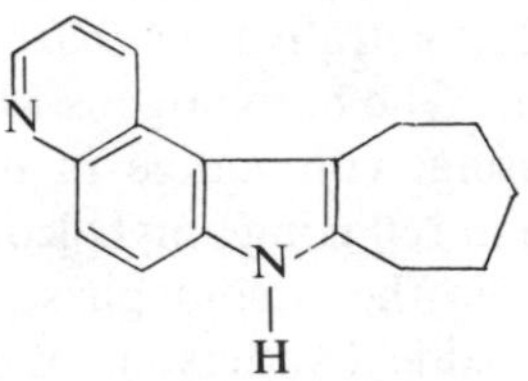 1267 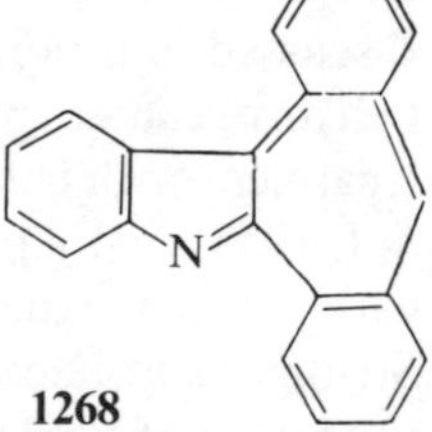 1268

Hoyle, 1964). It appears that dehydrogenation of **1207** in which $n > 5$ is not possible (Jacquignon and Buu-Hoï, 1957) since attempts to dehydrogenate **1207** ($R^1 = R^2 = H, n = 6$) using chloranil in refluxing xylene (Jacquignon and Buu-Hoï, 1957), **1207** (R^1 = H and 5-CH_3, $R^2 = H, n = 7$) and **1210** ($n = 7$) with chloranil in refluxing xylene or by heating with selenium at 350 °C (Buu-Hoï, Jacquignon and Loc, 1958), **1207** (R^1 = H and 5-CH_3, $R^2 = H, n = 10$), **1210** ($n = 10$) and **1212** with chloranil in refluxing xylene or by heating with palladium-on-carbon at 300–320 °C (Jacquignon, Buu-Hoï and Dufour, 1966) and **1214** with chloranil in refluxing xylene (Jacquignon and Buu-Hoï, 1957) were all unsuccessful.

Apart from the above syntheses of **1262**, **1264** and **1266**, benzazaazulines and dibenzazaazulenes, respectively, which only indirectly involved the Fischer indolization, several functionalized azaazulenes have been directly synthesized using a Fischer cyclization. Thus, treatment of **1164** with acid catalysts afforded the corresponding **1165**, the specific transformations which have been so effected being given in Table 25. In other related indolizations, **1183** [R = H,

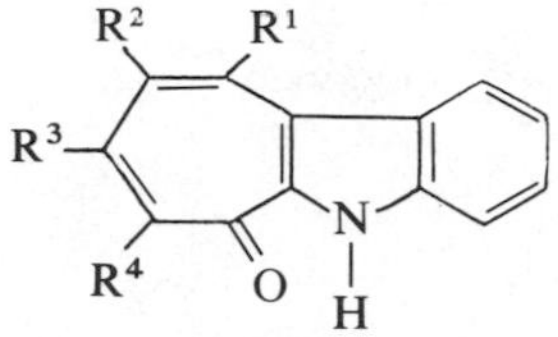

1269

CH_3, C_6H_5, $C_6H_4Cl(4)$ and $(CH_2)_2$-phthalimido], when heated in monoethylene glycol in the presence of concentrated sulphuric acid, afforded **1184** [$R^1 = COO(CH_2)_2OH$, $R^2 = H$ and CH_3) (Sunagawa and Sato, Y., 1962b), **1184** [$R^1 = COO(CH_2)_2OH$, $R^2 = C_6H_5$ and $C_6H_4Cl(4)$] (Sato, Y., 1963b) and **1184** [$R^1 = COO(CH_2)_2OH$, $R^2 = (CH_2)_2$-phthalimido] (Sankyo Co. Ltd, 1964; Sato, Y., 1963c), respectively. The last product was hydrolysed to the corresponding 3-(2-aminoethyl)-2-carboxylic acid (Sankyo Co. Ltd, 1964; Sato, Y., 1963c) and compounds **1184** [$R^1 = COO(CH_2)_2OH$, $R^2 = C_6H_5$ and $C_6H_4Cl(4)$] were sequentially hydrolysed by refluxing in ethanolic potassium hydroxide and decarboxylated by heating with copper powder in quinoline to afford **1184** [$R^1 = H$, $R^2 = C_6H_5$ and $C_6H_4Cl(4)$, respectively] (Sato, Y., 1963b). Attempts to similarly indolize **1183** ($R = COOC_2H_5$) (Sunagawa and Sato, Y., 1962b) and **1183** [$R = CH_2CN$ and $CH_2N(C_2H_5)_2$] (Sato, Y., 1963c) by heating with either a monoethylene glycol–concentrated sulphuric acid mixture or with polyphosphoric acid were unsuccessful. In an extension of these studies, the azines **1166** [$R^1 = R^2 = H$, $R^3 = CH_3$, C_2H_5, C_6H_5 and $(CH_2)_2$-phthalimido; $R^1 = H$, $R^2 = R^3 = CH_3$; $R^1 = H$, $R^2 + R^3 = (CH_2)_{3,4\text{ and }5}$ and $R^1 = H$, $R^2 = CH_3$, $R^3 = (CH_2)_2$-phthalimido] have been converted into **1270** [$R^1 = R^2 = H$, $R^3 = CH_3$, C_2H_5, C_6H_5 and $(CH_2)_2$-phthalimido; $R^1 = H$, $R^2 = R^3 = CH_3$; $R^1 = H$, $R^2 + R^3 = (CH_2)_{3,4\text{ and }5}$ and $R^1 = H$, $R^2 = CH_3$, $R_3 = (CH_2)_2$-phthalimido, respectively] (Sato, Y. and Sunagawa, 1967; Sunagawa and Sato, Y., 1967), **1166** [$R^1 = Br$, $R^2 = CH_3$, $R^3 = (CH_2)_2$-phthalimido] gave **1270** [$R^1 = Br$, $R^2 = CH_3$, $R^3 = (CH_2)_2$-phthalimido] (Sato, Y. and Sunagawa, 1967) and **1166** [$R^1 = R^3 = H$, $R^2 = COOCH_3$; $R^1 = Br$, $R^2 = H$, $R^3 = CH_3$ and $R^1 = Br$, $R^2 + R^3 = (CH_2)_3$] gave **1270** [$R^1 = R^3 = H$, $R^2 = COOCH_3$; $R^1 = Br$, $R^2 = H$, $R^3 = CH_3$ and $R^1 = Br$, $R^2 + R^3 = (CH_2)_3$, respectively] (Sunagawa and Sato, Y., 1967) when heated with polyphosphoric acid. Subsequent alkali

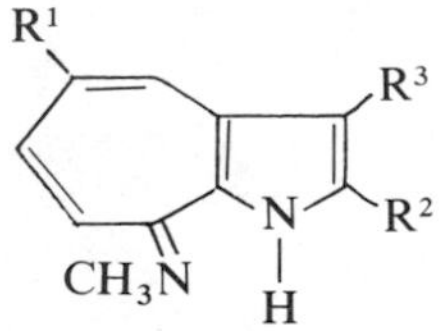

1270

Table 25. The Compounds **1164** which have been Converted under Conditions of Acid Catalysis into the Corresponding **1165**

Reactant **1164** (Unless otherwise stated, R^1–R^6 = H)

R^1	R^2	R^3	R^4	R^5	R^6	Catalyst(s)	Reference
						1, 2	1
					CH_3	1, 2	1
					C_2H_5	1, 2	1
					C_6H_5	3	2
				CH_3		(a)	2
				CH_3	CH_3	1, 2	1
				CH_3	$(CH_2)_2$-phthalimido	4(b)	3, 4
				C_6H_5		(a)	2
				$(CH_2)_3$		(c)	5
				$(CH_2)_4$		2(d)	5
				$(CH_2)_5$		1, 2	5
				$(CH_2)_2N(CH_3)CH_2$		1	6, 7
				$(CH_2)_2N(CH_2C_6H_5)CH_2$		1, 2	5
				$CH_2CH_2N^{\oplus}$(2-methylpyridinium) $Br^{\ominus}$		5(e)	6, 7
			Br			1, 2	1
			Br		CH_3	1, 2	1
			Br	CH_3	CH_3	1, 2	1

			Br	CH_3	$(CH_2)_2$-phthalimido	1(b)	3, 4
			Br		$(CH_2)_4$	2(d)	5
			OCH_3		CH_3	2	8
			OC_6H_5		CH_3	2	8
		$CH(CH_3)_2$			$(CH_2)_4$	2(d)	5
	Br		Br		CH_3	1, 2	1
	OCH_3				C_2H_5	1, 2	1
$CH(CH_3)_2$					$(CH_2)_4$	2(d)	5
Br			Br		CH_3	1, 2	1

Notes

a. The hydrazone was synthesized but was not indolized; b. The phthalimido group was removed, subsequent to indolization, by alkaline hydrolysis to afford the corresponding 2-aminoethyl substituted products; c. The hydrazone was synthesized but was recovered unchanged after heating with either dilute sulphuric acid or with 85% phosphoric acid; d. The product was dehydrogenated to the corresponding **1269** by treatment with either chloranil in refluxing xylene or 2,3-dichloro-5,6-dicyano-1,4-benzoquinone in refluxing benzene; e. The product was difficult to purify and was therefore subjected to immediate catalytic reduction.

Catalysts

1. Polyphosphoric acid; 2. Dilute sulphuric acid; 3. Methanolic hydrochloric acid; 4. Concentrated sulphuric acid in monoethylene glycol; 5. Ethanolic hydrobromic acid.

References

1. Sunagawa, Soma, Nakano and Matsumoto, 1961; 2. Nozoe, Kitahara and Arai, 1954; 3. Sato, Y. and Sunagawa, 1967; 4. Sankyo Co. Ltd, 1964; 5. Nozoe, Sin, Yamane and Fujimori, K., 1975; 6. Sunagawa and Sato, Y., 1962a; 7. Sankyo Co. Ltd, 1963; 8. Sato, Y., 1963a.

catalyzed hydrolysis of the phthalimido group in the appropriate three of these products afforded the corresponding 3-(2-aminoethyl)substituted compounds (Sato, Y. and Sunagawa, 1967). Attempts to similarly cyclize **1166** ($R^1 = R^3 = H$, $R^2 = H$ and CH_3) were unsuccessful (Sato, Y. and Sunagawa, 1967).

The 4-troponylhydrazones **1271** (R = H, CH_3 and $COOC_2H_5$) have been synthesized using diazotized 4-aminotropone in the appropriate Japp–Klingemann reactions. The first two of these hydrazones were subsequently heated with concentrated sulphuric acid in monoethylene glycol to form **1272** (R = H and CH_3, respectively) (transesterification also occurred under these reaction conditions – see Chapter V, H) (Sunagawa and Sato, Y., 1962b).

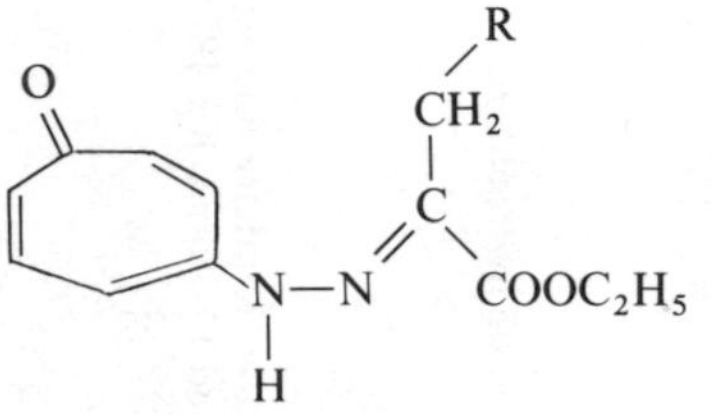

1271

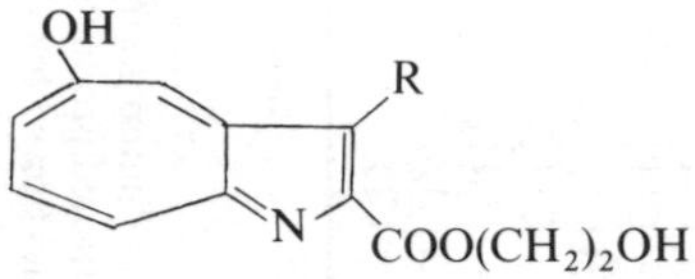

1272

E. Synthesis of Tryptophans

Although the Fischer indolization has often been used to prepare the necessary indoles for syntheses of tryptophans utilizing preformed indolic nuclei, the following tryptophan syntheses have more directly involved the indolization.

Several groups have observed that the indolization of the hydrazone **1273** (R^1–R^4 = H, $R^5 = R^6 = COOC_2H_5$, $R^7 = NHCOCH_3$), or a mixture of its

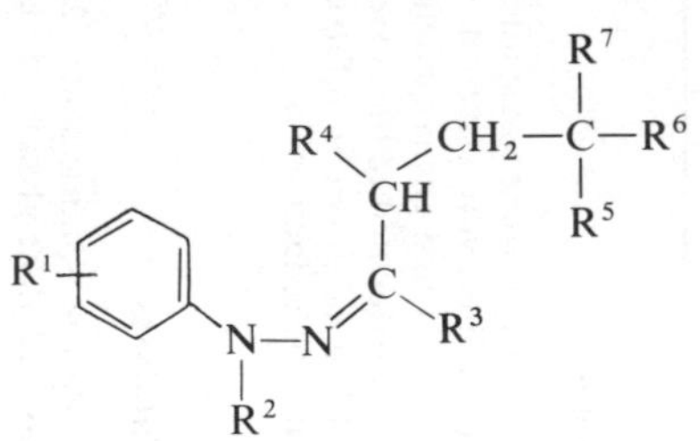

1273

component phenylhydrazine and aldehydic moiety, yielded **1274** (R^1–R^3 = H, $R^4 = R^5 = COOC_2H_5$, $R^6 = NHCOCH_3$) (Britton and Van der Weele, 1955; Cornforth, J. W., Cornforth, R. H. Dalgliesh and Neuberger, 1951, Grudzinski and Kotelko, 1955; Kidwai, A. R. and Khan' N. H., 1963; Novotny and Cerveny, 1974; Novotny, Cerveny, Kmonickova, Vlk and Blazkova, 1976; Warner and Moe, 1948; Yamada, S.' Chibata and Tsurui, 1953; Yamada, S. *et*

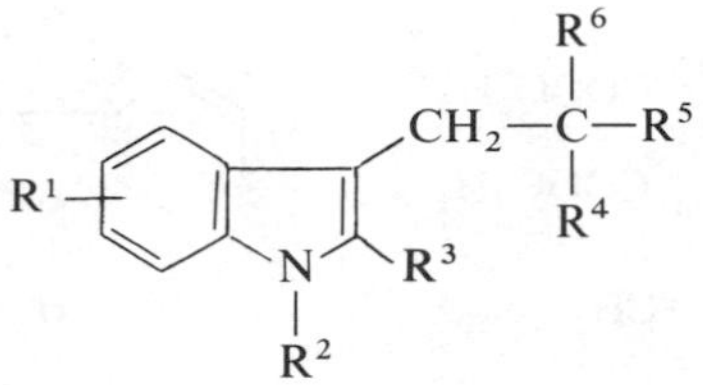

1274

al., 1954) which after hydrolysis of the ester groups, decarboxylation and deacetylation afforded tryptophan (**1274**; R^1–R^3 = R^4 = H, R^5 = COOH, R^6 = NH_2) (Cornforth, J. W., Cornforth, R. H., Dalgliesh, C. E. and Neuberger, 1951; Grudzinski and Kotelko, 1955; Novotny and Cerveny, 1974; Novotny, Cerveny, Kmonickova, Vlk and Blazkova, 1976; Warner and Moe, 1948). These last three reactions could be effected in 'one pot' under sequentially varying conditions without the isolation of intermediates (Opie, Warner and Moe, 1952) (cf. Snyder and Smith, C., 1944) and the final product could be isolated by chromatography upon an acidic ion exchange resin (Petrova, T. D., Savchenko, Ardyukova and Yakobson, 1971). This total reaction sequence has been extended to the syntheses of 4-, 5- and 6-methyl- and 4,6-dimethyl- (Snyder, Beilfuss and Williams, J. K., 1953), 5- and 7-ethyl- (the projected synthesis of the 4- and 6-ethyl isomers was curtailed after the initial indolization stage which afforded, from the corresponding 3-ethylphenylhydrazone, a mixture of both possible isomeric indoles which could not be separated) (Lingens and Weiler, 1963), 1-benzyl- (Cornforth, J. W., Cornforth, R. H., Dalgliesh and Neuberger, 1951), 5-fluoro- (Rinderknecht and Niemann, 1950), 5-nitro- (Cavallini and Ravenna, 1958), 7-nitro- (Hiremath and Siddappa, 1962), 7-methoxy- (Bergmann, E. D. and Hoffmann, E., 1962), 5-benzyloxy- (Suvorov, Morozovskaya and Sorokina, G. M., 1961), 5-sulphonamido- (Giuliano and Stein, M. L., 1957) and 4,5- and 6,7-benzo- (Rydon and Siddappa, 1951) tryptophans. However, for the synthesis of 5-ethoxytryptophan, the route via 5-ethoxyindole and 5-ethoxygramine was more satisfactory (Rydon and Siddappa, 1951) and failures attended early attempts to indolize **1273** (R^1 = 2- and 4-NO_2, R^2–R^4 = H, R^5 = R^6 = $COOC_2H_5$, R^7 = $NHCOCH_3$) (Rydon and Siddappa, 1951) and attempts to indolize **1273** (R^1 = 3-NO_2, R^2–R^4 = H, R^5 = R^6 = $COOC_2H_5$, R^7 = $NHCOCH_3$) (Rydon and Siddappa, 1951), the corresponding 2-methyl-5-nitrophenylhydrazone (Hiremath and Siddappa, 1965) and **1273** (R^1 = 4-OCH_3, R^2 = H and $CH_2C_6H_5$, R^3 = H and CH_3, R^4 − H, R^5 − R^6 = $COOC_2H_5$, R^7 = $NHCOCH_3$) (Leonard and Tschannen, 1965). In an interesting extension of this synthetic route, **1273** (R^1–R^3 = H, R^4 = CH_3, R^5 = R^6 = $COOC_2H_5$, R^7 = $NHCOCH_3$) was converted in refluxing acetic acid into **1276**, assumedly via the intermediate **1275** (Witkop and Hill, R. K., 1955), the acetyl group on the side chain nitrogen atom not preventing the participation of the nitrogen p-electron pair in the indicated ring closure.

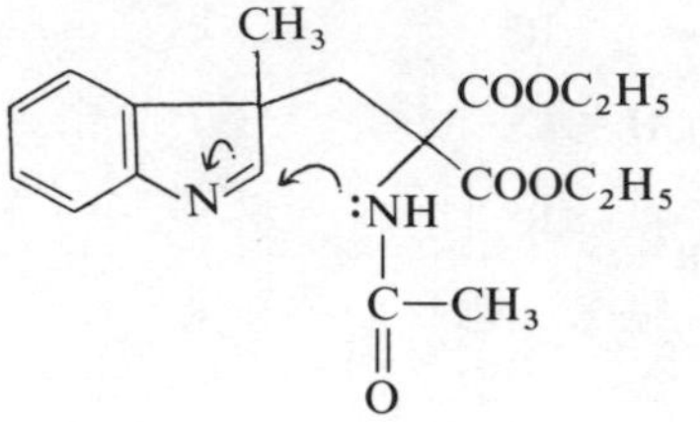

1275

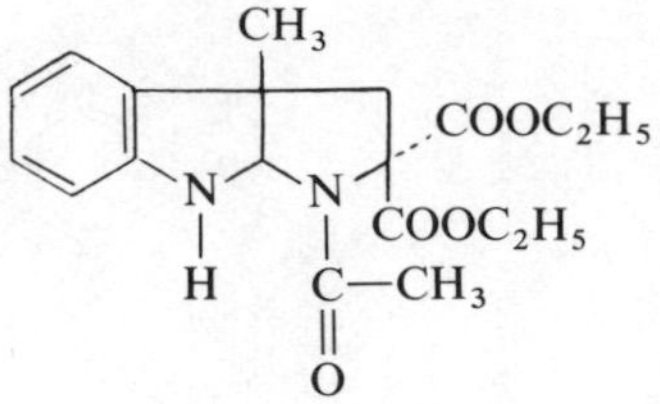

1276

Other related synthetic approaches to the tryptophan system involved (a) the acid catalysed conversion of **1277** (R = H) phenylhydrazone (Sakurai and Komachiya, 1964a, 1964b) into **1278** ($R^1 = R^2 = H$) (Ajinomoto Co., Inc., 1965) or tryptophan (Komachiya, Suzuki, S., Yamada, T., Miyayashiki and Sakurai, 1965) and the indolization of mixtures of **1277** ($R = CH_3$) with phenylhydrazine or 4-benzyloxyphenylhydrazine in concentrated hydrochloric acid to afford **1278** (R^1 = H and $OCH_2C_6H_5$, respectively, $R^2 = CH_3$) which after

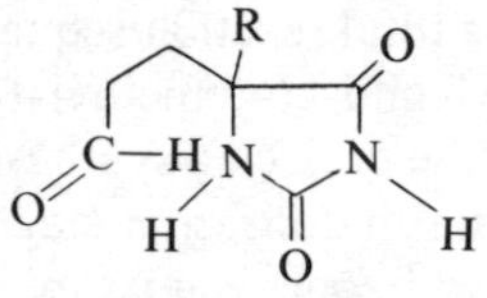

1277

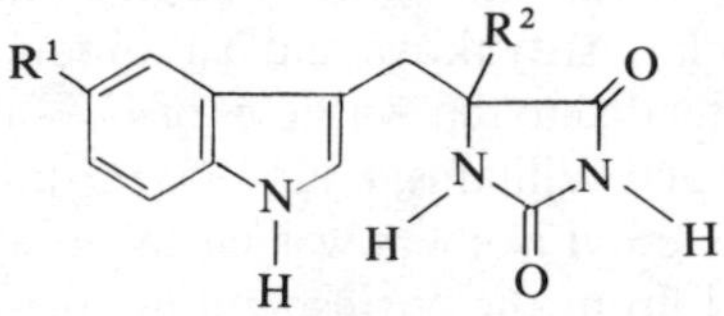

1278

hydrolysis with sodium hydroxide gave α-methyltryptophan and 5-benzyloxy-α-methyltryptophan, respectively (Sakuraba, Iwashita and Ninagawa, 1974), (b) the indolization of **1273** (R^1 = H, 4-F and 4-Cl, R^2–R^4 = H, $R^5 = CH_3$, $R^6 = COOC_2H_5$, $R^7 = NO_2$), by refluxing in dilute sulphuric acid solution, and of **1273** (R^1 = 4-CH_3 and 4-$OCH_2C_6H_5$, R^2–R^4 = H, $R^5 = CH_3$, $R^6 = COOC_2H_5$, $R^7 = NO_2$), by refluxing in aqueous sulphosalicylic acid solution, to afford **1274** (R^1 = H, 5-F, 5-Cl, 5-CH_3 and 5-$OCH_2C_6H_5$, respectively, $R^2 = R^3$ = H, $R^4 = CH_3$, $R^5 = COOC_2H_5$, $R^6 = NO_2$) which after sequential catalytic hydrogenation and hydrolysis gave α-methyltryptophan and 5-fluoro-, 5-chloro-, 5-methyl- and 5-benzyloxy-α-methyltryptophan, respectively (Suvorov, Morozovskaya and Ershova, 1962), (c) the indolization of a mixture of the dimethylacetal **1279** and 4-benzyloxyphenylhydrazine hydro-

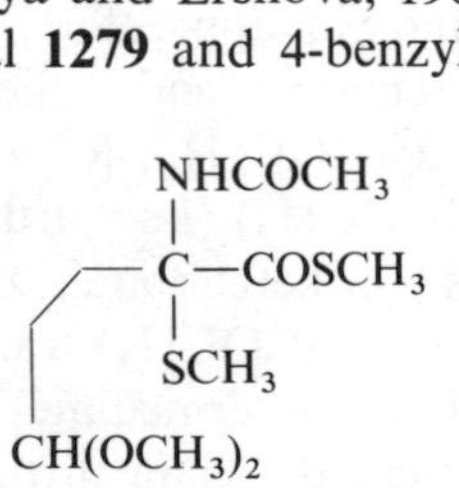

1279

chloride in aqueous acetic acid into **1274** ($R^1 = 5\text{-}OCH_2C_6H_5$, $R^2 = R^3 = H$, $R^4 = NHCOCH_3$, $R^5 = COSCH_3$, $R^6 = SCH_3$) which after sequential ester interchange with methyl alcohol and desulphurization with Raney nickel, which also caused O-debenzylation, gave the methyl ester of N-acetyl-5-hydroxy-tryptophan (Ogura and Tsuchihashi, 1974) and (d) the indolization of **1273** (R^1–R^5 = H, R^6 = COOH, $R^7 = NHCOCH_3$), as a mixture of its component aldehydic moiety and phenylhydrazine hydrochloride, in hot hydrochloric acid into **1274** (R^1–R^4 = H, R^5 = COOH, $R^6 = NHCOCH_3$) which upon de-acetylation yielded tryptophan (Iijima, K., Yamada, Y. and Homma, 1973). A slightly different synthesis of tryptophan involved the alcoholic hydrogen chloride catalysed indolization of **1273** ($R^1 = R^2 = R^4 = R^5 = H$, $R^3 = R^6 = COOC_2H_5$, $R^7 = NHCOOC_2H_5$) into **1274** ($R^1 = R^2 = R^4 = H$, $R^3 = R^5 = COOC_2H_5$, $R^6 = NHCOOC_2H_5$) which was converted into tryptophan via the sequential formations of **1274** ($R^1 = R^2 = R^4 = H$, $R^3 = R^5 = COOH$, $R^6 = NHCOOC_2H_5$), **1274** ($R^1 = R^2 = R^4 = H$, $R^3 = COOH$, $R^5 = COOC_2H_5$, $R^6 = NHCOOC_2H_5$) and **1274** (R^1–R^4 = H′ $R^5 = COOC_2H_5$, $R^6 = NHCOOC_2H_5$) (Plieninger, 1950a). A mixture of N_α-(4-chlorobenzoyl)-5-methoxyphenylhydrazine hydrochloride and **1280** has been indolized to afford 1-(4-chlorobenzoyl)-5-methoxy-2-methyltryptophan (Brenner, Clamkowski, Hinkley and Gal, 1969).

$$CH_3C(=O)CH_2CH_2CH(\overset{\oplus}{N}H_3)COO^{\ominus}$$

1280

F. Synthesis of Tryptamines and Related Systems

The first direct synthesis of tryptamines involving a Fischer indolization was effected when tryptamine (**1283**; R^1–R^4 = H) itself was produced by heating a mixture of 4-aminobutyraldehyde diethylacetal (**1282**; $R^3 = R^4 = H$) and phenylhydrazine (**1281**; $R^1 = R^2 = H$) with zinc chloride (Ewins, 1911; Ewins and Laidlaw, 1910), a synthesis which was later utilized by others

$$R^1C_6H_4N(R^2)NH_2 + NR^3R^4(CH_2)_3CH(OC_2H_5)_2 \longrightarrow$$ indole with R^1 on benzene ring, R^2 on N, and $CH_2CH_2NR^3R^4$ at C-3

1281 **1282** **1283**

Table 26. Tryptamines **1283** which have been formed by the Indolization of 4-Amino- or 4-Substitutedaminobutyraldehyde Arylhydrazones

Tryptamine **1283** (Unless otherwise stated, R^1–R^4 = H)

R^1	R^2	R^3	R^4	Reference(s) (Catalyst)
				1–3(1)
		$SO_2C_6H_4CH_3(4)$		4(1)
		$SO_2C_6H_4CH_3(4)$	CH_3	4(1)[a]
	CH_3			5(1)
	C_2H_5			6(1)
	C_6H_5			7(2)
	C_6H_5	CH_3		7(2)
	C_6H_5	$COCH_3$		7(2)
	$CH_2C_6H_5$			7(2), 8(1)
	$CH_2C_6H_5$	CH_3		7(2)
	$CH_2C_6H_5$	$COCH_3$		7(2)
	$CH_2C_6H_5$	COC_2H_5		7(2)
	$CH_2C_6H_5$	Nicotinoyl		7(2)
7-CH_3				9(1)
5-OCH_3				5(1), 10(3)
6-OCH_3[b]				11(1)
7-OCH_3				5(1), 12(3)
5-OCH_3		$COCH_3$		10(3)
5-OCH_3		CH_3	CH_3	13(3, 4)
5-OCH_3		CH_3	$CH_2C_6H_5$	13(3, 4)
5-OCH_3		C_2H_5	C_2H_5	13(3)
7-OCH_3		C_2H_5	C_2H_5	12(3)
5-OCH_3		$CH(CH_3)_2$	$CH(CH_3)_2$	13(3)
5-OCH_3		$(CH_2)_4$		13(3)
7-OCH_3		$(CH_2)_4$		12(3)
5-OCH_3		$(CH_2)_5$		13(3)
7-OCH_3		$(CH_2)_5$		12(3)
5-OCH_3		$(CH_2)_2O(CH_2)_2$		13(3)
5-OC_2H_5				12(3), 14(1)
5-OC_2H_5		$COCH_3$		12(3)
5-OC_2H_5		$SO_2C_6H_4CH_3(4)$		4(1)
5-OC_2H_5				
5-OC_2H_5		$SO_2C_6H_4CH_3(4)$	CH_3	4(1)[a]
5-$OCH_2C_6H_5$				10(3, 5), 15(1)
5-$OCH_2C_2H_5$		$COCH_3$		16(3)
5-$OCH_2C_2H_5$		COC_6H_5		12(6)
5-$OCH_2C_2H_5$		CH_3	CH_3	13(3, 4)
5-$OCH_2C_2H_5$		CH_3	$CH_2C_6H_5$	13(3, 4)
5-$OCH_2C_2H_5$		C_2H_5	C_2H_5	13(3, 4)
5-$OCH_2C_2H_5$		$CH(CH_3)_2$	$CH(CH_3)_2$	13(3, 4)
5-$OCH_2C_2H_5$		$(CH_2)_4$		13(4)
5-$OCH_2C_2H_5$		$(CH_2)_5$		13(4)
5-$OCH_2C_2H_5$		$(CH_2)_2O(CH_2)_2$		13(4)
5-$SCH_2C_6H_5$				17(7)
5-F				18, 19(1)
5,6-(OCH_3)		$CH(CH_2\text{-cyclohexyl})(CH_2)_4$		20(8)
5-SO_2NH_2				21(1)

Catalysts

1. Heat after admixture with excess zinc chloride; 2. Heat with concentrated hydrochloric acid; 3. Heat with 25% acetic acid[c]. 4. Heat with 25% acetic acid in concentrated hydrochloric acid; 5. Heat with 50% ethanolic 5% hydrochloric acid[c]; 6. Heat with 50% acetic acid in ethanol; 7. Heat with a mixture of acetic acid : ethanol : water (1 : 2 : 1) and concentrated hydrochloric acid; 8. Heat with ethanolic hydrogen chloride.

Notes

a. Heating the indolization product with a mixture of aniline and aniline hydrochloride in ethanol furnished the corresponding N-methyltryptamines; b. Only the 6-methoxy substituted product was isolated (as its N-acetyl derivative) but the 4-methoxy isomer appears also to have been formed (see also Chapter III, A); c. These catalysts were claimed to be preferable to zinc chloride for effecting these indolizations, although attempted indolization of a mixture of 3-benzyloxyphenylhydrazine hydrochloride with **1282** ($R^3 = R^4 = H$) or of 4-aminobutyraldehyde 2,4-dinitrophenylhydrazone by heating in 25% acetic acid or 5% ethanolic hydrochloric acid, respectively, were unsuccessful (Keglević, Stojanac and Desaty, 1961) and a mixture of phenylhydrazine hydrochloride with **1282** ($R^3 = R^4 = H$) afforded only a trace of tryptamine when heated in 25% acetic acid (Desaty and Keglević, 1965).

References

1. Ewins and Laidlaw, 1910; 2. Ewins, 1911 – later used by Scholz, C. (1935); 3. Manske, R. H. F., 1931b; 4. Hoshino and Kobayashi, T., 1935; 5. Späth and Lederer, 1930b; 6. Eiter and Svierak, 1952; 7. Duschinsky, 1953; 8. Hörlein, U., 1954; 9. Eiter and Nezval, 1950; 10. Keglević, Stojanac and Desaty, 1961; 11. Späth and Lederer, 1930a; 12. Desaty and Keglević, 1965; 13. Desaty and Keglević, 1964; 14. Hoshino, Kobayashi, T. and Kotake, Y., 1935; 15. Bernini, 1953 – see also Vejdelek, 1955; 16. Desaty, Hadžija, Iskerić, Keglević and Kveder, 1962; 17. Keglević and Goleš, 1970; 18. Quadbeck and Röhm, 1954; 19. Adlerová, Ernest, Hněvsová, Jílek, Novák, Pomykáček, Rajšner, Sova, Vejdělek and Protiva, 1960; 20. Zenitz, 1966; 21. De Bellis and Stein, M. L., 1961.

(Manske, R. H. F., 1931b; Scholz, C., 1935). This synthetic approach was subsequently extended by using other hydrazino moieties, often as their hydrochlorides, substituted aminoacetals and other catalysts (Table 26). As can be seen, although early studies utilized heating with zinc chloride as the indolization conditions, later investigators often found acetic acid or hydrochloric acid to be the preferred catalyst. However, in a more recent example in which a mixture of 1-amino-1,2,3,4-tetrahydroquinoline and **1282** ($R^3 = R^4 = H$) was indolized to afford **1284**, zinc chloride (in an equimolar amount) was again used as the catalyst but in this case the reaction was performed in boiling dry toluene (Steck, Fletcher and Carabateas, 1974). However, using 25% acetic acid as the potential indolization catalyst, a mixture of 3-benzyloxyphenylhydrazine hydrochloride and **1282** ($R^3 = R^4 = H$) failed to yield the corresponding tryptamine (Desaty and Keglević, 1965; Keglević, Stojanac and Desaty, 1961) and only traces of tryptamine, detected by paper chromatography, were formed from a mixture of phenylhydrazine hydrochloride

NH_2

N

1284

and **1282** ($R^3 = R^4 = H$) (Desaty and Keglević, 1965). In connection with this synthetic approach, a general method for the synthesis of **1282** (R^3 and R^4 = alkyl) was developed (Keglević and Leonhard, 1963) in which **1285**, paraformaldehyde and a secondary amine reacted together in the presence of copper(II) acetate to form **1286** which upon hydrogenation with a 10%

$$HC{\equiv}C{-}CH(OC_2H_5)_2 \quad \textbf{1285}$$

$$(R^3)(R^4)N{-}CH_2{-}C{\equiv}C{-}CH(OC_2H_5)_2 \quad \textbf{1286}$$

palladium-on-barium sulphate catalyst in ethanol at room temperature and atmospheric pressure gave the corresponding **1282**. By this method, compounds **1282** [$R^3 = R^4 = C_2H_5$, $CH(CH_3)_2$, $(CH_2)_2OH$ and $CH_2C_6H_5$ and $R^3 = CH_3$, $R^4 = CH_2C_6H_5$] have been prepared although hydrogenation of the acetylenic dibenzylamine **1286** ($R^3 = R^4 = CH_2C_6H_5$) also brought about the partial hydrogenolysis of the N-benzyl bond(s). Although nitrosobenzene reacted with ethylamine to afford acetaldehyde phenylhydrazone, it did not react similarly with 1,4-diaminobutane (Groves and Swan, 1952). This potential route to 4-aminobutyraldehyde arylhydrazones was therefore not further investigated.

The synthesis of **1287** (R = Cl) by a Fischer indolization has been effected but the synthesis of **1287** (R = F) by an analogous route was unsuccessful (Macander, 1969).

OCH_3 $N(CH_3)_2$ N H R

1287

This above synthetic approach to tryptamines has been extended to the synthesis of 3-(2-mercaptoethyl)indoles. Thus, reaction of phenylhydrazine, N_α-methylphenylhydrazine and 4-benzyloxy-, 4-ethoxy- and 4- and 2-methoxyphenylhydrazine with 4-benzylthiobutyraldehyde diethylacetal (**1288**; $n = 3$),

$S{-}CH_2C_6H_5$ $(CH_2)_n$ $CH(OC_2H_5)_2$

1288

R^1 R^2 R^3 R^4 $(CH_2)_nS{-}CH_2C_6H_5$ N R^5

1289

using aqueous ethanolic acetic acid as catalyst, afforded **1289** (R^1–R^5 = H; R^1–R^4 = H, R^5 = CH_3; R^1 = R^3–R^5 = H, R^2 = $OCH_2C_6H_5$, OC_2H_5 and OCH_3 and R^1–R^3 = R^5 = H, R^4 = OCH_3, respectively, $n = 2$). These, upon debenzylation with sodium in liquid ammonia, furnished the corresponding thiols, **1290** (O-debenzylation of the 5-benzyloxy substituent also occurred under these conditions) along with the corresponding disulphides **1291**, products with could be separated chromatographically (Keglević, Desaty, Goleš and Stančić, 1968). Similarly, 3-benzyloxyphenylhydrazine and 3,4-dimethylphenylhydrazine hydrochloride reacted with **1288** ($n = 3$) in aqueous ethanolic acetic acid to form low yields of **1289** (R^1 = $OCH_2C_6H_5$, R^2–R^5 = H and R^1 = R^2 = R^4 = R^5 = H, R^3 = $OCH_2C_6H_5$, $n = 2$) and **1289** (R^1 = R^2 = CH_3, R^3–R^5 = H and R^1 = R^4 = R^5 = H, R^2 = R^3 = CH_3, $n = 2$), respectively, the last two products being debenzylated into the corresponding thiols, **1290** (R^1 = R^2 = CH_3, R^3–R^5 = H and R^1 = R^4 = R^5 = H, R^2 = R^3 = CH_3, respectively) along with **1291** (R^1 = R^4 = R^5 = H, R^2 = R^3 = CH_3) in the latter case (Keglević and Goleš, 1970). Under similar indolization

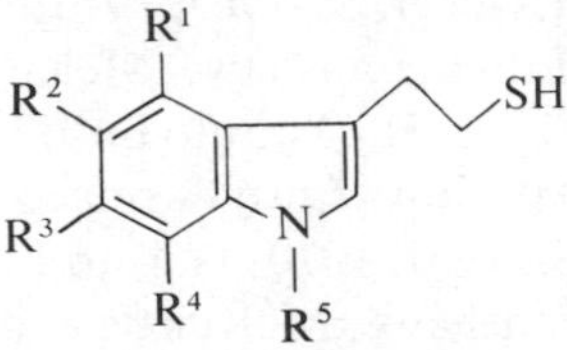

1290 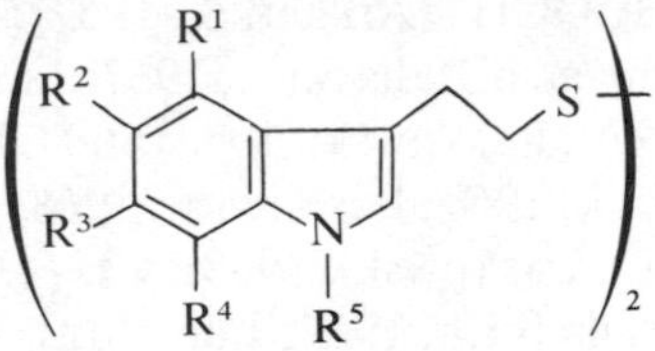 **1291**

conditions, **1288** ($n = 2$) reacted with phenylhydrazine and 4-benzyloxy- and 4-methoxyphenylhydrazine hydrochloride to give low yields of **1289** (R^1 = R^3–R^5 = H, R^2 = H, $OCH_2C_6H_5$ and OCH_3, respectively, $n = 1$) along with toluene-ω-thiol, dibenzyldisulphide and other indolic products, of unknown structures, formed by further reaction of the initially formed **1289**s under the indolization conditions (Keglević and Goleš, 1970).

A large number of 1-[(3-indolyl)alkyl]piperazines, **1293**, have been prepared by the indolization of the corresponding substituted phenylhydrazones, **1292**. The **1293** so synthesized had R^1 = H (Archer, S., 1967, 1970; Istituto Luso Farmaco d'Italia S.r.l., 1967; Laskowski, 1968), 6-CH_3 (Instituto Luso Farmaco d'Italia S.r.l., 1967 and Laskowski, 1968), 5-F (Lanzilotti, Littell,

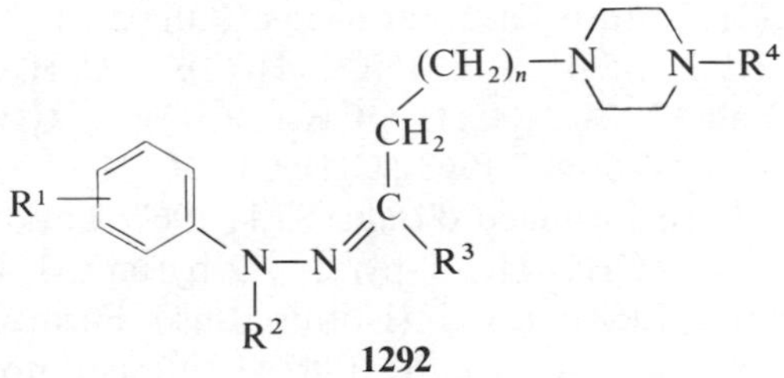

1292

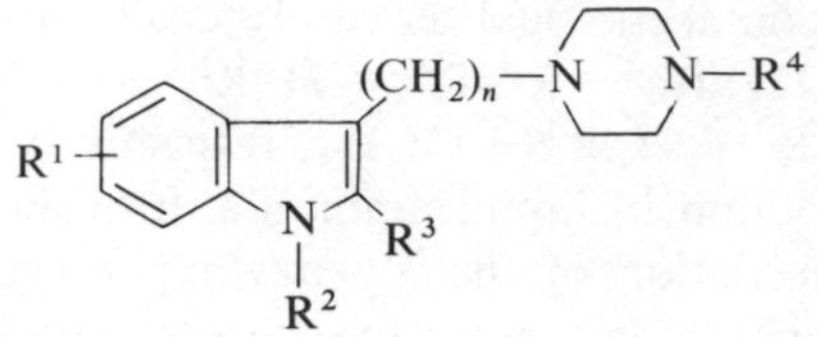

$$\mathrm{R^1}\text{-indole}(\mathrm{N{-}R^2})(2\text{-}\mathrm{R^3})\text{-}3\text{-}(\mathrm{CH_2})_n\text{-N}\langle\text{piperazine}\rangle\text{N-R}^4$$

1293

Fanshawe, McKenzie and Lovell, 1979; Laskowski, 1968), 4-Cl (Laskowski, 1968), 6-Cl (Instituto Luso Farmaco d'Italia S.r.l., 1967; Laskowski, 1968), 4- and 5-OCH_3 (Laskowski, 1968), 6-OCH_3 (Instituto Luso Farmaco d'Italia S.r.l., 1967; Laskowski, 1968), 7-OCH_3 (Laskowski, 1968), 5-$OCH_2C_6H_5$ (Istituto Luso Farmaco d'Italia S.r.l., 1967; Laskowski, 1968), 5-OH (Laskowski, 1968), 5,6-$(OCH_3)_2$ (Archer, S., 1970; Istituto Luso Farmaco d'Italia S.r.l., 1967; Laskowski, 1968), 5,6-$(OC_2H_5)_2$ and 5,6-OCH_2O (Instituto Luso Farmaco d'Italia S.r.l., 1967; Laskowski, 1968), 5-SCH_3, 5-NO_2, 5-NH_2 and 5-$NHCOOC_2H_5$, 5-OC_2H_5-6-OCH_3 and 5,6-$O(CH_2)_2O$ (Laskowski, 1968) and 5-$NHCOCH_3$ (Archer, S., 1967); R^2 = H (Archer, S., 1967, 1970; Instituto Luso Farmaco d'Italia S.r.l., 1967; Lanzilotti, Littell, Fanshawe, McKenzie and Lovell; 1979; Laskowski, 1968) and CH_3 and $CH_2C_6H_5$ (Istituto Luso Farmaco d'Italia S.r.l., 1967; Laskowski, 1968); R^3 = H (Istituto Luso Farmaco d'Italia, S.r.l., 1967; Laskowski, 1968), CH_3 (Archer, S., 1967, 1970; Istituto Luso Farmaco d'Italia S.r.l., 1967; Lanzilotti, Littell, Fanshawe, McKenzie and Lovell, 1979; Laskowski, 1968) and C_6H_5 (Laskowski, 1968); R^4 = H (Istituto Luso Farmaco d'Italia S.r.l., 1967), CH_3 (Istituto Luso Farmaco d'Italia S.r.l., 1967; Laskowski, 1968), C_6H_{11} (Archer' S., 1970; Laskowski, 1968), $(CH_2)_2OH$ (Istituto Luso Farmaco d'Italia S.r.l., 1967; Laskowski, 1968), $(CH_2)_2OCOCH_3$ (Istituto Luso Farmaco d'Italia S.r.l., 1967), C_6H_5 (Archer, S., 1968, 1971, Istituto Luso Farmaco d'Italia S.r.l., 1967; Lanzilotti, Littell, Fanshawe, McKenzie and Lovell, 1979; Laskowski, 1968), $C_6H_4CH_3$(2 and 3) (Laskowski, 1968), $C_6H_4CH_3$(4) (Istituto Luso Farmaco d'Italia S.r.l., 1967; Laskowski, 1968), C_6H_4F(4) (Archer, S., 1970), C_6H_4Cl(2) (Laskowski, 1968), C_6H_4Cl(3) (Archer, S., 1967, 1970; Laskowski, 1968), C_6H_4Cl(4) (Archer, S., 1967; Istituto Luso Farmaco d'Italia S.r.l., 1967; Laskowski, 1968), $C_6H_4C_2H_5$(2), $C_6H_3(CH_3)_2$(2, 2) (Laskowski, 1968), $C_6H_4(CH_3)_3$(2, 2, 4) (Archer, S., 1967), $C_6H_3Cl(3)CH_3(4)$ (Laskowski, 1968), $C_6H_2Cl(3)(CH_3)_2$(2, 4) (Archer, S., 1967), $C_6H_4OCH_3$(2) (Istituto Luso Farmaco d'Italia S.r.l., 1967; Laskowski, 1968), $C_6H_4OCH_3$(3 and 4), $C_6H_4OC_2H_5$(2), $C_6H_4O(CH_2)_2CH_3$(2), $C_6H_3(OCH_3)_2$(2, 3 and 2, 2), $C_6H_3OCH_3(2)Cl(3)$, C_6H_4OH(4) and $C_6H_4SCH_3$(2 and 4) (Laskowski, 1968), $C_6H_4COOC_2H_5$(4) (Archer, S., 1967), $CH_2C_6H_5$ (Istituto Luso Farmaco d'Italia S.r.l., 1967; Laskowski, 1968) and $(CH_2)_2C_6H_5$, $CH_2CH{=}CHC_6H_5$, 2-pyridyl, 2-pyrimidyl and 1-(4-methylpiperidyl) (Laskowski, 1968); $n = 1$ (Istituto Luso Farmaco d'Italia S.r.l., 1967; Laskowski, 1968), 2 (Archer, S., 1967, 1970; Istituto Luso Farmaco d'Italia S.r.l., 1967; Laskowski, 1968), 3 (Archer, S., 1970; Istituto Luso Farmaco d'Italia S.r.l., 1967; Laskowski, 1968) and 4 (Laskowski, 1968) and with $CH(CH_3)CH_2$ in place of $(CH_2)_2$ when $n = 2$ (Laskowski, 1968). The hydrazone formed between 1-amino-1,2,3,4-tetrahydroquinoline and the ketonic moiety of **1292** ($R^3 = R^4 = CH_3$, $n = 2$) has also been indolized (Steck, Fletcher and Carabateas, 1974).

The use of polyphosphoric acid to effect the claimed indolization of **1292** (R^1 = 4-F, R^2 = H, $R^3 = CH_3$, $R^4 = C_6H_5$, $n = 2$) into **1293** (R^1 = 5-F, R^2 = H, $R^3 = CH_3$, $R^4 = C_6H_5$, $n = 2$) (Lanzilotti, Littell, Fanshawe, McKenzie and Lovell, 1979) appears to be a particularly inappropriate choice of catalyst because it is well established (see Chapter III, G) that this catalyst favours new C—C bond formation at the methyl group carbon atom during the indolization of an alkyl methyl ketone. With respect to this, it is significant that the product from this reaction was an uncharacterized brown precipitate which was partially soluble in chloroform to afford a chloroform soluble, uncharacterized, brown glass (45% yield), observations which certainly do not eliminate the possible partial, or even exclusive, formation of the alternative isomeric 3-unsubstituted indolic product. Furthermore, it was also reported (Lanzilotti, Littell, Fanshawe, McKenzie and Lovell, 1979) that other hydrazones of the ketonic moiety of this hydrazone did not indolize under similar conditions, failures which may have been overcome by a more suitable choice of catalyst.

Indolization of the arylhydrazones **1294** with a variety of catalysts [boron trifluoride etherate in ethyl acetate, zinc chloride in anhydrous ethanol at 150 °C (oil bath), polyphosphoric acid or sulphuric acid] has afforded the corresponding 3-(2-aminoethyl)-3*H*-indoles **1295**. Many such 3*H*-indoles have been so produced and the reaction has been extended to the synthesis of the homologous 3-(2- and 3-aminopropyl)-3*H*-indoles (Table 27).

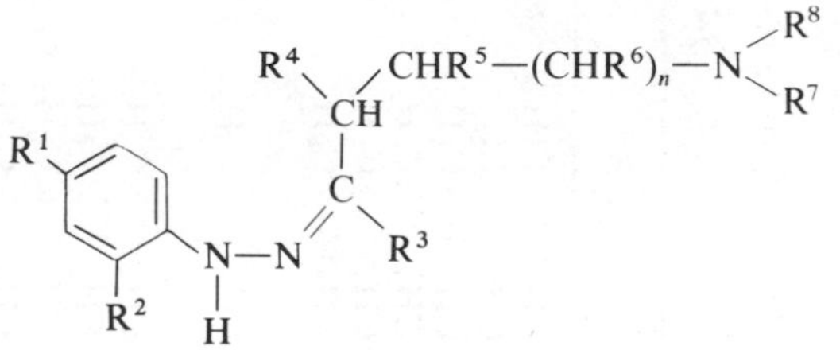

R^1, R^2, R^3, R^4, $CHR^5—(CHR^6)_n—N(R^7)R^8$, CH, C, N—N, H

1294

R^1, R^2, R^3, R^4, $CHR^5—(CHR^6)_n—N(R^7)R^8$, N

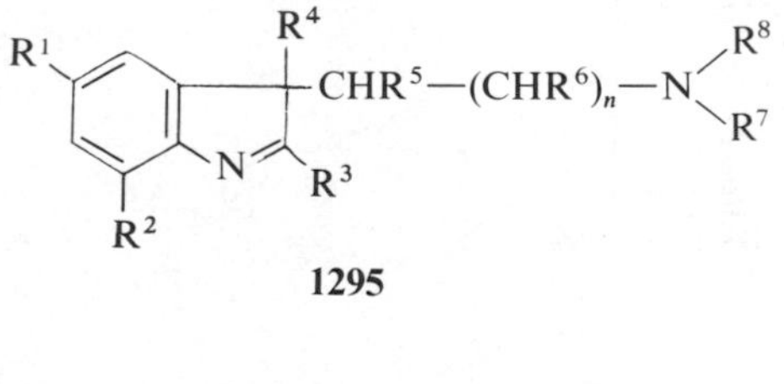

1295

Table 27. 3*H*-Indoles **1295** Produced by the Indolization of the Corresponding **1294**

		3*H*-Indoles **1295** (Unless otherwise stated, R^1–R^8 = H)							
R^1	R^2	R^{3a}	R^4	R^5	R^6	R^7	R^8	*n*	Reference(s)
		CH_3	CH_3				$CH_2CH{=}CH_2$	1	1
		CH_3	CH_3			CH_3	CH_3	1	2–4
		CH_3	C_6H_5			CH_3	CH_3	1	2–4
		CH_3	C_6H_5			CH_3	C_2H_5	1	5, 6
		CH_3	C_6H_5			CH_3	$CH_2CH{=}CH_2$	1	2–4
		CH_3	C_6H_5			CH_3	$CH_2CH{=}C(CH_3)_2$	1	3, 4
		CH_3	C_6H_5			CH_3	CH_2—◁	1	3, 4
		CH_3	C_6H_5			CH_3	$(CH_2)_2C_6H_5$	1	2–4
		CH_3	C_6H_5			C_2H_5	C_2H_5	1	2, 3, 5
		CH_3	C_6H_5			$(CH_2)_4$		1	2–4
		CH_3	C_6H_5			$(CH_2)_5$		1	2–4
		CH_3	C_6H_5			$(CH_2)_2O(CH_2)_2$		1	2–4
		CH_3	C_6H_5			$(CH_2)_2CH(C_6H_5)(CH_2)_2$		1	1, 3, 4
		CH_3	C_6H_5			$(CH_2)_2CH[C_6H_4CH_3(4)](CH_2)_2$		1	1
		CH_3	C_6H_5			$(CH_2)_2CH[C_6H_4OCH_3(4)](CH_2)_2$		1	1
		CH_3	C_6H_5			$(CH_2)_2C(C_6H_5){=}CHCH_2$		1	1, 3, 4

		CH_3	C_6H_5			$(CH_2)_2N(C_6H_5)(CH_2)_2$		1	1, 5, 6
		CH_3	$C_6H_4CH_3(4)$			CH_3	CH_3	1	2–4
		CH_3	$C_6H_4OCH_3(3)$			CH_3	CH_3	1	2–4
		CH_3	$C_6H_4OCH_3(4)$			CH_3	CH_3	1	2–4
		CH_3	$C_6H_4Cl(4)$			CH_3	CH_3	1	2–4
		CH_3	$C_6H_3(OCH_3)_2(3, 4)$			CH_3	CH_3	1	2–4
CH_3		CH_3	C_6H_5			CH_3	CH_3	1	2–4
OCH_3		CH_3	C_6H_5			CH_3	CH_3	1	2–4
	OCH_3	CH_3	C_6H_5			CH_3	CH_3	1	2–4
Cl		CH_3	C_6H_5			CH_3	CH_3	1	2–4
	Cl	CH_3	C_6H_5			CH_3	CH_3	1	2–4
		C_6H_5	C_6H_5			CH_3	CH_3	1	2–4
		C_6H_5	C_6H_5			C_2H_5	C_2H_5	1	2–4
		CH_3	CH_3			CH_3	CH_3	2	2–4
		CH_3	C_6H_5	CH_3		CH_3	CH_3	1	2, 4
		CH_3	C_6H_5		CH_3	CH_3	CH_3	1	2

Note

a. When $R^3 = CH_3$, the direction of indolization was potentially ambiguous (see Chapter III, G).

References

1. Thomae G.m.b.H., 1968b; 2. Thomae G.m.b.H., 1967a; 3. Thomae G.m.b.H., 1968a; 4. Thomae G.m.b.H., 1968c; 5. Thomae G.m.b.H., 1967b; 6. Landgraf and Seeger, 1968.

A widely applicable synthesis of tryptamines, first investigated in 1955 (Abramovitch and Shapiro, 1955), has been effected via the indolization of piperidin-2,3-dione 3-arylhydrazones **1296**. The 1,2,3,4-tetrahydro-β-carbolin-1-ones, structures **1297**, thus formed were then converted into the tryptamines **1299** by a method published earlier (Keimatsu, Sugasawa and Kasuya, 1928 – see also Frey, A. J., Ott, Bruderer and Stadler, 1960; Fujisawa Pharmaceutical Co. Ltd, 1965a, 1965b; Sandoz Patents Ltd, 1965) involving their hydrolysis in aqueous alcoholic alkali to ultimately yield the tryptamine-2-carboxylic acids **1298** which were then decarboxylated, usually by reacting in refluxing mineral acid solution. Using this route, the tryptamines listed in Table 28 and **1300** (R = H) (Nominé and Pénasse, 1959b), **1300** (R = CH_3) (Suvorov, Preobrazhenskaya and Uvarova, 1963; Suvorov, Preobrazhenskaya, Uvarova and Sheinker, Y. N., 1962), **1301** (X = N, Y = CH, n = 1) (Pietra and Tacconi, 1964; Yakhontov, Glushkov, Pronina, E. V. and Smirnova, 1973), **1301** (X = N, Y = CH, n = 2) and **1301** (X = CH, Y = N, n = 1 and 2) (Yakhontov, Glushkov, Pronina, E. V. and Smirnova, 1973) have been synthesized. Presumably, the syntheses of 6,7-dichloro-, 4-chloro-7-methoxy-, 7-chloro-6-methoxy- and 4,5,6-trimethoxytryptamine (Roussel-UCLAF, 1963a; Velluz, L., Muller, Joly,

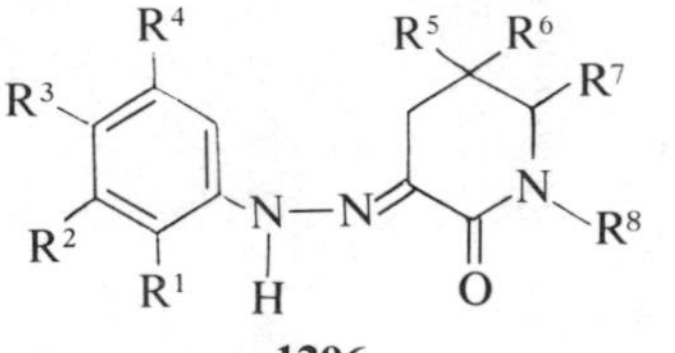

R⁴ R⁵ R⁶ R⁷ R³ R² R¹ N—N N R⁸ H O

1296

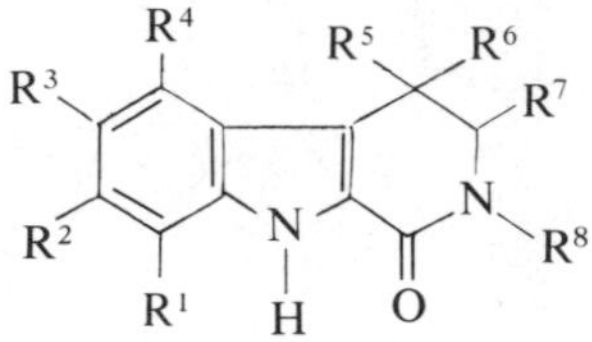

R⁴ R⁵ R⁶ R⁷ R³ R² R¹ N H N R⁸ O

1297

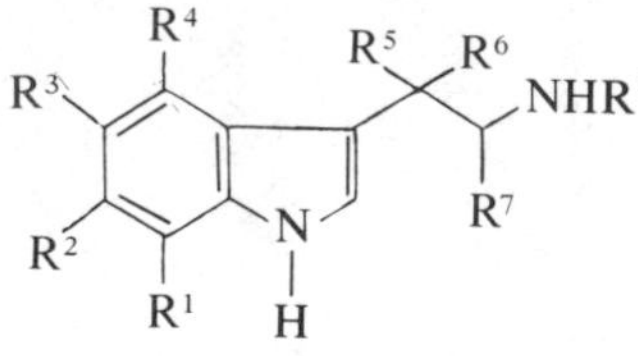

R⁴ R⁵ R⁶ NHR⁸ R³ R² R¹ N H R⁷

1299

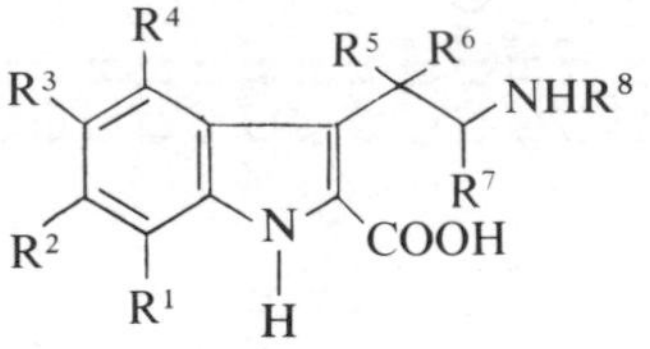

R⁴ R⁵ R⁶ NHR⁸ R³ R² R¹ N H R⁷ COOH

1298

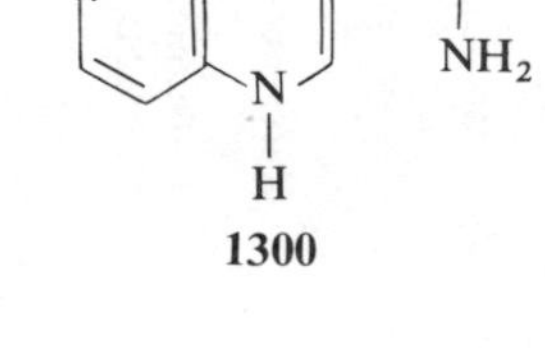

R NH_2 N H

1300

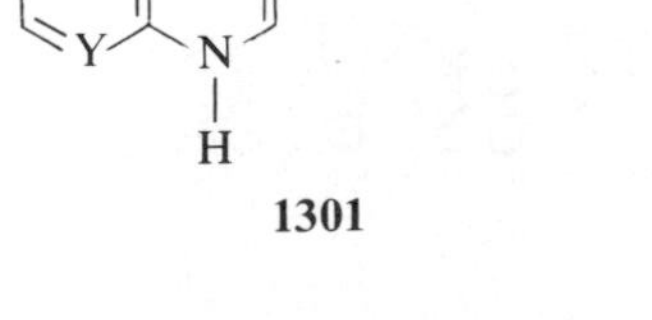

X Y N H $(CH_2)_nCH_2NH_2$

1301

Table 28. Tryptamines which have been Prepared via the Indolization of Piperidin-2,3-dione 3-(Phenylhydrazones and Benz-substituted Phenylhydrazones)

Tryptamines **1299** (Unless otherwise stated, R^1–R^8 = H)

R^1	R^2	R^3	R^4	R^5	R^6	R^7	R^8	Reference(s) [Indolization catalyst(s) or Note]
CH_3								1[1], 2[1, 2], 3[1, 2] 4[3], 5[4]
CH_3		OCH_3						6[5]
CH_3			F					5[4]
$(CH_2)_2CH_3$								7[6]
OCH_3	Cl							8[7], 9[4], 10[8]
OCH_3			Cl					9[8], 10[8], 11[7]
SCH_3								3[2], 12[2], 13[2]
Cl								10[8], 14[8]
Cl	OCH_3							8[7], 10[8]
Cl		OCH_3						15[2]
Cl	Cl							9[4], 10[8], 16[8]
Cl			Cl					8[7], 10[8], 17[8]
	OCH_3							4[2, 5], 18[a]
	OCH_3	OCH_3	OCH_3					10[8], 19[8], 20[8]
	F	OCH_3						21[2]
	F	OCH_3	F					21[2]
	$COCH_3$[b]							22[2], 23[2], 24[2]
	$COCH_3$[b]			CH_3				24[2]
	COC_6H_5							24[2]
		CH_3						3[2], 5[4]
		CH_3				CH_3		25[2]
		$C(CH_3)_3$						26[9]
		CF_3				CH_3		27[2]
		OCH_3						1[2], 2[2], 3[2]

(continued)

Table 28 (*continued*)

R^1	R^2	R^3	R^4	R^5	R^6	R^7	R^8	Reference(s) [Indolization catalyst(s) or Note]
		OCH_3		CH_3		CH_3		28[2]
		OCH_3				CH_3		29[2]
		OCH_3				CH_3	CH_3	30[5]
		OC_2H_5						3[2], 12[2]
		$OCH_2C_6H_5$						31[10]
		SCH_3						3[2], 12[2], 13[2]
		SCH_3		CH_3				32[2, 8]
		SCH_3		CH_3		CH_3		32[2, 8]
		SCH_3				CH_3		32[2, 8]
		$SCH_2C_6H_5$						26[9], 33[8]
		$SCH_2C_6H_5$		CH_3				32[2, 8]
		$SCH_2C_6H_5$		CH_3		CH_3		32[2, 8]
		$SCH_2C_6H_5$				CH_3		32[2, 8]
		F						3[2], 12[2], 13[2], 34[4]
		F		CH_3				35[8]
		F		CH_3	CH_3			35[8]
		F				CH_3		36[2], 37[11]
		Cl						4[2]
		Cl		CH_3				35[8]
		Cl		CH_3	CH_3			35[8]
		Cl				CH_3		27[2], 36[2], 37[11]
		Br		CH_3				35[8]
		Br		CH_3	CH_3			35[8]
		Br				CH_3		36[2], 37[11]
		$COCH_3$[b]						22[2], 23[2], 24[2]
		$COCH_3$[b]		CH_3				24[2]
		$COCH_3$[b]				CH_3		24[2], 38[2]
		COC_2H_5						23[2], 24[2]

COC_2H_5		CH_3			24[2]
$OCH_2C_6H_5$					23[2], 24[2]
$OCH_2C_6H_5$		CH_3			24[2]
$COC_6H_4Cl(4)$					23[2]
Isonicotinoyl					23[1]
COOH					39[12]
COOH		CH_3	CH_3		40[12]
COOH			CH_3		40[12]
SO_2NH_2					31[1], 41[1]
$SO_2N(CH_3)_2$					31[2]
	OCH_3				18[a], 42[c], 43[d]
		CH_3			44[2]
		CH_3	CH_3		28[2]
			CH_3		25[2], 29[2], 37[11], 45[10]
			CH_3	CH_3	30[5]

Notes

a. Product formed via the indolization of **1296** (R^1 = Br, $R^2 = R^3 = R^5$–R^8 = H, $R^4 = OCH_3$) in refluxing acetic acid–hydrochloric acid which yielded **1297** (R^1 = Br, $R^2 = R^3 = R^5$–R^8 = H, $R^4 = OCH_3$ and $R^1 = R^4 = R^5$–R^8 = H, $R^2 = OCH_3$, R^3 = Br) (see Chapters II, G1 and III, A1). These products were then converted into **1297** (R^1–$R^3 = R^5$–R^8 = H, $R^4 = OCH_3$ and $R^1 = R^3$–R^8 = H, $R^2 = OCH_3$, respectively) by treatment with hydrazine hydrate in the presence of 10% palladium-on-carbon in refluxing ethanol (see also notes c and d); b. Treatment of piperidin-2,3-dione 3-(2-acetylphenylhydrazones) with acids gave indazoles not indoles (Shavel, Strandtmann and Cohen, M. P., 1962; Strandtmann, Cohen, M. P. and Shavel, 1963) (see Chapter VII, O); c. Formed via the 'directed indolization' (see Chapter III, A1) of **1296** (R^1 = Br, $R^2 = R^3 = R^5$–R^8 = H, $R^4 = OCH_3$) in refluxing acetic acid–hydrochloric acid to afford **1297** (R^1 = Br, $R^2 = R^3 = R^5$–R^8 = H, $R^4 = OCH_3$) which was debrominated by heating with Raney nickel in ethanol containing hydrazine hydrate to furnish **1297** (R^1–$R^3 = R^5$–R^8 = H, $R^4 = OCH_3$). In a later publication, the formation of a further indolic product was also claimed from this reaction (see note a); d. Formed via the 'directed indolization' (see Chapter III, A1) of **1296** (R^1 = Cl, $R^2 = R^3 = R^5$–R^8 = H, $R^4 = OCH_3$) in refluxing acetic acid–hydrochloric acid to give **1297** (R^1 = Cl, $R^2 = R^3 = R^5$–R^8 = H, $R^4 = OCH_3$) which upon heating with palladium-on-carbon in refluxing ethanol containing hydrazine afforded **1297** (R^1–$R^3 = R^5$–R^8 = H, $R^4 = OCH_3$). A 0.5% yield of **1297** ($R^1 = R^3$–R^8 = H, $R^2 = OCH_3$) was also formed from the indolization (see Chapter II, G1).

Catalysts

1. Polyphosphoric acid; 2. Formic acid [this catalyst was found to be preferable to polyphosphoric acid which, when used in large scale indolizations, caused considerable charring (Abramovitch and Shapiro, 1956)]; 3. Acetic acid; 4. Concentrated hydrochloric acid in acetic acid; 5. Ethanolic hydrogen chloride; 6. Acetic acid–sulphuric acid mixture; 7. Methanolic hydrochloric acid; 8. Hydrogen chloride in acetic acid; 9. Reflux a solution of the hydrazone hydrochloride; 10. Sulphosalicylic acid; 11. Not stated; 12. Formic acid on the corresponding 4-carbomethoxyphenylhydrazone.

References

1. Abramovitch and Shapiro, 1955; 2. Abramovitch and Shapiro, 1956; 3. Adlerova, Ernest, Hněvsová, Jíkek, Novák, Pomykáček, Rajšner, Sova, Vejdělek and Protiva, 1960; 4. Abramovitch, 1956; 5. Pelchowicz and Bergmann, E. D., 1960; 6. Roussel-UCLAF, 1964a; 7. Roussel-UCLAF, 1964b; 8. Roussel-UCLAF, 1962; 9. Velluz, L., Muller, Nominé, Pénasse and Pierdet, 1964; 10. Roussel-UCLAF, 1963b; 11. Nominé and Pénasse, 1959a; 12. Protiva, Adlerová, Vejdělek, Novák, Rajšner and Ernest, 1959; 13. Protiva, Ernest, Hněvsova, Novák and Rajšner, 1960; 14. Allais, 1959a; 15. Magidson, Suvorov, Travin, Sorokina, N. P. and Novikova, 1965; 16. Nominé, Pénasse and Pierdet, 1959; 17. Nominé and Pénasse, 1959c; 18. Morozovskaya, Ogareva and Suvorov, 1969; 19. Allais, 1959b; 10. Velluz, L., Muller and Allais, 1962; 21. Kirk, 1976; 22. Shavel, Strandtmann and Cohen, M. P., 1962; 23. Strandtmann, Cohen, M. P. and Shavel, 1963; 24. Shavel, Strandtmann and Cohen, M. P., 1965; 25. Terzyan, Safrazbekyan, Sukasyan and Tatevosyan, 1961; 26. Colwell, Horner, J. K. and Skinner, 1964; 27. Fellows, 1965; 28. Mkhitaryan, Kogodovskaya, Terzyan and Tatevosyan, 1962; 29. Terzian, Safrasbekin, Sukasian and Tatevosian, 1961; 30. Heath-Brown and Philpott, P. G., 1965a; 31. Suvorov, Sorokina, N. P. and Tsvetkova, 1964; 32. Horner, J. K., DeGraw and Skinner, 1966; 33. Horner, J. K. and Skinner, 1964; 34. Pelchowicz and Bergmann, E. D., 1959; 35. DeGraw and Skinner, 1967; 36. Suvorov, Preobrazhenskaya, and Uvarova, 1963; 37. Suvorov, Preobrazhenskaya, Uvarova and Sheinker, Y. N., 1962; 38. Terzyan and Tatevosyan, 1962; 39. Terzyan, Kogodovskaya and Tatevosyan, 1964; 40. Terzyan, Aznauryan and Tatevosyan, 1965; 41. De Bellis and Stein, M. L., 1961; 42. Suvorov, Fedotova, M. V., Orlova, L. M. and Ogareva, 1962; 43. Morozovskaya, Ogareva and Suvorov, 1970; 44. Abramovitch and Muchowski, 1960a; 45. Suvorov, Preobrazhenskaya and Uvarova, 1962.

Nominé, Mathieu, Allais, Warnant, Valls, Bucourt and Jolly, 1958), of 5- and 7-chloro-, 4,7-dichloro-, 5-chloro-6-methoxy-, 6-chloro-7-methyl-, 7-methoxy- and 7-methyltryptamine (Roussel-UCLAF, 1963a) and of 4,5-benzo-, 5-chloro-6-methoxy- and 6-chloro-7-methoxytryptamine (Velluz, L. Muller, Joly, Nominé, Mathieu, Allais, Warnant, Valls, Bucourt and Jolly, 1958), details of which were not given, were also effected by this route in view of this group's other studies in this area (see Table 28). In connection with the syntheses of azatryptamines by this route it was observed that the decarboxylation stage required more vigorous conditions than did the corresponding reaction in the synthesis of tryptamines (Yakhontov, Glushkov, Pronina, E. V. and Smirnova, 1973). However, even in these latter syntheses, the decarboxylation stage could prove to be difficult and sometimes failed altogether. Thus, **1298** ($R^1 = R^2 = R^4$–R^8 = H, R^3 = Cl) was unaffected by refluxing 10% hydrochloric acid and was only decarboxylated under more vigorous conditions (i.e. fusion with resorcinol at 230 °C or refluxing with 20% hydrochloric acid in acetic acid) (Abramovitch, 1956). 5-Nitrotryptamine-2-carboxylic acid was resistant to decarboxylation under a variety of conditions but its 1-methyl derivative, formed by sequential indolization of **1296** ($R^1 = R^2 = R^4$–R^8 = H, $R^3 = NO_2$) using a polyphosphoric acid catalyst, N-1 methylation with dimethyl sulphate in the presence of potassium hydroxide and hydrolysis, afforded a small yield of 1-methyl-5-nitrotryptamine upon prolonged refluxing in mineral acid. Likewise, the decarboxylation of 5-chlorotryptamine-2-carboxylic acid was difficult (e.g. the compound was unaffected in refluxing 10% hydrochloric acid and yielded 6-chloro-1,2,3,4-tetrahydro-β-carbolin-1-one when fused with resorcinol) whereas the corresponding 1-methylated derivative readily decarboxylated in refluxing 20% hydrochloric acid–acetic acid. Although it could be readily converted into tryptamine by sequential hydrolysis and decarboxylation (Table 28), **1297** (R^1–R^8 = H) has also been methylated on the N-9 atom prior to these steps to ultimately afford 1-methyltryptamine (Abramovitch, 1956). Failure has also attended attempts to decarboxylate 5-benzyloxytryptamine-2-carboxylic acid (Abramovitch and Shapiro, 1956) although it was subsequently (Suvorov, Sorokina, N. P. and Tsvetkova, 1963, 1964) decarboxylated via the formation of its phthalimido derivative by heating in quinoline with phthalic anhydride and a copper–chrome catalyst, the phthalyl group then being removed by treatment with hydrazine hydrate in refluxing alcohol. Similarly, whereas initial attempts to decarboxylate 5-sulphonamidotryptamine-2-carboxylic acid by refluxing in hydrochloric acids of various concentrations yielded, after basification of the reaction mixture with sodium carbonate, a precipitate whose properties were not consistent with those expected of **1299** ($R^1 = R^2 = R^4$–R^8 = H, $R^3 = SO_2NH_2$) and whose structure remained unestablished (De Bellis and Stein, M. L., 1961), the decarboxylation of this amino acid into **1299** ($R^1 = R^2 = R^4$–R^8 = H, $R^3 = SO_2NH_2$) was again smoothly effected via formation of the phthalimido derivative, a technique which was also applied to the conversion of **1298** [$R^1 = R^2 = R^4–R^8 = H$, $R^3 = SO_2N(CH_3)_2$)] into **1299** [$R^1 = R^2 = R^4–R^8 = H$, $R^3 = SO_2N(CH_3)_2$] (Suvorov, Sorokina, N. P. and Tsvetkova, 1964).

5-Iodotryptamine could not be prepared by application of this synthetic route because, although **1297** ($R^1 = R^2 = R^4–R^8 = H$, $R^3 = I$) could be prepared by the formic acid catalysed indolization of **1296** ($R^1 = R^2 = R^4–R^8 = H$, $R^3 = I$), loss of iodine occurred during both the subsequent hydrolytic and decarboxylative stages (Preobrazhenskaya, Fedotova, M. N., Sorokina, N. P., Ogareva, Uvarova and Suvorov, 1964; Suvorov, Preobrazhenskaya and Uvarova, 1963). Care also has to be taken when decarboxylating **1298** ($R^1 = R^2 = R^4–R^6 = H$, $R^3 = OCH_3$, $R^7 = R^8 = CH_3$), for whereas after refluxing in 2.25–2.5 M sulphuric acid the corresponding 5-methoxytryptamine **1299** ($R^1 = R^2 = R^4–R^6 = H$, $R^3 = OCH_3$, $R^7 = R^8 = CH_3$) was obtained, refluxing in 3 M sulphuric acid also caused O-demethylation to yield **1299** ($R^1 = R^2 = R^4–R^6 = H$, $R^3 = OH$, $R^7 = R^8 = CH_3$) (Heath-Brown and Philpott, P. G., 1965a). Both the hydrolytic and decarboxylative stages in the above reaction sequence have also been effected in 'one-pot' by prolonged refluxing of the 1,2,3,4-tetrahydro-β-carbolin-1-one in a 1 : 1 mixture of concentrated hydrochloric acid and acetic acid. Under these conditions, **1297** ($R^1 = R^2 = R^4 = R^7 = R^8 = H$, $R^3 = F$, Cl and Br, $R^5 = CH_3$, $R^6 = H$ and CH_3) afforded **1299** ($R^1 = R^2 = R^4 = R^7 = R^8 = H$, $R^3 = F$, Cl and Br, $R^5 = CH_3$, $R^6 = H$ and CH_3, respectively). However, using similar conditions, **1297** ($R^1 = R^2 = R^4 = R^7 = R^8 = H$, $R^3 = OCH_3$, $R^5 = R^6 = CH_3$), which was not hydrolysed by treatment with alkali, afforded **1299** ($R^1 = R^2 = R^4 = R^7 = R^8 = H$, $R^3 = OH$, $R^5 = R^6 = CH_3$), O-demethylation again occurring under strongly acidic conditions (DeGraw and Skinner, 1967). Because **1297** ($R^1 = R^2 = R^4 = R^7 = R^8 = H$, $R^3 = SCH_3$, $R^5 = R^6 = CH_3$) was stable to alkaline hydrolysis, the synthesis of the corresponding tryptamine was not effected (Horner, J. K., DeGraw and Skinner, 1966). The decarboxylations of the 4-acyltryptamine-2-carboxylic acids **1298** ($R^1–R^3 = R^5–R^8 = H$, $R^4 = COCH_3$ and COC_6H_5; $R^1–R^3 = R^6–R^8 = H$, $R^4 = COCH_3$, $R^5 = CH_3$ and $R^1 = Cl$, $R^2 = R^3 = R^5–R^8 = H$, $R^4 = COCH_3$) in refluxing hydrochloric acid–acetic acid were accompanied in each case by a cyclodehydration, between the 4-acetyl or 4-benzoyl group and the primary amino function, to furnish **526** ($R^1 = COCH_3$, $R^2–R^4 = H$) (Shavel, Strandtmann and Cohen, M. P., 1962, 1965; Strandtmann, Cohen, M. P. and Shavel, 1963), **526** ($R^1 = C_6H_5$, $R^2–R^4 = H$) (Strandtmann, Cohen, M. P. and Shavel, 1963), **526** ($R^1 = R^4 = CH_3$, $R^2 = R^3 = H$) (Shavel, Strandtmann and Cohen, M. P., 1965) and **526** ($R^1 = CH_3$, $R^2 = Cl$, $R^3 = R^4 = H$) together with **526** ($R^1 = CH_3$, $R^2 = Cl$, $R^3 = COOH$, $R^4 = H$) (approximately 1 : 3 yield ratio) (Strandtmann, Cohen, M. P. and Shavel, 1965), respectively.

The attempted indolization of **1302** into **1303** ($R = H$) was unsuccessful. After refluxing in 90% acetic acid the hydrazone remained unchanged and upon

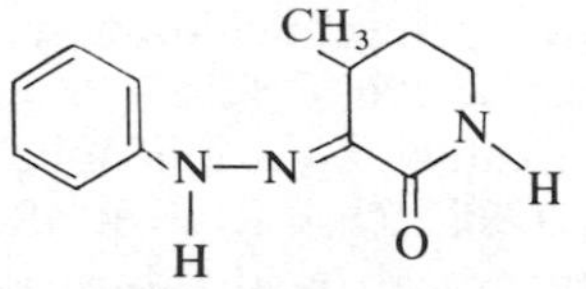

1302

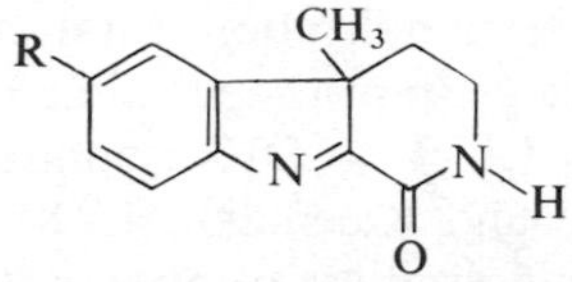

1303

heating with boron trifluoride etherate in acetic acid or with zinc chloride at 170 °C only an isomeric product, probably geometrical (see Chapter VII, I), was obtained in each case. In view of these failures, the projected sequential hydrolysis of the potential product **1303** (R = H) into **1304** (R = COOH), decarboxylation into **1304** (R = H) and cyclization into the eserine ring system **1305** could not be investigated (Abramovitch, 1958). However, the indolization of the related phenylhydrazone **1306** (probably the *trans* isomer) appears to have been successful. After refluxing in 90% formic acid, a product was obtained which was tentatively assigned the structure **1307**. Although i.r. and u.v. spectroscopic data supported this structure, the microanalytical data, while clearly indicating that one of the original nitrogen atoms in **1306** had been lost, did not correspond and, furthermore, attempted alkaline hydrolysis of the product simply gave rise to an isomer having a sharp melting point (Abramovitch and Muchowski, 1960b). Clearly, these reactions are worthy of further investigation.

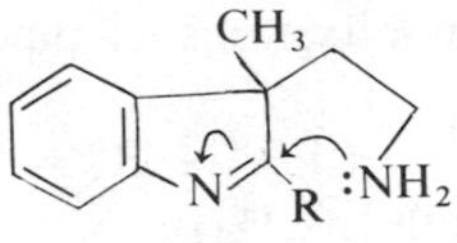

1304

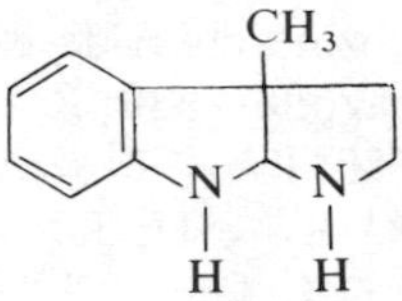

1305

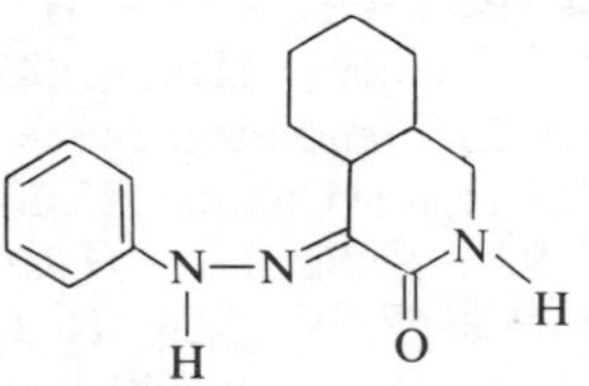

1306

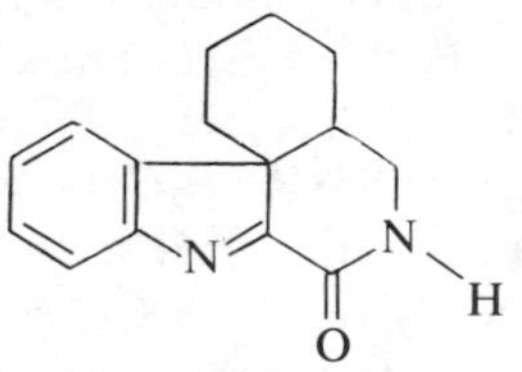

1307

The above reaction sequence ultimately leading to **1299** has also been arrested at intermediate stages, to afford **1298** or **1297** as the final products. Thus, **1298** ($R^1 = R^2 = R^4$–R^8 = H, R^3 = OH and $R^1 = R^5$–R^8 = H, $R^2 = R^4$ = Cl, R^3 = OH) (indolization catalysts were ethanolic hydrogen chloride or hydrogen chloride in acetic acid and hydrogen chloride in acetic acid, respectively) (Ried and Baumbach, 1969), **1298** ($R^1 = R^5$–R^8 = H, R^2–R^4 = OCH_3; R^1 = Cl, R^2–R^8 = H; R^1 = OCH_3, $R^2 = R^3 = R^5$–R^8 = H, R^4 = Cl; R^1 = Cl, R^2 = OCH_3, R^3–R^8 = H; $R^1 = R^2$ = Cl, R^3–R^8 = H and R^1 = OCH_3, R^2 = Cl,

R^3–R^8 = H) (indolization catalyst was hydrogen chloride in acetic acid) (Laboratories Francais de Chimiotherapie, 1962), **1298** (R^1 = R^5–R^8 = H, R^2 = R^4 = Cl, R^3 = OH) (indolization catalyst was ethanolic hydrochloric acid) (Ried and Kleemann, 1968) and **1298** (R^1 = R^2 = R^4–R^8 = H, R^3 = COOH) (indolization catalyst was formic acid on the 4-carbethoxyphenylhydrazone) (Fujisawa Pharmaceutical Co. Ltd, 1965a, 1965b) have been prepared from the corresponding piperidin-2,3-dione 3-arylhydrazones **1296**. The sequence was likewise arrested after the first stage when **1297** (R^1 = R^2 = R^4–R^6 = R^8 = H, R^3 = $COOCH_3$, R^7 = CH_3 and R^1 = R^2 = R^4 = R^5 = R^8 = H, R^3 = $COOCH_3$, R^6 = R^7 = CH_3) (Terzyan, Safrazbekyan, Sukasyan, Akopyran and Tatevosyan, 1964) and **1297** (R^1 = R^2 = R^4–R^8 = H, R^3 = $COOCH_3$) (Terzyan, Kogodovskaya and Tatevosyan, 1964) were obtained by refluxing the corresponding **1296** in formic acid, when compounds **1308** (R = H and OCH_3) were indolized into **1309** (R = H and OCH_3, respectively) by

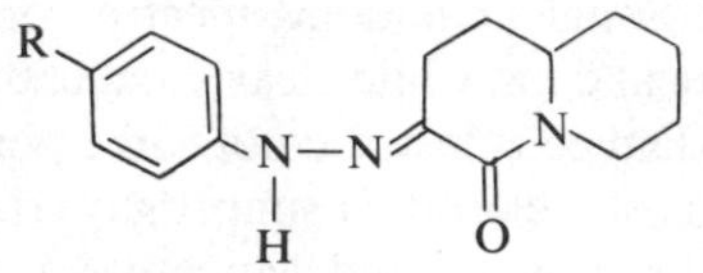

1308 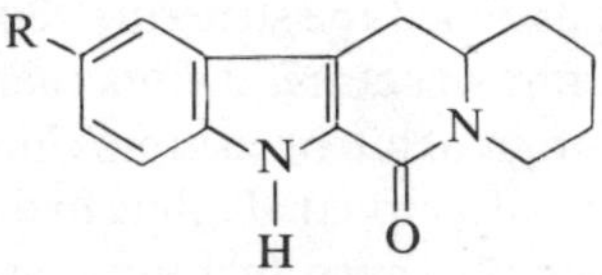 **1309**

treatment with hydrogen chloride in acetic acid (Arata, Sakai, M. and Soga, 1972), when the zinc chloride catalysed indolizations of piperidin-2,3-dione 3-(3-pyridylhydrazone) [**580**; R^1–R^3 = R^5 = H, R^4 + R^6 = $CONH(CH_2)_2$], its N-oxide and **580** [R^1 = R^3 = R^5 = H, R^2 = Cl, R^4 + R^6 = $CONH(CH_2)_2$] afforded **581** [R^1–R^3 = H, R^4 + R^5 = $CONH(CH_2)_2$, X = N and N → O, Y = CH] and **581** [R^1 = R^3 = H, R^2 = Cl, R^4 + R^5 = $CONH(CH_2)_2$, X = CH, Y = N], respectively (Tacconi and Perotti, 1965), when the zinc chloride catalysed indolizations of **584** (R^1 = CH_3, R^2 = H, R^3 = C_2H_5; R^1 = R^2 = H, R^3 = CH_3 and R^1 = CH_3, R^2 = R^3 = H) gave **585** (R^1 = CH_3, R^2 = H, R^3 = C_2H_5; R^1 = R^2 = H, R^3 = CH_3) and either **585** (R^1 = R^3 = H, R^2 = CH_3) or **585** (R^1 = CH_3, R^2 = R^3 = H) (see Chapter III, C), respectively (Tacconi and Pietra, 1965) and when **524** [R^1 = $COOCH_3$, $CH_2COOC_2H_5$ and $(CH_2)_2COOC_2H_5$, R^2 = Cl] gave **525** [R^1 = $COOCH_3$, $CH_2COOC_2H_5$ and $(CH_2)_2COOC_2H_5$, respectively, R^2 = Cl] upon treatment with boron trifluoride etherate in acetic acid (Nagasaka, Yuge and Ohki, 1977). Similar arrests occurred when **1310** ($n = 1$) afforded **1311** ($n = 1$) after treatment with hydrogen chloride in acetic acid (Glushkov, Zasosova, Ovcharova, Rudzit, Saratikov, Livshits and Kostyuchenko, 1978), when **524** (R^1 = OCH_3, R^2 = Br) produced **525** (R^1 = OCH_3, R^2 = Br) by treatment with a hydrochloric acid-acetic acid mixture (Morozovskaya, Ogareva and Suvorov, 1969; Suvorov, Fedotova, M. V., Orlova, L. M. and Ogareva, 1962), when **524** (R^1 = OCH_3, R^2 = Cl) similarly gave **525** (R^1 = OCH_3, R^2 = Cl) (Morozovskaya, Ogareva and Suvorov, 1969, 1970), when **1312** (R = CH_3 and $CH_2C_6H_5$, $n = 1$ and 2) yielded **1313** (R = CH_3 and

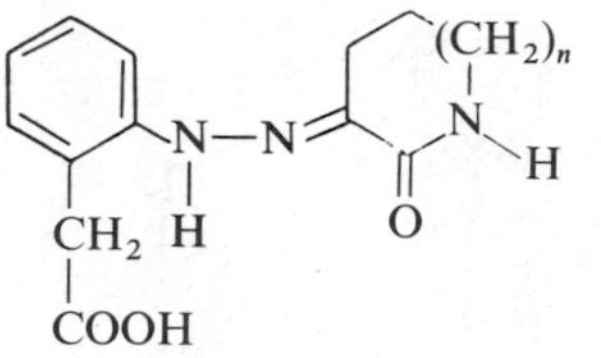

1310 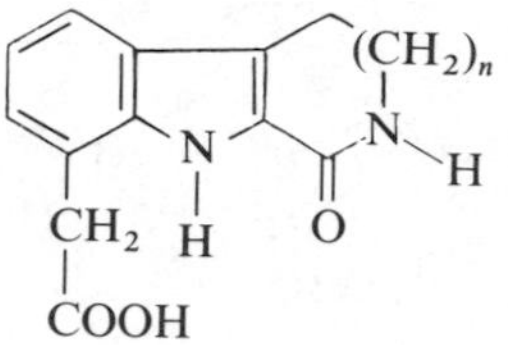 **1311**

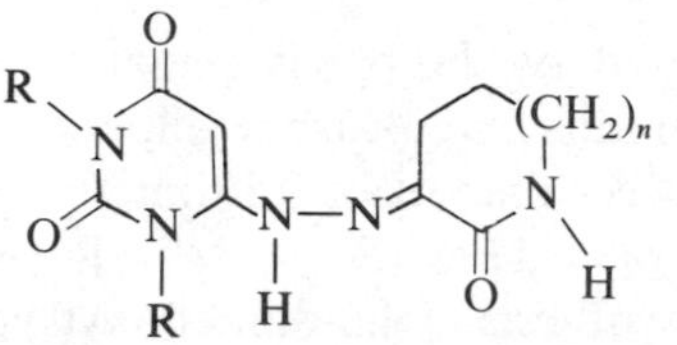

1312 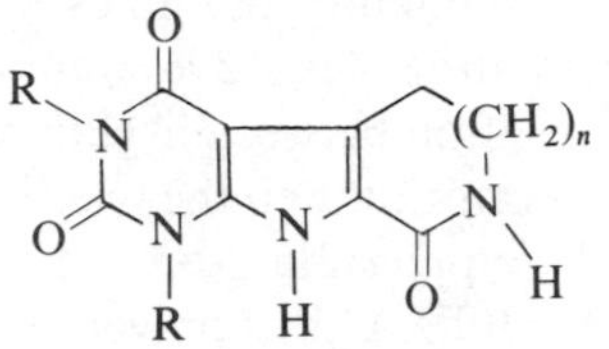 **1313**

$CH_2C_6H_5$, n = 1 and 2 respectively, by treatment with polyphosphoric acid or by refluxing in diethylene glycol (Glushkov, Zasasova and Ovcharova, 1977), when **565** afforded both **566** and **567** upon treatment with acoholic sulphuric acid (Sladkov, Anisimova and Suvorov, 1977), when **1314** (R = OCH_3 $OCH_2C_6H_5$) yielded **1297** (R^1 = R^3–R^8 = H, R^2 = OCH_3 and $OCH_2C_6H_5$, respectively) upon treatment with acetic acid–hydrochloric acid (Henecka, Timmler and Lorenz, 1959) and, when under similar conditions, **124** (R^1 = H and OCH_3, R^2 = H) gave **125** (R^1 = H and OCH_3, R^2 = H, respectively) (Henecka, Timmler, Lorenz and Geiger, 1957).

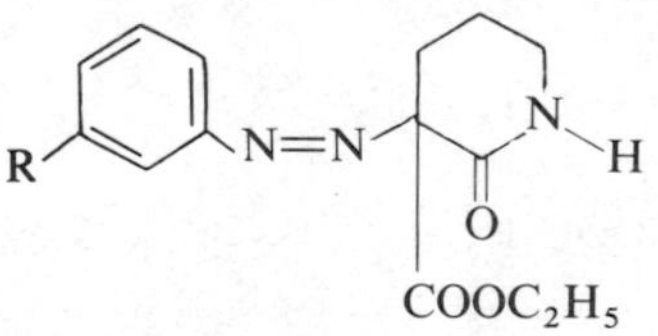

1314

A recent interesting synthesis of 1,2,3,4-tetrahydro-β-carbolin-1-ones involved the Fischer indolization of mixtures of arylhydrazines with 2-formyl-γ-lactams (Grandberg and Tokmakov, 1975b; Tokmakov and Grandberg, 1976, 1980). or 2-dimethylaminomethylene γ-lactams (Grandberg and Tokmakov, 1975b; Tokmakov and Grandberg, 1975, 1976).

Indolization of **1310** (n = 2) in acetic acid containing hydrogen chloride produced **1311** (n = 2) (Glushkov, Zasosova, Ovcharova, Rudzit, Saratikov, Livshits and Kostyuchenko, 1978) and of mixtures of **1315** with phenylhydrazine and 4-methoxyphenylhydrazine hydrochloride in ethanolic sulphuric acid and with 2-pyridylhydrazine in fused zinc chloride afforded **1316** (R = H and OCH_3, X = CH and R = H, X = N, respectively) (Glushkov, Yakhontov, Pronina,

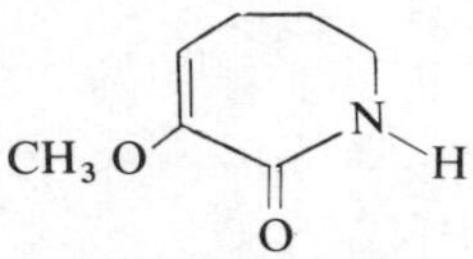

1315

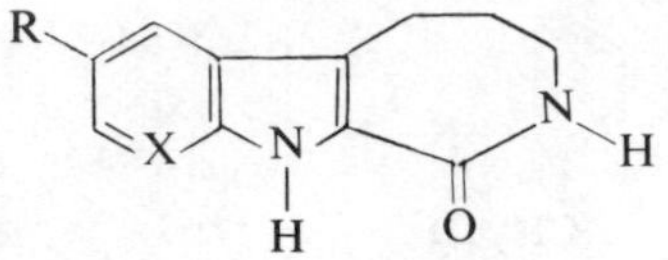

1316

E. V. and Magidson, 1969). Such compounds are potential sources of the corresponding homotryptamines.

The arylhydrazones **1296** were prepared by Japp–Klingemann reactions involving reaction between aryldiazonium salts and 3-carboxypiperidin-2-ones (or their esters), although the reaction between diazotized 3,5-dimethoxyaniline and 3-carboxypiperidin-2-one failed to yield **1296** ($R^1 = R^3 = R^5$–R^8 = H, $R^2 = R^4 = OCH_3$), thus preventing the synthesis of 4,6-dimethoxytryptamine via the indolization of this hydrazone (Brown, V. H., Skinner and DeGraw, 1969). By analogy with the earlier observation that diazotized aniline reacted with 2-carbethoxycyclopentanone to afford the half ethyl ester phenylhydrazone of 2-oxoadipic acid (Manske, R. H. F. and Robinson, R., 1927), it was anticipated (Shapiro and Abramovitch, 1955) when the reaction between diazonium salts and 3-carbethoxypiperidin-2-one (Albertson and Fillman, 1949) was first investigated that the product formed when starting with diazotized aniline would be **1317**, formed via the intermediate **1318**. In practice, only tars were produced from this reaction and coupling was only observed when the ester was first hydrolysed to give 3-carboxypiperidin-2-one. When this was reacted with benzenediazonium chloride, concomitant decarboxylation occurred to leave **1296** (R^1–R^8 = H) (Shapiro and Abramovitch, 1955).

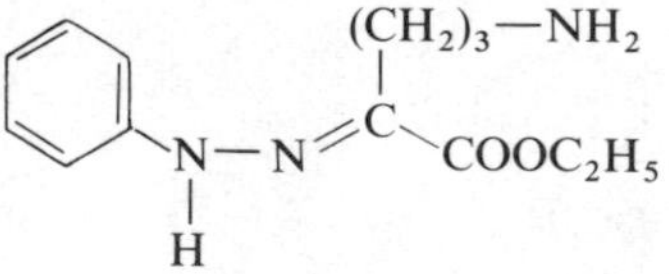

1317

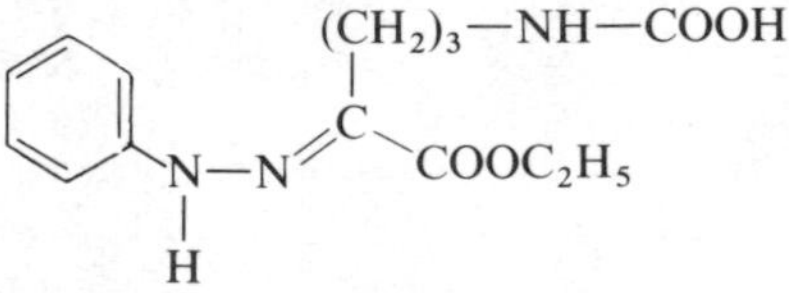

1318

The intermediate azo compounds **124** (R^1 = H and OCH_3, R^2 = H) have been isolated from these coupling reactions and have been directly indolized to afford **125** (R^1 = H and OCH_3, R^2 = H, respectively) (Henecka, Timmler, Lorenz and Geiger, 1957). Difficulty has, however, been experienced in coupling benzenediazonium chloride and 4-methoxybenzenediazonium chloride with 3-carbethoxy-1,6-dimethylpiperidin-2-one, slow coupling being successful in only the latter case, and then only by using the 3-ketoester in potassium carbonate rather than in sodium hydroxide solution, to furnish a low yield of **124** ($R^1 = OCH_3$, $R^2 = CH_3$) (Heath-Brown and Philpott, P. G., 1965b).

An alternative method of synthesizing **1296** involved the reaction of the arylhydrazine with an appropriate enamine. Thus, **134** (R = H, X = CH_2, m = 2,

$n = 1$ and 2) reacted with 2- and 4-hydrazinopyridine in the absence of acid catalysts to afford compounds **1319** ($R^1 = R^2 = H$; X = CH, Y = N, $n = 1$ and 2 and X = N, Y = CH, $n = 1$ and 2, respectively) (Yakhontov, Glushkov, Pronina, E. V. and Smirnova, 1973) which, as already mentioned, were converted via Fischer indolization into **1301** (X = CH, Y = N, $n = 1$ and 2 and X = N, Y = CH, $n = 1$ and 2, respectively), part of an analogous reaction sequence being used when **134** (R = H, X = CH_2, $m = 2$, $n = 1$ and 2) reacted with **1320** (R = CH_3 and $CH_2C_6H_5$) to give structures **1312** (R = CH_3 and $CH_2C_6H_5$, $n = 1$ and 2) (Glushkov, Zasosova and Ovcharova, 1977)

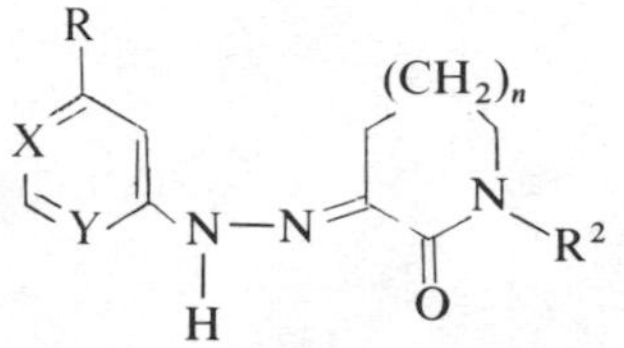

1319

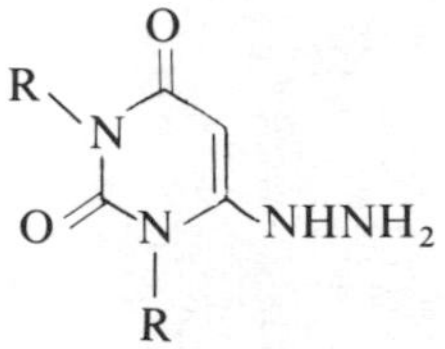

1320

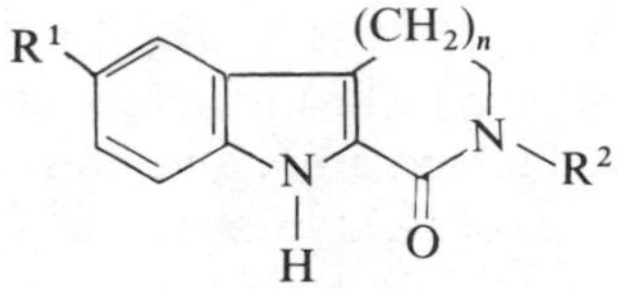

1321

which were then indolized as already described, when **134** (R = H and CH_3, X = CH_2, $m = n = 2$ and R = H, X = CH_2, $m = 2$, $n = 1$) reacted with phenylhydrazine in cold 20% sulphuric acid to give **1319** ($R^1 = H$, $R^2 = H$ and CH_3, X = Y = CH, $n = 2$ and $R^1 = R^2 = H$, X = Y = CH, $n = 1$, respectively) and when **134** (R = H, X = CH_2, $m = n = 2$) reacted with 4-hydrazinopyrimidine and 4-hydrazino-6-hydroxypyrimidine under similar conditions to afford **1319** ($R^1 = R^2 = H$, X = Y = N, $n = 2$ and $R^1 = OH$, $R^2 = H$, X = Y = N, $n = 2$, respectively). When **1319** ($R^1 = H$, $R^2 = CH_3$, X = Y = CH, $n = 2$) was refluxed in ethanol containing concentrated sulphuric acid it was converted into **1321** ($R^1 = H$, $R^2 = CH_3$, $n = 2$) (Glushkov, Smirnova, Zasosova and Ovcharova, 1975). If the initial hydrazone formation was effected under these last mentioned conditions, then it was formed *in situ* and the reaction product was the corresponding indole. Thus, **134** (R = H, X = CH_2, $m = 2$, $n = 1$) reacted with 4-methoxyphenylhydrazine hydrochloride in refluxing ethanolic sulphuric acid to furnish **1321** ($R^1 = CH_3O$, $R^2 = H$, $n = 1$) (Glushkov, Volskova, Smirnova and Magidson, 1969, 1972) and, under similar conditions, **134** (R = H, X = CH_2, $m = n = 1$ and 2 and X = O, $m = 2$, $n = 1$ and 2) reacted with phenylhydrazine to afford **1321** ($R^1 = R^2 = H$, $n = 1$ and 2, respectively) and **134** (R = H, X = CH_2,

$m = 1$, $n = 1$ and 2; R = H, X = CH_2, $m = n = 2$ and R = H, X = O, $m = 2$, $n = 1$ and 2) reacted with 4-methoxyphenylhydrazine to afford **1321** ($R^1 = OCH_3$, R^2 = H, $n = 1$ and 2, respectively) (Glushkov, Volskova, Smirnova and Magidson, 1969). In related reactions, **1322** reacted with phenylhydrazine and 4-methoxyphenylhydrazine under similar acidic conditions to give **1321** (R^1 = H and OCH_3, respectively, R^2 = H, $n = 1$) (Glushkov, Volskova, Smirnova and Magidson, 1969).

A further synthesis of **1296** involved the catalytic reduction, with subsequent simultaneous cyclization, of the arylhydrazones **1323** (Henecka, Timmler,

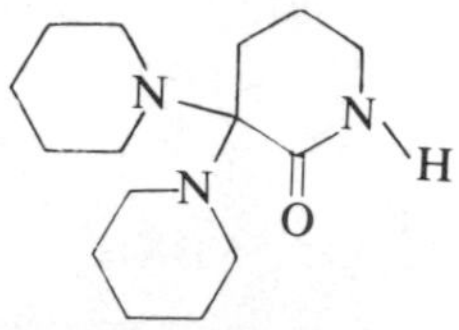

1322 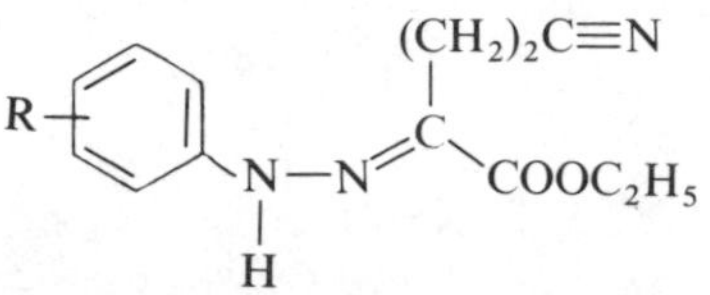 **1323**

Lorenz and Geiger, 1957). Using this route, compounds **1323** (R = H, 4-OCH_3, 4-$OCH_2C_6H_5$, 4-Cl and 3,4-OCH_2O) were converted into **1296** ($R^1 = R^2 = R^4$–R^8 = H, R^3 = H, OCH_3, $OCH_2C_6H_5$ and Cl; $R^1 = R^4$–R^8 = H, $R^2 + R^3 = OCH_2O$, respectively) when hydrogenated in methanolic solution over Raney nickel at elevated temperature and pressure (Timmler, 1957).

Treatment of arylhydrazones of 4-chlorobutyraldehyde (**1324**; $R^1 = R^2$ = H,

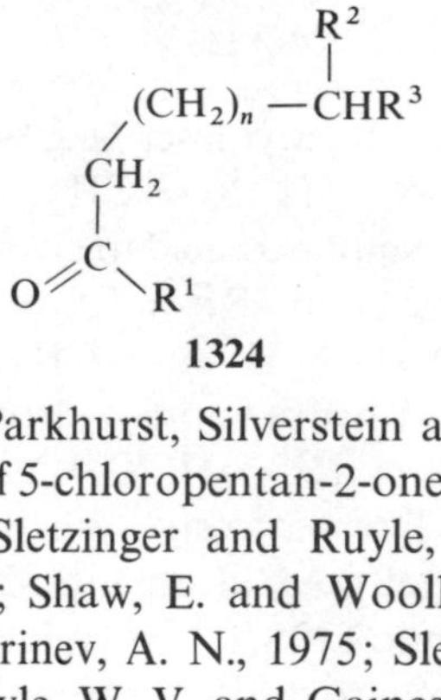

1324

R^3 = Cl, $n = 1$) (McKay, Parkhurst, Silverstein and Skinner, 1963; Shaw, E. and Woolley, D. W., 1953), of 5-chloropentan-2-one (**1324**; $R^1 = CH_3$, R^2 = H, R^3 = Cl, $n = 1$) (Gaines, Sletzinger and Ruyle, W., 1961; McCrea, 1966; Merck and Co., Inc., 1961; Shaw, E. and Woolley, D. W., 1953; Shvedov, Trofimkin, Vasil'eva and Grinev, A. N., 1975; Sletzinger, Gaines and Ruyle, W. V., 1957; Sletzinger, Ruyle, W. V. and Gaines, 1961; Steck, Fletcher and Carabateas, 1974) and of 5-chlorohexan-2-one (**1324**; $R^1 = R^2 = CH_3$, R^3 = Cl, $n = 1$) (Gaines, Sletzinger and Ruyle, W., 1961; Sletzinger, Ruyle, W. V. and Gaines, 1961) with hydrochloric acid as a catalyst under a variety of conditions and of arylhydrazones of **1324** [$R^1 = C_6H_5$, R^2 = H, R^3 = Br, $n = 2$; $R^1 = C_6H_5$, R^2 = H, R^3 = Cl, $n = 1$; $R^1 = C_6H_4Cl(4)$ and $C_6H_4Br(4)$, R^2 = H, R^3 = Cl, $n = 1$] with polyphosphoric acid (Julia, Melamed and Gombert, 1965) produced the corresponding indoles **1325** [$R^1 = R^2$ = H, R^3 = Cl, $n = 1$; $R^1 = CH_3$, R^2 = H, R^3 = Cl, $n = 1$; $R^1 = R^2 = CH_3$, R^3 = Cl, $n = 1$; $R^1 = C_6H_5$, R^2 = H, R^3 = Br, $n = 2$; $R^1 = C_6H_5$, R^2 = H,

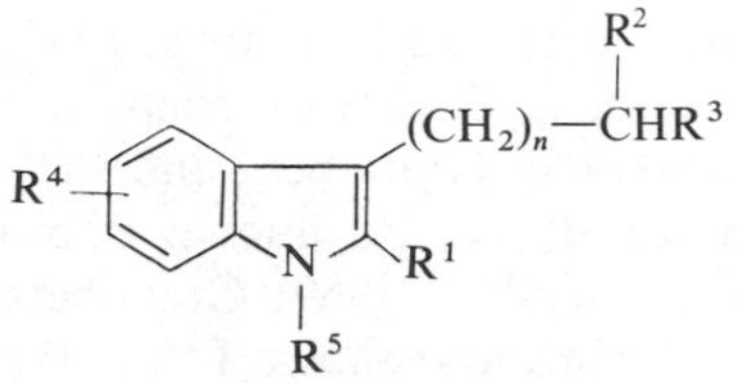

1325

R^3 = Cl, n = 1 and $R^1 = C_6H_4Cl(4)$ and $C_6H_4Br(4)$, R^2 = H, R^3 = Cl, n = 1, respectively]. These, upon treatment with ammonia (Gaines, Sletzinger and Ruyle, W., 1961; Julia, Melamed and Gombert, 1965; McKay, Parkhurst, Silverstein and Skinner, 1963; Shaw, E. and Woolley, D. W., 1953; Sletzinger, Gaines and Ruyle, W. V., 1957), with primary or secondary amines (Gaines, Sletzinger and Ruyle, W., 1961; Julia, Melamed and Gombert, 1965; McCrea, 1966; McKay, Parkhurst, Silverstein and Skinner, 1963; Merck and Co., Inc., 1961; Shaw, E. and Woolley, D. W., 1953; Sletzinger, Gaines and Ruyle, W. V., 1957) and with the sodium salt of β-alanine (McKay, Parkhurst, Silverstein and Skinner, 1963) gave the corresponding tryptamines and (2- and 3-aminopropyl)indoles. The halogen in the halogenoalkyl side chain could also be replaced by a primary amino group via formation of the phthalimido derivative (Sletzinger, Ruyle, W. V. and Gaines, 1961) and by a thiol group via reaction with thiourea (Steck, Fletcher and Carabateas, 1974). The corresponding 3-(2-hydroxyethyl) indoles (tryptophols), formed by hydrolysis of the chloro compound under the reaction conditions, have also been isolated from some of these indolizations (Shaw, E. and Woolley, D. W., 1953).

In the absence of acid catalysts, tryptamines **1333** ($R^2 = R^3$ = H, n = 1) were obtained directly by reacting arylhydrazines **1326** with 4-chloro-, 4-bromo-, 4-iodo- or 4-tosylbutyraldehyde **1327** ($R^3 = R^4$ = H, R^5 = Cl, Br, I or tosyl, respectively, n = 1) [optimum yields in these and the related reactions described below were obtained using the chloro compounds (Grandberg, Zuyanova, Afonina and Ivanova, T. A., 1967)] in refluxing aqueous alcoholic (usually methanolic or ethanolic) solution (Grandberg, 1966; Grandberg, Afonina, Zuyanova, 1967, 1968; Grandberg and Bobrova, 1973, 1974a; Grandberg, Bobrova and Zuyanova, 1972; Grandberg and Przheval'skii, 1972; Grandberg, Przheval'skii, Ivanova, T. A., Zuyanova, Bobrova, Dashkevich, S. I., Nikitina, Shcherbina and Yaryshev, 1969; Grandberg and Tokmakov, 1975b, 1975c; Grandberg and Zuyanova, 1967; Grandberg, Zuyanova, Afonina and Ivanova, T. A., 1967; Grandberg, Zuyanova and Bobrova, 1967). In analogous reactions under similar conditions, **1327** ($R^3 = CH_3$, R^4 = H, n = 1) afforded 2-methyltryptamines **1333** ($R^3 = CH_3$, n = 1) (Grandberg, Afonina and Zuyanova, 1968; Grandberg and Bobrova, 1974a; Grandberg and Zuyanova, 1967, 1968, 1971; Grandberg, Zuyanova, Afonina and Ivanova, T. A., 1967; Grandberg, Zuyanova and Bobrova, 1967; Grandberg, Zuyanova, Przheval'skii and Minkin, 1970), **1327** ($R^3 = C_6H_5$, R^4 = H, n = 1) afforded

2-phenyltryptamines (Grandberg and Zuyanova, 1967; Grandberg, Zuyanova and Bobrova, 1967; Grandberg, Zuyanova, Przheval'skii and Minkin, 1970), other 2-alkyl- (Grandberg, 1966; Grandberg and Przheval'skii, 1972; Grandberg, Przhevalskii, Ivanova, T. A., Zuyanova, Bobrova, Dashkevich, S. I., Nikitina, Shcherbina and Yaryshev, 1969; Grandberg and Zuyanova, 1967; Grandberg, Zuyanova, Afonina and Ivanova, T. A., 1967; Grandberg, Zuyanova and Bobrova, 1967) and 2-aryl- (Grandberg, 1966; Grandberg and Przheval'skii, 1972; Grandberg and Zuyanova, 1967; Grandberg, Zuyanova and Bobrova, 1967) tryptamines were similarly prepared and **1326** reacted with **1327** (R^3 = H, CH_3 and alkyl, R^4 = H, $n = 2$) to yield homotryptamines **1333** (R^3 = H, $n = 2$) (Grandberg, Afonina and Zuyanova, 1967), 2-methylhomotryptamines **1333** ($R^3 = CH_3$, $n = 2$) (Grandberg, Afonina and Zuyanova, 1967; Grandberg and Zuyanova, 1970, 1971) and 2-alkylhomotryptamines **1333** (R^3 = alkyl, $n = 2$) (Grandberg, Afonina and Zuyanova, 1967; Grandberg, Zuyanova, Afonina and Ivanova, T. A., 1967), respectively. The bisulphite derivatives of **1327** (R^3 = H and CH_3, R^4 = H, R^5 = Cl, $n = 1$) (Grandberg and Bobrova, 1974b) and of related 4-halogenocarbonyl compounds in general (Grandberg and Bobrova, 1974d) have also been similarly utilized in the preparation of tryptamines, 2-methyltryptamines and 2-alkyl- or 2-aryltryptamines, respectively. The use of these derivatives was particularly important in the synthesis of 2-unsubstituted tryptamines since 4-chlorobutyraldehyde, prepared by either the Rosenmund reduction of 4-chlorobutyryl chloride (Grandberg, Afonina and Zuyanova, 1968; Grandberg and Bobrova, 1973, 1974c; Loftfield, 1951; McKay, Parkhurst, Silverstein and Skinner, 1963) or from tetrahydrofurfuryl alcohol (Paul, R., 1941; Paul, R. and Tchelitcheff, 1948; Shaw, E. and Woolley, D. W., 1953), was an extremely unstable compound (Grandberg and Bobrova, 1974b). In a similar reaction, **1328** (R^1 = H and 4-OCH_3, $R^2 = R^4$ = H, $R^3 = COOC_2H_5$, R^5 = Cl, $n = 1$), formed *in situ* from the appropriate Japp–Klingemann reactions, when refluxed in butanol containing a trace of water, gave **1333** (R^1 = H and 5-OCH_3, respectively, R^2 = H, $R^3 = COOC_2H_5$, $n = 1$) which upon sequential hydrolysis with sodium hydroxide and decarboxylation in refluxing sulphuric acid formed **1333** (R^1 = H and 5-OCH_3, respectively, $R^2 = R^3$ = H, $n = 1$) (Szantay, Szabo and Kalaus, 1974a, 1974b).

It has been suggested (Grandberg and Przheval'skii, 1972; Grandberg, Przheval'skii, Ivanova, T. A., Zuyanova, Bobrova, Dashkevich, S. I., Nikitina, Shcherbina and Yaryshev, 1969; Grandberg, Przheval'skii, Vysotskii and Khmel'nitskii, 1972; Grandberg and Zuyanova, 1970; Grandberg, Zuyanova, Afonina and Ivanova, T. A., 1967) that these direct syntheses of tryptamines proceeded through the intermediates given in Scheme 42 by a mechanism related to the Fischer indolization. Verification for this scheme has been obtained from the following observations:

(i) 1-^{15}N-2-Methyltryptamine and 2-methyltryptamine with the ^{15}N label in the 2-aminoethyl side chain were obtained by reacting $^{15}N_\alpha$- and $^{15}N_\beta$-

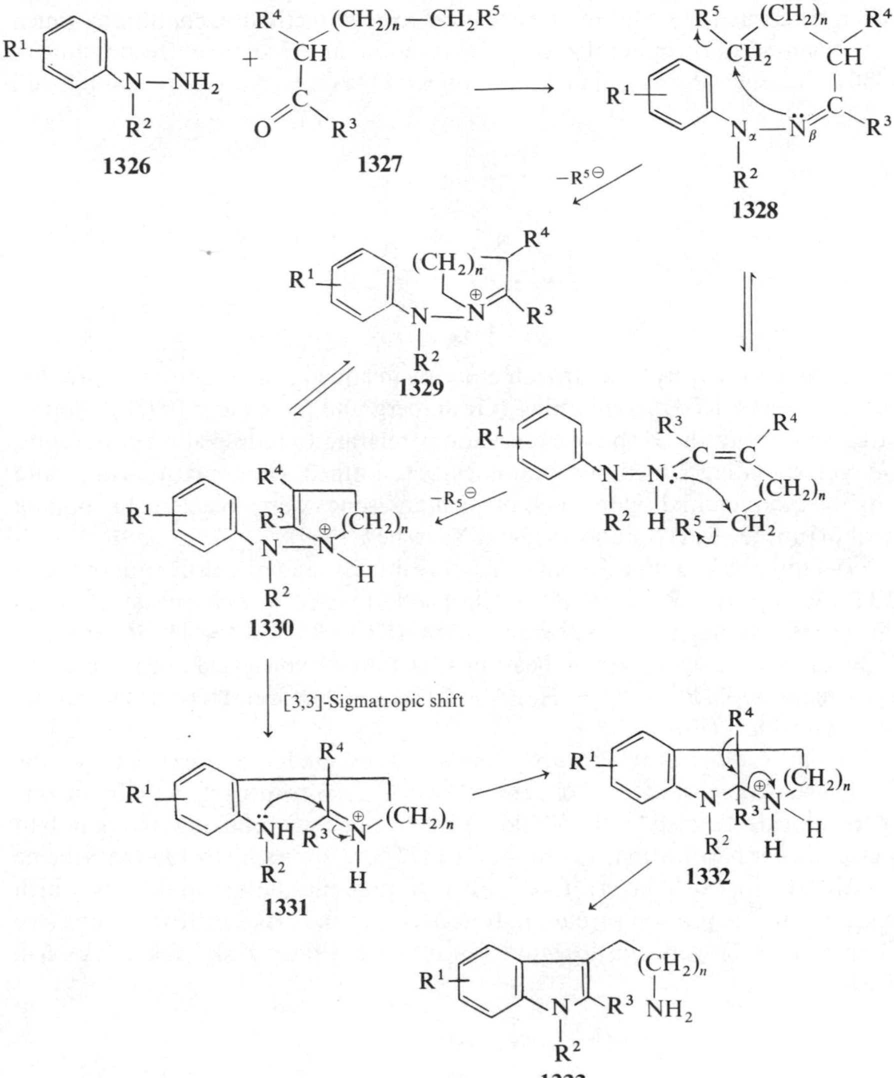

Scheme 42

phenylhydrazine, respectively, with 5-chloropentan-2-one in refluxing alcohol (Grandberg, Przheval'skii, Vysotskii and Khmel'nitskii, 1970).

(ii) The reaction of pyridylhydrazines with methyl 3-halogenopropyl ketones to afford azatryptamines (Grandberg and Yaryshev, 1968) required more forcing conditions than the corresponding reactions with phenylhydrazine or substituted phenylhydrazines. Thus, for 2-pyridylhydrazine and N_{α}-methyl-2-pyridylhydrazine to react with 5-chloropentan-2-one to give 2-methyl- and 1,2-dimethyl-7-azatryptamine, respectively, the reactants had to be heated in aqueous ethanol at 160 °C in an autoclave (Grandberg and Yaryshev, 1972a).

When the former reactant mixture was refluxed in methanol, conditions which were too mild to effect indolization, the reaction intermediate corresponding to **1330** in Scheme 42 was isolated as its dimer **1334** (X = N). Upon heating with

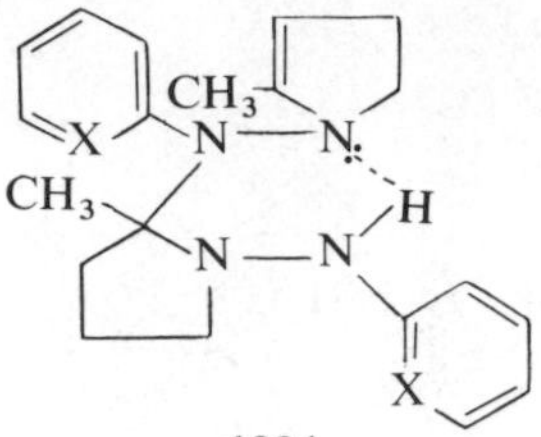

1334

an equimolar quantity of hydrogen chloride in aqueous methanol, this product formed 2-methyl-7-azatryptamine (Grandberg and Yaryshev, 1972b). Apparently contrary to these above observations relating to indolization conditions, the corresponding 2-methyltryptamines were formed when mixtures of 2- and 8-hydrazinoquinoline with 5-chloropentan-2-one were reacted in boiling alcohol (unspecified) (Grandberg and Yaryshev, 1969).

(iii) Equimolar quantities of phenylhydrazine and 5-chloropentan-2-one (**1327**; $R^3 = CH_3$, $R^4 = H$, $R^5 = Cl$, $n = 1$) reacted in benzene at 20 °C to afford the intermediate hydrazone **1328** ($R^1 = R^2 = R^4 = H$, $R^3 = CH_3$, $R^5 = Cl$, $n = 1$) which upon heating was rapidly converted into 2-methyltryptamine (**1333**, $R^1 = R^2 = H$, $R^3 = CH_3$, $n = 1$) (Grandberg, Przheval'skii and Vysotskii, 1970).

(iv) The *ortho* methyl groups sterically prevented the occurrence of the [3,3]-sigmatropic shift in Scheme 42 when 2,4,6-trimethylphenylhydrazine hydrochloride reacted with 5-chloropentan-2-one in refluxing methanol to produce, after basification, a compound **1335** corresponding to **1330** in Scheme 42. All attempts to convert **1335** into a tryptamine under conditions which might involve migration or elimination of one of the *ortho* methyl groups (see Chapter II, G2) were unsuccessful (Grandberg, Przheval'skii and Vysotskii, 1970).

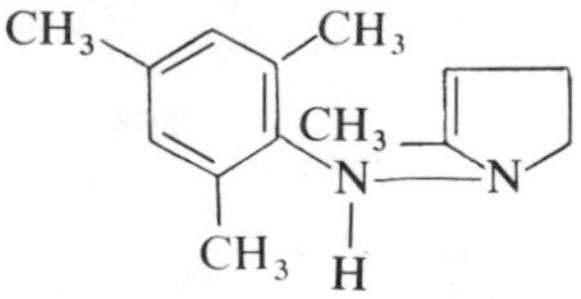

1335

(v) When N_α-acetyl- and N_α-benzoylphenylhydrazine reacted with 5-chloropentan-2-one in refluxing methanol, compounds **1336** ($R = CH_3$ and C_6H_5, respectively) were obtained, apparently being formed via hydrolysis of the immonium cationic tautomers of the intermediates **1330** ($R^1 = R^4 = H$, $R^2 = COCH_3$ and COC_6H_5, respectively, $R^3 = CH_3$, $n = 1$). The former of these products underwent ready dehydration when treated with picric acid, subsequent liberation of the base, with simultaneous deacetylation, by treatment with liquid ammonia affording the intermediate corresponding to **1330** ($R^1 = R^2 = R^4 = H$, $R^3 = CH_3$, $n = 2$), isolated as its dimer **1334** (X = CH). Refluxing a solution of **1336** ($R = CH_3$) in xylene, with removal of an equimolar amount of water, yielded a product corresponding to intermediate **1331** in Scheme 42 and having the tautomeric structure **1337**. Treatment of this product with refluxing acetic anhydride afforded **1338** which upon hydrolysis with methanolic hydrochloric acid gave 2-methyltryptamine (Grandberg, Zuyanova and Zhigulev, 1972).

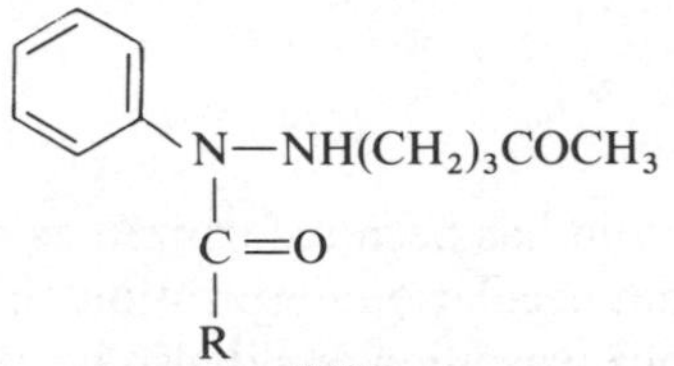

N—NH$(CH_2)_3COCH_3$
C=O
R

1336

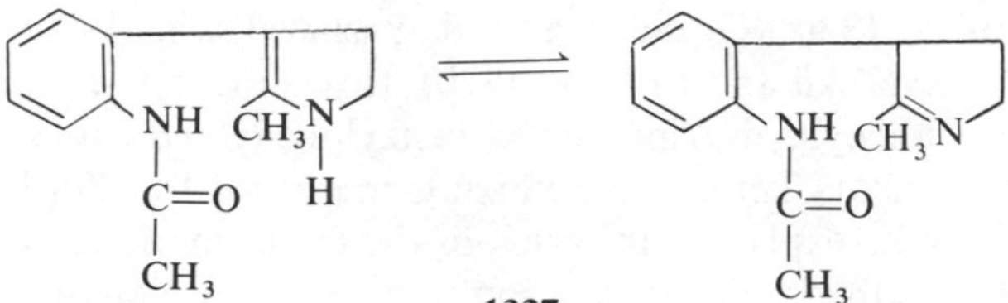

NH CH_3 N H C=O CH_3 ⇌ NH CH_3 N C=O CH_3

1337

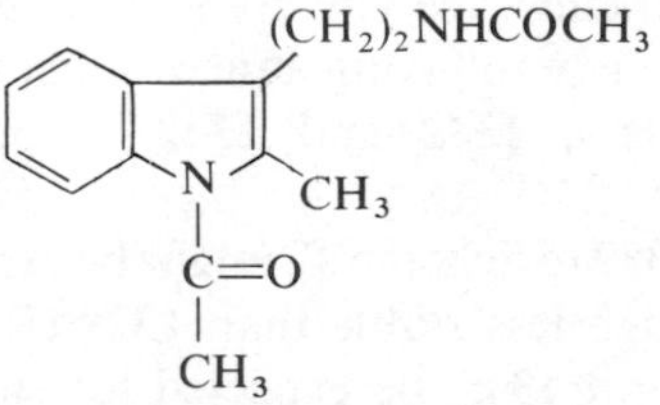

$(CH_2)_2NHCOCH_3$
N CH_3
C=O
CH_3

1338

(vi) Compounds corresponding to **1332** in Scheme 42 have been isolated with $R^4 \neq H$ (Grandberg and Ivanova, T. A., 1967; Grandberg, Zuyanova, Afonina and Ivanova, T. A., 1967). Thus, arylhydrazines reacted with **1327** ($R^3 = R^4 = CH_3$, R^5 = halogeno, $n = 1$) in refluxing alcoholic solution to afford the corresponding **1332** ($R^3 = R^4 = CH_3$, $n = 1$) (Grandberg and Ivanova, T. A., 1970a, 1970b, 1970d, 1975; Grandberg, Zuyanova, Afonina and Ivanova, T. A., 1967) and under similar conditions reacted with **1327** ($R^3 = R^4 = CH_3$, R^5 = halogeno, $n = 2$) to afford **1332** ($R^3 = R^4 = CH_3$, $n = 2$) (Grandberg and Ivanova, T. A., 1970c, 1970d) and with 2-(2-chloroethyl)-cyclohexanone [**1327**; $R^3 + R^4 = (CH_2)_4$, $R^5 = Cl$, $n = 1$] to give **1332**

$[R^3 + R^4 = (CH_2)_4,\ n = 1]$ (Grandberg and Ivanova, T. A., 1970a, 1970b, 1970d). Likewise, compounds **1339** (R = H, CH_3 and C_2H_5) have been prepared by reacting mixtures of 2-pyridylhydrazine and N_α-methyl- and N_α-ethyl-2-pyridylhydrazine with **1327** ($R^3 = R^4 = CH_3$, R^5 = Cl, n = 1) in refluxing aqueous ethanol (Grandberg and Yaryshev, 1972a).

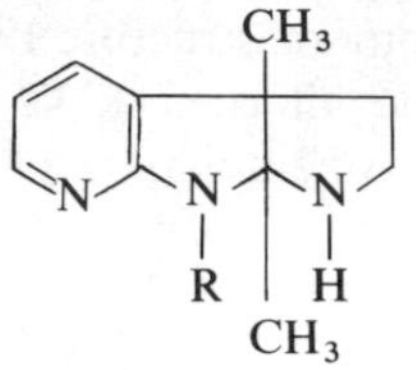

1339

(vii) In **1328** the N_β-atom has been recognized as being more basic than the N_α-atom. Thus, the intramolecular quaternization occurred at the former atom to furnish **1329** (analogous reasoning could also be applied to the formation of **1330** by the alternative route shown in Scheme 42) rather than at the N_α-atom to furnish **1340** (Grandberg and Przheval'skii, 1970; Grandberg, Zuyanova, Przheval'skii and Minkin, 1970). However, this was not invariably the case, since phenylhydrazine or N_α-benzylphenylhydrazine reacted with 3-chloropropyl phenyl ketone in refluxing methanol to afford **1341** ($R^1 = R^2 = C_6H_5$), results explained in terms of the conformational analysis of the hydrazones (Grandberg, Zuyanova, Przheval'skii and Minkin, 1970). These observations were substantiated by the analogous results obtained by studying the reactions of ethylhydrazine, acetylhydrazine and *asymm.*-dimethylhydrazine with 5-chloropentan-2-one in refluxing aqueous methanol which afforded **1341** ($R^1 = C_2H_5$, $R^2 = CH_3$), **1342** and **1343**, respectively (Grandberg and Przheval'skii, 1970).

The intermediate in **1329** in Scheme 42 might be expected to be in equilibrium with **1344** which, although less stable than **1329** ($R^3 \neq$ H) would, in its enehydrazine tautomeric form **1345**, be expected for steric reasons to undergo a

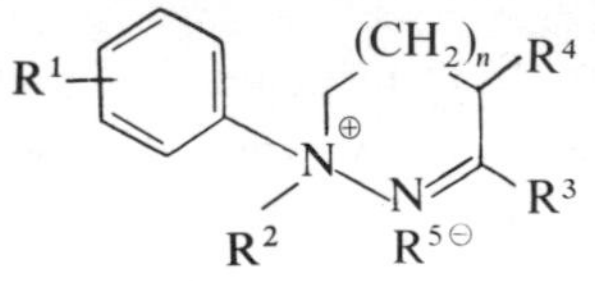

1340

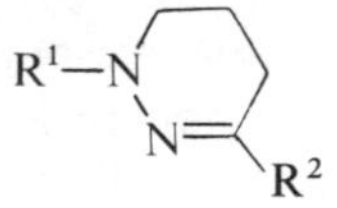

1341

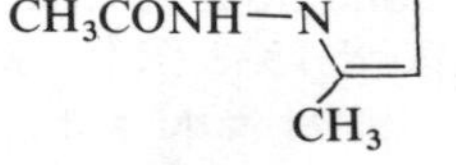

1342

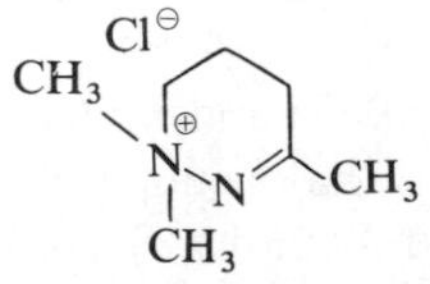

1343

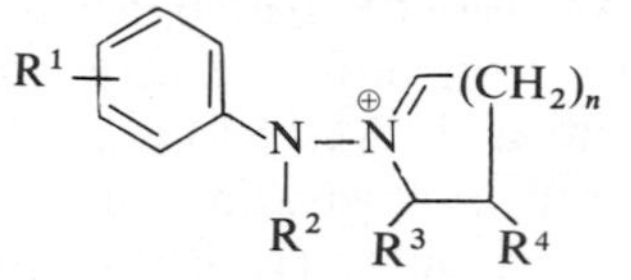

1344 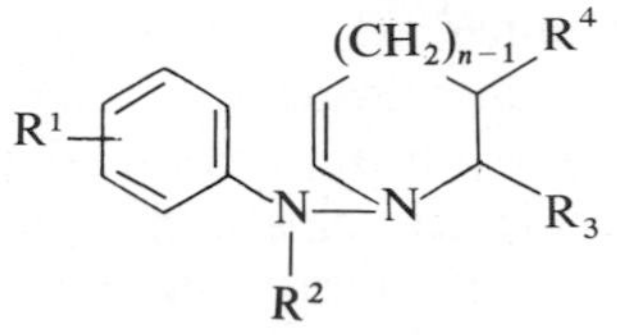 **1345**

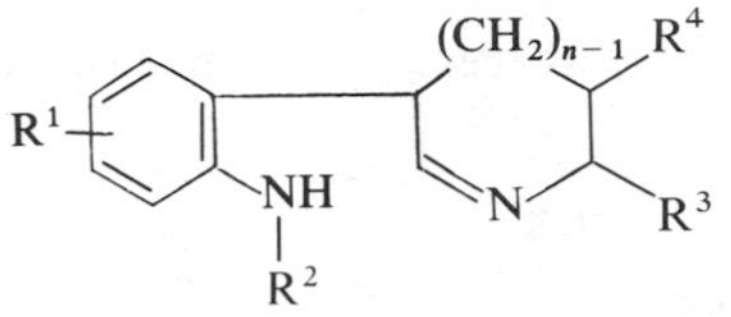

1346 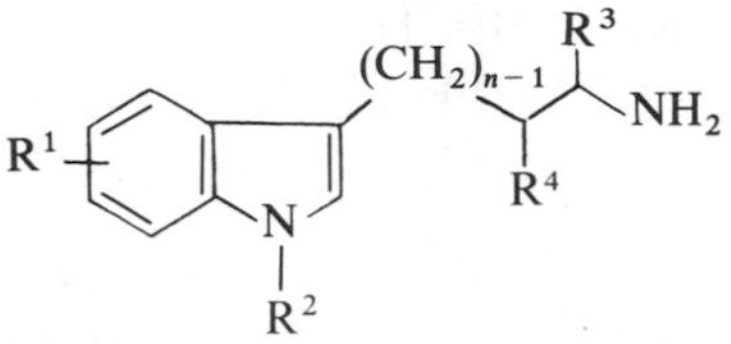 **1347**

[3,3]-sigmatropic shift to afford **1346** more readily than **1330** would give **1331**. Such a rearrangement would then ultimately yield **1347**. It would be interesting to investigate further the total reaction products from these above indolizations with a view to the possibility of isolating such products from reactions using **1327** (R^3 and/or $R^4 \neq$ H).

That under neutral conditions (i.e. refluxing aqueous alcohol) tryptamines were formed directly from **1328** according to Scheme 42 whereas under mineral acidic conditions the corresponding 3-halogenoalkylindoles were formed, as described previously (**1324** → **1325**) by 'normal' indolization has been attributed to the low basicities of arylhydrazones. In aqueous alcoholic solutions these were present largely in the unprotonated form which could undergo intramolecular quaternization (i.e. **1328** → **1329**) to yield ultimately tryptamines according to Scheme 42. However, in strong acid solution, the hydrazones **1328** would be largely protonated on N_β. This would therefore prevent intramolecular quaternization and, furthermore, would increase the rate of enehydrazine formation which ultimately leads to 'normal' indolization (Grandberg and Bobrova, 1974b).

This direct route to the synthesis of tryptamines and homotryptamines has been reviewed (Grandberg, 1974 – see also Grandberg and Tokmakov, 1975c).

In view of these above studies, the formulation as **1341** ($R^1 = C_6H_5$, $R^2 = CH_3$) of the product resulting from a reaction between phenylhydrazine and 5-chloropentan-2-one in refluxing alcohol reported earlier (Grandberg, Kost and Terentyev, 1957) is open to doubt. Although the product was obtained as an oil which could not be crystallized, even after prolonged standing, it had b.p. = 197–198 °C/6 mm, afforded a picrate, m.p. = 226.5 °C and had a nitrogen content corresponding to an empirical formula $C_{11}H_{14}N_2$. Clearly this product, the u.v. spectrum of which was unfortunately not recorded, could well be 2-methyltryptamine, m.p. 107 °C, b.p. = 203–208 °C/10 mm, picrate, m.p. = 218–219 °C (Grandberg and Zuyanova, 1968).

Tryptamines and homotryptamines have also been prepared directly by heating arylhydrazine salts with cyclic vinyl amines in dimethylformamide solution (Grandberg, Przheval'skii and Ivanova, T. A., 1969) or by the related hydrochloric acid catalysed reaction between arylhydrazines and the *in situ* formed salts of 1-pyrrolines **1349** ($R^3 = H, CH_3$ and $C_6H_5, n = 1$) (Grandberg and Nikitina, 1971b) and 1-piperideines **1349** ($R^3 = H, n = 2$) (Grandberg and Nikitina, 1971a, 1971b, 1971c) and **1349** ($R^3 = CH_3, n = 2$) (Grandberg and Nikitina, 1971b). The two parent heterocyclic bases were used in the form of their trimers **1348** ($n = 1$ and 2, respectively) as were the other bases which

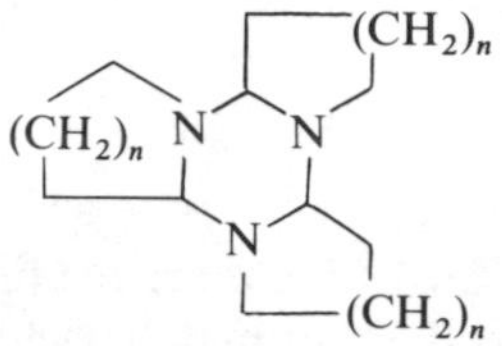

$(CH_2)_n$ N N $(CH_2)_n$ N $(CH_2)_n$

1348

detrimerized and protonated upon treatment with acid (Grandberg, Nikitina and Yaryshev, 1970). Reaction of 1-acetyl-2-piperideine with arylhydrazine salts in refluxing dimethylformamide likewise afforded 3-(3-acetylaminopropyl)-indoles (Grandberg and Nikitina, 1971a, 1971c). The mechanism proposed (Grandberg and Nikitina, 1971a, 1971b) for these reactions is given in Scheme 43. The initial reaction between the arylhydrazine and **1349** (formed *in situ* from the corresponding **1348**) afforded **1350** which underwent ring cleavage to produce the arylhydrazone **1351**, Fischer indolization of which, via tautomerism with **1352**, ultimately led to the tryptamine or homotryptamine **1333** ($n = 1$ and 2, respectively). In extensions of this reaction, phenylhydrazine and N_α-benzylphenylhydrazine hydrochloride were reacted with **1349** ($R^3 = C_2H_5$, $n = 1$) and N_α-benzylphenylhydrazine hydrochloride was reacted with **1349** ($R^3 = C_2H_5, n = 2$) in refluxing isopropanol in the presence of hydrogen chloride equimolar with the cyclic enamine. In each case, the product isolated was not the corresponding 2-ethyltryptamine or 2-ethylhomotryptamine **1333** ($R^1 =$ H, $R^2 = H$ and $CH_2C_6H_5$, $R^3 = C_2H_5$, $n = 1$ and $R^1 = H$, $R^2 = CH_2C_6H_5$, $R^3 = C_2H_5, n = 2$, respectively) but **1353** ($R = H$ and $CH_2C_6H_5, n = 3$) and $R = CH_2C_6H_5$, $n = 4$, respectively). The formation of these products was attributed to the tautomerism of the proposed intermediates **1351** ($R^1 = H$, $R^2 = H$ and $CH_2C_6H_5$, $R^3 = C_2H_5$, $n = 1$ and $R^1 = H$, $R^2 = CH_2C_6H_5$, $R^3 = C_2H_5$, $n = 2$) not with **1352** ($R^1 = H$, $R^2 = H$ and $CH_2C_6H_5$, $R^3 =$ C_2H_5, $n = 1$ and $R^1 = H$, $R^2 = CH_2C_6H_5$, $R^3 = C_2H_5$, $n = 2$, respectively) but with **1354** ($R = H$ and $CH_2C_6H_5$, $n = 3$ and $R = CH_2C_6H_5$, $n = 4$, respectively), which would ultimately give the isolated products (Grandberg and Nikitina, 1972). It would certainly be of interest to perform a very careful

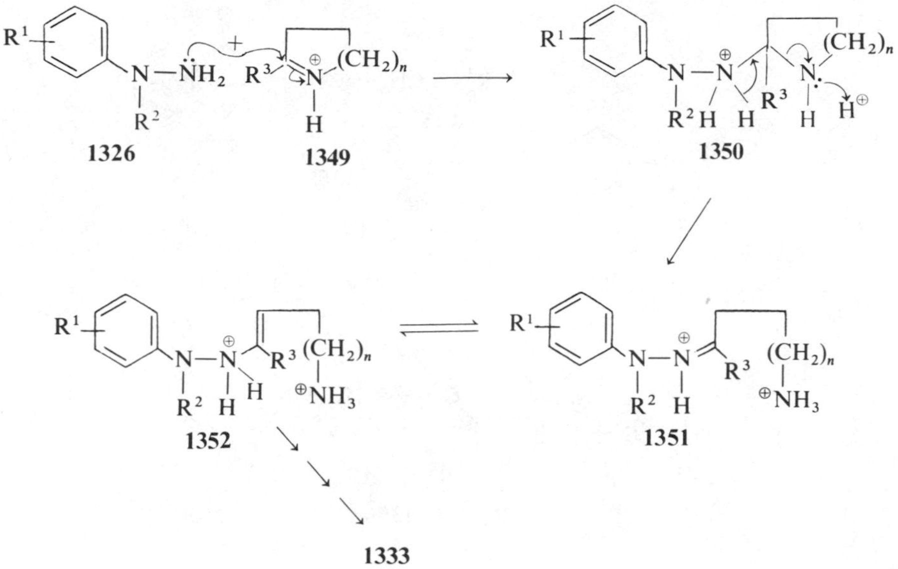

1326 1349 1350

1352 1351

1333

Scheme 43

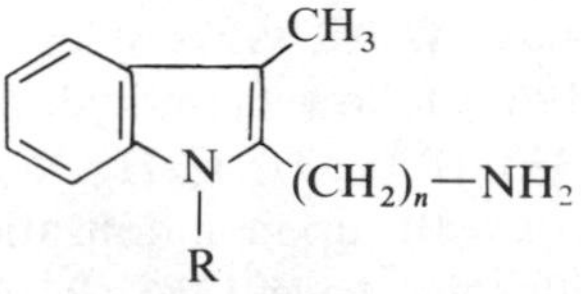

1353 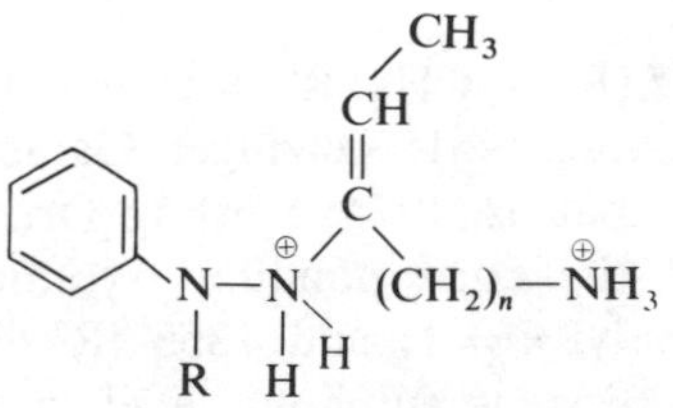 1354

analysis of the total reaction products resulting from these reactions since the concomitant formation of the corresponding 2-ethyl-3-aminoalkylindoles might be expected (see Chapter III, G).

The synthesis of tryptamines by this route has also been reviewed (Grandberg, 1974).

An indirect synthesis of tryptamines proceeded via the formation of 3-(2-phthalimidoethyl)indoles which were prepared by the appropriate Fischer indolizations. Thus, the arylhydrazones **1355** ($R^3 = COOC_2H_5$, $R^4 = H$, $n = 1$), prepared by the Japp–Klingemann reaction, have been indolized to form **1356** ($R^3 = COOC_2H_5$, $n = 1$) which upon sequential hydrolysis, decarboxylation and dephthalylation by treatment with hydrazine hydrate was converted into **1333** ($R^3 = H$, $n = 1$) (Bretherick, Gaimster and Wragg, W. R., 1961; Chilikin, Gorelova, Shvarts and Suvorov, 1979; Gordeev, Kobets and Suvorov, 1975; Keimatsu, Sugasawa and Kasuya, 1928). Likewise, indolizations of **1355** ($R^3 = H$ and C_6H_5, $R^4 = H$, $n = 1$) (Sterling Drug Inc., 1962)

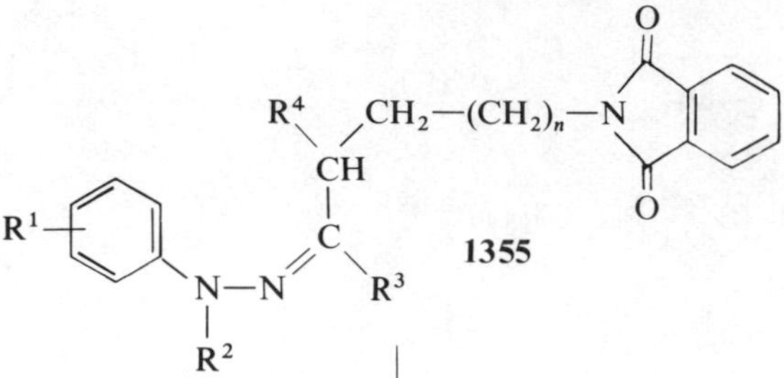

1355

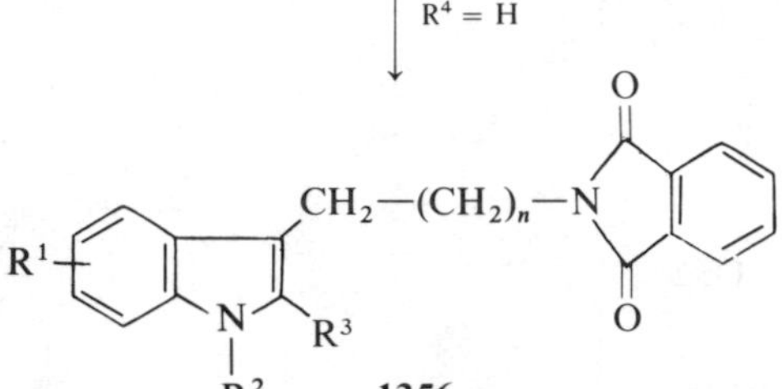

$\xrightarrow{R^4 = H}$

1356

↓

1333

and **1355** ($R^3 = CH_3, R^4 = H, n = 1$) (Archer, S., 1963; McCrea, 1966; Merck and Co., Inc., 1961; Sletzinger, Gaines and Ruyle, W. V., 1957; Steck, Fletcher and Carabateas, 1974; Sterling Drug Inc., 1962) followed by hydrazinolysis afforded the corresponding tryptamines **1333** (R^3 = H, C_6H_5 and CH_3, respectively, $n = 1$) and **1355** (R^1–R^4 = H, $n = 0$), upon indolization with 4-methylbenzene sulphonic acid in ethyl cellosolve gave **1356** (R^1–R^3 = H, $n = 0$) (Mamaev and Sedova, 1961) although this product was not dephthalylated. This synthetic approach could be varied by effecting hydrolysis of the ester group at the indolic 2-position and hydrazinolysis of the phthalyl group, but not decarboxylation, subsequent to the indolization. In these cases the intermediate tryptamine-2-carboxylic acids underwent intramolecular condensation to form 1,2,3,4-tetrahydro-*β*-carbolin-1-ones (Keimatsu, Sugasawa and Kasuya, 1928). A further variation of the post indolization hydrolysis–decarboxylation–hydrazinolysis sequence involved the base catalysed hydrolysis of **1356** ($R^2 = H, R^3 = COOC_2H_5, n = 1$) to give **1357**. This, upon pyrolysis, was converted into **1356** ($R^2 = R^3 = H, n = 1$) which upon hydrazinolysis

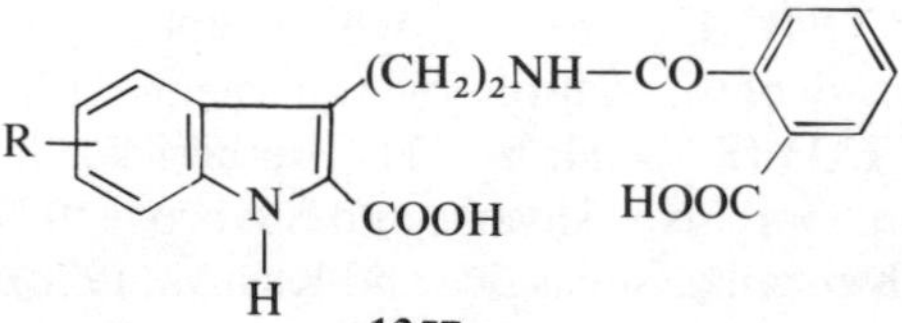

1357

afforded the tryptamines **1333** ($R^2 = R^3 = H$, $n = 1$). Using this approach, 5- and 6-benzyloxy-, 5-methoxy- and 4,5,6-trimethoxytryptamine (Gaimster, 1960) and tryptamine and 5-chloro-, 5-(2-hydroxyethoxy)-, 5-methoxy-, 5-methyl- and 5-nitrotryptamine (Suvorov, Gordeev and Vasin, 1974) were synthesized. In an attempt (Manske, R. H. F., Perkin and Robinson, R., 1927) to use this synthesis to prepare the 2-acetyl- and 2-benzoyltryptamine (**1356**; R^1 = H and 6-Cl, R^2 = H, $R^3 = COCH_3$, $n = 1$ and $R^1 = R^2 = H$, $R^3 = COC_6H_5$, $n = 1$ respectively), the conditions of the necessary Japp–Klingemann reactions also caused partial hydrolysis of the phthalimido group to give **1358** (R^1 = H and Cl, $R^2 = CH_3$ and R^1 = H, $R^2 = C_6H_5$). Treatment of these products, by passage of hydrogen chloride through their solutions in refluxing acetic acid, yielded **1359** (R^1 = H and Cl, $R^2 = CH_3$ and R^1 = H,

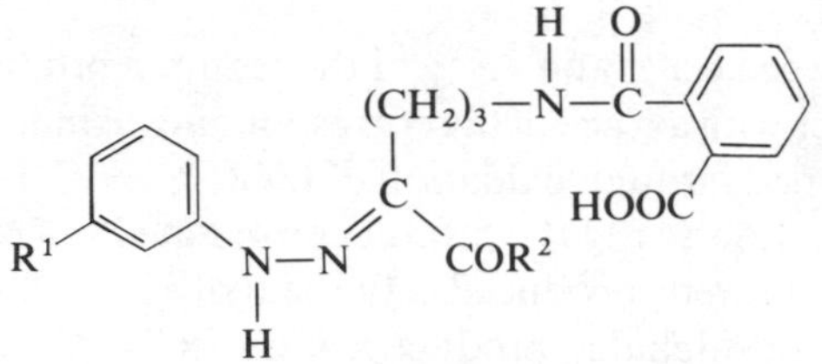

1358

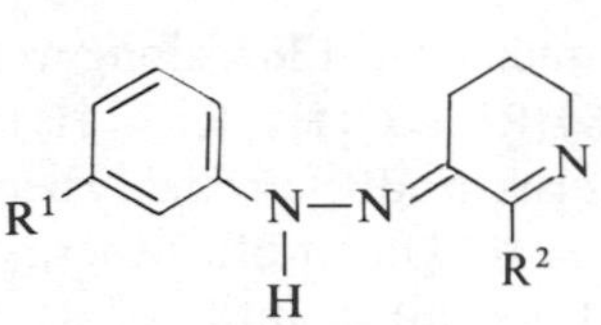

1359

$R^2 = C_6H_5$, respectively) and attempts to indolize these products were unsuccessful, probably because enehydrazine formation did not occur. However, when **1358** (R^1 = H and OCH_3, $R^2 = CH_3$) was heated in acetic anhydride, the phthalimido ring was reformed and indolization of the products so formed, by passage of hydrogen chloride through their refluxing alcoholic solutions, afforded **1356** (R^1 = H and 6-OCH_3, respectively, R^2 = H, $R^3 = COCH_3$, $n = 1$). Subsequent hydrazinolysis followed by dehydration formed 3,4-dihydro-1-methyl-*β*-carboline and its 7-methoxy derivative **535**. The phthalimido group could also be reduced by treatment with lithium aluminium hydride to afford the corresponding 3-(2-isoindolinylethyl)indoles (Steck, Fletcher and Carabateas, 1974). The syntheses of the 3-(2-phthalimidoethyl) derivatives of the parent heterocycles **1165**, **1184** and **1270**, together with studies carried out upon the further reactions of these products have been discussed in detail earlier in this chapter (see Table 25 and related studies). An interesting early application of the total synthetic sequence to the preparation of the corresponding 3*H*-indolic system ultimately afforded the eserine ring system. Treatment of **1355** (R^1 = 4-OC_2H_5, R^2 = H, $R^3 = COOC_2H_5$, $R^4 = CH_3$, $n = 1$) with refluxing ethanolic hydrogen chloride gave **1360** ($R^1 = C_2H_5$, $R^2 = COOC_2H_5$). However, attempts to decarbethoxylate this product were not very successful although eventually it was found that its hydrolysis, by treatment with ethanolic potassium hydroxide, afforded **1361** which upon pyrolysis as a suspension in refluxing xylene furnished as the major product a compound tentatively

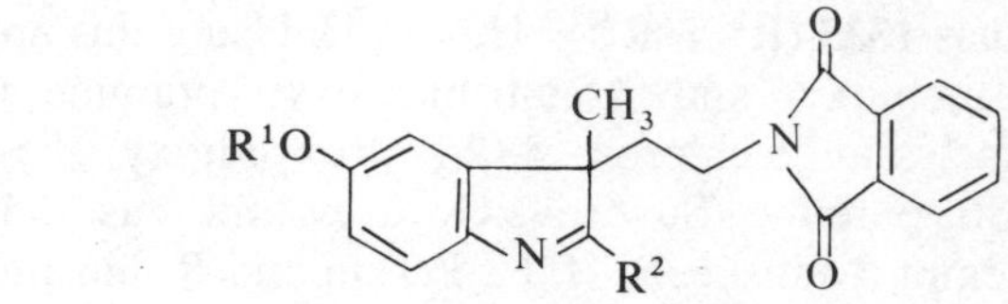

1360

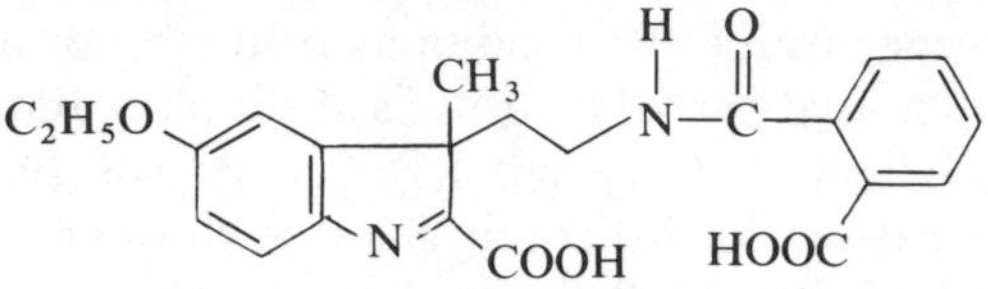

1361

formulated as **1362**, along with between 6.5% and 8.5% of the required product **1360** ($R^1 = C_2H_5, R^2 = H$) together with a trace of the corresponding oxindole, this last product probably being formed through oxidation of **1360** ($R^1 = C_2H_5$, $R^2 = H$). Under other thermal conditions, only the product formulated as **1362** and a small quantity of the oxindole were produced. Hydrazinolysis of **1360** ($R^1 = C_2H_5$, $R^2 = H$), as its methosulphate, produced **1363** ($R^1 = C_2H_5$, $R^2 = H$) which upon basification cyclized into **1364** ($R = C_2H_5$). Sequential

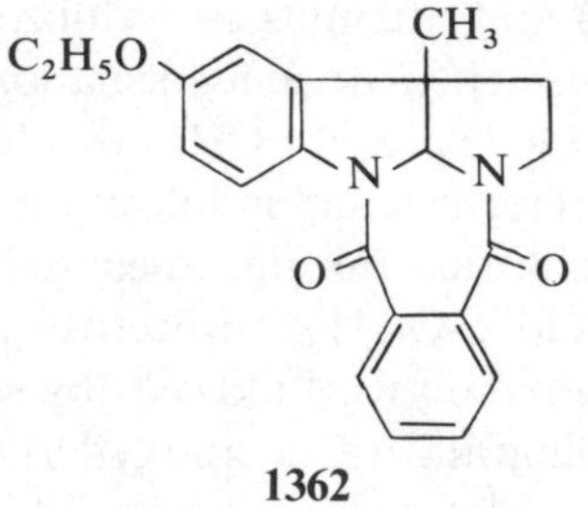

1362

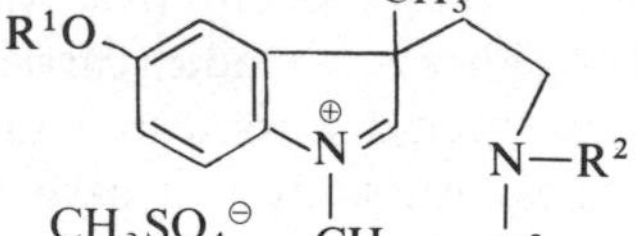

1363

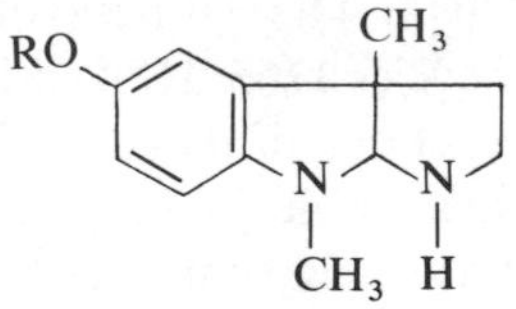

1364

hydrazinolysis and hydrolysis of **1360** ($R^1 = C_2H_5$, $R^2 = COOC_2H_5$) gave **1303** ($R = OC_2H_5$) (Robinson, R. and Suginome, 1932b). To overcome these problems associated with the removal of the 2-carboxylic ester group in this synthetic route, the 4-methoxy- and 4-ethoxyphenylhydrazone **1355** (R^1 = 4-OCH_3 and 4-OC_2H_5, respectively, $R^2 = R^3 = H$, $R^4 = CH_3$, $n = 1$) were treated with refluxing ethanolic hydrogen chloride to yield **1360** ($R^1 = CH_3$ and C_2H_5, respectively, $R^2 = H$) in good yields. These were converted into their methosulphates **1363** [$R^1 = CH_3$ and C_2H_5, respectively, $2R^2$ = phthalyl] which upon hydrazinolysis afforded, via the intermediates **1363** ($R^1 = CH_3$ and C_2H_5, respectively, $R^2 = H$), the eserine ring systems **1364** ($R = CH_3$ and C_2H_5, respectively) (King, F. E., Liguori and Robinson, R., 1933, 1934).

The eserine ring system was also synthesized via the tin(II) chloride catalysed indolization of **1365** ($R^1 = H$, $R^2 = CH_3$, $R^3 = C_2H_5$) into **1366** (isolated as a salt formed with the catalyst). This, upon sequential acidification with hydrochloric acid, removal of the tin(II) ion with hydrogen sulphide and basification to pH 9–10 afforded **1367** which upon warming with methanolic potassium hydroxide gave **1369** ($R^1 = H$, $R^2 = CH_3$, $X = Y = O$), probably via the intermediate **1368** [the mechanism of ring closure shown in **1367** is analogous to the proposed illustrated mechanism of ring closure of **1370**, by treatment with base, to form the corresponding optical isomer of **1369** ($R^1 = OC_2H_5$, $R^2 = CH_3, X = O, Y = H_2$) (Longmore and Robinson, B., 1966)]. The product **1369** ($R^1 = H$, $R^2 = CH_3$, $X = Y = O$) was then reacted with ammonia or methanolic methylamine to give **1369** ($R^1 = H$, $R^2 = CH_3$, $X = NH$ and

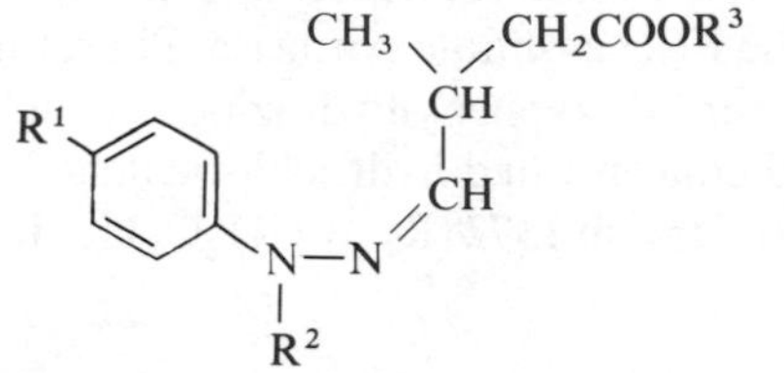

1365

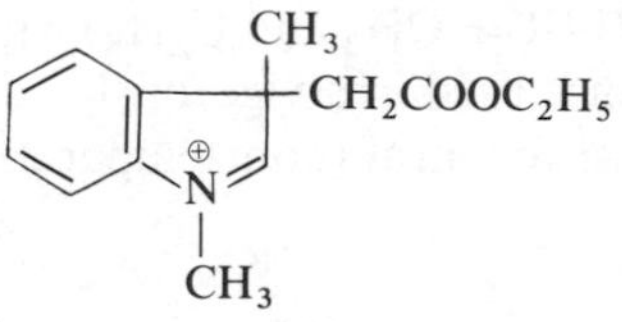

1366

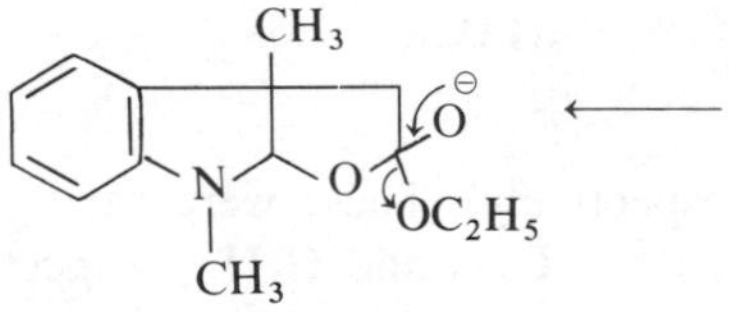

1368

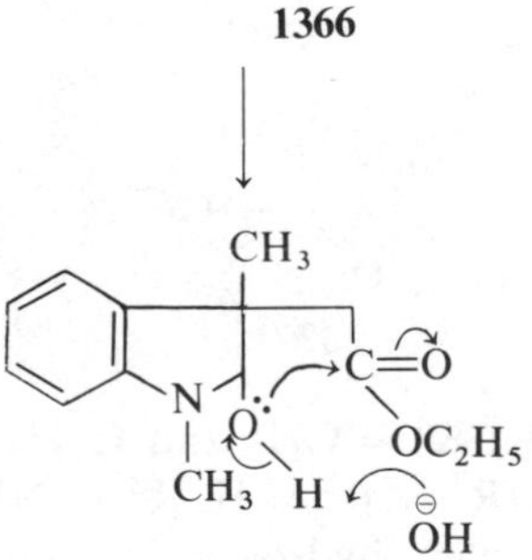

1367

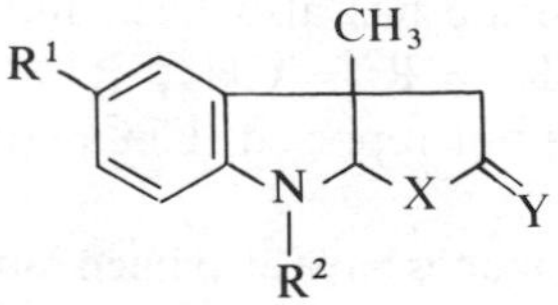

1369

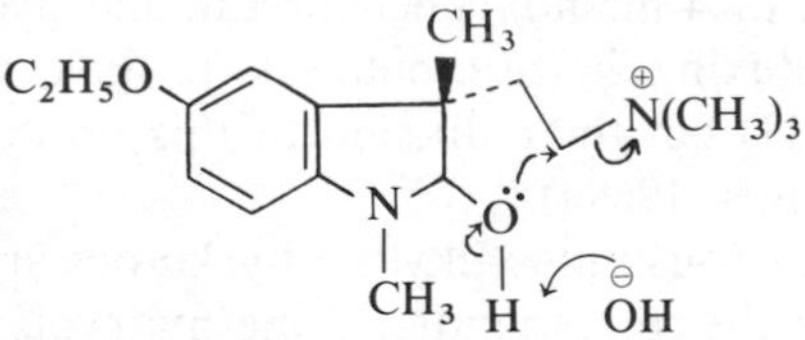

1370

NCH_3, respectively, Y = O) and the latter product, upon reduction with lithium aluminium hydride, was converted into **1369** (R^1 = H, R^2 = CH_3, X = NCH_3, Y = H_2) (Rosenmund and Sotiriou, 1964, 1975). In an extension of this reaction sequence, but using 4-methylbenzene sulphonic acid as the indolization catalyst, structures **1365** (R^1 = H, R^2 = $CH_2CH{=}CH_2$ and $CH_2C_6H_5$, R^3 = C_2H_5 and R^1 = OCH_3 and OC_2H_5, R^2 = CH_3, R^3 = C_2H_5) were converted into **1369** (R^1 = H, R^2 = $CH_2CH{=}CH_2$ and $CH_2C_6H_5$, respectively, X = Y = O and R^1 = OCH_3 and OC_2H_5, respectively, R^2 = CH_3, X = Y = O), some of which upon sequential reaction with various primary amines and reduction with lithium aluminium hydride produced the correspondingly substituted eserine systems (Rosenmund and Sadri, 1979). Related to this synthetic route is the observation that a mixture of phenylhydrazine and 3-methyllaevulinic acid was indolized in refluxing methanolic sulphuric acid to afford **647** (R^1 = H, R^3 = CH_3, R^4 = CH_2COOH) (18% yield) together with its methyl ester (52%), only the former product (62%) being produced when the indolization was effected in refluxing aqueous sulphuric acid (Rosenstock, 1966). However, no attempt was made in this study to effect a further cyclization to form a tetrahydrofuran ring.

Tryptamines have also been prepared by the reduction of 3-(2-nitroethyl)-indoles which could be prepared by the Fischer indole synthesis. Thus, mixtures of **1371** (R = CH_3 and C_2H_5) with 4-benzyloxyphenylhydrazine hydrochloride were indolized in a vigorously stirred concentrated hydrochloric acid–benzene biphase system at room temperature to furnish **1372** (R^1 = $CH_2C_6H_5$, R^2 = H,

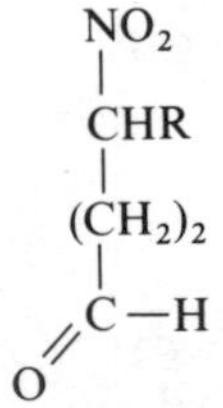

1371

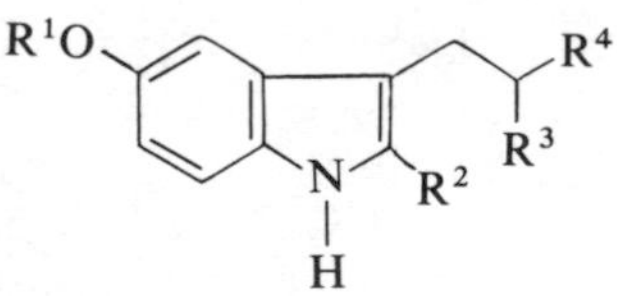

1372

R^3 = NO_2, R^4 = CH_3 and C_2H_5, respectively). These were then converted into **1372** (R^1 = R^2 = H, R^3 = NH_2, R^4 = CH_3 and C_2H_5, respectively) by treatment with hydrogen in the presence of a 10% palladium-on-carbon catalyst at elevated pressure (Lewis, A. D., 1962). A mixture of 5-nitropentan-2-one with 4-methoxyphenylhydrazine hydrochloride has also been indolized, by refluxing in *tert.*-butanol, to give **1372** (R^1 = R^2 = CH_3, R^3 = NO_2, R^4 = H) but the reduction of this product was not reported (Chemerda and Sletzinger, 1969a).

Numerous aminoalkyl methyl ketone arylhydrazones have also been indolized to give the corresponding 2-methyltryptamines and 2-methylhomotryptamines (see Chapter III, G).

3-Aminomethylindole derivatives have been synthesized by indolization of **1373** (R^1 = $COCH_3$, R^2 = $COOC_2H_5$, R^3 = CH_3) (Shavel and Strandtmann, 1965; Strandtmann, Puchalski and Shavel, 1964), **1373** (R^1 = R^2 = H, R^3 + R^3 = phthalyl) (Mamaev and Sedova, 1961), **1373** (R^1 = $NHCOCH_3$ and NO_2, R^2 = $COOC_2H_5$, R^3 = C_2H_5) (Berlin, A. Y. and Zaitseva, 1960) and **1373** (R^1 = H, R^2 = $COOC_2H_5$, R^3 = C_2H_5) (Boekelheide and Ainsworth, C., 1950b; Hegedüs, 1946).

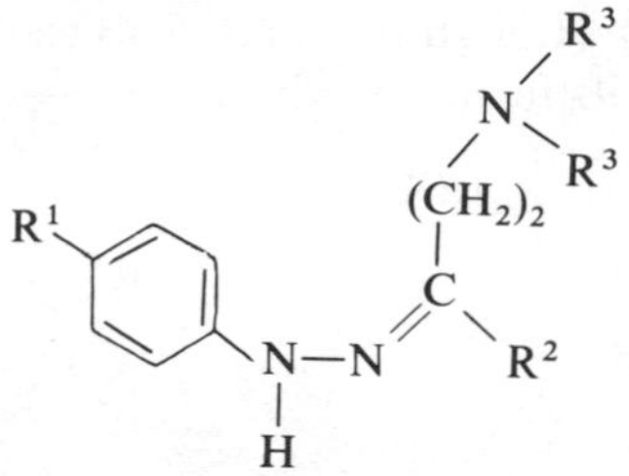

1373

G. Synthesis of Tryptophols and Related Systems

As already referred to (Chapter III, G), 2-methyltryptophols [3-(2-hydroxyethyl)-2-methylindoles] have been synthesized by the indolization of 5-hydroxypentan-2-one arylhydrazones (Brenner, Clamkowskii, Hinkley and Gal, 1969; Glamkowski, Gal and Sletzinger, 1973; Grandberg, Kost and Terentyev, 1957; Grandberg, Sibiryakova and Brovkin, 1969; Kost, Vinogradova, Daut and Terent'ev, A. P., 1962; Kost, Yudin, Dmitriev, B. A. and Terent'ev, A. P., 1959; Kost, Yudin and Popravko, 1962; Shvedov, Trofimkin, Vasil'eva and Grinev, A. N., 1975; Yamamoto, H., Nakamura, Y., Nakao and Kobayashi, T., 1970a; Yamamoto, H. and Nakao, 1968c), 3-(2-hydroxypropyl)-2-methylindoles have been synthesized by the indolization of 5-hydroxyhexan-2-one arylhydrazones (Chemerda and Sletzinger, 1968e) and 3-(2-phenoxyethyl)-2-methylindoles have been synthesized by the indolization of 5-phenoxypentan-2-one arylhydrazones (Boyd-Barrett and Robinson, R., 1932; King, F. E. and Robinson, R., 1932, 1933). 5-Methoxy-3-(2-phenoxyethyl)indole (**665**; R = H) has been synthesized by indolization of **664** into **665** (R = $COOC_2H_5$) with subsequent sequential hydrolysis and decarboxylation (King, F. E. and Robinson, R., 1932). A mixture of 4-methoxyphenylhydrazine hydrochloride with 4-benzyloxybutyraldehyde dimethylacetal in refluxing isopropanol afforded 3-(2-benzyloxyethyl)-5-methoxyindole which was catalytically debenzylated, upon treatment with hydrogen in the presence of a palladium-on-carbon catalyst, to give 5-methoxytryptophol (Shen, T. Y., 1968; Shen, T.-Y., 1969). 3-(3-Hydroxypropyl)indoles have been synthesized by indolization of 5-hydroxyvaleraldehyde arylhydrazones (Canas-Rodriguez and Leeming, 1971; Yamamoto, H. and Nakao, 1968c).

Indolization of **1374** (R = H and OCH_3) in refluxing alcoholic hydrogen chloride yielded **1375** (R = H) (King, F. E. and Robinson, R., 1933) and **1375** (R = OCH_3) (King, F. E., Liguori and Robinson, R., 1933), respectively. However, attempts to dephenylate the methochloride of the latter product with hydrobromic acid under a variety of conditions were unsuccessful (King, F. E., Liguori and Robinson, R., 1933). In view of this failure, the potential use of these compounds as intermediates in the synthesis of the ring system of the alkaloid physovenine (see below) (Longmore and Robinson, B., 1965, 1967, 1969; Robinson, B., 1964a) awaits further study.

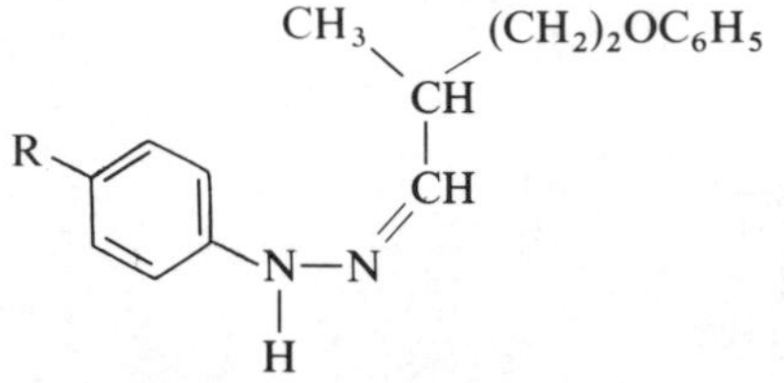

CH_3 $(CH_2)_2OC_6H_5$ CH CH R N—N H

1374

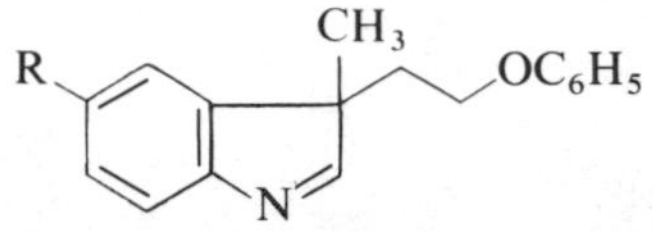

CH_3 R OC_6H_5 N

1375

The already discussed and widely used indolization of **1296** into **1297** has been applied to the indolization of analogous lactones. Thus, **1376** (R = H) has been indolized into **1377** by heating with hydrogen chloride in acetic acid (Plieninger, 1950b). Similarly, using ethanolic hydrogen chloride as the catalyst, a mixture of the component hydrazino (as its hydrochloride) and ketonic moieties of **1376** (R = OCH_3) and other similar mixtures, has been converted into **1378**, and other similar compounds, hydrolysis of the lactone group occurring under the reaction conditions (Sterling Drug Inc., 1970).

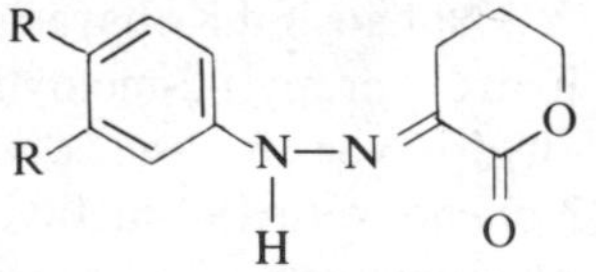

R R N—N H O O

1376

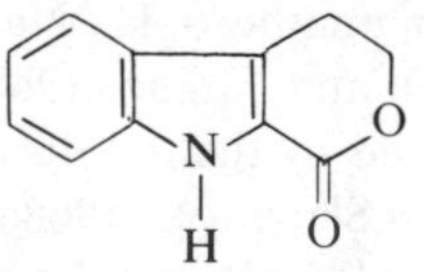

N H O O

1377

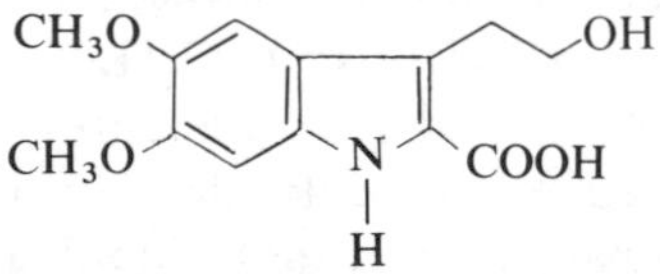

CH_3O CH_3O OH N H COOH

1378

A very useful synthesis of tryptophols and homotryptophols is analogous to the synthesis of tryptamines and homotryptamines from **1349** (n = 1 and 2, respectively). Thus, heating mixtures of an arylhydrazine **1326** salt (usually hydrochloride) with 2,3-dihydrofuran (**1379**; R^3–R^5 = H, n = 1) and 2,3-dihydro-4H-pyran (**1379**; R^3–R^5 = H, n = 2) in a solvent (dioxane containing approximately 5% of water gave the best results) produced tryptophols **1382** ($R^4 = R^5 = H$, $n = 1$) (Demerson, Humber, Philipp and Martel, 1976; Grandberg and Afonina, 1969; Grandberg and Moskvina, 1972) and homotryptophols **1382** ($R^4 = R^5 = H$, $n = 2$) (Grandberg and Afonina, 1969; Grandberg and Moskvina, 1970, 1971; Katsube, Sasajima, Ono, Nakao, Maruyama, Takayama, Katayama, Tanaka, Y., Inaba and Yamamoto, H., 1974). Similarly, N_α-benzylphenylhydrazine hydrochloride reacted with **1379** ($R^3 = R^4 = H$, $R^5 = C_6H_5$, $n = 2$ and $R^3 = R^5 = H$, $R^4 = CH_3$, $n = 2$) in aqueous dioxane to yield **1382** ($R^1 = R^4 = H$, $R^2 = CH_2C_6H_5$, $R^5 = C_6H_5$,

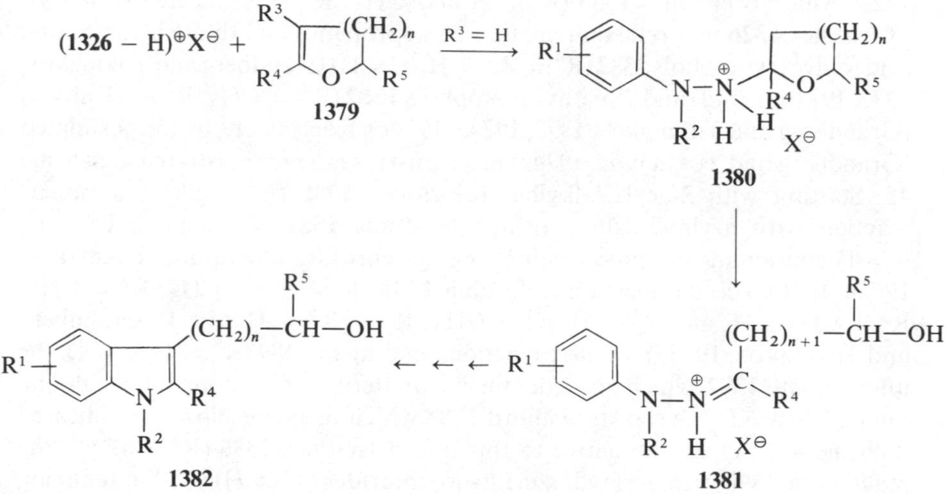

(1326 − H)$^\oplus$X$^\ominus$ + 1379 —(R^3 = H)→ 1380 → 1381 ← ← ← 1382

Scheme 44

$n = 2$ and $R^1 = R^5 = H$, $R^2 = CH_2C_6H_5$, $R^4 = CH_3$, $n = 2$, respectively) but attempts to react this hydrazine salt with **1379** ($R^3 = R^4 = CH_3$, $R^5 = H$, $n = 2$ and $R^3 = C_2H_5$, $R^4 = CH_3$, $R^5 = H$, $n = 2$) only gave resins from which no recognizable product could be isolated (Grandberg and Moskvina, 1973). Since these indolizations could occur in an anhydrous medium it was suggested that rather than **1379** ($R^3 = H$, $n = 1$ and 2) undergoing an initial hydrolysis, via formation of the cyclic hemiacetal **1383** ($n = 1$ and 2, respectively), prior to reaction with the arylhydrazine salt, they reacted directly with the salt to give **1380** followed by isomerization into **1381** which then indolized according to the Fischer indolization mechanism (Grandberg and Moskvina, 1970 – see also Grandberg and Moskvina, 1972) (Scheme 44). Cyclic hemiacetals **1383** ($R^4 = R^5 = H$, $n = 1$ and 2) did, however, react with arylhydrazines in refluxing benzene, with azeotropic removal of water, to afford **1384** ($n = 1$ and 2, respectively) which upon pyrolysis under reduced pressure yielded **1382** ($R^4 = R^5 = H$, $n = 1$ and 2, respectively). Presumably, this reaction occurred

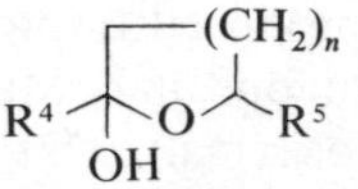

1383

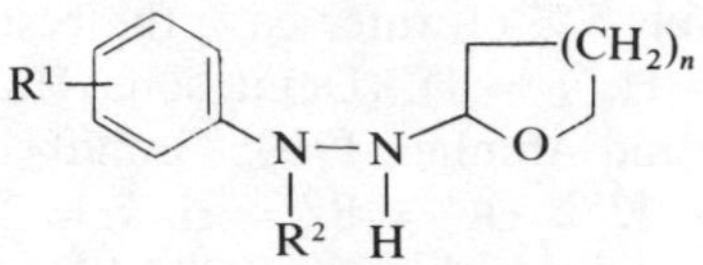

1384

via ring opening to yield the free bases of **1381** ($R^4 = R^5 = H$, $n = 1$ and 2, respectively) which then thermally indolized (Grandberg and Moskvina, 1974a, 1974b).

2-Acylbutyrolactones **1385** ($R^3 = H$ and CH_3, $R^4 = H$) reacted with arylhydrazines **1326** in a refluxing mixture of isopropanol and dilute hydrochloric acid to give tryptophols **1382** ($R^4 = R^5 = H, n = 1$) (Grandberg and Tokmakov, 1973, 1974b, 1974d) and 2-methyltryptophols **1382** ($R^4 = CH_3, R^5 = H, n = 1$) (Grandberg and Tokmakov, 1973, 1974a, 1974c), respectively, by the postulated (Grandberg and Tokmakov, 1974a) mechanistic sequence illustrated in Scheme 45. Starting with 2-acyl-2-alkylbutyrolactones **1385** (R^4 = alkyl), a similar reaction with arylhydrazines yielded products **1389** (R^4 = alkyl, $R^5 = H$, $n = 1$) containing the physovenine ring system (Grandberg and Tokmakov, 1975c, 1976), specific reactions affording **1389** ($R^3 = R^4 = CH_3$; $R^3 = CH_3$, $R^4 = CH_2C_6H_5$ and $R^3 = H$, $R^4 = CH_3$; $R^1 = R^5 = H, n = 1$) (Grandberg and Tokmakov, 1975a). In these reactions leading to **1389** ($R^5 = H, n = 1$), the intermediate **1387**, which now contained a quaternary C-3 atom in the indoline nucleus, lost $\overset{\ominus}{N}H_2$ as shown to afford **1388** which then ring closed as indicated (Scheme 45). As an alternative to this use of lactones **1385** (R^4 = alkyl), the reaction of **1390** with arylhydrazine hydrochlorides (**1326**-H)$^{\oplus}$Cl$^{\ominus}$ in refluxing dimethylformamide produced the corresponding **1389** ($R^5 = H$, $n = 1$) (Dashkevich and Grandberg, 1970; Grandberg and Dashkevich, 1970b), specific reactions giving **1389** ($R^3 = R^4 = CH_3$, $R^5 = H, n = 1$) (Grandberg and Dashkevich, 1970a, 1971a; Grandberg, Dashkevich, Markaryan and Nazaryan, 1975) and **1389** [$R^3 = CH_3$, $R^4 = (CH_2)_2CH_3$ and $CH_2C_6H_5$; $R^3 = (CH_2)_2CH_3$ and $CH_2C_6H_5$, $R^4 = CH_3$ and $R^3 + R^4 = (CH_2)_4$; $R^5 =$ H, $n = 1$] when starting with **1390** [$R^3 = R^4 = CH_3$; $R^3 = CH_3$, $R^4 =$ $(CH_2)_2CH_3$ and $CH_2C_6H_5$; $R^3 = (CH_2)_2CH_3$ and $CH_2C_6H_5$, $R^4 = CH_3$ and $R^3 + R^4 = (CH_2)_4$, respectively] (Grandberg and Dashkevich, 1971b). Similar intramolecular cyclizations of intermediate 3*H*-indolium cations occurred when mixtures of 2-(2-hydroxyethyl)cyclohexanone diethyl ketal with phenylhydrazine, 2-(2-hydroxyethyl)cyclohexanone ethylene ketal with N_α-methylphenylhydrazine, 2-(2-hydroxypropyl)cyclohexanone diethyl ketal with phenylhydrazine and 2-(3-hydroxypropyl)cyclohexanone diethyl ketal with phenylhydrazine and N_α-methylphenylhydrazine were refluxed in glacial acetic acid to afford **1389** [$R^1 = R^5 = H$, $R^3 + R^4 = (CH_2)_4$, $R^2 = H$ and CH_3, $n = 1$; $R^1 = R^2 = H$, $R^3 + R^4 = (CH_2)_4$, $R^5 = CH_3$, $n = 1$ and $R^1 = R^5 = H$, $R^2 = H$ and

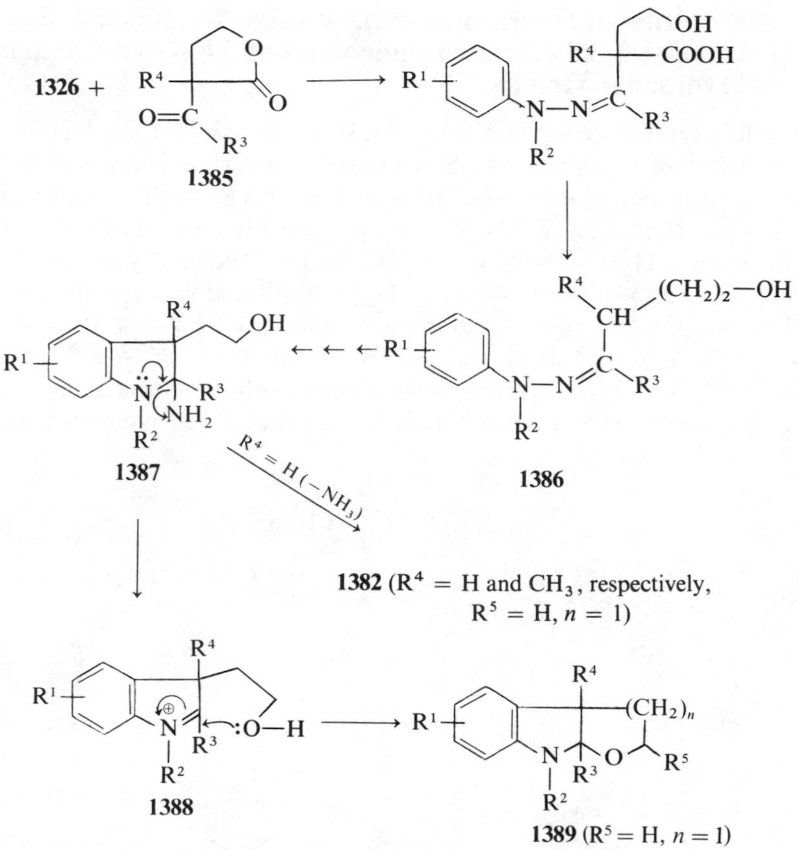

1326 + R⁴ ... 1385 → ... 1386 → → → ... 1387 → 1388 → 1389 ($R^5 = H$, $n = 1$)

$R^4 = H(-NH_3)$

1382 (R^4 = H and CH_3, respectively, R^5 = H, n = 1)

Scheme 45

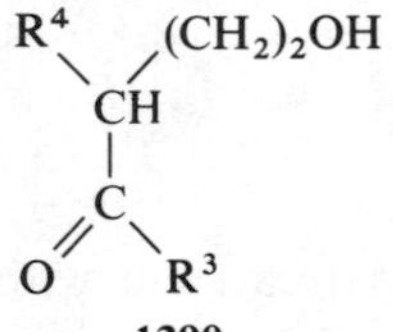

1390

CH_3, $R^3 + R^4 = (CH_2)_4$, $n = 2$, respectively] (Britten, A. Z., Bardsley and Hill, C. M., 1971).

The syntheses of tryptophols and homotryptophols from **1379** (n = 1 and 2, respectively) and of tryptophols and the physovenine ring system illustrated in Schemes 44 and 45 have been reviewed (Grandberg, 1974 – see also Grandberg and Tokmakov, 1975c).

H. Indolization of Hydrazones formed using 1-Aminoindolines, 1-Amino-1,2,3,4-tetrahydroquinolines and Related Compounds as Hydrazino Moieties

The first report (Perkin and Riley, 1923) of a reaction of this type involved the reduction of **1391** (X = NH, Y = O) with zinc dust in acetic acid, followed by the addition of cyclohexanone and warming. This led to the isolation of **1393** [$R^1 + R^2 = (CH_2)_4$, X = NH, Y = O], presumably via the formation of **1392** [$R^1 + R^2 = (CH_2)_4$, X = NH, Y = O]. In the following year (Clemo and Perkin, 1924a), **1391** X = C=O, Y = H_2] was reduced with zinc dust in acetic acid in the presence of cyclohexanone to afford **1392** [$R^1 + R^2 = (CH_2)_4$, X = C=O, Y = H_2] which was converted into **1393** [$R^1 + R^2 = (CH_2)_4$, X = C=O, Y = H_2] upon warming in dilute sulphuric acid. Similarly, the N-nitroso compounds **1394** and **1396** were reduced and converted into the

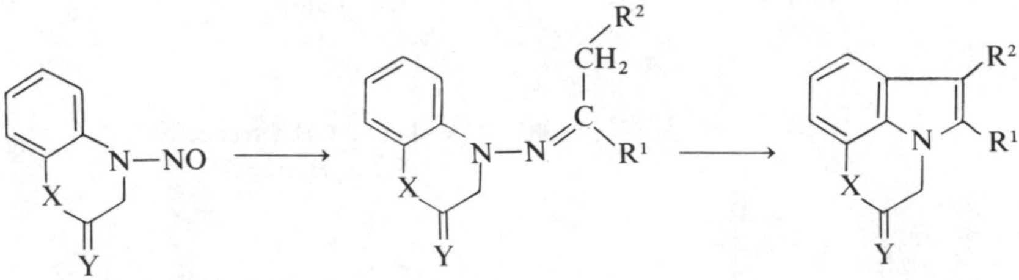

1391 **1392** **1393**

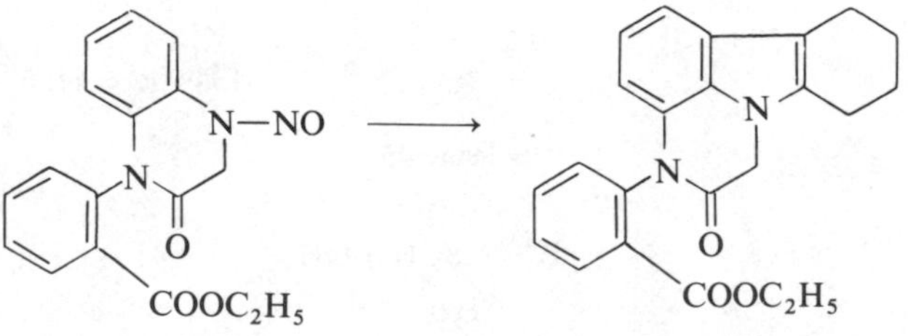

1394 **1395**

corresponding hydrazones by treatment with zinc dust in acetic acid in the presence of cyclohexanone, conditions which also then effected indolization to give **1395** and **1397**, respectively (Clemo, Perkin and Robinson, R., 1924). Subsequently (Manjunath, 1927), the reduction of 1,2,3,4,4a,9a-hexahydro-9-nitrosocarbazole and its 6-methyl homologue with zinc and acetic acid in the presence of cyclohexanone, with subsequent warming of the reaction mixture, was reported. Presumably, the intermediate hydrazone **1398** [R^1 = H, R^2 = H and CH_3, respectively, $R^3 + R^3 = (CH_2)_4$] was formed which would then afford **1399** [R^1 = H and CH_3, respectively, $R^2 + R^2 = (CH_2)_4$]. Unfortunately, the former product was incorrectly formulated as **1400** (R = H),

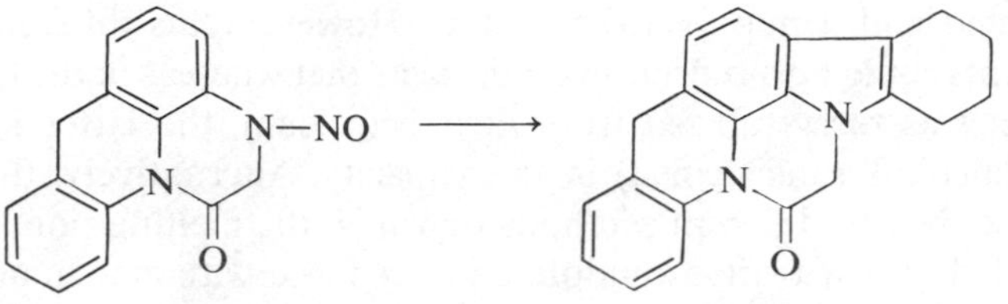

1396 **1397**

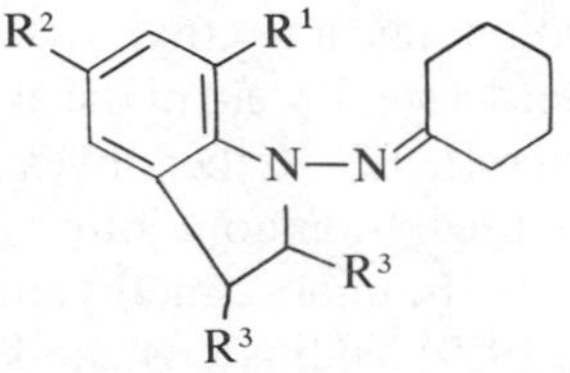

1398 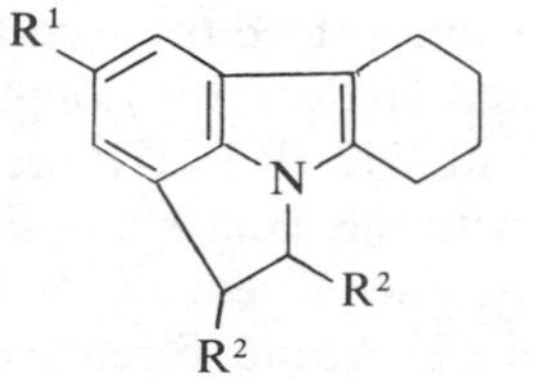 **1399**

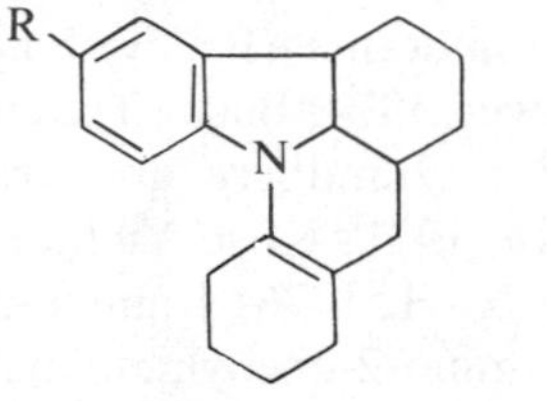

1400

which would imply the latter product had structure **1400** (R = CH_3) (Manjunath, 1927). However, there is little doubt that this was a typographical error, since the names, molecular formulae and chemical properties of the products as quoted corresponded to the correct structures **1399** [R^1 = H and CH_3, respectively, $R^2 + R^2 = (CH_2)_4$]. This error was later (Lions and Ritchie, 1939a, 1939b) corrected, Manjunath's observations were confirmed and the failure to apply the reaction to 1,2,3,4,4a,9a-hexahydro-8-methyl-9-nitrosocarbazole, presumably via the formation of the intermediate **1398** [R^1 = CH_3, R^2 = H, $R^3 + R^3 = (CH_2)_4$] was taken as further evidence for the occurrence of the previous indolizations although, in light of more recent studies (see Chapter II, G2), methyl group migration and/or elimination might occur with this arylhydrazone, thus permitting its indolization under the appropriate conditions. Extension of this synthetic route (Lions and Ritchie, 1939a, 1939b) by reduction of 1-nitrosoindoline with zinc dust in acetic acid in the presence of cyclohexanone furnished **1399** ($R^1 = R^2$ = H). Some doubt was later (Kost, Yudin, Berlin, Y. A. and Terent'ev, A. P., 1959) expressed concerning the identity of this product since it had a m.p. = 154 °C whereas the product of the same empirical formula obtained in this latter study by indolization of a mixture of 1-aminoindoline and cyclohexanone with 20%

sulphuric acid had a m.p. = 169.5–170 °C. However, this difference between melting points could be a reflection on the facts that whereas in the former study the product was recrystallized from light petroleum, the latter investigation employed 'alcohol' (unspecified) in this capacity. Alternatively, the difference could be ascribed to different methods by which the melting points may have been recorded. When 1-nitrosoindoline was reduced with zinc in acetic acid in the presence of pyruvic acid or ethyl pyruvate, in each case the only recoverable product was a small amount of 1-nitrosoindoline. However, the indolization of the hydrazone formed between 1-aminoindoline and ethyl pyruvate was later (Rapoport and Tretter, 1958) found to afford a small yield of the expected indole **313** along with **312** (R = H) (see Chapter II, E2). Other investigations contemporary with this later study similarly used 1-aminoindoline (Kost, Yudin, Berlin, Y. A. and Terent'ev, A. P., 1959), 1-amino-2-methylindoline (Kost, Yudin, Berlin, Y. A. and Terent'ev, A. P., 1959; Yudin, Kost, and Berlin, Y. A., 1958) and 1-amino-2-phenylindoline (Kost, Yudin, Berlin, Y. A. and Terent'ev, 1959) as the hydrazino moieties in indolizations. Later related indolizations have similarly utilized 1-aminoindoline (Bailey, Baldry and Scott, P. W., 1979; Bailey, Hill, P. A. and Seager, 1974; Blake, Tretter and Rapoport, 1965; Grandberg and Bobrova, 1973; Grandberg and Ivanova, 1970b, 1970c, 1970d; Grandberg and Zuyanova, 1971; Kost, Yudin and Popravko, 1962; Pachter, 1967; Welstead and Chen, Y.-H., 1976), 1-amino-2-methylindoline (Kost, Yudin and Popravko, 1962), 1-amino-2-phenylindoline (Kost, Yudin and Popravko, 1962; Pachter, 1967) and 1-amino-5-chloroindoline (Welstead and Chen, Y.-H., 1971) and the use of a large number of other substituted 1-aminoindolines has been claimed in a patent (Pachter, 1967). The attempted indolizations of the hydrazones formed by reacting propionaldehyde, butyraldehyde, isovaleraldehyde, phenylacetaldehyde, acetophenone and propiophenone with 1-amino-2-methylindoline were unsuccessful (Kost, Yudin, Berlin, Y. A. and Terent'ev, A. P., 1959) as were attempts using hydrazones with **1401** as the hydrazino moiety (Sharkova, N. M., Kucherova and Zagorevskii, 1969), these latter failures probably being the result of steric hindrance.

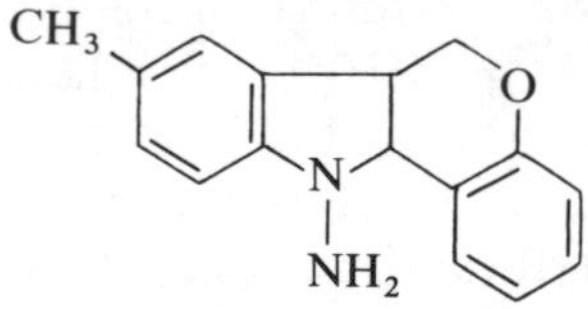

1401

A synthetic utility of this type of indolization was developed when compounds **1402** (R = CH_3 and C_6H_5) were indolized in refluxing ethanolic sulphuric acid to produce **1403** (R = CH_3 and C_6H_5, respectively) and the latter product was hydrolysed by refluxing in aqueous ethanolic sulphuric acid to yield **1404** (R = H), isolated as its N-acetyl derivative **1404** (R = $COCH_3$). This hydrolysis,

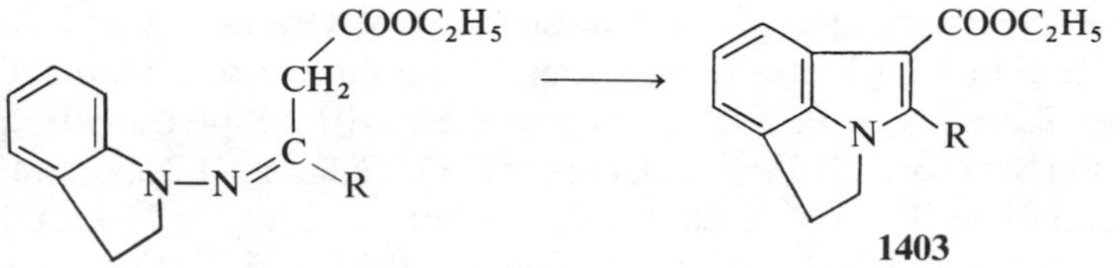

1402

1403

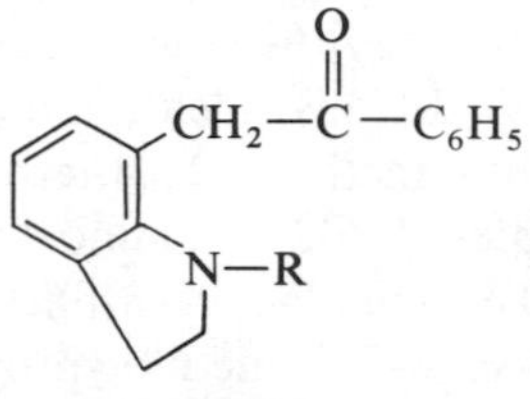

1404

which may be regarded as a partial reversal of the Fischer indolization, was presumed to be the result of the considerable steric strain in the 6,5,5-fused ring system in **1403** (Blake, Tretter and Rapoport, 1965).

The first indolization starting with a 1-amino-1,2,3,4-tetrahydroquinoline as the hydrazino moiety involved the indolizations of the pyruvic acid and acetophenone hydrazones (**473**; R^1–R^3 = R^5 = H, R^4 = COOH and C_6H_5, respectively) into **474** (R^1–R^3 = R^5 = H, R^4 = COOH and C_6H_5, respectively). However, attempted indolization failed when using the corresponding acetone hydrazone **473** (R^1–R^3 = R^5 = H, R^4 = CH_3) (Barger and Dyer, 1938), an observation later confirmed (Kost, Yudin and Terent'ev, A. N., 1959), and when using **1405** (Barger and Dyer, 1938). It was not until nearly ten years later

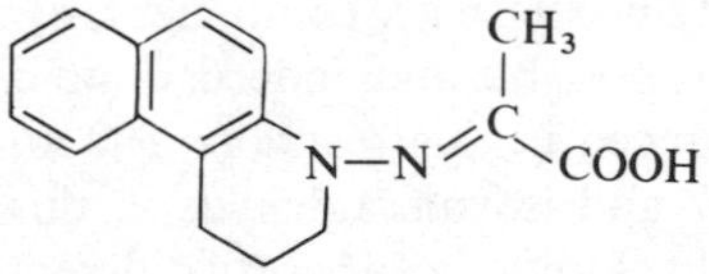

1405

(Dalgliesh and Mann, 1947) that the next report appeared utilizing 1-amino-1,2,3,4-tetrahydroquinoline as the hydrazino moiety in a Fischer indolization and a further ten years elapsed before the similar use of this moiety was next reported (Kost and Yudin, 1957). Since then, several other examples have appeared likewise using 1-amino-1,2,3,4-tetrahydroquinoline (Bahadur, Bailey and Baldry, 1977; Grandberg and Bobrova, 1973; Grandberg and Dashkevich, S. N., 1971b; Grandberg and Ivanova, T. A., 1970b, 1970c, 1970d; Illingsworth, Spencer, H. E., Mee and Haseltine, 1968; Kost, Yudin and Terent'ev, A. N., 1959; Pachter, 1967; Steck, Fletcher and Carabateas, 1974; Yudin, Budylin and Kost, 1964), 1-amino-1,2,3,4-tetrahydro-2-methylquinoline (Grandberg and

Bobrova, 1973; Grandberg and Nikitina, 1971a; Grandberg and Zuyanova, 1971), 1-amino-1,2,3,4-tetrahydro-6-methylquinoline (Kost, Yudin, Dmitriev, B. A. and Terent'ev, A. P., 1959), 1-amino-8-benzyl-1,2,3,4-tetrahydroquinoline (Fusco and Sannicolò, 1978c) (see Chapter II, G2) 1-amino-1,2,3,4-tetrahydro-8-methylquinoline (Fusco and Sannicolò, 1978c) (see Chapter II, G2), 1-amino-1,2,3,4-tetrahydro-5,7- (Fusco and Sannicolò, 1975a), -6,7- (Fusco and Sannicolò, 1975a, 1975b), -6,8- (Fusco and Sannicolò, 1975b, 1978c) and -7,8- (Fusco and Sannicolò, 1975a, 1975b, 1978c)dimethylquinoline (see Chapter II, G2), 1-amino-1,2,3,4-tetrahydro-8-methyl-6- (Fusco and Sannicolò, 1975b, 1978c) and -7-(Fusco and Sannicolò, 1975b)trideuteriomethylquinoline (see Chapter II, G2), 1-amino-8-diethylaminomethyl-1,2,3,4-tetrahydroquinoline (Fusco and Sannicolò, 1976a) (see Chapter II, G2), 1-amino-1,2,3,4-tetrahydro-2-(Pachter, 1967), -6- and -8-(Fusco and Sannicolò, 1976a)phenylquinoline (see Chapter II, G2), 1-amino-1,2,3,4-tetrahydro-7-(3-methoxypropyl)quinoline (Fusco and Sannicolò, 1975c), 1-amino-1,2,3,4-tetrahydro-6,7- and -7,8-trimethylenequinoline (Fusco and Sannicolò, 1975c) (see Chapter II, G2), 1-amino-1,2,3,4-tetrahydro-6,7-tetramethylenequinoline (Fusco and Sannicolò, 1973), 1-amino-1,2,3,4-tetrahydro-7,8-tetramethylenequinoline (Fusco and Sannicolò, 1973, 1974) (see Chapter II, G2), 1-amino-6-chloro-1,2,3,4-tetrahydroquinoline (Steck, Fletcher and Carabateas, 1974) and 1-amino-1,2,3,4-tetrahydro-6-methoxyquinoline (Pachter, 1967). The similar use of a large number of other substituted 1-amino-1,2,3,4-tetrahydroquinolines has been claimed in a patent (Pachter, 1967).

It is interesting that starting from 2-methylcyclohexanone and 1-amino-1,2,3,4-tetrahydroquinoline, only **474** [R^1–R^3 = H, $R^4 + R^5 = CH(CH_3)$-$(CH_2)_3$] was isolated from the reaction mixture (Kost, Yudin and Terent'ev, 1959). However, since this was isolated in significantly lower yields than were other related indoles and since the basic components of the final reaction mixture were not investigated it is possible and, indeed, quite likely (see Chapter III, H) that **1406** was also produced in this reaction. The attempted use of propionaldehyde, butyraldehyde and isovaleraldehyde in this reaction with 1-amino-1,2,3,4-tetrahydroquinolines under a variety of different catalytic conditions was unsuccessful (Kost, Yudin, Dmitriev, B. A. and Terent'ev, 1959).

Other indolizations of hydrazones **1392** (X = NH, Y = O) into **1393** (X = NH, Y = O) have been reported (Sellstedt and Wolf, M., 1974) and the indolizations of **1407** [R^1 = H and Cl, $R^2 + R^3 = (CH_2)_4$] gave **1408** [R^1 = H

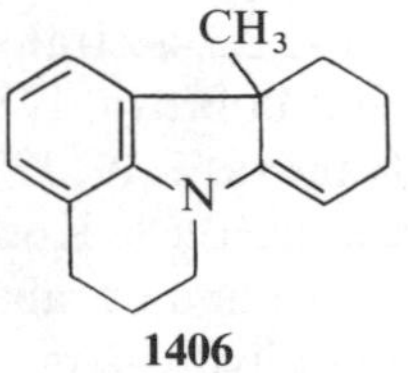

CH_3

N

1406

and Cl, respectively, $R^2 + R^3 = (CH_2)_4$] (Kim, 1975, 1976), of **1409** (R^1 = H, X = S, n = 1) gave **1410** (R^1 = H, X = S, n = 1) (Rajagopalan, 1974; Sharkova, N. M., Kucherova and Zagorevskii, 1964), of **1409** (R^1 = H, X = O, n = 1) gave **1410** (R^1 = H, X = O, n = 1) (Rajagopalan, 1974), of **1409** (R^1 = H, X = S, n = 2) gave **1410** (R^1 = H, X = S, n = 2) (Orlova, E. K., Sharkova, N. M., Meshcheryakova, Zagorevskii and Kucherova, 1975; Rajagopalan, 1974), of **1409** (R^1 = H, X = O, n = 2) gave **1410** (R^1 = H, X = O, n = 2) (Orlova, E. K., Sharkova, N. M., Meshcheryakova, Zagorevskii and Kucherova, 1975), of **1409** (R^1 = alkyl, aralkyl, substituted phenyl, Cl and OCH_3, X = S, n = 2) gave **1410** (R^1 = alkyl, aralkyl, substituted phenyl, Cl and OCH_3, respectively, X = S, n = 2) (Rajagopalan, 1974), of **1411** (X = O and S, n = 2 and 3) gave **1412** (X = O and S, n = 2 and 3, respectively) (Rajagopalan, 1974), of **1413** (n = 1 and 2) gave **1414** (n = 1 and 2, respectively)

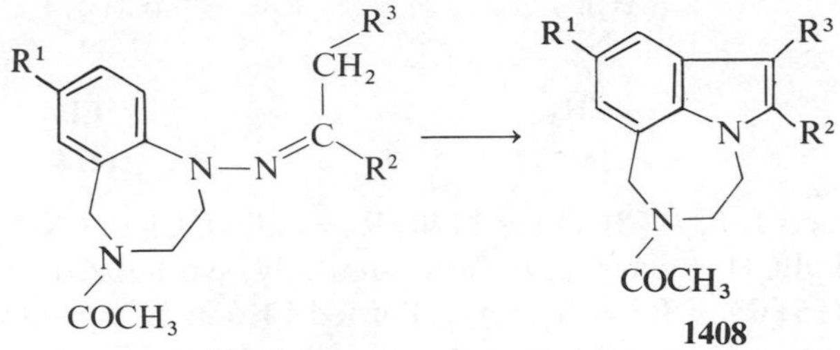

1407 → **1408**

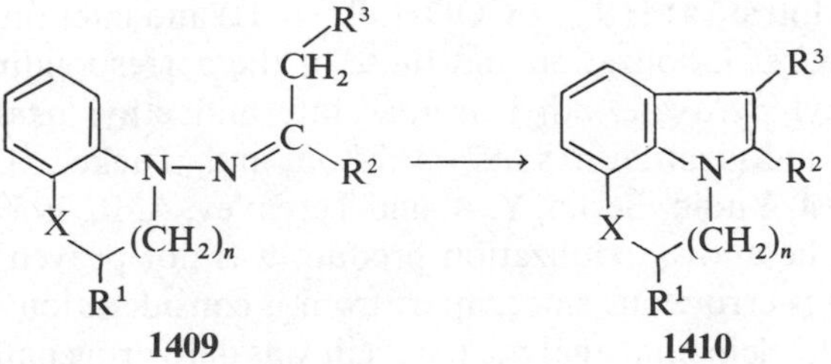

1409 → **1410**

(Sharkova, N. M., Kucherova and Zagorevskii, 1969), of **1139** (R^1 = H) gave **1140** (R^1 = H) (Adams, C. DeW., 1976; Cattanach, Cohen, A. and Heath-Brown, 1973; Cohen, A., Heath-Brown and Cattanach, 1968; F. Hoffmann-La Roche and Co. A. G., 1965; Heath-Brown, 1975; Mooradian, 1976a; Roche Products Ltd. 1965a, 1965c, 1974) (for the preparation of the parent hydrazine by the alkali metal hydride reduction of the corresponding N-nitroso compound, see also Roche Products Ltd., 1965b) and of **1139** ($R^1 = CH_3$) gave **1140** ($R^1 = CH_3$) (Cohen, A., Heath-Brown, Smithen and Cattanach, 1968; Roche Products Ltd, 1968a) (from the preparation of the parent hydrazine by the alkali metal hydride reduction of the corresponding N-nitroso compound, see also

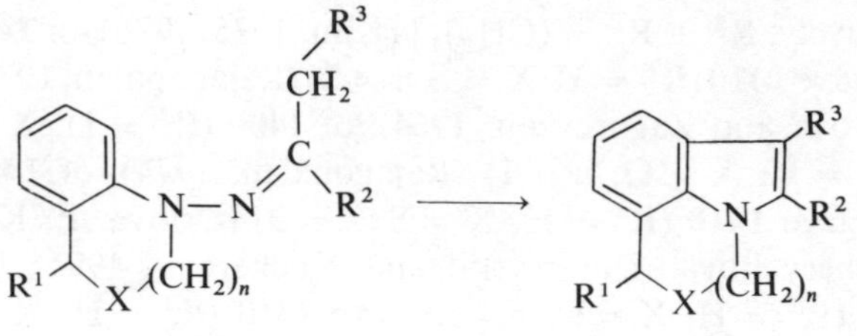

1411 → 1412

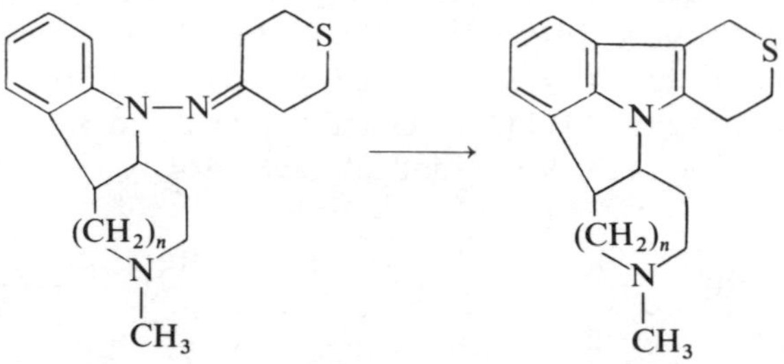

1413 → 1414

Roche Products Ltd, 1968). Other **1140** [R^1 = alkyl, Cl and NO_2, $R^2 + R^3$ = $(CH_2)_2N(alkyl)CH_2$] have also been similarly synthesized (Heath-Brown, 1978) and **1415** [$R^1 + R^2 = (CH_2)_4$] afforded **1416** in 21% yield when refluxed in tetralin containing dry hydrogen chloride (Preston and Tucker, 1943). Earlier attempts to achieve this last indolization had failed (Manjunath, 1927), as had attempts to indolize **1415** (R^1 = COOH, R^2 = H) and later further attempts to perform this latter indolization and those of the corresponding hydrazone of acetone, methyl pyruvate, ethyl acetoacetate and ethyl oxaloacetate led to 'discouraging indolization results' (Preston and Tucker, 1943). The later statement (Kost, Yudin, Berlin, Y. A. and Terent'ev, A. P., 1959) that the structure **1416** for the above indolization product was not proven by Preston and Tucker (1943) is erroneous since, apart from a consideration of its method of preparation and elemental analytical data, it was dehydrogenated, by treatment with sulphur in refluxing quinoline, to yield **1417** (Preston and Tucker, 1943)

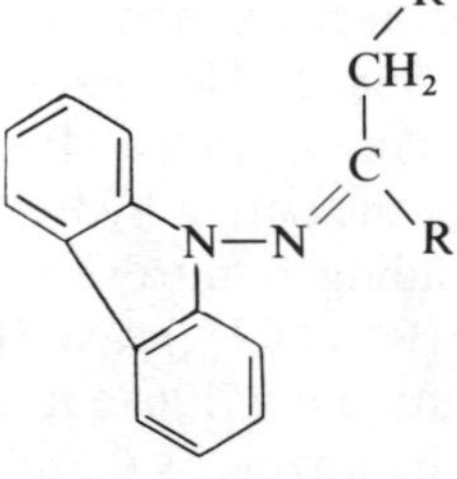

1415 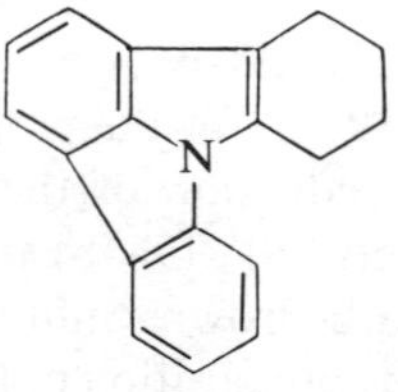 1416 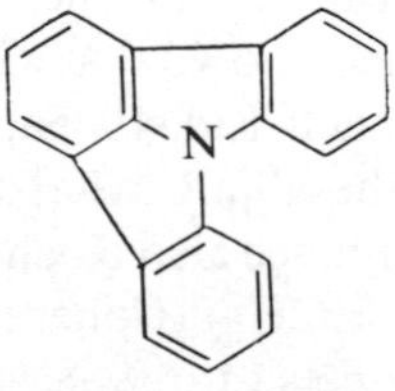 1417

which was compared with an authentic sample which had been synthesized by an independent route (Dunlop and Tucker, 1939).

A related but possibly ambiguous indolization involved the zinc dust in acetic acid reduction of **1418** in the presence of cyclohexanone, the *in situ* generated hydrazone apparently furnishing under these conditions **1419** as the only isolated product. This structure for the product was distinguished from the possible alternative, **1422**, by hydrolysis with alcoholic potassium hydroxide into **1420** which upon warming with concentrated sulphuric acid cyclized to form **1421**, a compound which gave a positive primary aromatic amine test. Such a cyclization would not be possible after hydrolysis of **1422** (Linnell and Perkin, 1924). However, in light of the mechanism of the indolization, it is unlikely that **1419** should be the only product from this reaction and reinvestigation with a view to isolating **1422** would be of interest.

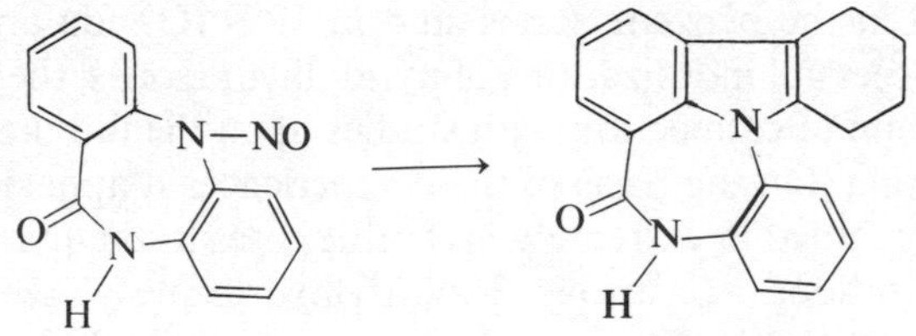

1418 → **1419** → **1420** → **1421**

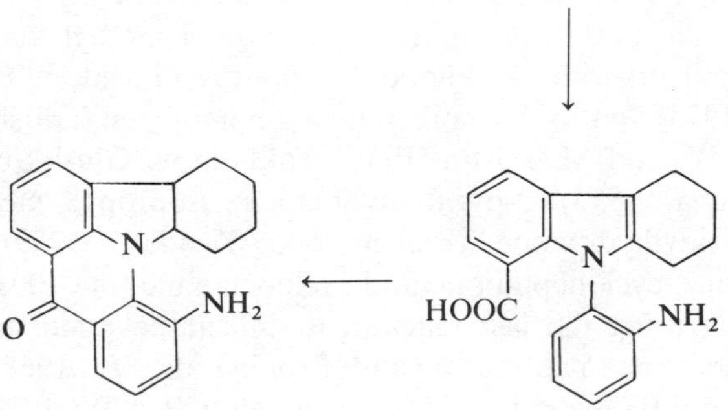

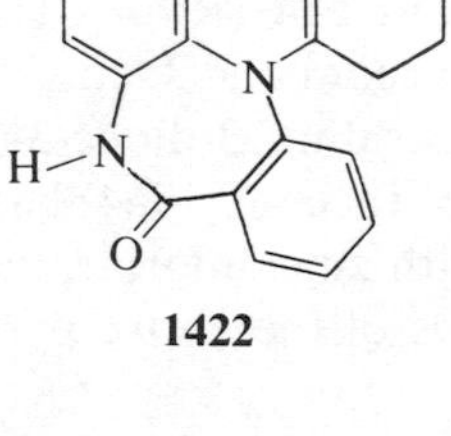

1422

I. Synthesis of Azaindoles and Related Studies

Pyridylhydrazones, when treated with a variety of acids, usually only indolized with difficulty or failed to do so at all, whereas the corresponding phenylhydrazones generally indolized relatively easily under similar conditions. Thus, although several pyridylhydrazones have been indolized under conditions of acid catalysis, the yields from many of these reactions were low, possibly because of the necessary severity of the reaction conditions relative to those required to effect indolization of the corresponding phenylhydrazones (Ficken and Kendall, 1959; Okuda and Robison, 1959). Indeed, in 1915 it was stated, with reference to the synthesis of 7-azaindoles and 7-aza-5,6-benzindoles by Fischer indolization of 2-pyridylhydrazones and 2-quinolylhydrazones, respectively (Fargher and Furness, 1915), that 'it appears unlikely that such substances can be obtained in the pyridine series by reactions analogous to those which work so well in the benzene series' and in 1959 (Okuda and Robison, 1959), after some successful indolizations of pyridylhydrazones, the guarded comment was made, again in connection with studies upon the indolization of 2-pyridylhydrazones, that 'On the basis of these experiences it appears that this cyclization is only practical in extremely favorable cases in the pyridine series'.

The indolizations of various 3-pyridylhydrazones have been referred to previously (Chapter III, C). Several 2- and 4-pyridylhydrazones have also been indolized under conditions of acid catalysis. Thus, cyclohexanone (Okuda and Robison, 1959; Yakhontov, Pronina, E. V. and Rubtsov, 1965), benzyl phenyl ketone (Okuda and Robison, 1959), isopropyl methyl ketone (Ficken and Kendall, 1959), piperidin-2,3-dione (Yakhontov, Glushkov, Pronina, R. V. and Smirnova, 1973) and its 7-membered ring homologue (Glushkov, Yakhontov, Pronina, E. V. and Magidson, 1969; Yakhontov, Glushkov, Pronina, E. V. and Smirnova, 1973) 2-pyridylhydrazone, isopropyl methyl ketone N_α-methyl-2-pyridylhydrazone (Ficken and Kendall, 1959), phenylacetone, cyclohexanone, cycloheptanone and cyclooctanone (the 100.0% product yield quoted when using the last reactant is difficult to understand) 2-(6-bromopyridyl)hydrazone (Yakhontov and Pronina, E. V., 1968), cyclohexanone 3-(2-methylpyridyl)hydrazone (Clemo and Holt, R. J. W., 1953), cyclohexanone (Mann, Prior and Willcox, 1959), isopropyl methyl ketone (Ficken 1963) and piperidin-2,3-dione and its 7-membered ring homologue (Yakhontov, Glushkov, Pronina, E. V. and Smirnova, 1973) 4-pyridylhydrazone, isopropyl methyl ketone 4-(2- and 4-methylpyridyl)hydrazone (Ficken, 1963), cyclohexanone 4-pyridylhydrazone 1-oxide and **584** ($R^1 = CH_3$, $R^2 = H$, $R^3 = C_2H_5$ and $R^1 = R^2 = H$, $R^3 = CH_3$) (Tacconi and Pietra, 1965 – see also Clark, B. A. J. and Parrick, 1974) and piperidin-2,3-dione-3-(4-pyridylhydrazone-1-oxide) (**584**; R^1–$R^3 = H$) (Pietra and Tacconi, 1964) have all been indolized, usually by fusion of their mixtures with zinc chloride, into the corresponding indolic products. The ketonic moieties of the above pyridylhydrazones gave phenylhydrazones which indolized under mild conditions in very high yields. In the case of other ketones, some of which did not give rise to such phenylhydrazones, it is well established that their pyridylhydrazones did not indolize in acidic conditions. Thus, cyclopentanone (Yakhontov and Pronina, E. V., 1968), not surprisingly acetaldehyde (see Chapter IV, A) (Fargher and Furness, 1915; Okuda and Robison, 1959), acetone (Fargher and Furness, 1915; Okuda and Robison, 1959; Yakhontov and Pronina, E. V., 1968), pyruvic acid (Fargher and Furness, 1915; Okuda and Robison, 1959), acetophenone (Fargher and Furness, 1915) and 1-methylpiperidin-4-one (Hörlein, U., 1954) 2-pyridylhydrazone, pyruvic acid 3-(2-methylpyridyl)hydrazone (Clemo and Holt, R. J. W., 1953), 1,2,3,4-tetrahydro-1-methyl- and 1,2,3,4-tetrahydro-1-phenylquinolin-4-one 4-pyridylhydrazone (Mann, Prior and Willcox, 1959), cyclohexan-1,2-dione bis-4-pyridylhydrazone (Mann, Prior and Willcox, 1959) and acetophenone, ethyl pyruvate and propionaldehyde 4-pyridylhydrazone-1-oxide (Tacconi and Pietra, 1965) all failed to indolize when treated with various acid catalysts. An interesting elimination occurred when a mixture of the 2-pyridylhydrazine **1423** and 1-methylpiperidin-4-one was heated with alcoholic hydrogen chloride to afford not the expected 8-aza-1,2,3,4-tetrahydro-γ-carboline **1424** but the hydrazone **1425** (Hörlein, U., 1954).

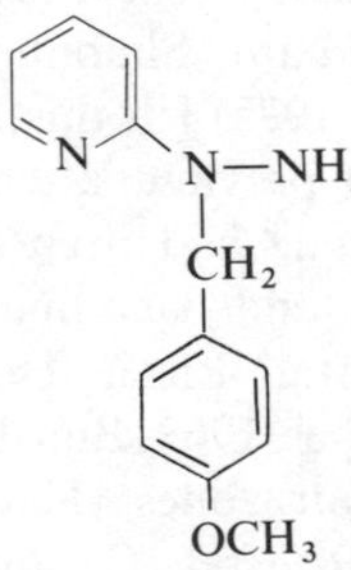

1423 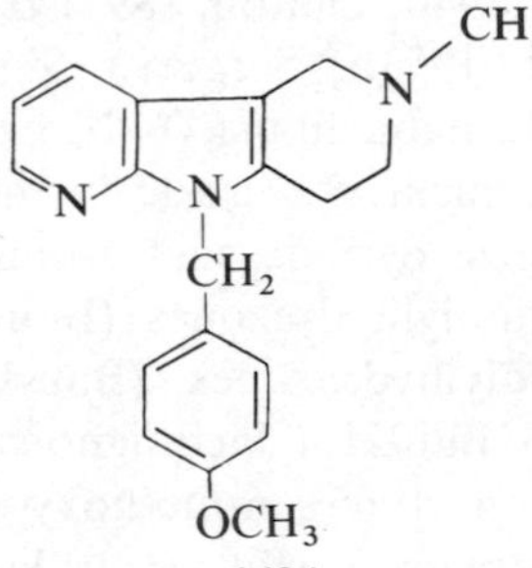 **1424**

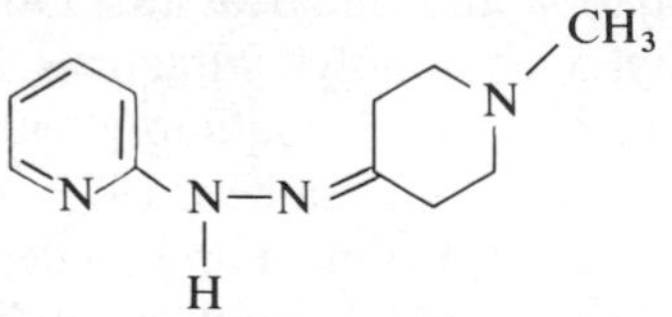

1425

Passage of acetaldehyde, acetone and ethyl methyl ketone 2-pyridylhydrazone over an alumina catalyst gave 7-azaindole, 7-aza-2-methylindole (Yakhontov, Suvorov, Kanterov, Podkhalyuzina, Pronina, E. V., Starostenko and Shkil'kova, 1972) and 7-aza-2,3-dimethylindole (Yakhontov, Suvorov, Pronina, E. V., Kanterov, Podkhalyuzina, Starostenko and Shkil'kova, 1972), respectively, along with 2-aminopyridine and **1167** (R = H) in each case.

By analogy with pyridylhydrazones, the indolization of quinolyl- and isoquinolylhydrazones might be expected to be hindered relative to the indolization of the corresponding phenylhydrazones. With the hydrazone moiety on the benzenoid ring of either of these heterocyclic systems, the presence of the nitrogen atom in the pyridinoid ring had no marked effect in this respect since it was too distant and such hydrazones indolized normally. Examples of such indolizations involving 6- and 7-quinolylhydrazones and 6- and 7-isoquinolylhydrazones have already been referred to (Chapter III, E and F, respectively) and other indolizations involving 5-quinolylhydrazones (Buu-Hoï, Bellavita, Ricci, Hoeffinger and Balucani, 1966; Buu-Hoï, Jacquignon and Hoeffinger, 1963; Buu-Hoï, Jacquignon and Périn, 1962; Buu-Hoï, Mangane and Jacquignon, 1966; Buu-Hoï, Périn and Jacquignon, 1960, 1962; Buu-Hoï and Saint-Ruf, 1963; Clemo and Felton, 1951a; Dewar, 1944; Dufton, 1892; Gryaznov, Akhvlediani, Volodina, Vasil'ev, A. M., Babushkina, T. A. and Suvorov, 1977; Huisgen, 1948a; Jacquignon, Buu-Hoï and Dufour, 1966; Roussel, O., Buu-Hoï and Jacquignon, 1965), 8-quinolylhydrazones (Buu-Hoï, Jacquignon and Hoeffinger, 1963; Buu-Hoï, Jacquignon, Ledésert, Ricci and Balucani, 1969; Buu-Hoï, Jacquignon and Périn, 1962; Buu-Hoï, Périn and Jacquignon, 1960, 1962; Buu-Hoï and Saint-Ruf, 1963; Clemo and Felton, 1951a; Dewar, 1944; Dufton, 1891; Grandberg and Yaryshev, 1969; Ockenden and Schofield, 1953b; Sergeeva, Z. F., Akhvlediani, Shabunova, Korolev, Vasil'ev, A. M., Babushkina, T. N. and Suvorov, 1975) [in this last paper the erroneous statement was made to the effect that previous attempts (Dufton, 1891) to indolize pyruvic acid 8-quinolylhydrazone had been unsuccessful], 2-methyl-8-quinolylhydrazones (Buu-Hoï, Périn and Jacquignon, 1962), 6-methyl-8-quinolylhydrazones (Buu-Hoï, Bellavita, Ricci, Hoeffinger and Balucani, 1966; Buu-Hoï, Jacquignon and Hoeffinger, 1963; Buu-Hoï, Mangane and Jacquignon, 1966), 6-methoxy-8-quinolylhydrazones (Nandi, 1940), 5-isoquinolylhydrazones (Buu-Hoï, Jacquignon, Roussel, O. and Hoeffinger, 1964; Buu-Hoï, Mangane and Jacquignon, 1966; Buu-Hoï and Saint-Ruf, 1963; Govindachari, Rajappa and Sudarsanam, 1966; Manske, R. H. F. and Kulka, 1949), 3-methyl-5-isoquinolylhydrazones (Buu-Hoï, Jacquignon, Roussel, O. and Hoeffinger, 1964), 8-isoquinolylhydrazones (Baradarani and Joule, 1978; Govindachari and Sudarsanam, 1971) and 5-ethyl-8-isoquinolylhydrazones (Cohylakis, Hignett, Lichman and Joule, 1974) have been effected. However, occasionally (Dewar, 1944), harsher conditions than those used to indolize the corresponding phenylhydrazones were required. The difficulty experienced (Buu-Hoï, Périn and Jacquignon, 1962) in indolizing 1-tetralone 8-quinolylhydrazone and 2-methyl-8-quinolylhydrazone and 1,2,3,4-tetrahydrophenanthren-1 and -4-one 8-quinolylhydrazone, which initially could not be effected even when 'drastic procedures' were used (Buu-Hoï, Périn and Jacquignon, 1960), was attributed (Buu-Hoï, Périn and Jacquignon, 1960, 1962) to the unfavourable steric conditions present in *cis* biangular azabenzindoles in the vicinity of *peri* nitrogen atoms as shown in **1426** (R = H and CH_3), **1427** and **1428**, respectively, the initial indolization products formed from these indolizations. Under the severe indolization conditions (i.e. pyrolysis with zinc chloride), dehydrogenation occurred to afford in each case the completely aromatized azabenzindolic products (Buu-Hoï, Périn and Jacquignon, 1962).

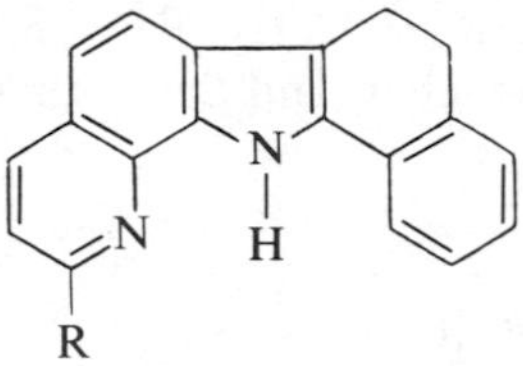

1426

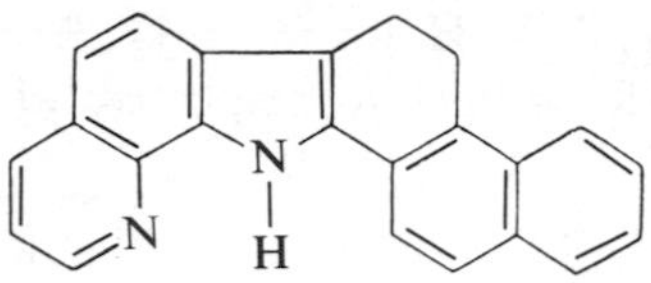

1427

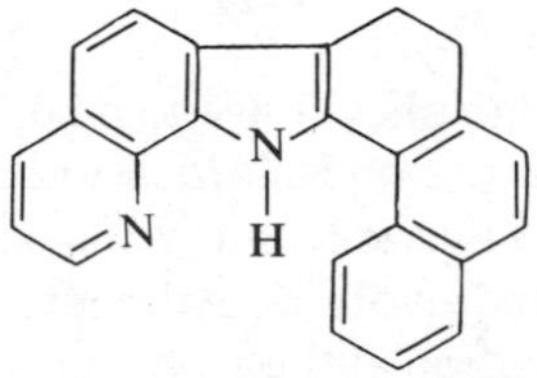

1428

When the hydrazino moiety was on the pyridinoid ring of quinolyl- and isoquinolylhydrazones, indolization was noticeably more difficult than it was with the corresponding phenylhydrazones. Thus, attempted indolizations, using acid catalysts, of acetone (Perkin and Robinson, R., 1913), pyruvic acid (Fargher and Furness, 1915) and apparently some other (Fargher and Furness, 1915) 2-quinolylhydrazones and 1,2,3,4-tetrahydro-1-methylquinolin-4-one 4-quinolylhydrazone (Mann, Prior and Willcox, 1959) were unsuccessful and attempts to indolize cyclopentanone, cyclohexanone, ethyl methyl ketone, benzyl methyl ketone, diethyl ketone, propionaldehyde and butyraldehyde 4-quinolylhydrazone and 2-, 6-, 7- and 8-methyl-, 6-, 7- and 8-chloro- and 6-, 7- and 8-methoxy-4-quinolylhydrazone using hydrochloric acid, polyphosphoric acid or sulphuric acid or a mixture of acetic acid and hydrochloric acid either failed or led to only traces of the desired 5-azaindolic products (Khan, M. A. and Rocha, 1978). However, under the appropriate acidic conditions, cyclohexanone 3-(2-methylquinolyl)hydrazone (Clemo and Holt, R. J. W., 1953; Robinson, G. M. and Robinson, R., 1924), other 3-quinolylhydrazones already referred to (Chapter III, C) and cyclohexanone 4-quinolylhydrazone (Mann, Prior and Willcox, 1959) have been successfully indolized. Although the phenylhydrazones of some of these ketones indolized very easily, in other examples (Govindachari, Rajappa and Sudarsanam, 1961) this was not the case and it could be that successful indolization of these 3-quinolylhydrazones was facilitated by the presence of the benzenoid nucleus or by the use of a more forcing catalyst. Thus, upon pyrolysis with zinc chloride, ethyl pyruvate, acetone, ethyl methyl ketone, acetophenone, propiophenone, phenylacetaldehyde, benzyl methyl ketone, benzyl phenyl ketone and cyclohexanone 4-isoquinolylhydrazone have been indolized to furnish **1429** [$R^1 = COOC_2H_5$, $R^2 = H$; $R^1 = CH_3$, $R^2 = H$; $R^1 = R^2 = CH_3$; $R^1 = C_6H_5$, $R^2 = H$; $R^1 = C_6H_5$, $R^2 = CH_3$; $R^1 = H$, $R^2 = C_6H_5$; $R^1 = CH_3$, $R^2 = C_6H_5$; $R^1 = R^2 = C_6H_5$ and $R^1 + R^2 = (CH_2)_4$, respectively] (Govindachari and Sudarsanam, 1967).

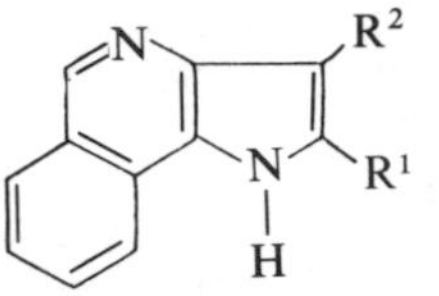

1429

The claim (Grandberg and Yaryshev, 1969) that boiling a mixture of 2-hydrazinoquinoline with 5-chloropentan-2-one afforded 3-(2-aminoethyl)-7-aza-2-methylindole has been regarded (Parrick and Willcox, 1976) as surprising, since, assuming that this reaction involved hydrazone formation and subsequent pyrrole ring synthesis, cyclization at the 3-position of the quinolyl nucleus would be expected to involve a higher energy transition state than that for reaction at the 1-position (see pages 527–528).

When it was still believed that the new C—C bond formation during indolization occurred via a concerted cyclic rearrangement involving an intramolecular electrophilic attack on the aromatic nucleus by the C-2 atom of the enehydrazine tautomer (see Scheme 10), it was suggested (Crooks and Robinson, B., 1967, 1969; Ficken and Kendall, 1959) that the deactivation, by the inductive effect of the nitrogen atom, of the pyridinoid nucleus relative to the benzenoid nucleus towards electrophilic attack may account for these above observations by a retardation or prevention of this rearrangement. However, it is now known that a [3,3]-sigmatropic shift, which would be largely insensitive to such changes, is responsible for this rearrangement (see Chapter II, F). It would appear, therefore, that the retardation or prevention of indolization by this nitrogen atom was caused by its inductive effect retarding or preventing the movement away of the π-electron pair in the hydrazone during the formation of the enehydrazine which is now known to be the rate determining step of the indolization mechanism. This also explains the detrimental effects which electron withdrawing substituents on the benzenoid ring of phenylhydrazones have upon their indolizations.

Under the conditions of acid catalysis, this effect of the pyridinoid nitrogen atom is more manifest since under these conditions it is protonated and thus carries a positive charge. In an analogous manner it can interact with Lewis acid catalysts (Yakhontov and Marshalkin, 1972). In order to prevent this protonation during indolization, the non-catalytic thermal indolization technique, first developed in 1957 (Fitzpatrick, J. T. and Hiser, 1957) for the indolization of phenylhydrazones, has been applied, with considerable success, to the indolization of 2-, 3- and 4-pyridylhydrazones. The results of these indolizations are summarized in Table 29. In generalized terms it can be seen that, when using carbonyl moieties whose phenylhydrazones are readily indolized, good yields

Table 29. Thermal Indolizations which have been Effected upon 2-, 3- and 4-Pyridylhydrazones

Pyridylhydrazone				
Hydrazine Moiety	Ketone or Aldehyde Moiety	Solvent	Yield of Azaindolic Product	Reference
2-Pyridyl-[a]	Propionaldehyde	DEG	25	1
2-Pyridyl-[a]	Butyraldehyde	DEG	37	1
2-Pyridyl-[a]	Phenylacetaldehyde	DEG	88	1
2-Pyridyl-[a]	Ethyl methyl ketone	DEG	43	1
2-Pyridyl-[a]	Isopropyl methyl ketone	DEG[b]	5	1
2-Pyridyl-[a]	Acetophenone	TEG	63	1
2-Pyridyl-[a]	Benzyl phenyl ketone	DEG	56	2
2-Pyridyl-[a]	Cyclopentanone	DEG	67	1
2-Pyridyl-[a]	Cyclohexanone	DEG	70	2
2-Pyridyl-[a]	1-Indanone	DEG	95	1
2-Pyridyl-[a]	1-Tetralone	DEG[b]	77	1
2-Pyridyl-[a]	Acetone	DEG or TEG	0	1
2-Pyridyl-[a]	Pyruvic acid	DEG or TEG	0	1
2-Pyridyl-[a]	Ethyl pyruvate	DEG or TEG	0	1
2-Pyridyl-[a]	2-Acetylpyridine	DEG or TEG	0	3
2-Pyridyl-[a]	4-Acetylpyridine	DEG or TEG	0	3
2-(4-Methylpyridyl)-	Propionaldehyde	DEG	25	3
2-(4-Methylpyridyl)-	Phenylacetaldehyde	DEG	72	3
2-(4-Methylpyridyl)-	Acetophenone	TEG	45	3
2-(4-Methylpyridyl)-	Benzyl phenyl ketone	DEG	82	3
2-(4-Methylpyridyl)-	Cyclohexanone	DEG	64	3
2-(4-Methylpyridyl)-	3-Acetylpropionic acid	DEG	0	3
3-Pyridyl-	Propionaldehyde	MEG[b]	21(6)[c,d]	4
3-Pyridyl-	Butyraldehyde	MEG[b]	48[d,e]	4
3-Pyridyl-	Phenacetaldehyde	MEG[b]	26(4)[c,d]	4
3-Pyridyl-	Ethyl methyl ketone	MEG[b]	20(3)[c,d]	4
3-Pyridyl-	Acetophenone	DEG[b]	54[d,e]	4
3-Pyridyl-	Benzyl phenyl ketone	DEG[b]	23(5)[c,d]	4
3-Pyridyl-	Cyclopentanone	MEG[b]	68(10)[c,d]	4
3-Pyridyl-	Cyclohexanone	MEG[b]	38(6)[c,d]	4
3-Pyridyl-	1-Tetralone	DEG[b]	53[d,e]	4
4-Pyridyl-	Propionaldehyde	DEG	34.5	5
4-Pyridyl-	Propionaldehyde	DEG(?)	0	4
4-Pyridyl-	Butyraldehyde	DEG	22	4

(*continued*)

Table 29 (*continued*)

Pyridylhydrazone				
Hydrazine Moiety	Ketone or Aldehyde Moiety	Solvent	Yield of Azaindolic Product	Reference
4-Pyridyl-	Phenylacetaldehyde	DEG	10	4
4-Pyridyl-	Ethyl methyl ketone	DEG	52	4, 5
4-Pyridyl-	Benzyl phenyl ketone	DEG	18	4
4-Pyridyl-	Cyclopentanone	DEG	77	4
4-Pyridyl-	Cyclohexanone	DEG	95	2
4-Pyridyl-	1-Tetralone	TEG	66	4
4-Pyridyl-	Acetone	MEG, DEG or TEG	0	4, 5
4-Pyridyl-	Pyruvic acid	MEG, DEG or TEG	0	5
4-Pyridyl-	Acetophenone	DEG(?)	0	4

Notes

a. Other investigators (Potts and Schneller, 1968) found that only 2-aminopyridine and unrecognizable tars resulted from boiling solutions of 2-pyridylhydrazones in tetralin; b. Reactions performed under an atmosphere of nitrogen; c. The figure outside the bracket is the yield of the 4-azaindole, that inside the bracket being the yield of 6-azaindole; d. Structures of 4- and 6-aza isomers were distinguished using i.r. and u.v. spectroscopy; e. No 6-azaindole isolated MEG = monoethylene glycol; DEG = diethylene glycol; TEG = triethylene glycol.

References

1. Kelly, A. H. and Parrick, 1966; 2. Kelly, A. H., McLeod and Parrick, 1965; 3. Herbert and Wibberley, 1969; 4. Kelly, A. H. and Parrick, 1970; 5. Crooks and Robinson, B., 1969.

of azaindoles were obtained whereas using carbonyl moieties whose phenylhydrazones are difficult to indolize led to only low yields of azaindoles or failed altogether, acetone, pyruvic acid and ethyl pyruvate pyridylhydrazones being notable among these failures. In an attempt (Kelly, A. H. and Parrick, 1970) to extend this synthetic technique to the indolization of pyridylhydrazone-1-oxides, boiling acetophenone 4-pyridylhydrazone-1-oxide in diethylene glycol led to the isolation of only acetophenone 4-pyridylhydrazone whereas similar treatment of the cyclohexanone analogue again caused reduction of the N-oxide function but also effected indolization to afford 6-aza-1,2,3,4-tetrahydrocarbazole. However, this was obtained in lower yield than when it was obtained by direct indolization of cyclohexanone 4-pyridylhydrazone. Azaindoles were not obtained by heating cyclohexanone and ethyl methyl ketone 2-(5-nitropyridyl)-hydrazone in diethylene or triethylene glycol (Kelly, A. H. and Parrick, 1966–see also Clark, B. A. J. and Parrick, 1974).

An interesting rearrangement occurred when isobutyraldehyde 4-pyridylhydrazone (**1430**; X = N, Y = CH, R = CH_3) was boiled under reflux in diethylene glycol. Instead of the expected 5-aza-3*H*-indole **1431** (X = N, Y = CH, R = CH_3), the only product isolated from this reaction was **1432**

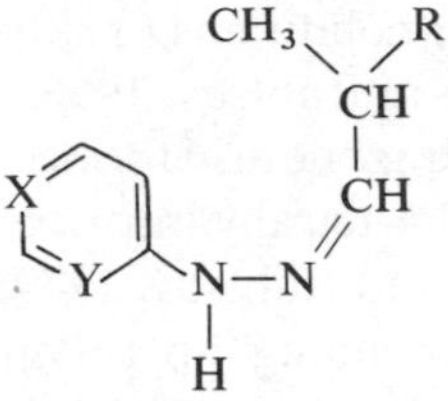

1430 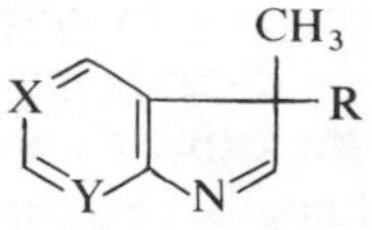 **1431** 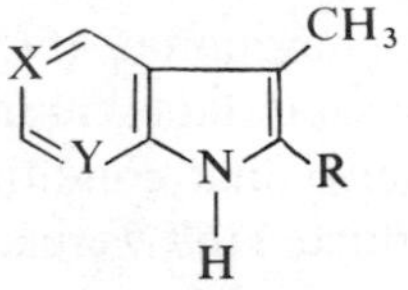 **1432**

(X = N, Y = CH, R = CH_3), presumably formed from the 3*H*-indole by a thermally induced Plancher rearrangement (Crooks and Robinson, B., 1967, 1969; Kelly, A. H. and Parrick, 1970). Likewise, 2-methylbutyraldehyde 4-pyridylhydrazone (**1430**; X = N, Y = CH, R = C_2H_5) gave **1432** (X = N, Y = CH, R = C_2H_5) (Crooks and Robinson, B., 1969), the exclusive formation of which was in accord with the observation (Jackson, A. H. and Smith, P., 1968) that an ethyl group has a higher migratory aptitude than a methyl group and migrated preferentially in a Plancher rearrangement. The earlier of these observations led to the suggestion (Crooks and Robinson, B., 1967, 1969) that the compound prepared by the non-catalytic thermal indolization of isobutyraldehyde 2-pyridylhydrazone (**1430**; X = CH, Y = N, R = CH_3) and formulated as **1431** (X = CH, Y = N, R = CH_3) (Kelly, A. H. and Parrick, 1966) should be structurally reassigned as **1432** (X = CH, Y = N, R = CH_3). Indeed, the physical properties quoted for this product were almost identical with those quoted for **1432** (X = CH, Y = N, R = CH_3) when prepared by the non-catalytic thermal indolization of ethyl methyl ketone 2-pyridylhydrazone (Kelly, A. H. and Parrick, 1966). This suggestion was later verified as being correct (Parrick, 1969).

Comparative studies of the reactions of cyclohexanone 2-pyridylhydrazone, several 5- and 6-substituted 2-pyridylhydrazones of cyclohexanone and other ketone 4-pyridylhydrazones and 2,6-disubstituted 4-pyridylhydrazones, under the action of various indolization catalysts, have been made. With the 2-pyridylhydrazones, an electron releasing group at the 5- or 6-position of the pyridine ring facilitated indolization and, usually, high yields of the corresponding 8-aza-1,2,3,4-tetrahydrocarbazoles were obtained using boron trifluoride (excess), polyphosphoric acid, sulphosalicylic acid, 4-methylbenzene sulphonic acid, copper(I) chloride or zinc chloride as catalysts (Yakhontov and Pronina, E. V., 1969; Yakhontov, Pronina, E. V. and Rubtsov, 1967, 1970) although the occasional claimed yields of 100% (Yakhontov and Pronina, E. V., 1969) are difficult to understand.

Using boron trifluoride etherate in acetic acid as the catalyst and cyclohexanone 2-pyridyl- and 2-(6-methylpyridyl)hydrazone as the reactants, the yield of indolization product in each case was lowered because of the concomitant formation of by-products formulated as **1167** (R = H and CH_3, respectively), although **1167** (R = Br) was not obtained by reacting cyclohexanone 2-(6-bromopyridyl)hydrazone under similar conditions (Yakhontov and Pronina, E. V., 1969; Yakhontov, Pronina, E. V. and Rubtsov, 1966, 1970). Cyclohexanone 2-(5-chloro- and -5-methylpyridyl)hydrazone also formed analogous products, along with the expected 8-aza-1,2,3,4-tetrahydrocarbazoles, under similar conditions (Yakhontov, Pronina, E. V. and Rubtsov, 1970). The products **1167** were also formed by heating the corresponding 2-pyridylhydrazines with boron trifluoride etherate in acetic acid and their formation simultaneously with the above indolization products no doubt arose by similar reactions, the required hydrazine being formed by hydrolysis of the hydrazone under the reaction conditions (Yakhontov, Pronina, E. V. and Rubtsov, 1966, 1970). Likewise, yields of 8-aza-1,2,3,4-tetrahydrocarbazoles were diminished because of by-product formation when cyclohexanone 2-pyridylhydrazone and 5- and 6-substituted 2-pyridylhydrazones were indolized using hydrochloric acid as catalyst. Thus, **1433** ($R^1 = R^2 = H$) (Yakhontov and Pronina, E. V., 1969; Yakhontov, Pronina, E. V., Rozynov and Rubtsov, 1968; Yakhontov, Pronina, E. V. and Rubtsov, 1966, 1967, 1970), **1433** ($R^1 = H$, $R^2 = CH_3$) (Yakhontov and Pronina, E. V., 1969; Yakhontov, Pronina, E. V., Rozynov and Rubtsov, 1968; Yakhontov, Pronina, E. V. and Rubtsov, 1966, 1967), **1433** ($R^1 = H$, $R^2 = Br$) (Yakhontov and Pronina, E. V., 1969) and **1433** ($R^1 = CH_3$ and Cl, $R^2 = H$) (Yakhontov, Pronina, E. V., Rozynov and Rubtsov, 1968; Yakhontov, Pronina, E. V. and Rubtsov, 1970) were converted into the corresponding **1434** ($R^1 = H$, $R^2 = H$, CH_3 and Br and $R^1 = CH_3$ and Cl, $R^2 = H$, respectively) and in each case another product which was formulated as **1435** ($R^1 = H$, $R^2 = H$, CH_3 and Br and $R^1 = CH_3$ and Cl, $R^2 = H$, respectively). Attempted indolization of **1433** ($R^1 = NO_2$, $R^2 = H$) was unsuccessful and the only product isolated after its acid treatment, even after heating with 4-methylbenzene sulphonic acid, a strong acid, was 2-amino-5-nitropyridine (Yakhontov,

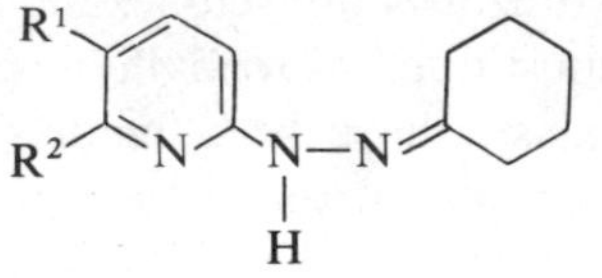

1433

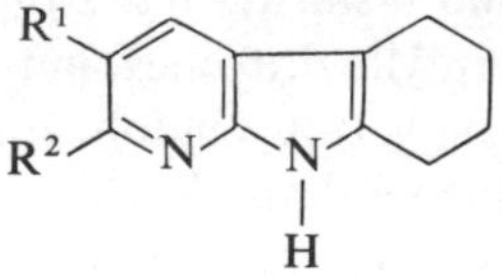

1434

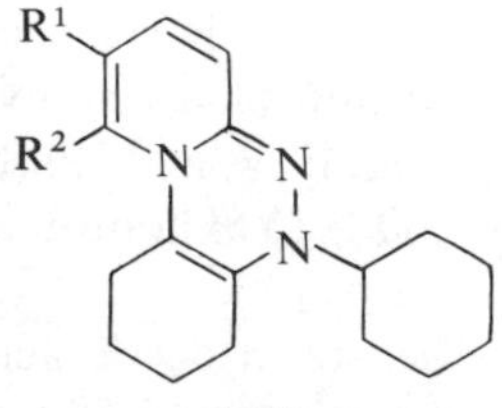

1435

Pronina, E. V. and Rubtsov, 1970). The postulated (Yakhontov, Pronina, E. V. and Rubtsov, 1966) mechanism of formation of **1435** from **1433** involved the addition of cyclohexanone, because of its labile methylene group, to the C=N group of the hydrazone **1433** to afford **1436** which then reacted further to give **1435**. Presumably, the reference to a labile methylene group implied the formation from cyclohexanone of the corresponding enol which then acted as a nucleophile. However, since the attack by a nucleophile at the nitrogen atom of the C=N group of **1433** is not feasible, intermediate **1436** could not be formed and the various structures **1435** for the above products therefore apparently require revision. That cyclohexanone was involved in the formation of these products was supported by the observation (Yakhontov, Pronina, E. V. and Rubtsov, 1966) that their yields were increased when cyclohexanone was added to the reaction mixture. It is possible that these products have structures which are related to **1437** which was assigned to the product formed when

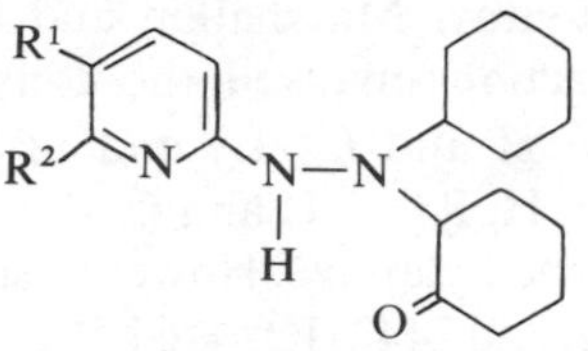

1436

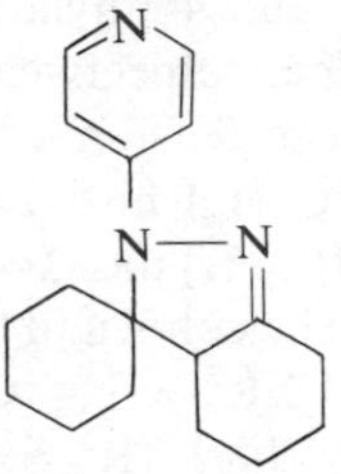

1437

cyclohexanone 4-pyridylhydrazone was treated with concentrated hydrochloric acid or sulphosalicylic acid and to **1446** (R^1 = H, R^2 = C_2H_5 and R^1 = Cl, R^2 = H) assigned to products formed from ethyl methyl ketone 4-pyridylhydrazone and acetaldehyde 2,6-dichloro-4-pyridylhydrazone, respectively, as discussed below. Such structures would at least be mechanistically feasible and such reformulations are certainly not at variance with the i.r., p.m.r. and mass spectral data quoted for these products (Yakhontov, Pronina, E. V., Rozynov and Rubtsov, 1968; Yakhontov, Pronina, E. V. and Rubtsov, 1966) nor with the chemical properties reported (Yakhontov, Pronina, E. V. and Rubtsov, 1966) for the product obtained from **1433** (R^1 = R^2 = H). Clearly, a ^{13}C n.m.r. spectroscopic analysis would solve this structural problem.

In an investigation of the dependence of its indolization upon the nature of the catalyst used, cyclohexanone 4-pyridylhydrazone was heated with zinc chloride, concentrated hydrochloric acid, sulphosalicylic acid, 4-methylbenzene sulphonic acid, copper(I) chloride, polyphosphoric acid and sodium ethoxide and was heated in the absence of a catalyst in refluxing diethylene glycol. In all cases 6-aza-1,2,3,4-tetrahydrocarbazole was obtained, the first catalyst leading to the highest yield of this product. The use of the second and third catalysts also led to the formation of compounds **1437** and **1438**, the latter product also being formed when using the fourth catalyst. From all reactions except that using zinc chloride as catalyst, 4-aminopyridine (**1439**; R = H) was also isolated (Yakhontov, Marshalkin and Anisimova, 1972 – see also Yakhontov and Marshalkin, 1972). From the indolization using sodium ethoxide as a catalyst, 4-ethylaminopyridine (**1439**; R = C_2H_5) was also isolated (Marshalkin and Yakhontov, 1972b; Yakhontov and Marshalkin, 1971, 1973; Yakhontov, Marshalkin and Anisimova, 1972). The isolation of this last product was developed into a general synthesis of 4-alkylaminopyridines when it was found that benzaldehyde and 4-isopropylbenzaldehyde 4-pyridylhydrazone, in which Fischer indolization was excluded, afforded high yields of 4-methyl-, 4-ethyl-, 4-isopropyl-, 4-butyl- and 4-heptylaminopyridine when heated with sodium methoxide, ethoxide, isopropoxide, butoxide and heptoxide, respectively, along with traces of 4-aminopyridine (Marshalkin and Yakhontov, 1972b; Yakhontov and Marshalkin, 1971, 1973). From the similar reactions of sodium ethoxide with benzaldehyde 2-, 3- and 4-pyridylhydrazone there were isolated 2-, 3- and 4-ethylaminopyridine, respectively, along with 2-, 3- and 4-aminopyridine, respectively (Yakhontov, Marshalkin and Pronina, E. V., 1972). Further extension of this reaction involved the conversion of **1440** (R = N=CHC_6H_5) into **1440** (R = H and C_2H_5) and of **1441** (R^1 = H, R^2 = N=CHC_6H_5) into **1441** (R^1 = H, R^2 = H and C_2H_5), dibenzylamine, dibenzyl, benzyl alcohol and benzylidenethylamine. However, similar treatment of **1441** (R^1 = Cl, R^2 = N=CHC_6H_5) and **1442** (R^1 = Cl, R^2 = N=CHC_6H_5) furnished only **1441** (R^1 = Cl, R^2 = H) and **1442** (R^1 = Cl, R^2 = H and R^1 = OC_2H_5, R^2 = N=CHC_6H_5), respectively, the last product being stable to treatment with sodium ethoxide at temperatures of up to 290 °C (Marshalkin, Azimov, Linberg and Yakhontov, 1978). The formations of **1440** (R = C_2H_5) and **1441** (R^1 = H, R^2 = C_2H_5) from the corresponding benzaldehyde hydrazones were facilitated and proceeded in higher yields when varying the ethoxide, in the order sodium ethoxide $<$ potassium ethoxide $<$ rubidium ethoxide (Marshalkin, Azimov, Gasparyants and Yakhontov, 1979).

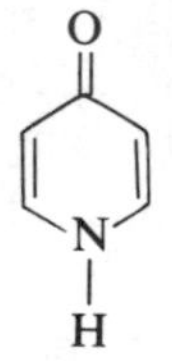

1438 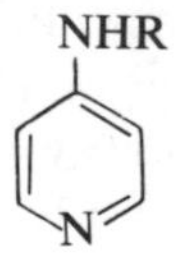 **1439** 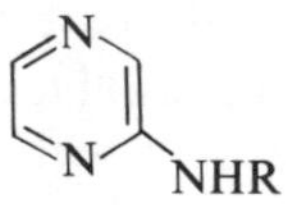 **1440**

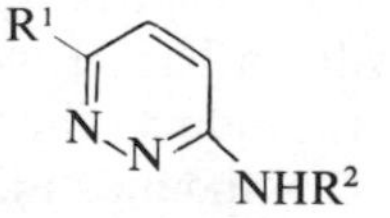

1441 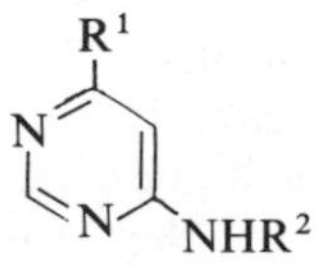 **1442**

The above formations of the alkylaminopyridines, alkylaminopyrazines and alkylaminopyridazines have been envisaged to occur, as exemplified by the formation of 4-alkylaminopyridines, as shown in **1443** to afford **1439** and **1444** (Yakhontov and Marshalkin, 1971, 1973; Yakhontov, Marshalkin and Anisimova, 1972). The concept of a four membered cyclic transition state in the formation of such compounds was subsequently further alluded to (Marshalkin, Azimov, Gasparyants and Yakhontov, 1979; Marshalkin, Azimov, Linberg and Yakhontov, 1978). However, it is difficult to envisage this cleavage of the C—O bond in the alkoxide anion which therefore invalidates this mechanistic proposal. It is very likely that, in the above reactions, the alkoxide anions are oxidized by the hydrazone moiety to the corresponding aldehyde or ketone which then condenses with the aminopyridines formed synchronously (the formation of the corresponding aromatic primary amines from arylhydrazones by pyrolysis,

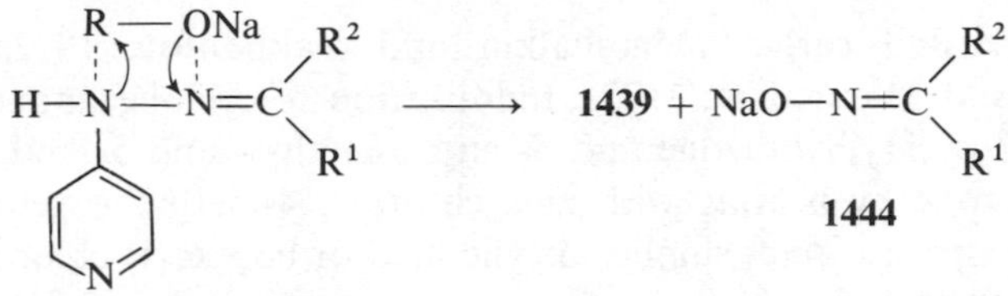

R—ONa, H—N—N=C(R^2)(R^1) ⟶ **1439** + NaO—N=C(R^2)(R^1)

1443 **1444**

often under indolization conditions, is well established – see Chapter VII, J) to afford the corresponding Schiff's bases (e.g. **1445** from the reaction of sodium ethoxide with cyclohexanone 4-pyridylhydrazone) which are then reduced to the alkylaminopyridines by the hydrazinopyridine formed during the initial oxidation.

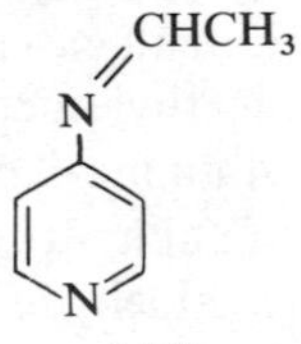

N=$CHCH_3$

1445

Further studies of the dependence of indolization of pyridylhydrazones upon the nature of the indolization catalyst have been undertaken. Propionaldehyde and ethyl methyl ketone 4-pyridylhydrazone have been indolized by heating with zinc chloride, copper(I) chloride, concentrated hydrochloric acid, polyphosphoric acid, 4-methylbenzene sulphonic acid, sulphosalicylic acid or sodium ethoxide or by refluxing in diethylene glycol. By far the highest yields of 5-aza-3-methylindole and 5-aza-2,3-dimethylindole, respectively, were obtained using zinc chloride and refluxing diethylene glycol, respectively. 4-Aminopyridine was also formed in all cases, 4-ethylaminopyridine was also formed when using sodium ethoxide and the indolization of ethyl methyl ketone 4-pyridylhydrazone with polyphosphoric acid, 4-methylbenzene sulphonic acid and sulphosalicylic acid also afforded a product formulated as **1446** (R^1 = H, R^2 = C_2H_5) whose formation was analogous to the formation

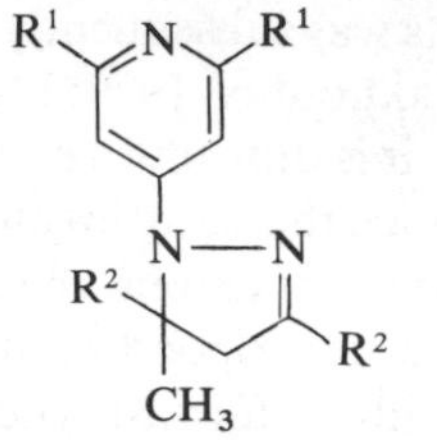

1446

of **1437** discussed earlier (Marshalkin and Yakhontov, 1972a – see also Yakhontov and Marshalkin, 1972). Indolization of cyclohexanone 4-pyridylhydrazone, 2-pyridylhydrazone and 5- and 6-methyl- and 5- and 6-chloro-2-pyridylhydrazone by heating with zinc chloride, 4-methylbenzene sulphonic acid, polyphosphoric acid, sulphosalicylic acid or copper(I) chloride afforded in all cases good yields of the corresponding 6- or 8-aza-1,2,3,4-tetrahydrocarbazoles although the claimed 100% yields from some of these reactions (Yakhontov, Marshalkin and Anisimova, 1972) are again difficult to understand. Catalytic conditions similar to those last mentioned, together with heating with concentrated hydrochloric acid and refluxing in diethylene glycol, have also been used to indolize propionaldehyde, ethyl methyl ketone and cyclohexanone 2,6-dichloro-4-pyridylhydrazone. Again, by far the highest yields of azaindoles, higher than from the corresponding 4-pyridylhydrazones, were obtained when using the zinc chloride catalyst and in all cases, except those involving the zinc chloride catalysed indolization of ethyl methyl ketone and cyclohexanone 2,6-dichloro-4-pyridylhydrazone, 4-amino-2,6-dichloropyridine was also isolated from the reaction mixtures (Yakhontov and Marshalkin, 1972). In an extension of these investigations, acetaldehyde 2,6-dichloro-4-pyridylhydrazone was heated with zinc chloride. As might have been expected (see Chapter IV, A), no 5-aza-4,6-dichloroindole, which would have been formed by indolization, could be isolated from this reaction which did, however, produce 4-amino-2,6-dichloropyridine, **1446** (R^1 = Cl, R^2 = H) and, surprisingly, acetone 2,6-dichloro-4-pyridylhydrazone. This last product was supposedly formed by alkylation of the starting hydrazone by methyl radicals formed by its pyrolysis (Yakhontov, Marshalkin, Anisimova and Kostyuchenko, 1973).

The application of the non-catalytic thermal indolization technique has been extended to 3- and 4-quinolylhydrazones to furnish very good yields of the corresponding pyrroloquinolines. Thus, cyclopentanone, cyclohexanone, ethyl methyl ketone and benzyl phenyl ketone 4-quinolylhydrazone, upon refluxing in diethylene glycol, afforded **1447** [$R^1 = R^2 = H$; $R^3 + R^4 = (CH_2)_{3 \text{ and } 4}$ and $R^3 = R^4 = CH_3$ and C_6H_5, respectively]. Similar treatment of cyclohexanone and ethyl methyl ketone 2-methyl-4-quinolylhydrazone hydrochloride gave **1447** [$R^1 = CH_3$, $R^2 = H$; $R^3 + R^4 = (CH_2)_4$ and $R^3 = R^4 = CH_3$, respectively] (Parrick and Wilcox, 1976). Later (Khan, M. A. and Rocha, 1978), the pyrroloquinolines **1447** [$R^1 = CH_3$, $R^2 = H$; $R^3 + R^4 = (CH_2)_4$, $R^3 = CH_3$

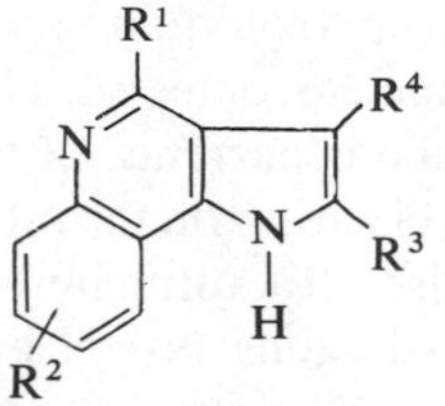

1447

and C_6H_5 and $R^4 = CH_3$; $R^1 = R^3 = H$, $R^2 = H$ and 6-, 7- and 8-Cl, 6-, 7- and 8-CH_3 and 6-, 7- and 8-OCH_3, $R^4 = CH_3$; $R^1 = H$, $R^2 = H$, 6-, 7- and 8-Cl, 6-, 7- and 8-CH_3 and 6-, 7- and 8-OCH_3, $R^3 = R^4 = CH_3$; $R^1 = H$, $R^2 =$ 8-Cl and 8-CH_3 and 6-OCH_3, $R^3 = C_2H_5$, $R^4 = CH_3$; $R^1 = R^3 = H$, $R^2 =$ 8-Cl, $R^4 = C_2H_5$; $R^1 = H$, $R^2 =$ 6- and 8-Cl, $R^3 + R^4 = (CH_2)_3$; $R^1 = H$, $R^2 = H$ and 6-, 7- and 8-Cl, 6-, 7- and 8-CH_3 and 6-, 7- and 8-OCH_3, $R^3 + R^4 = (CH_2)_4$ and $R^1 = H$, $R^2 = H$ and 6-, 7- and 8-Cl, 6-, 7- and 8-CH_3 and 6-, 7- and 8-OCH_3, $R^3 = CH_3$, $R^4 = C_6H_5$] were synthesized, in most cases in yields exceeding 70%, by refluxing solutions of the appropriate 4-quinolylhydrazones in diethylene glycol for between 10 min and 1 h depending upon the hydrazone being indolized. In four of the indolizations, refluxing Dowtherm was used in place of diethylene glycol. Likewise, refluxing solutions of cyclohexanone and ethyl methyl ketone 2-methyl-3-quinolylhydrazone in diethylene glycol furnished **593** [$R^1 = CH_3$, $R^2 + R^3 = (CH_2)_4$ and $R^2 = R^3 = CH_3$, respectively] (Parrick and Wilcox, 1976).

Initial studies (Potts and Schneller, 1968) reported that pyrolysis of 2-quinolylhydrazones yielded 2-aminoquinoline and unrecognizable tars. Further (Parrick and Wilcox, 1976) related studies of these hydrazones involved refluxing solutions of ethyl methyl ketone and benzyl phenyl ketone 2-quinolylhydrazone in diethylene glycol. From the former reaction were isolated three products, 2-aminoquinoline, **1448** and a compound $C_{18}H_{14}N_4$ which was likely to be an aminodiquinolylamine **1449** and from the latter reaction were isolated 2-aminoquinoline, 2,3,4,5-tetraphenylpyrrole and **1450**. In neither case

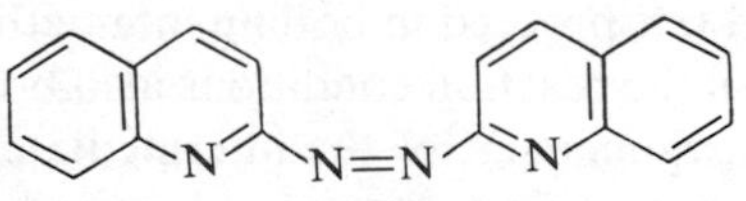

1448

$H_2N—C_9H_5N—NH—C_9H_5N$

1449

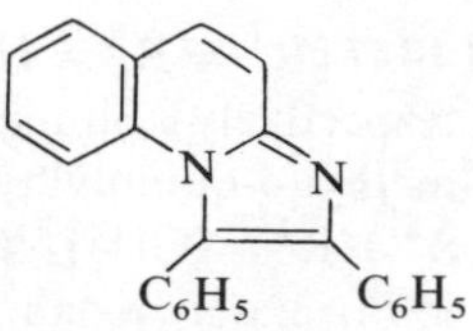

1450

was any of the corresponding pyrroloquinoline, resulting from indolization involving C-3 of the quinolyl nucleus, detected. The formation of these products was rationalized in terms of the occurrence of two competing processes, one involving an intramolecular cyclization on the nitrogen atom with a subsequent Fischer indolization mechanism (to ultimately afford **1450**) and the other involving homolysis of the hydrazone N—N bond (to ultimately afford the other reaction products).

Extensions of the non-catalytic thermal indolization technique to diazaphenylhydrazones have met with only limited success. Thus, cyclohexanone and ethyl methyl ketone 4-pyrimidylhydrazone [**1451**; $R^1 = R^2 = R^5 = H$, $R^3 + R^4 = (CH_2)_4$ and $R^3 = R^4 = CH_3$, respectively] (Crooks and Robinson, B., 1967, 1969) and cyclohexanone 4-(2-methyl- and -2,6-dimethylpyrimidyl)-hydrazone [**1451**; $R^1 = R^5 = H$, $R^2 = CH_3$, and $R^1 = R^2 = CH_3$, $R^5 = H$; $R^3 + R^4 = (CH_2)_4$, respectively] (Duffy and Wibberley, 1974) afforded

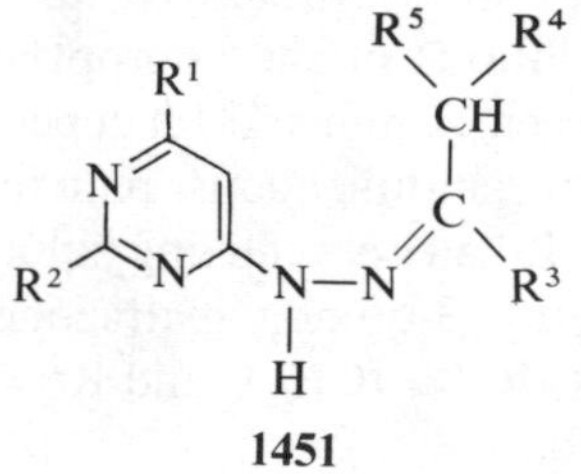

1451

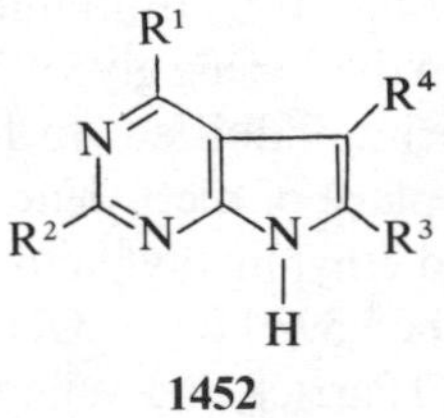

1452

1452 [$R^1 = R^2 = H$, $R^3 + R^4 = (CH_2)_4$ and $R^3 = R^4 = CH_3$; $R^1 = H$, $R^2 = CH_3$ and $R^1 = R^2 = CH_3$; $R^3 + R^4 = (CH_2)_4$, respectively] in 32%, 10.5%, 98% and 45% yields, respectively, after refluxing in diethylene glycol. However, the related pyrimidylhydrazones **1451** [$R^1 = R^2 = R^3 = CH_3$, $R^4 = C_6H_5$, $R^5 = H$; $R^1 = R^2 = CH_3$, $R^3 = C_6H_5$, $R^4 = R^5 = H$; $R^1 = R^2 = R^3 = R^4 = CH_3$, $R^5 = H$; $R^1 = R^2 = C_6H_5$, $R^3 + R^4 = (CH_2)_4$, $R^5 = H$ and $R^1 = R^2 = R^3 = C_6H_5$, $R^3 = R^5 = H$] could not be indolized under similar conditions (Duffy and Wibberley, 1974). Indolization of isobutyraldehyde 4-pyrimidylhydrazone did not proceed in boiling monoethylene glycol whereas in boiling diethylene glycol the reaction conditions also brought about a thermally induced Plancher rearrangement of the intermediate **1431** (X = Y = N, $R = CH_3$) which was presumably formed to afford **1452**

($R^1 = R^2 = H$, $R^3 = R^4 = CH_3$) as the only isolated product (Crooks and Robinson, B., 1967, 1969).

Application of the non-catalytic thermal indolization technique to pyrazinylhydrazones **1453** has met with considerable success and the 4,7-diazaindoles **1454** [$R^1 = H$, $R^2 = CH_3$; $R^1 = H$, $R^2 = C_6H_5$; $R^1 = R^2 = CH_3$; $R^1 = C_6H_5$, $R^2 = CH_3$; $R^1 = CH_3$, $R^2 = C_6H_5$; $R^1 = R^2 = C_6H_5$ and $R^1 + R^2 = (CH_2)_4$] and **1455** ($n = 1$ and 2) have been synthesized by boiling solutions of

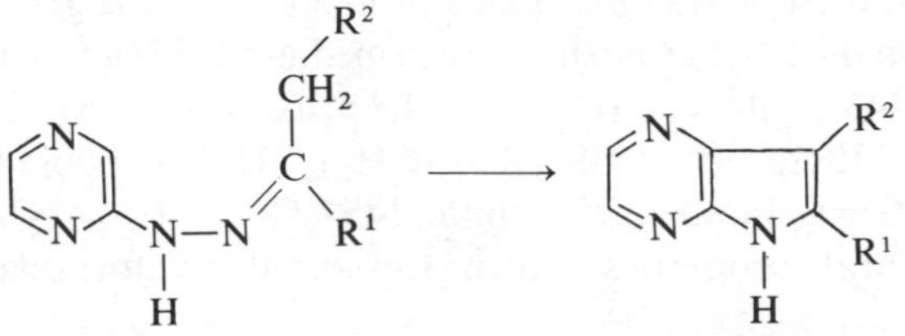

1453 **1454**

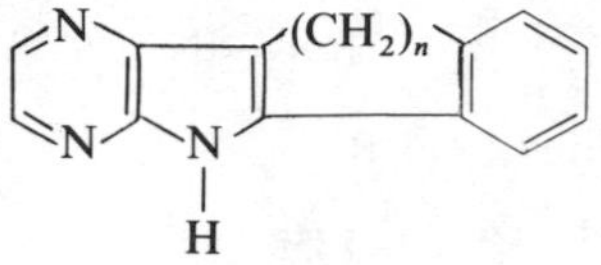

1455

the corresponding pyrazinylhydrazones in diethylene glycol under reflux. The product yields of 38% to 88% were higher than those obtained from the indolization of the corresponding pyrimidylhydrazones (Clark, B. A. J., Dorgan, R. J. and Parrick, 1975; Clark, B. A. J., Parrick and Dorgan, R. J. J., 1976).

Except in the formation of **1450**, in all the above reactions where indolization could theoretically involve either the carbon or nitrogen atom adjacent to the carbon atom carrying the hydrazino function, only products resulting from the former route have been obtained, although compounds resulting from the alternative direction of indolization might also have been expected (Kelly, A. H. and Parrick, 1970). However, in the case of the attempted thermal indolization of a series of 3-(6-methylpyridazinyl)hydrazones **1456** ($R^1 = CH_3$; $R^2 = C_2H_5$, $R^3 = CH_3$; $R^2 = CH_2C_6H_5$, $R^3 = C_6H_5$; $R^2 = R^3 = C_6H_5$; $R^2 = C_6H_5$, $R^3 = H$; $R^2 = C_6H_5$, $R^3 = CH_3$ and $R^2 = H$, $R^3 = C_6H_5$), the

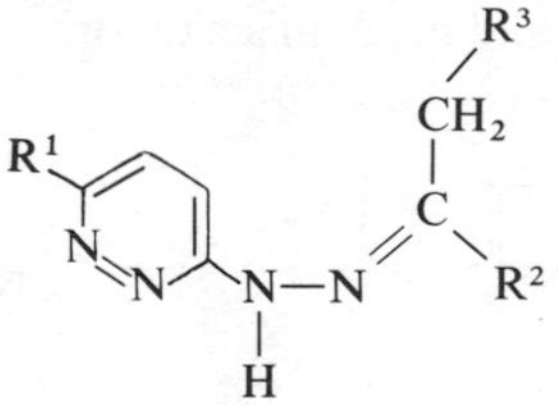

1456

reaction appeared to involve exclusively the adjacent nitrogen atom since the only products isolated were the corresponding 6-methyl-*s*-triazolo[4,3-*b*]-pyridazines **1458** (R = C_2H_5, $CH_2C_6H_5$, C_6H_5, C_6H_5, C_6H_5 and H, respectively), presumably formed by elimination of either R^2 or CH_2R^3 (benzyl more readily than phenyl or hydrogen and ethyl or methyl more readily than phenyl) from the intermediate **1457** (How and Parrick, 1976). Starting with **1456** ($R^1 = R^3 = CH_3$, $R^2 = H$), compounds **1458** (R = H and C_2H_5) were obtained, although since the latter was the major product the hydrogen atom was lost more readily than the ethyl group in the intermediate **1457** (R^2 = H, $R^3 = CH_3$). Starting with **1456** [$R^1 = CH_3$, $R^2 + R^3 = (CH_2)_4$], two products were isolated. One of these was **1458** [R = $(CH_2)_4CH_3$], supposedly formed by rupture of the C—C bond in the spiran **1457** [$R^2 + R^3 = (CH_2)_4$] and the other had spectral properties which supported its formulation as **1459**

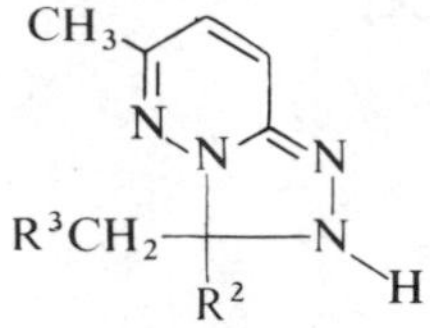

1457

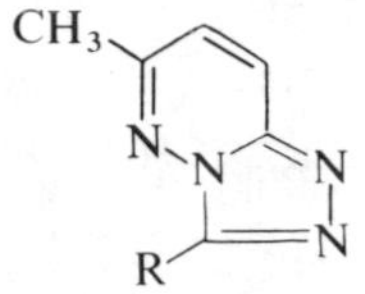

1458

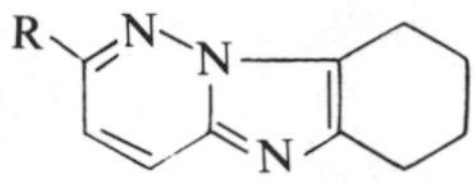

1459

(R = CH_3). This could be formed by operation of the Fischer indolization mechanism but the initial [3,3]-sigmatropic shift would again have involved attack at the nitrogen rather than carbon atom adjacent to the initial hydrazino moiety (How and Parrick, 1976). Similarly, when **1456** [R^1 = Cl, $R^2 + R^3 = (CH_2)_4$] was heated with zinc chloride, a product $C_{10}H_{10}ClN_3$ was formed which was initially (obviously incorrectly) formulated as **1460** and correspondingly named as '2-Chloro-5,6,7,8-tetrahydro-1,9a-diazacarbazole' (Yakhontov, Uritskaya and Glushkov, 1976). Subsequently (Kitaev and Troepol'skaya, 1978) this product was reformulated as **1461** which would represent a molecular formula of $C_{10}H_{12}ClN_3$ and is therefore also probably erroneous. Neither the formation of **1460** nor **1461** from **1456** [R^1 = Cl, $R^2 + R^3 = (CH_2)_4$] was mechanistically rationalized. However, the formulation of this reaction product as **1459** (R = Cl) would be in accordance with its molecular formula of $C_{10}H_{10}ClN_3$ and would be analogous to the product structure **1459** (R = CH_3) and mechanism of formation assigned to this product

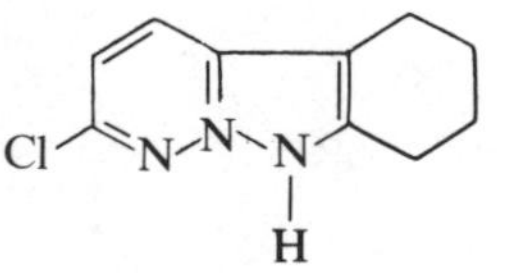

1460

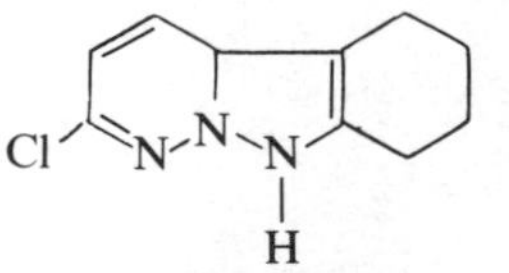

1461

obtained from the pyrolysis of **1456** [$R^1 = CH_3$, $R^2 + R^3 = (CH_2)_4$] (How and Parrick, 1976) as described above. When **1456** [$R^1 = H$, $R^2 + R^3 = (CH_2)_4$] was heated with zinc chloride, complete resinification occurred (Yakhontov, Uritskaya and Glushkov, 1976). The above formation of the compounds of general structure **1458** is mechanistically similar to the pyrolytic conversion of the hydrazones **1462** (R^1 = alkyl or aryl) into **1463** reported previously (Shiho and Tagami, 1960) although later workers (Potts and Schneller, 1968) could not repeat these observations.

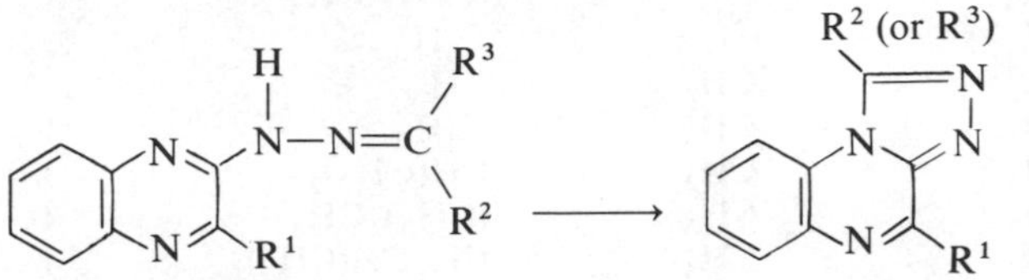

1462 → **1463**

The syntheses of the azaindoles **587–590** by the indolization of the corresponding **586** and the indolization of **256** into the corresponding **592** have already been discussed (see Chapter III, C). Analogously to these syntheses of **587–590** and **592**, the hydrazones **1465**, prepared by reaction of **1464** with the appropriate carbonyl moiety, have been indolized to the corresponding **1466**

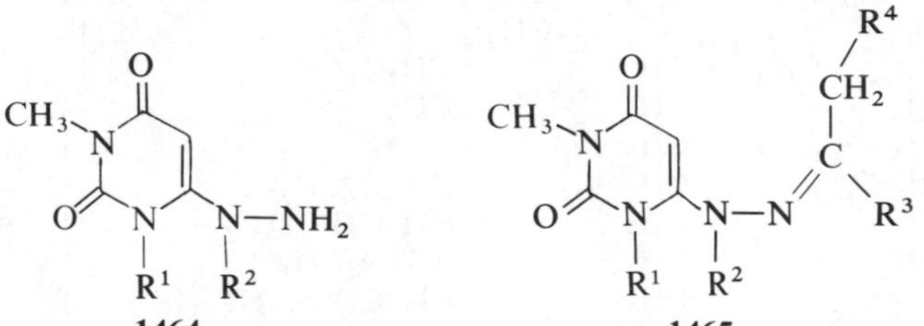

1464 **1465**

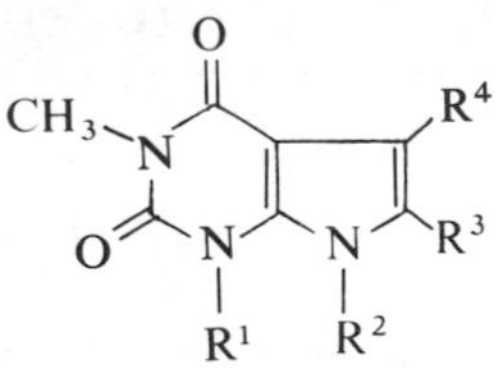

1466

(Table 30), using in some cases acid catalysis (Duffy and Wibberley, 1974) but more often the non-catalytic thermal technique using either the hydrazone **1465** or a mixture of its hydrazino and carbonyl moieties in refluxing ethylene glycol or tetralin (Senda and Hirota, 1972, 1974) or the hydrazone in refluxing

Table 30. The Products **1466** which have been formed from the Indolizations of Compounds **1465**

Product **1466** (Unless otherwise stated, $R^2 = R^3 = H$)

R^1	R^2	R^3	R^4	Reference(s)
CH_3		CH_3		1, 2
CH_3			CH_3	1, 2
CH_3			C_2H_5	1
CH_3			$(CH_2)_2CH_3$	1
CH_3			C_6H_5	1, 2
CH_3		CH_3	CH_3	1–3
CH_3		CH_3	C_2H_5	1
CH_3		CH_3	$CH(CH_3)_2$	1
CH_3		CH_3	$(CH_2)_2CH_3$	1
CH_3		CH_3	$CH_2CH(CH_3)_2$	1
CH_3		CH_3	$(CH_2)_4CH_3$	1
CH_3		CH_3	CH_2COOCH_3	1
CH_3		CH_3	C_6H_5	3
CH_3		C_2H_5	CH_3	1
CH_3		$(CH_2)_2CH_3$	C_2H_5	1
CH_3		C_6H_5	CH_3	1, 2
CH_3		C_6H_5	C_2H_5	1
CH_3		C_6H_5	$(CH_2)_3CH_3$	1
CH_3		C_6H_5	C_6H_5	1
CH_3		$(CH_2)_3$		1
CH_3		$(CH_2)_4$		1–3
CH_3	CH_3	CH_3	CH_3	1, 2
C_6H_5		CH_3	CH_3	1
C_6H_5		CH_3	C_2H_5	1
C_6H_5		CH_3	$(CH_2)_2CH_3$	1
C_6H_5		CH_3	$CH_2CH(CH_3)_2$	1
C_6H_5		CH_3	$(CH_2)_4CH_3$	1
C_6H_5		CH_3	$CH_2COOC_2H_5$	1
C_6H_5		$(CH_2)_3$		1
C_6H_5		$(CH_2)_4$		1
C_6H_5	CH_3	CH_3	CH_3	1
C_6H_5	CH_3	CH_3	$(CH_2)_3CH_3$	1

References

1. Senda and Hirota, 1974; 2. Senda and Hirota, 1972; 3. Duffy and Wibberley, 1974.

di- or triethylene glycol under an atmosphere of nitrogen (Duffy and Wibberley, 1974). Likewise, the hydrazones **1312** (R = CH_3 and $CH_2C_6H_5$, n = 1 and 2) were indolized into **1313** (R = CH_3 and $CH_2C_6H_5$, n = 1 and 2, respectively) using both polyphosphoric acid as catalyst and by refluxing in diethylene glycol solution, the latter technique affording the higher product yields (Glushkov, Zasosova and Ovcharova, 1977).

As with the previously described (see Chapter III, C) hydrazones **586** and **256**, the mechanistic role of the 1,2,3,4-tetrahydropyrimidin-2,4-dione moiety was reversed when compounds **254** were indolized to give **378** by either refluxing in tetralin (Senda, Hirota and Takahashi, M., 1975) or by heating with formic acid or 1 M hydrochloric acid (Wright, 1976). Starting with **254** (R^1–R^3 = H, $R^4 = R^5 = CH_3$, X = O) (Senda, Hirota and Takahashi, M., 1975; Wright, 1976) and **254** (R^1–R^5 = H; $R^1 = CH_3$, R^2–R^5 = H; R^1–R^4 = H, $R^5 = CH_3$; R^1–$R^3 = R^5$ = H, $R^4 = CH_3$; $R^1 = R^2 = R^4 = R^5$ = H, $R^3 = CH_3$ and R^1–R^3 = H, $R^4 = R^5 = CH_3$; X = O) (Wright, 1976) the corresponding indoles **378** were synthesized and likewise **254** ($R^1 = R^2 = CH_3$, R^3–R^5 = H, X = O) formed an approximately equimolar mixture of **378** ($R^1 = R^2 = CH_3$, R^3–R^5 = H, X = O) and its isomer **569**, the structural assignments being based upon p.m.r. spectroscopic data (Wright, 1976). Yields of the above products from the acid catalysed indolizations were low (Wright, 1976). In the case of the reactions of **254** (R^1–R^5 = H and $R^1 = CH_3$, R^2–R^5 = H; X = O) catalysed by 1 M hydrochloric acid, this was found to be caused by the concomitant hydrolysis of the starting materials to barbituric acid and phenylhydrazine and 4-methylphenylhydrazine, respectively (Wright, 1976). When these indolizations were effected using refluxing 98% formic acid as catalyst, then in each case, along with the expected indoles, a second product, **1470** (R^1 = H and CH_3, respectively, R^2 = H, $R^3 = NH_2$) was isolated in 48% (Wright, 1976) and 20% (Wright, 1977) yields, respectively. Likewise, **254** (R^1 = Br, R^2–R^5 = H, X = O) afforded **1470** (R^1 = Br, R^2 = H, $R^3 = NH_2$) in 69% yield (Wright, 1977). It was proposed (Wright, 1976) that these products arose as shown in Scheme 46, via sequential N-formylation of **254** (when R^3 = H) to give **1467**, cyclization by intramolecular nucleophilic attack (when R^4 = H) to give **1468**, elimination of water (when R^5 = H) to give **1469** and hydration, decarboxylation and isomerism to give **1470** (R^1 = H, CH_3 and Br, respectively, R^2 = H, $R^3 = NH_2$). Support for this scheme was derived from the observations that no triazole was isolated when **254** ($R^1 = R^2 = R^4 = R^5$ = H, $R^3 = CH_3$, X = O) was treated with refluxing 98% formic acid (blockage of the N-formylation step) [an 81% yield of **378** ($R^1 = R^2 = R^4 = R^5$ = H, $R^3 = CH_3$, X = O) was produced] (Wright, 1976) or when **254** (R^1–R^4 = H, $R^5 = CH_3$, X = O) was similarly treated (prevention of dehydration step) (Wright, 1977). Unfortunately, attempts to prepare the corresponding 5-methyltriazole by treating **254** (R^1–R^5 = H, X = O) with refluxing acetic acid gave only starting material after 1 h and a 42% yield of **378** (R^1–R^5 = H, X = O) after 21 h. However, the isolation of acetanilide from this reaction supported the occurrence of the initial N-acetylation, this product being formed by subsequent N—N bond cleavage (Wright, 1976).

Whereas **1471** was converted into **1472** by refluxing for 15 min in 1 M hydrochloric acid [refluxing for 1 h led to hydrolysis of the methylthio group to afford **378** ($R^1 = CH_3$, R^2–R^5 = H, X = O)], attempted indolization of **254**

254 (R^3 = H) ⟶ **1467**

(R^4 = H)

1468 ⟶ (R^5 = H) **1469**

1470 ($R^6 = NH_2$)

Scheme 46

1471 ⟶ **1472**

(R^1 = CH_3, R^2–R^5 = H, X = S) using both formic acid or hydrochloric acid failed to yield indolic products, the reaction products at the time (Wright, 1976) remaining unidentified. Subsequent (Wright and Gambino, 1979) further investigation of these reactions showed that when treated with refluxing 1 M hydrochloric acid, **254** (R^1 = CH_3, R^2–R^5 = H, X = S) gave 4-methylphenylhydrazine hydrochloride, by hydrolysis of the starting hydrazone, and **288** (R = H, X = S), which represented an intermediate in the indolization mechanistic pathway (see Chapter II, E2). When this latter product was refluxed with N,N-dimethylaniline, the expected indolic product, **289** (R^1 = CH_3, R^2–R^5 = H, X = S) was formed. In this later study, refluxing with 98% formic acid did, in fact, give the expected indolic product **289** (R^1 = CH_3, R^2–R^5 = H, X = S) in 17% yield, along with **288** (R = CHO, X = S) (see also Chapter II, E2) as the major product (31% yield) which was hydrolysed by sodium hydroxide into **288** (R = H, X = S). A 2% yield of **1470** (R^1 = CH_3, R^2 = H, R^3 = OH) was also isolated from this latter 98% formic acid catalysed reaction.

A review including the application of the Fischer indolization to the synthesis of azaindoles appeared in 1968 (Yakhontov, 1968) although this was restricted by the somewhat limited investigations which had until then been undertaken on this subject. The synthesis of pyrrolopyrimidines has also been reviewed (Amarnath and Madhav, 1974).

J. Indolization of 1,2-Diketone Mono- and Diarylhydrazones

1. Acyclic 1,2-Diketone Arylhydrazones

Compounds of this type, usually prepared by the Japp–Klingemann reaction, indolized as expected to furnish the corresponding 2-acyl- and 2-aroylindoles. Thus, when **1473** (R^1 = R^2 = H, R^3 = H, CH_3, OCH_3 and Cl, R^4 = CH_3; R^1 = R^4 = CH_3, R^2 = H, R^3 = H, CH_3, OCH_3 and Cl; R^1 = OCH_3, R^2 = H, R^3 = H, CH_3 and OCH_3, R^4 = CH_3; R^1 = Cl, R^2 = H, R^3 = H, OCH_3 and Cl, R^4 = CH_3 and R^1 = H and CH_3, R^2 = R^3 = OCH_3, R^4 = CH_3) were refluxed in formic acid solution, the corresponding compounds **1474** were produced (Shvedov, Alekseev, V. V. and Grinev, A. N., 1968a, 1968b).

R^4 CH_2 R^1 N–N=C C=O H R^3 R^2 ⟶ R^1 R^4 N H C=O R^3 R^2

1473 **1474**

Using hydrochloric acid as the catalyst, **1475** [R^1–R^3 = H, R^4 = CH_3, R^5 = C_6H_5 and $(CH_2)_2$-phthalimido] afforded **1476** [R^1–R^3 = H, R^4 = CH_3, R^5 = C_6H_5 and $(CH_2)_2$-phthalimido, respectively] (Manske, R. H. F., Perkin

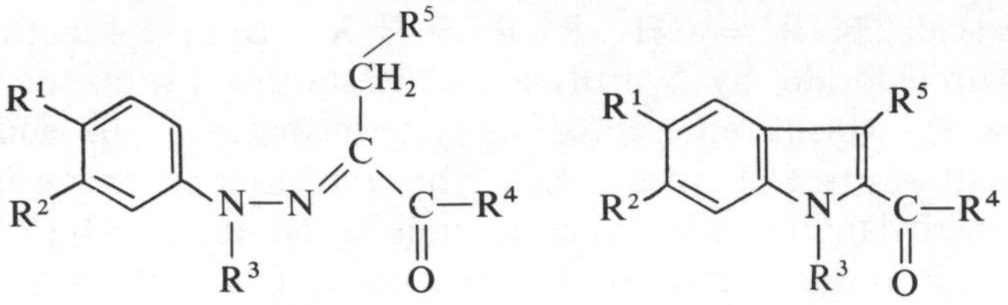

1475 **1476**

and Robinson, R., 1927), **1475** [R^1 = R^3 = H, R^2 = OCH_3, R^4 = CH_3, R^5 = $(CH_2)_2$-phthalimido] afforded **1476** [R^1 = R^3 = H, R^2 = OCH_3, R^4 = CH_3, R^5 = $(CH_2)_2$-phthalimido] (Manske, R. H. F., Perkin and Robinson, R., 1927), **1475** (R^1 = R^2 = OCH_3, R^3 = H, R^4 = CH_3, R^5 = C_6H_5) gave **1476** (R^1 = R^2 = OCH_3, R^3 = H, R^4 = CH_3, R^5 = C_6H_5) (Lions and Spruson, 1932), **1475** (R^1 = Cl, R^2 = R^3 = H, R^4 = R^5 = CH_3) gave **1476** (R^1 = Cl, R^2 = R^3 = H, R^4 = R^5 = CH_3) (Walser, Blount and Fryer, 1973) and **1475** (R^1 = Cl, R^2 = R^3 = H, R^4 = CH_3, R^5 = C_6H_5) gave **1476** (R^1 = Cl, R^2 = R^3 = H, R^4 = CH_3, R^5 = C_6H_5) (Inaba, Ishizumi, Okamoto and Yamamoto, H., 1975; Walser, Blount and Fryer, 1973). The hydrazone **1475** [R^1–R^3 = H, R^4 = CH_3, R^5 = $(CH_2)_2OC_6H_5$] has been synthesized by the Japp–Klingemann reaction but contemporary attempts to effect its indolization were not made (Manske, R. H. F., 1931a). Several arylhydrazones **1475** (R^4 = CH_3, R^5 = CH_2COOH) have been prepared by the Japp–Klingemann reaction and although their indolization by heating with 50% sulphuric acid appears to have been successfully effected, as evidenced by the separation of ammonium sulphate from the reaction mixture, no other characteristic products could be isolated (Feofilaktov and Semenova, N. K., 1953a).

In the syntheses of the above 2-acyl- and 2-aroylindoles, the indolic C-3 atom was quaternary. However, when this carbon atom was tertiary, migration of the 2-acyl or 2-aroyl group to the indolic 3-position can occur under 'forcing' indolization conditions although such migration has not been observed with the use of milder indolization conditions. Thus, treatment of butan-2,3-dione N_α-methylphenylhydrazone (**1475**; R^1 = R^2 = R^5 = H, R^3 = R^4 = CH_3) with dilute hydrochloric acid afforded **1476** (R^1 = R^2 = R^5 = H, R^3 = R^4 = CH_3) (Diels and Köllisch, 1911 – see also Chastrette, 1970), of butan-2,3-dione N_α-methyl-4-nitrophenylhydrazone (**1475**; R^1 = NO_2, R^2 = R^5 = H, R^3 = R^4 = CH_3) with concentrated sulphuric acid yielded **1476** (R^1 = NO_2, R^2 = R^5 = H, R^3 = R^4 = CH_3) (Diels and Durst, 1914), of butan-2,3-dione N_α-phenylphenylhydrazone (**1475**; R^1 = R^2 = R^5 = H, R^3 = C_6H_5, R^4 = CH_3) with dilute hydrochloric acid yielded **1476** (R^1 = R^2 = R^5 = H, R^3 = C_6H_5, R^4 = CH_3) (Chastrette, 1970), of **1475** (R^1 = OCH_3, R^2 = R^4 =

CH_3, $R^3 = R^5 = H$) with methanolic concentrated hydrochloric acid gave **1476** ($R^1 = OCH_3, R^2 = R^4 = CH_3, R^3 = R^5 = H$) (Remers, Roth and Weiss, M. J., 1964) and of **1475** [$R^1 = OCH_3$, $R^2 = R^5 = H$, $R^3 = CH_2C_6H_4Cl(4)$, $R^4 = CH_3$], as a mixture of its component ketonic moiety and the hydrochloride of its hydrazino moiety in refluxing 2-propanol, afforded **1476** [$R^1 = OCH_3$, $R^2 = R^5 = H$, $R^3 = CH_2C_6H_4Cl(4)$, $R^4 = CH_3$] (Walton, Jenkins, Nutt and Holly, 1968). However, butan-2,3-dione phenylhydrazone [for earlier syntheses of this hydrazone by the Japp–Klingemann reaction, see Abramovitch (1956) and Benary (1926)] was more difficult to indolize. No reaction was observed when it was treated with dilute hydrochloric acid (Chastrette, 1970; Diels and Köllisch, 1911), with ethanolic acetic acid or with hydrochloric acid in acetic acid and only extensive resinification was observed when using concentrated hydrochloric acid, zinc chloride or boron trifluoride in different solvents as catalysts (Chastrette, 1970). Using concentrated sulphuric acid in formic acid or in acetic acid as catalyst, a 1.7% yield of 2-acetylindole (**1476**; R^1–$R^3 = R^5 = H$, $R^4 = CH_3$) was produced (Chastrette, 1970).

Polyphosphoric acid is a catalyst particularly suitable for converting methyl alkyl ketones into the corresponding 2-substituted indoles (see Chapter III, G). In attempts to use heating with this catalyst to convert butan-2,3-dione phenylhydrazone (**1475**; R^1–$R^3 = R^5 = H$, $R^4 = CH_3$) into 2-acetylindole (**1476**; R^1–$R^3 = R^5 = H$, $R^4 = CH_3$), 3-acetylindole (**1477**; $R^1 = R^2 = H$) was the only product isolated (Chang, M. C., Hsing and Ho, T. S., 1966; Chastrette, 1970), similar catalytic conditions likewise achieving the converions of **1475** ($R^1 = R^2 = R^5 = H$, $R^3 = CH_3$ and C_6H_5, $R^4 = CH_3$ and $R^1 = COCH_3$, $R^2 = R^3 = R^5 = H$, $R^4 = CH_3$) into **1477** ($R^1 = H$, $R^2 = CH_3$ and C_6H_5) (Chastrette, 1970) and **1477** ($R^1 = COCH_3$, $R^2 = H$) (Avramenko, Mosina

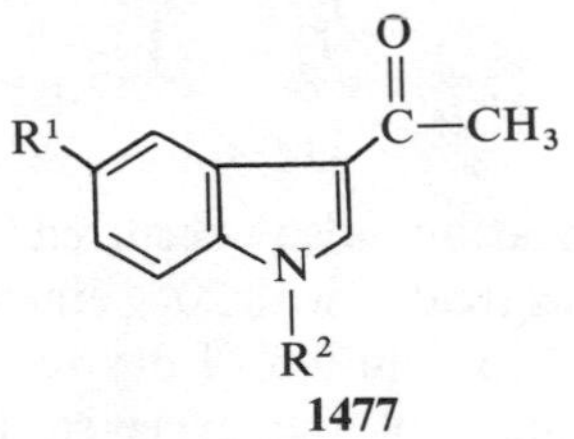

O
C—CH$_3$
R^1
N
R^2

1477

and Suvorov, 1970), respectively. Indolization of **1475** (R^1–$R^3 = R^5 = H$, $R^4 = CH_3$), apparently by heating with zinc chloride, appeared to afford 2-acetylindole (**1476**; R^1–$R^3 = R^5 = H$, $R^4 = CH_3$) [Aktieselskabet Dumex (Dumex Ltd), 1963] but verification for the orientation of the acetyl group was lacking. These formations of 3-acetylindoles result from acetyl group migration from the indolic $2 \rightarrow 3$ position under the indolization conditions (i.e. heating with polyphosphoric acid). Indeed, the isomerization of 2-acetylindoles into 3-acetylindoles has been observed when the former products were heated with polyphosphoric acid (Budylin, Kost, Matveeva and Minkin, 1972; Chang,

M. C., Hsing and Ho, T. S., 1966; Chastrette, 1970), with aluminium chloride (Chastrette, 1970) and with trifluoroacetic acid (Budylin, Kost, Matveeva and Minkin, 1972; Chastrette, 1970) and has been generalized by using other 2-acyl- or 2-aroyl- (e.g. formyl, acetoxyacetyl, benzoyl, 4-methylbenzoyl and 4-methoxybenzoyl) indoles which gave very high yields (although the example of a claimed 100% yield is difficult to understand) of the corresponding 3-acyl- or 3-aroylindoles when heated with polyphosphoric acid (Budylin, Kost, Matveeva and Minkin, 1972). Tentative mechanistic proposals have been made (Budylin, Kost, Matveeva and Minkin, 1972; Chastrette, 1970) to account for these acyl or aroyl group migrations.

In view of these observations, care must be taken when effecting the indolization of acyclic 1,2-dione arylhydrazones **1475** ($R^5 = H$) to use the mildest possible indolization conditions if the 2-acyl- or 2-aroylindolic product is required. As already mentioned, when the 3-position of the potential indolic product was quaternary, such precautions were unnecessary and thus treatment of **1475** ($R^1 = COCH_3, R^2 = R^3 = H, R^4 = R^5 = CH_3$) with polyphosphoric acid furnished 2,5-diacetyl-3-methylindole (**1476**; $R^1 = COCH_3, R^2 = R^3 = H, R^4 = R^5 = CH_3$) (Ishizumi, Shioiri and Yamada, S., 1967). Analogous migrations also failed to occur when compounds **137** (R = H and C_2H_5) were indolized into **1478** (R = H) (Jackson, A. and Joule, 1967; Jackson, A., Wilson, N. D. V., Gaskell and Joule, 1969) and **1478** ($R = C_2H_5$) (Jackson, A., Gaskell, Wilson, N. D. V. and Joule, 1968; Jackson, A., Wilson, N. D. V., Gaskell and Joule, 1969), respectively, by heating with syrupy phosphoric acid.

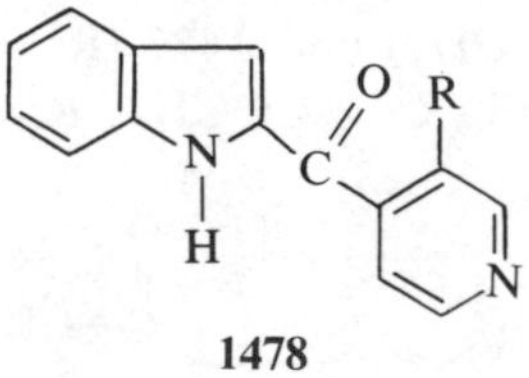

1478

An interesting isomerization also occurred when **1475** (R^1–R^3 = H, $R^4 = CH_3, R^5$ = alkyl) was treated with 20% ethanolic sulphuric acid. Whereas treatment with a saturated solution of dry hydrogen chloride in absolute ethanol afforded a low yield of the expected indoles **1476** (R^1–R^3 = H, $R^4 = CH_3$, R^5 = alkyl), together with large quantities of resinous materials, use of the former catalyst failed to effect indolization but led to the formation of the isomeric hydrazone **1479** (Yasuda, H., 1954).

When benzyl phenyl diketone reacted with phenylhydrazine in ethanolic solution in the presence of a trace of acetic acid, a monophenylhydrazone was formed which was formulated, on the basis of the C=O and C=N stretching frequencies in its i.r. spectrum, as **1475** (R^1–R^3 = H, $R^4 = R^5 = C_6H_5$) rather than that resulting from reaction involving the carbonyl group adjacent to the phenyl group. This hydrazone yielded **1476** (R^1–R^3 = H, $R^4 = R^5 = C_6H_5$)

when refluxed in acetic acid solution (Curtin and Poutsma, 1962). This same indole was produced when **1480** was stirred with zinc dust in glacial acetic acid at room temperature, a transformation which at some stage was suggested to involve a Fischer indolization although the intermediacy of **1475** (R^1–R^3 = H, $R^4 = R^5 = C_6H_5$) was eliminated since this hydrazone failed to indolize under similar acidic reductive conditions (Curtin and Poutsma, 1962).

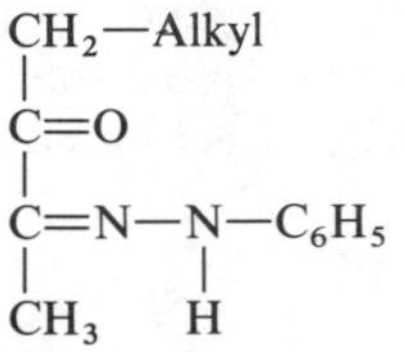

CH_2—Alkyl
C=O
C=N—N—C_6H_5
CH_3 H
1479

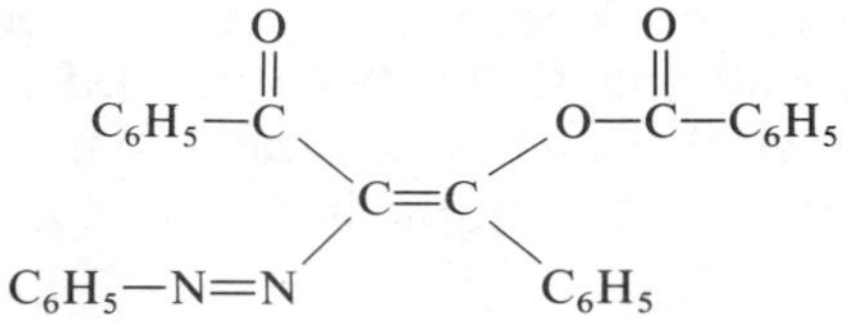

O O
C_6H_5—C O—C—C_6H_5
C=C
C_6H_5—N=N C_6H_5
1480

Indolization of mixtures of N_α-benzyl- and N_α-methylphenylhydrazine and 4-methoxy-N_α-methylphenylhydrazine with **1481**, formed by a Japp–Klingemann reaction between methyl 4-phenylacetoacetate and benzenediazonium chloride, in boiling methanolic sulphuric acid, yielded **1482** (R^1 = H, $R^2 = CH_3$ and $CH_2C_6H_5$ and $R^1 = OCH_3$, $R^2 = CH_3$, respectively) (Shvedov, Kurilo, Cherkasova and Grinev, A. N., 1973). Attempts to indolize the 2-ketoaldehyde arylhydrazones **1483** (R^1 = H, 4-CH_3, 4-OCH_3, 3-Cl and 2- and 4-NO_2, $R^2 = CH_3$ and R^1 = H, 4-CH_3, 4-OCH_3 and 3-Cl, $R^2 = C_2H_5$) under mild conditions were unsuccessful and under more drastic conditions the reactants were completely resinified (Shvedov, Altukhova and Grinev, A. N., 1966a). Pentan-2,3-dione 2-phenylhydrazone has been prepared but has not been indolized (Shvedov, Altukhova and Grinev, A. N., 1973).

Although 2-acetyl-1-methylindole was converted into the phenylhydrazone **1484** (Diels and Köllisch, 1911), no attempt was made to indolize this product. It would be interesting to attempt this indolization, polyphosphoric acid probably being the catalyst with the greatest potential in this respect (see Chapter III, G).

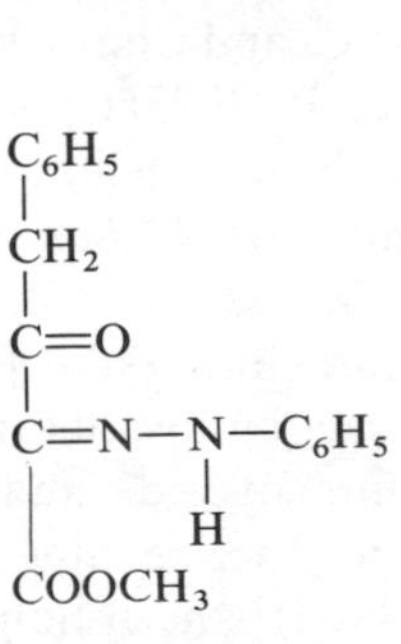

C_6H_5
CH_2
C=O
C=N—N—C_6H_5
H
$COOCH_3$
1481

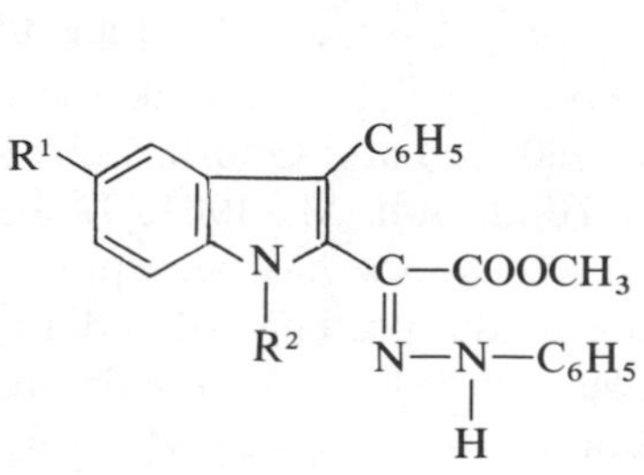

R^1 C_6H_5
N C—$COOCH_3$
R^2 N—N—C_6H_5
H
1482

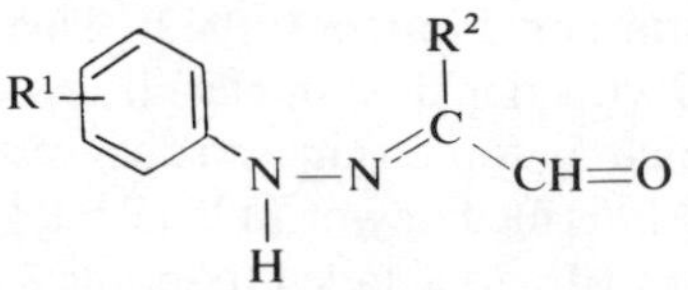

1483

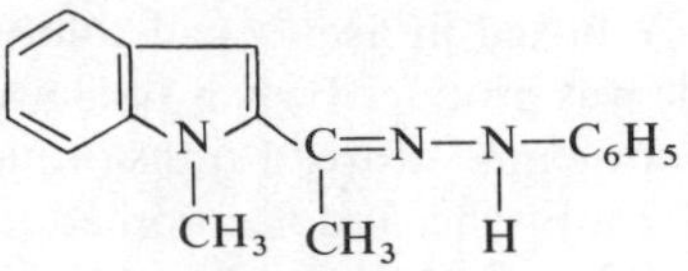

1484

Reduction of the carbonyl group to methylene in 2-acylindoles (Walton, Jenkins, Nutt and Holly, 1968) afforded a ready synthesis of the resulting products which would in most cases be unambiguously directly unobtainable by the Fischer indole synthesis.

2. Cyclic 1,2-Diketone Arylhydrazones

By far the most widely studied system in this respect is that of the cyclohexan-1,2-dione monoarylhydrazones **1485** (R^1 = NNHaryl, n = 1) which have been prepared by reaction between aryldiazonium salts and 2-hydroxymethylene-cyclohexanones **1485** (R^1 = CHOH, n = 1) (Altiparmakian and Braithwaite,

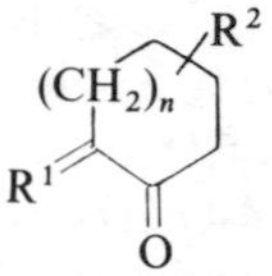

1485

1967; Anderson, G. and Campbell, 1950; Ballantine, Barrett, C. B., Beer, Boggiano, Eardley, Jennings and Robertson, 1957; Bhattacharyya, P., Basak, Islam and Chakraborty, 1975; Bhide, Tikotkar and Tilak, 1960; Brunelli, Fravolini, Grandolini and Strappaghetti, 1980; Chakraborty, Bhattacharyya, P., Roy, S., Bhattacharyya, S. P. and Biswas, A. K., 1978; Chakraborty, Chatterji and Ganguly, 1969; Chakraborty, Das, K. C. and Basak, 1968; Chakraborty and Chowdhury, 1966, 1968; Chakraborty, Das, K. C. and Chowdhury, 1966a, 1966b, 1969; Chakraborty, Islam and Bhattacharyya, P., 1973; Chowdhury and Chakraborty, 1974; Coffey, 1923; Elks, Elliott and Hems, 1944a; Kent, 1935; Kent and McNeil, 1938; Mears, Oakeshott and Plant, 1934; Mester, Choudhury and Reisch, 1980; Sen and Ghosh, 1927; Shah, G. D. and Patel, B. P. J., 1979; Soutter and Tomlinson, M., 1961; Teuber and Cornelius, 1965) [2-hydroxy-methylenecyclohexanone has been prepared by reacting cyclohexanone with ethyl formate in the presence of sodium or sodium hydride and a trace of ethanol. It was colourless liquid which began to polymerize after standing at room temperature for several days (Ainsworth, C., 1963)], the cyclohexanone-2-carboxylate anions **1486** (n = 1) (Altiparmakian and Braithwaite, 1967;

Braithwaite and Robinson, G. K., 1962; Douglas, B., Kirkpatrick, Moore, B. P. and Weisbach, 1964; Kent and McNeil, 1938; Linstead and Wang, A. B.-L., 1937; Lions, 1932; Teuber and Cornelius, 1965), the piperidino enamines **1487** ($X = CH_2$, $n = 1$) (Shvedov, Altukhova and Grinev, A. N., 1964, 1965a, 1966b) or the morpholino enamines **1487** ($X = O$, $n = 1$) (Bisagni, Ducrocq and Hung, 1980; Bisagni, Ducrocq, Lhoste, Rivalle and Civier, 1979), failure having attended the earlier attempted use of morpholino enamines in this type of reaction (Shvedov, Altukhova and Grinev, A. N., 1965a) (see also Chapter I, F). This formation of **1485** (R^1 = NNHaryl, $n = 1$) from enamines proceeded in better yields than those obtained by coupling the corresponding aryldiazonium salts with the potassium salt of **1486** ($R = H$, $n = 1$) (Bisagni, Ducrocq, Lhoste, Rivalle and Civier, 1979).

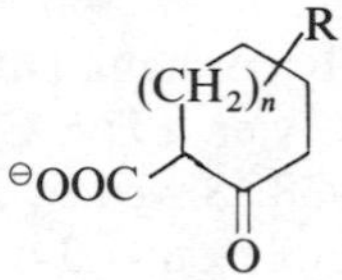

R (CH₂)ₙ ⊖OOC O

1486

R (CH₂)ₙ N X

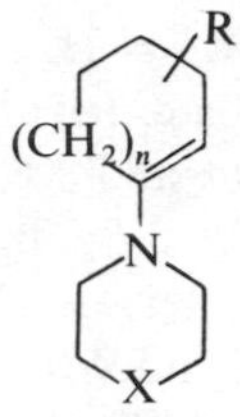

1487

Indolizations of the arylhydrazones **1488** ($n = 1$) into **1489** have been usually achieved with formic acid or aqueous or ethanolic hydrochloric acid or sulphuric acid as catalysts, although concentrated hydrochloric acid in acetic acid (Brunelli, Fravolini, Grandolini and Strappaghetti, 1980) or freshly prepared polyphosphoric acid (Shah, G. D. and Patel, B. P. J., 1979) also gave good yields of products, and with R^2–R^4 = H {Anderson, G. and Campbell, 1950; Ballantine, Barrett, C. B., Beer, Boggiano, Eardley, Jennings and Robertson, 1957 – see also Kent, 1935; Bhide, Tikotkar and Tilak, 1960; Bisagni, Ducrocq and Hung, 1980; Bisagni, Ducrocq, Lhoste, Rivalle and Civier, 1979; Chakraborty, Bhattacharyya, P., Roy, S., Bhattacharyya, S. P. and Biswas, A. K., 1978; Chakraborty, Chatterji and Ganguly, 1969; Chakraborty, Islam and Bhattacharyya, P. 1973; Coffey, 1923; Douglas, B., Kirkpatrick, Moore, B. P. and Weisbach, 1964; Hester, 1969b; Kent, 1935 [better than the original method

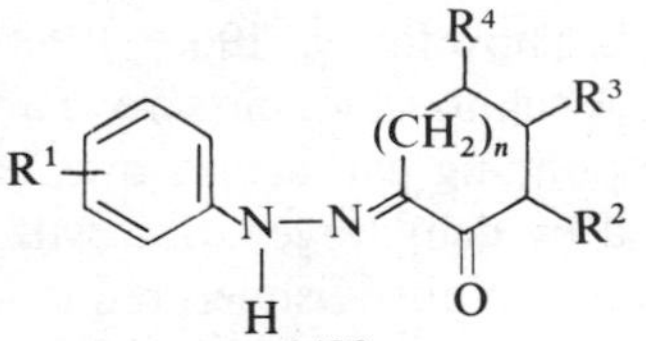

R⁴ R³ (CH₂)ₙ R¹ N—N R² H O

1488

R⁴ R³ R¹ N R² H O

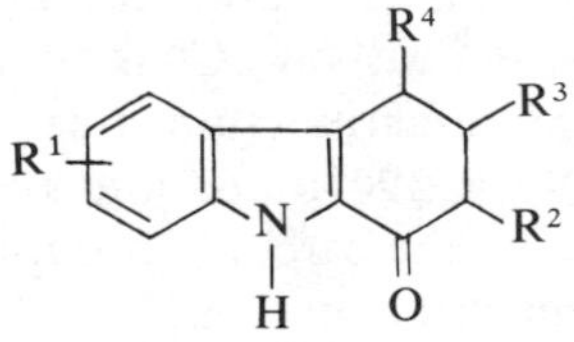

1489

(Coffey, 1923) for the synthesis of **1489** (R^1–R^4 = H)]; Lions, 1932; Mashkovskii, Grinev, A. N., Andreeva, Alkhova, Shvedov, Avrutskii and Gromova, 1974; Mears, Oakeshott and Plant, 1934; Mester, Choudhury and Reisch, 1980; Sen and Ghosh, 1927; Shah, G. D. and Patel, B. P. J., 1979; Shvedov, Altukhova and Grinev, A. N., 1965a, 1966b; Shvedov, Altukhova, Komissarova and Grinev, A. N., 1965a, 1965b; Soutter and Tomlinson, M., 1961} [presumably the 1,2,3,4-tetrahydrocarbazol-1-one, the source of which was not given, employed in other studies (Wenkert and Dave, K. G., 1962) could have been made by the indolization of **1488** (R^1–R^4 = H, $n = 1$)], with $R^2 = CH_3$, $R^3 = R^4$ = H (Chowdhury and Chakraborty, 1974; Chakraborty, Das and Basak, 1968; Kent and McNeil, 1938; Sen and Ghosh, 1927; Shvedov, Altukhova and Grinev, A. N., 1966b), with $R^2 = R^4$ = H, $R^3 = CH_3$ (Bhattacharyya, P., Basak, Islam and Chakraborty, 1975; Chakraborty and Chowdhury, 1966, 1968; Chakraborty, Das, K. C. and Chowdhury, 1966, 1969; Kent and McNeil, 1938) and with $R^2 = R^3$ = H, $R^4 = CH_3$ (Bisagni, Ducrocq and Hung, 1980; Bisagni, Ducrocq, Lhoste, Rivalle and Civier, 1979; Chakraborty, Das, K. C. and Basak, 1968; Kent and McNeil, 1938; Shvedov, Altukhova and Grinev, A. N., 1966b). Similarly, 3,6-dimethylcyclohexan-1,2-dione 2-phenylhydrazone and 3,5-dimethylcyclohexan-1,2-dione 2-(2-nitrophenyl)hydrazone afforded the corresponding 3*H*-indoles **1490** ($R^1 = R^3$ = H, $R^2 = CH_3$) (Teuber and Cornelius, 1965) and **1490** ($R^1 = NO_2$, R^2 = H, $R^3 = CH_3$) (Soutter and

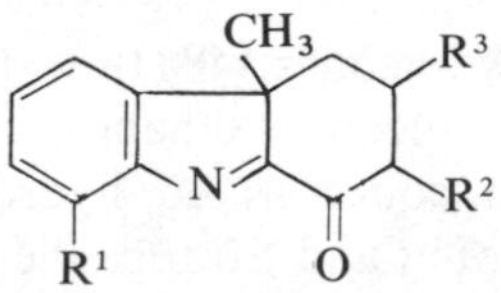

CH_3 R^3 N R^2 R^1 O

1490

Tomlinson, M., 1961) when treated with concentrated hydrochloric acid in acetic acid and polyphosphoric acid, respectively. No **1490** ($R^1 = R^2$ = H, $R^3 = CH_3$) was isolated after 3,5-dimethylcyclohexan-1,2-dione 2-phenylhydrazone in a mixture of concentrated hydrochloric acid and acetic acid was refluxed (Teuber and Cornelius, 1965). A large variety of substituted phenylhydrazines have been used in the preparation of **1488** which has led to a wide variation of R^1 in **1489**.

The 1,2,3,4-tetrahydrocarbazol-1-ones **1489** could be converted into the corresponding 1-hydroxycarbazoles by dehydrogenative pyrolysis with palladium-on-carbon (Chakraborty and Chowdhury, 1966, 1968) or with palladium-on-barium sulphate (Shvedov, Altukhova, Komissarova and Grinev, A. N., 1965a, 1965b), or into the corresponding carbazoles by Clemmensen reduction of the carbonyl group followed by dehydrogenation with chloranil (Anderson, G. and Campbell, 1950) or by Wolff–Kishner reduction of the carbonyl group followed by pyrolysis with palladium-on-carbon (Bhattacharyya,

P., Basak, Islam and Chakroborty, 1975; Chakraborty, Das, K. C. and Basak, 1968; Chakraborty, Das, K. C. and Chowdhury, 1966, 1969). When compounds **1489** were heated with pyridine hydrochloride, the corresponding 1-unsubstituted carbazoles were formed (Bisagni, Ducrocq and Hung, 1980) (i.e. the reaction involved elimination of the oxygen atom of the carbonyl function).

An occasional failure to effect an indolization of the above type has been noted as, for example, with cyclohexan-1,2-dione mono-2-bromophenylhydrazone (Mears, Oakeshott and Plant, 1934). However, a more general failure to achieve indolization occurred when cyclization into cinnolines (1,2-diazanaphthalenes) took place, an alternative reaction which was particularly favoured when treating the hydrazone with cold (0 °C) sulphuric acid, conditions which minimized indolization, sulphonation and hydrolysis (Altiparmakian and Braithwaite, 1967; Braithwaite and Robinson, G. K., 1962). Examples of this cyclization involved the conversion of **1488** (R^1–R^4 = H, n = 1) into **1491** [$R^1 + R^2 = (CH_2)_4$] by treatment with concentrated sulphuric acid which afforded only a 15% yield of the product, a large amount of unidentified amphoteric material also being produced (Moore, B. P., 1949), with sulphuric acid at 60 °C which afforded *ca.* 5% of the product (Soutter and Tomlinson, M., 1961), with polyphosphoric acid at 80 °C which afforded *ca.* 5% of the cinnoline together with 1,2,3,4-tetrahydrocarbazol-1-one (Soutter and Tomlinson, M., 1961) and with sulphuric acid at room temperature or 0 °C to give a low yield of product (Braithwaite and Robinson, G. K., 1962), of **1492** ($R^1 = C_6H_5$,

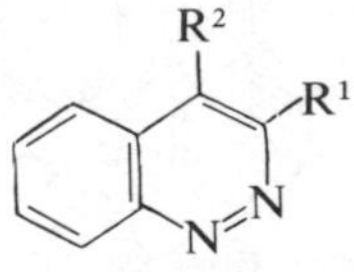

1491

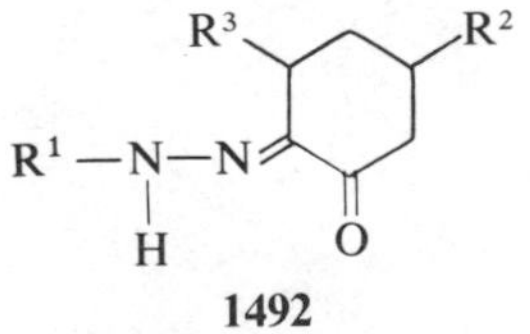

1492

$R^2 = R^3 = CH_3$) with concentrated sulphuric acid (Moore, B. P., 1949), with concentrated sulphuric acid at room temperature (Soutter and Tomlinson, M., 1961) or when dissolved in concentrated sulphuric acid without cooling (Teuber and Cornelius, 1965) to furnish good yields of **1491** [$R^1 + R^2 = CH(CH_3)CH_2CH(CH_3)CH_2$] and with polyphosphoric acid at 80 °C to produce *ca.* 7% yield of **1491** [$R^1 + R^2 = CH(CH_3)CH_2CH(CH_3)CH_2$] (Soutter and Tomlinson, M., 1961), of **1492** (R^1 = 1-naphthyl, $R^2 = R^3$ = H) with sulphuric acid at 0 °C to afford **1493** ($R^1 = R^2$ = H) (Braithwaite and Robinson, G. K., 1962), of **1492** [R^1 = 1-naphthyl, $R^2 = CH_3$, R^3 = H and R^1 = 1-(4-methylnaphthyl), $R^2 = R^3$ = H] with sulphuric acid at 0 °C to afford **1493** (R^1 = H, $R^2 = CH_3$ and $R^1 = CH_3$, R^2 = H, respectively) (Altiparmakian and Braithwaite, 1967), of **1492** (R^1 = 2-naphthyl, $R^2 = R^3$ = H and R^1 = 2-naphthyl, $R^2 = CH_3$, R^3 = H) with sulphuric acid at 0 °C to form

1494 (R = H) (Braithwaite and Robinson, G. K., 1962) and **1494** (R = CH_3) (Altiparmakian and Braithwaite, 1967) and of **1492** (R^1 = 5-acenaphthenyl, 9-phenanthryl, and 1- and 2-anthryl, $R^2 = R^3$ = H) with sulphuric acid at 0 °C to form **1495** (Altiparmakian and Braithwaite, 1967; Braithwaite and Robinson, G. K., 1962), **1496**, **1497** and **1498** (Altiparmakian and Braithwaite,

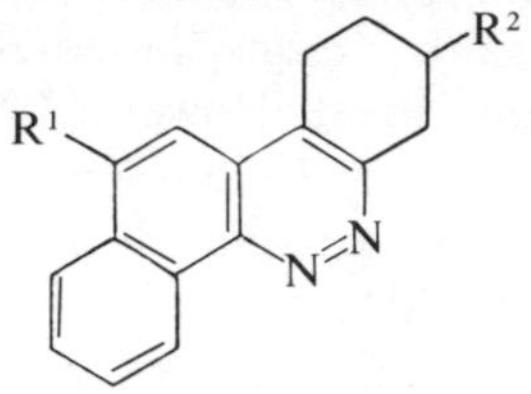

1493 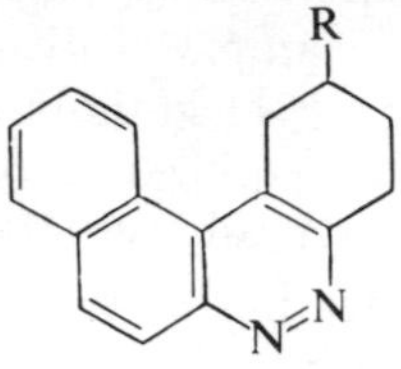 **1494**

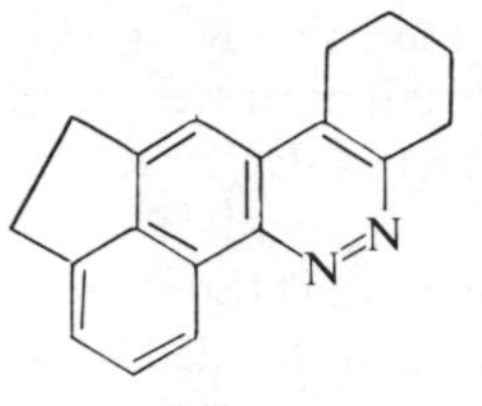

1495 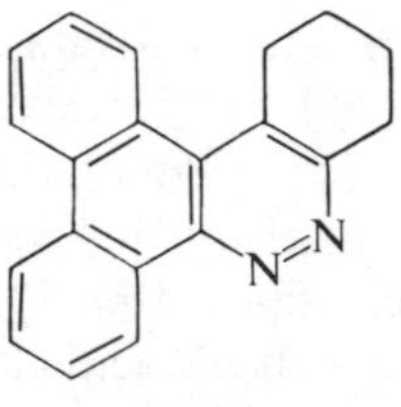 **1496**

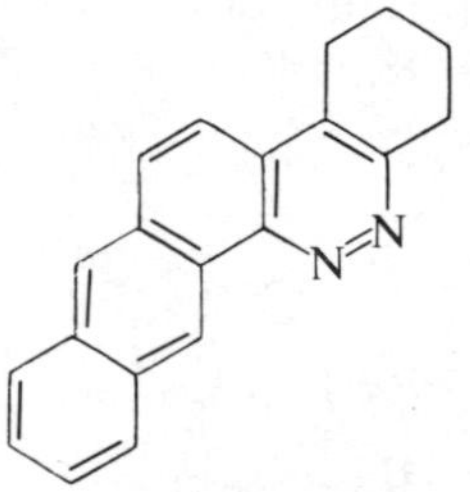

1497 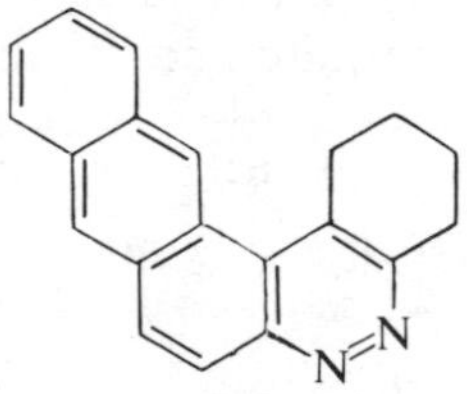 **1498**

1967), respectively. It was suggested (Altiparmakian and Braithwaite, 1967) that this formation of cinnolines involved sequential O-protonation of the carbonyl function in **1492**, intramolecular electrophilic attack on the *ortho* position of the aromatic nucleus by the carbonium ion and elimination of the elements of water. However, even under the favourable conditions of treatment with cold concentrated sulphuric acid, other attempted analogous cinnoline formations were unsuccessful. Thus, **1492** (R^1 = 5,6,7,8-tetrahydro-2-naphthyl, $R^2 = R^3$ = H) (Braithwaite and Robinson, G. K., 1962), **1492** (R^1 = 3-acenaphthenyl, 6-chrysenyl and 4-phenylazophenyl, $R^2 = R^3$ = H) (Altiparmakian and Braithwaite, 1967) and cyclopentan-1,2-dione monophenylhydrazone and mono-1-naphthylhydrazone (Altiparmakian and Braithwaite, 1967) afforded neither cinnolines nor indoles under these conditions and in other cases, using warm ethanolic sulphuric acid as the catalyst, exclusive indole formation was observed (Lions, 1932; Shvedov, Altukhova and Grinev, A. N., 1966b; Shvedov, Altukhova, Komissarova and Grinev, A. N., 1965a, 1965b).

Monoarylhydrazones of cycloalkan-1,2-diones **1485** ($R^1 = NNHC_6H_4R$, $R^2 = H$, $n = 0$, 2 and 3) have also been indolized, again usually using formic acid, hydrochloric acid or sulphuric acid as catalysts, to yield the corresponding **1499** ($n = 1$) (Elks, Elliott and Hems, 1944a; Manske, R. H. F., 1931a; Nagarajan, Pillai and Kulkarni, C. L., 1969; Rensen, 1959) [for the synthesis of cyclopentan-1,2-dione monophenylhydrazone and mono-2- and -4-nitrophenylhydrazone by the Japp–Klingemann reaction, see also Linstead and Wang, A. B.-L. (1937). When cyclopentan-1,2-dione monophenylhydrazone was first prepared, using the Japp–Klingemann reaction, indolization studies were not effected (Dieckmann, 1901)], **1499** ($n = 3$) (Bahadur, Bailey, Costello and Scott, P. W., 1979; Epstein and Goldman, 1973; Mühlstädt and Lichtmann, 1970; Mühlstädt and Treibs, 1957; Nagarajan, Pillai and Kulkarni, C. L., 1969; Shvedov, Altukhova and Grinev, A. N., 1965b, 1966b, 1966c; Yamane and Fujimori, K., 1976) and **1499** ($n = 4$) (Witkop, Patrick and Rosenblum, 1951).

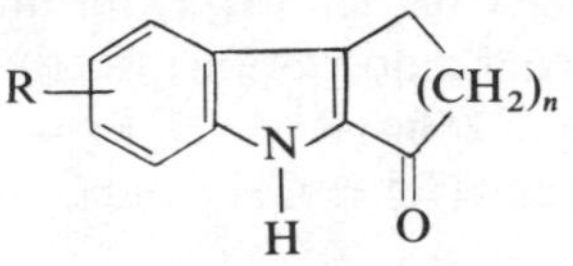

R $(CH_2)_n$ N H O

1499

The arylhydrazones **1500** (R = H, CH_3 and Br) yielded **1501** (R = H, CH_3 and Br, respectively, X = O) when treated in acetic acid solution with concentrated hydrochloric acid or zinc chloride. However, when **1500** (R = H) or the parent dione was treated with excess phenylhydrazine in acetic acid, **1501** (R = H, X = $NNHC_6H_5$) was formed although this was not produced by reacting **1501** (R = H, X = O) with phenylhydrazine (Yamane and Fujimori, K., 1972).

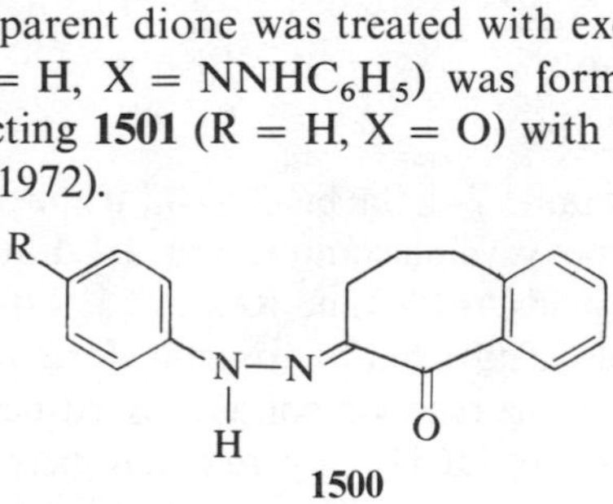

R N—N H O

1500

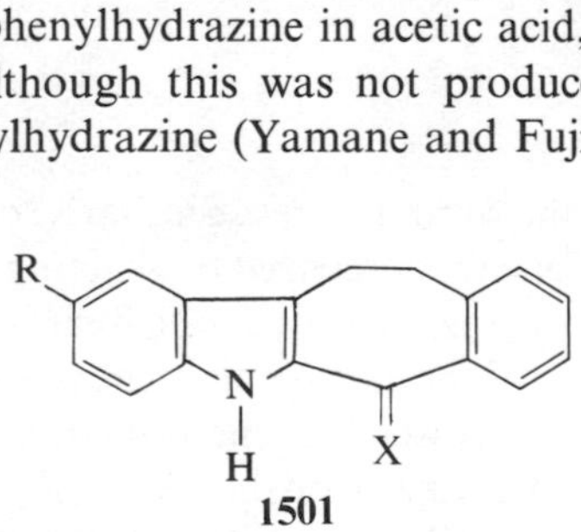

R N H X

1501

An interesting indolization occurred when cyclohexan-1,2-dione reacted with 1-phenylpyrazolidine hydrochloride [(**334**; $R^1 = R^2 = H, n = 1$) hydrochloride] to give **1502**. This was formed by dehydrative cyclization of the initial indolization product, **1503**, whose presence in the reaction was also detected (Eberle, Kahle and Talati, 1973). Other examples of similar indolizations, in which the nitrogen atom which was cleaved off upon forming the indole nucleus was still bound to the molecule, can be found in Chapter II, E3).

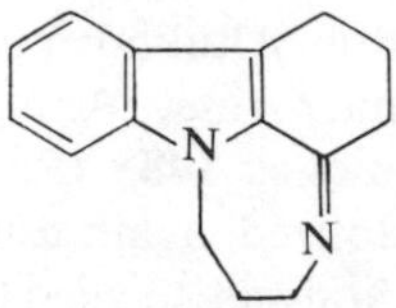

1502

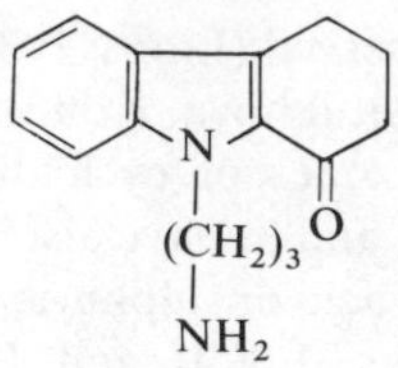

1503

Cyclohexan-1,2-dione bisphenylhydrazone (**1504**; $R^1 = R^2 = H$) was prepared by reacting cyclohexan-1,2-dione with a 2 × molar equivalent of phenylhydrazine in acetic acid solution (Wallach and Weissenborn, 1924), by reacting cyclohexan-1,2-dione monophenylhydrazone with phenylhydrazine (Coffey, 1923) and by reacting 2-chlorocyclohexanone or 2-hydroxycyclohexanone with phenylhydrazine in the presence of acetic acid (Bloink and Pausacker, 1950) or methanolic acetic acid and sodium acetate (Moldenhauer and Simon, 1969). Detailed studies of these last two reactions, osazone formations, have been effected (Bloink and Pausacker, 1950; Buckingham and Guthrie, 1968b; Simon and Moldenhauer, 1969 – see Chapter VII, Q for further discussion of osazone formation in relation to Fischer indolization) and their synthetic utility has been extended to the preparation of **1504** ($R^1 = H$, $R^2 = CH_3$, Cl and NO_2) (Bloink and Pausacker, 1950) and **1504** ($R^1 = CH_3$ and $CH_2C_6H_5$, $R^2 = H$) (Moldenhauer and Simon, 1969). Similarly, cyclohexan-1,2-dione bis-2,4-dinitrophenylhydrazone was prepared by reacting 2-methoxycyclohexanone with 2,4-dinitrophenylhydrazine in alcoholic hydrochloric acid (Adkins and Rossow, 1949), the presence of the nitro groups preventing indolization but not osazone formation after hydrolysis of the methoxy group under the reaction conditions. Attempts to prepare analogously **1504** ($R^1 = H$, $R^2 = OCH_3$) by reaction between 2-hydroxycyclohexanone and 4-methoxyphenylhydrazine in acetic acid at room temperature furnished only 1,2,3,4-tetrahydro-6-methoxycarbazol-1-one (Bloink and Pausacker, 1950). The reaction between 2-benzyloxycyclohexanone and phenylhydrazine in sulphuric acid gave 1,2,3,4-tetrahydrocarbazol-1-one and benzyl alcohol, between 2-acetoxy-1-tetralone (**1505**) and phenylhydrazine in boiling alcohol gave 1,2,3,4-tetrahydronaphthalen-1,2-dione bisphenylhydrazone (**1506**) and between **1505** and phenylhydrazine in acetic acid gave a product formulated as **1507** (Barnes, Pausacker and Badcock, 1951), although

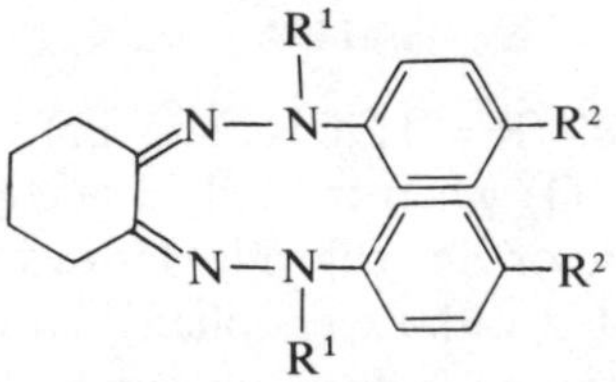

1504

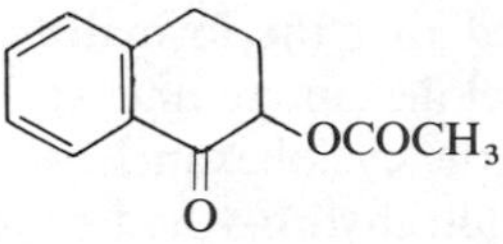

1505

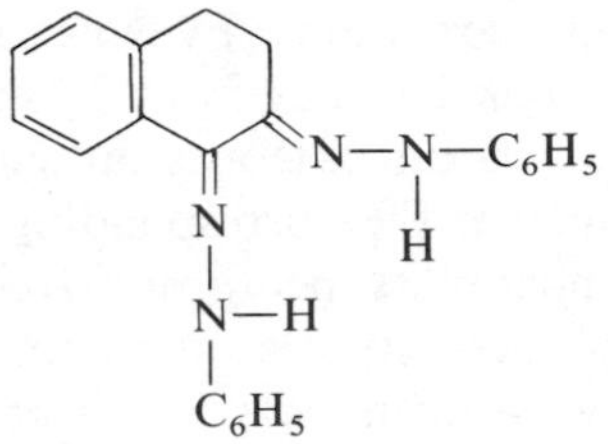

1506

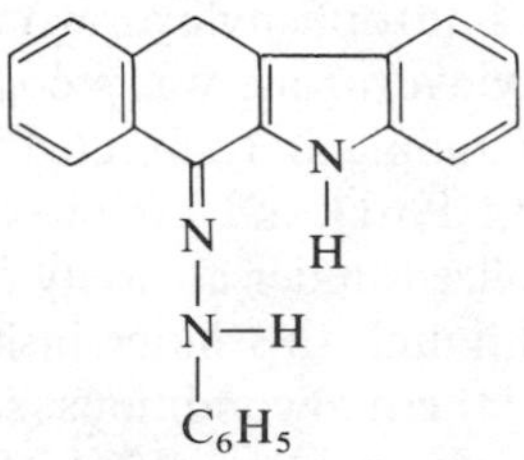

1507

the structure of this last product was not verified. The earlier observation that the reaction of 2-hydroxycyclohexanone with phenylhydrazine in 25% acetic acid afforded 2-hydroxycyclohexanone phenylhydrazone (Bergmann, M. and Gierth, 1926) could not be repeated by subsequent investigators, only **1504** ($R^1 = R^2 = H$) being isolated from the reaction. When equimolar amounts of 2-hydroxycyclohexanone and phenylhydrazine were heated *in vacuo*, a loss in weight corresponding to one mole equivalent of water was observed but the product was an uncrystallizable red oil. This, upon heating in aqueous alcoholic sulphuric acid, yielded aniline and 1,2,3,4-tetrahydrocarbazol-1-one (Bloink and Pausacker, 1950). Other investigations in this area have involved the formation of osazones by reacting arylhydrazones with 2-bromocyclohexanone (Buckingham and Guthrie, 1966, 1968b), 2-hydroxycyclohexanone (Hassner and Catsoulacos, 1967; Shemyakin, Maimind, Ermolaev and Bamdas, 1965) and 2-acetoxycyclohexanone (Caglioti, Rosini and Rossi, 1966; Hassner and Catsoulacos, 1967) systems.

Treatment of **1504** ($R^1 = R^2 = H$) in hot dilute sulphuric acid caused indolization of one of the hydrazone units and hydrolysis of the other to give 1,2,3,4-tetrahydrocarbazol-1-one (Bloink and Pausacker, 1950; Plancher and Ghigi, 1929). This product was also produced, along with aniline, when 2-hydroxycyclohexanone reacted with phenylhydrazine (1 mole) in refluxing ethanolic sulphuric acid, the yield of 1,2,3,4-tetrahydrocarbazol-1-one being doubled when a 2 × molar excess of phenylhydrazine was used. When a mixture of 2-hydroxycyclohexanone with phenylhydrazine in ethanol was refluxed, ammonia was evolved and **1504** ($R^1 = R^2 = H$) and **1489** (R^1–R^4 = H) were formed, an early example of a 'non-catalytic thermal indolization' but under relatively mild conditions. Evidence suggested that the formation of 1,2,3,4-tetrahydrocarbazol-1-ones from 2-hydroxycyclohexanone and arylhydrazines did not occur via the osazones but either via the cyclohexan-1,2-dione mono-arylhydrazones or the corresponding 1,2,3,4-tetrahydro-1-hydroxycarbazoles, the former possibility being preferred (Bloink and Pausacker, 1950). In an extension of these studies, a mixture of 2-hydroxycyclohexanone with N_{α}-methylphenylhydrazine was indolized in aqueous alcoholic sulphuric acid to afford 1,2,3,4-tetrahydro-9-methylcarbazol-1-one (Douglas, B., Kirkpatrick, Moore, B. P. and Weisbach, 1964).

Cyclohexan-1,2-dione bis-2-nitrophenylhydrazone was isolated after cyclohexan-1,2-dione 2-nitrophenylhydrazone was stood in sulphuric acid and the solution then basified with ammonia and subjected to column chromatography (Soutter and Tomlinson, M., 1961) and cyclohexan-1,2-dione bis-4-pyridylhydrazone could not be indolized under a variety of acidic conditions (Mann, Prior and Willcox, 1959) although this latter bishydrazone may well be indolizable under non-catalytic thermal conditions (see Chapter IV, I).

Treatment of the bisphenylhydrazone **1504** ($R^1 = R^2 = H$) with warm ethanolic hydrogen chloride yielded 1,2,3,4-tetrahydrocarbazol-1-one phenylhydrazone hydrochloride **1508** (R = H) (Mann and Willcox, 1958), with methanolic acetic acid under reflux it gave **1509** (R = H, X = O) and its phenylhydrazone **1509** (R = H, X = $NNHC_6H_5$) (Moldenhauer and Simon, 1969) and with acetic acid under reflux it afforded **1510** (R = H) (Moldenhauer

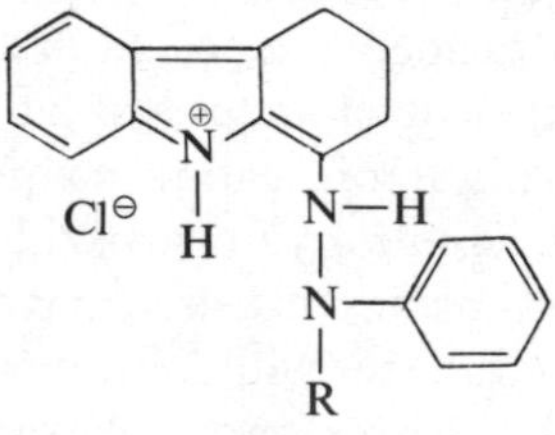

1508

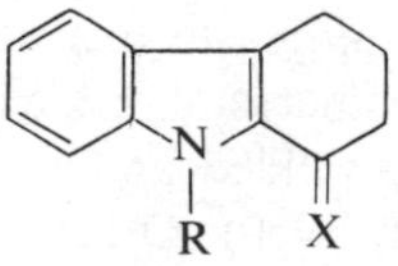

1509

and Simon, 1969). The bishydrazone **1504** ($R^1 = CH_3$, $R^2 = H$), upon treatment with methanolic acetic acid under reflux or with acetic acid at 100 °C, gave **1509** (R = CH_3, X = O) and **1510** (R = CH_3) or **1510** (R = CH_3), respectively, a mixture of cyclohexan-1,2-dione with N_{α}-phenylphenylhydrazine, upon heating with ethanolic acetic acid, yielded **1509** [R = C_6H_5, X = $NN(C_6H_5)C_6H_5$] which in refluxing acetic acid gave **1510** (R = C_6H_5) and **1504** ($R^1 = CH_2C_6H_5$, $R^2 = H$) in refluxing acetic acid afforded **1510** (R = $CH_2C_6H_5$), this product also being formed, along with **1509** (R = $CH_2C_6H_5$, X = O), when a mixture of cyclohexan-1,2-dione with N_{α}-benzylphenylhydrazine hydrochloride reacted in warm ethanolic acetic acid (Moldenhauer and Simon, 1969). A mixture of **1509** (R = CH_3, X = O) with phenylhydrazine or N_{α}-methylphenylhydrazine did not indolize in ethanolic acetic acid

(Buckingham and Guthrie, 1968b). However, mixtures of **1509** (R = H, X = O) with phenylhydrazine and N_α-methylphenylhydrazine in refluxing ethanolic hydrogen chloride yielded **1508** (R = H and CH_3, respectively). Upon pyrolysis of either of these products *in vacuo*, or upon treatment of the former product in refluxing acetic acid, compounds **1511** (R = H and CH_3, respectively) were produced, these reactions involving yet a further example of dehydrogenative aromatization concomitant with indolization (Mann and Willcox, 1958). Similarly, **1509** (R = H, X = O) reacted with a 2 × molar excess of phenylhydrazine in refluxing acetic acid to afford **1511** (R = H) which was also formed when a mixture of cyclohexan-1,2-dione phenylhydrazone with a 2 × molar excess of phenylhydrazine reacted in refluxing acetic acid (Bhide, Tikotkar and Tilak, 1957).

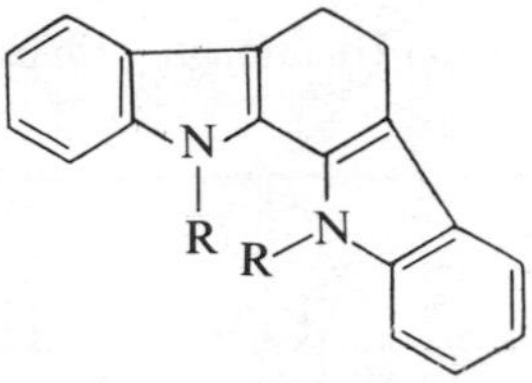

1510 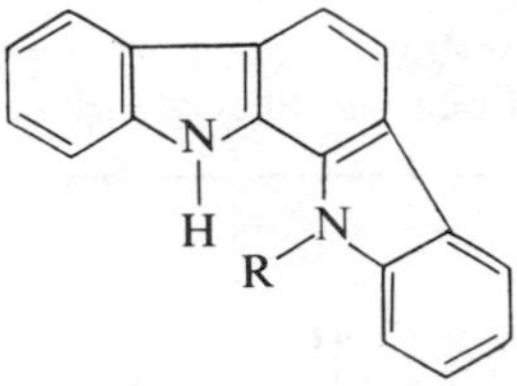 **1511**

Cyclopentan-1,2-dione bisphenylhydrazone has been prepared by reacting cyclopentan-1,2-dione with phenylhydrazine in warm acetic acid (Dieckmann, 1897) or by reacting cyclopentan-1,2-dione monophenylhydrazone with phenylhydrazine (Dieckmann, 1901, Linstead and Wang, A. B.-L., 1937; Lions, 1932) but indolization studies upon it were not carried out.

K. Other Arylhydrazine, Aldehyde and Ketone Moieties with Indolization Unambiguity which have been Employed in the Fischer Indole Synthesis

The indolizations of arylhydrazones in which the direction of indolization is potentially ambiguous because of the nature of either the ketonic or hydrazino moiety have already been considered in Chapter III, along with a discussion of the indolization of the arylhydrazones of heterocyclic ketones which were either directionally ambiguous or unambiguous with respect to the ketonic moiety. In the previous sections of the present chapter, other selected types of Fischer indolization which were directionally unambiguous have been elucidated. In Tables 31 and 32 are presented references to the use in Fischer indolizations of further examples of directionally unambiguous arylhydrazines and aldehydes or ketones, respectively, or their moieties. Other uses of some of these moieties in Fischer syntheses have also appeared elsewhere in this present study but it is intended that reference to the appropriate literature quoted in Tables 31 and 32 may expedite the choice of the catalyst and the reaction conditions in the design of arylhydrazone indolization.

Although in many instances the intermediate arylhydrazones were prepared by reaction of the corresponding arylhydrazine with the aldehyde or ketone listed in Tables 31 and 32, respectively, in some cases *in situ* hydrazone formation was achieved in the presence of an indolization catalyst. Furthermore, in many of the cases involving the use of the arylhydrazones of 2-ketoacids or 2-ketoesters (Table 32), the Japp–Klingemann reaction (see Chapter I, F), rather than the direct interaction of the corresponding arylhydrazines and ketones, was used to achieve the arylhydrazone preparation.

Table 31. Other Arylhydrazine Moieties with Indolization Unambiguity which have been Employed in the Fischer Indole Synthesis

---

Phenylhydrazine

Adkins and Coonradt, 1941; Ainsworth, D. P. and Suschitzky, 1967a; Aktieselskabet Dumex (Dumex Ltd), 1963, 1964; Al-Azawe and Sarkis, 1973; Ames, A. F., Ames, D. E., Coyne, Grey, Lockhart and Ralph, 1959; Anderson, R. M., Clemo and Swan, 1954; Arbusow, A. E., Saizew and Rasumov, 1935; Arbusow, A. E. and Tichwinsky, 1910b; Arbuzov, A. E. and Friauf, 1913; Arbuzov, A. E. and Rotermel, 1932; Arbuzov, A. E. and Wagner, R. E., 1913; Arbuzov, A. E. and Zaitzev, 1934; Arbuzov, A. E., Zaitzev and Rasumov, 1935; Armit and Robinson, R., 1922; Arnold, E., 1888; Atkinson, Simpson and Taylor, A., 1954; Auwers, Hilliger and Wulf, 1922; Avramenko, Plutitskii, Strekozov and Suvorov, 1979; Avanesova and Tatevosyan, 1974; Badische Anilin- & Soda-Fabrik, A.-G., 1965; Baeyer, 1894; Baeyer and Tutein, 1889; Bahadur, Bailey, Costello and Scott, P. W., 1979; Bailey and Bogle, 1977; Bailey, Scottergood and Warr, 1971; Ballantine, Barrett, C. B., Beer, Boggiano, Eardley, Jennings and Robertson, 1957; Bazile, Cointet and Pigerol, 1978; Bellamy and Guthrie, 1965b; Benary and Baravian, 1915; Bisagni, Ducrocq, Lhoste, Rivalle and Civier, 1979; Blades and Wilds, 1956; Bloch-Chaudé, Rumpf and Sadet, 1955; Boekelheide and Ainsworth, C., 1950b; Boido and Boido Canu, 1973; Borsche and Klein, 1941; Borsche, Witte, A. and Bothe, 1908; Bourdais and Lorre, 1975; Boyakhchyan, Rashidyan and Tatevosyan, 1966; Boyd-Barrett and Robinson, R., 1932; Braun and Bayer, 1925, 1929; Braun and Haensel, 1926; Bremner, and Browne, 1975; Brown, F. and Mann, 1948a, 1948b; Brown, R. K., Nelson, Sandin and Tanner, 1952; Bruce, 1960, 1962; Bullock and Fox, 1951; Bullock and Hand, J. J., 1956b; Buu-Hoï, 1949a, 1949b, 1954, 1958; Buu-Hoï, Cagniant, P., Hoán and Khôi, 1950; Buu-Hoï, Croisy, Jacquignon and Martani, 1971; Buu-Hoï, Delcey, Jacquignon and Périn, 1968; Buu-Hoï and Hoán, 1949, 1950, 1951, 1952; Buu-Hoï, Hoán and Jacquignon, 1949; Buu-Hoï, Hoán and Khôi, 1950a, 1950b, 1950c; Buu-Hoï, Hoán, Khôi and Xuong, 1950, 1951; Buu-Hoï, Hoán and Xuong, 1951; Buu-Hoï and Jacquignon, 1954, 1956; Buu-Hoï, Jacquignon and Lavit, 1956; Buu-Hoï, Jacquignon, and Ledésert, 1970; Buu-Hoï, Khôi and Xuong, 1950, 1951; Buu-Hoï, and Lavit, 1955; Buu-Hoï, Périn and Jacquignon, 1965, 1966; Buu-Hoï and Royer, 1950; Buu-Hoï and Saint-Ruf, 1962, 1963, 1965a, 1965b; Buu-Hoï, Saint-Ruf and Dufour, 1964; Buu-Hoï, Saint-Ruf, Jacquignon and Marty, 1963; Buu-Hoï and Xuong, 1952; Buu-Hoï, Xuong and Lavit, 1954; Buu-Hoï, Xuong and Thang, K. van, 1953; Campbell and Copper, 1935; Carson, D. F. and Mann, 1965; Chakraborty and Chowdhury, 1966, 1968; Chakraborty, Das, K. C. and Chowdhury, 1969; Ch'ang-pai, Evstigneeva and

Preobrazhenskii, N. A., 1958; Chapman, Clarke and Hughes, H., 1965; Chen, G.-S. J. and Gibson, 1975; Clemo and Perkin, 1924a; Clemo and Seaton, 1954; Colle, M.-A. and David, 1960; Cook, A. H. and Reed, K. J., 1945; Coombes, Harvey, D. J. and Reid, S. T., 1970; Cornforth, J. W., Hughes, G. K., Lions and Harradence, 1937–1938; Da Settimo, Biagi, Primofiore, Ferrarini and Livi, 1978; Da Settimo, Primofiore, Biagi and Santerini, 1976a, 1976b; Da Settimo and Saettone, 1965; Dave, V., 1972; David and Monnier, 1956, 1959; David and Régent, 1964; Davidge, 1959; Douglas, B., Kirkpatrick, Moore, B. P. and Weisbach, 1964; Duc and Fetizon, 1969; Dufour, Buu-Hoï, Jacquignon and Hien, 1972; Duncan and Boswell, 1973; Duncan, Helsley and Boswell, 1973; Ehrhart, 1953; Elgersma, 1969; Elks, Elliott and Hems, 1944b; Ellinger, 1904, 1905; Farbenfabriken Bayer, A.-G., 1954; Feofilaktov, 1947a, 1947b, 1952; Feofilaktov and Semenova, N. K., 1952a, 1952b, 1953b; Finkelstein and Lee, J., 1956; Fischer, Emil, 1886b, 1886c, 1886e, 1889b; Fischer, Emil and Jourdan, 1883; Fischer, Emil and Schmidt, T., 1888; Fischer, Emil and Schmitt, 1888; Fitzpatrick, J. T. and Hiser, 1957; Fox and Bullock, 1951a, 1955a, 1955b; Freed, Hertz and Rice, 1964; Freedman and Judd, 1972; Gale, Lin and Wilshire, 1976; Gardner, D. V., McOmie and Prabhu, 1970; Geigy, A.-G., 1966; Gesellschaft für Teerverwertung m.b.H., 1911; Ghigi, 1930; Gordon and Jackson, R. W., 1935; Grammaticakis, 1937, 1939; Grandberg and Bobrova, 1973, 1974b, 1978; Grandberg, Bobrova and Zuyanova, 1972; Grandberg and Dashkevich, S. N., 1970a, 1971a; Grandberg and Ivanova, 1970a, 1970c; Grandberg, Kost and Terentyev, 1957; Grandberg and Moskvina, 1970, 1971, 1972, 1974a, Grandberg and Nikitina, 1971a; Grandberg and Zuyanova, 1968, 1970; Grandberg, Zuyanova, Afonina and Ivanova, T. A., 1967; Grinev, Sadovskaya and Ufimtsev, 1963; Gupta and Ojha, 1971; Hahn, W. E., Bartnik and Zawadzka, 1966; Hai, Buu-Hoï and Xuong, 1958; Hamana and Kumadaki, 1967; Harradence and Lions, 1938, 1939; Harvey, D. J., 1968; Hausmann, 1889; Hegedüs, 1946; Henle, 1905; Herdieckerhoff and Tschunkur, 1933; Hino, Suzuki, T. and Nakagawa, 1973; Holla, B. S. and Ambekar, 1974; Hörlein, H. U., 1957; Horning, E. C., Horning, M. G. and Walker, G. N., 1948; Hoshino and Shimodaira, 1935; Hoshino and Takiura, 1936; Huffman, 1962; Hughes, G. K. and Lions, 1937–1938a; I. G. Farbenind, A.-G., 1930d; Jackson, A. H. and Smith, P., 1968; Jackson, R. W. and Manske, R. H., 1930, 1935; Janetzky and Verkade, 1945; Jardine and Brown, R. K., 1965; Jordaan and Arndt, 1968; Joshi, K. C., Pathak and Chand, 1978a, 1978b; Joshi, S. S. and Gambhir, 1956; Julia, Melamed and Gombert, 1965; Julian, Karpel, Magnani and Meyer, E. W., 1948; Kaji and Nagashima, 1952; Kalb, Schweizer and Schimpf, 1926; Kametani, Fukumoto and Masuko, 1963; Kanaoka, Ban, Miyashita, Irie and Yonemitsu, 1966; Kao and Robinson, R., 1955; Keglević, Desaty, Goleš and Stančić, 1968; Keimatsu and Sugasawa, 1928a, 1928b; Keimatsu, Sugasawa and Kasuya, 1928; Kent and McNeil, 1938; Kermack, Perkin and Robinson, R., 1921; Kiang, Mann, Prior and Topham, 1956; Kidwai, A. R. and Khan, N. H., 1963; King, F. E. and L'Ecuyer, 1934; Kipping, 1894; Kipping and Hill, A., 1899; Kissmanm Farnsworth and Witkop, 1952; Kögl and Kostermans, 1935; Korczynski and Kierzek, 1925; Korczyński and Kierzek, 1925; Kucherova, Petruchenko and Zagorevskii, 1962; Kuroda, 1923; LaForge, 1928; Landquist and Marsden, 1966; Laubmann, 1888b; Lee, T. B. and Swan, 1956; Lespagnol, A., Lespagnol, C. and Henichart, 1969; Leuchs, Heller, A. and Hoffmann, A., 1929; Leuchs and Kowalski, 1925; Leuchs and Winzer, 1925; Leuchs, Wulkow and Gerland, 1932; Lions, 1932; MacPhillamy, Dziemian, Lucas and Kuehne, 1958; Manske, R. H. F., 1931a; Manske, R. H. F. and Robinson, R., 1927; Marchand, Streffer and Jauer, 1961; Marion and Oldfield, 1947; Marshalkin and Yakhontov, 1972b; McClelland, 1929; McClelland and D'Silva, 1932; McLoughlin and Smith, L. H., 1965; Menichi, Bisagni and Royer, 1964; Merchant and Shah, N. J., 1975; Merck and Co., Inc., 1964c, 1965, 1967a; Mndzhoyan, Tatevosyan and Ekmekdzhayan, 1957; Mndzhoyan, Tatevosyan, Terzyan and Ekmekdzhayan, 1958; Moordian, 1976b;

(*continued*)

Table 31 (*continued*)
Phenylhydrazine (*continued*)
Morrison, Waite, R. O., Caro, A. N. and Shavel, 1967; Muhlstädt and Treibs, 1957; Nagarajan, Pillai and Kulkarni, C. L., 1969; Nakazaki, 1960a; Nakazaki and Isoe, 1955; Noland, Rush and Smith, L. R., 1966; Ockenden and Schofield, 1953a, 1953b; Okawa and Konishi, 1974; Oki and Nagasaka, 1973; Pacheco, 1951; Perkin, 1904a, 1904b; Perkin and Plant, 1921; Pfenninger, 1968; Pigerol, Chandavoine, De Cointet de Fillain and Nanthavong, 1975; Pigerol, De Cointet de Fillain, Nanthavong and Le Blay, 1976; Plancher, 1898e, 1898g; Plancher and Carrasco, 1905; Plancher and Forghieri, 1902; Plant and Tomlinson, M. L., 1933; Plieninger, 1950a, 1950b; Polaczkowa and Porowska, 1950; Prasad and Swan, 1958; Pretka and Lindwall, 1954; Rastogi, Bindra, Rai and Anand, 1972; Reynolds, B. and Carson, J., 1969; Robinson, B., 1963a; Robinson, B. and Zubair, 1971, 1973; Robinson, R. and Thornley, 1926; Rogers, C. U. and Corson, 1947, 1950, 1963; Rosenthal and Yalpani, 1965; Sakurai and Ito, T., 1957; Saleha, Khan, N. H., Siddiqui and Kidwai, M. M., 1978; Sarett and Shen, T.-Y., 1966a; Seka and Kellermann, 1942; Sempronj, 1938; Sen and Ghosh, 1927; Sharkova, N. M., Kucherova, Portnova and Zagorevskii, 1968; Shavel and Morrison, 1967; Shen, T-Y., 1964; Shimizu, J., Murakami, S., Oishi and Ban, 1971; Shriner, Ashley and Welch, E., 1942, 1955; Shukri, Alazawe and Al-Tai, 1970; Shvedov, Altukhova, Komissarova and Grinev, A. N., 1965a, 1965b; Shvedov, Kurilo and Grinev, A. N., 1969, 1970; Sibiryakova, Brovkin, Belyakova and Grandberg, 1969; Sircar and Gopalan, 1932; Skrabal, Steiger and Zollinger, 1975; Societe des Usines Chimiques Rhone-Poulenc, 1960; Spickett, 1966; Sternbach, Fryer, Metlesics, Sachs and Stempel, 1962; Sturm, Tritschler and Zeidler, 1972; Sugasawa and Takano, 1959; Sugasawa, Terashima and Kanaoka, 1956; Suvorov and Antonov, 1952; Suvorov, Antonov and Rokhlin, 1953; Suvorov, Mamaev and Shagalov, 1957; Suvorov and Murasheva, 1961; Suvorov, Samsoniya, Chilikin, Chikvaidze, I. S., Turchin, Efimova, Tret'yakova and Gverdtsiteli, I. M., 1978; Swaninathan and Ranganathan, 1957; Swaninathan, Ranganathan and Sulochana, 1958; Szmuszkovicz, 1967; Szmuszkovicz, Glenn, Heinzelman, Hester and Youngdale, 1966; Tanaka, 1940a, 1940b; Tenud, 1975; Teuber and Cornelius, 1965; Theilacker and Leichtle, 1951; Thomae G.m.b.H., 1967a; Thomas, R. C., 1975; Titley, 1928; Trenkler, 1888; Ufimtsev, Grineva and Sadovskaya, 1962; Unanyan and Tatevosyan, 1961; Upjohn Co., 1966; Verkade and Janetzky, 1943b; Verley and Beduwé, 1925; Walther and Clemen, 1900; Warner-Lambert Pharmaceutical Co., 1966; Watanabe, Yamamoto, M., Shim, Miyanaga and Mitsudo, 1980; Weisbach, Macko, De Sanctis, Cava and Douglas, B., 1964; Welch, 1977a, 1977b; Wieland, T. and Rühl, 1963; Wislicenus and Arnold, E., 1887; Wislicenus and Münzesheimer, 1898; Wislicenus and Schultz, F., 1924; Wislicenus and Waldmüller, 1911; Witkop, 1944, 1953b; Witkop and Patrick, 1952; Witkop, Patrick and Kissman, 1952; Witkop, Patrick and Rosenblum, 1951; Yakovenko, Mikhlina, Oganesyan and Yakhontov, 1975; Yamada, S., Chibata and Tsurui, 1953; Yamada, S. *et al.*, 1954; Yamamoto, H., Okamoto, Sasajima, Nakao, Maruyama and Katayama, 1971; Yoshimura, Sakamoto, and Matsumaga, 1969; Yoshitomi Pharmaceutical Industries Ltd, 1967; Zenitz, 1965, 1966.

1-^{14}C-Phenylhydrazine

Robinson, J. R., 1957 – see also Robinson, J. R. and Good, 1957.

2-Methylphenylhydrazine

Ainsworth, D. P. and Suschitzky, 1967b; Anderson, G. and Campbell, 1950; Andrisano and Vitali, 1957; Bajwa and Brown, R. K., 1968a, 1970; Barclay and Campbell, 1945;

Basangoudar and Siddappa, 1972; Bisagni, Ducrocq and Hung, 1980; Boss and Timberlake, 1963; Bullock and Hand, J. J., 1956a, 1956b; Buu-Hoï and Hoán, 1951; Buu-Hoï, Hoán and Khôi, 1950b; Buu-Hoï, Hoán, Khôi and Xuong, 1949; Buu-Hoï and Saint-Ruf, 1962, 1965a; Buu-Hoï, Saint-Ruf, Jacquignon and Marty, 1963; Carlin, Henley and Carlson, 1957; Chakraborty, Das, K. C. and Basak, 1968; Dambal and Siddappa, 1965; Da Settimo, Primofiore, Biagi and Santerini, 1976b; Feofilaktov and Semenova, N. K., 1953b; Ghigi, 1933b; Gilbert, J., Rousselle, Gansser and Viel, 1979; Grandberg and Bobrova, 1973, 1974b; Grandberg and Dashkevich, S. N., 1970a; Grandberg and Ivanova, T. A., 1970a, 1970c; Grandberg and Moskvina, 1970, 1971, 1972, 1974a; Grandberg and Nikitina, 1971b; Grandberg and Zuyanova, 1968, 1970; Hester, 1968; Hunsberger, Shaw, E. R., Fugger, Ketcham and Lednicer, 1956 (preparation); Kotov, Sagitullin and Gorbunov, 1970; Lakshmanan, 1960; Lions and Ritchie, 1939a, 1939b; Marion and Oldfield, 1947; Mears, Oakeshott and Plant, 1934; Mendlik and Wibaut, 1931; Mooradian, 1976b; Pausacker and Schubert, 1949b; Pecca and Albonico, 1970, 1971; Raschen, 1887; Shvedov, Altukhova, Komissarova and Grinev, A. N., 1965a, 1965b; Sibiryakova, Brovkin, Belyakova and Grandberg, 1969; Stevens, F. J. and Su., 1962; Suvorov and Antonov, 1952 – see also Suvorov, 1957; Suvorov and Murasheva, 1961; Szmuszkovicz, 1967; Szmuskovicz, Glenn, Heinzelman, Hester and Youngdale, 1966; Walker, C., 1894; Wieland, T. and Rühl, 1963; Yamane and Fujimori, K., 1976.

2,3-Dimethylphenylhydrazine

Anderson and Campbell, 1950; Buu-Hoï, Hoán and Khôi, 1950b; Buu-Hoï, Hoán, Khôi and Xuong, 1949; Buu-Hoï and Saint-Ruf, 1962, 1965a, 1965b; Buu-Hoï, Saint-Ruf, Jacquignon and Marty, 1963; Franzen, Onsager and Faerden, 1918 (preparation); Marion and Oldfield, 1947; Stevens, F. J. and Su, 1962; Thomas, T. J., 1975; Yamane and Fujimori, K., 1976.

4-Methylphenylhydrazine

Aksanova, Kucherova and Sharkova, L. M., 1969; Andrisano and Vitali, 1957; Arbusow, A. E. and Tichwinsky, 1910b; Arbuzov, A. E. and Tikhvinskii, 1913b; Badische Anilin- & Soda-Fabrik, A.-G., 1965; Barclay and Campbell, 1945; Basanagoudar and Siddappa, 1972; Boido and Boido Canu, 1973; Borsche and Klein, 1941; Borsche, Witte, A. and Bothe, 1908; Bremner and Browne, 1975; Brown, F. and Mann, 1948a, 1948b; Bullock and Hand, J. J., 1956a, 1956b; Buu-Hoï, Cagniant, P., Hoán and Khôi, 1950; Buu-Hoï and Hoán, 1949, 1951, 1952; Buu-Hoï, Hoán and Khôi, 1950b; Buu-Hoï, Hoán, Khôi and Xuong, 1949, 1950, 1951; Buu-Hoï and Jacquignon, 1956; Buu-Hoï, Khôi and Xuong, 1950, 1951; Buu-Hoï, Périn and Jacquignon, 1965; Buu-Hoï, Roussel, O. and Jacquignon, 1964; Buu-Hoï and Saint-Ruf, 1962, 1963, 1965a, 1965b; Buu-Hoï, Saint-Ruf, Jacquignon and Marty, 1963; Buu-Hoï and Xuong, 1952; Campbell and Cooper, 1935; Cardani, Piozzi and Casnati, 1955; Cattanach, Cohen, A. and Heath-Brown, 1968; Chakraborty, Das, K. C. and Basak, 1968; Chapman, Clarke and Hughes, H., 1965; Cornforth, J. W., Hughes, G. K., Lions and Haradence, 1937–1938; Cowper and Stevens, T. S., 1974; Dalgliesh and Mann, 1947; Dambal and Siddappa, 1965; Da Settimo, Biagi, Primofiore, Ferrarini and Livi, 1978; Da Settimo, Primofiore, Biagi and Santerini, 1976b; David and Régent, 1964; Duc and Fetizon, 1969; Duncan and Boswell, 1973; Farbenfabriken vorm. Friedr. Bayer and Co., 1901; Fitzpatrick, J. T. and Hiser, 1957; Fox and Bullock, 1951b, 1955b; Gadaginamath and Siddappa, 1975; Geigy A.-G., 1966; Grandberg and Bobrova,

(*continued*)

Table 31 (*continued*)
4-Methylphenylhydrazine (*continued*)

1973, 1974b; Grandberg and Dashkevich, 1970a, 1971a; Grandberg and Ivanova, T. A., 1970a, 1970c; Grandberg and Moskvina, 1970, 1971, 1972, 1974a; Grandberg and Zuyanova, 1968; Hester, 1968; Hunsberger, Shaw, E. R., Fugger, Ketcham and Lednicer, 1956 (preparation); Jackman and Archer, S., 1946; Janetzky and Verkade, 1945; Kaji and Nagashima, 1952; Kelly, A. H., McLeod and Parrick, 1965; Kempter, Schwalba, Stoss and Walter, 1962; Kent and McNeil, 1938; Kögl and Kostermans, 1935; Lakshmanan, 1960; Lions, 1932; Marion and Oldfield, 1947; Mashkovskii, Grinev, A. N., Andreeva, Altukhova, Shvedov, Avrutskii and Gromova, 1974; McClelland and D'Silva, 1932; Mears, Oakeshott and Plant, 1934; Mendlik and Wibaut, 1931; Merck and Co., Inc., 1962, 1963, 1964a, 1964b, 1965; Mooradian, 1976b; Morooka, Tamoto and Matuura, 1976; Mühlstädt and Lichtmann, 1970; Oakeshott and Plant, 1926; Ockenden and Schofield, 1953a, 1953b, 1957; Okawa and Konishi, 1974; Pausacker and Schubert, 1949b; Pecca and Albonico, 1970, 1971; Raschen, 1887; Rastogi, Bindra, Rai and Anand, 1972; Robson, 1924–1925; Rothstein and Feitelson, 1956; Sarett and Shen, T.-Y., 1965, 1966a, 1966b; Sen and Ghosh, 1927; Sharkova, N. M., Kucherova, Portnova and Zagorevskii, 1968; Shen, T.-Y., 1964, 1965, 1966, 1967b; Shvedov, Altukhova and Grinev, A. N., 1966b; Shvedov, Altukhova, Komissarova and Grinev, A. N., 1965a, 1965b; Shvedov, Kurilo and Grinev, A. N., 1969; Snyder, Beilfuss and Williams, J. K., 1953; Stevens, F. J., Ashby and Downey, 1957; Stevens, F. J. and Su, 1962; Suvorov and Antonov, 1952; Suvorov, Gordeev and Vasin, 1974; Suvorov and Murasheva, 1961; Szmuszkovicz, Glenn, Heinzelman, Hester and Youngdale, 1966; Thomae G.m.b.H., 1967a; Upjohn Co., 1966; Verkade and Janetsky, 1943b; Walker, C., 1894; Walton, Stammer, Nutt, Jenkins and Holly, 1965; Welch, W. M., 1977b; Wieland, T. and Rühl, 1963; Woolley, D. W. and Shaw, E., 1955; Yamamoto, H., Inaba, Okamoto, Hirohashi, T., Ishizumi, Yamamoto, M., Maruyama, Mori and Kobayashi, T., 1970; Yamamoto, H., Inaba, Okamoto, Hirohashi, T., Yamamoto, M., Mori, Ishigura, Maruyama and Kobayashi, T., 1972a, 1972b; Yamane and Fujimori, K., 1976; Yoshitomi Pharmaceutical Industries Ltd, 1967; Zenitz, 1966.

2,4-Dimethylphenylhydrazine

Anderson and Campbell, 1950; Borsche and Groth, 1941; Borsche, Witte, A. and Bothe, 1908; Bullock and Hand, J. J., 1956a, 1956b; Buu-Hoï and Saint-Ruf, 1962; Buu-Hoï, Saint-Ruf, Jacquignon and Marty, 1963; Cardani, Piozzi and Casnati, 1955; Carlin, Henley and Carlson, 1957; Casnati, Langella, Piozzi, Ricca and Umani-Ronchi, 1964b; Cattanach, Cohen, A. and Heath-Brown, 1968; David and Régent, 1964; Franzen, Onsager and Faerden, 1918 (preparation); Hunsberger, Shaw, E. R., Fugger, Ketcham and Lednicer, 1956; Kesswani, 1970; Merck and Co., Inc., 1964a; Stevens, F. J. and Su, 1962; Walser, Blount and Fryer, 1973; Yamane and Fujimori, K., 1976; Yoshitomi Pharmaceutical Industries Ltd, 1967.

2,3,4-Trimethylphenylhydrazine

Carlin and Moores, 1959, 1962; Shah, G. D. and Patel, B. P. J., 1979.

2,5-Dimethylphenylhydrazine

Anderson and Campbell, 1950; Bisagni, Ducrocq and Hung, 1980; Buu-Hoï, Cagniant, P., Hoán and Khôi, 1950; Buu-Hoï, Hoán and Khôi, 1950b; Buu-Hoï and Saint-Ruf, 1962, 1965a; Buu-Hoï, Saint-Ruf, Jacquignon and Marty, 1963; Carlin and Carlson, 1957, 1959; Carlin, Henley and Carlson, 1957; Cranwell and Saxton, 1962; Driver,

Matthews and Sainsbury, 1979; Franzen, Onsager and Faerden, 1918 (preparation); Marion and Oldfield, 1947; Neber, Knöller, Herbst and Trissler, 1929; Plancher and Caravaggi, 1905; Stevens, F. J. and Su, 1962; Thomas, T. J., 1975; Yoshitomi Pharmaceutical Industries Ltd, 1967.

3,5-Dimethylphenylhydrazine

Borsche and Groth, 1941; Stevens, F. J. and Su, 1962; Snyder, Beilfuss and Williams, J. K., 1953.

2,4,5-Trimethylphenylhydrazine

Borsche, Witte, A. and Bothe, 1908; Carlin and Harrison, J. W., 1965; Carlin and Moores, 1962; Cranwell and Sexton, 1962; Franzen, Onsager and Faerden, 1918 (preparation); Miller, B. and Matjeka, 1980.

2-Ethylphenylhydrazine

Bajwa and Brown, R. K., 1968a; Hunsberger, Shaw, E. R., Fugger, Ketcham and Lednicer, 1956 (preparation).

4-Ethylphenylhydrazine

Gaudion, Hook and Plant, 1947; Jones, G., 1975; Merck and Co., Inc., 1965; Plant, Rogers, K. M. and Williams, S. B. C., 1935; Plant and Williams, S. B. C., 1934; Willgerodt and Harter, 1905 (preparation).

2-(1-Hydroxyethyl)phenylhydrazine

Strandtmann, Cohen, M. P. and Shavel, 1963.

2-Carboxymethylphenylhydrazine

Neber, 1922 (preparation).

4-Carboxymethylphenylhydrazine

Hunsberger, Shaw, E. R., Fugger, Ketcham and Lednicer, 1956 (preparation).

4-Carbethoxymethylphenylhydrazine

Ishii, Murakami, Y., Furuse, Hosoya, Takeda and Ikeda, 1973.

4-Cyanomethylphenylhydrazine

Winchester and Popp, 1975.

(*continued*)

Table 31 (*continued*)

3-Cyanomethyl-2,5-dimethylphenylhydrazine

Winchester and Popp, 1975.

4-(2-Phthalimidoethyl)phenylhydrazine

Manske, R. H. F. and Kulka, 1947.

4-Propylphenylhydrazine

Jones, G., 1975.

4-Isopropylphenylhydrazine

Jones, G., 1975.

2-Cyclopropylphenylhydrazine

Demerson, Humber, Philipp and Martel, 1976.

4-(2-Carboxyethyl)phenylhydrazine

Manske, R. H. F. and Kulka, 1947.

4-(2-Carbethoxyethyl)phenylhydrazine

Shvedov, Altukhova and Grinev, A. N., 1966b.

4-Butylphenylhydrazine

Biere, Rufer, Ahrens, Schroeder, E., Losert, Loge and Schillinge, 1973; Colwell, Horner, J. K. and Skinner, 1964.

2- and 4-tert.-Butylphenylhydrazines

Hunsberger, Shaw, E. R., Fugger, Ketcham and Lednicer, 1956 (preparation).

4-(1,1-Dimethylpropyl)phenylhydrazine

Buu-Hoï, Hoán and Khôi, 1950b.

4-(1-Methylbutyl)phenylhydrazine

Hunsberger, Shaw, E. R., Fugger, Ketcham and Lednicer, 1956 (preparation).

4-(3-Methylbutyl)phenylhydrazine

I. G. Farbenind, A.-G., 1930d.

2,4-Di(3-methylbutyl)phenylhydrazine

Casnati, Langella, Piozzi, Ricca and Umani-Ronchi, 1964a, 1964b.

2,3-Tetramethylenephenylhydrazine

Fusco and Sannicolò, 1974; Illingsworth, Spencer, Mee and Heseltine, 1968; and Shagalov, Ostapchuk, Zlobina, Eraksina and Survorov, 1977.

2-Biphenylyhydrazine

Buu-Hoï and Hoán, 1951; Buu-Hoï, Hoán and Khôi, 1950b; Buu-Hoï, Hoán, Khôi and Xuong, 1949; Kost, Sugorova and Yakubov, 1965; Scherrer, 1969.

4-Biphenylyhydrazine

Buu-Hoï, Hoán, Khôi and Xuong, 1949; Buu-Hoï, Périn and Jacquignon, 1965; Jones, G., 1975; Jones, G. and Tringham, 1975; Kempter, Schwalba, Stoss and Walter, 1962; Stevens, F. J., Ashby and Downey, 1957; Suvorov, Antonov and Rokhlin, 1953.

4-Benzylphenylhydrazine

Woolley, D. and Shaw, E. N., 1959.

2-(α-Hydroxybenzyl)phenylhydrazine

Strandtmann, Cohen, M. P. and Shavel, 1963.

2-Methoxyphenylhydrazine

Ainsworth, D. P. and Suschitzky, 1967a; Aktieselskabet Dumex (Dumex Ltd), 1963; Barnes, Pausacker and Schubert, 1949; Bell and Lindwall, 1948; Bisagni, Ducrocq and Hung, 1980; Blaikie and Perkin, 1924; Borsche and Groth, 1941; Buu-Hoï, Hoán and Khôi, 1950b; Chalmers, J. R., Openshaw and Smith, G. F., 1957; Cook, J. W., Loudon and McCloskey, 1952; Cross, P. E. and Jones, E. R. H., 1964; Da Settimo, Biagi, Primofiore, Ferrarini and Livi, 1978; Da Settimo, Primofiore, Biagi and Santerini, 1976b; Desaty and Keglević, 1965; Douglas, B., Kirkpatrick, Moore, B. P. and Weisbach, 1964; Findlay, S. P. and Dougherty, 1948; Grandberg and Bobrova, 1973; Grandberg and Dashkevich, S. N., 1970a; Grandberg and Ivanova, T. A., 1970a, 1970c; Grandberg and Moskvina, 1970, 1971, 1972; Grandberg and Zuyanova, 1968, 1970; Hester, 1968; Hunsberger, Shaw, E. R., Fugger, Ketcham and Lednicer, 1956 (preparation); Kanaoka, Ban, Miyashita, Irie and Yonemitsu, 1966; Keglević, Desaty, Goleš and Stančić, 1968; Manske, R. H. F., 1931a; Pecca and Albonico, 1970, 1971; Perkin and Rubenstein, 1926; Rastogi, Bindra, Rai and Anand, 1972; Sakai, S., Wakabayashi and Nishina, 1969; Shen, T.-Y., 1964; Shvedov, Altukhova, Komissarova and Grinev, A. N., 1965a, 1965b; Szmuszkovicz, Glenn, Heinzelman, Hester and Youngdale, 1966; Thomae G.m.b.H., 1967a, 1967b; Upjohn Co., 1966; Vejdělek, 1957.

2-Methoxy-4-methylphenylhydrazine

Shen, T.-Y., 1964.

(*continued*)

Table 31 (*continued*)

5-Cyanomethyl-2-methoxyphenylhydrazine

Manske, R. H. F. and Kulka, 1950a.

4-Methoxyphenylhydrazine

Aktieselskabet Dumex (Dumex Ltd), 1964; Altschul, 1892 (preparation); Barrett, H. S. B., Perkin and Robinson, R., 1929; Basangoudar and Siddappa, 1972; Bell and Lindwall, 1948; Beer, Donavanik and Robertson, 1954; Blaikie and Perkin, 1924; Boido and Boido Canu, 1973; Borsche, Witte, A. and Bothe, 1908; Bremner and Browne, 1975; Buu-Hoï, Mangane and Jacquignon, 1966, 1967; Buu-Hoï, Périn and Jacquignon, 1966; Chakraborty, Das, K. C. and Chowdhury, 1966a, 1966b, 1969; Chapman, Clarke and Hughes, H., 1965; Chemerda and Sletzinger, 1968a, 1968b, 1968c, 1968d, 1968e, 1968f, 1969a, 1969b, 1969c; Chowdhury and Chakraborty, 1974; Ciba Ltd, 1960; Colwell, Horner, J. K. and Skinner, 1964; Coombes, Harvey, D. J. and Reid, S. T., 1970; Da Settimo, Biagi, Primofiore, Ferrarini and Livi, 1978; Da Settimo, Primofiore, Biagi and Santerini, 1976a, 1976b; Desaty and Keglević, 1965; Douglas, B., Kirkpatrick, Moore, B. P. and Weisbach, 1964; Duncan and Boswell, 1973; Epstein and Goldman, 1973; Fabbrica Italiana Sintetici S.p.A., 1968; Fernandez Alvarez and Monge Vega, 1975; Findlay, S. P. and Dougherty, 1948; Gabrielyan and Papayan, 1973; Gaimster, 1960; Geigy A.-G., 1966; Gilbert, J., Rousselle, Gansser and Viel, 1979; Grandberg and Bobrova, 1973; Grandberg and Dashkevich, S. N., 1970a, 1971a; Grandberg and Ivanova, T. A., 1970a, 1970c; Grandberg and Zuyanova, 1968, 1970; Grandberg, Zuyanova, Afonina and Ivanova, T. A., 1967; Harvey, D. J., 1968; Herdieckerhoff and Tschunkur, 1933; Hester, 1968; Hughes, G. K. and Lions, 1937–1938a; Hunsberger, Shaw, E. R., Fugger, Ketcham and Lednicer, 1956 (preparation); Johnson, R. P. and Oswald, 1968; Julia, Melamed and Gombert, 1965; Kaji and Nagashima, 1952; Keglević, Desaty, Goleš and Stančić, 1968; Kempter, Schwalba, Stoss and Walter, 1962; Kermack and Tebrich, 1940; Korczynski and Kierzek, 1925; Korczyński and Kierzek, 1925; Kost, Yurovskaya, Mel'nikova and Potanina, 1973; Kotov, Sagitullin and Gorbunov, 1970; Lee, T. B. and Swan, 1956; Leekning, 1964, 1967; Maki, Masugi, Hiramitsu and Ogiso, 1973; Merck and Co., Inc., 1961, 1962, 1963, 1964a, 1964b, 1964c, 1965, 1966, 1967b; Millson and Robinson, R., 1955; Milne and Tomlinson, M. L., 1952; Mooradian, 1976b; Morooka, Tamoto and Matuura, 1976; Mühlstädt and Lichtmann, 1970; Nagaraja and Sunthankar, 1958; Ockenden and Schofield, 1957; Pecca and Albonico, 1970, 1971; Perkin and Rubenstein, 1926; Rensen, 1959; Sakai, S., Wakabayashi and Nishina, 1969; Sarett and Shen, T.-.Y., 1965, 1966a, 1966b; Sharkova, N. M., Kucherova, Portnova and Zagorevskii, 1968; Shaw, E., 1955; Shen, T.Y., 1968; Shen, T.-Y., 1964, 1965, 1966, 1967a, 1967b, 1967c, 1969; Shvedov, Altukhova and Grinev, A. N., 1966b; Shvedov, Altukhova, Komissarova and Grinev, A. N., 1965a, 1965b; Shvedov, Kurilo and Grinev, A. N., 1969; Societe des Usines Chimiques Rhone-Poulenc, 1960; Späth and Brunner, O., 1925; Suvorov, Mamaev and Shagalov, 1955 – see also Suvorov, 1957; Szmuszkovicz, 1967; Szmuszkovicz, Glenn, Heinzelman, Hester and Youngdale, 1966; Thomae G.m.b.H., 1967a, 1967b; Unanyan and Tatevosyan, 1961; Upjohn Co., 1966; Vejdělek, 1957; Welch, 1977a, 1977b; Wheeler, 1970; Woolley, D. and Shaw, E. N., 1959; Yamamoto, H., Inaba, Okamoto, Hirohashi, T., Yamamoto, M., Mori, Ishigura, Maruyama and Kobayashi, T., 1972a, 1972b; Yoshitomi Pharmaceutical Industries Ltd, 1967; Zenitz, 1974.

3,4-Dimethoxy-2-propylphenylhydrazine

Beer, Brown, J. P. and Robertson, 1951.

2-Ethoxyphenylhydrazine

Aktieselskabet Dumex (Dumex Ltd), 1964; Barnes, Pausacker and Schubert, 1949; Basangoudar and Siddappa, 1972; Franzen and Schmidt, M., 1917 (preparation); Hughes, G. K. and Lions, 1937–1938a; Hunsberger, Shaw, E. R., Fugger, Ketcham and Lednicer, 1956 (preparation).

4-Ethoxyphenylhydrazine

Altschul, 1892 (preparation); Ballantine, Barrett, C. B., Beer, Boggiano, Eardley, Jennings and Robertson, 1957; Barclay and Campbell, 1945; Boido and Boido Canu, 1973; Borsche, Witte, A. and Bothe, 1908; Desaty and Keglević, 1965; Fernandez Alvarez and Monge Vega, 1975; Franzen and Schmidt, M., 1917 (preparation); Gordeev, Kobets and Suvorov, 1975; Hoshino and Takiura, 1936; Hughes, G. K. and Lions, 1937–1938a; Keglević, Desaty, Goleš and Stančić, 1968; Kempter, Schwalba, Stoss and Walter, 1962; Kochetkov, Kucherova and Evdakov, 1957; Merck and Co., Inc., 1967b; Nagaraja and Sunthankar, 1958; Shen, T.-Y., 1964, 1967c; Stolz, 1892 (preparation); Suvorov and Murasheva, 1961; Yoshitomi Pharmaceutical Industries Ltd, 1967.

4-(2-Hydroxyethoxy)phenylhydrazine

Suvorov, Gordeev and Vasin, 1974.

4-(2-Methoxyethoxy)phenylhydrazine

Gordeev, Kobets and Suvorov, 1975.

4-(2-Ethoxyethoxy)phenylhydrazine

Gordeev, Kobets and Suvorov, 1975.

4-Allyloxyphenylhydrazine

Shen, T.-Y., 1964.

4-Propoxyphenylhydrazine

Gordeev, Kobets and Suvorov, 1975; Merck and Co., Inc., 1967b; Suvorov and Murasheva, 1961.

4-Isopropoxyphenylhydrazine

Jones, G., 1975.

4-Butoxyphenylhydrazine

I. G. Farbenind, A.-G., 1930d; Suvorov and Murasheva, 1961.

(*continued*)

Table 31 (*continued*)

2-Phenoxyphenylhydrazine

Hunsberger, Shaw, E. R., Fugger, Ketcham and Lednicer, 1956 (preparation).

4-Phenoxyphenylhydrazine

Hunsberger, Shaw, E. R., Fugger, Ketcham and Lednicer, 1956 (preparation); Suvorov, Mamaev and Shagalov, 1955 – see also Suvorov, 1957.

2-Benzyloxyphenylhydrazine

Ek and Witkop, 1954 (the attempted synthesis was unsuccessful, treatment of 2-benzyloxybenzenediazonium chloride with either tin(II) chloride or sodium sulphite affording only 2-benzyloxyaniline hydrochloride).

4-Benzyloxyphenylhydrazine

Ahmed and Robinson, B., 1965; Beer, Broadhurst and Robertson, 1952; Bretherick, Gaimster and Wragg, W. R., 1961; Bisagni, Ducrocq, Lhoste, Rivalle and Civier, 1979; Cattanach, Cohen, A. and Heath-Brown, 1968; Coombes, Harvey, D. J. and Reid, S. T., 1970; Desaty and Keglević, 1965; Drogas Vacunas y Sueros, S. A. (Drovyssa), 1957; Gadaginamath and Siddappa, 1975b; Gaimster, 1960; Grandberg and Bobrova, 1974b; Grandberg and Zuyanova, 1968; Harvey, D. J., 1968; Jones, G. and Tringham, 1975; Justoni and Pessina, 1955, 1957a, 1960; Keglević, Desaty, Goleš and Stančić, 1968; Keglević, Stojanac and Desaty, 1961; Maki, Masugi, Hiramitsu and Ogiso, 1973; Marchand, Streffer and Jauer, 1961; Shaw, E., 1955; Societe des Usines Chimiques Rhone-Poulenc, 1960; Suvorov, Mamaev and Shagalov, 1955 – see also Suvorov, 1957; Suvorov and Murasheva, 1958, 1960; Zenitz, 1966.

4-Benzoyloxyphenylhydrazine

Lester, M. G., Petrow, V. and Stephenson, O., 1965.

2-Hydroxyphenylhydrazine

Clerc-Bory, 1954.

4-Hydroxyphenylhydrazine

Lester, M. G., Petrow, V. and Stephenson, O., 1965 (N_α-benzoyl derivative used); Mooradian, 1976b; and Ried and Kleemann, 1968.

4-Methylthiophenylhydrazine

Merck and Co., Inc., 1964a; Yoshitomi Pharmaceutical Industries Ltd, 1967. The preparation of mercaptophenylhydrazones has been stated to be 'quite unsatisfactory' (Piers, Haarsted, Cushley and Brown, R. K., 1962).

4-Phenylthiophenylhydrazine

Duncan and Andrako, 1968.

4-Benzylthiophenylhydrazine

Colwell, Horner, J. K. and Skinner, 1964; Keglević and Goleš, 1970.

2-Dimethylaminophenylhydrazine

Ainsworth, D. P. and Suschitzky, 1967a.

4-Dimethylaminophenylhydrazine

Merck and Co., Inc., 1965; Skrabal, Steiger and Zollinger, 1975 (claimed to be too unstable for use in the indolization process); Yamamoto, H., Inaba, Nakao and Koshiba, 1972.

2-Piperdinophenylhydrazine

Ainsworth, D. P. and Suschitzky, 1967a.

2-Morpholinophenylhydrazine

Ainsworth, D. P. and Suschitzky, 1967a.

2-[1-(4-Methylpiperazinyl)]phenylhydrazine

Ainsworth, D. P. and Suschitzky, 1967a.

4-Acetamidophenylhydrazine

Edwards and Plant, 1923; Perkin and Plant, 1921; Skrabal, Steiger and Zollinger, 1975.

1-Acetyl-5-hydrazino-7-hydroxy-6-methoxyindoline

Komoto, Enomoto, Miyagaki, Tanaka, Y., Nitanai and Umezawa, 1979.

1-Carbamoyl-5-hydrazino-7-hydroxy-6-methoxyindoline

Komoto, Enomoto, Tanaka, Y., Nitanai and Umazawa, 1979.

4-Dimethylthioaminophenylhydrazine

Merck and Co., Inc., 1965.

(*continued*)

Table 31 (*continued*)

2-Trifluoromethylphenylhydrazine

Forbes, E. J., Stacey, Tatlow and Wragg, R. T., 1960; Szmuszkovicz, 1970.

4-Trifluoromethylphenylhydrazine

Abbott Laboratories, 1967; Biere, Rufer, Ahrens, Schroeder, E., Losert, Loge and Schillinge, 1973; Cattanach, Cohen, A. and Heath-Brown, 1968; Forbes, E. J., Stacey, Tatlow and Wragg, R. T., 1960; Johnson, R. P. and Oswald, 1968; Merck and Co., Inc., 1965; Szmuszkovicz, 1970; Zenitz, 1966.

3,5-Ditrifluoromethylphenylhydrazine

Shen, T.-Y., 1964.

4-(2,2,2-Trifluoro-1-hydroxy-1-trifluoromethylethyl)phenylhydrazine

Dalton, Fahrenholtz and Silverzweig, 1979.

2-Fluorophenylhydrazine

Allen, F. L., Brunton and Suschitzky, 1955; Allen, F. L. and Suschitzky, 1953 – see also Suschitzky, 1953; Mooradian, 1973; Pecca and Albonico, 1970, 1971; Rastogi, Bindra, Rai and Anand, 1972; Sterling Drug Inc., 1975; Szmuszkovicz, Glenn, Heinzelman, Hester and Youngdale, 1966; Upjohn Co., 1966; Yamamoto, H., Okamoto, Sasajima, Nakao, Maruyama and Katayama, 1971.

2-Fluoro-4-methylphenylhydrazine

Shen, T.-Y., 1965.

4-Fluorophenylhydrazine

Allen, F. L., Brunton and Suschitzky, 1955; Allen, F. L. and Suschitzky, 1953 – see also Suschitzky, 1953; Biere, Rufer, Ahrens, Schroeder, E., Losert, Loge and Schillinge, 1973; Boido and Boido Canu, 1973; Bullock and Fox, 1951; Chapman, Clarke and Hughes, H., 1965; Da Settimo, Biagi, Primofiore, Ferrarini and Livi, 1978; Da Settimo, Primofiore, Biagi and Santerini, 1976a, 1976b; Duc and Fetizon, 1969; Duncan and Boswell, 1973; Finger, Gortatowski, Shiley and White, R. H., 1959; Fox and Bullock, 1955a; Harbert, Plattner, Welch, W. M., Weissman and Koe, 1980; Hester, 1968; Hunsberger, Shaw, E. R., Fugger, Ketcham and Lednicer, 1956 (preparation); Johnson, R. P. and Oswald, 1968; Joshi, K. C., Pathak and Chand, 1978a, 1978b; Lanzilotti, Littell, Fanshawe, McKenzie and Lovell, 1979; Merck and Co., Inc., 1965; Mooradian, 1973; Pecca and Albonico, 1970, 1971; Quadbeck and Röhm, 1954; Rastogi, Bindra, Rai and Anand, 1972; Rinderknecht and Niemann, 1950; Schiemann and Winkelmüller, 1933; Shen, T.-Y., 1967a; Sterling Drug Inc., 1975; Suvorov, Mamaev and Shagalov, 1953 – see also Suvorov, 1957; Szmuszkovicz, 1967; Szmuszkovicz, Glenn, Heinzelman, Hester and Youngdale, 1966; Upjohn Co., 1966; Welch, W. M., 1977a, 1977b; Welch, W. M. and Herbert, 1980; Yamamoto, H., Okamoto, Sasajima, Nakao, Maruyama and Katayama, 1971; Yoshitomi Pharmaceutical Industries Ltd, 1967; Zenitz, 1966, 1974.

4-Fluoro-2-methoxyphenylhydrazine

Shen, T.-Y., 1964.

2,4-Difluorophenylhydrazine

Mooradian, 1973; Zenitz, 1974.

3,5-Difluorophenylhydrazine

Shen, T.-Y., 1964.

2,3,4,5-Tetrafluorophenylhydrazine

Petrova, Mamaev and Yakobson, 1967, 1969.

2-Chlorophenylhydrazine

Ainsworth, D. P. and Suschitzky, 1967a; Aktieselskabet Dumex (Dumex Ltd), 1964; Barclay and Campbell, 1945; Borsche, Witte, A. and Bothe, 1908; Bullock and Hand, J. J., 1956b; Carlin, Wallace, J. G. and Fisher, 1952; Coldham, Lewis, J. W. and Plant, 1954; Da Settimo, Primofiore, Biagi and Santerini, 1976a, 1976b; Engvild, 1977; Feofilaktov and Semenova, N. K., 1953b; Fitzpatrick, J. T. and Hiser, 1957; Fox and Bullock, 1951b; Kotov, Sagitullin and Gorbunov, 1970; Mooradian, 1976b; Pappalardo and Vitali, 1958b; Pecca and Albonico, 1970, 1971; Robinson, F. P. and Brown, R. K., 1964; Stevens, F. J. and Fox, 1948; Suvorov, Mamaev and Shagalov, 1953 – see also Suvorov, 1957; Thomae G.m.b.H., 1967a, 1967b.

2-Chloro-5-methylphenylhydrazine

Coldham, Lewis, J. W. and Plant, 1954; Holla, S. and Ambekar, 1979b; Pausacker and Robinson, R., 1947.

2-Chloro-4,5-dimethylphenylhydrazine

Merck and Co., Inc., 1965.

2-Chloro-3-methoxyphenylhydrazine

Cattanach, Cohen, A. and Heath-Brown, 1968.

2-Chloro-4-methoxyphenylhydrazine

Merck and Co., Inc., 1964a; Suvorov, Mamaev and Shagalov, 1955 – see also Suvorov, 1957.

(continued)

Table 31 (*continued*)

3-Chloro-2-methylphenylhydrazine

Bullock and Hand, J. J., 1956a, 1956b; Hunsberger, Shaw, E. R., Fugger, Ketcham and Lednicer, 1956 (preparation).

2,3-Dichlorophenylhydrazine

Cattanach, Cohen, A. and Heath-Brown, 1968; Engvïld, 1977; Hunsberger, Shaw, E. R., Fugger, Ketcham and Lednicer, 1956 (preparation); Walser, Blount and Fryer, 1973.

4-Chlorophenylhydrazine

Abbott Laboratories, 1967; Basangoudar and Siddappa, 1972; Berger and Corraz, 1974, 1977; Boido and Boido Canu, 1973; Borsche, Witte, A. and Bothe, 1908; Bullock and Hand, 1956b; Buu-Hoï, Cagniant, P., Hoán and Khôi, 1950; Buu-Hoï, Hoán and Khôi, 1950b; Buu-Hoï, Hoán, Khôi and Xuong, 1949; Cattanach, Cohen, A. and Heath-Brown, 1968; Chapman, Clarke and Hughes, H., 1965; Ciba Ltd, 1960; Da Settimo, Primofiore, Biagi and Santerini, 1976a, 1976b; Duncan and Boswell, 1973; Engvïld, 1977; Farbenfabriken vorm. Friedr. Bayer and Co., 1901; Feofilaktov and Semenova, N. K., 1953b; Fox and Bullock, 1951b, 1955b; Geigy, A.-G., 1966; Gurien and Teitel, 1979, 1980; Inaba, Akatsu, Hirohashi, T. and Yamamoto, H., 1976; Inaba, Ishizumi, Okamoto and Yamamoto, H., 1972; Inaba, Ishizumi and Yamamoto, H., 1971; Ishizumi, Mori, Inaba and Yamamoto, H., 1973; Johnson, R. P. and Oswald, 1968; Kermack and Tebrich, 1940; Kotov, Sagitullin and Gorbunov, 1970; Lespagnol, A., Lespagnol, C. and Henichart, 1969; Massey and Plant, 1931; Merck and Co., Inc., 1964a; Mooradian, 1976b; Morooka, Tamoto and Matuura, 1976; Mühlstädt and Lichtmann, 1970; Nagaraja and Sunthankar, 1958; Ockenden and Schofield, 1957; Pappalardo and Vitali, 1958b; Pecca and Albonico, 1970, 1971; Pfenninger, 1968; Reynolds, B. and Carson, J., 1969; Rothstein and Feitelson, 1956; Sarett and Shen, T.-Y., 1966b; Shvedov, Altukhova and Grinev, A. N., 1966b; Societe des Usines Chimiques Rhone-Poulenc, 1960; Stevens, F. J. and Fox, 1948; Suvorov, Gordeev and Vasin, 1974; Suvorov, Mamaev and Shagalov, 1953 – see also Suvorov, 1957; Szmuszkovicz, Glenn, Heinzelman, Hester and Youngdale, 1966; Thomae G.m.b.H., 1967a, 1967b; Upjohn Co., 1966; Warner-Lambert Pharmaceutical Co., 1966; Welch, 1977a, 1977b; Welstead and Chen, Y.-H., 1976; Yamamoto, H., Inaba, Hirohashi, T., Ishizumi, Maruyama and Mori, 1970; Yamamoto, H., Inaba, Hirohashi, T., Mori, Ishizumi and Maruyama, 1969; Yamamoto, H., Inaba, Izumi, Hirohashi, T., Akatsu and Maruyama, 1970b; Yamamoto, H., Inaba, Okamoto, Hirohashi, T., Ishizumi, Akatsu, Kobayashi, T., Mori, Kume and Sato, H., 1973; Yamamoto, H., Inaba, Okamoto, Hirohashi, T., Ishizumi, Maruyama, Kobayashi, T., Yamamoto, M. and Mori, 1970; Yamamoto, H., Inaba, Okamoto, Hirohashi, T., Ishizumi, Yamamoto, M., Maruyama, Mori and Kobayashi, T., 1969a, 1969b; Yamamoto, H., Inaba, Okamoto, Hirohashi, T., Yamamoto, M., Mori, Ishigura, Maruyama and Kobayashi, T., 1972a, 1972c; Yoshitomi Pharmaceutical Industries Ltd. 1967.

4-Chloro-2-methylphenylhydrazine

Bullock and Hand, J. J., 1956a, 1956b; Cattanach, Cohen, A. and Heath-Brown, 1968; Engvïld, 1977.

2,4-Dichlorophenylhydrazine

Bullock and Fox, 1951; Bullock and Hand, J. J., 1956b; Bülow, 1918; Carlin and Fisher, 1948; Cattanach, Cohen, A. and Heath-Brown, 1968; Engvild, 1977; Fox and Bullock, 1951b, 1955a, 1955b; Hunsberger, Shaw, E. R., Fugger, Ketcham and Lednicer, 1956 (preparation); Morooka, Tamoto and Matuura, 1976; Pappalardo and Vitali, 1958b; Stevens, F. J. and Fox, 1948; Walser, Blount and Fryer, 1973.

5-Chloro-2-methylphenylhydrazine

Bullock and Hand, J. J., 1956a, 1956b.

5-Chloro-2-methoxyphenylhydrazine

Biere, Rufer, Ahrens, Schroeder, E., Losert, Loge and Schillinge, 1973; Bullock and Hand, J. J., 1956b; Hunsberger, Shaw, E. R., Fugger, Ketcham and Lednicer, 1956 (preparation); Stevens, F. J. and Higginbotham, 1954.

2,5-Dichlorophenylhydrazine

Barclay and Campbell, 1945; Carlin and Fisher, 1948; Chalmers, A. J. and Lions, 1933; Clemo and Seaton, 1954; Engvild, 1977; Fitzpatrick, J. T. and Hiser, 1957; Hansch and Muir, 1950; Pappalardo and Vitali, 1958b; Walser, Blount and Fryer, 1973.

3,5-Dichlorophenylhydrazine

Carlin and Fisher, 1948; Engvild, 1977; Pappalardo and Vitali, 1958b.

3,5-Dichloro-2-hydroxyphenylhydrazine

Ried and Kleemann, 1968.

3,5-Dichloro-4-hydroxyphenylhydrazine

Ried and Kleemann, 1968.

2-Bromophenylhydrazine

Barclay and Campbell, 1945; Barnes, Pausacker and Badcock, 1951; Carlin and Larson, 1957; Engvild, 1977; Grandberg and Bobrova, 1973; Grandberg and Dashkevich, S. N., 1970a; Hunsberger, Shaw, E. R., Fugger, Ketcham and Lednicer, 1956 (preparation); Kotov, Sagitullin and Gorbunov, 1970; Pappalardo and Vitali, 1958b; Pecca and Albonico, 1970, 1971; Plant and Tomlinson, M. L., 1931; Stevens, F. J. and Higginbotham, 1954.

2-Bromo-4-methylphenylhydrazine

Mears, Oakeshott and Plant, 1934.

(*continued*)

Table 31 (*continued*)

2-Bromo-5-methoxyphenylhydrazine

Grandberg and Zuyanova, 1968.

2-Bromo-4-chlorophenylhydrazine

Carlin and Larson, 1957.

4-Bromophenylhydrazine

Abbott Laboratories Ltd, 1967; Amorosa, 1956; Bannister and Plant, 1948; Barclay and Campbell, 1945; Basangoudar and Siddappa, 1972; Borsche, Witte, A. and Bothe, 1908; Bremner and Browne, 1975; Buu-Hoï; Cagniant, P., Hoán and Khôi, 1950; Buu-Hoï, Hoán and Khôi, 1950b; Buu-Hoï, Hoán, Khôi and Xuong, 1949, 1951; Buu-Hoï and Xuong, 1952; Cattanach, Cohen, A. and Heath-Brown, 1968; Chapman, Clarke and Hughes, H., 1965; Coló, Asero and Vercellone, 1954; Cornforth, J. W., Hughes, G. K., Lions and Harradence, 1937–1938; Dann, Volz, Demant, Pfeifer, Bergen, G., Fick and Walkenhorst, 1973; Da Settimo, Biagi, Primofiore, Ferrarini and Livi, 1978; Da Settimo, Primofiore, Biagi and Santerini, 1976a, 1976b; Elks, Elliott and Hems, 1944b; Farmaceutici Italia, S. A., 1954; Fox and Bullock, 1951; Grandberg and Bobrova, 1973; Hughes, G. K. and Lions, 1937–1938a; Inaba, Ishizumi, Okamoto and Yamamoto, H., 1972, 1975; Johnson, R. P. and Oswald, 1968; Kaji and Nagashima, 1952; Kipping and Hill, A., 1899; Korczyński and Kierzek, 1925; Kost, Yudin, Budylin and Yaryshev, 1965; Kotov, Sagitullin and Gorbunov, 1970; Lanzilotti, Littell, Fanshawe, McKenzie and Lovell, 1979; McClelland and D'Silva, 1932; Mears, Oakeshott and Plant, 1934; Morooka, Tamoto and Matuura, 1976; Nagaraja and Sunthankar, 1958; Pappalardo and Vitali, 1958b; Pecca and Albonico, 1970, 1971; Perkin and Plant, 1928; Plant and Tomlinson, M. L., 1931, 1933; Quadbeck and Röhm, 1954; Rothstein and Feitelson, 1956; Spickett, 1966; Stevens, F. J., Ashby and Downey, 1957; Stevens, F. J. and Higginbotham, 1954; Suvorov, Mamaev and Shagalov, 1953 – see also Suvorov, 1957; Welch, W. M., 1977b.

4-Bromo-2-methylphenylhydrazine

Mears, Oakeshott and Plant, 1934; Stevens, F. J. and Higginbotham, 1954.

4-Bromo-2-methoxyphenylhydrazine

Crum and Sprague, 1966.

4-Bromo-2-chlorophenylhydrazine

Carlin and Larson, 1957.

2,4-Dibromophenylhydrazine

Brunner, K., Wiedner and Kling, 1931; Carlin and Larson, 1957; Stevens, F. J. and Higginbotham, 1954.

2,5-Dibromophenylhydrazine

Brunner, K., Wiedner and Kling, 1931.

2-Iodophenylhydrazine

Hunsberger, Shaw, E. R., Fugger, Ketcham and Lednicer, 1956 (preparation); Pecca and Albonico, 1970, 1971.

4-Iodophenylhydrazine

Kalb, Schweizer, Zellner and Berthold, 1926; Pecca and Albonico, 1970, 1971; Suvorov, Mamaev and Shagalov, 1953 – see also Suvorov, 1957.

3,5-Diiodo-4-methoxyphenylhydrazine

Kalb, Schweizer, Zellner and Berthold, 1926.

3,4,5-Triiodophenylhydrazine

Kalb, Schweizer, Zellner and Berthold, 1926.

2-Nitrophenylhydrazine – see also Chapter VII, O

Ainsworth, D. P. and Suschitzky, 1967a; Atkinson, Simpson and Taylor, A., 1954; Bauer and Strauss, 1932; Bogat-skii, Ivanova, R. Y., Andronati and Zhilina, 1979; Borsche, Witte, A. and Bothe, 1908; Brunner, K., Wiedner and Kling, 1931; Deorha and Joshi, S. S., 1961; Fennell and Plant, 1932; Frasca, 1962a; Grammaticakis, 1960; Kinsley and Plant, 1956; Kotov, Sagitullin and Gorbunov, 1970; Müller, H., Montigel and Reichstein, 1937 (preparation); Ockenden and Schofield, 1953a, 1953b; Pappalardo and Vitali, 1958a; Parmerter, Cook, A. G. and Dixon, 1958; Plant, 1929; Plant and Rosser, 1928; Schofield and Theobald, 1949, 1950; Scriven, Suschitzky, Thomas, D. R. and Newton, 1979; Stevens, F. J. and Fox, 1948; Sych, 1953; Vejdělek, 1957; Yamane and Fujimori, K., 1976.

4-Methyl-2-nitrophenylhydrazine

Bogat-skii, Ivanova, R. Y., Andronati and Zhilina, 1979; Mears, Oakeshott and Plant, 1934.

5-Methyl-2-nitrophenylhydrazine

Gadaginamath, 1976.

4-Methoxy-2-nitrophenylhydrazine

Shvedov, Altukhova and Grinev, A. N., 1966b.

4-Ethoxy-2-nitrophenylhydrazine

Bogat-skii, Ivanova, R. Y., Andronati and Zhilina, 1979.

(*continued*)

Table 31 (*continued*)

4-Chloro-2-nitrophenylhydrazine

Deorha and Joshi, S. S., 1961; Bogat-skii, Ivanova, R. Y., Andronati and Zhilina, 1979; Plant and Rosser, 1928.

4-Bromo-2-nitrophenylhydrazine

Bogatt-skii, Ivanova, R. Y., Andronati and Zhilina, 1979.

5-Bromo-2-nitrophenylhydrazine

Moggridge and Plant, 1937.

4-Iodo-2-nitrophenylhydrazine

Deorha and Joshi, S. S., 1961.

4-Nitrophenylhydrazine – see also Chapter VII, O

Ainsworth, D. P. and Suschitzky, 1967a; Amorosa and Lipparini, 1956; Atkinson, Simpson and Taylor, A., 1954; Bauer and Strauss, 1932; Beyts and Plant, 1939; Biere, Rufer, Ahrens, Schroeder, E., Losert, Loge and Schillinge, 1973; Borsche, Witte, A. and Bothe, 1908; Brunner, K., Wiedner and Kling, 1931; Carey, Gal and Sletzinger, 1967; Carey, Gal, Sletzinger and Reinhold, 1968; Carry, Gal and Sletzinger, 1968; Chapman, Clarke and Hughes, H., 1965; Chemerda and Sletzinger, 1968c, 1969b; Colwell, Horner, J. K. and Skinner, 1964; Dalgliesh and Mann, 1947; Deorha and Joshi, S. S., 1961; Fennell and Plant, 1932; Feofilaktov and Semenova, N. K., 1953b; Frasca, 1962a; Grammaticakis, 1960; Inaba, Ishizumi, Mori and Yamamoto, H., 1971; Joshi, K. C., Pathak and Chand, 1978b; Kalb, Schweizer, Zellner and Berthold, 1926; Kucherova, Petruchenko and Zagorevskii, 1962; Merck and Co., Inc., 1962, 1964a, 1964b, 1965; Morooka, Tamoto and Matuura, 1976; Ockenden and Schofield, 1953b, 1957; Ogandzhanyan, Avanesova and Tatevosyan, 1968; Pappalardo and Vitali, 1958a; Parmerter, Cook, A. G. and Dixon, 1958; Plant, 1929; Plant and Rosser, 1928; Plant and Tomlinson, M. L., 1933; Rogers, C. U. and Corson, 1947; Rothstein and Feitelson, 1956; Sarett and Shen, T.-Y., 1966a, 1966b; Schofield and Theobald, 1949, 1950; Scriven, Suschitzky, Thomas, D. R. and Newton, 1979; Sen and Ghosh, 1927; Shaw, E. and Woolley, D. W., 1953; Shen, T.-Y., 1964, 1965, 1966, 1967b, 1967c; Shen, T.-Y. and Sarrett, 1966; Skrabal, Steiger and Zollinger, 1975; Suvorov, Gordeev and Vasin, 1974; Vejdělek, 1957; Walther and Clemen, 1900; Yamamoto, H., Inaba, Hirohashi, T., Ishizumi, Maruyama and Mori, 1970; Yamamoto, H., Inaba, Hirohashi, T., Mori, Ishizumi and Maruyama, 1969; Yamamoto, H., Inaba, Hirohashi, T., Okamoto, Ishizumi, Yamamoto, M., Maruyama, Mori and Kobayashi, T., 1970; Yamamoto, H., Inaba, Okamoto, Hirohashi, T., Ishizumi, Yamamoto, M., Maruyama, Mori and Kobayashi, T., 1969c; Yamamoto, H., Inaba, Okamoto, Hirohashi, T., Yamamoto, M., Mori, Ishigura, Maruyama and Kobayashi, T., 1972a, 1972b; Yamane and Fujimori, K., 1976.

2-Methyl-4-nitrophenylhydrazine

Carey, Gal and Sletzinger, 1967; Carry, Gal and Sletzinger, 1968.

2-Methoxy-4-nitrophenylhydrazine

Shen, T.-Y., 1964.

2-Hydroxy-4-nitrophenylhydrazine

Ried and Kleemann, 1968.

2-Fluoro-4-nitrophenylhydrazine

Carey, Gal and Sletzinger, 1967; Carry, Gal and Sletzinger, 1968.

2,3-Difluoro-4-nitrophenylhydrazine

Carry, Gal and Sletzinger, 1968.

2-Chloro-4-nitrophenylhydrazine

Ainsworth, D. P. and Suschitzky, 1967a; Deorha and Joshi, S. S., 1961; Shaw, E., 1954.

5-Bromo-2-hydroxy-4-nitrophenylhydrazine

Ried and Kleemann, 1968.

2-Iodo-4-nitrophenylhydrazine

Deorha and Joshi, S. S., 1961.

2,4-Dinitrophenylhydrazine—see also Chapter VII, O

Deorha and Joshi, S. S., 1961; Joshi, S. S. and Gambhir, 1956; Saleha, Siddiqui and Khan, N. H., 1979.

5-Methyl-2,4-dinitrophenylhydrazine

Deorha and Joshi, S. S., 1961.

2-Chloro-5-nitrophenylhydrazine

Ainsworth, D. P. and Suschitzky, 1967b; Bannister and Plant, 1948; Moggridge and Plant, 1937; Perkin and Plant, 1921; Sych, 1953.

3-Chloro-2-hydroxy-5-nitrophenylhydrazine

Ried and Kleemann, 1968.

(*continued*)

Table 31 (*continued*)

2-Carboxyphenylhydrazine

Avanesova, Astvatsatryan, Sarkisyan, Garibyan and Tatevosyan, 1975; Avanesova, Musaelyan and Tatevosyan, 1972; Barclay and Campbell, 1945; Beitz, Stroh, H.-H. and Fiebig, 1967; Brimblecombe, Downing and Hunt, R. R., 1966; Brown, U. M., Carter, P. H. and Tomlinson, M., 1958; Coldham, Lewis, J. W. and Plant, 1954; Collar and Plant, 1926; Feofilaktov and Semenova, N. K., 1953a, 1953b; Ogandzhanyan and Tatevosyan, 1970; Terzyan and Tatevosyan, 1960.

2-Carboxy-5-methylphenylhydrazine

Coldham, Lewis, J. W. and Plant, 1954.

4-Carboxyphenylhydrazine

Akopyan and Tatevosyan, 1974; Avanesova, Astvatsatryan, Sarkisyan, Garibyan and Tatevosyan, 1975; Avanesova, Musaelyan and Tatevosyan, 1972; Barclay and Campbell, 1945; Beitz, Stroh, H.-H. and Fiebig, 1967; Biere, Rufer, Ahrens, Schroeder, E., Losert, Loge and Schillinge, 1973; Brimblecombe, Downing and Hunt, R. R., 1966; Brown, U. M., Carter, P. H. and Tomlinson, M., 1958; Coldham, Lewis, J. W. and Plant, 1954; Collar and Plant, 1926; Esayan, Terzyan, Astratyan, Dzhanpoladyan and Tatevosyan, 1968; I. G. Farbenindustrie Akt.-Ges., 1934; Lindwall and Mantell, 1953; Rashidyan, Asratyan, Karagezyan, Mkrtchyan, Sedrakyan and Tatevosyan, 1968; Sharkova, N. M., Kucherova, Portnova and Zagorevskii, 1968; Terzyan, Akopyan and Tatevosyan, 1961; Terzyan and Tatevosyan, 1960.

4-Carboxy-2-hydroxyphenylhydrazine

Chakraborty, Bhattacharyya, P., Roy, S., Bhattacharyya, S. P. and Biswas, A. K., 1978.

2-Carbomethoxyphenylhydrazine

Ainsworth, D. P. and Suschitzky, 1967a.

2-Carbethoxyphenylhydrazine

Aksanova, Pidevich, Sharkova, L. M. and Kucherova, 1968; Avanesova and Tatevosyan, 1970; Beitz, Stroh, H.-H. and Fiebig, 1967.

4-Carbethoxyphenylhydrazine

Aksanova, Pidevich, Sharkova, L. M. and Kucherova, 1968; Avanesova and Tatevosyan, 1970; Beitz, Stroh, H.-H. and Fiebig, 1967; Esayan and Tatevosyan, 1972; Hughes, G. K. and Lions, 1937–1938a; Kochetkov, Kucherova, Pronina, L. P. and Petruchenko, 1959; Kucherova, Petruchenko and Zagorevskii, 1962; Lindwall and Mantell, 1953; Rothstein and Feitelson, 1956; Shvedov, Altukhova and Grinev, A. N., 1966b; Szmuszkovicz, 1971; Youngdale, Glenn, Lednicer and Szmuszkovicz, 1969.

4-Carbethoxy-2,5-dimethylphenylhydrazine

Govindachari, Rajappa and Sudarsanam, 1963.

5-Carbethoxy-2-ethylphenylhydrazine

Kesswani, 1967.

4-(2-Carbethoxy-3-indolyl)phenylhydrazine

Chilikin, Gorelova, Shvarts and Suvorov, 1979; Suvorov, Samsoniya, Chilikin, Chikvaidze, I. S., Turchin, Efimova, Tret'yakova and Gverdtsiteli, 1978.

2-Cyanophenylhydrazine

Morooka, Tamoto and Matuura, 1976.

3-Cyano-2-methylphenylhydrazine

Mosher, Crews, Acton and Goodman, 1966; Schmutz and Wittwer, 1960.

3-Cyano-2,5-dimethylphenylhydrazine

Govindachari, Rajappa and Sudarsanam, 1963.

4-Cyanophenylhydrazine

Abbott Laboratories, 1967; Johnson, R. P. and Oswald, 1968; Lindwall and Mantell, 1953; Merck and Co., Inc., 1962, 1965; Shaw, E. and Woolley, D. W., 1957; Shen, T.-Y., 1964, 1967b, 1967c; Shen, T.-Y. and Sarett, 1966.

4-Cyano-2,5-dimethylphenylhydrazine

Govindachari, Rajappa and Sudarsanam, 1963.

4-Carbamoylphenylhydrazine

Brimblecombe, Downing and Hunt, R. R., 1966 (hydrolysis of the carbamoyl group to a carboxyl group occurred under the indolization conditions); Shaw, E. and Woolley, D. W., 1957.

4-Amidinophenylhydrazine

Shaw, E. and Woolley, D. W., 1957.

2-Acetylphenylhydrazine

Strandtmann, Cohen, M. P. and Shavel, 1963.

(*continued*)

Table 31 (*continued*)

4-Acetylphenylhydrazine

Avramenko, Mosina and Suvorov, 1970; Borsche and Groth, 1941; Ishizumi, Shiori and Yamada, S., 1967; Shavel and Strandtmann, 1965; Strandtmann, Cohen, M. P. and Shavel, 1963.

4-Propionylphenylhydrazine

Strandtmann, Cohen, M. P. and Shavel, 1963.

2-Benzoylphenylhydrazine

Strandtmann, Cohen, M. P. and Shavel, 1963.

2-Benzoyl-4-methylphenylhydrazine

Ivanova, R. Y., Bogatskii, Andronati and Zhilina, 1979.

4-Benzoylphenylhydrazine

Plant and Tomlinson, M. L., 1932a; Strandtmann, Cohen, M. P. and Shavel, 1963.

4-(4-Chlorobenzoyl)phenylhydrazine

Strandtmann, Cohen, M. P. and Shavel, 1963.

4-Isonicotinoylphenylhydrazine

Strandtmann, Cohen, M. P. and Shavel, 1963.

2-Sulphophenylhydrazine

General Aniline Works, Inc., 1932; I. G. Farbenindustrie, Akt.-Ges., 1932.

4-Sulphophenylhydrazine

Feofilaktov and Semenova, N. K., 1953b (sodium salt); I. G. Farbenindustrie Akt.-Ges., 1932; Johnson, M., 1921 (preparation); Pfülf, 1887.

4-Sulphamoylphenylhydrazine

De Bellis and Stein, M. L., 1961.

N_α-Methylphenylhydrazine

Adkins and Coonradt, 1941; Armit and Robinson, R., 1922; Bailey, Hill, P. A. and Seager, 1974; Ballantine, Barrett, C. B., Beer, Boggiano, Eardley, Jennings and Robertson, 1957; Bellamy and Guthrie, 1965b; Berlin, K. D., Clark, P. E., Schroeder, J. T. and Hopper, 1968; Berti and Da Settimo, 1960; Bloch-Chaudé, Rumpf and Sadet, 1955; Borsche and Klein, 1941; Bradsher and Litzinger, 1964; Braun and Schörnig, 1925; Brown, F. and Mann,

1948a, 1948b; Bullock and Fox, 1951; Buu-Hoï and Jacquignon, 1956; Campbell and Cooper, 1935; Chalmers, A. J. and Lions, 1933; Coombes, Harvey, D. J. and Reid, S. T., 1970; Cornforth, J. W., Hughes, G. K., Lions and Harradence, 1937–1938; Degen, 1886; Duuren, 1961; Ehrhart, 1953; Elgersma, 1969; Fischer, Emil, 1886b, 1886d (preparation), 1886e; Fischer, Emil and Hess, 1884; Fischer, Emil and Jourdan, 1883; Fitzpatrick, J. T. and Hiser, 1957; Fox and Bullock, 1955a; Gal'bershtam and Samoilova, 1973; Gale, Lin and Wilshire, 1976; Ghigi, 1930; Grammaticakis, 1940a (preparation); Grandberg, Afonina and Zuyanova, 1968; Grandberg and Bobrova, 1973, 1974b; Grandberg and Ivanova, T. A., 1970b, 1970c; Grandberg and Moskvina, 1970, 1971, 1972, 1974a; Grandberg and Nikitina, 1971a; Grandberg, Sibiryakova and Brovkin, 1969; Grandberg and Tokmakov, 1973, 1974a, 1974d, 1975a; Grandberg and Zuyanova, 1970; Hamana and Kumadaki, 1967; Hanna and Schueler, 1952 (preparation); Hartman and Roll, 1943 (preparation); Harvey, D. J., 1968; Hegel, 1886; Horning, E. C., Horning, M. G. and Walker, G. N., 1948; Illingsworth, Spencer; Mee and Haseltine, 1968; Janetzky and Verkade, 1945, 1946; Janetzky, Verkade and Lieste, 1946; Julia and Lenzi, 1971; Julian and Pikl, 1933; Kanaoka, Ban, Miyashita, Irie and Yonemitsu, 1966; Keglević, Desaty, Goleš and Stančić, 1968; Kermack, Perkin and Robinson, R., 1921; Kermack and Slater, 1928; Kiang, Mann, Prior and Topham, 1956; King, F. E. and L'Ecuyer, 1934; Kissman, Farnsworth and Witkop, 1952; Komzolova, Kucherova and Zagorevskii, 1967; Kuroda, 1923; Mann and Haworth, 1944; Marion and Oldfield, 1947; McClelland and D'Silva, 1932; Millson and Robinson, R., 1955; Mooradian, 1976b; Neber, Knöller, Herbst and Trissler, 1929; Noland and Robinson, D. N., 1958; Perkin and Plant, 1921, 1923b; Plancher, 1902; Plancher, Cecchetti and Ghigi, 1929; Plancher and Forghieri, 1902; Prasad and Swan, 1958; Reif, 1909; Rice and Scott, K. R., 1970; Rice, Sheth and Wheeler, 1971; Robinson, R. and Thornley, 1924; Rosenmund, Meyer, G. and Hansal, 1975; Sharkova, N. M., Kucherova, Portnova and Zagorevskii, 1968; Shirley and Roussel, P. A., 1953; Sibiryakova, Brovkin, Belyakova and Grandberg, 1969; Sircar and Gopalan, 1932; Snyder and Cook, P. L., 1956; Snyder and Eliel, 1948; Snyder, Eliel and Carnahan, 1951; Snyder and Smith, C. W., 1943; Spickett, 1966; Sumitomo, Hayakawa and Tsubojima, 1969; Suvorov and Antonov, 1952 – see also Suvorov, 1957; Suvorov, Sorokina, N. P. and Sheinker, I. N., 1959; Szmuszkovicz, 1967; Szmuszkovicz, Glenn, Heinzelman, Hester and Youngdale, 1966; Teotino, 1959; Thyagarajan, Hillard, Reddy, K. V. and Majumder, 1974; Trofimov, Ryabchenko and Grinev, A. N., 1975; Upjohn Co., 1966; Utley and Yeboah, 1978; Verkade and Janetzky, 1943b; Woolley, D. W. and Shaw, E., 1955; Wuyts and Lacourt, 1935; Yoshitomi Pharmaceutical Industries Ltd, 1967.

N_α,2-Dimethylphenylhydrazine

Bloss and Timberlake, 1963; Marion and Oldfield, 1947; Suvorov and Antonov, 1952.

N_α,4-Dimethylphenylhydrazine

Bloss and Timberlake, 1963; Braun and Schörnig, 1925; Farbenfabriken vorm. Friedr. Bayer and Co., 1902; Kost, Yudin and Zinchenko, 1973; Suvorov and Antonov, 1952; Trofimov, Ryabchenko and Grinev, A. N., 1975.

N_α-2,4-Trimethylphenylhydrazine

Desaty and Keglević, 1965.

(*continued*)

Table 31 (*continued*)

N_α,2,5-Trimethylphenylhydrazine

Kricka and Vernon, 1974.

4-Ethyl-N_α-methylphenylhydrazine

Suvorov and Sorokina, N. P., 1960.

N_α-Methyl-2-biphenylylhydrazine

Little, Taylor, W. I. and Thomas, B. R., 1954.

2-Methoxy-N_α-methylphenylhydrazine

Bell and Lindwall, 1948; Millson and Robinson, R., 1955.

4-Methoxy-N_α-methylphenylhydrazine

Bell and Lindwall, 1948; Gardner, J. H. and Stevens, J. R., 1947; Kermack and Tebrich, 1940; King, F. E. and Robinson, R., 1933; Mann and Tetlow, 1957; Millson and Robinson, R., 1955; Schlittler, Burckhardt and Gellert, 1953; Shaw, E., 1955; Späth and Brunner, O., 1925; Trofimov, Ryabchenko and Grinev, A. N., 1975; Woolley, D. and Shaw, E. N., 1959; Woolley, D. W. and Shaw, E., 1955.

4-Ethoxy-N_α-methylphenylhydrazine

Stedman, 1924.

4-Acetamido-N_α-methylphenylhydrazine

Perkin and Plant, 1921.

4-Chloro-N_α-methylphenylhydrazine

Farbenfabriken vorm. Friedr. Bayer and Co., 1902; Trofimov, Ryabchenko and Grinev, A. N., 1975; Yamamoto, H., Inaba, Okamoto, Hirohashi, T., Yamamoto, M., Mori, Ishigura, Maruyama and Kobayashi, T., 1972a, 1972b.

N_α-Methyl-2-nitrophenylhydrazine

Millson and Robinson, R., 1955.

N_α-Methyl-4-nitrophenylhydrazine

Amorosa and Lipparini, 1959; Diels and Dürst, 1914; Millson and Robinson, R., 1955.

N_α-Ethylphenylhydrazine

Brown, F. and Mann, 1948a; Crowther, Mann and Purdie, 1943; Eiter and Svierak, 1952; Elgersma, 1969; Farbenfabriken vorm. Friedr. Bayer and Co., 1902; Fischer, Emil, 1875c (preparation); Fischer, Emil and Hess, 1884; Grammaticakis, 1940a (preparation); Grandberg, Afonina and Zuyanova, 1968; Grandberg, Zuyanova, Afonina and Ivanova, T. A., 1967; Hegel, 1886; Janetzky, Verkade and Lieste, 1946; Michaelis, 1886, 1897; Neber, Knöller, Herbst and Trissler, 1929; Philips, 1889 (preparation); Rosenmund, Meyer, G. and Hansal, 1975; Singh, B. K., 1913 (preparation); Szmuskovicz, Glenn, Heinzelman, Hester and Youngdale, 1966.

N_α-Ethyl-4-methylphenylhydrazine

Crowther, Mann and Purdie, 1943; Farbenfabriken vorm. Friedr. Bayer and Co., 1902.

4-Ethoxy-N_α-ethylphenylhydrazine

Mann and Haworth, 1944.

4-Chloro-N_α-ethylphenylhydrazine

Farbenfabriken vorm. Friedr. Bayer and Co., 1902.

N_α-Allylphenylhydrazine

Michaelis and Luxembourg, 1893 (preparation); Mooradian, 1976b; Rosenmund, Meyer, G. and Hansal, 1975; Singh, B. K., 1920 (preparation).

N_α-Allyl-4-methylphenylhydrazine

Michaelis and Luxembourg, 1893 (preparation).

N_α-Propylphenylhydrazine

Crowther, Mann and Purdie, 1943; Elgersma, 1969; Michaelis, 1897; Singh, B. K., 1920 (preparation).

N_α-Isopropylphenylhydrazine

Elgersma, 1969; Grammaticakis, 1940a (preparation); Grandberg and Bobrova, 1973; Grandberg and Nikitina, 1971b; Grandberg and Tokmakov, 1973, 1974a, 1974d, 1975a; Michaelis, 1897; Philips, 1889 (preparation).

N_α-Butylphenylhydrazine

Farbenfabriken Bayer A.-G., 1954; Hörlein, H. U., 1957; Spickett, 1966.

(*continued*)

Table 31 (*continued*)

N_α-Butyl-4-carbethoxyphenylhydrazine

Kucherova, Petruchenko and Zagorevskii, 1962.

N_α-Isobutylphenylhydrazine

Crowther, Mann and Purdie, 1943; Elgersma, 1969; Michaelis, 1897; Philips, 1889 (preparation).

4-Chloro-N_α-cyclopropylmethylphenylhydrazine

Yamamoto, H., Inaba, Hirohashi, T., Ishigura, Maruyama and Mori, 1971; Yamamoto, H., Inaba, Ishigura, Maruyama, Hirohashi, T. and Mori, 1971; Yamamoto, H., Inaba, Okamoto, Hirohashi, T., Ishizumi, Yamamoto, M., Maruyama, Mori and Kobayashi, T., 1969a; Yamamoto, H., Inaba, Okamoto, Hirohashi, T., Yamamoto, M., Mori, Ishigura, Maruyama and Kobayashi, T., 1972a, 1972b.

N_α-(3-Methylbutyl)phenylhydrazine

Michaelis, 1897; Philips, 1889 (preparation).

N_α-Cyclohexylphenylhydrazine

Elgersma, 1969.

N_α-(2-Hydroxyethyl)phenylhydrazine

Clemo and Perkin, 1924b.

N_α-(2-Ethoxyethyl)phenylhydrazine

Farbenfabriken Bayer A.-G., 1955a; Hörlein, U., 1954, 1956.

N_α-(2-Propoxyethyl)phenylhydrazine

Hörlein, U., 1954.

N_α-(2-Isopropoxyethyl)phenylhydrazine

Farbenfabriken Bayer A.-G., 1955a; Hörlein, U., 1954, 1956.

N_α-(2-Butoxyethyl)phenylhydrazine

Farbenfabriken Bayer A.-G., 1955a; Hörlein, U., 1954, 1956.

N_α-(2-Isobutoxyethyl)phenylhydrazine

Hörlein, U., 1954.

N_α-(2-*sec*.-Butoxyethyl)phenylhydrazine

Farbenfabriken Bayer, A.-G., 1955a; Hörlein, U., 1956.

N_α-(2-*tert*.-Butoxyethyl)phenylhydrazine

Farbenfabriken Bayer A.-G., 1955a; Hörlein, U., 1956.

N_α-(2-Cyclohexoxyethyl)phenylhydrazine

Hörlein, U., 1954.

N_α-(2-Phenoxyethyl)phenylhydrazine

Hörlein, U., 1954.

N_α-(2-Ethylthioethyl)phenylhydrazine

Farbenfabriken Bayer A.-G., 1954; Hörlein, U., 1954, 1956.

4-Chloro-N_α-(2-ethylthioethyl)phenylhydrazine

Farbenfabriken Bayer A.-G., 1954; Hörlein, U., 1957.

N_α-(2-Isopropylthioethyl)phenylhydrazine

Farbenfabriken Bayer A.-G., 1955a; Hörlein, U., 1956.

N_α-(2-Dimethylaminoethyl)phenylhydrazine

Aksanova, Sharkova, N. M., Baranova, Kucherova and Zagorevskii, 1966.

4-Chloro-N_α-(2-dimethylaminoethyl)phenylhydrazine

Farbenfabriken Bayer A.-G., 1955b.

N_α-(2-Diethylaminoethyl)phenylhydrazine

Ciba Ltd, 1960; Mooradian, 1976b.

N_α-(3-Dimethylaminopropyl)phenylhydrazine

Mooradian, 1976b; Yamamoto, H. and Atami, 1976.

N_α-(3-Dimethylamino-2-methylpropyl)phenylhydrazine

Yamamoto, H. and Atami, 1976.

(*continued*)

Table 31 (*continued*)

N_α-(4-Diethylamino-1-methylbutyl)phenylhydrazine

Ciba Ltd, 1960.

N_α-(2-Chloroethyl)phenylhydrazine

Clemo and Perkin, 1924b.

N_α-(3-Chlorobut-2-enyl)phenylhydrazine

Akopyan and Tatevosyan, 1971.

N_α-(Carbethoxymethyl)phenylhydrazine

Mooradian, 1976b; Perkin and Riley, 1923; Smith, W. S. and Moir, 1952.

N_α-[2-(3,4,5-Trimethoxybenzoyloxy)ethyl]phenylhydrazine

Mooradian, 1976b.

N_α-(1-Carboxycyclopentyl)phenylhydrazine

Plant and Rippon, 1928.

N_α-(Cyanomethyl)phenylhydrazine

Hahn, W. E., Bartnik, Zawadzka and Nowaczyk, W., 1968; Hahn, W. E. and Zawadzka, 1967; Hahn, W., Nowaczyk, M., Bartnik and Zawadzka, 1968.

N_α-(2-Cyanoethyl)phenylhydrazine

Hahn, W. E. and Bartnik, 1972; Hahn, W. E., Bartnik and Zawadzka, 1966; Hahn, W. E., Bartnik, Zawadzka and Nowaczyk, W., 1968; Hahn, W. E., Nowaczyk, M. and Bartnik, 1968; Hahn, W. E., Zawadzka and Szwedowska, 1969; Hahn, W., Nowaczyk, M., Bartnik and Zawadzka, 1968.

N_α-(2-Cyanoethyl)-4-methoxyphenylhydrazine

Hahn, W., Kryczka and Bartnik, 1975.

N_α-Phenylphenylhydrazine

Arbuzov, A. E. and Valitova, 1957 (preparation); Canas–Rodriguez and Leeming, 1971; Drapkina, Grigor'eva and Doroshina, 1974; Farbenfabriken Bayer A.-G., 1954; Fischer, Emil, 1876b (preparation); Fischer, Emil, 1887; Fischer, Emil and Hess, 1884; Grandberg and Bobrova, 1973; Grandberg and Moskvina, 1971, 1972, 1973; Grandberg and Nikitina, 1971a, 1971b; Grandberg and Tokmakov, 1973, 1974a, 1974d, 1975a; Grandberg and Zuyanova, 1970; Hörlein, H. U., 1957; Huang-Hsinmin and Mann, 1949; Kollenz, 1978; Kollenz, Ziegler, Eder and Prewedourakis, 1970; Linnell and Perkin,

1924; Mann and Haworth, 1944; Neber, Knöller, Herbst and Trissler, 1929; Pfülf, 1887; Sandoz Ltd, 1980; Scherrer, 1969; Spickett, 1966; Sumitomo, Hayakawa and Tsubojima, 1969; Suvorov, Antonov and Rokhlin, 1953; Welch, W. M., 1977a, 1977b; Wittig and Reichel, 1963.

4-Methyl-N_α-(4-methylphenyl)phenylhydrazine

Gordian and Gal'bershtam, 1971; Sandoz Ltd, 1980.

4-Methoxy-N_α-(4-methoxyphenyl)phenylhydrazine

Sandoz Ltd, 1980.

4-Fluoro-N_α-(4-fluorophenyl)phenylhydrazine

Gordian and Gal'bershtam, 1971; Sandoz Ltd, 1980.

4-Chloro-N_α-(4-chlorophenyl)phenylhydrazine

Sandoz Ltd, 1980.

N_α-Benzylphenylhydrazine

Alexander and Mooradian, 1977b; Antrick, 1885; Cornforth, J. W., Cornforth, R. H., Dalgliesh and Neuberger, 1951; Farbenfabriken Bayer A.-G., 1954; Geigy A.-G., 1966, 1968; Grammaticakis, 1940a (preparation); Grandberg, Afonina and Zuyanova, 1968; Grandberg and Bobrova, 1973, 1974b; Grandberg and Ivanova, T. A., 1970b; Grandberg and Moskvina, 1970, 1971, 1972; Grandberg and Nikitina, 1971a, 1971b; Grandberg, Sibiryakova and Brovkin, 1969; Grandberg and Tokmakov, 1973; 1974a, 1974d, 1975a; Grandberg and Zuyanova, 1970; Grandberg, Zuyanova, Afonina and Ivanova, T. A., 1967; Harvey, D. J., 1968; Hörlein, H. U., 1957; Hörlein, U., 1954; Kochetkov, Kucherova and Evdakov, 1956; Mooradian, 1976b; Namis, Cortes, Collera and Walls, F., 1966; Neber, Knöller, Herbst and Trissler, 1929; Philips, 1889 (preparation); Rosenmund, Meyer, G. and Hansal, 1975; Schut, Ward, Lorenzetti and Hong, 1970; Spickett, 1966; Suvorov and Murasheva, 1961; Trofimov, Ryabchenko and Grinev, A. N., 1975; Yoneda, Miyamae and Nitta, 1967.

N_α-Benzyl-4-methylphenylhydrazine

Geigy, A.-G., 1966, 1968; Kochetkov, Kucherova and Evdakov, 1956; Trofimov, Ryabchenko and Grinev, A. N., 1975.

N_α-Benzyl-4-methoxyphenylhydrazine

Gaines, Sletzinger and Ruyle, M., 1961; Grandberg, Afonina and Zuyanova, 1968; Grandberg and Bobrova, 1973; Grandberg, Zuyanova, Afonina and Ivanova, T. A., 1967; Shaw, E., 1955; Shen, T.-Y., 1967c; Sletzinger, Gaines and Ruyle, W. V., 1957; Sletzinger, Ruyle and Gaines, 1961; Trofimov, Ryabchenko and Grinev, A. N., 1975; Woolley, D. and Shaw, E. N., 1959.

(continued)

Table 31 (*continued*)

N_α-Benzyl-4-ethoxyphenylhydrazine

Kochetkov, Kucherova and Evdakov, 1956; Kucherova, Petruchenko and Zagorevskii, 1962.

N_α-Benzyl-4-benzyloxyphenylhydrazine

Harvey, D. J., 1968.

N_α-Benzyl-4-methylthiophenylhydrazine

Archer, S., 1963; Sterling Drug Inc., 1962.

N_α-Benzyl-4-chlorophenylhydrazine

Geigy, A.-G., 1966, 1968; Hörlein, U., 1954; Trofimov, Ryabchenko and Grinev, A. N., 1975.

N_α-Benzyl-4-nitrophenylhydrazine

Shen, T.-Y. and Sarett, 1966.

N_α-Benzyl-4-carbethoxyphenylhydrazine

Kochetkov, Kucherova and Evdakov, 1956; Kucherova, Petruchenko and Zagorevskii, 1962; Sharkova, N. M., Kucherova, Portnova and Zagorevskii, 1968.

N_α-(3-Methylbenzyl)phenylhydrazine

Geigy A.-G., 1966.

N_α-(4-Methylbenzyl)-4-methylthiophenylhydrazine

Sterling Drug Inc., 1962.

N_α-(3-Methoxybenzyl)phenylhydrazine

Geigy A.-G., 1968.

4-Methoxy-N_α-(4-methoxybenzyl)phenylhydrazine

Merck and Co., Inc., 1961.

4-Benzyloxy-N_α-(4-methoxybenzyl)phenylhydrazine

Merck and Co., Inc., 1961.

4-Chloro-N_α-(4-methoxybenzyl)phenylhydrazine

Farbenfabriken Bayer A.-G., 1954; Hörlein, H. U., 1957.

4-Methoxy-N_α-(2,4-dimethoxybenzyl)phenylhydrazine

Sarett and Shen, T.-Y., 1966b.

N_α-(3,4-Methylenedioxybenzyl)-4-methylthiophenylhydrazine

Archer, S., 1963; Sterling Drug Inc., 1962.

4-Methoxy-N_α-(4-methylthiobenzyl)phenylhydrazine

Sarett and Shen, T.-Y., 1966b.

N_α-(2-Chlorobenzyl)phenylhydrazine

Geigy A.-G., 1966, 1968.

N_α-(2-Chlorobenzyl)-4-methylthiophenylhydrazine

Archer, S., 1963; Sterling Drug Inc., 1962.

N_α-(4-Chlorobenzyl)phenylhydrazine

Farbenfabriken Bayer A.-G., 1954; Geigy A.-G., 1966, 1968; Hörlein, U., 1954; Mooradian, 1976b; Yoneda, Miyamae and Nitta, 1967; Yoshitomi Pharmaceutical Industries Ltd, 1967.

N_α-(4-Chlorobenzyl)-4-methylphenylhydrazine

Walton, Stammer, Nutt, Jenkins and Holly, 1965.

N_α-(4-Chlorobenzyl)-4-methoxyphenylhydrazine

Sarett and Shen, T.-Y., 1966b.

N_α-(4-Chlorobenzyl)-4-methylthiophenylhydrazine

Archer, S., 1963; Sarett and Shen, T.-Y., 1965, 1966a, 1966b; Sterling Drug Inc., 1962.

2-Chloro-N_α-(4-chlorobenzyl)phenylhydrazine

Farbenfabriken Bayer A.-G., 1954.

(*continued*)

Table 31 (*continued*)

4-Chloro-N_α-(4-chlorobenzyl)phenylhydrazine

Farbenfabriken Bayer A.-G., 1954; Hörlein, H. U., 1957; Hörlein, U., 1954.

N_α-(2,4-Dichlorobenzyl)-4-methylthiophenylhydrazine

Archer, S., 1963; Sterling Drug Inc., 1962.

N_α-(3,4-Dichlorobenzyl)-4-methylthiophenylhydrazine

Archer, S., 1963; Sterling Drug Inc., 1962.

4-Bromo-N_α-(4-bromobenzyl)phenylhydrazine

Gordian and Gal'bershtam, 1971.

4-Methoxy-N_α-(4-nitrobenzyl)phenylhydrazine

Sarett and Shen, T.-Y., 1966b.

N_α-(1-Phenylethyl)phenylhydrazine

Geigy A.-G., 1966, 1968.

N_α-(2-Phenylethyl)phenylhydrazine

Hörlein, U., 1954.

N_α-(2-Pyridylmethyl)phenylhydrazine

Berger, L. and Corraz, 1973; Farbenfabriken Bayer A.-G., 1954; Geigy A.-G., 1966; Hörlein, H. U., 1957; Hörlein, U., 1954; Yamamoto, H., Atsumi, Aono and Kuwazima, 1970.

4-Methyl-N_α-(2-pyridylmethyl)phenylhydrazine

Berger, L. and Corraz, 1973.

4-Chloro-N_α-(2-pyridylmethyl)phenylhydrazine

Berger, L. and Corraz, 1973; Farbenfabriken Bayer A.-G., 1954.

N_α-(3-Pyridylmethyl)phenylhydrazine

Berger, L. and Corraz, 1973.

4-Methyl-N_{α}-(3-pyridylmethyl)phenylhydrazine

Berger, L. and Corraz, 1973.

4-Chloro-N_{α}-(3-pyridylmethyl)phenylhydrazine

Berger, L. and Corraz, 1973.

N_{α}-[(2-Methyl-5-pyridyl)methyl]phenylhydrazine

Berger, L. and Corraz, 1973.

4-Methyl-N_{α}-[(2-methyl-5-pyridyl)methyl]phenylhydrazine

Berger, L. and Corraz, 1973.

4-Chloro-N_{α}-[(2-methyl-5-pyridyl)methyl]phenylhydrazine

Berger, L. and Corraz, 1973.

N_{α}-[(2-Methyl-5-pyridyl-N-oxide)methyl]phenylhydrazine

Berger, L. and Corraz, 1973.

4-Methyl-N_{α}-[(2-methyl-5-pyridyl-N-oxide)methyl]phenylhydrazine

Berger, L. and Corraz, 1973.

4-Chloro-N_{α}-[(2-methyl-5-pyridyl-N-oxide)methylphenylhydrazine

Berger, L. and Corraz, 1973.

N_{α}-(4-Pyridylmethyl)phenylhydrazine

Berger, L. and Corraz, 1973.

4-Methyl-N_{α}-(4-pyridylmethyl)phenylhydrazine

Berger, L. and Corraz, 1973.

4-Chloro-N_{α}-(4-pyridylmethyl)phenylhydrazine

Berger, L. and Corraz, 1973.

N_{α}-[1-(2-Pyridyl)ethyl]phenylhydrazine

Yamamoto, H., Atsumi, Aono and Kuwazima, 1970.

(*continued*)

Table 31 (*continued*)

N$_{\alpha}$-[2-(2-Pyridyl)ethyl]phenylhydrazine

Berger, L. and Corraz, 1973; Grandberg and Bobrova, 1973.

4-Methyl-N$_{\alpha}$-[2-(2-pyridyl)ethyl]phenylhydrazine

Berger, L. and Corraz, 1973.

4-Chloro-N$_{\alpha}$-[2-(2-pyridyl)ethyl]phenylhydrazine

Berger, L. and Corraz, 1973.

N$_{\alpha}$-[2-(3-Pyridyl)ethyl]phenylhydrazine

Berger, L. and Corraz, 1973.

4-Methyl-N$_{\alpha}$-[2-(3-pyridyl)ethyl]phenylhydrazine

Berger, L. and Corraz, 1973.

4-Chloro-N$_{\alpha}$-[1-(3-pyridyl)ethyl]phenylhydrazine

Berger, L. and Corraz, 1972b.

4-Chloro-N$_{\alpha}$-[2-(3-pyridyl)ethyl]phenylhydrazine

Berger, L. and Corraz, 1968, 1970b, 1973.

N$_{\alpha}$-[2-(2-Methyl-5-pyridyl)ethyl]phenylhydrazine

Berger, L. and Corraz, 1973; Kost, Terent'ev, A. P., Vinogradova, Terent'ev, P. B. and Ershov, 1960; Kost, Vinogradova, Daut and Terent'ev, A. P., 1962; Kost, Vinogradova, Trofimov, Mukhanova, Nozdrich and Shadurskii, 1967; Vinogradova, Daut, Kost and Terent'ev, A. P., 1962.

4-Methyl-N$_{\alpha}$-[2-(2-methyl-5-pyridyl)ethyl]phenylhydrazine

Berger, L. and Corraz, 1968, 1973; Kost, Vinogradova, Daut and Terent'ev, A. P., 1962; Kost, Vinogradova, Trofimov, Mukhanova, Nozdrich and Shadurskii, 1967; Vinogradova, Daut, Kost and Terent'ev, A. P., 1962.

4-Methoxy-N$_{\alpha}$-[2-(2-methyl-5-pyridyl)ethyl]phenylhydrazine

Kost, Vinogradova, Daut and Terent'ev, A. P., 1962; Kost, Vinogradova, Trofimov, Mukhanova, Nozdrich and Shadurskii, 1967; Vinogradova, Daut, Kost and Terent'ev, A. P., 1962.

4-Chloro-N_α-[2-(2-methyl-5-pyridyl)ethyl]phenylhydrazine

Berger, L. and Corraz, 1968, 1973.

4-Chloro-2-methyl-N_α-[2-(2-methyl-5-pyridyl)ethyl]phenylhydrazine

Berger, L. and Corraz, 1968.

4-Bromo-N_α-[2-(2-methyl-5-pyridyl)ethyl]phenylhydrazine

Berger, L. and Corraz, 1968.

N_α-[2-(2-Methyl-5-pyridyl-N-oxide)ethyl]phenylhydrazine

Berger, L. and Corraz, 1973; Kost, Vinogradova, Trofimov, Mukhanova, Nozdrich and Shadurskii, 1967.

4-Methyl-N_α-[2-(2-methyl-5-pyridyl-N-oxide)ethyl]phenylhydrazine

Berger, L. and Corraz, 1973.

4-Chloro-N_α-[2-(2-methyl-5-pyridyl-N-oxide)ethyl]phenylhydrazine

Berger, L. and Corraz, 1973.

N_α-[2-(4-Pyridyl)ethyl]phenylhydrazine

Berger, L. and Corraz, 1972a, 1973; Trofimov, Ryabchenko and Grinev, A. N., 1975.

2-Methyl-N_α-[2-(4-pyridyl)ethyl]phenylhydrazine

Berger, L. and Corraz, 1972a.

4-Methyl-N_α-[2-(4-pyridyl)ethyl]phenylhydrazine

Berger, L. and Corraz, 1973; Kost, Trofimov, Tsyshkova and Shadurskii, 1969; Trofimov, Ryabchenko and Grinev, A. N., 1975.

2,4-Dimethyl-N_α-[2-(4-pyridyl)ethyl]phenylhydrazine

Berger, L. and Corraz, 1972a.

4-Methoxy-N_α-[2-(4-pyridyl)ethyl]phenylhydrazine

Berger, L. and Corraz, 1972a; Trofimov, Ryabchenko and Grinev, A. N., 1975.

(*continued*)

Table 31 (*continued*)

4-Methoxy-2-methyl-N_α-[2-(4-pyridyl)ethyl]phenylhydrazine

Berger, L. and Corraz, 1972a.

N_α-[2-(4-Pyridyl)ethyl]-4-trifluoromethylphenylhydrazine

Berger, L. and Corraz, 1972a.

2-Methyl-N_α-[2-(4-pyridyl)ethyl]-4-trifluoromethylphenylhydrazine

Berger, L. and Corraz, 1972a.

4-Chloro-N_α-[2-(4-pyridyl)ethyl]phenylhydrazine

Berger, L. and Corraz, 1972a, 1973; Trofimov, Ryabchenko and Grinev, A. N., 1975.

4-Chloro-2-methyl-N_α-[2-(4-pyridyl)ethyl]phenylhydrazine

Berger, L. and Corraz, 1972a.

N_α-[3-(2-Pyridyl)propyl]phenylhydrazine

Berger, L. and Corraz, 1973.

4-Methyl-N_α-[3-(2-pyridyl)propyl]phenylhydrazine

Berger, L. and Corraz, 1973.

4-Chloro-N_α-[3-(2-pyridyl)propyl]phenylhydrazine

Berger, L. and Corraz, 1973.

N_α-[3-(3-Pyridyl)propyl]phenylhydrazine

Berger, L. and Corraz, 1973.

4-Methyl-N_α-[4-(2-pyridyl)butyl]hydrazine

Berger, L. and Corraz, 1973.

4-Chloro-N_α-[3-(3-pyridyl)propyl]phenylhydrazine

Berger, L. and Corraz, 1973.

N_α-[3-(2-Methyl-5-pyridyl)propyl]phenylhydrazine

Berger, L. and Corraz, 1973.

4-Methyl-N$_{\alpha}$-[3-(2-methyl-5-pyridyl)propyl]phenylhydrazine

Berger, L. and Corraz, 1973.

4-Chloro-N$_{\alpha}$-[3-(2-methyl-5-pyridyl)propyl]phenylhydrazine

Berger, L. and Corraz, 1973.

N$_{\alpha}$-[3-(2-Methyl-5-pyridyl-N-oxide)propyl]phenylhydrazine

Berger, L. and Corraz, 1973.

4-Methyl-N$_{\alpha}$-[3-(2-methyl-5-pyridyl-N-oxide)propyl]phenylhydrazine

Berger, L. and Corraz, 1973.

4-Chloro-N$_{\alpha}$-[3-(2-methyl-5-pyridyl-N-oxide)propyl]phenylhydrazine

Berger, L. and Corraz, 1973.

N$_{\alpha}$-[3-(4-Pyridyl)propyl]phenylhydrazine

Berger, L. and Corraz, 1973.

4-Methyl-N$_{\alpha}$-[3-(4-pyridyl)propyl]phenylhydrazine

Berger, L. and Corraz, 1973.

4-Chloro-N$_{\alpha}$-[3-(4-pyridyl)propyl]phenylhydrazine

Berger, L. and Corraz, 1973.

N$_{\alpha}$-[4-(2-Pyridyl)butyl]phenylhydrazine

Berger, L. and Corraz, 1973.

4-Methyl-N$_{\alpha}$-[4-(2-pyridyl)butyl]hydrazine

Berger, L. and Corraz, 1973.

4-Chloro-N$_{\alpha}$-[4-(2-pyridyl)butyl]phenylhydrazine

Berger, L. and Corraz, 1973.

N$_{\alpha}$-[4-(3-Pyridyl)butyl]phenylhydrazine

Berger, L. and Corraz, 1973.

(*continued*)

Table 31 (*continued*)

4-Methyl-N_α-[4-(3-pyridyl)butyl]phenylhydrazine

Berger, L. and Corraz, 1973.

4-Chloro-N_α-[4-(3-pyridyl)butyl]phenylhydrazine

Berger, L. and Corraz, 1973.

N_α-[4-(2-Methyl-5-pyridyl)butyl]phenylhydrazine

Berger, L and Corraz, 1973.

4-Methyl-N_α-[4-(2-methyl-5-pyridyl)butyl]phenylhydrazine

Berger, L. and Corraz, 1973.

4-Chloro-N_α-[4-(2-methyl-5-pyridyl)butyl]phenylhydrazine

Berger, L. and Corraz, 1973.

N_α-[4-(2-Methyl-5-pyridyl-N-oxide)butyl]phenylhydrazine

Berger, L. and Corraz, 1973.

4-Methyl-N_α-[4-(2-methyl-5-pyridyl-N-oxide)butyl]phenylhydrazine

Berger, L. and Corraz, 1973.

4-Chloro-N_α-[4-(2-methyl-5-pyridyl-N-oxide)butyl]phenylhydrazine

Berger, L. and Corraz, 1973.

N_α-[4-(4-Pyridyl)butyl]phenylhydrazine

Berger, L. and Corraz, 1973.

4-Methyl-N_α-[4-(4-pyridyl)butyl]phenylhydrazine

Berger, L. and Corraz, 1973.

4-Chloro-N_α-[4-(4-pyridyl)butyl]phenylhydrazine

Berger, L. and Corraz, 1973.

N_α-[5-(2-Pyridyl)pentyl]phenylhydrazine

Berger, L. and Corraz, 1973.

4-Methyl-N_α-[5-(2-pyridyl)pentyl]phenylhydrazine

Berger, L. and Corraz, 1973.

4-Chloro-N_α-[5-(2-pyridyl)pentyl]phenylhydrazine

Berger, L. and Corraz, 1973.

N_α-[5-(3-Pyridyl)pentyl]phenylhydrazine

Berger, L. and Corraz, 1973.

4-Methyl-N_α-[5-(3-pyridyl)pentyl]phenylhydrazine

Berger, L. and Corraz, 1973.

4-Chloro-N_α-[5-(3-pyridyl)pentyl]phenylhydrazine

Berger, L. and Corraz, 1973.

N_α-[5-(2-Methyl-5-pyridyl)pentyl]phenylhydrazine

Berger, L. and Corraz, 1973.

4-Methyl-N_α-[5-(2-methyl-5-pyridyl)pentyl]phenylhydrazine

Berger, L. and Corraz, 1973.

4-Chloro-N_α-[5-(2-methyl-5-pyridyl)pentyl]phenylhydrazine

Berger, L. and Corraz, 1973.

N_α-[5-(2-Methyl-5-pyridyl-N-oxide)pentyl]phenylhydrazine

Berger, L. and Corraz, 1973.

4-Methyl-N_α-[5-(2-methyl-5-pyridyl-N-oxide)pentyl]phenylhydrazine

Berger, L. and Corraz, 1973.

4-Chloro-N_α-[5-(2-methyl-5-pyridyl-N-oxide)pentyl]phenylhydrazine

Berger, L. and Corraz, 1973.

N_α-[5-(4-Pyridyl)pentyl]phenylhydrazine

Berger, L. and Corraz, 1973.

(*continued*)

Table 31 (*continued*)

4-Methyl-N_α-[5-(4-pyridyl)pentyl]phenylhydrazine

Berger, L. and Corraz, 1973.

4-Chloro-N_α-[5-(4-pyridyl)pentyl]phenylhydrazine

Berger, L. and Corraz, 1973.

N_α-(4-Quinazolinyl)phenylhydrazine

Birchall, G. R., Hepworth and Smith, S. C., 1973; Doyle and Smith, S. C., 1974.

4-Methyl-N_α-(4-quinazolinyl)phenylhydrazine

Doyle and Smith, S. C., 1974.

4-Methoxy-N_α-(4-quinazolinyl)phenylhydrazine

Birchall, G. R., Hepworth and Smith, S. C., 1973; Doyle and Smith, S. C., 1974.

N_α-(2-Methyl-4-quinazolinyl)phenylhydrazine

Doyle and Smith, S. C., 1974.

4-Methyl-N_α-(2-methyl-4-quinazolinyl)phenylhydrazine

Doyle and Smith, S. C., 1974.

4-Methoxy-N_α-(2-methyl-4-quinazolinyl)phenylhydrazine

Doyle and Smith, S. C., 1974.

N_α-(2-Methylthio-4-quinazolinyl)phenylhydrazine

Doyle and Smith, S. C., 1974; I.C.I. Ltd, 1975.

4-Methyl-N_α-(2-methylthio-4-quinazolinyl)phenylhydrazine

Doyle and Smith, S. C., 1974; I.C.I. Ltd, 1975.

4-Methoxy-N_α-(2-methylthio-4-quinazolinyl)phenylhydrazine

Doyle and Smith, S. C., 1974; and I.C.I. Ltd, 1975.

N_α-(2-Ethylthio-4-quinazolinyl)-4-methoxyphenylhydrazine

I.C.I. Ltd, 1975.

N_α-(7-Chloro-4-quinazolinyl)phenylhydrazine

Doyle and Smith, S. C., 1974.

4-Methyl-N_α-(7-chloro-4-quinazolinyl)phenylhydrazine

Doyle and Smith, S. C., 1974.

4-Methoxy-N_α-(7-chloro-4-quinazolinyl)phenylhydrazine

Doyle and Smith, S. C., 1974.

N_α-(2-Quinoxalinyl)phenylhydrazine

Birchall, G. R., Hepworth and Smith, S. C., 1973.

4-Methoxy-N_α-(2-quinoxalinyl)phenylhydrazine

Birchall, G. R., Hepworth and Smith, S. C., 1973.

N_α-(2-Benzothiazolyl)phenylhydrazine

Birchall, G. R., Hepworth and Smith, S. C., 1973.

N_α-(2-Benzothiazolyl)-4-methoxyphenylhydrazine

Birchall, G. R., Hepworth and Smith, S. C., 1973.

4,4′-Dihydrazinobiphenyl

Arheidt, 1887; Suvorov, Samsoniya, Chilikin, Chikvaidze, I. S., Turchin, Efimova, Tret'yakova and Gverdtsiteli, I. M., 1978.

4,4′-Dihydrazinodiphenylmethane

Borsche and Kienitz, 1910; Borsche and Manteuffel, 1934; Bruck, 1970; Samsoniya, Chikvaidze, S. I., Suvorov and Gverdtsiteli, N., 1978.

4,4′-Di(N_α-methylhydrazino)diphenylmethane

Braun, 1908a (preparation); Braun, 1908b.

1-Naphthylhydrazine

Borghero and Finsterle, 1955; Borsche, Witte, A. and Bothe, 1908; Bryant and Plant, 1931; Buu-Hoï, Cagniant, P., Hoán and Khôi, 1950; Buu-Hoï, Croisy, Jacquignon and Martani, 1971; Buu-Hoï and Hoán, 1951, 1952; Buu-Hoï, Hoán and Khôi, 1949, 1950a,

(*continued*)

Table 31 (*continued*)

1-Naphthylhydrazine (*continued*)

1950b, 1950c; Buu-Hoï, Hoán, Khôi and Xuong, 1949, 1950, 1951; Buu-Hoï and Jacquignon, 1954; Buu-Hoï, Jacquignon and Ledésert, 1970; Buu-Hoï, Khôi and Xuong, 1950, 1951; Buu-Hoï, Mangane and Jacquignon, 1966, 1967; Buu-Hoï, Roussel, O. and Jacquignon, 1964; Buu-Hoï and Saint-Ruf, 1962, 1963, 1965a, 1965b; Buu-Hoï, Saint-Ruf and Dufour, 1964; Buu-Hoï and Xuong, 1952; Clemo and Felton, 1952; Dufour, Buu-Hoï, Jacquignon and Hien, 1972; Felton, 1952; Feofilaktov, 1947a; Feofilaktov and Semenova, N. K., 1953b; Findlay, S. P. and Dougherty, 1948; Fischer, Emil, 1886a (preparation); Freed, Hertz and Rice, 1964; Herdieckerhoff and Tschunkur, 1933; Hughes, G. K. and Lions, 1937–1938a; Korczynski, Brydowna and Kierzek, 1926; Lions, 1932: Menichi, Bisagni and Royer, 1964; Newberry, 1974; Oakeshott and Plant, 1928; Patel, H. P. and Tedder, 1963; Roussel, O., Buu-Hoï and Jacquignon, 1965; Schlieper, 1887; Shagalov, Eraksina and Suvorov, 1970; Shagalov, Eraksina, Tkachenko and Suvorov, 1973.

4-Chloro-1-naphthylhydrazine

Barnes, Pausacker and Badcock, 1951.

4-Bromo-1-naphthylhydrazine

Hunsberger, Shaw, E. R., Fugger, Ketcham and Lednicer, 1956 (preparation), Plant and Tomlinson, M. L., 1932b.

1-Hydrazinonaphthalene-4-sulphonic acid

Hunsberger, Shaw, E. R., Fugger, Ketcham and Lednicer, 1956 (preparation).

N_α-Methyl-1-naphthalhydrazine

Illingsworth, Spencer, Mee and Heseltine, 1968.

1-Chloro-2-naphthylhydrazine

Hunsberger, Shaw, E. R., Fugger, Ketcham and Lednicer, 1956 (preparation).

1,5-Dihydrazinonaphthalene

Samsoniya, Trapaidze, Gverdtsiteli, I. M. and Suvorov, 1977; Samsoniya, Trapaidze, Suvorov and Gverdtsiteli, I. M., 1978.

9-Phenanthrylhydrazine

Bayanov, Mirzametova, Tret'yakova, Efimova and Suvorov, 1978; Hunsberger, Shaw, E. R., Fugger, Ketcham and Lednicer, 1956 (preparation).

2-Fluorenylhydrazine

Diels, 1901; Hunsberger, Shaw, E. R., Fugger, Ketcham and Lednicer, 1956 (preparation).

3-Pyrenylhydrazine

Hunsberger, Shaw, E. R., Fugger, Ketcham and Lednicer, 1956 (preparation).

3-Coumarinylhydrazine

Khan, M. A. and Morley, 1978.

4-(9-Acridanonyl)hydrazine

Clemo, Perkin and Robinson, R., 1924.

2*H*-[1,4]benzothiazin-3(4*H*)one-8-ylhydrazine

Brunelli, Fravolini, Grandolini and Strappaghetti, 1980.

2-Methyl- and 2-Ethyl-3-thienylhydrazine

Binder, Habison and Noe, 1977 (the free hydrazines are unstable and are thus produced *in situ* under indolization conditions from their N_α-carbo-*tert.*-butoxy derivatives).

3-Carbethoxy-2-methyl-4-thienylhydrazine

Borisova and Kartashova, 1979; Shvedov, Trofimkin, Vasil'eva and Grinev, A. N., 1975 (also prepared *in situ* under indolization conditions from the corresponding N-formyl or N-acetyl derivatives); Shvedov, Trofimkin, Vasil'eva, Vlasova and Grinev, A. N., 1975 (preparation); Shvedov, Vasil'eva, Grinev, A. N. and Trofimkin, 1975 (prepared *in situ* under indolization conditions from the corresponding N-formyl or N-acetyl derivatives).

Table 32. Other Aldehyde and Ketonic Moieties with Indolization Unambiguity which have been Employed in the Fischer Indole Synthesis

---

Phenylacetaldehyde

Aktieselskabet Dumex (Dumex Ltd), 1963, 1964; Fischer, Emil, 1886b; Fischer, Emil and Schmidt, 1888; Fischer, Emil and Schmitt, 1888; Govindachari and Sudarsanam, 1967; Henle, 1905; Ince, 1889; Jackson, A. H. and Smith, P., 1968; Kanaoka, Ban, Miyashita, Irie and Yonemitsu, 1966; Kost, Sugorova and Yakubov, 1965.

2-Methoxyphenylacetaldehyde

Saleha, Khan, N. H., Siddiqui and Kidwai, M. M., 1978.

(*continued*)

Table 32 (*continued*)

4-Methoxyphenylacetaldehyde

Hino, Suzuki, T. and Nakagawa, 1973; Szmuszkovicz, Glenn, Heinzelman, Hester and Youngdale, 1966.

3,4-Dimethoxyphenylacetaldehyde

Kanaoka, Ban, Miyashita, Irie and Yonemitsu, 1966.

2-Hydroxyphenylacetaldehyde

Saleha, Khan, N. H., Siddiqui and Kidwai, M. M., 1978.

3-Chlorophenylacetaldehyde

Aktieselskabet Dumex (Dumex Ltd), 1964.

4-Bromophenylacetaldehyde

Hino, Suzuki, T. and Nakagawa, 1973.

3-Indolylacetaldehyde

Saleha, Khan, N. H., Siddiqui and Kidwai, M. M., 1978.

Ethylthioacetaldehyde

Jardine and Brown, R. K., 1965.

Carbethoxymethylthioacetaldehyde

Wieland, T. and Rühl, 1963.

Phenylthioacetaldehyde

Wieland, T. and Rühl, 1963.

Propionaldehyde

Ainsworth, D. P. and Suschitzky, 1967a; Arbusow, A. E. and Tichwinsky, 1910b; Arbuzov and Tikhvinskii, 1913b; Bauer and Strauss, 1932; Cook, J. W., Loudon and McCloskey, 1952; De Bellis and Stein, M. L., 1961; Degen, 1886; Fischer, Emil, 1886b, 1886c, 1886e; Fitzpatrick, J. T. and Hiser, 1957; Gesellschaft für Teerverwertung m.b.H., 1911; Govindachari, Rajappa and Sudarsanam, 1966; Grandberg and Bobrova, 1978; Grandberg, Sibiryakova and Brovkin, 1969; Janetzky, Verkade and Lieste, 1946; Kanaoka, Ban, Miyashita, Irie and Yonemitsu, 1966; Keglević, Stojanac and Desaty, 1961 (as the diethyl acetal); Khan, M. A. and Morley, 1978, 1979; Khan, M. A. and Rocha, 1978; King, F. E. and Robinson, R., 1933; Kost, Sugorova and Yakubov, 1965; Marion and Oldfield, 1947; Mendlik and Wibaut, 1931; Noland and Robinson, D. N., 1958; Saleha, Khan, N. H., Siddiqui and Kidwai, M. M., 1978; Sibiryakova, Brovkin, Belyakova and Grandberg, 1969; Späth and Brunner, O., 1925; Swaminathan and Ranganathan, 1957; Verley and Beduwé, 1925; Vinogradova, Daut, Kost and Terent'ev, A. P., 1962; Watanabe, Y., Yamamoto, M., Shim, Miyanaga and Mitsudo, 1980.

3-Phenylpropionaldehyde

Aktieselskabet Dumex (Dumex Ltd), 1963, 1964; Julian and Pikl, 1933; Saleha, Khan, N. H., Siddiqui and Kidwai, M. M., 1978; Swaminathan, Ranganathan and Sulochana, 1958.

3-(4-Methyl-2-pyridyl)propionaldehyde Diethylacetal

Morrison, Waite, R. O., Caro, A. N. and Shavel, 1967; Shavel and Morrison, 1967; Warner-Lambert Pharmaceutical Co., 1966.

3-Carboxypropionaldehyde

Amorosa and Lipparini, 1956; Borghero and Finsterle, 1955; Clerc-Bory, 1954; Engvild, 1977; Fox and Bullock, 1951a, 1951b, 1955b; Jackson, R. W. and Manske, R. H., 1935; Kögl and Kostermans, 1935; Leekning, 1964, 1967; Mentzer, Beaudet and Bory, 1953; Merck and Co., Inc., 1964a; Sarett and Shen, T.-Y., 1966a; Shaw, E., 1955; Shen, T.-Y., 1964; Suvorov and Murasheva, 1958, 1960, 1961; Woolley, D. and Shaw, E. N., 1959; Woolley, D. W. and Shaw, E., 1955.

3-Carbomethoxypropionaldehyde

Amorosa and Lipparini, 1959; Ellinger, 1904; Shen, T.-Y., 1964; Walton, Stammer, Nutt, Jenkins and Holly, 1965.

3-Carbethoxypropionaldehyde

Fox and Bullock, 1955a (as the dimethyl acetal); Sarett and Shen, T.-Y., 1966a, 1966b (as the dimethyl acetal); Shen, T.-Y., 1965 (as the dimethyl acetal); Stevens, F. J. and Su, 1962.

3,3-Dicarbethoxypropionaldehyde

De Bellis and Stein, M. L., 1961.

3-Carbopropoxypropionaldehyde

Shen, T.-Y., 1964.

(*continued*)

Table 32 (*continued*)

3-Carbobenzoxypropionaldehyde Diethylacetal

Shen, T.-Y., 1969.

3-Carbalkoxypropionaldehyde

Yoshimura, Sakamoto and Matsunaga, 1969.

3-Cyanopropionaldehyde Diethylacetal

Desaty and Keglević, 1965; Tanaka, Z., 1940b.

Butyraldehyde

Adkins and Coonradt, 1941; Ainsworth, D. P. and Suschitzky, 1967a; Braun and Bayer, 1925; Bullock and Fox, 1951; Desaty and Keglević, 1965; Fitzpatrick, J. T. and Hiser, 1957; Jackson, A. H. and Smith, P., 1968; Kanaoka, Ban, Miyashita, Irie and Yonemitsu, 1966; Keglević, Stojanac and Desaty, 1961 (as the diethyl acetal); Khan, M. A. and Rocha, 1978; Korczynski, Brydowna and Kierzek, 1926; Namis, Cortes, Collera and Walls, F., 1966; Plancher and Carrasco, 1905; Saleha, Khan, N. H., Siddiqui and Kidwai, M. M., 1978; Snyder, Eliel and Carnahan, 1951; Vinogradova, Daut, Kost and Terent'ev, A. P., 1962; Watanabe, Y., Yamamoto, M., Shim, Miyanaga and Mitsudo, 1980.

Isobutyraldehyde

Abramovitch and Brown, R. A., 1969; Ahmed and Robinson, B., 1965; Brunner, K., Wiedner and Kling, 1931; Grammaticakis, 1940b; Kanaoka, Ban, Miyashita, Irie and Yonemitsu, 1966.

3-Methylbutyraldehyde

Aksanova, Kucherova and Sharkova, L. M., 1969; Snyder and Smith, C. W., 1943; Trenkler, 1888.

2-Ethylbutyraldehyde

Desaty and Keglević, 1965.

2-Methyl-4-phenoxybutyraldehyde

King, F. E. and Robinson, R., 1933.

4-Benzyloxybutyraldehyde Dimethylacetal

Shen, T.Y., 1968.

4-Benzylthiobutyraldehyde

Keglević, Desaty, Goleš and Stančić, 1968 (as the diethyl acetal); Keglević and Goleš, 1970.

3-Carboxybutyraldehyde

Ellinger, 1905; Kögl and Kostermans, 1935.

3-Carbomethoxybutyraldehyde

Shen, T.-Y., 1964.

4-Carboxybutyraldehyde

Ellinger, 1905; Suvorov and Murasheva, 1961.

4-Carbomethoxybutyraldehyde

Merck and Co., Inc., 1964c.

4,4-Dicarbomethoxybutyraldehyde

Sumitomo Chemical Co. Ltd, 1966.

4-Carbobenzoxybutyraldehyde Diethyl Acetal

Shen, T.-Y., 1969.

Valeraldehyde

Saleha, Khan, N. H., Siddiqui and Kidwai, M. M., 1978; Watanabe, Y., Yamamoto, M., Shim, Miyanaga and Mitsudo, 1980.

3-Carbo-*tert*.-butoxyvaleraldehyde

Merck and Co., Inc., 1965.

4-Carbomethoxyvaleraldehyde

Merck and Co., Inc., 1964c.

5-Carboxyvaleraldehyde

Suvorov and Murasheva, 1961.

(*continued*)

Table 32 (*continued*)

5-Carbomethoxyvaleraldehyde

Bullock and Hand, J. J., 1956b; Shagalov, Eraksina, Turchin and Suvorov, 1970; Sumitomo Chemical Co., Ltd, 1966.

5-Carbethoxyvaleraldehyde

Bullock and Hand, J. J., 1956b; Shagalov, Eraksina and Suvorov, 1970; Shagalov, Eraksina, Turchin and Suvorov, 1970; Suvorov and Antonov, 1952; Suvorov, Antonov and Rokhlin, 1953; Suvorov, Mamaev and Shagalov, 1953, 1955, 1957 – see also Suvorov, 1957.

3-Carbo-*tert*.-butoxycaproaldehyde

Merck and Co., Inc., 1965.

Heptaldehyde

Korczyński and Kierzek, 1925; Trenkler, 1888.

3-Carbo-*tert*.-butoxyheptaldehyde

Merck and Co., Inc., 1965.

Octaldehyde

Korczynski, Brydowna and Kierzek, 1926 [although the product obtained from the indolization of the phenylhydrazone of this aldehyde was incorrectly named 2-n-Essilindolo in this paper, it was correctly formulated as 3-hexylindole. Unfortunately, the former erroneous nomenclature was extrapolated to the use of the corresponding incorrect formulation in a later review (Meyer, H., 1940)].

Nonaldehyde

Korczynski, Brydowna and Kierzek, 1926.

Decaldehyde

Desaty and Keglević, 1965.

Acetone

Adkins and Coonradt, 1941; Aktieselskabet Dumex (Dumex Ltd), 1963; Arheidt, 1887; Badische Anilin- & Soda-Fabrik A.-G., 1965; Bell and Lindwall, 1948; Berti and Da Settimo, 1960; Bogat-skii, Ivanova, R. Y., Andronati and Zhilina, 1979; Bülow, 1918; Cardani, Piozzi and Casnati, 1955; Carlin and Fisher, 1948; Chapman, Clarke and Hughes,

H., 1965; Coló, Asero and Vercellone, 1954; Cook, J. W., Loudon and McCloskey, 1952; Degen, 1886; Dewar, 1944; Duuren, 1961; Farbenfabriken vorm. Friedr. Bayer and Co., 1901, 1902; Farmaceutici Italia, S. A., 1954; Fischer, Emil, 1886b, 1886c, 1886e; 1889b; Fitzpatrick, J. T. and Hiser, 1957; Fusco and Sannicolò, 1974; Gesellschaft für Teerverwertung m.b.H., 1911; Gordon and Jackson, R. W., 1935; Govindachari, Rajappa and Sudarsanam, 1966; Govindachari and Sudarsanam, 1967, 1971; Grammaticakis, 1937; Herdieckerhoff and Tschunkur, 1933; I. G. Farbenindustrie Akt.-Ges., 1932; Kissman, Farnsworth and Witkop, 1952; Korczyński and Kierzek, 1925; Kost, Sugorova and Yakubov, 1965; Kost, Vinogradova, Daut and Terent'ev, A. P., 1962; Mann and Haworth, 1944; Mann and Tetlow, 1957; Marion and Oldfield, 1947; Pretka and Lindwall, 1954; Quadbeck and Röhm, 1954; Raschen, 1887; Schiemann and Winkelmüller, 1933; Schlieper, 1886, 1887; Sergeeva, Z. F., Akhvlediani, Shabunova, Korolev, Vasil'ev, A. M., Babushkina, T. N. and Suvorov, 1975; Sharkova, Kucherova and Zagorevskii, 1972b; Shvedov, Trofimkin, Vasil'eva and Grinev, A. N., 1975; Späth and Brunner, O., 1925; Venuto and Landis, 1968; Vinogradova, Daut, Kost and Terent'ev, A. P., 1962; Walther and Clemen, 1900; Welch, W. M., 1977a; Yamada, S., Chibata and Tsurui, 1953; Yamada, S. *et al.*, 1954; Yamamoto, H., Nakamura, Y., Atami, Nakao and Kobayashi, T., 1970e.

Methyl *tert*.-Butyl Ketone

Casnati, Langella, Piozzi, Ricca and Umani-Ronchi, 1964b; Colle, M.-A. and David, 1960; Plancher and Forghieri, 1902.

Methyl 1,1-Dimethylpropyl Ketone

Casnati, Langella, Piozzi, Ricca and Umani-Ronchi, 1964a, 1964b; David and Monnier, 1956, 1959.

Acetophenone (Methyl Phenyl Ketone)

Adkins and Coonradt, 1941; Al-Azawe and Sarkis, 1973; Bruce, 1960; Campbell and Cooper, 1935; Carlin and Carlson, 1957, 1959; Carlin and Fisher, 1948; Carlin and Harrison, J. W., 1965; Carlin and Larson, 1957; Carlin, Wallace, J. G. and Fisher, 1952; Crowther, Mann and Purdie, 1943; Degen, 1886; Farbenfabriken vorm. Friedr. Bayer and Co., 1901, 1902; Fischer, Emil, 1886b, 1886c, 1886e, 1887; Fitzpatrick, J. T. and Hiser, 1957; Gale, Lin and Wilshire, 1976; Gesellschaft für Teerverwertung m.b.H., 1911; Govindachari, Rajappa and Sudarsanam, 1966; Govindachari and Sudarsanam, 1967, 1971; Grammaticakis, 1937; Grandberg and Bobrova, 1978; Grinev, N. I., Sadovskaya and Ufimtsev, 1963; Herdieckerhoff and Tschunkur, 1933; Huang-Hsinmin and Mann, 1949; I. G. Farbenindustrie Akt.-Ges., 1932; Illingsworth, Spencer, Mee and Heseltine, 1968; Ince, 1889; Khan, M. A. and Morley, 1979; Kissman, Farnsworth and Witkop, 1952; Korczyński and Kierzek, 1925; Kost, Sugorova and Yakubov, 1965; MacPhillamy, Dziemian, Lucas and Kuehne, 1958; Mann and Haworth, 1944; Noland, Rush and Smith, L. R., 1966; Okawa and Konishi, 1974; Patel, H. P. and Tedder, 1963; Petrova, T. D., Mamaev and Yakobson, 1969; Pfülf, 1887; Pretka and Lindwall, 1954; Ried and Kleemann, 1968; Saleha, Khan, N. H., Siddiqui and Kidwai, M. M., 1978; Shriner, Ashley and Welch, E., 1942, 1955; Shredov, Trofimkin, Vasil'eva and Grinev, A. N., 1975; Snyder and Smith, C. W., 1943; Theilacker and Leichtle, 1951; Ufimtsev, Grineva and Salovskaya, 1962.

(*continued*)

Table 32 (*continued*)

4-Methylacetophenone

Brown, F. and Mann, 1948a; Calvaire and Pallaud, 1960; I. G. Farbenindustrie Akt.-Ges., 1932; Kaji and Nagashima, 1952.

2,5-Dimethylacetophenone

Buu-Hoï and Hoán, 1949.

4-Ethylacetophenone

Buu-Hoï and Hoán, 1949; Calvaire and Pallaud, 1960.

3-Carbethoxymethylacetophenone

Duncan and Boswell, 1973.

4-Carbethoxymethylacetophenone

Duncan and Boswell, 1973.

4-Propylacetophenone

Calvaire and Pallaud, 1960.

4-Isopropylacetophenone

Buu-Hoï and Hoán, 1949.

4-Cyclopentylacetophenone

Hai, Buu-Hoï and Xuong, 1958.

4-Cyclohexylacetophenone

Buu-Hoï, Binh, Loc, Xuong and Jacquignon, 1957.

4-Phenylacetophenone

Brown, F. and Mann, 1948a; Carlin and Fisher, 1948; Carlin, Wallace, J. G. and Fisher, 1952; Kissman, Farnsworth and Witkop, 1952; Pigerol, De Cointet de Fillain, Nanthavong and Le Blay, 1976; Shukri, Alazawe and Al-Tai, 1970.

2,3-Dimethoxyacetophenone

Bruce, 1960.

4-Methoxyacetophenone

Gale, Lin and Wilshire, 1976; Kaji and Nagashima, 1952; Korczynski and Kierzek, 1925, Korczyński and Kierzek, 1925; Szmuszkovicz, Glenn, Heinzelman, Hester and Youngdale, 1966.

4-Methoxy-3-methylacetophenone

Gale, Lin and Wilshire, 1976.

3,4-Dimethoxyacetophenone

Bruce, 1960; Korczynski, Brydowna and Kierzek, 1926; Shvedov, Kurilo and Grinev, A. N., 1970; Woodward, Cava, Ollis, Hunger, Daeniker and Schenker, 1954, 1963.

2,5-Dimethoxyacetophenone

Bruce, 1960.

4-Dodecoxyacetophenone

Pigerol, De Cointet de Fillain, Nanthavong and Le Blay, 1976.

4-Phenoxyacetophenone

Korczynski, Brydowna and Kierzek, 1926.

2-Hydroxyacetophenone

Bourdais and Lorre, 1975; Calvaire and Pallaud, 1960; McLoughlin and Smith, L. H., 1965.

2-Hydroxy-3,4-dimethylacetophenone

Merchant and Shah, N. J., 1975.

2-Hydroxy-4,6-dimethylacetophenone

Merchant and Shah, N. J., 1975.

2,3-Dihydroxyacetophenone

Bruce, 1960.

4-Hydroxyacetophenone

Ames, A. F., Ames, D. E., Coyne, Grey, Lockhart and Ralph, 1959; Korczynski and Kierzek, 1925; Korczyński and Kierzek, 1925; Pigerol, De Cointet de Fillain, Nanthavong and Le Blay, 1976.

(*continued*)

Table 32 (*continued*)

4-Hydroxy-2,5-dimethylacetophenone

Merchant and Shah, N. J., 1975.

4-Hydroxy-3,5-dimethylacetophenone

Merchant and Shah, N. J., 1975.

4-Hydroxy-3-methoxyacetophenone

Bazile, Cointet and Pigerol, 1978; Pigerol, Chandavoine, De Cointet de Fillain and Nanthavong, 1975.

3,4-Dihydroxyacetophenone

Bruce, 1960.

2,5-Dihydroxyacetophenone

Bruce, 1960.

4,5-Dihydroxy-2-methylacetophenone

Bruce, 1960.

4-Mercaptoacetophenone

Pigerol, De Cointet de Fillain, Nanthavong and Le Blay, 1976.

4-Methylthioacetophenone

Pigerol, De Cointet de Fillain, Nanthavong and Le Blay, 1976.

2-Aminoacetophenone

Duncan, Helsley and Boswell, 1973; Kiang, Mann, Prior and Topham, 1956.

2-Amino-3-ethyl-5-methylacetophenone

MacPhillamy, Dziemian, Lucas and Kuehne, 1958.

2-Amino-5-chloroacetophenone

Duncan, Helsley and Boswell, 1973.

2-Amino-5-bromoacetophenone

Duncan, Helsley and Boswell, 1973.

4-Dimethylaminoacetophenone

Gale, Lin and Wilshire, 1976.

4-Acetylaminoacetophenone

Pigerol, De Cointet de Fillain, Nanthavong and Le Blay, 1976.

2-Fluoro-5-methylacetophenone

Joshi, K. C., Pathak, Arya and Chand, 1979; Joshi, K. C., Pathak and Chand, 1978b.

4-Fluoroacetophenone

Joshi, K. C., Pathak, Arya and Chand, 1979; Joshi, K. C., Pathak and Chand, 1978a, 1978b.

4-Fluoro-2-methylacetophenone

Joshi, K. C., Pathak, Arya and Chand, 1979; Joshi, K. C., Pathak and Chand, 1978b.

4-Fluoro-3-methylacetophenone

Joshi, K. C., Pathak, Arya and Chand, 1979; Joshi, K. C., Pathak and Chand, 1978b.

3,4-Difluoroacetophenone

Joshi, K. C., Pathak and Chand, 1978b.

2-Chloroacetophenone

Ames, A. F., Ames, D. E., Coyne, Grey, Lockhart and Ralph, 1959.

3-Chloro-4-fluoroacetophenone

Joshi, K. C., Pathak, Arya and Chand, 1979; Joshi, K. C., Pathak and Chand, 1978b.

4-Chloroacetophenone

Aktieselskabet Dumex (Dumex Ltd), 1963; Brown, F. and Mann, 1948a; Carlin and Fisher, 1948; Carlin, Wallace, J. G. and Fisher, 1952; Crowther, Mann and Purdie, 1943; Gale, Lin and Wilshire, 1976; I. G. Farbenindustrie Akt.-Ges., 1932; Okawa and Konishi, 1974; Weisbach, Macko, De Sanctis, Cava and Douglas, B., 1964.

(*continued*)

Table 32 (*continued*)

5-Chloro-2-hydroxyacetophenone

Gupta and Ojha, 1971.

3-Bromoacetophenone

Dann, Bergen, G., Demant and Volz, 1971.

3-Bromo-2-hydroxy-5-methylacetophenone

Gupta and Ojha, 1971.

3-Bromo-5-chloro-2-hydroxyacetophenone

Gupta and Ohja, 1971.

4-Bromoacetophenone

Al-Azawe and Sarkis, 1973; Dann, Bergen, G., Demant and Volz, 1971; Gale, Lin and Wilshire, 1976; Kaji and Nagashima, 1952.

5-Bromo-2-hydroxyacetophenone

Gupta and Ojha, 1971.

5-Bromo-3-chloro-2-hydroxyacetophenone

Gupta and Ojha, 1971.

3,5-Dibromo-2-hydroxyacetophenone

Gupta and Ojha, 1971.

3,5-Dibromo-4-hydroxyacetophenone

Buu-Hoï, Xuong and Lavit, 1954.

2-Iodoacetophenone

Bruce, 1962.

4-Iodoacetophenone

Calvaire and Pallaud, 1960.

2-Nitroacetophenone

MacPhillamy, Dziemian, Lucas and Kuehne, 1958.

3-Nitroacetophenone

Calvaire and Pallaud, 1960.

4-Nitroacetophenone

Al-Azawe and Sarkis, 1973; Da Settimo and Saettone, 1965.

4-Carboxyacetophenone

Pigerol, De Cointet de Fillain, Nanthavong and Le Blay, 1976.

4-Carbamoylacetophenone

Pigerol, De Cointet de Fillain, Nanthavong and Le Blay, 1976.

4-Cyanoacetophenone

Pigerol, De Cointet de Fillain, Nanthavong and Le Blay, 1976.

3-Methoxy-4-methylsulphonyloxyacetophenone

Bazile, Cointet and Pigerol, 1978.

3-Methoxy-4-phenylsulphonyloxyacetophenone

Bazile, Cointet and Pigerol, 1978.

1,4-Diacetylbenzene

Buu-Hoï; Périn and Jacquignon, 1965.

1,3,5-Triacetylbenzene

Buu-Hoï, Périn and Jacquignon, 1965.

4,4′-Diacetyldiphenyl Ether

Buu-Hoï, Xuong and Lavit, 1954.

2-Acetylfuran

Calvaire and Pallaud, 1960.

(*continued*)

Table 32 (*continued*)

3-Acetyldibenzofuran

Buu-Hoï and Royer, 1950.

2-Acetylthiophen

Brunck, 1893; Holla, B. S. and Ambekar, 1974.

2-Acetyl-5-methylthiophen

Buu-Hoï and Hoán, 1949.

3-Acetyl-2,5-dimethylthiophen

Buu-Hoï and Hoán, 1949.

2-Acetyl-5-ethylthiophen

Buu-Hoï, 1958.

2-Acetyl-5-(4-methylphenyl)thiophen

Buu-Hoï and Hoán, 1950.

5-Acetyl-1-methylimidazole

Jordaan and Arndt, 1968.

5-Acetyl-2-ethylthio-1-methylimidazole

Jordaan and Arndt, 1968.

2-Acetylpyridine

Bradsher and Litzinger, 1964; Lakshmanan, 1960; Prasad and Swan, 1958; Schut, Ward, Lorenzetti and Hong, 1970; Shukri, Alazawe and Al-Tai, 1970; Societe des Usines Chimiques Rhone-Poulenc, 1960; Sugasawa, Terashima and Kanaoka, 1956.

3-Acetylpyridine

Duc and Fetizon, 1969; Gray and Archer, W. L., 1957; Huffman, 1962; Illingsworth, Spencer, Mee and Haseltine, 1968; Schut, Ward, Lorenzetti and Hong, 1970; Shukri, Alazawe and Al-Tai, 1970; Sugusawa, Terashima and Kanaoka, 1956.

4-Acetylpyridine

Gray and Archer, W. L., 1957; Lakshmanan, 1960; Shukri, Alazawe and Al-Tai, 1970; Sugusawa, Terashima and Kanaoka, 1956.

2,6-Diacetylpyridine

Buu-Hoï, Périn and Jacquignon, 1965.

1-Acetylnaphthalene

Blades and Wilds, 1956 [the earlier claim (Brunck, 1893) for the use of this ketonic moiety in a Fischer indolization was shown to be incorrect when the reactant which had been used was identified with the 2-substituted isomer (Blades and Wilds, 1956)].

2-Acetylnaphthalene

Blades and Wilds, 1956; Carlin, Wallace, J. G. and Fisher, 1952; Shukri, Alazawe and Al-Tai, 1970.

3-Acetylindole

Shukri, Alazawe and Al-Tai, 1970.

3-Acetyl-5-bromobenzofuran

Dann, Volz, Demant, Pfeifer, Bergen, G., Fick and Walkenhorst, 1973.

2-Acetylquinoline

Hamana and Kumadaki, 1967; Sugusawa and Takano, 1959.

1-Acetylisoquinoline

Sugusawa and Takano, 1959.

3-Acetylisoquinoline

Bradsher and Litzinger, 1964.

5-Acetylindane

Shukri, Alazawe and Al-Tai, 1970.

2-Acetylbiphenylene

Gardner, D. V., McOmie and Prabhu, 1970.

2-Acetylfluorene

Shukri, Alazawe and Al-Tai, 1970.

(*continued*)

Table 32 (*continued*)

2-Acetylphenanthrene

Buu-Hoï and Jacquignon, 1954.

2-Acetyltriphenylene

Buu-Hoï and Jacquignon, 1953.

Methoxyacetylbenzene

Robinson, R. and Thornley, 1926.

Hydroxyacetylbenzene

Laubmann, 1888b – see also Chapter I, A.

Benzamidoacetylbenzene

Robinson, R. and Thornley, 1926.

5,5,5-Trifluorolaevulinic Acid

Sarett and Shen, T.-Y., 1966a, 1966b.

Diethyl Ketone

Arbuzov, A. E. and Rotermel, 1932; Cattanach, Cohen, A. and Heath-Brown, 1973; Dave, V., 1976; Deorha and Joshi, S. S., 1961; Herdieckerhoff and Tschunkur, 1933; I. G. Farbenindustrie Akt.-Ges., 1932, 1934; Jackson, A. H. and Smith, P., 1968; Janetzky and Verkade, 1946; Khan, M. A. and Rocha, 1978; Kost and Yudin, 1957; Kost, Yudin, Berlin, Y. A. and Terent'ev, A. P., 1959; Kost, Yudin and Terent'ev, A. N., 1959; Miller, B. and Matjeka, 1980; Plancher, 1898f, 1902; Plancher and Forghieri, 1902; Watanabe, Yamamoto, M., Shim, Miyanaga and Mitsudo, 1980; Yudin, Kost and Berlin, Y. A., 1958.

Ethyl *tert*.-Butyl Ketone

David and Régent, 1964 and Robinson, B., 1963b.

Propiophenone (Ethyl Phenyl Ketone)

Aktieselskabet Dumex (Dumex Ltd), 1964; Arbusow, A. E., Saizew and Rasumov, 1935; Arbuzov, A. E., Zaitzev and Rasumov, 1935; Atkinson, Simpson and Taylor, A., 1954; Beer, Donavanik and Robertson, 1954; Blades and Wilds, 1956; Buu-Hoï and Xuong, 1952; Govindachari, Rajappa and Sudarsanam, 1966; Govindachari and Sudarsanam, 1967, 1971; Jackson, A. H. and Smith, P., 1968; Kanaoka, Ban, Miyashita, Irie and Yonemitsu, 1966; Khan, M. A. and Morley, 1978, 1979; Kissman, Farnsworth and

Witkop, 1952; Korczynski, Brydowna and Kierzek, 1926; Kost, Sugorova and Yakubov, 1965; Kost, Yudin, Dmitriev, B. A. and Terent'ev, A. P., 1959; Nakazaki, 1960a; Neber, Knöller, Herbst and Trissler, 1929; Plant and Tomlinson, M. L., 1933; Robinson, B., 1963b; Saleha, Khan, N. H., Siddiqui and Kidwai, M. M., 1978; Shvedov, Trofimkin, Vasil'eva and Grinev, A. N., 1975; Snyder and Smith, C. W., 1943; Verkade and Janetzky, 1943b; Witkop and Patrick, 1952; Yamada, S., Chibata and Tsurui, 1953; Yamamoto, H., Nakao and Kobayashi, T., 1969.

4-Methylpropiophenone

Buu-Hoï and Hoán, 1949.

2,4-Dimethylpropiophenone

Buu-Hoï and Hoán, 1949.

2,5-Dimethylpropiophenone

Buu-Hoï and Hoán, 1949.

2,4,5-Trimethylpropiophenone

Buu-Hoï, 1949a.

4-Ethylpropiophenone

Buu-Hoï, 1949a.

4-Isopropylpropiophenone

Buu-Hoï, 1949a.

4-*tert*.-Butylpropiophenone

Buu-Hoï, 1949a.

4-Phenylpropiophenone

Buu-Hoï and Hoán, 1949.

1,2,3,4-Tetrahydro-6-propionylnaphthalene

Buu-Hoï, 1949a.

2-Methoxy-5-methylpropiophenone

Buu-Hoï, 1949a.

(*continued*)

Table 32 (*continued*)

2-Methoxy-3,4-dimethylpropiophenone

Buu-Hoï, 1949a.

4-Methoxypropiophenone

Bailey and Bogle, 1977; Buu-Hoï, 1949b; Mentzer, 1946; Szmuszkovicz, Glenn, Heinzelman, Hester and Youngdale, 1966; Witkop, Patrick and Kissman, 1952.

4-(4-Methoxyphenyl)propiophenone

Buu-Hoï, Xuong and Thang, K. van, 1953.

4-Methoxy-2-methylpropiophenone

Buu-Hoï, 1949a.

4-Methoxy-3-methylpropiophenone

Buu-Hoï, 1949a.

2,4-Dimethoxypropiophenone

Buu-Hoï, 1949a.

3,4-Dimethoxypropiophenone

Buu-Hoï, 1949a.

4-Ethoxy-5-isopropyl-2-methylpropiophenone

Buu-Hoï, 1949a.

2,5-Dimethyl-4-propoxypropiophenone

Buu-Hoï, 1949a.

2-Benzoyloxy-5-methylpropiophenone

Auwers, Hilliger and Wulf, 1922 (the phenylhydrazone of the corresponding phenol could not be indolized).

2-Hydroxypropiophenone

Robinson, B. and Zubair, 1971.

4-Hydroxypropiophenone

Buu-Hoï, 1949b; Yamamoto, H., Nakao and Atsumi, 1968a; Yamamoto, H., Nakao and Kobayashi, T., 1969.

3,4-Dihydroxypropiophenone

Bruce, 1960.

3-Fluoro-4-hydroxypropiophenone

Buu-Hoï, Xuong and Lavit, 1954.

4-Fluoropropiophenone

Buu-Hoï, Hoán and Jacquignon, 1949.

3-Chloro-4-methoxypropiophenone

Buu-Hoï, Xuong and Lavit, 1954.

3-Chloro-4-hydroxypropiophenone

Buu-Hoï, Xuong and Lavit, 1954.

4-Chloropropiophenone

Blades and Wilds, 1956.

3-Bromo-4-methoxypropiophenone

Buu-Hoï, Xuong and Lavit, 1954.

3-Bromo-4-hydroxypropiophenone

Buu-Hoï, Xuong and Lavit, 1954.

3-Bromo-5-chloro-4-hydroxypropiophenone

Buu-Hoï, Xuong and Lavit, 1954.

4-Bromopropiophenone

Buu-Hoï and Hoán, 1949; Dann, Bergen, G., Demant and Volz, 1971.

Table 32 (*continued*)

3,5-Dibromo-4-hydroxypropiophenone

Buu-Hoï, Xuong and Lavit, 1954.

2-Aminopropiophenone

Robinson, B. and Zubair, 1973.

3-Aminopropiophenone

Robinson, B. and Zubair, 1973.

4-Aminopropiophenone

Robinson, B. and Zubair, 1973.

3-Nitropropiophenone

Buu-Hoï, 1949a.

2-Propionylthiophen

Buu-Hoï and Hoán, 1949.

3-Methyl-2-propionylthiophen

Buu-Hoï and Hoán, 1949.

5-Methyl-2-propionylthiophen

Buu-Hoï and Hoán, 1949.

2,5-Dimethyl-3-propionylthiophen

Buu-Hoï and Hoán, 1949.

2-Phenyl-5-propionylthiophen

Buu-Hoï and Hoán, 1949.

2-(4-Methylphenyl)-5-propionylthiophen

Buu-Hoï and Hoán, 1950.

2-Chloro-5-propionylthiophen

Buu-Hoï and Hoán, 1949.

2-(4-Chlorophenyl)-5-propionylthiophen

Buu-Hoï and Hoán, 1950.

2-Propionylpyridine

Bradsher and Litzinger, 1964; Kao and Robinson, R., 1955.

3-Propionylpyridine

Pfenninger, 1968.

4-Propionylpyridine

Pfenninger, 1968.

1-Methoxy-4-propionylnaphthalene

Buu-Hoï, 1949b.

2-Methoxy-6-propionylnaphthalene

Buu-Hoï, 1949b.

3-Propionylisoquinoline

Elderfield and Wythe, 1954.

1,4-Dimethoxy-2,5-dipropionylbenzene

Noland and Baude, 1966.

4,4′-Dipropionyldiphenylether

Buu-Hoï, Xuong and Lavit, 1954.

3-Propionylpyrene

Buu-Hoï, 1949a.

Dipropyl Ketone

Arbuzov, A. E. and Wagner, R. E., 1913; Grandberg, Sibiryakova and Brovkin, 1969.

(continued)

Table 32 (*continued*)

Butyrophenone (Phenyl Propyl Ketone)

Buu-Hoï and Xuong, 1952; Fitzpatrick, J. T. and Hiser, 1957; Korozynski, Brydowna and Kierzek, 1926.

4-Phenylbutyrophenone

Buu-Hoï and Hoán, 1949.

6-Butyryl-1,2,3,4-tetrahydronaphthalene

Buu-Hoï, 1949a.

4-Methoxybutyrophenone

Blades and Wilds, 1956.

2,4-Dimethoxybutyrophenone

Buu-Hoï, 1949a.

3,4-Dimethoxybutyrophenone

Buu-Hoï, 1949a.

4-Hydroxybutyrophenone

Buu-Hoï, Xuong and Lavit, 1954.

4-Fluorobutyrophenone

Buu-Hoï, Hoán and Jacquignon, 1949.

4-Chlorobutyrophenone

Buu-Hoï and Hoán, 1949.

4-Bromobutyrophenone

Buu-Hoï and Hoán, 1949.

3,5-Dibromo-4-hydroxybutyrophenone

Buu-Hoï, Xuong and Lavit, 1954.

2-Butyrylthiophen

Buu-Hoï and Hoán, 1949.

2-Butyryl-3-methylthiophen

Buu-Hoï and Hoán, 1949.

2-Butyryl-5-methylthiophen

Buu-Hoï and Hoán, 1949.

3-Butyryl-2,5-dimethylthiophen

Buu-Hoï and Hoán, 1949.

2-Butyryl-5-phenylthiophen

Buu-Hoï and Hoán, 1949.

2-Butyryl-5-(4-methylphenyl)thiophen

Buu-Hoï and Hoán, 1950.

2-Butyryl-5-chlorothiophen

Buu-Hoï and Hoán, 1949.

2-Butyrylpyridine

Kao and Robinson, R., 1955; Societe des Usines Chimiques Rhone-Poulenc, 1960.

2-Butyryl-4-methylpyridine

Societe des Usines Chimiques Rhone-Poulenc, 1960.

2-Butyryl-5-ethylpyridine

Anderson, R. M., Clemo and Swan, 1954; Lee, T. B. and Swan, 1956.

2-Butyryl-4,5-diethylpyridine

Lee, T. B. and Swan, 1956.

2-Butyryl-5-ethyl-4-isopropylpyridine

Prasad and Swan, 1958.

3-Butyryl-5,6,7,8-tetrahydroisoquinoline

Julian, Karpel, Magnani and Meyer, E. W., 1948.

(continued)

Table 32 (*continued*)

3-Butyrylpyridine

LaForge, 1928.

2-Butyryl-6-methoxynaphthalene

Buu-Hoï, 1949b.

3-Butyrylisoquinoline

Julian, Karpel, Magnani and Meyer, E. W., 1948.

5-Butyrylisoquinoline

Witkop, 1953b.

3-Butyrylpyrene

Buu-Hoï and Jacquignon, 1954.

Diisopropyl Ketone

Bloch-Chaudé, Rumpf and Sadet, 1955; Plancher, 1898a, 1898b, 1898g.

Isobutyrophenone (Isopropyl Phenyl Ketone)

Evans, F. J., Lyle, G. G., Watkins and Lyle, R. E., 1962; Evans, F. J. and Lyle, R. E., 1960; Kissman, Farnsworth and Witkop, 1952; Leuchs, Heller, A. and Hoffman, A., 1929 – see also Haselbach and Heilbronner, 1968; Nakazaki, 1960a; Skrabal, Steiger and Zollinger, 1975; Witkop, Patrick and Kissman, 1952.

2-Aminoisobutyrophenone

Robinson, B. and Zubair, 1973.

4-Dimethylaminoisobutyrophenone

Carson, D. F. and Mann, 1965; Skrabal, Steiger and Zollinger, 1975.

3-Nitroisobutyrophenone

Skrabal, Steiger and Zollinger, 1975.

3-Hydroxypropyl 2-Pyridyl Ketone

Ch'ang-pai, Evstigneeva and Preobrazhenskii, N. A., 1958.

2-Carboxyethyl Phenyl Ketone

Merck and Co., Inc., 1964a; Saleha, Khan, N. H., Siddiqui and Kidwai, M. M., 1978; Sarett and Shen, T.-Y., 1966a; Shen, T.-Y., 1967a; Walton, Stammer, Nutt, Jenkins and Holly, 1965; Yamamoto, H., Nakao and Atsumi, 1968a.

2-Carboxyethyl 4-Methylphenyl Ketone

Merck and Co., Inc., 1966.

2-Carboxyethyl 4-Methoxyphenyl Ketone

Merck and Co., Inc., 1966; Yamamoto, H., Nakao and Atsumi, 1968a.

2-Carboxyethyl 2,4-Dimethoxyphenyl Ketone

Chalmers, A. J. and Lions, 1933.

2-Carboxyethyl 3,4-Dimethoxyphenyl Ketone

Chalmers, A. J. and Lions, 1933.

2-Carboxyethyl 4-Hydroxyphenyl Ketone

Merck and Co., Inc., 1966.

2-Carboxyethyl 4-Chlorophenyl Ketone

Yamamoto, H., Nakao and Atsumi, 1968a.

2-Carbomethoxyethyl Phenyl Ketone

Shen, T.-Y., 1967c.

2-Carbethoxyethyl Phenyl Ketone

Julia, Melamed and Gombert, 1965; Saleha, Khan, N. H., Siddiqui and Kidwai, M. M., 1978; Sarett and Shen, T.-Y., 1966b.

2-Carbethoxyethyl 4-Fluorophenyl Ketone

Merck and Co., Inc., 1966; Newberry, 1974.

2-Carbethoxyethyl 4-Chlorophenyl Ketone

Newberry, 1974.

(*continued*)

Table 32 (*continued*)

Butyl Phenyl Ketone

Buu-Hoï and Royer, 1947; Buu-Hoï and Xuong, 1952.

2,5-Dimethyl-3-valerylthiophen

Buu-Hoï and Hoán, 1949.

3-Valerydibenzofuran

Buu-Hoï, 1949a.

2-Carbethoxypropyl Phenyl Ketone

Shen, T.-Y. and Sarett, 1966.

3-Carbethoxypropyl Phenyl Ketone

Yoshitomi Pharmaceutical Industries Ltd, 1967.

2,5-Dimethyl-3-(3-methylbutyryl)thiophen

Buu-Hoï and Hoán, 1949.

2-Caproylnaphthalene

Buu-Hoï and Hoán, 1949.

2-Caproyl-6-methoxynaphthalene

Buu-Hoï, 1949b.

3-Caproyldibenzofuran

Buu-Hoï, 1949a.

3-Caproylbenzothiophen

Buu-Hoï and Hoán, 1949.

Cyclopentyl Phenyl Ketone

Nakazaki, Yamamoto, K. and Yamagami, 1960.

4-Carboxybutyl Phenyl Ketone

Saleha, Khan, N. H., Siddiqui and Kidwai, M. M., 1978.

4-Carbethoxybutyl Phenyl Ketone

Suvorov, Antonov and Rokhlin, 1953.

4-Carbethoxybutyl 4-Methylphenyl Ketone

Suvorov, Antonov and Rokhlin, 1953 – see also Suvorov, 1957.

2-Heptanoyl-6-methoxynaphthalene

Buu-Hoï, 1949b.

9-Ethyl-3-heptanoylcarbazole

Buu-Hoï, 1949a.

9-Ethyl-3-heptanoyl-6-nitrocarbazole

Buu-Hoï, 1949a.

3-Heptanoylbenzothiophen

Buu-Hoï and Hoán, 1949.

Cyclohexyl Phenyl Ketone

Hughes, G. K. and Lions, 1937–1938b.

Cyclohexyl 4-Methoxyphenyl Ketone

Hughes, G. K. and Lions, 1937–1938b.

Cyclohexyl 3,4-Dimethoxyphenyl Ketone

Hughes, G. K. and Lions, 1937–1938b.

3-Caprylyl-2,5-dimethylthiophen

Buu-Hoï and Hoán, 1949.

2-Caprylylnaphthalene

Buu-Hoï and Hoán, 1949.

3-Caprylylbenzothiophen

Buu-Hoï and Hoán, 1949.

(continued)

Table 32 (*continued*)

3-Nonanoylbenzothiophen

Buu-Hoï and Hoán, 1949.

3,5-Dibromo-4-hydroxyphenyl Nonyl Ketone

Buu-Hoï, Xuong and Lavit, 1954.

2-Caprylnaphthalene

Buu-Hoï and Hoán, 1949.

3-Caprylbenzothiophen

Buu-Hoï and Hoán, 1949.

2-Undecanoylnaphthalene

Buu-Hoï and Hoán, 1949.

3-Undecanoylbenzothiophen

Buu-Hoï and Hoán, 1949.

2-Laurylnaphthalene

Buu-Hoï and Hoán, 1949.

3-Laurylnaphthalene

Buu-Hoï and Hoán, 1949.

2-Myristylnaphthalene

Buu-Hoï and Hoán, 1949.

3-Myristylbenzothiophen

Buu-Hoï and Hoán, 1949.

Dipentadecyl Ketone

Buu-Hoï and Royer, 1947.

2,3,4-Trihydroxyphenyl Pentadecyl Ketone

Buu-Hoï, 1954.

2,5-Dimethyl-3-palmitylthiophen

Buu-Hoï and Hoán, 1949.

2-Palmitylnaphthalene

Buu-Hoï and Hoán, 1949.

3-Palmitylbenzothiophen

Buu-Hoï and Hoán, 1949.

2-Stearylnaphthalene

Buu-Hoï and Hoán, 1949.

Heneicosyl 2,4-Dihydroxyphenyl Ketone

Buu-Hoï, 1954.

Heneicosyl 3,4-Dihydroxyphenyl Ketone

Buu-Hoï, 1954.

Heneicosyl 2,3,4-Trihydroxyphenyl Ketone

Buu-Hoï, 1954.

Dibenzyl Ketone

Korczynski, Brydowna and Kierzek, 1926; Neber, Knöller, Herbst and Trissler, 1929; Rosenthal and Yalpani, 1965; Trenkler, 1888.

1-Hydroxyimino-1,3-diphenylpropan-2-one

Neber, Knöller, Herbst and Trissler, 1929.

Benzyl Phenyl Ketone (Desoxybenzoin)

Aktieselskabet Dumex (Dumex Ltd), 1963; Beer, Donavanik and Robertson, 1954; Coldham, Lewis, J. W. and Plant, 1954; Dave, V., 1972; Fennell and Plant, 1932; Fischer, Emil, 1886b, 1886c, 1886e; Govindachari, Rajappa and Sudarsanam, 1966; Govindachari and Sudarsanam, 1967, 1971; Herdieckerhoff and Tschunkur, 1933; Kametani, Fukumoto and Masuko, 1963; Korczynski, Brydowna and Kierzek, 1926; Kost, Sugorova and Yakubov, 1965; Kricka and Vernon, 1974; Ockenden and Schofield, 1953a, 1957; Plant and Thompson, 1950; Plant and Tomlinson, M. L., 1933; Rosenthal and Yalpani, 1965; Schofield and Theobald, 1950; Shvedov, Trofimkin, Vasil'eva and Grinev, A. N., 1975; Steck, Fletcher and Carabateas, 1974.

(*continued*)

Table 32 (*continued*)

Benzyl 4-Methylphenyl Ketone

Brown, F. and Mann, 1948b.

Benzyl 4-Biphenylyl Ketone

Buu-Hoï and Hoán, 1949.

Benzyl 4-Methoxyphenyl Ketone

Cowper and Stevens, T. S., 1947; Landquist and Marsden, 1966; Szmuszkovicz, Glenn, Heinzelman, Hester and Youngdale, 1966.

Benzyl 3,4-Dimethoxyphenyl Ketone

Chalmers, A. J. and Lions, 1933; Szmuszkovicz, Glenn, Heinzelman, Hester and Youngdale, 1966.

Benzyl 4-Hydroxyphenyl Ketone

Buu-Hoï, Xuong and Lavit, 1954; Council of Scientific and Industrial Research (India), 1975.

Benzyl 4-Fluorophenyl Ketone

Buu-Hoï, Hoán and Jacquignon, 1949.

Benzyl 3-Fluoro-4-methoxyphenyl Ketone

Buu-Hoï, Xuong and Lavit, 1954.

Benzyl 3-Fluoro-4-hydroxyphenyl Ketone

Buu-Hoï, Xuong and Lavit, 1954.

Benzyl 4-Chlorophenyl Ketone

Buu-Hoï and Hoán, 1949.

Benzyl 4-Bromophenyl Ketone

Buu-Hoï and Hoán, 1949.

Benzyl 3-Bromo-5-fluoro-4-hydroxyphenyl Ketone

Buu-Hoï, Xuong and Lavit, 1954.

2-Phenylacetylthiophen

Buu-Hoï and Hoan, 1949.

2,5-Dimethyl-3-phenylacetylthiophen

Buu-Hoï and Hoán, 1949.

2-Phenyl-5-phenylacetylthiophen

Buu-Hoï and Hoán, 1949.

2-(4-Methylphenyl)-5-phenylacetylthiophen

Buu-Hoï and Hoán, 1950.

2-Chloro-5-phenylacetylthiophen

Buu-Hoï and Hoán, 1949.

2-Bromo-5-phenylacetylthiophen

Buu-Hoï and Hoán, 1949.

1-Methoxy-4-phenylacetylnaphthalene

Buu-Hoï and Lavit, 1955.

2-Methoxy-6-phenylacetylnaphthalene

Buu-Hoï, 1949b.

3-Phenylacetylbenzothiophen

Buu-Hoï, 1949a.

Benzyl Methylthio Ketone

Wuyts and Lacourt, 1935.

3-Methylbenzyl Methylthio Ketone

Wuyts and Lacourt, 1935.

4-Methylbenzyl Phenyl Ketone

Brown, F. and Mann, 1948b.

(*continued*)

Table 32 (*continued*)

4-Methoxybenzyl Phenyl Ketone

Cowper and Stevens, T. S., 1947; Szmuszkovicz, Glenn, Heinzelman, Hester and Youngdale, 1966.

4-Hydroxybenzyl Phenyl Ketone

Council of Scientific and Industrial Research (India), 1975.

2-Fluorobenzyl Phenyl Ketone

Sternbach, Fryer, Metlesics, Sach and Stempel, 1962.

2-Chlorobenzyl Phenyl Ketone

Chen, G.-S. J. and Gibson, 1975.

3-Chlorobenzyl Phenyl Ketone

Chen, G.-S. J. and Gibson, 1975.

4-Nitrobenzyl Phenyl Ketone

Da Settimo and Saettone, 1965.

2,2′-Dimethoxydesoxybenzoin

Szmuszkovicz, Glenn, Heinzelman, Hester and Youngdale, 1966.

3,3′-Dimethoxydesoxybenzoin

Szmuszkovicz, Glenn, Heinzelman, Hester and Youngdale, 1966.

4,4′-Dimethoxydesoxybenzoin

Szmuszkovicz, 1967, 1970, 1971; Szmuszkovicz, Glenn, Heinzelman, Hester and Youngdale,1966; Thomas, R. C., 1975 (with carbonyl-^{14}C); Upjohn Co., 1966; Yamamoto, H., Nakao and Atami, 1969; Youngdale, Glenn, Lednicer and Szmuszkovicz, 1969.

4,4′-Dihydroxydesoxybenzoin

Szmuszkovicz, 1967.

4,4′-Dichlorodesoxybenzoin

Szmuszkovicz, Glenn, Heinzelman, Hester and Youngdale, 1966.

4,4′-Dinitrodesoxybenzoin

Szmuszkovicz, Glenn, Heinzelman, Hester and Youngdale, 1966.

2-Phenacylpyridine

Ockenden and Schofield, 1953b.

2-Phenacylquinoline

Hamana and Kumadaki, 1967.

Phenyl 2-Phenylethyl Ketone

Yakovenko, Mikhlina, Oganesyan and Yakhontov, 1975.

2-(4-Methoxyphenyl)ethyl Phenyl Ketone

Yakovenko, Mikhlina, Oganesyan and Yakhontov, 1975.

2-(4-Dimethylaminophenyl)ethyl Phenyl Ketone

Yakovenko, Mikhlina, Oganesyan and Yakhontov, 1975.

2-(3-Chlorophenyl)ethyl Phenyl Ketone

Yakovenko, Mikhlina, Oganesyan and Yakhontov, 1975.

2-(4-Chlorophenyl)ethyl Phenyl Ketone

Buu-Hoï, Hoán and Xuong, 1951.

1-Benzyl-2-phenylethyl Phenyl Ketone

Leuchs, Wulkow and Gerland, 1932.

2-(4-Chlorophenyl)ethyl 4-Ethylphenyl Ketone

Buu-Hoï, Hoán and Xuong, 1951.

2-(4-Chlorophenyl)ethyl 4-Biphenylyl Ketone

Buu-Hoï, Hoán and Xuong, 1951.

4-Methoxyphenyl 2-Phenylethyl Ketone

Buu-Hoï, Hoán and Xuong, 1951.

(*continued*)

Table 32 (*continued*)

2-(4-Chlorophenyl)ethyl 4-Methoxyphenyl Ketone

Buu-Hoï, Hoán and Xuong, 1951.

2-Carbomethoxy-2-phenylethyl Phenyl Ketone

Lespagnol, A., Lespagnol, C. and Henichart, 1969.

2-Carbomethoxy-2-(4-methoxyphenyl)ethyl Phenyl Ketone

Lespagnol, A., Lespagnol, C. and Henichart, 1969.

2-Carbomethoxy-2-(4-methoxyphenyl)ethyl 4-Methoxyphenyl Ketone

Lespagnol, A., Lespagnol, C. and Henichart, 1969.

4(?)-Methoxyphenyl 3-Phenylpropyl Ketone

Buu-Hoï, Hoán and Xuong, 1951.

3-Phenylpropyl 2-Thienyl Ketone

Buu-Hoï, Hoán and Xuong, 1951.

Pyruvic Acid

Andrisano and Vitali, 1957; Bell and Lindwall, 1948; Cattanach, Cohen, A. and Heath-Brown, 1973; Cornforth, J. W., Cornforth, R. H., Dalgliesh and Neuberger, 1951; Cortes and Walls, F., 1964; Dufton, 1891, 1892; Fischer, Emil, 1886b, 1886c, 1886e; Fischer, Emil and Hess, 1884; Fischer, Emil and Jourdan, 1883; Jackman and Archer, S., 1946; Kermack and Tebrich, 1940; Kögl and Kostermans, 1935; Kost, Yudin, Dmitriev, B. A. and Terent'ev, A. P., 1959; Michaelis, 1897; Nandi, 1940; Raschen, 1887; Robson, 1924–1925; Samsoniya, Trapaidze, Gverdtsiteli, I. M. and Suvorov, 1977; Samsoniya, Trapaidze, Suvorov and Gverdtsiteli, I. M., 1978; Schlieper, 1886, 1887; Sergeeva, Z. F., Akhvlediani, Shabunova, Korolev, Vasil'ev, A. M., Babushkina, T. N. and Suvorov, 1975; Shagalov, Ostapchuk, Zlobina, Eraksina, Babushkina, T. A., Vasil'ev, A. M., Ogorad-nikova and Suvorov, 1978; Shirley and Roussel, P. A., 1953; Snyder and Cook, P. L., 1956; Steck, Fletcher and Carabateas, 1974.

Phenylpyruvic Acid

Borsche and Klein, 1941; Inaba, Ishizumi, Mori and Yamamoto, H., 1971 (prepared *in situ*); Kidwai, A. R. and Khan, N. H., 1963; Manske, R. H. F., Perkin and Robinson, R., 1927; Yamamoto, H., Inaba, Hirohashi, T., Ishizumi, Maruyama and Mori, 1970; Yamamoto, H., Inaba, Hirohashi, T., Mori, Ishizumi and Maruyama, 1969; Yamamoto, H., Inaba, Hirohashi, T., Okamoto, Ishizumi, Yamamoto, M., Maruyama, Mori and Kobayashi, T., 1970; Yamamoto, H., Inaba, Okamoto, Hirohashi, T., Ishizumi,

Yamamoto, M., Maruyama, Mori and Kobayashi, T., 1969a, 1969c, 1970; Yamamoto, H., Inaba, Okamoto, Hirohashi, T., Yamamoto, M., Mori, Ishigura, Maruyama and Kobayashi, T., 1972a, 1972b (prepared *in situ*).

4-Methoxyphenylpyruvic Acid

Steck, Fletcher and Carabateas, 1974.

3,4-Dimethoxyphenylpyruvic Acid

Chalmers, A. J. and Lions, 1933.

2-Fluorophenylpyruvic Acid

Yamamoto, H., Inaba, Hirohashi, T., Okamoto, Ishizumi, Yamamoto, M., Maruyama, Mori and Kobayashi, T., 1970.

2-Chlorophenylpyruvic Acid

Yamamoto, H., Inaba, Hirohashi, T., Okamoto, Ishizumi, Yamamoto, M., Maruyama, Mori and Kobayashi, T., 1970.

4-Chlorophenylpyruvic Acid

Yamamoto, H., Inaba, Hirohashi, T., Okamoto, Ishizumi, Yamamoto, M., Maruyama, Mori and Kobayashi, T., 1970.

2-Nitrophenylpyruvic Acid

Kermack and Slater, 1928; Kermack and Tebrich, 1940; Neber, Knöller, Herbst and Trissler, 1929.

5-Chloro-2-nitrophenylpyruvic Acid

Freedman and Judd, 1972.

1-Bromo-2-naphthylpyruvic Acid

Sempronj, 1938.

2-Oxobutyric Acid

Arnold, E., 1888; Blaikie and Perkin, 1924; Kermack, Perkin and Robinson, R., 1921; Kollenz, Ziegler, Eder and Prewedourakis, 1970; Nakao, Katayama and Yamamoto, H., 1972a; Nakao, Katayama and Yamamoto, H., 1972b (the corresponding enol methyl ether was used); Sandoz Ltd, 1980; Wislicensus and Arnold, E., 1887.

(*continued*)

Table 32 (*continued*)

3-Methyl-2-oxobutyric Acid

Millson and Robinson, R., 1955.

2-Oxo-4-phenylbutyric Acid

Borsche and Klein, 1941; Wislicenus and Münzesheimer, 1898.

4-Hydroxy-2-oxobutyric Acid Lactone

Harradence and Lions, 1938; Snyder and Smith, C. W., 1943.

2-Oxoglutaric Acid

Feofilaktov and Semenova, N. K., 1953a, 1953b; Fox and Bullock, 1951b; Kermack, Perkin and Robinson, R., 1921; King, F. E. and L'Ecuyer, 1934; Merck and Co., Inc., 1967a; Perkin and Rubenstein, 1926; Robinson, J. R., 1957; Rosenmund, Meyer, G. and Hansal, 1975; Saleha, Khan, N. H., Siddiqui and Kidwai, M. M., 1978; Sarett and Shen, T.-Y., 1966a, 1966b; Steck, Fletcher and Carabateas, 1974; Stedman, 1924; Stevens, F. J. and Su, 1962; Wislicenus and Waldmüller, 1911; Yamamoto, H., and Nakao, 1968d, 1970a, 1970e, 1970h.

4-Methyl-2-oxoglutaric Acid

Shen, T.-Y. and Sarett, 1966.

2-Oxoadipic Acid

Drogas Vacunas y Sueros, S. A. (Drovyssa), 1957; Justoni and Pessina, 1955, 1957a, 1960; Kalb, Schweizer and Schimpf, 1926; Kalb, Schweizer, Zellner and Berthold, 1926; Manske, R. H. F., 1931b; Yamamoto, H. and Nakao, 1968g.

5-Hydroxy-2-oxoadipic Acid

Tenud, 1975.

2-Oxosuberic Acid

Manske, R. H. F. and Leitch, 1936.

Methyl Pyruvate

Govindachari, Rajappa and Sudarsanam, 1966; Kost, Sugorova and Yakubov, 1965; Snyder and Smith, C. W., 1943.

Dimethyl 2-Oxoadipate

Justoni and Pessina, 1955.

Methyl 2-Oxopalmitate

Saleha, Khan, N. H., Siddiqui and Kidwai, M. M., 1978.

Ethyl Pyruvate

Ainsworth, D. P. and Suschitzky, 1967a; Beer, Brown, J. P. and Robertson, 1951; Bell and Lindwall, 1948; Bogat-skii, Ivanova, R. Y., Andronati and Zhilina, 1979; Borsche and Groth, 1941; Carlin, Henley and Carlson, 1957; Dambal and Siddappa, 1965; Duncan and Andrako, 1968; Elks, Elliott and Hems, 1944b; Fernandez Alvarez and Monge Vega, 1975; Fischer, Emil, 1886b; Gadaginamath, 1976; Gesellschaft für Teerverwertung m.b.H., 1911; Goldsmith and Lindwall, 1953; Govindachari and Sudarsanam, 1967, 1971; Gryaznov, Akhvlediani, Volodina, Vasil'ev, A. M., Babushkina, T. A. and Suvorov, 1977; Holla, B. S. and Ambekar, 1979b; Hughes, G. K. and Lions, 1937–1938a; Ishii, Murakami, Y., Furuse, Hosoya, Takeda and Ikeda, 1973; Ivanova, R. Y., Bogatskii, Andronati and Zhilina, 1979; Komoto, Enomoto, Tanaka, Y., Nitanai and Umezawa, 1979; Komoto, Enomoto, Miyagaki, Tanaka, Y., Nitanai and Umezawa, 1979; Lindwall and Mantell, 1953; Lions and Spruson, 1932; Marchand, Streffer and Jauer, 1961; Marion and Oldfield, 1947; Namis, Cortes, Collera and Walls, F., 1966; Pappalardo and Vitali, 1958b; Raschen, 1887; Ried and Kleemann, 1968; Roder, 1886; Samsoniya, Chikvaidze, S. I., Suvorov and Gverdtsiteli, N., 1978; Schlieper, 1887; Scriven, Suschitzky, Thomas, D. R. and Newton, 1979; Smith, W. S. and Moir, 1952; Steck, Fletcher and Carabateas, 1974; Suvorov, Gordeev and Velezheva, 1975; Suvorov, Samsoniya, Chilikin, Chikvaidze, I. S., Turchin, Efimova, Tret'yakova and Gverdtsiteli, I. M., 1978.

Ethyl Phenylpyruvate

Chalmers, A. J. and Lions, 1933; Dalton, Fahrenholtz and Silverzweig, 1979; Gadaginamath and Siddappa, 1975; Hughes, G. K. and Lions, 1937–1938a; Inaba, Ishizumi, Mori and Yamamoto, H., 1971; Inaba, Ishizumi and Yamamoto, H., 1971; Lions and Spruson, 1932; Ried and Kleemann, 1968; Walser, Blount and Fryer, 1973; Yamamoto, H., Inaba, Hirohashi, T., Ishigura, Maruyama and Mori, 1971; Yamamoto, H., Inaba, Hirohashi, T., Ishizumi, Maruyama and Mori, 1970; Yamamoto, H., Inaba, Hirohashi, T., Mori, Ishizumi and Maruyama, 1969; Yamamoto, H., Inaba, Okamoto, Hirohashi, T., Yamamoto, M., Mori, Ishigura, Maruyama and Kobayashi, T., 1972c.

Ethyl 2-Fluorophenylpyruvate

Inaba, Ishizumi and Yamamoto, H., 1971; Yamamoto, H., Inaba, Hirohashi, T., Mori, Ishizumi and Maruyama, 1969; Yamamoto, H., Inaba, Okamoto, Hirohashi, T., Yamamoto, M., Mori, Ishigura, Maruyama and Kobayashi, T., 1972c.

Ethyl 2-Chlorophenylpyruvate

Inaba, Ishizumi and Yamamoto, H., 1971; Yamamoto, H., Inaba, Hirohashi, T., Ishizumi, Maruyama and Mori, 1970; Yamamoto, H., Inaba, Hirohashi, T., Mori, Ishizumi and Maruyama, 1969; Yamamoto, H., Inaba, Okamoto, Hirohashi, T., Ishizumi, Yamamoto, M., Maruyama, Mori and Kobayashi, T., 1969b.

(*continued*)

Table 32 (*continued*)

Ethyl 2,6-Dichlorophenylpyruvate

Yamamoto, H., Inaba, Okamoto, Hirohashi, T., Ishizumi, Akatsu, Kobayashi, T., Mori, Kume and Sato, H., 1973 (other analogues, unspecified in the abstract, were also employed with the chlorine atoms replaced by other halogen atoms and also in other positions upon the benzenoid nucleus).

Ethyl 2-Bromophenylpyruvate

Chalmers, A. J. and Lions, 1933.

Ethyl 2-Nitrophenylpyruvate

Clemo and Felton, 1951a; Kermack and Tebrich, 1940.

Ethyl 4-Nitrophenylpyruvate

Suvorov, Samsoniya, Chilikin, Chikvaidze, I. S., Turchin, Efimova, Tret'yakova and Gverdtsiteli, I. M., 1978; Wislicenus and Schultz, F., 1924.

Ethyl 2-Furylpyruvate

Gabrielyan and Papayan, 1973.

Ethyl 2-Pyridylpyruvate

Inaba, Ishizumi, Okamoto and Yamamoto, H., 1972; Yamamoto, H., Inaba, Okamoto, Hirohashi, T., Ishizumi, Maruyama, Kobayashi, T., Yamamoto, M. and Mori, 1970.

Ethyl 2-Pyridylpyruvate N-Oxide

Inaba, Ishizumi, Okamoto and Yamamoto, H., 1975.

Ethyl 2-Oxobutyrate

Beer, Brown, J. P. and Robertson, 1951; Bogat-skii, Ivanova, R. Y., Andronati and Zhilina, 1979; Carlin, Henley and Carlson, 1957; Dambal and Siddappa, 1965; Elks, Elliott and Hems, 1949b; Hughes, G. K. and Lions, 1937–1938a; Inaba, Akatsu, Hirohashi, T. and Yamamoto, H., 1976; Lions and Spruson, 1932; Marchand, Streffer and Jauer, 1961; Reynolds, B. and Carson, J., 1969; Ried and Kleemann, 1968; Smith, W. S. and Moir, 1952; Yamamoto, H., Inaba, Izumi, Hirohashi, T., Akatsu and Maruyama, 1970b.

Ethyl 2-Oxo-4-phenylbutyrate

Inaba, Akatsu, Hirohashi, T. and Yamamoto, H., 1976; Shvedov, Kurilo and Grinev, A. N., 1969; Yamamoto, H., Inaba, Izumi, Hirohashi, T., Akatsu and Maruyama, 1970b.

Ethyl 2-Oxo-4-(2-pyridyl)butyrate

Clemo and Seaton, 1954; Finkelstein and Lee, J., 1956.

Ethyl 2-Oxovalerate

Bogat-skii, Ivanova, R. Y., Andronati and Zhilina, 1979.

Ethyl 2-Oxo-5-phenoxyvalerate

Manske, R. H. F., 1931a.

Diethyl 2-Oxoglutarate

De Bellis and Stein, M. L., 1961; Feofilaktov and Semenova, N. K., 1953a, 1953b; Findlay, S. P. and Dougherty, 1948; Finger, Gortatowski, Shiley and White, R. H., 1959; Kalb, Schweizer, Zellner and Berthold, 1926; King, F. E. and L'Ecuyer, 1934; Plieninger, 1954; Ried and Kleemann, 1968; Sakurai and Ito, T., 1957; Saleha, Khan, N. H., Siddiqui and Kidwai, M. M., 1978; Tanaka, Z., 1940a.

Ethyl 4-Cyano-2-oxobutyrate

Feofilaktov and Semenova, N. K., 1952a; Keimatsu and Sugasawa, 1928b; Snyder and Smith, C. W., 1943.

Ethyl 2-Oxocarproate

Hughes, G. K. and Lions, 1937–1938a; Ried and Kleeman, 1968.

Ethyl 4-Carboxy-2-oxovalerate

Barrett, H. S. B., Perkin and Robinson, R., 1929; Feofilaktov, 1947a, 1952; Jackson, R. W. and Manske, R. H., 1930; Kametani, Ohsawa and Ihara, 1979; Koelsch, 1943; Manske, R. H. F. and Robinson, R., 1927; Oki and Nagasaka, 1973 (also used other esters but these were unspecified in the abstract); Rensen, 1959; Zenitz, 1966.

Diethyl 2-Oxoadipate

Basanagoudar and Siddappa, 1972; Feofilaktov, 1947a; Kalb, Schweizer, Zellner and Berthold, 1926; Manske, R. H. F., 1931a; Ried and Kleemann, 1968; Unanyan and Tatevosyan, 1961.

Ethyl 5-Cyano-2-oxovalerate

Keimatsu and Sugasawa, 1928b.

(*continued*)

Table 32 (*continued*)

Ethyl 6-Carboxy-2-oxoheptanoate

Feofilaktov, 1947b; Feofilaktov and Semenova, N. K., 1952b; Jackson, R. W. and Manske, R. H., 1930; Lions and Spruson, 1932; Manske, R. H. and Boos, 1960; Oki and Nagasaka, 1973 (also used other esters but these were unspecified in the abstract); Pacheco, 1951; Polaczkowa and Porowska, 1950; Zenitz, 1966.

Diethyl 2-Oxopimelate

Ried and Kleemann, 1968; Yoshitomi Pharmaceutical Industries Ltd, 1967.

Ethyl 9-Cyano-2-oxocaprate

Avramenko, Plutitskii, Strekozov and Suvorov, 1979.

Pyruvonitrile

Robinson, B., 1963c.

Phenylpyruvonitrile and 2-Chloro-, 4-cyano-, 2-fluoro-, 4-methoxy-, 4-methyl-, 4-dimethylamino- and 3-nitrophenylpyruvonitrile

Morooka, Tamoto and Matuura, 1976.

Diethyl 3-Oxoglutarate

King, F. E. and L'Ecuyer, 1934; Trofimov, Ryabchenko and Grinev, A. N., 1975.

L. Catalysts which have been Employed in the Fischer Indole Synthesis

A large number of Brønsted and Lewis acid catalysts have been employed in the Fischer indole synthesis (Table 33). Usually, the arylhydrazone or, where appropriate, a mixture of its ketonic or aldehydic and hydrazino moieties, was heated with the catalyst, often at the boiling point of any solvent which may have been present. Often, when using Lewis acid catalysts, the use of an inert dry solvent, as opposed to fusing the mixture of the arylhydrazone and catalyst, had beneficial effects upon product yield and quality by reducing undesired thermal decompositions. The first application of such a technique involved the use of a zinc chloride catalyst in 'solvent naphtha' or 'neutral tar oils' with the reaction temperature kept to the minimum possible (Gesellschaft für Teerverwertung m.b.H., 1911). Likewise, the yields of indolic products could often be improved by conducting indolizations under nitrogen atmospheres (e.g. Bajwa and Brown, R. K., 1968a; Carlin and Moores, 1959). Of the large number of compounds which are listed in Table 33, only a few have been found to be sufficiently useful for general adoption, as is evidenced by the number of their related references. Probably the most widely used and most vigorous indolization catalyst is zinc chloride which is closely followed with respect to both these properties by polyphosphoric acid. Less vigorous but still very widely used catalysts are hydrochloric acid, phosphoric acid and sulphuric acid in their various dilutions with a variety of solvents, and likewise hydrogen chloride. Among the most often used mildest indolization catalysts are to be found acetic acid, boron trifluoride, cation exchange resins, formic acid and sulphosalicyclic acid. In the appropriate sections of the present study, the relationship between the choice of catalyst and the type of indolization has been discussed as have cases in which the choice of catalyst can exert specific directional influences upon directionally ambiguous indolizations. A very useful account which has attempted to relate the choice of catalyst to the type of indolization has also been presented (Brown, R. K., 1972a). Very often, however, this choice of catalyst appeared to be arbitrary, optimum product yields and quality from the indolization of related arylhydrazones could usually only be obtained by the use of a wide spectrum of catalysts and even when indolizing very closely related arylhydrazones using the same catalytic procedure, some modifications of conditions were often necessary in attempts to optimize these factors (e.g. see Chapman, Clarke and Hughes, H., 1965; Stevens, F. J. and Su, 1962). From Table 33 it can also be seen that arylhydrazones have also been indolized by heating with reagents as diverse as sodium ethoxide, zinc granules, ceramic powder and Grignard reagents. Indeed, in view of the now well established 'non-catalytic' thermal indolization technique (see below), it could be that Fischer indolizations effected by pyrolysis with some of these diverse reagents would have occurred regardless of the presence of the reagent and, indeed, in others it could be that indolization occurred not because of but in spite of the presence of the reagent.

Some Fischer indolizations have been unsuccessful because of the sensitivity of reaction intermediates or products to acidic conditions. In many such cases the 'non-catalytic' thermal indolization technique may be of use but drawbacks would be apparent when using this technique when reaction intermediates or products were thermally labile. To overcome these problems, the use of a pyridine hydrochloride catalyst in pyridine as solvent has been introduced. Such reactions were effected by reacting equimolar mixtures of ketones with arylhydrazine hydrochlorides in refluxing pyridine (Welch, W. M., 1977a). Another technique which could be attempted when product thermal and/or acid lability presents problems is the use of a biphasic indolization system at room temperature. Using this technique (i.e. stirring the arylhydrazone in a mixture of concentrated hydrochloric acid with benzene at room temperature for several hours), mixtures of **1371** ($R = CH_3$ and C_2H_5) with 4-benzyloxyphenylhydrazine hydrochloride were converted into **1372** ($R^1 = CH_2C_6H_5$, $R^2 = H$, $R^3 = NO_2$, $R^4 = CH_3$ and C_2H_5, respectively) (Lewis, A. D., 1962), **1591** afforded **1592** ($R = Cl$) along with some **1592** ($R = OH$) (Shaw, E. and Woolley,

D. W., 1953) and **508** [$R^1 = NO_2$, $R^2 = R^3 = H$, $R^4 = C_2H_5$ and $(CH_2)_2Cl$] gave a mixture of **509** [$R^1 = NO_2$, $R^2 = R^3 = H$, $R^4 = C_2H_5$ and $(CH_2)_2Cl$, respectively] and **510** [$R^1 = NO_2$, $R^2 = R^3 = H$, $R^4 = C_2H_5$ and $(CH_2)_2Cl$ respectively] (McKay, Parkhurst, Silverstein and Skinner, 1963). Flow-through indolization techniques can also be used to produce indoles that would otherwise be unstable under the indolization conditions (see Chapter IV, A).

Prior to 1957, several examples of the occurrence of Fischer indolizations under thermal conditions in the absence of an acid catalyst had been reported. When a mixture of 5_β-cholestan-3-one [**873**; R^1–R^5 = H, $R^6 = CH(CH_3)$-$(CH_2)_3CH(CH_3)_2$] with phenylhydrazine was refluxed, a low yield of the corresponding indolosteroid was formed (Dorée and Gardner, J. A., 1908) although the product was not, in fact, recognized as being indolic until the following year (Dorée, 1909) and the structure of the product was not established until many years later (see Chapter III, J2). Distillation of acetophenone phenylhydrazone afforded a low yield of 2-phenylindole (Wolff and Mayen, 1912). The evolution of ammonia occurred when a mixture of 2-benzyl-1-indanone (**1550**; $R = CH_2C_6H_5$) with an excess of phenylhydrazine was heated in xylenol at 130 °C (Leuchs, Wutke and Gieseler, 1913) and a product was formed which was later shown to result from the reaction between phenylhydrazine and the expected 3*H*-indolic product (Leuchs, Philpott, D., Sander, Heller, A. and Köhler, 1928; Leuchs and Winzer, 1925) (see Chapter V, C). The evolution of ammonia was also observed when **815** was heated, which suggested the occurrence of a Fischer indolization although indolic products from this reaction were not isolated (Perkin and Titley, 1922). When a mixture of phenyl propyl ketone with 2-naphthylhydrazine was heated and then distilled *in vacuo*, 4,5-benzo-3-ethyl-2-phenylindole was obtained (Korczynski, Brydowna and Kierzek, 1926). When a mixture of phenylhydrazine with 2-(1-piperidylmethyl)-cyclohexanone was heated, there were formed ammonia, aniline, piperidine and 1,2,3,4-tetrahydrocarbazole (Mannich and Hönig, 1927), the formation of the last two products resulting from the occurrence of a retro Mannich reaction in either the starting hydrazone (Grandberg, Kost and Yaguzhinskii, 1960) or in the 3*H*-indolic product (see Fritz, Losacker, Stock and Gerber, 1971 and Chapter III, H). When mixtures of **992** (R^1 = H, X = O) with phenylhydrazine and 2-naphthylhydrazine were heated at 120 °C and 130 °C, respectively, until most of the water from the formation of the hydrazone had boiled away, effervescence of ammonia suddenly occurred and the reactions yielded **1075** (R^1–R^3 = H, X = O) and **1136** (X = O) (Cawley and Plant, 1938), a reaction analogous to the formation of **87** (R = H) from the heating at 200 °C of a mixture of **86** with an excess of phenylhydrazine (Japp and Findlay, A., 1897). Heating 4-methylcyclohexanone phenylhydrazone furnished 3-methyl-1,2,3,4-tetrahydrocarbazole (Grammaticakis, 1939). When a mixture of 2-hydroxy-cyclohexanone and phenylhydrazine was refluxed in ethanolic solution, a copious evolution of ammonia occurred and a mixture of cyclohexan-1,2-dione bisphenylhydrazone and 1,2,3,4-tetrahydrocarbazol-1-one (**1489**; R^1–R^4 = H) was formed, resulting from osazone formation (see Chapters IV, J2 and VII, Q) and indolization (Bloink and Pausacker, 1950).

In 1957 (Fitzpatrick, J. T. and Hiser, 1957), after the second (Wolff and Mayen, 1912) of these above observations had been noted, the general realization that Fischer indolizations could be effected under thermal conditions in the absence of an acid catalyst became apparent and was developed into a synthetically useful route. Thus, refluxing solutions of ethyl methyl ketone phenylhydrazone, N_{α}-methylphenylhydrazone and 2,5-dichlorophenylhydrazone, butyraldehyde phenylhydrazone, propionaldehyde 4-methylphenylhydrazone and N_{α}-methylphenylhydrazone and acetophenone and butyrophenone phenylhydrazone in monoethylene glycol afforded 2,3-dimethylindole (70%), 1,2,3-trimethylindole (65%), 4,7-dichloro-2,3-dimethylindole (66%), 3-ethylindole (44%), 3,5-dimethylindole (44%), 1,3-dimethylindole (70%), 2-phenylindole (54%) and 3-ethyl-2-phenylindole (50%), respectively. In some cases, the reaction rate was accelerated by using the higher boiling diethylene glycol and in this manner acetone phenylhydrazone and ethyl methyl ketone 2-chlorophenylhydrazone gave 2-methylindole (36%) and 7-chloro-2,3-dimethylindole (55%), respectively. The polarity of the refluxing solvent was not critical and refluxing tetralin was comparable with refluxing glycols in effecting indolizations. Thus, ethyl methyl ketone phenylhydrazone was converted in refluxing tetralin into 2,3-dimethylindole (48%), a conversion which could also be achieved in 68% yield by refluxing in ethylene glycol containing 2% sodium hydroxide. Although not referred to by J. T. Fitzpatrick and Hiser (1957), Fischer indolizations had previously been effected under basic conditions when acetophenone phenylhydrazone furnished 2-phenylindole in the presence of sodium ethoxide (Wolff and Mayen, 1912) and when acetone phenylhydrazone afforded aniline and 2-methylindole when heated with sodium hydroxide at 240–260 °C (Seibert, 1948). The term 'non-catalytic' for the above thermal indolizations has been questioned in as much as the catalytic effect of the hydroxylic solvent (cf. Shine, 1956) should be considered (Yakhontov, Pronina, E. V. and Rubtsov, 1967), as should the difficulty in precluding trace amounts of acid (Sucrow and Chondromatidis, 1970), although this latter difficulty would obviously not be apparent in indolizations performed under basic conditions.

Fischer indolizations have also been achieved in a chemical ionization source and the products subsequently detected by mass spectrometric analysis, conditions under which acetaldehyde, acetone, ethyl methyl ketone and isopropyl methyl ketone phenylhydrazone afforded indole, 2-methylindole, 2,3-dimethylindole and 2,3,3-trimethyl-3*H*-indole, respectively, by a mechanism which was suggested as being analogous to that operating in 'solution' indolizations (Glish and Cooks, 1978). Indolization of arylhydrazones in the metallic injection chamber during their v.p.c. was also apparent when acetaldehyde, acetone and propionaldehyde phenylhydrazone and acetaldehyde N_{α}-methylphenylhydrazone exhibited peaks in their v.p. chromatograms attributed to the corresponding indoles (Nakazaki and Yamamoto, K., 1976).

Table 33. Catalysts/Conditions which have been Utilized to Effect the Fischer Indolization

Acetic Acid (Glacial)

Ainsworth, D. P. and Suschitzky, 1967a; Ban, Sato, Y., Inoue, Nagai, M., Oishi, Terashima, Yonemitsu and Kanaoka, 1965; Barnes, Pausacker and Badcock, 1951; Beer, Broadhurst and Robertson, 1952; Borghero and Finsterle, 1955; Borsche and Kienitz, 1910; Carlin and Moores, 1959, 1962; Cetenko and Morrison, 1979; Chalmers, A. J. and Lions, 1933; Ciba Ltd, 1960; Coombes, Harvey, D. J. and Reid, S. T., 1970; Dalgliesh and Mann, 1947; Green and Ritchie, 1949; Harvey, D. J., 1968; Hinman and Whipple, 1962; Hughes, G. K. and Lions, 1937–1938b; Huisgen, 1948a; Inoue and Ban, 1970; Kadzyauskas, Butkus, Vasyulite, Averina and Zefirov, 1979; Keglevic, Stojanac and Desaty, 1961; Klioze and Darmory, 1975; Maki, Masugi, Hiramitsu and Ogiso, 1973; Merck and Co. Inc., 1961; Neber, Knöller, Herbst and Trissler, 1929; Newberry, 1974; Perkin and Plant, 1921; Sakai, S., Wakabayashi and Nishina, 1969; Sandoz Ltd, 1980; Saxton, Smith, A. J. and Lawton, 1975; Sen and Ghosh, 1927; Shukri, Alazawe and Al-Tai, 1970; Steck, Fletcher and Carabateas, 1974; Stork and Dolfini, 1963; Walton, Stammer, Nutt, Jenkins and Holly, 1965; Witkop and Hill, R. K., 1955; Witkop, Patrick and Rosenblum, 1951; Yamamoto, H., Inaba, Hirohashi, T., Ishigura, Maruyama and Mori, 1971; Yamamoto, H., Inaba, Okamoto, Hirohashi, T., Ishizumi, Yamamoto, M., Maruyama, Mori and Kobayashi, T., 1970.

Acetic Acid (Glacial?)

Ahmed and Robinson, B., 1967; Benary and Baravian, 1915; Binder, Habison and Noe, 1977; Bloss and Timberlake, 1963; Borsche and Kienitz, 1910; Cattanach, Cohen, A. and Heath-Brown, 1973; Chalmers, A. J. and Lions, 1933; Clemo and Perkin, 1924b; Clemo, Perkin and Robinson, R., 1924; Cohen, A., Heath-Brown and Cattanach, 1968; Cornforth, J. W., Hughes, G. K., Lions and Harradence, 1937–1938; Davidge, 1959; Evans, F. J. and Lyle, R. E., 1963; Evans, F. J., Lyle, G. G., Watkins and Lyle, R. E., 1962; Gale, Lin and Wilshire, 1976; Georgian, 1957, 1958; Horning, E. C., Horning, M. G. and Walker, G. N., 1948; Huisgen, 1948a; Kempter, Schwalba, Stoss and Walter, 1962; Kost, Yudin and Popravko, 1962; Linnell and Perkin, 1924; Lions and Ritchie, 1939a, 1939b; Lohr, 1977; McClelland, 1929; McClelland and D'Silva, 1932; Miller, F. M. and Lohr, 1978; Nakazaki, 1960a; Neber, Knöller, Herbst and Trissler, 1929; Oliver, 1968; Perkin and Riley, 1923; Rice and Scott, K. R., 1970; Robinson, R. and Thornley, 1926; Rogers, C. U. and Corson, 1947, 1950, 1963; Schut, Ward, Lorenzetti and Hong, 1970; Shukri, Alazawe and Al-Tai, 1970; Sircar and Gopalan, 1932; Thomas, T. J. 1975; Winchester and Popp, 1975.

Acetic Acid (Aqueous)

Crum and Sprague, 1966, Desaty and Keglevic, 1965; Herdieckerhoff and Tschunkur, 1933; King, F. E. and L'Ecuyer, 1934; Teotino, 1959.

Acetic Acid (50%)

Kempter, Schwalba, Stoss and Walter, 1962; Steck, Fletcher and Carabateas, 1974.

Acetic Acid in Ethanol

Duncan and Boswell, 1973; Keglević, Desaty, Goleš and Stančić, 1968 (aqueous ethanol); Sumitomo, Hayakawa and Tsubojima, 1969 (and in methanol).

Acetyl Chloride in Dioxane–Carbon Tetrachloride

Grandberg, Kost and Terentyev, 1957.

Alkylating Agents

Allyl bromide, benzyl chloride, dimethyl sulphate (Grandberg, Sibiryakova and Brovkin, 1969 – see also Grandberg and Sibiryakova, 1967; Grandberg, Przheval'skii, Ivanova, T. A., Zuyanova, Bobrova, Dashkevich, S. I., Nikitina, Shcherbina and Yaryshev, 1969; methyl iodide (Posvic, Dombro, Ito, H. and Telinski, 1974).

Aluminium Powder

Korczynski and Kierzek, 1925.

Aluninium Oxide, Other Oxides and Zinc Chloride using Flow-through Techniques

(see Chapter IV, A).

Aniline Hydrochloride

Zaitsev, 1938.

Beryllium Chloride

Zaitsev, 1938.

Boric Acid in Strong Mineral Acid

Badische Anilin- & Soda-Fabrik A.-G., 1965.

Boron Trifluoride Etherate in Acetic Acid

Atkinson, Simpson and Taylor, A., 1954; Bailey and Bogle, 1977; Ballantine, Barrett, C. B., Beer, Boggiano, Eardley, Jennings and Robertson, 1957; Bellamy and Guthrie, 1965b; Blades and Wilds, 1956; Brown, R. K., Nelson, Sandin and Tanner, 1952; Council of Scientific and Industrial Research (India), 1975; Grandberg, Belyaeva and Dmitriev, L. B., 1971a, 1971b, 1971c, 1973; Hino, Suzuki, T. and Nakagawa, 1973; Ishii, Murakami, Y., Furuse, Hosoya and Ikeda, 1973; Jackson, A. H. and Smith, P., 1968; Jardine and Brown, R. K., 1965; Kesswani, 1967, 1970; Nakazaki, 1960a; Noland and Robinson, D. N., 1958; Ockenden and Schofield, 1953a, 1953b; Robinson, F. P. and Brown, R. K., 1964; Shen, T.-Y. and Sarrett, 1966; Shukri, Alazawe and Al-Tai, 1970; Snyder, Eliel and

(*continued*)

Table 33 (*continued*)

Boron Trifluoride Etherate in Acetic Acid (*continued*)

Carnahan, 1951; Snyder and Smith, C. W., 1943 (initial use of this indolization catalyst which failed to indolize acetaldehyde and acetone phenylhydrazone); Witkop and Patrick, 1952; Witkop, Patrick and Kissman, 1952; Yakhontov and Pronina, E. V., 1969.

Boron Trifluoride Etherate in Ethanol

Lohr, 1977; Miller, F. M. and Schinske, 1978.

Boron Trifluoride Etherate in Ethyl Acetate

Ishii, Murakami, Y., Furuse, Hosoya and Ikeda, 1973.

Boron Trifluoride Etherate in Benzene

Bullock and Fox, 1951.

Boron Trifluoride–Anisole Complex

Nesmeyanov and Golovnya, 1961.

Boron Trifluoride (heat in a seated tube)

Yakhontov and Pronina, E. V., 1969.

Cadmium Chloride (Catalytic Quantity)

Kitaev, Troepol'skaya and Arbuzov, A. E., 1964, 1966a, 1966b.

Cation Exchange Resin in refluxing Water

Bergmann, E. D. and Hoffman, E., 1962; Yamada, S., Chibata and Tsurui, 1953; Yamada, S. *et al.*, 1954.

Cation Exchange Resin in Ethanol

Suzuki, N. and Sato, Y., 1955.

Cation Exchange Resin in refluxing Toluene

Posvic, Dombro, Ito, H. and Telinskii, 1974.

Cation Exchange Resin in a Continuous System

Sibiryakova, Brovkin, Belyakova and Grandberg, 1969.

Ceramic Powder

Sumitomo Chemical Co. Ltd, 1969a.

Chromium(II) Chloride

Korczynski and Kierzek, 1925; Korczyński and Kierzek, 1925.

Cobalt Powder

Korczynski and Kierzek, 1925; Korczyński and Kiersek, 1925.

Cobalt(II) Chloride

Korczynski and Kierzek, 1925; Korczyński and Kierzek, 1925.

Copper Acetylacetone

Although it was indicated (Kitaev, 1959) that this compound had been used earlier as a catalyst for Fischer indolization, this only one attempted such use (Korczynski, Brydowna and Kierzek, 1926) involved its failure to indolize methyl phenyl ketone phenylhydrazone in refluxing benzene or chloroform.

Copper Powder

Korczynski and Kierzek, 1925; Korczyński and Kierzek, 1925.

Copper(I) Bromide (Catalytic Quantity)

Arbusow, A. E. and Tichwinsky, 1910b; Arbuzov, A. E. and Tikhvinskii, 1913b.

Copper(II) Chloride

Korczyński and Kierzek, 1925.

Copper(I) Chloride (Catalytic Quantity)

Arbusow, A. E., Saizew and Rasumow, 1935; Arbusow, A. E. and Tichwinsky, 1910b; Arbuzŏv, A. E. and Friauf, 1913; Arbuzov, A. E. and Rotermel, 1932; Arbuzov, A. E. and Tikhvinskii, 1913b; Arbuzov, A. E. and Wagner, R. E., 1913; Arbuzov, A. E. and Zaitev, 1934; Arbuzov, A. E., Zaitev and Rasumov, 1935; Grandberg, Belyaeva and Dmitriev, L. B., 1971a, 1971c; Grandberg, Kost and Terentyev, 1957; Hoshino, 1933; Janetzky and Verkade, 1945, 1946; Kitaev, Troepol'skaya and Arbuzov, A. E., 1966a, 1966b; Marshalkin and Yakhontov, 1972a; Verkade and Janetsky, 1943b; Yakhontov and Marshalkin, 1972; Yakhontov, Marshalkin and Anisimova, 1972; Yakhontov and Pronina, E. V., 1969; Yakhontov, Pronina, E. V. and Rubtsov, 1967, 1970.

(*continued*)

Table 33 (*continued*)

Copper (I) Cyanide (Catalytic Quantity)

Kitaev, Troepol'skaya and Arbuzov, A. E., 1964.

Formic Acid

Akagi, Oishi and Ban, 1969; Bahadur, Bailey, Costello and Scott, P. W., 1979; Ban and Iijima, I., 1969; Ban, Oishi, Kishio and Iijima, I., 1967; Hester, 1968; Inoue and Ban, 1970; Kidwai, A. R. and Khan, N. H., 1963; Kirk, 1976; Magidson, Suvorov, Travin, Sorkina and Novikova, 1965; Protiva, Ernest, Hněvsová, Novák and Rajšner, 1960; Rogers, C. U. and Corson, 1947; Shimizu, J., Murakami, S., Oishi and Ban, 1971; Strandtmann, Cohen, M. P. and Shavel, 1963; Wright and Gambino, 1979; Yamamoto, H., Inaba, Okamoto, Hirohashi, T., Ishizumi, Yamamoto, M., Maruyama, Mori and Kobayashi, T., 1969c.

Formic Acid–Water (1:1)

Saleha, Khan, N. H., Siddiqui and Kidwai, M. M., 1978 [alcoholic formic acid (1:1) necessitated shorter reaction times and led to increased product yields].

Formic Acid in Ethanol

Trofimov, Tsyshkova, Garnova and Grinev, A. N., 1975.

Formic Acid in Ethanol–Carbon Tetrachloride

Saleha, Siddiqui and Khan, N. H., 1979.

Grignard Reagents

Benzyl magnesium chloride (Grammaticakis, 1939); Ethyl magnesium bromide (Grammaticakis, 1937, 1939); Methyl magnesium iodide (Grammaticakis, 1939, 1940b); Phenyl magnesium bromide (Grammaticakis, 1939, 1940b).

Heat a Mixture of the Arylhydrazine Hydrochloride with an Aldehyde or Ketone in Solution (unless otherwise stated, ethanolic)

Alexander and Mooradian, 1977b (ketal used); Archer, S., 1963; Baradarani and Joule, 1978; Borisova and Kartashova, 1979; Bullock and Hand, J. J., 1956b; Campaigne, Ergener, Hallum and Lake, 1959 (stir in aqueous methanol at room temperature); Cattanach, Cohen, A. and Heath-Brown, 1973; Chalmers, A. J. and Lions, 1933; Chemerda and Sletzinger, 1968a, 1968b (dry benzene); Chemerda and Sletzinger, 1968c, 1968d, 1968e, 1968f, 1969a, 1969c (*tert.*-butanol); Cohylakis, Hignett, Lichmann and Joule, 1974; Colwell, Horner, J. K. and Skinner, 1964; Grandberg, Belyaeva and Dmitriev, L. B., 1971a, 1971b (dimethylformamide); Grandberg, Zuyanova, Afonina and Ivanova, T. A., 1969 (neutral anhydrous solvent); Harbert, Plattner, Welch, W. M., Weissmann and Koe, 1980; Herdieckerhoff and Tschunkur, 1933 (water); I. G. Farbenindustrie Akt.-Ges., 1934 (water); Katsube, Sasajima, Ono, Nakao, Maruyama, Takayama, Katayama,

Tanaka, Y., Inaba and Yamamoto, H., 1974 (aldehyde precursor in aqueous dioxane); Kempter, Schwalba, Stoss and Walter, 1962; Merck and Co. Inc., 1965 (*tert.*-butanol); Orlova, E. K., Sharkova, N. M., Meshcheryakova, Zagorevskii and Kucherova, 1975; Sarett and Shen, T.-Y., 1966a; Sharkova, Kucherova and Zagorevskii, 1972a, 1972b, 1972c; Shen, T. Y., 1968 and Shen, T.-Y., 1969 (aldehyde dimethyl acetal in isopropanol); Shen, T.-Y., 1969 (isopropanol); Shvedov, Trofimkin, Vasil'eva and Grinev, A. N., 1975; Sterling Drug Inc., 1962; Titley, 1928 (trace of concentrated hydrochloric acid); Walton, Jenkins, Nutt and Holly, 1968; Walton, Jenkins, Nutt and Holly, 1968 (isopropanol); Wheeler, 1970 (dry methanol); Yamamoto, H. and Atami, 1976; Yamamoto, H., Hirohashi, A., Izumi and Koshiba, 1973a, 1973b, 1973c; Yamamoto, H., Misaki, N., Izumi and Koshiba, 1972.

Heat a Mixture of the Arylhydrazine Hydrochloride with a Ketoacid

Avanesova and Tatevosyan, 1974; Walton, Stammer, Nutt, Jenkins and Holly, 1965 (methanol); Yamamoto, H. and Nakao, 1969c.

Heat a Mixture of Arylhydrazine with a Ketoacid in Ethanol

Chalmers, A. J. and Lions, 1933.

Heat the Arylhydrazone Perchlorate in Bromobenzene

Theilacker and Leichtle, 1951.

Hydrochloric Acid (Concentrated)

Antrick, 1885; Armit and Robinson, R., 1922; Atkinson, Simpson and Taylor, A., 1954; Auwers, Hilliger and Wulf, 1922; Bannister and Plant, 1948; Bauer and Strauss, 1932; Braun and Bayer, 1929; Carey, Gal and Sletzinger, 1967; Carey, Gal, Sletzinger and Reinhold, 1968; Carry, Gal and Sletzinger, 1968; Cohen, A., Heath-Brown and Cattanach, 1968; Colwell, Horner, J. K. and Skinner, 1965; Dufton, 1891, 1892; Grandberg, Belyaeva and Dmitriev, L. B. 1973; Hahn, W., Nowaczyk, Bartnik and Zawadzka, 1968; Hausmann, 1889; Hughes, G. K., Lions and Ritchie, 1939; Kakurina, Kucherova and Zagorevskii, 1965b; Kermack, Perkin and Robinson, R., 1921; Kermack and Tebrich, 1940; Kipping and Hill, A., 1899; LaForge, 1928; Leuchs and Kowalski, 1925; Leuchs and Winzer, 1925; Manske, R. H. F., Perkin and Robinson, R., 1927; Michaelis and Luxembourg, 1893; Ockenden and Schofield, 1953b; Ogandzhanyan, Avanesova and Tatevosyan, 1968; Pappalardo and Vitali, 1958a; Perkin, 1904a, 1904b; Petrova, T. D., Mamaev and Yakobson, 1967, 1969; Scherrer, 1969; Schofield and Theobald, 1950; Sergeeva, Z. F., Akhvlediani, Shabunova, Korolev, Vasil'ev, A. M., Babushkina, T. N. and Suvorov, 1975; Shaw, E. and Woolley, D. W., 1957; Shen, T.-Y., 1967b; Steck, Fletcher and Carabateas, 1974; Walser, Blount and Fryer, 1973; Yakhontov and Marshalkin, 1972; Yakhontov, Marshalkin and Anisimova, 1972; Yakhontov and Pronina, E. V., 1969.

Hyfrochloric Acid (50% Concentrated)

Rosenmund, Meyer, G. and Hansal, 1975.

(*continued*)

Table 33 (*continued*)

Hydrochloric Acid (20%)

Driver, Matthews and Sainsbury, 1979.

Hydrochloric Acid (Dilute)

Bell and Lindwall, 1948; Borsche and Klein, 1941; Diels and Köllisch, 1911; Fischer, Emil and Hess, 1884; Fischer, Emil and Jourdan, 1883; Hegel, 1886; Herdieckerhoff and Tschunkur, 1933; Julia, Melamed and Gombert, 1965; Kermack, Perkin and Robinson, R., 1921; Kost, Yudin, Dmitriev, B. A. and Terent'ev, 1959; Kulka and Manske, R. H. F., 1952; Michaelis, 1897; Rogers, C. U. and Corson, 1947, 1950; Snyder and Cook, P. L., 1956; Steck, Fletcher and Carabateas, 1974; Szmuszkovicz, 1967; Wittig and Reichel, 1963.

Hydrochloric Acid in Alcohol (Ethanol?)

Ainsworth, D. P. and Suschitzky, 1967b; Berger, L. and Corraz, 1968; Carlin and Fisher, 1948; Cowper and Stevens, T. S., 1947; Duncan and Boswell, 1973; Fischer, Emil and Schmidt, 1888; Fischer, Emil and Schmitt, 1888; Govindachari, Rajappa and Sudarsanam, 1963; Hahn, W. E., Nowaczyk and Bartnik, 1968; Kametani, Fukumoto and Masuko, 1963; Kempter, Schwalba, Stoss and Walter, 1962; Kipping, 1894; Komzolova, Kucherova and Zagorevskii, 1967; Korczynski, Brydowna and Kierzek, 1926; Marquez, Cranston, Ruddon, Kier and Burckhalter, 1972; Mosher, Crews, Acton and Goodman, 1966; Neber, Knöller, Herbst and Trissler, 1929; Pappalardo and Vitali, 1958; Perkin and Rubenstein, 1926; Rogers, C. U. and Corson, 1947; Rosenmund and Sotiriou, 1975; Schmutz and Wittwer, 1960; Steck, Fletcher and Carabateas, 1974; Sterling Drug Inc., 1975; Szmuszkovicz, 1967, 1970; Trenkler, 1888; Upjohn Co., 1966; Walton, Stammer, Nutt, Jenkins and Holly, 1965; Wieland, T. and Rühl, 1963; Yamamoto, H., Inaba, Hirohashi, T., Ishigura, Murayama and Mori, 1971; Yamamoto, H., Inaba, Ishigura, Maruyama, Hirohashi, T. and Mori, 1971; Youngdale, Glenn, Lednicer and Szmuszkovicz, 1969.

Hydrochloric Acid (Concentrated) in Methanol

Chen, G.-S. J. and Gibson, 1975; Fusco and Sannicolò, 1973; Seka and Kellermann, 1942; Shvedov, Trofimkin, Vasil'eva and Grinev, A. N., 1975; Sternbach, Fryer, Metlesics, Sach and Stempel, 1962; Wang, T. S. T., 1975; Wuyts and Lacourt, 1935.

Hydrochloric Acid (Concentrated) in Ethanol

Bannister and Plant, 1948; Da Settimo, Biagi, Primofiore, Ferrarini and Livi, 1978; Da Settimo, Primofiore, Biagi and Santerini, 1976b; Ellinger, 1904; Gadaginamath and Siddappa, 1975; Gaines, Sletzinger and Ruyle, W., 1961; Hahn, W. E. and Zawadzka, 1967, 1969; Ince, 1889; Joshi, S. S. and Gambhir, 1956; Keimatsu and Sugasawa, 1928b; Kögl and Kostermans, 1935; Kollenz, 1978; Kollenz, Ziegler, Eder and Prewedourakis, 1970; Manske, R. H. F., 1931b; Manske, R. H. F., Perkin and Robinson, R., 1927; Marchand, Streffer and Jauer, 1961; Neber, Knöller, Herbst and Trissler, 1929; Perkin and Rubenstein, 1926; Rogers, C. U. and Corson, 1947; Sandoz Ltd, 1980.

Hydrochloric Acid (Concentrated) in Isopropanol

Yamamoto, H., Inaba, Okamoto, Hirohashi, T., Ishigumi, Yamamoto, M., Maruyama, Mori and Kobayashi, T., 1969c.

Hydrochloric Acid (Concentrated) in *tert*.-Butanol

Merck and Co., Inc., 1967b.

Hydrochloric Acid (Concentrated) in Dioxane

Drogas Vacunas y Sueros, S. A. (Drovyssa), 1957.

Hydrochloric Acid (Concentrated) in Acetic Acid

Armit and Robinson, R., 1922; Ballantine, Barrett, C. B., Beer, Boggiano, Eardley, Jennings and Robertson, 1957; Coldham, Lewis, J. W. and Plant, 1954; Dalton, Fahrenholtz and Silverzweig, 1979; Da Settimo and Saettone, 1965; Fennell and Plant, 1932; Harradence and Lions, 1938; 1939; Inaba, Ishizumi, Mori and Yamamoto, H., 1971; Kinsley and Plant, 1956; Kipping, 1894; Kulka and Manske, R. H. F., 1952; Pelchowicz and Bergmann, E. D., 1960; Schofield and Theobald, 1950; Shvedov, Trofimkin, Vasil'eva and Grinev, A. N., 1975; Velluz, L., Muller, Nominé, Pénasse and Pierdet, 1964; Yamamoto, H., Inaba, Hirohashi, T., Ishizumi, Maruyama and Mori, 1970; Yamamoto, H., Inaba, Hirohashi, T., Mori, Ishizumi and Maruyama, 1969; Yamamoto, H., Inaba, Okamoto, Hirohashi, T., Ishizumi, Yamamoto, M., Maruyama, Mori and Kobayashi, T., 1969c; Yamane and Fujimori, K., 1972.

Hydrochloric Acid (Concentrated)–85% Phosphoric Acid–Pyridine

Engvild, 1977; Robinson, J. R., 1957.

Hydrochloric Acid (Concentrated)–Acetic Acid (50:50)

Plant and Tomlinson, 1933 [failed to indolize acetone phenylhydrazone (Bell and Lindwall, 1948)].

Hydrochloric Acid (Dilute) in Methanol

Borsche and Klein, 1941.

Hydrochloric Acid (Dilute) in Ethanol (?)

Borsche and Klein, 1941; Cornforth, J. W., Cornforth, R. H., Dalgliesh and Neuberger, 1951; Kadzyauskas, Butkus, Vasyulite, Averina and Zefirov, 1979; Merck and Co., Inc., 1961, 1962; Ried and Kleemann, 1968.

Hydrogen Bromide in Aqueous Ethanol

Sunagawa and Sato, Y., 1962a.

(*continued*)

Table 33 (*continued*)

Hydrogen Bromide in Ethanol or Methanol

Elderfield, Lagowski, McCurdy and Wythe, 1958.

Hydrogen Bromide in Acetic Acid

Hughes, G. K., Lions and Ritchie, 1939; Ried and Kleemann, 1968.

Hydrogen Chloride in Methanol

Bullock and Hand, J. J., 1956b; Fabbrica Italiana Sintetici S.p.A., 1968; Fusco and Sannicolò, 1978c; Merck and Co., Inc., 1966; Nozoe, Kitahara and Arai, 1954; Shagalov, Ostapchuk, Zlobina, Eraksina, Babushkina, T. A., Vasil'ev, A. M., Ogoradnikova and Suvorov, 1978; Stillwell, 1964; Yamamoto, H., Inaba, Okamoto, Hirohashi, T., Ishizumi, Yamamoto, M., Maruyama, Mori and Kobayashi, T., 1969b.

Hydrogen Chloride (Dry) in Methanol

Justoni and Pessina, 1957a; Merck and Co., Inc., 1964c.

Hydrogen Chloride in Ethanol

Beer, Donavanik and Robertson, 1954; Bell and Lindwall, 1948; Borsche and Klein, 1941; Brown, F. and Mann, 1948b; Carson, D. F. and Mann, 1965; Cattanach, Cohen, A. and Heath-Brown, 1973; Clemo and Felton, 1951a; Cohen, A., Heath-Brown, Smithen and Cattanach, 1968; Crowther, Mann and Purdie, 1943; Dambal and Siddappa, 1965; Da Settimo, Biagi, Primofiore, Ferrarini and Livi, 1978; Da Settimo, Primofiore, Biagi and Santerini, 1976b; Ehrhart, 1953; Fabbrica Italiana Sintetici S.p.A., 1968; Farbenfabriken Bayer A.-G., 1954, 1955a; Feofilaktov and Semenova, N. K., 1953b; Findlay, S. P. and Dougherty, 1948; Gaimster, 1960; Geigy A.-G., 1966; Gilbert, J., Rousselle, Ganser and Viel, 1979; Goldsmith and Lindwall, 1953; Grandberg, Belyaeva and Dmitriev, L. B., 1971a, 1971b; Hahn, W. E., Bartnik and Zawadzka, 1966; I. G. Farbenind. A.-G., 1930d; Ishii, Hagiwara, Ishikawa, Ikeda and Murakami, Y., 1975; Ishii, Murakami, Y., Furuse, Hosoya and Ikeda, 1973; Ishii, Murakami, Y., Furuse, Hosoya, Takeda and Ikeda, 1973; Ishii, Murakami, Y., Hosoya, Takeda, Suzuki, Y. and Ikeda, 1973; Kakurina, Kucherova and Zagorevkii, 1965b; Kao and Robinson, R., 1955; Kermack, Perkin and Robinson, R., 1921; Kermack and Tebrich, 1940; Kiang, Mann, Prior and Topham, 1956; King, F. E., and Robinson, R., 1933; Korczynski, Brydowna and Kierzek, 1926; Kornet, Thio, A. P. and Tolbert, 1980; Kost, Trofimov, Tsyshkova and Shadurskii, 1969; Mann and Tetlow, 1957; Merck and Co., Inc., 1962, 1963, 1964a, 1964b, 1967b; Orlova, E. K., Sharkova, N. M., Meshcheryakova, Zagorevskii and Kucherova, 1975; Plant and Tomlinson, M. L., 1933; Robinson, R. and Suginome, 1932b; Roussel-UCLAF, 1965; Sarett and Shen, T.-Y., 1965, 1966a, 1966b; Sempronj, 1938; Shen, T.-Y., 1964, 1965, 1966, 1967a, 1967b, 1967c; Shen, T.-Y. and Sarrett, 1966; Skrabal, Steiger and Zollinger, 1975; Spickett, 1966; Suvorov, Gordeev and Vasin, 1974; Szmuszkovicz, Glenn, Heinzelman, Hester and Youngdale, 1966; Tanaka, Z., 1940a; Thomas, R. C., 1975; Trofimov, Ryabchenko and Grinev, A. N., 1975; Walton, Stammer, Nutt, Jenkins and Holly, 1956; Wislicenus and Waldmüller, 1911; Woolley, D. and Shaw, E. N., 1959; Yakovenko, Mikhlina, Oganesyan and Yakhontov, 1955; Yamamoto, H., Inaba, Hirohashi, T., Ishizumi, Maruyama and Mori, 1970; Yamamoto, H., Inaba, Hirohashi, T., Mori, Ishizumi and Maruyama, 1969; Zenitz, 1966.

Hydrogen Chloride (Dry) in Dry Ethanol

Anderson, R. M., Clemo and Swan, 1954; Andrisano and Vitali, 1957; Beer, Brown, J. P. and Robertson, 1951; Bell and Lindwall, 1948 (failed to indolize acetone phenylhydrazone); Chalmers, A. J. and Lions, 1933; Clemo and Seaton, 1954; Clemo and Swan, 1946; Elderfield and Wythe, 1954; Feofilaktov and Semenova, N. K., 1952a, 1953b; Finger, Gortatowski, Shiley and White, R. H., 1959; Harradence and Lions, 1939; Hughes, G. K. and Lions, 1937–1938a; Hughes, G. K., Lions and Ritchie, 1939; Julian, Karpel, Magnani and Meyer, E. W., 1948; Kametani, Ohsawa and Ihara, 1979; Kao and Robinson, R., 1955; Keimatsu and Sugasawa, 1928b; King, F. E. and L'Ecuyer, 1934; Komoto, Enomoto, Miyagaki, Tanaka, Y., Nitanai and Umezawa, 1979; Komoto, Enomoto, Tanaka, Y., Nitanai and Umezawa, 1979; Kost, Vinogradova, Daut and Terent'ev, A. P., 1962; Kost, Vinogradova, Trofimov, Mukhanova, Nozdrich and Shadurskii, 1967; Kost, Yudin and Propravko, 1962; Kost, Yurovskaya, Mel'nikova and Potanina, 1973; Lee, T. B. and Swan, 1956; Nakazaki and Maeda, 1962; Nakazaki, Yamamoto, K. and Yamagami, 1960; Namis, Cortes, Collera and Walls, F., 1966; Pfenninger, 1968; Prasad and Swan, 1958; Ried and Kleemann, 1968; Robinson, R. and Suginome, 1932a; Sharkova, N. M., Kucherova, Portnova and Zagorevskii, 1968; Shaw, E., 1955; Smith, W. S. and Moir, 1952; Suvorov, Samsoniya, Chilikin, Chikvaidze, I. S., Turchin, Efimova, Tret'yakova and Gverdtsiteli, I. M., 1978; Trofimov, Ryabchenko and Grinev, A. N., 1975; Walser, Blount and Fryer, 1973; Witkop, 1953b; Witkop, Patrick and Rosenblum, 1951; Yamamoto, H., Inaba, Okamoto, Hirohashi, T., Ishizumi, Yamamoto, M., Maruyama, Mori and Kobayashi, T., 1969a.

Hydrogen Chloride in Isopropanol

Morooka, Tamoto and Matuura, 1976.

Hydrogen Chloride (Dry) in Butanol

Kulka and Manske, R. H. F., 1952; Manske, R. H. F., Perkin and Robinson, R., 1927.

Hydrogen Chloride in Benzyl Alcohol

Fabbrica Italiana Sintetici S.p.A., 1968; Shen, T.-Y., 1969.

Hydrogen Chloride (Dry) in Absolute Ether

Kost, Yudin and Zinchenko, 1973.

Hydrogen Chloride (Dry, 5%) in Anhydrous Dioxane

Drogas Vacunas y Sueros, S. A. (Drovyssa), 1957; Justoni and Pessina, 1955, 1957a.

(*continued*)

Table 33 (*continued*)

Hydrogen Chloride in Benzene

Fusco and Sannicolò, 1978c; Kakurina, Kucherova and Zagorevskii, 1965a; Pausacker, 1950a.

Hydrogen Chloride (Dry) in Xylene

Chalmers, A. J. and Lions, 1933.

Hydrogen Chloride in Acetic Acid

Blades and Wilds, 1956; Bloss and Timberlake, 1963; Bremner and Browne, 1975; Buu-Hoï, 1949a, 1949b, 1954; Buu-Hoï, Cagniant, Hoán, and Khôi, 1950; Buu-Hoï, Croisy, Jacquignon and Martani, 1971; Buu-Hoï and Hoán, 1949, 1950, 1951, 1952; Buu-Hoï, Hoán and Jacquignon, 1949; Buu-Hoï, Hoán and Khôi, 1949, 1950a, 1950b, 1950c; Buu-Hoï, Hoán, Khôi and Xuong, 1949, 1950, 1951; Buu-Hoï, Hoán and Xuong, 1951; Buu-Hoï and Jacquignon, 1951, 1954, 1956; Buu-Hoï; Jacquignon and Lavit, 1956; Buu-Hoï, Jacquignon and Ledésert, 1970; Buu-Hoï, Jacquignon, Ledésert, Ricci and Balucani, 1969; Buu-Hoï, Khôi and Xuong, 1950, 1951; Buu-Hoï and Lavit, 1955; Buu-Hoï, Mangane and Jacquignon, 1966; Buu-Hoï, Martani, Ricci, Dufour, Jacquignon and Saint-Ruf, 1966; Buu-Hoï, Périn and Jacquignon, 1966; Buu-Hoï, Roussel, O., and Jacquignon, 1964; Buu-Hoï, Royer, Eckert and Jacquignon, 1952; Buu-Hoï and Saint-Ruf, 1962, 1963, 1965a, 1965b; Buu-Hoï, Saint-Ruf, Deschamps, Bigot and Hieu, 1971; Buu-Hoï, Saint-Ruf and Dufour, 1964; Buu-Hoï, Saint-Ruf, Jacquignon and Marty, 1963; Buu-Hoï, Saint-Ruf, Martani, Ricci and Balucani, 1968; Buu-Hoï and Xuong, 1952; Buu-Hoï, Xuong and Lavit, 1954; Buu-Hoï, Xuong and Thang, K. van, 1953; Croisy, Ricci, Janćevska, Jacquignon and Balucani, 1976; David and Régent, 1964; Dufour, Buu-Hoï, Jacquignon and Hien, 1972; Huisgen and Ugi, 1957; Kempter, Schwalba, Stoss and Walter, 1962; Maréchal, Christaens, Renson and Jacquignon, 1978; Menichi, Bisagni and Royer, 1964; Plieninger, 1950b; Rice and Scott, K. R., 1970; Robinson, B. and Zubair, 1971, 1973; Thang, D. C., Kossoff, Jacquignon and Dufour, 1976; Velluz, L., Muller and Allais, 1962; Yamamoto, H., Inaba, Okamoto, Hirohashi, T., Ishizumi, Yamamoto, M., Maruyama, Mori and Kobayashi, T., 1970.

Hydrogen Chloride over the Molten Arylhydrazone

Robinson, G. M. and Robinson, R., 1918.

Hydrogen Iodide in Absolute Ethanol

Brunner, K., 1898b, 1900; Plancher, 1900a, 1902; Plancher and Bonavia, 1902; Plangger, 1905.

Hydrogen Iodide in Acetic Acid

Kalb, Schweizer, Zellner and Berthold, 1926.

Iron Powder

Korczynski and Kierzek, 1925.

Magnesium Chloride

Zaitsev, 1938.

4-Methylbenzene Sulphonic Acid

Yakhontov and Marshalkin, 1972; Yakhontov, Marshalkin and Anisimova, 1972; Yakhontov and Pronina, E. V., 1968, 1969; Yakhontov, Pronina, E. V. and Rubtsov, 1965, 1967, 1970.

4-Methylbenzene Sulphonic Acid in Benzene

Ishii, Ikeda and Murakami, Y., 1972; Ishii, and Murakami, Y., 1975; Ishii, Murakami, Y., Furuse and Hosoya, 1979; Ishii, Murakami, Y., Furuse, Hosoya, Takeda and Ikeda, 1973; Ishii, Murakami, Y., Furuse, Takeda and Ikeda, 1973; Ishii, Murakami, Y., Hosoya, Furuse, Takeda and Ikeda, 1972; Ishii, Murakami, Y., Takeda and Furuse, 1974; Ishii, Murakami, Y. and Ishikawa, 1977a.

4-Methylbenzene Sulphonic Acid in Toluene

Borch and Newell, R. G., 1973 and Newell, R., 1975.

4-Methylbenzene Sulphonic Acid in Methanol

Rosenmund and Sadri, 1979; Shvedov, Trofimkin, Vasil'eva and Grinev, A. N., 1975; Sumitomo, Hayakawa and Tsubojima, 1969.

4-Methylbenzene Sulphonic Acid in Ethyl Cellosolve

Mamaev and Sedova, 1961.

4-Methylbenzene Sulphonic Acid in Acetic Anhydride

Suvorov, Sorokina, N. P. and Sheinker, I. N., 1959.

Molybdenum Powder

Korczynski and Kierzek, 1925.

Naphthalene-1,5-disulphonic Acid

Herdieckerhoff and Tschunkur, 1933.

Nickel Powder

Korczynski and Kierzek, 1925; Korczyński and Kierzek, 1925.

(*continued*)

Table 33 (*continued*)

Nickel (II) Chloride

Huisgen, 1948a; Korczynski, Brydowna and Kierzek (far better than zinc chloride and was successful when the use of ethanolic hydrogen chloride failed); Korczynski and Kierzek, 1925; Korczyński and Kierzek, 1925; Witkop, 1944.

'Non-Catalytic' Thermal Methods

(*i*) *Heat in Benzene Solution at 235 °C in an Autoclave*
Rosenthal and Yalpani, 1965.

(ii) *Pyrolysis*
Shvedov, Kurilo and Grinev, A. N., 1972.

(iii) *Reflux in 1,2-Dichlorobenzene*
Baldwin and Tzodikov, 1977.

(iv) *Reflux in Diethylene Glycol*

Fitzpatrick, J. T. and Hiser, 1957; Lohr, 1977; Miller, F. M. and Schinske, 1978; Yakhontov and Marshalkin, 1972. See also Chapter IV, I, for other applications of this technique.

(v) *Reflux in Monoethylene Glycol*
Bailey, Scottergood and Warr, 1971; Baldwin and Tzodikov, 1977; Casnati, Langella, Piozzi, Ricci and Umani-Rochi, 1964b; Dashkevich, S. N., 1978; Fitzpatrick, J. T. and Hiser, 1957; Gale and Wilshire, 1974; Robinson, B., 1963c; Robinson, B. and Zubair, 1973; Schut, Ward, Lorenzetti and Hong, 1970 (using N_α-methyl- or N_α-benzylarylhydrazones, this method was preferred over those using an acid catalyst); Szmuszkovicz, 1970, 1971; Youngdale, Glenn, Lednicer and Szmuszkovicz, 1969. See also Chapter IV, I, for other applications of this technique.

(vi) *Reflux in tert.-Butanol*
Chemerda and Sletzinger, 1969b.

(vii) *Reflux in Naphthalene*
Shukri, Alazawe and Al-Tai, 1970.

(viii) *Reflux in Tetralin*
How and Parrick, 1976; Fitzpatrick, J. T. and Hiser, 1957; Mills, Al Khawaja, Al-Saleh and Joule, 1981.

(ix) *Reflux in Triethylene Glycol*
See Chapter IV, I for examples.

Orthophosphoric Acid

Shagalov, Eraksina and Suvorov, 1970 (in cellosolve); Shagalov, Eraksina, Turchin and Suvorov, 1970 (in ethyl cellosolve).

Oxalic Acid in Water

Henle, 1905 (concentrated solution); Kost, Sugorova and Yakubov, 1965 (saturated solution).

Perchloric Acid in Anhydrous Acetic Acid

Fusco and Sannicolò, 1976a.

Phenol

Zaitsev, 1938.

Phosphoric Acid (Orthophosphoric Acid?)

Borghero and Finsterle, 1955 (in ethanol); Clerc-Bory, 1954 (in methanol); Fischer, Emil, 1886b; Jackson, A., Gaskell, Wilson, N. D. V. and Joule, 1968; Jackson, A. and Joule, 1967; Jackson, A., Wilson, N. D. V., Gaskell and Joule, 1969; Kadzyauskas, Butkus, Vasyulite, Averina and Zefirov, 1979; Lakshmanan, 1960; Leekning, 1964, 1967; Mentzer, Beaudet and Bory, 1953; Pigerol, Chandavoine, De Cointet de Fillain and Nanthavong, 1975; Rogers, C. U. and Corson, 1947; Shvedov, Kurilo and Grinev, A. N., 1970; Suvorov, Mamaev and Shagalov, 1955 in methanol – see also Suvorov, 1957; Suvorov, Mamaev and Shagalov, 1957 (in cellosolve); Ufimtsev, Grineva and Sadovskaya, 1962.

Phosphorus Oxychloride

Kermack and Smith, J. F., 1930 (in toluene); Manhas, Brown, J. W. and Pandit, 1975; Manhas, Brown, J. W., Pandit and Houdewind, 1975 (at room temperature).

Platinum(II) Chloride

Arbusow, A. E. and Tichwinsky, 1910b; Arbuzov, A. E. and Tikhvinskii, 1913b.

Polyphosphate Ester

Inoue and Ban, 1970; Kanaoka, Ban, Miyashita, Irie and Yonemitsu, 1966; Kanaoka, Ban, Yonemitsu, Irie and Miyashita, 1965; Krutak, 1975.

Polyphosphoric Acid

Ainsworth, D. P. and Suschitzky, 1967b; Al-Azawe and Sarkis, 1973; Amorosa and Lipparini, 1959; Andrisano and Vitali, 1957; Avramenko, Plutitskii, Strekozov and Suvorov, 1979; Bazile, Cointet and Pigerol, 1978; Berlin, A. Y. and Zaitseva, 1960; Berti and Da Settimo, 1960; Bogat-skii, Ivanova, R. Y., Andronati and Zhilina, 1979; Bourdais and Lorre, 1975; Bradsher and Litzinger, 1964; Bruce, 1960 (better to use 88–90% polyphosphoric acid with dihydroxyphenyl methyl ketone phenylhydrazones); Bruce, 1962 (zinc chloride was preferable); Buu-Hoï, Delcey, Jacquignon and Périn, 1968; Buu-Hoï, Périn and Jacquignon, 1965; Calvaire and Pallaud, 1960; Casnati, Langella, Piozzi, Ricci and Umani-Ronchi, 1964a, 1964b; Cavallini and Ravenna, 1958; Chapman, Clarke and Hughes, H., 1965 (not as good as zinc chloride for acetone phenylhydrazone); Colle, M.-A. and David, 1960; Collins, I, Roberts and Suschitzky, 1971; Da Settimo, Primofiore, Biagi and Santerini, 1976a; De Bellis and Stein, M. L., 1961; Duc and Fétizon, 1969; Duncan

(*continued*)

Table 33 (*continued*)

Polyphosphoric Acid (*continued*)

and Andrako, 1968; Duncan and Boswell, 1973; Duncan, Helsley and Boswell, 1973; Duuren, 1961; Evans, F. J., Lyle, G. G., Watkins and Lyle, R. E., 1962; Fusco and Sannicolò, 1974; Gadaginamath, 1976; Gale, Lin and Wilshire, 1976; Gardner, D. V., McOmie and Prabhu, 1970; Grandberg, Belyaeva and Dmitriev, L. B., 1971a, 1971b, 1971c, 1973; Gray and Archer, W. L., 1957; Grinev, N. I., Sadovskaya and Ufimtsev, 1963; Hamana and Kumadaki, 1967; Hiremath and Siddappa, 1962; Holla, S. and Ambekar, 1974; Holla, B. S. and Ambekar, 1979b; Huffman, 1962; Huisgen and Ugi, 1957; Ishizumi, Shioiri and Yamada, S., 1967; Joshi, K. C., Pathak and Chand, 1978a, 1978b; Julia and Lenzi, 1971; Julia, Melamed and Gombert, 1965; Kadzyauskas, Butkus, Vasyulite, Averina, and Zefirov, 1979; Kempter, Schwalba, Stoss and Walter, 1962; Kissman, Farnsworth and Witkop, 1952 (initial use of this indolization catalyst); Kost, Sugorova and Yakubov, 1965; Kricka and Vernon, 1974; Lakshmanan, 1960; Leekning, 1964, 1967; Lohr, 1977; MacPhillamy, Dziemian, Lucas and Kuehne, 1958; Marshalkin and Yakhontov, 1972a; Miller, F. M. and Schinske, 1978; Noland and Baude, 1966; Noland, Rush and Smith, L. R., 1966 [better than zinc chloride for the indolization of methyl phenyl ketone phenylhydrazone but modified by the gradual addition of the phenylhydrazone to the catalyst to prevent an otherwise uncontrolled eruptive reaction using a large scale (1 × mole) reaction – see also Joshi, K. C., Pathak, Arya and Chand, 1979]; Nozoe, Sin, Yamane and Fujimori, K., 1975; Pappalardo and Vitali, 1958b; Prasad and Swan, 1958 (indolization failed using ethanolic hydrogen chloride); Sato, Y., and Sunagawa, 1967; Sergeeva, Z. F., Akhvlediani, Shabunova, Korolev, Vasil'ev, A. M., Babushkina, T. N. and Suvorov, 1975; Shavel and Strandtmann, 1965; Shvedov, Kurilo and Grinev, A. N., 1972; Strandtmann, Cohen, M. P. and Shavel, 1963; Strandtmann, Puchalski and Shavel, 1964; Sugasawa and Takano, 1959; Sugasawa, Terashima and Kanaoka, 1956; Sunagawa and Sato, Y., 1962b, 1967; Sunagawa, Soma, Nakao and Matsumoto, 1961; Suvorov, Gordeev and Vasin, 1974; Szmuszkovicz, Glenn, Heinzelman, Hester and Youngdale, 1966; Trofimov, Tsyshkova, Garnova and Grinev, A. N., 1975; Vinogradova, D'aut, Kost and Terent'ev, A. P., 1962; Weng, Wang, C.-T., and Wang, T., 1962; Witkop, Patrick and Kissman, 1952; Yakhontov and Marshalkin, 1972; Yakhontov and Pronina, E. V., 1969; Yakhontov, Pronina, E. V. and Rubtsov, 1967, 1970; Yamamoto, H., Inaba, Okamoto, Hirohashi, T., Ishizumi, Akatsu, Kobayashi, T., Mori, Kume and Sato, H., 1973 (much improved yields than when using 85% formic acid); Yamamoto, H., Misaki, Izumi and Koshiba, 1971.

Polyphosphoric Acid in Absolute Ethanol Saturated with Dry Hydrogen Chloride

Robinson, J. R. and Good, 1957.

Propionic Acid

Rogers, C. U. and Corson, 1947.

Pyridine Hydrochloride in Pyridine (Ketone + Arylhydrazine Hydrochloride in Refluxing Pyridine)

Although successful when using 1-carbethoxypiperidin-4-ones and other non-basic ketones, this method failed when using a mixture of 1-methylpiperidin-4-one or its hydrochloride with phenylhydrazine hydrochloride in refluxing pyridine. This observation was explained by assuming a facile 1,3-transfer of a pseudoaxial proton from N-1 to C-3 in the enehydrazine tautomer of the corresponding hydrazone (the hydrazone was readily formed) with reversion to the hydrazone, a process energetically favoured over indolization (Welch, W. M., 1977a).

Quaternary Salts

N,N,2,6-Tetramethylanilinium Iodide (Posvic, Dombro, Ito, H. and Telinski, 1974; Tetra-*tert*.-butylammonium bromide and benzyl trimethylammonium bromide in the presence of potassium carbonate which afforded the corresponding 1-*tert*.-butyl and 1-benzylindoles, respectively starting with N_α-unsubstituted arylhydrazones (Suvorov, Plutitskii and Smushkevich, 1980).

Rhodium Complexes in Ethanol (i.e. neutral conditions)

Watanabe, Y., Yamamoto, M., Shim, Miyanaga and Mitsudo, 1980.

Sodium Ethoxide

Marshalkin and Yakhontov, 1972a, 1972b; Wolff and Mayen, 1912; Yakhontov and Marshalkin, 1971; Yakhontov, Marshalkin and Anisimova, 1972.

Sodium Hydride

Grandberg and Bobrova, 1978.

Sodium Hydroxide

Fitzpatrick, J. T. and Hiser, 1957 (2% in refluxing ethylene glycol); Seibert, 1948 – see also Ioffe, Sergeeva, Z. I. and Stopskii, 1966.

Sulphanilic Acid

Kitaev, Troepol'skaya and Arbuzov, A. E., 1964, 1966a, 1966b.

Sulphosalicylic Acid (usually in ethanol or methanol)

Grandberg, Belyaeva and Dmitriev, L. B., 1971a, 1971b, 1971c; Shagalov, Eraksina and Suvorov, 1970; Shagalov, Eraksina, Turchin and Suvorov, 1970; Suvorov, Morozovskaya and Sorokina, G. M., 1961; Suvorov and Murasheva, 1958, 1960, 1961; Yakhontov and Marshalkin, 1972; Yakhontov, Marshalkin and Anisimova, 1972; Yakhontov and Pronina, E. V., 1969; Yakhontov, Pronina, E. V. and Rubtsov, 1967, 1970. Basanagoudar and Siddappa, 1972; Borch and Newell, R. G., 1973; Colwell, Horner, J. K. and Skinner, 1965; Cook, A. H. and Reed, K. J., 1945; Diels and Durst, 1914; Hughes, G. K., Lions and Ritchie, 1939; Nef, 1891; Nozoe, Sin, Yamane and Fujimori, K., 1975; Rensen, 1959; Walker, C., 1894; Yoshimura, Sakamoto and Matsunaga, 1969; Yudin, Budylin and Kost, 1964.

Sulphuric Acid (Concentrated) in Methanol

Bullock and Hand, J. J., 1956b; Fusco and Sannicolò, 1975a, 1975c, 1976a, 1978c; Shagalov, Ostapchuk, Zlobina, Eraksina, Babushkina, T. A., Vasil'ev, A. M., Ogoradnikova and

(*continued*)

Table 33 (*continued*)

Sulphuric Acid (Concentrated)

Suvorov, 1978; Shvedov, Kurilo and Grinev, A. N., 1972; Suvorov, Mamaev and Shagalov, 1953 – see also Suvorov, 1957.

Sulphuric Acid (Concentrated) in Absolute Methanol

Rosenstock, 1966.

Sulphuric Acid (Concentrated) in Ethanol

Akopyan and Tatevosyan, 1971; Amorosa, 1955, 1956 (fails); Amorosa and Lipparini, 1956; Bisagni, Ducrocq, Lhoste, Rivalle and Civier, 1979; Blaikie and Perkin, 1924; Boyd-Barrett and Robinson, R., 1932; Brunner, K., 1894; Bullock and Fox, 1951; Bullock and Hand, J. J., 1956a, 1956b; Carlin, Henley and Carlson, 1957; Colwell, Horner, J. K. and Skinner, 1964; Cook, J. W., Loudon and McCloskey, 1952; Dambal and Siddappa, 1965; Elks, Elliott and Hems, 1944b; Ellinger, 1905; Feofilaktov, 1947a, 1952; Feofilaktov and Semenova, N. K., 1952b, 1953b; Fischer, Emil, 1886b; Fox and Bullock, 1951a, 1951b, 1955a, 1955b; Inaba, Akatsu, Hirohashi, T. and Yamamoto, H., 1976; Jackson, R. W. and Manske, R. H., 1930, 1935; Kalb, Schweizer and Schimpf, 1926; Kalb, Schweizer, Zellner and Berthold, 1926; Keimatsu and Sugasawa, 1928b; King, F. E. and Robinson, R., 1932; Lanzilotti, Littell, Fanshawe, McKenzie and Lovell, 1979; Lohr, 1977; Manske, R. H. F., 1931b; Miller, F. M. and Schinske, 1978; Mndzhoyan, Tatevosyan and Ekmekdzhayan, 1957; Mndzhoyan, Tatevosyan, Terzyan and Ekmekdzhyan, 1958; Pacheco, 1951; Perkin and Rubenstein, 1926; Plancher and Carrasco, 1904; Razumov and Gurevich, 1967a; Robinson, R. and Suginome, 1932a; Sakurai and Ito, T. 1957; Sato, Y., 1963a; Scherrer, 1969; Shvedov, Kurilo and Grinev, A. N. 1969; Stevens, F. J., Ashby and Downey, 1957; Stevens, F. J. and Su, 1962; Suvorov and Antonov, 1952; Suvorov, Antonov and Rokhlin, 1953; Suvorov, Mamaev and Shagalov, 1953 – see also Suvorov, 1957; Suvorov and Sorokina, N. P., 1960; Suvorov, Sorokina, N. P. and Sheinker, I. N., 1959; Terzyan, Akopyan and Tatevosyan, 1961; Terzyan and Tatevosyan, 1960; Trofimov, Tsyshkova, Garnova and Grinev, A. N., 1975; Unanyan and Tatevosyan, 1961.

Sulphuric Acid (Concentrated?) in Alcohol (Ethanol?)

Andrisano and Vitali, 1957; Arnold, E., 1888; Barrett, H. S. B., Perkin and Robinson, R., 1929; Blaikie and Perkin, 1924; Boyakhchyan, Rashidyan and Tatevosyan, 1966; Ellinger, 1905; Gabrielyan and Papayan, 1973; Kalb, Schweizer and Schimpf, 1926; King, F. E. and Robinson, R., 1933; Kost, Yudin and Popravko, 1962; Manske, R. H. F., 1931b; Manske, R. H. F. and Robinson, R., 1927; Marion and Oldfield, 1947; Perkin and Rubenstein, 1926; Plant and Thompson, 1950; Shvedov, Kurilo and Grinev, A. N., 1969; Wislicenus and Arnold, E., 1887; Wislicenus and Münzesheimer, 1898; Wislicenus and Schultz, F., 1924.

Sulphuric Acid (5%) in Ethanol

Orlova, E. K., Sharkova, N. M., Meshcheryakova, Zagoreveskii and Kucherova, 1975.

Sulphuric Acid (10%) in Ethanol

Bell and Lindwall (failed to indolize acetone phenylhydrazone); Grandberg, Belyaeva and Dmitriev, L. B., 1971a; Pachter, 1967.

Sulphuric Acid (15%) in Ethanol

Gardner, J. H. and Stevens, J. R., 1947.

Sulphuric Acid (33%) in Ethanol

Suvorov, Samsoniya, Chilikin, Chikvaidze, I. S., Turchin, Efimova, Tretyak'ova and Gverdtsiteli, I. M., 1978.

Sulphuric Acid (Concentrated) in Propanol

Yamamoto, H., Inaba, Hirohashi, T., Mori, Ishizumi and Maruyama, 1969.

Sulphuric Acid (Concentrated) in Isopropanol

Yamamoto, H., Inaba, Hirohashi, T., Ishizumi, Maruyama and Mori, 1970.

Sulphuric Acid (Concentrated) in Monoethylene Glycol

Sankyo Co. Ltd, 1964, Sato, Y. 1963b, 1963c; Sunagawa and Sato, Y., 1962b.

Sulphuric Acid (Concentrated) in Acetic Acid

Andrisano and Vitali, 1957; Boekelheide and Ainsworth, C., 1950b; Buu-Hoï, Jacquignon and Hoeffinger, 1963; Buu-Hoï, Jacquignon, Ledésert, Ricci and Balucani, 1969; Buu-Hoï, Jacquignon, Roussel, O. and Hoeffinger, 1964; Buu-Hoï, Mangane and Jacquignon, 1966; Buu-Hoï, Périn and Jacquignon, 1962; Buu-Hoï and Saint-Ruf, 1963; Clemo and Felton, 1951a; Elks, Elliott and Hems, 1944b, 1944c; Finkelstein and Lee, J., 1956; Gryaznov, Akhvediani, Volodina, Vasil'ev, A. M., Babushkina, T. A., and Suvorov, 1977; Hegedüs, 1946; Inaba, Ishizumi, Okamoto and Yamamoto, H., 1972, 1975; Ishii, Murakami, Y., Furuse, Hosaya and Ikeda, 1973; Kost, Sugorova and Yakubov, 1965; Samsoniya, Trapaidze, Gverdtsiteli, I. M. and Suvorov, 1977; Sergeeva, Z. F., Akhvlediani, Shabunova, Korolev, Vasil'ev, A. M., Babushkina, T. N. and Suvorov, 1975; Yamamoto, H., Inaba, Okamoto, Hirohashi, T., Ishizumi, Maruyama, Kobayashi, T., Yamamoto, M. and Mori, 1970.

Sulphuric Acid (Dilute)

Bahadur, Bailey and Baldry, 1977; Bailey, Baldry and Scott, P. W., 1979; Bailey, Hill, P. A. and Seager, 1974; Bajwa and Brown, R. K., 1968a, 1970; Baldwin and Tzodikov, 1977; Ballantine, Barrett, C. B., Beer, Boggiano, Eardley, Jennings and Robertson, 1957; Barclay and Campbell, 1945; Barnes, Pausacker and Badcock, 1951; Barnes, Pausacker

(*continued*)

Table 33 (*continued*)

Sulphuric Acid (Concentrated) in Acetic Acid (*continued*)

and Schubert, 1949; Bergmann, E. D. and Hoffmann, E., 1962; Borsche, Witte, A. and Bothe, 1908; Braun, 1908b; Braun and Bayer, 1929; Braun and Haensel, 1926; Braun and Schörnig, 1925; Clemo and Felton, 1951b; Cook, A. H. and Reed, K. J., 1945; Cornforth, J. W., Cornforth, R. H., Dalgliesh and Neuberg, 1951; Da Settimo, Primofiore, Biagi and Santerini, 1976a; Douglas, B., Kirkpatrick, Moore, B. P. and Weisbach, 1964; Elks, Elliott and Hems, 1944a; Ghigi, 1930; Grammaticakis, 1939, 1940b; Grandberg, Belyaeva and Dmitriev, L. B., 1973; Hörlein, U., 1954; Hoshino and Takiura, 1936; Huisgen, 1948b; Julia and Lenzi, 1971; Kochetkov, Kucherova and Evdakov, 1956, 1957; Kost, Sugorova and Yakubov, 1965; Kost, Terent'ev, A. P., Vinogradova, Terent'ev, P. B. and Ershov, 1960; Kost, Vinogradova, Trofimov, Mukhanova, Nozdrich and Shadurskii, 1967; Kost and Yudin, 1957; Kost, Yudin, Berlin, Y. A. and Terent'ev, A. P., 1959; Kost, Yudin, Dmitriev, B. A. and Terent'ev, A. P., 1959; Kost, Yudin and Terent'ev, A. N., 1959; Lingens, and Weiler, 1963; Manske, R. H. F. and Kulka, 1947, 1950a; Milne and Tomlinson, M. L., 1952; Morrison, Waite, R. O., Caro, A. N., and Shavel, 1967; Nagaraja and Sunthankar, 1958; Nozoe, Sin, Yamane and Fujimori, K., 1975; Oakeshott and Plant, 1926; Perkin and Plant, 1923b (the use of stronger acid or hydrochloric or acetic acid led to considerable hydrolysis and the formation of much tar); Plancher, Cecchetti and Ghigi, 1929; Plancher and Ghigi, 1929; Plant, Rogers, K. M. and Williams, S. B. C., 1935; Plant and Tomlinson, M. L., 1932a, 1932b; Robinson, R. and Thornley, 1924; Rosenstock, 1966; Sato, Y., 1963a; Shavel and Morrison, 1967; Sparatore and Cerri, 1968; Sturm, Tritschler and Zeidler, 1972; Sunagawa, Soma, Nakano and Matsumoto, 1961; Tenud, 1975; Teuber and Cornelius, 1965; Vinogradova, Daut, Kost and Terent'ev, A. P., 1962; Warner-Lambert Pharmaceutical Co., 1966; Wright and Gambino, 1979; Yudin, Kost and Berlin, Y. A., 1958.

Tin(II) Chloride (using various conditions)

Brunner, K., 1900 (in hydrochloric acid); Carlin and Amoros-Marin, 1959; Carlin, Wallace, J. G. and Fisher, 1952 (heat with the dihydrate, to remove the water, then fuse); Carson, D. F. and Mann, 1965 (in an ethanol–concentrated hydrochloric acid mixture. Indolization failed using polyphosphoric acid, dilute sulphuric acid or zinc chloride as catalyst); Council of Scientific and Industrial Research (India), 1975 (in an organic acid); Jenisch, 1906 (in an ethanol–concentrated hydrochloric acid mixture); Kögl and Kostermans, 1935 (cf. Fischer, Emil, 1886c); Robinson, G. M. and Robinson, R., 1924 (crystallized catalyst in an acetic acid–concentrated hydrochloric acid mixture); Rosenmund and Sotiriou, 1975 (in an ethanol–concentrated hydrochloric acid mixture); Zaitsev, 1938.

Titanium Powder

Korczynski and Kierzek, 1925.

Trifluoracetic Acid

Zinnes, 1975.

Tungsten Powder

Korczynski and Kierzek, 1925.

Uranium Powder

Korczynski and Kierzek, 1925.

Uranium Hexachloride

Korczynski and Kierzek, 1925; Korczyński and Kierzek, 1925.

Zeolite Catalysts in a Continuous Flow System

Venuto and Landis, 1968.

Zinc

Korczynski and Kierzek, 1925 (powder); McLoughlin and Smith, L. H., 1965 (granules).

Zinc Bromide in Nitrobenzene

Carlin and Larson, 1957.

Zinc Chloride (Excess) [for the preparation of this catalyst, see Pray (1957)]

Adkins and Coonradt, 1941; Aktieselskabet Dumex (Dumex Ltd), 1963, 1964; Ames, A. F., Ames, D. E., Coyne, Grey, Lockhart and Ralph, 1959; Amorosa, 1956; Arheidt, 1887; Auwers, Hilliger and Wulf, 1922; Beitz, Stroh, H.-H. and Fiebig, 1967; Bell and Lindwall, 1948; Blades and Wilds, 1956; Bornstein, Leone, Sullivan and Bennett, O. F., 1957 (indolizations failed or gave only very low yields of products when using only small amounts of zinc chloride or when using several other catalysts); Borsche and Groth, 1941; Brown, F. and Mann, 1948a; Bruce, 1962 (preferred to polyphosphoric acid); Brunck, 1893; Brunner, K., 1895a; Bullock and Hand, J. J., 1956a; Bülow, 1918; Buu-Hoï, 1958; Buu-Hoï, Binh, Loc, Xuong and Jacquignon, 1957; Buu-Hoï and Hoán, 1949, 1950; Buu-Hoï and Jacquignon, 1953, 1954; Buu-Hoï, Jacquignon and Hoeffinger, 1963; Buu-Hoï, Jacquignon, Roussel, O. and Hoeffinger, 1964; Buu-Hoï, Périn and Jacquignon, 1962, 1965; Buu-Hoï and Royer, 1950; Campbell and Cooper, 1935; Cardani, Piozzi and Casnati, 1955; Carlin and Fisher, 1948; Carlin, Henley and Carlson, 1957; Carlin and Larson, 1957; Carlin, Wallace, J. G. and Fisher, 1952; Coló, Asero and Vercellone, 1954; Cook, J. W., Loudon and McCloskey, 1952; Crowther, Mann and Purdie, 1943, David and Monnier, 1959; David and Régent, 1964; Degen, 1886; Ehrhart, 1953; Eiter and Svierak, 1952; Farbenfabriken vorm. Friedr. Bayer and Co., 1901, 1902; Farmaceutici Italia, S. A., 1954; Fischer, Emil, 1886b, 1886c, 1886e, 1887, 1889b; Fischer, Emil and Schmidt, T., 1888; Fischer, Emil and Schmitt, 1888; Fox and Bullock, 1951a (a saturated solution in concentrated hydrochloric acid failed to indolize 3-carboxypropionaldehyde phenylhydrazone); General Aniline Works, Inc., 1932; Gordon and Jackson, R. W., 1935; Govindachari, Rajappa and Sudarsanam, 1966; Gupta and Ojha, 1971; Hai, Buu-Hoï and Xuong, 1958; Hamana and Kumadaki, 1967 (conditions not stated); Hansch and Muir, 1950; Hörlein, U., 1954; Hoshino and Shimodaira, 1935; Huang-Hsinmin and Mann,

(*continued*)

Table 33 (*continued*)

Zinc Chloride (Excess) [for the preparation of this catalyst, see Pray (1957)] (*continued*)

1949; Ince, 1889; Janetsky, Verkade and Lieste, 1946; Julian and Pikl, 1933; Kaji and Nagashima, 1952; Kametani, Fukumoto and Masuko, 1963; Kiang, Mann, Prior and Topham, 1956; Kohlrausch, 1889; Korczynski and Kierzek, 1925; Kost, Sugorova and Yakubov, 1965; Kuroda, 1923; Laubmann, 1888b; Lespagnol, A., Lespagnol, C. and Henichart, 1969; Lindwall and Mantell, 1953; Mann and Haworth, 1944; Marion and Oldfield, 1947; Marshalkin and Yakhontov, 1972a; Mendlik and Wibaut, 1931; Merchant and Shah, N. J., 1975; Michaelis and Luxembourg, 1893; Nakazaki, 1960a; Ockenden and Schofield, 1953a; Patel, H. P. and Tedder, 1963; Petrova, T. D., Mamaev and Yakobson, 1969; Pfülf, 1887; Piper and Stevens, F. J., 1962; Plancher, 1898e, 1898f; Plancher and Caravaggi, 1905; Plancher and Forghieri, 1902; Pretka and Lindwall, 1954; Raschen, 1887; Reif, 1909; Robinson, B. and Zubair, 1971; Roder, 1886; Schiemann and Winkelmüller, 1933; Schlieper, 1886, 1887; Shriner, Ashley and Welch, E., 1942, 1955 (using equal parts of the hydrazone with zinc chloride, the yield of indolic product was lowered compared with that obtained when using an excess of zinc chloride); Späth and Brunner, O., 1925; Steche, 1887; Stevens, F. J. and Fox, 1948; Stevens, F. J. and Higginbotham, 1954; Stevens, F. J. and Su, 1962; Swaminathan, Ranganathan and Sulochana, 1958; Tanaka, Z., 1940b; Trenkler, 1888; Walther and Clemen, 1900; Weisbach, Macko, De Sanctis, Cava and Douglas, B., 1964; Yakhontov and Marshalkin, 1972; Yakhontov, Marshalkin and Anisimova, 1972.

Zinc Chloride (Catalytic Quantity)

Arbusow, A. E. and Tichwinsky, 1910b; Arbuzov, A. E. and Rotermel, 1932; Arbuzov, A. E. and Tikhvinskii, 1913b; Grandberg, Belyaeva and Dmitriev, L. B., 1971a, 1971b, 1971c, 1973; Jackson, A. H. and Smith, P., 1968; Kost, Vinogradova, Trofimov, Mukhanova, Nozdrich and Shadurskii, 1967; Swaminathan and Ranganathan, 1957; Terent'ev, P. B., Kost, Shchegolev and Terent'ev, A. P., 1961; Yakhontov and Pronina, E. V., 1969; Yakhontov, Pronina, E. V. and Rubtsov, 1967, 1970.

Zinc Chloride in Dry Ethanol

Amorosa, 1956; Amorosa and Lipparini, 1956; Braun and Bayer, 1925; Brunner, K., 1895b, 1896a, 1898a, 1900; Clemo and Seaton, 1954; Gal'bershtam and Samoilova, 1973; Ghigi, 1933a, 1933b; Grgin, 1906; Hoshino, 1933; Hoshino and Kobayashi, T., 1935; Landgraf and Seeger, 1968; Leuchs, Heller, A. and Hoffmann, A., 1929; Leuchs and Overberg, 1931; Leuchs, Wulkow and Gerland, 1932; Lohr, 1977; Merck and Co., Inc., 1962, 1964b; Miller, F. M. and Lohr, 1978; Miller, F. M. and Schinske, 1978; Nakazaki and Maeda, 1962; Plancher, 1898a, 1898b, 1898c, 1898d, 1898g, 1900a, 1902; Plancher and Bettinelli, 1899; Plancher and Bonavia, 1900, 1902; Plancher and Carrasco, 1905; Plancher and Testoni, 1900; Sarett and Shen, T.-Y., 1966a, 1966b; Shen, T.-Y., 1967a, 1967b; Shen, T.-Y. and Sarett, 1966; Witkop, Patrick and Kissman, 1952.

Zinc Chloride (Saturated Solution in Concentrated Hydrochloric Acid)

Stevens, F. J. and Fox, 1948.

Zinc Chloride in Acetic Acid

Ishii, Hagiwara, Ishikawa, Ikeda and Murakami, Y., 1975; Ishii, Murakami, Y., Furuse, Hosoya and Ikeda, 1973; Ishii, Murakami, Y., Takeda, and Furuse, 1974; Mann and Haworth, 1944; Oki and Nagasaka, 1973; Yamane and Fujimori, K., 1972.

Zinc Chloride in Alcoholic Concentrated Hydrochloric Acid

Konschegg, 1905.

Zinc Chloride in an Organic Acid

Council for Scientific and Industrial Research (India), 1975.

Zinc Chloride in Cumene

Chapman, Clarke and Hughes, H., 1965 (better than hydrogen chloride, hydrochloric acid, boron trifluoride, polyphosphoric acid or copper(I) chloride); Hughes, G. K., Lions and Ritchie, 1939.

Zinc Chloride in 2-Methylnaphthalene

Jordaan and Arndt, 1968 [indolization failures have been reported (Carlin and Fisher, 1948) when using refluxing 1-methylnaphthalene or tetralin as the solvent].

Zinc Chloride in *para*-Cymene

Dewar, 1944; Govindachari and Sudarsanam, 1971.

Zinc Chloride in Nitrobenzene

Carlin and Carlson, 1959; Carlin and Fisher, 1948; Carlin and Harrison, J. W., 1965; Carlin and Larson, 1957.

Zinc Chloride in 2-Nitrotoluene or Phenol

Carlin and Fisher, 1948.

Zinc Chloride in *para*-Cresol

Carlin and Fisher, 1948; Carlin, Wallace, J. G. and Fisher, 1952.

Zinc Chloride in 'Solvent Naphtha' or 'Tar Oils'

Gesellschaft für Teerverwertung m.b.H., 1911.

Zinc Chloride in the presence of Sulpholane (Yamamoto, T., 1970)

Okawa and Konishi, 1975.

(continued)

Table 33 (*continued*)

Zinc Chloride in Toluene

Doyle and Smith, S. C., 1974.

Zinc Chloride in Xylene

Bullock and Hand, J. J., 1956a; De Bellis and Stein, M. L., 1961; Giuliano and Stein, M. L., 1957.

Zinc Chloride-Calcium Chloride Mixture

Tanaka, Z., 1940b.

Zinc Chloride–Sand Mixture

Carlin and Larson, 1957.

CHAPTER V

Successful Fischer Indolizations with Subsequent Product Modification Under the Reaction Conditions

A. Group Migrations, Group Eliminations and Rearrangements

The migration of alkyl and aryl groups in 2-unsubstituted indoles from the 3- to the 2-position of the indole nucleus under forcing indolization conditions is well documented. Indeed, the first such observation was made by Fischer and one of his students, Theodor Schmidt (Schmitt?) who found that whereas indolization of phenylacetaldehyde phenylhydrazone (**1**; $R^1 = R^2 = H$, $R^3 = C_6H_5$) with alcoholic hydrochloric acid afforded the expected 3-phenylindole (**2**; $R^1 = R^2 = H$, $R^3 = C_6H_5$), use of zinc chloride at 180–185 °C to effect this indolization gave 2-phenylindole (**2**; $R^1 = R^3 = H$, $R^2 = C_6H_5$) which had earlier (Fischer, Emil, 1886b) been synthesized by indolization of acetophenone phenylhydrazone with a zinc chloride catalyst (Fischer, Emil and Schmidt, T., 1888; Fischer, Emil and Schmitt, 1888). The first of these reactions was later (Henle, 1905) repeated using oxalic acid as the catalyst and although the product, m.p. 88–89 °C, was referred to as 'α-Phenylindol' it was clearly recognized as being identical with Fischer and Schmidt's (Schmitt's) (1888) Pr.3.Phenylindol (3-phenylindole), of the same melting point. Upon treatment with a zinc chloride catalyst, propionaldehyde phenylhydrazone furnished only 3-methylindole, no trace of 2-methylindole being detected (Fischer, Emil and Schmitt, 1888). The isomerization of 3- to 2-phenylindole was, in fact, found to be effected by zinc chloride at 170 °C (Fischer, Emil and Schmidt, T., 1888), a reaction subsequently (Crowther, Mann and Purdie, 1943) achieved in a claimed quantitative yield. Subsequent to the initial observations, another of Fischer's students (Ince, 1889) observed further similar migrations, using heated zinc chloride, when 1-methyl-3-phenylindole, formed by indolization of phenylacetaldehyde N_α-methylphenylhydrazone with alcoholic hydrochloric acid, was converted into 1-methyl-2-phenylindole and the 4,5-benzo-3-phenylindole, obtained by alcoholic hydrochloric acid catalysed indolization of phenylacetaldehyde 2-naphthylhydrazone, was likewise converted into

4,5-benzo-2-phenylindole. Both these 2-phenylindoles were synthesized from the N_α-methylphenylhydrazone and 2-naphthylhydrazone, respectively, of acetophenone using zinc chloride as catalyst (see also Verkade and Janetzky, 1943a). Other analogous examples of such group migrations are known. Thus, indolization of **1** [$R^1 = R^2 = H$, $R^3 = C(CH_3)_3$] with either a polyphosphoric acid (Colle, M. A., 1960) or a zinc chloride (David, 1960) catalyst afforded 2-*tert*.-butylindole [**2**; $R^1 = R^3 = H$, $R^2 = C(CH_3)_3$] and fusion of 1-methyl-3-(4-methylphenyl)indole and 1-methyl-3-(4-biphenylyl)indole with zinc chloride yielded 1-methyl-2-(4-methylphenyl)indole and 1-methyl-2-(4-biphenyl)indole, respectively, 1-methyl-3-(4-chlorophenyl)indole, 1-ethyl-3-(4-methylphenyl)indole and 1-ethyl-3-(4-chlorophenyl)indole remaining unchanged under these conditions (Brown, F. and Mann, 1948a). 3-Benzylindole gave 2-benzylindole when heated with aluminium chloride (Biswas, K. M. and Jackson, A. H., 1969; Clemo and Seaton, 1954) and similar conditions isomerized 3-methyl- and 3-(2-pyridylmethyl)indole into 2-methyl- and 2-(2-pyridylmethyl)indole, respectively (Clemo and Seaton, 1954) although under these conditions 1-acetyl-3-benzyl- and 1-acetyl-3-(2-pyridylmethyl)indole were not isomerized (Clemo and Seaton, 1954). Likewise 3-benzyl-, 3-*tert*.-butyl-, 3-methyl- and 3-phenylindole, 1,3-dimethylindole, 1-methyl-3-phenylindole and 3,7-diphenylindole gave 2-benzyl-, 2-*tert*.-butyl-, 2-methyl- and 2-phenylindole, 1,2-dimethylindole, 1-methyl-2-phenylindole and 2,7-diphenylindole, respectively, when heated with polyphosphoric acid, in yields of between 50% and 90% in all but one case. Starting from 3-methylindole, the yield of 2-methylindole was only 10%, probably because of competing acid catalysed polymerization which would account for the considerable resinification observed during this reaction and also when using 1,3-dimethylindole, which afforded only a 50% yield of 1,2-dimethylindole (Budylin, Kost and Matveeva, 1969, 1972; Kost, Budylin, Matveeva and Sterligov, 1970). Similarly, 3-benzylindole furnished 2-benzylindole (30% yield) upon refluxing in trifluoroacetic acid and 3-phenylindole furnished 2-phenylindole (10% yield) upon standing in concentrated sulphuric acid at room temperature (Budylin, Kost and Matveeva, 1972). Following the initial observation (Davies, W. and Middleton, 1958) that, upon heating with polyphosphoric acid, 3-phenylbenzofuran was converted into 2-phenylbenzofuran in 80% yield, other similar isomerizations were also observed when 3-*tert*.-butyl- and 3-phenylbenzofuran and 3-methyl- and 3-phenylbenzothiophen were similarly treated to yield the corresponding 2-substituted isomers in good yields (Kost, Budylin, Matveeva and Sterligov, 1970). The possible $3 \rightarrow 2$ migration of alkyl and aryl substituents must therefore not only be considered when effecting the appropriate Fischer indolizations but also when synthesizing the appropriate benzofurans by the extension to the Fischer indole synthesis discussed in Chapter VI, C).

These acid catalysed migrations have been suggested to occur by the mechanism shown in Scheme 47 (Budylin, Kost and Matveeva, 1973 – see also

Kost, Budylin, Matveeva and Sterligov, 1970), it being possible that the species **1515** and **1516** were intermediates between **1512** and **1513** and **1513** and **1514**, respectively and that migration of R^2 and the elimination of $H^{\oplus}$ was a synchronous process. The observation that when a mixture of 3-*tert*.-butylindole and 3-phenylbenzofuran was heated with polyphosphoric acid, only 2-*tert*.-butylindole and 2-phenylbenzofuran were formed, the reaction products being analysed by g.l.c., supported the intramolecular nature of the isomerization (Budylin, Kost and Matveeva, 1972; Kost, Budylin, Matveeva and Sterligov, 1970). This conclusion was also reached earlier (Buu-Hoï and Jacquignon, 1967) from the observation that when 3-phenylindole was heated with aluminium chloride in toluene, 2-phenylindole but no 2-tolylindole was isolated.

R^2 N H R^1 $\xrightarrow{H^{\oplus}}$ H R^2 ⊕ N H R^1 **1512**

H ⊕ N R^2 H R^1 **1513** $\xrightarrow{-H^{\oplus}}$ H N R^2 R^1 **1514**

Scheme 47

H ⊕R^2 N H R^1 **1515** H ⊕H N R^2 R^1 **1516**

It has been claimed (Razumov and Gurevich, 1967a, 1967b; Razumov, Gurevich and Baigil'dina, 1976 – see also S. M. Kirov Chemical Engineering Institute, 1965) that the indolization of the phenylhydrazones **1517** [$R^1 = H$, $R^2 = R^3 = C_2H_5$, $CH(CH_3)_2$, $(CH_2)_2CH_3$, $n = 1$; $R^1 = H$, $R^2 = C_2H_5$, $R^3 = C_6H_5$, $n = 1$; $R^1 = H$, CH_3, $R^2 = R^3 = C_2H_5$, $n = 0$; $R^1 = H$, $R^2 = C_2H_5$, $R^3 = C_6H_5$, $n = 0$ and $R^1 = R^2 = H$, $R^3 = C_6H_5$, $n = 0$ and 1] yielded

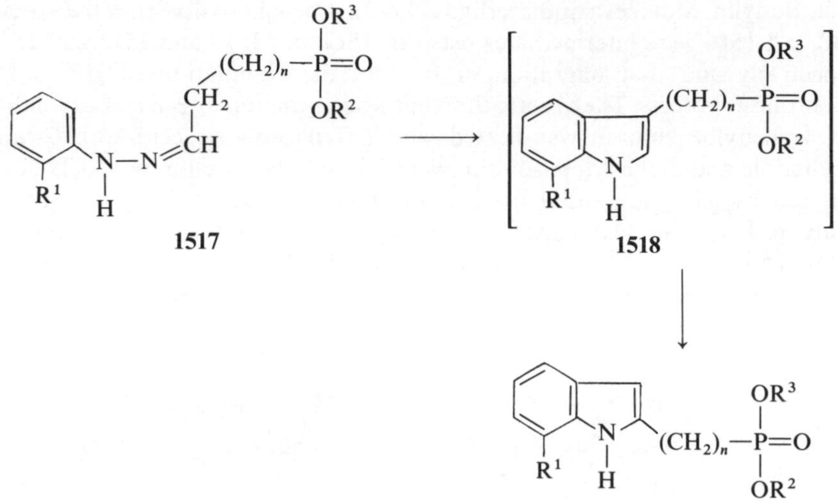

1517 1518 1519

the corresponding 2-substituted indoles **1519** [R^1 = H, R^2 = R^3 = C_2H_5, $CH(CH_3)_2$, $(CH_2)_2CH_3$, $n = 1$; R^1 = H, R^2 = C_2H_5, R^3 = C_6H_5, $n = 1$; R^1 = H, CH_3, R^2 = R^3 = C_2H_5 $n = 0$; R^1 = H, R^2 = C_2H_5, R^3 = C_6H_5, $n = 0$ and R^1 = R^2 = H, R^3 = C_6H_5, $n = 0$ and 1, respectively] which were formed by rearrangement of the initial indolization products, **1518**, under the indolization conditions. In support of this it was found that authentic samples of **1518** [R^1 = H, R^2 = R^3 = C_2H_5 and $CH(CH_3)_2$, $n = 1$], synthesized by reacting gramine methiodide with triethylphosphite or sodium diethylphosphite and triisopropylphosphite or sodium diisopropylphosphite, respectively (Torralba and Myers, 1957), were isomerized by heating with excess zinc chloride in a sealed tube at 175–180 °C for 20 min to afford high yields of products identical with those obtained from the above indolizations and presumed to have structures **1519** [R^1 = H, R^2 = R^3 = C_2H_5 and $CH(CH_3)_2$, respectively, $n = 1$]. It is unfortunate that this possible isomerization of these authentic 3-substituted indoles was not investigated using refluxing alcoholic concentrated sulphuric acid, the conditions used to achieve the indolizations of **1517**, although these observations appear to establish that the proposed rearrangement did occur during the indolizations to afford the corresponding **1519** ($n = 1$). However, in the cases of the lower homologues, structural assignments were based (Razumov and Gurevich, 1967a) upon rather tenuous u.v. and i.r. spectroscopic assignments. Conclusive characterization was lacking and the products of such syntheses could well be (Redmore, 1971) the 3-substituted isomers **1518** (R^1 = H, CH_3, R^2 = R^3 = C_2H_5, $n = 0$ and R^1 = H, R^2 = H and C_2H_5, R^3 = C_6H_5, $n = 0$) and not the corresponding 2-substituted isomers as was claimed (Razumov and Gurevich, 1967a, 1967b; Razumov, Gurevich and Baigil'dina, 1976).

A further example of group migration occurred when the hydrazone **1520** was indolized by refluxing in ethylene glycol solution under a nitrogen atmosphere. In this case, two products were isolated. One of these resulted from the migration of the dimethylallyl group in the supposed intermediate 3*H*-indole **1521** from the 3- to the 1-position and was formulated as **1522** (9% yield). The other product was shown to be 3-methylindole (4% yield) which was probably formed by thermal elimination of isoprene from **1521** (Baldwin and Tzodikov, 1977).

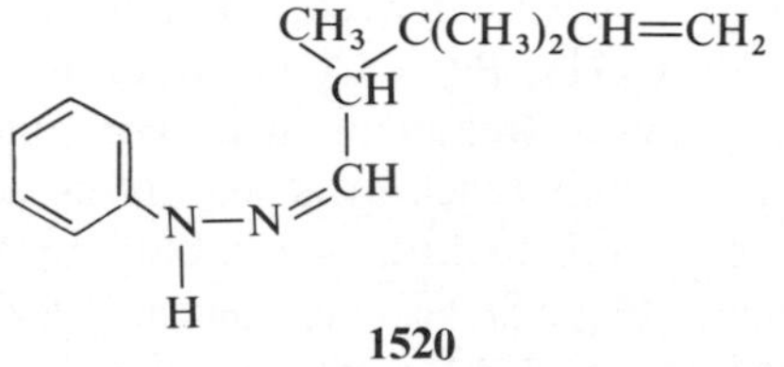

1520 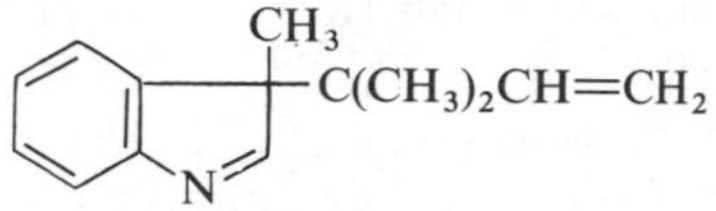 **1521**

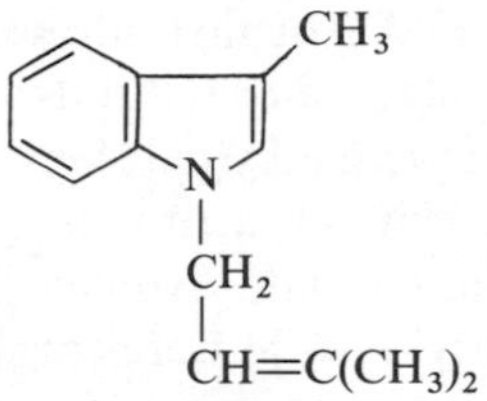

1522

The elimination of 3-substituents under indolization conditions from initially formed 3*H*-indoles has also been observed in other cases. Thus, treatment of **44** (R^1–R^4 = H, $R^5 = R^6 = CH_3$) with zinc chloride in ethanol afforded 3,3-dimethyl-3*H*-indole and 3-methylindole and treatment of **44** (R^1–R^3 = H, R^4–R^6 = CH_3) with Grignard reagents gave 2,3,3-trimethyl-3*H*-indole and 2,3-dimethylindole (Grammaticakis, 1940b). The zinc chloride catalysed indolization of **646** [$R^1 = R^2$ = H, $R^3 = CH_3$, $R^4 = (CH_2)_2CN$] to give 2,3-dimethylindole {with 20% sulphuric acid as catalyst, the product was **647** [R^1 = H, $R^3 = CH_3$, $R^4 = (CH_2)_2CN$] which also afforded 2,3-dimethylindole upon heating with zinc chloride}, the zinc chloride catalysed indolization of **736** [$R^1 = R^3$ = H, $R^2 = (CH_2)_2CN$, $n = 2$] at 200–220 °C to afford 1,2,3,4-tetrahydrocarbazole (44%) along with traces of an unidentified product, at 170–190 °C to yield 1,2,3,4-tetrahydrocarbazole (13.5%) and **738** [$R^1 = R^3$ = H, $R^2 = (CH_2)_2CN$, $n = 2$] (32%) {with sulphuric acid as catalyst, the products were **737** [$R^1 = R^3$ = H, $R^2 = (CH_2)_2CN$, $n = 2$] (18%) and **738** [$R^1 = R^3$ = H, $R^2 = (CH_2)_2CN$, $n = 2$] (50%), treatment of the former product with zinc chloride affording 1,2,3,4-tetrahydrocarbazole} (Kost, Yudin and Yü-Chou, 1964) and the formation of 1,2,3,4-tetrahydrocarbazole by heating the picrate of **737** [$R^1 = R^3$ = H, $R^2 = C(CH_3)_3$, $n = 2$] in boiling ethanol or at its m.p. (121 °C) or by heating the free base in 2 M hydrochloric acid on a water-bath for a few minutes and by heating **737** [$R^1 = R^3 = H$, $R^2 = CH(CH_3)_2$ and $CH_2C_6H_5$] in 4 M hydrochloric acid at 170–180 °C for 2 h (Nakazaki, Isoe and Tanno, 1955) have already been referred to in Chapter III, H). Likewise, boiling a solution of 2,3-dimethyl-3-*tert*.-butyl-3*H*-indole in 4 M hydrochloric acid for 10 min or heating 3-benzyl-2,3-dimethyl-3*H*-indole in 4 M hydrochloric acid at 140 °C for 6 h gave, in both cases, 2,3-dimethylindole, it being suggested that these above facile removals of the *tert*.-butyl and benzyl groups were caused by the simultaneous relief of considerable steric strain (Nakazaki, Isoe and Tanno, 1955). A particularly interesting example of a related elimination has been claimed when **646** ($R^1 = R^2 = H$, $R^3 = C_2H_5$, $R^4 = COOC_2H_5$) furnished **644** ($R^1 = R^2 = H$, $R^3 = COOC_2H_5$) upon treatment with concentrated sulphuric acid (Walker, C., 1894). Clearly, this reaction is worthy of further investigation. An elimination of a *tert*.-butyl substituent, as *tert*.-butyl bromide, occurred when 3-*tert*.-butylindole (David, 1960; David and Régent, 1964) and 2-methyl-3-*tert*.-butylindole (David and Régent, 1964) were treated with hydrobromic acid to afford indole and 2-methylindole, respectively, although under similar conditions 2-*tert*.-butyl, 2-*tert*.-butyl-3-methyl, 2-*tert*.-butyl-3,5-dimethyl- and 2-*tert*.-butyl-3,5,7-trimethylindole were stable (David and Régent, 1964). When 2-*tert*.-butylindole was heated with methyl iodide in a sealed tube, 1,2,3,3-tetramethyl-3*H*-indolium iodide, hydrogen iodide and 2-methylpentene (from the eliminated *tert*.-butyl group) were formed (Plancher and Forghieri, 1902).

Exchange reactions between the 2- and 3-substituents on an indole nucleus under conditions appertaining to vigorous indolization conditions have also been observed. Thus, heating 2-methyl-3-phenylindole with a mixture of aluminium chloride and sodium chloride led to the formation of 3-methyl-2-phenylindole (Nakazaki, 1960a) [an observation also reported a decade later (Kost, Budylin, Matveeva and Sterligov, 1970) using an aluminium chloride catalyst] and the similar treatment of 3,4-benzo-1,2-dihydrocarbazole (**868**; R^1–R^3 = H) gave 1,2-benzocarbazole (**62**), possibly formed by dehydrogenation of the rearrangement product, 1,2-benzo-3,4-dihydrocarbazole, under the reaction conditions (Nakazaki, 1960a). Alternatively, dehydrogenation could have preceded rearrangement to afford 3,4-benzocarbazole (**609**; $R^1 = R^2 = H$, $X = Y = CH$) which could have then rearranged into 1,2-benzocarbazole (**62**) (Zander and Franke, W. H., 1967) (see later in this section). A related observation involved the formation of 1,2-benzocarbazole (**62**) [identified with the product previously (Borsche, Witte, A. and Bothe, 1908) obtained from cyclohexanone 1-naphthylhydrazine by sequential indolization and dehydrogenation] by heating 2-tetralone phenylhydrazone **866** (R^1–R^3 = H) with zinc chloride in dry ethanol followed by distillation of the product under reduced pressure. These conditions also caused dehydrogenation either of the initially produced 3,4-benzo-1,2-dihydrocarbazole (**868**; R^1–R^3 = H) or of its rearrangement product, 1,2-benzo-3,4-dihydrocarbazole (see Chapter III, L, for other examples of aromatization concomitantly with indolization). The former dihydro compound was the isolated reaction product when **866** (R^1–R^3 = H) was indolized by heating in dilute sulphuric acid (Ghigi, 1930, 1931). The exchange reactions between 2- and 3-substituents on an indole nucleus were suggested to proceed via electrophilic attack by the Lewis acid at the indole 3-position, followed by a twofold Wagner–Meerwein rearrangement and finally the loss of the Lewis acid as shown in Scheme 48 (A = $AlCl_3$) (Nakazaki, 1960a).

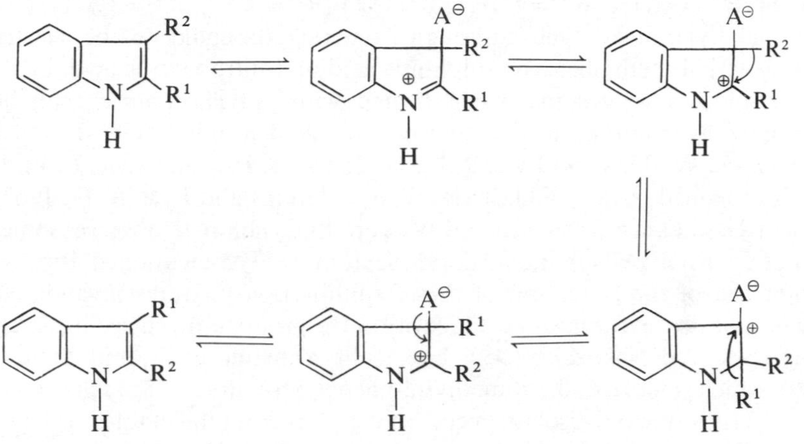

Scheme 48

Exchange reactions between 2- and 3-substituents on a 3*H*-indole nucleus can also occur under indolization conditions. Thus, although it has been reported that 3,3-dimethyl-2-phenyl-3*H*-indole (**1524**; $R^1 = C_6H_5$, $R^2 = CH_3$) was the only product resulting from the treatment of isobutyrophenone phenylhydrazone (**1523**; $R^1 = C_6H_5$, $R^2 = CH_3$) with polyphosphoric acid (Kissmann, Farnsworth and Witkop, 1952), with acetic acid (Evans, F. J., and Lyle, R. E., 1960; Evans, F. J., Lyle, G. G., Watkins and Lyle, R. E., 1962) or with zinc chloride (Leuchs, Heller, A. and Hoffmann, A., 1929) as catalyst, subsequent repeat of the indolization using polyphosphoric acid as catalyst yielded a product consisting of **1524** ($R^1 = C_6H_5$, $R^2 = CH_3$) and **1524** ($R^1 = CH_3$, $R^2 = C_6H_5$). Treatment of the former isomer with polyphosphoric acid at 150 °C was found to establish an equilibrium between it and the latter isomer in the

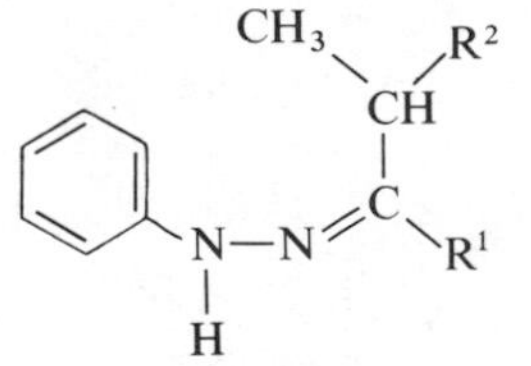

1523 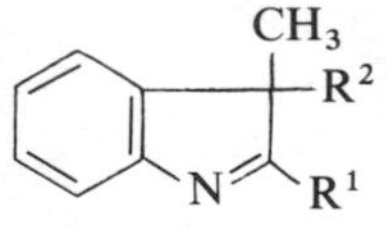 **1524**

ratio of 3 : 7, respectively (Evans, F. J. and Lyle, R. E., 1960; Evans, F. J., Lyle, G. G., Watkins and Lyle, R. E., 1962). The same mixture also resulted from similar treatment of the latter isomer which was synthesized, without rearrangement during indolization, by treatment of **1523** ($R^1 = CH_3$, $R^2 = C_6H_5$) with refluxing acetic acid (Evans, F. J. and Lyle, R. E., 1960; Evans, F. J., Lyle, G. G., Watkins and Lyle, R. E., 1962). Heating with boron trifluoride, aluminium chloride or iron(III) chloride at 150 °C also brought about the partial conversion of **1524** ($R^1 = C_6H_5$, $R^2 = CH_3$) into **1524** ($R^1 = CH_3$, $R^2 = C_6H_5$) (Evans, F. J. and Lyle, R. E., 1960) although isomerization could not be effected by heating with 4-methylbenzene sulphonic acid or 4-nitrobenzoic acid. Likewise, the 3*H*-indoles were stable when heated with polyphosphoric acid, boron trifluoride, hydrobromic acid or hydrochloric acid at only 100 °C (Evans, F. J., Lyle, G. G., Watkins and Lyle, R. E., 1962; Evans, F. J. and Lyle, R. E., 1960) which explained (Evans, F. J., Lyle, G. G., Watkins and Lyle, R. E., 1962) the earlier (Kissmann, Farnsworth and Witkop, 1952) failure to observe isomerization using a polyphosphoric acid indolization catalyst mentioned above. This recognition of the formation of these equilibration mixtures invalidated the conclusions drawn from, and the identity of the products produced in, an earlier study (Isoe and Nakazaki, 1959; Nakazaki, Yamamoto, K., and Yamagami, 1960) which reported 3,3-dimethyl-2-phenyl-3*H*-indole (**1524**; $R^1 = C_6H_5$, $R^2 = CH_3$) as the only isolated product resulting from the indolization of **1523** ($R^1 = CH_3$, $R^2 = C_6H_5$) with zinc chloride, boron trifluoride etherate or dry ethanolic hydrogen chloride. Furthermore, by analogy with this, doubt was cast (Evans, F. J., Lyle, G. G., Watkins and Lyle, R. E., 1962) upon the characterization of other products obtained in this earlier study (Nakazaki, Yamamoto, K. and Yamagami, 1960). These reported that the ethanolic hydrogen chloride catalysed treatment of cyclopentyl phenyl ketone phenylhydrazone [**44**; R^1–R^3 = H, $R^4 = C_6H_5$, $R^5 + R^6 = (CH_2)_4$] gave **45** [$R^1 = R^2$ = H, $R^3 = C_6H_5$, $R^4 + R^5 = (CH_2)_4$] which with polyphosphoric acid was isomerized into **737** ($R^1 = R^3$ = H, $R^2 = C_6H_5$, $n = 2$) (Isoe and Nakazaki, 1959; Nakazaki, Yamamoto, K. and Yamagami, 1960). This was synthesized by ethanolic hydrogen chloride catalysed indolization of 2-phenylcyclohexanone phenylhydrazone. Likewise, the ethanolic hydrogen chloride catalysed indolization of **743** was reported to give **744**, polyphosphoric acid isomerization of which afforded **1525** (Isoe and Nakazaki, 1959; Nakazaki, Yamamoto, K. and Yamagami, 1960) which was also synthesized by treatment of 2-methyl-1-

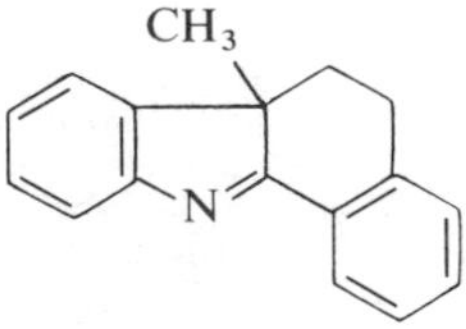

CH$_3$
N

1525

tetralone phenylhydrazone with polyphosphoric acid. Clearly these doubts were valid, since the potential for the establishment of equilibrium mixtures of 3*H*-indolic products during some, if not all, of these reactions is very likely and further investigation of these possibilities would be of interest. In related studies upon 1-methyl-3*H*-indolium iodide rearrangements (Nakazaki, 1960b), similar equilibrium mixtures may also have been obtained. The mechanism of these above acid catalysed 3*H*-indole isomerizations involves a twofold Wagner–Meerwein rearrangement of the protonated 3*H*-indole (Evans, F. and Lyle, R. E., 1963; Evans, F. J., Lyle, G. G., Watkins and Lyle, R. E., 1962; Nakazaki, Yamamoto, K. and Yamagami, 1960).

Isomerization of initially formed 3*H*-indoles into indoles by the well established Plancher rearrangement can also occur under indolization conditions. Thus, heating a mixture of isobutyraldehyde phenylhydrazone with zinc chloride at 140–150 °C afforded 2,3-dimethylindole (Brunner, K., 1895a – see also Grammaticakis, 1940b). Likewise, whereas treatment of **44** [R^1–R^4 = H, $R^5 + R^6 = (CH_2)_5$] with zinc chloride in refluxing ethanol afforded the expected 3*H*-indole **45** [R^1–R^3 = H, $R^4 + R^5 = (CH_2)_5$], treatment by heating with powdered zinc chloride gave **147** [R^1 = H, $R^2 + R^3 = (CH_2)_5$] (Jones, G. and Stevens, T. S., 1953). The hydrazone **1526** was indolized with 20% methanolic hydrochloric acid at room temperature to yield **1527** which upon heating with zinc chloride in an atmosphere of nitrogen gave **1528** (Wang, T. S. T., 1975). Similar rearrangements have also been observed during the indolization of the appropriate 2- and 4-pyridylhydrazones and 4-pyrimidylhydrazones under non-catalytic thermal conditions (Crooks and Robinson, B., 1969; Kelly, A. H. and Parrick, 1966, 1970; Parrick, 1969), reactions already thoroughly discussed

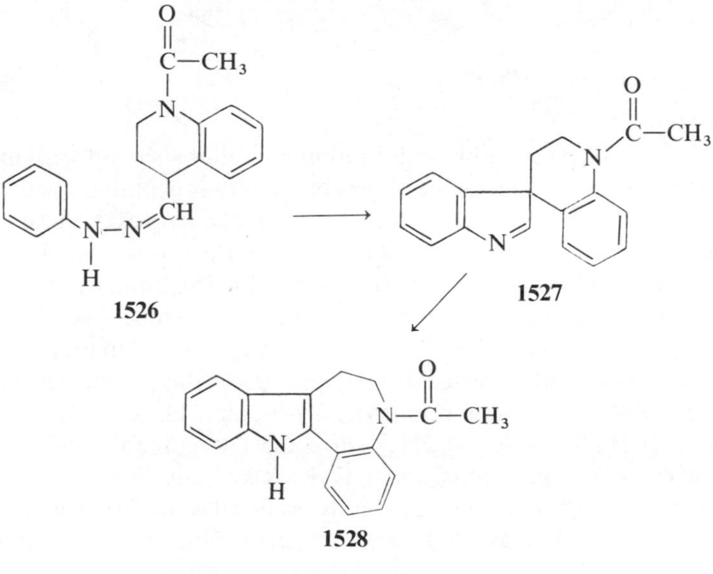

1526 **1527**

1528

(see Chapter IV, I). The rearrangement of 3,3-disubstituted 3*H*-indoles into 2,3-disubstituted indoles has been thoroughly investigated, with particular reference to the migratory aptitudes of the substituents (Jackson, A. H. and Smith, P., 1968 – see also Jackson, A. H. and Smith, A. E., 1965). In view of the potential occurrence of Plancher rearrangements in appropriate cases under indolization conditions, both basic and neutral fractions resulting from such reactions should be investigated.

The rearrangement of 2-acylindoles, produced by indolization of 1,2-diketone monoarylhydrazones, into 3-acylindoles under vigorous indolization conditions has already been discussed (see Chapter IV, J1).

It has been well established (Buu-Hoï and Lavit, 1960; Buu-Hoï and Lavit-Lamy, 1961, 1962; Zander, 1962) that polynuclear aromatic hydrocarbons isomerized when heated with aluminium chloride. Similar observations were also made when 3,4,5,6-dibenzocarbazole (**88**) yielded 1,2,5,6-dibenzocarbazole (**71**) upon treatment with aluminium chloride in boiling benzene or with an aluminium chloride–sodium chloride melt (Zander and Franke, W. H., 1964, 1967) and when the latter conditions caused the conversions of 3,4-benzocarbazole (**609**; $R^1 = R^2 = H$, X = Y = CH) into 1,2-benzocarbazole (**62**) and of **1529** into **1530** (Zander and Franke, W. H., 1967). The possible occurrence of such isomerizations must therefore be considered when using Lewis acids to effect indolizations which would lead to such types of polynuclear aromatic indolic products.

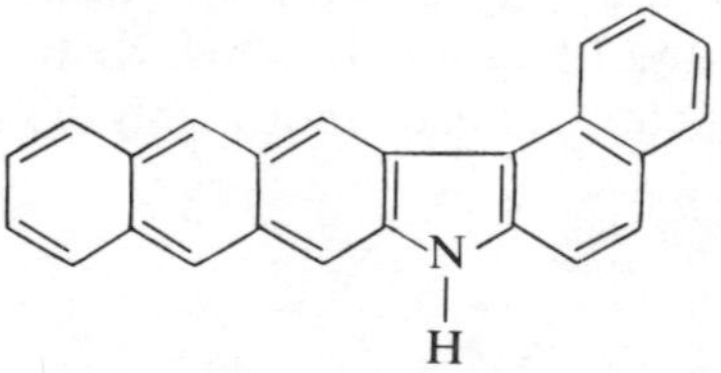

N–H **1529** N–H **1530**

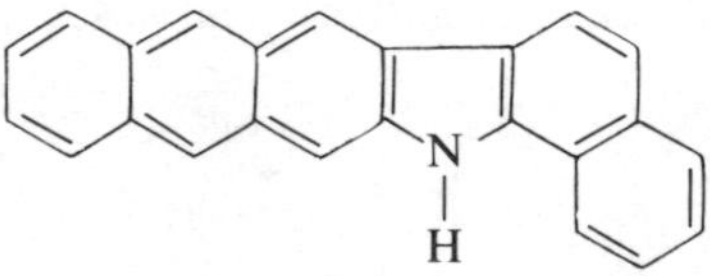

Clearly, the mildest possible indolization conditions should be employed if the occurrences of rearrangements of the above types concomitant with indolization are to be avoided. Thus, for example, alcoholic hydrogen chloride rather than zinc chloride has been used as the catalyst in the synthesis of 3-arylindoles by the Fischer method (Brown, F. and Mann, 1948b). Similarly, treatment of isopropyl 4-dimethylaminophenylketone phenylhydrazone [**44**; R^1–R^3 = H, $R^4 = C_6H_4N(CH_3)_2(4)$, $R^5 = R^6 = CH_3$] with either ethanolic hydrogen chloride or acetic acid afforded similar products which were formulated as 3,3-dimethyl-2-(4-dimethylaminophenyl)-3*H*-indole [**45**; $R^1 = R^2 = H$, $R^3 = C_6H_4N(CH_3)_2(4)$, $R^4 = R^5 = CH_3$] since the latter 'mild' catalyst was unlikely to effect rearrangement (Carson, D. F. and Mann, 1965).

An interesting rearrangement apparently occurred when **1531** was heated with zinc chloride in absolute alcohol, conditions which induced the elimination of the elements of ammonia. However, the product was not the 3*H*-indole **1532** since it was non-basic, being unextractable from ethereal solution into 5 M hydrochloric acid and forming neither a picrate nor a perchlorate salt. Neither does it appear to be the alternative indole **1533** [R^1 = H, R^2 = $CH(CH_2C_6H_5)_2$] since it remained unchanged upon heating with acetic anhydride. It was suggested that a 3 to 1 migration of a benzyl group had occurred in **1532** under the indolization conditions to afford **1533** [R^1 = $CH_2C_6H_5$, R^2 = $(CH_2)_2C_6H_5$] (Leuchs and Overberg, 1931). Further investigation of the structure of this reaction product using spectroscopic techniques, with a subsequent unambiguous synthesis, would be of interest.

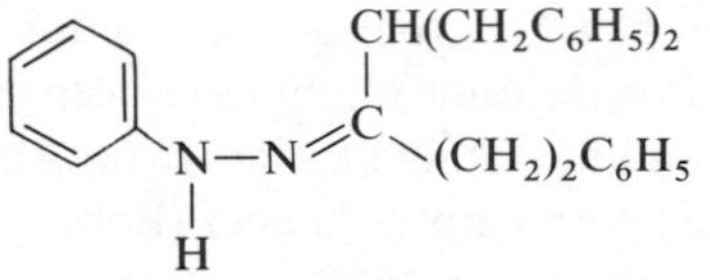

$CH(CH_2C_6H_5)_2$ C N—N H $(CH_2)_2C_6H_5$

1531

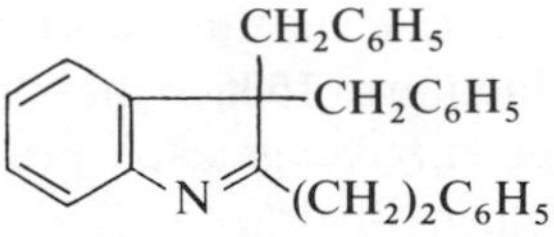

$CH_2C_6H_5$ $CH_2C_6H_5$ N $(CH_2)_2C_6H_5$

1532

$CH_2C_6H_5$ N R^2 R^1

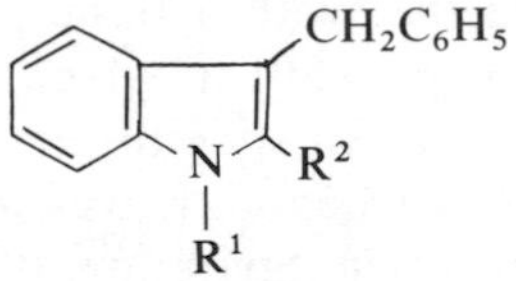

1533

Another 3*H*-indole rearrangement under indolization conditions was suggested to occur when **1534** (X = O) was reacted with phenylhydrazine in boiling aqueous acetic acid to furnish a compound $C_{26}H_{33}ON$, inactive towards alcoholic potassium hydroxide, acetic anhydride, bromine in acetic acid, concentrated hydrochloric acid and lithium aluminium hydride in ether and which, upon the basis of its u.v. spectrum and the presence of amide C=O and N—H stretching bands in its i.r. spectrum, was formulated as **1536**. It was presumed that **1534** (X = O) reacted with the phenylhydrazine to give **1534** (X = $NNHC_6H_5$) which then indolized into **1535**, Wagner–Meerwein rearrangement then affording **1536** (Clark, K. J., 1957). The stability of a compound of structure **1536** towards lithium aluminium hydride is surprising but could possibly be ascribed to the low boiling point of the ethereal reaction solvent or

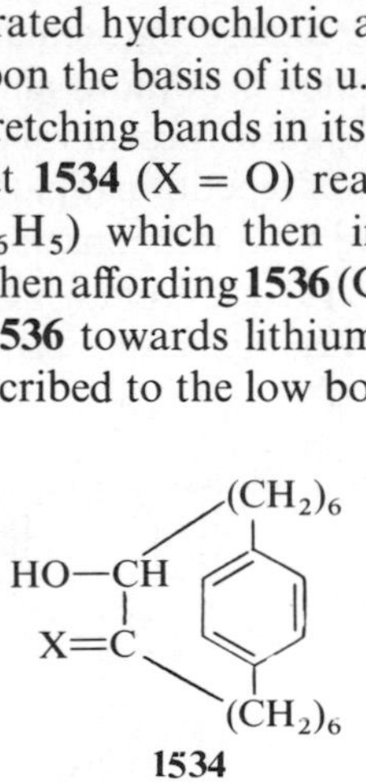

$(CH_2)_6$ HO—CH X=C $(CH_2)_6$

1534

OH $(CH_2)_6$ N $(CH_2)_6$

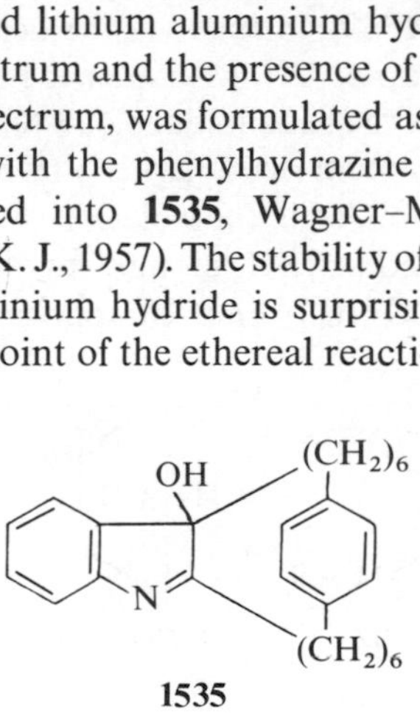

1535

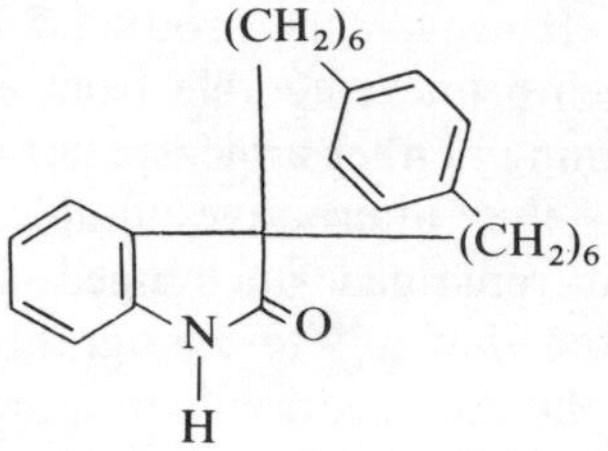

(CH2)6 (CH2)6 N H O

1536

to the insolubility in this solvent of the anion of **1536** formed by removal of a proton by the lithium aluminium hydride from the N—H grouping. It is unfortunate that dioxane was not used in place of ether as the reaction solvent, when reduction of **1536** to the corresponding indoline might have been expected.

Although no recognizable products could be obtained after treating camphor 4-bromophenylhydrazone (**1537**; R = Br) with sulphuric acid (Borsche, Witte, A. and Bothe, 1908), camphor phenylhydrazone (**1537**; R = H) was subsequently (Kuroda, 1923) indolized to yield an indolic product, 'camphorindole', $C_{16}H_{19}N$, m.p. = 94 °C, together with a basic product, $C_{16}H_{22}N_2$, m.p. = 207 °C, and a third product, m.p. = 201 °C which upon boiling with alkali was converted into the basic product. In spite of the observation that the indolic product gave a strong Ehrlich reaction, it was formulated as the 2,3-disubstituted indole **1538**. Later repetition (Sparatore, 1958) of this indolization using powdered zinc chloride as the catalyst at 180 °C led to the isolation of several compounds from the reaction product. The neutral fraction gave 'camphorindole' which was again formulated as **1538**, along with two other compounds, m.p. = 153–154 °C and 215–218 °C, the structures of which were not established. From the basic fraction were separated aniline, a not uncommon by-product from indolizations of phenylhydrazones (see Chapter VII, J), *ortho*-phenylenediamine (**1539**), formed via the occurrence of an *ortho*-semidine rearrangement (see Chapter VII, M) under the indolization conditions and a compound formulated as the benzimidazole **1540** which was also formed via the *ortho*-semidine rearrangement (see Chapter VII, M). A later repetition (Sparatore, 1962) of this indolization led to the isolation of **1540** and yet another basic product which was formulated as **1541**, along with a neutral fraction from which was isolated **1542** (R = $CONH_2$) and 'camphorindole'. Yet again this last product was

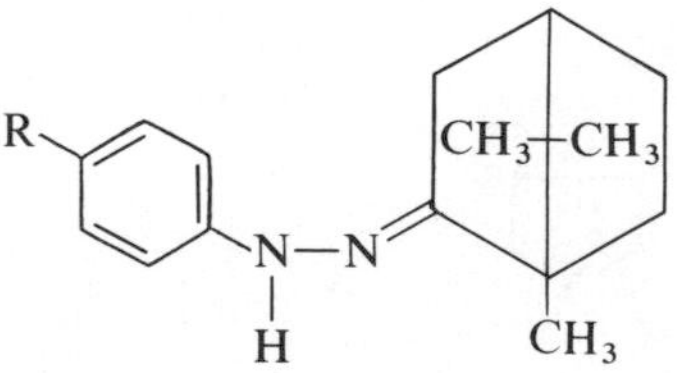

R N—N H CH3 CH3 CH3

1537

CH3 CH3 N H CH3

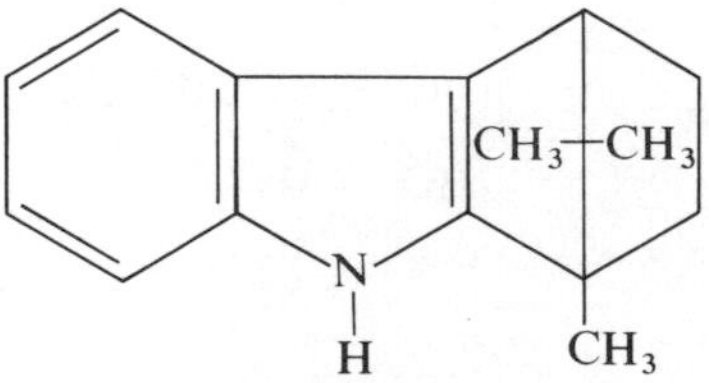

1538

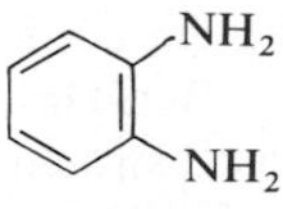

1539 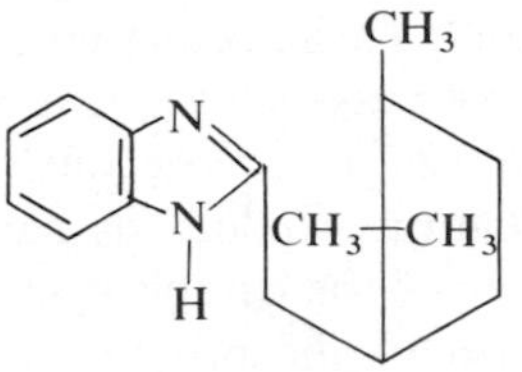 **1540**

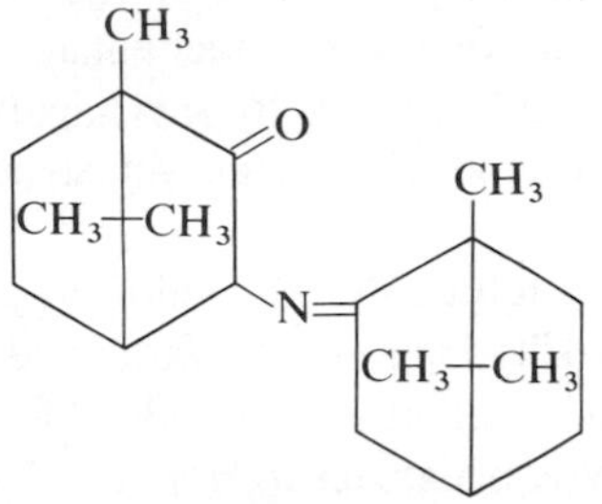

1541 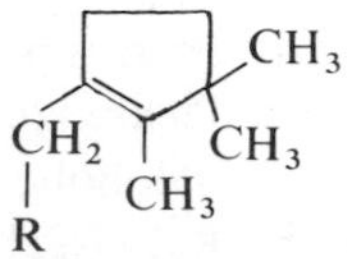 **1542**

formulated as **1538**, in spite of the fact that it was again reported to give an intense red-violet coloration in the Ehrlich test and, even more surprisingly, in view of its u.v. spectrum which was some 20 nm bathochromically shifted relative to, and of significantly higher absorbance than that of 1,2,3,4-tetrahydrocarbazole with which it was actually compared. It was later (Beck, Schenker, Stuber and Zürcher, 1965) recognized that these u.v. spectroscopic data suggested that 'camphorindole' contained an indole with a conjugated C=C as a chromophore. This was confirmed by catalytic hydrogenation to a dihydro derivative which showed typical indolic u.v. absorption. P.m.r. spectral examination of 'camphorindole' then clearly indicated the presence of the moiety **1543** which was supposed to be substituted at the 2- rather than the 3-position of an indole nucleus, to afford **1544**, because of the strong red–violet colour produced by the product in the Ehrlich reaction and because of the signal at $\tau = 3.55$ (multiplet) in its p.m.r. spectrum which was assigned to the presence of a 3-indolic proton rather than a 2-indolic proton. However, a contemporary report (Sparatore and Pirisino, 1965) showed that this assignment was incorrect when 'camphorindole' was shown to have structure **188** (R = H) from its unambiguous synthesis by indolization of the phenylhydrazone of **1542** (R = CHO), synthesized from camphor oxime via **1542** (R = $CONH_2$ and COOH, respectively) (Cornubert, 1933 – see also Blanc, 1898; Bouveault, 1898; Bredt, Rochussen and Monheim, 1901; Buchman and Sargent, 1942;

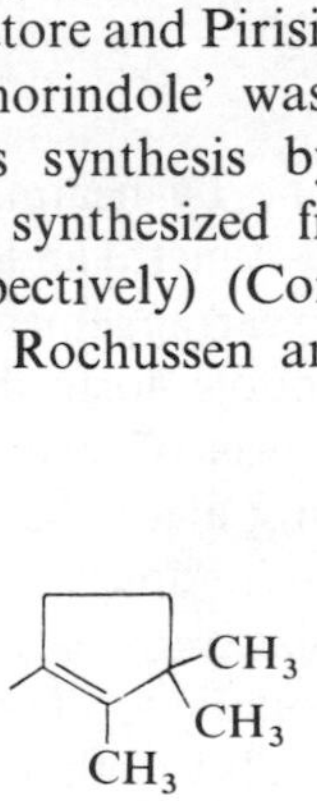

1543 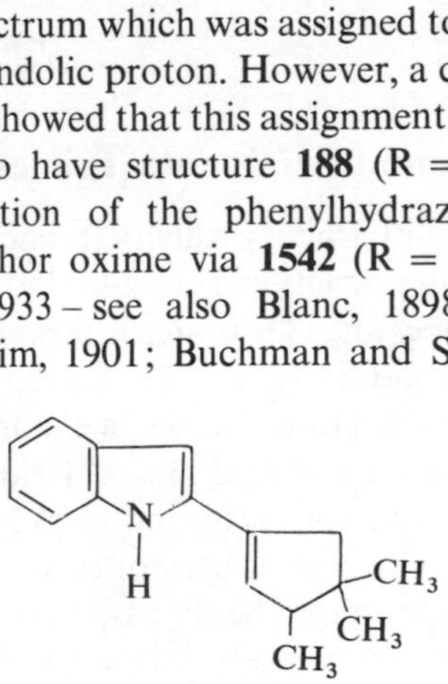 **1544**

Simonsen, 1949; Yurina, Goryaev and Dembitskii, 1969). The authors of a paper published three years subsequent to this reformulation (Buu-Hoï, Jacquignon and Béranger, 1968) were obviously unaware of this revised structure when they referred to 'camphorindole' as having structure **1544**. The mechanism of the rearrangement of the camphor moiety in the formation of **188** (R = H) and its relationship to the progress of the indolization mechanism still require clarification. Furthermore, it would be of interest to investigate the other products formed during this indolization which are as yet structurally unassigned.

It has recently been reported (Saleha, Khan, N. H., Siddiqui and Kidwai, M. M., 1978) that 'fenchylaldehyde' phenylhydrazone was indolized in refluxing 50% aqueous or alcoholic formic acid to produce an indole formulated as **2** ($R^1 = R^2 = H$, $R^3 = C_8H_{14}$). However, no structure for the C_8H_{14} indolic 3-substituent was proposed and, furthermore, the structure of 'fenchylaldehyde' is difficult to understand since fenchol is the secondary alcohol **1545** ($R^1 = H$, $R^2 = OH$) and the corresponding oxidation product is fenchone (**1545**; $R^1 + R^2 = O$). If the indolization referred to involved the phenylhydrazone of this ketone, in which both C-2 atoms are quaternary, then rearrangement of the monoterpenoid moiety must also have occurred. Certainly the published data (Saleha, Khan, N. H., Siddiqui and Kidwai, M. M., 1978) left much to be desired and these problems were not further clarified by one of the authors (Saleha, 1979).

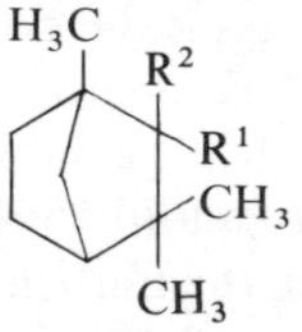

H_3C R^2 R^1 CH_3 CH_3

1545

B. Formation of 3*H*-Indole Trimers

Indolization of isobutyraldehyde phenylhydrazone by heating its mixture with powdered zinc chloride at 140–150 °C gave 2,3-dimethylindole (Brunner, K., 1895a) – see also Grammaticakis, 1940b) by rearrangement, under the indolization conditions, of the 3,3-dimethyl-3*H*-indole formed initially, as already discussed. However, when the indolization was effected under milder conditions, using powdered zinc chloride in refluxing alcoholic solution, this rearrangement did not occur and the 3,3-dimethyl-3*H*-indole was isolated as a zinc salt **1546**. Upon treatment of this with alkali, a steam volatile oil was obtained which upon rapid heating formed a steam involatile solid of m.p. = 215–216 °C. This was shown to be a trimer of the expected 3,3-dimethyl-3*H*-indole by the determination of a molecular formula of $C_{30}H_{33}N_3$ (Brunner, K., 1895b;

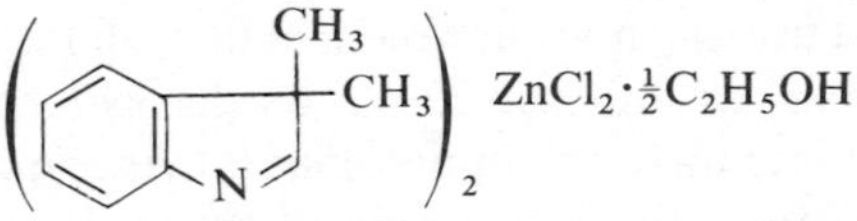

$\left(\text{3,3-dimethyl-3}H\text{-indole} \right)_2 ZnCl_2 \cdot \frac{1}{2}C_2H_5OH$

1546

Leuchs, Heller, A. and Hoffmann, A., 1929). Subsequent (Reinshagen, 1960) u.v. and i.r. spectroscopic examination of this product showed it to contain an indoline chromophore and no N—H group, respectively, and structure **1547** (R^1 = H, R^2 = CH_3) was therefore proposed for it. This structure was later (Fritz and Pfaender, 1965; Jackson, A. H. and Smith, A. E., 1965) confirmed by p.m.r. spectral investigation of a solution of the compound in deuterochloroform which in particular showed the three N—CH—N protons as non-equivalent singlets between $\tau = 5$–6 (see also Berti, Da Settimo and Nannipieri, 1969). These data, together with a molecular model study, allowed the expansion of the trimer structure into **1548** (Fritz and Pfaender, 1965). U.v. and p.m.r. spectral studies undertaken in acidic solutions showed that, under these conditions, detrimerization occurred to form the corresponding 3,3-dimethyl-3*H*-indolium cation (Fritz and Pfaender, 1965; Jackson, A. H. and Smith, A. E., 1965) and, furthermore, p.m.r. examination in deuterochloroform solution showed that at 30 °C about 15% of the monomeric 3*H*-indole was present and that at 120 °C, detrimerization to the monomer was complete. Subsequent similar studies (Grandberg, Nikitina and Yaryshev, 1970) supported these observations by showing that in perchloric acid solution or upon heating in solution, detrimerization, with formation of the monomer perchlorate or free base, respectively, occurred. Indolization of a mixture of 4-methoxyphenylhydrazine and isobutyraldehyde in refluxing glacial acetic acid led to the isolation of both 5-methoxy-3,3-dimethyl-3*H*-indole and the trimer **1547** (R^1 = OCH_3, R^2 = CH_3) (Ahmed and Robinson, B., 1967). This trimer was less stable than **1547** (R^1 = H, R^2 = CH_3) since u.v. and p.m.r. spectral investigation in ethanol and deuterochloroform solutions, respectively, showed that under these conditions at room temperature for a few hours, complete detrimerization to the 3*H*-indolic

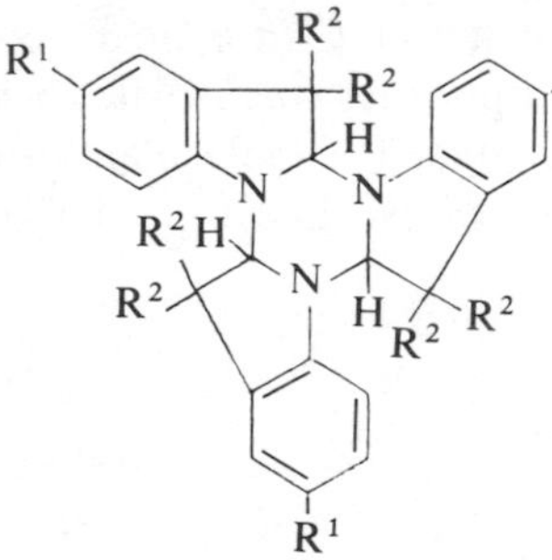

1547

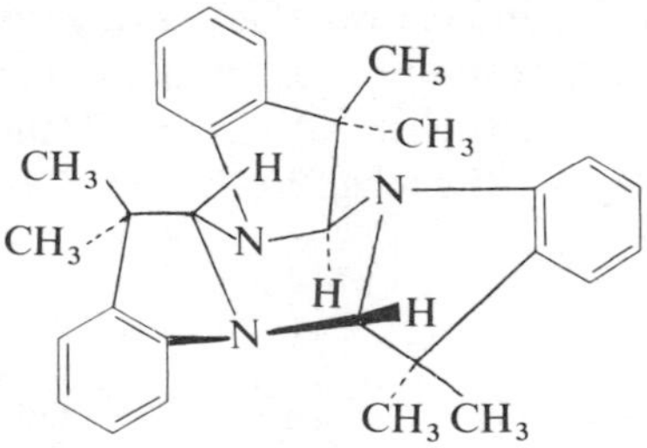

1548

compound occurred, a transformation which, in fact, almost went to completion immediately the trimer was dissolved in deuterochloroform (Ahmed and Robinson, B., 1967). Obviously, when indolizing hydrazones of structure **1549** ($R^2 = H$), the possibility of the above type of trimer formation must be considered. However, using **1549** ($R^2 \neq H$; for example, $R^2 = CH_3$), trimer formation did not appear to occur, possibly for steric reasons. Thus, for example, isopropyl methyl ketone phenylhydrazone (**1549**; $R^1 = H$, R^2–$R^4 = CH_3$),

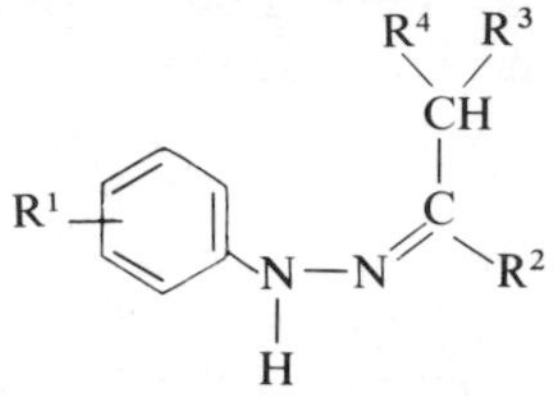

1549

upon indolization with zinc chloride in absolute ethanol, afforded 2,3,3-trimethyl-3*H*-indole, an oil which solidified at *ca.* 6–8 °C, as the only isolated product (Berti, Da Settimo and Nannipieri, 1969; Plancher, 1898a, 1898b, 1898c, 1898d, 1898g, 1901, 1905; Plancher and Bettinelli, 1899). Steric reasons are also likely to be responsible for the observations that whereas 3,3-dimethyl-, 3,3-diethyl-, 3-ethyl-3-methyl-, 3-methyl-3-propyl-, 3-isopropyl-3-methyl- and 3-allyl-3-methyl-3*H*-indole all partially existed as the corresponding trimer in deuterochloroform solution at 30 °C, under similar conditions 3-benzyl-3-methyl-3*H*-indole existed solely as the monomer (as would be expected, all seven of these 3*H*-indoles existed solely as the corresponding monomeric cations in trifluoroacetic acid solution) (Jackson, A. H. and Smith, P., 1968). Compound **45** [R^1–$R^3 = H$, $R^4 + R^5 = (CH_2)_4$] was converted into **1547** [$R^1 = H$, $R^2 + R^2 = (CH_2)_4$] after four days' storage (Golubev and Suvorov, 1970).

C. Other Reactions of 3*H*-Indoles

When 2-benzyl-1-indanone (**1550**; $R = CH_2C_6H_5$) and an excess of phenylhydrazine were heated together in xylenol at 130 °C (non-acidic conditions), an evolution of ammonia occurred and a product was isolated which, upon the evidence of elemental analysis, was assigned a molecular formula $C_{28}H_{25}N_3$ (Leuchs, Wutke and Gieseler, 1913 – see also Leuchs and Winzer,

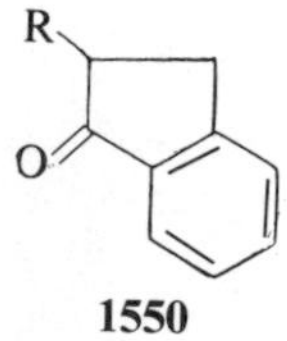

1550

1925). Subsequently (Leuchs and Winzer, 1925) it was found that when this product was heated with concentrated hydrochloric acid, phenylhydrazine was eliminated and a basic product, $C_{22}H_{17}N$, was produced. This latter compound was also produced directly when **1550** (R = $CH_2C_6H_5$) and phenylhydrazine were converted into the corresponding hydrazone **327** (R = $CH_2C_6H_5$, n = 1) and this was then treated with 12 M hydrochloric acid. Furthermore, it could also be reconverted into the compound $C_{28}H_{25}N_3$ by heating with phenylhydrazine (Leuchs and Winzer, 1925). Tentative structures were proposed for both these compounds $C_{28}H_{25}N_3$ (Leuchs, Wutke and Gieseler, 1913) and $C_{22}H_{17}N$ (Leuchs and Winzer, 1925) which, even at that time, would probably have been far from feasible although the relationship between the loss of ammonia and the Fischer indole synthesis was noted (Leuchs and Winzer, 1925). From further studies (Leuchs, Philpott, D., Sander, Heller, A. and Köhler, 1928) it was suggested that the compound $C_{28}H_{25}N_3$ was formed by the addition of phenylhydrazine across a C=N group in the compound $C_{22}H_{17}N$ and although structures were not actually given for these compounds, by analogy with other structures given for related compounds it might be inferred that they should be formulated as **1551** (R = $CH_2C_6H_5$) and **328** (R = $CH_2C_6H_5$, n = 1), respectively. It is unfortunate that in a later review (Julian, Meyer, E. W. and Printy, 1952) which included these studies, the starting ketone was erroneously referred to as '1-benzyl-2-hydrindone' (1-benzyl-2-indanone) and formulated accordingly as **1552** (Nakazaki and Maeda, 1962) which also led to related incorrect structures being given corresponding to **1551** (R = $CH_2C_6H_5$) and **328** (R = $CH_2C_6H_5$, n = 1). As might be expected, the direct formation from 2-substituted 1-indanones and arylhydrazines of corresponding com-

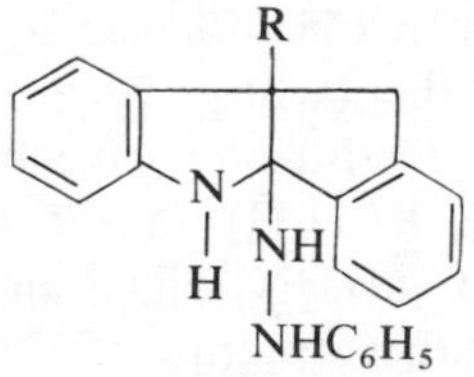

R
N
H
NH
NHC_6H_5
1551

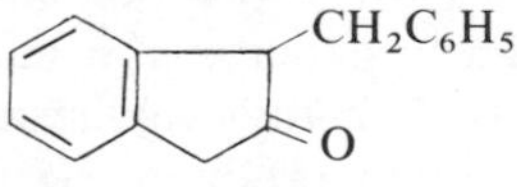

$CH_2C_6H_5$
O
1552

pounds related to **1551** appeared to be confined to reactions effected in the presence of excess arylhydrazine under non-acidic thermal conditions. Thus, an equimolar mixture of 2-ethyl-1-indanone (**1550**; R = C_2H_5) with phenylhydrazine yielded **328** (R = C_2H_5, n = 1) upon heating with zinc chloride (Leuchs, Philpott, D., Sander, Heller, A. and Köhler, 1928), 2-methyl-1-indanone phenylhydrazone (**327**; R = CH_3, n = 1) afforded **328** (R = CH_3, n = 1) upon heating with concentrated hydrochloric acid (Leuchs, Wulkow and Gerland, 1932), 2-phenyl-1-indanone phenylhydrazone (**327**; R = C_6H_5, n = 1) gave **328** (R = C_6H_5, n = 1) when its solution in chloroform was shaken with 12 M hydrochloric acid and similar treatment of 2-benzyl-1-tetralone and 2-(2-phenylethyl)-1-tetralone phenylhydrazone **327** [R = $CH_2C_6H_5$, $(CH_2)_2C_6H_5$, respectively, $n = 2$] gave **328** [R = $CH_2C_6H_5$, $(CH_2)_2C_6H_5$, respectively, $n = 2$]. Unfortunately, reactions of these phenylhydrazones or their ketonic moieties with excess phenylhydrazine were not investigated, although **328** (R = C_6H_5, $n = 1$) was heated with excess phenylhydrazine to afford **1551** (R = C_6H_5) (Leuchs, Philpott, D., Sander, Heller, A. and Köhler, 1928). It was later (Leuchs, Heller, A. and Hoffmann, A., 1929) found that neither 3,3-dimethyl-2-phenyl-3*H*-indole nor **328** (R = $CH_2C_6H_5$ and C_2H_5, $n = 2$), all prepared by the appropriate Fischer indolisation, would react with phenylhydrazine, from which it was concluded that for such addition to occur across the imine group of a 3*H*-indole, the 2,3-bond of this moiety had to be fused with a 5-membered ring. Furthermore, for such addition to occur during Fischer indolization, excess phenylhydrazine must be present and non-catalytic thermal indolization conditions must be used. Thus, indolization of **327** (R = C_2H_5, $n = 1$) using zinc chloride in ethanol, of **327** (R = CH_3, $n = 3$ and 4) using absolute ethanol saturated with hydrogen chloride (Nakazaki and Maeda, 1962) and of **327** (R = CH_3, $n = 2$) using absolute ethanol saturated with hydrogen chloride (Nakazaki, Yamamoto, K. and Yamagami, 1960) afforded **328** (R = C_2H_5, $n = 1$; R = CH_3, $n = 3$, 4 and 2, respectively). However, several attempts to convert **327** (R = CH_3, $n = 1$) into **328** (R = CH_3, $n = 1$) failed (Nakazaki and Maeda, 1962, footnote 11) although the ethyl homologue could be indolized as described above. A further addition across the imine group of a 3*H*-indole under the conditions of its formation occurred when using Grignard reagents as indolization catalysts. Thus, whereas the phenyl magnesium bromide catalysed indolization of cyclohexanone phenylhydrazone yielded 1,2,3,4-tetrahydrocarbazole together with **1553**, formed by addition of the Grignard reagent across the imine group of the starting hydrazone, a similar indolization of 2-methylcyclohexanone phenylhydrazone gave both the expected indolization products, **54** ($R^1 = CH_3$, $R^2 = H$) and **55** ($R^1 = CH_3$, $R^2 = H$), along with a further basic product, $C_{19}H_{21}N$, which on the basis of its u.v. spectroscopic properties was assigned either structure **1554** ($R^1 = CH_3$, $R^2 = H$) or **1554** ($R^1 = H$, $R^2 = CH_3$) (Grammaticakis, 1939). Clearly, since compound **55** ($R^1 = CH_3$, $R^2 = H$) is unlikely to exist in the form of its 3*H*-indolic tautomer, structure **1554** ($R^1 = H$, $R^2 = CH_3$) is highly favoured for this further basic product. In an analogous manner, indolization of 2-ethyl-

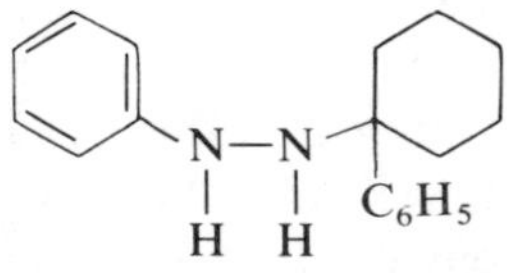

N—N, H H, C_6H_5

1553

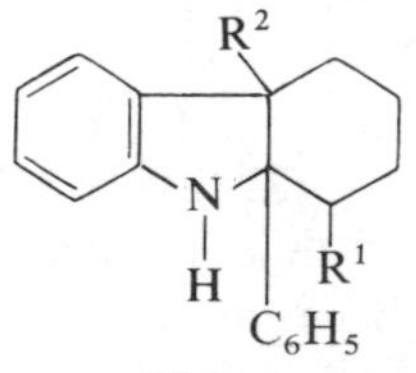

R^2, N, H, R^1, C_6H_5

1554

cyclohexanone in refluxing formic acid gave **54** ($R^1 = C_2H_5$, $R^2 = H$) (3.3%), possibly a trace of **55** ($R^1 = C_2H_5$, $R^2 = H$), although this could not be isolated, and **747** (63.7%) (Ban, Oishi, Kishio and Iijima, I., 1967).

D. Indolization of Arylhydrazones of Optically Active Ketones with Synchronous Racemization

When S-3-methylpentanal phenylhydrazone (**1555**) was indolized by heating with boron trifluoride etherate in acetic acid, the expected S-3-*sec*.-butylindole (**1556**) was formed along with some (±)-3-*sec*.-butylindole. The formation of the racemic product was attributed to the influence of the acid indolization conditions upon **1556** (Menicagli, Malanga and Lardicci, 1979). Analogies for such a racemization were found in the observations that (+)-2-phenylpentane (Burwell and Shields, 1955) and (+)-2-phenylbutane (Eliel, Wilken and Fang, 1957) were both rapidly racemized when treated with aluminium chloride at room temperature and 0 °C respectively, and other similar but unspecified observations under the influence of Lewis acids have also been made (Menicagli, Malanga and Lardicci, 1979). Indolization of **1557**, by heating with ethanolic hydrochloric acid at 85–90 °C for 1.5 h under pressure, afforded a product formulated as **1195** ($R^1 = R^2 = R^4 = R^5 = H$, $R^3 = OCH_3$, $R^6 = COOCH_3$, $R^7 = R^8 = CH_3$, $n = 1$) (Frey, A. J., Ott, Bruderer and Stadler, 1960; Sandoz Patents Ltd, 1965). In spite of the statement in the former publication (Frey, A. J., Ott, Bruderer and Stadler, 1960) that the indolization took place without racemization, the optical purity of the product might be questioned in the light of the above observations.

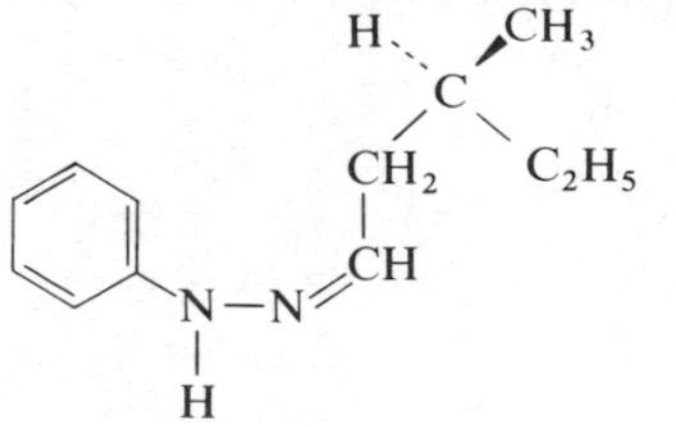

1555

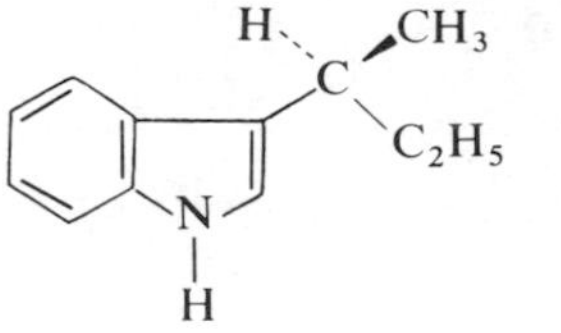

1556

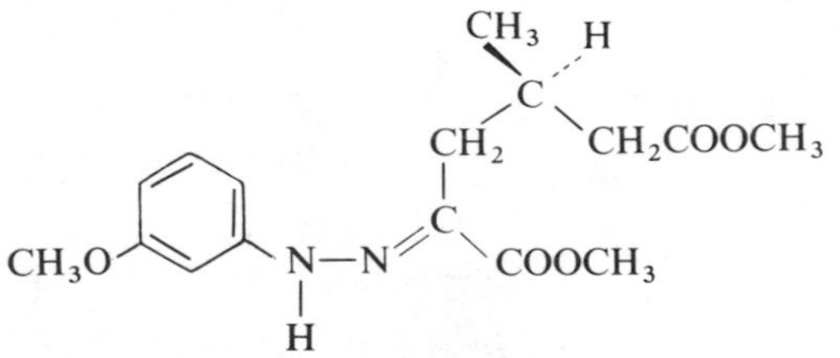

1557

E. Reactions of Aldehyde and Ketonic Moieties formed from Arylhydrazones under Indolization Conditions

Arylhydrazones often hydrolyse under indolization conditions. In the cases of aldehyde arylhydrazones, condensation products resulting from the reaction between one molecule of the liberated aldehyde and two molecules of the 2-unsubstituted indolization products have been isolated. The first reported (Bauer and Strauss, 1932) isolation of such products was observed when propionaldehyde 2-, 3- and 4-nitrophenylhydrazone (**1558**; $R^1 = R^2 = H$, $R^3 = NO_2$; $R^1 = R^3 = H$, $R^2 = NO_2$ and $R^1 = NO_2$, $R^2 = R^3 = H$, respectively; $n = 2$) were treated with concentrated hydrochloric acid to yield **1559** (R^1–$R^3 = H$, $R^4 = NO_2$, $R^5 = CH_3$; $R^1 = NO_2$ (or H), $R^2 = R^4 = H$, $R^3 = H$ (or NO_2), $R^5 = CH_3$ and $R^1 = R^3 = R^4 = H$, $R^2 = NO_2$, $R^5 = CH_3$, respectively; $n = 2$). The structures of these products were based upon their elemental analyses. Subsequently (Shaw, E. and Woolley, D. W., 1953), it was found that butyraldehyde 4-nitrophenylhydrazone (**1558**; $R^1 = NO_2$, $R^2 = R^3 = H$, $n = 3$), upon indolization in a stirred benzene–concentrated hydrochloric acid (1 : 1 v/v) biphase system, furnished both 3-ethyl-5-nitroindole and **1559** ($R^1 = R^3 = R^4 = H$, $R^2 = NO_2$, $R^5 = C_2H_5$, $n = 3$), separation being achieved by chromatography and structural assignments being based upon elemental analytical data and u.v. spectroscopic data. Nearly two decades later, other examples of analogous by-product formation were reported. Thus, the indolization of the 2-naphthylhydrazone **1560** ($R = C_2H_5$), which was assumed to involve the 1-position and not the 3-position of the naphthalene ring (see Chapter III, D), with orthophosphoric acid in ethanol and with sulphosalicylic acid in both ethanol and methanol gave the expected 4,5-benzindolic product **608** [R^1–$R^3 = H$, $R^4 = (CH_2)_3COOC_2H_5$, X–Z = CH] (64%) and

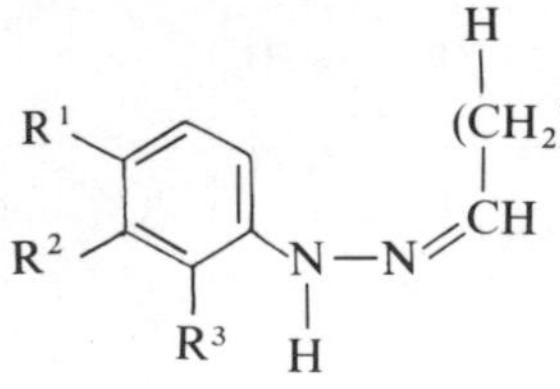

1558

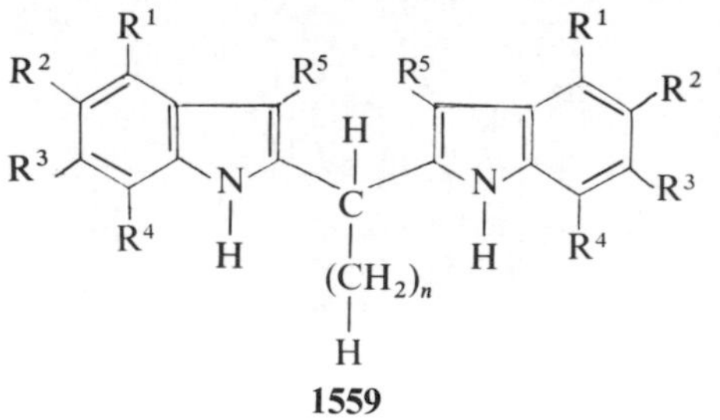

1559

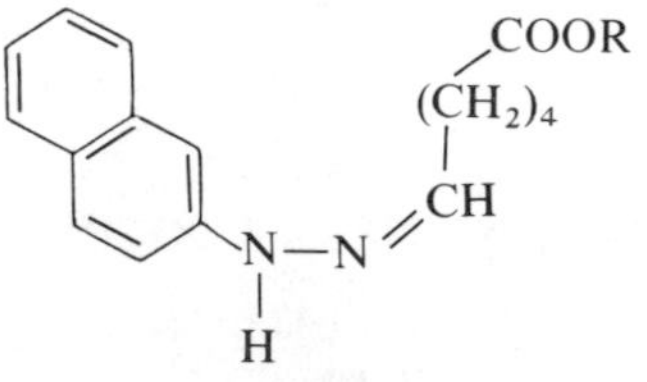

1560

1561 (R = C_2H_5) (5.2%), **608** [R^1–R^3 = H, R^4 = $(CH_2)_3COOC_2H_5$, X–Z = CH] (17%) and **1561** (R = C_2H_5) (22%) and **608** [R^1–R^3 = H, R^4 = $(CH_2)_3COOCH_3$, X–Z = CH] (23%) and **1561** (R = CH_3) (45%), respectively, in the latter reaction a transesterification also occurring. However, using orthophosphoric acid in ethyl cellosolve as catalyst, **1560** (R = CH_3) afforded **608** [R^1–R^3 = H, R^4 = $(CH_2)_3COOCH_3$, X–Z = CH] (88%) as the only isolated product (Shagalov, Eraksina, Turchin and Suvorov, 1970). The structure of the products **1561** (R = CH_3 and C_2H_5) was based mainly upon spectroscopic data, of which the p.m.r. spectra indicated the absence of a signal attributable to a proton at a 2-position of an indole nucleus in the products, the products resulting from their lithium aluminium hydride reduction and the acetyl derivatives of the latter. Furthermore, reaction of **608** [R^1–R^3 = H, R^4 = $(CH_2)_3COOCH_3$ and $(CH_2)_3COOC_2H_5$, X–Z = CH] with ethyl 5-formylvalerate in heated methanolic sulphosalicylic acid gave **1561** (R = CH_3 and C_2H_5, respectively) (Shagalov, Eraksina, Turchin and Suvorov, 1970) although the selectivity of the transesterifications which also occurred concomitantly with these condensations is difficult to understand. Likewise, the corresponding 1-naphthylhydrazone **1562** furnished **1563** (R = C_2H_5) (58%) as the only isolated product when orthophosphoric acid in ethyl cellosolve was used as the indolization catalyst but when using sulphosalicylic acid as the catalyst in either ethanolic or methanolic solution, both **1563** (R = C_2H_5) (76%) and **1564** (R = C_2H_5) (5.4%) or **1563** (R = CH_3) (49%) and **1564** (R = CH_3) (19.7%), respectively, were isolated (Shagalov, Eraksina and Suvorov, 1970). The structures of these condensation products were further supported by the results of earlier investigations, by one of Emil Fischer's students (Wenzing, 1887 – see also Fischer, Emil, 1888), of the reaction between 3-methylindole and benzaldehyde which afforded a product, the formulation

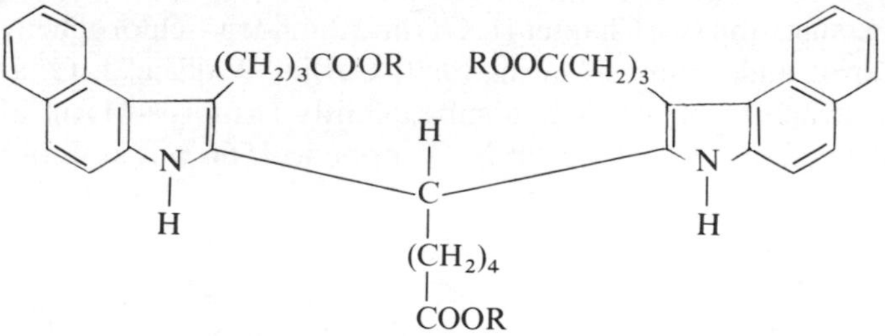

1561

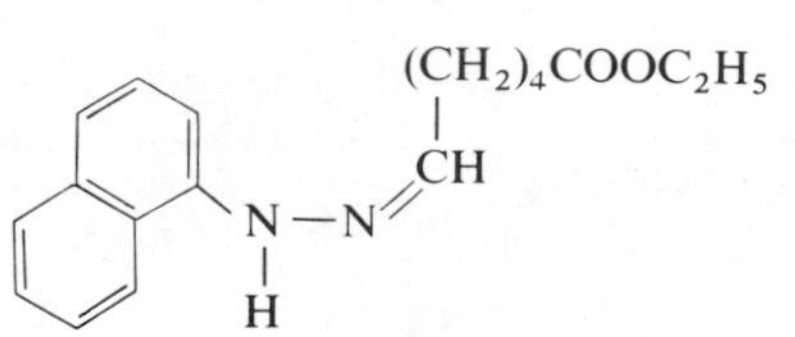

1562

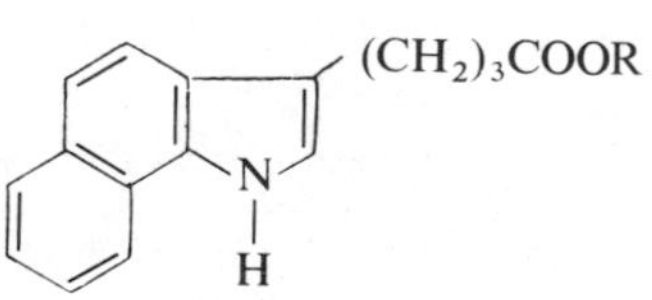

1563

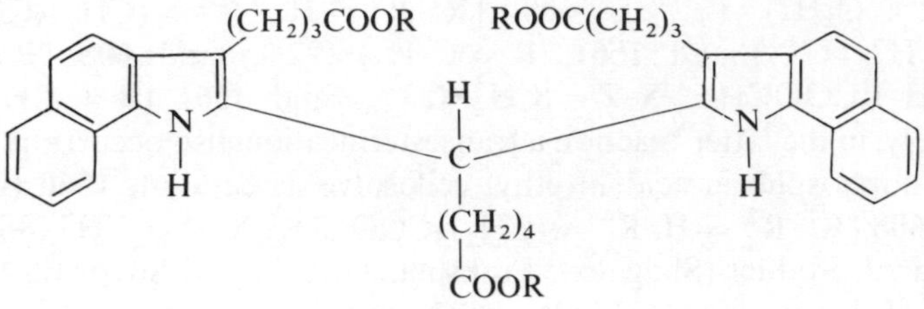

(CH₂)₃COOR ROOC(CH₂)₃

1564

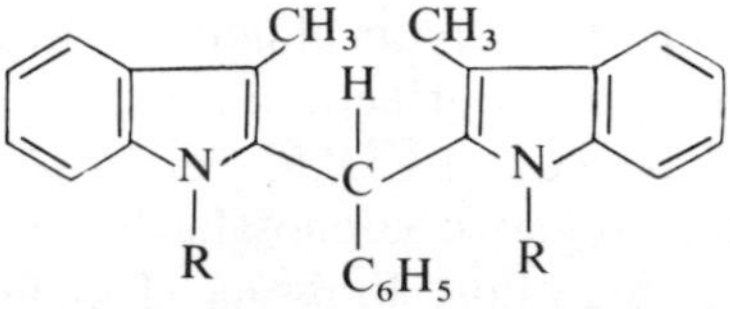

1565

of which as **1565** (R = H) (Dostál, 1938; Étienne and Heymès, 1948; Passerini and Bonciani, 1933; Dobeneck and Maresch, 1952) was subsequently (Noland and Robinson, D. N., 1958) confirmed by i.r. and u.v. spectroscopic examination and by the observations that 1,3-dimethylindole similarly reacted with benzaldehyde to give **1565** (R = CH_3) whereas 2,3-dimethylindole was recovered unchanged after similar treatment.

An interesting reaction, under the indolization conditions, of the ketonic moiety liberated by hydrolysis during an indolization has been observed. Treatment of 4-chloroacetophenone 2,6-dichlorophenylhydrazone with molten tin(II) chloride produced 7-chloro-2-(4-chlorophenyl)indole (formed by halogen elimination (see Chapter II, G1), 5,7-dichloro-2-(4-chlorophenyl)indole (formed by halogen migration (see Chapter II, G1) and 2,4,6-tri(4-chlorophenyl)pyridine (**1566**) (Carlin and Amoros-Marin, 1959; Carlin, Wallace, J. G. and Fisher, 1952), the structure of which was subsequently (Amoros-Marin and Carlin, 1959) confirmed by a direct synthesis. Compound **1566** was also formed, along

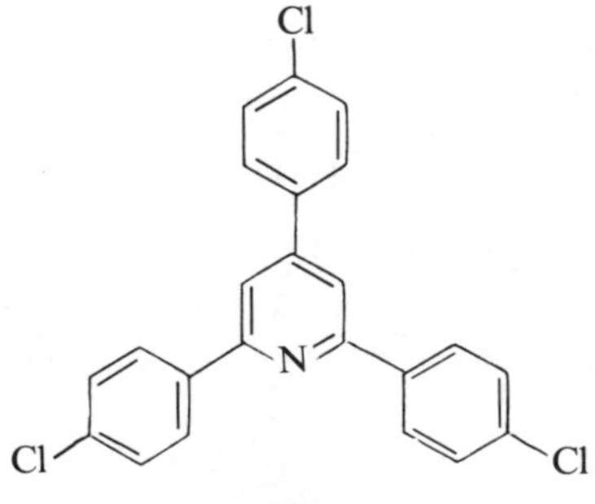

1566

with the expected 7-chloro-2-(4-chlorophenyl)indole, when 4-chloroacetophenone 2-chlorophenylhydrazone was similarly treated with tin(II) chloride or zinc chloride (Carlin and Amoros-Marin, 1959). Although the mechanism by which **1566** was formed concomitantly with the above indolizations remained obscure, it was likely that it resulted from the reaction, under the indolization conditions, between the 4-chloroacetophenone, formed by hydrolysis of the hydrazone, and the ammonia produced during indolization (Amoros-Marin and Carlin, 1959; Carlin and Amoros-Marin, 1959).

Ketonic functions liberated from their arylhydrazones during indolization of the latter can also react with a 2-methylene group in a 3*H*-indolic product. Thus, isopropyl methyl ketone 2-pyridylhydrazone, upon heating with a catalytic quantity of zinc chloride, afforded 7-aza-2,3,3-trimethyl-3*H*-indole together with a second product which appeared to have structure **1567** (Ficken and Kendall, 1959).

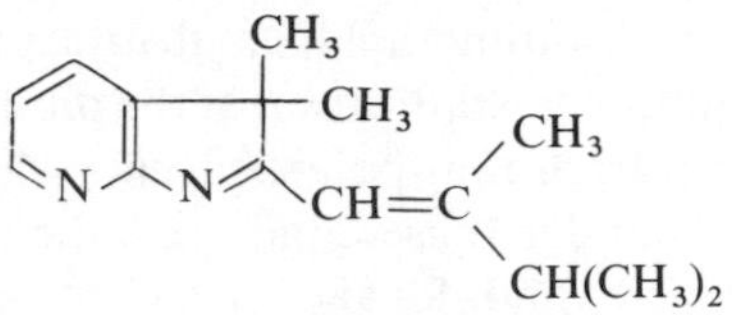

CH_3 CH_3 CH_3 N N $CH{=}C$ $CH(CH_3)_2$

1567

F. Oxidation of Indolic Products

It is well established that 2,3-disubstituted indoles undergo aerial oxidation to form the corresponding 2,3-disubstituted 3*H*-indole-3-hydroperoxides (Beer, Broadhurst and Robertson, 1952, 1953; Beer, Donavanik and Robertson, 1954; Beer, McGrath and Robertson, 1950a, 1950b; Jackson, A. H. and Smith, A. E., 1965; Leete, 1961b; Leete and Chen, F. Y.-H., 1963; McCapra and Chang, Y. C., 1966; Taylor, W. I., 1962; Wasserman and Floyd, 1963; Witkop and Patrick, 1951a, 1951b, 1952; Witkop, Patrick and Rosenblum, 1951) which can further react to give dioxindoles or the corresponding 3-hydroxy-3*H*-indoles or undergo cleavage of their heterocyclic ring. Such hydroperoxide formations have been used to characterize 1,2,3,4-tetrahydrocarbazoles (Campaigne and Lake, 1959). These oxidative reactions may also occur subsequent to Fischer indolization during the reaction product 'work up'. Thus, when cyclohexanone 2-bromophenylhydrazone was indolized by refluxing in dilute sulphuric acid solution there were isolated, along with the expected 8-bromo-1,2,3,4-tetrahydrocarbazole, three products whose molecular formulae were proposed to be $C_{12}H_{10}O_2N$ Br, $C_{12}H_{10}O_2N$ Br and $C_{12}H_8O$ N Br although Rast molecular weight determinations were inconsistent with the last two formulae. Although the structures of these last three products were not established, they were suggested as arising from the 2-bromoanilino radical (**1568**), formed as a result of homolysis of the N—N bond of the hydrazone, thought at the time by these workers to represent the first stage of the indolization mechanism (Barnes, Pausacker and Badcock, 1951) (see Chapter II, A). However, subsequent studies (Beer, Broadhurst and Robertson, 1953) showed that the first of these products was **1569** and that the other two were artefacts derived from it for which the structures **1570** and **1571**, respectively, were established, the molecular formulae originally proposed requiring slight modification to accommodate these structures. When the ultimate required products are the corresponding carbazoles, it has been found that to prevent such peroxidations of the initially formed 1,2,3,4-tetrahydrocarbazole, immediate dehydrogenation of the indolization products should be effected. This has usually been carried out with 5% palladium-on-carbon in refluxing xylene or mesitylene but occasionally with chloranil (Fusco and Sannicolò, 1978c). Attempted prevention of 3-hydroperoxide formation during indolization product 'work up' was unsuccessful when ethyl methyl ketone 2,4-dimethylphenylhydrazone was indolized using boron trifluoride etherate and crystallization of the product was attempted from polar solvents rather than from non-polar solvents which would facilitate the oxidation. However, even under these conditions the product proved to be a mixture of 2,3,5,7-tetramethylindole and the corresponding hydroperoxide **1572** (Kesswani, 1970).

Indolization of ethyl 2-hydroxyphenyl ketone phenylhydrazone with either hydrogen chloride in acetic acid or zinc chloride as catalyst afforded the indolic product **1573** (R^1 = OH, R^2 = H, R^3 = CH_3) as an oil. Attempts to crystallize this from light petroleum led to the separation from the solution of **1574** (R = OH, X = O) which was most likely produced via the formation of **1574** (R = OOH, X = O) and stabilized by intramolecular hydrogen bonding as shown (Robinson, B. and Zubair, 1971). By analogy with an earlier suggestion (Hino, Nakagawa and Akaboshi, 1967) it is possible that the 3*H*-indole hydro-

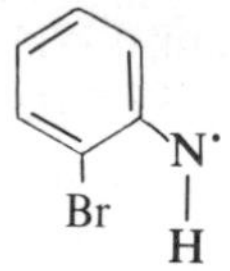

1568 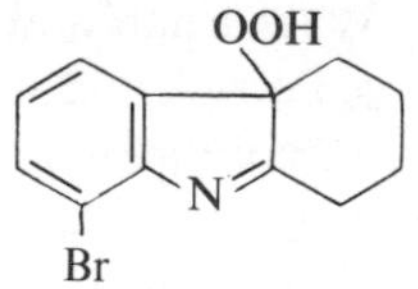 **1569** 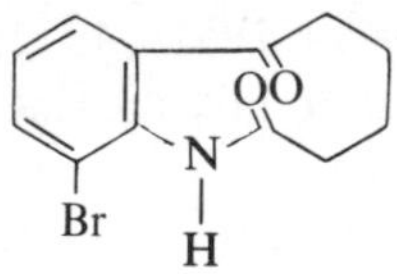 **1570**

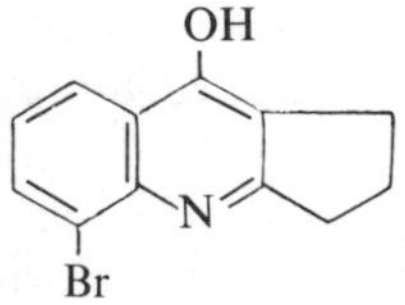

1571 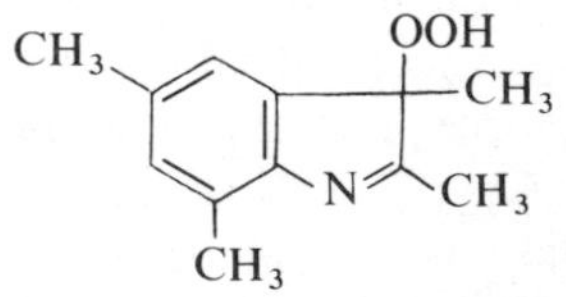 **1572**

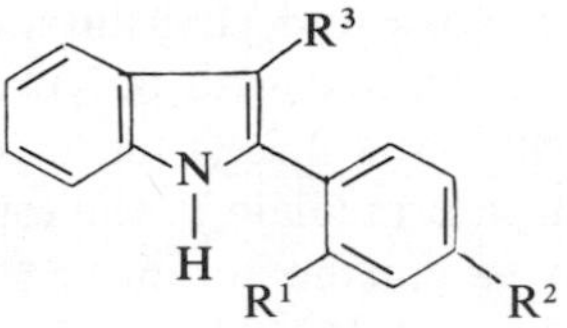

R

1573

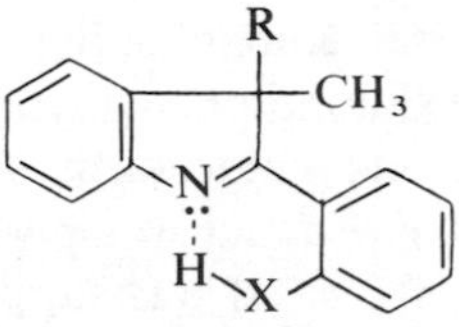

1574

peroxide was formed by oxidation of the species **1574** (R = H, X = O), the transient formation of which was possible because of its stabilization by intramolecular hydrogen bonding as shown in **1574**. Such hydrogen bonding certainly appeared to be required in connection with this oxidation since both **1573** ($R^1 = OCH_3$, $R^2 = H$, $R^3 = CH_3$ and $R^1 = H$, $R^2 = OH$, $R^3 = CH_3$) remained unchanged when boiled in light petroleum exposed to the atmosphere for 16 h. Furthermore, no analogous oxidation was observed during earlier Fischer syntheses of **1573** ($R^1 = OH$, $R^2 = R^3 = H$) from 2-hydroxyphenyl methyl ketone phenylhydrazone, either using a polyphosphoric acid catalyst, to afford the indolic product of m.p. = 161–162 °C (Calvaire and Pallaud, 1960), or by pyrolysis of the hydrazone with granulated zinc followed by digestion of the reaction mixture in hydrochloric acid to afford the indolic product of m.p. = 175–176 °C (McLoughlin and Smith, L. H., 1965). It is therefore possible that for such oxidation to occur, the indolic 3-position needed to be substituted (Robinson, B. and Zubair, 1971). When 2-(2-aminophenyl)-3-methylindole (**1573**; $R^1 = NH_2$, $R^2 = H$, $R^3 = CH_3$) was prepared by refluxing 2-aminophenyl ethyl ketone phenylhydrazone in either glacial acetic acid saturated with hydrogen chloride or in monoethylene glycol, the product could be isolated without the occurrence of oxidation analogous to that described above. Only after prolonged boiling in light petroleum was **1574** (R = OH, X = NH) formed, a transformation which did not occur when the 3- and 4-aminophenyl isomers were similarly treated. These observations again supported the suggestion that stabilization by intramolecular hydrogen bonding, stronger with an OH than with an NH_2 group, was necessary in connection with such oxidations (Robinson, B. and Zubair, 1973).

G. Further Reaction when Using Polyphosphate Ester as Indolization Catalyst

Fischer indolizations can be effected by treating arylhydrazones with polyphosphate ester in refluxing chloroform (Inoue and Ban, 1970; Kanaoka, Ban, Miyashita, Irie and Yonemitsu, 1966; Kanaoka, Ban, Yonemitsu, Irie and Miyashita, 1965; Krutak, 1975). However, at higher reaction temperatures, alkylation occurred of the indoles initially produced. Thus, when cyclohexanone phenylhydrazone was heated with polyphosphate ethyl ester at 150 °C, **54** ($R^1 = C_2H_5$, $R^2 = H$) was produced (Kanaoka, Ban, Miyashita, Irie and

Yonemitsu, 1966; Kanaoka, Ban, Yonemitsu, Irie and Miyashita, 1965) whereas when cyclohexanone phenylhydrazone was treated with polyphosphate ethyl ester in refluxing chloroform, the product was 1,2,3,4-tetrahydrocarbazole. That this latter product was probably an intermediate in the formation of **54** ($R^1 = C_2H_5$, $R^2 = H$) was supported by its conversion into **54** ($R^1 = C_2H_5$, $R^2 = H$) (Kanaoka, Ban, Yonemitsu, Irie and Miyashita, 1965; Yonemitsu, Miyashita, Ban and Kanaoka, 1969) and its ethiodide and 9-ethyl-1,2,3,4-tetrahydrocarbazole (Yonemitsu, Miyashita, Ban and Kanaoka, 1969) by heating with polyphosphate ethyl ester at 160 °C. Indeed, at such elevated temperatures, polyphosphate ester acts as a general alkylating agent for converting 2,3-disubstituted indoles into 2,3,3-trisubstituted-3*H*-indoles (Kanaoka, Ban, Yonemitsu, Irie and Miyashita, 1965; Yonemitsu, Miyashita, Ban and Kanaoka, 1969). Since no 2-ethyl-3,3-dimethyl-3*H*-indole could be detected, by p.m.r. spectroscopic analysis, after such treatment of 2,3-dimethylindole which afforded 3-ethyl-2,3-dimethyl-3*H*-indole and 1,3-diethyl-2,3-dimethyl-3*H*-indolium iodide, the possibility of the occurrence of a subsequent Plancher rearrangement was initially ruled out (Yonemitsu, Miyashita, Ban and Kanaoka, 1969). However, subsequent chromatographic investigation of this reaction mixture, and others related to it, detected the formation of such Plancher rearrangement products under these conditions (Kanaoka, Miyashita and Yonemitsu, 1969).

H. Alcoholyses of Nitrile Groups, Esterifications and Transesterifications

When using refluxing monoethylene glycol to effect 'non-catalytic' thermal indolizations, the solvent can react, concomitantly with indolization, with other functional groups present. Thus, *cis*-pyruvonitrile phenylhydrazone in refluxing monoethylene glycol afforded **2**[$R^1 = R^3 = H$, $R^2 = COO(CH_2)_2OH$], the nitrile group having undergone alcoholysis under the indolization conditions (Robinson, B., 1963c). Under similar conditions, transesterification also occurred when a mixture of 4-carbethoxyphenylhydrazine with **1575** gave **1576** (Szmuszkovicz, 1971; Youngdale, Glenn, Lednicer and Smuszkovicz, 1969). Further simultaneous alcoholysis and hydrolysis of the nitrile group have been observed when compounds **1577** ($n = 2$ and 3) were treated with ethanolic hydrochloric or sulphuric acid or dry hydrogen chloride in absolute ethanol to give **1578** ($R = H, n = 1$ and $R = C_2H_5, n = 2$, respectively) (Keimatsu and Sugasawa, 1928b). Later investigators (Feofilaktov and Semenova, N. K., 1952a, 1953b) also observed the occurrence of ethanolysis of a nitrile group synchronously with indolization when using heated ethanolic hydrogen chloride as the catalyst.

Using sulphuric acid in monoethylene glycol to carry out the indolization of ethyl pyruvate arylhydrazones, transesterification occurred, resulting in the

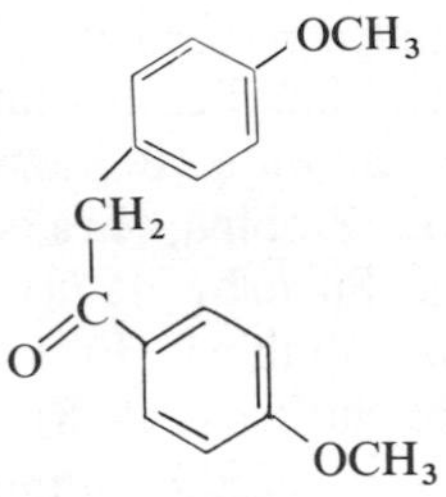

1575 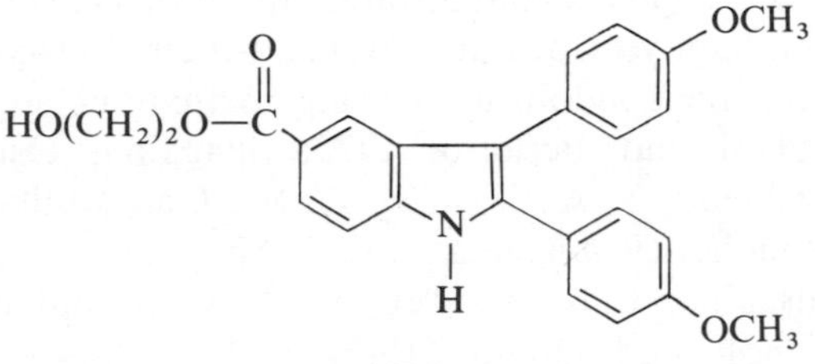 **1576**

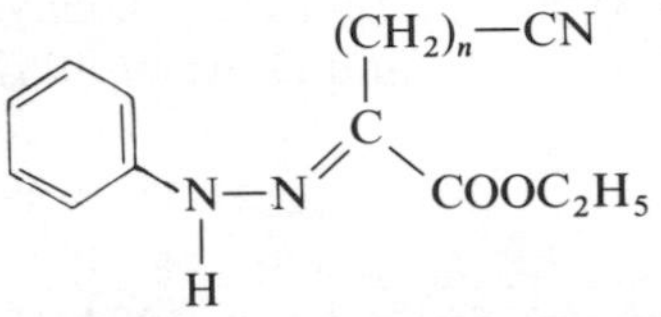

1577 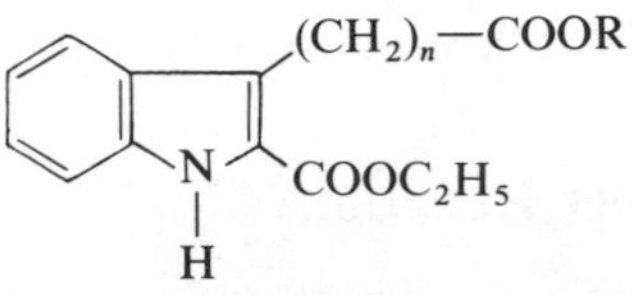 **1578**

formation of indole-2-carboxylic acid 2-hydroxyethyl esters (Sato, Y., 1963b, 1963c; Sunagawa and Sato, Y., 1962b). Similar transesterification was observed when indolization of ethyl pyruvate arylhydrazones with methanolic sulphuric acid afforded the corresponding 2-carbomethoxyindoles (Suvorov, Samsoniya, Chilikin, Chikvaidze, I. S., Turchin, Efimova, Tret'yakova and Gverdtsiteli, I. M., 1978). Likewise, transesterification accompanied the indolization of aldehydoester arylhydrazones using sulphosalicylic acid as the catalyst in alcoholic solutions (Shagalov, Eraksina and Suvorov, 1970) and also occurred when using methyl laevulinate as the ketonic moiety and 2 M ethanolic hydrogen chloride as the catalyst (Woolley, D. and Shaw, E. N., 1959) and when using dimethyl-2-oxoadipate as the ketonic moiety and ethanolic hydrogen chloride as the catalyst (Justoni and Pessina, 1955).

Esterification of carboxylic acid groups present in the original arylhydrazone, to produce the corresponding carbethoxyindoles, has also occurred during indolizations using as catalyst concentrated sulphuric acid in reflxuing ethanol (Akopyan and Tatevosyan, 1971, 1974, 1976; Arnold, E., 1888; Bailey and Seager, 1974; Feofilaktov, 1947a, 1952; Feofilaktov and Semenova, N. K., 1953b; Fox and Bullock, 1955a; Jackson, R. W. and Manske, R. H., 1930, 1935; Kalb, Schweizer and Schimpf, 1926; Koelsch, 1943; Manske, R. H. and Boos, 1960; Manske, R. H. F., 1931a; Manske, R. H. F. and Robinson, R., 1927; Mndzhoyan, Tatevosyan, Terzyan and Ekmekdzhyan, 1958; Pacheco, 1951; Sato, Y., 1963b; Scherrer, 1969; Terzyan, Akopyan and Tatevosyan, 1961; Terzyan and Tatevosyan, 1960; Wislicenus and Arnold, E., 1887) or dry hydrogen chloride in ethanol (Feofilaktov and Semenova, 1953b; Lions and Spruson, 1932; Sterling Drug Inc., 1970; Wislicenus and Waldmüller, 1911; Yamamoto, H., Inaba, Okamoto, Hirohashi, T., Ishizumi, Yamamoto, M., Maruyama, Mori and Kobayashi, T., 1969a). Likewise, 2-hydroxyethyl esterification accompanied indolization using concentrated sulphuric acid as the catalyst in monoethylene glycol (Sankyo Co. Ltd, 1964) and methyl, isopropyl and butyl esterification accompanied indolization using methanolic hydrogen chloride (Bertazzoni, Bartoletti and Perlotto, 1970; Shagalov, Ostapchuk, Zlobina, Eraksina, Babushkina, T. A., Vasil'ev, A. M., Ogoradnikova and Suvorov, 1978) and methanolic (Rosenstock, 1966; Shagalov, Ostapchuk, Zlobina, Eraksina, Babushkina, T. A., Vasil'ev, A. M., Ogoradnikova and Suvorov, 1978), isopropanolic and butanolic (Stevens, F. J., Ashby and Downey, 1957) sulphuric acid under reflux as catalysts. These conditions can be useful for the direct synthesis of the appropriate indole carboxylic acid ester from aldehydoacid or ketoacid arylhydrazones (e.g. Stevens, F. J., Ashby and Downey, 1957).

I. Other Reactions

Selective O-demethylation was found to occur when **1004** (R^1 = 3,4-$(OCH_3)_2$, R^2 = H, X = Br] was treated with ethanolic hydrogen bromide to afford **1005** (R^1 = 3-OCH_3-4-OH, R^2 = H, X = Br). Similar treatment of **1004** (R^1 = 3- and 4-OCH_3, R^2 = H, X = Br) gave **1005** (R^1 = 3-OCH_3 and 4-OH, respectively, R^2 = H, X = Br), again indicating the selective nature of the O-demethylation under these conditions (Elderfield, Lagowski, McCurdy and Wythe, 1958). The further selectivity of this side reaction was noted when treatment of **1006** [R = 3,4-$(OCH_3)_2$ and 3- and 4-OCH_3, X = Br] with methanolic hydrogen bromide gave **1007** [R = 3,4-$(OCH_3)_2$ and 3- and 4-OCH_3, respectively, X = Br] (Elderfield, Lagowski, McCurdy and Wythe, 1958).

After **1579** (R^1 = R^3 = H, R^2 = CH_3, R^4 = $CH_2C_6H_5$) was subjected to indolization by treatment with methanolic hydrochloric acid, **1580** was isolated along with a trace of what was apparently **1581** (R = C_6H_5). The structure of this latter product, which was assumed to arise by methylation of **1580** under the indolization conditions, was based upon a comparison of its m.p. with that of the product obtained by the indolization of **1579** (R^1 = H, R^2 = R^3 = CH_3, R^4 = $CH_2C_6H_5$) with a methanolic hydrochloric acid catalyst (Wuyts and Lacourt, 1935).

Dehalogenations have been found to occur under indolization conditions. Thus, when **1582** (R^1 = I, R^2 = OH, R^3 = C_2H_5, R^4 = C_6H_5) was indolized with ethanolic hydrochloric acid, concomitant displacement of one of the iodine atoms occurred to afford **1583** (R^1 = H, R^2 = OH, R^3 = I, R^4 = C_2H_5, R^5 = C_6H_5). This structure was distinguished from the possible alternative, **1583** (R^1 = I, R^2 = OH, R^3 = H, R^4 = C_2H_5, R^5 = C_6H_5) by the appearance in the p.m.r. spectrum of the product of two 1H aromatic singlets, indicative of the two *para* protons (Ried and Kleemann, 1968). However, indolization of **1582** (R^1 = Cl, R^2 = OH, R^3 = C_2H_5, R^4 = C_6H_5) with either ethanolic hydrogen chloride or 40% hydrogen bromide in acetic acid occurred without

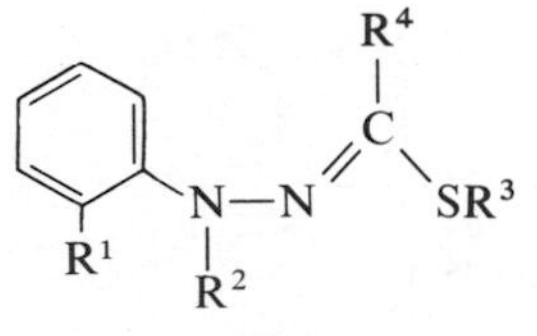

1579 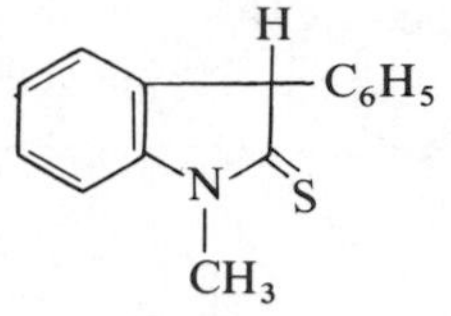 1580 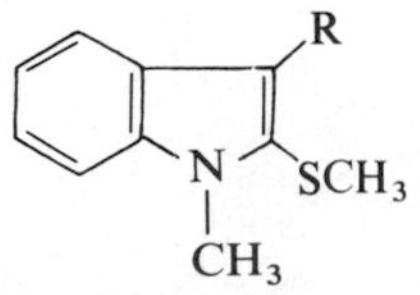 1581

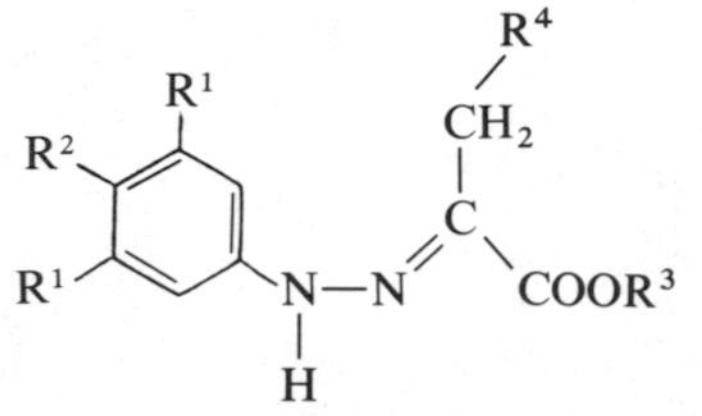

1582 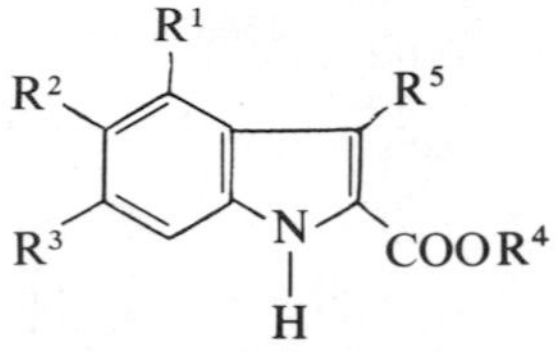 1583

halogen displacement to furnish **1583** ($R^1 = R^3 = Cl$, $R^2 = OH$, $R^4 = C_2H_5$, $R^5 = C_6H_5$) (Ried and Kleemann, 1968) and likewise **1582** [$R^1 = R^2 = I$, $R^3 = H$, $R^4 = (CH_2)_2COOH$] and **1582** [$R^1 = I$, $R^2 = OCH_3$, $R^3 = H$, $R^4 = (CH_2)_2COOH$], upon indolization with 30% alcoholic sulphuric acid, followed by base catalysed hydrolysis, yielded **1583** [R^1–$R^3 = I$, $R^4 = H$, $R^5 = (CH_2)_2COOH$ and $R^1 = R^3 = I$, $R^2 = OCH_3$, $R^4 = H$, $R^5 = (CH_2)_2COOH$, respectively] (Kalb, Schweizer, Zellner and Berthold, 1926). When the hydrazones **1584** ($R^1 = CH_3$, C_2H_5 and C_6H_5, $R^2 = OH$ and OCH_3, $R^3 = F$, Cl and Br) were indolized in acetic acid saturated with hydrogen chloride, the expected indoles **1585** were formed together with some dehalogenated products, although details of these latter products have not been given (Buu-Hoï, Xuong and Lavit, 1954).

Dehydration has also been observed to accompany Fischer indolization. Thus, whereas indolization of **1586** ($R = CH_3$), by heating with catalytic amounts of zinc chloride, afforded **1587** ($R = CH_3$), similar treatment of **1586** [$R + R = (CH_2)_5$] afforded **1587** [$R + R = (CH_2)_5$] along with **1588**, these two products being separated by column chromatography on alumina (Terent'ev, P. B., Kost, Shchegolev and Terent'ev, A. P., 1961). Nucleophilic replacement of a hydroxy group by a chloro group was observed to accompany indolization

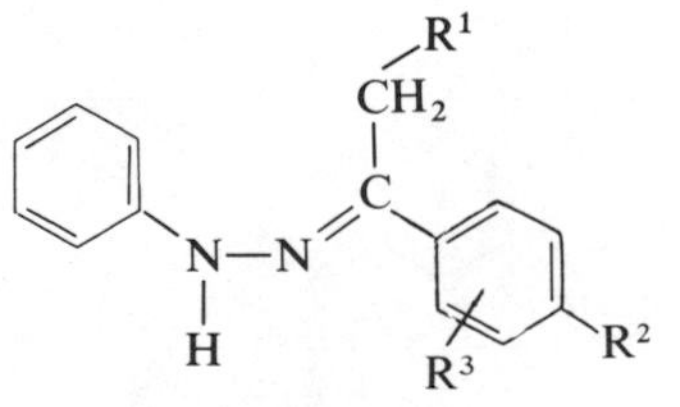

1584 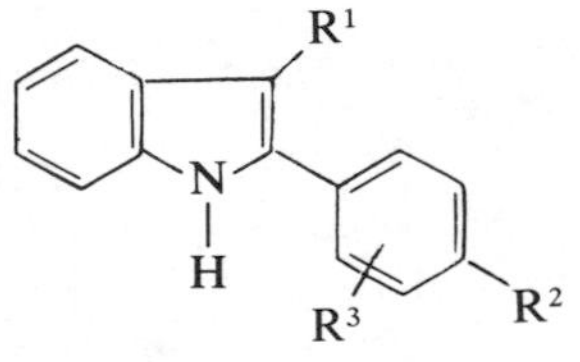 1585

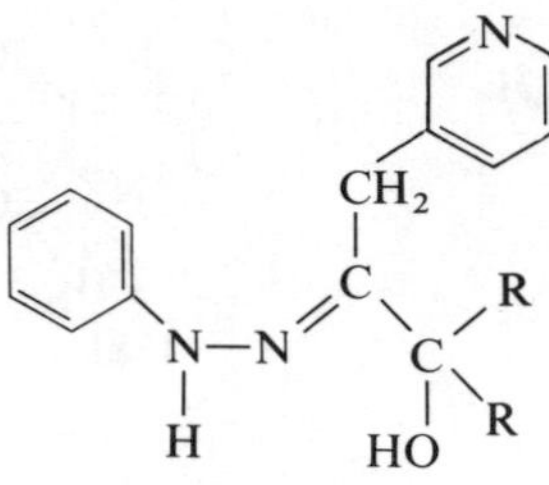

1586

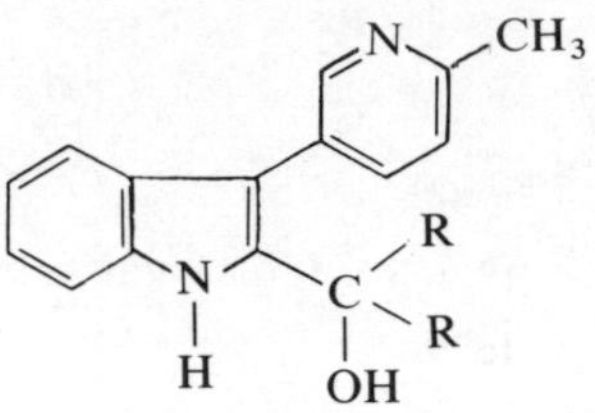

1587

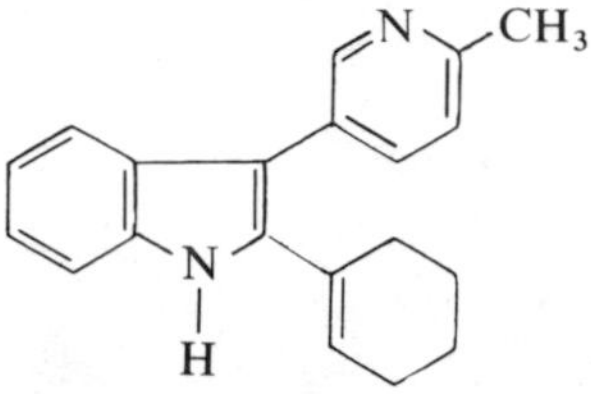

1588

when the arylhydrazone **1589** was converted into **1590** by treatment in refluxing alcohol saturated with hydrogen chloride (Kost, Vinogradova, Daut and Terent'ev, A. P., 1962). The reverse of this displacement reaction occurred when **1591** was converted, upon stirring with concentrated hydrochloric acid and benzene, into a mixture of **1592** (R = Cl and OH) which were separated by column chromatography on alumina (Shaw, E. and Woolley, D. W., 1953).

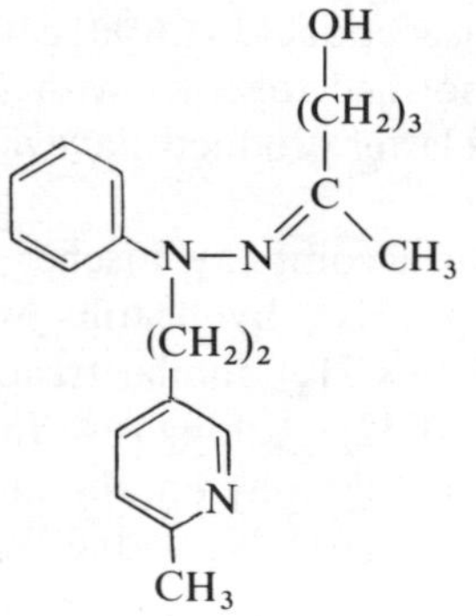

1589

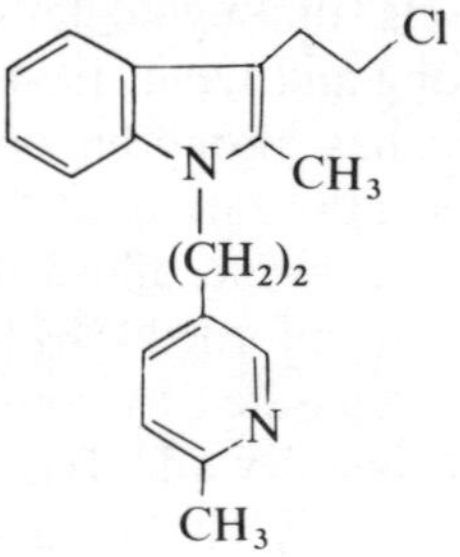

1590

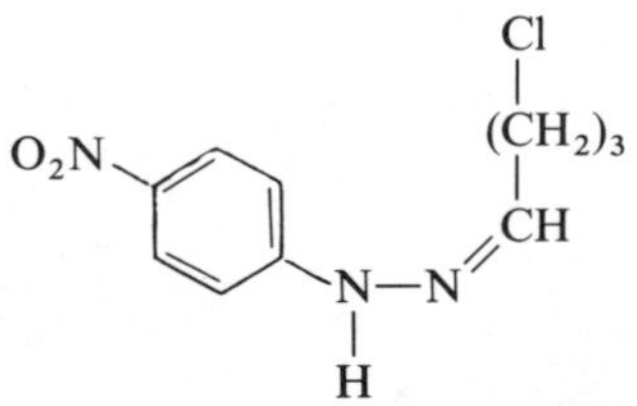

1591

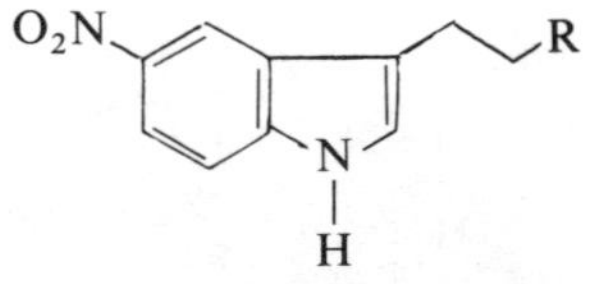

1592

Benzacetylation has also been shown to occur when using 4-methylbenzene sulphonic acid in refluxing acetic anhydride to effect indolization. In this way, ethyl methyl ketone N_α-methylphenylhydrazone was converted into a separable mixture of N-methylacetanilide, N_β-acetyl-N_α-methylphenylhydrazine and three monoacetyl-1,2,3-trimethylindoles. These same three monoacetyl-1,2,3-trimethylindoles were also prepared by treating 1,2,3-trimethylindole, synthesized by reacting the hydrazone in refluxing ethanol containing concentrated sulphuric acid, with 4-methylbenzene sulphonic acid in boiling acetic anhydride. This indicated that the acetyl substituents were probably introduced into the benzenoid ring subsequent to indolization. Although these three products were converted into the corresponding ethyl derivatives by the Huang–Minlon modification of the Wolff–Kishner reduction, the orientations of these ethyl groups on the benzenoid ring were not established (Suvorov, Sorokina, N. P. and Sheinker, I. N., 1959).

Subsequent to indolization, further spontaneous cyclization under the reaction conditions can occur. Thus, **1593** (R = CH_3), upon treatment with phosphoryl chloride in boiling toluene, gave **1595** (R = CH_3) (Kermack and Smith, J. F., 1930), via the intermediacy of **1594** (R = CH_3). The yield and quality of the product **1595** (R = CH_3) were subsequently (Kiang, Mann, Prior and Topham, 1956) much improved using ethanolic hydrogen chloride as the indolization catalyst, this catalyst also being used to convert **1593** (R = C_6H_5) into **1595** (R = C_6H_5) (Kiang, Mann, Prior and Topham, 1956).

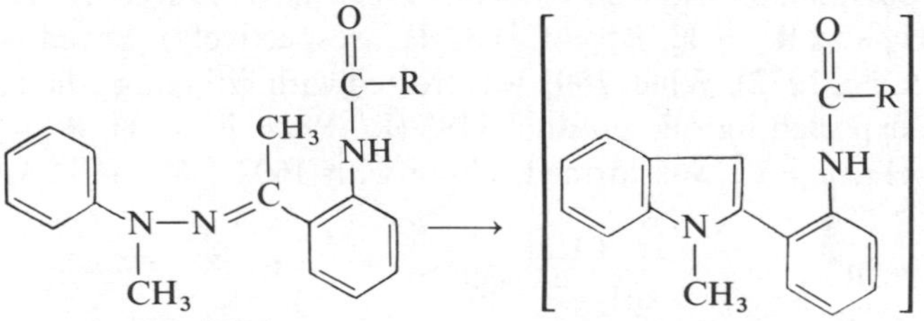

O C—R CH₃ NH N—N=C CH₃

1593

O C—R NH N CH₃

1594

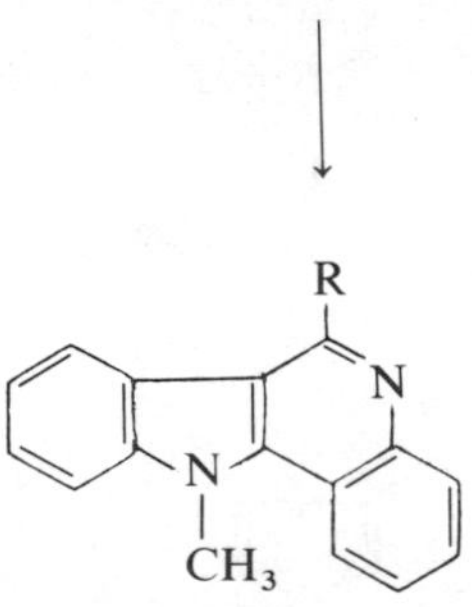

R N N CH₃

1595

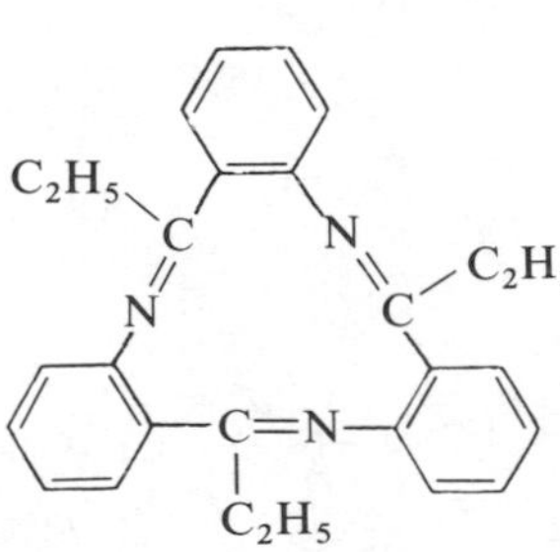

C_2H_5 C N N C C_2H_5 C=N C_2H_5

1596

When a mixture of 2-aminophenyl ethyl ketone with phenylhydrazine was heated, the only isolated product was **1596**. However, the corresponding phenylhydrazone was prepared by treating the reactants with refluxing ethanol containing acetic acid. Subsequent indolization of this hydrazone in refluxing acetic acid saturated with hydrogen chloride afforded the expected 2-(2-aminophenyl)-3-methylindole along with **996** (R^1–R^3 = H, R^4 = CH_3). This latter product appeared to arise by cyclization of 2-acetamidophenyl-3-methylindole, formed by acetylation of the initial indolic product under the indolization conditions. Indeed, when the indolization was brought about by refluxing the hydrazone in monoethylene glycol, the formation of **996** (R^1–R^3 = H, R^4 = CH_3) was prevented and only 2-(2-aminophenyl)-3-methylindole was isolated (Robinson, B. and Zubair, 1973). Indolization of **1597** (R = H) with ethanolic hydrogen chloride produced **1598** (R^1 = H, R^2 = $COOC_2H_5$) which upon hydrolysis gave **1598** (R^1 = H, R^2 = COOH). This was then cyclodehydrated by heating to yield **1599** (R = H). However, attempts to indolize **1597** (R = Cl) to **1598** (R^1 = Cl, R^2 = $COOC_2H_5$) by heating in alcoholic sulphuric acid or in alcohol with zinc chloride at 125 °C were unsuccessful whereas heating at 150 °C with zinc chloride in dry ethanol also caused a further spontaneous cyclization of the intermediate **1598** (R^1 = Cl, R^2 = $COOC_2H_5$), analogous to that above, to give **1599** (R = Cl) (Clemo and Seaton, 1954). A similar simultaneous cyclization occurred during the indolization of **1600** (R^1 = H and OCH_3, R^2 = CH_3 and R^1 = H, R^2 = $CH_2C_6H_5$), using a polyphosphoric acid catalyst, which gave **710** (R^1 = H and OCH_3, R^2 = CH_3 and R^1 = H, R^2 = $CH_2C_6H_5$, respectively) (Shvedov, Kurilo and Grinev, A. N., 1972). When **1601** was treated with refluxing ethanolic sulphuric acid, the expected indolic product **1195** (R^1–R^5 = R^7 = H, R^6 = $COOC_2H_5$, R^8 = C_2H_5, n = 2) was formed, along with **1602** [R = $(CH_2)_3COOC_2H_5$].

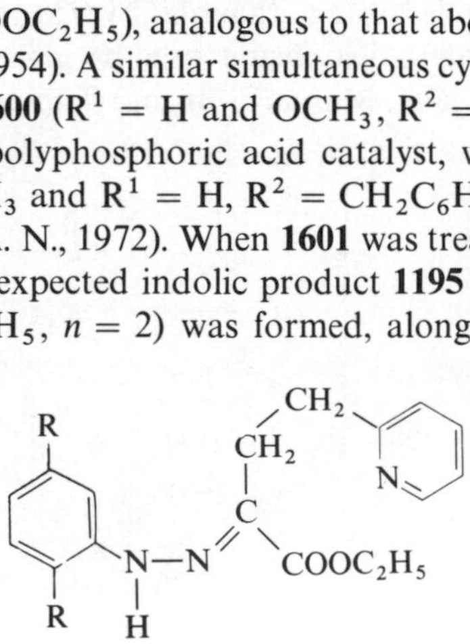

1597

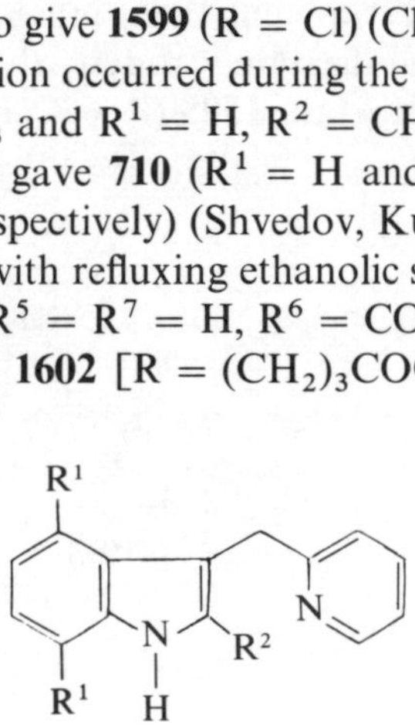

1598

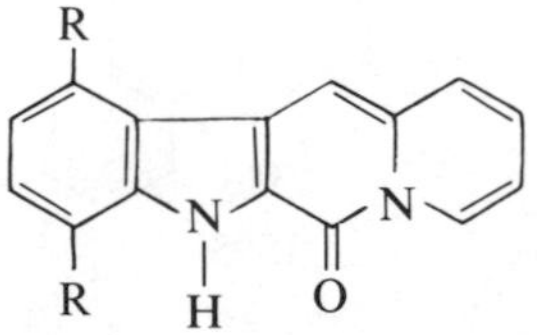

1599

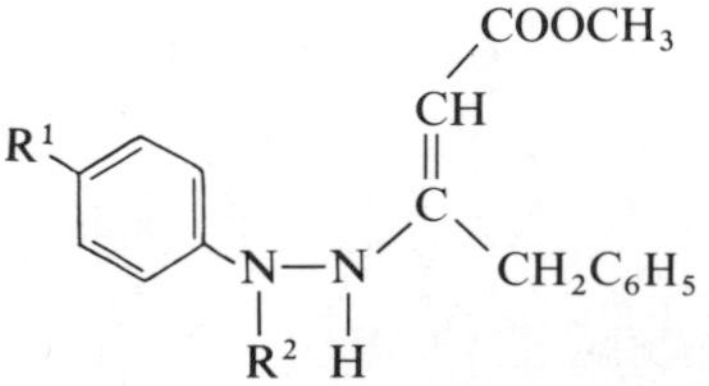

1600

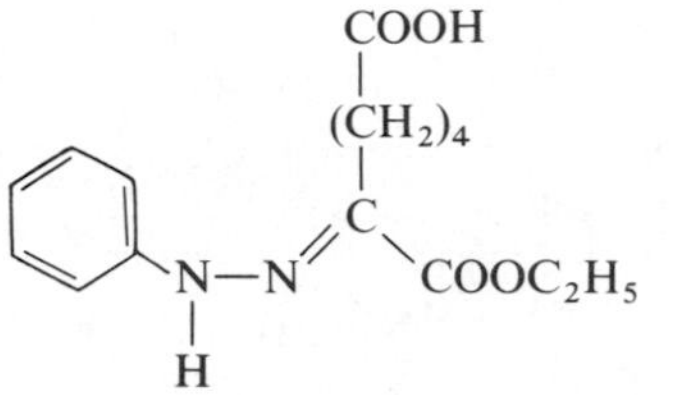

COOH
$(CH_2)_4$
N—N=C
H
$COOC_2H_5$

1601

R
O
N
N
O
R

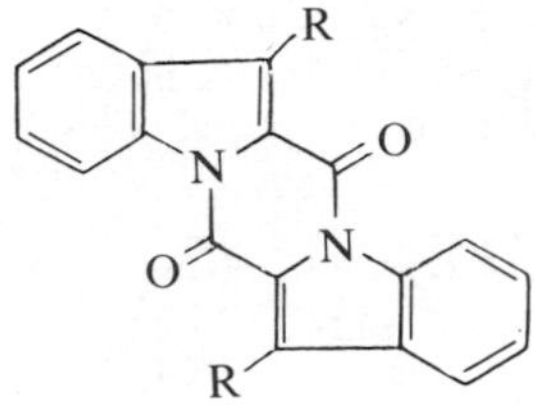

1602

The structure of the latter product was based upon C, H and N elemental analytical data, a Rast molecular weight determination and its hydrolysis, in refluxing propanolic potassium hydroxide, into **1195** (R^1–R^5 = R^7 = R^8 = H, R^6 = COOH, $n = 2$) (Manske, R. H. and Boos, 1960). Nearly forty years earlier (Kermack, Perkin and Robinson, R., 1921) a similar alkali catalysed hydrolysis of **1602** (R = H), obtained by the action of acetic anhydride upon indole-2-carboxylic acid, into potassium indole-2-carboxylate had been effected.

When **1603** [R = $NN(CH_3)C_6H_5$] was refluxed in 2 M ethanolic hydrogen chloride, the expected indole **1604** was produced. However, when **1603** [R = $NN(CH_3)C_6H_5$] was heated with polyphosphoric acid, compound **1606** was produced. This same product was also formed when **1603** (R = O) was similarly treated, a reaction which must initially have involved the transformation of the ketone hydrazide into the hydrazone hydrazide. The structure **1606** was based upon p.m.r. spectroscopic data and upon the observations that the product was also formed when **1604** or a mixture of ethyl 1,2-dimethylindole-3-carboxylate with N_α-methylphenylhydrazine was heated with polyphosphoric acid. It was suggested that **1606** was formed, as shown in Scheme 49, from **1603** [R = $NN(CH_3)C_6H_5$] via **1604**. This underwent a rearrangement reminiscent of a benzidine rearrangement to afford **1605** which was then cyclized by an acid catalysed intramolecular acylation (Kornet, Thio, A. P. and Tolbert, 1980).

Decarboxylation accompanied indolization when mixtures of arylhydrazines with **1607** were indolized to furnish the corresponding 2-methyl-3-indolylacetic acids (Yamamoto, H. and Nakao, 1968g, 1970i). Whether the decarboxylations preceded, occurred during or followed the cyclization was not established.

Heating mixtures of phenylhydrazine and 4-benzyloxy- and 4-methoxyphenylhydrazine hydrochloride with 3-benzylthiopropionaldehyde diethyl acetal **1288** ($n = 2$) in aqueous ethanolic acetic acid gave low yields of the corresponding 3-benzylthiomethylindoles along with, in each case, toluene-ω-thiol, dibenzyldisulphide and other indolic products, of unknown structures, formed by further reaction of the 3-benzylthiomethylindoles under the indolization conditions (Keglević and Goleš, 1970).

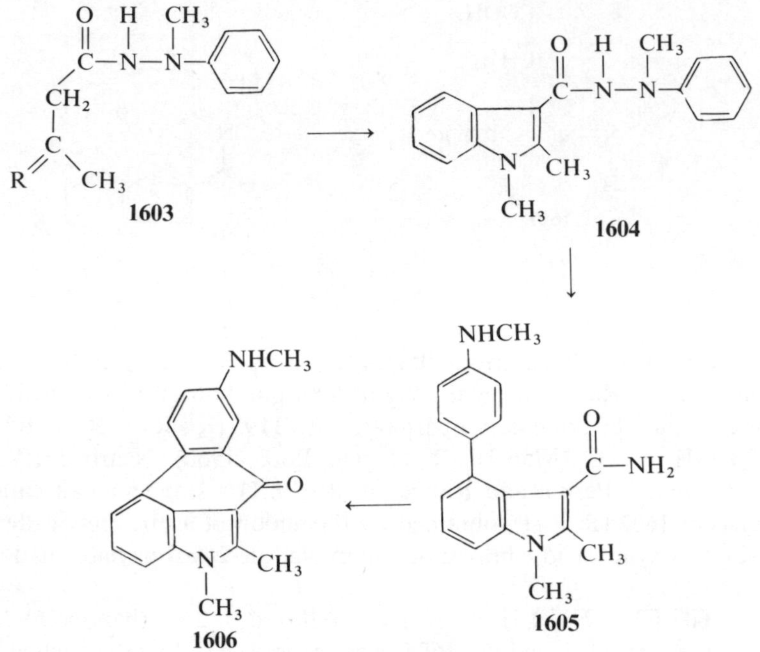

1603 → 1604 → 1605 → 1606

Scheme 49

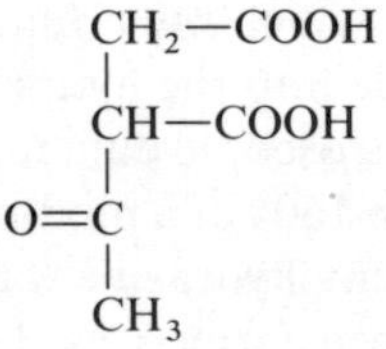

1607

Indolization of acetone 4-chlorophenylhydrazone, using a polyphosphoric acid catalyst, produced a low yield of a white crystalline solid, thought to be 2,2′-dimethyl-5,5′-bisindolyl (Chapman, Clarke and Hughes, H., 1965). However, neither experimental details for the reaction nor structural verification of the product were given.

Many examples have been recorded in which spontaneous dehydrogenation occurred subsequent to indolization (see Chapters I, D; III, L3; IV, I and J2 and V, A).

CHAPTER VI

Extensions to the Fischer Indole Synthesis

A. The Brunner Synthesis of Oxindoles

The Fischer indole synthesis can be indirectly applied to the synthesis of oxindoles by using, for example, isobutyraldehyde N_α-methylphenylhydrazone (**44**; $R^1 = R^2 = R^4 = H, R^3 = R^5 = R^6 = CH_3$) which afforded 2-hydroxy-1,3,3-trimethylindoline (**48**; $R^1 = CH_3, R^2 = H$). Upon oxidation with alcoholic ammoniacal silver nitrate, this product yielded 1,3,3-trimethyloxindole (**49**; $R^1 = H, R^2$–$R^4 = CH_3$) (Brunner, K., 1896a), subsequent (Brunner, K., Wiedner and Kling, 1931) similar oxidations being effected by alkaline potassium permanganate. The oxindole **49** ($R^1 = H, R^2$–$R^4 = CH_3$) was also obtained by Brunner, K. (1896b) by heating isobutyryl N_α-methylphenylhydrazide (**50**; $R^1 = H, R^2$–$R^4 = CH_3$) with lime, this reaction being the prototype of the synthetic route to oxindoles which was to carry Brunner's name. The synthesis was included in an early review (Sumpter, 1945) upon the chemistry of oxindole.

The mechanistic analogy between the Brunner oxindole synthesis and the Fischer indole synthesis, first referred to in 1924 (Robinson, G. M. and Robinson, R., 1924), is apparent if the species **50** is considered to enolize, under the

Scheme 50

conditions of base catalysis, to give **1608** which then rearranges, with formation of the new C—C bond, to form **1609**, loss of ammonia then leading to **49** (Scheme 50). An alternative mechanism for the Brunner oxindole synthesis was envisaged to be as shown in Scheme 51 involving the intermediacy of an anion **1610** formed under basic conditions by proton extraction from the carbonyl C-2 atom in the arylhydrazide (Wenkert, Bhattacharyya, N. K., Reid, T. L. and Stevens, T. E. (1956).

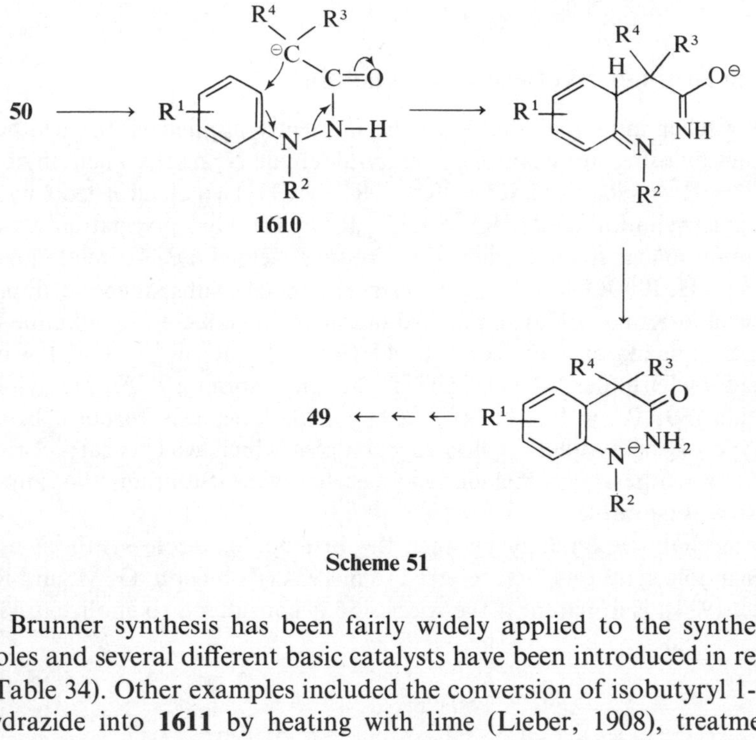

1610

Scheme 51

The Brunner synthesis has been fairly widely applied to the synthesis of oxindoles and several different basic catalysts have been introduced in relation to it (Table 34). Other examples included the conversion of isobutyryl 1-naphthylhydrazide into **1611** by heating with lime (Lieber, 1908), treatment of isobutyryl 2-naphthylhydrazide under similar conditions affording only one product, **1612** or **1613** (Lieber, 1908), the former structure being preferred by analogy with the known direction of indolization of 2-naphthylhydrazones (see Chapter III, D).

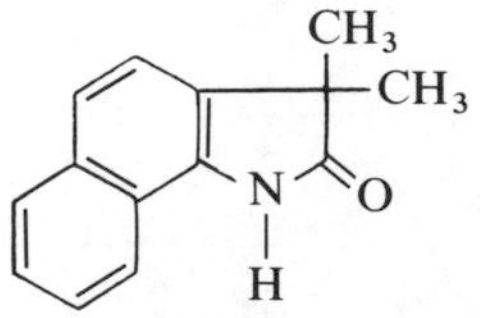

1611 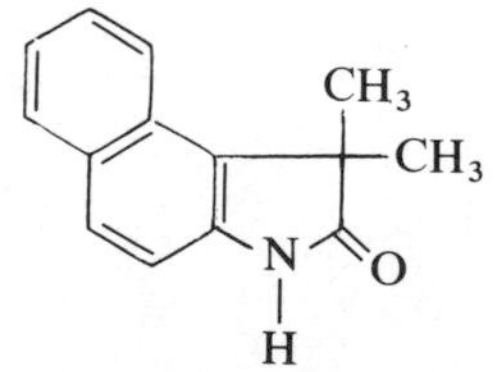 **1612** 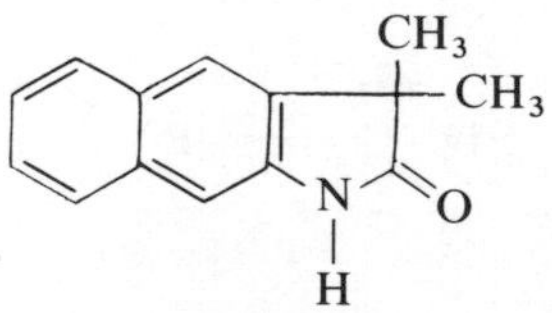 **1613**

Table 34. Oxindoles **49** which have been Prepared by the Brunner Oxindole Synthesis

Oxindole **49** (Unless otherwise stated, R^1–R^4 = H)

R^1	R^2	R^3	R^4	Reference (Catalyst)
				1, 4(1), 2(2), 3(3)
			CH_3	1, 4(1), 3(4), 5(5), 6(6)
			C_2H_5	1, 4(1)
			$CH(CH_3)_2$	7(1)[a], 4(1)
			$CH(CH_3)CH_2CH_3$[b]	8(1)[c]
			C_6H_5	1, 4(1)
			$CH_2C_6H_5$	9(1)
		CH_3	CH_3	4, 10–12(1)
		CH_3	C_6H_5	13(1)
		$(CH_2)_4$		14(1)
		$(CH_2)_5$		14, 15(1)
	CH_3			4, 16(1)
	CH_3		CH_3	4, 16(1)
	CH_3		$CH(CH_3)_2$	4, 7(1)
	CH_3		$CH_2C_6H_5$	9(1)
	CH_3	CH_3	CH_3	4, 16, 17(1), 18(7)
	CH_3	CH_3	C_6H_5	19(1)
4-CH_3			CH_3	3(4), 20(1)
6-CH_3			CH_3	3(4), 20(1)
4(6)-CH_3		CH_3	$CH_2C_6H_5$	9(1)[d]
5-CH_3			$CH_2C_6H_5$	9(1)
7-CH_3			$CH_2C_6H_5$	9(1)
4(6)-CH_3		CH_3	CH_3	17, 21(1)[e]
5-CH_3		CH_3	CH_3	4(1)
7-CH_3		CH_3	CH_3	4, 17(1)
5-CH_3		$(CH_2)_5$		14(1)
7-CH_3		$(CH_2)_5$		14(1)
5-OC_2H_5			$(CH_2)_2CH_3$	22(1)
5-OC_2H_5			$CH(CH_3)_2$	22(1)
5-OCH_3		CH_3	CH_3	23(1)
7-OCH_3		CH_3	CH_3	23(1)
5-OCH_3		$(CH_2)_5$		14(1)
5-Cl		$(CH_2)_5$		14(1)
5-OC_2H_5	CH_3		$CH(CH_3)_2$	22(1)
5-OC_2H_5	CH_3	CH_3	C_2H_5	22(1)

Notes

a. Subsequent (Wenkert, Bhattacharyya, N. K., Reid, T. L. and Stevens, T. E., 1956) attempts to repeat this reaction led only to the isolation of valeranilide. It was suggested that this was formed under the reaction conditions by cleavage of the N—N bond in the hydrazide followed by transamidation between the aniline thus produced and the remaining hydrazide. Analogous N—N bond cleavages in arylhydrazones to afford the

(*continued*)

Table 34. (*continued*)

corresponding anilines under Fischer indolization conditions are well known (see Chapter VII, J). b. (+)-3-Methylvaleroyl-N_β-phenylhydrazide, obtained from S-3-methylvaleric acid, was used in this reaction which yielded (+)-3-*sec*.-butyloxindole; c. The use of sodium methoxide in refluxing naphthalene as the catalyst afforded only unreacted hydrazide (quantitatively?). Whereas the yield of the oxindolic product from the calcium oxide catalysed reaction was 65%, use of calcium hydride in dimethylbenzamide as the catalyst gave only a 53% yield of total product consisting of a mixture of the desired oxindolic product together with the corresponding hydroxindole in a yield ratio of 3 : 1; d. Only one product was obtained but the orientation of the methyl group was not established; e. Both products were apparently formed but were neither clearly identified nor differentiated.

Catalysts

1. Lime; 2. Sodamide; 3. Heat the potassium compound of the hydrazide in quinoline; 4. Alkali metal alcoholate (e.g. sodium methoxide) or alkali metals; 5. Calcium hydride; 6. Sodium methoxide; 7. Calcium hydride [heating with sodamide, recommended by Staněk and Rybář (1946), failed to afford the oxindole].

References

1. Brunner, K., 1897b; 2. Staněk and Rybář, 1946; 3. C. F. Boehringer and Söhne, 1908b; 4. Brunner, K., 1906; 4. Endler and Becker, E. I., 1963; 6. Wieland, T., Weiberg, Fischer, Edgar and Hörlein, G., 1954; 7. Schwarz, 1903; 8. Menicagli, Malanga and Lardicci, 1979; 9. Tomicek, 1922; 10. Brunner, K., 1897a; 11. Kates and Marion, 1951; 12. Döpp and Weiler, 1979; 13. Bruce and Sutcliffe, 1957; 14. Moore, R. F. and Plant, 1951; 15. Jones, G. and Stevens, T. S., 1953; 16. Brunner, K., 1896b; 17. Brunner, K., Mikoss and Riedl, 1933; 18. Carson, D. F. and Mann, 1965; 19. Boyd-Barrett, 1932; 20. C. F. Boehringer and Söhne, 1908a; 21. Brunner, K., Wiedner and Kling, 1931; 22. Brunner, K. and Moser, 1932; 23. Wahl, 1917.

However, the Brunner synthesis failed when using hydroxyphenylhydrazides, the required hydroxyoxindoles being indirectly prepared using the synthesis via their corresponding methoxy derivatives (Wahl, G., 1917). Nitrooxindoles (Brunner, K., Wiedner and Kling, 1931), halogenooxindoles (Brunner, K., 1897b, 1906; Brunner, K., Wiedner and Kling, 1931; Schwarz, 1903) and oxindole-5-sulphonic acids (Brunner, K., Mikoss and Riedl, 1933) also appeared to be better prepared indirectly by nitration, halogenation and sulphonation, respectively, of their parent oxindoles. However, **50** [R^1 = 4-Cl, R^2 = H, $R^3 + R^4 = (CH_2)_5$] afforded **49** [R^1 = 5-Cl, R^2 = H, $R^3 + R^4 = (CH_2)_5$] upon heating with lime although the yield of product was only 25% and considerable difficulties were encountered in the reaction 'work up' (Moore, R. F. and Plant, 1951).

Although the yields of oxindoles obtained from phenylhydrazides were usually higher than those of the corresponding 1-methyloxindoles obtained from N_α-methylphenylhydrazides, it was changes in the degree of 2-alkylation on the acyl moiety which had the more marked effect upon product yields. Thus, although oxindole and 1-methyloxindole were obtained in only poor yields (approximately equal) (Brunner, 1896b), 3-methyloxindole (Brunner, 1897b), 1,3-dimethyloxindole (Brunner, 1896b), 3,3-dimethyloxindole (Brunner, 1897a) and 1,3,3-trimethyloxindole (Brunner, 1896b) were obtained in yields of 78%, 56%, 63% and 67%, respectively. It appears from these results and those listed in Table 34 that the Brunner synthesis is best adopted for the synthesis of 3-alkyl- and 3,3-dialkyloxindoles.

The Brunner synthesis has been applied to the conversions of **50** (R^1–R^3 = H, R^4 = CH_3 and C_6H_5) into 3-methyloxindole (Endler and Becker, E. I., 1963) and 3-phenyloxindole (Kornet, Ong, T. H. and Thio, P. A., 1971), respectively, using calcium hydride without a solvent and in N,N-dimethylbenzamide, respectively. However, using dimethylformamide as solvent, 3-pyrazolin-5-ones **1616** were exclusively produced when similar hybrazides were treated with calcium hydride. It is suggested in these cases that C or N formylation of the hydrazide **50** (R^2 = R^3 = H) occurred to give intermediates **1614** or **1615**,

50 (R^2 = R^3 = H) ⟶ **1614** or **1615** ⟶ **1616**

respectively, which then cyclized accordingly (Kornet, Ong, T. H. and Thio, P. A., 1971), either by nucleophilic attack of the N_α p-electron pair on the formyl carbonyl group in **1614** or by an intramolecular Claisen-type condensation in **1615** (Kornet, Ong, T. H. and Thio, P. A., 1971).

The interesting 'Kost' (Stradyn', 1979) reaction, which bears a formal analogy to the Brunner oxindole synthesis, occurred when acyl arylhydrazides **1617** (R^5 = H) were reacted with phosphorus oxychloride in refluxing absolute ether to afford 2-aminoindole hydrochlorides **1619** (R^5 = H)$\rightleftharpoons$**1620** which are formed by the postulated mechanism shown in Scheme 52. This mechanism

R[5] HC—R[4] R[1] N—N C O R[2] R[3]
1617

R[5] HC—R[4] R[1] N—N ⊕C $OPOCl_2$ R[2] R[3] Cl⊖

R[5] C—R[4] R[1] N—N C $OPOCl_2$ R[2] R[3]

R[5] HC—R[4] R[1] N—N⊕=C $OPOCl_2$ R[2] R[3] Cl⊖

(i) [3,3]-Sigmatropic shift
(ii) Ring closure

R[1] R[5] R[4] O PCl_2 N R[2] N⊕ H O R[3] H
1618

R[1] R[5] R[4] N ⊕NH R[2] R[3] Cl⊖ ⟷ R[1] R[5] R[4] ⊕N NH Cl⊖ R[2] R[3]
1619

(R^5 = H)

R[1] R[4] Cl⊖ N ⊕NH_2 R[2] R[3]
1620

Scheme 52

parallels that of the Fischer indolization except that in the intermediate **1618** the elimination of the more polar PO_2Cl_2 group, rather than the amino group, occurs (Portnov, Golubeva and Kost, 1972a). In addition to phosphorus oxychloride, phosphorus trichloride or phosphorus tribromide could be used to effect the reaction, the order of reactivity in this respect being phosphorus oxychloride < phosphorus trichloride < phosphorus tribromide. With phosphorus trichloride, the reaction was complete after 1–2 h at room temperature whereas with phosphorus tribromide the reaction proceeded exothermically a few minutes after mixing the reagents in absolute ether (Portnov, Golubeva and Kost, 1972a, 1972b). 2-Aminoindole hydrochlorides **1619** (R^5 = H) ⇌ **1620** which have been prepared by this route by Kost and his co-workers are given in Table 35. In a more recent related study (Kost, Zabrodnyaya, Portnov and Voronin, 1980), compounds **1617** [$R^1 = R^3 = R^4$ = H, $R^2 = CH_3$ and C_6H_5, $R^5 = (CH_2)_{1-5}$-phthalimido] were converted into **2** [$R^1 = CH_3$ and C_6H_5, $R^2 = NH_2$, $R^3 = (CH_2)_{1-5}$-phthalimido, respectively] by treatment with phosphorus oxychloride, subsequent dephthalylation of **2** [$R^1 = CH_3$, R^2 = H, $R^3 = (CH_2)_{2-5}$-phthalimido], by treatment with hydrochloric acid, giving **2**[$R^1 = CH_3$, $R^2 = NH_2$, $R^3 = (CH_2)_{2-5}NH_2$].

Table 35. Compounds **1619** (R^5 = H) ⇌ **1620** which have been synthesized by Kost and his co-workers from **1617** (R^5 = H)

Compound **1619** (R^5 = H) ⇌ **1620** (Unless otherwise stated, R^1–R^4 = H)

R^1	$R^{2\,a}$	$R^{3\,a}$	R^4	References
			C_6H_5	1
		CH_3	CH_3	2
		CH_3	C_6H_5	1
	CH_3		CH_3	1, 2
	CH_3		C_2H_5	1
	CH_3		C_6H_5	1, 2
	CH_3	CH_3		1
	C_6H_5		CH_3	1
	$CH_2C_6H_5$		CH_3	1
	$CH_2C_6H_5$		C_2H_5	1, 2
		$(CH_2)_3$	CH_3	2, 3
	$CH_2CH(CH_3)CH_2$		CH_3	2–4
	$CH_2CH(CH_3)CH_2$		C_7H_5	2, 3
Br	CH_3		CH_3	1, 2
Br	$CH_2CH(CH_3)CH_2$		CH_3	2, 3

Note

a. Arylhydrazides have been converted into N_α-alkylarylhydrazides upon treatment with an equimolar quantity of sodium ethoxide and an excess of alkyl halide and into N_α, N_β-dialkylarylhydrazides when similarly treated with a 2 × molar excess of sodium ethoxide (Kost, Ostrovskii, Goluebea and Tikhonov, 1974).

References

1. Golubeva, Portnov and Kost, 1973; 2. Portnov, Golubeva and Kost, 1972a; 3. Kost, Golubeva and Portnov, 1971; 4. Kost, Golubeva, Zabrodnyaya and Portnov, 1975.

In many cases when acetyl arylhydrazides were reacted with phosphorus oxychloride in refluxing absolute ether, considerable resinification occurred and the required products could not be isolated. However, this could be overcome by heating these hydrazides with phosphorus oxychloride in absolute methylene chloride at 90 °C in a sealed ampoule under argon. Under these conditions, **1617** ($R^1 = R^3$–$R^5 = H$, $R^2 = CH_3$) afforded **1619** ($R^5 = H$) $\rightleftharpoons$ **1620** ($R^1 = R^3 = R^4 = H$, $R^2 = CH_3$), a reaction which could also be achieved, without significant change in product yield, using phosphorus trichloride in place of phosphorus oxychloride under similar conditions, and **1617** ($R^1 = R^2 = R^4 = R^5 = H$, $R^3 = CH_3$) afforded **1619** ($R^5 = H$) $\rightleftharpoons$ **1620** ($R^1 = R^2 = R^4 = H$, $R^3 = CH_3$) (Portnov, Golubeva, Kost and Volkov, 1973). However, using phosphorus oxychloride in refluxing absolute ether, **1617** [$R^1 = R^4 = R^5 = H$, $R^2 = R^3 = CH_3$; $R^1 = R^4 = R^5 = H$, $R^2 + R^3 = (CH_2)_3$ and $CH_2CH(CH_3)CH_2$ and $R^1 = Br$, $R^2 + R^3 = CH_2CH(CH_3)CH_2$, $R^4 = R^5 = H$], in all of which both N_α and N_β were alkylated, gave the corresponding **1619** ($R^5 = H$) $\rightleftharpoons$ **1620** [$R^1 = R^4 = H$, $R^2 = R^3 = CH_3$; $R^1 = R^4 = H$, $R^2 + R^3 = (CH_2)_3$ and $CH_2CH(CH_3)CH_2$; $R^1 = Br$, $R^2 + R^3 = CH_2CH(CH_3)CH_2$, $R^4 = H$, respectively] (Portnov, Golubeva, Kost and Volkov, 1973). The melting point (decomposition above 240 °C) for the product **1619** ($R^5 = H$) $\rightleftharpoons$ **1620** ($R^1 = R^4 = H$, $R^2 = R^3 = CH_3$) in this report differed appreciably from that recorded (m.p. = 278–280 °C) for the same product obtained by a similar reaction in a related investigation (Golubeva, Portnov and Kost, 1973). All attempts to similarly react N_β-acetylphenylhydrazine (**1617**; R^1–$R^5 = H$) were unsuccessful, a result reminiscent of the many early failures to indolize acetaldehyde phenylhydrazone (Portnov, Golubeva, Kost and Volkov, 1973) (see Chapter IV, A).

Attempts to liberate the free bases from the salts **1619** ($R^5 = H$) $\rightleftharpoons$ **1620** resulted in the formation of the corresponding **1621** ($R^5 = OH$), formed by aerial oxidation (Golubeva, Portnov and Kost, 1973; Kost, Golubeva, Zabrodnyaya and Portnov, 1975, Kost, Portnov and Glubeva, 1972; Portnov, Golubeva and Kost, 1972a). Reduction of **1621** [$R^2 + R^3 = CH_2CH(CH_3)CH_2$, $R^5 = OH$] with a sodium–potassium alloy in butyl alcohol gave the corresponding **1622** (Portnov, Golubeva and Kost, 1972a), a ring opening reminiscent of the formation of **339** from the intermediate **338**. Treatment with base of the salts **1619** ($R^4 = R^5 = H$) $\rightleftharpoons$ **1620** ($R^4 = H$) furnished the corresponding **1621** ($R^4 + R^5 = O$), again formed by aerial oxidation (Portnov, Golubeva, Kost and Volkov, 1973). However, the former type of oxidation was not observed during the later isolation of **1621** ($R^1 = R^3 = R^4 = H$, $R^2 = CH_3$, $R^5 = C_6H_5$) after treatment of **1619** ($R^1 = R^3 = R^4 = H$, $R^2 = CH_3$, $R^5 = C_6H_5$) with 1 M sodium hydroxide solution, even though oxygen was not specifically excluded during this procedure (Kost, Golubeva and Popova, 1979). Similarly, **1623** has afforded **1624** (Kost, Ostrovskii and Golubeva, 1973). In extensions of the above cyclization by reacting **1617** (R^4 and $R^5 = H$)

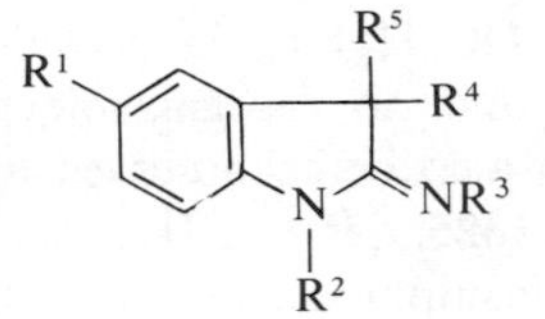

1621

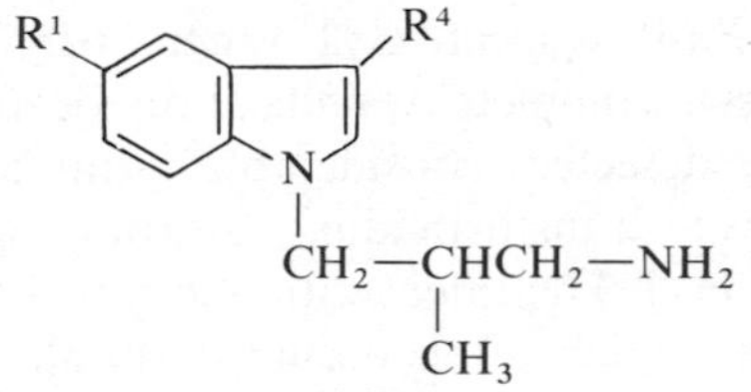

1622

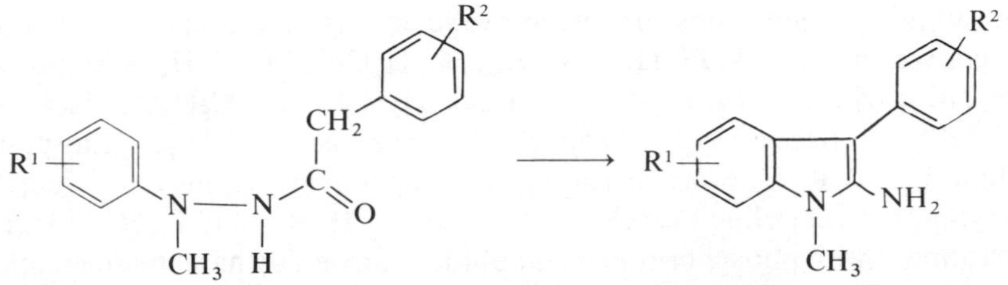

1623 → 1624

with phosphorus oxychloride in refluxing absolute benzene, compounds **1617** [$R^1 = R^3 = H$ and $R^2 = R^4 = R^5 = CH_3$; $R^2 = R^4 = CH_3$, $R^5 = C_6H_5$; $R^2 = CH_2C_6H_5$, $R^4 = R^5 = CH_3$ and $R^2 = CH_3$, $R^4 + R^5 = (CH_2)_5$] were converted into the corresponding **1619** from which the free bases **1621** [$R^1 = R^3 = H$ and $R^2 = R^4 = R^5 = CH_3$; $R^2 = R^4 = CH_3$, $R^5 = C_6H_5$; $R^2 = CH_2C_6H_5$, $R^4 = R^5 = CH_3$ and $R^2 = CH_3$, $R^4 + R^5 = (CH_2)_5$, respectively] were isolated by treatment of the salts with 2 M sodium hydroxide (Kost, Golubeva, Zabrodnyaya and Portnov, 1975).

Concomitantly with the above studies of Kost and his co-workers, another group performed a series of similar studies. Thus, the hydrazides **1617** [$R^1 = R^3 = H$, $R^2 = CH_3$ and substituted CH_3, $R^4 = CH_3$, aminoalkyl, alkoxycarbonylalkyl, chloroalkyl and substituted phenyl, $R^5 = H$, CH_3 and C_6H_5 and $R^4 + R^5 = (CH_2)_4$ and $(CH_2)_5$] were converted into **1621**, or the corresponding indoles (when $R^5 = H$) (Winters and Di Mola, 1975), **1623** ($R^1 = H$, $R^2 = 4\text{-Cl}$) afforded **1624** ($R^1 = H$, $R^2 = 4\text{-Cl}$) by warming its suspension in benzene or toluene in the presence of phosphorus pentachloride and **1623** [$R^1 = H$, $R^2 = 4\text{-}CH_3$, $4\text{-}CH(CH_3)_2$, $4\text{-}OCH_3$ and $3,4\text{-}(OCH_3)_2$] gave **1624** [$R^1 = H$, $R^2 = 4\text{-}CH_3$, $4\text{-}CH(CH_3)_2$, $4\text{-}OCH_3$ and $3,4\text{-}(OCH_3)_2$, respectively] when reacted with refluxing phosphorus oxychloride (Winters, Di Mola, Berti and Arioli, 1979).

In an expansion (Kost, Golubeva and Popova, 1979) of these synthetic studies, compounds **1625** [$R = CH_3$, C_2H_5, $(CH_2)_2CH_3$, C_6H_5, $C_6H_4CH_3(2)$, $C_6H_4OCH_3(4)$ and $C_6H_4NO_2(4)$] were treated with phosphorus oxychloride or phosphorus trichloride which gave both the disulphides **1626** [$R = CH_3$, C_2H_5, $(CH_2)_2CH_3$, C_6H_5, $C_6H_4CH_3(2)$, $C_6H_4OCH_3(4)$ and $C_6H_4NO_2(4)$, respectively] and traces of **1619** [$R^1 = R^3 = R^4 = H$, $R^2 = CH_3$, $R^5 = CH_3$, C_2H_5, $(CH_2)_2CH_3$, C_6H_5, $C_6H_4CH_3(2)$, $C_6H_4OCH_3(4)$ and

$C_6H_4NO_2$(4), respectively]. When **1625** (R = OC_6H_5) was subjected to these conditions, complete resinification occurred and no recognizable products could be detected, considerable resinification also being observed alongside the above transformations starting with **1625** [R = CH_3, C_2H_5 and $(CH_2)_2CH_3$]. Treatment with phosphorus pentasulphide also caused the above reactions, which often occurred simultaneously during the synthesis of the thiohydrazides **1625** by reacting the corresponding hydrazides with phosphorus pentasulphide. Limitations on these above reactions were apparent in that only the conversions of **1625** [R = C_6H_5, $C_6H_4CH_3$(2), $C_6H_4OCH_3$(4) and $C_6H_4NO_2$(4)] into **1626** [R = C_6H_5, $C_6H_4CH_3$(2), $C_6H_4OCH_3$(4) and $C_6H_4NO_2$(4), respectively], using phosphorus oxychloride or phosphorus trichloride and of phenylacetic acid N_α-methylphenylhydrazide into 2-imino-1-methyl-3-phenylindoline **1621** ($R^1 = R^3 = R^4$ = H, $R^2 = CH_3$, $R^5 = C_6H_5$), by treatment with phosphorus pentasulphide followed by base treatment of the crude **1619** ($R^1 = R^3 = R^4$ = H, $R^2 = CH_3$, $R^5 = C_6H_5$) initially formed, could be effected upon a preparative scale. Furthermore, the formation of the disulphides remained unexplained although the formation of an intermediate **1627**, from which in each case both reaction products could be derived, was postulated (Kost, Golubeva and Popova, 1979).

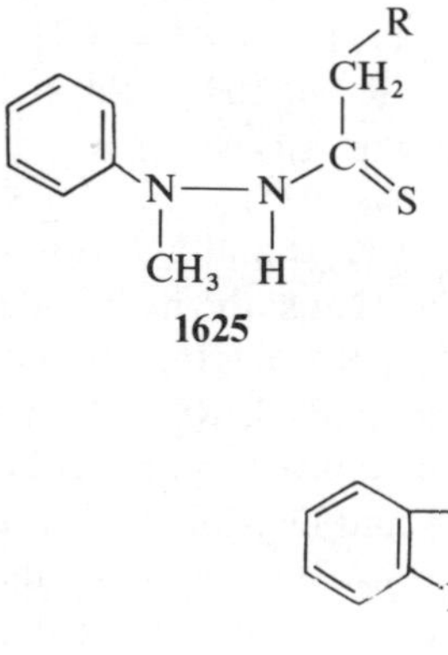

1625 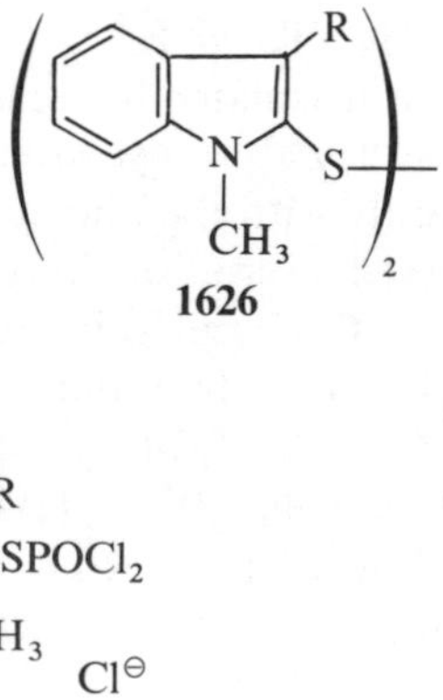 **1626**

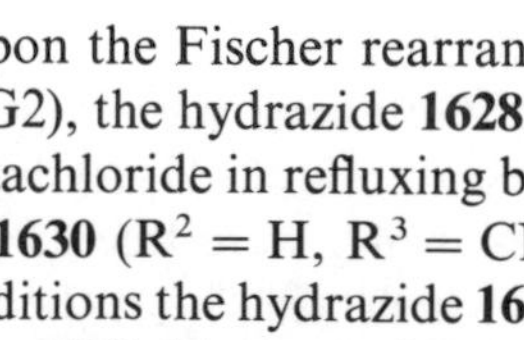

1627

By analogy with studies upon the Fischer rearrangement of 2,6-dialkylarylhydrazones (see Chapter II, G2), the hydrazide **1628** ($R^1 = R^2$ = H) has been treated with phosphorus pentachloride in refluxing benzene to afford a mixture of **1629** ($R^1 = R^2$ = H) and **1630** (R^2 = H, R^3 = CN) (Fusco and Sannicolò, 1976b) and under similar conditions the hydrazide **1628** (R^1 = H, $R^2 = C_6H_5$) gave **1630** ($R^2 = C_6H_5$, R^3 = CN) (Fusco and Sannicolò, 1978d). Compound **1629** ($R^1 = R^2$ = H) was converted into **1630** (R^2 = H, R^3 = COOH) by treatment with concentrated hydrobromic acid which suggested its inter-

1628

$-HPO_2Cl_2$

1629

(R^1 = H)
1,3-Shift

1630

Scheme 53

mediacy in the formation of **1630** (R^2 = H, R^3 = CN) from **1628** (R^1 = R^2 = H) and led to the postulation of the following mechanistic pathway (Scheme 53) which in its initial stages resembled Scheme 52 (Fusco and Sannicolò, 1976b). However, the possibility that the reaction proceeded through the alternative cyclohexadieneoneimine intermediate **1631** which could undergo a similar 1,3-shift to produce **1630** (R^2 = H, R^3 = CN) could not be excluded (Fusco and Sannicolò, 1978d). An analogy has also been drawn (Fusco and Sannicolò, 1976b) between the final stage of Scheme 53 and the earlier reported [Miller, B., 1974 – this reference is incorrectly quoted in the literature (Fusco and Sannicolò, 1976b)] 1,3-shift of a benzyl group during the acid catalysed rearrangement of **1632** which afforded, along with other products, **1633**. By contrast with **1628** (R^1 = R^2 = H), compound **1634** (R = H) did not react with phosphorus pentachloride, even on heating the mixture to 130 °C without a solvent. However, **1634** (R = CH_3) under similar conditions afforded **1635** as the only isolated product, this probably being produced via intermediates analogous to those given in Scheme 53 (Fusco and Sannicolò, 1976b). In a subsequent extension (Fusco and Sannicolò, 1978d) of these studies, **1628** ($R^1 = CH_3$, $R^2 = H$) was treated with phosphorus pentachloride in refluxing dry benzene to afford **1629** ($R^1 = CH_3$, $R^2 = H$) and **1636–1640**. Compound **1638** was also apparently (Fusco and Sannicolò, 1978d) formed as a third product in the decomposition of **1628** ($R^1 = R^2 = H$) described above, although it was not referred to in the original publication (Fusco and Sannicolò, 1976b). The five products **1636–1640** were all apparently formed from **1629** ($R^1 = CH_3$, $R^2 = H$) as was evidenced by the treatment of this compound in refluxing 20% hydrochloric acid to give **1636**, **1637**, **1638**, phenylacetic acid and benzaldehyde, these last two compounds being further transformation products of **1639** and

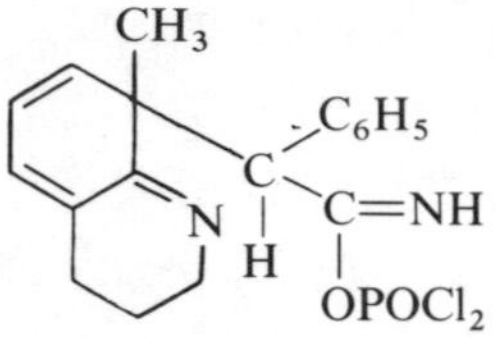

1631 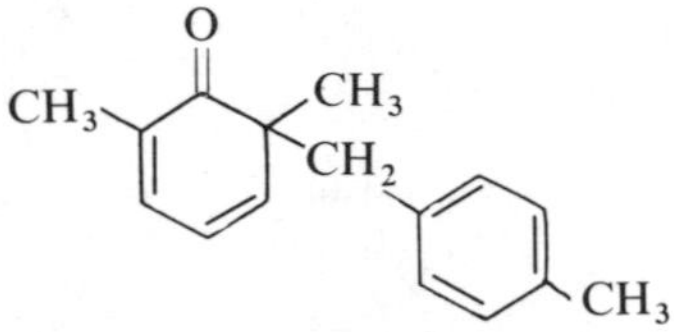 **1632** 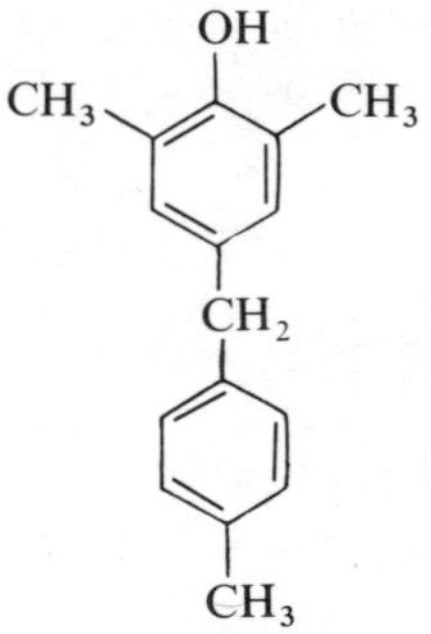 **1633**

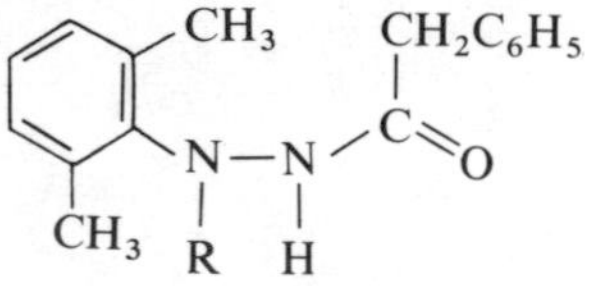

1634 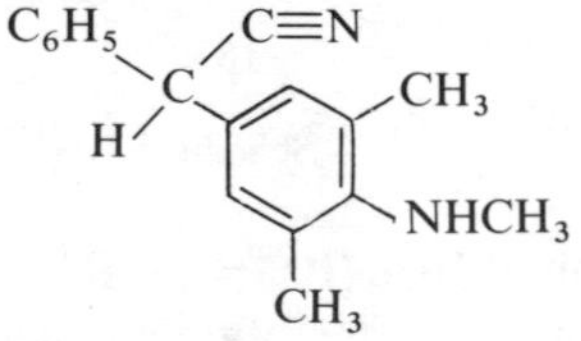 **1635** 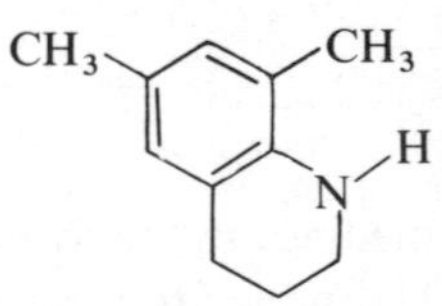 **1636**

$C_6H_5CH_2C{\equiv}N$

1639

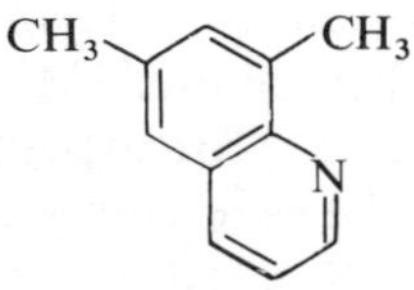

1637 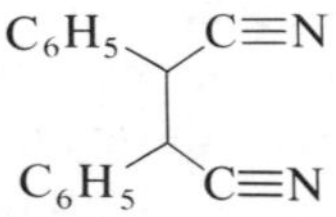 **1638**

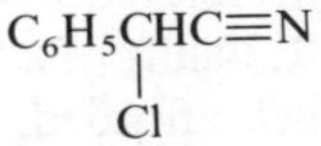

$C_6H_5CH(Cl)C{\equiv}N$

1640

1640, respectively. Evidently, if phenylacetonitrile was eliminated from **1629** ($R^1 = CH_3$, $R^2 = H$), the resulting dihydroquinoline underwent disproportionation to afford **1636** and **1637** and the product **1638** possibly resulted from a homolysis of **1629** ($R^1 = CH_3$, $R^2 = H$) (Fusco and Sannicolò, 1978d). These reactions showed completely different courses to those observed in the analogous arylhydrazone rearrangements (see Chapter II, G2). The above studies of Fusco and Sannicolò have been briefly reviewed (Fusco and Sannicolò, 1980).

Thiooxindoles have also been prepared by the Fischer indole synthesis. Thus, when **1579** ($R^1 = R^3 = H$, $R^2 = CH_3$, $R^4 = CH_2C_6H_5$) was treated with methanolic hydrochloric acid, a product formulated as **1580** was obtained along with a small amount of indolic material which probably resulted from the methylation of **1580**. However, similar treatment of **1579** (R^1–$R^3 = H$, $R^4 = CH_2C_6H_5$) furnished mainly phenylhydrazine hydrochloride along with a trace of yellow oil which was probably the corresponding indole (Wuyts and Lacourt, 1935). Indeed, indoles **1581** [$R = C_6H_5$ and $C_6H_4CH_3(3)$] were produced when structures **1579** [$R^1 = H$, $R^2 = R^3 = CH_3$, $R^4 = CH_2C_6H_5$ and $CH_2C_6H_4CH_3(3)$, respectively] were treated with methanolic hydrochloric acid at either elevated or room temperature, respectively. However, similar treatment of **1579** [$R^1 = H$, $R^2 = R^3 = CH_3$, $R^4 = CH(CH_2)_5$] was only observed to cause hydrolysis of the hydrazone group, as it did with **1579** [$R^1 = R^3 = CH_3$, $R^2 = H$, $R^4 = C_6H_4CH_3(2)$] (Wuyts and Lacourt, 1935).

B. The Hugershoff Synthesis of 2-Aminobenzothiazoles

The hydrochloric acid catalysed cyclization of a 1-phenylthiosemicarbazide was first effected by Fischer and one of his co-workers (Fischer, Emil and Besthorn, 1882) by heating the parent compound, **1641** (R^1–$R^6 = H$), in 20% hydrochloric acid at 125–130 °C in a sealed tube for 12 h to afford a product which was formulated as **1642**. Other later studies (Harries, C. D. and Loewenstein, 1894) upon the reaction of **1641** ($R^1 = R^2 = R^4 = H$, $R^3 = R^5 = CH_3$, $R^6 = C_6H_5$) in alcoholic hydrochloric acid afforded N-methylaniline and a product which was given structure **1643** ($R = CH_3$). At the same time, the product obtained earlier from **1641** (R^1–$R^6 = H$) was reformulated as **1643** ($R = H$). Subsequently, the analogy between these reactions and the Fischer indole synthesis was recognized and the above two

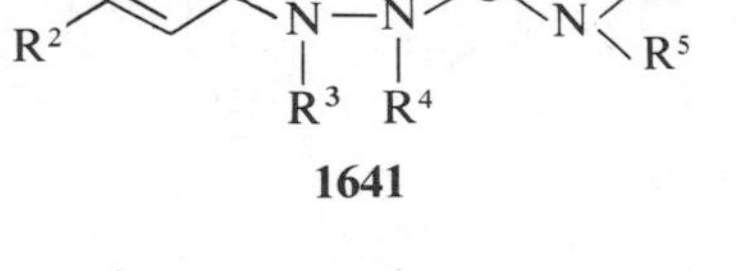

R1 R2 N—N C S N R6 R5 R3 R4

1641

C_6H_5—N—N—H C S

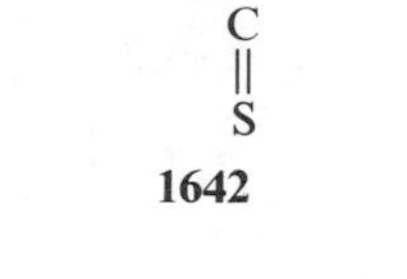

1642

products were consequently reformulated as **1644** (R^1 = H and CH_3, respectively, R^2 = H) by Hugershoff (1903). This synthesis subsequently found a wide application in the preparation of 2-aminobenzothiazoles (Sprague and Land, 1957) and Hugershoff's pioneering studies which clarified the reaction have been recognized by the fact that the reaction now carries his name.

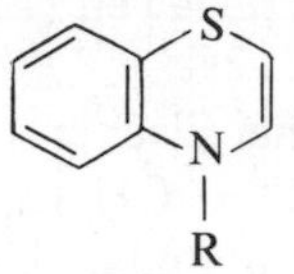

S N R

1643

S N NR^2 R^1

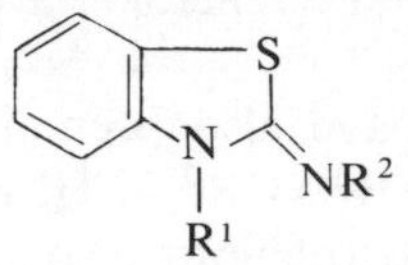

1644

The mechanistic analogy with the Fischer indole synthesis has been supported by the observation that, starting from $^{15}N_{\beta}$-phenylhydrazine hydrochloride, the parent cyclization in the Hugershoff synthesis, effected by heating **1645** with with 50% hydrochloric acid, led to an equal division of the ^{15}N label between the resulting 2-aminobenzothiazole and ammonia. These results were consistent with the formation of the tautomeric intermediate diamine **1646** in the mechanism shown in Scheme 54 (Clusius and Weisser, 1952). Further support for the

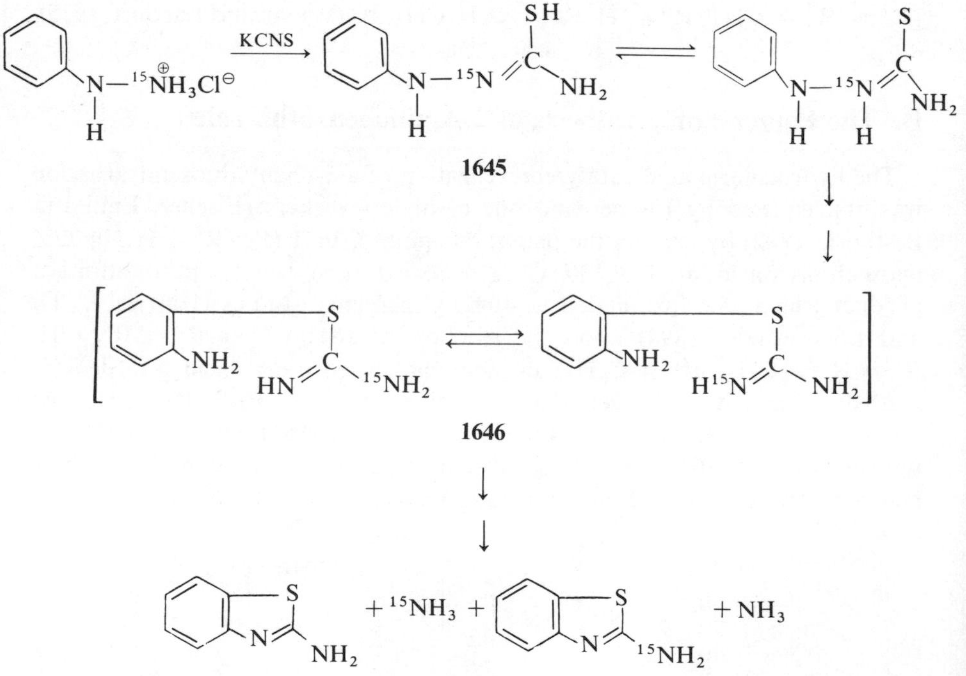

$N-{}^{15}\overset{\oplus}{N}H_3Cl^{\ominus}$ H $\xrightarrow{KCNS}$ $N-{}^{15}N=C(SH)NH_2$ H $\rightleftharpoons$ $N-{}^{15}N-C(=S)NH_2$ H H

1645

[NH_2 S C HN ${}^{15}NH_2$ $\rightleftharpoons$ NH_2 S C $H^{15}N$ NH_2]

1646

S N NH_2 $+ {}^{15}NH_3 +$ S N ${}^{15}NH_2$ $+ NH_3$

Scheme 54

intermediacy of species **1646** arose from the observations that **1647** (R = C_6H_5, $C_6H_4OCH_3(2)$ and $C_6H_4Br(4)$], prepared by reaction of 2-aminothiophenol with diphenyl-, di-2-methoxyphenyl- and di-4-bromophenylcarbodiimide, respectively, gave **1648** [R^1–R^3 = H, R^4 = C_6H_5, $C_6H_4OCH_3(2)$ and $C_6H_4Br(4)$, respectively] when refluxed in ethanolic concentrated hydrochloric acid and **1647** (R = C_6H_5) again furnished **1648** (R^1–R^3 = H, R^4 = C_6H_5) when refluxed in glacial acetic acid or when pyrolysed just above its m.p. at 165–170 °C (Kurzer and Sanderson, 1962).

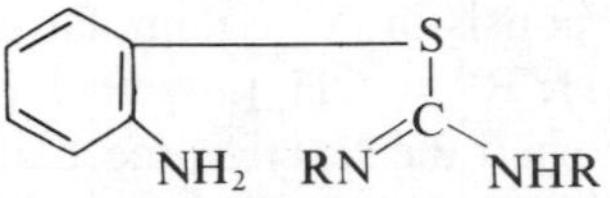

1647 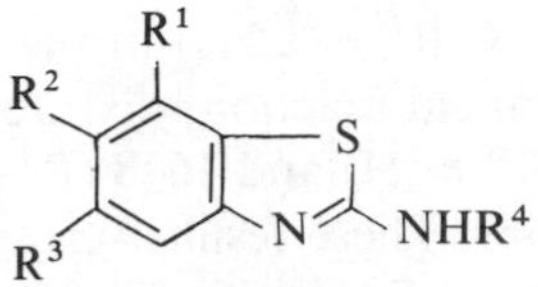 **1648**

Although 4-methylbenzene sulphonic acid in heated benzene or toluene had no effect upon arylthiosemicarbazides and both dilute and concentrated acetic or sulphuric acid yielded only cleavage products, the use of polyphosphoric acid was found to be preferable to hydrochloric acid in catalysing the reaction of **1641** (R^1–R^6 = H) into **1648** (R^1–R^4 = H) and was also found to catalyse the conversion of **1641** (R^1 = CH_3 and Cl, R^2–R^6 = H) into **1648** (R^1 = R^3 = R^4 = H, R^2 = CH_3) (65% yield) and **1648** (R^1 = R^3 = R^4 = H, R^2 = Cl) (70% yield), respectively. The similar conversion of **1641** (R^1 = OCH_3, R^2–R^6 = H) into **1648** (R^1 = R^3 = R^4 = H, R^2 = OCH_3) occurred in only 14% yield becuase of the occurrence of an alternative reaction involving N—N bond cleavage leading to the formation of a large amount of 4-methoxyaniline (the corresponding anilines have often been isolated as by-products in Fischer indolizations – see Chapter VII, J). The conversion of **1641** (R^1 = NO_2, R^2–R^6 = H) into **1648** (R^1 = R^3 = R^4 = H, R^2 = NO_2) could not be effected, the only products from such an attempted reaction being 4-nitrophenylhydrazine and 4-nitroaniline (Kost, Lebedenko and Sviridova, 1976).

Both **1648** (R^1 = Cl, R^2–R^4 = H) and **1648** (R^1 = R^2 = R^4 = H, R^3 = Cl), in a yield ratio of 1 : 1, were formed when **1641** (R^1 = R^3–R^6 = H, R^2 = Cl) was heated with polyphosphoric acid and similarly both **1648** (R^1 = CH_3, R^2–R^4 = H) and **1648** (R^1 = R^2 = R^4 = H, R^3 = CH_3), but in a yield ratio of 2 : 5, respectively, were formed from **1641** (R^1 = R^3–R^6 = H, R^2 = CH_3) whereas only **1648** (R^1 = R^2 = R^4 = H, R^3 = OCH_3) was isolated after cyclization of **1641** (R^1 = R^3–R^6 = H, R^2 = OCH_3) (Kost, Lebedenko, Sviridova and Torocheshnikov, 1978).

Contrary to the equivalent observations that, in the Fischer indolization, N_α-methylation of the arylhydrazone facilitated the reaction (Chastrette, 1970); Chalmers, A. J. and Lions, 1933; Diels and Köllisch, 1911; Fischer, Emil, 1886b; Harvey, D. J. and Reid, S. T., 1972; Ishii, Murakami, Y., Hosaya, Takeda, Suzuki, Y. and Ikeda, 1973; Kermack, Perkin and Robinson, R., 1921;

Kermack and Slater, 1928; Mann and Wilkinson, 1957; Padfield and Tomlinson, M. L., 1950; Perkin and Plant, 1923b; Robinson, R. and Thornley, 1924; Woolley, D. W. and Shaw, E., 1955 – see also Michaelis, 1897), the product yield was diminished by comparison with that obtained in the synthesis of the parent compound when **1641** ($R^1 = R^2 = R^4$–R^6 = H, $R^3 = CH_3$) was converted into **1644** ($R^1 = CH_3$, R^2 = H) using a polyphosphoric acid catalyst (Kost, Lebedenko, Sviridova and Torocheshnikov, 1978).

The reaction of either **1641** (R^1–$R^3 = R^5 = R^6$ = H, $R^4 = CH_3$) or **1641** (R^1–R^5 = H, $R^6 = CH_3$) in the presence of a polyphosphoric acid catalyst yielded identical reaction mixtures, each consisting of equimolar amounts of **1648** (R^1–R^4 = H) and **1648** (R^1–R^3 = H, $R^4 = CH_3$) together with the disulphide **1649**. These results further supported the reaction mechanism postulated in Scheme 54, in this particular case the tautomeric intermediate diamine from both reactants being **1650**. This could cyclize into **1651**, loss of ammonia

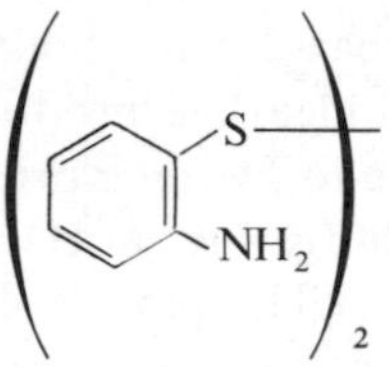

S
NH2
2
1649

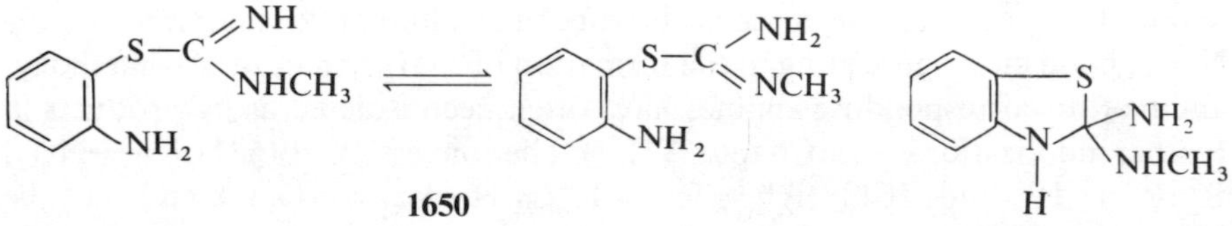

NH
S—C
NHCH3
NH2
NH2
S—C
NCH3
NH2
1650

S
NH2
N
NHCH3
H
1651

or methylamine from which gave the benzothiazoles **1648** (R^1–R^3 = H, $R^4 = CH_3$) and **1648** (R^1–R^4 = H), respectively. The disulphide **1649** was probably formed during the strong alkalization of the reaction mixture during 'work up' as a result of hydrolysis of uncyclized **1650** followed by oxidation of the 2-aminothiophenol thus produced. Likewise, **1641** (R^1 = H, CH_3 and Cl, R^2–R^5 = H, $R^6 = C_6H_5$) gave both **1648** ($R^1 = R^3 = R^4$ = H, R^2 = H, CH_3 and Cl, respectively) and **1648** ($R^1 = R^3$ = H, R^2 = H, CH_3 and Cl, respectively, $R^4 = C_6H_5$) (Kost, Lebedenko, Sviridova and Torocheshnikov, 1978). When **1641** ($R^1 = R^2 = R^4 = R^5$ = H, $R^3 = CH_3$ and $R^6 = C_6H_5$) was treated with polyphosphoric acid at 120 °C, p.m.r. spectral investigation of the reaction mixture indicated the formation of both **1644** ($R^1 = CH_3$, $R^2 = C_6H_5$) and **1644** ($R^1 = CH_3$, R^2 = H) in a yield ratio of 5 : 1, respectively (Kost, Lebedenko, Sviridova and Torocheshnikov, 1978).

Attempts to vary the catalyst using **1641** (R^1–R^5 = H, R^6 = C_6H_5) as the reactant led to some interesting observations. Treatment with an acidic ion exchange resin brought about no change, heating in glacial acetic acid or trifluoroacetic acid afforded complex mixtures, using 85% formic acid resulted in the formation of the mesoionic compound **1652**, heating with polyphosphoric acid at 120 °C, 180 °C or 205 °C gave in each case mixtures of **1648** (R^1–R^4 = H) and **1648** (R^1–R^3 = H, R^4 = C_6H_5) in the yield ratio of 1 : 1 and pyrolysis above the m.p. without solvent or catalyst yielded **1648** (R^1–R^3 = H, R^4 = C_6H_5) [**1648** (R^1–R^4 = H) decomposed under these thermal conditions] (Kost, Lebedenko, Sviridova and Torocheshnikov, 1978).

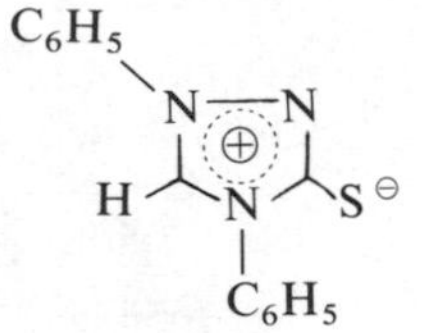

1652

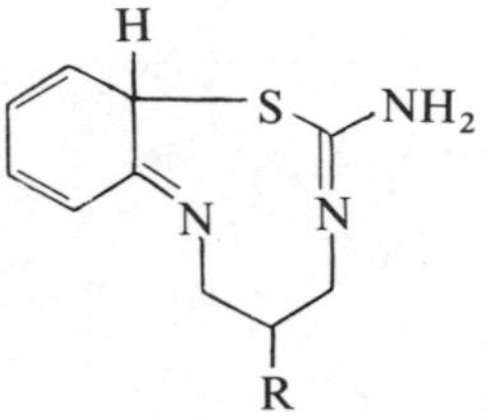

1653

An interesting extension of the Hugershoff synthesis involved the treatment of **1641** [R^1 = R^2 = R^5 = R^6 = H, R^3 + R^4 = $(CH_2)_3$ and $CH_2CH(CH_3)CH_2$] by heating with methanolic hydrogen chloride which yielded **1644** [R^1 + R^2 = $(CH_2)_3$] (63% yield) and **1644** [R = $CH_2CH(CH_3)CH_2$] (55% yield), respectively. The suggested intermediate in these transformations was **1653** (R = H and CH_3, respectively) (Kost, Golubeva and Sviridova, 1973).

C. Synthesis of Benzofurans from O-Aryloximes

This reaction, involving the formation of benzofurans **1660** by the treatment of O-aryloximes **1654** with acid catalysts, was first reported in 1966 (Sheradsky, 1966) and since then has been developed by many investigators into a very useful synthesis for both simple and ring fused benzofurans. It has been proposed (Dupont, 1968; Mooradian, 1967; Mooradian and Dupont, 1967b; Sheradsky, 1966; Sheradsky and Elgavi, 1968; Sheradsky and Salemnick, 1971a) that the mechanism of this reaction is as shown in Scheme 55. This is analogous to the mechanism of the Fischer indolizations but involved a 3,4-oxaza-Cope rearrangement in the formation of **1657** from **1656** instead of a 3,4-diaza-Cope rearrangement which operated in the corresponding step of the indolization (Heimgartner, Hansen and Schmid, H., 1979; Winterfeldt, 1970). Other reactions which appear to involve [3,3]-sigmatropic rearrangements of the C=C—O—N—C=C system have been developed (Coates and Said,

Scheme 55

1977; Makisumi and Takada, 1976), including reactions analogous to the benzidine rearrangement (Sheradsky and Salemnick, 1971b, 1972), and a similar rearrangement of the O=C—O—N—C=C system has also been suggested (House and Richey, 1969).

Deuterium labelling experiments have shown that this 3,4-oxaza-Cope rearrangement occurred in the mechanism of O-aryloxime benzofuranization after the rate determining step (Dupont, 1968) which appeared to be the step involving the formation of **1655** from **1654** (Grandberg and Sorokin, 1973). Furthermore, although intermediates corresponding to **1655** could not be isolated, support for the initial tautomerism in Scheme 55 was apparent from the observation that the rate of the rearrangment of O-aryloximes into benzofurans was directly related to the enolizability of the parent ketones (Dupont, 1968). 'Crossover' experiments have shown that the transformation of O-aryloximes into benzofurans was of an intramolecular nature (Dupont, 1968).

The mechanism as shown in Scheme 55 has received further support by the isolation of intermediates or of products resulting from the arrest of the reaction at an intermediate stage in the mechanistic pathway. Several of these relate to intermediate **1658**. Thus, when **1661** was treated with hydrogen chloride in glacial acetic acid at 25 °C, compound **1662** was isolated in good yield. This, upon heating with the same catalyst, afforded 5-cyano-2-methylbenzofuran (**1663**) which was also formed directly by heating **1661** in a hydrogen chloride–glacial acetic acid solution. Arresting this latter reaction after all the starting material had been consumed also led to the isolated of compound **1662** as the major product (Mooradian and Dupont, 1967b). Treatment of **1664** with refluxing 4% ethanolic hydrogen chloride for 5 min afforded **1665**. Under similar conditions for 30–60 min, either **1664** or **1665** gave **1666**, corresponding to intermediate **1659** in Scheme 55. Treatment of **1664**, **1665** or **1666** in refluxing 10% hydrogen chloride in acetic acid effected elimination of the elements of ammonia to yield **1667** (Sheradsky and Elgavi, 1968). Prolonged heating of **1668** (R = H, X = SO_2) in ethanolic hydrogen chloride gave **1669**, formed by hydrolysis of the imino group in the intermediate corresponding to **1658**. However, the product **1669** could not be converted into the corresponding benzofuran under a variety of acidic conditions, probably because of steric strain in the potential product, and it is possibly for this reason that it was isolated (Aksanova, Sharkova, L. M., Kucherova and Zagorevskii, 1973). When compounds **1668** (R = H and NO_2, X = O) were heated in ethanolic hydrogen chloride, compounds **1670** (R = H and NO_2, respectively) were formed (Zagorevskii, Samarina and Sharkova, L. M., 1978). Boiling solutions in dimethylsulphoxide (non-catalytic thermal conditions) of cyclopentane, cyclohexanone, acetophenone, 1-indanone and 1-tetralone O-(2-pyridyl)oxime led to the isolation of low yields of **1671** [$R^1 + R^2 = (CH_2)_3$ and $(CH_2)_4$ and $R^1 = C_6H_5$, R^2 = H] and moderate yields of **1672** (n = 1 and 2), respectively. These compounds resulted from interruption of the benzofuranization mechanism at intermediate **1657** of Scheme 55. This was possible since 2-hydroxypyridines were known to exist predominantly as the 2-pyridones, the known reactivity of which suggested it was unlikely that the hydroxyl group of the minor tautomer would exhibit the necessary nucleophilicity required to effect cyclization. Furthermore the 2-pyridone imidic carbonyl group should be inert to possible nucleophilic attack by the imine group which under reaction 'work up' conditions was hydrolysed (Sheradsky and Salemnick, 1971a). Cyclization of intermediate **1671** [$R^1 + R^2 = (CH_2)_4$] and **1672** (n = 2) into **1673** (n = 2) and **1674** (n = 2), respectively, was achieved in both sulphuric acid and

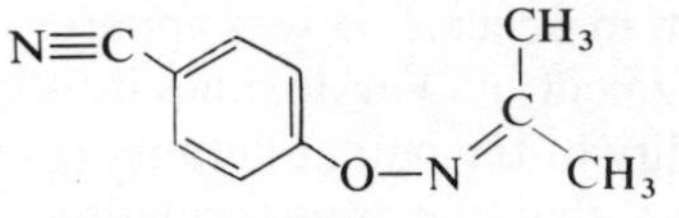

1661 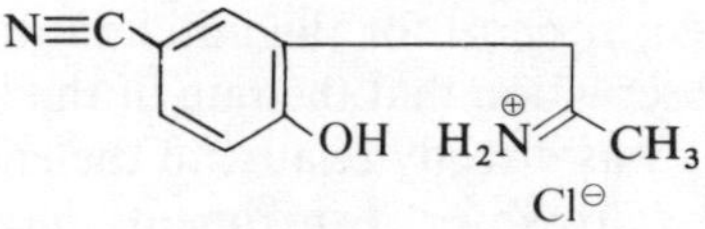 **1662**

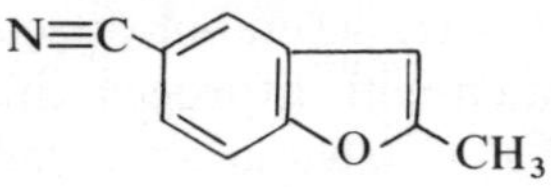

1663 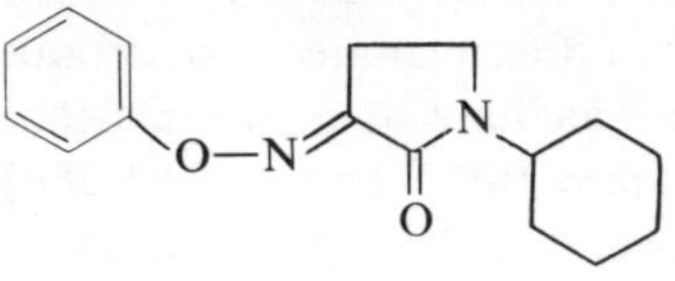 **1664**

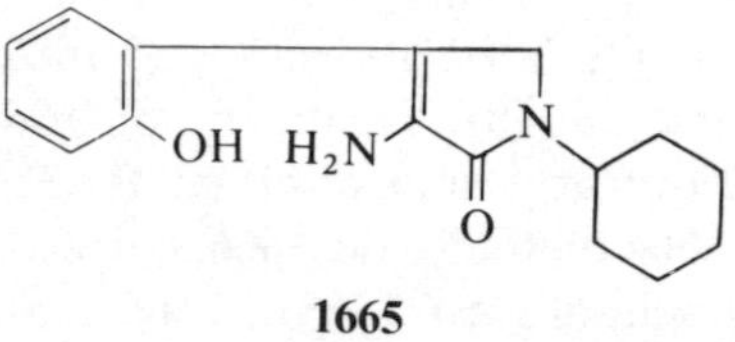

1665 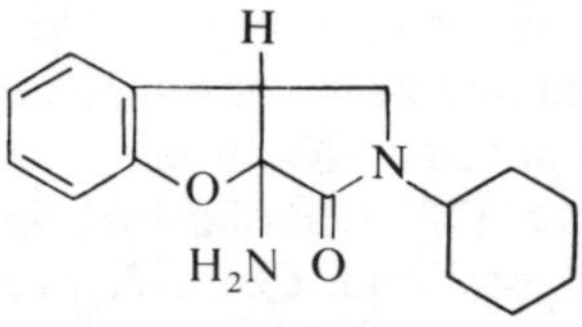 **1666**

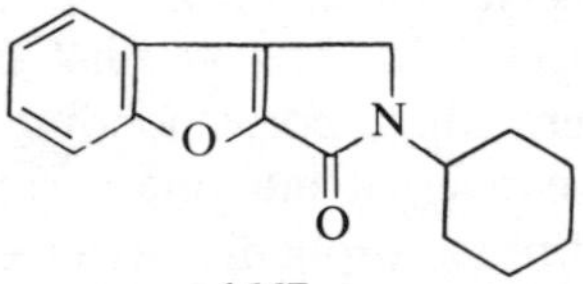

1667 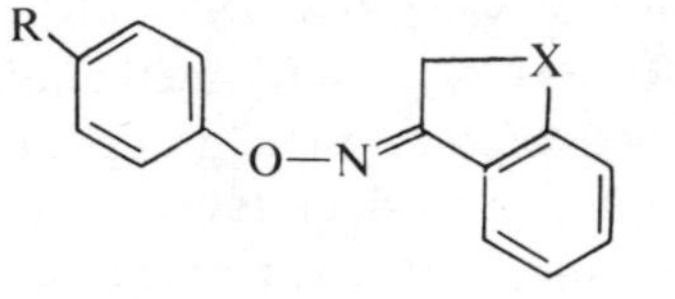 **1668**

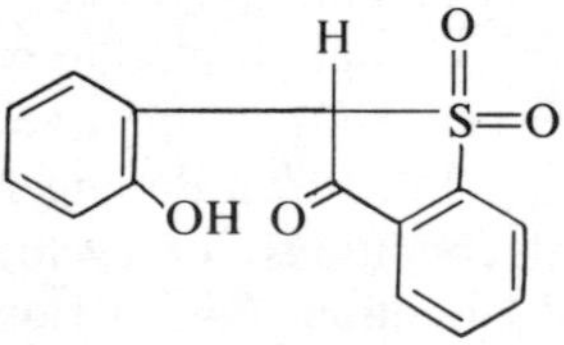

1669 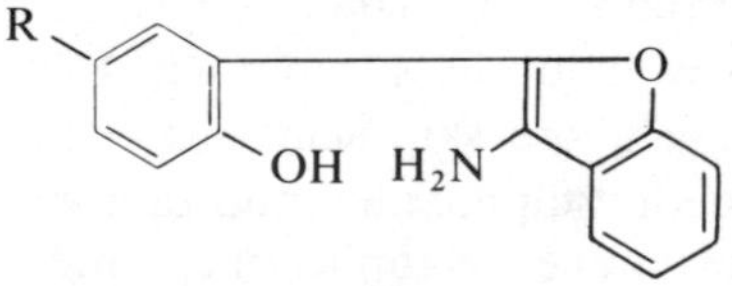 **1670**

polyphosphoric acid but attempts to effect similar cyclizations of compounds **1671** [$R^1 + R^2 = (CH_2)_3$] and **1672** (n = 1) into compounds **1673** (n = 1) and **1674** (n = 1), respectively, failed, probably because of the inaccessibility of the two carbonyl groups in these two five-membered ring ketones (Sheradsky, and Salemnick, 1971a). Reaction of O-phenylhydroxylamine hydrochloride with **1675** ($R = CH_3$) in boiling absolute alcohol, followed by basification using sodium bicarbonate, yielded **1676** (Zagorevskii, Samarina, Sharkova, L. M. and Aksanova, 1979). Reaction of **1677** in refluxing alcoholic hydrogen chloride gave **1678** which in refluxing trifluoroacetic acid yielded **1679** (Samarina, Sharkova, L. M. and Zagorevskii, 1979).

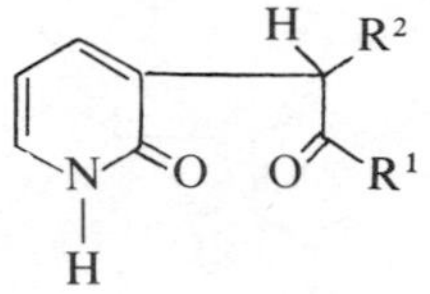

1671

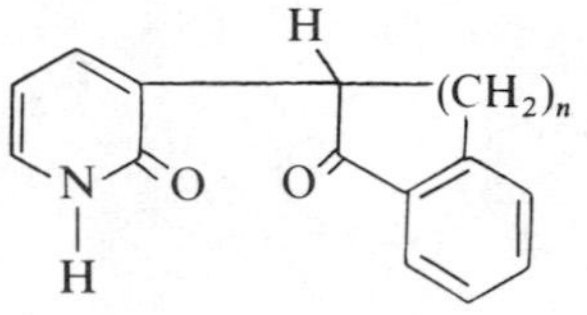

1672

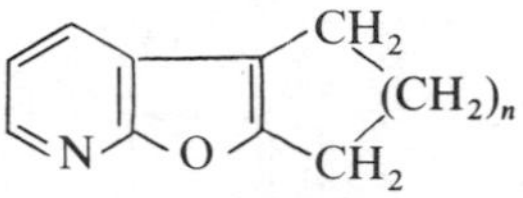

1673

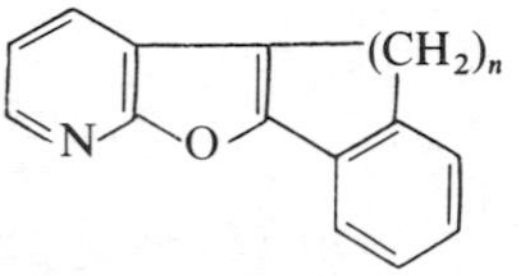

1674

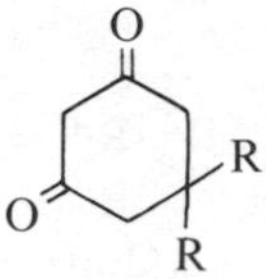

1675

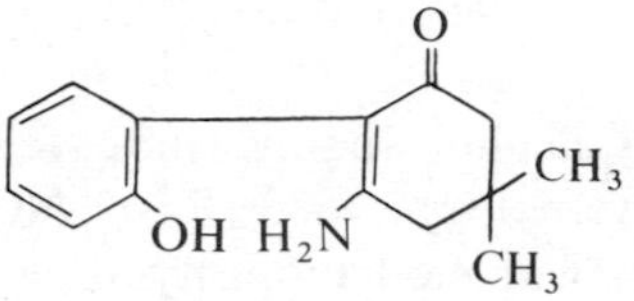

1676

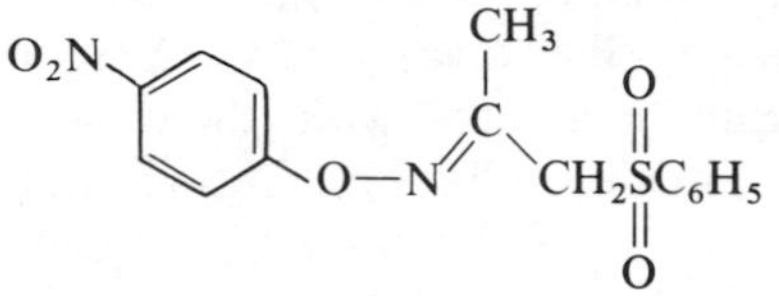

1677

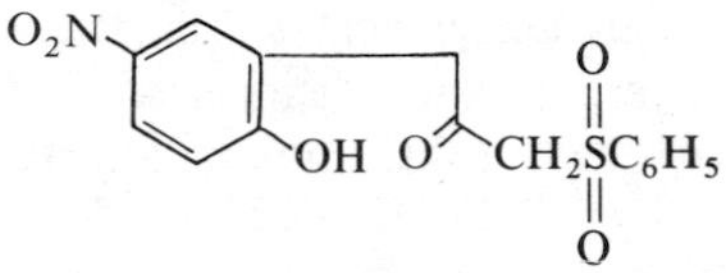

1678

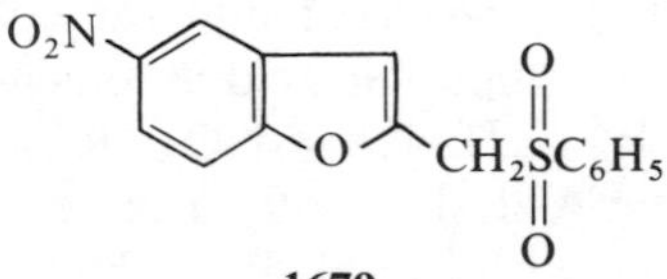

1679

Whereas the oximes **1680** (R^1 = NO_2 and $COOC_2H_5$, R^2–R^4 = R^6 = H, R^5 = CH_3) were converted into the corresponding benzofurans **1681** (R^1 = NO_2 and $COOC_2H_5$, respectively, R^2–R^4 = R^6 = H, R^5 = CH_3) upon treatment with an acetic acid–sulphuric acid mixture, treatment with methanolic hydrogen chloride yielded a mixture of **1682** (R^1 = NO_2 and $COOC_2H_5$, respectively, R^2–R^4 = R^6 = H, R^5 = CH_3) and **1683** (R^1 = NO_2 and $COOC_2H_5$, respectively, R^2–R^4 = R^6 = H, R^5 = R^7 = CH_3). The former products resulted from the hydrolysis of the imine group subsequent to the occurrence of the formation of the new C—C bond and the latter products resulted from ketalization of the **1682** involving the phenolic hydroxyl group

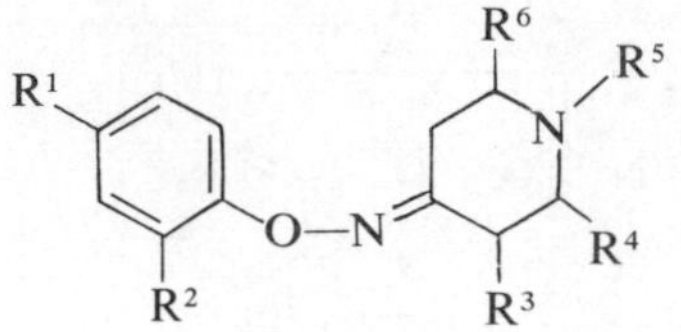

1680 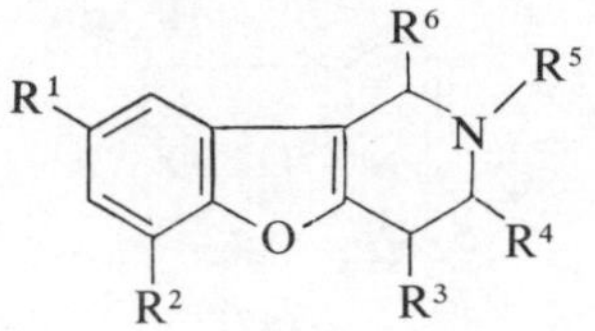 **1681**

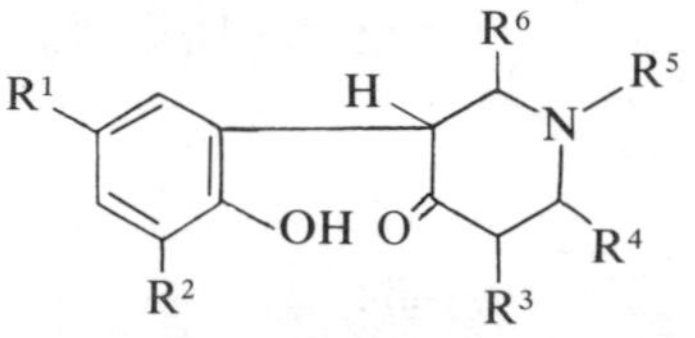

1682 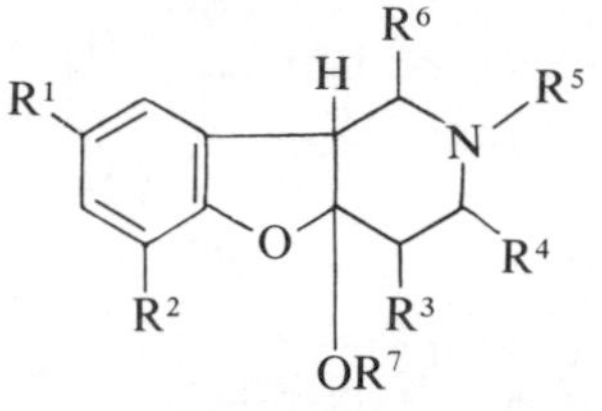 **1683**

and the solvent alcohol. Indeed, **1682** ($R^1 = NO_2$, R^2–$R^4 = R^6 = H$, $R^5 = CH_3$) was converted into **1683** ($R^1 = NO_2$, R^2–$R^4 = R^6 = H$, $R^5 = R^7 = CH_3$) by heating with methanolic hydrogen chloride and both **1682** ($R^1 = NO_2$ and $COOC_2H_5$, R^2–$R^4 = R^6 = H$, $R^5 = CH_3$) and **1683** ($R^1 = NO_2$ and $COOC_2H_5$, R^2–$R^4 = R^6 = H$, $R^5 = R^7 = CH_3$) afforded **1681** ($R^1 = NO_2$ and $COOC_2H_5$, respectively, R^2–$R^4 = R^6 = H$, $R^5 = CH_3$) upon treatment with an acetic acid–sulphuric acid mixture (Sharkova, L. M., Aksanova, Kucherova and Zagorevskii, 1971b). Simultaneously with these studies, another group (Cattanach and Rees, 1971) made similar observations when a series of **1680** ($R^1 = NO_2$, $R^2 = H$ and $R^1 = H$, $R^2 = NO_2$) yielded the corresponding **1683** ($R^7 = CH_3$ and C_2H_5) when treated with hot methanolic and ethanolic hydrogen chloride, respectively. These products could be distilled unchanged but were converted into the corresponding **1681** when heated with 4-methylbenzene sulphonic acid in toluene. Whereas treatment of **1680** ($R^1 = R^3 = R^4 = R^6 = H$, $R^2 = NO_2$, $R^5 = CH_3$ and C_2H_5) with hydrogen chloride in acetic acid at 100 °C furnished products believed to be **1683** ($R^1 = R^3 = R^4 = R^6 = R^7 = H$, $R^2 = NO_2$, $R^5 = CH_3$ and C_2H_5, respectively), treatment of **1680** ($R^1 = R^3 = R^4 = R^6 = H$, $R^2 = NO_2$, $R^5 = CH_3$) with hydrogen chloride in acetic acid at room temperature yielded **1684** ($R^1 = R^4 = H$, $R^2 = NO_2$, $R^3 = CH_3$, $X = NH$) which was readily hydrolysed to form **1684** ($R^1 = R^4 = H$, $R^2 = NO_2$, $R^3 = CH_3$, $X = O$), both these products giving **1683** ($R^1 = R^3 = R^4 = R^6 = H$, $R^2 = NO_2$, $R^5 = CH_3$, $R^7 = C_2H_5$) when treated with ethanolic hydrogen chloride. In a similar manner structures **1684** ($R^1 = H$, $R^2 = NO_2$, $R^3 = R^4 = CH_3$ and $R^3 = CH_2C_6H_5$, $R^4 = CH_3$, $X = O$ and $R^1 = NO_2$, $R^2 = R^4 = H$, $R^3 = CH_3$, $X = O$) were prepared. Treatment of **1680** (R^1–$R^4 = R^6 = H$, $R^5 = CH_3$) with ethanolic hydrogen chloride gave low yields of **1683** (R^1–$R^4 = R^6 = H$, $R^5 = CH_3$, $R^7 = C_2H_5$) and **1681** (R^1–$R^4 = R^6 = H$, $R^5 = CH_3$).

Although the former product, **1683** (R^1–R^4 = R^6 = H, R^5 = CH_3, R^7 = C_2H_5), could be distilled unchanged under reduced pressure, upon pyrolysis at 120 °C at atmospheric pressure it afforded the latter product, **1681** (R^1–R^4 = R^6 = H, R^5 = CH_3), whereas in order to effect the above mentioned similar reactions with the benz-nitro substituted analogues, an acid catalyst was required. Similar observations are made starting with 1,3-dimethylpiperidin-4-one and *cis*-2,6-dimethylpiperidin-4-one O-phenyloximes which afforded **1683** (R^1 = R^2 = R^4 = R^6 = H, R^3 = R^5 = CH_3, R^7 = C_2H_5) (isolated) and **1681** (R^1 = R^2 = R^4 = R^6 = H, R^3 = R^5 = CH_3) (detected by t.l.c.) and a mixture containing **1681** (R^1–R^3 = R^5 = H, R^4 = R^6 = CH_3), respectively. Attempts to convert **1681** into **1683** (R^7 = C_2H_5) by treatment of the former with refluxing ethanolic hydrogen chloride were unsuccessful, showing that in these above reactions, **1681** was not essential precursor of **1683**. Whereas the rearrangement of **1685** afforded **1686**, the isolation of which supported the direct cyclization of **1658** into **1659** as shown in Scheme 55 (Dupont, 1968), the rearrangement of **1687** gave **1688**, the isolation of which suggested hydrolysis of the intermediate **1658** into the corresponding ketone prior to cyclization to ultimately produce **1660** (Dupont, 1968).

R^1 R^4 R^3 N OH X R^2

1684

CH_3 CH—CH_3 O_2N C O—N $CH(CH_3)_2$ → O_2N CH_3 CH_3 O $CH(CH_3)_2$ NH_2

1685 **1686**

O C_6H_5—S O CH_3 C O—N CH_3 → O C_6H_5—S O O CH_3 OC_2H_5

1687 **1688**

The intermediacy of **1658** in Scheme 55 found further support from the observations that treatment of a mixture of O-(4-nitrophenyl)hydroxylamine hydrochloride with **1689** (R = C_6H_5) in a boiling acetic acid–concentrated sulphuric acid (9 : 1) mixture gave **1690** (R^1 = NO_2, R^2 = C_6H_5) (74%), reaction of similar reactants in boiling alcoholic hydrogen chloride afforded

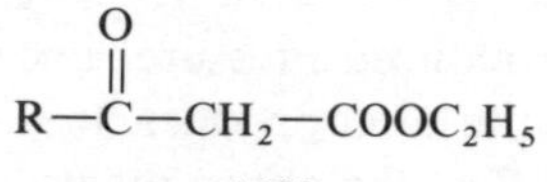

$$R-\overset{O}{\overset{\|}{C}}-CH_2-COOC_2H_5$$

1689

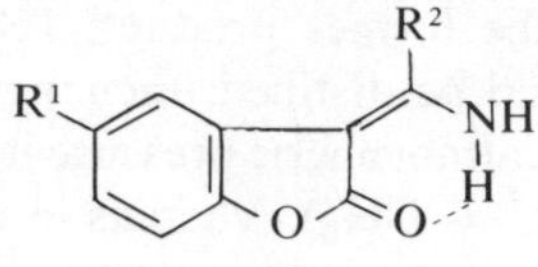

1690

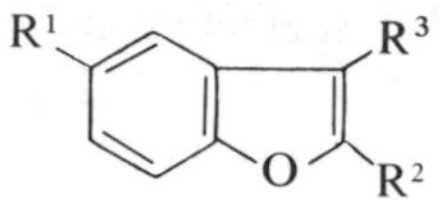

1691

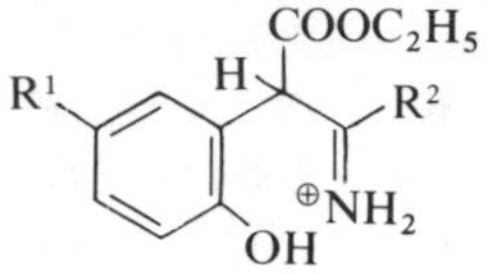

1692

1690 ($R^1 = NO_2$, $R^2 = C_6H_5$) (4% yield) and **1691** ($R^1 = NO_2$, $R^2 = C_6H_5$, $R^3 = H$ and $COOC_2H_5$), treatment of a mixture of O-phenylhydroxylamine hydrochloride with **1689** ($R = C_6H_5$) under the latter conditions afforded **1690** ($R^1 = H$, $R^2 = C_6H_5$) (45% yield) and treatment of a mixture of O-(4-nitrophenyl)hydroxylamine hydrochloride with **1689** ($R = CH_3$) in boiling alcoholic hydrogen chloride furnished **1690** ($R^1 = NO_2$, $R^2 = CH_3$) (9% yield) and **1691** ($R^1 = NO_2$, $R^2 = CH_3$, $R^3 = COOC_2H_5$) (52% yield). Clearly, these products were formed via **1692** ($R^1 = H$ and NO_2, $R^2 = C_6H_5$ and $R^1 = NO_2$, $R^2 = CH_3$), subsequent lactonization or loss of ammonium ion leading to the corresponding **1690** or **1691**, respectively (Zagorevskii, Samarina, Sharkova, L. M. and Aksanova, 1979).

Benzofuranizations of O-aryloximes have been effected in which the direction of the reaction was ambiguous, this direction being in many cases in sharp contrast to the direction(s) of indolization of the analogous arylhydrazones. The observation (Mooradian, 1967, 1969a, 1969b; Mooradian and Dupont, 1967a) that ethyl methyl ketone O-(4-nitrophenyl)oxime afforded both possible isomeric products, 2,3-dimethyl-5-nitrobenzofuran (isolated) and 2-ethyl-5-nitrobenzofuran (detected by p.m.r. examination of the mother liquors) in a 3 : 1 ratio has been extended (Grandberg and Sorokin, 1973) to cover a series of alkyl methyl ketone O-phenyloximes. Benzofuranization of **1693** [$R = CH_3$, C_2H_5, C_6H_5, $CH_2C_6H_5$ and $(CH_2)_2OH$], prepared *in situ* from O-phenylhydroxylamine and the appropriate ketone, in refluxing isopropanolic hydrochloric acid gave both possible products, **1694** [$R = CH_3$, C_2H_5, C_6H_5, $CH_2C_6H_5$ and $(CH_2)_2OH$, respectively] and **1695** [$R = CH_3$, C_2H_5, C_6H_5, $CH_2C_6H_5$ and $(CH_2)_2OH$, respectively]. However, similar reaction of **1693** ($R = COOC_2H_5$, $COCH_3$ and CN), formed *in situ*, gave only **1694** ($R = COOC_2H_5$, $COCH_3$ and $CONH_2$, respectively). Contrary to what might have been expected from these observations, benzofuranization of **1693** ($R = SO_2C_6H_5$), formed *in situ*, in refluxing alcoholic hydrogen chloride afforded **1694** ($R = SO_2C_6H_5$) (19% yield) and **1695** ($R = SO_2C_6H_5$) (36%

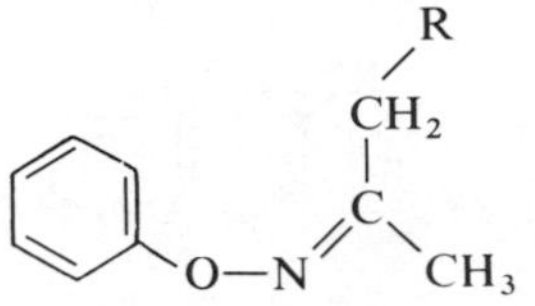

R
CH$_2$
C
O—N CH$_3$
1693

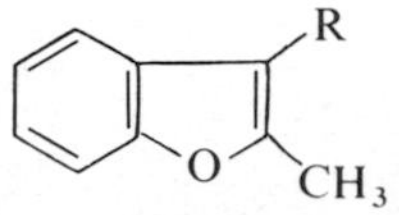

R
O CH$_3$
1694

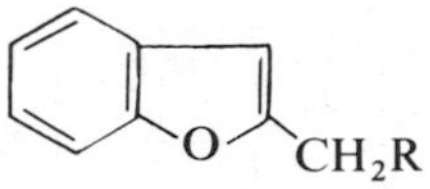

O CH$_2$R
1695

yield), the 4-nitro analogue, **1677**, yielded **1679**, via the isolation of **1678**, as the only isolated product (Samarina, Sharkova, L. M. and Zagorevskii, 1979) and only 3-carbethoxymethyl-2-methyl-5-nitrobenzofuran was isolated after the benzofuranization of ethyl laevulinate O-(4-nitrophenyl)oxime (Mooradian, 1969a, 1969b). The isomer ratio **1694** ($R = CH_3$) : **1695** ($R = CH_3$) obtained from **1693** ($R = CH_3$) has been found to depend upon the catalyst employed to effect the reaction and upon the conditions under which it is used (Sorokin, 1973) (Table 36). Benzofuranization of **1696** ($R^1 = R^3 = H$, R^2 = alkyl) has been reported to afford only **1697** ($R^1 = R^3 = H$, R^2 = alkyl), none of the possible alternative products **1698** ($R^1 = R^3 = H$, R^2 = alkyl) being detected

Table 36. Ratio of the Isomeric Products **1694** ($R = CH_3$) : **1695** ($R = CH_3$) obtained by the Benzofuranization of **1693** ($R = CH_3$) under a Variety of Acid Catalytic Conditions

Acid Catalyst[a] [Acid : **1693** ($R = CH_3$) = 1 : 1]	Solvent	Isomer Ratio of Products **1694** ($R = CH_3$) : **1695** ($R = CH_3$)
HCl	Methanol	22:78
HCl	2-Propanol	45:55
HCl	*tert.*-Butanol	63:37
HCl	Nitromethane	67:33
HCl	Dimethylsulphoxide	70:30
HCl	Dimethylformamide	75:25
HCl	Dioxane	88:12
HCl	Tetrahydrofuran	92:8
HCl	Benzene	78:22
H_2SO_4	Methanol	18:82
H_2SO_4	2-Propanol	15:75[b]
H_2SO_4	Dioxane	42:58
$CH_3C_6H_4SO_3H(4)$	2-Propanol	35:65
$HClO_4$	2-Propanol	19:81

Notes

a. All reactions were carried out at 100 °C; b. This ratio is obviously misquoted in (Grandberg and Sorokin, 1974) and should read 25 : 75 or 15 : 85.

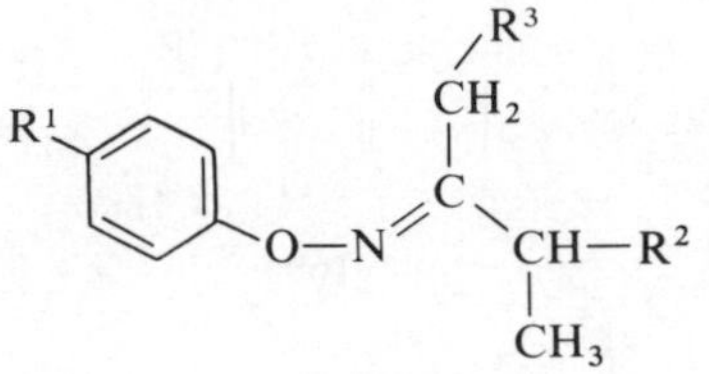

1696

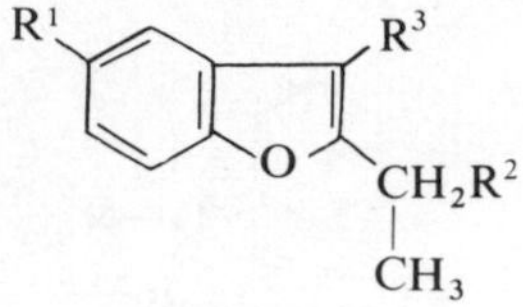

1697

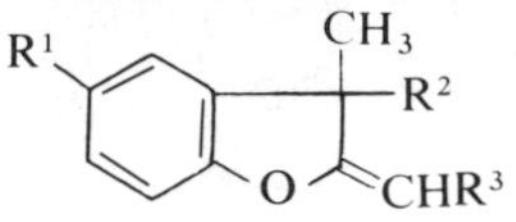

1698

by g.l.c. and p.m.r. analysis of the reaction mixtures (Sorokin, 1973). Similarly, the formation of **1698** [$R^1 = NO_2$, $R^2 + R^3 = (CH_2)_3$] could not be detected from the reaction of **1696** [$R^1 = NO_2$, $R^2 + R^3 = (CH_2)_3$] in a refluxing concentrated hydrochloric acid–absolute ethanol mixture which gave a nearly quantitative yield of **1697** [$R^1 = NO_2$, $R^2 + R^3 = (CH_2)_3$] (Kaminsky, Shavel and Meltzer, 1967). However, treatment of **1696** ($R^1 = NO_2$, $R^2 = COOC_2H_5$, $R^3 = H$), formed *in situ*, with refluxing ethanolic hydrogen chloride yielded only **1699** (31% yield) (Zagorevskii, Samarina, Sharkova, L. M. and Aksanova, 1979).

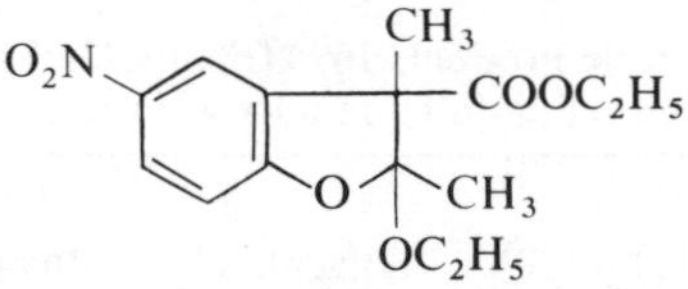

1699

A series of 3-substituted cycloalkanone O-aryloximes **1700** have been treated under either thermal or acid catalytic conditions to yield ultimately **1701** [$R^1 = NO_2$, NH_2, Cl, F, CH_3, $NHCOCH_3$, OCH_3, CN, $CONH_2$, COOH and $COCH_3$, $R^2 = H$, CH_3, C_2H_5 and $(CH_2)_2N(CH_3)_2$, $n = 1$ and 2] (Berger, L., Leimgruber and Schenker, 1972). The 2-benzyloxy-1,2,3,4-tetrahydro-8-nitrodibenzofuran quoted in the literature (Mooradian, 1969b; Mooradian

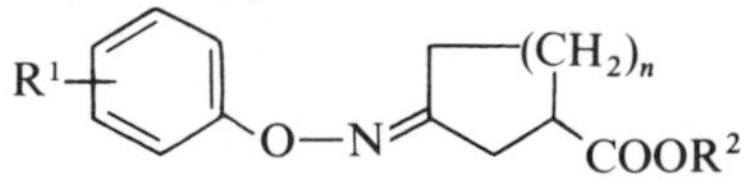

1700

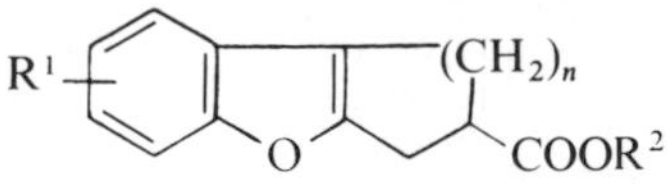

1701

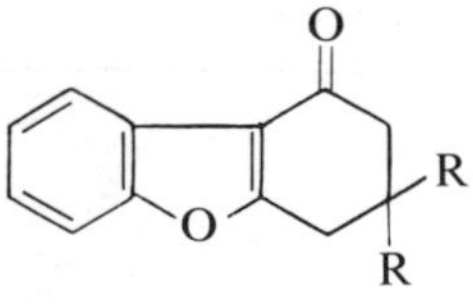

1702

and Dupont, 1967a) may also have been similarly synthesized but its origin was not given.

Reaction of a mixture of **1675** (R = H) with O-phenylhydroxylamine in refluxing methanolic hydrogen chloride and of a mixture of **1675** (R = CH_3) with O-phenylhydroxylamine hydrochloride in refluxing alcoholic hydrogen chloride led in each case to the isolation of only one product, **1702** (R = H and CH_3, respectively) (Kucherova, Aksanova, Sharkova, L. M. and Zagorevskii, 1973a).

The formation of benzofurans from heterocyclic ketone O-aryloximes has also been investigated and those involving the acid catalysed conversions of **1703** into **1704** are given in Table 37. In none of these reactions in which

Table 37. Compounds **1704** which have been Synthesized by Benzofuranization of the Aryloximes **1703**

1704 (Unless otherwise stated, R^1–R^6 = H)							Reference
R^1	R^2	R^3	R^4	R^5	R^6	X	(Catalyst)
						S	1(1)
$COOC_2H_5$						S	2(1)
NO_2						S	2(1), 3(1)
						SO_2	4(1)
						NH	5(1), 6(1)
	CH_3					NH	5(1), 6(1)
$COOC_2H_5$		CH_3	CH_3	CH_3	CH_3	NH	7(2)
NO_2		CH_3	CH_3	CH_3	CH_3	NH	7(2)
						NCH_3	1(1)
CH_3						NCH_3	1(1)
$COOC_2H_5$						NCH_3	7(2)
NO_2						NCH_3	3(2, 3), 7(2)
$COOC_2H_5$	CH_3			CH_3		NCH_3	7(2)
NO_2	CH_3			CH_3		NCH_3	3(2, 3), 7(2)
	CH_3			CH_3		NCH_3	1(1)

Catalysts

1. Hydrogen chloride in absolute alcohol under reflux; 2. A refluxing mixture of acetic acid and sulphuric acid; 3. Boron trifluoride etherate in refluxing acetic acid.

References

1. Aksanova, Kucherova, Sharkova, L. M. and Zagorevskii, 1972; 2. Sharkova, L. M., Aksanova, Kucherova and Zagorevskii, 1971a; 3. Aksanova, Sharkova, L. M., Kucherova and Zagorevskii, 1970; 4. Aksanova, Sharkova, L. M., Kucherova and Zagorevskii, 1973; 5. Kucherova, Aksanova, Sharkova, L. M. and Zagorevskii, 1973b; 6. Aksanova, Barkov, Zagorevskii, Kucherova and Sharkova, L. M., 1975; 7. Kucherova, Aksanova, Sharkova, L. M. and Zagorevskii, 1971.

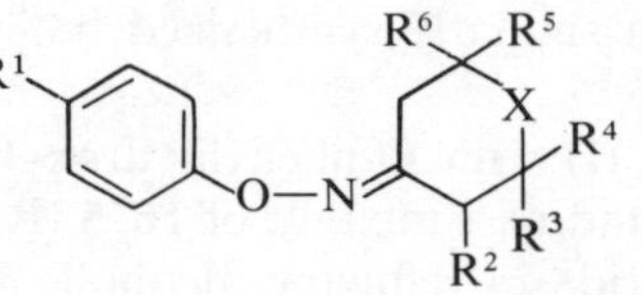

1703

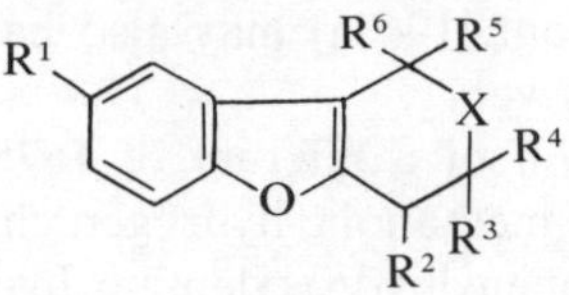

1704

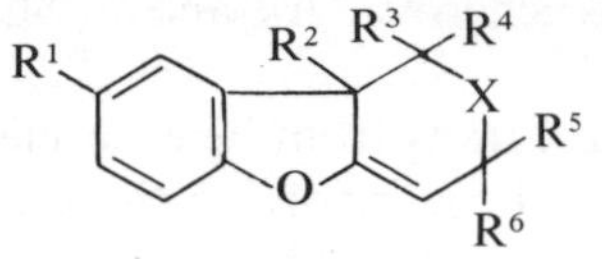

1705

$R^2 \neq H$ (i.e. $R^2 = CH_3$) were products **1705** isolated. These might have resulted from the alternative direction of benzofuranization in accord with the conversions of **1680** ($R^3 \neq H$) and **1696** [$R^2 + R^3 = (CH_2)_3$] into **1681** ($R^3 \neq H$) (Sharkova, L. M., Aksanova, Kucherova and Zagorevskii, 1971b) and **1697** [$R^2 + R^3 = (CH_2)_3$] (Kaminsky, Shavel and Meltzer, 1967), respectively, these being obtained as the only isolated products. Treatment of **1706** (R = H and NO_2) with refluxing alcoholic hydrogen chloride furnished **1707** (R = H) (Aksanova, Kucherova, Sharkova, L. M. and Zagorevskii, 1972) and **1707** (R = NO_2) (Aksanova, Sharkova, L. M., Kucherova and Zagorevskii, 1970; Sharkova, L. M., Aksanova, Kucherova and Zagorevskii, 1971a), respectively. Treatment of **1708** (X = NCH_3) with a refluxing acetic acid–sulphuric acid mixture and of the *in situ* formed **1708** (X = S and SO_2) with refluxing alcoholic hydrogen chloride afforded **1709** (X = NCH_3 and S, Y = CH_2) and **1709** (X = CH_2, Y = SO_2), respectively, as the only isolated products (Aksanova, Sharkova, L. M., Kucherova and Zagorevskii, 1973). Reaction of a mixture of O-phenylhydroxylamine with **1710** in refluxing alcoholic hydrogen chloride gave **1711** as the only isolated product (Aksanova, Sharkova, L. M., Kucherova and Zagorevskii, 1973). Mixtures of **1712** (R = H and CH_3) with O-phenylhydroxylamine hydrochloride in refluxing absolute ethanolic hydrogen chloride

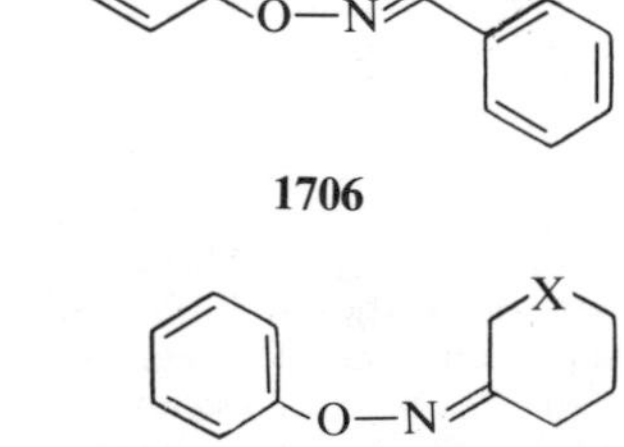

1706 → 1707

1708

1709

gave **1713** (R = H and CH_3, respectively) along with, in the latter reaction, a second product to which structure **1714** was assigned. The formation of **1714** was assumed to involve alcoholysis of the seven-membered lactam ring in one of the intermediate stages of the reaction (Kucherova, Aksanova, Sharkova, L. M. and Zagorevskii, 1973a). In a related reaction, **1715** was treated with refluxing alcoholic hydrogen chloride to yield **1716** and **1718**. It was suggested that these arose as shown in Scheme 56 in which the intermediate **1717** underwent cyclization to yield **1716** or amide group hydrolysis followed by bicyclization to yield **1718**. When **1716** was stood in trifluoroacetic solution for three days at 20 °C, dehydration occurred to give **1719** (Glushkov, Zasosova, Ovcharova, Solov'eva, Anisimova and Sheinker, Y. N., 1978).

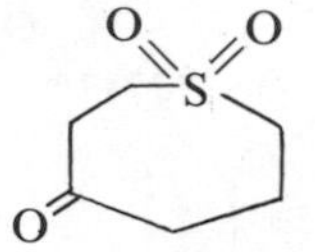

1710

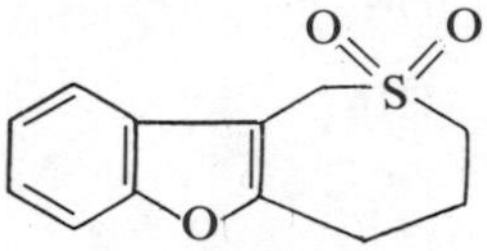

1711

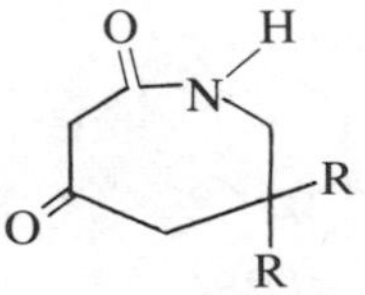

1712

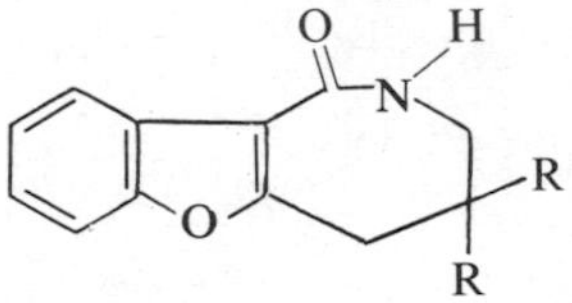

1713

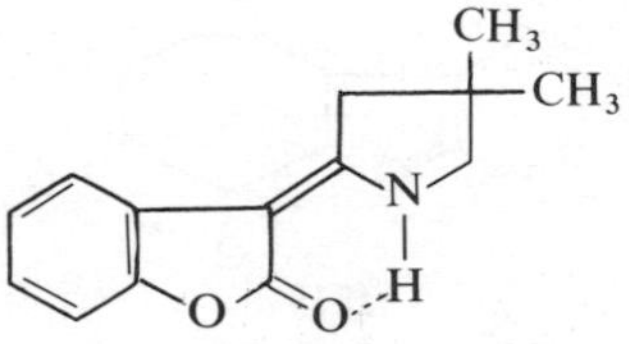

1714

Other benzofurans **1721** which have been unambiguously synthesized from the corresponding O-aryloximes **1720** by treatment with acid catalysts are given in Table 38. Still further examples involved the conversion of **1722** into **1723** by heating in alcoholic mineral acid (Mooradian, 1969a, 1969b) and the conversion of **1724** [R^1 = CH_3 and C_6H_5, R^2 = H and $R^1 + R^2 = (CH_2)_4$] into **1725** [R^1 = CH_3 and C_6H_5, R^2 = H and $R^1 + R^2 = (CH_2)_4$, respectively] by refluxing in a sulphuric acid–acetic acid (1 : 9) mixture. The attempted use of 30% ethanolic hydrogen chloride as the catalyst in the latter reactions led to the formation of 4-ethoxycoumarin (Aksanova, Sharkova, L. M., Kucherova and Zagorevskii, 1971).

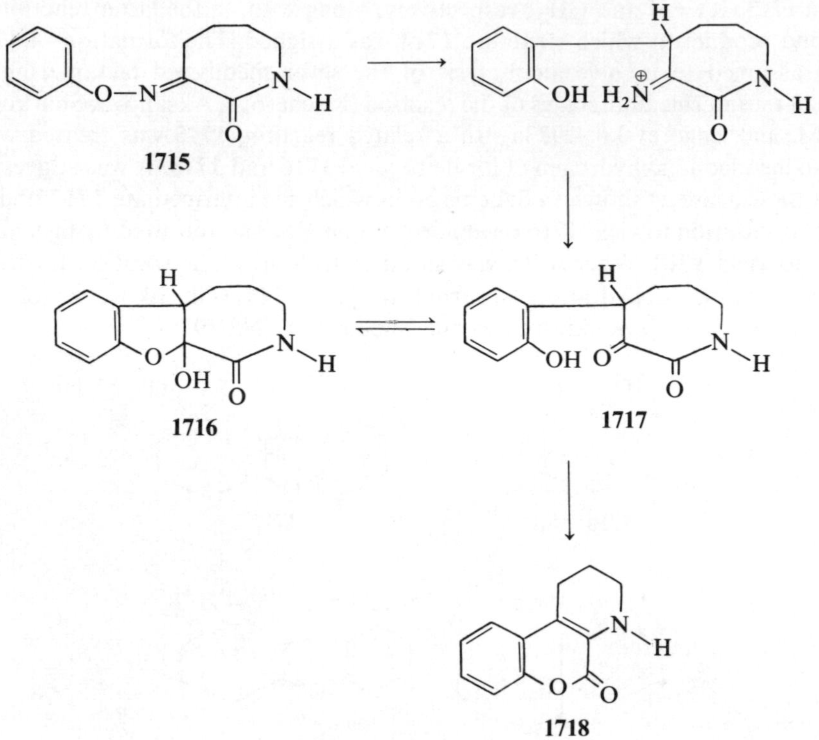

1715

1716

1717

1718

Scheme 56

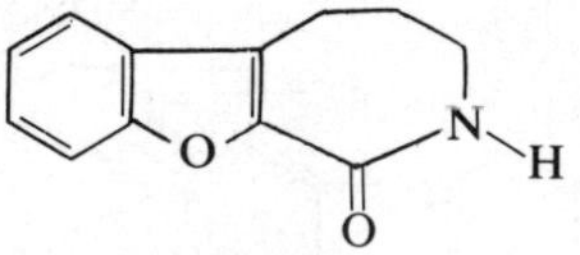

1719

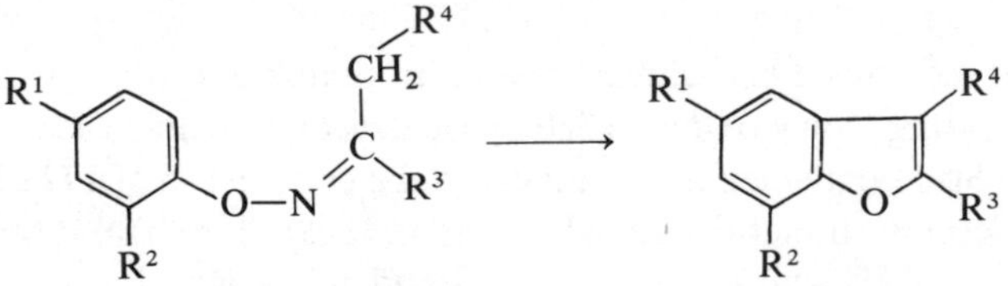

1720

1721

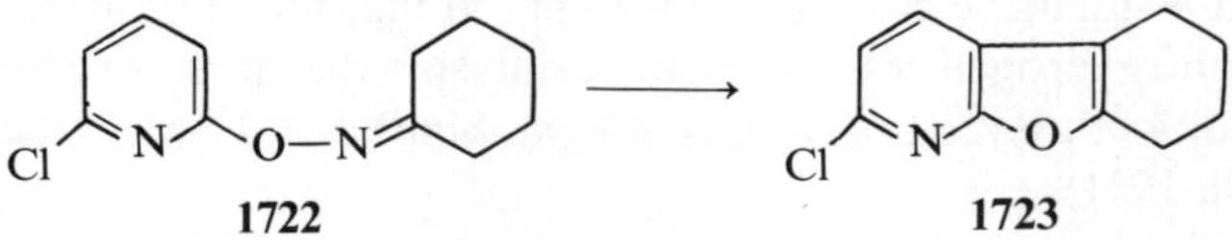

1722

1723

Table 38. Benzofurans which have been Prepared by the Acid catalysed Cyclization of O-Aryloximes

Benzofuran **1721** (Unless otherwise stated, R^1, R^2 and R^4 = H)			Reference (Catalysts) and Reaction Conditions
R^1	R^2	R^3	
		CH_3	1(1)
CF_3		CH_3	2(2), 3(3), 4(4), 5(4, 5)
CN		CH_3	2(2), 3(3), 5(4, 5)
$COCH_3$		CH_3	2(2), 5(4, 5)
NO_2		CH_3	2(2), 4(4), 5(4, 5), 6(6)
$SO_2C_6H_5$		CH_3	2(2), 3(3), 5(4, 5)
CF_3	NO_2	CH_3	2(2), 3(3), 5(4, 5)
NO_2	CF_3	CH_3	2(2), 3(3), 5(4, 5)
NO_2	$COOC_2H_5$[a]	CH_3	2(2), 3(3), 5(4, 5), 7(6)
NO_2	Cl	CH_3	2(2), 3(3), 5(4, 5)
NO_2	NO_2	CH_3	2(2), 3(3), 4(4), 5(4, 5), 8(7)
	NO_2	CH_3	2(2), 4(4), 5(4, 5), 6(6)
		C_6H_5	1(1)
CF_3		C_6H_5	2(2), 3(3), 5(4, 5)
$COOC_2H_5$		C_6H_5	2(2), 3(3), 5(4, 5)
NO_2		C_6H_5	2(2), 3(3), 5(4, 5)
$SO_2N(CH_3)_2$		C_6H_5	2(2), 3(3), 5(4, 5)
	CF_3	C_6H_5	2(2), 3(3), 5(4, 5)
NO_2	NO_2	C_6H_5	8(7)
$COOC_2H_5$		$C_6H_4OCH_3(2)$	2(2), 3(3), 5(4, 5)
$SO_2N(CH_3)_2$		$C_6H_4OCH_3(2)$	2(2), 3(3), 5(4, 5)
	NO_2	$C_6H_4OCH_3(2)$	2(2), 3(3), 5(4, 5)
	$(CH_2)_4$		1(1)
NO_2	$(CH_2)_4$		2(2), 6(6)
NO_2	NO_2	$(CH_2)_4$	8(7)
	NO_2	$(CH_2)_4$	4(4), 5(4, 5), 6(6)

Note

a. The ester was obtained from the corresponding O-(2-carboxy-4-nitrophenyl)oxime under the benzofuranization conditions.

Catalysts and Reaction Conditions

1. Boron trifluoride etherate in acetic acid at 100 °C; 2. Heat with strong mineral acid in alcoholic solution; 3. Refluxing alcoholic hydrogen chloride or heat with acetic acid containing hydrogen chloride: the attempted thermal non-catalytic benzofuranization afforded only intractable tars, observations which were attributed to the thermal instability of O-phenyloximes; 4. Refluxing alcoholic hydrogen chloride; 5. Heat with hydrogen bromide in acetic acid or with trifluoroacetic acid; 6. Reflux with concentrated hydrochloric acid in absolute ethanol; 7. Concentrated sulphuric acid in refluxing acetic acid.

References

1. Sheradsky, 1966; 2. Mooradian, 1969a; 3. Mooradian and Dupont, 1967a; 4. Mooradian, 1967; 5. Mooradian, 1969b; 6. Kaminsky, Shavel and Meltzer, 1967; 7. Mooradian, 1971; 8. Sheradsky, 1967.

1724 → 1725

Attempts to extend this reaction to the synthesis of 2-unsubstituted benzofurans were unsuccessful. When equimolar mixtures of O-phenylhydroxylamine hydrochloride with propionaldehyde or 4-aminobutyraldehyde diethylacetal were heated in ethanolic hydrogen chloride, no traces of the expected benzofurans were detectable, the only recognizable product being 2-aminophenol, isolated in 39% and 45% yields, respectively. This product was also isolated in *ca.* 5% yield, along with the expected 2-phenylbenzofuran (71% yield), when an equimolar mixture of O-phenylhydroxylamine hydrochloride and acetophenone was heated in ethanolic hydrogen chloride, and in 13% yield when O-phenylhydroxylamine hydrochloride alone was subjected to similar conditions. In none of these reactions was 4-aminophenol detectable in the reaction product (Carter, M. A. and Robinson, B., 1974). These formations of 2-aminophenol were probably the result of intramolecular *ortho* rearrangements, many examples of which are well established (Ingold, 1969a), and are analogous to the formation of *ortho*-phenylenediamine concomitantly with Fischer indolization (see Chapter VII, M).

In the formation of benzofurans from O-aryloximes, the acid catalysts are only necessary to effect the initial tautomerism of the oximes **1654** into **1655** (Scheme 55). This has been clearly established by reacting **1726** (R^1 = H, CH_3, Cl and Br, R^2 = H and R^1 = H, R^2 = CH_3) with dimethyl acetylenedicarboxylate (**1727**) in ethanolic solution at room temperature, or even at −30 °C, to afford high yields of **1732** (R^1 = H, CH_3, Cl and Br, R^2 = H and R^1 = H, R^2 = CH_3, respectively). Presumably, the initial adducts **1728** underwent spontaneous rearrangement into **1729** and the intermediate **1730** lactonized to furnish **1731** (Scheme 57). Under similar conditions, **1726** (R^1 = NO_2, R^2 = H) reacted with **1727** to produce **1732** (R^1 = NO_2, R^2 = H) (10%) together with **1733**. The latter product remained unchanged in refluxing ethanol but upon treatment with refluxing ethanolic hydrogen chloride it was rapidly converted into **1732** (R^1 = NO_2, R^2 = H) (Sheradsky, Nov, Segal and Frank, 1977). In a preliminary communication of some of these studies (Sheradsky and Lewinter, 1972) it was reported that **1726** (R^1 = H, CH_3 and NO_2, R^2 = H) reacted with **1727** in ethanolic or ethereal solutions at −30 °C to afford **1732** (R^1 = H and CH_3, R^2 = H) and **1732** (R^1 = NO_2, R^2 = H) and **1733**, respectively. Clearly, these formations of **1732** from **1726** are related to the transformations of **314** into **316** (Chapter II, E2). In a related study, **1734**

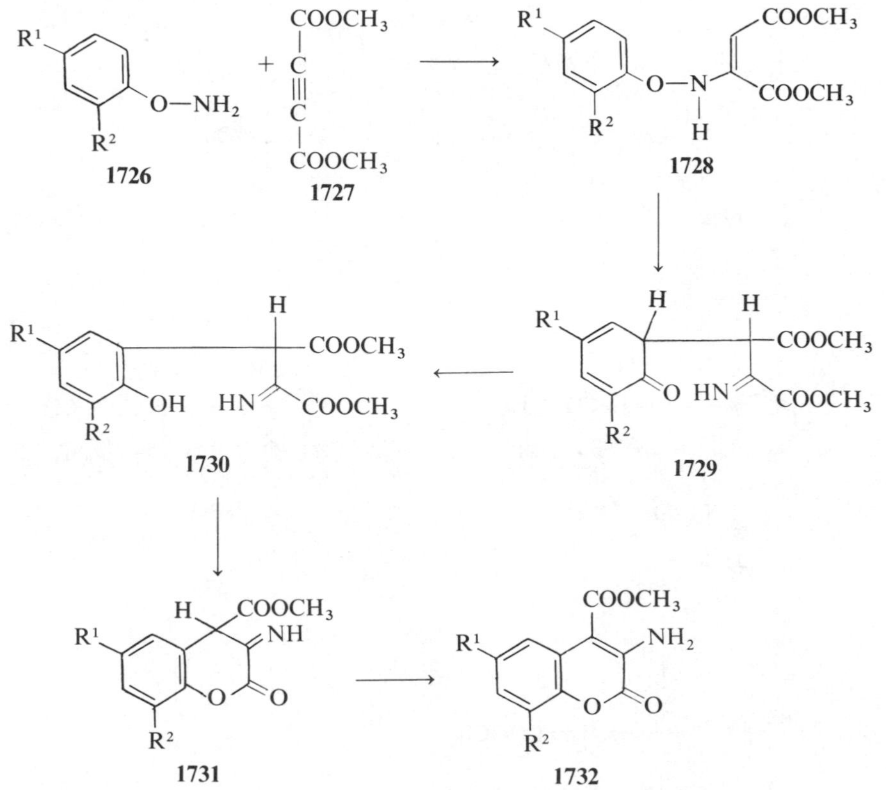

Scheme 57

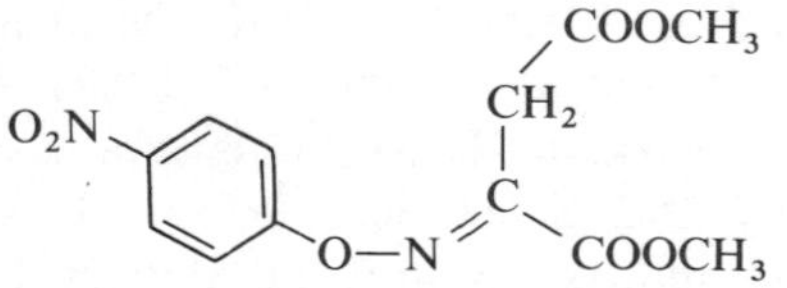

(R^1 = H, R^2 = C_6H_5 and R^1 = H, CH_3 and Cl, R^2 = $OCH_2C_6H_5$) reacted with **1727** in the presence of potassium *tert*.-butoxide in dry dimethyl sulphoxide at room temperature to afford the corresponding **1737**s, presumably via formation of the adducts **1735** which spontaneously rearranged into **1736** which then afforded **1737**. When compounds **1734** (R^1 = H, R^2 = C_6H_5 and R^1 = H, CH_3 and Cl, R^2 = $OCH_2C_6H_5$) were reacted with **1727** in refluxing ethanolic potassium hydroxide or sodium ethoxide, the only product from the first reaction was **1738** (R^1 = H, R^2 = C_6H_5) and the major product from the other three reactions was **1738** (R^1 = H, CH_3 and Cl, respectively, R^2 = $OCH_2C_6H_5$), formed by retro Claisen cleavage of the corresponding **1737**s as shown. A minor product isolated from each of the last three reactions was **1740** (R^1 = H, CH_3 and Cl, respectively, R^2 = $OCH_2C_6H_5$), probably formed via the dehydration of the corresponding intermediates **1739** which were produced when compounds **1737** cyclized by the attack of their amide anions upon their ketonic carbonyl groups (Sheradsky, Nov, Segal and Frank, 1977).

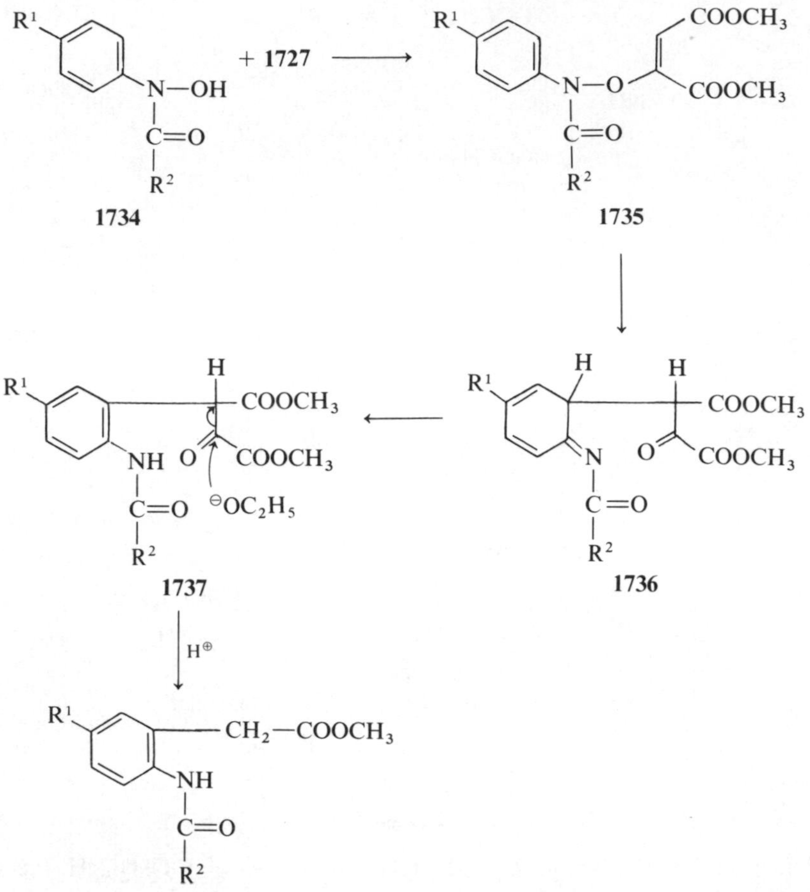

O-Aryloximes have been synthesized by reacting the sodium or potassium salt of an oxime with an aryl halide (usually fluoride) substituted *ortho* and/or *para* with electron withdrawing groups (Cattanach and Rees, 1971; Dupont, 1968; Kaminsky, Shavel and Meltzer, 1967; Kucherova, Aksanova, Sharkova, L. M. and Zagorevskii, 1971; Mooradian, 1967, 1969a, 1969b; Mooradian and Dupont, 1967a, 1967b; Sharkova, L. M., Aksanova, Kucherova and Zagorevskii, 1971a; Sheradsky and Salemnick, 1971a). This synthetic route was particularly favoured (Kaminsky, Shavel and Meltzer, 1967) in view of the low yields reported in the preparation of O-phenylhydroxylamine (Bumgardner

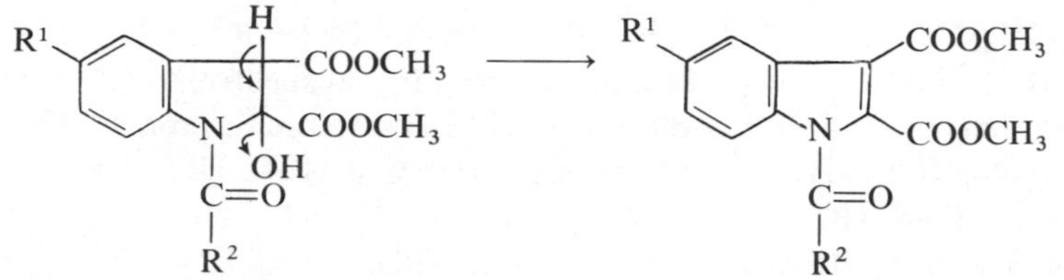

1739 → 1740

and Lilly, 1962 – see also Nicholson and Peak, 1962) and of its instability, properties which were presumably extrapolated to O-phenylhydroxylamines bearing electron withdrawing substituents on the benzenoid ring. O-Aryloximes have also been prepared by the acid catalysed reaction between O-arylhydroxylamines or their salts and ketones (Aksanova, Sharkova, L. M., Kucherova and Zagorevskii, 1973; Dupont, 1968; Sheradsky, 1966, 1967) and have been prepared *in situ* from similar reactants during the above syntheses of benzofurans (Aksanova, Barkov, Zagorevskii, Kucherova and Sharkova, L. M., 1975; Aksanova, Kucherova, Sharkova, L. M. and Zagorevskii, 1972; Aksanova, Sharkova, L. M., Kucherova and Zagorevskii, 1973; Grandberg and Sorokin, 1973; Kucherova, Aksanova, Sharkova, L. M. and Zagorevskii, 1973a, 1973b; Zagorevskii, Samarina, Sharkova, L. M. and Aksanova, 1979).

The somewhat limited number of syntheses of O-arylhydroxylamines was reviewed during the introduction of a new synthetic approach (Sheradsky, Salemnick and Nir, 1972). This involved the reaction of the aryl halides **1741** ($R^1 = NO_2$, $R^2 = R^3 = H$; $R^1 = R^3 = H$, $R^2 = NO_2$ and $R^1 = R^2 = NO_2$, $R^3 = H$, X = F and $R^1 = R^3 = NO_2$, $R^2 = H$ and R^1–$R^3 = NO_2$, X = Cl) with *tert.*-butyl N-hydroxycarbamate (**1742**; $R^4 = H$) to afford the corresponding structures **1743** which were readily hydrolysed and decarboxylated, by brief treatment with trifluoroacetic acid, to give **1744** ($R^1 = NO_2$,

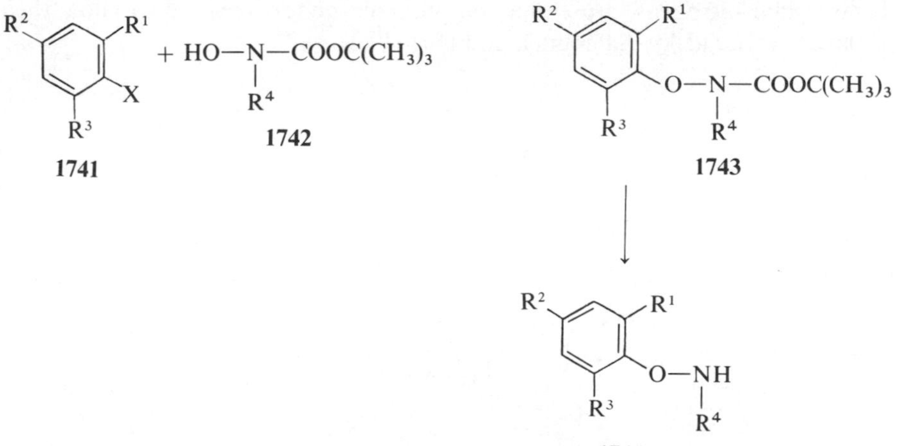

1741 + HO—N(R^4)—COOC$(CH_3)_3$ **1742** → **1743** → **1744**

$R^2 = R^3 = H$; $R^1 = R^3 = H$, $R^2 = NO_2$; $R^1 = R^2 = NO_2$, $R^3 = H$; $R^1 = R^3 = NO_2$, $R^2 = H$ and R^1–$R^3 = NO_2$, respectively, $R^4 = H$). In an earlier (Sheradsky, 1967) application of this synthetic approach, **1741** ($R^1 = R^2 = NO_2$, $R^3 = H$, $X = Cl$) was reacted with **1742** ($R^4 = H$) to furnish ultimately **1744** ($R^1 = R^2 = NO_2$, $R^3 = R^4 = H$). However, the above method starting with the fluoro analogue **1741** ($R^1 = R^2 = NO_2$, $R^3 = H$, $X = F$) was far superior (Sheradsky, Salemnick and Nir, 1972; Sheradsky and Nir, 1969). Use of 2,4-dinitrofluorobenzene (**1741**; $R^1 = R^2 = NO_2$, $R^3 = H$, $X = F$) and **1742** ($R^4 = CH_3$) as the initial reactants in a similar reaction sequence ultimately gave **1744** ($R^1 = R^2 = NO_2$, $R^3 = H$, $R^4 = CH_3$) (Sheradsky, Salemnick and Nir, 1972). Such N-alkylated O-arylhydroxylamines could well be useful when benzofuranization under mild and possibly non-acidic conditions is desired since their reaction with ketones should, by analogy with the use of N_β-alkylated arylhydrazines in the Fischer indolization, form N-alkylated **1655**, thus permitting benzofuranization, during which the elements of the corresponding alkyl amine would be liberated, to proceed under mild conditions. Unfortunately, attempts to extend the reaction sequence **1741** → **1744** to the synthesis of O-(N-heteroaryl)hydroxylamines were unsuccessful. Reaction of 2-chloro-5-nitropyridine with **1742** ($R^4 = H$) afforded **1745** but acidolysis of this product gave **1746** and **1747**, the latter product appearing to be formed via the desired O-(5-nitro-2-pyridyl)hydroxylamine which was unstable under the reaction conditions, yielding **1746**, as was also reported elsewhere during another attempted synthesis (Sheradsky, Salemnick and Frankel, 1971). The corresponding carbamates of other N-heteroaromatic systems have been synthesized but attempted acidolysis to the hydroxylamines was again unsuccessful, either yielding the corresponding hydroxy compound or causing decomposition. From these results it was concluded 'that materials bearing aminooxy groups on carbons adjacent to electron-withdrawing heterocyclic nitrogens, are either too unstable or too reactive to allow their isolation' (Sheradsky, Salemnick and Nir, 1972).

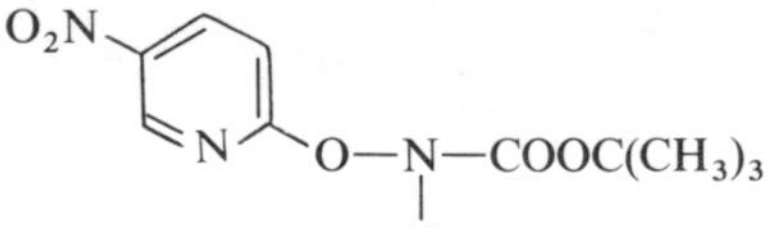

1745

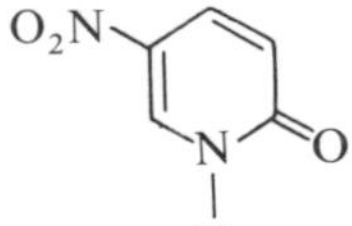

1746

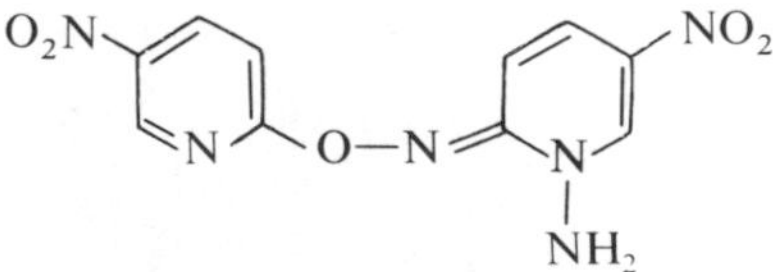

1747

A review of benzofurans (Mustafa, 1974) included their synthesis by the acid catalysed cyclization of O-aryloximes. Unfortunately, this aspect of this review was very brief and, furthermore, quoted only four original literature references, three of which were quoted erroneously. However, more thorough reviews, encompassing the majority of the early publications in this area, have appeared (Cagniant, P. and Cagniant, D., 1975; Grandberg and Sorokin, 1974).

Attempts which have been made to convert S-arylthiooximes **1748** into benzothiophens **1749** have been unsuccessful. Treatment of **1748** [$R^1 = NO_2$, $R^2 + R^3 = (CH_2)_4$, X = CH and $R^1 = H$, $R^2 + R^3 = (CH_2)_2N(CH_3)CH_2$, X = N] in refluxing ethanolic concentrated hydrochloric acid (Kaminsky, Shavel and Meltzer, 1967) and of **1748** ($R^1 = R^3 = H$, $R^2 = CH_3$, X = CH) in refluxing acetic or hydrochloric acids (Davis, F. A. and Skibo, 1974) afforded **1750** ($R = NO_2$) and **1751** [$R^1 + R^1 = (CH_2)_3$, $R^2 = C_6H_4NO_2(4)$], **1751** [$R^1 + R^1 = CH_2N(CH_3)CH_2$, R^2 = 2-pyridyl) and **1750** (R = H) and **1751** ($R^1 = H$, $R^2 = C_6H_5$), respectively. When the third thiooxime was treated

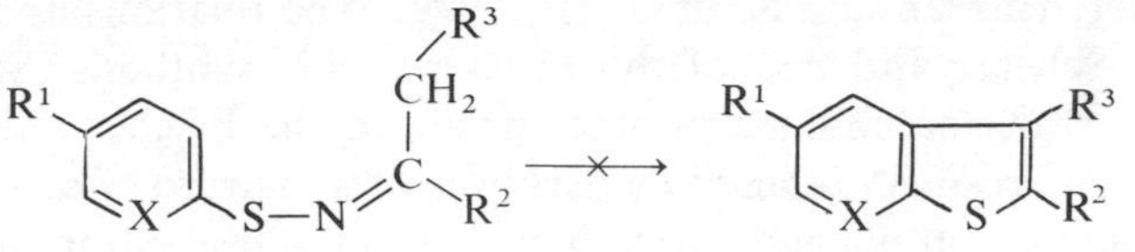

1748 **1749**

1750 **1751**

with boron trifluoride etherate at −78 °C, the only isolated product was **1750** (R = H) and other related attempts to convert S-arylthiooximes into benzothiophens were also unsuccessful (Davis, F. A. and Skibo, 1974). The failure of all these reactions probably reflected the lack of formation of the enesulphenamide tautomer of **1748**, required for benzothiophenization, the major reaction of **1748** with acids involving cleavage of the N—S bond (Davis, F. A. and Skibo, 1974).

D. The Piloty Synthesis of Pyrroles and Related Reactions

In 1910 it was observed (Piloty, 1910) that heating bisdiethyl ketone azine (**154**; $R^1 = C_2H_5$, $R^2 = CH_3$) with zinc chloride led to the formation of 2,5-diethyl-3,4-dimethylpyrrole (**157**; $R^1 = C_2H_5$, $R^2 = CH_3$, $R^3 = H$). This observation was not referred to when, eight years later (Robinson, G. M. and Robinson, R., 1918), bisbenzyl phenyl ketone azine (**154**; $R^1 = R^2 = C_6H_5$) was converted into 2,3,4,5-tetraphenylpyrrole (**157**; $R^1 = R^2 = C_6H_5$,

R^3 = H) either by heating with zinc chloride or, better, by passing dry hydrogen chloride over the molten azine. Piloty's publication of 1910 was again overlooked (Perkin and Plant, 1925) when reference was made to the Robinsons' conversion of **154** ($R^1 = R^2 = C_6H_5$) into **157** ($R^1 = R^2 = C_6H_5$, R^3 = H) and also when it was later stated (King, F. E. and Paterson, 1936) that G. M. and R. Robinson had introduced this synthetic route to pyrroles in 1918.

During the pioneering studies (Piloty, 1910) upon this synthesis, the analogy with the Fischer indole synthesis was drawn, an analogy which was later (Robinson, G. M. and Robinson, R., 1918) extended to embrace related mechanistic proposals for both syntheses. After subsequent refinement, the Robinsons' mechanistic proposal for the Piloty pyrrole synthesis has been represented (see, for example, Chapelle, Elguero, Jacquier and Tarrago, 1970b, 1971; Fritz and Uhrhan, 1971a; Posvic, Dombro, Ito, H. and Telinski, 1974) as shown in Scheme 58, involving a 3,4-diaza-Cope rearrangement of the intermediate **1753** ($R^3 = R^4$ = H) into the intermediate **1754** ($R^3 = R^4$ = H) (Heimgartner, Hansen and Schmid, H., 1979). The relationship between this mechanistic scheme and that of the Fischer indole synthesis (Scheme 10) is obvious. As with analogous experimentation on the Fischer indole synthesis (Chapter II, A), results obtained by pyrolysis of a mixture of bisacetophenone and bis-4-methylacetophenone azine have led to the suggestion that pyrrolization occurs via homolysis of the azine followed by sequential rearrangement

154 ⇌ **1752** ($R^3 = R^4$ = H) → **1753** ($R^3 = R^4$ = H) → **1754** ($R^3 = R^4$ = H) → ($R^3 = R^4$ = H) $-NH_3$ → **157** (R^3 = H)

Scheme 58

and recombination of the radicals formed (Tsuge, Tashiro, Hokama and Yamada, K., 1968; Tsuge, Watanabe and Hokama, 1971). However, simple transazinization, analogous to the transhydrazonization suggested in the equivalent indolization studies, could explain what was presumably the formation of 'crossed pyrrolization' products. The bisenamines **1752** [$R^1 = H$, $R^2 = H$, CH_3 and $CH(CH_3)_2$, $R^3 = R^4 = CH_3$] have been postulated as intermediates in the thermolysis of **1755** [R = H, CH_3 and $CH(CH_3)_2$, respectively] into **157** [$R^1 = H$, $R^2 = H$, CH_3 and $CH(CH_3)_2$, respectively, $R^3 = CH_3$] (Sucrow, Bethke and Chondromatidis, 1971).

Starting with a variety of azines, the bisenamines **1752** (R^3 and $R^4 \neq H$), the monoenaminoazines **1756** ($R^3 \neq H$) and mixtures of the ketone with hydrazine, the Piloty pyrrole synthesis has been fairly widely applied to the preparation of the symmetrically substituted pyrroles **157** shown in Table 39 and to **1757** (R = H, $n = 1$) (Cornforth, J. W., Hughes, G. K., Lions and Harradence, 1937–1938), **1757** (R = H, $n = 2$) (Baumes, Jacquier and Tarrago, 1973; Tsuge, Tashiro and Hokama, 1968), **1757** (R = CH_3, $n = 2$) (Baumes, Jacquier and Tarrago, 1973; Fritz and Uhrhan, 1971a), **1758** (R = H, $n = 1$) (Cornforth, J. W., Hughes, G. K., Lions and Harradence, 1937–1938), **1758** (R = CH_3, $n = 2$) (Chappelle, Elguero, Jacquier and Tarrago, 1970b; Baumes, Jacquier and Tarrago, 1973; Sucrow and Chondromatidis, 1970), **1759** (Cornforth, J. W., Hughes, G. K., Lions and Harradence, 1937–1938), **1760** (from *trans*-2-decalone), **1761** and **1762** (Sucrow and Condromatidis, 1970).

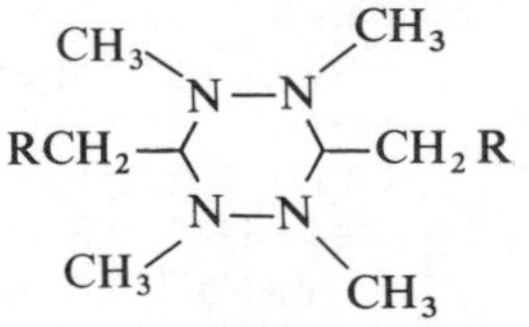

1755

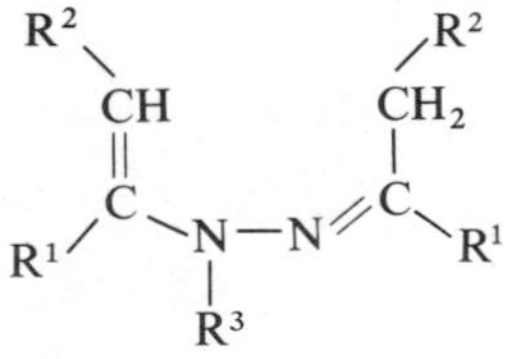

1756

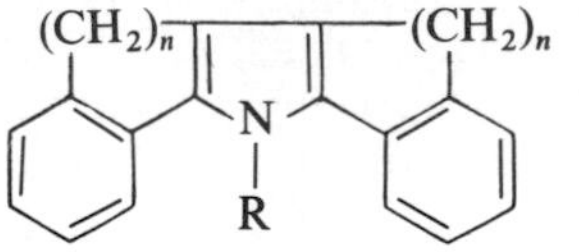

1757

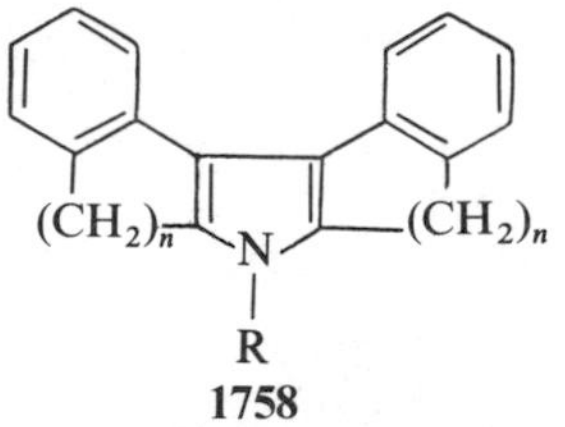

1758

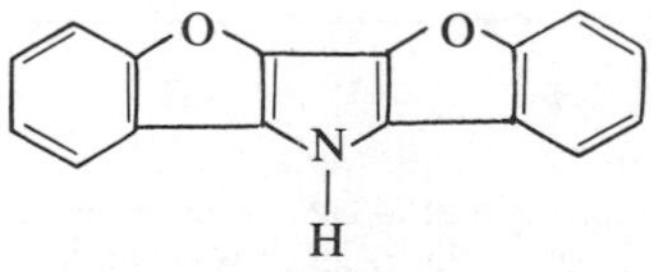

1759

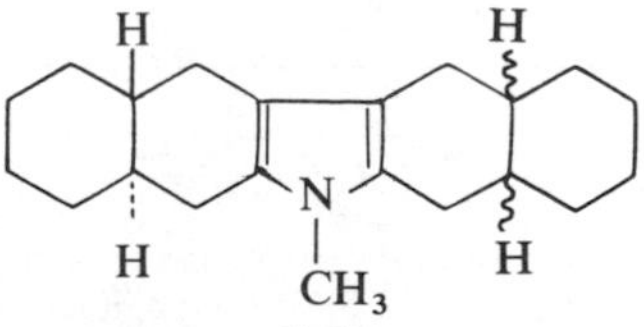

1760

Table 39. Pyrroles which have been Prepared by the Piloty Synthesis

Pyrrole **157** (Unless otherwise stated, R^1–R^3 = H)			
R^1	R^2	R^3	Reference (Catalyst)[a]
C_6H_5			1(1)
$C_6H_4CH_3(4)$			1(1)
$C_6H_4OCH_3(4)$			1(1)
$C_6H_4Cl(4)$			1(1)
	CH_3		2(2)
C_2H_5	CH_3		2(3), 3(4), 4(5)
	$(CH_2)_3$		3(4)
	$(CH_2)_4$		2(6), 5(7), 6(5, 8)
C_6H_5	C_6H_5		7(5, 9), 8(1)
C_6H_5	$C_6H_4CH_3(4)$		8(1)
C_6H_5	$C_6H_4Cl(4)$		8(1)
C_6H_5	2-Pyridyl		9(1)
C_6H_5	4-Pyridyl		9(1)
$C_6H_4CH_3(4)$	C_6H_5		8(1)
$C_6H_4Cl(4)$	C_6H_5		8(1)
$C_6H_4OCH_3(4)$	$C_6H_4OCH_3(4)$		10(9)
CH_3	$CH(CH_3)_2$	CH_3	11(10)
CH_3	$CH_2CH(CH_3)_2$	CH_3	11(10)
CH_3	C_6H_5	CH_3	3(11)[b], 11(10), 12(10), 13(1), 14(10)
C_2H_5	CH_3	CH_3	3(4)
C_2H_5	C_6H_5	CH_3	11(10), 12(10), 13(1), 14(10)
C_6H_5	C_6H_5	CH_3	15(1)
$CH_2C_6H_5$	C_6H_5	CH_3	3(11), 11(10), 15(1)
	$(CH_2)_4$	CH_3	3(4, 11, 12), 11(10), 12(10), 13(1), 14(10), 15(1), 16(1), 17(10), 18(13), 19(1), 20(1), 21(10)
	$(CH_2)_4$	C_2H_5	21(10)
	$(CH_2)_4$	$(CH_2)_3CH_3$	16(10)
	$(CH_2)_4$	$CH_2C_6H_5$	16(10)
	$(CH_2)_4$	$COCH_3$	6(14), 22(14)[c]
	$(CH_2)_4$	$COCH_2Cl$	22(15)[c]
	$(CH_2)_5$	CH_3	18(13)
	$(CH_2)_2N(CH_3)CH_2$	CH_3	18(13)[d]
	CH_3	CH_3	12(10)
	C_2H_5	CH_3	12(10)
	$CH(CH_3)_2$	CH_3	12(10)
	C_6H_5	CH_3	18(13)

Notes

a. The corresponding pyrroles were also produced by passage of aldazines or ketazines through heated activated alumina although, unfortunately, the scope of this reaction was not defined in the abstract of the report of these studies (Suvorov, Zamyshlyaeva and Tupikina, 1969). A similar technique has also been used to effect Fischer indolizations, including that of acetaldehyde phenylhydrazone into indole (see Chapter IV, A); b. The product from this reaction was incorrectly named as 1,3,4-trimethyl-2,5-diphenylpyrrole in this report; c. The products from these reactions were formulated in this report as **1763** (R = $COCH_3$ and $COCH_2Cl$, respectively) but later studies (see below) suggested that these should be reformulated as the corresponding pyrroles; d. When a mixture of 1-methylpiperidin-4-one with N_α,N_β-dimethylhydrazine was heated in an acetic acid–benzene solution, **157** [$R^1 + R^2 = (CH_2)_2N(CH_3)CH_2$, $R^3 = CH_3$] was formed. However, when a mixture of the same reactants was heated in benzene alone, a mixture of this pyrrolic product along with **1764** [$R^1 + R^2 = (CH_2)_2N(CH_3)(CH_2)_2$, $R^3 = CH_3$] was formed. Upon heating with or without an acid catalyst, the latter product was converted into the former product (Sucrow and Chondromatidis, 1970). This formation of **1764** [$R^1 + R^2 = (CH_2)_2N(CH_3)(CH_2)_2$, $R^3 = CH_3$] was analogous to the formation of **1764** [$R^1 = H$, $R^2 = CH_3$, $R^3 = C_2H_5$ and $R^1 = R^2 = H$, $R^3 = CH(CH_3)_2$] from the reaction of N_α,N_β-diethylamine with acetaldehyde (Eberson and Persson, 1964) and of N_α,N_β-diisopropylhydrazine with formaldehyde (Zinner, Kliegel, Ritter, W. and Böhlke, 1966) respectively (for another similar reaction, see Grashey and Adelsberger, 1962). The latter hydrazine also reacted with isobutyraldehyde and 2-ethylbutyraldehyde to afford **1765** [R = $CH(CH_3)_2$ and $CH(C_2H_5)_2$, respectively] via **1764** [$R^1 = H$, $R^2 = CH(CH_3)_2$ and $CH(C_2H_5)_2$, respectively, $R^3 = CH(CH_3)_2$] (Zinner, Kliegel, Ritter, W. and Böhlke, 1966). Compounds of the type **1765** were converted into the corresponding pyrroles by thermolysis in a mass spectrometer sample chamber (Sucrow, Bethke and Chondromatidis, 1971); e. When 1,2-dialkylhydrazines reacted with aldehydes or ketones bearing at least one hydrogen atom on their C-2 atom(s), the bisenamine **1752** (R^3 and $R^4 \neq H$) structure was formed. This removed the first step of the pyrrolization mechanism (Scheme 58) which was apparently the stage dependent upon acid catalysis since the bisenamines **1752** (R^3 and $R^4 \neq H$) could be pyrrolized under fairly mild non-catalytic thermal conditions (e.g. in refluxing toluene). Many similar observations have been noted in connection with the Fischer indole synthesis (see Chapter II, E1).

Catalysts

1. Non-catalytic pyrolysis[e]; 2. Nickel(II) or cobalt(II) halides (e.g. cobalt(II) iodide or nickel(II) chloride); 3. Cobalt(II) iodide; 4. Pyrolysis of the azine methiodide; 5. Zinc chloride; 6. Nickel(II) chloride; 7. Hydrogen chloride; 8. Aniline hydrochloride or quinoline hydrochloride; 9. Dry hydrogen chloride passed over the molten azine; 10. 4-Methylbenzene sulphonic acid; 1. Methylamine hydrochloride; 12. Methylamine acetate; 13. Acetic acid; 14. Acetyl chloride; 15. Chloroacetyl chloride.

References

1. Tsuge, Tashiro, Hokama and Yamada, K., 1968; 2. Stapfer and D'Andrea, 1970; 3 Posvic, Dombro, Ito, H. and Telinski, 1974; 4. Piloty, 1910; 5. Perkin and Plant, 1924; 6. Kost and Grandberg, 1956a; 7. Robinson, G. M. and Robinson, R., 1918; 8. Tsuge, Hokama and Watanabe, 1969b; 9. Tsuge, Hokama and Watanabe, 1969a; 10. King, F. E. and Paterson, 1936; 11. Chapelle, Elguero, Jacquier and Tarrago, 1971; 12. Chapelle, Elguero, Jacquier and Tarrago, 1970b; 13. Jacquier, Chapelle, Elguero and Tarrago, 1969; 14. Chapelle, Elguero, Jacquier and Tarrago, 1969; 15. Fritz and Uhrhan, 1971a; 16. Sucrow, 1969; 17. Baumes, Jacquier and Tarrago, 1973; 18. Sucrow and Chondromatidis, 1970; 19. Tsuge, Tashiro and Hokama, 1968; 20. Baumes, Jacquier and Tarrago, 1976; 21. Schmitz and Fechner, 1969; 22. Benary, 1934.

When potential ambiguity existed in the direction of pyrrolization because of the nature of the ketonic moiety, reaction occurred in the same direction as the enolization of the corresponding ketone and therefore parallels the reaction directions observed in the indolization of the arylhydrazones of similar ketones (see Chapter III, G–L). A brief discussion of such pyrrolization ambiguity has been published (Chapelle, Elguero, Jacquier and Tarrago, 1971). Examples of

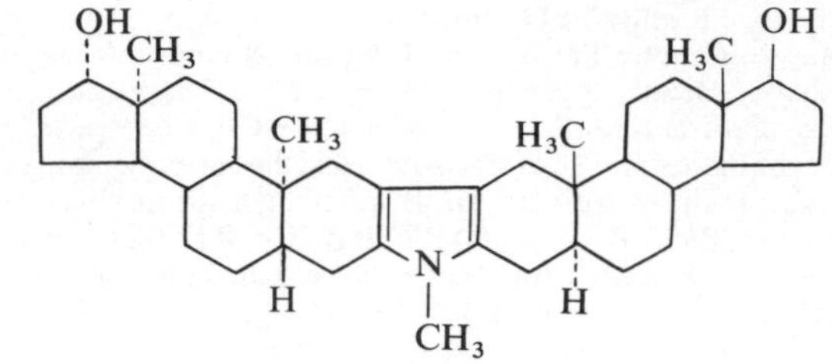

1761

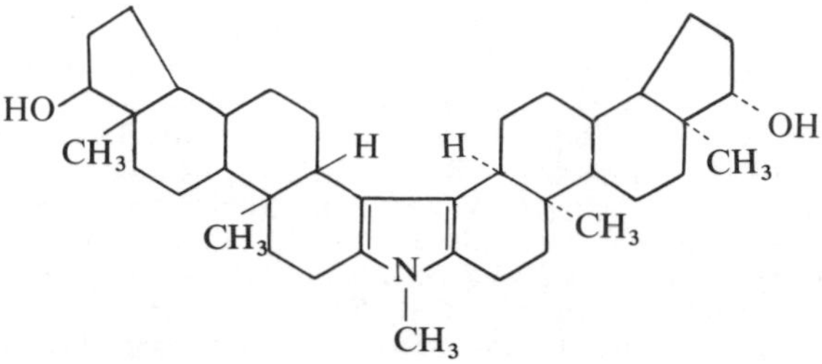

1762

such pyrrolizations are to be found in Table 39 and in the above described formations of **1758** (R = CH_3, n = 2), **1760**, **1761** and **1762** although, by analogy with the indolization of 3-ketosteroid arylhydrazones (Harvey, D. J. and Reid, S. T., 1972) (see Chapter III, J2), the formation of the last two products might be expected to be accompanied by the formation of small quantities of the corresponding isomers (two in each case) formed by pyrrolization occurring in either one or both possible alternative directions.

Starting with the monoenaminoazines **1756** [$R^3 \neq$ H (usually = CH_3)] the possibility arose that both N-unsubstituted pyrroles **157** (R^3 = H) and N-alkylated pyrroles **157** ($R^3 \neq$ H) may be formed from pyrrolization because of the two possible modes of cyclization, shown in **1766** and **1767**, respectively, the intermediates resulting from the rearrangement involving new C—C bond formation. Cases are reported (see Table 39) from such reactions in which the 1-alkylated pyrroles were formed (Baumes, Jacquier and Tarrago, 1973, 1974; Chapelle, Elguero, Jacquier and Tarrago, 1971; Fritz and Uhrhan, 1971a), in which the 1-unsubstituted pyrroles were formed (Baumes, Jacquier and Tarrago, 1974; Posvic, Dombro, Ito, H. and Telinski, 1974) and in which both possible pyrroles were formed (Baumes, Jacquier and Tarrago, 1973, 1974; Posvic, Dombro, Ito, H. and Telinski, 1974). It appeared that in cases which were governed by electronic effects, the more nucleophilic N-alkylated nitrogen atom attacked the more electrophilic imine function to produce ultimately the N-alkylated pyrrole, whereas when adverse steric effects were operating the additional steric crowding which an N-alkyl substituent would have imposed directed the reaction towards the formation of the N-unsubstituted pyrrole

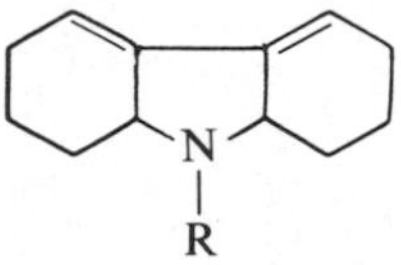

1763

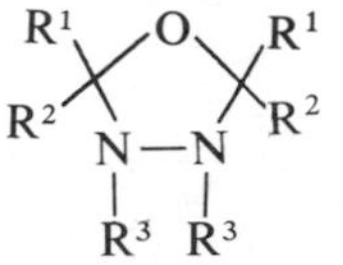

1764

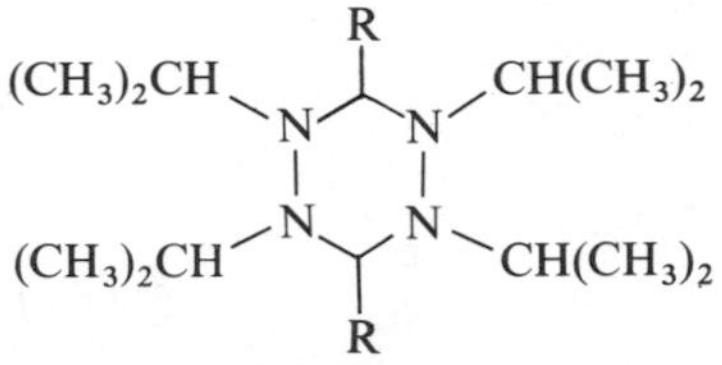

1765

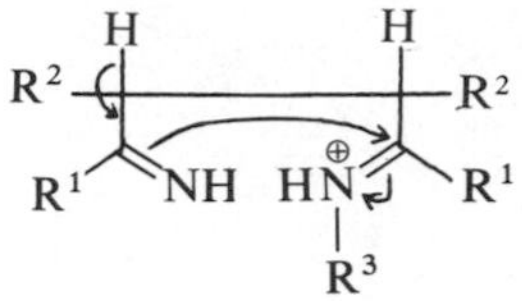

1766 $\downarrow$ $-CH_3NH_2$ **157** (R^3 = H)

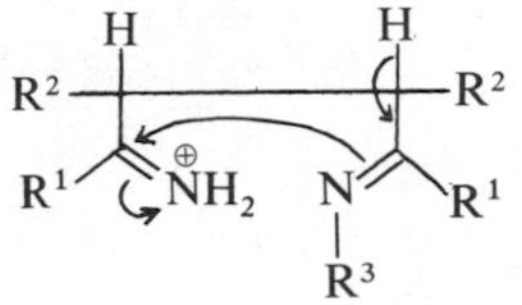

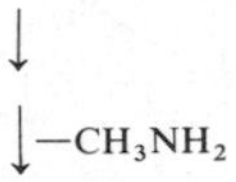

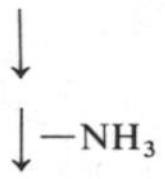

1767 $\downarrow$ $-NH_3$ **157**

(Baumes, Jacquier and Tarrago, 1974; Posvic, Dombro, Ito, H. and Telinski, 1974).

By heating with 4-methylbenzene sulphonic acid in toluene solution, a mixture of **1768** (R^1 = H, $R^2 = R^3 = CH_3$) (formed by heating cyclohexan-1,3-dione with N_α,N_β-dimethylhydrazine in benzene solution) with cyclohexanone afforded **1769** (R^1 = H, $R^2 = CH_3$) (Sucrow, and Wiese, 1970), of **1768** (R^1–$R^3 = CH_3$) [formed by heating 5,5-dimethylcyclohexan-1,3-dione (dimedone) with N_α,N_β-dimethylhydrazine in benzene solution] with cyclohexanone afforded **1769** (R^1 = H, $R^2 = CH_3$) (Sucrow, and Wiese, 1970), of Wiese, 1970), of **1768** (R^1 = H and CH_3, $R^2 = R^3 = CH_3$) with **874** (R^1–R^7 = H, R^8 = OH) furnished **1770** (R^1 = H and CH_3, respectively, R^2 = OH) (Sucrow and Wiese, 1970), of **1768** (R^1–$R^3 = CH_3$) with **874** [R^1–R^7 = H, $R^8 = CH(CH_3)(CH_2)_3CH(CH_3)_2$] furnished **1770** [$R^1 = CH_3$, $R^2 = CH(CH_3)(CH_2)_3CH(CH_3)_2$] (Sucrow and Wiese, 1970) [traces of products isomeric with the **1770**s obtained in these pyrrolizations might also be expected by analogy with the corresponding indolizations (see Chapter III, J2)], of **1768** (R^1 = H and CH_3, $R^2 = R^3 = CH_2C_6H_5$) with cyclohexanone yielded **1769** (R^1 = H and CH_3, respectively, $R^2 = CH_2C_6H_5$), of **1768** ($R^1 = R^2 = CH_3$,

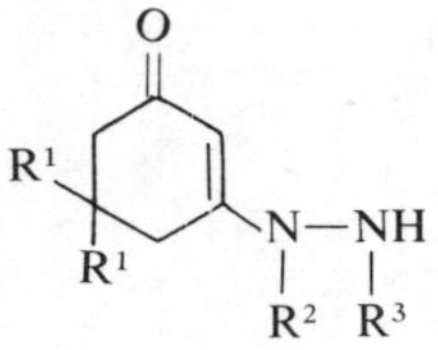

1768 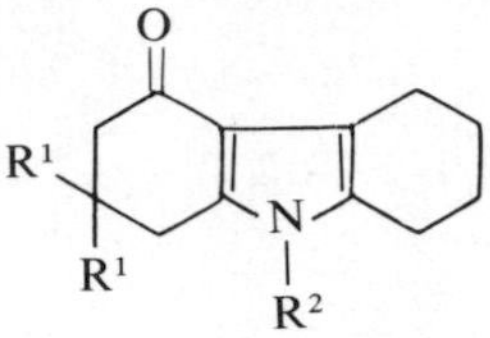 **1769**

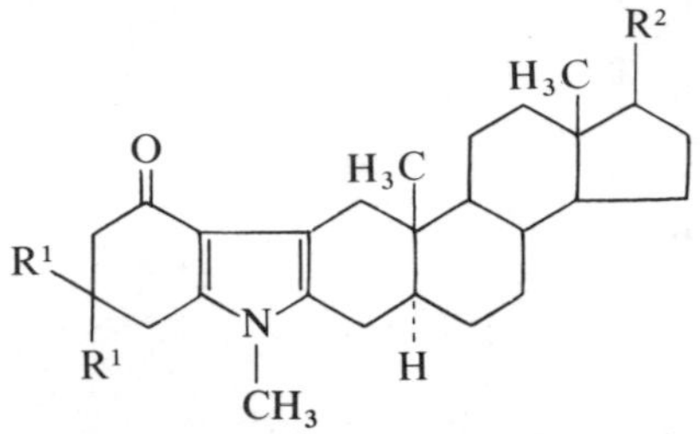

1770

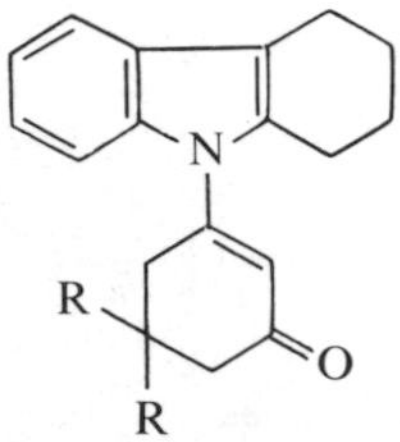

1771 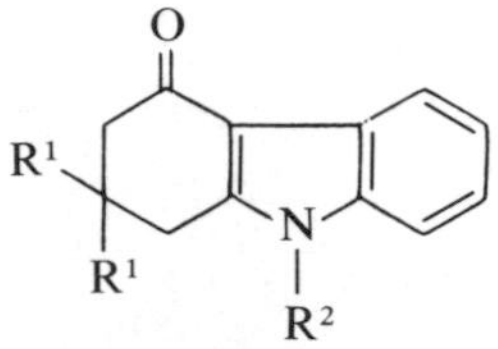 **1772**

$R^3 = CH_2C_6H_5$) with cyclohexanone gave only **1769** ($R^1 = R^2 = CH_3$), as evidenced by t.l.c. analysis of the total reaction mixture, and benzylamine, indicating the specificity of this potentially ambiguous pyrrolization, and of **1768** (R^1 = H and CH_3, $R^2 = C_6H_5$, $R^3 = CH_2C_6H_5$) with cyclohexanone furnished both **1769** (R^1 = H, $R^2 = C_6H_5$) and **1771** (R = H) and both **1769** ($R^1 = CH_3$, $R^2 = C_6H_5$) and **1771** ($R = CH_3$), respectively (Sucrow, Slopianka and Mentzel, 1973). Dehydrogenation of **1769** (R^1 = H and CH_3, $R^2 = CH_3$ and C_6H_5), by heating with chloranil in xylene solution, gave **1772** (R^1 = H, $R^2 = CH_3$) (Sucrow and Wiese, 1970), **1772** ($R^1 = R^2 = CH_3$) (Sucrow, 1969; Sucrow and Wiese, 1970) and **1772** (R^1 = H and CH_3, $R^2 = C_6H_5$) (Sucrow, Slopianka and Mentzel, 1973), respectively. This therefore led to a synthetic approach to 1,2,3,4-tetrahydrocarbazol-4-ones alternative to that involving the Fischer indolization of cyclohexan-1,3-dione monoarylhydrazones (see Chapter III, K2).

Addition of dimethyl acetylenedicarboxylate to cyclopentanone and cyclohexanone N_α-methylhydrazones [**1773**; $R^1 + R^2 = (CH_2)_4$ and $(CH_2)_5$,

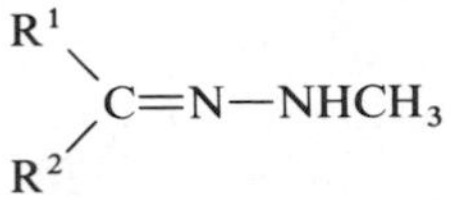

$R^1R^2C{=}N{-}NHCH_3$

1773

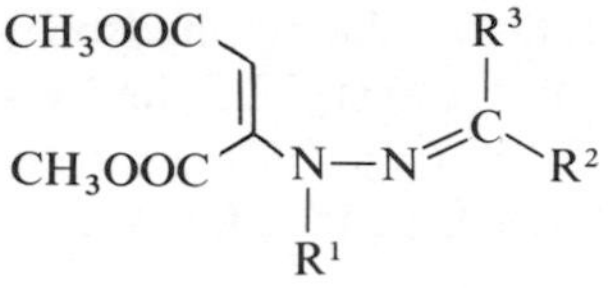

1774

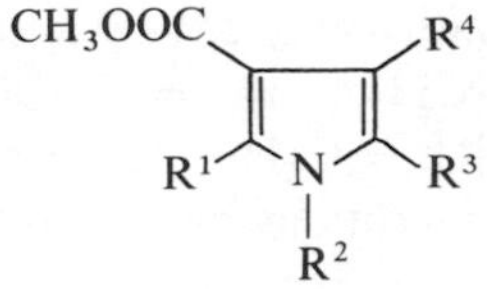

1775

$CH_3OOC{-}C{\equiv}CH$

1776

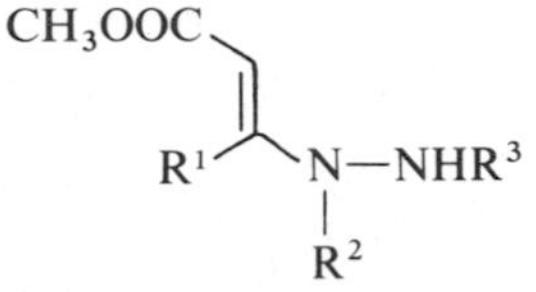

1777

respectively] gave **1774** [$R^1 = CH_3$, $R^2 + R^3 = (CH_2)_4$ and $(CH_2)_5$, respectively] which could be cyclized by heating in xylene or acetic acid to yield **1775** [$R^1 = COOCH_3$, $R^2 = CH_3$, $R^3 + R^4 = (CH_2)_3$] and **1775** [$R^1 = COOCH_3$, $R^2 = H$ and CH_3, $R^3 + R^4 = (CH_2)_4$], respectively (Bardakos, Sucrow and Fehlauer, 1975). The generation of the species **1774** was effected *in situ* when dimethyl acetylenedicarboxylate reacted with **1773** [$R^1 = CH_3$, $R^2 = C_2H_5$ and $CH_2C_6H_5$; $R^1 = C_2H_5$, $R^2 = C_2H_5$ and $CH_2C_6H_5$ and $R^1 + R^2 = (CH_2)_4$ and $(CH_2)_5$] in refluxing xylene to yield **1775** [$R^1 = COOCH_3$; R^2–$R^4 = CH_3$; $R^2 = H$ and CH_3, $R^3 = CH_3$, $R^4 = C_6H_5$; $R^2 = H$ and CH_3, $R^3 = C_2H_5$, $R^4 = CH_3$; $R^2 = H$ and CH_3, $R^3 = C_2H_5$, $R^4 = C_6H_5$ and $R^2 = H$ and CH_3, $R^3 + R^4 = (CH_2)_3$ and $(CH_2)_4$, respectively] (Baumes, Jacquier and Tarrago, 1974). In a similar manner, **1776** reacted with **1773** [$R^1 = CH_3$, $R^2 = C_2H_5$ and $CH_2C_6H_5$; $R^1 = C_2H_5$, $R^2 = C_2H_5$ and $CH_2C_6H_5$ and $R^1 + R^2 = (CH_2)_4$ and $(CH_2)_5$] to afford **1775** [$R^1 = H$; R^2–$R^4 = CH_3$; $R^2 = H$ and CH_3, $R^3 = CH_3$, $R^4 = C_6H_5$; $R^2 = H$ and CH_3, $R^3 = C_2H_5$, $R^4 = CH_3$; $R^2 = H$, $R^3 = C_2H_5$, $R^4 = C_6H_5$ and $R^2 = CH_3$, $R^3 + R^4 = (CH_2)_3$ and $(CH_2)_4$, respectively] (Baumes, Jacquier and Tarrago, 1974). The last mentioned product was also formed when **1777** ($R^1 = H$, $R^2 = R^3 = CH_3$) was heated with cyclohexanone in a toluene–acetic acid mixture (Sucrow and Grosz, 1976). Compounds **1774** [$R^1 = CH_3$, $R^2 = H$, $R^3 = C_6H_5$; R^1–$R^3 = CH_3$; $R^1 = CH_3$, $R^2 + R^3 = (CH_2)_5$] have also been prepared by reacting benzaldehyde, acetone and cyclohexanone, respectively, with **1777** ($R^1 = COOCH_3$, $R^2 = CH_3$, $R^3 = H$) which was synthesized by reaction of dimethyl acetylenedicarboxylate with methyl hydrazine. Similarly, the reaction of dimethyl acetylenedicarboxylate with N_α, N_β-dimethylhydrazine gave **1777** ($R^1 = COOCH_3$, $R^2 = R^3 = CH_3$) and of **1776** with benzylhydrazine afforded **1777** ($R^1 = R^3 = H$, $R^2 = C_6H_5CH_2$) (Fehlauer, Grosz, Slopianka, Sucrow, Lockley and Lwowski, 1976).

Treatment of biscyclohexanone azine [**154**; $R^1 + R^2 = (CH_2)_4$] and bis-diethyl ketone azine (**154**; $R^1 = C_2H_5$, $R^2 = CH_3$] with boiling acetic anhydride gave 9-acetyl-1,2,3,4,5,6,7,8-octahydrocarbazole [**157**; $R^1 + R^2 = (CH_2)_4$, $R^3 = COOCH_3$] and 1-acetyl-2,5-diethyl-3,4-dimethylpyrrole (**157**; $R^1 = C_2H_5$, $R^2 = CH_3$, $R^3 = COCH_3$), respectively, whereas under these conditions biscyclopentanone azine [**154**; $R^1 + R^2 = (CH_2)_3$] afforded **1778**.

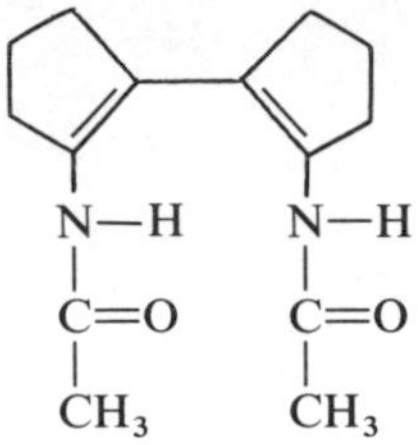

N—H N—H
C=O C=O
CH_3 CH_3

1778

Although this compound could not be converted into **157** [$R^1 + R^2 = (CH_2)_3$, R^3 = H or $COCH_3$], it can be regarded as a trapped intermediate in the pyrrolization mechanistic pathway corresponding to **1754** in Scheme 58, such trapping being possible because of the strain in the 5-5-5 ring system of the potential pyrrolic product which thus prevented its formation (Posvic, Dombro, Ito, H. and Telinski, 1974).

When biscyclohexanone azine [**154**; $R^1 + R^2 = (CH_2)_4$] was treated with hydrogen chloride, a product was obtained which was formulated as **157** [$R^1 + R^2 = (CH_2)_4$, R^3 = H] (Perkin and Plant, 1924). Since methylation of this product afforded a compound which was not identical with the compound previously obtained by catalytic hydrogenation of 9-methylcarbazole and formulated as **157** [$R^1 + R^2 = (CH_2)_4$, $R^3 = CH_3$] (Braun and Ritter, 1922), this hydrogenation product was reformulated as **1779** or **1780** (Perkin and Plant, 1924). After further investigations, these suggestions were rejected by others (Braun and Bayer, 1925; Braun and Schörnig, 1925) who in turn suggested structure **1763** (R = H) as a possible structure for the product obtained by acid catalysed deaminative cyclization of the azine. After the advancement of further counter argument (Perkin and Plant, 1925), compound **157** [$R^1 + R^2 = (CH_2)_4$, $R^3 = CH_3$] was unambiguously synthesized and was found to be identical with the catalytic hydrogenation product of 9-methylcarbazole, observations which led to the repeated reformulation of the Piloty cyclization product as **1763** (R = H). Later, using acetyl chloride and chloroacetyl chloride, biscyclohexanone azine was converted into products which,

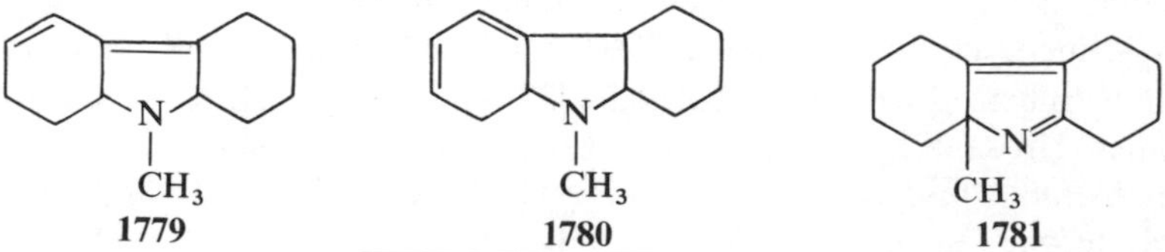

N
CH_3
1779

N
CH_3
1780

N
CH_3
1781

by analogy with these earlier related conclusions, were formulated as **1763** (R = $COCH_3$ and $COCH_2Cl$, respectively) (Benary, 1934). In a much later repeat of the original cyclization, using zinc chloride instead of hydrogen chloride as the catalyst, the product was formulated as **157** [$R^1 + R^2 = (CH_2)_4$, R^3 = H] without experimental verification (Kost and Grandberg, 1956a) although this was later (Robinson, B., 1964c) obtained by u.v., i.r. and p.m.r. spectroscopic studies which also showed that the methylation product obtained earlier (Perkin and Plant, 1924) had structure **1781**. A later (Stapfer and D'Andrea, 1970) formulation as **1763** (R = H), rather than **157** [$R^1 + R^2 = (CH_2)_4$, R^3 = H], of the product resulting from the nickel(II) chloride catalysed transformation of biscyclohexanone azine, is obviously erroneous.

Biscyclohexanone azine has also been suggested as an intermediate in the decomposition of cyclohexanone phenylsemicarbazone and semicarbazone by heating with zinc chloride. Under these conditions the former reactant gave aniline (47%), 1,2,3,4-tetrahydrocarbazole (10%) and 1,2,3,4,5,6,7,8-octahydrocarbazole (24%) and the latter reactant gave 1,2,3,4-tetrahydrocarbazole (3%) and 1,2,3,4,5,6,7,8-octahydrocarbazole (54%). Pyrolysis of the latter reactant in the absence of a catalyst yielded **1782** and biscyclohexanone azine. Furthermore, heating biscyclohexanone azine with zinc chloride yielded 1,2,3,4,5,6,7,8-octahydrocarbazole (74%) and 1,2,3,4-tetrahydrocarbazole (12%) and the former product afforded the latter product upon pyrolysis. These above data led to the suggestion that the zinc chloride catalysed decomposition of cyclohexanone phenylsemicarbazone and semicarbazone proceeded by N—N bond homolysis to afford **1783** and **1784** and **1783** and **1785**, respectively. In both cases radical pairing of the **1783**s afforded biscyclohexanone azine and ultimately the hydrocarbazoles, radical **1784** was stabilized by conversion into aniline and radical pairing of **1785** produced **1782** (Marshalkin and Yakhontov, 1978).

$H_2NC(=O)NHNHC(=O)NH_2$ **1782** $\dot{N}$=cyclohexylidene **1783** $C_6H_5NHC(=O)\dot{N}H$ **1784** $H_2NC(=O)\dot{N}H$ **1785**

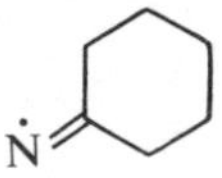

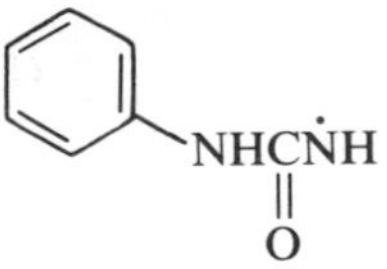

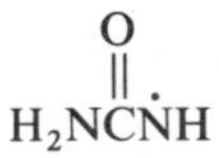

The Piloty pyrrole synthesis has been extended to a synthesis of pyrrolin-5-ones **1788** (R^5 = H and CH_3) by treating acylhydrazones **1786** and acylenehydrazines **1787**, respectively, with sodium methylate, reactions which are analogous to the extension of the Fischer indole synthesis to the Brunner oxindole synthesis (see Chapter VI, A). By this method, starting with the appropriate **1786** or **1787**, the pyrrolin-5-ones **1788** [R^1 = H, $R^2 = C_6H_5$, $R^3 + R^4 = (CH_2)_4$, $R^5 = CH_3$] (Fritz and Uhrhan, 1971b, 1972), **1788**

[R^1 = H, R^2–R^4 = C_6H_5, R^5 = CH_3; R^1 = R^2 = R^5 = CH_3, R^3 = R^4 = C_6H_5; R^1 = R^2 = R^5 = CH_3, R^3 = H, R^4 = C_6H_5 and R^1 = R^5 = H, R^2 = C_6H_5, $R^3 + R^4$ = $(CH_2)_4$], **1788** [R^1 = R^2 = R^5 = CH_3, $R^3 + R^4$ = $(CH_2)_4$], together with its isomer **1789**, and **1790**, formed via the intermediate **1788** (R^1 = R^3 = H, R^2 = R^4 = C_6H_5, R^5 = CH_3), have been prepared (Fritz and Uhrhan, 1972). An obvious mechanism which encompasses the basic principles of the mechanisms of both the Fischer indole synthesis and the Brunner oxindole synthesis was proposed (Fritz and Uhrhan, 1972) for these reactions.

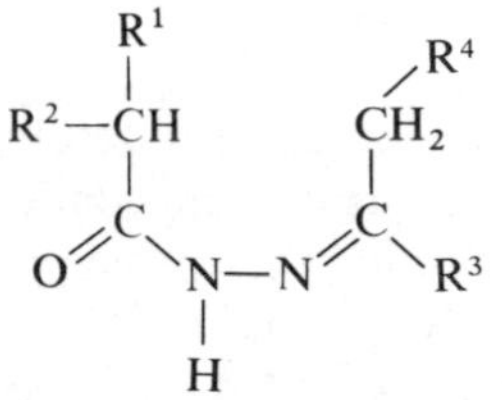

1786

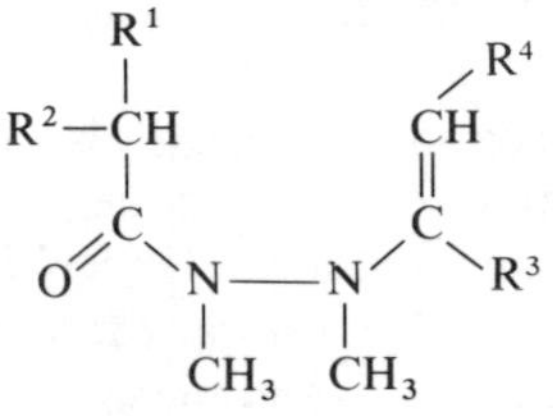

1787

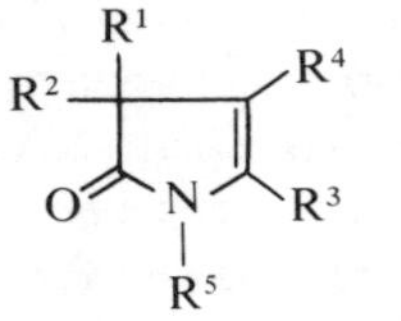

1788 (R^5 = H or CH_3)

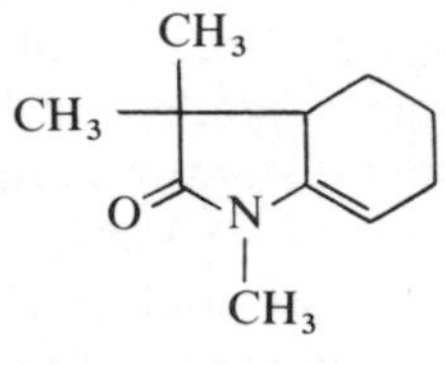

1789

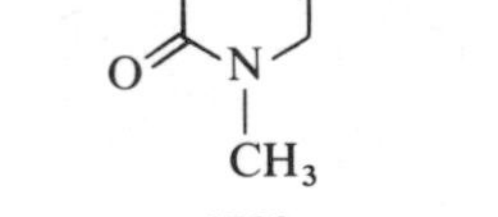

1790

By analogy with the benzofuranization of O-aryloximes (see Chapter VI, C), the O-vinyloximes **1791** [R^1 = C_6H_5, R^2 = H and $R^1 + R^2$ = $(CH_2)_4$], formed by reacting dimethyl acetylenedicarboxylate with acetophenone and cyclohexanone oximes, respectively, in boiling methanolic sodium methoxide, afforded **1792** [R^1 = C_6H_5, R^2 = H and $R^1 + R^2$ = $(CH_2)_4$, respectively] upon heating. The mechanism postulated for these reactions is given in Scheme 59 (Sheradsky, 1970), the similarity with that of the Piloty pyrrole synthesis (Scheme 58) being obvious. The formation of the pyrroles rather than the

Scheme 59

corresponding furans from these two reactions is of interest, especially in relation to the benzofuranization of O-aryloximes discussed earlier (see this chapter, C).

In a review of recent synthetic methods for pyrroles (Patterson, 1976), brief references have been made to some of the above studies.

Several failures to pyrrolize azines have been observed. Thus, attempted acid catalysed pyrrolization of **1793** was unsuccessful (Harradence, Hughes, G. K. and Lions, 1938) and failure also attended the attempted thermal pyrrolization of **154** (R^1 = H, R^2 = H and CH_3; R^1 = CH_3 and C_2H_5, R^2 = CH_3 and R^1 = CH_3, R^2 = H) (Tsuge, Tashiro and Hokama, 1968), **1794** (R = H, X = S) (Tsuge, Watanabe and Hokama, 1971), **1794** (R = C_6H_5, X = S) (Tsuge, Hokama and Watanabe, 1969a), **1794** (R = H, X = O) (Tsuge, Watanabe and Hokama, 1971) and **1795** (Tsuge, Hokama and Koga, 1969) although several other products from these reactions were isolated and characterized. The reported (Kost and Grandberg, 1956a) failures to pyrrolize

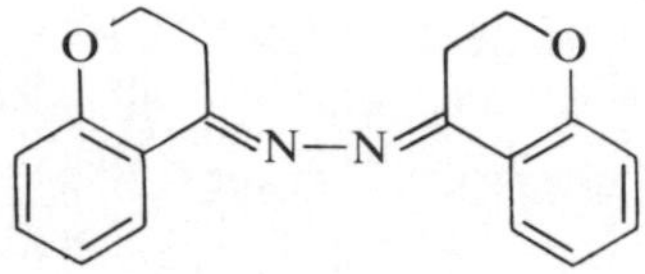

1793

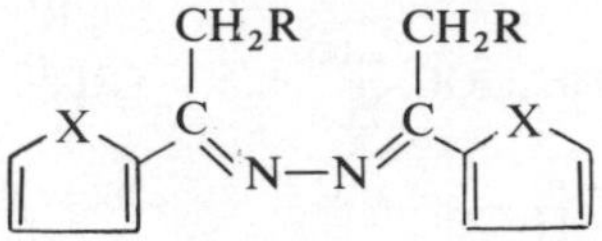

1794

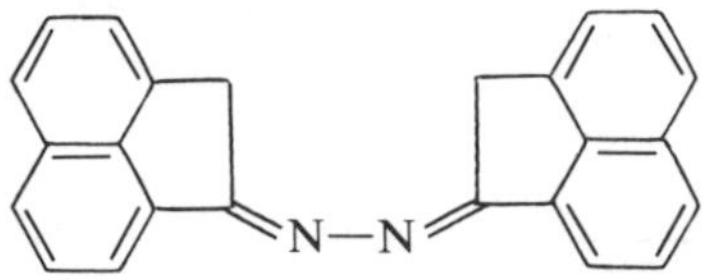

1795

bisacetophenone (see also Fritz and Uhrhan, 1971a), bisbutyraldehyde, bisisovaleraldehyde and biscyclopentanone azine using zinc chloride as the catalyst were no doubt the result of the unsuitability of the catalyst since the first of these azines, which in an acetic acid–hydrochloric acid mixture at room temperature afforded acetophenone (Tsuge, Samura and Tashiro, 1972), has been pyrrolized upon pyrolysis (see Table 39) and the corresponding N-methylpyrroles have been directly obtained by Piloty syntheses using the ketonic moieties of the next two with the appropriate hydrazine (see Table 39). Treatment of **154** [$R^1 + R^2 = (CH_2)_3$] with hydrogen chloride, conditions which pyrrolized the cyclohexanone homologue, yielded only **1796** and attempts to pyrrolize cyclohexanone cyclopentanone azine failed to yield any crystalline product (Perkin and Plant, 1925).

1796

Under the appropriate acidic conditions, bisketone and bisaldehyde azines can also undergo a reaction alternative to pyrrolization when they are isomerized to afford pyrazolines. The first azine to be so isomerized was bisacetone azine **154** ($R^1 = CH_3$, $R^2 = H$) which, upon heating with maleic acid, gave **1797** ($R^1 = R^3 = R^4 = CH_3$, $R^2 = R^5 = R^6 = H$) (Curtius and Försterling, 1894), a transformation which was subsequently effected by heating the azine with oxalic acid (Frey, K. W. and Hofmann, R., 1901; Kost and Grandberg, 1965b), with aqueous tartaric acid, boric acid, hydrochloric acid, metaphosphoric acid, orthophosphoric acid, succinic acid or tartaric acid or with dry hydrogen chloride in toluene (Frey, K. W. and Hofmann, R., 1901), with cobalt(II) bromide (Stapfer and D'Andrea, 1970) or using unspecified conditions (Robinson, G. M. and Robinson, R., 1918). Heating with maleic acid likewise converted the azines **154** [$R^1 = C_2H_5$, $(CH_2)_2CH_3$ and $(CH_2)_5CH_3$, $R^2 = H$] into **1797** [$R^1 = R^3 = C_2H_5$, $(CH_2)_2CH_3$ and $(CH_2)_5CH_3$, respectively, $R^4 = CH_3$, $R^2 = R^5 = R^6 = H$] although the possibility that cyclization had involved the 2-methylene groups rather than the 2-methyl groups was not eliminated (Curtius and Zinkeisen, 1898). Similarly, heating bisethyl methyl ketone azine with succinic acid afforded a product which was formulated as **1797** ($R^1 = R^3 = C_2H_5$, $R^4 = CH_3$, $R^2 = R^5 = R^6 = H$) rather than **1797** ($R^1 = R^3 = R^5 = CH_3$, $R^4 = C_2H_5$, $R^2 = R^6 = H$) (Kizhner, 1912). This postulation was subsequently (Kost and Grandberg, 1956b) verified when this latter transformation was effected by heating with oxalic acid [a later group (Stapfer and D'Andrea, 1970) used nickel(II) or cobalt(II) halides as catalysts] and the resulting product was then unambiguously synthesized. In other analogous transformations, **154** ($R^1 = R^2 = H$) gave **1797** (R^1–$R^3 = R^5 = R^6 = H$, $R^4 = CH_3$) upon heating with nickel(II)

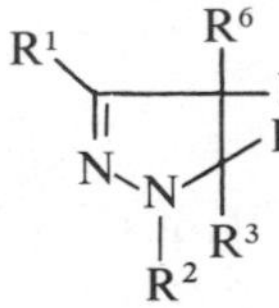

or cobalt(II) halides (Stapfer and D'Andrea, 1970), a transformation which was alluded to earlier (Robinson, G. M. and Robinson, R., 1918) without any experimental or structural details and which was also effected, as was the conversion of **154** ($R^1 = H$, $R^2 = CH_3$) into **1797** (R^1–$R^3 = R^5 = H$, $R^4 = C_2H_5$, $R^6 = CH_3$), though neither very satisfactorily, using a maleic acid catalyst (Curtius, 1916; Curtius and Zinkeisen, 1898). However, with this last catalyst or with hydrochloric acid or hydriodic acid, bisisobutyraldehyde azine readily furnished **1797** [R^1–$R^3 = H$, $R^4 = CH(CH_3)_2$, $R^5 = R^6 = CH_3$] (Curtius and Zinkeisen, 1898; Franke, A., 1899). Similarly, bismethyl propyl ketone azine (**154**; $R^1 = CH_3$, $R^2 = C_2H_5$) and acetone cyclohexanone azine (**1798**) yielded **1797** [$R^1 = R^3 = (CH_2)_2CH_3$, $R^4 = CH_3$, $R^2 = R^5 = R^6 = H$ and $R^1 = CH_3$, $R^3 + R^4 = (CH_2)_5$, $R^2 = R^5 = R^6 = H$, respectively] upon heating with oxalic acid (Kost and Grandberg, 1956b) [the structure of the latter product was established by unambiguous synthesis (Grandberg, Kost and Terentyev, 1956)] although attempts to similarly cyclize bisisopropyl methyl ketone azine, by heating with either oxalic acid or formic acid, only caused extensive resinification or led to recovery of the azine (Kost, Grandberg and Evreinova, 1958).

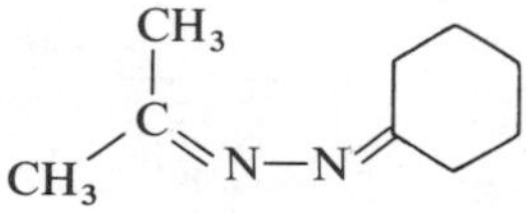

It has been suggested (Stapfer and D'Andrea, 1970) that such pyrazoline formation was favoured over pyrrolization when a methyl and hydrogen or methylene group was the substituent unit on the nitrogen-bonding carbon atoms of the azine. The reaction pathway may also be dependent upon the nature of the catalyst. Clearly, the former suggestion is not in agreement with the above mentioned conversions of bisaldehyde azines into pyrazolines. However, in agreement with the postulation are the observations that heating mixtures of acetone and methyl ethyl, methyl propyl, butyl methyl, isobutyl methyl, isopentyl methyl and methyl neopentyl ketone with 1,2-dimethylhydrazine in the presence of trace amounts of 4-methylbenzene sulphonic acid gave the pyrazolines **1799** [R = CH_3, C_2H_5, $(CH_2)_2CH_3$, $(CH_2)_3CH_3$, $CH_2CH(CH_3)_2$, $(CH_2)_2CH(CH_3)_2$ and $CH_2C(CH_3)_3$, respectively] whereas, under similar conditions, cyclohexanone, benzyl methyl ketone, benzyl ethyl ketone, propionaldehyde and 2-tetralone afforded only the corresponding pyrrolization products [see Table 39 and **1758** (R = CH_3, n = 2)]. Similarly in agreement, the pyrazoline **1799** (R = C_6H_5), the pyrazole **1800** and the pyrazolinium salt **1801** (isolated as the picrate) were formed when mixtures of

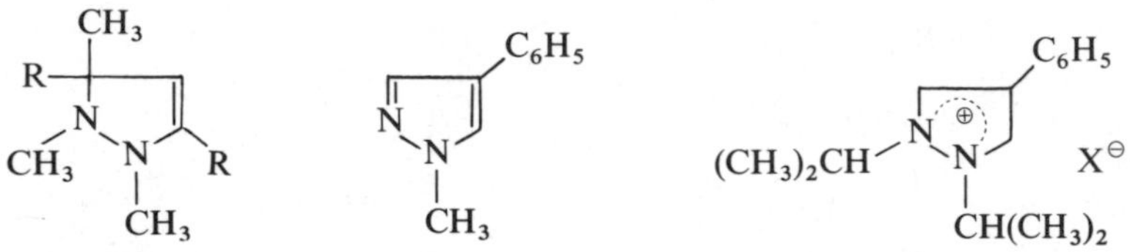

1799 **1800** **1801**

acetophenone with 1,2-dimethylhydrazine (Chapelle, Elguero, Jacquier and Tarrago, 1970b; Fritz and Uhrhan, 1971a; Jacquier, Chapelle, Elguero and Tarrago, 1969) and of phenylacetaldehyde with methylhydrazine and 1,2-diisopropylhydrazine (Fritz and Uhrhan, 1971a), respectively, were heated and butyraldehyde and isovaleraldehyde afforded mixtures of the expected pyrroles (see Table 39) and 4-ethyl-1,2-dimethyl-5-propyl-3-pyrazoline and 5-isobutyl-4-isopropyl-1,2-dimethyl-3-pyrazoline, respectively (Chapelle, Elguero, Jacquier and Tarrago, 1970b). However, further observations contrary to the above former postulation were made when biscyclohexanone azine [**154**; $R^1 + R^2 = (CH_2)_4$] was heated with oxalic acid (other catalysts afforded lower yields of product) to give the pyrazoline **1797** [$R^1 + R^5 = (CH_2)_4$, $R^3 + R^4 = (CH_2)_5$, $R^2 = R^6$ = H] which could be isolated or generated *in situ*, losing nitrogen to afford **1802** (n = 2) which ultimately gave rise to 1-cyclohexylcyclohexene (Kost and Grandberg, 1955b, 1956a – see also Kost and Grandberg, 1955a). Further studies of this type of reaction have been made (Grandberg, Kost and Terentyev, 1956). The pyrazoline **1797** [$R^1 + R^5 = (CH_2)_4$, $R^3 + R^4 = (CH_2)_5$, $R^2 = R^6$ = H] was also formed as the major product, along with some aniline and some **157** [$R^1 + R^2 = (CH_2)_4$,

R^3 = H], the Piloty cyclization product, when **154** [$R^1 + R^2 = (CH_2)_4$] was heated with aniline hydrochloride (Kost and Grandberg, 1956a). Biscyclopentanone azine **154** [$R^1 + R^2 = (CH_2)_3$] reacted similarly to its homologue when heated with cobalt(II) chloride to afford **1802** (n = 1) although this was probably not formed via the corresponding pyrazoline **1797** [$R^1 + R^5 = (CH_2)_3$, $R^3 + R^4 = (CH_2)_4$, $R^2 = R^6$ = H] but via **1803** (Stapfer and

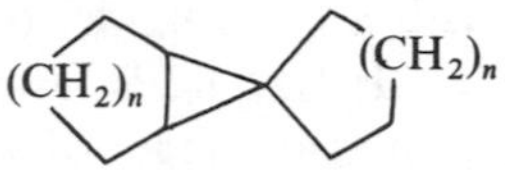

$(CH_2)_n$ $(CH_2)_n$ NNH_2

1802 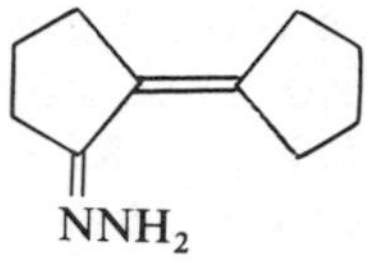 **1803**

D'Andrea, 1970). However, previous attempts to prepare **1797** [$R^1 + R^5 = (CH_2)_3$, $R^3 + R^4 = (CH_2)_4$, $R^2 = R^6$ = H] by treating biscyclopentanone azine [**154**; $R^1 + R^2 = (CH_2)_3$] with oxalic acid (Kupletskaya, Kost and Grandberg, 1956) or formic acid (Kost, Grandberg and Evreinova, 1958) [but not with (Kost, Grandberg and Evreinova, 1958) acetic acid as earlier (Kupletskaya, Kost and Grandberg, 1956) claimed] appeared to have been successful. The reaction product, m.p. 140 °C, was, upon the evidence of an empirical formula determination of $C_{15}H_{22}N_2$ and u.v. spectroscopic data, formulated as either **1804** or **1805** and was suggested as being formed via **1797** [$R^1 + R^5 = (CH_2)_3$, $R^3 + R^4 = (CH_2)_4$, $R^2 = R^6$ = H] (Kost, Grandberg and Evreinova, 1958). Clearly, neither of these structural proposals was verified and further investigation would therefore be of interest.

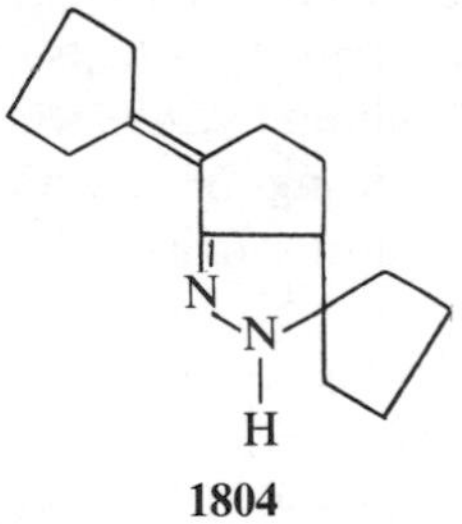

N N H N N H

1804 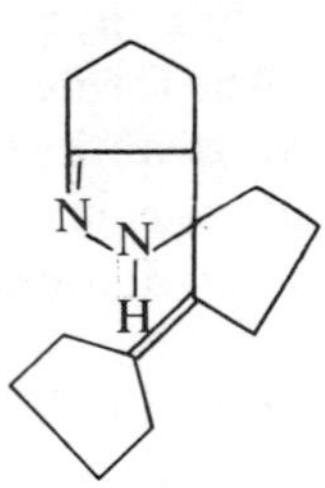 **1805**

Treatment with formic acid has also been used to convert **154** [$R^1 + R^2 = (CH_2)_4$] (Kost and Grandberg, 1955b), **154** ($R^1 = CH_3$, $R^2 = CH_3$ and C_2H_5) (Kost and Grandberg, 1956b), **154** [R^1 = H, R^2 = H, CH_3, C_2H_5, $CH(CH_3)_2$ and $(CH_2)_3CH_3$] and bisisobutyraldehyde azine (Kost and Grandberg, 1956c) into the corresponding 1-formylpyrazolines from which the pyrazolines can be obtained by hydrolysis. Similarly, further reaction at the N-1 atom of the pyrazoline nucleus occurred when the cyclization was carried out by reacting azines with alkyl halides. Thus, when bisacetone azine (**154**; $R^1 = CH_3$, $R^2 = H$) was heated with butyl bromide or benzyl chloride or reacted with allyl bromide or chloride at room temperature, the bromide or chloride of 3,5,5-trimethylpyrazoline was formed in each case along with its 1-butyl, 1-benzyl and 1-allyl derivatives, respectively (Kost, Golubeva and Grandberg, 1956). An analogous reaction also occurred with bisacetone azine in refluxing ethyl iodide (Kost, Grandberg and Golubeva, 1956). Treatment with alkyl halides also effected the cyclization of bisaldehyde azines. Thus, when bisisobutyraldehyde azine was reacted with allyl bromide, benzyl chloride or methyl iodide, **1797** [R^1–R^3 = H, $R^4 = CH(CH_3)_2$, $R^5 = R^6 = CH_3$] was produced in each case along with the corresponding 1-allyl, 1-benzyl or 1-methyl derivatives, respectively, and likewise bisisovaleraldehyde azine [**154**; R^1 = H, $R^2 = CH(CH_3)_2$], upon refluxing with methyl iodide, afforded **1797** [R^1–$R^3 = R^5$ = H, $R^4 = CH_2CH(CH_3)_2$, $R^6 = CH(CH_3)_2$] and its 1-methyl homologue (Kost, Grandberg and Golubeva, 1956). Acid chlorides also effected similar conversions when bisbutyraldehyde azine was refluxed with benzoyl chloride to furnish **1797** [$R^1 = R^3 = R^5$ = H, $R^2 = COC_6H_5$, $R^4 = (CH_2)_2CH_3$, $R^6 = C_2H_5$] and 1,2-dibenzoylhydrazine and when bis-acetone azine was refluxed with acetyl chloride to furnish **1797** ($R^1 = R^3 = R^4 = CH_3$, $R^2 = COCH_3$, $R^5 = R^6$ = H). However, treatment of bis-cyclopentanone azine with acetyl chloride or benzoyl chloride under similar conditions gave only 1,2-diacetyl- and 1,2-dibenzoylhydrazine, respectively, and likewise only 1,2-dibenzoylhydrazine was isolated when bisisopropyl methyl ketone azine was refluxed with benzoyl chloride (Kost, Grandberg and Evre-inova, 1958). When bisphenylacetaldehyde and bisbenzyl methyl, bisbenzyl ethyl, bisbenzyl propyl, bisbenzyl isopropyl and bisbenzyl phenyl ketone azine were heated with sodium hydride, both the corresponding pyrroles and pyra-zoles were formed. The formation of the former products was favoured by the use of small catalytic quantitites of the sodium hydride whereas the formation of the latter products was favoured when using the larger alkyl benzyl ketone azines and equimolar amounts of sodium hydride (Sorokin, Larshin and Grandberg, 1977).

The Piloty synthesis also failed using 3-ketoesters. Thus, when ethyl aceto-acetate was heated with 1,2-dimethylhydrazine in benzene solution, **1806** (R^1 = H, $R^2 = CH_3$) was produced, along with **1807** and 2-carbethoxycyclo-hexanone likewise reacted to afford **1806** [$R^1 + R^2 = (CH_2)_4$] (Sucrow and Wiese, 1970), analogous products having been similarly obtained earlier from ethyl acetoacetate with hydrazine, methylhydrazine and 1,2-dimethylhydrazine (Auwers and Niemeyer, 1925). Pyrazoles and related compounds, and not the corresponding pyrroles, were also formed when compounds **1808** [R = C_2H_5 and $CH(CH_3)_2$] were heated in xylene (Bardakos, Sucrow and Fehlauer, 1975; Sucrow, Mentzel and Slopianka, 1974), as they were from related hydrazones of formaldehyde (Sucrow and Grosz, 1976), benzaldehyde (Sucrow and

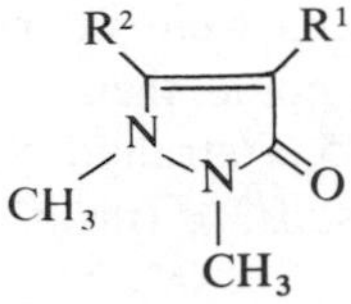

1806 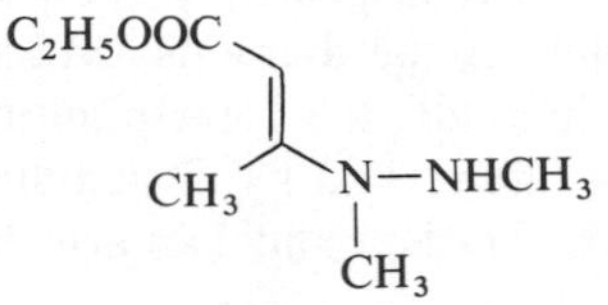 **1807** 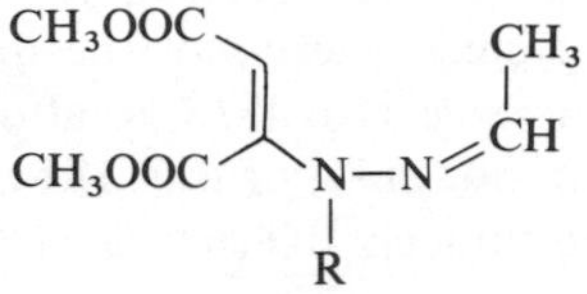 **1808**

Slopianka, 1972; Sucrow, Slopianka and Neophytou, 1972) and substituted benzaldehydes (Sucrow, Mentzel and Slopianka, 1973; Sucrow, Slopianka and Neophytou, 1972) in which pyrrolization was not possible. These and related cyclizations have also been extensively studied by another group (Aubagnac, Elguero and Jacquier, 1967, 1969; Baumes, Jacquier and Tarrago, 1976; Chapelle, Elguero, Jacquier and Tarrago, 1970a; Coispeau and Elguero, 1970; Coispeau, Elguero, Jacquier and Tizane, 1970).

Unsymmetrical pyrroles have also be prepared using the Piloty synthetic approach but symmetrical pyrroles sometimes resulted because of intermediate transazinization under the reaction conditions. Thus, when refluxed in xylene in the presence of 4-methylbenzene sulphonic acid with concomitant removal of water, a mixture of **1809** ($R^1 = CH_3$, $R^2 = C_6H_5$ with cyclohexanone gave **1810** but mixtures of **1809** ($R^1 = C_2H_5$, $R^2 = C_6H_5$) with cyclohexanone and

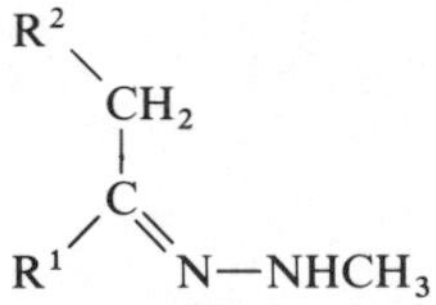

1809 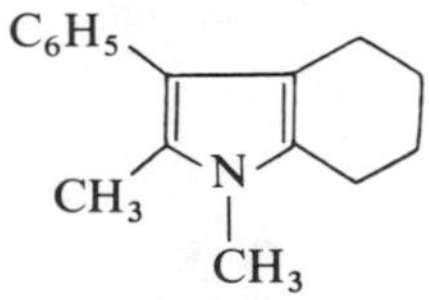 **1810**

of **1809** [$R^1 + R^2 = (CH_2)_4$] with benzyl ethyl ketone both afforded **157** [$R^1 + R^2 = (CH_2)_4$, $R^3 = CH_3$], benzyl ethyl ketone and **1809** ($R^1 = C_2H_5$, $R^2 = C_6H_5$) and a mixture of **1809** ($R^1 = H$, $R^2 = CH_3$) with benzyl ethyl ketone yielded **1809** ($R^1 = H$, $R^2 = CH_3$ and $R^1 = C_2H_5$, $R^2 = C_6H_5$), **157** ($R^1 = C_2H_5$, $R^2 = C_6H_5$, $R^3 = CH_3$) and **1797** ($R^1 = R^3 = R^5 = H$, $R^2 = R^6 = CH_3$, $R^4 = C_2H_5$) (Chapelle, Elguero, Jacquier and Tarrago, 1971).

In contrast to the Piloty synthesis which, except for the examples mentioned above using a sodium hydride catalyst (Sorokin, Larshin and Grandberg, 1977), is catalysed by acid, the reaction of bisalkyl aryl ketone azines in the presence of lithium diisopropylamide in tetrahydrofuran afforded either pyrroles or 2,3,4,5-tetrahydropyridazines, depending upon the nature of the ketonic moiety. Thus, **154** ($R^1 = C_6H_5$ and 2-naphthyl, $R^2 = H$) yielded **1811** ($R = C_6H_5$ and 2-naphthyl, respectively) whereas **154** [$R^1 = C_6H_5$, $R^2 = CH_3$ and $R^1 = C_6H_4CH_3(4)$, $R^2 = H$] gave **157** [$R^1 = C_6H_5$, $R^2 = CH_3$, $R^3 = H$ and

$R^1 = C_6H_4CH_3(4)$, $R^2 = R^3 = H$, respectively]. These results were interpreted in terms of the formation of the dianions **1812** which could then rearrange into **1813**, to afford ultimatedly the corresponding 1,4,5,6-tetrahydropyridazines, or into **1814**, most likely via a [3,3]-sigmatropic shift, to furnish ultimately the pyrroles (Yoshida, Harada and Tamaru, 1976).

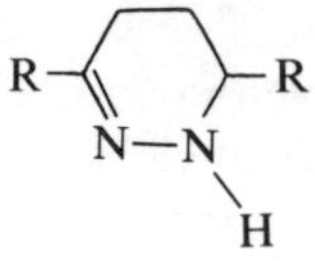

R— —R N—N H

1811

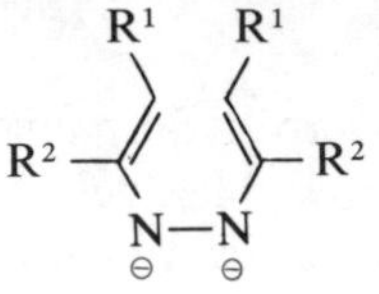

R^1 R^1 R^2— —R^2 N—N ⊖ ⊖

1812

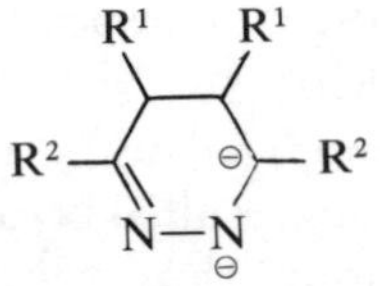

R^1 R^1 R^2— ⊖ —R^2 N—N ⊖

1813

R^1 R^1 R^2— —R^2 N N ⊖ ⊖

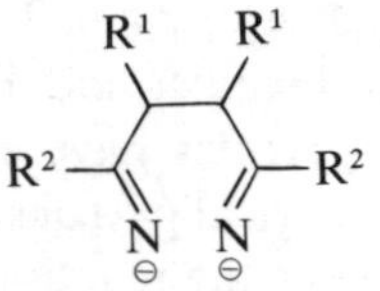

1814

CHAPTER VII

Limitations, Exceptions and Alternative Reactions to Fischer Indolization

A. Inhibition of Indolization by *Ortho* Substituents on the Arylhydrazine Moiety

Many examples of this phenomenon have been recorded. In some cases the attempted indolization of a 2-substituted phenylhydrazone failed altogether, whereas under similar conditions the corresponding 3- and/or 4-substituted isomer indolized, and in other cases the product yields from the indolization of the 2-substituted isomers were much lower than the product yields obtained from the 3- and/or 4-substituted isomers under similar conditions.

Several examples of the former type, involving complete inhibition of indolization, have been observed. 2-Indanone 3- and 4-bromophenylhydrazone were both indolized by refluxing in dilute sulphuric acid whereas, under similar conditions, attempted indolization of 2-indanone 2-bromophenylhydrazone was unsuccessful (Plant and Tomlinson, M. L., 1931). Acetophenone 3- and 4-methylphenylhydrazone were indolized by fusing with zinc chloride whereas, under similar conditions, the 2-methyl isomer failed to indolize at fusion temperatures of between 120 °C and 400 °C (Campbell and Cooper, 1935). Benzyl phenyl ketone 3- and 4-nitrophenylhydrazone were indolized in a refluxing mixture of acetic and concentrated hydrochloric acids, conditions which did not indolize benzyl phenyl ketone 2-nitrophenylhydrazone (Fennell and Plant, 1932). Failures also attended the attempts to indolize mixtures of **992** (R^1 = H, X = S) with 2,4-dinitrophenylhydrazine, 2-methylphenylhydrazine or 1-naphthylhydrazine or of **1076** with 2-methylphenylhydrazine whereas mixtures of either of these two ketones and other related compounds with substituted phenylhydrazines bearing substituents not in the 2-positions indolized readily (Dalgliesh and Mann, 1947). 2-Tetralone phenylhydrazone and 4-bromo-, 4-chloro-, 4-fluoro-, 4-iodo-, 4-methoxy- and 4-methylphenylhydrazone all readily indolized, in high yields, in refluxing acetic acid saturated with hydrogen chloride but under these conditions the corresponding 2-substituted phenylhydrazones gave either only relatively poor yields, or none of the corresponding indoles which were eventually prepared in approximately

50 % yields under more vigorous conditions (i.e. heating in a sealed ampoule at 140 °C under nitrogen in a solution of acetic acid saturated with hydrogen chloride and containing a trace of di-*tert*.-butyl-4-cresol) (Pecca and Albonico, 1971). Only starting material was recovered from attempts to indolize ethyl pyruvate 2-tosyloxyphenylhydrazone (Ek and Witkop, 1954). The statement (Dalgliesh and Mann, 1947) to the effect that previous studies (Barclay and Campbell, 1945) had found that cyclohexanone 2-methylphenylhydrazone decomposed rapidly, even at room temperature, without detectable conversion into 1,2,3,4-tetrahydro-8-methylcarbazole, is misleading since the expected indolization was carried out by refluxing the hydrazone in dilute sulphuric acid for 6 h (Barclay and Campbell, 1945).

Examples of indolizations of the latter type, in which only a partial inhibition of indolization occurred, have been observed. Mixtures of **1070** (R = OH) with 2-, 3- and 4-methylphenylhydrazine hydrochloride in refluxing acetic acid yielded the corresponding 7-, 6- and 5-methylindoles in 30 %, almost 100 % and 98 % yields, respectively (Buu-Hoï, Hoan, Khôi and Xuong, 1949). Pyruvic acid 4-methoxy-N_α-methylphenylhydrazone in an acetic acid–concentrated hydrochloric acid mixture afforded 5-methoxy-1-methylindole-2-carboxylic acid (33 % yield) whereas similar treatment of the corresponding 2-substituted phenylhydrazone produced 7-methoxy-1-methylindole-2-carboxylic acid in only trace amounts (89 mg from 12.5 g of the hydrazine) (Bell and Lindwall, 1948). Diethyl 2-oxosuccinate 4-methoxy-N_α-methylphenylhydrazone afforded 5-methoxy-1-methylindole-2-carboxylic acid (11.5 % yield) after indolization in hot alcohol saturated with hydrogen chloride followed by base catalysed hydrolysis and decarboxylation whereas similar treatment of the corresponding 2-substituted phenylhydrazone gave only 54 mg of 7-methoxy-1-methylindole-2-carboxylic acid after starting from 12.5 g of the hydrazine (Bell and Lindwall, 1948). Likewise, the yields of indolic products obtained from other 4-methoxyphenylhydrazones were significantly greater than those obtained from corresponding 2-methoxyphenylhydrazones under similar indolization conditions (Grandberg and Zuyanova, 1970), analogous yield differences being observed upon indolization of other 2- and 4-methoxy-, 2- and 4-bromo- and 2- and 4-methylphenylhydrazones (Grandberg and Bobrova, 1973). A comparison of the indolizations of 2-oxoglutaric acid 2-, 3- and 4-chlorophenylhydrazone showed that the 2-isomer was the most difficult and that the 3-isomer was the easiest to indolize (Feofilaktov and Semenova, 1953c). The 25 % acetic acid catalysed indolization of mixtures of 4-methoxyphenylhydrazine hydrochloride with several diethyl acetals gave noticeably higher product yields than when indolizing mixtures of 2-methoxyphenylhydrazine hydrochloride with the corresponding diethyl acetals (Desaty and Kegliević, 1965). The polyphosphoric acid catalysed indolization of ethyl pyruvate 4-methyl-3-nitrophenylhydrazone occurred in higher yield than the indolization of ethyl pyruvate 5-methyl-2-nitrophenylhydrazone (Gadaginamath, 1976). Other examples in which the presence of a 2-substituent appeared to hinder indolization include the conversions of cyclohexanone 2,5-dichloro- and 2-methylphenylhydrazone into 5,8-dichloro-1,2,3,4-tetrahydrocarbazole and 1,2,3,4-tetrahydro-8-methylcarbazole, respectively, which necessitated refluxing in dilute sulphuric acid for 4.5 and 6 h, respectively, whereas other cyclohexanone phenylhydrazones unsubstituted in the 2-position were indolized simply by shaking in dilute sulphuric acid solution at room temperature (Barclay and Campbell, 1945). The recovery of 33% unchanged starting material, along with the formation of the expected indole and a further product (see Chapter V, E) was achieved after 4-chlorophenyl methyl ketone 2-chlorophenylhydrazone was treated with excess tin(II) chloride at 260 °C (Carlin and Amoros-Marin, 1959).

Failures to indolize pyruvic acid 2-nitrophenylhydrazone have also been reported (Hughes, G. K., Lions and Ritchie, 1939; Rydon and Siddappa, 1951) although in this respect the general reluctance of nitrophenylhydrazones to indolize should be noted (see Chapter VII, O).

N_α-Methylated arylhydrazones are known to indolize more readily than the corresponding indolizable arylhydrazones (Chalmers, A. J. and Lions, 1933; Chastrette, 1970; Diels and Köllisch, 1911; Fischer, Emil, 1886b; Harvey, D. J. and Reid, 1972; Ishii, Murakami, Y., Hosoya, Takeda, Suzuki, Y. and Ikeda, 1973; Kermack, Perkin and Robinson, R., 1921; Kermack and Slater, 1928; Mann and Wilkinson, 1957; Padfield and Tomlinson, M. L., 1950; Perkin and Plant, 1923b; Robinson, R. and Thornley, 1924; Woolley, D. W. and Shaw, E., 1955) and other N_α-alkyl groups also facilitate indolization (Michaelis, 1897). In furtherance of this generalization and that of indolization inhibition by *ortho* substituents, it has been observed that whereas indolization of pyruvic acid 2-chlorophenylhydrazone was only achieved by refluxing in ethanol saturated with hydrogen chloride, the indolization of the corresponding N_α-methylphenylhydrazone into 7-chloro-1-methylindole-2-carboxylic acid could be effected with a similar catalyst at room temperature and whereas attempted indolization under similar conditions of ethyl pyruvate 2-chlorophenylhydrazone was unsuccessful, the corresponding N_α-methylphenylhydrazone gave ethyl 7-chloro-1-methylindole-2-carboxylate (Ishii, Murakami, Y., Hosoya, Takeda, Suzuki, Y. and Ikeda, 1973).

Although it has been reported (Marion and Oldfield, 1947; Mendlik and Wibaut, 1931) that propionaldehyde 2-methylphenylhydrazone was indolized to afford 3,7-dimethylindole, subsequent attempts (Bajwa and Brown R. K., 1969) to repeat this synthesis using the published conditions were unsuccessful. However, by forcing the formation of the enehydrazine, by refluxing N_β,2-dimethylphenylhydrazine hydrochloride with propionaldehyde in dry benzene, with azeotropic distillation of the water produced, the desired 3,7-dimethylindole was produced, along with the expected methylamine hydrochloride (Bajwa and Brown, R. K., 1969). As mentioned earlier (Chapter II, E1), this technique should certainly be considered for use where indolizations using phenylhydrazones or N_α-substituted phenylhydrazones only proceed with difficulty or are unsuccessful.

The inhibition of Fischer indolizations by the presence of 2-substituents on the phenylhydrazine moiety of a phenylhydrazone has been suggested (Buu-Hoï, Hoán, Khôi and Xuong, 1949) to be caused by steric hindrance exerted by the 2-substituent, a view which was apparently further supported when it was noted (Dalgliesh and Mann, 1947; Pecca and Albonico, 1971) that electronic factors did not appear to play a major role in this inhibition since it was observed irrespective of the nature of the 2-substituent. Steric hindrance certainly appeared to be responsible for the failure of the attempted indolization of 2,6-dimethylcyclohexanone 2,6-dimethylphenylhydrazone (Bajwa and Brown, R. K., 1970). A related steric inhibition, now between the 3- and 4-substituents of the potential indole nucleus, appeared to be responsible for the failure to indolize **1815** (R^1 = Cl, R^2 = Br) by treatment with dry ethanolic hydrogen chloride, since under similar conditions **1815** (R^1 = H, R^2 = Br and R^1 = Cl, R^2 = H) yielded **1816** (R^1 = H, R^2 = Br and R^1 = Cl, R^2 = H, respectively) (Chalmers, A. J. and Lions, 1933). However, factors other than steric hindrance may be involved in the inhibition of indolization by *ortho* substituents, although such circumstances were not stated (Dalgliesh and Mann, 1947).

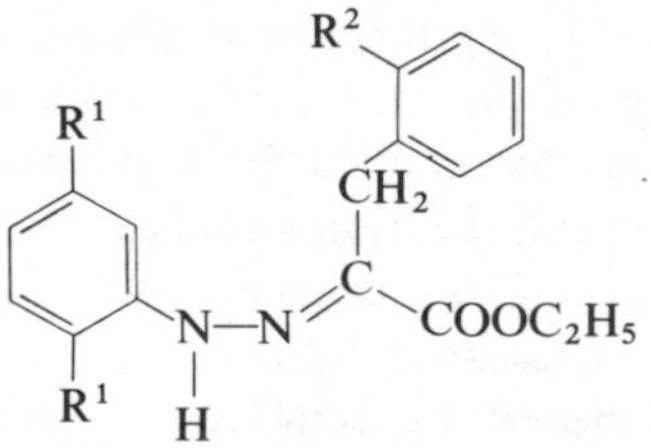

R^2 R^1 CH_2 C N—N $COOC_2H_5$ R^1 H

1815

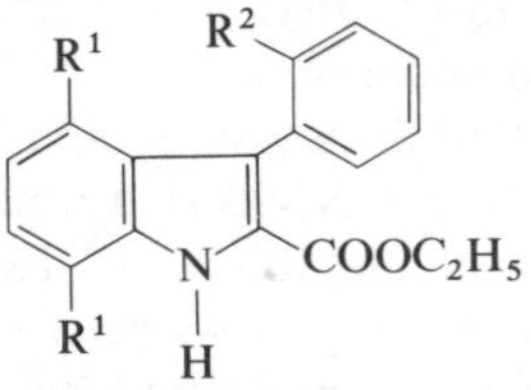

R^1 R^2 N $COOC_2H_5$ R^1 H

1816

It has been suggested that the presence of a 2-nitro substituent in benzyl phenyl ketone 2-nitrophenylhydrazone might inhibit the *ortho*-benzidine rearrangement (i.e. the indolization process) which could then permit an alternative *ortho*-semidine rearrangement, ultimately accounting for the formation of osazones from this and related attempted indolizations (Kinsley and Plant, 1956) (see Chapter VII, Q).

Because of these above failures or low yields, it must not be assumed that all 2-substituted phenylhydrazones react analogously. Indeed, a large number of such hydrazones have been readily indolized, many of them furnishing good yields of products (see Chapter IV, K, Table 31).

B. Indolization Failures

Failures to indolize several acetone arylhydrazones have been recorded. Thus, the phenylhydrazone with tin(II) chloride in hydrochloric acid (Brunner, K., 1900) or with Amberlite 1R-120 (a cation exchange resin) in boiling water

(Yamada, S., Chibata and Tsurui, 1953) failed to give 2-methylindole and the 2- and 4-methoxyphenylhydrazone could be indolized only by heating with zinc chloride at 110 °C under vacuum. Other conditions, such as heating with zinc chloride at 180 °C with or without 1,2,3,4-tetrahydronaphthalene as solvent, with ethanol saturated with hydrogen chloride, with ethanol containing 10% sulphuric acid or with a 1 : 1 mixture of acetic acid–concentrated hydrochloric acid, all proved to be unsuccessful (Bell and Lindwall, 1948; Späth and Brunner, O., 1925). No 2-methyl-5-nitroindole, for which m.p.s. of 176–176.5 °C (Noland, Smith, L. R. and Johnson, D. C., 1963), 171.5–172.5 °C (Terent'ev, A. P., Preobrazhenskaya, Bobkov and Sorokina, G. M., 1959) and 170 °C (Walther and Clemen, 1900) were subsequently recorded, could be isolated from the attempted indolization of acetone 4-nitrophenylhydrazone with a zinc chloride catalyst. However, a crystalline product, m.p. = 61–62 °C (0.5 g starting from 10 g of the hydrazone), was isolated from this reaction but, apart from an elemental analytical determination of C = 45.76%, H = 4% and N = 9.07% and 9.30%, the small amount of product available precluded further investigation of its structure at the time (Bamberger and Sternitzki, 1893 – see also Hyde, 1899). Clearly, structural studies using modern spectroscopic techniques should now be undertaken. Failures also attended the attempted indolization of acetone 2-quinolylhydrazone under a variety of conditions (e.g. heating with zinc chloride or concentrated sulphuric, hydrochloric or hydrobromic acids), all of which simply led to hydrolysis of the hydrazone to form 2-quinolylhydrazine and acetone (Perkin and Robinson, R., 1913), of acetone 4-bromo- or 3-nitrophenylhydrazone (Korczynski, Brydowna, and Kierzek, 1926) and of acetone 6-bromo-2-pyridylhydrazone by heating with 4-methylbenzene sulphonic acid (Yakhontov and Pronina, E. V., 1968). Although this last failure was ascribed to the relative difficulty in the enolization of acetone, which is extrapolatable to the corresponding difficulty in enehydrazine formation in the hydrazone and hence to indolization (Yakhontov and Pronina, E. V., 1968), a relationship postulated fifty years earlier (Robinson, G. M. and Robinson, R., 1918), a further determinant factor could be that in many of these cases the aryl nucleus carried electron withdrawing functions. Many attempts to indolize **1139** ($R^1 = R^3 = H, R^2 = CH_3$) failed, although other ketonic hydrazones of this hydrazino moiety could be readily indolized (Cattanach, Cohen, A., and Heath-Brown, 1973) and attempts to indolize **1070** [$R = NHN{=}C(CH_3)_2$] were also unsuccessful whereas similar hydrazones of many other ketones were readily indolized (Shvedov, Trofimkin, Vasil'eva and Grinev, A. N., 1975). Even though, as mentioned in the previous section of this chapter, N_α-methylated derivatives of indolizable arylhydrazones are more readily indolized than are the parent arylhydrazones, failure attended the attempted indolization of acetone N_α-methylphenylhydrazone using tin(II) chloride in hydrochloric acid as the catalyst (Brunner, K., 1900). These above indolization failures were most probably caused by the incorrect choice of catalyst, since many acetone arylhydrazones have been successfully indolized (see Chapter IV, K, Table 32), often using zinc chloride or polyphosphoric acid as the catalyst. Indeed, whereas acetone 4-bromophenylhydrazone has been indolized into 5-bromo-2-methylindole by several groups (Chapman, Clarke and Hughes, H., 1965; Colo, Asero and Vercellone, 1954; Farmaceutici Italia, S. A., 1954; Quadbeck and Röhm, 1954) in yields ranging between 8% and 45% depending upon the reaction conditions [the highest yield was obtained when the indolization was carried out with a zinc chloride catalyst in refluxing cumene under a nitrogen atmosphere (Chapman, Clarke and Hughes, H., 1965)], subsequent attempts to effect the indolization of this arylhydrazone either by treatment with boron trifluoride etherate in refluxing acetic acid or by heating with polyphosphoric acid were unsuccessful (Noland and Reich, 1967). Likewise, the choice of catalyst was critical for the indolization of ethyl pyruvate phenylhydrazone and 4-chlorophenylhydrazone into ethyl indole-2-carboxylate and ethyl 5-chloroindole-2-carboxylate, respectively. The former reaction could be effected by warming with an acetic acid–concentrated sulphuric acid mixture whereas treatment with alcoholic hydrogen chloride or sulphuric acid, aqueous sulphuric acid, zinc chloride in alcohol or hydrogen bromide in acetic acid under a variety of conditions led only to reactant decomposition (Elks, Elliott and Hems, 1944c) and the latter reaction failed to occur using either acetic acid or ethanolic hydrogen chloride as the catalyst (Young, E. H. P., 1958) but was achieved using a polyphosphoric acid catalyst (Rydon and Tweddle, 1955). Other cases have been reported in which indolization has failed but where an alternative choice of catalyst may well have satisfactorily effected the reaction. Thus, attempts to indolize a mixture of the ketone **1817** with phenylhydrazine to afford **1818**

CH_3OOC, CH_3—CH, $COOCH_3$, O, $COOCH_3$, CH—$COOCH_3$, CH_3 ⟶✕⟶ CH_3OOC, CH_3—CH, $COOCH_3$, N, $COOCH_3$, CH—$COOCH_3$, CH_3

1817 **1818**

failed (Levy and Robinson, R., 1962), as did attempts to perform the double indolization of **1819** into **1820**, or its Plancher rearrangement product, by refluxing in ethanolic hydrogen chloride or glacial acetic acid or by heating with polyphosphoric acid, no recognizable product being obtained from any of this latter group of reactions (Noland and Baude, 1966). The reported failure (McClelland and D'Silva, 1932) to effect the indolization of a mixture of **1821** with phenylhydrazine in boiling glacial acetic acid into **1822** is not surprising. Success would probably be achieved by heating the hydrazone with

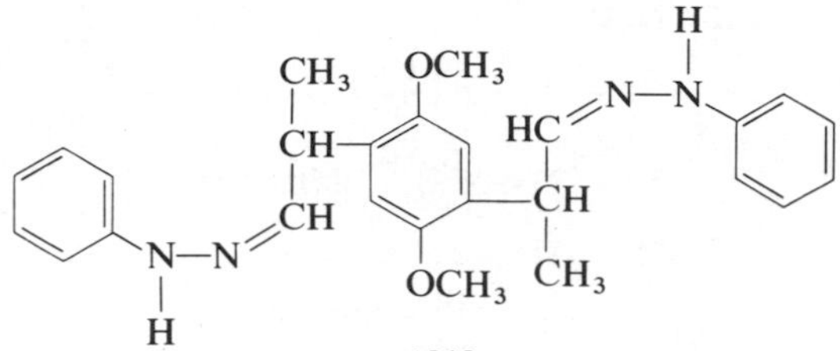

1819

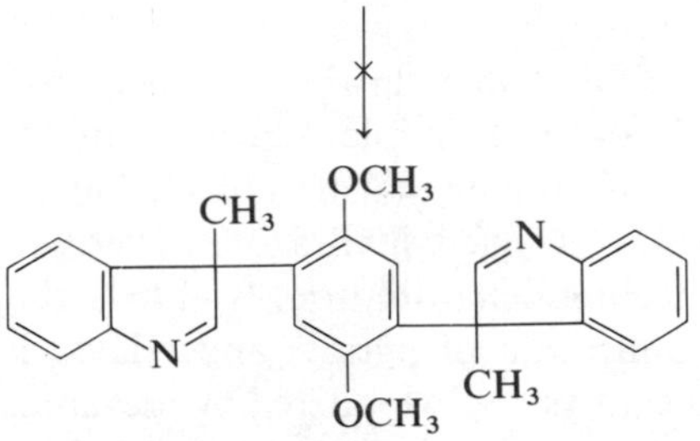

1820

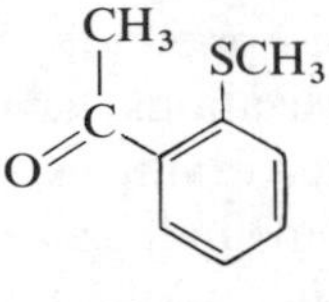

1821

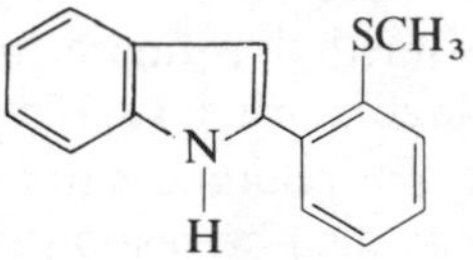

1822

zinc chloride or polyphosphoric acid. Likewise, acetophenone phenylhydrazone was not indolized using a polyphosphate ester catalyst (Kanaoka, Ban, Miyashita, Irie and Yonemitsu, 1966) and some benz substituted acetophenone arylhydrazones could not be indolized by pyrolysis with nickel(II) salts, conditions which brought about the indolization of other such hydrazones (Korczynski and Kierzek, 1925). Whereas the indolization of **1239** phenylhydrazone using dry hydrogen chloride in acetic acid proceeded in good yield and those of the 4-bromophenylhydrazone and 2-naphthylhydrazone proceeded in only poor yield, no indolic products could be isolated after treatment of the 4-methoxy- and 4-methylphenylhydrazone under similar conditions. However, other catalysts or the non-catalytic thermal indolization technique were not investigated with respect to these last two reactions. Furthermore, although it might be expected that substituents in the phenylhydrazine moiety such as methoxy and methyl would facilitate indolization (see Chapter II, F), these could also likewise increase the rate of side reactions which might thus occur preferentially (Bremner and Browne, 1975).

Attempts to indolize the ethyl pyruvate arylhydrazones **1823** (R = piperidino or morpholino), using acetic acid, sulphuric acid or polyphosphoric acid as catalysts, all failed, undoubtedly owing to protonation of the basic piperidino or

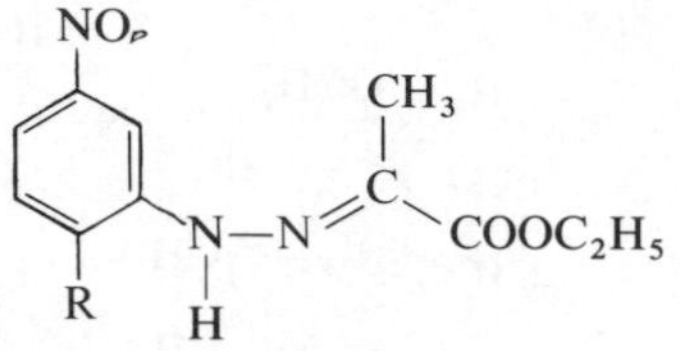

1823

morpholino substituents under the reaction conditions which would deactivate the benzenoid ring accordingly. This was illustrated by the observation that ethyl pyruvate 2-chloro-5-nitrophenylhydrazone indolized in almost quantitative yield to ethyl 7-chloro-4-nitroindole-2-carboxylate (Ainsworth, D. P. and Suschitzky, 1967b). It is unfortunate that attempts to indolize the piperidino and morpholino substituted nitrophenylhydrazones were not made using the non-catalytic thermal technique (Robinson, B., 1969), the uses of which have been proven in the indolization of pyridyl and related hydrazones in which indolization using acid catalysts was hindered or prevented by the protonation of the hetero nitrogen atom under the reaction conditions (see Chapter IV, I). The syntheses of 5-acetamido-, 5-amino-, 5-di(2-hydroxyethyl)amino-, 5-methylamino-, 5-(4-methyl-1-piperazinyl)-, 5-(morpholino)-, 5-(1-piperazinyl)- and 5-(1-pyrrolidinyl)indole via Fischer indolizations have been reported (Shen, T.-Y., 1964). For references to other related examples, see Table 31.

Steric reasons may also be responsible for some indolization failures, other than their suggested involvement in unsuccessful indolizations of *ortho* substituted phenylhydrazones discussed in the previous section. Thus, the failure to indolize 2-methylcyclohexanone 2-naphthylhydrazone was apparently caused by steric hindrance since the corresponding hydrazones of 3- and 4-methylcyclohexanone readily yielded indolic products (Campaigne, Ergener, Hallum and Lake, 1959). Similarly, the failure to indolize **1824** ($R^1 = CH_3$, $R^2 = C_2H_5$, $R^3 = NO_2$) by heating with polyphosphoric acid, which produced

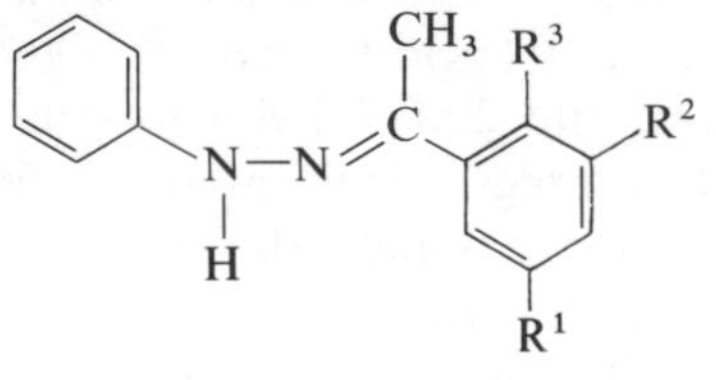

1824

only black tars, was suggested (Wenkert, 1958) as being caused by steric crowding because of the *diortho* substitution of the nitro group which might be sufficient to have caused it to undergo interaction, possibly in an oxidative–reductive manner, with the neighbouring hydrazino intermediates in the Fischer reaction. In support of this, **1824** ($R^1 = R^2 = H$, $R^3 = NO_2$) and **1824** ($R^1 = CH_3$, $R^2 = C_2H_5$, $R^3 = NH_2$) indolized, when heated with polyphosphoric acid, to furnish **1825** ($R^1 = R^2 = H$, $R^3 = NO_2$ and $R^1 = CH_3$, $R^2 = C_2H_5$, $R^3 = NH_2$), respectively (MacPhillamy, Dziemian, Lucas and Kuehne, 1958).

Using polyphosphoric acid as catalyst, compounds **679** (R = H, X = O and S) were indolized into **680** (X = O) (Holla, B. S. and Ambekar, 1976) and **680** (X = S) (Holla, B. S. and Ambekar, 1974), respectively. However, **679** ($R = NO_2$, X = O) and apparently **679** ($R = NO_2$, X = S) could not be indolized under varying (unspecified) conditions (Holla, B. S. and Ambekar, 1978). Whereas benzyl 3,4-dimethoxyphenyl ketone reacted with phenylhydrazine to give the expected hydrazone which could be indolized, attempts to prepare the corresponding N_α-methyl hydrazone, by reacting the ketone with N_α-methylphenylhydrazine, were unsuccessful (Chalmers, A. J. and Lions, 1933). Heating with formic acid failed to indolize pyruvic acid phenylhydrazone (Kidwai, A. R. and Khan, N. H., 1963) although this catalyst causes a large variety of indolizations (see Chapter IV, L, Table 33). Likewise, treatment with dry ethanolic hydrogen chloride at room temperature or with refluxing acetic acid failed to indolize ethyl pyruvate 4-carbethoxyphenylhydrazone whereas under the former conditions, **1826** [$R^1 = COOC_2H_5$, $R^2 = H$, $R^3 = CH_3$, $(CH_2)_2CH_3$ and C_6H_5; $R^1 = OCH_3$, OC_2H_5 and Br, $R^2 = R^3 = H$ and $R^1 = R^3 = H$, $R^2 = OC_2H_5$] and ethyl pyruvate 1- and 2-naphthylhydrazone afforded the

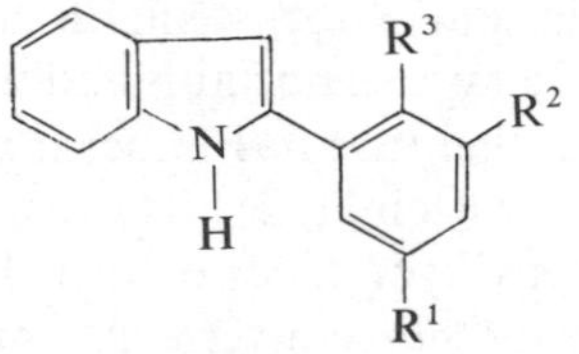

1825 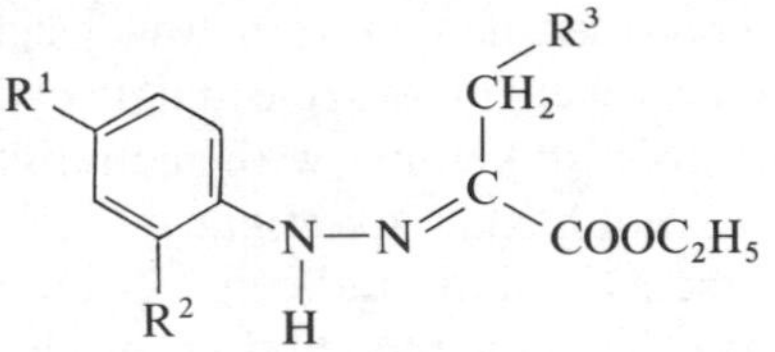 **1826**

expected indolic products (Hughes, G. K. and Lions, 1937–1938a). Although a large number of diethyl 2-oxoglutarate arylhydrazones have been successfully indolized (see Chapter IV, K, Table 32), characterizable crystalline indolic products could not be obtained from several attempted indolizations of the 2-, 3- and 4-chlorophenylhydrazone under a variety of catalytic conditions (Findlay, S. P. and Dougherty, 1948).

Other indolization failures have been discussed elsewhere. Thus, many pyridylhydrazones failed to indolize using acid catalysts (see Chapter IV, I), the failure to indolize **1240** ($R^1 = H$, $R^2 = CH_3$, $n = 1$) (Nakazaki and Maeda, 1962) (see Chapter IV, D) is difficult to understand in view of the successful indolizations of several related and homologous hydrazones and further attempts should be made to indolize compounds **1579** [$R^1 = H$, $R^2 = R^3 = CH_3$, $R^4 = CH(CH_2)_5$ and $R^1 = R^3 = CH_3$, $R^2 = H$, $R^4 = C_6H_4CH_3(2)$] which as yet have only been observed to hydrolyse under acidic conditions (ethanolic hydrogen chloride) (Wuyts and Lacourt, 1935) (see Chapter VI, A).

C. Preparation and Indolization of Dimethoxyphenylhydrazones and Aminophenylhydrazines

Because of their instability, a property of many arylhydrazines (Terent'ev, A. P. and Preobrazhenskaia, 1958), difficulty was experienced when preparing 2,5- and 3,4-dimethoxyphenylhydrazine by sequential diazotization and reduction of 2,5- and 3,4-dimethoxyaniline, respectively. Probably related to this was the difficulty experienced in utilizing these hydrazines in Fischer indolization. Thus, refluxing mixtures of 3,4-dimethoxyphenylhydrazine hydrochloride with cyclohexanone or ethyl pyruvate, respectively, in ethanol containing sodium acetate afforded the corresponding hydrazones as oils which upon boiling with alcoholic hydrochloric acid gave 1,2,3,4-tetrahydro-6,7-dimethoxycarbazole [**1827**; $R^1 + R^2 = (CH_2)_4$] and ethyl 5,6-dimethoxyindole-2-carboxylate (**1827**; $R^1 = COOC_2H_5$, $R^2 = H$), respectively, but these products

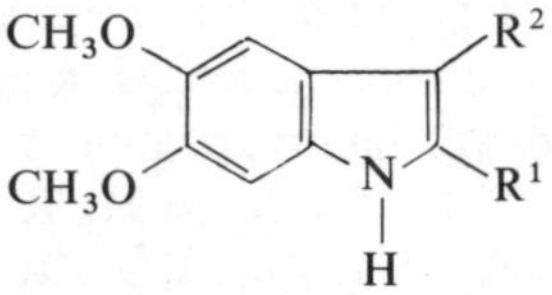

1827

were only obtained in very low yields (Perkin and Rubenstein, 1926). 2,5-Dimethoxyphenylhydrazine and its hydrochloride were so unstable and thereby condensed with ketones with such difficulty that their use in Fischer indolizations was not further investigated (Perkin and Rubenstein, 1926). The instability of 2-dimethylamino- and other 2-*tert*.-aminophenylhydrazines or their hydrochlorides likewise rendered unsuccessful their *in situ* conversions into the corresponding hydrazones (Ainsworth, D. P. and Suschitzky, 1967a) and 4-dimethylaminophenylhydrazine was too unstable for it to be used in the Fischer indolization (Skrabal, Steiger and Zollinger, 1975). However, see Table 31.

The problem of hydrazine instability was overcome in the case of 3,4-dimethoxyphenylhydrazones by carrying out their syntheses using the Japp–Klingemann reaction. In this manner, the 3,4-dimethoxyphenylhydrazones **1828** [$R^1 = COOC_2H_5$, $R^2 = H$, CH_3, C_6H_5 and $(CH_2)_3COOH$ and $R^1 = COCH_3$, $R^2 = C_6H_5$] were prepared by coupling 3,4-dimethoxybenzenediazonium chloride with **114** [$R^2 = C_2H_5$, $R^3 = H$, CH_3 and C_6H_5, $R^4 = CH_3$;

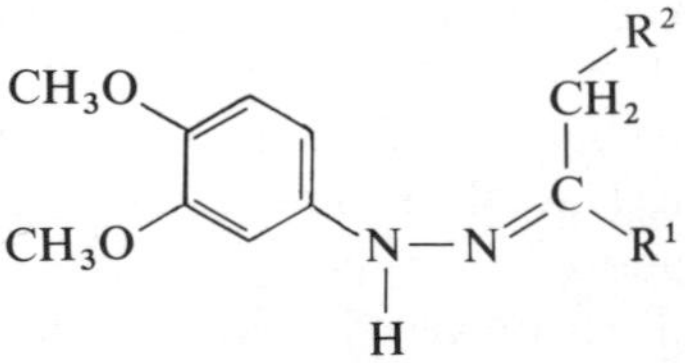

1828

$R^2 = C_2H_5, R^3 + R^4 = (CH_2)_3$ and $R^2 = H, R^3 = C_6H_5, R^4 = CH_3$, respectively]. All these hydrazones were readily indolized to give reasonable yields of **1827** [$R^1 = COOC_2H_5$, $R^2 = H$, CH_3, C_6H_5 and $(CH_2)_3COOH$ and $R^1 = COCH_3$, $R^2 = C_6H_5$, respectively] by treatment with dry hydrogen chloride in ethanol (Lions and Spruson, 1932), the syntheses of **1827** ($R^1 = COOC_2H_5$, $R^2 = CH_3$ and C_6H_5) by this route being subsequently (Beer, McGrath, Robertson and Woodier, 1949) repeated. However, difficulties arose with this synthetic approach when using the free acid hydrazones **1828** ($R^1 = COOH$), formed by hydrolysis of the 2-ketoester hydrazones with alcoholic potassium hydroxide, which upon similar treatment with acid afforded only tars from which indolic products could not be isolated (Lions and Spruson, 1932). This was probably caused by the instability of 5,6-dimethoxyindole-2-carboxylic acids to mineral acids (Lions and Spruson, 1932), as noted earlier (Oxford and Raper, 1927). Subsequent to these studies, 3,4-dimethoxyphenylhydrazine hydrochloride has been successfully employed in Fischer indolizations. Thus, by reaction with ethyl methyl ketone in ethanolic acetic acid in the presence of sodium acetate it yielded the hydrazone **1828** ($R^1 = R^2 = CH_3$) which was immediately indolized, by refluxing in ethanol saturated with dry hydrogen chloride, to give **1827** ($R^1 = R^2 = CH_3$) [Beer, McGrath, Robertson and Woodier (with Holker), 1949]. Under basic conditions it also formed the free hydrazine, immediate treatment of which with 2-oxo-5-hydroxyvalerolactone yielded the corresponding hydrazone which, without purification, was converted into **1827** [$R^1 = COOH$, $R^2 = (CH_2)_2OH$] by treatment with ethanolic hydrogen chloride (Sterling Drug Inc., 1970). Ethyl phenylpyruvate 2,5-dimethoxyphenylhydrazone has also been prepared by the Japp–Klingemann reaction and has been successfully subjected to Fischer indolization (Ishii, Murakami, Y., Tani, S., Abe, K. and Ikeda, 1970) and 2,4- and 3,5-dimethoxy-, 2,4-dimethoxy-3-methyl and 3,4,5-trimethoxyphenylhydrazones have also been synthesized and converted into the corresponding 5,7-dimethoxyindoles (Shen, T.-Y., 1964), 4,6-dimethoxyindoles (Bhattacharyya, P., Basak, Islam and Chakraborty, 1975; Douglas, B., Kirkpatrick, Moore, B. P. and Weisbach, 1964), 5,7-dimethoxy-6-methylindoles (Kametani, Ohsawa and Ihara, 1979) and 4,5,6-trimethoxyindoles (Douglas, B., Kirkpatrick, Moore, B. P. and Weisbach, 1964; Gaimster, 1960).

D. Reaction of Acyclic 1,3- and 1,4-Diketones with Arylhydrazines

The reaction between N_α-unsubstituted arylhydrazines and acyclic 1,3-diketones has already been referred to (see Chapter III, K1), the products from such reactions being not the arylhydrazones **922** ($R^2 = H$) but the pyrazoles **923** (Barry and McClelland, 1935; Bülow, 1918; Coispeau and Elguero, 1970; Dewar, 1944; Fowkes and McClelland, 1941; Fusco, 1967; Gorbunova and

Suvorov, 1973; Jacobs, 1957a; Kost and Grandberg, 1966; Laubmann, 1888a; Lee, W. Y., 1976; McClelland and Smith, P. W., 1945; Phillips, 1959; Rull and Strat, 1975; Schank and Eistert, 1966; Schoutissen, 1934, 1935; Shvedov, Trofimkin, Vasil'eva and Grinev, A. N., 1975; Tämnefors, Claesson and Karlsson, 1975; Vul'fson and Zhurin, 1961a, 1961b; Wieland, H., Juchum and Maier, 1931). As already discussed in Chapter III, K1, attempts to prepare 3-formylindoles, by treatment under indolization conditions of N_α-unsubstituted arylhydrazones of 3-ketoaldehydes with the aldehyde function protected, again afforded only the corresponding pyrazoles, although N_α-substituted arylhydrazones of 1,3-diketones, in which pyrazole formation was thus prevented, indolized as expected.

The reaction between acyclic 1,4-diketones and arylhydrazines furnished either the corresponding bisarylhydrazone or 1-arylaminopyrrole, or a mixture of both these products, depending upon the reactants and reaction conditions. Hexane-2,5-dione bisarylhydrazones or mixtures of the corresponding ketonic and hydrazino moieties have been indolized to yield the corresponding 3,3-bisindolyls or, in the latter cases, the corresponding 3-acetylmethylindoles resulting from monoindolization. Studies carried out in this area have already been discussed in detail (see Chapter III, K3).

E. Reaction of 3- and 4-Ketoacids and their Esters and 2-Cyanoketones with Arylhydrazines

When 3-ketoester arylhydrazones **922** (R^2 = H, R^4 = O-alkyl) or mixtures of the component 3-ketoesters and arylhydrazines reacted, usually in neutral or weakly acidic conditions, pyrazol-3-ones **924** were formed to the exclusion of the corresponding indoles (Ainsworth, D. P. and Suschitzky, 1967b; Altschul, 1892; Arnold, E., 1888; Bell and Lindwall, 1948; Bülow, 1918; Coispeau and Elguero, 1970; Croisy, Ricci, Jančevska, Jacquignon and Balucani, 1976; Dieckmann, 1901; Dufton, 1892; Eistert, Haupter and Schank, 1963; Feofilaktov and Ivanova, 1955; Fusco, 1967; Julian, Meyer, E. W. and Printy, 1952; Knorr, 1883; Knorr and Klotz, 1887; Le Count and Greer, 1974; Lespagnol, A., Bar, Erb-Debruyne, Delhomenie-Sauvage, Labiau and Tudo, 1960; Michael, 1892; Miller, S. J. and Tanaka, R., 1970; Neber, Knöller, Herbst and Trissler, 1929; Nef, 1891; Rodionov and Suvorov, 1950; Schiemann and Winkelmüller, 1933; Schoutissen, 1934; Stolz, 1892; Walker, C., 1892, 1894; Wiley, R. H. and Wiley, P., 1964; Wislicenus, 1888a; Wislicenus, Böklen and Reuthe, 1908; Wislicenus, Butterfass and Koken, 1924; Wislicenus and Münzesheimer, 1898; Wislicenus and Riethmuller, 1924; Wislicenus and Waldmüller, 1911). Indeed, the reaction between 3-ketoesters and hydrazines has become well established as a major synthetic route to pyrazol-3-ones (Jacobs, 1957a; Johnson, A. W., 1950) although the corresponding phenylhydrazones have also been isolated from such reactions using phenylhydrazine (Wislicenus, 1888a, 1888b).

Although treatment of ethyl acetoacetate 4-nitrophenylhydrazone with concentrated sulphuric acid afforded the pyrazol-3-one **924** ($R^1 = 4\text{-}NO_2$, $R^3 = CH_3$) (Bauer and Strauss, 1932), indolization rather than pyrazol-3-one formation occurred when ethyl acetoacetate phenylhydrazone was treated with concentrated sulphuric acid to give an indolic product (Nef, 1891) which was later (Walker, C., 1892) recognized as ethyl 2-methylindole-3-carboxylate (**2**; $R^1 = H$, $R^2 = CH_3$, $R^3 = COOC_2H_5$) and, likewise, ethyl acetoacetate 4-methylphenylhydrazone afforded ethyl 2,5-dimethylindole-3-carboxylate (Walker, C., 1894). It was later (Mills, Al Khawaja, Al-Saleh and Joule, 1981) presummed that pyrazol-3-one formation under these strongly acid conditions would be discouraged by protonation of the N_α atom, the yield of ethyl 2-methylindole-3-carboxylate from the first reaction was considerably improved (60%) by gradual addition of the hydrazone, either alone or in acetic acid solution, to vigorously stirred concentrated sulphuric acid at −10 °C and the reaction was extended to the conversion of **922** ($R^1 = R^2 = H$, $R^3 = CH_2COOC_2H_5$, $R^4 = OC_2H_5$) into **2** ($R^1 = H$, $R^2 = CH_2COOC_2H_5$, $R^3 = COOC_2H_5$) but pyrazol-3-one formation was so fast when using ethyl benzoylacetate that even the isolation of the phenylhydrazone was not possible.

N_α-Substituted arylhydrazones of 3-ketoesters (usually ethyl acetoacetate), in which the N_α-substituent prevented pyrazol-3-one formation, underwent indolization without complication (Bender, 1887; Binder, Habison and Noe, 1977; Borsche and Klein, 1941; Degen, 1886; Findlay, S. P. and Dougherty, 1948; Fusco and Sannicolò, 1973, 1975a, 1975c, 1976a, 1978c; King, F. E. and L'Ecuyer, 1934; Kornet, Thio, A. P. and Tolbert, 1980; Kost, Vinogradova, Daut and Terent'ev, A. P., 1962; Kost, Yudin, Dmitriev and Terent'ev, A. P., 1959; Kost, Yudin and Popravko, 1962; Reif, 1909; Steck, Fletcher and Carabateas, 1974; Thyagarajan, Hillard, Reddy, K. V. and Majumdar, 1974). In an approach to the synthesis of 1-unsubstituted indole-3-carboxylates, 3-ketoester N_α-benzylphenylhydrazones were analogously converted into 1-benzylindole-3-carboxylates but attempts to debenzylate these products either failed or effected other further reaction. Similar attempts to utilize an aroyl group to protect the N_α-atom failed at the indolization stage (Mills, Al Khawaja, Al-Saleh and Joule, 1981).

When a mixture of **1829** ($n = 2$) with phenylhydrazine was warmed, **1830** was formed (Ghigi, 1930). However, when mixtures of **1829** ($n = 1$ and 2) with phenylhydrazine hydrochloride were each warmed with a small amount of concentrated hydrochloric acid, indolization, hydrolysis and decarboxylation occurred in each case to give **1241** ($R = H$, $n = 1$ and 2, respectively) (Titley, 1928). By contrast with these observations, phenylhydrazine reacted with 1-carbethoxy-2-indanone to afford **815** which could not be converted into the corresponding pyrazol-3-one but which, upon heating, lost ammonia, suggesting

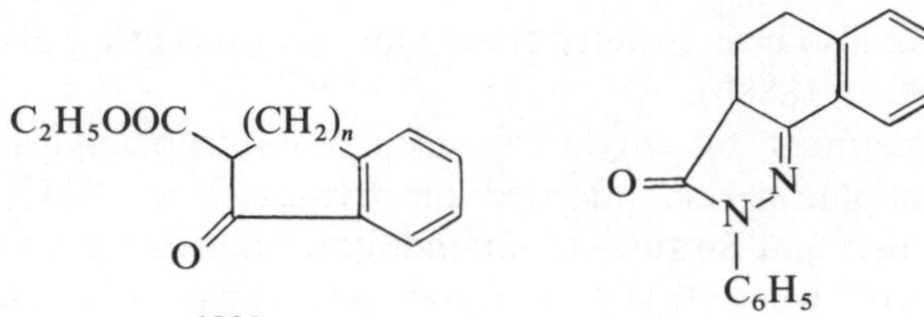

C_2H_5OOC $(CH_2)_n$ O

1829

O N N C_6H_5

1830

the occurrence of indolization, a reaction which was carried out by heating in acetic acid to afford **816** along with another structurally unidentified product (Perkin and Titley, 1922).

Studies involving the indolization of 3-ketoester arylhydrazones derived from the reaction between acetylenedicarboxylic esters and N_α-substituted arylhydrazines have already been described (see Chapter II, E2).

3-Ketonitriles **1831** reacted with hydrazines, including arylhydrazines **139**, in hydrochloric acid to give 3-aminopyrazoles **1833**, presumably via formation of the hydrazones **1832** (Grandberg, Wei-pi and Kost, 1961).

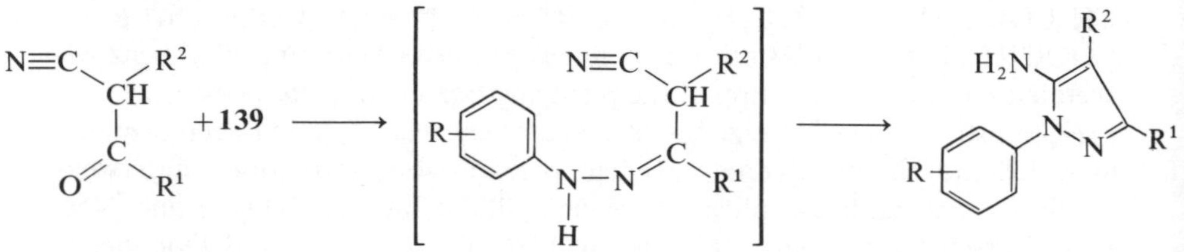

N≡C R^2 CH C O R^1 +139 ⟶ [N≡C R^2 CH C R N—N R^1 H] ⟶ R^2 H_2N N N R^1 R

1831 **1832** **1833**

4-Aldehydo- and 4-ketoacids underwent cyclodehydration and their esters underwent cyclodealcoholization when reacted with N_α-unsubstituted arylhydrazines to afford 2-aryl-4,5-dihydro-3(2*H*)-pyridazinone (De Bellis and Stein, 1961; Eraksina, Kost, Khazanova and Vinogradova, 1966; Farbwerke vorm. Meister, Lucius und Brüning, 1885; Feofilaktov and Semenova, N. K., 1953a, 1953b, 1953c; Fischer, Emil, 1886c; Fischer, Emil and Ach, 1889; Lespagnol, A., Bar, Erb-Debruyne, Delhomenie-Sauvage, Labiau and Tudo, 1960; Lespagnol, A., Lespagnol, C. and Henichart, 1969; Overend and Wiggins, 1947; Sakurai and Komachiya, 1961; Shaw, E., 1955; Steche, 1887; Stevens, F. J. and Fox, 1948, Suvorov, Morazovskaya and Sorokina, G. M., 1961; Wislicenus, Böklen and Reuthe, 1908; Wislicenus and Waldmüller, 1911 – see also Brown, P., Burdon, Smith, T. J. and Tatlow 1960). However, such products, upon heating with either ethanolic hydrochloric acid (Amorosa and Lipparini, 1957; Feofilaktov and Semenova, N. K., 1953b; Wislicenus and Waldmüller, 1911) or ethanolic sulphuric acid (Feofilaktov and Semenova, N. K., 1953a, 1953c; Sakurai and Komachiya, 1961) furnished, via the *in situ* formation of their parent hydrazones, the corresponding 3-indolylacetic acids or ethyl esters. Of course, the use of N_α-substituted arylhydrazines in analogous reactions gave

1-substituted 3-indolylacetic acids or esters without complication and indeed, in many other cases using N_α-unsubstituted arylhydrazines the expected indolization products were isolated. Thus, whereas treatment of the 2-nitrophenylhydrazone of **1834** (R^1–R^3 = H) in concentrated sulphuric acid at room temperature afforded only **1835** (R^1 = R^3 = H, R^2 = NO_2) (Stevens, F. J. and Fox, 1948), treatment of the phenylhydrazone of **1834** (R^1–R^3 = H) with refluxing absolute ethanolic concentrated sulphuric acid gave a good yield of

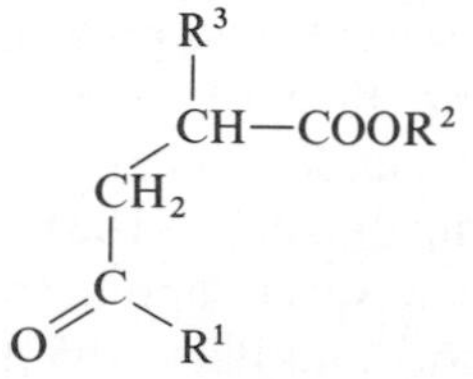

1834

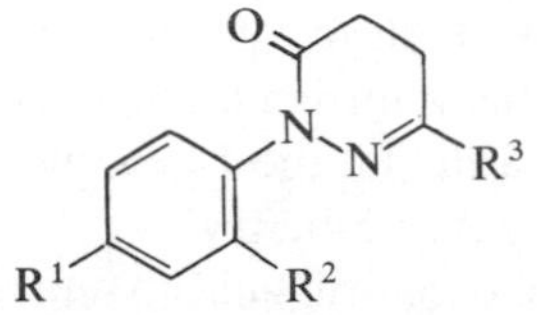

1835

3-indolylacetic acid (the reaction failed using boron trifluoride, boron trifluoride in acetic acid or zinc chloride in concentrated hydrochloric acid as catalysts) (Fox and Bullock, 1951a), treatment of the 4-bromophenylhydrazone of **1834** (R^1 = CH_3, R^2 = R^3 = H) with zinc chloride, with zinc chloride in ethanol, with 5% dry hydrogen chloride in dioxane, with 12.5% ethanolic concentrated sulphuric acid and with 10% methanolic phosphoric acid produced 5-bromo-2-methyl-3-indolylacetic acid in 2% or 22% yields, this indole (11%) together with 4-bromophenylhydrazine (12%) and **1835** (R^1 = Br, R^2 = H, R^3 = CH_3) (54%) and the ethyl and methyl esters of the starting hydrazone, respectively (Amorosa, 1956), treatment of the 4-nitrophenylhydrazone of **1834** (R^1–R^3 = H) with ethanolic zinc chloride, 12.5% ethanolic concentrated sulphuric acid and 5% hydrogen chloride in dioxane yielded 5-nitro-3-indolylacetic acid (33%), its ethyl ester (9–10%) and the indolic acid (1%) together with **1835** (R^1 = NO_2, R^2 = R^3 = H) (17%), respectively (Amorosa and Lipparini, 1956) [treatment of the 4-nitrophenylhydrazone of **1834** (R^1 = R^3 = H, R^2 = CH_3) gave only **1835** (R^1 = NO_2, R^2 = R^3 = H) (Amorosa and Lipparini, 1959)], treatment of the 4-benzyloxyphenylhydrazone of **1834** (R^1–R^3 = H) with sulphosalicylic acid in refluxing dry ethanol gave 5-benzyloxy-3-indolylacetic acid (Suvorov and Murasheva, 1960), heating the 4-sulphonamidophenylhydrazones of **1834** (R^1 = CH_3, R^2 = C_2H_5, R^3 = H; R^1 = CH_3, R^2 = R^3 = H, R^1 = $COOC_2H_5$, R^2 = C_2H_5, R^3 = H and R^1 = H, R^2 = C_2H_5, R^3 = $COOC_2H_5$) with zinc chloride, zinc chloride in xylene, polyphosphoric acid and without a catalyst (?), respectively, yielded ethyl 2-methyl-5-sulphonamido -3-indolylacetate, 2-methyl-5-sulphonamido-3-indolylacetic acid, ethyl 2-carbethoxy-5-sulphonamido-3-indolylacetate and diethyl 5-sulphonamido-3-indolylmalonate, respectively (De Bellis and Stein, 1961) and a series of N_α-unsubstituted arylhydrazones of **1834** (R^1 = COOH, R^2 = R^3 = H) have been converted into the corresponding ethyl 2-carbethoxy-3-indolylacetates by heating with ethanolic hydrogen chloride or sulphuric acid (Feofilaktov and Semenova, N. K., 1953b). In other related studies, heating mixtures of 4-ketoacids with phenylhydrazines with or without acetic acid led to the formation of only the corresponding **1835** whereas mixtures of the corresponding 4-ketoesters with phenylhydrazine, when heated with zinc chloride, gave the corresponding indoles (Lespagnol, A., Lespagnol, C. and Henichart, 1969) and the expected 3-indolylacetic acids were obtained by acid treatment of **1834** (R^1–R^3 = H) arylhydrazones although the product yields were low, probably because of competitive formation of the corresponding **1835** (R^3 = H) since the N_α-alkylhydrazones afforded high yields of indoles under similar conditions (Shaw, E., 1955). A claim that the 4-benzyloxyphenylhydrazone of **1834** ($R^1 = R^3$ = H, $R^2 = CH_3$) gave 5-benzyloxy-3-indolylacetic acid upon treatment with methanolic phosphoric acid (Mentzer, Beaudet and Bory, 1953) was later not substantiated when only **1835** ($R^1 = OCH_2C_6H_5$, $R^2 = R^3$ = H) could be isolated from the reaction (Sorokina, N. P., 1960). Based upon an elemental analysis, the product isolated after reacting **1834** (R^1–R^3 = H) with a 2 × molar equivalent of phenylhydrazine at 150 °C was formulated as the corresponding phenylhydrazone phenylhydrazide (Perkin and Sprankling, 1899).

F. Reactions of 2,3-Unsaturated and 3-Substituted (other than 3-keto) Aldehydes and Ketones with Arylhydrazines

Attempts to indolize 1,5-dimethylcyclohexen-3-one (**1836**), carvone (**1837**) and pulegone (**1838**) 3-nitrophenylhydrazone failed, although the catalyst used was limited to dilute sulphuric acid (Borsche, Witte, A. and Bothe, 1908). Similarly, failure attended attempts to indolize a mixture of **1836** with **1839** (Braun,

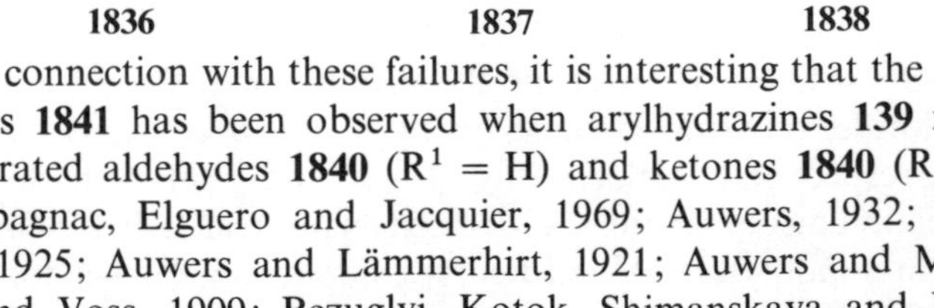

CH_3 O CH_3 CH_3 CH_3 O CH_3 CH_3 O CH_3 CH_3

1836 **1837** **1838**

1908b). In connection with these failures, it is interesting that the formation of pyrazolines **1841** has been observed when arylhydrazines **139** reacted with 2,3-unsaturated aldehydes **1840** (R^1 = H) and ketones **1840** (R^1 = alkyl or aryl) (Aubagnac, Elguero and Jacquier, 1969; Auwers, 1932; Auwers and Kreuder, 1925; Auwers and Lämmerhirt, 1921; Auwers and Müller, 1908; Auwers and Voss, 1909; Bezuglyi, Kotok, Shimanskaya and Bondarenko, 1969; Ferres, Hamdam and Jackson, W. R., 1971; Fischer, Emil, 1886e; Fischer,

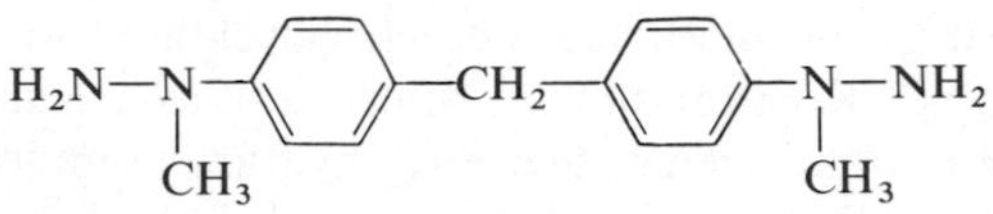

1839

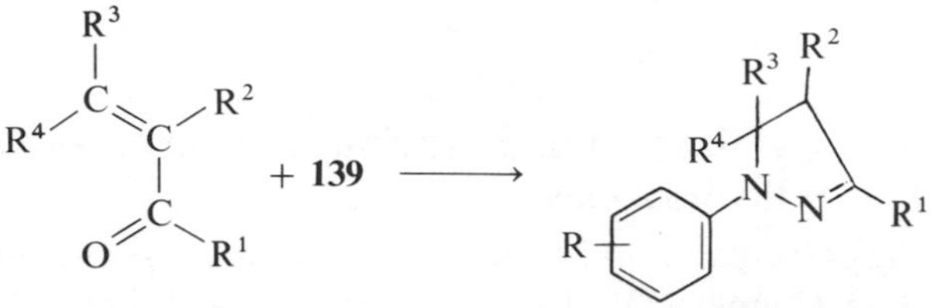

1840

1841

Emil and Knoevenagel, 1887; Gardner, D. V., McOmie and Prabhu, 1970; Grandberg and Kost, 1959; Kohler, E. P., 1909; Laubmann, 1888a; Leuchs and Overberg, 1931; Murakami, K., 1929; Onoda and Koshinaka, 1957; Raiford and Davis, H. L., 1928; Raiford and Entrikin, 1933; Raiford and Gundy, 1938; Raiford and Manley, 1940; Raiford and Peterson, 1937; Raiford and Tanzer, 1941; Ried and Dankert, 1957; Schäfer, H. and Tollens, 1906; Wiley, R., Jarboe, C. and Aayer, 1958 – see also Fischer, Emil, 1884). This synthetic approach to pyrazolines has been reviewed (Jacobs, 1957a, Jarboe, C. H., 1967; Wagner, A., Schellhammer and Petersen, 1966). Pyrazoles are formed by reacting 2,3-acetylenic aldehydes and ketones with hydrazines, including arylhydrazines (Jacobs, 1957a; Tämnefors, Claesson and Karlsson, 1975).

Contrary to the above indolization failures, steroidal 2,3-unsaturated ketone arylhydrazones have been successfully indolized (see Chapter III, J3).

Pyrazolines **1841** were also readily formed when 3-substituted ketones **1842**,

1842

in which the 3-substituent, R^5, was a readily displaceable group, were reacted with arylhydrazines **139**. Such displaceable 3-substituents included dialkylamino, halogeno, hydroxyl, mercapto and arylseleno. These reactions have been reviewed (Jacobs, 1957a; Wagner, A., Schellhammer and Petersen, 1966), as has the further use of Mannich bases in such reactions (Grandberg and Kost, 1959 – see also Scott, F. L., Houlihan and Fenton, 1970). Whereas the pyrazoline **1841** [$R = R^3 = R^4 = H, R^1 + R^2 = (CH_2)_4$] was formed when a mixture of phenylhydrazine with 2-dimethylaminomethylcyclohexanone was refluxed in ethanol (Grandberg, Kost and Yaguzhinskii, 1959), Mannich bases **1842** ($R^2 \neq H, R^3$ = *tert.*-amino) reacted with arylhydrazines under acidic conditions to give the corresponding indoles and 3*H*-indoles, the latter products undergoing a retro Mannich reaction under the indolization conditions (see Chapter III, H).

G. Alternative Cyclization of 2-Acylphenylhydrazones and 2-Aroylphenylhydrazones

Although the corresponding 3- and 4-isomer indolized normally in refluxing formic acid (see Chapter IV, F), when piperidin-2,3-dione 2-acetylphenylhydrazone (**1843**; R = CH_3) (Shavel, Strandtmann and Cohen, M. P., 1962; Strandtmann, Cohen, M. P. and Shavel, 1963) and piperidin-2,3-dione 2-benzoylphenylhydrazone (**1843**; R = COC_6H_5) (Strandtmann, Cohen, M. P. and Shavel, 1963) were refluxed in glacial acetic acid, an alternative cyclization, believed to involve intramolecular nucleophilic attack of the more basic N_β-atom on the C=O group in the enehydrazine tautomers **1844**, occurred to furnish the indazoles **1845** (R = CH_3 and C_6H_5, respectively) (Scheme 60).

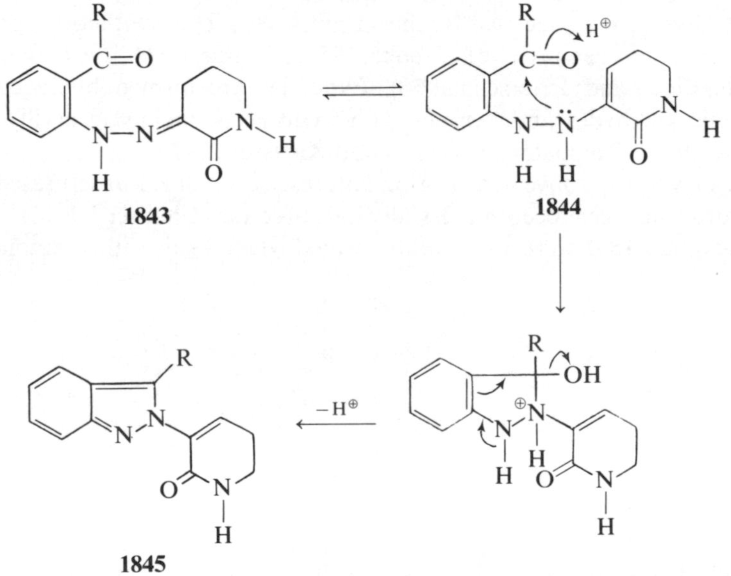

1843

1844

1845

Scheme 60

In an attempt to circumvent this indazole formation, structures **1846** (R = CH_3 and C_6H_5) were refluxed with 88% formic acid, but again the occurrence of indolization was not detected, the reaction products being **1848** (R = CH_3 and C_6H_5, respectively). In these reactions, it was postulated that the initial hydrazones lost hydroxyl ion under the acidic conditions and the azo tautomer,

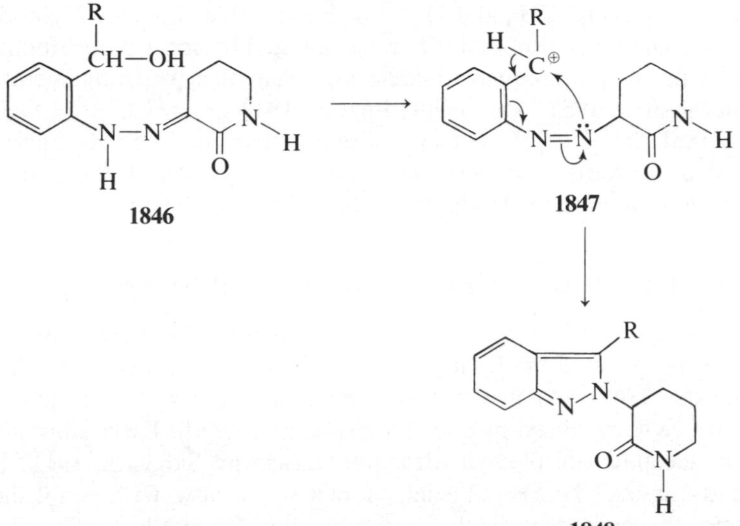

Scheme 61

1847, of the resulting cation intramolecularly cyclized (Scheme 61) (Strandtmann, Cohen, M. P. and Shavel, 1963).

H. Indolization of Ethyl 3-(2-furanyl)-2-oxopropionate Arylhydrazones

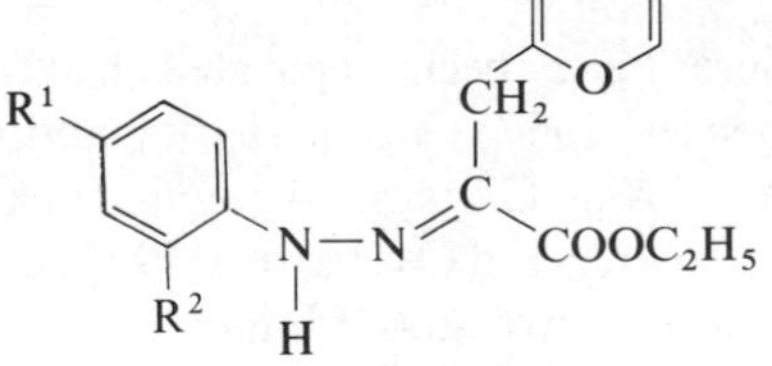

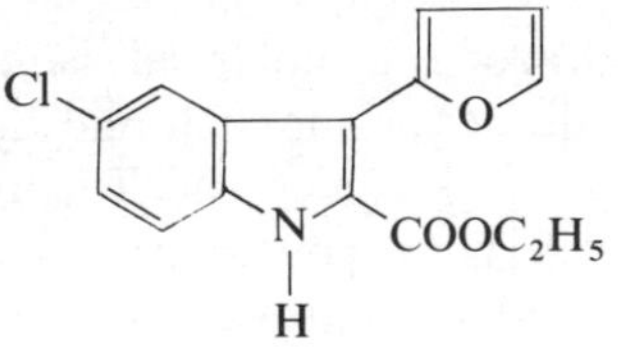

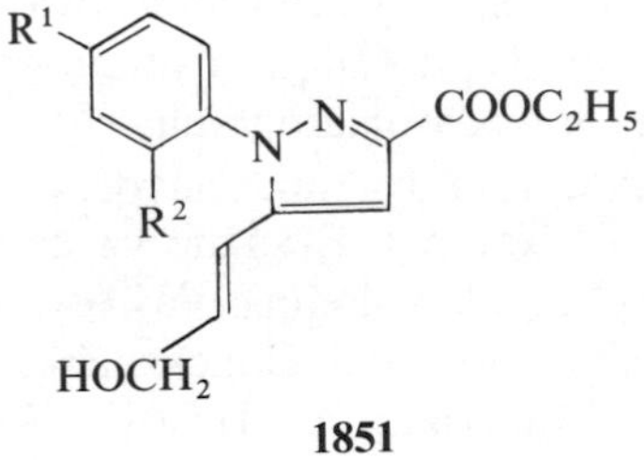

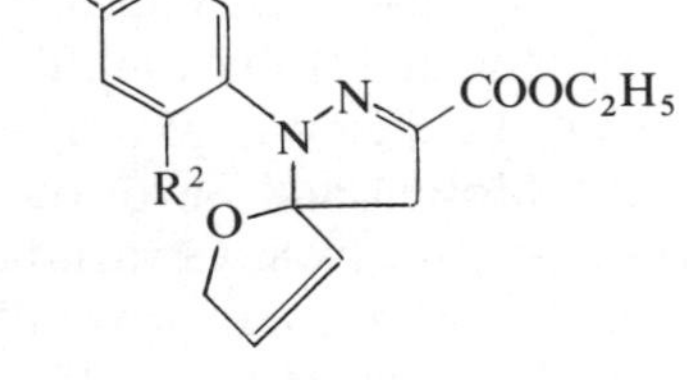

When **1849** (R^1 = Cl, R^2 = H) was refluxed with ethanolic sulphuric acid, the expected indole, **1850**, was obtained in only 10% yield, the major product from the reaction, obtained in 40% yield, being **1851** (R^1 = Cl, R^2 = H). Likewise, structures **1849** (R^1 = NO_2, CH_3 and H, R^2 = H) were converted into **1851** (R^1 = NO_2, CH_3 and H, R^2 = H) in yields of 32%, 22% and 24%, respectively. The formation of **1851** was suggested to occur by protonation of the furan ring with subsequent formation of and then rearrangement of the spiro intermediate **1852**. The arylhydrazone **1849** (R^1 = Cl, R^2 = CH_3) did not yield **1851** (R^1 = Cl, R^2 = CH_3) under similar conditions but afforded a 48% yield of a product, isomeric with the starting material, the structure of which was not established (Holla, B. S. and Ambekar, 1979a).

I. Indolization of Arylhydrazone Geometrical Isomers

Geometrical isomerism in arylhydrazones has been thoroughly investigated and has been discussed briefly in a review of the chemistry of arylhydrazones (Buckingham, 1969) and in far more detail in a review of the structure of hydrazones (Kitaev, Buzykin and Troepol'skaya, 1970). Early studies in the area used acetaldehyde phenylhydrazone (Laws and Sidgwick, 1911; Lockemann and Liesche, 1905) and such isomerism is now well established for hydrazones in general, particularly, though not essentially (e.g. Ch'ang-pai, Evstigneeva and Preobrazhenskii, N. A., 1958; Ramirez and Kirby, 1954), those with a carbonyl group adjacent to their hydrazino moiety. The geometrical isomers of the latter hydrazones are represented by the *syn* and *anti* structures **1853** and **1854**, respectively.

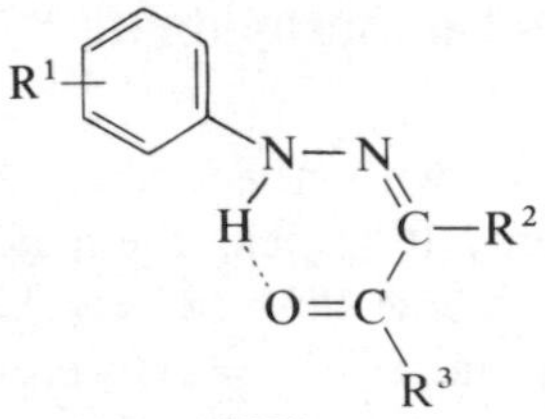

1853 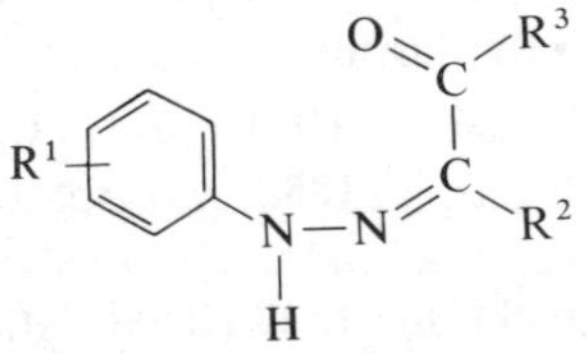 **1854**

Arylhydrazone geometrical isomers have been separated by fractional crystallization (Abramovitch and Spenser, 1957; Carlin, Henley and Carlson, 1957; Elguero, Jacquier and Tarrago, 1966; Gannon, Benigni, Dickson and Minnis, 1969; Glushkov, Smirnova, Zasosova and Ovcharova, 1975; Glushkov, Zasosova and Ovcharova, 1977; Heath-Brown and Philpott, P. G., 1965a; Morozovskaya, Ogareva and Suvorov, 1969; Ramirez and Kirby, 1952; Sempronj, 1938; Yakhontov, Glushkov, Pronina, E. V. and Smirnova, 1973; Yamamoto, H., Inaba, Okamoto, Hirohashi, T., Ishizumi, Yamamoto, M., Maruyama, Mori and Kobayashi, T., 1969c) or by chromatography (Alyab'eva, Khoshtariya, Vasil'ev, A. M., Tret'yakova, Efimova and Suvorov, 1979; Avramenko, Mosina and Suvorov, 1970; Ch'ang-pai, Evstigneeva and Preobrazhenskii, N. A., 1958; Gryaznov, Akhvlediani, Volodina, Vasil'ev, A. M., Babushkina, T. A. and Suvorov, 1977; Isherwood and Cruickshank, 1954; Isherwood and Jones, R. L., 1955; Ishii, Murakami, Y., Hosoya, Takeda,

Suzuki, Y. and Ikeda, 1973; Ishii, Murakami, Y., Suzuki, Y. and Ikeda, 1970; Ishii, Murakami, Y., Takeda and Furuse, 1974; Khoshtariya, Sikharulidze, Tret'yakova, Efimova and Suvorov, 1979; Nagasaka, Yuge and Ohki, 1977; Shagalov, Eraksina, Tkachenko, Mamonov and Suvorov, 1972; Shagalov, Ostapchuk, Zlobina, Eraksina, Babushkina, T. A., Vasil'ev, A. M., Ogaradnikova and Suvorov, 1978; Suvorov, Samsoniya, Chilikin, Chikvaidze, I. S., Turchin, Efimova, Tret'yakova and Gverdtsiteli, L. G., 1978; Yakhontov, Glushkov, Pronina, E. V. and Smirnova, 1973) of the isomeric mixture or by selective formation from the component hydrazine and carbonyl moieties under specified differing conditions (Abramovitch and Spenser, 1957; Clemo and Seaton, 1954; Laws and Sidgwick, 1911; Lockemann and Liesche, 1905) and under unspecified conditions (Suvorov, Preobrazhenskaya, Uvarova and Sheinker, Y. N., 1962).

The structural assignments **1853** and **1854** have been based upon the relevant spectroscopic differences that exist between the geometrical isomers. The *syn* isomers **1853** exhibited u.v. spectra which were bathochromically shifted relative to those of the corresponding *anti* isomers **1854** (Abramovitch and Spenser, 1957; Alyab'eva, Khoshtariya, Vasil'ev, A. M., Tret'yakova, Efimova and Suvorov, 1979; Elguero, Jacquier and Tarrago, 1966; Glushkov, Smirnova, Zasosova and Ovcharova, 1975; Glushkov, Zasosova and Ovcharova, 1977; Gryaznov, Akhvlediani, Volodina, Vasil'ev, A. M., Babushkina, T. A. and Suvorov, 1977; Heath-Brown and Philpott, P. G., 1965a, 1965b; Heneka, Timmler, Lorenz and Geiger, 1957; Inaba, Ishizumi, Mori and Yamamoto, H., 1971; Isherwood and Cruickshank, 1954; Khoshtariya, Sikharulidze, Tret'yakova, Efimova and Suvorov, 1979; Nagasaka, Yuge and Ohki, 1977; Preobrazhenskaya, Uvarova, Sheinker, Y. N. and Suvorov, 1963; Ramirez and Kirby, 1954; Shagalov, Eraksina, Tkachenko, Mamonov and Suvorov, 1972; Shagalov, Ostapchuk, Zlobina, Eraksina, Babushkina, T. A., Vasil'ev, A. M., Ogorodnikova and Suvorov, 1978; Sladkov, Shner, Anisimova and Suvorov, 1972; Suvorov, Samsoniya, Chilikin, Chikvaidze, I. S., Turchin, Efimova, Tret'yakova and Gverdtsiteli, L. G., 1978; Yakhontov, Glushkov, Pronina, E. V. and Smirnova, 1973). Apparently contrary to this generalization, in one example of chromatographically separated geometrical isomeric products, the structural assignments **1855** and **1856** were made although it was the former isomer which exhibited the shorter wavelength u.v. absorption spectrum.

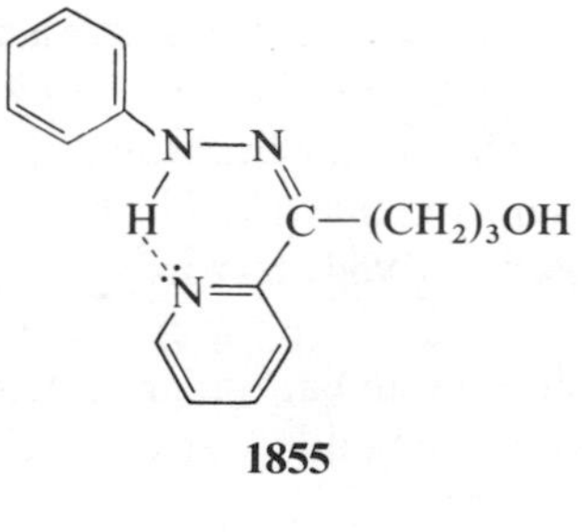

N—N
H C—$(CH_2)_3OH$
N=

1855

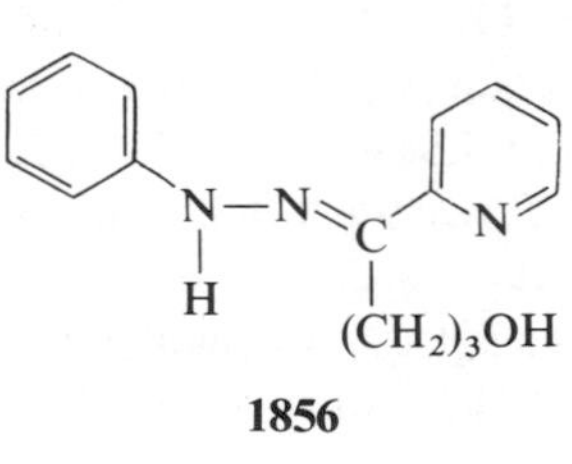

N—N=C N
H $(CH_2)_3OH$

1856

Evidence for these structural assignments was based upon the lower basicity of the former isomer, attributed to the presence of the illustrated intramolecular hydrogen bond, and upon the observation that the latter isomer, upon heating *in vacuo*, underwent dehydration to afford a product formulated as **1857**

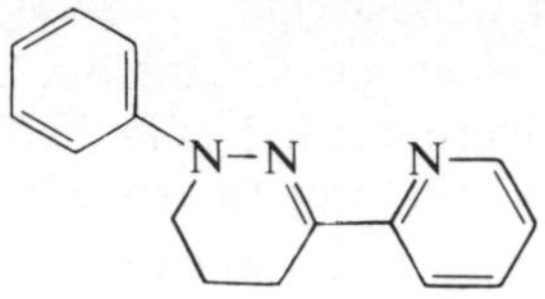

1857

(Ch'ang-pai, Evstigneeva and Preobrazhenskii, N. A. 1958). However, neither experimental verification for the structure of this product nor experimental details for its formation were presented although these structural assignments were supported by the observations that whereas the former isomer could not be indolized using a variety of catalysts, the latter isomer was indolized without difficulty to give the expected product (Ch'ang-pai, Evstigneeva and Preobrazhenskii, N. A., 1958). Clarification of this problem awaits further study.

The i.r. spectra of the isomers **1853** showed N—H and C=O stretching absorption bands at lower frequencies than the corresponding bands of the corresponding isomers **1854** (Abramovitch, 1956; Abramovitch and Muchowski, 1958; Abramovitch and Spenser, 1957; Alyab'eva, Khoshtariya, Vasil'ev, A. M., Tret'yakova, Efimova and Suvorov, 1979; Arata, Sakai, M. and Yasuda, S., 1972; Avramenko, Mosina and Suvorov, 1970; Elguero, Jacquier and Tarrago, 1966; Gannon, Benigni, Dickson and Minnis, 1969; Heath-Brown and Philpott, P. G., 1965a; Heindel, Kennewell and Pfau, 1970; Henecka, Timmler, Lorenz and Geiger, 1957; Inaba, Ishizumi, Mori and Yamamoto, H., 1971; Isherwood and Jones, R. L., 1955; Ishii, Murakami, Y., Hosoya, Takeda, Suzuki, Y. and Ikeda, 1973; Ishii, Murakami, Y., Suzuki, Y. and Ikeda, 1970; Ishii, Murakami, Y., Takeda and Furuse, 1974; Khoshtariya, Sikharulidze, Tret'yakova, Efimova and Suvorov, 1979; Morozovskaya, Ogareva and Suvorov, 1969; Nagasaka, Yuge and Ohki, 1977; Shagalov, Ostapchuk, Zlobina, Eraksina, Babushkina, T. A., Vasil'ev, A. M., Ogorodnikova and Suvorov, 1978; Sladkov, Shner, Anisimova and Suvorov, 1972; Suvorov, Gordeev and Vasin, 1974). Furthermore, in the i.r. spectra of the *syn* isomers of 2-ketoester arylhydrazones, these absorptions gave rise to split bands because of the rotational isomerism between **1853** ($R^3 = OR^4$) and **1858** (Avramenko, Mosina and Suvorov, 1970; Morozovskaya, Ogareva and Suvorov, 1969; Shagalov, Eraksina, Tkachenko, Mamonov and Suvorov, 1972 – see also Preobrazhenskaya, Uvarova, Sheinker, Y. N. and Suvorov, 1963; Ramirez and Kirby, 1954; Suvorov, Gordeev and Vasin, 1974).

The p.m.r. spectra of the isomers **1853** showed, in particular, a signal caused by their N—H proton (intramolecularly hydrogen bonded) at a much lower field

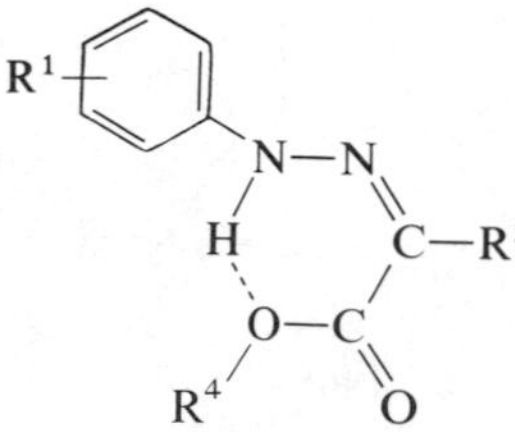

1858

than the analogous signal in the spectra of the corresponding isomers **1854** (Arata, Sakai, M. and Yasuda, S., 1972; Elguero, Jacquier and Tarrago, 1966; Inaba, Ishizumi, Mori and Yamamoto, H., 1971; Ishii, Murakami, Y., Hosoya, Takeda, Suzuki, Y. and Ikeda, 1973; Ishii, Murakami, Y., Suzuki, Y, and Ikeda, 1970; Ishii, Murakami, Y., Takeda and Furuse, 1974; Nagasaka, Yuge and Ohki, 1977; Suvorov, Samsoniya, Chilikin, Chikvaidze, I. S., Turchin, Efimova, Tret'yakova and Gverdtsiteli, L. G., 1978). Furthermore, in the p.m.r. spectra of the geometrical isomers of pyruvic ester arylhydrazones, the signal in each spectrum attributed the methyl group protons of the *syn* isomers was 3–5 Hz downfield relative to that of the corresponding *anti* isomers (Elguero, Jacquier and Tarrago, 1966; Shagalov, Eraksina, Tkachenko, Mamonov and Suvorov, 1974). P.m.r. spectroscopic analysis has also been used to determine the relative amounts of *syn* and *anti* isomers in a wide range of N_α-methylphenylhydrazones (Karabatsos and Krumel, 1967).

A combination of use of these above comparative spectroscopic properties has allowed the assignment of the *syn-syn*, *syn-anti* and *anti-anti* structures **1859**, **1860** and **1861**, respectively, to the three geometrical isomers which were separated by a combination of fractional crystallization and chromatography of bisethylpyruvate *meta*-phenylenedihydrazone (Samsoniya, Targamadze, Tret'yakova, Efimova, Turchin, Gverdtsiteli, I. M. and Suvorov, 1977).

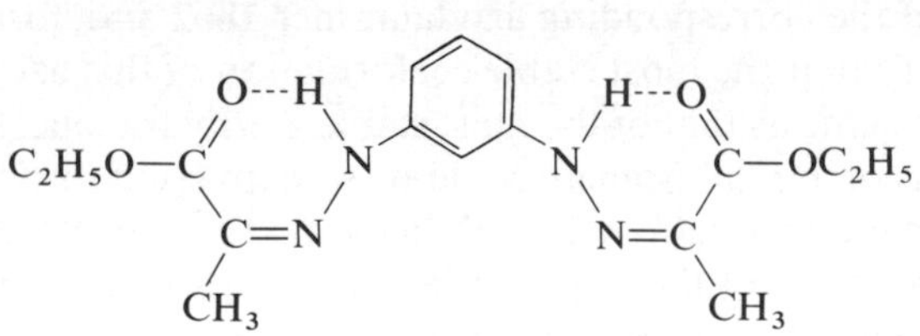

1859

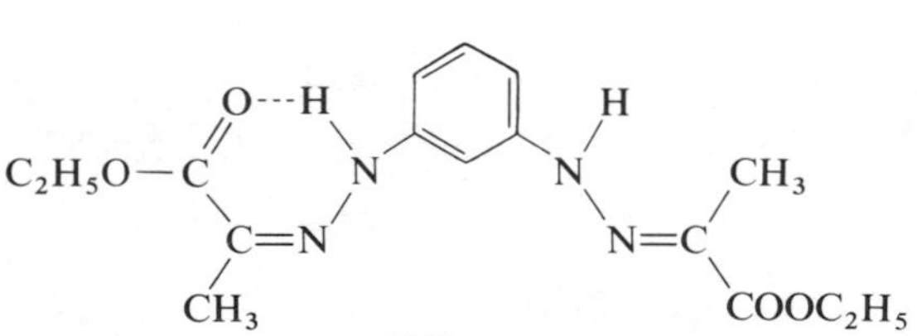

1860

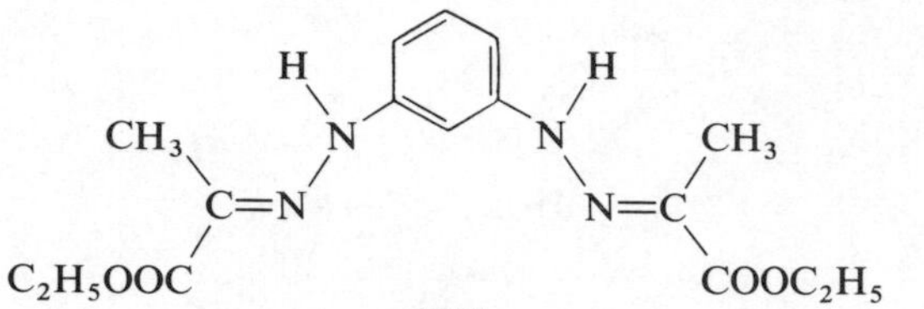

$$\begin{array}{c} \text{CH}_3\diagdown \quad \overset{\text{H}}{\text{N}}\text{—C}_6\text{H}_4\text{—}\overset{\text{H}}{\text{N}} \quad \diagup\text{CH}_3 \\ \text{C}{=}\text{N} \qquad\qquad \text{N}{=}\text{C} \\ \text{C}_2\text{H}_5\text{OOC}\diagup \qquad\qquad\qquad \diagdown\text{COOC}_2\text{H}_5 \end{array}$$

1861

The *anti* isomers **1854** have been converted into the corresponding *syn* isomers **1853** by heating (Arata, Sakai, M. and Yasuda, S., 1972 – see also Heindel, Kennewell and Pfau, 1970), by prolonged standing in solution (Arata, Sakai, M. and Yasuda, S., 1972), by treatment with sodium alcoholate or sodium hydride (Preobrazhenskaya, Uvarova, Sheinker, Y. N. and Suvorov, 1963) or by treatment with acid (indolization conditions?) (Avramenko, Mosina and Suvorov, 1970; Henecka, Timmler, Lorenz and Geiger, 1957; Inaba, Ishizumi, Mori and Yamamoto, H., 1971; Ishii, Murakami, Y., Furuse and Hosoya, 1979; Sladkov, Shner, Anisimova and Suvorov, 1972; Yamamoto, H., Inaba, Okamoto, Hirohashi, T., Ishizumi, Yamamoto, M., Maruyama, Mori and Kobayashi, T., 1969c – see also Heindel, Kennewell and Pfau, 1970), the *syn* isomers **1853** have been converted into the corresponding *anti* isomers **1854** under indolization conditions (Abramovitch, 1958; Abramovitch and Muchowski, 1958) or under basic conditions (Abramovitch and Spenser, 1957) and unspecified geometrical isomerization of arylhydrazones has been effected by their treatment under indolization conditions (Hughes, G. K., Lions and Ritchie, 1939; Ockenden and Schofield, 1957; Ried and Kleemann, 1968; Rydon and Siddappa, 1951; Snyder and Smith, C. W., 1943) or via formation of the hydrochloride salt (Clemo and Seaton, 1954). By-products isolated from successful indolizations may also be geometrical isomers of the starting arylhydrazones (Ockenden and Schofield, 1957). It has been suggested that these isomerizations may have occurred via the formation of the corresponding azo tautomer **1862** and, furthermore, since it is not necessary that the most stable conformation of this azo tautomer (as a complex) be the same as that of the most stable arylhydrazone, the more stable arylhydrazone geometrical isomers could be so converted into their less stable isomers (Snyder and Smith, C. W., 1943). In the case of base catalysed isomerizations, the intermediacy of the species **1863** has been suggested (Preobrazhenskaya, Uvarova, Sheinker, Y. N. and Suvorov, 1963).

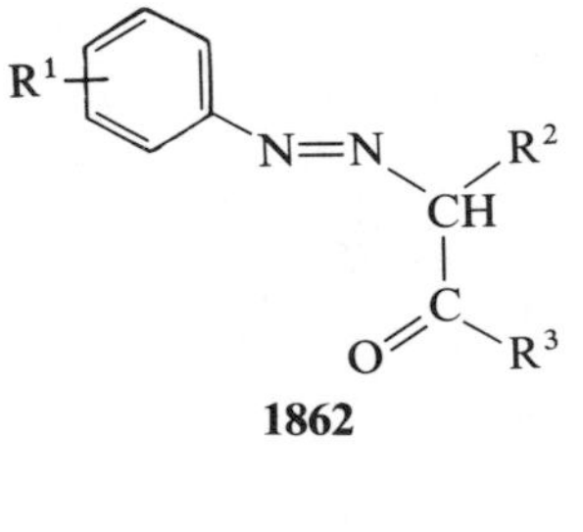

R^1–C$_6$H$_4$–N=N–CH(R^2)–C(=O)–R^3

1862

R^1–C$_6$H$_4$–N=N–C(R^2)=C($O^{\ominus}$)–R^3

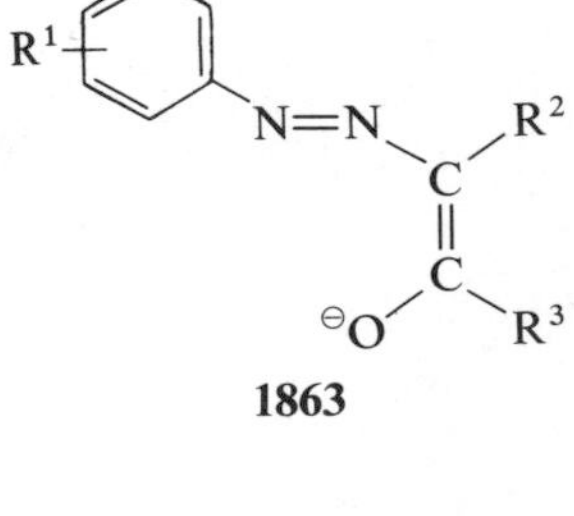

1863

Although in many cases, little or no difference has been observed between the indolizability of geometrically isomeric pairs of arylhydrazones (Arata, Sakai, M. and Yasuda, S., 1972; Gannon, Benigni, Dickson and Minnis, 1969; Glushkov, Smirnova, Zasosova and Ovcharova, 1975; Heath-Brown and Philpott, P. G., 1965a; Hughes, G. K., Lions and Ritchie, 1939; Ishii, Murakami, Y., Hosoya, Takeda, Suzuki, Y. and Ikeda, 1973; Nagasaka, Yuge and Ohki, 1977; Suvorov, Preobrazhenskaya and Uvarova, 1963; Yakhontov, Glushkov, Pronina, E. V. and Smirnova, 1973), probably because of the forcing indolization conditions which also enabled ready geometrical isomerization of the arylhydrazones to occur (Ishii, Murakami, Y., Hosoya, Takeda, Suzuki, Y. and Ikeda, 1973; Suvorov, Samsoniya, Chilikin, Chikvaidze, I. S., Turchin, Efimova, Tret'yakova and Gverdtsiteli, L. G., 1978), in other cases the *anti* isomer afforded a considerably higher yield of indolic product than the corresponding *syn* isomer (Arata, Sakai, M. and Yasuda, S., 1972; Glushkov, Zasosova and Ovcharova, 1977) and occasionally the *syn* isomer failed to indolize under conditions which readily indolized the *anti* isomer (Ch'ang-pai, Evstigneeva and Preobrazhenskii, N., 1958a; Sladkov, Shner, Anisimova and Suvorov, 1972). These differences were probably caused by the presence of the very strong intramolecular hydrogen bond in the *syn* isomer **1853** (Ch'ang-pai, Evstigneeva and Preobrazhenskii, N., 1958a; Glushkov, Smirnova, Zasosova and Ovcharova, 1975; Glushkov, Zasosova and Ovcharova, 1977; Sladkov, Shner, Anisimova and Suvorov, 1972). However, lower yields of indolization products have also been obtained from the *anti* isomers by comparison with those from the *syn* isomers of arylhydrazones (Morozovskaya, Ogareva and Suvorov, 1969).

The probable existence of ethyl pyruvate 2-chloro-3-pyridylhydrazone as *syn* and *anti* geometrical isomers **1864** and **1865** has been used to explain the

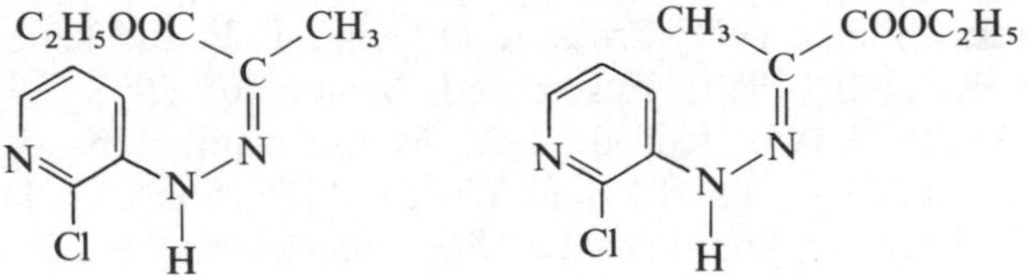

1864 **1865**

observation that its treatment with polyphosphoric acid resulted in the isolation of only a 20% yield of the azacinnoline **1866**. It was suggested (Kochhar, 1969)

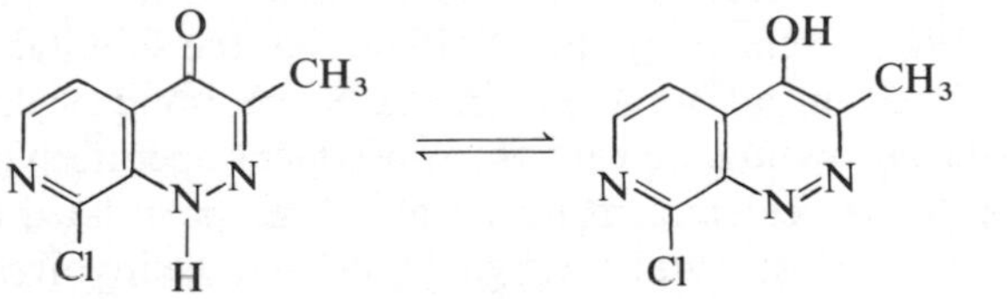

1866

that the isomer **1864** led to the formation of **1866** whereas the isomer **1865** would afford the corresponding indolic product. Unfortunately, none of this latter product was isolated but, because of the very low yield of **1866** from the above reaction, its possible formation is not excluded. This possibility should be investigated.

Geometrical isomerism in a blocked enehydrazine is also possible. Thus, dibenzylketone reacted with N_α,N_β,2,5-tetramethylphenylhydrazine in acetic acid to yield the corresponding enehydrazine **249** ($R^1 = R^3$–$R^5 = CH_3$, $R^2 = R^7 = H$, $R^6 = C_6H_5CH_2$, $R^8 = C_6H_5$), isolated in two forms, m.p. = 86 °C and 104 °C, which are possibly stereoisomers. Both these isomers yielded 2-benzyl-1,4,7-trimethyl-2-phenylindole (**250**; $R^1 = R^3 = R^4 = CH_3$, $R^2 = H$, $R^5 = C_6H_5CH_2$, $R^6 = C_6H_5$) when treated with ethanolic hydrochloric acid (Neber, Knöller, Herbst and Trissler, 1929). However, subsequent attempts to repeat this preparation of the enehydrazine led only to the isolation of the indolic product (Elgersma, 1969) (see Chapter II, E1).

J. Formation of the Corresponding Substituted Anilines and/or Substituted Benzenes Concomitantly with Indolization

There are numerous reports in which formation of the corresponding aniline accompanied the indolization of an arylhydrazone (Alyab'eva, Khoshtariya, Vasil'ev, A. M., Tret'yakova, Efimova and Suvorov, 1979; Arbuzov, A. E. and Friauf, 1913; Arbuzov, A. E. and Wagner, R. E., 1913; Barnes, Pausacker and Badcock, 1951; Barnes, Pausacker and Schubert, 1949; Berlin, A. Y. and Zaitseva, 1960; Boido and Boido Canu, 1977; Carlin and Carlson, 1959; Carlin and Harrison, J. W., 1965; Carlin, Magistro and Mains, 1964; Collins, I., Roberts and Suschitzky, 1971; Cross, A. D., King, F. E. and King, T. J., 1961; Ficken and Kendall, 1959; Fusco and Sannicolò, 1973, 1976a, 1978c; Grammaticakis, 1937, 1939; Kuroda, 1923; Mannich and Hönig, 1927; Millson and Robinson, R., 1955; Parrick and Wilcox, 1976; Shah, G. D. and Patel, B. P. J., 1979; Suvorov, Dmitrevskaya, Smushkevich and Pozdnyakov, 1972; Szmuszkovicz, Glenn, Heinzelman, Hester and Youngdale, 1966; Yakhontov and Pronina, E. V., 1968; Yakhontov, Pronina, E. V. and Rubtsov, 1970 – see also Mann, Prior and Willcox, 1959). Although the indolization process does not involve the intermediacy of free radicals (see Chapter II, A), it has been suggested (Berlin, A. Y. and Zaitseva, 1960; Parrick and Wilcox, 1976, Pausacker and Schubert, 1949b; Schiess and Sendi, 1978) that the simultaneous formation of anilines involved initial homolysis of the N—N bond of the enehydrazine tautomer of the arylhydrazone with the subsequent operation of a free radical mechanism. A Neber rearrangement has also been postulated (Szmuszkovicz, Glenn, Heinzelman, Hester and Youngdale, 1966) to be involved in this formation of anilino by-products (see Chapter VII, L).

Cleavages between the aromatic C-1 atom and the N_α-atom in the starting arylhydrazone, to afford the parent substituted benzene concomitantly with indolization, have been reported. Thus, **11**, when indolized by heating with zinc chloride, furnished the expected 5,5′-bi(2-methylindolyl) (**12**) along with biphenyl and possibly 2-methyl-5-phenylindole (**13**) (Arheidt, 1887), **1180** ($R = C_2H_5$, $n = 0$), when indolized by refluxing in ethanolic sulphuric acid, afforded **1181** ($R = COOC_2H_5$, $n = 0$) along with **1182** ($R = COOC_2H_5$, $n = 0$) (Suvorov, Samsoniya, Chilikin, Chikvaidze, I. S., Turchin, Efimova, Tret'yakova and Gverdtsiteli, I. M., 1978) and **1180** ($R = C_2H_5$, $n = 1$) similarly gave **1181** ($R = COOC_2H_5$, $n = 1$) and **1182** ($R = COOC_2H_5$, $n = 1$) (Samsoniya, Chikvaidze, S. I., Suvorov and Gverdtsiteli, N., 1978). It was assumed (Suvorov, Samsoniya, Chilikin, Chikvaidze, I. S., Turchin, Efimova, Tret'yakova and Gverdtsiteli, I. M., 1978) that this C-1—N_α bond cleavage occurred by thermal decomposition of the azo tautomer of the arylhydrazone. Likewise, when acetone, propiophenone or cyclohexanone pentachlorophenylhydrazone were heated with polyphosphoric acid, zinc chloride or acetic acid, indolic products could not be detected but pentachlorobenzene was sometimes isolated, along with almost invariable occurrence of tar and pentachloroaniline formations (Collins, I., Roberts and Suschitzky, 1971). When cyclohexanone 2,4,6-tribromophenylhydrazone was refluxed with dilute sulphuric acid, 1,3,5-tribromobenzene was formed along with indolization products (see Chapter II, G1) (Barnes, Pausacker and Schubert, 1949) and only 2-hydroxy-4,5,8-trimethylquinoline was isolated from the attempted indolization of **638** [$R^1 = CH_3$, $R^2 = NHN{=}C(CH_3)C_2H_5$) (Huisgen, 1948a). Related to the above studies is the observation that pyrolysis of 2-chlorophenylhydrazine produced 2-chloroaniline and ammonia (Hewitt, 1891) and the later observations that, under similar conditions, phenylhydrazine, 4-bromo- and 4-methylphenylhydrazine and 1- and 2-naphthylhydrazine yielded in each case nitrogen and ammonia along with aniline and benzene, 4-bromoaniline and bromobenzene, 4-methylaniline and toluene, and 1- and 2-aminonaphthalene and naphthalene, respectively, N_α-methyl- and N_α-phenylphenylhydrazine afforded in each case nitrogen and ammonia along with methylphenylamine and diphenylamine, respectively, and N_β-phenylphenylhydrazine afforded aniline and azobenzene (Chattaway and Aldridge, 1911). Indeed, because of their decomposition into the corresponding anilines, methoxyphenylhydrazones should not be distilled, even *in vacuo*. Such decompositions are facilitated by electron donating groups but are hindered by the presence of electron withdrawing, especially chloro, groups. These relationships also apply to the storage of the hydrazones. Wherever possible, methoxy substituted phenylhydrazines should be used immediately or stored as their hydrochloride or sulphate salts (Dufresne, 1972). Similar decomposition of phenylhydrazine has also been found to be effected by pyrolysis with mercury(II) or copper(I) salts. Thus, reaction of phenylhydrazine with mercury(II) cyanide afforded a compound, formulated as HgC_2N_2, $2C_6H_5NHNH_2$, which upon pyrolysis gave aniline, benzene, ammonia, nitrogen, hydrogen cyanide, mercury and phenylhydrazine and likewise the reaction with copper(I) cyanide afforded an isolated intermediate compound CuCN, $C_6H_5NHNH_2$ which upon pyrolysis gave aniline, benzene, ammonia, nitrogen and copper(I) cyanide. In the latter reaction, the catalytic nature of the copper(I) cyanide was also established (Struthers, 1905). Copper(I) chloride, copper(I) bromide and copper(I) iodide also effected the catalytic decomposition of phenylhydrazine into aniline, ammonia and nitrogen, the chloride being the most active and the iodide being the least active in this respect. A solid intermediate, formulated as CuI, $C_6H_5NHNH_2$, was isolated by reacting copper(I) iodide with phenylhydrazine (Arbusow, A. E. and Tikhwinsky, 1910a; Arbuzov, A. E. and Tichvinskii, 1913a).

More recently, pyrolysis of pentafluorophenylhydrazine (**1867**; R = $NHNH_2$) furnished pentafluoroaniline (**1867**; R = NH_2), pentafluorobenzene (**1867**; R = H), nitrogen and ammonia in 44%, 39%, 49% and 42% yields, respectively,

F, F, F, F, F, R

$$\longrightarrow \quad \dot{N}H_2 + C_6F_5\dot{N}H$$

1867 (R = $NHNH_2$)

$$C_6F_5\dot{N}H + \mathbf{1867}\ (R = NHNH_2) \longrightarrow \mathbf{1867}\ (R = NH_2) + C_6F_5\dot{N}NH_2$$

$$\dot{N}H_2 + \mathbf{1867}\ (R = NHNH_2) \longrightarrow NH_3 + C_6F_5\dot{N}NH_2$$

$$C_6F_5\dot{N}NH_2 + R\cdot \longrightarrow C_6F_5N{=}NH + RH$$

(where $R\cdot = C_6F_5\dot{N}NH_2$, $C_6F_5\dot{N}H$ or $\dot{N}H_2$)

$$C_6F_5N{=}NH \longrightarrow \mathbf{1867}\ (R = H) + N_2$$

Scheme 62

a decomposition which was postulated to proceed via the free radical process as shown in Scheme 62 (Birchall, J. M., Haseldine and Parkinson, 1962). However, from the observations that 2,4,6-trichlorophenylhydrazine was rapidly oxidized to 1,3,5-trichlorobenzene and nitrogen with either alkaline permanganate or Fehling's solution (Chattaway and Irving, 1931) and that treatment of cyclohexanone 2,4,6-tribromophenylhydrazone with refluxing dilute sulphuric acid produced comparatively high yields of 2,4,6-tribromoaniline and 1,3,5-tribromobenzene (Barnes, Pausacker and Schubert, 1949), it was suggested that the operation of oxidation–reduction processes could be responsible for the production of the corresponding substituted anilines and benzenes from arylhydrazones and arylhydrazines as described above (Gore, Hughes, G. R.

and Ritchie, 1950). Alternatively, radical cations have been suggested as intermediates in the polyphosphoric acid decomposition of several benzophenone arylhydrazones. Thus, whereas under these conditions **1868** ($R^1 = R^2 = H$) afforded *ortho*-phenylenediamine, probably via an *ortho*-semidine rearrangement (see Chapter VII, M), and **1868** ($R^1 = CH_3$, $R^2 = H$) afforded **1869** by the postulated mechanism shown in Scheme 63, **1868** ($R^1 = H$, $R^2 = CH_3$)

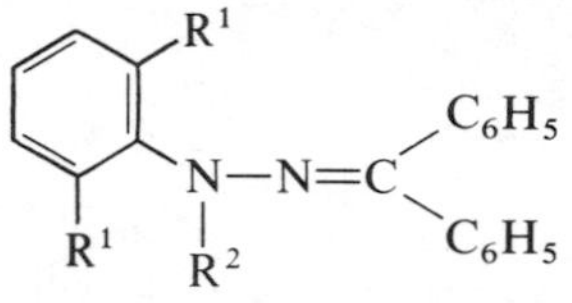

1868

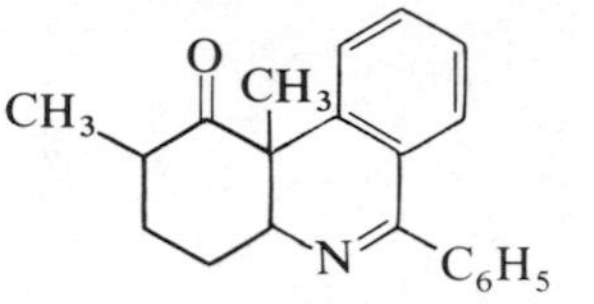

1869

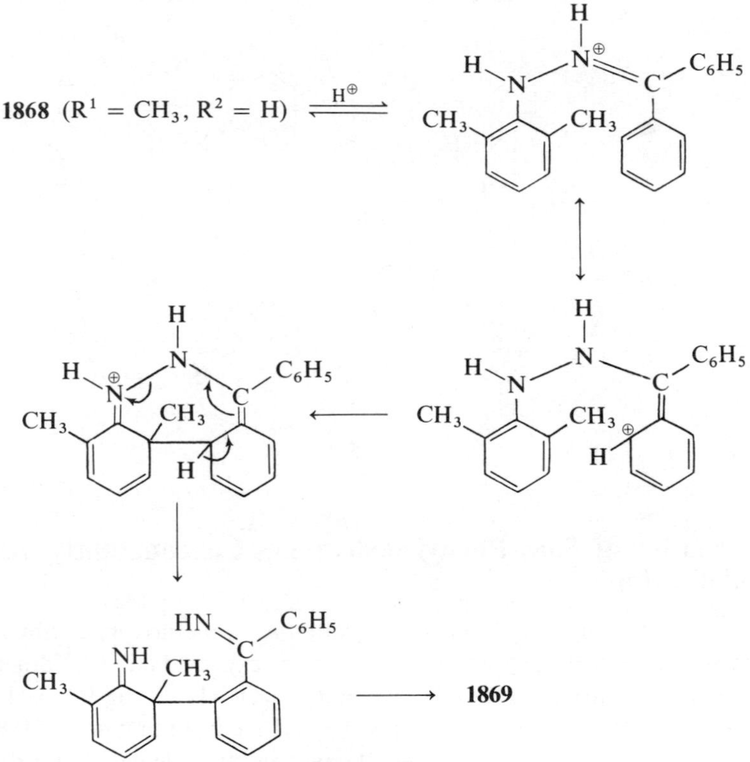

1868 ($R^1 = CH_3$, $R^2 = H$) $\overset{H^{\oplus}}{\rightleftharpoons}$

→ **1869**

Scheme 63

and **1870** afforded **1871** and **1872**, respectively, along with benzophenone in each case. It was proposed that these products arose by homolysis of the diprotonated hydrazones to give, in each case, **1873** which yielded benzophenone, along with **1874** and **1875**, respectively, which underwent radical dimerization to yield **1871** and **1872**, respectively (Fusco 1978; Fusco and Sannicolò, 1980).

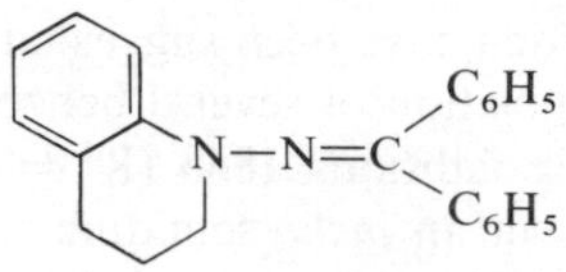

1870

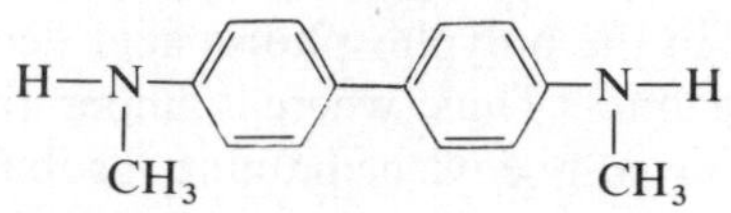

1871

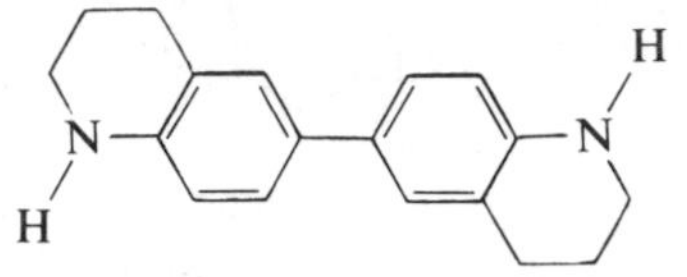

1872

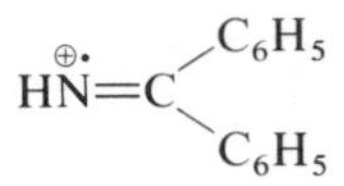

1873

1874

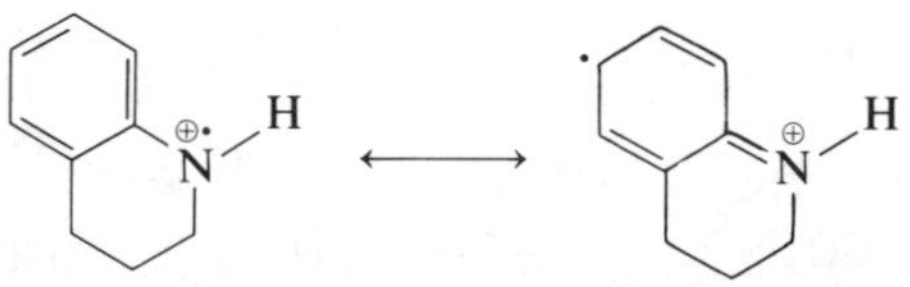

1875

K. Formation of *Para*-Phenylenediamines Concomitantly with Indolization

Treatment of *anti*-ethyl phenylpyruvate 2,5-dimethoxyphenylhydrazone (**1876**) with ethanolic hydrogen chloride gave the expected ethyl 4,7-dimethoxy-3-phenylindole-2-carboxylate (**495**; $R^1 = R^4 = OCH_3$, $R^2 = R^3 = H$, $R^5 = C_6H_5$) as only a minor product, along with ethyl 2-ethoxycinnamate (**1880**) and the major reaction product, 2,5-dimethoxy-*para*-phenylenediamine dihydrochloride (**1881**). It was suggested that these products arose as shown in Scheme 64 via the intermediate **1877**. This could afford **495** ($R^1 = R^4 = OCH_3$, $R^2 = R^3 = H$, $R^5 = C_6H_5$) by the operation of the normal indolization mechanism or undergo sequential rearrangement into **1878**, isomerization into **1879** and ethanolysis to furnish **1880** and **1881**. Alternatively, the formations of **1880** and **1881** might have occurred via the ethanolysis of **1876** into **1880** and the corresponding phenylhydrazine since 2,5-dimethoxyphenylhydrazine

CH_3O C_6H_5 CH_2 C N—N $COOC_2H_5$ CH_3O H

1876

→

$COOC_2H_5$ $H_2\overset{\oplus}{N}$=C CH_3O CH—C_6H_5 H NH CH_3O

1877

CH_3O C_6H_5 N $COOC_2H_5$ CH_3O H

495 ($R^1 = R^4 = OCH_3$, $R^2 = R^3 = H$, $R^5 = C_6H_5$)

C_6H_5CH $COOC_2H_5$ C OCH_3 $H_2N^{\oplus}$ NH_2 CH_3O

1878

$C_6H_5CH_2$ $COOC_2H_5$ C OCH_3 $HN^{\oplus}$ NH_2 CH_3O

1879

CH_3O $H_3\overset{\oplus}{N}$ $2Cl^{\ominus}$ $\overset{\oplus}{N}H_3$ CH_3O

1881

+ C_6H_5CH=C OC_2H_5 $COOC_2H_5$

1880

Scheme 64

afforded 2,5-dimethoxy-*para*-phenylenediamine dihydrochloride (**1881**) when treated with ethanolic hydrogen chloride (Ishii, Murakami, Y., Tani, S., Abe, K. and Ikeda, 1970).

L. Indolization with Simultaneous Neber Rearrangement

When desoxyanisoin phenylhydrazone (**1882**) was refluxed in ethanolic hydrogen chloride, then along with the expected 2,3-di(4-methoxyphenyl)indole, isolated as the major product, there was isolated aniline hydrochloride (4.3 % – the actual yield appears to have been incorrectly quoted: the corresponding anilines have often been isolated as by-products during the indolization of arylhydrazones – e.g. Chapter VII, J) and **1884** (2.7% – the actual yield appears to have been incorrectly quoted). The formation of these two by-products was postulated (Szmuszkovicz, Glenn, Heinzelmann, Hester and Youngdale, 1966) to occur by a Neber rearrangement (O'Brien, 1964) of the hydrazone (**1882**) as shown in Scheme 65. However, it should be noted that the Neber rearrangement is catalysed by base and also that Scheme 65 involved the protonation of **1882** on the less basic N_α-atom to afford the intermediate **1883**. Furthermore, an

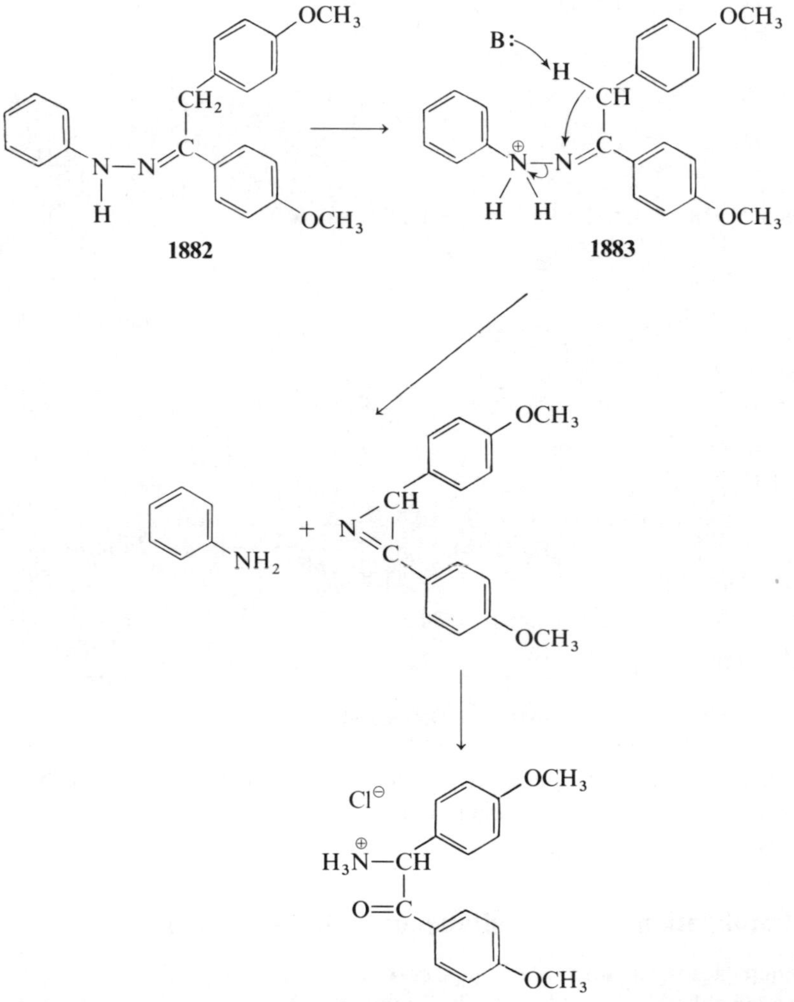

1882 **1883** **1884**

Scheme 65

alternative mechanistic source for the formation of the corresponding anilines during Fischer indolization has been suggested, involving homolysis of the N—N bond of the enehydrazine tautomer of the arylhydrazone with subsequent operation of a free radical mechanism (Berlin, A. Y. and Zaitseva, 1960; Parrick and Wilcox, 1976; Pausacker and Schubert, 1949b; Schiess and Sendi, 1978) (see Chapter VII, J).

M. Indolization with Concomitant *ortho*-Semidine Rearrangement

When camphor phenylhydrazone (**1537**; R = H) was subjected to Fischer indolization, then along with an indolic product (see Chapter V, A) there was isolated *ortho*-phenylenediamine (**1539**) and a product which was formulated as **1540**. Both these products were suggested as resulting from intermediate **1885**, formed by an *ortho*-semidine rearrangement (Banthorpe, Hughes, E. D.,

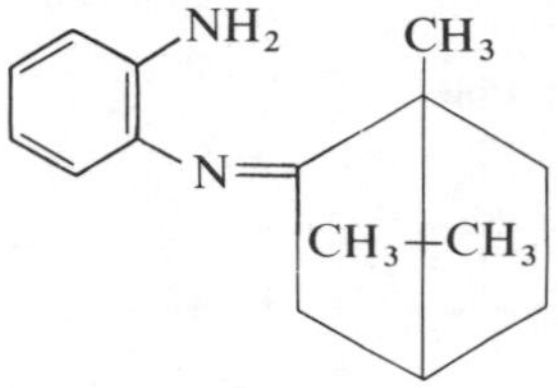

1885

Ingold, Bramley and Thomas, J. A., 1964; Ingold, 1969c; Robinson, R., 1941, Shine, 1969; Večeřa, Gasparič and Petránek, 1957) of the starting hydrazone. Hydrolysis of **1885** would give **1539** and cyclization with simultaneous opening of the camphor ring would give **1540** (Sparatore, 1958, 1962). The suggestion (Sparatore, 1962) that **1539** arose from **1537** (R = H) by initial homolysis of the N—N bond followed by free radical coupling to give **1885** is interesting but unlikely in view of the results of studies upon a wide range of *ortho* rearrangements (Ingold, 1969a). The formation of **1540** from **1885** is related to other cyclizations which have been reported, the gain in resonance stabilization accompanying formation of the imidazole nucleus being suggested as providing the driving force for the cleavage of the C—C bond (Elderfield and Kreysa, 1948). Furthermore, factors governing which of the two possible C—C bond cleavages preferentially occurred have been investigated (Elderfield and McCarthy, 1951).

Further related examples of such *ortho*-semidine rearrangements have been observed when the phenylhydrazones **1886** [R = CH_3, C_2H_5, $CH(CH_3)_2$, $C(CH_3)_3$, C_6H_5 and **1887**] were mixed with powdered zinc chloride and heated at 180–190 °C under an atmosphere of nitrogen. The products isolated from these reactions were aniline and 2-methylbenzimidazole (**1888**; R = CH_3), aniline and **1888** (R = CH_3), 2,3,3-trimethyl-3*H*-indole, 3-(4-aminophenyl)-3-methylbutanone (**190**) (see Chapter II, A), aniline and **1888** (R = CH_3), aniline and **1888** ($R = CH_3$), aniline, *ortho*-phenylenediamine (**1539**) and **1888** ($R = C_6H_5$) and aniline and **1888** ($R = CH_3$), respectively. Again the intermediacy of a species **1889**, resulting from an *ortho*-semidine rearrangement of **1886**, was suggested, hydrolysis then affording *ortho*-phenylenediamine (**1539**) or cyclization affording **1888**, after elimination of the appropriate hydrocarbon (cf. Elderfield and McCarthy, 1951; Theilacker and Leichtle, 1951) (Boido and Boido Canu, 1977). When benzophenone phenylhydrazone was heated with polyphosphoric acid, benzophenone and *ortho*-phenylenediamine (**1539**) were produced (Fusco, 1978).

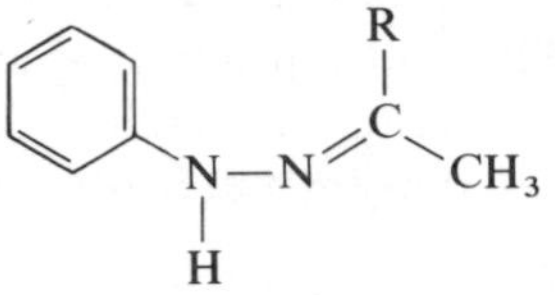

1886

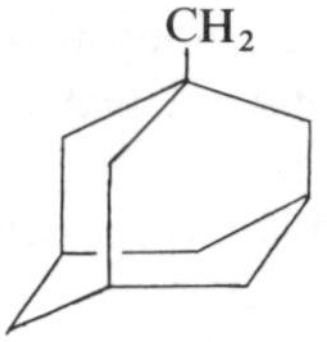

1887

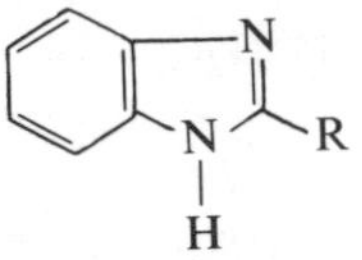

1888

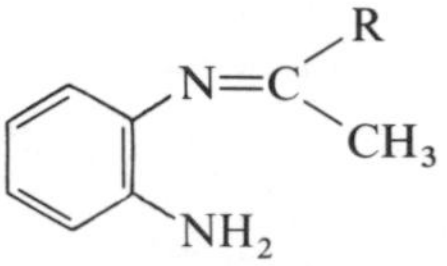

1889

Another related rearrangement occurred when 2-aminophenol was formed in 39%, 45%, *ca.* 5% (along with a 71% yield of 2-phenylbenzofuran) and 13% yields by heating mixtures of O-phenylhydroxylamine hydrochloride with propionaldehyde, 4-aminobutyraldehyde diethyl acetal and acetophenone and O-phenylhydroxylamine hydrochloride itself in ethanolic hydrogen chloride, respectively (Carter, M. A. and Robinson, B., 1974) (see Chapter VI, C).

Although in early investigations of the benzidine rearrangement under thermal non-catalytic conditions (Shine, 1956; Shine, Huang and Snell, 1961; Shine and Snell, 1957; Shine and Trisler, 1960), products resulting from *ortho*- and *para*-semidine rearrangements (for a review, see Ingold, 1969c) were not isolated, such products were later found to be formed, along with products resulting from *ortho*- and *para*-benzidine rearrangements and along with the corresponding benzocarbazoles, when N_α,N_β-diarylhydrazines underwent intramolecular (Banthorpe and Hughes, E. D., 1964b; Shine, Huang and Snell, 1961; Večeřa, Gasparič and Petránek, 1957; Večeřa, Petránek and Gasparič, 1957) rearrangement under such conditions (Banthorpe, 1964; Banthorpe and Hughes, E. D., 1964a; Večeřa, Gasparič and Petránek, 1957; Večeřa, Petránek and Gasparič, 1957). Although from the treatment of similar reactants under acidic conditions (Banthorpe, 1972 and references quoted therein), *ortho*- and *para*-semidine rearrangement products were not structurally identified but may have been formed in small amounts (Banthorpe, 1962a; Banthorpe, Ingold, Roy and Sommerville, 1962), *ortho*-semidine rearrangement products have been isolated after the acid treatment of other N_{α},N_{β}-diarylhydrazines (Carlin and Wich, 1958; Shine and Chamness, 1963a, 1963b, 1967; Shine and Stanley, 1965, 1967). However, in other acid catalysed reactions of such hydrazines, only the corresponding benzidine rearrangement products, and often the benzocarbazoles, were formed (Banthorpe, 1962a, 1962b, 1962c; Banthorpe and Hughes, E. D., 1962a, 1962b; Banthorpe, Hughes, E. D. and Ingold, 1962a, 1962b; Banthorpe, Hughes, E. D., Ingold and Humberlin, 1962; Banthorpe, Ingold, Roy and Somerville, 1962 – see also Shine, 1956; Shine and Chamness, 1963a; Shine, Huang and Snell, 1961; Shine and Snell, 1957; Shine and Trisler, 1960). Only products resulting from *ortho*- and *para*-semidine rearrangements, along with diphenylamine, were isolated from the thermal decomposition of tetraphenylhydrazine (Welzel, 1970).

N. Indolization with Synchronous Formation of Other By-products

Many examples of such reactions have been discussed in various chapters elsewhere in the present study but others are referred to here and are of particular interest in that the structures of some of these by-products remain to be elucidated.

When **992** (R^1 = H, X = O) and phenylhydrazine were heated together (cf. Cawley and Plant, 1938 and other related work in Chapter III, L6 and 10), an exothermic reaction ensured at approximately 100 °C and, as well as the expected indolic product **1075** (R^1–R^3 = H, X = O), other products were separated from the reaction mixture by column chromatography. These other products were mainly formed when using large quantities of reactants and when the reaction temperature was allowed to rise. Of the three such by-products isolated, the molecular formula of one, consisting of yellow needles, m.p. = 201–202 °C, was established as $C_{22}H_{13}NO_3$ and, on the basis of the absence of absorption in its i.r. spectrum characteristic of OH and NH groups and the presence of an absorption band in this spectrum at 1635–1645 cm^{-1} (possibly indicating the presence of the O N-grouping), it was tentatively formulated as **1890** (R = H). Another by-product, m.p. = 216–217 °C, had a molecular formula of $C_{44}H_{32}N_4O_3$ but no structure was postulated for it and the third by-product was isolated as an amorphous green powder (Schroeder, D. C., Corcoran, Holden and Mulligan, 1962). Clearly, further structural investigations of these products are required. Similarly, **992** (R^1 = H, X = O) and 3-chlorophenylhydrazine afforded a product $C_{22}H_{12}ClNO_3$ which, again on the basis of i.r. spectroscopic data (no OH or NH absorption bands but absorption at 1635–1645 cm^{-1}), was tentatively formulated as **1890** (R = Cl). Again, further

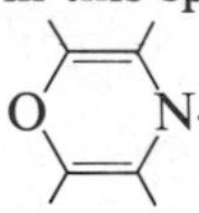

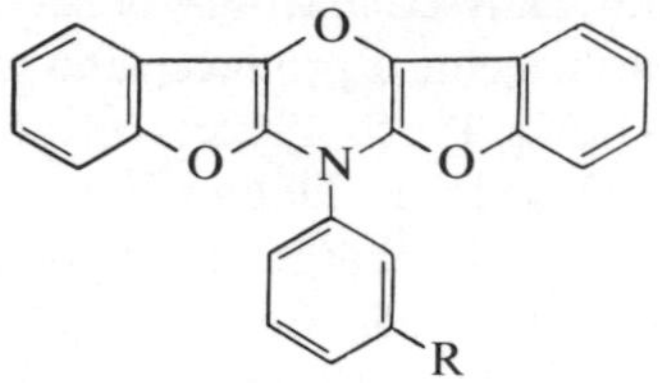

1890

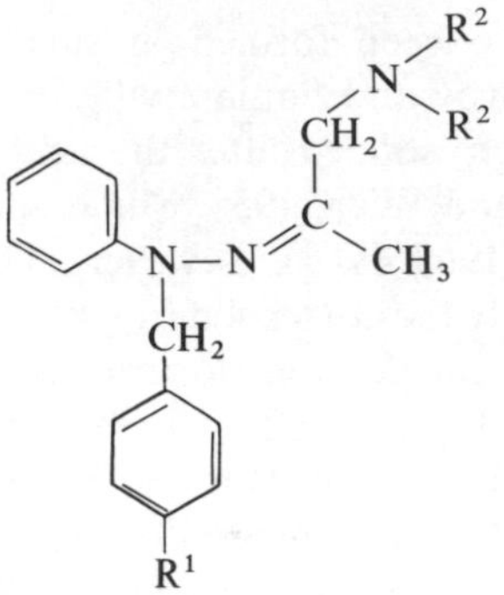

1891

structural studies are required for verification. Contrary to these observations, the corresponding reaction with 3-nitrophenylhydrazine furnished only the expected indolic products (Schroeder, D. C., Corcoran, Holden and Mulligan, 1962).

When the hydrazones **1891** [R^1 = H and Cl, $R^2 = CH_3$, C_2H_5 and $R^2 + R^2 = (CH_2)_5$ and $(CH_2)_2O(CH_2)_2$] were heated in the presence of a trace of anhydrous acetic acid, the corresponding indolic products **1892** [R^1 = H and Cl, $R^2 = CH_3$, C_2H_5 and $R^2 + R^2 = (CH_2)_5$ and $(CH_2)_2O(CH_2)_2$, respectively] were formed, no products resulting from indolization involving the CH_3 group on the carbonyl moiety being isolated (see Chapter III, G). Similar indolic products resulted when the separate carbonyl and hydrazino moieties were likewise treated without the isolation of the corresponding hydrazones, although in these cases small amounts of **1893** (R = H) [when starting with N_α-benzylphenylhydrazine (**1894**, R^1 = H)] and **1893** (R = Cl) [when starting with N_α-4-chlorobenzylphenylhydrazine (**1894**, R^1 = Cl)] were formed. When these reactions were attempted in ethanolic or methanolic solution with or

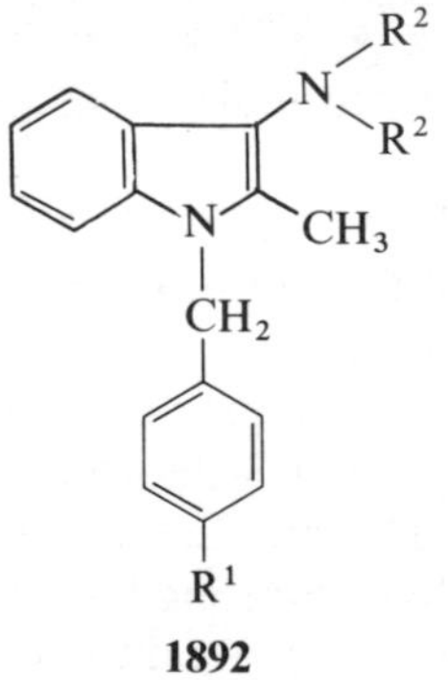

1892

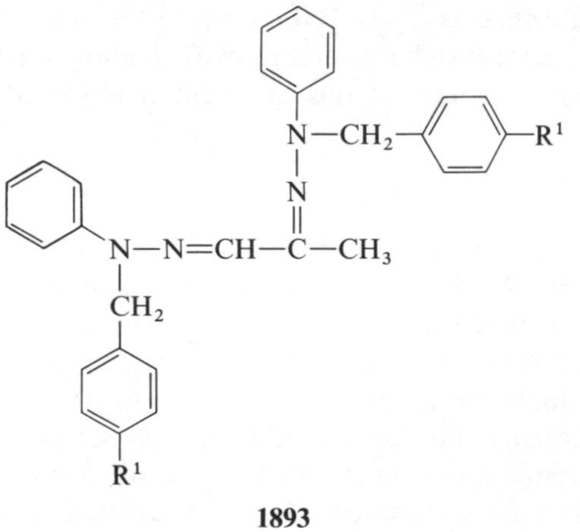

1893

$$CH_3-\overset{O}{\overset{\|}{C}}-CH_2-N(R^2)_2 + \underset{\mathbf{1894}}{C_6H_5-N(NH_2)-CH_2-C_6H_4-R^1} \xrightarrow{\text{In } C_2H_5OH} C_6H_5-N(CH_2C_6H_4R^1)-NH-CH_2-\overset{O}{\overset{\|}{C}}-CH_3$$

$$\rightleftharpoons C_6H_5-N(CH_2C_6H_4R^1)-NH-CH=C(OH)-CH_3 \longrightarrow C_6H_5-NH-CH_2-C_6H_4-R^1 + HN=CH-\overset{O}{\overset{\|}{C}}-CH_3$$

$$\xrightarrow{\text{(i) } +\mathbf{1894};\ \text{(ii) } -NH_3} C_6H_5-N(CH_2C_6H_4R^1)-N=CH-\overset{O}{\overset{\|}{C}}-CH_3 \xrightarrow{+\mathbf{1894}} \mathbf{1893}$$

Scheme 66

without acetic acid, indolization did not occur and moderate yields of **1893** (R^1 = H or Cl, respectively) were the only isolated products. It was proposed, by analogy with related published studies, that these two products arose by the mechanism shown in Scheme 66 (Yoneda, Miyamae and Nitta, 1967).

When 4-methoxyphenylacetaldehyde phenylhydrazone was refluxed with boron trifluoride etherate in acetic acid and the total product was subjected to column chromatography on silica gel, two compounds were isolated. One of these was the expected 3-(4-methoxyphenyl)indole and the other, isolated in small quantity, was obtained as colourless plates, m.p. 139–141 °C (Hino, Suzuki, T. and Nakagawa, 1973). Neither analytical nor spectroscopic data for this latter product were recorded and its structural elucidation awaits further study.

O. Indolization of Nitrophenylhydrazones and the Formation of Indazoles by the Polyphosphoric Acid Catalysed Cyclization of Aryl Methyl Ketone 4-Nitrophenylhydrazones

Several examples are recorded in which nitrophenylhydrazones have failed to indolize (Aksanova, Kucherova and Zagorevskii, 1964a; Bannister and Plant, 1948; Bennett, G. and Waddington, 1929; Cawley and Plant, 1938; Da Settimo, Primofiore, Biagi and Santerini, 1976a; Fennell and Plant, 1932; Hiremath and Siddappa, 1965; Hughes, G. K., Lions and Ritchie, 1939; Kadzyauskas, Butkus, Vasyulite, Averina and Zefirov, 1979; Kakurina, Kucherova and Zagorevskii, 1965a; Kelly, A. H. and Parrick, 1966; Kinsley and Plant, 1956; Massey and Plant, 1931; Ockenden and Schofield, 1953b; Rydon and Siddappa, 1951; Sharkova, N. M., Kucherova and Zagorevskii, 1962; Yakontov, Pronina, E. V. and Rubtsov, 1970; Yevich, 1970) or have only indolized with some difficulty (Kakurina, Kucherova and Zagorevskii, 1965b; Kelly, A. H., McLeod and Parrick, 1965; Rogers, C. U. and Corson, 1947) whereas the corresponding phenylhydrazones indolized normally. It has also been stated, although neither references nor experimental details were given, that aldehyde nitrophenylhydrazones indolize with difficulty (Terent'ev, A. P. and Preobrazhenskaia, 1958). However, contrary to an earlier observation reporting failure (Cawley and Plant, 1938), benzofuran-3(2*H*)-one 2-nitrophenylhydrazone was subsequently (Kinsley and Plant, 1956) found to indolize. No doubt some of the other indolization failures in the above reports should also be reinvestigated under different conditions of catalysis. Indeed, many nitrophenylhydrazones have been indolized (see Chapter IV, K, Table 31), often using hydrochloric acid or an acetic acid–hydrochloric acid mixture as the catalyst. Earlier claims (Hughes, G. K., Lions and Ritchie, 1939) that ethyl pyruvate 2-nitrophenylhydrazone was indolized to ethyl 7-nitroindole-2-carboxylate were not substantiated in later studies (Rydon and Siddappa, 1951) but subsequently it was found (Parmerter, Cook, A. G. and Dixon, 1958) that, using polyphosphoric acid as catalyst, this indolization, together with those of the corresponding 3- and 4-nitrophenylhydrazone, could be readily achieved in good yields. Likewise using a polyphosphoric acid catalyst, ethyl pyruvate 2-methyl-4-nitro, 2-methyl-5-nitro-, 4-methyl-2-nitro- and 4-methoxy-2-nitrophenylhydrazone were converted into **1895** ($R^1 = H$, $R^2 = NO_2$, $R^3 = CH_3$; $R^1 = NO_2$, $R^2 = H$, $R^3 = CH_3$; $R^1 = H$, $R^2 = CH_3$, $R^3 = NO_2$; $R^1 = H$, $R^2 = OCH_3$, $R^3 = NO_2$, respectively) (Hiremath and Siddappa, 1964).

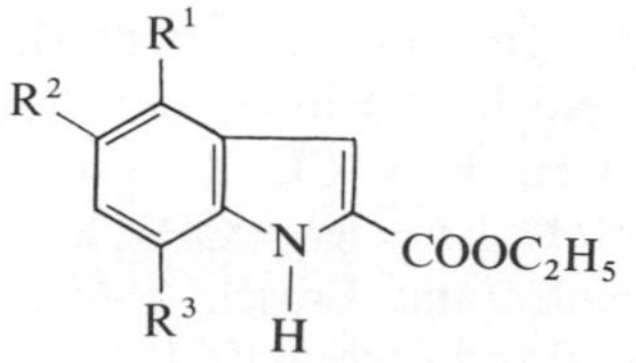

R^1, R^2, R^3, $COOC_2H_5$, N, H

1895

O_2N, R^2, R^1, N, H, NO_2

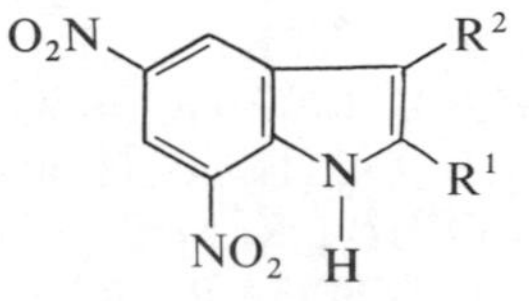

1896

However, attempts to use this catalyst in later related attempted indolizations were unsuccessful (Frasca, 1962a).

It has been suggested (Arbuzov, A. E. and Kitaev, 1957a) that 5,7-dinitroindoles cannot be prepared by indolization of 2,4-dinitrophenylhydrazones and, indeed, such hydrazones have failed to indolize under conditions which have indolized the corresponding phenylhydrazones (Barclay and Campbell, 1945; Dalgliesh and Mann, 1947; Keglević, Stojanac and Desaty, 1961). However, the choice of catalyst obviously plays a critical role in such reactions, since, using a concentrated sulphuric acid–acetic acid catalyst, ethyl methyl, diethyl and methyl propyl ketone, cyclohexanone and 4-methylcyclohexanone 2,4-dinitrophenylhydrazone underwent indolization to yield **1896** [$R^1 = R^2 = CH_3$; $R^1 = C_2H_5$, $R^2 = CH_3$; $R^1 = CH_3$, $R^2 = C_2H_5$; $R^1 + R^2 = (CH_2)_4$ and $(CH_2)_2CH(CH_3)CH_2$, respectively] (Deorha, and Joshi, S. S., 1961) and other 2,4-dinitrophenylhydrazones have also been indolized (Joshi, S. S. and Gambhir, 1956). Some limitations using 2,4-dinitrophenylhydrazones still however occurred in the former investigation since attempted indolizations of isopropyl methyl ketone and 2-methylcyclohexanone 2,4-dinitrophenylhydrazone failed, although the mononitrophenylhydrazones of these two ketones indolized using concentrated hydrochloric acid as the catalyst (Deorha and Joshi, S. S., 1961). Recent studies (Saleha, Siddiqi and Khan, N. H., 1979) have shown that alcoholic formic acid in refluxing carbon tetrachloride or chloroform is the method of choice for indolizing 2,4-dinitrophenylhydrazones. With this catalyst in the former solvent, the 2,4-dinitrophenylhydrazones of acetone, ethyl methyl ketone, laevulinic acid, ethyl laevulinate, acetophenone, propiophenone, 3-benzoylpropionic acid, ethyl 3-benzoylpropionate, 2-oxoglutaric acid and ethyl 2-oxoglutarate were converted into **1896** ($R^1 = CH_3$, $R^2 = H$, CH_3, CH_2COOH and $CH_2COOC_2H_5$; $R^1 = C_6H_5$, $R^2 = H$, CH_3, CH_2COOH and $CH_2COOC_2H_5$, $R^1 = COOH$, $R^2 = CH_2COOH$ and $R^1 = COOC_2H_5$ and $R^2 = CH_2COOC_2H_5$, respectively) and with this catalyst in the latter solvent, acetone and laevulinic acid 2,4-dinitrophenylhydrazone afforded **1896** ($R^1 = CH_3$, $R^2 = H$ and CH_2COOH, respectively), the yields from all these reactions being *ca.* 70%.

The attempted indolization of several phenyl substituted acetophenone 4-nitrophenylhydrazones, using either sulphuric acid or hydrochloric acid as catalyst, failed (Dennler and Frasca, 1966b; Frasca, 1962b). However, treatment of these hydrazones with polyphosphoric acid as catalyst furnished the corresponding 1-(4-nitrophenyl)indazoles. In this way **1897** (R^1–R^4 = H; $R^1 = R^2 = R^4$ = H, R^3 = CH_3, OCH_3 and Br) (Dennler and Frasca, 1966b; Frasca, 1962b), **1897** ($R^2 = R^3 = R^4$ = H, R^1 = Cl; $R^1 = R^2 = R^4$ = H, R^3 = Cl, C_2H_5, OC_2H_5, C_6H_5 and $OCOCH_3$; $R^1 = R^3 = CH_3$, $R^2 = R^4$ = H; $R^1 = R^3 = OCH_3$, $R^2 = R^4$ = H) (Dennler and Frasca, 1966b) and **1897** (R^1 = CH_3 and OCH_3, R^2–R^4 = H) (Portal and Frasca, 1971b) were converted into the corresponding **1898**. Whereas **1897** ($R^1 = R^4$ = H, $R^2 = R^3 = OCH_3$) gave only **1898** ($R^1 = R^4$ = H, $R^2 = R^3 = OCH_3$) in this reaction (Dennler

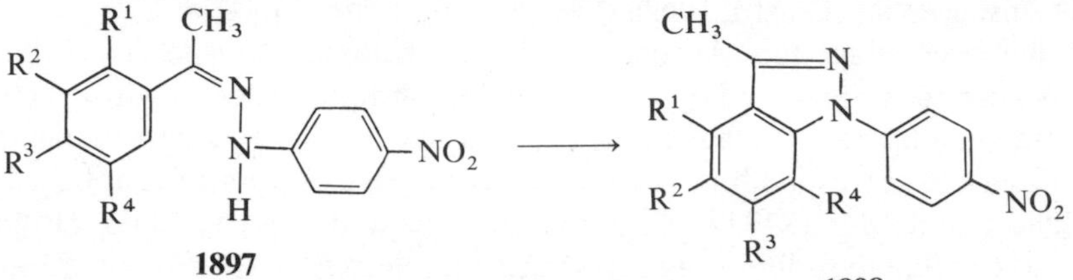

1897 → **1898**

and Frasca, 1966b), similar hydrazones in which the ring closure was potentially ambiguous yielded both possible cyclization products. Thus **1897** [$R^1 = R^4$ = H, $R^2 + R^3 = (CH_2)_3$] gave **1898** [$R^1 = R^4$ = H, $R^2 + R^3 = (CH_2)_3$ and $R^1 = R^2$ = H, $R^3 + R^4 = (CH_2)_3$] (Portal, Dennler and Frasca, 1967) and **1897** ($R^1 = R^3 = R^4$ = H, R^2 = Cl, Br, I, CH_3, OCH_3; $R^1 = R^4$ = H, $R^2 = R^3 = CH_3$) gave both **1898** ($R^1 = R^3 = R^4$ = H, R^2 = Cl, Br, I, CH_3, OCH_3; $R^1 = R^4$ = H, $R^2 = R^3 = CH_3$, respectively) and **1898** (R^1–R^3 = H, R^4 = Cl, Br, I, CH_3, OCH_3; $R^1 = R^2$ = H, $R^3 = R^4 = CH_3$, respectively) (Portal and Frasca, 1971b). In this latter study the isomers in which R^4 = H were always obtained in higher yield than those where $R^4 \neq$ H, this being attributed (Portal and Frasca, 1971b) to the steric repulsion between the indazolic 1-(4-nitrophenyl) substituent and C-7 substituent which had previously (Dennler, Portal and Frasca, 1967) been observed. This reaction has been extended to the 4-nitrophenylhydrazones of several other acetylated aromatic hydrocarbon and heterocyclic systems. Thus, **1899**, **1901** and **1903** (R = CH_3)

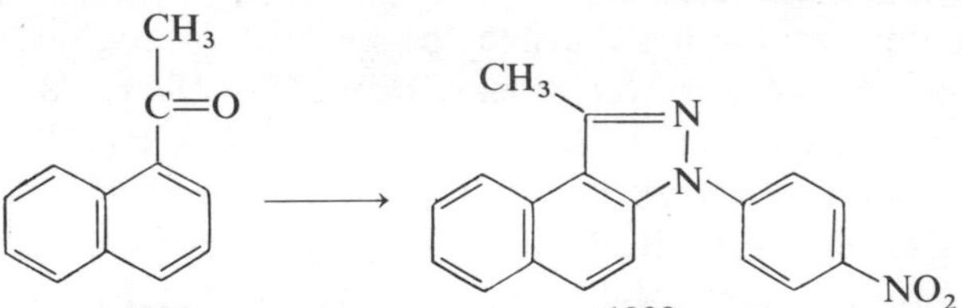

1899 → **1900**

(Dennler and Frasca, 1967) and **1905** (Portal and Frasca, 1971a) afforded **1900**, **1902**, **1904** (R = CH_3) and **1906**, respectively. In other cases in which the direction of cyclization was potentially ambiguous, only one product was obtained [i.e. **1903** (R = C_6H_5) → **1904** (R = C_6H_5), **1907** → **1908**, **1909** → **1910**, **1911** → **1912** (Dennler and Frasca, 1967) and **1913** (X = S) → **1914**

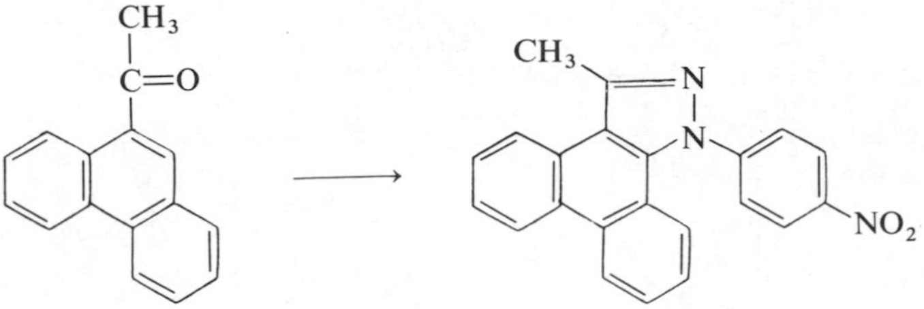

1901 **1902**

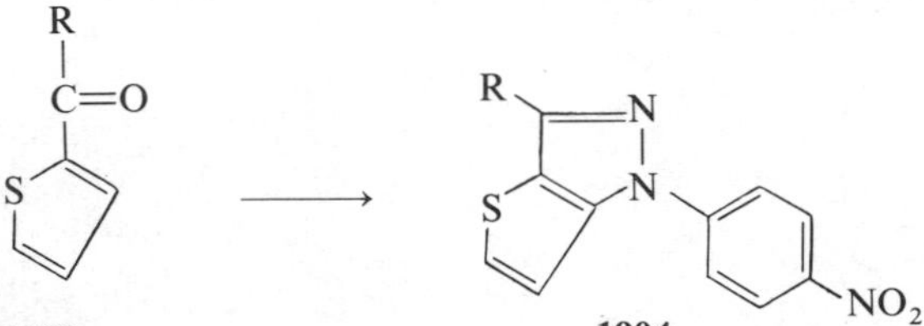

1903 **1904**

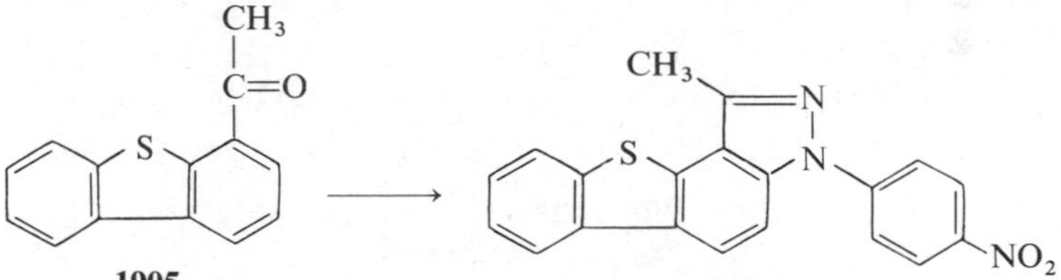

1905 **1906**

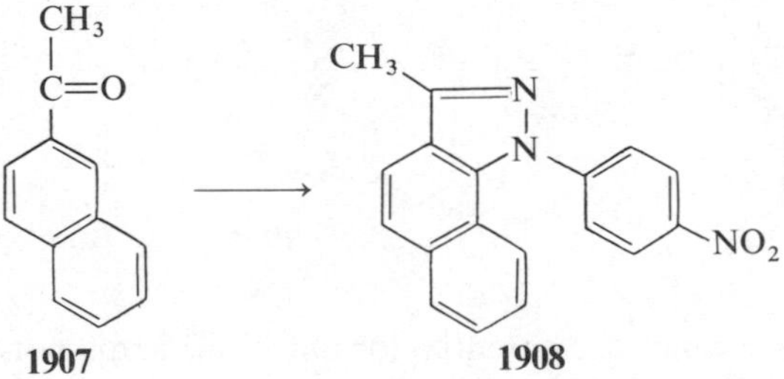

1907 **1908**

(X = S) (Portal and Frasca, 1971a)] although both possible products were formed when **1915** was cyclized into **1916** and **1917** (Dennler and Frasca, 1967) and when **1913** (X = O) was cyclized into **1914** (X = O) and **1918** (Portal and Frasca, 1971a). However, 2-, 3- and 4-acetylpyridine could not be similarly converted into the corresponding 1-(4-nitrophenyl)azaindazoles (Dennler and Frasca, 1967).

The 1-(4-nitrophenyl)indazoles **1920** [$R^1 = R^2 = H$, $R^3 = OCH_3$; $R^1 = C_6H_5$, $R^2 = R^3 = H$; $R^1 = C_6H_4OCH_3(4)$, $R^2 = H$, $R^3 = OCH_3$] were similarly prepared from 4-methoxybenzaldehyde, diphenyl ketone and di(4-methoxyphenyl) ketone 4-nitrophenylhydrazone, respectively (Dennler and

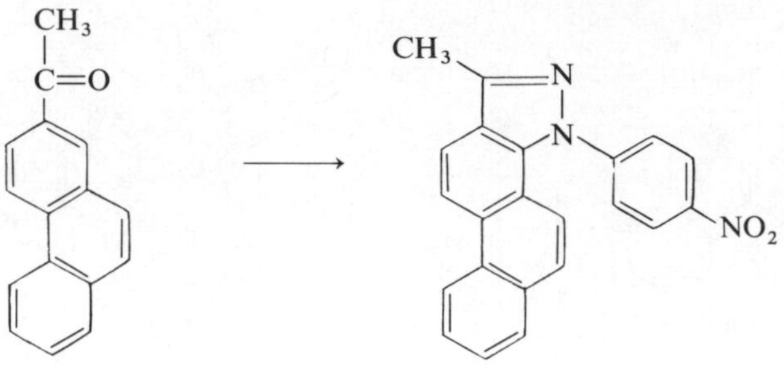

1909 → 1910

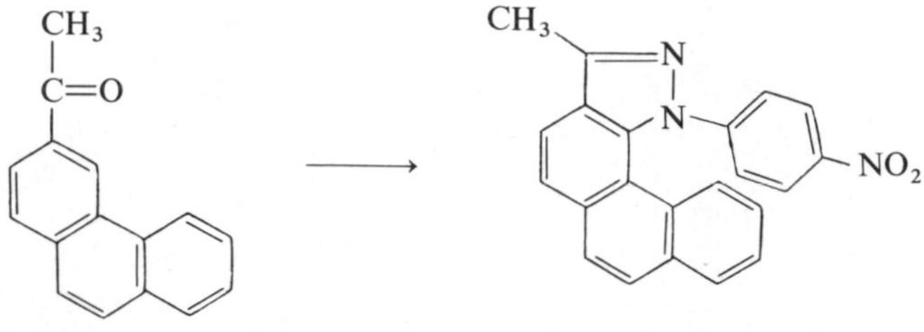

1911 → 1912

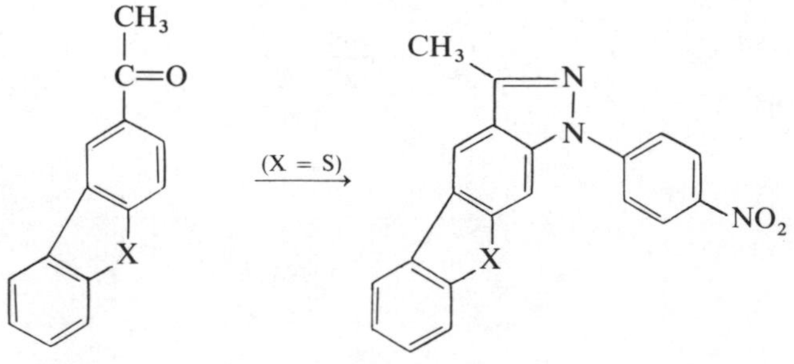

1913 —(X = S)→ 1914

Frasca, 1966b) in which the potential for indolization did not exist. In two cases where the direction of other similar cyclizations was ambiguous, only one of the two possible products was obtained from each reaction. Thus **1919** (R^1 = H, R^2 = OCH_3, R^3 = OH and R^1 = H, R^2 = R^3 = OCH_3) afforded **1920** (R^1 = H, R^2 = OCH_3, R^3 = OH and R^1 = H, R^2 = R^3 = OCH_3, respectively) (Dennler and Frasca, 1966b). When **1897** (R^1 = R^4 = CH_3, R^2 = R^3 = H) was treated with polyphosphoric acid, a mixture (*ca.* 1 : 1) of **1898** (R^1 = R^4 = CH_3, R^2 = R^3 = H) and **1898** (R^2 = R^4 = CH_3, R^1 = R^3 = H) was produced, the latter product resulting from a methyl group migration. When **1897** (R^1 = R^3 = R^4 = CH_3, R^2 = H) was similarly treated, unrearranged, rearranged and demethylated 1-(4-nitrophenyl)indazoles resulted. These migrations and eliminations were rationalized in terms of the steric

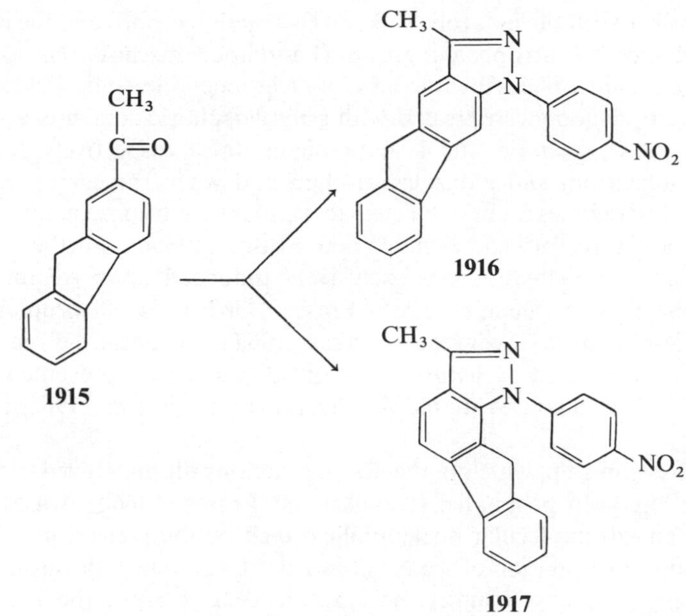

1915 → 1916 + 1917

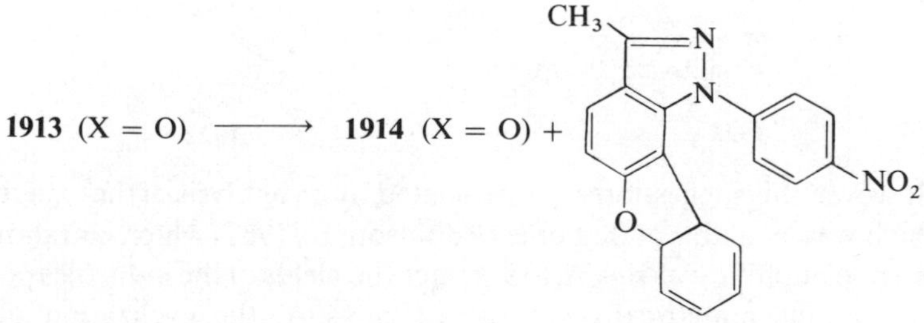

1913 (X = O) ⟶ **1914** (X = O) + **1918**

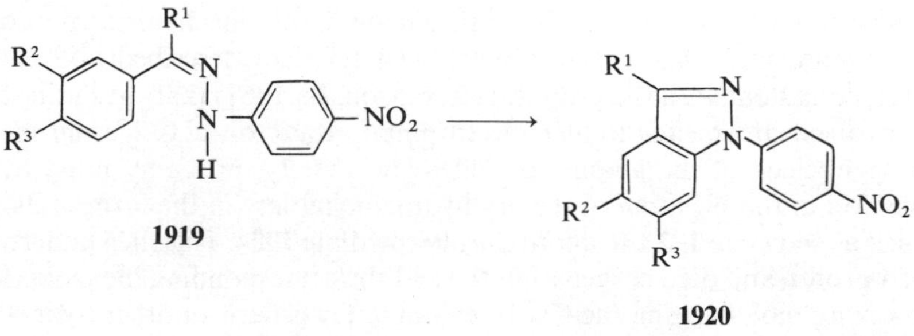

1919 ⟶ **1920**

interactions between the indazolic C-3 and C-4 methyl groups and the indazolic C-7 methyl and 1-(4-nitrophenyl) groups (Land and Frasca, 1971).

When the 3-nitrophenylhydrazones of acetophenone and 4-chloro- and 4-methylacetophenone were treated with polyphosphoric acid, mixtures of the corresponding 4-, 4- and 6- and 4- and 6-nitroindoles, respectively, formed by Fischer indolization, and indazoles resulted and when the corresponding 2-nitrophenylhydrazones were subjected to similar conditions, neither indoles nor indazoles were formed (Dennler and Frasca, 1966b). Furthermore, the corresponding phenylhydrazones exclusively indolized upon treatment with polyphosphoric acid (Dennler, 1965; Frasca, 1962b), as did propiophenone 2-, 3- and 4-nitrophenylhydrazone (Frasca, 1962a), whereas the replacement of polyphosphoric acid by hydrochloric or sulphuric acid prevented the formation of either indazoles or indoles (Dennler and Frasca, 1966b; Frasca, 1962b).

The cyclization step, which in the above reactions ultimately led to indazole formation, has been postulated (Dennler and Frasca, 1966b; Frasca, 1962b) to involve an intramolecular nucleophilic attack by the p-electron pair of the N_α-atom, after protonation of the N_β-atom, at a C-2 atom of the aromatic ring of the ketonic or aldehydic moiety, as shown in **1921**, to afford the intermediate

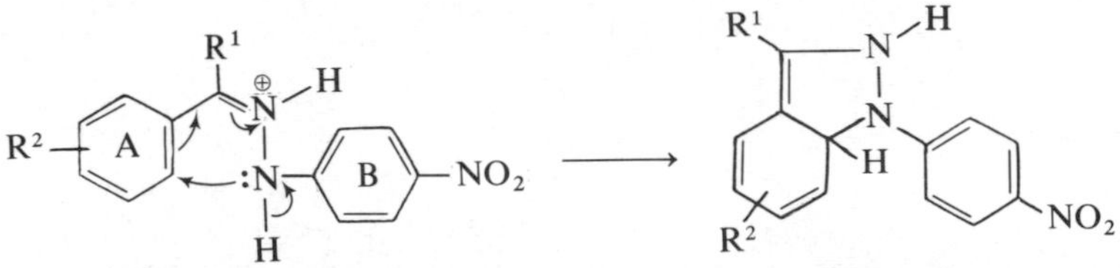

1921 **1922**

1922. However, this suggestion was invalidated by an analysis of the experimental data which was available at that time (Robinson, B., 1967) which had shown that the more nucleophilic was ring A, the greater the yields of the indazoles produced and that in unsymmetrically substituted rings A, the cyclization occurred exclusively or predominantly at the more nucleophilic of the two *ortho* carbon atoms. Furthermore, the necessity for a 3- or 4-nitro group on ring B for indazole formation, which would retard the mechanism shown in **1921** relative to the ring B unsubstituted analogue, was inconsistent with this mechanistic proposal. These experimental data, together with the later data described above, are, however, consistent with the proposal (Robinson, B., 1967) that the cyclization stage involved an intramolecular electrophilic attack on a C-2 atom of the aromatic nucleus of the ketonic or aldehydic moiety, probably initiated by protonation of the N_β-atom of the arylhydrazine moiety in the form of its azo tautomer as shown in **1923**, to afford the intermediate **1924**. This then underwent loss of a proton and dehydrogenation to yield the corresponding indazole **1925**, the oxidizing moiety being the C=N and/or NO_2 groups of other hydrazone

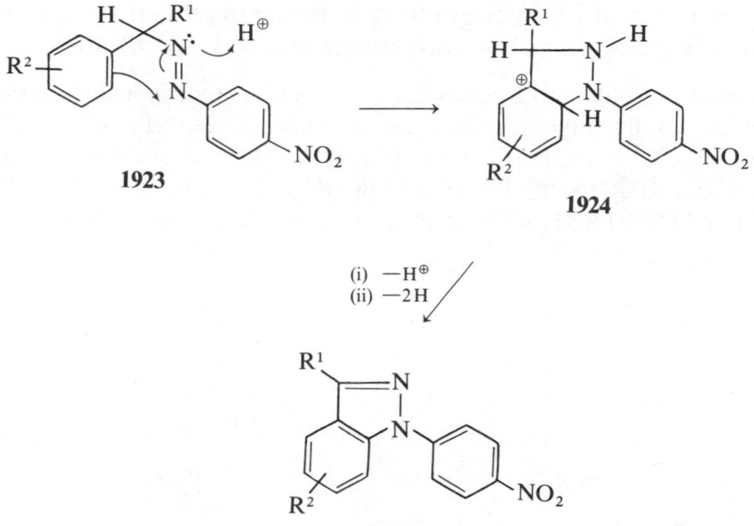

1923

1924

(i) $-H^{\oplus}$
(ii) $-2H$

1925

Scheme 67

molecules (Dennler and Frasca, 1966b; Frasca, 1962b) (Scheme 67). In the case of 3-nitrophenylhydrazones, the hydrazine moiety would be more basic (Stroh and Westphal, 1963) which it was suggested (Robinson, B., 1967) would permit the competing equilibrium between the hydrazine and its enehydrazine tautomer to become established and thus ultimately lead to some indolization. It was also suggested (Robinson, B., 1967) that in a 2-nitrophenylhydrazone, the inhibition of indazole formation was caused by the intramolecular hydrogen bonding as shown in **1926** which prevented the formation of the necessary azo tautomer. Support for this was later forthcoming (Gale and Wilshire, 1973) by the observation of similar hydrogen bonding shown in **1927** (R = NO_2 and H, respectively) which also led to stabilization. Furthermore, this cyclization of arylhydrazones into indazoles was later shown to resemble the initial stage in the mass spectral fragmentation of the former compounds (Das, K. G., Kulkarni, P. S. and Chinchwadkar, 1969).

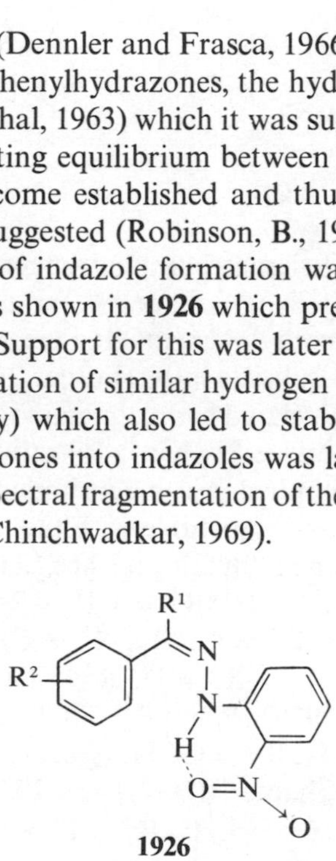

1926

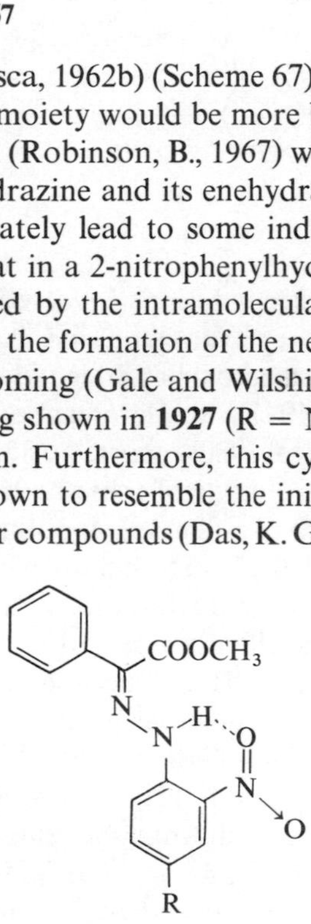

1927

P. Occurrence of [5,5]-Sigmatropic Rearrangement instead of [3,3]-Sigmatropic Rearrangement and Related Studies

Treatment of 4-methoxybenzaldehyde, 4,4′-dimethoxybenzophenone and fluorenone 2,6-dimethylphenylhydrazone, **1928** [$R^1 = CH_3$, $R^2 = R^3 = H$, $R^4 = OCH_3$, $R^5 = H$ and $C_6H_4OCH_3(4)$] and **1929**, respectively, with polyphosphoric acid at 100 °C afforded **1930** [$R^1 = CH_3$, $R^2 = H$, $R^3 = OCH_3$, $R^4 = H$ and $C_6H_4OCH_3(4)$] and **1931**, respectively. The mechanism proposed

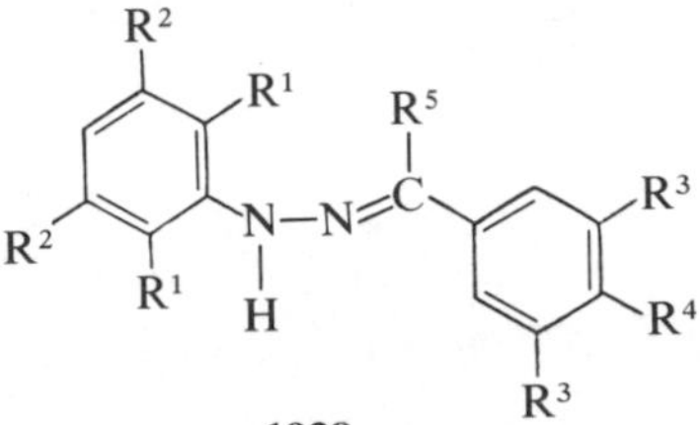

1928

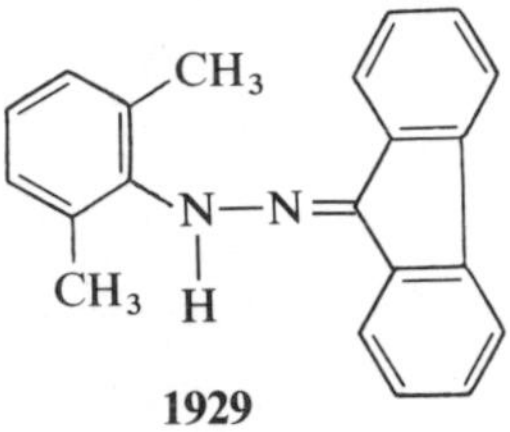

1929

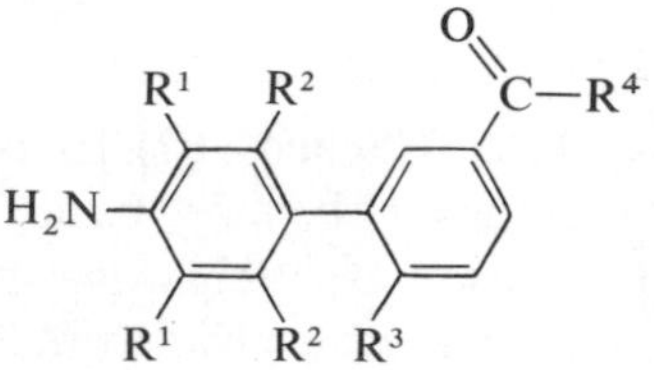

1930

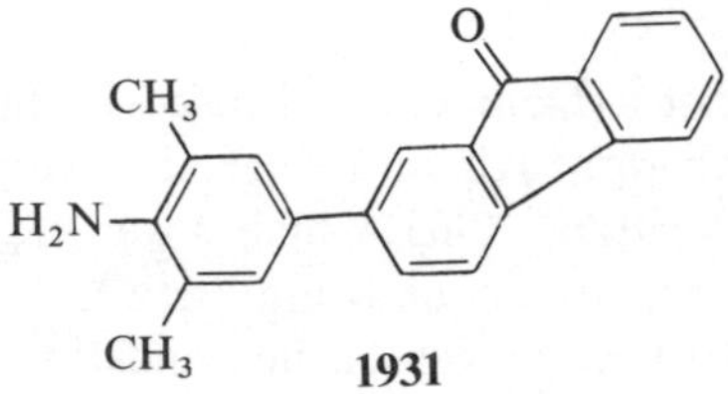

1931

for these three reactions involved a [5,5]-sigmatropic rearrangement of the protonated arylhydrazone and is illustrated in Scheme 68 for the reaction involving 4-methoxybenzaldehyde 2,6-dimethylphenylhydrazone (Fusco and Sannicolò, 1977). Similar reactions were subsequently (Fusco, 1978; Fusco and Sannicolò, 1980) reported using closely related hydrazones and hydrazides. Although with none of these hydrazones was a Fischer indolization possible, the treatment of potentially indolizable acetophenones under similar conditions showed that the [5-5]-sigmatropic shift was competitive with the indolization process. Thus, 4-methoxyacetophenone phenylhydrazone and 2,6- and 3,5-dimethylphenylhydrazone (**1928**; $R^1 = R^2 = H$; $R^1 = CH_3$, $R^2 = H$ and $R^1 = H$, $R^2 = CH_3$, respectively, $R^3 = H$, $R^4 = OCH_3$, $R^5 = CH_3$) in each case gave both possible products, **1930** ($R^1 = R^2 = H$, $R^3 = OCH_3$, $R^4 = CH_3$) and 2-(4-methoxyphenyl)indole, **1930** ($R^1 = R^4 = CH_3$, $R^2 = H$, $R^3 = OCH_3$) and **1932** ($R^1 = H$, $R^2 = CH_3$, $R^3 = OCH_3$) [formed by methyl group migration during indolization (see Chapter II, G2)] and **1930** ($R^1 = H$, $R^2 = R^4 = CH_3$, $R^3 = OCH_3$) and **1932** ($R^1 = CH_3$, $R^2 = H$, $R^3 = OCH_3$), respectively (Fusco and Sannicolò, 1978a, 1980). Competitive reactions also occurred when **1928** ($R^1 = R^5 = CH_3$, $R^2 = R^3 = H$, $R^4 = NH_2$) was heated

1928 ($R^1 = CH_3, R^2 = R^3 = R^5 = H, R^4 = OCH_3$)

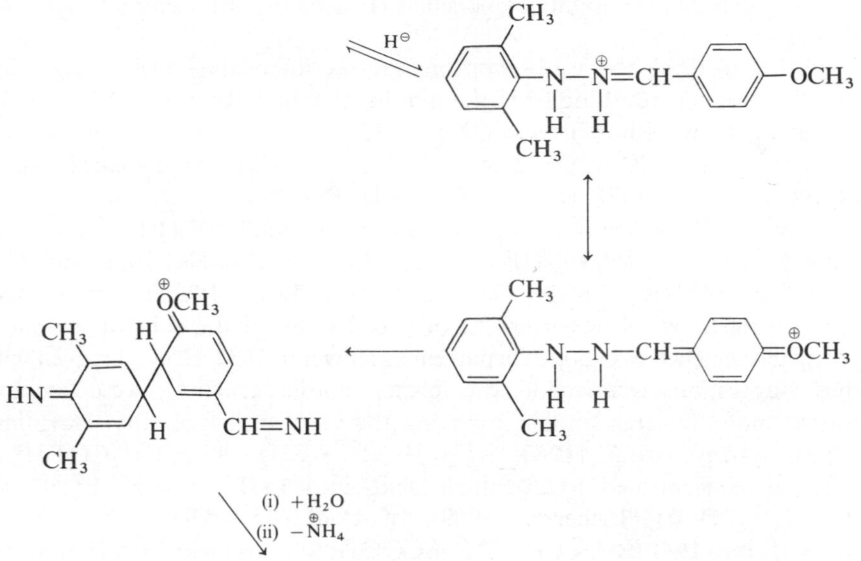

1930 ($R^1 = CH_3, R^2 = R^4 = H, R^3 = OCH_3$)

Scheme 68

with polyphosphoric acid to furnish **1930** ($R^1 = R^4 = CH_3$, $R^2 = H$, $R^3 = NH_2$), **1932** ($R^1 = H$, $R^2 = CH_3$, $R^3 = NH_2$), **1933** and **1934** (Fusco, 1978; Fusco and Sannicolò, 1980) but similar treatment of **1928** ($R^1 = R^3 = R^5 = CH_3$, $R^2 = H$, $R^4 = OH$) yielded **1935**, rearrangements analogous to the latter also being observed when starting with 4-hydroxy-3,5-dimethylacetophenone and 4-hydroxy-3,5-dimethylbenzophenone 3,5-dimethylphenylhydrazone (Fusco and Sannicolò, 1980). The formation of all these reaction products was

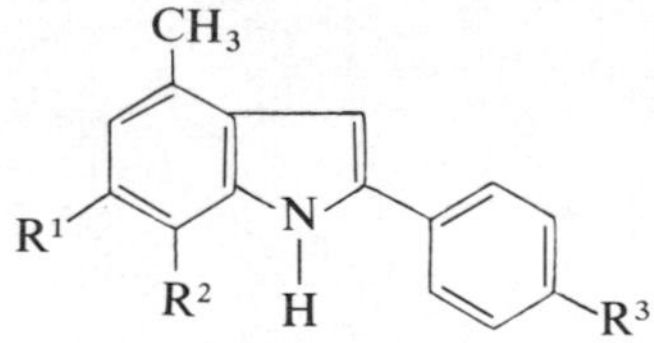

1932

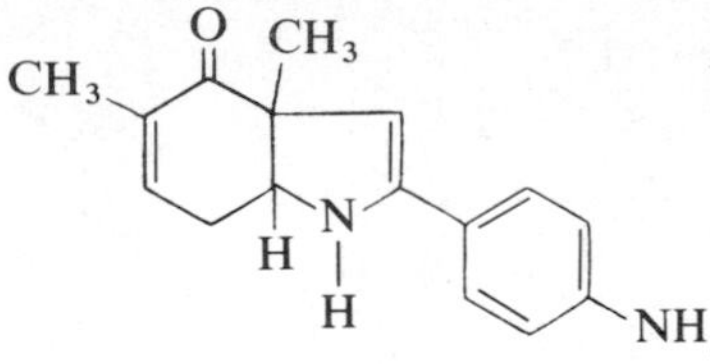

1933

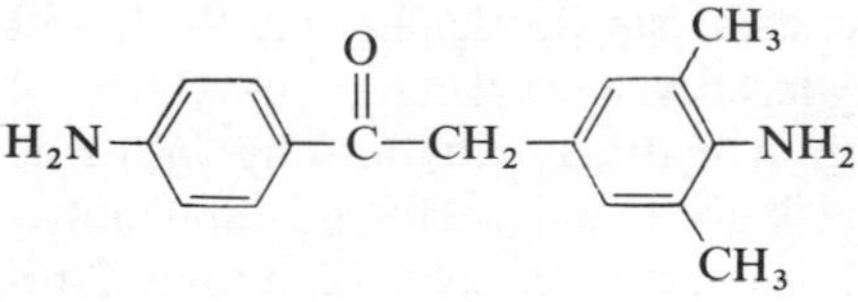

1934

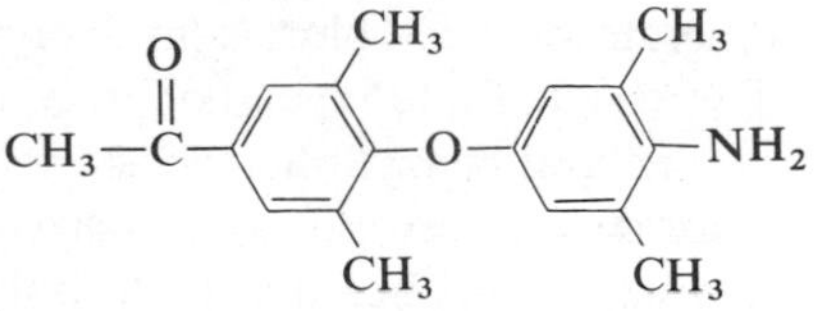

1935

mechanistically rationalized via the occurrence of the appropriate [3,3]-, [3,5]-, [5,5]- or [5,7]-sigmatropic shifts (Fusco, 1978; Fusco and Sannicolò, 1980).

Related to these above observations are the interesting earlier suggestions that the products resulting from the indolization of a mixture of 2-biphenylylhydrazine (**1936**) with **1070** (R = OH) and **1228** ($R^1 = R^4$–R^6 = H, $R^2 = NO_2$, R^3 = H and $R^1 = R^4$–R^6 = H, $R^2 = SCH_3$, $R^3 = CH_3$) were possibly not the expected indoles **1071** (R = C_6H_5) and **1937** ($R^1 = NO_2$, R^2 = H and $R^1 = SCH_3$, $R^2 = CH_3$), respectively, but compounds **1072** (Buu-Hoï, Hoán, Khôi and Xuong, 1949), **1938** ($R^1 = NO_2$, R^2 = H) (Buu-Hoï, Hoán and Khôi, 1950b) and **1938** ($R^1 = SCH_3$, $R^2 = CH_3$) (Buu-Hoï and Hoán, 1951), respectively, formed by the involvement of a C-2 atom of the substituent phenyl group in the new C—C bond formation as shown in **1939**. However, even when these suggestions were made, the former indolic structures were favoured. Furthermore, in later studies involving the indolization of ethyl laevulinate 2-biphenylylhydrazone (**1940**; R^1 = H, $R^2 = CH_3$, $R^3 = CH_2COOH$), in refluxing concentrated hydrochloric acid, into **1941** (R^1 = H, $R^2 = CH_3$, $R^3 = CH_2COOH$) (Scherrer, 1969), of **1940** ($R^1 = CH_3$, R^2 = COOH, R^3 = H) into **1941** ($R^1 = CH_3$, R^2 = COOH, R^3 = H) with boron trifluoride

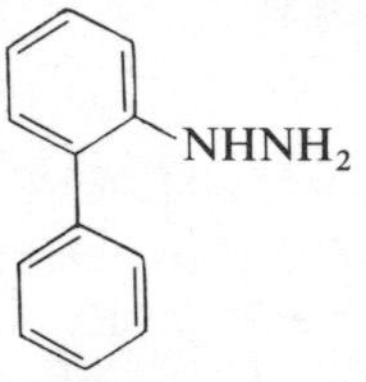

1936 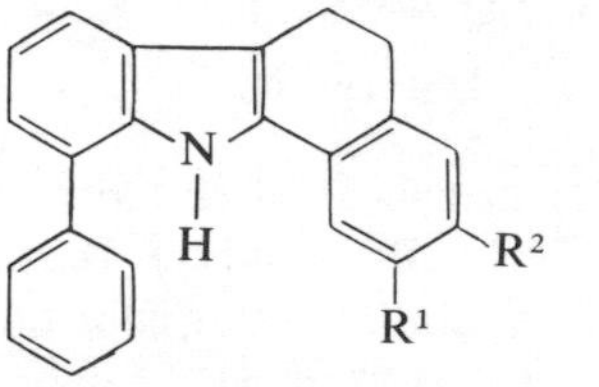 **1937** 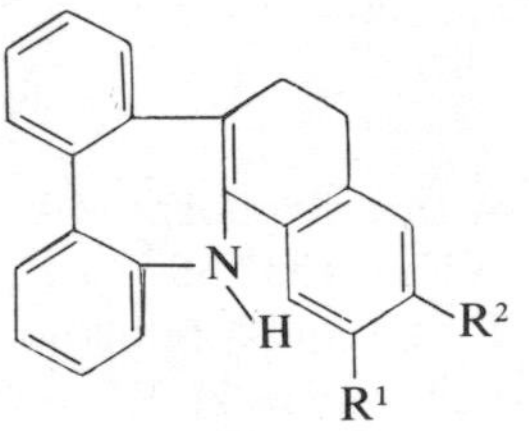 **1938**

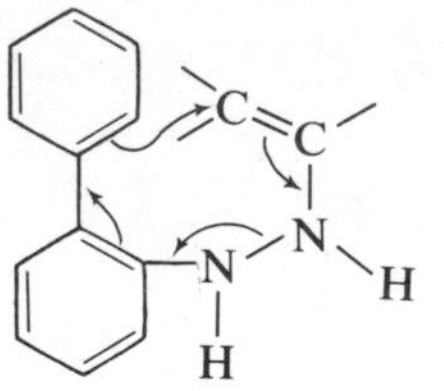

1939 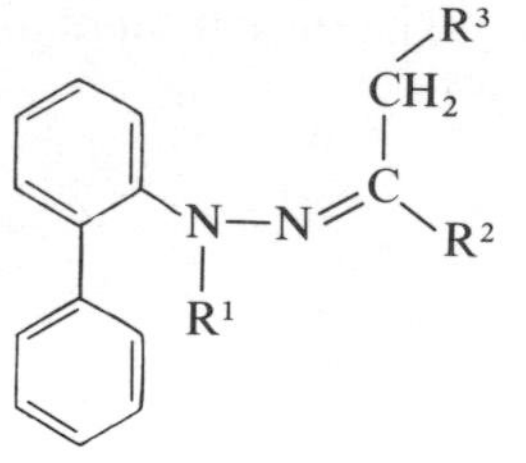 **1940** 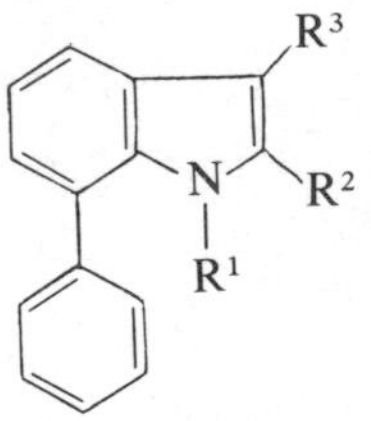 **1941**

etherate or zinc chloride in acetic acid as catalysts (Little, Taylor, W. I. and Thomas, B. R., 1954) and of propionaldehyde, ethyl methyl ketone, ethyl phenyl ketone and benzyl phenyl ketone 2-biphenylylhydrazone by heating with zinc chloride, of acetophenone, benzyl methyl ketone and acetone 2-biphenylylhydrazone by heating with polyphosphoric acid, of cyclohexanone 2-biphenylylhydrazone by heating with 20% sulphuric acid, of phenylacetaldehyde

2-biphenylylhydrazone by boiling with a saturated solution of oxalic acid and of methyl pyruvate 2-biphenylylhydrazone by boiling in an acetic acid–sulphuric acid mixture into **1941** [$R^1 = R^2 = H, R^3 = CH_3, R^1 = H, R^2 = R^3 = CH_3$; $R^1 = H, R^2 = C_6H_5, R^3 = CH_3$; $R^1 = H, R^2 = R^3 = C_6H_5, R^1 = R^3 = H, R^2 = C_6H_5$; $R^1 = H, R^2 = CH_3, R^3 = C_6H_5$; $R^1 = R^3 = H, R^2 = CH_3$; $R^1 = H, R^2 + R^3 = (CH_2)_4$; $R^1 = R^2 = H, R^3 = C_6H_5$ and $R^1 = R^3 = H, R^2 = COOH$, respectively] (Kost, Sugorova and Yakubov, 1965), the possible occurrence of this type of alternative cyclization was not even considered.

Q. Osazone Formation during Attempted Indolization

Heating benzyl phenyl ketone 2-nitrophenylhydrazone (**1942**) in an acetic acid–concentrated hydrochloric acid mixture did not produce the corresponding 7-nitro-2,3-diphenylindole but gave the osazone **1943**. It was possible that this resulted from the oxidation of **1942** by free 2-nitrophenylhydrazine, formed by hydrolysis of **1942** under the reaction conditions, since the yield of **1943** was increased when 2-nitrophenylhydrazine was added to the reaction mixture and 2-nitroaniline, a reduction product of 2-nitrophenylhydrazine, was also isolated from the reaction mixture (Kinsley and Plant, 1956). A probable mechanism for this reaction has been suggested (Robinson, M. J. T., 1956) in which the ene-hydrazine tautomer of **1942** underwent an *ortho*-semidine related rearrangement to afford **1944**, the *ortho*-benzidine rearrangement, leading ultimately to indolization, being inhibited by the *ortho* nitro group (see Chapter VII, A). Hydrolysis of **1944** would then afford **1945** which, by analogy with earlier

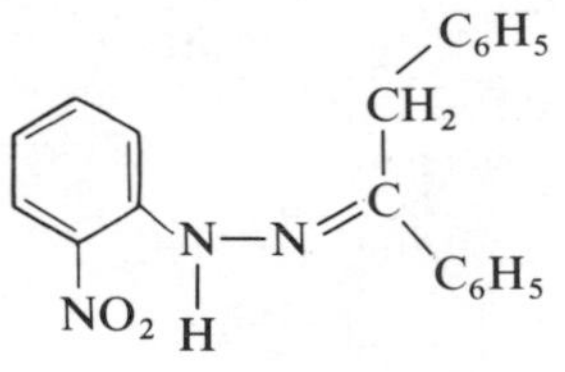

C_6H_5 CH_2 N—N=C C_6H_5 NO_2 H

1942

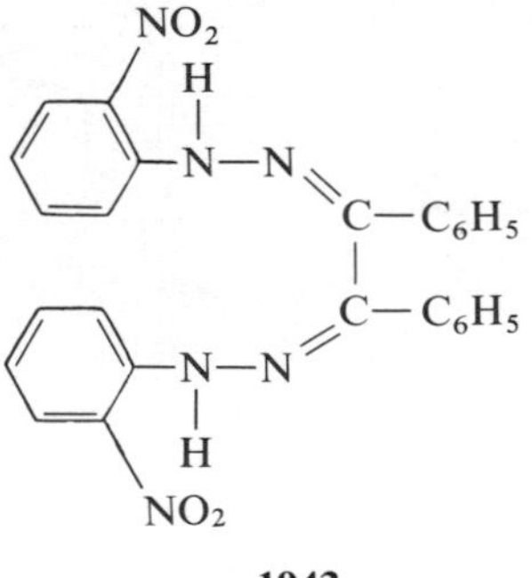

NO_2 H N—N=C—C_6H_5 C—C_6H_5 N—N H NO_2

1943

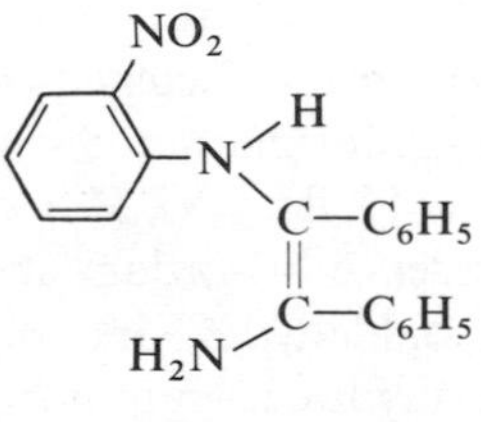

NO_2 N H C—C_6H_5 C—C_6H_5 H_2N

1944

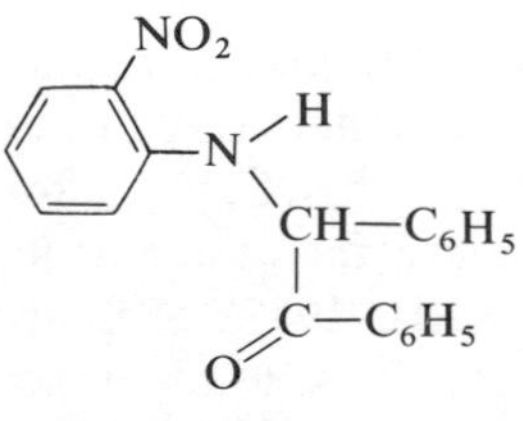

NO_2 N H CH—C_6H_5 O=C—C_6H_5

1945

observations (Wolf, 1953), could further react with 2-nitrophenylhydrazine, either directly or via the formation of benzoin, to furnish **1943**. In these latter reactions the 2-nitrophenylhydrazine would be reduced to 2-nitroaniline. Similar osazone formation occurred using other 2-nitrophenylhydrazones which were resistant to indolization. Thus, when a mixture of acenaphthenone (**1225**) with 2-nitrophenylhydrazine was heated in a mixture of acetic acid and hydrochloric acids, **1946** was formed [**1225** 2-nitrophenylhydrazone does not indolize (Bannister and Plant, 1948)] and when a mixture of 9-hydroxyphenanthrene (**86**) with 2-nitrophenylhydrazine was similarly treated, both indolization and oxidation occurred to afford **87** (R = NO_2) and **1947**, respectively, along with 2-nitroaniline (Kinsley and Plant, 1956). However, similar treatment of 2-indanone 2-nitrophenylhydrazone (**1948**) yielded only **1949** and 1-indanone and propiophenone 2-nitrophenylhydrazone yielded neither indoles nor osazones (Kinsley and Plant, 1956).

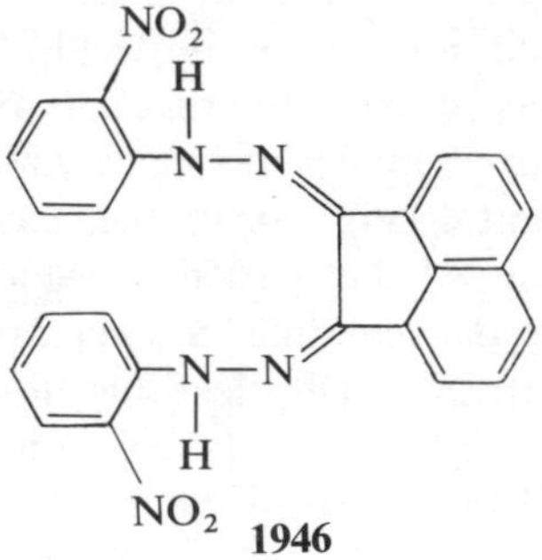

1946 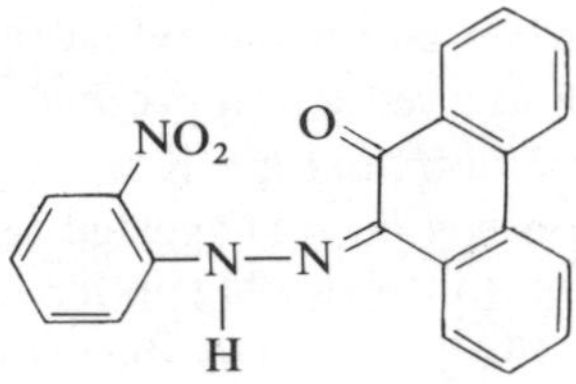 **1947**

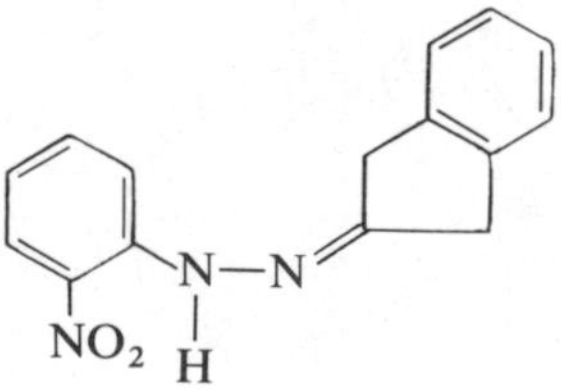

1948 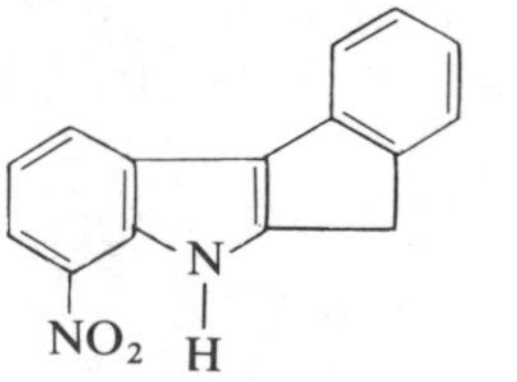 **1949**

Other examples of osazone formation, involving the reacting of acyclic 2-amino- and 2-hydroxyketones and 2-chloro-, 2-hydroxy- and 2-methoxycyclohexanone with arylhydrazines, have already been discussed (see Chapters I, A and IV, J2, respectively). In extensions of the former studies, **17** [R^1 = H, R^2 = $N(CH_3)_2$, R^3 = C_6H_5] afforded **22** (R^1 = H, R^2 = C_6H_5) when reacted with phenylhydrazine in acetic acid in the presence of sodium acetate (Jacob and Madinaveitia, 1937). Similarly, 2-alkylthioaldehydes, and therefore presumably 2-alkylthioketones, reacted with arylhydrazines to form osazones. Thus, **320** [R = C_2H_5 and $(CH_2)_3CH_3$] reacted with phenylhydrazine to give, via the supposed intermediates **321** [R = C_2H_5 and $(CH_2)_3CH_3$, respectively], the indoles **322** [R = C_2H_5 and $(CH_2)_3CH_3$, respectively] and the osazones **22** [R^1 = H, R^2 = $COOC_2H_5$ and $COO(CH_2)_3CH_3$, respectively], these latter products being formed by reaction of the corresponding intermediates **321** with a further two moles of phenylhydrazine. In both reactions, the formation of ethanethiol was detected by odour (Bonnema and Arens, 1960a). The analogy was made (Bonnema and Arens, 1960a) between these osazone formations and the formation of glyoxal 2,4-dinitrophenylosazone by heating ethylthioacetaldehyde 2,4-dinitrophenylhydrazone with 2,4-dinitrophenylhydrazine (Bonnema and Arens, 1960b).

R. Alternative Decomposition, including Nitrile Formation, of Arylhydrazones using Copper(I) Chloride, Sulphanilic Acid and other Catalysts

Concomitantly with the observations that the copper(I) chloride catalysed indolization of propionaldehyde phenylhydrazone and 4-methylphenylhydrazone gave 3-methyl- and 3,5-dimethylindole, respectively (Arbuzov, A. E. and Tikhvinski, 1913b), it was found (Arbuzov, A. E., 1913) that nitriles were produced corresponding to the aldehydic moiety in such hydrazones, the yields of such nitriles being greater the larger the aldehydic moiety. Thus, when heated with catalytic amounts of copper(I) chloride, **1950** [R^1 = H, R^2 = $CH(CH_3)_2$, $CH_2CH(CH_3)_2$, $(CH_2)_2CH(CH_3)_2$ and $(CH_2)_5CH_3$] afforded the nitriles **1952** [R^2 = $CH(CH_3)_2$] (Arbusow, A. E., 1910), **1952** [R^2 = $CH_2CH(CH_3)_2$] (Arbusow, A. E. 1910; Arbuzov, A. E., 1913), **1952** [R^2 = $(CH_2)_2CH(CH_3)_2$] (Arbuzov, A. E., 1913) and **1952** [R^2 = $(CH_2)_5CH_3$] (Arbusow, A. E., 1910; Arbuzov, A. E., 1913), respectively, along with aniline (**1951**; R^1 = H). This

$$C_6H_5-N(R^1)-N{=}CHR^2 \longrightarrow C_6H_5-NH(R^1) + N{\equiv}C-R^2$$

1950 → **1951** + **1952**

type of reaction, which had been observed initially (Anselmino, 1903) when salicylaldehyde phenylhydrazone was dry distilled to furnish salicylonitrile, aniline, benzene and ammonia, was suggested (Arbuzov, A. E., 1913) as being a valuable method of synthesizing alkyl nitriles containing at least four carbon atoms in the molecule. Indeed, it was later (Grandberg, Kost and Naumov, 1963) extended when benzaldehyde phenylhydrazone (**1950**; R^1 = H, R^2 = C_6H_5) and benzaldehyde, propionaldehyde and isovaleraldehyde N_α-phenylphenylhydrazone [**1950**; R^1 = C_6H_5, R^2 = C_6H_5, C_2H_5 and $CH_2CH(CH_3)_2$] were heated with catalytic amounts of copper(I) chloride to give **1952**

($R^2 = C_6H_5$) and aniline and **1952** [$R^2 = C_6H_5$, C_2H_5 and $CH_2CH(CH_3)_2$] and diphenylamine (**1951**; $R^2 = C_6H_5$), respectively. Furthermore, alkylhydrazones and benzaldehyde benzoylhydrazone were also found to undergo analogous reactions. Other catalysts such as zinc chloride or platinum(II) chloride also effected this reaction, it being suggested that the catalyst (e.g. zinc chloride) first added to the more basic nitrogen atom of the hydrazone **1950** to give **1953** which, as a result of a redistribution of the electron density, broke down

H—C—R^2 ‖ N—N(:ZnCl$_2$)(R^1)(C_6H_5) ⟶ **1952** + $C_6H_5N(R^1)H \cdot ZnCl_2$

1953

Scheme 69

by rupture of the N—N bond and migration of a hydrogen atom by a process proposed to be as shown in Scheme 69 (Grandberg, Kost and Naumov, 1963). The reaction was thus analogous (Grandberg, Kost and Naumov, 1963) to the thermal decomposition, in the presence of colloidal platinum, of pyrazolines **1954** (X = NH) (equivalent to cyclic hydrazones) (Kost, Golubeva, Terent'ev, A. P. and Grandberg, 1962) or their hydrochlorides **1954** ($X = \overset{\oplus}{N}H_2Cl^{\ominus}$) (Grandberg and Golubeva, 1963; Grandberg and Potanova, 1962) into 3-aminonitriles **1955** as shown in Scheme 70. This rearrangement and the above formation of nitriles from hydrazones have been included in a review (Naumov and Grandberg, 1966). Heating hydrazones with bases (e.g. sodamide or potassium hydroxide) also caused a similar decomposition (Naumov, Kost and Grandberg, 1965).

1954 $\xrightarrow[\overset{\oplus}{N}H_2Cl^{\ominus}]{X\ =\ NH\ or}$ $N{\equiv}C{-}C(R^3)(R^2){-}CH(R^1){-}NH_2$ **1955**

Scheme 70

When acetone phenylhydrazone was heated with a catalytic quantity of copper(I) chloride, evolution of ammonia occurred and aniline and an unidentified product, for which the molecular formula $C_{12}H_{14}N_2$ was proposed, were isolated. This latter product afforded acetone upon hydrolysis but no structure for it was postulated (Arbuzov, A. E. and Shapshinskaya, 1954–1955). Subsequently (Arbuzov, A. E. and Kitaev, 1957b, 1957e) it was also found that sulphanilic acid would similarly effect this decomposition of acetone phenylhydrazone and also that methyl isopropyl ketone phenylhydrazone was likewise converted into aniline and an oil which was formulated, on the evidence of only a nitrogen analysis, as $C_{16}H_{22}N_2$ and which liberated isopropyl methyl ketone upon hydrolysis. On the basis of these somewhat exiguous data, together with the observation that an analogous product could not be obtained from acetone N_α-methylphenylhydrazone, these two products were formulated as **1956** (R^1–R^3 = CH_3, R^4 = H) and **1957**, respectively (Arbuzov, A. E. and Kitaev, 1957b, 1957e). Analogous products could not be obtained by the similar treatment of pinacolone, pyruvic acid, ethyl pyruvate, acetophenone and 4-chloroacetophenone phenylhydrazone (Arbuzov, A. E. and Kitaev, 1957b, 1957e). To account for the above formations of **1956** (R^1–R^3 = CH_3, R^4 = H) and **1957** and for the formation of nitriles from the phenylhydrazones already

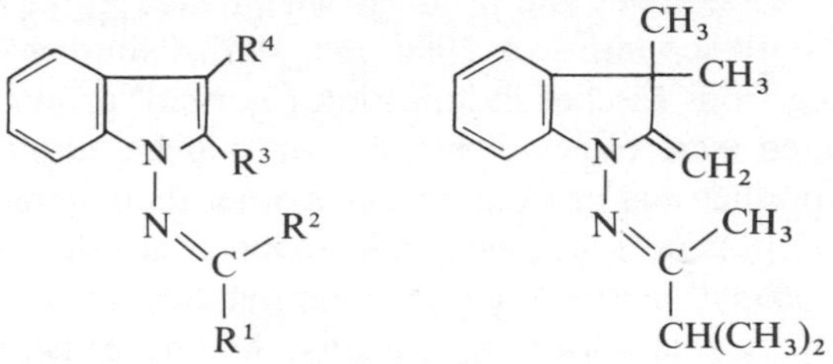

1956 **1957**

discussed, two related mechanisms were proposed. These were similar to the mechanisms proposed for the catalytic decomposition of phenylhydrazine into nitrogen, ammonia, benzene and aniline and for the disproportionation of hydrazobenzene into azobenzene and aniline (Arbuzov, A. E. and Kitaev, 1957b, 1957e). However, the necessity for these mechanistic postulations was subsequently found to be non-existent when it was found (Kitaev, Troepol'skaya and Arbuzov, A. E., 1966b) that the product supposedly having the molecular formula $C_{12}H_{14}N_2$ and resulting from the decomposition of acetone phenylhydrazone was, in fact, an azeotropic mixture composed mainly of 3,5-dimethyl-1-phenylpyrazole along with some 2-methylindole and starting hydrazone, this last component obviously being responsible for the formation of acetone upon hydrolysis of the product mixture. Similarly, **923** (R^1 = R^3 = H, R^4 = CH_3) was formed when acetaldehyde phenylhydrazone was heated with copper(I) chloride (Kitaev and Troepol'skaya, 1978) (see also Chapter IV, A). Clearly, the structural proposal **1957** is, by analogy with these studies, also in doubt. Even by comparison with the structural formulations **1956** (R^1–R^3 = CH_3, R^4 = H) and **1957**, the statement to the effect that a compound **1956** (R^1 = R^4 = C_6H_5, R^2 = R^3 = H) was prepared by an analogous reaction (Arbuzov, A. E. and Kitaev, 1957e) is difficult to understand, involving as it would the cyclization of phenylacetaldehyde phenylhydrazone but with the ultimate formation of a benzaldehyde condensation product. With the above mentioned discovery of the heterogeneous nature of the acetone phenylhydrazone decomposition product, it is even more difficult to understand the inference (Arbuzov, A. E. and Kitaev, 1957e) that this phenylacetaldehyde (?) phenylhydrazone decomposition product was the same as the specimen of **1956** ($R^1 = R^4 = C_6H_5, R^2 = R^3 = H$) which had been prepared earlier (Alder and Niklas, 1954) by condensing 1-amino-3-phenylindole with benzaldehyde. However, experimental details in relation to this later work (Arbuzov, A. E. and Kitaev, 1957e) were not given.

The catalytic effects of sulphanilic acid upon acetone, ethyl methyl ketone and isopropyl methyl ketone phenylhydrazone, acetone 2-, 3- and 4-methylphenylhydrazone and 1- and 2-naphthylhydrazone and isopropyl methyl ketone 2- and 3-methylphenylhydrazone have been investigated. In all these reactions the ratio of anilines and products which were at the time thought to have structures corresponding to **1956** and **1957** ('abnormal' products) to products resulting from Fischer indolization ('normal' products) were determined (also formed were traces of benzene or toluene, aliphatic imines and amines, azo compounds and gaseous products other than ammonia). This ratio decreased in the order acetone phenylhydrazone > acetone 2-methylphenylhydrazone > isopropyl methyl ketone 2-methylphenylhydrazone > acetone 3-methylphenylhydrazone > isopropyl methyl ketone 3-methylphenylhydrazone > ethyl methyl ketone phenylhydrazone (Kitaev, Troepol'skaya and Arbuzov, A. E., 1964). Unfortunately, no recognizable products were isolated when starting with acetone 1- and 2-naphthylhydrazone. Cadmium chloride was similarly investigated as a catalyst and, with the exception of isopropyl methyl ketone phenylhydrazone, favoured the occurrence of Fischer indolization rather than the formation of 'abnormal' products by comparison with the use of sulphanilic acid (Kitaev, Troepol'skaya and Arbuzov, A. E., 1964). A preliminary investigation upon the decomposition of acetone and ethyl methyl ketone phenylhydrazone, representing the extremes in the above 'normal' : 'abnormal' product ratio studies was also effected using copper(I) cyanide. In the former case, evolution of ammonia occurred and benzene and an aliphatic imine were formed along with much tar and unchanged hydrazone. In the latter reaction, ammonia was again evolved and unrecognizable oils, together with 2,3-dimethylindole and tar, were formed (Kitaev, Troepol'skaya and Arbuzov, A. E., 1964). These results are far from conclusive. Indeed, subsequent studies upon the decomposition of various ketone arylhydrazones in the presence of copper(I) chloride, cadmium chloride and sulphanilic acid resulted in the formation of the expected indolic products along with aromatic and aliphatic amines, ketimines, ketone anils and aromatic hydrocarbons (Kitaev, Troepol'skaya and Arbuzov, A. E., 1966a).

Pyrolysis of the phenylhydrazones **1958** ($R = H, CH_3, C_2H_5$ and C_6H_5) in the vapour phase yielded in each case benzonitrile and aniline along with

N-methyl-, N-ethyl- and N-phenylaniline, respectively, from the last three hydrazones. The intermediacy of a four centre mechanism shown in **1958**

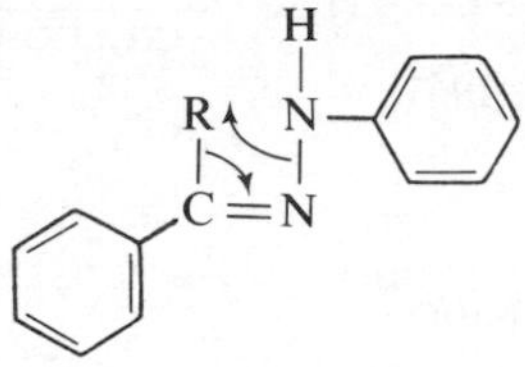

1958

(R = H, CH_3, C_2H_5 and C_6H_5) along with a competing free radical process was suggested in order to account for these observations (Crow and Solly, 1966). Treatment of aldehyde phenylhydrazones **1959** with potassium *tert.*-butoxide

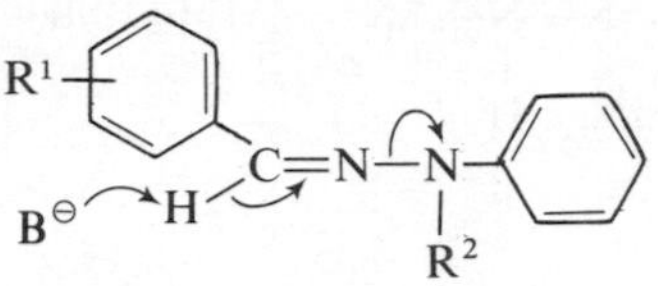

1959

in heated toluene or diglyme afforded the corresponding nitriles and anilines. In the absence of oxygen, these reactions were suggested to proceed via the attack of *tert.*-butoxide ion indicated in **1959** whereas, in the presence of oxygen, the sequential intermediacy of free radicals, their pairing and similar nucleophilic attack by *tert.*-butoxide ion was suggested (Grundon and Scott, M. D., 1964). Both acid and base catalysed decomposition of arylhydrazones into nitriles and anilines appear to proceed heterolytically (Naumov, Kost and Grandberg, 1965).

The conversion of arylhydrazones into the corresponding nitriles and anilines was reviewed at the time the conversion of arylaldehyde arylhydrazones into amidines, by treatment with sodamide in refluxing benzene or xylene (the Robev rearrangement), was investigated (Grandberg, Naumov and Kost, 1965).

S. Reaction of 2-Heterylhydrazines with Aryl Bromomethyl Ketones

Reaction of 2-pyridylhydrazine (**1960**) with bromomethyl phenyl ketone (**1961**) (2:1 molar ratio) in ethanol in the presence of sodium bicarbonate furnished **1962**. However, in methanol in the presence of acetic acid, these two reactants (10:1 molar ratio) produced **1963** which was also formed by refluxing a methanolic acetic acid solution of **1962**. The formation of **1963** was suggested to occur as shown in Scheme 71. Products similar to **1963** were also

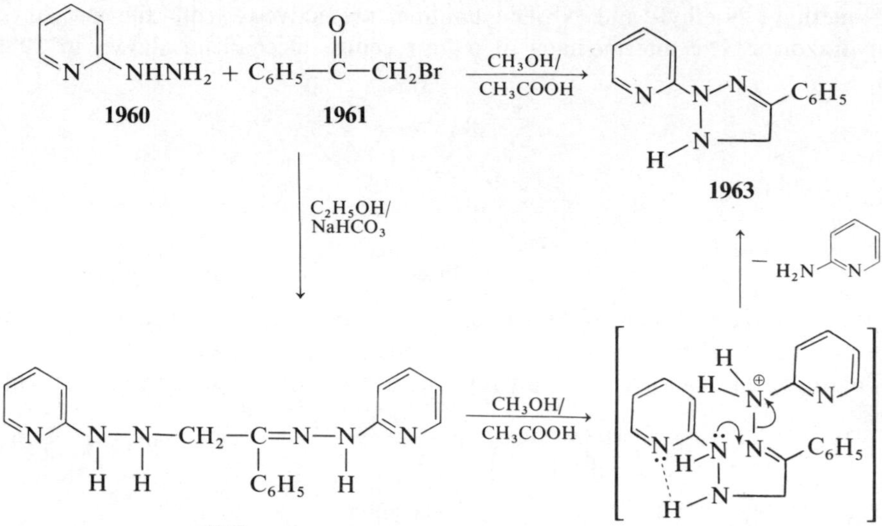
O
NHNH2 + C6H5—C—CH2Br
1960
1961
CH3OH/
CH3COOH
C6H5
1963
C2H5OH/
NaHCO3
− H2N
N—N—CH2—C=N—N
H H C6H5 H
1962
CH3OH/
CH3COOH

Scheme 71

obtained by reacting 2-benzothiazolyl-, 1-isoquinolyl- and 2-quinolylhydrazine with bromomethyl phenyl ketone, 2-pyridylhydrazine with bromomethyl 4-methylphenyl ketone and 2-quinolylhydrazine with bromomethyl 4-bromo- and 4-methoxyphenyl ketone (Reddy, P. A., Singh, S. and Srinivasan, 1976).

References

Abbott Laboratories (1967). Brit. Pat. 1,149,442.
Abramenko, P. I. and Zhiryakov, V. G. (1971). *Chem. Het. Comps.*, **7**, 171.
Abramovitch, R. A. (1956). *J. Chem. Soc.*, 4593.
Abramovitch, R. A. (1958). *Canad. J. Chem.*, **36**, 354.
Abramovitch, R. A. and Adams, K. A. H. (1962). *Canad. J. Chem.*, **40**, 864.
Abramovitch, R. A. and Brown, R. A. (1969). *Org. Prep. Proc.*, **1**, 39.
Abramovitch, R. A. and Muchowski, J. M. (1958). *Canad. J. Chem.*, **36**, 354.
Abramovitch, R. A. and Muchowski, J. M. (1960a). *Canad. J. Chem.*, **38**, 554.
Abramovitch, R. A. and Muchowski, J. M. (1960b). *Canad. J. Chem.*, **38**, 557.
Abramovitch, R. A. and Shapiro, D. (1955). *Chem. Ind.*, 1255.
Abramovitch, R. A. and Shapiro, D. (1956). *J. Chem. Soc.*, 4589.
Abramovitch, R. A. and Spenser, I. D. (1957). *J. Chem. Soc.*, 3767.
Acheson, R. M. and Vernon, J. M. (1962). *J. Chem. Soc.*, 1148.
Adams, C. DeW. (1976). US Pat. 3,983,123 (*Chem. Abs.* 1977, **86**, 72613a).
Adkins, H. and Coonradt, H. L. (1941). *J. Amer. Chem. Soc.*, **63**, 1563.
Adkins, H. and Lundsted, L. G. (1949). *J. Amer. Chem. Soc.*, **71**, 2964.
Adkins, H. and Rossow, A. G. (1949). *J. Amer. Chem. Soc.*, **71**, 3836.
Adlerová, E., Ernest, I., Hněvsová, V., Jílek, J. O., Novák, L., Pomykáček, J., Rasjner, M., Sova, J., Vejdělek, Z. J. and Protiva, M. (1960). *Collect. Czech. Chem. Communs.*, **25**, 784 (*Chem. Abs.*, 1960, **54**, 1309a).
Agosta, W. C. (1961). *J. Org. Chem.*, **26**, 1724.
Ahlbrecht, H. (1971). *Tetrahedron Letts.*, 545.
Ahlbrecht, H. and Henk, H. (1975). *Chem. Ber.*, **108**, 1659.
Ahlbrecht, H. and Henk, H. (1976). *Chem. Ber.*, **109**, 1516.
Ahmed, M. (1966). M.Sc. Thesis, University of Manchester, England (quoted in Robinson, B., 1969).
Ahmed, M. and Robinson, B. (1965). *J. Pharm. Pharmacol.*, **17**, 728.
Ahmed, M. and Robinson, B. (1967). *J. Chem. Soc.* (*C*), 411.
Ainsworth, C. (1963). *Org. Syns.*, Coll. Vol. 4, 536.
Ainsworth, D. P. and Suschitzky, H. (1967a). *J. Chem. Soc.* (*C*), 315.
Ainsworth, D. P. and Suschitzky, H. (1967b). *J. Chem. Soc.* (*C*), 1003.
Ajinomoto C., Inc., (1965). Brit. Pat. 982, 727 (*Chem. Abs.*, 1965, **62**, 14821e).
Akabori, S. and Saito, K. (1930a). *Ber.*, **63**, 2245.
Akabori, S. and Saito, K. (1930b). *Proc. Imp. Acad., Tokyo*, **6**, 236 (*Chem. Zentr.*, 1930, **II**, 3257).
Akagi, M. T., Oishi, T. and Ban, Y. (1969). *Tetrahedron Letts.*, 2063.
Akhren, A. A. and Titov, Y. A. (1967). *Russian Chem. Revs.*, **36**, 311.

Akiba, M. and Ohki, S. (1970). *Chem. Pharm. Bull.*, **18**, 2195.
Akopyan, Z. G. and Tatevosyan, G. T. (1971). *Arm. Khim. Zh.*, **24**, 1025 (*Chem. Abs.*, 1972, **76**, 140447y).
Akopyan, Z. G. and Tatevosyan, G. T. (1974). *Arm. Khim. Zh.*, **27**, 604 (*Chem. Abs.*, 1974, **81**, 169387q).
Akopyan, Z. G. and Tatevosyan, G. T. (1976). *Arm. Khim. Zh.*, **29**, 1039 (*Chem. Abs.*, 1977, **87**, 22946m).
Aksanova, L. A., Barkov, N. K., Zagorevskii, V. A., Kucherova, N. F. and Sharkova, L. M. (1975). *Khim.-Farm. Zh.*, **9**, 7 (*Chem. Abs.*, 1975, **82**, 139982h).
Aksanova, L. A., Kucherova, N. F., Portnova, S. L. and Zagorevskii, V. A. (1967). *Chem. Het. Comps.*, **3**, 825.
Aksanova, L. A., Kucherova, N. F. and Sharkova, L. M. (1969). *Chem. Het. Comps.*, **5**, 746.
Aksanova, L. A., Kucherova, N. F., Sharkova, L. M. and Zagorevskii, V. A. (1972). *Chem. Het. Comps.*, **8**, 669.
Aksanova, L. A., Kucherova, N. F. and Zagorevskii, V. A. (1963). *J. Gen. Chem. USSR*, **33**, 213.
Aksanova, L. A., Kucherova, N. F. and Zagorevskii, V. A. (1964a). *J. Gen. Chem. USSR* **34**, 1619.
Aksanova, L. A., Kucherova, N. F. and Zagorevskii, V. A. (1964b). *J. Gen. Chem. USSR*, **34**, 3417.
Aksanova, L. A., Kucherova, N. F. and Zagorevskii, V. A. (1965). *J. Org. Chem. USSR*, **1**, 2254.
Aksanova, L. A., Pidevich, I. N., Sharkova, L. M. and Kucherova, N. F. (1968). *Khim.-Farm. Zh.*, **2**, 3 (*Chem. Abs.*, 1969, **70**, 68209a).
Aksanova, L. A., Sharkova, N. M., Baranova, M. A., Kucherova, N. F. and Zagorevskii, V. A. (1966). *J. Org. Chem. USSR*, **2**, 159.
Aksanova, L. A., Sharkova, L. M. and Kucherova, N. F. (1970). *Chem. Het. Comps.*, **6**, 864.
Aksanova, L. A., Sharkova, L. M., Kucherova, N. F. and Zagorevskii, V. A. (1970). *Chem. Het. Comps.*, **6**, 1478.
Aksanova, L. A., Sharkova, L. M., Kucherova, N. F. and Zagorevskii, V. A. (1971). *Chem. Het. Comps.*, **7**, 529.
Aksanova, L. A., Sharkova, L. M., Kucherova, N. F. and Zagorevskii, V. A. (1973). *Chem. Het. Comps.*, **9**, 289.
Aktieselskabet Dumex (Dumex Ltd) (1963). Dan. Pat. 96, 884 (*Chem. Abs.*, 1964, **60**, 7999h).
Aktieselskabet Dumex (Dumex Ltd) (1964). Brit. Pat. 959, 203 (*Chem. Abs.*, 1964, **61**, 10659g).
Al-Azawe, S. and Sarkis, G. Y. (1973). *J. Chem. Eng. Data*, **18**, 109.
Albertson, N. F. and Fillman, J. L. (1949). *J. Amer. Chem. Soc.*, **71**, 2818.
Albonico, S. M. and Gallo Pecca, J. (1971). *J. Med. Chem.*, **14**, 448.
Alder, K. and Niklas, H. (1954). *Liebig's Ann. Chem.*, **585**, 97.
Alexander, E. J. and Mooradian, A. (1972). Ger. Pat. 2,127,352 (*Chem. Abs.*, 1972, **76**, 126778x).
Alexander, E. J. and Mooradian, A. (1975). US Pat. 3,905,998 (*Chem. Abs.*, 1976, **84**, 74101q).
Alexander, E. J. and Mooradian, A. (1976). US Pat. 3,948,939 (*Chem. Abs.*, 1976, **85**, 46382j).
Alexander, E. J. and Mooradian, A. (1977a). US Pat. 4,001,270 (*Chem. Abs.*, 1977, **87**, 39275q).
Alexander, E. J. and Mooradian, A. (1977b). US Pat. 4,028,382 (*Chem. Abs.*, 1977, **87**, 84814a).

Allais, A. (1959a). Fr. Pat. 1,180,512 (*Chem. Abs.*, 1961, **55**, 14480f).
Allais, A. (1959b). Fr. Pat. 1,186,258 (*Chem. Abs.*, 1961, **55**, 23562g).
Allen, C. F. H. and Allen, J. A. van (1951). *J. Amer. Chem. Soc.*, **73**, 5850.
Allen, C. F. H. and Wilson, C. V. (1943). *J. Amer. Chem. Soc.*, **65**, 611.
Allen, C. F. H., Young, D. M. and Gilbert, M. R. (1937). *J. Org. Chem.*, **2**, 235.
Allen, F. L., Brunton, J. C. and Suschitzky, H. (1955). *J. Chem. Soc.*, 1283.
Allen, F. L. and Suschitzky, H. (1953). *J. Chem. Soc.*, 3845.
Allen, G. R. (1970). *J. Het. Chem.*, **7**, 239.
Allen, G. R., Poletto, J. F. and Weiss, M. J. (1965). *J. Org. Chem.*, **30**, 2897.
Almond, C. Y. and Mann, F. G. (1951). *J. Chem. Soc.*, 1906.
Almond, C. Y. and Mann, F. G. (1952). *J. Chem. Soc.*, 1870.
Altiparmakian, A. H. and Braithwaite, R. S. W. (1967). *J. Chem. Soc.* (*C*), 1973.
Altschul, J. (1892). *Ber.*, **25**, 1842.
Alyab'eva, T. M., Khoshtariya, T. E., Vasil'ev, A. M., Tret'yakova, L. G., Efimova, T. K. and Suvorov, N. N. (1979). *Chem. Het. Comps.*, **15**, 894.
Amarnath, V. and Madhav, R. (1974). *Synthesis*, 848.
Ambekar, S. Y. and Siddappa, S. (1964-5). *J. Karnatak Univ.*, **9-10**, 5 (*Chem. Abs.*, 1966, **64**, 9814b).
Ambekar, S. Y. and Siddappa, S. (1967). *Monat. Chem.*, **98**, 798.
Ames, A. F., Ames, D. E., Coyne, C. R., Grey, T. F., Lockhart, I. M. and Ralph, R. S. (1959). *J. Chem. Soc.*, 3388.
Ames, D. E. and Novitt, B. (1970). *J. Chem. Soc.* (*C*), 1700.
Ames, D. E., Novitt, B., Waite, D. and Lund, H. (1969). *J. Chem. Soc.* (*C*), 796.
Amorosa, M. (1955). *Gazz. Chim. Ital.*, **85**, 1445.
Amorosa, M. (1956). *Annali Chim.*, **46**, 335.
Amorosa, M. and Lipparini, L. (1956). *Annali Chim.*, **46**, 451.
Amorosa, M. and Lipparini, L. (1957). *Annali Chim.*, **47**, 722.
Amorosa, M. and Lipparini, L. (1959). *Annali Chim.*, **49**, 322.
Amoros-Marin, L. and Carlin, R. B. (1959). *J. Amer. Chem. Soc.*, **81**, 733.
Anderson, A. G. and Tazuma, J. (1952). *J. Amer. Chem. Soc.*, **74**, 3455.
Anderson, G. and Campbell, N. (1950). *J. Chem. Soc.*, 2855.
Anderson, J. A. and Kohler, R. D. (1974). *Synthesis*, 277.
Anderson, R. M., Clemo, G. R. and Swan, G. A. (1954). *J. Chem. Soc.*, 2962.
Andrisano, R. and Vitali, T. (1957). *Gazz. Chim. Ital.*, **87**, 949.
Anselmino, O. (1903). *Ber.*, **36**, 580.
Antaki, H. and Petrow, V. (1951). *J. Chem. Soc.*, 901.
Antrick, O. (1885). *Liebig's Ann. Chem.*, **227**, 360.
Arata, Y., Sakai, M. and Soga, M. (1972). Jap. Pat. 25,080 (*Chem. Abs.*, 1972, **77**, 114381v).
Arata, Y., Sakai, M. and Yasuda, S. (1972). *Chem. Pharm. Bull.*, **20**, 1745.
Arbusow, A. and Chrutzki, N. (1913). *J. Russ. Phys. Chem. Soc.*, **45**, 699 (*Chem. Zentr.*, 1913, II, 1474).
Arbusow, A. E. (1910). *Ber.*, **43**, 2296.
Arbusow, A. E., Saizew, J. A. and Rasumov, A. J. (1935). *Ber.*, **68**, 1792.
Arbusow, A. E. and Tichwinsky, W. M. (1910a). *Ber.*, **43**, 2295.
Arbusow, A. E. and Tichwinsky, W. M. (1910b). *Ber.*, **43**, 2301.
Arbuzov, A. E. (1913). *J. Russ. Phys. Chem. Soc.*, **45**, 74 (*Chem. Abs.*, 1913, **7**, 2225).
Arbuzov, A. E. (1977). Birthday centenary, *Russian Chem. Revs.*, **46**, 789.
Arbuzov, A. E. and Friauf, A. P. (1913). *J. Russ. Phys. Chem. Soc.*, **45**, 694 (*Chem. Abs.*, 1913, **7**, 3599) (*Chem. Zentr.*, 1913, II, 1474).
Arbuzov, A. E. and Khrutzkii, N. E. (1913). *J. Russ. Phys. Chem. Soc.*, **45**, 699 (*Chem. Abs.*, 1913, **7**, 3599).

Arbuzov, A. E. and Kitaev, Y. P. (1957a). *Proc. Acad. Sci. USSR*, **113**, 243.
Arbuzov, A. E. and Kitaev, Y. P. (1957b). *Proc. Acad. Sci. USSR*, **113**, 303.
Arbuzov, A. E. and Kitaev, Y. P. (1957c). *Trudy Kazan. Khim. Tekhnol. Inst. im. S.M. Kirova*, **23**, 60 (*Chem. Abs.*, 1958, **52**, 9980b).
Arbuzov, A. E. and Kitaev, Y. P. (1957d). *J. Gen. Chem. USSR*, **27**, 2388.
Arbuzov, A. E. and Kitaev, Y. P. (1957e). *J. Gen. Chem. USSR*, **27**, 2401.
Arbuzov, A. E., Kitaev, Y. P., Shagidullin, R. R. and Petrova, L. E. (1967). *Bull. Acad. Sci. USSR, Div. Chem. Sci.*, 1822.
Arbuzov, A. E. and Rotermel, V. A. (1932). *J. Gen. Chem. USSR*, **2**, 397 (*Chem. Abs.*, 1933, **27**, 291) (*Chem. Zentr.*, 1933, I, 2935).
Arbuzov, A. E. and Shapshinskaya, O. M. (1954–55). *Trans. Kazan. Khim. Tekhnol. Inst. im. S. M. Kirova*, **19–20**, 27 (*Chem. Abs.*, 1957, **51**, 11240b).
Arbuzov, A. E. and Tikhvinskii, V. M. (1913a). *J. Russ. Phys. Chem. Soc.*, **45**, 69 (*Chem. Abs.*, 1913, **7**, 2225).
Arbuzov, A. E. and Tikhvinskii, V. M. (1913b). *J. Russ. Phys. Chem. Soc.*, **45**, 70 (*Chem. Abs.*, 1913, **7**, 2225).
Arbuzov, A. E. and Valitova, F. G. (1957). *J. Gen. Chem. USSR*, **27**, 2413.
Arbuzov, A. E. and Wagner, R. E. (1913). *J. Russ. Phys. Chem. Soc.*, **45**, 697 (*Chem. Abs.*, 1913, **7**, 3599).
Arbuzov, A. E. and Zaitzev, I. A. (1934). *Trans. Butlerov Inst. Chem. Tech. Kazan*, **1**, 33 (*Chem. Abs.*, 1935, **29**, 4006q) (*Chem. Zentr.*, 1936, II, 1925).
Arbuzov, A. E., Zaitzev, I. A. and Rasumov, A. J. (1935). *Zhur. Obs. Khim.*, 1935, **6**, 289 (*Chem. Zentr.*, 1936, II, 3668).
Arbuzov, B. A., Samitov, Y. Y. and Kitaev, Y. P. (1966). *Bull. Acad. Sci. USSR, Div. Chem. Sci.*, 41.
Archer, S. (1963). US Pat. 3,074,960 (*Chem. Abs.*, 1963, **59**, 578a).
Archer, S. (1967). US Pat. 3,328,407 (*Chem. Abs.*, 1968, **68**, 49652g).
Archer, S. (1970). US Pat. 3,547,922 (*Chem. Abs.*, 1971, **74**, 125731y).
Arheidt, R. (1887). *Liebig's Ann. Chem.*, **239**, 206.
Armit, J. W. and Robinson, R. (1922). *J. Chem. Soc.*, **121**, 827.
Arnold, E. (1888). *Liebig's Ann. Chem.*, **246**, 329.
Arnold, R. T. and Collins, C. J. (1939). *J. Amer. Chem. Soc.*, **61**, 1407.
Arnold, R. T., Collins, C. and Zenk, W. (1940). *J. Amer. Chem. Soc.*, **62**, 983.
Aron, M. A. and Elvidge, J. A. (1958). *Chem. Ind.*, 1234.
Ash, A. S. F. and Wragg, W. R. (1958). *J. Chem. Soc.*, 3887.
Asselin, A. A., Humber, L. G. and Dobson, T. A. (1976). Ger. Pat. 2,526,966 (*Chem. Abs.*, 1976, **84**, 135464j).
Asselin, A. A., Humber, L. G. and Dobson, T. A. (1977). US Pat. 4,057,559 (*Chem. Abs.*, 1978, **88**, 120981j).
Atami, T., Izumi, T. and Yamamoto, H. (1970). Jap. Pat. 70 38,048 (*Chem. Abs.*, 1971, **74**, 87827z).
Atkinson, C. M. and Simpson, J. C. E. (1947). *J. Chem. Soc.*, 1649.
Atkinson, C. M., Simpson, J. C. E. and Taylor, A. (1954). *J. Chem. Soc.*, 165.
Aubagnac, J.-L., Elguero, J. and Jacquier, R. (1967). *Bull. Soc. Chim. France*, 3516.
Aubagnac, J.-L., Elguero, J. and Jacquier, R. (1969). *Bull. Soc. Chim. France*, 3292.
Audrieth, L. F., Weisiger, J. R. and Carter, H. E. (1941). *J. Org. Chem.*, **6**, 417.
Augustine, R. L. and Pierson, W. G. (1969). *J. Org. Chem.*, **34**, 1070.
Auwers, K. v. (1932). *Ber.*, **65**, 831.
Auwers, K. v., Hilliger, E. and Wulf, E. (1922). *Liebig's Ann. Chem.*, **429**, 190.
Auwers, K. v. and Kreuder, A. (1925). *Ber.*, **58**, 1974.
Auwers, K. v. and Lämmerhirt, E. (1921). *Ber.*, **54**, 1000.

Auwers, K. v. and Müller, K. (1908). *Ber.*, **41**, 4230.
Auwers, K. v. and Niemeyer, F. (1925). *J. Prakt. Chem.*, **110**, 153.
Auwers, K. v. and Voss, H. (1909). *Ber.*, **42**, 4411.
Auwers, K. v. and Wunderling, H. (1931). *Ber.*, **64**, 2758.
Avanesova, D. A., Astvatsatryan, S. T., Sarkisyan, T. S., Garibyan, D. K. and Tatevosyan, G. T. (1975). *Arm. Khim. Zh.*, **28**, 720 (*Chem. Abs.*, 1976, **84**, 30801p).
Avanesova, D. A., Musaelyan, A. G. and Tatevosyan, G. T. (1972). *Arm. Khim. Zh.*, **25**, 531 (*Chem. Abs.*, 1972, **77**, 139713t).
Avanesova, D. A. and Tatevosyan, G. T. (1970). *Arm. Khim. Zh.*, **23**, 280 (*Chem. Abs.*, 1970, **73**, 76980x).
Avanesova, D. A. and Tatevosyan, G. T. (1974). *Arm. Khim. Zh.*, **27**, 143 (*Chem. Abs.*, 1974, **81**, 37453y).
Avanesova, D. A. and Tatevosyan, G. T. (1979). *Arm. Khim. Zh.*, **32**, 322 (*Chem. Abs.*, 1980, **92**, 6443t).
Avramenko, V. G., Mosina, G. S. and Suvorov, N. N. (1970). *Chem. Het. Comps.*, **6**, 1131.
Avramenko, V. G., Plutitskii, D. N., Strekozov, A. N. and Suvorov, N. N. (1979). *Izv. Vyssh. Uchebn. Zaved., Khim. Khim. Tekhnol.*, **22**, 885 (*Chem. Abs.*, 1979, **91**, 193102v).
Badcock, W. E. and Pausacker, K. H. (1951). *J. Chem. Soc.*, 1373.
Badger, G. M., Cook, J. W., Hewett, C. L., Kennaway, E. L., Kennaway, N. M. and Martin, R. H. (1942–1943). *Proc. Roy. Soc.*, **131B**, 170.
Badische Anilin- & Soda-Fabrik A.-G. (1965). Belg. Pat. 663,712 (*Chem. Abs.*, 1966, **64**, 19563d).
Baeyer, A. (1866). *Liebig's Ann. Chem. Pharm.*, **140**, 295.
Baeyer, A. (1868). *Ber.*, **1**, 17.
Baeyer, A. (1879). *Ber.*, **12**, 456.
Baeyer, A. (1883). *Ber.*, **16**, 2188.
Baeyer, A. (1894). *Liebig's Ann. Chem.*, **278**, 88.
Baeyer, A. and Caro, H. (1877). *Ber.*, **10**, 1262.
Baeyer, A. and Emmerling, A. (1869). *Ber.*, **2**, 679.
Baeyer, A. and Kochendoerfer, E. (1889). *Ber.*, **22**, 2189.
Baeyer, A. and Knop, C. A. (1866). *Liebig's Ann. Chem. Parm.*, **140**, 1.
Baeyer, A. and Tutein, F. (1889). *Ber.*, **22**, 2178.
Bahadur, G. A., Bailey, A. S. and Baldry, P. A. (1977). *J. Chem. Soc., Perkin I*, 1619.
Bahadur, G. A., Bailey, A. S., Costello, G. and Scott, P. W. (1979). *J. Chem. Soc., Perkin I*, 2154.
Bailey, A. S., Baldry, P. A. and Scott, P. W. (1979). *J. Chem. Soc., Perkin I.*, 2387.
Bailey, A. S. and Bogle, P. H. (1977). *J. Chem. Res.* (*M*), 2447.
Bailey, A. S., Hill, P. A. and Seager, J. F. (1974). *J. Chem. Soc., Perkin I*, 967.
Bailey, A. S., Peach, J. M. and Vandrevala, M. H. (1978). *J. Chem. Soc., Chem. Communs.*, 845.
Bailey, A. S., Scott, P. W. and Vandrevala, M. H. (1980). *J. Chem. Soc., Perkin I*, 97.
Bailey, A. S., Scottergood, R. and Warr, W. A. (1971). *J. Chem. Soc.* (*C*), 3769.
Bailey, A. S. and Seager, J. F. (1974). *J. Chem. Soc., Perkin I*, 763.
Bajwa, G. S. and Brown, R. K. (1968a). *Canad. J. Chem.*, **46**, 1927.
Bajwa, G. S. and Brown, R. K. (1968b). *Canad. J. Chem.*, **46**, 3105.
Bajwa, G. S. and Brown, R. K. (1969). *Canad. J. Chem.*, **47**, 785.
Bajwa, G. S. and Brown, R. K. (1970). *Canad. J. Chem.*, **48**, 2293.
Bakke, J., Heikman, H. and Hellgren, E. B. (1974). *Acta Chem. Scand., Ser. B.*, **28**, 393.
Baldwin, J. E. and Tzodikov, N. R. (1977). *J. Org. Chem.*, **42**, 1878.
Ballantine, J. A., Barrett, C. B., Beer, R. J. S., Boggiano, B. G., Eardley, S., Jennings, B. E. and Robertson, A. (1957). *J. Chem. Soc.*, 2227.

Ballauf, F. and Schmetzer, A. (1929). Ger. Pat. 544,621 (*Chem. Abs.*, 1932, **26**, 3522); US Pat. 1,834,015 (*Chem. Abs.*, 1932, **26**, 1000).
Baly, E. C. C. and Tuck, W. B. (1906). *J. Chem. Soc.* (*Trans.*), **89**, 982.
Bamberger, E. and Landau, A. (1919). *Ber.*, **52**, 1093.
Bamberger, E. and Sternitzki, H. (1893). *Ber.*, **26**, 1291.
Ban, Y. and Iijima, I. (1969). *Tetrahedron Letts.*, 2523.
Ban, Y., Iijima, I., Inoue, I., Akagi, M. and Oishi, T. (1969). *Tetrahedron Letts.*, 2067.
Ban, Y., Oishi, T., Kishio, Y. and Iijima, I. (1967). *Chem. Pharm. Bull.*, **15**, 531.
Ban, Y. and Sato, Y. (1965). *Chem. Pharm. Bull.*, **13**, 1073.
Ban, Y., Sato, Y., Inoue, I., Nagai, M., Oishi, T., Terashima, M., Yonemitsu, O. and Kanoaka, Y. (1965). *Tetrahedron Letts.*, 2261.
Ban, Y., Wakamatsu, T., Fujimoto, Y., and Oishi, T. (1968). *Tetrahedron Letts.*, 3383.
Bannister, B. and Plant, S. G. P. (1948). *J. Chem. Soc.*, 1247.
Banthorpe, D. V. (1962a). *J. Chem. Soc.*, 2407.
Banthorpe, D. V. (1962b). *J. Chem. Soc.*, 2413.
Banthorpe, D. V. (1962c). *J. Chem. Soc.*, 2429.
Banthorpe, D. V. (1964). *J. Chem. Soc.*, 2854.
Banthorpe, D. V. (1969). In *Topics in Carbocyclic Chemistry*, ed. by Lloyd, D., Logos Press, London, Vol. 1, p. 1.
Banthorpe, D. V. (1972). *Tetrahedron Letts.*, 2707.
Banthorpe, D. V. and Hughes, E. D. (1962a). *J. Chem. Soc.*, 2402.
Banthorpe, D. V. and Hughes, E. D. (1962b). *J. Chem. Soc.*, 3314.
Banthorpe, D. V. and Hughes, E. D. (1964a). *J. Chem. Soc.*, 2849.
Banthorpe, D. V. and Hughes, E. D. (1964b). *J. Chem. Soc.*, 2860.
Banthorpe, D. V., Hughes, E. D. and Ingold, C. (1962a). *J. Chem. Soc.*, 2386.
Banthorpe, D. V., Hughes, E. D. and Ingold, C. (1962b). *J. Chem. Soc.*, 2418.
Banthorpe, D. V., Hughes, E. D., Ingold, C., Bramley, R. and Thomas, J. A. (1964). *J. Chem. Soc.*, 2864.
Banthorpe, D. V., Hughes, E. D., Ingold, C. and Humberlin, R. (1962). *J. Chem. Soc.*, 3299.
Banthorpe, D. V., Ingold, C., Roy, J. and Somerville, S. M. (1962). *J. Chem. Soc.*, 2436.
Baradarani, M. M. and Joule, J. A. (1978). *J. Chem. Soc., Chem. Communs.*, 309.
Barbulescu, N., Bornaz, C. and Greff, C. (1971). *Rev. Chim.* (*Bucharest*), **22**, 269 (*Chem. Abs.*, 1971, **75**, 118253m).
Barclay, B. M. and Campbell, N. (1945). *J. Chem. Soc.*, 530.
Barclay, B. M., Campbell, N. and Gow, R. S. (1946). *J. Chem. Soc.*, 997.
Bard, R. R. and Bunnett, J. F. (1980). *J. Org. Chem.*, **45**, 1546.
Bardakos, V., Sucrow, W. and Fehlauer, A. (1975). *Chem. Ber.*, **108**, 2161.
Barger, G. and Dyer, E. (1938). *J. Amer. Chem. Soc.*, **60**, 2414.
Barltrop, J. A., Acheson, R. M., Philpot, P. G., MacPhee, K. E. and Hunt, J. S. (1956). *J. Chem. Soc.*, 2928.
Barltrop, J. A. and Taylor, D. A. H. (1954). *J. Chem. Soc.*, 3399.
Barnes, C. S., Pausacker, K. H. and Badcock, W. E. (1951). *J. Chem. Soc.*, 730.
Barnes, C. S., Pausacker, K. H. and Schubert, C. I. (1949). *J. Chem. Soc.*, 1381.
Barrett, H. S. B., Perkin, W. H. and Robinson, R. (1929). *J. Chem. Soc.*, 2942.
Barry, W. J. and McClelland, E. W. (1935). *J. Chem. Soc.*, 471.
Basanagoudar, L. D. and Siddappa, S. (1972). *J. Karnatak. Univ.*, **17**, 33 (*Chem. Abs.*, 1975, **82**, 111898m).
Bauer, H. and Strauss, E. (1932). *Ber.*, **65**, 308.
Baumes, R., Jacquier, R. and Tarrago, G. (1973). *Bull. Soc. Chim. France*, 317.
Baumes, R., Jacquier, R. and Tarrago, G. (1974). *Bull. Soc. Chim. France*, 1147.

Baumes, R., Jacquier, R. and Tarrago, G. (1976). *Bull. Soc. Chim. France*, 260.
Baumgarten, H. E. and Furnas, J. L. (1961). *J. Org. Chem.*, **26**, 1536.
Bayanov, V. N., Mirzametova, R. M., Tret'yakova, L. G., Efimova, T. K. and Suvorov, N. N. (1978). *Khim. Get. Soedin.*, **14**, 1061 (*Chem. Abs.*, 1978, **89**, 215170d).
Bazile, Y., Cointet, P. de and Pigerol, C. (1978). *J. Het. Chem.*, **15**, 859.
Beck, D., Schenker, K., Stuber, F. and Zürcher, R. (1965). *Tetrahedron Letts.*, 2285.
Beckmann, (1920). *Sitzber. Preuss. Akad.*, 698 (*Chem. Abs.*, 1921, **15**, 1029).
Beer, R. J. S., Broadhurst, T. and Robertson, A. (1952). *J. Chem. Soc.*, 4946.
Beer, R. J. S., Broadhurst, T. and Robertson, A. (1953). *J. Chem. Soc.*, 2440.
Beer, R. J. S., Brown, J. P. and Robertson, A. (1951). *J. Chem. Soc.*, 2426.
Beer, R. J. S., Clarke, K., Davenport, N. F. and Robertson, A. (1951). *J. Chem. Soc.*, 2029.
Beer, R. J. S., Clarke, K., Khorana, H. G. and Robertson, A. (1948). *J. Chem. Soc.*, 1605.
Beer, R. J. S., Donavanik, T. and Robertson, A. (1954). *J. Chem. Soc.*, 4139.
Beer, R. J. S., McGrath, L. and Robertson, A. (1950a). *J. Chem. Soc.*, 2118.
Beer, R. J. S., McGrath, L. and Robertson, A. (1950b). *J. Chem. Soc.*, 3283.
Beer, R. J. S., McGrath, L., Robertson, A. and Woodier, A. B. (1949). *J. Chem. Soc.*, 2061.
Beer, R. J. S., McGrath, L., Robertson, A. and Woodier, A. B. (with Holker, J. S. E.), (1949). *J. Chem. Soc.*, 2061 (2066).
Beitz, H., Stroh, H.-H. and Fiebig, H.-J. (1967). *J. Prakt. Chem.*, **36**, 304.
Bell, J. B. and Lindwall, H. G. (1948). *J. Org. Chem.*, **13**, 547.
Bellamy, A. J. and Guthrie, R. D. (1964). *Chem. Ind.*, 1575.
Bellamy, A. J. and Guthrie, R. D. (1965a). *J. Chem. Soc.*, 2788.
Bellamy, A. J. and Guthrie, R. D. (1965b). *J. Chem. Soc.*, 3528.
Bellamy, A. J. and Guthrie, R. D. (1968). *J. Chem. Soc.* (*C*), 2090.
Bellamy, A. J., Guthrie, R. D. and Chittenden, G. J. F. (1966). *J. Chem. Soc.* (*C*), 1989.
Benary, E. (1926). *Ber.*, **59**, 2198.
Benary, E. (1934). *Ber.*, **67**, 708.
Benary, E. and Baravian, A. (1915). *Ber.*, **48**, 593.
Bender, G. (1887). *Ber.*, **20**, 2747.
Bennett, G. and Waddington, N. (1929). *J. Chem. Soc.*, 2829.
Berger, L. and Corraz, A. J. (1968). US Pat. 3,409,628 (*Chem. Abs.*, 1969, **71**, 38939s) [see also US Pat. 3,484,449 (*Chem. Abs.*, 1970, **72**, 121507r)].
Berger, L. and Corraz, A. J. (1970a). US Pat. 3,502,688 (*Chem. Abs.*, 1970, **73**, 3905s).
Berger, L. and Corraz, A. J. (1970b). US Pat. 3,522,262 (*Chem. Abs.*, 1970, **73**, 120600z).
Berger, L. and Corraz, A. J. (1972a). US Pat. 3,646,045 (*Chem. Abs.*, 1972, **77**, 5354t).
Berger, L. and Corraz, A. J. (1972b). US Pat. 3,654,290 (*Chem. Abs.*, 1972, **77**, 5342n).
Berger, L. and Corraz, A. J. (1972c). Ger. Pat. 2,125,926 (*Chem. Abs.*, 1972, **76**, 140519v).
Berger, L. and Corraz, A. J. (1973). Ger. Pat. 2,239,648 (*Chem. Abs.* 1973, **78**, 124439e).
Berger, L. and Corraz, A. J. (1974). Ger. Pat. 2,337,340 (*Chem. Abs.*, 1974, **80**, 108366q).
Berger, L. and Corraz, A. J. (1977). US Pat. 4,009,181 (*Chem. Abs.*, 1977, **86**, 171258v).
Berger, L., Leimgruber, W. and Schenker, F. E. (1972). Ger. Pat. 2,214,501 (*Chem. Abs.*, 1973, **78**, 29609b).
Berger, L. and Scott, J. W. (1979). US Pat. 4,146,542 (*Chem. Abs.*, 1979, **91**, 20319t).
Bergman, J. and Erdtman, H. (1969). *Acta Chem. Scand.*, **23**, 2578.
Bergmann, E. D. and Hoffmann, E. (1962). *J. Chem. Soc.*, 2827.
Bergmann, E. D. and Pelchowicz, Z. (1959). *J. Chem. Soc.*, 1913.
Bergmann, M. (1924). *J. Amer. Chem. Soc.*, **46**, 2133.
Bergmann, M. and Gierth, M. (1926). *Liebig's Ann. Chem.*, **448**, 48.
Berlage, F. and Karrer, P. (1957). *Helv. Chim. Acta*, **40**, 736.
Berlin, A. Y. and Zaitseva, V. N. (1960). *J. Gen. Chem. USSR*, **30**, 2349.

Berlin, K. D., Clark, P. E., Schroeder, J. T. and Hopper, D. G. (1968). *Proc. Okla. Acad. Sci.*, **47**, 215 (*Chem. Abs.*, 1969, **70**, 96537z).
Bernini, G. (1953). *Annali Chim.*, **43**, 559.
Bertazzoni, D., Bartoletti, B. and Perlotto, T. (1970). *Boll. Chim. Farm.*, **109**, 60 (*Chem. Abs.*, 1970, **73**, 14610t).
Berti, G. and Da Settimo, A. (1960). *Gazz. Chim. Ital.*, **90**, 525.
Berti, G., Da Settimo, A. and Nannipieri, E. (1969). *Gazz. Chim. Ital.*, **99**, 1236.
Berti, G., Da Settimo, A. and Segnini, D. (1960). *Gazz. Chim. Ital.*, **90**, 539.
Besford, L. S., Allen, G. and Bruce, J. M. (1963). *J. Chem. Soc.*, 2867.
Besford, L. S. and Bruce, J. M. (1964). *J. Chem. Soc.*, 4037.
Bettembourg, M.-C. and David, S. (1962). *Bull. Soc. Chim. France*, 772.
Beyts, N. M. and Plant, S. G. P. (1939). *J. Chem. Soc.*, 1534.
Bezuglyi, V. D., Kotok, L. A., Shimanskaya, N. P. and Bondarenko, V. E. (1969). *J. Gen. Chem. USSR*, **39**, 2116.
Bhatnagar, I. and George, M. V. (1967). *J. Org. Chem.*, **32**, 2252.
Bhattacharyya, P., Basak, S. P., Islam, A. and Chakraborty, D. P. (1975). *J. Indian Chem. Soc.*, **53**, 861.
Bhide, G. V., Pai, N. R., Tikotkar, N. L. and Tilak, B. D. (1958). *Tetrahedron*, **4**, 420.
Bhide, G. V., Tikotkar, N. L. and Tilak, B. D. (1957). *Chem. Ind.*, 363.
Bhide, G. V., Tikotkar, N. L. and Tilak, B. D. (1960). *Tetrahedron*, **10**, 230.
Biere, H., Rufer, C., Ahrens, H., Schroeder, E., Losert, W., Loge, O. and Schillinge, E. (1973). Ger. Pat. 2,226,703 (*Chem. Abs.*, 1974, **80**, 59861m).
Bikova, N., Vitev, M., Dyankova, L. and Ilarionov, I. (1972). *Tr. Nauchnoizsled. Khim.-Farm. Inst.*, **8**, 101 (*Chem. Abs.*, 1973, **78**, 147728n).
Bikova, N., Vitev, M. and Ilarionov, I. (1974). *Tr. Nauchnoizsled. Khim.-Farm. Inst.*, **9**, 155 (*Chem. Abs.*, 1975, **83**, 58591y).
Binder, D., Habison, G. and Noe, C. R. (1977). *Synthesis*, 487.
Biniecki, S. and Jakubowski, J. (1974). *Rocz. Chem.*, **48**, 1599 (*Chem. Abs.*, 1975, **82**, 72717e).
Binns, T. D. and Brettle, R. (1966). *J. Chem. Soc. (C)*, 341.
Birchall, G. R., Hepworth, W. and Smith, S. C. (1973). Ger. Pat. 2,253,927 (*Chem. Abs.*, 1973, **79**, 31865w).
Birchall, J. M., Haseldine, R. N. and Parkinson, A. R. (1962). *J. Chem. Soc.*, 4966.
Bisagni, E., Ducrocq, C. and Civier, A. (1976). *Tetrahedron*, **32**, 1383.
Bisagni, E., Ducrocq, C. and Hung, N. C. (1980). *Tetrahedron*, **36**, 1327.
Bisagni, E., Ducrocq, C., Lhoste, J.-M., Rivalle, C. and Civier, A. (1979). *J. Chem. Soc., Perkin I*, 1706.
Bistochi, G. A., De Meo, G., Ricci, A., Croisy, A. and Jacquignon, P. (1978). *Heterocycles*, **9**, 247.
Biswas, K. M. and Jackson, A. H. (1969). *Tetrahedron*, **25**, 227.
Blades, C. E. and Wilds, A. L. (1956). *J. Org. Chem.*, **21**, 1013.
Blaikie, K. G. and Perkin, W. H. (1924). *J. Chem. Soc.*, **125**, 296.
Blair, H. S. and Roberts, G. A. F. (1967). *J. Chem. Soc. (C)*, 2425.
Blake, J., Tretter, J. R. and Rapoport, H. (1965). *J. Amer. Chem. Soc.*, **87**, 1397.
Blanc, M. G. (1898). *Bull. Soc. Chim. France*, **19**, 350.
Bloch-Chaudé, O., Rumpf, M. P. and Sadet, J. (1955). *Compt. Rend. Acad. Sci.*, **240**, 1426.
Bloink, G. J. and Pausacker, K. H. (1950). *J. Chem. Soc.*, 1328.
Bloss, K. H. and Timberlake, C. E. (1963). *J. Org. Chem.*, **28**, 267.
Bodforss, S. (1925). *Ber.*, **58**, 775.
Boehme, W. R. (1953). *J. Amer. Chem. Soc.*, **75**, 2502.
Boekelheide, V. and Ainsworth, C. (1950a). *J. Amer. Chem. Soc.*, **72**, 2132.
Boekelheide, V. and Ainsworth, C. (1950b). *J. Amer. Chem. Soc.*, **72**, 2134.

Bogat-skii, A. V., Ivanova, R. Y., Andronati, S. A. and Zhilina, Z. I. (1979). *Chem. Het. Comps.*, **15**, 688.
Boido, V. and Boido Canu, C. (1973). *Annali Chim.*, **63**, 593.
Boido, V. and Boido Canu, C. (1977). *Chim. Ind.*, **59**, 300.
Boido, V. and Sparatore, F. (1968). *Studi Sassaresi*, **46** (Suppl.), 337.
Boido, V. and Sparatore, F. (1977). *Chim. Ind.*, **59**, 300.
Bokii, N. G., Babushkina, T. A., Vasil'ev, A. M., Volodina, T. A., Kozik, T. A., Struchkov, Y. T. and Suvorov, N. N. (1975). *J. Org. Chem. USSR*, **11**, 984.
Boltze, K. H., Brendler, O., Dell, H. D. and Jacobi, H. (1974). Ger. Pat. 2,234,651 (*Chem. Abs.*, 1974, **81**, 3765n).
Boltze, K. H., Opitz, W., Raddatz, S., Seidel, P. K., Jacobi, H., Dell, H. D. and Schoelln-hammer, G. (1979). Ger. Pat. 2,740,836 (*Chem. Abs.*, 1979, **91**, 91500f).
Bonnema, J. and Arens, J. F. (1960a). *Rec. Trav. Chim.*, **79**, 1137.
Bonnema. J. and Arens, J. F. (1960b). Unpublished observations quoted as reference 9 in Bonnema and Arens, 1960a.
Borch, R. F. and Newell, R. G. (1973). *J. Org. Chem.*, **38**, 2729.
Borghero, S. and Finsterle, O. (1955). *Gazz. Chim. Ital.*, **85**, 651.
Borisova, L. N. and Kartashova, T. A. (1979). *Chem. Het. Comps.*, **15**, 162.
Borisova, L. N., Kucherova, N. F., Kartashova, T. A. and Zagorevskii, V. A. (1972). *Chem. Het. Comps.*, **8**, 584.
Borisova, L. N., Kucherova, N. F. and Zagorevskii, V. A. (1970a). *Chem. Het. Comps.*, **6**, 860.
Borisova, L. N., Kucherova, N. F. and Zagorevskii, V. A. (1970b). *Chem. Het. Comps.*, **6**, 868.
Bornstein, J., Leone, S. A., Sullivan, W. F. and Bennett, O. F. (1957). *J. Amer. Chem. Soc.*, **79**, 1745.
Borsche, W. and Groth, H. (1941). *Liebig's Ann. Chem.*, **549**, 238.
Borsche, W. and Kienitz, G. A. (1910). *Ber.*, **43**, 2333.
Borsche, W. and Klein, A. (1941). *Liebig's Ann. Chem.*, **548**, 64.
Borsche, W. and Manteuffel, R. (1934). *Ber.*, **67B**, 144.
Borsche, W., Witte, A. and Bothe, W. (1908). *Liebig's Ann. Chem.*, **359**, 49.
Bourdais, J. and Germain, C. (1970). *Tetrahedron Letts.*, 195.
Bourdais, J. and Lorre, A. (1975). *J. Het. Chem.*, **12**, 1111.
Bouveault, M. L. (1898). *Bull. Soc. Chim. France*, **19**, 565.
Bowman, R. E., Goodburn, T. G. and Reynolds, A. A. (1972). *J. Chem. Soc., Perkin I*, 1121.
Bowman, R. E. and Islip, P. J. (1971). *Chem. Ind.*, 154.
Boyakhchyan, A. P., Rashidyan, L. G. and Tatevosyan, G. T. (1966). *Arm. Khim. Zh.*, **19**, 636 (*Chem. Abs.*, 1967, **66**, 65374g).
Boyd, W. J. and Robson, W. (1935). *Biochem. J.*, **29**, 555.
Boyd-Barrett, H. S. (1932). *J. Chem. Soc.*, 321.
Boyd-Barrett, H. S. and Robinson, R. (1932). *J. Chem. Soc.*, 317.
Bozzini, S., Gratton, S., Pellizer, G., Risaliti, A. and Stener, A. (1979). *J. Chem. Soc., Perkin I*, 869.
Bradsher, C. K. and Litzinger, E. F. (1964). *J. Org. Chem.*, **29**, 3584.
Braithwaite, R. S. W. and Robinson, G. K. (1962). *J. Chem. Soc.*, 3671.
Bramely, R. K., Caldwell, J. and Grigg, R. (1973). *J. Chem. Soc., Perkin I*, 1913.
Braun, J. v. (1908a). *Ber.*, **41**, 2169.
Braun, J. v. (1908b). *Ber.*, **41**, 2604.
Braun, J. v. and Bayer, O. (1925). *Ber.*, **58**, 387.
Braun, J. v. and Bayer, O. (1929). *Liebig's Ann. Chem.*, **472**, 90.
Braun, J. v. and Haensel, W. (1926). *Ber.*, **59**, 1999.

Braun, J. v. and Kruber, O. (1912). *Ber.*, **45**, 2977.
Braun, J. v., Manz, G. and Reinsch, E. (1929). *Liebig's Ann. Chem.*, **468**, 277.
Braun, J. v. and Ritter, H. (1922). *Ber.*, **55**, 3792.
Braun, J. v. and Schörnig, L. (1925). *Ber.*, **58**, 2156.
Braunholtz, J. T. and Mann, F. G. (1955a). *J. Chem. Soc.*, 381.
Braunholtz, J. T. and Mann, F. G. (1955b). *J. Chem. Soc.*, 393.
Braunholtz, J. T. and Mann, F. G. (1958). *J. Chem. Soc.*, 3377.
Bredt, J., Rochussen, F. and Monheim, J. (1901). *Liebig's Ann. Chem.*, **314**, 369.
Bremner, J. B. and Browne, E. J. (1975). *J. Het. Chem.*, **12**, 301.
Brenner, G. S., Clamkowski, E. J., Hinkley, D. F. and Gal, G. (1969). Ger. Pat. 1,917,128 (*Chem. Abs.*, 1970, **72**, 12565n).
Brenner, G. S., Glamkowski, E. J., Hinkley, D. F. and Russ, W. K. (1969). Ger. Pat. 1,917,126 (*Chem. Abs.*, 1970, **72**, 31610q).
Bretherick, L., Gaimster, K. and Wragg, W. R. (1961). *J. Chem. Soc.*, 2919.
Brimblecombe, R. W., Downing, D. F. and Hunt, R. R. (1966). *J. Med. Chem.*, **9**, 345.
Briscoe, E. F. and Plant, S. G. P. (1928). *J. Chem. Soc.*, 1990.
Britten, A. and Lockwood, G. (1974). *J. Chem. Soc., Perkin I*, 1824.
Britten, A. Z., Bardsley, W. G. and Hill, C. M. (1971). *Tetrahedron*, **27**, 5631.
Britton, E. C. and Van der Weele, J. C. (1955). US Pat. 2,701,800 (*Chem. Abs.*, 1956, **50**, 409i).
Brooke, G. M., Musgrave, W. K. R., Rutherford, R. J. D. and Smith, T. W. (1971). *Tetrahedron*, **27**, 5653.
Brooke, G. M., Musgrave, W. K. R. and Thomas, T. R. (1971). *J. Chem. Soc.* (*C*), 3596.
Brown, B. R. (1951). *Quarterly Revs.*, **5**, 131.
Brown, F. and Mann, F. G. (1948a). *J. Chem. Soc.*, 847.
Brown, F. and Mann, F. G. (1948b). *J. Chem. Soc.*, 858.
Brown, P., Burdon, J., Smith, T. J. and Tatlow, J. C. (1960). *Tetrahedron*, **10**, 164.
Brown, R. K. (1972a). In *The Chemistry of Heterocyclic Compounds*, ed. by Houlihan, W. J., Wiley-Interscience, New York, London, Sydney and Toronto, Vol. 25, Part I, Chapter 2, p. 232.
Brown, R. K. (1972b). In *The Chemistry of Heterocyclic Compounds*, ed. by Houlihan, W. J., Wiley-Interscience, New York, London, Sydney and Toronto, Vol. 25, Part I, Chapter 2, p. 234.
Brown, R. K. (1972c). In *The Chemistry of Heterocyclic Compounds*, ed. by Houlihan, W. J., Wiley-Interscience, New York, London, Sydney and Toronto, Vol. 25, Part I, Chapter 2, p. 479.
Brown, R. K. (1972d). In *The Chemistry of Heterocyclic Compounds*, ed. by Houlihan, W. J., Wiley-Interscience, New York, London, Sydney and Toronto, Vol. 25, Part I, Chapter 2, p. 484.
Brown, R. K. (1972e). In *The Chemistry of Heterocyclic Compounds*, ed. by Houlihan, W. J., Wiley-Interscience, New York, London, Sydney and Toronto, Vol. 25, Part I, Chapter 2, p. 492.
Brown, R. K. (1972f). In *The Chemistry of Heterocyclic Compounds*, ed. by Houlihan, W. J., Wiley-Interscience, New York, London, Sydney and Toronto, Vol. 25, Part I, Chapter 2, p. 496.
Brown, R. K., Nelson, N. A., Sandin, R. B. and Tanner, K. G. (1952). *J. Amer. Chem. Soc.*, **74**, 3934.
Brown, U. M., Carter, P. H. and Tomlinson, M. (1958). *J. Chem. Soc.*, 1843.
Brown, V. H., Skinner, W. A. and DeGraw, J. I. (1969). *J. Het. Chem.*, **5**, 539.
Bruce, J. M. (1959). *J. Chem. Soc.*, 2366.
Bruce, J. M. (1960). *J. Chem. Soc.*, 360.

Bruce, J. M. (1962). *J. Chem. Soc.*, 1514.
Bruce, J. M. and Knowles, P. (1964). *J. Chem. Soc.*, 4046.
Bruce, J. M., Knowles, P. and Besford, L. S. (1964). *J. Chem. Soc.*, 4044.
Bruce, J. M. and Sutcliffe, F. K. (1957). *J. Chem. Soc.*, 4789.
Bruck, P. (1970). *J. Org. Chem.*, **35**, 2222.
Brunck, R. (1893). *Liebig's Ann. Chem.*, **272**, 201.
Brunelli, C., Fravolini, A., Grandolini, G. and Strappaghetti (1980). *J. Het. Chem.*, **17**, 199.
Brunner, K. (1894). *Monat. Chem.*, **15**, 747.
Brunner, K. (1895a). *Monat. Chem.*, **16**, 183.
Brunner, K. (1895b). *Monat. Chem.*, **16**, 849.
Brunner, K. (1896a). *Monat. Chem.*, **17**, 253.
Brunner, K. (1896b). *Monat. Chem.*, **17**, 479.
Brunner, K. (1897a). *Monat. Chem.*, **18**, 95.
Brunner, K. (1897b). *Monat. Chem.*, **18**, 527.
Brunner, K. (1898a). *Ber.*, **31**, 612.
Brunner, K. (1898b). *Ber.*, **31**, 1943.
Brunner, K. (1900). *Monat. Chem.*, **21**, 156.
Brunner, K. (1906). *Monat. Chem.*, **27**, 1183.
Brunner, K., Mikoss, M. von and Riedl, J. (1933). *Monat. Chem.*, **62**, 373.
Brunner, K. and Moser, H. (1932). *Monat. Chem.*, **61**, 15.
Brunner, K., Wiedner, K. and Kling, W. (1931). *Monat. Chem.*, **58**, 369.
Bryant, S. A. and Plant, S. G. P. (1931). *J. Chem. Soc.*, 93.
Bucherer, H. T. and Brandt, W. (1934). *J. Prakt. Chem.*, **140**, 129.
Bucherer, H. T. and Schmidt, M. (1909). *J. Prakt. Chem.*, **79**, 369.
Bucherer, H. T. and Seyde, F. (1908). *J. Prakt. Chem.*, **77**, 403.
Bucherer, H. T. and Seyde, F. (1909). Ger. Pat. 208,960 (*Chem. Zentr.*, 1909, **I**, 1951).
Bucherer, H. T. and Sonnenburg, E. F. (1910). *J. Prakt. Chem.*, **81**, 1.
Bucherer, H. T. and Wahl, R. (1922). *J. Prakt. Chem.*, **103**, 253.
Bucherer, H. T. and Zimmerman, W. (1922). *J. Prakt. Chem.*, **103**, 294.
Buchi, G. and Warnhoff, E. W. (1959). *J. Amer. Chem. Soc.*, **81**, 4433.
Buchman, E. R. and Sargent, H. (1942). *J. Org. Chem.*, **7**, 140.
Buchmann, G., Rehor, H. and Wegwart, H. (1968). *Pharmazie*, **23**, 557 (*Chem. Abs.*, 1969, **70**, 47236s).
Buckingham, J. (1969). *Quarterly Revs.*, **23**, 37.
Buckingham, J. and Guthrie, R. D. (1966). *Chem. Communs.*, 781.
Buckingham, J. and Guthrie, R. D. (1967). *J. Chem. Soc.* (*C*), 1700.
Buckingham, J. and Guthrie, R. D. (1968a). *J. Chem. Soc.* (*C*), 1445.
Buckingham, J. and Guthrie, R. D. (1968b). *J. Chem. Soc.* (*C*), 3079.
Budylin, V. A., Kost, A. N. and Matveeva, E. D. (1969). *Vestn. Mosk. Univ., Khim.*, **24**, 121 (*Chem. Abs.*, 1970, **72**, 55148w).
Budylin, V. A., Kost, A. N. and Matveeva, E. D. (1972). *Chem. Het. Comps.*, **8**, 52.
Budylin, V. A., Kost, A. N., Matveeva, E. D. and Minkin, V. I. (1972). *Chem. Het. Comps.*, **8**, 63.
Bullock, M. W. and Fox, S. W. (1951). *J. Amer. Chem. Soc.*, **73**, 5155.
Bullock, M. W. and Hand, J. J. (1956a). *J. Amer. Chem. Soc.*, **78**, 5852.
Bullock, M. W. and Hand, J. J. (1956b). *J. Amer. Chem. Soc.*, **78**, 5854.
Bülow, C. (1886). *Liebig's Ann. Chem.*, **236**, 184.
Bülow, C. (1918). *Ber.*, **51**, 399.
Bumgardner, C. L. and Lilly, R. L. (1962). *Chem. Ind.*, 559.
Burton, H. and Duffield, J. A. (1949). *J. Chem. Soc.*, 78.
Burton, H. and Leong, M. (1953). *Chem. Ind.*, 1035.

Burton, H. and Stoves, J. L. (1937). *J. Chem. Soc.*, 1726.
Burwell, R. L. and Shields, A. D. (1955). *J. Amer. Chem. Soc.*, **77**, 2766.
Busch, M. and Dietz, W. (1914). *Ber.*, **47**, 3277.
Buu-Hoï, N. P. (1949a). *J. Chem. Soc.*, 2882.
Buu-Hoï, N. P. (1949b). *Rec. Trav. Chim.*, **68**, 759.
Buu-Hoï, N. P. (1954). *J. Org. Chem.*, **19**, 1770.
Buu-Hoï, N. P. (1958). *J. Chem. Soc.*, 2418.
Buu-Hoï, N. P., Bellavita, V., Ricci, A., Hoeffinger, J. P. and Balucani, D. (1966). *J. Chem. Soc.* (*C*), 47.
Buu-Hoï, N. P., Binh, L. C., Loc, T. B., Xuong, N. D. and Jacquignon, P. (1957). *J. Chem. Soc.*, 3126.
Buu-Hoï, N. P., Cagniant, P., Hoán, N. and Khôi, N. H. (1950). *J. Org. Chem.*, **15**, 950.
Buu-Hoï, N. P., Croisy, A., Jacquignon, P., Hien, D.-P., Martani, A. and Ricci, A. (1972). *J. Chem. Soc., Perkin I*, 1266.
Buu-Hoï, N. P., Croisy, A., Jacquignon, P. and Martani, A. (1971). *J. Chem. Soc.* (*C*), 1109.
Buu-Hoï, N. P., Croisy, A., Jacquignon, P., Renson, M. and Ruwet, A. (1970). *J. Chem. Soc.* (*C*), 1058.
Buu-Hoï, N. P., Croisy, A., Ricci, A., Jacquignon, P. and Périn, F. (1966). *Chem. Communs.*, 269.
Buu-Hoï, N. P., Delcey, M., Jacquignon, P. and Perin, F. (1968). *J. Het. Chem.*, **5**, 259.
Buu-Hoï, N. P. and Hoán, N. (1949). *Rec. Trav. Chim.*, **68**, 441.
Buu-Hoï, N. P. and Hoán, N. (1950). *Rec. Trav. Chim.*, **69**, 1455.
Buu-Hoï, N. P. and Hoán, N. (1951). *J. Chem. Soc.*, 2868.
Buu-Hoï, N. P. and Hoán, N. (1952). *J. Chem. Soc.*, 3745.
Buu-Hoï, N. P., Hoán, N. and Jacquignon, P. (1949). *Rec. Trav. Chim.*, **68**, 781.
Buu-Hoï, N. P., Hoán, N. and Khôi, N. H. (1949). *J. Org. Chem.*, **14**, 492.
Buu-Hoï, N. P., Hoán, N. and Khôi, N. H. (1950a). *J. Org. Chem.*, **15**, 131.
Buu-Hoï, N. P., Hoán, N. and Khôi, N. H. (1950b). *Rec. Trav. Chim.*, **69**, 1053.
Buu-Hoï, N. P., Hoán, N. and Khôi, N. H. (1950c). *J. Org. Chem.*, **15**, 957.
Buu-Hoï, N. P., Hoán, N., Khôi, N. H. and Xuong, N. D. (1949). *J. Org. Chem.*, **14**, 802.
Buu-Hoï, N. P., Hoán, N., Khôi, N. H. and Xuong, N. D. (1950). *J. Org. Chem.*, **15**, 962.
Buu-Hoï, N. P., Hoán, N., Khôi, N. H. and Xuong, N. D. (1951). *J. Org. Chem.*, **16**, 309.
Buu-Hoï, N. P., Hoán, N. and Xuong, N. D. (1951). *J. Chem. Soc.*, 3499.
Buu-Hoï, N. P. and Jacquignon, P. (1951). *J. Chem. Soc.*, 2964.
Buu-Hoï, N. P. and Jacquignon, P. (1953). *J. Chem. Soc.*, 941.
Buu-Hoï, N. P. and Jacquignon, P. (1954). *J. Chem. Soc.*, 513.
Buu-Hoï, N. P. and Jacquignon, P. (1956). *J. Chem. Soc.*, 1515.
Buu-Hoï, N. P. and Jacquignon, P. (1960). *Compt. Rend. Acad. Sci.*, **251**, 1297.
Buu-Hoï, N. P. and Jacquignon, P. (1967). *Bull. Soc. Chim. France*, 1104.
Buu-Hoï, N. P., Jacquignon, P. and Béranger, S. (1968). *Bull. Soc. Chim. France*, 2476.
Buu-Hoï, N. P., Jacquignon, P., Croisy, A., Loiseau, A., Périn, F., Ricci, A. and Martani, A. (1969). *J. Chem. Soc.* (*C*), 1422.
Buu-Hoï, N. P., Jacquignon, P., Croisy, A., and Ricci, A. (1968). *J. Chem. Soc.* (*C*), 45.
Buu-Hoï, N. P., Jacquignon, P. and Hoeffinger, J. P. (1963). *J. Chem. Soc.*, 4754.
Buu-Hoï, N. P., Jacquignon, P. and Lavit, D. (1956). *J. Chem. Soc.*, 2593.
Buu-Hoï, N. P., Jacquignon, P. and Ledésert, L. (1970). *Bull. Soc. Chim. France*, 628.
Buu-Hoï, N. P., Jacquignon, P., Ledésert, L., Ricci, A. and Balucani, D. (1969). *J. Chem. Soc.* (*C*), 2196.
Buu-Hoï, N. P., Jacquignon, P. and Loc, T. B. (1958). *J. Chem. Soc.*, 738.
Buu-Hoï, N. P., Jacquignon, P. and Long, C. T. (1957). *J. Chem. Soc.*, 4994.

Buu-Hoï, N. P., Jacquignon, P. and Périn, F. (1962). *Bull. Soc. Chim. France*, **29**, 109.
Buu-Hoï, N. P., Jacquignon, P. and Périn-Roussel, O. (1965). *Bull. Soc. Chim. France*, 2849.
Buu-Hoï, N. P., Jacquignon, P., Roussel, O. and Hoeffinger, J. P. (1964). *J. Chem. Soc.*, 3924.
Buu-Hoï, N. P., Khôi, N. H. and Xuong, N. D. (1950). *J. Org. Chem.*, **15**, 511.
Buu-Hoï, N. P., Khôi, N. H. and Xuong, N. D. (1951). *J. Org. Chem.*, **16**, 315.
Buu-Hoï, N. P. and Lavit, D. (1955). *J. Org. Chem.*, **20**, 823.
Buu-Hoï, N. P. and Lavit, D. (1960). *Proc. Chem. Soc.*, 120.
Buu-Hoï, N. P. and Lavit-Lamy, D. (1961). *Bull. Soc. Chim. France*, 1657.
Buu-Hoï, N. P. and Lavit-Lamy, D. (1962). *Bull. Soc. Chim. France*, 1398.
Buu-Hoï, N. P., Mangane, M. and Jacquignon, P. (1966). *J. Chem. Soc. (C)*, 50.
Buu-Hoï, N. P., Mangane, M. and Jacquignon, P. (1967). *J. Chem. Soc. (C)*, 662.
Buu-Hoï, N. P., Martani, A., Croisy, A., Jacquignon, P. and Périn, F. (1966). *J. Chem. Soc. (C)*, 1789.
Buu-Hoï, N. P., Martani, A., Ricci, A., Dufour, M., Jacquignon, P. and Saint-Ruf, G. (1966). *J. Chem. Soc. (C)*, 1790.
Buu-Hoï, N. P., Périn, F. and Jacquignon, P. (1960). *J. Chem. Soc.*, 4500.
Buu-Hoï, N. P., Périn, F. and Jacquignon, P. (1962). *J. Chem. Soc.*, 146.
Buu-Hoï, N. P., Périn, F. and Jacquignon, P. (1965). *J. Het. Chem.*, **2**, 7.
Buu-Hoï, N. P., Périn, F. and Jacquignon, P. (1966). *Bull. Soc. Chim. France*, 584.
Buu-Hoï, N. P., Roussel, O. and Jacquignon, P. (1964). *J. Chem. Soc.*, 708.
Buu-Hoï, N. P. and Royer, R. (1947). *Rec. Trav. Chim.*, **66**, 305.
Buu-Hoï, N. P. and Royer, R. (1950). *Rec. Trav. Chim.*, **69**, 861.
Buu-Hoï, N. P., Royer, R., Eckert, B. and Jacquignon, P. (1952). *J. Chem. Soc.*, 4867.
Buu-Hoï, N. P. and Saint-Ruf, G. (1962). *J. Chem. Soc.*, 2630.
Buu-Hoï, N. P. and Saint-Ruf, G. (1963). *Israel J. Chem.*, **1**, 369.
Buu-Hoï, N. P. and Saint-Ruf, G. (1965a). *J. Chem. Soc.*, 2642.
Buu-Hoï, N. P. and Saint-Ruf, G. (1965b). *J. Chem. Soc.*, 5464.
Buu-Hoï, N. P., Saint-Ruf, G., Deschamps, D., Bigot, P. and Hieu, H.-T. (1971). *J. Chem. Soc. (C)*, 2606.
Buu-Hoï, N. P., Saint-Ruf, G. and Dufour, M. (1964). *J. Chem. Soc.*, 5433.
Buu-Hoï, N. P., Saint-Ruf, G., Jacquignon, P. and Barrett, G. C. (1958). *J. Chem. Soc.*, 4308.
Buu-Hoï, N. P., Saint-Ruf, G., Jacquignon, P. and Marty, M. (1963). *J. Chem. Soc.*, 2274.
Buu-Hoï, N. P., Saint-Ruf, G., Martani, A., Ricci, A. and Balucani, D. (1968). *J. Chem. Soc. (C)*, 609.
Buu-Hoï, N. P. and Xuong, N. D. (1952). *J. Chem. Soc.*, 2225.
Buu-Hoï, N. P., Xuong, N. D. and Lavit, D. (1954). *J. Chem. Soc.*, 1034.
Buu-Hoï, N. P., Xuong, N. D. and Thang, K. van (1953). *Rec. Trav. Chim.*, **72**, 774.
Buyanov, V. N., Mirzametova, R. M., Tret'yakova, L. G., Efimova, T. K. and Suvorov, N. N. (1978). *Chem. Het. Comps.*, **14**, 851.
Cadogan, J. I. G., Cameron-Wood, M., Mackie, R. K. and Searle, R. J. G. (1965). *J. Chem. Soc.*, 4831.
Caglioti, L., Rosini, G. and Rossi, F. (1966). *J. Amer. Chem. Soc.*, **88**, 3865.
Cagniant, P. and Cagniant, D. (1975). In *Advances in Heterocyclic Chemistry*, ed. by Katritzky, A. R. and Boulton, A. J., Academic Press, New York, San Francisco and London, Vol. 18, p. 337.
Calvaire, A. and Pallaud, R. (1960). *Compt. Rend. Acad. Sci.*, **250**, 3194.
Campaigne, E., Ergener, L., Hallum, J. V. and Lake, R. D. (1959). *J. Org. Chem.*, **24**, 487.
Campaigne, E. and Lake, R. D. (1959). *J. Org. Chem.*, **24**, 478.

Campbell, N. and Barclay, B. M. (1947). *Chem. Revs.*, **40**, 361.
Campbell, N. and Cooper, R. C. (1935). *J. Chem. Soc.*, 1208.
Canas-Rodriguez, A. and Leeming, P. R. (1971). Brit. Pat. 1,220,628 (*Chem. Abs.*, 1971, **75**, 5690h).
Cardani, C., Piozzi, F. and Casnati, G. (1955). *Gazz. Chim. Ital.*, **85**, 263.
Carelli, V., Marchini, P., Cardellini, M., Micheletti Moracci, F. and Liso, G. (1969). *Annali Chim.*, **59**, 163.
Carey, D. J., Gal, G. and Sletzinger, M. (1967). Fr. Pat. 1,488,841 (*Chem. Abs.*, 1968, **69**, 67216t).
Carey, D. J., Gal, G., Sletzinger, M. and Reinhold, D. F. (1968). Fr. Pat. 1,544,381 (*Chem. Abs.*, 1969, **71**, 124232n).
Carlin, R. B. (1952). *J. Amer. Chem. Soc.*, **74**, 1077.
Carlin, R. B. and Amoros-Marin, L. (1959). *J. Amer. Chem. Soc.*, **81**, 730.
Carlin, R. B. and Carlson, D. P. (1957). *J. Amer. Chem. Soc.*, **79**, 3605.
Carlin, R. B. and Carlson, D. P. (1959). *J. Amer. Chem. Soc.*, **81**, 4673.
Carlin, R. B. and Fisher, E. E. (1948). *J. Amer. Chem. Soc.*, **70**, 3421.
Carlin, R. B. and Forshey, W. O. (1950). *J. Amer. Chem. Soc.*, **72**, 793.
Carlin, R. B. and Harrison, J. W. (1965). *J. Org. Chem.*, **30**, 563.
Carlin, R. B., Henley, W. O. and Carlson, D. P. (1957). *J. Amer. Chem. Soc.*, **79**, 5712.
Carlin, R. B. and Larson, G. W. (1957). *J. Amer. Chem. Soc.*, **79**, 934.
Carlin, R. B., Magistro, A. J. and Mains, G. J. (1964). *J. Amer. Chem. Soc.*, **86**, 5300.
Carlin, R. B. and Moores, M. S. (1959). *J. Amer. Chem. Soc.*, **81**, 1259.
Carlin, R. B. and Moores, M. S. (1962). *J. Amer. Chem. Soc.*, **84**, 4107.
Carlin, R. B., Wallace, J. G. and Fisher, E. E. (1952). *J. Amer. Chem. Soc.*, **74**, 990.
Carlin, R. B. and Wich, G. S. (1958). *J. Amer. Chem. Soc.*, **80**, 4023.
Carry, D. J., Gal, G. and Sletzinger, M. (1968). Brit. Pat. 1,121,798 (*Chem. Abs.*, 1968, **69**, 86812a).
Carson, D. F. and Mann, F. G. (1965). *J. Chem. Soc.*, 5819.
Carter, P. H., Plant, S. G. P. and Tomlinson, M. (1957). *J. Chem. Soc.*, 2210.
Carter, M. A. and Robinson, B. (1974). *Chem. Ind.*, 304.
Casini, G. and Goodman, L. (1964). *Canad. J. Chem.*, **42**, 1235.
Casnati, G., Langella, M. R., Piozzi, F., Ricca, A. and Umani-Ronchi, A. (1964a). *Tetrahedron Letts.*, 1597.
Casnati, G., Langella, M. R., Piozzi, F., Ricca, A. and Umani-Ronchi, A. (1964b). *Gazz. Chim. Ital.*, **94**, 1221.
Cassebaum. H., Dierbach, K. and Hilger, H. (1970). Ger.(East) Pat. 77,974 (*Chem. Abs.*, 1972, **76**, 85694w).
Catsoulacos, P. and Papadopoulos, B. (1976). *J. Het. Chem.*, **13**, 159.
Cattanach, C. J., Cohen, A. and Heath-Brown, B. (1968). *J. Chem. Soc.* (*C*), 1235.
Cattanach, C. J., Cohen, A. and Heath-Brown, B. (1971). *J. Chem. Soc.* (*C*), 359.
Cattanach, C. J., Cohen, A. and Heath-Brown, B. (1973). *J. Chem. Soc., Perkin I*, 1041.
Cattanach, C. J. and Rees, R. G. (1971). *J. Chem. Soc.* (*C*), 53.
Cavallini, G. and Ravenna, V. (1958). *Il Farmaco* (*Pavia*), *Ed. Sci.*, **13**, 105 (*Chem. Abs.*, 1958, **52**, 20126b).
Cavallini, G., Ravenna, F. and Grasso, I. (1958). *Il Farmaco* (*Pavia*), *Ed. Sci.*, **13**, 113 (*Chem. Abs.*, 1958, **52**, 20126i).
Cawley, S. R. and Plant, S. G. P. (1938). *J. Chem. Soc.*, 1214.
Cecchetti, B. and Ghigi, E. (1930). *Gazz. Chim. Ital.*, **60**, 185.
Cetenko, W. A. and Morrison, G. C. (1979). US Pat. 4,144,349 (*Chem. Abs.*, 1979, **91**, 5217q).
C. F. Boehringer and Söhne (1908a). Ger. Pat. 218,477 (*Chem. Zentr.* 1910, **I**, 781) (*Chem. Abs.*, 1910, **4**, 2028).

C. F. Boehringer and Söhne (1908b). Ger. Pat. 218,727 (*Chem. Zentr.*, 1910, I, 876) (*Chem. Abs.*, 1910, **4**, 2028).

Chakraborty, D. P. (1966). *Tetrahedron Letts.*, 661.

Chakraborty, D. P., Bhattacharyya, P., Roy, S., Bhattacharyya, S. P. and Biswas, A. K. (1978). *Phytochem.*, **17**, 834.

Chakraborty, D. P., Chatterji, D. and Ganguly, S. N. (1969). *Chem. Ind.*, 1662.

Chakraborty, D. P. and Chowdhury, B. K. (1966). *Sci. Cult.* (*Calcutta*), **32**, 590 (*Chem. Abs.*, 1967, **67**, 53975m).

Chakraborty, D. P. and Chowdhury, B. K. (1968). *J. Org. Chem.*, **33**, 1265.

Chakraborty, D. P., Das, K. C. and Basak, S. P. (1968). *J. Indian Chem. Soc.*, **45**, 84.

Chakraborty, D. P., Das, K. C. and Chowdhury, B. K. (1966a). *Chem. Ind.*, 1684.

Chakraborty, D. P., Das, K. C. and Chowdhury, B. K. (1966b). *Sci. Cult.* (*Calcutta*), **32**, 245 (*Chem. Abs.*, 1967, **66**, 28609s).

Chakraborty, D. P., Das, K. C. and Chowdhury, B. K. (1969). *Phytochem.*, **8**, 773.

Chakraborty, D. P., Islam, A. and Bhattacharyya, P. (1973). *J. Org. Chem.*, **38**, 2728.

Chalmers, A. J. and Lions, F. (1933). *J. Proc. Roy. Soc. NSW*, **67**, 178.

Chalmers, J. R., Openshaw, H. T. and Smith, G. F. (1957). *J. Chem. Soc.*, 1115.

Chang, M. C., Hsing, C. I. and Ho, T. S. (1966). *Hua Hsueh Hseuh Pao*, **32**, 64 (*Chem. Abs.*, 1966, **65**, 10554g).

Ch'ang-pai, C., Evstigneeva, R. P. and Preobrazhenskii, N. (1958). *Proc. Acad. Sci. USSR*, **123**, 927.

Ch'ang-pai, C., Evstigneeva, R. P. and Preobrazhenskii, N. A. (1958). *J. Gen. Chem. USSR*, **28**, 3116.

Ch'ang-pai, C., Evstigneeva, R. P. and Preobrazhenskii, N. A. (1960). *J. Gen. Chem. USSR*, **30**, 2066.

Chapelle, J.-P., Elguero, J., Jacquier, R. and Tarrago, G. (1969). *Bull. Soc. Chim. France*, 4464.

Chapelle, J.-P., Elguero, J., Jacquier, R. and Tarrago, G. (1970a). *Bull. Soc. Chim. France*, 240.

Chapelle, J.-P., Elguero, J., Jacquier, R. and Tarrago, G. (1970b). *Bull. Soc. Chim. France*, 3147.

Chapelle, J.-P., Elguero, J., Jacquier, R. and Tarrago, G. (1971). *Bull. Soc. Chim. France*, 1971, 280.

Chaplin, A. F., Hey, D. H. and Honeyman, J. (1959). *J. Chem. Soc.*, 3194.

Chapman, N. B., Clarke, K. and Hughes, H. (1965). *J. Chem. Soc.*, 1424.

Charles, M., Descotes, G., Martin, J. C. and Querou, Y. (1968). *Bull. Soc. Chim. France*, 4159.

Chastrette, F. (1970). *Bull. Soc. Chim. France*, 1151.

Chattaway, F. D. (1906). *J. Chem. Soc., Trans.*, **89**, 462.

Chattaway, F. D. and Aldridge, M. (1911). *J. Chem. Soc., Trans.*, **99**, 404.

Chattaway, F. D. and Humphrey, W. G. (1927). *J. Chem. Soc.*, 1323.

Chattaway, F. D. and Irving, H. (1931). *J. Chem. Soc.*, 1740.

Chemerda, J. M. and Sletzinger, M. (1968a). Fr. Pat. 1,534,198 (*Chem. Abs.*, 1970, **72**, 3371u).

Chemerda, J. M. and Sletzinger, M. (1968b). Fr. Pat. 1,534,326 (*Chem. Abs.*, 1969, **71**, 81168b).

Chemerda, J. M. and Sletzinger, M. (1968c). Fr. Pat. 1,534,375 (*Chem. Abs.*, 1969, **71**, 81166z).

Chemerda, J. M. and Sletzinger, M. (1968d). Fr. Pat. 1,534,376 (*Chem. Abs.*, 1969, **71**, 81173z).

Chemerda, J. M. and Sletzinger, M. (1968e). Fr. Pat. 1,534,459 (*Chem. Abs.*, 1969, **71**, 81158y).

Chemerda, J. M. and Sletzinger, M. (1968f). Fr. Pat. 1,534,487 (*Chem. Abs.*, 1969, **71**, 61209n).
Chemerda, J. M. and Sletzinger, M. (1969a). US Pat. 3,449,364 (*Chem. Abs.*, 1969, **71**, 61204g).
Chemerda, J. M. and Sletzinger, M. (1969b). US Pat. 3,457,276 (*Chem. Abs.*, 1970, **72**, 3368y).
Chemerda, J. M. and Sletzinger, M. (1969c). US Pat. 3,467,669 (*Chem. Abs.*, 1969, **71**, 124235r).
Chen, F. M. F. and Forrest, T. P. (1973). *Canad. J. Chem.*, **51**, 881.
Chen, G.-S. J. and Gibson, M. S. (1975). *J. Chem. Soc., Perkin I*, 1138.
Chernova, A. V., Shagidullin, R. R. and Kitaev, Y. P. (1964). *Bull. Acad. Sci. USSR, Div. Chem. Sci.*, 1470.
Chernova, A. V., Shagidullin, R. R. and Kitaev, Y. P. (1967). *J. Org. Chem. USSR*, **3**, 882.
Chilikin, L. G., Gorelova, N. V., Shvarts, G. Y. and Suvorov, N. N. (1979). *Khim.-Farm. Zh.*, **13**, 48 (*Chem. Abs.*, 1979, **91**, 39245n).
Chippendale, K. E., Iddon, B. and Suschitzky, H. (1972). *J. Chem. Soc., Perkin I*, 2023.
Chittenden, G. J. F. and Guthrie, R. D. (1964). Unpublished result quoted as ref. 11 in Bellamy and Guthrie, 1964.
Chowdhury, B. K. and Chakraborty, D. P. (1974). *Trans. Bose Res. Inst. Calcutta*, **37**, 57 (*Chem. Abs.*, 1977, **86**, 121578d).
Ciamician, G. and Zatti, C. (1889). *Ber.*, **22**, 1976.
Ciba, Ltd (1960). Brit. Pat. 830,223 (*Chem. Abs.*, 1960, **54**, 18550c).
Clark, B. A. J., Dorgan, R. J. and Parrick, J. (1975). *Chem. Ind.*, 215.
Clark, B. A. J. and Parrick, J. (1974). *J. Chem. Soc., Perkin I*, 2270.
Clark, B. A. J., Parrick, J. and Dorgan, R. J. J. (1976). *J. Chem. Soc., Perkin I*, 1361.
Clark, B. A. J., Parrick, J., West, P. J. and Kelly, A. H. (1970). *J. Chem. Soc. (C)*, 498.
Clark, K. J. (1957). *J. Chem. Soc.*, 2202.
Clemo, G. R. and Felton, D. G. I. (1951a). *J. Chem. Soc.*, 671.
Clemo, G. R. and Felton, D. G. I. (1951b). *J. Chem. Soc.*, 700.
Clemo, G. R. and Felton, D. G. I. (1952). *J. Chem. Soc.*, 1658.
Clemo, G. R. and Holt, R. J. W. (1953). *J. Chem. Soc.*, 1313.
Clemo, G. R. and Perkin, W. H. (1924a). *J. Chem. Soc.*, **125**, 1608.
Clemo, G. R. and Perkin, W. H. (1924b). *J. Chem. Soc.*, **125**, 1804.
Clemo, G. R., Perkin, W. H. and Robinson, R. (1924). *J. Chem. Soc.*, **125**, 1751.
Clemo, G. R. and Seaton, J. C. (1954). *J. Chem. Soc.*, 2582.
Clemo, G. R. and Swan, G. A. (1946). *J. Chem. Soc.*, 617.
Clemo, G. R. and Swan, G. A. (1949). *J. Chem. Soc.*, 487.
Clemo, G. R. and Weiss, J. (1945). *J. Chem. Soc.*, 702.
Clerc-Bory, M. (1954). *Bull. Soc. Chim. France*, 337.
Clifford, B., Nixon, P., Salt, C. and Tomlinson, M. (1961). *J. Chem. Soc.*, 3516.
Clifton, P. V. and Plant, S. G. P. (1951). *J. Chem. Soc.*, 461.
Clusius, K. and Barsh, M. (1954). *Helv. Chim. Acta*, **37**, 2013.
Clusius, K. and Weisser, H. R. (1952). *Helv. Chim. Acta*, **35**, 400.
Coates, R. M. and Said, I. M. (1977). *J. Amer. Chem. Soc.*, **99**, 2355.
Cockerill, A. F., Davies, G. L. O., Harden, R. C. and Rackham, D. M. (1973). *Chem. Revs.*, **73**, 553.
Coffey, S. (1923). *Rec. Trav. Chim.*, **42**, 528.
Cohen, A. and Cattanach, C. J. (1967). US Pat. 3,316,271 (*Chem. Abs.*, 1967, **67**, 21900s).
Cohen, A., Heath-Brown, B. and Cattanach, C. J. (1968). US Pat. 3,373,153 (*Chem. Abs.*, 1968, **69**, 43898r).
Cohen, A., Heath-Brown, B., Smithen, C. E. and Cattanach, C. J. (1968). S. African Pat. 68 00,169 (*Chem. Abs.*, 1969, **70**, 77929n).

Cohn, G. (1919). *Die Carbazolgruppe*, Thieme, Leipzig.

Cohylakis, D., Hignett, G. J., Lichman, K. V. and Joule, J. A. (1974). *J. Chem. Soc., Perkin I*, 1518.

Coispeau, G. and Elguero, J. (1970). *Bull. Soc. Chim. France*, 2717.

Coispeau, G., Elguero, J., Jacquier, R. and Tizane, D. (1970). *Bull. Soc. Chim. France*, 1581.

Coldham, M. W. G., Lewis, J. W. and Plant, S. G. P. (1954). *J. Chem. Soc.*, 4528.

Collar, W. M. and Plant, S. G. P. (1926). *J. Chem. Soc.*, 808.

Colle, M. A. (1960). Diplome d'Etudes Supierieures, Nancy, France (quoted in Bettembourg and David, 1962).

Colle, M.-A. and David, S. (1960). *Compt. Rend. Acad. Sci.*, **250**, 2226.

Collins, I., Roberts, S. M. and Suschitzky, H. (1971). *J. Chem. Soc.* (*C*), 167.

Coló, V., Asero, B. and Vercellone, A. (1954). *Il Farmaco, Ed. Sci.*, **9**, 611 (*Chem. Abs.*, 1955, **49**, 14732i).

Colwell, W. T., Horner, J. K. and Skinner, W. A. (1964). *US Dept. Com., Office Tech. Serv.* AD 435,889 (*Chem. Abs.*, 1965, **62**, 11763a).

Connor, R., Folkers, K. and Adkins, H. (1932). *J. Amer. Chem. Soc.*, **54**, 1138.

Conroy, H. and Firestone, R. A. (1953). *J. Amer. Chem. Soc.*, **75**, 2530.

Conroy, H. and Firestone, R. A. (1956). *J. Amer. Chem. Soc.*, **78**, 2290.

Cook, A. H. and Reed, K. J. (1945). *J. Chem. Soc.*, 399.

Cook, J. W., Loudon, J. D. and McCloskey, P. (1951). *J. Chem. Soc.*, 1203.

Cook, J. W., Loudon, J. D. and McCloskey, P. (1952). *J. Chem. Soc.*, 3904.

Cooke, G. W. and Gulland, J. M. (1939). *J. Chem. Soc.*, 872.

Cookson, R. C. and Mann, F. G. (1949). *J. Chem. Soc.*, 67.

Coombes, G. E. A., Harvey, D. J. and Reid, S. T. (1970). *J. Chem. Soc.* (*C*), 325.

Corbett, J. F. and Holt, P. F. (1960). *J. Chem. Soc.*, 3646.

Cornforth, J. W., Cornforth, R. H., Dalgliesh, C. E. and Neuberger, A. (1951). *Biochem. J.*, **48**, 591.

Cornforth, J. W., Hughes, G. K., Lions, F. and Harradence, R. H. (1937–1938). *J. Proc. Roy. Soc. NSW*, **71**, 486.

Cornubert, R. (1933). In *Le Camphre et ses Dérivés*, Masson, Paris, p. 214.

Cortes, E. and Walls, F. (1964). *Bol. Inst. Quim. Univ. Nacl. Auton. Mex.*, 163 (*Chem. Abs.*, 1965, **63**, 564c).

Council of Scientific and Industrial Research (India) (1975). Indian Pat. 103,093 (*Chem. Abs.*, 1980, **92**, 128720b).

Cowper, R. M. and Stevens, T. S. (1947). *J. Chem. Soc.*, 1041.

Coxworth, E. (1965). In *The Alkaloids*, ed. by Manske, R. H. F., Academic Press, New York and London, Vol. 8, p. 27.

Cranwell, P. A. and Saxton, J. E. (1962). *J. Chem. Soc.*, 3482.

Crary, J. W., Quayle, O. R. and Lester, C. T. (1956). *J. Amer. Chem. Soc.*, **78**, 5584.

CRC Compagnia di Ricerca Chimica S.A. (1979). Fr. Pat. 2,400,016 (*Chem. Abs.*, 1980, **92**, 58606t).

Criegee, R. and Lohaus, G. (1951). *Chem. Ber.*, **84**, 219.

Croce, P. D. (1973). *Annali Chim.*, **63**, 29.

Croisy, A., Jacquignon, P. and Fravolini, A. (1974). *J. Het. Chem.*, **11**, 113.

Croisy, A., Ricci, A., Janćevska, M., Jacquignon, P. and Balucani, D. (1976). *Chem. Letts.*, 5.

Crooks, P. A. and Robinson, B. (1967). *Chem. Ind.*, 547.

Crooks, P. A. and Robinson, B. (1969). *Canad. J. Chem.*, **47**, 2061.

Cross, P. E. and Jones, E. R. H. (1964). *J. Chem. Soc.*, 5919.

Cross, A. D., King, F. E. and King, T. J. (1961). *J. Chem. Soc.*, 2714.

Crow, W. D. and Solly, R. K. (1966). *Aust. J. Chem.*, **19**, 2119.

Crowther, A. F., Mann, F. G. and Purdie, D. (1943). *J. Chem. Soc.*, 58.

Crum, J. D. and Sprague, P. W. (1966). *Chem. Communs.*, 417.

Culmann, J. (1888). *Ber.*, **21**, 2595.

Cummins, J. A., Kaye, B. F. and Tomlinson, M. L. (1954). *J. Chem. Soc.*, 1414.

Cummins, J. A. and Tomlinson, M. L. (1955). *J. Chem. Soc.*, 3475.

Curtin, D. Y. and Crawford, R. J. (1957). *J. Amer. Chem. Soc.*, **79**, 3156.

Curtin, D. Y. and Poutsma, M. L. (1962). *J. Amer. Chem. Soc.*, **84**, 4887.

Curtius, T. (1916). *J. Prakt. Chem.*, **94**, 273.

Curtius, T. and Försterling, H. A. (1894). *Ber.*, **27**, 770.

Curtius, T. and Zinkeisen, E. (1898). *J. Prakt. Chem.*, **58**, 310.

Dalgliesh, C. E. and Mann, F. G. (1947). *J. Chem. Soc.*, 653.

Dalton, C., Fahrehnoltz, K. E. and Silverzweig, M. Z. (1979). US Pat. 4,156,016 (*Chem. Abs.*, 1979, **91**, 74467h).

Dambal, S. B. and Siddappa, S. (1965). *J. Indian Chem. Soc.*, **42**, 112.

Dann, O., Bergen, G., Demant, E. and Volz, G. (1971). *Liebig's Ann. Chem.*, **749**, 68.

Dann, O., Fick, H., Pietzner, B., Walkenhorst, E., Fernbach, R. and Feh, D. (1975). *Liebig's Ann. Chem.*, 160.

Dann, O., Volz, G., Demant, E., Pfeifer, W., Bergen, G., Fick, H. and Walkenhorst, E. (1973). *Liebig's Ann. Chem.*, 1112.

Das, K. G., Kulkarni, P. S. and Chinchwadkar, C. A. (1969). *Indian J. Chem.*, **7**, 140.

Da Settimo, A., Biagi, G., Primofiore, G., Ferrarini, P. L. and Livi, O. (1978). *Il Farmaco, Ed. Sci.*, **33**, 770 (*Chem. Abs.*, 1979, **90**, 72088r).

Da Settimo, A., Primofiore, G., Biagi, G. and Santerini, V. (1976a). *J. Het. Chem.*, **13**, 97.

Da Settimo, A., Primofiore, G., Biagi, G. and Santerini, V. (1976b). *Il Farmaco, Ed. Sci.*, **31**, 587.

Da Settimo, A. and Saettone, M. F. (1965). *Tetrahedron*, **21**, 823.

Dashkevich, S. N. (1978). *Khim. Geter. Soedin.*, **14**, 132 (*Chem. Abs.*, 1978, **88**, 152348z).

Dashkevich, S. N. and Grandberg, I. I. (1970). *Dokl. Timiryazev Sel'skakhoz. Akad.*, 160, 243 (*Chem. Abs.*, 1971, **74**, 125884a).

Dave, V. (1972). *Canad. J. Chem.*, **50**, 3397.

Dave, V. (1976). *Org. Prep. Proced.*, **8**, 41 (*Chem. Abs.*, 1976, **86**, 32754v).

David, S. (1960). *Compt. Rend. Acad. Sci.*, **251**, 2549.

David, S. and Monnier, J. (1956). *Compt. Rend. Acad. Sci.*, **243**, 597.

David, S. and Monnier, J. (1959). *Bull. Soc. Chim. France*, 1333.

David, S. and Régent, P. (1964). *Bull. Soc. Chim. France*, 101.

Davidge, H. (1959). *J. Appl. Chem.*, **9**, 241.

Davies, W. and Middleton, S. (1958). *J. Chem. Soc.*, 822.

Davis, F. A. and Skibo, E. B. (1974). *J. Org. Chem.*, **39**, 807.

De Bellis, L. and Stein, M. L. (1961). *Annali Chim.*, **51**, 663 (*Chem. Abs.*, 1962, **56**, 11544d).

Degen, J. (1886). *Liebig's Ann. Chem.*, **236**, 151.

DeGraw, J. I. and Skinner, W. A. (1967). *Canad. J. Chem.*, **45**, 63.

De Jong, J. and Boyer, J. H. (1972). *J. Org. Chem.*, **37**, 3571.

Demerson, C. A., Humber, L. G., Philipp, A. H. and Martel, R. R., (1976). *J. Med. Chem.*, **19**, 391.

Dennler, E. B. (1965). Doctoral Thesis, University of Buenos Aires (quoted in Dennler and Frasca, 1966b).

Dennler, E. B. and Frasca, A. R. (1966a). *Anales Asoc. Quim. Argent.*, **54**, 1.

Dennler, E. B. and Frasca, A. R. (1966b). *Tetrahedron*, **22**, 3131.

Dennler, E. B. and Frasca, A. R. (1967). *Canad. J. Chem.*, **45**, 697.

Dennler, E. B., Portal, C. R. and Frasca, A. R. (1967). *Spectrochemica Acta*, **23A**, 2243.

Deorha, D. S. and Joshi, S. S. (1961). *J. Org. Chem.*, **26**, 3527.
Desaty, D., Hadžija, O., Iskrić, S., Keglević, D. and Kveder, S. (1962). *Biochim. Biophys. Acta*, **62**, 179.
Desaty, D. and Keglević, D. (1964). *Croat. Chem. Acta*, **36**, 103.
Desaty, D. and Keglević, D. (1965). *Croat. Chem. Acta*, **37**, 25.
Deulofeu, V. (1938). *Rev. Brasil. Chim.*, (*São Paulo*), **5**, 270 (*Chem. Abs.*, 1938, **32**, 8412_5).
Deutsche Gold-und Silberscheideanstalt (1925). Brit. Pat. 259,982 (quoted in Ficken and Kendall, 1959) [cf. Takahashi, T., Saikachi, H., Goto, H. and Shimamura, S. (1944). *J. Pharm. Soc. Japan*, **64**, 7 (*Chem. Abs.*, 1951, **45**, 8529)].
Dewar, M. J. S. (1944). *J. Chem. Soc.*, 615.
Dewar, M. J. S. (1963). In *Molecular Rearrangements*, ed. by de Mayo, P., Wiley-Interscience, New York and London, Chapter 5, p. 303.
Dickel, D. F. and DeStevens, G. (1970). US Pat. 3,538,223 (*Chem. Abs.*, 1971, **74**, 22694j).
Dickel, D. F. and DeStevens, G. (1972). US Pat. 3,686,211 (*Chem. Abs.*, 1972, **77**, 151925w).
Dieckmann, W. (1897). *Ber.*, **30**, 1470.
Dieckmann, W. (1901). *Liebig's Ann. Chem.*, **317**, 27.
Diels, O. (1901). *Ber.*, **34**, 1758.
Diels, O. and Dürst, W. (1914). *Ber.*, **47**, 284.
Diels, O. and Köllisch, A. (1911). *Ber.*, **44**, 263.
Diels, O. and Reese, J. (1934). *Liebig's Ann. Chem.*, **511**, 168.
Diels, O. and Reese, J. (1935). *Liebig's Ann. Chem.*, **519**, 147.
Djerassi, C. and Nakano, T. (1960). *Chem. Ind.*, 1385.
Djerassi, C. and Scholz, C. R. (1948). *J. Amer. Chem. Soc.*, **70**, 417.
Dmitrevskaya, L. I., Smushkevich, Y. I., Pozdnyakov, A. D. and Suvurov, N. N. (1973). *Chem. Het. Comps.*, **9**, 476.
Dmitrevskaya, L. I., Smushkevich, Y. I. and Suvorov, N. N. (1973). *Zh. Vses. Khim. Obshchest.*, **18**, 117 (*Chem. Abs.*, 1973, **78**, 135995j).
Dobeneck, H. von and Maresch, G. (1952). *Hoppe-Seyler's Zeit. Physiol. Chem.*, **289**, 271.
Dokunikhin, N. S. and Bystritskii, G. I. (1963). *J. Gen. Chem. USSR*, **33**, 962.
Dolby, L. J. and Esfandiari, Z. (1972). *J. Org. Chem.*, **37**, 43.
Doorenbos, N. J. and Wu, M. T. (1968). *J. Org. Chem.*, **11**, 158.
Dopp, D. and Weiler, H. (1979). *Chem. Ber.*, **112**, 3950.
Dorée, C. (1909). *J. Chem. Soc.*, **95**, 653.
Dorée, C. and Gardner, J. A. (1908). *J. Chem. Soc.*, **93**, 1625.
Dorée, C. and Petrow, V. A. (1935). *J. Chem. Soc.*, 1391.
Dostál, V. (1938). *Chem. Listy*, **32**, 13 (*Chem. Abs.*, 1938, **32**, 5399_7, 6242_8).
Douglas, A. W. (1978). *J. Amer. Chem. Soc.*, **100**, 6463.
Douglas, A. W. (1979). *J. Amer. Chem. Soc.*, **101**, 5676.
Douglas, B., Kirkpatrick, J. L., Moore, B. P. and Weisbach, J. A. (1964). *Aust. J. Chem.*, **17**, 246.
Doyle, M. and Smith, S. C. (1974). Ger. Pat. 2,410,699 (*Chem. Abs.*, 1975, **82**, 4289s).
Drapkina, D. A., Grigor'eva, N. M. and Doroshina, N. I. (1974). USSR Pat. 455,658 (*Chem. Abs.*, 1975, **82**, 125274s).
Drechsel, E. (1888). *J. Prakt. Chem.*, **38**, 65.
Driver, M., Matthews, I. T. and Sainsbury, M. (1979). *J. Chem. Soc.*, *Perkin I*, 2506.
Drogas Vacunas, y Sueros, S. A. (Drovyssa) (1957). Span. Pat. 227,606 (*Chem. Abs.*, 1958, **52**, 2923b).
Duc, D. K. M. and Fetizon, M. (1969). *Bull. Soc. Chem. France*, 4154.
Ducrocq, C., Bisagni, E., Lhoste, J.-M., Mispelter, J. and Defaye, J. (1976). *Tetrahedron*, **32**, 773.
Ducrocq, C., Civier, A., André-Louisfert, J. and Bisagni, E. (1975). *J. Het. Chem.*, **12**, 963.

Duffy, T. D. and Wibberley, D. G. (1974). *J. Chem. Soc., Perkin I*, 1921.
Dufour, M., Buu-Hoï, N. P., Jacquignon, P. and Hien, D.-P. (1972). *J. Chem. Soc., Perkin I*, 527.
Dufresne, R. F. (1972). Ph.D. Thesis, Carnegie-Mellon University, USA (*Diss. Abs.*, 1973, **33**, 4724B) (*Chem. Abs.*, 1973, **79**, 18523p).
Dufton, S. F. (1891). *J. Chem. Soc.* (*Trans.*), **59**, 756.
Dufton, S. F. (1892). *J. Chem. Soc.* (*Trans.*), **61**, 782.
Duisberg, C. (1919). *Ber.*, **52A**, 149.
Duncan, R. L. and Andrako, J. (1968). *J. Pharm. Sci.*, **57**, 979.
Duncan, R. L. and Boswell, R. F. (1973). Ger. Pat. 2,241,269 (*Chem. Abs.*, 1973, **78**, 136058t).
Duncan, R. L., Helsley, G. C. and Boswell, R. F. (1973). *J. Het. Chem.*, **10**, 65.
Dunlop, H. G. and Tucker, S. H. (1939). *J. Chem. Soc.*, 1945.
Dupont, P. E. (1968). Ph.D. Thesis, Rensselaer Polytechnic Institute, USA (*Diss. Abs.*, 1969, **29**, 4092B) (*Chem. Abs.*, 1969, **71**, 112147a).
Duschinsky, R. (1953). US Pat. 2,642,438 (*Chem. Abs.*, 1954, **48**, 5230c).
Duuren, B. L. van (1961). *J. Org. Chem.*, **26**, 2954.
Eberle, M. K. (1971). Ger. Pat. 2,049,559 (*Chem. Abs.*, 1971, **75**, 35731q).
Eberle, M. K. (1972a). US Pat. 3,634,426 (*Chem. Abs.*, 1972, **76**, 99699p).
Eberle, M. K. (1972b). US Pat. 3,696,118 (*Chem. Abs.*, 1973, **78**, 29807q).
Eberle, M. K. and Brzechffa, L. (1976). *J. Org. Chem.*, **41**, 3775.
Eberle, M. K. and Kahle, G. G. (1973). *Tetrahedron*, **29**, 4049.
Eberle, M. K., Kahle, G. G. and Talati, S. M. (1973). *Tetrahedron*, **29**, 4045.
Eberson, L. and Persson, K. (1964). *Acta Chem. Scand.*, **18**, 721.
Ebnöther, A., Niklaus, P. and Süess, R. (1968). *Chimia*, **22**, 493.
Ebnöther, A., Niklaus, P. and Süess, R. (1969). *Helv. Chim. Acta*, **52**, 629.
Edwards, G. A. and Plant, S. G. P. (1923). *J. Chem. Soc.*, **123**, 2393.
Ehrhart, G. (1953). Ger. Pat. 878,802 (*Chem. Abs.*, 1958, **52**, 10202g).
Eistert, B., Haupter, F. and Schank, K. (1963). *Liebig's Ann. Chem.*, **665**, 55.
Eiter, K. and Nagy, M. (1949). *Monat. Chem.*, **80**, 607.
Eiter, K. and Nezval, A. (1950). *Monat. Chem.*, **81**, 404.
Eiter, K. and Svierak, O. (1952). *Monat. Chem.*, **83**, 1453.
Ek, A. and Witkop, B. (1954). *J. Amer. Chem. Soc.*, **76**, 5579.
Ekmekdzhyan, S. P. and Tatevosyan, G. T. (1960). *Izv. Akad. Nauk. Arm. SSR, Khim. Nauki*, **13**, 201 (*Chem. Abs.*, 1961, **55**, 9374g).
Elderfield, R. C. and Kreysa, F. J. (1948). *J. Amer. Chem. Soc.*, **70**, 44.
Elderfield, R. C., Lagowski, J. M., McCurdy, O. L. and Wythe, S. L. (1958). *J. Org. Chem.*, **23**, 435.
Elderfield, R. C. and McCarthy, J. R. (1951). *J. Amer. Chem. Soc.*, **73**, 975.
Elderfield, R. C. and Wythe, S. L. (1954). *J. Org. Chem.*, **19**, 693.
Elgersma, R. H. C. (1969). Thesis, University of Leiden, Holland.
Elgersma, R. H. C. and Havinga, E. (1969). *Tetrahedron Letts.*, 1735.
Elguero, J., Jacquier, R. and Tarrago, G. (1966). *Bull. Soc. Chim. France*, 2981.
Eliel, E. L., Wilken, P. H. and Fang, F. T. (1957). *J. Org. Chem.*, **22**, 231.
Elks, J., Elliott, D. F. and Hems, B. A. (1944a). *J. Chem. Soc.*, 624.
Elks, J., Elliott, D. F. and Hems, B. A. (1944b). *J. Chem. Soc.*, 626.
Elks, J., Elliott, D. F. and Hems, B. A. (1944c). *J. Chem. Soc.*, 629.
Ellinger, A. (1904). *Ber.*, **37**, 1801.
Ellinger, A. (1905). *Ber.*, **38**, 2884.
Ellsworth, R. L., Gatto, G. J., Meriwether, H. T. and Mertel, H. E. (1978). *J. Labelled Compounds Radiopharm.*, **15**, 613.

Ender, F., Moisar, E., Schafer, K. and Teuber, H.-J. (1959). *Z. Elektrochem.*, **63**, 349.
Endler, A. S. and Becker, E. I. (1963). *Org. Syns.*, Coll. Vol. 4, 657.
Engvild, K. C. (1977). *Acta Chem. Scand., Ser. B.*, **B31**, 338.
Epstein, J. W. and Goldman, L. (1973). US Pat. 3,772,325 (*Chem. Abs.*, 1974, **80**, 59939t).
Eraksina, V. N., Kost, A. N., Khazanova, T. S. and Vinogradova, E. V. (1966). *Chem. Het. Comps.*, **2**, 160.
Eraksina, V. N., Maslennikova, L. V., Shagalov, L. B. and Suvorov, N. N. (1979). *Chem. Het. Comps.*, **15**, 1259.
Erdmann, O. L. (1841). *J. Prakt. Chem.*, **24**, 1.
Esayan, Z. V. and Tatevosyan, G. T. (1972). *Arm. Khim. Zh.*, **25**, 969 (*Chem. Abs.*, 1973, **78**, 136174c).
Esayan, Z. V., Terzyan, A. G., Astratyan, S. N., Dzhanpoladyan, E. G. and Tatevosyan, G. T. (1968). *Arm. Khim. Zh.*, **21**, 348 (*Chem. Abs.*, 1969, **70**, 11476f).
Étienne, A. and Heymès, R. (1948). *Bull. Soc. Chim. France*, 841.
Evans, F. and Lyle, R. E. (1963). *Chem. Ind.*, 986.
Evans, F. J., Lyle, G. G., Watkins, J. and Lyle, R. E. (1962). *J. Org. Chem.*, **27**, 1553.
Evans, F. J. and Lyle, R. E. (1960). *Chem. Ind.*, 497.
Ewins, A. J. (1911). *J. Chem. Soc.*, **99**, 270.
Ewins, A. J. and Laidlaw, P. P. (1910). *Proc. Chem. Soc.*, **26**, 343.
Fabbrica Italiana Sintetici S.p.A. (1968). Fr. Pat. 1,512,023 (*Chem. Abs.*, 1969, **70**, 106379r).
Farbenfabriken Bayer A.-G. (1954). Brit. Pat. 721,171 (*Chem. Abs.*, 1956, **50**, 2685e) [also as US Pat. 2,786,059 (*Chem. Abs.*, 1957, **51**, 8147b) and Ger. Pat. 930,444 (*Chem. Abs.*, 1958, **52**, 20208b)].
Farbenfabriken Bayer A.-G. (1955a). Brit. Pat. 733,123 (*Chem. Abs.*, 1956, **50**, 10799f).
Farbenfabriken Bayer A.-G. (1955b). Ger. Pat. 930,988 (*Chem. Abs.*, 1958, **52**, 17288g) (*Chem. Zentr.*, 1957, **128**, 999).
Farbenfabriken vorm. Friedr. Bayer and Co. (1901). Ger. Pat. 127,245 (*Chem. Zentr.* 1902, **I**, 154).
Farbenfabriken vorm. Friedr. Bayer and Co. (1902). Ger. Pat. 128,660 (*Chem. Zentr.*, 1902, **I**, 610).
Farber, E. (1963). In *Nobel Prize Winners in Chemistry, 1901–1961*, Abelard-Schuman, London, New York, Toronto, p. 5.
Farber, E. (1972). In *Dictionary of Scientific Biography*, ed. by Gillispie, C. C., Charles Scribner's Sons, New York, Vol. 5, p. 1.
Farbwerke vorm. Meister, Lucius und Brüning (1885). Ger. Pat. 37,727 (*Chem. Tech. Jahresbericht*, 1886, 464).
Farbwerke vorm. Meister, Lucius und Brüning (1886a). Ger. Pat. 38,784 (*Fortschritte der Theerfarbenfabrikation und verwandter Industriezweige*, 1877–1887, Springer, Berlin, 1888, p. 151) [*Chem. Zentr.*, 1887, **18** (Ser. 3), 617] (*Chem. Tech. Jahresbericht*, 1887, 732).
Farbwerke vorm. Meister, Lucius und Brüning (1886b). Eng. Pat. 7137 (Provisional Application) (*Fortschritte der Theerfarbenfabrikation und verwandter Industriezweige*, 1877–1887, Springer, Berlin, 1888, p. 154).
Farbwerke vorm. Meister, Lucius und Brüning (1886c). Fr. Pat. 176,701 (*Fortschritte der Theerfarbenfabrikation und verwandter Industriezweige*, 1877–1887, Springer, Berlin, 1888, p. 154).
Fargher, R. G. and Furness, R. (1915). *J. Chem. Soc.*, **57**, 688.
Farmaceutici Italia, S. A. (1954). Brit. Pat. 773,440 (*Chem. Abs.*, 1957, **51**, 12147d).
Fehlauer, A., Grosz, K.-P., Slopianka, M., Sucrow, W., Lockley, W. J. S. and Lwowski, W. (1976). *Chem. Ber.*, **109**, 253.
Fellows, E. J. (1965). US Pat. 3,216,898 (*Chem. Abs.*, 1966, **64**, 3489g).
Felton, D. G. I. (1952). *J. Chem. Soc.*, 1668.

Fennell, R. C. G. and Plant, S. G. P. (1932). *J. Chem. Soc.*, 2872.
Feofilaktov, V. V. (1947a). *J. Gen. Chem.* USSR, **17**, 993 (*Chem. Abs.*, 1948, **42**, 4537i).
Feofilaktov, V. V. (1947b). USSR Pat. 68,316 (*Chem. Abs.*, 1949, **43**, 5424f).
Feofilaktov, V. V. (1952). *Akad. Nauk. SSSR, Inst. Org. Khim., Sintezy Org. Soedin., Sbornik*, **2**, 103 (*Chem. Abs.*, 1954, **48**, 666g).
Feofilaktov, V. V. and Ivanova, T. N. (1955). *J. Gen. Chem. USSR*, **25**, 111.
Feofilaktov, V. V. and Semenova, N. K. (1952a). *Akad. Nauk SSSR, Inst. Org. Khim. Sintezy Org. Soedin., Sbornik*, **2**, 63 (*Chem. Abs.*, 1954, **48**, 666c).
Feofilaktov, V. V. and Semenova, N. K. (1952b). *Akad. Nauk SSSR, Inst. Org. Khim. Sintezy Org. Soedin., Sbornik*, **2**, 98 (*Chem. Abs.*, 1954, **48**, 668i).
Feofilaktov, V. V. and Semenova, N. K. (1953a). *J. Gen. Chem. USSR*, **23**, 463.
Feofilaktov, V. V. and Semenova, N. K. (1953b). *J. Gen. Chem. USSR*, **23**, 669.
Feofilaktov, V. V. and Semenova, N. K. (1953c). *J. Gen. Chem. USSR*, **23**, 889.
Fernandez Alvarez, E. and Monge Vega, A. (1975). Span. Pat. 400,436 (*Chem. Abs.*, 1975, **83**, 114205q).
Ferres, H., Hamdam, M. S. and Jackson, W. R. (1971). *J. Chem. Soc. (B)*, 1892.
F. Hoffmann-La Roche and Co., A.-G. (1963). Belg. Pat. 626,589 (*Chem. Abs.*, 1964, **60**, 10691f).
F. Hoffmann-La Roche and Co., A.-G. (1965). Neth. Pat. 6,506,929 (*Chem. Abs.*, 1966, **64**, 17603h).
Ficken, G. E. (1963). (quoted in Robinson, B., 1963a).
Ficken, G. E. and Kendall, J. D. (1959). *J. Chem. Soc.*, 3202.
Ficken, G. E. and Kendall, J. D. (1961). *J. Chem. Soc.*, 584.
Fieser, L. F. and Fieser, M. (1959). In *Steroids*, Reinhold Publishing Co., New York; Chapman and Hall Ltd., London, p. 205.
Findlay, S. P. and Dougherty, G. (1948). *J. Org. Chem.*, **13**, 560.
Finger, G. C., Gortatowski, M. J., Shiley, R. H. and White, R. H. (1959). *J. Amer. Chem. Soc.*, **81**, 94.
Finkelstein, J. and Lee, J. (1956). US Pat. 2,773,875 (*Chem. Abs.*, 1957, **51**, 6702d).
Finsterle, O. (1955). *Il Farmaco, Ed. Sci.*, **10**, 432 (*Chem. Abs.*, 1956, **50**, 4114i).
Fioshin, M. Y., Ginna, G. P. and Mamaev, V. P. (1956). *J. Gen. Chem. USSR*, **26**, 2585.
Firestone, R. A. and Sletzinger, M. (1970). Ger. Pat. 2,009,724 (*Chem. Abs.*, 1970, **73**, 109677j).
Fischer, Emil (1874). *Ber.*, **7**, 1211.
Fischer, Emil (1875a). *Ber.*, **8**, 589.
Fischer, Emil (1875b). *Ber.*, **8**, 1005.
Fischer, Emil (1875c). *Ber.*, **8**, 1641.
Fischer, Emil (1876a). *Ber.*, **9**, 880.
Fischer, Emil (1876b). *Ber.*, **9**, 1840.
Fischer, Emil (1877). *Ber.*, **10**, 1331.
Fischer, Emil (1878). *Liebig's Ann. Chem.*, **190**, 67.
Fischer, Emil (1884). *Ber.*, **17**, 572
Fischer, Emil (1886a). *Liebig's Ann. Chem.*, **232**, 236.
Fischer, Emil (1886b). *Liebig's Ann. Chem.*, **236**, 116.
Fischer, Emil (1886c). *Liebig's Ann. Chem.*, **236**, 126.
Fischer, Emil (1886d). *Liebig's Ann. Chem.*, **236**, 198.
Fischer, Emil (1886e). *Ber.*, **19**, 1563.
Fischer, Emil (1887). *Liebig's Ann. Chem.*, **239**, 220.
Fischer, Emil (1888). *Liebig's Ann. Chem.*, **242**, 372.
Fischer, Emil (1889a). In *Exercises in the Preparation of Organic Compounds*, translated by Kling, A., Hodge and Co., Glasgow; Williams and Norgate, London and Edinburgh, p. 21.

Fischer, Emil (1889b). In *Exercises in the Preparation of Organic Compounds*, translated by Kling, A., Hodge and Co., Glasgow; Williams and Norgate, London and Edinburgh, p. 59.
Fischer, Emil (1893). *Ber.*, **26**, 92.
Fischer, Emil (1894). *Amer. Chem. J.*, **16**, 159.
Fischer, Emil (1896). *Ber.*, **29**, 793.
Fischer, Emil (1902). Nobel Lecture, in *Nobel Lectures – Chemistry – 1901–1921*, published for the Nobel Foundation by Elsevier Publishing Co., Amsterdam, London and New York, 1966, p. 21.
Fischer, Emil (1907). In *Synthetical Chemistry in its Relation to Biology*, Faraday Lecture, 18 October, in Lectures delivered before the Chemical Society – Faraday Lectures, 1869–1928. The Chemical Society, London.
Fischer, Emil (1922). In *Aus Meinem Leben*, Julius Springer, Berlin.
Fischer, Emil and Ach, F. (1889). *Liebig's Ann. Chem.*, **253**, 57.
Fischer, Emil and Besthorn, E. (1882). *Liebig's Ann. Chem.*, **212**, 316.
Fischer, Emil and Hess, O. (1884). *Ber.*, **17**, 559.
Fischer, Emil and Jourdan, F. (1883). *Ber.*, **16**, 2241.
Fischer, Emil and Knoevenagel, O. (1887). *Liebig's Ann. Chem.*, **239**, 194.
Fischer, Emil and Schmidt, T. (1888). *Ber.*, **21**, 1811.
Fischer, Emil and Schmitt, T. (1888). *Ber.*, **21**, 1071.
Fisnerova, L. and Nemecek, O. (1976). Czech. Pat. 162,230 (*Chem. Abs.*, 1977, **86**, 106373q).
Fitzpatrick, J. T. and Hiser, R. D. (1957). *J. Org. Chem.*, **22**, 1703.
Fliedner, L. J. (1979). Eur. Pat. Appl. 4,342 (*Chem. Abs.*, 1980, **92**, 128717f).
Fodor, G. V. and Szarvas, P. (1943). *Ber.*, **76**, 334.
Forbes, E. J., Stacey, M., Tatlow, J. C. and Wragg, R. T. (1960). *Tetrahedron*, **8**, 67.
Forbes, E. J., Tatlow, J. C. and Wragg, R. T. (1960). *Tetrahedron*, **8**, 73.
Forrest, T. P. and Chen, F. M. F. (1972). *J. Chem. Soc., Chem. Communs.*, 1067.
Forster, M. O. (1920). *J. Chem. Soc.*, **117**, 1157.
Forster, M. O. (1933). Emil Fischer Memorial Lecture in *Memorial Lectures Delivered Before the Chemical Society*, The Chemical Society, Burlington House, Piccadilly, London, W.1.
Forster, M. O. (1921–1922). *Proc. Roy. Soc.*, **98A**, 1.
Fowkes, F. S. and McClelland, E. W. (1941). *J. Chem. Soc.*, 187.
Fox, S. W. and Bullock, M. W. (1951a). *J. Amer. Chem. Soc.*, **73**, 2754.
Fox, S. W. and Bullock, M. W. (1951b). *J. Amer. Chem. Soc.*, **73**, 2756.
Fox, S. W. and Bullock, M. W. (1955a). US Pat. 2,701,250 (*Chem. Abs.*, 1956, **50**, 1922c).
Fox, S. W. and Bullock, M. W. (1955b). US Pat. 2,701,251 (*Chem. Abs.*, 1956, **50**, 1922a).
Francia Barra, E. and Carmelo Marin Moga, A. (1979). Span. Pat. 471,436 (*Chem. Abs.*, 1980, **92**, 110849r).
Franke, A. (1899). *Monat. Chem.*, **20**, 847.
Franzen, H., Onsager, A. and Faerden, G. (1918). *J. Prakt. Chem.*, **97**, 336.
Franzen, H. and Schmidt, M. (1917). *J. Prakt. Chem.*, **96**, 1.
Frasca, A. R. (1962a). *Anales Asoc. Quim. Argent.*, **50**, 162 (*Chem. Abs.*, 1963, **59**, 12743g).
Frasca, A. R. (1962b). *Tetrahedron Letts.*, 1115.
Fravolini, A., Croisy, A. and Jacquignon, P. (1976). *Il Farmaco (Pavia), Ed. Sci.*, **31**, 418.
Freed, M. E., Hertz, E. and Rice, L. M. (1964). *J. Med. Chem.*, **7**, 628.
Freedman, J. and Judd, C. I. (1972). US Pat. 3,637,747 (*Chem. Abs.*, 1972, **76**, 113059p).
Freer, P. C. (1892). *J. Prakt. Chem.*, **45**, 414.
Freer, P. C. (1893). *J. Prakt. Chem.*, **47**, 236.
Freer, P. C. (1894). *Liebig's Ann. Chem.*, **283**, 380.
Freer, P. C. (1899). *Amer. Chem. J.*, **21**, 14.

Freudenberg, W. (1952). In *Heterocyclic Compounds*, ed. by Elderfield, R. C., Wiley, New York, Vol. 3, Chapter 3.
Frey, A. J., Ott, H., Bruderer, H. and Stadler, P. (1960). *Chimia*, **14**, 423.
Frey, K. W. and Hofmann, R. (1901). *Monat. Chem.*, **22**, 760.
Friedländer, P. (1916). *Chem. Ztg.*, **40**, 918.
Friedländer, P. (1921). *Ber.*, **54B**, 620.
Fritz, H. and Fischer, O. (1964). *Tetrahedron*, **20**, 1737.
Fritz, H. and Losacker, P. (1967). *Liebig's Ann. Chem.*, **709**, 135.
Fritz, H., Losacker, P., Stock, E. and Gerber, H.-G. (1971). *Liebig's Ann. Chem.*, **749**, 168.
Fritz, H. and Pfaender, P. (1965). *Chem. Ber.*, **98**, 989.
Fritz, H. and Rubach, G. (1968). *Liebig's Ann. Chem.*, **715**, 135.
Fritz, H. and Schenk, S. (1972). *Liebig's Ann. Chem.*, **762**, 121.
Fritz, H. and Stock, E. (1969). *Liebig's Ann. Chem.*, **721**, 82.
Fritz, H. and Stock, E. (1970). *Tetrahedron*, **26**, 5821.
Fritz, H. and Uhrhan, P. (1971a). *Liebig's Ann. Chem.*, **744**, 81.
Fritz, H. and Uhrhan, P. (1971b). *Tetrahedron Letts.*, 4183.
Fritz, H. and Uhrhan, P. (1972). *Liebig's Ann. Chem.*, **763**, 198.
Fuchs, W. and Niszel, F. (1927). *Ber.*, **60**, 209.
Fugger, J., Tien, J. M. and Hunsberger, I. M. (1955). *J. Amer. Chem. Soc.*, **77**, 1843.
Fujimori, H. and Yamane, K. (1978). *Bull. Chem. Soc. Japan.*, **51**, 3579.
Fujisawa Pharmaceutical Co., Ltd (1965a). Jap. Pat. 6,240 (*Chem. Abs.*, 1965, **63**, 1789e).
Fujisawa Pharmaceutical Co., Ltd (1965b). Jap. Pat. 19,336 ('65) (*Chem. Abs.*, 1965, **63**, 16309f).
Fukui, K. and Fujimoto, H. (1966). *Tetrahedron Letts.*, 251.
Fusco, R. (1967). In *The Chemistry of Heterocyclic Compounds*, ed. by Wiley, R. H., Wiley-Interscience, New York, London and Sydney, Vol. 22, Part 1, Chapter 3, p. 10.
Fusco, R. (1978). *Chim. Ind.*, **60**, 903.
Fusco, R. and Sannicolò, F. (1973). *Gazz. Chim. Ital.*, **103**, 197.
Fusco, R. and Sannicolò, F. (1974). *Gazz. Chim. Ital.*, **104**, 813.
Fusco, R. and Sannicolò, F. (1975a). *Gazz. Chim. Ital.*, **105**, 465.
Fusco, R. and Sannicolò, F. (1975b). *Tetrahedron Letts.*, 3351.
Fusco, R. and Sannicolò, F. (1975c). *Gazz. Chim. Ital.*, **105**, 1105.
Fusco, R. and Sannicolò, F. (1976a). *Gazz. Chim. Ital.*, **106**, 85.
Fusco, R. and Sannicolò, F. (1976b). *Tetrahedron Letts.*, 3991.
Fusco, R. and Sannicolò, F. (1977). *Tetrahedron Letts.*, 3163.
Fusco, R. and Sannicolò, F. (1978a). *Tetrahedron Letts.*, 1233.
Fusco, R. and Sannicolò, F. (1978b). *Tetrahedron Letts.*, 4827.
Fusco, R. and Sannicolò, F. (1978c). *Chem. Het. Comps.*, **14**, 157.
Fusco, R. and Sannicolò, F. (1978d). *Chem. Het. Comps.*, **14**, 413.
Fusco, R. and Sannicolò, F. (1980). *Tetrahedron*, **36**, 161.
Gabriel, S. and Pinkus, G. (1893). *Ber.*, **26**, 2197.
Gabrielyan, G. E. and Papayan, G. L. (1973). *Arm. Khim. Zh.*, **26**, 768 (*Chem. Abs.*, 1974, **80**, 70638q).
Gadaginamath, G. S. (1976). *Curr. Sci.*, **45**, 507.
Gadaginamath, G. S. and Siddappa, S. (1975). *Indian J. Chem.*, **13**, 1251.
Gaimster, K. (1960). Brit. Pat. 841,524 (*Chem. Abs.*, 1962, **56**, 1431g).
Gaines, W. A., Sletzinger, M. and Ruyle, W. (1961). US Pat. 3,014,043 (*Chem. Abs.*, 1962, **56**, 15486e).
Gal'bershtam, M. A. and Samoilova, N. P. (1973). *Chem. Het. Comps.*, **9**, 1098.
Gale, D. J., Lin, J. and Wilshire, J. F. K. (1976). *Aust. J. Chem.*, **29**, 2747.
Gale, D. J. and Wilshire, J. F. K. (1973). *Aust. J. Chem.*, **26**, 2683.

Gale, D. J. and Wilshire, J. F. K. (1974). *Aust. J. Chem.*, **27**, 1295.
Gallagher, M. J. and Mann, F. G. (1962). *J. Chem. Soc.*, 5110.
Gallagher, M. J. and Mann, F. G. (1963). *J. Chem. Soc.*, 4855.
Gannon, W. F., Benigni, J. D., Dickson, D. E. and Minnis, R. L. (1969). *J. Org. Chem.*, **34**, 3002.
Gardner, D. V., McOmie, J. F. W. and Prabhu, T. P. (1970). *J. Chem. Soc.* (*C*), 2500.
Gardner, J. H. and Stevens, J. R. (1947). *J. Amer. Chem. Soc.*, **69**, 3086.
Gassman, P. G. (1975a). US Pat. 3,901,899 (*Chem. Abs.*, 1975, **83**, 193079c).
Gassman, P. G. (1975b). US Pat. 3,897,451 (*Chem. Abs.*, 1975, **83**, 193077a).
Gassman, P. G. (1976). US Pat. 3,992,392 (*Chem. Abs.*, 1977, **86**, 139849c).
Gassman, P. G. and Bergen, T. J. van (1973a). *J. Amer. Chem. Soc.*, **95**, 590.
Gassman, P. G. and Bergen, T. J. van (1973b). *J. Amer. Chem. Soc.*, **95**, 591.
Gassman, P. G. and Bergen, T. J. van (1973c). *J. Amer. Chem. Soc.*, **95**, 2718.
Gassman, P. G. and Bergen, T. J. van (1974). *J. Amer. Chem. Soc.*, **96**, 5508.
Gassman, P. G., Bergen, T. J. van, Gilbert, D. P. and Cue, B. W. (1974). *J. Amer. Chem. Soc.*, **96**, 5495.
Gassman, P. G., Bergen, T. J. van and Gruetzmacher, G. D. (1973). *J. Amer. Chem. Soc.*, **95**, 6508.
Gassman, P. G., Gilbert, D. P. and Bergen, T. J. van (1974). *J. Chem. Soc., Chem. Communs.*, 201.
Gassman, P. G. and Gruetzmacher, G. D. (1976). US Pat. 3,960,926 (*Chem. Abs.*, 1976, **85**, 142979e).
Gassman, P. G., Gruetzmacher, G. and Bergen, T. J. van (1974). *J. Amer. Chem. Soc.*, **96**, 5512.
Gaudion, W. J., Hook, W. H. and Plant, S. G. P. (1947). *J. Chem. Soc.*, 1631.
Geigy, A.-G., J. R. (1966). Neth. Pat. 6,515,701 (*Chem. Abs.*, 1966, **65**, 13714b).
Geigy, A.-G., J. R. (1968). Fr. Pat. 1,524,830 (*Chem. Abs.*, 1970, **72**, 21675a).
Geller, B. A. (1978). *Russian Chem. Revs.*, **47**, 297.
Geller, B. A. and Skrunts, L. K. (1964). *J. Gen. Chem. USSR*, **34**, 663.
General Aniline Works, Inc. (1932). US Pat. 1,866,956 (*Chem. Zentr.*, 1932, **II**, 1977).
Genkina, N. K., Gordeev, E. N. and Suvorov, N. N. (1976). *J. Org. Chem. USSR*, **12**, 1446.
Georgian, V. (1957). *Chem. Ind.*, 1124.
Georgian, V. (1958). US Pat. 2,858,314 (*Chem. Abs.*, 1959, **53**, 10258e).
Georgian, V. (1962). US Pat. 3,015,661 (*Chem. Abs.*, 1962, **57**, 4636i).
Germain, C. and Bourdais, J. (1976). *J. Het. Chem.*, **13**, 1209.
Gerszberg, S., Cueva, P. and Frasca, A. R. (1972). *Anales Asoc. Quim. Argent.*, **60**, 331 (*Chem. Abs.*, 1973, **78**, 42775x).
Gesellschaft für Teerverwertung m.b.H. (1911). Ger. Pat. 238,138 (*Chem. Zentr.*, 1911, **15**, ii, 1080) (*Chem. Abs.*, 1912, **6**, 1659).
Ghigi, E. (1930). *Gazz. Chim. Ital.*, **60**, 194.
Ghigi, E. (1931). *Gazz. Chim. Ital.*, **61**, 43.
Ghigi, E. (1933a). *Gazz. Chim. Ital.*, **63**, 698.
Ghigi, E. (1933b). *Gazz. Chim. Ital.*, **63**, 701.
Gilbert, B. (1965). In *The Alkaloids*, ed. by Manske, R. H. F., Academic Press, New York and London, Vol. 8, p. 335.
Gilbert, B. (1968). In *The Alkaloids*, ed. by Manske, R. H. F., Academic Press, New York and London, Vol. 11, p. 222.
Gilbert, J., Rousselle, D., Gansser, C. and Viel, C. (1979). *J. Het. Chem.*, **16**, 7.
GilChrist, T. L. and Storr, R. C. (1972), In *Organic Reactions and Orbital Symmetry*, Cambridge University Press, p. 204.
Giuliano, R. and Leonardi, G. (1957). *Ricerca Sci.*, **27**, 1843 (*Chem. Abs.*, 1959, **53**, 18937i).

Giuliano, R. and Stein, M. L. (1957). *Atti Accad. Nazl. Lincei, Rend. Classe sci. fis. mat. e nat.*, **22**, 626 (*Chem. Abs.*, 1958, **52**, 2835a).
Glamkowski, E. J., Gal, G. and Sletzinger, M. (1973). *J. Med. Chem.*, **16**, 176.
Glish, G. L. and Cooks, R. G. (1978). *J. Amer. Chem. Soc.*, **100**, 6720.
Glover, E. E. and Jones, G. (1958). *J. Chem. Soc.*, 1750.
Glushkov, R. G., Smirnova, V. G., Zasosova, I. M. and Ovcharova, I. M. (1975). *Chem. Het. Comps.*, **11**, 696.
Glushkov, R. G., Volskova, V. A., Smirnova, V. G. and Magidson, O. Y. (1969). *Proc. Acad. Sci. USSR*, **187**, 532.
Glushkov, R. G., Volskova, V. A., Smirnova, V. G. and Magidson, O. Y. (1972). USSR Pat. 332,084 (*Chem. Abs.*, 1972, **77**, 140013w).
Glushkov, R. G., Yakhontov, L. N., Pronina, E. V. and Magidson, O. Y. (1969). *Chem. Het. Comps.*, **5**, 421.
Glushkov, R. G., Zasosova, I. M. and Ovcharova, I. M. (1977). *Chem. Het. Comps.* **13**, 1122.
Glushkov, R. G., Zasosova, I. M., Ovcharova, I. M., Rudzit, E. A., Saratikov, A. S., Livshits, N. S. and Kostyuchenko, N. P. (1978). *Khim. Farm. Zh.*, **12**, 48 (*Chem. Abs.*, 1978, **89**, 43178u).
Glushkov, R. G., Zasosova, I. M., Ovcharova, I. M., Solov'eva, N. P., Anisimova, O. S. and Sheinker, Y. N. (1978). *Chem. Het. Comps.*, **14**, 1223.
Goel, O. P. (1966). Ph.D. Thesis, Carnegie Institute of Technology, USA.
Goldsmith, E. A. and Lindwall, H. G. (1953). *J. Org. Chem.*, **18**, 507.
Golubev, V. E. and Suvorov, N. N. (1970). *Chem. Het. Comps.*, **6**, 701.
Golubeva, G. A., Portnov, Y. N. and Kost, A. N. (1973). *Chem. Het. Comps.*, **9**, 471.
Gorbunova, V. P. and Suvorov, N. N. (1973). *Chem. Het. Comps.*, **9**, 1374.
Gordeev, E. N., Kobets, N. S. and Suvorov, N. N. (1975). *Tr. Mosk. Khim. - Tekhnol. Inst.*, **86**, 38 (*Chem. Abs.*, 1976, **85**, 142945r).
Gordian, M. B. and Gal'bershtam, M. A. (1971). *Anilino-Krasoch. Prom.*, [3], 11 (*Chem. Abs.*, 1973, **78**, 16006c).
Gordon, W. G. and Jackson, R. W. (1935). *J. Biol. Chem.*, **110**, 151.
Gore, P. H., Hughes, G. K. and Ritchie, E. (1949). *Nature*, **164**, 835.
Gore, P. H., Hughes, G. K. and Ritchie, E. (1950). *J. Proc. Roy. Soc. NSW*, **84**, 59.
Gould, W. A. and Larsen, A. A. (1967). US Pat. 3,297,717 (*Chem. Abs.*, 1967, **67**, 99994v).
Goutarel, R., Janot, M.-M., Le Hir, A., Corrodi, H. and Prelog. V. (1954). *Helv. Chim. Acta*, **37**, 1805.
Govindachari, T. R., Rajappa, S. and Sudarsanam, V. (1961). *Tetrahedron*, **16**, 1.
Govindachari, T. R., Rajappa, S. and Sudarsanam, V. (1963). *Indian J. Chem.*, **1**, 247.
Govindachari, T. R., Rajappa, S. and Sudarsanam, V. (1966). *Indian J. Chem.*, **4**, 118.
Govindachari, T. R. and Sudarsanam, V. (1967). *Indian J. Chem.*, **5**, 16.
Govindachari, T. R. and Sudarsanam, V. (1971). *Indian J. Chem.*, **9**, 402.
Graebe, C. and Glaser, C. (1872). *Liebig's Ann. Chem. Pharm.*, **163**, 343.
Grammaticakis, P. (1937). *Compt. Rend. Acad. Sci.*, **204**, 502.
Grammaticakis, P. (1939). *Compt. Rend. Acad. Sci.*, **209**, 317.
Grammaticakis, P. (1940a). *Compt. Rend. Acad. Sci.*, **210**, 303.
Grammaticakis, P. (1940b). *Compt. Rend. Acad. Sci.*, **210**, 569.
Grammaticakis, P. (1946). *Compt. Rend. Acad. Sci.*, **223**, 804.
Grammaticakis, P. (1947). *Bull. Soc. Chim. France*, **14**, 438.
Grammaticakis, P. (1960). *Compt. Rend. Acad. Sci.*, **251**, 2728.
Grandberg, I. I. (1966). Paper delivered to the General Meeting of the Division of General and Technical Chemistry of the Academy of Sciences of the USSR, 31 May (*Bull. Acad. Sci. USSR, Div. Chem. Sci.*, 1633).

Grandberg, I. I. (1972). *Izv. Timiryazev Sel'skokhoz. Akad.*, [5], 188 (*Chem. Abs.*, 1973, **78**, 84125r).
Grandberg, I. I. (1974). *Chem. Het. Comps.*, **10**, 501.
Grandberg, I. I. and Afonina, N. I. (1969). USSR Pat. 239,341 (*Chem. Abs.*, 1969, **71**, 49765h).
Grandberg, I. I., Afonina, N. I. and Zuyanova, T. I. (1967). USSR Pat. 201,412 (*Chem. Abs.*, 1968, **69**, 19021d).
Grandberg, I. I., Afonina, N. I. and Zuyanova, T. I. (1968). *Chem. Het. Comps.*, **4**, 753.
Grandberg, I. I., Belyaeva, L. D. and Dmitriev, L. B. (1971a). *Chem. Het. Comps.*, **7**, 54.
Grandberg, I. I., Belyaeva, L. D. and Dmitriev, L. B. (1971b). *Chem. Het. Comps.*, **7**, 1131.
Grandberg, I. I., Belyaeva, L. D. and Dmitriev, L. B. (1971c). *Dokl. Timiryazev Sel'skokhoz. Akad.*, [162], 398 (*Chem. Abs.*, 1971, **75**, 140618f).
Grandberg, I. I., Belyaeva, L. D. and Dmitriev, L. B. (1973). *Chem. Het. Comps.*, **9**, 31.
Grandberg, I. I. and Bobrova, N. I. (1973). *Chem. Het. Comps.*, **9**, 196.
Grandberg, I. I. and Bobrova, N. I. (1974a). USSR Pat. 433,145 (*Chem. Abs.*, 1974, **81**, 91348f).
Grandberg, I. I. and Bobrova, N. I. (1974b). *Chem. Het. Comps.*, **10**, 943.
Grandberg, I. I. and Bobrova, N. I. (1974c). *Izv. Timiryazev Sel'skokhoz. Akad.*, [6], 198 (*Chem. Abs.*, 1975, **82**, 30905j).
Grandberg, I. I. and Bobrova, N. I. (1974d). USSR Pat. 451,696 (*Chem. Abs.*, 1975, **82**, 72782x).
Grandberg, I. I. and Bobrova, N. I. (1978). *Izv. Timiryazev Sel'skokhoz. Akad.*, [6], 220 (*Chem. Abs.*, 1979, **90**, 54766f).
Grandberg, I. I., Bobrova, N. I. and Zuyanova, T. I. (1972). *Sin. Geter. Soedin.*, [9], 18 (*Chem. Abs.*, 1973, **79**, 146321h).
Grandberg, I. I. and Dashkevich, S. N. (1970a). *Chem. Het. Comps.*, **6**, 1522.
Grandberg, I. I. and Dashkevich, S. N. (1970b). USSR Pat. 276,062 (*Chem. Abs.*, 1971, **74**, 100017g).
Grandberg, I. I. and Dashkevich, S. N. (1971a). *Chem. Het. Comps.*, **7**, 316.
Grandberg, I. I. and Dashkevich, S. N. (1971b). *Chem. Het. Comps.*, **7**, 729.
Grandberg, I. I., Dashkevich, S. N., Markaryan, E. A. and Nazaryan, V. A. (1975). *Vsb. Sintezy Geter. Soedin. Erevan*, [10], 62 (*Chem. Abs.*, 1976, **84**, 135520z).
Grandberg, I. I. and Golubeva, G. A. (1963). *J. Gen. Chem. USSR*, **33**, 237.
Grandberg, I. I. and Ivanova, T. A. (1967). USSR Pat. 201,411 (*Chem. Abs.*, 1968, **69**, 27584e).
Grandberg, I. I. and Ivanova, T. A. (1970a). *Chem. Het. Comps.*, **6**, 444.
Grandberg, I. I. and Ivanova, T. A. (1970b). *Chem. Het. Comps.*, **6**, 872.
Grandberg, I. I. and Ivanova, T. A. (1970c). *Chem. Het. Comps.*, **6**, 1388.
Grandberg, I. I. and Ivanova, T. A. (1970d). Dokl. *Timiryazev Sel'skokhoz. Akad.*, 1970, [160], 232 (*Chem. Abs.*, 1971, **75**, 20732s).
Grandberg, I. I. and Ivanova, T. A. (1975). *Sin. Geter. Soedin.*, **10**, 31 (*Chem. Abs.*, 1976, **85**, 94582w).
Grandberg, I. I. and Kost, A. N. (1959). *J. Gen. Chem. USSR*, **29**, 650.
Grandberg, I. I. and Kost, A. N. (1966). *Izv. Timiryazev Sel'skokhoz. Akad.*, [5], 210 (*Chem. Abs.*, 1967, **66**, 55415g).
Grandberg, I. I., Kost, A. N. and Naumov, Y. A. (1963). *Proc. Acad. Sci. USSR*, **149**, 278.
Grandberg, I. I., Kost, A. N. and Terentyev, A. P. (1956). *J. Gen. Chem. USSR*, **26**, 3839.
Grandberg, I. I., Kost, A. N. and Terentyev, A. P. (1957). *J. Gen. Chem. USSR*, **27**, 3378.
Grandberg, I. I., Kost, A. N. and Yaguzhinskii, L. S. (1959). *J. Gen. Chem. USSR*, **29**, 2499.
Grandberg, I. I., Kost, A. N. and Yaguzhinskii, L. S. (1960). *J. Gen. Chem. USSR*, **30**, 3082.
Grandberg, I. I. and Moskvina, T. P. (1970). *Chem. Het. Comps.*, **6**, 875.

Grandberg, I. I. and Moskvina, T. P. (1971). *Dokl. Timiryazev Sel'skokhoz. Akad.*, [162], 380 (*Chem. Abs.*, 1971, **75**, 76512n).
Grandberg, I. I. and Moskvina, T. P. (1972). *Chem. Het. Comps.*, **8**, 1235.
Grandberg, I. I. and Moskvina, T. P. (1973). *Izv. Timiryazev Sel'skokhoz. Akad.*, [4], 167 (*Chem. Abs.*, 1973, **79**, 91889s).
Grandberg, I. I. and Moskvina, T. P. (1974a). *Chem. Het. Comps.*, **10**, 78.
Grandberg, I. I. and Moskvina, T. P. (1974b). USSR Pat. 431,164 (*Chem. Abs.*, 1974, **81**, 91344b).
Grandberg, I. I., Naumov, Y. A. and Kost, A. N. (1965). *J. Org. Chem. USSR*, **1**, 809.
Grandberg, I. I. and Nikitina, S. B. (1971a). *Chem. Het. Comps.*, **7**, 50.
Grandberg, I. I. and Nikitina, S. B. (1971b). *Chem. Het. Comps.*, **7**, 1128.
Grandberg, I. I. and Nikitina, S. B. (1971c). *Dokl. Timiryazev Sel'skokhoz. Akad.*, [162], 374 (*Chem. Abs.*, 1971, **75**, 76503k).
Grandberg, I. I. and Nikitina, S. B. (1972). *Chem. Het. Comps.*, **8**, 1099.
Grandberg, I. I., Nikitina, S. B. and Yaryshev, N. G. (1970). *Izv. Timiryazev Sel'skokhoz. Akad.*, [2], 196 (*Chem. Abs.*, 1970, **73**, 87721q).
Grandberg, I. I. and Potanova, A. V. (1962). *J. Gen. Chem. USSR*, **32**, 644.
Grandberg, I. I. and Przheval'skii, N. M. (1969). *Chem. Het. Comps.*, **5**, 704.
Grandberg, I. I. and Przheval'skii, N. M. (1970). *Chem. Het. Comps.*, **6**, 1189.
Grandberg, I. I. and Przheval'skii, N. M. (1972). *Izv. Timiryazev Sel'skokhoz. Akad.*, [2], 192 (*Chem. Abs.*, 1972, **76**, 139696u).
Grandberg, I. I. and Przheval'skii, N. M. (1974). *Izv. Timiryazev Sel'skokhoz. Akad.*, [2], 177 (*Chem. Abs.*, 1974, **80**, 145944x).
Grandberg, I. I., Przheval'skii, N. M. and Ivanova, T. A. (1969). USSR Pat. 248,689 (*Chem. Abs.*, 1970, **72**, 90288p).
Grandberg, I. I., Przheval'skii, N. M., Ivanova, T. A., Zuyanova, T. I., Bobrova, N. I., Dashkevich, S. I., Nikitina, S. B., Shcherbina, T. P. and Yaryshev, N. G. (1969). *Izv. Timiryazev Sel'skokhoz. Akad.*, [5], 208 (*Chem. Abs.*, 1970, **72**, 78794t).
Grandberg, I. I., Przheval'skii, N. M. and Vysotskii, V. I. (1970). *Chem. Het. Comps.*, **6**, 1398.
Grandberg, I. I., Przheval'skii, N. M., Vysotskii, V. I. and Khmel'nitskii, R. A. (1970). *Chem. Het. Comps.*, **6**, 441.
Grandberg, I. I. and Sibiryakova, D. V. (1967). USSR Pat. 199,892 (*Chem. Abs.*, 1968, **68**, 95677s).
Grandberg, I. I., Sibiryakova, D. V. and Brovkin, L. V. (1969). *Chem. Het. Comps.*, **5**, 75.
Grandberg, I. I. and Sorokin, V. I. (1973). *Chem. Het. Comps.*, **9**, 26.
Grandberg, I. I. and Sorokin, V. I. (1974). *Russian Chem. Revs.*, **43**, 115.
Grandberg, I. I. and Tokmakov, G. P. (1973). *Tezisy Dokl.-Simp. Khim. Tekhnol. Geter. Soedin. Goryuch. Iskop., 2nd.*, 53 (*Chem. Abs.*, 1977, **86**, 16493z).
Grandberg, I. I. and Tokmakov, G. P. (1974a). *Chem. Het. Comps.*, **10**, 179.
Grandberg, I. I. and Tokmakov, G. P. (1974b). USSR Pat. 445,659 (*Chem. Abs.*, 1975, **82**, 57561f).
Grandberg, I. I. and Tokmakov, G. P. (1974c). USSR Pat. 445,660 (*Chem. Abs.*, 1975, **82**, 57560e).
Grandberg, I. I. and Tokmakov, G. P. (1974d). *Chem. Het. Comps.*, **10**, 941.
Grandberg, I. I. and Tokmakov, G. P. (1975a). *Chem. Het. Comps.*, **11**, 176.
Grandberg, I. I. and Tokmakov, G. P. (1975b). *V. sb., Khimiya i Farmakol. Indol'n Soedin.*, 52 (*Chem. Abs.*, 1976, **84**, 121689z).
Grandberg, I. I. and Tokmakov, G. P. (1975c). *Khimiya I Farmacol. Indol'n Soedin.*, 40 (*Chem. Abs.*, 1976, **84**, 90067p).
Grandberg, I. I. and Tokmakov, G. P. (1976). USSR Pat. 502,883 (*Chem. Abs.*, 1976, **84**, 180191d).

Grandberg, I. I., Wei-pi, D. and Kost, A. N. (1961). *J. Gen. Chem. USSR*, **31**, 2153.
Grandberg, I. I. and Yaryshev, N. G. (1968). USSR Pat. 221,709 (*Chem. Abs.*, 1969, **70**, 4115j).
Grandberg, I. I. and Yaryshev, N. G. (1969). USSR Pat. 241,441 (*Chem. Abs.*, 1969, **71**, 81325a).
Grandberg, I. I. and Yaryshev, N. G. (1972a). *Chem. Het. Comps.*, **8**, 966.
Grandberg, I. I. and Yaryshev, N. G. (1972b). *Chem. Het. Comps.*, **8**, 1087.
Grandberg, I. I. and Zuyanova, T. I. (1967). USSR Pat. 196,852 (*Chem. Abs.*, 1968, **68**, 29592j).
Grandberg, I. I. and Zuyanova, T. I. (1968). *Chem. Het. Comps.*, **4**, 632.
Grandberg, I. I. and Zuyanova, T. I. (1970). *Chem. Het. Comps.*, **6**, 1394.
Grandberg, I. I. and Zuyanova, T. I. (1971). *Chem. Het. Comps.*, **7**, 47.
Grandberg, I. I., Zuyanova, T. I., Afonina, N. I. and Ivanova, T. A. (1967). *Proc. Acad. Sci. USSR*, **176**, 828.
Grandberg, I. I., Zuyanova, T. I., Afonia, N. I. and Ivanova, T. A. (1969). USSR Pat. 242,901 (*Chem. Abs.*, 1969, **71**, 70487d).
Grandberg, I. I., Zuyanova, T. I. and Bobrova, N. I. (1967). USSR Pat. 192,818 (*Chem. Abs.*, 1968, **69**, 27788z).
Grandberg, I. I., Zuyanova, T. I., Przheval'skii, N. M. and Minkin, V. I. (1970). *Chem. Het. Comps.*, **6**, 693.
Grandberg, I. I. Zuyanova, T. I. and Zhigulev, K. K. (1972). *Chem. Het. Comps.*, **8**, 1092.
Grashey, R. and Adelsberger, K. (1962). *Angnew. Chem. (Intern. Edit.)*, **1**, 267.
Gray, A. P. and Archer, W. L. (1957). *J. Amer. Chem. Soc.*, **79**, 3554.
Green, K. H. B. and Ritchie, E. (1949). *J. Proc. Roy. Soc. NSW*, **83**, 120.
Grgin, D. J. (1906). *Monat. Chem.*, **27**, 731.
Grieder, A. (1970). Thesis, University of Basel, Switzerland.
Grieder, A. and Schiess, P. (1970). *Chimia*, **24**, 25.
Grinev, N. I., Sadovskaya, V. L. and Ufimtsev, V. N. (1963). *J. Gen. Chem. USSR*, **33**, 545.
Grotta, H. M., Riggle, C. J. and Bearse, A. E. (1961). *J. Org. Chem.*, **26**, 1509.
Groves, L. H. and Swan, G. A. (1952). *J. Chem. Soc.*, 650.
Grudziński, S. and Kotelko, A. (1955). *Acta Polon. Pharm.*, **12**, 201 (*Chem. Abs.*, 1956, **50**, 12978f).
Grundon, M. F. and Scott, M. D. (1964). *J. Chem. Soc.*, 5674.
Gryaznov, A. P., Akhvlediani, R. N., Volodina, T. A., Vasil'ev, A. M., Babushkina, T. A. and Suvorov, N. N. (1977). *Chem. Het. Comps.*, **13**, 298.
Gupta, D. R. and Ojha, A. C. (1971). *J. Indian Chem. Soc.*, **48**, 295.
Gupta, S. P., Jetley, U. K., Rani, S. and Malik, O. P. (1975). *Acta Cienc. Indica*, **1**, 290 (*Chem. Abs.*, 1976, **85**, 21006b).
Gurien, H. and Teitel, S. (1979). US Pat. 4,158,007 (*Chem. Abs.*, 1979, **91**, 107889c).
Gurien, H. and Teitel, S. (1980). Eur. Pat. Appl. 8,446 (*Chem. Abs.*, 1980, **93**, 46417q).
Gurowitz, W. D. and Joseph, M. A. (1965). *Tetrahedron Letts.*, 4433.
Gurowitz, W. D. and Joseph, M. A. (1967). *J. Org. Chem.*, **32**, 3289.
Hadáček, J. and Švehla, J. (1954). *Publs. Fac. Sci. Univ. Masaryk Číslo*, **357**, 257 (*Chem. Abs.*, 1955, **49**, 13978g).
Hahn, W. E. and Bartnik, R. (1972). *Soc. Sci. Lodz., Acta Chim.*, **17**, 181 (*Chem. Abs.*, 1973, **78**, 43184r).
Hahn, W. E., Bartnik, R. and Zawadzka, H. (1966). *Lodz. Tow. Nauk Wydz. III, Acta Chim.*, **11**, 83 (*Chem. Abs.*, 1967, **66**, 75925b).
Hahn, W. E., Bartnik, R., Zawadzka, H. and Nowaczyk, W. (1968). *Lodz. Tow. Nauk Wydz. III, Acta Chim.*, **13**, 73 (*Chem. Abs.*, 1969, **71**, 81085x).

Hahn, W. E., Kryczka, B. and Bartnik, R. (1974). *Soc. Sci. Lodz., Acta Chim.*, **18**, 175 (*Chem. Abs.*, 1975, **82**, 43129j).
Hahn, W. E., Nowaczyk, M. and Bartnik, R. (1968). *Lodz. Tow. Nauk Wydz. III, Acta Chim.*, **13**, 59 (*Chem. Abs.*, 1969, **71**, 81084w).
Hahn, W. E. and Zawadzka, H. (1967). *Lodz. Tow. Nauk Wydz. III, Acta Chim.*, **12**, 69 (*Chem. Abs.*, 1969, **71**, 112734q).
Hahn, W. E. and Zawadzka, H. (1969). *Lodz. Tow. Nauk Wydz. III, Acta Chim.*, **14**, 53 (*Chem. Abs.*, 1970, **72**, 100419y).
Hahn, W. E., Zawadzka, H. and Szwedowska, A. (1969). *Lodz. Tow. Nauk Wydz. III, Acta Chim.*, **14**, 61 (*Chem. Abs.*, 1970, **72**, 100418x).
Hahn, W., Kryczka, B. and Bartnik, R. (1975). Pol. Pat. 79,057 (*Chem. Abs.*, 1977, **86**, 55281y).
Hahn, W., Nowaczyk, M., Bartnik, R. and Zawadzka, H. (1968). Pol. Pat. 54,536 (*Chem. Abs.*, 1969, **70**, 19931w).
Hai, P. V., Buu-Hoï, N. P. and Xuong, N. D. (1958). *J. Org. Chem.*, **23**, 39.
Hall, J. A. and Plant, S. G. P. (1953). *J. Chem. Soc.*, 116.
Hamana, M. and Kumadaki, I. (1967). *Chem. Pharm. Bull.*, **15**, 363.
Hand, E. S. and Cohen, T. (1967). *Tetrahedron*, **23**, 2911.
Hanna, C. and Schueler, F. W. (1952). *J. Amer. Chem. Soc.*, **74**, 3693.
Hansch, C. and Muir, R. M. (1950). *Plant Physiol.*, **25**, 389.
Hansen, H.-J., Sutter, B. and Schmid, H. (1968). *Helv. Chim. Acta*, **51**, 828.
Harbert, C. A. Plattner, J. J., Welch, W. M., Weissman, A. and Koe, B. K. (1980). *J. Med. Chem.*, **23**, 635.
Hardegger, E. and Corrodi, H. (1956). *Helv. Chim. Acta*, **39**, 514.
Harley-Mason, J. and Pavri, E. H. (1963). *J. Chem. Soc.*, 2504.
Harradence, R. H., Hughes, G. K. and Lions, F. (1938). *J. Proc. Roy. Soc. NSW*, **72**, 273.
Harradence, R. H. and Lions, F. (1938). *J. Proc. Roy. Soc. NSW*, **72**, 221.
Harradence, R. H. and Lions, F. (1939). *J. Proc. Roy. Soc. NSW*, **73**, 14.
Harries, C., Abderhalden, E., Weinberg, A. von, Trendelenburg, E. and Lewin, L. (1919). *Naturwissenschaften*, **46**, 843.
Harries, C. D. and Loewenstein, E. (1894). *Ber.*, **27**, 861.
Harrison, J. W. (1962). Ph.D. Thesis, Carnegie Institute of Technology, USA.
Harrison, M. J., Norman, R. O. C. and Gladstone, W. A. F. (1967). *J. Chem. Soc.* (*C*), 735.
Harrow, B. (1919). *Science*, **50**, [1285], 150.
Hart, P. A. (1963). In *Steroid Reactions. An Outline for Organic Chemists*, ed. Djerassi, C., Holden-Day Inc., San Francisco, p. 182.
Hartman, W. W. and Roll, L. R. (1943). *Org. Syns.*, Coll. Vol. 2, 418.
Harvey, D. G. (1955). *J. Chem. Soc.*, 2536.
Harvey, D. G. (1958). *J. Chem. Soc.*, 3760.
Harvey, D. G. (1959). *J. Chem. Soc.*, 473.
Harvey, D. G. and Robson, W. (1938). *J. Chem. Soc.*, 97.
Harvey, D. J. (1968). Ph.D. Thesis, University of Glasgow, Scotland.
Harvey, D. J., Laurie, W. A. and Reed, R. I. (1971a). *Org. Mass Spectrom.*, **5**, 1183.
Harvey, D. J., Laurie, W. A. and Reed, R. I. (1971b). *Org. Mass Spectrom.*, **5**, 1189.
Harvey, D. J. and Reid, S. T. (1970). *J. Chem. Soc.* (*C*), 2074.
Harvey, D. J. and Reid, S. T. (1972). *Tetrahedron*, **28**, 2489.
Haselbach, E. and Heilbronner, E. (1968). *Helv. Chim. Acta*, **51**, 16.
Hassner, A. and Catsoulacos, P. (1967). *Chem. Communs.*, 121.
Hausmann, J. (1889). *Ber.*, **22**, 2019.
Haynes, R. K. and Hewgill, F. R. (1972). *J. Chem. Soc.*, *Perkin I*, 396.
H. E. A. (1919). *Nature*, **103**, [2596], 430.

Heath-Brown, B. (1975). Ger. Pat. 2,513,197 (*Chem. Abs.*, 1976, **84**, 30932g).
Heath-Brown, B. (1978). Roche Products Ltd, Welwyn Garden City, Hertfordshire, England, Personal Communication.
Heath-Brown, B. and Philpott, P. G. (1965a). *J. Chem. Soc.*, 7165.
Heath-Brown, B. and Philpott, P. G. (1965b). *J. Chem. Soc.*, 7185.
Hegedüs, B. (1946). *Helv. Chim. Acta*, **29**, 1499.
Hegel, S. (1886). *Liebig's Ann. Chem.*, **232**, 214.
Heidt, J., Gombos, E. and Tudos, F. (1966). *Kozlemen*, **14**, 183 (*Chem. Abs.*, 1967, **66**, 18552d).
Heilbron, I. M., Kennedy, T., Spring, F. S. and Swain, G. (1938). *J. Chem. Soc.*, 869.
Heimgartner, H., Hansen, H.-J. and Schmid, H. (1979). In *Advances in Organic Chemistry, Methods and Results*, ed. by Taylor, E. C.; *Iminium Salts in Organic Chemistry*, ed. by Böhme, H. and Viehe, H. G., John Wiley and Sons, New York, Sydney and Toronto, Vol. 9, Part II, p. 655.
Heindel, N. D., Kennewell, P. D. and Pfau, M. (1969). *Chem. Communs.*, 757.
Heindel, N. D., Kennewell, P. D. and Pfau, M. (1970). *J. Org. Chem.*, **35**, 80.
Heine, H. W., Hoye, T. R., Williard, P. G. and Hoye, R. C. (1973). *J. Org. Chem.*, **38**, 2984.
Helferich, B. (1953). *Angew. Chem.*, **65**, 45.
Helferich, B. (1961). In *Great Chemists*, ed. by Farber, E., Interscience, New York and London, p. 983.
Heller, G. (1917). *Ber.*, **50**, 1202.
Hendrickson, J. B. (1974). *Angew. Chem.* (*Intern. Edit.*), **13**, 47.
Henecka, H., Timmler, H. and Lorenz, R. (1959). US Pat. 2,888,451 (*Chem. Abs.*, 1959, **53**, 20097g).
Henecka, H., Timmler, H., Lorenz, R. and Geiger, W. (1957). *Chem. Ber.*, **90**, 1060.
Henle, F. (1905). *Ber.*, **38**, 1362.
Herbert, R. and Wibberley, D. G. (1969). *J. Chem. Soc.* (*C*), 1505.
Herdieckerhoff, E. and Tschunkur, E. (1933). Ger. Pat. 574,840 (*Chem. Abs.*, 1933, **27**, 4541) (*Chem. Zentr.*, 1933, II, 622).
Hester, J. B. (1967). *J. Org. Chem.*, **32**, 3804.
Hester, J. B. (1968). Fr. Pat. 1,524,495 (*Chem. Abs.*, 1970, **72**, 55430g).
Hester, J. B. (1969a). Fr. Pat. 1,566,173 (*Chem. Abs.*, 1970, **72**, 90425f).
Hester, J. B. (1969b). Fr. Pat. 1,566,174 (*Chem. Abs.*, 1970, **72**, 90426g).
Hester, J. B., Tang, A. H., Keasling, H. H. and Veldkamp, W. (1968). *J. Med. Chem.*, **11**, 101.
Hewitt, J. T. (1891). *J. Chem. Soc.* (*Trans.*), **59**, 209.
Heyl, F. W. and Herr, M. E. (1953). *J. Amer. Chem. Soc.*, **75**, 1918.
Hinman, R. L. (1968). *Tetrahedron*, **24**, 185.
Hinman, R. L. and Whipple, E. B. (1962). *J. Amer. Chem. Soc.*, **84**, 2534.
Hino, T., Nakagawa, M. and Akaboshi, S. (1967). *Chem. Communs.*, 656.
Hino, T., Suzuki, T. and Nakagawa, M. (1973). *Chem. Pharm. Bull.*, **21**, 2786.
Hinshaw, J. C. (1975). *J. Org. Chem.*, **40**, 47.
Hiremath, S. P. and Hosame, R. S. (1973), In *Advances in Heterocyclic Chemistry*, ed. by Katritzky, A. R. and Boulton, A. J., Academic Press, New York and London, Vol. 15, p. 277.
Hiremath, S. P. and Siddappa, S. (1962). *J. Karnatak Univ.*, **6**, 1 (*Chem. Abs.*, 1963, **59**, 8855f).
Hiremath, S. P. and Siddappa, S. (1964). *J. Indian. Chem. Soc.*, **41**, 357.
Hiremath, S. P. and Siddappa, S. (1965). *J. Indian Chem. Soc.*, **42**, 836.
Hoerlein, H. U. (1956). Brit. Pat. 752,668 (*Chem. Abs.*, 1957, **51**, 5844i).
Hoesch, K. (1921). In *Emil Fischer, sein Leben und sein Werk*, *Ber.*, **54**, 1 (reviewed in *Chem. Weekblad.*, 1922, **19**, 98).

Hoffman, R. and Woodward, R. B. (1965a). *J. Amer. Chem. Soc.*, **87**, 2511.
Hoffman, R. and Woodward, R. B. (1965b). *J. Amer. Chem. Soc.*, **87**, 4389.
Hofmann, K. A. (1919). *Ber.*, **52A**, 125.
Holla, B. S. and Ambekar, S. Y. (1976). *Indian J. Chem.* (*B*), **14B**, 579 (*Chem. Abs.*, 1977, **86**, 72358w).
Holla, B. S. and Ambekar, S. Y. (1978). *Indian J. Chem.*, **16B**, 240.
Holla, B. S. and Ambekar, S. Y. (1979a). *J. Chem. Soc.*, *Chem. Communs.*, 221.
Holla, B. S. and Ambekar, S. Y. (1979b). *Indian J. Chem.*, **17B**, 66.
Holla, S. and Ambekar, S. Y. (1974). *J. Indian Chem. Soc.*, **51**, 965.
Hollins, C. (1922). *J. Amer. Chem. Soc.*, **44**, 1598.
Hollins, C. (1924a). In *The Synthesis of Nitrogen Ring Compounds Containing a Single Hetero-Atom* (*Nitrogen*), Ernest Benn, London, p. 80.
Hollins, C. (1924b). In *The Synthesis of Nitrogen Ring Compounds Containing a Single Hetero-Atom* (*Nitrogen*), Ernest Benn, London, p. 92.
Hollins, C. (1924c). In *The Synthesis of Nitrogen Ring Compounds Containing a Single Hetero-Atom* (*Nitrogen*), Ernest Benn, London, p. 168.
Holt, P. F. and McNae, C. J. (1964). *J. Chem. Soc.*, 1759.
Hörlein, H. U. (1955). Ger. Pat. 930,444 (*Chem. Abs.*, 1958, **52**, 20208b).
Hörlein, H. U. (1957). US Pat. 2,786,059 (*Chem. Abs.*, 1957, **51**, 8147b).
Hörlein, U. (1954). *Chem. Ber.*, **87**, 463.
Hörlein, U. (1956). US Pat. 2,759,943 (*Chem. Abs.*, 1957, **51**, 8148h).
Horner, J. K., DeGraw, J. I. and Skinner, W. A. (1966). *Canad. J. Chem.*, **44**, 307.
Horner, J. K. and Skinner, W. A. (1964). *Canad. J. Chem.*, **42**, 2904.
Horner, L. (1939). *Liebig's Ann. Chem.*, **540**, 73.
Horning, E. C., Horning, M. G. and Walker, G. N. (1948). *J. Amer. Chem. Soc.*, **70**, 3935.
Horning, E. C., Sweeley, C. C., Dalgliesh, C. E. and Kelly, W. (1959). *Biochim. Biophys. Acta*, **32**, 566.
Hoshino, T. (1933). *Liebig's Ann. Chem.*, **500**, 35.
Hoshino, T. and Kobayashi, T. (1935). *Liebig's Ann. Chem.*, **520**, 11.
Hoshino, T., Kobayashi, T. and Kotake, Y. (1935). *Liebig's Ann. Chem.*, **516**, 81.
Hoshino, T. and Shimodaira, K. (1935). *Liebig's Ann. Chem.*, **520**, 19.
Hoshino, T. and Takiura, K. (1936). *Bull. Chem. Soc. Japan*, **11**, 218 (*Chem. Abs.*, 1936, **30**, 5985_2) (*Chem. Zentr.*, 1936, II, 1925).
Hosmane, R. S., Hiremath, S. P. and Schneller, S. W. (1973). *J. Chem. Soc.*, *Perkin I*, 2450.
Hosmane, R. S., Hiremath, S. P. and Schneller, S. W. (1974). *J. Het. Chem.*, **11**, 29.
House, H. O. and Richey, F. A. (1969). *J. Org. Chem.*, **34**, 1430.
How, P.-Y. and Parrick, J. (1976). *J. Chem. Soc.*, *Perkin I*, 1363.
Huang-Hsinmin and Mann, F. G. (1949). *J. Chem. Soc.*, 2903.
Hudson, B. J. F. and Robinson, R. (1942). *J. Chem. Soc.*, 691.
Hudson, C. S. (1953). *J. Chem. Educ.*, **30**, 120–1 (*Chem. Abs.*, 1953, **47**, 4145h).
Huffman, J. W. (1962). *J. Org. Chem.*, **27**, 503.
Hugershoff, A. (1903). *Ber.*, **36**, 3134.
Hughes, G. K. and Lions, F. (1937–1938a). *J. Proc. Roy. Soc. NSW*, **71**, 475.
Hughes, G. K. and Lions, F. (1937–1938b). *J. Proc. Roy. Soc. NSW*, **71**, 494 (*Chem. Abs.*, 1939, **33**, 588).
Hughes, G. K., Lions, F. and Ritchie, E. (1939). *J. Proc. Roy. Soc. NSW*, **72**, 209 (*Chem. Abs.*, 1939, **33**, 6837).
Huisgen, R. (1948a). *Liebig's Ann. Chem.*, **559**, 101.
Huisgen, R. (1948b). *Liebig's Ann. Chem.*, **559**, 174.
Huisgen, R., Jakob, F., Siegel, W. and Cadus, A. (1954). *Liebig's Ann. Chem.*, **590**, 1.

Huisgen, R. and Ugi, I. (1957). *Liebig's Ann. Chem.*, **610**, 57.
Hunig, S. and Steinmetzer, H. C. (1976). *Liebig's Ann. Chem.*, 1090.
Hunsberger, I. M., Shaw, E. R., Fugger, J., Ketcham, R. and Lednicer, D. (1956). *J. Org. Chem.*, **21**, 394.
Huntress, E. H. (1952). *Proc. Amer. Acad. Arts and Sci.*, **81**, 50.
Huntress, E. H., Bornstein, J. and Hearon, W. M. (1956). *J. Amer. Chem. Soc.*, **78**, 2225.
Huntress, E. H. and Hearon, W. M. (1941). *J. Amer. Chem. Soc.*, **63**, 2762.
Huntress, E. H., Lesslie, T. E. and Hearon, W. M. (1956). *J. Amer. Chem. Soc.*, **78**, 419.
Hutton, R. F. and Steel, C. (1964). *J. Amer. Chem. Soc.*, **86**, 745.
Hyde, E. (1899). *Ber.*, **32**, 1810.
Ibrahim, H. G. and Rippie, E. G. (1976). *J. Pharm. Sci.*, **65**, 1640.
I.C.I. Ltd (1975). Fr. Pat. 2,259,613 (*Chem. Abs.*, 1976, **84**, 135715s).
Iffland, D. C., Salisbury, L. and Schafer, W. R. (1961). *J. Amer. Chem. Soc.*, **83**, 747.
I. G. Farbenind. A.-G. (1929). Brit. Pat. 337,821 (*Chem. Abs.*, 1931, **25**, 2302).
I. G. Farbenind. A.-G. (1930a). Fr. Pat. 698,148 (*Chem. Abs.*, 1931, **25**, 3012).
I. G. Farbenind. A.-G. (1930b). Ger. Pat. 533,470 (*Chem. Abs.*, 1932, **26**, 479).
I. G. Farbenind. A.-G. (1930c). Ger. Pat. 542,422 (*Chem. Abs.*, 1932, **26**, 2469).
I. G. Farbenind. A.-G. (1930d). Ger. Pat. 548,818 (*Chem. Abs.*, 1932, **26**, 4068).
I. G. Farbenind. A.-G. (1930e). Ger. Pat. 548,819 (*Chem. Abs.*, 1932, **26**, 4068).
I. G. Farbenind. A.-G. (1931a). Fr. Pat. 713,500 (*Chem. Abs.*, 1932, **26**, 1620).
I. G. Farbenind. A.-G. (1931b). Fr. Pat. 716,158 (*Chem. Abs.*, 1932, **26**, 1943).
I. G. Farbenind. A.-G. (1933). Fr. Pat. 744,595 (*Chem. Abs.*, 1933, **27**, 4239).
I. G. Farbenind. A.-G. (1935). Fr. Pat. 778,861 (*Chem. Abs.*, 1935, **29**, 4774).
I. G. Farbenindustrie A.-G. (1935). Brit. Pat 436,110 (*Chem. Abs.*, 1936, **30**, 1391).
I. G. Farbenindustrie Akt.-Ges. (1932). Ger. Pat. 555,935 and Swiss Pat. 150,300 (*Chem. Zentr.*, 1932, II, 1976).
I. G. Farbenindustrie Akt.-Ges. (1934). E.P. 409,350 (*Chem. Zentr.*, 1934, II, 1532).
Iijima, K., Yamada, Y. and Homma, M. (1973). Jap. Pat. 73 14,661 (*Chem. Abs.*, 1973, **78**, 147793e).
Ikezaki, M., Wakamatsu, T. and Ban, Y. (1969). *Chem. Communs.*, 88.
Illingsworth, B. D., Spencer, H. E., Mee, J. D. and Heseltine, D. W. (1968). Fr. Pat. 1,520,818 (*Chem. Abs.*, 1969, **71**, 103183g).
Illy, H. and Funderburk, L. (1968). *J. Org. Chem.*, **33**, 4283.
Inaba, S., Akatsu, M., Hirohashi, T. and Yamamoto, H. (1976). *Chem. Pharm. Bull.*, **24**, 1076.
Inaba, S., Ishizumi, K., Mori, K. and Yamamoto, H. (1971). *Chem. Pharm. Bull.*, 1971, **19**, 722.
Inaba, S., Ishizumi, K., Okamoto, T. and Yamamoto, H. (1972). *Chem. Pharm. Bull.*, **20**, 1628.
Inaba, S., Ishizumi, K., Okamoto, T. and Yamamoto, H. (1975). *Chem. Pharm. Bull.*, **23**, 3279.
Inaba, S., Ishizumi, K. and Yamamoto, H. (1971). *Chem. Pharm. Bull.*, **19**, 263.
Ince, W. H. (1889). *Liebig's Ann. Chem.*, **253**, 35.
Ingold, C. K. (1969a). *Structure and Mechanism in Organic Chemistry*, 2nd Edition, G. Bell and Sons, London, p. 891.
Ingold, C. K. (1969b). *Structure and Mechanism in Organic Chemistry*, 2nd Edition, G. Bell and Sons, London, p. 916.
Ingold, C. K. (1969c). *Structure and Mechanism in Organic Chemistry*, 2nd Edition, G. Bell and Sons, London, pp. 934 and 938.
Inoue, I. and Ban, Y. (1970). *J. Chem. Soc.* (*C*), 602.
Ioffe, B. V. and Gershtein, L. M. (1969). *J. Org. Chem. USSR*, **5**, 257.

Ioffe, B. V., Sergeeva, Z. I. and Stopskii, V. S. (1966). *Proc. Acad. Sci., USSR*, **167**, 393.
Ioffe, B. V. and Stopskii, V. S. (1967). *Proc. Acad. Sci. USSR*, **175**, 712.
Ioffe, B. V. and Stopskii, V. S. (1968). *J. Org. Chem. USSR*, **4**, 1446.
Ioffe, B. V. and Stopskii, V. S. (1968). *Tetrahedron Letts.*, 1333.
Irismetov, M. P., Goryaev, M. I. and Rivkina, T. V. (1979). *Izv. Akad. Nauk Kaz. SSR, Ser. Khim.*, 60 (*Chem. Abs.*, 1979, **91**, 193516h).
Isherwood, F. A. and Cruickshank, D. H. (1954). *Nature*, **173**, 121.
Isherwood, F. A. and Jones, R. L. (1955). *Nature*, 1955, **175**, 419.
Ishii, H., Hagiwara, T., Ishikawa, T., Ikeda, N. and Murakami, Y. (1975). *Heterocycles*, **3**, [1], 71.
Ishii, H., Harada, K., Abe, K., Doki, K. and Ikeda, N. (1971). *Yakugaku Zasshi*, **91**, 947 (*Chem. Abs.*, 1971, **75**, 151471d).
Ishii, H., Ikeda, N. and Murakami, Y. (1972). Jap. Pat. 72 30,663 (*Chem. Abs.*, 1973, **78**, 29621z).
Ishii, H. and Murakami, Y. (1975). *Tetrahedron*, **31**, 933.
Ishii, H. and Murakami, Y. (1979). *Yakagaku Zasshi*, **99**, 413 (*Chem. Abs.*, 1979, **91**, 38438r).
Ishii, H., Murakami, Y., Furuse, T. and Hosoya, K. (1979). *Chem. Pharm. Bull.*, **27**, 346.
Ishii, H., Murakami, Y., Furuse, T., Hosoya, K. and Ikeda, N. (1973). *Chem. Pharm. Bull.*, **21**, 1495.
Ishii, H., Murakami, Y., Furuse, T., Hosoya, K., Takeda, H. and Ikeda, N. (1971). Third International Congress of Heterocyclic Chemistry, B, Sendai, Japan, p. 446.
Ishii, H., Murakami, Y., Furuse, T., Hosoya, K., Takeda, H. and Ikeda, N. (1973). *Tetrahedron*, **29**, 1991.
Ishii, H., Murakami, Y., Furuse, T., Takeda, H. and Ikeda, N. (1973). *Tetrahedron Letts.*, 355.
Ishii, H., Murakami, Y., Hosoya, K., Furuse, T., Takeda, H. and Ikeda, N. (1972). *Chem. Pharm. Bull.*, **20**, 1088.
Ishii, H., Murakami, Y., Hosoya, K., Takeda, H., Suzuki, Y. and Ikeda, N. (1973). *Chem. Pharm. Bull.*, **21**, 1481.
Ishii, H., Murakami, Y. and Ishikawa, T. (1977a). *Heterocycles*, **6**, 1686.
Ishii, H., Murakami, Y. and Ishikawa, T. (1977b). *Symp. Heterocycl.*, 188 (*Chem. Abs.*, 1978, **89**, 179749p).
Ishii, H., Murakami, K., Murakami, Y. and Hosoya, K. (1977). *Chem. Pharm. Bull.*, **25**, 3122.
Ishii, H., Murakami, Y., Suzuki, Y. and Ikeda, N. (1970). *Tetrahedron Letts.*, 1181.
Ishii, H., Murakami, Y., Takeda, H. and Furuse, T. (1974). *Chem. Pharm. Bull.*, **22**, 1981.
Ishii, H., Murakami, Y., Takeda, H. and Ikeda, N. (1974). *Heterocycles*, **2**, 229.
Ishii, H., Murakami, Y., Tani, S., Abe, K. and Ikeda, N. (1970). *Yakugaku Zasshi.*, **90**, 724 (*Chem. Abs.*, 1970, **73**, 55728f).
Ishizumi, K. and Mori, K. (1976). Jap. Pat. 76, 149,271 (*Chem. Abs.*, 1977, **87**, 68147e).
Ishizumi, K., Mori, K., Inaba, S. and Yamamoto, H. (1973). *Chem. Pharm. Bull.*, **21**, 1027.
Ishizumi, K., Shioiri, T. and Yamada, S. (1967). *Chem. Pharm. Bull.*, **15**, 863.
Isoe, S. and Nakazaki, M. (1959). *Chem. Ind.*, 1574.
Istituto Luso Farmaco d'Italia S.r.l. (1967). Brit. Pat. 1,075,156 (*Chem. Abs.*, 1968, **68**, 69041k).
Ivanova, R. Y., Bogatskii, A. V., Andronati, S. A. and Zhilina, Z. I. (1979). *Dopov. Akad. Nauk Ukr. RSR, Ser. B: Geol., Khim. Biol. Nauki*, 31 (*Chem. Abs.*, 1979, **90**, 151918q).
Iyer, R., Jackson, A. H. and Shannon, P. V. R. (1973). *J. Chem. Soc., Perkin II*, 878.
Iyer, R., Jackson, A. H., Shannon, P. V. R. and Naidoo, B. (1972). *J. Chem. Soc., Chem. Communs.*, 461.

Iyer, R., Jackson, A. H., Shannon, P. V. R. and Naidoo, B. (1973). *J. Chem. Soc., Perkin II*, 872.
Jackman, M. E. and Archer, S. (1946). *J. Amer. Chem. Soc.*, **68**, 2105.
Jackson, A., Gaskell, A. J., Wilson, N. D. V. and Joule, J. A. (1968). *Chem. Communs.*, 364.
Jackson, A. and Joule, J. A. (1967). *Chem. Communs.*, 459.
Jackson, A., Wilson, N. D. V., Gaskell, A. J. and Joule, J. A. (1969). *J. Chem. Soc.* (*C*), 2738.
Jackson, A. H. and Smith, A. E. (1965). *Tetrahedron*, **21**, 989.
Jackson, A. H. and Smith, P. (1968). *Tetrahedron*, **24**, 2227.
Jackson, R. W. and Manske, R. H. (1930). *J. Amer. Chem. Soc.*, **52**, 5029.
Jackson, R. W. and Manske, R. H. (1935). *Canad. J. Res.*, **13B**, 170.
Jacob, A. and Madinaveitia, J. (1937). *J. Chem. Soc.*, 1929.
Jacobs, T. L. (1957a). In *Heterocyclic Compounds*, ed. by Elderfield, R. C., John Wiley, New York and Chapman and Hall, London, Vol. 5, Chapter 2, p. 45.
Jacobs, T. L. (1957b). In *Heterocyclic Compounds*, ed. by Elderfield, R. C., John Wiley New York and Chapman and Hall, London, Vol. 6, Chapter 5, p. 159.
Jacobson, P. (1919). *Chem.-Ztg.*, **43**, 565 (*Chem. Abs.*, 1920, **14**, 399).
Jacquier, R., Chapelle, J.-P., Elguero, J. and Tarrago, G. (1969). *Chem. Communs.*, 752.
Jacquignon, P. and Buu-Hoï, N. P. (1957). *J. Org. Chem.*, **22**, 72.
Jacquignon, P. and Buu-Hoï, N. P. (1961). *Bull. Soc. Chim. France*, 663.
Jacquignon, P., Buu-Hoï, N. P. and Dufour, M. (1966). *Bull. Soc. Chim. France*, 2765.
Jacquignon, P., Croisy-Delcey, M. and Croisy, A. (1972). *Bull. Soc. Chim. France*, 4251.
Jacquignon, P., Fravolini, A., Feron, A. and Croisy, A. (1974). *Experientia*, **30**, 452.
Janetzky, E. F. J. and Verkade, P. E. (1945). *Rec. Trav. Chim.*, **64**, 129.
Janetzky, E. F. J. and Verkade, P. E. (1946). *Rec. Trav. Chim.*, **65**, 905.
Janetzky, E. F. J., Verkade, P. E. and Lieste, J. (1946). *Rec. Trav. Chim.*, **65**, 193.
Japp, F. R. and Findlay, A. (1897). *J. Chem. Soc.*, **71**, 1115.
Japp, F. R. and Klingemann, F. (1887a). *Ber.*, **20**, 2942.
Japp, F. R. and Klingemann, F. (1887b). *Ber.*, **20**, 3284.
Japp, F. R. and Klingemann, F. (1887c). *Ber.*, **20**, 3398.
Japp, F. R. and Klingemann, F. (1887d). *Proc. Chem. Soc.*, 140.
Japp, F. R. and Klingemann, F. (1888a). *Liebig's Ann. Chem.*, **247**, 190.
Japp, F. R. and Klingemann, F. (1888b). *Ber.*, **21**, 549.
Japp, F. R. and Klingemann, F. (1888c). *J. Chem. Soc.* (*Trans.*), **53**, 519.
Japp, F. R. and Klingemann, F. (1888d). *Proc. Chem. Soc.*, 11.
Japp, F. R. and Maitland, W. (1901). *Proc. Chem. Soc.*, **17**, 176.
Japp, F. R. and Maitland, W. (1903a). *Proc. Chem. Soc.*, **19**, 19.
Japp, F. R. and Maitland, W. (1903b). *J. Chem. Soc.*, **83**, 267.
Japp, F. R. and Meldrum, A. N. (1899). *J. Chem. Soc.*, **75**, 1044.
Jarboe, C. H. (1967). In *The Chemistry of Heterocyclic Compounds*, ed. by Wiley, R. H., Wiley-Interscience, New York, London and Sydney, Vol. 22, Part 2, Chapter 7, p. 180.
Jardine, R. V. and Brown, R. K. (1965). *Canad. J. Chem.*, **43**, 1293.
Jefferson, A. and Scheinmann, F. (1968). *Quarterly Revs.*, **22**, 391.
Jenisch, G. (1906). *Monat. Chem.*, **27**, 1223.
Jogdeo, P. S. and Bhide, G. V. (1980). *Steroids*, **35**, 599.
Johnson, A. W. (1950). In *The Chemistry of the Acetylenic Compounds*, Edward Arnold and Co., London, Vol. 2, p. 132.
Johnson, F. Malhotra, S. K. (1965). *J. Amer. Chem. Soc.*, **87**, 5492.
Johnson, F. and Whitehead, A. (1964). *Tetrahedron Letts.*, 3825.
Johnson, M. (1921). *J. Soc. Chem. Ind.*, **40**, 176T.

Johnson, R. P. and Oswald, J. P. (1968). US Pat. 3,419,568 (*Chem. Abs.*, 1969, **70**, 68175m).
Johnston, K. M. and Shotter, R. G. (1974). *Tetrahedron*, **30**, 4059.
Jones, G. (1975). Ger. Pat. 2,438,413 (*Chem. Abs.*, 1975, **83**, 43364c).
Jones, G. and Stevens, T. S. (1953). *J. Chem. Soc.*, 2344.
Jones, G. and Tringham, G. T. (1975). *J. Chem. Soc., Perkin I*, 1280.
Jones, N. A. and Tomlinson, M. L. (1953). *J. Chem. Soc.*, 4114.
Jordaan, A. and Arndt, R. R. (1968). *J. Het. Chem.*, **5**, 723.
Joshi, K. C., Pathak, V. N., Arya, P. and Chand, P. (1979). *Agric. Biol. Chem.*, **43**, 171.
Joshi, K. C., Pathak, V. N. and Chand, P. (1978a). *Indian J. Chem.*, **16B**, 933.
Joshi, K. C., Pathak, V. N. and Chand, P. (1978b). *J. Prakt. Chem.*, **320**, 701.
Joshi, S. S. and Gambhir, I. R. (1956). *J. Amer. Chem. Soc.*, **78**, 2222.
Julia, M. and Lenzi, J. (1971). *Bull. Soc. Chim. France*, 4084.
Julia, M. and Manoury, P. (1965). *Bull. Soc. Chim. France*, 1411.
Julia, M., Melamed, R. and Gombert, R. (1965). *Ann. Inst. Pasteur*, **109**, 343 (*Chem. Abs.*, 1966, **64**, 677c).
Julia, M. and Nickel, P. (1966). *Medd. Norsk. Farm. Selsk.*, **28**, 153 (*Chem. Abs.*, 1967, **66**, 85664e).
Julian, P. L. (1931). Dissertation, Vienna (quoted in Julian and Pikl, 1933).
Julian, P. L., Karpel, W. J., Magnani, A. and Meyer, E. W. (1948). *J. Amer. Chem. Soc.*, **70**, 180.
Julian, P. L. and Magnani, A. (1949). *J. Amer. Chem. Soc.*, **71**, 3207.
Julian, P. L., Meyer, E. W. and Printy, H. C. (1952). In *Heterocyclic Compounds*, ed. by Elderfield, R. C., Wiley, New York, Vol. 3, Chapter 1.
Julian, P. L. and Pikl, J. (1933). *J. Amer. Chem. Soc.*, **55**, 2105.
Julian, P. L. and Pikl, J. (1935). *Proc. Indiana Acad. Sci.*, **45**, 145 (*Chem. Abs.*, 1937, **31**, 1026_8).
Justoni, R. and Pessina, R. (1955). *Il Farmaco, Ed. Sci.*, **10**, 356 (*Chem. Abs.*, 1955, **49**, 13968d).
Justoni, R. and Pessina, R. (1957a). Brit. Pat. 766,036 (*Chem. Abs.*, 1957, **51**, 13931h).
Justoni, R. and Pessina, R. (1957b). Brit. Pat. 770,370 (*Chem. Abs.*, 1957, **51**, 14822a).
Justoni, R. and Pessina, R. (1960). US Pat. 2,945,046 (*Chem. Abs.*, 1961, **55**, 567e).
Kadzyauskas, P. P., Butkus, E. P., Vasyulite, Y. A., Averina, N. V. and Zefirov, N. S. (1979). *Chem. Het. Comps.*, **15**, 258.
Kajfez, F. and Mihalic, M. (1978). Swiss Pat. 605,744 (*Chem. Abs.*, 1979, **90**, 54820u).
Kaji, K. and Nagashima, H. (1952). *J. Pharm. Soc. Japan*, **72**, 1589 (*Chem. Abs.*, 1953, **47**, 9317f).
Kakurina, L. N., Kucherova, N. F. and Zagorevskii, V. A. (1964). *J. Gen. Chem. USSR*, **34**, 2829.
Kakurina, L. N., Kucherova, N. F. and Zagorevskii, V. A. (1965a). *J. Gen. Chem. USSR*, **35**, 311.
Kakurina, L. N., Kucherova, N. F. and Zagorevskii, V. A. (1965b). *J. Org. Chem. USSR*, **1**, 1118.
Kalb, L., Schweizer, F. and Schimpf, G. (1926). *Ber.*, **59**, 1858.
Kalb, L., Schweizer, F., Zellner, H. and Berthold, E. (1926). *Ber.*, **59**, 1860.
Kalir, A. and Pelah, Z. (1966). *Israel J. Chem.*, **4**, 155.
Kametani, T., Fukumoto, K. and Masuko, K. (1963). *Yakugaku Zasshi*, **83**, 1052 (*Chem. Abs.*, 1964, **60**, 10732h).
Kametani, T., Ohsawa, T. and Ihara, M. (1979). *J. Chem. Res. (S)*, 364.
Kametani, T., Seino, C., Noguchi, I., Shibuya, S., Fukumoto, K., Kohno, T. and Takano, S. (1971). *J. Chem. Soc. (C)*, 1803.
Kaminsky, D., Shavel, J. and Meltzer, R. I. (1967). *Tetrahedron Letts.*, 859.

Kanaoka, Y., Ban, Y., Miyashita, K., Irie, K. and Yonemitsu, O. (1966). *Chem. Pharm. Bull.*, **14**, 934.

Kanaoka, Y., Ban, Y., Yonemitsu, O., Irie, K. and Miyashita, K. (1965). *Chem. Ind.*, 473.

Kanaoka, Y., Miyashita, K. and Yonemitsu, O. (1969). *Tetrahedron*, **25**, 2757.

Kaneko, H. (1960). *Yakugaku Zasshi*, **80**, 1374 (*Chem. Abs.*, 1961, **55**, 6512f).

Kanterov, V. Y., Starostenko, N. E., Oleinikov, Y. G. and Suvorov, N. N. (1970). *Tr. Mosk. Khim.-Tekhnol. Inst.*, [66], 113 (*Chem. Abs.*, 1971, **75**, 75650u).

Kanterov, V. Y., Starostenko, N. E. and Suvorov, N. N. (1970). *Tr. Mosk. Khim.-Tekhnol. Inst.*, [66], 117 (*Chem. Abs.*, 1971, **75**, 75651v).

Kanterov, V. Y., Suvorov, N. N., Starostenko, N. E. and Oleinikov, Y. G. (1970). *Russian J. Phys. Chem.*, **44**, 889.

Kao, Y.-S. and Robinson, R. (1955). *J. Chem. Soc.*, 2865.

Karabatsos, G. J., Graham, J. D. and Vane, F. M. (1962). *J. Amer. Chem. Soc.*, **84**, 753.

Karabatsos, G. J. and Krumel, K. L. (1967). *Tetrahedron*, **23**, 1097.

Karabatsos, G. J. and Taller, R. A. (1963). *J. Amer. Chem. Soc.*, **85**, 3624.

Karabatsos, G. J., Vane, F. M., Taller, R. A. and Hsi, N. (1964). *J. Amer. Chem. Soc.*, **86**, 3351.

Karady, S., Ly, M. G., Pines, S. H., Chemerda, J. M. and Sletzinger, M. (1973). *Synthesis*, 50.

Kates, M. and Marion, L. (1951). *Canad. J. Chem.*, **29**, 37.

Katritzky, A. R., Dennis, N., Sabongi, G. J. and Turker, L. (1979). *J. Chem. Soc., Perkin I*, 1525.

Katsube, J., Sasajima, K., Ono, K., Nakao, M., Maruyama, I., Takayama, M., Katayama, S., Tanaka, Y., Inaba, S. and Yamamoto, H. (1974). Jap. Pat. 74 07,272 (*Chem. Abs.*, 1974, **81**, 13381x).

Keglević, D., Desaty, D., Goleš, D. and Stančić, L. (1968). *Croat. Chem. Acta*, **40**, 7.

Keglević, D. and Goleš, D. (1970). *Croat. Chem. Acta*, **42**, 513.

Keglević, D. and Leonhard, B. (1963). *Croat. Chem. Acta*, **35**, 175.

Keglević, D., Stojanac, N. and Desaty, D. (1961). *Croat. Chem. Acta*, **33**, 83.

Keimatsu, S. and Sugasawa, S. (1928a). *J. Pharm. Soc. Japan*. **48**, 348 (*Chem. Abs.*, 1928, **22**, 3163). *J. Pharm. Soc. Japan*, **48**, 63 (*Chem. Zentr.* 1928, II, 48).

Keimatsu, S. and Sugasawa, S. (1928b). *J. Pharm. Soc. Japan*, **48**, 755 (*Chem. Abs.*, 1929, **23**, 834). *J. Pharm. Soc. Japan*, **48**, 101 (*Chem. Zentr.*, 1928, II. 1881).

Keimatsu, S., Sugasawa, S. and Kasuya, G. (1928). *J. Pharm. Soc. Japan*, **48**, 762 (*Chem. Abs.*, 1929, **23**, 834). *J. Pharm. Soc. Japan*, **48**, 105 (*Chem. Zentr.* 1928, II, 1882).

Kelly, A. H., McLeod, D. H. and Parrick, J. (1965). *Canad. J. Chem.*, **43**, 296.

Kelly, A. H. and Parrick, J. (1966). *Canad. J. Chem.*, **44**, 2455.

Kelly, A. H. and Parrick, J. (1970). *J. Chem. Soc.* (*C*), 303.

Kempter, G., Schwalba, M., Stoss, W. and Walter, K. (1962). *J. Prakt. Chem.*, **18**, 39.

Kent, A. (1935). *J. Chem. Soc.*, 976.

Kent, A. and McNeil, D. (1938). *J. Chem. Soc.*, 8.

Kermack, W. O. (1924). *J. Chem. Soc.*, **125**, 2285.

Kermack, W. O., Perkin, W. H. and Robinson, R. (1921). *J. Chem. Soc.*, **119**, 1602.

Kermack, W. O., Perkin, W. H. and Robinson, R. (1922). *J. Chem. Soc.*, **121**, 1872.

Kermack, W. O. and Slater, R. H. (1928). *J. Chem. Soc.*, 32.

Kermack, W. O. and Smith, J. F. (1930). *J. Chem. Soc.*, 1999.

Kermack, W. O. and Tebrich, W. (1940). *J. Chem. Soc.*, 314.

Kesswani, H. I. (1967). Ph.D. Thesis, University of Birmingham, England (quoted in Kesswani, 1970).

Kesswani, H. I. (1970). *Canad. J. Chem.*, **48**, 689.

Keufer, J. (1950). *Ann. Pharm. Franc.*, **8**, 816 (*Chem. Abs.*, 1951, **45**, 10246).

Khan, M. A. and Morley, M. L. de B. (1978). *J. Het. Chem.*, **15**, 1399.
Khan, M. A. and Morley, M. L. de B. (1979). *J. Het. Chem.*, **16**, 997.
Khan, M. A. and Rocha, J. F. da (1978). *J. Het. Chem.*, **15**, 913.
Khoshtariya, T. E., Sikharulidze, M. I., Tret'yakova, L. G., Efimova, T. K. and Suvorov, N. N. (1979). *Chem. Het. Comps.*, **15**, 642.
Kiang, A. K. and Mann, F. G. (1951). *J. Chem. Soc.*, 1909.
Kiang, A. K., Mann, F. G., Prior, A. F. and Topham, A. (1956). *J. Chem. Soc.*, 1319.
Kidwai, A. R. and Khan, N. H. (1963). *Compt. Rend. Acad. Sci.*, **256**, 3709.
Kim, D. H. (1975). US Pat. 3,914,250 (*Chem. Abs.*, 1976, **84**, 31150n).
Kim, D. H. (1976). *J. Het. Chem.*, **13**, 1187.
Kim, D. H. and Santilli, A. A. (1969). *J. Het. Chem.*, **6**, 819.
Kimura, M., Inaba, S. and Yamamoto, H. (1972). Ger. Pat. 2,142,196 (*Chem. Abs.*, 1972, **77**, 34323n).
Kimura, M., Inaba, S. and Yamamoto, H. (1973a). Jap. Pat. 73 32,864 (*Chem. Abs.*, 1973, **79**, 31864v).
Kimura, M., Inaba, S. and Yamamoto, H. (1973b). Jap. Pat. 73 34,872 (*Chem. Abs.*, 1973, **79**, 53178b).
Kimura, M., Inaba, S. and Yamamoto, H. (1973c). Jap. Pat. 73 34,873 (*Chem. Abs.*, 1973, **79**, 42338c).
Kimura, M., Inaba, S. and Yamamoto, H. (1973d). Jap. Pat. 73 39,474 (*Chem. Abs.*, 1973, **79**, 78607d).
King, F. E. and King, T. J. (1945). *J. Chem. Soc.*, 824.
King, F. E. and L'Ecuyer, P. (1934). *J. Chem. Soc.*, 1901.
King, F. E., Liguori, M. and Robinson, R. (1933). *J. Chem. Soc.*, 1475.
King, F. E., Liguori, M. and Robinson, R. (1934). *J. Chem. Soc.*, 1416.
King, F. E. and Paterson, G. D. (1936). *J. Chem. Soc.*, 400.
King, F. E. and Robinson, R. (1932). *J. Chem. Soc.*, 326.
King, F. E. and Robinson, R. (1933). *J. Chem. Soc.*, 270.
Kinsley, D. A. and Plant, S. G. P. (1956). *J. Chem. Soc.*, 4814.
Kinsley, D. A. and Plant, S. G. P. (1958). *J. Chem. Soc.*, 1.
Kipping, F. S. (1894). *J. Chem. Soc.*, **65**, 480.
Kipping, F. S. and Hill, A. (1899). *J. Chem. Soc.*, **75**, 144.
Kirk, K. L. (1976). *J. Het. Chem.*, **13**, 1253.
Kissman, H. M., Farnsworth, D. W. and Witkop, B. (1952). *J. Amer. Chem. Soc.*, **74**, 3948.
Kitaev, Y. P. (1959). *Usp. Khim.*, **28**, 336 (*Chem. Abs.*, 1959, **53**, 14913).
Kitaev, Y. P. (1971). *Probl. Org. Fiz. Khim. Mater. Nauch. Sess., Posvyashch. 25–letizu Inst. Org. Fiz. Khim. Akad. Nauk SSSR.*, 75 (*Chem. Abs.*, 1973, **79**, 105132d).
Kitaev, Y. P. and Arbuzov, A. E. (1957). *Bull. Acad. Sci. USSR, Div. Chem. Sci.*, 1068.
Kitaev, Y. P. and Arbuzov, A. E. (1960). *Bull. Acad. Sci. USSR, Div. Chem. Sci.*, 1306.
Kitaev, Y. P. and Buzykin, B. I. (1972). *Russian Chem. Revs.*, **41**, 495.
Kitaev, Y. P., Buzykin, B. I. and Troepol'skaya, T. V. (1970). *Russian Chem. Revs.*, **39**, 441.
Kitaev, Y. P., Flegontov, S. A. and Troepol'skaya, T. V. (1966). *Bull. Acad. Sci. USSR, Div. Chem. Sci.*, 2023.
Kitaev, Y. P. and Troepol'skaya, T. V. (1963a). *Bull. Acad. Sci. USSR, Div. Chem. Sci.*, 408.
Kitaev, Y. P. and Troepol'skaya, T. V. (1963b). *Bull. Acad. Sci. USSR, Div. Chem. Sci.*, 418.
Kitaev, Y. P. and Troepol'skaya, T. V. (1967). *Bull. Acad. Sci. USSR, Div. Chem. Sci.*, 1828.
Kitaev, Y. P. and Troepol'skaya, T. V. (1978). *Chem. Het. Comps.*, **14**, 807.
Kitaev, Y. P. and Troepol'skaya, T. V. (1979). *Pyatichlen. Aromat. Getorotsikly., Riga*, 102 (*Chem. Abs.*, 1980, **93**, 70210n).

Kitaev, Y. P., Troepol'skaya, T. V. and Arbuzov, A. E. (1964). *J. Gen. Chem. USSR*, **34**, 1848.
Kitaev, Y. P., Troepol.skaya, T. V. and Arbuzov, A. E. (1966a). *J. Org. Chem. USSR*, **2**, 333.
Kitaev, Y. P., Troepol'skaya, T. V. and Arbuzov, A. E. (1966b). *Bull. Acad. Sci. USSR, Div. Chem. Sci.*, 910.
Kizhner, N. M. (1912). *J. Russ. Phys. Chem. Soc.*, **44**, 165 (*Chem. Abs.*, 1912, **6**, 1431).
Klare, H. (1980). *Vopr. Istorii Estestovzn. i Tekhn., Moskva*, 79 (*Chem. Abs.* 1980, **93**, 113220k).
Klioze, S. S. and Darmory, F. P. (1975). *J. Org. Chem.*, **40**, 1588.
Knorr, L. (1883). *Ber.*, **16**, 2597.
Knorr, L. (1887). *Liebig's Ann. Chem.*, **238**, 137.
Knorr, L. (1919). *Ber.*, **52A**, 132.
Knorr, L. and Klotz, C. (1887). *Ber.*, **20**, 2545.
Knueppel, C. (1900). *Liebig's Ann. Chem.*, **310**, 75.
Kobayashi, T. (1939). *Liebig's Ann. Chem.*, **539**, 213.
Kochetkov, N. K., Kucherova, N. F. and Evdakov, V. P. (1956). *J. Gen. Chem. USSR*, **26**, 3505.
Kochetkov, N. K., Kucherova, N. F. and Evdakov, V. P. (1957). *J. Gen. Chem. USSR*, **27**, 283.
Kochetkov, N. K., Kucherova, N. F., Pronina, L. P. and Petruchenko, M. I. (1959). *J. Gen. Chem. USSR*, **29**, 3581.
Kochetkov, N. K., Kucherova, N. F. and Zhukova, I. G. (1961). *J. Gen. Chem. USSR*, **31**, 853.
Kochhar, M. M. (1969). *J. Het. Chem.*, **6**, 977.
Koelsch, C. F. (1943). *J. Org. Chem.*, **8**, 295.
Kögl, F. and Kostermans, D. G. F. R. (1935). *Hoppe-Seyler's Zeit. Physiol. Chem.*, **235**, 201.
Kohler, E. P. (1909). *Amer. Chem. J.*, **42**, 375.
Kohlrausch, K. (1889). *Liebig's Ann. Chem.*, **253**, 15.
Kollenz, G. (1971). *Monat. Chem.*, **102**, 108.
Kollenz, G. (1972a). *Liebig's Ann. Chem.*, **762**, 23.
Kollenz, G. (1972b). *Monat. Chem.*, **103**, 947.
Kollenz, G. (1978). *Monat. Chem.*, **109**, 249.
Kollenz, G. and Labes, C. (1975). *Liebig's Ann. Chem.*, 1979.
Kollenz, G. and Labes, C. (1976). *Liebig's Ann. Chem.*, 174.
Kollenz, G., Ziegler, E., Eder, M. and Prewedourakis, E. (1970). *Monat. Chem.*, **101**, 1597.
Kolosov, M. N. and Preobrazhensky, N. A. (1953). *J. Gen. Chem. USSR*, **23**, 1641.
Komachiya, Y., Suzuki, S., Yamada, T., Miyayashiki, H., and Sakurai, S. (1965). *Nippon Kagaku Zasshi*, **86**, 856 (*Chem. Abs.*, 1966, **65**, 13818c).
Komoto, N., Enomoto, Y., Miyagaki, M., Tanaka, Y., Nitanai, K. and Umezawa, H. (1979). *Agric. Biol. Chem.*, **43**, 555.
Komoto, N., Enomoto, Y., Tanaka, Y., Nitanai, K. and Umezawa, H. (1979). *Agric. Biol. Chem.*, **43**, 559.
Komzolova, N. N., Kucherova, N. F. and Zagorevskii, V. A. (1964). *J. Gen. Chem. USSR*, **34**, 2396.
Komzolova, N. N., Kucherova, N. F. and Zagorevskii, V. A. (1967). *Chem. Het. Comps.*, **3**, 556.
König, W. and Haller, H. (1920). *J. Prakt. Chem.*, **101**, 38.
Konschegg, A. (1905). *Monat. Chem.*, **26**, 931.
Konschegg, A. (1906). *Monat. Chem.*, **27**, 247.
Korczynski, A., Brydowna, W. and Kierzek, L. (1926). *Gazz. Chim. Ital.*, **56**, 903.

Korczynski, A. and Kierzek, L. (1925). *Gazz. Chim. Ital.*, **55**, 361.
Korczyński, A. and Kierzek, L. (1925). *Roczniki Chemji*, **5**, 23 (*J. Chem. Soc., Abs.*, 1925, **128i**, 973).
Kornet, M. J., Ong, T. H. and Thio, P. A. (1971). *J. Het. Chem.*, **8**, 999.
Kornet, M. J., Thio, A. P. and Tolbert, L. M. (1980). *J. Org. Chem.*, **45**, 30.
Kosa, I. and Kovacs, V. (1970). Ger. Pat. 2,009,474 (*Chem. Abs.*, 1971, **74**, 22693h).
Kosa, I. and Kovacs, G. (1972). Hung. Pat. 4,889 (*Chem. Abs.*, 1973, **78**, 43269x).
Kosa, I. and Kovacs, G. (1975). Hung. Pat. 9,689 (*Chem. Abs.*, 1976, **84**, 74102r).
Kost, A. N., Budylin, V. A., Matveeva, E. D. and Sterligov, D. O. (1970). *J. Org. Chem. USSR*, **6**, 1516.
Kost, A. N., Golubeva, G. A. and Grandberg, I. I. (1956). *J. Gen. Chem. USSR*, **26**, 2201.
Kost, A. N., Golubeva, G. A. and Popova, A. G. (1979). *Chem. Het. Comps.*, **15**, 282.
Kost, A. N., Golubeva, G. A. and Portnov, Y. N. (1971). *Proc. Acad. Sci. USSR*, **200**, 772.
Kost, A. N., Golubeva, G. A. and Sviridova, L. A. (1973). *Chem. Het. Comps.*, **9**, 456.
Kost, A. N., Golubeva, G. A., Terent'ev, A. P. and Grandberg, I. I. (1962). *Proc. Acad. Sci. USSR*, **144**, 431.
Kost, A. N., Golubeva, G. A., Zabrodnyaya, V. G. and Portnov, Y. N. (1975). *Chem. Het. Comps.*, **11**, 1383.
Kost, A. N. and Grandberg, I. I. (1955a). *J. Gen. Chem. USSR*, **25**, 1673.
Kost, A. N. and Grandberg, I. I. (1955b). *J. Gen. Chem. USSR*, **25**, 2017.
Kost, A. N. and Grandberg, I. I. (1956a). *J. Gen. Chem. USSR*, **26**, 607.
Kost, A. N. and Grandberg, I. I. (1956b). *J. Gen. Chem. USSR*, **26**, 1925.
Kost, A. N. and Grandberg, I. I. (1956c). *J. Gen. Chem. USSR*, **26**, 2593.
Kost, A. N. and Grandberg, I. I. (1966). In *Advances in Heterocyclic Chemistry*, ed. by Katritzky, A. R. and Boulton, A. J., Academic Press, New York and London, Vol. 6, p. 347.
Kost, A. N., Grandberg, I. I. and Evreinova, E. B. (1958). *J. Gen. Chem. USSR*, **28**, 503.
Kost, A. N., Grandberg, I. I. and Golubeva, G. A. (1956). *J. Gen. Chem. USSR*, **26**, 2905.
Kost, A. N., Lebedenko, N. Y. and Sviridova, L. A. (1976). *J. Org. Chem. USSR*, **12**, 2374.
Kost, A. N., Lebedenko, N. Y., Sviridova, L. A. and Torocheshnikov, V. N. (1978). *Chem. Het. Comps.*, **14**, 380.
Kost, A. N., Ostrovskii, M. K. and Golubeva, G. A. (1973). *Tezisy Dokl.-Simp. Khim. Tekhnol. Get. Soedin. Goryunch. Iskop., 2nd.*, 191 (*Chem. Abs.*, 1977, **86**, 5259b).
Kost, A. N., Ostrovskii, M. K., Goluebea, G. A. and Tikhonov, V. E. (1974). *J. Org. Chem. USSR*, **10**, 2629.
Kost, A. N., Portnov, Y. N. and Golubeva, G. A. (1972). USSR Pat. 334,218 (*Chem. Abs.*, 1972, **77**, 48505k).
Kost, A. N., Sugorova, I. P. and Yakubov, A. P. (1965). *J. Org. Chem. USSR*, **1**, 121.
Kost, A. N., Sviridova, L. A., Golubeva, G. A. and Portnov, Y. N. (1970a). *Chem. Het. Comps.*, **6**, 344.
Kost, A. N., Sviridova, L. A., Golubeva, G. A. and Portnov, Y. N. (1970b). USSR Pat. 281,473 (*Chem. Abs.*, 1971, **74**, 99868b).
Kost, A. N., Terent'ev, A. P., Vinogradova, E. V., Terent'ev, P. B. and Ershov, V. V. (1960). *J. Gen. Chem. USSR*, **30**, 2538.
Kost, A. N., Trofimov, F. A., Tsyshkova, N. G. and Shadurskii, K. S. (1969). USSR Pat. 259,888 (*Chem. Abs.*, 1970, **73**, 14827u).
Kost, A. N., Vinogradova, E. V., Daut, K. and Terent'ev, A. P. (1962). *J. Gen. Chem. USSR*, **32**, 2030.
Kost, A. N., Vinogradova, E. V., Trofimov, F. A., Mukhanova, T. I., Nozdrich, V. I. and Shadurskii, K. (1967). *Khim.-Farm. Zh.*, **1**, 25 (*Chem. Abs.*, 1968, **68**, 39516b).

Kost, A. N. and Yudin, L. G. (1957). *Khim. Nauka i Prom.*, **2**, 800 (*Chem. Abs.*, 1958, **52**, 11833d).
Kost, A. N., Yudin, L. G., Berlin, Y. A. and Terent'ev, A. P. (1959). *J. Gen. Chem. USSR*, **29**, 3782.
Kost, A. N., Yudin, L. G., Budylin, V. A. and Yaryshev, N. G. (1965). *Chem. Het. Comps.*, **1**, 426.
Kost, A. N., Yudin, L. G. and Chernyshova, N. B. (1968). *Proc. Acad. Sci. USSR*, **183**, 962.
Kost, A. N., Yudin, L. G., Dmitriev, B. A. and Terent'ev, A. P. (1959). *J. Gen. Chem. USSR*, **29**, 3937.
Kost, A. N., Yudin, L. G. and Popravko, S. A. (1962). *J. Gen. Chem. USSR*, **32**, 1530.
Kost, A. N., Yudin, L. G. and Terent'ev, A. N. (1959). *J. Gen. Chem. USSR*, **29**, 1920.
Kost, A. N., Yudin, L. G. and Yü-Chou, C. (1964). *J. Gen. Chem. USSR*, **34**, 3487.
Kost, A. N., Yudin, L. G. and Zinchenko, E. Y. (1973). *Chem. Het. Comps.*, **9**, 306.
Kost, A. N., Yurovskaya, M. A., Mel'nikova, T. V. and Potanina, O. I. (1973). *Chem. Het. Comps.*, **9**, 191.
Kost, A. N., Yurovskaya, M. A. and Trofimov, F. A. (1973). *Chem. Het. Comps.*, **9**, 267.
Kost, A. N., Zabrodnyaya, V. G., Portnov, Y. N. and Voronin, V. G. (1980). *Khim. Get. Soedin.*, 484 (*Chem. Abs.*, 1980, **93**, 114247m).
Kotov, A. L., Sagitullin, R. S. and Gorbunov, V. I. (1970). *Regul. Rosta Rast. Khim. Sredstvami*, 111 (*Chem. Abs.*, 1971, **74**, 64156r).
Kralt, T., Asma, W. J., Haeck, H. H. and Moed, H. D. (1961). *Rec. Trav. Chim.*, **80**, 313.
Kratzl, K. and Berger, K. P. (1958). *Monat. Chem.*, **89**, 83.
Kricka, L. J. (1974). *Chem. Revs.*, **74**, 117.
Kricka, L. J. and Vernon, J. M. (1974). *Canad. J. Chem.*, **52**, 299.
Kruber, O. (1926). *Ber.*, **59**, 2752.
Krutak, J. J. (1975). US Pat. 3,865,837 (*Chem. Abs.*, 1975, **83**, 58649y).
Kucherova, N. F., Aksanova, L. A., Sharkova, N. M. and Zagorevskii, V. A. (1963). *J. Gen. Chem. USSR*, **33**, 3593.
Kucherova, N. F., Aksanova, L. A., Sharkova, L. M. and Zagorevskii, V. A. (1971). *Chem. Het. Comps.*, **7**, 1368.
Kucherova, N. F., Aksanova, L. A., Sharkova, L. M. and Zagorevskii, V. A. (1973a). *Chem. Het. Comps.*, **9**, 137.
Kucherova, N. F., Aksanova, L. A., Sharkova, L. M. and Zagorevskii, V. A. (1973b). *Chem. Het. Comps.*, **9**, 835.
Kucherova, N. F., Aksanova, L. A., Sharkova, L. M. and Zagorevskii, V. A. (1978). *Chem. Het. Comps.*, **14**, 137.
Kucherova, N. F., Aksanova, L. A. and Zagorevskii, V. A. (1963). *J. Gen. Chem. USSR*, **33**, 3331.
Kucherova, N. F., Borisova, L. N., Sharkova, N. M. and Zagorevskii, V. A. (1970). *Chem. Het. Comps*, **6**, 1138.
Kucherova, N. F. and Kochetkov, N. K. (1956). *J. Gen. Chem. USSR*, **26**, 3511.
Kucherova, N. F., Petruchenko, M. I. and Zagorevskii, V. A. (1962). *J. Gen. Chem. USSR*, **32**, 3577.
Kucherova, N. F., Zhukova, I. G., Kamzolova, N. N., Petruchenko, M. I., Sharkova, N. M. and Kochetkov, N. K. (1961). *J. Gen. Chem. USSR*, **31**, 858.
Kuchländer, U. (1975). *Tetrahedron*, **31**, 1631.
Kuehne, M. E. (1962). *J. Amer. Chem. Soc.*, **84**, 837.
Kuehne, M. E. (1969). In *Enamines: Synthesis, Structure and Reactions*, ed. by Cook, A. G., Dekker, New York and London, Chapter 8, p. 414.
Kuehne, M. E. and Bayha, C. (1966). *Tetrahedron Letts.*, 1311.
Kulka, M. and Manske, R. H. F. (1952). *Canad. J. Chem.*, **30**, 711.

Kumar, Y. and Jain, P. C. (1979). *Indian J. Chem.*, **17B**, 623.
Kunori, M. (1962). *Nippon Kagaku Zasshi*, **83**, 839 (*Chem. Abs.*, 1963, **59**, 1573e).
Kupletskaya, N. B., Kost, A. N. and Grandberg, I. I. (1956). *J. Gen. Chem. USSR*, **26**, 3495.
Kurilo, G. N., Ryabova, S. Y. and Grinev, A. N. (1979a). *Chem. Het. Comps.*, **15**, 681.
Kurilo, G. N., Ryabova, S. Y. and Grinev, A. N. (1979b). USSR Pat. 690,017 (*Chem. Abs.*, 1980, **92**, 76545b).
Kuroda, S. (1923). *J. Pharm. Soc. Japan*, **493**, 13 (*Chem. Abs.*, 1923, **17**, 3031) (*Chem. Zentr.*, 1923, III, 142).
Kurzer, F. and Sanderson, P. M. (1962). *J. Chem. Soc.*, 230.
Kyziol, J. B. and Lyzniak, A. (1980). *Tetrahedron*, **36**, 3017.
Laboratories Francais de Chimiotherapie (1962). Brit. Pat. 888,426 (*Chem. Abs.*, 1962, **57**, 15075g).
Lacoume, B. (1972a). Ger. Pat. 2,141,640 (*Chem. Abs.*, 1972, **76**, 153590m).
Lacoume, B. (1972b). S. African Pat. 71 05,217 (*Chem. Abs.*, 1972, **77**, 164474y).
Lacoume, B. Milcent, G. and Olivier, A. (1972). *Tetrahedron*, **28**, 667.
LaForge, F. B. (1928). *J. Amer. Chem. Soc.*, **50**, 2477.
Lakshmanan, V. K. (1960). *Proc. Indian Acad. Sci.*, *Sect. A.*, **51**, 41 (*Chem. Abs.*, 1961, **55**, 528a).
Lamb, A. B. (1924). *J. Amer. Chem. Soc.*, **46**, 2133.
Land, H. B. and Frasca, A. R. (1971). *Anales Asoc. Quim. Argent.*, **59**, 251 (*Chem. Abs.*, 1972, **76**, 58664e).
Landgraf, C. A. and Seeger, E. (1968). S. African Pat. 67 00,516 (*Chem. Abs.*, 1969, **70**, 77779p).
Landor, S. R., Landor, P. D., Fomun, Z. T. and Mpango, G. M. (1977). *Tetrahedron Letts.*, 3743.
Landquist, J. K. and Marsden, C. J. (1966). *Chem. Ind.*, 1032.
Langlois, Y., Langlois, N. and Potier, P. (1975). *Tetrahedron Letts.*, 955.
Lanzilotti, A. E., Littell, R., Fanshawe, W. J., McKenzie, T. C. and Lovell, F. M. (1979). *J. Org. Chem.*, **44**, 4809.
Laskowski, S. C. (1968). Fr. Pat. 1,551,082 (*Chem. Abs.*, 1970, **72**, 43733v) (as in US Pat. 3,468,882).
Laubmann, H. (1888a). *Ber.*, **21**, 1212.
Laubmann, H. (1888b). *Liebig's Ann. Chem.*, **243**, 244.
Laufer, R. J. (1958). Ph.D. Thesis, Carnegie Institute of Technology, USA.
Laurent, A. (1841). *J. Prakt. Chem.*, **24**, 1.
Laurent, A. (1842). *Ann. Chim. Phys.*, **3**, 371.
Laws, E. G. and Sidgwick, N. V. (1911). *J. Chem. Soc.*, **99**, 2085.
Lazier, W. A. and Arnold, H. R. (1943). *Org. Syns.*, Coll. Vol. 2, 142.
LeCount, D. J. and Greer, A. T. (1974). *J. Chem. Soc.*, *Perkin I*, 297.
Lee, F. G. H., Dickson, D. E., Suzuki, J., Zirnis, A. and Manian, A. A. (1973). *J. Het. Chem.*, **10**, 649.
Lee, T. B. and Swan, G. A. (1956). *J. Chem. Soc.*, 771.
Lee, W. Y. (1974). *Daehan Hwahak Hwoejee*, **18**, 50 (*Chem. Abs.*, 1974, **81**, 77586a).
Lee, W. Y. (1976). *Taehan Hwahak Hoechi*, **20**, 500 (*Chem. Abs.*, 1977, **87**, 5858s).
Lee, W. Y. and Lee, Y. Y. (1969). *Soul Taehokkyo Kyoyang Kwajongbu Nonmunjip, Chayon Kwahak-Pyon*, **1**, 21 (*Chem. Abs.*, 1971, **74**, 87522w).
Leekning, M. E. (1964). *Rev. Fac. Farm. Bioquim.*, *Univ. Sao Paulo*, **2**, 45 (*Chem. Abs.*, 1965, **63**, 13190b).
Leekning, M. E. (1967). *Rev. Fac. Farm. Odontol. Araraquara*, **1**, 219 (*Chem. Abs.*, 1968, **69**, 27164t).

Leete, E. (1961a). *Tetrahedron*, **14**, 35.
Leete, E. (1961b). *J. Amer. Chem. Soc.*, **83**, 3645.
Leete, E. and Chen, F. Y.-H. (1963). *Tetrahedron Letts.*, 2013.
Le Fevre, G. and Hamelin, J. (1979). *Tetrahedron Letts.*, 1757.
Leggetter, B. E. and Brown, R. K. (1960). *Canad. J. Chem.*, **38**, 1467.
Leonard, F. and Tschannen, W. (1965). *J. Med. Chem.*, **8**, 287.
Lespagnol, A., Bar, D., Erb-Debruyne, Delhomenie-Sauvage, Labiau and Tudo (1960). *Bull. Soc. Chim. France*, 1166.
Lespagnol, A., Lespagnol, C. and Henichart, J. P. (1969). *Compt. Rend. Acad. Sci., Ser. C.*, **268**, 1528.
Lester, M. G., Petrow, V. and Stephenson, O. (1965). *Tetrahedron*, **21**, 1761.
Leuchs, H., Heller, A. and Hoffmann, A. (1929). *Ber.*, **62**, 871.
Leuchs, H. and Kowalski, G. (1925). *Ber.*, **58**, 2822.
Leuchs, H. and Overberg, H. S. (1931). *Ber.*, **64**, 1896.
Leuchs, H., Philpott, D., Sander, P., Heller, H. and Köhler, H. (1928). *Liebig's Ann. Chem.*, **461**, 27.
Leuchs, H., and Winzer, K. (1925). *Ber.*, **58**, 1520.
Leuchs, H., Wulkow, G. and Gerland, H. (1932). *Ber.*, **65**, 1586.
Leuchs, H., Wutke, J. and Gieseler, E. (1913). *Ber.*, **46**, 2200.
Levy, P. R. and Robinson, R. (1962). *J.S. African Chem. Inst.*, **15**, 85 (*Chem. Abs.*, 1963, **59**, 9953f).
Lewis, A. D. (1962). US Pat. 3,037,031 (*Chem. Abs.*, 1962, **57**, 12439e).
Lieber, D. (1908). *Monat. Chem.*, **29**, 421.
Limpricht, H. (1888). *Ber.*, **21**, 3409.
Lindwall, H. G. and Mantell, G. J. (1953). *J. Org. Chem.*, **18**, 345.
Lingens, F. and Weiler, K. H. (1963). *Liebig's Ann. Chem.*, **662**, 139.
Linnell, W. H. and Perkin, W. H. (1924). *J. Chem. Soc.*, **125**, 2451.
Linstead, R. P. and Wang, A. B.-L. (1937). *J. Chem. Soc.*, 807.
Lions, F. (1932). *J. Proc. Roy. Soc. NSW*, **66**, 516.
Lions, F. (1937–1938). *J. Proc. Roy. Soc. NSW*, **71**, 192.
Lions, F. and Ritchie, E. (1939a). *J. Proc. Roy. Soc. NSW*, **73**, 125.
Lions, F. and Ritchie, E. (1939b). *J. Amer. Chem. Soc.*, **61**, 1927.
Lions, F. and Spruson, M. J. (1932). *J. Proc. Roy. Soc. NSW*, **66**, 171.
Liston, A. J. (1966). *J. Org. Chem.*, **31**, 2105.
Liston, A. J. and Howarth, M. (1967). *J. Org. Chem.*, **32**, 1034.
Little, J. S., Taylor, W. I. and Thomas, B. R. (1954). *J. Chem. Soc.*, 4036.
Livingstone, R. (1973). In *Rodd's Chemistry of Carbon Compounds*, ed. by Coffey, S., Elsevier, Amsterdam, London and New York, Vol. 4A, Chapter 5, pp. 398–401 and 487–489, together with references to several examples throughout the chapter.
Lockemann, G. and Liesche, O. (1905). *Liebig's Ann. Chem.*, **342**, 14.
Loev, B. (1969). US Pat. 3,427,311 (*Chem. Abs.*, 1969, **70**, 68391d).
Loev, B. and Snader, K. M. (1967). *J. Het. Chem.*, **4**, 403.
Loftfield, R. B. (1951). *J. Amer. Chem. Soc.*, **73**, 1365.
Lohr, R. A. (1977). Ph.D. Thesis, Northern Illinois University, USA (*Diss. Abs.*, 1978, **38**, [8], 3697B) (*Chem. Abs.*, 1978, **88**, 120920p).
Long, R. S. (1956). US Pat. 2,731,474 (*Chem. Abs.*, 1956, **50**, 13096d).
Longmore, R. B. and Robinson, B. (1965). *Chem. Ind.*, 1297.
Longmore, R. B. and Robinson, B. (1966). *Chem. Ind.*, 1638.
Longmore, R. B. and Robinson, B. (1967). *Collect. Czech. Chem. Communs.*, **32**, 2184.
Longmore, R. B. and Robinson, B. (1969). *J. Pharm. Pharmacol.*, **21S**, 118.
Losacker, P. (1966). Dissertation, Frankfurt am Main, West Germany (quoted in Fritz and Stock, 1970).

Lyle, R. E. and Skarlos, L. (1966). *Chem. Communs.*, 644.
Macander, R. F. (1969). Ph.D. Thesis, Wayne State University, USA. (*Diss. Abs.*, 1971, **32**, 2606B) (*Chem. Abs.*, 1972, **76**, 85641b).
MacPhillamy, H. B., Dziemian, R. L., Lucas, R. A. and Kuehne, M. E. (1958). *J. Amer. Chem. Soc.*, **80**, 2172.
Maeda, M. and Nakazaki, M. (1960). *Chem. Ind.*, 719.
Magidson, O. Y., Suvorov, N. N., Travin, A. I., Sorkina, N. P. and Novikova, S. P. (1965). *Biol. Aktivn. Soedin., Akad. Nauk SSSR*, 5 (*Chem. Abs.*, 1965, **63**, 18192b).
Maiti, B. C., Thomson, R. H. and Mahendran, M. (1978). *J. Chem. Res. (S)*, 126.
Majima, R. and Kotake, M. (1930). *Ber.*, **63**, 2237.
Maki, Y., Masugi, T., Hiramitsu, T. and Ogiso, T. (1973). *Chem. Pharm. Bull.*, **21**, 2460.
Makisumi, Y. and Takada, S. (1976). *Chem. Pharm. Bull.*, **24**, 770.
Maksimov, N. Y., Chetverikov, V. P. and Kost, A. N. (1979). USSR Pat. 685,664 (*Chem. Abs.*, 1980, **92**, 41938e).
Malhotra, S. K. and Johnson, F. (1965). *J. Amer. Chem. Soc.*, **87**, 5493.
Malhotra, S. K., Moackley, D. F. and Johnson, F. (1967). *Chem. Communs.*, 448.
Mamaev, V. P. and Sedova, V. F. (1961). *Izv. Sibirsk. Otd. Akad. Nauk SSSR*, 142 (*Chem. Abs.*, 1962, **57**, 5872i).
Manhas, M. S., Brown, J. W. and Pandit, U.K. (1975). *Heterocycles*, **3**, 117.
Manhas, M. S., Brown, J. W., Pandit, U. K. and Houdewind, P. (1975). *Tetrahedron*, **31**, 1325.
Manjunath, B. L. (1927). *Quarterly J. Indian Chem. Soc.*, **4**, 271.
Mann, F. G. (1949). *J. Chem. Soc.*, 2816.
Mann, F. G. and Haworth, R. C. (1944). *J. Chem. Soc.*, 670.
Mann, F. G., Prior, A. F. and Willcox, T. J. (1959). *J. Chem. Soc.*, 3830.
Mann, F. G. and Smith, B. B. (1951). *J. Chem. Soc.*, 1898.
Mann, F. G. and Tetlow, A. J. (1957). *J. Chem. Soc.*, 3352.
Mann, F. G. and Wilkinson, A. J. (1957). *J. Chem. Soc.*, 3346.
Mann, F. G. and Willcox, T. J. (1958). *J. Chem. Soc.*, 1525.
Mannich, C. and Hönig, D. (1927). *Arch. Pharm.*, **265**, 598 (*Chem. Abs.*, 1928, **22**, 591.
Manske, R. H. and Boos, W. R. (1960). *Canad. J. Chem.*, **38**, 620.
Manske, R. H. F. (1931a). *Canad. J. Res.*, **4**, 591.
Manske, R. H. F. (1931b). *Canad. J. Res.*, **5**, 592.
Manske, R. H. F. and Kulka, M. (1947). *Canad. J. Res.*, **25B**, 376.
Manske, R. H. F. and Kulka, M. (1949). *Canad. J. Res.*, **27B**, 291.
Manske, R. H. F. and Kulka, M. (1950a). *Canad. J. Res.*, **28B**, 443.
Manske, R. H. F. and Kulka, M. (1950b). *J. Amer. Chem. Soc.*, **72**, 4997.
Manske, R. H. F. and Leitch, L. C. (1936). *Canad. J. Res.*, **14B**, 1.
Manske, R. H. F., Perkin, W. H. and Robinson, R. (1927). *J. Chem. Soc.*, 1.
Manske, R. H. F. and Robinson, R. (1927). *J. Chem. Soc.*, 240.
Marchand, B., Streffer, C. and Jauer, H. (1961). *J. Prakt. Chem.*, **13**, 54.
Marchetti, L. and Tosi, G. (1969a). *Annali Chim.*, **59**, 315.
Marchetti, L. and Tosi, G. (1969b). *Annali Chim.*, **59**, 328.
Maréchal, G., Christiaens, L., Renson, M. and Jacquignon, P. (1978). *Collect. Czech. Chem. Communs.*, **43**, 2916.
Marion, L. and Oldfield, C. W. (1947). *Canad. J. Res.*, **25B**, 1.
Märkl, G., Habel, G. and Baier, H. (1979). *Phosphorus and Sulfur*, **5**, 257.
Marquez, V. E., Cranston, J. W., Ruddon, R. W., Kier, L. B. and Burckalter, J. H. (1972). *J. Med. Chem.*, **15**, 36.
Marshalkin, M. F., Azimov, V. A., Gasparyants, A. S. and Yakhontov, L. N. (1979). *Chem. Het. Comps.*, **15**, 1037.

Marshalkin, M. F., Azimov, V. A., Lindberg, L. F. and Yakhontov, L. N. (1978). *Chem. Het. Comps.*, **14**, 905.
Marshalkin, M. F. and Yakhontov, L. N. (1972a). *Chem. Het. Comps.*, **8**, 590.
Marshalkin, M. F. and Yakhontov, L. N. (1972b). *Chem. Het. Comps.*, **8**, 1426.
Marshalkin, M. F. and Yakhontov, L. N. (1973). *Russian Chem. Revs.*, **42**, 725.
Marshalkin, M. F. and Yakhontov, L. N. (1978). *Chem. Het. Comps.*, **14**, 1208.
Martynov, V. F. and Martynova, V. F. (1954). *Zh. Obs. Khim.*, **24**, 2146 (*Chem. Abs.*, 1956, **50**, 287f.
Marumo, S., Abe, H., Hattori, H. and Munakata, K. (1968). *Agric. Biol. Chem.*, **32**, 117.
Marvell, E. N., Stephenson, J. L. and Ong, J. (1965). *J. Amer. Chem. Soc.*, **87**, 1267.
Mashkovskii, M. D., Grinev, A. N., Andreeva, N. I., Altukhova, L. B., Shvedov, V. I., Avrutskii, G. Y. and Gromova, V. V. (1974). *Khim.-Farm. Zh.*, **8**, 60 (*Chem. Abs.*, 1974, **81**, 3878a).
Massey, J. P. and Plant, S. G. P. (1931). *J. Chem. Soc.*, 1990.
McCapra, F. and Chang, Y. C. (1966). *Chem. Communs.*, 522.
McClelland, E. W. (1929). *J. Chem. Soc.*, 1588.
McClelland, E. W. and D'Silva, J. L. (1932). *J. Chem. Soc.*, 227.
McClelland, E. W. and Smith, P. W. (1945). *J. Chem. Soc.*, 408.
McCrea, P. A. (1966). Brit. Pat. 1,021,422 (*Chem. Abs.*, 1966, **64**, 19563h).
McKay, J. B., Parkhurst, R. M., Silverstein, R. M. and Skinner, W. A. (1963). *Canad. J. Chem.*, **41**, 2585.
McKusick, B. C., Heckert, R. E., Cairns, T. L., Coffman, D. D. and Mower, H. F. (1958). *J. Amer. Chem. Soc.*, **80**, 2806.
McLean, J., McLean, S. and Reed, R. I. (1955). *J. Chem. Soc.*, 2519.
McLoughlin, B. J. and Smith, L. H. (1965). Belg. Pat. 660,800 (*Chem. Abs.*, 1966, **64**, 2091g).
Mears, A. J., Oakeshott, S. H. and Plant, S. G. P. (1934). *J. Chem. Soc.*, 272.
Meisenheimer, J. and Witte, K. (1903). *Ber.*, **36**, 4153.
Mendlik, F. and Wibaut, J. P. (1931). *Rec. Trav. Chim.*, **50**, 91.
Menicagli, R., Malanga, C. and Lardicci, L. (1979). *J. Het. Chem.*, **16**, 667.
Menichi, G., Bisagni, E. and Royer, R. (1964). *Compt. Rend. Acad. Sci.*, **259**, 2258.
Mentzer, C. (1946). *Compt. Rend. Acad. Sci.*, **222**, 1176.
Mentzer, C., Beaudet, C. and Bory, M. (1953). *Bull. Soc. Chim. France*, 421.
Merchant, J. R. and Salgar, S. S. (1963). *J. Indian Chem. Soc.*, **40**, 23.
Merchant, J. R. and Shah, N. J. (1975). *Curr. Sci.*, **44**, 12.
Merck and Co., Inc. (1961). Brit. Pat. 859,223 (*Chem. Abs.*, 1961, **55**, 15510d).
Merck and Co., Inc. (1962). Belg. Pat. 615,395 (*Chem. Abs.*, 1963, **59**, 8707e).
Merck and Co., Inc. (1963). Belg. Pat. 621,313 (*Chem. Abs.*, 1963, **59**, 11434e).
Merck and Co., Inc. (1964a). Brit. Pat. 948,460 (*Chem. Abs.*, 1964, **60**, 14477g).
Merck and Co., Inc. (1964b). Neth. Pat. 6,404,781 (*Chem. Abs.*, 1965, **62**, 16198f).
Merck and Co., Inc. (1964c). Neth. Pat. 6,405,591 (*Chem. Abs.*, 1965, **62**, 16198b).
Merck and Co., Inc. (1965). Fr. Pat. 1,384,248 (*Chem. Abs.*, 1965, **62**, 16197a).
Merck and Co., Inc. (1966). Neth. Pat. 6,517,262 (*Chem. Abs.*, 1966, **65**, 20102b).
Merck and Co., Inc. (1967a). Neth. Pat. 6,609,058 (*Chem. Abs.*, 1967, **67**, 90670d).
Merck and Co., Inc. (1967b). Neth. Pat. 6,609,138 (*Chem. Abs.*, 1967, **67**, 32591t).
Mester, I., Choudhury, M. K. and Reisch, J. (1980). *Liebig's Ann. Chem.*, 241.
Meyer, H. (1940). In *Synthese der Kohlenstoffverbindungen*. Julius Springer, Vienna, Part 2 (i).
Meyer, K. (1964). *Zeit. Chem.*, **4**, 147.
Meyer, V. (1877). *Ber.*, **10**, 2075.
Meyer, V. and Lecco, M. T. (1883). *Ber.*, **16**, 2976.

Michael, A. (1892). *Amer. J. Chem.*, **14**, 481.
Michaelis, A. (1886). *Ber.*, **19**, 2448.
Michaelis, A. (1889). *Liebig's Ann. Chem.*, **252**, 266.
Michaelis, A. (1897). *Ber.*, **30**, 2809.
Michaelis, A. and Luxembourg, K. (1893). *Ber.*, **26**, 2174.
Miller, B. (1974). *J. Amer. Chem. Soc.*, **96**, 7155.
Miller, B. and Matjeka, E. R. (1977). *Tetrahedron Letts.*, 131.
Miller, B. and Matjeka, E. R. (1980). *J. Amer. Chem. Soc.*, **102**, 4772.
Miller, F. M. and Lohr, R. A. (1978). *J. Org. Chem.*, **43**, 3388.
Miller, F. M. and Schinske, W. N. (1978). *J. Org. Chem.*, **43**, 3384.
Miller, S. A. (1958). Brit. Pat. 806, 493 (*Chem. Abs.*, 1959, **53**, 11409d).
Miller, S. I. and Tanaka, R. (1970). In *Selective Organic Transformations*, ed. by Thyagarajan, B. S., Wiley-Interscience, New York, London, Sydney and Toronto, Vol. 1, p. 143.
Mills, K. (1969). M.Sc. Thesis, University of Manchester, England.
Mills, K., Al Khawaja, I. K., Al-Saleh, F. S. and Joule, J. A. (1981). *J. Chem. Soc., Perkin I*, 636.
Millson, M. F. and Robinson, R. (1955). *J. Chem. Soc.*, 3362.
Milne, A. H. and Tomlinson, M. L. (1952). *J. Chem. Soc.*, 2789.
Mispelter, J., Croisy, A., Jacquignon, P., Ricci, A., Rossi, C. and Schiaffela, F. (1977). *Tetrahedron*, **33**, 2383.
Mkhitaryan, A. V., Kogodovskaya, A. A., Terzyan, A. G. and Tatevosyan, G. T. (1962). *Izv. Akad. Nauk. Arm. SSR, Khim. Nauki*, **15**, 379 (*Chem. Abs.*, 1963, **59**, 2753e).
Mndzhoyan, A. L., Tatevosyan, G. T. and Ekmekdzhayan, S. P. (1957). *Izv. Akad. Nauk Arm. SSR, Khim. Nauki*, **10**, 291 (*Chem. Abs.*, 1960, **54**, 12105h).
Mndzhoyan, A. L., Tatevosyan, G. T., Terzyan, A. G. and Ekmekdzhyan, S. P. (1958). *Izv. Akad. Nauk Arm. SSR, Khim. Nauki*, **11**, 127 (*Chem. Abs.*, 1959, **53**, 9196g).
Moggridge, R. C. G. and Plant, S. G. P. (1937). *J. Chem. Soc.*, 1125.
Möhlau, R. (1881). *Ber.*, **14**, 171.
Moldenhauer, W. and Simon, H. (1969). *Chem. Ber.*, **102**, 1198.
Mooradian, A. (1967). *Tetrahedron Letts.*, 407.
Mooradian, A. (1969a). US Pat. 3,452,033 (*Chem. Abs.*, 1969, **71**, 91284a).
Mooradian, A. (1969b). US Pat. 3,481,944 (*Chem. Abs.*, 1970, **72**, 55237z).
Mooradian, A. (1971). US Pat. 3,558,667 (*Chem. Abs.*, 1971, **75**, 20176p).
Mooradian, A. (1973). Ger. Pat. 2,240,211 (*Chem. Abs.*, 1973, **78**, 136069x).
Mooradian, A. (1975). Belg. Pat. 817,970 (*Chem. Abs.*, 1975, **83**, 97014g).
Mooradian, A. (1976a). US Pat. 3,931,223 (*Chem. Abs.*, 1976, **84**, 90026z).
Mooradian, A. (1976b). US Pat. 3,959,309 (*Chem. Abs.*, 1976, **85**, 123759s).
Mooradian, A. (1976c). Ger. Pat. 2,435,394 (*Chem. Abs.*, 1976, **84**, 164605m).
Mooradian, A. and Dupont, P. E. (1967a). *J. Het. Chem.*, **4**, 441.
Mooradian, A. and Dupont, P. E. (1967b). *Tetrahedron Letts.*, 2867.
Moore, P. B. (1949). *Nature*, **163**, 918.
Moore, R. F. and Plant, S. G. P. (1951). *J. Chem. Soc.*, 3475.
Morgan, G. T. and Walls, L. P. (1930). *J. Chem. Soc.*, 1502.
Morita, K., Noguchi, S., Kishimoto, S., Agata, I. and Otsuka, K. (1972). Jap. Pat. 72 10,707 (*Chem. Abs.*, 1972, **77**, 75125t).
Morooka, S., Tamoto, K. and Matuura, A. (1976). Ger. Pat. 2,625,029 (*Chem. Abs.*, 1977, **86**, 106382s).
Morosawa, S. (1960). *Bull. Chem. Soc. Japan*, **33**, 1113 (*Chem. Abs.*, 1961, **55**, 27363f).
Morozovskaya, L. M., Ogareva, O. B. and Suvorov, N. N. (1969). *Chem. Het. Comps.*, **5**, 749.
Morozovskaya, L. M., Ogareva, O. B. and Suvorov, N. N. (1970). *Zh. Vses. Khim. Obshchest.*, **15**, 712 (*Chem. Abs.*, 1971, **74**, 53583b).

Morrison, G. C., Waite, R. O., Caro, A. N. and Shavel, J. (1967). *J. Org. Chem.*, **32**, 3691.
Morton, A. A. (1946). In *The Chemistry of Heterocyclic Compounds*, McGraw-Hill, New York and London, Chapter 6.
Morton, A. A. and Slaunwhite, W. R. (1949). *J. Biol. Chem.*, **179**, 259.
Mosher, C. W., Crews, O. P., Acton, E. M. and Goodman, L. (1966). *J. Med. Chem.*, **9**, 237.
Moskovkina, T. V. and Tilichenko, M. N. (1976). *Chem. Het. Comps.*, **12**, 541.
Mühlstädt, M. and Treibs, W. (1957). *Liebig's Ann. Chem.*, **608**, 38.
Mühlstädt, M. and Lichtmann, H. (1970). *J. Prakt. Chem.*, **312**, 466.
Müller, H., Montigel, C. and Reichstein, T. (1937). *Helv. Chim. Acta*, **20**, 1468.
Murakami, K. (1929). *Science Repts. Tôhoku Imp. Univ., 1st Ser.*, **18**, 651. (*Chem. Abs.*, 1930, **24**, 2445).
Murakami, Y. and Morimoto, K. (1974). Jap. Pat. 74 11,875 (*Chem. Abs.*, 1974, **80**, 108367r).
Muramatsu, M., Ishizumi, K. and Katsube, J. (1978). Jap. Pat. 78 71,066 (*Chem. Abs.*, 1978, **89**, 163396k).
Murphy, H. W. (1964). *J. Pharm. Sci.*, **53**, 272.
Mustafa, A. (1974). In *The Chemistry of Heterocyclic Compounds*, ed. by Weissberger, A. and Taylor, E. C., Wiley, New York, London, Sydney and Toronto, Vol. 29, p. 24.
Muth, C. W. and Hoyle, R. E. (1964). *Proc. West Va. Acad. Sci.*, **36**, 93 (*Chem. Abs.*, 1964, **61**, 14624d).
Muth, C. W., Steiniger, D. O. and Papanastassiou, Z. B. (1955). *J. Amer. Chem. Soc.*, **77**, 1006.
Nagai, Y., Uno, H., Shimizu, M. and Karasawa, T. (1977). Jap. Pat. 77 89,699 (*Chem. Abs.*, 1978, **88**,, 62374a).
Nagaraja, S. N. and Sunthankar, S. V. (1958). *J. Sci. Ind. Research* (*India*), **17B**, 457 (*Chem. Abs.*, 1959, **53**, 11340f).
Nagarajan, K., Pillai, P. M. and Kulkarni, C. L. (1969). *Indian J. Chem.*, **7**, 319.
Nagasaka, T., Yuge, T. and Ohki, S. (1977). *Heterocycles*, **8**, 371.
Nakao, M., Katayama, S. and Yamamoto, H. (1972a). Jap. Pat. 72 31,301 (*Chem. Abs.*, 1972, **77**, 164478c).
Nakao, M., Katayama, S. and Yamamoto, H. (1972b). Jap. Pat. 72 37,625 (*Chem. Abs.*, 1972, **77**, 164463u).
Nakao, M, Takahashi, K. and Yamamoto, H. (1971). Jap. Pat. 71 14,899 (*Chem. Abs.*, 1971, **75**, 35729v).
Nakatsuka, I., Hazue, M., Makari, Y., Kawahara, K., Endo, M. and Yoshitake, A. (1976). *J. Labelled Compounds Radiopharm.*, **12**, 395 (*Chem. Abs.*, 1977, **86**, 55232h).
Nakazaki, M. (1960a). *Bull. Chem. Soc. Japan*, **33**, 461.
Nakazaki, M. (1960b). *Bull. Chem. Soc. Japan*, **33**, 472.
Nakazaki, M. and Isoe, S. (1955). *Nippon Kagaku Zasshi*, **76**, 1159 (*Chem. Abs.*, 1957, **51**, 17877f).
Nakazaki, M., Isoe, S. and Tanno, K. (1955). *Nippon Kagaku Zasshi*, **76**, 1262 (*Chem. Abs.*, 1957, **51**, 17878b).
Nakazaki, M. and Maeda, M. (1962). *Bull. Chem. Soc. Japan*, **35**, 1380.
Nakazaki, M. and Yamamoto, K. (1976). *J. Org. Chem.*, **41**, 1877.
Nakazaki, M., Yamamoto, K. and Yamagami, K. (1960). *Bull. Chem. Soc. Japan*, **33**, 466.
Namis, A. J., Cortes, E., Collera, O. and Walls, F. (1966). *Bol. Inst. Quim. Univ. Nacl. Auton. Mex.*, **18**, 64 (*Chem. Abs.*, 1967, **67**, 73028h).
Nandi, B. K. (1940). *J. Indian Chem. Soc.*, **17**, 449.
Nastvogel, O. (1888). *Liebig's Ann. Chem.*, **248**, 85.
National Cash Register Co. (1961). Brit. Pat. 883,803 (*Chem. Abs.*, 1962, **57**, 4638b).
Naumov, Y. A. and Grandberg, I. I. (1966). *Russian Chem. Revs.*, **35**, 9.

Naumov, Y. A., Kost, A. N. and Grandberg, I. I. (1965). *Vestn. Mosk. Univ., Khim.*, **20**, 46 (*Chem. Abs.*, 1965, **62**, 14447b).
Neber, P. W. (1922). *Ber.*, **55**, 826.
Neber, P. W. (1925). *Chem.-Ztg.*, **101**, 709.
Neber, P. W., Knöller, G., Herbst, K. and Trissler, A. (1929). *Liebig's Ann. Chem.*, **471**, 113.
Nef, J. U. (1891). *Liebig's Ann. Chem.*, **266**, 52.
Nesmeyanov, A. N. and Golovnya, R. V. (1960). *Proc. Acad. Sci. USSR*, **133**, 961.
Nesmeyanov, A. N. and Golovnya, R. V. (1961). *Proc. Acad. Sci. USSR*, **136**, 147.
Neuberg, C. and Federer, M. (1905). *Ber.*, **38**, 866.
Neuss, N., Boaz, H. E. and Forbes, J. W. (1954). *J. Amer. Chem. Soc.*, **76**, 2463.
Newberry, R. A. (1974). Brit. Pat. 1,356,431 (*Chem. Abs.*, 1975, **82**, 16712n).
Newell, R. (1975). Ph.D. Thesis, University of Minnesota, USA (*Diss. Abs.*, 1976, **36**, 3392B).
Nicholson, J. S. and Peak, D. A. (1962). *Chem. Ind.*, 1244.
Niementowskii, S. V. (1901). *Ber.*, **34**, 3325.
Nietzki, R. and Goll, O. (1885). *Ber.*, **18**, 3252.
Noda, K., Nakagawa, A., Miyata, S., Nakajima, Y. and Ide, H. (1979). Jap. Pat. 79 03,060 (*Chem. Abs.*, 1979, **90**, 168454h).
Noland, W. E. and Baude, F. J. (1966). *J. Org. Chem.*, **31**, 3321.
Noland, W. E. and Reich, C. (1967). *J. Org. Chem.*, **32**, 828.
Noland, W. E. and Robinson, D. N. (1958). *Tetrahedron*, **3**, 68.
Noland, W. E., Rush, K. R. and Smith, L. R. (1966). *J. Org. Chem.*, **31**, 65.
Noland, W. E., Smith, L. R. and Johnson, D. C. (1963). *J. Org. Chem.*, **28**, 2262.
Noland, W. E., Smith, L. R. and Rush, K. R. (1965). *J. Org. Chem.*, **30**, 3457.
Nominé, G. and Pénasse, L. (1959a). Fr. Pat. 1,181,214 (*Chem. Abs.*, 1961, **55**, 14481b).
Nominé, G. and Pénasse, L. (1959b). Fr. Pat. 1,184,706 (*Chem. Abs.*, 1961, **55**, 14480h).
Nominé, G. and Pénasse, L. (1959c). Fr. Pat. 1,188,326 (*Chem. Abs.*, 1960, **54**, 24800i).
Nominé, G., Pénasse, L. and Pierdet, A. (1959). Fr. Pat. 1,189,456 (*Chem. Abs.*, 1962, **56**, 2429b).
Novikova, N. N., Silenko, I. D., Kucherova, N. F., Rozenberg, S. G. and Zagorevskii, V. A. (1971). *Chem. Het. Comps.*, **7**, 1002.
Novotny, A. and Cerveny, L. (1974). Czech. Pat. 152,565 (*Chem. Abs.*, 1974, **81**, 105274t).
Novotny, A., Cerveny, L., Kmonickova, S., Vlk, J. and Blazkova, H. (1976). *Chem. Prum.*, **26**, 458 (*Chem. Abs.*, 1977, **86**, 155914a).
Nozoe, T., Kitahara, Y. and Arai, T. (1954). *Proc. Japan. Acad.*, **30**, 478 (*Chem. Abs.*, 1956, **50**, 4900f).
Nozoe, T., Sin, J.-K., Yamane, K. and Fujimori, K. (1975). *Bull. Chem. Soc. Japan*, **48**, 314.
Oakeshott, S. H. and Plant, S. G. P. (1926). *J. Chem. Soc.*, 1210.
Oakeshott, S. H. and Plant, S. G. P. (1928). *J. Chem. Soc.*, 1840.
O'Brien, C. (1964). *Chem. Revs.*, **64**, 81.
Ockenden, D. W. and Schofield, K. (1953a). *J. Chem. Soc.*, 612.
Ockenden, D. W. and Schofield, K. (1953b). *J. Chem. Soc.*, 3440.
Ockenden, D. W. and Schofield, K. (1957). *J. Chem. Soc.*, 3175.
O'Connor, R. (1961). *J. Org. Chem.*, **26**, 4375.
O'Connor, R. and Henderson, G. (1965). *Chem. Ind.*, 850.
O'Connor, R. and Rosenbrook, W. (1961). *J. Org. Chem.*, **26**, 5208.
Ogandzhanyan, N. M., Avanesova, D. A. and Tatevosyan, G. T. (1968). *Arm. Khim. Zh.*, **21**, 730 (*Chem. Abs.*, 1969, **70**, 106294j).
Ogandzhanyan, N. M. and Tatevosyan, G. T. (1970). *Arm. Khim. Zh.*, **23**, 458 (*Chem. Abs.*, 1971, **74**, 3457t).
Ogura, K. and Tsuchihashi, G. (1974). *J. Amer. Chem. Soc.*, **96**, 1960.

Oikawa, Y. and Yonemitsu, O. (1977). *J. Org. Chem.*, **42**, 1213.
Okamoto, T., Kobayashi, T. and Yamamoto, H. (1970). Ger. Pat. 1,944,759 (*Chem. Abs.*, 1970, **72**, 121365t).
Okamoto, T., Kobayashi, T. and Yamamoto, H. (1974). Jap. Pat. 74 20,591 (*Chem. Abs.*, 1975, **82**, 140106g).
Okamoto, T., Niizaki, M., Kobayashi, T., Izumi, T., Yamamoto, H., Inaba, S., Nakamura, Y. and Nakao, M. (1971). Jap. Pat. 71 40,621 (*Chem. Abs.*, 1972, **76**, 59447y).
Okamoto, T., Niizaki, S., Kobayashi, T., Izumi, T. and Yamamoto, H. (1972). Jap. Pat. 72 13,266 (*Chem. Abs.*, 1972, **77**, 34329u).
Okamoto, T. and Shudo, K. (1973). *Tetrahedron Letts.*, 4533.
Okawa, M. and Konishi, S. (1974). Jap. Pat. 74 87,660 (*Chem. Abs.*, 1975, **82**, 139958e).
Oki, S. and Nagasaka, T. (1973). Jap. Pat. 73 22,495 (*Chem. Abs.*, 1973, **79**, 92190u).
Okuda, S. and Robison, M. M. (1959). *J. Amer. Chem. Soc.*, **81**, 740.
Okuno, H. (1952). *Kagaku no Ryôiki* (*J. Japan. Chem.*), **6**, 646 (*Chem. Abs.*, 1954, **48**, 13289f).
Oliver, G. L. (1968). Brit. Pat. 1,135,207 (*Chem. Abs.*, 1969, **70**, 68153c).
Onoda, R. and Koshinaka, K. (1957). *Mem. Gakugei Fac., Akita Univ.*, **7**, 67 (*Chem. Abs.*, 1959, **53**, 4255h).
Openshaw, H. T. and Robinson, R. (1937). *J. Chem. Soc.*, 941.
Opie, J. W., Warner, D. T. and Moe, O. A. (1952). US Pat. 2,583,010 (*Chem. Abs.*, 1952, **46**, 8680b).
Order, R. B. van and Lindwall, H. G. (1942). *Chem. Revs.*, **30**, 78.
Organon, N. V. (1972). Neth. Pat. 71 00,213, (*Chem. Abs.*, 1972, **77**, 164479d).
Orlova, E. K., Sharkova, N. M., Meshcheryakova, L. M., Zagorevskii, V. A. and Kucherova, N. F. (1975). *Chem. Het. Comps.*, **11**, 1099.
Orr, J. C. and Bowers, A. (1962). US Pat. 3,032,551 (*Chem. Abs.*, 1963, **58**, 8006a).
Overend, W. G. and Wiggins, L. F. (1947). *J. Chem. Soc.*, 549.
Owellen, R. J., Fitzgerald, J. A., Fitzgerald, B. M., Welsh, D. A., Walker, D. M. and Southwick, P. L. (1967). *Tetrahedron Letts.*, 1741.
Oxford, A. E. and Raper, H. S. (1927). *J. Chem. Soc.*, 417.
Pacheco, H. (1951). *Bull. Soc. Chim. France*, 633.
Pachter, I. J. (1967). US Pat. 3,299,078 (*Chem. Abs.*, 1968, **68**, 21919a).
Padfield, E. M. and Tomlinson, M. L. (1950). *J. Chem. Soc.*, 2272.
Pakula, R., Wojciechowski, J., Poslinska, H., Pichnej, L., Ptaszynski, L., Przepalkowski, A. and Logwinienko, R. (1969). Ger. Pat. 1,816,993 (*Chem. Abs.*, 1970, **72**, 12563k).
Pakula, R., Wojciechowski, J., Poslinska, H., Pichnej, L., Ptaszynski, L., Przepalkowski, A. and Logwinienko, R. (1971). Pol. Pat. 62,464 (*Chem. Abs.*, 1972, **76**, 113058n).
Pakula, R., Wojciechowski, J., Poslinska, H., Pichnej, L., Ptaszynski, L., Przepalkowski, A. and Logwinienko, R. (1975). US Pat. 3,919,247 (*Chem. Abs.*, 1976, **84**, 59187d).
Palazzo, G. and Baiocchi, L. (1965). *Annali Chim.*, **55**, 935 (*Chem. Abs.*, 1965, **63**, 16335g).
Palmer, M. H. and McIntyre, P. S. (1969). *J. Chem. Soc.*, (*B*), 446.
Pappalardo, G. and Vitali, T. (1958a). *Gazz. Chim. Ital.*, **88**, 564.
Pappalardo, G. and Vitali, T. (1958b). *Gazz. Chim. Ital.*, **88**, 574.
Pappalardo, G. and Vitali, T. (1958c). *Gazz. Chim. Ital.*, **88**, 1147.
Paragamian, V. (1973). US Pat. 3,723,459 (*Chem. Abs.*, 1973, **79**, 42471r).
Paragamian, V. (1974a). US Pat. 3,818,038 (*Chem. Abs.*, 1974, **81**, 105327n).
Paragamian, V. (1974b). US Pat. 3,818,039 (*Chem. Abs.*, 1974, **81**, 105326n).
Parmerter, S. M., Cook, A. G. and Dixon, W. B. (1958). *J. Amer. Chem. Soc.*, **80**, 4621.
Parrick, J. (1969) (quoted in Crooks and Robinson, B., 1969).
Parrick, J. and Wilcox, R. (1976). *J. Chem. Soc., Perkin I*, 2121.

Passerini, M. and Bonciani, T. (1933). *Gazz. Chim. Ital.*, **63**, 138.
Patel, H. P. and Tedder, J. M. (1963). *J. Chem. Soc.*, 4593.
Patterson, J. M. (1976). *Synthesis*, 281.
Paul, H. and Weise, A. (1963). *Tetrahedron Letts.*, 163.
Paul, R. (1941). *Bull. Soc. Chim. France*, **8**, 911.
Paul, R. and Tchelitcheff, S. (1948). *Bull. Soc. Chim. France*, 197.
Pausacker, K. H. (1949). *Nature*, **163**, 602.
Pausacker, K. H. (1950a). *J. Chem. Soc.*, 621.
Pausacker, K. H. (1950b). *J. Chem. Soc.*, 3478.
Pausacker, K. H. and Robinson, R. (1947). *J. Chem. Soc.*, 1557.
Pausacker, K. H. and Schubert, C. I. (1949a). *Nature*, **163**, 289.
Pausacker, K. H. and Schubert, C. I. (1949b). *J. Chem. Soc.*, 1384.
Pausacker, K. H. and Schubert, C. I. (1950). *J. Chem. Soc.*, 1814.
Pecca, J. G. and Albonico, S. M. (1970). *J. Med. Chem.*, **13**, 327.
Pecca, J. G. and Albonico, S. M. (1971). *J. Med. Chem.*, **14**, 448.
Pelchowicz, Z. and Bergmann, E. D. (1959). *J. Chem. Soc.*, 847.
Pelchowicz, Z. and Bergmann, E. D. (1960). *J. Chem. Soc.*, 4699.
Pellicciari, R., Natalini, B., Ricci, A., Alunni-Bistocchi, G. and De Meo, G. (1978). *J. Het. Chem.*, **15**, 927.
Pennington, F. C., Jellinek, M. and Thurn, R. D. (1959). *J. Org. Chem.*, **24**, 565.
Pennington, F. C., Martin, L. J., Reid, R. E. and Lapp, T. W. (1959). *J. Org. Chem.*, **24**, 2030.
Pennington, F. C., Tritle, G. L., Boyd, S. D., Bowersox, W. and Aniline, O. (1965). *J. Org. Chem.*, **30**, 2801.
Périn-Roussel, O., Buu-Hoï, N. P. and Jacquignon, P. (1972). *J. Chem. Soc., Perkin I*, 531.
Perkin, W. H. (1904a). *Proc. Chem. Soc.*, **20**, 51.
Perkin, W. H. (1904b). *J. Chem. Soc.*, **85**, 416.
Perkin, W. H. and Plant, S. G. P. (1921). *J. Chem. Soc.*, **119**, 1825.
Perkin, W. H. and Plant, S. G. P. (1923a). *J. Chem. Soc.*, **123**, 676.
Perkin, W. H. and Plant, S. G. P. (1923b). *J. Chem. Soc.*, **123**, 3242.
Perkin, W. H. and Plant, S. G. P. (1924). *J. Chem. Soc.*, **125**, 1503.
Perkin, W. H. and Plant, S. G. P. (1925). *J. Chem. Soc.*, **127**, 1138.
Perkin, W. H. and Plant, S. G. P. (1928). *J. Chem. Soc.*, 2583.
Perkin, W. H. and Riley, G. C. (1923). *J. Chem. Soc.*, **123**, 2399.
Perkin, W. H. and Robinson, R. (1913). *J. Chem. Soc.*, **103**, 1973.
Perkin, W. H. and Rubenstein, L. (1926). *J. Chem. Soc.*, 357.
Perkin, W. H. and Sprankling, C. H. G. (1899). *J. Chem. Soc.*, **75**, 11.
Perkin, W. H. and Titley, A. F. (1922). *J. Chem. Soc.*, **121**, 1562.
Petrova, T. D., Mamaev, V. P. and Yakobson, G. G. (1967). *Izv. Akad. Nauk SSSR, Ser. Khim.*, 1633 (*Chem. Abs.*, 1968, **68**, 49394z).
Petrova, T. D., Mamaev, V. P. and Yakobson, G. G. (1969). *Izv. Akad. Nauk SSSR, Ser. Khim.*, 679 (*Chem. Abs.*, 1969, **71**, 30318t).
Petrova, T. D., Savchenko, T. I., Ardyukova, T. F. and Yakobson, G. G. (1971). *Chem. Het. Comps.*, **7**, 195.
Pfenninger, H. A. (1968). S. African Pat. 67 7,171 (*Chem. Abs.*, 1969, **70**, 77795r) (also as US Pat. 3,468,894).
Pfülf, A. (1887). *Liebig's Ann. Chem.*, **239**, 215.
Philips, B. (1889). *Liebig's Ann. Chem.*, **252**, 270.
Phillips, R. R. (1959). In *Organic Reactions*, John Wiley and Sons, Inc., New York and Chapman and Hall, London, Vol. 10, p. 143.
Piers, E. and Brown, R. K. (1962). *Canad. J. Chem.*, **40**, 559.

Piers, E. and Brown, R. K. (1963). *Canad. J. Chem.*, **41**, 329.
Piers, E., Haarsted, V. B., Cushley, R. J. and Brown, R. K. (1962). *Canad. J. Chem.*, **40**, 511.
Pietra, S. and Tacconi, G. (1964). *Il Farmaco (Pavia), Ed. Sci.*, **19**, 741 (*Chem. Abs.*, 1965, **62**, 1637e).
Pigerol, C., Chandavoine, M. M., De Cointet de Fillain, P. and Nanthavong, S. (1975). Ger. Pat. 2,524,659 (*Chem. Abs.*, 1976, **84**, 122827e).
Pigerol, C., De Cointet de Fillain, P., Nanthavong, S. and Le Blay, J. (1976). Ger. Pat. 2,526,317 (*Chem. Abs.*, 1976, **84**, 164604k).
Piloty, O. (1910). *Ber.*, **43**, 489.
Piper, J. R. and Stevens, F. J. (1962). *J. Org. Chem.*, **27**, 3134.
Piper, J. R. and Stevens, F. J. (1966). *J. Het. Chem.*, **3**, 95.
Plancher, G. (1898a). *Atti R. Accad. dei Lincei Roma*, [5], 7, I, 275 and 316 (*Chem. Zentr.*, 1898, II, 297).
Plancher, G. (1898b). *Ber.*, **31**, 1488.
Plancher, G. (1898c). *Chem.-Ztg.*, **22**, 37 (*Chem. Zentr.*, 1898, I, 463).
Plancher, G. (1898d). *Gazz. Chim. Ital.*, **28**, II, 30.
Plancher, G. (1898e). *Gazz. Chim. Ital.*, **28**, II, 333.
Plancher, G. (1898f). *Gazz. Chim. Ital.*, **28**, II, 374.
Plancher, G. (1898g). *Gazz. Chim. Ital.*, **28**, II, 418.
Plancher, G. (1900a). *Gazz. Chim. Ital.*, **30**, II, 548.
Plancher, G. (1900b). *Gazz. Chim. Ital.*, **30**, II, 558.
Plancher, G. (1901). *Gazz. Chim. Ital.*, **31**, I, 280.
Plancher, G. (1902). *Gazz. Chim. Ital.*, **32**, II, 398.
Plancher, G. (1905). *Rend. Soc. Chim. Roma*, 152.
Plancher, G. and Bettinelli, D. (1899). *Gazz. Chim. Ital.*, **29**, I, 106.
Plancher, G. and Bonavia, A. (1900). *Atti R. Accad. dei Lincei Roma*, [5], **9**, I, 115 (*Chem. Zentr.*, 1903, I, 867) (*J. Chem. Soc., Abs.*, 1900, **78**, i, 560).
Plancher, G. and Bonavia, A. (1902). *Gazz. Chim. Ital.*, **32**, II, 414.
Plancher, G. and Caravaggi, A. (1905). *Atti R. Accad. dei Lincei Roma*, [5], **14**, I, 157 (*Chem. Zentr.*, 1905, I, 1154) (*J. Chem. Soc., Abs.*, 1905, **88**, i, 298).
Plancher, G. and Carrasco, O. (1904). *Atti R. Accad. dei Lincei Roma*, [5], **13**, I, 632 (*Chem. Zentr.*, 1904, II, 342) (*J. Chem. Soc., Abs.*, 1904, **86**, i, 777).
Plancher, G. and Carrasco, O. (1905). *Atti R. Accad. dei Lincei Roma*, **14**, 31 (*Chem. Zentr.*, 1905, II, 676) (*J. Chem. Soc., Abs.*, 1905, **88**, i, 719).
Plancher, G. and Carrasco, O. (1909). *Atti R. Accad. dei Lincei Roma*, [5], **18**, II, 274 (*Chem. Abs.*, 1910, **4**, 2297) (*J. Chem. Soc., Abs.*, 1909, **96**, i, 959).
Plancher, G., Cecchetti, B. and Ghigi, E. (1929). *Gazz. Chim. Ital.*, **59**, 334.
Plancher, G. and Ciusa, R. (1906). *Atti R. Accad. dei Lincei Roma*, **15**, 447, (*J. Chem. Soc., Abs.*, 1907, **92**, i, 80).
Plancher, G. and Forghieri, L. (1902). *Atti R. Accad. dei Lincei Roma*, [5], **11**, II, 182 (*Chem. Zentr.*, 1902, II, 1322) (*J. Chem. Soc., Abs.*, 1903, **84**, i, 114).
Plancher, G. and Ghigi, E. (1929). *Gazz. Chim. Ital.*, **59**, 339).
Plancher, G. and Testoni, G. (1900). *Atti R. Accad. dei Lincei Roma*, [5], **9**, I, 218 (*Chem. Zentr.*, 1900, I, 1027).
Plancher, G., Testoni, G. and Olivari, F. (1920). *Giorn. Chim. Ind. Applicata*, **2**, 458 (*Chem. Abs.*, 1921, **15**, 3103).
Plangger, A. (1905). *Monat. Chem.*, **26**, 833.
Plant, S. G. P. (1929). *J. Chem. Soc.*, 2493.
Plant, S. G. P. (1930). *J. Chem. Soc.*, 1595.
Plant, S. G. P. (1936). *J. Chem. Soc.*, 899.

Plant, S. G. P. and Rippon, D. M. L. (1928). *J. Chem. Soc.*, 1906.
Plant, S. G. P. and Rogers, K. M. (1936). *J. Chem. Soc.*, 40.
Plant, S. G. P., Rogers, K. M. and Williams, S. B. C. (1935). *J. Chem. Soc.*, 741.
Plant, S. G. P. and Rosser, R. J. (1928). *J. Chem. Soc.*, 2454.
Plant, S. G. P. and Thompson, M. W. (1950). *J. Chem. Soc.*, 1066.
Plant, S. G. P. and Tomlinson, M. L. (1931). *J. Chem. Soc.*, 3324.
Plant, S. G. P. and Tomlinson, M. L. (1932a). *J. Chem. Soc.*, 2188.
Plant, S. G. P. and Tomlinson, M. L. (1932b). *J. Chem. Soc.*, 2192.
Plant, S. G. P. and Tomlinson, M. L. (1933). *J. Chem. Soc.*, 955.
Plant, S. G. P. and Whitaker, W. A. (1940). *J. Chem. Soc.*, 283.
Plant, S. G. P. and Williams, S. B. C. (1934). *J. Chem. Soc.*, 1142.
Plant, S. G. P. and Wilson, A. E. J. (1939). *J. Chem. Soc.*, 237.
Plattner, J. J., Harbert, C. A., Tretter, J. R. and Welch, W. M. (1975). Ger. Pat. 2,514,084 (*Chem. Abs.*, 1976, **84**, 44008x).
Plieninger, H. (1950a). *Chem. Ber.*, **83**, 268.
Plieninger, H. (1950b). *Chem. Ber.*, **83**, 271.
Plieninger, H. (1950c). *Chem. Ber.*, **83**, 273.
Plieninger, H. (1954). *Chem. Ber.*, **87**, 228.
Plieninger, H. and Nógrádi, I. (1955a). *Chem. Ber.*, **88**, 1961.
Plieninger, H. and Nógrádi, I. (1955b). *Chem. Ber.*, **88**, 1964.
Plieninger, H., Suehiro, T., Suhr, K. and Decker, M. (1955). *Chem. Ber.*, **88**, 370.
Poierier, R. H. and Benington, F. (1952). *J. Amer. Chem. Soc.*, **74**, 3192.
Polaczkowa, W. and Porowska, N. (1950). *Przem. Chem.*, **6**, 340 (*Chem. Abs.*, 1952, **46**, 3039a).
Pope, W. J. (1922). *J. Soc. Chem. Ind.*, **41**, 495R.
Poraĭ-Koshits, B. A. and Salyamon, G. S. (1944). *J. Gen. Chem. USSR*, **14**, 1019 (*Chem. Abs.*, 1945, **39**, 4599_8).
Porcher, C. and Hervieux, C. (1907). *Compt. Rend. Acad. Sci.*, **145**, 345.
Portal, C. R., Dennler, E. B. and Frasca, A. R. (1967). *Anales Asoc. Quim. Argent.*, **55**, 245 (*Chem. Abs.*, 1968, **69**, 96557k).
Portal, C. R. and Frasca, A. R. (1971a). *Anales Asoc. Quim. Argent.*, **59**, 69 (*Chem. Abs.*, 1971, **75**, 35885t).
Portal, C. R. and Frasca, A. R. (1971b). *Anales Asoc. Quim. Argent.*, **59**, 77 (*Chem. Abs.*, 1971, **75**, 35872m).
Portnov, Y. N., Golubeva, G. A. and Kost, A. N. (1972a). *Chem. Het. Comps.*, **8**, 57.
Portnov, Y. N., Golubeva, G. A. and Kost, A. N. (1972b). USSR Pat. 331, 060 (*Chem. Abs.*, 1972, **77**, 19527p).
Portnov, Y. N., Golubeva, G. A., Kost, A. N. and Volkov, V. S. (1973). *Chem. Het. Comps.*, **9**, 598.
Poslinska, H., Pakula, R., Wojciechowski, J. and Pichnej, L. (1972). Pol. Pat. 65,781 (*Chem. Abs.*, 1972, **77**, 151928z).
Posvic, H., Dombro, R., Ito, H. and Telinski, T. (1974). *J. Org. Chem.*, **39**, 2575.
Potts, K. T. and Schneller, S. W. (1968). *J. Het. Chem.*, **5**, 485.
Prasad, K. B. and Swan, G. A. (1958). *J. Chem. Soc.*, 2024.
Pray, A. R. (1957). *Inorg. Syns.*, **5**, 153.
Preobrazhenskaya, M. N., Fedotova, M. N., Sorokina, N. P., Ogareva, O. B., Uvarova, N. V. and Suvorov, N. N. (1964). *J. Gen. Chem. USSR*, **34**, 1310.
Preobrazhenskaya, M. N., Uvarova, N. V., Sheinker, Y. N. and Suvorov, N. N. (1963). *Proc. Acad. Sci. USSR*, **148**, 149.
Preston, R. W. G. and Tucker, S. H. (1943). *J. Chem. Soc.*, 659.
Pretka, J. E. and Lindwall, H. G. (1954). *J. Org. Chem.*, **19**, 1080.

Protiva, M., Adlerová, E., Vejdělek, Z. J., Novák, L., Rajšner, M. and Ernest, I. (1959). *Naturwissenschaften*, **46**, 263.
Protiva, M., Ernest, I., Hněvsova, V., Novák, L. and Rajšner, M. (1960). Czech. Pat. 96,101 (*Chem. Abs.*, 1961, **55**, 10478i).
Przheval'skii, N. M. and Grandberg, I. I. (1974). *Chem. Het. Comps.*, **10**, 1395.
Przheval'skii, N. M., Grandberg, I. I. and Klyuev, N. A. (1976). *Chem. Het. Comps.*, **12**, 880.
Przheval'skii, N. M., Grandberg, I. I., Klyuev, N. A. and Belikov, A. B. (1978). *Chem. Het. Comps.*, **14**, 1093.
Pudovik, A. N. (1957). *J. Gen. Chem. USSR*, **27**, 2375.
Quadbeck, G. and Röhm, E. (1954). *Hoppe Seyler's Zeit. Physiol. Chem.*, **297**, 229.
Raiford, L. C. and Davis, H. L. (1928). *J. Amer. Chem. Soc.*, **50**, 156.
Raiford, L. C. and Entrikin, J. B. (1933). *J. Amer. Chem. Soc.*, **55**, 1125.
Raiford, L. C. and Gundy, G. V. (1938). *J. Org. Chem.*, **3**, 265.
Raiford, L. C. and Manley, R. H. (1940). *J. Org. Chem.*, **5**, 590.
Raiford, L. C. and Peterson, W. J. (1937). *J. Org. Chem.*, **1**, 544.
Raiford, L. C. and Tanzer, L. K. (1941). *J. Org. Chem.*, **6**, 722.
Rajagopalan, P. (1974). Ger. Pat. 2,330,719 (*Chem. Abs.*, 1974, **80**, 96047w).
Ramart-Lucas, P., Hoch, J. and Martynoff, M. (1937). *Bull. Soc. Chim. France*, **4**, 481.
Ramirez, F. and Kirby, A. F. (1952). *J. Amer. Chem. Soc.*, **74**, 6026.
Ramirez, F. and Kirby, A. F. (1954). *J. Amer. Chem. Soc.*, **76**, 1037.
Rapoport, H. and Tretter, J. R. (1958). *J. Amer. Chem. Soc.*, **80**, 5574.
Raschen, J. (1887). *Liebig's Ann. Chem.*, **239**, 223.
Rashidyan, L. G., Asratyan, S. N., Karagezyan, K. S., Mkrtchyan, A. R., Sedrakyan, R. O. and Tatevosyan, G. T. (1968). *Arm. Khim. Zh.*, **21**, 793 (*Chem. Abs.*, 1969, **71**, 21972z).
Rastogi, S. N., Bindra, J. S., Rai, S. N. and Anand, N. (1972). *Indian J. Chem.*, **10**, 673.
Rausser, R., Weber, L., Hershberg, E. B. and Oliveto, E. P. (1966). *J. Org. Chem.*, **31**, 1342.
Razumov, A. I. and Gurevich, P. A. (1967a). *J. Gen. Chem. USSR*, **37**, 1532.
Razumov, A. I. and Gurevich, P. A. (1967b). *Trans. Kazan. Khim. Tekhnol. Inst.*, 480 (*Chem. Abs.*, 1969, **70**, 20160a).
Razumov, A. I., Gurevich, P. A. and Baigil'dina, S. Y. (1976). *Chem. Het. Comps.*, **12**, 723.
Reckhow, W. A. and Tarbell, D. S. (1952). *J. Amer. Chem. Soc.*, **74**, 4960.
Reddelien, G. (1912). *Liebig's Ann. Chem.*, **388**, 165.
Reddy, P. A., Singh, S. and Srinivasan, V. R. (1976). *Indian J. Chem.*, **14B**, 793.
Redies, F., Redies, B., Tuerk, D. and Gille, C. (1969). Ger. Pat. 1,906,832 (*Chem. Abs.*, 1970, **72**, 55251z).
Redmore, D. (1971). *Chem. Revs.*, **71**, 329, 330, 334, 335.
Reed, G. W. B. (1975). Ph.D. Thesis, University of Toronto, Canada (*Diss. Abs.*, 1977, **37**, 5092B).
Regis Chemical Company (1973). Morton Grove, Illinois 60053, USA. Unpublished results quoted as reference 6 in Lee, F. G. H., Dickson, Suzuki, J., Zirnis and Manian, 1973.
Reif, G. (1909). *Ber.*, **42**, 3036.
Reinshagen, H. (1960). Dissertation, University of Frankfurt am Main, Germany (quoted in Fritz and Pfaender, 1965).
Reissert, A. and Heller, H. (1904). *Ber.*, **37**, 4364.
Remers, W. A., Roth, R. H., Gibs, G. J. and Weiss, M. J. (1971). *J. Org. Chem.*, **36**, 1232.
Remers, W. A., Roth, R. H. and Weiss, M. J. (1964). *J. Amer. Chem. Soc.*, **86**, 4612.
Remers, W. A. and Weiss, M. J. (1971). *J. Org. Chem.*, **36**, 1241.
Rensen, M. (1959). *Bull. Soc. Chim. Belges*, **68**, 258 (*Chem. Abs.*, 1960, **54**, 494d).
Reynolds, B. and Carson, J. (1969). Ger. Pat. 1,928,726 (*Chem. Abs.*, 1970, **72**, 55528v) (see also McNeil Laboratories, Brit. Pat. 1,278,787).

Rhoads, S. J. (1963). In *Molecular Rearrangements*, ed. by de Mayo, P., Wiley Interscience, New York and London, Chapter 11, p. 655.
Rice, L. M., Hertz, E. and Freed, M. E. (1964). *J. Med. Chem.*, **7**, 313.
Rice, L. M. and Scott, K. R. (1970). *J. Med. Chem.*, **13**, 308.
Rice, L. M., Sheth, B. S. and Wheeler, J. W. (1971). *J. Het. Chem.*, **8**, 751.
Rieche, A. and Seeboth, H. (1958a). *Angew. Chem.*, **70**, 52.
Rieche, A. and Seeboth, H. (1958b). *Angew. Chem.*, **70**, 312.
Rieche, A. and Seeboth, H. (1960a). *Liebig's Ann. Chem.*, **638**, 43.
Rieche, A. and Seeboth, H. (1960b). *Liebig's Ann. Chem.*, **638**, 76.
Rieche, A. and Seeboth, H. (1960c). *Liebig's Ann. Chem.*, **638**, 81.
Ried, W. and Baumbach, E. A. (1969). *Liebig's Ann. Chem.*, **726**, 81.
Ried, W. and Dankert, G. (1957). *Chem. Ber.*, **90**, 2707.
Ried, W. and Kleemann, A. (1968). *Liebig's Ann. Chem.*, **713**, 127.
Rinderknecht, H. and Niemann, C. (1950). *J. Amer. Chem. Soc.*, **72**, 2296.
Robinson, B. (1962). *Chem. Ind.*, 1291.
Robinson, B. (1963a). *Chem. Revs.*, **63**, 373.
Robinson, B. (1963b). *J. Chem. Soc.*, 586.
Robinson, B. (1963c). *J. Chem. Soc.*, 2417.
Robinson, B. (1963d). *J. Chem. Soc.*, 3097.
Robinson, B. (1964a). *J. Chem. Soc.*, 1503.
Robinson, B. (1964b). *Canad. J. Chem.*, **42**, 2900.
Robinson, B. (1964c). *Tetrahedron*, **20**, 515.
Robinson, B. (1967). *Tetrahedron Letts.*, 5085.
Robinson, B. (1969). *Chem. Revs.*, **69**, 227.
Robinson, B. and Smith, G. F. (1960). *J. Chem. Soc.*, 4574.
Robinson, B. and Zubair, M. U. (1971). *J. Chem. Soc. (C)*, 976.
Robinson, B. and Zubair, M. U. (1973). *Tetrahedron*, **29**, 1429.
Robinson, F. P. and Brown, R. K. (1964). *Canad. J. Chem.*, **42**, 1940.
Robinson, G. M. and Robinson, R. (1918). *J. Chem. Soc.*, **113**, 639.
Robinson, G. M. and Robinson, R. (1924). *J. Chem. Soc.*, **125**, 827.
Robinson, J. R. (1957). *Canad. J. Chem.*, **35**, 1570.
Robinson, J. R. and Good, N. E. (1957). *Canad. J. Chem.*, **35**, 1578.
Robinson, M. J. T. (1956) (quoted in Kinsley and Plant, 1956).
Robinson, P. and Slaytor, M. (1961). *Aust. J. Chem.*, **14**, 606.
Robinson, R. (1941). *J. Chem. Soc.*, 220.
Robinson, R. and Suginome, H. (1932a). *J. Chem. Soc.*, 298.
Robinson, R. and Suginome, H. (1932b). *J. Chem. Soc.*, 304.
Robinson, R. and Thornley, S. (1924). *J. Chem. Soc.*, **125**, 2169.
Robinson, R. and Thornley, S. (1926). *J. Chem. Soc.*, 3144.
Robson, W. (1924–1925). *J. Biol. Chem.*, **62**, 495.
Roche Products Ltd (1965a). Brit. Pat. 1,035,448.
Roche Products Ltd (1965b). Brit. Pat. 1,035,449.
Roche Products Ltd (1965c). Brit. Pat. 1,035,450.
Roche Products Ltd (1968a). Brit. Pat. 1,149,507.
Roche Products Ltd (1968b). Brit. Pat. 1,149,508.
Roche Products Ltd (1974). Brit. Pat. 1,464,432.
Roder, A. (1886). *Liebig's Ann. Chem.*, **236**, 164.
Rodionov, V. M. and Suvorov, N. N. (1950). *Zh. Obs. Khim.*, **20**, 1273 (*Chem. Abs.*, 1951, **45**, 1543g).
Rogers, C. U. and Corson, B. B. (1947). *J. Amer. Chem. Soc.*, **69**, 2910.
Rogers, C. U. and Corson, B. B. (1950). *Org. Syns.*, **30**, 90.

Rogers, C. U. and Corson, B. B. (1963). *Org. Syns.*, Coll. Vol. 4, 884.
Roosmalen, F. L. W. van (1934). *Rec. Trav. Chim.*, **53**, 359.
Rosenmund, P., Meyer, G. and Hansal, I. (1975). *Chem. Ber.*, **108**, 3538.
Rosenmund, P. and Sadri, E. (1979). *Liebig's Ann. Chem.*, 927.
Rosenmund, P. and Sotiriou, A. (1964). *Angew. Chem.* (*Intern. Edit.*), **3**, 641; *Angew. Chem.*, **76**, 787.
Rosenmund, P. and Sotiriou, A. (1975). *Chem. Ber.*, **108**, 208.
Rosenstock, P. D. (1966). *J. Het. Chem.*, **3**, 537.
Rosenthal, A. and Yalpani, M. (1965). *Canad. J. Chem.*, **43**, 3449.
Rosnati, V. and Palazzo, G. (1954). *Gazz. Chim. Ital.*, **84**, 644.
Rossner, W. (1937). *Hoppe-Seyler's Zeit. Physiol. Chem.*, **249**, 267.
Rothstein, R. and Feitelson, B. N. (1956). *Compt. Rend. Acad. Sci.*, **242**, 1042.
Roussel, O., Buu-Hoï, N. P. and Jacquignon, P. (1965). *J. Chem. Soc.*, 5458.
Roussel, P. A. (1953). *J. Chem. Educ.*, **30**, 122.
Roussel-UCLAF (1962). Brit. Pat. 888,413 (*Chem. Abs.*, 1966, **64**, 15890d).
Roussel-UCLAF (1963a). Ger. Pat. 1,143,823 (*Chem. Abs.*, 1964, **61**, 10729c).
Roussel-UCLAF (1963b). Swiss Pat. 368,492 (*Chem. Zentr.*, 1966, **137**, [11], 2627) [also in Ger. Pat. 1,219,030 (*Chem. Zentr.*, 1967, **138**, [15], 2746)].
Roussel-UCLAF (1964a). Fr. Pat. 1,366,721 (*Chem. Abs.*, 1965, **62**, 2802g).
Roussel-UCLAF (1964b). Fr. Pat. 1,366,726 (*Chem. Abs.*, 1965, **62**, 1700c).
Rousselle, D., Gilbert, J. and Viel, C. (1977). *Compt. Rend. Acad. Sci.*, **284**, 377.
Royer, R. (1946). *Ann. Chim.*, **1**, *Ser.* 12, 395.
Ruff, O. and Stein, V. (1901). *Ber.*, **34**, 1668.
Ruggli, P. (1917). *Ber.*, **50**, 883.
Ruggli, P. and Petitjean, C. (1936). *Helv. Chim. Acta*, **19**, 928.
Ruggli, P. and Straub, O. (1938). *Helv. Chim. Acta*, **21**, 1084.
Rull, T. and Le Strat, G. (1975). *Bull. Soc. Chim. France*, 1371.
Rydon, H. N. (1948). *J. Chem. Soc.*, 705.
Rydon, H. N. and Long, C. A. (1949). *Nature*, **164**, 575.
Rydon, H. N. and Siddappa, S. (1951). *J. Chem. Soc.*, 2462.
Rydon, H. N. and Tweddle, J. C. (1955). *J. Chem. Soc.*, 3499.
Sagitullin, R. S. and Koronelli, T. V. (1964). *Vestn. Mosk. Univ., Ser. II: Khim.*, **19**, 68 (*Chem. Abs.*, 1964, **61**, 3055g).
Sainsbury, M. (1977). *Synthesis*, 437.
Sakai, S., Wakabayashi, M. and Nishina, M. (1969). *Yakugaku Zasshi.*, **89**, 1061 (*Chem. Abs.*, 1970, **72**, 3308d).
Sako, S. (1934). *Bull. Chem. Soc. Japan*, **9**, 55 (*Chem. Abs.*, 1934, **28**, 3730q).
Sako, S. (1936). *Bull. Chem. Soc. Japan*, **11**, 144 (*Chem. Abs.*, 1936, **30**, 5984_3).
Sakuraba, M., Iwashita, Y. and Ninagawa, S. (1974). Jap. Pat. 74 66,678 (*Chem. Abs.*, 1975, **82**, 43415z).
Sakurai, S. and Ito, T. (1957). *Nippon Kagaku Zasshi*, **78**, 1665 (*Chem. Abs.*, 1960, **54**, 1488f).
Sakurai, S. and Komachiya, Y. (1961). *Nippon Kagaku Zasshi*, **82**, 490, (*Chem. Abs.*, 1962, **56**, 10266d).
Sakurai, S. and Komachiya, Y. (1964a). Jap. Pat. 19,805 (*Chem. Abs.*, 1965, **62**, 9144g). S
Sakurai, S. and Komachiya, Y. (1964b). Jap. Pat. 19,806 (*Chem. Abs.*, 1965, **62**, 9144g).
Saleha, S. (1979). Personal communication.
Saleha, S., Khan, N. H., Siddiqui, A. A. and Kidwai, M. M. (1978). *Indian J. Chem.*, **16B**, 1122.
Saleha, S., Siddiqi, A. A. and Khan, N. H. (1979). *Indian J. Chem.*, **17B**, 636.
Sallay, S. I. (1964). *Tetrahedron Letts.*, 2443.

Sallay, S. I. (1966). US Pat. 3,294, 817 (*Chem. Abs.*, 1967, **67**, 90787x).
Sallay, S. I. (1967a). *J. Amer. Chem. Soc.*, **89**, 6762.
Sallay, S. I. (1967b). US Pat. 3,329,686 (*Chem. Abs.*, 1968, **68**, 49449w).
Samarina, L. A., Sharkova, L. M. and Zagorevskii, V. A. (1979). *Chem. Het., Comps.*, **15**, 955.
Samsoniya, S. A., Chikvaidze, S. I., Suvorov, N. N. and Gverdtsiteli, N. (1978). *Soobschch. Akad. Nauk Graz. SSR*, **91**, 609 (*Chem. Abs.*, 1979, **90**, 137617j).
Samsoniya, S. A., Targamadze, N. L. and Suvorov, N. N. (1980). *Khim. Get. Soedin.*, **16**, 849 (*Chem. Abs.*, 1980, **93**, 204500t).
Samsoniya, S. A., Targamadze, N. L., Tret'yakova, L. G., Efimova, T. K., Turchin, K. F., Gverdtsiteli, I. M. and Suvorov, N. N. (1977). *Chem. Het. Comps.*, **13**, 758.
Samsoniya, S. A., Trapaidze, M. V., Gverdtsiteli, I. M. and Suvorov, N. N. (1977). *Chem. Het. Comps.*, **13**, 1035.
Samsoniya, S. A., Trapaidze, M. V., Suvorov, N. N. and Gverdtsiteli, I. M. (1978). *Soobschch. Akad. Nauk Graz. SSR*, **91**, 361 (*Chem. Abs.*, 1979, **90**, 54860g).
Sandoz Ltd (1980). Ger. Pat. 2,933,636 (*Chem. Abs.*, 1980, **93**, 71555r) (also in Brit. Pat. 2,032,423).
Sandoz Patents Ltd (1965). Brit. Pat. 1,004,661 (*Chem. Abs.*, 1966, **64**, 2058h) (also in US Pat. 3,211,744 and Belg. Pat. 607,751).
Sankyo Co. Ltd (1963). Jap. Pat. 17,348 ('63) (*Chem. Abs.*, 1964, **60**, 3027e).
Sankyo Co. Ltd (1964). Neth. Pat. 6,406,914 (*Chem. Abs.*, 1965, **63**, 4261c).
Sarett, L. H. and Shen, T.-Y. (1965). US Pat. 3,196,162 (*Chem. Abs.*, 1965, **63**, 16308a).
Sarett, L. H. and Shen, T.-Y. (1966a). US Pat. 3,242,162 (*Chem. Abs.*, 1966, **65**, 688d).
Sarett, L. H. and Shen, T.-Y. (1966b). US Pat. 3,242, 193 (*Chem. Abs.*, 1966, **65**, 3840e) [also in US Pat. 3,242,163 (*Chem. Abs.*, 1966, **65**, 3843e) but this reference limits the 5-substituents on the indolic product to CN, $COOC_2H_5$ and COOAlkyl].
Sato, Y. (1963a). *Sankyo Kenkyusho Nempo*, **15**, 47 (*Chem. Abs.*, 1964, **60**, 11979a).
Sato, Y. (1963b). *Chem. Pharm. Bull.*, **11**, 1431.
Sato, Y. (1963c). *Chem. Pharm. Bull.*, **11**, 1440.
Sato, Y. and Sunagawa, G. (1967). *Chem. Pharm. Bull.*, **15**, 634.
Savelli, F., Sparatore, F. and Cordella, G. (1977). *Chim. Ind.*, **59**, 300.
Sawa, Y. and Miyamoto, T. (1970). Jap. Pat. 70 25,300 (*Chem. Abs.*, 1970, **73**, 120798v).
Saxton, J. E., Smith, A. J. and Lawton, G. (1975). *Tetrahedron Letts.*, 4161.
Schäfer, H. and Tollens, B. (1906). *Ber.*, **39**, 2181.
Schank, K. and Eistert, B. (1966). *Chem. Ber.*, **99**, 1414.
Scheltus, P. I. T. (1959). Thesis, Leiden, Holland.
Scherrer, R. A. (1969). US Pat. 3,476,770 (*Chem. Abs.*, 1970, **72**, 43445c).
Schiemann, G. and Winkelmüller, W. (1933). *Ber.*, **66**, 727.
Schiess, P. and Grieder, A. (1969). *Tetrahedron Letts.*, 2097.
Schiess, P. and Grieder, A. (1974). *Helv. Chim. Acta*, **57**, 2643.
Schiess, P. and Sendi, E. (1978). *Helv. Chim. Acta*, **61**, 1364.
Schindler, W. and Häfligers, F. (1957). Swiss Pat. 324,080 (*Chem. Abs.*, 1959, **53**, 10251f).
Schlieper, A. (1886). *Liebig's Ann. Chem.*, **236**, 174.
Schlieper, A. (1887). *Liebig's Ann. Chem.*, **239**, 229.
Schlittler, E., Burckhardt, C. A. and Gellert, E. (1953). *Helv. Chim. Acta*, **36**, 1337.
Schlittler, E. and Weber, N. (1972). *J. Prakt. Chem.*, **314**, 669.
Schmid, M., Hansen, H.-J. and Schmid, H. (1971). *Helv. Chim. Acta*, **54**, 937.
Schmitz, E. and Fechner, H. (1969). *Org. Prep. Proced.*, **1**, 253.
Schmutz, J. and Wittwer, H. (1960). *Helv. Chim. Acta*, **43**, 793.
Schofield, K. and Theobald, R. S. (1949). *J. Chem. Soc.*, 796.
Schofield, K. and Theobald, R. S. (1950). *J. Chem. Soc.*, 1505.
Scholz, C. (1935). *Helv. Chim. Acta*, **18**, 923.

Schöpff, M. (1896). *Ber.*, **29**, 265.
Schoutissen, H. A. J. (1933). *J. Amer. Chem. Soc.*, **55**, 4545.
Schoutissen, H. A. J. (1934). *Rec. Trav. Chim.*, **53**, 561.
Schoutissen, H. A. J. (1935). *Rec. Trav. Chim.*, **54**, 253.
Schroeder, D. C., Corcoran, P. O., Holden, C. A. and Mulligan, M. C. (1962). *J. Org. Chem.*, **27**, 586.
Schueler, F. W. and Hanna, C. (1951). *J. Amer. Chem. Soc.*, **73**, 4996.
Schultz, A. G. and Hagmann, W. K. (1976). *J. Chem. Soc., Chem. Communs.*, 726.
Schulz, M. and Somogyi, L. (1967). *Angew. Chem.*, **6**, 168.
Schut, R. N. (1968). Fr. Pat. 1,515,629 (*Chem. Abs.*, 1969, **70**, 106382m).
Schut, R. N., Ward, F. E., Lorenzetti, O. J. and Hong, E., (1970). *J. Med. Chem.*, **13**, 394.
Schwarz, H. (1903) *Monat. Chem.*, **24**, 568.
Schwenk, E. and Whitman, B. (1937). *J. Amer. Chem. Soc.*, **59**, 949.
Scott, F. L., Houlihan, S. A. and Fenton, D. F. (1970). *Tetrahedron Letts.*, 1991.
Scriven, E. F. V., Suschitzky, H., Thomas, D. R. and Newton, R. F. (1979). *J. Chem. Soc., Perkin I*, 53.
Seeboth, H. (1967). *Angew. Chem.* (*Intern. Edit.*), **6**, 307.
Seeboth, H., Bärwolff, D. and Becker, B. (1965). *Liebig's Ann. Chem.*, **683**, 85.
Seeboth, H., Neumann, H. and Görsch, H. (1965). *Liebig's Ann. Chem.*, **683**, 93.
Seibert, W. (1948). *Chem. Ber.*, **81**, 266.
Seka, R. and Kellermann, W. (1942). *Ber.*, **75B**, 1730.
Sellstedt, J. H. and Wolf, M. (1974). US Pat. 3,813,392 (*Chem. Abs.*, 1974, **81**, 49700q).
Sempronj, A. (1938). *Gazz. Chim. Ital.*, **68**, 263.
Sen, H. K. and Ghosh, S. K. (1927). *Quarterly J. Indian Chem. Soc.*, **4**, 477 (*Chem. Abs.*, 1928, **22**, 1145).
Senda, S. and Hirota, K. (1972). *Chem. Letts.*, 367.
Senda, S. and Hirota, K. (1974). *Chem. Pharm. Bull.*, **22**, 1459.
Senda, S., Hirota, K. and Takahashi, M. (1975). *J. Chem. Soc., Perkin I*, 503.
Sergeeva, Z. F., Akhvlediani, R. N., Shabunova, V. P., Korolev, B. A., Vasil'ev, A. M., Babushkina, T. N. and Suvorov, N. N. (1975). *Chem. Het. Comps.*, **12**, 1402.
Shagalov, L. B., Eraksina, V. N. and Suvorov, N. N. (1970). *Chem. Het. Comps.*, **6**, 884.
Shagalov, L. B., Eraksina, V. N., Tkachenko, T. A., Mamonov, V. I. and Suvorov, N. N. (1972). *J. Org. Chem. USSR*, **8**, 2357.
Shagalov, L. B., Eraksina, V. N., Tkachenko, T. A. and Suvorov, N. N. (1973). USSR Pat. 371,223 (*Chem. Abs.*, 1973, **79**, 31866x).
Shagalov, L. B., Eraksina, V. N., Turchin, K. F. and Suvorov, N. N. (1970). *Chem. Het. Comps.*, **6**, 878.
Shaglov, L. B., Ostapchuk, G. M., Zlobina, A. D., Eraksina, V. N., Babushkina, T. A., Vasil'ev, A. M., Ogorodnikova, V. V. and Suvorov, N. N. (1978). *Chem. Het. Comps.*, **14**, 518.
Shagalov, L. B., Ostapchuk, G. M., Zlobina, A. D., Eraksina, V. N. and Suvorov, N. N. (1977). *Tr. Mosk. Khim.-Tekhnol. Inst.*, **94**, 32 (*Chem. Abs.*, 1980, **92**, 110768p).
Shagalov, L. B., Sorokina, N. P. and Suvorov, N. N. (1964). *J. Gen. Chem. USSR*, **34**, 1602.
Shagalov, L. B., Tkachenko, T. A., Eraksina, V. N. and Suvorov, N. N. (1973). *Tr. Mosk. Khim.-Tekhnol. Inst.*, **74**, 61 (*Chem. Abs.*, 1975, **82**, 31203r).
Shagidullin, R. R., Sattarova, F. K., Semenova, N. V., Troepol'skaya, T. V. and Kitaev, Y. P. (1963). *Bull. Acad. Sci. USSR, Div. Chem. Sci.*, 568.
Shagidullin, R. R., Sattarova, F. K., Troepol'skaya, T. V. and Kitaev, Y. P. (1963a). *Bull. Acad. Sci. USSR, Div. Chem. Sci.*, 347.
Shagidullin, R. R., Sattarova, F. K., Troepol'skaya, T. V. and Kitaev, Y. P. (1963b). *Bull. Acad. Sci. USSR, Div. Chem. Sci.*, 425.

Shah, G. D. and Patel, B. P. J. (1979). *Indian J. Chem.*, **18B**, 451.
Shapiro, D. and Abramovitch, R. A. (1955). *J. Amer. Chem. Soc.*, **77**, 6690.
Sharkova, L. M., Aksanova, L. A. and Kucherova, N. F. (1971). *Chem. Het. Comps.*, **7**,62.
Sharkova, L. M., Aksanova, L. A., Kucherova, N. F. and Zagorevskii, V. A. (1971a). *Chem. Het. Comps.*, **7**, 710.
Sharkova, L. M., Aksanova, L. A., Kucherova, N. F. and Zagorevskii, V. A. (1971b). *Chem. Het. Comps.*, **7**, 1482.
Sharkova, N. M., Kucherova, N. F., Aksanova, L. A. and Zagorevskii, V. A. (1969). *Chem. Het. Comps.*, **5**, 66.
Sharkova, N. M., Kucherova, N. F., Portnova, S. L. and Zagorevskii, V. A. (1968). *Chem. Het. Comps.*, **4**, 101.
Sharkova, N. M., Kucherova, N. F. and Zagorevskii, V. A. (1962). *J. Gen. Chem. USSR*, **32**, 3572.
Sharkova, N. M., Kucherova, N. F. and Zagorevskii, V. A. (1964). *J. Gen. Chem. USSR*, **34**, 1623.
Sharkova, N. M., Kucherova, N. F. and Zagorevskii, V. A. (1969). *Chem. Het. Comps.*, **5**, 71.
Sharkova, N. M., Kucherova, N. F. and Zagorevskii, V. A. (1972a). *Chem. Het. Comps.*, **8**, 75.
Sharkova, N. M., Kucherova, N. F. and Zagorevskii, V. A. (1972b). *Chem. Het. Comps.*, **8**, 259.
Sharkova, N. M., Kucherova, N. F. and Zagorevskii, V. A. (1972c). *Chem. Het. Comps.*, **8**, 970.
Sharkova, N. M., Kucherova, N. F. and Zagorevskii, V. A. (1974). *Chem. Het. Comps.*, **10**, 1393.
Shavel, J. and Morrison, G. C. (1967). US Pat. 3,359,273 (*Chem. Abs.*, 1968, **69**, 2872j).
Shavel, J. and Strandtmann, M. von (1965). US Pat. 3,217,029 (*Chem. Abs.*, 1966, **64**, 2057h).
Shavel, J., Strandtmann, M. von and Cohen, M. P. (1962). *J. Amer. Chem. Soc.*, **84**, 881.
Shavel, J., Strandtmann, M. von and Cohen, M. P. (1965). US Pat. 3,182,071 (*Chem. Abs.*, 1965, **63**, 11509f) [see also US Pat. 3,215,699 (*Chem. Abs.*, 1966, **64**, 12648c)].
Shaw, E. (1954). *J. Amer. Chem. Soc.*, **76**, 1384.
Shaw, E. (1955). *J. Amer. Chem. Soc.*, **77**, 4319.
Shaw, E. and Woolley, D. W. (1953). *J. Amer. Chem. Soc.*, **75**, 1877.
Shaw, E. and Woolley, D. W. (1957). *J. Amer. Chem. Soc.*, **79**, 3561.
Shemyakin, M. M., Maimind, V. I., Ermolaev, K. M. and Bamdas, E. M. (1965). *Tetrahedron*, **21**, 2771.
Shen, T.-Y. (1968). Fr. Pat. 1,529,368 (*Chem. Abs.*, 1969, **71**, 30357e).
Shen, T.-Y. (1964). US Pat. 3,161,654 (*Chem. Abs.*, 1965, **63**, 2957b).
Shen, T.-Y. (1965). US Pat. 3,201,414 (*Chem. Abs.*, 1966, **64**, 2059d).
Shen, T.-Y. (1966). US Pat. 3,242,185 (*Chem. Abs.*, 1966, **64**, 17555b).
Shen, T.-Y. (1967a). US Pat. 3,316,260 (*Chem. Abs.*, 1968, **68**, 49447u).
Shen, T.-Y. (1967b). US Pat. 3,316,267 (*Chem. Abs.*, 1968, **68**, 95683r).
Shen, T.-Y. (1967c). US Pat. 3,336,194 (*Chem. Abs.*, 1968, **68**, 29596p).
Shen, T.-Y. (1969). US Pat. 3,462,450 (*Chem. Abs.*, 1970, **72**, 12566p).
Shen, T.-Y., Gal, G. and Utne, T. (1970). Ger. Pat. 1,943,156 (*Chem. Abs.*, 1970, **73**, 14686x).
Shen, T.-Y. and Sarett, L. H. (1966). US Pat. 3,271,416 (*Chem. Abs.*, 1967, **66**, 18668w).
Shen, T.-Y., Windholz, T. B., Rosegay, A., Witzel, B. E., Wilson, A. N., Willett, J. D., Holtz, W. J., Ellis, R. L., Matzuk, A. R., Lucas, S., Stammer, C. H., Holly, F. W., Sarett, L. H., Risley, E. A., Nuss, G. W. and Winter, C. A. (1963). *J. Amer. Chem. Soc.*, **85**, 488.
Sheradsky, T. (1966). *Tetrahedron Letts.*, 5225.

Sheradsky, T. (1967). *J. Het. Chem.*, **4**, 413.
Sheradsky, T. (1970). *Tetrahedron Letts.*, 25.
Sheradsky, T. and Elgavi, A. (1968). *Israel J. Chem.*, **6**, 895.
Sheradsky, T. and Lewinter, S. (1972). *Tetrahedron Letts.*, 3941.
Sheradsky, T. and Nir, Z. (1969). *Tetrahedron Letts.*, 77.
Sheradsky, T., Nov, E., Segal, S. and Frank, A. (1977). *J. Chem. Soc., Perkin I*, 1827.
Sheradsky, T. and Salemnick, G. (1971a). *J. Org. Chem.*, **36**, 1061.
Sheradsky, T. and Salemnick, G. (1971b). *Tetrahedron Letts.*, 645.
Sheradsky, T. and Salemnick, G. (1972). *Israel J. Chem.*, **10**, 857.
Sheradsky, T., Salemnick, G. and Frankel, M. (1971). *Israel J. Chem.*, **9**, 263.
Sheradsky, T., Salemnick, G. and Nir, Z. (1972). *Tetrahedron*, **28**, 3833.
Shiho, D.-I. and Tagami, S. (1960). *J. Amer. Chem. Soc.*, **82**, 4044.
Shimizu, J., Murakami, S., Oishi, T. and Ban, Y. (1971). *Chem. Pharm. Bull.*, **19**, 2561.
Shine, H. J. (1956). *J. Amer. Chem. Soc.*, **78**, 4807.
Shine, H. J. (1967a). *Aromatic Rearrangements*, Elsevier, Amsterdam, London and New York, p. 126.
Shine, H. J. (1967b). *Aromatic Rearrangements*, Elsevier, Amsterdam, London and New York, p. 190.
Shine, H. J. (1969). In *Mechanisms of Molecular Migration*, ed. by Thyagarajan, B. S., Wiley-Interscience, New York, London, Sydney and Toronto, Vol. 2, p. 191.
Shine, H. J. and Chamness, J. T. (1963a). *J. Org. Chem.*, **28**, 1232.
Shine, H. J. and Chamness, J. T. (1963b). *Tetrahedron Letts.*, 641.
Shine, H. J. and Chamness, J. T. (1967). *J. Org. Chem.*, **32**, 901.
Shine, H. J., Huang, F.-T. and Snell, R. L. (1961). *J. Org. Chem.*, **26**, 380.
Shine, H. J. and Snell, R. L. (1957). *Chem. Ind.*, 706.
Shine, H. J. and Stanley, J. P. (1965). *Chem. Communs.*, 294.
Shine, H. J. and Stanley, J. P. (1967). *J. Org. Chem.*, **32**, 905.
Shine, H. J. and Trisler, J. C. (1960). *J. Amer. Chem. Soc.*, **82**, 4054.
Shirley, D. A. and Roussel, P. A. (1953). *J. Amer. Chem. Soc.*, **75**, 375.
Shorygin, P. P. and Polyakova, K. S. (1939). *Sintezy Dushistykh Veshchestv, Sbornik Stateĭ*, 130 (*Chem. Abs.*, 1942, **36**, 3802_4).
Shorygin, P. P. and Polyakova, K. S. (1940). *Khim. Referat. Zhur.*, [4], 114 (*Chem. Abs.*, 1942, **36**, 3802_4).
Shostakovskii, M. F., Komarov, N. V. and Roman, V. K. (1968). *Chem. Het. Comps.*, **4**, 827.
Shriner, R. L., Ashley, W. C. and Welch, E. (1942). *Org. Syns.*, **22**, 98.
Shriner, R. L., Ashley, W. C. and Welch, E. (1955). *Org. Syns.*, Coll. Vol. 3, 725.
Shukri, J., Alazawe, S. and Al-Tai, A. S. (1970). *J. Indian Chem. Soc.*, **47**, 123.
Shvedov, V. I., Alekseev, V. V. and Grinev, A. N. (1968a). USSR Pat. 215,217 (*Chem. Abs.*, 1968, **69**, 106547y).
Shvedov, V. I., Alekseev, V. V. and Grinev, A. N. (1968b). *Khim.-Farm. Zh.*, **2**, 8 (*Chem. Abs.*, 1969, **70**, 11469f).
Shvedov, V. I., Altukhova, L. B. and Grinev, A. N. (1964). USSR Pat. 165,464 (*Chem. Abs.*, 1965, **62**, 5208e).
Shvedov, V. I., Altukhova, L. B. and Grinev, A. N. (1965a). *J. Org. Chem. USSR*, **1**, 882.
Shvedov, V. I., Altukhova, L. B. and Grinev, A. N. (1965b). USSR Pat. 169,508 (*Chem. Abs.*, 1965, **63**, 2928g).
Shvedov, V. I., Altukhova, L. B. and Grinev, A. N. (1966a). *J. Org. Chem. USSR*, **2**, 387.
Shvedov, V. I., Altukhova, L. B. and Grinev, A. N. (1966b). *J. Org. Chem. USSR*, **2**, 1586.
Shvedov, V. I., Kurilo, G. N. and Grinev, A. N. (1969). *Khim.-Farm. Zh.*, **3**, 10 (*Chem. Abs.*, 1970, **72**, 12471d).

Shvedov, V. I., Altukhova, L. B. and Grinev, A. N. (1973). USSR Pat. 396,323 (*Chem. Abs.*, 1974, **80**, 14743h).
Shvedov, V. I., Altukhova, L. B., Komissarova, E. K. and Grinev, A. N. (1965a). *Chem. Het. Comps.*, **1**, 241.
Shvedov, V. I., Altukhova, L. B., Komissarova, E. K. and Grinev, A. N. (1965b). *Khim. Get. Soedin. Akad. Nauk Latv. SSR*, 365 (*Chem. Abs.*, 1965, **63**, 14800d).
Shvedov, V. I., Kurilo, G. N., Cherkasova, A. A. and Grinev, A. N. (1973). *Chem. Het. Comps.*, **9**, 965.
Shvedov, V. I., Kurilo, G. N., Cherkasova, A. A. and Grinev, A. N. (1975). *Chem. Het. Comps.*, **11**, 956.
Shvedov, V. I., Kurilo, G. N., Cherkasova, A. A. and Grinev, A. N. (1977). *Chem. Het. Comps.*, **13**, 305.
Shvedov, V. I., Kurilo, G. N. and Grinev, A. N. (1969). *Khim.-Farm. Zh.*, **3**, 10 (*Chem. Abs.*, 1970, **72**, 12471d).
Shvedov, V. I., Kurilo, G. N. and Grinev, A. N. (1970). *Khim.-Farm. Zh.*, **4**, 11 (*Chem. Abs.*, 1970, **73**, 3734k).
Shvedov, V. I., Kurilo, G. N. and Grinev, A. N. (1972). *Chem. Het. Comps.*, **8**, 974.
Shvedov, V. I., Trofimkin, Y. I., Vasil'eva, V. K. and Grinev, A. N. (1975). *Chem. Het. Comps.*, **11**, 1133.
Shvedov, V. I., Trofimkin, Y. I., Vasil'eva, V. K., Vlasova, T. F. and Grinev, A. N. (1975). *Chem. Het. Comps.*, **11**, 802.
Shvedov, V. I., Vasil'eva, V. K., Grinev, A. N. and Trofimkin, Y. I. (1975). USSR Pat. 478,833 (*Chem. Abs.*, 1975, **83**, 193277r).
Sibiryakova, D. V., Brovkin, L. V., Belyakova, T. A. and Grandberg, I. I. (1969). *Chem. Het. Comps.*, **5**, 77.
Sikharulidze, M. I., Khoshtariya, T. E., Kurkovskaya, L. N., Tret'yakova, L. G., Efimova, T. K. and Suvorov, N. N. (1979). *Chem. Het. Comps.*, **15**, 1097.
Simon, H. and Moldenhauer, W. (1967). *Chem. Ber.*, **100**, 1949.
Simon, H. and Moldenhauer, W. (1968). *Chem. Ber.*, **101**, 2124.
Simon, H. and Moldenhauer, W. (1969). *Chem. Ber.*, **102**, 1191.
Simonsen, J. L. (1949). In *The Terpenes*, Second Edition (revised by Simonsen, J. and Owen, L. N.), University Press, Cambridge, Vol. 2, pp. 438–440.
Singer, H. and Shive, W. (1955a). *J. Org. Chem.*, **20**, 1458.
Singer, H. and Shive, W. (1955b). *J. Amer. Chem. Soc.*, **77**, 5700.
Singer, H. and Shive, W. (1957). *J. Org. Chem.*, **22**, 84.
Singh, B. K. (1913). *J. Chem. Soc.*, **103**, 604.
Singh, B. K. (1920). *J. Chem. Soc.*, **117**, 1202.
Sircar, A. C. and Gopalan, M. D. R. (1932). *J. Indian Chem. Soc.*, **9**, 297.
Skrabal, P., Steiger, J. and Zollinger, H. (1975). *Helv. Chim. Acta*, **58**, 800.
Sladkov, V. I., Anisimova, O. S. and Suvorov, N. N. (1977). *Chem. Het. Comps.*, **13**, 875.
Sladkov, V. I., Shner, V. F., Alekseeva, L. M., Turchin, K. F., Anisimova, O. S., Sheinker, Y. N. and Suvorov, N. N. (1971). *Proc. Acad. Sci. USSR*, **198**, 443.
Sladkov, V. I., Shner, V. F., Anisimova, O. S., Alekseeva, L. M., Lisitsa, L. I., Terekhina, A. I. and Suvorov, N. N. (1974). *J. Org. Chem. USSR*, **10**, 1295.
Sladkov, V. I., Shner, V. F., Anisimova, O. S. and Suvorov, N. N. (1972). *J. Org. Chem. USSR*, **8**, 1291.
Sletzinger, M., Gaines, W. A. and Ruyle, W. V. (1957). *Chem. Ind.*, 1215.
Sletzinger, M. and Gal, G. (1968a). Fr. Pat. 1,538,296 (*Chem. Abs.*, 1970, **72**, 3374x).
Sletzinger, M. and Gal, G. (1968b). Fr. Pat. 1,540,724 (*Chem. Abs.*, 1969, **71**, 81159z).
Sletzinger, M., Gal, G. and Chemerda, J. M. (1968). Fr. Pat. 1,540,725 (*Chem. Abs.*, 1969, **71**, 81161u).

Sletzinger, M., Ruyle, W. V. and Gaines, W. A. (1961). US Pat. 2,995,566 (*Chem. Abs.*, 1962, **56**, 1431b).
Smith, A. (1896). *Liebig's Ann. Chem.*, **289**, 310.
Smith, A. and McCoy, H. N. (1902). *Ber.*, **35**, 2169.
Smith, A. and Utley, J. H. P. (1970). *J. Chem. Soc.* (*C*), 1.
Smith, C. R. (1930). *J. Amer. Chem. Soc.*, **52**, 397.
Smith, G. F. and Wróbel, J. T. (1960). *J. Chem. Soc.*, 792.
Smith, L. I. and Sogn, A. W. (1945). *J. Amer. Chem. Soc.*, **67**, 822.
Smith, W. S. and Moir, R. Y. (1952). *Canad. J. Chem.*, **30**, 411.
S. M. Kirov Chemical Engineering Institute (1965). USSR Pat. 172,796 (*Chem. Abs.*, 1966, **64**, 707b).
Snyder, H. R., Beilfuss, H. R. and Williams, J. K. (1953). *J. Amer. Chem. Soc.*, **75**, 1873.
Snyder, H. R. and Cook, P. L. (1956). *J. Amer. Chem. Soc.*, **78**, 969.
Snyder, H. R. and Eliel, E. L. (1948). *J. Amer. Chem. Soc.*, **70**, 1703.
Snyder, H. R., Eliel, E. L. and Carnahan, R. E. (1951). *J. Amer. Chem. Soc.*, **73**, 970.
Snyder, H. R., Merica, E. P., Force, C. G. and White, E. G. (1958). *J. Amer. Chem. Soc.*, **80**, 4622.
Snyder, H. R. and Smith, C. (1944). *J. Amer. Chem. Soc.*, **66**, 350.
Snyder, H. R. and Smith, C. W. (1943). *J. Amer. Chem. Soc.*, **65**, 2452.
Societe des Usines Chimiques Rhone-Poulenc (1960). Fr. Pat. 71,930 (*Chem. Abs.*, 1962, **57**, 3419d).
Sorokin, V. I. (1973). Candidate's Thesis, Timiryazev Agricultural Academy, Moscow, USSR, 1973 (quoted as ref. 52 in Grandberg and Sorokin, 1974).
Sorokin, V. I., Larshin, Y. A. and Grandberg, I. I. (1977). *Dokl. Timiryazev Sel'skokhoz. Akad.*, 228, 140 (*Chem. Abs.*, 1979, **90**, 22461m).
Sorokina, N. P. (1960) (quoted in Suvorov and Murasheva, 1960).
Sorrentino, P. D. (1968). S. African Pat. 68 01,254 (*Chem. Abs.*, 1969, **70**, 77783k).
Southwick, P. L. (1978). Personal communication.
Southwick, P. L., McGrew, B., Engel, R. R., Milliman, G. E. and Owellen, R. J. (1963). *J. Org. Chem.*, **28**, 3058.
Southwick, P. L. and Owellen, R. J. (1960). *J. Org. Chem.*, **25**, 1133.
Southwick, P. L., Vida, J. A., FitzGerald, B. M. and Lee, S. K. (1968). *J. Org. Chem.*, **33**, 2051.
Soutter, R. A. and Tomlinson, M. (1961). *J. Chem. Soc.*, 4256.
Sparatore, F. (1958). *Gazz. Chim. Ital.*, **88**, 755.
Sparatore, F. (1962). *Gazz. Chim. Ital.*, **92**, 596.
Sparatore, F., Boido, V. and Pirisino, G. (1974). *Tetrahedron Letts.*, 2371.
Sparatore, F. and Cerri, R. (1968). *Annali Chim.*, **58**, 1477 (*Chem. Abs.*, 1969, **70**, 96538a).
Sparatore, F. and Pirisino, G. (1965). *Gazz. Chim. Ital.*, **95**, 546.
Späth, E. and Brunner, O. (1925). *Ber.*, **58**, 518.
Späth, E. and Lederer, E. (1930a). *Ber.*, **63**, 120.
Späth, E. and Lederer, E. (1930b). *Ber.*, **63**, 2102.
Spickett, R. G. W. (1966). *J. Med. Chem.*, **9**, 436.
Sprague, J. M. and Land, A. H. (1957). In *Heterocyclic Compounds*, ed. by Elderfield, R. C., John Wiley and Sons, Inc., New York and Chapman and Hall, Ltd, London, Vol. 5, Chapter 8, pp. 512 and 581.
Srivastava, K. C. and Berlin, K. D. (1972). *J. Org. Chem.*, **37**, 4487.
Staněk, J. and Rybář, D. (1946). *Chem. Listy*, **40**, 173 (*Chem. Abs.*, 1951, **45**, 5147e).
Stapfer, C. H. and D'Andrea, R. W. (1970). *J. Het. Chem.*, **7**, 651.
Staunton, R. S. and Topham, A. (1953). *J. Chem. Soc.*, 1889.
Steche, A. (1887). *Liebig's Ann. Chem.*, **242**, 367.

Steck, E. A., Fletcher, L. T. and Carabateas, C. D. (1974). *J. Het. Chem.*, **11**, 387.
Stedman, E. (1924). *J. Chem. Soc.*, **125**, 1373.
Steinman, M. and Tahbaz, P. (1980). *Eur. Pat. Appl.*, 10,617 (*Chem. Abs.*, 1980, **93**, 239223g).
Sterling Drug Inc. (1962). Brit. Pat. 895,430 (*Chem. Abs.*, 1962, **57**, 13725e).
Sterling Drug Inc. (1970). Brit. Pat. 1,189,064 (*Chem. Abs.*, 1970, **73**, 14684v).
Sterling Drug Inc. (1975). Brit. Pat. 1,386,391 (*Chem. Abs.*, 1975, **83**, 58652u).
Sternbach, L. H., Fryer, R. I., Metlesics, W., Sach, G. and Stempel, A. (1962). *J. Org. Chem.*, **27**, 3781.
Stevens, F. J. (1962). Personal communication (quoted as ref. 113 in Robinson, B., 1969).
Stevens, F. J., Ashby, E. C. and Downey, W. E. (1957). *J. Amer. Chem. Soc.*, **79**, 1680.
Stevens, F. J. and Fox, S. W. (1948). *J. Amer. Chem. Soc.*, **70**, 2263.
Stevens, F. J. and Higginbotham, D. H. (1954). *J. Amer. Chem. Soc.*, **76**, 2206.
Stevens, F. J. and Su, H. C.-F. (1962). *J. Org. Chem.*, **27**, 500.
Stevens, R. V., Fitzpatrick, J. M., Kaplan, M. and Zimmerman, R. L. (1971). *Chem. Communs.*, 857.
Stillwell, R. N. (1964). Ph.D. Thesis, Harvard University, USA (quoted in Sainsbury, 1977 in which the relevant experimental details are also given) (*Diss. Abs.*, 1964, **25**, 2769).
Stobbe, H. and Nowak, R. (1913). *Ber.*, **46**, 2887.
Stoll, A., Troxler, F., Peyer, J. and Hofmann, A. (1955). *Helv. Chim. Acta*, **38**, 1452.
Stolz, F. (1892). *Ber.*, **25**, 1663.
Stork, G. (1964). In *Special Lectures presented at the Third International Symposium on the Chemistry of Natural Products*, Kyoto, Japan (April 1964), Butterworths, London, p. 131.
Stork, G. and Dolfini, J. E. (1963). *J. Amer. Chem. Soc.*, **85**, 2872.
Stradyn', Y. P. (1979). *Chem. Het. Comps.*, **15**, 1262.
Strandtmann, M. von, Cohen, M. P. and Shavel, J. (1963). *J. Med. Chem.*, **6**, 719.
Strandtmann, M. von, Cohen, M. P. and Shavel, J. (1965). *J. Med. Chem.*, **8**, 200.
Strandtmann, M. von, Puchalski, C. and Shavel, J. (1964). *J. Med. Chem.*, **7**, 141.
Strecker, A. (1871). *Ber.*, **4**, 784.
Stroh, H. H. and Westphal, G. (1963). *Chem. Ber.*, **96**, 184.
Struthers, R. de J. F. (1905). *Proc. Chem. Soc.*, **21**, 95.
Sturm, H. J., Tritschler, C. and Zeidler, A. (1972). Ger. Pat. 2,104,377 (*Chem. Abs.*, 1972, **77**, 151929a).
Sucrow, W. (1969). *Chimia*, **23**, 36.
Sucrow, W., Bethke, H. and Chondromatidis, G. (1971). *Tetrahedron Letts.*, 1481.
Sucrow, W. and Chondromatidis, G. (1970). *Chem. Ber.*, **103**, 1759.
Sucrow, W. and Grosz, K.-P. (1976). *Chem. Ber.*, **109**, 2154.
Sucrow, W., Mentzel, C. and Slopianka, M. (1973). *Chem. Ber.*, **106**, 450.
Sucrow, W., Mentzel, C. and Slopianka, M. (1974). *Chem. Ber.*, **107**, 1318.
Sucrow, W. and Slopianka, M. (1972). *Chem. Ber.*, **105**, 3807.
Sucrow, W. and Slopianka, M. (1978). *Chem. Ber.*, **111**, 780.
Sucrow, W., Slopianka, M. and Mentzel, C. (1973). *Chem. Ber.*, **106**, 745.
Sucrow, W., Slopianka, M. and Neophytou, A. (1972). *Chem. Ber.*, **105**, 2143.
Sucrow, W. and Wiese, E. (1970). *Chem. Ber.*, **103**, 1767.
Sugasawa, S. and Nakamura, S. (1953). *J. Pharm. Soc. Japan*, **73**, 647 (*Chem. Abs.*, 1954, **48**, 5181b).
Sugasawa, S. and Takano, S. (1959). *Chem. Pharm. Bull.*, **7**, 417.
Sugasawa, S., Terashima, M. and Kanaoka, Y. (1956). *Chem. Pharm. Bull.*, **4**, 16.
Sumitomo, Y. (1969). Jap. Pat. 69 31,342 (*Chem. Abs.*, 1970, **72**, 121364s).
Sumitomo, Y., Hayakawa, A. and Tsubojima, K. (1969). Jap. Pat. 69 31,342 (*Chem. Abs.*, 1970, **72**, 121364s).

Sumitomo Chemical Co. Ltd (1966). Neth. Pat. 6,605,169 (*Chem. Abs.*, 1967, **67**, 90668j).
Sumitomo Chemical Co. Ltd (1967). Jap. Pat. 15,092 (*Chem. Abs.*, 1968, **68**, 104976t).
Sumitomo Chemical Co. Ltd (1968a). Fr. Pat. 1,543,321 (*Chem. Abs.*, 1970, **72**, 121366u).
Sumitomo Chemical Co. Ltd (1968b). Fr. Pat. 1,545,576 (*Chem. Abs.*, 1969, **71**, 91294d).
Sumitomo Chemical Co. Ltd (1968c). Fr. Pat. 1,551,429 (*Chem. Abs.*, 1969, **71**, 124231m).
Sumitomo Chemical Co. Ltd (1969a). Fr. Pat. 1,583,552 (*Chem. Abs.*, 1970, **73**, 98797v).
Sumitomo Chemical Co. Ltd (1969b). Fr. Pat. 6774 (*Chem. Abs.*, 1971, **74**, 125418b).
Sumpter, W. C. (1945). *Chem. Revs.*, **37**, 443.
Sumpter, W. C. and Miller, F. M. (1954). In *Heterocyclic Compounds with Indole and Carbazole Systems*, ed. by Weissberger, A., Interscience, New York and London, Chapter 1 and 2.
Sunagawa, G. and Sato, Y. (1962a). *Yakugaku Zasshi*, **82**, 408 (*Chem. Abs.*, 1963, **58**, 6773d).
Sunagawa, G. and Sato, Y. (1962b). *Yakugaku Zasshi*, **82**, 414 (*Chem. Abs.*, 1963, **58**, 6773h).
Sunagawa, G. and Sato, Y. (1967). Jap. Pat. 67 21,828 (*Chem. Abs.*, 1968, **69**, 19012b).
Sunagawa, G., Soma, N., Nakano, H. and Matsumoto, Y. (1961). *Yakugaku Zasshi*, **81**, 1799 (*Chem. Abs.*, 1962, **57**, 16534b).
Sundberg, R. J. (1970). In *The Chemistry of Indoles*, ed. by Blomquist, A. T., Academic Press, New York and London, pp. 142, 217, 228, 234, 251 and 417.
Suschitzky, H. (1953). *J. Chem. Soc.*, 3326.
Suvorov, N. N. (1957). *J. Gen. Chem. USSR*, **27**, 2372.
Suvorov, N. N., Alyab'eva, T. M. and Khoshtariya, T. E. (1978). *Chem. Het. Comps.*, **14**, 1036.
Suvorov, N. N. and Antonov, V. K. (1952). *Dokl. Akad. Nauk SSSR*, **84**, 971 (*Chem. Abs.*, 1953, **47**, 3294e).
Suvorov, N. N., Antonov, V. K. and Rokhlin, E. M. (1953). *Dokl. Akad. Nauk SSSR*, **91**, 1345 (*Chem. Abs.*, 1954, **48**, 12078b).
Suvorov, N. N., Avramenko, V. G. and Shkil'kova, V. N. (1970). USSR Pat. 262,904 (*Chem. Abs.*, 1970, **73**, 45332h).
Suvorov, N. N., Avramenko, V. G., Shkil'kova, V. N. and Zamyshlyaeva, L. I. (1969a). Brit. Pat. 1,174,034 (*Chem. Abs.*, 1970, **72**, 66814m).
Suvorov, N. N., Avramenko, V. G., Shkil'kova, V. N. and Zamyshlyaeva, L. I. (1969b). Fr. Pat. 1,562,253 (*Chem. Abs.*, 1970, **72**, 90281f).
Suvorov, N. N., Avramenko, V. G., Shkil'kova, V. N. and Zamyshlyaeva, L. I. (1974). US Pat. 3,790,596 (*Chem. Abs.*, 1974, **80**, 82647e).
Suvorov, N. N., Bykhovskii, M. Y., Dmitrevskaya, L. I., Starostenko, N. E., Smushkevich, Y. I. and Shkil'kova, V. N. (1977). *Tr. Mosk. Khim.-Tekhnol. Inst.*, **94**, 46 (*Chem. Abs.*, 1980, **92**, 110078p).
Suvorov, N. N., Bykhovskii, M. Y. and Podkhalyuzina, N.Y. (1977). *J. Org. Chem. USSR*, **13**, 382.
Suvorov, N. N., Bykhovskii, M. Y., Pozdnyakov, A. D. and Sadovnikov, V. V. (1975). *J. Org. Chem. USSR*, **11**, 2684.
Suvorov, N. N., Dmitrevskaya, L. I., Smushkevich, Y. I. and Petrovskaya, L. Y. (1975). *J. Gen. Chem. USSR*, **45**, 1565.
Suvorov, N. N., Dmitrevskaya, L. I., Smushkevich, Y. I. and Pozdnyakov, A. D. (1972). *J. Gen. Chem. USSR*, **42**, 2736.
Suvorov, N. N., Dmitrevskaya, L. I., Smushkevich, Y. I. and Przhiyalgovskaya, N. M. (1976). *J. Org. Chem. USSR*, **12**, 872.
Suvorov, N. N., Fedotova, M. V., Orlova, L. M. and Ogareva, O. D. (1962). *J. Gen. Chem. USSR*, **32**, 2325.
Suvorov, N. N., Gordeev, E. N. and Vasin, M. V. (1974). *Chem. Het. Comps.*, **10**, 1316.

Suvorov, N. N., Gordeev, E. N. and Velezheva, V. S. (1975). USSR Pat. 463,667 (*Chem. Abs.*, 1975, **83**, 28100v).
Suvorov, N. N., Mamaev, V. P. and Rodinov, V. M. (1959). *Reaktsii Metody Issledovan. Org. Soedin*, **9** (Rodinov *et al.*, Editors, Moscow: Gosudarst. Nauch-Tekh. Izdatel. Khim. Lit.), p. 7 (*Chem. Abs.*, 1960, **54**, 17368h).
Suvorov, N. N., Mamaev, V. P. and Shagalov, L. B. (1953). *Dokl. Akad. Nauk SSSR*, **93**, 835 (*Chem. Abs.*, 1955, **49**, 1006h).
Suvorov, N. N., Mamaev, V. P. and Shagalov, L. B. (1955). *Dokl. Akad. Nauk SSSR*, **101**, 103 (*Chem. Abs.*, 1956, **50**, 2543c).
Suvorov, N. N., Mamaev, V. P. and Shagalov, L. B. (1957). USSR Pat. 105,124 (*Chem. Abs.*, 1957, **51**, 12982b).
Suvorov, N. N., Morozovskaya, L. M. and Ershova, L. I. (1962). *J. Gen. Chem. USSR*, **32**, 2521.
Suvorov, N. N., Morozovskaya, L. M. and Sorokina, G. M. (1961). *J. Gen. Chem. USSR*, **31**, 864.
Suvorov, N. N. and Murasheva, V. S. (1958). USSR. Pat. 116,106 (*Chem. Abs.*, 1959, **53**, 20090d).
Suvorov, N. N. and Murasheva, V. S. (1960). *J. Gen. Chem. USSR*, **30**, 3086.
Suvorov, N. N. and Murasheva, V. S. (1961). *Med. Prom. SSSR*, **15**, 6 (*Chem. Abs.*, 1962, **57**, 15056e).
Suvorov, N. N., Plutitskii, D. N. and Smushkevich, Y. I. (1980). *Zh. Org. Khim.*, **16**, 872 (*Chem. Abs.*, 1980, **93**, 46318h).
Suvorov, N. N., Preobrazhenskaya, M. N. and Uvarova, N. V. (1962). *J. Gen. Chem. USSR*, **32**, 1552.
Suvorov, N. N., Preobrazhenskaya, M. N. and Uvarova, N. V. (1963). *J. Gen. Chem. USSR*, **33**, 3672.
Suvorov, N. N., Preobrazhenskaya, M. N., Uvarova, N. V. and Sheinker, Y. N. (1962). *Izv. Akad. Nauk SSSR, Otdel. Khim. Nauk*, 729 (*Chem. Abs.*, 1962, **57**, 15057i).
Suvorov, N. N., Samsoniya, S. A., Chilikin, L. G., Chikvaidze, I. S., Turchin, K. F., Efimova, T. K., Tret'yakova, L. G. and Gverdtsiteli, I. M. (1978). *Chem. Het. Comps.*, **14**, 173.
Suvorov, N. N., Sergeeva, Z. F., Gryaznev, A. P., Shabunova, V. P., Tret'yakova, L. G., Efimova, T. K., Volodina, T. A., Morozova, I. A., Akhvlediani, R. N. *et al.* (1977). *Tr. Mosk. Khim.-Tekhnol. Inst.*, **94**, 23 (*Chem. Abs.*, 1980, **92**, 163872z).
Suvorov, N. N., Shkil'kova, V. N., Avramenko, V. G. and Zamyshlyaeva, L. I. (1970). USSR Pat. 279,619 (*Chem. Abs.*, 1971, **74**, 53518j).
Suvorov, N. N., Smushkevich, Y. I., Pozdnyakov, A. D. and Shteinpress, A. B. (1974). *Tr. Mosk. Khim.-Tekhnol. Inst.*, **80**, 63 (*Chem. Abs.*, 1976, **85**, 62905n).
Suvorov, N. N. and Sorokina, N. P. (1960). *J. Gen. Chem. USSR*, **30**, 2036.
Suvorov, N. N. and Sorokina, N. P. (1961). *Proc. Acad. Sci. USSR*, **136**, 151.
Suvorov, N. N., Sorokina, N. P. and Sheinker, Y. N. (1957). *Khim. Nauka i Prom.* **2**, 394 (*Chem. Abs.*, 1958, **52**, 355d).
Suvorov, N. N., Sorokina, N. P. and Sheinker, I. N. (1958). *J. Gen. Chem. USSR*, **28**, 1058.
Suvorov, N. N., Sorokina, N. P. and Sheinker, I. N. (1959). *J. Gen. Chem. USSR*, **29**, 962.
Suvorov, N. N., Sorokina, N. P. and Tsvetkova, G. N. (1963). USSR Pat. 146,311 (*Chem. Abs.*, 1963, **59**, 11430f).
Suvorov, N. N., Sorokina, N. P. and Tsvetkova, G. N. (1964). *J. Gen. Chem. USSR*, **34**, 1605.
Suvorov, N. N., Starostenko, N. E., Antipina, T. V., Kanterov, V. Y., Podkhalyuzina, N. Y. and Bulgakov, O. V. (1972). *Russian J. Phys. Chem.*, **46**, 1160.
Suvorov, N. N., Starostenko, N. E., Kanterov, V. Y. and Podkhalyuzina, N. Y. (1976). *Katalitich. Sintez. i Prevrashcheniya Geterotsikl. Soedin. Geterogen. Kataliz.*, 74 (*Chem. Abs.*, 1977, **87**, 134917j).

Suvorov, N. N., Zamyshlyaeva, L. I. and Tupikina, N. A. (1969). USSR Pat. 258,313 (*Chem. Abs.*, 1970, **72**, 132506r).

Suzuki, N. and Sato, Y. (1955). *Nagoya Shiritsu Diagaku Kyoyobu Kujo*, **1**, 35 (*Chem. Abs.*, 1958, **52**, 11818d).

Swaminathan, S. and Ranganathan, S. (1957). *J. Org. Chem.*, **22**, 70.

Swaminathan, S., Ranganathan, S. and Sulochana, S. (1958). *J. Org. Chem.*, **23**, 707.

Swan, G. A. (1950). *J. Chem. Soc.*, 1534.

Swan, G. A. (1958). *J. Chem. Soc.*, 2038.

Swan, G. A. and Thomas, P. R. (1963). *J. Chem. Soc.*, 3440.

Sych, E. D. (1953). *Ukrain. Khim. Zh.*, **19**, 643 (*Chem. Abs.*, 1955, **49**, 12429e).

Szantay, C., Szabo, L. and Kalaus, G. (1974a). *Synthesis*, 354.

Szantay, C., Szabo, L. and Kalaus, G. (1974b). Ger. Pat. 2,344,919 (*Chem. Abs.*, 1974, **80**, 146016h).

Szmant, H. H. and Planinsek, H. J. (1950). *J. Amer. Chem. Soc.*, **72**, 4042.

Szmuszkovicz, J. (1967). Fr. Pat. 1,505,197 (*Chem. Abs.*, 1969, **70**, 47296m).

Szmuszkovicz, J. (1970). US Pat. 3,551,451 (*Chem. Abs.*, 1971, **74**, 141523n).

Szmuszkovicz, J. (1971). US Pat. 3,565,912 (*Chem. Abs.*, 1971, **75**, 35734t).

Szmuszkovicz, J., Glenn, E. M., Heinzelman, R. V., Hester, J. B. and Youngdale, G. A. (1966). *J. Med. Chem.*, **9**, 527.

Taber, D., Becker, E. I. and Spoerri, P. E. (1954). *J. Amer. Chem. Soc.*, **76**, 776.

Tacconi, G. and Perotti, A. (1965). *Annali Chim.*, **55**, 1223.

Tacconi, G. and Pietra, S. (1965). *Annali Chim.*, **55**, 810.

Tafel, J. (1885). *Ber.*, **18**, 1739.

Takahashi, T., Saikachi, H., Goto, H. and Shimamura, S. (1944). *J. Pharm. Soc. Japan*, **64** [8A], 7 (*Chem. Abs.*, 1951, **45**, 8529i).

Takase, K., Asao, T. and Hirata, N. (1968). *Bull. Chem. Soc. Japan.*, **41**, 3027.

Takayama, M., Nakao, M., Kimura, M., Inaba, S. and Yamamoto, H. (1974). Jap. Pat. 74 72,247 (*Chem. Abs.*, 1975, **83**, 114201k).

Tämnefors, I., Claesson, A. and Karlsson, M. (1975). *Acta Pharm. Suec.*, **12**, 435.

Tanaka, Z. (1940a). *J. Pharm. Soc. Japan*, **60**, 74 (*Chem. Abs.*, 1940, **34**, 3735_3).

Tanaka, Z. (1940b). *J. Pharm. Soc. Japan*, **60**, 219 (*Chem. Abs.*, 1940, **34**, 5446_7).

Tani, H., Otani, M., Mizutani, N. and Mashimo, K. (1969). Jap. Pat. 69 05,224 (*Chem. Abs.*, 1969, **70**, 115005t).

Tarbell, D. S. (1944). *Org. Reactions*, **2**, 1.

Täuber, E. (1890). *Ber.*, **23**, 3266.

Täuber, E. (1891). *Ber.*, **24**, 197.

Täuber, E. (1892). *Ber.*, **25**, 128.

Täuber, E. and Loewenherz, R. (1891). *Ber.*, **24**, 1033.

Taylor, E. C. and Sowinski, F. (1975). *J. Org. Chem.*, **40**, 2321.

Taylor, W. F., Weiss, H. A. and Wallace, T. J. (1968). *Chem. Ind.*, 1226.

Taylor, W. F., Weiss, H. A. and Wallace, T. J. (1969). *J. Org. Chem.*, **34**, 1759.

Taylor, W. I. (1962). *Proc. Chem. Soc.*, 247.

Tenud, L. (1975). Ger. Pat. 2,440,419 (*Chem. Abs.*, 1975, **83**, 43190t).

Teotino, U. M. (1959). *Gazz. Chim. Ital.*, **89**, 1853.

Teotino, U. M. and Maffii, G. (1962). US Pat. 3,005,827 (*Chem. Abs.*, 1962, **56**, 3460c).

Terent'ev, A. P. and Preobrazhenskaia, M. N. (1958). *Proc. Acad. Sci. USSR*, **118**, 49.

Terent'ev, A. P., Preobrazhenskaya, M. N., Bobkov, A. S. and Sorokina, G. M. (1959). *J. Gen. Chem. USSR*, **29**, 2504.

Terent'ev, P. B., Kost, A. N., Shchegolev, A. A. and Terent'ev, A. P. (1961). *Proc. Acad. Sci. USSR*, **141**, 1099.

Terzian, A. G., Safrasbekian, R. R., Sukasian, R. S. and Tatevosian, G. T. (1961). *Experientia*, **17**, 493.

Terzyan, A. G., Akopyan, Z. G. and Tatevosyan, G. T. (1961). *Izv. Akad. Nauk Arm. SSR, Khim. Nauki*, **14**, 71 (*Chem. Abs.*, 1961, **55**, 27266f).
Terzyan, A. G., Aznauryan, N. V. and Tatevosyan, G. T. (1965). *Izv. Akad. Nauk Arm. SSR, Khim. Nauki*, **18**, 88 (*Chem. Abs.*, 1965, **63**, 6948e).
Terzyan, A. G., Kogadovskaya, A. A. and Tatevosyan, G. T. (1964). *Izv. Akad. Nauk Arm. SSR, Khim. Nauki*, **17**, 230 (*Chem. Abs.*, 1964, **61**, 8356a).
Terzyan, A. G., Safrazbekyan R. R., Sukasyan, R. S., Akopyran, Z. G. and Tatevosyan, G. T. (1964). *Izv. Akad. Nauk Arm. SSR, Khim. Nauki*, **17**, 567 (*Chem. Abs.*, 1965, **62**, 11868c).
Terzyan, A. G., Safrazbekyan, R. R., Sukasyan, R. S. and Tatevosyan, G. T. (1961). *Izv. Akad. Nauk Arm. SSR, Khim. Nauki*, **14**, 261 (*Chem. Abs.*, 1962, **57**, 8531i).
Terzyan, A. G. and Tatevosyan, G. T. (1960). *Izv. Akad. Nauk Arm. SSR, Khim. Nauki*, **13**, 193 (*Chem. Abs.*, 1961, **55**, 7384i).
Terzyan, A. G. and Tatevosyan, G. T. (1962). *Izv. Akad. Nauk Arm. SSR, Khim. Nauki*, **15**, 563 (*Chem. Abs.*, 1963, **59**, 6343d).
Teuber, H.-J. and Cornelius, D. (1964). *Liebig's Ann. Chem.*, **671**, 127.
Teuber, H.-J. and Cornelius, D. (1965). *Chem. Ber.*, **98**, 2111.
Teuber, H.-J., Cornelius, D. and Wölcke, U. (1966). *Liebig's Ann. Chem.*, **696**, 116.
Teuber, H.-J., Cornelius, D. and Worbs, E. (1964). *Tetrahedron Letts.*, 331.
Teuber, H.-J., Gholami, A., Reinehr, U. and Bader, H.-J. (1979). *Liebig's Ann. Chem.*, 1048.
Teuber, H.-J. and Vogel, L. (1970a). *Chem. Ber.*, **103**, 3302.
Teuber, H.-J. and Vogel, L. (1970b). *Chem. Ber.*, **103**, 3319.
Teuber, H.-J., Worbs, E and Cornelius, D. (1968). *Chem. Ber.*, **101**, 3918.
Thang, D. C., Can, C. X., Buu-Hoï, N. P. and Jacquignon, P. (1972). *J. Chem. Soc., Perkin I*, 1932.
Thang, D. C., Kossoff, E. H., Jacquignon, P. and Dufour, M. (1976). *Collect. Czech. Chem. Communs.*, **41**, 1212.
Théel, H. (1966). In *Nobel Lectures – Chemistry – 1901 to 1921*, published for the Nobel Foundation by Elsevier Publishing Co., Amsterdam, London and New York, p. 17.
Theilacker, W. and Leichtle, O. R. (1951). *Liebig's Ann. Chem.*, **572**, 121.
Thiele, J. and Heuser, K. (1896). *Liebig's Ann. Chem.*, **290**, 1.
Thomae G.m.b.H., Dr Karl (1967a). Fr. Pat. 1,469,468 (*Chem. Abs.*, 1967, **67**, 99995w) (also as Brit. Pat. 1,093,912; Fr. Pat. M 4423; Ger. Pat. 1,470,370; US Pat. 3,475,437).
Thomae G.m.b.H., Dr Karl (1967b). Belg. Pat. 693,450.
Thomae G.m.b.H., Dr Karl (1968a). Neth. Pat. 6,511,253 (*Chem. Abs.*, 1968, **68**, 39469p).
Thomae G.m.b.H., Dr Karl (1968b). Brit. Pat. 1,111,489 (*Chem. Abs.*, 1968, **69**, 51996m).
Thomae G.m.b.H., Dr Karl (1968c). Belg. Pat. 671,440 (*Chem. Abs.*, 1968, **68**, 12856a).
Thomas, R. C. (1975). *J. Labelled Compds.*, **11**, 355.
Thomas, T. J. (1975). Ph.D. Thesis, Carnegie Institute of Technology, USA.
Thyagarajan, B. S., Hillard, J. B., Reddy, K. V. and Majumdar, K. C. (1974). *Tetrahedron Letts.*, 1999.
Tien, J. M. and Hunsberger, I. M. (1955a). *Chem. Ind.*, 119.
Tien, J. M. and Hunsberger, I. M. (1955b). *J. Amer. Chem. Soc.*, **77**, 6604.
Tien, J. M. and Hunsberger, I. M. (1955c). *J. Amer. Chem. Soc.*, **77**, 6696.
Timmler, H. (1957). Ger. Pat. 964,047 (*Chem. Abs.*, 1959, **53**, 13065e).
Titley, A. F. (1928). *J. Chem. Soc.*, 2571.
Tokmakov, G. P. and Grandberg, I. I. (1975). *Khim. Dikar. Soedin., Tezisy Dokl. Vses. Konf., 4th.*, 165 (*Chem. Abs.*, 1977, **87**, 53119y).
Tokmakov, G. P. and Grandberg, I. I. (1976). USSR Pat. 523,096 (*Chem. Abs.*, 1977, **86**, 106553y).
Tokmakov, G. P. and Grandberg, I. I. (1980). *Khim. Get. Soedin.*, 331 (*Chem. Abs.*, 1980, **93**, 114352s).

Tomicek, O. (1922). *Chem. Listy*, **16**, 1, 35 (*Chem. Abs.*, 1923, **17**, 1467) (*J. Chem. Soc., Abs.*, 1922, **122**i, 679).
Tomlinson, M. L. (1951). *J. Chem. Soc.*, 809.
Torralba, A. F. and Myers, T. C. (1957). *J. Org. Chem.*, **22**, 972.
Toth, G., Szabo, G., Eibel, G. and Somfai, E. (1972). Ger. Pat. 2,135,145 (*Chem. Abs.*, 1972, **77**, 5329p).
Toth, G., Szaba, G., Eibel, G. and Somfai, E. (1973). Hung. Pat. 6608 (*Chem. Abs.*, 1974, **80**, 47835t).
Treibs, W. (1952). *Liebig's Ann. Chem.*, **576**, 110.
Treibs, W. (1959). *Naturwissenschaften*, **46**, 170.
Treibs, W., Steinert, R. and Kirchhof, W. (1953). *Liebig's Ann. Chem.*, **581**, 54.
Trenkler, B. (1888). *Liebig's Ann. Chem.*, **248**, 106.
Tret'yakova, L. G., Suvorov, N. N., Efimova, T. K., Vasil'ev, A. M., Shagalov, L. B. and Babushkina, T. A. (1978). *Chem. Het. Comps.*, **14**, 847.
Troepol'skaya, T. V. (1967). Candidate's Thesis, Kazan State University, USSR (quoted as ref. 13 in Kitaev, Buzykin and Troepol'skaya, 1960).
Trofimov, F. A., Garnova, V. I., Grinev, A. N. and Tsyshkova, N. G. (1979). *Chem. Het. Comps.*, **15**, 63.
Trofimov, F. A., Ryabchenko, V. I. and Grinev, A. N. (1975). *Chem. Het. Comps.*, **11**, 1147.
Trofimov, F. A., Tsyshkova, N. G., Garnova, V. I. and Grinev, A. N. (1975). *Chem. Het. Comps.*, **11**, 1091 [see also *Vsb., Khimiya i Farmakol. Indol'n Soedinenii*, 1975, 39 (*Chem. Abs.*, 1976, **84**, 135412r)].
Trofimov, F. A., Tsyshkova, N. G. and Grenev, A. N. (1975). USSR Pat. 457,698 (*Chem. Abs.*, 1975, **83**, 9785f).
Troxler, F., Bormann, G. and Seemann, F. (1968). *Helv. Chim. Acta*, **51**, 1203.
Tsuge, O., Hokama, K. and Koga, M. (1969). *Yakugaku Zasshi*, **89**, 798 (*Chem. Abs.*, 1969, **71**, 81292n).
Tsuge, O., Hokama, K. and Watanabe, H. (1969a). *Yakugaku Zasshi*, **89**, 783 (*Chem. Abs.*, 1969, **71**, 81263d).
Tsuge, O., Hokama, K. and Watanabe, H. (1969b). *Kogyo Kagaku Zasshi*, **72**, 1107 (*Chem. Abs.*, 1969, **71**, 81264e).
Tsuge, O., Samura, H. and Tashiro, M. (1972). *Chem. Letts.*, 1185.
Tsuge, O., Tashiro, M. and Hokama, K. (1968). *Kogyo Kagaku Zasshi*, **71**, 1203 (*Chem. Abs.*, 1969, **70**, 37083a).
Tsuge, O., Tashiro, M., Hokama, K. and Yamada, K. (1968). *Kogyo Kagaku Zasshi*, **71**, 1667 (*Chem. Abs.*, 1969, **70**, 37587t).
Tsuge, O., Watanabe, H. and Hokama, K. (1971). *Bull. Chem. Soc. Japan*, **44**, 505.
Uemura, T. and Yasuo, I. (1935). *Bull. Chem. Soc. Japan*, **10**, 169 (*Chem. Abs.*, 1935, **29**, 6217_6).
Ufimtsev, V. N., Grineva, N. I. and Sadovskaya, V. L. (1962). USSR Pat. 149,432 (*Chem. Abs.*, 1963, **58**, 10174b).
Uhle, F. C. (1949). *J. Amer. Chem. Soc.*, **71**, 761.
Ullmann, F. (1898). *Ber.*, **31**, 1697.
Ullmann, F. (1904). *Liebig's Ann. Chem.*, **332**, 82.
Ulrich, K. (1952). *Zucker*, **5**, 494 (*Chem. Abs.*, 1953, **47**, 2551i).
Unanyan, M. P. and Tatevosyan, G. T. (1961). *Izv. Akad. Nauk Arm. SSR, Khim. Nauki*, **14**, 387 (*Chem. Abs.*, 1962, **57**, 16531f).
Upjohn Co. (1966). Neth. Pat. 6,512,085 (*Chem. Abs.*, 1966, **65**, 8879d).
Ushakov, A. P., Timofeev, V. E. and Tyuryaev, I. Y. (1979). *Zh. Prikl. Khim.* (*Leningrad*), **52**, 2458 (*Chem. Abs.*, 1980, **92**, 197671y).
Utley, J. H. P. and Yeboah, S. O. (1978). *J. Chem. Soc., Perkin I*, 888.

Van de Velde, A. J. J. (1952). *Mededel. Koninkl. Vlaam. Acad. Wetenschap. Belg., Kl. Wetenschap.*, **14** [6] (*Chem. Abs.*, 1954, **48**, 3733b).
Večeřa, M., Gasparič, J. and Petránek, J. (1957). *Chem. Ind.*, 299.
Večeřa, M., Petránek, J. and Gasparič, J. (1957). *Collect. Czech. Chem. Communs.*, **22**, 1603.
Vejdelek, Z. (1955). *Ceskoslov. Farm.*, **4**, 510 (quoted as part of ref. 5 in Suvorov and Murasheva, 1960).
Vejdělek, Z. J. (1957). *Chem. Listy*, **51**, 1338 (*Chem. Abs.*, 1957, **51**, 17874i).
Velluz, L., Muller, G. and Allais, A. (1962). US Pat. 3,047,578 (*Chem. Abs.*, 1963, **58**, 11421h).
Velluz, L., Muller, G., Joly, R., Nominé, G., Mathieu, J., Allais, A., Warnant, J., Valls, J., Bucourt, R. and Jolly, J. (1958). *Bull. Soc. Chim. France*, 673.
Velluz, L., Muller, G., Nominé, G., Pénasse, L. and Pierdet, A. (1964). US Pat. 3,148,192 (*Chem. Abs.*, 1966, **64**, 15940c).
Venuto, P. B. and Landis, P. S. (1968). *Advances in Catalysis*, **18**, 259.
Verkade, P. E. and Janetzky, E. F. J. (1943a). *Rec. Trav. Chim.*, **62**, 763.
Verkade, P. E. and Janetzky, E. F. J. (1943b). *Rec. Trav. Chim.*, **62**, 775.
Verkade, P. E., Janetzky, E. F. J., Werner, E. G. G. and Lieste, J. (1943). *Verslag Gewone Vergader. Afdeel. Natuurk. Nederland. Akad. Wetenschap.*, **52**, 295 (*Chem. Abs.*, 1947, **41**, 756i).
Verkade, P. E. and Lieste, J. (1946). *Rec. Trav. Chim.*, **65**, 912.
Verley, M. A. and Beduwé, J. (1925). *Bull. Soc. Chim. France*, **37**, 189.
Vesely, V. (1905). *Ber.*, **38**, 136.
Vinogradova, E. V., Daut, K., Kost, A. N. and Terent'ev, A. P. (1962). *J. Gen. Chem. USSR*, **32**, 1536.
Vogel, A. I. (1951). In *Practical Organic Chemistry*, 2nd edition, Longmans, Green and Co., London, p. 808.
Vul'fson, N. S. and Zhurin, R. B. (1961a). *J. Gen. Chem. USSR*, **31**, 3151.
Vul'fson, N. S. and Zhurin, R. B. (1961b). *Zhur. Vsesoyuz. Khim. Obshchestva im D.I. Mendeleeva*, **6**, 239 (*Chem. Abs.*, 1961, **55**, 19912f).
Wagner, A., Schellhammer, C.-W. and Petersen, S. (1966). *Angew. Chem.* (*Intern. Edit.*), **5**, 699.
Wahl, G. (1917). *Monat. Chem.*, **38**, 525.
Wakamatsu, T., Hara, H. and Ban, Y. (1977). *Heterocycles*, **8**, 335.
Walker, C. (1892). *Amer. Chem. J.*, **14**, 576.
Walker, C. (1894). *Amer. Chem. J.*, **16**, 430.
Walker, G. N. (1955). *J. Amer. Chem. Soc.*, **77**, 3844.
Wallach, O. and Weissenborn, A. (1924). *Liebig's Ann. Chem.*, **437**, 148.
Walser, A., Blount, J. F. and Fryer, R. I. (1973). *J. Org. Chem.*, **38**, 3077.
Walther, R. von and Clemen, J. (1900). *J. Prakt. Chem.*, **61**, 249.
Walton, E., Jenkins, S. R., Nutt, R. F. and Holly, F. W. (1968). *J. Med. Chem.*, **11**, 1252.
Walton, E., Stammer, C. H., Nutt, R. F., Jenkins, S. R. and Holly, F. W. (1965). *J. Med. Chem.*, **8**, 204.
Wang, T. S. T. (1975). *Tetrahedron Letts.*, 1637.
Warner, D. T. and Moe, O. A. (1948). *J. Amer. Chem. Soc.*, **70**, 2765.
Warner-Lambert Pharmaceutical Co. (1966). Brit. Pat. 1,035,176 (*Chem. Abs.*, 1966, **65**, 13713e).
Warnhoff, E. W. and NaNonggai, P. (1962). *J. Org. Chem.*, **27**, 1186.
Warren, F. L. (1942). Unpublished experiment (quoted in Badger, Cook, J. W., Hewett, Kennaway, E. L., Kennaway, N. M. and Martin, R. H., 1942–1943).
Wassermann, H. H. and Floyd, M. B. (1963). *Tetrahedron Letts.*, 2009.

Watanabe, Y., Yamamoto, M., Shim, S. C., Miyanaga, S. and Mitsudo, T. (1980). *Chem. Letts.*, 603.
Weisbach, J. A., Macko, E., De Sanctis, N. J., Cava, M. P. and Douglas, B. (1964). *J. Med. Chem.*, **7**, 735.
Welch, W. H. and Harbert, C. A. (1976). US Pat. 3,968,231 (*Chem. Abs.*, 1977, **86**, 5434e).
Welch, W. M. (1977a). *Synthesis*, 645.
Welch, W. M. (1977b). US Pat. 4,014,890 (*Chem. Abs.*, 1977, **87**, 53078j).
Welch, W. M. and Harbert, C. A. (1977). US Pat. 4,006,164 (*Chem. Abs.*, 1977, **86**, 189906q).
Welch, W. M. and Harbert, C. A. (1980). *J. Med. Chem.*, **23**, 704.
Wells, J. E., Babcock, D. E. and France, W. G. (1936). *J. Amer. Chem. Soc.*, **58**, 2630.
Welstead, W. J. and Chen, Y.-H. (1971). Ger. Pat. 2,117,116 (*Chem. Abs.*, 1972, **76**, 59669x).
Welstead, W. J. and Chen, Y.-H. (1976). US Pat. 3,941,806 (*Chem. Abs.*, 1976, **85**, 46379p).
Welzel, P. (1970). *Chem. Ber.*, **103**, 1318.
Weng, T.-Y., Wang, C.-T. and Wang, T. (1962). *Hua Hseuh Hseuh Pao*, **28**, 108 (*Chem. Abs.*, 1963, **59**, 12743d).
Wenkert, E. (1958). (quoted as ref. 8 in MacPhillamy, Dziemian, Lucas and Kuehne, 1958).
Wenkert, E., Bhattacharyya, N. K., Reid, T. L. and Stevens, T. E. (1956). *J. Amer. Chem. Soc.*, **78**, 797.
Wenkert, E. and Dave, K. G. (1962). *J. Amer. Chem. Soc.*, **84**, 94.
Wenzing, M. (1887). *Liebig's Ann. Chem.*, **239**, 239.
Werner, L. H. (1962). US Pat. 3,024,248 (*Chem. Abs.*, 1962, **57**, 8580f).
Werner, L. H., Schroeder, D. C. and Ricca, S. (1957). *J. Amer. Chem. Soc.*, **79**, 1675.
Wheeler, W. J. (1970). Ph.D. Thesis, Purdue University, USA (*Diss. Abs.*, 1971, **31**, 4605B).
Wichelhaus, H. (1919). *Ber.*, **52A**, 129.
Widmer, U., Zsindely, J., Hansen, H.-J. and Schmid, H. (1973). *Helv. Chim. Acta*, **56**, 75.
Wieland, H. (1913). In *Die Hydrazine*. Ferdinand Enke, pp. 123–125.
Wieland, H. and Horner, L. (1938). *Liebig's Ann. Chem.*, **536**, 89.
Wieland, H., Juchum, D. and Maier, J. (1931). *Ber.*, **64**, 2513.
Wieland, T. and Grimm, D. (1965). *Chem. Ber.*, **98**, 1727.
Wieland, T. and Rühl, K. (1963). *Chem. Ber.*, **96**, 260.
Wieland, T., Weiberg, O., Fischer, Edgar and Hörlein, G. (1954). *Liebig's Ann. Chem.*, **587**, 146.
Wiley, R., Jarboe, C. and Aayer, F. (1958). *J. Org. Chem.*, **29**, 732.
Wiley, R. H. and Wiley, P. (1964). In *The Chemistry of Heterocyclic Compounds*, ed. by by Weissberger, A., Wiley-Interscience, New York, London and Sydney, Vol. 20, Part 1, Chapter 2, p. 13.
Willgerodt, C. and Harter, H. (1905). *J. Prakt. Chem.*, **71**, 409.
Winchester, M. J. and Popp, F. D. (1975). *J. Het. Chem.*, **12**, 547.
Winterfeldt, E. (1970). *Fortschr. Chem. Forsch.*, **16**, 75.
Winterfeldt, E., Krohn, W. and Stracke, H. U. (1969). *Chem. Ber.*, **102**, 2346.
Winters, G. and Di Mola, N. (1975). Ger. Pat. 2,442,667 (*Chem. Abs.*, 1975, **83**, 28096y).
Winters, G., Di Mola, N., Berti, M. and Arioli, V. (1979). *Il Farmaco, Ed. Sci.*, **34**, 507.
Wislicenus, W. (1888a). *Liebig's Ann. Chem.*, **246**, 306.
Wislicenus, W. (1888b). *Liebig's Ann. Chem.*, **246**, 339.
Wislicenus, W. and Arnold, E. (1887). *Ber.*, **20**, 3394.
Wislicenus, W., Böklen, E. and Reuthe, E. (1908). *Liebig's Ann. Chem.*, **363**, 340.
Wislicenus, W., Butterfass, G. and Koken, G. (1924). *Liebig's Ann. Chem.*, **436**, 69.
Wislicenus, W. and Münzesheimer, M. (1898). *Ber.*, **31**, 551.
Wislicenus, W. and Riethmüller, H. (1924). *Liebig's Ann. Chem.*, **436**, 82.
Wislicenus, W. and Schultz, F. (1924). *Liebig's Ann. Chem.*, **436**, 55.
Wislicenus, W. and Waldmüller, M. (1911). *Ber.*, **44**, 1564.

Witkop, B. (1944). *Liebig's Ann. Chem.*, **556**, 103.
Witkop, B. (1953a) [quoted in Ginsberg, D. and Pappo, R. (1953). *J. Amer. Chem. Soc.*, **75**, 1094].
Witkop, B. (1953b). *J. Amer. Chem. Soc.*, **75**, 3361.
Witkop, B. and Hill, R. K. (1955). *J. Amer. Chem. Soc.*, **77**, 6592 [see Witkop, B. and Hill, R. K. (1956). *J. Amer. Chem. Soc.*, **78**, 6421 for a minor correction in the original paper which is, however, unconnected with the Fischer indolization].
Witkop, B. and Patrick, J. B. (1951a). *J. Amer. Chem. Soc.*, **73**, 1558.
Witkop, B. and Patrick, J. B. (1951b). *J. Amer. Chem. Soc.*, **73**, 2188.
Witkop, B. and Patrick, J. B. (1951c). *J. Amer. Chem. Soc.*, **73**, 2196.
Witkop, B. and Patrick, J. B. (1952). *J. Amer. Chem. Soc.*, **74**, 3855.
Witkop, B. and Patrick, J. B. (1953). *J. Amer. Chem. Soc.*, **75**, 2572.
Witkop, B., Patrick, J. B. and Kissman, H. M. (1952). *Chem. Ber.*, **85**, 949.
Witkop, B., Patrick, J. B. and Rosenblum, M. (1951). *J. Amer. Chem. Soc.*, **73**, 2641.
Wittekind, R. R. and Lazarus, S. (1970). *J. Het. Chem.*, **7**, 1241.
Wittig, G. and Reichel, B. (1963). *Chem. Ber.*, **96**, 2851.
Wolf, V. (1953). *Chem. Ber.*, **86**, 840.
Wolff, L. and Mayen, H. (1912). *Liebig's Ann. Chem.*, **394**, 105.
Woodward, R. B., Cava, M. P., Ollis, W. D., Hunger, A., Daeniker, H. U. and Schenker, K. (1954). *J. Amer. Chem. Soc.*, **76**, 4749.
Woodward, R. B., Cava, M. P., Ollis, W. D., Hunger, A., Daeniker, H. U. and Schenker, K. (1963). *Tetrahedron*, **19**, 247.
Woolley, D. and Shaw, E. N. (1959). US Pat. 2,890,223 (*Chem. Abs.*, 1959, **53**, 22018a).
Woolley, D. W. and Shaw, E. (1955). *Fed. Proc.*, **14**, 307.
Wright, G. E. (1976). *J. Het. Chem.*, **13**, 539.
Wright, G. E. (1977). *J. Het. Chem.*, **14**, 701.
Wright, G. E. and Brown, N. C. (1974). *J. Med. Chem.*, **17**, 1277.
Wright, G. E. and Gambino, J. (1979). *J. Het. Chem.*, **16**, 401.
Wuyts, H. and Lacourt, A. (1935). *Bull. Sci. Acad. Roy. Belg.*, **21**, 736 (*Chem. Abs.*, 1935, **29**, 7952_5).
Yakhontov, L. N. (1968). *Russian Chem. Revs.*, **37**, 551.
Yakhontov, L. N., Glushkov, R. G., Pronina, E. V. and Smirnova, V. G. (1973). *Proc. Acad. Sci. USSR*, **212**, 751.
Yakhontov, L. N. and Marshalkin, M. F. (1971). *Proc. Acad. Sci. USSR*, **199**, 633.
Yakhontov, L. N. and Marshalkin, M. F. (1972). *Chem. Het. Comps.*, **8**, 1486.
Yakhontov, L. N. and Marshalkin, M. F. (1973). *Tetrahedron Letts.*, 2807.
Yakhontov, L. N., Marshalkin, M. F. and Anisimova, O. S. (1972). *Chem. Het. Comps.*, **8**, 463.
Yakhontov, L. N., Marshalkin, M. F., Anisimova, O. S. and Kostyuchenko, N. P. (1973). *Chem. Het. Comps.*, **9**, 345.
Yakhontov, L. N., Marshalkin, M. F. and Pronina, E. V. (1972). *Chem. Het. Comps.*, **8**, 318.
Yakhontov, L. N. and Pronina, E. V. (1968). *J. Org. Chem. USSR*, **4**, 1608.
Yakhontov, L. N. and Pronina, E. V. (1969). *Chem. Het. Comps.*, **5**, 851.
Yakhontov, L. N., Pronina, E. V., Rozynov, B. V. and Rubtsov, M. V. (1968). *Proc. Acad. Sci. USSR*, **178**, 25.
Yakhontov, L. N., Pronina, E. V. and Rubtsov, M. V. (1965). USSR Pat. 176,900 (*Chem. Abs.*, 1966, **64**, 12678h).
Yakhontov, L. N., Pronina, E. V. and Rubtsov, M. V. (1966). *Proc. Acad. Sci. USSR*, **169**, 705.
Yakhontov, L. N., Pronina, E. V. and Rubtsov, M. V. (1967). *Chem. Het. Comps.*, **3**, 549.
Yakhontov, L. N., Pronina, E. V. and Rubtsov, M. V. (1970). *Chem. Het. Comps.*, **6**, 170.

Yakhontov, L. N., Suvorov, N. N., Kanterov, V. Y., Podkhalyuzina, N. Y., Pronina, E, V., Starostenko, N. E. and Shkil'kova, V. N. (1972). *Chem. Het. Comps.*, **8**, 594.
Yakhontov, L. N., Suvorov, N. N., Pronina, E. V., Kanterov, V. Y., Podkhalyuzina, N. Y., Starostenko, N. E. and Shkil'kova, V. N. (1972). *Chem. Het. Comps.*, **8**, 1037.
Yakhontov, L. N., Uritskaya, M. Y. and Glushkov, R. G. (1976). *Chem. Het. Comps.*, **12**, 931.
Yakovenko, V. I., Mikhlina, E. E., Oganesyan, E. T. and Yakhontov, L. N. (1975). *Chem. Het. Comps.*, **11**, 1104.
Yamada, S., Chibata, I. and Tsurui, R. (1953). *Chem. Pharm. Bull.*, **1**, 14.
Yamada, S. *et al.*, (1954). Jap. Pat. 1284 ('54) (*Chem. Abs.*, 1955, **49**, 11720f).
Yamada, S. and Kunieda, T. (1967). *Chem. Pharm. Bull.*, **15**, 499.
Yamamoto, H. (1967a). *J. Org. Chem.*, **32**, 3693.
Yamamoto, H. (1967b). *Bull. Chem. Soc. Japan*, **40**, 425.
Yamamoto, H. (1968). *Chem. Pharm. Bull.*, **16**, 17.
Yamamoto, H. and Atami, T. (1969). Jap. Pat. 69 10,785 (*Chem. Abs.*, 1969, **71**, 112806q).
Yamamoto, H. and Atami, T. (1970). Jap. Pat. 70 27,966 (*Chem. Abs.*, 1971, **74**, 3501c).
Yamamoto, H. and Atami, T. (1976). Jap. Pat. 76 15,040 (*Chem. Abs.*, 1977, **86**, 89784e).
Yamamoto, H., Atsuko, M. and Takuhiro, I. (1970). Jap. Pat. 70 13,100 (*Chem. Abs.*, 1970, **73**, 66416d).
Yamamoto, H. and Atsumi, T. (1968a). *Chem. Pharm. Bull.*, **16**, 1831.
Yamamoto, H. and Atsumi, T. (1968b). *Bull. Chem. Soc. Japan*, **41**, 2431.
Yamamoto, H., Atsumi, T., Aono, S. and Kuwazima, H. (1970). Ger. Pat. 1,813,229 (*Chem. Abs.*, 1970, **73**, 87907e).
Yamamoto, H., Hirohashi, A., Izumi, T. and Koshiba, M. (1973a). Ger. Pat. 2,144,569 (*Chem. Abs.*, 1973, **78**, 147797j).
Yamamoto, H., Hirohashi, A., Izumi, T. and Koshiba, M. (1973b). Brit. Pat. 1,324,494 (*Chem. Abs.*, 1973, **79**, 126304b).
Yamamoto, H., Hirohashi, A., Izumi, T. and Koshiba, M. (1973c). Jap. Pat. 73 28,466 (*Chem. Abs.*, 1973, **78**, 159425z).
Yamamoto, H., Inaba, S., Hirohashi, T., Ishigura, K., Maruyama, I. and Mori, K. (1971). Jap. Pat. 71 12,444 (*Chem. Abs.*, 1971, **75**, 35736v).
Yamamoto, H., Inaba, S., Hirohashi, T., Ishizumi, K., Maruyama, I. and Mori, K. (1970). Ger. Pat. 1,817,757 (*Chem. Abs.*, 1971, **74**, 53528n).
Yamamoto, H., Inaba, S., Hirohashi, T., Mori, K., Ishizumi, K. and Maruyama, I. (1969). Ger. Pat. 1,812,205 (*Chem. Abs.*, 1969, **71**, 124521f).
Yamamoto, H., Inaba, S., Hirohashi, T., Okamoto, T., Ishizumi, K., Yamamoto, M., Maruyama, I., Mori, K. and Kobayashi, T. (1970). Ger. Pat. 1,817,761 (*Chem. Abs.*, 1971, **74**, 3671h).
Yamamoto, H., Inaba, S., Ishigura, K., Maruyama, I., Hirohashi, T. and Mori, K. (1971). Jap. Pat. 71 09,578 (*Chem. Abs.*, 1971, **75**, 35737w).
Yamamoto, H., Inaba, S., Izumi, T., Hirohashi, T., Akatsu, M and Maruyama, I. (1970a). Ger. Pat. 1,920,207 (*Chem. Abs.*, 1970, **72**, 90540q).
Yamamoto, H., Inaba, S., Izumi, T., Hirohashi, T., Akatsu, M. and Maruyama, I. (1970b). Ger. Pat. 1,935,671 (*Chem. Abs.*, 1970, **73**, 14676u).
Yamamoto, H., Inaba, S., Izumi, T., Okamoto, T., Hirohashi, T., Ishizumi, K., Maruyama, I., Kobayashi, T., Yamamoto, M. and Mori, K. (1971). Ger. Pat. 2,016,793 (*Chem. Abs.*, 1971, **74**, 100120k).
Yamamoto, H., Inaba, S., Nakamura, Y., Nakao, M., Niizaki, M. and Atami, T. (1970). Jap. Pat. 70 31,285 (*Chem. Abs.*, 1971, **74**, 53523g).
Yamamoto, H., Inaba, S., Nakao, M. and Koshiba, M. (1972). Jap. Pat. 72 47,384 (*Chem. Abs.*, 1973, **78**, 124441z).

Yamamoto, H., Inaba, S., Nakao, M. and Niizaki, M. (1970). Jap. Pat. 70 31,286 (*Chem. Abs.*, 1971, **74**, 53524h).

Yamamoto, H., Inaba, S., Okamoto, T., Hirohashi, T., Ishizumi, K., Akatsu, M., Kobayashi, T., Mori, K., Kume, R. and Sato, H. (1973). Jap. Pat. 73 35,073 (*Chem. Abs.*, 1974, **80**, 82652c).

Yamamoto, H., Inaba, S., Okamoto, T., Hirohashi, T., Ishizumi, K., Maruyama, I., Kobayashi, T., Yamamoto, M. and Mori, K. (1970). Ger. Pat. 2,005,845 (*Chem. Abs.*, 1970, **73**, 131048b).

Yamamoto, H., Inaba, S., Okamoto, T., Hirohashi, T., Ishizumi, K., Yamamoto, M., Maruyama, I., Mori, K. and Kobayashi, T. (1969a). S. African Pat. 68 03,041 (*Chem. Abs.*, 1969, **71**, 124519m).

Yamamoto, H., Inaba, S., Okamoto, T., Hirohashi, T., Ishizumi, K., Yamamoto, M., Maruyama, I., Mori, K. and Kobayashi, T. (1969b). S. African Pat. 68 06,061 (*Chem. Abs.*, 1970, **72**, 90541r).

Yamamoto, H., Inaba, S., Okamoto, T., Hiroashi, T., Ishizumi, K., Yamamoto, M., Maruyama, I., Mori, K. and Kobayashi, T. (1969c). Ger. Pat. 1,806,106 (*Chem. Abs.*, 1969, **71**, 70659m).

Yamamoto, H., Inaba, S., Okamoto, T., Hiroashi, T., Ishizumi, K., Yamamoto, M., Maruyama, I., Mori, K. and Kobayashi, T. (1970). Ger. Pat. 1,811,831 (*Chem. Abs.*, 1970, **73**, 35419c).

Yamamoto, H., Inaba, S., Okamoto, T., Hirohashi, T., Yamamoto, M., Mori, K., Ishigura, K., Maruyama, I. and Kobayashi, T. (1972a). Jap. Pat. 72 11,070 (*Chem. Abs.*, 1972, **77**, 34313j).

Yamamoto, H., Inaba, S., Okamoto, T., Hiroashi, T., Yamamoto, M., Mori, K., Ishigura, K., Maruyama, I. and Kobayashi, T. (1972b). Jap. Pat. 72 11,071 (*Chem. Abs.*, 1972, **77**, 48237z).

Yamamoto, H., Inaba, S., Okamoto, T., Hirohashi, T., Yamamoto, M., Mori, K., Ishigura, K., Maruyama, I. and Kobayashi, T. (1972c). Jap. Pat. 72 30,700 (*Chem. Abs.*, 1972, **77**, 139801v).

Yamamoto, H. and Kimura, M. (1974). Jap. Pat. 74 20,777 (*Chem. Abs.*, 1975, **82**, 111940u).

Yamamoto, H., Misaki, A. and Imanaka, M. (1968). *Chem. Pharm. Bull.*, **16**, 2313.

Yamamoto, H., Misaki, A. and Izumi, T. (1968). S. African Pat. 67 07,250 (*Chem. Abs.*, 1969, **70**, 87788e).

Yamamoto, H., Misaki, A. and Izumi, T. (1970). Jap. Pat. 70 13,100 (*Chem. Abs.*, 1970, **73**, 66416d).

Yamamoto, H., Misaki, A., Izumi, T. and Koshiba, M. (1971). Jap. Pat. 71 01,408 (*Chem. Abs.*, 1971, **74**, 87819y).

Yamamoto, H., Misaki, N., Izumi, T. and Koshiba, M. (1972). Jap. Pat. 72 13,267 (*Chem. Abs.*, 1972, **77**, 34304g).

Yamamoto, H., Nakamura, Y., Atami, T., Kobayashi, T. and Nakao, M. (1971). Jap. Pat. 71 38,784 (*Chem. Abs.*, 1972, **76**, 25103t).

Yamamoto, H., Nakamura, Y., Atami, T., Nakao, M. and Kobayashi, T. (1970a). Jap. Pat. 70 20,896 (*Chem. Abs.*, 1970, **73**, 87788s).

Yamamoto, H., Nakamura, Y., Atami, T., Nakao, M. and Kobayashi, T. (1970b). Jap. Pat. 70 20,898 (*Chem. Abs.*, 1970, **73**, 87787r).

Yamamoto, H., Nakamura, Y., Atami, T., Nakao, M. and Kobayashi, T., (1970c). Jap. Pat. 70 28,778 (*Chem. Abs.*, 1971, **74**, 42276c).

Yamamoto, H., Nakamura, Y., Atami, T., Nakao, M. and Kobayashi, T. (1970d). Jap. Pat. 70 28,982 (*Chem. Abs.*, 1971, **74**, 12997w).

Yamamoto, H., Nakamura, Y., Atami, T., Nakao, M. and Kobayashi, T. (1970e). Jap. Pat. 70 35,308 (*Chem. Abs.*, 1971, **74**, 125419c).

Yamamoto, H., Nakamura, Y., Atami, T., Nakao, M. and Kobayashi, T. (1970f). Jap. Pat. 70 40,284 (*Chem. Abs.*, 1971, **74**, 141527s).
Yamamoto, H., Nakamura, Y., Atami, T., Nakao, M. and Kobayashi, T. (1970g). Jap. Pat. 70 41,381 (*Chem. Abs.*, 1971, **75**, 20189v).
Yamamoto, H., Nakamura, Y., Atami, T., Nakao, M. and Kobayashi, T. (1970h). Jap. Pat. 70 41,382 (*Chem. Abs.*, 1971, **75**, 20192r).
Yamamoto, H., Nakamura, Y., Atami, T., Nakao, M., Kobayashi, T., Saito, C. and Awata, H. (1970). Jap. Pat. 70 38,045 (*Chem. Abs.*, 1971, **74**, 125422y).
Yamamoto, H., Nakamura. Y., Atami, T., Nakao, M., Kobayashi, T., Saito, C. and Kurita, H. (1969). Jap. Pat. 69 08,663 (*Chem. Abs.*, 1970, **72**, 43444b).
Yamamoto, H., Nakamura, Y. and Kobayashi, T. (1970). Jap. Pat. 70 36,745 (*Chem. Abs.*, 1971, **75**, 5692k).
Yamamoto, H., Nakamura, Y., Nakao, M., Atsumi, T. and Kobayashi, T. (1970). US Pat. 3,535,326 (*Chem. Abs.*, 1971, **74**, 125428e).
Yamamoto, H., Nakamura, Y., Nakao, M. and Kobayashi, T. (1969a). Jap. Pat. 69 05,223 (*Chem. Abs.*, 1969, **71**, 13018c).
Yamamoto, H., Nakamura, Y., Nakao, M. and Kobayashi, T. (1969b). Jap. Pat. 69 26,867 (*Chem. Abs.*, 1970, **72**, 78872s).
Yamamoto, H., Nakamura, Y., Nakao, M. and Kobayashi, T. (1970a). Jap. Pat. 70 16,944 (*Chem. Abs.*, 1970, **73**, 55967h).
Yamamoto, H., Nakamura, Y., Nakao, M. and Kobayashi, T. (1970b). Jap. Pat. 70 18,856 (*Chem. Abs.*, **73**, 87791n).
Yamamoto, H. and Nakao, M. (1968a). Jap. Pat. 68 19,949 (*Chem. Abs.*, 1969, **71**, 3270q).
Yamamoto, H. and Nakao, M. (1968b). Jap. Pat. 68 19,950 (*Chem. Abs.*, 1969, **71**, 3271r).
Yamamoto, H. and Nakao, M. (1968c). Jap. Pat. 68 19,952 (*Chem. Abs.*, 1969, **71**, 3269w).
Yamamoto, H. and Nakao, M. (1968d). Jap. Pat. 68 21,424 (*Chem. Abs.*, 1969, **70**, 68148e).
Yamamoto, H. and Nakao, M. (1968e). Jap. Pat. 68 21,425 (*Chem. Abs.*, 1969, **70**, 68146c).
Yamamoto, H. and Nakao, M. (1968f). Jap. Pat. 68 24,418 (*Chem. Abs.*, 1969, **70**, 57643m).
Yamamoto, H. and Nakao, M. (1968g). S. African Pat. 67 02,683 (*Chem. Abs.*, 1969, **70**, 68149f) (as in Sumitomo Chemical Co. Ltd, Ger. Pat. 1,620,441).
Yamamoto, H. and Nakao, M. (1969a). Jap. Pat. 69 08,662 (*Chem. Abs.*, 1971, **75**, 140693b).
Yamamoto, H. and Nakao, M. (1969b). Jap. Pat. 69 10,264 (*Chem. Abs.*, 1971, **75**, 140694c).
Yamamoto, H. and Nakao, M. (1969c). Jap. Pat. 69 12,140 (*Chem. Abs.*, 1969, **71**, 81171x).
Yamamoto, H. and Nakao, M. (1969d). Jap. Pat. 69 20,340 (*Chem. Abs.*, 1969, **71**, 112804n).
Yamamoto, H. and Nakao, M. (1969e). *J. Med. Chem.*, **12**, 176.
Yamamoto, H. and Nakao, M. (1970a). Jap. Pat. 70 02,380 (*Chem. Abs.*, 1970, **72**, 111296g).
Yamamoto, H. and Nakao, M. (1970b). Jap. Pat. 70 10,140 (*Chem. Abs.*, 1970, **73**, 45333j).
Yamamoto, H. and Nakao, M. (1970c). Jap. Pat. 70 10,141 (*Chem. Abs.*, 1970, **73**, 45334k).
Yamamoto, H. and Nakao, M. (1970d). Jap. Pat. 70 10,142 (*Chem. Abs.*, 1970, **73**, 45335m).
Yamamoto, H. and Nakao, M. (1970e). Jap. Pat. 70 10,143 (*Chem. Abs.*, 1970, **73**, 45338q).
Yamamoto, H. and Nakao, M. (1970f). Jap. Pat. 70 11,145 (*Chem. Abs.*, 1970, **73**, 25291r).
Yamamoto, H. and Nakao, M. (1970g). Jap. Pat. 70 11,896 (*Chem. Abs.*, 1970, **73**, 45336n).
Yamamoto, H. and Nakao, M. (1970h). Jap. Pat. 70 18,655 (*Chem. Abs.*, 1970, **73**, 77048m).
Yamamoto, H. and Nakao, M. (1970i). Jap. Pat. 70 37,528 (*Chem. Abs.*, 1971, **74**, 125421x).
Yamamoto, H. and Nakao, M. (1970j). Jap. Pat. 70 03,774 (*Chem. Abs.*, 1970, **72**, 132516u).
Yamamoto, H. and Nakao, M. (1971a). US Pat. 3,564,011 (*Chem. Abs.*, 1971, **75**, 35727t).
Yamamoto, H. and Nakao, M. (1971b). US Pat. 3,629,284 (*Chem. Abs.*, 1972, **76**, 113060g).
Yamamoto, H. and Nakao, M. (1971c). Jap. Pat. 71 27,191 (*Chem. Abs.*, 1971, **75**, 110187x).
Yamamoto, H. and Nakao, M. (1973). US Pat. 3,770,752 (*Chem. Abs.*, 1975, **82**, 125273r).
Yamamoto, H. and Nakao, M. (1974a). US Pat. 3,810,906 (*Chem. Abs.*, 1974, **81**, 49563x).
Yamamoto, H. and Nakao, M. (1974b). US Pat. 3,822,275 (*Chem. Abs.*, 1975, **82**, 125272q).

Yamamoto, H., Nakao, M. and Atami, T. (1969). Jap. Pat. 69 10,265 (*Chem. Abs.*, 1970, **72**, 66818r).
Yamamoto, H., Nakao, M. and Atsumi, T. (1968a). Jap. Pat. 68 21,426 (*Chem. Abs.*, 1969, **70**, 68154d).
Yamamoto, H., Nakao, M. and Atsumi, T. (1968b). Jap. Pat. 68 21,428 (*Chem. Abs.*, 1969, **70**, 68155e).
Yamamoto, H., Nakao, M. and Kobayashi, T. (1968). *Chem. Pharm. Bull.*, **16**, 647.
Yamamoto, H., Nakao, M. and Kobayashi, T. (1969). Jap. Pat. 69 08,664 (*Chem. Abs.*, 1969, **71**, 49766j).
Yamamoto, H., Nakao, M. and Kurita, H. (1968a). Jap. Pat. 68 21,427 (*Chem. Abs.*, 1969, **70**, 68147d).
Yamamoto, H., Nakao, M. and Kurita, H. (1968b). Jap. Pat. 68 21,429 (*Chem. Abs.*, 1969, **70**, 68145b).
Yamamoto, H., Nakao, M. and Okamoto, T. (1968). *Chem. Pharm. Bull.*, **16**, 1927.
Yamamoto, H., Okamoto, T. and Kobayashi, T. (1970). Ger. Pat. 1,944,758 (*Chem. Abs.*, 1970, **72**, 121361p).
Yamamoto, H., Okamoto, T., Sasajima, K., Nakao, M., Maruyama, I. and Katayama, S. (1971). Ger. Pat. 2,033,909 (*Chem. Abs.*, 1971, **74**, 99879f).
Yamamoto, H., Okamoto, T., Sasajima, K., Nakao, M., Maruyama, I. and Katayama, S. (1975). US Pat. 3,907,812 (*Chem. Abs.*, 1976, **84**, 17153j).
Yamamoto, H., Saito, C., Okamoto, T., Awata, H., Inukai, T., Hirohashi, A. and Yukawa, Y. (1969). *Arzneim.-Forsch.*, **19**, 981.
Yamamoto, T. (1970). *Yuki Gosei Kagaku Kyokai Shi*, **28**, 853 (*Chem. Abs.*, 1970, **73**, 109587e).
Yamane, K. and Fujimori, K. (1972). *Bull. Chem. Soc. Japan*, **45**, 269.
Yamane, K. and Fujimori, K. (1976). *Bull. Chem. Soc. Japan*, **49**, 1101.
Yamauchi, T., Yada, S. and Kudo, S. (1973). Jap. Pat. 73 76,864 (*Chem. Abs.*, 1974, **80**, 14842q).
Yanagita, M. and Yamakawa, K. (1956). *J. Org. Chem.*, **21**, 500.
Yanagita, M. and Yamakawa, K. (1957). *J. Org. Chem.*, **22**, 291.
Yao, H. C. (1964). *J. Org. Chem.*, **29**, 2959.
Yao, H. C. and Resnick, P. (1962). *J. Amer. Chem. Soc.*, **84**, 3514.
Yao, H. C. and Resnick, P. (1965). *J. Org. Chem.*, **30**, 2832.
Yardley, J. P. and Smith, H. (1972). US Pat. 3,637,744 (*Chem. Abs.*, 1972, **76**, 113194d).
Yasuda, H. (1954). *Repts. Sci. Research Inst.* (*Japan*), **30**, 139 (*Chem. Abs.*, 1955, **49**, 6832g).
Yevich, J. P. (1970). Ph.D. Thesis, Carnegie-Mellon University, Pittsburgh, USA (*Diss. Abs.*, 1970, **31**, [4], 1851B) (*Chem. Abs.*, 1971, **75**, 35592v).
Yevich, J. P., Murphy, J. R., Dufresne, R. F. and Southwick, P. L. (1978). *J. Het. Chem.*, **15**, 1463.
Yoneda, F., Miyamae, T. and Nitta, Y. (1967). *Chem. Pharm. Bull.*, **15**, 8.
Yonemitsu, O., Miyashita. K., Ban, Y. and Kanaoka, Y. (1969). *Tetrahedron*, **25**, 95.
Yoshida, Z., Harada, T. and Tamaru, Y. (1976). *Tetrahedron Letts.*, 3823.
Yoshimura, I., Sakamoto, H. and Matsunaga, T. (1969). Jap. Pat. 69 32,780 (*Chem. Abs.*, 1970, **72**, 66813k).
Yoshitomi Pharmaceutical Industries Ltd (1967). Fr. Pat. 1,477,152 (*Chem. Abs.*, 1968, **68**, 29609v).
Young, E. H. P. (1958). *J. Chem. Soc.*, 3495.
Young, T. E. and Ohnmacht, C. J. (1968). *J. Org. Chem.*, **33**, 1306.
Young, T. E., Ohnmacht, C. J. and Hamel, C. R. (1967). *J. Org. Chem.*, **32**, 3622.
Young, T. E. and Scott, P. H. (1965). *J. Org. Chem.*, **30**, 3613.

Young, T. E. and Scott, P. H. (1966). *J. Org. Chem.*, **31**, 343.
Young, T. E. and Scott, P. H. (1967). US Pat. 3,314,972 (*Chem. Abs.*, 1968, **68**, 39608h).
Young, T. E. and Scott, P. H. (1968a). US Pat. 3,388,133 (*Chem. Abs.*, 1968, **69**, 59213z).
Young, T. E. and Scott, P. H. (1968b). US Pat. 3,388,134 (*Chem. Abs.*, 1968, **69**, 67357q).
Youngdale, G. A., Glenn, E. M., Lednicer, D. and Szmuszkovicz, J. (1969). *J. Med. Chem.*, **12**, 948.
Yudin, L. G., Budylin, V. A. and Kost, A. N. (1964). *Metody Polucheniya Khim. Reaktivov i Preparatov*, [11], 65 (*Chem. Abs.*, 1966, **64**, 19551f).
Yudin, L. G., Chernyshova, N. B. and Kost, A. N. (1968). USSR Pat. 229,522 (*Chem. Abs.*, 1969, **70**, 87786c).
Yudin, L. G., Kost, A. N. and Berlin, Y. A. (1958). *Khim. Nauka i Prom.*, **3**, 406 (*Chem. Abs.*, 1958, **52**, 20163f).
Yudin, L. G., Kost, A. N. and Chernyshova, N. B. (1970). *Chem. Het. Comps.*, **6**, 447.
Yudin, L. G., Papravko, S. A. and Kost, A. N. (1962). *J. Gen. Chem. USSR*, **32**, 3519.
Yurina, R. A., Goryaev, M. I. and Dembitskii, A. D. (1969). *Izv. Akad. Nauk Kaz. SSR, Ser. Khim.*, **19**, 27 (*Chem. Abs.*, 1970, **72**, 21784k).
Zagorevskii, V. A., Kucherova, N. F. and Sharkova, N. M. (1977). *Chem. Het. Comps.*, **13**, 862.
Zagorevskii, V. A., Kucherova, N. F., Sharkova, N. M., Ivanova, T. I. and Klyuev, S. M. (1975). *Chem. Het. Comps.*, **11**, 1156.
Zagorevskii, V. A., Samarina, L. A. and Sharkova, L. M. (1978). *Chem. Het. Comps.*, **14**, 702.
Zagorevskii, V. A., Samarina, L. A., Sharkova, L. M. and Aksanova, L. A. (1979). *Chem. Het. Comps.*, **15**, 842.
Zaitsev, I. A. (1938). Dissertation, Kazan (quoted in Kitaev, 1959).
Zander, M. (1962). *Naturwissenschaften*, **49**, 300.
Zander, M. and Franke, W. (1963). *Chem. Ber.*, **96**, 699.
Zander, M. and Franke, W. H. (1964). *Angew. Chem.* (*Intern. Edit*), **3**, 755.
Zander, M. and Franke, W. H. (1967). *Chem. Ber.*, **100**, 2649.
Zander, M. and Franke, W. H. (1969). *Chem. Ber.*, **102**, 2728.
Zee, S. and Ho, Y.-S. (1970). *J. Chin. Chem. Soc.* (*Taipei*), **17**, 179 (*Chem. Abs.*, 1971, **74**, 53404u).
Zenitz, B. L. (1965). US Pat. 3,193,235 (*Chem. Abs.*, 1965, **63**, 18047d) [correction in US Pat. 3,183,235 (*Chem. Abs.*, 1966, **64**, 5049c)].
Zenitz, B. L. (1966). US Pat. 3,238,215 (*Chem. Abs.*, 1966, **65**, 7145h).
Zenitz, B. L. (1974). Ger. Pat. 2,329,430 (*Chem. Abs.*, 1974, **80**, 95725d).
Zenitz, B. L. (1977). US Pat. 4,021,431 (*Chem. Abs.*, 1977, **87**, 102164v).
Zhungietu, G. I. and Dorofeenko, G. N. (1967). *Russian Chem. Revs.*, **36**, 24.
Zinner, G., Kliegel, W., Ritter, W. and Böhlke, H. (1966). *Chem. Ber.*, **99**, 1678.
Zinnes, H. (1975). US Pat. 3,892,766 (*Chem. Abs.*, 1975, **83**, 178814v).

Index

2-Aminoketones, acyclic, 798